2012 INTERNATIONAL RESIDENTIAL CODE®

FOR ONE- AND TWO-FAMILY DWELLINGS

A Member of the International Code Family®

IRC®

KING CO
CLIMATE ZONE 4C

GRANT CO
CLIMATE ZONE 5B

2012 International Residential Code®

First Printing: May 2011

ISBN: 978-1-60983-042-7 (soft-cover edition)
ISBN: 978-1-60983-041-0 (loose-leaf edition)

PRINTED IN THE U.S.A.

PREFACE

Introduction

Internationally, code officials recognize the need for a modern, up-to-date residential code addressing the design and construction of one- and two-family dwellings and townhouses. The *International Residential Code®*, in this 2012 edition, is designed to meet these needs through model code regulations that safeguard the public health and safety in all communities, large and small.

This comprehensive, stand-alone residential code establishes minimum regulations for one- and two-family dwellings and townhouses using prescriptive provisions. It is founded on broad-based principles that make possible the use of new materials and new building designs. This 2012 edition is fully compatible with all of the *International Codes®* (I-Codes®) published by the International Code Council® (ICC)®, including the *International Building Code®, International Energy Conservation Code®, International Existing Building Code®, International Fire Code®, International Fuel Gas Code®, International Green Construction Code™* (to be available March 2012), *International Mechanical Code®*, ICC *Performance Code®, International Plumbing Code®, International Private Sewage Disposal Code®, International Property Maintenance Code®, International Swimming Pool and Spa Code™* (to be available March 2012), *International Wildland-Urban Interface Code®* and *International Zoning Code®*.

The *International Residential Code* provisions provide many benefits, among which is the model code development process that offers an international forum for residential construction professionals to discuss prescriptive code requirements. This forum provides an excellent arena to debate proposed revisions. This model code also encourages international consistency in the application of provisions.

Development

The first edition of the *International Residential Code* (2000) was the culmination of an effort initiated in 1996 by ICC and consisting of representatives from the three statutory members of the International Code Council at the time, including: Building Officials and Code Administrators International, Inc. (BOCA), International Conference of Building Officials (ICBO) and Southern Building Code Congress International (SBCCI), and representatives from the National Association of Home Builders (NAHB). The intent was to draft a stand-alone residential code consistent with and inclusive of the scope of the existing model codes. Technical content of the 1998 *International One- and Two-Family Dwelling Code* and the latest model codes promulgated by BOCA, ICBO, SBCCI and ICC was used as the basis for the development, followed by public hearings in 1998 and 1999 to consider proposed changes. This 2012 edition represents the code as originally issued, with changes reflected in the 2009 edition, and further changes developed through the ICC Code Development Process through 2010. Residential electrical provisions are based on the 2011 *National Electrical Code®* (NFPA 70). A new edition such as this is promulgated every three years.

Energy provisions in Chapter 11 are duplicated from the *International Energy Conservation Code®*—Residential Provisions applicable to residential buildings which fall under the scope of this code.

Fuel gas provisions have been included through an agreement with the American Gas Association (AGA). Electrical provisions have been included through an agreement with the National Fire Protection Association (NFPA).

This code is founded on principles intended to establish provisions consistent with the scope of a residential code that adequately protects public health, safety and welfare; provisions that do not unnecessarily increase construction costs; provisions that do not restrict the use of new materials, products or methods of construction; and provisions that do not give preferential treatment to particular types or classes of materials, products or methods of construction.

Adoption

The *International Residential Code* is available for adoption and use by jurisdictions internationally. Its use within a governmental jurisdiction is intended to be accomplished through adoption by reference in accordance with proceedings establishing the jurisdiction's laws. At the time of adoption, jurisdictions should insert the appropriate information in provisions requiring specific local information, such as the name of the adopting jurisdiction. These locations are shown in bracketed words in small capital letters in the code and in the sample ordinance. The sample adoption ordinance on page xvii addresses several key elements of a code adoption ordinance, including the information required for insertion into the code text.

Maintenance

The *International Residential Code* is kept up-to-date through the review of proposed changes submitted by code enforcing officials, industry representatives, design professionals and other interested parties. Proposed changes are carefully considered through an open code development process in which all interested and affected parties may participate.

The contents of this work are subject to change both through the Code Development Cycles and the governmental body that enacts the code into law. For more information regarding the code development process, contact the Codes and Standards Development Department of the International Code Council.

The maintenance process for the fuel gas provisions is based upon the process used to maintain the *International Fuel Gas Code*, in conjunction with the American Gas Association. The maintenance process for the electrical provisions is undertaken by the National Fire Protection Association.

While the development procedure of the *International Residential Code* assures the highest degree of care, ICC, the founding members of ICC, its members and those participating in the development of this code do not accept any liability resulting from compliance or noncompliance with the provisions because ICC and its founding members do not have the power or authority to police or enforce compliance with the contents of this code. Only the governmental body that enacts the code into law has such authority.

Code Development Committee Responsibilities

In each code development cycle, proposed changes to the code are considered at the Code Development Hearings by the applicable International Code Development Committee as follows:

[RB] = IRC—Building Code Development Committee

[RE] = Residential Energy Code Development Committee

[RMP] = IRC—Mechanical/Plumbing Code Development Committee

The [RE] committee is also responsible for the IECC—Residential Provisions.

Note that, for the development of the 2015 edition of the I-Codes, there will be two groups of code development committees and they will meet in separate years. The groupings are as follows:

Group A Codes (Heard in 2012, Code Change Proposals Deadline: January 3, 2012)	Group B Codes (Heard in 2013, Code Change Proposals Deadline: January 3, 2013)
International Building Code	Administrative Provisions (Chapter 1 all codes except the IECC, IRC and ICCPC, administrative updates to currently referenced standards, and designated definitions)
International Fuel Gas Code	International Energy Conservation Code
International Mechanical Code	International Existing Building Code
International Plumbing Code	International Fire Code
International Private Sewage Disposal Code	International Green Construction Code
	ICC Performance Code
	International Property Maintenance Code
	International Residential Code
	International Swimming Pool and Spa Code
	International Wildland-Urban Interface Code
	International Zoning Code

The International Residential Code is included in the Group B Codes. Therefore, any code change proposals to the IRC will be heard in the 2013 code cycle. The deadline for proposed changes to the IRC is January 3, 2013.

Marginal Markings

Solid vertical lines in the margins within the body of the code indicate a technical change from the requirements of the 2009 edition. Deletion indicators in the form of an arrow (➡) are provided in the margin where an entire section, paragraph, exception or table has been deleted or an item in a list of items or a table has been deleted.

A single asterisk [*] placed in the margin indicates that text or a table has been relocated within the code. A double asterisk [**] placed in the margin indicates that the text or table immediately following it has been relocated there from elsewhere in the code. The following table indicates such relocations in the 2012 Edition of the *International Residential Code*.

2012 LOCATION	2009 LOCATION
R312.2	R612.2
R507	R502.2.2
R602.3.5	R602.10.1.2.1
R602.10.6.5	R602.12
R702.7	R601.3

Italicized Terms

Selected terms set forth in Chapter 2, Definitions, are italicized where they appear in code text. Such terms are not italicized where the definition set forth in Chapter 2 does not impart the intended meaning in the use of the term. The terms selected have definitions which the user should read carefully to facilitate better understanding of the code.

Effective Use of the International Residential Code

Effective Use of the International Residential Code

The *International Residential Code*® (IRC®) was created to serve as a complete, comprehensive code regulating the construction of single-family houses, two-family houses (duplexes) and buildings consisting of three or more townhouse units. All buildings within the scope of the IRC are limited to three stories above grade plane. For example, a four-story single-family house would fall within the scope of the *International Building Code*® (IBC®), not the IRC. The benefits of devoting a separate code to residential construction include the fact that the user need not navigate through a multitude of code provisions that do not apply to residential construction in order to locate that which is applicable. A separate code also allows for residential and nonresidential code provisions to be distinct and tailored to the structures that fall within the appropriate code's scopes.

The IRC contains coverage for all components of a house or townhouse, including structural components, fireplaces and chimneys, thermal insulation, mechanical systems, fuel gas systems, plumbing systems and electrical systems.

The IRC is a prescriptive-oriented (specification) code with some examples of performance code language. It has been said that the IRC is the complete cookbook for residential construction. Section R301.1, for example, is written in performance language, but states that the prescriptive requirements of the code will achieve such performance.

It is important to understand that the IRC contains coverage for what is conventional and common in residential construction practice. While the IRC will provide all of the needed coverage for most residential construction, it might not address construction practices and systems that are atypical or rarely encountered in the industry. Sections such as R301.1.3, R301.2.1, R301.2.2, R320.1, R322.1, M1301.1, G2401.1 and P2601.1 refer to other codes either as an alternative to the provisions of the IRC or where the IRC lacks coverage for a particular type of structure, design, system, appliance or method of construction. In other words, the IRC is meant to be all inclusive for typical residential construction and it relies on other codes only where alternatives are desired or where the code lacks coverage for the uncommon aspect of residential construction. Of course, the IRC constantly evolves to address new technologies and construction practices that were once uncommon, but now common.

The IRC is unique in that much of it, including Chapters 3 through 9 and Chapters 34 through 43, is presented in an ordered format that is consistent with the normal progression of construction, starting with the design phase and continuing through the final trim-out phase. This is consistent with the "cookbook" philosophy of the IRC.

The IRC is divided into eight main parts, specifically, Part I—Administration, Part II—Definitions, Part III—Building Planning and Construction, Part IV—Energy Conservation, Part V—Mechanical, Part VI—Fuel Gas, Part VII—Plumbing and Part VIII—Electrical.

The following provides a brief description of the content of each chapter and appendix of the IRC:

Chapter 1 Scope and Administration. This chapter contains provisions for the application, enforcement and administration of subsequent requirements of the code. In addition to establishing the scope of the code, Chapter 1 identifies which buildings and structures come under its purview. Chapter 1 is largely concerned with maintaining "due process of law" in enforcing the building criteria contained in the body of the code. Only through careful observation of the administrative provisions can the building official reasonably expect to demonstrate that "equal protection under the law" has been provided.

Chapter 2 Definitions. Terms defined in the code are listed alphabetically in Chapter 2. It is important to note that two chapters have their own definitions sections: Chapter 24 for the defined terms that are unique to fuel gas and Chapter 35 containing terms that are applicable to electrical Chapters 34 through 43. In the case where Chapter 2 and another chapter both define the same term differently, the definition found in Chapter 24 and/or 35 is intended to prevail where the term is used in Chapter 24 and/or 35 and the definition contained in Chapter 2 is intended to prevail where the term is used in all other locations in the code. Except where Chapter 24 or 35 has a definition that will prevail therein, the definitions in Chapter 2 are applicable throughout the code.

Additional definitions regarding skylights that are not listed in Chapter 2 are found in Section R308.6.1.

Where understanding a term's definition is key to or necessary for understanding a particular code provision, the term is shown in italics where it appears in the code. This is true only for those terms that have a meaning that is unique to the code. In other words, the generally understood meaning of a term or phrase might not be sufficient or consistent with the meaning prescribed by the code; therefore, it is essential that the code-defined meaning be known.

Guidance regarding not only tense, gender and plurality of defined terms, but also terms not defined in this code, is provided.

Chapter 3 Building Planning. Chapter 3 provides guidelines for a minimum level of structural integrity, life safety, fire safety and livability for inhabitants of dwelling units regulated by this code. Chapter 3 is a compilation of the code requirements specific to the building planning sector of the design and construction process. This chapter sets forth code requirements dealing with light, ventilation, sanitation, minimum room size, ceiling height and environmental comfort. Chapter 3 establishes life-safety provisions including limitations on glazing used in hazardous areas, specifications on stairways, use of guards at elevated surfaces, window and fall protection, and rules for means of egress. Snow, wind and seismic design and flood-resistant construction, as well as live and dead loads, are addressed in this chapter.

Chapter 4 Foundations. Chapter 4 provides the requirements for the design and construction of foundation systems for buildings regulated by this code. Provisions for seismic load, flood load and frost protection are contained in this chapter. A foundation system consists of two interdependent components: the foundation structure itself and the supporting soil.

The prescriptive provisions of this chapter provide requirements for constructing footings and walls for foundations of wood, masonry, concrete and precast concrete. In addition to a foundation's ability to support the required design loads, this chapter addresses several other factors that can affect foundation performance. These include controlling surface water and subsurface drainage, requiring soil tests where conditions warrant and evaluating proximity to slopes and minimum depth requirements. The chapter also provides requirements to minimize adverse effects of moisture, decay and pests in basements and crawl spaces.

Chapter 5 Floors. Chapter 5 provides the requirements for the design and construction of floor systems that will be capable of supporting minimum required design loads. This chapter covers four different types: wood floor framing, wood floors on the ground, cold-formed steel floor framing and concrete slabs on the ground. Allowable span tables are provided that greatly simplify the determination of joist, girder and sheathing sizes for raised floor systems of wood framing and cold-formed steel framing. This chapter also contains prescriptive requirements for attaching a deck to the main building.

Chapter 6 Wall Construction. Chapter 6 contains provisions that regulate the design and construction of walls. The wall construction covered in Chapter 6 consists of five different types: wood framed, cold-formed steel framed, masonry, concrete and structural insulated panel (SIP). The primary concern of this chapter is the structural integrity of wall construction and transfer of all imposed loads to the supporting structure. This chapter provides the requirements for the design and construction of wall systems that are capable of supporting the minimum design vertical loads (dead, live and snow loads) and lateral loads (wind or seismic loads). This chapter contains the prescriptive requirements for wall bracing and/or shear walls to resist the imposed lateral loads due to wind and seismic.

Chapter 6 also regulates exterior windows and doors installed in walls. The chapter contains criteria for the performance of exterior windows and doors and includes provisions for window sill height, testing and labeling, vehicular access doors, wind-borne debris protection and anchorage details.

Chapter 7 Wall Covering. Chapter 7 contains provisions for the design and construction of interior and exterior wall coverings. This chapter establishes the various types of materials, materials standards and methods of application permitted for use as interior coverings, including interior plaster, gypsum board, ceramic tile, wood veneer paneling, hardboard paneling, wood shakes and wood shingles. Chapter 7 also contains requirements for the use of vapor retarders for moisture control in walls.

Exterior wall coverings provide the weather-resistant exterior envelope that protects the building's interior from the elements. Chapter 7 provides the requirements for wind resistance and water-resistive barrier for exterior wall coverings. This chapter prescribes the exterior wall coverings as well as the water-resistive barrier required beneath the exterior materials. Exterior wall coverings regulated by this section include aluminum, stone and masonry veneer, wood, hardboard, particleboard, wood structural panel siding, wood shakes and shingles, exterior plaster, steel, vinyl, fiber cement and exterior insulation finish systems.

Chapter 8 Roof-ceiling Construction. Chapter 8 regulates the design and construction of roof-ceiling systems. This chapter contains two roof-ceiling framing systems: wood framing and cold-formed steel framing. Allowable span tables are provided to simplify the selection of rafter and ceiling joist size for wood roof framing and cold-formed steel framing. Chapter 8 also provides requirements for the application of ceiling finishes, the proper ventilation of concealed spaces in roofs (e.g., enclosed attics and rafter spaces), unvented attic assemblies and attic access.

Chapter 9 Roof Assemblies. Chapter 9 regulates the design and construction of roof assemblies. A roof assembly includes the roof deck, vapor retarder, substrate or thermal barrier, insulation, vapor retarder and roof covering. This chapter provides the requirement for wind resistance of roof coverings.

The types of roof covering materials and installation regulated by Chapter 9 are: asphalt shingles, clay and concrete tile, metal roof shingles, mineral-surfaced roll roofing, slate and slate-type shingles, wood shakes and shingles, built-up roofs, metal roof panels, modified bitumen roofing, thermoset and thermoplastic single-ply roofing, sprayed polyurethane foam roofing, liquid applied coatings and photovoltaic modules/shingles. Chapter 9 also provides requirements for roof drainage, flashing, above deck thermal insulation and recovering or replacing an existing roof covering.

Chapter 10 Chimneys and Fireplaces. Chapter 10 contains requirements for the safe construction of masonry chimneys and fireplaces and establishes the standards for the use and installation of factory-built chimneys, fireplaces and masonry heaters. Chimneys and fireplaces constructed of masonry rely on prescriptive requirements for the details of their construction; the factory-built type relies on the listing and labeling method of approval. Chapter 10 provides the requirements for seismic reinforcing and anchorage of masonry fireplaces and chimneys.

Chapter 11 Energy Efficiency. The purpose of Chapter 11 is to provide minimum design requirements that will promote efficient utilization of energy in buildings. The requirements are directed toward the design of building envelopes with adequate thermal resistance and low air leakage, and toward the design and selection of mechanical, water heating, electrical and illumination systems that promote effective use of depletable energy resources. The provisions of Chapter 11 are duplicated from the *International Energy Conservation Code—Residential Provisions*, as applicable for buildings which fall under the scope of the IRC.

For ease of use and coordination of provisions, the corresponding IECC—Residential Provisions section number is indicated following the IRC section number [e.g. N1102.1 (R402.1)].

Chapter 12 Mechanical Administration. Chapter 12 establishes the limits of applicability of the code and describes how the code is to be applied and enforced. A mechanical code, like any other code, is intended to be adopted as a legally enforceable document and it cannot be effective without adequate provisions for its administration and enforcement. The provisions of Chapter 12 establish the authority and duties of the code official appointed by the jurisdiction having authority and also establish the rights and privileges of the design professional, contractor and property owner. It also relates this chapter to the administrative provisions in Chapter 1.

Chapter 13 General Mechanical System Requirements. Chapter 13 contains broadly applicable requirements related to appliance listing and labeling, appliance location and installation, appliance and systems access, protection of structural elements and clearances to combustibles, among others.

Chapter 14 Heating and Cooling Equipment. Chapter 14 is a collection of requirements for various heating and cooling appliances, dedicated to single topics by section. The common theme is that all of these types of appliances use energy in one form or another, and the improper installation of such appliances would present a hazard to the occupants of the dwellings, due to either the potential for fire or the accidental release of refrigerants. Both situations are undesirable in dwellings that are covered by this code.

Chapter 15 Exhaust Systems. Chapter 15 is a compilation of code requirements related to residential exhaust systems, including kitchens and bathrooms, clothes dryers and range hoods. The code regulates the materials used for constructing and installing such duct systems. Air brought into the building for ventilation, combustion or makeup purposes is protected from contamination by the provisions found in this chapter.

Chapter 16 Duct Systems. Chapter 16 provides requirements for the installation of ducts for supply, return and exhaust air systems. This chapter contains no information on the design of these systems from the standpoint of air movement, but is concerned with the structural integrity of the systems and the overall impact of the systems on the fire-safety performance of the building. This chapter regulates the materials and methods of construction which affect the performance of the entire air distribution system.

Chapter 17 Combustion Air. Complete combustion of solid and liquid fuel is essential for the proper operation of appliances, control of harmful emissions and achieving maximum fuel efficiency. If insufficient quantities of oxygen are supplied, the combustion process will be incomplete, creating dangerous byproducts and wasting energy in the form of unburned fuel (hydrocarbons). The byproducts of incomplete combustion are poisonous, corrosive and combustible, and can cause serious appliance or equipment malfunctions that pose fire or explosion hazards.

The combustion air provisions in this code from previous editions have been deleted from Chapter 17 in favor of a single section that directs the user to NFPA 31 for oil-fired appliance combustion air requirements and the manufacturer's installation instructions for solid fuel-burning appliances. If fuel gas appliances are used, the provisions of Chapter 24 must be followed.

Chapter 18 Chimneys and Vents. Chapter 18 regulates the design, construction, installation, maintenance, repair and approval of chimneys, vents and their connections to fuel-burning appliances. A properly designed chimney or vent system is needed to conduct the flue gases produced by a fuel-burning appliance to the outdoors. The provisions of this chapter are intended to minimize the hazards associated with high temperatures and potentially toxic and corrosive combustion gases. This chapter addresses factory-built and masonry chimneys, vents and venting systems used to vent oil-fired and solid fuel-burning appliances.

Chapter 19 Special Fuel-burning Equipment. Chapter 19 regulates the installation of fuel-burning appliances that are not covered in other chapters, such as ranges and ovens, sauna heaters, fuel cell power plants and hydrogen systems. Because the subjects in this chapter do not contain the volume of text necessary to warrant individual chapters, they have been combined into a single chapter. The only commonality is that the subjects use energy to perform some task or function. The intent is to provide a reasonable level of protection for the occupants of the dwelling.

Chapter 20 Boilers and Water Heaters. Chapter 20 regulates the installation of boilers and water heaters. Its purpose is to protect the occupants of the dwelling from the potential hazards associated with such appliances. A water heater is any appliance that heats potable water and supplies it to the plumbing hot water distribution system. A boiler either heats water or generates steam for space heating and is generally a closed system.

Chapter 21 Hydronic Piping. Hydronic piping includes piping, fittings and valves used in building space conditioning systems. Applications include hot water, chilled water, steam, steam condensate, brines and water/antifreeze mixtures. Chapter 21 regulates installation, alteration and repair of all hydronic piping systems to insure the reliability, serviceability, energy efficiency and safety of such systems.

Chapter 22 Special Piping and Storage Systems. Chapter 22 regulates the design and installation of fuel oil storage and piping systems. The regulations include reference to construction standards for above-ground and underground storage tanks, material standards for piping systems (both above-ground and underground) and extensive requirements for the proper assembly of system piping and components. The purpose of this chapter is to prevent fires, leaks and spills involving fuel oil storage and piping systems, whether inside or outside structures and above or underground.

Chapter 23 Solar Systems. Chapter 23 contains requirements for the construction, alteration and repair of all systems and components of solar energy systems used for space heating or cooling, and domestic hot water heating or processing. The provisions of this chapter are limited to those necessary to achieve installations that are relatively hazard free.

A solar energy system can be designed to handle 100 percent of the energy load of a building, although this is rarely accomplished. Because solar energy is a low-intensity energy source and dependent on the weather, it is usually necessary to supplement a solar energy system with traditional energy sources.

As our world strives to find alternate means of producing power for the future, the requirements of this chapter will become more and more important over time.

Chapter 24 Fuel Gas. Chapter 24 regulates the design and installation of fuel gas distribution piping and systems, appliances, appliance venting systems and combustion air provisions. The definition of "Fuel gas" includes natural, liquefied petroleum and manufactured gases and mixtures of these gases.

The purpose of this chapter is to establish the minimum acceptable level of safety and to protect life and property from the potential dangers associated with the storage, distribution and use of fuel gases and the byproducts of combustion of such fuels. This code also protects the personnel who install, maintain, service and replace the systems and appliances addressed herein.

Chapter 24 is composed entirely of text extracted from the IFGC; therefore, whether using the IFGC or the IRC, the fuel gas provisions will be identical. Note that to avoid the potential for confusion and conflicting definitions, Chapter 24 has its own definition section.

Chapter 25 Plumbing Administration. The requirements of Chapter 25 do not supersede the administrative provisions of Chapter 1. Rather, the administrative guidelines of Chapter 25 pertain to plumbing installations that are best referenced and located within the plumbing chapters. This chapter addresses how to apply the plumbing provisions of this code to specific types or phases of construction. This chapter also outlines the responsibilities of the applicant, installer and inspector with regard to testing plumbing installations.

Chapter 26 General Plumbing Requirements. The content of Chapter 26 is often referred to as "miscellaneous," rather than general plumbing requirements. This is the only chapter of the plumbing chapters of the code whose requirements do not interrelate. If a requirement cannot be located in another plumbing chapter, it should be located in this chapter. Chapter 26 contains safety requirements for the installation of plumbing systems and includes requirements for the identification of pipe, pipe fittings, traps, fixtures, materials and devices used in plumbing systems. If specific provisions do not demand that a requirement be located in another chapter, the requirement is located in this chapter.

Chapter 27 Plumbing Fixtures. Chapter 27 requires fixtures to be of the proper type, approved for the purpose intended and installed properly to promote usability and safe, sanitary conditions. This chapter regulates the quality of fixtures and faucets by requiring those items to comply with nationally recognized standards. Because fixtures must be properly installed so that they are usable by the occupants of the building, this chapter contains the requirements for the installation of fixtures.

Chapter 28 Water Heaters. Chapter 28 regulates the design, approval and installation of water heaters and related safety devices. The intent is to minimize the hazards associated with the installation and operation of water heaters. Although this chapter does not regulate the size of a water heater, it does regulate all other aspects of the water heater installation such as temperature and pressure relief valves, safety drip pans and connections. Where a water heater also supplies water for space heating, this chapter regulates the maximum water temperature supplied to the water distribution system.

Chapter 29 Water Supply and Distribution. This chapter regulates the supply of potable water from both public and individual sources to every fixture and outlet so that it remains potable and uncontaminated by cross connections. Chapter 29 also regulates the design of the water distribution system, which will allow fixtures to function properly. Because it is critical that the potable water supply system remain free of actual or potential sanitary hazards, this chapter has the requirements for providing backflow protection devices.

Chapter 30 Sanitary Drainage. The purpose of Chapter 30 is to regulate the materials, design and installation of sanitary drainage piping systems as well as the connections made to the system. The intent is to design and install sanitary drainage systems that will function reliably, are neither undersized nor oversized and are constructed from materials, fittings and connections whose quality is regulated by this section. This chapter addresses the proper use of fittings for directing the flow into and within the sanitary drain piping system. Materials and provisions necessary for servicing the drainage system are also included in this chapter.

Chapter 31 Vents. Venting protects the trap seal of each trap. The vents are designed to limit differential pressures at each trap to 1 inch of water column (249 Pa). Because waste flow in the drainage system creates pressure fluctuations that can negatively affect traps, the sanitary drainage system must have a properly designed venting system. Chapter 31 covers the requirements for vents and venting. All of the provisions set forth in this chapter are intended to limit the pressure differentials in the drainage system to a maximum of 1 inch of water column (249 Pa) above or below atmospheric pressure (i.e., positive or negative pressures).

Chapter 32 Traps. Traps prevent sewer gas from escaping from the drainage piping into the building. Water seal traps are the simplest and most reliable means of preventing sewer gas from entering the interior environment. This chapter lists prohibited trap types as well as specifies the minimum trap size for each type of fixture.

Chapter 33 Storm Drainage. Rainwater infiltration into the ground adjacent to a building can cause the interior of foundation walls to become wet. The installation of a subsoil drainage system prevents the build-up of rainwater on the exterior of the foundation walls. This chapter provides the specifications for subsoil drain piping. Where the discharge of the subsoil drain system is to a sump, this chapter also provides coverage for sump construction, pumps and discharge piping.

Chapter 34 General Requirements. This chapter contains broadly applicable, general and miscellaneous requirements including scope, listing and labeling, equipment locations and clearances for conductor materials and connections and conductor identification.

Chapter 35 Electrical Definitions. Chapter 35 is the repository of the definitions of terms used in the body of Part VIII of the code. To avoid the potential for confusion and conflicting definitions, Part VIII, Electrical, has its own definition chapter.

Codes are technical documents and every word, term and punctuation mark can impact the meaning of the code text and the intended results. The code often uses terms that have a unique meaning in the code, which can differ substantially from the ordinarily understood meaning of the term as used outside of the code.

The terms defined in Chapter 35 are deemed to be of prime importance in establishing the meaning and intent of the electrical code text that uses the terms. The user of the code should be familiar with and consult this chapter because the definitions are essential to the correct interpretation of the code and because the user may not be aware that a term is defined.

Chapter 36 Services. This chapter covers the design, sizing and installation of the building's electrical service equipment and grounding electrode system. It includes an easy-to-use load calculation method and service conductor sizing table. The electrical service is generally the first part of the electrical system to be designed and installed.

Chapter 37 Branch Circuit and Feeder Requirements. Chapter 37 addresses the requirements for designing the power distribution system which consists of feeders and branch circuits emanating from the service equipment. This chapter dictates the ratings of circuits and the allowable loads, the number and types of branch circuits required, the wire sizing for such branch circuits and feeders and the requirements for protection from overcurrent for conductors. A load calculation method specific to feeders is also included. This chapter is used to design the electrical system on the load side of the service.

Chapter 38 Wiring Methods. Chapter 38 specifies the allowable wiring methods, such as cable, conduit and raceway systems, and provides the installation requirements for the wiring methods. This chapter is primarily applicable to the "rough-in" phase of construction.

Chapter 39 Power and Lighting Distribution. This chapter mostly contains installation requirements for the wiring that serves the lighting outlets, receptacle outlets, appliances and switches located throughout the building. The required distribution and spacing of receptacle outlets and lighting outlets is prescribed in this chapter, as well as the requirements for ground-fault and arc-fault circuit interrupter protection.

Chapter 40 Devices and Luminaires. This chapter focuses on the devices, including switches and receptacles, and lighting fixtures that are typically installed during the final phase of construction.

Chapter 41 Appliance Installation. Chapter 41 addresses the installation of appliances including HVAC appliances, water heaters, fixed space-heating equipment, dishwashers, garbage disposals, range hoods and suspended paddle fans.

Chapter 42 Swimming Pools. This chapter covers the electrical installation requirements for swimming pools, storable swimming pools, wading pools, decorative pools, fountains, hot tubs, spas and hydromassage bathtubs. The allowable wiring methods are specified along with the required clearances between electrical system components and pools, spas and tubs. This chapter includes the special grounding requirements related to pools, spas and tubs, and also prescribes the equipotential bonding requirements that are unique to pools, spas and tubs.

Chapter 43 Class 2 Remote-control, Signaling and Power-limited Circuits. This chapter covers the power supplies, wiring methods and installation requirements for the Class 2 circuits found in dwellings. Such circuits include thermostat wiring, alarm systems, security systems, automated control systems and doorbell systems.

Chapter 44 Referenced Standards. The code contains numerous references to standards that are used to regulate materials and methods of construction. Chapter 44 contains a comprehensive list of all standards that are referenced in the code. The standards are part of the code to the extent of the reference to the standard. Compliance with the referenced standard is necessary for compliance with this code. By providing specifically adopted standards, the construction and installation requirements necessary for compliance with the code can be readily determined. The basis for code compliance is, therefore, established and available on an equal basis to the code official, contractor, designer and owner.

Chapter 44 is organized in a manner that makes it easy to locate specific standards. It lists all of the referenced standards, alphabetically, by acronym of the promulgating agency of the standard. Each agency's standards are then listed in either alphabetical or numeric order based upon the standard identification. The list also contains the title of the standard; the edition (date) of the standard referenced; any addenda included as part of the ICC adoption; and the section or sections of this code that reference the standard.

Appendix A Sizing and Capacities of Gas Piping. This appendix is informative and not part of the code. It provides design guidance, useful facts and data and multiple examples of how to apply the sizing tables and sizing methodologies of Chapter 24.

Appendix B Sizing of Venting Systems Serving Appliances Equipped with Draft Hoods, Category I Appliances and Appliances Listed for Use with Type B Vents. This appendix is informative and not part of the code. It contains multiple examples of how to apply the vent and chimney tables and methodologies of Chapter 24.

Appendix C Exit Terminals of Mechanical Draft and Direct-vent Venting Systems. This appendix is informative and not part of the code. It consists of a figure and notes that visually depict code requirements from Chapter 24 for vent terminals with respect to the openings found in building exterior walls.

Appendix D Recommended Procedure for Safety Inspection of an Existing Appliance Installation. This appendix is informative and not part of the code. It provides recommended procedures for testing and inspecting an appliance installation to determine if the installation is operating safely and if the appliance is in a safe condition.

Appendix E Manufactured Housing Used as Dwellings. The criteria for the construction of manufactured homes are governed by the National Manufactured Housing Construction and Safety Act. While this act may seem to cover the bulk of the construction of manufactured housing, it does not cover those areas related to the placement of the housing on the property. The provisions of Appendix E are not applicable to the design and construction of manufactured homes. Appendix E provides a complete set of regulations in conjunction with federal law for the installation of manufactured housing. This appendix also contains provisions for existing manufactured home installations.

Appendix F Radon Control Methods. Radon comes from the natural (radioactive) decay of the element radium in soil, rock and water and finds its way into the air. Appendix F contains requirements to mitigate the transfer of radon gases from the soil into the dwelling. The provisions of this appendix regulate the design and construction of radon-resistant measures intended to reduce the entry of radon gases into the living space of residential buildings.

Appendix G Swimming Pool, Spas and Hot Tubs. Appendix G provides the regulations for swimming pools, hot tubs and spas installed in or on the lot of a one- or two-family dwelling. This appendix contains provisions for an effective barrier surrounding the water area and entrapment protection for suction outlets to reduce the potential for drowning of young children.

Appendix H Patio Covers. Appendix H sets forth the regulations and limitations for patio covers. The provisions address those uses permitted in patio cover structures, the minimum design loads to be assigned for structural purposes, and the effect of the patio cover on egress and emergency escape or rescue from sleeping rooms. This appendix also contains the special provisions for aluminum screen enclosures in hurricane-prone regions.

Appendix I Private Sewage Disposal. Appendix I simply provides the opportunity to utilize the International Private Sewage Disposal Code for the design and installation of private sewage disposal in one- and two-family dwellings.

Appendix J Existing Buildings and Structures. Appendix J contains the provisions for the repair, renovation, alteration and reconstruction of existing buildings and structures that are within the scope of this code. To accomplish this objective and to make the rehabilitation process more available, this appendix allows for a controlled departure from full code compliance without compromising minimum life safety, fire safety, structural and environmental features of the rehabilitated existing building or structure.

Appendix K Sound Transmission. Appendix K regulates the sound transmission of wall and floor-ceiling assemblies separating dwelling units and townhouse units. Air-borne sound insulation is required for walls. Air-borne sound insulation and impact sound insulation are required for floor-ceiling assemblies. The provisions in Appendix K set forth a minimum Sound Transmission Class (STC) rating for common walls and floor-ceiling assemblies between dwelling units. In addition, a minimum Impact Insulation Class (IIC) rating is also established to limit structure-borne sound through common floor-ceiling assemblies separating dwelling units.

Appendix L Permit Fees. Appendix L provides guidance to jurisdictions for setting appropriate permit fees. This appendix will aid many jurisdictions to assess permit fees that will assist to fairly and properly administer the code. This appendix can be used for informational purposes only or may be adopted when specifically referenced in the adopting ordinance.

Appendix M Home Day Care–R-3 Occupancy. Appendix M provides means of egress and smoke detection requirements for a Group R-3 Occupancy that is to be used as a home day care for more than five children who receive custodial care for less than 24 hours. This appendix is strictly for guidance and/or adoption by those jurisdictions that have Licensed Home Care Provider laws and statutes that allow more than five children to be cared for in a person's home. When a jurisdiction adopts this appendix, the provisions for day care and child care facilities in the IBC should be considered also.

Appendix N Venting Methods. Because venting of sanitary drainage systems is perhaps the most difficult concept to understand, and Chapter 31 uses only words to describe venting requirements, illustrations can offer greater insight into what the words mean. Appendix N has a number of illustrations for commonly installed sanitary drainage systems in order for the reader to gain a better understanding of this code's venting requirements.

Appendix O Automatic Vehicular Gates. Appendix O provides the requirements for the design and construction of automatic vehicular gates. The provisions are for where automatic gates are installed for use at a vehicular entrance or exit on the lot of a one- or two-family dwelling. The requirements provide protection for individuals from potential entrapment between an automatic gate and a stationary object or surface.

Appendix P Sizing of Water Piping System. Appendix P provides two recognized methods for sizing the water service and water distribution piping for a building. The method under Section AP103 provides friction loss diagrams that require the user to "plot" points and read values from the diagrams in order to perform the required calculations and necessary checks. This method is the most accurate of the two presented in this appendix. The method under Section AP201 is known to be conservative; however, very few calculations are necessary in order to determine a pipe size that satisfies the flow requirements of any application.

Appendix Q ICC International Residential Code Electrical Provisions/National Electrical Code Cross Reference. This cross reference allows the code user to trace the code sections in Chapters 34 through 43 back to their source: the *National Electrical Code*. See the introduction to Chapter 34 for more information on the relationship between Part VIII of this code and the NEC, NFPA 70.

LEGISLATION

The *International Codes* are designed and promulgated to be adopted by reference by legislative action. Jurisdictions wishing to adopt the 2012 *International Residential Code* as an enforceable regulation governing one- and two-family dwellings and townhouses should ensure that certain factual information is included in the adopting legislation at the time adoption is being considered by the appropriate governmental body. The following sample adoption legislation addresses several key elements, including the information required for insertion into the code text.

SAMPLE LEGISLATION FOR ADOPTION OF THE *INTERNATIONAL RESIDENTIAL CODE* ORDINANCE NO.________

A**[N] [ORDINANCE/STATUTE/REGULATION]** of the **[JURISDICTION]** adopting the 2012 edition of the *International Residential Code*, regulating and governing the construction, alteration, movement, enlargement, replacement, repair, equipment, location, removal and demolition of detached one- and two-family dwellings and multiple single-family dwellings (townhouses) not more than threes stories in height with separate means of egress in the **[JURISDICTION]**; providing for the issuance of permits and collection of fees therefor; repealing **[ORDINANCE/STATUTE/REGULATION]** No. ______ of the **[JURISDICTION]** and all other ordinances or parts of laws in conflict therewith.

The **[GOVERNING BODY]** of the **[JURISDICTION]** does ordain as follows:

Section 1. That a certain document, three (3) copies of which are on file in the office of the **[TITLE OF JURISDICTION'S KEEPER OF RECORDS]** of **[NAME OF JURISDICTION]**, being marked and designated as the *International Residential Code*, 2012 edition, including Appendix Chapters **[FILL IN THE APPENDIX CHAPTERS BEING ADOPTED]** (see *International Residential Code* Section R102.5, 2012 edition), as published by the International Code Council, be and is hereby adopted as the Residential Code of the **[JURISDICTION]**, in the State of **[STATE NAME]** for regulating and governing the construction, alteration, movement, enlargement, replacement, repair, equipment, location, removal and demolition of detached one- and two-family dwellings and multiple single-family dwellings (townhouses) not more than threes stories in height with separate means of egress as herein provided; providing for the issuance of permits and collection of fees therefor; and each and all of the regulations, provisions, penalties, conditions and terms of said Residential Code on file in the office of the **[JURISDICTION]** are hereby referred to, adopted, and made a part hereof, as if fully set out in this ordinance, with the additions, insertions, deletions and changes, if any, prescribed in Section 2 of this ordinance.

Section 2. The following sections are hereby revised:

Section R101.1. Insert: **[NAME OF JURISDICTION]**

Table R301.2 (1) Insert: **[APPROPRIATE DESIGN CRITERIA]**

Section P2603.6.1 Insert: **[NUMBER OF INCHES IN TWO LOCATIONS]**

Section 3. That **[ORDINANCE/STATUTE/REGULATION]** No. ______ of **[JURISDICTION]** entitled **[FILL IN HERE THE COMPLETE TITLE OF THE LEGISLATION OR LAWS IN EFFECT AT THE PRESENT TIME SO THAT THEY WILL BE REPEALED BY DEFINITE MENTION]** and all other ordinances or parts of laws in conflict herewith are hereby repealed.

Section 4. That if any section, subsection, sentence, clause or phrase of this legislation is, for any reason, held to be unconstitutional, such decision shall not affect the validity of the remaining portions of this ordinance. The **[GOVERNING BODY]** hereby declares that it would have passed this law, and each section, subsection, clause or phrase thereof, irrespective of the fact that any one or more sections, subsections, sentences, clauses and phrases be declared unconstitutional.

Section 5. That nothing in this legislation or in the Residential Code hereby adopted shall be construed to affect any suit or proceeding impending in any court, or any rights acquired, or liability incurred, or any cause or causes of action acquired or existing, under any act or ordinance hereby repealed as cited in Section 3 of this law; nor shall any just or legal right or remedy of any character be lost, impaired or affected by this legislation.

Section 6. That the **[JURISDICTION'S KEEPER OF RECORDS]** is hereby ordered and directed to cause this legislation to be published. (An additional provision may be required to direct the number of times the legislation is to be published and to specify that it is to be in a newspaper in general circulation. Posting may also be required.)

Section 7. That this law and the rules, regulations, provisions, requirements, orders and matters established and adopted hereby shall take effect and be in full force and effect **[TIME PERIOD]** from and after the date of its final passage and adoption.

TABLE OF CONTENTS

Part I—Administrative 1

CHAPTER 1 SCOPE AND ADMINISTRATION 1

PART 1—SCOPE AND APPLICATION 1

Section

R101 General 1

R102 Applicability 1

PART 2—ADMINISTRATION AND ENFORCEMENT 2

R103 Department of Building Safety 2

R104 Duties and Powers of the Building Official 2

R105 Permits 3

R106 Construction Documents 5

R107 Temporary Structures and Uses 6

R108 Fees 6

R109 Inspections 6

R110 Certificate of Occupancy 7

R111 Service Utilities 7

R112 Board of Appeals 8

R113 Violations 8

R114 Stop Work Order 9

Part II—Definitions 11

CHAPTER 2 DEFINITIONS 11

Section

R201 General 11

R202 Definitions 11

Part III—Building Planning and Construction 25

CHAPTER 3 BUILDING PLANNING 25

Section

R301 Design Criteria 25

R302 Fire-resistant Construction 48

R303 Light, Ventilation and Heating 53

R304 Minimum Room Areas 54

R305 Ceiling Height 54

R306 Sanitation 54

R307 Toilet, Bath and Shower Spaces 54

R308 Glazing 55

R309 Garages and Carports 58

R310 Emergency Escape and Rescue Openings 58

R311 Means of Egress 59

R312 Guards and Window Fall Protection 62

R313 Automatic Fire Sprinkler Systems 62

R314 Smoke Alarms 63

R315 Carbon Monoxide Alarms 63

R316 Foam Plastic 64

R317 Protection of Wood and Wood Based Products Against Decay 65

R318 Protection Against Subterranean Termites 67

R319 Site Address 68

R320 Accessibility 68

R321 Elevators and Platform Lifts 68

R322 Flood-resistant Construction 68

R323 Storm Shelters 71

CHAPTER 4 FOUNDATIONS 73

Section

R401 General 73

R402 Materials 73

R403 Footings 74

R404 Foundation and Retaining Walls 91

R405 Foundation Drainage 109

R406 Foundation Waterproofing and Dampproofing 110

R407 Columns 111

R408 Under-floor Space 111

CHAPTER 5 FLOORS 113

Section

R501 General 113

R502 Wood Floor Framing 113

R503 Floor Sheathing 124

R504 Pressure Preservatively Treated-wood Floors (On Ground) 125

R505 Steel Floor Framing 126

R506 Concrete Floors (On Ground) 144

R507 Decks 146

CHAPTER 6 WALL CONSTRUCTION 149

Section

R601 General 149
R602 Wood Wall Framing 149
R603 Steel Wall Framing 192
R604 Wood Structural Panels 264
R605 Particleboard 264
R606 General Masonry Construction 264
R607 Unit Masonry 272
R608 Multiple-wythe Masonry 273
R609 Grouted Masonry 274
R610 Glass Unit Masonry 276
R611 Exterior Concrete Wall Construction 277
R612 Exterior Windows and Doors 347
R613 Structural Insulated Panel Wall Construction 350

CHAPTER 7 WALL COVERING 359

Section

R701 General 359
R702 Interior Covering 359
R703 Exterior Covering 363

CHAPTER 8 ROOF-CEILING CONSTRUCTION 377

Section

R801 General 377
R802 Wood Roof Framing 377
R803 Roof Sheathing 403
R804 Steel Roof Framing 403
R805 Ceiling Finishes 436
R806 Roof Ventilation 436
R807 Attic Access 437

CHAPTER 9 ROOF ASSEMBLIES 439

Section

R901 General 439
R902 Roof Classification 439
R903 Weather Protection 439
R904 Materials 439
R905 Requirements for Roof Coverings 440
R906 Roof Insulation 450
R907 Reroofing 450

CHAPTER 10 CHIMNEYS AND FIREPLACES . . . 453

Section

R1001 Masonry Fireplaces 453
R1002 Masonry Heaters 456
R1003 Masonry Chimneys 457
R1004 Factory-built Fireplaces 461
R1005 Factory-built Chimneys 462
R1006 Exterior Air Supply 462

Part IV—Energy Conservation 463

CHAPTER 11 ENERGY EFFICIENCY 463

Section

N1101 General 463
N1102 Building Thermal Envelope 482
N1103 Systems 485
N1104 Electrical Power and Lighting Systems (Mandatory) 487
N1105 Simulated Performance Alternative (Performance) 487

Part V—Mechanical 493

CHAPTER 12 MECHANICAL ADMINISTRATION 493

Section

M1201 General 493
M1202 Existing Mechanical Systems 493

CHAPTER 13 GENERAL MECHANICAL SYSTEM REQUIREMENTS 495

Section

M1301 General 495
M1302 Approval 495
M1303 Labeling of Appliances 495
M1304 Type of Fuel 495
M1305 Appliance Access 495
M1306 Clearances from Combustible Construction . . . 496
M1307 Appliance Installation 497
M1308 Mechanical Systems Installation 500

CHAPTER 14 HEATING AND COOLING EQUIPMENT AND APPLIANCES 501

Section

M1401 General 501
M1402 Central Furnaces 501
M1403 Heat Pump Equipment 501
M1404 Refrigeration Cooling Equipment 501
M1405 Baseboard Convectors 501
M1406 Radiant Heating Systems 501
M1407 Duct Heaters 502
M1408 Vented Floor Furnaces 502
M1409 Vented Wall Furnaces 502
M1410 Vented Room Heaters 503
M1411 Heating and Cooling Equipment 503
M1412 Absorption Cooling Equipment 504
M1413 Evaporative Cooling Equipment 504
M1414 Fireplace Stoves 504
M1415 Masonry Heaters 505

CHAPTER 15 EXHAUST SYSTEMS 507

Section

M1501 General 507
M1502 Clothes Dryer Exhaust 507
M1503 Range Hoods 508
M1504 Installation of Microwave Ovens 508
M1505 Overhead Exhaust Hoods 508
M1506 Exhaust Ducts and Exhaust Openings 508
M1507 Mechanical Ventilation 508

CHAPTER 16 DUCT SYSTEMS 511

Section

M1601 Duct Construction 511
M1602 Return Air 513

CHAPTER 17 COMBUSTION AIR 515

Section

M1701 General 515

CHAPTER 18 CHIMNEYS AND VENTS 517

Section

M1801 General 517
M1802 Vent Components 517
M1803 Chimney and Vent Connectors 518
M1804 Vents 519
M1805 Masonry and Factory-built Chimneys 519

CHAPTER 19 SPECIAL APPLIANCES, EQUIPMENT AND SYSTEMS 521

Section

M1901 Ranges and Ovens 521
M1902 Sauna Heaters 521
M1903 Stationary Fuel Cell Power Plants 521
M1904 Gaseous Hydrogen Systems 521

CHAPTER 20 BOILERS AND WATER HEATERS 523

Section

M2001 Boilers 523
M2002 Operating and Safety Controls 523
M2003 Expansion Tanks 523
M2004 Water Heaters Used for Space Heating 524
M2005 Water Heaters 524
M2006 Pool Heaters 524

CHAPTER 21 HYDRONIC PIPING 525

Section

M2101 Hydronic Piping Systems Installation 525
M2102 Baseboard Convectors 525
M2103 Floor Heating Systems 525
M2104 Low Temperature Piping 527
M2105 Ground Source Heat Pump System Loop Piping 527

CHAPTER 22 SPECIAL PIPING AND STORAGE SYSTEMS 529

Section

M2201 Oil Tanks 529
M2202 Oil Piping, Fitting and Connections 529
M2203 Installation 529
M2204 Oil Pumps and Valves 530

CHAPTER 23 SOLAR ENERGY SYSTEMS 531

Section

M2301 Thermal Solar Energy Systems 531
M2302 Photovoltaic Solar Energy Systems 531

Part VI—Fuel Gas *533*

CHAPTER 24 FUEL GAS 533

Section

G2401 General 533
G2402 General 533
G2403 General Definitions 533
G2404 General 539
G2405 Structural Safety 540
G2406 Appliance Location 540
G2407 Combustion, Ventilation and Dilution Air 540
G2408 Installation 544
G2409 Clearance Reduction 545
G2410 Electrical 548
G2411 Electrical Bonding 548
G2412 General 548
G2413 Pipe Sizing 549
G2414 Piping Materials 550
G2415 Piping System Installation 573
G2416 Piping Bends and Changes in Direction 575
G2417 Inspection, Testing and Purging 575
G2418 Piping Support 578
G2419 Drips and Sloped Piping 578
G2420 Gas Shutoff Valves 578
G2421 Flow Controls 579
G2422 Appliance Connections 580
G2423 CNG Gas-dispensing Systems 581
G2424 Piping Support Intervals 581
G2425 General 581
G2426 Vents 582
G2427 Venting of Appliances 583
G2428 Sizing of Category I Appliance Venting Systems 592
G2429 Direct-vent, Integral Vent, Mechanical Vent and Ventilation/Exhaust Hood Venting 596
G2430 Factory-built Chimneys 596
G2431 General 596
G2432 Decorative Appliances for Installation in Fireplaces 596
G2433 Log Lighters 596
G2434 Vented Gas Fireplaces (Decorative Appliances) 605
G2435 Vented Gas Fireplace Heaters 605
G2436 Vented Wall Furnaces 605
G2437 Floor Furnaces 605
G2438 Clothes Dryers 606
G2439 Clothes Dryer Exhaust 606
G2440 Sauna Heaters 607
G2441 Pool and Spa Heaters 607
G2442 Forced-air Warm-air Furnaces 607
G2443 Conversion Burners 608
G2444 Unit Heaters 608
G2445 Unvented Room Heaters 608
G2446 Vented Room Heaters 609
G2447 Cooking Appliances 609
G2448 Water Heaters 609
G2449 Air Conditioning Appliances 609
G2450 Illuminating Appliances 610
G2451 Infrared Radiant Heaters 610
G2452 Boilers 610
G2453 Chimney Damper Opening Area 610
G2454 Outdoor Decorative Appliances 610

Part VII—Plumbing *613*

CHAPTER 25 PLUMBING ADMINISTRATION 613

Section

P2501 General 613
P2502 Existing Plumbing Systems 613
P2503 Inspection and Tests 613

CHAPTER 26 GENERAL PLUMBING REQUIREMENTS 615

Section

P2601 General 615
P2602 Individual Water Supply and Sewage Disposal 615
P2603 Structural and Piping Protection 615
P2604 Trenching and Backfilling 616
P2605 Support 616
P2606 Penetrations 616
P2607 Waterproofing of Openings 616
P2608 Workmanship 616
P2609 Materials Evaluation and Listing 617

CHAPTER 27 PLUMBING FIXTURES 619

Section

P2701 Fixtures, Faucets and Fixture Fittings 619
P2702 Fixture Accessories 619
P2703 Tail Pieces 619

P2704 Access to Connections 619
P2705 Installation 619
P2706 Waste Receptors 619
P2707 Directional Fittings 621
P2708 Showers 621
P2709 Shower Receptors 621
P2710 Shower Walls 622
P2711 Lavatories 622
P2712 Water Closets 622
P2713 Bathtubs 623
P2714 Sinks 623
P2715 Laundry Tubs 623
P2716 Food Waste Grinder 623
P2717 Dishwashing Machines 623
P2718 Clothes Washing Machine 623
P2719 Floor Drains 623
P2720 Whirlpool Bathtubs 623
P2721 Bidet Installations 624
P2722 Fixture Fitting 624
P2723 Macerating Toilet Systems 624
P2724 Speciality Temperature Control Devices and Valves 624

CHAPTER 28 WATER HEATERS 625

Section

P2801 General 625
P2802 Water Heaters Used for Space Heating 625
P2803 Relief Valves 625

CHAPTER 29 WATER SUPPLY AND DISTRIBUTION 627

Section

P2901 General 627
P2902 Protection of Potable Water Supply 627
P2903 Water-supply System 629
P2904 Dwelling Unit Fire Sprinkler Systems 632
P2905 Materials, Joints and Connections 645
P2906 Changes in Direction 649
P2907 Support 649
P2908 Drinking Water Treatment Units 649

CHAPTER 30 SANITARY DRAINAGE 651

Section

P3001 General 651
P3002 Materials 651
P3003 Joints and Connections 653
P3004 Determining Drainage Fixture Units 656
P3005 Drainage System 656
P3006 Sizing of Drain Pipe Offsets 659
P3007 Sumps and Ejectors 659
P3008 Backwater Valves 660
P3009 Gray Water Recycling Systems 660

CHAPTER 31 VENTS 665

Section

P3101 Vent Systems 665
P3102 Vent Stacks and Stack Vents 665
P3103 Vent Terminals 665
P3104 Vent Connections and Grades 665
P3105 Fixture Vents 666
P3106 Individual Vent 666
P3107 Common Vent 666
P3108 Wet Venting 666
P3109 Waste Stack Vent 667
P3110 Circuit Venting 667
P3111 Combination Waste and Vent System 667
P3112 Island Fixture Venting 668
P3113 Vent Pipe Sizing 668
P3114 Air Admittance Valves 668

CHAPTER 32 TRAPS 671

Section

P3201 Fixture Traps 671

CHAPTER 33 STORM DRAINAGE 673

Section

P3301 General 673
P3302 Subsoil Drains 673
P3303 Sumps and Pumping Systems 673

Part VIII—Electrical 675

CHAPTER 34 GENERAL REQUIREMENTS 675

Section

E3401 General 675
E3402 Building Structure Protection 676
E3403 Inspection and Approval 676
E3404 General Equipment Requirements 676

E3405 Equipment Location and Clearances 677
E3406 Electrical Conductors and Connections 678
E3407 Conductor and Terminal Identification 680

CHAPTER 35 ELECTRICAL DEFINITIONS 683
Section
E3501 General 683

CHAPTER 36 SERVICES 687
Section
E3601 General Services 687
E3602 Service Size and Rating 687
E3603 Service, Feeder and Grounding Electrode Conductor Sizing 688
E3604 Overhead Service and Service-entrance Conductor Installation 689
E3605 Service-entrance Conductors 691
E3606 Service Equipment—General 691
E3607 System Grounding 692
E3608 Grounding Electrode System 692
E3609 Bonding 694
E3610 Grounding Electrode Conductors 695
E3611 Grounding Electrode Conductor Connection to the Grounding Electrodes 695

CHAPTER 37 BRANCH CIRCUIT AND FEEDER REQUIREMENTS 697
Section
E3701 General 697
E3702 Branch Circuit Ratings 697
E3703 Required Branch Circuits 698
E3704 Feeder Requirements 699
E3705 Conductor Sizing and Overcurrent Protection 700
E3706 Panelboards 703

CHAPTER 38 WIRING METHODS 705
Section
E3801 General Requirements 705
E3802 Above-ground Installation Requirements 705
E3803 Underground Installation Requirements 706

CHAPTER 39 POWER AND LIGHTING DISTRIBUTION 711
Section
E3901 Receptacle Outlets 711
E3902 Ground-fault and Arc-fault Circuit-interrupter Protection 713
E3903 Lighting Outlets 714
E3904 General Installation Requirements 715
E3905 Boxes, Conduit Bodies and Fittings 715
E3906 Installation of Boxes, Conduit Bodies and Fittings 729
E3907 Cabinets and Panelboards 731
E3908 Grounding 732
E3909 Flexible Cords 735

CHAPTER 40 DEVICES AND LUMINAIRES 737
Section
E4001 Switches 737
E4002 Receptacles 738
E4003 Fixtures 739
E4004 Luminaire Installation 741
E4005 Track Lighting 741

CHAPTER 41 APPLIANCE INSTALLATION 743
Section
E4101 General 743

CHAPTER 42 SWIMMING POOLS 747
Section
E4201 General 747
E4202 Wiring Methods for Pools, Spas, Hot Tubs and Hydromassage Bathtubs 747
E4203 Equipment Location and Clearances 749
E4204 Bonding 750
E4205 Grounding 753
E4206 Equipment Installation 754
E4207 Storable Swimming Pools 756
E4208 Spas and Hot Tubs 757
E4209 Hydromassage Bathtubs 757

CHAPTER 43 CLASS 2 REMOTE-CONTROL, SIGNALING AND POWER-LIMITED CIRCUITS 759
Section
E4301 General 759
E4302 Power Sources 759
E4303 Wiring Methods 759
E4304 Installation Requirements 759

Part IX—Referenced Standards *761*

CHAPTER 44 REFERENCED STANDARDS 761

APPENDIX A SIZING AND CAPACITIES OF GAS PIPING 785

APPENDIX B SIZING OF VENTING SYSTEMS SERVING APPLIANCES EQUIPPED WITH DRAFT HOODS, CATEGORY I APPLIANCES, AND APPLIANCES LISTED FOR USE WITH TYPE B VENTS 797

APPENDIX C EXIT TERMINALS OF MECHANICAL DRAFT AND DIRECT-VENT VENTING SYSTEMS 807

APPENDIX D RECOMMENDED PROCEDURE FOR SAFETY INSPECTION OF AN EXISTING APPLIANCE INSTALLATION 809

APPENDIX E MANUFACTURED HOUSING USED AS DWELLINGS. 811

Section

AE101 Scope 811

AE102 Application to Existing Manufactured Homes and Building Service Equipment 811

AE201 Definitions 812

AE301 Permits 812

AE302 Application for Permit 812

AE303 Permits Issuance 813

AE304 Fees 814

AE305 Inspections 814

AE306 Special Inspections 815

AE307 Utility Service 815

AE401 Occupancy Classification 815

AE402 Location on Property 815

AE501 Design 815

AE502 Foundation Systems 816

AE503 Skirting and Perimeter Enclosures 816

AE504 Structural Additions 816

AE505 Building Service Equipment. 817

AE506 Exits 817

AE507 Occupancy, Fire Safety and Energy Conservation Standards 817

AE600 Special Requirements for Foundation Systems 817

AE601 Footings and Foundations 817

AE602 Pier Construction 817

AE603 Height of Piers 817

AE604 Anchorage Installations 818

AE605 Ties, Materials and Installation 818

AE606 Referenced Standards. 819

APPENDIX F RADON CONTROL METHODS . . . 821

Section

AF101 Scope 821

AF102 Definitions 821

AF103 Requirements 821

APPENDIX G SWIMMING POOLS, SPAS AND HOT TUBS 829

Section

AG101 General. 829

AG102 Definitions 829

AG103 Swimming Pools 829

AG104 Spas and Hot Tubs 829

AG105 Barrier Requirements 829

AG106 Entrapment Protection for Swimming Pool and Spa Suction Outlets. 830

AG107 Abbreviations. 831

AG108 Referenced Standards. 831

APPENDIX H PATIO COVERS 833

Section

AH101 General. 833

AH102 Definition 833

AH103 Exterior Walls and Openings 833

AH104 Height. 833

AH105 Structural Provisions 833

AH106 Special Provisions for Aluminum Screen Enclosures in Hurricane-prone Regions 833

APPENDIX I PRIVATE SEWAGE DISPOSAL . . . 835

Section

AI101 General. 835

APPENDIX J EXISTING BUILDINGS AND STRUCTURES 837

Section

AJ101 Purpose and Intent 837

AJ102 Compliance 837
AJ103 Preliminary Meeting 837
AJ104 Evaluation of an Existing Building 838
AJ105 Permit 838
AJ201 Definitions 838
AJ301 Repairs 838
AJ401 Renovations 839
AJ501 Alterations 839
AJ601 Reconstruction 840

APPENDIX K SOUND TRANSMISSION 841

Section
AK101 General 841
AK102 Air-borne Sound 841
AK103 Structural-borne Sound 841
AK104 Referenced Standards 841

APPENDIX L PERMIT FEES 843

APPENDIX M HOME DAY CARE—R-3 OCCUPANCY 845

Section
AM101 General 845
AM102 Definition 845
AM103 Means of Egress 845
AM104 Smoke Detection 846

APPENDIX N VENTING METHODS 847

APPENDIX O AUTOMATIC VEHICULAR GATES 853

Section
AO101 General 853
AO102 Definitions 853
AO103 Automatic Vehicular Gates 853

APPENDIX P SIZING OF WATER PIPING SYSTEM 855

Section
AP101 General 855
AP102 Information Required 855
AP103 Selection of Pipe Size 855
AP201 Selection of Pipe Size 872

APPENDIX Q ICC INTERNATIONAL RESIDENTIAL CODE ELECTRICAL PROVISIONS/NATIONAL ELECTRICAL CODE CROSS REFERENCE 875

INDEX 891

Part I—Administrative

CHAPTER 1

SCOPE AND ADMINISTRATION

PART 1—SCOPE AND APPLICATION

SECTION R101 GENERAL

R101.1 Title. These provisions shall be known as the *Residential Code for One- and Two-family Dwellings* of **[NAME OF JURISDICTION]**, and shall be cited as such and will be referred to herein as "this code."

R101.2 Scope. The provisions of the *International Residential Code for One- and Two-family Dwellings* shall apply to the construction, *alteration*, movement, enlargement, replacement, repair, equipment, use and occupancy, location, removal and demolition of detached one- and two-family dwellings and townhouses not more than three stories above *grade plane* in height with a separate means of egress and their *accessory structures*.

Exceptions:

1. Live/work units complying with the requirements of Section 419 of the *International Building Code* shall be permitted to be built as one- and two-family *dwellings* or townhouses. Fire suppression required by Section 419.5 of the *International Building Code* when constructed under the *International Residential Code for One- and Two-family Dwellings* shall conform to Section P2904.
2. Owner-occupied lodging houses with five or fewer guestrooms shall be permitted to be constructed in accordance with the *International Residential Code for One- and Two-family Dwellings* when equipped with a fire sprinkler system in accordance with Section P2904.

R101.3 Intent. The purpose of this code is to establish minimum requirements to safeguard the public safety, health and general welfare through affordability, structural strength, means of egress facilities, stability, sanitation, light and ventilation, energy conservation and safety to life and property from fire and other hazards attributed to the built environment and to provide safety to fire fighters and emergency responders during emergency operations.

SECTION R102 APPLICABILITY

R102.1 General. Where there is a conflict between a general requirement and a specific requirement, the specific requirement shall be applicable. Where, in any specific case, different sections of this code specify different materials, methods of construction or other requirements, the most restrictive shall govern.

R102.2 Other laws. The provisions of this code shall not be deemed to nullify any provisions of local, state or federal law.

R102.3 Application of references. References to chapter or section numbers, or to provisions not specifically identified by number, shall be construed to refer to such chapter, section or provision of this code.

R102.4 Referenced codes and standards. The codes and standards referenced in this code shall be considered part of the requirements of this code to the prescribed extent of each such reference and as further regulated in Sections R102.4.1 and R102.4.2.

Exception: Where enforcement of a code provision would violate the conditions of the *listing* of the *equipment* or *appliance*, the conditions of the *listing* and manufacturer's instructions shall apply.

R102.4.1 Differences. Where differences occur between provisions of this code and referenced codes and standards, the provisions of this code shall apply.

R102.4.2 Provisions in referenced codes and standards. Where the extent of the reference to a referenced code or standard includes subject matter that is within the scope of this code, the provisions of this code, as applicable, shall take precedence over the provisions in the referenced code or standard.

R102.5 Appendices. Provisions in the appendices shall not apply unless specifically referenced in the adopting ordinance.

R102.6 Partial invalidity. In the event any part or provision of this code is held to be illegal or void, this shall not have the effect of making void or illegal any of the other parts or provisions.

R102.7 Existing structures. The legal occupancy of any structure existing on the date of adoption of this code shall be permitted to continue without change, except as is specifically covered in this code, the *International Property Maintenance Code* or the *International Fire Code*, or as is deemed necessary by the *building official* for the general safety and welfare of the occupants and the public.

R102.7.1 Additions, alterations or repairs. *Additions, alterations* or repairs to any structure shall conform to the requirements for a new structure without requiring the existing structure to comply with all of the requirements of this code, unless otherwise stated. *Additions, alterations* or repairs shall not cause an existing structure to become unsafe or adversely affect the performance of the building.

PART 2—ADMINISTRATION AND ENFORCEMENT

SECTION R103
DEPARTMENT OF BUILDING SAFETY

R103.1 Creation of enforcement agency. The department of building safety is hereby created and the official in charge thereof shall be known as the *building official*.

R103.2 Appointment. The *building official* shall be appointed by the chief appointing authority of the *jurisdiction*.

R103.3 Deputies. In accordance with the prescribed procedures of this *jurisdiction* and with the concurrence of the appointing authority, the *building official* shall have the authority to appoint a deputy *building official*, the related technical officers, inspectors, plan examiners and other employees. Such employees shall have powers as delegated by the *building official*.

SECTION R104
DUTIES AND POWERS OF THE BUILDING OFFICIAL

R104.1 General. The *building official* is hereby authorized and directed to enforce the provisions of this code. The *building official* shall have the authority to render interpretations of this code and to adopt policies and procedures in order to clarify the application of its provisions. Such interpretations, policies and procedures shall be in conformance with the intent and purpose of this code. Such policies and procedures shall not have the effect of waiving requirements specifically provided for in this code.

R104.2 Applications and permits. The *building official* shall receive applications, review *construction documents* and issue permits for the erection and alteration of buildings and structures, inspect the premises for which such permits have been issued and enforce compliance with the provisions of this code.

R104.3 Notices and orders. The *building official* shall issue all necessary notices or orders to ensure compliance with this code.

R104.4 Inspections. The *building official* is authorized to make all of the required inspections, or the *building official* shall have the authority to accept reports of inspection by *approved agencies* or individuals. Reports of such inspections shall be in writing and be certified by a responsible officer of such *approved* agency or by the responsible individual. The *building official* is authorized to engage such expert opinion as deemed necessary to report upon unusual technical issues that arise, subject to the approval of the appointing authority.

R104.5 Identification. The *building official* shall carry proper identification when inspecting structures or premises in the performance of duties under this code.

R104.6 Right of entry. Where it is necessary to make an inspection to enforce the provisions of this code, or where the *building official* has reasonable cause to believe that there exists in a structure or upon a premises a condition which is contrary to or in violation of this code which makes the structure or premises unsafe, dangerous or hazardous, the *building official* or designee is authorized to enter the structure or premises at reasonable times to inspect or to perform the duties imposed by this code, provided that if such structure or premises be occupied that credentials be presented to the occupant and entry requested. If such structure or premises be unoccupied, the *building official* shall first make a reasonable effort to locate the owner or other person having charge or control of the structure or premises and request entry. If entry is refused, the *building official* shall have recourse to the remedies provided by law to secure entry.

R104.7 Department records. The *building official* shall keep official records of applications received, permits and certificates issued, fees collected, reports of inspections, and notices and orders issued. Such records shall be retained in the official records for the period required for the retention of public records.

R104.8 Liability. The *building official*, member of the board of appeals or employee charged with the enforcement of this code, while acting for the *jurisdiction* in good faith and without malice in the discharge of the duties required by this code or other pertinent law or ordinance, shall not thereby be rendered liable personally and is hereby relieved from personal liability for any damage accruing to persons or property as a result of any act or by reason of an act or omission in the discharge of official duties. Any suit instituted against an officer or employee because of an act performed by that officer or employee in the lawful discharge of duties and under the provisions of this code shall be defended by legal representative of the *jurisdiction* until the final termination of the proceedings. The *building official* or any subordinate shall not be liable for cost in any action, suit or proceeding that is instituted in pursuance of the provisions of this code.

R104.9 Approved materials and equipment. Materials, *equipment* and devices *approved* by the *building official* shall be constructed and installed in accordance with such approval.

R104.9.1 Used materials and equipment. Used materials, *equipment* and devices shall not be reused unless *approved* by the *building official.*

R104.10 Modifications. Wherever there are practical difficulties involved in carrying out the provisions of this code, the *building official* shall have the authority to grant modifications for individual cases, provided the *building official* shall first find that special individual reason makes the strict letter of this code impractical and the modification is in compliance with the intent and purpose of this code and that such modification does not lessen health, life and fire safety or structural requirements. The details of action granting modifications shall be recorded and entered in the files of the department of building safety.

R104.10.1 Flood hazard areas. The *building official* shall not grant modifications to any provision related to flood hazard areas as established by Table R301.2(1) without the granting of a variance to such provisions by the board of appeals.

R104.11 Alternative materials, design and methods of construction and equipment. The provisions of this code are not intended to prevent the installation of any material or to prohibit any design or method of construction not specifically prescribed by this code, provided that any such alternative has been *approved*. An alternative material, design or method of construction shall be *approved* where the *building official* finds that the proposed design is satisfactory and complies with the intent of the provisions of this code, and that the material, method or work offered is, for the purpose intended, at least the equivalent of that prescribed in this code. Compliance with the specific performance-based provisions of the International Codes in lieu of specific requirements of this code shall also be permitted as an alternate.

R104.11.1 Tests. Whenever there is insufficient evidence of compliance with the provisions of this code, or evidence that a material or method does not conform to the requirements of this code, or in order to substantiate claims for alternative materials or methods, the *building official* shall have the authority to require tests as evidence of compliance to be made at no expense to the *jurisdiction*. Test methods shall be as specified in this code or by other recognized test standards. In the absence of recognized and accepted test methods, the *building official* shall approve the testing procedures. Tests shall be performed by an *approved* agency. Reports of such tests shall be retained by the *building official* for the period required for retention of public records.

SECTION R105 PERMITS

R105.1 Required. Any owner or authorized agent who intends to construct, enlarge, alter, repair, move, demolish or change the occupancy of a building or structure, or to erect, install, enlarge, alter, repair, remove, convert or replace any electrical, gas, mechanical or plumbing system, the installation of which is regulated by this code, or to cause any such work to be done, shall first make application to the *building official* and obtain the required *permit*.

R105.2 Work exempt from permit. *Permits* shall not be required for the following. Exemption from *permit* requirements of this code shall not be deemed to grant authorization for any work to be done in any manner in violation of the provisions of this code or any other laws or ordinances of this *jurisdiction*.

Building:

1. One-story detached *accessory structures* used as tool and storage sheds, playhouses and similar uses, provided the floor area does not exceed 200 square feet (18.58 m^2).
2. Fences not over 7 feet (2134 mm) high.
3. Retaining walls that are not over 4 feet (1219 mm) in height measured from the bottom of the footing to the top of the wall, unless supporting a surcharge.
4. Water tanks supported directly upon *grade* if the capacity does not exceed 5,000 gallons (18 927 L) and the ratio of height to diameter or width does not exceed 2 to 1.
5. Sidewalks and driveways.
6. Painting, papering, tiling, carpeting, cabinets, counter tops and similar finish work.
7. Prefabricated swimming pools that are less than 24 inches (610 mm) deep.
8. Swings and other playground equipment.
9. Window awnings supported by an exterior wall which do not project more than 54 inches (1372 mm) from the exterior wall and do not require additional support.
10. Decks not exceeding 200 square feet (18.58 m^2) in area, that are not more than 30 inches (762 mm) above *grade* at any point, are not attached to a *dwelling* and do not serve the exit door required by Section R311.4.

Electrical:

1. *Listed* cord-and-plug connected temporary decorative lighting.
2. Reinstallation of attachment plug receptacles but not the outlets therefor.
3. Replacement of branch circuit overcurrent devices of the required capacity in the same location.
4. Electrical wiring, devices, *appliances,* apparatus or *equipment* operating at less than 25 volts and not capable of supplying more than 50 watts of energy.
5. Minor repair work, including the replacement of lamps or the connection of *approved* portable electrical *equipment* to *approved* permanently installed receptacles.

Gas:

1. Portable heating, cooking or clothes drying *appliances*.
2. Replacement of any minor part that does not alter approval of *equipment* or make such *equipment* unsafe.
3. Portable-fuel-cell *appliances* that are not connected to a fixed piping system and are not interconnected to a power grid.

Mechanical:

1. Portable heating *appliances*.
2. Portable ventilation *appliances*.
3. Portable cooling units.
4. Steam, hot- or chilled-water piping within any heating or cooling *equipment* regulated by this code.
5. Replacement of any minor part that does not alter approval of *equipment* or make such *equipment* unsafe.
6. Portable evaporative coolers.

7. Self-contained refrigeration systems containing 10 pounds (4.54 kg) or less of refrigerant or that are actuated by motors of 1 horsepower (746 W) or less.
8. Portable-fuel-cell *appliances* that are not connected to a fixed piping system and are not interconnected to a power grid.

The stopping of leaks in drains, water, soil, waste or vent pipe; provided, however, that if any concealed trap, drainpipe, water, soil, waste or vent pipe becomes defective and it becomes necessary to remove and replace the same with new material, such work shall be considered as new work and a *permit* shall be obtained and inspection made as provided in this code.

The clearing of stoppages or the repairing of leaks in pipes, valves or fixtures, and the removal and reinstallation of water closets, provided such repairs do not involve or require the replacement or rearrangement of valves, pipes or fixtures.

R105.2.1 Emergency repairs. Where *equipment* replacements and repairs must be performed in an emergency situation, the *permit* application shall be submitted within the next working business day to the *building official*.

R105.2.2 Repairs. Application or notice to the *building official* is not required for ordinary repairs to structures, replacement of lamps or the connection of *approved* portable electrical *equipment* to *approved* permanently installed receptacles. Such repairs shall not include the cutting away of any wall, partition or portion thereof, the removal or cutting of any structural beam or load-bearing support, or the removal or change of any required means of egress, or rearrangement of parts of a structure affecting the egress requirements; nor shall ordinary repairs include *addition* to, *alteration* of, replacement or relocation of any water supply, sewer, drainage, drain leader, gas, soil, waste, vent or similar piping, electric wiring or mechanical or other work affecting public health or general safety.

R105.2.3 Public service agencies. A *permit* shall not be required for the installation, alteration or repair of generation, transmission, distribution, metering or other related *equipment* that is under the ownership and control of public service agencies by established right.

R105.3 Application for permit. To obtain a *permit*, the applicant shall first file an application therefor in writing on a form furnished by the department of building safety for that purpose. Such application shall:

1. Identify and describe the work to be covered by the *permit* for which application is made.
2. Describe the land on which the proposed work is to be done by legal description, street address or similar description that will readily identify and definitely locate the proposed building or work.
3. Indicate the use and occupancy for which the proposed work is intended.
4. Be accompanied by *construction documents* and other information as required in Section R106.1.
5. State the valuation of the proposed work.
6. Be signed by the applicant or the applicant's authorized agent.
7. Give such other data and information as required by the *building official*.

R105.3.1 Action on application. The *building official* shall examine or cause to be examined applications for permits and amendments thereto within a reasonable time after filing. If the application or the *construction documents* do not conform to the requirements of pertinent laws, the *building official* shall reject such application in writing stating the reasons therefor. If the *building official* is satisfied that the proposed work conforms to the requirements of this code and laws and ordinances applicable thereto, the *building official* shall issue a *permit* therefor as soon as practicable.

R105.3.1.1 Determination of substantially improved or substantially damaged existing buildings in flood hazard areas. For applications for reconstruction, rehabilitation, *addition* or other improvement of existing buildings or structures located in a flood hazard area as established by Table R301.2(1), the *building official* shall examine or cause to be examined the *construction documents* and shall prepare a finding with regard to the value of the proposed work. For buildings that have sustained damage of any origin, the value of the proposed work shall include the cost to repair the building or structure to its predamaged condition. If the *building official* finds that the value of proposed work equals or exceeds 50 percent of the market value of the building or structure before the damage has occurred or the improvement is started, the finding shall be provided to the board of appeals for a determination of substantial improvement or substantial damage. Applications determined by the board of appeals to constitute substantial improvement or substantial damage shall require all existing portions of the entire building or structure to meet the requirements of Section R322.

R105.3.2 Time limitation of application. An application for a *permit* for any proposed work shall be deemed to have been abandoned 180 days after the date of filing unless such application has been pursued in good faith or a *permit* has been issued; except that the *building official* is authorized to grant one or more extensions of time for additional periods not exceeding 180 days each. The extension shall be requested in writing and justifiable cause demonstrated.

R105.4 Validity of permit. The issuance or granting of a *permit* shall not be construed to be a *permit* for, or an *approval* of, any violation of any of the provisions of this code or of any other ordinance of the *jurisdiction*. Permits presuming to give authority to violate or cancel the provisions of this code or other ordinances of the *jurisdiction* shall not be valid. The issuance of a *permit* based on *construction documents* and other data shall not prevent the *building official* from requiring the correction of errors in the *construction documents* and other data. The *building official* is also authorized to prevent occupancy or use of a structure where in violation of this code or of any other ordinances of this *jurisdiction*.

R105.5 Expiration. Every *permit* issued shall become invalid unless the work authorized by such *permit* is commenced within 180 days after its issuance, or if the work authorized by such *permit* is suspended or abandoned for a period of 180 days after the time the work is commenced. The *building official* is authorized to grant, in writing, one or more extensions of time, for periods not more than 180 days each. The extension shall be requested in writing and justifiable cause demonstrated.

R105.6 Suspension or revocation. The *building official* is authorized to suspend or revoke a *permit* issued under the provisions of this code wherever the *permit* is issued in error or on the basis of incorrect, inaccurate or incomplete information, or in violation of any ordinance or regulation or any of the provisions of this code.

R105.7 Placement of permit. The building *permit* or copy thereof shall be kept on the site of the work until the completion of the project.

R105.8 Responsibility. It shall be the duty of every person who performs work for the installation or repair of building, structure, electrical, gas, mechanical or plumbing systems, for which this code is applicable, to comply with this code.

R105.9 Preliminary inspection. Before issuing a *permit,* the *building official* is authorized to examine or cause to be examined buildings, structures and sites for which an application has been filed.

SECTION R106
CONSTRUCTION DOCUMENTS

R106.1 Submittal documents. Submittal documents consisting of *construction documents*, and other data shall be submitted in two or more sets with each application for a *permit.* The *construction documents* shall be prepared by a registered *design professional* where required by the statutes of the *jurisdiction* in which the project is to be constructed. Where special conditions exist, the *building official* is authorized to require additional *construction documents* to be prepared by a registered *design professional.*

Exception: The *building official* is authorized to waive the submission of *construction documents* and other data not required to be prepared by a registered *design professional* if it is found that the nature of the work applied for is such that reviewing of *construction documents* is not necessary to obtain compliance with this code.

R106.1.1 Information on construction documents. *Construction documents* shall be drawn upon suitable material. Electronic media documents are permitted to be submitted when *approved* by the *building official. Construction documents* shall be of sufficient clarity to indicate the location, nature and extent of the work proposed and show in detail that it will conform to the provisions of this code and relevant laws, ordinances, rules and regulations, as determined by the *building official.* Where required by the *building official*, all braced wall lines, shall be identified on the *construction documents* and all pertinent information including, but not limited to, bracing methods, location and length of braced wall panels, foundation requirements of braced wall panels at top and bottom shall be provided.

R106.1.2 Manufacturer's installation instructions. Manufacturer's installation instructions, as required by this code, shall be available on the job site at the time of inspection.

R106.1.3 Information for construction in flood hazard areas. For buildings and structures located in whole or in part in flood hazard areas as established by Table R301.2(1), *construction documents* shall include:

1. Delineation of flood hazard areas, floodway boundaries and flood zones and the design flood elevation, as appropriate;
2. The elevation of the proposed lowest floor, including *basement*; in areas of shallow flooding (AO Zones), the height of the proposed lowest floor, including *basement*, above the highest adjacent *grade*;
3. The elevation of the bottom of the lowest horizontal structural member in coastal high hazard areas (V Zone); and
4. If design flood elevations are not included on the community's Flood Insurance Rate Map (FIRM), the *building official* and the applicant shall obtain and reasonably utilize any design flood elevation and floodway data available from other sources.

R106.2 Site plan or plot plan. The *construction documents* submitted with the application for *permit* shall be accompanied by a site plan showing the size and location of new construction and existing structures on the site and distances from *lot lines*. In the case of demolition, the site plan shall show construction to be demolished and the location and size of existing structures and construction that are to remain on the site or plot. The *building official* is authorized to waive or modify the requirement for a site plan when the application for permit is for alteration or repair or when otherwise warranted.

R106.3 Examination of documents. The *building official* shall examine or cause to be examined *construction documents* for code compliance.

R106.3.1 Approval of construction documents. When the *building official* issues a *permit*, the *construction documents* shall be *approved* in writing or by a stamp which states "REVIEWED FOR CODE COMPLIANCE." One set of *construction documents* so reviewed shall be retained by the *building official*. The other set shall be returned to the applicant, shall be kept at the site of work and shall be open to inspection by the *building official* or his or her authorized representative.

R106.3.2 Previous approvals. This code shall not require changes in the *construction documents*, construction or designated occupancy of a structure for which a lawful *permit* has been heretofore issued or otherwise lawfully authorized, and the construction of which has been pursued in good faith within 180 days after the effective date of this code and has not been abandoned.

R106.3.3 Phased approval. The *building official* is authorized to issue a *permit* for the construction of foundations or any other part of a building or structure before the *construction documents* for the whole building or structure have been submitted, provided that adequate information and detailed statements have been filed complying with pertinent requirements of this code. The holder of such *permit* for the foundation or other parts of a building or structure shall proceed at the holder's own risk with the building operation and without assurance that a *permit* for the entire structure will be granted.

R106.4 Amended construction documents. Work shall be installed in accordance with the *approved construction documents*, and any changes made during construction that are not in compliance with the *approved construction documents* shall be resubmitted for approval as an amended set of *construction documents*.

R106.5 Retention of construction documents. One set of *approved construction documents* shall be retained by the *building official* for a period of not less than 180 days from date of completion of the permitted work, or as required by state or local laws.

SECTION R107 TEMPORARY STRUCTURES AND USES

R107.1 General. The *building official* is authorized to issue a *permit* for temporary structures and temporary uses. Such permits shall be limited as to time of service, but shall not be permitted for more than 180 days. The *building official* is authorized to grant extensions for demonstrated cause.

R107.2 Conformance. Temporary structures and uses shall conform to the structural strength, fire safety, means of egress, light, ventilation and sanitary requirements of this code as necessary to ensure the public health, safety and general welfare.

R107.3 Temporary power. The *building official* is authorized to give permission to temporarily supply and use power in part of an electric installation before such installation has been fully completed and the final certificate of completion has been issued. The part covered by the temporary certificate shall comply with the requirements specified for temporary lighting, heat or power in NFPA 70.

R107.4 Termination of approval. The *building official* is authorized to terminate such *permit* for a temporary structure or use and to order the temporary structure or use to be discontinued.

SECTION R108 FEES

R108.1 Payment of fees. A *permit* shall not be valid until the fees prescribed by law have been paid. Nor shall an amendment to a *permit* be released until the additional fee, if any, has been paid.

R108.2 Schedule of permit fees. On buildings, structures, electrical, gas, mechanical and plumbing systems or *alterations* requiring a *permit*, a fee for each *permit* shall be paid as required, in accordance with the schedule as established by the applicable governing authority.

R108.3 Building permit valuations. Building *permit* valuation shall include total value of the work for which a *permit* is being issued, such as electrical, gas, mechanical, plumbing equipment and other permanent systems, including materials and labor.

R108.4 Related fees. The payment of the fee for the construction, alteration, removal or demolition for work done in connection with or concurrently with the work authorized by a building *permit* shall not relieve the applicant or holder of the *permit* from the payment of other fees that are prescribed by law.

R108.5 Refunds. The *building official* is authorized to establish a refund policy.

R108.6 Work commencing before permit issuance. Any person who commences work requiring a *permit* on a building, structure, electrical, gas, mechanical or plumbing system before obtaining the necessary permits shall be subject to a fee established by the applicable governing authority that shall be in addition to the required *permit* fees.

SECTION R109 INSPECTIONS

R109.1 Types of inspections. For onsite construction, from time to time the *building official*, upon notification from the *permit* holder or his agent, shall make or cause to be made any necessary inspections and shall either approve that portion of the construction as completed or shall notify the *permit* holder or his or her agent wherein the same fails to comply with this code.

R109.1.1 Foundation inspection. Inspection of the foundation shall be made after poles or piers are set or trenches or *basement* areas are excavated and any required forms erected and any required reinforcing steel is in place and supported prior to the placing of concrete. The foundation inspection shall include excavations for thickened slabs intended for the support of bearing walls, partitions, structural supports, or *equipment* and special requirements for wood foundations.

R109.1.2 Plumbing, mechanical, gas and electrical systems inspection. Rough inspection of plumbing, mechanical, gas and electrical systems shall be made prior to covering or concealment, before fixtures or *appliances* are set or installed, and prior to framing inspection.

Exception: Backfilling of ground-source heat pump loop systems tested in accordance with Section M2105.1 prior to inspection shall be permitted.

R109.1.3 Floodplain inspections. For construction in flood hazard areas as established by Table R301.2(1), upon placement of the lowest floor, including *basement*, and prior to further vertical construction, the *building official* shall require submission of documentation, prepared and sealed by a registered *design professional*, of the elevation of the lowest floor, including *basement*, required in Section R322.

R109.1.4 Frame and masonry inspection. Inspection of framing and masonry construction shall be made after the roof, masonry, all framing, firestopping, draftstopping and bracing are in place and after the plumbing, mechanical and electrical rough inspections are *approved*.

R109.1.5 Other inspections. In addition to the called inspections above, the *building official* may make or require any other inspections to ascertain compliance with this code and other laws enforced by the *building official*.

R109.1.5.1 Fire-resistance-rated construction inspection. Where fire-resistance-rated construction is required between *dwelling units* or due to location on property, the *building official* shall require an inspection of such construction after all lathing and/or wallboard is in place, but before any plaster is applied, or before wallboard joints and fasteners are taped and finished.

R109.1.6 Final inspection. Final inspection shall be made after the permitted work is complete and prior to occupancy.

R109.1.6.1 Elevation documentation. If located in a flood hazard area, the documentation of elevations required in Section R322.1.10 shall be submitted to the *building official* prior to the final inspection.

R109.2 Inspection agencies. The *building official* is authorized to accept reports of *approved* agencies, provided such agencies satisfy the requirements as to qualifications and reliability.

R109.3 Inspection requests. It shall be the duty of the *permit* holder or their agent to notify the *building official* that such work is ready for inspection. It shall be the duty of the person requesting any inspections required by this code to provide access to and means for inspection of such work.

R109.4 Approval required. Work shall not be done beyond the point indicated in each successive inspection without first obtaining the approval of the *building official*. The *building official* upon notification, shall make the requested inspections and shall either indicate the portion of the construction that is satisfactory as completed, or shall notify the *permit* holder or an agent of the *permit* holder wherein the same fails to comply with this code. Any portions that do not comply shall be corrected and such portion shall not be covered or concealed until authorized by the *building official*.

SECTION R110 CERTIFICATE OF OCCUPANCY

R110.1 Use and occupancy. No building or structure shall be used or occupied, and no change in the existing occupancy classification of a building or structure or portion thereof shall be made until the *building official* has issued a certificate of occupancy therefor as provided herein. Issuance of a certificate of occupancy shall not be construed as an approval of a violation of the provisions of this code or of other ordinances of the *jurisdiction*. Certificates presuming to give authority to violate or cancel the provisions of this code or other ordinances of the *jurisdiction* shall not be valid.

Exceptions:

1. Certificates of occupancy are not required for work exempt from permits under Section R105.2.
2. Accessory buildings or structures.

R110.2 Change in use. Changes in the character or use of an existing structure shall not be made except as specified in Sections 3408 and 3409 of the *International Building Code*.

R110.3 Certificate issued. After the *building official* inspects the building or structure and finds no violations of the provisions of this code or other laws that are enforced by the department of building safety, the *building official* shall issue a certificate of occupancy which shall contain the following:

1. The building *permit* number.
2. The address of the structure.
3. The name and address of the owner.
4. A description of that portion of the structure for which the certificate is issued.
5. A statement that the described portion of the structure has been inspected for compliance with the requirements of this code.
6. The name of the *building official*.
7. The edition of the code under which the *permit* was issued.
8. If an automatic sprinkler system is provided and whether the sprinkler system is required.
9. Any special stipulations and conditions of the building *permit*.

R110.4 Temporary occupancy. The *building official* is authorized to issue a temporary certificate of occupancy before the completion of the entire work covered by the *permit*, provided that such portion or portions shall be occupied safely. The *building official* shall set a time period during which the temporary certificate of occupancy is valid.

R110.5 Revocation. The *building official* shall, in writing, suspend or revoke a certificate of occupancy issued under the provisions of this code wherever the certificate is issued in error, or on the basis of incorrect information supplied, or where it is determined that the building or structure or portion thereof is in violation of any ordinance or regulation or any of the provisions of this code.

SECTION R111 SERVICE UTILITIES

R111.1 Connection of service utilities. No person shall make connections from a utility, source of energy, fuel or power to any building or system that is regulated by this code for which a *permit* is required, until *approved* by the *building official*.

R111.2 Temporary connection. The *building official* shall have the authority to authorize and approve the temporary connection of the building or system to the utility, source of energy, fuel or power.

R111.3 Authority to disconnect service utilities. The *building official* shall have the authority to authorize disconnection of utility service to the building, structure or system regulated by this code and the referenced codes and standards set forth in Section R102.4 in case of emergency where necessary to eliminate an immediate hazard to life or property or when such utility connection has been made without the approval required by Section R111.1 or R111.2. The *building official* shall notify the serving utility and whenever possible the owner and occupant of the building, structure or service system of the decision to disconnect prior to taking such action if not notified prior to disconnection. The owner or occupant of the building, structure or service system shall be notified in writing as soon as practical thereafter.

SECTION R112 BOARD OF APPEALS

R112.1 General. In order to hear and decide appeals of orders, decisions or determinations made by the *building official* relative to the application and interpretation of this code, there shall be and is hereby created a board of appeals. The *building official* shall be an ex officio member of said board but shall have no vote on any matter before the board. The board of appeals shall be appointed by the governing body and shall hold office at its pleasure. The board shall adopt rules of procedure for conducting its business, and shall render all decisions and findings in writing to the appellant with a duplicate copy to the *building official.*

R112.2 Limitations on authority. An application for appeal shall be based on a claim that the true intent of this code or the rules legally adopted thereunder have been incorrectly interpreted, the provisions of this code do not fully apply, or an equally good or better form of construction is proposed. The board shall have no authority to waive requirements of this code.

R112.2.1 Determination of substantial improvement in flood hazard areas. When the *building official* provides a finding required in Section R105.3.1.1, the board of appeals shall determine whether the value of the proposed work constitutes a substantial improvement. A substantial improvement means any repair, reconstruction, rehabilitation, *addition* or improvement of a building or structure, the cost of which equals or exceeds 50 percent of the market value of the building or structure before the improvement or repair is started. If the building or structure has sustained substantial damage, all repairs are considered substantial improvement regardless of the actual repair work performed. The term does not include:

1. Improvements of a building or structure required to correct existing health, sanitary or safety code violations identified by the *building official* and which are the minimum necessary to assure safe living conditions; or
2. Any alteration of an historic building or structure, provided that the alteration will not preclude the continued designation as an historic building or structure. For the purpose of this exclusion, an historic building is:
    - 2.1. *Listed* or preliminarily determined to be eligible for *listing* in the National Register of Historic Places; or
    - 2.2. Determined by the Secretary of the U.S. Department of Interior as contributing to the historical significance of a registered historic district or a district preliminarily determined to qualify as an historic district; or
    - 2.3. Designated as historic under a state or local historic preservation program that is *approved* by the Department of Interior.

R112.2.2 Criteria for issuance of a variance for flood hazard areas. A variance shall be issued only upon:

1. A showing of good and sufficient cause that the unique characteristics of the size, configuration or topography of the site render the elevation standards in Section R322 inappropriate.
2. A determination that failure to grant the variance would result in exceptional hardship by rendering the *lot* undevelopable.
3. A determination that the granting of a variance will not result in increased flood heights, additional threats to public safety, extraordinary public expense, cause fraud on or victimization of the public, or conflict with existing local laws or ordinances.
4. A determination that the variance is the minimum necessary to afford relief, considering the flood hazard.
5. Submission to the applicant of written notice specifying the difference between the design flood elevation and the elevation to which the building is to be built, stating that the cost of flood insurance will be commensurate with the increased risk resulting from the reduced floor elevation, and stating that construction below the design flood elevation increases risks to life and property.

R112.3 Qualifications. The board of appeals shall consist of members who are qualified by experience and training to pass on matters pertaining to building construction and are not employees of the *jurisdiction.*

R112.4 Administration. The *building official* shall take immediate action in accordance with the decision of the board.

SECTION R113 VIOLATIONS

R113.1 Unlawful acts. It shall be unlawful for any person, firm or corporation to erect, construct, alter, extend, repair, move, remove, demolish or occupy any building, structure or *equipment* regulated by this code, or cause same to be done, in conflict with or in violation of any of the provisions of this code.

R113.2 Notice of violation. The *building official* is authorized to serve a notice of violation or order on the person responsible for the erection, construction, alteration, extension, repair, moving, removal, demolition or occupancy of a building or structure in violation of the provisions of this code, or in violation of a detail statement or a plan *approved* thereunder, or in violation of a *permit* or certificate issued under the provisions of this code. Such order shall direct the discontinuance of the illegal action or condition and the abatement of the violation.

R113.3 Prosecution of violation. If the notice of violation is not complied with in the time prescribed by such notice, the *building official* is authorized to request the legal counsel of the *jurisdiction* to institute the appropriate proceeding at law or in equity to restrain, correct or abate such violation, or to require the removal or termination of the unlawful occupancy of the building or structure in violation of the provisions of this code or of the order or direction made pursuant thereto.

R113.4 Violation penalties. Any person who violates a provision of this code or fails to comply with any of the requirements thereof or who erects, constructs, alters or repairs a building or structure in violation of the *approved construction documents* or directive of the *building official,* or of a *permit* or certificate issued under the provisions of this code, shall be subject to penalties as prescribed by law.

SECTION R114
STOP WORK ORDER

R114.1 Notice to owner. Upon notice from the *building official* that work on any building or structure is being prosecuted contrary to the provisions of this code or in an unsafe and dangerous manner, such work shall be immediately stopped. The stop work order shall be in writing and shall be given to the owner of the property involved, or to the owner's agent or to the person doing the work and shall state the conditions under which work will be permitted to resume.

R114.2 Unlawful continuance. Any person who shall continue any work in or about the structure after having been served with a stop work order, except such work as that person is directed to perform to remove a violation or unsafe condition, shall be subject to penalties as prescribed by law.

Part II—Definitions

CHAPTER 2
DEFINITIONS

SECTION R201
GENERAL

R201.1 Scope. Unless otherwise expressly stated, the following words and terms shall, for the purposes of this code, have the meanings indicated in this chapter.

R201.2 Interchangeability. Words used in the present tense include the future; words in the masculine gender include the feminine and neuter; the singular number includes the plural and the plural, the singular.

R201.3 Terms defined in other codes. Where terms are not defined in this code such terms shall have meanings ascribed to them as in other code publications of the International Code Council.

R201.4 Terms not defined. Where terms are not defined through the methods authorized by this section, such terms shall have ordinarily accepted meanings such as the context implies.

SECTION R202
DEFINITIONS

ACCESSIBLE. Signifies access that requires the removal of an access panel or similar removable obstruction.

ACCESSIBLE, READILY. Signifies access without the necessity for removing a panel or similar obstruction.

ACCESSORY STRUCTURE. A structure not greater than 3,000 square feet (279 m^2) in floor area, and not over two stories in height, the use of which is customarily accessory to and incidental to that of the dwelling(s) and which is located on the same *lot*.

ADDITION. An extension or increase in floor area or height of a building or structure.

ADHERED STONE OR MASONRY VENEER. Stone or masonry veneer secured and supported through the adhesion of an *approved* bonding material applied to an *approved* backing.

AIR ADMITTANCE VALVE. A one-way valve designed to allow air into the plumbing drainage system when a negative pressure develops in the piping. This device shall close by gravity and seal the terminal under conditions of zero differential pressure (no flow conditions) and under positive internal pressure.

AIR BARRIER. See Section N1101.9 for definition applicable in Chapter 11.

AIR BREAK (DRAINAGE SYSTEM). An arrangement in which a discharge pipe from a fixture, *appliance* or device drains indirectly into a receptor below the flood-level rim of the receptor, and above the trap seal.

AIR CIRCULATION, FORCED. A means of providing space conditioning utilizing movement of air through ducts or plenums by mechanical means.

AIR-CONDITIONING SYSTEM. A system that consists of heat exchangers, blowers, filters, supply, exhaust and return-air systems, and shall include any apparatus installed in connection therewith.

AIR GAP, DRAINAGE SYSTEM. The unobstructed vertical distance through free atmosphere between the outlet of a waste pipe and the flood-level rim of the fixture or receptor into which it is discharging.

AIR GAP, WATER-DISTRIBUTION SYSTEM. The unobstructed vertical distance through free atmosphere between the lowest opening from a water supply discharge to the flood-level rim of a plumbing fixture.

AIR-IMPERMEABLE INSULATION. An insulation having an air permanence equal to or less than 0.02 L/s-m^2 at 75 Pa pressure differential tested according to ASTM E 2178 or E 283.

ALTERATION. Any construction or renovation to an existing structure other than repair or addition that requires a *permit.* Also, a change in a mechanical system that involves an extension, addition or change to the arrangement, type or purpose of the original installation that requires a *permit.*

ANCHORED STONE OR MASONRY VENEER. Stone or masonry veneer secured with *approved* mechanical fasteners to an approved backing.

ANCHORS. See "Supports."

ANTISIPHON. A term applied to valves or mechanical devices that eliminate siphonage.

APPLIANCE. A device or apparatus that is manufactured and designed to utilize energy and for which this code provides specific requirements.

APPROVED. Acceptable to the *building official.*

APPROVED AGENCY. An established and recognized agency regularly engaged in conducting tests or furnishing inspection services, when such agency has been *approved* by the building official.

ASPECT RATIO. The ratio of longest to shortest perpendicular dimensions, or for wall sections, the ratio of height to length.

ATTIC. The unfinished space between the ceiling assembly of the top *story* and the roof assembly.

ATTIC, HABITABLE. A finished or unfinished area, not considered a *story*, complying with all of the following requirements:

1. The occupiable floor area is at least 70 square feet (17 m^2), in accordance with Section R304,
2. The occupiable floor area has a ceiling height in accordance with Section R305, and
3. The occupiable space is enclosed by the roof assembly above, knee walls (if applicable) on the sides and the floor-ceiling assembly below.

BACKFLOW, DRAINAGE. A reversal of flow in the drainage system.

BACKFLOW PREVENTER. A device or means to prevent backflow.

BACKFLOW PREVENTER, REDUCED-PRESSURE-ZONE TYPE. A backflow-prevention device consisting of two independently acting check valves, internally force loaded to a normally closed position and separated by an intermediate chamber (or zone) in which there is an automatic relief means of venting to atmosphere internally loaded to a normally open position between two tightly closing shutoff valves and with means for testing for tightness of the checks and opening of relief means.

BACKFLOW, WATER DISTRIBUTION. The flow of water or other liquids into the potable water-supply piping from any sources other than its intended source. Backsiphonage is one type of backflow.

BACKPRESSURE. Pressure created by any means in the water distribution system, which by being in excess of the pressure in the water supply mains causes a potential backflow condition.

BACKPRESSURE, LOW HEAD. A pressure less than or equal to 4.33 psi (29.88 kPa) or the pressure exerted by a 10-foot (3048 mm) column of water.

BACKSIPHONAGE. The flowing back of used or contaminated water from piping into a potable water-supply pipe due to a negative pressure in such pipe.

BACKWATER VALVE. A device installed in a drain or pipe to prevent backflow of sewage.

BASEMENT. A *story* that is not a *story above grade plane*. (see "*Story above grade plane*").

BASEMENT WALL. The opaque portion of a wall that encloses one side of a *basement* and has an average below *grade* wall area that is 50 percent or more of the total opaque and non-opaque area of that enclosing side.

BASIC WIND SPEED. Three-second gust speed at 33 feet (10 058 mm) above the ground in Exposure C (see Section R301.2.1) as given in Figure R301.2(4)A.

BATHROOM GROUP. A group of fixtures, including or excluding a bidet, consisting of a water closet, lavatory, and bathtub or shower. Such fixtures are located together on the same floor level.

BEND. A drainage fitting, designed to provide a change in direction of a drain pipe of less than the angle specified by the amount necessary to establish the desired slope of the line (see "Elbow" and "Sweep").

BOILER. A self-contained *appliance* from which hot water is circulated for heating purposes and then returned to the boiler, and which operates at water pressures not exceeding 160 pounds per square inch gage (psig) (1102 kPa gauge) and at water temperatures not exceeding 250°F (121°C).

BOND BEAM. A horizontal grouted element within masonry in which reinforcement is embedded.

BRACED WALL LINE. A straight line through the building plan that represents the location of the lateral resistance provided by the wall bracing.

BRACED WALL LINE, CONTINUOUSLY SHEATHED. A *braced wall line* with structural sheathing applied to all sheathable surfaces including the areas above and below openings.

BRACED WALL PANEL. A full-height section of wall constructed to resist in-plane shear loads through interaction of framing members, sheathing material and anchors. The panel's length meets the requirements of its particular bracing method, and contributes toward the total amount of bracing required along its *braced wall line* in accordance with Section R602.10.1.

BRANCH. Any part of the piping system other than a riser, main or stack.

BRANCH, FIXTURE. See "Fixture branch, drainage."

BRANCH, HORIZONTAL. See "Horizontal branch, drainage."

BRANCH INTERVAL. A vertical measurement of distance, 8 feet (2438 mm) or more in *developed length*, between the connections of horizontal branches to a drainage stack. Measurements are taken down the stack from the highest horizontal branch connection.

BRANCH, MAIN. A water-distribution pipe that extends horizontally off a main or riser to convey water to branches or fixture groups.

BRANCH, VENT. A vent connecting two or more individual vents with a vent stack or stack vent.

BTU/H. The *listed* maximum capacity of an *appliance*, absorption unit or burner expressed in British thermal units input per hour.

BUILDING. Building shall mean any one- and two-family dwelling or portion thereof, including *townhouses*, that is used, or designed or intended to be used for human habitation, for living, sleeping, cooking or eating purposes, or any combination thereof, and shall include accessory structures thereto.

BUILDING DRAIN. The lowest piping that collects the discharge from all other drainage piping inside the house and extends 30 inches (762 mm) in *developed length* of pipe, beyond the *exterior walls* and conveys the drainage to the *building sewer*.

BUILDING, EXISTING. Existing building is a building erected prior to the adoption of this code, or one for which a legal building *permit* has been issued.

BUILDING LINE. The line established by law, beyond which a building shall not extend, except as specifically provided by law.

BUILDING OFFICIAL. The officer or other designated authority charged with the administration and enforcement of this code.

BUILDING SEWER. That part of the drainage system that extends from the end of the *building drain* and conveys its discharge to a public sewer, private sewer, individual sewage-disposal system or other point of disposal.

BUILDING THERMAL ENVELOPE. The *basement walls*, *exterior walls*, floor, roof and any other building element that enclose *conditioned spaces*.

BUILT-UP ROOF COVERING. Two or more layers of felt cemented together and surfaced with a cap sheet, mineral aggregate, smooth coating or similar surfacing material.

CAP PLATE. The top plate of the double top plates used in structural insulated panel (SIP) construction. The cap plate is cut to match the panel thickness such that it overlaps the wood structural panel facing on both sides.

CEILING HEIGHT. The clear vertical distance from the finished floor to the finished ceiling.

CEMENT PLASTER. A mixture of portland or blended cement, portland cement or blended cement and hydrated lime, masonry cement or plastic cement and aggregate and other *approved* materials as specified in this code.

CHIMNEY. A primary vertical structure containing one or more flues, for the purpose of carrying gaseous products of combustion and air from a fuel-burning *appliance* to the outside atmosphere.

CHIMNEY CONNECTOR. A pipe that connects a fuel-burning *appliance* to a chimney.

CHIMNEY TYPES.

Residential-type appliance. An *approved* chimney for removing the products of combustion from fuel-burning, residential-type *appliances* producing combustion gases not in excess of 1,000°F (538°C) under normal operating conditions, but capable of producing combustion gases of 1,400°F (760°C) during intermittent forces firing for periods up to 1 hour. All temperatures shall be measured at the *appliance* flue outlet. Residential-type *appliance* chimneys include masonry and factory-built types.

CIRCUIT VENT. A vent that connects to a horizontal drainage branch and vents two traps to a maximum of eight traps or trapped fixtures connected into a battery.

CLADDING. The exterior materials that cover the surface of the building envelope that is directly loaded by the wind.

CLEANOUT. An accessible opening in the drainage system used for the removal of possible obstruction.

CLOSET. A small room or chamber used for storage.

COMBINATION WASTE AND VENT SYSTEM. A specially designed system of waste piping embodying the horizontal wet venting of one or more sinks, lavatories or floor drains by means of a common waste and vent pipe adequately sized to provide free movement of air above the flow line of the drain.

COMBUSTIBLE MATERIAL. Any material not defined as noncombustible.

COMBUSTION AIR. The air provided to fuel-burning *equipment* including air for fuel combustion, draft hood dilution and ventilation of the *equipment* enclosure.

[CE] COMMERCIAL, BUILDING. See Section N1101.9.

COMMON VENT. A single pipe venting two trap arms within the same *branch interval*, either back-to-back or one above the other.

CONDENSATE. The liquid that separates from a gas due to a reduction in temperature, e.g., water that condenses from flue gases and water that condenses from air circulating through the cooling coil in air conditioning *equipment*.

CONDENSING APPLIANCE. An *appliance* that condenses water generated by the burning of fuels.

CONDITIONED AIR. Air treated to control its temperature, relative humidity or quality.

CONDITIONED AREA. That area within a building provided with heating and/or cooling systems or *appliances* capable of maintaining, through design or heat loss/gain, 68°F (20°C) during the heating season and/or 80°F (27°C) during the cooling season, or has a fixed opening directly adjacent to a conditioned area.

CONDITIONED FLOOR AREA. The horizontal projection of the floors associated with the *conditioned space*.

CONDITIONED SPACE. For energy purposes, space within a building that is provided with heating and/or cooling *equipment* or systems capable of maintaining, through design or heat loss/gain, 50°F (10°C) during the heating season and 85°F (29°C) during the cooling season, or communicates directly with a *conditioned space*. For mechanical purposes, an area, room or space being heated or cooled by any *equipment* or *appliance*.

CONSTRUCTION DOCUMENTS. Written, graphic and pictorial documents prepared or assembled for describing the design, location and physical characteristics of the elements of a project necessary for obtaining a building *permit*. Construction drawings shall be drawn to an appropriate scale.

CONTAMINATION. An impairment of the quality of the potable water that creates an actual hazard to the public health through poisoning or through the spread of disease by sewage, industrial fluids or waste.

CONTINUOUS WASTE. A drain from two or more similar adjacent fixtures connected to a single trap.

CONTROL, LIMIT. An automatic control responsive to changes in liquid flow or level, pressure, or temperature for limiting the operation of an *appliance*.

CONTROL, PRIMARY SAFETY. A safety control responsive directly to flame properties that senses the presence or absence of flame and, in event of ignition failure or unintentional flame extinguishment, automatically causes shutdown of mechanical equipment.

CONVECTOR. A system-incorporating heating element in an enclosure in which air enters an opening below the heating element, is heated and leaves the enclosure through an opening located above the heating element.

CORE. The light-weight middle section of the structural insulated panel composed of foam plastic insulation, which provides the link between the two facing shells.

CORROSION RESISTANCE. The ability of a material to withstand deterioration of its surface or its properties when exposed to its environment.

COURT. A space, open and unobstructed to the sky, located at or above *grade* level on a *lot* and bounded on three or more sides by walls or a building.

CRIPPLE WALL. A framed wall extending from the top of the foundation to the underside of the floor framing of the first *story above grade plane*.

CROSS CONNECTION. Any connection between two otherwise separate piping systems whereby there may be a flow from one system to the other.

CURTAIN WALL. See Section N1101.9 for definition applicable in Chapter 11.

DALLE GLASS. A decorative composite glazing material made of individual pieces of glass that are embedded in a cast matrix of concrete or epoxy.

DAMPER, VOLUME. A device that will restrict, retard or direct the flow of air in any duct, or the products of combustion of heat-producing *equipment*, vent connector, vent or chimney.

DEAD END. A branch leading from a DWV system terminating at a *developed length* of 2 feet (610 mm) or more. Dead ends shall be prohibited except as an *approved* part of a rough-in for future connection.

DEAD LOADS. The weight of all materials of construction incorporated into the building, including but not limited to walls, floors, roofs, ceilings, stairways, built-in partitions, finishes, cladding, and other similarly incorporated architectural and structural items, and fixed service *equipment*.

DECORATIVE GLASS. A carved, leaded or Dalle glass or glazing material whose purpose is decorative or artistic, not functional; whose coloring, texture or other design qualities or components cannot be removed without destroying the glazing material; and whose surface, or assembly into which it is incorporated, is divided into segments.

DEMAND RECIRCULATION WATER SYSTEM. See Section N1101.9 for definition applicable in Chapter 11.

DESIGN PROFESSIONAL. See "*Registered design professional.*"

DEVELOPED LENGTH. The length of a pipeline measured along the center line of the pipe and fittings.

DIAMETER. Unless specifically stated, the term "diameter" is the nominal diameter as designated by the *approved* material standard.

DIAPHRAGM. A horizontal or nearly horizontal system acting to transmit lateral forces to the vertical resisting elements. When the term "*diaphragm*" is used, it includes horizontal bracing systems.

DILUTION AIR. Air that enters a draft hood or draft regulator and mixes with flue gases.

DIRECT-VENT APPLIANCE. A fuel-burning *appliance* with a sealed combustion system that draws all air for combustion from the outside atmosphere and discharges all flue gases to the outside atmosphere.

DRAFT. The pressure difference existing between the *appliance* or any component part and the atmosphere, that causes a continuous flow of air and products of combustion through the gas passages of the *appliance* to the atmosphere.

Induced draft. The pressure difference created by the action of a fan, blower or ejector, that is located between the *appliance* and the chimney or vent termination.

Natural draft. The pressure difference created by a vent or chimney because of its height, and the temperature difference between the flue gases and the atmosphere.

DRAFT HOOD. A device built into an *appliance*, or a part of the vent connector from an *appliance*, which is designed to provide for the ready escape of the flue gases from the *appliance* in the event of no draft, backdraft or stoppage beyond the draft hood; prevent a backdraft from entering the *appliance*; and neutralize the effect of stack action of the chimney or gas vent on the operation of the *appliance*.

DRAFT REGULATOR. A device that functions to maintain a desired draft in the *appliance* by automatically reducing the draft to the desired value.

DRAFT STOP. A material, device or construction installed to restrict the movement of air within open spaces of concealed areas of building components such as crawl spaces, floor-ceiling assemblies, roof-ceiling assemblies and *attics*.

DRAIN. Any pipe that carries soil and water-borne wastes in a building drainage system.

DRAINAGE FITTING. A pipe fitting designed to provide connections in the drainage system that have provisions for establishing the desired slope in the system. These fittings are made from a variety of both metals and plastics. The methods of coupling provide for required slope in the system (see "Durham fitting").

DUCT SYSTEM. A continuous passageway for the transmission of air which, in addition to ducts, includes duct fittings, dampers, plenums, fans and accessory air-handling *equipment* and *appliances*. For definition applicable in Chapter 11, see Section N1101.9.

DURHAM FITTING. A special type of drainage fitting for use in the Durham systems installations in which the joints are made with recessed and tapered threaded fittings, as opposed to bell and spigot lead/oakum or solvent/cemented or soldered joints. The tapping is at an angle (not 90 degrees) to provide for proper slope in otherwise rigid connections.

DURHAM SYSTEM. A term used to describe soil or waste systems where all piping is of threaded pipe, tube or other such rigid construction using recessed drainage fittings to correspond to the types of piping.

DWELLING. Any building that contains one or two *dwelling units* used, intended, or designed to be built, used, rented, leased, let or hired out to be occupied, or that are occupied for living purposes.

DWELLING UNIT. A single unit providing complete independent living facilities for one or more persons, including permanent provisions for living, sleeping, eating, cooking and sanitation.

DWV. Abbreviated term for drain, waste and vent piping as used in common plumbing practice.

EFFECTIVE OPENING. The minimum cross-sectional area at the point of water-supply discharge, measured or expressed in terms of diameter of a circle and if the opening is not circular, the diameter of a circle of equivalent cross-sectional area. (This is applicable to air gap.)

ELBOW. A pressure pipe fitting designed to provide an exact change in direction of a pipe run. An elbow provides a sharp turn in the flow path (see "Bend" and "Sweep").

EMERGENCY ESCAPE AND RESCUE OPENING. An operable exterior window, door or similar device that provides for a means of escape and access for rescue in the event of an emergency.

EQUIPMENT. All piping, ducts, vents, control devices and other components of systems other than *appliances* that are permanently installed and integrated to provide control of environmental conditions for buildings. This definition shall also include other systems specifically regulated in this code.

EQUIVALENT LENGTH. For determining friction losses in a piping system, the effect of a particular fitting equal to the friction loss through a straight piping length of the same nominal diameter.

ESCARPMENT. With respect to topographic wind effects, a cliff or steep slope generally separating two levels or gently sloping areas.

ESSENTIALLY NONTOXIC TRANSFER FLUIDS. Fluids having a Gosselin rating of 1, including propylene glycol; mineral oil; polydimethy oil oxane; hydrochlorofluorocarbon, chlorofluorocarbon and hydrofluorocarbon refrigerants; and FDA-*approved* boiler water additives for steam boilers.

ESSENTIALLY TOXIC TRANSFER FLUIDS. Soil, water or gray water and fluids having a Gosselin rating of 2 or more including ethylene glycol, hydrocarbon oils, ammonia refrigerants and hydrazine.

EVAPORATIVE COOLER. A device used for reducing air temperature by the process of evaporating water into an airstream.

EXCESS AIR. Air that passes through the combustion chamber and the *appliance* flue in excess of that which is theoretically required for complete combustion.

EXHAUST HOOD, FULL OPENING. An exhaust hood with an opening at least equal to the diameter of the connecting vent.

EXISTING INSTALLATIONS. Any plumbing system regulated by this code that was legally installed prior to the effective date of this code, or for which a *permit* to install has been issued.

EXTERIOR INSULATION AND FINISH SYSTEMS (EIFS). EIFS are nonstructural, nonload-bearing *exterior wall* cladding systems that consist of an insulation board attached either adhesively or mechanically, or both, to the substrate; an integrally reinforced base coat; and a textured protective finish coat.

EXTERIOR INSULATION AND FINISH SYSTEMS (EIFS) WITH DRAINAGE. An EIFS that incorporates a means of drainage applied over a water-resistive barrier.

EXTERIOR WALL. An above-*grade* wall that defines the exterior boundaries of a building. Includes between-floor spandrels, peripheral edges of floors, roof and *basement* knee walls, dormer walls, gable end walls, walls enclosing a mansard roof and *basement walls* with an average below-*grade* wall area that is less than 50 percent of the total opaque and nonopaque area of that enclosing side.

EXTERIOR WALL COVERING. A material or assembly of materials applied on the exterior side of exterior walls for the purpose of providing a weather-resistive barrier, insulation or for aesthetics, including but not limited to, veneers, siding, exterior insulation and finish systems, architectural trim and embellishments such as cornices, soffits, and fascias.

FACING. The wood structural panel facings that form the two outmost rigid layers of the structural insulated panel.

FACTORY-BUILT CHIMNEY. A *listed* and *labeled* chimney composed of factory-made components assembled in the field in accordance with the manufacturer's instructions and the conditions of the listing.

FENESTRATION. Skylights, roof windows, vertical windows (whether fixed or moveable); opaque doors; glazed doors; glass block; and combination opaque/glazed doors. For definition applicable in Chapter 11, see Section N1101.9

FIBER-CEMENT SIDING. A manufactured, fiber-reinforcing product made with an inorganic hydraulic or calcium silicate binder formed by chemical reaction and reinforced with discrete organic or inorganic nonasbestos fibers, or both. Additives which enhance manufacturing or product performance are permitted. Fiber-cement siding products have either smooth or textured faces and are intended for *exterior wall* and related applications.

FIREBLOCKING. Building materials or materials *approved* for use as fireblocking, installed to resist the free passage of flame to other areas of the building through concealed spaces.

FIREPLACE. An assembly consisting of a hearth and fire chamber of noncombustible material and provided with a chimney, for use with solid fuels.

Factory-built fireplace. A *listed* and *labeled* fireplace and chimney system composed of factory-made components, and assembled in the field in accordance with manufacturer's instructions and the conditions of the listing.

Masonry chimney. A field-constructed chimney composed of solid masonry units, bricks, stones or concrete.

Masonry fireplace. A field-constructed fireplace composed of solid masonry units, bricks, stones or concrete.

FIREPLACE STOVE. A free-standing, chimney-connected solid-fuel-burning heater designed to be operated with the fire chamber doors in either the open or closed position.

FIREPLACE THROAT. The opening between the top of the firebox and the smoke chamber.

FIRE-RETARDANT-TREATED WOOD. Pressure-treated lumber and plywood that exhibit reduced surface burning characteristics and resist propagation of fire.

Other means during manufacture. A process where the wood raw material is treated with a fire-retardant formulation while undergoing creation as a finished product.

Pressure process. A process for treating wood using an initial vacuum followed by the introduction of pressure above atmospheric.

FIRE SEPARATION DISTANCE. The distance measured from the building face to one of the following:

1. To the closest interior *lot line*; or
2. To the centerline of a street, an alley or public way; or
3. To an imaginary line between two buildings on the *lot*.

The distance shall be measured at a right angle from the face of the wall.

FIXTURE. See "Plumbing fixture."

FIXTURE BRANCH, DRAINAGE. A drain serving two or more fixtures that discharges into another portion of the drainage system.

FIXTURE BRANCH, WATER-SUPPLY. A water-supply pipe between the fixture supply and a main water-distribution pipe or fixture group main.

FIXTURE DRAIN. The drain from the trap of a fixture to the junction of that drain with any other drain pipe.

FIXTURE FITTING.

Supply fitting. A fitting that controls the volume and/or directional flow of water and is either attached to or accessible from a fixture or is used with an open or atmospheric discharge.

Waste fitting. A combination of components that conveys the sanitary waste from the outlet of a fixture to the connection of the sanitary drainage system.

FIXTURE GROUP, MAIN. The main water-distribution pipe (or secondary branch) serving a plumbing fixture grouping such as a bath, kitchen or laundry area to which two or more individual fixture branch pipes are connected.

FIXTURE SUPPLY. The water-supply pipe connecting a fixture or fixture fitting to a fixture branch.

FIXTURE UNIT, DRAINAGE (d.f.u.). A measure of probable discharge into the drainage system by various types of plumbing fixtures, used to size DWV piping systems. The drainage fixture-unit value for a particular fixture depends on its volume rate of drainage discharge, on the time duration of a single drainage operation and on the average time between successive operations.

FIXTURE UNIT, WATER-SUPPLY (w.s.f.u.). A measure of the probable hydraulic demand on the water supply by various types of plumbing fixtures used to size water-piping systems. The water-supply fixture-unit value for a particular fixture depends on its volume rate of supply, on the time duration of a single supply operation and on the average time between successive operations.

FLAME SPREAD. The propagation of flame over a surface.

FLAME SPREAD INDEX. A comparative measure, expressed as a dimensionless number, derived from visual measurements of the spread of flame versus time for a material tested in accordance with ASTM E 84 or UL 723.

FLIGHT. A continuous run of rectangular treads or winders or combination thereof from one landing to another.

FLOOD-LEVEL RIM. The edge of the receptor or fixture from which water overflows.

FLOOR DRAIN. A plumbing fixture for recess in the floor having a floor-level strainer intended for the purpose of the collection and disposal of waste water used in cleaning the floor and for the collection and disposal of accidental spillage to the floor.

FLOOR FURNACE. A self-contained furnace suspended from the floor of the space being heated, taking air for combustion from outside such space, and with means for lighting the *appliance* from such space.

FLOW PRESSURE. The static pressure reading in the water-supply pipe near the faucet or water outlet while the faucet or water outlet is open and flowing at capacity.

FLUE. See "Vent."

FLUE, APPLIANCE. The passages within an *appliance* through which combustion products pass from the combustion chamber to the flue collar.

FLUE COLLAR. The portion of a fuel-burning *appliance* designed for the attachment of a draft hood, vent connector or venting system.

FLUE GASES. Products of combustion plus excess air in *appliance* flues or heat exchangers.

FLUSH VALVE. A device located at the bottom of a flush tank that is operated to flush water closets.

FLUSHOMETER TANK. A device integrated within an air accumulator vessel that is designed to discharge a predetermined quantity of water to fixtures for flushing purposes.

FLUSHOMETER VALVE. A flushometer valve is a device that discharges a predetermined quantity of water to fixtures for flushing purposes and is actuated by direct water pressure.

FOAM BACKER BOARD. Foam plastic used in siding applications where the foam plastic is a component of the siding.

FOAM PLASTIC INSULATION. A plastic that is intentionally expanded by the use of a foaming agent to produce a reduced-density plastic containing voids consisting of open or closed cells distributed throughout the plastic for thermal insulating or acoustic purposes and that has a density less than 20 pounds per cubic foot (320 kg/m^3) unless it is used as interior trim.

FOAM PLASTIC INTERIOR TRIM. Exposed foam plastic used as picture molds, chair rails, crown moldings, baseboards, handrails, ceiling beams, door trim and window trim and similar decorative or protective materials used in fixed applications.

FUEL-PIPING SYSTEM. All piping, tubing, valves and fittings used to connect fuel utilization *equipment* to the point of fuel delivery.

FULLWAY VALVE. A valve that in the full open position has an opening cross-sectional area equal to a minimum of 85 percent of the cross-sectional area of the connecting pipe.

FURNACE. A vented heating *appliance* designed or arranged to discharge heated air into a *conditioned space* or through a duct or ducts.

GLAZING AREA. The interior surface area of all glazed fenestration, including the area of sash, curbing or other framing elements, that enclose *conditioned space.* Includes the area of glazed fenestration assemblies in walls bounding conditioned *basements.*

GRADE. The finished ground level adjoining the building at all *exterior walls.*

GRADE FLOOR OPENING. A window or other opening located such that the sill height of the opening is not more than 44 inches (1118 mm) above or below the finished ground level adjacent to the opening.

GRADE, PIPING. See "Slope."

GRADE PLANE. A reference plane representing the average of the finished ground level adjoining the building at all *exterior walls.* Where the finished ground level slopes away from the *exterior walls*, the reference plane shall be established by the lowest points within the area between the building and the *lot line* or, where the *lot line* is more than 6 feet (1829 mm) from the building between the structure and a point 6 feet (1829 mm) from the building.

GRAY WATER. Waste discharged from lavatories, bathtubs, showers, clothes washers and laundry trays.

GRIDDED WATER DISTRIBUTION SYSTEM. A water distribution system where every water distribution pipe is interconnected so as to provide two or more paths to each fixture supply pipe.

GROSS AREA OF EXTERIOR WALLS. The normal projection of all *exterior walls*, including the area of all windows and doors installed therein.

GROUND-SOURCE HEAT PUMP LOOP SYSTEM. Piping buried in horizontal or vertical excavations or placed in a body of water for the purpose of transporting heat transfer liquid to and from a heat pump. Included in this definition are closed loop systems in which the liquid is recirculated and open loop systems in which the liquid is drawn from a well or other source.

GUARD. A building component or a system of building components located near the open sides of elevated walking surfaces that minimizes the possibility of a fall from the walking surface to the lower level.

GUESTROOM. Any room or rooms used or intended to be used by one or more guests for living or sleeping purposes.

HABITABLE SPACE. A space in a building for living, sleeping, eating or cooking. Bathrooms, toilet rooms, closets, halls, storage or utility spaces and similar areas are not considered *habitable spaces.*

HANDRAIL. A horizontal or sloping rail intended for grasping by the hand for guidance or support.

HANGERS. See "Supports."

HAZARDOUS LOCATION. Any location considered to be a fire hazard for flammable vapors, dust, combustible fibers or other highly combustible substances.

HEAT PUMP. An *appliance* having heating or heating/cooling capability and that uses refrigerants to extract heat from air, liquid or other sources.

HEATING DEGREE DAYS (HDD). The sum, on an annual basis, of the difference between 65°F (18°C) and the mean temperature for each day as determined from "NOAA Annual Degree Days to Selected Bases Derived from the 1960-1990 Normals" or other weather data sources acceptable to the code official.

HEIGHT, BUILDING. The vertical distance from *grade plane* to the average height of the highest roof surface.

HEIGHT, STORY. The vertical distance from top to top of two successive tiers of beams or finished floor surfaces; and, for the topmost *story*, from the top of the floor finish to the top of the ceiling joists or, where there is not a ceiling, to the top of the roof rafters.

HIGH-EFFICACY LAMPS. See Section N1101.9 for definition applicable in Chapter 11.

HIGH-TEMPERATURE (H.T.) CHIMNEY. A high temperature chimney complying with the requirements of UL 103. A Type H.T. chimney is identifiable by the markings "Type H.T." on each chimney pipe section.

HILL. With respect to topographic wind effects, a land surface characterized by strong relief in any horizontal direction.

HORIZONTAL BRANCH, DRAINAGE. A drain pipe extending laterally from a soil or waste stack or *building drain*, that receives the discharge from one or more *fixture drains.*

HORIZONTAL PIPE. Any pipe or fitting that makes an angle of less than 45 degrees (0.79 rad) with the horizontal.

HOT WATER. Water at a temperature greater than or equal to 110°F (43°C).

HURRICANE-PRONE REGIONS. Areas vulnerable to hurricanes, defined as the U.S. Atlantic Ocean and Gulf of Mexico coasts where the basic wind speed is greater than 90 miles per hour (40 m/s), and Hawaii, Puerto Rico, Guam, Virgin Islands, and America Samoa.

HYDROGEN GENERATING APPLIANCE. A self-contained package or factory-matched packages of integrated systems for generating gaseous hydrogen. Hydrogen generating *appliances* utilize electrolysis, reformation, chemical, or other processes to generate hydrogen.

IGNITION SOURCE. A flame, spark or hot surface capable of igniting flammable vapors or fumes. Such sources include *appliance* burners, burner ignitions and electrical switching devices.

INDIRECT WASTE PIPE. A waste pipe that discharges into the drainage system through an air gap into a trap, fixture or receptor.

INDIVIDUAL SEWAGE DISPOSAL SYSTEM. A system for disposal of sewage by means of a septic tank or mechanical treatment, designed for use apart from a public sewer to serve a single establishment or building.

INDIVIDUAL VENT. A pipe installed to vent a single-*fixture drain* that connects with the vent system above or terminates independently outside the building.

INDIVIDUAL WATER SUPPLY. A supply other than an *approved* public water supply that serves one or more families.

INSULATING CONCRETE FORM (ICF). A concrete forming system using stay-in-place forms of rigid foam plastic insulation, a hybrid of cement and foam insulation, a hybrid of cement and wood chips, or other insulating material for constructing cast-in-place concrete walls.

INSULATING SHEATHING. An insulating board having a minimum thermal resistance of R-2 of the core material. For definition applicable in Chapter 11, see Section N1101.9.

JURISDICTION. The governmental unit that has adopted this code under due legislative authority.

KITCHEN. Kitchen shall mean an area used, or designated to be used, for the preparation of food.

LABEL. An identification applied on a product by the manufacturer which contains the name of the manufacturer, the function and performance characteristics of the product or material, and the name and identification of an *approved agency* and that indicates that the representative sample of the product or material has been tested and evaluated by an *approved agency*. (See also "Manufacturer's designation" and "Mark.")

LABELED. *Equipment*, materials or products to which have been affixed a *label*, seal, symbol or other identifying *mark* of a nationally recognized testing laboratory, inspection agency or other organization concerned with product evaluation that maintains periodic inspection of the production of the above-*labeled* items and whose labeling indicates either that the *equipment*, material or product meets identified standards or has been tested and found suitable for a specified purpose.

LIGHT-FRAME CONSTRUCTION. A type of construction whose vertical and horizontal structural elements are primarily formed by a system of repetitive wood or cold-formed steel framing members.

LISTED. *Equipment*, materials, products or services included in a list published by an organization acceptable to the code official and concerned with evaluation of products or services that maintains periodic inspection of production of *listed equipment* or materials or periodic evaluation of services and whose listing states either that the *equipment*, material, product or service meets identified standards or has been tested and found suitable for a specified purpose.

LIVE LOADS. Those loads produced by the use and occupancy of the building or other structure and do not include construction or environmental loads such as wind load, snow load, rain load, earthquake load, flood load or dead load.

LIVING SPACE. Space within a *dwelling unit* utilized for living, sleeping, eating, cooking, bathing, washing and sanitation purposes.

LOCAL EXHAUST. An exhaust system that uses one or more fans to exhaust air from a specific room or rooms within a dwelling.

LODGING HOUSE. A one-family dwelling where one or more occupants are primarily permanent in nature, and rent is paid for guestrooms.

LOT. A portion or parcel of land considered as a unit.

LOT LINE. A line dividing one *lot* from another, or from a street or any public place.

MACERATING TOILET SYSTEMS. A system comprised of a sump with macerating pump and with connections for a water closet and other plumbing fixtures, that is designed to accept, grind and pump wastes to an *approved* point of discharge.

MAIN. The principal pipe artery to which branches may be connected.

MAIN SEWER. See "Public sewer."

MANIFOLD WATER DISTRIBUTION SYSTEMS. A fabricated piping arrangement in which a large supply main is fitted with multiple branches in close proximity in which water is distributed separately to fixtures from each branch.

MANUFACTURED HOME. *Manufactured home* means a structure, transportable in one or more sections, which in the traveling mode is 8 body feet (2438 body mm) or more in width or 40 body feet (12 192 body mm) or more in length, or, when erected on site, is 320 square feet (30 m^2) or more, and which is built on a permanent chassis and designed to be used as a *dwelling* with or without a permanent foundation when connected to the required utilities, and includes the plumbing, heating, air-conditioning and electrical systems contained therein; except that such term shall include any structure that meets all the requirements of this paragraph except the size requirements and with respect to which the manufacturer voluntarily files a certification required by the secretary (HUD) and complies with the standards established under this title. For mobile homes built prior to June 15, 1976, a *label* certifying compliance to the Standard for Mobile Homes, NFPA 501, in effect at the time of manufacture is required. For the purpose of these provisions, a mobile home shall be considered a *manufactured home*.

MANUFACTURER'S DESIGNATION. An identification applied on a product by the manufacturer indicating that a product or material complies with a specified standard or set of rules. (See also "*Mark*" and "*Label*.")

MANUFACTURER'S INSTALLATION INSTRUCTIONS. Printed instructions included with *equipment* as part of the conditions of listing and labeling.

MARK. An identification applied on a product by the manufacturer indicating the name of the manufacturer and the function of a product or material. (See also "Manufacturer's designation" and "*Label*.")

MASONRY CHIMNEY. A field-constructed chimney composed of solid masonry units, bricks, stones or concrete.

MASONRY HEATER. A masonry heater is a solid fuel burning heating *appliance* constructed predominantly of concrete or solid masonry having a mass of at least 1,100 pounds (500 kg), excluding the chimney and foundation. It is designed to absorb and store a substantial portion of heat from a fire built in the firebox by routing exhaust gases through internal heat exchange channels in which the flow path downstream of the firebox includes at least one 180-degree (3.14-rad) change in flow direction before entering the chimney and which deliver heat by radiation through the masonry surface of the heater.

MASONRY, SOLID. Masonry consisting of solid masonry units laid contiguously with the joints between the units filled with mortar.

MASONRY UNIT. Brick, tile, stone, glass block or concrete block conforming to the requirements specified in Section 2103 of the *International Building Code*.

Clay. A building unit larger in size than a brick, composed of burned clay, shale, fire clay or mixtures thereof.

Concrete. A building unit or block larger in size than 12 inches by 4 inches by 4 inches (305 mm by 102 mm by 102 mm) made of cement and suitable aggregates.

Glass. Nonload-bearing masonry composed of glass units bonded by mortar.

Hollow. A masonry unit whose net cross-sectional area in any plane parallel to the loadbearing surface is less than 75 percent of its gross cross-sectional area measured in the same plane.

Solid. A masonry unit whose net cross-sectional area in every plane parallel to the loadbearing surface is 75 percent or more of its cross-sectional area measured in the same plane.

MASS WALL. Masonry or concrete walls having a mass greater than or equal to 30 pounds per square foot (146 kg/m^2), solid wood walls having a mass greater than or equal to 20 pounds per square foot (98 kg/m^2), and any other walls having a heat capacity greater than or equal to 6 Btu/ft^2 · °F [266 J/(m^2 · K)].

MEAN ROOF HEIGHT. The average of the roof eave height and the height to the highest point on the roof surface, except that eave height shall be used for roof angle of less than or equal to 10 degrees (0.18 rad).

MECHANICAL DRAFT SYSTEM. A venting system designed to remove flue or vent gases by mechanical means, that consists of an induced draft portion under nonpositive static pressure or a forced draft portion under positive static pressure.

Forced-draft venting system. A portion of a venting system using a fan or other mechanical means to cause the removal of flue or vent gases under positive static pressure.

Induced draft venting system. A portion of a venting system using a fan or other mechanical means to cause the removal of flue or vent gases under nonpositive static vent pressure.

Power venting system. A portion of a venting system using a fan or other mechanical means to cause the removal of flue or vent gases under positive static vent pressure.

MECHANICAL EXHAUST SYSTEM. A system for removing air from a room or space by mechanical means.

MECHANICAL SYSTEM. A system specifically addressed and regulated in this code and composed of components, devices, *appliances* and *equipment*.

METAL ROOF PANEL. An interlocking metal sheet having a minimum installed weather exposure of at least 3 square feet (0.28 m^2) per sheet.

METAL ROOF SHINGLE. An interlocking metal sheet having an installed weather exposure less than 3 square feet (0.28 m^2) per sheet.

MEZZANINE, LOFT. An intermediate level or levels between the floor and ceiling of any *story* with an aggregate floor area of not more than one-third of the area of the room or space in which the level or levels are located.

MODIFIED BITUMEN ROOF COVERING. One or more layers of polymer modified asphalt sheets. The sheet materials shall be fully adhered or mechanically attached to the substrate or held in place with an *approved* ballast layer.

MULTIPLE STATION SMOKE ALARM. Two or more single station alarm devices that are capable of interconnection such that actuation of one causes all integral or separate audible alarms to operate.

NATURAL DRAFT SYSTEM. A venting system designed to remove flue or vent gases under nonpositive static vent pressure entirely by natural draft.

NATURALLY DURABLE WOOD. The heartwood of the following species with the exception that an occasional piece with corner sapwood is permitted if 90 percent or more of the width of each side on which it occurs is heartwood.

Decay resistant. Redwood, cedar, black locust and black walnut.

Termite resistant. Alaska yellow cedar, redwood, Eastern red cedar and Western red cedar including all sapwood of Western red cedar.

NONCOMBUSTIBLE MATERIAL. Materials that pass the test procedure for defining noncombustibility of elementary materials set forth in ASTM E 136.

NONCONDITIONED SPACE. A space that is not a *conditioned space* by insulated walls, floors or ceilings.

NOSING. The leading edge of treads of stairs and of landings at the top of stairway flights.

OCCUPIED SPACE. The total area of all buildings or structures on any *lot* or parcel of ground projected on a horizontal plane, excluding permitted projections as allowed by this code.

OFFSET. A combination of fittings that makes two changes in direction bringing one section of the pipe out of line but into a line parallel with the other section.

OWNER. Any person, agent, firm or corporation having a legal or equitable interest in the property.

PAN FLASHING. Corrosion-resistant flashing at the base of an opening that is integrated into the building exterior wall to direct water to the exterior and is premanufactured, fabricated, formed or applied at the job site.

PANEL THICKNESS. Thickness of core plus two layers of structural wood panel facings.

PELLET FUEL-BURNING APPLIANCE. A closed combustion, vented *appliance* equipped with a fuel feed mechanism for burning processed pellets of solid fuel of a specified size and composition.

PELLET VENT. A vent *listed* and *labeled* for use with a *listed* pellet fuel-burning *appliance.*

PERFORMANCE CATEGORY. A designation of wood structural panels as related to the panel performance used in Chapters 4, 5, 6 and 8.

PERMIT. An official document or certificate issued by the authority having *jurisdiction* that authorizes performance of a specified activity.

PERSON. An individual, heirs, executors, administrators or assigns, and also includes a firm, partnership or corporation, its or their successors or assigns, or the agent of any of the aforesaid.

PHOTOVOLTAIC MODULES/SHINGLES. A roof covering composed of flat-plate photovoltaic modules fabricated into shingles.

PITCH. See "Slope."

PLATFORM CONSTRUCTION. A method of construction by which floor framing bears on load bearing walls that are not continuous through the *story* levels or floor framing.

PLENUM. A chamber that forms part of an air-circulation system other than the *occupied space* being conditioned.

PLUMBING. For the purpose of this code, plumbing refers to those installations, repairs, maintenance and *alterations* regulated by Chapters 25 through 33.

PLUMBING APPLIANCE. An energized household *appliance* with plumbing connections, such as a dishwasher, food-waste grinder, clothes washer or water heater.

PLUMBING APPURTENANCE. A device or assembly that is an adjunct to the basic plumbing system and demands no additional water supply nor adds any discharge load to the system. It is presumed that it performs some useful function in the operation, maintenance, servicing, economy or safety of the plumbing system. Examples include filters, relief valves and aerators.

PLUMBING FIXTURE. A receptacle or device that is connected to a water supply system or discharges to a drainage system or both. Such receptacles or devices require a supply of water; or discharge liquid waste or liquid-borne solid waste; or require a supply of water and discharge waste to a drainage system.

PLUMBING SYSTEM. Includes the water supply and distribution pipes, plumbing fixtures, supports and appurtenances; soil, waste and vent pipes; sanitary drains and *building sewers* to an *approved* point of disposal.

POLLUTION. An impairment of the quality of the potable water to a degree that does not create a hazard to the public health but that does adversely and unreasonably affect the aesthetic qualities of such potable water for domestic use.

PORTABLE-FUEL-CELL APPLIANCE. A fuel cell generator of electricity, which is not fixed in place. A portable-fuel-cell *appliance* utilizes a cord and plug connection to a grid-isolated load and has an integral fuel supply.

POSITIVE ROOF DRAINAGE. The drainage condition in which consideration has been made for all loading deflections of the roof deck, and additional slope has been provided to ensure drainage of the roof within 48 hours of precipitation.

POTABLE WATER. Water free from impurities present in amounts sufficient to cause disease or harmful physiological effects and conforming in bacteriological and chemical quality to the requirements of the public health authority having *jurisdiction.*

PRECAST CONCRETE. A structural concrete element cast elsewhere than its final position in the structure.

PRECAST CONCRETE FOUNDATION WALLS. Preengineered, precast concrete wall panels that are designed to withstand specified stresses and used to build below-*grade* foundations.

PRESSURE-RELIEF VALVE. A pressure-actuated valve held closed by a spring or other means and designed to automatically relieve pressure at the pressure at which it is set.

PUBLIC SEWER. A common sewer directly controlled by public authority.

PUBLIC WATER MAIN. A water-supply pipe for public use controlled by public authority.

PUBLIC WAY. Any street, alley or other parcel of land open to the outside air leading to a public street, which has been deeded, dedicated or otherwise permanently appropriated to the public for public use and that has a clear width and height of not less than 10 feet (3048 mm).

PURGE. To clear of air, gas or other foreign substances.

QUICK-CLOSING VALVE. A valve or faucet that closes automatically when released manually or controlled by mechanical means for fast-action closing.

***R*-VALUE, THERMAL RESISTANCE.** The inverse of the time rate of heat flow through a *building thermal envelope* element from one of its bounding surfaces to the other for a unit temperature difference between the two surfaces, under steady state conditions, per unit area ($h \cdot ft^2 \cdot °F/Btu$).

RAMP. A walking surface that has a running slope steeper than 1 unit vertical in 20 units horizontal (5-percent slope).

RECEPTOR. A fixture or device that receives the discharge from indirect waste pipes.

REFRIGERANT. A substance used to produce refrigeration by its expansion or evaporation.

REFRIGERANT COMPRESSOR. A specific machine, with or without accessories, for compressing a given refrigerant vapor.

REFRIGERATING SYSTEM. A combination of interconnected parts forming a closed circuit in which refrigerant is circulated for the purpose of extracting, then rejecting, heat. A direct refrigerating system is one in which the evaporator or condenser of the refrigerating system is in direct contact with the air or other substances to be cooled or heated. An indirect refrigerating system is one in which a secondary coolant cooled or heated by the refrigerating system is circulated to the air or other substance to be cooled or heated.

REGISTERED DESIGN PROFESSIONAL. An individual who is registered or licensed to practice their respective design profession as defined by the statutory requirements of the professional registration laws of the state or *jurisdiction* in which the project is to be constructed.

RELIEF VALVE, VACUUM. A device to prevent excessive buildup of vacuum in a pressure vessel.

REPAIR. The reconstruction or renewal of any part of an existing building for the purpose of its maintenance. For definition applicable in Chapter 11, see Section N1101.9

REROOFING. The process of recovering or replacing an existing roof covering. See "Roof recover."

RETURN AIR. Air removed from an *approved conditioned space* or location and recirculated or exhausted.

RIDGE. With respect to topographic wind effects, an elongated crest of a hill characterized by strong relief in two directions.

RISER.

1. The vertical component of a *step* or *stair*.
2. A water pipe that extends vertically one full *story* or more to convey water to branches or to a group of fixtures.

ROOF ASSEMBLY. A system designed to provide weather protection and resistance to design loads. The system consists of a roof covering and roof deck or a single component serving as both the roof covering and the roof deck. A roof assembly includes the roof deck, vapor retarder, substrate or thermal barrier, insulation, vapor retarder, and roof covering.

ROOF COVERING. The covering applied to the roof deck for weather resistance, fire classification or appearance.

ROOF COVERING SYSTEM. See "Roof assembly."

ROOF DECK. The flat or sloped surface not including its supporting members or vertical supports.

ROOF RECOVER. The process of installing an additional roof covering over a prepared existing roof covering without removing the existing roof covering.

ROOF REPAIR. Reconstruction or renewal of any part of an existing roof for the purposes of its maintenance.

ROOFTOP STRUCTURE. An enclosed structure on or above the roof of any part of a building.

ROOM HEATER. A freestanding heating *appliance* installed in the space being heated and not connected to ducts.

ROUGH-IN. The installation of all parts of the plumbing system that must be completed prior to the installation of fixtures. This includes DWV, water supply and built-in fixture supports.

RUNNING BOND. The placement of masonry units such that head joints in successive courses are horizontally offset at least one-quarter the unit length.

SANITARY SEWER. A sewer that carries sewage and excludes storm, surface and groundwater.

SCUPPER. An opening in a wall or parapet that allows water to drain from a roof.

SEISMIC DESIGN CATEGORY (SDC). A classification assigned to a structure based on its occupancy category and the severity of the design earthquake ground motion at the site.

SEPTIC TANK. A water-tight receptor that receives the discharge of a building sanitary drainage system and is constructed so as to separate solids from the liquid, digest organic matter through a period of detention, and allow the liquids to discharge into the soil outside of the tank through a system of open joint or perforated piping or a seepage pit.

SEWAGE. Any liquid waste containing animal matter, vegetable matter or other impurity in suspension or solution.

SEWAGE PUMP. A permanently installed mechanical device for removing sewage or liquid waste from a sump.

SHALL. The term, when used in the code, is construed as mandatory.

SHEAR WALL. A general term for walls that are designed and constructed to resist racking from seismic and wind by use of masonry, concrete, cold-formed steel or wood framing in accordance with Chapter 6 of this code and the associated limitations in Section R301.2 of this code.

SIDE VENT. A vent connecting to the drain pipe through a fitting at an angle less than 45 degrees (0.79 rad) to the horizontal.

SINGLE PLY MEMBRANE. A roofing membrane that is field applied using one layer of membrane material (either homogeneous or composite) rather than multiple layers.

SINGLE STATION SMOKE ALARM. An assembly incorporating the detector, control *equipment* and alarm sounding device in one unit that is operated from a power supply either in the unit or obtained at the point of installation.

SKYLIGHT. See Section N1101.9 for definition applicable in Chapter 11.

SKYLIGHT AND SLOPED GLAZING. See Section R308.6.1.

SKYLIGHT, UNIT. See Section R308.6.1.

SLEEPING UNIT. See Section N1101.9 for definition applicable in Chapter 11.

SLIP JOINT. A mechanical-type joint used primarily on fixture traps. The joint tightness is obtained by compressing a friction-type washer such as rubber, nylon, neoprene, lead or special packing material against the pipe by the tightening of a (slip) nut.

SLOPE. The fall (pitch) of a line of pipe in reference to a horizontal plane. In drainage, the slope is expressed as the fall in units vertical per units horizontal (percent) for a length of pipe.

SMOKE-DEVELOPED INDEX. A comparative measure, expressed as a dimensionless number, derived from measurements of smoke obscuration versus time for a material tested in accordance with ASTM E 84 or UL 723.

SOIL STACK OR PIPE. A pipe that conveys sewage containing fecal material.

SOLAR HEAT GAIN COEFFICIENT (SHGC). The solar heat gain through a fenestration or glazing assembly relative to the incident solar radiation (Btu/h · ft^2 · °F).

SOLID MASONRY. Load-bearing or nonload-bearing construction using masonry units where the net cross-sectional area of each unit in any plane parallel to the bearing surface is not less than 75 percent of its gross cross-sectional area. Solid masonry units shall conform to ASTM C 55, C 62, C 73, C 145 or C 216.

SPLINE. A strip of wood structural panel cut from the same material used for the panel facings, used to connect two structural insulated panels. The strip (spline) fits into a groove cut into the vertical edges of the two structural insulated panels to be joined. Splines are used behind each facing of the structural insulated panels being connected as shown in Figure R613.8.

STACK. Any main vertical DWV line, including offsets, that extends one or more stories as directly as possible to its vent terminal.

STACK BOND. The placement of masonry units in a bond pattern is such that head joints in successive courses are vertically aligned. For the purpose of this code, requirements for stack bond shall apply to all masonry laid in other than running bond.

STACK VENT. The extension of soil or waste stack above the highest horizontal drain connected.

STACK VENTING. A method of venting a fixture or fixtures through the soil or waste stack without individual fixture vents.

STAIR. A change in elevation, consisting of one or more risers.

STAIRWAY. One or more flights of stairs, either interior or exterior, with the necessary landings and platforms connecting them to form a continuous and uninterrupted passage from one level to another within or attached to a building, porch or deck.

STANDARD TRUSS. Any construction that does not permit the roof/ceiling insulation to achieve the required *R*-value over the *exterior walls*.

STATIONARY FUEL CELL POWER PLANT. A self-contained package or factory-matched packages which constitute an automatically-operated assembly of integrated systems for generating useful electrical energy and recoverable thermal energy that is permanently connected and fixed in place.

STORM SEWER, DRAIN. A pipe used for conveying rainwater, surface water, subsurface water and similar liquid waste.

STORY. That portion of a building included between the upper surface of a floor and the upper surface of the floor or roof next above.

STORY ABOVE GRADE PLANE. Any *story* having its finished floor surface entirely above *grade plane*, or in which the finished surface of the floor next above is:

1. More than 6 feet (1829 mm) *above grade plane; or*
2. More than 12 feet (3658 mm) above the finished ground level at any point.

STRUCTURAL COMPOSITE LUMBER. Structural members manufactured using wood elements bonded together with exterior adhesives.

Examples of structural composite lumber are:

Laminated veneer lumber (LVL). A composite of wood veneer elements with wood fibers primarily oriented along the length of the member, where the veneer element thicknesses are 0.25 inches (6.4 mm) or less.

Parallel strand lumber (PSL). A composite of wood strand elements with wood fibers primarily oriented along the length of the member, where the least dimension of the wood strand elements is 0.25 inch (6.4 mm) or less and their average lengths are a minimum of 300 times the least dimension of the wood strand elements.

Laminated strand lumber (LSL). A composite of wood strand elements with wood fibers primarily oriented along the length of the member, where the least dimension of the wood strand elements is 0.10 inch (2.54 mm) or less and their average lengths are a minimum of 150 times the least dimension of the wood strand elements.

Oriented strand lumber (OSL). A composite of wood strand elements with wood fibers primarily oriented along the length of the member, where the least dimension of the wood strand elements is 0.10 inch (2.54 mm) or less and their average lengths are a minimum of 75 times and less than 150 times the least dimension of the wood strand elements.

STRUCTURAL INSULATED PANEL (SIP). A structural sandwich panel that consists of a light-weight foam plastic core securely laminated between two thin, rigid wood structural panel facings.

STRUCTURE. That which is built or constructed.

SUBSOIL DRAIN. A drain that collects subsurface water or seepage water and conveys such water to a place of disposal.

SUMP. A tank or pit that receives sewage or waste, located below the normal *grade* of the gravity system and that must be emptied by mechanical means.

SUMP PUMP. A pump installed to empty a sump. These pumps are used for removing storm water only. The pump is selected for the specific head and volume of the load and is usually operated by level controllers.

SUNROOM. A one-story structure attached to a *dwelling* with a *glazing area* in excess of 40 percent of the gross area of the structure's *exterior walls* and roof. For definition applicable in Chapter 11, see Section N1101.9.

SUPPLY AIR. Air delivered to a *conditioned space* through ducts or plenums from the heat exchanger of a heating, cooling or ventilating system.

SUPPORTS. Devices for supporting, hanging and securing pipes, fixtures and *equipment*.

SWEEP. A drainage fitting designed to provide a change in direction of a drain pipe of less than the angle specified by the amount necessary to establish the desired slope of the line. Sweeps provide a longer turning radius than bends and a less turbulent flow pattern (see "Bend" and "Elbow").

TEMPERATURE- AND PRESSURE-RELIEF (T AND P) VALVE. A combination relief valve designed to function as both a temperature-relief and pressure-relief valve.

TEMPERATURE-RELIEF VALVE. A temperature-actuated valve designed to discharge automatically at the temperature at which it is set.

TERMITE-RESISTANT MATERIAL. Pressure-preservative treated wood in accordance with the AWPA standards in Section R318.1, naturally durable termite-resistant wood, steel, concrete, masonry or other *approved* material.

THERMAL ISOLATION. Physical and space conditioning separation from *conditioned space(s)* consisting of existing or new walls, doors and/or windows. The *conditioned space(s)* shall be controlled as separate zones for heating and cooling or conditioned by separate *equipment*. For definition applicable in Chapter 11, see Section N1101.9.

THERMAL RESISTANCE, *R*-VALUE. The inverse of the time rate of heat flow through a body from one of its bounding surfaces to the other for a unit temperature difference between the two surfaces, under steady state conditions, per unit area (h · ft^2 · °F/Btu) (m^2 · K)/W.

THERMAL TRANSMITTANCE, *U*-FACTOR. The coefficient of heat transmission (air to air) through a building envelope component or assembly, equal to the time rate of heat flow per unit area and unit temperature difference between the warm side and cold side air films (Btu/h · ft^2 · °F) W/(m^2 · K).

THIRD-PARTY CERTIFICATION AGENCY. An approved agency operating a product or material certification system that incorporates initial product testing, assessment and surveillance of a manufacturer's quality control system.

THIRD PARTY CERTIFIED. Certification obtained by the manufacturer indicating that the function and performance characteristics of a product or material have been determined by testing and ongoing surveillance by an approved third-party certification agency. Assertion of certification is in the form of identification in accordance with the requirements of the third-party certification agency.

THIRD-PARTY TESTED. Procedure by which an approved testing laboratory provides documentation that a product material or system conforms to specified requirements.

TOWNHOUSE. A single-family *dwelling unit* constructed in a group of three or more attached units in which each unit extends from foundation to roof and with a *yard* or public way on at least two sides.

TRAP. A fitting, either separate or built into a fixture, that provides a liquid seal to prevent the emission of sewer gases without materially affecting the flow of sewage or waste water through it.

TRAP ARM. That portion of a *fixture drain* between a trap weir and the vent fitting.

TRAP PRIMER. A device or system of piping to maintain a water seal in a trap, typically installed where infrequent use of the trap would result in evaporation of the trap seal, such as floor drains.

TRAP SEAL. The trap seal is the maximum vertical depth of liquid that a trap will retain, measured between the crown weir and the top of the dip of the trap.

TRIM. Picture molds, chair rails, baseboards, handrails, door and window frames, and similar decorative or protective materials used in fixed applications.

TRUSS DESIGN DRAWING. The graphic depiction of an individual truss, which describes the design and physical characteristics of the truss.

TYPE L VENT. A *listed* and *labeled* vent conforming to UL 641 for venting oil-burning *appliances listed* for use with Type L vents or with gas *appliances listed* for use with Type B vents.

***U*-FACTOR, THERMAL TRANSMITTANCE.** See Section N1101.9 for definition applicable in Chapter 11.

UNDERLAYMENT. One or more layers of felt, sheathing paper, nonbituminous saturated felt, or other *approved* material over which a roof covering, with a slope of 2 to 12 (17-percent slope) or greater, is applied.

VACUUM BREAKERS. A device which prevents back-siphonage of water by admitting atmospheric pressure through ports to the discharge side of the device.

VAPOR PERMEABLE. The property of having a moisture vapor permeance rating of 5 perms (2.9 x 10^{-10} kg/Pa · s · m^2) or greater, when tested in accordance with the desiccant method using Procedure A of ASTM E 96. A vapor permeable material permits the passage of moisture vapor.

VAPOR RETARDER CLASS. A measure of the ability of a material or assembly to limit the amount of moisture that passes through that material or assembly. Vapor retarder class shall be defined using the desiccant method with Procedure A of ASTM E 96 as follows:

Class I: 0.1 perm or less

Class II: 0.1 < perm ≤ 1.0 perm

Class III: 1.0 < perm ≤ 10 perm

VENT. A passageway for conveying flue gases from fuel-fired *appliances*, or their vent connectors, to the outside atmosphere.

VENT COLLAR. See "Flue collar."

VENT CONNECTOR. That portion of a venting system which connects the flue collar or draft hood of an *appliance* to a vent.

VENT DAMPER DEVICE, AUTOMATIC. A device intended for installation in the venting system, in the outlet of an individual, automatically operated fuel burning *appliance* and that is designed to open the venting system automatically when the *appliance* is in operation and to close off the venting system automatically when the *appliance* is in a standby or shutdown condition.

VENT GASES. Products of combustion from fuel-burning *appliances*, plus excess air and dilution air, in the venting system above the draft hood or draft regulator.

VENT STACK. A vertical vent pipe installed to provide circulation of air to and from the drainage system and which extends through one or more stories.

VENT SYSTEM. Piping installed to equalize pneumatic pressure in a drainage system to prevent trap seal loss or blow-back due to siphonage or back pressure.

VENTILATION. The natural or mechanical process of supplying conditioned or unconditioned air to, or removing such air from, any space. For definition applicable in Chapter 11, see Section N1101.9.

VENTING. Removal of combustion products to the outdoors.

VENTING SYSTEM. A continuous open passageway from the flue collar of an *appliance* to the outside atmosphere for the purpose of removing flue or vent gases. A venting system is usually composed of a vent or a chimney and vent connector, if used, assembled to form the open passageway.

VERTICAL PIPE. Any pipe or fitting that makes an angle of 45 degrees (0.79 rad) or more with the horizontal.

VINYL SIDING. A shaped material, made principally from rigid polyvinyl chloride (PVC), that is used to cover exterior walls of buildings.

WALL, RETAINING. A wall not laterally supported at the top, that resists lateral soil load and other imposed loads.

WALLS. Walls shall be defined as follows:

Load-bearing wall. A wall supporting any vertical load in addition to its own weight.

Nonbearing wall. A wall which does not support vertical loads other than its own weight.

WASTE. Liquid-borne waste that is free of fecal matter.

WASTE PIPE OR STACK. Piping that conveys only liquid sewage not containing fecal material.

WATER DISTRIBUTION SYSTEM. Piping which conveys water from the service to the plumbing fixtures, *appliances*, appurtenances, *equipment*, devices or other systems served, including fittings and control valves.

WATER HEATER. Any heating *appliance* or *equipment* that heats potable water and supplies such water to the potable hot water distribution system.

WATER MAIN. A water supply pipe for public use.

WATER OUTLET. A valved discharge opening, including a hose bibb, through which water is removed from the potable water system supplying water to a plumbing fixture or plumbing *appliance* that requires either an air gap or backflow prevention device for protection of the supply system.

WATER-RESISTIVE BARRIER. A material behind an *exterior wall* covering that is intended to resist liquid water that has penetrated behind the exterior covering from further intruding into the *exterior wall* assembly.

WATER SERVICE PIPE. The outside pipe from the water main or other source of potable water supply to the water distribution system inside the building, terminating at the service valve.

WATER SUPPLY SYSTEM. The water service pipe, the water-distributing pipes and the necessary connecting pipes, fittings, control valves and all appurtenances in or adjacent to the building or premises.

WET VENT. A vent that also receives the discharge of wastes from other fixtures.

WHOLE-HOUSE MECHANICAL VENTILATION SYSTEM. An exhaust system, supply system, or combination thereof that is designed to mechanically exchange indoor air for outdoor air when operating continuously or through a programmed intermittent schedule to satisfy the whole-house ventilation rate. For definition applicable in Chapter 11, see Section N1101.9.

WIND-BORNE DEBRIS REGION. Areas within *hurricane-prone regions* as designated in accordance with Figure R302.1(4)C.

WINDER. A tread with nonparallel edges.

WOOD/PLASTIC COMPOSITE. A composite material made primarily from wood or cellulose-based materials and plastic.

WOOD STRUCTURAL PANEL. A panel manufactured from veneers; or wood strands or wafers; bonded together with waterproof synthetic resins or other suitable bonding systems. Examples of wood structural panels are plywood, OSB or composite panels.

YARD. An open space, other than a court, unobstructed from the ground to the sky, except where specifically provided by this code, on the *lot* on which a building is situated.

Part III—Building Planning and Construction

CHAPTER 3
BUILDING PLANNING

SECTION R301
DESIGN CRITERIA

R301.1 Application. Buildings and structures, and all parts thereof, shall be constructed to safely support all loads, including dead loads, live loads, roof loads, flood loads, snow loads, wind loads and seismic loads as prescribed by this code. The construction of buildings and structures in accordance with the provisions of this code shall result in a system that provides a complete load path that meets all requirements for the transfer of all loads from their point of origin through the load-resisting elements to the foundation. Buildings and structures constructed as prescribed by this code are deemed to comply with the requirements of this section.

R301.1.1 Alternative provisions. As an alternative to the requirements in Section R301.1 the following standards are permitted subject to the limitations of this code and the limitations therein. Where engineered design is used in conjunction with these standards, the design shall comply with the *International Building Code*.

1. AF&PA *Wood Frame Construction Manual* (WFCM).
2. AISI *Standard for Cold-Formed Steel Framing—Prescriptive Method for One- and Two-Family Dwellings* (AISI S230).
3. ICC *Standard on the Design and Construction of Log Structures* (ICC 400).

R301.1.2 Construction systems. The requirements of this code are based on platform and balloon-frame construction for light-frame buildings. The requirements for concrete and masonry buildings are based on a balloon framing system. Other framing systems must have equivalent detailing to ensure force transfer, continuity and compatible deformations.

R301.1.3 Engineered design. When a building of otherwise conventional construction contains structural elements exceeding the limits of Section R301 or otherwise not conforming to this code, these elements shall be designed in accordance with accepted engineering practice. The extent of such design need only demonstrate compliance of nonconventional elements with other applicable provisions and shall be compatible with the performance of the conventional framed system. Engineered design in accordance with the *International Building Code* is permitted for all buildings and structures, and parts thereof, included in the scope of this code.

R301.2 Climatic and geographic design criteria. Buildings shall be constructed in accordance with the provisions of this code as limited by the provisions of this section. Additional criteria shall be established by the local *jurisdiction* and set forth in Table R301.2(1).

R301.2.1 Wind design criteria. Buildings and portions thereof shall be constructed in accordance with the wind provisions of this code using the basic wind speed in Table R301.2(1) as determined from Figure R301.2(4)A. The structural provisions of this code for wind loads are not permitted where wind design is required as specified in Section R301.2.1.1. Where different construction methods and structural materials are used for various portions of a building, the applicable requirements of this section for each portion shall apply. Where not otherwise specified, the wind loads listed in Table R301.2(2) adjusted for height and exposure using Table R301.2(3) shall be used to determine design load performance requirements for wall coverings, curtain walls, roof coverings, exterior windows, skylights, garage doors and exterior doors. Asphalt shingles shall be designed for wind speeds in accordance with Section R905.2.4. A continuous load path shall be provided to transmit the applicable uplift forces in Section R802.11.1 from the roof assembly to the foundation.

R301.2.1.1 Wind limitations and wind design required. The wind provisions of this code shall not apply to the design of buildings where wind design is required in accordance with Figure R301.2(4)B or where the basic wind speed from Figure R301.2(4)A equals or exceeds 110 miles per hour (49 m/s).

Exceptions:

1. For concrete construction, the wind provisions of this code shall apply in accordance with the limitations of Sections R404 and R611.
2. For structural insulated panels, the wind provisions of this code shall apply in accordance with the limitations of Section R613.

In regions where wind design is required in accordance with Figure R301.2(4)B or where the basic wind speed shown on Figure R301.2(4)A equals or exceeds 110 miles per hour (49 m/s), the design of buildings for wind loads shall be in accordance with one or more of the following methods:

1. AF&PA *Wood Frame Construction Manual* (WFCM); or
2. ICC *Standard for Residential Construction in High-Wind Regions* (ICC 600); or

3. ASCE *Minimum Design Loads for Buildings and Other Structures* (ASCE 7); or
4. AISI *Standard for Cold-Formed Steel Framing—Prescriptive Method For One- and Two-Family Dwellings* (AISI S230); or
5. *International Building Code.*

The elements of design not addressed by the methods in Items 1 through 5 shall be in accordance with the provisions of this code. When ASCE 7 or the *International Building Code* is used for the design of the building, the wind speed map and exposure category requirements as specified in ASCE 7 and the *International Building Code* shall be used.

R301.2.1.2 Protection of openings. Exterior glazing in buildings located in windborne debris regions shall be protected from windborne debris. Glazed opening protection for windborne debris shall meet the requirements of the Large Missile Test of ASTM E 1996 and ASTM E 1886 referenced therein. The applicable wind zones for establishing misile types in ASTM E 1996 are shown on Figure R301.2(4)C. Garage door glazed opening protection for windborne debris shall meet the requirements of an *approved* impact-resisting standard or ANSI/DASMA 115.

Exception: Wood structural panels with a minimum thickness of $^7/_{16}$ inch (11 mm) and a maximum span of 8 feet (2438 mm) shall be permitted for opening protection in one- and two-story buildings. Panels shall be precut and attached to the framing surrounding the opening containing the product with the glazed opening. Panels shall be predrilled as required for the anchorage method and shall be secured with the attachment hardware provided. Attachments shall be designed to resist the component and cladding loads determined in accordance with either Table R301.2(2) or ASCE 7, with the permanent corrosion-resistant attachment hardware provided and anchors permanently installed on the building. Attachment in accordance with Table R301.2.1.2 is permitted for buildings with a mean roof height of 33 feet (10 058 mm) or less where located in Wind Zones 1 and 2 in accordance with Figure R301.2(4)C.

TABLE R301.2(1)
CLIMATIC AND GEOGRAPHIC DESIGN CRITERIA

GROUND SNOW LOAD	WIND DESIGN		SEISMIC DESIGN CATEGORY[f]	SUBJECT TO DAMAGE FROM			WINTER DESIGN TEMP[e]	ICE BARRIER UNDERLAYMENT REQUIRED[h]	FLOOD HAZARDS[g]	AIR FREEZING INDEX[i]	MEAN ANNUAL TEMP[j]
	Speed[d] (mph)	Topographic effects[k]		Weathering[a]	Frost line depth[b]	Termite[c]					

For SI: 1 pound per square foot = 0.0479 kPa, 1 mile per hour = 0.447 m/s.

a. Weathering may require a higher strength concrete or grade of masonry than necessary to satisfy the structural requirements of this code. The weathering column shall be filled in with the weathering index (i.e., "negligible," "moderate" or "severe") for concrete as determined from the Weathering Probability Map [Figure R301.2(3)]. The grade of masonry units shall be determined from ASTM C 34, C 55, C 62, C 73, C 90, C 129, C 145, C 216 or C 652.

b. The frost line depth may require deeper footings than indicated in Figure R403.1(1). The jurisdiction shall fill in the frost line depth column with the minimum depth of footing below finish grade.

c. The jurisdiction shall fill in this part of the table to indicate the need for protection depending on whether there has been a history of local subterranean termite damage.

d. The jurisdiction shall fill in this part of the table with the wind speed from the basic wind speed map [Figure R301.2(4)A]. Wind exposure category shall be determined on a site-specific basis in accordance with Section R301.2.1.4.

e. The outdoor design dry-bulb temperature shall be selected from the columns of $97^1/_2$-percent values for winter from Appendix D of the *International Plumbing Code*. Deviations from the Appendix D temperatures shall be permitted to reflect local climates or local weather experience as determined by the building official.

f. The jurisdiction shall fill in this part of the table with the seismic design category determined from Section R301.2.2.1.

g. The jurisdiction shall fill in this part of the table with (a) the date of the jurisdiction's entry into the National Flood Insurance Program (date of adoption of the first code or ordinance for management of flood hazard areas), (b) the date(s) of the Flood Insurance Study and (c) the panel numbers and dates of all currently effective FIRMs and FBFMs or other flood hazard map adopted by the authority having jurisdiction, as amended.

h. In accordance with Sections R905.2.7.1, R905.4.3.1, R905.5.3.1, R905.6.3.1, R905.7.3.1 and R905.8.3.1, where there has been a history of local damage from the effects of ice damming, the jurisdiction shall fill in this part of the table with "YES." Otherwise, the jurisdiction shall fill in this part of the table with "NO."

i. The jurisdiction shall fill in this part of the table with the 100-year return period air freezing index (BF-days) from Figure R403.3(2) or from the 100-year (99 percent) value on the National Climatic Data Center data table "Air Freezing Index-USA Method (Base 32°F)" at www.ncdc.noaa.gov/fpsf.html.

j. The jurisdiction shall fill in this part of the table with the mean annual temperature from the National Climatic Data Center data table "Air Freezing Index-USA Method (Base 32°F)" at www.ncdc.noaa.gov/fpsf.html.

k. In accordance with Section R301.2.1.5, where there is local historical data documenting structural damage to buildings due to topographic wind speed-up effects, the jurisdiction shall fill in this part of the table with "YES." Otherwise, the jurisdiction shall indicate "NO" in this part of the table.

TABLE R301.2(2)
COMPONENT AND CLADDING LOADS FOR A BUILDING WITH A MEAN ROOF HEIGHT OF 30 FEET LOCATED IN EXPOSURE B (psf)[a, b, c, d, e]

	ZONE	EFFECTIVE WIND AREA (feet²)	BASIC WIND SPEED (mph-3-second gust)																							
			85		90		100		105		110		120		125		130		140		145		150		170	
Roof > 0 to 10 degrees	1	10	10.0	-13.0	10.0	-14.6	10.0	-18.0	10.0	-19.8	10.0	-21.8	10.5	-25.9	11.4	-28.1	12.4	-30.4	14.3	-35.3	15.4	-37.8	16.5	-40.5	21.1	-52.0
	1	20	10.0	-12.7	10.0	-14.2	10.0	-17.5	10.0	-19.3	10.0	-21.2	10.0	-25.2	10.7	-27.4	11.6	-29.6	13.4	-34.4	14.4	-36.9	15.4	-39.4	19.8	-50.7
	1	50	10.0	-12.2	10.0	-13.7	10.0	-16.9	10.0	-18.7	10.0	-20.5	10.0	-24.4	10.0	-26.4	10.6	-28.6	12.3	-33.2	13.1	-35.6	14.1	-38.1	18.1	-48.9
	1	100	10.0	-11.9	10.0	-13.3	10.0	-18.5	10.0	-18.2	10.0	-19.9	10.0	-23.7	10.0	-25.7	10.0	-27.8	11.4	-32.3	12.2	-34.6	13.0	-37.0	16.7	-47.6
	2	10	10.0	-21.8	10.0	-24.4	10.0	-30.2	10.0	-33.3	10.0	-36.5	10.5	-43.5	11.4	-47.2	12.4	-51.0	14.3	-59.2	15.4	-63.5	16.5	-67.9	21.1	-87.2
	2	20	10.0	-19.5	10.0	-21.8	10.0	-27.0	10.0	-29.7	10.0	-32.6	10.0	-38.8	10.7	-42.1	11.6	-45.6	13.4	-52.9	14.4	-56.7	15.4	-60.7	19.8	-78.0
	2	50	10.0	-16.4	10.0	-18.4	10.0	-22.7	10.0	-25.1	10.0	-27.5	10.0	-32.7	10.0	-35.5	10.6	-38.4	12.3	-44.5	13.1	-47.8	14.1	-51.1	18.1	-65.7
	2	100	10.0	-14.1	10.0	-15.8	10.0	-19.5	10.0	-21.5	10.0	-23.6	10.0	-28.1	10.0	-30.5	10.0	-33.0	11.4	-38.2	12.2	-41.0	13.0	-43.9	16.7	-56.4
	3	10	10.0	-32.8	10.0	-36.8	10.0	-45.4	10.0	-50.1	10.0	-55.0	10.5	-65.4	11.4	-71.0	12.4	-76.8	14.3	-89.0	15.4	-95.5	16.5	102.2	21.1	131.3
	3	20	10.0	-27.2	10.0	-30.5	10.0	-37.6	10.0	-41.5	10.0	-45.5	10.0	-54.2	10.7	-58.8	11.6	-63.6	13.4	-73.8	14.4	-79.1	15.4	-84.7	19.8	108.7
	3	50	10.0	-19.7	10.0	-22.1	10.0	-27.3	10.0	-30.1	10.0	-33.1	10.0	-39.3	10.0	-42.7	10.6	-46.2	12.3	-53.5	13.1	-57.4	14.1	-61.5	18.1	-78.9
	3	100	10.0	-14.1	10.0	-15.8	10.0	-19.5	10.0	-21.5	10.0	-23.6	10.0	-28.1	10.0	-30.5	10.0	-33.0	11.4	-38.2	12.2	-41.0	13.0	-43.9	16.7	-56.4
Roof > 10 to 30 degrees	1	10	10.0	-11.9	10.0	-13.3	10.4	-16.5	11.4	-18.2	12.5	-19.9	14.9	-23.7	16.2	-25.7	17.5	-27.8	20.3	-32.3	21.8	-34.6	23.3	-37.0	30.0	-47.6
	1	20	10.0	-11.6	10.0	-13.0	10.0	-16.0	10.4	-17.6	11.4	-19.4	13.6	-23.0	14.8	-25.0	16.0	-27.0	18.5	-31.4	19.9	-33.7	21.3	-36.0	27.3	-46.3
	1	50	10.0	-11.1	10.0	-12.5	10.0	-15.4	10.0	-17.0	10.0	-18.6	11.9	-22.2	12.9	-24.1	13.9	-26.0	16.1	-30.2	17.3	-32.4	18.5	-34.6	23.8	-44.5
	1	100	10.0	-10.8	10.0	-12.1	10.0	-14.9	10.0	-16.5	10.0	-18.1	10.5	-21.5	11.4	-23.3	12.4	-25.2	14.3	-29.3	15.4	-31.4	16.5	-33.6	21.1	-43.2
	2	10	10.0	-25.1	10.0	-28.2	10.4	-34.8	11.4	-38.3	12.5	-42.1	14.9	-50.1	16.2	-54.3	17.5	-58.7	20.3	-68.1	21.8	-73.1	23.3	-78.2	30.0	100.5
	2	20	10.0	-22.8	10.0	-25.6	10.0	-31.5	10.4	-34.8	11.4	-38.2	13.6	-45.4	14.8	-49.3	16.0	-53.3	18.5	-61.8	19.9	-66.3	21.3	-71.0	27.3	-91.2
	2	50	10.0	-19.7	10.0	-22.1	10.0	-27.3	10.0	-30.1	10.0	-33.0	11.9	-39.3	12.9	-42.7	13.9	-46.1	16.1	-53.5	17.3	-57.4	18.5	-61.4	23.8	-78.9
	3	20	10.0	-22.8	10.0	-25.6	10.0	-31.5	10.4	-34.8	11.4	-38.2	13.6	-45.4	14.8	-49.3	16.0	-53.3	18.5	-61.8	19.9	-66.3	21.3	-71.0	27.3	-91.2
	3	50	10.0	-19.7	10.0	-22.1	10.0	-27.3	10.0	-30.1	10.0	-33.0	11.9	-39.3	12.9	-42.7	13.9	-46.1	16.1	-53.5	17.3	-57.4	18.5	-61.4	23.8	-78.9
	3	100	10.0	-17.4	10.0	-19.5	10.0	-24.1	10.0	-26.6	10.0	-29.1	10.5	-34.7	11.4	-37.6	12.4	-40.7	14.3	-47.2	15.4	-50.6	16.5	-54.2	21.1	-69.6
Roof > 30 to 45 degrees	1	10	11.9	-13.0	13.3	-14.6	16.5	-18.0	18.2	-19.8	19.9	-21.8	23.7	-25.9	25.7	-28.1	27.8	-30.4	32.3	-35.3	34.6	-37.8	37.0	-40.5	47.6	-52.0
	1	20	11.6	-12.3	13.0	-13.8	16.0	-17.1	17.6	-18.8	19.4	-20.7	23.0	-24.6	25.0	-26.7	27.0	-28.9	31.4	-33.5	33.7	-35.9	36.0	-38.4	46.3	-49.3
	1	50	11.1	-11.5	12.5	-12.8	15.4	-15.9	17.0	-17.5	18.6	-19.2	22.2	-22.8	24.1	-24.8	26.0	-25.8	30.2	-31.1	32.4	-33.3	34.6	-35.7	44.5	-45.8
	1	100	10.8	-10.8	12.1	-12.1	14.9	-14.9	16.5	-16.5	18.1	-18.1	21.5	-21.5	23.3	-23.3	25.2	-25.2	29.3	-29.3	31.4	-31.4	33.6	-33.6	43.2	-43.2
	2	10	11.9	-15.2	13.3	-17.0	16.5	-21.0	18.2	-23.2	19.9	-25.5	23.7	-30.3	25.7	-32.9	27.8	-35.6	32.3	-41.2	34.6	-44.2	37.0	-47.3	47.6	-60.8
	2	20	11.6	-14.5	13.0	-16.3	16.0	-20.1	17.6	-22.2	19.4	-24.3	23.0	-29.0	25.0	-31.4	27.0	-34.0	31.4	-39.4	33.7	-42.3	36.0	-45.3	46.3	-58.1
	2	50	11.1	-13.7	12.5	-15.3	15.4	-18.9	17.0	-20.8	18.6	-22.9	22.2	-27.2	24.1	-29.5	26.0	-32.0	30.2	-37.1	32.4	-39.8	34.6	-42.5	44.5	-54.6
	2	100	10.8	-13.0	12.1	-14.6	14.9	-18.0	16.5	-19.8	18.1	-21.8	21.5	-25.9	23.3	-28.1	25.2	-30.4	29.3	-35.3	31.4	-37.8	33.6	-40.5	43.2	-52.0
	3	10	11.9	-15.2	13.3	-17.0	16.5	-21.0	18.2	-23.2	19.9	-25.5	23.7	-30.3	25.7	-32.9	27.8	-35.6	32.3	-41.2	34.6	-44.2	37.0	-47.3	47.6	-60.8
	3	20	11.6	-14.5	13.0	-16.3	16.0	-20.1	17.6	-22.2	19.4	-24.3	23.0	-29.0	25.0	-31.4	27.0	-34.0	31.4	-39.4	33.7	-42.3	36.0	-45.3	46.3	-58.1
	3	50	11.1	-13.7	12.5	-15.3	15.4	-18.9	17.0	-20.8	18.6	-22.9	22.2	-27.2	24.1	-29.5	26.0	-32.0	30.2	-37.1	32.4	-39.8	34.6	-42.5	44.5	-54.5
	3	100	10.8	-13.0	12.1	-14.6	14.9	-18.0	16.5	-19.8	18.1	-21.8	21.5	-25.9	23.3	-28.1	25.2	-30.4	29.3	-35.3	31.4	-37.8	33.6	-40.5	43.2	-52.0
Wall	4	10	13.0	-14.1	14.6	-15.8	18.0	-19.5	19.8	-21.5	21.8	-23.6	25.9	-28.1	28.1	-30.5	30.4	-33.0	35.3	-38.2	37.8	-41.0	40.5	-43.9	52.0	-56.4
	4	20	12.4	-13.5	13.9	-15.1	17.2	-18.7	18.9	-20.6	20.8	-22.6	24.7	-26.9	26.8	-29.2	29.0	-31.6	33.7	-36.7	36.1	-39.3	38.7	-42.1	49.6	-54.1
	4	50	11.6	-12.7	13.0	-14.3	16.1	-17.6	17.8	-19.4	19.5	-21.3	23.2	-25.4	25.2	-27.5	27.2	-29.8	31.6	-34.6	33.9	-37.1	36.2	-39.7	46.6	-51.0
	4	100	11.1	-12.2	12.4	-13.6	15.3	-16.8	16.9	-18.5	18.5	-20.4	22.0	-24.2	23.9	-26.3	25.9	-28.4	30.0	-33.0	32.2	-35.4	34.4	-37.8	44.2	-48.6
	5	10	13.0	-17.4	14.6	-19.5	18.0	-24.1	19.8	-26.6	21.8	-29.1	25.9	-34.7	28.1	-37.6	30.4	-40.7	35.3	-47.2	37.8	-50.6	40.5	-54.2	52.0	-69.6
	5	20	12.4	-16.2	13.9	-18.2	17.2	-22.5	18.9	-24.8	20.8	-27.2	24.7	-32.4	26.8	-35.1	29.0	-38.0	33.7	-44.0	36.1	-47.2	38.7	-50.5	49.6	-64.9
	5	50	11.6	-14.7	13.0	-16.5	16.1	-20.3	17.8	-22.4	19.5	-24.6	23.2	-29.3	25.2	-31.8	27.2	-34.3	31.6	-39.8	33.9	-42.7	36.2	-45.7	46.6	-58.7
	5	100	11.1	-13.5	12.4	-15.1	15.3	-18.7	16.9	-20.6	18.5	-22.6	22.0	-26.9	23.9	-29.2	25.9	-31.6	30.0	-36.7	32.2	-39.3	34.4	-42.1	44.2	-54.1

For SI: 1 foot = 304.8 mm, 1 square foot = 0.0929 m², 1 mile per hour = 0.447 m/s, 1 pound per square foot = 0.0479 kPa.

a. The effective wind area shall be equal to the span length multiplied by an effective width. This width shall be permitted to be not be less than one-third the span length. For cladding fasteners, the effective wind area shall not be greater than the area that is tributary to an individual fastener.

b. For effective areas between those given above, the load may be interpolated; otherwise, use the load associated with the lower effective area.

c. Table values shall be adjusted for height and exposure by multiplying by the adjustment coefficient in Table R301.2(3).

d. See Figure R301.2(7) for location of zones.

e. Plus and minus signs signify pressures acting toward and away from the building surfaces.

TABLE R301.2(3)
HEIGHT AND EXPOSURE ADJUSTMENT COEFFICIENTS FOR TABLE R301.2(2)

MEAN ROOF HEIGHT	EXPOSURE		
	B	C	D
15	1.00	1.21	1.47
20	1.00	1.29	1.55
25	1.00	1.35	1.61
30	1.00	1.40	1.66
35	1.05	1.45	1.70
40	1.09	1.49	1.74
45	1.12	1.53	1.78
50	1.16	1.56	1.81
55	1.19	1.59	1.84
60	1.22	1.62	1.87

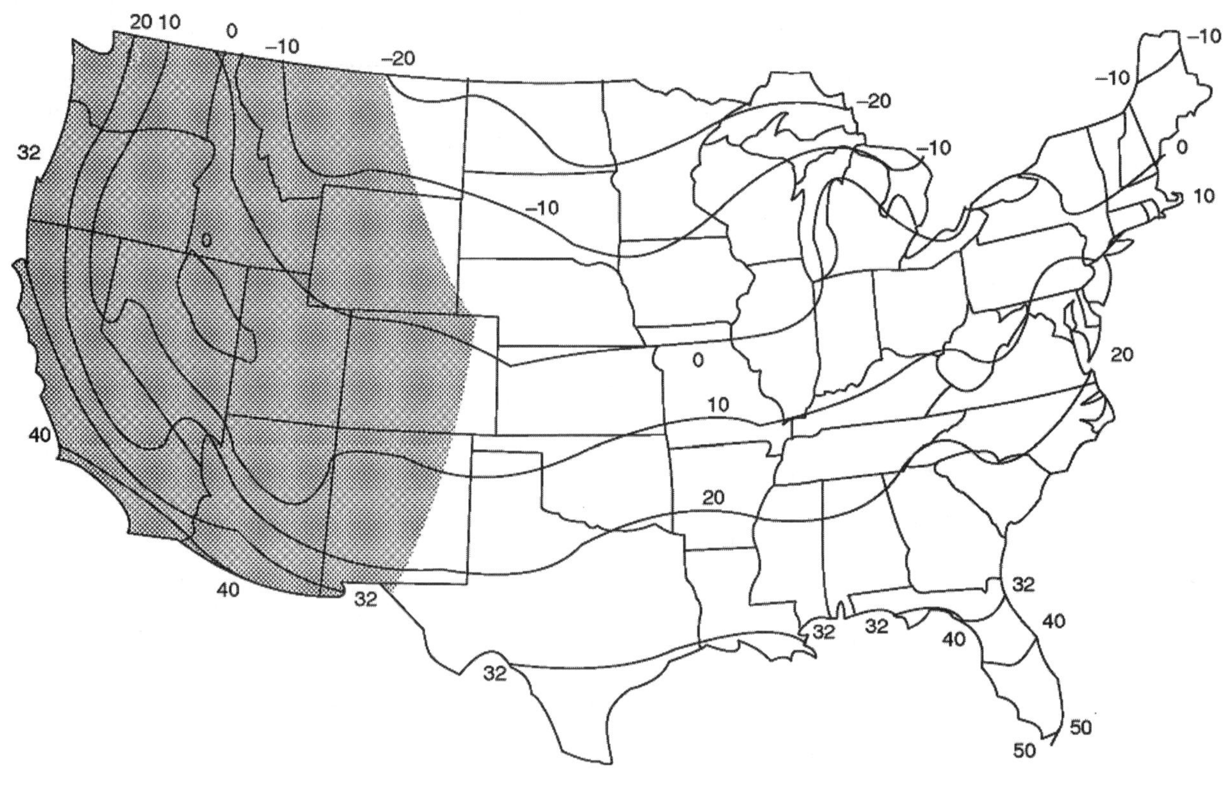

For SI: °C = [(°F)-32]/1.8.

FIGURE R301.2(1)
ISOLINES OF THE 97¹/₂ PERCENT WINTER (DECEMBER, JANUARY AND FEBRUARY) DESIGN TEMPERATURES (°F)

Explanation

%g		Seismic Design Category
		E
125		D_2
83		D_1
67		D_0
50		C
33		B
17		A
0		

REFERENCES

Building Seismic Safety Council, 2009, NEHRP Recommended Provisions for Seismic Regulations for New Buildings and Other Structures: FEMA P750/2009 Edition, Federal Emergency Management Agency, Washington, DC

Huang, Yin-Nan, Whittaker, A.S., and Luco, Nicolas, 2008, Maximum spectral demands in the near-fault region, Earthquake Spectra Volume 24, Issue 1, pp. 319-341.

Luco, Nicolas, Ellingwood, B.R., Hamburger, R.O., Hooper, J.D., Kimball, J.K., and Kircher, C.A., 2007, Risk-Targeted versus Current Seismic Design Maps for the Conterminous United States, Structural Engineers Association of California 2007 Convention Proceedings, pp. 163-175.

Wesson, Robert L., Boyd, Oliver S., Mueller, Charles S., Bufe, Charles G., Frankel, Arthur D., Petersen, Mark D., 2007, Revision of time-Independent probabilistic seismic hazard maps for Alaska U.S. Geological Survey Open-File Report 2007-1043.

250 0 250 500 MILES

250 0 250 500 KILOMETERS

Map prepared by U.S. Geological Survey in collaboration with the Federal Emergency Management Agency (FEMA) funded Building Seismic Safety Council's (BSSC) Code Resource Support Committee (CRSC).

FIGURE R301.2(2)
SEISMIC DESIGN CATEGORIES—SITE CLASS D

(continued)

REFERENCES

Building Seismic Safety Council, 2009, NEHRP Recommended Provisions for Seismic Regulations for New Buildings and Other Structures: FEMA P750/2009 Edition, Federal Emergency Management Agency, Washington, DC.

Huang, Yin-Nan, Whittaker, A.S., and Luco, Nicolas, 2008, Maximum spectral demands in the near-fault region, Earthquake Spectra Volume 24, Issue 1, pp. 319-341.

Klein, F., Frankel, A., Mueller, C., Wesson, R., and Okubo, P., 2001, Seismic Hazard in Hawaii: High Rate of Large Earthquakes and Probabilistic Ground Motion Maps, Bulletin of the Seismological Society of America, v. 91, pp. 479-498.

Klein, F., Frankel, A., Mueller, C., Wesson, R., and Okubo, P., 2000, Seismic Hazard Maps for Hawaii: U.S. Geological Survey Miscellaneous Investigations Series I-2724, 2 sheets, scale 1:2,000,000.

Luco, Nicolas, Ellingwood, B.R., Hamburger, R.O., Hooper, J.D., Kimball, J.K., and Kircher, C.A., 2007, Risk-Targeted versus Current Seismic Design Maps for the Conterminous United States, Structural Engineers Association of California 2007 Convention Proceedings, pp. 163-175.

Map prepared by U.S. Geological Survey in collaboration with the Federal Emergency Management Agency (FEMA)-funded Building Seismic Safety Council's (BSSC) Code Resource Support Committee (CRSC).

Explanation

%g	Seismic Design Category
125	E
83	D_2
67	D_1
50	D_0
33	C
17	B
0	A

FIGURE R301.2(2)—continued
SEISMIC DESIGN CATEGORIES—SITE CLASS D

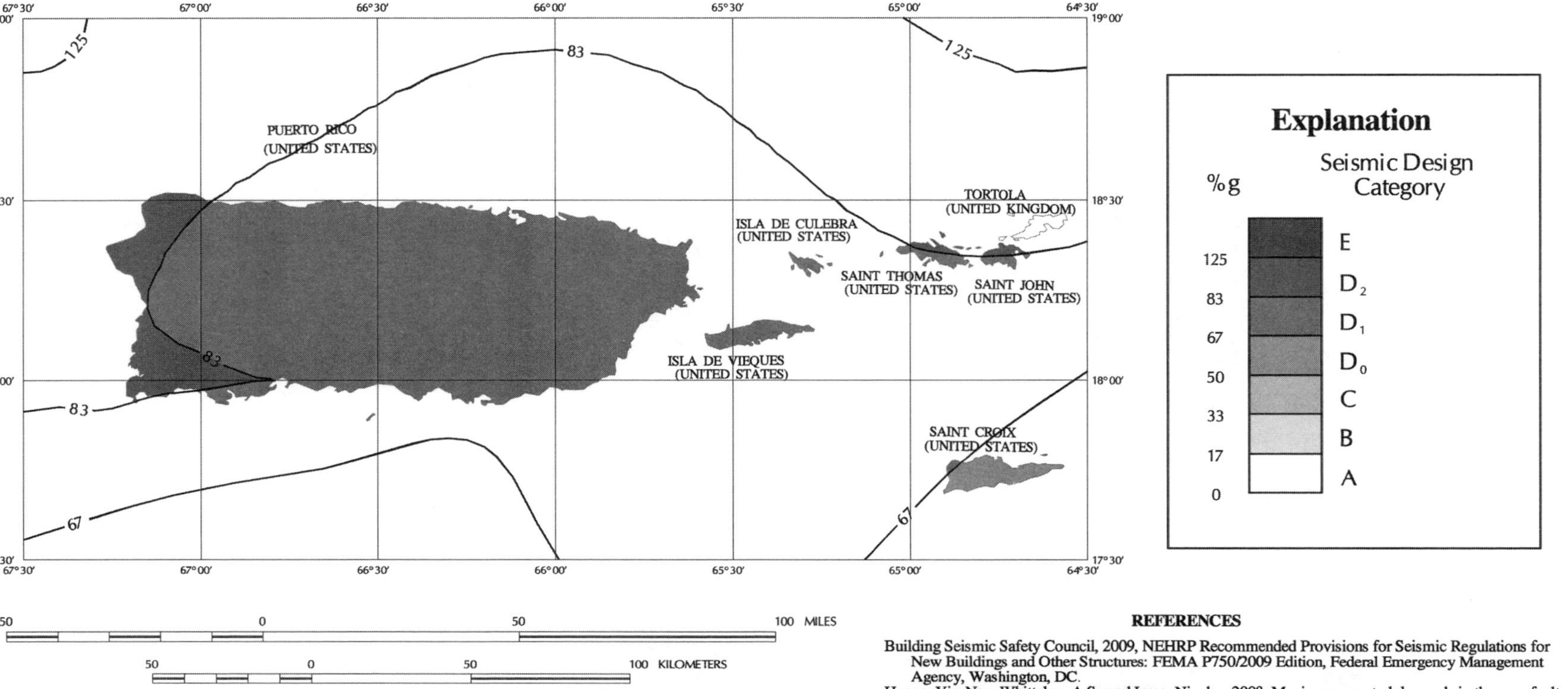

REFERENCES

Building Seismic Safety Council, 2009, NEHRP Recommended Provisions for Seismic Regulations for New Buildings and Other Structures: FEMA P750/2009 Edition, Federal Emergency Management Agency, Washington, DC.

Huang, Yin-Nan, Whittaker, A.S., and Luco, Nicolas, 2008, Maximum spectral demands in the near-fault region, Earthquake Spectra Volume 24, Issue 1, pp. 319-341.

Luco, Nicolas, Ellingwood, B.R., Hamburger, R.O., Hooper, J.D., Kimball, J.K., and Kircher, C.A., 2007, Risk-Targeted versus Current Seismic Design Maps for the Conterminous United States, Structural Engineers Association of California 2007 Convention Proceedings, pp. 163-175.

Mueller, C., Frankel, A., Petersen, M., and Leyendecker, E., 2003, Documentation for 2003 USGS Seismic Hazard Maps for Puerto Rico and the U.S. Virgin Islands, U.S. Geological Survey Open-File Report 03-379.

Map prepared by U.S. Geological Survey in collaboration with the Federal Emergency Management Agency (FEMA)-funded Building Seismic Safety Council's (BSSC) Code Resource Support Committee (CRSC).

FIGURE R301.2(2)—continued
SEISMIC DESIGN CATEGORIES—SITE CLASS D

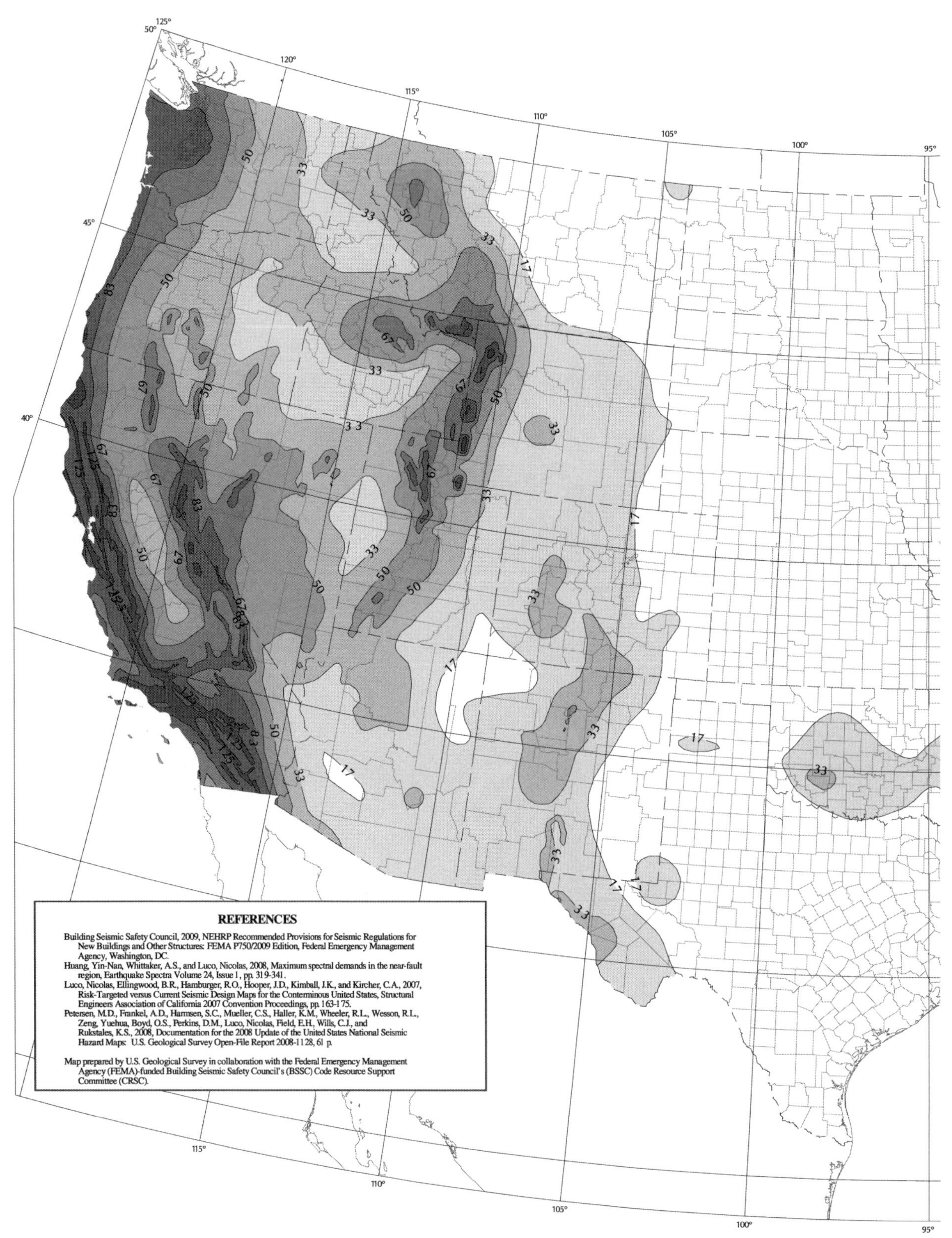

FIGURE R301.2(2)—continued
SEISMIC DESIGN CATEGORIES—SITE CLASS D

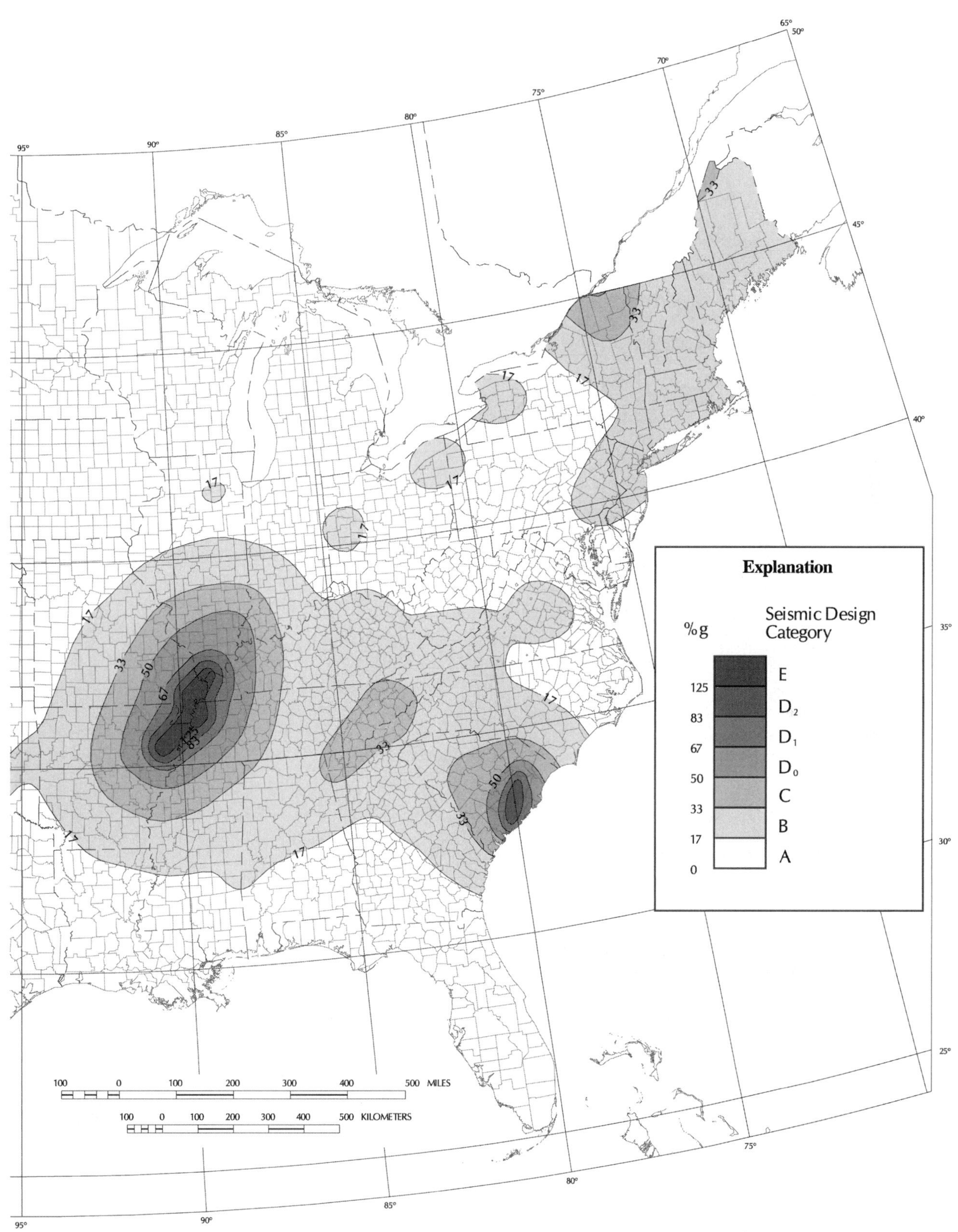

FIGURE R301.2(2)—continued
SEISMIC DESIGN CATEGORIES—SITE CLASS D

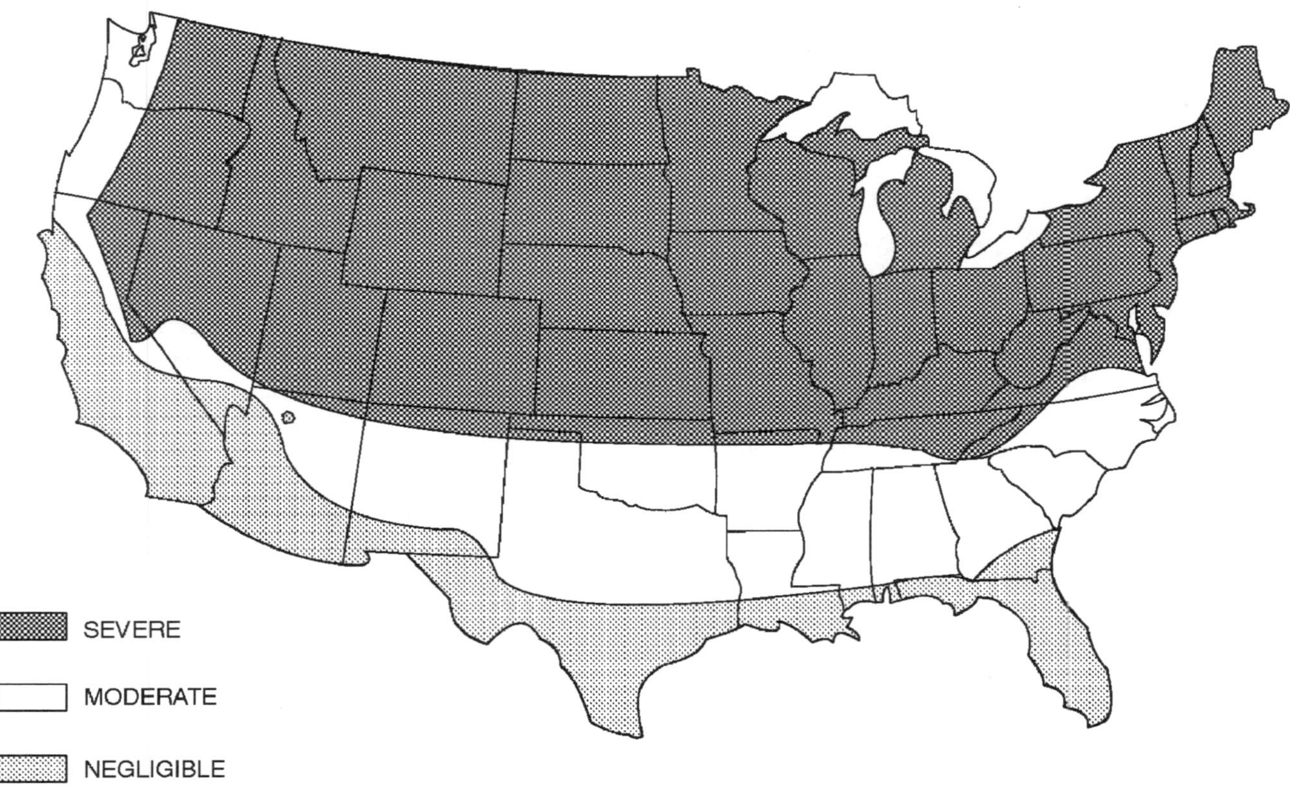

a. Alaska and Hawaii are classified as severe and negligible, respectively.

b. Lines defining areas are approximate only. Local conditions may be more or less severe than indicated by region classification. A severe classification is where weather conditions result in significant snowfall combined with extended periods during which there is little or no natural thawing causing deicing salts to be used extensively.

FIGURE R301.2(3)
WEATHERING PROBABILITY MAP FOR CONCRETE[a, b]

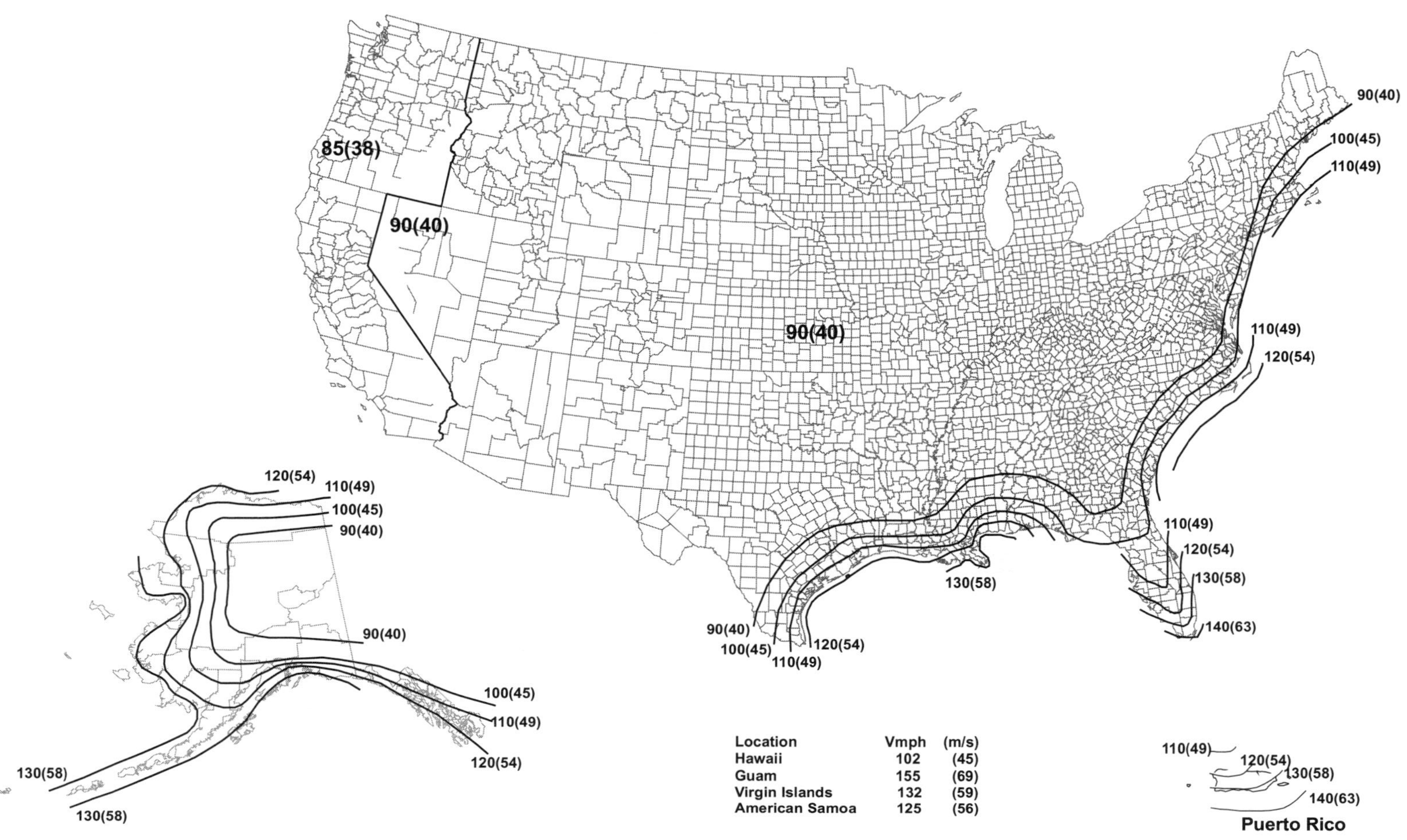

Location	Vmph	(m/s)
Hawaii	102	(45)
Guam	155	(69)
Virgin Islands	132	(59)
American Samoa	125	(56)

Notes:
1. Values are nominal design 3-second gust wind speeds in miles per hour (m/s) at 33 ft (10 m) above ground for Exposure C category.
2. Linear interpolation between contours is permitted.
3. Islands and coastal areas outside the last contour shall use the last wind speed contour of the coastal area.
4. Mountainous terrain, gorges, ocean promontories, and special wind regions shall be examined for unusual wind conditions.

FIGURE R301.2(4)A
BASIC WIND SPEEDS

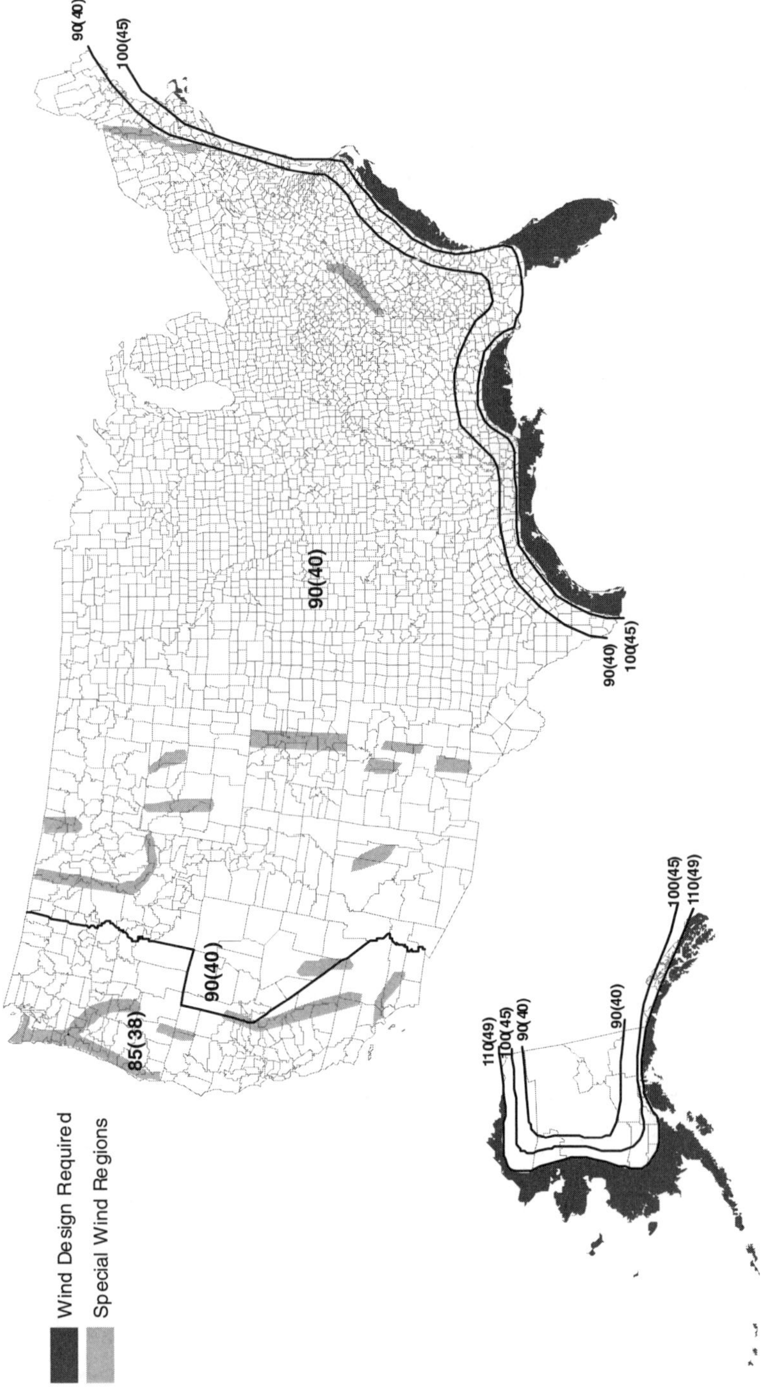

Wind Design Required
Special Wind Regions

85(38)
90(40)
90(40)
90(40)
100(45)
90(40)
100(45)
110(49)
100(45)
90(40)
90(40)
100(45)
110(49)

Other Locations where Wind Design Required
Puerto Rico
Guam
Virgin Islands
American Samoa
Hawaii - Special Wind Regions

FIGURE R301.2(4)B
REGIONS WHERE WIND DESIGN IS REQUIRED

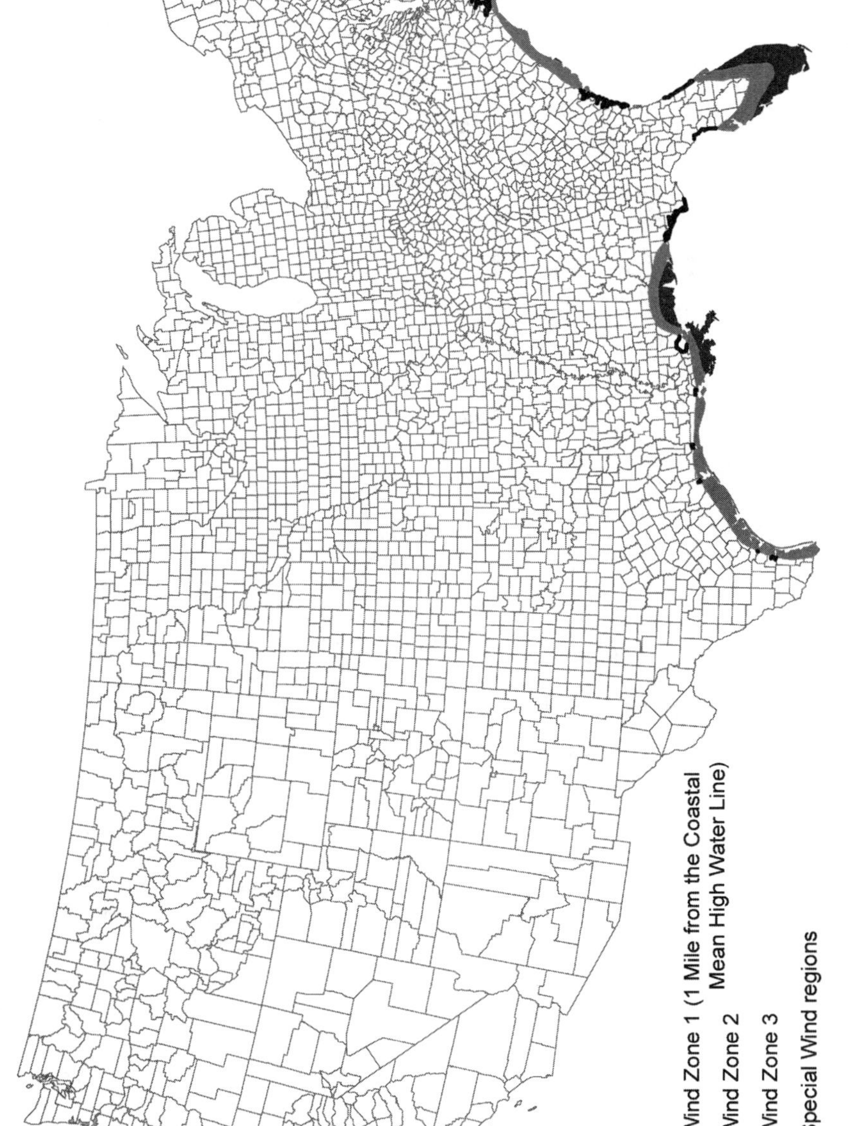

Note:
Wind Zone 3 applies for:
Guam
Virgin Islands
Amerigan Samoa
Puerto Rico

Note: Wind Zone 3 applies in Wind Zone 2 areas that are within a mile of the Coastal Mean High Water Line.

Note: Wind Zone 1 applies in Hawaii - Special Wind Regions.

FIGURE R301.2(4)C
WIND-BORNE DEBRIS REGIONS

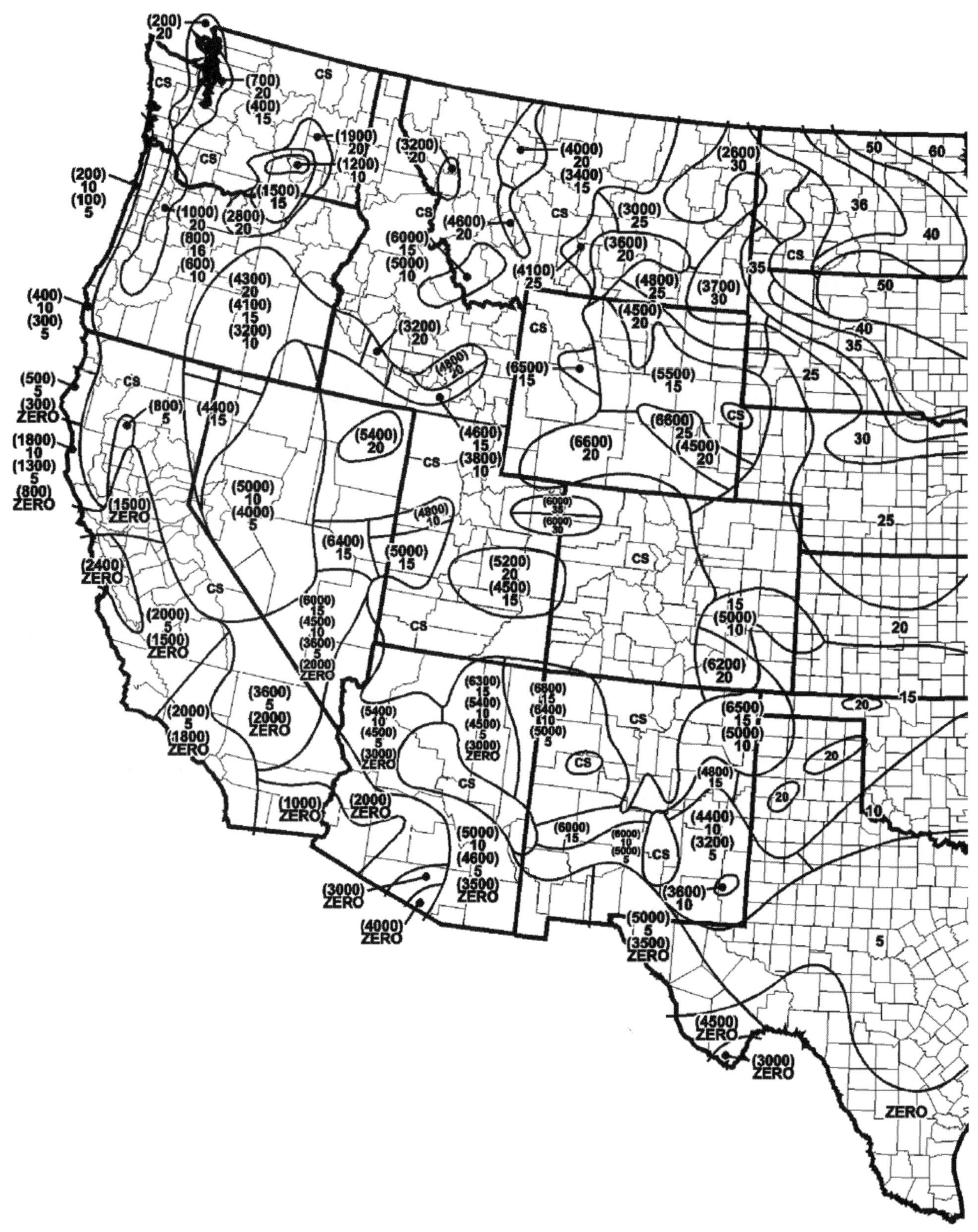

FIGURE R301.2(5)
GROUND SNOW LOADS, P_g, FOR THE UNITED STATES (lb/ft^2)
(continued)

For SI: 1 foot = 304.8 mm, 1 pound per square foot = 0.0479 kPa.

FIGURE R301.2(5)—continued
GROUND SNOW LOADS, P_g, FOR THE UNITED STATES (lb/ft^2)

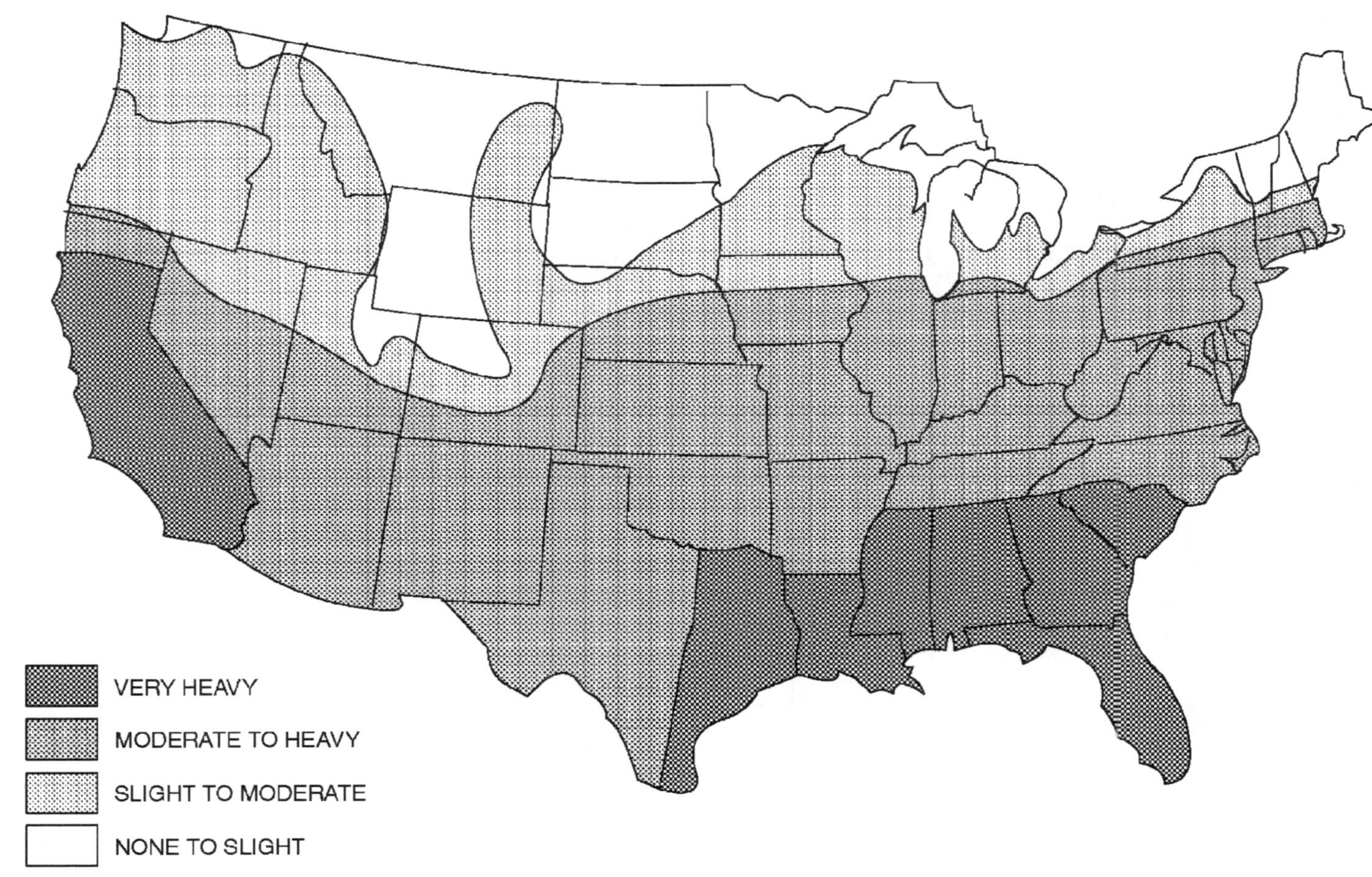

Note: Lines defining areas are approximate only. Local conditions may be more or less severe than indicated by the region classification.

FIGURE R301.2(6)
TERMITE INFESTATION PROBABILITY MAP

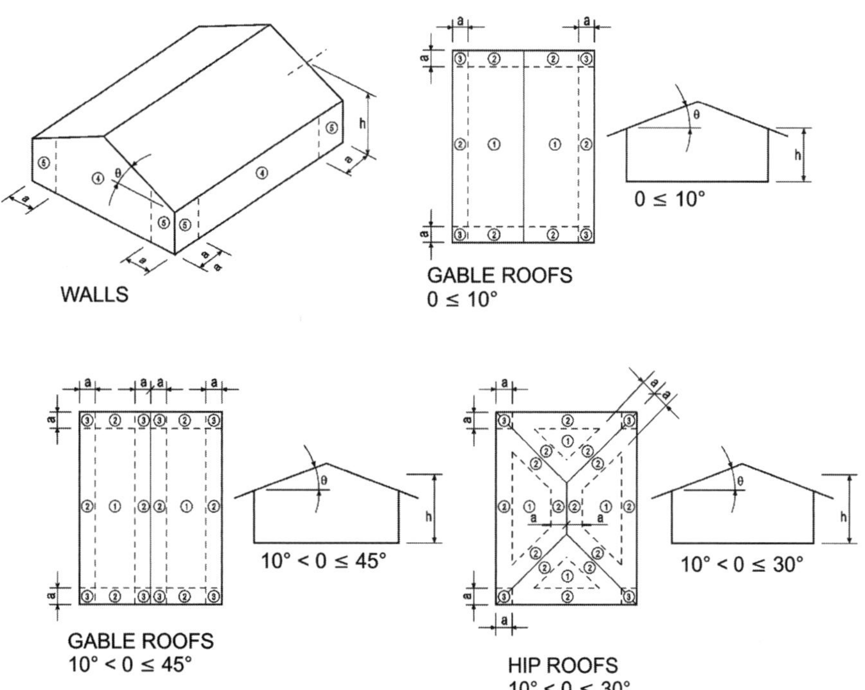

For SI: 1 foot = 304.8 mm, 1 degree = 0.0175 rad.
Note: a = 4 feet in all cases.

**FIGURE R301.2(7)
COMPONENT AND CLADDING PRESSURE ZONES**

**TABLE R301.2.1.2
WINDBORNE DEBRIS PROTECTION FASTENING SCHEDULE FOR WOOD STRUCTURAL PANELS[a, b, c, d]**

FASTENER TYPE	FASTENER SPACING (inches)[a, b]		
	Panel span ≤ 4 feet	4 feet < panel span ≤ 6 feet	6 feet < panel span ≤ 8 feet
No. 8 wood screw based anchor with 2-inch embedment length	16	10	8
No. 10 wood screw based anchor with 2-inch embedment length	16	12	9
$^1/_4$-inch lag screw based anchor with 2-inch embedment length	16	16	16

For SI: 1 inch = 25.4 mm, 1 foot = 304.8 mm, 1 pound = 4.448 N, 1 mile per hour = 0.447 m/s.

a. This table is based on 130 mph wind speeds and a 33-foot mean roof height.
b. Fasteners shall be installed at opposing ends of the wood structural panel. Fasteners shall be located a minimum of 1 inch from the edge of the panel.
c. Anchors shall penetrate through the exterior wall covering with an embedment length of 2 inches minimum into the building frame. Fasteners shall be located a minimum of $2^1/_2$ inches from the edge of concrete block or concrete.
d. Where panels are attached to masonry or masonry/stucco, they shall be attached using vibration-resistant anchors having a minimum ultimate withdrawal capacity of 1500 pounds.

R301.2.1.3 Wind speed conversion. When referenced documents are based on fastest mile wind speeds, the three-second gust basic wind speeds, V_{3s}, of Figure R301.2(4) shall be converted to fastest mile wind speeds, V_{fm}, using Table R301.2.1.3.

R301.2.1.4 Exposure category. For each wind direction considered, an exposure category that adequately reflects the characteristics of ground surface irregularities shall be determined for the site at which the building or structure is to be constructed. For a site located in the transition zone between categories, the category resulting in the largest wind forces shall apply. Account shall be taken of variations in ground surface roughness that arise from natural topography and vegetation as well as from constructed features. For a site where multiple detached one- and two-family dwellings, *townhouses* or other structures are to be constructed as part of a subdivision, master-planned community, or otherwise designated as a developed area by the authority having jurisdiction, the exposure category for an individual structure shall be based upon the site conditions that will exist at the time when all adjacent structures on the site have been constructed, provided their construction is expected to begin within one year of the start of construction for the structure for which the

**TABLE R301.2.1.3
EQUIVALENT BASIC WIND SPEEDS[a]**

3-second gust, V_{3s}	85	90	100	105	110	120	125	130	140	145	150	160	170
Fastest mile, V_{fm}	71	76	85	90	95	104	109	114	123	128	133	142	152

For SI: 1 mile per hour = 0.447 m/s.
a. Linear interpolation is permitted.

exposure category is determined. For any given wind direction, the exposure in which a specific building or other structure is sited shall be assessed as being one of the following categories:

1. Exposure A. Large city centers with at least 50 percent of the buildings having a height in excess of 70 feet (21 336 mm). Use of this exposure category shall be limited to those areas for which terrain representative of Exposure A prevails in the upwind direction for a distance of at least 0.5 mile (0.8 km) or 10 times the height of the building or other structure, whichever is greater. Possible channeling effects or increased velocity pressures due to the building or structure being located in the wake of adjacent buildings shall be taken into account.
2. Exposure B. Urban and suburban areas, wooded areas, or other terrain with numerous closely spaced obstructions having the size of single-family dwellings or larger. Exposure B shall be assumed unless the site meets the definition of another type exposure.
3. Exposure C. Open terrain with scattered obstructions, including surface undulations or other irregularities, having heights generally less than 30 feet (9144 mm) extending more than 1,500 feet (457 m) from the building site in any quadrant. This exposure shall also apply to any building located within Exposure B type terrain where the building is directly adjacent to open areas of Exposure C type terrain in any quadrant for a distance of more than 600 feet (183 m). This category includes flat, open country and grasslands.
4. Exposure D. Flat, unobstructed areas exposed to wind flowing over open water for a distance of at least 1 mile (1.61 km). Shorelines in Exposure D include inland waterways, the Great Lakes, and coastal areas of California, Oregon, Washington and Alaska. This exposure shall apply only to those buildings and other structures exposed to the wind coming from over the water. Exposure D extends inland from the shoreline a distance of 1500 feet (457 m) or 10 times the height of the building or structure, whichever is greater.

R301.2.1.5 Topographic wind effects. In areas designated in Table R301.2(1) as having local historical data documenting structural damage to buildings caused by wind speed-up at isolated hills, ridges and escarpments that are abrupt changes from the general topography of the area, topographic wind effects shall be considered in the design of the building in accordance with Section R301.2.1.5.1 or in accordance with the provisions of ASCE 7. See Figure R301.2.1.5.1(1) for topographic features for wind speed-up effect.

In these designated areas, topographic wind effects shall apply only to buildings sited on the top half of an isolated hill, ridge or escarpment where all of the following conditions exist:

1. The average slope of the top half of the hill, ridge or escarpment is 10 percent or greater.
2. The hill, ridge or escarpment is 60 feet (18 288 mm) or greater in height for Exposure B, 30 feet (9144 mm) or greater in height for Exposure C, and 15 feet (4572 mm) or greater in height for Exposure D.
3. The hill, ridge or escarpment is isolated or unobstructed by other topographic features of similar height in the upwind direction for a distance measured from its high point of 100 times its height or 2 miles, whichever is less. See Figure R301.2.1.5.1(3) for upwind obstruction.
4. The hill, ridge or escarpment protrudes by a factor of two or more above the height of other upwind topographic features located in any quadrant within a radius of 2 miles measured from its high point.

R301.2.1.5.1 Simplified topographic wind speed-up method. As an alternative to the ASCE 7 topographic wind provisions, the provisions of Section R301.2.1.5.1 shall be permitted to be used to design for wind speed-up effects, where required by Section R301.2.1.5.

Structures located on the top half of isolated hills, ridges or escarpments meeting the conditions of Section R301.2.1.5 shall be designed for an increased basic wind speed as determined by Table R301.2.1.5.1. On the high side of an escarpment, the increased basic wind speed shall extend horizontally downwind from the edge of the escarpment 1.5

TABLE R301.2.1.5.1
BASIC WIND MODIFICATION FOR TOPOGRAPHIC WIND EFFECT

BASIC WIND SPEED FROM FIGURE R301.2(4) (mph)	AVERAGE SLOPE OF THE TOP HALF OF HILL, RIDGE OR ESCARPMENT (percent)						
	0.10	0.125	0.15	0.175	0.20	0.23	0.25 or greater
	Required basic wind speed-up, modified for topographic wind speed up (mph)						
85	100	100	100	110	110	110	120
90	100	100	110	110	120	120	120
100	110	120	120	130	130	130	140
110	120	130	130	140	140	150	150
120	140	140	150	150	N/A	N/A	N/A
130	150	N/A	N/A	N/A	N/A	N/A	N/A

For SI: 1 mile per hour = 0.447 m/s.

times the horizontal length of the upwind slope (1.5L) or 6 times the height of the escarpment (6H), whichever is greater. See Figure R301.2.1.5.1(2) for where wind speed increase is applied.

R301.2.2 Seismic provisions. The seismic provisions of this code shall apply as follows:

1. Townhouses in Seismic Design Categories C, D_0, D_1 and D_2.
2. Detached one- and two-family dwellings in Seismic Design Categories, D_0, D_1 and D_2.

R301.2.2.1 Determination of seismic design category. Buildings shall be assigned a seismic design category in accordance with Figure R301.2(2).

R301.2.2.1.1 Alternate determination of seismic design category. The seismic design categories and corresponding short period design spectral response

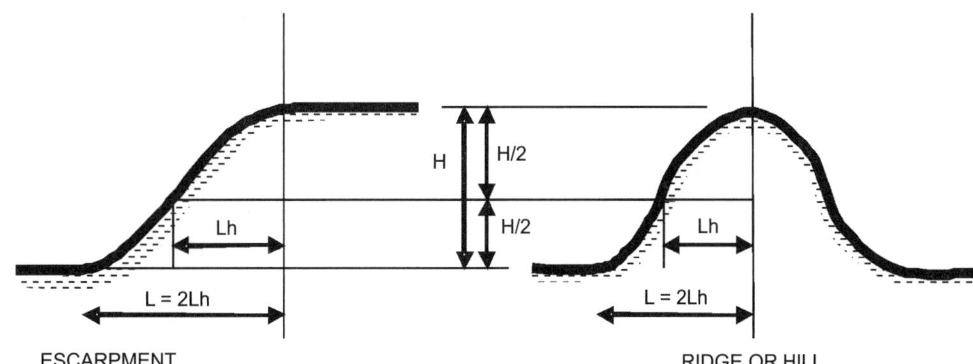

Note: H/2 determines the measurement point for Lh. L is twice Lh.

FIGURE R301.2.1.5.1(1)
TOPOGRAPHIC FEATURES FOR WIND SPEED-UP EFFECT

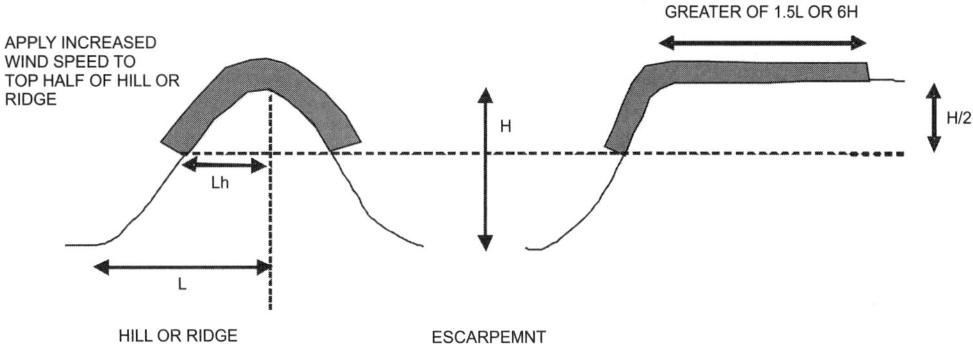

FIGURE R301.2.1.5.1(2)
ILLUSTRATION OF WHERE ON A TOPOGRAPHIC FEATURE, WIND SPEED INCREASE IS APPLIED

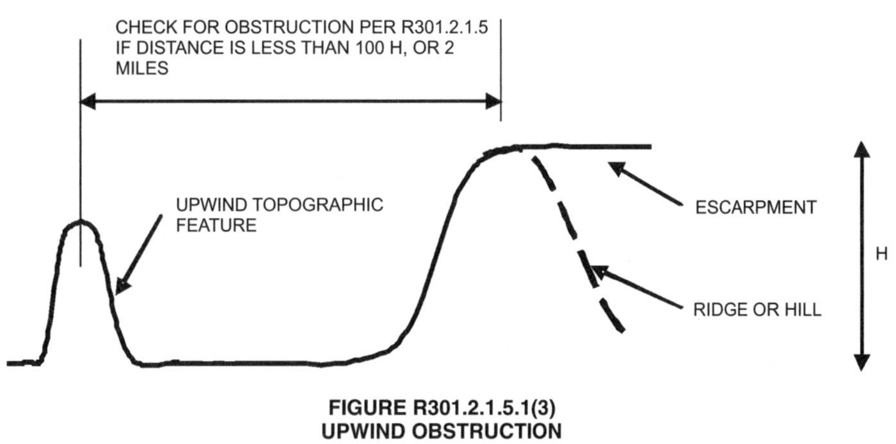

FIGURE R301.2.1.5.1(3)
UPWIND OBSTRUCTION

accelerations, S_{DS} shown in Figure R301.2(2) are based on soil Site Class D, as defined in Section 1613.5.2 of the *International Building Code*. If soil conditions are other than Site Class D, the short period design spectral response accelerations, S_{DS}, for a site can be determined according to Section 1613.5 of the *International Building Code*. The value of S_{DS} determined according to Section 1613.5 of the *International Building Code* is permitted to be used to set the seismic design category according to Table R301.2.2.1.1, and to interpolate between values in Tables R602.10.1.2(2), R603.9.2(1) and other seismic design requirements of this code.

TABLE R301.2.2.1.1
SEISMIC DESIGN CATEGORY DETERMINATION

CALCULATED S_{DS}	SEISMIC DESIGN CATEGORY
$S_{DS} \leq 0.17g$	A
$0.17g < S_{DS} \leq 0.33g$	B
$0.33g < S_{DS} \leq 0.50g$	C
$0.50g < S_{DS} \leq 0.67g$	D_0
$0.67g < S_{DS} \leq 0.83g$	D_1
$0.83g < S_{DS} \leq 1.17g$	D_2
$1.17g < S_{DS}$	E

R301.2.2.1.2 Alternative determination of Seismic Design Category E. Buildings located in Seismic Design Category E in accordance with Figure R301.2(2) are permitted to be reclassified as being in Seismic Design Category D_2 provided one of the following is done:

1. A more detailed evaluation of the seismic design category is made in accordance with the provisions and maps of the *International Building Code*. Buildings located in Seismic Design Category E per Table R301.2.2.1.1, but located in Seismic Design Category D per the *International Building Code*, may be designed using the Seismic Design Category D_2 requirements of this code.
2. Buildings located in Seismic Design Category E that conform to the following additional restrictions are permitted to be constructed in accordance with the provisions for Seismic Design Category D_2 of this code:
   - 2.1. All exterior shear wall lines or *braced wall panels* are in one plane vertically from the foundation to the uppermost story.
   - 2.2. Floors shall not cantilever past the exterior walls.
   - 2.3. The building is within all of the requirements of Section R301.2.2.2.5 for being considered as regular.

R301.2.2.2 Seismic Design Category C. Structures assigned to Seismic Design Category C shall conform to the requirements of this section.

R301.2.2.2.1 Weights of materials. Average dead loads shall not exceed 15 pounds per square foot (720 Pa) for the combined roof and ceiling assemblies (on a horizontal projection) or 10 pounds per square foot (480 Pa) for floor assemblies, except as further limited by Section R301.2.2. Dead loads for walls above *grade* shall not exceed:

1. Fifteen pounds per square foot (720 Pa) for exterior light-frame wood walls.
2. Fourteen pounds per square foot (670 Pa) for exterior light-frame cold-formed steel walls.
3. Ten pounds per square foot (480 Pa) for interior light-frame wood walls.
4. Five pounds per square foot (240 Pa) for interior light-frame cold-formed steel walls.
5. Eighty pounds per square foot (3830 Pa) for 8-inch-thick (203 mm) masonry walls.
6. Eighty-five pounds per square foot (4070 Pa) for 6-inch-thick (152 mm) concrete walls.
7. Ten pounds per square foot (480 Pa) for SIP walls.

Exceptions:

1. Roof and ceiling dead loads not exceeding 25 pounds per square foot (1190 Pa) shall be permitted provided the wall bracing amounts in Chapter 6 are increased in accordance with Table R301.2.2.2.1.
2. Light-frame walls with stone or masonry veneer shall be permitted in accordance with the provisions of Sections R702.1 and R703.
3. Fireplaces and chimneys shall be permitted in accordance with Chapter 10.

TABLE R301.2.2.2.1
WALL BRACING ADJUSTMENT FACTORS BY ROOF COVERING DEAD LOAD[a]

WALL SUPPORTING	ROOF/CEILING DEAD LOAD	
	15 psf or less	25 psf
Roof only	1.0	1.2
Roof plus one or two stories	1.0	1.1

For SI: 1 pound per square foot = 0.0479 kPa.
a. Linear interpolation shall be permitted.

R301.2.2.2.2 Stone and masonry veneer. Anchored stone and masonry veneer shall comply with the requirements of Sections R702.1 and R703.

R301.2.2.2.3 Masonry construction. Masonry construction shall comply with the requirements of Section R606.12.

R301.2.2.2.4 Concrete construction. Detached one- and two-family *dwellings* with exterior above-*grade* concrete walls shall comply with the requirements of Section R611, PCA 100 or shall be designed in accordance with ACI 318. *Townhouses* with above-*grade* exterior concrete walls shall comply with the requirements of PCA 100 or shall be designed in accordance with ACI 318.

R301.2.2.2.5 Irregular buildings. The seismic provisions of this code shall not be used for irregular structures located in Seismic Design Categories C, D_0, D_1 and D_2. Irregular portions of structures shall be designed in accordance with accepted engineering practice to the extent the irregular features affect the performance of the remaining structural system. When the forces associated with the irregularity are resisted by a structural system designed in accordance with accepted engineering practice, design of the remainder of the building shall be permitted using the provisions of this code. A building or portion of a building shall be considered to be irregular when one or more of the following conditions occur:

1. When exterior shear wall lines or *braced wall panels* are not in one plane vertically from the foundation to the uppermost *story* in which they are required.

   Exception: For wood light-frame construction, floors with cantilevers or setbacks not exceeding four times the nominal depth of the wood floor joists are permitted to support *braced wall panels* that are out of plane with *braced wall panels* below provided that:

   1. Floor joists are nominal 2 inches by 10 inches (51 mm by 254 mm) or larger and spaced not more than 16 inches (406 mm) on center.
   2. The ratio of the back span to the cantilever is at least 2 to 1.
   3. Floor joists at ends of *braced wall panels* are doubled.
   4. For wood-frame construction, a continuous rim joist is connected to ends of all cantilever joists. When spliced, the rim joists shall be spliced using a galvanized metal tie not less than 0.058 inch (1.5 mm) (16 gage) and $1^1/_2$ inches (38 mm) wide fastened with six 16d nails on each side of the splice or a block of the same size as the rim joist of sufficient length to fit securely between the joist space at which the splice occurs fastened with eight 16d nails on each side of the splice; and
   5. Gravity loads carried at the end of cantilevered joists are limited to uniform wall and roof loads and the reactions from headers having a span of 8 feet (2438 mm) or less.

2. When a section of floor or roof is not laterally supported by shear walls or *braced wall lines* on all edges.

   Exception: Portions of floors that do not support shear walls or *braced wall panels* above, or roofs, shall be permitted to extend no more than 6 feet (1829 mm) beyond a shear wall or *braced wall line.*

3. When the end of a *braced wall panel* occurs over an opening in the wall below and ends at a horizontal distance greater than 1 foot (305 mm) from the edge of the opening. This provision is applicable to shear walls and *braced wall panels* offset in plane and to *braced wall panels* offset out of plane as permitted by the exception to Item 1 above.

   Exception: For wood light-frame wall construction, one end of a *braced wall panel* shall be permitted to extend more than 1 foot (305 mm) over an opening not more than 8 feet (2438 mm) wide in the wall below provided that the opening includes a header in accordance with the following:

   1. The building width, loading condition and framing member species limitations of Table R502.5(1) shall apply; and
   2. Not less than one 2 × 12 or two 2 × 10 for an opening not more than 4 feet (1219 mm) wide; or
   3. Not less than two 2 × 12 or three 2 × 10 for an opening not more than 6 feet (1829 mm) wide; or
   4. Not less than three 2 × 12 or four 2 × 10 for an opening not more than 8 feet (2438 mm) wide; and
   5. The entire length of the *braced wall panel* does not occur over an opening in the wall below.

4. When an opening in a floor or roof exceeds the lesser of 12 feet (3658 mm) or 50 percent of the least floor or roof dimension.
5. When portions of a floor level are vertically offset.

    Exceptions:

    1. Framing supported directly by continuous foundations at the perimeter of the building.
    2. For wood light-frame construction, floors shall be permitted to be vertically offset when the floor framing is lapped or tied together as required by Section R502.6.1.

6. When shear walls and *braced wall lines* do not occur in two perpendicular directions.
7. When stories above *grade* plane partially or completely braced by wood wall framing in accordance with Section R602 or steel wall framing in accordance with Section R603 include masonry or concrete construction.

    Exception: Fireplaces, chimneys and masonry veneer as permitted by this code. When this irregularity applies, the entire *story* shall be designed in accordance with accepted engineering practice.

R301.2.2.3 Seismic Design Categories D_0, D_1 and D_2. Structures assigned to Seismic Design Categories D_0, D_1 and D_2 shall conform to the requirements for Seismic Design Category C and the additional requirements of this section.

R301.2.2.3.1 Height limitations. Wood-framed buildings shall be limited to three stories above *grade* plane or the limits given in Table R602.10.3(3). Cold-formed, steel-framed buildings shall be limited to less than or equal to three stories above *grade* plane in accordance with AISI S230. Mezzanines as defined in Section R202 shall not be considered as stories. Structural insulated panel buildings shall be limited to two stories above *grade* plane.

R301.2.2.3.2 Stone and masonry veneer. Anchored stone and masonry veneer shall comply with the requirements of Sections R702.1 and R703.

R301.2.2.3.3 Masonry construction. Masonry construction in Seismic Design Categories D_0 and D_1 shall comply with the requirements of Section R606.12.1. Masonry construction in Seismic Design Category D_2 shall comply with the requirements of Section R606.12.4.

R301.2.2.3.4 Concrete construction. Buildings with exterior above-*grade* concrete walls shall comply with PCA 100 or shall be designed in accordance with ACI 318.

R301.2.2.3.5 Cold-formed steel framing in Seismic Design Categories D_0, D_1 and D_2. In Seismic Design Categories D_0, D_1 and D_2 in addition to the requirements of this code, cold-formed steel framing shall comply with the requirements of AISI S230.

R301.2.2.3.6 Masonry chimneys. Masonry chimneys shall be reinforced and anchored to the building in accordance with Sections R1003.3 and R1003.4.

R301.2.2.3.7 Anchorage of water heaters. Water heaters shall be anchored against movement and overturning in accordance with Section M1307.2.

R301.2.2.4 Seismic Design Category E. Buildings in Seismic Design Category E shall be designed to resist seismic loads in accordance with the *International Building Code*, except when the seismic design category is reclassified to a lower seismic design category in accordance with Section R301.2.2.1. Components of buildings not required to be designed to resist seismic loads shall be constructed in accordance with the provisions of this code.

R301.2.3 Snow loads. Wood-framed construction, cold-formed, steel-framed construction and masonry and concrete construction, and structural insulated panel construction in regions with ground snow loads 70 pounds per square foot (3.35 kPa) or less, shall be in accordance with Chapters 5, 6 and 8. Buildings in regions with ground snow loads greater than 70 pounds per square foot (3.35 kPa) shall be designed in accordance with accepted engineering practice.

R301.2.4 Floodplain construction. Buildings and structures constructed in whole or in part in flood hazard areas (including A or V Zones) as established in Table R301.2(1) shall be designed and constructed in accordance with Section R322. Buildings and structures located in whole or in part in identified floodways shall be designed and constructed in accordance with ASCE 24.

R301.2.4.1 Alternative provisions. As an alternative to the requirements in Section R322.3 for buildings and structures located in whole or in part in coastal high-hazard areas (V Zones) and coastal A Zones, if delineated, ASCE 24 is permitted subject to the limitations of this code and the limitations therein.

R301.3 Story height. The wind and seismic provisions of this code shall apply to buildings with story heights not exceeding the following:

1. For wood wall framing, the laterally unsupported bearing wall stud height permitted by Table R602.3(5) plus a height of floor framing not to exceed 16 inches (406 mm).

    Exception: For wood-framed wall buildings with bracing in accordance with Tables R602.10.3(1) and R602.10.3(3), the wall stud clear height used to determine the maximum permitted *story height* may be increased to 12 feet (3658 mm) without requiring an engineered design for the building wind and seismic force-resisting systems provided that the length of bracing required by Table R602.10.3(1) is increased by multiplying by a factor of 1.10 and the length of bracing required by Table R602.10.3(3) is increased by multiplying by a factor of 1.20. Wall studs are still subject to the requirements of this section.

2. For steel wall framing, a stud height of 10 feet (3048 mm), plus a height of floor framing not to exceed 16 inches (406 mm).
3. For masonry walls, a maximum bearing wall clear height of 12 feet (3658 mm) plus a height of floor framing not to exceed 16 inches (406 mm).

   Exception: An additional 8 feet (2438 mm) is permitted for gable end walls.
4. For insulating concrete form walls, the maximum bearing wall height per *story* as permitted by Section R611 tables plus a height of floor framing not to exceed 16 inches (406 mm).
5. For structural insulated panel (SIP) walls, the maximum bearing wall height per *story* as permitted by Section R613 tables shall not exceed 10 feet (3048 mm) plus a height of floor framing not to exceed 16 inches (406 mm).

Individual walls or walls studs shall be permitted to exceed these limits as permitted by Chapter 6 provisions, provided *story heights* are not exceeded. Floor framing height shall be permitted to exceed these limits provided the *story height* does not exceed 11 feet 7 inches (3531 mm). An engineered design shall be provided for the wall or wall framing members when they exceed the limits of Chapter 6. Where the *story height* limits of this section are exceeded, the design of the building, or the noncompliant portions thereof, to resist wind and seismic loads shall be in accordance with the *International Building Code*.

R301.4 Dead load. The actual weights of materials and construction shall be used for determining dead load with consideration for the dead load of fixed service *equipment*.

R301.5 Live load. The minimum uniformly distributed live load shall be as provided in Table R301.5.

R301.6 Roof load. The roof shall be designed for the live load indicated in Table R301.6 or the snow load indicated in Table R301.2(1), whichever is greater.

TABLE R301.6
MINIMUM ROOF LIVE LOADS IN POUNDS-FORCE PER SQUARE FOOT OF HORIZONTAL PROJECTION

ROOF SLOPE	TRIBUTARY LOADED AREA IN SQUARE FEET FOR ANY STRUCTURAL MEMBER		
	0 to 200	201 to 600	Over 600
Flat or rise less than 4 inches per foot (1:3)	20	16	12
Rise 4 inches per foot (1:3) to less than 12 inches per foot (1:1)	16	14	12
Rise 12 inches per foot (1:1) and greater	12	12	12

For SI: 1 square foot = 0.0929 m^2, 1 pound per square foot = 0.0479 kPa, 1 inch per foot = 83.3 mm/m.

TABLE R301.5
MINIMUM UNIFORMLY DISTRIBUTED LIVE LOADS (in pounds per square foot)

USE	LIVE LOAD
Uninhabitable attics without storage[b]	10
Uninhabitable attics with limited storage[b, g]	20
Habitable attics and attics served with fixed stairs	30
Balconies (exterior) and decks[e]	40
Fire escapes	40
Guardrails and handrails[d]	200[h]
Guardrail in-fill components[f]	50[h]
Passenger vehicle garages[a]	50[a]
Rooms other than sleeping room	40
Sleeping rooms	30
Stairs	40[c]

For SI: 1 pound per square foot = 0.0479 kPa, 1 square inch = 645 mm^2, 1 pound = 4.45 N.

a. Elevated garage floors shall be capable of supporting a 2,000-pound load applied over a 20-square-inch area.

b. Uninhabitable attics without storage are those where the maximum clear height between joists and rafters is less than 42 inches, or where there are not two or more adjacent trusses with web configurations capable of accommodating an assumed rectangle 42 inches high by 24 inches in width, or greater, within the plane of the trusses. This live load need not be assumed to act concurrently with any other live load requirements.

c. Individual stair treads shall be designed for the uniformly distributed live load or a 300-pound concentrated load acting over an area of 4 square inches, whichever produces the greater stresses.

d. A single concentrated load applied in any direction at any point along the top.

e. See Section R502.2.2 for decks attached to exterior walls.

f. Guard in-fill components (all those except the handrail), balusters and panel fillers shall be designed to withstand a horizontally applied normal load of 50 pounds on an area equal to 1 square foot. This load need not be assumed to act concurrently with any other live load requirement.

g. Uninhabitable attics with limited storage are those where the maximum clear height between joists and rafters is 42 inches or greater, or where there are two or more adjacent trusses with web configurations capable of accommodating an assumed rectangle 42 inches in height by 24 inches in width, or greater, within the plane of the trusses.

The live load need only be applied to those portions of the joists or truss bottom chords where all of the following conditions are met:

1. The attic area is accessible from an opening not less than 20 inches in width by 30 inches in length that is located where the clear height in the attic is a minimum of 30 inches.
2. The slopes of the joists or truss bottom chords are no greater than 2 inches vertical to 12 units horizontal.
3. Required insulation depth is less than the joist or truss bottom chord member depth.

The remaining portions of the joists or truss bottom chords shall be designed for a uniformly distributed concurrent live load of not less than 10 lb/ft^2.

h. Glazing used in handrail assemblies and guards shall be designed with a safety factor of 4. The safety factor shall be applied to each of the concentrated loads applied to the top of the rail, and to the load on the in-fill components. These loads shall be determined independent of one another, and loads are assumed not to occur with any other live load.

R301.7 Deflection. The allowable deflection of any structural member under the live load listed in Sections R301.5 and R301.6 or wind loads determined by Section R301.2.1 shall not exceed the values in Table R301.7.

TABLE R301.7
ALLOWABLE DEFLECTION OF STRUCTURAL MEMBERS[b, c]

STRUCTURAL MEMBER	ALLOWABLE DEFLECTION
Rafters having slopes greater than 3:12 with no finished ceiling attached to rafters	L/180
Interior walls and partitions	H/180
Floors/ceilings with plaster or stucco finish	L/360
All other structural members	L/240
Exterior walls—wind loads[a] with plaster or stucco finish	H/360
Exterior walls with other brittle finishes	H/240
Exterior walls with flexible finishes	H/120[d]
Lintels supporting masonry veneer walls[e]	L/600

Note: *L* = span length, *H* = span height.

a. The wind load shall be permitted to be taken as 0.7 times the Component and Cladding loads for the purpose of the determining deflection limits herein.

b For cantilever members, *L* shall be taken as twice the length of the cantilever.

c. For aluminum structural members or panels used in roofs or walls of sunroom additions or patio covers, not supporting edge of glass or sandwich panels, the total load deflection shall not exceed *L*/60. For continuous aluminum structural members supporting edge of glass, the total load deflection shall not exceed *L*/175 for each glass lite or *L*/60 for the entire length of the member, whichever is more stringent. For sandwich panels used in roofs or walls of sunroom additions or patio covers, the total load deflection shall not exceed *L*/120.

d. Deflection for exterior walls with interior gypsum board finish shall be limited to an allowable deflection of *H*/180.

e. Refer to Section R703.7.2.

R301.8 Nominal sizes. For the purposes of this code, where dimensions of lumber are specified, they shall be deemed to be nominal dimensions unless specifically designated as actual dimensions.

SECTION R302
FIRE-RESISTANT CONSTRUCTION

R302.1 Exterior walls. Construction, projections, openings and penetrations of *exterior walls* of *dwellings* and accessory buildings shall comply with Table R302.1(1); or *dwellings* equipped throughout with an *automatic sprinkler system* installed in accordance with Section P2904 shall comply with Table R302.1(2).

Exceptions:

1. Walls, projections, openings or penetrations in walls perpendicular to the line used to determine the *fire separation distance*.
2. Walls of *dwellings* and *accessory structures* located on the same *lot*.
3. Detached tool sheds and storage sheds, playhouses and similar structures exempted from permits are not required to provide wall protection based on location on the *lot*. Projections beyond the *exterior wall* shall not extend over the *lot line*.
4. Detached garages accessory to a *dwelling* located within 2 feet (610 mm) of a *lot line* are permitted to have roof eave projections not exceeding 4 inches (102 mm).
5. Foundation vents installed in compliance with this code are permitted.

R302.2 Townhouses. Each *townhouse* shall be considered a separate building and shall be separated by fire-resistance-rated wall assemblies meeting the requirements of Section R302.1 for exterior walls.

Exception: A common 1-hour fire-resistance-rated wall assembly tested in accordance with ASTM E 119 or UL 263 is permitted for townhouses if such walls do not contain plumbing or mechanical equipment, ducts or vents in the cavity of the common wall. The wall shall be rated for fire exposure from both sides and shall extend to and be tight against exterior walls and the underside of the roof sheathing. Electrical installations shall be installed in accordance with Chapters 34 through 43. Penetrations of electrical outlet boxes shall be in accordance with Section R302.4.

R302.2.1 Continuity. The fire-resistance-rated wall or assembly separating *townhouses* shall be continuous from the foundation to the underside of the roof sheathing, deck or slab. The fire-resistance rating shall extend the full length of the wall or assembly, including wall extensions

TABLE R302.1(1)
EXTERIOR WALLS

EXTERIOR WALL ELEMENT		MINIMUM FIRE-RESISTANCE RATING	MINIMUM FIRE SEPARATION DISTANCE
Walls	Fire-resistance rated	1 hour—tested in accordance with ASTM E 119 or UL 263 with exposure from both sides	< 5 feet
	Not fire-resistance rated	0 hours	≥ 5 feet
Projections	Fire-resistance rated	1 hour on the underside	≥ 2 feet to < 5 feet
	Not fire-resistance rated	0 hours	≥ 5 feet
Openings in walls	Not allowed	N/A	< 3 feet
	25% maximum of wall area	0 hours	3 feet
	Unlimited	0 hours	5 feet
Penetrations	All	Comply with Section R302.4	< 5 feet
		None required	5 feet

For SI: 1 foot = 304.8 mm.
N/A = Not Applicable.

through and separating attached enclosed *accessory structures*.

R302.2.2 Parapets. Parapets constructed in accordance with Section R302.2.3 shall be constructed for *townhouses* as an extension of exterior walls or common walls in accordance with the following:

1. Where roof surfaces adjacent to the wall or walls are at the same elevation, the parapet shall extend not less than 30 inches (762 mm) above the roof surfaces.
2. Where roof surfaces adjacent to the wall or walls are at different elevations and the higher roof is not more than 30 inches (762 mm) above the lower roof, the parapet shall extend not less than 30 inches (762 mm) above the lower roof surface.

   Exception: A parapet is not required in the two cases above when the roof is covered with a minimum class C roof covering, and the roof decking or sheathing is of noncombustible materials or *approved* fire-retardant-treated wood for a distance of 4 feet (1219 mm) on each side of the wall or walls, or one layer of $^5/_8$-inch (15.9 mm) Type X gypsum board is installed directly beneath the roof decking or sheathing, supported by a minimum of nominal 2-inch (51 mm) ledgers attached to the sides of the roof framing members, for a minimum distance of 4 feet (1219 mm) on each side of the wall or walls and there are no openings or penetrations in the roof within 4 feet (1219 mm) of the common walls.
3. A parapet is not required where roof surfaces adjacent to the wall or walls are at different elevations and the higher roof is more than 30 inches (762 mm) above the lower roof. The common wall construction from the lower roof to the underside of the higher roof deck shall have not less than a 1-hour fire-resistance rating. The wall shall be rated for exposure from both sides.

R302.2.3 Parapet construction. Parapets shall have the same fire-resistance rating as that required for the supporting wall or walls. On any side adjacent to a roof surface, the parapet shall have noncombustible faces for the uppermost 18 inches (457 mm), to include counterflashing and coping materials. Where the roof slopes toward a parapet at slopes greater than 2 units vertical in 12 units horizontal (16.7-percent slope), the parapet shall extend to the same height as any portion of the roof within a distance of 3 feet (914 mm), but in no case shall the height be less than 30 inches (762 mm).

R302.2.4 Structural independence. Each individual *townhouse* shall be structurally independent.

Exceptions:

1. Foundations supporting *exterior walls* or common walls.
2. Structural roof and wall sheathing from each unit may fasten to the common wall framing.
3. Nonstructural wall and roof coverings.
4. Flashing at termination of roof covering over common wall.
5. *Townhouses* separated by a common 1-hour fire-resistance-rated wall as provided in Section R302.2.

R302.3 Two-family dwellings. *Dwelling units* in two-family dwellings shall be separated from each other by wall and/or floor assemblies having not less than a 1-hour fire-resistance rating when tested in accordance with ASTM E 119 or UL 263. Fire-resistance-rated floor/ceiling and wall assemblies shall extend to and be tight against the *exterior wall*, and wall assemblies shall extend from the foundation to the underside of the roof sheathing.

Exceptions:

1. A fire-resistance rating of $^1/_2$ hour shall be permitted in buildings equipped throughout with an automatic sprinkler system installed in accordance with NFPA 13.

TABLE R302.1(2)
EXTERIOR WALLS—DWELLINGS WITH FIRE SPRINKLERS

EXTERIOR WALL ELEMENT		MINIMUM FIRE-RESISTANCE RATING	MINIMUM FIRE SEPARATION DISTANCE
Walls	Fire-resistance rated	1 hour—tested in accordance with ASTM E 119 or UL 263 with exposure from the outside	0 feet
	Not fire-resistance rated	0 hours	3 feet[a]
Projections	Fire-resistance rated	1 hour on the underside	2 feet[a]
	Not fire-resistance rated	0 hours	3 feet
Openings in walls	Not allowed	N/A	< 3 feet
	Unlimited	0 hours	3 feet[a]
Penetrations	All	Comply with Section R302.4	< 3 feet
		None required	3 feet[a]

For SI: 1 foot = 304.8 mm.

N/A = Not Applicable

a. For residential subdivisions where all dwellings are equipped throughout with an automatic sprinkler systems installed in accordance with Section P2904, the fire separation distance for nonrated exterior walls and rated projections shall be permitted to be reduced to 0 feet, and unlimited unprotected openings and penetrations shall be permitted, where the adjoining lot provides an open setback yard that is 6 feet or more in width on the opposite side of the property line.

2. Wall assemblies need not extend through *attic* spaces when the ceiling is protected by not less than $^5/_8$-inch (15.9 mm) Type X gypsum board and an *attic* draft stop constructed as specified in Section R302.12.1 is provided above and along the wall assembly separating the *dwellings*. The structural framing supporting the ceiling shall also be protected by not less than $^1/_2$-inch (12.7 mm) gypsum board or equivalent.

R302.3.1 Supporting construction. When floor assemblies are required to be fire-resistance rated by Section R302.3, the supporting construction of such assemblies shall have an equal or greater fire-resistance rating.

R302.4 Dwelling unit rated penetrations. Penetrations of wall or floor/ceiling assemblies required to be fire-resistance rated in accordance with Section R302.2 or R302.3 shall be protected in accordance with this section.

R302.4.1 Through penetrations. Through penetrations of fire-resistance-rated wall or floor assemblies shall comply with Section R302.4.1.1 or R302.4.1.2.

Exception: Where the penetrating items are steel, ferrous or copper pipes, tubes or conduits, the annular space shall be protected as follows:

1. In concrete or masonry wall or floor assemblies, concrete, grout or mortar shall be permitted where installed to the full thickness of the wall or floor assembly or the thickness required to maintain the fire-resistance rating, provided:
   - 1.1. The nominal diameter of the penetrating item is a maximum of 6 inches (152 mm); and
   - 1.2. The area of the opening through the wall does not exceed 144 square inches (92 900 mm^2).
2. The material used to fill the annular space shall prevent the passage of flame and hot gases sufficient to ignite cotton waste where subjected to ASTM E 119 or UL 263 time temperature fire conditions under a minimum positive pressure differential of 0.01 inch of water (3 Pa) at the location of the penetration for the time period equivalent to the fire-resistance rating of the construction penetrated.

R302.4.1.1 Fire-resistance-rated assembly. Penetrations shall be installed as tested in the *approved* fire-resistance-rated assembly.

R302.4.1.2 Penetration firestop system. Penetrations shall be protected by an *approved* penetration firestop system installed as tested in accordance with ASTM E 814 or UL 1479, with a minimum positive pressure differential of 0.01 inch of water (3 Pa) and shall have an F rating of not less than the required fire-resistance rating of the wall or floor/ceiling assembly penetrated.

R302.4.2 Membrane penetrations. Membrane penetrations shall comply with Section R302.4.1. Where walls are required to have a fire-resistance rating, recessed fixtures shall be installed so that the required fire-resistance rating will not be reduced.

Exceptions:

1. Membrane penetrations of maximum 2-hour fire-resistance-rated walls and partitions by steel electrical boxes that do not exceed 16 square inches (0.0103 m^2) in area provided the aggregate area of the openings through the membrane does not exceed 100 square inches (0.0645 m^2) in any 100 square feet $(9.29\ m)^2$ of wall area. The annular space between the wall membrane and the box shall not exceed $^1/_8$ inch (3.1 mm). Such boxes on opposite sides of the wall shall be separated by one of the following:
   - 1.1. By a horizontal distance of not less than 24 inches (610 mm) where the wall or partition is constructed with individual noncommunicating stud cavities;
   - 1.2. By a horizontal distance of not less than the depth of the wall cavity when the wall cavity is filled with cellulose loose-fill, rockwool or slag mineral wool insulation;
   - 1.3. By solid fire blocking in accordance with Section R302.11;
   - 1.4. By protecting both boxes with listed putty pads; or
   - 1.5. By other listed materials and methods.
2. Membrane penetrations by listed electrical boxes of any materials provided the boxes have been tested for use in fire-resistance-rated assemblies and are installed in accordance with the instructions included in the listing. The annular space between the wall membrane and the box shall not exceed $^1/_8$ inch (3.1 mm) unless listed otherwise. Such boxes on opposite sides of the wall shall be separated by one of the following:
   - 2.1. By the horizontal distance specified in the listing of the electrical boxes;
   - 2.2. By solid fireblocking in accordance with Section R302.11;
   - 2.3. By protecting both boxes with listed putty pads; or
   - 2.4. By other listed materials and methods.
3. The annular space created by the penetration of a fire sprinkler provided it is covered by a metal escutcheon plate.

R302.5 Dwelling/garage opening/penetration protection. Openings and penetrations through the walls or ceilings separating the *dwelling* from the garage shall be in accordance with Sections R302.5.1 through R302.5.3.

R302.5.1 Opening protection. Openings from a private garage directly into a room used for sleeping purposes shall not be permitted. Other openings between the garage and residence shall be equipped with solid wood doors not less than $1^3/_8$ inches (35 mm) in thickness, solid or honeycomb-core steel doors not less than $1^3/_8$ inches (35 mm) thick, or 20-minute fire-rated doors, equipped with a self-closing device.

R302.5.2 Duct penetration. Ducts in the garage and ducts penetrating the walls or ceilings separating the *dwelling* from the garage shall be constructed of a minimum No. 26 gage (0.48 mm) sheet steel or other *approved* material and shall have no openings into the garage.

R302.5.3 Other penetrations. Penetrations through the separation required in Section R302.6 shall be protected as required by Section R302.11, Item 4.

R302.6 Dwelling/garage fire separation. The garage shall be separated as required by Table R302.6. Openings in garage walls shall comply with Section R302.5. This provision does not apply to garage walls that are perpendicular to the adjacent *dwelling unit* wall.

R302.7 Under-stair protection. Enclosed accessible space under stairs shall have walls, under-stair surface and any soffits protected on the enclosed side with $^1/_2$-inch (12.7 mm) gypsum board.

R302.8 Foam plastics. For requirements for foam plastics see Section R316.

R302.9 Flame spread index and smoke-developed index for wall and ceiling finishes. Flame spread and smoke index for wall and ceiling finishes shall be in accordance with Sections R302.9.1 through R302.9.4.

R302.9.1 Flame spread index. Wall and ceiling finishes shall have a flame spread index of not greater than 200.

Exception: Flame spread index requirements for finishes shall not apply to trim defined as picture molds, chair rails, baseboards and handrails; to doors and windows or their frames; or to materials that are less than $^1/_{28}$ inch (0.91 mm) in thickness cemented to the surface of walls or ceilings if these materials exhibit flame spread index values no greater than those of paper of this thickness cemented to a noncombustible backing.

R302.9.2 Smoke-developed index. Wall and ceiling finishes shall have a smoke-developed index of not greater than 450.

R302.9.3 Testing. Tests shall be made in accordance with ASTM E 84 or UL 723.

R302.9.4 Alternative test method. As an alternative to having a flame spread index of not greater than 200 and a smoke-developed index of not greater than 450 when tested in accordance with ASTM E 84 or UL 723, wall and ceiling finishes shall be permitted to be tested in accordance with NFPA 286. Materials tested in accordance with NFPA 286 shall meet the following criteria:

The interior finish shall comply with the following:

1. During the 40 kW exposure, flames shall not spread to the ceiling.
2. The flame shall not spread to the outer extremity of the sample on any wall or ceiling.
3. Flashover, as defined in NFPA 286, shall not occur.
4. The peak heat release rate throughout the test shall not exceed 800 kW.
5. The total smoke released throughout the test shall not exceed 1,000 m^2.

R302.10 Flame spread index and smoke-developed index for insulation. Flame spread and smoke-developed index for insulation shall be in accordance with Sections R302.10.1 through R302.10.5.

R302.10.1 Insulation. Insulation materials, including facings, such as vapor retarders and vapor-permeable membranes installed within floor/ceiling assemblies, roof/ceiling assemblies, wall assemblies, crawl spaces and *attics* shall have a flame spread index not to exceed 25 with an accompanying smoke-developed index not to exceed 450 when tested in accordance with ASTM E 84 or UL 723.

Exceptions:

1. When such materials are installed in concealed spaces, the flame spread index and smoke-developed index limitations do not apply to the facings, provided that the facing is installed in substantial contact with the unexposed surface of the ceiling, floor or wall finish.
2. Cellulose loose-fill insulation, which is not spray applied, complying with the requirements of Section R302.10.3, shall only be required to meet the smoke-developed index of not more than 450.
3. Foam plastic insulation shall comply with Section R316.

TABLE R302.6
DWELLING/GARAGE SEPARATION

SEPARATION	MATERIAL
From the residence and attics	Not less than $^1/_2$-inch gypsum board or equivalent applied to the garage side
From all habitable rooms above the garage	Not less than $^5/_8$-inch Type X gypsum board or equivalent
Structure(s) supporting floor/ceiling assemblies used for separation required by this section	Not less than $^1/_2$-inch gypsum board or equivalent
Garages located less than 3 feet from a dwelling unit on the same lot	Not less than $^1/_2$-inch gypsum board or equivalent applied to the interior side of exterior walls that are within this area

For SI: 1 inch = 25.4 mm, 1 foot = 304.8 mm.

R302.10.2 Loose-fill insulation. Loose-fill insulation materials that cannot be mounted in the ASTM E 84 or UL 723 apparatus without a screen or artificial supports shall comply with the flame spread and smoke-developed limits of Section R302.10.1 when tested in accordance with CAN/ULC S102.2.

Exception: Cellulose loose-fill insulation shall not be required to be tested in accordance with CAN/ULC S102.2, provided such insulation complies with the requirements of Section R302.10.1 and Section R302.10.3.

R302.10.3 Cellulose loose-fill insulation. Cellulose loose-fill insulation shall comply with CPSC 16 CFR, Parts 1209 and 1404. Each package of such insulating material shall be clearly *labeled* in accordance with CPSC 16 CFR, Parts 1209 and 1404.

R302.10.4 Exposed attic insulation. All exposed insulation materials installed on *attic* floors shall have a critical radiant flux not less than 0.12 watt per square centimeter.

R302.10.5 Testing. Tests for critical radiant flux shall be made in accordance with ASTM E 970.

R302.11 Fireblocking. In combustible construction, fireblocking shall be provided to cut off all concealed draft openings (both vertical and horizontal) and to form an effective fire barrier between stories, and between a top *story* and the roof space.

Fireblocking shall be provided in wood-frame construction in the following locations:

1. In concealed spaces of stud walls and partitions, including furred spaces and parallel rows of studs or staggered studs, as follows:
   1.1. Vertically at the ceiling and floor levels.
   1.2. Horizontally at intervals not exceeding 10 feet (3048 mm).
2. At all interconnections between concealed vertical and horizontal spaces such as occur at soffits, drop ceilings and cove ceilings.
3. In concealed spaces between stair stringers at the top and bottom of the run. Enclosed spaces under stairs shall comply with Section R302.7.
4. At openings around vents, pipes, ducts, cables and wires at ceiling and floor level, with an *approved* material to resist the free passage of flame and products of combustion. The material filling this annular space shall not be required to meet the ASTM E 136 requirements.
5. For the fireblocking of chimneys and fireplaces, see Section R1003.19.
6. Fireblocking of cornices of a two-family *dwelling* is required at the line of *dwelling unit* separation.

R302.11.1 Fireblocking materials. Except as provided in Section R302.11, Item 4, fireblocking shall consist of the following materials.

1. Two-inch (51 mm) nominal lumber.
2. Two thicknesses of 1-inch (25.4 mm) nominal lumber with broken lap joints.
3. One thickness of $^{23}/_{32}$-inch (18.3 mm) wood structural panels with joints backed by $^{23}/_{32}$-inch (18.3 mm) wood structural panels.
4. One thickness of $^{3}/_{4}$-inch (19.1 mm) particleboard with joints backed by $^{3}/_{4}$-inch (19.1 mm) particleboard.
5. One-half-inch (12.7 mm) gypsum board.
6. One-quarter-inch (6.4 mm) cement-based millboard.
7. Batts or blankets of mineral wool or glass fiber or other *approved* materials installed in such a manner as to be securely retained in place.
8. Cellulose insulation installed as tested for the specific application.

R302.11.1.1 Batts or blankets of mineral or glass fiber. Batts or blankets of mineral or glass fiber or other *approved* nonrigid materials shall be permitted for compliance with the 10-foot (3048 mm) horizontal fireblocking in walls constructed using parallel rows of studs or staggered studs.

R302.11.1.2 Unfaced fiberglass. Unfaced fiberglass batt insulation used as fireblocking shall fill the entire cross section of the wall cavity to a minimum height of 16 inches (406 mm) measured vertically. When piping, conduit or similar obstructions are encountered, the insulation shall be packed tightly around the obstruction.

R302.11.1.3 Loose-fill insulation material. Loose-fill insulation material shall not be used as a fireblock unless specifically tested in the form and manner intended for use to demonstrate its ability to remain in place and to retard the spread of fire and hot gases.

R302.11.2 Fireblocking integrity. The integrity of all fireblocks shall be maintained.

R302.12 Draftstopping. In combustible construction where there is usable space both above and below the concealed space of a floor/ceiling assembly, draftstops shall be installed so that the area of the concealed space does not exceed 1,000 square feet (92.9 m^2). Draftstopping shall divide the concealed space into approximately equal areas. Where the assembly is enclosed by a floor membrane above and a ceiling membrane below, draftstopping shall be provided in floor/ceiling assemblies under the following circumstances:

1. Ceiling is suspended under the floor framing.
2. Floor framing is constructed of truss-type open-web or perforated members.

R302.12.1 Materials. Draftstopping materials shall not be less than $^{1}/_{2}$-inch (12.7 mm) gypsum board, $^{3}/_{8}$-inch (9.5 mm) wood structural panels or other *approved* materials adequately supported. Draftopping shall be installed parallel to the floor framing members unless otherwise *approved* by the *building official*. The integrity of the draftstops shall be maintained.

R302.13 Combustible insulation clearance. Combustible insulation shall be separated a minimum of 3 inches (76 mm) from recessed luminaires, fan motors and other heat-producing devices.

Exception: Where heat-producing devices are listed for lesser clearances, combustible insulation complying with the listing requirements shall be separated in accordance with the conditions stipulated in the listing.

Recessed luminaires installed in the *building thermal envelope* shall meet the requirements of Section N1102.4.4 of this code.

SECTION R303
LIGHT, VENTILATION AND HEATING

R303.1 Habitable rooms. All habitable rooms shall have an aggregate glazing area of not less than 8 percent of the floor area of such rooms. Natural *ventilation* shall be through windows, doors, louvers or other *approved* openings to the outdoor air. Such openings shall be provided with ready access or shall otherwise be readily controllable by the building occupants. The minimum openable area to the outdoors shall be 4 percent of the floor area being ventilated.

Exceptions:

1. The glazed areas need not be openable where the opening is not required by Section R310 and a whole-house mechanical *ventilation* system is installed in accordance with Section M1507.
2. The glazed areas need not be installed in rooms where Exception 1 above is satisfied and artificial light is provided capable of producing an average illumination of 6 footcandles (65 lux) over the area of the room at a height of 30 inches (762 mm) above the floor level.
3. Use of sunroom and patio covers, as defined in Section R202, shall be permitted for natural *ventilation* if in excess of 40 percent of the exterior sunroom walls are open, or are enclosed only by insect screening.

R303.2 Adjoining rooms. For the purpose of determining light and *ventilation* requirements, any room shall be considered as a portion of an adjoining room when at least one-half of the area of the common wall is open and unobstructed and provides an opening of not less than one-tenth of the floor area of the interior room but not less than 25 square feet (2.3 m^2).

Exception: Openings required for light and/or *ventilation* shall be permitted to open into a sunroom with thermal isolation or a patio cover, provided that there is an openable area between the adjoining room and the sunroom or patio cover of not less than one-tenth of the floor area of the interior room but not less than 20 square feet (2 m^2). The minimum openable area to the outdoors shall be based upon the total floor area being ventilated.

R303.3 Bathrooms. Bathrooms, water closet compartments and other similar rooms shall be provided with aggregate glazing area in windows of not less than 3 square feet (0.3 m^2), one-half of which must be openable.

Exception: The glazed areas shall not be required where artificial light and a local exhaust system are provided. The minimum local exhaust rates shall be determined in accordance with Section M1507. Exhaust air from the space shall be exhausted directly to the outdoors.

R303.4 Mechanical ventilation. Where the air infiltration rate of a dwelling unit is less than 5 air changes per hour when tested with a blower door at a pressure of 0.2 inch w.c (50 Pa) in accordance with Section N1102.4.1.2, the dwelling unit shall be provided with whole-house mechanical ventilation in accordance with Section M1507.3.

R303.5 Opening location. Outdoor intake and exhaust openings shall be located in accordance with Sections R303.5.1 and R303.5.2.

R303.5.1 Intake openings. Mechanical and gravity outdoor air intake openings shall be located a minimum of 10 feet (3048 mm) from any hazardous or noxious contaminant, such as vents, chimneys, plumbing vents, streets, alleys, parking lots and loading docks, except as otherwise specified in this code. Where a source of contaminant is located within 10 feet (3048 mm) of an intake opening, such opening shall be located a minimum of 3 feet (914 mm) below the contaminant source.

For the purpose of this section, the exhaust from *dwelling* unit toilet rooms, bathrooms and kitchens shall not be considered as hazardous or noxious.

R303.5.2 Exhaust openings. Exhaust air shall not be directed onto walkways.

R303.6 Outside opening protection. Air exhaust and intake openings that terminate outdoors shall be protected with corrosion-resistant screens, louvers or grilles having a minimum opening size of $^1/_4$ inch (6 mm) and a maximum opening size of $^1/_2$ inch (13 mm), in any dimension. Openings shall be protected against local weather conditions. Outdoor air exhaust and intake openings shall meet the provisions for *exterior wall* opening protectives in accordance with this code.

R303.7 Stairway illumination. All interior and exterior stairways shall be provided with a means to illuminate the stairs, including the landings and treads. Interior stairways shall be provided with an artificial light source located in the immediate vicinity of each landing of the stairway. For interior stairs the artificial light sources shall be capable of illuminating treads and landings to levels not less than 1 footcandle (11 lux) measured at the center of treads and landings. Exterior stairways shall be provided with an artificial light source located in the immediate vicinity of the top landing of the stairway. Exterior stairways providing access to a *basement* from the outside *grade* level shall be provided with an artificial light source located in the immediate vicinity of the bottom landing of the stairway.

Exception: An artificial light source is not required at the top and bottom landing, provided an artificial light source is located directly over each stairway section.

R303.7.1 Light activation. Where lighting outlets are installed in interior stairways, there shall be a wall switch at each floor level to control the lighting outlet where the stairway has six or more risers. The illumination of exterior stairways shall be controlled from inside the *dwelling* unit.

Exception: Lights that are continuously illuminated or automatically controlled.

R303.8 Required glazed openings. Required glazed openings shall open directly onto a street or public alley, or a *yard* or court located on the same *lot* as the building.

Exceptions:

1. Required glazed openings may face into a roofed porch where the porch abuts a street, *yard* or court and the longer side of the porch is at least 65 percent unobstructed and the ceiling height is not less than 7 feet (2134 mm).
2. Eave projections shall not be considered as obstructing the clear open space of a *yard* or court.
3. Required glazed openings may face into the area under a deck, balcony, bay or floor cantilever provided a clear vertical space at least 36 inches (914 mm) in height is provided.

R303.8.1 Sunroom additions. Required glazed openings shall be permitted to open into sunroom *additions* or patio covers that abut a street, *yard* or court if in excess of 40 percent of the exterior sunroom walls are open, or are enclosed only by insect screening, and the ceiling height of the sunroom is not less than 7 feet (2134 mm).

R303.9 Required heating. When the winter design temperature in Table R301.2(1) is below 60°F (16°C), every *dwelling unit* shall be provided with heating facilities capable of maintaining a minimum room temperature of 68°F (20°C) at a point 3 feet (914 mm) above the floor and 2 feet (610 mm) from exterior walls in all habitable rooms at the design temperature. The installation of one or more portable space heaters shall not be used to achieve compliance with this section.

SECTION R304
MINIMUM ROOM AREAS

R304.1 Minimum area. Every *dwelling* unit shall have at least one habitable room that shall have not less than 120 square feet (11 m^2) of gross floor area.

R304.2 Other rooms. Other habitable rooms shall have a floor area of not less than 70 square feet (6.5 m^2).

Exception: Kitchens.

R304.3 Minimum dimensions. Habitable rooms shall not be less than 7 feet (2134 mm) in any horizontal dimension.

Exception: Kitchens.

R304.4 Height effect on room area. Portions of a room with a sloping ceiling measuring less than 5 feet (1524 mm) or a furred ceiling measuring less than 7 feet (2134 mm) from the finished floor to the finished ceiling shall not be considered as contributing to the minimum required habitable area for that room.

SECTION R305
CEILING HEIGHT

R305.1 Minimum height. *Habitable space*, hallways, bathrooms, toilet rooms, laundry rooms and portions of *basements* containing these spaces shall have a ceiling height of not less than 7 feet (2134 mm).

Exceptions:

1. For rooms with sloped ceilings, at least 50 percent of the required floor area of the room must have a ceiling height of at least 7 feet (2134 mm) and no portion of the required floor area may have a ceiling height of less than 5 feet (1524 mm).
2. Bathrooms shall have a minimum ceiling height of 6 feet 8 inches (2032 mm) at the center of the front clearance area for fixtures as shown in Figure R307.1. The ceiling height above fixtures shall be such that the fixture is capable of being used for its intended purpose. A shower or tub equipped with a showerhead shall have a minimum ceiling height of 6 feet 8 inches (2032 mm) above a minimum area 30 inches (762 mm) by 30 inches (762 mm) at the showerhead.

R305.1.1 Basements. Portions of *basements* that do not contain *habitable space*, hallways, bathrooms, toilet rooms and laundry rooms shall have a ceiling height of not less than 6 feet 8 inches (2032 mm).

Exception: Beams, girders, ducts or other obstructions may project to within 6 feet 4 inches (1931 mm) of the finished floor.

SECTION R306
SANITATION

R306.1 Toilet facilities. Every *dwelling* unit shall be provided with a water closet, lavatory, and a bathtub or shower.

R306.2 Kitchen. Each *dwelling* unit shall be provided with a kitchen area and every kitchen area shall be provided with a sink.

R306.3 Sewage disposal. All plumbing fixtures shall be connected to a sanitary sewer or to an *approved* private sewage disposal system.

R306.4 Water supply to fixtures. All plumbing fixtures shall be connected to an *approved* water supply. Kitchen sinks, lavatories, bathtubs, showers, bidets, laundry tubs and washing machine outlets shall be provided with hot and cold water.

SECTION R307
TOILET, BATH AND SHOWER SPACES

R307.1 Space required. Fixtures shall be spaced in accordance with Figure R307.1, and in accordance with the requirements of Section P2705.1.

R307.2 Bathtub and shower spaces. Bathtub and shower floors and walls above bathtubs with installed shower heads and in shower compartments shall be finished with a nonabsorbent surface. Such wall surfaces shall extend to a height of not less than 6 feet (1829 mm) above the floor.

SECTION R308 GLAZING

R308.1 Identification. Except as indicated in Section R308.1.1 each pane of glazing installed in hazardous locations as defined in Section R308.4 shall be provided with a manufacturer's designation specifying who applied the designation, designating the type of glass and the safety glazing standard with which it complies, which is visible in the final installation. The designation shall be acid etched, sandblasted, ceramic-fired, laser etched, embossed, or be of a type which once applied cannot be removed without being destroyed. A *label* shall be permitted in lieu of the manufacturer's designation.

Exceptions:

1. For other than tempered glass, manufacturer's designations are not required provided the *building official* approves the use of a certificate, affidavit or other evidence confirming compliance with this code.
2. Tempered spandrel glass is permitted to be identified by the manufacturer with a removable paper designation

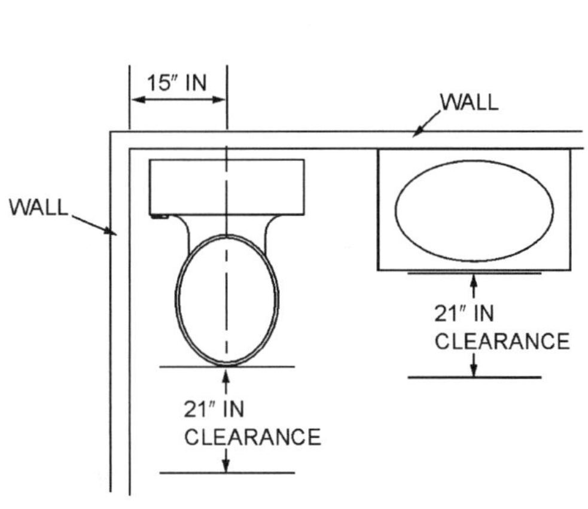

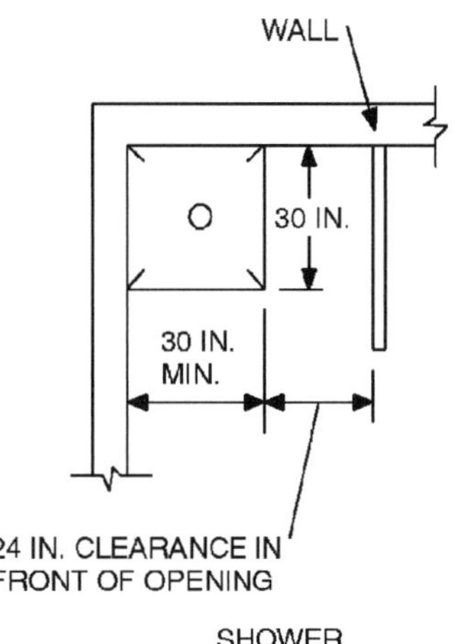

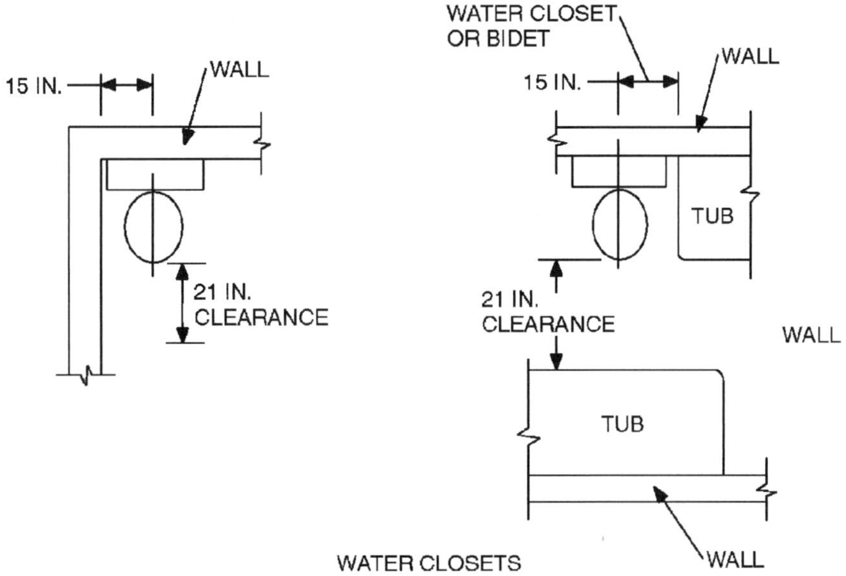

For SI: 1 inch = 25.4 mm.

FIGURE R307.1 MINIMUM FIXTURE CLEARANCES

R308.1.1 Identification of multiple assemblies. Multipane assemblies having individual panes not exceeding 1 square foot (0.09 m^2) in exposed area shall have at least one pane in the assembly identified in accordance with Section R308.1. All other panes in the assembly shall be *labeled* "CPSC 16 CFR 1201" or "ANSI Z97.1" as appropriate.

R308.2 Louvered windows or jalousies. Regular, float, wired or patterned glass in jalousies and louvered windows shall be no thinner than nominal $^3/_{16}$ inch (5 mm) and no longer than 48 inches (1219 mm). Exposed glass edges shall be smooth.

R308.2.1 Wired glass prohibited. Wired glass with wire exposed on longitudinal edges shall not be used in jalousies or louvered windows.

R308.3 Human impact loads. Individual glazed areas, including glass mirrors in hazardous locations such as those indicated as defined in Section R308.4, shall pass the test requirements of Section R308.3.1.

Exceptions:

1. Louvered windows and jalousies shall comply with Section R308.2.
2. Mirrors and other glass panels mounted or hung on a surface that provides a continuous backing support.
3. Glass unit masonry complying with Section R610.

R308.3.1 Impact test. Where required by other sections of the code, glazing shall be tested in accordance with CPSC 16 CFR 1201. Glazing shall comply with the test criteria for Category II unless otherwise indicated in Table R308.3.1(1).

Exception: Glazing not in doors or enclosures for hot tubs, whirlpools, saunas, steam rooms, bathtubs and showers shall be permitted to be tested in accordance with ANSI Z97.1. Glazing shall comply with the test criteria for Class A unless indicated in Table R308.3.1 (2).

R308.4 Hazardous locations. The locations specified in Sections R308.4.1 through R308.4.7 shall be considered specific hazardous locations for the purposes of glazing.

R308.4.1 Glazing in doors. Glazing in all fixed and operable panels of swinging, sliding and bifold doors shall be considered a hazardous location.

Exceptions:

1. Glazed openings of a size through which a 3-inch-diameter (76 mm) sphere is unable to pass.
2. Decorative glazing.

R308.4.2 Glazing adjacent doors. Glazing in an individual fixed or operable panel adjacent to a door where the nearest vertical edge of the glazing is within a 24-inch (610 mm) arc of either vertical edge of the door in a closed position and where the bottom exposed edge of the glazing is less than 60 inches (1524 mm) above the floor or walking surface shall be considered a hazardous location.

Exceptions:

1. Decorative glazing.
2. When there is an intervening wall or other permanent barrier between the door and the glazing.
3. Glazing in walls on the latch side of and perpendicular to the plane of the door in a closed position.
4. Where access through the door is to a closet or storage area 3 feet (914 mm) or less in depth. Glazing in this application shall comply with section R308.4.3.
5. Glazing that is adjacent to the fixed panel of patio doors.

R308.4.3 Glazing in windows. Glazing in an individual fixed or operable panel that meets all of the following conditions shall be considered a hazardous location:

1. The exposed area of an individual pane is larger than 9 square feet (0.836 m^2);

TABLE R308.3.1(1)
MINIMUM CATEGORY CLASSIFICATION OF GLAZING USING CPSC 16 CFR 1201

EXPOSED SURFACE AREA OF ONE SIDE OF ONE LITE	GLAZING IN STORM OR COMBINATION DOORS (Category Class)	GLAZING IN DOORS (Category Class)	GLAZED PANELS REGULATED BY SECTION R308.4.3 (Category Class)	GLAZED PANELS REGULATED BY SECTION R308.4.2 (Category Class)	GLAZING IN DOORS AND ENCLOSURES REGULATED BY SECTION 308.4.5 (Category Class)	SLIDING GLASS DOORS PATIO TYPE (Category Class)
9 square feet or less	I	I	NR	I	II	II
More than 9 square feet	II	II	II	II	II	II

For SI: 1 square foot = 0.0929 m^2.
NR means "No Requirement."

TABLE R308.3.1(2)
MINIMUM CATEGORY CLASSIFICATION OF GLAZING USING ANSI Z97.1

EXPOSED SURFACE AREA OF ONE SIDE OF ONE LITE	GLAZED PANELS REGULATED BY SECTION R308.4.3 (Category Class)	GLAZED PANELS REGULATED BY SECTION R308.4.2 (Category Class)	DOORS AND ENCLOSURES REGULATED BY SECTION R308.4.5[a] (Category Class)
9 square feet or less	No requirement	B	A
More than 9 square feet	A	A	A

For SI: 1 square foot = 0.0929 m^2.
a. Use is permitted only by the exception to Section R308.3.1.

2. The bottom edge of the glazing is less than 18 inches (457 mm) above the floor;
3. The top edge of the glazing is more than 36 inches (914 mm) above the floor; and
4. One or more walking surfaces are within 36 inches (914 mm), measured horizontally and in a straight line, of the glazing.

Exceptions:

1. Decorative glazing.
2. When a horizontal rail is installed on the accessible side(s) of the glazing 34 to 38 inches (864 to 965 mm) above the walking surface. The rail shall be capable of withstanding a horizontal load of 50 pounds per linear foot (730 N/m) without contacting the glass and be a minimum of $1^1/_2$ inches (38 mm) in cross sectional height.
3. Outboard panes in insulating glass units and other multiple glazed panels when the bottom edge of the glass is 25 feet (7620 mm) or more above *grade*, a roof, walking surfaces or other horizontal [within 45 degrees (0.79 rad) of horizontal] surface adjacent to the glass exterior.

R308.4.4 Glazing in guards and railings. Glazing in guards and railings, including structural baluster panels and nonstructural in-fill panels, regardless of area or height above a walking surface shall be considered a hazardous location.

R308.4.5 Glazing and wet surfaces. Glazing in walls, enclosures or fences containing or facing hot tubs, spas, whirlpools, saunas, steam rooms, bathtubs, showers and indoor or outdoor swimming pools where the bottom exposed edge of the glazing is less than 60 inches (1524 mm) measured vertically above any standing or walking surface shall be considered a hazardous location. This shall apply to single glazing and all panes in multiple glazing.

Exception: Glazing that is more than 60 inches (1524 mm), measured horizontally and in a straight line, from the water's edge of a bathtub, hot tub, spa, whirlpool, or swimming pool.

R308.4.6 Glazing adjacent stairs and ramps. Glazing where the bottom exposed edge of the glazing is less than 36 inches (914 mm) above the plane of the adjacent walking surface of stairways, landings between flights of stairs and ramps shall be considered a hazardous location.

Exceptions:

1. When a rail is installed on the accessible side(s) of the glazing 34 to 38 inches (864 to 965 mm) above the walking surface.The rail shall be capable of withstanding a horizontal load of 50 pounds per linear foot (730 N/m) without contacting the glass and be a minimum of $1^1/_2$ inches (38 mm) in cross sectional height.
2. Glazing 36 inches (914 mm) or more measured horizontally from the walking surface.

R308.4.7 Glazing adjacent to the bottom stair landing. Glazing adjacent to the landing at the bottom of a stairway where the glazing is less than 36 inches (914 mm) above the landing and within 60 inches (1524 mm) horizontally of the bottom tread shall be considered a hazardous location.

Exception: The glazing is protected by a guard complying with Section R312 and the plane of the glass is more than 18 inches (457 mm) from the guard.

R308.5 Site built windows. Site built windows shall comply with Section 2404 of the *International Building Code*.

R308.6 Skylights and sloped glazing. Skylights and sloped glazing shall comply with the following sections.

R308.6.1 Definitions.

SKYLIGHT, UNIT. A factory assembled, glazed fenestration unit, containing one panel of glazing material, that allows for natural daylighting through an opening in the roof assembly while preserving the weather-resistant barrier of the roof.

SKYLIGHTS AND SLOPED GLAZING. Glass or other transparent or translucent glazing material installed at a slope of 15 degrees (0.26 rad) or more from vertical. Glazing materials in skylights, including unit skylights, tubular daylighting devices, solariums, sunrooms, roofs and sloped walls are included in this definition.

TUBULAR DAYLIGHTING DEVICE (TDD). A nonoperable fenestration unit primarily designed to transmit daylight from a roof surface to an interior ceiling via a tubular conduit. The basic unit consists of an exterior glazed weathering surface, a light-transmitting tube with a reflective interior surface, and an interior-sealing device such as a translucent ceiling panel. The unit may be factory assembled, or field assembled from a manufactured kit.

R308.6.2 Permitted materials. The following types of glazing may be used:

1. Laminated glass with a minimum 0.015-inch (0.38 mm) polyvinyl butyral interlayer for glass panes 16 square feet (1.5 m^2) or less in area located such that the highest point of the glass is not more than 12 feet (3658 mm) above a walking surface or other accessible area; for higher or larger sizes, the minimum interlayer thickness shall be 0.030 inch (0.76 mm).
2. Fully tempered glass.
3. Heat-strengthened glass.
4. Wired glass.
5. *Approved* rigid plastics.

R308.6.3 Screens, general. For fully tempered or heat-strengthened glass, a retaining screen meeting the requirements of Section R308.6.7 shall be installed below the glass, except for fully tempered glass that meets either condition listed in Section R308.6.5.

R308.6.4 Screens with multiple glazing. When the inboard pane is fully tempered, heat-strengthened or wired glass, a retaining screen meeting the requirements of Section R308.6.7 shall be installed below the glass, except for either condition listed in Section R308.6.5. All other panes in the multiple glazing may be of any type listed in Section R308.6.2.

R308.6.5 Screens not required. Screens shall not be required when fully tempered glass is used as single glazing or the inboard pane in multiple glazing and either of the following conditions are met:

1. Glass area 16 square feet (1.49 m^2) or less. Highest point of glass not more than 12 feet (3658 mm) above a walking surface or other accessible area, nominal glass thickness not more than $^3/_{16}$ inch (4.8 mm), and (for multiple glazing only) the other pane or panes fully tempered, laminated or wired glass.
2. Glass area greater than 16 square feet (1.49 m^2). Glass sloped 30 degrees (0.52 rad) or less from vertical, and highest point of glass not more than 10 feet (3048 mm) above a walking surface or other accessible area.

R308.6.6 Glass in greenhouses. Any glazing material is permitted to be installed without screening in the sloped areas of greenhouses, provided the greenhouse height at the ridge does not exceed 20 feet (6096 mm) above *grade*.

R308.6.7 Screen characteristics. The screen and its fastenings shall be capable of supporting twice the weight of the glazing, be firmly and substantially fastened to the framing members, and have a mesh opening of no more than 1 inch by 1 inch (25 mm by 25 mm).

R308.6.8 Curbs for skylights. All unit skylights installed in a roof with a pitch flatter than three units vertical in 12 units horizontal (25-percent slope) shall be mounted on a curb extending at least 4 inches (102 mm) above the plane of the roof unless otherwise specified in the manufacturer's installation instructions.

R308.6.9 Testing and labeling. Unit skylights and tubular daylighting devices shall be tested by an *approved* independent laboratory, and bear a *label* identifying manufacturer, performance *grade* rating and *approved* inspection agency to indicate compliance with the requirements of AAMA/WDMA/CSA 101/I.S.2/A440.

SECTION R309
GARAGES AND CARPORTS

R309.1 Floor surface. Garage floor surfaces shall be of *approved* noncombustible material.

The area of floor used for parking of automobiles or other vehicles shall be sloped to facilitate the movement of liquids to a drain or toward the main vehicle entry doorway.

R309.2 Carports. Carports shall be open on at least two sides. Carport floor surfaces shall be of *approved* noncombustible material. Carports not open on at least two sides shall be considered a garage and shall comply with the provisions of this section for garages.

Exception: Asphalt surfaces shall be permitted at ground level in carports.

The area of floor used for parking of automobiles or other vehicles shall be sloped to facilitate the movement of liquids to a drain or toward the main vehicle entry doorway.

R309.3 Flood hazard areas. For buildings located in flood hazard areas as established by Table R301.2(1), garage floors shall be:

1. Elevated to or above the design flood elevation as determined in Section R322; or
2. Located below the design flood elevation provided they are at or above *grade* on at least one side, are used solely for parking, building access or storage, meet the requirements of Section R322 and are otherwise constructed in accordance with this code.

R309.4 Automatic garage door openers. Automatic garage door openers, if provided, shall be listed and labeled in accordance with UL 325.

R309.5 Fire sprinklers. Private garages shall be protected by fire sprinklers where the garage wall has been designed based on Table R302.1(2), Footnote a. Sprinklers in garages shall be connected to an automatic sprinkler system that complies with Section P2904. Garage sprinklers shall be residential sprinklers or quick-response sprinklers, designed to provide a density of 0.05 gpm/ft^2. Garage doors shall not be considered obstructions with respect to sprinkler placement.

SECTION R310
EMERGENCY ESCAPE AND RESCUE OPENINGS

R310.1 Emergency escape and rescue required. *Basements,* habitable attics and every sleeping room shall have at least one operable emergency escape and rescue opening. Where *basements* contain one or more sleeping rooms, emergency egress and rescue openings shall be required in each sleeping room. Where emergency escape and rescue openings are provided they shall have a sill height of not more than 44 inches (1118 mm) measured from the finished floor to the bottom of the clear opening. Where a door opening having a threshold below the adjacent ground elevation serves as an emergency escape and rescue opening and is provided with a bulkhead enclosure, the bulkhead enclosure shall comply with Section R310.3. The net clear opening dimensions required by this section shall be obtained by the normal operation of the emergency escape and rescue opening from the inside. Emergency escape and rescue openings with a finished sill height below the adjacent ground elevation shall be provided with a window well in accordance with Section R310.2. Emergency escape and rescue openings shall open directly into a public way, or to a *yard* or court that opens to a public way.

Exception: *Basements* used only to house mechanical *equipment* and not exceeding total floor area of 200 square feet (18.58 m^2).

R310.1.1 Minimum opening area. All emergency escape and rescue openings shall have a minimum net clear opening of 5.7 square feet (0.530 m^2).

Exception: *Grade* floor openings shall have a minimum net clear opening of 5 square feet (0.465 m^2).

R310.1.2 Minimum opening height. The minimum net clear opening height shall be 24 inches (610 mm).

R310.1.3 Minimum opening width. The minimum net clear opening width shall be 20 inches (508 mm).

R310.1.4 Operational constraints. Emergency escape and rescue openings shall be operational from the inside of the room without the use of keys, tools or special knowledge.

R310.2 Window wells. The minimum horizontal area of the window well shall be 9 square feet (0.9 m^2), with a minimum horizontal projection and width of 36 inches (914 mm). The area of the window well shall allow the emergency escape and rescue opening to be fully opened.

Exception: The ladder or steps required by Section R310.2.1 shall be permitted to encroach a maximum of 6 inches (152 mm) into the required dimensions of the window well.

R310.2.1 Ladder and steps. Window wells with a vertical depth greater than 44 inches (1118 mm) shall be equipped with a permanently affixed ladder or steps usable with the window in the fully open position. Ladders or steps required by this section shall not be required to comply with Sections R311.7 and R311.8. Ladders or rungs shall have an inside width of at least 12 inches (305 mm), shall project at least 3 inches (76 mm) from the wall and shall be spaced not more than 18 inches (457 mm) on center vertically for the full height of the window well.

R310.2.2 Drainage. Window wells shall be designed for proper drainage by connecting to the building's foundation drainage system required by Section R405.1 or by an approved alternative method.

Exception: A drainage system for window wells is not required when the foundation is on well-drained soil or sand-gravel mixture soils according to the United Soil Classification System, Group I Soils, as detailed in Table R405.1.

R310.3 Bulkhead enclosures. Bulkhead enclosures shall provide direct access to the *basement*. The bulkhead enclosure with the door panels in the fully open position shall provide the minimum net clear opening required by Section R310.1.1. Bulkhead enclosures shall also comply with Section R311.7.8.2.

R310.4 Bars, grilles, covers and screens. Bars, grilles, covers, screens or similar devices are permitted to be placed over emergency escape and rescue openings, bulkhead enclosures, or window wells that serve such openings, provided the minimum net clear opening size complies with Sections R310.1.1 to R310.1.3, and such devices shall be releasable or removable from the inside without the use of a key, tool, special knowledge or force greater than that which is required for normal operation of the escape and rescue opening.

R310.5 Emergency escape windows under decks and porches. Emergency escape windows are allowed to be installed under decks and porches provided the location of the deck allows the emergency escape window to be fully opened and provides a path not less than 36 inches (914 mm) in height to a *yard* or court.

SECTION R311
MEANS OF EGRESS

R311.1 Means of egress. All *dwellings* shall be provided with a means of egress as provided in this section. The means of egress shall provide a continuous and unobstructed path of vertical and horizontal egress travel from all portions of the *dwelling* to the exterior of the *dwelling* at the required egress door without requiring travel through a garage.

R311.2 Egress door. At least one egress door shall be provided for each *dwelling* unit. The egress door shall be side-hinged, and shall provide a minimum clear width of 32 inches (813 mm) when measured between the face of the door and the stop, with the door open 90 degrees (1.57 rad). The minimum clear height of the door opening shall not be less than 78 inches (1981 mm) in height measured from the top of the threshold to the bottom of the stop. Other doors shall not be required to comply with these minimum dimensions. Egress doors shall be readily openable from inside the *dwelling* without the use of a key or special knowledge or effort.

R311.3 Floors and landings at exterior doors. There shall be a landing or floor on each side of each exterior door. The width of each landing shall not be less than the door served. Every landing shall have a minimum dimension of 36 inches (914 mm) measured in the direction of travel. Exterior landings shall be permitted to have a slope not to exceed $^1/_4$ unit vertical in 12 units horizontal (2-percent).

Exception: Exterior balconies less than 60 square feet (5.6 m^2) and only accessible from a door are permitted to have a landing less than 36 inches (914 mm) measured in the direction of travel.

R311.3.1 Floor elevations at the required egress doors. Landings or finished floors at the required egress door shall not be more than $1^1/_2$ inches (38 mm) lower than the top of the threshold.

Exception: The landing or floor on the exterior side shall not be more than $7^3/_4$ inches (196 mm) below the top of the threshold provided the door does not swing over the landing or floor.

Where exterior landings or floors serving the required egress door are not at *grade*, they shall be provided with access to *grade* by means of a ramp in accordance with Section R311.8 or a stairway in accordance with Section R311.7.

R311.3.2 Floor elevations for other exterior doors. Doors other than the required egress door shall be provided with landings or floors not more than $7^3/_4$ inches (196 mm) below the top of the threshold.

Exception: A landing is not required where a stairway of two or fewer risers is located on the exterior side of the door, provided the door does not swing over the stairway.

R311.3.3 Storm and screen doors. Storm and screen doors shall be permitted to swing over all exterior stairs and landings.

R311.4 Vertical egress. Egress from habitable levels including habitable attics and *basements* not provided with an egress door in accordance with Section R311.2 shall be by a ramp in accordance with Section R311.8 or a stairway in accordance with Section R311.7.

R311.5 Construction.

R311.5.1 Attachment. Exterior landings, decks, balconies, stairs and similar facilities shall be positively anchored to the primary structure to resist both vertical and lateral forces or shall be designed to be self-supporting. Attachment shall not be accomplished by use of toenails or nails subject to withdrawal.

R311.6 Hallways. The minimum width of a hallway shall be not less than 3 feet (914 mm).

R311.7 Stairways.

R311.7.1 Width. Stairways shall not be less than 36 inches (914 mm) in clear width at all points above the permitted handrail height and below the required headroom height. Handrails shall not project more than 4.5 inches (114 mm) on either side of the stairway and the minimum clear width of the stairway at and below the handrail height, including treads and landings, shall not be less than $31^1/_2$ inches (787 mm) where a handrail is installed on one side and 27 inches (698 mm) where handrails are provided on both sides.

Exception: The width of spiral stairways shall be in accordance with Section R311.7.9.1.

R311.7.2 Headroom. The minimum headroom in all parts of the stairway shall not be less than 6 feet 8 inches (2032 mm) measured vertically from the sloped line adjoining the tread nosing or from the floor surface of the landing or platform on that portion of the stairway.

Exception: Where the nosings of treads at the side of a flight extend under the edge of a floor opening through which the stair passes, the floor opening shall be allowed to project horizontally into the required headroom a maximum of $4^3/_4$ inches (121 mm).

R311.7.3 Vertical rise. A flight of stairs shall not have a vertical rise larger than 12 feet (3658 mm) between floor levels or landings.

R311.7.4 Walkline. The walkline across winder treads shall be concentric to the curved direction of travel through the turn and located 12 inches (305 mm) from the side where the winders are narrower. The 12-inch (305 mm) dimension shall be measured from the widest point of the clear stair width at the walking surface of the winder. If winders are adjacent within the flight, the point of the widest clear stair width of the adjacent winders shall be used.

R311.7.5 Stair treads and risers. Stair treads and risers shall meet the requirements of this section. For the purposes of this section all dimensions and dimensioned surfaces shall be exclusive of carpets, rugs or runners.

R311.7.5.1 Risers. The maximum riser height shall be $7^3/_4$ inches (196 mm). The riser shall be measured vertically between leading edges of the adjacent treads. The greatest riser height within any flight of stairs shall not exceed the smallest by more than $^3/_8$ inch (9.5 mm). Risers shall be vertical or sloped from the underside of the nosing of the tread above at an angle not more than 30 degrees (0.51 rad) from the vertical. Open risers are permitted provided that the opening between treads does not permit the passage of a 4-inch-diameter (102 mm) sphere.

Exception: The opening between adjacent treads is not limited on stairs with a total rise of 30 inches (762 mm) or less.

R311.7.5.2 Treads. The minimum tread depth shall be 10 inches (254 mm). The tread depth shall be measured horizontally between the vertical planes of the foremost projection of adjacent treads and at a right angle to the tread's leading edge. The greatest tread depth within any flight of stairs shall not exceed the smallest by more than $^3/_8$ inch (9.5 mm).

R311.7.5.2.1 Winder treads. Winder treads shall have a minimum tread depth of 10 inches (254 mm) measured between the vertical planes of the foremost projection of adjacent treads at the intersections with the walkline. Winder treads shall have a minimum tread depth of 6 inches (152 mm) at any point within the clear width of the stair. Within any flight of stairs, the largest winder tread depth at the walkline shall not exceed the smallest winder tread by more than $^3/_8$ inch (9.5 mm). Consistently shaped winders at the walkline shall be allowed within the same flight of stairs as rectangular treads and do not have to be within $^3/_8$ inch (9.5 mm) of the rectangular tread depth.

R311.7.5.3 Nosings. The radius of curvature at the nosing shall be no greater than $^9/_{16}$ inch (14 mm). A nosing not less than $^3/_4$ inch (19 mm) but not more than $1^1/_4$ inches (32 mm) shall be provided on stairways with solid risers. The greatest nosing projection shall not exceed the smallest nosing projection by more than $^3/_8$ inch (9.5 mm) between two stories, including the nosing at the level of floors and landings. Beveling of nosings shall not exceed $^1/_2$ inch (12.7 mm).

Exception: A nosing is not required where the tread depth is a minimum of 11 inches (279 mm).

R311.7.5.4 Exterior wood/plastic composite stair treads. Wood/plastic composite stair treads shall comply with the provisions of Section R507.3.

R311.7.6 Landings for stairways. There shall be a floor or landing at the top and bottom of each stairway. The minimum width perpendicular to the direction of travel shall be no less than the width of the flight served. Landings of shapes other than square or rectangular shall be permitted provided the depth at the walk line and the total area is not less than that of a quarter circle with a radius equal to the required landing width. Where the stairway has a straight run, the minimum depth in the direction of travel shall be not less than 36 inches (914 mm).

Exception: A floor or landing is not required at the top of an interior flight of stairs, including stairs in an enclosed garage, provided a door does not swing over the stairs.

R311.7.7 Stairway walking surface. The walking surface of treads and landings of stairways shall be sloped no steeper than one unit vertical in 48 inches horizontal (2-percent slope).

R311.7.8 Handrails. Handrails shall be provided on at least one side of each continuous run of treads or flight with four or more risers.

R311.7.8.1 Height. Handrail height, measured vertically from the sloped plane adjoining the tread nosing, or finish surface of ramp slope, shall be not less than 34 inches (864 mm) and not more than 38 inches (965 mm).

Exceptions:

1. The use of a volute, turnout or starting easing shall be allowed over the lowest tread.
2. When handrail fittings or bendings are used to provide continuous transition between flights, transitions at winder treads, the transition from handrail to guardrail, or used at the start of a flight, the handrail height at the fittings or bendings shall be permitted to exceed the maximum height.

R311.7.8.2 Continuity. Handrails for stairways shall be continuous for the full length of the flight, from a point directly above the top riser of the flight to a point directly above the lowest riser of the flight. Handrail ends shall be returned or shall terminate in newel posts or safety terminals. Handrails adjacent to a wall shall have a space of not less than $1^1/_2$ inch (38 mm) between the wall and the handrails.

Exceptions:

1. Handrails shall be permitted to be interrupted by a newel post at the turn.
2. The use of a volute, turnout, starting easing or starting newel shall be allowed over the lowest tread.

R311.7.8.3 Grip-size. All required handrails shall be of one of the following types or provide equivalent graspability.

1. Type I. Handrails with a circular cross section shall have an outside diameter of at least $1^1/_4$ inches (32 mm) and not greater than 2 inches (51 mm). If the handrail is not circular, it shall have a perimeter dimension of at least 4 inches (102 mm) and not greater than $6^1/_4$ inches (160 mm) with a maximum cross section of dimension of $2^1/_4$ inches (57 mm). Edges shall have a minimum radius of 0.01 inch (0.25 mm).
2. Type II. Handrails with a perimeter greater than $6^1/_4$ inches (160 mm) shall have a graspable finger recess area on both sides of the profile. The finger recess shall begin within a distance of $^3/_4$ inch (19 mm) measured vertically from the tallest portion of the profile and achieve a depth of at least $^5/_{16}$ inch (8 mm) within $^7/_8$ inch (22 mm) below the widest portion of the profile. This required depth shall continue for at least $^3/_8$ inch (10 mm) to a level that is not less than $1^3/_4$ inches (45 mm) below the tallest portion of the profile. The minimum width of the handrail above the recess shall be $1^1/_4$ inches (32 mm) to a maximum of $2^3/_4$ inches (70 mm). Edges shall have a minimum radius of 0.01 inch (0.25 mm).

R311.7.8.4 Exterior wood/plastic composite handrails. Wood/plastic composite handrails shall comply with the provisions of Section R507.3.

R311.7.9 Illumination. All stairs shall be provided with illumination in accordance with Section R303.6.

R311.7.10 Special stairways. Spiral stairways and bulkhead enclosure stairways shall comply with all requirements of Section R311.7 except as specified below.

R311.7.10.1 Spiral stairways. Spiral stairways are permitted, provided the minimum clear width at and below the handrail shall be 26 inches (660 mm) with each tread having a $7^1/_2$-inch (190 mm) minimum tread depth at 12 inches (914 mm) from the narrower edge. All treads shall be identical, and the rise shall be no more than $9^1/_2$ inches (241 mm). A minimum headroom of 6 feet 6 inches (1982 mm) shall be provided.

R311.7.10.2 Bulkhead enclosure stairways. Stairways serving bulkhead enclosures, not part of the required building egress, providing access from the outside *grade* level to the *basement* shall be exempt from the requirements of Sections R311.3 and R311.7 where the maximum height from the *basement* finished floor level to *grade* adjacent to the stairway does not exceed 8 feet (2438 mm) and the *grade* level opening to the stairway is covered by a bulkhead enclosure with hinged doors or other *approved* means.

R311.8 Ramps.

R311.8.1 Maximum slope. Ramps shall have a maximum slope of 1 unit vertical in 12 units horizontal (8.3-percent slope).

Exception: Where it is technically infeasible to comply because of site constraints, ramps may have a maximum slope of one unit vertical in eight horizontal (12.5-percent slope).

R311.8.2 Landings required. A minimum 3-foot-by-3-foot (914 mm by 914 mm) landing shall be provided:

1. At the top and bottom of ramps.
2. Where doors open onto ramps.
3. Where ramps change direction.

R311.8.3 Handrails required. Handrails shall be provided on at least one side of all ramps exceeding a slope of one unit vertical in 12 units horizontal (8.33-percent slope).

R311.8.3.1 Height. Handrail height, measured above the finished surface of the ramp slope, shall be not less than 34 inches (864 mm) and not more than 38 inches (965 mm).

R311.8.3.2 Grip size. Handrails on ramps shall comply with Section R311.7.8.3.

R311.8.3.3 Continuity. Handrails where required on ramps shall be continuous for the full length of the ramp. Handrail ends shall be returned or shall terminate in newel posts or safety terminals. Handrails adjacent to a wall shall have a space of not less than $1^1/_2$ inches (38 mm) between the wall and the handrails.

SECTION R312
GUARDS AND WINDOW FALL PROTECTION

R312.1 Guards. Guards shall be provided in accordance with Sections R312.1.1 through R312.1.4.

R312.1.1 Where required. *Guards* shall be located along open-sided walking surfaces, including stairs, ramps and landings, that are located more than 30 inches (762 mm) measured vertically to the floor or *grade* below at any point within 36 inches (914 mm) horizontally to the edge of the open side. Insect screening shall not be considered as a *guard.*

R312.1.2 Height. Required *guards* at open-sided walking surfaces, including stairs, porches, balconies or landings, shall be not less than 36 inches (914 mm) high measured vertically above the adjacent walking surface, adjacent fixed seating or the line connecting the leading edges of the treads.

Exceptions:

1. *Guards* on the open sides of stairs shall have a height not less than 34 inches (864 mm) measured vertically from a line connecting the leading edges of the treads.
2. Where the top of the *guard* also serves as a handrail on the open sides of stairs, the top of the *guard* shall not be less than 34 inches (864 mm) and not more than 38 inches (965 mm) measured vertically from a line connecting the leading edges of the treads.

R312.1.3 Opening limitations. Required *guards* shall not have openings from the walking surface to the required *guard* height which allow passage of a sphere 4 inches (102 mm) in diameter.

Exceptions:

1. The triangular openings at the open side of stair, formed by the riser, tread and bottom rail of a guard, shall not allow passage of a sphere 6 inches (153 mm) in diameter.
2. *Guards* on the open side of stairs shall not have openings which allow passage of a sphere $4^3/_8$ inches (111 mm) in diameter.

R312.1.4 Exterior woodplastic composite guards. Woodplastic composite *guards* shall comply with the provisions of Section R317.4.

R312.2 Window fall protection. Window fall protection shall be provided in accordance with Sections R312.2.1 and R312.2.2. **

R312.2.1 Window sills. In dwelling units, where the opening of an operable window is located more than 72 inches (1829 mm) above the finished grade or surface below, the lowest part of the clear opening of the window shall be a minimum of 24 inches (610 mm) above the fininshed floor of the room in which the window is located. Operable sections of windows shall not permit openings that allow passage of a 4-inch-diameter (102 mm) sphere where such openings are located within 24 inches (610 mm) of the finished floor.

Exceptions:

1. Windows whose openings will not allow a 4-inch-diameter (102 mm) sphere to pass through the opening when the opening is in its largest opened position.
2. Openings that are provided with window fall prevention devices that comply with ASTM F 2090.
3. Windows that are provided with window opening control devices that comply with Section R312.2.2.

R312.2.2 Window opening control devices. Window opening control devices shall comply with ASTM F 2090. The window opening control device, after operation to release the control device allowing the window to fully open, shall not reduce the minimum net clear opening area of the window unit to less than the area required by Section R310.1.1.

SECTION R313
AUTOMATIC FIRE SPRINKLER SYSTEMS

R313.1 Townhouse automatic fire sprinkler systems. An automatic residential fire sprinkler system shall be installed in *townhouses*.

Exception: An automatic residential fire sprinkler system shall not be required when *additions* or *alterations* are made to existing *townhouses* that do not have an automatic residential fire sprinkler system installed.

R313.1.1 Design and installation. Automatic residential fire sprinkler systems for *townhouses* shall be designed and installed in accordance with Section P2904.

R313.2 One- and two-family dwellings automatic fire systems. An automatic residential fire sprinkler system shall be installed in one- and two-family *dwellings*.

Exception: An automatic residential fire sprinkler system shall not be required for *additions* or *alterations* to existing buildings that are not already provided with an automatic residential sprinkler system.

R313.2.1 Design and installation. Automatic residential fire sprinkler systems shall be designed and installed in accordance with Section P2904 or NFPA 13D.

SECTION R314 SMOKE ALARMS

R314.1 Smoke detection and notification. All smoke alarms shall be listed and labeled in accordance with UL 217 and installed in accordance with the provisions of this code and the household fire warning *equipment* provisions of NFPA 72.

R314.2 Smoke detection systems. Household fire alarm systems installed in accordance with NFPA 72 that include smoke alarms, or a combination of smoke detector and audible notification device installed as required by this section for smoke alarms, shall be permitted. The household fire alarm system shall provide the same level of smoke detection and alarm as required by this section for smoke alarms. Where a household fire warning system is installed using a combination of smoke detector and audible notification device(s), it shall become a permanent fixture of the occupancy and owned by the homeowner. The system shall be monitored by an *approved* supervising station and be maintained in accordance with NFPA 72.

Exception: Where smoke alarms are provided meeting the requirements of Section R314.4.

R314.3 Location. Smoke alarms shall be installed in the following locations:

1. In each sleeping room.
2. Outside each separate sleeping area in the immediate vicinity of the bedrooms.
3. On each additional *story* of the *dwelling*, including *basements* and habitable attics but not including crawl spaces and uninhabitable *attics*. In *dwellings* or *dwelling units* with split levels and without an intervening door between the adjacent levels, a smoke alarm installed on the upper level shall suffice for the adjacent lower level provided that the lower level is less than one full *story* below the upper level.

R314.3.1 Alterations, repairs and additions. When *alterations*, repairs or *additions* requiring a *permit* occur, or when one or more sleeping rooms are added or created in existing *dwellings*, the individual *dwelling unit* shall be equipped with smoke alarms located as required for new *dwellings*.

Exceptions:

1. Work involving the exterior surfaces of *dwellings*, such as the replacement of roofing or siding, or the *addition* or replacement of windows or doors, or the *addition* of a porch or deck, are exempt from the requirements of this section.
2. Installation, *alteration* or repairs of plumbing or mechanical systems are exempt from the requirements of this section.

R314.4 Power source. Smoke alarms shall receive their primary power from the building wiring when such wiring is served from a commercial source, and when primary power is interrupted, shall receive power from a battery. Wiring shall be permanent and without a disconnecting switch other than those required for overcurrent protection.

Exceptions:

1. Smoke alarms shall be permitted to be battery operated when installed in buildings without commercial power.
2. Hard wiring of smoke alarms in existing areas shall not be required where the *alterations* or repairs do not result in the removal of interior wall or ceiling finishes exposing the structure, unless there is an *attic*, crawl space or *basement* available which could provide access for hard wiring without the removal of interior finishes.

R314.5 Interconnection. Where more than one smoke alarm is required to be installed within an individual dwelling unit in accordance with Section R314.3, the alarm devices shall be interconnected in such a manner that the actuation of one alarm will activate all of the alarms in the individual unit. Physical interconnection of smoke alarms shall not be required where listed wireless alarms are installed and all alarms sound upon activation of one alarm.

Exception: Interconnection of smoke alarms in existing areas shall not be required where alterations or repairs do not result in removal of interior wall or ceiling finishes exposing the structure, unless there is an attic, crawl space or basement available which could provide access for interconnection without the removal of interior finishes.

SECTION R315 CARBON MONOXIDE ALARMS

R315.1 Carbon monoxide alarms. For new construction, an approved carbon monoxide alarm shall be installed outside of each separate sleeping area in the immediate vicinity of the bedrooms in *dwelling units* within which fuel-fired *appliances* are installed and in dwelling units that have attached garages.

R315.2 Carbon monoxide detection systems. Carbon monoxide detection systems that include carbon monoxide detectors and audible notification appliances, installed and maintained in accordance with this section for carbon monoxide alarms and NFPA 720, shall be permitted. The carbon monoxide detectors shall be listed as complying with UL 2075. Where a household carbon monoxide detection system is installed, it shall become a permanent fixture of the occupancy, owned by the homeowner and shall be monitored by an approved supervising station.

Exception: Where carbon monoxide alarms are installed meeting the requirements of Section R315.1, compliance with Section 315.2 is not required.

R315.3 Where required in existing dwellings. Where work requiring a *permit* occurs in existing *dwellings* that have attached garages or in existing dwellings within which fuel-fired *appliances* exist, carbon monoxide alarms shall be provided in accordance with Section R315.1.

R315.4 Alarm requirements. Single-station carbon monoxide alarms shall be listed as complying with UL 2034 and shall be installed in accordance with this code and the manufacturer's installation instructions.

SECTION R316
FOAM PLASTIC

R316.1 General. The provisions of this section shall govern the materials, design, application, construction and installation of foam plastic materials.

R316.2 Labeling and identification. Packages and containers of foam plastic insulation and foam plastic insulation components delivered to the job site shall bear the *label* of an *approved agency* showing the manufacturer's name, the product listing, product identification and information sufficient to determine that the end use will comply with the requirements.

R316.3 Surface burning characteristics. Unless otherwise allowed in Section R316.5 or R316.6, all foam plastic or foam plastic cores used as a component in manufactured assemblies used in building construction shall have a flame spread index of not more than 75 and shall have a smoke-developed index of not more than 450 when tested in the maximum thickness intended for use in accordance with ASTM E 84 or UL 723. Loose-fill-type foam plastic insulation shall be tested as board stock for the flame spread index and smoke-developed index.

Exception: Foam plastic insulation more than 4 inches (102 mm) thick shall have a maximum flame spread index of 75 and a smoke-developed index of 450 where tested at a minimum thickness of 4 inches (102 mm), provided the end use is *approved* in accordance with Section R316.6 using the thickness and density intended for use.

R316.4 Thermal barrier. Unless otherwise allowed in Section R316.5 or Section R316.6, foam plastic shall be separated from the interior of a building by an *approved* thermal barrier of minimum $^{1}/_{2}$ inch (12.7 mm) gypsum wallboard or a material that is tested in accordance with and meets the acceptance criteria of both the Temperature Transmission Fire Test and the Integrity Fire Test of NFPA 275.

R316.5 Specific requirements. The following requirements shall apply to these uses of foam plastic unless specifically *approved* in accordance with Section R316.6 or by other sections of the code or the requirements of Sections R316.2 through R316.4 have been met.

R316.5.1 Masonry or concrete construction. The thermal barrier specified in Section R316.4 is not required in a masonry or concrete wall, floor or roof when the foam plastic insulation is separated from the interior of the building by a minimum 1-inch (25 mm) thickness of masonry or concrete.

R316.5.2 Roofing. The thermal barrier specified in Section R316.4 is not required when the foam plastic in a roof assembly or under a roof covering is installed in accordance with the code and the manufacturer's installation instructions and is separated from the interior of the building by tongue-and-groove wood planks or wood structural panel sheathing in accordance with Section R803, not less than $^{15}/_{32}$ inch (11.9 mm) thick bonded with exterior glue and identified as Exposure 1, with edges supported by blocking or tongue-and-groove joints or an equivalent material. The smoke-developed index for roof applications shall not be limited.

R316.5.3 Attics. The thermal barrier specified in Section R316.4 is not required where all of the following apply:

1. *Attic* access is required by Section R807.1.
2. The space is entered only for purposes of repairs or maintenance.
3. The foam plastic insulation is protected against ignition using one of the following ignition barrier materials:
   - 3.1. $1^{1}/_{2}$-inch-thick (38 mm) mineral fiber insulation;
   - 3.2. $^{1}/_{4}$-inch-thick (6.4 mm) wood structural panels;
   - 3.3. $^{3}/_{8}$-inch (9.5 mm) particleboard;
   - 3.4. $^{1}/_{4}$-inch (6.4 mm) hardboard;
   - 3.5. $^{3}/_{8}$-inch (9.5 mm) gypsum board; or
   - 3.6. Corrosion-resistant steel having a base metal thickness of 0.016 inch (0.406 mm);
   - 3.7. $1^{1}/_{2}$-inch-thick (38 mm) cellulose insulation.

The above ignition barrier is not required where the foam plastic insulation has been tested in accordance with Section R316.6.

R316.5.4 Crawl spaces. The thermal barrier specified in Section R316.4 is not required where all of the following apply:

1. Crawlspace access is required by Section R408.4
2. Entry is made only for purposes of repairs or maintenance.
3. The foam plastic insulation is protected against ignition using one of the following ignition barrier materials:
   - 3.1. $1^{1}/_{2}$-inch-thick (38 mm) mineral fiber insulation;
   - 3.2. $^{1}/_{4}$-inch-thick (6.4 mm) wood structural panels;

3.3. $^3/_8$-inch (9.5 mm) particleboard;

3.4. $^1/_4$-inch (6.4 mm) hardboard;

3.5. $^3/_8$-inch (9.5 mm) gypsum board; or

3.6. Corrosion-resistant steel having a base metal thickness of 0.016 inch (0.406 mm).

The above ignition barrier is not required where the foam plastic insulation has been tested in accordance with Section R316.6.

R316.5.5 Foam-filled exterior doors. Foam-filled exterior doors are exempt from the requirements of Sections R316.3 and R316.4.

R316.5.6 Foam-filled garage doors. Foam-filled garage doors in attached or detached garages are exempt from the requirements of Sections R316.3 and R316.4.

R316.5.7 Foam backer board. The thermal barrier specified in Section R316.4 is not required where siding backer board foam plastic insulation has a maximum thickness of 0.5 inch (12.7 mm) and a potential heat of not more than 2000 Btu per square foot (22 720 kJ/m^2) when tested in accordance with NFPA 259 provided that:

1. The foam plastic insulation is separated from the interior of the building by not less than 2 inches (51 mm) of mineral fiber insulation;
2. The foam plastic insulation is installed over existing *exterior wall* finish in conjunction with re-siding; or
3. The foam plastic insulation has been tested in accordance with Section R316.6.

R316.5.8 Re-siding. The thermal barrier specified in Section R316.4 is not required where the foam plastic insulation is installed over existing *exterior wall* finish in conjunction with re-siding provided the foam plastic has a maximum thickness of 0.5 inch (12.7 mm) and a potential heat of not more than 2000 Btu per square foot (22 720 kJ/m^2) when tested in accordance with NFPA 259.

R316.5.9 Interior trim. The thermal barrier specified in Section R316.4 is not required for exposed foam plastic interior trim, provided all of the following are met:

1. The minimum density is 20 pounds per cubic foot (320 kg/m^3).
2. The maximum thickness of the trim is 0.5 inch (12.7 mm) and the maximum width is 8 inches (204 mm).
3. The interior trim shall not constitute more than 10 percent of the aggregate wall and ceiling area of any room or space.
4. The flame spread index does not exceed 75 when tested per ASTM E 84 or UL 723. The smoke-developed index is not limited.

R316.5.10 Interior finish. Foam plastics shall be permitted as interior finish where *approved* in accordance with Section R316.6 Foam plastics that are used as interior finish shall also meet the flame spread index and smoke-developed index requirements of Sections R302.9.1 and R302.9.2.

R316.5.11 Sill plates and headers. Foam plastic shall be permitted to be spray applied to a sill plate and header without the thermal barrier specified in Section R316.4 subject to all of the following:

1. The maximum thickness of the foam plastic shall be $3^1/_4$ inches (83 mm).
2. The density of the foam plastic shall be in the range of 0.5 to 2.0 pounds per cubic foot (8 to 32 kg/m^3).
3. The foam plastic shall have a flame spread index of 25 or less and an accompanying smoke-developed index of 450 or less when tested in accordance with ASTM E 84 or UL 723.

R316.5.12 Sheathing. Foam plastic insulation used as sheathing shall comply with Section R316.3 and Section R316.4. Where the foam plastic sheathing is exposed to the *attic* space at a gable or kneewall, the provisions of Section R316.5.3 shall apply.

R316.5.13 Floors. The thermal barrier specified in Section R316.4 is not required to be installed on the walking surface of a structural floor system that contains foam plastic insulation when the foam plastic is covered by a minimum nominal $^1/_2$-inch-thick (12.7 mm) wood structural panel or equivalent. The thermal barrier specified in Section R316.4 is required on the underside of the structural floor system that contains foam plastic insulation when the underside of the structural floor system is exposed to the interior of the building.

R316.6 Specific approval. Foam plastic not meeting the requirements of Sections R316.3 through R316.5 shall be specifically *approved* on the basis of one of the following *approved* tests: NFPA 286 with the acceptance criteria of Section R302.9.4, FM4880, UL 1040, or UL 1715, or fire tests related to actual end-use configurations. Approval shall be based on the actual end use configuration and shall be performed on the finished foam plastic assembly in the maximum thickness intended for use. Assemblies tested shall include seams, joints and other typical details used in the installation of the assembly and shall be tested in the manner intended for use.

R316.7 Termite damage. The use of foam plastics in areas of "very heavy" termite infestation probability shall be in accordance with Section R318.4.

SECTION R317
PROTECTION OF WOOD AND WOOD BASED PRODUCTS AGAINST DECAY

R317.1 Location required. Protection of wood and wood based products from decay shall be provided in the following locations by the use of naturally durable wood or wood that is preservative-treated in accordance with AWPA U1 for the species, product, preservative and end use. Preservatives shall be listed in Section 4 of AWPA U1.

1. Wood joists or the bottom of a wood structural floor when closer than 18 inches (457 mm) or wood girders when closer than 12 inches (305 mm) to the exposed

ground in crawl spaces or unexcavated area located within the periphery of the building foundation.

2. All wood framing members that rest on concrete or masonry exterior foundation walls and are less than 8 inches (203 mm) from the exposed ground.

3. Sills and sleepers on a concrete or masonry slab that is in direct contact with the ground unless separated from such slab by an impervious moisture barrier.

4. The ends of wood girders entering exterior masonry or concrete walls having clearances of less than $^1/_2$ inch (12.7 mm) on tops, sides and ends.

5. Wood siding, sheathing and wall framing on the exterior of a building having a clearance of less than 6 inches (152 mm) from the ground or less than 2 inches (51 mm) measured vertically from concrete steps, porch slabs, patio slabs, and similar horizontal surfaces exposed to the weather.

6. Wood structural members supporting moisture-permeable floors or roofs that are exposed to the weather, such as concrete or masonry slabs, unless separated from such floors or roofs by an impervious moisture barrier.

7. Wood furring strips or other wood framing members attached directly to the interior of exterior masonry walls or concrete walls below *grade* except where an *approved* vapor retarder is applied between the wall and the furring strips or framing members.

R317.1.1 Field treatment. Field-cut ends, notches and drilled holes of preservative-treated wood shall be treated in the field in accordance with AWPA M4.

R317.1.2 Ground contact. All wood in contact with the ground, embedded in concrete in direct contact with the ground or embedded in concrete exposed to the weather that supports permanent structures intended for human occupancy shall be *approved* pressure-preservative-treated wood suitable for ground contact use, except untreated wood may be used where entirely below groundwater level or continuously submerged in fresh water.

R317.1.3 Geographical areas. In geographical areas where experience has demonstrated a specific need, *approved* naturally durable or pressure-preservative-treated wood shall be used for those portions of wood members that form the structural supports of buildings, balconies, porches or similar permanent building appurtenances when those members are exposed to the weather without adequate protection from a roof, eave, overhang or other covering that would prevent moisture or water accumulation on the surface or at joints between members. Depending on local experience, such members may include:

1. Horizontal members such as girders, joists and decking.

2. Vertical members such as posts, poles and columns.

3. Both horizontal and vertical members.

R317.1.4 Wood columns. Wood columns shall be *approved* wood of natural decay resistance or *approved* pressure-preservative-treated wood.

Exceptions:

1. Columns exposed to the weather or in *basements* when supported by concrete piers or metal pedestals projecting 1 inch (25.4 mm) above a concrete floor or 6 inches (152 mm) above exposed earth and the earth is covered by an *approved* impervious moisture barrier.

2. Columns in enclosed crawl spaces or unexcavated areas located within the periphery of the building when supported by a concrete pier or metal pedestal at a height more than 8 inches (203 mm) from exposed earth and the earth is covered by an impervious moisture barrier.

R317.1.5 Exposed glued-laminated timbers. The portions of glued-laminated timbers that form the structural supports of a building or other structure and are exposed to weather and not properly protected by a roof, eave or similar covering shall be pressure treated with preservative, or be manufactured from naturally durable or preservative-treated wood.

R317.2 Quality mark. Lumber and plywood required to be pressure-preservative-treated in accordance with Section R318.1 shall bear the quality *mark* of an *approved* inspection agency that maintains continuing supervision, testing and inspection over the quality of the product and that has been *approved* by an accreditation body that complies with the requirements of the American Lumber Standard Committee treated wood program.

R317.2.1 Required information. The required quality *mark* on each piece of pressure-preservative-treated lumber or plywood shall contain the following information:

1. Identification of the treating plant.
2. Type of preservative.
3. The minimum preservative retention.
4. End use for which the product was treated.
5. Standard to which the product was treated.
6. Identity of the *approved* inspection agency.
7. The designation "Dry," if applicable.

Exception: Quality *marks* on lumber less than 1 inch (25.4 mm) nominal thickness, or lumber less than nominal 1 inch by 5 inches (25.4 mm by 127 mm) or 2 inches by 4 inches (51 mm by 102 mm) or lumber 36 inches (914 mm) or less in length shall be applied by stamping the faces of exterior pieces or by end labeling not less than 25 percent of the pieces of a bundled unit.

R317.3 Fasteners and connectors in contact with preservative-treated and fire-retardant-treated wood. Fasteners, including nuts and washers, and connectors in contact with preservative-treated wood and fire-retardant-treated wood shall be in accordance with this section. The coating weights for zinc-coated fasteners shall be in accordance with ASTM A 153.

R317.3.1 Fasteners for preservative-treated wood. Fasteners, including nuts and washers, for preservative-treated wood shall be of hot-dipped, zinc-coated galvanized steel, stainless steel, silicon bronze or copper. Coating types and weights for connectors in contact with preservative-treated wood shall be in accordance with the connector manufacturer's recommendations. In the absence of manufacturer's recommendations, a minimum of ASTM A 653 type G185 zinc-coated galvanized steel, or equivalent, shall be used.

Exceptions:

1. One-half-inch-diameter (12.7 mm) or greater steel bolts.
2. Fasteners other than nails and timber rivets shall be permitted to be of mechanically deposited zinc-coated steel with coating weights in accordance with ASTM B 695, Class 55 minimum.
3. Plain carbon steel fasteners in SBX/DOT and zinc borate preservative-treated wood in an interior, dry environment shall be permitted.

R317.3.2 Fastenings for wood foundations. Fastenings, including nuts and washers, for wood foundations shall be as required in AF&PA PWF.

R317.3.3 Fasteners for fire-retardant-treated wood used in exterior applications or wet or damp locations. Fasteners, including nuts and washers, for fire-retardant-treated wood used in exterior applications or wet or damp locations shall be of hot-dipped, zinc-coated galvanized steel, stainless steel, silicon bronze or copper. Fasteners other than nails and timber rivets shall be permitted to be of mechanically deposited zinc-coated steel with coating weights in accordance with ASTM B 695, Class 55 minimum.

R317.3.4 Fasteners for fire-retardant-treated wood used in interior applications. Fasteners, including nuts and washers, for fire-retardant-treated wood used in interior locations shall be in accordance with the manufacturer's recommendations. In the absence of the manufacturer's recommendations, Section R317.3.3 shall apply.

R317.4 Wood/plastic composites. Wood/plastic composites used in exterior deck boards, stair treads, handrails and guardrail systems shall bear a *label* indicating the required performance levels and demonstrating compliance with the provisions of ASTM D 7032.

R317.4.1 Labeling. Deck boards and stair treads shall bear a label that indicates compliance to ASTM D 7032 and includes the allowable load and maximum allowable span. Handrails and guardrail systems or their packaging shall bear a label that indicates compliance to ASTM D 7032 and includes the maximum allowable span.

R317.4.2 Installation. Wood/plastic composites shall be installed in accordance with the manufacturer's instructions.

SECTION R318
PROTECTION AGAINST SUBTERRANEAN TERMITES

R318.1 Subterranean termite control methods. In areas subject to damage from termites as indicated by Table R301.2(1), methods of protection shall be one of the following methods or a combination of these methods:

1. Chemical termiticide treatment, as provided in Section R318.2.
2. Termite baiting system installed and maintained according to the *label*.
3. Pressure-preservative-treated wood in accordance with the provisions of Section R317.1.
4. Naturally durable termite-resistant wood.
5. Physical barriers as provided in Section R318.3 and used in locations as specified in Section R317.1.
6. Cold-formed steel framing in accordance with Sections R505.2.1 and R603.2.1.

R318.1.1 Quality mark. Lumber and plywood required to be pressure-preservative-treated in accordance with Section R318.1 shall bear the quality *mark* of an *approved* inspection agency which maintains continuing supervision, testing and inspection over the quality of the product and which has been *approved* by an accreditation body which complies with the requirements of the American Lumber Standard Committee treated wood program.

R318.1.2 Field treatment. Field-cut ends, notches, and drilled holes of pressure-preservative-treated wood shall be retreated in the field in accordance with AWPA M4.

R318.2 Chemical termiticide treatment. Chemical termiticide treatment shall include soil treatment and/or field applied wood treatment. The concentration, rate of application and method of treatment of the chemical termiticide shall be in strict accordance with the termiticide *label*.

R318.3 Barriers. *Approved* physical barriers, such as metal or plastic sheeting or collars specifically designed for termite prevention, shall be installed in a manner to prevent termites from entering the structure. Shields placed on top of an exterior foundation wall are permitted to be used only if in combination with another method of protection.

R318.4 Foam plastic protection. In areas where the probability of termite infestation is "very heavy" as indicated in Figure R301.2(6), extruded and expanded polystyrene, polyisocyanurate and other foam plastics shall not be installed on the exterior face or under interior or exterior foundation walls or slab foundations located below *grade*. The clearance between foam plastics installed above *grade* and exposed earth shall be at least 6 inches (152 mm).

Exceptions:

1. Buildings where the structural members of walls, floors, ceilings and roofs are entirely of noncombustible materials or pressure-preservative-treated wood.

2. When in *addition* to the requirements of Section R318.1, an *approved* method of protecting the foam plastic and structure from subterranean termite damage is used.
3. On the interior side of *basement walls*.

SECTION R319
SITE ADDRESS

R319.1 Address numbers. Buildings shall have *approved* address numbers, building numbers or *approved* building identification placed in a position that is plainly legible and visible from the street or road fronting the property. These numbers shall contrast with their background. Address numbers shall be Arabic numbers or alphabetical letters. Numbers shall be a minimum of 4 inches (102 mm) high with a minimum stroke width of $^1/_2$ inch (12.7 mm). Where access is by means of a private road and the building address cannot be viewed from the public way, a monument, pole or other sign or means shall be used to identify the structure.

SECTION R320
ACCESSIBILITY

R320.1 Scope. Where there are four or more *dwelling* units or sleeping units in a single structure, the provisions of Chapter 11 of the *International Building Code* for Group R-3 shall apply.

SECTION R321
ELEVATORS AND PLATFORM LIFTS

R321.1 Elevators. Where provided, passenger elevators, limited-use/limited-application elevators or private residence elevators shall comply with ASME A17.1.

R321.2 Platform lifts. Where provided, platform lifts shall comply with ASME A18.1.

R321.3 Accessibility. Elevators or platform lifts that are part of an accessible route required by Chapter 11 of the *International Building Code*, shall comply with ICC A117.1.

SECTION R322
FLOOD-RESISTANT CONSTRUCTION

R322.1 General. Buildings and structures constructed in whole or in part in flood hazard areas (including A or V Zones) as established in Table R301.2(1) shall be designed and constructed in accordance with the provisions contained in this section. Buildings and structures located in whole or in part in identified floodways shall be designed and constructed in accordance with ASCE 24.

R322.1.1 Alternative provisions. As an alternative to the requirements in Section R322.3 for buildings and structures located in whole or in part in coastal high-hazard areas (V Zones) and Coastal A Zones, if delineated, ASCE 24 is permitted subject to the limitations of this code and the limitations therein.

R322.1.2 Structural systems. All structural systems of all buildings and structures shall be designed, connected and anchored to resist flotation, collapse or permanent lateral movement due to structural loads and stresses from flooding equal to the design flood elevation.

R322.1.3 Flood-resistant construction. All buildings and structures erected in areas prone to flooding shall be constructed by methods and practices that minimize flood damage.

R322.1.4 Establishing the design flood elevation. The design flood elevation shall be used to define flood hazard areas. At a minimum, the design flood elevation is the higher of:

1. The base flood elevation at the depth of peak elevation of flooding (including wave height) which has a 1 percent (100-year flood) or greater chance of being equaled or exceeded in any given year; or
2. The elevation of the design flood associated with the area designated on a flood hazard map adopted by the community, or otherwise legally designated.

R322.1.4.1 Determination of design flood elevations. If design flood elevations are not specified, the *building official* is authorized to require the applicant to:

1. Obtain and reasonably use data available from a federal, state or other source; or
2. Determine the design flood elevation in accordance with accepted hydrologic and hydraulic engineering practices used to define special flood hazard areas. Determinations shall be undertaken by a registered *design professional* who shall document that the technical methods used reflect currently accepted engineering practice. Studies, analyses and computations shall be submitted in sufficient detail to allow thorough review and approval.

R322.1.4.2 Determination of impacts. In riverine flood hazard areas where design flood elevations are specified but floodways have not been designated, the applicant shall demonstrate that the effect of the proposed buildings and structures on design flood elevations, including fill, when combined with all other existing and anticipated flood hazard area encroachments, will not increase the design flood elevation more than 1 foot (305 mm) at any point within the jurisdiction.

R322.1.5 Lowest floor. The lowest floor shall be the floor of the lowest enclosed area, including *basement*, but excluding any unfinished flood-resistant enclosure that is useable solely for vehicle parking, building access or limited storage provided that such enclosure is not built so as to render the building or structure in violation of this section.

R322.1.6 Protection of mechanical and electrical systems. Electrical systems, *equipment* and components; heating, ventilating, air conditioning; plumbing *appliances* and plumbing fixtures; *duct systems*; and other service *equipment* shall be located at or above the elevation required in Section R322.2 (flood hazard areas including A Zones) or R322.3 (coastal high-hazard areas including V Zones). If replaced as part of a substantial improvement, electrical systems, *equipment* and components; heating, ventilating, air conditioning and plumbing *appliances* and plumbing fixtures; *duct systems*; and other service *equipment* shall meet the requirements of this section. Systems, fixtures, and *equipment* and components shall not be mounted on or penetrate through walls intended to break away under flood loads.

Exception: Locating electrical systems, *equipment* and components; heating, ventilating, air conditioning; plumbing *appliances* and plumbing fixtures; *duct systems*; and other service *equipment* is permitted below the elevation required in Section R322.2 (flood hazard areas including A Zones) or R322.3 (coastal high-hazard areas including V Zones) provided that they are designed and installed to prevent water from entering or accumulating within the components and to resist hydrostatic and hydrodynamic loads and stresses, including the effects of buoyancy, during the occurrence of flooding to the design flood elevation in accordance with ASCE 24. Electrical wiring systems are permitted to be located below the required elevation provided they conform to the provisions of the electrical part of this code for wet locations.

R322.1.7 Protection of water supply and sanitary sewage systems. New and replacement water supply systems shall be designed to minimize or eliminate infiltration of flood waters into the systems in accordance with the plumbing provisions of this code. New and replacement sanitary sewage systems shall be designed to minimize or eliminate infiltration of floodwaters into systems and discharges from systems into floodwaters in accordance with the plumbing provisions of this code and Chapter 3 of the *International Private Sewage Disposal Code.*

R322.1.8 Flood-resistant materials. Building materials used below the elevation required in Section R322.2 (flood hazard areas including A Zones) or R322.3 (coastal high-hazard areas including V Zones) shall comply with the following:

1. All wood, including floor sheathing, shall be pressure-preservative-treated in accordance with AWPA U1 for the species, product, preservative and end use or be the decay-resistant heartwood of redwood, black locust or cedars. Preservatives shall be listed in Section 4 of AWPA U1.
2. Materials and installation methods used for flooring and interior and *exterior walls* and wall coverings shall conform to the provisions of FEMA/FIA-TB-2.

R322.1.9 Manufactured homes. New or replacement *manufactured homes* shall be elevated in accordance with Section R322.2 (flood hazard areas including A Zones) or Section R322.3 in coastal high-hazard areas (V Zones). The anchor and tie-down requirements of Sections AE604 and AE605 of Appendix E shall apply. The foundation and anchorage of *manufactured homes* to be located in identified floodways shall be designed and constructed in accordance with ASCE 24.

R322.1.10 As-built elevation documentation. A registered *design professional* shall prepare and seal documentation of the elevations specified in Section R322.2 or R322.3.

R322.2 Flood hazard areas (including A Zones). All areas that have been determined to be prone to flooding but not subject to high-velocity wave action shall be designated as flood hazard areas. Flood hazard areas that have been delineated as subject to wave heights between $1^1/_2$ feet (457 mm) and 3 feet (914 mm) shall be designated as Coastal A Zones. All building and structures constructed in whole or in part in flood hazard areas shall be designed and constructed in accordance with Sections R322.2.1 through R322.2.3.

R322.2.1 Elevation requirements.

1. Buildings and structures in flood hazard areas not designated as Coastal A Zones shall have the lowest floors elevated to or above the design flood elevation.
2. Buildings and structures in flood hazard areas designated as Coastal A Zones shall have the lowest floors elevated to or above the base flood elevation plus 1 foot (305 mm), or to the design flood elevation, whichever is higher.
3. In areas of shallow flooding (AO Zones), buildings and structures shall have the lowest floor (including *basement*) elevated at least as high above the highest adjacent *grade* as the depth number specified in feet on the FIRM, or at least 2 feet (610 mm) if a depth number is not specified.
4. Basement floors that are below *grade* on all sides shall be elevated to or above the design flood elevation.

Exception: Enclosed areas below the design flood elevation, including *basements* whose floors are not below *grade* on all sides, shall meet the requirements of Section R322.2.2.

R322.2.2 Enclosed area below design flood elevation. Enclosed areas, including crawl spaces, that are below the design flood elevation shall:

1. Be used solely for parking of vehicles, building access or storage.
2. Be provided with flood openings that meet the following criteria:
   - 2.1. There shall be a minimum of two openings on different sides of each enclosed area; if a building has more than one enclosed area below the design flood elevation, each area shall have openings on exterior walls.
   - 2.2. The total net area of all openings shall be at least 1 square inch (645 mm^2) for each square foot (0.093 m^2) of enclosed area, or the openings shall be designed and the *construction documents* shall include a state-

ment by a registered *design professional* that the design of the openings will provide for equalization of hydrostatic flood forces on exterior walls by allowing for the automatic entry and exit of floodwaters as specified in Section 2.6.2.2 of ASCE 24.

2.3. The bottom of each opening shall be 1 foot (305 mm) or less above the adjacent ground level.

2.4. Openings shall be not less than 3 inches (76 mm) in any direction in the plane of the wall.

2.5. Any louvers, screens or other opening covers shall allow the automatic flow of floodwaters into and out of the enclosed area.

2.6. Openings installed in doors and windows, that meet requirements 2.1 through 2.5, are acceptable; however, doors and windows without installed openings do not meet the requirements of this section.

R322.2.3 Foundation design and construction. Foundation walls for all buildings and structures erected in flood hazard areas shall meet the requirements of Chapter 4.

Exception: Unless designed in accordance with Section R404:

1. The unsupported height of 6-inch (152 mm) plain masonry walls shall be no more than 3 feet (914 mm).
2. The unsupported height of 8-inch (203 mm) plain masonry walls shall be no more than 4 feet (1219 mm).
3. The unsupported height of 8-inch (203 mm) reinforced masonry walls shall be no more than 8 feet (2438 mm).

For the purpose of this exception, unsupported height is the distance from the finished *grade* of the under-floor space to the top of the wall.

R322.3 Coastal high-hazard areas (including V Zones). Areas that have been determined to be subject to wave heights in excess of 3 feet (914 mm) or subject to high-velocity wave action or wave-induced erosion shall be designated as coastal high-hazard areas. Buildings and structures constructed in whole or in part in coastal high-hazard areas shall be designed and constructed in accordance with Sections R322.3.1 through R322.3.6.

R322.3.1 Location and site preparation.

1. New buildings and buildings that are determined to be substantially improved pursuant to Section R105.3.1.1, shall be located landward of the reach of mean high tide.
2. For any alteration of sand dunes and mangrove stands the *building official* shall require submission of an engineering analysis which demonstrates that the proposed *alteration* will not increase the potential for flood damage.

R322.3.2 Elevation requirements.

1. All buildings and structures erected within coastal high-hazard areas shall be elevated so that the lowest portion of all structural members supporting the lowest floor, with the exception of piling, pile caps, columns, grade beams and bracing, is:

   1.1. Located at or above the design flood elevation, if the lowest horizontal structural member is oriented parallel to the direction of wave approach, where parallel shall mean less than or equal to 20 degrees (0.35 rad) from the direction of approach, or

   1.2. Located at the base flood elevation plus 1 foot (305 mm), or the design flood elevation, whichever is higher, if the lowest horizontal structural member is oriented perpendicular to the direction of wave approach, where perpendicular shall mean greater than 20 degrees (0.35 rad) from the direction of approach.

2. Basement floors that are below *grade* on all sides are prohibited.
3. The use of fill for structural support is prohibited.
4. Minor grading, and the placement of minor quantities of fill, shall be permitted for landscaping and for drainage purposes under and around buildings and for support of parking slabs, pool decks, patios and walkways.

   Exception: Walls and partitions enclosing areas below the design flood elevation shall meet the requirements of Sections R322.3.4 and R322.3.5.

R322.3.3 Foundations. Buildings and structures erected in coastal high-hazard areas shall be supported on pilings or columns and shall be adequately anchored to such pilings or columns. The space below the elevated building shall be either free of obstruction or, if enclosed with walls, the walls shall meet the requirements of Section R322.3.4. Pilings shall have adequate soil penetrations to resist the combined wave and wind loads (lateral and uplift). Water-loading values used shall be those associated with the design flood. Wind-loading values shall be those required by this code. Pile embedment shall include consideration of decreased resistance capacity caused by scour of soil strata surrounding the piling. Pile systems design and installation shall be certified in accordance with Section R322.3.6. Spread footing, mat, raft or other foundations that support columns shall not be permitted where soil investigations that are required in accordance with Section R401.4 indicate that soil material under the spread footing, mat, raft or other foundation is subject to scour or erosion from wave-velocity flow conditions. If permitted, spread footing, mat, raft or other foundations that support columns shall be designed in accordance with ASCE 24. Slabs, pools, pool decks and walkways shall be located and constructed to be structurally independent of buildings and structures and their foundations to prevent transfer of flood loads to the buildings and structures during conditions of flooding, scour or erosion from wave-velocity flow conditions, unless the buildings and structures and their foundation are designed to resist the additional flood load.

R322.3.4 Walls below design flood elevation. Walls and partitions are permitted below the elevated floor, provided that such walls and partitions are not part of the structural support of the building or structure and:

1. Electrical, mechanical, and plumbing system components are not to be mounted on or penetrate through walls that are designed to break away under flood loads; and
2. Are constructed with insect screening or open lattice; or
3. Are designed to break away or collapse without causing collapse, displacement or other structural damage to the elevated portion of the building or supporting foundation system. Such walls, framing and connections shall have a design safe loading resistance of not less than 10 (479 Pa) and no more than 20 pounds per square foot (958 Pa); or
4. Where wind loading values of this code exceed 20 pounds per square foot (958 Pa), the *construction documents* shall include documentation prepared and sealed by a registered *design professional* that:
   - 4.1. The walls and partitions below the design flood elevation have been designed to collapse from a water load less than that which would occur during the design flood.
   - 4.2. The elevated portion of the building and supporting foundation system have been designed to withstand the effects of wind and flood loads acting simultaneously on all building components (structural and nonstructural). Water-loading values used shall be those associated with the design flood. Wind-loading values shall be those required by this code.

R322.3.5 Enclosed areas below design flood elevation. Enclosed areas below the design flood elevation shall be used solely for parking of vehicles, building access or storage.

R322.3.6 Construction documents. The *construction documents* shall include documentation that is prepared and sealed by a registered *design professional* that the design and methods of construction to be used meet the applicable criteria of this section.

SECTION R323
STORM SHELTERS

R323.1 General. This section applies to the construction of storm shelters when constructed as separate detached buildings or when constructed as safe rooms within buildings for the purpose of providing safe refuge from storms that produce high winds, such as tornados and hurricanes. In addition to other applicable requirements in this code, storm shelters shall be constructed in accordance with ICC/NSSA-500.

CHAPTER 4

FOUNDATIONS

SECTION R401 GENERAL

R401.1 Application. The provisions of this chapter shall control the design and construction of the foundation and foundation spaces for all buildings. In addition to the provisions of this chapter, the design and construction of foundations in flood hazard areas as established by Table R301.2(1) shall meet the provisions of Section R322. Wood foundations shall be designed and installed in accordance with AF&PA PWF.

Exception: The provisions of this chapter shall be permitted to be used for wood foundations only in the following situations:

1. In buildings that have no more than two floors and a roof.
2. When interior *basement* and foundation walls are constructed at intervals not exceeding 50 feet (15 240 mm).

Wood foundations in Seismic Design Category D_0, D_1 or D_2 shall be designed in accordance with accepted engineering practice.

R401.2 Requirements. Foundation construction shall be capable of accommodating all loads according to Section R301 and of transmitting the resulting loads to the supporting soil. Fill soils that support footings and foundations shall be designed, installed and tested in accordance with accepted engineering practice. Gravel fill used as footings for wood and precast concrete foundations shall comply with Section R403.

R401.3 Drainage. Surface drainage shall be diverted to a storm sewer conveyance or other *approved* point of collection that does not create a hazard. *Lots* shall be graded to drain surface water away from foundation walls. The *grade* shall fall a minimum of 6 inches (152 mm) within the first 10 feet (3048 mm).

Exception: Where *lot lines*, walls, slopes or other physical barriers prohibit 6 inches (152 mm) of fall within 10 feet (3048 mm), drains or swales shall be constructed to ensure drainage away from the structure. Impervious surfaces within 10 feet (3048 mm) of the building foundation shall be sloped a minimum of 2 percent away from the building.

R401.4 Soil tests. Where quantifiable data created by accepted soil science methodologies indicate expansive, compressible, shifting or other questionable soil characteristics are likely to be present, the *building official* shall determine whether to require a soil test to determine the soil's characteristics at a particular location. This test shall be done by an *approved agency* using an *approved* method.

R401.4.1 Geotechnical evaluation. In lieu of a complete geotechnical evaluation, the load-bearing values in Table R401.4.1 shall be assumed.

TABLE R401.4.1
PRESUMPTIVE LOAD-BEARING VALUES OF FOUNDATION MATERIALS[a]

CLASS OF MATERIAL	LOAD-BEARING PRESSURE (pounds per square foot)
Crystalline bedrock	12,000
Sedimentary and foliated rock	4,000
Sandy gravel and/or gravel (GW and GP)	3,000
Sand, silty sand, clayey sand, silty gravel and clayey gravel (SW, SP, SM, SC, GM and GC)	2,000
Clay, sandy clay, silty clay, clayey silt, silt and sandy silt (CL, ML, MH and CH)	1,500[b]

For SI: 1 pound per square foot = 0.0479 kPa.

a. When soil tests are required by Section R401.4, the allowable bearing capacities of the soil shall be part of the recommendations.

b. Where the building official determines that in-place soils with an allowable bearing capacity of less than 1,500 psf are likely to be present at the site, the allowable bearing capacity shall be determined by a soils investigation.

R401.4.2 Compressible or shifting soil. Instead of a complete geotechnical evaluation, when top or subsoils are compressible or shifting, they shall be removed to a depth and width sufficient to assure stable moisture content in each active zone and shall not be used as fill or stabilized within each active zone by chemical, dewatering or presaturation.

SECTION R402 MATERIALS

R402.1 Wood foundations. Wood foundation systems shall be designed and installed in accordance with the provisions of this code.

R402.1.1 Fasteners. Fasteners used below *grade* to attach plywood to the exterior side of exterior *basement* or crawlspace wall studs, or fasteners used in knee wall construction, shall be of Type 304 or 316 stainless steel. Fasteners used above *grade* to attach plywood and all lumber-to-lumber fasteners except those used in knee wall construction shall be of Type 304 or 316 stainless steel, silicon bronze, copper, hot-dipped galvanized (zinc coated) steel nails, or hot-tumbled galvanized (zinc coated) steel nails. Electro-galvanized steel nails and galvanized (zinc coated) steel staples shall not be permitted.

R402.1.2 Wood treatment. All lumber and plywood shall be pressure-preservative treated and dried after treatment in accordance with AWPA U1 (Commodity Specification A, Use Category 4B and Scction 5.2), and shall bear the *label* of an accredited agency. Where lumber and/or plywood is cut or drilled after treatment, the treated surface shall be field treated with copper naphthenate, the concentration of which shall contain a minimum of 2 percent copper metal, by repeated brushing, dipping or soaking until the wood absorbs no more preservative.

R402.2 Concrete. Concrete shall have a minimum specified compressive strength of f'_c, as shown in Table R402.2. Concrete subject to moderate or severe weathering as indicated in Table R301.2(1) shall be air entrained as specified in Table R402.2. The maximum weight of fly ash, other pozzolans, silica fume, slag or blended cements that is included in concrete mixtures for garage floor slabs and for exterior porches, carport slabs and steps that will be exposed to deicing chemicals shall not exceed the percentages of the total weight of cementitious materials specified in Section 4.2.3 of ACI 318. Materials used to produce concrete and testing thereof shall comply with the applicable standards listed in Chapter 3 of ACI 318 or ACI 332.

R402.3 Precast concrete. Precast concrete foundations shall be designed in accordance with Section R404.5 and shall be installed in accordance with the provisions of this code and the manufacturer's installation instructions.

R402.3.1 Precast concrete foundation materials. Materials used to produce precast concrete foundations shall meet the following requirements.

1. All concrete used in the manufacture of precast concrete foundations shall have a minimum compressive strength of 5,000 psi (34 470 kPa) at 28 days. Concrete exposed to a freezing and thawing environment shall be air entrained with a minimum total air content of 5 percent.
2. Structural reinforcing steel shall meet the requirements of ASTM A 615, A 706 or A 996. The minimum yield strength of reinforcing steel shall be 40,000 psi (Grade 40) (276 MPa). Steel reinforcement for precast concrete foundation walls shall have a minimum concrete cover of $^3/_4$ inch (19.1 mm).
3. Panel-to-panel connections shall be made with Grade II steel fasteners.
4. The use of nonstructural fibers shall conform to ASTM C 1116.
5. Grout used for bedding precast foundations placed upon concrete footings shall meet ASTM C 1107.

SECTION R403 FOOTINGS

R403.1 General. All exterior walls shall be supported on continuous solid or fully grouted masonry or concrete footings, crushed stone footings, wood foundations, or other *approved* structural systems which shall be of sufficient design to accommodate all loads according to Section R301 and to transmit the resulting loads to the soil within the limitations as determined from the character of the soil. Footings shall be supported on undisturbed natural soils or engineered fill. Concrete footing shall be designed and constructed in accordance with the provisions of Section R403 or in accordance with ACI 332.

R403.1.1 Minimum size. Minimum sizes for concrete and masonry footings shall be as set forth in Table R403.1 and Figure R403.1(1). The footing width, W, shall be based on the load-bearing value of the soil in accordance with Table R401.4.1. Spread footings shall be at least 6 inches (152 mm) in thickness, T. Footing projections, P, shall be at least 2 inches (51 mm) and shall not exceed the thickness of the footing. The size of footings supporting piers and columns shall be based on the tributary load and allowable soil pressure in accordance with Table R401.4.1. Footings for wood foundations shall be in accordance with the details set forth in Section R403.2, and Figures R403.1(2) and R403.1(3).

TABLE R402.2
MINIMUM SPECIFIED COMPRESSIVE STRENGTH OF CONCRETE

TYPE OR LOCATION OF CONCRETE CONSTRUCTION	MINIMUM SPECIFIED COMPRESSIVE STRENGTH[a] (f'_c)		
	Weathering Potential[b]		
	Negligible	Moderate	Severe
Basement walls, foundations and other concrete not exposed to the weather	2,500	2,500	2,500[c]
Basement slabs and interior slabs on grade, except garage floor slabs	2,500	2,500	2,500[c]
Basement walls, foundation walls, exterior walls and other vertical concrete work exposed to the weather	2,500	3,000[d]	3,000[d]
Porches, carport slabs and steps exposed to the weather, and garage floor slabs	2,500	3,000[d, e, f]	3,500[d, e, f]

For SI: 1 pound per square inch = 6.895 kPa.

a. Strength at 28 days psi.
b. See Table R301.2(1) for weathering potential.
c. Concrete in these locations that may be subject to freezing and thawing during construction shall be air-entrained concrete in accordance with Footnote d.
d. Concrete shall be air-entrained. Total air content (percent by volume of concrete) shall be not less than 5 percent or more than 7 percent.
e. See Section R402.2 for maximum cementitious materials content.
f. For garage floors with a steel-troweled finish, reduction of the total air content (percent by volume of concrete) to not less than 3 percent is permitted if the specified compressive strength of the concrete is increased to not less than 4,000 psi

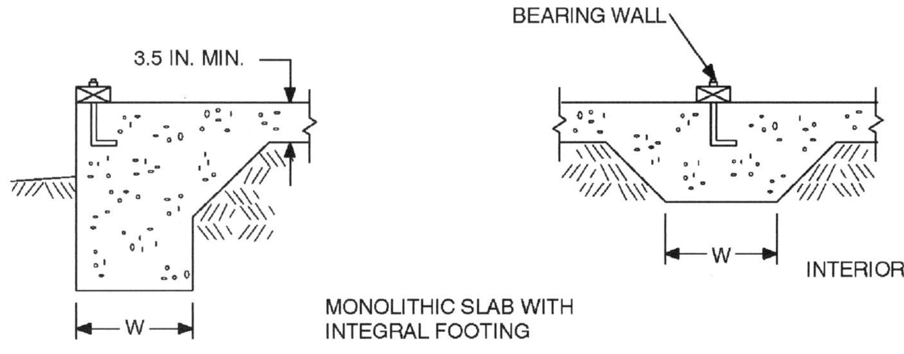

MONOLITHIC SLAB WITH INTEGRAL FOOTING

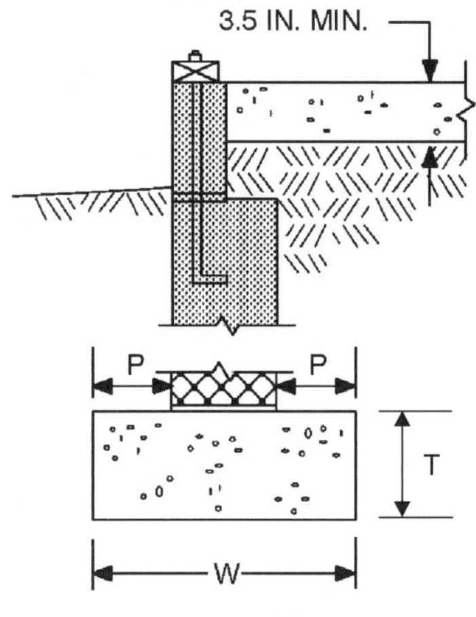

GROUND SUPPORT SLAB WITH MASONRY WALL AND SPREAD FOOTING

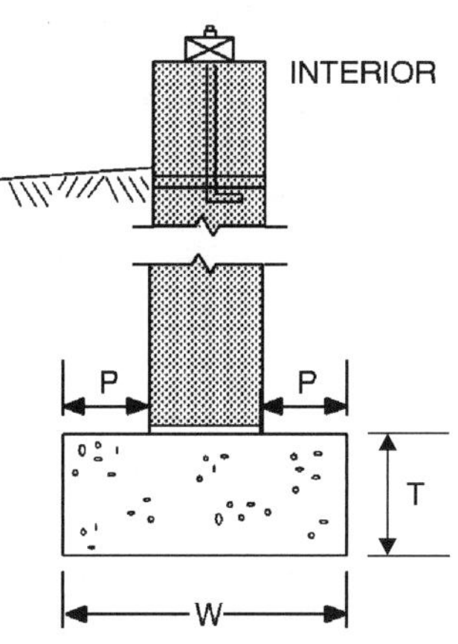

BASEMENT OR CRAWL SPACE WITH MASONRY WALL AND SPREAD FOOTING

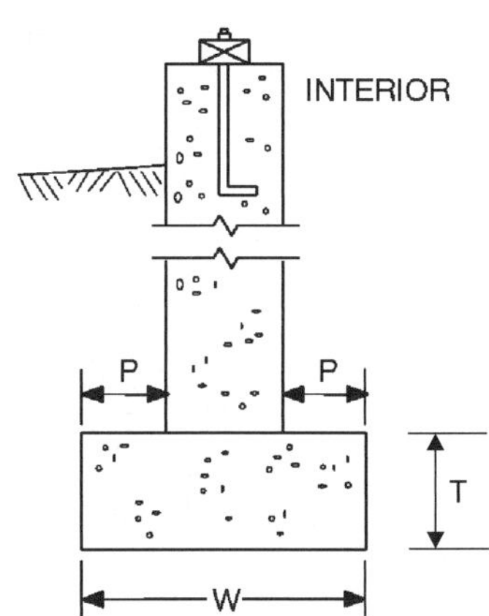

BASEMENT OR CRAWL SPACE WITH CONCRETE WALL AND SPREAD FOOTING

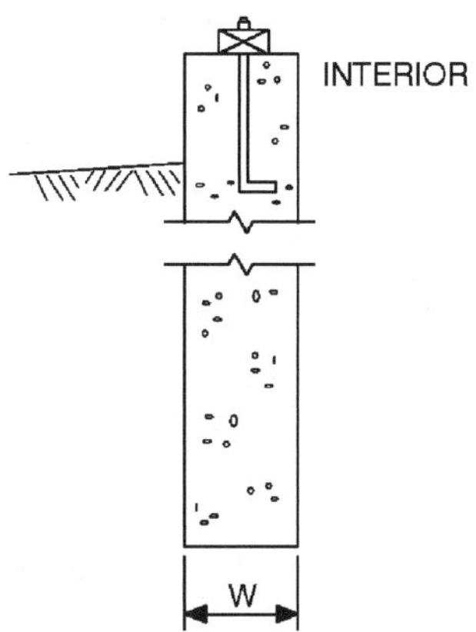

BASEMENT OR CRAWL SPACE WITH FOUNDATION WALL BEARING DIRECTLY ON SOIL

For SI: 1 inch = 25.4 mm.

FIGURE R403.1(1)
CONCRETE AND MASONRY FOUNDATION DETAILS

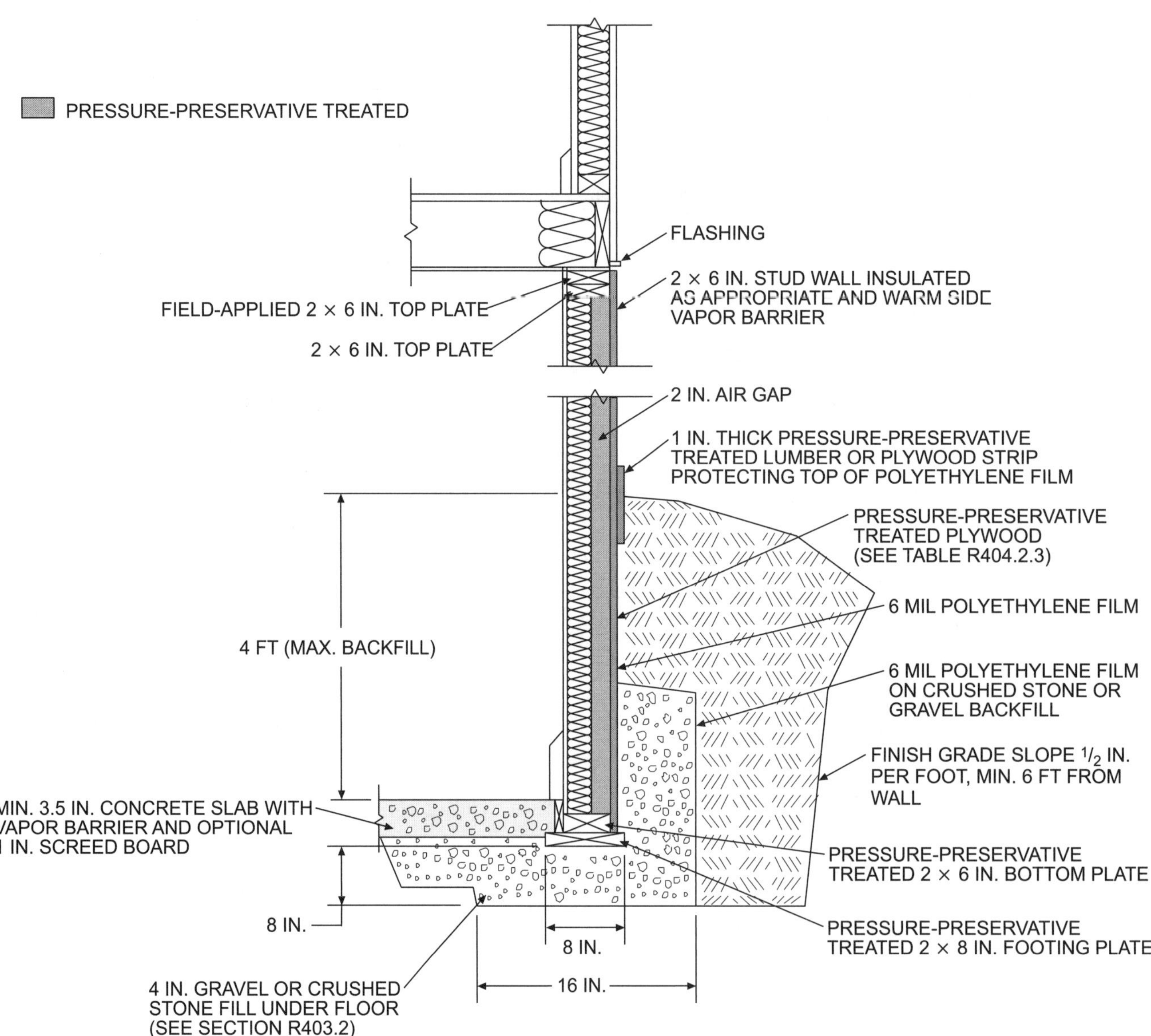

For SI: 1 inch = 25.4 mm, 1 foot = 304.8 mm, 1 mil = 0.0254.

FIGURE R403.1(2)
PERMANENT WOOD FOUNDATION BASEMENT WALL SECTION

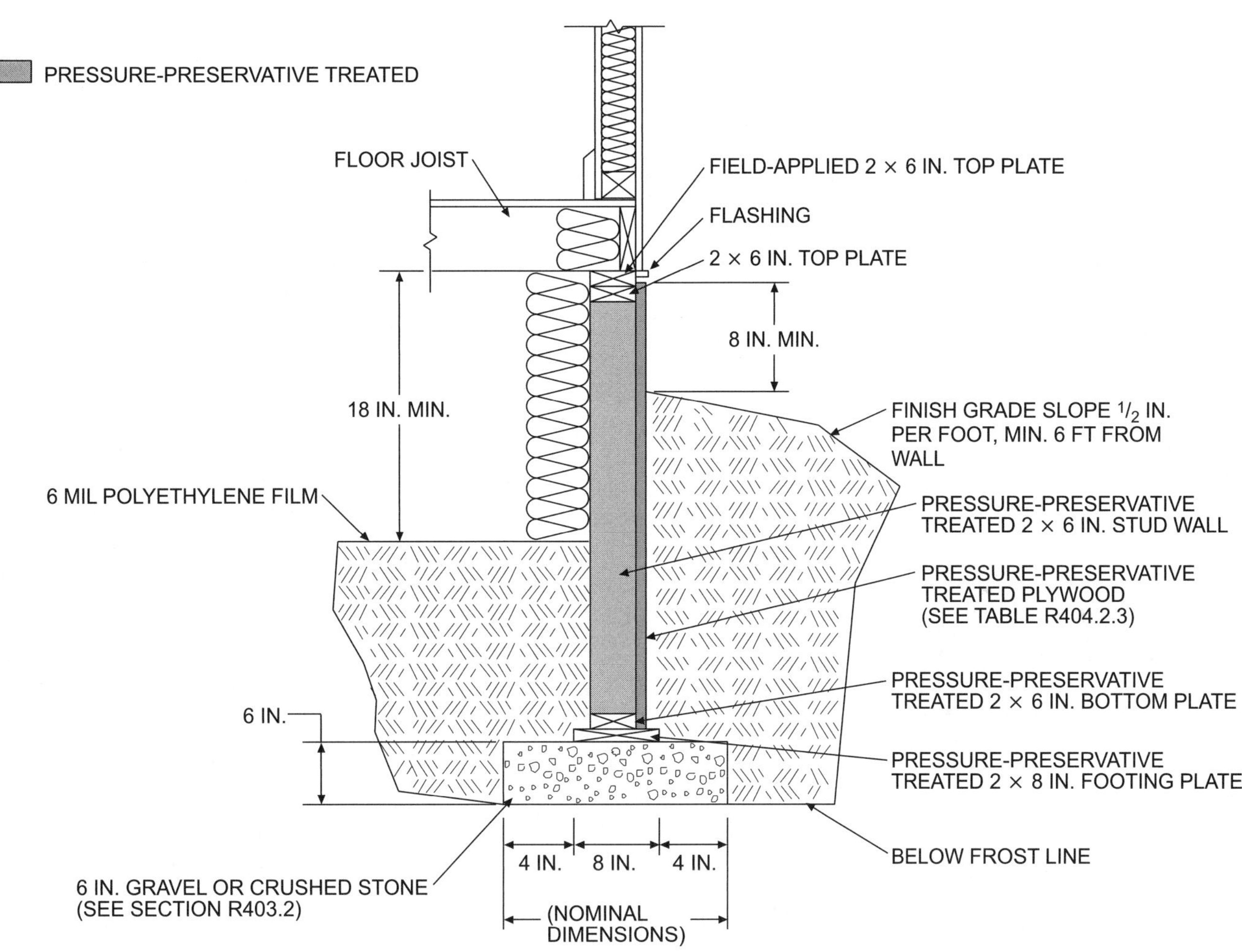

For SI: 1 inch = 25.4 mm, 1 foot = 304.8 mm, 1 mil = 0.0254 mm.

FIGURE R403.1(3)
PERMANENT WOOD FOUNDATION CRAWL SPACE SECTION

R403.1.2 Continuous footing in Seismic Design Categories D_0, D_1 and D_2. The *braced wall panels* at exterior walls of buildings located in Seismic Design Categories D_0, D_1 and D_2 shall be supported by continuous footings. All required interior *braced wall panels* in buildings with plan dimensions greater than 50 feet (15 240 mm) shall also be supported by continuous footings.

TABLE R403.1
MINIMUM WIDTH OF CONCRETE, PRECAST OR MASONRY FOOTINGS (inches)[a]

	LOAD-BEARING VALUE OF SOIL (psf)			
	1,500	2,000	3,000	≥ 4,000
Conventional light-frame construction				
1-story	12	12	12	12
2-story	15	12	12	12
3-story	23	17	12	12
4-inch brick veneer over light frame or 8-inch hollow concrete masonry				
1-story	12	12	12	12
2-story	21	16	12	12
3-story	32	24	16	12
8-inch solid or fully grouted masonry				
1-story	16	12	12	12
2-story	29	21	14	12
3-story	42	32	21	16

For SI: 1 inch = 25.4 mm, 1 pound per square foot = 0.0479 kPa.

a. Where minimum footing width is 12 inches, use of a single wythe of solid or fully grouted 12-inch nominal concrete masonry units is permitted.

R403.1.3 Seismic reinforcing. Concrete footings located in Seismic Design Categories D_0, D_1 and D_2, as established in Table R301.2(1), shall have minimum reinforcement. Bottom reinforcement shall be located a minimum of 3 inches (76 mm) clear from the bottom of the footing.

In Seismic Design Categories D_0, D_1 and D_2 where a construction joint is created between a concrete footing and a stem wall, a minimum of one No. 4 bar shall be installed at not more than 4 feet (1219 mm) on center. The vertical bar shall extend to 3 inches (76 mm) clear of the bottom of the footing, have a standard hook and extend a minimum of 14 inches (357 mm) into the stem wall.

In Seismic Design Categories D_0, D_1 and D_2 where a grouted masonry stem wall is supported on a concrete footing and stem wall, a minimum of one No. 4 bar shall be installed at not more than 4 feet (1219 mm) on center. The vertical bar shall extend to 3 inches (76 mm) clear of the bottom of the footing and have a standard hook.

In Seismic Design Categories D_0, D_1 and D_2 masonry stem walls without solid grout and vertical reinforcing are not permitted.

Exception: In detached one- and two-family *dwellings* which are three stories or less in height and constructed with stud bearing walls, isolated plain concrete footings, supporting columns or pedestals are permitted.

R403.1.3.1 Foundations with stemwalls. Foundations with stem walls shall have installed a minimum of one No. 4 bar within 12 inches (305 mm) of the top of the wall and one No. 4 bar located 3 inches (76 mm) to 4 inches (102 mm) from the bottom of the footing.

R403.1.3.2 Slabs-on-ground with turned-down footings. Slabs on ground with turned down footings shall have a minimum of one No. 4 bar at the top and the bottom of the footing.

Exception: For slabs-on-ground cast monolithically with the footing, locating one No. 5 bar or two No. 4 bars in the middle third of the footing depth shall be permitted as an alternative to placement at the footing top and bottom.

Where the slab is not cast monolithically with the footing, No. 3 or larger vertical dowels with standard hooks on each end shall be provided in accordance with Figure R403.1.3.2. Standard hooks shall comply with Section R611.5.4.5.

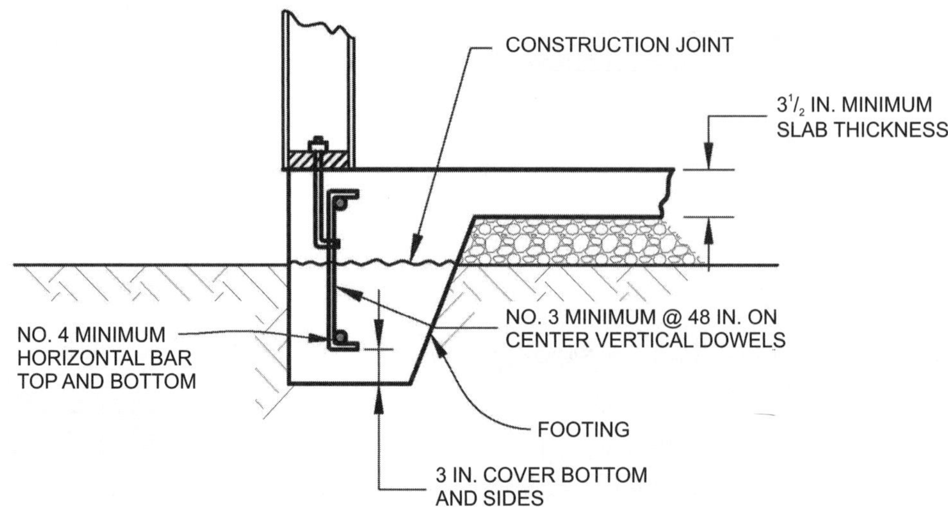

For SI: 1 inch = 25.4 mm.

FIGURE R403.1.3.2
DOWELS FOR SLABS-ON-GROUND WITH TURNED-DOWN FOOTINGS

R403.1.4 Minimum depth. All exterior footings shall be placed at least 12 inches (305 mm) below the undisturbed ground surface. Where applicable, the depth of footings shall also conform to Sections R403.1.4.1 through R403.1.4.2.

R403.1.4.1 Frost protection. Except where otherwise protected from frost, foundation walls, piers and other permanent supports of buildings and structures shall be protected from frost by one or more of the following methods:

1. Extended below the frost line specified in Table R301.2.(1);
2. Constructing in accordance with Section R403.3;
3. Constructing in accordance with ASCE 32; or
4. Erected on solid rock.

Exceptions:

1. Protection of freestanding *accessory structures* with an area of 600 square feet (56 m^2) or less, of light-frame construction, with an eave height of 10 feet (3048 mm) or less shall not be required.
2. Protection of freestanding *accessory structures* with an area of 400 square feet (37 m^2) or less, of other than light-frame construction, with an eave height of 10 feet (3048 mm) or less shall not be required.
3. Decks not supported by a dwelling need not be provided with footings that extend below the frost line.

Footings shall not bear on frozen soil unless the frozen condition is permanent.

R403.1.4.2 Seismic conditions. In Seismic Design Categories D_0, D_1 and D_2, interior footings supporting bearing or bracing walls and cast monolithically with a slab on *grade* shall extend to a depth of not less than 12 inches (305 mm) below the top of the slab.

R403.1.5 Slope. The top surface of footings shall be level. The bottom surface of footings shall not have a slope exceeding one unit vertical in 10 units horizontal (10-percent slope). Footings shall be stepped where it is necessary to change the elevation of the top surface of the footings or where the slope of the bottom surface of the footings will exceed one unit vertical in ten units horizontal (10-percent slope).

R403.1.6 Foundation anchorage. Sill plates and walls supported directly on continuous foundations shall be anchored to the foundation in accordance with this section.

Wood sole plates at all exterior walls on monolithic slabs, wood sole plates of *braced wall panels* at building interiors on monolithic slabs and all wood sill plates shall be anchored to the foundation with anchor bolts spaced a maximum of 6 feet (1829 mm) on center. Bolts shall be at least $^1/_2$ inch (12.7 mm) in diameter and shall extend a minimum of 7 inches (178 mm) into concrete or grouted cells of concrete masonry units. A nut and washer shall be tightened on each anchor bolt. There shall be a minimum of two bolts per plate section with one bolt located not more than 12 inches (305 mm) or less than seven bolt diameters from each end of the plate section. Interior bearing wall sole plates on monolithic slab foundation that are not part of a *braced wall panel* shall be positively anchored with *approved* fasteners. Sill plates and sole plates shall be protected against decay and termites where required by Sections R317 and R318. Cold-formed steel framing systems shall be fastened to wood sill plates or anchored directly to the foundation as required in Section R505.3.1 or R603.3.1.

Exceptions:

1. Foundation anchorage, spaced as required to provide equivalent anchorage to $^1/_2$-inch-diameter (12.7 mm) anchor bolts.
2. Walls 24 inches (610 mm) total length or shorter connecting offset *braced wall panels* shall be anchored to the foundation with a minimum of one anchor bolt located in the center third of the plate section and shall be attached to adjacent *braced wall panels* at corners as shown in item 8 of Table R602.3(1).
3. Connection of walls 12 inches (305 mm) total length or shorter connecting offset *braced wall panels* to the foundation without anchor bolts shall be permitted. The wall shall be attached to adjacent *braced wall panels* at corners as shown in item 8 of Table R602.3(1).

R403.1.6.1 Foundation anchorage in Seismic Design Categories C, D_0, D_1 and D_2. In addition to the requirements of Section R403.1.6, the following requirements shall apply to wood light-frame structures in Seismic Design Categories D_0, D_1 and D_2 and wood light-frame townhouses in Seismic Design Category C.

1. Plate washers conforming to Section R602.11.1 shall be provided for all anchor bolts over the full length of required *braced wall lines* except where *approved* anchor straps are used. Properly sized cut washers shall be permitted for anchor bolts in wall lines not containing *braced wall panels.*
2. Interior braced wall plates shall have anchor bolts spaced at not more than 6 feet (1829 mm) on center and located within 12 inches (305 mm) of the ends of each plate section when supported on a continuous foundation.
3. Interior bearing wall sole plates shall have anchor bolts spaced at not more than 6 feet (1829 mm) on center and located within 12 inches (305 mm) of the ends of each plate section when supported on a continuous foundation.
4. The maximum anchor bolt spacing shall be 4 feet (1219 mm) for buildings over two stories in height.
5. Stepped cripple walls shall conform to Section R602.11.2.

6. Where continuous wood foundations in accordance with Section R404.2 are used, the force transfer shall have a capacity equal to or greater than the connections required by Section R602.11.1 or the *braced wall panel* shall be connected to the wood foundations in accordance with the *braced wall panel*-to-floor fastening requirements of Table R602.3(1).

R403.1.7 Footings on or adjacent to slopes. The placement of buildings and structures on or adjacent to slopes steeper than one unit vertical in three units horizontal (33.3-percent slope) shall conform to Sections R403.1.7.1 through R403.1.7.4.

R403.1.7.1 Building clearances from ascending slopes. In general, buildings below slopes shall be set a sufficient distance from the slope to provide protection from slope drainage, erosion and shallow failures. Except as provided in Section R403.1.7.4 and Figure R403.1.7.1, the following criteria will be assumed to provide this protection. Where the existing slope is steeper than one unit vertical in one unit horizontal (100-percent slope), the toe of the slope shall be assumed to be at the intersection of a horizontal plane drawn from the top of the foundation and a plane drawn tangent to the slope at an angle of 45 degrees (0.79 rad) to the horizontal. Where a retaining wall is constructed at the toe of the slope, the height of the slope shall be measured from the top of the wall to the top of the slope.

R403.1.7.2 Footing setback from descending slope surfaces. Footings on or adjacent to slope surfaces shall be founded in material with an embedment and setback from the slope surface sufficient to provide vertical and lateral support for the footing without detrimental settlement. Except as provided for in Section R403.1.7.4 and Figure R403.1.7.1, the following setback is deemed adequate to meet the criteria. Where the slope is steeper than one unit vertical in one unit horizontal (100-percent slope), the required setback shall be measured from an imaginary plane 45 degrees (0.79 rad) to the horizontal, projected upward from the toe of the slope.

R403.1.7.3 Foundation elevation. On graded sites, the top of any exterior foundation shall extend above the elevation of the street gutter at point of discharge or the inlet of an *approved* drainage device a minimum of 12 inches (305 mm) plus 2 percent. Alternate elevations are permitted subject to the approval of the *building official*, provided it can be demonstrated that required drainage to the point of discharge and away from the structure is provided at all locations on the site.

R403.1.7.4 Alternate setback and clearances. Alternate setbacks and clearances are permitted, subject to the approval of the *building official*. The *building official* is permitted to require an investigation and recommendation of a qualified engineer to demonstrate that the intent of this section has been satisfied. Such an investigation shall include consideration of material, height of slope, slope gradient, load intensity and erosion characteristics of slope material.

R403.1.8 Foundations on expansive soils. Foundation and floor slabs for buildings located on expansive soils shall be designed in accordance with Section 1808.6 of the *International Building Code*.

Exception: Slab-on-ground and other foundation systems which have performed adequately in soil conditions similar to those encountered at the building site are permitted subject to the approval of the *building official*.

R403.1.8.1 Expansive soils classifications. Soils meeting all four of the following provisions shall be considered expansive, except that tests to show compliance with Items 1, 2 and 3 shall not be required if the test prescribed in Item 4 is conducted:

1. Plasticity Index (PI) of 15 or greater, determined in accordance with ASTM D 4318.

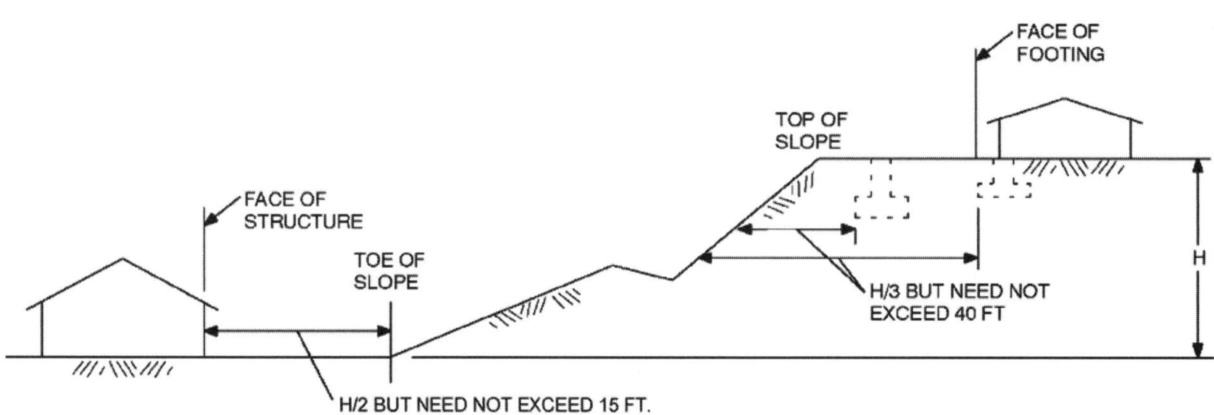

For SI: 1 foot = 304.8 mm.

FIGURE R403.1.7.1
FOUNDATION CLEARANCE FROM SLOPES

2. More than 10 percent of the soil particles pass a No. 200 sieve (75 μm), determined in accordance with ASTM D 422.
3. More than 10 percent of the soil particles are less than 5 micrometers in size, determined in accordance with ASTM D 422.
4. Expansion Index greater than 20, determined in accordance with ASTM D 4829.

R403.2 Footings for wood foundations. Footings for wood foundations shall be in accordance with Figures R403.1(2) and R403.1(3). Gravel shall be washed and well graded. The maximum size stone shall not exceed $^3/_4$ inch (19.1 mm). Gravel shall be free from organic, clayey or silty soils. Sand shall be coarse, not smaller than $^1/_{16}$-inch (1.6 mm) grains and shall be free from organic, clayey or silty soils. Crushed stone shall have a maximum size of $^1/_2$ inch (12.7 mm).

R403.3 Frost-protected shallow foundations. For buildings where the monthly mean temperature of the building is maintained at a minimum of 64°F (18°C), footings are not required to extend below the frost line when protected from frost by insulation in accordance with Figure R403.3(1) and Table R403.3(1). Foundations protected from frost in accordance with Figure R403.3(1) and Table R403.3(1) shall not be used for unheated spaces such as porches, utility rooms, garages and carports, and shall not be attached to basements or crawl spaces that are not maintained at a minimum monthly mean temperature of 64°F (18°C).

Materials used below *grade* for the purpose of insulating footings against frost shall be *labeled* as complying with ASTM C 578.

R403.3.1 Foundations adjoining frost-protected shallow foundations. Foundations that adjoin frost-protected shallow foundations shall be protected from frost in accordance with Section R403.1.4.

R403.3.1.1 Attachment to unheated slab-on-ground structure. Vertical wall insulation and horizontal insulation of frost protected shallow foundations that adjoin a slab-on-ground foundation that does not have a monthly mean temperature maintained at a minimum of 64°F (18°C) shall be in accordance with Figure R403.3(3) and Table R403.3(1). Vertical wall insulation shall extend between the frost protected shallow foundation and the adjoining slab foundation. Required horizontal insulation shall be continuous under the adjoining slab foundation and through any foundation walls adjoining the frost protected shallow foundation. Where insulation passes through a foundation wall, it shall either be of a type complying with this section and having bearing capacity equal to or greater than the structural loads imposed by the building, or the building shall be designed and constructed using beams, lintels, cantilevers or other means of transferring building loads such that the structural loads of the building do not bear on the insulation.

R403.3.1.2 Attachment to heated structure. Where a frost-protected shallow foundation abuts a structure that has a monthly mean temperature maintained at a minimum of 64°F (18°C), horizontal insulation and vertical wall insulation shall not be required between the frost-protected shallow foundation and the adjoining structure. Where the frost-protected shallow foundation abuts the heated structure, the horizontal insulation and vertical wall insulation shall extend along the adjoining foundation in accordance with Figure R403.3(4) a distance of not less than Dimension A in Table R403.3(1).

Exception: Where the frost-protected shallow foundation abuts the heated structure to form an inside corner, vertical insulation extending along the adjoining foundation is not required.

R403.3.2 Protection of horizontal insulation below ground. Horizontal insulation placed less than 12 inches (305 mm) below the ground surface or that portion of horizontal insulation extending outward more than 24 inches (610 mm) from the foundation edge shall be protected against damage by use of a concrete slab or asphalt paving on the ground surface directly above the insulation or by cementitious board, plywood rated for below-ground use, or other *approved* materials placed below ground, directly above the top surface of the insulation.

R403.3.3 Drainage. Final *grade* shall be sloped in accordance with Section R401.3. In other than Group I Soils, as detailed in Table R405.1, gravel or crushed stone beneath horizontal insulation below ground shall drain to daylight or into an *approved* sewer system.

R403.3.4 Termite damage. The use of foam plastic in areas of "very heavy" termite infestation probability shall be in accordance with Section R318.4.

R403.4 Footings for precast concrete foundations. Footings for precast concrete foundations shall comply with Section R403.4.

R403.4.1 Crushed stone footings. Clean crushed stone shall be free from organic, clayey or silty soils. Crushed stone shall be angular in nature and meet ASTM C 33, with the maximum size stone not to exceed $^1/_2$ inch (12.7 mm) and the minimum stone size not to be smaller than $^1/_{16}$-inch (1.6 mm). Crushed stone footings for precast foundations shall be installed in accordance with Figure R403.4(1) and Table R403.4. Crushed stone footings shall be consolidated using a vibratory plate in a maximum of 8-inch lifts. Crushed stone footings shall be limited to Seismic Design Categories A, B and C.

R403.4.2 Concrete footings. Concrete footings shall be installed in accordance with Section R403.1 and Figure R403.4(2).

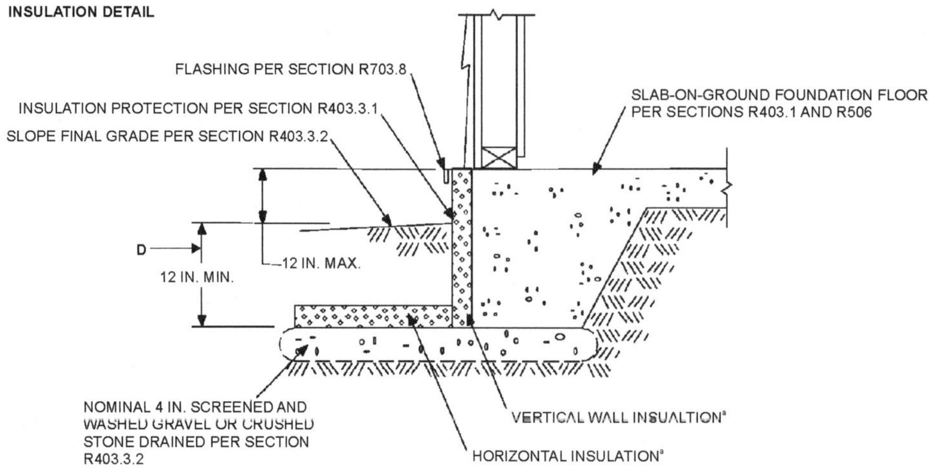

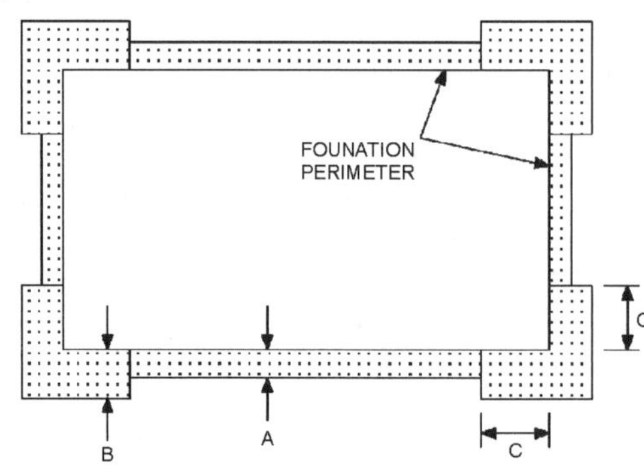

For SI: 1 inch = 25.4 mm.

a. See Table R403.3(1) for required dimensions and *R*-values for vertical and horizontal insulation and minimum footing depth.

FIGURE R403.3(1)
INSULATION PLACEMENT FOR FROST PROTECTED FOOTINGS IN HEATED BUILDINGS

TABLE R403.3(1)
MINIMUM FOOTING DEPTH AND INSULATION REQUIREMENTS FOR FROST-PROTECTED FOOTINGS IN HEATED BUILDINGS[a]

AIR FREEZING INDEX (°F-days)[b]	MINIMUM FOOTING DEPTH, D (inches)	VERTICAL INSULATION R-VALUE[c, d]	HORIZONTAL INSULATION R-VALUE[c, e]		HORIZONTAL INSULATION DIMENSIONS PER FIGURE R403.3(1) (inches)		
			Along walls	At corners	A	B	C
1,500 or less	12	4.5	Not required	Not required	Not required	Not required	Not required
2,000	14	5.6	Not required	Not required	Not required	Not required	Not required
2,500	16	6.7	1.7	4.9	12	24	40
3,000	16	7.8	6.5	8.6	12	24	40
3,500	16	9.0	8.0	11.2	24	30	60
4,000	16	10.1	10.5	13.1	24	36	60

For SI: 1 inch = 25.4 mm, °C = [(°F) - 32]/1.8.

a. Insulation requirements are for protection against frost damage in heated buildings. Greater values may be required to meet energy conservation standards.

b. See Figure R403.3(2) or Table R403.3(2) for Air Freezing Index values.

c. Insulation materials shall provide the stated minimum *R*-values under long-term exposure to moist, below-ground conditions in freezing climates. The following *R*-values shall be used to determine insulation thicknesses required for this application: Type II expanded polystyrene-2.4*R* per inch; Type IV extruded polystyrene-4.5*R* per inch; Type VI extruded polystyrene-4.5*R* per inch; Type IX expanded polystyrene-3.2*R* per inch; Type X extruded polystyrene-4.5*R* per inch.

d. Vertical insulation shall be expanded polystyrene insulation or extruded polystyrene insulation.

e. Horizontal insulation shall be extruded polystyrene insulation.

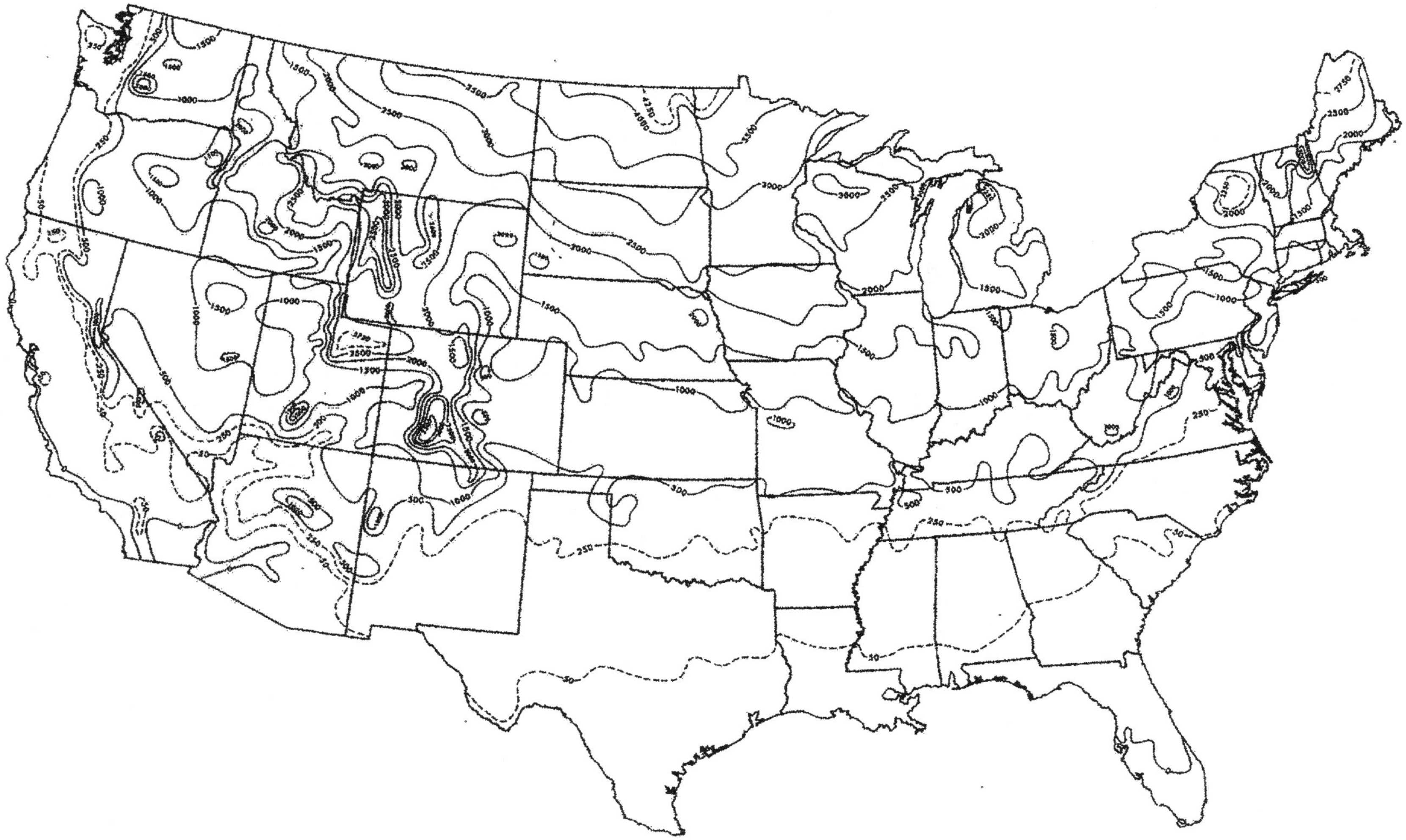

For SI: °C = [(°F) -32]/1.8.

Note: The air-freezing index is defined as cumulative degree days below 32°F. It is used as a measure of the combined magnitude and duration of air temperature below freezing. The index was computed over a 12-month period (July-June) for each of the 3,044 stations used in the above analysis. Date from the 1951-80 period were fitted to a Weibull probability distribution to produce an estimate of the 100-year return period.

FIGURE R403.3(2)
AIR-FREEZING INDEX AN ESTIMATE OF THE 100-YEAR RETURN PERIOD

TABLE R403.3(2)
AIR-FREEZING INDEX FOR U.S. LOCATIONS BY COUNTY

STATE	AIR-FREEZING INDEX					
	1500 or less	2000	2500	3000	3500	4000
Alabama	All counties	—	—	—	—	—
Alaska	Ketchikan Gateway, Prince of Wales-Outer Ketchikan (CA), Sitka, Wrangell-Petersburg (CA)	—	Aleutians West (CA), Haines, Juneau, Skagway-Hoonah-Angoon (CA), Yakutat	—	—	All counties not listed
Arizona	All counties	—	—	—	—	—
Arkansas	All counties	—	—	—	—	—
California	All counties not listed	Nevada, Sierra	—	—	—	—
Colorado	All counties not listed	Archuleta, Custer, Fremont, Huerfano, Las Animas, Ouray, Pitkin, San Miguel	Clear Creek, Conejos, Costilla, Dolores, Eagle, La Plata, Park, Routt, San Juan, Summit	Alamosa, Grand, Jackson, Larimer, Moffat, Rio Blanco, Rio Grande	Chaffee, Gunnison, Lake, Saguache	Hinsdale, Mineral
Connecticut	All counties not listed	Hartford, Litchfield	—	—	—	—
Delaware	All counties	—	—	—	—	—
District of Columbia	All counties	—	—	—	—	—
Florida	All counties	—	—	—	—	—
Georgia	All counties	—	—	—	—	—
Hawaii	All counties	—	—	—	—	—
Idaho	All counties not listed	Adams, Bannock, Blaine, Clearwater, Idaho, Lincoln, Oneida, Power, Valley, Washington	Bingham, Bonneville, Camas, Caribou, Elmore, Franklin, Jefferson, Madison, Teton	Bear Lake, Butte, Custer, Fremont, Lemhi	Clark	—
Illinois	All counties not listed	Boone, Bureau, Cook, Dekalb, DuPage, Fulton, Grundy, Henderson, Henry, Iroquois, Jo Daviess, Kane, Kankakee, Kendall, Knox, La Salle, Lake, Lee, Livingston, Marshall, Mason, McHenry, McLean, Mercer, Peoria, Putnam, Rock Island, Stark, Tazewell, Warren, Whiteside, Will, Woodford	Carroll, Ogle, Stephenson, Winnebago	—	—	—
Indiana	All counties not listed	Allen, Benton, Cass, Fountain, Fulton, Howard, Jasper, Kosciusko, La Porte, Lake, Marshall, Miami, Newton, Porter, Pulaski, Starke, Steuben, Tippecanoe, Tipton, Wabash, Warren, White	—	—	—	—

(continued)

TABLE R403.3(2)—continued
AIR-FREEZING INDEX FOR U.S. LOCATIONS BY COUNTY

STATE	AIR-FREEZING INDEX					
	1500 or less	2000	2500	3000	3500	4000
Iowa	Appanoose, Davis, Fremont, Lee, Van Buren	All counties not listed	Allamakee, Black Hawk, Boone, Bremer, Buchanan, Buena Vista, Butler, Calhoun, Cerro Gordo, Cherokee, Chickasaw, Clay, Clayton, Delaware, Dubuque, Fayette, Floyd, Franklin, Grundy, Hamilton, Hancock, Hardin, Humboldt, Ida, Jackson, Jasper, Jones, Linn, Marshall, Palo Alto, Plymouth, Pocahontas, Poweshiek, Sac, Sioux, Story, Tama, Webster, Winnebago, Woodbury, Worth, Wright	Dickinson, Emmet, Howard, Kossuth, Lyon, Mitchell, O'Brien, Osceola, Winneshiek	—	—
Kansas	All counties	—	—	—	—	—
Kentucky	All counties	—	—	—	—	—
Louisiana	All counties	—	—	—	—	—
Maine	York	Knox, Lincoln, Sagadahoc	Androscoggin, Cumberland, Hancock, Kennebec, Waldo, Washington	Aroostook, Franklin, Oxford, Penobscot, Piscataquis, Somerset	—	—
Maryland	All counties	—	—	—	—	—
Massachusetts	All counties not listed	Berkshire, Franklin, Hampden, Worcester	—	—	—	—
Michigan	Berrien, Branch, Cass, Kalamazoo, Macomb, Ottawa, St. Clair, St. Joseph	All counties not listed	Alger, Charlevoix, Cheboygan, Chippewa, Crawford, Delta, Emmet, Iosco, Kalkaska, Lake, Luce, Mackinac, Menominee, Missaukee, Montmorency, Ogemaw, Osceola, Otsego, Roscommon, Schoolcraft, Wexford	Baraga, Dickinson, Iron, Keweenaw, Marquette	Gogebic, Houghton, Ontonagon	—
Minnesota	—	—	Houston, Winona	All counties not listed	Aitkin, Big Stone, Carlton, Crow Wing, Douglas, Itasca, Kanabec, Lake, Morrison, Pine, Pope, Stearns, Stevens, Swift, Todd, Wadena	Becker, Beltrami, Cass, Clay, Clearwater, Grant, Hubbard, Kittson, Koochiching, Lake of the Woods, Mahnomen, Marshall, Norman, Otter Tail, Pennington, Polk, Red Lake, Roseau, St. Louis, Traverse, Wilkin

(continued)

TABLE R403.3(2)—continued
AIR-FREEZING INDEX FOR U.S. LOCATIONS BY COUNTY

STATE	AIR-FREEZING INDEX					
	1500 or less	2000	2500	3000	3500	4000
Mississippi	All counties	—	—	—	—	—
Missouri	All counties not listed	Atchison, Mercer, Nodaway, Putnam	—	—	—	—
Montana	Mineral	Broadwater, Golden Valley, Granite, Lake, Lincoln, Missoula, Ravalli, Sanders, Sweet Grass	Big Horn, Carbon, Jefferson, Judith Basin, Lewis and Clark, Meagher, Musselshell, Powder River, Powell, Silver Bow, Stillwater, Westland	Carter, Cascade, Deer Lodge, Falcon, Fergus, Flathead, Gallanting, Glacier, Madison, Park, Petroleum, Ponder, Rosebud, Teton, Treasure, Yellowstone	Beaverhead, Blaine, Chouteau, Custer, Dawson, Garfield, Liberty, McCone, Prairie, Toole, Wibaux	Daniels, Hill, Phillips, Richland, Roosevelt, Sheridan, Valley
Nebraska	Adams, Banner, Chase, Cheyenne, Clay, Deuel, Dundy, Fillmore, Franklin, Frontier, Furnas, Gage, Garden, Gosper, Harlan, Hayes, Hitchcock, Jefferson, Kimball, Morrill, Nemaha, Nuckolls, Pawnee, Perkins, Phelps, Red Willow, Richardson, Saline, Scotts Bluff, Seward, Thayer, Webster	All counties not listed	Boyd, Burt, Cedar, Cuming, Dakota, Dixon, Dodge, Knox, Thurston	—	—	—
Nevada	All counties not listed	Elko, Eureka, Nye, Washoe, White Pine	—	—	—	—
New Hampshire	—	All counties not listed	—	—	—	Carroll, Coos, Grafton
New Jersey	All counties	—	—	—	—	—
New Mexico	All counties not listed	Rio Arriba	Colfax, Mora, Taos	—	—	—
New York	Albany, Bronx, Cayuga, Columbia, Cortland, Dutchess, Genessee, Kings, Livingston, Monroe, Nassau, New York, Niagara, Onondaga, Ontario, Orange, Orleans, Putnam, Queens, Richmond, Rockland, Seneca, Suffolk, Wayne, Westchester, Yates	All counties not listed	Clinton, Essex, Franklin, Hamilton, Herkimer, Jefferson, Lewis, St. Lawrence, Warren	—	—	—
North Carolina	All counties	—	—	—	—	—

(continued)

TABLE R403.3(2)—continued
AIR-FREEZING INDEX FOR U.S. LOCATIONS BY COUNTY

STATE	AIR-FREEZING INDEX					
	1500 or less	2000	2500	3000	3500	4000
North Dakota	—	—	—	Billings, Bowman	Adams, Dickey, Golden Valley, Hettinger, LaMoure, Oliver, Ransom, Sargent, Sioux, Slope, Stark	All counties not listed
Ohio	All counties not listed	Ashland, Crawford, Defiance, Holmes, Huron, Knox, Licking, Morrow, Paulding, Putnam, Richland, Seneca, Williams	—	—	—	—
Oklahoma	All counties	—	—	—	—	—
Oregon	All counties not listed	Baker, Crook, Grant, Harney	—	—	—	—
Pennsylvania	All counties not listed	Berks, Blair, Bradford, Cambria, Cameron, Centre, Clarion, Clearfield, Clinton, Crawford, Elk, Forest, Huntingdon, Indiana, Jefferson, Lackawanna, Lycoming, McKean, Pike, Potter, Susquehanna, Tioga, Venango, Warren, Wayne, Wyoming	—	—	—	—
Rhode Island	All counties	—	—	—	—	—
South Carolina	All counties	—	—	—	—	—
South Dakota	—	Bennett, Custer, Fall River, Lawrence, Mellette, Shannon, Todd, Tripp	Bon Homme, Charles Mix, Davison, Douglas, Gregory, Jackson, Jones, Lyman	All counties not listed	Beadle, Brookings, Brown, Campbell, Codington, Corson, Day, Deuel, Edmunds, Faulk, Grant, Hamlin, Kingsbury, Marshall, McPherson, Perkins, Roberts, Spink, Walworth	—
Tennessee	All counties	—	—	—	—	—
Texas	All counties	—	—	—	—	—
Utah	All counties not listed	Box Elder, Morgan, Weber	Garfield, Salt Lake, Summit	Carbon, Daggett, Duchesne, Rich, Sanpete, Uintah, Wasatch	—	—

(continued)

TABLE R403.3(2)—continued
AIR-FREEZING INDEX FOR U.S. LOCATIONS BY COUNTY

STATE	AIR-FREEZING INDEX					
	1500 or less	2000	2500	3000	3500	4000
Vermont	—	Bennington, Grand Isle, Rutland, Windham	Addison, Chittenden, Franklin, Orange, Washington, Windsor	Caledonia, Essex, Lamoille, Orleans	—	—
Virginia	All counties	—	—	—	—	—
Utah	All counties not listed	Box Elder, Morgan, Weber	Garfield, Salt Lake, Summit	Carbon, Daggett, Duchesne, Rich, Sanpete, Uintah, Wasatch	—	—
West Virginia	All counties	—	—	—	—	—
Wisconsin	—	Kenosha, Kewaunee, Racine, Sheboygan, Walworth	All counties not listed	Ashland, Barron, Burnett, Chippewa, Clark, Dunn, Eau Claire, Florence, Forest, Iron, Jackson, La Crosse, Langlade, Marathon, Monroe, Pepin, Polk, Portage, Price, Rust, St. Croix, Taylor, Trempealeau, Vilas, Wood	Bayfield, Douglas, Lincoln, Oneida, Sawyer, Washburn	—
Wyoming	Goshen, Platte	Converse, Crook, Laramie, Niobrara	Campbell, Carbon, Hot Springs, Johnson, Natrona, Sheridan, Uinta, Weston	Albany, Big Horn, Park, Washakie	Fremont, Teton	Lincoln, Sublette, Sweetwater

INSULATION DETAIL

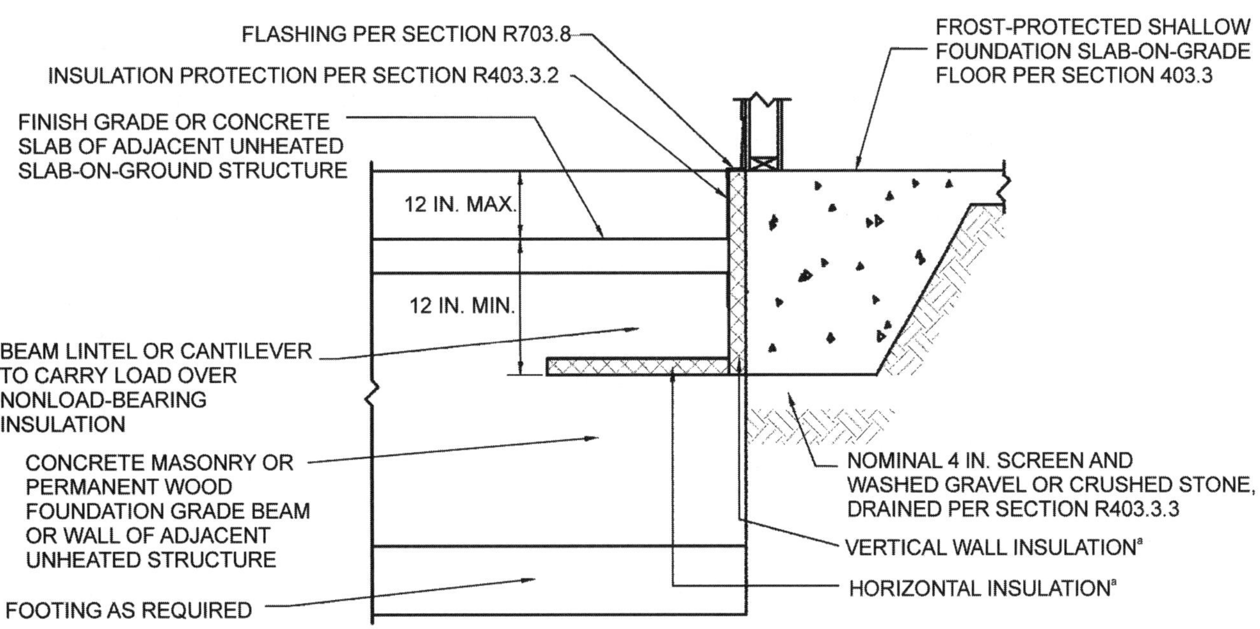

HORIZONTAL INSULATION PLAN

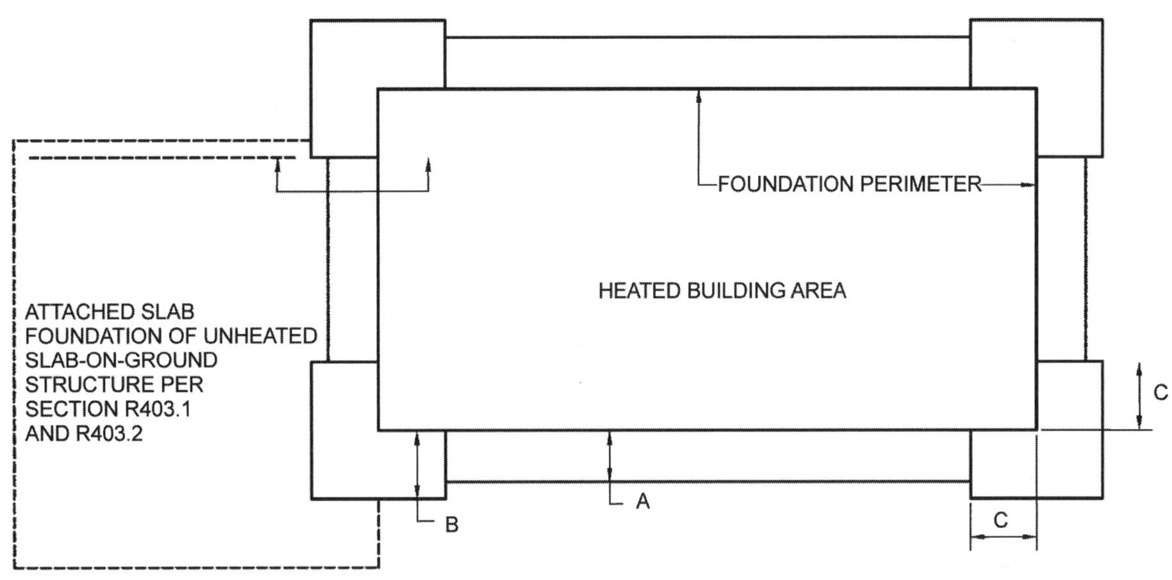

For SI: 1 inch = 25.4 mm.

a. See Table R403.3(1) for required dimensions and *R*-values for vertical and horizontal insulation.

FIGURE R403.3(3)
INSULATION PLACEMENT FOR FROST-PROTECTED FOOTINGS ADJACENT TO UNHEATED SLAB-ON-GROUND STRUCTURE

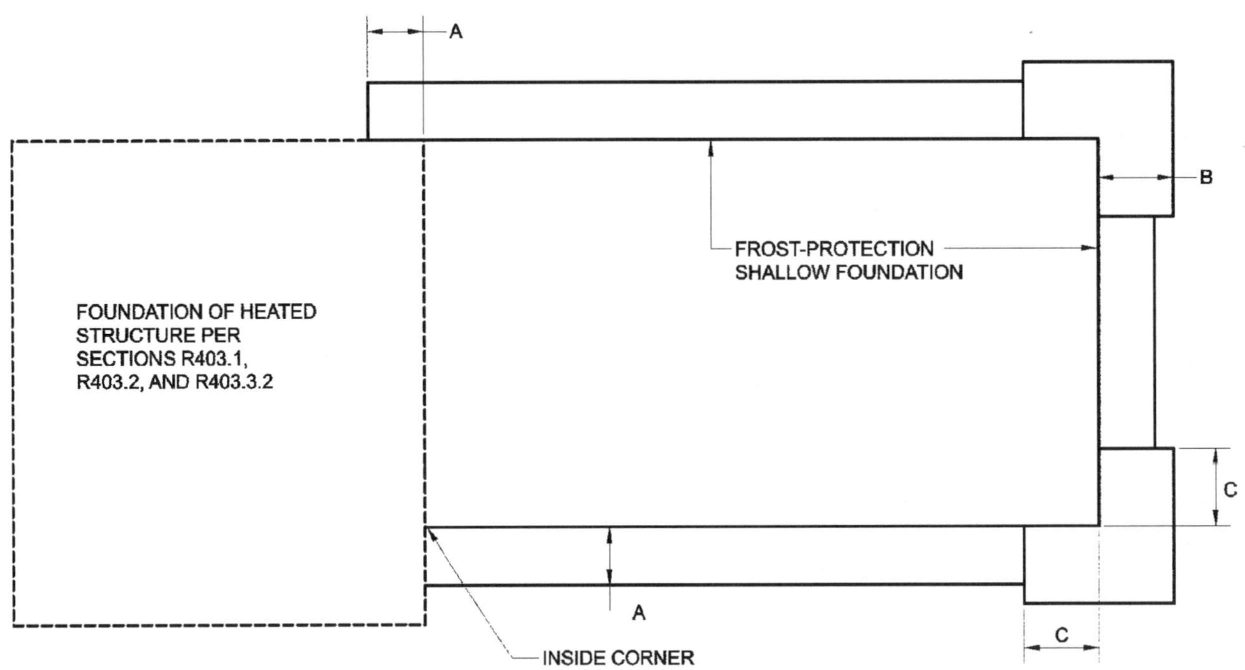

FIGURE R403.3(4)
INSULATION PLACEMENT FOR FROST-PROTECTED FOOTINGS ADJACENT TO HEATED STRUCTURE

TABLE R403.4
MINIMUM DEPTH OF CRUSHED STONE FOOTINGS (D), (inches)

		LOAD-BEARING VALUE OF SOIL (psf)															
		1500				2000				3000				4000			
		MH, CH, CL, ML				SC, GC, SM, GM, SP, SW				GP, GW							
		Wall width (inches)				Wall width (inches)				Wall width (inches)				Wall width (inches)			
		6	8	10	12	6	8	10	12	6	8	10	12	6	8	10	12
Conventional light-frame construction																	
1-story	1100 plf	6	4	4	4	6	4	4	4	6	4	4	4	6	4	4	4
2-story	1800 plf	8	6	4	4	6	4	4	4	6	4	4	4	6	4	4	4
3-story	2900 plf	16	14	12	10	10	8	6	6	6	4	4	4	6	4	4	4
4-inch brick veneer over light-frame or 8-inch hollow concrete masonry																	
1-story	1500 plf	6	4	4	4	6	4	4	4	6	4	4	4	6	4	4	4
2-story	2700 plf	14	12	10	8	10	8	6	4	6	4	4	4	6	4	4	4
3-story	4000 plf	22	22	20	18	16	14	12	10	10	8	6	4	6	4	4	4
8-inch solid or fully grouted masonry																	
1-story	2000 plf	10	8	6	4	6	4	4	4	6	4	4	4	6	4	4	4
2-story	3600 plf	20	18	16	16	14	12	10	8	8	6	4	4	6	4	4	4
3-story	5300 plf	32	30	28	26	22	22	20	18	14	12	10	8	10	8	6	4

For SI: 1 inch = 25.4 mm, 1 pound per square inch = 6.89 pounds per linear foot, 1 plf = 2.44 N/m, 1 pounds per square foot = 47.9 N/m^2.

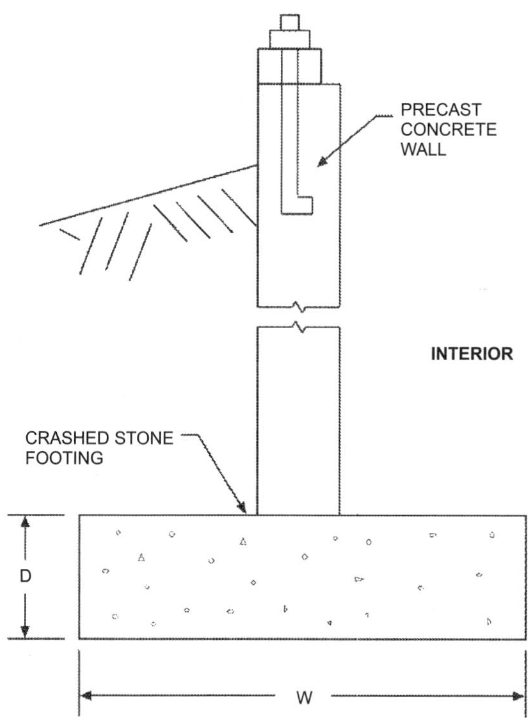

FIGURE R403.4(1)
BASEMENT OR CRAWL SPACE WITH PRECAST FOUNDATION WALL BEARING ON CRUSHED STONE

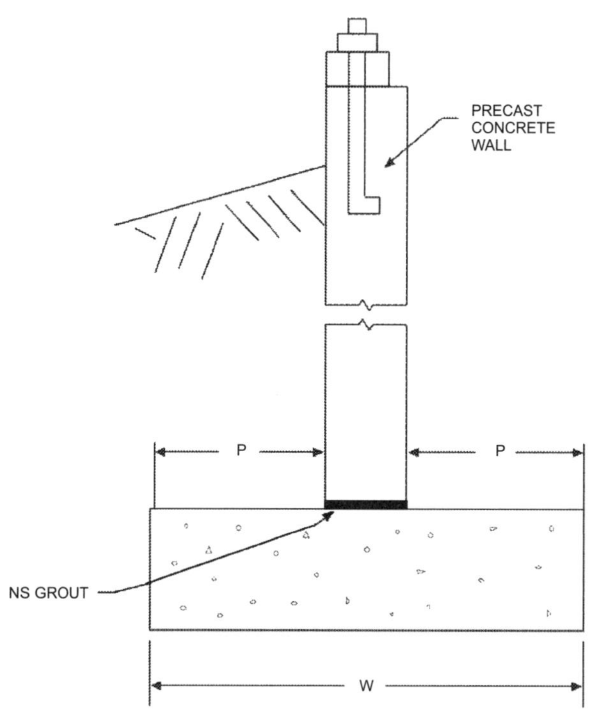

FIGURE R403.4(2)
BASEMENT OR CRAWL SPACE WITH PRECAST FOUNDATION WALL ON SPREAD FOOTING

SECTION R404
FOUNDATION AND RETAINING WALLS

R404.1 Concrete and masonry foundation walls. Concrete foundation walls shall be selected and constructed in accordance with the provisions of Section R404.1.2. Masonry foundation walls shall be selected and constructed in accordance with the provisions of Section R404.1.1.

R404.1.1 Design of masonry foundation walls. Masonry foundation walls shall be designed and constructed in accordance with the provisions of this section or in accordance with the provisions of TMS 402/ACI 530/ASCE 5 or NCMA TR68-A. When TMS 402/ACI 530/ASCE 5, NCMA TR68-A or the provisions of this section are used to design masonry foundation walls, project drawings, typical details and specifications are not required to bear the seal of the architect or engineer responsible for design, unless otherwise required by the state law of the *jurisdiction* having authority.

R404.1.1.1 Masonry foundation walls. Concrete masonry and clay masonry foundation walls shall be constructed as set forth in Table R404.1.1(1), R404.1.1(2), R404.1.1(3) or R404.1.1(4) and shall also comply with applicable provisions of Sections R606, R607 and R608. In buildings assigned to Seismic Design Categories D_0, D_1 and D_2, concrete masonry and clay masonry foundation walls shall also comply with Section R404.1.4.1. Rubble stone masonry foundation walls shall be constructed in accordance with Sections R404.1.8 and R607.2.2. Rubble stone masonry walls shall not be used in Seismic Design Categories D_0, D_1 and D_2.

R404.1.2 Concrete foundation walls. Concrete foundation walls that support light-frame walls shall be designed and constructed in accordance with the provisions of this section, ACI 318, ACI 332 or PCA 100. Concrete foundation walls that support above-grade concrete walls that are within the applicability limits of Section R611.2 shall be designed and constructed in accordance with the provisions of this section, ACI 318, ACI 332 or PCA 100. Concrete foundation walls that support above-grade concrete walls that are not within the applicability limits of Section R611.2 shall be designed and constructed in accordance with the provisions of ACI 318, ACI 332 or PCA 100. When ACI 318, ACI 332, PCA 100 or the provisions of this section are used to design concrete foundation walls, project drawings, typical details and specifications are not required to bear the seal of the architect or engineer responsible for design, unless otherwise required by the state law of the *jurisdiction* having authority.

R404.1.2.1 Concrete cross-section. Concrete walls constructed in accordance with this code shall comply with the shapes and minimum concrete cross-sectional dimensions required by Table R611.3. Other types of forming systems resulting in concrete walls not in compliance with this section and Table R611.3 shall be designed in accordance with ACI 318.

TABLE R404.1.1(1)
PLAIN MASONRY FOUNDATION WALLS

MAXIMUM WALL HEIGHT (feet)	MAXIMUM UNBALANCED BACKFILL HEIGHT[c] (feet)	PLAIN MASONRY[a] MINIMUM NOMINAL WALL THICKNESS (inches)		
		Soil classes[b]		
		GW, GP, SW and SP	GM, GC, SM, SM-SC and ML	SC, MH, ML-CL and inorganic CL
5	4	6 solid[d] or 8	6 solid[d] or 8	6 solid[d] or 8
	5	6 solid[d] or 8	8	10
6	4	6 solid[d] or 8	6 solid[d] or 8	6 solid[d] or 8
	5	6 solid[d] or 8	8	10
	6	8	10	12
7	4	6 solid[d] or 8	8	8
	5	6 solid[d] or 8	10	10
	6	10	12	10 solid[d]
	7	12	10 solid[d]	12 solid[d]
8	4	6 solid[d] or 8	6 solid[d] or 8	8
	5	6 solid[d] or 8	10	12
	6	10	12	12 solid[d]
	7	12	12 solid[d]	Footnote e
	8	10 solid[d]	12 solid[d]	Footnote e
9	4	6 solid[d] or 8	6 solid[d] or 8	8
	5	8	10	12
	6	10	12	12 solid[d]
	7	12	12 solid[d]	Footnote e
	8	12 solid[d]	Footnote e	Footnote e
	9	Footnote e	Footnote e	Footnote e

For SI: 1 inch = 25.4 mm, 1 foot = 304.8 mm, 1 pound per square inch = 6.895 Pa.

a. Mortar shall be Type M or S and masonry shall be laid in running bond. Ungrouted hollow masonry units are permitted except where otherwise indicated.

b. Soil classes are in accordance with the Unified Soil Classification System. Refer to Table R405.1.

c. Unbalanced backfill height is the difference in height between the exterior finish ground level and the lower of the top of the concrete footing that supports the foundation wall or the interior finish ground level. Where an interior concrete slab-on-grade is provided and is in contact with the interior surface of the foundation wall, measurement of the unbalanced backfill height from the exterior finish ground level to the top of the interior concrete slab is permitted.

d. Solid grouted hollow units or solid masonry units.

e. Wall construction shall be in accordance with either Table R404.1.1(2), Table R404.1.1(3), Table R404.1.1(4), or a design shall be provided.

TABLE R404.1.1(2)
8-INCH MASONRY FOUNDATION WALLS WITH REINFORCING WHERE d > 5 INCHES[a, c]

WALL HEIGHT	HEIGHT OF UNBALANCED BACKFILL[e]	MINIMUM VERTICAL REINFORCEMENT AND SPACING (INCHES)[b, c]		
		Soil classes and lateral soil load[d] (psf per foot below grade)		
		GW, GP, SW and SP soils 30	GM, GC, SM, SM-SC and ML soils 45	SC, ML-CL and inorganic CL soils 60
6 feet 8 inches	4 feet (or less)	#4 at 48	#4 at 48	#4 at 48
	5 feet	#4 at 48	#4 at 48	#4 at 48
	6 feet 8 inches	#4 at 48	#5 at 48	#6 at 48
7 feet 4 inches	4 feet (or less)	#4 at 48	#4 at 48	#4 at 48
	5 feet	#4 at 48	#4 at 48	#4 at 48
	6 feet	#4 at 48	#5 at 48	#5 at 48
	7 feet 4 inches	#5 at 48	#6 at 48	#6 at 40
8 feet	4 feet (or less)	#4 at 48	#4 at 48	#4 at 48
	5 feet	#4 at 48	#4 at 48	#4 at 48
	6 feet	#4 at 48	#5 at 48	#5 at 48
	7 feet	#5 at 48	#6 at 48	#6 at 40
	8 feet	#5 at 48	#6 at 48	#6 at 32
8 feet 8 inches	4 feet (or less)	#4 at 48	#4 at 48	#4 at 48
	5 feet	#4 at 48	#4 at 48	#5 at 48
	6 feet	#4 at 48	#5 at 48	#6 at 48
	7 feet	#5 at 48	#6 at 48	#6 at 40
	8 feet 8 inches	#6 at 48	#6 at 32	#6 at 24
9 feet 4 inches	4 feet (or less)	#4 at 48	#4 at 48	#4 at 48
	5 feet	#4 at 48	#4 at 48	#5 at 48
	6 feet	#4 at 48	#5 at 48	#6 at 48
	7 feet	#5 at 48	#6 at 48	#6 at 40
	8 feet	#6 at 48	#6 at 40	#6 at 24
	9 feet 4 inches	#6 at 40	#6 at 24	#6 at 16
10 feet	4 feet (or less)	#4 at 48	#4 at 48	#4 at 48
	5 feet	#4 at 48	#4 at 48	#5 at 48
	6 feet	#4 at 48	#5 at 48	#6 at 48
	7 feet	#5 at 48	#6 at 48	#6 at 32
	8 feet	#6 at 48	#6 at 32	#6 at 24
	9 feet	#6 at 40	#6 at 24	#6 at 16
	10 feet	#6 at 32	#6 at 16	#6 at 16

For SI: 1 inch = 25.4 mm, 1 foot = 304.8 mm, 1 pound per square foot per foot = 0.157 kPa/mm.

a. Mortar shall be Type M or S and masonry shall be laid in running bond.

b. Alternative reinforcing bar sizes and spacings having an equivalent cross-sectional area of reinforcement per lineal foot of wall shall be permitted provided the spacing of the reinforcement does not exceed 72 inches.

c. Vertical reinforcement shall be Grade 60 minimum. The distance, *d*, from the face of the soil side of the wall to the center of vertical reinforcement shall be at least 5 inches.

d. Soil classes are in accordance with the Unified Soil Classification System and design lateral soil loads are for moist conditions without hydrostatic pressure. Refer to Table R405.1.

e. Unbalanced backfill height is the difference in height between the exterior finish ground level and the lower of the top of the concrete footing that supports the foundation wall or the interior finish ground level. Where an interior concrete slab-on-grade is provided and is in contact with the interior surface of the foundation wall, measurement of the unbalanced backfill height from the exterior finish ground level to the top of the interior concrete slab is permitted.

TABLE R404.1.1(3)
10-INCH MASONRY FOUNDATION WALLS WITH REINFORCING WHERE d > 6.75 INCHES[a, c]

WALL HEIGHT	HEIGHT OF UNBALANCED BACKFILL[e]	MINIMUM VERTICAL REINFORCEMENT AND SPACING (INCHES)[b, c]		
		Soil classes and later soil load[d] (psf per foot below grade)		
		GW, GP, SW and SP soils 30	GM, GC, SM, SM-SC and ML soils 45	SC, ML-CL and inorganic CL soils 60
6 feet 8 inches	4 feet (or less)	#4 at 56	#4 at 56	#4 at 56
	5 feet	#4 at 56	#4 at 56	#4 at 56
	6 feet 8 inches	#4 at 56	#5 at 56	#5 at 56
7 feet 4 inches	4 feet (or less)	#4 at 56	#4 at 56	#4 at 56
	5 feet	#4 at 56	#4 at 56	#4 at 56
	6 feet	#4 at 56	#4 at 56	#5 at 56
	7 feet 4 inches	#4 at 56	#5 at 56	#6 at 56
8 feet	4 feet (or less)	#4 at 56	#4 at 56	#4 at 56
	5 feet	#4 at 56	#4 at 56	#4 at 56
	6 feet	#4 at 56	#4 at 56	#5 at 56
	7 feet	#4 at 56	#5 at 56	#6 at 56
	8 feet	#5 at 56	#6 at 56	#6 at 48
8 feet 8 inches	4 feet (or less)	#4 at 56	#4 at 56	#4 at 56
	5 feet	#4 at 56	#4 at 56	#4 at 56
	6 feet	#4 at 56	#4 at 56	#5 at 56
	7 feet	#4 at 56	#5 at 56	#6 at 56
	8 feet 8 inches	#5 at 56	#6 at 48	#6 at 32
9 feet 4 inches	4 feet (or less)	#4 at 56	#4 at 56	#4 at 56
	5 feet	#4 at 56	#4 at 56	#4 at 56
	6 feet	#4 at 56	#5 at 56	#5 at 56
	7 feet	#4 at 56	#5 at 56	#6 at 56
	8 feet	#5 at 56	#6 at 56	#6 at 40
	9 feet 4 inches	#6 at 56	#6 at 40	#6 at 24
10 feet	4 feet (or less)	#4 at 56	#4 at 56	#4 at 56
	5 feet	#4 at 56	#4 at 56	#4 at 56
	6 feet	#4 at 56	#5 at 56	#5 at 56
	7 feet	#5 at 56	#6 at 56	#6 at 48
	8 feet	#5 at 56	#6 at 48	#6 at 40
	9 feet	#6 at 56	#6 at 40	#6 at 24
	10 feet	#6 at 48	#6 at 32	#6 at 24

For SI: 1 inch = 25.4 mm, 1 foot = 304.8 mm, 1 pound per square foot per foot = 0.157 kPa/mm.

a. Mortar shall be Type M or S and masonry shall be laid in running bond.

b. Alternative reinforcing bar sizes and spacings having an equivalent cross-sectional area of reinforcement per lineal foot of wall shall be permitted provided the spacing of the reinforcement does not exceed 72 inches.

c. Vertical reinforcement shall be Grade 60 minimum. The distance, *d*, from the face of the soil side of the wall to the center of vertical reinforcement shall be at least 6.75 inches.

d. Soil classes are in accordance with the Unified Soil Classification System and design lateral soil loads are for moist conditions without hydrostatic pressure. Refer to Table R405.1.

e. Unbalanced backfill height is the difference in height between the exterior finish ground level and the lower of the top of the concrete footing that supports the foundation wall or the interior finish ground level. Where an interior concrete slab-on-grade is provided and is in contact with the interior surface of the foundation wall, measurement of the unbalanced backfill height from the exterior finish ground level to the top of the interior concrete slab is permitted.

TABLE R404.1.1(4)
12-INCH MASONRY FOUNDATION WALLS WITH REINFORCING WHERE d > 8.75 INCHES[a, c]

WALL HEIGHT	HEIGHT OF UNBALANCED BACKFILL[e]	MINIMUM VERTICAL REINFORCEMENT AND SPACING (INCHES)[b, c]		
		Soil classes and lateral soil load[d] (psf per foot below grade)		
		GW, GP, SW and SP soils 30	GM, GC, SM, SM-SC and ML soils 45	SC, ML-CL and inorganic CL soils 60
6 feet 8 inches	4 feet (or less)	#4 at 72	#4 at 72	#4 at 72
	5 feet	#4 at 72	#4 at 72	#4 at 72
	6 feet 8 inches	#4 at 72	#4 at 72	#5 at 72
7 feet 4 inches	4 feet (or less)	#4 at 72	#4 at 72	#4 at 72
	5 feet	#4 at 72	#4 at 72	#4 at 72
	6 feet	#4 at 72	#4 at 72	#5 at 72
	7 feet 4 inches	#4 at 72	#5 at 72	#6 at 72
8 feet	4 feet (or less)	#4 at 72	#4 at 72	#4 at 72
	5 feet	#4 at 72	#4 at 72	#4 at 72
	6 feet	#4 at 72	#4 at 72	#5 at 72
	7 feet	#4 at 72	#5 at 72	#6 at 72
	8 feet	#5 at 72	#6 at 72	#6 at 64
8 feet 8 inches	4 feet (or less)	#4 at 72	#4 at 72	#4 at 72
	5 feet	#4 at 72	#4 at 72	#4 at 72
	6 feet	#4 at 72	#4 at 72	#5 at 72
	7 feet	#4 at 72	#5 at 72	#6 at 72
	8 feet 8 inches	#5 at 72	#7 at 72	#6 at 48
9 feet 4 inches	4 feet (or less)	#4 at 72	#4 at 72	#4 at 72
	5 feet	#4 at 72	#4 at 72	#4 at 72
	6 feet	#4 at 72	#5 at 72	#5 at 72
	7 feet	#4 at 72	#5 at 72	#6 at 72
	8 feet	#5 at 72	#6 at 72	#6 at 56
	9 feet 4 inches	#6 at 72	#6 at 48	#6 at 40
10 feet	4 feet (or less)	#4 at 72	#4 at 72	#4 at 72
	5 feet	#4 at 72	#4 at 72	#4 at 72
	6 feet	#4 at 72	#5 at 72	#5 at 72
	7 feet	#4 at 72	#6 at 72	#6 at 72
	8 feet	#5 at 72	#6 at 72	#6 at 48
	9 feet	#6 at 72	#6 at 56	#6 at 40
	10 feet	#6 at 64	#6 at 40	#6 at 32

For SI: 1 inch = 25.4 mm, 1 foot = 304.8 mm, 1 pound per square foot per foot = 0.157 kPa/mm.

a. Mortar shall be Type M or S and masonry shall be laid in running bond.

b. Alternative reinforcing bar sizes and spacings having an equivalent cross-sectional area of reinforcement per lineal foot of wall shall be permitted provided the spacing of the reinforcement does not exceed 72 inches.

c. Vertical reinforcement shall be Grade 60 minimum. The distance, *d*, from the face of the soil side of the wall to the center of vertical reinforcement shall be at least 8.75 inches.

d. Soil classes are in accordance with the Unified Soil Classification System and design lateral soil loads are for moist conditions without hydrostatic pressure. Refer to Table R405.1.

e. Unbalanced backfill height is the difference in height between the exterior finish ground level and the lower of the top of the concrete footing that supports the foundation wall or the interior finish ground levels. Where an interior concrete slab-on-grade is provided and in contact with the interior surface of the foundation wall, measurement of the unbalanced backfill height is permitted to be measured from the exterior finish ground level to the top of the interior concrete slab is permitted.

R404.1.2.2 Reinforcement for foundation walls. Concrete foundation walls shall be laterally supported at the top and bottom. Horizontal reinforcement shall be provided in accordance with Table R404.1.2(1). Vertical reinforcement shall be provided in accordance with Table R404.1.2(2), R404.1.2(3), R404.1.2(4), R404.1.2(5), R404.1.2(6), R404.1.2(7) or R404.1.2(8). Vertical reinforcement for flat *basement* walls retaining 4 feet (1219 mm) or more of unbalanced backfill is permitted to be determined in accordance with Table R404.1.2(9). For *basement* walls supporting above-grade concrete walls, vertical reinforcement shall be the greater of that required by Tables R404.1.2(2) through R404.1.2(8) or by Section R611.6 for the above-grade wall. In buildings assiged to Seismic Design Category D_0, D_1 or D_2, concrete foundation walls shall also comply with Section R404.1.4.2.

R404.1.2.2.1 Concrete foundation stem walls supporting above-grade concrete walls. Foundation stem walls that support above-grade concrete walls shall be designed and constructed in accordance with this section.

1. Stem walls not laterally supported at top.Concrete stem walls that are not monolithic with slabs-on-ground or are not otherwise laterally supported by slabs-on-ground shall comply with this section. Where unbalanced backfill retained by the stem wall is less than or equal to 18 inches (457 mm), the stem wall and above-grade wall it supports shall be provided with vertical reinforcement in accordance with Section R611.6 and Table R611.6(1), R611.6(2) or R611.6(3) for above-grade walls. Where unbalanced backfill retained by the stem wall is greater than 18 inches (457 mm), the stem wall and above-grade wall it supports shall be provided with vertical reinforcement in accordance with Section R611.6 and Table R611.6(4).
2. Stem walls laterally supported at top.Concrete stem walls that are monolithic with slabs-on-ground or are otherwise laterally supported by slabs-on-ground shall be vertically reinforced in accordance with Section R611.6 and Table R611.6(1), R611.6(2) or R611.6(3) for above-grade walls. Where the unbalanced backfill retained by the stem wall is greater than 18 inches (457 mm), the connection between the stem wall and the slab-on-ground, and the portion of the slab-on-ground providing lateral support for the wall shall be designed in accordance with PCA 100 or in accordance with accepted engineering practice. Where the unbalanced backfill retained by the stem wall is greater than 18 inches (457 mm), the minimum nominal thickness of the wall shall be 6 inches (152 mm).

R404.1.2.2.2 Concrete foundation stem walls supporting light-frame above-grade walls. Concrete foundation stem walls that support light-frame above-grade walls shall be designed and constructed in accordance with this section.

1. Stem walls not laterally supported at top.Concrete stem walls that are not monolithic with slabs-on-ground or are not otherwise laterally supported by slabs-on-ground and retain 48 inches (1219 mm) or less of unbalanced fill, measured from the top of the wall, shall be constructed in accordance with Section R404.1.2. Foundation stem walls that retain more than 48 inches (1219 mm) of unbalanced fill, measured from the top of the wall, shall be designed in accordance with Sections R404.1.3 and R404.4.
2. Stem walls laterally supported at top.Concrete stem walls that are monolithic with slabs-on-ground or are otherwise laterally supported by slabs-on-ground shall be constructed in accordance with Section R404.1.2. Where the unbalanced backfill retained by the stem wall is greater than 48 inches (1219 mm), the connection between the stem wall and the slab-on-ground, and the portion of the slab-on-ground providing lateral support for the wall shall be designed in accordance with PCA 100 or in accordance with accepted engineering practice.

R404.1.2.3 Concrete, materials for concrete, and forms. Materials used in concrete, the concrete itself and forms shall conform to requirements of this section or ACI 318.

R404.1.2.3.1 Compressive strength. The minimum specified compressive strength of concrete, f'_c, shall comply with Section R402.2 and shall be not less

TABLE R404.1.2(1)
MINIMUM HORIZONTAL REINFORCEMENT FOR CONCRETE BASEMENT WALLS[a, b]

MAXIMUM UNSUPPORTED HEIGHT OF BASEMENT WALL (feet)	LOCATION OF HORIZONTAL REINFORCEMENT
≤ 8	One No. 4 bar within 12 inches of the top of the wall story and one No. 4 bar near mid-height of the wall story.
> 8	One No. 4 bar within 12 inches of the top of the wall story and one No. 4 bar near third points in the wall story.

For SI: 1 inch = 25.4 mm, 1 foot = 304.8 mm, 1 pound per square inch = 6.895 kPa.

a. Horizontal reinforcement requirements are for reinforcing bars with a minimum yield strength of 40,000 psi and concrete with a minimum concrete compressive strength 2,500 psi.

b. See Section R404.1.2.2 for minimum reinforcement required for foundation walls supporting above-grade concrete walls.

than 2,500 psi (17.2 MPa) at 28 days in buildings assigned to Seismic Design Category A, B or C and 3000 psi (20.5 MPa) in buildings assigned to Seismic Design Category D_0, D_1 or D_2.

R404.1.2.3.2 Concrete mixing and delivery. Mixing and delivery of concrete shall comply with ASTM C 94 or ASTM C 685.

R404.1.2.3.3 Maximum aggregate size. The nominal maximum size of coarse aggregate shall not exceed one-fifth the narrowest distance between sides of forms, or three-fourths the clear spacing between reinforcing bars or between a bar and the side of the form.

Exception: When *approved*, these limitations shall not apply where removable forms are used and workability and methods of consolidation permit concrete to be placed without honeycombs or voids.

R404.1.2.3.4 Proportioning and slump of concrete. Proportions of materials for concrete shall be established to provide workability and consistency to permit concrete to be worked readily into forms and around reinforcement under conditions of placement to be employed, without segregation or excessive bleeding. Slump of concrete placed in removable forms shall not exceed 6 inches (152 mm).

Exception: When *approved*, the slump is permitted to exceed 6 inches (152 mm) for concrete mixtures that are resistant to segregation, and are in accordance with the form manufacturer's recommendations.

TABLE R404.1.2(2)
MINIMUM VERTICAL REINFORCEMENT FOR 6-INCH NOMINAL FLAT CONCRETE BASEMENT WALLS[b, c, d, e, g, h, i, j]

MAXIMUM UNSUPPORTED WALL HEIGHT (feet)	MAXIMUM UNBALANCED BACKFILL HEIGHT[f] (feet)	MINIMUM VERTICAL REINFORCEMENT-BAR SIZE AND SPACING (inches)		
		Soil classes[a] and design lateral soil (psf per foot of depth)		
		GW, GP, SW, SP 30	GM, GC, SM, SM-SC and ML 45	SC, ML-CL and inorganic CL 60
8	4	NR	NR	NR
	5	NR	6 @ 39	6 @ 48
	6	5 @ 39	6 @ 48	6 @ 35
	7	6 @ 48	6 @ 34	6 @ 25
	8	6 @ 39	6 @ 25	6 @ 18
9	4	NR	NR	NR
	5	NR	5 @ 37	6 @ 48
	6	5 @ 36	6 @ 44	6 @ 32
	7	6 @ 47	6 @ 30	6 @ 22
	8	6 @ 34	6 @ 22	6 @ 16
	9	6 @ 27	6 @ 17	DR
10	4	NR	NR	NR
	5	NR	5 @ 35	6 @ 48
	6	6 @ 48	6 @ 41	6 @ 30
	7	6 @ 43	6 @ 28	6 @ 20
	8	6 @ 31	6 @ 20	DR
	9	6 @ 24	6 @ 15	DR
	10	6 @ 19	DR	DR

For SI: 1 foot = 304.8 mm; 1 inch = 25.4 mm; 1 pound per square foot per foot = 0.1571 kPa^2/m, 1 pound per square inch = 6.895 kPa.

NR = Not required.

a. Soil classes are in accordance with the Unified Soil Classification System. Refer to Table R405.1.

b. Table values are based on reinforcing bars with a minimum yield strength of 60,000 psi concrete with a minimum specified compressive strength of 2,500 psi and vertical reinforcement being located at the centerline of the wall. See Section R404.1.2.3.7.2.

c. Vertical reinforcement with a yield strength of less than 60,000 psi and/or bars of a different size than specified in the table are permitted in accordance with Section R404.1.2.3.7.6 and Table R404.1.2(9).

d. Deflection criterion is *L*/240, where *L* is the height of the basement wall in inches.

e. Interpolation is not permitted.

f. Where walls will retain 4 feet or more of unbalanced backfill, they shall be laterally supported at the top and bottom before backfilling.

g. NR indicates no vertical wall reinforcement is required, except for 6-inch-nominal walls formed with stay-in-place forming systems in which case vertical reinforcement shall be No. 4@48 inches on center.

h. See Section R404.1.2.2 for minimum reinforcement required for basement walls supporting above-grade concrete walls.

i. See Table R611.3 for tolerance from nominal thickness permitted for flat walls.

j. DR means design is required in accordance with the applicable building code, or where there is no code, in accordance with ACI 318.

Slump of concrete placed in stay-in-place forms shall exceed 6 inches (152 mm). Slump of concrete shall be determined in accordance with ASTM C 143.

R404.1.2.3.5 Consolidation of concrete. Concrete shall be consolidated by suitable means during placement and shall be worked around embedded items and reinforcement and into corners of forms. Where stay-in-place forms are used, concrete shall be consolidated by internal vibration.

Exception: When *approved* for concrete to be placed in stay-in-place forms, self-consolidating concrete mixtures with slumps equal to or greater than 8 inches (203 mm) that are specifically designed for placement without internal vibration need not be internally vibrated.

R404.1.2.3.6 Form materials and form ties. Forms shall be made of wood, steel, aluminum, plastic, a composite of cement and foam insulation, a composite of cement and wood chips, or other *approved* material suitable for supporting and containing concrete. Forms shall provide sufficient strength to contain concrete during the concrete placement operation.

Form ties shall be steel, solid plastic, foam plastic, a composite of cement and wood chips, a composite of cement and foam plastic, or other suitable material capable of resisting the forces created by fluid pressure of fresh concrete.

R404.1.2.3.6.1 Stay-in-place forms. Stay-in-place concrete forms shall comply with this section.

1. Surface burning characteristics. The flame-spread index and smoke-developed index of forming material, other than foam plastic, left exposed on the interior shall comply with Section R302. The surface burning characteristics of foam plastic used in insu-

TABLE R404.1.2(3)
MINIMUM VERTICAL REINFORCEMENT FOR 8-INCH (203 mm) NOMINAL FLAT CONCRETE BASEMENT WALLS[b, c, d, e, f, h, i]

MAXIMUM UNSUPPORTED WALL HEIGHT (feet)	MAXIMUM UNBALANCED BACKFILL HEIGHT[g] (feet)	MINIMUM VERTICAL REINFORCEMENT-BAR SIZE AND SPACING (inches)		
		Soil classes[a] and design lateral soil (psf per foot of depth)		
		GW, GP, SW, SP 30	GM, GC, SM, SM-SC and ML 45	SC, ML-CL and inorganic CL 60
8	4	NR	NR	NR
	5	NR	NR	NR
	6	NR	NR	6 @ 37
	7	NR	6 @ 36	6 @ 35
	8	6 @ 41	6 @ 35	6 @ 26
9	4	NR	NR	NR
	5	NR	NR	NR
	6	NR	NR	6 @ 35
	7	NR	6 @ 35	6 @ 32
	8	6 @ 36	6 @ 32	6 @ 23
	9	6 @ 35	6 @ 25	6 @ 18
10	4	NR	NR	NR
	5	NR	NR	NR
	6	NR	NR	6 @ 35
	7	NR	6 @ 35	6 @ 29
	8	6 @ 35	6 @ 29	6 @ 21
	9	6 @ 34	6 @ 22	6 @ 16
	10	6 @ 27	6 @ 17	6 @ 13

For SI: 1 foot = 304.8 mm; 1 inch = 25.4 mm; 1 pound per square foot per foot = 0.1571 kPa^2/m, 1 pound per square inch = 6.895 kPa.

NR = Not required.

a. Soil classes are in accordance with the Unified Soil Classification System. Refer to Table R405.1.

b. Table values are based on reinforcing bars with a minimum yield strength of 60,000 psi, concrete with a minimum specified compressive strength of 2,500 psi and vertical reinforcement being located at the centerline of the wall. See Section R404.1.2.3.7.2.

c. Vertical reinforcement with a yield strength of less than 60,000 psi and/or bars of a different size than specified in the table are permitted in accordance with Section R404.1.2.3.7.6 and Table R404.1.2(9).

d. NR indicates no vertical reinforcement is required.

e. Deflection criterion is *L*/240, where *L* is the height of the basement wall in inches.

f. Interpolation is not permitted.

g. Where walls will retain 4 feet or more of unbalanced backfill, they shall be laterally supported at the top and bottom before backfilling.

h. See Section R404.1.2.2 for minimum reinforcement required for basement walls supporting above-grade concrete walls.

i. See Table R611.3 for tolerance from nominal thickness permitted for flat walls.

lating concrete forms shall comply with Section R316.3.

2. Interior covering. Stay-in-place forms constructed of rigid foam plastic shall be protected on the interior of the building as required by Section R316. Where gypsum board is used to protect the foam plastic, it shall be installed with a mechanical fastening system. Use of adhesives in addition to mechanical fasteners is permitted.

3. Exterior wall covering. Stay-in-place forms constructed of rigid foam plastics shall be protected from sunlight and physical damage by the application of an *approved* exterior wall covering complying with this code. Exterior surfaces of other stay-in-place forming systems shall be protected in accordance with this code.

4. Termite hazards. In areas where hazard of termite damage is very heavy in accordance with Figure R301.2(6), foam plastic insulation shall be permitted below *grade* on foundation walls in accordance with one of the following conditions:

   4.1. Where in addition to the requirements in Section R318.1, an *approved* method of protecting the foam plastic and structure from subterranean termite damage is provided.

   4.2. The structural members of walls, floors, ceilings and roofs are entirely of noncombustible materials or pressure-preservative-treated wood.

   4.3. On the interior side of *basement* walls.

TABLE R404.1.2(4)
MINIMUM VERTICAL REINFORCEMENT FOR 10-INCH NOMINAL FLAT CONCRETE BASEMENT WALLS[b, c, d, e, f, h, i]

MAXIMUM UNSUPPORTED WALL HEIGHT (feet)	MAXIMUM UNBALANCED BACKFILL HEIGHT[g] (feet)	MINIMUM VERTICAL REINFORCEMENT-BAR SIZE AND SPACING (inches)		
		Soil classes[a] and design lateral soil (psf per foot of depth)		
		GW, GP, SW, SP 30	GM, GC, SM, SM-SC and ML 45	SC, ML-CL and inorganic CL 60
8	4	NR	NR	NR
	5	NR	NR	NR
	6	NR	NR	NR
	7	NR	NR	NR
	8	6 @ 48	6 @ 35	6 @ 28
9	4	NR	NR	NR
	5	NR	NR	NR
	6	NR	NR	NR
	7	NR	NR	6 @ 31
	8	NR	6 @ 31	6 @ 28
	9	6 @ 37	6 @ 28	6 @ 24
10	4	NR	NR	NR
	5	NR	NR	NR
	6	NR	NR	NR
	7	NR	NR	6 @ 28
	8	NR	6 @ 28	6 @ 28
	9	6 @ 33	6 @ 28	6 @ 21
	10	6 @ 28	6 @ 23	6 @ 17

For SI: 1 foot = 304.8 mm; 1 inch = 25.4 mm; 1 pound per square foot per foot = 0.1571 kPa^2/m, 1 pound per square inch = 6.895 kPa.

NR = Not required.

a. Soil classes are in accordance with the Unified Soil Classification System. Refer to Table R405.1.

b. Table values are based on reinforcing bars with a minimum yield strength of 60,000 psi concrete with a minimum specified compressive strength of 2,500 psi and vertical reinforcement being located at the centerline of the wall. See Section R404.1.2.3.7.2.

c. Vertical reinforcement with a yield strength of less than 60,000 psi and/or bars of a different size than specified in the table are permitted in accordance with Section R404.1.2.3.7.6 and Table R404.1.2(9).

d. NR indicates no vertical reinforcement is required.

e. Deflection criterion is *L*/240, where *L* is the height of the basement wall in inches.

f. Interpolation is not permitted.

g. Where walls will retain 4 feet or more of unbalanced backfill, they shall be laterally supported at the top and bottom before backfilling.

h. See Section R404.1.2.2 for minimum reinforcement required for basement walls supporting above-grade concrete walls.

i. See Table R611.3 for tolerance from nominal thickness permitted for flat walls.

5. Flat ICF wall system forms shall conform to ASTM E 2634.

R404.1.2.3.7 Reinforcement.

R404.1.2.3.7.1 Steel reinforcement. Steel reinforcement shall comply with the requirements of ASTM A 615, A 706, or A 996. ASTM A 996 bars produced from rail steel shall be Type R. In buildings assigned to Seismic Design Category A, B or C, the minimum yield strength of reinforcing steel shall be 40,000 psi (Grade 40) (276 MPa). In buildings assigned to Seismic Design Category D_0, D_1 or D_2, reinforcing steel shall comply with the requirements of ASTM A 706 for low-alloy steel with a minimum yield strength of 60,000 psi (Grade 60) (414 MPa).

R404.1.2.3.7.2 Location of reinforcement in wall. The center of vertical reinforcement in *basement* walls determined from Tables R404.1.2(2) through R404.1.2(7) shall be located at the centerline of the wall. Vertical reinforcement in *basement* walls determined from Table R404.1.2(8) shall be located to provide a maximum cover of 1.25 inches (32 mm) measured from the inside face of the wall. Regardless of the table used to determine vertical wall reinforcement, the center of the steel shall not vary from the specified location by more than the greater of 10 percent of the wall thickness and $^3/_8$-inch (10 mm). Horizontal and vertical reinforcement shall be located in foundation walls to provide the minimum cover required by Section R404.1.2.3.7.4.

R404.1.2.3.7.3 Wall openings. Vertical wall reinforcement required by Section R404.1.2.2 that is interrupted by wall openings shall have additional vertical reinforcement of the same size placed within 12 inches (305 mm) of each side of the opening.

R404.1.2.3.7.4 Support and cover. Reinforcement shall be secured in the proper location in the forms with tie wire or other bar support system to prevent displacement during the concrete placement operation. Steel reinforcement in concrete

TABLE R404.1.2(5)
MINIMUM VERTICAL WALL REINFORCEMENT FOR 6-INCH WAFFLE-GRID BASEMENT WALLS[b, c, d, e, g, h, i]

MAXIMUM UNSUPPORTED WALL HEIGHT (feet)	MAXIMUM UNBALANCED BACKFILL HEIGHT[f] (feet)	MINIMUM VERTICAL REINFORCEMENT-BAR SIZE AND SPACING (inches) — Soil classes[a] and design lateral soil (psf per foot of depth): GW, GP, SW, SP 30	GM, GC, SM, SM-SC and ML 45	SC, ML-CL and inorganic CL 60
8	4	4 @ 48	4 @ 46	6 @ 39
	5	4 @ 45	5 @ 46	6 @ 47
	6	5 @ 45	6 @ 40	DR
	7	6 @ 44	DR	DR
	8	6 @ 32	DR	DR
9	4	4 @ 48	4 @ 46	4 @ 37
	5	4 @ 42	5 @ 43	6 @ 44
	6	5 @ 41	6 @ 37	DR
	7	6 @ 39	DR	DR
	> 8	DR[i]	DR	DR
10	4	4 @ 48	4 @ 46	4 @ 35
	5	4 @ 40	5 @ 40	6 @ 41
	6	5 @ 38	6 @ 34	DR
	7	6 @ 36	DR	DR
	> 8	DR	DR	DR

For SI: 1 foot = 304.8 mm; 1 inch = 25.4 mm; 1 pound per square foot per foot = 0.1571 kPa^2/m, 1 pound per square inch = 6.895 kPa.

a. Soil classes are in accordance with the Unified Soil Classification System. Refer to Table R405.1.

b. Table values are based on reinforcing bars with a minimum yield strength of 60,000 psi concrete with a minimum specified compressive strength of 2,500 psi and vertical reinforcement being located at the centerline of the wall. See Section R404.1.2.3.7.2.

c. Maximum spacings shown are the values calculated for the specified bar size. Where the bar used is Grade 60 and the size specified in the table, the actual spacing in the wall shall not exceed a whole-number multiple of 12 inches (i.e., 12, 24, 36 and 48) that is less than or equal to the tabulated spacing. Vertical reinforcement with a yield strength of less than 60,000 psi and/or bars of a different size than specified in the table are permitted in accordance with Section R404.1.2.3.7.6 and Table R404.1.2(9).

d. Deflection criterion is *L*/240, where *L* is the height of the basement wall in inches.

e. Interpolation is not permitted.

f. Where walls will retain 4 feet or more of unbalanced backfill, they shall be laterally supported at the top and bottom before backfilling.

g. See Section R404.1.2.2 for minimum reinforcement required for basement walls supporting above-grade concrete walls.

h. See Table R611.3 for thicknesses and dimensions of waffle-grid walls.

i. DR means design is required in accordance with the applicable building code, or where there is no code, in accordance with ACI 318.

cast against the earth shall have a minimum cover of 3 inches (75 mm). Minimum cover for reinforcement in concrete cast in removable forms that will be exposed to the earth or weather shall be $1^1/_2$ inches (38 mm) for No. 5 bars and smaller, and 2 inches (50 mm) for No. 6 bars and larger. For concrete cast in removable forms that will not be exposed to the earth or weather, and for concrete cast in stay-in-place forms, minimum cover shall be $^3/_4$ inch (19 mm). The minus tolerance for cover shall not exceed the smaller of one-third the required cover or $^3/_8$ inch (10 mm).

R404.1.2.3.7.5 Lap splices. Vertical and horizontal wall reinforcement shall be the longest lengths practical. Where splices are necessary in reinforcement, the length of lap splice shall be in accordance with Table R611.5.4.(1) and Figure R611.5.4(1). The maximum gap between non-contact parallel bars at a lap splice shall not exceed the smaller of one-fifth the required lap length and 6 inches (152 mm). See Figure R611.5.4(1).

R404.1.2.3.7.6 Alternate grade of reinforcement and spacing. Where tables in Section R404.1.2.2 specify vertical wall reinforcement based on minimum bar size and maximum spacing, which are based on Grade 60 (414 MPa) steel reinforcement, different size bars and/or bars made from a different grade of steel are permitted provided an equivalent area of steel per linear foot of wall is provided. Use of Table R404.1.2(9) is permitted to determine the maximum bar spacing for different bar sizes than

TABLE R404.1.2(6)
MINIMUM VERTICAL REINFORCEMENT FOR 8-INCH WAFFLE-GRID BASEMENT WALLS[b, c, d, e, f, h, i, j]

MAXIMUM UNSUPPORTED WALL HEIGHT (feet)	MAXIMUM UNBALANCED BACKFILL HEIGHT[g] (feet)	MINIMUM VERTICAL REINFORCEMENT-BAR SIZE AND SPACING (inches)		
		Soil classes[a] and design lateral soil (psf per foot of depth)		
		GW, GP, SW, SP 30	GM, GC, SM, SM-SC and ML 45	SC, ML-CL and inorganic CL 60
8	4	NR	NR	NR
	5	NR	5 @ 48	5 @ 46
	6	5 @ 48	5 @ 43	6 @ 45
	7	5 @ 46	6 @ 43	6 @ 31
	8	6 @ 48	6 @ 32	6 @ 23
9	4	NR	NR	NR
	5	NR	5 @ 47	5 @ 46
	6	5 @ 46	5 @ 39	6 @ 41
	7	5 @ 42	6 @ 38	6 @ 28
	8	6 @ 44	6 @ 28	6 @ 20
	9	6 @ 34	6 @ 21	DR
10	4	NR	NR	NR
	5	NR	5 @ 46	5 @ 44
	6	5 @ 46	5 @ 37	6 @ 38
	7	5 @ 38	6 @ 35	6 @ 25
	8	6 @ 39	6 @ 25	DR
	9	6 @ 30	DR	DR
	10	6 @ 24	DR	DR

For SI: 1 foot = 304.8 mm; 1 inch = 25.4 mm; 1 pound per square foot per foot = 0.1571 kPa^2/m, 1 pound per square inch = 6.895 kPa.

NR = Not required.

a. Soil classes are in accordance with the Unified Soil Classification System. Refer to Table R405.1.

b. Table values are based on reinforcing bars with a minimum yield strength of 60,000 psi concrete with a minimum specified compressive strength of 2,500 psi and vertical reinforcement being located at the centerline of the wall. See Section R404.1.2.3.7.2.

c. Maximum spacings shown are the values calculated for the specified bar size. Where the bar used is Grade 60 (420 MPa) and the size specified in the table, the actual spacing in the wall shall not exceed a whole-number multiple of 12 inches (i.e., 12, 24, 36 and 48) that is less than or equal to the tabulated spacing. Vertical reinforcement with a yield strength of less than 60,000 psi and/or bars of a different size than specified in the table are permitted in accordance with Section R404.1.2.3.7.6 and Table R404.1.2(9).

d. NR indicates no vertical reinforcement is required.

e. Deflection criterion is *L*/240, where *L* is the height of the basement wall in inches.

f. Interpolation shall not be permitted.

g. Where walls will retain 4 feet or more of unbalanced backfill, they shall be laterally supported at the top and bottom before backfilling.

h. See Section R404.1.2.2 for minimum reinforcement required for basement walls supporting above-grade concrete walls.

i. See Table R611.3 for thicknesses and dimensions of waffle-grid walls.

j. DR means design is required in accordance with the applicable building code, or where there is no code, in accordance with ACI 318.

specified in the tables and/or bars made from a different grade of steel. Bars shall not be spaced less than one-half the wall thickness, or more than 48 inches (1219 mm) on center.

R404.1.2.3.7.7 Standard hooks. Where reinforcement is required by this code to terminate with a standard hook, the hook shall comply with Section R611.5.4.5 and Figure R611.5.4(3).

R404.1.2.3.7.8 Construction joint reinforcement. Construction joints in foundation walls shall be made and located to not impair the strength of the wall. Construction joints in plain concrete walls, including walls required to have not less than No. 4 bars at 48 inches (1219 mm) on center by Sections R404.1.2.2 and R404.1.4.2, shall be located at points of lateral support, and a minimum of one No. 4 bar shall extend across the construction joint at a spacing not to exceed 24 inches (610 mm) on center. Construction joint reinforcement shall have a minimum of 12 inches (305 mm) embedment on both sides of the joint. Construction joints in reinforced concrete walls shall be located in the middle third of the span between lateral supports, or located and constructed as required for joints in plain concrete walls.

Exception: Use of vertical wall reinforcement required by this code is permitted in lieu of construction joint reinforcement provided the spacing does not exceed 24 inches (610 mm), or the combination of wall reinforcement and No.4 bars described above does not exceed 24 inches (610 mm).

R404.1.2.3.8 Exterior wall coverings. Requirements for installation of masonry veneer, stucco and other wall coverings on the exterior of concrete walls and other construction details not covered in this section shall comply with the requirements of this code.

TABLE R404.1.2(7)
MINIMUM VERTICAL REINFORCEMENT FOR 6-INCH (152 mm) SCREEN-GRID BASEMENT WALLS[b, c, d, e, g, h, i]

MAXIMUM UNSUPPORTED WALL HEIGHT (feet)	MAXIMUM UNBALANCED BACKFILL HEIGHT[f] (feet)	MINIMUM VERTICAL REINFORCEMENT-BAR SIZE AND SPACING (inches)		
		Soil classes[a] and design lateral soil (psf per foot of depth)		
		GW, GP, SW, SP 30	GM, GC, SM, SM-SC and ML 45	SC, ML-CL and inorganic CL 60
8	4	4 @ 48	4 @ 48	5 @ 43
	5	4 @ 48	5 @ 48	5 @ 37
	6	5 @ 48	6 @ 45	6 @ 32
	7	6 @ 48	DR	DR
	8	6 @ 36	DR	DR
9	4	4 @ 48	4 @ 48	4 @ 41
	5	4 @ 48	5 @ 48	6 @ 48
	6	5 @ 45	6 @ 41	DR
	7	6 @ 43	DR	DR
	> 8	DR	DR	DR
10	4	4 @ 48	4 @ 48	4 @ 39
	5	4 @ 44	5 @ 44	6 @ 46
	6	5 @ 42	6 @ 38	DR
	7	6 @ 40	DR	DR
	> 8	DR	DR	DR

For SI: 1 foot = 304.8 mm; 1 inch = 25.4 mm; 1 pound per square foot per foot = 0.1571 kPa^2/m, 1 pound per square inch = 6.895 kPa.

a. Soil classes are in accordance with the Unified Soil Classification System. Refer to Table R405.1.

b. Table values are based on reinforcing bars with a minimum yield strength of 60,000 psi, concrete with a minimum specified compressive strength of 2,500 psi and vertical reinforcement being located at the centerline of the wall. See Section R404.1.2.3.7.2.

c. Maximum spacings shown are the values calculated for the specified bar size. Where the bar used is Grade 60 and the size specified in the table, the actual spacing in the wall shall not exceed a whole-number multiple of 12 inches (i.e., 12, 24, 36 and 48) that is less than or equal to the tabulated spacing. Vertical reinforcement with a yield strength of less than 60,000 psi and/or bars of a different size than specified in the table are permitted in accordance with Section R404.1.2.3.7.6 and Table R404.1.2(9).

d. Deflection criterion is $L/240$, where L is the height of the basement wall in inches.

e. Interpolation is not permitted.

f. Where walls will retain 4 feet or more of unbalanced backfill, they shall be laterally supported at the top and bottom before backfilling.

g. See Sections R404.1.2.2 for minimum reinforcement required for basement walls supporting above-grade concrete walls.

h. See Table R611.3 for thicknesses and dimensions of screen-grid walls.

i. DR means design is required in accordance with the applicable building code, or where there is no code, in accordance with ACI 318.

TABLE R404.1.2(8)
MINIMUM VERTICAL REINFORCEMENT FOR 6-, 8-, 10-INCH AND 12-INCH NOMINAL FLAT BASEMENT WALLS[b, c, d, e, f, h, i, k, n]

MAXIMUM WALL HEIGHT (feet)	MAXIMUM UNBALANCED BACKFILL HEIGHT[g] (feet)	MINIMUM VERTICAL REINFORCEMENT-BAR SIZE AND SPACING (inches)											
		Soil classes[a] and design lateral soil (psf per foot of depth)											
		GW, GP, SW, SP 30				GM, GC, SM, SM-SC and ML 45				SC, ML-CL and inorganic CL 60			
		Minimum nominal wall thickness (inches)											
		6	8	10	12	6	8	10	12	6	8	10	12
5	4	NR	NR	NR	NR	NR	NR	NR	NR	NR	NR	NR	NR
	5	NR	NR	NR	NR	NR	NR	NR	NR	NR	NR	NR	NR
6	4	NR	NR	NR	NR	NR	NR	NR	NR	NR	NR	NR	NR
	5	NR	NR	NR	NR	NR	NR[l]	NR	NR	4 @ 35	NR[l]	NR	NR
	6	NR	NR	NR	NR	5 @ 48	NR	NR	NR	5 @ 36	NR	NR	NR
7	4	NR	NR	NR	NR	NR	NR	NR	NR	NR	NR	NR	NR
	5	NR	NR	NR	NR	NR	NR	NR	NR	5 @ 47	NR	NR	NR
	6	NR	NR	NR	NR	5 @ 42	NR	NR	NR	6 @ 43	5 @ 48	NR[l]	NR
	7	5 @ 46	NR	NR	NR	6 @ 42	5 @ 46	NR[l]	NR	6 @ 34	6 @ 48	NR	NR
8	4	NR	NR	NR	NR	NR	NR	NR	NR	NR	NR	NR	NR
	5	NR	NR	NR	NR	4 @ 38	NR[l]	NR	NR	5 @ 43	NR	NR	NR
	6	4 @ 37	NR[l]	NR	NR	5 @ 37	NR	NR	NR	6 @ 37	5 @ 43	NR[l]	NR
	7	5 @ 40	NR	NR	NR	6 @ 37	5 @ 41	NR[l]	NR	6 @ 34	6 @ 43	NR	NR
	8	6 @ 43	5 @ 47	NR[l]	NR	6 @ 34	6 @ 43	NR	NR	6 @ 27	6 @ 32	6 @ 44	NR
9	4	NR	NR	NR	NR	NR	NR	NR	NR	NR	NR	NR	NR
	5	NR	NR	NR	NR	4 @ 35	NR[l]	NR	NR	5 @ 40	NR	NR	NR
	6	4 @ 34	NR[l]	NR	NR	6 @ 48	NR	NR	NR	6 @ 36	6 @ 39	NR[l]	NR
	7	5 @ 36	NR	NR	NR	6 @ 34	5 @ 37	NR	NR	6 @ 33	6 @ 38	5 @ 37	NR[l]
	8	6 @ 38	5 @ 41	NR[l]	NR	6 @ 33	6 @ 38	5 @ 37	NR[l]	6 @ 24	6 @ 29	6 @ 39	4 @ 48[m]
	9	6 @ 34	6 @ 46	NR	NR	6 @ 26	6 @ 30	6 @ 41	NR	6 @ 19	6 @ 23	6 @ 30	6 @ 39
10	4	NR	NR	NR	NR	NR	NR	NR	NR	NR	NR	NR	NR
	5	NR	NR	NR	NR	4 @ 33	NR[l]	NR	NR	5 @ 38	NR	NR	NR
	6	5 @ 48	NR[l]	NR	NR	6 @ 45	NR	NR	NR	6 @ 34	5 @ 37	NR	NR
	7	6 @ 47	NR	NR	NR	6 @ 34	6 @ 48	NR	NR	6 @ 30	6 @ 35	6 @ 48	NR[l]
	8	6 @ 34	5 @ 38	NR	NR	6 @ 30	6 @ 34	6 @ 47	NR[l]	6 @ 22	6 @ 26	6 @ 35	6 @ 45[m]
	9	6 @ 34	6 @ 41	4 @ 48	NR[l]	6 @ 23	6 @ 27	6 @ 35	4 @ 48[m]	DR	6 @ 22	6 @ 27	6 @ 34
	10	6 @ 28	6 @ 33	6 @ 45	NR	DR[j]	6 @ 23	6 @ 29	6 @ 38	DR	6 @ 22	6 @ 22	6 @ 28

For SI: 1 foot = 304.8 mm; 1 inch = 25.4 mm; 1 pound per square foot per foot = 0.1571 kPa^2/m, 1 pound per square inch = 6.895 kPa.

NR = Not required.

a. Soil classes are in accordance with the Unified Soil Classification System. Refer to Table R405.1.

b. Table values are based on reinforcing bars with a minimum yield strength of 60,000 psi.

c. Vertical reinforcement with a yield strength of less than 60,000 psi and/or bars of a different size than specified in the table are permitted in accordance with Section R404.1.2.3.7.6 and Table R404.1.2(9).

d. NR indicates no vertical wall reinforcement is required, except for 6-inch nominal walls formed with stay-in-place forming systems in which case vertical reinforcement shall be #4@48 inches on center.

e. Allowable deflection criterion is *L*/240, where *L* is the unsupported height of the basement wall in inches.

f. Interpolation is not permitted.

g. Where walls will retain 4 feet or more of unbalanced backfill, they shall be laterally supported at the top and bottom before backfilling.

h. Vertical reinforcement shall be located to provide a cover of 1.25 inches measured from the inside face of the wall. The center of the steel shall not vary from the specified location by more than the greater of 10 percent of the wall thickness or $^3/_8$-inch.

i. Concrete cover for reinforcement measured from the inside face of the wall shall not be less than $^3/_4$-inch. Concrete cover for reinforcement measured from the outside face of the wall shall not be less than $1^1/_2$ inches for No. 5 bars and smaller, and not less than 2 inches for larger bars.

j. DR means design is required in accordance with the applicable building code, or where there is no code in accordance with ACI 318.

k. Concrete shall have a specified compressive strength, f'_c, of not less than 2,500 psi at 28 days, unless a higher strength is required by footnote l or m.

l. The minimum thickness is permitted to be reduced 2 inches, provided the minimum specified compressive strength of concrete, f'_c, is 4,000 psi.

m. A plain concrete wall with a minimum nominal thickness of 12 inches is permitted, provided minimum specified compressive strength of concrete, f'_c, is 3,500 psi.

n. See Table R611.3 for tolerance from nominal thickness permitted for flat walls.

TABLE R404.1.2(9) MINIMUM SPACING FOR ALTERNATE BAR SIZE AND/OR ALTERNATE GRADE OF STEEL[a, b, c]

BAR SPACING FROM APPLICABLE TABLE IN SECTION R404.1.2.2 (inches)	BAR SIZE FROM APPLICABLE TABLE IN SECTION R404.1.2.2															
	#4					#5					#6					
	Alternate bar size and/or alternate grade of steel desired															
	Grade 60		Grade 40			Grade 60		Grade 40			Grade 60		Grade 40			
	#5	#6	#4	#5	#6	#4	#6	#4	#5	#6	#4	#5	#4	#5	#6	
	Maximum spacing for alternate bar size and/or alternate grade of steel (inches)															
8	12	18	5	8	12	5	11	3	5	8	4	6	2	4	5	
9	14	20	6	9	13	6	13	4	6	9	4	6	3	4	6	
10	16	22	7	10	15	6	14	4	7	9	5	7	3	5	7	
11	17	24	7	11	16	7	16	5	7	10	5	8	3	5	7	
12	19	26	8	12	18	8	17	5	8	11	5	8	4	6	8	
13	20	29	9	13	19	8	18	6	9	12	6	9	4	6	9	
14	22	31	9	14	21	9	20	6	9	13	6	10	4	7	9	
15	23	33	10	16	22	10	21	6	10	14	7	11	5	7	10	
16	25	35	11	17	23	10	23	7	11	15	7	11	5	8	11	
17	26	37	11	18	25	11	24	7	11	16	8	12	5	8	11	
18	28	40	12	19	26	12	26	8	12	17	8	13	5	8	12	
19	29	42	13	20	28	12	27	8	13	18	9	13	6	9	13	
20	31	44	13	21	29	13	28	9	13	19	9	14	6	9	13	
21	33	46	14	22	31	14	30	9	14	20	10	15	6	10	14	
22	34	48	15	23	32	14	31	9	15	21	10	16	7	10	15	
23	36	48	15	24	34	15	33	10	15	22	10	16	7	11	15	
24	37	48	16	25	35	15	34	10	16	23	11	17	7	11	16	
25	39	48	17	26	37	16	35	11	17	24	11	18	8	12	17	
26	40	48	17	27	38	17	37	11	17	25	12	18	8	12	17	
27	42	48	18	28	40	17	38	12	18	26	12	19	8	13	18	
28	43	48	19	29	41	18	40	12	19	26	13	20	8	13	19	
29	45	48	19	30	43	19	41	12	19	27	13	20	9	14	19	
30	47	48	20	31	44	19	43	13	20	28	14	21	9	14	20	
31	48	48	21	32	45	20	44	13	21	29	14	22	9	15	21	
32	48	48	21	33	47	21	45	14	21	30	15	23	10	15	21	
33	48	48	22	34	48	21	47	14	22	31	15	23	10	16	22	
34	48	48	23	35	48	22	48	15	23	32	15	24	10	16	23	
35	48	48	23	36	48	23	48	15	23	33	16	25	11	16	23	
36	48	48	24	37	48	23	48	15	24	34	16	25	11	17	24	
37	48	48	25	38	48	24	48	16	25	35	17	26	11	17	25	
38	48	48	25	39	48	25	48	16	25	36	17	27	12	18	25	
39	48	48	26	40	48	25	48	17	26	37	18	27	12	18	26	
40	48	48	27	41	48	26	48	17	27	38	18	28	12	19	27	
41	48	48	27	42	48	26	48	18	27	39	19	29	12	19	27	
42	48	48	28	43	48	27	48	18	28	40	19	30	13	20	28	
43	48	48	29	44	48	28	48	18	29	41	20	30	13	20	29	
44	48	48	29	45	48	28	48	19	29	42	20	31	13	21	29	
45	48	48	30	47	48	29	48	19	30	43	20	32	14	21	30	
46	48	48	31	48	48	30	48	20	31	44	21	32	14	22	31	
47	48	48	31	48	48	30	48	20	31	44	21	33	14	22	31	
48	48	48	32	48	48	31	48	21	32	45	22	34	15	23	32	

For SI: 1 inch = 25.4 mm, 1 pound per square inch = 6.895 kPa.

a. This table is for use with tables in Section R404.1.2.2 that specify the minimum bar size and maximum spacing of vertical wall reinforcement for foundation walls and above-grade walls. Reinforcement specified in tables in Sections R404.1.2.2 is based on Grade 60 steel reinforcement.

b. Bar spacing shall not exceed 48 inches on center and shall not be less than one-half the nominal wall thickness.

c. For Grade 50 steel bars (ASTM A 996, Type R), use spacing for Grade 40 bars or interpolate between Grades 40 and 60.

R404.1.2.4 Requirements for Seismic Design Category C. Concrete foundation walls supporting above-grade concrete walls in townhouses assigned to Seismic Design Category C shall comply with ACI 318, ACI 332 or PCA 100 (see Section R404.1.2).

R404.1.3 Design required. Concrete or masonry foundation walls shall be designed in accordance with accepted engineering practice when either of the following conditions exists:

1. Walls are subject to hydrostatic pressure from groundwater.
2. Walls supporting more than 48 inches (1219 mm) of unbalanced backfill that do not have permanent lateral support at the top or bottom.

R404.1.4 Seismic Design Category D_0, D_1 or D_2.

R404.1.4.1 Masonry foundation walls. In addition to the requirements of Table R404.1.1(1) plain masonry foundation walls in buildings assigned to Seismic Design Category D_0, D_1 or D_2, as established in Table R301.2(1), shall comply with the following.

1. Wall height shall not exceed 8 feet (2438 mm).
2. Unbalanced backfill height shall not exceed 4 feet (1219 mm).
3. Minimum nominal thickness for plain masonry foundation walls shall be 8 inches (203 mm).
4. Masonry stem walls shall have a minimum vertical reinforcement of one No. 3 (No. 10) bar located a maximum of 4 feet (1219 mm) on center in grouted cells. Vertical reinforcement shall be tied to the horizontal reinforcement in the footings.

Foundation walls in buildings assigned to Seismic Design Category D_0, D_1 or D_2, as established in Table R301.2(1), supporting more than 4 feet (1219 mm) of unbalanced backfill or exceeding 8 feet (2438 mm) in height shall be constructed in accordance with Table R404.1.1(2), R404.1.1(3) or R404.1.1(4). Masonry foundation walls shall have two No. 4 (No. 13) horizontal bars located in the upper 12 inches (305 mm) of the wall.

R404.1.4.2 Concrete foundation walls. In buildings assigned to Seismic Design Category D_0, D_1 or D_2, as established in Table R301.2(1), concrete foundation walls that support light-frame walls shall comply with this section, and concrete foundation walls that support above-grade concrete walls shall comply with ACI 318, ACI 332 or PCA 100 (see Section R404.1.2). In addition to the horizontal reinforcement required by Table R404.1.2(1), plain concrete walls supporting light-frame walls shall comply with the following.

1. Wall height shall not exceed 8 feet (2438 mm).
2. Unbalanced backfill height shall not exceed 4 feet (1219 mm).
3. Minimum thickness for plain concrete foundation walls shall be 7.5 inches (191 mm) except that 6 inches (152 mm) is permitted where the maximum wall height is 4 feet, 6 inches (1372 mm).

Foundation walls less than 7.5 inches (191 mm) in thickness, supporting more than 4 feet (1219 mm) of unbalanced backfill or exceeding 8 feet (2438 mm) in height shall be provided with horizontal reinforcement in accordance with Table R404.1.2(1), and vertical reinforcement in accordance with Table R404.1.2(2), R404.1.2(3), R404.1.2(4), R404.1.2(5), R404.1.2(6), R404.1.2(7) or R404.1.2(8). Where Tables R404.1.2(2) through R404.1.2(8) permit plain concrete walls, not less than No. 4 (No. 13) vertical bars at a spacing not exceeding 48 inches (1219 mm) shall be provided.

R404.1.5 Foundation wall thickness based on walls supported. The thickness of masonry or concrete foundation walls shall not be less than that required by Section R404.1.5.1 or R404.1.5.2, respectively.

R404.1.5.1 Masonry wall thickness. Masonry foundation walls shall not be less than the thickness of the wall supported, except that masonry foundation walls of at least 8-inch (203 mm) nominal thickness shall be permitted under brick veneered frame walls and under 10-inch-wide (254 mm) cavity walls where the total height of the wall supported, including gables, is not more than 20 feet (6096 mm), provided the requirements of Section R404.1.1 are met.

R404.1.5.2 Concrete wall thickness. The thickness of concrete foundation walls shall be equal to or greater than the thickness of the wall in the *story* above. Concrete foundation walls with corbels, brackets or other projections built into the wall for support of masonry veneer or other purposes are not within the scope of the tables in this section.

Where a concrete foundation wall is reduced in thickness to provide a shelf for the support of masonry veneer, the reduced thickness shall be equal to or greater than the thickness of the wall in the *story* above. Vertical reinforcement for the foundation wall shall be based on Table R404.1.2(8) and located in the wall as required by Section R404.1.2.3.7.2 where that table is used. Vertical reinforcement shall be based on the thickness of the thinner portion of the wall.

Exception: Where the height of the reduced thickness portion measured to the underside of the floor assembly or sill plate above is less than or equal to 24 inches (610 mm) and the reduction in thickness does not exceed 4 inches (102 mm), the vertical reinforcement is permitted to be based on the thicker portion of the wall.

R404.1.5.3 Pier and curtain wall foundations. Use of pier and curtain wall foundations shall be permitted to support light-frame construction not more than two stories in height, provided the following requirements are met:

1. All load-bearing walls shall be placed on continuous concrete footings placed integrally with the exterior wall footings.
2. The minimum actual thickness of a load-bearing masonry wall shall be not less than 4 inches (102 mm) nominal or $3^3/_8$ inches (92 mm) actual thickness, and shall be bonded integrally with piers spaced in accordance with Section R606.9.
3. Piers shall be constructed in accordance with Section R606.6 and Section R606.6.1, and shall be bonded into the load-bearing masonry wall in accordance with Section R608.1.1 or R608.1.1.2.
4. The maximum height of a 4-inch (102 mm) load-bearing masonry foundation wall supporting

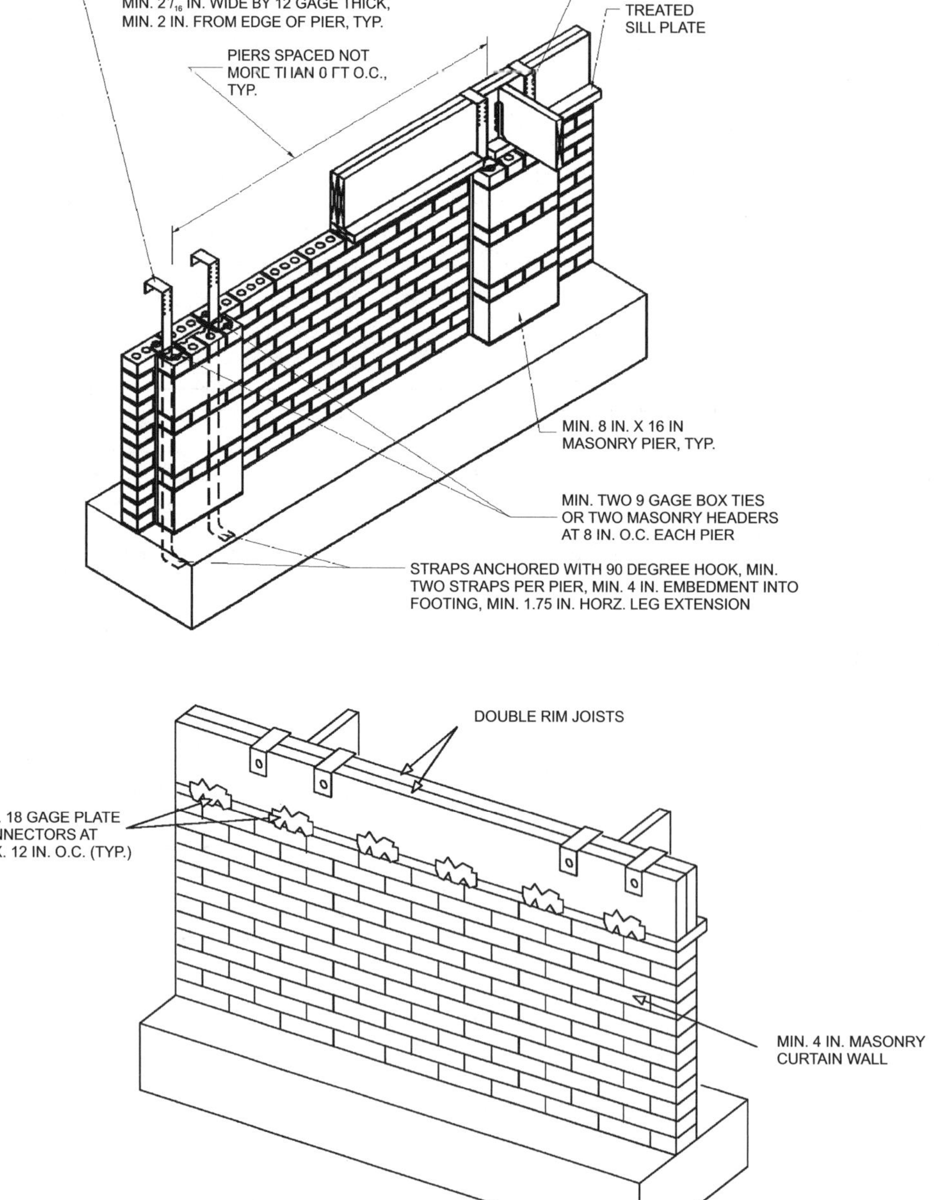

For SI: 1 inch = 25.4 mm, 1 foot = 304.8 mm, 1 degree = 0.0175 rad.

FIGURE R404.1.5(1)
FOUNDATION WALL CLAY MASONRY CURTAIN WALL WITH CONCRETE MASONRY PIERS

wood-frame walls and floors shall not be more than 4 feet (1219 mm).

5. Anchorage shall be in accordance with Section R403.1.6, Figure R404.1.5(1), or as specified by engineered design accepted by the *building official.*
6. The unbalanced fill for 4-inch (102 mm) foundation walls shall not exceed 24 inches (610 mm) for solid masonry or 12 inches (305 mm) for hollow masonry.
7. In Seismic Design Categories D_0, D_1 and D_2, prescriptive reinforcement shall be provided in the horizontal and vertical direction. Provide minimum horizontal joint reinforcement of two No. 9 gage wires spaced not less than 6 inches (152 mm) or one $^1/_4$ inch (6.4 mm) diameter wire at 10 inches (254 mm) on center vertically. Provide minimum vertical reinforcement of one No. 4 bar at 48 inches (1220 mm) on center horizontally grouted in place.

R404.1.6 Height above finished grade. Concrete and masonry foundation walls shall extend above the finished *grade* adjacent to the foundation at all points a minimum of 4 inches (102 mm) where masonry veneer is used and a minimum of 6 inches (152 mm) elsewhere.

R404.1.7 Backfill placement. Backfill shall not be placed against the wall until the wall has sufficient strength and has been anchored to the floor above, or has been sufficiently braced to prevent damage by the backfill.

Exception: Bracing is not required for walls supporting less than 4 feet (1219 mm) of unbalanced backfill.

R404.1.8 Rubble stone masonry. Rubble stone masonry foundation walls shall have a minimum thickness of 16 inches (406 mm), shall not support an unbalanced backfill exceeding 8 feet (2438 mm) in height, shall not support a soil pressure greater than 30 pounds per square foot per foot (4.71 kPa/m), and shall not be constructed in Seismic Design Categories D_0, D_1, D_2 or townhouses in Seismic Design Category C, as established in Figure R301.2(2).

R404.1.9 Isolated masonry piers. Isolated masonry piers shall be constructed in accordance with this section and the general masonry construction requirements of Section R606. Hollow masonry piers shall have a minimum nominal thickness of 8 inches (203 mm), with a nominal height not exceeding four times the nominal thickness and a nominal length not exceeding three times the nominal thickness. Where hollow masonry units are solidly filled with concrete or grout, piers shall be permitted to have a nominal height not exceeding ten times the nominal thickness. Footings for isolated masonry piers shall be sized in accordance with Section R403.1.1.

R404.1.9.1 Pier cap. Hollow masonry piers shall be capped with 4 inches (102 mm) of solid masonry or concrete, a masonry cap block, or shall have cavities of the top course filled with concrete or grout. Where required, termite protection for the pier cap shall be provided in accordance with Section R318.

R404.1.9.2 Masonry piers supporting floor girders. Masonry piers supporting wood girders sized in accordance with Tables R502.5(1) and R502.5(2) shall be permitted in accordance with this section. Piers supporting girders for interior bearing walls shall have a minimum nominal dimension of 12 inches (305 mm) and a maximum height of 10 feet (3048 mm) from top of footing to bottom of sill plate or girder. Piers supporting girders for exterior bearing walls shall have a minimum nominal dimension of 12 inches (305 mm) and a maximum height of 4 feet (1220 mm) from top of footing to bottom of sill plate or girder. Girders and sill plates shall be anchored to the pier or footing in accordance with Section R403.1.6 or Figure R404.1.5(1). Floor girder bearing shall be in accordance with Section R502.6.

R404.1.9.3 Masonry piers supporting braced wall panels. Masonry piers supporting *braced wall panels* shall be designed in accordance with accepted engineering practice.

R404.1.9.4 Seismic design of masonry piers. Masonry piers in all *dwellings* located in Seismic Design Category D_0, D_1 or D_2, and townhouses in Seismic Design Category C, shall be designed in accordance with accepted engineering practice.

R404.1.9.5 Masonry piers in flood hazard areas. Masonry piers for *dwellings* in flood hazard areas shall be designed in accordance with Section R322.

R404.2 Wood foundation walls. Wood foundation walls shall be constructed in accordance with the provisions of Sections R404.2.1 through R404.2.6 and with the details shown in Figures R403.1(2) and R403.1(3).

R404.2.1 Identification. All load-bearing lumber shall be identified by the grade *mark* of a lumber grading or inspection agency which has been *approved* by an accreditation body that complies with DOC PS 20. In lieu of a grade *mark*, a certificate of inspection issued by a lumber grading or inspection agency meeting the requirements of this section shall be accepted. Wood structural panels shall conform to DOC PS 1 or DOC PS 2 and shall be identified by a grade *mark* or certificate of inspection issued by an *approved agency.*

R404.2.2 Stud size. The studs used in foundation walls shall be 2-inch by 6-inch (51 mm by 152 mm) members. When spaced 16 inches (406 mm) on center, a wood species with an F_b value of not less than 1,250 pounds per square inch (8619 kPa) as listed in AF&PA/NDS shall be used. When spaced 12 inches (305 mm) on center, an F_b of not less than 875 psi (6033 kPa) shall be required.

R404.2.3 Height of backfill. For wood foundations that are not designed and installed in accordance with AF&PA PWF, the height of backfill against a foundation wall shall not exceed 4 feet (1219 mm). When the height of fill is more than 12 inches (305 mm) above the interior *grade* of a crawl space or floor of a *basement*, the thickness of the plywood sheathing shall meet the requirements of Table R404.2.3.

R404.2.4 Backfilling. Wood foundation walls shall not be backfilled until the *basement* floor and first floor have been constructed or the walls have been braced. For crawl space construction, backfill or bracing shall be installed on the interior of the walls prior to placing backfill on the exterior.

R404.2.5 Drainage and dampproofing. Wood foundation basements shall be drained and dampproofed in accordance with Sections R405 and R406, respectively.

R404.2.6 Fastening. Wood structural panel foundation wall sheathing shall be attached to framing in accordance with Table R602.3(1) and Section R402.1.1.

R404.3 Wood sill plates. Wood sill plates shall be a minimum of 2-inch by 4-inch (51 mm by 102 mm) nominal lumber. Sill plate anchorage shall be in accordance with Sections R403.1.6 and R602.11.

R404.4 Retaining walls. Retaining walls that are not laterally supported at the top and that retain in excess of 24 inches (610 mm) of unbalanced fill shall be designed to ensure stability against overturning, sliding, excessive foundation pressure and water uplift. Retaining walls shall be designed for a safety factor of 1.5 against lateral sliding and overturning.

R404.5 Precast concrete foundation walls.

R404.5.1 Design. Precast concrete foundation walls shall be designed in accordance with accepted engineering practice. The design and manufacture of precast concrete foundation wall panels shall comply with the materials requirements of Section R402.3 or ACI 318. The panel design drawings shall be prepared by a registered design professional where required by the statutes of the *jurisdiction* in which the project is to be constructed in accordance with Section R106.1.

R404.5.2 Precast concrete foundation design drawings. Precast concrete foundation wall design drawings shall be submitted to the *building official* and *approved* prior to installation. Drawings shall include, at a minimum, the information specified below:

1. Design loading as applicable;
2. Footing design and material;

TABLE R404.2.3
PLYWOOD GRADE AND THICKNESS FOR WOOD FOUNDATION CONSTRUCTION (30 pcf equivalent-fluid weight soil pressure)

HEIGHT OF FILL (inches)	STUD SPACING (inches)	FACE GRAIN ACROSS STUDS			FACE GRAIN PARALLEL TO STUDS		
		Grade[a]	Minimum thickness (inches)	Span rating	Grade[a]	Minimum thickness (inches)[b, c]	Span rating
24	12	B	$^{15}/_{32}$	32/16	A	$^{15}/_{32}$	32/16
					B	$^{15}/_{32}$[c]	32/16
	16	B	$^{15}/_{32}$	32/16	A	$^{15}/_{32}$[c]	32/16
					B	$^{19}/_{32}$[c] (4, 5 ply)	40/20
36	12	B	$^{15}/_{32}$	32/16	A	$^{15}/_{32}$	32/16
					B	$^{15}/_{32}$[c] (4, 5 ply)	32/16
					B	$^{19}/_{32}$ (4, 5 ply)	40/20
	16	B	$^{15}/_{32}$[c]	32/16	A	$^{19}/_{32}$	40/20
					B	$^{23}/_{32}$	48/24
48	12	B	$^{15}/_{32}$	32/16	A	$^{15}/_{32}$[c]	32/16
					B	$^{19}/_{32}$[c](4, 5 ply)	40/20
	16	B	$^{19}/_{32}$	40/20	A	$^{19}/_{32}$[c]	40/20
					A	$^{23}/_{32}$	48/24

For SI: 1 inch = 25.4 mm, 1 foot = 304.8 mm, 1 pound per cubic foot = 0.1572 kN/m^3.

a. Plywood shall be of the following minimum grades in accordance with DOC PS 1 or DOC PS 2:

1. DOC PS 1 Plywood grades marked:
   - 1.1. Structural I C-D (Exposure 1).
   - 1.2. C-D (Exposure 1).
2. DOC PS 2 Plywood grades marked:
   - 2.1. Structural I Sheathing (Exposure 1).
   - 2.2. Sheathing (Exposure 1).
3. Where a major portion of the wall is exposed above ground and a better appearance is desired, the following plywood grades marked exterior are suitable:
   - 3.1. Structural I A-C, Structural I B-C or Structural I C-C (Plugged) in accordance with DOC PS 1.
   - 3.2. A-C Group 1, B-C Group 1, C-C (Plugged) Group 1 or MDO Group 1 in accordance with DOC PS 1.
   - 3.3. Single Floor in accordance with DOC PS 1 or DOC PS 2.

b. Minimum thickness $^{15}/_{32}$ inch, except crawl space sheathing may be $^{3}/_{8}$ inch for face grain across studs 16 inches on center and maximum 2-foot depth of unequal fill.

c. For this fill height, thickness and grade combination, panels that are continuous over less than three spans (across less than three stud spacings) require blocking 16 inches above the bottom plate. Offset adjacent blocks and fasten through studs with two 16d corrosion-resistant nails at each end.

3. Concentrated loads and their points of application;
4. Soil bearing capacity;
5. Maximum allowable total uniform load;
6. Seismic design category; and
7. Basic wind speed.

R404.5.3 Identification. Precast concrete foundation wall panels shall be identified by a certificate of inspection *label* issued by an *approved* third party inspection agency.

SECTION R405
FOUNDATION DRAINAGE

R405.1 Concrete or masonry foundations. Drains shall be provided around all concrete or masonry foundations that retain earth and enclose habitable or usable spaces located below *grade*. Drainage tiles, gravel or crushed stone drains, perforated pipe or other *approved* systems or materials shall be installed at or below the area to be protected and shall discharge by gravity or mechanical means into an *approved* drainage system. Gravel or crushed stone drains shall extend at least 1 foot (305 mm) beyond the outside edge of the footing and 6 inches (152 mm) above the top of the footing and be covered with an *approved* filter membrane material. The top of open joints of drain tiles shall be protected with strips of building paper. Perforated drains shall be surrounded with an *approved* filter membrane or the filter membrane shall cover the washed gravel or crushed rock covering the drain. Drainage tiles or perforated pipe shall be placed on a minimum of 2 inches (51 mm) of washed gravel or crushed rock at least one sieve size larger than the tile joint opening or perforation and covered with not less than 6 inches (152 mm) of the same material.

Exception: A drainage system is not required when the foundation is installed on well-drained ground or sand-gravel mixture soils according to the Unified Soil Classification System, Group I Soils, as detailed in Table R405.1.

R405.1.1 Precast concrete foundation. Precast concrete walls that retain earth and enclose habitable or useable space located below-*grade* that rest on crushed stone footings shall have a perforated drainage pipe installed below the base of the wall on either the interior or exterior side of the wall, at least one foot (305 mm) beyond the edge of the wall. If the exterior drainage pipe is used, an *approved* filter membrane material shall cover the pipe. The drainage system shall discharge into an *approved* sewer system or to daylight.

R405.2 Wood foundations. Wood foundations enclosing habitable or usable spaces located below *grade* shall be adequately drained in accordance with Sections R405.2.1 through R405.2.3.

TABLE R405.1
PROPERTIES OF SOILS CLASSIFIED ACCORDING TO THE UNIFIED SOIL CLASSIFICATION SYSTEM

SOIL GROUP	UNIFIED SOIL CLASSIFICATION SYSTEM SYMBOL	SOIL DESCRIPTION	DRAINAGE CHARACTERISTICS[a]	FROST HEAVE POTENTIAL	VOLUME CHANGE POTENTIAL EXPANSION[b]
Group I	GW	Well-graded gravels, gravel sand mixtures, little or no fines	Good	Low	Low
	GP	Poorly graded gravels or gravel sand mixtures, little or no fines	Good	Low	Low
	SW	Well-graded sands, gravelly sands, little or no fines	Good	Low	Low
	SP	Poorly graded sands or gravelly sands, little or no fines	Good	Low	Low
	GM	Silty gravels, gravel-sand-silt mixtures	Good	Medium	Low
	SM	Silty sand, sand-silt mixtures	Good	Medium	Low
Group II	GC	Clayey gravels, gravel-sand-clay mixtures	Medium	Medium	Low
	SC	Clayey sands, sand-clay mixture	Medium	Medium	Low
	ML	Inorganic silts and very fine sands, rock flour, silty or clayey fine sands or clayey silts with slight plasticity	Medium	High	Low
	CL	Inorganic clays of low to medium plasticity, gravelly clays, sandy clays, silty clays, lean clays	Medium	Medium	Medium to Low
Group III	CH	Inorganic clays of high plasticity, fat clays	Poor	Medium	High
	MH	Inorganic silts, micaceous or diatomaceous fine sandy or silty soils, elastic silts	Poor	High	High
Group IV	OL	Organic silts and organic silty clays of low plasticity	Poor	Medium	Medium
	OH	Organic clays of medium to high plasticity, organic silts	Unsatisfactory	Medium	High
	Pt	Peat and other highly organic soils	Unsatisfactory	Medium	High

For SI: 1 inch = 25.4 mm.

a. The percolation rate for good drainage is over 4 inches per hour, medium drainage is 2 inches to 4 inches per hour, and poor is less than 2 inches per hour.

b. Soils with a low potential expansion typically have a plasticity index (PI) of 0 to 15, soils with a medium potential expansion have a PI of 10 to 35 and soils with a high potential expansion have a PI greater than 20.

R405.2.1 Base. A porous layer of gravel, crushed stone or coarse sand shall be placed to a minimum thickness of 4 inches (102 mm) under the *basement* floor. Provision shall be made for automatic draining of this layer and the gravel or crushed stone wall footings.

R405.2.2 Vapor retarder. A 6-mil-thick (0.15 mm) polyethylene vapor retarder shall be applied over the porous layer with the *basement* floor constructed over the polyethylene.

R405.2.3 Drainage system. In other than Group I soils, a sump shall be provided to drain the porous layer and footings. The sump shall be at least 24 inches (610 mm) in diameter or 20 inches square (0.0129 m^2), shall extend at least 24 inches (610 mm) below the bottom of the *basement* floor and shall be capable of positive gravity or mechanical drainage to remove any accumulated water. The drainage system shall discharge into an *approved* sewer system or to daylight.

SECTION R406 FOUNDATION WATERPROOFING AND DAMPPROOFING

R406.1 Concrete and masonry foundation dampproofing. Except where required by Section R406.2 to be waterproofed, foundation walls that retain earth and enclose interior spaces and floors below *grade* shall be dampproofed from the top of the footing to the finished *grade*. Masonry walls shall have not less than $^3/_8$ inch (9.5 mm) portland cement parging applied to the exterior of the wall. The parging shall be dampproofed in accordance with one of the following:

1. Bituminous coating.
2. Three pounds per square yard (1.63 kg/m^2) of acrylic modified cement.
3. One-eighth inch (3.2 mm) coat of surface-bonding cement complying with ASTM C 887.
4. Any material permitted for waterproofing in Section R406.2.
5. Other *approved* methods or materials.

Exception: Parging of unit masonry walls is not required where a material is *approved* for direct application to the masonry.

Concrete walls shall be dampproofed by applying any one of the above listed dampproofing materials or any one of the waterproofing materials listed in Section R406.2 to the exterior of the wall.

R406.2 Concrete and masonry foundation waterproofing. In areas where a high water table or other severe soil-water conditions are known to exist, exterior foundation walls that retain earth and enclose interior spaces and floors below *grade* shall be waterproofed from the top of the footing to the finished *grade*. Walls shall be waterproofed in accordance with one of the following:

1. Two-ply hot-mopped felts.
2. Fifty-five-pound (25 kg) roll roofing.
3. Six-mil (0.15 mm) polyvinyl chloride.
4. Six-mil (0.15 mm) polyethylene.
5. Forty-mil (1 mm) polymer-modified asphalt.
6. Sixty-mil (1.5 mm) flexible polymer cement.
7. One-eighth-inch (3 mm) cement-based, fiber-reinforced, waterproof coating.
8. Sixty-mil (0.22 mm) solvent-free liquid-applied synthetic rubber.

Exception: Organic-solvent-based products such as hydrocarbons, chlorinated hydrocarbons, ketones and esters shall not be used for ICF walls with expanded polystyrene form material. Use of plastic roofing cements, acrylic coatings, latex coatings, mortars and pargings to seal ICF walls is permitted. Cold-setting asphalt or hot asphalt shall conform to type C of ASTM D 449. Hot asphalt shall be applied at a temperature of less than 200°F (93°C).

All joints in membrane waterproofing shall be lapped and sealed with an adhesive compatible with the membrane.

R406.3 Dampproofing for wood foundations. Wood foundations enclosing habitable or usable spaces located below *grade* shall be dampproofed in accordance with Sections R406.3.1 through R406.3.4.

R406.3.1 Panel joint sealed. Plywood panel joints in the foundation walls shall be sealed full length with a caulking compound capable of producing a moisture-proof seal under the conditions of temperature and moisture content at which it will be applied and used.

R406.3.2 Below-grade moisture barrier. A 6-mil-thick (0.15 mm) polyethylene film shall be applied over the below-*grade* portion of exterior foundation walls prior to backfilling. Joints in the polyethylene film shall be lapped 6 inches (152 mm) and sealed with adhesive. The top edge of the polyethylene film shall be bonded to the sheathing to form a seal. Film areas at *grade* level shall be protected from mechanical damage and exposure by a pressure preservatively treated lumber or plywood strip attached to the wall several inches above finish *grade* level and extending approximately 9 inches (229 mm) below *grade*. The joint between the strip and the wall shall be caulked full length prior to fastening the strip to the wall. Other coverings appropriate to the architectural treatment may also be used. The polyethylene film shall extend down to the bottom of the wood footing plate but shall not overlap or extend into the gravel or crushed stone footing.

R406.3.3 Porous fill. The space between the excavation and the foundation wall shall be backfilled with the same material used for footings, up to a height of 1 foot (305 mm) above the footing for well-drained sites, or one-half the total back-fill height for poorly drained sites. The porous fill shall be covered with strips of 30-pound (13.6 kg) asphalt paper or 6-mil (0.15 mm) polyethylene to permit water seepage while avoiding infiltration of fine soils.

R406.3.4 Backfill. The remainder of the excavated area shall be backfilled with the same type of soil as was removed during the excavation.

R406.4 Precast concrete foundation system dampproofing. Except where required by Section R406.2 to be waterproofed, precast concrete foundation walls enclosing habitable or useable spaces located below *grade* shall be dampproofed in accordance with Section R406.1.

R406.4.1 Panel joints sealed. Precast concrete foundation panel joints shall be sealed full height with a sealant meeting ASTM C 920, Type S or M, *Grade* NS, Class 25, Use NT, M or A. Joint sealant shall be installed in accordance with the manufacturer's installation instructions.

SECTION R407 COLUMNS

R407.1 Wood column protection. Wood columns shall be protected against decay as set forth in Section R317.

R407.2 Steel column protection. All surfaces (inside and outside) of steel columns shall be given a shop coat of rust-inhibitive paint, except for corrosion-resistant steel and steel treated with coatings to provide corrosion resistance.

R407.3 Structural requirements. The columns shall be restrained to prevent lateral displacement at the bottom end. Wood columns shall not be less in nominal size than 4 inches by 4 inches (102 mm by 102 mm). Steel columns shall not be less than 3-inch-diameter (76 mm) Schedule 40 pipe manufactured in accordance with ASTM A 53 Grade B or *approved* equivalent.

Exception: In Seismic Design Categories A, B and C, columns no more than 48 inches (1219 mm) in height on a pier or footing are exempt from the bottom end lateral displacement requirement within under-floor areas enclosed by a continuous foundation.

SECTION R408 UNDER-FLOOR SPACE

R408.1 Ventilation. The under-floor space between the bottom of the floor joists and the earth under any building (except space occupied by a *basement*) shall have ventilation openings through foundation walls or exterior walls. The minimum net area of ventilation openings shall not be less than 1 square foot (0.0929 m^2) for each 150 square feet (14 m^2) of under-floor space area, unless the ground surface is covered by a Class 1 vapor retarder material. When a Class 1 vapor retarder material is used, the minimum net area of ventilation openings shall not be less than 1 square foot (0.0929 m^2) for each 1,500 square feet (140 m^2) of under-floor space area. One such ventilating opening shall be within 3 feet (914 mm) of each corner of the building.

R408.2 Openings for under-floor ventilation. The minimum net area of ventilation openings shall not be less than 1 square foot (0.0929 m^2) for each 150 square feet (14 m^2) of under-floor area. One ventilation opening shall be within 3 feet (915 mm) of each corner of the building. Ventilation openings shall be covered for their height and width with any of the following materials provided that the least dimension of the covering shall not exceed $^1/_4$ inch (6.4 mm):

1. Perforated sheet metal plates not less than 0.070 inch (1.8 mm) thick.
2. Expanded sheet metal plates not less than 0.047 inch (1.2 mm) thick.
3. Cast-iron grill or grating.
4. Extruded load-bearing brick vents.
5. Hardware cloth of 0.035 inch (0.89 mm) wire or heavier.
6. Corrosion-resistant wire mesh, with the least dimension being $^1/_8$ inch (3.2 mm) thick.

Exception: The total area of ventilation openings shall be permitted to be reduced to $^1/_{1,500}$ of the under-floor area where the ground surface is covered with an *approved* Class I vapor retarder material and the required openings are placed to provide cross ventilation of the space. The installation of operable louvers shall not be prohibited.

R408.3 Unvented crawl space. Ventilation openings in under-floor spaces specified in Sections R408.1 and R408.2 shall not be required where:

1. Exposed earth is covered with a continuous Class I vapor retarder. Joints of the vapor retarder shall overlap by 6 inches (152 mm) and shall be sealed or taped. The edges of the vapor retarder shall extend at least 6 inches (152 mm) up the stem wall and shall be attached and sealed to the stem wall or insulation; and
2. One of the following is provided for the under-floor space:
   - 2.1. Continuously operated mechanical exhaust ventilation at a rate equal to 1 cubic foot per minute (0.47 L/s) for each 50 square feet (4.7m^2) of crawlspace floor area, including an air pathway to the common area (such as a duct or transfer grille), and perimeter walls insulated in accordance with Section N1103.2.1 of this code;
   - 2.2. *Conditioned air* supply sized to deliver at a rate equal to 1 cubic foot per minute (0.47 L/s) for each 50 square feet (4.7 m^2) of under-floor area, including a return air pathway to the common area (such as a duct or transfer grille), and perimeter walls insulated in accordance with Section N1102.2 of this code;
   - 2.3. Plenum in existing structures complying with Section M1601.5, if under-floor space is used as a plenum.

R408.4 Access. Access shall be provided to all under-floor spaces. Access openings through the floor shall be a minimum of 18 inches by 24 inches (457 mm by 610 mm). Openings through a perimeter wall shall be not less than 16 inches by 24 inches (407 mm by 610 mm). When any portion of the through-wall access is below *grade*, an areaway not less than 16 inches by 24 inches (407 mm by 610 mm) shall be provided. The bottom of the areaway shall be below the threshold of the access opening. Through wall access openings shall not be located under a door to the residence. See Section M1305.1.4 for access requirements where mechanical *equipment* is located under floors.

R408.5 Removal of debris. The under-floor *grade* shall be cleaned of all vegetation and organic material. All wood forms used for placing concrete shall be removed before a building is occupied or used for any purpose. All construction materials shall be removed before a building is occupied or used for any purpose.

R408.6 Finished grade. The finished *grade* of under-floor surface may be located at the bottom of the footings; however, where there is evidence that the groundwater table can rise to within 6 inches (152 mm) of the finished floor at the building perimeter or where there is evidence that the surface water does not readily drain from the building site, the *grade* in the under-floor space shall be as high as the outside finished *grade*, unless an *approved* drainage system is provided.

R408.7 Flood resistance. For buildings located in flood hazard areas as established in Table R301.2(1):

1. Walls enclosing the under-floor space shall be provided with flood openings in accordance with Section R322.2.2.
2. The finished ground level of the under-floor space shall be equal to or higher than the outside finished ground level on at least one side.

   Exception: Under-floor spaces that meet the requirements of FEMA/FIA TB 11-1.

CHAPTER 5
FLOORS

SECTION R501
GENERAL

R501.1 Application. The provisions of this chapter shall control the design and construction of the floors for all buildings including the floors of *attic* spaces used to house mechanical or plumbing fixtures and *equipment*.

R501.2 Requirements. Floor construction shall be capable of accommodating all loads according to Section R301 and of transmitting the resulting loads to the supporting structural elements.

R501.3 Fire protection of floors. Floor assemblies, not required elsewhere in this code to be fire-resistance rated, shall be provided with a $^{1}/_{2}$-inch (12.7 mm) gypsum wallboard membrane, $^{5}/_{8}$-inch (16 mm) wood structural panel membrane, or equivalent on the underside of the floor framing member.

Exceptions:

1. Floor assemblies located directly over a space protected by an automatic sprinkler system in accordance with Section P2904, NFPA13D, or other approved equivalent sprinkler system.
2. Floor assemblies located directly over a crawl space not intended for storage or fuel-fired appliances.
3. Portions of floor assemblies can be unprotected when complying with the following:
   3.1. The aggregate area of the unprotected portions shall not exceed 80 square feet per story
   3.2. Fire blocking in accordance with Section R302.11.1 shall be installed along the perimeter of the unprotected portion to separate the unprotected portion from the remainder of the floor assembly.
4. Wood floor assemblies using dimension lumber or structural composite lumber equal to or greater than 2-inch by 10-inch (50.8 mm by 254 mm) nominal dimension, or other approved floor assemblies demonstrating equivalent fire performance.

SECTION R502
WOOD FLOOR FRAMING

R502.1 Identification. Load-bearing dimension lumber for joists, beams and girders shall be identified by a grade *mark* of a lumber grading or inspection agency that has been *approved* by an accreditation body that complies with DOC PS 20. In lieu of a grade *mark*, a certificate of inspection issued by a lumber grading or inspection agency meeting the requirements of this section shall be accepted.

R502.1.1 Preservative-treated lumber. Preservative treated dimension lumber shall also be identified as required by Section R317.2.

R502.1.2 Blocking and subflooring. Blocking shall be a minimum of utility grade lumber. Subflooring may be a minimum of utility grade lumber or No. 4 common grade boards.

R502.1.3 End-jointed lumber. *Approved* end-jointed lumber identified by a grade *mark* conforming to Section R502.1 may be used interchangeably with solid-sawn members of the same species and grade. End-jointed lumber used in an assembly required elsewhere in this code to have a fire-resistance rating shall have the designation "Heat Resistant Adhesive" or "HRA" included in its grade mark.

R502.1.4 Prefabricated wood I-joists. Structural capacities and design provisions for prefabricated wood I-joists shall be established and monitored in accordance with ASTM D 5055.

R502.1.5 Structural glued laminated timbers. Glued laminated timbers shall be manufactured and identified as required in ANSI/AITC A190.1 and ASTM D 3737.

R502.1.6 Structural log members. Stress grading of structural log members of nonrectangular shape, as typically used in log buildings, shall be in accordance with ASTM D 3957. Such structural log members shall be identified by the grade *mark* of an *approved* lumber grading or inspection agency. In lieu of a grade *mark* on the material, a certificate of inspection as to species and grade issued by a lumber-grading or inspection agency meeting the requirements of this section shall be permitted to be accepted.

R502.1.7 Structural composite lumber. Structural capacities for structural composite lumber shall be established and monitored in accordance with ASTM D 5456.

R502.2 Design and construction. Floors shall be designed and constructed in accordance with the provisions of this chapter, Figure R502.2 and Sections R317 and R318 or in accordance with AF&PA/NDS.

R502.2.1 Framing at braced wall lines. A load path for lateral forces shall be provided between floor framing and *braced wall panels* located above or below a floor, as specified in Section R602.10.8.

*

R502.3 Allowable joist spans. Spans for floor joists shall be in accordance with Tables R502.3.1(1) and R502.3.1(2). For other grades and species and for other loading conditions, refer to the AF&PA Span Tables for Joists and Rafters.

R502.3.1 Sleeping areas and attic joists. Table R502.3.1(1) shall be used to determine the maximum allowable span of floor joists that support sleeping areas and *attics* that are accessed by means of a fixed stairway in accordance with Section R311.7 provided that the design live load does not exceed 30 pounds per square foot (1.44 kPa) and the design dead load does not exceed 20 pounds per square foot (0.96 kPa). The allowable span of ceiling joists that support *attics* used for limited storage or no storage shall be determined in accordance with Section R802.4.

R502.3.2 Other floor joists. Table R502.3.1(2) shall be used to determine the maximum allowable span of floor joists that support all other areas of the building, other than sleeping rooms and *attics*, provided that the design live load does not exceed 40 pounds per square foot (1.92 kPa) and the design dead load does not exceed 20 pounds per square foot (0.96 kPa).

R502.3.3 Floor cantilevers. Floor cantilever spans shall not exceed the nominal depth of the wood floor joist. Floor cantilevers constructed in accordance with Table R502.3.3(1) shall be permitted when supporting a light-frame bearing wall and roof only. Floor cantilevers supporting an exterior balcony are permitted to be constructed in accordance with Table R502.3.3(2)

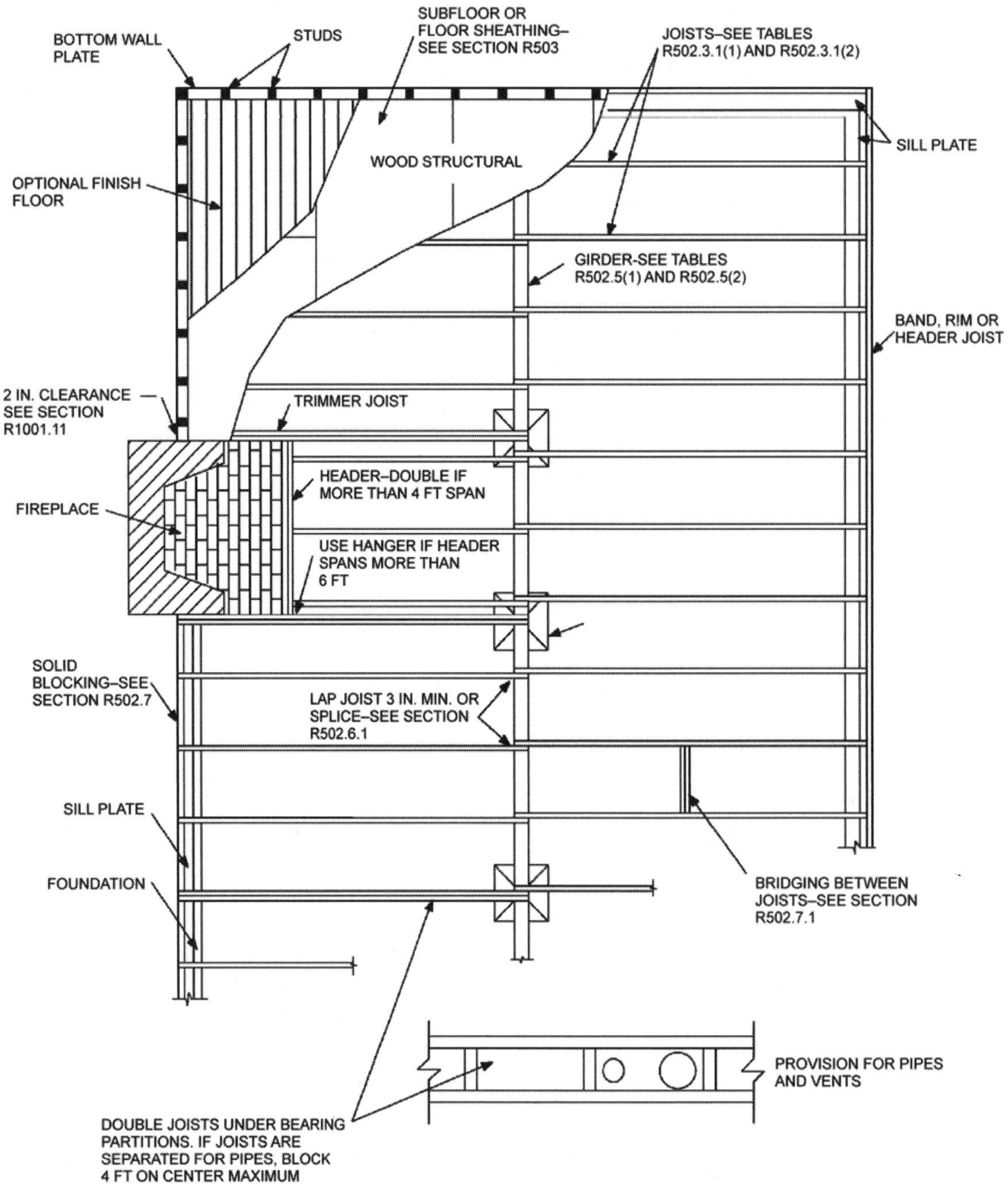

For SI: 1 inch = 25.4 mm, 1 foot = 304.8 mm.

FIGURE R502.2
FLOOR CONSTRUCTION

TABLE R502.3.1(1)
FLOOR JOIST SPANS FOR COMMON LUMBER SPECIES
(Residential sleeping areas, live load = 30 psf, L/Δ = 360)[a]

JOIST SPACING (inches)	SPECIES AND GRADE		DEAD LOAD = 10 psf				DEAD LOAD = 20 psf			
			2 × 6	2 × 8	2 × 10	2 × 12	2 × 6	2 × 8	2 × 10	2 × 12
			Maximum floor joist spans							
			(ft - in.)	(ft - in.)	(ft - in.)	(ft - in.)	(ft - in.)	(ft - in.)	(ft - in.)	(ft - in.)
12	Douglas fir-larch	SS	12-6	16-6	21-0	25-7	12-6	16-6	21-0	25-7
	Douglas fir-larch	#1	12-0	15-10	20-3	24-8	12-0	15-7	19-0	22-0
	Douglas fir-larch	#2	11-10	15-7	19-10	23-0	11-6	14-7	17-9	20-7
	Douglas fir-larch	#3	9-8	12-4	15-0	17-5	8-8	11-0	13-5	15-7
	Hem-fir	SS	11-10	15-7	19-10	24-2	11-10	15-7	19-10	24-2
	Hem-fir	#1	11-7	15-3	19-5	23-7	11-7	15-2	18-6	21-6
	Hem-fir	#2	11-0	14-6	18-6	22-6	11-0	14-4	17-6	20-4
	Hem-fir	#3	9-8	12-4	15-0	17-5	8-8	11-0	13-5	15-7
	Southern pine	SS	12-3	16-2	20-8	25-1	12-3	16-2	20-8	25-1
	Southern pine	#1	12-0	15-10	20-3	24-8	12-0	15-10	20-3	24-8
	Southern pine	#2	11-10	15-7	19-10	24-2	11-10	15-7	18-7	21-9
	Southern pine	#3	10-5	13-3	15-8	18-8	9-4	11-11	14-0	16-8
	Spruce-pine-fir	SS	11-7	15-3	19-5	23-7	11-7	15-3	19-5	23-7
	Spruce-pine-fir	#1	11-3	14-11	19-0	23-0	11-3	14-7	17-9	20-7
	Spruce-pine-fir	#2	11-3	14-11	19-0	23-0	11-3	14-7	17-9	20-7
	Spruce-pine-fir	#3	9-8	12-4	15-0	17-5	8-8	11-0	13-5	15-7
16	Douglas fir-larch	SS	11-4	15-0	19-1	23-3	11-4	15-0	19-1	23-0
	Douglas fir-larch	#1	10-11	14-5	18-5	21-4	10-8	13-6	16-5	19-1
	Douglas fir-larch	#2	10-9	14-1	17-2	19-11	9-11	12-7	15-5	17-10
	Douglas fir-larch	#3	8-5	10-8	13-0	15-1	7-6	9-6	11-8	13-6
	Hem-fir	SS	10-9	14-2	18-0	21-11	10-9	14-2	18-0	21-11
	Hem-fir	#1	10-6	13-10	17-8	20-9	10-4	13-1	16-0	18-7
	Hem-fir	#2	10-0	13-2	16-10	19-8	9-10	12-5	15-2	17-7
	Hem-fir	#3	8-5	10-8	13-0	15-1	7-6	9-6	11-8	13-6
	Southern pine	SS	11-2	14-8	18-9	22-10	11-2	14-8	18-9	22-10
	Southern pine	#1	10-11	14-5	18-5	22-5	10-11	14-5	17-11	21-4
	Southern pine	#2	10-9	14-2	18-0	21-1	10-5	13-6	16-1	18-10
	Southern pine	#3	9-0	11-6	13-7	16-2	8-1	10-3	12-2	14-6
	Spruce-pine-fir	SS	10-6	13-10	17-8	21-6	10-6	13-10	17-8	21-4
	Spruce-pine-fir	#1	10-3	13-6	17-2	19-11	9-11	12-7	15-5	17-10
	Spruce-pine-fir	#2	10-3	13-6	17-2	19-11	9-11	12-7	15-5	17-10
	Spruce-pine-fir	#3	8-5	10-8	13-0	15-1	7-6	9-6	11-8	13-6

(continued)

TABLE R502.3.1(1)—continued
FLOOR JOIST SPANS FOR COMMON LUMBER SPECIES
(Residential sleeping areas, live load = 30 psf, L/Δ = 360)[a]

JOIST SPACING (inches)	SPECIES AND GRADE		DEAD LOAD = 10 psf				DEAD LOAD = 20 psf			
			2 × 6	2 × 8	2 × 10	2 × 12	2 × 6	2 × 8	2 × 10	2 × 12
			Maximum floor joist spans							
			(ft - in.)	(ft - in.)	(ft - in.)	(ft - in.)	(ft - in.)	(ft - in.)	(ft - in.)	(ft - in.)
19.2	Douglas fir-larch	SS	10-8	14-1	18-0	21-10	10-8	14-1	18-0	21-0
	Douglas fir-larch	#1	10-4	13-7	16-9	19-6	9-8	12-4	15-0	17-5
	Douglas fir-larch	#2	10-1	12-10	15-8	18-3	9-1	11-6	14-1	16-3
	Douglas fir-larch	#3	7-8	9-9	11-10	13-9	6-10	8-8	10-7	12-4
	Hem-fir	SS	10-1	13-4	17-0	20-8	10-1	13-4	17-0	20-7
	Hem-fir	#1	9-10	13-0	16-4	19-0	9-6	12-0	14-8	17-0
	Hem-fir	#2	9-5	12-5	15-6	17-1	8-11	11-4	13-10	16-1
	Hem-fir	#3	7-8	9-9	11-10	13-9	6-10	8-8	10-7	12-4
	Southern pine	SS	10-6	13-10	17-8	21-6	10-6	13-10	17-8	21-6
	Southern pine	#1	10-4	13-7	17-4	21-1	10-4	13-7	16-4	19-6
	Southern pine	#2	10-1	13-4	16-5	19-3	9-6	12-4	14-8	17-2
	Southern pine	#3	8-3	10-6	12-5	14-9	7-4	9-5	11-1	13-2
	Spruce-pine-fir	SS	9-10	13-0	16-7	20-2	9-10	13-0	16-7	19-6
	Spruce-pine-fir	#1	9-8	12-9	15-8	18-3	9-1	11-6	14-1	16-3
	Spruce-pine-fir	#2	9-8	12-9	15-8	18-3	9-1	11-6	14-1	16-3
	Spruce-pine-fir	#3	7-8	9-9	11-10	13-9	6-10	8-8	10-7	12-4
24	Douglas fir-larch	SS	9-11	13-1	16-8	20-3	9-11	13-1	16-2	18-9
	Douglas fir-larch	#1	9-7	12-4	15-0	17-5	8-8	11-0	13-5	15-7
	Douglas fir-larch	#2	9-1	11-6	14-1	16-3	8-1	10-3	12-7	14-7
	Douglas fir-larch	#3	6-10	8-8	10-7	12-4	6-2	7-9	9-6	11-0
	Hem-fir	SS	9-4	12-4	15-9	19-2	9-4	12-4	15-9	18-5
	Hem-fir	#1	9-2	12-0	14-8	17-0	8-6	10-9	13-1	15-2
	Hem-fir	#2	8-9	11-4	13-10	16-1	8-0	10-2	12-5	14-4
	Hem-fir	#3	6-10	8-8	10-7	12-4	6-2	7-9	9-6	11-0
	Southern pine	SS	9-9	12-10	16-5	19-11	9-9	12-10	16-5	19-11
	Southern pine	#1	9-7	12-7	16-1	19-6	9-7	12-4	14-7	17-5
	Southern pine	#2	9-4	12-4	14-8	17-2	8-6	11-0	13-1	15-5
	Southern pine	#3	7-4	9-5	11-1	13-2	6-7	8-5	9-11	11-10
	Spruce-pine-fir	SS	9-2	12-1	15-5	18-9	9-2	12-1	15-0	17-5
	Spruce-pine-fir	#1	8-11	11-6	14-1	16-3	8-1	10-3	12-7	14-7
	Spruce-pine-fir	#2	8-11	11-6	14-1	16-3	8-1	10-3	12-7	14-7
	Spruce-pine-fir	#3	6-10	8-8	10-7	12-4	6-2	7-9	9-6	11-0

For SI: 1 inch = 25.4 mm, 1 foot = 304.8 mm, 1 pound per square foot = 0.0479 kPa.

Note: Check sources for availability of lumber in lengths greater than 20 feet.

a. Dead load limits for townhouses in Seismic Design Category C and all structures in Seismic Design Categories D_0, D_1 and D_2 shall be determined in accordance with Section R301.2.2.2.1.

TABLE R502.3.1(2)
FLOOR JOIST SPANS FOR COMMON LUMBER SPECIES
(Residential living areas, live load = 40 psf, L/Δ = 360)[b]

JOIST SPACING (inches)	SPECIES AND GRADE		DEAD LOAD = 10 psf				DEAD LOAD = 20 psf			
			2 × 6	2 × 8	2 × 10	2 × 12	2 × 6	2 × 8	2 × 10	2 × 12
			Maximum floor joist spans							
			(ft - in.)	(ft - in.)	(ft - in.)	(ft - in.)	(ft - in.)	(ft - in.)	(ft - in.)	(ft - in.)
12	Douglas fir-larch	SS	11-4	15-0	19-1	23-3	11-4	15-0	19-1	23-3
	Douglas fir-larch	#1	10-11	14-5	18-5	22-0	10-11	14-2	17-4	20-1
	Douglas fir-larch	#2	10-9	14-2	17-9	20-7	10-6	13-3	16-3	18-10
	Douglas fir-larch	#3	8-8	11-0	13-5	15-7	7-11	10-0	12-3	14-3
	Hem-fir	SS	10-9	14-2	18-0	21-11	10-9	14-2	18-0	21-11
	Hem-fir	#1	10-6	13-10	17-8	21-6	10-6	13-10	16-11	19-7
	Hem-fir	#2	10-0	13-2	16-10	20-4	10-0	13-1	16-0	18-6
	Hem-fir	#3	8-8	11-0	13-5	15-7	7-11	10-0	12-3	14-3
	Southern pine	SS	11-2	14-8	18-9	22-10	11-2	14-8	18-9	22-10
	Southern pine	#1	10-11	14-5	18-5	22-5	10-11	14-5	18-5	22-5
	Southern pine	#2	10-9	14-2	18-0	21-9	10-9	14-2	16-11	19-10
	Southern pine	#3	9-4	11-11	14-0	16-8	8-6	10-10	12-10	15-3
	Spruce-pine-fir	SS	10-6	13-10	17-8	21-6	10-6	13-10	17-8	21-6
	Spruce-pine-fir	#1	10-3	13-6	17-3	20-7	10-3	13-3	16-3	18-10
	Spruce-pine-fir	#2	10-3	13-6	17-3	20-7	10-3	13-3	16-3	18-10
	Spruce-pine-fir	#3	8-8	11-0	13-5	15-7	7-11	10-0	12-3	14-3
16	Douglas fir-larch	SS	10-4	13-7	17-4	21-1	10-4	13-7	17-4	21-0
	Douglas fir-larch	#1	9-11	13-1	16-5	19-1	9-8	12-4	15-0	17-5
	Douglas fir-larch	#2	9-9	12-7	15-5	17-10	9-1	11-6	14-1	16-3
	Douglas fir-larch	#3	7-6	9-6	11-8	13-6	6-10	8-8	10-7	12-4
16	Hem-fir	SS	9-9	12-10	16-5	19-11	9-9	12-10	16-5	19-11
	Hem-fir	#1	9-6	12-7	16-0	18-7	9-6	12-0	14-8	17-0
	Hem-fir	#2	9-1	12-0	15-2	17-7	8-11	11-4	13-10	16-1
	Hem-fir	#3	7-6	9-6	11-8	13-6	6-10	8-8	10-7	12-4
	Southern pine	SS	10-2	13-4	17-0	20-9	10-2	13-4	17-0	20-9
	Southern pine	#1	9-11	13-1	16-9	20-4	9-11	13-1	16-4	19-6
	Southern pine	#2	9-9	12-10	16-1	18-10	9-6	12-4	14-8	17-2
	Southern pine	#3	8-1	10-3	12-2	14-6	7-4	9-5	11-1	13-2
	Spruce-pine-fir	SS	9-6	12-7	16-0	19-6	9-6	12-7	16-0	19-6
	Spruce-pine-fir	#1	9-4	12-3	15-5	17-10	9-1	11-6	14-1	16-3
	Spruce-pine-fir	#2	9-4	12-3	15-5	17-10	9-1	11-6	14-1	16-3
	Spruce-pine-fir	#3	7-6	9-6	11-8	13-6	6-10	8-8	10-7	12-4

(continued)

TABLE R502.3.1(2)—continued
FLOOR JOIST SPANS FOR COMMON LUMBER SPECIES
(Residential living areas, live load = 40 psf, L/Δ = 360)[b]

JOIST SPACING (inches)	SPECIES AND GRADE		DEAD LOAD = 10 psf				DEAD LOAD = 20 psf			
			2 × 6	2 × 8	2 × 10	2 × 12	2 × 6	2 × 8	2 × 10	2 × 12
			Maximum floor joist spans							
			(ft - in.)	(ft - in.)	(ft - in.)	(ft - in.)	(ft - in.)	(ft - in.)	(ft - in.)	(ft - in.)
19.2	Douglas fir-larch	SS	9-8	12-10	16-4	19-10	9-8	12-10	16-4	19-2
	Douglas fir-larch	#1	9-4	12-4	15-0	17-5	8-10	11-3	13-8	15-11
	Douglas fir-larch	#2	9-1	11-6	14-1	16-3	8-3	10-6	12-10	14-10
	Douglas fir-larch	#3	6-10	8-8	10-7	12-4	6-3	7-11	9-8	11-3
	Hem-fir	SS	9-2	12-1	15-5	18-9	9-2	12-1	15-5	18-9
	Hem-fir	#1	9-0	11-10	14-8	17-0	8-8	10-11	13-4	15-6
	Hem-fir	#2	8-7	11-3	13-10	16-1	8-2	10-4	12-8	14-8
	Hem-fir	#3	6-10	8-8	10-7	12-4	6-3	7-11	9-8	11-3
	Southern pine	SS	9-6	12-7	16-0	19-6	9-6	12-7	16-0	19-6
	Southern pine	#1	9-4	12-4	15-9	19-2	9-4	12-4	14-11	17-9
	Southern pine	#2	9-2	12-1	14-8	17-2	8-8	11-3	13-5	15-8
	Southern pine	#3	7-4	9-5	11-1	13-2	6-9	8-7	10-1	12-1
	Spruce-pine-fir	SS	9-0	11-10	15-1	18-4	9-0	11-10	15-1	17-9
	Spruce-pine-fir	#	8-9	11-6	14-1	16-3	8-3	10-6	12-10	14-10
	Spruce-pine-fir	#2	8-9	11-6	14-1	16-3	8-3	10-6	12-10	14-10
	Spruce-pine-fir	#3	6-10	8-8	10-7	12-4	6-3	7-11	9-8	11-3
24	Douglas fir-larch	SS	9-0	11-11	15-2	18-5	9-0	11-11	14-9	17-1
	Douglas fir-larch	#1	8-8	11-0	13-5	15-7	7-11	10-0	12-3	14-3
	Douglas fir-larch	#2	8-1	10-3	12-7	14-7	7-5	9-5	11-6	13-4
	Douglas fir-larch	#3	6-2	7-9	9-6	11-0	5-7	7-1	8-8	10-1
	Hem-fir	SS	8-6	11-3	14-4	17-5	8-6	11-3	14-4	16-10[a]
	Hem-fir	#1	8-4	10-9	13-1	15-2	7-9	9-9	11-11	13-10
	Hem-fir	#2	7-11	10-2	12-5	14-4	7-4	9-3	11-4	13-1
	Hem-fir	#3	6-2	7-9	9-6	11-0	5-7	7-1	8-8	10-1
	Southern pine	SS	8-10	11-8	14-11	18-1	8-10	11-8	14-11	18-1
	Southern pine	#1	8-8	11-5	14-7	17-5	8-8	11-3	13-4	15-11
	Southern pine	#2	8-6	11-0	13-1	15-5	7-9	10-0	12-0	14-0
	Southern pine	#3	6-7	8-5	9-11	11-10	6-0	7-8	9-1	10-9
	Spruce-pine-fir	SS	8-4	11-0	14-0	17-0	8-4	11-0	13-8	15-11
	Spruce-pine-fir	#1	8-1	10-3	12-7	14-7	7-5	9-5	11-6	13-4
	Spruce-pine-fir	#2	8-1	10-3	12-7	14-7	7-5	9-5	11-6	13-4
	Spruce-pine-fir	#3	6-2	7-9	9-6	11-0	5-7	7-1	8-8	10-1

For SI: 1 inch = 25.4 mm, 1 foot = 304.8 mm, 1 pound per square foot = 0.0479 kPa.

Note: Check sources for availability of lumber in lengths greater than 20 feet.

a. End bearing length shall be increased to 2 inches.

b. Dead load limits for townhouses in Seismic Design Category C and all structures in Seismic Design Categories D_0, D_1, and D_2 shall be determined in accordance with Section R301.2.2.2.1.

TABLE R502.3.3(1)
CANTILEVER SPANS FOR FLOOR JOISTS SUPPORTING LIGHT-FRAME EXTERIOR BEARING WALL AND ROOF ONLY[a, b, c, f, g, h]
(Floor Live Load ≤ 40 psf, Roof Live Load ≤ 20 psf)

Member & Spacing	Maximum Cantilever Span (Uplift Force at Backspan Support in Lbs.)[d, e]											
	Ground Snow Load											
	≤ 20 psf			30 psf			50 psf			70 psf		
	Roof Width			Roof Width			Roof Width			Roof Width		
	24 ft	32 ft	40 ft	24 ft	32 ft	40 ft	24 ft	32 ft	40 ft	24 ft	32 ft	40 ft
2 × 8 @ 12″	20″ (177)	15″ (227)	—	18″ (209)	—	—	—	—	—	—	—	—
2 × 10 @ 16″	29″ (228)	21″ (297)	16″ (364)	26″ (271)	18″ (354)	—	20″ (375)	—	—	—	—	—
2 × 10 @ 12″	36″ (166)	26″ (219)	20″ (270)	34″ (198)	22″ (263)	16″ (324)	26″ (277)	—	—	19″ (356)	—	—
2 × 12 @ 16″	—	32″ (287)	25″ (356)	36″ (263)	29″ (345)	21″ (428)	29″ (367)	20″ (484)	—	23″ (471)	—	—
2 × 12 @ 12″	—	42″ (209)	31″ (263)	—	37″ (253)	27″ (317)	36″ (271)	27″ (358)	17″ (447)	31″ (348)	19″ (462)	—
2 × 12 @ 8″	—	48″ (136)	45″ (169)	—	48″ (164)	38″ (206)	—	40″ (233)	26″ (294)	36″ (230)	29″ (304)	18″ (379)

For SI: 1 inch = 25.4 mm, 1 foot = 304.8 mm, 1 pound per square foot = 0.0479 kPa.

a. Tabulated values are for clear-span roof supported solely by exterior bearing walls.

b. Spans are based on No. 2 Grade lumber of Douglas fir-larch, hem-fir, southern pine and spruce-pine-fir for repetitive (three or more) members.

c. Ratio of backspan to cantilever span shall be at least 3:1.

d. Connections capable of resisting the indicated uplift force shall be provided at the backspan support.

e. Uplift force is for a backspan to cantilever span ratio of 3:1. Tabulated uplift values are permitted to be reduced by multiplying by a factor equal to 3 divided by the actual backspan ratio provided (3/backspan ratio).

f. See Section R301.2.2.2.5, Item 1, for additional limitations on cantilevered floor joists for detached one- and two-family dwellings in Seismic Design Category D_0, D_1, or D_2 and townhouses in Seismic Design Category C, D_0, D_1 or D_2.

g. A full-depth rim joist shall be provided at the unsupported end of the cantilever joists. Solid blocking shall be provided at the supported end.

h. Linear interpolation shall be permitted for building widths and ground snow loads other than shown.

TABLE R502.3.3(2)
CANTILEVER SPANS FOR FLOOR JOISTS SUPPORTING EXTERIOR BALCONY[a, b, e, f]

Member Size	Spacing	Maximum Cantilever Span (Uplift Force at Backspan Support in lb)[c, d]		
		Ground Snow Load		
		≤ 30 psf	50 psf	70 psf
2 × 8	12″	42″ (139)	39″ (156)	34″ (165)
2 × 8	16″	36″ (151)	34″ (171)	29″ (180)
2 × 10	12″	61″ (164)	57″ (189)	49″ (201)
2 × 10	16″	53″ (180)	49″ (208)	42″ (220)
2 × 10	24″	43″ (212)	40″ (241)	34″ (255)
2 × 12	16″	72″ (228)	67″ (260)	57″ (268)
2 × 12	24″	58″ (279)	54″ (319)	47″ (330)

For SI: 1 inch = 25.4 mm, 1 pound per square foot = 0.0479 kPa.

a. Spans are based on No. 2 Grade lumber of Douglas fir-larch, hem-fir, southern pine and spruce-pine-fir for repetitive (three or more) members.

b. Ratio of backspan to cantilever span shall be at least 2:1.

c. Connections capable of resisting the indicated uplift force shall be provided at the backspan support.

d. Uplift force is for a backspan to cantilever span ratio of 2:1. Tabulated uplift values are permitted to be reduced by multiplying by a factor equal to 2 divided by the actual backspan ratio provided (2/backspan ratio).

e. A full-depth rim joist shall be provided at the unsupported end of the cantilever joists. Solid blocking shall be provided at the supported end.

f. Linear interpolation shall be permitted for ground snow loads other than shown.

R502.4 Joists under bearing partitions. Joists under parallel bearing partitions shall be of adequate size to support the load. Double joists, sized to adequately support the load, that are separated to permit the installation of piping or vents shall be full depth solid blocked with lumber not less than 2 inches (51 mm) in nominal thickness spaced not more than 4 feet (1219 mm) on center. Bearing partitions perpendicular to joists shall not be offset from supporting girders, walls or partitions more than the joist depth unless such joists are of sufficient size to carry the additional load.

R502.5 Allowable girder spans. The allowable spans of girders fabricated of dimension lumber shall not exceed the values set forth in Tables R502.5(1) and R502.5(2).

R502.6 Bearing. The ends of each joist, beam or girder shall have not less than 1.5 inches (38 mm) of bearing on wood or metal and not less than 3 inches (76 mm) on masonry or concrete except where supported on a 1-inch by 4-inch (25.4 mm by 102 mm) ribbon strip and nailed to the adjacent stud or by the use of approved joist hangers. The bearing on masonry or concrete shall be direct, or a sill plate of 2-inch-minimum (51 mm) nominal thickness shall be provided under the joist, beam or girder. The sill plate shall provide a minimum nominal bearing area of 48 square inches (30 865 square mm).

R502.6.1 Floor systems. Joists framing from opposite sides over a bearing support shall lap a minimum of 3 inches (76 mm) and shall be nailed together with a minimum three 10d face nails. A wood or metal splice with strength equal to or greater than that provided by the nailed lap is permitted.

R502.6.2 Joist framing. Joists framing into the side of a wood girder shall be supported by *approved* framing anchors or on ledger strips not less than nominal 2 inches by 2 inches (51 mm by 51 mm).

R502.7 Lateral restraint at supports. Joists shall be supported laterally at the ends by full-depth solid blocking not less than 2 inches (51 mm) nominal in thickness; or by attachment to a full-depth header, band or rim joist, or to an adjoining stud or shall be otherwise provided with lateral support to prevent rotation.

Exceptions:

1. Trusses, structural composite lumber, structural glued-laminated members and I-joists shall be supported laterally as required by the manufacturer's recommendations.
2. In Seismic Design Categories D_0, D_1 and D_2, lateral restraint shall also be provided at each intermediate support.

R502.7.1 Bridging. Joists exceeding a nominal 2 inches by 12 inches (51 mm by 305 mm) shall be supported laterally by solid blocking, diagonal bridging (wood or metal), or a continuous 1 inch by 3 inch (25.4 mm by 76 mm) strip nailed across the bottom of joists perpendicular to joists at intervals not exceeding 8 feet (2438 mm).

Exception: Trusses, structural composite lumber, structural glued-laminated members and I-joists shall be supported laterally as required by the manufacturer's recommendations.

R502.8 Cutting, drilling and notching. Structural floor members shall not be cut, bored or notched in excess of the limitations specified in this section. See Figure R502.8.

R502.8.1 Sawn lumber. Notches in solid lumber joists, rafters and beams shall not exceed one-sixth of the depth of the member, shall not be longer than one-third of the depth of the member and shall not be located in the middle one-third of the span. Notches at the ends of the member shall not exceed one-fourth the depth of the member. The tension side of members 4 inches (102 mm) or greater in nominal thickness shall not be notched except at the ends of the members. The diameter of holes bored or cut into members shall not exceed one-third the depth of the member. Holes shall not be closer than 2 inches (51 mm) to the top or bottom of the member, or to any other hole located in the member. Where the member is also notched, the hole shall not be closer than 2 inches (51 mm) to the notch.

R502.8.2 Engineered wood products. Cuts, notches and holes bored in trusses, structural composite lumber, structural glue-laminated members or I-joists are prohibited except where permitted by the manufacturer's recommendations or where the effects of such alterations are specifically considered in the design of the member by a *registered design professional*.

R502.9 Fastening. Floor framing shall be nailed in accordance with Table R602.3(1). Where posts and beam or girder construction is used to support floor framing, positive connections shall be provided to ensure against uplift and lateral displacement.

R502.10 Framing of openings. Openings in floor framing shall be framed with a header and trimmer joists. When the header joist span does not exceed 4 feet (1219 mm), the header joist may be a single member the same size as the floor joist. Single trimmer joists may be used to carry a single header joist that is located within 3 feet (914 mm) of the trimmer joist bearing. When the header joist span exceeds 4 feet (1219 mm), the trimmer joists and the header joist shall be doubled and of sufficient cross section to support the floor joists framing into the header. *Approved* hangers shall be used for the header joist to trimmer joist connections when the header joist span exceeds 6 feet (1829 mm). Tail joists over 12 feet (3658 mm) long shall be supported at the header by framing anchors or on ledger strips not less than 2 inches by 2 inches (51 mm by 51 mm).

R502.11 Wood trusses.

R502.11.1 Design. Wood trusses shall be designed in accordance with *approved* engineering practice. The design and manufacture of metal plate connected wood trusses shall comply with ANSI/TPI 1. The truss design drawings shall be prepared by a registered professional where required by the statutes of the *jurisdiction* in which the project is to be constructed in accordance with Section R106.1.

TABLE R502.5(1)
GIRDER SPANS[a] AND HEADER SPANS[a] FOR EXTERIOR BEARING WALLS
(Maximum spans for Douglas fir-larch, hem-fir, southern pine and spruce-pine-fir[b] and required number of jack studs)

GIRDERS AND HEADERS SUPPORTING	SIZE	GROUND SNOW LOAD (psf)[e]																	
		30						50						70					
		Building width[c] (feet)																	
		20		28		36		20		28		36		20		28		36	
		Span	NJ[d]	Span	NJ[d]	Span	NJ[d]	Span	NJ[d]	Span	NJ[d]	Span	NJ[d]	Span	NJ[d]	Span	NJ[d]	Span	NJ[d]
Roof and ceiling	2-2 × 4	3-6	1	3-2	1	2-10	1	3-2	1	2-9	1	2-6	1	2-10	1	2-6	1	2-3	1
	2-2 × 6	5-5	1	4-8	1	4-2	1	4-8	1	4-1	1	3-8	2	4-2	1	3-8	2	3-3	2
	2-2 × 8	6-10	1	5-11	2	5-4	2	5-11	2	5-2	2	4-7	2	5-4	2	4-7	2	4-1	2
	2-2 × 10	8-5	2	7-3	2	6-6	2	7-3	2	6-3	2	5-7	2	6-6	2	5-7	2	5-0	2
	2-2 × 12	9-9	2	8-5	2	7-6	2	8-5	2	7-3	2	6-6	2	7-6	2	6-6	2	5-10	3
	3-2 × 8	8-4	1	7-5	1	6-8	1	7-5	1	6-5	2	5-9	2	6-8	1	5-9	2	5-2	2
	3-2 × 10	10-6	1	9-1	2	8-2	2	9-1	2	7-10	2	7-0	2	8-2	2	7-0	2	6-4	2
	3-2 × 12	12-2	2	10-7	2	9-5	2	10-7	2	9-2	2	8-2	2	9-5	2	8-2	2	7-4	2
	4-2 × 8	9-2	1	8-4	1	7-8	1	8-4	1	7-5	1	6-8	1	7-8	1	6-8	1	5-11	2
	4-2 × 10	11-8	1	10-6	1	9-5	2	10-6	1	9-1	2	8-2	2	9-5	2	8-2	2	7-3	2
	4-2 × 12	14-1	1	12-2	2	10-11	2	12-2	2	10-7	2	9-5	2	10-11	2	9-5	2	8-5	2
Roof, ceiling and one center-bearing floor	2-2 × 4	3-1	1	2-9	1	2-5	1	2-9	1	2-5	1	2-2	1	2-7	1	2-3	1	2-0	1
	2-2 × 6	4-6	1	4-0	1	3-7	2	4-1	1	3-7	2	3-3	2	3-9	2	3-3	2	2-11	2
	2-2 × 8	5-9	2	5-0	2	4-6	2	5-2	2	4-6	2	4-1	2	4-9	2	4-2	2	3-9	2
	2-2 × 10	7-0	2	6-2	2	5-6	2	6-4	2	5-6	2	5-0	2	5-9	2	5-1	2	4-7	3
	2-2 × 12	8-1	2	7-1	2	6-5	2	7-4	2	6-5	2	5-9	3	6-8	2	5-10	3	5-3	3
	3-2 × 8	7-2	1	6-3	2	5-8	2	6-5	2	5-8	2	5-1	2	5-11	2	5-2	2	4-8	2
	3-2 × 10	8-9	2	7-8	2	6-11	2	7-11	2	6-11	2	6-3	2	7-3	2	6-4	2	5-8	2
	3-2 × 12	10-2	2	8-11	2	8-0	2	9-2	2	8-0	2	7-3	2	8-5	2	7-4	2	6-7	2
	4-2 × 8	8-1	1	7-3	1	6-7	1	7-5	1	6-6	1	5-11	2	6-10	1	6-0	2	5-5	2
	4-2 × 10	10-1	1	8-10	2	8-0	2	9-1	2	8-0	2	7-2	2	8-4	2	7-4	2	6-7	2
	4-2 × 12	11-9	2	10-3	2	9-3	2	10-7	2	9-3	2	8-4	2	9-8	2	8-6	2	7-7	2
Roof, ceiling and one clear span floor	2-2 × 4	2-8	1	2-4	1	2-1	1	2-7	1	2-3	1	2-0	1	2-5	1	2-1	1	1-10	1
	2-2 × 6	3-11	1	3-5	2	3-0	2	3-10	2	3-4	2	3-0	2	3-6	2	3-1	2	2-9	2
	2-2 × 8	5-0	2	4-4	2	3-10	2	4-10	2	4-2	2	3-9	2	4-6	2	3-11	2	3-6	2
	2-2 × 10	6-1	2	5-3	2	4-8	2	5-11	2	5-1	2	4-7	3	5-6	2	4-9	2	4-3	3
	2-2 × 12	7-1	2	6-1	3	5-5	3	6-10	2	5-11	3	5-4	3	6-4	2	5-6	3	5-0	3
	3-2 × 8	6-3	2	5-5	2	4-10	2	6-1	2	5-3	2	4-8	2	5-7	2	4-11	2	4-5	2
	3-2 × 10	7-7	2	6-7	2	5-11	2	7-5	2	6-5	2	5-9	2	6-10	2	6-0	2	5-4	2
	3-2 × 12	8-10	2	7-8	2	6-10	2	8-7	2	7-5	2	6-8	2	7-11	2	6-11	2	6-3	2
	4-2 × 8	7-2	1	6-3	2	5-7	2	7-0	1	6-1	2	5-5	2	6-6	1	5-8	2	5-1	2
	4-2 × 10	8-9	2	7-7	2	6-10	2	8-7	2	7-5	2	6-7	2	7-11	2	6-11	2	6-2	2
	4-2 × 12	10-2	2	8-10	2	7-11	2	9-11	2	8-7	2	7-8	2	9-2	2	8-0	2	7-2	2
Roof, ceiling and two center-bearing floors	2-2 × 4	2-7	1	2-3	1	2-0	1	2-6	1	2-2	1	1-11	1	2-4	1	2-0	1	1-9	1
	2-2 × 6	3-9	2	3-3	2	2-11	2	3-8	2	3-2	2	2-10	2	3-5	2	3-0	2	2-8	2
	2-2 × 8	4-9	2	4-2	2	3-9	2	4-7	2	4-0	2	3-8	2	4-4	2	3-9	2	3-5	2
	2-2 × 10	5-9	2	5-1	2	4-7	3	5-8	2	4-11	2	4-5	3	5-3	2	4-7	3	4-2	3
	2-2 × 12	6-8	2	5-10	3	5-3	3	6-6	2	5-9	3	5-2	3	6-1	3	5-4	3	4-10	3
	3-2 × 8	5-11	2	5-2	2	4-8	2	5-9	2	5-1	2	4-7	2	5-5	2	4-9	2	4-3	2
	3-2 × 10	7-3	2	6-4	2	5-8	2	7-1	2	6-2	2	5-7	2	6-7	2	5-9	2	5-3	2
	3-2 × 12	8-5	2	7-4	2	6-7	2	8-2	2	7-2	2	6-5	3	7-8	2	6-9	2	6-1	3
	4-2 × 8	6-10	1	6-0	2	5-5	2	6-8	1	5-10	2	5-3	2	6-3	2	5-6	2	4-11	2
	4-2 × 10	8-4	2	7-4	2	6-7	2	8-2	2	7-2	2	6-5	2	7-7	2	6-8	2	6-0	2
	4-2 × 12	9-8	2	8-6	2	7-8	2	9-5	2	8-3	2	7-5	2	8-10	2	7-9	2	7-0	2
Roof, ceiling, and two clear span floors	2-2 × 4	2-1	1	1-8	1	1-6	2	2-0	1	1-8	1	1-5	2	2-0	1	1-8	1	1-5	2
	2-2 × 6	3-1	2	2-8	2	2-4	2	3-0	2	2-7	2	2-3	2	2-11	2	2-7	2	2-3	2
	2-2 × 8	3-10	2	3-4	2	3-0	3	3-10	2	3-4	2	2-11	3	3-9	2	3-3	2	2-11	3

(continued)

TABLE R502.5(1)—continued
GIRDER SPANS[a] AND HEADER SPANS[a] FOR EXTERIOR BEARING WALLS
(Maximum spans for Douglas fir-larch, hem-fir, southern pine and spruce-pine-fir[b] and required number of jack studs)

GIRDERS AND HEADERS SUPPORTING	SIZE	GROUND SNOW LOAD (psf)[e]																	
		30						50						70					
		Building width[c] (feet)																	
		20		28		36		20		28		36		20		28		36	
		Span	NJ[d]	Span	NJ[d]	Span	NJ[d]	Span	NJ[d]	Span	NJ[d]	Span	NJ[d]	Span	NJ[d]	Span	NJ[d]	Span	NJ[d]
Roof, ceiling, and two clear span floors	2-2 × 10	4-9	2	4-1	3	3-8	3	4-8	2	4-0	3	3-7	3	4-7	3	4-0	3	3-6	3
	2-2 × 12	5-6	3	4-9	3	4-3	3	5-5	3	4-8	3	4-2	3	5-4	3	4-7	3	4-1	4
	3-2 × 8	4-10	2	4-2	2	3-9	2	4-9	2	4-1	2	3-8	2	4-8	2	4-1	2	3-8	2
	3-2 × 10	5-11	2	5-1	2	4-7	3	5-10	2	5-0	2	4-6	3	5-9	2	4-11	2	4-5	3
	3-2 × 12	6-10	2	5-11	3	5-4	3	6-9	2	5-10	3	5-3	3	6-8	2	5-9	3	5-2	3
	4-2 × 8	5-7	2	4-10	2	4-4	2	5-6	2	4-9	2	4-3	2	5-5	2	4-8	2	4-2	2
	4-2 × 10	6-10	2	5-11	2	5-3	2	6-9	2	5-10	2	5-2	2	6-7	2	5-9	2	5-1	2
	4-2 × 12	7-11	2	6-10	2	6-2	3	7-9	2	6-9	2	6-0	3	7-8	2	6-8	2	5-11	3

For SI: 1 inch = 25.4 mm, 1 pound per square foot = 0.0479 kPa.

a. Spans are given in feet and inches.

b. Tabulated values assume #2 grade lumber.

c. Building width is measured perpendicular to the ridge. For widths between those shown, spans are permitted to be interpolated.

d. NJ - Number of jack studs required to support each end. Where the number of required jack studs equals one, the header is permitted to be supported by an approved framing anchor attached to the full-height wall stud and to the header.

e. Use 30 psf ground snow load for cases in which ground snow load is less than 30 psf and the roof live load is equal to or less than 20 psf.

TABLE R502.5(2)
GIRDER SPANS[a] AND HEADER SPANS[a] FOR INTERIOR BEARING WALLS
(Maximum spans for Douglas fir-larch, hem-fir, southern pine and spruce-pine-fir[b] and required number of jack studs)

HEADERS AND GIRDERS SUPPORTING	SIZE	BUILDING Width[c] (feet)					
		20		28		36	
		Span	NJ[d]	Span	NJ[d]	Span	NJ[d]
One floor only	2-2 × 4	3-1	1	2-8	1	2-5	1
	2-2 × 6	4-6	1	3-11	1	3-6	1
	2-2 × 8	5-9	1	5-0	2	4-5	2
	2-2 × 10	7-0	2	6-1	2	5-5	2
	2-2 × 12	8-1	2	7-0	2	6-3	2
	3-2 × 8	7-2	1	6-3	1	5-7	2
	3-2 × 10	8-9	1	7-7	2	6-9	2
	3-2 × 12	10-2	2	8-10	2	7-10	2
	4-2 × 8	9-0	1	7-8	1	6-9	1
	4-2 × 10	10-1	1	8-9	1	7-10	2
	4-2 × 12	11-9	1	10-2	2	9-1	2
Two floors	2-2 × 4	2-2	1	1-10	1	1-7	1
	2-2 × 6	3-2	2	2-9	2	2-5	2
	2-2 × 8	4-1	2	3-6	2	3-2	2
	2-2 × 10	4-11	2	4-3	2	3-10	3
	2-2 × 12	5-9	2	5-0	3	4-5	3
	3-2 × 8	5-1	2	4-5	2	3-11	2
	3-2 × 10	6-2	2	5-4	2	4-10	2
	3-2 × 12	7-2	2	6-3	2	5-7	3
	4-2 × 8	6-1	1	5-3	2	4-8	2
	4-2 × 10	7-2	2	6-2	2	5-6	2
	4-2 × 12	8-4	2	7-2	2	6-5	2

For SI: 1 inch = 25.4 mm, 1 foot = 304.8 mm.

a. Spans are given in feet and inches.

b. Tabulated values assume #2 grade lumber.

c. Building width is measured perpendicular to the ridge. For widths between those shown, spans are permitted to be interpolated.

d. NJ - Number of jack studs required to support each end. Where the number of required jack studs equals one, the header is permitted to be supported by an approved framing anchor attached to the full-height wall stud and to the header.

R502.11.2 Bracing. Trusses shall be braced to prevent rotation and provide lateral stability in accordance with the requirements specified in the *construction documents* for the building and on the individual truss design drawings. In the absence of specific bracing requirements, trusses shall be braced in accordance with accepted industry practices, such as, the SBCA *Building Component Safety Information (BCSI) Guide to Good Practice for Handling, Installing & Bracing of Metal Plate Connected Wood Trusses.*

R502.11.3 Alterations to trusses. Truss members and components shall not be cut, notched, spliced or otherwise altered in any way without the approval of a registered *design professional. Alterations* resulting in the addition of load (e.g., HVAC *equipment*, water heater, etc.), that exceed the design load for the truss, shall not be permitted without verification that the truss is capable of supporting the additional loading.

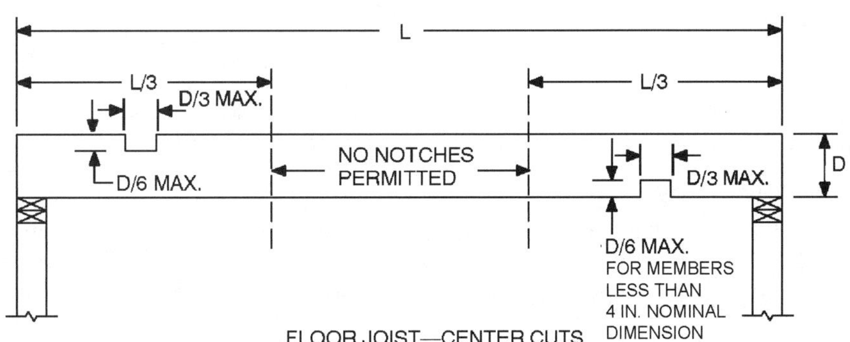

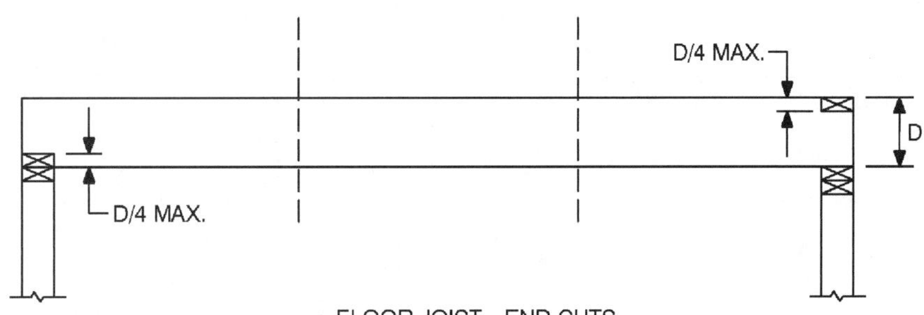

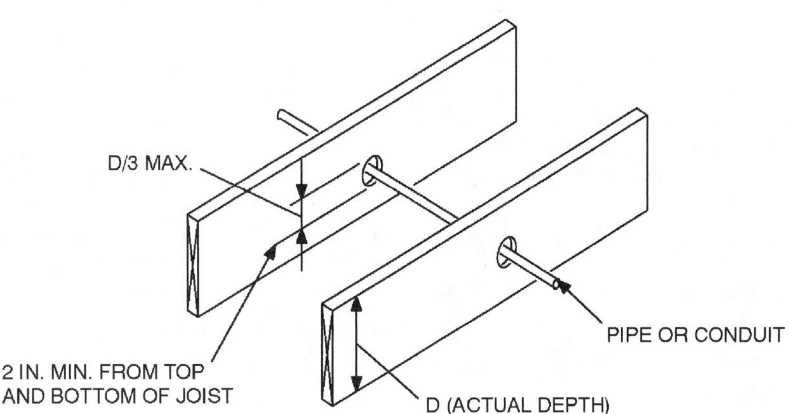

For SI: 1 inch = 25.4 mm.

FIGURE R502.8
CUTTING, NOTCHING AND DRILLING

R502.11.4 Truss design drawings. Truss design drawings, prepared in compliance with Section R502.11.1, shall be submitted to the *building official* and *approved* prior to installation. Truss design drawings shall be provided with the shipment of trusses delivered to the job site. Truss design drawings shall include, at a minimum, the information specified below:

1. Slope or depth, span and spacing.
2. Location of all joints.
3. Required bearing widths.
4. Design loads as applicable:
   4.1. Top chord live load;
   4.2. Top chord dead load;
   4.3. Bottom chord live load;
   4.4. Bottom chord dead load;
   4.5. Concentrated loads and their points of application; and
   4.6. Controlling wind and earthquake loads.
5. Adjustments to lumber and joint connector design values for conditions of use.
6. Each reaction force and direction.
7. Joint connector type and description, e.g., size, thickness or gauge, and the dimensioned location of each joint connector except where symmetrically located relative to the joint interface.
8. Lumber size, species and grade for each member.
9. Connection requirements for:
   9.1. Truss-to-girder-truss;
   9.2. Truss ply-to-ply; and
   9.3. Field splices.
10. Calculated deflection ratio and/or maximum description for live and total load.
11. Maximum axial compression forces in the truss members to enable the building designer to design the size, connections and anchorage of the permanent continuous lateral bracing. Forces shall be shown on the truss drawing or on supplemental documents.
12. Required permanent truss member bracing location.

R502.12 Draftstopping required. Draftstopping shall be provided in accordance with Section R302.12.

R502.13 Fireblocking required. Fireblocking shall be provided in accordance with Section R302.11.

SECTION R503
FLOOR SHEATHING

R503.1 Lumber sheathing. Maximum allowable spans for lumber used as floor sheathing shall conform to Tables R503.1, R503.2.1.1(1) and R503.2.1.1(2).

TABLE R503.1
MINIMUM THICKNESS OF LUMBER FLOOR SHEATHING

JOIST OR BEAM SPACING (inches)	MINIMUM NET THICKNESS	
	Perpendicular to joist	Diagonal to joist
24	$^{11}/_{16}$	$^{3}/_{4}$
16	$^{5}/_{8}$	$^{5}/_{8}$
48[a]	$1^{1}/_{2}$ T & G	N/A
54[b]		
60[c]		

For SI: 1 inch = 25.4 mm, 1 pound per square inch = 6.895 kPa.

N/A = Not applicable.

a. For this support spacing, lumber sheathing shall have a minimum F_b of 675 and minimum E of 1,100,000 (see AF&PA/NDS).

b. For this support spacing, lumber sheathing shall have a minimum F_b of 765 and minimum E of 1,400,000 (see AF&PA/NDS).

c. For this support spacing, lumber sheathing shall have a minimum F_b of 855 and minimum E of 1,700,000 (see AF&PA/NDS).

R503.1.1 End joints. End joints in lumber used as subflooring shall occur over supports unless end-matched lumber is used, in which case each piece shall bear on at least two joists. Subflooring may be omitted when joist spacing does not exceed 16 inches (406 mm) and a 1-inch (25.4 mm) nominal tongue-and-groove wood strip flooring is applied perpendicular to the joists.

R503.2 Wood structural panel sheathing.

R503.2.1 Identification and grade. Wood structural panel sheathing used for structural purposes shall conform to DOC PS 1, DOC PS 2 or, when manufactured in Canada, CSA 0437 or CSA 0325. All panels shall be identified for grade, bond classification, and Performance Category by a grade *mark* or certificate of inspection issued by an *approved agency*. The Performance Category value shall be used as the "nominal panel thickness" or "panel thickness" whenever referenced in this code.

R503.2.1.1 Subfloor and combined subfloor underlayment. Where used as subflooring or combination subfloor underlayment, wood structural panels shall be of one of the grades specified in Table R503.2.1.1(1). When sanded plywood is used as combination subfloor underlayment, the grade, bond classification, and Performance Category shall be as specified in Table R503.2.1.1(2).

TABLE R503.2.1.1(2)
ALLOWABLE SPANS FOR SANDED PLYWOOD COMBINATION SUBFLOOR UNDERLAYMENT[a]

IDENTIFICATION	SPACING OF JOISTS (inches)		
	16	20	24
Species group[b]	—	—	—
1	$^{1}/_{2}$	$^{5}/_{8}$	$^{3}/_{4}$
2, 3	$^{5}/_{8}$	$^{3}/_{4}$	$^{7}/_{8}$
4	$^{3}/_{4}$	$^{7}/_{8}$	1

For SI: 1 inch = 25.4 mm, 1 pound per square foot = 0.0479 kPa.

a. Plywood continuous over two or more spans and face grain perpendicular to supports. Unsupported edges shall be tongue-and-groove or blocked except where nominal $^{1}/_{4}$-inch-thick underlayment or $^{3}/_{4}$-inch wood finish floor is used. Allowable uniform live load at maximum span based on deflection of $^{1}/_{360}$ of span is 100 psf.

b. Applicable to all grades of sanded exterior-type plywood.

R503.2.2 Allowable spans. The maximum allowable span for wood structural panels used as subfloor or combination subfloor underlayment shall be as set forth in Table R503.2.1.1(1), or APA E30. The maximum span for sanded plywood combination subfloor underlayment shall be as set forth in Table R503.2.1.1(2).

R503.2.3 Installation. Wood structural panels used as subfloor or combination subfloor underlayment shall be attached to wood framing in accordance with Table R602.3(1) and shall be attached to cold-formed steel framing in accordance with Table R505.3.1(2).

R503.3 Particleboard.

R503.3.1 Identification and grade. Particleboard shall conform to ANSI A208.1 and shall be so identified by a grade *mark* or certificate of inspection issued by an *approved agency*.

R503.3.2 Floor underlayment. Particleboard floor underlayment shall conform to Type PBU and shall not be less than $^1/_4$ inch (6.4 mm) in thickness.

R503.3.3 Installation. Particleboard underlayment shall be installed in accordance with the recommendations of the manufacturer and attached to framing in accordance with Table R602.3(1).

SECTION R504 PRESSURE PRESERVATIVELY TREATED-WOOD FLOORS (ON GROUND)

R504.1 General. Pressure preservatively treated-wood *basement* floors and floors on ground shall be designed to withstand axial forces and bending moments resulting from lateral soil pressures at the base of the exterior walls and floor live

TABLE R503.2.1.1(1) ALLOWABLE SPANS AND LOADS FOR WOOD STRUCTURAL PANELS FOR ROOF AND SUBFLOOR SHEATHING AND COMBINATION SUBFLOOR UNDERLAYMENT[a, b, c]

SPAN RATING	MINIMUM NOMINAL PANEL THICKNESS (inch)	ALLOWABLE LIVE LOAD (psf)[h, l]		MAXIMUM SPAN (inches)		LOAD (pounds per square foot, at maximum span)		MAXIMUM SPAN (inches)
		SPAN @ 16″ o.c.	SPAN @ 24″ o.c.	With edge support[d]	Without edge support	Total load	Live load	
Sheathing[e]				Roof[f]				Subfloor[j]
16/0	$^3/_8$	30	—	16	16	40	30	0
20/0	$^3/_8$	50	—	20	20	40	30	0
24/0	$^3/_8$	100	30	24	20[g]	40	30	0
24/16	$^7/_{16}$	100	40	24	24	50	40	16
32/16	$^{15}/_{32}$, $^1/_2$	180	70	32	28	40	30	16[h]
40/20	$^{19}/_{32}$, $^5/_8$	305	130	40	32	40	30	20[h, i]
48/24	$^{23}/_{32}$, $^3/_4$	—	175	48	36	45	35	24
60/32	$^7/_8$	—	305	60	48	45	35	32
Underlayment, C-C plugged, single floor[e]				Roof[f]				Combination subfloor underlayment[k]
16 o.c.	$^{19}/_{32}$, $^5/_8$	100	40	24	24	50	40	16[i]
20 o.c.	$^{19}/_{32}$, $^5/_8$	150	60	32	32	40	30	20[i, j]
24 o.c.	$^{23}/_{32}$, $^3/_4$	240	100	48	36	35	25	24
32 o.c.	$^7/_8$	—	185	48	40	50	40	32
48 o.c.	$1^3/_{32}$, $1^1/_8$	—	290	60	48	50	40	48

For SI: 1 inch = 25.4 mm, 1 pound per square foot = 0.0479 kPa.

a. The allowable total loads were determined using a dead load of 10 psf. If the dead load exceeds 10 psf, then the live load shall be reduced accordingly.

b. Panels continuous over two or more spans with long dimension (strength axis) perpendicular to supports. Spans shall be limited to values shown because of possible effect of concentrated loads.

c. Applies to panels 24 inches or wider.

d. Lumber blocking, panel edge clips (one midway between each support, except two equally spaced between supports when span is 48 inches), tongue-and-groove panel edges, or other approved type of edge support.

e. Includes Structural 1 panels in these grades.

f. Uniform load deflection limitation: $^1/_{180}$ of span under live load plus dead load, $^1/_{240}$ of span under live load only.

g. Maximum span 24 inches for $^{15}/_{32}$-and $^1/_2$-inch panels.

h. Maximum span 24 inches where $^3/_4$-inch wood finish flooring is installed at right angles to joists.

i. Maximum span 24 inches where 1.5 inches of lightweight concrete or approved cellular concrete is placed over the subfloor.

j. Unsupported edges shall have tongue-and-groove joints or shall be supported with blocking unless minimum nominal $^1/_4$-inch thick underlayment with end and edge joints offset at least 2 inches or 1.5 inches of lightweight concrete or approved cellular concrete is placed over the subfloor, or $^3/_4$-inch wood finish flooring is installed at right angles to the supports. Allowable uniform live load at maximum span, based on deflection of $^1/_{360}$ of span, is 100 psf.

k. Unsupported edges shall have tongue-and-groove joints or shall be supported by blocking unless nominal $^1/_4$-inch-thick underlayment with end and edge joints offset at least 2 inches or $^3/_4$-inch wood finish flooring is installed at right angles to the supports. Allowable uniform live load at maximum span, based on deflection of $^1/_{360}$ of span, is 100 psf, except panels with a span rating of 48 on center are limited to 65 psf total uniform load at maximum span.

l. Allowable live load values at spans of 16" o.c. and 24" o.c taken from reference standard APA E30, APA Engineered Wood Construction Guide. Refer to reference standard for allowable spans not listed in the table.

and dead loads. Floor framing shall be designed to meet joist deflection requirements in accordance with Section R301.

R504.1.1 Unbalanced soil loads. Unless special provision is made to resist sliding caused by unbalanced lateral soil loads, wood *basement* floors shall be limited to applications where the differential depth of fill on opposite exterior foundation walls is 2 feet (610 mm) or less.

R504.1.2 Construction. Joists in wood *basement* floors shall bear tightly against the narrow face of studs in the foundation wall or directly against a band joist that bears on the studs. Plywood subfloor shall be continuous over lapped joists or over butt joints between in-line joists. Sufficient blocking shall be provided between joists to transfer lateral forces at the base of the end walls into the floor system.

R504.1.3 Uplift and buckling. Where required, resistance to uplift or restraint against buckling shall be provided by interior bearing walls or properly designed stub walls anchored in the supporting soil below.

R504.2 Site preparation. The area within the foundation walls shall have all vegetation, topsoil and foreign material removed, and any fill material that is added shall be free of vegetation and foreign material. The fill shall be compacted to assure uniform support of the pressure preservatively treated-wood floor sleepers.

R504.2.1 Base. A minimum 4-inch-thick (102 mm) granular base of gravel having a maximum size of $^3/_4$ inch (19.1 mm) or crushed stone having a maximum size of $^1/_2$ inch (12.7 mm) shall be placed over the compacted earth.

R504.2.2 Moisture barrier. Polyethylene sheeting of minimum 6-mil (0.15 mm) thickness shall be placed over the granular base. Joints shall be lapped 6 inches (152 mm) and left unsealed. The polyethylene membrane shall be placed over the pressure preservatively treated-wood sleepers and shall not extend beneath the footing plates of the exterior walls.

R504.3 Materials. All framing materials, including sleepers, joists, blocking and plywood subflooring, shall be pressure-preservative treated and dried after treatment in accordance with AWPA U1 (Commodity Specification A, Use Category 4B and Section 5.2), and shall bear the *label* of an accredited agency.

SECTION R505
STEEL FLOOR FRAMING

R505.1 Cold-formed steel floor framing. Elements shall be straight and free of any defects that would significantly affect structural performance. Cold-formed steel floor framing members shall comply with the requirements of this section.

R505.1.1 Applicability limits. The provisions of this section shall control the construction of cold-formed steel floor framing for buildings not greater than 60 feet (18 288 mm) in length perpendicular to the joist span, not greater than 40 feet (12 192 mm) in width parallel to the joist span, and less than or equal to three stories above *grade* plane. Cold-formed steel floor framing constructed in accordance with the provisions of this section shall be limited to sites subjected to a maximum design wind speed of 110 miles per hour (49 m/s), Exposure B or C, and a maximum ground snow load of 70 pounds per square foot (3.35 kPa).

R505.1.2 In-line framing. When supported by cold-formed steel framed walls in accordance with Section R603, cold-formed steel floor framing shall be constructed with floor joists located in-line with load-bearing studs located below the joists in accordance with Figure R505.1.2 and the tolerances specified as follows:

1. The maximum tolerance shall be $^3/_4$ inch (19.1 mm) between the centerline of the horizontal framing member and the centerline of the vertical framing member.
2. Where the centerline of the horizontal framing member and bearing stiffener are located to one side of the centerline of the vertical framing member, the maximum tolerance shall be $^1/_8$ inch (3 mm) between the web of the horizontal framing member and the edge of the vertical framing member.

R505.1.3 Floor trusses. Cold-formed steel trusses shall be designed, braced and installed in accordance with AISI S100, Section D4. In the absence of specific bracing requirements, trusses shall be braced in accordance with accepted industry practices, such as the SBCA *Cold-Formed Steel Building Component Safety Information (CFSBCSI), Guide to Good Practice for Handling, Installing & Bracing of Cold-Formed Steel Trusses*. Truss members shall not be notched, cut or altered in any manner without an *approved* design.

TABLE R505.2(1)
COLD-FORMED STEEL JOIST SIZES

MEMBER DESIGNATION[a]	WEB DEPTH (inches)	MINIMUM FLANGE WIDTH (inches)	MAXIMUM FLANGE WIDTH (inches)	MINIMUM LIP SIZE (inches)
550S162-t	5.5	1.625	2	0.5
800S162-t	8	1.625	2	0.5
1000S162-t	10	1.625	2	0.5
1200S162-t	12	1.625	2	0.5

For SI: 1 inch = 25.4 mm, 1 mil = 0.0254 mm.

a. The member designation is defined by the first number representing the member depth in 0.01 inch, the letter "S" representing a stud or joist member, the second number representing the flange width in 0.01 inch, and the letter "t" shall be a number representing the minimum base metal thickness in mils [See Table R505.2(2)].

R505.2 Structural framing. Load-bearing cold-formed steel floor framing members shall comply with Figure R505.2(1) and with the dimensional and minimum thickness requirements specified in Tables R505.2(1) and R505.2(2). Tracks shall comply with Figure R505.2(2) and shall have a minimum flange width of $1^1/_4$ inches (32 mm).

TABLE R505.2(2)
MINIMUM THICKNESS OF COLD-FORMED STEEL MEMBERS

DESIGNATION THICKNESS (mils)	MINIMUM BASE STEEL THICKNESS (inches)
33	0.0329
43	0.0428
54	0.0538
68	0.0677
97	0.0966

For SI: inch = 25.4 mm, 1 mil = 0.0254 mm.

R505.2.1 Material. Load-bearing cold-formed steel framing members shall be cold formed to shape from structural quality sheet steel complying with the requirements of one of the following:

1. ASTM A 653: Grades 33 and 50 (Class 1 and 3).
2. ASTM A 792: Grades 33 and 50A.
3. ASTM A 1003: Structural Grades 33 Type H and 50 Type H.

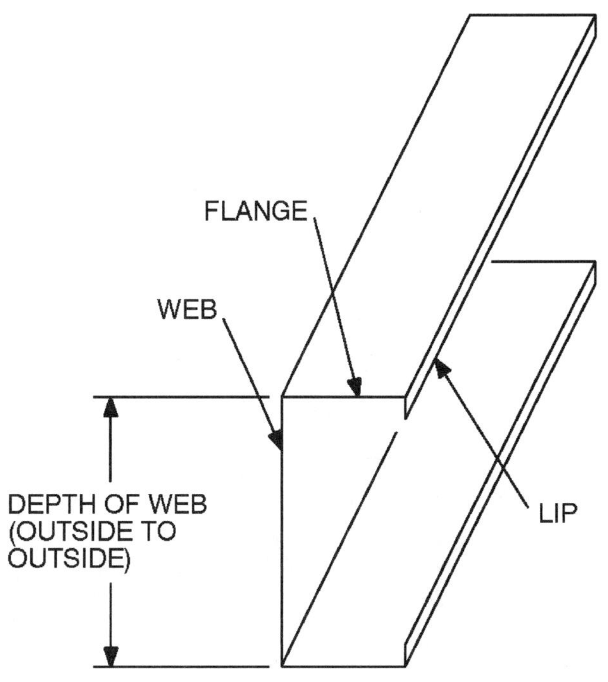

FIGURE R505.2(1)
C-SHAPED SECTION

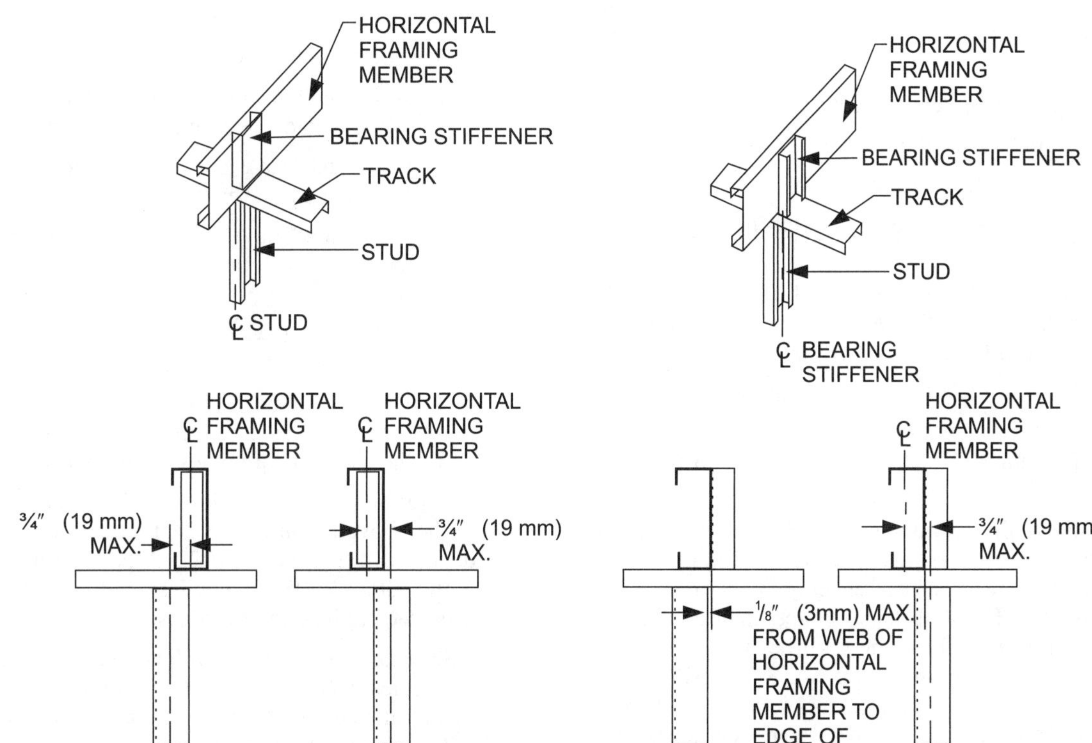

For SI: 1 inch = 25.4 mm.

FIGURE R505.1.2
IN-LINE FRAMING

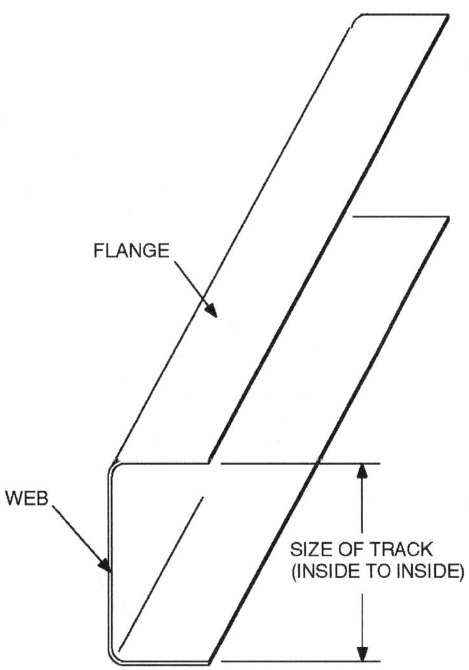

FIGURE R505.2(2)
TRACK SECTION

R505.2.2 Identification. Load-bearing cold-formed steel framing members shall have a legible *label*, stencil, stamp or embossment with the following information as a minimum:

1. Manufacturer's identification.
2. Minimum base steel thickness in inches (mm).
3. Minimum coating designation.
4. Minimum yield strength, in kips per square inch (ksi) (MPa).

R505.2.3 Corrosion protection. Load-bearing cold-formed steel framing shall have a metallic coating complying with ASTM A 1003 and one of the following:

1. A minimum of G 60 in accordance with ASTM A 653.
2. A minimum of AZ 50 in accordance with ASTM A 792.

R505.2.4 Fastening requirements. Screws for steel-to-steel connections shall be installed with a minimum edge distance and center-to-center spacing of $^1/_2$ inch (12.7 mm), shall be self-drilling tapping, and shall conform to ASTM C 1513. Floor sheathing shall be attached to cold-formed steel joists with minimum No. 8 self-drilling tapping screws that conform to ASTM C 1513. Screws attaching floor-sheathing to cold-formed steel joists shall have a minimum head diameter of 0.292 inch (7.4 mm) with countersunk heads and shall be installed with a minimum edge distance of $^3/_8$ inch (9.5 mm). Gypsum board ceilings shall be attached to cold-formed steel joists with minimum No. 6 screws conforming to ASTM C 954 or ASTM C 1513 with a bugle head style and shall be installed in accordance with Section R702. For all connections, screws shall extend through the steel a minimum of three exposed threads. All fasteners shall have rust inhibitive coating suitable for the installation in which they are being used, or be manufactured from material not susceptible to corrosion.

Where No. 8 screws are specified in a steel-to-steel connection, the required number of screws in the connection is permitted to be reduced in accordance with the reduction factors in Table R505.2.4 when larger screws are used or when one of the sheets of steel being connected is thicker than 33 mils (0.84 mm). When applying the reduction factor, the resulting number of screws shall be rounded up.

TABLE R505.2.4
SCREW SUBSTITUTION FACTOR

SCREW SIZE	THINNEST CONNECTED STEEL SHEET (mils)	
	33	43
#8	1.0	0.67
#10	0.93	0.62
#12	0.86	0.56

For SI: 1 mil = 0.0254 mm.

R505.2.5 Web holes, web hole reinforcing and web hole patching. Web holes, web hole reinforcing, and web hole patching shall be in accordance with this section.

R505.2.5.1 Web holes. Web holes in floor joists shall comply with all of the following conditions:

1. Holes shall conform to Figure R505.2.5.1;
2. Holes shall be permitted only along the centerline of the web of the framing member;
3. Holes shall have a center-to-center spacing of not less than 24 inches (610 mm);
4. Holes shall have a web hole width not greater than 0.5 times the member depth, or $2^1/_2$ inches (64.5 mm);
5. Holes shall have a web hole length not exceeding $4^1/_2$ inches (114 mm); and
6. Holes shall have a minimum distance between the edge of the bearing surface and the edge of the web hole of not less than 10 inches (254 mm).

Framing members with web holes not conforming to the above requirements shall be reinforced in accordance with Section R505.2.5.2, patched in accordance with Section R505.2.5.3 or designed in accordance with accepted engineering practices.

R505.2.5.2 Web hole reinforcing. Reinforcement of web holes in floor joists not conforming to the requirements of Section R505.2.5.1 shall be permitted if the hole is located fully within the center 40 percent of the span and the depth and length of the hole does not exceed 65 percent of the flat width of the web. The reinforcing shall be a steel plate or C-shape section with a hole that does not exceed the web hole size limitations of Section R505.2.5.1 for the member being reinforced. The steel reinforcing shall be the same thickness as the receiving member and shall extend at least 1 inch (25.4 mm) beyond all edges of the hole. The steel reinforcing shall be fastened to the web of the receiving member with No. 8 screws spaced no more than 1 inch (25.4 mm) center-to-center along the edges of the patch with minimum edge distance of $^1/_2$ inch (12.7 mm).

R505.2.5.3 Hole patching. Patching of web holes in floor joists not conforming to the requirements in Section R505.2.5.1 shall be permitted in accordance with either of the following methods:

1. Framing members shall be replaced or designed in accordance with accepted engineering practices where web holes exceed the following size limits:
   1.1. The depth of the hole, measured across the web, exceeds 70 percent of the flat width of the web; or
   1.2. The length of the hole measured along the web, exceeds 10 inches (254 mm) or the depth of the web, whichever is greater.
2. Web holes not exceeding the dimensional requirements in Section R505.2.5.3, Item 1, shall be patched with a solid steel plate, stud section, or track section in accordance with Figure R505.2.5.3. The steel patch shall, as a minimum, be of the same thickness as the receiving member and shall extend at least 1 inch (25 mm) beyond all edges of the hole. The steel patch shall be fastened to the web of the receiving member with No. 8 screws spaced no more than 1 inch (25 mm) center-to-center along the edges of the patch with minimum edge distance of $^1/_2$ inch (13 mm).

R505.3 Floor construction. Cold-formed steel floors shall be constructed in accordance with this section.

R505.3.1 Floor to foundation or load-bearing wall connections. Cold-formed steel framed floors shall be anchored to foundations, wood sills or load-bearing walls in accordance with Table R505.3.1(1) and Figure R505.3.1(1), R505.3.1(2), R505.3.1(3), R505.3.1(4), R505.3.1(5) or R505.3.1(6). Anchor bolts shall be located not more than 12 inches (305 mm) from corners or the termination of bottom tracks. Continuous cold-formed steel joists supported by interior load-bearing walls shall be constructed in accordance with Figure R505.3.1(7). Lapped cold-formed steel joists shall be constructed in accordance with Figure R505.3.1(8). End floor joists constructed on foundation walls parallel to the joist span shall be doubled unless a C-shaped bearing stiffener, sized in accordance with Section R505.3.4, is installed web-to-web with the floor joist beneath each supported wall stud, as shown in Figure R505.3.1(9). Fastening of cold-formed steel joists to other framing members shall be in accordance with Section R505.2.4 and Table R505.3.1(2).

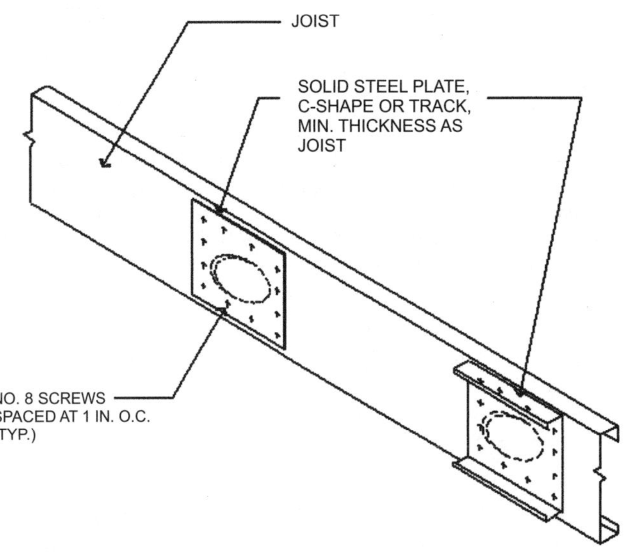

For SI: 1 inch = 25.4 mm.

FIGURE R505.2.5.3
WEB HOLE PATCH

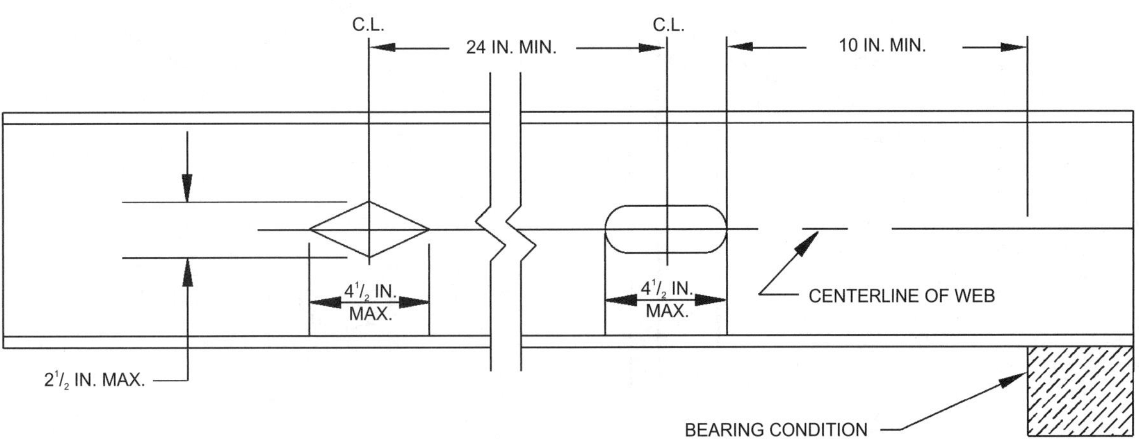

For SI: 1 inch = 25.4 mm.

FIGURE R505.2.5.1
FLOOR JOIST WEB HOLES

TABLE R505.3.1(1)
FLOOR TO FOUNDATION OR BEARING WALL CONNECTION REQUIREMENTS[a, b]

FRAMING CONDITION	BASIC WIND SPEED (mph) AND EXPOSURE	
	85 mph Exposure C or less than 110 mph Exposure B	Less than 110 mph Exposure C
Floor joist to wall track of exterior wall per Figure R505.3.1(1)	2-No. 8 screws	3-No. 8 screws
Rim track or end joist to load-bearing wall top track per Figure R505.3.1(1)	1-No. 8 screw at 24 inches o.c.	1-No. 8 screw at 24 inches o.c.
Rim track or end joist to wood sill per Figure R505.3.1(2)	Steel plate spaced at 4 feet o.c. with 4-No. 8 screws and 4-10d or 6-8d common nails	Steel plate spaced at 2 feet o.c. with 4-No. 8 screws and 4-10d or 6-8d common nails
Rim track or end joist to foundation per Figure R505.3.1(3)	$^1/_2$ inch minimum diameter anchor bolt and clip angle spaced at 6 feet o.c. with 8-No. 8 screws	$^1/_2$ inch minimum diameter anchor bolt and clip angle spaced at 4 feet o.c. with 8-No. 8 screws
Cantilevered joist to foundation per Figure R505.3.1(4)	$^1/_2$ inch minimum diameter anchor bolt and clip angle spaced at 6 feet o.c. with 8-No. 8 screws	$^1/_2$ inch minimum diameter anchor bolt and clip angle spaced at 4 feet o.c. with 8-No. 8 screws
Cantilevered joist to wood sill per Figure R505.3.1(5)	Steel plate spaced at 4 feet o.c. with 4-No. 8 screws and 4-10d or 6-8d common nails	Steel plate spaced at 2 feet o.c. with 4-No. 8 screws and 4-10d or 6-8d common nails
Cantilevered joist to exterior load-bearing wall track per Figure R505.3.1(6)	2-No. 8 screws	3-No. 8 screws

For SI: 1 inch = 25.4 mm, 1 pound per square foot = 0.0479 kPa, 1 mile per hour = 0.447 m/s, 1 foot = 304.8 mm.

a. Anchor bolts are to be located not more than 12 inches from corners or the termination of bottom tracks (e.g., at door openings or corners). Bolts extend a minimum of 15 inches into masonry or 7 inches into concrete. Anchor bolts connecting cold-formed steel framing to the foundation structure are to be installed so that the distance from the center of the bolt hole to the edge of the connected member is not less than one and one-half bolt diameters.

b. All screw sizes shown are minimum.

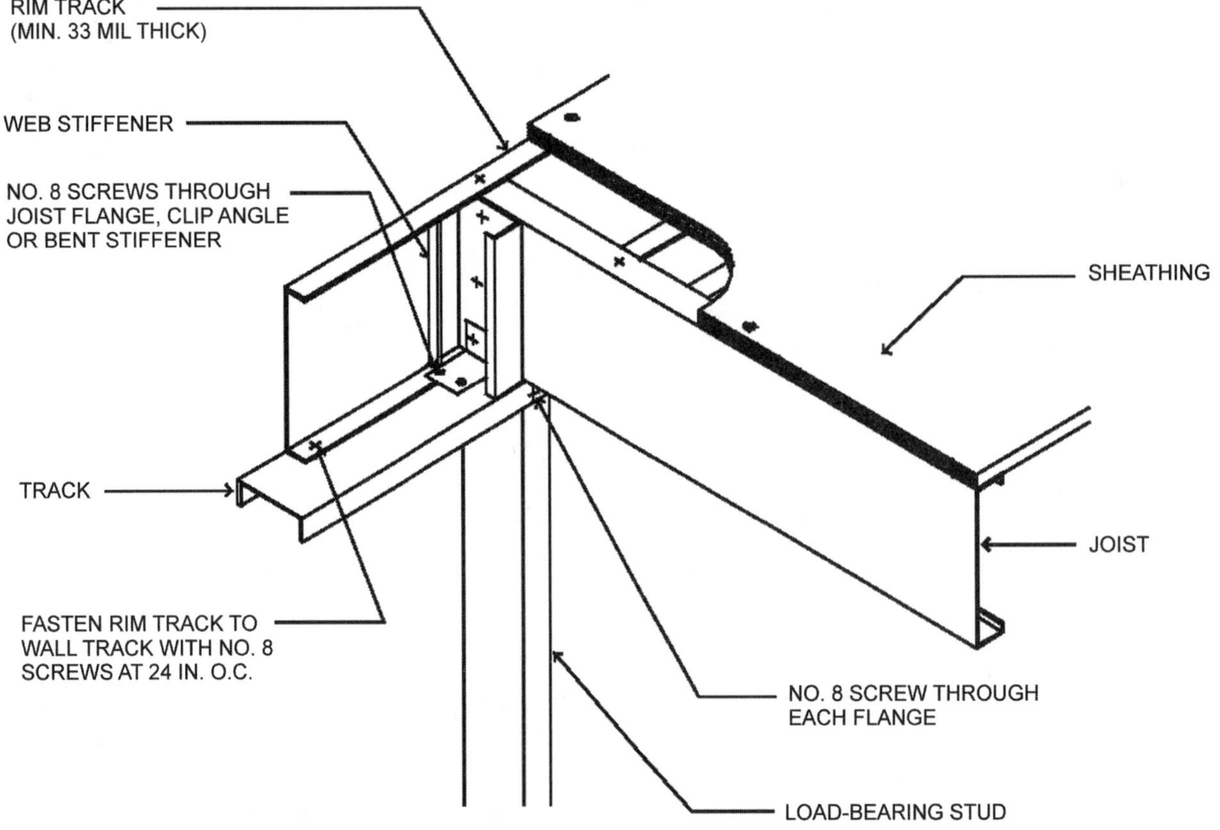

For SI: 1 mil = 0.0254 mm, 1 inch = 25.4 mm.

FIGURE 505.3.1(1)
FLOOR TO EXTERIOR LOAD-BEARING WALL STUD CONNECTION

**TABLE R505.3.1(2)
FLOOR FASTENING SCHEDULE[a]**

DESCRIPTION OF BUILDING ELEMENTS	NUMBER AND SIZE OF FASTENERS	SPACING OF FASTENERS
Floor joist to track of an interior load-bearing wall per Figures R505.3.1(7) and R505.3.1(8)	2 No. 8 screws	Each joist
Floor joist to track at end of joist	2 No. 8 screws	One per flange or two per bearing stiffener
Subfloor to floor joists	No. 8 screws	6 in. o.c. on edges and 12 in. o.c. at intermediate supports

For SI: 1 inch = 25.4 mm.

a. All screw sizes shown are minimum.

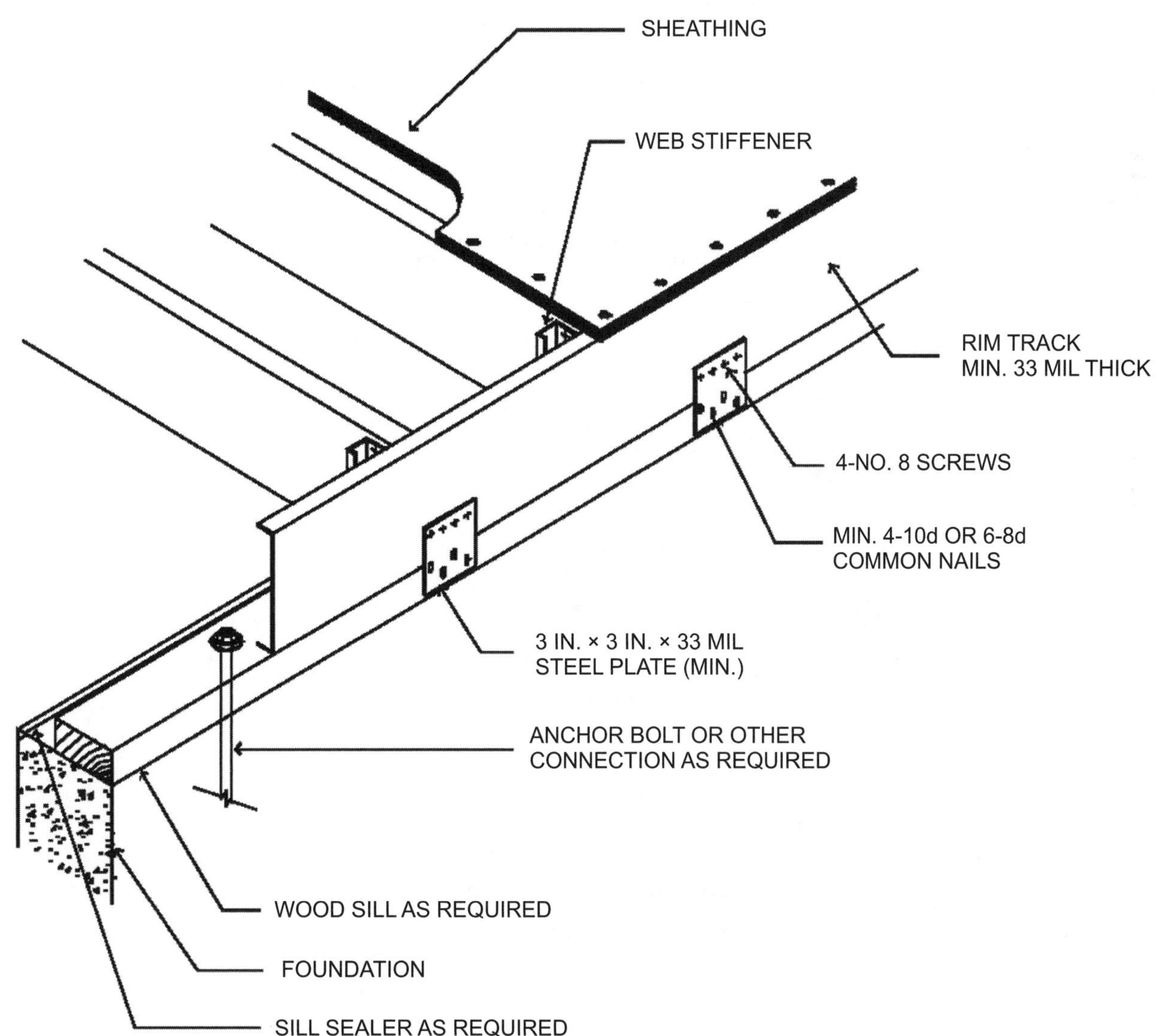

For SI: 1 mil = 0.0254 mm, 1 inch = 25.4 mm.

**FIGURE R505.3.1(2)
FLOOR TO WOOD SILL CONNECTION**

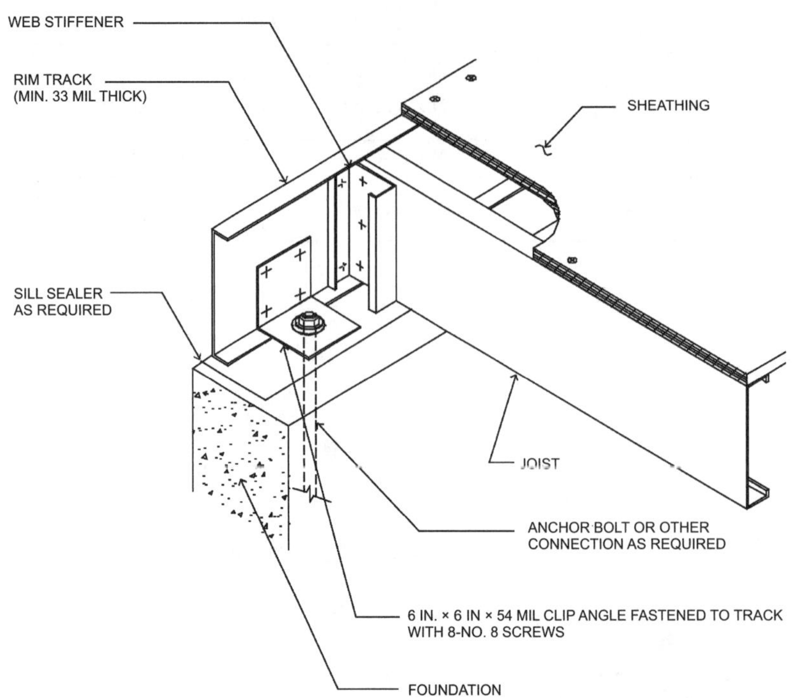

For SI: 1 mil = 0.0254 mm, 1 inch = 25.4 mm.

FIGURE R505.3.1(3)
FLOOR TO FOUNDATION CONNECTION

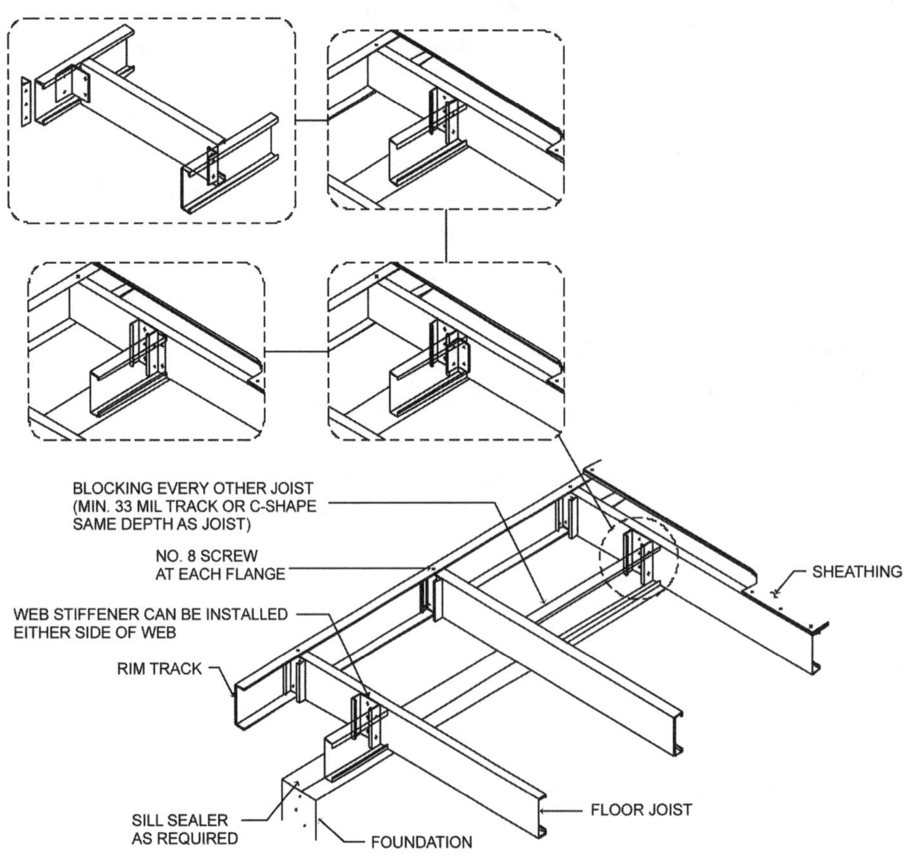

For SI: 1 mil = 0.0254 mm.

FIGURE R505.3.1(4)
CANTILEVERED FLOOR TO FOUNDATION CONNECTION

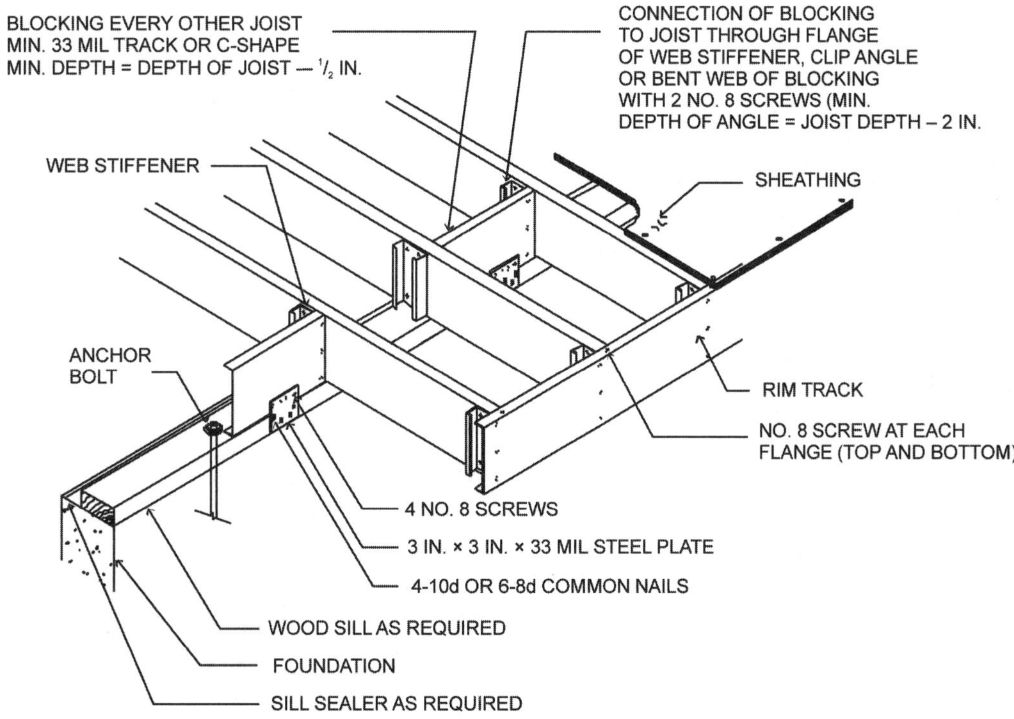

For SI: 1 mil = 0.0254 mm, 1 inch = 25.4 mm.

FIGURE R505.3.1(5)
CANTILEVERED FLOOR TO WOOD SILL CONNECTION

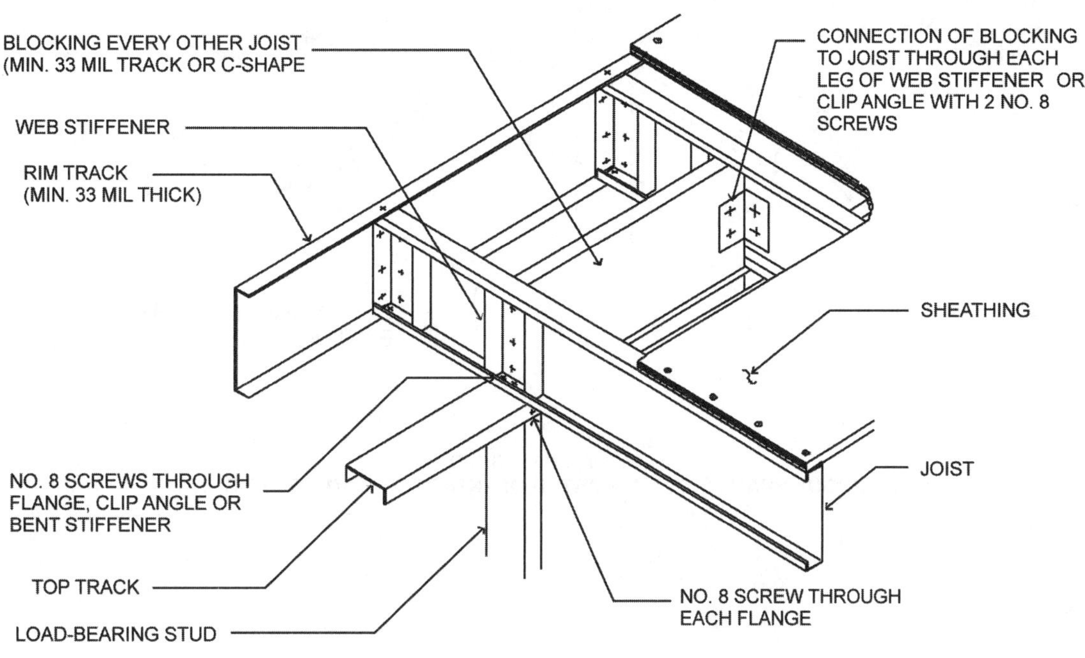

For SI: 1 mil = 0.0254 mm.

FIGURE R505.3.1(6)
CANTILEVERED FLOOR TO EXTERIOR LOAD-BEARING WALL CONNECTION

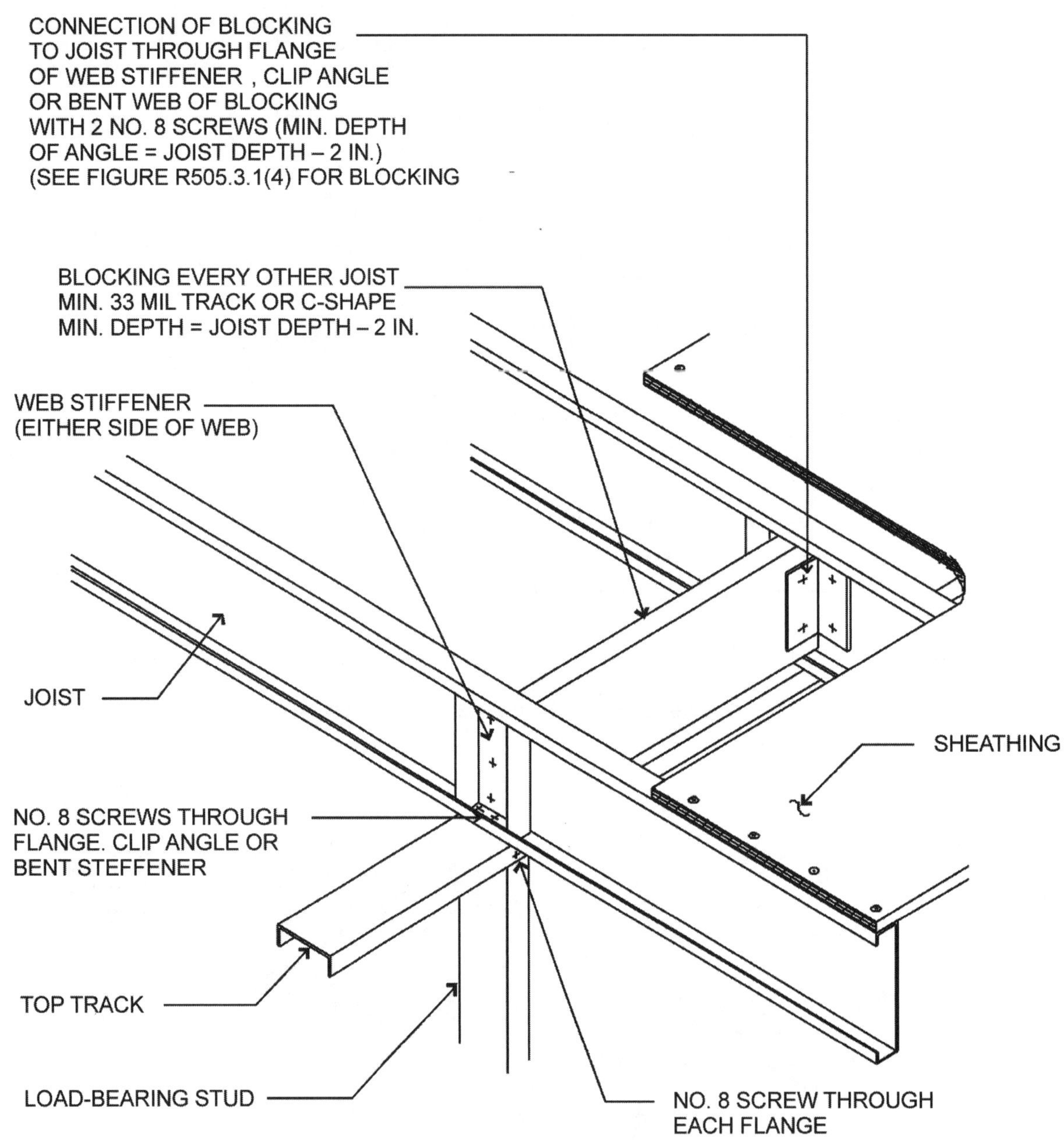

For SI: 1 mil = 0.0254 mm, 1 inch = 25.4 mm.

FIGURE R505.3.1(7)
CONTINUOUS SPAN JOIST SUPPORTED ON INTERIOR LOAD-BEARING WALL

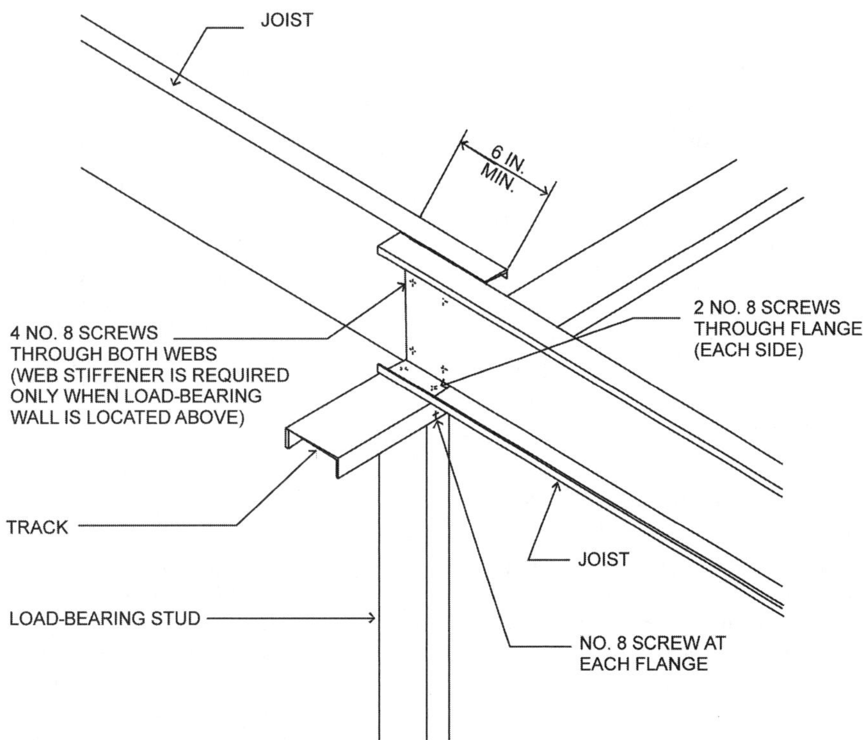

For SI: 1 inch = 25.4 mm.

FIGURE R505.3.1(8)
LAPPED JOISTS SUPPORTED ON INTERIOR LOAD-BEARING WALL

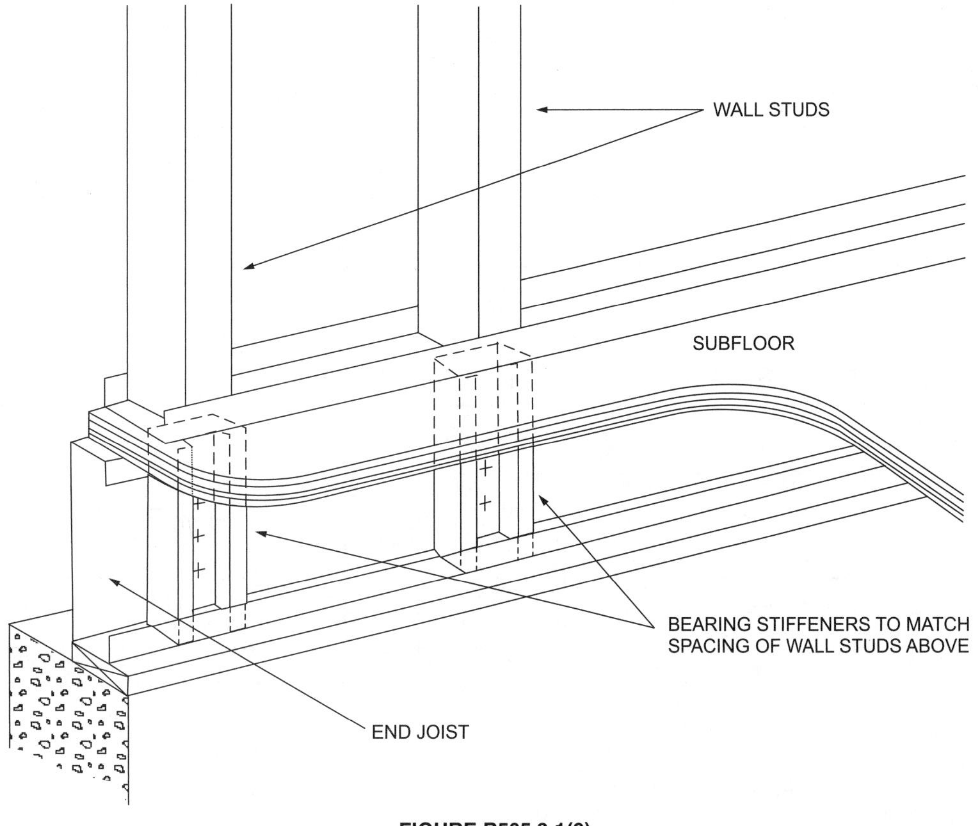

FIGURE R505.3.1(9)
BEARING STIFFENERS FOR END JOISTS

R505.3.2 Minimum floor joist sizes. Floor joist size and thickness shall be determined in accordance with the limits set forth in Table R505.3.2(1) for single spans, and Tables R505.3.2(2) and R505.3.2(3) for multiple spans. When continuous joist members are used, the interior bearing supports shall be located within 2 feet (610 mm) of mid-span of the cold-formed steel joists, and the individual spans shall not exceed the spans in Table R505.3.2(2) or R505.3.2(3), as applicable. Floor joists shall have a bearing support length of not less than $1^1/_2$ inches (38 mm) for exterior wall supports and $3^1/_2$ inches (89 mm) for interior wall supports. Tracks shall be a minimum of 33 mils (0.84 mm) thick except when used as part of a floor header or trimmer in accordance with Section R505.3.8. Bearing stiffeners shall be installed in accordance with Section R505.3.4.

R505.3.3 Joist bracing and blocking. Joist bracing and blocking shall be in accordance with this section.

R505.3.3.1 Joist top flange bracing. The top flanges of cold-formed steel joists shall be laterally braced by the application of floor sheathing fastened to the joists in accordance with Section R505.2.4 and Table R505.3.1(2).

R505.3.3.2 Joist bottom flange bracing/blocking. Floor joists with spans that exceed 12 feet (3658 mm) shall have the bottom flanges laterally braced in accordance with one of the following:

1. Gypsum board installed with minimum No. 6 screws in accordance with Section R702.
2. Continuous steel straps installed in accordance with Figure R505.3.3.2(1). Steel straps shall be spaced at a maximum of 12 feet (3658 mm) on center and shall be at least $1^1/_2$ inches (38 mm) in width and 33 mils (0.84 mm) in thickness. Straps shall be fastened to the bottom flange of each joist with one No. 8 screw, fastened to blocking with two No. 8 screws, and fastened at each end (of strap) with two No. 8 screws. Blocking in accordance with Figure R505.3.3.2(1) or Figure R505.3.3.2(2) shall be installed between joists at each end of the continuous strapping and at a maximum spacing of 12 feet (3658 mm) measured along the continuous strapping (perpendicular to the joist run). Blocking shall also be located at the termination of all straps. As an alternative to blocking at the ends, anchoring the strap to a stable building component with two No. 8 screws shall be permitted.

TABLE R505.3.2(1)
ALLOWABLE SPANS FOR COLD-FORMED STEEL JOISTS—SINGLE SPANS[a, b, c, d] 33 ksi STEEL

JOIST DESIGNATION	30 PSF LIVE LOAD				40 PSF LIVE LOAD			
	Spacing (inches)				Spacing (inches)			
	12	16	19.2	24	12	16	19.2	24
550S162-33	11′-7″	10′-7″	9′-6″	8′-6″	10′-7″	9′-3″	8′-6″	7′-6″
550S162-43	12′-8″	11′-6″	10′-10″	10′-2″	11′-6″	10′-5″	9′-10″	9′-1″
550S162-54	13′-7″	12′-4″	11′-7″	10′-9″	12′-4″	11′-2″	10′-6″	9′-9″
550S162-68	14′-7″	13′-3″	12′-6″	11′-7″	13′-3″	12′-0″	11′-4″	10′-6″
550S162-97	16′-2″	14′-9″	13′-10″	12′-10″	14′-9″	13′-4″	12′-7″	11′-8″
800S162-33	15′-8″	13′-11″	12′-9″	11′-5″	14′-3″	12′-5″	11′-3″	9′-0″
800S162-43	17′-1″	15′-6″	14′-7″	13′-7″	15′-6″	14′-1″	13′-3″	12′-4″
800S162-54	18′-4″	16′-8″	15′-8″	14′-7″	16′-8″	15′-2″	14′-3″	13′-3″
800S162-68	19′-9″	17′-11″	16′-10″	15′-8″	17′-11″	16′-3″	15′-4″	14′-2″
800S162-97	22′-0″	20′-0″	16′-10″	17′-5″	20′-0″	18′-2″	17′-1″	15′-10″
1000S162-43	20′-6″	18′-8″	17′-6″	15′-8″	18′-8″	16′-11″	15′-6″	13′-11″
1000S162-54	22′-1″	20′-0″	18′-10″	17′-6″	20′-0″	18′-2″	17′-2″	15′-11″
1000S162-68	23′-9″	21′-7″	20′-3″	18′-10″	21′-7″	19′-7″	18′-5″	17′-1″
1000S162-97	26′-6″	24′-1″	22′-8″	21′-0″	24′-1″	21′-10″	20′-7″	19′-1″
1200S162-43	23′-9″	20′-10″	19′-0″	16′-8″	21′-5″	18′-6″	16′-6″	13′-2″
1200S162-54	25′-9″	23′-4″	22′-0″	20′-1″	23′-4″	21′-3″	20′-0″	17′-10″
1200S162-68	27′-8″	25′-1″	23′-8″	21′-11″	25′-1″	22′-10″	21′-6″	21′-1″
1200S162-97	30′-11″	28′-1″	26′-5″	24′-6″	28′-1″	25′-6″	24′-0″	22′-3″

For SI: 1 inch = 25.4 mm, 1 foot = 304.8 mm, 1 pound per square foot = 0.0479 kPa.

a. Deflection criteria: *L*/480 for live loads, *L*/240 for total loads.

b. Floor dead load = 10 psf.

c. Table provides the maximum clear span in feet and inches.

d. Bearing stiffeners are to be installed at all support points and concentrated loads.

**TABLE R505.3.2(2)
ALLOWABLE SPANS FOR COLD-FORMED STEEL JOISTS—MULTIPLE SPANS[a, b, c, d, e, f] 33 ksi STEEL**

JOIST DESIGNATION	30 PSF LIVE LOAD				40 PSF LIVE LOAD			
	Spacing (inches)				Spacing (inches)			
	12	16	19.2	24	12	16	19.2	24
550S162-33	12′-1″	10′-5″	9′-6″	8′-6″	10′-9″	9′-3″	8′-6″	7′-6″
550S162-43	14′-5″	12′-5″	11′-4″	10′-2″	12′-9″	11′-11″	10′-1″	9′-0″
550S162-54	16′-3″	14′-1″	12′-10″	11′-6″	14′-5″	12′-6″	11′-5″	10′-2″
550S162-68	19′-7″	17′-9″	16′-9″	15′-6″	17′-9″	16′-2″	15′-2″	14′-1″
550S162-97	21′-9″	19′-9″	18′-7″	17′-3″	19′-9″	17′-11″	16′-10″	15′-4″
800S162-33	14′-8″	11′-10″	10′-4″	8′-8″	12′-4″	9′-11″	8′-7″	7′-2″
800S162-43	20′-0″	17′-4″	15′-9″	14′-1″	17′-9″	15′-4″	14′-0″	12′-0″
800S162-54	23′-7″	20′-5″	18′-8″	16′-8″	21′-0″	18′-2″	16′-7″	14′-10″
800S162-68	26′-5″	23′-1″	21′-0″	18′-10″	23′-8″	20′-6″	18′-8″	16′-9″
800S162-97	29′-6″	26′-10″	25′-3″	22′-8″	26′-10″	24′-4″	22′-6″	20′-2″
1000S162-43	22′-2″	18′-3″	16′-0″	13′-7″	18′-11″	15′-5″	13′-6″	11′-5″
1000S162-54	26′-2″	22′-8″	20′-8″	18′-6″	23′-3″	20′-2″	18′-5″	16′-5″
1000S162-68	31′-5″	27′-2″	24′-10″	22′-2″	27′-11″	24′-2″	22′-1″	19′-9″
1000S162-97	35′-6″	32′-3″	29′-11″	26′-9″	32′-3″	29′-2″	26′-7″	23′-9″
1200S162-43	21′-8″	17′-6″	15′-3″	12′-10″	18′-3″	14′-8″	12′-8″	10′-6″
1200S162-54	28′-5″	24′-8″	22′-6″	19′-6″	25′-3″	21′-11″	19′-4″	16′-6″
1200S162-68	33′-7″	29′-1″	26′-6″	23′-9″	29′-10″	25′-10″	23′-7″	21′-1″
1200S162-97	41′-5″	37′-8″	34′-6″	30′-10″	37′-8″	33′-6″	30′-7″	27′-5″

For SI: 1 inch = 25.4 mm, 1 foot = 304.8 mm, 1 pound per square foot = 0.0479 kPa.

a. Deflection criteria: *L*/480 for live loads, *L*/240 for total loads.

b. Floor dead load = 10 psf.

c. Table provides the maximum clear span in feet and inches to either side of the interior support.

d. Interior bearing supports for multiple span joists consist of structural (bearing) walls or beams.

e. Bearing stiffeners are to be installed at all support points and concentrated loads.

f. Interior supports shall be located within 2 feet of mid-span provided that each of the resulting spans does not exceed the appropriate maximum span shown in the table above.

TABLE R505.3.2(3)
ALLOWABLE SPANS FOR COLD-FORMED STEEL JOISTS—MULTIPLE SPANS[a, b, c, d, e, f] 50 ksi STEEL

JOIST DESIGNATION	30 PSF LIVE LOAD				40 PSF LIVE LOAD			
	Spacing (inches)				Spacing (inches)			
	12	16	19.2	24	12	16	19.2	24
550S162-33	13′-11″	12′-0″	11′-0″	9′-3″	12′-3″	10′-8″	9′-7″	8′-4″
550S162-43	16′-3″	14′-1″	12′-10″	11′-6″	14′-6″	12′-6″	11′-5″	10′-3″
550S162-54	18′-2″	16′-6″	15′-4″	13′-8″	16′-6″	14′-11″	13′-7″	12′-2″
550S162-68	19′-6″	17′-9″	16′-8″	15′-6″	17′-9″	16′-1″	15′-2″	14′-0″
550S162-97	21′-9″	19′-9″	18′-6″	17′-2″	19′-8″	17′-10″	16′-8″	15′-8″
800S162-33	15′-6″	12′-6″	10′-10″	9′-1″	13′-0″	10′-5″	8′-11″	6′-9″
800S162-43	22′-0″	19′-1″	17′-5″	15′-0″	19′-7″	16′-11″	14′-10″	12′-8″
800S162-54	24′-6″	22′-4″	20′-6″	17′-11″	22′-5″	19′-9″	17′-11″	15′-10″
800S162-68	26′-6″	24′-1″	22′-8″	21′-0″	24′-1″	21′-10″	20′-7″	19′-2″
800S162-97	29′-9″	26′-8″	25′-2″	23′-5″	26′-8″	24′-3″	22′-11″	21′-4″
1000S162-43	23′-6″	19′-2″	16′-9″	14′-2″	19′-11″	16′-2″	14′-0″	11′-9″
1000S162-54	28′-2″	23′-10″	21′-7″	18′-11″	24′-8″	20′-11″	18′-9″	18′-4″
1000S162-68	31′-10″	28′-11″	27′-2″	25′-3″	28′-11″	26′-3″	24′-9″	22′-9″
1000S162-97	35′-4″	32′-1″	30′-3″	28′-1″	32′-1″	29′-2″	27′-6″	25′-6″
1200S162-43	22′-11″	18′-5″	16′-0″	13′-4″	19′-2″	15′-4″	13′-2″	10′-6″
1200S162-54	32′-8″	28′-1″	24′-9″	21′-2″	29′-0″	23′-10″	20′-11″	17′-9″
1200S162-68	37′-1″	32′-5″	29′-4″	25′-10″	33′-4″	28′-6″	25′-9″	22′-7″
1200S162-97	41′-2″	37′-6″	35′-3″	32′-9″	37′-6″	34′-1″	32′-1″	29′-9″

For SI: 1 inch = 25.4 mm, 1 foot = 304.8 mm, 1 pound per square foot = 0.0479 kPa.

a. Deflection criteria: *L*/480 for live loads, *L*/240 for total loads.

b. Floor dead load = 10 psf.

c. Table provides the maximum clear span in feet and inches to either side of the interior support.

d. Interior bearing supports for multiple span joists consist of structural (bearing) walls or beams.

e. Bearing stiffeners are to be installed at all support points and concentrated loads.

f. Interior supports shall be located within 2 feet of mid-span provided that each of the resulting spans does not exceed the appropriate maximum span shown in the table above.

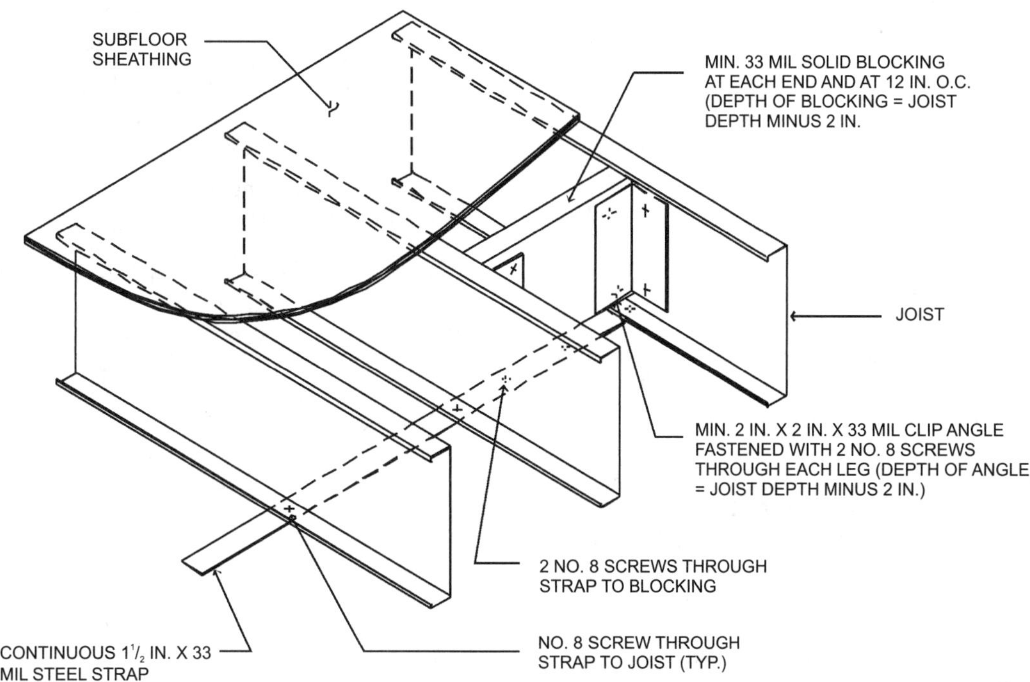

For SI: 1 mil = 0.0254, 1 inch = 25.4 mm.

FIGURE R505.3.3.2(1)
JOIST BLOCKING (SOLID)

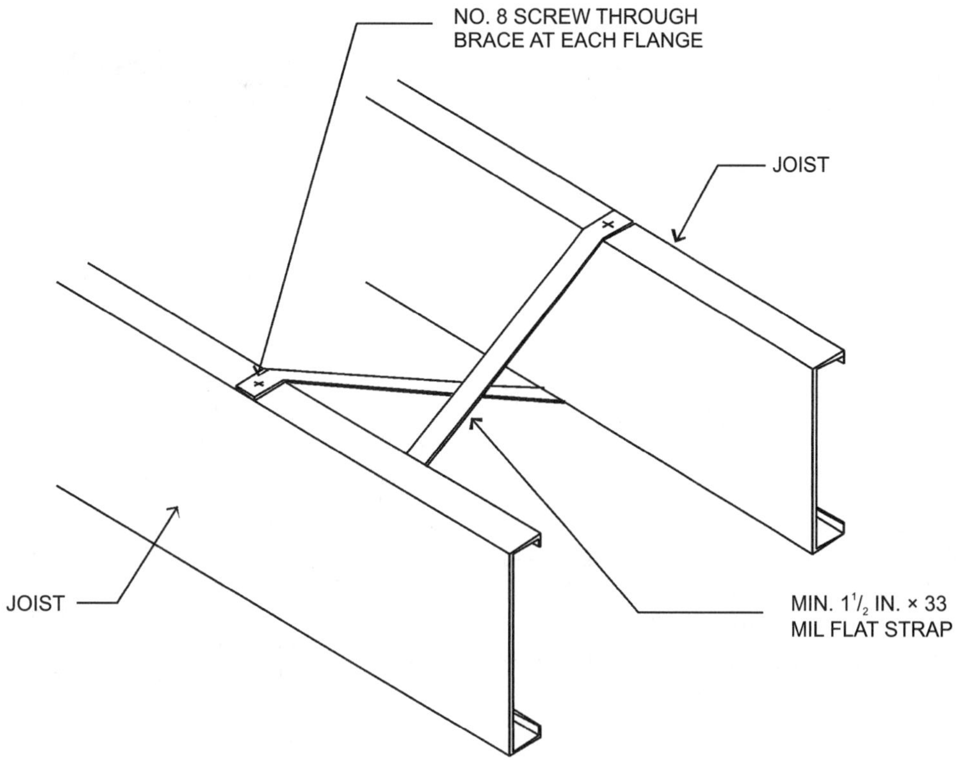

For SI: 1 mil = 0.0254, 1 inch = 25.4 mm.

FIGURE R505.3.3.2(2)
JOIST BLOCKING (STRAP)

R505.3.3.3 Blocking at interior bearing supports. Blocking is not required for continuous back-to-back floor joists at bearing supports. Blocking shall be installed between every other joist for single continuous floor joists across bearing supports in accordance with Figure R505.3.1(7). Blocking shall consist of C-shape or track section with a minimum thickness of 33 mils (0.84 mm). Blocking shall be fastened to each adjacent joist through a 33-mil (0.84 mm) clip angle, bent web of blocking or flanges of web stiffeners with two No. 8 screws on each side. The minimum depth of the blocking shall be equal to the depth of the joist minus 2 inches (51 mm). The minimum length of the angle shall be equal to the depth of the joist minus 2 inches (51 mm).

R505.3.3.4 Blocking at cantilevers. Blocking shall be installed between every other joist over cantilever bearing supports in accordance with Figure R505.3.1(4), R505.3.1(5) or R505.3.1(6). Blocking shall consist of C-shape or track section with minimum thickness of 33 mils (0.84 mm). Blocking shall be fastened to each adjacent joist through bent web of blocking, 33 mil clip angle or flange of web stiffener with two No. 8 screws at each end. The depth of the blocking shall be equal to the depth of the joist. The minimum length of the angle shall be equal to the depth of the joist minus 2 inches (51 mm). Blocking shall be fastened through the floor sheathing and to the support with three No. 8 screws (top and bottom).

R505.3.4 Bearing stiffeners. Bearing stiffeners shall be installed at each joist bearing location in accordance with this section, except for joists lapped over an interior support not carrying a load-bearing wall above. Floor joists supporting jamb studs with multiple members shall have two bearing stiffeners in accordance with Figure R505.3.4(1). Bearing stiffeners shall be fabricated from a C-shaped, track or clip angle member in accordance with the one of following:

1. C-shaped bearing stiffeners:
   - 1.1. Where the joist is not carrying a load-bearing wall above, the bearing stiffener shall be a minimum 33 mil (0.84 mm) thickness.
   - 1.2. Where the joist is carrying a load-bearing wall above, the bearing stiffener shall be at least the same designation thickness as the wall stud above.
2. Track bearing stiffeners:
   - 2.1. Where the joist is not carrying a load-bearing wall above, the bearing stiffener shall be a minimum 43 mil (1.09 mm) thickness.
   - 2.2. Where the joist is carrying a load-bearing wall above, the bearing stiffener shall be at least one designation thickness greater than the wall stud above.
3. Clip angle bearing stiffeners: Where the clip angle bearing stiffener is fastened to both the web of the member it is stiffening and an adjacent rim track using the fastener pattern shown in Figure R505.3.4(2), the bearing stiffener shall be a minimum 2 inch by 2 inch (51 mm by 51 mm) angle sized in accordance with Tables R505.3.4(1), R505.3.4(2), R505.3.4(3), and R505.3.4(4).

The minimum length of a bearing stiffener shall be the depth of member being stiffened minus $^3/_8$ inch (9.5 mm). Each bearing stiffener shall be fastened to the web of the member it is stiffening as shown in Figure R505.3.4(2). Each clip angle bearing stiffener shall also be fastened to the web of the adjacent rim track using the fastener pattern shown in Figure R505.3.4(2). No. 8 screws shall be used for C-shaped and track members of any thickness and for clip angle members with a designation thickness less than or equal to 54. No. 10 screws shall be used for clip angle members with a designation thickness greater than 54.

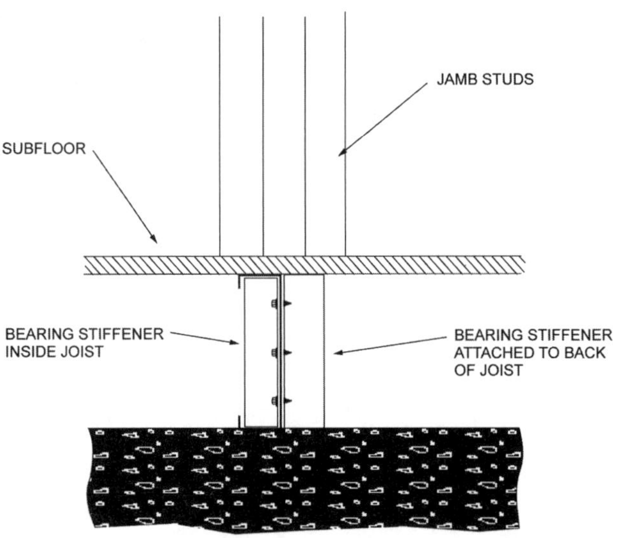

FIGURE R505.3.4(1)
BEARING STIFFENERS UNDER JAMB STUDS

R505.3.5 Cutting and notching. Flanges and lips of load-bearing cold-formed steel floor framing members shall not be cut or notched.

R505.3.6 Floor cantilevers. Floor cantilevers for the top floor of a two- or three-story building or the first floor of a one-story building shall not exceed 24 inches (610 mm). Cantilevers, not exceeding 24 inches (610 mm) and supporting two stories and roof (i.e., first floor of a two-story building), shall also be permitted provided that all cantilevered joists are doubled (nested or back-to-back). The doubled cantilevered joists shall extend a minimum of 6 feet (1829 mm) toward the inside and shall be fastened with a minimum of two No. 8 screws spaced at 24 inches (610 mm) on center through the webs (for back-to-back) or flanges (for nested joists).

R505.3.7 Splicing. Joists and other structural members shall not be spliced. Splicing of tracks shall conform to Figure R505.3.7.

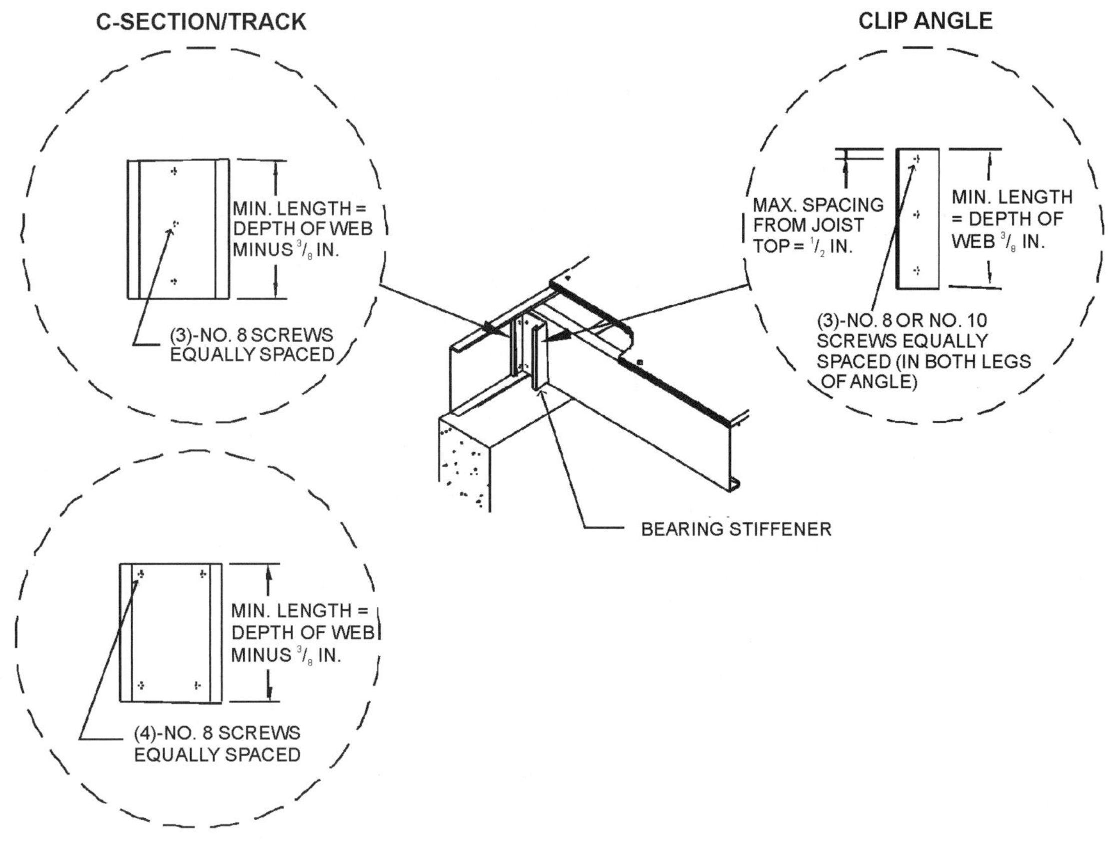

For SI: 1 inch = 25.4 mm.

FIGURE R505.3.4(2)
BEARING STIFFENER

TABLE R505.3.4(1)
CLIP ANGLE BEARING STIFFENERS
(20 psf equivalent snow load)

JOIST DESIGNATION	MINIMUM THICKNESS (mils) OF 2 INCH × 2 INCH CLIP ANGLE											
	Top floor				Bottom floor in 2 story Middle floor in 3 story				Bottom floor in 3 story			
	Joist spacing (inches)				Joist spacing (inches)				Joist spacing (inches)			
	12	16	19.2	24	12	16	19.2	24	12	16	19.2	24
800S162-33	43	43	43	43	43	54	68	68	68	97	97	—
800S162-43	43	43	43	43	54	54	68	68	97	97	97	97
800S162-54	43	43	43	43	43	54	68	68	68	97	97	—
800S162-68	43	43	43	43	43	43	54	68	54	97	97	—
800S162-97	43	43	43	43	43	43	43	43	43	43	54	97
1000S162-43	43	43	43	43	54	68	97	97	97	—	—	—
1000S162-54	43	43	43	43	54	68	68	97	97	97	—	—
1000S162-68	43	43	43	43	54	68	97	97	97	—	—	—
1000S162-97	43	43	43	43	43	43	43	54	43	68	97	—
1200S162-43	43	54	54	54	97	97	97	97	—	—	—	—
1200S162-54	54	54	54	54	97	97	97	97	—	—	—	—
1200S162-68	43	43	54	54	68	97	97	97	—	—	—	—
1200S162-97	43	43	43	43	43	54	68	97	97	—	—	—

For SI: 1 mil = 0.254 mm, 1 inch = 25.4 mm, 1 pound per square foot = 0.0479 kPa.

TABLE R505.3.4(2)
CLIP ANGLE BEARING STIFFENERS
(30 psf equivalent snow load)

JOIST DESIGNATION	MINIMUM THICKNESS (mils) OF 2 INCH × 2 INCH CLIP ANGLE											
	Top floor				Bottom floor in 2 story Middle floor in 3 story				Bottom floor in 3 story			
	Joist spacing (inches)				Joist spacing (inches)				Joist spacing (inches)			
	12	16	19.2	24	12	16	19.2	24	12	16	19.2	24
800S162-33	43	43	43	43	54	68	68	97	97	97	97	—
800S162-43	43	43	43	54	68	68	68	97	97	97	97	—
800S162-54	43	43	43	43	54	68	68	97	97	97	—	—
800S162-68	43	43	43	43	43	54	68	97	68	97	97	—
800S162-97	43	43	43	43	43	43	43	43	43	43	68	97
1000S162-43	54	54	54	54	68	97	97	97	97	—	—	—
1000S162-54	54	54	54	54	68	97	97	97	97	—	—	—
1000S162-68	43	43	54	68	68	97	97	—	97	—	—	—
1000S162-97	43	43	43	43	43	43	54	68	54	97	—	—
1200S162-43	54	68	68	68	97	97	97	—	—	—	—	—
1200S162-54	68	68	68	68	97	97	—	—	—	—	—	—
1200S162-68	68	68	68	68	97	97	97	—	—	—	—	—
1200S162-97	43	43	43	43	54	68	97	—	97	—	—	—

For SI: 1 mil = 0.0254 mm, 1 inch = 25.4 mm, 1 pound per square foot = 0.0479 kPa.

TABLE R505.3.4(3)
CLIP ANGLE BEARING STIFFENERS
(50 psf equivalent snow load)

JOIST DESIGNATION	MINIMUM THICKNESS (mils) OF 2 INCH × 2 INCH CLIP ANGLE											
	Top floor				Bottom floor in 2 story Middle floor in 3 story				Bottom floor in 3 story			
	Joist spacing (inches)				Joist spacing (inches)				Joist spacing (inches)			
	12	16	19.2	24	12	16	19.2	24	12	16	19.2	24
800S162-33	54	54	54	54	68	97	97	97	97	—	—	—
800S162-43	68	68	68	68	97	97	97	97	—	—	—	—
800S162-54	54	68	68	68	97	97	97	97	—	—	—	—
800S162-68	43	43	54	54	68	97	97	97	97	—	—	—
800S162-97	43	43	43	43	43	43	43	54	54	68	97	—
1000S162-43	97	68	68	68	97	97	97	97	—	—	—	—
1000S162-54	97	97	68	68	97	97	97	—	—	—	—	—
1000S162-68	68	97	97	97	97	—	—	—	—	—	—	—
1000S162-97	43	43	43	43	54	68	97	97	—	—	—	—
1200S162-43	97	97	97	97	—	—	—	—	—	—	—	—
1200S162-54	—	97	97	97	—	—	—	—	—	—	—	—
1200S162-68	97	97	97	97	—	—	—	—	—	—	—	—
1200S162-97	54	68	68	97	97	—	—	—	—	—	—	—

For SI: 1 mil = 0.0254 mm, 1 inch = 25.4 mm, 1 pound per square foot = 0.0479 kPa.

TABLE R505.3.4(4)
CLIP ANGLE BEARING STIFFENERS
(70 psf equivalent snow load)

JOIST DESIGNATION	MINIMUM THICKNESS (mils) OF 2 INCH × 2 INCH CLIP ANGLE											
	Top floor				Bottom floor in 2 story Middle floor in 3 story				Bottom floor in 3 story			
	Joist spacing (inches)				Joist spacing (inches)				Joist spacing (inches)			
	12	16	19.2	24	12	16	19.2	24	12	16	19.2	24
800S162-33	68	68	68	68	97	97	97	97	—	—	—	—
800S162-43	97	97	97	97	97	97	97	—	—	—	—	—
800S162-54	97	97	97	97	97	—	—	—	—	—	—	—
800S162-68	68	68	68	97	97	97	97	—	—	—	—	—
800S162-97	43	43	43	43	43	54	68	97	97	97	—	—
1000S162-43	97	97	97	97	—	—	—	—	—	—	—	—
1000S162-54	—	97	97	97	—	—	—	—	—	—	—	—
1000S162-68	97	97	—	—	—	—	—	—	—	—	—	—
1000S162-97	68	68	68	68	97	97	—	—	—	—	—	—
1200S162-43	97	97	97	97	—	—	—	—	—	—	—	—
1200S162-54	—	—	—	—	—	—	—	—	—	—	—	—
1200S162-68	—	—	—	—	—	—	—	—	—	—	—	—
1200S162-97	97	97	97	—	—	—	—	—	—	—	—	—

For SI: 1 mil = 0.0254 mm, 1 inch = 25.4 mm, 1 pound per square foot = 0.0479 kPa.

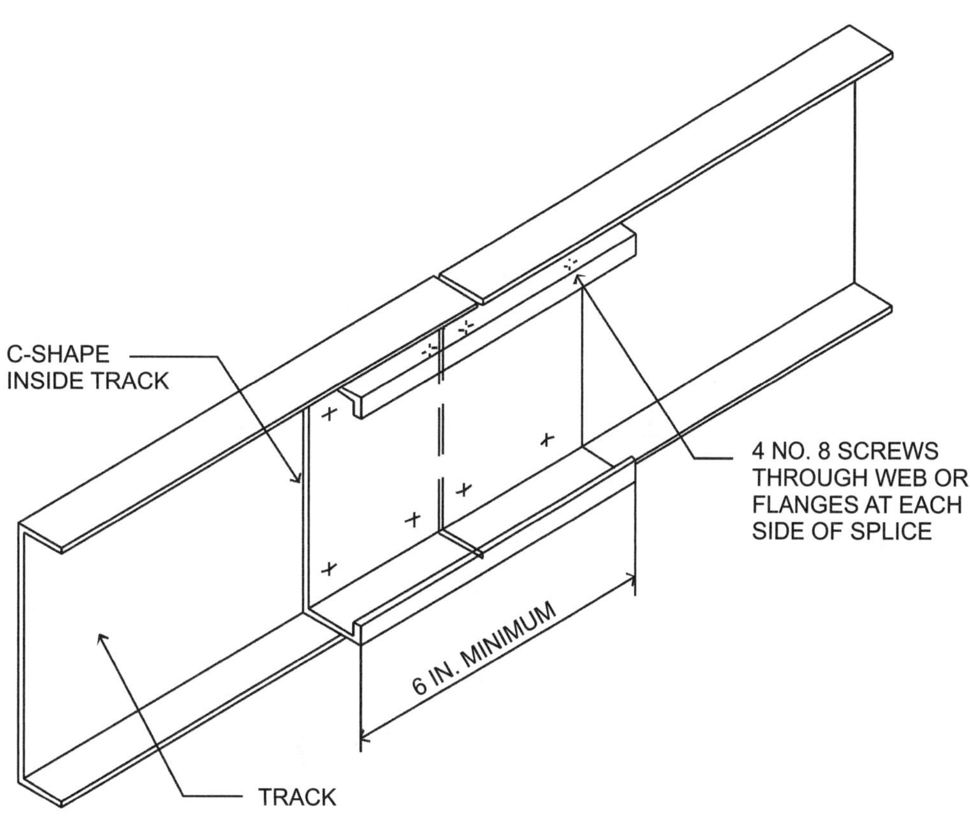

For SI: 1 inch = 25.4 mm.

FIGURE R505.3.7
TRACK SPLICE

R505.3.8 Framing of floor openings. Openings in floors shall be framed with header and trimmer joists. Header joist spans shall not exceed 6 feet (1829 mm) or 8 feet (2438 mm) in length in accordance with Figure R505.3.8(1) or R505.3.8(2), respectively. Header and trimmer joists shall be fabricated from joist and track members, having a minimum size and thickness at least equivalent to the adjacent floor joists and shall be installed in accordance with Figures R505.3.8(1), R505.3.8(2), R505.3.8(3), and R505.3.8(4). Each header joist shall be connected to trimmer joists with four 2 inch by 2 inch (51 mm by 51 mm) clip angles. Each clip angle shall be fastened to both the header and trimmer joists with four No. 8 screws, evenly spaced, through each leg of the clip angle. The clip angles shall have a thickness not less than that of the floor joist. Each track section for a built-up header or trimmer joist shall extend the full length of the joist (continuous).

SECTION R506 CONCRETE FLOORS (ON GROUND)

R506.1 General. Concrete slab-on-ground floors shall be designed and constructed in accordance with the provisions of this section or ACI 332. Floors shall be a minimum 3.5 inches (89 mm) thick (for expansive soils, see Section R403.1.8). The specified compressive strength of concrete shall be as set forth in Section R402.2.

R506.2 Site preparation. The area within the foundation walls shall have all vegetation, top soil and foreign material removed.

R506.2.1 Fill. Fill material shall be free of vegetation and foreign material. The fill shall be compacted to assure uniform support of the slab, and except where *approved*, the fill depths shall not exceed 24 inches (610 mm) for clean sand or gravel and 8 inches (203 mm) for earth.

R506.2.2 Base. A 4-inch-thick (102 mm) base course consisting of clean graded sand, gravel, crushed stone or crushed blast-furnace slag passing a 2-inch (51 mm) sieve shall be placed on the prepared subgrade when the slab is below *grade*.

Exception: A base course is not required when the concrete slab is installed on well-drained or sand-gravel mixture soils classified as Group I according to the United Soil Classification System in accordance with Table R405.1.

R506.2.3 Vapor retarder. A 6-mil (0.006 inch; 152 μm) polyethylene or *approved* vapor retarder with joints lapped not less than 6 inches (152 mm) shall be placed between the concrete floor slab and the base course or the prepared subgrade where no base course exists.

Exception: The vapor retarder may be omitted:

1. From garages, utility buildings and other unheated *accessory structures*.
2. For unheated storage rooms having an area of less than 70 square feet (6.5 m^2) and carports.
3. From driveways, walks, patios and other flatwork not likely to be enclosed and heated at a later date.
4. Where *approved* by the *building official*, based on local site conditions.

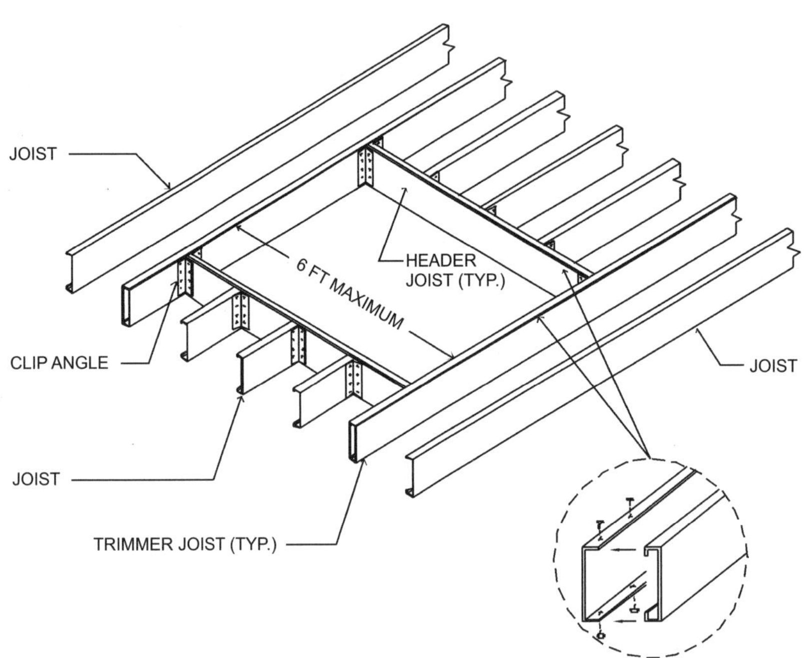

For SI: 1 foot = 304.8 mm.

FIGURE R505.3.8(1)
COLD-FORMED STEEL FLOOR CONSTRUCTION: 6-FOOT FLOOR OPENING

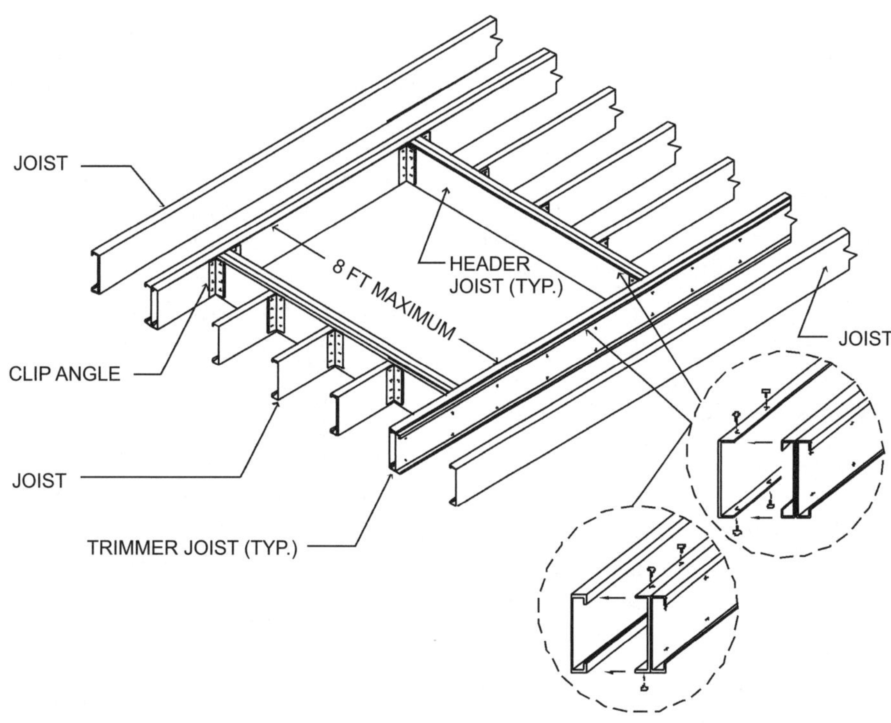

For SI: 1 foot = 304.8 mm.

FIGURE R505.3.8(2)
COLD-FORMED STEEL FLOOR CONSTRUCTION—8-FOOT FLOOR OPENING

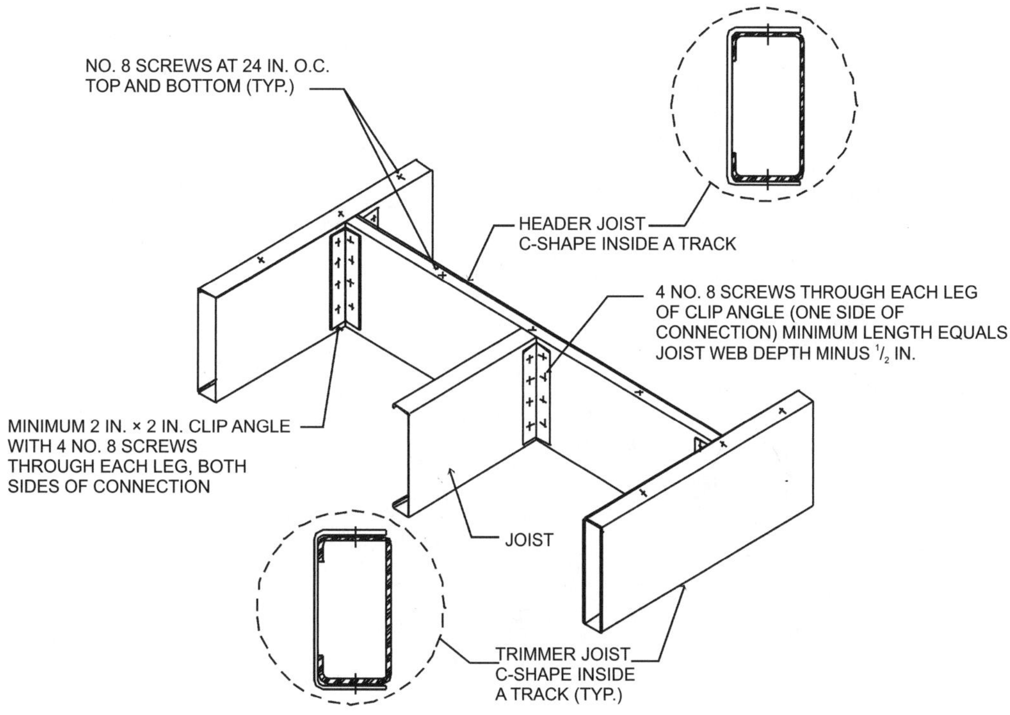

For SI: 1 inch = 25.4 mm, 1 foot = 304.8 mm.

FIGURE R505.3.8(3)
COLD-FORMED STEEL FLOOR CONSTRUCTION: FLOOR HEADER TO TRIMMER CONNECTION—6-FOOT OPENING

R506.2.4 Reinforcement support. Where provided in slabs on ground, reinforcement shall be supported to remain in place from the center to upper one third of the slab for the duration of the concrete placement.

**

SECTION R507
DECKS

R507.1 Decks. Where supported by attachment to an exterior wall, decks shall be positively anchored to the primary structure and designed for both vertical and lateral loads. Such attachment shall not be accomplished by the use of toenails or nails subject to withdrawal. Where positive connection to the primary building structure cannot be verified during inspection, decks shall be self-supporting. For decks with cantilevered framing members, connections to exterior walls or other framing members, shall be designed and constructed to resist uplift resulting from the full live load specified in Table R301.5 acting on the cantilevered portion of the deck.

R507.2 Deck ledger connection to band joist. For decks supporting a total design load of 50 pounds per square foot (2394 Pa) [40 pounds per square foot (1915 Pa) live load plus 10 pounds per square foot (479 Pa) dead load], the connection between a deck ledger of pressure-preservative-treated Southern Pine, incised pressure-preservative-treated Hem-Fir or *approved* decay-resistant species, and a 2-inch (51 mm) nominal lumber band joist bearing on a sill plate or wall plate shall be constructed with $^{1}/_{2}$-inch (12.7 mm) lag screws or bolts with washers in accordance with Table R507.2. Lag screws, bolts and washers shall be hot-dipped galvanized or stainless steel.

R507.2.1 Placement of lag screws or bolts in deck ledgers and band joists. The lag screws or bolts in deck ledgers and band joists shall be placed in accordance with Table R507.2.1 and Figures R507.2.1(1) and R507.2.1(2).

R507.2.2 Alternate deck ledger connections. Deck ledger connections not conforming to Table R507.2 shall be designed in accordance with accepted engineering practice. Girders supporting deck joists shall not be supported on deck ledgers or band joists. Deck ledgers shall not be supported on stone or masonry veneer.

R507.2.3 Deck lateral load connection. The lateral load connection required by Section R507.1 shall be permitted to be in accordance with Figure R507.2.3. Where the lateral load connection is provided in accordance with Figure 507.2.3, hold-down tension devices shall be installed in not less than two locations per deck, and each device shall have an allowable stress design capacity of not less than 1500 pounds (6672 N).

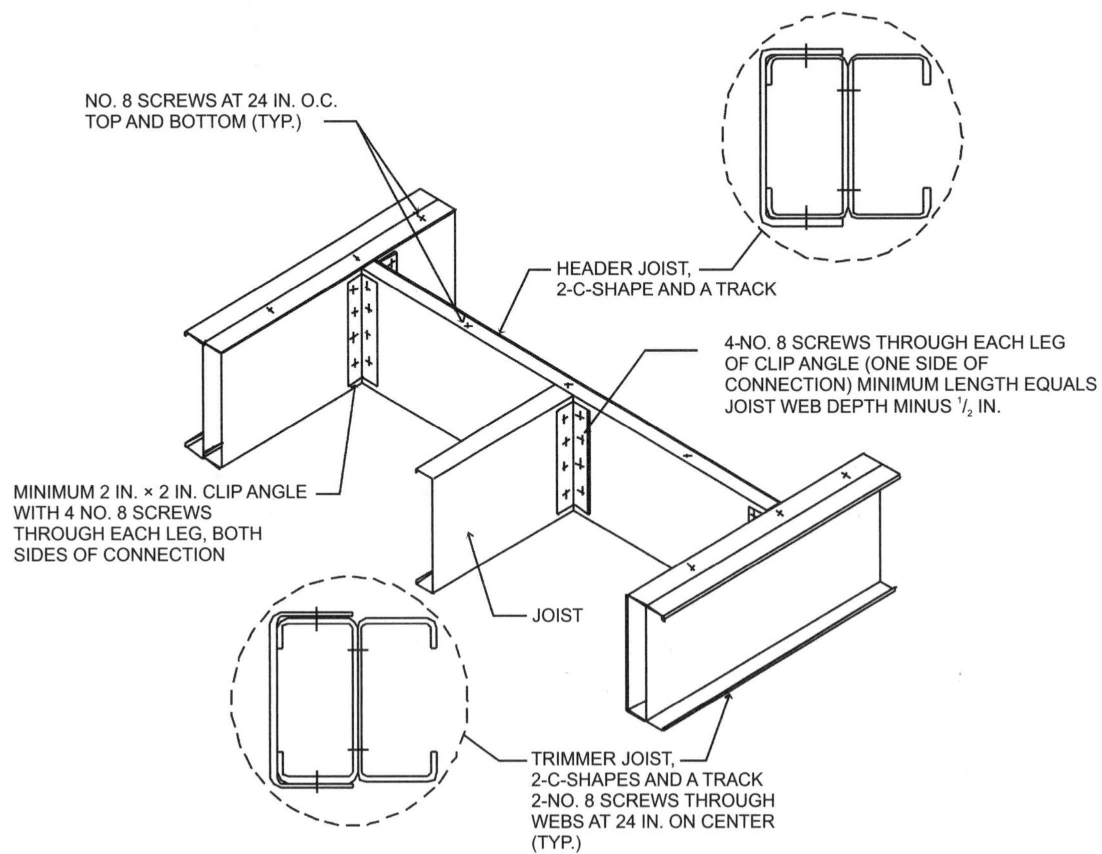

For SI: 1 inch = 25.4 mm, 1 foot = 304.8 mm.

FIGURE R505.3.8(4)
COLD-FORMED STEEL FLOOR CONSTRUCTION: FLOOR HEADER TO TRIMMER CONNECTION—8-FOOT OPENING

TABLE R507.2
FASTENER SPACING FOR A SOUTHERN PINE OR HEM-FIR DECK LEDGER AND A 2-INCH-NOMINAL SOLID-SAWN SPRUCE-PINE-FIR BAND JOIST[c, f, g]
(Deck live load = 40 psf, deck dead load = 10 psf)

JOIST SPAN	6′ and less	6′1″ to 8′	8′1″ to 10′	10′1″ to 12′	12′1″ to 14′	14′1″ to 16′	16′1″ to 18′
Connection details	On-center spacing of fasteners[d, e]						
$^1/_2$ inch diameter lag screw with $^{15}/_{32}$ inch maximum sheathing[a]	30	23	18	15	13	11	10
$^1/_2$ inch diameter bolt with $^{15}/_{32}$ inch maximum sheathing	36	36	34	29	24	21	19
$^1/_2$ inch diameter bolt with $^{15}/_{32}$ inch maximum sheathing and $^1/_2$ inch stacked washers[b, h]	36	36	29	24	21	18	16

For SI: 1 inch = 25.4 mm, 1 foot = 304.8 mm. 1 pound per square foot = 0.0479 kPa.

a. The tip of the lag screw shall fully extend beyond the inside face of the band joist.
b. The maximum gap between the face of the ledger board and face of the wall sheathing shall be $^1/_2$ inch.
c. Ledgers shall be flashed to prevent water from contacting the house band joist.
d. Lag screws and bolts shall be staggered in accordance with Section R507.2.1.
e. Deck ledger shall be minimum 2 × 8 pressure-preservative-treated No. 2 grade lumber, or other approved materials as established by standard engineering practice.
f. When solid-sawn pressure-preservative-treated deck ledgers are attached to a minimum 1-inch-thick engineered wood product (structural composite lumber, laminated veneer lumber or wood structural panel band joist), the ledger attachment shall be designed in accordance with accepted engineering practice.
g. A minimum 1 × 9$^1/_2$ Douglas Fir laminated veneer lumber rimboard shall be permitted in lieu of the 2-inch nominal band joist.
h. Wood structural panel sheathing, gypsum board sheathing or foam sheathing not exceeding 1 inch in thickness shall be permitted. The maximum distance between the face of the ledger board and the face of the band joist shall be 1 inch.

TABLE 507.2.1
PLACEMENT OF LAG SCREWS AND BOLTS IN DECK LEDGERS AND BAND JOISTS

	MINIMUM END AND EDGE DISTANCES AND SPACING BETWEEN ROWS			
	TOP EDGE	BOTTOM EDGE	ENDS	ROW SPACING
Ledger[a]	2 inches[d]	$^1/_4$ inch	2 inches[b]	1$^5/_8$ inches[b]
Band Joist[c]	$^3/_4$ inch	2 inches	2 inches[b]	1$^5/_8$ inches[b]

For SI: 1 inch = 25.4 mm.

a. Lag screws or bolts shall be staggered from the top to the bottom along the horizontal run of the deck ledger in accordance with Figure R507.2.1(1).
b. Maximum 5 inches.
c. For engineered rim joists, the manufacturer's recommendations shall govern.
d. The minimum distance from bottom row of lag screws or bolts to the top edge of the ledger shall be in accordance with Figure R507.2.1(1).

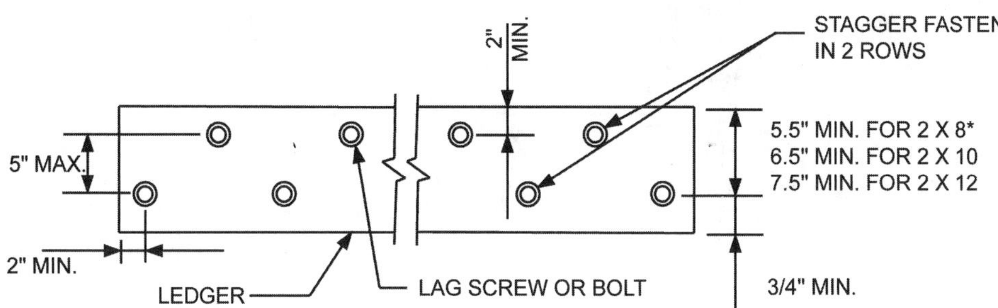

*DISTANCE SHALL BE PERMITTED TO BE REDUCED TO 4.5" IF LAG SCREWS ARE USED OR BOLT SPACING IS REDUCED TO THAT OF LAG SCREWS TO ATTACH 2 X 8 LEDGERS TO 2 X 8 BAND JOISTS.

For SI: 1 inch = 25.4 mm.

FIGURE R507.2.1(1)
PLACEMENT OF LAG SCREWS AND BOLTS IN LEDGERS

R507.3 Wood/plastic composites. Wood/plastic composites used in exterior deck boards, stair treads, handrails and guardrail systems shall bear a label indicating the required performance levels and demonstrating compliance with the provisions of ASTM D 7032.

R507.3.1 Installation of wood/plastic composites. Wood/plastic composites shall be installed in accordance with the manufacturer's instructions.

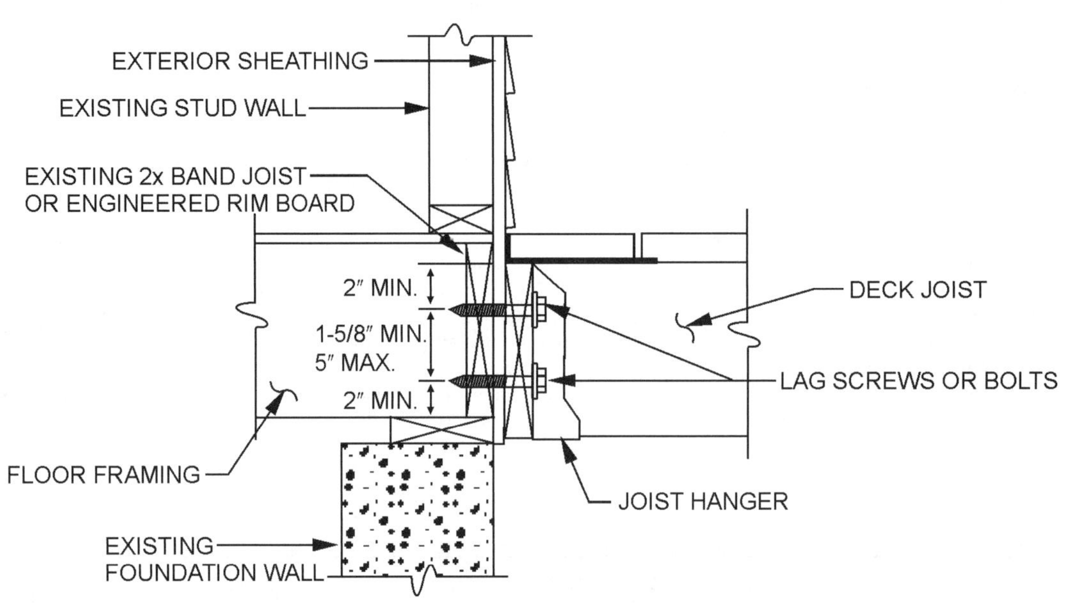

For SI: 1 inch = 25.4 mm.

FIGURE R507.2.1(2)
PLACEMENT OF LAG SCREWS AND BOLTS IN BAND JOISTS

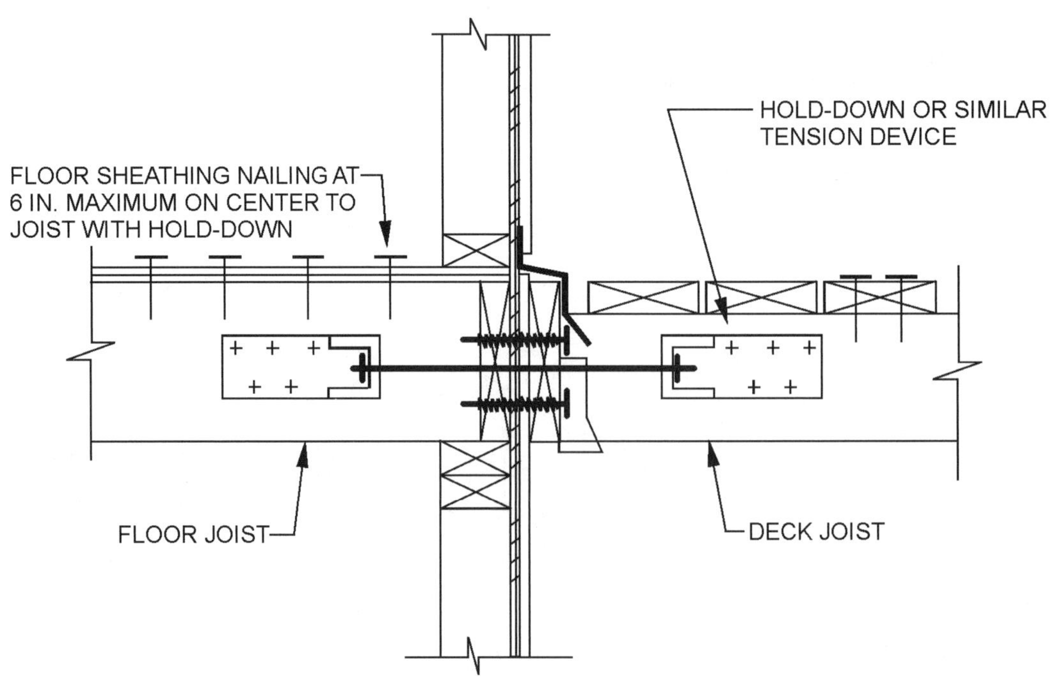

For SI: 1 inch = 25.4 mm.

FIGURE 507.2.3
DECK ATTACHMENT FOR LATERAL LOADS

CHAPTER 6
WALL CONSTRUCTION

SECTION R601
GENERAL

R601.1 Application. The provisions of this chapter shall control the design and construction of all walls and partitions for all buildings.

R601.2 Requirements. Wall construction shall be capable of accommodating all loads imposed according to Section R301 and of transmitting the resulting loads to the supporting structural elements.

R601.2.1 Compressible floor-covering materials. Compressible floor-covering materials that compress more than $^{1}/_{32}$ inch (0.8 mm) when subjected to 50 pounds (23 kg) applied over 1 inch square (645 mm) of material and are greater than $^{1}/_{8}$ inch (3 mm) in thickness in the uncompressed state shall not extend beneath walls, partitions or columns, which are fastened to the floor.

*

SECTION R602
WOOD WALL FRAMING

R602.1 Identification. Load-bearing dimension lumber for studs, plates and headers shall be identified by a grade mark of a lumber grading or inspection agency that has been *approved* by an accreditation body that complies with DOC PS 20. In lieu of a grade mark, a certification of inspection issued by a lumber grading or inspection agency meeting the requirements of this section shall be accepted.

R602.1.1 End-jointed lumber. Approved end-jointed lumber identified by a grade mark conforming to Section R602.1 may be used interchangeably with solid-sawn members of the same species and grade. End-jointed lumber used in an assembly required elsewhere in this code to have a fire-resistance rating shall have the designation "Heat Resistant Adhesive" or "HRA" included in its grade mark.

R602.1.2 Structural glued laminated timbers. Glued laminated timbers shall be manufactured and identified as required in ANSI/AITC A190.1 and ASTM D 3737.

R602.1.3 Structural log members. Stress grading of structural log members of nonrectangular shape, as typically used in log buildings, shall be in accordance with ASTM D 3957. Such structural log members shall be identified by the grade mark of an *approved* lumber grading or inspection agency. In lieu of a grade mark on the material, a certificate of inspection as to species and grade, issued by a lumber-grading or inspection agency meeting the requirements of this section, shall be permitted to be accepted.

R602.1.4 Structural composite lumber. Structural capacities for structural composite lumber shall be established and monitored in accordance with ASTM D 5456.

R602.2 Grade. Studs shall be a minimum No. 3, standard or stud grade lumber.

Exception: Bearing studs not supporting floors and nonbearing studs may be utility grade lumber, provided the studs are spaced in accordance with Table R602.3(5).

R602.3 Design and construction. Exterior walls of wood-frame construction shall be designed and constructed in accordance with the provisions of this chapter and Figures R602.3(1) and R602.3(2) or in accordance with AF&PA's NDS. Components of exterior walls shall be fastened in accordance with Tables R602.3(1) through R602.3(4). Wall sheathing shall be fastened directly to framing members and, when placed on the exterior side of an exterior wall, shall be capable of resisting the wind pressures listed in Table R301.2(2) adjusted for height and exposure using Table R301.2(3). Wood structural panel sheathing used for exterior walls shall conform to DOC PS 1, DOC PS 2 or, when manufactured in Canada, CSA 0437 or CSA 0325. All panels shall be identified for grade, bond classification, and Performance Category by a grade mark or certificate of inspection issued by an approved agency and shall conform to the requirements of Table R602.3(3). Wall sheathing used only for exterior wall covering purposes shall comply with Section R703.

Studs shall be continuous from support at the sole plate to a support at the top plate to resist loads perpendicular to the wall. The support shall be a foundation or floor, ceiling or roof diaphragm or shall be designed in accordance with accepted engineering practice.

Exception: Jack studs, trimmer studs and cripple studs at openings in walls that comply with Tables R502.5(1) and R502.5(2).

R602.3.1 Stud size, height and spacing. The size, height and spacing of studs shall be in accordance with Table R602.3.(5).

Exceptions:

1. Utility grade studs shall not be spaced more than 16 inches (406 mm) on center, shall not support more than a roof and ceiling, and shall not exceed 8 feet (2438 mm) in height for exterior walls and load-bearing walls or 10 feet (3048 mm) for interior nonload-bearing walls.
2. Studs more than 10 feet (3048 mm) in height which are in accordance with Table R602.3.1.

TABLE R602.3(1)
FASTENER SCHEDULE FOR STRUCTURAL MEMBERS

ITEM	DESCRIPTION OF BUILDING ELEMENTS	NUMBER AND TYPE OF FASTENER[a, b, c]	SPACING OF FASTENERS
		Roof	
1	Blocking between joists or rafters to top plate, toe nail	3-8d (2$^1/_2$″ × 0.113″)	—
2	Ceiling joists to plate, toe nail	3-8d (2$^1/_2$″ × 0.113″)	—
3	Ceiling joists not attached to parallel rafter, laps over partitions, face nail	3-10d	—
4	Collar tie to rafter, face nail or 1$^1/_4$″ × 20 gage ridge strap	3-10d (3″ × 0.128″)	—
5	Rafter or roof truss to plate, toe nail	3-16d box nails (3$^1/_2$″ × 0.135″) or 3-10d common nails (3″ × 0.148″)	2 toe nails on one side and 1 toe nail on opposite side of each rafter or truss[j]
6	Roof rafters to ridge, valley or hip rafters: toe nail face nail	4-16d (3$^1/_2$″ × 0.135″) 3-16d (3$^1/_2$″ × 0.135″)	—
		Wall	
7	Built-up studs-face nail	10d (3″ × 0.128″)	24″ o.c.
8	Abutting studs at intersecting wall corners, face nail	16d (3 ½" x 0.135")	12″ o.c.
9	Built-up header, two pieces with $^1/_2$″ spacer	16d (3$^1/_2$″ × 0.135″)	16″ o.c. along each edge
10	Continued header, two pieces	16d (3$^1/_2$″ × 0.135″)	16″ o.c. along each edge
11	Continuous header to stud, toe nail	4-8d (2$^1/_2$″ × 0.113″)	—
12	Double studs, face nail	10d (3″ × 0.128″)	24″ o.c.
13	Double top plates, face nail	10d (3″ × 0.128″)	24″ o.c.
14	Double top plates, minimum 24-inch offset of end joints, face nail in lapped area	8-16d (3$^1/_2$″ × 0.135″)	—
15	Sole plate to joist or blocking, face nail	16d (3$^1/_2$″ × 0.135″)	16″ o.c.
16	Sole plate to joist or blocking at braced wall panels	3-16d (3$^1/_2$″ × 0.135″)	16″ o.c.
17	Stud to sole plate, toe nail	3-8d (2$^1/_2$″ × 0.113″) or 2-16d (3$^1/_2$″ × 0.135″)	— —
18	Top or sole plate to stud, end nail	2-16d (3$^1/_2$″ × 0.135″)	—
19	Top plates, laps at corners and intersections, face nail	2-10d (3″ × 0.128″)	—
20	1″ brace to each stud and plate, face nail	2-8d (2$^1/_2$″ × 0.113″) 2 staples 1$^3/_4$″	— —
21	1″ × 6″ sheathing to each bearing, face nail	2-8d (2$^1/_2$″ × 0.113″) 2 staples 1$^3/_4$″	— —
22	1″ × 8″ sheathing to each bearing, face nail	2-8d (2$^1/_2$″ × 0.113″) 3 staples 1$^3/_4$″	— —
23	Wider than 1″ × 8″ sheathing to each bearing, face nail	3-8d (2$^1/_2$″ × 0.113″) 4 staples 1$^3/_4$″	— —
		Floor	
24	Joist to sill or girder, toe nail	3-8d (2$^1/_2$″ × 0.113″)	—
25	Rim joist to top plate, toe nail (roof applications also)	8d (2$^1/_2$″ × 0.113″)	6″ o.c.
26	Rim joist or blocking to sill plate, toe nail	8d (2 ½″ × 0.113″)	6″ o.c.
27	1″ × 6″ subfloor or less to each joist, face nail	2-8d (2$^1/_2$″ × 0.113″) 2 staples 1$^3/_4$″	— —
28	2″ subfloor to joist or girder, blind and face nail	2-16d (3$^1/_2$″ × 0.135″)	—
29	2″ planks (plank & beam - floor & roof)	2-16d (3$^1/_2$″ × 0.135″)	at each bearing
30	Built-up girders and beams, 2-inch lumber layers	10d (3″ × 0.128″)	Nail each layer as follows: 32″ o.c. at top and bottom and staggered. Two nails at ends and at each splice.
31	Ledger strip supporting joists or rafters	3-16d (3$^1/_2$″ × 0.135″)	At each joist or rafter

(continued)

TABLE R602.3(1)—continued
FASTENER SCHEDULE FOR STRUCTURAL MEMBERS

ITEM	DESCRIPTION OF BUILDING MATERIALS	DESCRIPTION OF FASTENER[b, c, e]	SPACING OF FASTENERS	
			Edges (inches)[i]	Intermediate supports[c, e] (inches)
Wood structural panels, subfloor, roof and interior wall sheathing to framing and particleboard wall sheathing to framing				
32	$^3/_8$″ - $^1/_2$″	6d common (2″ × 0.113″) nail (subfloor wall)[j] 8d common (2$^1/_2$″ × 0.131″) nail (roof)[f]	6	12[g]
33	$^{19}/_{32}$″ - 1″	8d common nail (2$^1/_2$″ × 0.131″)	6	12[g]
34	1$^1/_8$″ - 1$^1/_4$″	10d common (3″ × 0.148″) nail or 8d (2$^1/_2$″ × 0.131″) deformed nail	6	12
Other wall sheathing[h]				
35	$^1/_2$″ structural cellulosic fiberboard sheathing	$^1/_2$″ galvanized roofing nail, $^7/_{16}$″ crown or 1″ crown staple 16 ga., 1$^1/_4$″ long	3	6
36	$^{25}/_{32}$″ structural cellulosic fiberboard sheathing	1$^3/_4$″ galvanized roofing nail, $^7/_{16}$″ crown or 1″ crown staple 16 ga., 1$^1/_2$″ long	3	6
37	$^1/_2$″ gypsum sheathing[d]	1$^1/_2$″ galvanized roofing nail; staple galvanized, 1$^1/_2$″ long; 1$^1/_4$ screws, Type W or S	7	7
38	$^5/_8$″ gypsum sheathing[d]	1$^3/_4$″ galvanized roofing nail; staple galvanized, 1$^5/_8$″ long; 1$^5/_8$″ screws, Type W or S	7	7
	Wood structural panels, combination subfloor underlayment to framing			
39	$^3/_4$″ and less	6d deformed (2″ × 0.120″) nail or 8d common (2$^1/_2$″ × 0.131″) nail	6	12
40	$^7/_8$″ - 1″	8d common (2$^1/_2$″ × 0.131″) nail or 8d deformed (2$^1/_2$″ × 0.120″) nail	6	12
41	1$^1/_8$″ - 1$^1/_4$″	10d common (3″ × 0.148″) nail or 8d deformed (2$^1/_2$″ × 0.120″) nail	6	12

For SI: 1 inch = 25.4 mm, 1 foot = 304.8 mm, 1 mile per hour = 0.447 m/s; 1 Ksi = 6.895 MPa.

a. All nails are smooth-common, box or deformed shanks except where otherwise stated. Nails used for framing and sheathing connections shall have minimum average bending yield strengths as shown: 80 ksi for shank diameter of 0.192 inch (20d common nail), 90 ksi for shank diameters larger than 0.142 inch but not larger than 0.177 inch, and 100 ksi for shank diameters of 0.142 inch or less.

b. Staples are 16 gage wire and have a minimum $^7/_{16}$-inch on diameter crown width.

c. Nails shall be spaced at not more than 6 inches on center at all supports where spans are 48 inches or greater.

d. Four-foot by 8-foot or 4-foot by 9-foot panels shall be applied vertically.

e. Spacing of fasteners not included in this table shall be based on Table R602.3(2).

f. For regions having basic wind speed of 110 mph or greater, 8d deformed (2$^1/_2$″ × 0.120) nails shall be used for attaching plywood and wood structural panel roof sheathing to framing within minimum 48-inch distance from gable end walls, if mean roof height is more than 25 feet, up to 35 feet maximum.

g. For regions having basic wind speed of 100 mph or less, nails for attaching wood structural panel roof sheathing to gable end wall framing shall be spaced 6 inches on center. When basic wind speed is greater than 100 mph, nails for attaching panel roof sheathing to intermediate supports shall be spaced 6 inches on center for minimum 48-inch distance from ridges, eaves and gable end walls; and 4 inches on center to gable end wall framing.

h. Gypsum sheathing shall conform to ASTM C 1396 and shall be installed in accordance with GA 253. Fiberboard sheathing shall conform to ASTM C 208.

i. Spacing of fasteners on floor sheathing panel edges applies to panel edges supported by framing members and required blocking and at all floor perimeters only. Spacing of fasteners on roof sheathing panel edges applies to panel edges supported by framing members and required blocking. Blocking of roof or floor sheathing panel edges perpendicular to the framing members need not be provided except as required by other provisions of this code. Floor perimeter shall be supported by framing members or solid blocking.

j. Where a rafter is fastened to an adjacent parallel ceiling joist in accordance with this schedule, provide two toe nails on one side of the rafter and toe nails from the ceiling joist to top plate in accordance with this schedule. The toe nail on the opposite side of the rafter shall not be required.

**TABLE R602.3(2)
ALTERNATE ATTACHMENTS TO TABLE R602.3(1)**

NOMINAL MATERIAL THICKNESS (inches)	DESCRIPTION[a, b] OF FASTENER AND LENGTH (inches)	SPACING[c] OF FASTENERS	
		Edges (inches)	Intermediate supports (inches)
Wood structural panels subfloor, roof[g] and wall sheathing to framing and particleboard wall sheathing to framing[f]			
Up to $^1/_2$	Staple 15 ga. $1^3/_4$	4	8
	0.097 - 0.099 Nail $2^1/_4$	3	6
	Staple 16 ga. $1^3/_4$	3	6
$^{19}/_{32}$ and $^5/_8$	0.113 Nail 2	3	6
	Staple 15 and 16 ga. 2	4	8
	0.097 - 0.099 Nail $2^1/_4$	4	8
$^{23}/_{32}$ and $^3/_4$	Staple 14 ga. 2	4	8
	Staple 15 ga. $1^3/_4$	3	6
	0.097 - 0.099 Nail $2^1/_4$	4	8
	Staple 16 ga. 2	4	8
1	Staple 14 ga. $2^1/_4$	4	8
	0.113 Nail $2^1/_4$	3	6
	Staple 15 ga. $2^1/_4$	4	8
	0.097 - 0.099 Nail $2^1/_2$	4	8
NOMINAL MATERIAL THICKNESS (inches)	DESCRIPTION[a,b] OF FASTENER AND LENGTH (inches)	SPACING[c] OF FASTENERS	
		Edges (inches)	Body of panel[d] (inches)
Floor underlayment; plywood-hardboard-particleboard[f]			
Plywood			
$^1/_4$ and $^5/_{16}$	$1^1/_4$ ring or screw shank nail-minimum $12^1/_2$ ga. (0.099″) shank diameter	3	6
	Staple 18 ga., $^7/_8$, $^3/_{16}$ crown width	2	5
$^{11}/_{32}$, $^3/_8$, $^{15}/_{32}$, and $^1/_2$	$1^1/_4$ ring or screw shank nail-minimum $12^1/_2$ ga. (0.099″) shank diameter	6	8[e]
$^{19}/_{32}$, $^5/_8$, $^{23}/_{32}$ and $^3/_4$	$1^1/_2$ ring or screw shank nail-minimum $12^1/_2$ ga. (0.099″) shank diameter	6	8
	Staple 16 ga. $1^1/_2$	6	8
Hardboard[f]			
0.200	$1^1/_2$ long ring-grooved underlayment nail	6	6
	4d cement-coated sinker nail	6	6
	Staple 18 ga., $^7/_8$ long (plastic coated)	3	6
Particleboard			
$^1/_4$	4d ring-grooved underlayment nail	3	6
	Staple 18 ga., $^7/_8$ long, $^3/_{16}$ crown	3	6
$^3/_8$	6d ring-grooved underlayment nail	6	10
	Staple 16 ga., $1^1/_8$ long, $^3/_8$ crown	3	6
$^1/_2$, $^5/_8$	6d ring-grooved underlayment nail	6	10
	Staple 16 ga., $1^5/_8$ long, $^3/_8$ crown	3	6

For SI: 1 inch = 25.4 mm.

a. Nail is a general description and may be T-head, modified round head or round head.

b. Staples shall have a minimum crown width of $^7/_{16}$-inch on diameter except as noted.

c. Nails or staples shall be spaced at not more than 6 inches on center at all supports where spans are 48 inches or greater. Nails or staples shall be spaced at not more than 12 inches on center at intermediate supports for floors.

d. Fasteners shall be placed in a grid pattern throughout the body of the panel.

e. For 5-ply panels, intermediate nails shall be spaced not more than 12 inches on center each way.

f. Hardboard underlayment shall conform to CPA/ANSI A135.4

g. Specified alternate attachments for roof sheathing shall be permitted for windspeeds less than 100 mph. Fasteners attaching wood structural panel roof sheathing to gable end wall framing shall be installed using the spacing listed for panel edges.

TABLE R602.3(3) REQUIREMENTS FOR WOOD STRUCTURAL PANEL WALL SHEATHING USED TO RESIST WIND PRESSURES[a, b, c]

MINIMUM NAIL		MINIMUM WOOD STRUCTURAL PANEL SPAN RATING	MINIMUM NOMINAL PANEL THICKNESS (inches)	MAXIMUM WALL STUD SPACING (inches)	PANEL NAIL SPACING		MAXIMUM WIND SPEED (mph)		
Size	Penetration (inches)				Edges (inches o.c.)	Field (inches o.c.)	Wind exposure category		
							B	C	D
6d Common (2.0″ × 0.113″)	1.5	24/0	$^{3}/_{8}$	16	6	12	110	90	85
8d Common (2.5″ × 0.131″)	1.75	24/16	$^{7}/_{16}$	16	6	12	130	110	105
				24	6	12	110	90	85

For SI: 1 inch = 25.4 mm, 1 mile per hour = 0.447 m/s.

a. Panel strength axis parallel or perpendicular to supports. Three-ply plywood sheathing with studs spaced more than 16 inches on center shall be applied with panel strength axis perpendicular to supports.

b. Table is based on wind pressures acting toward and away from building surfaces per Section R301.2. Lateral bracing requirements shall be in accordance with Section R602.10.

c. Wood structural panels with span ratings of Wall-16 or Wall-24 shall be permitted as an alternate to panels with a 24/0 span rating. Plywood siding rated 16 o.c. or 24 o.c. shall be permitted as an alternate to panels with a 24/16 span rating. Wall-16 and Plywood siding 16 o.c. shall be used with studs spaced a maximum of 16 inches on center.

TABLE R602.3(4) ALLOWABLE SPANS FOR PARTICLEBOARD WALL SHEATHING[a]

THICKNESS (inch)	GRADE	STUD SPACING (inches)	
		When siding is nailed to studs	When siding is nailed to sheathing
$^{3}/_{8}$	M-1 Exterior glue	16	—
$^{1}/_{2}$	M-2 Exterior glue	16	16

For SI: 1 inch = 25.4 mm.

a. Wall sheathing not exposed to the weather. If the panels are applied horizontally, the end joints of the panel shall be offset so that four panels corners will not meet. All panel edges must be supported. Leave a $^{1}/_{16}$-inch gap between panels and nail no closer than $^{3}/_{8}$ inch from panel edges.

TABLE R602.3(5) SIZE, HEIGHT AND SPACING OF WOOD STUDS[a]

STUD SIZE (inches)	BEARING WALLS					NONBEARING WALLS	
	Laterally unsupported stud height[a] (feet)	Maximum spacing when supporting a roof-ceiling assembly or a habitable attic assembly, only (inches)	Maximum spacing when supporting one floor, plus a roof-ceiling assembly or a habitable attic assembly (inches)	Maximum spacing when supporting two floors, plus a roof-ceiling assembly or a habitable attic assembly (inches)	Maximum spacing when supporting one floor height[a] (feet)	Laterally unsupported stud height[a] (feet)	Maximum spacing (inches)
2 × 3[b]	—	—	—	—	—	10	16
2 × 4	10	24[c]	16[c]	—	24	14	24
3 × 4	10	24	24	16	24	14	24
2 × 5	10	24	24	—	24	16	24
2 × 6	10	24	24	16	24	20	24

For SI: 1 inch = 25.4 mm, 1 foot = 304.8 mm, 1 square foot = 0.093 m^2.

a. Listed heights are distances between points of lateral support placed perpendicular to the plane of the wall. Increases in unsupported height are permitted where justified by analysis.

b. Shall not be used in exterior walls.

c. A habitable attic assembly supported by 2 × 4 studs is limited to a roof span of 32 feet. Where the roof span exceeds 32 feet, the wall studs shall be increased to 2 × 6 or the studs shall be designed in accordance with accepted engineering practice.

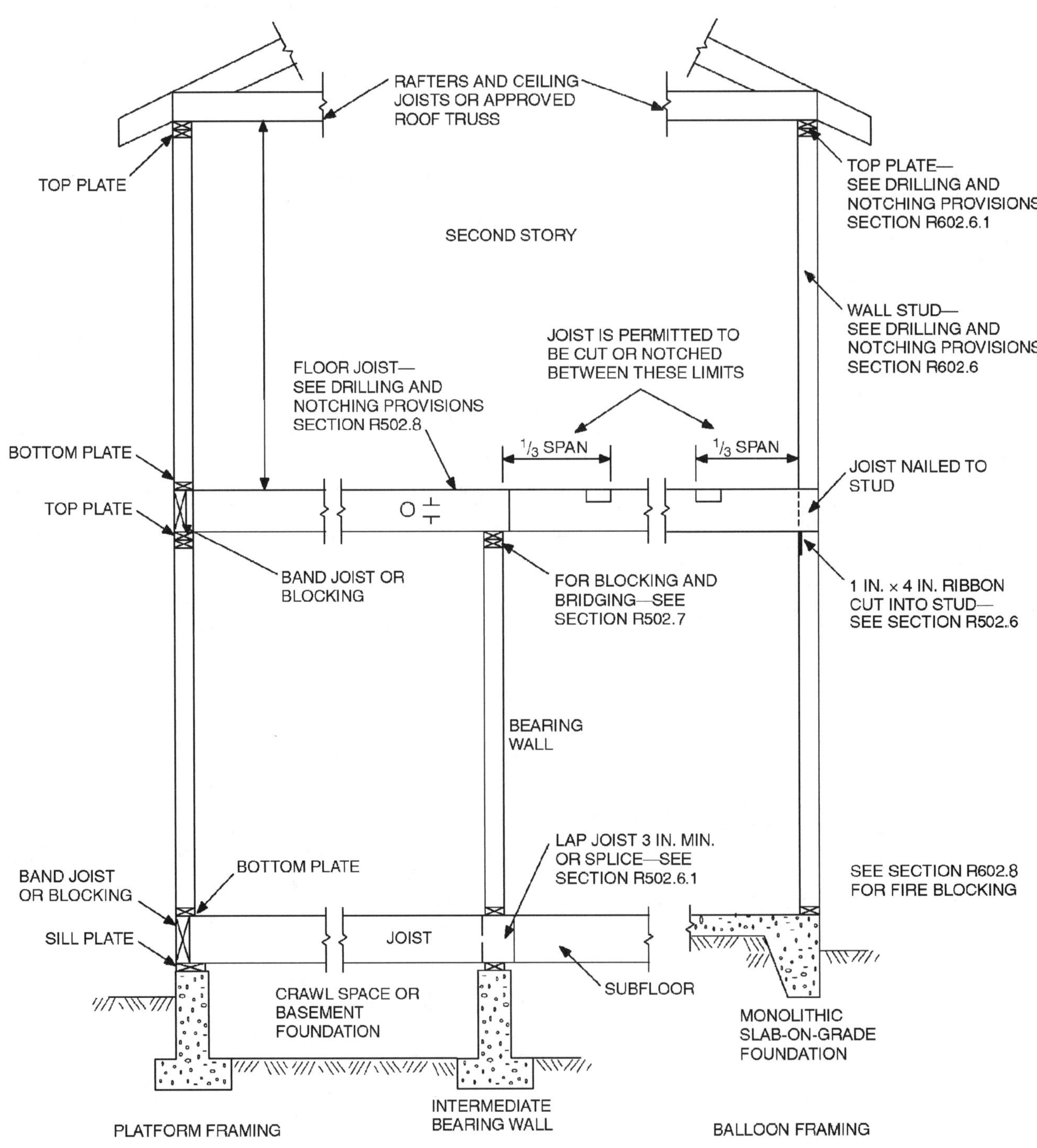

For SI: 1 inch = 25.4 mm, 1 foot = 304.8 mm.

FIGURE R602.3(1)
TYPICAL WALL, FLOOR AND ROOF FRAMING

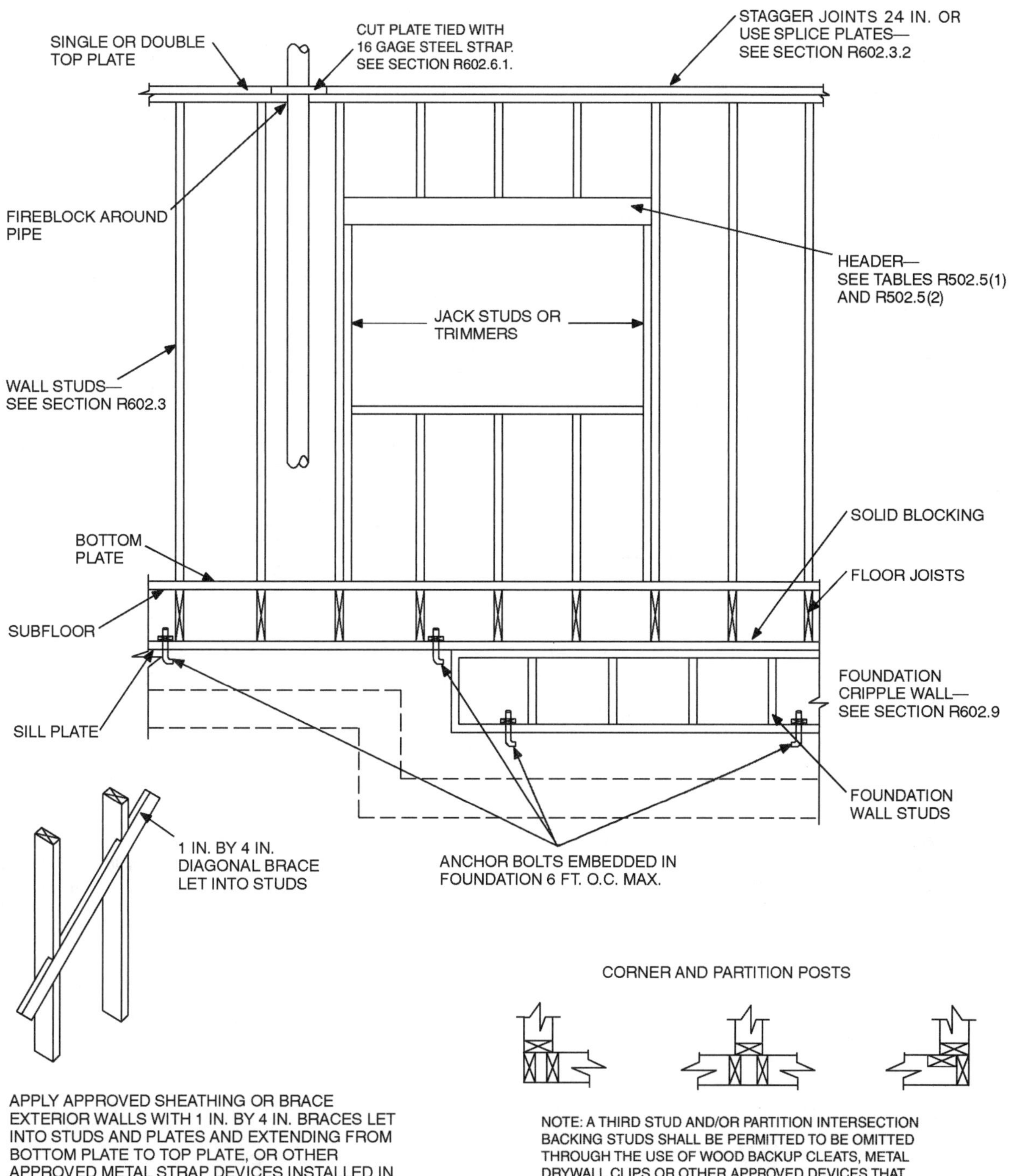

For SI: 1 inch = 25.4 mm, 1 foot = 304.8 mm.

FIGURE R602.3(2)
FRAMING DETAILS

TABLE R602.3.1
MAXIMUM ALLOWABLE LENGTH OF WOOD WALL STUDS EXPOSED TO WIND SPEEDS OF 100 MPH OR LESS IN SEISMIC DESIGN CATEGORIES A, B, C, D_0, D_1 and D_2[b, c]

HEIGHT (feet)	ON-CENTER SPACING (inches)			
	24	16	12	8
Supporting a roof only				
> 10	2 × 4	2 × 4	2 × 4	2 × 4
12	2 × 6	2 × 4	2 × 4	2 × 4
14	2 × 6	2 × 6	2 × 6	2 × 4
16	2 × 6	2 × 6	2 × 6	2 × 4
18	NA[a]	2 × 6	2 × 6	2 × 6
20	NA[a]	NA[a]	2 × 6	2 × 6
24	NA[a]	NA[a]	NA[a]	2 × 6
Supporting one floor and a roof				
> 10	2 × 6	2 × 4	2 × 4	2 × 4
12	2 × 6	2 × 6	2 × 6	2 × 4
14	2 × 6	2 × 6	2 × 6	2 × 6
16	NA[a]	2 × 6	2 × 6	2 × 6
18	NA[a]	2 × 6	2 × 6	2 × 6
20	NA[a]	NA[a]	2 × 6	2 × 6
24	NA[a]	NA[a]	NA[a]	2 × 6
Supporting two floors and a roof				
> 10	2 × 6	2 × 6	2 × 4	2 × 4
12	2 × 6	2 × 6	2 × 6	2 × 6
14	2 × 6	2 × 6	2 × 6	2 × 6
16	NA[a]	NA[a]	2 × 6	2 × 6
18	NA[a]	NA[a]	2 × 6	2 × 6
20	NA[a]	NA[a]	NA[a]	2 × 6
22	NA[a]	NA[a]	NA[a]	NA[a]
24	NA[a]	NA[a]	NA[a]	NA[a]

For SI: 1 inch = 25.4 mm, 1 foot = 304.8 mm, 1 pound per square foot = 0.0479 kPa
1 pound per square inch = 6.895 kPa, 1 mile per hour = 0.447 m/s.

a. Design required.

b. Applicability of this table assumes the following: Snow load not exceeding 25 psf, f_b not less than 1310 psi determined by multiplying the AF&PA NDS tabular base design value by the repetitive use factor, and by the size factor for all species except southern pine, E not less than 1.6×10^6 psi, tributary dimensions for floors and roofs not exceeding 6 feet, maximum span for floors and roof not exceeding 12 feet, eaves not over 2 feet in dimension and exterior sheathing. Where the conditions are not within these parameters, design is required.

c. Utility, standard, stud and No. 3 grade lumber of any species are not permitted.

(continued)

TABLE R602.3.1—continued
MAXIMUM ALLOWABLE LENGTH OF WOOD WALL STUDS EXPOSED TO WIND SPEEDS OF 100 MPH OR LESS IN SEISMIC DESIGN CATEGORIES A, B, C, D_0, D_1 and D_2

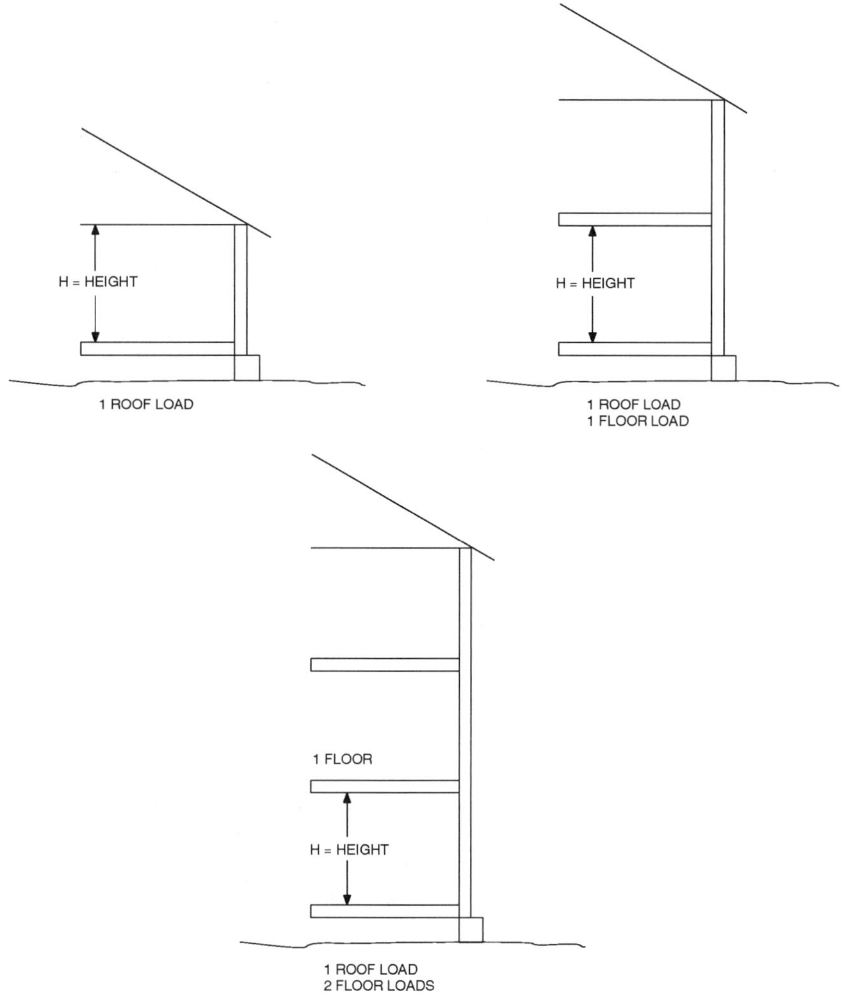

R602.3.2 Top plate. Wood stud walls shall be capped with a double top plate installed to provide overlapping at corners and intersections with bearing partitions. End joints in top plates shall be offset at least 24 inches (610 mm). Joints in plates need not occur over studs. Plates shall be not less than 2-inches (51 mm) nominal thickness and have a width at least equal to the width of the studs.

Exception: A single top plate may be installed in stud walls, provided the plate is adequately tied at joints, corners and intersecting walls by a minimum 3-inch by 6-inch by a 0.036-inch-thick (76 mm by 152 mm by 0.914 mm) galvanized steel plate that is nailed to each wall or segment of wall by six 8d nails on each side, provided the rafters or joists are centered over the studs with a tolerance of no more than 1 inch (25 mm). The top plate may be omitted over lintels that are adequately tied to adjacent wall sections with steel plates or equivalent as previously described.

R602.3.3 Bearing studs. Where joists, trusses or rafters are spaced more than 16 inches (406 mm) on center and the bearing studs below are spaced 24 inches (610 mm) on center, such members shall bear within 5 inches (127 mm) of the studs beneath.

Exceptions:

1. The top plates are two 2-inch by 6-inch (38 mm by 140 mm) or two 3-inch by 4-inch (64 mm by 89 mm) members.
2. A third top plate is installed.
3. Solid blocking equal in size to the studs is installed to reinforce the double top plate.

R602.3.4 Bottom (sole) plate. Studs shall have full bearing on a nominal 2-by (51 mm) or larger plate or sill having a width at least equal to the width of the studs. **

R602.3.5 Braced wall panel uplift load path. *Braced wall panels* located at exterior walls that support roof rafters or trusses (including stories below top *story*) shall have the framing members connected in accordance with one of the following:

1. Fastening in accordance with Table R602.3(1) where:
   1.1. The basic wind speed does not exceed 90 mph (40 m/s), the wind exposure category is B, the roof pitch is 5:12 or greater, and the roof span is 32 feet (9754 mm) or less, or
   1.2. The net uplift value at the top of a wall does not exceed 100 plf. The net uplift value shall be determined in accordance with Section R802.11 and shall be permitted to be reduced by 60 plf (86 N/mm) for each full wall above.
2. Where the net uplift value at the top of a wall exceeds 100 plf (146 N/mm), installing *approved* uplift framing connectors to provide a continuous load path from the top of the wall to the foundation or to a point where the uplift force is 100 plf (146 N/mm) or less. The net uplift value shall be as determined in Item 1.2 above.
3. Wall sheathing and fasteners designed in accordance with accepted engineering practice to resist combined uplift and shear forces.

R602.4 Interior load-bearing walls. Interior load-bearing walls shall be constructed, framed and fireblocked as specified for exterior walls.

R602.5 Interior nonbearing walls. Interior nonbearing walls shall be permitted to be constructed with 2 inch by 3 inch (51 mm by 76 mm) studs spaced 24 inches (610 mm) on center or, when not part of a *braced wall line*, 2 inch by 4 inch (51 mm by 102 mm) flat studs spaced at 16 inches (406 mm) on center. Interior nonbearing walls shall be capped with at least a single top plate. Interior nonbearing walls shall be fireblocked in accordance with Section R602.8.

R602.6 Drilling and notching of studs. Drilling and notching of studs shall be in accordance with the following:

1. Notching. Any stud in an exterior wall or bearing partition may be cut or notched to a depth not exceeding 25 percent of its width. Studs in nonbearing partitions may be notched to a depth not to exceed 40 percent of a single stud width.
2. Drilling. Any stud may be bored or drilled, provided that the diameter of the resulting hole is no more than 60 percent of the stud width, the edge of the hole is no more than $^5/_8$ inch (16 mm) to the edge of the stud, and the hole is not located in the same section as a cut or notch. Studs located in exterior walls or bearing partitions drilled over 40 percent and up to 60 percent shall also be doubled with no more than two successive doubled studs bored. See Figures R602.6(1) and R602.6(2).

   Exception: Use of *approved* stud shoes is permitted when they are installed in accordance with the manufacturer's recommendations.

R602.6.1 Drilling and notching of top plate. When piping or ductwork is placed in or partly in an exterior wall or interior load-bearing wall, necessitating cutting, drilling or notching of the top plate by more than 50 percent of its width, a galvanized metal tie not less than 0.054 inch thick (1.37 mm) (16 ga) and $1^1/_2$ inches (38 mm) wide shall be fastened across and to the plate at each side of the opening with not less than eight 10d (0.148 inch diameter) having a minimum length of $1^1/_2$ inches (38 mm) at each side or equivalent. The metal tie must extend a minimum of 6 inches past the opening. See Figure R602.6.1.

Exception: When the entire side of the wall with the notch or cut is covered by wood structural panel sheathing.

R602.7 Headers. For header spans see Tables R502.5(1), R502.5(2) and R602.7.1.

R602.7.1 Single member headers. Single headers shall be framed with a single flat 2-inch-nominal (51 mm) member or wall plate not less in width than the wall studs on the top and bottom of the header in accordance with Figures R602.7.1(1) and R602.7.1(2).

R602.7.2 Wood structural panel box headers. Wood structural panel box headers shall be constructed in accordance with Figure R602.7.2 and Table R602.7.2.

R602.7.3 Nonbearing walls. Load-bearing headers are not required in interior or exterior nonbearing walls. A single flat 2-inch by 4-inch (51 mm by 102 mm) member may be used as a header in interior or exterior nonbearing walls for openings up to 8 feet (2438 mm) in width if the vertical distance to the parallel nailing surface above is not more than 24 inches (610 mm). For such nonbearing headers, no cripples or blocking are required above the header.

R602.8 Fireblocking required. Fireblocking shall be provided in accordance with Section R302.11.

R602.9 Cripple walls. Foundation cripple walls shall be framed of studs not smaller than the studding above. When exceeding 4 feet (1219 mm) in height, such walls shall be framed of studs having the size required for an additional *story*.

Cripple walls with a stud height less than 14 inches (356 mm) shall be continuously sheathed on one side with wood structural panels fastened to both the top and bottom plates in accordance with Table R602.3(1), or the cripple walls shall be constructed of solid blocking.

All cripple walls shall be supported on continuous foundations.

R602.10 Wall bracing. Buildings shall be braced in accordance with this section or, when applicable, Section R602.12. Where a building, or portion thereof, does not comply with one or more of the bracing requirements in this section, those portions shall be designed and constructed in accordance with Section R301.1.

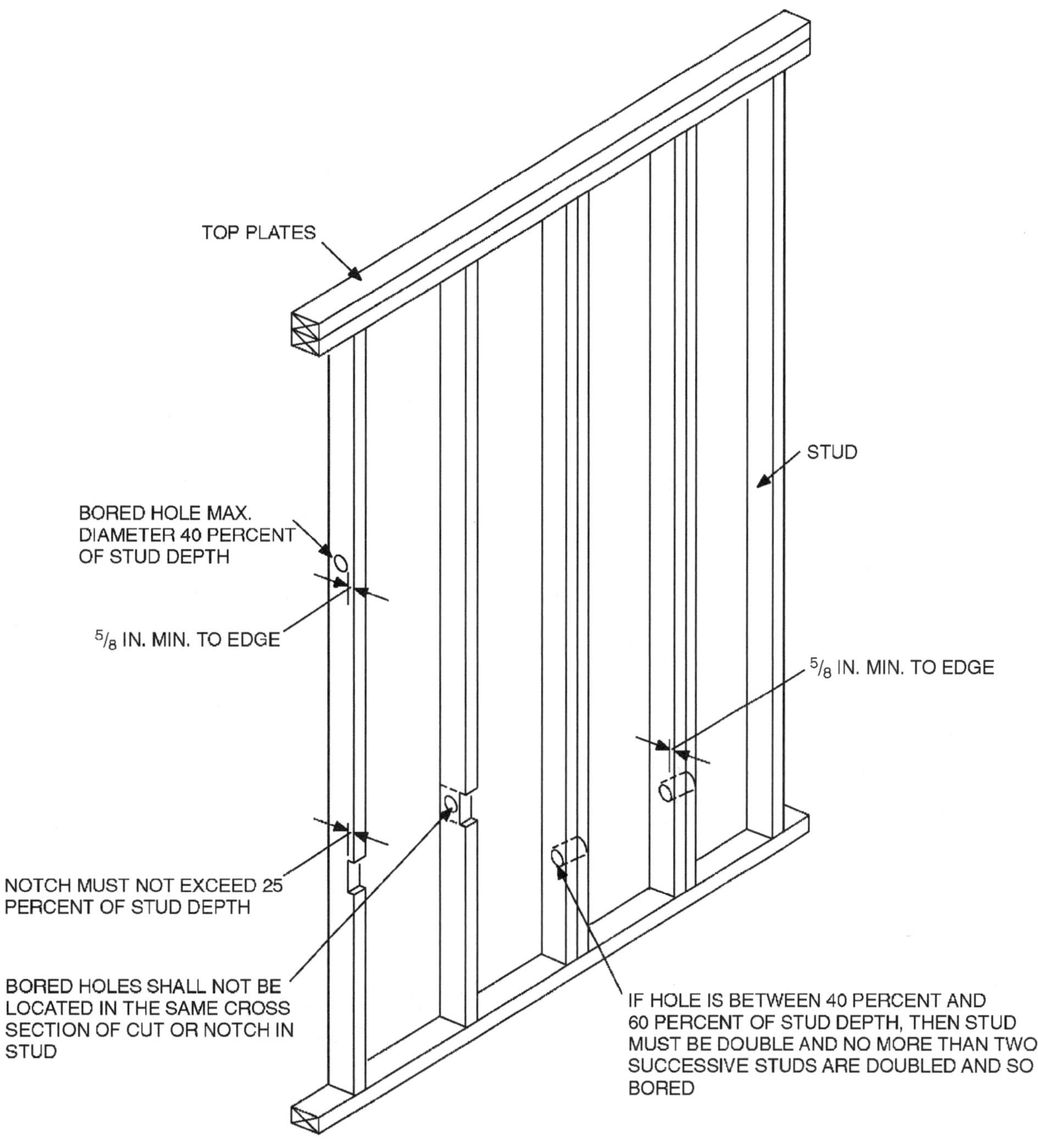

For SI: 1 inch = 25.4 mm.
Note: Condition for exterior and bearing walls.

FIGURE R602.6(1)
NOTCHING AND BORED HOLE LIMITATIONS FOR EXTERIOR WALLS AND BEARING WALLS

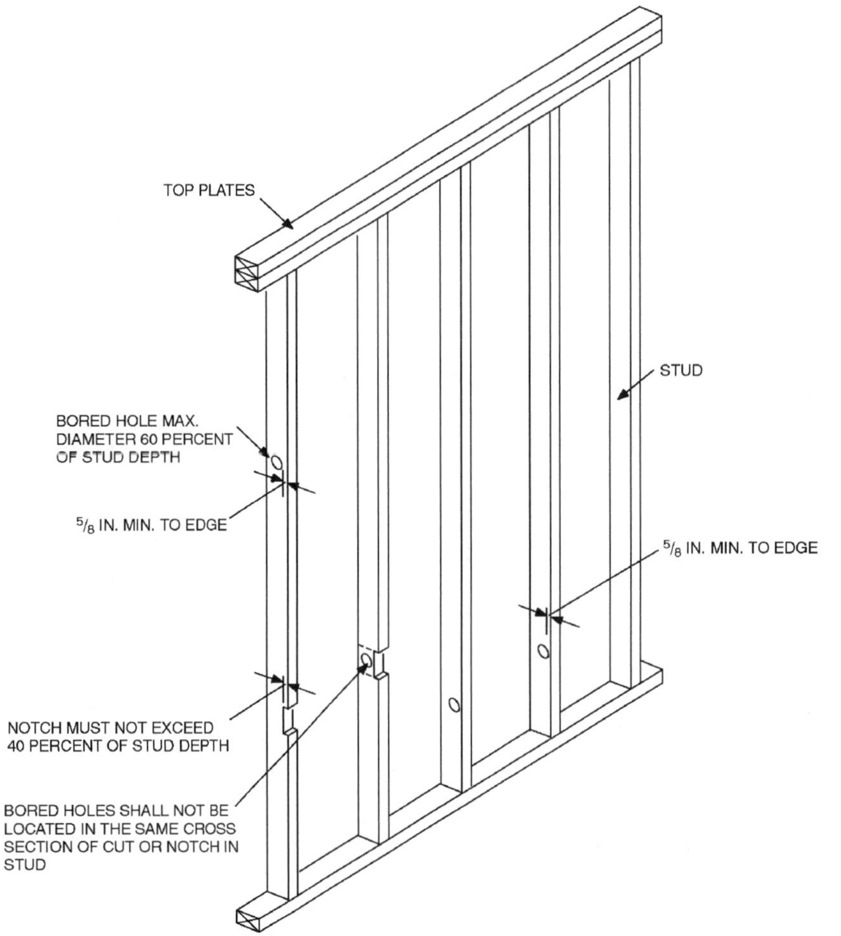

For SI: 1 inch = 25.4 mm.

FIGURE R602.6(2)
NOTCHING AND BORED HOLE LIMITATIONS FOR INTERIOR NONBEARING WALLS

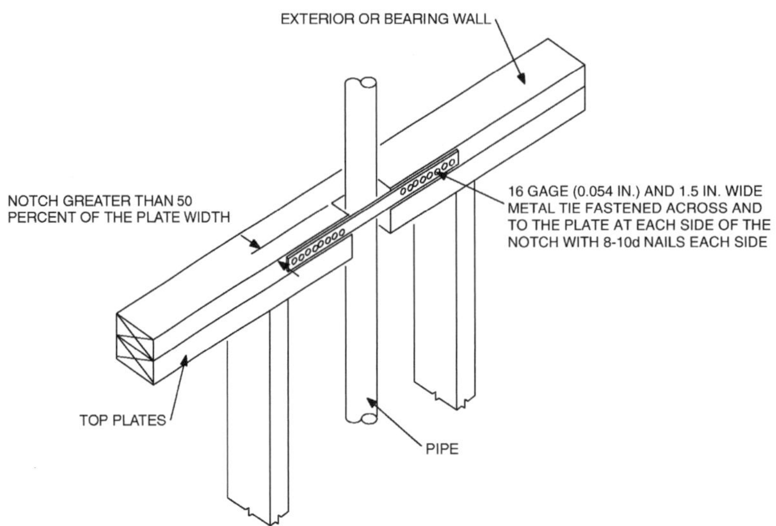

For SI: 1 inch = 25.4 mm.

FIGURE R602.6.1
TOP PLATE FRAMING TO ACCOMMODATE PIPING

TABLE R602.7.1
SPANS FOR MINIMUM No.2 GRADE SINGLE HEADER[a, b, c, f]

SINGLE HEADERS SUPPORTING	SIZE	WOOD SPECIES	GROUND SNOW LOAD (psf)								
			≤ 20[d]			30			50		
			Building Width (feet)[e]								
			20	28	36	20	28	36	20	28	36
Roof and ceiling	2 × 8	Spruce-Pine-Fir	4-10	4-2	3-8	4-3	3-8	3-3	3-7	3-0	2-8
		Hem-Fir	5-1	4-4	3-10	4-6	3-10	3-5	3-9	3-2	2-10
		Douglas-Fir or Southern Pine	5-3	4-6	4-0	4-7	3-11	3-6	3-10	3-3	2-11
	2 × 10	Spruce-Pine-Fir	6-2	5-3	4-8	5-5	4-8	4-2	4-6	3-11	3-1
		Hem-Fir	6-6	5-6	4-11	5-8	4-11	4-4	4-9	4-1	3-7
		Douglas-Fir or Southern Pine	6-8	5-8	5-1	5-10	5-0	4-6	4-11	4-2	3-9
	2 × 12	Spruce-Pine-Fir	7-6	6-5	5-9	6-7	5-8	4-5	5-4	3-11	3-1
		Hem-Fir	7-10	6-9	6-0	6-11	5-11	5-3	5-9	4-8	3-8
		Douglas-Fir or Southern Pine	8-1	6-11	6-2	7-2	6-1	5-5	5-11	5-1	4-6
Roof, ceiling and one center-bearing floor	2 × 8	Spruce-Pine-Fir	3-10	3-3	2-11	3-9	3-3	2-11	3-5	2-11	2-7
		Hem-Fir	4-0	3-5	3-1	3-11	3-5	3-0	3-7	3-0	2-8
		Douglas-Fir or Southern Pine	4-1	3-7	3-2	4-1	3-6	3-1	3-8	3-2	2-9
	2 × 10	Spruce-Pine-Fir	4-11	4-2	3-8	4-10	4-1	3-6	4-4	3-7	2-10
		Hem-Fir	5-1	4-5	3-11	5-0	4-4	3-10	4-6	3-11	3-4
		Douglas-Fir or Southern Pine	5-3	4-6	4-1	5-2	4-5	4-0	4-8	4-0	3-7
	2 × 12	Spruce-Pine-Fir	5-8	4-2	3-4	5-5	4-0	3-6	4-9	3-6	2-10
		Hem-Fir	5-11	4-11	3-11	5-10	4-9	4-2	5-5	4-2	3-4
		Douglas-Fir or Southern Pine	6-1	5-3	4-8	6-0	5-2	4-10	5-7	4-10	4-3
Roof, ceiling and one clear span floor	2 × 8	Spruce-Pine-Fir	3-5	2-11	2-7	3-4	2-11	2-7	3-3	2-10	2-6
		Hem-Fir	3-7	3-1	2-9	3-6	3-0	2-8	3-5	2-11	2-7
		Douglas-Fir or Southern Pine	3-8	3-2	2-10	3-7	3-1	2-9	3-6	3-0	2-9
	2 × 10	Spruce-Pine-Fir	4-4	3-7	2-10	4-3	3-6	2-9	4-2	3-4	2-7
		Hem-Fir	4-7	3-11	3-5	4-6	3-10	3-3	4-4	3-9	3-1
		Douglas-Fir or Southern Pine	4-8	4-0	3-7	4-7	4-0	3-6	4-6	3-10	3-5
	2 × 12	Spruce-Pine-Fir	4-11	3-7	2-10	4-9	3-6	2-9	4-6	3-4	2-7
		Hem-Fir	5-6	4-3	3-5	5-6	4-2	3-3	5-4	3-11	3-1
		Douglas-Fir or Southern Pine	5-8	4-11	4-4	5-7	4-10	4-3	5-6	4-8	4-2

For SI: 1 inch=25.4 mm, 1 pound per square foot = 0.0479 kPa.

a. Spans are given in feet and inches.

b. Table is based on a maximum roof-ceiling dead load of 15 psf.

c. The header is permitted to be supported by an approved framing anchor attached to the full-height wall stud and to the header in lieu of the required jack stud.

d. The 20 psf ground snow load condition shall apply only when the roof pitch is 9:12 or greater. In conditions where the ground snow load is 30 psf or less and the roof pitch is less than 9:12, use the 30 psf ground snow load condition.

e. Building width is measured perpendicular to the ridge. For widths between those shown, spans are permitted to be interpolated.

f. The header shall bear on a minimum of one jack stud at each end.

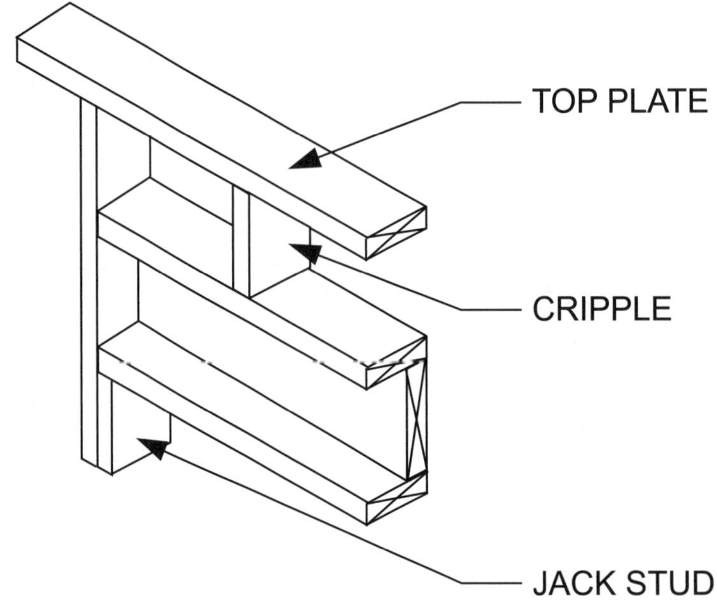

FIGURE R602.7.1(1)
SINGLE MEMBER HEADER IN EXTERIOR BEARING WALL

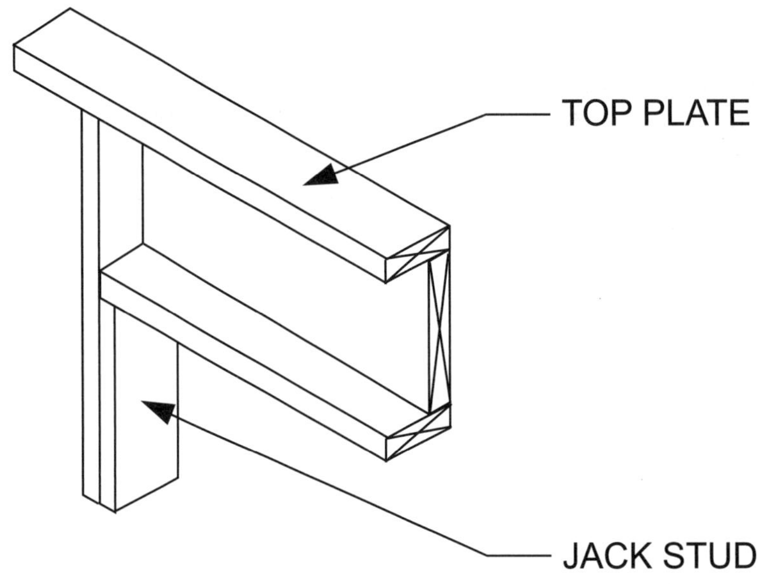

FIGURE R602.7.1(2)
ALTERNATIVE SINGLE MEMBER HEADER WITHOUT CRIPPLE

TABLE R602.7.2
MAXIMUM SPANS FOR WOOD STRUCTURAL PANEL BOX HEADERS[a]

HEADER CONSTRUCTION[b]	HEADER DEPTH (inches)	HOUSE DEPTH (feet)				
		24	26	28	30	32
Wood structural panel–one side	9 15	4 5	4 5	3 4	3 3	— 3
Wood structural panel–both sides	9 15	7 8	5 8	5 7	4 7	3 6

For SI: 1 inch = 25.4 mm, 1 foot = 304.8 mm.

a. Spans are based on single story with clear-span trussed roof or two-story with floor and roof supported by interior-bearing walls.

b. See Figure R602.7.2 for construction details.

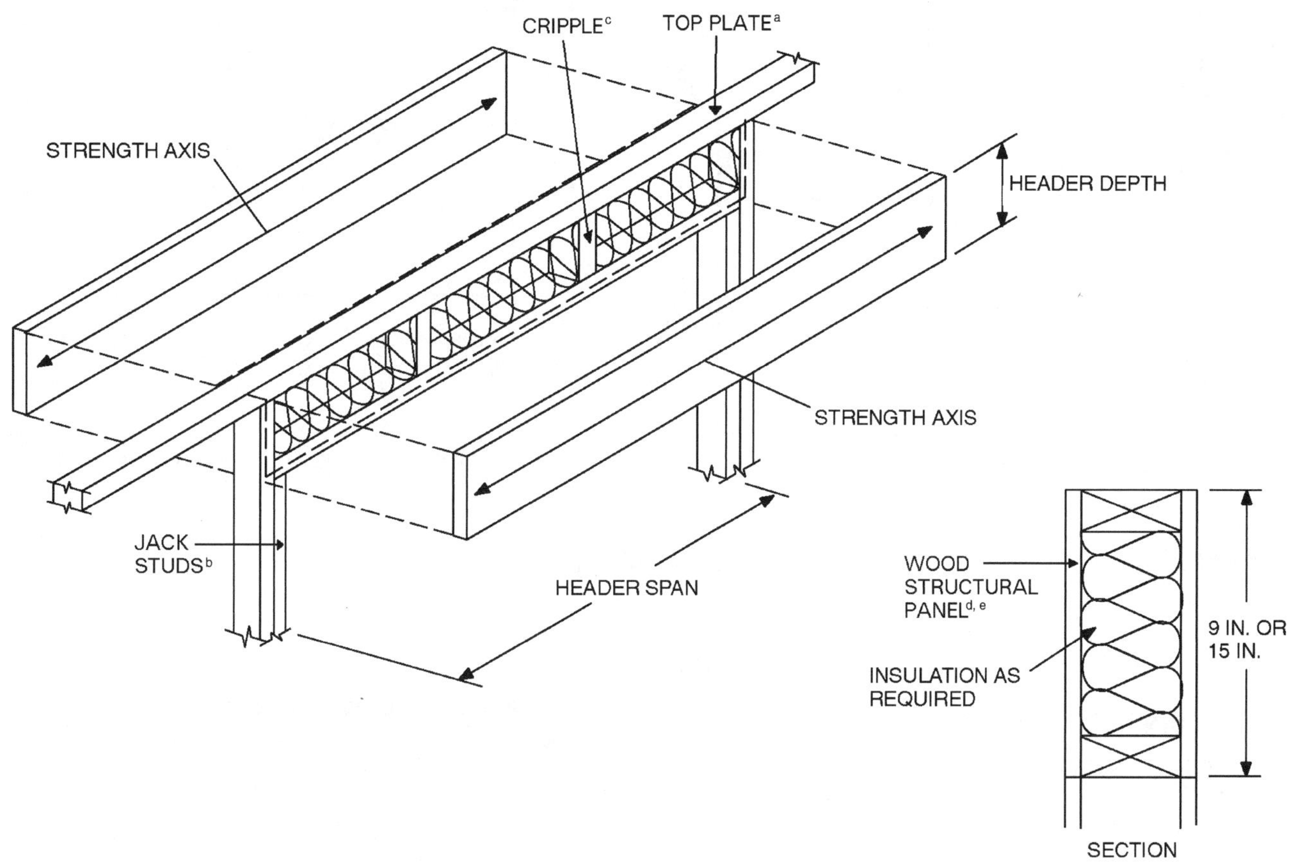

For SI: 1 inch = 25.4 mm, 1 foot = 304.8 mm.

NOTES:

a. The top plate shall be continuous over header.

b. Jack studs shall be used for spans over 4 feet.

c. Cripple spacing shall be the same as for studs.

d. Wood structural panel faces shall be single pieces of $^{15}/_{32}$-inch-thick Exposure 1 (exterior glue) or thicker, installed on the interior or exterior or both sides of the header.

e. Wood structural panel faces shall be nailed to framing and cripples with 8d common or galvanized box nails spaced 3 inches on center, staggering alternate nails $^{1}/_{2}$ inch. Galvanized nails shall be hot-dipped or tumbled.

FIGURE R602.7.2
TYPICAL WOOD STRUCTURAL PANEL BOX HEADER CONSTRUCTION

R602.10.1 Braced wall lines. For the purpose of determining the amount and location of bracing required in each story level of a building, *braced wall lines* shall be designated as straight lines in the building plan placed in accordance with this section.

R602.10.1.1 Length of a braced wall line. The length of a *braced wall line* shall be the distance between its ends. The end of a *braced wall line* shall be the intersection with a perpendicular *braced wall line*, an angled *braced wall line* as permitted in Section R602.10.1.4 or an exterior wall as shown in Figure R602.10.1.1.

R602.10.1.2 Offsets along a braced wall line. All exterior walls parallel to a *braced wall line* shall be offset not more than 4 feet (1219 mm) from the designated *braced wall line* location as shown Figure R602.10.1.1. Interior walls used as bracing shall be offset not more than 4 feet (1219 mm) from a *braced wall line* through the interior of the building as shown in Figure R602.10.1.1.

R602.10.1.3 Spacing of braced wall lines. The spacing between parallel *braced wall lines* shall be in accordance with Table R602.10.1.3. Intermediate *braced*

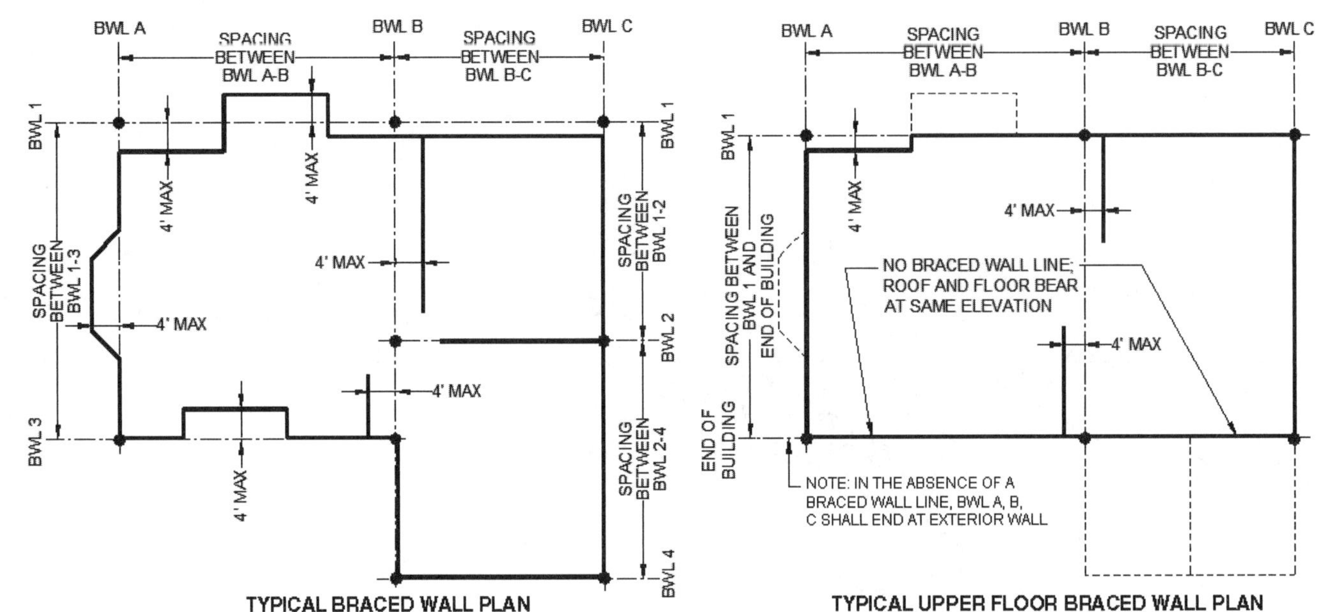

For SI: 1 foot = 304.8 mm.

FIGURE R602.10.1.1
BRACED WALL LINES

TABLE R602.10.1.3
BRACED WALL LINE SPACING

APPLICATION	CONDITION	BUILDING TYPE	BRACED WALL LINE SPACING CRITERIA	
			Maximum Spacing	Exception to Maximum Spacing
Wind bracing	85 mph to < 110 mph	Detached, townhouse	60 feet	None
Seismic bracing	SDC A – C	Detached	Use wind bracing	
	SDC A – B	Townhouse	Use wind bracing	
	SDC C	Townhouse	35 feet	Up to 50 feet when length of required bracing per Table R602.10.3(3) is adjusted in accordance with Table R602.10.3(4).
	SDC D_0, D_1, D_2	Detached, townhouses, one- and two-story only	25 feet	Up to 35 feet to allow for a single room not to exceed 900 square feet. Spacing of all other braced wall lines shall not exceed 25 feet.
	SDC D_0, D_1, D_2	Detached, townhouse	25 feet	Up to 35 feet when length of required bracing per Table R602.10.3(3) is adjusted in accordance with Table R602.10.3(4).

For SI: 1 foot = 304.8 mm, 1 square foot = 0.0929 m^2, 1 mile per hour = 0.447 m/s.

wall lines through the interior of the building shall be permitted.

R602.10.1.4 Angled walls. Any portion of a wall along a *braced wall line* shall be permitted to angle out of plane for a maximum diagonal length of 8 feet (2438 mm). Where the angled wall occurs at a corner, the length of the *braced wall line* shall be measured from the projected corner as shown in Figure R602.10.1.4. Where the diagonal length is greater than 8 feet (2438 mm), it shall be considered a separate *braced wall line* and shall be braced in accordance with Section R602.10.1.

R602.10.2 Braced wall panels. *Braced wall panels* shall be full-height sections of wall that shall have no vertical or horizontal offsets. *Braced wall panels* shall be constructed and placed along a *braced wall line* in accordance with this section and the bracing methods specified in Section R602.10.4.

R602.10.2.1 Braced wall panel uplift load path. The bracing lengths in Table R602.10.3(1) apply only when uplift loads are resisted in accordance with Section R602.3.5.

R602.10.2.2 Locations of braced wall panels. A *braced wall panel* shall begin within 10 feet (3810 mm) from each end of a *braced wall line* as determined in Section R602.10.1.1. The distance between adjacent edges of *braced wall panels* along a *braced wall line* shall be no greater than 20 feet (6096 mm) as shown in Figure R602.10.2.2.

R602.10.2.2.1 Location of braced wall panels in Seismic Design Categories D_0, D_1 and D_2. *Braced wall panels* shall be located at each end of a *braced wall line*.

Exception: *Braced wall panels* constructed of Methods WSP or BV-WSP and continuous sheathing methods as specified in Section R602.10.4 shall be permitted to begin no more than 10 feet (3048 mm) from each end of a *braced wall line* provided each end complies with one of the following.

1. A minimum 24-inch-wide (610 mm) panel for Methods WSP, BV-WSP, CS-WSP, CS-G, and CS-PF, and 32-inch-wide (813 mm) panel for Method CS-SFB is applied to each side of the building corner as shown in Condition 4 of Figure R602.10.7.
2. The end of each *braced wall panel* closest to the end of the *braced wall line* shall have an 1,800 lb (8 kN) hold-down device fastened to the stud at the edge of the *braced wall panel* closest to the corner and to the foundation or framing below as shown in Condition 5 of Figure R602.10.7.
3. For Method BV-WSP, hold-down devices shall be provided in accordance with Table R602.10.6.5 at the ends of each *braced wall panel.*

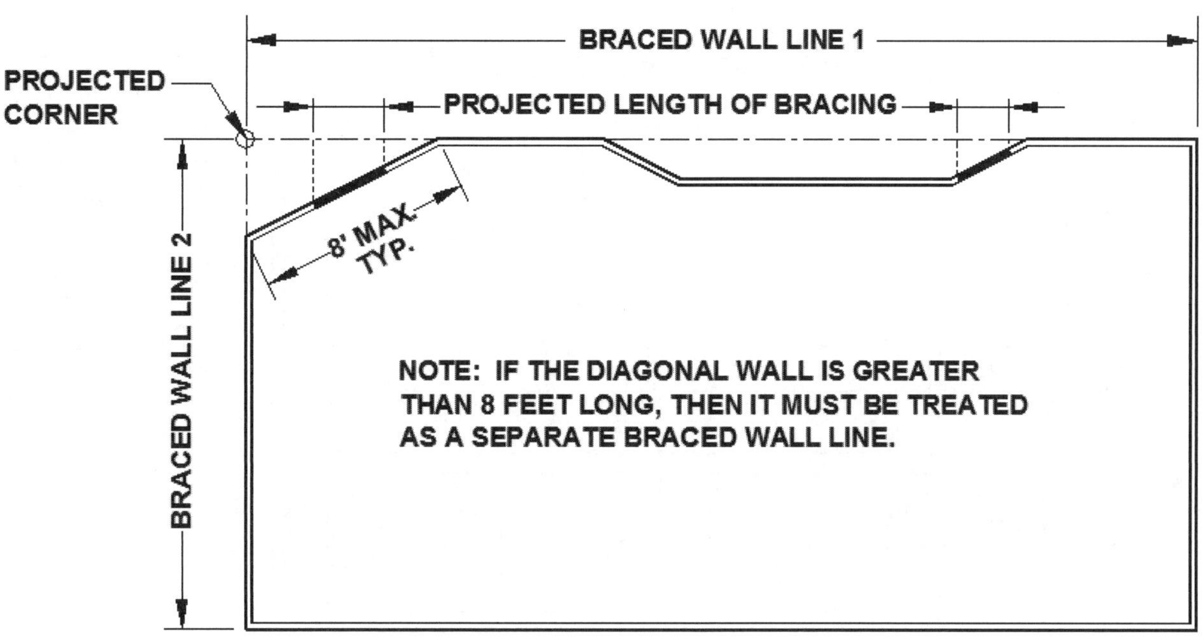

For SI: 1 foot = 304.8 mm.

**FIGURE R602.10.1.4
ANGLED WALLS**

R602.10.2.3 Minimum number of braced wall panels. *Braced wall lines* with a length of 16 feet (4877 mm) or less shall have a minimum of two *braced wall panels* of any length or one *braced wall panel* equal to 48 inches (1219 mm) or more. *Braced wall lines* greater than 16 feet (4877 mm) shall have a minimum of two *braced wall panels*.

R602.10.3 Required length of bracing. The required length of bracing along each *braced wall line* shall be determined as follows.

1. All buildings in Seismic Design Categories A and B shall use Table R602.10.3(1) and the applicable adjustment factors in Table R602.10.3(2).
2. Detached buildings in Seismic Design Category C shall use Table R602.10.3(1) and the applicable adjustment factors in Table R602.10.3(2).
3. Townhouses in Seismic Design Category C shall use the greater value determined from Table R602.10.3(1) or R602.10.3(3) and the applicable adjustment factors in Table R602.10.3(2) or R602.10.3(4) respectively.
4. All buildings in Seismic Design Categories D_0, D_1 and D_2 shall use the greater value determined from Table R602.10.3(1) or R602.10.3(3) and the applicable adjustment factors in Table R602.10.3(2) or R602.10.3(4) respectively.

Only *braced wall panels* parallel to the *braced wall line* shall contribute toward the required length of bracing of that *braced wall line*. *Braced wall panels* along an angled wall meeting the minimum length requirements of Tables R602.10.5 and R602.10.5.2 shall be permitted to contribute its projected length toward the minimum required length of bracing for the *braced wall line* as shown in Figure R602.10.1.4. Any *braced wall panel* on an angled wall at the end of a *braced wall line* shall contribute its projected length for only one of the *braced wall lines* at the projected corner.

Exception: The length of wall bracing for dwellings in Seismic Design Categories D_0, D_1 and D_2 with stone or masonry veneer installed per Section R703.7 and exceeding the first-story height shall be in accordance with Section R602.10.6.5.

R602.10.4 Construction methods for braced wall panels. Intermittent and continuously sheathed *braced wall panels* shall be constructed in accordance with this section and the methods listed in Table R602.10.4.

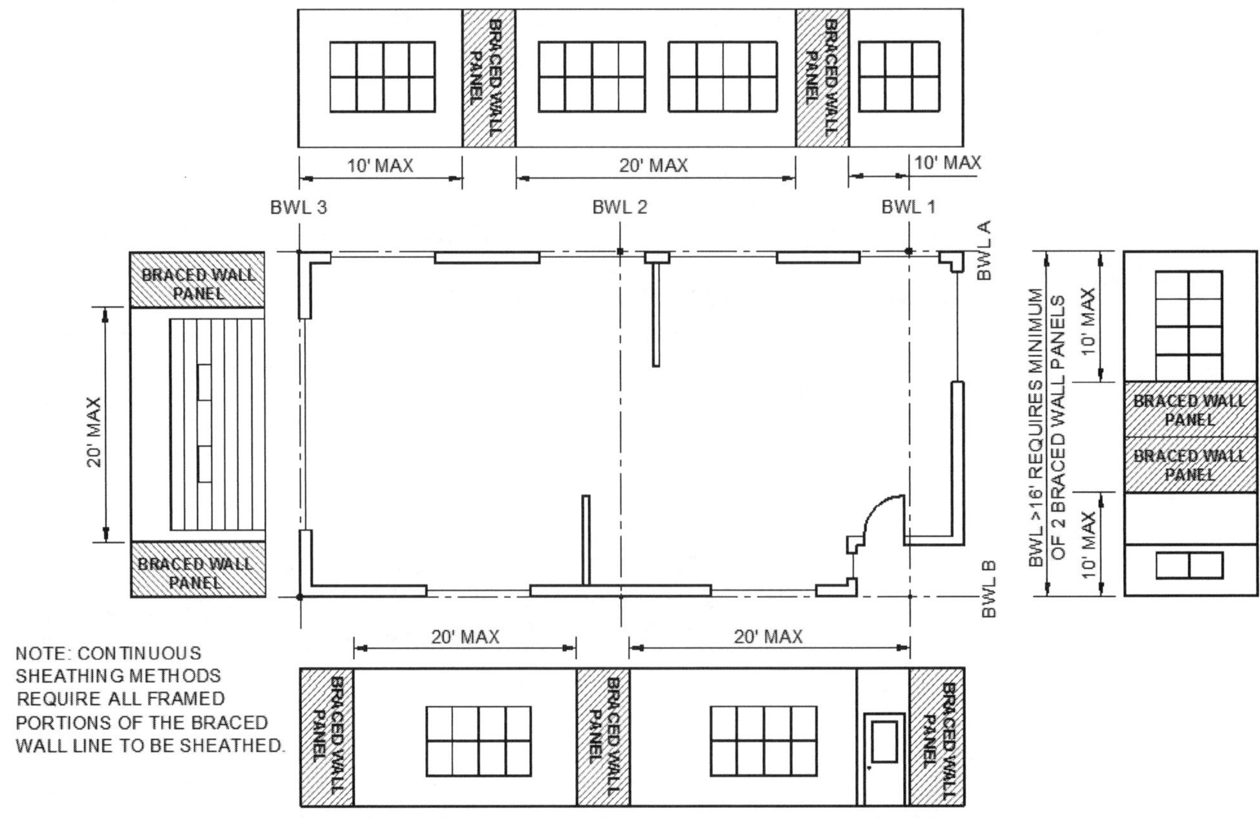

For SI: 1 foot = 304.8 mm.

FIGURE R602.10.2.2
LOCATION OF BRACED WALL PANELS

TABLE R602.10.3(1)
BRACING REQUIREMENTS BASED ON WIND SPEED

• EXPOSURE CATEGORY B • 30 FOOT MEAN ROOF HEIGHT • 10 FOOT EAVE-TO-RIDGE HEIGHT • 10 FOOT WALL HEIGHT • 2 BRACED WALL LINES			MINIMUM TOTAL LENGTH (FEET) OF BRACED WALL PANELS REQUIRED ALONG EACH BRACED WALL LINE[a]			
Basic Wind Speed (mph)	**Story Location**	**Braced Wall Line Spacing (feet)**	**Method LIB[b]**	**Method GB**	**Methods DWB, WSP, SFB, PBS, PCP, HPS, CS-SFB[c]**	**Methods CS-WSP, CS-G, CS-PF**
≤ 85		10	3.5	3.5	2.0	1.5
		20	6.0	6.0	3.5	3.0
		30	8.5	8.5	5.0	4.5
		40	11.5	11.5	6.5	5.5
		50	14.0	14.0	8.0	7.0
		60	16.5	16.5	9.5	8.0
		10	6.5	6.5	3.5	3.0
		20	11.5	11.5	6.5	5.5
		30	16.5	16.5	9.5	8.0
		40	21.5	21.5	12.5	10.5
		50	26.5	26.5	15.0	13.0
		60	31.5	31.5	18.0	15.5
		10	NP	9.0	5.5	4.5
		20	NP	17.0	10.0	8.5
		30	NP	24.5	14.0	12.0
		40	NP	32.0	18.0	15.5
		50	NP	39.0	22.5	19.0
		60	NP	46.5	26.5	22.5
≤ 90		10	3.5	3.5	2.0	2.0
		20	7.0	7.0	4.0	3.5
		30	9.5	9.5	5.5	5.0
		40	12.5	12.5	7.5	6.0
		50	15.5	15.5	9.0	7.5
		60	18.5	18.5	10.5	9.0
		10	7.0	7.0	4.0	3.5
		20	13.0	13.0	7.5	6.5
		30	18.5	18.5	10.5	9.0
		40	24.0	24.0	14.0	12.0
		50	29.5	29.5	17.0	14.5
		60	35.0	35.0	20.0	17.0
		10	NP	10.5	6.0	5.0
		20	NP	19.0	11.0	9.5
		30	NP	27.5	15.5	13.5
		40	NP	35.5	20.5	17.5
		50	NP	44.0	25.0	21.5
		60	NP	52.0	30.0	25.5

(continued)

TABLE R602.10.3(1)—continued
BRACING REQUIREMENTS BASED ON WIND SPEED

• EXPOSURE CATEGORY B • 30 FOOT MEAN ROOF HEIGHT • 10 FOOT EAVE-TO-RIDGE HEIGHT • 10 FOOT WALL HEIGHT • 2 BRACED WALL LINES			MINIMUM TOTAL LENGTH (FEET) OF BRACED WALL PANELS REQUIRED ALONG EACH BRACED WALL LINE[a]			
Basic Wind Speed (mph)	**Story Location**	**Braced Wall Line Spacing** (feet)	**Method LIB[b]**	**Method GB**	**Methods** DWB, WSP, SFB, PBS, PCP, HPS, CS-SFB[c]	**Methods** CS-WSP, CS-G, CS-PF
≤ 100		10	4.5	4.5	2.5	2.5
		20	8.5	8.5	5.0	4.0
		30	12.0	12.0	7.0	6.0
		40	15.5	15.5	9.0	7.5
		50	19.0	19.0	11.0	9.5
		60	22.5	22.5	13.0	11.0
		10	8.5	8.5	5.0	4.5
		20	16.0	16.0	9.0	8.0
		30	23.0	23.0	13.0	11.0
		40	29.5	29.5	17.0	14.5
		50	36.5	36.5	21.0	18.0
		60	43.5	43.5	25.0	21.0
		10	NP	12.5	7.5	6.0
		20	NP	23.5	13.5	11.5
		30	NP	34.0	19.5	16.5
		40	NP	44.0	25.0	21.5
		50	NP	54.0	31.0	26.5
		60	NP	64.0	36.5	31.0
< 110[c]		10	5.5	5.5	3.0	3.0
		20	10.0	10.0	6.0	5.0
		30	14.5	14.5	8.5	7.0
		40	18.5	18.5	11.0	9.0
		50	23.0	23.0	13.0	11.5
		60	27.5	27.5	15.5	13.5
		10	10.5	10.5	6.0	5.0
		20	19.0	19.0	11.0	9.5
		30	27.5	27.5	16.0	13.5
		40	36.0	36.0	20.5	17.5
		50	44.0	44.0	25.5	21.5
		60	52.5	52.5	30.0	25.5
		10	NP	15.5	9.0	7.5
		20	NP	28.5	16.5	14.0
		30	NP	41.0	23.5	20.0
		40	NP	53.0	30.5	26.0
		50	NP	65.5	37.5	32.0
		60	NP	77.5	44.5	37.5

For SI: 1 inch = 25.4 mm, 1 foot = 305 mm, 1 mile per hour = 0.447 m/s.

a. Linear interpolation shall be permitted.

b. Method LIB shall have gypsum board fastened to at least one side with nails or screws in accordance withTable R602.3(1) for exterior sheathing or Table R702.3.5 for interior gypsum board. Spacing of fasteners at panel edges shall not exceed 8 inches.

c. Method CS-SFB does not apply where the wind speed is greater than 100 mph.

TABLE R602.10.3(2)
WIND ADJUSTMENT FACTORS TO THE REQUIRED LENGTH OF WALL BRACING

ADJUSTMENT BASED ON	STORY/ SUPPORTING	CONDITION	ADJUSTMENT FACTOR[a, b] [multiply length from Table R602.10.3(1) by this factor]	APPLICABLE METHODS
Exposure category	One-story structure	B	1.00	All methods
		C	1.20	
		D	1.50	
	Two-story structure	B	1.00	
		C	1.30	
		D	1.60	
	Three-story structure	B	1.00	
		C	1.40	
		D	1.70	
Roof eave-to-ridge height	Roof only	≤ 5 feet	0.70	
		10 feet	1.00	
		15 feet	1.30	
		20 feet	1.60	
	Roof + 1 floor	≤ 5 feet	0.85	
		10 feet	1.00	
		15 feet	1.15	
		20 feet	1.30	
	Roof + 2 floors	≤ 5 feet	0.90	
		10 feet	1.00	
		15 feet	1.10	
		20 feet	Not permitted	
Wall height adjustment	Any story	8 feet	0.90	
		9 feet	0.95	
		10 feet	1.00	
		11 feet	1.05	
		12 feet	1.10	
Number of braced wall lines (per plan direction)[c]	Any story	2	1.00	
		3	1.30	
		4	1.45	
		≤ 5	1.60	
Additional 800-pound hold-down device	Top story only	Fastened to the end studs of each braced wall panel and to the foundation or framing below	0.80	DWB, WSP, SFB, PBS, PCP, HPS
Interior gypsum board finish (or equivalent)	Any story	Omitted from inside face of braced wall panels	1.40	DWB, WSP, SFB,PBS, PCP, HPS, CS-WSP, CS-G, CS-SFB
Gypsum board fastening	Any story	4 inches o.c. at panel edges, including top and bottom plates, and all horizontal joints blocked	0.7	GB

For SI: 1 inch = 25.4 mm, 1 foot = 305 mm, 1 pound = 4.48 N.

a. Linear interpolation shall be permitted.

b. The total adjustment factor is the product of all applicable adjustment factors.

c. The adjustment factor is permitted to be 1.0 when determining bracing amounts for intermediate braced wall lines provided the bracing amounts on adjacent braced wall lines are based on a spacing and number that neglects the intermediate braced wall line.

TABLE R602.10.3(3)
BRACING REQUIREMENTS BASED ON SEISMIC DESIGN CATEGORY

• SOIL CLASS D[b] • WALL HEIGHT = 10 FEET • 10 PSF FLOOR DEAD LOAD • 15 PSF ROOF/CEILING DEAD LOAD • BRACED WALL LINE SPACING ≤ 25 FEET			MINIMUM TOTAL LENGTH (FEET) OF BRACED WALL PANELS REQUIRED ALONG EACH BRACED WALL LINE[a]				
Seismic Design Category	**Story Location**	**Braced Wall Line Length (feet)**	**Method LIB[c]**	**Method GB**	**Methods DWB, SFB, PBS, PCP, HPS, CS-SFB[d]**	**Method WSP**	**Methods CS-WSP, CS-G**
C (townhouses only)		10	2.5	2.5	2.5	1.6	1.4
		20	5.0	5.0	5.0	3.2	2.7
		30	7.5	7.5	7.5	4.8	4.1
		40	10.0	10.0	10.0	6.4	5.4
		50	12.5	12.5	12.5	8.0	6.8
		10	NP	4.5	4.5	3.0	2.6
		20	NP	9.0	9.0	6.0	5.1
		30	NP	13.5	13.5	9.0	7.7
		40	NP	18.0	18.0	12.0	10.2
		50	NP	22.5	22.5	15.0	12.8
		10	NP	6.0	6.0	4.5	3.8
		20	NP	12.0	12.0	9.0	7.7
		30	NP	18.0	18.0	13.5	11.5
		40	NP	24.0	24.0	18.0	15.3
		50	NP	30.0	30.0	22.5	19.1
D_0		10	NP	2.8	2.8	1.8	1.6
		20	NP	5.5	5.5	3.6	3.1
		30	NP	8.3	8.3	5.4	4.6
		40	NP	11.0	11.0	7.2	6.1
		50	NP	13.8	13.8	9.0	7.7
		10	NP	5.3	5.3	3.8	3.2
		20	NP	10.5	10.5	7.5	6.4
		30	NP	15.8	15.8	11.3	9.6
		40	NP	21.0	21.0	15.0	12.8
		50	NP	26.3	26.3	18.8	16.0
		10	NP	7.3	7.3	5.3	4.5
		20	NP	14.5	14.5	10.5	9.0
		30	NP	21.8	21.8	15.8	13.4
		40	NP	29.0	29.0	21.0	17.9
		50	NP	36.3	36.3	26.3	22.3

(continued)

TABLE R602.10.3(3)—continued
BRACING REQUIREMENTS BASED ON SEISMIC DESIGN CATEGORY

• SOIL CLASS D[b] • WALL HEIGHT = 10 FEET • 10 PSF FLOOR DEAD LOAD • 15 PSF ROOF/CEILING DEAD LOAD • BRACED WALL LINE SPACING ≤ 25 FEET			MINIMUM TOTAL LENGTH (FEET) OF BRACED WALL PANELS REQUIRED ALONG EACH BRACED WALL LINE[a]				
Seismic Design Category	Story Location	Braced Wall Line Length (feet)	Method LIB[c]	Method GB	Methods DWB, SFB, PBS, PCP, HPS, CS-SFB[d]	Method WSP	Methods CS-WSP, CS-G
D_1		10	NP	3.0	3.0	2.0	1.7
		20	NP	6.0	6.0	4.0	3.4
		30	NP	9.0	9.0	6.0	5.1
		40	NP	12.0	12.0	8.0	6.8
		50	NP	15.0	15.0	10.0	8.5
		10	NP	6.0	6.0	4.5	3.8
		20	NP	12.0	12.0	9.0	7.7
		30	NP	18.0	18.0	13.5	11.5
		40	NP	24.0	24.0	18.0	15.3
		50	NP	30.0	30.0	22.5	19.1
		10	NP	8.5	8.5	6.0	5.1
		20	NP	17.0	17.0	12.0	10.2
		30	NP	25.5	25.5	18.0	15.3
		40	NP	34.0	34.0	24.0	20.4
		50	NP	42.5	42.5	30.0	25.5
D_2		10	NP	4.0	4.0	2.5	2.1
		20	NP	8.0	8.0	5.0	4.3
		30	NP	12.0	12.0	7.5	6.4
		40	NP	16.0	16.0	10.0	8.5
		50	NP	20.0	20.0	12.5	10.6
		10	NP	7.5	7.5	5.5	4.7
		20	NP	15.0	15.0	11.0	9.4
		30	NP	22.5	22.5	16.5	14.0
		40	NP	30.0	30.0	22.0	18.7
		50	NP	37.5	37.5	27.5	23.4
		10	NP	NP	NP	NP	NP
		20	NP	NP	NP	NP	NP
		30	NP	NP	NP	NP	NP
		40	NP	NP	NP	NP	NP
		50	NP	NP	NP	NP	NP
	Cripple wall below one- or two-story dwelling	10	NP	NP	NP	7.5	6.4
		20	NP	NP	NP	15.0	12.8
		30	NP	NP	NP	22.5	19.1
		40	NP	NP	NP	30.0	25.5
		50	NP	NP	NP	37.5	31.9

For SI: 1 inch = 25.4 mm, 1 foot = 305 mm, 1 pound per square foot = 0.0479 kPa.

a. Linear interpolation shall be permitted.

b. Wall bracing lengths are based on a soil site class "D." Interpolation of bracing length between the S_{ds} values associated with the Seismic Design Categories shall be permitted when a site-specific S_{ds} value is determined in accordance with Section 1613.3 of the *International Building Code*.

c. Method LIB shall have gypsum board fastened to at least one side with nails or screws per Table R602.3(1) for exterior sheathing or Table R702.3.5 for interior gypsum board. Spacing of fasteners at panel edges shall not exceed 8 inches.

d. Method CS-SFB applies in SDC C only.

TABLE R602.10.3(4)
SEISMIC ADJUSTMENT FACTORS TO THE REQUIRED LENGTH OF WALL BRACING

ADJUSTMENT BASED ON:	STORY/SUPPORTING	CONDITION	ADJUSTMENT FACTOR[a, b] [Multiply length from Table R602.10.3(1) by this factor]	APPLICABLE METHODS
Story height (Section 301.3)	Any story	≤ 10 feet	1.0	All methods
		> 10 feet and ≤ 12 feet	1.2	
Braced wall line spacing, townhouses in SDC C	Any story	≤ 35 feet	1.0	
		> 35 feet and ≤ 50 feet	1.43	
Braced wall line spacing, in SDC D_0, D_1, D_2[c]	Any story	> 25 feet and ≤ 30 feet	1.2	
		> 30 feet and ≤ 35 feet	1.4	
Wall dead load	Any story	> 8 psf and < 15 psf	1.0	
		< 8 psf	0.85	
Roof/ceiling dead load for wall supporting	Roof only or roof plus one or two stories	≤15 psf	1.0	
	Roof plus one or two stories	> 15 psf and ≤ 25 psf	1.1	
	Roof only	> 15 psf and ≤ 25 psf	1.2	
Walls with stone or masonry veneer, townhouses in SDC[d, e]		1.0		All intermittent and continuous methods
	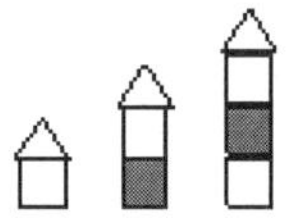	1.5		
	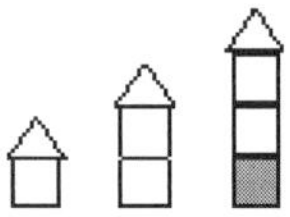	1.5		
Walls with stone or masonry veneer, detached one-and two-family dwellings in SDC $D_0 - D_2$[d]	Any story	See Table R602.10.6.5		BV-WSP
Interior gypsum board finish (or equivalent)	Any story	Omitted from inside face of braced wall panels	1.5	DWB, WSP, SFB, PBS, PCP, HPS, CS-WSP, CS-G, CS-SFB

For SI: 1 foot = 304.8 mm, 1 pound per square foot = 0.0479 kPa.

a. Linear interpolation shall be permitted.

b. The total length of bracing required for a given wall line is the product of all applicable adjustment factors.

c. The length-to-width ratio for the floor/roof *diaphragm* shall not exceed 3:1. The top plate lap splice nailing shall be a minimum of 12-16d nails on each side of the splice.

d. Applies to stone or masonry veneer exceeding the first story height. See Section R602.10.6.5 for requirements when stone or masonry veneer does not exceed the first story height.

e. The adjustment factor for stone or masonry veneer shall be applied to all exterior *braced wall lines* and all *braced wall lines* on the interior of the building, backing or perpendicular to and laterally supported veneered walls.

TABLE R602.10.4
BRACING METHODS

METHODS, MATERIAL		MINIMUM THICKNESS	FIGURE	CONNECTION CRITERIA[a]	
				Fasteners	Spacing
Intermittent Bracing Method	**LIB** Let-in-bracing	1 × 4 wood or approved metal straps at 45° to 60° angles for maximum 16″ stud spacing		Wood: 2-8d common nails or 3-8d ($2^1/_2$″ long x 0.113″ dia.) nails	Wood: per stud and top and bottom plates
				Metal strap: per manufacturer	Metal: per manufacturer
	DWB Diagonal wood boards	$^3/_4$″(1″ nominal) for maximum 24″ stud spacing		2-8d ($2^1/_2$″ long × 0.113″ dia.) nails or 2 - $1^3/_4$″ long staples	Per stud
	WSP Wood structural panel (See Section R604)	$^3/_8$″		Exterior sheathing per Table R602.3(3)	6″ edges 12″ field
				Interior sheathing per Table R602.3(1) or R602.3(2)	Varies by fastener
	BV-WSP[e] Wood Structural Panels with Stone or Masonry Veneer (See Section R602.10.6.5)	$^7/_{16}$″	See Figure R602.10.6.5	8d common ($2^1/_2$″ × 0.131) nails	4″ at panel edges 12″ at intermediate supports 4″ at braced wall panel end posts
	SFB Structural fiberboard sheathing	$^1/_2$″ or $^{25}/_{32}$″ for maximum 16″ stud spacing		$1^1/_2$″ long × 0.12″ dia. (for $^1/_2$″ thick sheathing) $1^3/_4$″ long × 0.12″ dia. (for $^{25}/_{32}$″ thick sheathing) galvanized roofing nails or 8d common ($2^1/_2$″ long × 0.131″ dia.) nails	3″ edges 6″ field
	GB Gypsum board	$^1/_2$″		Nails or screws per Table R602.3(1) for exterior locations	For all braced wall panel locations: 7″ edges (including top and bottom plates) 7″ field
				Nails or screws per Table R702.3.5 for interior locations	
	PBS Particleboard sheathing (See Section R605)	$^3/_8$″ or $^1/_2$″ for maximum 16″ stud spacing		For $^3/_8$″, 6d common (2″ long × 0.113″ dia.) nails For $^1/_2$″, 8d common ($2^1/_2$″ long × 0.131″ dia.) nails	3″ edges 6″ field
	PCP Portland cement plaster	See Section R703.6 for maximum 16″ stud spacing		$1^1/_2$″ long, 11 gage, $^7/_{16}$″ dia. head nails or $^7/_8$″ long, 16 gage staples	6″ o.c. on all framing members
	HPS Hardboard panel siding	$^7/_{16}$″ for maximum 16″ stud spacing		0.092″ dia., 0.225″ dia. head nails with length to accommodate $1^1/_2$″ penetration into studs	4″ edges 8″ field
	ABW Alternate braced wall	$^3/_8$″		See Section R602.10.6.1	See Section R602.10.6.1

(continued)

TABLE R602.10.4—continued BRACING METHODS

METHODS, MATERIAL		MINIMUM THICKNESS	FIGURE	CONNECTION CRITERIA[a]	
				Fasteners	Spacing
Intermittent Bracing Methods	**PFH** Portal frame with hold-downs	3/8"		See Section R602.10.6.2	See Section R602.10.6.2
	PFG Portal frame at garage	7/16"		See Section R602.10.6.3	See Section R602.10.6.3
Continuous Sheathing Methods	**CS-WSP** Continuously sheathed wood structural panel	3/8"		Exterior sheathing per Table R602.3(3)	6" edges 12" field
				Interior sheathing per Table R602.3(1) or R602.3(2)	Varies by fastener
	CS-G[b, c] Continuously sheathed wood structural panel adjacent to garage openings	3/8"		See Method CS-WSP	See Method CS-WSP
	CS-PF Continuously sheathed portal frame	7/16"		See Section R602.10.6.4	See Section R602.10.6.4
	CS-SFB[d] Continuously sheathed structural fiberboard	1/2" or 25/32" for maximum 16" stud spacing		1 1/2" long × 0.12" dia. (for 1/2" thick sheathing) 1 3/4" long × 0.12" dia. (for 25/32" thick sheathing) galvanized roofing nails or 8d common (2 1/2" long × 0.131" dia.) nails	3" edges 6" field

For SI: 1 inch = 25.4 mm, 1 foot = 305 mm, 1 degree = 0.0175 rad, 1 pound per square foot = 47.8 N/m^2, 1 mile per hour = 0.447 m/s.

a. Adhesive attachment of wall sheathing, including Method GB, shall not be permitted in Seismic Design Categories C, D_0, D_1 and D_2.

b. Applies to panels next to garage door opening when supporting gable end wall or roof load only. May only be used on one wall of the garage. In Seismic Design Categories D_0, D_1 and D_2 roof covering dead load may not exceed 3 psf.

c. Garage openings adjacent to a Method CS-G panel shall be provided with a header in accordance with Table R502.5(1). A full height clear opening shall not be permitted adjacent to a Method CS-G panel.

d. Method CS-SFB does not apply in Seismic Design Categories D_0, D_1 and D_2 and in areas where the wind speed exceeds 100 mph.

e. Method applies to detached one- and two-family dwellings in Seismic Design Categories D_0 through D_2 only.

R602.10.4.1 Mixing methods. Mixing of bracing methods shall be permitted as follows:

1. Mixing intermittent bracing and continuous sheathing methods from story to story shall be permitted.
2. Mixing intermittent bracing methods from *braced wall line* to *braced wall line* within a story shall be permitted. Within Seismic Design Categories A, B and C or in regions where the basic wind speed is less than or equal to 100 mph (45 m/s), mixing of intermittent bracing and continuous sheathing methods from *braced wall line* to *braced wall line* within a story shall be permitted.
3. Mixing intermittent bracing methods along a *braced wall line* shall be permitted in Seismic Design Categories A and B, and detached dwellings in Seismic Design Category C provided the length of required bracing in accordance with Table R602.10.3(1) or R602.10.3(3) is the highest value of all intermittent bracing methods used.
4. Mixing of continuous sheathing methods CS-WSP, CS-G and CS-PF along a *braced wall line* shall be permitted.
5. In Seismic Design Categories A and B, and for detached one- and two-family dwellings in Seismic Design Category C, mixing of intermittent bracing methods along the interior portion of a *braced wall line* with continuous sheathing methods CS-WSP, CS-G and CS-PF along the exterior portion of the same *braced wall line* shall be permitted. The length of required bracing shall be the highest value of all intermittent bracing methods used in accordance with Table R602.10.3(1) or R602.10.3(3) as adjusted by Tables R602.10.3(2) and R602.10.3(4), respectively. The requirements of Section R602.10.7 shall apply to each end of the continuously sheathed portion of the *braced wall line*.

R602.10.4.2 Continuous sheathing methods. Continuous sheathing methods require structural panel sheathing to be used on all sheathable surfaces on one side of a *braced wall line* including areas above and below openings and gable end walls and shall meet the requirements of Section R602.10.7.

R602.10.4.3 Braced wall panel interior finish material. *Braced wall panels* shall have gypsum wall board installed on the side of the wall opposite the bracing material. Gypsum wall board shall be not less than $^1/_2$ inch (12.7 mm) in thickness and be fastened with nails or screws in accordance with Table R602.3(1) for exterior sheathing or Table R702.3.5 for interior gypsum wall board. Spacing of fasteners at panel edges for gypsum wall board opposite Method LIB bracing shall not exceed 8 inches (203 mm). Interior finish material shall not be glued in Seismic Design Categories D_0, D_1 and D_2.

Exceptions:

1. Interior finish material is not required opposite wall panels that are braced in accordance with Methods GB, BV-WSP, ABW, PFH, PFG and CS-PF, unless otherwise required by Section R302.6.
2. An approved interior finish material with an in-plane shear resistance equivalent to gypsum board shall be permitted to be substituted, unless otherwise required by Section R302.6.
3. Except for Method LIB, gypsum wall board is permitted to be omitted provided the required length of bracing in Tables R602.10.3(1) and R602.10.3(3) is multiplied by the appropriate adjustment factor in Tables R602.10.3(2) and R602.10.3(4) respectively, unless otherwise required by Section R302.6.

R602.10.5 Minimum length of a braced wall panel. The minimum length of a *braced wall panel* shall comply with Table R602.10.5. For Methods CS-WSP and CS-SFB, the minimum panel length shall be based on the adjacent clear opening height in accordance with Table R602.10.5 and Figure R602.10.5. When a panel has an opening on either side of differing heights, the taller opening height shall be used to determine the panel length.

R602.10.5.1 Contributing length. For purposes of computing the required length of bracing in Tables R602.10.3(1) and R602.10.3(3), the contributing length of each *braced wall panel* shall be as specified in Table R602.10.5.

R602.10.5.2 Partial credit. For Methods DWB, WSP, SFB, PBS, PCP and HPS in Seismic Design Categories A, B and C, panels between 36 inches and 48 inches (914 mm and 1219 mm)) in length shall be considered a *braced wall panel* and shall be permitted to partially contribute toward the required length of bracing in Tables R602.10.3(1) and R602.10.3(3), and the contributing length shall be determined from Table R602.10.5.2.

R602.10.6 Construction of Methods ABW, PFH, PFG, CS-PF and BV-WSP. Methods ABW, PFH, PFG, CS-PF and BV-WSP shall be constructed as specified in Sections R602.10.6.1 through R602.10.6.5.

R602.10.6.1 Method ABW: Alternate braced wall panels. Method ABW *braced wall panels* shall be constructed in accordance with Figure R602.10.6.1. The hold-down force shall be in accordance with Table R602.10.6.1.

R602.10.6.2 Method PFH: Portal frame with hold-downs. Method PFH *braced wall panels* shall be constructed in accordance with Figure R602.10.6.2.

TABLE R602.10.5
MINIMUM LENGTH OF BRACED WALL PANELS

METHOD (See Table R602.10.4)		MINIMUM LENGTH[a] (inches)					CONTRIBUTING LENGTH (inches)
		Wall Height					
		8 feet	9 feet	10 feet	11 feet	12 feet	
DWB, WSP, SFB, PBS, PCP, HPS, BV-WSP		48	48	48	53	58	Actual[b]
GB		48	48	48	53	58	Double sided = Actual Single sided = 0.5 × Actual
LIB		55	62	69	NP	NP	Actual[b]
ABW	SDC A, B and C, wind speed < 110 mph	28	32	34	38	42	48
ABW	SDC D_0, D_1 and D_2, wind speed < 110 mph	32	32	34	NP	NP	48
PFH	Supporting roof only	16	16	16	18[c]	20[c]	48
PFH	Supporting one story and roof	24	24	24	27[c]	29[c]	48
PFG		24	27	30	33[d]	36[d]	1.5 × Actual[b]
CS-G		24	27	30	33	36	Actual[b]
CS-PF		16	18	20	22[e]	24[e]	Actual[b]
CS-WSP, CS-SFB	Adjacent clear opening height (inches)						
CS-WSP, CS-SFB	≤ 64	24	27	30	33	36	Actual[b]
CS-WSP, CS-SFB	68	26	27	30	33	36	Actual[b]
CS-WSP, CS-SFB	72	27	27	30	33	36	Actual[b]
CS-WSP, CS-SFB	76	30	29	30	33	36	Actual[b]
CS-WSP, CS-SFB	80	32	30	30	33	36	Actual[b]
CS-WSP, CS-SFB	84	35	32	32	33	36	Actual[b]
CS-WSP, CS-SFB	88	38	35	33	33	36	Actual[b]
CS-WSP, CS-SFB	92	43	37	35	35	36	Actual[b]
CS-WSP, CS-SFB	96	48	41	38	36	36	Actual[b]
CS-WSP, CS-SFB	100	—	44	40	38	38	Actual[b]
CS-WSP, CS-SFB	104	—	49	43	40	39	Actual[b]
CS-WSP, CS-SFB	108	—	54	46	43	41	Actual[b]
CS-WSP, CS-SFB	112	—	—	50	45	43	Actual[b]
CS-WSP, CS-SFB	116	—	—	55	48	45	Actual[b]
CS-WSP, CS-SFB	120	—	—	60	52	48	Actual[b]
CS-WSP, CS-SFB	124	—	—	—	56	51	Actual[b]
CS-WSP, CS-SFB	128	—	—	—	61	54	Actual[b]
CS-WSP, CS-SFB	132	—	—	—	66	58	Actual[b]
CS-WSP, CS-SFB	136	—	—	—	—	62	Actual[b]
CS-WSP, CS-SFB	140	—	—	—	—	66	Actual[b]
CS-WSP, CS-SFB	144	—	—	—	—	72	Actual[b]

For SI: 1 inch = 25.4 mm, 1 foot = 304.8 mm, 1 mile per hour = 0.447 m/s.

NP = Not Permitted.

a. Linear interpolation shall be permitted.

b. Use the actual length when it is greater than or equal to the minimum length.

c. Maximum header height for PFH is 10 feet in accordance with Figure R602.10.6.2, but wall height may be increased to 12 feet with pony wall.

d. Maximum opening height for PFG is 10 feet in accordance with Figure R602.10.6.3, but wall height may be increased to 12 feet with pony wall.

e. Maximum opening height for CS-PF is 10 feet in accordance with Figure R602.10.6.4, but wall height may be increased to 12 feet with pony wall.

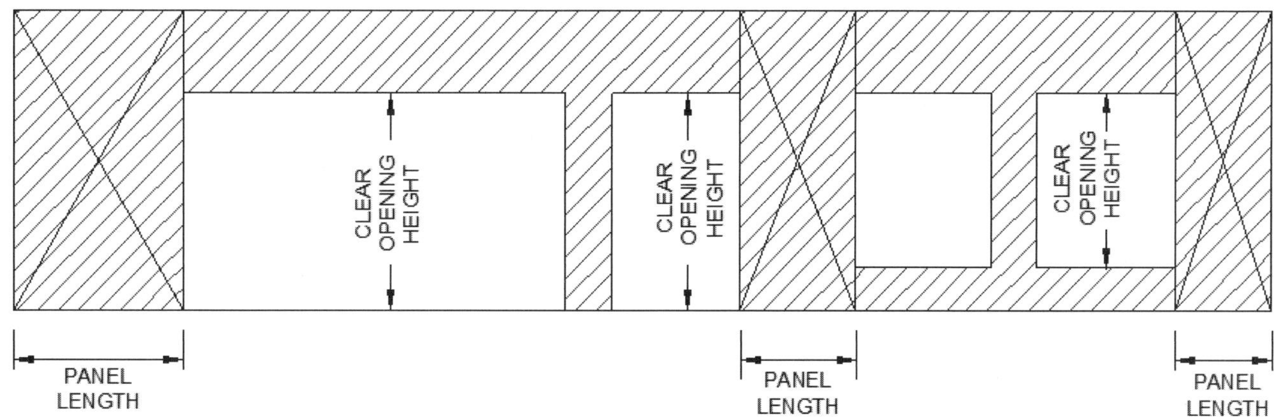

FIGURE R602.10.5
BRACED WALL PANELS WITH CONTINUOUS SHEATHING

TABLE R602.10.5.2
PARTIAL CREDIT FOR BRACED WALL PANELS LESS THAN 48 INCHES IN ACTUAL LENGTH

ACTUAL LENGTH OF BRACED WALL PANEL (inches)	CONTRIBUTING LENGTH OF BRACED WALL PANEL (inches)[a]	
	8-foot Wall Height	9-foot Wall Height
48	48	48
42	36	36
36	27	N/A

For SI: 1 inch = 25.4 mm, 1 foot = 304.8 mm.

N/A = Not Applicable.

a. Linear interpolation shall be permitted.

TABLE R602.10.6.1
MINIMUM HOLD-DOWN FORCES FOR METHOD ABW BRACED WALL PANELS

SEISMIC DESIGN CATEGORY AND WIND SPEED	SUPPORTING/STORY	HOLD DOWN FORCE (pounds)				
		Height of Braced Wall Panel				
		8 feet	9 feet	10 feet	11 feet	12 feet
SDC A, B and C Wind speed < 110 mph	One story	1,800	1,800	1,800	2,000	2,200
	First of two stories	3,000	3,000	3,000	3,300	3,600
SDC D_o, D_1 and D_2 Wind speed < 110 mph	One story	1,800	1,800	1,800	NP	NP
	First of two stories	3,000	3,000	3,000	NP	NP

For SI: 1 inch = 25.4 mm, 1 foot = 304.8 mm, 1 pound = 4.45 N, 1 mile per hour = 0.447 m/s.

NP = Not Permitted.

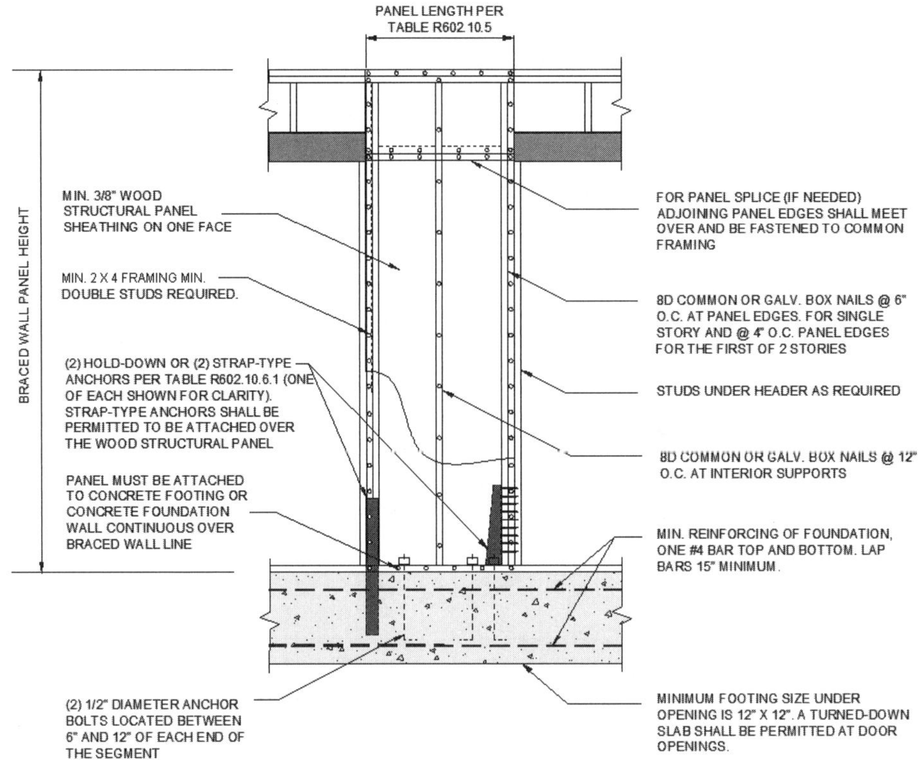

For SI: 1 inch = 25.4 mm.

FIGURE R602.10.6.1
METHOD ABW—ALTERNATE BRACED WALL PANEL

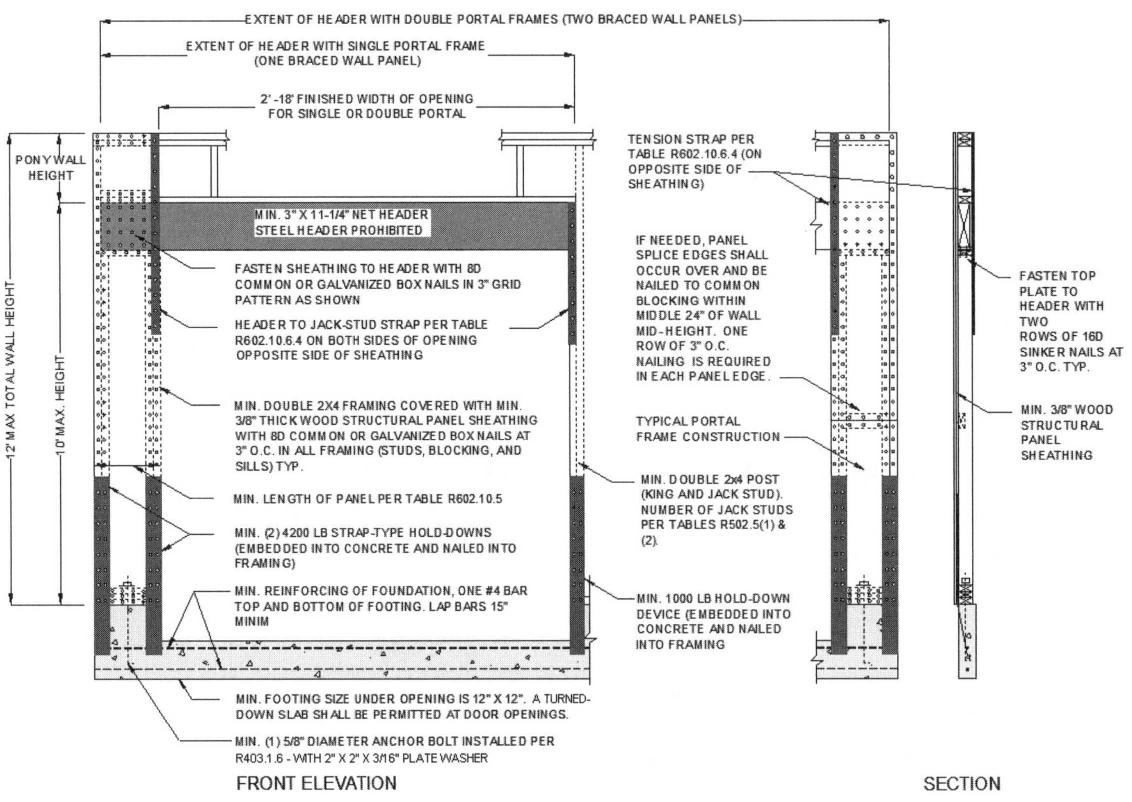

For SI: 1 inch = 25.4 mm, 1 foot = 304.8 mm.

FIGURE R602.10.6.2
METHOD PFH—PORTAL FRAME WITH HOLD-DOWNS

R602.10.6.3 Method PFG: Portal frame at garage door openings in Seismic Design Categories A, B and C. Where supporting a roof or one story and a roof, a Method PFG *braced wall panel* constructed in accordance with Figure R602.10.6.3 shall be permitted on either side of garage door openings.

R602.10.6.4 Method CS-PF: Continuously sheathed portal frame. Continuously sheathed portal frame *braced wall panels* shall be constructed in accordance with Figure R602.10.6.4 and Table R602.10.6.4. The number of continuously sheathed portal frame panels in a single *braced wall line* shall not exceed four.

R602.10.6.5 Wall bracing for dwellings with stone and masonry veneer in Seismic Design Categories D_0, D_1 and D_2. Where stone and masonry veneer are installed in accordance with Section R703.7, wall bracing on exterior *braced wall lines* and *braced wall lines* on the interior of the building, backing or perpendicular to and laterally supporting veneered walls shall comply with this section.

Where dwellings in Seismic Design Categories D_0, D_1 and D_2 have stone or masonry veneer installed in accordance with Section R703.7, and the veneer does not exceed the first-story height, wall bracing shall be in accordance with Section R602.10.3.

Where detached one- or two-family dwellings in Seismic Design Categories D_0, D_1 and D_2 have stone or masonry veneer installed in accordance with Section R703.7, and the veneer exceeds the first-*story height*, wall bracing at exterior *braced wall lines* and *braced wall lines* on the interior of the building shall be constructed using Method BV-WSP in accordance with this section and Figure R602.10.6.5. Cripple walls shall not be permitted, and required interior *braced wall lines* shall be supported on continuous foundations.

Townhouses in Seismic Design Categories D_0, D_1 and D_2 with stone or masonry veneer exceeding the first-story height shall be designed in accordance with accepted engineering practice.

R602.10.6.5.1 Length of bracing. The length of bracing along each *braced wall line* shall be the greater of that required by the design wind speed and *braced wall line* spacing in accordance with Table R602.10.3(1) as adjusted by the factors in the Table R602.10.3(2) or the Seismic Design Category and *braced wall line* length in accordance with Table R602.10.6.5. Angled walls shall be permitted to be counted in accordance with Section R602.10.1.4, and *braced wall panel* location shall be in accordance with Section R602.10.2.2. The seismic adjustment factors in Table R602.10.3(4) shall not be applied to the length of bracing determined using Table R602.10.6.5. In no case shall the minimum total length of bracing in a *braced wall line*, after all adjustments have been taken, be less than 48 inches (1219 mm) total.

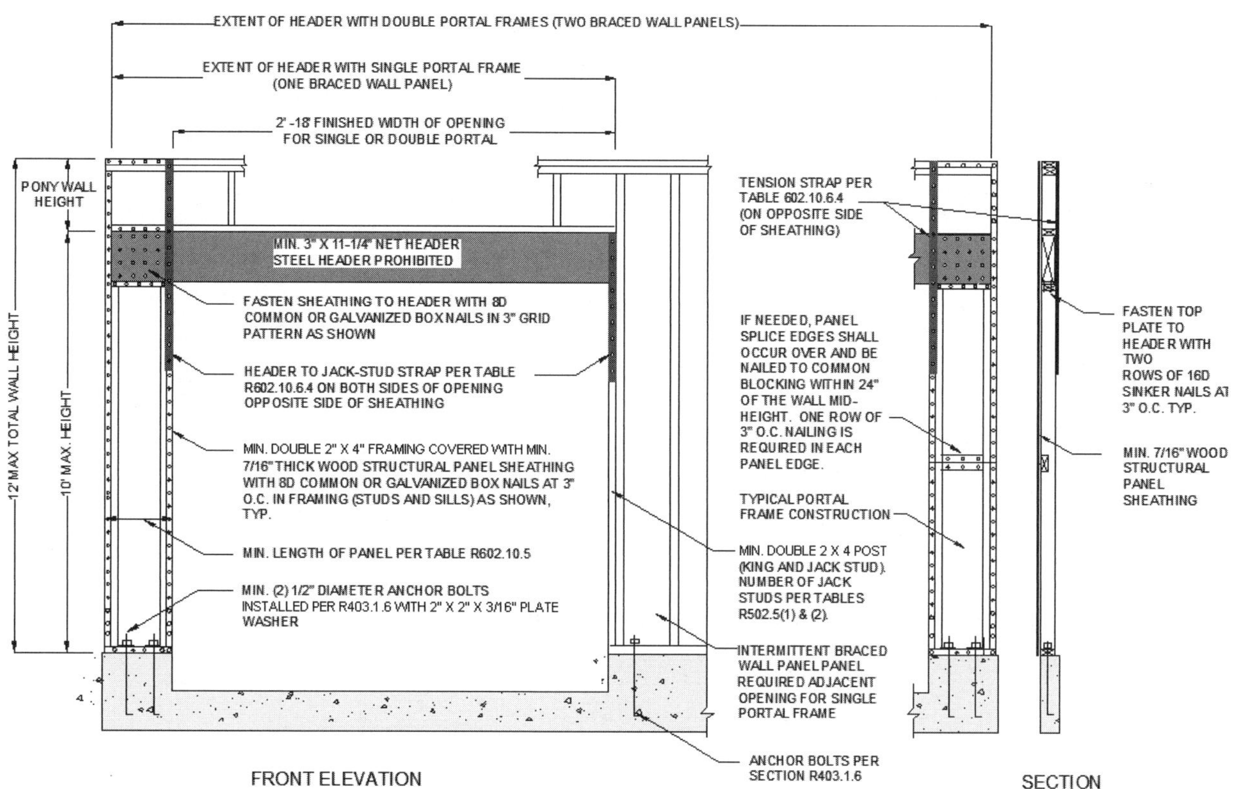

For SI: 1 inch = 25.4 mm, 1 foot = 304.8 mm.

FIGURE R602.10.6.3
METHOD PFG—PORTAL FRAME AT GARAGE DOOR OPENINGS IN SEISMIC DESIGN CATEGORIES A, B AND C

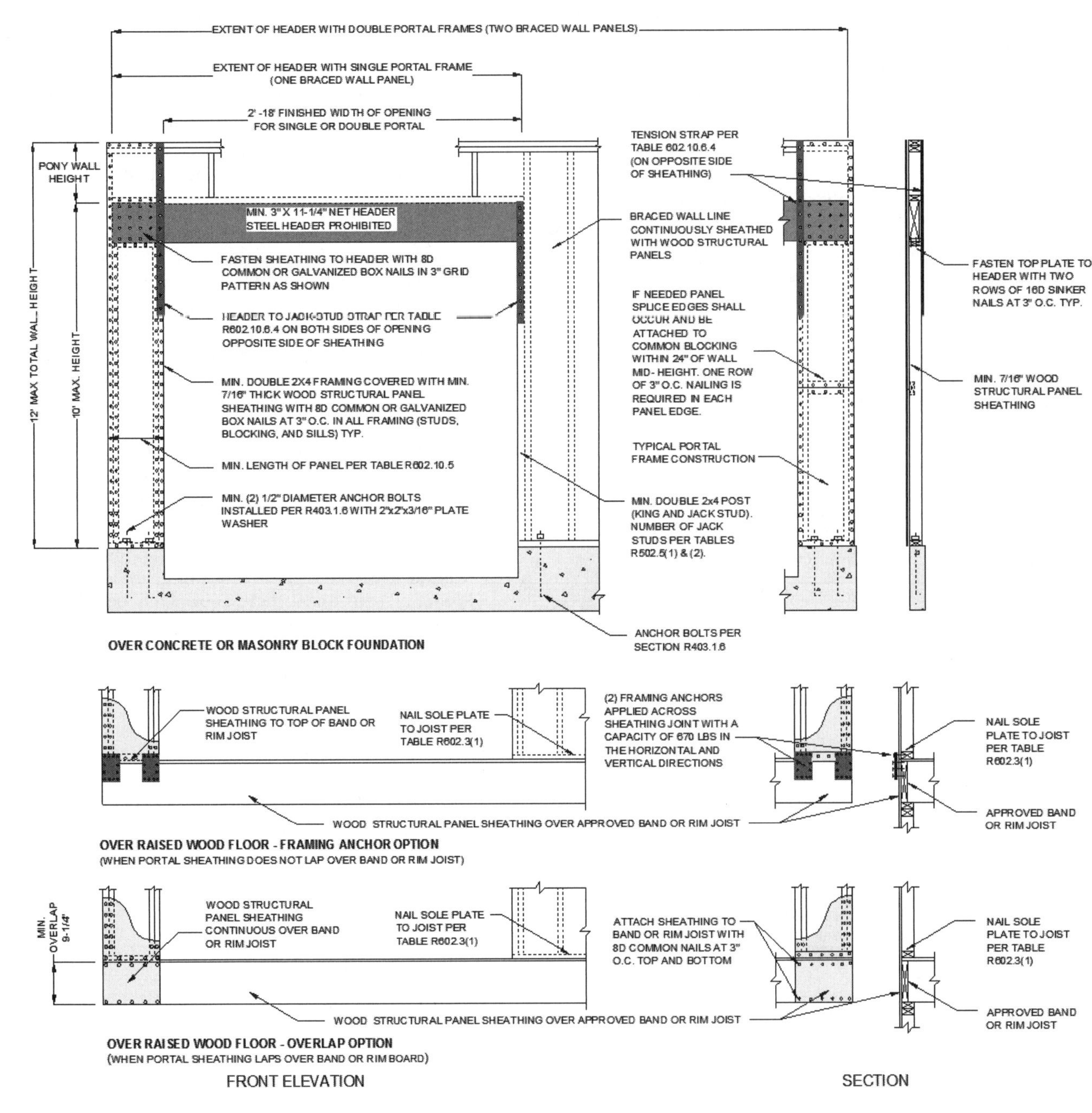

For SI: 1 inch = 25.4 mm, 1 foot = 304.8 mm.

FIGURE R602.10.6.4
METHOD CS-PF—CONTINUOUSLY SHEATHED PORTAL FRAME PANEL CONSTRUCTION

**TABLE R602.10.6.4
TENSION STRAP CAPACITY REQUIRED FOR RESISTING WIND PRESSURES
PERPENDICULAR TO METHOD PFH, PFG AND CS-PF BRACED WALL PANELS**

MINIMUM WALL STUD FRAMING NOMINAL SIZE AND GRADE	MAXIMUM PONY WALL HEIGHT (feet)	MAXIMUM TOTAL WALL HEIGHT (feet)	MAXIMUM OPENING WIDTH (feet)	TENSION STRAP CAPACITY REQUIRED (pounds)[a, b]					
				Basic Wind Speed (mph)					
				85	90	100	85	90	100
				Exposure B			Exposure C		
2 × 4 No. 2 Grade	0	10	18	1,000	1,000	1,000	1,000	1,000	1,000
	1	10	9	1,000	1,000	1,000	1,000	1,000	1,275
			16	1,000	1,000	1,750	1,800	2,325	3,500
			18	1,000	1,200	2,100	2,175	2,725	DR
	2	10	9	1,000	1,000	1,025	1,075	1,550	2,500
			16	1,525	2,025	3,125	3,200	3,900	DR
			18	1,875	2,400	3,575	3,700	DR	DR
	2	12	9	1,000	1,200	2,075	2,125	2,750	4,000
			16	2,600	3,200	DR	DR	DR	DR
			18	3,175	3,850	DR	DR	DR	DR
	4	12	9	1,775	2,350	3,500	3,550	DR	DR
			16	4,175	DR	DR	DR	DR	DR
2 × 6 Stud Grade	2	12	9	1,000	1,000	1,325	1,375	1,750	2,550
			16	1,650	2,050	2,925	3,000	3,550	DR
			18	2,025	2,450	3,425	3,500	4,100	DR
	4	12	9	1,125	1,500	2,225	2,275	2,775	3,800
			16	2,650	3,150	DR	DR	DR	DR
			18	3,125	3,675	DR	DR	DR	DR

For SI: 1 inch = 25.4 mm, 1 foot = 304.8 mm, 1 pound = 4.45 N.
a. DR = design required.
b. Strap shall be installed in accordance with manufacturer's recommendations.

TABLE R602.10.6.5
METHOD BV-WSP WALL BRACING REQUIREMENTS

SEISMIC DESIGN CATEGORY	STORY	BRACED WALL LINE LENGTH (FEET)					SINGLE-STORY HOLD-DOWN FORCE (pounds)[a]	CUMULATIVE HOLD-DOWN FORCE (pounds)[b]
		10	20	30	40	50		
		MINIMUM TOTAL LENGTH (FEET) OF BRACED WALL PANELS REQUIRED ALONG EACH BRACED WALL LINE						
D_0		4.0	7.0	10.5	14.0	17.5	N/A	—
		4.0	7.0	10.5	14.0	17.5	1900	—
		4.5	9.0	13.5	18.0	22.5	3500	5400
		6.0	12.0	18.0	24.0	30.0	3500	8900
D_1		4.5	9.0	13.5	18.0	22.5	2100	—
		4.5	9.0	13.5	18.0	22.5	3700	5800
		6.0	12.0	18.0	24.0	30.0	3700	9500
D_2		5.5	11.0	16.5	22.0	27.5	2300	—
		5.5	11.0	16.5	22.0	27.5	3900	6200
		NP	NP	NP	NP	NP	N/A	N/A

For SI: 1 inch = 25.4 mm, 1 foot = 304.8 mm, 1 pound per square foot = 0.479 kPa, 1 pound-force = 4.448 N.

NP = Not Permitted.

N/A = Not Applicable.

a. Hold-down force is minimum allowable stress design load for connector providing uplift tie from wall framing at end of braced wall panel at the noted story to wall framing at end of braced wall panel at the story below, or to foundation or foundation wall. Use single-story hold-down force where edges of braced wall panels do not align; a continuous load path to the foundation shall be maintained.

b. Where hold-down connectors from stories above align with stories below, use cumulative hold-down force to size middle- and bottom-story hold-down connectors.

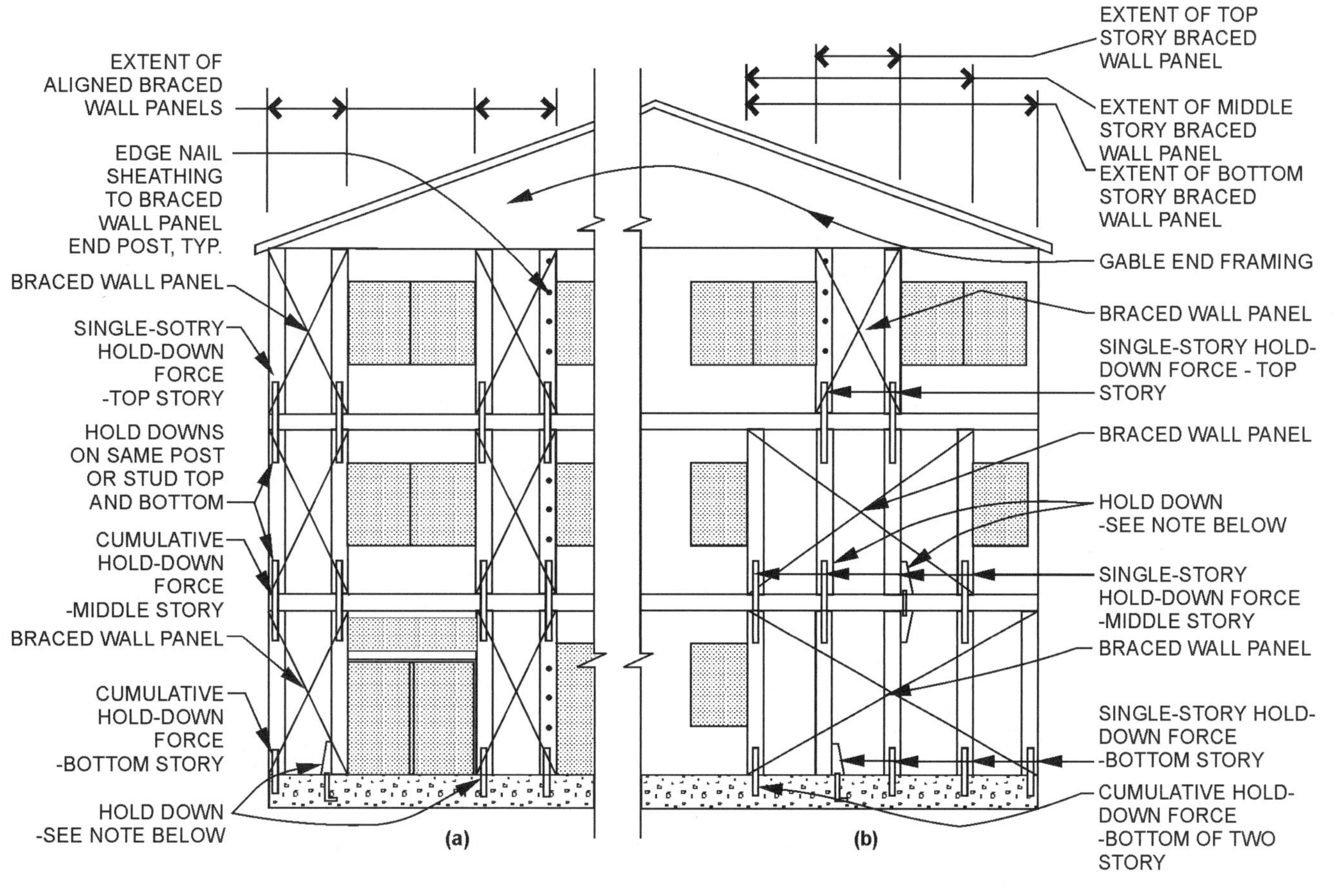

FIGURE R602.10.6.5
METHOD BV-WSP—WALL BRACING FOR DWELLINGS WITH STONE AND MASONRY VENEER IN SEISMIC DESIGN CATEGORIES D_0, D_1 and D_2

R602.10.7 Ends of braced wall lines with continuous sheathing. Each end of a *braced wall line* with continuous sheathing shall have one of the conditions shown in Figure R602.10.7.

R602.10.8 Braced wall panel connections. *Braced wall panels* shall be connected to floor framing or foundations as follows:

1. Where joists are perpendicular to a *braced wall panel* above or below, a rim joist, band joist or blocking shall be provided along the entire length of the *braced wall panel* in accordance with Figure R602.10.8(1). Fastening of top and bottom wall plates to framing, rim joist, band joist and/or blocking shall be in accordance with Table R602.3(1).
2. Where joists are parallel to a *braced wall panel* above or below, a rim joist, end joist or other parallel framing member shall be provided directly above and below the *braced wall panel* in accordance with Figure R602.10.8(2). Where a parallel framing member cannot be located directly above and below the panel, full-depth blocking at 16-inch (406 mm) spacing shall be provided between the parallel framing members to each side of the *braced wall panel* in accordance with Figure R602.10.8(2). Fastening of blocking and wall plates shall be in accordance with Table R602.3(1) and Figure R602.10.8(2).
3. Connections of *braced wall panels* to concrete or masonry shall be in accordance with Section R403.1.6.

R602.10.8.1 Braced wall panel connections for Seismic Design Categories D_0, D_1 and D_2. *Braced wall panels* shall be fastened to required foundations in accordance with Section R602.11.1, and top plate lap splices shall be face-nailed with at least eight 16d nails on each side of the splice.

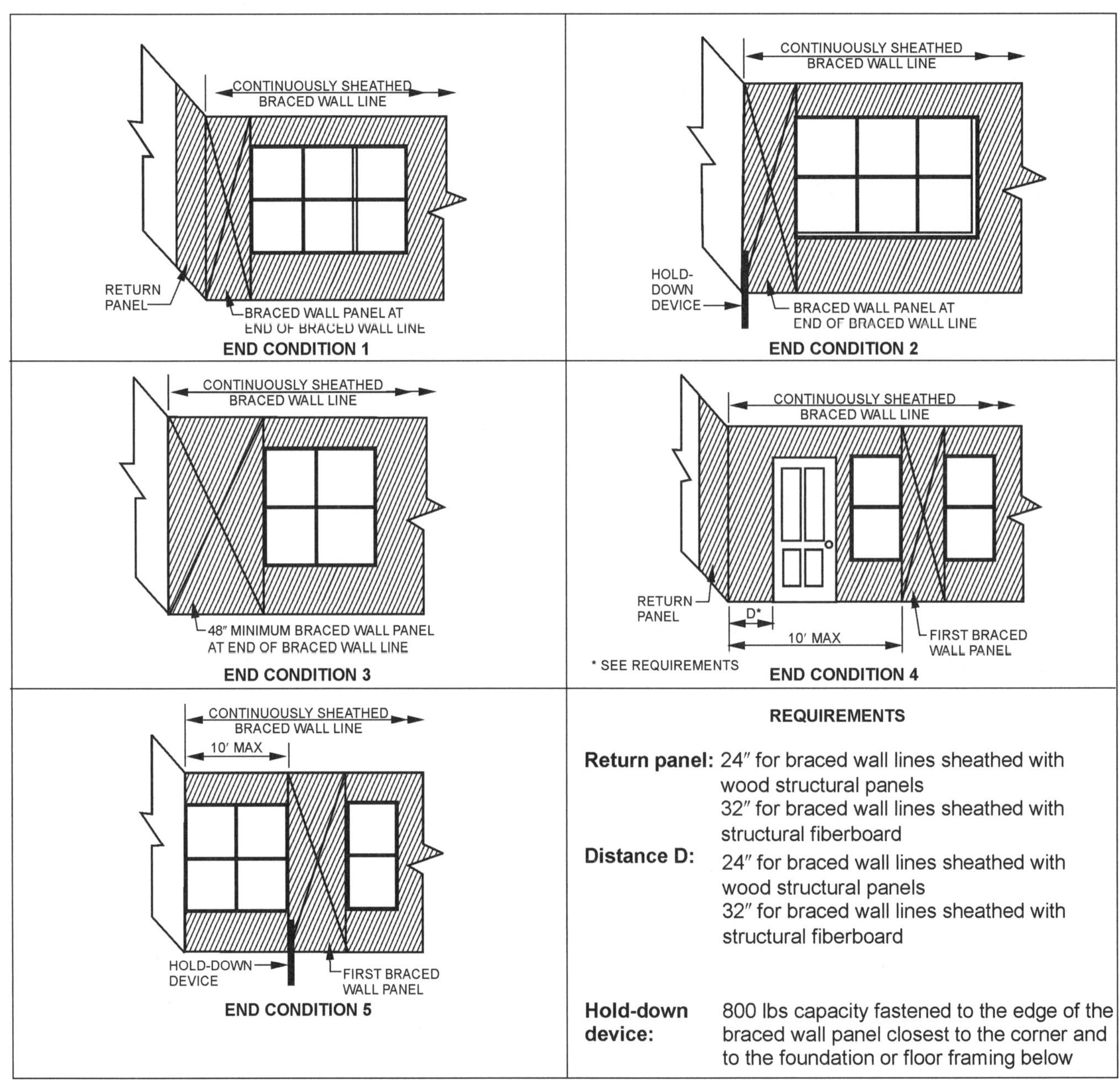

For SI: 1 inch = 25.4 mm, 1 foot = 304.8 mm, 1 pound = 4.45 N.

FIGURE R602.10.7
END CONDITIONS FOR BRACED WALL LINES WITH CONTINUOUS SHEATHING

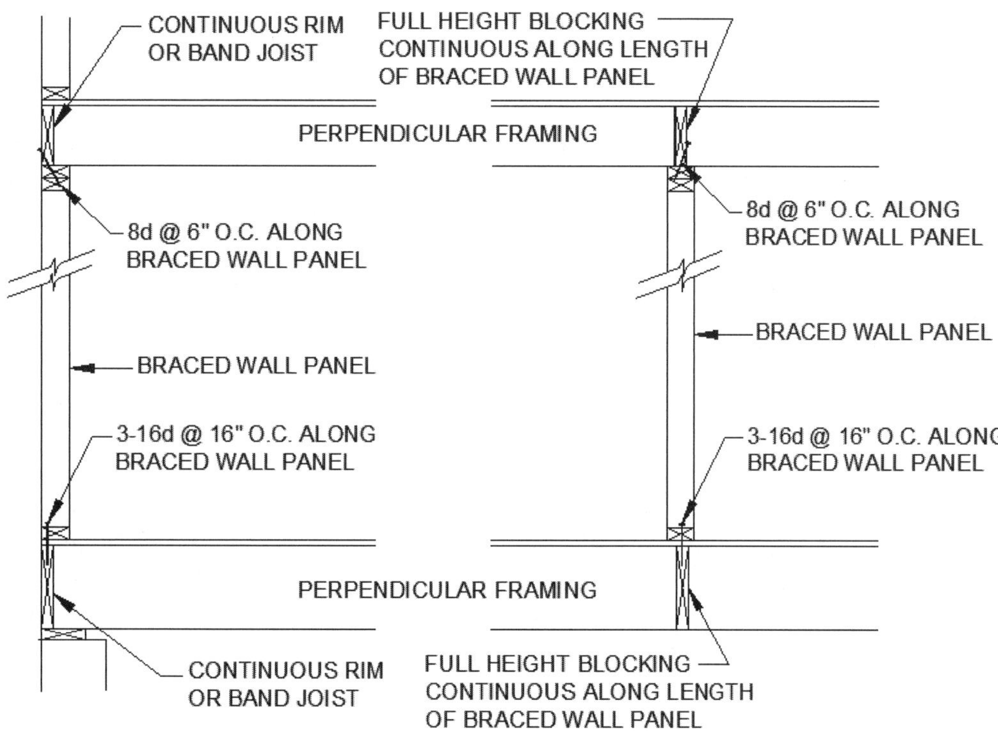

For SI: 1 inch = 25.4 mm.

FIGURE R602.10.8(1)
BRACED WALL PANEL CONNECTION WHEN PERPENDICULAR TO FLOOR/CEILING FRAMING

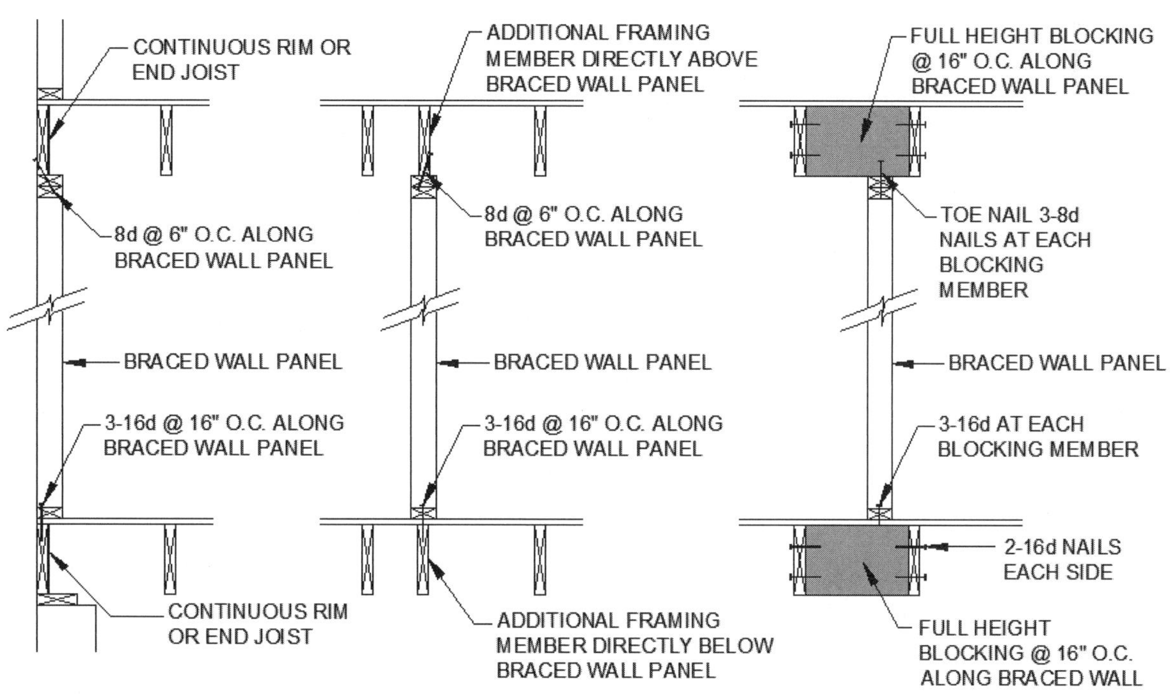

For SI: 1 inch = 25.4 mm.

FIGURE R602.10.8(2)
BRACED WALL PANEL CONNECTION WHEN PARALLEL TO FLOOR/CEILING FRAMING

R602.10.8.2 Connections to roof framing. Top plates of exterior *braced wall panels* shall be attached to rafters or roof trusses above in accordance with Table R602.3(1) and this section. Where required by this section, blocking between rafters or roof trusses shall be attached to top plates of *braced wall panels* and to rafters and roof trusses in accordance with Table R602.3(1). A continuous band, rim, or header joist or roof truss parallel to the *braced wall panels* shall be permitted to replace the blocking required by this section. Blocking shall not be required over openings in continuously-sheathed *braced wall lines*. In addition to the requirements of this section, lateral support shall be provided for rafters and ceiling joists in accordance with Section R802.8 and for trusses in accordance with Section R802.10.3. Roof ventilation shall be provided in accordance with Section R806.1.

1. For Seismic Design Categories A, B and C and wind speeds less than 100 mph (45 m/s) where the distance from the top of the *braced wall panel* to the top of the rafters or roof trusses above is $9^1/_4$ inches (235 mm) or less, blocking between rafters or roof trusses shall not be required. Where the distance from the top of the *braced wall panel* to the top of the rafters or roof trusses above is between $9^1/_4$ inches (235 mm) and $15^1/_4$ inches (387 mm), blocking between rafters or roof trusses shall be provided above the *braced wall panel* in accordance with Figure R602.10.8.2(1).
2. For Seismic Design Categories D_0, D_1 and D_2 or wind speeds of 100 mph (45 m/s) or greater, where the distance from the top of the *braced wall panel* to the top of the rafters or roof trusses is $15^1/_4$ inches (387 mm) or less, blocking between rafters or roof trusses shall be provided above the *braced wall panel* in accordance with Figure R602.10.8.2(1).
3. Where the distance from the top of the *braced wall panel* to the top of rafters or roof trusses exceeds $15^1/_4$ inches (387 mm), the top plates of the *braced wall panel* shall be connected to perpendicular rafters or roof trusses above in accordance with one or more of the following methods:
    3.1. Soffit blocking panels constructed in accordance with Figure R602.10.8.2(2);
    3.2. Vertical blocking panels constructed in accordance with Figure R602.10.8.2(3);
    3.3. Full-height engineered blocking panels designed in accordance with the AF&PA WFCM; or
    3.4. Blocking, blocking panels, or other methods of lateral load transfer designed in accordance with accepted engineering practice.

R602.10.9 Braced wall panel support. *Braced wall panel* support shall be provided as follows:

1. Cantilevered floor joists complying with Section R502.3.3 shall be permitted to support *braced wall panels*.
2. Elevated post or pier foundations supporting *braced wall panels* shall be designed in accordance with accepted engineering practice.
3. Masonry stem walls with a length of 48 inches (1219 mm) or less supporting *braced wall panels* shall be reinforced in accordance with Figure R602.10.9. Masonry stem walls with a length greater than 48 inches (1219 mm) supporting *braced wall panels* shall be constructed in accordance with Section R403.1 Methods ABW and PFH shall not be permitted to attach to masonry stem walls.
4. Concrete stem walls with a length of 48 inches (1219 mm) or less, greater than 12 inches (305 mm) tall and less than 6 inches (152 mm) thick shall have reinforcement sized and located in accordance with Figure R602.10.9.

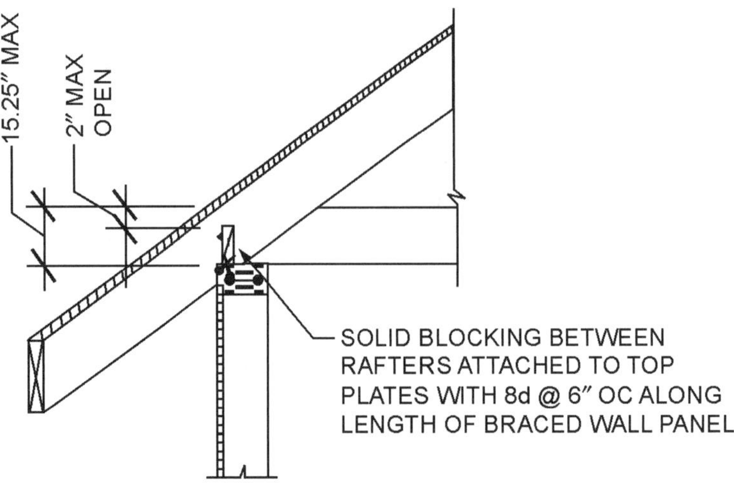

For SI: 1 inch = 25.4 mm.

FIGURE R602.10.8.2(1)
BRACED WALL PANEL CONNECTION TO PERPENDICULAR RAFTERS

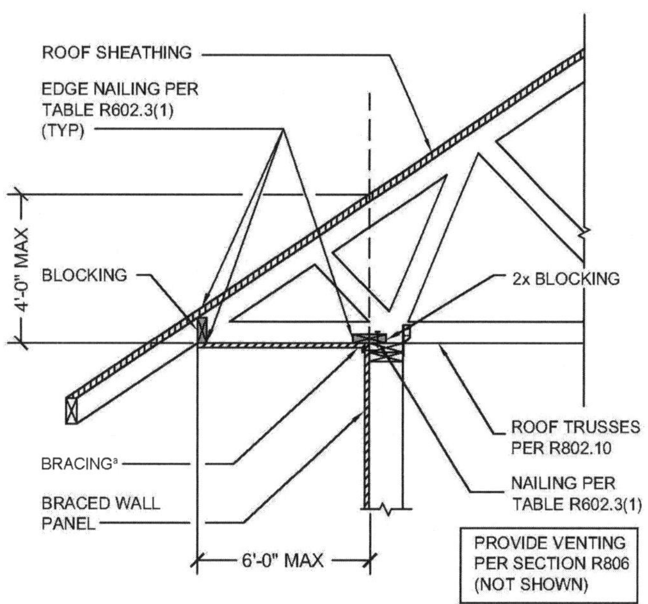

For SI: 1 inch = 25.4 mm, 1 foot = 304.8 mm.

a. Methods of bracing shall be as described in Section R602.10.4.

FIGURE R602.10.8.2(2)
BRACED WALL PANEL CONNECTION OPTION TO PERPENDICULAR RAFTERS OR ROOF TRUSSES

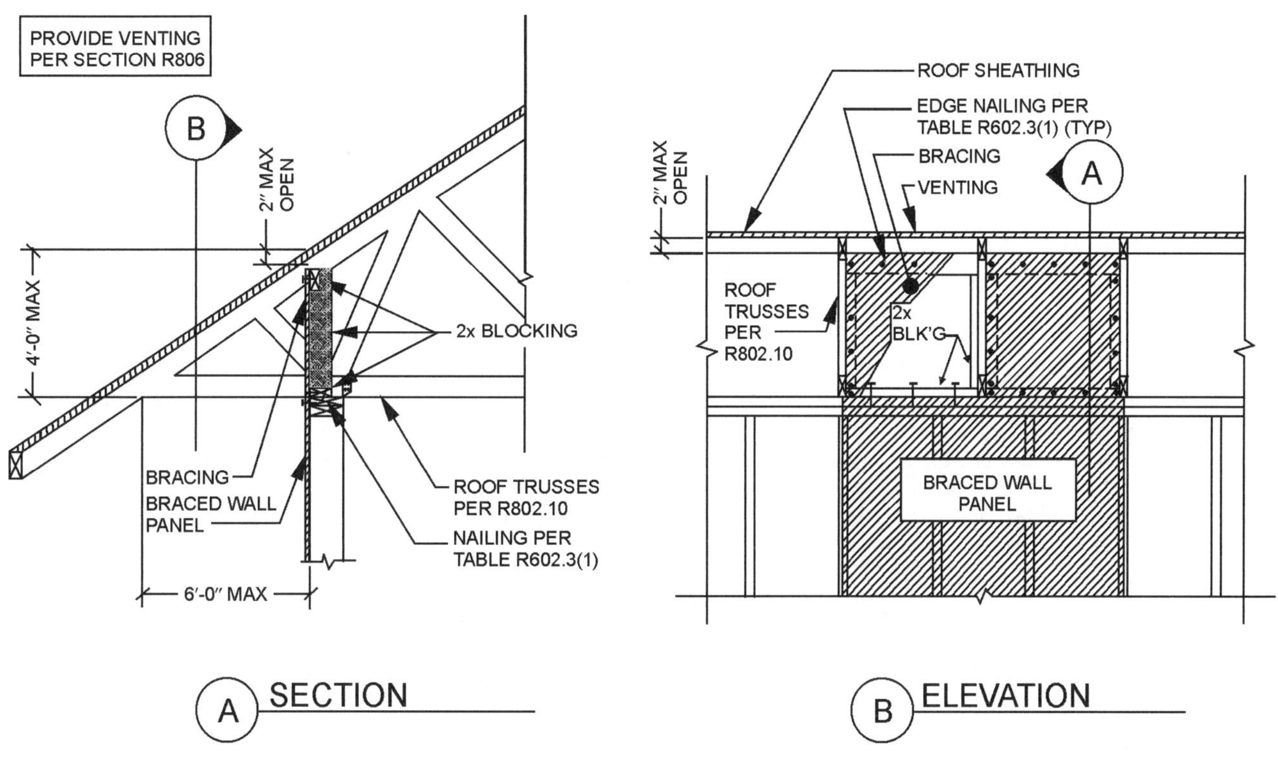

For SI: 1 inch = 25.4 mm, 1 foot = 304.8 mm.

a. Methods of bracing shall be as described in Section R602.10.4.

FIGURE R602.10.8.2(3)
BRACED WALL PANEL CONNECTION OPTION TO PERPENDICULAR RAFTERS OR ROOF TRUSSES

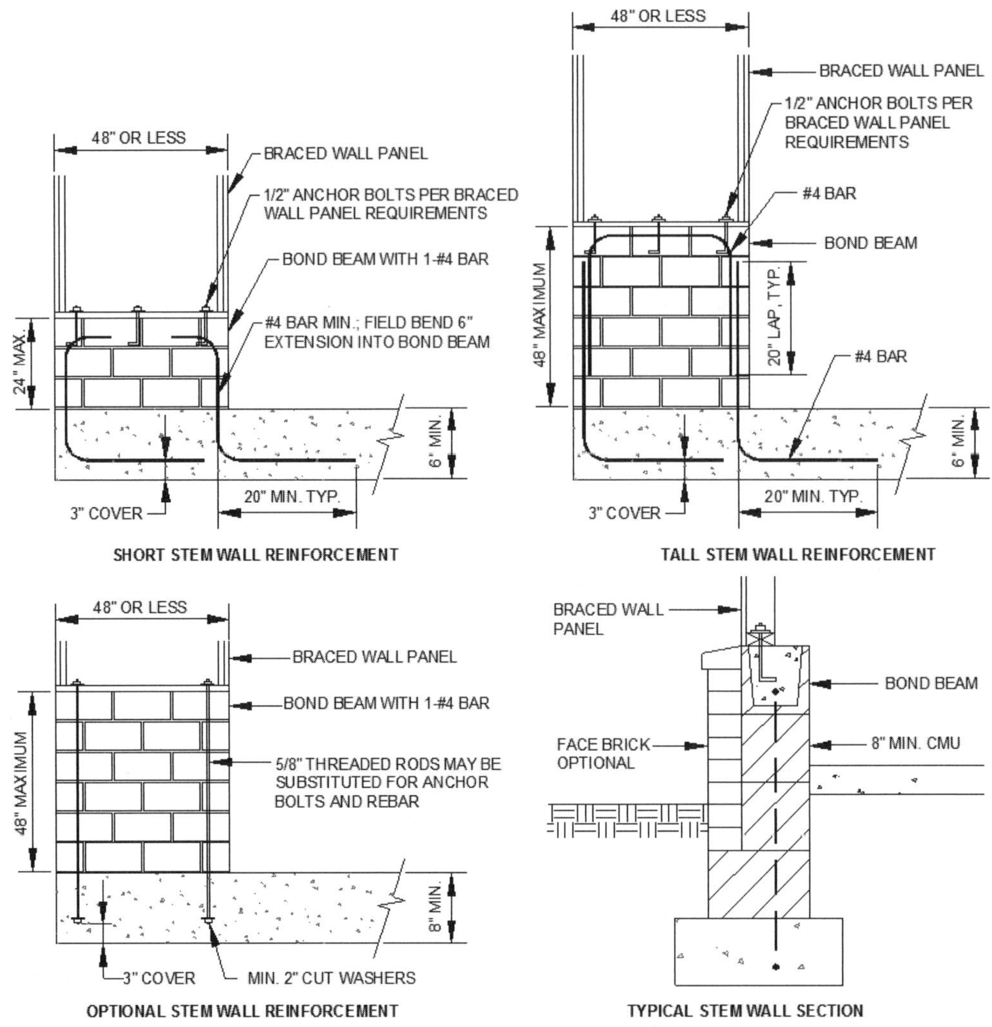

For SI: 1 inch = 25.4 mm.

FIGURE R602.10.9
MASONRY STEM WALLS SUPPORTING BRACED WALL PANELS

R602.10.9.1 Braced wall panel support for Seismic Design Category D_2. In one-story buildings located in Seismic Design Category D_2, *braced wall panels* shall be supported on continuous foundations at intervals not exceeding 50 feet (15 240 mm). In two-story buildings located in Seismic Design Category D_2, all *braced wall panels* shall be supported on continuous foundations.

Exception: Two-story buildings shall be permitted to have interior *braced wall panels* supported on continuous foundations at intervals not exceeding 50 feet (15 240 mm) provided that:

1. The height of cripple walls does not exceed 4 feet (1219 mm).
2. First-floor *braced wall panels* are supported on doubled floor joists, continuous blocking or floor beams.
3. The distance between bracing lines does not exceed twice the building width measured parallel to the *braced wall line*.

R602.10.10 Panel joints. All vertical joints of panel sheathing shall occur over, and be fastened to, common studs. Horizontal joints in *braced wall panels* shall occur over, and be fastened to, common blocking of a minimum $1^1/_2$ inch (38 mm) thickness.

Exceptions:

1. Vertical joints of panel sheathing shall be permitted to occur over double studs, where adjoining panel edges are attached to separate studs with the required panel edge fastening schedule, and the adjacent studs are attached together with two rows of 10d box nails [3 inches by 0.128 inch (76.2 mm by 3.25 mm)] at 10 inches o.c. (254 mm).

2. Blocking at horizontal joints shall not be required in wall segments that are not counted as *braced wall panels*.
3. Where the bracing length provided is at least twice the minimum length required by Table R602.10.3(1) and Table R602.10.3(3) blocking at horizontal joints shall not be required in *braced wall panels* constructed using Methods WSP, SFB, GB, PBS or HPS.
4. When Method GB panels are installed horizontally, blocking of horizontal joints is not required.

R602.10.11 Cripple wall bracing. Cripple walls shall be constructed in accordance with Section R602.9 and braced in accordance with this section. Cripple walls shall be braced with the length and method of bracing used for the wall above in accordance with Tables R602.10.3(1) and R602.10.3(3), and the applicable adjustment factors in Table R602.10.3(2) or R602.10.3(4), repspectivley, except that the length of cripple wall bracing shall be multiplied by a factor of 1.15. The distance between adjacent edges of *braced wall panels* shall be reduced from 20 feet (6096 mm) to 14 feet (4267 mm).

R602.10.11.1 Cripple wall bracing for Seismic Design Categories D_0 and D_1 and townhouses in Seismic Design Category C. In addition to the requirements in Section R602.10.11, the distance between adjacent edges of *braced wall panels* for cripple walls along a *braced wall line* shall be 14 feet (4267 mm) maximum.

Where *braced wall lines* at interior walls are not supported on a continuous foundation below, the adjacent parallel cripple walls, where provided, shall be braced with Method WSP or Method CS-WSP in accordance with Section R602.10.4. The length of bracing required in accordance with Table R602.10.3(3) for the cripple walls shall be multiplied by 1.5. Where the cripple walls do not have sufficient length to provide the required bracing, the spacing of panel edge fasteners shall be reduced to 4 inches (102 mm) on center and the required bracing length adjusted by 0.7. If the required length can still not be provided, the cripple wall shall be designed in accordance with accepted engineering practice.

R602.10.11.2 Cripple wall bracing for Seismic Design Category D_2. In Seismic Design Category D_2, cripple walls shall be braced in accordance with Tables R602.10.3(3) and R602.10.3(4).

R602.10.11.3 Redesignation of cripple walls. Where all cripple wall segments along a *braced wall line* do not exceed 48 inches (1219 mm) in height, the cripple walls shall be permitted to be redesignated as a first-*story* wall for purposes of determining wall bracing requirements. Where any cripple wall segment in a *braced wall line* exceeds 48 inches (1219 mm) in height, the entire cripple wall shall be counted as an additional *story*. If the cripple walls are redesignated, the stories above the redesignated *story* shall be counted as the second and third stories, respectively.

R602.11 Wall anchorage. *Braced wall line* sills shall be anchored to concrete or masonry foundations in accordance with Sections R403.1.6 and R602.11.1.

R602.11.1 Wall anchorage for all buildings in Seismic Design Categories D_0, D_1 and D_2 and townhouses in Seismic Design Category C. Plate washers, a minimum of 0.229 inch by 3 inches by 3 inches (5.8 mm by 76 mm by 76 mm) in size, shall be provided between the foundation sill plate and the nut except where *approved* anchor

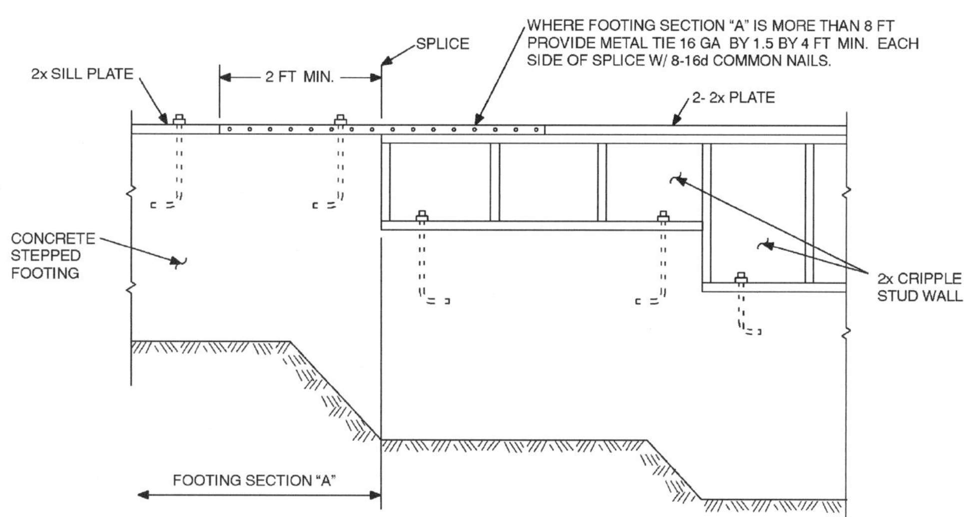

For SI: 1 inch = 25.4 mm, 1 foot = 304.8 mm.

Note: Where footing Section "A" is less than 8 feet long in a 25-foot-long wall, install bracing at cripple stud wall.

FIGURE R602.11.2
STEPPED FOUNDATION CONSTRUCTION

straps are used. The hole in the plate washer is permitted to be diagonally slotted with a width of up to $^3/_{16}$ inch (5 mm) larger than the bolt diameter and a slot length not to exceed $1^3/_4$ inches (44 mm), provided a standard cut washer is placed between the plate washer and the nut.

R602.11.2 Stepped foundations in Seismic Design Categories D_0, D_1 and D_2. In all buildings located in Seismic Design Categories D_0, D_1 or D_2, where the height of a required *braced wall line* that extends from foundation to floor above varies more than 4 feet (1219 mm), the *braced wall line* shall be constructed in accordance with the following:

1. Where the lowest floor framing rests directly on a sill bolted to a foundation not less than 8 feet (2440 mm) in length along a line of bracing, the line shall be considered as braced. The double plate of the cripple stud wall beyond the segment of footing that extends to the lowest framed floor shall be spliced by extending the upper top plate a minimum of 4 feet (1219 mm) along the foundation. Anchor bolts shall be located a maximum of 1 foot and 3 feet (305 and 914 mm) from the step in the foundation. See Figure R602.11.2.
2. Where cripple walls occur between the top of the foundation and the lowest floor framing, the bracing requirements of Sections R602.10.11, R602.10.11.1 and R602.10.11.2 shall apply.
3. Where only the bottom of the foundation is stepped and the lowest floor framing rests directly on a sill bolted to the foundations, the requirements of Sections R403.1.6 and R602.11.1 shall apply.

* **R602.12 Simplified wall bracing.** Buildings meeting all of the conditions listed in items 1-8 shall be permitted to be braced in accordance with this section as an alternative to the requirements of Section R602.10. The entire building shall be braced in accordance with this section; the use of other bracing provisions of R602.10, except as specified herein, shall not be permitted.

1. There shall be no more than two stories above the top of a concrete or masonry foundation or basement wall. Permanent wood foundations shall not be permitted.
2. Floors shall not cantilever more than 24 inches (607 mm) beyond the foundation or bearing wall below.
3. Wall height shall not be greater than 10 feet (2743 mm).
4. The building shall have a roof eave-to-ridge height of 15 feet (4572 mm) or less.
5. All exterior walls shall have gypsum board with a minimum thickness of $^1/_2$ inch (12.7 mm) installed on the interior side fastened in accordance with Table R702.3.5.
6. The structure shall be located where the basic wind speed is less than or equal to 90 mph (40 m/s), and the Exposure Category is A or B.
7. The structure shall be located in Seismic Design Category A, B or C for detached one- and two-family dwellings or Seismic Design Category A or B for townhouses.
8. Cripple walls shall not be permitted in two-story buildings.

R602.12.1 Circumscribed rectangle. The bracing required for each building shall be determined by circumscribing a rectangle around the entire building on each floor as shown in Figure R602.12.1. The rectangle shall surround all enclosed offsets and projections such as sunrooms and attached garages. Open structures, such as carports and decks, shall be permitted to be excluded. The rectangle shall have no side greater than 60 feet (18 288 mm), and the ratio between the long side and short side shall be a maximum of 3:1.

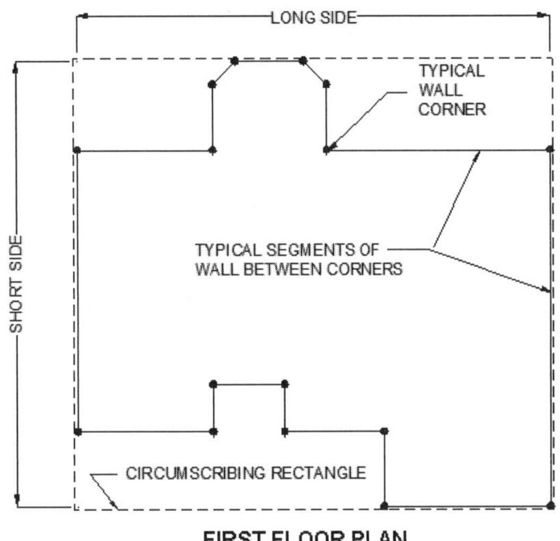

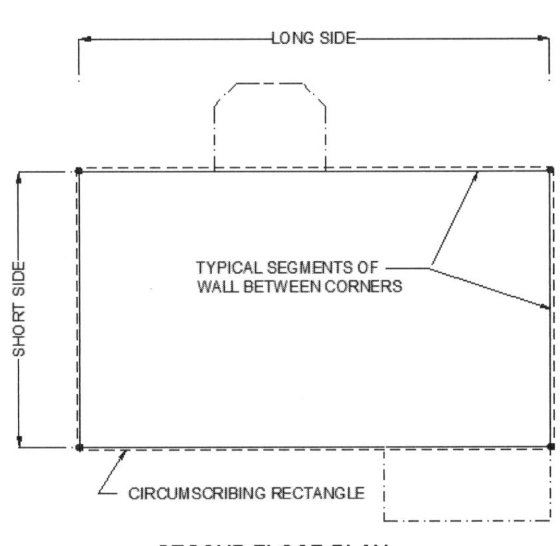

FIGURE R602.12.1
RECTANGLE CIRCUMSCRIBING AN ENCLOSED BUILDING

R602.12.2 Sheathing materials. The following sheathing materials installed on the exterior side of exterior walls shall be used to construct a bracing unit as defined in Section R602.12.3. Mixing materials is prohibited.

1. Wood structural panels with a minimum thickness of $^{3}/_{8}$ inch (9.5 mm) fastened in accordance with Table R602.3(3).
2. Structural fiberboard sheathing with a minimum thickness of $^{1}/_{2}$ inch (12.7 mm) fastened in accordance with Table R602.3(1).

R602.12.3 Bracing unit. A bracing unit shall be a full-height sheathed segment of the exterior wall with no openings or vertical or horizontal offsets and a minimum length as specified herein. Interior walls shall not contribute toward the amount of required bracing. Mixing of Items 1 and 2 is prohibited on the same story.

1. Where all framed portions of all exterior walls are sheathed in accordance with Section R602.12.2, including wall areas between bracing units, above and below openings and on gable end walls, the minimum length of a bracing unit shall be 3 feet (914 mm).
2. Where the exterior walls are braced with sheathing panels in accordance with Section R602.12.2 and areas between bracing units are covered with other materials, the minimum length of a bracing unit shall be 4 feet (1219 mm).

R602.12.3.1 Multiple bracing units. Segments of wall compliant with Section R602.12.3 and longer than the minimum bracing unit length shall be considered as multiple bracing units. The number of bracing units shall be determined by dividing the wall segment length by the minimum bracing unit length. Full-height sheathed segments of wall narrower than the minimum bracing unit length shall not contribute toward a bracing unit except as specified in Section R602.12.6.

R602.12.4 Number of bracing units. Each side of the circumscribed rectangle, as shown in Figure R602.12.1, shall have, at a minimum, the number of bracing units in accordance with Table R602.12.4 placed on the parallel exterior walls facing the side of the rectangle. Bracing units shall then be placed using the distribution requirements specified in Section R602.12.5.

R602.12.5 Distribution of bracing units. The placement of bracing units on exterior walls shall meet all of the following requirements as shown in Figure R602.12.5.

1. A bracing unit shall begin no more than 12 feet (3658 mm) from any wall corner.
2. The distance between adjacent edges of bracing units shall be no greater than 20 feet (6096 mm).
3. Segments of wall greater than 8 feet (2438 mm) in length shall have a minimum of one bracing unit.

R602.12.6 Narrow panels. The bracing methods referenced in Section R602.10 and specified in Sections R602.12.6.1 through R602.12.6.3 shall be permitted when using simplified wall bracing.

R602.12.6.1 Method CS-G. *Braced wall panels* constructed as Method CS-G in accordance with Tables R602.10.4 and R602.10.5 shall be permitted for one-story garages when all framed portions of all exterior walls are sheathed with wood structural panels. Each CS-G panel shall be equivalent to 0.5 of a bracing unit. Segments of wall which include a Method CS-G panel shall meet the requirements of Section R602.10.4.2.

R602.12.6.2 Method CS-PF. *Braced wall panels* constructed as Method CS-PF in accordance with Section R602.10.6.4 shall be permitted when all framed por-

TABLE R602.12.4
MINIMUM NUMBER OF BRACING UNITS ON EACH SIDE OF THE CIRCUMSCRIBED RECTANGLE

STORY LEVEL	EAVE-TO-RIDGE HEIGHT (feet)	MINIMUM NUMBER OF BRACING UNITS ON EACH LONG SIDE[a, b]						MINIMUM NUMBER OF BRACING UNITS ON EACH SHORT SIDE[a, b]					
		Length of short side (feet)[c]						Length of long side (feet)[c]					
		10	20	30	40	50	60	10	20	30	40	50	60
	10	1	2	2	2	3	3	1	2	2	2	3	3
		2	3	3	4	5	6	2	3	3	4	5	6
	15	1	2	3	3	4	4	1	2	3	3	4	4
		2	3	4	5	6	7	2	3	4	5	6	7

For SI: 1 inch = 25.4 mm, 1 foot = 304.8 mm.

a. Interpolation shall not be permitted.

b. Cripple walls or wood-framed basement walls in a walk-out condition of a one-story structure shall be designed as the first floor of a two-story house.

c. Actual lengths of the sides of the circumscribed rectangle shall be rounded to the next highest unit of 10 when using this table.

tions of all exterior walls are sheathed with wood structural panels. Each CS-PF panel shall equal 0.5 bracing units. A maximum of four CS-PF panels shall be permitted on all segments of walls parallel to each side of the circumscribed rectangle. Segments of wall which include a Method CS-PF panel shall meet the requirements of Section R602.10.4.2.

R602.12.6.3 Methods PFH and PFG. *Braced wall panels* constructed as Method PFH and PFG shall be permitted when bracing units are constructed using wood structural panels. Each PFH panel shall equal one bracing unit and each PFG panel shall be equal to 0.75 bracing units.

R602.12.7 Lateral support. For bracing units located along the eaves, the vertical distance from the outside edge of the top wall plate to the roof sheathing above shall not exceed 9.25 inches (235 mm) at the location of a bracing unit unless lateral support is provided in accordance with Section R602.10.8.2.

R602.12.8 Stem walls. Masonry stem walls with a height and length of 48 inches (1219 mm) or less supporting a bracing unit or a Method CS-G, CS-PF or PFG *braced wall panel* shall be constructed in accordance with Figure R602.10.9. Concrete stem walls with a length of 48 inches (1219 mm) or less, greater than 12 inches (305 mm) tall and less than 6 inches (152 mm) thick shall be reinforced sized and located in accordance with Figure R602.10.9.

SECTION R603
STEEL WALL FRAMING

R603.1 General. Elements shall be straight and free of any defects that would significantly affect structural performance. Cold-formed steel wall framing members shall comply with the requirements of this section.

R603.1.1 Applicability limits. The provisions of this section shall control the construction of exterior cold-formed steel wall framing and interior load-bearing cold-formed steel wall framing for buildings not more than 60 feet (18 288 mm) long perpendicular to the joist or truss span, not more than 40 feet (12 192 mm) wide parallel to the joist or truss span, and less than or equal to three stories above *grade plane.* All exterior walls installed in accordance with the provisions of this section shall be considered as load-bearing walls. Cold-formed steel walls constructed in accordance with the provisions of this section shall be limited to sites subjected to a maximum design wind speed of 110 miles per hour (49 m/s) Exposure B or C and a maximum ground snow load of 70 pounds per square foot (3.35 kPa).

R603.1.2 In-line framing. Load-bearing cold-formed steel studs constructed in accordance with Section R603 shall be located in-line with joists, trusses and rafters in accordance with Figure R603.1.2 and the tolerances specified as follows:

1. The maximum tolerance shall be $^3/_4$ inch (19 mm) between the centerline of the horizontal framing member and the centerline of the vertical framing member.
2. Where the centerline of the horizontal framing member and bearing stiffener are located to one side of the centerline of the vertical framing member, the maximum tolerance shall be $^1/_8$ inch (3 mm) between the web of the horizontal framing member and the edge of the vertical framing member.

R603.2 Structural framing. Load-bearing cold-formed steel wall framing members shall comply with Figure R603.2(1) and with the dimensional and minimum thickness requirements specified in Tables R603.2(1) and R603.2(2). Tracks shall comply with Figure R603.2(2) and shall have a minimum flange width of $1^1/_4$ inches (32 mm).

R603.2.1 Material. Load-bearing cold-formed steel framing members shall be cold-formed to shape from structural quality sheet steel complying with the requirements of one of the following:

1. ASTM A 653: Grades 33 and 50 (Class 1 and 3).
2. ASTM A 792: Grades 33 and 50A.
3. ASTM A 1003: Structural Grades 33 Type H, and 50 Type H.

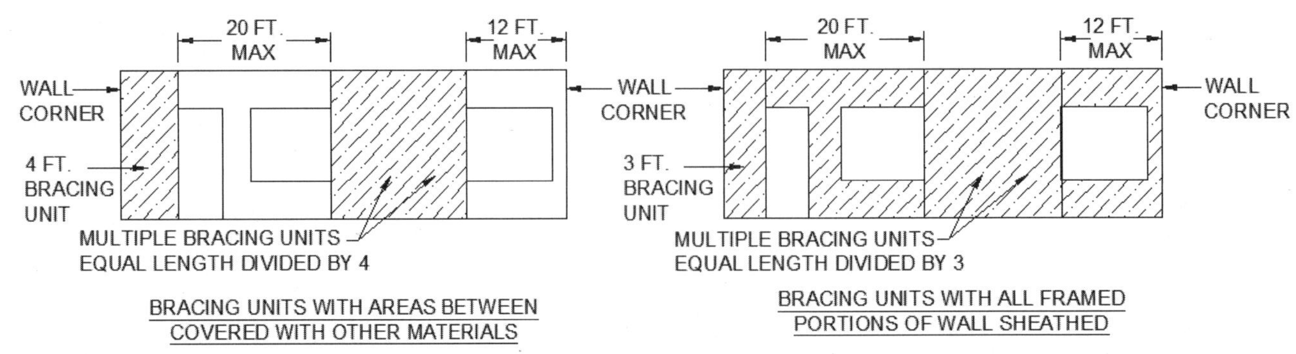

For SI: 1 foot = 304.8 mm.

FIGURE R602.12.5
BRACING UNIT DISTRIBUTION

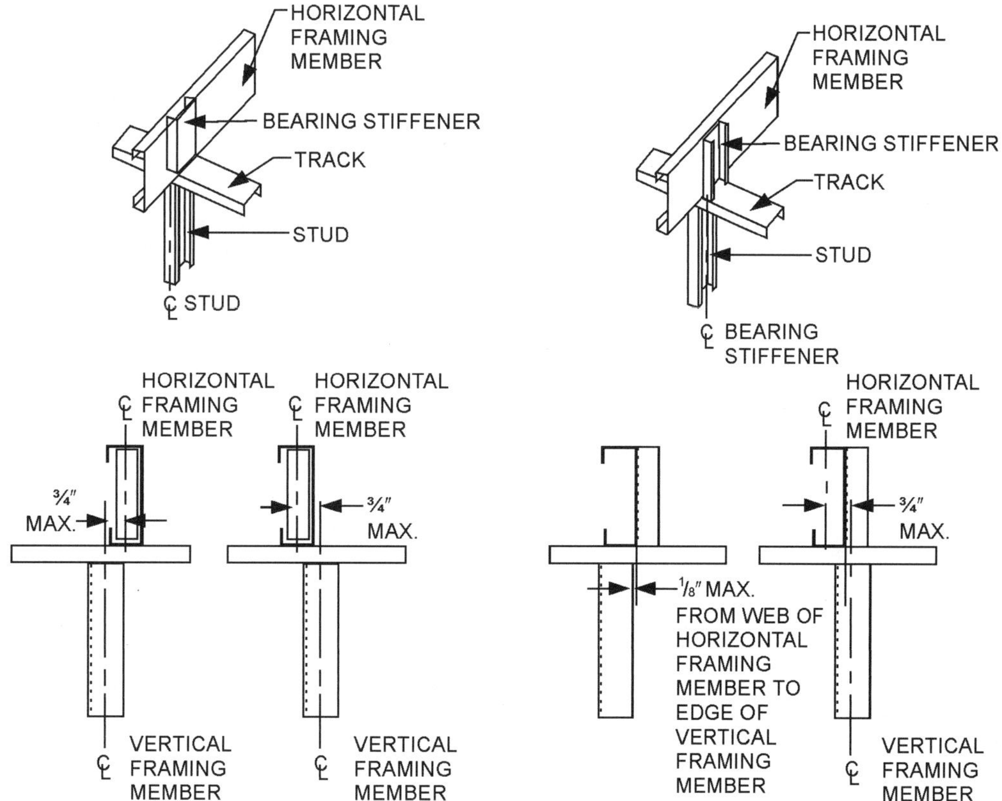

For SI: 1 inch = 25.4 mm,

FIGURE R603.1.2
IN-LINE FRAMING

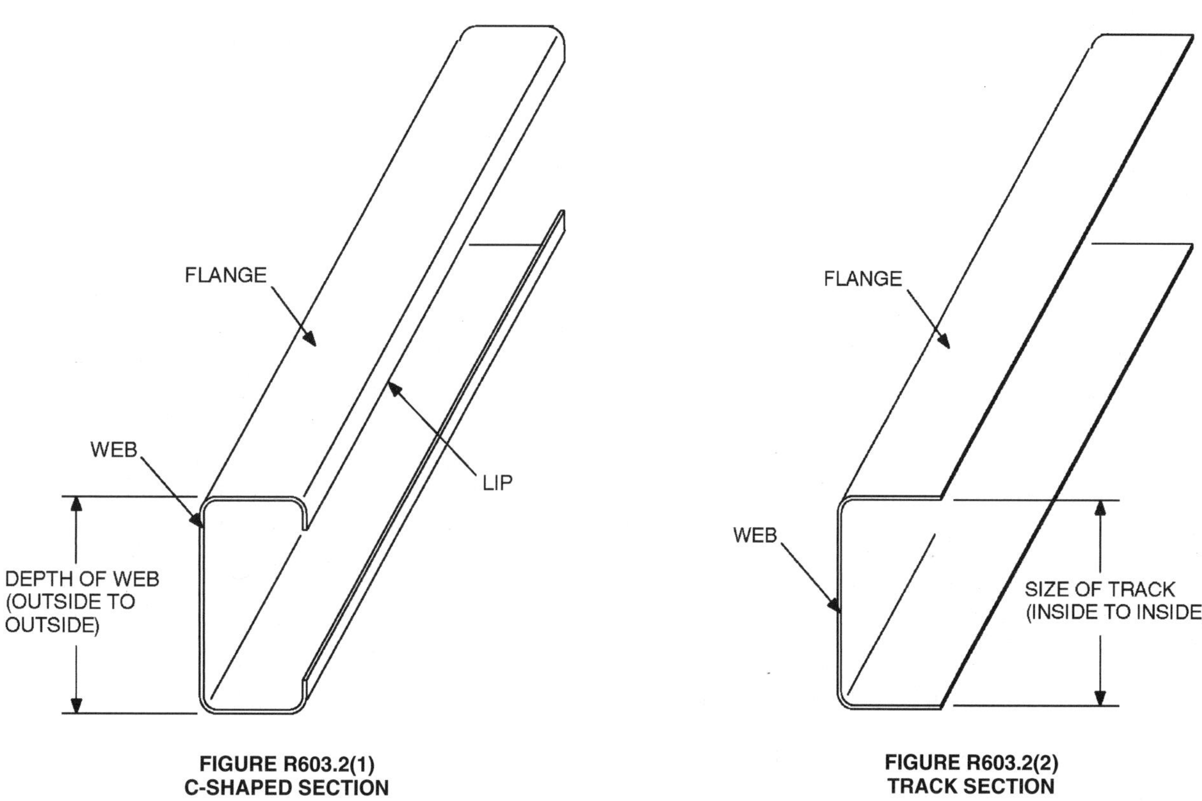

FIGURE R603.2(1)
C-SHAPED SECTION

FIGURE R603.2(2)
TRACK SECTION

**TABLE R603.2(1)
LOAD-BEARING COLD-FORMED STEEL STUD SIZES**

MEMBER DESIGNATION[a]	WEB DEPTH (inches)	MINIMUM FLANGE WIDTH (inches)	MAXIMUM FLANGE WIDTH (inches)	MINIMUM LIP SIZE (inch)
350S162-t	3.5	1.625	2	0.5
550S162-t	5.5	1.625	2	0.5

For SI: 1 inch = 25.4 mm; 1 mil = 0.0254 mm.

a. The member designation is defined by the first number representing the member depth in hundredths of an inch "S" representing a stud or joist member, the second number representing the flange width in hundredths of an inch, and the letter "t" shall be a number representing the minimum base metal thickness in mils [See Table R603.2(2)].

**TABLE R603.2(2)
MINIMUM THICKNESS OF COLD-FORMED STEEL MEMBERS**

DESIGNATION THICKNESS (mils)	MINIMUM BASE STEEL THICKNESS (inch)
33	0.0329
43	0.0428
54	0.0538
68	0.0677
97	0.0966

For SI: 1 mil = 0.0254 mm, 1 inch = 25.4 mm.

R603.2.2 Identification. Load-bearing cold-formed steel framing members shall have a legible *label*, stencil, stamp or embossment with the following information as a minimum:

1. Manufacturer's identification.
2. Minimum base steel thickness in inches (mm).
3. Minimum coating designation.
4. Minimum yield strength, in kips per square inch (ksi) (MPa).

R603.2.3 Corrosion protection. Load-bearing cold-formed steel framing shall have a metallic coating complying with ASTM A 1003 and one of the following:

1. A minimum of G 60 in accordance with ASTM A 653.
2. A minimum of AZ 50 in accordance with ASTM A 792.

R603.2.4 Fastening requirements. Screws for steel-to-steel connections shall be installed with a minimum edge distance and center-to-center spacing of $^1/_2$ inch (12.7 mm), shall be self-drilling tapping and shall conform to ASTM C 1513. Structural sheathing shall be attached to cold-formed steel studs with minimum No. 8 self-drilling tapping screws that conform to ASTM C 1513. Screws for attaching structural sheathing to cold-formed steel wall framing shall have a minimum head diameter of 0.292 inch (7.4 mm) with countersunk heads and shall be installed with a minimum edge distance of $^3/_8$ inch (9.5 mm). Gypsum board shall be attached to cold-formed steel wall framing with minimum No. 6 screws conforming to ASTM C 954 or ASTM C 1513 with a bugle head style and shall be installed in accordance with Section R702. For all connections, screws shall extend through the steel a minimum of three exposed threads. All fasteners shall have rust inhibitive coating suitable for the installation in which they are being used, or be manufactured from material not susceptible to corrosion.

Where No. 8 screws are specified in a steel-to-steel connection, the required number of screws in the connection is permitted to be reduced in accordance with the reduction factors in Table R603.2.4, when larger screws are used or when one of the sheets of steel being connected is thicker than 33 mils (0.84 mm). When applying the reduction factor, the resulting number of screws shall be rounded up.

**TABLE R603.2.4
SCREW SUBSTITUTION FACTOR**

SCREW SIZE	THINNEST CONNECTED STEEL SHEET (mils)	
	33	43
#8	1.0	0.67
#10	0.93	0.62
#12	0.86	0.56

For SI: 1 mil = 0.0254 mm.

R603.2.5 Web holes, web hole reinforcing and web hole patching. Web holes, web hole reinforcing and web hole patching shall be in accordance with this section.

R603.2.5.1 Web holes. Web holes in wall studs and other structural members shall comply with all of the following conditions:

1. Holes shall conform to Figure R603.2.5.1;
2. Holes shall be permitted only along the centerline of the web of the framing member;
3. Holes shall have a center-to-center spacing of not less than 24 inches (610 mm);
4. Holes shall have a web hole width not greater than 0.5 times the member depth, or $1^1/_2$ inches (38 mm);

5. Holes shall have a web hole length not exceeding $4^1/_2$ inches (114 mm); and

6. Holes shall have a minimum distance between the edge of the bearing surface and the edge of the web hole of not less than 10 inches (254 mm).

Framing members with web holes not conforming to the above requirements shall be reinforced in accordance with Section R603.2.5.2, patched in accordance with Section R603.2.5.3 or designed in accordance with accepted engineering practice.

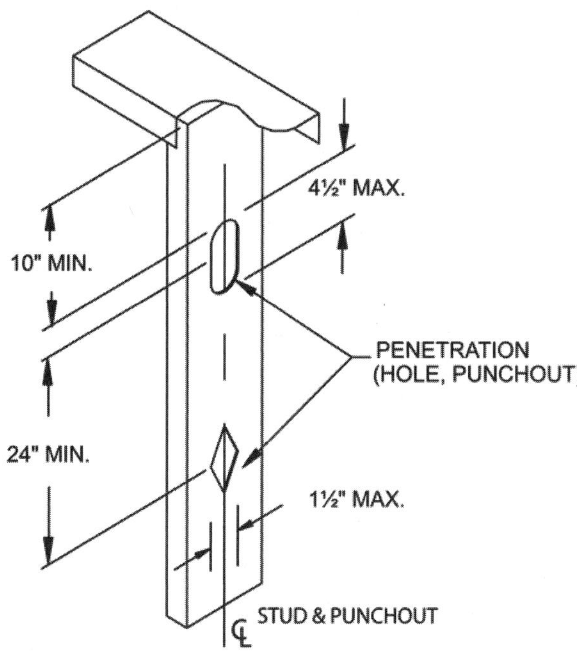

For SI: 1 inch = 25.4 mm.

FIGURE R603.2.5.1
WEB HOLES

R603.2.5.2 Web hole reinforcing. Web holes in gable endwall studs not conforming to the requirements of Section R603.2.5.1 shall be permitted to be reinforced if the hole is located fully within the center 40 percent of the span and the depth and length of the hole does not exceed 65 percent of the flat width of the web. The reinforcing shall be a steel plate or C-shape section with a hole that does not exceed the web hole size limitations of Section R603.2.5.1 for the member being reinforced. The steel reinforcing shall be the same thickness as the receiving member and shall extend at least 1 inch (25.4 mm) beyond all edges of the hole. The steel reinforcing shall be fastened to the web of the receiving member with No.8 screws spaced no more than 1 inch (25.4 mm) center-to-center along the edges of the patch with minimum edge distance of $^1/_2$ inch (12.7 mm).

R603.2.5.3 Hole patching. Web holes in wall studs and other structural members not conforming to the requirements in Section R603.2.5.1 shall be permitted to be patched in accordance with either of the following methods:

1. Framing members shall be replaced or designed in accordance with accepted engineering practice when web holes exceed the following size limits:
   1.1. The depth of the hole, measured across the web, exceeds 70 percent of the flat width of the web; or
   1.2. The length of the hole measured along the web exceeds 10 inches (254 mm) or the depth of the web, whichever is greater.
2. Web holes not exceeding the dimensional requirements in Section R603.2.5.3, Item 1 shall be patched with a solid steel plate, stud section or track section in accordance with Figure R603.2.5.3. The steel patch shall, as a minimum, be the same thickness as the receiving member and shall extend at least 1 inch (25.4 mm) beyond all edges of the hole. The steel patch shall be fastened to the web of the receiving member with No. 8 screws spaced no more than 1 inch (25.4 mm) center-to-center along the edges of the patch with a minimum edge distance of $^1/_2$ inch (12.7 mm).

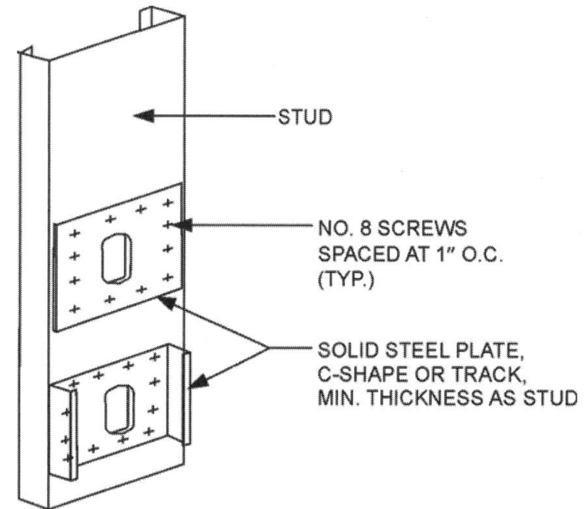

For SI: 1 inch = 25.4 mm.

FIGURE R603.2.5.3
STUD WEB HOLE PATCH

R603.3 Wall construction. All exterior cold-formed steel framed walls and interior load-bearing cold-formed steel framed walls shall be constructed in accordance with the provisions of this section.

R603.3.1 Wall to foundation or floor connection. Cold-formed steel framed walls shall be anchored to foundations or floors in accordance with Table R603.3.1 and Figure R603.3.1(1), R603.3.1(2) or R603.3.1(3). Anchor bolts shall be located not more than 12 inches (305 mm) from corners or the termination of bottom tracks. Anchor bolts shall extend a minimum of 15 inches (381 mm) into masonry or 7 inches (178 mm) into concrete. Foundation anchor straps shall be permitted, in lieu of anchor bolts, if spaced as required to provide equivalent anchorage to the required anchor bolts and installed in accordance with manufacturer's requirements.

R603.3.1.1 Gable endwalls. Gable endwalls with heights greater than 10 feet (3048 mm) shall be anchored to foundations or floors in accordance with Tables R603.3.1.1(1) or R603.3.1.1(2).

R603.3.2 Minimum stud sizes. Cold-formed steel walls shall be constructed in accordance with Figure R603.3.1(1), R603.3.1(2) or R603.3.1(3), as applicable. Exterior wall stud size and thickness shall be determined in accordance with the limits set forth in Tables R603.3.2(2) through R603.3.2(31). Interior load-bearing wall stud size and thickness shall be determined in accordance with the limits set forth in Tables R603.3.2(2) through R603.3.2(31) based upon an 85 miles per hour (38 m/s) Exposure A/B wind value and the building width, stud spacing and snow load, as appropriate. Fastening requirements shall be in accordance with Section R603.2.4 and Table R603.3.2(1). Top and bottom tracks shall have the same minimum thickness as the wall studs.

Exterior wall studs shall be permitted to be reduced to the next thinner size, as shown in Tables R603.3.2(2)

TABLE R603.3.1
WALL TO FOUNDATION OR FLOOR CONNECTION REQUIREMENTS[a, b]

FRAMING CONDITION	WIND SPEED (MPH) AND EXPOSURE					
	85 B	90 B	100 B 85 C	110 B 90 C	100 C	< 110 C
Wall bottom track to floor per Figure R603.3.1(1)	1-No. 8 screw at 12″ o.c.	1-No. 8 screw at 12″ o.c.	1-No. 8 screw at 12″ o.c.	1-No. 8 screw at 12″ o.c.	2-No. 8 screws at 12″ o.c.	2 No. 8 screws at 12″ o.c.
Wall bottom track to foundation per Figure R603.3.1(2)[d]	$^1/_2$″ minimum diameter anchor bolt at 6″ o.c.	$^1/_2$″ minimum diameter anchor bolt at 6″ o.c.	$^1/_2$″ minimum diameter anchor bolt at 4″ o.c.	$^1/_2$″ minimum diameter anchor bolt at 4″ o.c.	$^1/_2$″ minimum diameter anchor bolt at 4″ o.c.	$^1/_2$″ minimum diameter anchor bolt at 4″ o.c.
Wall bottom track to wood sill per Figure R603.3.1(3)	Steel plate spaced at 4″ o.c., with 4-No. 8 screws and 4-10d or 6-8d common nails	Steel plate spaced at 4″ o.c., with 4-No. 8 screws and 4-10d or 6-8d common nails	Steel plate spaced at 3″ o.c., with 4-No. 8 screws and 4-10d or 6-8d common nails	Steel plate spaced at 3″ o.c., with 4-No. 8 screws and 4-10d or 6-8d common nails	Steel plate spaced at 2″ o.c., with 4-No. 8 screws and 4-10d or 6-8d common nails	Steel plate spaced at 2″ o.c., with 4-No. 8 screws and 4-10d or 6-8d common nails
Wind uplift connector strength to 16″ stud spacing[c]	NR	NR	NR	NR	NR	65 lb per foot of wall length
Wind uplift connector strength for 24″ stud spacing[c]	NR	NR	NR	NR	NR	100 lb per foot of wall length

For SI: 1 inch = 25.4 mm, 1 mile per hour = 0.447 m/s, 1 foot = 304.8 mm, 1 pound = 4.45 N.

a. Anchor bolts are to be located not more than 12 inches from corners or the termination of bottom tracks (e.g., at door openings or corners). Bolts are to extend a minimum of 15 inches into masonry or 7 inches into concrete.

b. All screw sizes shown are minimum.

c. NR = uplift connector not required.

d. Foundation anchor straps are permitted in place of anchor bolts, if spaced as required to provide equivalent anchorage to the required anchor bolts and installed in accordance with manufacturer's requirements.

TABLE R603.3.1.1(1)
GABLE ENDWALL TO FLOOR CONNECTION REQUIREMENTS[a, b, c]

BASIC WIND SPEED (mph)		WALL BOTTOM TRACK TO FLOOR JOIST OR TRACK CONNECTION		
Exposure		Stud height, *h* (feet)		
B	C	10 < *h* ≤ 14	14 < *h* ≤ 18	18 < *h* ≤ 22
85	—	1-No. 8 screw @ 12″ o.c.	1-No. 8 screw @ 12″ o.c.	1-No. 8 screw @ 12″ o.c.
90	—	1-No. 8 screw @ 12″ o.c.	1-No. 8 screw @ 12″ o.c.	1-No. 8 screw @ 12″ o.c.
100	85	1-No. 8 screw @ 12″ o.c.	1-No. 8 screw @ 12″ o.c.	1-No. 8 screw @ 12″ o.c.
110	90	1-No. 8 screw @ 12″ o.c.	1-No. 8 screw @ 12″ o.c.	2-No. 8 screws @ 12″ o.c.
—	100	1-No. 8 screw @ 12″ o.c.	2-No. 8 screws @ 12″ o.c.	1-No. 8 screw @ 8″ o.c.
—	110	2-No. 8 screws @ 12″ o.c.	1-No. 8 screw @ 8″ o.c.	2-No. 8 screws @ 8″ o.c.

For SI: 1 inch = 25.4 mm, 1 mile per hour = 0.447 m/s, 1 foot = 304.8 mm.

a. Refer to Table R603.3.1.1(2) for gable endwall bottom track to foundation connections.

b. Where attachment is not given, special design is required.

c. Stud height, *h*, is measured from wall bottom track to wall top track or brace connection height.

through R603.3.2(31), but not less than 33 mils (0.84 mm), where both of the following conditions exist:

1. Minimum of $^1/_2$ inch (12.7 mm) gypsum board is installed and fastened in accordance with Section R702 on the interior surface.
2. Wood structural sheathing panels of minimum $^7/_{16}$-inch-thick (11 mm) oriented strand board or $^{15}/_{32}$-inch-thick (12 mm) plywood is installed and fastened in accordance with Section R603.9.1 and Table R603.3.2(1) on the outside surface.

Interior load-bearing walls shall be permitted to be reduced to the next thinner size, as shown in Tables R603.3.2(2) through R603.3.2(31), but not less than 33 mils (0.84 mm), where a minimum of $^1/_2$-inch (12.7 mm) gypsum board is installed and fastened in accordance with Section R702 on both sides of the wall. The tabulated stud thickness for load- bearing walls shall be used when the *attic* load is 10 pounds per square feet (480 Pa) or less. A limited *attic* storage load of 20 pounds per square feet (960 Pa) shall be permitted provided that the next higher snow load column is used to select the stud size from Tables R603.3.2(2) through R603.3.2(31).

For two-story buildings, the tabulated stud thickness for walls supporting one floor, roof and ceiling shall be used when second floor live load is 30 pounds per

TABLE R603.3.1.1(2)
GABLE ENDWALL BOTTOM TRACK TO FOUNDATION CONNECTION REQUIREMENTS[a, b, c]

BASIC WIND SPEED (mph)		MINIMUM SPACING FOR $^1/_2$-INCH-DIAMETER ANCHOR BOLTS[d]		
Exposure		Stud height, *h* (feet)		
B	C	$10 < h \leq 14$	$14 < h \leq 18$	$18 < h \leq 22$
85	—	6′- 0″ o.c.	6′- 0″ o.c.	6′- 0″ o.c.
90	—	6′- 0″ o.c.	5′- 7″ o.c.	6′- 0″ o.c.
100	85	5′- 10″ o.c.	6′- 0″ o.c.	6′- 0″ o.c.
110	90	4′- 10″ o.c.	5′- 6″ o.c.	6′- 0″ o.c.
—	100	4′- 1″ o.c.	6′- 0″ o.c.	6′- 0″ o.c.
—	110	5′- 1″ o.c.	6′- 0″ o.c.	5′- 2″ o.c.

For SI: 1 inch = 25.4 mm, 1 mile per hour = 0.447 m/s, 1 foot = 304.8 mm.

a. Refer to Table R603.3.1.1(1) for gable endwall bottom track to floor joist or track connection connections.

b. Where attachment is not given, special design is required.

c. Stud height, *h*, is measured from wall bottom track to wall top track or brace connection height.

d. Foundation anchor straps are permitted in place of anchor bolts if spaced as required to provide equivalent anchorage to the required anchor bolts and installed in accordance with manufacturer's requirements.

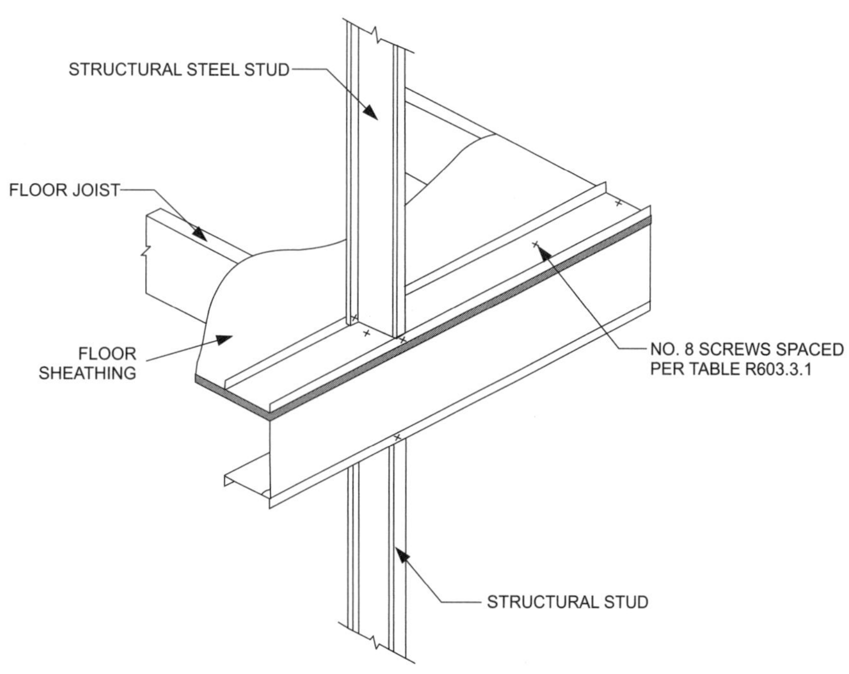

FIGURE R603.3.1(1)
WALL TO FLOOR CONNECTION

square feet (1440 Pa). Second floor live loads of 40 psf (1920 pounds per square feet) shall be permitted provided that the next higher snow load column is used to select the stud size from Tables R603.3.2(2) through R603.3.2(21).

For three-story buildings, the tabulated stud thickness for walls supporting one or two floors, roof and ceiling shall be used when the third floor live load is 30 pounds per square feet (1440 Pa). Third floor live loads of 40 pounds per square feet (1920 Pa) shall be permitted provided that the next higher snow load column is used to select the stud size from Tables R603.3.2(22) through R603.3.2(31).

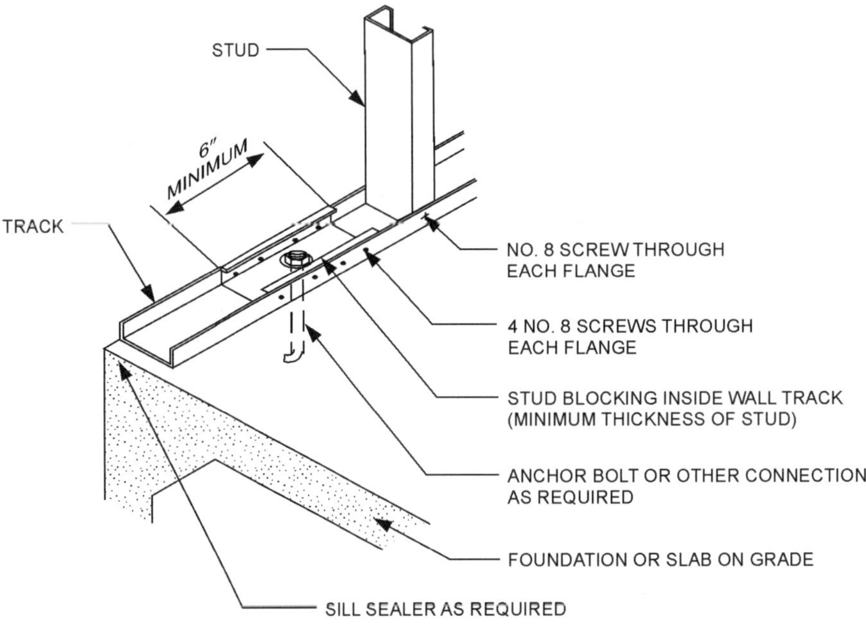

For SI: 1 inch = 25.4 mm.

FIGURE R603.3.1(2)
WALL TO FOUNDATION CONNECTION

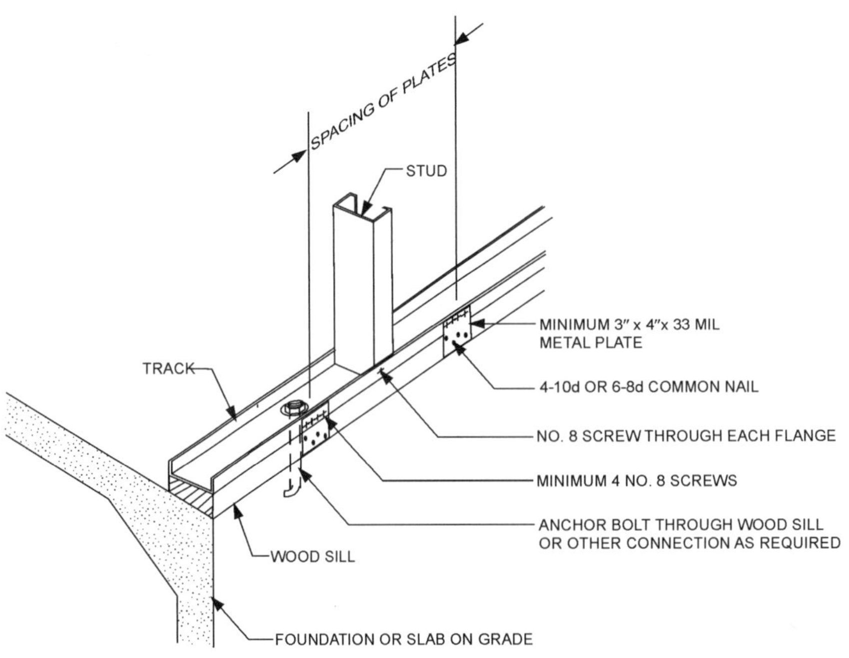

For SI: 1 mil = 0.0254 mm, 1 inch = 25.4 mm.

FIGURE R603.3.1(3)
WALL TO WOOD SILL CONNECTION

TABLE R603.3.2(1)
WALL FASTENING SCHEDULE[a]

DESCRIPTION OF BUILDING ELEMENT	NUMBER AND SIZE OF FASTENERS[a]	SPACING OF FASTENERS
Floor joist to track of load-bearing wall	2-No. 8 screws	Each joist
Wall stud to top or bottom track	2-No. 8 screws	Each end of stud, one per flange
Structural sheathing to wall studs	No. 8 screws[b]	6″ o.c. on edges and 12″ o.c. at intermediate supports
Roof framing to wall	Approved design or tie down in accordance with Section R802.11.	

For SI: 1 inch = 25.4 mm.

a. All screw sizes shown are minimum.

b. Screws for attachment of structural sheathing panels are to be bugle-head, flat-head, or similar head styles with a minimum head diameter of 0.29 inch.

TABLE R603.3.2(2)
24-FOOT-WIDE BUILDING SUPPORTING ROOF AND CEILING ONLY[a, b, c]
33 KSI STEEL

WIND SPEED		MEMBER SIZE	STUD SPACING (inches)	MINIMUM STUD THICKNESS (mils)											
				8-foot Studs				9-foot Studs				10-foot Studs			
				Ground Snow Load (psf)											
Exp. B	Exp. C			20	30	50	70	20	30	50	70	20	30	50	70
85 mph	—	350S162	16	33	33	33	33	33	33	33	33	33	33	33	33
			24	33	33	33	43	33	33	33	43	33	33	43	43
		550S162	16	33	33	33	33	33	33	33	33	33	33	33	33
			24	33	33	33	33	33	33	33	33	33	33	33	33
90 mph	—	350S162	16	33	33	33	33	33	33	33	33	33	33	33	33
			24	33	33	33	43	33	33	33	43	33	33	43	43
		550S162	16	33	33	33	33	33	33	33	33	33	33	33	33
			24	33	33	33	33	33	33	33	33	33	33	33	33
100 mph	85 mph	350S162	16	33	33	33	33	33	33	33	33	33	33	33	33
			24	33	33	33	43	33	33	33	43	43	43	43	43
		550S162	16	33	33	33	33	33	33	33	33	33	33	33	33
			24	33	33	33	43	33	33	33	33	33	33	33	43
110 mph	90 mph	350S162	16	33	33	33	33	33	33	33	33	33	33	33	33
			24	33	33	33	43	43	43	43	43	43	43	43	54
		550S162	16	33	33	33	33	33	33	33	33	33	33	33	33
			24	33	33	33	43	33	33	33	33	43	43	43	43
—	100 mph	350S162	16	33	33	33	33	33	33	33	33	43	43	43	43
			24	43	43	43	43	43	43	43	43	54	54	54	54
		550S162	16	33	33	33	33	33	33	33	33	33	33	33	33
			24	33	33	33	43	43	43	43	43	43	43	43	43
—	110 mph	350S162	16	33	33	33	33	43	43	43	43	43	43	43	43
			24	43	43	43	43	54	54	54	54	68	68	68	68
		550S162	16	33	33	33	33	33	33	33	33	33	33	33	33
			24	33	43	43	43	43	43	43	43	43	43	43	43

For SI: 1 inch = 25.4 mm, 1 foot = 304.8 mm, 1 mil = 0.0254 mm, 1 mile per hour = 0.447 m/s, 1 pound per square foot = 0.0479 kPa, 1 Ksi = 1,000 psi = 6.895 MPa.

a. Deflection criterion: *L*/240.

b. Design load assumptions:
Second floor dead load is 10 psf.
Second floor live load is 30 psf.
Roof/ceiling dead load is 12 psf.
Attic live load is 10 psf.

c. Building width is in the direction of horizontal framing members supported by the wall studs.

TABLE R603.3.2(3)
24-FOOT-WIDE BUILDING SUPPORTING ROOF AND CEILING ONLY[a, b, c]
50 KSI STEEL

WIND SPEED		MEMBER SIZE	STUD SPACING (inches)	MINIMUM STUD THICKNESS (mils)											
				8-foot Studs				9-foot Studs				10-foot Studs			
				Ground Snow Load (psf)											
Exp. B	Exp. C			20	30	50	70	20	30	50	70	20	30	50	70
85 mph	—	350S162	16	33	33	33	33	33	33	33	33	33	33	33	33
			24	33	33	33	43	33	33	33	33	33	33	33	43
		550S162	16	33	33	33	33	33	33	33	33	33	33	33	33
			24	33	33	33	33	33	33	33	33	33	33	33	33
90 mph	—	350S162	16	33	33	33	33	33	33	33	33	33	33	33	33
			24	33	33	33	43	33	33	33	33	33	33	33	43
		550S162	16	33	33	33	33	33	33	33	33	33	33	33	33
			24	33	33	33	33	33	33	33	33	33	33	33	33
100 mph	85 mph	350S162	16	33	33	33	33	33	33	33	33	33	33	33	33
			24	33	33	33	43	33	33	33	33	33	33	33	43
		550S162	16	33	33	33	33	33	33	33	33	33	33	33	33
			24	33	33	33	33	33	33	33	33	33	33	33	33
110 mph	90 mph	350S162	16	33	33	33	33	33	33	33	33	33	33	33	33
			24	33	33	33	43	33	33	33	43	43	43	43	43
		550S162	16	33	33	33	33	33	33	33	33	33	33	33	33
			24	33	33	33	33	33	33	33	33	33	33	33	33
—	100 mph	350S162	16	33	33	33	33	33	33	33	33	33	33	33	33
			24	33	33	33	43	43	43	43	43	43	43	43	43
		550S162	16	33	33	33	33	33	33	33	33	33	33	33	33
			24	33	33	33	33	33	33	33	33	33	33	33	33
—	110 mph	350S162	16	33	33	33	33	33	33	33	33	33	33	33	33
			24	33	33	33	43	43	43	43	43	54	54	54	54
		550S162	16	33	33	33	33	33	33	33	33	33	33	33	33
			24	33	33	33	33	33	33	33	33	33	33	33	33

For SI: 1 inch = 25.4 mm, 1 foot = 304.8 mm, 1 mil = 0.0254 mm, 1 mile per hour = 0.447 m/s, 1 pound per square foot = 0.0479 kPa, 1 Ksi = 1,000 psi = 6.895 MPa.

a. Deflection criterion: *L*/240.

b. Design load assumptions:
 Second floor dead load is 10 psf.
 Second floor live load is 30 psf.
 Roof/ceiling dead load is 12 psf.
 Attic live load is 10 psf.

c. Building width is in the direction of horizontal framing members supported by the wall studs.

TABLE R603.3.2(4) 28-FOOT-WIDE BUILDING SUPPORTING ROOF AND CEILING ONLY[a, b, c] 33 KSI STEEL

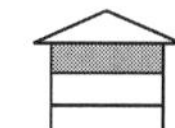

WIND SPEED		MEMBER SIZE	STUD SPACING (inches)	MINIMUM STUD THICKNESS (mils)											
				8-foot Studs				9-foot Studs				10-foot Studs			
				Ground Snow Load (psf)											
Exp. B	Exp. C			20	30	50	70	20	30	50	70	20	30	50	70
85 mph	—	350S162	16	33	33	33	33	33	33	33	33	33	33	33	33
			24	33	33	43	43	33	33	43	43	33	33	43	54
		550S162	16	33	33	33	33	33	33	33	33	33	33	33	33
			24	33	33	33	43	33	33	33	43	33	33	33	43
90 mph	—	350S162	16	33	33	33	33	33	33	33	33	33	33	33	33
			24	33	33	43	43	33	33	43	43	33	33	43	54
		550S162	16	33	33	33	33	33	33	33	33	33	33	33	33
			24	33	33	33	43	33	33	33	43	33	33	33	43
100 mph	85 mph	350S162	16	33	33	33	33	33	33	33	33	33	33	33	33
			24	33	33	43	43	33	33	43	43	43	43	43	54
		550S162	16	33	33	33	33	33	33	33	33	33	33	33	33
			24	33	33	33	43	33	33	33	43	33	33	33	43
110 mph	90 mph	350S162	16	33	33	33	33	33	33	33	33	33	33	33	43
			24	33	33	43	43	43	43	43	43	43	43	43	54
		550S162	16	33	33	33	33	33	33	33	33	33	33	33	33
			24	33	33	33	43	33	33	33	43	33	33	33	43
—	100 mph	350S162	16	33	33	33	33	33	33	33	33	43	43	43	43
			24	43	43	43	54	43	43	43	54	54	54	54	54
		550S162	16	33	33	33	33	33	33	33	33	33	33	33	33
			24	33	33	33	43	33	33	33	43	33	33	33	43
—	110 mph	350S162	16	33	33	33	33	43	43	43	43	43	43	43	43
			24	43	43	43	54	54	54	54	54	68	68	68	68
		550S162	16	33	33	33	33	33	33	33	33	33	33	33	33
			24	33	33	33	43	33	33	33	43	43	43	43	43

For SI: 1 inch = 25.4 mm, 1 foot = 304.8 mm, 1 mil = 0.0254 mm, 1 mile per hour = 0.447 m/s, 1 pound per square foot = 0.0479 kPa, 1 Ksi = 1,000 psi = 6.895 MPa.

a. Deflection criterion: *L*/240.

b. Design load assumptions:
  Second floor dead load is 10 psf.
  Second floor live load is 30 psf.
  Roof/ceiling dead load is 12 psf.
  Attic live load is 10 psf.

c. Building width is in the direction of horizontal framing members supported by the wall studs.

TABLE R603.3.2(5)
28-FOOT-WIDE BUILDING SUPPORTING ROOF AND CEILING ONLY[a, b, c]
50 KSI STEEL

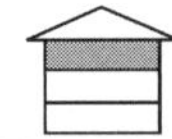

WIND SPEED		MEMBER SIZE	STUD SPACING (inches)	MINIMUM STUD THICKNESS (mils)											
				8-foot Studs				9-foot Studs				10-foot Studs			
				Ground Snow Load (psf)											
Exp. B	Exp. C			20	30	50	70	20	30	50	70	20	30	50	70
85 mph	—	350S162	16	33	33	33	33	33	33	33	33	33	33	33	33
			24	33	33	33	43	33	33	33	43	33	33	33	43
		550S162	16	33	33	33	33	33	33	33	33	33	33	33	33
			24	33	33	33	33	33	33	33	33	33	33	33	33
90 mph	—	350S162	16	33	33	33	33	33	33	33	33	33	33	33	33
			24	33	33	33	43	33	33	33	43	33	33	33	43
		550S162	16	33	33	33	33	33	33	33	33	33	33	33	33
			24	33	33	33	33	33	33	33	33	33	33	33	33
100 mph	85 mph	350S162	16	33	33	33	33	33	33	33	33	33	33	33	33
			24	33	33	33	43	33	33	33	43	33	33	43	43
		550S162	16	33	33	33	33	33	33	33	33	33	33	33	33
			24	33	33	33	33	33	33	33	33	33	33	33	33
110 mph	90 mph	350S162	16	33	33	33	33	33	33	33	33	33	33	33	33
			24	33	33	33	43	33	33	33	43	43	43	43	43
		550S162	16	33	33	33	33	33	33	33	33	33	33	33	33
			24	33	33	33	33	33	33	33	33	33	33	33	33
—	100 mph	350S162	16	33	33	33	33	33	33	33	33	33	33	33	33
			24	33	33	33	43	43	43	43	43	43	43	43	43
		550S162	16	33	33	33	33	33	33	33	33	33	33	33	33
			24	33	33	33	43	33	33	33	33	33	33	33	33
—	110 mph	350S162	16	33	33	33	33	33	33	33	33	33	33	33	33
			24	33	33	43	43	43	43	43	43	54	54	54	54
		550S162	16	33	33	33	33	33	33	33	33	33	33	33	33
			24	33	33	33	33	33	33	33	33	33	33	33	43

For SI: 1 inch = 25.4 mm, 1 foot = 304.8 mm, 1 mil = 0.0254 mm, 1 mile per hour = 0.447 m/s, 1 pound per square foot = 0.0479 kPa, 1 Ksi = 1,000 psi = 6.895 MPa.

a. Deflection criterion: *L*/240.

b. Design load assumptions:
Second floor dead load is 10 psf.
Second floor live load is 30 psf.
Roof/ceiling dead load is 12 psf.
Attic live load is 10 psf.

c. Building width is in the direction of horizontal framing members supported by the wall studs.

TABLE R603.3.2(6)
32-FOOT-WIDE BUILDING SUPPORTING ROOF AND CEILING ONLY[a, b, c]
33 KSI STEEL

WIND SPEED		MEMBER SIZE	STUD SPACING (inches)	MINIMUM STUD THICKNESS (mils)											
				8-foot Studs				9-foot Studs				10-foot Studs			
				Ground Snow Load (psf)											
Exp. B	Exp. C			20	30	50	70	20	30	50	70	20	30	50	70
85 mph	—	350S162	16	33	33	33	33	33	33	33	33	33	33	33	43
			24	33	33	43	54	33	33	43	43	33	33	43	54
		550S162	16	33	33	33	33	33	33	33	33	33	33	33	33
			24	33	33	33	43	33	33	33	43	33	33	33	43
90 mph	—	350S162	16	33	33	33	33	33	33	33	33	33	33	33	43
			24	33	33	43	54	33	33	43	43	33	33	43	54
		550S162	16	33	33	33	33	33	33	33	33	33	33	33	33
			24	33	33	33	43	33	33	33	43	33	33	33	43
100 mph	85 mph	350S162	16	33	33	33	33	33	33	33	33	33	33	33	43
			24	33	33	43	54	33	33	43	54	43	43	43	54
		550S162	16	33	33	33	33	33	33	33	33	33	33	33	33
			24	33	33	33	43	33	33	33	43	33	33	33	43
110 mph	90 mph	350S162	16	33	33	33	43	33	33	33	33	33	33	33	43
			24	33	33	43	54	43	43	43	54	43	43	43	54
		550S162	16	33	33	33	33	33	33	33	33	33	33	33	33
			24	33	33	33	43	33	33	33	43	33	33	43	43
—	100 mph	350S162	16	33	33	33	43	33	33	33	43	43	43	43	43
			24	43	43	43	54	43	43	43	54	54	54	54	54
		550S162	16	33	33	33	33	33	33	33	33	33	33	33	33
			24	33	33	43	43	33	33	33	43	33	33	43	43
—	110 mph	350S162	16	33	33	33	43	43	43	43	43	43	43	43	43
			24	43	43	43	54	54	54	54	54	68	68	68	68
		550S162	16	33	33	33	33	33	33	33	33	33	33	33	33
			24	33	33	43	43	33	33	43	43	43	43	43	43

For SI: 1 inch = 25.4 mm, 1 foot = 304.8 mm, 1 mil = 0.0254 mm, 1 mile per hour = 0.447 m/s, 1 pound per square foot = 0.0479 kPa, 1 Ksi = 1,000 psi = 6.895 MPa.

a. Deflection criterion: *L*/240.

b. Design load assumptions:
Second floor dead load is 10 psf.
Second floor live load is 30 psf.
Roof/ceiling dead load is 12 psf.
Attic live load is 10 psf.

c. Building width is in the direction of horizontal framing members supported by the wall studs.

TABLE R603.3.2(7) 32-FOOT-WIDE BUILDING SUPPORTING ROOF AND CEILING ONLY[a, b, c] 50 KSI STEEL

WIND SPEED		MEMBER SIZE	STUD SPACING (inches)	MINIMUM STUD THICKNESS (mils)											
				8-foot Studs				9-foot Studs				10-foot Studs			
				Ground Snow Load (psf)											
Exp. B	Exp. C			20	30	50	70	20	30	50	70	20	30	50	70
85 mph	—	350S162	16	33	33	33	33	33	33	33	33	33	33	33	33
			24	33	33	33	43	33	33	33	43	33	33	43	43
		550S162	16	33	33	33	33	33	33	33	33	33	33	33	33
			24	33	33	33	43	33	33	33	33	33	33	33	43
90 mph	—	350S162	16	33	33	33	33	33	33	33	33	33	33	33	33
			24	33	33	33	43	33	33	33	43	33	33	43	43
		550S162	16	33	33	33	33	33	33	33	33	33	33	33	33
			24	33	33	33	43	33	33	33	33	33	33	33	43
100 mph	85 mph	350S162	16	33	33	33	33	33	33	33	33	33	33	33	33
			24	33	33	43	43	33	33	33	43	33	33	43	43
		550S162	16	33	33	33	33	33	33	33	33	33	33	33	33
			24	33	33	33	43	33	33	33	33	33	33	33	43
110 mph	90 mph	350S162	16	33	33	33	33	33	33	33	33	33	33	33	33
			24	33	33	43	43	33	33	33	43	43	43	43	54
		550S162	16	33	33	33	33	33	33	33	33	33	33	33	33
			24	33	33	33	43	33	33	33	33	33	33	33	43
—	100 mph	350S162	16	33	33	33	33	33	33	33	33	33	33	33	33
			24	33	33	43	43	43	43	43	43	43	43	43	54
		550S162	16	33	33	33	33	33	33	33	33	33	33	33	33
			24	33	33	33	43	33	33	33	43	33	33	33	43
—	110 mph	350S162	16	33	33	33	33	33	33	33	33	33	33	33	43
			24	33	33	43	43	43	43	43	43	54	54	54	54
		550S162	16	33	33	33	33	33	33	33	33	33	33	33	33
			24	33	33	33	43	33	33	33	43	33	33	33	43

For SI: 1 inch = 25.4 mm, 1 foot = 304.8 mm, 1 mil = 0.0254 mm, 1 mile per hour = 0.447 m/s, 1 pound per square foot = 0.0479 kPa, 1 Ksi = 1,000 psi = 6.895 MPa.

a. Deflection criterion: *L*/240.

b. Design load assumptions:
Second floor dead load is 10 psf.
Second floor live load is 30 psf.
Roof/ceiling dead load is 12 psf.
Attic live load is 10 psf.

c. Building width is in the direction of horizontal framing members supported by the wall studs.

TABLE R603.3.2(8) 36-FOOT-WIDE BUILDING SUPPORTING ROOF AND CEILING ONLY[a, b, c] 33 KSI STEEL

WIND SPEED		MEMBER SIZE	STUD SPACING (inches)	MINIMUM STUD THICKNESS (mils)											
				8-foot Studs				9-foot Studs				10-foot Studs			
				Ground Snow Load (psf)											
Exp. B	Exp. C			20	30	50	70	20	30	50	70	20	30	50	70
85 mph	—	350S162	16	33	33	33	43	33	33	33	43	33	33	33	43
			24	33	33	43	54	33	33	43	54	33	43	43	54
		550S162	16	33	33	33	33	33	33	33	33	33	33	33	33
			24	33	33	43	43	33	33	43	43	33	33	43	43
90 mph	—	350S162	16	33	33	33	43	33	33	33	43	33	33	33	43
			24	33	33	43	54	33	33	43	54	33	43	43	54
		550S162	16	33	33	33	33	33	33	33	33	33	33	33	33
			24	33	33	43	43	33	33	43	43	33	33	43	43
100 mph	85 mph	350S162	16	33	33	33	43	33	33	33	43	33	33	33	43
			24	33	33	43	54	33	33	43	54	43	43	54	54
		550S162	16	33	33	33	33	33	33	33	33	33	33	33	33
			24	33	33	43	43	33	33	43	43	33	33	43	43
110 mph	90 mph	350S162	16	33	33	33	43	33	33	33	33	33	33	33	43
			24	33	33	43	54	43	43	43	43	43	43	54	68
		550S162	16	33	33	33	33	33	33	33	33	33	33	33	33
			24	33	33	43	43	33	33	43	43	33	33	43	43
—	100 mph	350S162	16	33	33	33	43	33	33	33	43	43	43	43	43
			24	43	43	43	54	43	43	43	54	54	54	54	68
		550S162	16	33	33	33	33	33	33	33	33	33	33	33	33
			24	33	33	43	43	33	33	43	43	33	33	43	43
—	110 mph	350S162	16	33	33	33	43	43	43	43	43	43	43	43	43
			24	43	43	54	54	54	54	54	54	68	68	68	68
		550S162	16	33	33	33	33	33	33	33	33	33	33	33	33
			24	33	33	43	54	33	33	43	43	43	43	43	54

For SI: 1 inch = 25.4 mm, 1 foot = 304.8 mm, 1 mil = 0.0254 mm, 1 mile per hour = 0.447 m/s, 1 pound per square foot = 0.0479 kPa, 1 Ksi = 1,000 psi = 6.895 MPa.

a. Deflection criterion: *L*/240.

b. Design load assumptions:
Second floor dead load is 10 psf.
Second floor live load is 30 psf.
Roof/ceiling dead load is 12 psf.
Attic live load is 10 psf.

c. Building width is in the direction of horizontal framing members supported by the wall studs.

TABLE R603.3.2(9)
36-FOOT-WIDE BUILDING SUPPORTING ROOF AND CEILING ONLY[a, b, c]
50 KSI STEEL

WIND SPEED		MEMBER SIZE	STUD SPACING (inches)	MINIMUM STUD THICKNESS (mils)											
				8-foot Studs				9-foot Studs				10-foot Studs			
				Ground Snow Load (psf)											
Exp. B	Exp. C			20	30	50	70	20	30	50	70	20	30	50	70
85 mph	—	350S162	16	33	33	33	33	33	33	33	33	33	33	33	33
			24	33	33	43	43	33	33	43	43	33	33	43	54
		550S162	16	33	33	33	33	33	33	33	33	33	33	33	33
			24	33	33	33	43	33	33	33	43	33	33	33	43
90 mph	—	350S162	16	33	33	33	33	33	33	33	33	33	33	33	33
			24	33	33	43	43	33	33	43	43	33	33	43	54
		550S162	16	33	33	33	33	33	33	33	33	33	33	33	33
			24	33	33	33	43	33	33	33	43	33	33	33	43
100 mph	85 mph	350S162	16	33	33	33	33	33	33	33	33	33	33	33	33
			24	33	33	43	43	33	33	43	43	33	33	43	54
		550S162	16	33	33	33	33	33	33	33	33	33	33	33	33
			24	33	33	33	43	33	33	33	43	33	33	33	43
110 mph	90 mph	350S162	16	33	33	33	33	33	33	33	33	33	33	33	43
			24	33	33	43	54	33	33	33	43	43	43	43	54
		550S162	16	33	33	33	33	33	33	33	33	33	33	33	33
			24	33	33	33	43	33	33	33	43	33	33	33	43
—	100 mph	350S162	16	33	33	33	33	33	33	33	33	33	33	33	43
			24	33	33	33	54	43	43	43	43	43	43	43	54
		550S162	16	33	33	33	33	33	33	33	33	33	33	33	33
			24	33	33	33	43	33	33	33	43	33	33	33	43
—	110 mph	350S162	16	33	33	33	43	33	33	33	33	33	33	33	43
			24	33	33	43	54	43	43	43	54	54	54	54	54
		550S162	16	33	33	33	33	33	33	33	33	33	33	33	33
			24	33	33	33	43	33	33	33	43	33	33	33	43

For SI: 1 inch = 25.4 mm, 1 foot = 304.8 mm, 1 mil = 0.0254 mm, 1 mile per hour = 0.447 m/s, 1 pound per square foot = 0.0479 kPa, 1 Ksi = 1,000 psi = 6.895 MPa.

a. Deflection criterion: *L*/240.

b. Design load assumptions:
   Second floor dead load is 10 psf.
   Second floor live load is 30 psf.
   Roof/ceiling dead load is 12 psf.
   Attic live load is 10 psf.

c. Building width is in the direction of horizontal framing members supported by the wall studs.

TABLE R603.3.2(10) 40-FOOT-WIDE BUILDING SUPPORTING ROOF AND CEILING ONLY[a, b, c] 33 KSI STEEL

WIND SPEED		MEMBER SIZE	STUD SPACING (inches)	MINIMUM STUD THICKNESS (mils)											
				8-foot Studs				9-foot Studs				10-foot Studs			
				Ground Snow Load (psf)											
Exp. B	Exp. C			20	30	50	70	20	30	50	70	20	30	50	70
85 mph	—	350S162	16	33	33	33	43	33	33	33	43	33	33	33	43
			24	33	33	43	54	33	33	43	54	43	43	54	68
		550S162	16	33	33	33	33	33	33	33	33	33	33	33	33
			24	33	33	43	54	33	33	43	43	33	33	43	54
90 mph	—	350S162	16	33	33	33	43	33	33	33	43	33	33	33	43
			24	33	33	43	54	33	33	43	54	43	43	54	68
		550S162	16	33	33	33	33	33	33	33	33	33	33	33	33
			24	33	33	43	54	33	33	43	43	33	33	43	54
100 mph	85 mph	350S162	16	33	33	33	43	33	33	33	43	33	33	33	43
			24	33	43	43	54	33	43	43	54	43	43	54	68
		550S162	16	33	33	33	43	33	33	33	33	33	33	33	33
			24	33	33	43	54	33	33	43	43	33	33	43	54
110 mph	90 mph	350S162	16	33	33	33	43	33	33	33	43	33	33	43	43
			24	33	43	43	54	43	43	43	54	43	43	54	68
		550S162	16	33	33	33	43	33	33	33	33	33	33	33	43
			24	33	33	43	54	33	33	43	43	33	33	43	54
—	100 mph	350S162	16	33	33	33	43	33	33	33	43	43	43	43	43
			24	43	43	54	68	43	43	54	54	54	54	54	68
		550S162	16	33	33	33	43	33	33	33	33	33	33	33	43
			24	33	33	43	54	33	33	43	54	33	33	43	54
—	110 mph	350S162	16	33	33	43	43	43	43	43	43	43	43	43	54
			24	43	43	54	68	54	54	54	68	68	68	68	68
		550S162	16	33	33	33	43	33	33	33	43	33	33	33	43
			24	33	33	43	54	33	33	43	54	43	43	43	54

For SI: 1 inch = 25.4 mm, 1 foot = 304.8 mm, 1 mil = 0.0254 mm, 1 mile per hour = 0.447 m/s, 1 pound per square foot = 0.0479 kPa, 1 Ksi = 1,000 psi = 6.895 MPa.

a. Deflection criterion: *L*/240.

b. Design load assumptions:

Second floor dead load is 10 psf.

Second floor live load is 30 psf.

Roof/ceiling dead load is 12 psf.

Attic live load is 10 psf.

c. Building width is in the direction of horizontal framing members supported by the wall studs.

TABLE R603.3.2(11)
40-FOOT-WIDE BUILDING SUPPORTING ROOF AND CEILING ONLY[a, b, c]
50 KSI STEEL

WIND SPEED		MEMBER SIZE	STUD SPACING (inches)	MINIMUM STUD THICKNESS (mils)											
				8-foot Studs				9-foot Studs				10-foot Studs			
				Ground Snow Load (psf)											
Exp. B	Exp. C			20	30	50	70	20	30	50	70	20	30	50	70
85 mph	—	350S162	16	33	33	33	33	33	33	33	33	33	33	33	43
			24	33	33	43	54	33	33	43	43	33	33	43	54
		550S162	16	33	33	33	33	33	33	33	33	33	33	33	33
			24	33	33	33	43	33	33	33	43	33	33	33	43
90 mph	—	350S162	16	33	33	33	33	33	33	33	33	33	33	33	43
			24	33	33	43	54	33	33	43	43	33	33	43	54
		550S162	16	33	33	33	33	33	33	33	33	33	33	33	33
			24	33	33	33	43	33	33	33	43	33	33	33	43
100 mph	85 mph	350S162	16	33	33	33	43	33	33	33	33	33	33	33	43
			24	33	33	43	54	33	33	43	54	33	33	43	54
		550S162	16	33	33	33	33	33	33	33	33	33	33	33	33
			24	33	33	33	43	33	33	33	43	33	33	33	43
110 mph	90 mph	350S162	16	33	33	33	43	33	33	33	33	33	33	33	43
			24	33	33	43	54	33	33	43	54	43	43	43	54
		550S162	16	33	33	33	33	33	33	33	33	33	33	33	33
			24	33	33	33	43	33	33	33	43	33	33	33	43
—	100 mph	350S162	16	33	33	33	43	33	33	33	43	33	33	33	43
			24	33	33	43	54	43	43	43	54	43	43	54	54
		550S162	16	33	33	33	33	33	33	33	33	33	33	33	33
			24	33	33	43	43	33	33	33	43	33	33	43	43
—	110 mph	350S162	16	33	33	33	43	33	33	33	43	33	33	33	43
			24	33	33	43	54	43	43	43	54	54	54	54	68
		550S162	16	33	33	33	33	33	33	33	33	33	33	33	33
			24	33	33	43	43	33	33	33	43	33	33	43	43

For SI: 1 inch = 25.4 mm, 1 foot = 304.8 mm, 1 mil = 0.0254 mm, 1 mile per hour = 0.447 m/s, 1 pound per square foot = 0.0479 kPa, 1 Ksi = 1,000 psi = 6.895 MPa.

a. Deflection criterion: *L*/240.

b. Design load assumptions:

Second floor dead load is 10 psf.

Second floor live load is 30 psf.

Roof/ceiling dead load is 12 psf.

Attic live load is 10 psf.

c. Building width is in the direction of horizontal framing members supported by the wall studs.

TABLE R603.3.2(12) 24-FOOT-WIDE BUILDING SUPPORTING ONE FLOOR, ROOF AND CEILING[a, b, c] 33 KSI STEEL

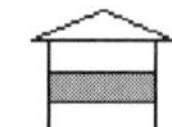

WIND SPEED		MEMBER SIZE	STUD SPACING (inches)	MINIMUM STUD THICKNESS (mils)											
				8-foot Studs				9-foot Studs				10-foot Studs			
				Ground Snow Load (psf)											
Exp. B	Exp. C			20	30	50	70	20	30	50	70	20	30	50	70
85 mph	—	350S162	16	33	33	33	33	33	33	33	33	33	33	33	43
			24	33	33	43	43	33	43	43	43	43	43	43	54
		550S162	16	33	33	33	33	33	33	33	33	33	33	33	33
			24	33	33	33	43	33	33	33	43	33	33	33	43
90 mph	—	350S162	16	33	33	33	33	33	33	33	33	33	33	33	43
			24	33	33	43	43	33	43	43	43	43	43	43	54
		550S162	16	33	33	33	33	33	33	33	33	33	33	33	33
			24	33	33	33	43	33	33	33	43	33	33	33	43
100 mph	85 mph	350S162	16	33	33	33	33	33	33	33	33	33	33	33	43
			24	33	43	43	43	43	43	43	43	43	43	43	54
		550S162	16	33	33	33	33	33	33	33	33	33	33	33	33
			24	33	33	33	43	33	33	33	43	33	33	33	43
110 mph	90 mph	350S162	16	33	33	33	43	33	33	33	33	33	33	43	43
			24	43	43	43	43	43	43	43	43	54	54	54	54
		550S162	16	33	33	33	33	33	33	33	33	33	33	33	33
			24	33	33	33	43	33	33	33	43	43	43	43	43
—	100 mph	350S162	16	33	33	33	43	33	33	33	43	43	43	43	43
			24	43	43	43	54	43	43	54	54	54	54	54	54
		550S162	16	33	33	33	33	33	33	33	33	33	33	33	33
			24	33	33	33	43	43	43	43	43	43	43	43	43
—	110 mph	350S162	16	33	33	33	43	43	43	43	43	43	43	43	43
			24	43	43	43	54	54	54	54	54	68	68	68	68
		550S162	16	33	33	33	33	33	33	33	33	33	33	33	33
			24	43	43	43	43	43	43	43	43	43	43	43	43

For SI: 1 inch = 25.4 mm, 1 foot = 304.8 mm, 1 mil = 0.0254 mm, 1 mile per hour = 0.447 m/s, 1 pound per square foot = 0.0479 kPa, 1 Ksi = 1,000 psi = 6.895 MPa.

a. Deflection criterion: *L*/240.

b. Design load assumptions:

Second floor dead load is 10 psf.

Second floor live load is 30 psf.

Roof/ceiling dead load is 12 psf.

Attic live load is 10 psf.

c. Building width is in the direction of horizontal framing members supported by the wall studs.

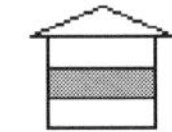

TABLE R603.3.2(13)
24-FOOT-WIDE BUILDING SUPPORTING ONE FLOOR, ROOF AND CEILING[a, b, c]
50 KSI STEEL

WIND SPEED		MEMBER SIZE	STUD SPACING (inches)	MINIMUM STUD THICKNESS (mils)											
				8-foot Studs				9-foot Studs				10-foot Studs			
				Ground Snow Load (psf)											
Exp. B	Exp. C			20	30	50	70	20	30	50	70	20	30	50	70
85 mph	—	350S162	16	33	33	33	33	33	33	33	33	33	33	33	33
			24	33	33	33	43	33	33	33	43	33	33	43	43
		550S162	16	33	33	33	33	33	33	33	33	33	33	33	33
			24	33	33	33	33	33	33	33	33	33	33	33	33
90 mph	—	350S162	16	33	33	33	33	33	33	33	33	33	33	33	33
			24	33	33	33	43	33	33	33	43	33	33	43	43
		550S162	16	33	33	33	33	33	33	33	33	33	33	33	33
			24	33	33	33	33	33	33	33	33	33	33	33	33
100 mph	85 mph	350S162	16	33	33	33	33	33	33	33	33	33	33	33	33
			24	33	33	33	43	33	33	33	43	43	43	43	43
		550S162	16	33	33	33	33	33	33	33	33	33	33	33	33
			24	33	33	33	33	33	33	33	33	33	33	33	33
110 mph	90 mph	350S162	16	33	33	33	33	33	33	33	33	33	33	33	33
			24	33	33	43	43	33	33	43	43	43	43	43	43
		550S162	16	33	33	33	33	33	33	33	33	33	33	33	33
			24	33	33	33	33	33	33	33	33	33	33	33	33
—	100 mph	350S162	16	33	33	33	33	33	33	33	33	33	33	33	33
			24	33	33	43	43	43	43	43	43	43	43	43	54
		550S162	16	33	33	33	33	33	33	33	33	33	33	33	33
			24	33	33	33	43	33	33	33	33	33	33	33	43
—	110 mph	350S162	16	33	33	33	33	33	33	33	33	33	33	43	43
			24	43	43	43	43	43	43	43	43	54	54	54	54
		550S162	16	33	33	33	33	33	33	33	33	33	33	33	33
			24	33	33	33	43	33	33	33	33	33	33	33	43

For SI: 1 inch = 25.4 mm, 1 foot = 304.8 mm, 1 mil = 0.0254 mm, 1 mile per hour = 0.447 m/s, 1 pound per square foot = 0.0479 kPa, 1 Ksi = 1,000 psi = 6.895 MPa.

a. Deflection criterion: *L*/240.

b. Design load assumptions:
Second floor dead load is 10 psf.
Second floor live load is 30 psf.
Roof/ceiling dead load is 12 psf.
Attic live load is 10 psf.

c. Building width is in the direction of horizontal framing members supported by the wall studs.

TABLE R603.3.2(14)
28-FOOT-WIDE BUILDING SUPPORTING ONE FLOOR, ROOF AND CEILING[a, b, c]
33 KSI STEEL

WIND SPEED		MEMBER SIZE	STUD SPACING (inches)	MINIMUM STUD THICKNESS (mils)											
				8-foot Studs				9-foot Studs				10-foot Studs			
				Ground Snow Load (psf)											
Exp. B	Exp. C			20	30	50	70	20	30	50	70	20	30	50	70
85 mph	—	350S162	16	33	33	33	43	33	33	33	43	33	33	33	43
			24	43	43	43	54	43	43	43	54	43	43	43	54
		550S162	16	33	33	33	33	33	33	33	33	33	33	33	33
			24	33	33	43	43	33	33	43	43	33	33	43	43
90 mph	—	350S162	16	33	33	33	43	33	33	33	43	33	33	33	43
			24	43	43	43	54	43	43	43	54	43	43	43	54
		550S162	16	33	33	33	33	33	33	33	33	33	33	33	33
			24	33	33	43	43	33	33	43	43	33	33	43	43
100 mph	85 mph	350S162	16	33	33	33	43	33	33	33	43	33	33	43	43
			24	43	43	43	54	43	43	43	54	43	43	54	54
		550S162	16	33	33	33	33	33	33	33	33	33	33	33	33
			24	33	33	43	43	33	33	43	43	33	33	43	43
110 mph	90 mph	350S162	16	33	33	33	43	33	33	33	43	43	43	43	43
			24	43	43	43	54	43	43	43	54	54	54	54	54
		550S162	16	33	33	33	33	33	33	33	33	33	33	33	33
			24	33	33	43	43	33	33	43	43	43	43	43	43
—	100 mph	350S162	16	33	33	33	43	33	33	43	43	43	43	43	43
			24	43	43	43	54	54	54	54	54	54	54	54	68
		550S162	16	33	33	33	33	33	33	33	33	33	33	33	33
			24	33	33	43	43	43	43	43	43	43	43	43	43
—	110 mph	350S162	16	33	33	43	43	43	43	43	43	43	43	43	54
			24	43	43	54	54	54	54	54	54	68	68	68	68
		550S162	16	33	33	33	33	33	33	33	33	33	33	33	33
			24	43	43	43	43	43	43	43	43	43	43	43	43

For SI: 1 inch = 25.4 mm, 1 foot = 304.8 mm, 1 mil = 0.0254 mm, 1 mile per hour = 0.447 m/s, 1 pound per square foot = 0.0479 kPa, 1 Ksi = 1,000 psi = 6.895 MPa.

a. Deflection criterion: *L*/240.

b. Design load assumptions:
Second floor dead load is 10 psf.
Second floor live load is 30 psf.
Roof/ceiling dead load is 12 psf.
Attic live load is 10 psf.

c. Building width is in the direction of horizontal framing members supported by the wall studs.

TABLE R603.3.2(15)
28-FOOT-WIDE BUILDING SUPPORTING ONE FLOOR, ROOF AND CEILING[a, b, c]
50 KSI STEEL

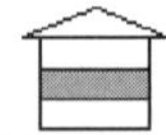

WIND SPEED		MEMBER SIZE	STUD SPACING (inches)	MINIMUM STUD THICKNESS (mils)											
				8-foot Studs				9-foot Studs				10-foot Studs			
				Ground Snow Load (psf)											
Exp. B	Exp. C			20	30	50	70	20	30	50	70	20	30	50	70
85 mph	—	350S162	16	33	33	33	33	33	33	33	33	33	33	33	33
			24	33	33	43	43	33	33	43	43	43	43	43	54
		550S162	16	33	33	33	33	33	33	33	33	33	33	33	33
			24	33	33	33	43	33	33	33	43	33	33	33	43
90 mph	—	350S162	16	33	33	33	33	33	33	33	33	33	33	33	33
			24	33	33	43	43	33	33	43	43	43	43	43	54
		550S162	16	33	33	33	33	33	33	33	33	33	33	33	33
			24	33	33	33	43	33	33	33	43	33	33	33	43
100 mph	85 mph	350S162	16	33	33	33	33	33	33	33	33	33	33	33	43
			24	33	33	43	43	33	33	43	43	43	43	43	54
		550S162	16	33	33	33	33	33	33	33	33	33	33	33	33
			24	33	33	33	43	33	33	33	43	33	33	33	43
110 mph	90 mph	350S162	16	33	33	33	33	33	33	33	33	33	33	33	43
			24	33	33	43	43	43	43	43	43	43	43	43	54
		550S162	16	33	33	33	33	33	33	33	33	33	33	33	33
			24	33	33	33	43	33	33	33	43	33	33	33	43
—	100 mph	350S162	16	33	33	33	33	33	33	33	33	33	33	33	43
			24	43	43	43	54	43	43	43	43	43	43	54	54
		550S162	16	33	33	33	33	33	33	33	33	33	33	33	33
			24	33	33	33	43	33	33	33	43	33	33	33	43
—	110 mph	350S162	16	33	33	33	43	33	33	33	33	43	43	43	43
			24	43	43	43	54	43	43	43	43	54	54	54	54
		550S162	16	33	33	33	33	33	33	33	33	33	33	33	33
			24	33	33	33	43	33	33	33	43	33	33	33	43

For SI: 1 inch = 25.4 mm, 1 foot = 304.8 mm, 1 mil = 0.0254 mm, 1 mile per hour = 0.447 m/s, 1 pound per square foot = 0.0479 kPa, 1 Ksi = 1,000 psi = 6.895 MPa.

a. Deflection criterion: *L*/240.

b. Design load assumptions:

Second floor dead load is 10 psf.

Second floor live load is 30 psf.

Roof/ceiling dead load is 12 psf.

Attic live load is 10 psf.

c. Building width is in the direction of horizontal framing members supported by the wall studs.

TABLE R603.3.2(16)
32-FOOT-WIDE BUILDING SUPPORTING ONE FLOOR, ROOF AND CEILING[a, b, c]
33 KSI STEEL

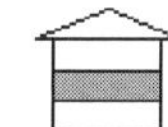

WIND SPEED		MEMBER SIZE	STUD SPACING (inches)	MINIMUM STUD THICKNESS (mils)											
				8-foot Studs				9-foot Studs				10-foot Studs			
				Ground Snow Load (psf)											
Exp. B	Exp. C			20	30	50	70	20	30	50	70	20	30	50	70
85 mph	—	350S162	16	33	33	33	43	33	33	33	43	33	33	43	43
			24	43	43	43	54	43	43	43	54	43	43	54	54
		550S162	16	33	33	33	43	33	33	33	33	33	33	33	43
			24	33	43	43	54	33	33	43	43	33	33	43	43
90 mph	—	350S162	16	33	33	33	43	33	33	33	43	33	33	43	43
			24	43	43	43	54	43	43	43	54	43	43	54	54
		550S162	16	33	33	33	43	33	33	33	33	33	33	33	43
			24	33	43	43	54	33	33	43	43	33	33	43	43
100 mph	85 mph	350S162	16	33	33	33	43	33	33	33	43	33	43	43	43
			24	43	43	43	54	43	43	43	54	54	54	54	68
		550S162	16	33	33	33	43	33	33	33	33	33	33	33	43
			24	33	43	43	54	33	33	43	43	33	33	43	43
110 mph	90 mph	350S162	16	33	33	43	43	33	33	33	43	43	43	43	43
			24	43	43	54	54	43	43	54	54	54	54	54	68
		550S162	16	33	33	33	43	33	33	33	33	33	33	33	43
			24	33	43	43	54	33	33	43	43	43	43	43	54
—	100 mph	350S162	16	33	33	43	43	43	43	43	43	43	43	43	43
			24	43	43	54	54	54	54	54	54	54	54	54	54
		550S162	16	33	33	33	43	33	33	33	33	33	33	33	43
			24	33	43	43	54	43	43	43	43	43	43	43	54
—	110 mph	350S162	16	43	43	43	43	43	43	43	43	43	43	54	54
			24	54	54	54	68	54	54	54	68	68	68	68	68
		550S162	16	33	33	33	43	33	33	33	43	33	33	33	43
			24	43	43	43	54	43	43	43	43	43	43	43	54

For SI: 1 inch = 25.4 mm, 1 foot = 304.8 mm, 1 mil = 0.0254 mm, 1 mile per hour = 0.447 m/s, 1 pound per square foot = 0.0479 kPa, 1 Ksi = 1,000 psi = 6.895 MPa.

a. Deflection criterion: *L*/240.

b. Design load assumptions:

Second floor dead load is 10 psf.

Second floor live load is 30 psf.

Roof/ceiling dead load is 12 psf.

Attic live load is 10 psf.

c. Building width is in the direction of horizontal framing members supported by the wall studs.

TABLE R603.3.2(17)
32-FOOT-WIDE BUILDING SUPPORTING ONE FLOOR, ROOF AND CEILING[a, b, c]
50 KSI STEEL

WIND SPEED		MEMBER SIZE	STUD SPACING (inches)	MINIMUM STUD THICKNESS (mils)											
				8-foot Studs				9-foot Studs				10-foot Studs			
				Ground Snow Load (psf)											
Exp. B	Exp. C			20	30	50	70	20	30	50	70	20	30	50	70
85 mph	—	350S162	16	33	33	33	43	33	33	33	33	33	33	33	43
			24	33	33	43	54	33	33	43	43	43	43	43	54
		550S162	16	33	33	33	33	33	33	33	33	33	33	33	33
			24	33	33	43	43	33	33	33	43	33	33	33	43
90 mph	—	350S162	16	33	33	33	43	33	33	33	33	33	33	33	43
			24	33	33	43	54	33	33	43	43	43	43	43	54
		550S162	16	33	33	33	33	33	33	33	33	33	33	33	33
			24	33	33	43	43	33	33	33	43	33	33	33	43
100 mph	85 mph	350S162	16	33	33	33	43	33	33	33	33	33	33	33	43
			24	33	33	43	54	33	33	43	43	43	43	43	54
		550S162	16	33	33	33	33	33	33	33	33	33	33	33	33
			24	33	33	43	43	33	33	33	43	33	33	33	43
110 mph	90 mph	350S162	16	33	33	33	43	33	33	33	33	33	33	33	43
			24	43	43	43	54	43	43	43	54	43	43	54	54
		550S162	16	33	33	33	33	33	33	33	33	33	33	33	33
			24	33	33	43	43	33	33	33	43	33	33	33	43
—	100 mph	350S162	16	33	33	33	43	33	33	33	43	33	33	43	43
			24	43	43	43	54	43	43	43	54	54	54	54	54
		550S162	16	33	33	33	33	33	33	33	33	33	33	33	33
			24	33	33	43	43	33	33	33	43	33	33	43	43
—	110 mph	350S162	16	33	33	33	43	33	33	33	43	43	43	43	43
			24	43	43	43	54	43	43	43	54	54	54	54	54
		550S162	16	33	33	33	33	33	33	33	33	33	33	33	33
			24	33	33	43	43	33	33	33	43	33	33	43	43

For SI: 1 inch = 25.4 mm, 1 foot = 304.8 mm, 1 mil = 0.0254 mm, 1 mile per hour = 0.447 m/s, 1 pound per square foot = 0.0479 kPa, 1 Ksi = 1,000 psi = 6.895 MPa.

a. Deflection criterion: *L*/240.

b. Design load assumptions:
  Second floor dead load is 10 psf.
  Second floor live load is 30 psf.
  Roof/ceiling dead load is 12 psf.
  Attic live load is 10 psf.

c. Building width is in the direction of horizontal framing members supported by the wall studs.

**TABLE R603.3.2(18)
36-FOOT-WIDE BUILDING SUPPORTING ONE FLOOR, ROOF AND CEILING[a, b, c]
33 KSI STEEL**

WIND SPEED		MEMBER SIZE	STUD SPACING (inches)	MINIMUM STUD THICKNESS (mils)											
				8-foot Studs				9-foot Studs				10-foot Studs			
				Ground Snow Load (psf)											
Exp. B	Exp. C			20	30	50	70	20	30	50	70	20	30	50	70
85 mph	—	350S162	16	33	33	43	43	33	33	43	43	33	33	43	43
			24	43	43	54	54	43	43	54	54	54	54	54	68
		550S162	16	33	33	33	43	33	33	33	43	33	33	33	43
			24	43	43	43	54	43	43	43	54	43	43	43	54
90 mph	—	350S162	16	33	33	43	43	33	33	43	43	33	33	43	43
			24	43	43	54	54	43	43	54	54	54	54	54	68
		550S162	16	33	33	33	43	33	33	33	43	33	33	33	43
			24	43	43	43	54	43	43	43	54	43	43	43	54
100 mph	85 mph	350S162	16	33	33	43	43	33	33	43	43	43	43	43	43
			24	43	43	54	68	43	43	54	54	54	54	54	68
		550S162	16	33	33	33	43	33	33	33	43	33	33	33	43
			24	43	43	43	54	43	43	43	54	43	43	43	54
110 mph	90 mph	350S162	16	33	33	43	43	33	33	43	43	43	43	43	54
			24	43	43	54	68	54	54	54	54	54	54	54	68
		550S162	16	33	33	33	43	33	33	33	43	33	33	33	43
			24	43	43	43	54	43	43	43	54	43	43	43	54
—	100 mph	350S162	16	33	33	43	43	43	43	43	43	43	43	43	54
			24	54	54	54	68	54	54	54	68	54	68	68	68
		550S162	16	33	33	33	43	33	33	33	43	33	33	33	43
			24	43	43	43	54	43	43	43	54	43	43	43	54
—	110 mph	350S162	16	43	43	43	43	43	43	43	43	43	54	54	54
			24	54	54	54	68	54	54	54	68	68	68	68	68
		550S162	16	33	33	33	43	33	33	33	43	33	33	33	43
			24	43	43	43	54	43	43	43	54	43	43	43	54

For SI: 1 inch = 25.4 mm, 1 foot = 304.8 mm, 1 mil = 0.0254 mm, 1 mile per hour = 0.447 m/s, 1 pound per square foot = 0.0479 kPa, 1 Ksi = 1,000 psi = 6.895 MPa.

a. Deflection criterion: *L*/240.

b. Design load assumptions:
Second floor dead load is 10 psf.
Second floor live load is 30 psf.
Roof/ceiling dead load is 12 psf.
Attic live load is 10 psf.

c. Building width is in the direction of horizontal framing members supported by the wall studs.

TABLE R603.3.2(19)
36-FOOT-WIDE BUILDING SUPPORTING ONE FLOOR, ROOF AND CEILING[a, b, c]
50 KSI STEEL

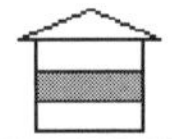

WIND SPEED		MEMBER SIZE	STUD SPACING (inches)	MINIMUM STUD THICKNESS (mils)											
				8-foot Studs				9-foot Studs				10-foot Studs			
				Ground Snow Load (psf)											
Exp. B	Exp. C			20	30	50	70	20	30	50	70	20	30	50	70
85 mph	—	350S162	16	33	33	33	43	33	33	33	43	33	33	33	43
			24	43	43	43	54	33	33	43	54	43	43	43	54
		550S162	16	33	33	33	33	33	33	33	33	33	33	33	33
			24	33	33	43	43	33	33	43	43	33	33	43	43
90 mph	—	350S162	16	33	33	33	43	33	33	33	43	33	33	33	43
			24	43	43	43	54	33	33	43	54	43	43	43	54
		550S162	16	33	33	33	33	33	33	33	33	33	33	33	33
			24	33	33	43	43	33	33	43	43	33	33	43	43
100 mph	85 mph	350S162	16	33	33	33	43	33	33	33	43	33	33	33	43
			24	43	43	43	54	43	43	43	54	43	43	54	54
		550S162	16	33	33	33	33	33	33	33	33	33	33	33	33
			24	33	33	43	43	33	33	43	43	33	33	43	43
110 mph	90 mph	350S162	16	33	33	33	43	33	33	33	43	33	33	43	43
			24	43	43	43	54	43	43	43	54	43	43	54	54
		550S162	16	33	33	33	33	33	33	33	33	33	33	33	33
			24	33	33	43	43	33	33	43	43	33	33	43	43
—	100 mph	350S162	16	33	33	33	43	33	33	33	43	43	43	43	43
			24	43	43	43	54	43	43	43	54	54	54	54	68
		550S162	16	33	33	33	33	33	33	33	33	33	33	33	33
			24	33	33	43	43	33	33	43	43	33	33	43	43
—	110 mph	350S162	16	33	33	43	43	33	33	33	43	43	43	43	43
			24	43	43	54	54	43	43	54	54	54	54	54	68
		550S162	16	33	33	33	33	33	33	33	33	33	33	33	33
			24	33	33	43	43	33	33	43	43	43	43	43	43

For SI: 1 inch = 25.4 mm, 1 foot = 304.8 mm, 1 mil = 0.0254 mm, 1 mile per hour = 0.447 m/s, 1 pound per square foot = 0.0479 kPa, 1 Ksi = 1,000 psi = 6.895 MPa.

a. Deflection criterion: *L*/240.

b. Design load assumptions:
Second floor dead load is 10 psf.
Second floor live load is 30 psf.
Roof/ceiling dead load is 12 psf.
Attic live load is 10 psf.

c. Building width is in the direction of horizontal framing members supported by the wall studs.

TABLE R603.3.2(20)
40-FOOT-WIDE BUILDING SUPPORTING ONE FLOOR, ROOF AND CEILING[a, b, c]
33 KSI STEEL

WIND SPEED		MEMBER SIZE	STUD SPACING (inches)	MINIMUM STUD THICKNESS (mils)											
				8-foot Studs				9-foot Studs				10-foot Studs			
				Ground Snow Load (psf)											
Exp. B	Exp. C			20	30	50	70	20	30	50	70	20	30	50	70
85 mph	—	350S162	16	33	33	43	43	33	33	43	43	43	43	43	54
			24	43	43	54	68	43	43	54	68	54	54	54	68
		550S162	16	33	33	33	43	33	33	33	43	33	33	33	43
			24	43	43	54	54	43	43	43	54	43	43	43	54
90 mph	—	350S162	16	33	33	43	43	33	33	43	43	43	43	43	54
			24	43	43	54	68	43	43	54	68	54	54	54	68
		550S162	16	33	33	33	43	33	33	33	43	33	33	33	43
			24	43	43	54	54	43	43	43	54	43	43	43	54
100 mph	85 mph	350S162	16	33	33	43	43	33	33	43	43	43	43	43	54
			24	43	43	54	68	43	43	54	68	54	54	54	68
		550S162	16	33	33	33	43	33	33	33	43	33	33	33	43
			24	43	43	54	54	43	43	43	54	43	43	43	54
110 mph	90 mph	350S162	16	33	33	43	43	43	43	43	43	43	43	43	54
			24	43	43	54	68	54	54	54	68	54	54	68	68
		550S162	16	33	33	43	43	33	33	33	43	33	33	33	43
			24	43	43	54	54	43	43	43	54	43	43	43	54
—	100 mph	350S162	16	43	43	43	54	43	43	43	54	43	43	54	54
			24	54	54	54	68	54	54	54	68	68	68	68	97
		550S162	16	33	33	43	43	33	33	33	43	33	33	43	43
			24	43	43	54	54	43	43	43	54	43	43	54	54
—	110 mph	350S162	16	43	43	43	54	43	43	43	54	54	54	54	54
			24	54	54	54	68	54	54	68	68	68	68	68	97
		550S162	16	33	33	43	43	33	33	33	43	33	33	43	43
			24	43	43	54	54	43	43	43	54	43	43	54	54

For SI: 1 inch = 25.4 mm, 1 foot = 304.8 mm, 1 mil = 0.0254 mm, 1 mile per hour = 0.447 m/s, 1 pound per square foot = 0.0479 kPa, 1 Ksi = 1,000 psi = 6.895 MPa.

a. Deflection criterion: *L*/240.

b. Design load assumptions:
 Second floor dead load is 10 psf.
 Second floor live load is 30 psf.
 Roof/ceiling dead load is 12 psf.
 Attic live load is 10 psf.

c. Building width is in the direction of horizontal framing members supported by the wall studs.

TABLE R603.3.2(21)
40-FOOT-WIDE BUILDING SUPPORTING ONE FLOOR, ROOF AND CEILING[a, b, c]
50 KSI STEEL

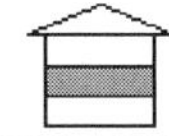

WIND SPEED		MEMBER SIZE	STUD SPACING (inches)	MINIMUM STUD THICKNESS (mils)											
				8-foot Studs				9-foot Studs				10-foot Studs			
				Ground Snow Load (psf)											
Exp. B	Exp. C			20	30	50	70	20	30	50	70	20	30	50	70
85 mph	—	350S162	16	33	33	33	43	33	33	33	43	33	33	43	43
			24	43	43	43	54	43	43	43	54	43	43	54	54
		550S162	16	33	33	33	43	33	33	33	33	33	33	33	33
			24	33	43	43	54	33	33	43	43	33	33	43	43
90 mph	—	350S162	16	33	33	33	43	33	33	33	43	33	33	43	43
			24	43	43	43	54	43	43	43	54	43	43	54	54
		550S162	16	33	33	33	43	33	33	33	33	33	33	33	33
			24	33	43	43	54	33	33	43	43	33	33	43	43
100 mph	85 mph	350S162	16	33	33	33	43	33	33	33	43	33	33	43	43
			24	43	43	54	54	43	43	43	54	43	43	54	68
		550S162	16	33	33	33	43	33	33	33	33	33	33	33	33
			24	33	43	43	54	33	33	43	43	33	33	43	43
110 mph	90 mph	350S162	16	33	33	43	43	33	33	33	43	33	33	43	43
			24	43	43	54	54	43	43	43	54	54	54	54	68
		550S162	16	33	33	33	43	33	33	33	33	33	33	33	43
			24	33	43	43	54	33	33	43	43	33	33	43	43
—	100 mph	350S162	16	33	33	43	43	33	33	33	43	43	43	43	43
			24	43	43	54	54	43	43	54	54	54	54	54	68
		550S162	16	33	33	33	43	33	33	33	33	33	33	33	43
			24	33	43	43	54	33	33	43	43	33	43	43	43
—	110 mph	350S162	16	33	33	43	43	33	33	43	43	43	43	43	54
			24	43	43	54	68	54	54	54	54	54	54	54	68
		550S162	16	33	33	33	43	33	33	33	33	33	33	33	43
			24	33	43	43	54	33	33	43	43	43	43	43	54

For SI: 1 inch = 25.4 mm, 1 foot = 304.8 mm, 1 mil = 0.0254 mm, 1 mile per hour = 0.447 m/s, 1 pound per square foot = 0.0479 kPa, 1 Ksi = 1,000 psi = 6.895 MPa.

a. Deflection criterion: *L*/240.

b. Design load assumptions:
Second floor dead load is 10 psf.
Second floor live load is 30 psf.
Roof/ceiling dead load is 12 psf.
Attic live load is 10 psf.

c. Building width is in the direction of horizontal framing members supported by the wall studs.

TABLE R603.3.2(22) 24-FOOT-WIDE BUILDING SUPPORTING TWO FLOORS, ROOF AND CEILING[a, b, c] 33 KSI STEEL

WIND SPEED		MEMBER SIZE	STUD SPACING (inches)	MINIMUM STUD THICKNESS (mils)											
				8-foot Studs				9-foot Studs				10-foot Studs			
				Ground Snow Load (psf)											
Exp. B	Exp. C			20	30	50	70	20	30	50	70	20	30	50	70
85 mph	—	350S162	16	43	43	43	43	33	33	33	43	43	43	43	43
			24	54	54	54	54	43	43	54	54	54	54	54	54
		550S162	16	33	33	43	43	33	33	33	33	33	33	33	43
			24	43	43	54	54	43	43	43	43	43	43	43	54
90 mph	—	350S162	16	43	43	43	43	33	33	33	43	43	43	43	43
			24	54	54	54	54	43	43	54	54	54	54	54	54
		550S162	16	33	33	43	43	33	33	33	33	33	33	33	43
			24	43	43	54	54	43	43	43	43	43	43	43	54
100 mph	85 mph	350S162	16	43	43	43	43	33	33	33	43	43	43	43	43
			24	54	54	54	54	54	54	54	54	54	54	54	68
		550S162	16	33	33	43	43	33	33	33	33	33	33	33	43
			24	43	43	54	54	43	43	43	43	43	43	43	54
110 mph	90 mph	350S162	16	43	43	43	43	43	43	43	43	43	43	43	43
			24	54	54	54	54	54	54	54	54	54	54	68	68
		550S162	16	33	33	43	43	33	33	33	33	33	33	33	43
			24	43	43	54	54	43	43	43	43	43	43	43	54
—	100 mph	350S162	16	43	43	43	43	43	43	43	43	43	43	43	54
			24	54	54	54	54	54	54	54	54	68	68	68	68
		550S162	16	33	33	43	43	33	33	33	33	33	33	33	43
			24	43	43	54	54	43	43	43	43	43	43	43	54
—	110 mph	350S162	16	43	43	43	43	43	43	43	43	54	54	54	54
			24	54	54	54	68	54	54	68	68	68	68	68	97
		550S162	16	33	33	43	43	33	33	33	33	33	33	33	43
			24	43	43	54	54	43	43	43	43	43	43	43	54

For SI: 1 inch = 25.4 mm, 1 foot = 304.8 mm, 1 mil = 0.0254 mm, 1 mile per hour = 0.447 m/s, 1 pound per square foot = 0.0479 kPa, 1 Ksi = 1,000 psi = 6.895 MPa.

a. Deflection criterion: *L*/240.

b. Design load assumptions:

Top and middle floor dead load is 10 psf.

Top floor live load is 30 psf.

Middle floor live load is 40 psf.

Roof/ceiling dead load is 12 psf.

Attic live load is 10 psf.

c. Building width is in the direction of horizontal framing members supported by the wall studs.

TABLE R603.3.2(23)
24-FOOT-WIDE BUILDING SUPPORTING TWO FLOORS, ROOF AND CEILING[a, b, c]
33 KSI STEEL

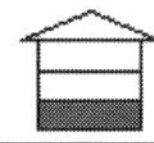

WIND SPEED		MEMBER SIZE	STUD SPACING (inches)	MINIMUM STUD THICKNESS (mils)											
				8-foot Studs				9-foot Studs				10-foot Studs			
				Ground Snow Load (psf)											
Exp. B	Exp. C			20	30	50	70	20	30	50	70	20	30	50	70
85 mph	—	350S162	16	33	33	33	43	33	33	33	33	33	33	33	33
			24	43	43	54	54	43	43	43	43	43	43	43	54
		550S162	16	33	33	33	33	33	33	33	33	33	33	33	33
			24	43	43	43	43	43	43	43	43	43	43	43	43
90 mph	—	350S162	16	33	33	33	43	33	33	33	33	33	33	33	33
			24	43	43	54	54	43	43	43	43	43	43	43	54
		550S162	16	33	33	33	33	33	33	33	33	33	33	33	33
			24	43	43	43	43	43	43	43	43	43	43	43	43
100 mph	85 mph	350S162	16	33	33	33	43	33	33	33	33	33	33	33	33
			24	43	43	54	54	43	43	43	43	43	43	54	54
		550S162	16	33	33	33	33	33	33	33	33	33	33	33	33
			24	43	43	43	43	43	43	43	43	43	43	43	43
110 mph	90 mph	350S162	16	33	33	33	43	33	33	33	33	33	33	43	43
			24	43	43	54	54	43	43	43	43	54	54	54	54
		550S162	16	33	33	33	33	33	33	33	33	33	33	33	33
			24	43	43	43	43	43	43	43	43	43	43	43	43
—	100 mph	350S162	16	33	33	33	43	33	33	33	33	43	43	43	43
			24	43	43	54	54	43	43	54	54	54	54	54	54
		550S162	16	33	33	33	33	33	33	33	33	33	33	33	33
			24	43	43	43	43	43	43	43	43	43	43	43	43
—	110 mph	350S162	16	33	33	33	43	33	33	33	43	43	43	43	43
			24	54	54	54	54	54	54	54	54	54	54	54	68
		550S162	16	33	33	33	33	33	33	33	33	33	33	33	33
			24	43	43	43	43	43	43	43	43	43	43	43	43

For SI: 1 inch = 25.4 mm, 1 foot = 304.8 mm, 1 mil = 0.0254 mm, 1 mile per hour = 0.447 m/s, 1 pound per square foot = 0.0479 kPa, 1 Ksi = 1,000 psi = 6.895 MPa.

a. Deflection criterion: *L*/240.

b. Design load assumptions:

Top and middle floor dead load is 10 psf.

Top floor live load is 30 psf.

Middle floor live load is 40 psf.

Attic live load is 10 psf.

c. Building width is in the direction of horizontal framing members supported by the wall studs.

TABLE R603.3.2(24)
28-FOOT-WIDE BUILDING SUPPORTING TWO FLOORS, ROOF AND CEILING[a, b, c]
33 KSI STEEL

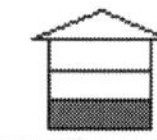

WIND SPEED		MEMBER SIZE	STUD SPACING (inches)	MINIMUM STUD THICKNESS (mils)											
				8-foot Studs				9-foot Studs				10-foot Studs			
				Ground Snow Load (psf)											
Exp. B	Exp. C			20	30	50	70	20	30	50	70	20	30	50	70
85 mph	—	350S162	16	43	43	43	43	43	43	43	43	43	43	43	43
			24	54	54	54	68	54	54	54	54	54	54	54	68
		550S162	16	43	43	43	43	43	43	43	43	43	43	43	43
			24	54	54	54	54	54	54	54	54	54	54	54	54
90 mph	—	350S162	16	43	43	43	43	43	43	43	43	43	43	43	43
			24	54	54	54	68	54	54	54	54	54	54	54	68
		550S162	16	43	43	43	43	43	43	43	43	43	43	43	43
			24	54	54	54	54	54	54	54	54	54	54	54	54
100 mph	85 mph	350S162	16	43	43	43	43	43	43	43	43	43	43	43	43
			24	54	54	54	68	54	54	54	54	54	54	68	68
		550S162	16	43	43	43	43	43	43	43	43	43	43	43	43
			24	54	54	54	54	54	54	54	54	54	54	54	54
110 mph	90 mph	350S162	16	43	43	43	43	43	43	43	43	43	43	43	43
			24	54	54	54	68	54	54	54	54	68	68	68	68
		550S162	16	43	43	43	43	43	43	43	43	43	43	43	43
			24	54	54	54	54	54	54	54	54	54	54	54	54
—	100 mph	350S162	16	43	43	43	43	43	43	43	43	43	43	54	54
			24	54	54	54	68	54	54	68	68	68	68	68	97
		550S162	16	43	43	43	43	43	43	43	43	43	43	43	43
			24	54	54	54	54	54	54	54	54	54	54	54	54
—	110 mph	350S162	16	43	43	43	43	43	43	43	43	54	54	54	54
			24	54	68	68	68	68	68	68	68	68	68	97	97
		550S162	16	43	43	43	43	43	43	43	43	43	43	43	43
			24	54	54	54	54	54	54	54	54	54	54	54	54

For SI: 1 inch = 25.4 mm, 1 foot = 304.8 mm, 1 mil = 0.0254 mm, 1 mile per hour = 0.447 m/s, 1 pound per square foot = 0.0479 kPa, 1 Ksi = 1,000 psi = 6.895 MPa.

a. Deflection criterion: *L*/240.

b. Design load assumptions:
 Top and middle floor dead load is 10 psf.
 Top floor live load is 30 psf.
 Middle floor live load is 40 psf.
 Roof/ceiling dead load is 12 psf.
 Attic live load is 10 psf.

c. Building width is in the direction of horizontal framing members supported by the wall studs.

TABLE R603.3.2(25)
28-FOOT-WIDE BUILDING SUPPORTING TWO FLOORS, ROOF AND CEILING[a, b, c]
50 KSI STEEL

WIND SPEED		MEMBER SIZE	STUD SPACING (inches)	MINIMUM STUD THICKNESS (mils)											
				8-foot Studs				9-foot Studs				10-foot Studs			
				Ground Snow Load (psf)											
Exp. B	Exp. C			20	30	50	70	20	30	50	70	20	30	50	70
85 mph	—	350S162	16	43	43	43	43	33	33	33	43	43	43	43	43
			24	54	54	54	54	43	43	54	54	54	54	54	54
		550S162	16	33	33	33	43	33	33	33	33	33	33	33	33
			24	43	43	43	54	43	43	43	43	43	43	43	43
90 mph	—	350S162	16	43	43	43	43	33	33	33	43	43	43	43	43
			24	54	54	54	54	43	43	54	54	54	54	54	54
		550S162	16	33	33	33	43	33	33	33	33	33	33	33	33
			24	43	43	43	54	43	43	43	43	43	43	43	43
100 mph	85 mph	350S162	16	43	43	43	43	33	33	33	43	43	43	43	43
			24	54	54	54	54	43	43	54	54	54	54	54	54
		550S162	16	33	33	33	43	33	33	33	33	33	33	33	33
			24	43	43	43	54	43	43	43	43	43	43	43	43
110 mph	90 mph	350S162	16	43	43	43	43	33	33	33	43	43	43	43	43
			24	54	54	54	54	43	43	54	54	54	54	54	54
		550S162	16	33	33	33	43	33	33	33	33	33	33	33	33
			24	43	43	43	54	43	43	43	43	43	43	43	43
—	100 mph	350S162	16	43	43	43	43	33	33	33	43	43	43	43	43
			24	54	54	54	54	54	54	54	54	54	54	54	68
		550S162	16	33	33	33	43	33	33	33	33	33	33	33	33
			24	43	43	43	54	43	43	43	43	43	43	43	43
—	110 mph	350S162	16	43	43	43	43	43	43	43	43	43	43	43	43
			24	54	54	54	54	54	54	54	54	68	68	68	68
		550S162	16	33	33	33	43	33	33	33	33	33	33	33	33
			24	43	43	43	54	43	43	43	43	43	43	43	43

For SI: 1 inch = 25.4 mm, 1 foot = 304.8 mm, 1 mil = 0.0254 mm, 1 mile per hour = 0.447 m/s, 1 pound per square foot = 0.0479 kPa, 1 Ksi = 1,000 psi = 6.895 MPa.

a. Deflection criterion: *L*/240.

b. Design load assumptions:

Top and middle floor dead load is 10 psf.

Top floor live load is 30 psf.

Middle floor live load is 40 psf.

Roof/ceiling dead load is 12 psf.

Attic live load is 10 psf.

c. Building width is in the direction of horizontal framing members supported by the wall studs.

TABLE R603.3.2(26) 32-FOOT-WIDE BUILDING SUPPORTING TWO FLOORS, ROOF AND CEILING[a, b, c] 33 KSI STEEL

WIND SPEED		MEMBER SIZE	STUD SPACING (inches)	MINIMUM STUD THICKNESS (mils)											
				8-foot Studs				9-foot Studs				10-foot Studs			
				Ground Snow Load (psf)											
Exp. B	Exp. C			20	30	50	70	20	30	50	70	20	30	50	70
85 mph	—	350S162	16	43	43	43	54	43	43	43	43	43	43	43	54
			24	68	68	68	68	54	54	68	68	68	68	68	68
		550S162	16	43	43	43	43	43	43	43	43	43	43	43	43
			24	54	54	54	68	54	54	54	54	54	54	54	54
90 mph	—	350S162	16	43	43	43	54	43	43	43	43	43	43	43	54
			24	68	68	68	68	54	54	68	68	68	68	68	68
		550S162	16	43	43	43	43	43	43	43	43	43	43	43	43
			24	54	54	54	68	54	54	54	54	54	54	54	54
100 mph	85 mph	350S162	16	43	43	43	54	43	43	43	43	43	43	43	54
			24	68	68	68	68	54	54	68	68	68	68	68	68
		550S162	16	43	43	43	43	43	43	43	43	43	43	43	43
			24	54	54	54	68	54	54	54	54	54	54	54	54
110 mph	90 mph	350S162	16	43	43	43	54	43	43	43	43	43	43	54	54
			24	68	68	68	68	54	54	68	68	68	68	68	68
		550S162	16	43	43	43	43	43	43	43	43	43	43	43	43
			24	54	54	54	68	54	54	54	54	54	54	54	54
—	100 mph	350S162	16	43	43	43	54	43	43	43	43	54	54	54	54
			24	68	68	68	68	68	68	68	68	68	68	97	97
		550S162	16	43	43	43	43	43	43	43	43	43	43	43	43
			24	54	54	54	68	54	54	54	54	54	54	54	54
—	110 mph	350S162	16	43	43	43	54	43	43	54	54	54	54	54	54
			24	68	68	68	68	68	68	68	68	97	97	97	97
		550S162	16	43	43	43	43	43	43	43	43	43	43	43	43
			24	54	54	54	68	54	54	54	54	54	54	54	54

For SI: 1 inch = 25.4 mm, 1 foot = 304.8 mm, 1 mil = 0.0254 mm, 1 mile per hour = 0.447 m/s, 1 pound per square foot = 0.0479 kPa, 1 Ksi = 1,000 psi = 6.895 MPa.

a. Deflection criterion: *L*/240.

b. Design load assumptions:

Top and middle floor dead load is 10 psf.

Top floor live load is 30 psf.

Middle floor live load is 40 psf.

Roof/ceiling dead load is 12 psf.

Attic live load is 10 psf.

c. Building width is in the direction of horizontal framing members supported by the wall studs.

TABLE R603.3.2(27)
32-FOOT-WIDE BUILDING SUPPORTING TWO FLOORS, ROOF AND CEILING[a, b, c]
50 KSI STEEL

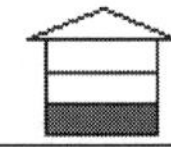

WIND SPEED		MEMBER SIZE	STUD SPACING (inches)	MINIMUM STUD THICKNESS (mils)											
				8-foot Studs				9-foot Studs				10-foot Studs			
				Ground Snow Load (psf)											
Exp. B	Exp. C			20	30	50	70	20	30	50	70	20	30	50	70
85 mph	—	350S162	16	43	43	43	43	43	43	43	43	43	43	43	43
			24	54	54	54	68	54	54	54	54	54	54	54	68
		550S162	16	43	43	43	43	33	33	33	43	33	33	43	43
			24	54	54	54	54	43	43	43	54	43	43	54	54
90 mph	—	350S162	16	43	43	43	43	43	43	43	43	43	43	43	43
			24	54	54	54	68	54	54	54	54	54	54	54	68
		550S162	16	43	43	43	43	33	33	33	43	33	33	43	43
			24	54	54	54	54	43	43	43	54	43	43	54	54
100 mph	85 mph	350S162	16	43	43	43	43	43	43	43	43	43	43	43	43
			24	54	54	54	68	54	54	54	54	54	54	54	68
		550S162	16	43	43	43	43	33	33	33	43	33	33	43	43
			24	54	54	54	54	43	43	43	54	43	43	54	54
110 mph	90 mph	350S162	16	43	43	43	43	43	43	43	43	43	43	43	43
			24	54	54	54	68	54	54	54	54	54	54	54	68
		550S162	16	43	43	43	43	33	33	33	43	33	33	43	43
			24	54	54	54	54	43	43	43	54	43	43	54	54
—	100 mph	350S162	16	43	43	43	43	43	43	43	43	43	43	43	43
			24	54	54	54	68	54	54	54	54	68	68	68	68
		550S162	16	43	43	43	43	33	33	33	43	33	33	43	43
			24	54	54	54	54	43	43	43	54	43	43	54	54
—	110 mph	350S162	16	43	43	43	43	43	43	43	43	43	43	43	54
			24	54	54	54	68	54	54	54	54	68	68	68	68
		550S162	16	43	43	43	43	33	33	33	43	33	33	43	43
			24	54	54	54	54	43	43	43	54	43	43	54	54

For SI: 1 inch = 25.4 mm, 1 foot = 304.8 mm, 1 mil = 0.0254 mm, 1 mile per hour = 0.447 m/s, 1 pound per square foot = 0.0479 kPa, 1 Ksi = 1,000 psi = 6.895 MPa.

a. Deflection criterion: $L/240$.

b. Design load assumptions:

Top and middle floor dead load is 10 psf.

Top floor live load is 30 psf.

Middle floor live load is 40 psf.

Roof/ceiling dead load is 12 psf.

Attic live load is 10 psf.

c. Building width is in the direction of horizontal framing members supported by the wall studs.

TABLE R603.3.2(28)
36-FOOT-WIDE BUILDING SUPPORTING TWO FLOORS, ROOF AND CEILING[a, b, c]
33 KSI STEEL

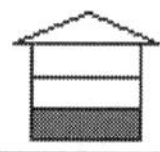

WIND SPEED		MEMBER SIZE	STUD SPACING (inches)	MINIMUM STUD THICKNESS (mils)											
				8-foot Studs				9-foot Studs				10-foot Studs			
				Ground Snow Load (psf)											
Exp. B	Exp. C			20	30	50	70	20	30	50	70	20	30	50	70
85 mph	—	350S162	16	54	54	54	54	43	43	43	54	54	54	54	54
			24	68	68	68	97	68	68	68	68	68	68	68	97
		550S162	16	43	43	43	54	43	43	43	43	43	43	43	43
			24	68	68	68	68	54	54	54	68	54	54	68	68
90 mph	—	350S162	16	54	54	54	54	43	43	43	54	54	54	54	54
			24	68	68	68	97	68	68	68	68	68	68	68	97
		550S162	16	43	43	43	54	43	43	43	43	43	43	43	43
			24	68	68	68	68	54	54	54	68	54	54	68	68
100 mph	85 mph	350S162	16	54	54	54	54	43	43	43	54	54	54	54	54
			24	68	68	68	97	68	68	68	68	68	68	68	97
		550S162	16	43	43	43	54	43	43	43	43	43	43	43	43
			24	68	68	68	68	54	54	54	68	54	54	68	68
110 mph	90 mph	350S162	16	54	54	54	54	43	43	43	54	54	54	54	54
			24	68	68	68	97	68	68	68	68	68	68	97	97
		550S162	16	43	43	43	54	43	43	43	43	43	43	43	43
			24	68	68	68	68	54	54	54	68	54	54	68	68
—	100 mph	350S162	16	54	54	54	54	43	43	54	54	54	54	54	54
			24	68	68	68	97	68	68	68	68	97	97	97	97
		550S162	16	43	43	43	54	43	43	43	43	43	43	43	43
			24	68	68	68	68	54	54	54	68	54	54	68	68
—	110 mph	350S162	16	54	54	54	54	54	54	54	54	54	54	54	68
			24	68	68	68	97	68	68	68	97	97	97	97	97
		550S162	16	43	43	43	54	43	43	43	43	43	43	43	43
			24	68	68	68	68	54	54	54	68	54	54	68	68

For SI: 1 inch = 25.4 mm, 1 foot = 304.8 mm, 1 mil = 0.0254 mm, 1 mile per hour = 0.447 m/s, 1 pound per square foot = 0.0479 kPa, 1 Ksi = 1,000 psi = 6.895 MPa.

a. Deflection criterion: *L*/240.

b. Design load assumptions:

Top and middle floor dead load is 10 psf.

Top floor live load is 30 psf.

Middle floor live load is 40 psf.

Roof/ceiling dead load is 12 psf.

Attic live load is 10 psf.

c. Building width is in the direction of horizontal framing members supported by the wall studs.

TABLE R603.3.2(29)
36-FOOT-WIDE BUILDING SUPPORTING TWO FLOORS, ROOF AND CEILING[a, b, c]
50 KSI STEEL

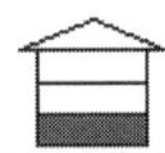

WIND SPEED		MEMBER SIZE	STUD SPACING (inches)	MINIMUM STUD THICKNESS (mils)											
				8-foot Studs				9-foot Studs				10-foot Studs			
				Ground Snow Load (psf)											
Exp. B	Exp. C			20	30	50	70	20	30	50	70	20	30	50	70
85 mph	—	350S162	16	43	43	43	54	43	43	43	43	43	43	43	43
			24	68	68	68	68	54	54	54	68	68	68	68	68
		550S162	16	43	43	43	43	43	43	43	43	43	43	43	43
			24	54	54	54	54	54	54	54	54	54	54	54	54
90 mph	—	350S162	16	43	43	43	54	43	43	43	43	43	43	43	43
			24	68	68	68	68	54	54	54	68	68	68	68	68
		550S162	16	43	43	43	43	43	43	43	43	43	43	43	43
			24	54	54	54	54	54	54	54	54	54	54	54	54
100 mph	85 mph	350S162	16	43	43	43	54	43	43	43	43	43	43	43	43
			24	68	68	68	68	54	54	54	68	68	68	68	68
		550S162	16	43	43	43	43	43	43	43	43	43	43	43	43
			24	54	54	54	54	54	54	54	54	54	54	54	54
110 mph	90 mph	350S162	16	43	43	43	54	43	43	43	43	43	43	43	43
			24	68	68	68	68	54	54	54	68	68	68	68	68
		550S162	16	43	43	43	43	43	43	43	43	43	43	43	43
			24	54	54	54	54	54	54	54	54	54	54	54	54
—	100 mph	350S162	16	43	43	43	54	43	43	43	43	43	43	43	54
			24	68	68	68	68	54	54	54	68	68	68	68	68
		550S162	16	43	43	43	43	43	43	43	43	43	43	43	43
			24	54	54	54	54	54	54	54	54	54	54	54	54
—	110 mph	350S162	16	43	43	43	54	43	43	43	43	43	54	54	54
			24	68	68	68	68	54	54	68	68	68	68	68	68
		550S162	16	43	43	43	43	43	43	43	43	43	43	43	43
			24	54	54	54	54	54	54	54	54	54	54	54	54

For SI: 1 inch = 25.4 mm, 1 foot = 304.8 mm, 1 mil = 0.0254 mm, 1 mile per hour = 0.447 m/s, 1 pound per square foot = 0.0479 kPa, 1 Ksi = 1,000 psi = 6.895 MPa.

a. Deflection criterion: *L*/240.

b. Design load assumptions:
   Top and middle floor dead load is 10 psf.
   Top floor live load is 30 psf.
   Middle floor live load is 40 psf.
   Roof/ceiling dead load is 12 psf.
   Attic live load is 10 psf.

c. Building width is in the direction of horizontal framing members supported by the wall studs.

TABLE R603.3.2(30)
40-FOOT-WIDE BUILDING SUPPORTING TWO FLOORS, ROOF AND CEILING[a, b, c]
33 KSI STEEL

WIND SPEED		MEMBER SIZE	STUD SPACING (inches)	MINIMUM STUD THICKNESS (mils)											
				8-foot Studs				9-foot Studs				10-foot Studs			
				Ground Snow Load (psf)											
Exp. B	Exp. C			20	30	50	70	20	30	50	70	20	30	50	70
85 mph	—	350S162	16	54	54	54	54	54	54	54	54	54	54	54	54
			24	97	97	97	97	68	68	68	97	97	97	97	97
		550S162	16	54	54	54	54	43	43	54	54	43	43	54	54
			24	68	68	68	68	68	68	68	68	68	68	68	68
90 mph	—	350S162	16	54	54	54	54	54	54	54	54	54	54	54	54
			24	97	97	97	97	68	68	68	97	97	97	97	97
		550S162	16	54	54	54	54	43	43	54	54	43	43	54	54
			24	68	68	68	68	68	68	68	68	68	68	68	68
100 mph	85 mph	350S162	16	54	54	54	54	54	54	54	54	54	54	54	54
			24	97	97	97	97	68	68	68	97	97	97	97	97
		550S162	16	54	54	54	54	43	43	54	54	43	43	54	54
			24	68	68	68	68	68	68	68	68	68	68	68	68
110 mph	90 mph	350S162	16	54	54	54	54	54	54	54	54	54	54	54	54
			24	97	97	97	97	68	68	68	97	97	97	97	97
		550S162	16	54	54	54	54	43	43	54	54	43	43	54	54
			24	68	68	68	68	68	68	68	68	68	68	68	68
—	100 mph	350S162	16	54	54	54	54	54	54	54	54	54	54	54	54
			24	97	97	97	97	68	68	68	97	97	97	97	97
		550S162	16	54	54	54	54	43	43	54	54	43	43	54	54
			24	68	68	68	68	68	68	68	68	68	68	68	68
—	110 mph	350S162	16	54	54	54	54	54	54	54	54	54	54	68	68
			24	97	97	97	97	68	68	97	97	97	97	97	97
		550S162	16	54	54	54	54	43	43	54	54	43	43	54	54
			24	68	68	68	68	68	68	68	68	68	68	68	68

For SI: 1 inch = 25.4 mm, 1 foot = 304.8 mm, 1 mil = 0.0254 mm, 1 mile per hour = 0.447 m/s, 1 pound per square foot = 0.0479 kPa, 1 Ksi = 1,000 psi = 6.895 MPa.

a. Deflection criterion: *L*/240.

b. Design load assumptions:

Top and middle floor dead load is 10 psf.

Top floor live load is 30 psf.

Middle floor live load is 40 psf.

Roof/ceiling dead load is 12 psf.

Attic live load is 10 psf.

c. Building width is in the direction of horizontal framing members supported by the wall studs.

TABLE R603.3.2(31)
40-FOOT-WIDE BUILDING SUPPORTING TWO FLOORS, ROOF AND CEILING[a, b, c]
50 KSI STEEL

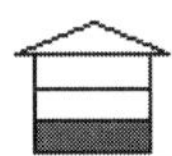

WIND SPEED		MEMBER SIZE	STUD SPACING (inches)	MINIMUM STUD THICKNESS (mils)											
				8-foot Studs				9-foot Studs				10-foot Studs			
				Ground Snow Load (psf)											
Exp. B	Exp. C			20	30	50	70	20	30	50	70	20	30	50	70
85 mph	—	350S162	16	54	54	54	54	43	43	43	43	43	54	54	54
			24	68	68	68	68	68	68	68	68	68	68	68	68
		550S162	16	43	43	43	43	43	43	43	43	43	43	43	43
			24	54	54	54	68	54	54	54	54	54	54	54	54
90 mph	—	350S162	16	54	54	54	54	43	43	43	43	43	54	54	54
			24	68	68	68	68	68	68	68	68	68	68	68	68
		550S162	16	43	43	43	43	43	43	43	43	43	43	43	43
			24	54	54	54	68	54	54	54	54	54	54	54	54
100 mph	85 mph	350S162	16	54	54	54	54	43	43	43	43	43	54	54	54
			24	68	68	68	68	68	68	68	68	68	68	68	68
		550S162	16	43	43	43	43	43	43	43	43	43	43	43	43
			24	54	54	54	68	54	54	54	54	54	54	54	54
110 mph	90 mph	350S162	16	54	54	54	54	43	43	43	43	43	54	54	54
			24	68	68	68	68	68	68	68	68	68	68	68	68
		550S162	16	43	43	43	43	43	43	43	43	43	43	43	43
			24	54	54	54	68	54	54	54	54	54	54	54	54
—	100 mph	350S162	16	54	54	54	54	43	43	43	43	43	54	54	54
			24	68	68	68	68	68	68	68	68	68	68	68	68
		550S162	16	43	43	43	43	43	43	43	43	43	43	43	43
			24	54	54	54	68	54	54	54	54	54	54	54	54
—	110 mph	350S162	16	54	54	54	54	43	43	43	43	54	54	54	54
			24	68	68	68	68	68	68	68	68	68	68	68	97
		550S162	16	43	43	43	43	43	43	43	43	43	43	43	43
			24	54	54	54	68	54	54	54	54	54	54	54	54

For SI: 1 inch = 25.4 mm, 1 foot = 304.8 mm, 1 mil = 0.0254 mm, 1 mile per hour = 0.447 m/s, 1 pound per square foot = 0.0479 kPa, 1 Ksi = 1,000 psi = 6.895 MPa.

a. Deflection criterion: *L*/240.

b. Design load assumptions:

Top and middle floor dead load is 10 psf.

Top floor live load is 30 psf.

Middle floor live load is 40 psf.

Roof/ceiling dead load is 12 psf.

Attic live load is 10 psf.

c. Building width is in the direction of horizontal framing members supported by the wall studs.

R603.3.2.1 Gable endwalls. The size and thickness of gable endwall studs with heights less than or equal to 10 feet (3048 mm) shall be permitted in accordance with the limits set forth in Table R603.3.2.1(1) or R603.3.2.1(2). The size and thickness of gable endwall studs with heights greater than 10 feet (3048 mm) shall be determined in accordance with the limits set forth in Table R603.3.2.1(3) or R603.3.2.1(4).

R603.3.3 Stud bracing. The flanges of cold-formed steel studs shall be laterally braced in accordance with one of the following:

1. Gypsum board on both sides, structural sheathing on both sides, or gypsum board on one side and structural sheathing on the other side of load-bearing walls with gypsum board installed with minimum No. 6 screws in accordance with Section R702 and

structural sheathing installed in accordance with Section R603.9.1 and Table R603.3.2(1).

2. Horizontal steel straps fastened in accordance with Figure R603.3.3(1) on both sides at mid-height for 8-foot (2438 mm) walls, and at one-third points for 9-foot and 10-foot (2743 mm and 3048 mm) walls. Horizontal steel straps shall be at least 1.5 inches in width and 33 mils in thickness (38 mm by 0.84 mm). Straps shall be attached to the flanges of studs with one No. 8 screw. In-line blocking shall be installed between studs at the termination of all straps and at 12 foot (3658 mm) intervals along the strap. Straps shall be fastened to the blocking with two No. 8 screws.

3. Sheathing on one side and strapping on the other side fastened in accordance with Figure R603.3.3(2). Sheathing shall be installed in accordance with Item 1. Steel straps shall be installed in accordance with Item 2.

TABLE R603.3.2.1(1)
ALL BUILDING WIDTHS GABLE ENDWALLS 8, 9 OR 10 FEET IN HEIGHT[a, b, c]
33 KSI STEEL

WIND SPEED		MEMBER SIZE	STUD SPACING (inches)	MINIMUM STUD THICKNESS (Mils)		
Exp. B	Exp. C			8-foot Studs	9-foot Studs	10-foot Studs
85 mph	—	350S162	16	33	33	33
			24	33	33	33
		550S162	16	33	33	33
			24	33	33	33
90 mph	—	350S162	16	33	33	33
			24	33	33	33
		550S162	16	33	33	33
			24	33	33	33
100 mph	85 mph	350S162	16	33	33	33
			24	33	33	43
		550S162	16	33	33	33
			24	33	33	33
110 mph	90 mph	350S162	16	33	33	33
			24	33	33	43
		550S162	16	33	33	33
			24	33	33	33
—	100 mph	350S162	16	33	33	43
			24	43	43	54
		550S162	16	33	33	33
			24	33	33	33
—	110 mph	350S162	16	33	43	43
			24	43	54	54
		550S162	16	33	33	33
			24	33	33	43

For SI: 1 inch = 25.4 mm, 1 foot = 304.8 mm, 1 mil = 0.0254 mm, 1 mile per hour = 0.447 m/s, 1 pound per square foot = 0.0479 kPa, 1 Ksi = 1,000 psi = 6.895 MPa.

a. Deflection criterion *L*/240.

b. Design load assumptions:
   Ground snow load is 70 psf.
   Roof/ceiling dead load is 12 psf.
   Floor dead load is 10 psf.
   Floor live load is 40 psf.
   Attic dead load is 10 psf.

c. Building width is in the direction of horizontal framing members supported by the wall studs.

TABLE R603.3.2.1(2)
ALL BUILDING WIDTHS GABLE ENDWALLS 8, 9 OR 10 FEET IN HEIGHT[a, b, c]
50 KSI STEEL

WIND SPEED		MEMBER SIZE	STUD SPACING (inches)	MINIMUM STUD THICKNESS (Mils)		
Exp. B	Exp. C			8-foot Studs	9-foot Studs	10-foot Studs
85 mph	—	350S162	16	33	33	33
			24	33	33	33
		550S162	16	33	33	33
			24	33	33	33
90 mph	—	350S162	16	33	33	33
			24	33	33	33
		550S162	16	33	33	33
			24	33	33	33
100 mph	85 mph	350S162	16	33	33	33
			24	33	33	33
		550S162	16	33	33	33
			24	33	33	33
110 mph	90 mph	350S162	16	33	33	33
			24	33	33	43
		550S162	16	33	33	33
			24	33	33	33
—	100 mph	350S162	16	33	33	33
			24	33	33	43
		550S162	16	33	33	33
			24	33	33	33
—	110 mph	350S162	16	33	33	33
			24	33	43	54
		550S162	16	33	33	33
			24	33	33	33

For SI: 1 inch = 25.4 mm, 1 foot = 304.8 mm, 1 mil = 0.0254 mm, 1 mile per hour = 0.447 m/s, 1 pound per square foot = 0.0479 kPa, 1 Ksi = 1,000 psi = 6.895 MPa.

a. Deflection criterion *L*/240.

b. Design load assumptions:
Ground snow load is 70 psf.
Roof/ceiling dead load is 12 psf.
Floor dead load is 10 psf.
Floor live load is 40 psf.
Attic dead load is 10 psf.

c. Building width is in the direction of horizontal framing members supported by the wall studs.

TABLE R603.3.2.1(3)
ALL BUILDING WIDTHS GABLE ENDWALLS OVER 10 FEET IN HEIGHT[a, b, c]
33 KSI STEEL

WIND SPEED		MEMBER SIZE	STUD SPACING (inches)	MINIMUM STUD THICKNESS (mils)					
				Stud Height, *h* (feet)					
Exp. B	Exp. C			10 < *h* ≤ 12	12 < *h* ≤ 14	14 < *h* ≤ 16	16 < *h* ≤ 18	18 < *h* ≤ 20	20 < *h* ≤ 22
85 mph	—	350S162	16	33	43	54	97	—	—
			24	43	54	97	—	—	—
		550S162	16	33	33	33	43	43	54
			24	33	33	43	54	68	97
90 mph	—	350S162	16	33	43	68	97	—	—
			24	43	68	97	—	—	—
		550S162	16	33	33	33	43	54	54
			24	33	33	43	54	68	97
100 mph	85 mph	350S162	16	43	54	97	—	—	—
			24	54	97	—	—	—	—
		550S162	16	33	33	43	54	54	68
			24	33	43	54	68	97	97
110 mph	90 mph	350S162	16	43	68	—	—	—	—
			24	68	—	—	—	—	—
		550S162	16	33	43	43	54	68	97
			24	43	54	68	97	97	—
—	100 mph	350S162	16	54	97	—	—	—	—
			24	97	—	—	—	—	—
		550S162	16	33	43	54	68	97	—
			24	43	68	97	97	—	—
—	110 mph	350S162	16	68	97	—	—	—	—
			24	97	—	—	—	—	—
		550S162	16	43	54	68	97	97	—
			24	54	68	97	—	—	—

For SI: 1 inch = 25.4 mm, 1 foot = 304.8 mm, 1 mil = 0.0254 mm, 1 mile per hour = 0.447 m/s, 1 pound per square foot = 0.0479 kPa, 1 Ksi = 1,000 psi = 6.895 MPa.

a. Deflection criterion *L*/240.

b. Design load assumptions:

Ground snow load is 70 psf.

Roof/ceiling dead load is 12 psf.

Floor dead load is 10 psf.

Floor live load is 40 psf.

Attic dead load is 10 psf.

c. Building width is in the direction of horizontal framing members supported by the wall studs.

TABLE R603.3.2.1(4)
ALL BUILDING WIDTHS GABLE ENDWALLS OVER 10 FEET IN HEIGHT[a, b, c]
50 KSI STEEL

WIND SPEED		MEMBER SIZE	STUD SPACING (inches)	MINIMUM STUD THICKNESS (mils)					
				Stud Height, *h* (feet)					
Exp. B	Exp. C			10 < *h* ≤ 12	12 < *h* ≤ 14	14 < *h* ≤ 16	16 < *h* ≤ 18	18 < *h* ≤ 20	20 < *h* ≤ 22
85 mph	—	350S162	16	33	43	54	97	—	—
			24	33	54	97	—	—	—
		550S162	16	33	33	33	33	43	54
			24	33	33	33	43	54	97
90 mph	—	350S162	16	33	43	68	97	—	—
			24	43	68	97	—	—	—
		550S162	16	33	33	33	33	43	54
			24	33	33	43	43	68	97
100 mph	85 mph	350S162	16	33	54	97	—	—	—
			24	54	97	—	—	—	—
		550S162	16	33	33	33	43	54	68
			24	33	33	43	54	97	97
110 mph	90 mph	350S162	16	43	68	—	—	—	—
			24	68	—	—	—	—	—
		550S162	16	33	33	43	43	68	97
			24	33	43	54	68	97	—
—	100 mph	350S162	16	54	97	—	—	—	—
			24	97	—	—	—	—	—
		550S162	16	33	33	43	54	97	—
			24	43	54	54	97	—	—
—	110 mph	350S162	16	54	97	—	—	—	—
			24	97	—	—	—	—	—
		550S162	16	33	43	54	68	97	—
			24	43	54	68	97	—	—

For SI: 1 inch = 25.4 mm, 1 foot = 304.8 mm, 1 mil = 0.0254 mm, 1 mile per hour = 0.447 m/s, 1 pound per square foot = 0.0479 kPa, 1 Ksi = 1,000 psi = 6.895 MPa.

a. Deflection criterion *L*/240.

b. Design load assumptions:
   Ground snow load is 70 psf.
   Roof/ceiling dead load is 12 psf.
   Floor dead load is 10 psf.
   Floor live load is 40 psf.
   Attic dead load is 10 psf.

c. Building width is in the direction of horizontal framing members supported by the wall studs.

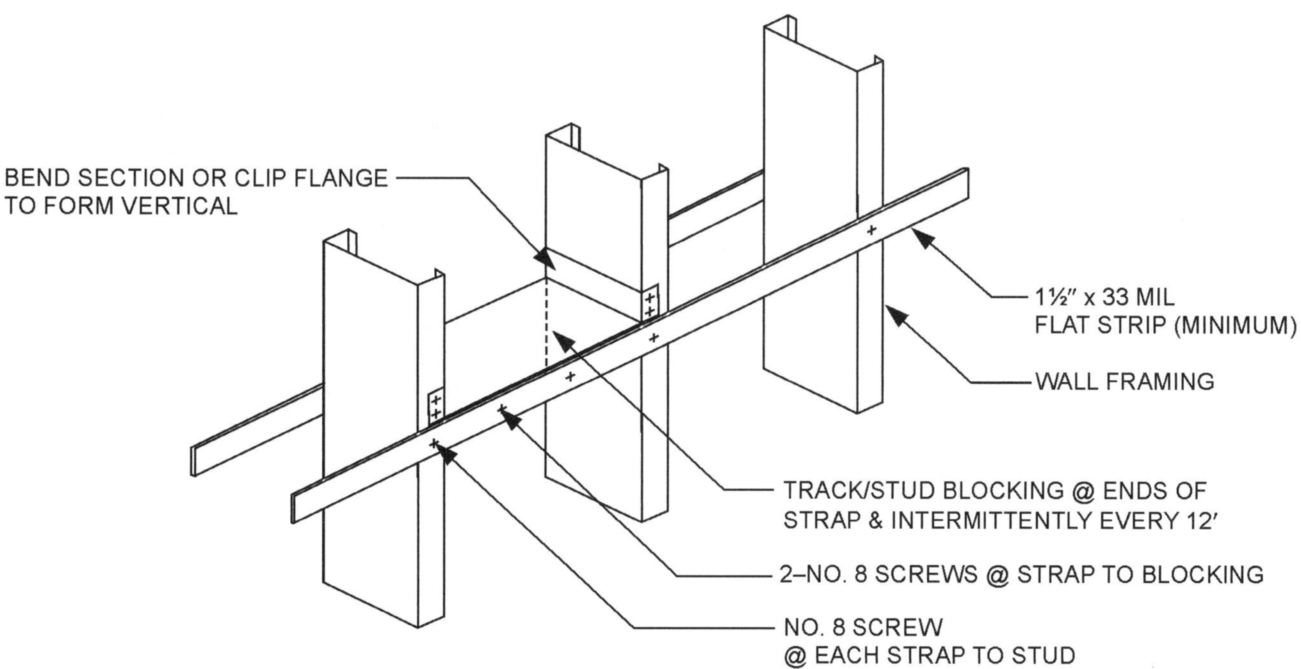

For SI: 1 mil = 0.0254 mm, 1 inch = 25.4 mm, 1 foot = 304.8 mm.

FIGURE R603.3.3(1)
STUD BRACING WITH STRAPPING ONLY

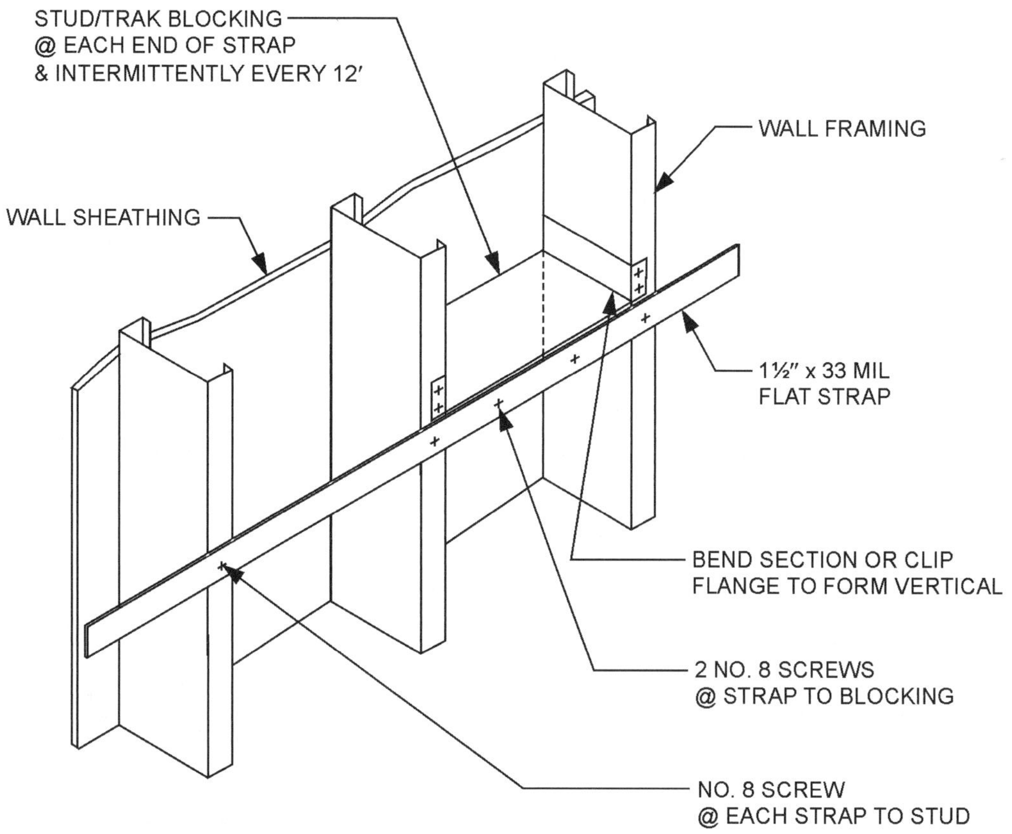

For SI: 1 mil = 0.0254 mm, 1 inch = 25.4 mm, 1 foot = 304.8 mm.

FIGURE R603.3.3(2)
STUD BRACING WITH STRAPPING AND SHEATHING MATERIAL

R603.3.4 Cutting and notching. Flanges and lips of cold-formed steel studs and headers shall not be cut or notched.

R603.3.5 Splicing. Steel studs and other structural members shall not be spliced. Tracks shall be spliced in accordance with Figure R603.3.5.

R603.4 Corner framing. In exterior walls, corner studs and the top tracks shall be installed in accordance with Figure R603.4.

R603.5 Exterior wall covering. The method of attachment of exterior wall covering materials to cold-formed steel stud wall framing shall conform to the manufacturer's installation instructions.

R603.6 Headers. Headers shall be installed above all wall openings in exterior walls and interior load-bearing walls. Box beam headers and back-to-back headers each shall be formed from two equal sized C-shaped members in accordance with Figures R603.6(1) and R603.6(2), respectively, and Tables R603.6(1) through R603.6(24). L-shaped headers shall be permitted to be constructed in accordance with AISI S230. Alternately, headers shall be permitted to be designed and constructed in accordance with AISI S100, Section D4.

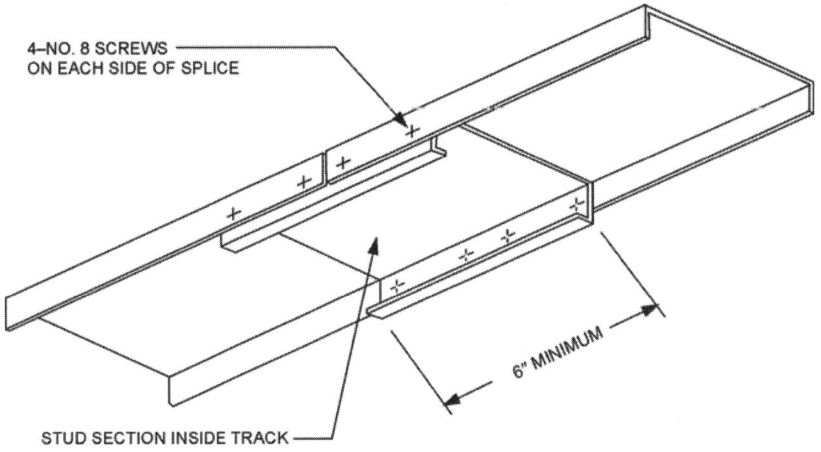

For SI: 1 inch = 25.4 mm.

FIGURE R603.3.5
TRACK SPLICE

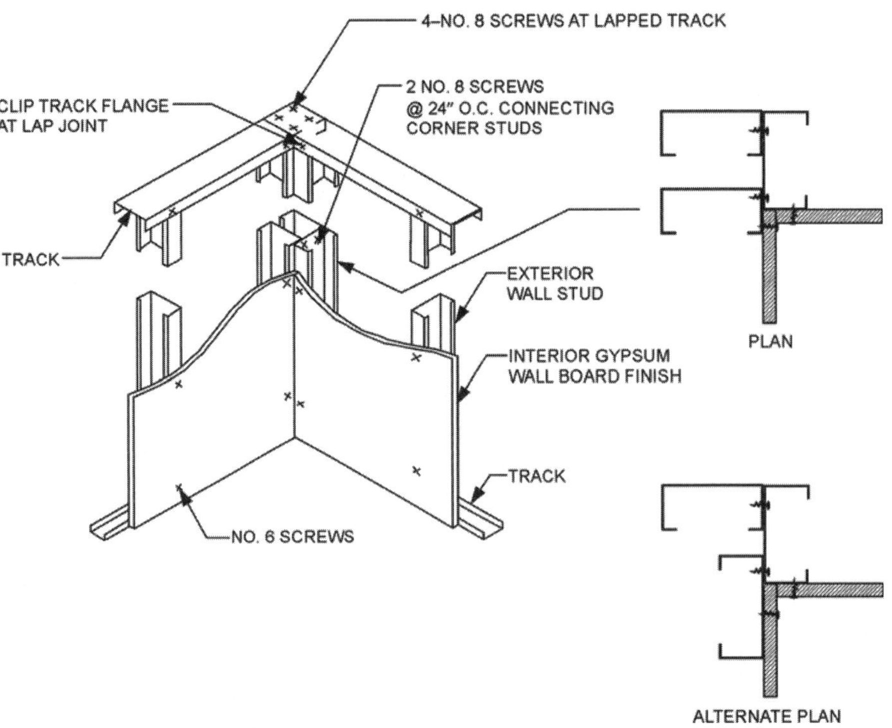

For SI: 1 inch = 25.4 mm.

FIGURE R603.4
CORNER FRAMING

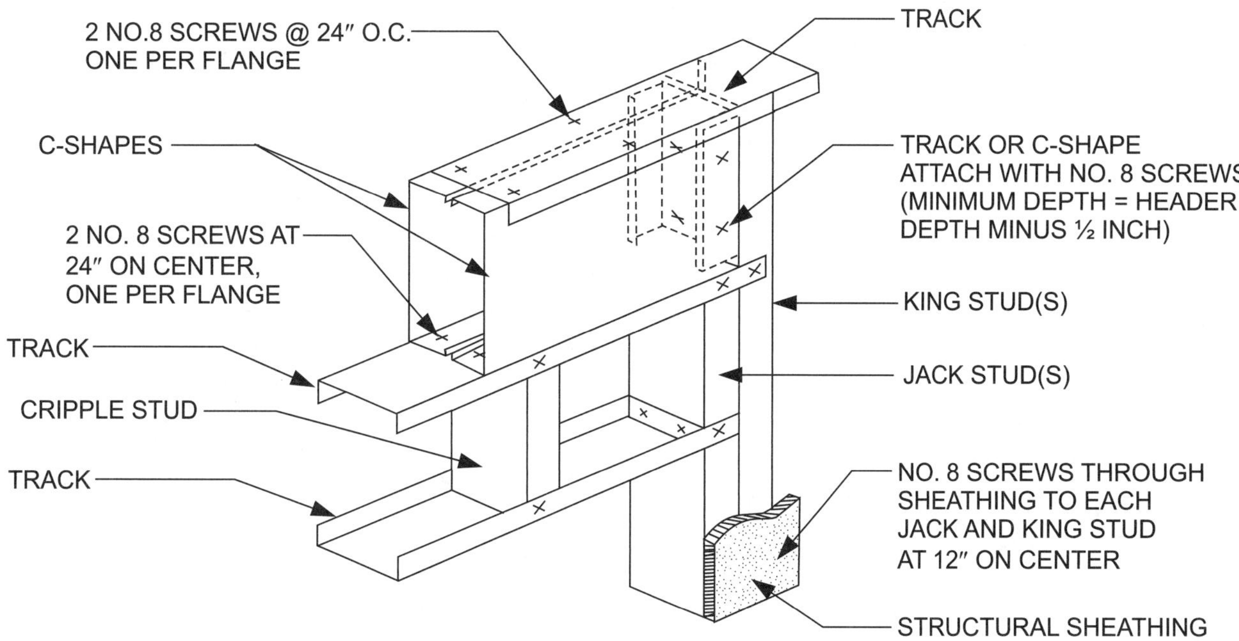

For SI: 1 inch = 25.4 mm.

**FIGURE R603.6(1)
BOX BEAM HEADER**

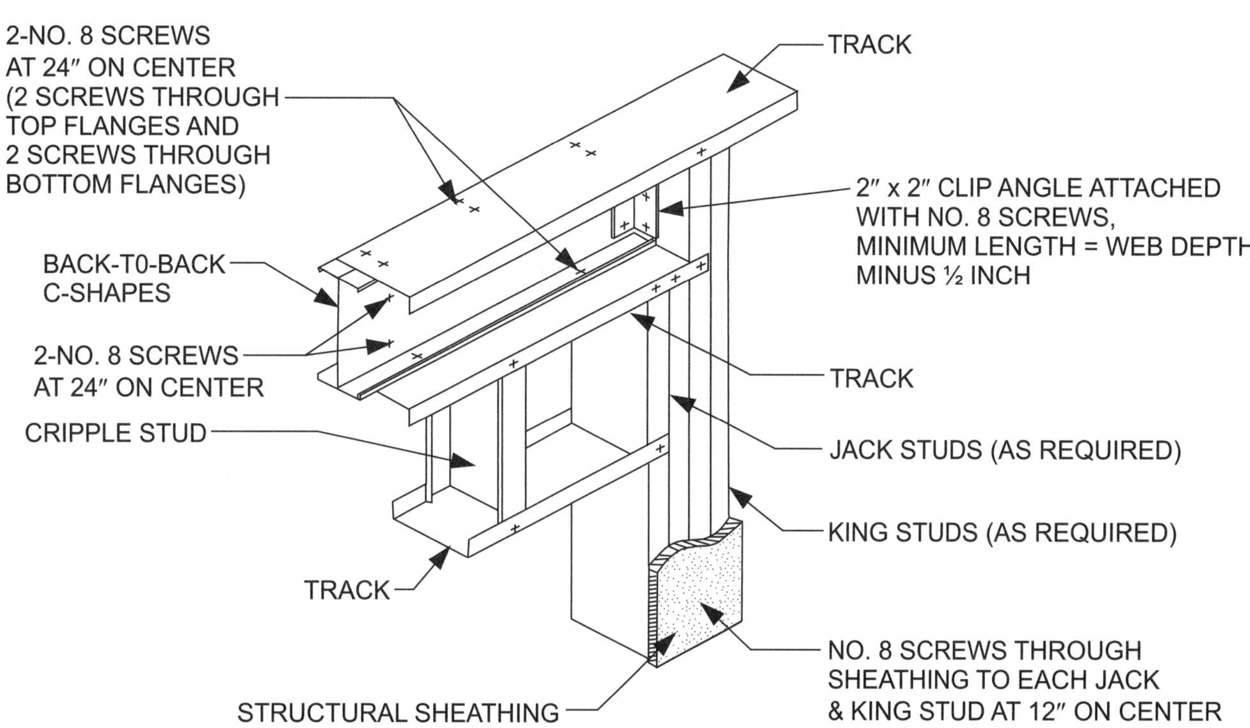

For SI: 1 inch = 25.4 mm.

**FIGURE R603.6(2)
BACK-TO-BACK HEADER**

**TABLE R603.6(1)
BOX-BEAM HEADER SPANS
Headers Supporting Roof and Ceiling Only
(33 Ksi steel)[a, b]**

MEMBER DESIGNATION	GROUND SNOW LOAD (20 psf)					GROUND SNOW LOAD (30 psf)				
	Building width[c] (feet)					Building width[c] (feet)				
	24	28	32	36	40	24	28	32	36	40
2-350S162-33	3′-3″	2′-8″	2′-2″	—	—	2′-8″	2′-2″	—	—	—
2-350S162-43	4′-2″	3′-9″	3′-4″	2′-11″	2′-7″	3′-9″	3′-4″	2′-11″	2′-7″	2′-2″
2-350S162-54	5′-0″	4′-6″	4′-1″	3′-8″	3′-4″	4′-6″	4′-1″	3′-8″	3′-3″	3′-0″
2-350S162-68	5′-7″	5′-1″	4′-7″	4′-3″	3′-10″	5′-1″	4′-7″	4′-2″	3′-10″	3′-5″
2-350S162-97	7′-1″	6′-6″	6′-1″	5′-8″	5′-3″	6′-7″	6′-1″	5′-7″	5′-3″	4′-11″
2-550S162-33	4′-8″	4′-0″	3′-6″	3′-0″	2′-6″	4′-1″	3′-6″	3′-0″	2′-6″	—
2-550S162-43	6′-0″	5′-4″	4′-10″	4′-4″	3′-11″	5′-5″	4′-10″	4′-4″	3′-10″	3′-5″
2-550S162-54	7′-0″	6′-4″	5′-9″	5′-4″	4′-10″	6′-5″	5′-9″	5′-3″	4′-10″	4′-5″
2-550S162-68	8′-0″	7′-4″	6′-9″	6′-3″	5′-10″	7′-5″	6′-9″	6′-3″	5′-9″	5′-4″
2-550S162-97	9′-11″	9′-2″	8′-6″	8′-0″	7′-6″	9′-3″	8′-6″	8′-0″	7′-5″	7′-0″
2-800S162-33	4′-5″	3′-11″	3′-5″	3′-1″	2′-10″	3′-11″	3′-6″	3′-1″	2′-9″	2′-3″
2-800S162-43	7′-3″	6′-7″	5′-11″	5′-4″	4′-10″	6′-7″	5′-11″	5′-4″	4′-9″	4′-3″
2-800S162-54	8′-10″	8′-0″	7′-4″	6′-9″	6′-2″	8′-1″	7′-4″	6′-8″	6′-1″	5′-7″
2-800S162-68	10′-5″	9′-7″	8′-10″	8′-2″	7′-7″	9′-8″	8′-10″	8′-1″	7′-6″	7′-0″
2-800S162-97	13′-1″	12′-1″	11′-3″	10′-7″	10′-0″	12′-2″	11′-4″	10′-6″	10′-0″	9′-4″
2-1000S162-43	7′-10″	6′-10″	6′-1″	5′-6″	5′-0″	6′-11″	6′-1″	5′-5″	4′-11″	4′-6″
2-1000S162-54	10′-0″	9′-1″	8′-3″	7′-7″	7′-0″	9′-2″	8′-4″	7′-7″	6′-11″	6′-4″
2-1000S162-68	11′-11″	10′-11″	10′-1″	9′-4″	8′-8″	11′-0″	10′-1″	9′-3″	8′-7″	8′-0″
2-1000S162-97	15′-3″	14′-3″	13′-5″	12′-6″	11′-10″	14′-4″	13′-5″	12′-6″	11′-9″	11′-0″
2-1200S162-54	11′-1″	10′-0″	9′-2″	8′-5″	7′-9″	10′-1″	9′-2″	8′-4″	7′-7″	7′-0″
2-1200S162-68	13′-3″	12′-1″	11′-2″	10′-4″	9′-7″	12′-3″	11′-2″	10′-3″	9′-6″	8′-10″
2-1200S162-97	16′-8″	15′-7″	14′-8″	13′-11″	13′-3″	15′-8″	14′-8″	13′-11″	13′-2″	12′-6″

For SI: 1 inch = 25.4 mm, 1 foot = 304.8 mm, 1 pound per square foot = 0.0479 kPa, 1 pound per square inch = 6.895 kPa, 1 Ksi = 1,000 psi = 6.895 MPa.

a. Deflection criterion: *L*/360 for live loads, *L*/240 for total loads.

b. Design load assumptions:

Roof/ceiling dead load is 12 psf.

Attic dead load is 10 psf.

c. Building width is in the direction of horizontal framing members supported by the header.

TABLE R603.6(2)
BOX-BEAM HEADER SPANS
Headers Supporting Roof and Ceiling Only
(50 Ksi steel)[a, b]

MEMBER DESIGNATION	GROUND SNOW LOAD (20 psf)					GROUND SNOW LOAD (30 psf)				
	Building width[c] (feet)					Building width[c] (feet)				
	24	28	32	36	40	24	28	32	36	40
2-350S162-33	4′-4″	3′-11″	3′-6″	3′-2″	2′-10″	3′-11″	3′-6″	3′-1″	2′-9″	2′-5″
2-350S162-43	5′-6″	5′-0″	4′-7″	4′-2″	3′-10″	5′-0″	4′-7″	4′-2″	3′-10″	3′-6″
2-350S162-54	6′-2″	5′-10″	5′-8″	5′-3″	4′-10″	5′-11″	5′-8″	5′-2″	4′-10″	4′-6″
2-350S162-68	6′-7″	6′-3″	6′-0″	5′-10″	5′-8″	6′-4″	6′-1″	5′-10″	5′-8″	5′-6″
2-350S162-97	7′-3″	6′-11″	6′-8″	6′-5″	6′-3″	7′-0″	6′-8″	6′-5″	6′-3″	6′-0″
2-550S162-33	6′-2″	5′-6″	5′-0″	4′-7″	4′-2″	5′-7″	5′-0″	4′-6″	4′-1″	3′-8″
2-550S162-43	7′-9″	7′-2″	6′-7″	6′-1″	5′-8″	7′-3″	6′-7″	6′-1″	5′-7″	5′-2″
2-550S162-54	8′-9″	8′-5″	8′-1″	7′-9″	7′-3″	8′-6″	8′-1″	7′-8″	7′-2″	6′-8″
2-550S162-68	9′-5″	9′-0″	8′-8″	8′-4″	8′-1″	9′-1″	8′-8″	8′-4″	8′-1″	7′-10″
2-550S162-97	10′-5″	10′-0″	9′-7″	9′-3″	9′-0″	10′-0″	9′-7″	9′-3″	8′-11″	8′-8″
2-800S162-33	4′-5″	3′-11″	3′-5″	3′-1″	2′-10″	3′-11″	3′-6″	3′-1″	2′-9″	2′-6″
2-800S162-43	9′-1″	8′-5″	7′-8″	6′-11″	6′-3″	8′-6″	7′-8″	6′-10″	6′-2″	5′-8″
2-800S162-54	10′-10″	10′-2″	9′-7″	9′-0″	8′-5″	10′-2″	9′-7″	8′-11″	8′-4″	7′-9″
2-800S162-68	12′-8″	11′-10″	11′-2″	10′-7″	10′-1″	11′-11″	11′-2″	10′-7″	10′-0″	9′-6″
2-800S162-97	14′-2″	13′-6″	13′-0″	12′-7″	12′-2″	13′-8″	13′-1″	12′-7″	12′-2″	11′-9″
2-1000S162-43	7′-10″	6′-10″	6′-1″	5′-6″	5′-0″	6′-11″	6′-1″	5′-5″	4′-11″	4′-6″
2-1000S162-54	12′-3″	11′-5″	10′-9″	10′-2″	9′-6″	11′-6″	10′-9″	10′-1″	9′-5″	8′-9″
2-1000S162-68	14′-5″	13′-5″	12′-8″	12′-0″	11′-6″	13′-6″	12′-8″	12′-0″	11′-5″	10′-10″
2-1000S162-97	17′-1″	16′-4″	15′-8″	14′-11″	14′-3″	16′-5″	15′-9″	14′-10″	14′-1″	13′-6″
2-1200S162-54	12′-11″	11′-3″	10′-0″	9′-0″	8′-2″	11′-5″	10′-0″	9′-0″	8′-1″	7′-4″
2-1200S162-68	15′-11″	14′-10″	14′-0″	13′-4″	12′-8″	15′-0″	14′-0″	13′-3″	12′-7″	11′-11″
2-1200S162-97	19′-11″	18′-7″	17′-6″	16′-8″	15′-10″	18′-9″	17′-7″	16′-7″	15′-9″	15′-0″

For SI: 1 inch = 25.4 mm, 1 foot = 304.8 mm, 1 pound per square foot = 0.0479 kPa, 1 pound per square inch = 6.895 kPa, 1 Ksi = 1,000 psi = 6.895 MPa.

a. Deflection criterion: *L*/360 for live loads, *L*/240 for total loads.

b. Design load assumptions:
Roof/ceiling dead load is 12 psf.
Attic dead load is 10 psf.

c. Building width is in the direction of horizontal framing members supported by the header.

TABLE R603.6(3)
BOX-BEAM HEADER SPANS
Headers Supporting Roof and Ceiling Only
(33 Ksi steel)[a, b]

MEMBER DESIGNATION	GROUND SNOW LOAD (50 psf)					GROUND SNOW LOAD (70 psf)				
	Building width[c] (feet)					Building width[c] (feet)				
	24	28	32	36	40	24	28	32	36	40
2-350S162-33	—	—	—	—	—	—	—	—	—	—
2-350S162-43	2′-4″	—	—	—	—	—	—	—	—	—
2-350S162-54	3′-1″	2′-8″	2′-3″	—	—	2′-1″	—	—	—	—
2-350S162-68	3′-7″	3′-2″	2′-8″	2′-3″	—	2′-6″	—	—	—	—
2-350S162-97	5′-1″	4′-7″	4′-3″	3′-11″	3′-7″	4′-1″	3′-8″	3′-4″	3′-0″	2′-8″
2-550S162-33	2′-2″	—	—	—	—	—	—	—	—	—
2-550S162-43	3′-8″	3′-1″	2′-6″	—	—	2′-3″	—	—	—	—
2-550S162-54	4′-7″	4′-0″	3′-6″	3′-0″	2′-6″	3′-3″	2′-8″	2′-1″	—	—
2-550S162-68	5′-6″	4′-11″	4′-5″	3′-11″	3′-6″	4′-3″	3′-8″	3′-1″	2′-7″	2′-1″
2-550S162-97	7′-3″	6′-7″	6′-1″	5′-8″	5′-3″	5′-11″	5′-4″	4′-11″	4′-6″	4′-1″
2-800S162-33	2′-7″	—	—	—	—	—	—	—	—	—
2-800S162-43	4′-6″	3′-9″	3′-1″	2′-5″	—	2′-10″	—	—	—	—
2-800S162-54	5′-10″	5′-1″	4′-6″	3′-11″	3′-4″	4′-3″	3′-6″	2′-9″	—	—
2-800S162-68	7′-2″	6′-6″	5′-10″	5′-3″	4′-8″	5′-7″	4′-10″	4′-2″	3′-7″	2′-11″
2-800S162-97	9′-7″	8′-9″	8′-2″	7′-7″	7′-0″	7′-11″	7′-2″	6′-7″	6′-0″	5′-7″
2-1000S162-43	4′-8″	4′-1″	3′-6″	2′-9″	—	3′-3″	2′-2″	—	—	—
2-1000S162-54	6′-7″	5′-10″	5′-1″	4′-5″	3′-9″	4′-10″	4′-0″	3′-2″	2′-3″	—
2-1000S162-68	8′-3″	7′-5″	6′-8″	6′-0″	5′-5″	6′-5″	5′-7″	4′-9″	4′-1″	3′-5″
2-1000S162-97	11′-4″	10′-5″	9′-8″	9′-0″	8′-5″	9′-5″	8′-6″	7′-10″	7′-2″	6′-7″
2-1200S162-54	7′-3″	6′-5″	5′-7″	4′-10″	4′-2″	5′-4″	4′-4″	3′-5″	2′-5″	—
2-1200S162-68	9′-2″	8′-2″	7′-5″	6′-8″	6′-0″	7′-1″	6′-2″	5′-4″	4′-6″	3′-9″
2-1200S162-97	12′-10″	11′-9″	10′-11″	10′-2″	9′-6″	10′-7″	9′-8″	8′-10″	8′-2″	7′-6″

For SI: 1 inch = 25.4 mm, 1 foot = 304.8 mm, 1 pound per square foot = 0.0479 kPa, 1 pound per square inch = 6.895 kPa, 1 Ksi = 1,000 psi = 6.895 MPa.

a. Deflection criterion: *L*/360 for live loads, *L*/240 for total loads.

b. Design load assumptions:

Roof/ceiling dead load is 12 psf.

Attic dead load is 10 psf.

c. Building width is in the direction of horizontal framing members supported by the header.

TABLE R603.6(4)
BOX-BEAM HEADER SPANS
Headers Supporting Roof and Ceiling Only
(50 Ksi steel)[a, b]

MEMBER DESIGNATION	GROUND SNOW LOAD (50 psf)					GROUND SNOW LOAD (70 psf)				
	Building width[c] (feet)					Building width[c] (feet)				
	24	28	32	36	40	24	28	32	36	40
2-350S162-33	2′-7″	2′-2″	—	—	—	—	—	—	—	—
2-350S162-43	3′-8″	3′-3″	2′-10″	2′-6″	2′-1″	2′-8″	2′-3″	—	—	—
2-350S162-54	4′-8″	4′-2″	3′-9″	3′-5″	3′-1″	3′-7″	3′-2″	2′-9″	2′-5″	2′-0″
2-350S162-68	5′-7″	5′-2″	4′-9″	4′-4″	3′-11″	4′-7″	4′-1″	3′-7″	3′-2″	2′-10″
2-350S162-97	6′-2″	5′-11″	5′-8″	5′-6″	5′-4″	5′-8″	5′-5″	5′-3″	4′-11″	4′-7″
2-550S162-33	3′-11″	3′-4″	2′-10″	2′-4″	—	2′-7″	—	—	—	—
2-550S162-43	5′-4″	4′-10″	4′-4″	3′-10″	3′-5″	4′-2″	3′-7″	3′-1″	2′-7″	2′-1″
2-550S162-54	6′-11″	6′-3″	5′-9″	5′-3″	4′-9″	5′-6″	4′-11″	4′-5″	3′-11″	3′-5″
2-550S162-68	8′-0″	7′-6″	6′-11″	6′-5″	5′-11″	6′-9″	6′-1″	5′-6″	5′-0″	4′-7″
2-550S162-97	8′-11″	8′-6″	8′-2″	7′-11″	7′-8″	8′-1″	7′-9″	7′-6″	7′-1″	6′-7″
2-800S162-33	2′-8″	2′-4″	2′-1″	1′-11″	1′-9″	2′-0″	1′-9″	—	—	—
2-800S162-43	5′-10″	5′-2″	4′-7″	4′-2″	3′-10″	4′-5″	3′-11″	3′-6″	3′-0″	2′-6″
2-800S162-54	8′-0″	7′-3″	6′-8″	6′-1″	5′-7″	6′-5″	5′-9″	5′-1″	4′-7″	4′-0″
2-800S162-68	9′-9″	9′-0″	8′-3″	7′-8″	7′-1″	8′-0″	7′-3″	6′-7″	6′-0″	5′-6″
2-800S162-97	12′-1″	11′-7″	11′-2″	10′-8″	10′-2″	11′-0″	10′-4″	9′-9″	9′-2″	8′-7″
2-1000S162-43	4′-8″	4′-1″	3′-8″	3′-4″	3′-0″	3′-6″	3′-1″	2′-9″	2′-6″	2′-3″
2-1000S162-54	9′-1″	8′-2″	7′-3″	6′-7″	6′-0″	7′-0″	6′-2″	5′-6″	5′-0″	4′-6″
2-1000S162-68	11′-1″	10′-2″	9′-5″	8′-8″	8′-1″	9′-1″	8′-3″	7′-6″	6′-10″	6′-3″
2-1000S162-97	13′-9″	12′-11″	12′-2″	11′-7″	11′-1″	11′-11″	11′-3″	10′-7″	9′-11″	9′-4″
2-1200S162-54	7′-8″	6′-9″	6′-1″	5′-6″	5′-0″	5′-10″	5′-1″	4′-7″	4′-1″	3′-9″
2-1200S162-68	12′-3″	11′-3″	10′-4″	9′-7″	8′-11″	10′-1″	9′-1″	8′-3″	7′-6″	6′-10″
2-1200S162-97	15′-4″	14′-5″	13′-7″	12′-11″	12′-4″	13′-4″	12′-6″	11′-10″	11′-1″	10′-5″

For SI: 1 inch = 25.4 mm, 1 foot = 304.8 mm, 1 pound per square foot = 0.0479 kPa, 1 pound per square inch = 6.895 kPa, 1 Ksi = 1,000 psi = 6.895 MPa.

a. Deflection criterion: *L*/360 for live loads, *L*/240 for total loads.

b. Design load assumptions:
Roof/ceiling dead load is 12 psf.
Attic dead load is 10 psf.

c. Building width is in the direction of horizontal framing members supported by the header.

TABLE R603.6(5)
BOX-BEAM HEADER SPANS
Headers Supporting One Floor, Roof and Ceiling
(33 Ksi steel)[a, b]

MEMBER DESIGNATION	GROUND SNOW LOAD (20 psf)					GROUND SNOW LOAD (30 psf)				
	Building width[c] (feet)					Building width[c] (feet)				
	24	28	32	36	40	24	28	32	36	40
2-350S162-33	—	—	—	—	—	—	—	—	—	—
2-350S162-43	2′-2″	—	—	—	—	2′-1″	—	—	—	—
2-350S162-54	2′-11″	2′-5″	—	—	—	2′-10″	2′-4″	—	—	—
2-350S162-68	3′-8″	3′-2″	2′-9″	2′-4″	-	3′-7″	3′-1″	2′-8″	2′-3″	—
2-350S162-97	4′-11″	4′-5″	4′-2″	3′-8″	3′-5″	4′-10″	4′-5″	4′-0″	3′-8″	3′-4″
2-550S162-33	—	—	—	—	—	—	—	—	—	—
2-550S162-43	3′-5″	2′-9″	2′-1″	—	—	3′-3″	2′-7″	—	—	—
2-550S162-54	4′-4″	3′-9″	3′-2″	2′-7″	2′-1″	4′-3″	3′-7″	3′-1″	2′-6″	—
2-550S162-68	5′-3″	4′-8″	4′-1″	3′-7″	3′-2″	5′-2″	4′-7″	4′-0″	3′-6″	3′-1″
2-550S162-97	7′-0″	6′-5″	5′-10″	5′-5″	5′-0″	6′-11″	6′-4″	5′-9″	5′-4″	4′-11″
2-800S162-33	2′-1″	—	—	—	—	—	—	—	—	—
2-800S162-43	4′-2″	3′-4″	2′-7″	—	—	4′-0″	3′-3″	2′-5″	—	—
2-800S162-54	5′-6″	4′-9″	4′-1″	3′-5″	2′-9″	5′-5″	4′-8″	3′-11″	3′-3″	2′-8″
2-800S162-68	6′-11″	6′-2″	5′-5″	4′-10″	4′-3″	6′-9″	6′-0″	5′-4″	4′-8″	4′-1″
2-800S162-97	9′-4″	8′-6″	7′-10″	7′-3″	6′-8″	9′-2″	8′-4″	7′-8″	7′-1″	6′-7″
2-1000S162-43	4′-4″	3′-9″	2′-11″	—	—	4′-3″	3′-8″	2′-9″	—	—
2-1000S162-54	6′-3″	5′-5″	4′-7″	3′-11″	3′-2″	6′-1″	5′-3″	4′-6″	3′-9″	3′-0″
2-1000S162-68	7′-11″	7′-0″	6′-3″	5′-6″	4′-10″	7′-9″	6′-10″	6′-1″	5′-4″	4′-9″
2-1000S162-97	11′-0″	10′-1″	9′-3″	8′-7″	8′-0″	10′-11″	9′-11″	9′-2″	8′-5″	7′-10″
2-1200S162-54	6′-11″	5′-11″	5′-1″	4′-3″	3′-5″	6′-9″	5′-9″	4′-11″	4′-1″	3′-3″
2-1200S162-68	8′-9″	7′-9″	6′-11″	6′-1″	5′-4″	8′-7″	7′-7″	6′-9″	5′-11″	5′-3″
2-1200S162-97	12′-4″	11′-5″	10′-6″	9′-8″	9′-0″	12′-3″	11′-3″	10′-4″	9′-6″	8′-10″

For SI: 1 inch = 25.4 mm, 1 foot = 304.8 mm, 1 pound per square foot = 0.0479 kPa, 1 pound per square inch = 6.895 kPa, 1 Ksi = 1,000 psi = 6.895 MPa.

a. Deflection criterion: *L*/360 for live loads, *L*/240 for total loads.

b. Design load assumptions:
   Second floor dead load is 10 psf.
   Roof/ceiling dead load is 12 psf.
   Second floor live load is 30 psf.
   Attic dead load is 10 psf.

c. Building width is in the direction of horizontal framing members supported by the header.

TABLE R603.6(6)
BOX-BEAM HEADER SPANS
Headers Supporting One Floor, Roof and Ceiling
(50 Ksi steel)[a, b]

MEMBER DESIGNATION	GROUND SNOW LOAD (20 psf)					GROUND SNOW LOAD (30 psf)				
	Building width[c] (feet)					Building width[c] (feet)				
	24	28	32	36	40	24	28	32	36	40
2-350S162-33	2′-4″	—	—	—	—	2′-3″	—	—	—	—
2-350S162-43	3′-4″	2′-11″	2′-6″	2′-1″	—	3′-3″	2′-10″	2′-5″	2′-0″	—
2-350S162-54	4′-4″	3′-10″	3′-5″	3′-1″	2′-9″	4′-3″	2′-9″	3′-4″	3′-0″	2′-8″
2-350S162-68	5′-0″	4′-9″	4′-7″	4′-2″	3′-9″	4′-11″	4′-8″	4′-6″	4′-1″	3′-9″
2-350S162-97	5′-6″	5′-3″	5′-1″	4′-11″	2′-9″	5′-5″	5′-2″	5′-0″	4′-10″	4′-8″
2-550S162-33	3′-6″	2′-11″	2′-4″	—	—	3′-5″	2′-10″	2′-3″	—	—
2-550S162-43	5′-0″	4′-5″	3′-11″	3′-5″	3′-0″	4′-11″	4′-4″	3′-10″	3′-4″	2′-11″
2-550S162-54	6′-6″	5′-10″	5′-3″	4′-9″	4′-4″	6′-4″	5′-9″	5′-2″	4′-8″	4′-3″
2-550S162-68	7′-2″	6′-10″	6′-5″	5′-11″	5′-6″	7′-0″	6′-9″	6′-4″	5′-10″	5′-4″
2-550S162-97	7′-11″	7′-7″	7′-3″	7′-0″	6′-10″	7′-9″	7′-5″	7′-2″	6′-11″	6′-9″
2-800S162-33	2′-5″	2′-2″	1′-11″	1′-9″	—	2′-5″	2′-1″	1′-10″	1′-8″	—
2-800S162-43	5′-5″	4′-9″	4′-3″	3′-9″	3′-5″	5′-3″	4′-8″	4′-1″	3′-9″	3′-5″
2-800S162-54	7′-6″	6′-9″	6′-2″	5′-7″	5′-0″	7′-5″	6′-8″	6′-0″	5′-5″	4′-11″
2-800S162-68	9′-3″	8′-5″	7′-8″	7′-1″	6′-6″	9′-1″	8′-3″	7′-7″	7′-0″	6′-5″
2-800S162-97	10′-9″	10′-3″	9′-11″	9′-7″	9′-3″	10′-7″	10′-1″	9′-9″	9′-5″	9′-1″
2-1000S162-43	4′-4″	3′-9″	3′-4″	3′-0″	2′-9″	4′-3″	3′-8″	3′-3″	2′-11″	2′-8″
2-1000S162-54	8′-6″	7′-6″	6′-8″	6′-0″	5′-5″	8′-4″	7′-4″	6′-6″	5′-10″	5′-4″
2-1000S162-68	10′-6″	9′-7″	8′-9″	8′-0″	7′-5″	10′-4″	9′-5″	8′-7″	7′-11″	7′-3″
2-1000S162-97	12′-11″	12′-4″	11′-8″	11′-1″	10′-6″	12′-9″	12′-2″	11′-6″	10′-11″	10′-5″
2-1200S162-54	7′-1″	6′-2″	5′-6″	5′-0″	4′-6″	6′-11″	6′-1″	5′-5″	4′-10″	4′-5″
2-1200S162-68	11′-7″	10′-7″	9′-8″	8′-11″	8′-2″	11′-5″	10′-5″	9′-6″	8′-9″	8′-0″
2-1200S162-97	14′-9″	13′-9″	13′-0″	12′-4″	11′-9″	14′-7″	13′-8″	12′-10″	12′-3″	11′-8″

For SI: 1 inch = 25.4 mm, 1 foot = 304.8 mm, 1 pound per square foot = 0.0479 kPa, 1 pound per square inch = 6.895 kPa, 1 Ksi = 1,000 psi = 6.895 MPa.

a. Deflection criterion: *L*/360 for live loads, *L*/240 for total loads.

b. Design load assumptions:
Second floor dead load is 10 psf.
Roof/ceiling dead load is 12 psf.
Second floor live load is 30 psf.
Attic live load is 10 psf.

c. Building width is in the direction of horizontal framing members supported by the header.

TABLE R603.6(7)
BOX-BEAM HEADER SPANS
Headers Supporting One Floor, Roof and Ceiling
(33 Ksi steel)[a, b]

MEMBER DESIGNATION	GROUND SNOW LOAD (50 psf)					GROUND SNOW LOAD (70 psf)				
	Building width[c] (feet)					Building width[c] (feet)				
	24	28	32	36	40	24	28	32	36	40
2-350S162-33	—	—	—	—	—	—	—	—	—	—
2-350S162-43	—	—	—	—	—	—	—	—	—	—
2-350S162-54	—	—	—	—	—	—	—	—	—	—
2-350S162-68	2′-8″	2′-3″	—	—	—	—	—	—	—	—
2-350S162-97	4′-0″	3′-7″	3′-3″	2′-11″	2′-7″	3′-4″	2′-11″	2′-6″	2′-2″	—
2-550S162-33	—	—	—	—	—	—	—	—	—	—
2-550S162-43	2′-0″	—	—	—	—	—	—	—	—	—
2-550S162-54	3′-1″	2′-6″	—	—	—	—	—	—	—	—
2-550S162-68	4′-1″	3′-6″	2′-11″	2′-5″	—	3′-1″	2′-5″	—	—	—
2-550S162-97	5′-10″	5′-3″	4′-10″	4′-5″	4′-0″	4′-11″	4′-5″	3′-11″	3′-6″	3′-2″
2-800S162-33	—	—	—	—	—	—	—	—	—	—
2-800S162-43	2′-6″	—	—	—	—	—	—	—	—	—
2-800S162-54	4′-0″	3′-3″	2′-6″	—	—	2′-8″	—	—	—	—
2-800S162-68	5′-5″	4′-8″	4′-0″	3′-4″	2′-8″	4′-2″	3′-4″	2′-6″	—	—
2-800S162-97	7′-9″	7′-1″	6′-6″	5′-11″	5′-5″	6′-7″	5′-11″	5′-4″	4′-10″	4′-4″
2-1000S162-43	2′-10″	—	—	—	—	—	—	—	—	—
2-1000S162-54	4′-7″	3′-8″	2′-9″	—	—	3′-0″	—	—	—	—
2-1000S162-68	6′-2″	5′-4″	4′-7″	3′-10″	3′-1″	4′-9″	3′-10″	2′-11″	—	—
2-1000S162-97	9′-3″	8′-5″	7′-8″	7′-1″	6′-6″	7′-10″	7′-1″	6′-5″	5′-9″	5′-2″
2-1200S162-54	5′-0″	4′-0″	3′-1″	—	—	3′-4″	—	—	—	—
2-1200S162-68	6′-10″	5′-11″	5′-0″	4′-3″	3′-5″	5′-3″	4′-3″	3′-2″	—	—
2-1200S162-97	10′-5″	9′-6″	8′-8″	8′-0″	7′-4″	8′-10″	8′-0″	7′-3″	6′-6″	5′-10″

For SI: 1 inch = 25.4 mm, 1 foot = 304.8 mm, 1 pound per square foot = 0.0479 kPa, 1 pound per square inch = 6.895 kPa, 1 Ksi = 1,000 psi = 6.895 MPa.

a. Deflection criterion: *L*/360 for live loads, *L*/240 for total loads.

b. Design load assumptions:

Second floor dead load is 10 psf.

Roof/ceiling dead load is 12 psf.

Second floor live load is 30 psf.

Attic live load is 10 psf.

c. Building width is in the direction of horizontal framing members supported by the header.

TABLE R603.6(8)
BOX-BEAM HEADER SPANS
Headers Supporting One Floor, Roof and Ceiling
(50 Ksi steel)[a, b]

MEMBER DESIGNATION	GROUND SNOW LOAD (50 psf)					GROUND SNOW LOAD (70 psf)				
	Building width[c] (feet)					Building width[c] (feet)				
	24	28	32	36	40	24	28	32	36	40
2-350S162-33	—	—	—	—	—	—	—	—	—	—
2-350S162-43	2′-8″	—	—	—	—	—	—	—	—	—
2-350S162-54	3′-5″	3′-0″	2′-7″	2′-2″	-	2′-8″	2′-2″	—	—	—
2-350S162-68	4′-6″	4′-1″	3′-8″	3′-3″	2′-11″	3′-9″	3′-3″	2′-10″	2′-5″	2′-1″
2-350S162-97	5′-1″	4′-10″	4′-8″	4′-6″	4′-5″	4′-10″	4′-7″	4′-4″	4′-0″	3′-8″
2-550S162-33	2′-4″	—	—	—	—	—	—	—	—	—
2-550S162-43	3′-10″	3′-4″	2′-9″	2′-3″	—	2′-11″	2′-3″	—	—	—
2-550S162-54	5′-3″	3′-8″	4′-1″	3′-8″	3′-2″	4′-3″	3′-8″	3′-1″	2′-7″	2′-0″
2-550S162-68	6′-5″	5′-10″	5′-3″	4′-9″	4′-4″	5′-5″	4′-9″	4′-3″	3′-9″	3′-4″
2-550S162-97	7′-4″	7′-0″	6′-9″	6′-6″	6′-4″	6′-11″	6′-8″	6′-3″	5′-10″	5′-5″
2-800S162-33	1′-11″	1′-8″	—	—	—	—	—	—	—	—
2-800S162-43	4′-2″	3′-8″	3′-4″	2′-9″	2′-2″	3′-5″	2′-9″	—	—	—
2-800S162-54	6′-1″	5′-5″	4′-10″	4′-3″	3′-9″	4′-11″	4′-3″	3′-8″	3′-0″	2′-5″
2-800S162-68	7′-8″	6′-11″	6′-3″	5′-9″	5′-2″	6′-5″	5′-9″	5′-1″	4′-6″	4′-0″
2-800S162-97	9′-11″	9′-6″	9′-2″	8′-10″	8′-3″	9′-5″	8′-10″	8′-2″	7′-7″	7′-0″
2-1000S162-43	3′-4″	2′-11″	2′-7″	2′-5″	2′-2″	2′-8″	2′-5″	2′-2″	—	—
2-1000S162-54	6′-7″	5′-10″	5′-3″	4′-9″	4′-3″	5′-4″	4′-9″	4′-1″	3′-5″	2′-9″
2-1000S162-68	8′-8″	7′-10″	7′-2″	6′-6″	5′-11″	7′-4″	6′-6″	5′-9″	5′-1″	4′-6″
2-1000S162-97	11′-7″	10′-11″	10′-3″	9′-7″	9′-0″	10′-5″	9′-7″	8′-10″	8′-2″	7′-8″
2-1200S162-54	5′-6″	4′-10″	4′-4″	3′-11″	3′-7″	4′-5″	3′-11″	3′-6″	3′-2″	2′-11″
2-1200S162-68	9′-7″	8′-8″	7′-11″	7′-2″	6′-6″	8′-1″	7′-2″	6′-4″	5′-8″	5′-0″
2-1200S162-97	12′-11″	12′-2″	11′-6″	10′-8″	10′-0″	11′-8″	10′-9″	9′-11″	9′-2″	8′-6″

For SI: 1 inch = 25.4 mm, 1 foot = 304.8 mm, 1 pound per square foot = 0.0479 kPa, 1 pound per square inch = 6.895 kPa, 1 Ksi = 1,000 psi = 6.895 MPa.

a. Deflection criterion: *L*/360 for live loads, *L*/240 for total loads.

b. Design load assumptions:
Second floor dead load is 10 psf.
Roof/ceiling dead load is 12 psf.
Second floor live load is 30 psf.
Attic live load is 10 psf.

c. Building width is in the direction of horizontal framing members supported by the header.

TABLE R603.6(9)
BOX-BEAM HEADER SPANS
Headers Supporting Two Floors, Roof and Ceiling
(33 Ksi steel)[a, b]

MEMBER DESIGNATION	GROUND SNOW LOAD (20 psf)					GROUND SNOW LOAD (30 psf)				
	Building width[c] (feet)					Building width[c] (feet)				
	24	28	32	36	40	24	28	32	36	40
2-350S162-33	—	—	—	—	—	—	—	—	—	—
2-350S162-43	—	—	—	—	—	—	—	—	—	—
2-350S162-54	—	—	—	—	—	—	—	—	—	—
2-350S162-68	—	—	—	—	—	—	—	—	—	—
2-350S162-97	3′-1″	2′-8″	2′-3″	—	—	3′-1″	2′-7″	2′-2″	—	—
2-550S162-33	—	—	—	—	—	—	—	—	—	—
2-550S162-43	—	—	—	—	—	—	—	—	—	—
2-550S162-54	—	—	—	—	—	—	—	—	—	—
2-550S162-68	2′-9″	—	—	—	—	2′-8″	—	—	—	—
2-550S162-97	4′-8″	4′-1″	3′-7″	3′-2″	2′-9″	4′-7″	4′-0″	3′-6″	3′-1″	2′-8″
2-800S162-33	—	—	—	—	—	—	—	—	—	—
2-800S162-43	—	—	—	—	—	—	—	—	—	—
2-800S162-54	2′-1″	—	—	—	—	—	—	—	—	—
2-800S162-68	3′-8″	2′-9″	—	—	—	3′-7″	2′-8″	—	—	—
2-800S162-97	6′-3″	5′-6″	4′-11″	4′-4″	3′-9″	6′-2″	5′-5″	4′-10″	4′-3″	3′-9″
2-1000S162-43	—	—	—	—	—	—	—	—	—	—
2-1000S162-54	2′-5″	—	—	—	—	2′-3″	—	—	—	—
2-1000S162-68	4′-3″	3′-2″	2′-0″	—	—	4′-2″	3′-1″	—	—	—
2-1000S162-97	7′-5″	6′-7″	5′-10″	5′-2″	4′-7″	7′-4″	6′-6″	5′-9″	5′-1″	4′-6″
2-1200S162-54	2′-7″	—	—	—	—	2′-6″	—	—	—	—
2-1200S162-68	4′-8″	3′-6″	2′-2″	—	—	4′-7″	3′-5″	2′-0″	—	—
2-1200S162-97	8′-5″	7′-5″	6′-7″	5′-10″	5′-2″	8′-3″	7′-4″	6′-6″	5′-9″	5′-1″

For SI: 1 inch = 25.4 mm, 1 foot = 304.8 mm, 1 pound per square foot = 0.0479 kPa, 1 pound per square inch = 6.895 kPa, 1 Ksi = 1,000 psi = 6.895 MPa.

a. Deflection criterion: *L*/360 for live loads, *L*/240 for total loads.

b. Design load assumptions:
   Second floor dead load is 10 psf.
   Roof/ceiling dead load is 12 psf.
   Second floor live load is 40 psf.
   Third floor live load is 30 psf.
   Attic live load is 10 psf.

c. Building width is in the direction of horizontal framing members supported by the header.

TABLE R603.6(10)
BOX-BEAM HEADER SPANS
Headers Supporting Two Floors, Roof and Ceiling
(50 Ksi steel)[a, b]

MEMBER DESIGNATION	GROUND SNOW LOAD (20 psf)					GROUND SNOW LOAD (30 psf)				
	Building width[c] (feet)					Building width[c] (feet)				
	24	28	32	36	40	24	28	32	36	40
2-350S162-33	—	—	—	—	—	—	—	—	—	—
2-350S162-43	—	—	—	—	—	—	—	—	—	—
2-350S162-54	2′-5″	—	—	—	—	2′-4″	—	—	—	—
2-350S162-68	3′-6″	3′-0″	2′-6″	2′-1″	—	3′-5″	2′-11″	2′-6″	2′-0″	—
2-350S162-97	4′-9″	4′-6″	4′-1″	3′-8″	3′-4″	4′-8″	4′-5″	4′-0″	3′-8″	3′-4″
2-550S162-33	—	—	—	—	—	—	—	—	—	—
2-550S162-43	2′-7″	—	—	—	—	2′-6″	—	—	—	—
2-550S162-54	3′-11″	3′-3″	2′-8″	2′-0″	—	3′-10″	3′-3″	2′-7″	—	—
2-550S162-68	5′-1″	4′-5″	3′-10″	3′-3″	2′-9″	5′-0″	4′-4″	3′-9″	3′-3″	2′-9″
2-550S162-97	6′-10″	6′-5″	5′-10″	5′-5″	4′-11″	6′-9″	6′-4″	5′-10″	5′-4″	4′-11″
2-800S162-33	—	—	—	—	—	—	—	—	—	—
2-800S162-43	3′-1″	2′-3″	—	—	—	3′-0″	2′-2″	—	—	—
2-800S162-54	4′-7″	3′-10″	3′-1″	2′-5″	—	4′-6″	3′-9″	3′-0″	2′-4″	—
2-800S162-68	6′-0″	5′-3″	4′-7″	3′-11″	3′-4″	6′-0″	5′-2″	4′-6″	3′-11″	3′-3″
2-800S162-97	9′-2″	8′-4″	7′-8″	7′-0″	6′-6″	9′-1″	8′-3″	7′-7″	7′-0″	6′-5″
2-1000S162-43	2′-6″	2′-2″	—	—	—	2′-6″	2′-2″	—	—	—
2-1000S162-54	5′-0″	4′-4″	3′-6″	2′-9″	—	4′-11″	4′-3″	3′-5″	2′-7″	—
2-1000S162-68	6′-10″	6′-0″	5′-3″	4′-6″	3′-10″	6′-9″	5′-11″	5′-2″	4′-5″	3′-9″
2-1000S162-97	10′-0″	9′-1″	8′-3″	7′-8″	7′-0″	9′-10″	9′-0″	8′-3″	7′-7″	7′-0″
2-1200S162-54	4′-2″	3′-7″	3′-3″	2′-11″	—	4′-1″	3′-7″	3′-2″	2′-10″	—
2-1200S162-68	7′-7″	6′-7″	5′-9″	5′-0″	4′-2″	7′-6″	6′-6″	5′-8″	4′-10″	4′-1″
2-1200S162-97	11′-2″	10′-1″	9′-3″	8′-6″	7′-10″	11′-0″	10′-0″	9′-2″	9′-2″	7′-9″

For SI: 1 inch = 25.4 mm, 1 foot = 304.8 mm, 1 pound per square foot = 0.0479 kPa, 1 pound per square inch = 6.895 kPa, 1 Ksi = 1,000 psi = 6.895 MPa.

a. Deflection criterion: *L*/360 for live loads, *L*/240 for total loads.

b. Design load assumptions:
   Second floor dead load is 10 psf.
   Roof/ceiling dead load is 12 psf.
   Second floor live load is 40 psf.
   Third floor live load is 30 psf.
   Attic live load is 10 psf.

c. Building width is in the direction of horizontal framing members supported by the header.

TABLE R603.6(11)
BOX-BEAM HEADER SPANS
Headers Supporting Two Floors, Roof and Ceiling
(33 Ksi steel)[a, b]

MEMBER DESIGNATION	GROUND SNOW LOAD (50 psf)					GROUND SNOW LOAD (70 psf)				
	Building width[c] (feet)					Building width[c] (feet)				
	24	28	32	36	40	24	28	32	36	40
2-350S162-33	—	—	—	—	—	—	—	—	—	—
2-350S162-43	—	—	—	—	—	—	—	—	—	—
2-350S162-54	—	—	—	—	—	—	—	—	—	—
2-350S162-68	—	—	—	—	—	—	—	—	—	—
2-350S162-97	2′-11″	2′-5″	2′-0″	—	—	2′-7″	2′-2″	—	—	—
2-550S162-33	—	—	—	—	—	—	—	—	—	—
2-550S162-43	—	—	—	—	—	—	—	—	—	—
2-550S162-54	—	—	—	—	—	—	—	—	—	—
2-550S162-68	2′-5″	—	—	—	—	—	—	—	—	—
2-550S162-97	4′-4″	3′-10″	3′-4″	2′-10″	2′-5″	4′-0″	3′-6″	3′-1″	2′-7″	2′-2″
2-800S162-33	—	—	—	—	—	—	—	—	—	—
2-800S162-43	—	—	—	—	—	—	—	—	—	—
2-800S162-54	—	—	—	—	—	—	—	—	—	—
2-800S162-68	3′-3″	2′-3″	—	—	—	2′-8″	—	—	—	—
2-800S162-97	5′-11″	5′-2″	4′-6″	4′-0″	3′-5″	5′-6″	4′-10″	4′-3″	3′-8″	3′-2″
2-1000S162-43	—	—	—	—	—	—	—	—	—	—
2-1000S162-54	—	—	—	—	—	—	—	—	—	—
2-1000S162-68	3′-9″	2′-7″	—	—	—	3′-1″	—	—	—	—
2-1000S162-97	7′-0″	6′-2″	5′-5″	4′-9″	4′-2″	6′-6″	5′-9″	5′-1″	4′-5″	3′-10″
2-1200S162-54	—	—	—	—	—	—	—	—	—	—
2-1200S162-68	4′-2″	2′-10″	—	—	—	3′-5″	2′-0″	—	—	—
2-1200S162-97	7′-11″	7′-0″	6′-2″	5′-5″	4′-8″	7′-4″	6′-6″	5′-9″	5′-0″	4′-4″

For SI: 1 inch = 25.4 mm, 1 foot = 304.8 mm, 1 pound per square foot = 0.0479 kPa, 1 pound per square inch = 6.895 kPa, 1 Ksi = 1,000 psi = 6.895 MPa.

a. Deflection criterion: *L*/360 for live loads, *L*/240 for total loads.

b. Design load assumptions:
Second floor dead load is 10 psf.
Roof/ceiling dead load is 12 psf.
Second floor live load is 40 psf.
Third floor live load is 30 psf.
Attic live load is 10 psf.

c. Building width is in the direction of horizontal framing members supported by the header.

TABLE R603.6(12)
BOX-BEAM HEADER SPANS
Headers Supporting Two Floors, Roof and Ceiling
(50 Ksi steel)[a,b]

MEMBER DESIGNATION	GROUND SNOW LOAD (50 psf)					GROUND SNOW LOAD (70 psf)				
	Building width[c] (feet)					Building width[c] (feet)				
	24	28	32	36	40	24	28	32	36	40
2-350S162-33	—	—	—	—	—	—	—	—	—	—
2-350S162-43	—	—	—	—	—	—	—	—	—	—
2-350S162-54	2′-2″	—	—	—	—	—	—	—	—	—
2-350S162-68	3′-3″	2′-9″	2′-3″	—	—	2′-11″	2′-5″	—	—	—
2-350S162-97	4′-6″	4′-3″	3′-10″	3′-6″	3′-2″	4′-3″	4′-0″	3′-7″	3′-3″	3′-0″
2-550S162-33	—	—	—	—	—	—	—	—	—	—
2-550S162-43	2′-3″	—	—	—	—	—	—	—	—	—
2-550S162-54	3′-7″	2′-11″	2′-3″	—	—	3′-3″	2′-7″	—	—	—
2-550S162-68	4′-9″	2′-1″	3′-6″	3′-0″	2′-5″	4′-4″	3′-9″	3′-2″	2′-8″	2′-1″
2-550S162-97	6′-5″	6′-1″	5′-7″	5′-1″	4′-8″	6′-3″	5′-10″	5′-4″	4′-10″	4′-5″
2-800S162-33	—	—	—	—	—	—	—	—	—	—
2-800S162-43	2′-8″	—	—	—	—	2′-2″	—	—	—	—
2-800S162-54	4′-3″	3′-5″	2′-8″	—	—	3′-9″	3′-0″	2′-3″	—	—
2-800S162-68	5′-8″	4′-11″	4′-2″	3′-7″	2′-11″	5′-3″	4′-6″	3′-10″	3′-3″	2′-7″
2-800S162-97	8′-9″	8′-0″	7′-3″	6′-8″	6′-2″	8′-4″	7′-7″	6′-11″	6′-4″	5′-10″
2-1000S162-43	2′-4″	2′-0″	—	—	—	2′-2″	—	—	—	—
2-1000S162-54	4′-8″	3′-11″	3′-1″	2′-2″	—	4′-3″	3′-5″	2′-7″	—	—
2-1000S162-68	6′-5″	5′-7″	4′-9″	4′-1″	3′-4″	5′-11″	5′-1″	4′-5″	3′-8″	2′-11″
2-1000S162-97	9′-6″	8′-8″	7′-11″	7′-3″	6′-8″	9′-0″	8′-3″	7′-6″	6′-11″	6′-4″
2-1200S162-54	3′-11″	3′-5″	3′-0″	2′-4″	—	3′-7″	3′-2″	2′-10″	—	—
2-1200S162-68	7′-1″	6′-2″	5′-3″	4′-6″	3′-8″	6′-6″	5′-8″	4′-10″	4′-0″	3′-3″
2-1200S162-97	10′-8″	9′-8″	8′-10″	8′-1″	7′-5″	10′-1″	9′-2″	8′-5″	7′-9″	7′-1″

For SI: 1 inch = 25.4 mm, 1 foot = 304.8 mm, 1 pound per square foot = 0.0479 kPa, 1 pound per square inch = 6.895 kPa, 1 Ksi = 1,000 psi = 6.895 MPa.

a. Deflection criterion: *L*/360 for live loads, *L*/240 for total loads.

b. Design load assumptions:
Second floor dead load is 10 psf.
Roof/ceiling dead load is 12 psf.
Second floor live load is 40 psf.
Third floor live load is 30 psf.
Attic live load is 10 psf.

c. Building width is in the direction of horizontal framing members supported by the header.

TABLE R603.6(13)
BACK-TO-BACK HEADER SPANS
Headers Supporting Roof and Ceiling Only
(33 Ksi steel)[a,b]

MEMBER DESIGNATION	GROUND SNOW LOAD (20 psf)					GROUND SNOW LOAD (30 psf)				
	Building width[c] (feet)					Building width[c] (feet)				
	24	28	32	36	40	24	28	32	36	40
2-350S162-33	2′-11″	2′-4″	—	—	—	2′-5″	—	—	—	—
2-350S162-43	4′-8″	3′-10″	3′-5″	3′-1″	2′-9″	3′-11″	3′-5″	3′-0″	2′-8″	2′-4″
2-350S162-54	5′-3″	4′-9″	4′-4″	4′-1″	3′-8″	4′-10″	4′-4″	4′-0″	3′-8″	3′-4″
2-350S162-68	6′-1″	5′-7″	5′-2″	4′-10″	4′-6″	5′-8″	5′-3″	4′-10″	4′-6″	4′-2″
2-350S162-97	7′-3″	6′-10″	6′-5″	6′-0″	5′-8″	6′-11″	6′-5″	6′-0″	5′-8″	5′-4″
2-550S162-33	4′-5″	3′-9″	3′-1″	2′-6″	—	3′-9″	3′-2″	2′-6″	—	—
2-550S162-43	6′-2″	5′-7″	5′-0″	4′-7″	4′-2″	5′-7″	5′-0″	4′-6″	4′-1″	3′-8″
2-550S162-54	7′-5″	6′-9″	6′-3″	5′-9″	5′-4″	6′-10″	6′-3″	5′-9″	5′-4″	4′-11″
2-550S162-68	6′-7″	7′-11″	7′-4″	6′-10″	6′-5″	8′-0″	7′-4″	6′-10″	6′-5″	6′-0″
2-550S162-97	10′-5″	9′-8″	9′-0″	8′-6″	8′-0″	9′-9″	9′-0″	8′-6″	8′-0″	7′-7″
2-800S162-33	4′-5″	3′-11″	3′-5″	3′-1″	2′-4″	3′-11″	3′-6″	3′-0″	2′-3″	—
2-800S162-43	7′-7″	6′-10″	6′-2″	5′-8″	5′-2″	6′-11″	6′-2″	5′-7″	5′-1″	4′-7″
2-800S162-54	9′-3″	8′-7″	7′-11″	7′-4″	6′-10″	8′-8″	7′-11″	7′-4″	6′-9″	6′-3″
2-800S162-68	10′-7″	9′-10″	9′-4″	8′-10″	8′-5″	9′-11″	9′-4″	8′-10″	8′-4″	7′-11″
2-800S162-97	13′-9″	12′-9″	12′-0″	11′-3″	10′-8″	12′-10″	12′-0″	11′-3″	10′-7″	10′-0″
2-1000S162-43	7′-10″	6′-10″	6′-1″	5′-6″	5′-0″	6′-11″	6′-1″	5′-5″	4′-11″	4′-6″
2-1000S162-54	10′-5″	9′-9″	9′-0″	8′-4″	7′-9″	9′-10″	9′-0″	8′-4″	7′-9″	7′-2″
2-1000S162-68	12′-1″	11′-3″	10′-8″	10′-1″	9′-7″	11′-4″	10′-8″	10′-1″	9′-7″	9′-1″
2-1000S162-97	15′-3″	14′-3″	13′-5″	12′-9″	12′-2″	14′-4″	13′-5″	12′-8″	12′-1″	11′-6″
2-1200S162-54	11′-6″	10′-9″	10′-0″	9′-0″	8′-2″	10′-10″	10′-0″	9′-0″	8′-1″	7′-4″
2-1200S162-68	13′-4″	12′-6″	11′-9″	11′-2″	10′-8″	12′-7″	11′-10″	11′-2″	10′-7″	10′-1″
2-1200S162-97	16′-8″	15′-7″	14′-8″	13′-11″	13′-3″	15′-8″	14′-8″	13′-11″	13′-2″	12′-7″

For SI: 1 inch = 25.4 mm, 1 foot = 304.8 mm, 1 pound per square foot = 0.0479 kPa, 1 pound per square inch = 6.895 kPa, 1 Ksi = 1,000 psi = 6.895 MPa.

a. Deflection criterion: *L*/360 for live loads, *L*/240 for total loads.

b. Design load assumptions:
   Second floor dead load is 12 psf.
   Attic live load is 10 psf.

c. Building width is in the direction of horizontal framing members supported by header.

TABLE R603.6(14)
BACK-TO-BACK HEADER SPANS
Headers Supporting Roof and Ceiling Only
(50 Ksi steel)[a,b]

MEMBER DESIGNATION	GROUND SNOW LOAD (20 psf)					GROUND SNOW LOAD (30 psf)				
	Building width[c] (feet)					Building width[c] (feet)				
	24	28	32	36	40	24	28	32	36	40
2-350S162-33	4′-2″	3′-8″	3′-3″	2′-10″	2′-6″	3′-8″	3′-3″	2′-10″	2′-5″	2′-1″
2-350S162-43	5′-5″	5′-0″	4′-6″	4′-2″	3′-10″	5′-0″	4′-7″	4′-2″	3′-10″	3′-6″
2-350S162-54	6′-2″	5′-10″	5′-8″	5′-4″	5′-0″	5′-11″	5′-8″	5′-4″	5′-0″	4′-8″
2-350S162-68	6′-7″	6′-3″	6′-0″	5′-10″	5′-8″	6′-4″	6′-1″	5′-10″	5′-8″	5′-6″
2-350S162-97	7′-3″	6′-11″	6′-8″	6′-5″	6′-3″	7′-0″	6′-8″	6′-5″	6′-3″	6′-0″
2-550S162-33	5′-10″	5′-3″	4′-8″	4′-3″	3′-9″	5′-3″	4′-9″	4′-2″	3′-9″	3′-3″
2-550S162-43	7′-9″	7′-2″	6′-7″	6′-1″	5′-8″	7′-3″	6′-7″	6′-1″	5′-8″	5′-3″
2-550S162-54	8′-9″	8′-5″	8′-1″	7′-9″	7′-5″	8′-6″	8′-1″	7′-9″	7′-5″	6′-11″
2-550S162-68	9′-5″	9′-0″	8′-8″	8′-4″	8′-1″	9′-1″	8′-8″	8′-4″	8′-1″	7′-10″
2-550S162-97	10′-5″	10′-0″	9′-7″	9′-3″	9′-0″	10′-0″	9′-7″	9′-3″	8′-11″	8′-8″
2-800S162-33	4′-5″	3′-11″	3′-5″	3′-1″	2′-10″	3′-11″	3′-6″	3′-1″	2′-9″	2′-6″
2-800S162-43	9′-1″	8′-5″	7′-8″	6′-11″	6′-3″	8′-6″	7′-8″	6′-10″	6′-2″	5′-8″
2-800S162-54	10′-10″	10′-2″	9′-7″	9′-1″	8′-8″	10′-2″	9′-7″	9′-0″	8′-7″	8′-1″
2-800S162-68	12′-8″	11′-10″	11′-2″	10′-7″	10′-1″	11′-11″	11′-2″	10′-7″	10′-0″	9′-7″
2-800S162-97	14′-2″	13′-6″	13′-0″	12′-7″	12′-2″	13′-8″	13′-1″	12′-7″	12′-2″	11′-9″
2-1000S162-43	7′-10″	6′-10″	6′-1″	5′-6″	5′-0″	6′-11″	6′-1″	5′-5″	4′-11″	4′-6″
2-1000S162-54	12′-3″	11′-5″	10′-9″	10′-3″	9′-9″	11′-6″	10′-9″	10′-2″	9′-8″	8′-11″
2-1000S162-68	14′-5″	13′-5″	12′-8″	12′-0″	11′-6″	13′-6″	12′-8″	12′-0″	11′-5″	10′-11″
2-1000S162-97	17′-1″	16′-4″	15′-8″	14′-11″	14′-3″	16′-5″	15′-9″	14′-10″	14′-1″	13′-6″
2-1200S162-54	12′-11″	11′-3″	10′-0″	9′-0″	8′-2″	11′-5″	10′-0″	9′-0″	8′-1″	7′-4″
2-1200S162-68	15′-11″	14′-10″	14′-0″	13′-4″	12′-8″	15′-0″	14′-0″	13′-3″	12′-7″	12′-0″
2-1200S162-97	19′-11″	18′-7″	17′-6″	16′-8″	15′-10″	18′-9″	17′-7″	16′-7″	15′-9″	15′-0″

For SI: 1 inch = 25.4 mm, 1 foot = 304.8 mm, 1 pound per square foot = 0.0479 kPa, 1 pound per square inch = 6.895 kPa, 1 Ksi = 1,000 psi = 6.895 MPa.

a. Deflection criterion: *L*/360 for live loads, *L*/240 for total loads.

b. Design load assumptions:
Roof/ceiling dead load is 12 psf.
Attic live load is 10 psf.

c. Building width is in the direction of horizontal framing members supported by the header.

TABLE R603.6(15)
BACK-TO-BACK HEADER SPANS
Headers Supporting Roof and Ceiling Only
(33 Ksi steel)[a, b]

MEMBER DESIGNATION	GROUND SNOW LOAD (50 psf)					GROUND SNOW LOAD (70 psf)				
	Building width[c] (feet)					Building width[c] (feet)				
	24	28	32	36	40	24	28	32	36	40
2-350S162-33	—	—	—	—	—	—	—	—	—	—
2-350S162-43	2′-6″	—	—	—	—	—	—	—	—	—
2-350S162-54	3′-6″	3′-1″	2′-8″	2′-4″	2′-0″	2′-7″	2′-1″	—	—	—
2-350S162-68	4′-4″	3′-11″	3′-7″	3′-3″	2′-11″	3′-5″	3′-0″	2′-8″	2′-4″	2′-1″
2-350S162-97	5′-5″	5′-0″	4′-8″	4′-6″	4′-1″	4′-6″	4′-2″	3′-10″	3′-6″	3′-3″
2-550S162-33	—	—	—	—	—	—	—	—	—	—
2-550S162-43	3′-10″	3′-3″	2′-9″	2′-2″	—	2′-6″	—	—	—	—
2-550S162-54	5′-1″	4′-7″	4′-1″	3′-8″	3′-4″	3′-11″	3′-5″	2′-11″	2′-6″	2′-0″
2-550S162-68	6′-2″	5′-8″	5′-2″	4′-9″	4′-5″	5′-0″	4′-6″	4′-1″	3′-9″	3′-4″
2-550S162-97	7′-9″	7′-2″	6′-8″	6′-3″	5′-11″	6′-6″	6′-0″	5′-7″	5′-2″	4′-10″
2-800S162-33	—	—	—	—	—	—	—	—	—	—
2-800S162-43	4′-10″	4′-1″	3′-6″	2′-11″	2′-3″	3′-3″	2′-5″	—	—	—
2-800S162-54	6′-6″	5′-10″	5′-3″	4′-9″	4′-4″	5′-1″	4′-6″	3′-11″	3′-4″	2′-10″
2-800S162-68	8′-1″	7′-5″	6′-10″	6′-4″	5′-11″	6′-8″	6′-1″	5′-6″	5′-0″	4′-7″
2-800S162-97	10′-3″	9′-7″	8′-11″	8′-5″	7′-11″	8′-8″	8′-0″	7′-6″	7′-0″	6′-7″
2-1000S162-43	4′-8″	4′-1″	3′-8″	3′-4″	2′-8″	3′-6″	2′-10″	—	—	—
2-1000S162-54	7′-5″	6′-8″	6′-1″	5′-6″	5′-0″	5′-10″	5′-1″	4′-6″	3′-11″	3′-4″
2-1000S162-68	9′-4″	8′-7″	7′-11″	7′-4″	6′-10″	7′-8″	7′-0″	6′-4″	5′-10″	5′-4″
2-1000S162-97	11′-9″	11′-0″	10′-5″	9′-11″	9′-5″	10′-3″	9′-7″	8′-11″	8′-4″	7′-10″
2-1200S162-54	7′-8″	6′-9″	6′-1″	5′-6″	5′-0″	5′-10″	5′-1″	4′-7″	4′-1″	3′-9″
2-1200S162-68	10′-4″	9′-6″	8′-10″	8′-2″	7′-7″	8′-7″	7′-9″	7′-1″	6′-6″	6′-0″
2-1200S162-97	12′-10″	12′-1″	11′-5″	10′-10″	10′-4″	11′-2″	10′-6″	9′-11″	9′-5″	9′-0″

For SI: 1 inch = 25.4 mm, 1 foot = 304.8 mm, 1 pound per square foot = 0.0479 kPa, 1 pound per square inch = 6.895 kPa, 1 Ksi = 1,000 psi = 6.895 MPa.

a. Deflection criterion: *L*/360 for live loads, *L*/240 for total loads.

b. Design load assumptions:
   Roof/ceiling dead load is 12 psf.
   Attic live load is 10 psf.

c. Building width is in the direction of horizontal framing members supported by the header.

TABLE R603.6(16)
BACK-TO-BACK HEADER SPANS
Headers Supporting Roof and Ceiling Only
(50 Ksi steel)[a, b]

MEMBER DESIGNATION	GROUND SNOW LOAD (50 psf)					GROUND SNOW LOAD (70 psf)				
	Building width[c] (feet)					Building width[c] (feet)				
	24	28	32	36	40	24	28	32	36	40
2-350S162-33	2′-3″	—	—	—	—	—	—	—	—	—
2-350S162-43	3′-8″	3′-3″	2′-10″	2′-6″	2′-2″	2′-8″	2′-3″	—	—	—
2-350S162-54	4′-9″	4′-4″	4′-0″	3′-8″	3′-8″	3′-10″	3′-5″	3′-1″	2′-9″	2′-5″
2-350S162-68	5′-7″	5′-4″	5′-2″	4′-11″	4′-7″	5′-1″	4′-8″	4′-3″	3′-11″	3′-8″
2-350S162-97	6′-2″	5′-11″	5′-8″	5′-6″	5′-4″	5′-8″	5′-5″	5′-3″	5′-0″	4′-11″
2-550S162-33	3′-6″	2′-10″	2′-3″	—	—	2′-0″	—	—	—	—
2-550S162-43	5′-5″	4′-10″	4′-4″	3′-11″	3′-6″	4′-2″	3′-8″	3′-2″	2′-8″	2′-3″
2-550S162-54	7′-2″	6′-6″	6′-0″	5′-7″	5′-2″	5′-10″	5′-3″	4′-10″	4′-5″	4′-0″
2-550S162-68	8′-0″	7′-8″	7′-3″	6′-11″	6′-6″	7′-2″	6′-7″	6′-1″	5′-8″	5′-4″
2-550S162-97	8′-11″	8′-6″	8′-2″	7′-11″	7′-8″	8′-1″	7′-9″	7′-6″	7′-2″	6′-11″
2-800S162-33	2′-8″	2′-4″	2′-1″	1′-11″	—	2′-0″	—	—	—	—
2-800S162-43	5′-10″	5′-2″	4′-7″	4′-2″	3′-10″	4′-5″	3′-11″	3′-6″	3′-2″	2′-9″
2-800S162-54	8′-4″	7′-8″	7′-1″	6′-7″	6′-1″	6′-10″	6′-3″	5′-8″	5′-2″	4′-9″
2-800S162-68	9′-9″	9′-2″	8′-8″	8′-3″	7′-10″	8′-6″	7′-11″	7′-4″	6′-10″	6′-5″
2-800S162-97	12′-1″	11′-7″	11′-2″	10′-8″	10′-2″	11′-0″	10′-4″	9′-9″	9′-3″	8′-10″
2-1000S162-43	4′-8″	4′-1″	2′-8″	3′-4″	3′-0″	3′-6″	10′-1″	2′-9″	2′-6″	2′-3″
2-1000S162-54	9′-3″	8′-2″	7′-3″	6′-7″	6′-0″	7′-0″	6′-2″	5′-6″	5′-0″	4′-6″
2-1000S162-68	11′-1″	10′-5″	9′-10″	9′-4″	8′-11″	9′-8″	9′-1″	8′-5″	7′-10″	7′-4″
2-1000S162-97	13′-9″	12′-11″	12′-2″	11′-7″	11′-1″	11′-11″	11′-3″	10′-7″	10′-1″	9′-7″
2-1200S162-54	7′-8″	6′-9″	6′-1″	5′-6″	5′-0″	5′-10″	5′-1″	4′-7″	4′-1″	3′-9″
2-1200S162-68	12′-3″	11′-6″	10′-11″	10′-4″	9′-11″	10′-8″	10′-0″	9′-2″	8′-4″	7′-7″
2-1200S162-97	15′-4″	14′-5″	13′-7″	12′-11″	12′-4″	13′-4″	12′-6″	11′-10″	11′-3″	10′-9″

For SI: 1 inch = 25.4 mm, 1 foot = 304.8 mm, 1 pound per square foot = 0.0479 kPa, 1 pound per square inch = 6.895 kPa, 1 Ksi = 1,000 psi = 6.895 MPa.

a. Deflection criterion: *L*/360 for live loads, *L*/240 for total loads.

b. Design load assumptions:
Roof/ceiling dead load is 12 psf.
Attic live load is 10 psf.

c. Building width is in the direction of horizontal framing members supported by the header.

TABLE R603.6(17)
BACK-TO-BACK HEADER SPANS
Headers Supporting One Floor, Roof and Ceiling
(33 Ksi steel)[a, b]

MEMBER DESIGNATION	GROUND SNOW LOAD (20 psf)					GROUND SNOW LOAD (30 psf)				
	Building width[c] (feet)					Building width[c] (feet)				
	24	28	32	36	40	24	28	32	36	40
2-350S162-33	—	—	—	—	—	—	—	—	—	—
2-350S162-43	2′-2″	—	—	—	—	2′-1″	—	—	—	—
2-350S162-54	3′-3″	2′-9″	2′-5″	2′-0″	—	3′-2″	2′-9″	2′-4″	—	—
2-350S162-68	4′-4″	3′-8″	3′-3″	2′-11″	2′-8″	4′-0″	3′-7″	3′-2″	2′-11″	2′-7″
2-350S162-97	5′-2″	4′-9″	4′-4″	4′-1″	3′-9″	5′-1″	4′-8″	4′-4″	4′-0″	3′-9″
2-550S162-33	—	—	—	—	—	—	—	—	—	—
2-550S162-43	3′-6″	2′-10″	2′-3″	—	—	3′-5″	2′-9″	2′-2″	—	—
2-550S162-54	4′-9″	4′-2″	3′-9″	3′-3″	2′-10″	4′-8″	4′-1″	3′-8″	3′-2″	2′-9″
2-550S162-68	5′-10″	5′-3″	4′-10″	4′-5″	4′-1″	5′-9″	5′-3″	4′-9″	4′-4″	4′-0″
2-550S162-97	7′-4″	6′-9″	6′-4″	5′-11″	5′-6″	7′-3″	6′-9″	6′-3″	5′-10″	5′-5″
2-800S162-33	—	—	—	—	—	—	—	—	—	—
2-800S162-43	4′-4″	3′-8″	2′-11″	2′-3″	—	4′-3″	3′-6″	2′-10″	2′-1″	—
2-800S162-54	6′-1″	5′-5″	4′-10″	4′-4″	3′-10″	6′-0″	5′-4″	4′-9″	4′-3″	3′-9″
2-800S162-68	7′-8″	7′-0″	6′-5″	5′-11″	5′-5″	7′-7″	6′-11″	6′-4″	5′-10″	5′-4″
2-800S162-97	9′-10″	9′-1″	8′-5″	7′-11″	7′-5″	9′-8″	8′-11″	8′-4″	7′-10″	7′-4″
2-1000S162-43	4′-4″	3′-9″	3′-4″	2′-8″	—	4′-3″	3′-8″	3′-3″	2′-6″	—
2-1000S162-54	6′-11″	6′-2″	5′-6″	5′-0″	4′-5″	6′-10″	6′-1″	5′-5″	4′-10″	4′-4″
2-1000S162-68	8′-10″	8′-1″	7′-5″	6′-10″	6′-4″	8′-8″	7′-11″	7′-3″	6′-8″	6′-2″
2-1000S162-97	11′-3″	10′-7″	9′-11″	9′-5″	8′-10″	11′-2″	10′-5″	9′-10″	9′-3″	8′-9″
2-1200S162-54	7′-1″	6′-2″	5′-6″	5′-0″	4′-6″	6′-11″	6′-1″	5′-5″	4′-10″	4′-5″
2-1200S162-68	9′-10″	9′-0″	8′-3″	7′-7″	7′-0″	9′-8″	8′-10″	8′-1″	7′-6″	6′-11″
2-1200S162-97	12′-4″	11′-7″	10′-11″	10′-4″	9′-10″	12′-3″	11′-5″	10′-9″	10′-3″	9′-9″

For SI: 1 inch = 25.4 mm, 1 foot = 304.8 mm, 1 pound per square foot = 0.0479 kPa, 1 pound per square inch = 6.895 kPa, 1 Ksi = 1,000 psi = 6.895 MPa.

a. Deflection criterion: $L/360$ for live loads, $L/240$ for total loads.

b. Design load assumptions:
Second floor dead load is 10 psf.
Roof/ceiling dead load is 12 psf.
Second floor live load is 30 psf.
Attic live load is 10 psf.

c. Building width is in the direction of horizontal framing members supported by the header.

TABLE R603.6(18)
BACK-TO-BACK HEADER SPANS
Headers Supporting One Floor, Roof and Ceiling
(50 Ksi steel)[a, b]

MEMBER DESIGNATION	GROUND SNOW LOAD (20 psf)					GROUND SNOW LOAD (30 psf)				
	Building width[c] (feet)					Building width[c] (feet)				
	24	28	32	36	40	24	28	32	36	40
2-350S162-33	—	—	—	—	—	—	—	—	—	—
2-350S162-43	3′-4″	2′-11″	2′-6″	2′-2″	—	3′-3″	2′-10″	2′-5″	2′-1″	—
2-350S162-54	4′-6″	4′-1″	3′-8″	3′-4″	3′-0″	4′-5″	4′-0″	3′-7″	3′-3″	2′-11″
2-350S162-68	5′-0″	4′-9″	4′-7″	4′-5″	4′-3″	4′-11″	4′-8″	4′-6″	4′-4″	4′-2″
2-350S162-97	5′-6″	5′-3″	5′-1″	4′-11″	4′-9″	5′-5″	5′-2″	5′-0″	4′-10″	4′-8″
2-550S162-33	3′-1″	2′-5″	—	—	—	3′-0″	2′-3″	—	—	—
2-550S162-43	5′-1″	4′-6″	4′-0″	3′-6″	3′-1″	4′-11″	4′-5″	3′-11″	3′-5″	3′-0″
2-550S162-54	6′-8″	6′-2″	5′-7″	5′-2″	4′-9″	6′-6″	6′-0″	5′-6″	5′-1″	4′-8″
2-550S162-68	7′-2″	6′-10″	6′-7″	6′-4″	6′-1″	7′-0″	6′-9″	6′-6″	6′-3″	6′-0″
2-550S162-97	7′-11″	7′-7″	7′-3″	7′-0″	6′-10″	7′-9″	7′-5″	7′-2″	6′-11″	6′-9″
2-800S162-33	2′-5″	2′-2″	1′-11″	—	—	2′-5″	2′-1″	1′-10″	—	—
2-800S162-43	5′-5″	4′-9″	4′-3″	3′-9″	3′-5″	5′-3″	4′-8″	4′-1″	3′-9″	3′-5″
2-800S162-54	7′-11″	7′-2″	6′-7″	6′-1″	5′-7″	7′-9″	7′-1″	6′-6″	6′-0″	5′-6″
2-800S162-68	9′-5″	8′-9″	8′-3″	7′-9″	7′-4″	9′-3″	8′-8″	8′-2″	7′-8″	7′-3″
2-800S162-97	10′-9″	10′-3″	9′-11″	9′-7″	9′-3″	10′-7″	10′-1″	9′-9″	9′-5″	9′-1″
2-1000S162-43	4′-4″	3′-9″	3′-4″	3′-0″	2′-9″	4′-3″	3′-8″	3′-3″	2′-11″	2′-8″
2-1000S162-54	8′-6″	7′-5″	6′-8″	6′-0″	5′-5″	8′-4″	7′-4″	6′-6″	5′-10″	5′-4″
2-1000S162-68	10′-8″	10′-0″	9′-5″	8′-11″	8′-4″	10′-7″	9′-10″	9′-4″	8′-9″	8′-3″
2-1000S162-97	12′-11″	12′-4″	11′-8″	11′-1″	10′-6″	12′-9″	12′-2″	11′-6″	10′-11″	10′-5″
2-1200S162-54	7′-1″	6′-2″	5′-6″	5′-0″	4′-6″	6′-11″	6′-1″	5′-5″	4′-10″	4′-5″
2-1200S162-68	11′-9″	11′-0″	10′-5″	9′-10″	9′-1″	11′-8″	10′-11″	10′-3″	9′-9″	8′-11″
2-1200S162-97	14′-9″	13′-9″	13′-0″	12′-4″	11′-9″	14′-7″	13′-8″	12′-10″	12′-3″	11′-8″

For SI: 1 inch = 25.4 mm, 1 foot = 304.8 mm, 1 pound per square foot = 0.0479 kPa, 1 pound per square inch = 6.895 kPa, 1 Ksi = 1,000 psi = 6.895 MPa.

a. Deflection criterion: *L*/360 for live loads, *L*/240 for total loads.

b. Design load assumptions:
   Second floor dead load is 10 psf.
   Roof/ceiling dead load is 12 psf.
   Second floor live load is 30 psf.
   Attic live load is 10 psf.

c. Building width is in the direction of horizontal framing members supported by the header.

TABLE R603.6(19)
BACK-TO-BACK HEADER SPANS
Headers Supporting One Floor, Roof and Ceiling
(33 Ksi steel)[a, b]

MEMBER DESIGNATION	GROUND SNOW LOAD (50 psf)					GROUND SNOW LOAD (70 psf)				
	Building width[c] (feet)					Building width[c] (feet)				
	24	28	32	36	40	24	28	32	36	40
2-350S162-33	—	—	—	—	—	—	—	—	—	—
2-350S162-43	—	—	—	—	—	—	—	—	—	—
2-350S162-54	2′-4″	—	—	—	—	—	—	—	—	—
2-350S162-68	3′-3″	2′-10″	2′-6″	2′-2″	—	2′-7″	2′-2″	—	—	—
2-350S162-97	4′-4″	4′-0″	3′-8″	3′-4″	3′-1″	3′-9″	3′-4″	3′-1″	2′-9″	2′-6″
2-550S162-33	—	—	—	—	—	—	—	—	—	—
2-550S162-43	2′-2″	—	—	—	—	—	—	—	—	—
2-550S162-54	3′-8″	3′-2″	2′-8″	2′-3″	—	2′-10″	2′-3″	—	—	—
2-550S162-68	4′-9″	4′-4″	3′-11″	3′-6″	3′-2″	4′-0″	3′-6″	3′-1″	2′-9″	2′-4″
2-550S162-97	6′-3″	5′-9″	5′-4″	5′-0″	4′-8″	5′-6″	5′-0″	4′-7″	4′-3″	3′-11″
2-800S162-33	—	—	—	—	—	—	—	—	—	—
2-800S162-43	2′-11″	2′-0″	—	—	—	—	—	—	—	—
2-800S162-54	4′-9″	4′-2″	3′-7″	3′-1″	2′-7″	3′-9″	3′-1″	2′-5″	—	—
2-800S162-68	6′-4″	5′-9″	5′-3″	4′-9″	4′-4″	5′-4″	4′-9″	4′-3″	3′-10″	3′-4″
2-800S162-97	8′-5″	7′-9″	7′-3″	6′-9″	6′-4″	7′-4″	6′-9″	6′-3″	5′-10″	5′-5″
2-1000S162-43	3′-4″	2′-5″	—	—	—	—	—	—	—	—
2-1000S162-54	5′-6″	4′-10″	4′-2″	3′-7″	3′-0″	4′-4″	3′-7″	2′-11″	2′-2″	—
2-1000S162-68	7′-4″	6′-8″	6′-1″	5′-7″	5′-1″	6′-3″	5′-7″	5′-0″	4′-5″	4′-0″
2-1000S162-97	9′-11″	8′-3″	8′-7″	8′-1″	7′-7″	8′-9″	8′-1″	7′-6″	7′-0″	6′-6″
2-1200S162-54	5′-6″	4′-10″	4′-4″	3′-11″	3′-5″	4′-5″	3′-11″	3′-3″	2′-6″	—
2-1200S162-68	8′-2″	7′-5″	6′-9″	6′-3″	5′-8″	6′-11″	6′-3″	5′-7″	5′-0″	4′-6″
2-1200S162-97	10′-10″	10′-2″	9′-8″	9′-2″	8′-7″	9′-9″	9′-2″	8′-6″	7′-11″	7′-5″

For SI: 1 inch = 25.4 mm, 1 foot = 304.8 mm, 1 pound per square foot = 0.0479 kPa, 1 pound per square inch = 6.895 kPa, 1 Ksi = 1,000 psi = 6.895 MPa.

a. Deflection criterion: *L*/360 for live loads, *L*/240 for total loads.

b. Design load assumptions:
   Second floor dead load is 10 psf.
   Roof/ceiling dead load is 12 psf.
   Second floor live load is 30 psf.
   Attic live load is 10 psf.

c. Building width is in the direction of horizontal framing members supported by the header.

TABLE R603.6(20)
BACK-TO-BACK HEADER SPANS
Headers Supporting One Floor, Roof and Ceiling
(50 Ksi steel)[a, b]

MEMBER DESIGNATION	GROUND SNOW LOAD (50 psf)					GROUND SNOW LOAD (70 psf)				
	Building width[c] (feet)					Building width[c] (feet)				
	24	28	32	36	40	24	28	32	36	40
2-350S162-33	—	—	—	—	—	—	—	—	—	—
2-350S162-43	2′-6″	2′-0″	—	—	—	—	—	—	—	—
2-350S162-54	3′-8″	3′-3″	2′-11″	2′-7″	2′-3″	3′-0″	2′-7″	2′-2″	—	—
2-350S162-68	4′-7″	4′-5″	4′-1″	3′-9″	3′-6″	4′-2″	3′-9″	3′-5″	3′-1″	2′-10″
2-350S162-97	5′-1″	4′-10″	4′-8″	4′-6″	4′-5″	4′-10″	4′-7″	4′-5″	4′-3″	4′-1″
2-550S162-33	—	—	—	—	—	—	—	—	—	—
2-550S162-43	3′-11″	3′-5″	2′-11″	2′-5″	—	3′-0″	2′-5″	—	—	—
2-550S162-54	5′-7″	5′-0″	4′-7″	4′-2″	3′-9″	4′-8″	4′-2″	3′-8″	3′-3″	2′-11″
2-550S162-68	6′-7″	6′-4″	5′-11″	5′-6″	5′-1″	6′-0″	5′-6″	5′-0″	4′-7″	4′-3″
2-550S162-97	7′-4″	7′-0″	6′-9″	6′-6″	6′-4″	6′-11″	6′-8″	6′-5″	6′-2″	6′-0″
2-800S162-33	1′-11″	—	—	—	—	—	—	—	—	—
2-800S162-43	4′-2″	3′-8″	3′-4″	3′-0″	2′-6″	3′-5″	3′-0″	2′-4″	—	—
2-800S162-54	6′-7″	5′-11″	5′-5″	4′-11″	4′-6″	5′-6″	4′-11″	4′-5″	3′-11″	3′-6″
2-800S162-68	8′-3″	7′-8″	7′-1″	6′-8″	6′-2″	7′-3″	6′-7″	6′-1″	5′-7″	5′-2″
2-800S162-97	9′-11″	9′-6″	9′-2″	8′-10″	8′-7″	9′-5″	9′-0″	8′-7″	8′-2″	7′-9″
2-1000S162-43	3′-4″	2′-11″	2′-7″	2′-5″	2′-2″	2′-8″	2′-5″	2′-2″	1′-11″	—
2-1000S162-54	6′-7″	5′-10″	5′-3″	4′-9″	4′-4″	5′-4″	4′-9″	4′-3″	3′-10″	3′-6″
2-1000S162-68	9′-4″	8′-9″	8′-1″	7′-7″	7′-1″	8′-3″	7′-7″	6′-11″	6′-5″	5′-11″
2-1000S162-97	11′-7″	10′-11″	10′-4″	9′-10″	9′-5″	10′-5″	9′-10″	9′-3″	8′-10″	8′-5″
2-1200S162-54	5′-6″	4′-10″	4′-4″	3′-11″	3′-7″	4′-5″	3′-11″	3′-6″	3′-2″	2′-11″
2-1200S162-68	10′-4″	9′-8″	8′-8″	7′-11″	7′-2″	8′-11″	7′-11″	7′-1″	6′-5″	5′-10″
2-1200S162-97	12′-11″	12′-2″	11′-6″	11′-0″	10′-6″	11′-8″	11′-0″	10′-5″	9′-10″	9′-5″

For SI: 1 inch = 25.4 mm, 1 foot = 304.8 mm, 1 pound per square foot = 0.0479 kPa, 1 pound per square inch = 6.895 kPa, 1 Ksi = 1,000 psi = 6.895 MPa.

a. Deflection criterion: *L*/360 for live loads, *L*/240 for total loads.

b. Design load assumptions:
Second floor dead load is 10 psf.
Roof/ceiling dead load is 12 psf.
Second floor live load is 30 psf.
Attic live load is 10 psf.

c. Building width is in the direction of horizontal framing members supported by the header.

TABLE R603.6(21)
BACK-TO-BACK HEADER SPANS
Headers Supporting Two Floors, Roof and Ceiling
(33 Ksi steel)[a, b]

MEMBER DESIGNATION	GROUND SNOW LOAD (20 psf)					GROUND SNOW LOAD (30 psf)				
	Building width[c] (feet)					Building width[c] (feet)				
	24	28	32	36	40	24	28	32	36	40
2-350S162-33	—	—	—	—	—	—	—	—	—	—
2-350S162-43	—	—	—	—	—	—	—	—	—	—
2-350S162-54	—	—	—	—	—	—	—	—	—	—
2-350S162-68	2′-5″	—	—	—	—	2′-4″	—	—	—	—
2-350S162-97	3′-6″	3′-2″	2′-10″	2′-6″	2′-3″	3′-6″	3′-1″	2′-9″	2′-6″	2′-3″
2-550S162-33	—	—	—	—	—	—	—	—	—	—
2-550S162-43	—	—	—	—	—	—	—	—	—	—
2-550S162-54	2′-6″	—	—	—	—	2′-5″	—	—	—	—
2-550S162-68	3′-9″	3′-3″	2′-9″	2′-4″	—	3′-8″	3′-2″	2′-9″	2′-4″	—
2-550S162-97	5′-3″	4′-9″	4′-4″	3′-11″	3′-8″	5′-2″	4′-8″	4′-3″	3′-11″	3′-7″
2-800S162-33	—	—	—	—	—	—	—	—	—	—
2-800S162-43	—	—	—	—	—	—	—	—	—	—
2-800S162-54	3′-5″	2′-8″	—	—	—	3′-4″	2′-7″	—	—	—
2-800S162-68	5′-1″	4′-5″	3′-11″	3′-4″	2′-11″	5′-0″	4′-4″	3′-10″	3′-4″	2′-10″
2-800S162-97	7′-0″	6′-5″	5′-11″	5′-5″	5′-0″	7′-0″	6′-4″	5′-10″	5′-5″	5′-0″
2-1000S162-43	—	—	—	—	—	—	—	—	—	—
2-1000S162-54	3′-11″	3′-1″	2′-3″	—	—	3′-10″	3′-0″	2′-2″	—	—
2-1000S162-68	5′-10″	5′-2″	4′-6″	4′-0″	3′-5″	5′-9″	5′-1″	4′-6″	3′-11″	3′-4″
2-1000S162-97	8′-5″	7′-8″	7′-1″	6′-6″	6′-1″	8′-4″	7′-7″	7′-0″	6′-6″	6′-0″
2-1200S162-54	4′-2″	3′-6″	2′-7″	—	—	4′-1″	3′-5″	2′-6″	—	—
2-1200S162-68	6′-6″	5′-9″	5′-1″	4′-6″	3′-11″	6′-6″	5′-8″	5′-0″	4′-5″	3′-10″
2-1200S162-97	9′-5″	8′-8″	8′-0″	7′-5″	6′-11″	9′-5″	8′-7″	7′-11″	7′-4″	6′-10″

For SI: 1 inch = 25.4 mm, 1 foot = 304.8 mm, 1 pound per square foot = 0.0479 kPa, 1 pound per square inch = 6.895 kPa, 1 Ksi = 1,000 psi = 6.895 MPa.

a. Deflection criterion: *L*/360 for live loads, *L*/240 for total loads.

b. Design load assumptions:

Second floor dead load is 10 psf.

Roof/ceiling dead load is 12 psf.

Second floor live load is 40 psf.

Third floor live load is 30 psf.

Attic live load is 10 psf.

c. Building width is in the direction of horizontal framing members supported by the header.

TABLE R603.6(22)
BACK-TO-BACK HEADER SPANS
Headers Supporting Two Floors, Roof and Ceiling
(50 Ksi steel)[a, b]

MEMBER DESIGNATION	GROUND SNOW LOAD (20 psf)					GROUND SNOW LOAD (30 psf)				
	Building width[c] (feet)					Building width[c] (feet)				
	24	28	32	36	40	24	28	32	36	40
2-350S162-33	—	—	—	—	—	—	—	—	—	—
2-350S162-43	—	—	—	—	—	—	—	—	—	—
2-350S162-54	2′-9″	2′-3″	—	—	—	2′-8″	2′-3″	—	—	—
2-350S162-68	3′-11″	3′-6″	3′-2″	2′-10″	2′-6″	3′-11″	3′-6″	3′-1″	2′-9″	2′-6″
2-350S162-97	4′-9″	4′-6″	4′-4″	4′-1″	3′-10″	4′-8″	4′-6″	4′-4″	4′-1″	3′-9″
2-550S162-33	—	—	—	—	—	—	—	—	—	—
2-550S162-43	2′-9″	2′-0″	—	—	—	2′-8″	—	—	—	—
2-550S162-54	4′-5″	3′-10″	3′-4″	2′-11″	2′-5″	4′-4″	3′-9″	3′-3″	2′-10″	2′-5″
2-550S162-68	5′-8″	5′-2″	4′-8″	4′-3″	3′-11″	5′-8″	5′-1″	4′-8″	4′-3″	3′-10″
2-550S162-97	6′-10″	6′-6″	6′-3″	6′-0″	5′-7″	6′-9″	6′-5″	6′-3″	5′-11″	5′-6″
2-800S162-33	—	—	—	—	—	—	—	—	—	—
2-800S162-43	3′-2″	2′-7″	—	—	—	3′-1″	2′-6″	—	—	—
2-800S162-54	5′-2″	4′-7″	4′-0″	3′-6″	3′-0″	5′-2″	4′-6″	3′-11″	3′-5″	2′-11″
2-800S162-68	6′-11″	6′-3″	5′-8″	5′-2″	4′-9″	6′-10″	6′-2″	5′-7″	5′-2″	4′-8″
2-800S162-97	9′-3″	8′-8″	8′-3″	7′-9″	7′-4″	9′-2″	8′-8″	8′-2″	7′-9″	7′-4″
2-1000S162-43	2′-6″	2′-2″	2′-0″	—	—	2′-6″	2′-2″	1′-11″	—	—
2-1000S162-54	5′-0″	4′-4″	3′-11″	3′-6″	3′-2″	4′-11″	4′-4″	3′-10″	3′-6″	3′-2″
2-1000S162-68	7′-10″	7′-2″	6′-6″	5′-11″	5′-6″	7′-9″	7′-1″	6′-5″	5′-11″	5′-5″
2-1000S162-97	10′-1″	9′-5″	8′-11″	8′-6″	8′-0″	10′-0″	9′-5″	8′-10″	8′-5″	7′-11″
2-1200S162-54	—	—	—	—	—	—	—	—	—	—
2-1200S162-68	7′-4″	6′-8″	6′-1″	5′-6″	5′-1″	7′-3″	6′-7″	6′-0″	5′-6″	5′-0″
2-1200S162-97	9′-5″	8′-8″	8′-1″	7′-6″	7′-1″	9′-4″	8′-8″	8′-0″	7′-6″	7′-0″

For SI: 1 inch = 25.4 mm, 1 foot = 304.8 mm, 1 pound per square foot = 0.0479 kPa, 1 pound per square inch = 6.895 kPa, 1 Ksi = 1,000 psi = 6.895 MPa.

a. Deflection criterion: *L*/360 for live loads, *L*/240 for total loads.

b. Design load assumptions:

Second floor dead load is 10 psf.

Roof/ceiling dead load is 12 psf.

Second floor live load is 40 psf.

Third floor live load is 30 psf.

Attic live load is 10 psf.

c. Building width is in the direction of horizontal framing members supported by the header.

TABLE R603.6(23)
BACK-TO-BACK HEADER SPANS
Headers Supporting Two Floors, Roof and Ceiling
(33 Ksi steel)[a, b]

MEMBER DESIGNATION	GROUND SNOW LOAD (50 psf)					GROUND SNOW LOAD (70 psf)				
	Building width[c] (feet)					Building width[c] (feet)				
	24	28	32	36	40	24	28	32	36	40
2-350S162-33	—	—	—	—	—	—	—	—	—	—
2-350S162-43	—	—	—	—	—	—	—	—	—	—
2-350S162-54	—	—	—	—	—	—	—	—	—	—
2-350S162-68	2′-2″	—	—	—	—	—	—	—	—	—
2-350S162-97	3′-3″	3′-0″	2′-8″	2′-4″	2′-1″	3′-1″	2′-9″	2′-6″	2′-2″	—
2-550S162-33	—	—	—	—	—	—	—	—	—	—
2-550S162-43	—	—	—	—	—	—	—	—	—	—
2-550S162-54	2′-2″	—	—	—	—	—	—	—	—	—
2-550S162-68	3′-6″	3′-0″	2′-6″	2′-1″	—	3′-2″	2′-9″	2′-3″	—	—
2-550S162-97	5′-0″	4′-6″	4′-1″	3′-9″	3′-5″	4′-8″	4′-3″	3′-11″	3′-7″	3′-3″
2-800S162-33	—	—	—	—	—	—	—	—	—	—
2-800S162-43	—	—	—	—	—	—	—	—	—	—
2-800S162-54	3′-0″	2′-3″	—	—	—	2′-7″	—	—	—	—
2-800S162-68	4′-9″	4′-2″	3′-7″	3′-1″	2′-7″	4′-5″	3′-10″	3′-3″	2′-9″	2′-3″
2-800S162-97	6′-9″	6′-1″	5′-7″	5′-2″	4′-9″	6′-4″	5′-10″	5′-4″	4′-11″	4′-7″
2-1000S162-43	—	—	—	—	—	—	—	—	—	—
2-1000S162-54	3′-6″	2′-8″	—	—	—	3′-1″	2′-2″	—	—	—
2-1000S162-68	5′-6″	4′-10″	4′-2″	3′-7″	3′-1″	5′-1″	4′-6″	3′-10″	3′-4″	2′-9″
2-1000S162-97	8′-0″	7′-4″	6′-9″	6′-3″	5′-9″	7′-7″	7′-0″	6′-5″	5′-11″	5′-6″
2-1200S162-54	3′-11″	3′-0″	2′-0″	—	—	3′-5″	2′-6″	—	—	—
2-1200S162-68	6′-2″	5′-5″	4′-9″	4′-1″	3′-6″	5′-9″	5′-0″	4′-4″	3′-9″	3′-2″
2-1200S162-97	9′-1″	8′-4″	7′-8″	7′-1″	6′-7″	8′-8″	7′-11″	7′-4″	6′-9″	6′-3″

For SI: 1 inch = 25.4 mm, 1 foot = 304.8 mm, 1 pound per square foot = 0.0479 kPa, 1 pound per square inch = 6.895 kPa, 1 Ksi = 1,000 psi = 6.895 MPa.

a. Deflection criterion: *L*/360 for live loads, *L*/240 for total loads.

b. Design load assumptions:
  Second floor dead load is 10 psf.
  Roof/ceiling dead load is 12 psf.
  Second floor live load is 40 psf.
  Third floor live load is 30 psf.
  Attic live load is 10 psf.

c. Building width is in the direction of horizontal framing members supported by the header.

TABLE R603.6(24)
BACK-TO-BACK HEADER SPANS
Headers Supporting Two Floors, Roof and Ceiling
(50 Ksi steel)[a, b]

MEMBER DESIGNATION	GROUND SNOW LOAD (50 psf)					GROUND SNOW LOAD (70 psf)				
	Building width[c] (feet)					Building width[c] (feet)				
	24	28	32	36	40	24	28	32	36	40
2-350S162-33	—	—	—	—	—	—	—	—	—	—
2-350S162-43	—	—	—	—	—	—	—	—	—	—
2-350S162-54	2′-6″	2′-1″	—	—	—	2′-3″	—	—	—	—
2-350S162-68	3′-9″	3′-4″	2′-11″	2′-7″	2′-4″	3′-6″	3′-1″	2′-9″	2′-5″	2′-2″
2-350S162-97	4′-6″	4′-4″	4′-2″	3′-11″	3′-8″	4′-4″	4′-2″	4′-0″	3′-9″	3′-6″
2-550S162-33	—	—	—	—	—	—	—	—	—	—
2-550S162-43	2′-5″	—	—	—	—	—	—	—	—	—
2-550S162-54	4′-1″	3′-7″	3′-1″	2′-7″	2′-2″	3′-10″	3′-3″	2′-10″	2′-4″	—
2-550S162-68	5′-5″	4′-11″	4′-5″	4′-0″	3′-8″	5′-1″	4′-7″	4′-2″	3′-10″	3′-5″
2-550S162-97	6′-5″	6′-2″	5′-11″	5′-9″	5′-4″	6′-3″	6′-0″	5′-9″	5′-6″	5′-2″
2-800S162-33	—	—	—	—	—	—	—	—	—	—
2-800S162-43	2′-11″	2′-2″	—	—	—	2′-6″	—	—	—	—
2-800S162-54	4′-11″	4′-3″	3′-8″	3′-2″	2′-8″	4′-6″	3′-11″	3′-5″	2′-11″	2′-4″
2-800S162-68	6′-7″	5′-11″	5′-4″	4′-11″	4′-6″	6′-2″	5′-7″	5′-1″	4′-8″	4′-3″
2-800S162-97	8′-9″	8′-5″	7′-11″	7′-6″	7′-0″	8′-5″	8′-1″	7′-9″	7′-3″	6′-10″
2-1000S162-43	2′-4″	2′-1″	—	—	—	2′-2″	1′-11″	—	—	—
2-1000S162-54	4′-8″	4′-1″	3′-8″	3′-3″	3′-0″	4′-4″	3′-10″	3′-5″	3′-1″	2′-9″
2-1000S162-68	7′-6″	6′-9″	6′-2″	5′-8″	5′-2″	7′-1″	6′-5″	5′-10″	5′-4″	4′-11″
2-1000S162-97	9′-9″	9′-2″	8′-7″	8′-2″	7′-8″	9′-5″	8′-10″	8′-5″	7′-11″	7′-5″
2-1200S162-54	—	—	—	—	—	—	—	—	—	—
2-1200S162-68	7′-0″	6′-4″	5′-9″	5′-3″	4′-9″	6′-7″	6′-0″	5′-5″	5′-0″	4′-6″
2-1200S162-97	9′-1″	8′-4″	7′-9″	7′-3″	6′-9″	8′-8″	8′-0″	7′-6″	7′-0″	6′-7″

For SI: 1 inch = 25.4 mm, 1 foot = 304.8 mm, 1 pound per square foot = 0.0479 kPa, 1 pound per square inch = 6.895 kPa, 1 Ksi = 1,000 psi = 6.895 MPa.

a. Deflection criterion: *L*/360 for live loads, *L*/240 for total loads.

b. Design load assumptions:

Second floor dead load is 10 psf.

Roof/ceiling dead load is 12 psf.

Second floor live load is 40 psf.

Third floor live load is 30 psf.

Attic live load is 10 psf.

c. Building width is in the direction of horizontal framing members supported by the header.

R603.6.1 Headers in gable endwalls. Box beam and back-to-back headers in gable endwalls shall be permitted to be constructed in accordance with Section R603.6 or with the header directly above the opening in accordance with Figures R603.6.1(1) and R603.6.1(2) and the following provisions:

1. Two 362S162-33 for openings less than or equal to 4 feet (1219 mm).
2. Two 600S162-43 for openings greater than 4 feet (1219 mm) but less than or equal to 6 feet (1830 mm).
3. Two 800S162-54 for openings greater than 6 feet (1829 mm) but less than or equal to 9 feet (2743 mm).

R603.7 Jack and king studs. The number of jack and king studs installed on each side of a header shall comply with Table R603.7(1). King, jack and cripple studs shall be of the same dimension and thickness as the adjacent wall studs. Headers shall be connected to king studs in accordance with Table R603.7(2) and the following provisions:

1. For box beam headers, one-half of the total number of required screws shall be applied to the header and one half to the king stud by use of C-shaped or track member in accordance with Figure R603.6(1). The track or C-shape sections shall extend the depth of the header minus $^1/_2$ inch (12.7 mm) and shall have a minimum thickness not less than that of the wall studs.
2. For back-to-back headers, one-half the total number of screws shall be applied to the header and one-half to the king stud by use of a minimum 2-inch by 2-inch (51 mm by 51 mm) clip angle in accordance with Figure R603.6(2). The clip angle shall extend the depth of the header minus $^1/_2$ inch (12.7 mm) and shall have a minimum thickness not less than that of the wall studs. Jack and king studs shall be interconnected with structural sheathing in accordance with Figures R603.6(1) and R603.6(2).

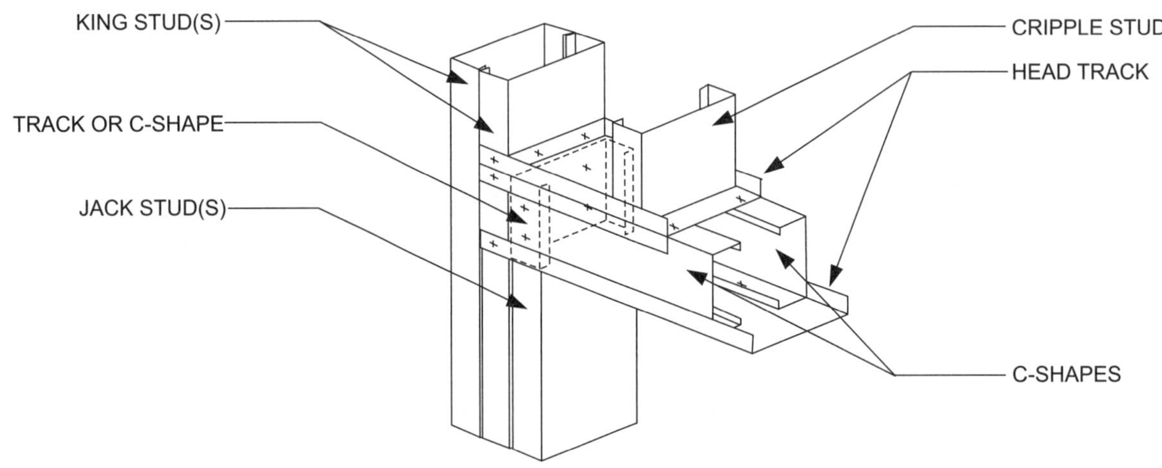

FIGURE R603.6.1(1)
BOX BEAM HEADER IN GABLE ENDWALL

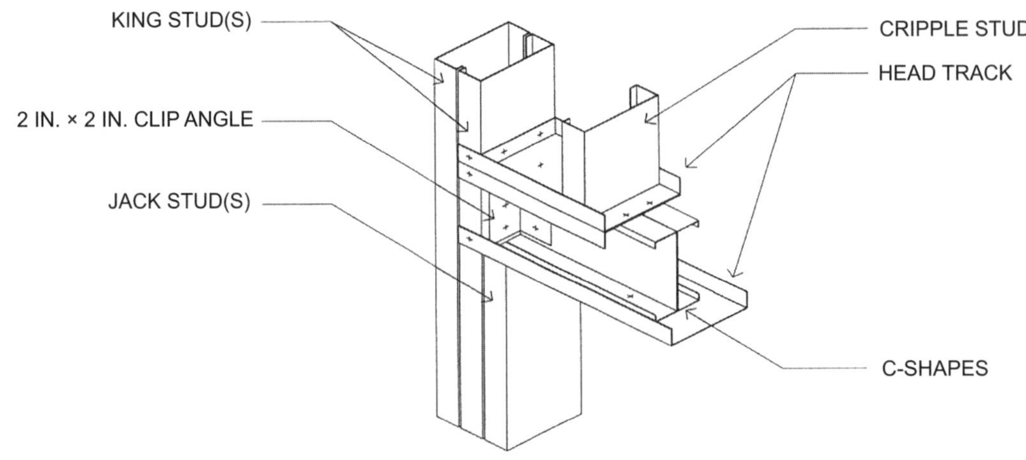

For SI: 1 inch = 25.4 mm.

FIGURE R603.6.1(2)
BACK-TO-BACK HEADER IN GABLE ENDWALL

**TABLE R603.7(1)
TOTAL NUMBER OF JACK AND KING STUDS REQUIRED AT EACH END OF AN OPENING**

SIZE OF OPENING (feet-inches)	24-INCH O.C. STUD SPACING		16-INCH O.C. STUD SPACING	
	No. of jack studs	No. of king studs	No. of jack studs	No. of king studs
Up to 3′-6″	1	1	1	1
> 3′-6″ to 5′-0″	1	2	1	2
> 5′-0″ to 5′-6″	1	2	2	2
> 5′-6″ to 8′-0″	1	2	2	2
> 8′-0″ to 10′-6″	2	2	2	3
> 10′-6″ to 12′-0″	2	2	3	3
> 12′-0″ to 13′-0″	2	3	3	3
> 13′-0″ to 14′-0″	2	3	3	4
> 14′-0″ to 16′-0″	2	3	3	4
> 16′-0″ to 18′-0″	3	3	4	4

For SI: 1 inch = 25.4 mm, 1 foot = 304.8 mm.

**TABLE R603.7(2)
HEADER TO KING STUD CONNECTION REQUIREMENTS[a, b, c, d]**

HEADER SPAN (feet)	BASIC WIND SPEED (mph), EXPOSURE		
	85 B or Seismic Design Categories A, B, C, D_0, D_1 and D_2	85 C or less than 110 B	Less than 110 C
≤ 4′	4-No. 8 screws	4-No. 8 screws	6-No. 8 screws
> 4′ to 8′	4-No. 8 screws	4-No. 8 screws	8-No. 8 screws
> 8′ to 12′	4-No. 8 screws	6-No. 8 screws	10-No. 8 screws
> 12′ to 16′	4-No. 8 screws	8-No. 8 screws	12-No. 8 screws

For SI: 1 inch = 25.4 mm, 1 foot = 304.8 mm, 1 mile per hour = 0.447 m/s, 1 pound = 4.448 N.

a. All screw sizes shown are minimum.

b. For headers located on the first floor of a two-story building or the first or second floor of a three-story building, the total number of screws is permitted to be reduced by 2 screws, but the total number of screws shall be no less than 4.

c. For roof slopes of 6:12 or greater, the required number of screws may be reduced by half, but the total number of screws shall be no less than four.

d. Screws can be replaced by an uplift connector which has a capacity of the number of screws multiplied by 164 pounds (e.g., 12-No. 8 screws can be replaced by an uplift connector whose capacity exceeds 12 × 164 pounds = 1,968 pounds).

R603.8 Head and sill track. Head track spans above door and window openings and sill track spans beneath window openings shall comply with Table R603.8. For openings less than 4 feet (1219 mm) in height that have both a head track and a sill track, multiplying the spans by 1.75 shall be permitted in Table R603.8. For openings less than or equal to 6 feet (1829 mm) in height that have both a head track and a sill track, multiplying the spans in Table R603.8 by 1.50 shall be permitted.

R603.9 Structural sheathing. Structural sheathing shall be installed in accordance with Figure R603.9 and this section on all sheathable exterior wall surfaces, including areas above and below openings.

R603.9.1 Sheathing materials. Structural sheathing panels shall consist of minimum $^{7}/_{16}$-inch-thick (11 mm) oriented strand board or $^{15}/_{32}$-inch-thick (12 mm) plywood.

R603.9.2 Determination of minimum length of full height sheathing. The minimum length of full height sheathing on each *braced wall line* shall be determined by multiplying the length of the *braced wall line* by the percentage obtained from Table R603.9.2(1) and by the plan aspect-ratio adjustment factors obtained from Table

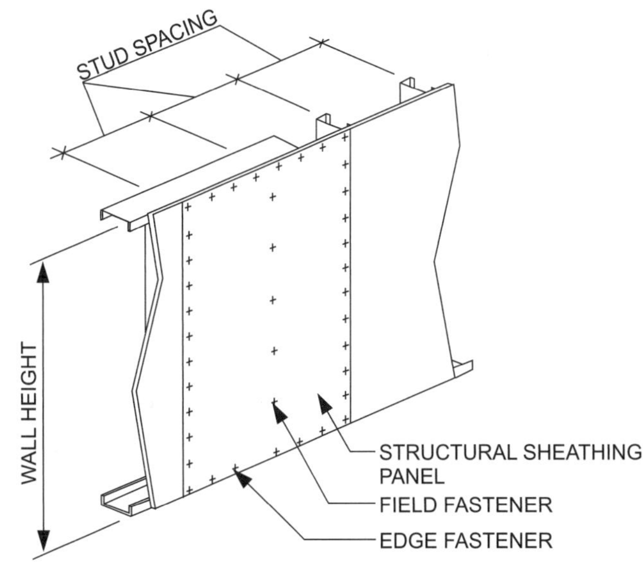

**FIGURE R603.9
STRUCTURAL SHEATHING FASTENING PATTERN**

R603.9.2(2). The minimum length of full height sheathing shall not be less than 20 percent of the *braced wall line* length.

To be considered full height sheathing, structural sheathing shall extend from the bottom to the top of the wall without interruption by openings. Only sheathed, full height wall sections, uninterrupted by openings, which are a minimum of 48 inches (1219 mm) wide, shall be counted toward meeting the minimum percentages in Table R603.9.2(1). In addition, structural sheathing shall comply with all of the following requirements:

1. Be installed with the long dimension parallel to the stud framing (i.e., vertical orientation) and shall cover the full vertical height of wall from the bottom of the bottom track to the top of the top track of each *story*. Installing the long dimension perpendicular to the stud framing or using shorter segments shall be permitted provided that the horizontal joint is blocked as described in Item 2.
2. Be blocked when the long dimension is installed perpendicular to the stud framing (i.e., horizontal orientation). Blocking shall be a minimum of 33 mil (0.84 mm) thickness. Each horizontal structural sheathing panel shall be fastened with No. 8 screws spaced at 6 inches (152 mm) on center to the blocking at the joint.

TABLE R603.8
HEAD AND SILL TRACK SPAN F_y = 33 KSI

BASIC WIND SPEED (mph)		ALLOWABLE HEAD AND SILL TRACK SPAN[a, b, c] (feet-inches)					
EXPOSURE		TRACK DESIGNATION					
B	C	350T125-33	350T125-43	350T125-54	550T125-33	550T125-43	550T125-54
85	—	5′-0″	5′-7″	6′-2″	5′-10″	6′-8″	7′-0″
90	—	4′-10″	5′-5″	6′-0″	5′-8″	6′-3″	6′-10″
100	85	4′-6″	5′-1″	5′-8″	5′-4″	5′-11″	6′-5″
110	90	4′-2″	4′-9″	5′-4″	5′-1″	5′-7″	6′-1″
120	100	3′-11″	4′-6″	5′-0″	4′-10″	5′-4″	5′-10″
130	110	3′-8″	4′-2″	4′-9″	4′-1″	5′-1″	5′-7″
140	120	3′-7″	4′-1″	4′-7″	3′-6″	4′-11″	5′-5″
150	130	3′-5″	3′-10″	4′-4″	2′-11″	4′-7″	5′-2″
—	140	3′-1″	3′-6″	4′-1″	2′-3″	4′-0″	4′-10″
—	150	2′-9″	3′-4″	3′-10″	2′-0″	3′-7″	4′-7″

For SI: 1 inch = 25.4 mm, 1 foot = 304.8 mm, 1 mile per hour = 0.447 m/s.

a. Deflection limit: *L*/240.

b. Head and sill track spans are based on components and cladding wind speeds and 48-inch tributary span.

c. For openings less than 4 feet in height that have both a head track and sill track, the above spans are permitted to be multiplied by 1.75. For openings less than or equal to 6 feet in height that have both a head track and a sill track, the above spans are permitted to be multiplied by a factor of 1.5.

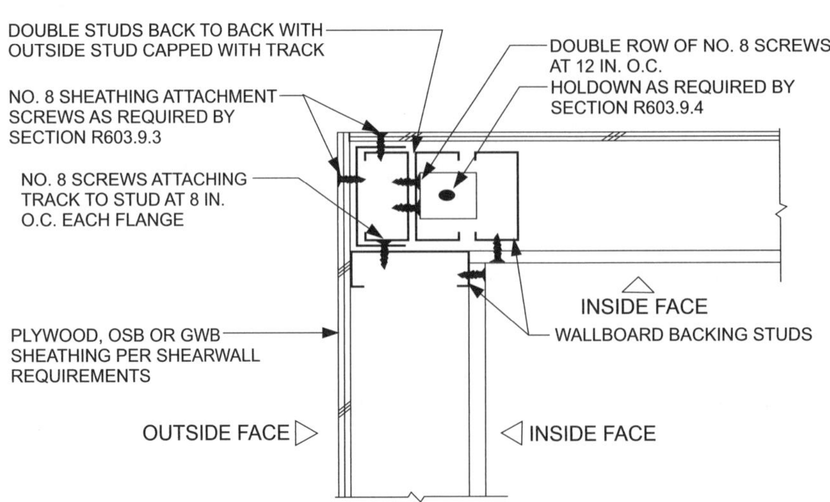

For SI: 1 inch = 25.4 mm.

FIGURE R603.9.2
CORNER STUD HOLD-DOWN DETAIL

3. Be applied to each end (corners) of each of the exterior walls with a minimum 48-inch-wide (1219 mm) panel.

R603.9.2.1 Full height sheathing. The minimum percentage of full-height structural sheathing shall be multiplied by 1.10 for 9-foot-high (2743 mm) walls and multiplied by 1.20 for 10-foot-high (3048 mm) walls.

R603.9.2.2 Full height sheathing in hip roof homes. For hip roofed homes, the minimum percentages of full height sheathing in Table R603.9.2(1), based upon wind, shall be permitted to be multiplied by a factor of 0.95 for roof slopes not exceeding 7:12 and a factor of 0.9 for roof slopes greater than 7:12.

R603.9.2.3 Full height sheathing in lowest story. In the lowest *story* of a *dwelling*, multiplying the percentage of full height sheathing required in Table R603.9.2(1) by 0.6, shall be permitted provided hold down anchors are provided in accordance with Section R603.9.4.2.

R603.9.3 Structural sheathing fastening. All edges and interior areas of structural sheathing panels shall be fastened to framing members and tracks in accordance with Figure R603.9 and Table R603.3.2(1). Screws for attachment of structural sheathing panels shall be bugle-head, flat-head, or similar head style with a minimum head diameter of 0.29 inch (8 mm).

For continuously-sheathed *braced wall lines* using wood structural panels installed with No. 8 screws spaced 4-inches (102 mm) on center at all panel edges and 12 inches (304.8 mm) on center on intermediate framing members, the following shall apply:

1. Multiplying the percentages of full height sheathing in Table R603.9.2(1) by 0.72 shall be permitted.
2. For bottom track attached to foundations or framing below, the bottom track anchor or screw connection spacing in Table R505.3.1(1) and Table R603.3.1 shall be multiplied by two-thrids.

R603.9.4 Uplift connection requirements. Uplift connections shall be provided in accordance with this section.

R603.9.4.1 Wind speeds greater than 100 mph. Where wind speeds are in excess of 100 miles per hour (45 m/s), Exposure C, walls shall be provided wind direct uplift connections in accordance with AISI S230,

TABLE R603.9.2(1)
MINIMUM PERCENTAGE OF FULL HEIGHT STRUCTURAL SHEATHING ON EXTERIOR WALLS[a, b]

WALL SUPPORTING	ROOF SLOPE	BASIC WIND SPEED AND EXPOSURE (mph)					
		85 B	90 B	100 B / 85 C	< 110 B / 90 C	100 C	< 110 C
Roof and ceiling only (one story or top floor of two- or three-story building).	3:12	8	9	9	12	16	20
	6:12	12	13	15	20	26	35
	9:12	21	23	25	30	50	58
	12:12	30	33	35	40	66	75
One story, roof and ceiling (first floor of a two-story building or second floor of a three-story building).	3:12	24	27	30	35	50	66
	6:12	25	28	30	40	58	74
	9:12	35	38	40	55	74	91
	12:12	40	45	50	65	100	115
Two story, roof and ceiling (first floor of a three-story building).	3:12	40	45	51	58	84	112
	6:12	38	43	45	60	90	113
	9:12	49	53	55	80	98	124
	12:12	50	57	65	90	134	155

For SI: 1 mile per hour = 0.447 m/s.

a. Linear interpolation is permitted.

b. For hip-roofed homes the minimum percentage of full height sheathing, based upon wind, is permitted to be multiplied by a factor of 0.95 for roof slopes not exceeding 7:12 and a factor of 0.9 for roof slopes greater than 7:12.

TABLE R603.9.2(2)
FULL HEIGHT SHEATHING LENGTH ADJUSTMENT FACTORS

PLAN ASPECT RATIO	LENGTH ADJUSTMENT FACTORS	
	Short wall	Long wall
1:1	1.0	1.0
1.5:1	1.5	0.67
2:1	2.0	0.50
3:1	3.0	0.33
4:1	4.0	0.25

Section E13.3, and AISI S230, Section F7.2, as required for 110 miles per hour (49 m/s), Exposure C.

R603.9.4.2 Hold-down anchor. Where the percentage of full height sheathing is adjusted in accordance with Section R603.9.2.3, a hold-down anchor, with a strength of 4,300 pounds (19 kN), shall be provided at each end of each full-height sheathed wall section used to meet the minimum percent sheathing requirements of Section R603.9.2. Hold-down anchors shall be attached to back-to-back studs; structural sheathing panels shall have edge fastening to the studs, in accordance with Section R603.9.3 and AISI S230, Table E11-1.

A single hold-down anchor, installed in accordance with Figure R603.9.2, shall be permitted at the corners of buildings.

R603.9.5 Structural sheathing for stone and masonry veneer. In Seismic Design Category C, where stone and masonry veneer is installed in accordance with Section R703.7, the length of structural sheathing for walls supporting one *story*, roof and ceiling shall be the greater of the amount required by Section R603.9.2 or 36 percent, modified by Section R603.9.2 except Section R603.9.2.2 shall not be permitted.

SECTION R604 WOOD STRUCTURAL PANELS

R604.1 Identification and grade. Wood structural panels shall conform to DOC PS 1, DOC PS 2 or ANSI/APA PRP 210 or, when manufactured in Canada, CSA 0437 or CSA 0325. All panels shall be identified by a grade mark or certificate of inspection issued by an *approved* agency.

R604.2 Allowable spans. The maximum allowable spans for wood structural panel wall sheathing shall not exceed the values set forth in Table R602.3(3).

R604.3 Installation. Wood structural panel wall sheathing shall be attached to framing in accordance with Table R602.3(1) or R602.3(3). Wood structural panels marked Exposure 1 or Exterior are considered water-repellent sheathing under the code.

SECTION R605 PARTICLEBOARD

R605.1 Identification and grade. Particleboard shall conform to ANSI A208.1 and shall be so identified by a grade mark or certificate of inspection issued by an *approved* agency. Particleboard shall comply with the grades specified in Table R602.3(4).

SECTION R606 GENERAL MASONRY CONSTRUCTION

R606.1 General. Masonry construction shall be designed and constructed in accordance with the provisions of this section, TMS 403 or in accordance with the provisions of TMS 402/ACI 530/ASCE 5.

R606.1.1 Professional registration not required. When the empirical design provisions of Chapter 5 of TMS 402/ACI 530/ASCE 5, the provisions of TMS 403, or the provisions of this section are used to design masonry, project drawings, typical details and specifications are not required to bear the seal of the architect or engineer responsible for design, unless otherwise required by the state law of the *jurisdiction* having authority.

R606.2 Thickness of masonry. The nominal thickness of masonry walls shall conform to the requirements of Sections R606.2.1 through R606.2.4.

R606.2.1 Minimum thickness. The minimum thickness of masonry bearing walls more than one *story* high shall be 8 inches (203 mm). *Solid masonry* walls of one-story *dwellings* and garages shall not be less than 6 inches (152 mm) in thickness when not greater than 9 feet (2743 mm) in height, provided that when gable construction is used, an additional 6 feet (1829 mm) is permitted to the peak of the gable. Masonry walls shall be laterally supported in either the horizontal or vertical direction at intervals as required by Section R606.9.

R606.2.2 Rubble stone masonry wall. The minimum thickness of rough, random or coursed rubble stone masonry walls shall be 16 inches (406 mm).

R606.2.3 Change in thickness. Where walls of masonry of hollow units or masonry-bonded hollow walls are decreased in thickness, a course of *solid masonry* shall be constructed between the wall below and the thinner wall above, or special units or construction shall be used to transmit the loads from face shells or wythes above to those below.

R606.2.4 Parapet walls. Unreinforced *solid masonry* parapet walls shall not be less than 8 inches (203 mm) thick and their height shall not exceed four times their thickness. Unreinforced hollow unit masonry parapet walls shall be not less than 8 inches (203 mm) thick, and their height shall not exceed three times their thickness. Masonry parapet walls in areas subject to wind loads of 30 pounds per square foot (1.44 kPa) located in Seismic Design Category D_0, D_1 or D_2, or on townhouses in Seismic Design Category C shall be reinforced in accordance with Section R606.12.

R606.3 Corbeled masonry. Corbeled masonry shall be in accordance with Sections R606.3.1 through R606.3.3.

R606.3.1 Units. *Solid masonry* units or masonry units filled with mortar or grout shall be used for corbeling.

R606.3.2 Corbel projection. The maximum projection of one unit shall not exceed one-half the height of the unit or one-third the thickness at right angles to the wall. The maximum corbeled projection beyond the face of the wall shall not exceed:

1. One-half of the wall thickness for multiwythe walls bonded by mortar or grout and wall ties or masonry headers, or
2. One-half the wythe thickness for single wythe walls, masonry-bonded hollow walls, multiwythe walls with open collar joints and veneer walls.

R606.3.3 Corbeled masonry supporting floor or roof-framing members. When corbeled masonry is used to support floor or roof-framing members, the top course of the corbel shall be a header course or the top course bed joint shall have ties to the vertical wall.

R606.4 Support conditions. Bearing and support conditions shall be in accordance with Sections R606.4.1 and R606.4.2.

R606.4.1 Bearing on support. Each masonry wythe shall be supported by at least two-thirds of the wythe thickness.

R606.4.2 Support at foundation. Cavity wall or masonry veneer construction may be supported on an 8-inch (203 mm) foundation wall, provided the 8-inch (203 mm) wall is corbeled to the width of the wall system above with masonry constructed of *solid masonry* units or masonry units filled with mortar or grout. The total horizontal projection of the corbel shall not exceed 2 inches (51 mm) with individual corbels projecting not more than one-third the thickness of the unit or one-half the height of the unit. The hollow space behind the corbeled masonry shall be filled with mortar or grout.

R606.5 Allowable stresses. Allowable compressive stresses in masonry shall not exceed the values prescribed in Table R606.5. In determining the stresses in masonry, the effects of all loads and conditions of loading and the influence of all forces affecting the design and strength of the several parts shall be taken into account.

R606.5.1 Combined units. In walls or other structural members composed of different kinds or grades of units, materials or mortars, the maximum stress shall not exceed the allowable stress for the weakest of the combination of units, materials and mortars of which the member is composed. The net thickness of any facing unit that is used to resist stress shall not be less than 1.5 inches (38 mm).

R606.6 Piers. The unsupported height of masonry piers shall not exceed ten times their least dimension. When structural clay tile or hollow concrete masonry units are used for isolated piers to support beams and girders, the cellular spaces shall be filled solidly with concrete or Type M or S mortar, except that unfilled hollow piers may be used if their unsupported height is not more than four times their least dimension. Where hollow masonry units are solidly filled with concrete or Type M, S or N mortar, the allowable compressive stress shall be permitted to be increased as provided in Table R606.5.

R606.6.1 Pier cap. Hollow piers shall be capped with 4 inches (102 mm) of *solid masonry* or concrete, a masonry cap block, or shall have cavities of the top course filled with concrete or grout.

R606.7 Chases. Chases and recesses in masonry walls shall not be deeper than one-third the wall thickness, and the maximum length of a horizontal chase or horizontal projection shall not exceed 4 feet (1219 mm), and shall have at least 8 inches (203 mm) of masonry in back of the chases and recesses and between adjacent chases or recesses and the jambs of openings. Chases and recesses in masonry walls shall be designed and constructed so as not to reduce the required strength or required fire resistance of the wall and in no case shall a chase or recess be permitted within the required area of a pier. Masonry directly above chases or recesses wider than 12 inches (305 mm) shall be supported on noncombustible lintels.

TABLE R606.5
ALLOWABLE COMPRESSIVE STRESSES FOR EMPIRICAL DESIGN OF MASONRY

CONSTRUCTION; COMPRESSIVE STRENGTH OF UNIT, GROSS AREA	ALLOWABLE COMPRESSIVE STRESSES[a] GROSS CROSS-SECTIONAL AREA[b]	
	Type M or S mortar	Type N mortar
Solid masonry of brick and other solid units of clay or shale; sand-lime or concrete brick:		
8,000 + psi	350	300
4,500 psi	225	200
2,500 psi	160	140
1,500 psi	115	100
Grouted[c] masonry, of clay or shale; sand-lime or concrete:		
4,500 + psi	225	200
2,500 psi	160	140
1,500 psi	115	100
Solid masonry of solid concrete masonry units:		
3,000 + psi	225	200
2,000 psi	160	140
1,200 psi	115	100
Masonry of hollow load-bearing units:		
2,000 + psi	140	120
1,500 psi	115	100
1,000 psi	75	70
700 psi	60	55
Hollow walls (cavity or masonry bonded[d]) solid units:		
2,500 + psi	160	140
1,500 psi	115	100
Hollow units	75	70
Stone ashlar masonry:		
Granite	720	640
Limestone or marble	450	400
Sandstone or cast stone	360	320
Rubble stone masonry:		
Coarse, rough or random	120	100

For SI: 1 pound per square inch = 6.895 kPa.

a. Linear interpolation shall be used for determining allowable stresses for masonry units having compressive strengths that are intermediate between those given in the table.

b. Gross cross-sectional area shall be calculated on the actual rather than nominal dimensions.

c. See Section R608.

d. Where floor and roof loads are carried upon one wythe, the gross cross-sectional area is that of the wythe under load; if both wythes are loaded, the gross cross-sectional area is that of the wall minus the area of the cavity between the wythes. Walls bonded with metal ties shall be considered as cavity walls unless the collar joints are filled with mortar or grout.

R606.8 Stack bond. In unreinforced masonry where masonry units are laid in stack bond, longitudinal reinforcement consisting of not less than two continuous wires each with a minimum aggregate cross-sectional area of 0.017 square inch (11 mm^2) shall be provided in horizontal bed joints spaced not more than 16 inches (406 mm) on center vertically.

R606.9 Lateral support. Masonry walls shall be laterally supported in either the horizontal or the vertical direction. The maximum spacing between lateral supports shall not exceed the distances in Table R606.9. Lateral support shall be provided by cross walls, pilasters, buttresses or structural frame members when the limiting distance is taken horizontally, or by floors or roofs when the limiting distance is taken vertically.

**TABLE R606.9
SPACING OF LATERAL SUPPORT FOR MASONRY WALLS**

CONSTRUCTION	MAXIMUM WALL LENGTH TO THICKNESS OR WALL HEIGHT TO THICKNESS[a, b]
Bearing walls:	
Solid or solid grouted	20
All other	18
Nonbearing walls:	
Exterior	18
Interior	36

For SI: 1 foot = 304.8 mm.

a. Except for cavity walls and cantilevered walls, the thickness of a wall shall be its nominal thickness measured perpendicular to the face of the wall. For cavity walls, the thickness shall be determined as the sum of the nominal thicknesses of the individual wythes. For cantilever walls, except for parapets, the ratio of height to nominal thickness shall not exceed 6 for solid masonry, or 4 for hollow masonry. For parapets, see Section R606.2.4.

b. An additional unsupported height of 6 feet is permitted for gable end walls.

R606.9.1 Horizontal lateral support. Lateral support in the horizontal direction provided by intersecting masonry walls shall be provided by one of the methods in Section R606.9.1.1 or Section R606.9.1.2.

R606.9.1.1 Bonding pattern. Fifty percent of the units at the intersection shall be laid in an overlapping masonry bonding pattern, with alternate units having a bearing of not less than 3 inches (76 mm) on the unit below.

R606.9.1.2 Metal reinforcement. Interior nonloadbearing walls shall be anchored at their intersections, at vertical intervals of not more than 16 inches (406 mm) with joint reinforcement of at least 9 gage [0.148 inch (4mm)], or $^1/_4$-inch (6 mm) galvanized mesh hardware cloth. Intersecting masonry walls, other than interior nonloadbearing walls, shall be anchored at vertical intervals of not more than 8 inches (203 mm) with joint reinforcement of at least 9 gage and shall extend at least 30 inches (762 mm) in each direction at the intersection. Other metal ties, joint reinforcement or anchors, if used, shall be spaced to provide equivalent area of anchorage to that required by this section.

R606.9.2 Vertical lateral support. Vertical lateral support of masonry walls in Seismic Design Category A, B or C shall be provided in accordance with one of the methods in Section R606.9.2.1 or Section R606.9.2.2.

R606.9.2.1 Roof structures. Masonry walls shall be anchored to roof structures with metal strap anchors spaced in accordance with the manufacturer's instructions, $^1/_2$-inch (13 mm) bolts spaced not more than 6 feet (1829 mm) on center, or other *approved* anchors. Anchors shall be embedded at least 16 inches (406 mm) into the masonry, or be hooked or welded to bond beam reinforcement placed not less than 6 inches (152 mm) from the top of the wall.

R606.9.2.2 Floor diaphragms. Masonry walls shall be anchored to floor *diaphragm* framing by metal strap anchors spaced in accordance with the manufacturer's instructions, $^1/_2$-inch-diameter (13 mm) bolts spaced at intervals not to exceed 6 feet (1829 mm) and installed as shown in Figure R606.11(1), or by other *approved* methods.

R606.10 Lintels. Masonry over openings shall be supported by steel lintels, reinforced concrete or masonry lintels or masonry arches, designed to support load imposed.

R606.11 Anchorage. Masonry walls shall be anchored to floor and roof systems in accordance with the details shown in Figure R606.11(1), R606.11(2) or R606.11(3). Footings may be considered as points of lateral support.

R606.12 Seismic requirements. The seismic requirements of this section shall apply to the design of masonry and the construction of masonry building elements located in Seismic Design Category D_0, D_1 or D_2. Townhouses in Seismic Design Category C shall comply with the requirements of Section R606.12.2. These requirements shall not apply to glass unit masonry conforming to Section R610 or masonry veneer conforming to Section R703.7.

R606.12.1 General. Masonry structures and masonry elements shall comply with the requirements of Sections R606.12.2 through R606.12.4 based on the seismic design category established in Table R301.2(1). Masonry structures and masonry elements shall comply with the requirements of Section R606.12 and Figures R606.11(1), R606.11(2) and R606.11(3) or shall be designed in accordance with TMS 402/ACI 530/ASCE 5 or TMS 403.

R606.12.1.1 Floor and roof diaphragm construction. Floor and roof *diaphragms* shall be constructed of wood structural panels attached to wood framing in accordance with Table R602.3(1) or to cold-formed steel floor framing in accordance with Table R505.3.1(2) or to cold-formed steel roof framing in accordance with Table R804.3. Additionally, sheathing panel edges perpendicular to framing members shall be backed by blocking, and sheathing shall be connected to the blocking with fasteners at the edge spacing. For Seismic Design Categories C, D_0, D_1 and D_2, where the width-to-thickness dimension of the *diaphragm* exceeds 2-to-1, edge spacing of fasteners shall be 4 inches (102 mm) on center.

R606.12.2 Seismic Design Category C. Townhouses located in Seismic Design Category C shall comply with the requirements of this section.

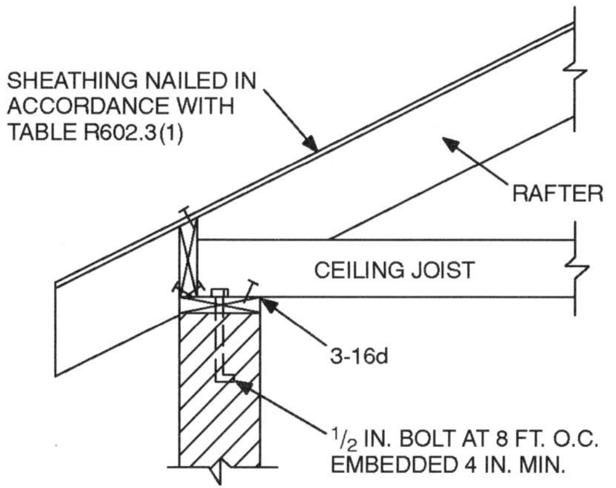

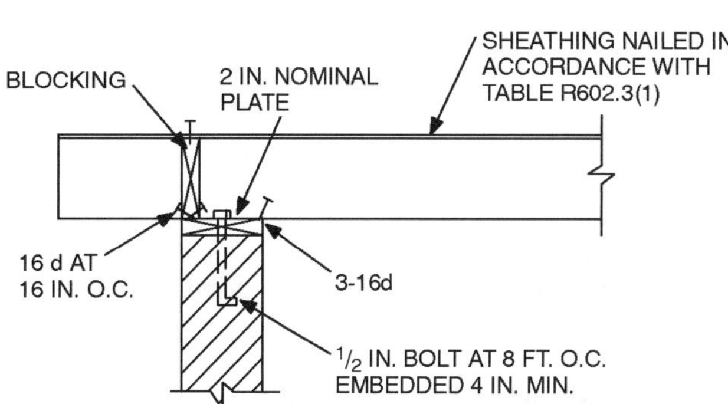

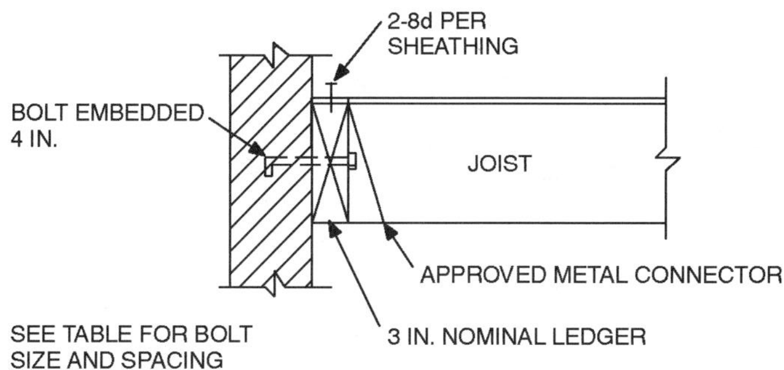

LEDGER BOLT SIZE AND SPACING

JOIST SPAN	BOLT SIZE AND SPACING	
	ROOF	FLOOR
10 FT.	1/2 AT 2 FT. 6 IN. 7/8 AT 3 FT. 6 IN.	1/2 AT 2 FT. 0 IN. 7/8 AT 2 FT. 9 IN.
10–15 FT.	1/2 AT 1 FT. 9 IN. 7/8 AT 2 FT. 6 IN.	1/2 AT 1 FT. 4 IN. 7/8 AT 2 FT. 0 IN.
15-20 FT.	1/2 AT 1 FT. 3 IN. 7/8 AT 2 FT. 0 IN.	1/2 AT 1 FT. 0 IN. 7/8 AT 1 FT. 6 IN.

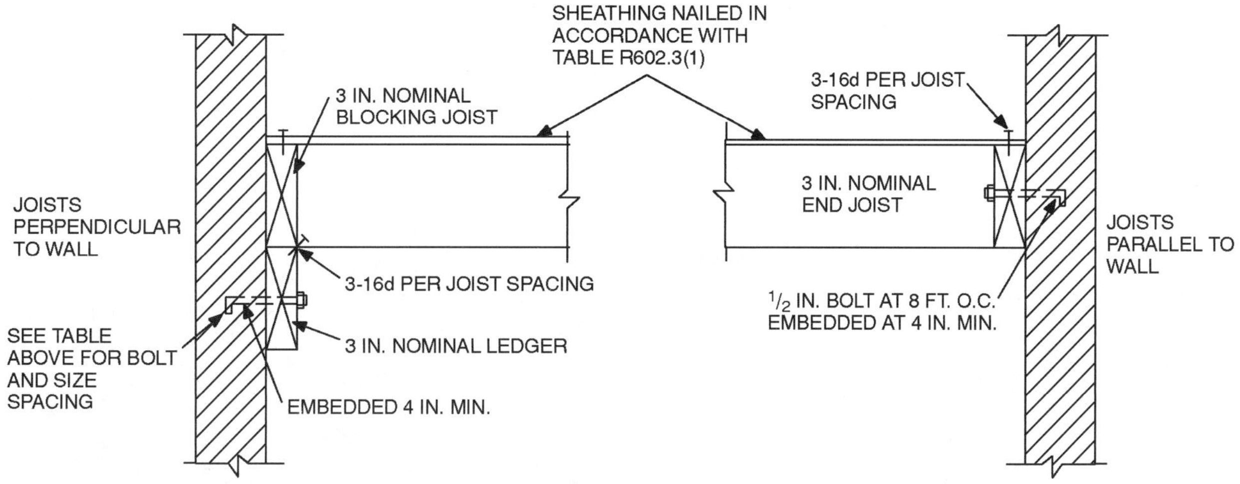

For SI: 1 inch = 25.4 mm, 1 foot = 304.8 mm, 1 pound per square foot = 0.0.479 kPa.

Note: Where bolts are located in hollow masonry, the cells in the courses receiving the bolt shall be grouted solid.

FIGURE R606.11(1)
ANCHORAGE REQUIREMENTS FOR MASONRY WALLS LOCATED IN SEISMIC DESIGN CATEGORY A, B OR C AND WHERE WIND LOADS ARE LESS THAN 30 PSF

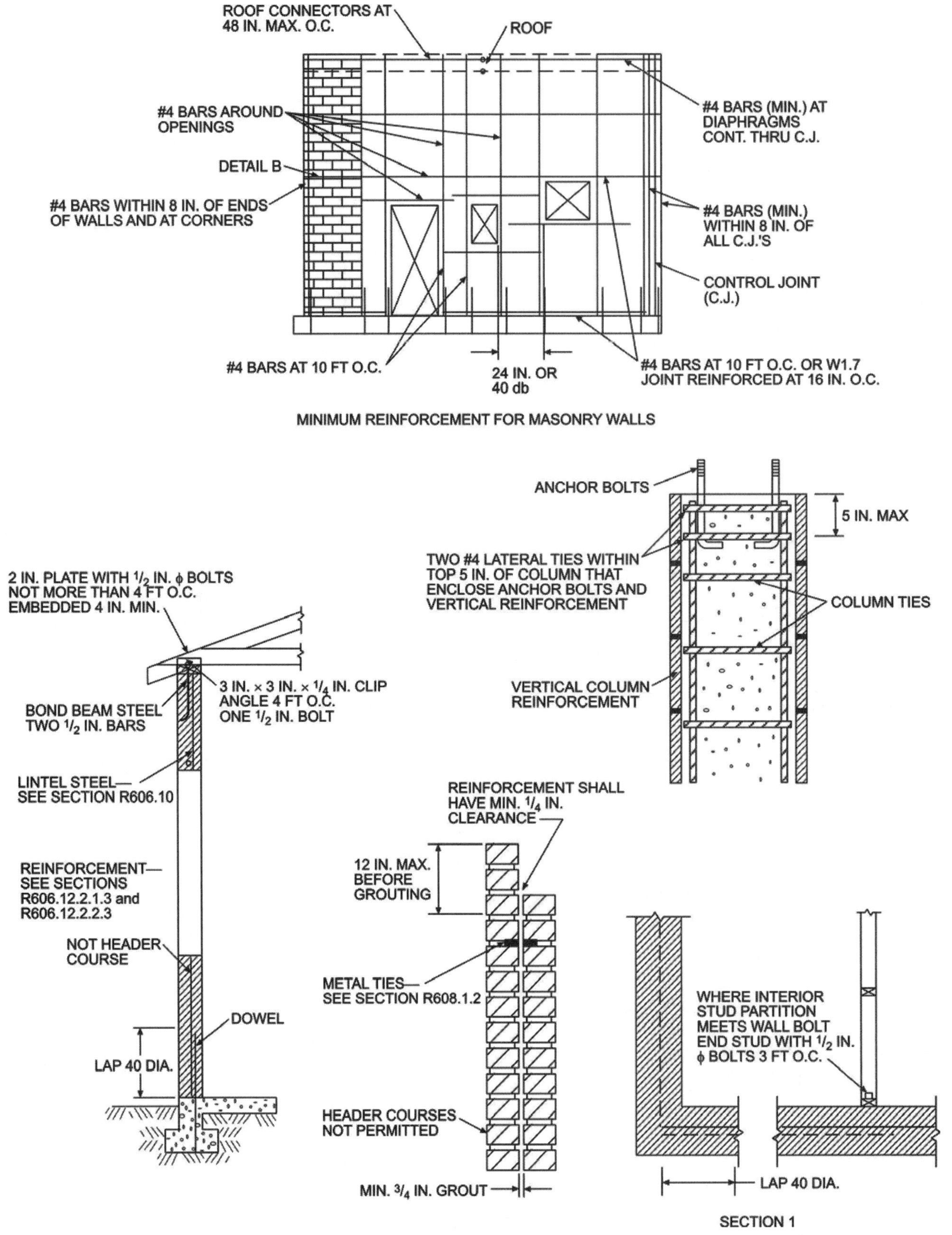

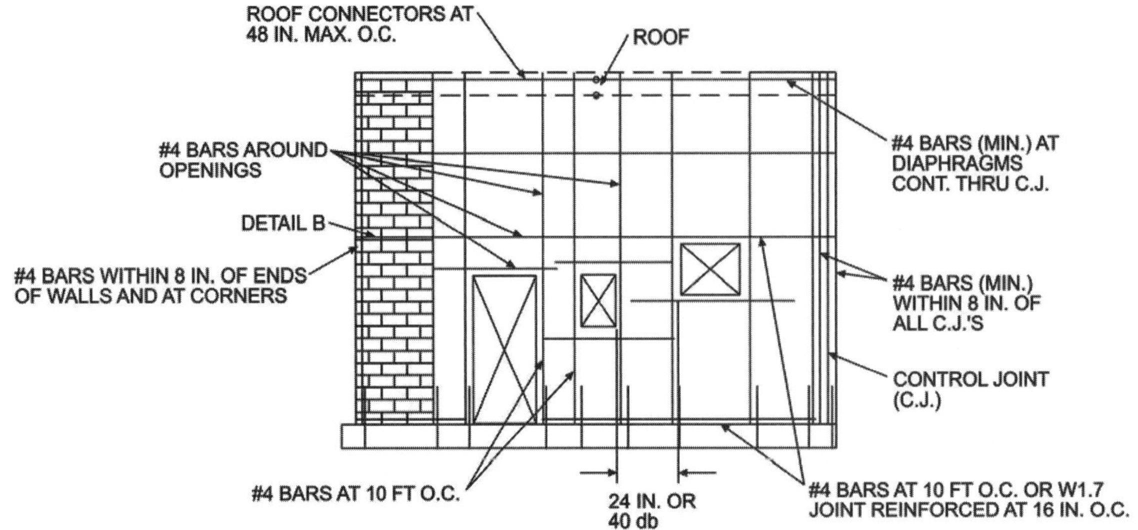

MINIMUM REINFORCEMENT FOR MASONRY WALLS

For SI: 1 inch = 25.4 mm, 1 foot = 304.8 mm.

FIGURE R606.11(2)
REQUIREMENTS FOR REINFORCED GROUTED MASONRY CONSTRUCTION IN SEISMIC DESIGN CATEGORY C

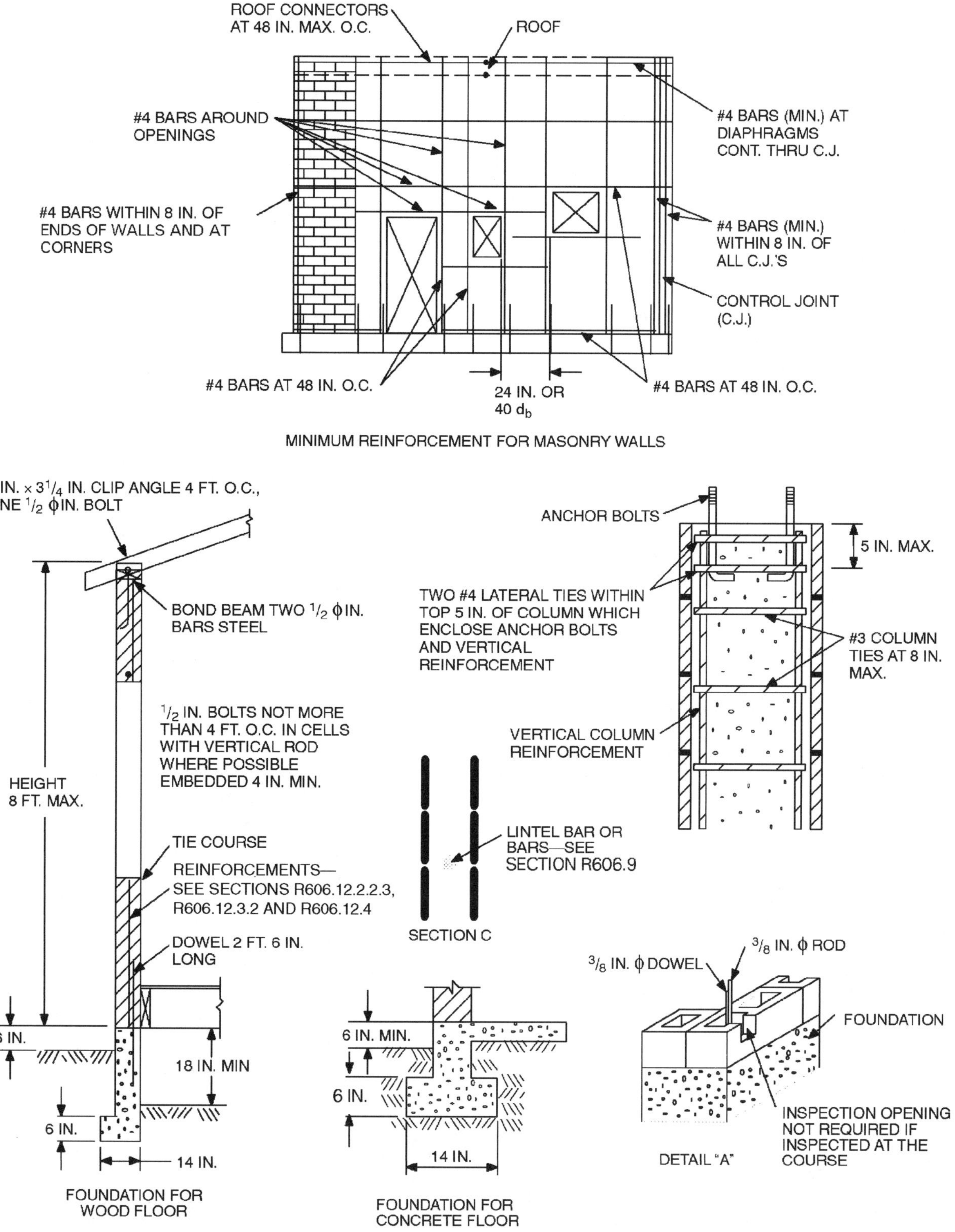

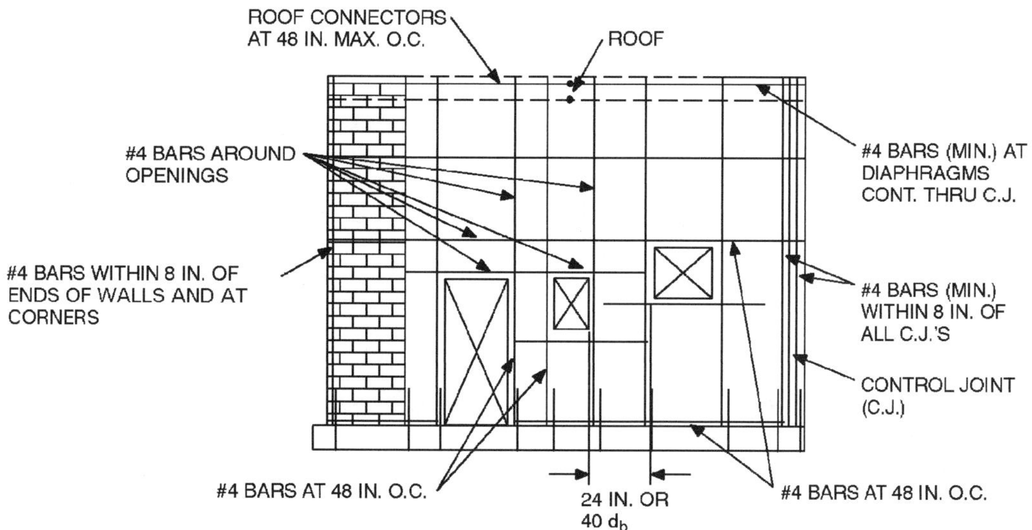

For SI: 1 inch = 25.4 mm, 1 foot = 304.8 mm.

Note: A full bed joint must be provided. All cells containing vertical bars are to be filled to the top of wall and provide inspection opening as shown on detail "A."

Horizontal bars are to be laid as shown on detail "B." Lintel bars are to be laid as shown on Section C.

FIGURE R606.11(3)
REQUIREMENTS FOR REINFORCED MASONRY CONSTRUCTION IN SEISMIC DESIGN CATEGORY D_0, D_1, OR D_2

R606.12.2.1 Minimum length of wall without openings. Table R606.12.2.1 shall be used to determine the minimum required solid wall length without openings at each masonry exterior wall. The provided percentage of solid wall length shall include only those wall segments that are 3 feet (914 mm) or longer. The maximum clear distance between wall segments included in determining the solid wall length shall not exceed 18 feet (5486 mm). Shear wall segments required to meet the minimum wall length shall be in accordance with Section R606.12.2.2.3.

R606.12.2.2 Design of elements not part of the lateral force-resisting system.

R606.12.2.2.1 Load-bearing frames or columns. Elements not part of the lateral force-resisting system shall be analyzed to determine their effect on the response of the system. The frames or columns shall be adequate for vertical load carrying capacity and induced moment caused by the design *story* drift.

R606.12.2.2.2 Masonry partition walls. Masonry partition walls, masonry screen walls and other masonry elements that are not designed to resist vertical or lateral loads, other than those induced by their own weight, shall be isolated from the structure so that vertical and lateral forces are not imparted to these elements. Isolation joints and connectors between these elements and the structure shall be designed to accommodate the design *story* drift.

R606.12.2.2.3 Reinforcement requirements for masonry elements. Masonry elements listed in Section R606.12.2.2.2 shall be reinforced in either the horizontal or vertical direction as shown in Figure R606.11(2) and in accordance with the following:

1. Horizontal reinforcement. Horizontal joint reinforcement shall consist of at least two longitudinal W1.7 wires spaced not more than 16 inches (406 mm) for walls greater than 4 inches (102 mm) in width and at least one longitudinal W1.7 wire spaced not more than 16 inches (406 mm) for walls not exceeding 4 inches (102 mm) in width; or at least one No. 4 bar spaced not more than 48 inches (1219 mm). Where two longitudinal wires of joint reinforcement are used, the space between these wires shall be the widest that the mortar joint will accommodate. Horizontal reinforcement shall be provided within 16 inches (406 mm) of the top and bottom of these masonry elements.
2. Vertical reinforcement. Vertical reinforcement shall consist of at least one No. 4 bar spaced not more than 48 inches (1219 mm). Vertical reinforcement shall be located within 16 inches (406 mm) of the ends of masonry walls.

R606.12.2.3 Design of elements part of the lateral force-resisting system.

R606.12.2.3.1 Connections to masonry shear walls. Connectors shall be provided to transfer forces between masonry walls and horizontal elements in accordance with the requirements of Section 1.7.4 of TMS 402/ACI 530/ASCE 5. Connectors shall be designed to transfer horizontal design forces acting either perpendicular or parallel to the wall, but not less than 200 pounds per linear foot (2919 N/m) of wall. The maximum spacing between connectors shall be 4 feet (1219 mm). Such anchorage mechanisms shall not induce tension stresses perpendicular to grain in ledgers or nailers.

R606.12.2.3.2 Connections to masonry columns. Connectors shall be provided to transfer forces between masonry columns and horizontal elements in accordance with the requirements of Section 1.7.4 of TMS 402/ACI 530/ASCE 5. Where anchor bolts are used to connect horizontal elements to the tops of columns, the bolts shall be placed within lateral ties. Lateral ties shall enclose both the vertical bars in the column and the anchor bolts. There shall be a minimum of two No. 4 lateral ties provided in the top 5 inches (127 mm) of the column.

R606.12.2.3.3 Minimum reinforcement requirements for masonry shear walls. Vertical reinforcement of at least one No. 4 bar shall be provided at corners, within 16 inches (406 mm) of each side of openings, within 8 inches (203 mm) of each side of movement joints, within 8 inches (203 mm) of the ends of walls, and at a maximum spacing of 10 feet (3048 mm).

Horizontal joint reinforcement shall consist of at least two wires of W1.7 spaced not more than 16 inches (406 mm); or bond beam reinforcement of at least one No. 4 bar spaced not more than 10 feet

TABLE R606.12.2.1
MINIMUM SOLID WALL LENGTH ALONG EXTERIOR WALL LINES

SESIMIC DESIGN CATEGORY	MINIMUM SOLID WALL LENGTH (percent)[a]		
	One story or top story of two story	Wall supporting light-framed second story and roof	Wall supporting masonry second story and roof
Townhouses in C	20	25	35
D_0 or D_1	25	NP	NP
D_2	30	NP	NP

NP = Not permitted, except with design in accordance with the *International Building Code*.

a. For all walls, the minimum required length of solid walls shall be based on the table percent multiplied by the dimension, parallel to the wall direction under consideration, of a rectangle inscribing the overall building plan.

(3048 mm) shall be provided. Horizontal reinforcement shall also be provided at the bottom and top of wall openings and shall extend not less than 24 inches (610 mm) nor less than 40 bar diameters past the opening; continuously at structurally connected roof and floor levels; and within 16 inches (406 mm) of the top of walls.

R606.12.3 Seismic Design Category D_0 or D_1. Structures in Seismic Design Category D_0 or D_1 shall comply with the requirements of Seismic Design Category C and the additional requirements of this section.

R606.12.3.1 Design requirements. Masonry elements other than those covered by Section R606.12.2.2.2 shall be designed in accordance with the requirements of Chapter 1 and Sections 2.1 and 2.3 of TMS 402, ACI 530/ASCE 5 and shall meet the minimum reinforcement requirements contained in Sections R606.12.3.2 and R606.12.3.2.1. Otherwise, masonry shall be designed in accordance with TMS 403.

Exception: Masonry walls limited to one *story* in height and 9 feet (2743 mm) between lateral supports need not be designed provided they comply with the minimum reinforcement requirements of Sections R606.12.3.2 and R606.12.3.2.1.

R606.12.3.2 Minimum reinforcement requirements for masonry walls. Masonry walls other than those covered by Section R606.12.2.2.3 shall be reinforced in both the vertical and horizontal direction. The sum of the cross-sectional area of horizontal and vertical reinforcement shall be at least 0.002 times the gross cross-sectional area of the wall, and the minimum cross-sectional area in each direction shall be not less than 0.0007 times the gross cross-sectional area of the wall. Reinforcement shall be uniformly distributed. Table R606.12.3.2 shows the minimum reinforcing bar sizes required for varying thicknesses of masonry walls. The maximum spacing of reinforcement shall be 48 inches (1219 mm) provided that the walls are solid grouted and constructed of hollow open-end units, hollow units laid with full head joints or two wythes of solid units. The maximum spacing of reinforcement shall be 24 inches (610 mm) for all other masonry.

R606.12.3.2.1 Shear wall reinforcement requirements. The maximum spacing of vertical and horizontal reinforcement shall be the smaller of one-third the length of the shear wall, one-third the height of the shear wall, or 48 inches (1219 mm). The minimum cross-sectional area of vertical reinforcement shall be one-third of the required shear reinforcement. Shear reinforcement shall be anchored around vertical reinforcing bars with a standard hook.

R606.12.3.3 Minimum reinforcement for masonry columns. Lateral ties in masonry columns shall be spaced not more than 8 inches (203 mm) on center and shall be at least $^3/_8$-inch (9.5 mm) diameter. Lateral ties shall be embedded in grout.

R606.12.3.4 Material restrictions. Type N mortar or masonry cement shall not be used as part of the lateral-force-resisting system.

R606.12.3.5 Lateral tie anchorage. Standard hooks for lateral tie anchorage shall be either a 135-degree (2.4 rad) standard hook or a 180-degree (3.2 rad) standard hook.

R606.12.4 Seismic Design Category D_2. All structures in Seismic Design Category D_2 shall comply with the requirements of Seismic Design Category D_1 and to the additional requirements of this section.

R606.12.4.1 Design of elements not part of the lateral force-resisting system. Stack bond masonry that is not part of the lateral force-resisting system shall have a horizontal cross-sectional area of reinforcement of at least 0.0015 times the gross cross-sectional area of masonry. Table R606.12.4.1 shows minimum reinforcing bar sizes for masonry walls. The maximum spacing of horizontal reinforcement shall be 24 inches (610 mm). These elements shall be solidly grouted and shall be constructed of hollow open-end units or two wythes of solid units.

TABLE R606.12.4.1
MINIMUM REINFORCING FOR STACKED BONDED MASONRY WALLS IN SEISMIC DESIGN CATEGORY D_2

NOMINAL WALL THICKNESS (inches)	MINIMUM BAR SIZE SPACED AT 24 INCHES
6	#4
8	#5
10	#5
12	#6

For SI: 1 inch = 25.4 mm.

TABLE R606.12.3.2
MINIMUM DISTRIBUTED WALL REINFORCEMENT FOR BUILDING ASSIGNED TO SEISMIC DESIGN CATEGORY D_0 or D_1

NOMINAL WALL THICKNESS (inches)	MINIMUM SUM OF THE VERTICAL AND HORIZONTAL REINFORCEMENT AREAS[a] (square inches per foot)	MINIMUM REINFORCEMENT AS DISTRIBUTED IN BOTH HORIZONTAL AND VERTICAL DIRECTIONS[b] (square inches per foot)	MINUMUM BAR SIZE FOR REINFORCEMENT SPACED AT 48 INCHES
6	0.135	0.047	#4
8	0.183	0.064	#5
10	0.231	0.081	#6
12	0.279	0.098	#6

For SI: 1 inch = 25.4 mm, 1 foot = 304.8 mm, 1 square inch per foot = 2064 mm²/m.

a. Based on the minimum reinforcing ratio of 0.002 times the gross cross-sectional area of the wall.

b. Based on the minimum reinforcing ratio each direction of 0.0007 times the gross cross-sectional area of the wall.

R606.12.4.2 Design of elements part of the lateral force-resisting system. Stack bond masonry that is part of the lateral force-resisting system shall have a horizontal cross-sectional area of reinforcement of at least 0.0025 times the gross cross-sectional area of masonry. Table R606.12.4.2 shows minimum reinforcing bar sizes for masonry walls. The maximum spacing of horizontal reinforcement shall be 16 inches (406 mm). These elements shall be solidly grouted and shall be constructed of hollow open-end units or two wythes of solid units.

TABLE R606.12.4.2
MINIMUM REINFORCING FOR STACKED BONDED MASONRY WALLS IN SEISMIC DESIGN CATEGORY D_2

NOMINAL WALL THICKNESS (inches)	MINIMUM BAR SIZE SPACED AT 16 INCHES
6	#4
8	#5
10	#5
12	#6

For SI: 1 inch = 25.4 mm.

R606.13 Protection for reinforcement. Bars shall be completely embedded in mortar or grout. Joint reinforcement embedded in horizontal mortar joints shall not have less than $^5/_8$-inch (15.9 mm) mortar coverage from the exposed face. All other reinforcement shall have a minimum coverage of one bar diameter over all bars, but not less than $^3/_4$ inch (19 mm), except where exposed to weather or soil, in which case the minimum coverage shall be 2 inches (51 mm).

R606.14 Beam supports. Beams, girders or other concentrated loads supported by a wall or column shall have a bearing of at least 3 inches (76 mm) in length measured parallel to the beam upon *solid masonry* not less than 4 inches (102 mm) in thickness, or upon a metal bearing plate of adequate design and dimensions to distribute the load safely, or upon a continuous reinforced masonry member projecting not less than 4 inches (102 mm) from the face of the wall.

R606.14.1 Joist bearing. Joists shall have a bearing of not less than $1^1/_2$ inches (38 mm), except as provided in Section R606.14, and shall be supported in accordance with Figure R606.11(1).

R606.15 Metal accessories. Joint reinforcement, anchors, ties and wire fabric shall conform to the following: ASTM A 82 for wire anchors and ties; ASTM A 36 for plate, headed and bent-bar anchors; ASTM A 510 for corrugated sheet metal anchors and ties; ASTM A 951 for joint reinforcement; ASTM B 227 for copper-clad steel wire ties; or ASTM A 167 for stainless steel hardware.

R606.15.1 Corrosion protection. Minimum corrosion protection of joint reinforcement, anchor ties and wire fabric for use in masonry wall construction shall conform to Table R606.15.1.

TABLE R606.15.1
MINIMUM CORROSION PROTECTION

MASONRY METAL ACCESSORY	STANDARD
Joint reinforcement, interior walls	ASTM A 641, Class 1
Wire ties or anchors in exterior walls completely embedded in mortar or grout	ASTM A 641, Class 3
Wire ties or anchors in exterior walls not completely embedded in mortar or grout	ASTM A 153, Class B-2
Joint reinforcement in exterior walls or interior walls exposed to moist environment	ASTM A 153, Class B-2
Sheet metal ties or anchors exposed to weather	ASTM A 153, Class B-2
Sheet metal ties or anchors completely embedded in mortar or grout	ASTM A 653, Coating Designation G60
Stainless steel hardware for any exposure	ASTM A 167, Type 304

SECTION R607
UNIT MASONRY

R607.1 Mortar. Mortar for use in masonry construction shall comply with ASTM C 270. The type of mortar shall be in accordance with Sections R607.1.1, R607.1.2 and R607.1.3 and shall meet the proportion specifications of Table R607.1 or the property specifications of ASTM C 270.

R607.1.1 Foundation walls. Masonry foundation walls constructed as set forth in Tables R404.1.1(1) through R404.1.1(4) and mortar shall be Type M or S.

R607.1.2 Masonry in Seismic Design Categories A, B and C. Mortar for masonry serving as the lateral-force-resisting system in Seismic Design Categories A, B and C shall be Type M, S or N mortar.

R607.1.3 Masonry in Seismic Design Categories D_0, D_1 and D_2. Mortar for masonry serving as the lateral-force-resisting system in Seismic Design Categories D_0, D_1 and D_2 shall be Type M or S portland cement-lime or mortar cement mortar.

R607.2 Placing mortar and masonry units.

R607.2.1 Bed and head joints. Unless otherwise required or indicated on the project drawings, head and bed joints shall be $^3/_8$ inch (10 mm) thick, except that the thickness of the bed joint of the starting course placed over foundations shall not be less than $^1/_4$ inch (7 mm) and not more than $^3/_4$ inch (19 mm).

R607.2.1.1 Mortar joint thickness tolerance. Mortar joint thickness for load-bearing masonry shall be within the following tolerances from the specified dimensions:

1. Bed joint: + $^1/_8$ inch (3 mm).

2. Head joint: - $^1/_4$ inch (7 mm), + $^3/_8$ inch (10 mm).
3. Collar joints: - $^1/_4$ inch (7 mm), + $^3/_8$ inch (10 mm).

R607.2.2 Masonry unit placement. The mortar shall be sufficiently plastic and units shall be placed with sufficient pressure to extrude mortar from the joint and produce a tight joint. Deep furrowing of bed joints that produces voids shall not be permitted. Any units disturbed to the extent that initial bond is broken after initial placement shall be removed and relaid in fresh mortar. Surfaces to be in contact with mortar shall be clean and free of deleterious materials.

R607.2.2.1 Solid masonry. *Solid masonry* units shall be laid with full head and bed joints and all interior vertical joints that are designed to receive mortar shall be filled.

R607.2.2.2 Hollow masonry. For hollow masonry units, head and bed joints shall be filled solidly with mortar for a distance in from the face of the unit not less than the thickness of the face shell.

R607.3 Installation of wall ties. The installation of wall ties shall be as follows:

1. The ends of wall ties shall be embedded in mortar joints. Wall ties shall have a minimum of $^5/_8$-inch (15.9 mm) mortar coverage from the exposed face.
2. Wall ties shall not be bent after being embedded in grout or mortar.
3. For solid masonry units, solid grouted hollow units, or hollow units in anchored masonry veneer, wall ties shall be embedded in mortar bed at least $1^1/_2$ inches (38 mm).
4. For hollow masonry units in other than anchored masonry veneer, wall ties shall engage outer face shells by at least $^1/_2$ inch (13 mm).

SECTION R608 MULTIPLE-WYTHE MASONRY

R608.1 General. The facing and backing of multiple-wythe masonry walls shall be bonded in accordance with Section R608.1.1, R608.1.2 or R608.1.3. In cavity walls, neither the facing nor the backing shall be less than 3 inches (76 mm) nominal in thickness and the cavity shall not be more than 4 inches (102 mm) nominal in width. The backing shall be at least as thick as the facing.

Exception: Cavities shall be permitted to exceed the 4-inch (102 mm) nominal dimension provided tie size and tie spacing have been established by calculation.

R608.1.1 Bonding with masonry headers. Bonding with solid or hollow masonry headers shall comply with Sections R608.1.1.1 and R608.1.1.2.

R608.1.1.1 Solid units. Where the facing and backing (adjacent wythes) of *solid masonry* construction are bonded by means of masonry headers, no less than 4 percent of the wall surface of each face shall be com-

TABLE R607.1
MORTAR PROPORTIONS[a, b]

		PROPORTIONS BY VOLUME (cementitious materials)									
MORTAR	TYPE	Portland cement or blended cement	Mortar cement			Masonry cement			Hydrated lime[c] or lime putty	Aggregate ratio (measured in damp, loose conditions)
			M	S	N	M	S	N		
Cement-lime	M	1	—	—	—	—	—	—	$^1/_4$	Not less than $2^1/_4$ and not more than 3 times the sum of separate volumes of lime, if used, and cement
	S	1	—	—	—	—	—	—	over $^1/_4$ to $^1/_2$	
	N	1	—	—	—	—	—	—	over $^1/_2$ to $1^1/_4$	
	O	1	—	—	—	—	—	—	over $1^1/_4$ to $2^1/_2$	
Mortar cement	M	1	—	—	1	—	—	—	—	
	M	—	1	—	—	—	—	—		
	S	$^1/_2$	—	—	1	—	—	—		
	S	—	—	1	—	—	—	—		
	N	—	—	—	1	—	—	—		
	O	—	—	—	1	—	—	—		
Masonry cement	M	1				—	—	1	—	
	M	—				1	—	—		
	S	$^1/_2$				—	—	1		
	S	—				—	1	—		
	N	—				—	—	1		
	O	—				—	—	1		

For SI: 1 cubic foot = 0.0283 m^3, 1 pound = 0.454 kg.

a. For the purpose of these specifications, the weight of 1 cubic foot of the respective materials shall be considered to be as follows:

Portland Cement	94 pounds	Masonry Cement	Weight printed on bag
Mortar Cement	Weight printed on bag	Hydrated Lime	40 pounds
Lime Putty (Quicklime)	80 pounds	Sand, damp and loose	80 pounds of dry sand

b. Two air-entraining materials shall not be combined in mortar.

c. Hydrated lime conforming to the requirements of ASTM C 207.

posed of headers extending not less than 3 inches (76 mm) into the backing. The distance between adjacent full-length headers shall not exceed 24 inches (610 mm) either vertically or horizontally. In walls in which a single header does not extend through the wall, headers from the opposite sides shall overlap at least 3 inches (76 mm), or headers from opposite sides shall be covered with another header course overlapping the header below at least 3 inches (76 mm).

R608.1.1.2 Hollow units. Where two or more hollow units are used to make up the thickness of a wall, the stretcher courses shall be bonded at vertical intervals not exceeding 34 inches (864 mm) by lapping at least 3 inches (76 mm) over the unit below, or by lapping at vertical intervals not exceeding 17 inches (432 mm) with units that are at least 50 percent thicker than the units below.

R608.1.2 Bonding with wall ties or joint reinforcement. Bonding with wall ties or joint reinforcement shall comply with Sections R608.1.2.1 through R608.1.2.3.

R608.1.2.1 Bonding with wall ties. Bonding with wall ties, except as required by Section R610, where the facing and backing (adjacent wythes) of masonry walls are bonded with $^3/_{16}$-inch-diameter (5 mm) wall ties embedded in the horizontal mortar joints, there shall be at least one metal tie for each 4.5 square feet (0.418 m^2) of wall area. Ties in alternate courses shall be staggered. The maximum vertical distance between ties shall not exceed 24 inches (610 mm), and the maximum horizontal distance shall not exceed 36 inches (914 mm). Rods or ties bent to rectangular shape shall be used with hollow masonry units laid with the cells vertical. In other walls, the ends of ties shall be bent to 90-degree (0.79 rad) angles to provide hooks no less than 2 inches (51 mm) long. Additional bonding ties shall be provided at all openings, spaced not more than 3 feet (914 mm) apart around the perimeter and within 12 inches (305 mm) of the opening.

R608.1.2.2 Bonding with adjustable wall ties. Where the facing and backing (adjacent wythes) of masonry are bonded with adjustable wall ties, there shall be at least one tie for each 2.67 square feet (0.248 m^2) of wall area. Neither the vertical nor the horizontal spacing of the adjustable wall ties shall exceed 24 inches (610 mm). The maximum vertical offset of bed joints from one wythe to the other shall be 1.25 inches (32 mm). The maximum clearance between connecting parts of the ties shall be $^1/_{16}$ inch (2 mm). When pintle legs are used, ties shall have at least two $^3/_{16}$-inch-diameter (5 mm) legs.

R608.1.2.3 Bonding with prefabricated joint reinforcement. Where the facing and backing (adjacent wythes) of masonry are bonded with prefabricated joint reinforcement, there shall be at least one cross wire serving as a tie for each 2.67 square feet (0.248 m^2) of wall area. The vertical spacing of the joint reinforcement shall not exceed 16 inches (406 mm). Cross wires on prefabricated joint reinforcement shall not be smaller than No. 9 gage. The longitudinal wires shall be embedded in the mortar.

R608.1.3 Bonding with natural or cast stone. Bonding with natural and cast stone shall conform to Sections R608.1.3.1 and R608.1.3.2.

R608.1.3.1 Ashlar masonry. In ashlar masonry, bonder units, uniformly distributed, shall be provided to the extent of not less than 10 percent of the wall area. Such bonder units shall extend not less than 4 inches (102 mm) into the backing wall.

R608.1.3.2 Rubble stone masonry. Rubble stone masonry 24 inches (610 mm) or less in thickness shall have bonder units with a maximum spacing of 3 feet (914 mm) vertically and 3 feet (914 mm) horizontally, and if the masonry is of greater thickness than 24 inches (610 mm), shall have one bonder unit for each 6 square feet (0.557 m^2) of wall surface on both sides.

R608.2 Masonry bonding pattern. Masonry laid in running and stack bond shall conform to Sections R608.2.1 and R608.2.2.

R608.2.1 Masonry laid in running bond. In each wythe of masonry laid in running bond, head joints in successive courses shall be offset by not less than one-fourth the unit length, or the masonry walls shall be reinforced longitudinally as required in Section R608.2.2.

R608.2.2 Masonry laid in stack bond. Where unit masonry is laid with less head joint offset than in Section R608.2.1, the minimum area of horizontal reinforcement placed in mortar bed joints or in bond beams spaced not more than 48 inches (1219 mm) apart, shall be 0.0007 times the vertical cross-sectional area of the wall.

SECTION R609
GROUTED MASONRY

R609.1 General. Grouted multiple-wythe masonry is a form of construction in which the space between the wythes is solidly filled with grout. It is not necessary for the cores of masonry units to be filled with grout. Grouted hollow unit masonry is a form of construction in which certain cells of hollow units are continuously filled with grout.

R609.1.1 Grout. Grout shall consist of cementitious material and aggregate in accordance with ASTM C 476 and the proportion specifications of Table R609.1.1. Type M or Type S mortar to which sufficient water has been added to produce pouring consistency can be used as grout.

R609.1.2 Grouting requirements. Maximum pour heights and the minimum dimensions of spaces provided for grout placement shall conform to Table R609.1.2. If the work is stopped for one hour or longer, the horizontal construction joints shall be formed by stopping all tiers at the same elevation and with the grout 1 inch (25 mm) below the top.

R609.1.3 Grout space (cleaning). Provision shall be made for cleaning grout space. Mortar projections that project more than $^1/_2$ inch (13 mm) into grout space and any other foreign matter shall be removed from grout space prior to inspection and grouting.

R609.1.4 Grout placement. Grout shall be a plastic mix suitable for pumping without segregation of the constituents and shall be mixed thoroughly. Grout shall be placed by pumping or by an *approved* alternate method and shall be placed before any initial set occurs and in no case more than $1^1/_2$ hours after water has been added. Grouting shall be done in a continuous pour, in lifts not exceeding 5 feet (1524 mm). It shall be consolidated by puddling or mechanical vibrating during placing and reconsolidated after excess moisture has been absorbed but before plasticity is lost.

R609.1.4.1 Grout pumped through aluminum pipes. Grout shall not be pumped through aluminum pipes.

R609.1.5 Cleanouts. Where required by the *building official*, cleanouts shall be provided as specified in this section. The cleanouts shall be sealed before grouting and after inspection.

R609.1.5.1 Grouted multiple-wythe masonry. Cleanouts shall be provided at the bottom course of the exterior wythe at each pour of grout where such pour exceeds 5 feet (1524 mm) in height.

R609.1.5.2 Grouted hollow unit masonry. Cleanouts shall be provided at the bottom course of each cell to be grouted at each pour of grout, where such pour exceeds 4 feet (1219 mm) in height.

R609.2 Grouted multiple-wythe masonry. Grouted multiple-wythe masonry shall conform to all the requirements specified in Section R609.1 and the requirements of this section.

R609.2.1 Bonding of backup wythe. Where all interior vertical spaces are filled with grout in multiple-wythe construction, masonry headers shall not be permitted. Metal wall ties shall be used in accordance with Section R608.1.2 to prevent spreading of the wythes and to maintain the vertical alignment of the wall. Wall ties shall be installed in accordance with Section R608.1.2 when the backup wythe in multiple-wythe construction is fully grouted.

R609.2.2 Grout spaces. Fine grout shall be used when interior vertical space to receive grout does not exceed 2 inches (51 mm) in thickness. Interior vertical spaces exceeding 2 inches (51 mm) in thickness shall use coarse or fine grout.

R609.2.3 Grout barriers. Vertical grout barriers or dams shall be built of *solid masonry* across the grout space the entire height of the wall to control the flow of the grout horizontally. Grout barriers shall not be more than 25 feet (7620 mm) apart. The grouting of any section of a wall between control barriers shall be completed in one day with no interruptions greater than one hour.

R609.3 Reinforced grouted multiple-wythe masonry. Reinforced grouted multiple-wythe masonry shall conform to all the requirements specified in Sections R609.1 and R609.2 and the requirements of this section.

R609.3.1 Construction. The thickness of grout or mortar between masonry units and reinforcement shall not be less

TABLE R609.1.1
GROUT PROPORTIONS BY VOLUME FOR MASONRY CONSTRUCTION

TYPE	PORTLAND CEMENT OR BLENDED CEMENT SLAG CEMENT	HYDRATED LIME OR LIME PUTTY	AGGREGATE MEASURED IN A DAMP, LOOSE CONDITION	
			Fine	Coarse
Fine	1	0 to 1/10	$2^1/_4$ to 3 times the sum of the volume of the cementitious materials	—
Coarse	1	0 to 1/10	$2^1/_4$ to 3 times the sum of the volume of the cementitious materials	1 to 2 times the sum of the volumes of the cementitious materials

TABLE R609.1.2
GROUT SPACE DIMENSIONS AND POUR HEIGHTS

GROUT TYPE	GROUT POUR MAXIMUM HEIGHT (feet)	MINIMUM WIDTH OF GROUT SPACES[a, b] (inches)	MINIMUM GROUT[b, c] SPACE DIMENSIONS FOR GROUTING CELLS OF HOLLOW UNITS (inches × inches)
Fine	1	0.75	1.5 × 2
	5	2	2 × 3
	12	2.5	2.5 × 3
	24	3	3 × 3
Coarse	1	1.5	1.5 × 3
	5	2	2.5 × 3
	12	2.5	3 × 3
	24	3	3 × 4

For SI: 1 inch = 25.4 mm, 1 foot = 304.8 mm.

a. For grouting between masonry wythes.

b. Grout space dimension is the clear dimension between any masonry protrusion and shall be increased by the horizontal projection of the diameters of the horizontal bars within the cross section of the grout space.

c. Area of vertical reinforcement shall not exceed 6 percent of the area of the grout space.

than $^1/_4$ inch (7 mm), except that $^1/_4$-inch (7 mm) bars may be laid in horizontal mortar joints at least $^1/_2$ inch (13 mm) thick, and steel wire reinforcement may be laid in horizontal mortar joints at least twice the thickness of the wire diameter.

R609.4 Reinforced hollow unit masonry. Reinforced hollow unit masonry shall conform to all the requirements of Section R609.1 and the requirements of this section.

R609.4.1 Construction. Requirements for construction shall be as follows:

1. Reinforced hollow-unit masonry shall be built to preserve the unobstructed vertical continuity of the cells to be filled. Walls and cross webs forming cells to be filled shall be full-bedded in mortar to prevent leakage of grout. Head and end joints shall be solidly filled with mortar for a distance in from the face of the wall or unit not less than the thickness of the longitudinal face shells. Bond shall be provided by lapping units in successive vertical courses.
2. Cells to be filled shall have vertical alignment sufficient to maintain a clear, unobstructed continuous vertical cell of dimensions prescribed in Table R609.1.2.
3. Vertical reinforcement shall be held in position at top and bottom and at intervals not exceeding 200 diameters of the reinforcement.
4. Cells containing reinforcement shall be filled solidly with grout. Grout shall be poured in lifts of 8-foot (2438 mm) maximum height. When a total grout pour exceeds 8 feet (2438 mm) in height, the grout shall be placed in lifts not exceeding 5 feet (1524 mm) and special inspection during grouting shall be required.
5. Horizontal steel shall be fully embedded by grout in an uninterrupted pour.

SECTION R610 GLASS UNIT MASONRY

R610.1 General. Panels of glass unit masonry located in load-bearing and nonload-bearing exterior and interior walls shall be constructed in accordance with this section.

R610.2 Materials. Hollow glass units shall be partially evacuated and have a minimum average glass face thickness of $^3/_{16}$ inch (5 mm). The surface of units in contact with mortar shall be treated with a polyvinyl butyral coating or latex-based paint. The use of reclaimed units is prohibited.

R610.3 Units. Hollow or solid glass block units shall be standard or thin units.

R610.3.1 Standard units. The specified thickness of standard units shall be at least $3^7/_8$ inches (98 mm).

R610.3.2 Thin units. The specified thickness of thin units shall be at least $3^1/_8$ inches (79 mm) for hollow units and at least 3 inches (76 mm) for solid units.

R610.4 Isolated panels. Isolated panels of glass unit masonry shall conform to the requirements of this section.

R610.4.1 Exterior standard-unit panels. The maximum area of each individual standard-unit panel shall be 144 square feet (13.4 m^2) when the design wind pressure is 20 psf (958 Pa). The maximum area of such panels subjected to design wind pressures other than 20 psf (958 Pa) shall be in accordance with Figure R610.4.1. The maximum panel dimension between structural supports shall be 25 feet (7620 mm) in width or 20 feet (6096 mm) in height.

R610.4.2 Exterior thin-unit panels. The maximum area of each individual thin-unit panel shall be 85 square feet (7.9 m^2). The maximum dimension between structural supports shall be 15 feet (4572 mm) in width or 10 feet (3048 mm) in height. Thin units shall not be used in applications where the design wind pressure as stated in Table R301.2(1) exceeds 20 psf (958 Pa).

R610.4.3 Interior panels. The maximum area of each individual standard-unit panel shall be 250 square feet (23.2 m^2). The maximum area of each thin-unit panel shall be 150 square feet (13.9 m^2). The maximum dimension between structural supports shall be 25 feet (7620 mm) in width or 20 feet (6096 mm) in height.

R610.4.4 Curved panels. The width of curved panels shall conform to the requirements of Sections R610.4.1, R610.4.2 and R610.4.3, except additional structural supports shall be provided at locations where a curved section joins a straight section, and at inflection points in multicurved walls.

R610.5 Panel support. Glass unit masonry panels shall conform to the support requirements of this section.

R610.5.1 Deflection. The maximum total deflection of structural members that support glass unit masonry shall not exceed $^1/_{600}$.

R610.5.2 Lateral support. Glass unit masonry panels shall be laterally supported along the top and sides of the panel. Lateral supports for glass unit masonry panels shall be designed to resist a minimum of 200 pounds per lineal feet (2918 N/m) of panel, or the actual applied loads, whichever is greater. Except for single unit panels, lateral support shall be provided by panel anchors along the top and sides spaced a maximum of 16 inches (406 mm) on center or by channel-type restraints. Single unit panels shall be supported by channel-type restraints.

Exceptions:

1. Lateral support is not required at the top of panels that are one unit wide.
2. Lateral support is not required at the sides of panels that are one unit high.

R610.5.2.1 Panel anchor restraints. Panel anchors shall be spaced a maximum of 16 inches (406 mm) on center in both jambs and across the head. Panel anchors shall be embedded a minimum of 12 inches (305 mm) and shall be provided with two fasteners so as to resist the loads specified in Section R610.5.2.

R610.5.2.2 Channel-type restraints. Glass unit masonry panels shall be recessed at least 1 inch (25 mm) within channels and chases. Channel-type restraints shall be oversized to accommodate expansion material in the opening, packing and sealant between the framing restraints, and the glass unit masonry perimeter units.

R610.6 Sills. Before bedding of glass units, the sill area shall be covered with a water base asphaltic emulsion coating. The coating shall be a minimum of $^1/_8$ inch (3 mm) thick.

R610.7 Expansion joints. Glass unit masonry panels shall be provided with expansion joints along the top and sides at all structural supports. Expansion joints shall be a minimum of $^3/_8$ inch (10 mm) in thickness and shall have sufficient thickness to accommodate displacements of the supporting structure. Expansion joints shall be entirely free of mortar and other debris and shall be filled with resilient material.

R610.8 Mortar. Glass unit masonry shall be laid with Type S or N mortar. Mortar shall not be retempered after initial set. Mortar unused within $1^1/_2$ hours after initial mixing shall be discarded.

R610.9 Reinforcement. Glass unit masonry panels shall have horizontal joint reinforcement spaced a maximum of 16 inches (406 mm) on center located in the mortar bed joint. Horizontal joint reinforcement shall extend the entire length of the panel but shall not extend across expansion joints. Longitudinal wires shall be lapped a minimum of 6 inches (152 mm) at splices. Joint reinforcement shall be placed in the bed joint immediately below and above openings in the panel. The reinforcement shall have not less than two parallel longitudinal wires of size W1.7 or greater, and have welded cross wires of size W1.7 or greater.

R610.10 Placement. Glass units shall be placed so head and bed joints are filled solidly. Mortar shall not be furrowed. Head and bed joints of glass unit masonry shall be $^1/_4$ inch (6.4 mm) thick, except that vertical joint thickness of radial panels shall not be less than $^1/_8$ inch (3 mm) or greater than $^5/_8$ inch (16 mm). The bed joint thickness tolerance shall be minus $^1/_{16}$ inch (1.6 mm) and plus $^1/_8$ inch (3 mm). The head joint thickness tolerance shall be plus or minus $^1/_8$ inch (3 mm).

SECTION R611
EXTERIOR CONCRETE WALL CONSTRUCTION

R611.1 General. Exterior concrete walls shall be designed and constructed in accordance with the provisions of this section or in accordance with the provisions of PCA 100 or ACI 318. When PCA 100, ACI 318 or the provisions of this section are used to design concrete walls, project drawings, typical details and specifications are not required to bear the seal of the architect or engineer responsible for design, unless otherwise required by the state law of the jurisdiction having authority.

R611.1.1 Interior construction. These provisions are based on the assumption that interior walls and partitions, both load-bearing and nonload-bearing, floors and roof/

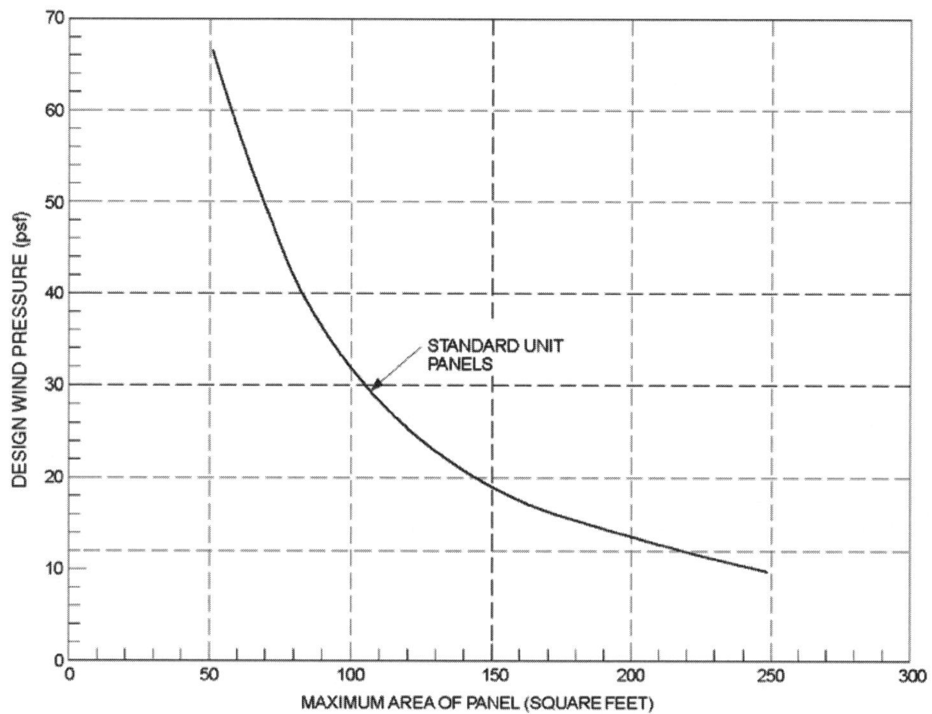

For SI: 1 square foot = 0.0929 m^2, 1 pound per square foot = 0.0479 kPa.

FIGURE R610.4.1
GLASS UNIT MASONRY DESIGN WIND LOAD RESISTANCE

ceiling assemblies are constructed of *light-framed construction* complying with the limitations of this code and the additional limitations of Section R611.2. Design and construction of light-framed assemblies shall be in accordance with the applicable provisions of this code. Where second-story exterior walls are of *light-framed construction*, they shall be designed and constructed as required by this code.

Aspects of concrete construction not specifically addressed by this code, including interior concrete walls, shall comply with ACI 318.

R611.1.2 Other concrete walls. Exterior concrete walls constructed in accordance with this code shall comply with the shapes and minimum concrete cross-sectional dimensions of Table R611.3. Other types of forming systems resulting in concrete walls not in compliance with this section shall be designed in accordance with ACI 318.

R611.2 Applicability limits. The provisions of this section shall apply to the construction of exterior concrete walls for buildings not greater than 60 feet (18 288 mm) in plan dimensions, floors with clear spans not greater than 32 feet (9754 mm) and roofs with clear spans not greater than 40 feet (12 192 mm). Buildings shall not exceed 35 feet (10 668 mm) in mean roof height or two stories in height above-grade. Floor/ceiling dead loads shall not exceed 10 pounds per square foot (479 Pa), roof/ceiling dead loads shall not exceed 15 pounds per square foot (718 Pa) and *attic* live loads shall not exceed 20 pounds per square foot (958 Pa). Roof overhangs shall not exceed 2 feet (610 mm) of horizontal projection beyond the exterior wall and the dead load of the overhangs shall not exceed 8 pounds per square foot (383 Pa).

Walls constructed in accordance with the provisions of this section shall be limited to buildings subjected to a maximum design wind speed of 130 miles per hour (58 m/s) Exposure B, 110 miles per hour (49 m/s) Exposure C and 100 miles per hour (45 m/s) Exposure D. Walls constructed in accordance with the provisions of this section shall be limited to detached one- and two-family *dwellings* and townhouses assigned to Seismic Design Category A or B, and detached one- and two-family *dwellings* assigned to Seismic Design Category C.

Buildings that are not within the scope of this section shall be designed in accordance with PCA 100 or ACI 318.

R611.3 Concrete wall systems. Concrete walls constructed in accordance with these provisions shall comply with the shapes and minimum concrete cross-sectional dimensions of Table R611.3.

R611.3.1 Flat wall systems. Flat concrete wall systems shall comply with Table R611.3 and Figure R611.3(1) and have a minimum nominal thickness of 4 inches (102 mm).

R611.3.2 Waffle-grid wall systems. Waffle-grid wall systems shall comply with Table R611.3 and Figure R611.3(2). and shall have a minimum nominal thickness of 6 inches (152 mm) for the horizontal and vertical concrete members (cores). The core and web dimensions shall comply with Table R611. 3. The maximum weight of waffle-grid walls shall comply with Table R611.3.

TABLE R611.3
DIMENSIONAL REQUIREMENTS FOR WALLS[a, b]

WALL TYPE AND NOMINAL THICKNESS	MAXIMUM WALL WEIGHT[c] (psf)	MINIMUM WIDTH, W, OF VERTICAL CORES (inches)	MINIMUM THICKNESS, T, OF VERTICAL CORES (inches)	MAXIMUM SPACING OF VERTICAL CORES (inches)	MAXIMUM SPACING OF HORIZONTAL CORES (inches)	MINIMUM WEB THICKNESS (inches)
4″ Flat[d]	50	N/A	N/A	N/A	N/A	N/A
6″ Flat[d]	75	N/A	N/A	N/A	N/A	N/A
8″ Flat[d]	100	N/A	N/A	N/A	N/A	N/A
10″ Flat[d]	125	N/A	N/A	N/A	N/A	N/A
6″ Waffle-grid	56	8[e]	5.5[e]	12	16	2
8″ Waffle-grid	76	8[f]	8[f]	12	16	2
6″ Screen-grid	53	6.25[g]	6.25[g]	12	12	N/A

For SI: 1 inch = 25.4 mm; 1 pound per square foot = 0.0479 kPa, 1 pound per cubic foot = 2402.77 kg/m^3, 1 square inch = 645.16 mm^2, 1 inch^4 = 42 cm^4.

a. Width "W," thickness "T," spacing and web thickness, refer to Figures R611.3(2) and R611.3(3).

b. N/A indicates not applicable.

c. Wall weight is based on a unit weight of concrete of 150 pcf. For flat walls the weight is based on the nominal thickness. The tabulated values do not include any allowance for interior and exterior finishes.

d. Nominal wall thickness. The actual as-built thickness of a flat wall shall not be more than $^1/_2$-inch less or more than $^1/_4$-inch more than the nominal dimension indicated.

e. Vertical core is assumed to be elliptical-shaped. Another shape core is permitted provided the minimum thickness is 5 inches, the moment of inertia, I, about the centerline of the wall (ignoring the web) is not less than 65 inch^4, and the area, A, is not less than 31.25 in^2. The width used to calculate A and I shall not exceed 8 inches.

f. Vertical core is assumed to be circular. Another shape core is permitted provided the minimum thickness is 7 inches, the moment of inertia, I, about the centerline of the wall (ignoring the web) is not less than 200 in^4, and the area, A, is not less than 49 square inch. The width used to calculate A and I shall not exceed 8 inches.

g. Vertical core is assumed to be circular. Another shape core is permitted provided the minimum thickness is 5.5 inches, the moment of inertia, I, about the centerline of the wall is not less than 76 inch^4, and the area, A, is not less than 30.25 square inch. The width used to calculate A and I shall not exceed 6.25 inches.

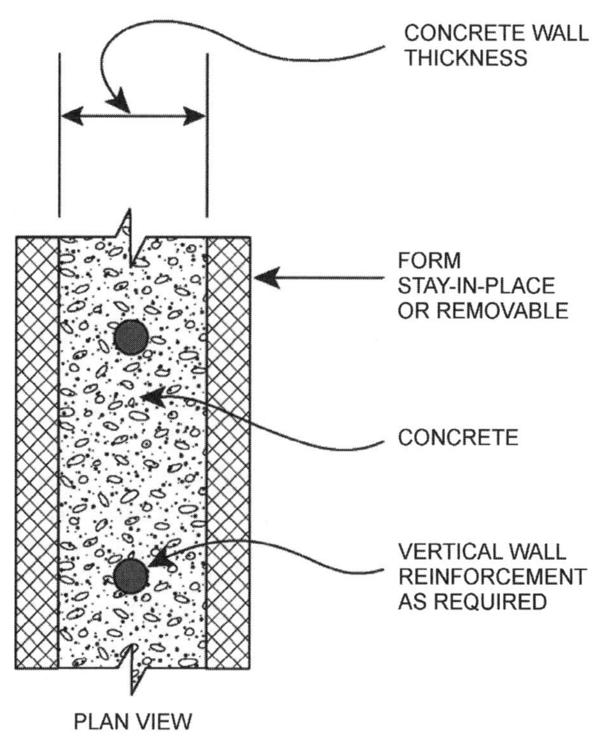

FIGURE R611.3(1)
FLAT WALL SYSTEM

R611.3.3 Screen-grid wall systems. Screen-grid wall systems shall comply with Table R611.3 and Figure R611.3(3) and shall have a minimum nominal thickness of 6 inches (152 mm) for the horizontal and vertical concrete members (cores). The core dimensions shall comply with Table R611.3. The maximum weight of screen-grid walls shall comply with Table R611.3.

R611.4 Stay-in-place forms. Stay-in-place concrete forms shall comply with this section.

R611.4.1 Surface burning characteristics. The flame spread index and smoke-developed index of forming material, other than foam plastic, left exposed on the interior shall comply with Section R302.9. The surface burning characteristics of foam plastic used in insulating concrete forms shall comply with Section R316.3.

R611.4.2 Interior covering. Stay-in-place forms constructed of rigid foam plastic shall be protected on the interior of the building as required by Sections R316.4 and R702.3.4. Where gypsum board is used to protect the foam plastic, it shall be installed with a mechanical fastening system. Use of adhesives is permitted in addition to mechanical fasteners.

R611.4.3 Exterior wall covering. Stay-in-place forms constructed of rigid foam plastics shall be protected from sunlight and physical damage by the application of an *approved* exterior wall covering complying with this code. Exterior surfaces of other stay-in-place forming systems shall be protected in accordance with this code.

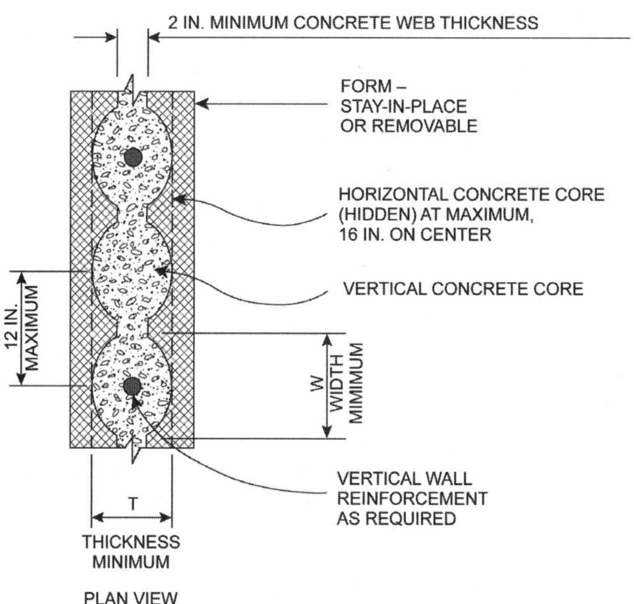

For SI: 1 inch = 25.4 mm.

FIGURE R611.3(2)
WAFFLE-GRID WALL SYSTEM

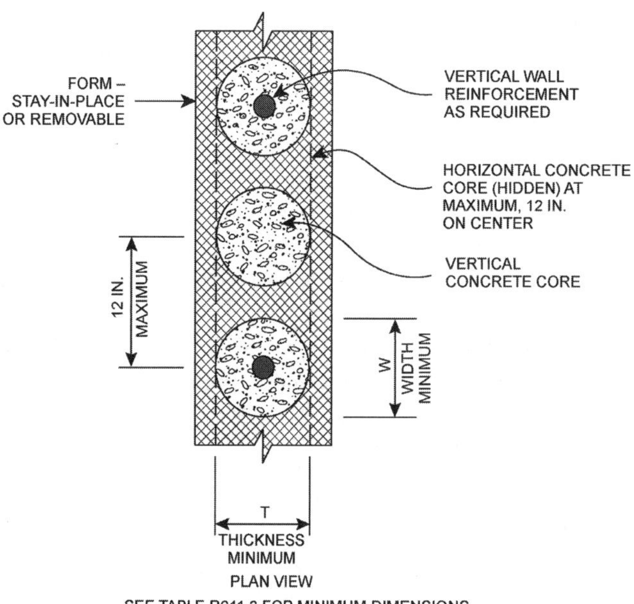

For SI: 1 inch = 25.4 mm.

FIGURE R611.3(3)
SCREEN-GRID WALL SYSTEM

Requirements for installation of masonry veneer, stucco and other finishes on the exterior of concrete walls and other construction details not covered in this section shall comply with the requirements of this code.

R611.4.4 Flat ICF wall systems. Flat ICF wall system forms shall conform to ASTM E 2634.

R611.5 Materials. Materials used in the construction of concrete walls shall comply with this section.

R611.5.1 Concrete and materials for concrete. Materials used in concrete, and the concrete itself, shall conform to requirements of this section, or ACI 318.

R611.5.1.1 Concrete mixing and delivery. Mixing and delivery of concrete shall comply with ASTM C 94 or ASTM C 685.

R611.5.1.2 Maximum aggregate size. The nominal maximum size of coarse aggregate shall not exceed one-fifth the narrowest distance between sides of forms, or three-fourths the clear spacing between reinforcing bars or between a bar and the side of the form.

Exception: When *approved*, these limitations shall not apply where removable forms are used and workability and methods of consolidation permit concrete to be placed without honeycombs or voids.

R611.5.1.3 Proportioning and slump of concrete. Proportions of materials for concrete shall be established to provide workability and consistency to permit concrete to be worked readily into forms and around reinforcement under conditions of placement to be employed, without segregation or excessive bleeding. Slump of concrete placed in removable forms shall not exceed 6 inches (152 mm).

Exception: When *approved*, the slump is permitted to exceed 6 inches (152 mm) for concrete mixtures that are resistant to segregation, and are in accordance with the form manufacturer's recommendations.

Slump of concrete placed in stay-in-place forms shall exceed 6 inches (152 mm). Slump of concrete shall be determined in accordance with ASTM C 143.

R611.5.1.4 Compressive strength. The minimum specified compressive strength of concrete, f'_c, shall comply with Section R402.2 and shall be not less than 2,500 pounds per square inch (17.2 MPa) at 28 days.

R611.5.1.5 Consolidation of concrete. Concrete shall be consolidated by suitable means during placement and shall be worked around embedded items and reinforcement and into corners of forms. Where stay-in-place forms are used, concrete shall be consolidated by internal vibration.

Exception: When *approved*, self-consolidating concrete mixtures with slumps equal to or greater than 8 inches (203 mm) that are specifically designed for placement without internal vibration need not be internally vibrated.

R611.5.2 Steel reinforcement and anchor bolts.

R611.5.2.1 Steel reinforcement. Steel reinforcement shall comply with ASTM A 615, ASTM A 706, or ASTM A 996. ASTM A 996 bars produced from rail steel shall be Type R.

R611.5.2.2 Anchor bolts. Anchor bolts for use with connection details in accordance with Figures R611.9(1) through R611.9(12) shall be bolts with heads complying with ASTM A 307 or ASTM F 1554. ASTM A 307 bolts shall be Grade A (i.e., with heads). ASTM F 1554 bolts shall be Grade 36 minimum. Instead of bolts with heads, it is permissible to use rods with threads on both ends fabricated from steel complying with ASTM A 36. The threaded end of the rod to be embedded in the concrete shall be provided with a hex or square nut.

R611.5.2.3 Sheet steel angles and tension tie straps. Angles and tension tie straps for use with connection details in accordance with Figures R611.9(1) through R611.9(12) shall be fabricated from sheet steel complying with ASTM A 653 SS, ASTM A 792 SS, or ASTM A 875 SS. The steel shall be minimum Grade 33 unless a higher grade is required by the applicable figure.

R611.5.3 Form materials and form ties. Forms shall be made of wood, steel, aluminum, plastic, a composite of cement and foam insulation, a composite of cement and wood chips, or other *approved* material suitable for supporting and containing concrete. Forms shall provide sufficient strength to contain concrete during the concrete placement operation.

Form ties shall be steel, solid plastic, foam plastic, a composite of cement and wood chips, a composite of cement and foam plastic, or other suitable material capable of resisting the forces created by fluid pressure of fresh concrete.

R611.5.4 Reinforcement installation details.

R611.5.4.1 Support and cover. Reinforcement shall be secured in the proper location in the forms with tie wire or other bar support system such that displacement will not occur during the concrete placement operation. Steel reinforcement in concrete cast against the earth shall have a minimum cover of 3 inches (76 mm). Minimum cover for reinforcement in concrete cast in removable forms that will be exposed to the earth or weather shall be $1^1/_2$ inches (38 mm) for No. 5 bars and smaller, and 2 inches (50 mm) for No. 6 bars and larger. For concrete cast in removable forms that will not be exposed to the earth or weather, and for concrete cast in stay-in-place forms, minimum cover shall be $^3/_4$ inch (19 mm). The minus tolerance for cover shall not exceed the smaller of one-third the required cover and $^3/_8$ inch (10 mm). See Section R611.5.4.4 for cover requirements for hooks of bars developed in tension.

R611.5.4.2 Location of reinforcement in walls. For location of reinforcement in foundation walls and above-grade walls, see Sections R404.1.2.3.7.2 and R611.6.5, respectively.

R611.5.4.3 Lap splices. Vertical and horizontal wall reinforcement required by Sections R611.6 and R611.7 shall be the longest lengths practical. Where splices are necessary in reinforcement, the length of lap splices shall be in accordance with Table R611.5.4(1) and Figure R611.5.4 (1). The maximum gap between noncontact parallel bars at a lap splice shall not exceed the smaller of one-fifth the required lap length and 6 inches (152 mm). See Figure R611.5.4(1).

R611.5.4.4 Development of bars in tension. Where bars are required to be developed in tension by other provisions of this code, development lengths and cover for hooks and bar extensions shall comply with Table R611.5.4(1) and Figure R611.5.4 (2). The development lengths shown in Table R611.5.4(1) also apply to bundled bars in lintels installed in accordance with Section R611.8.2.2.

TABLE R611.5.4(1)
LAP SPLICE AND TENSION DEVELOPMENT LENGTHS

	BAR SIZE NO.	YIELD STRENGTH OF STEEL, f_y- psi (MPa)	
		40,000 (280)	60,000 (420)
		Splice length or tension development length (inches)	
Lap splice length-tension	4	20	30
	5	25	38
	6	30	45
Tension development length for straight bar	4	15	23
	5	19	28
	6	23	34
Tension development length for: a. 90-degree and 180-degree standard hooks with not less than $2^1/_2$ inches of side cover perpendicular to plane of hook, and b. 90-degree standard hooks with not less than 2 inches of cover on the bar extension beyond the hook.	4	6	9
	5	7	11
	6	8	13
Tension development length for bar with 90-degree or 180-degree standard hook having less cover than required above.	4	8	12
	5	10	15
	6	12	18

For SI: 1 inch = 25.4 mm.

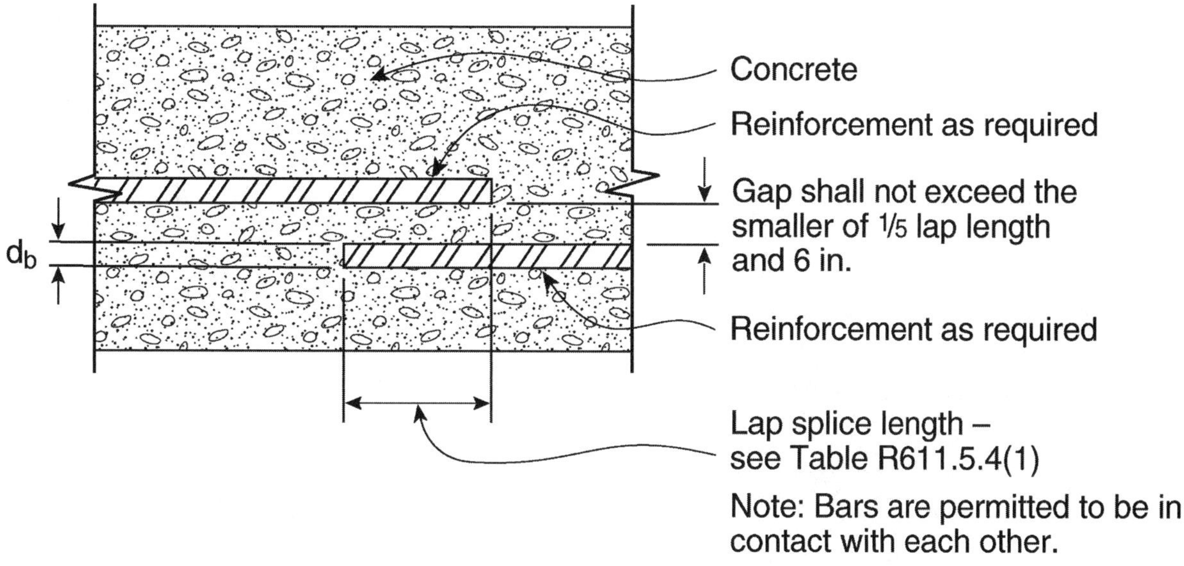

For SI: 1 inch = 25.4 mm.

FIGURE R611.5.4(1)
LAP SPLICES

R611.5.4.5 Standard hooks. Where reinforcement is required by this code to terminate with a standard hook, the hook shall comply with Figure R611.5.4(3).

R611.5.4.6 Webs of waffle-grid walls. Reinforcement, including stirrups, shall not be placed in webs of waffle-grid walls, including lintels. Webs are permitted to have form ties.

R611.5.4.7 Alternate grade of reinforcement and spacing. Where tables in Sections R404.1.2 and R611.6 specify vertical wall reinforcement based on minimum bar size and maximum spacing, which are based on Grade 60 (420 MPa) steel reinforcement, different size bars and/or bars made from a different grade of steel are permitted provided an equivalent area of steel per linear foot of wall is provided. Use of Table R611.5.4(2) is permitted to determine the maximum bar spacing for different bar sizes than specified in the tables and/or bars made from a different grade of steel. Bars shall not be spaced less than one-half the wall thickness, or more than 48 inches (1219 mm) on center.

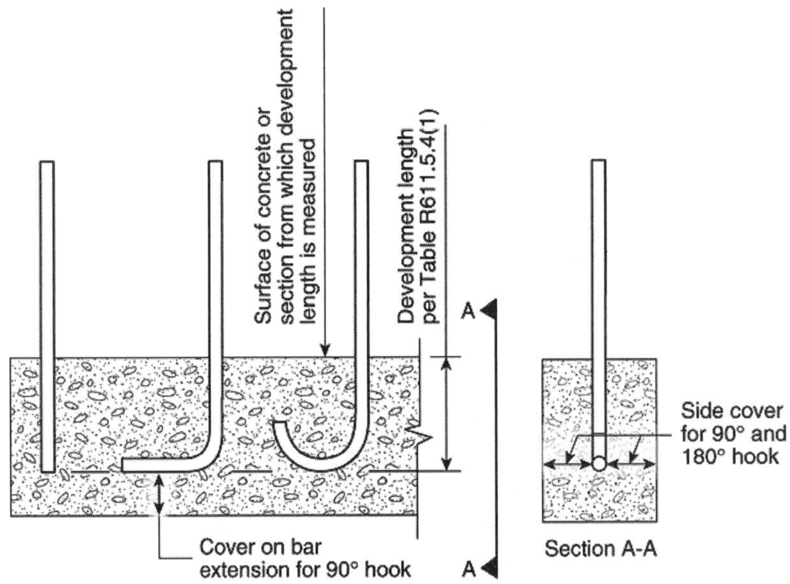

For SI: 1 degree = 0.0175 rad.

FIGURE R611.5.4(2)
DEVELOPMENT LENGTH AND COVER FOR HOOKS AND BAR EXTENSION

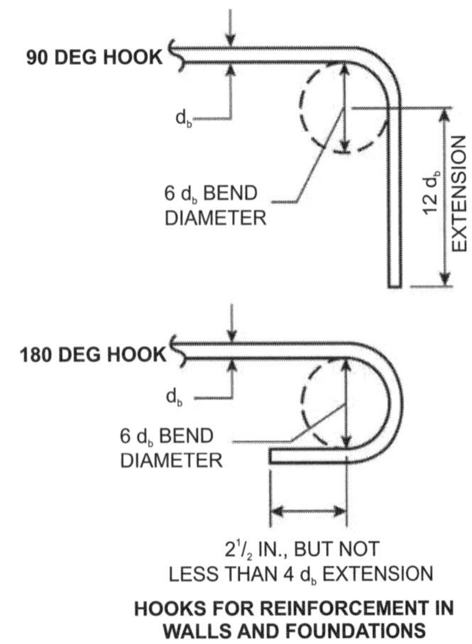

HOOKS FOR REINFORCEMENT IN WALLS AND FOUNDATIONS

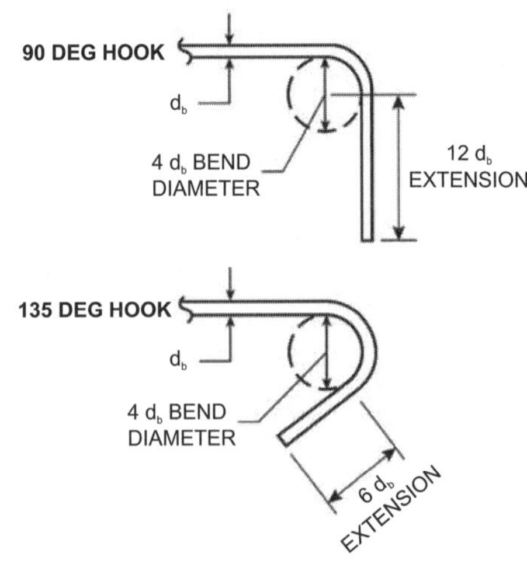

HOOKS FOR STIRRUPS IN LINTELS

For SI: 1 inch = 25.4 mm, 1 degree = 0.0175 rad.

FIGURE R611.5.4(3)
STANDARD HOOKS

TABLE R611.5.4(2)
MAXIMUM SPACING FOR ALTERNATIVE BAR SIZE AND/OR ALTERNATIVE GRADE OF STEEL[a, b, c]

BAR SPACING FROM APPLICABLE TABLE IN SECTION R611.6 (inches)	BAR SIZE FROM APPLICABLE TABLE IN SECTION R611.6														
	#4					#5					#6				
	Alternate bar size and/or alternate grade of steel desired														
	Grade 60		Grade 40			Grade 60		Grade 40			Grade 60		Grade 40		
	#5	#6	#4	#5	#6	#4	#6	#4	#5	#6	#4	#5	#4	#5	#6
	Maximum spacing for alternate bar size and/or alternate grade of steel (inches)														
8	12	18	5	8	12	5	11	3	5	8	4	6	2	4	5
9	14	20	6	9	13	6	13	4	6	9	4	6	3	4	6
10	16	22	7	10	15	6	14	4	7	9	5	7	3	5	7
11	17	24	7	11	16	7	16	5	7	10	5	8	3	5	7
12	19	26	8	12	18	8	17	5	8	11	5	8	4	6	8
13	20	29	9	13	19	8	18	6	9	12	6	9	4	6	9
14	22	31	9	14	21	9	20	6	9	13	6	10	4	7	9
15	23	33	10	16	22	10	21	6	10	14	7	11	5	7	10
16	25	35	11	17	23	10	23	7	11	15	7	11	5	8	11
17	26	37	11	18	25	11	24	7	11	16	8	12	5	8	11
18	28	40	12	19	26	12	26	8	12	17	8	13	5	8	12
19	29	42	13	20	28	12	27	8	13	18	9	13	6	9	13
20	31	44	13	21	29	13	28	9	13	19	9	14	6	9	13
21	33	46	14	22	31	14	30	9	14	20	10	15	6	10	14
22	34	48	15	23	32	14	31	9	15	21	10	16	7	10	15
23	36	48	15	24	34	15	33	10	15	22	10	16	7	11	15
24	37	48	16	25	35	15	34	10	16	23	11	17	7	11	16
25	39	48	17	26	37	16	35	11	17	24	11	18	8	12	17
26	40	48	17	27	38	17	37	11	17	25	12	18	8	12	17
27	42	48	18	28	40	17	38	12	18	26	12	19	8	13	18
28	43	48	19	29	41	18	40	12	19	26	13	20	8	13	19
29	45	48	19	30	43	19	41	12	19	27	13	20	9	14	19
30	47	48	20	31	44	19	43	13	20	28	14	21	9	14	20
31	48	48	21	32	45	20	44	13	21	29	14	22	9	15	21
32	48	48	21	33	47	21	45	14	21	30	15	23	10	15	21
33	48	48	22	34	48	21	47	14	22	31	15	23	10	16	22
34	48	48	23	35	48	22	48	15	23	32	15	24	10	16	23
35	48	48	23	36	48	23	48	15	23	33	16	25	11	16	23
36	48	48	24	37	48	23	48	15	24	34	16	25	11	17	24
37	48	48	25	38	48	24	48	16	25	35	17	26	11	17	25
38	48	48	25	39	48	25	48	16	25	36	17	27	12	18	25
39	48	48	26	40	48	25	48	17	26	37	18	27	12	18	26
40	48	48	27	41	48	26	48	17	27	38	18	28	12	19	27
41	48	48	27	42	48	26	48	18	27	39	19	29	12	19	27
42	48	48	28	43	48	27	48	18	28	40	19	30	13	20	28
43	48	48	29	44	48	28	48	18	29	41	20	30	13	20	29
44	48	48	29	45	48	28	48	19	29	42	20	31	13	21	29
45	48	48	30	47	48	29	48	19	30	43	20	32	14	21	30
46	48	48	31	48	48	30	48	20	31	44	21	32	14	22	31
47	48	48	31	48	48	30	48	20	31	44	21	33	14	22	31
48	48	48	32	48	48	31	48	21	32	45	22	34	15	23	32

For SI: 1 inch = 25.4 mm.

a. This table is for use with tables in Section R611.6 that specify the minimum bar size and maximum spacing of vertical wall reinforcement for foundation walls and above-grade walls. Reinforcement specified in tables in Section R611.6 is based on Grade 60 (420 MPa) steel reinforcement.

b. Bar spacing shall not exceed 48 inches on center and shall not be less than one-half the nominal wall thickness.

c. For Grade 50 (350 MPa) steel bars (ASTM A 996, Type R), use spacing for Grade 40 (280 MPa) bars or interpolate between Grade 40 (280 MPa) and Grade 60 (420 MPa).

R611.5.5 Construction joints in walls. Construction joints shall be made and located to not impair the strength of the wall. Construction joints in plain concrete walls, including walls required to have not less than No. 4 bars at 48 inches (1219 mm) on center by Section R611.6, shall be located at points of lateral support, and a minimum of one No. 4 bar shall extend across the construction joint at a spacing not to exceed 24 inches (610 mm) on center. Construction joint reinforcement shall have a minimum of 12 inches (305 mm) embedment on both sides of the joint. Construction joints in reinforced concrete walls shall be located in the middle third of the span between lateral supports, or located and constructed as required for joints in plain concrete walls.

Exception: Vertical wall reinforcement required by this code is permitted to be used in lieu of construction joint reinforcement, provided the spacing does not exceed 24 inches (610 mm), or the combination of wall reinforcement and No. 4 bars described above does not exceed 24 inches (610 mm).

R611.6 Above-grade wall requirements.

R611.6.1 General. The minimum thickness of load-bearing and nonload-bearing above-grade walls and reinforcement shall be as set forth in the appropriate table in this section based on the type of wall form to be used. Where the wall or building is not within the limitations of Section R611.2, design is required by the tables in this section, or the wall is not within the scope of the tables in this section, the wall shall be designed in accordance with ACI 318.

Above-grade concrete walls shall be constructed in accordance with this section and Figure R611.6(1), R611.6(2), R611.6(3) or R611.6(4). Above-grade concrete walls that are continuous with stem walls and not laterally supported by the slab-on-ground shall be designed and constructed in accordance with this section. Concrete walls shall be supported on continuous foundation walls or slabs-on-ground that are monolithic with the footing in accordance with Section R403. The minimum length of solid wall without openings shall be in accordance with Section R611.7. Reinforcement around openings, including lintels, shall be in accordance with Section R611.8. Lateral support for above-grade walls in the out-of-plane direction shall be provided by connections to the floor framing system, if applicable, and to ceiling and roof framing systems in accordance with Section R611.9. The wall thickness shall be equal to or greater than the thickness of the wall in the *story* above.

LIGHT-FRAMED ROOF
SEE SECTION R611.9.3
HORIZONTAL WALL REINFORCEMENT AS REQUIRED
FIRST-STORY UNSUPPORTED WALL HEIGHT 10 FT MAXIMUM
WALL—STAY-IN-PLACE OR REMOVABLE FORM
LIGHT-FRAMED FLOOR (OR CONCRETE SLAB-ON-GROUND)
SEE SECTION R611.9.2
VERTICAL WALL REINFORCEMENT AS REQUIRED
BASEMENT, CRAWLSPACE OR STEM WALL. fOR SLAB-ON-GROUND FOOTING, SEE FIGURE R611.6(4)

SECTION CUT THROUGH FLAT WALL OR VERTICAL CORE OF AWAFFLE—OR SCREEN-GRID WALL

For SI: 1 foot = 304.8 mm.

FIGURE R611.6(1)
ABOVE-GRADE CONCRETE WALL CONSTRUCTION ONE

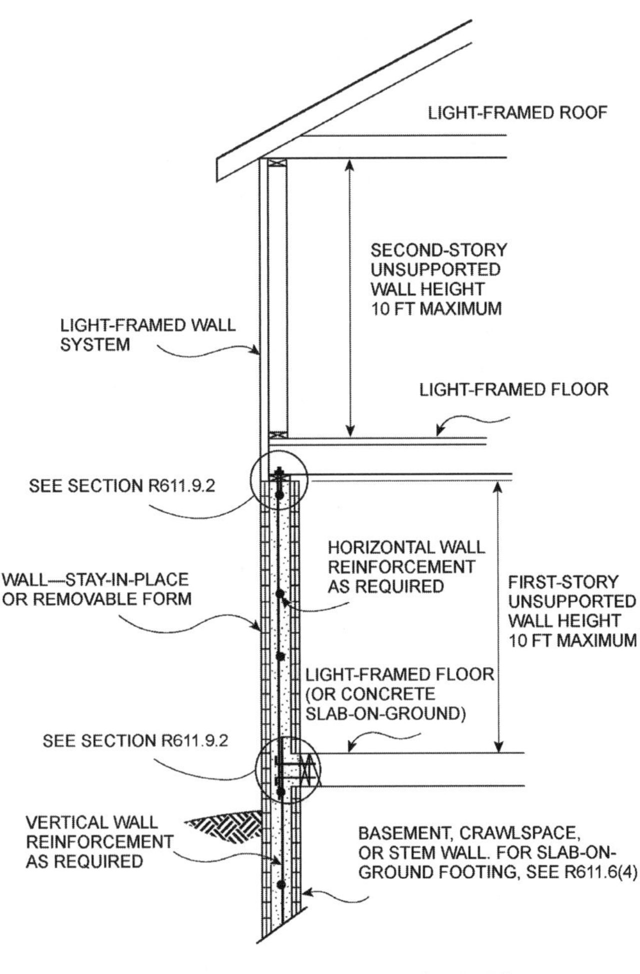

For SI: 1 foot = 304.8 mm.

FIGURE R611.6(2)
ABOVE-GRADE CONCRETE WALL CONSTRUCTION CONCRETE FIRST-STORY AND LIGHT-FRAMED SECOND-STORY

R611.6.2 Wall reinforcement for wind. Vertical wall reinforcement for resistance to out-of-plane wind forces shall be determined from Table R611.6(1), R611.6(2), R611.6(3) or R611.6(4). Also, see Sections R611.7.2.2.2 and R611.7.2.2.3. There shall be a vertical bar at all corners of exterior walls. Unless more horizontal reinforcement is required by Section R611.7.2.2.1, the minimum horizontal reinforcement shall be four No. 4 bars [Grade 40 (280 MPa)] placed as follows: top bar within 12 inches (305 mm) of the top of the wall, bottom bar within 12 inches (305 mm) of the finish floor, and one bar each at approximately one-third and two-thirds of the wall height.

R611.6.3 Continuity of wall reinforcement between stories. Vertical reinforcement required by this section shall be continuous between elements providing lateral support for the wall. Reinforcement in the wall of the *story* above shall be continuous with the reinforcement in the wall of the *story* below, or the foundation wall, if applicable. Lap splices, where required, shall comply with Section R611.5.4.3 and Figure R611.5.4(1). Where the above-grade wall is supported by a monolithic slab-on-ground and footing, dowel bars with a size and spacing to match the vertical above-grade concrete wall reinforcement shall be embedded in the monolithic slab-on-ground and footing the distance required to develop the dowel bar in tension in accordance with Section R611.5.4.4 and Figure R611.5.4(2) and lap-spliced with the above-grade wall reinforcement in accordance with Section R611.5.4.3 and Figure R611.5.4(1).

Exception: Where reinforcement in the wall above cannot be made continuous with the reinforcement in the wall below, the bottom of the reinforcement in the wall above shall be terminated in accordance with one of the following:

1. Extend below the top of the floor the distance required to develop the bar in tension in accordance with Section R611.5.4.4 and Figure R611.5.4(2).
2. Lap-spliced in accordance with Section R611.5.4.3 and Figure R611.5.4(1) with a dowel bar that extends into the wall below the distance required to develop the bar in tension in accordance with Section R611.5.4.4 and Figure R611.5.4(2).

Where a construction joint in the wall is located below the level of the floor and less than the distance required to develop the bar in tension, the distance required to develop the bar in tension shall be measured from the top of the concrete below the joint. See Section R611.5.5.

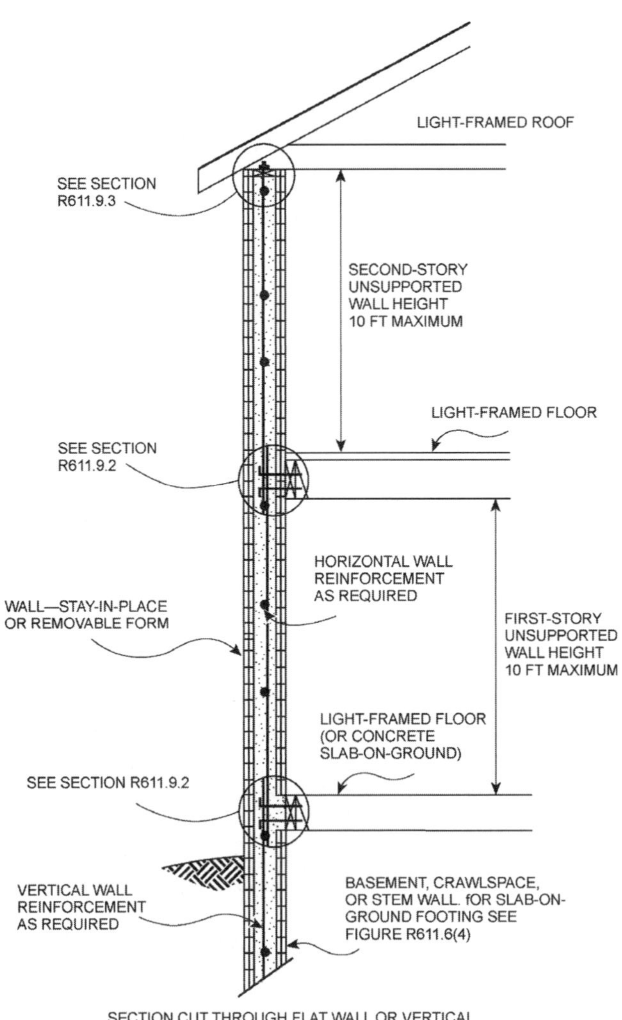

For SI: 1 foot = 304.8 mm.

FIGURE R611.6(3)
ABOVE-GRADE CONCRETE WALL CONSTRUCTION TWO-STORY

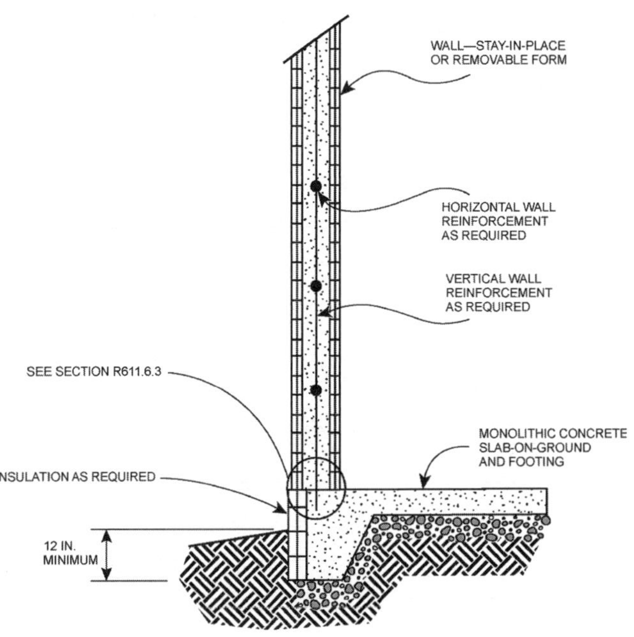

For SI: 1 inch = 25.4 mm.

FIGURE R611.6(4)
ABOVE-GRADE CONCRETE WALL SUPPORTED ON MONOLITHIC SLAB-ON-GROUND FOOTING

TABLE R611.6(1)
MINIMUM VERTICAL REINFORCEMENT FOR FLAT ABOVE-GRADE WALLS[a, b, c, d, e]

MAXIMUM WIND SPEED (mph)			MAXIMUM UNSUPPORTED WALL HEIGHT PER STORY (feet)	MINIMUM VERTICAL REINFORCEMENT-BAR SIZE AND SPACING (inches)[f, g]							
Exposure Category				Nominal[h] wall thickness (inches)							
				4		6		8		10	
B	C	D		Top[i]	Side[i]	Top[i]	Side[i]	Top[i]	Side[i]	Top[i]	Side[i]
85	—	—	8	4@48	4@48	4@48	4@48	4@48	4@48	4@48	4@48
			9	4@48	4@43	4@48	4@48	4@48	4@48	4@48	4@48
			10	4@47	4@36	4@48	4@48	4@48	4@48	4@48	4@48
90	—	—	8	4@48	4@47	4@48	4@48	4@48	4@48	4@48	4@48
			9	4@48	4@39	4@48	4@48	4@48	4@48	4@48	4@48
			10	4@42	4@34	4@48	4@48	4@48	4@48	4@48	4@48
100	85	—	8	4@48	4@40	4@48	4@48	4@48	4@48	4@48	4@48
			9	4@42	4@34	4@48	4@48	4@48	4@48	4@48	4@48
			10	4@34	4@34	4@48	4@48	4@48	4@48	4@48	4@48
110	90	85	8	4@44	4@34	4@48	4@48	4@48	4@48	4@48	4@48
			9	4@34	4@34	4@48	4@48	4@48	4@48	4@48	4@48
			10	4@34	4@31	4@48	4@37	4@48	4@48	4@48	4@48
120	100	90	8	4@36	4@34	4@48	4@48	4@48	4@48	4@48	4@48
			9	4@34	4@32	4@48	4@38	4@48	4@48	4@48	4@48
			10	4@30	4@27	4@48	5@48	4@48	4@48	4@48	4@48
130	110	100	8	4@34	4@34	4@48	4@48	4@48	4@48	4@48	4@48
			9	4@32	4@28	4@48	4@33	4@48	4@48	4@48	4@48
			10	4@26	4@23	4@48	5@43	4@48	4@48	4@48	4@48

For SI: 1 inch = 25.4 mm, 1 foot = 304.8 mm, 1 mile per hour = 0.447 m/s, 1 pound per square inch = 1.895 kPa, 1 square foot = 0.0929 m^2.

a. Table is based on ASCE 7 components and cladding wind pressures for an enclosed building using a mean roof height of 35 feet, interior wall area 4, an effective wind area of 10 square feet, and topographic factor, K_{zt}, and importance factor, *I*, equal to 1.0.

b. Table is based on concrete with a minimum specified compressive strength of 2,500 psi.

c. See Section R611.6.5 for location of reinforcement in wall.

d. Deflection criterion is *L*/240, where *L* is the unsupported height of the wall in inches.

e. Interpolation is not permitted.

f. Where No. 4 reinforcing bars at a spacing of 48 inches are specified in the table, use of bars with a minimum yield strength of 40,000 psi or 60,000 psi is permitted.

g. Other than for No. 4 bars spaced at 48 inches on center, table values are based on reinforcing bars with a minimum yield strength of 60,000 psi. Vertical reinforcement with a yield strength of less than 60,000 psi and/or bars of a different size than specified in the table are permitted in accordance with Section R611.5.4.7 and Table R611.5.4(2).

h. See Table R611.3 for tolerances on nominal thicknesses.

i. Top means gravity load from roof and/or floor construction bears on top of wall. Side means gravity load from floor construction is transferred to wall from a wood ledger or cold-formed steel track bolted to side of wall. Where floor framing members span parallel to the wall, use of the top bearing condition is permitted.

**TABLE R611.6(2)
MINIMUM VERTICAL REINFORCEMENT FOR WAFFLE-GRID ABOVE-GRADE WALLS[a, b, c, d, e]**

MAXIMUM WIND SPEED (mph)			MAXIMUM UNSUPPORTED WALL HEIGHT PER STORY (feet)	MINIMUM VERTICAL REINFORCEMENT-BAR SIZE AND SPACING (inches)[f, g]			
Exposure Category				Nominal[h] wall thickness (inches)			
				6		8	
B	C	D		Top[i]	Side[i]	Top[i]	Side[i]
85	—	—	8	4@48	4@36, 5@48	4@48	4@48
			9	4@48	4@30, 5@47	4@48	4@45
			10	4@48	4@26, 5@40	4@48	4@39
90	—	—	8	4@48	4@33, 5@48	4@48	4@48
			9	4@48	4@28, 5@43	4@48	4@42
			10	4@31, 5@48	4@24, 5@37	4@48	4@36
100	85	—	8	4@48	4@28, 5@44	4@48	4@43
			9	4@31, 5@48	4@24, 5@37	4@48	4@36
			10	4@25, 5@39	4@24, 5@37	4@48	4@31, 5@48
110	90	85	8	4@33, 5@48	4@25, 5@38	4@48	4@38
			9	4@26, 5@40	4@24, 5@37	4@48	4@31, 5@48
			10	4@24, 5@37	4@23, 5@35	4@48	4@27, 5@41
120	100	90	8	4@27, 5@42	4@24, 5@37	4@48	4@33, 5@48
			9	4@24, 5@37	4@23, 5@36	4@48	4@27, 5@43
			10	4@23, 5@35	4@19, 5@30	4@48	4@23, 5@36
130	110	100	8	4@24, 5@37	4@24, 5@37	4@48	4@29, 5@45
			9	4@24, 5@37	4@20, 5@32	4@48	4@24, 5@37
			10	4@19, 5@30	4@17, 5@26	4@23, 5@36	4@20, 5@31

For SI: 1 inch = 25.4 mm, 1 foot = 304.8 mm, 1 mile per hour = 0.447 m/s, 1 pound per square inch = 6.895 kPa, 1 square foot = 0.0929 m^2.

a. Table is based on ASCE 7 components and cladding wind pressures for an enclosed building using a mean roof height of 35 feet, interior wall area 4, an effective wind area of 10 square feet, and topographic factor, K_{zt}, and importance factor, I, equal to 1.0.

b. Table is based on concrete with a minimum specified compressive strength of 2,500 psi.

c. See Section R611.6.5 for location of reinforcement in wall.

d. Deflection criterion is $L/240$, where L is the unsupported height of the wall in inches.

e. Interpolation is not permitted.

f. Where No. 4 reinforcing bars at a spacing of 48 inches are specified in the table, use of bars with a minimum yield strength of 40,000 psi or 60,000 psi is permitted.

g. Other than for No. 4 bars spaced at 48 inches on center, table values are based on reinforcing bars with a minimum yield strength of 60,000 psi. Maximum spacings shown are the values calculated for the specified bar size. Where the bar used is Grade 60 and the size specified in the table, the actual spacing in the wall shall not exceed a whole-number multiple of 12 inches (i.e., 12, 24, 36 and 48) that is less than or equal to the tabulated spacing. Vertical reinforcement with a yield strength of less than 60,000 psi and/or bars of a different size than specified in the table are permitted in accordance with Section R611.5.4.7 and Table R611.5.4(2).

h. See Table R611.3 for minimum core dimensions and maximum spacing of horizontal and vertical cores.

i. Top means gravity load from roof and/or floor construction bears on top of wall. Side means gravity load from floor construction is transferred to wall from a wood ledger or cold-formed steel track bolted to side of wall. Where floor framing members span parallel to the wall, the top bearing condition is permitted to be used.

TABLE R611.6(3)
MINIMUM VERTICAL REINFORCEMENT FOR 6-INCH SCREEN-GRID ABOVE-GRADE WALLS[a, b, c, d, e]

MAXIMUM WIND SPEED (mph)			MAXIMUM UNSUPPORTED WALL HEIGHT PER STORY (feet)	MINIMUM VERTICAL REINFORCEMENT-BAR SIZE AND SPACING (inches)[f, g]	
Exposure Category				Nominal[h] wall thickness (inches)	
				6	
B	C	D		Top[i]	Side[i]
85	—	—	8	4@48	4@34, 5@48
			9	4@48	4@29, 5@45
			10	4@48	4@25, 5@39
90	—	—	8	4@48	4@31, 5@48
			9	4@48	4@27, 5@41
			10	4@30, 5@47	4@23, 5@35
100	85	—	8	4@48	4@27, 5@42
			9	4@30, 5@47	4@23, 5@35
			10	4@24, 5@38	4@22, 5@34
110	90	85	8	4@48	4@24, 5@37
			9	4@25, 5@38	4@22, 5@34
			10	4@22, 5@34	4@22, 5@34
120	100	90	8	4@26, 5@41	4@22, 5@34
			9	4@22, 5@34	4@22, 5@34
			10	4@22, 6@34	4@19, 5@26
130	110	100	8	4@22, 5@35	4@22, 5@34
			9	4@22, 5@34	4@20, 5@30
			10	4@19, 5@29	4@16, 5@25

For SI: 1 inch = 25.4 mm, 1 foot = 304.8 mm, 1 mile per hour = 0.447 m/s, 1 pound per square inch = 6.895 kPa, 1 square foot = 0.0929 m^2.

a. Table is based on ASCE 7 components and cladding wind pressures for an enclosed building using a mean roof height of 35 feet, interior wall area 4, an effective wind area of 10 square feet, and topographic factor, K_{zt}, and importance factor, I, equal to 1.0.

b. Table is based on concrete with a minimum specified compressive strength of 2,500 psi.

c. See Section R611.6.5 for location of reinforcement in wall.

d. Deflection criterion is $L/240$, where L is the unsupported height of the wall in inches.

e. Interpolation is not permitted.

f. Where No. 4 reinforcing bars at a spacing of 48 inches are specified in the table, use of bars with a minimum yield strength of 40,000 psi or 60,000 psi is permitted.

g. Other than for No. 4 bars spaced at 48 inches on center, table values are based on reinforcing bars with a minimum yield strength of 60,000 psi. Maximum spacings shown are the values calculated for the specified bar size. Where the bar used is Grade 60 and the size specified in the table, the actual spacing in the wall shall not exceed a whole-number multiple of 12 inches (i.e., 12, 24, 36 and 48) that is less than or equal to the tabulated spacing. Vertical reinforcement with a yield strength of less than 60,000 psi and/or bars of a different size than specified in the table are permitted in accordance with Section R611.5.4.7 and Table R611.5.4(2).

h. See Table R611.3 for minimum core dimensions and maximum spacing of horizontal and vertical cores.

i. Top means gravity load from roof and/or floor construction bears on top of wall. Side means gravity load from floor construction is transferred to wall from a wood ledger or cold-formed steel track bolted to side of wall. Where floor framing members span parallel to the wall, use of the top bearing condition is permitted.

TABLE R611.6(4)
MINIMUM VERTICAL REINFORCEMENT FOR FLAT, WAFFLE- AND SCREEN-GRID ABOVE-GRADE WALLS DESIGNED CONTINUOUS WITH FOUNDATION STEM WALLS[a, b, c, d, e, k, l]

MAXIMUM WIND SPEED (mph)			HEIGHT OF STEM WALL[h, i] (feet)	MAXIMUM DESIGN LATERAL SOIL LOAD (psf/ft)	MAXIMUM UNSUPPORTED HEIGHT OF ABOVE-GRADE WALL (feet)	MINIMUM VERTICAL REINFORCEMENT-BAR SIZE AND SPACING (inches)[f, g]						
Exposure Category						Wall type and nominal thickness[j] (inches)						
						Flat				Waffle		Screen
B	C	D				4	6	8	10	6	8	6
85	—	—	3	30	8	4@33	4@39	4@48	4@48	4@24	4@28	4@22
					10	4@26	5@48	4@41	4@48	4@19	4@22	4@18
				60	10	4@21	5@40	5@48	4@44	4@16	4@19	4@15
			6	30	10	DR	5@22	6@35	6@43	DR	4@11	DR
				60	10	DR	DR	6@26	6@28	DR	DR	DR
90	—	—	3	30	8	4@30	4@36	4@48	4@48	4@22	4@26	4@21
					10	4@24	5@44	4@38	4@48	4@17	4@21	4@17
				60	10	4@20	5@37	4@48	4@41	4@15	4@18	4@14
			6	30	10	DR	5@21	6@35	6@41	DR	4@10	DR
				60	10	DR	DR	6@26	6@28	DR	DR	DR
100	85	—	3	30	8	4@26	5@48	4@42	4@48	4@19	4@23	4@18
					10	4@20	5@37	4@33	4@41	4@15	4@18	4@14
			6	60	10	4@17	5@34	5@44	4@36	4@13	4@17	4@12
				30	10	DR	5@20	6@35	6@38	DR	4@9	DR
				60	10	DR	DR	6@24	6@28	DR	DR	DR
110	90	85	3	30	8	4@22	5@42	4@37	4@46	4@16	4@20	4@16
					10	4@17	5@34	5@44	4@35	4@12	4@17	4@12
				60	10	4@15	5@34	5@39	5@48	4@11	4@17	4@11
			6	30	10	DR	5@18	6@35	6@35	DR	4@9	DR
				60	10	DR	DR	6@23	6@28	DR	DR	DR
120	100	90	3	30	8	4@19	5@37	5@48	4@40	4@14	4@17	4@14
					10	4@14	5@34	5@38	5@48	4@11	4@17	4@10
				60	10	4@13	5@33	6@48	5@43	4@10	4@16	4@9
			6	30	10	DR	5@16	6@33	6@32	DR	4@8	DR
				60	10	DR	DR	6@22	6@28	DR	DR	DR
130	110	100	3	30	8	4@17	5@34	5@44	4@36	4@12	4@17	4@10
					10	DR	5@32	6@47	5@42	4@9	4@15	DR
				60	10	DR	5@29	6@43	5@39	DR	4@14	DR
			6	30	10	DR	5@15	6@30	6@29	DR	4@7	DR
				60	10	DR	DR	6@21	6@27	DR	DR	DR

For SI: 1 inch = 25.4 mm, 1 foot = 304.8 mm, 1 mile per hour = 0.447 m/s, 1 pound per square inch = 6.895 kPa, 1 square foot = 0.0929 m^2.

a. Table is based on ASCE 7 components and cladding wind pressures for an enclosed building using a mean roof height of 35 feet, interior wall area 4, an effective wind area of 10 square feet, and topographic factor, K_{zt}, and importance factor, I, equal to 1.0.

b. Table is based on concrete with a minimum specified compressive strength of 2,500 psi.

c. See Section R611.6.5 for location of reinforcement in wall.

d. Deflection criterion is $L/240$, where L is the height of the wall in inches from the exterior finish ground level to the top of the above-grade wall.

e. Interpolation is not permitted. For intermediate values of basic wind speed, heights of stem wall and above-grade wall, and design lateral soil load, use next higher value.

f. Where No. 4 reinforcing bars at a spacing of 48 inches are specified in the table, use of bars with a minimum yield strength of 40,000 psi or 60,000 psi is permitted.

g. Other than for No. 4 bars spaced at 48 inches on center, table values are based on reinforcing bars with a minimum yield strength of 60,000 psi. Maximum spacings shown are the values calculated for the specified bar size. In waffle and screen-grid walls where the bar used is Grade 60 and the size specified in the table, the actual spacing in the wall shall not exceed a whole-number multiple of 12 inches (i.e., 12, 24, 36 and 48) that is less than or equal to the tabulated spacing. Vertical reinforcement with a yield strength of less than 60,000 psi and/or bars of a different size than specified in the table are permitted in accordance with Section R611.5.4.7 and Table R611.5.4(2).

h. Height of stem wall is the distance from the exterior finish ground level to the top of the slab-on-ground.

i. Where the distance from the exterior finish ground level to the top of the slab-on-ground is equal to or greater than 4 feet, the stem wall shall be laterally supported at the top and bottom before backfilling. Where the wall is designed and constructed to be continuous with the above-grade wall, temporary supports bracing the top of the stem wall shall remain in place until the above-grade wall is laterally supported at the top by floor or roof construction.

j. See Table R611.3 for tolerances on nominal thicknesses, and minimum core dimensions and maximum spacing of horizontal and vertical cores for waffle- and screen-grid walls.

k. Tabulated values are applicable to construction where gravity loads bear on top of wall, and conditions where gravity loads from floor construction are transferred to wall from a wood ledger or cold-formed steel track bolted to side of wall. See Tables R611.6(1), R611.6(2) and R611.6(3).

l. DR indicates design required.

R611.6.4 Termination of reinforcement. Where indicated in Items 1 through 3, vertical wall reinforcement in the top-most *story* with concrete walls shall be terminated with a 90-degree (1.57 rad) standard hook complying with Section R611.5.4.5 and Figure R611.5.4(3).

1. Vertical bars adjacent to door and window openings required by Section R611.8.1.2.
2. Vertical bars at the ends of required solid wall segments. See Section R611.7.2.2.2.
3. Vertical bars (other than end bars, see Item 2) used as shear reinforcement in required solid wall segments where the reduction factor for design strength, R_3, used is based on the wall having horizontal and vertical shear reinforcement. See Section R611.7.2.2.3.

The bar extension of the hook shall be oriented parallel to the horizontal wall reinforcement and be within 4 inches (102 mm) of the top of the wall.

Horizontal reinforcement shall be continuous around the building corners by bending one of the bars and lap-splicing it with the bar in the other wall in accordance with Section R611.5.4.3 and Figure R611.5.4(1).

Exception: In lieu of bending horizontal reinforcement at corners, separate bent reinforcing bars shall be permitted provided that the bent bar is lap-spliced with the horizontal reinforcement in both walls in accordance with Section R611.5.4.3 and Figure R611.5.4(1).

In required solid wall segments where the reduction factor for design strength, R_3, is based on the wall having horizontal and vertical shear reinforcement in accordance with Section R611.7.2.2.1, horizontal wall reinforcement shall be terminated with a standard hook complying with Section R611.5.4.5 and Figure R611.5.4(3) or in a lap-splice, except at corners where the reinforcement shall be continuous as required above.

R611.6.5 Location of reinforcement in wall. Except for vertical reinforcement at the ends of required solid wall segments, which shall be located as required by Section R611.7.2.2.2, the location of the vertical reinforcement shall not vary from the center of the wall by more than the greater of 10 percent of the wall thickness and $^3/_8$-inch (10 mm). Horizontal and vertical reinforcement shall be located to provide not less than the minimum cover required by Section R611.5.4.1.

R611.7 Solid walls for resistance to lateral forces.

R611.7.1 Length of solid wall. Each exterior wall line in each *story* shall have a total length of solid wall required by Section R611.7.1.1. A solid wall is a section of flat, waffle-grid or screen-grid wall, extending the full *story height* without openings or penetrations, except those permitted by Section R611.7.2. Solid wall segments that contribute to the total length of solid wall shall comply with Section R611.7.2.

R611.7.1.1 Length of solid wall for wind. All buildings shall have solid walls in each exterior endwall line (the side of a building that is parallel to the span of the roof or floor framing) and sidewall line (the side of a building that is perpendicular to the span of the roof or floor framing) to resist lateral in-plane wind forces. The site-appropriate basic wind speed and exposure category shall be used in Tables R611.7(1A) through (1C) to determine the unreduced total length, *UR*, of solid wall required in each exterior endwall line and sidewall line. For buildings with a mean roof height of less than 35 feet (10 668 mm), the unreduced values determined from Tables R611.7(1A) though (1C) is permitted by multiplying by the applicable factor, R_1, from Table R611.7(2); however, reduced values shall not be less than the minimum values in Tables R611.7(1A) through (1C). Where the floor-to-ceiling height of a *story* is less than 10 feet (3048 mm), the unreduced values determined from Tables R611.7(1A) through (C), including minimum values, is permitted to be reduced by multiplying by the applicable factor, R_2, from Table R611.7(3). To account for different design strengths than assumed in determining the values in Tables R611.7(1A) through (1C), the unreduced lengths determined from Tables R611.7(1A) through (1C), including minimum values, are permitted to be reduced by multiplying by the applicable factor, R_3, from Table R611.7(4). The reductions permitted by Tables R611.7(2), R611.7(3) and R611.7(4) are cumulative.

The total length of solid wall segments, *TL*, in a wall line that comply with the minimum length requirements of Section R611.7.2.1 [see Figure R611.7(1)] shall be equal to or greater than the product of the unreduced length of solid wall from Tables R611.7(1A) through (1C), *UR* and the applicable reduction factors, if any, from Tables R611.7(2), R611.7(3) and R611.7(4) as indicated by Equation R6-1.

$$TL \geq R_1 \cdot R_2 \cdot R_3 \cdot UR \qquad \textbf{(Equation R6-1)}$$

where:

TL = Total length of solid wall segments in a wall line that comply with Section R611.7.2.1 [see Figure R611.7(1)];

R_1 = 1.0 or reduction factor for mean roof height from Table R611.7(2);

R_2 = 1.0 or reduction factor for floor-to-ceiling wall height from Table R611.7(3);

R_3 = 1.0 or reduction factor for design strength from Table R611.7(4), and

UR = Unreduced length of solid wall from Tables R611.7(1A) through (1C).

The total length of solid wall in a wall line, *TL*, shall not be less than that provided by two solid wall segments complying with the minimum length requirements of Section R611.7.2.1.

To facilitate determining the required wall thickness, wall type, number and *grade* of vertical bars at the each end of each solid wall segment, and whether shear reinforcement is required, use of Equation R6-2 is permitted.

$$R \leq \frac{TL}{R_1 \cdot R_2 \cdot UR} \qquad \textbf{(Equation R6-2)}$$

After determining the maximum permitted value of the reduction factor for design strength, R_3, in accordance with Equation R6-2, select a wall type from Table R611.7(4) with R_3 less than or equal to the value calculated.

TABLE R611.7(1A)
UNREDUCED LENGTH, *UR*, OF SOLID WALL REQUIRED IN EACH EXTERIOR ENDWALL FOR WIND PERPENDICULAR TO RIDGE ONE STORY OR TOP STORY OF TWO STORY[a, c, d, e, f, g]

SIDEWALL LENGTH (feet)	ENDWALL LENGTH (feet)	ROOF SLOPE	UNREDUCED LENGTH, *UR*, OF SOLID WALL REQUIRED IN ENDWALLS FOR WIND PERPENDICULAR TO RIDGE (feet)						
			Basic Wind Speed (mph) Exposure						
			85B	90B	100B	110B	120B	130B	Minimum[b]
					85C	90C	100C	110C	
						85D	90D	100D	
15	15	< 1:12	0.90	1.01	1.25	1.51	1.80	2.11	0.98
		5:12	1.25	1.40	1.73	2.09	2.49	2.92	1.43
		7:12	1.75	1.96	2.43	2.93	3.49	4.10	1.64
		12:12	2.80	3.13	3.87	4.68	5.57	6.54	2.21
	30	< 1:12	0.90	1.01	1.25	1.51	1.80	2.11	1.09
		5:12	1.25	1.40	1.73	2.09	2.49	2.92	2.01
		7:12	2.43	2.73	3.37	4.08	4.85	5.69	2.42
		12:12	4.52	5.07	6.27	7.57	9.01	10.58	3.57
	45	< 1:12	0.90	1.01	1.25	1.51	1.80	2.11	1.21
		5:12	1.25	1.40	1.73	2.09	2.49	2.92	2.59
		7:12	3.12	3.49	4.32	5.22	6.21	7.29	3.21
		12:12	6.25	7.00	8.66	10.47	12.45	14.61	4.93
	60	< 1:12	0.90	1.01	1.25	1.51	1.80	2.11	1.33
		5:12	1.25	1.40	1.73	2.09	2.49	2.92	3.16
		7:12	3.80	4.26	5.26	6.36	7.57	8.89	3.99
		12:12	7.97	8.94	11.05	13.36	15.89	18.65	6.29
30	15	< 1:12	1.61	1.80	2.23	2.70	3.21	3.77	1.93
		5:12	2.24	2.51	3.10	3.74	4.45	5.23	2.75
		7:12	3.15	3.53	4.37	5.28	6.28	7.37	3.12
		12:12	4.90	5.49	6.79	8.21	9.77	11.46	4.14
	30	< 1:12	1.61	1.80	2.23	2.70	3.21	3.77	2.14
		5:12	2.24	2.51	3.10	3.74	4.45	5.23	3.78
		7:12	4.30	4.82	5.96	7.20	8.57	10.05	4.52
		12:12	7.79	8.74	10.80	13.06	15.53	18.23	6.57
	45	< 1:12	1.61	1.80	2.23	2.70	3.21	3.77	2.35
		5:12	2.24	2.51	3.10	3.74	4.45	5.23	4.81
		7:12	5.44	6.10	7.54	9.12	10.85	12.73	5.92
		12:12	10.69	11.98	14.81	17.90	21.30	25.00	9.00
	60	< 1:12	1.61	1.80	2.23	2.70	3.21	3.77	2.56
		5:12	2.24	2.51	3.10	3.74	4.45	5.23	5.84
		7:12	6.59	7.39	9.13	11.04	13.14	15.41	7.32
		12:12	13.58	15.22	18.82	22.75	27.07	31.77	11.43

(continued)

TABLE R611.7(1A)—continued
UNREDUCED LENGTH, UR, OF SOLID WALL REQUIRED IN EACH EXTERIOR ENDWALL FOR WIND PERPENDICULAR TO RIDGE ONE STORY OR TOP STORY OF TWO STORY[a, c, d, e, f, g]

SIDEWALL LENGTH (feet)	ENDWALL LENGTH (feet)	ROOF SLOPE	UNREDUCED LENGTH, UR, OF SOLID WALL REQUIRED IN ENDWALLS FOR WIND PERPENDICULAR TO RIDGE (feet)						
			Basic Wind Speed (mph) Exposure						
			85B	90B	100B	110B	120B	130B	Minimum[b]
					85C	90C	100C	110C	
						85D	90D	100D	
60	15	< 1:12	2.99	3.35	4.14	5.00	5.95	6.98	3.83
		5:12	4.15	4.65	5.75	6.95	8.27	9.70	5.37
		7:12	5.91	6.63	8.19	9.90	11.78	13.83	6.07
		12:12	9.05	10.14	12.54	15.16	18.03	21.16	8.00
	30	< 1:12	2.99	3.35	4.14	5.00	5.95	6.98	4.23
		5:12	4.15	4.65	5.75	6.95	8.27	9.70	7.31
		7:12	7.97	8.94	11.05	13.36	15.89	18.65	8.71
		12:12	14.25	15.97	19.74	23.86	28.40	33.32	12.57
	45	< 1:12	3.11	3.48	4.30	5.20	6.19	7.26	4.63
		5:12	4.31	4.84	5.98	7.23	8.60	10.09	9.25
		7:12	10.24	11.47	14.19	17.15	20.40	23.84	11.35
		12:12	19.84	22.24	27.49	33.23	39.54	46.40	17.14
	60	< 1:12	3.22	3.61	4.46	5.39	6.42	7.53	5.03
		5:12	4.47	5.01	6.19	7.49	8.91	10.46	11.19
		7:12	12.57	14.09	17.42	21.05	25.05	29.39	13.99
		12:12	25.61	28.70	35.49	42.90	51.04	59.90	21.71

For SI: 1 inch = 25.4 mm, 1 foot = 304.8 mm, 1 mile per hour = 0.447 m/s, 1 pound-force per linear foot = 0.146 kN/m, 1 pound per square foot = 47.88 Pa.

a. Tabulated lengths were derived by calculating design wind pressures in accordance with Figure 6-10 of ASCE 7 for a building with a mean roof height of 35 feet. For wind perpendicular to the ridge, the effects of a 2-foot overhang on each endwall are included. The design pressures were used to calculate forces to be resisted by solid wall segments in each endwall [Table R611.7(1A) or R611.7(1B) or sidewall (Table R611.7(1C)], as appropriate. The forces to be resisted by each wall line were then divided by the default design strength of 840 pounds per linear foot of length to determine the required solid wall length. The actual mean roof height of the building shall not exceed the least horizontal dimension of the building.

b. Tabulated lengths in the "minimum" column are based on the requirement of Section 6.1.4.1 of ASCE 7 that the main windforce-resisting system be designed for a minimum service level force of 10 psf multiplied by the area of the building projected onto a vertical plane normal to the assumed wind direction. Tabulated lengths in shaded cells are less than the "minimum" value. Where the minimum controls, it is permitted to be reduced in accordance with Notes c, d and e. See Section R611.7.1.1.

c. For buildings with a mean roof height of less than 35 feet, tabulated lengths are permitted to be reduced by multiplying by the appropriate factor, R_1, from Table R611.7(2). The reduced length shall not be less than the "minimum" value shown in the table.

d. Tabulated lengths for "one story or top story of two story" are based on a floor-to-ceiling height of 10 feet. Tabulated lengths for "first story of two story" are based on floor-to-ceiling heights of 10 feet each for the first and second story. For floor-to-ceiling heights less than assumed, use the lengths in Table R611.7(1A), (1B) or (1C), or multiply the value in the table by the reduction factor, R_2, from Table R611.7(3).

e. Tabulated lengths are based on the default design shear strength of 840 pounds per linear foot of solid wall segment. The tabulated lengths are permitted to be reduced by multiplying by the applicable reduction factor for design strength, R_3, from Table R611.7(4).

f. The reduction factors, R_1, R_2 and R_3, in Tables R611.7(2), R611.7(3), and R611.7(4), respectively, are permitted to be compounded, subject to the limitations of Note b. However, the minimum number and minimum length of solid walls segments in each wall line shall comply with Sections R611.7.1 and R611.7.2.1, respectively.

g. For intermediate values of sidewall length, endwall length, roof slope and basic wind speed, use the next higher value, or determine by interpolation.

TABLE R611.7(1B)
UNREDUCED LENGTH, *UR*, OF SOLID WALL REQUIRED IN EACH EXTERIOR ENDWALL FOR WIND PERPENDICULAR TO RIDGE FIRST STORY OF TWO STORY[a, c, d, e, f, g]

SIDEWALL LENGTH (feet)	ENDWALL LENGTH (feet)	ROOF SLOPE	UNREDUCED LENGTH, *UR*, OF SOLID WALL REQUIRED IN ENDWALLS FOR WIND PERPENDICULAR TO RIDGE (feet)						
			Basic Wind Speed (mph) Exposure						
			85B	90B	100B	110B	120B	130B	Minimum[b]
					85C	90C	100C	110C	
						85D	90D	100D	
			Velocity pressure (psf)						
			11.51	12.90	15.95	19.28	22.94	26.92	
15	15	< 1:12	2.60	2.92	3.61	4.36	5.19	6.09	2.59
		5:12	3.61	4.05	5.00	6.05	7.20	8.45	3.05
		7:12	3.77	4.23	5.23	6.32	7.52	8.82	3.26
		12:12	4.81	5.40	6.67	8.06	9.60	11.26	3.83
	30	< 1:12	2.60	2.92	3.61	4.36	5.19	6.09	2.71
		5:12	3.61	4.05	5.00	6.05	7.20	8.45	3.63
		7:12	4.45	4.99	6.17	7.46	8.88	10.42	4.04
		12:12	6.54	7.33	9.06	10.96	13.04	15.30	5.19
	45	< 1:12	2.60	2.92	3.61	4.36	5.19	6.09	2.83
		5:12	3.61	4.05	5.00	6.05	7.20	8.45	4.20
		7:12	5.14	5.76	7.12	8.60	10.24	12.01	4.83
		12:12	8.27	9.27	11.46	13.85	16.48	19.34	6.55
	60	< 1:12	2.60	2.92	3.61	4.36	5.19	6.09	2.95
		5:12	3.61	4.05	5.00	6.05	7.20	8.45	4.78
		7:12	5.82	6.52	8.06	9.75	11.60	13.61	5.61
		12:12	9.99	11.20	13.85	16.74	19.92	23.37	7.90
30	15	< 1:12	4.65	5.21	6.45	7.79	9.27	10.88	5.16
		5:12	6.46	7.24	8.95	10.82	12.87	15.10	5.98
		7:12	6.94	7.78	9.62	11.62	13.83	16.23	6.35
		12:12	8.69	9.74	12.04	14.55	17.32	20.32	7.38
	30	< 1:12	4.65	5.21	6.45	7.79	9.27	10.88	5.38
		5:12	6.46	7.24	8.95	10.82	12.87	15.10	7.01
		7:12	8.09	9.06	11.21	13.54	16.12	18.91	7.76
		12:12	11.58	12.98	16.05	19.40	23.08	27.09	9.81
	45	< 1:12	4.65	5.21	6.45	7.79	9.27	10.88	5.59
		5:12	6.46	7.24	8.95	10.82	12.87	15.10	8.04
		7:12	9.23	10.35	12.79	15.46	18.40	21.59	9.16
		12:12	14.48	16.22	20.06	24.25	28.85	33.86	12.24
	60	< 1:12	4.65	5.21	6.45	7.79	9.27	10.88	5.80
		5:12	6.46	7.24	8.95	10.82	12.87	15.10	9.08
		7:12	10.38	11.63	14.38	17.38	20.69	24.27	10.56
		12:12	17.37	19.47	24.07	29.10	34.62	40.63	14.67

(continued)

TABLE R611.7(1B)—continued
UNREDUCED LENGTH, *UR*, OF SOLID WALL REQUIRED IN EACH EXTERIOR ENDWALL FOR WIND PERPENDICULAR TO RIDGE FIRST STORY OF TWO STORY[a, c, d, e, f, g]

SIDEWALL LENGTH (feet)	ENDWALL LENGTH (feet)	ROOF SLOPE	UNREDUCED LENGTH, *UR*, OF SOLID WALL REQUIRED IN ENDWALLS FOR WIND PERPENDICULAR TO RIDGE (feet)						
			Basic Wind Speed (mph) Exposure						
			85B	90B	100B	110B	120B	130B	Minimum[b]
					85C	90C	100C	110C	
						85D	90D	100D	
			Velocity pressure (psf)						
			11.51	12.90	15.95	19.28	22.94	26.92	
60	15	< 1:12	8.62	9.67	11.95	14.45	17.19	20.17	10.30
		5:12	11.98	13.43	16.61	20.07	23.88	28.03	11.85
		7:12	13.18	14.78	18.27	22.08	26.28	30.83	12.54
		12:12	16.32	18.29	22.62	27.34	32.53	38.17	14.48
	30	< 1:12	8.62	9.67	11.95	14.45	17.19	20.17	10.70
		5:12	11.98	13.43	16.61	20.07	23.88	28.03	13.79
		7:12	15.25	17.09	21.13	25.54	30.38	35.66	15.18
		12:12	21.52	24.12	29.82	36.05	42.89	50.33	19.05
	45	< 1:12	8.97	10.06	12.43	15.03	17.88	20.99	11.10
		5:12	12.46	13.97	17.27	20.88	24.84	29.15	15.73
		7:12	17.67	19.80	24.48	29.59	35.21	41.32	17.82
		12:12	27.27	30.56	37.79	45.68	54.35	63.78	23.62
	60	< 1:12	9.30	10.43	12.89	15.58	18.54	21.76	11.50
		5:12	12.91	14.47	17.90	21.63	25.74	30.20	17.67
		7:12	20.14	22.58	27.91	33.74	40.15	47.11	20.46
		12:12	33.19	37.19	45.99	55.59	66.14	77.62	28.19

For SI: 1 inch = 25.4 mm, 1 foot = 304.8 mm, 1 mile per hour = 0.447 m/s, 1 pound force per linear foot = 0.146 kN/m, 1 pound per square foot = 47.88 Pa.

a. Tabulated lengths were derived by calculating design wind pressures in accordance with Figure 6-10 of ASCE 7 for a building with a mean roof height of 35 feet. For wind perpendicular to the ridge, the effects of a 2-foot overhang on each endwall are included. The design pressures were used to calculate forces to be resisted by solid wall segments in each endwall [Table R611.7(1A) or R611.7(1B)] or sidewall [Table R611.7(1C)], as appropriate. The forces to be resisted by each wall line were then divided by the default design strength of 840 pounds per linear foot of length to determine the required solid wall length. The actual mean roof height of the building shall not exceed the least horizontal dimension of the building.

b. Tabulated lengths in the "minimum" column are based on the requirement of Section 6.1.4.1 of ASCE 7 that the main windforce-resisting system be designed for a minimum service level force of 10 psf multiplied by the area of the building projected onto a vertical plane normal to the assumed wind direction. Tabulated lengths in shaded cells are less than the "minimum" value. Where the minimum controls, it is permitted to be reduced in accordance with Notes c, d and e. See Section R611.7.1.1.

c. For buildings with a mean roof height of less than 35 feet, tabulated lengths are permitted to be reduced by multiplying by the appropriate factor, R_1, from Table R611.7(2). The reduced length shall not be less than the "minimum" value shown in the table.

d. Tabulated lengths for "one story or top story of two story" are based on a floor-to-ceiling height of 10 feet. Tabulated lengths for "first story of two story" are based on floor-to-ceiling heights of 10 feet each for the first and second story. For floor-to-ceiling heights less than assumed, use the lengths in Table R611.7(1A), (1B) or (1C), or multiply the value in the table by the reduction factor, R_2, from Table R611.7(3).

e. Tabulated lengths are based on the default design shear strength of 840 pounds per linear foot of solid wall segment. The tabulated lengths are permitted to be reduced by multiplying by the applicable reduction factor for design strength, R_3, from Table R611.7(4).

f. The reduction factors, R_1, R_2 and R_3, in Tables R611.7(2), R611.7(3), and R611.7(4), respectively, are permitted to be compounded, subject to the limitations of Note b. However, the minimum number and minimum length of solid walls segments in each wall line shall comply with Sections R611.7.1 and R611.7.2.1, respectively.

g. For intermediate values of sidewall length, endwall length, roof slope and basic wind speed, use the next higher value, or determine by interpolation.

TABLE R611.7(1C)
UNREDUCED LENGTH, *UR*, OF SOLID WALL REQUIRED IN EACH EXTERIOR SIDEWALL FOR WIND PARALLEL TO RIDGE[a, c, d, e, f, g]

SIDEWALL LENGTH (feet)	ENDWALL LENGTH (feet)	ROOF SLOPE	UNREDUCED LENGTH, *UR*, OF SOLID WALL REQUIRED IN ENDWALLS FOR WIND PERPENDICULAR TO RIDGE (feet)						
			Basic Wind Speed (mph) Exposure						
			85B	90B	100B	110B	120B	130B	Minimum[b]
					85C	90C	100C	110C	
						85D	90D	100D	
			One story or top story of two story						
< 30	15	< 1:12	0.95	1.06	1.31	1.59	1.89	2.22	0.90
		5:12	1.13	1.26	1.56	1.88	2.24	2.63	1.08
		7:12	1.21	1.35	1.67	2.02	2.40	2.82	1.17
		12:12	1.43	1.60	1.98	2.39	2.85	3.34	1.39
	30	< 1:12	1.77	1.98	2.45	2.96	3.53	4.14	1.90
		5:12	2.38	2.67	3.30	3.99	4.75	5.57	2.62
		7:12	2.66	2.98	3.69	4.46	5.31	6.23	2.95
		12:12	3.43	3.85	4.76	5.75	6.84	8.03	3.86
	45	< 1:12	2.65	2.97	3.67	4.43	5.27	6.19	2.99
		5:12	3.98	4.46	5.51	6.66	7.93	9.31	4.62
		7:12	4.58	5.14	6.35	7.68	9.14	10.72	5.36
		12:12	6.25	7.01	8.67	10.48	12.47	14.63	7.39
	60	< 1:12	3.59	4.03	4.98	6.02	7.16	8.40	4.18
		5:12	5.93	6.65	8.22	9.93	11.82	13.87	7.07
		7:12	6.99	7.83	9.69	11.71	13.93	16.35	8.38
		12:12	9.92	11.12	13.75	16.62	19.77	23.21	12.00
60	45	< 1:12	2.77	3.11	3.84	4.65	5.53	6.49	2.99
		5:12	4.15	4.66	5.76	6.96	8.28	9.72	4.62
		7:12	4.78	5.36	6.63	8.01	9.53	11.18	5.36
		12:12	6.51	7.30	9.03	10.91	12.98	15.23	7.39
	60	< 1:12	3.86	4.32	5.35	6.46	7.69	9.02	4.18
		5:12	6.31	7.08	8.75	10.57	12.58	14.76	7.07
		7:12	7.43	8.32	10.29	12.44	14.80	17.37	8.38
		12:12	10.51	11.78	14.56	17.60	20.94	24.57	12.00

(continued)

TABLE R611.7(1C)—continued UNREDUCED LENGTH, *UR*, OF SOLID WALL REQUIRED IN EACH EXTERIOR SIDEWALL FOR WIND PARALLEL TO RIDGE[a, c, d, e, f, g]

SIDEWALL LENGTH (feet)	ENDWALL LENGTH (feet)	ROOF SLOPE	UNREDUCED LENGTH, *UR*, OF SOLID WALL REQUIRED IN ENDWALLS FOR WIND PERPENDICULAR TO RIDGE (feet)						
			Basic Wind Speed (mph) Exposure						
			85B	90B	100B	110B	120B	130B	Minimum[b]
					85C	90C	100C	110C	
						85D	90D	100D	
			One story or top story of two story						
First story of two story									
< 30	15	< 1:12	2.65	2.97	3.67	4.44	5.28	6.20	2.52
		5:12	2.83	3.17	3.92	4.74	5.64	6.62	2.70
		7:12	2.91	3.26	4.03	4.87	5.80	6.80	2.79
		12:12	3.13	3.51	4.34	5.25	6.24	7.32	3.01
	30	< 1:12	4.81	5.39	6.67	8.06	9.59	11.25	5.14
		5:12	5.42	6.08	7.52	9.09	10.81	12.69	5.86
		7:12	5.70	6.39	7.90	9.55	11.37	13.34	6.19
		12:12	6.47	7.25	8.97	10.84	12.90	15.14	7.10
	45	< 1:12	6.99	7.83	9.69	11.71	13.93	16.35	7.85
		5:12	8.32	9.33	11.53	13.94	16.59	19.47	9.48
		7:12	8.93	10.01	12.37	14.95	17.79	20.88	10.21
		12:12	10.60	11.88	14.69	17.75	21.13	24.79	12.25
	60	< 1:12	9.23	10.35	12.79	15.46	18.40	21.59	10.65
		5:12	11.57	12.97	16.03	19.38	23.06	27.06	13.54
		7:12	12.63	14.15	17.50	21.15	25.17	29.54	14.85
		12:12	15.56	17.44	21.56	26.06	31.01	36.39	18.48
60	45	< 1:12	7.34	8.22	10.17	12.29	14.62	17.16	7.85
		5:12	8.72	9.77	12.08	14.60	17.37	20.39	9.48
		7:12	9.34	10.47	12.95	15.65	18.62	21.85	10.21
		12:12	11.08	12.41	15.35	18.55	22.07	25.90	12.25
	60	< 1:12	9.94	11.14	13.77	16.65	19.81	23.25	10.65
		5:12	12.40	13.89	17.18	20.76	24.70	28.99	13.54
		7:12	13.51	15.14	18.72	22.63	26.92	31.60	14.85
		12:12	16.59	18.59	22.99	27.79	33.06	38.80	18.48

For SI: 1 inch = 25.4 mm, 1 foot = 304.8 mm, 1 mile per hour = 0.447 m/s, 1 pound force per linear foot = 0.146 kN/m, 1 pound per square foot = 47.88 Pa.

a. Tabulated lengths were derived by calculating design wind pressures in accordance with Figure 6-10 of ASCE 7 for a building with a mean roof height of 35 feet. For wind perpendicular to the ridge, the effects of a 2-foot overhang on each endwall are included. The design pressures were used to calculate forces to be resisted by solid wall segments in each endwall [Table R611.7(1A) or R611.7(1B)] or sidewall [(Table R611.7(1C)], as appropriate. The forces to be resisted by each wall line were then divided by the default design strength of 840 pounds per linear foot of length to determine the required solid wall length. The actual mean roof height of the building shall not exceed the least horizontal dimension of the building.

b. Tabulated lengths in the "minimum" column are based on the requirement of Section 6.1.4.1 of ASCE 7 that the main windforce-resisting system be designed for a minimum service level force of 10 psf multiplied by the area of the building projected onto a vertical plane normal to the assumed wind direction. Tabulated lengths in shaded cells are less than the "minimum" value. Where the minimum controls, it is permitted to be reduced in accordance with Notes c, d and e. See Section R611.7.1.1.

c. For buildings with a mean roof height of less than 35 feet, tabulated lengths are permitted to be reduced by multiplying by the appropriate factor, R_1, from Table R611.7(2). The reduced length shall not be less than the "minimum" value shown in the table.

d. Tabulated lengths for "one story or top story of two story" are based on a floor-to-ceiling height of 10 feet. Tabulated lengths for "first story of two story" are based on floor-to-ceiling heights of 10 feet each for the first and second story. For floor-to-ceiling heights less than assumed, use the lengths in Table R611.7(1A), (1B) or (1C), or multiply the value in the table by the reduction factor, R_2, from Table R611.7(3).

e. Tabulated lengths are based on the default design shear strength of 840 pounds per linear foot of solid wall segment. The tabulated lengths are permitted to be reduced by multiplying by the applicable reduction factor for design strength, R_3, from Table R611.7(4).

f. The reduction factors, R_1, R_2 and R_3, in Tables R611.7(2), R611.7(3), and R611.7(4), respectively, are permitted to be compounded, subject to the limitations of Note b. However, the minimum number and minimum length of solid walls segments in each wall line shall comply with Sections R611.7.1 and R611.7.2.1, respectively.

g. For intermediate values of sidewall length, endwall length, roof slope and basic wind speed, use the next higher value, or determine by interpolation.

TABLE R611.7(2)
REDUCTION FACTOR, R_1, FOR BUILDINGS WITH MEAN ROOF HEIGHT LESS THAN 35 FEET[a]

MEAN ROOF HEIGHT[b, c] (feet)	REDUCTION FACTOR R_1, FOR MEAN ROOF HEIGHT		
	Exposure category		
	B	C	D
< 15	0.96	0.84	0.87
20	0.96	0.89	0.91
25	0.96	0.93	0.94
30	0.96	0.97	0.98
35	1.00	1.00	1.00

For SI: 1 foot = 304.8 mm, 1 degree = 0.0175 rad.

a. See Section R611.7.1.1 and Note c to Table R611.7(1A) for application of reduction factors in this table. This reduction is not permitted for "minimum" values.

b. For intermediate values of mean roof height, use the factor for the next greater height, or determine by interpolation.

c. Mean roof height is the average of the roof eave height and height of the highest point on the roof surface, except that for roof slopes of less than or equal to $2^1/_8$:12 (10 degrees), the mean roof height is permitted to be taken as the roof eave height.

TABLE R611.7(3)
REDUCTION FACTOR, R_2, FOR FLOOR-TO-CEILING WALL HEIGHTS LESS THAN 10 FEET[a, b]

STORY UNDER CONSIDERATION	FLOOR-TO-CEILING HEIGHT[c] (feet)	ENDWALL LENGTH (feet)	ROOF SLOPE	REDUCTION FACTOR, R_2
Endwalls—for wind perpendicular to ridge				
One story or top story of two story	8	15	< 5:12	0.83
			7:12	0.90
			12:12	0.94
		60	< 5:12	0.83
			7:12	0.95
			12:12	0.98
First story of two story	16 combined first and second story	15	< 5:12	0.83
			7:12	0.86
			12:12	0.89
		60	< 5:12	0.83
			7:12	0.91
			12:12	0.95
Sidewalls—for wind parallel to ridge				
One story or top story of two story	8	15	< 1:12	0.84
			5:12	0.87
			7:12	0.88
			12:12	0.89
		60	< 1:12	0.86
			5:12	0.92
			7:12	0.93
			12:12	0.95
First story of two story	16 combined first and second story	15	< 1:12	0.83
			5:12	0.84
			7:12	0.85
			12:12	0.86
		60	< 1:12	0.84
			5:12	0.87
			7:12	0.88
			12:12	0.90

For SI: 1 foot = 304.8 mm.

a. See Section R611.7.1.1 and Note d to Table R611.7(1A) for application of reduction factors in this table.

b. For intermediate values of endwall length, and/or roof slope, use the next higher value, or determine by interpolation.

c. Tabulated values in Table R611.7(1A) and (1C) for "one story or top story of two story" are based on a floor-to-ceiling height of 10 feet. Tabulated values in Table R611.7(1B) and (1C) for "first story of two story" are based on floor-to-ceiling heights of 10 feet each for the first and second story. For floor to ceiling heights between those shown in this table and those assumed in Table R611.7(1A), (1B) or (1C), use the solid wall lengths in Table R611.7(1A), (1B) or (1C), or determine the reduction factor by interpolating between 1.0 and the factor shown in this table.

TABLE R611.7(4)
REDUCTION FACTOR FOR DESIGN STRENGTH, R_3, FOR FLAT, WAFFLE- AND SCREEN-GRID WALLS[a, c]

NOMINAL THICKNESS OF WALL (inches)	VERTICAL BARS AT EACH END OF SOLID WALL SEGMENT		VERTICAL REINFORCEMENT LAYOUT DETAIL [see Figure R611.7(2)]	REDUCTION FACTOR, R_3, FOR LENGTH OF SOLID WALL			
				Horizontal and vertical shear reinforcement provided			
				No		Yes[d]	
	Number of bars	Bar size		40,000[b]	60,000[b]	40,000[b]	60,000[b]
Flat walls							
4	2	4	1	0.74	0.61	0.74	0.50
	3	4	2	0.61	0.61	0.52	0.27
	2	5	1	0.61	0.61	0.48	0.25
	3	5	2	0.61	0.61	0.26	0.18
6	2	4	3	0.70	0.48	0.70	0.48
	3	4	4	0.49	0.38	0.49	0.33
	2	5	3	0.46	0.38	0.46	0.31
	3	5	4	0.38	0.38	0.32	0.16
8	2	4	3	0.70	0.47	0.70	0.47
	3	4	5	0.47	0.32	0.47	0.32
	2	5	3	0.45	0.31	0.45	0.31
	4	4	6	0.36	0.28	0.36	0.25
	3	5	5	0.31	0.28	0.31	0.16
	4	5	6	0.28	0.28	0.24	0.12
10	2	4	3	0.70	0.47	0.70	0.47
	2	5	3	0.45	0.30	0.45	0.30
	4	4	7	0.36	0.25	0.36	0.25
	6	4	8	0.25	0.22	0.25	0.13
	4	5	7	0.24	0.22	0.24	0.12
	6	5	8	0.22	0.22	0.12	0.08
Waffle-grid walls[e]							
6	2	4	3	0.78	0.78	0.70	0.48
	3	4	4	0.78	0.78	0.49	0.25
	2	5	3	0.78	0.78	0.46	0.23
	3	5	4	0.78	0.78	0.24	0.16
8	2	4	3	0.78	0.78	0.70	0.47
	3	4	5	0.78	0.78	0.47	0.24
	2	5	3	0.78	0.78	0.45	0.23
	4	4	6	0.78	0.78	0.36	0.18
	3	5	5	0.78	0.78	0.23	0.16
	4	5	6	0.78	0.78	0.18	0.13
Screen-grid walls[e]							
6	2	4	3	0.93	0.93	0.70	0.48
	3	4	4	0.93	0.93	0.49	0.25
	2	5	3	0.93	0.93	0.46	0.23
	3	5	4	0.93	0.93	0.24	0.16

For SI: 1 inch = 25.4 mm, 1,000 pounds per square inch = 6.895 MPa.

a. See Note e to Table R611.7(1A) for application of adjustment factors in this table.

b. Yield strength in pounds per square inch of vertical wall reinforcement at ends of solid wall segments.

c. Values are based on concrete with a specified compressive strength, f'_c, of 2,500 psi. Where concrete with f'_c of not less than 3,000 psi is used, values in shaded cells are permitted to be decreased by multiplying by 0.91.

d. Horizontal and vertical shear reinforcement shall be provided in accordance with Section R611.7.2.2.

e. Each end of each solid wall segment shall have rectangular flanges. In the through-the-wall dimension, the flange shall not be less than $5^1/_2$ inches for 6-inch-nominal waffle- and screen-grid walls, and not less than $7^1/_2$ inches for 8-inch-nominal waffle-grid walls. In the in-plane dimension, flanges shall be long enough to accommodate the vertical reinforcement required by the layout detail selected from Figure R611.7(2) and provide the cover required by Section R611.5.4.1. If necessary to achieve the required dimensions, form material shall be removed or use of flat wall forms is permitted.

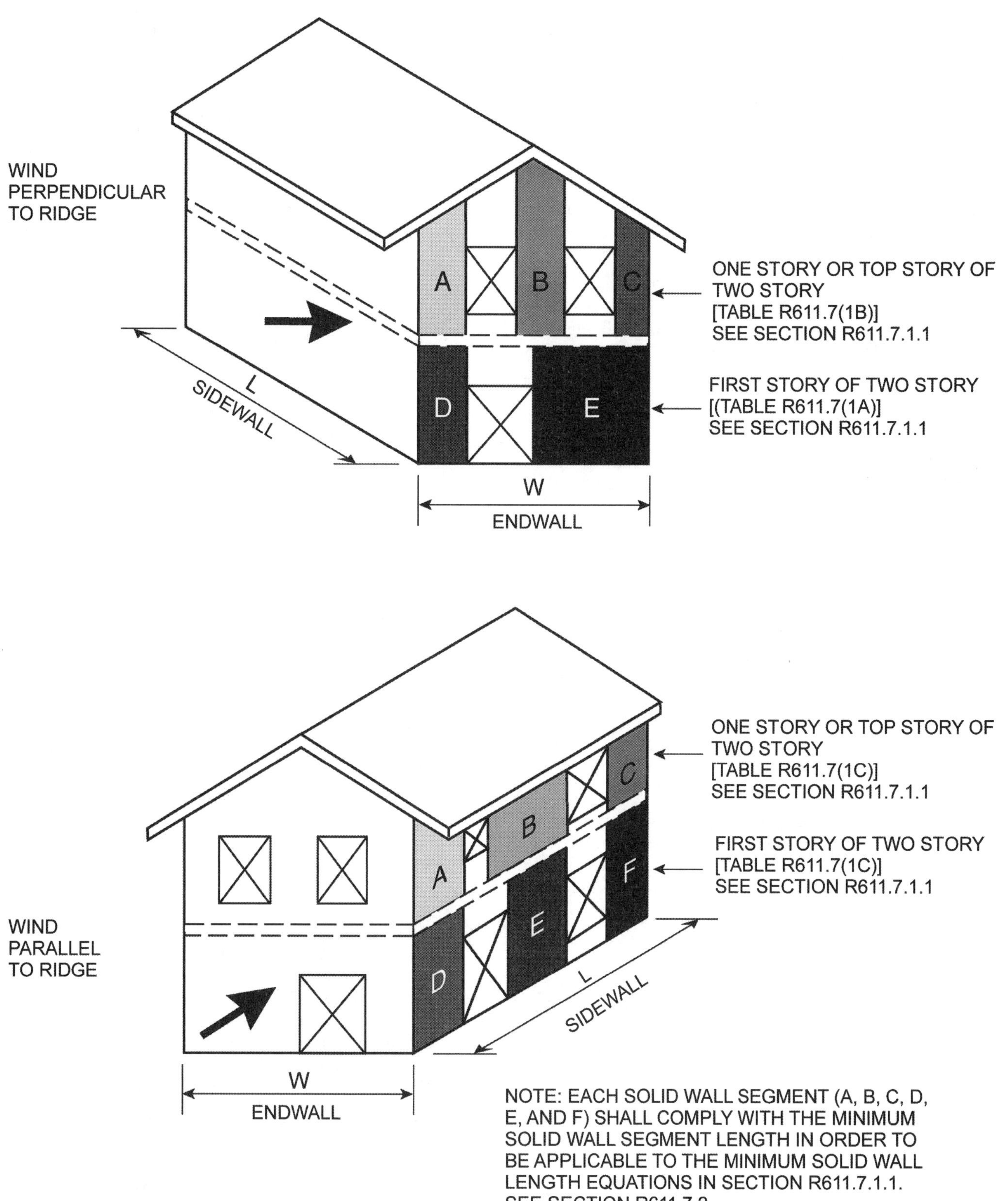

FIGURE R611.7(1)
MINIMUM SOLID WALL LENGTH

R611.7.2 Solid wall segments. Solid wall segments that contribute to the required length of solid wall shall comply with this section. Reinforcement shall be provided in accordance with Section R611.7.2.2 and Table R611.7(4). Solid wall segments shall extend the full story-height without openings, other than openings for the utilities and other building services passing through the wall. In flat walls and waffle-grid walls, such openings shall have an area of less than 30 square inches (19 355 mm^2) with no dimension exceeding $6^1/_4$ inches (159 mm), and shall not be located within 6 inches (152 mm) of the side edges of the solid wall segment. In screen-grid walls, such openings shall be located in the portion of the solid wall segment between horizontal and vertical cores of concrete and opening size and location are not restricted provided no concrete is removed.

R611.7.2.1 Minimum length of solid wall segment and maximum spacing. Only solid wall segments equal to or greater than 24 inches (610 mm) in length shall be included in the total length of solid wall required by Section R611.7.1. In addition, no more than two solid wall segments equal to or greater than 24 inches (610 mm) in length and less than 48 inches (1219 mm) in length shall be included in the required total length of solid wall. The maximum clear opening width shall be 18 feet (5486 mm). See Figure R611.7(1).

R611.7.2.2 Reinforcement in solid wall segments.

R611.7.2.2.1 Horizontal shear reinforcement. Where reduction factors for design strength, R_3, from Table R611.7(4) based on horizontal and vertical shear reinforcement being provided are used, solid wall segments shall have horizontal reinforcement consisting of minimum No. 4 bars. Horizontal shear reinforcement shall be the same grade of steel required for the vertical reinforcement at the ends of solid wall segments by Section R611.7.2.2.2.

The spacing of horizontal reinforcement shall not exceed the smaller of one-half the length of the solid wall segment, minus 2 inches (51 mm), and 18 inches (457 mm). Horizontal shear reinforcement shall terminate in accordance with Section R611.6.4.

R611.7.2.2.2 Vertical reinforcement. Vertical reinforcement applicable to the reduction factor(s) for design strength, R_3, from Table R611.7(4) that is used, shall be located at each end of each solid wall segment in accordance with the applicable detail in Figure R611.7(2). The No. 4 vertical bar required on each side of an opening by Section R611.8.1.2 is permitted to be used as reinforcement at the ends of solid wall segments where installed in accordance with the applicable detail in Figure R611.7(2). There shall be not less than two No. 4 bars at each end of solid wall segments located as required by the applicable detail in Figure R611.7(2). One of the bars at each end of solid wall segments shall be deemed to meet the requirements for vertical wall reinforcement required by Section R611.6.

The vertical wall reinforcement at each end of each solid wall segment shall be developed below the bottom of the adjacent wall opening [see Figure R611.7(3)] by one of the following methods:

1. Where the wall height below the bottom of the adjacent opening is equal to or greater than 22 inches (559 mm) for No. 4 or 28 inches (711 mm) for No. 5 vertical wall reinforcement, reinforcement around openings in accordance with Section R611.8.1 shall be sufficient, or
2. Where the wall height below the bottom of the adjacent opening is less than required by Item 1 above, the vertical wall reinforcement adjacent to the opening shall extend into the footing far enough to develop the bar in tension in accordance with Section R611.5.4.4 and Figure R611.5.4(2), or shall be lap-spliced with a dowel that is embedded in the footing far enough to develop the dowel-bar in tension.

R611.7.2.2.3 Vertical shear reinforcement. Where reduction factors for design strength, R_3, from Table R611.7(4) based on horizontal and vertical shear reinforcement being provided are used, solid wall segments shall have vertical reinforcement consisting of minimum No. 4 bars. Vertical shear reinforcement shall be the same grade of steel required by Section R611.7.2.2.2 for the vertical reinforcement at the ends of solid wall segments. The spacing of vertical reinforcement throughout the length of the segment shall not exceed the smaller of one third the length of the segment, and 18 inches (457 mm). Vertical shear reinforcement shall be continuous between stories in accordance with Section R611.6.3, and shall terminate in accordance with Section R611.6.4. Vertical shear reinforcement required by this section is permitted to be used for vertical reinforcement required by Table R611.6(1), R611.6(2), R611.6(3) or R611.6(4), whichever is applicable.

R611.7.2.3 Solid wall segments at corners. At all interior and exterior corners of exterior walls, a solid wall segment shall extend the full height of each wall *story*. The segment shall have the length required to develop the horizontal reinforcement above and below the adjacent opening in tension in accordance with Section R611.5.4.4. For an exterior corner, the limiting dimension is measured on the outside of the wall, and for an interior corner the limiting dimension is measured on the inside of the wall. See Section R611.8.1. The length of a segment contributing to the required length of solid wall shall comply with Section R611.7.2.1.

The end of a solid wall segment complying with the minimum length requirements of Section R611.7.2.1 shall be located no more than 6 feet (1829 mm) from each corner.

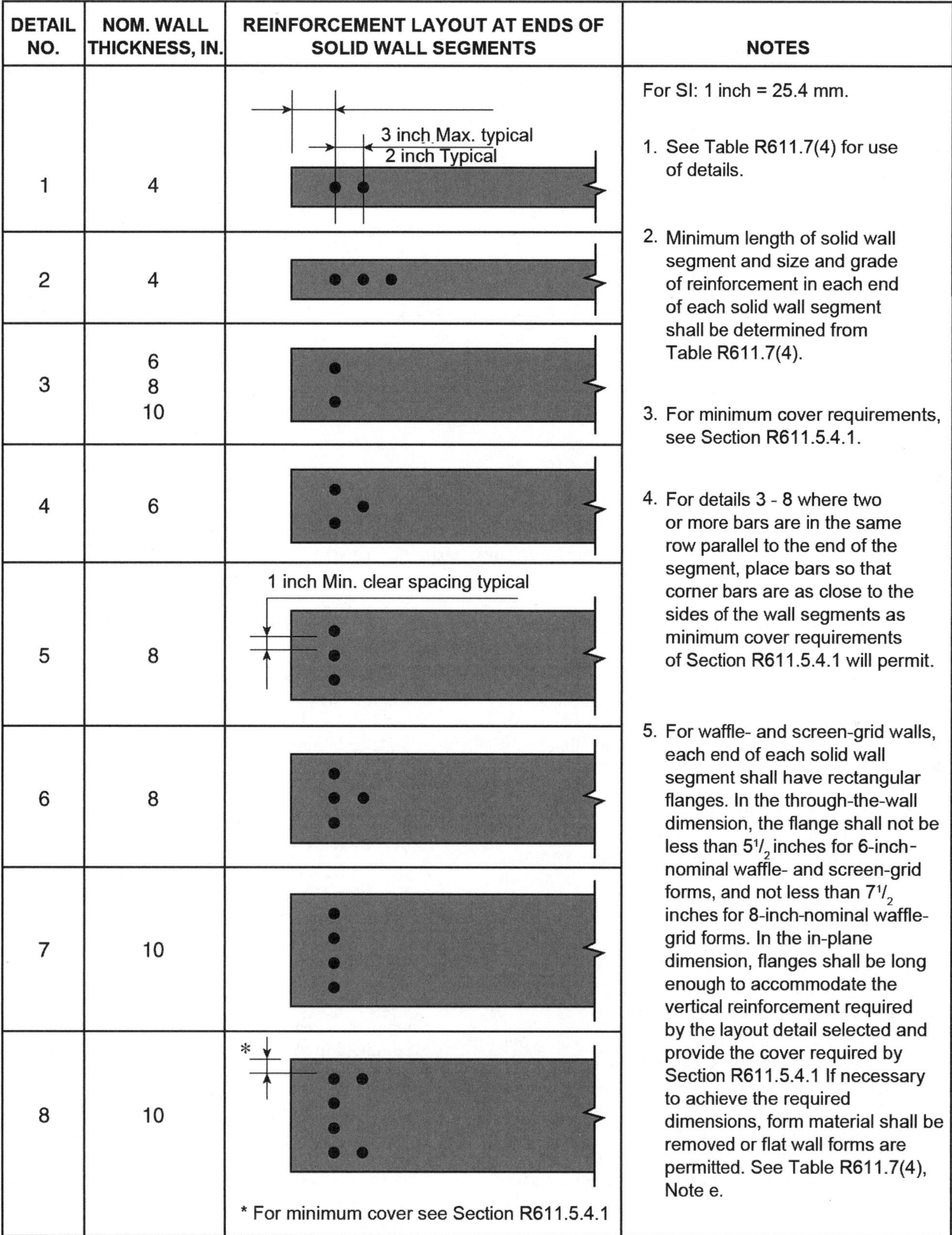

DETAIL NO.	NOM. WALL THICKNESS, IN.	REINFORCEMENT LAYOUT AT ENDS OF SOLID WALL SEGMENTS
1	4	3 inch Max. typical 2 inch Typical
2	4	
3	6 8 10	
4	6	
5	8	1 inch Min. clear spacing typical
6	8	
7	10	
8	10	*

* For minimum cover see Section R611.5.4.1

NOTES

For SI: 1 inch = 25.4 mm.

1. See Table R611.7(4) for use of details.
2. Minimum length of solid wall segment and size and grade of reinforcement in each end of each solid wall segment shall be determined from Table R611.7(4).
3. For minimum cover requirements, see Section R611.5.4.1.
4. For details 3 - 8 where two or more bars are in the same row parallel to the end of the segment, place bars so that corner bars are as close to the sides of the wall segments as minimum cover requirements of Section R611.5.4.1 will permit.
5. For waffle- and screen-grid walls, each end of each solid wall segment shall have rectangular flanges. In the through-the-wall dimension, the flange shall not be less than $5^1/_2$ inches for 6-inch-nominal waffle- and screen-grid forms, and not less than $7^1/_2$ inches for 8-inch-nominal waffle-grid forms. In the in-plane dimension, flanges shall be long enough to accommodate the vertical reinforcement required by the layout detail selected and provide the cover required by Section R611.5.4.1 If necessary to achieve the required dimensions, form material shall be removed or flat wall forms are permitted. See Table R611.7(4), Note e.

FIGURE R611.7(2)
VERTICAL REINFORCEMENT LAYOUT DETAIL

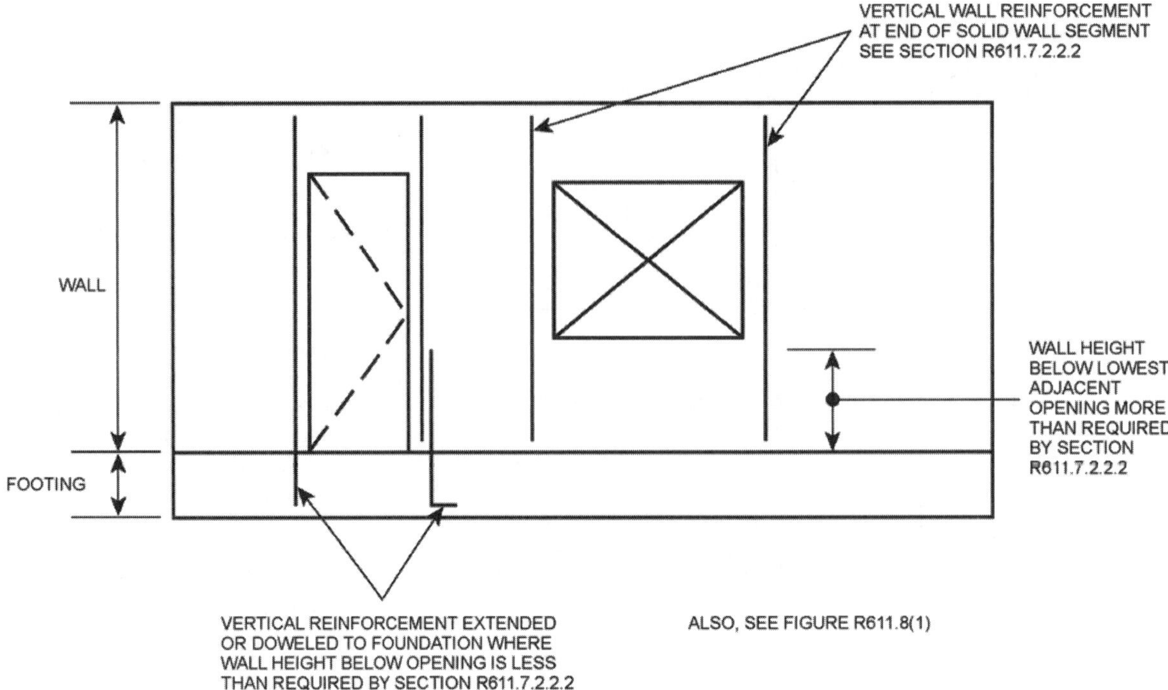

FIGURE R611.7(3)
VERTICAL WALL REINFORCEMENT ADJACENT TO WALL OPENINGS

R611.8 Requirements for lintels and reinforcement around openings.

R611.8.1 Reinforcement around openings. Reinforcement shall be provided around openings in walls equal to or greater than 2 feet (610 mm) in width in accordance with this section and Figure R611.8(1), in addition to the minimum wall reinforcement required by Sections R404.1.2, R611.6 and R611.7. Vertical wall reinforcement required by this section is permitted to be used as reinforcement at the ends of solid wall segments required by Section R611.7.2.2.2 provided it is located in accordance with Section R611.8.1.2. Wall openings shall have a minimum depth of concrete over the width of the opening of 8 inches (203 mm) in flat walls and waffle-grid walls, and 12 inches (305 mm) in screen-grid walls. Wall openings in waffle-grid and screen-grid walls shall be located such that not less than one-half of a vertical core occurs along each side of the opening.

R611.8.1.1 Horizontal reinforcement. Lintels complying with Section R611.8.2 shall be provided above wall openings equal to or greater than 2 feet (610 mm) in width.

Exception: Continuous horizontal wall reinforcement placed within 12 inches (305 mm) of the top of the wall *story* as required in Sections R404.1.2.2 and R611.6.2 is permitted in lieu of top or bottom lintel reinforcement required by Section R611.8.2 provided that the continuous horizontal wall reinforcement meets the location requirements specified in Figures R611.8(2), R611.8(3), and R611.8(4) and the size requirements specified in Tables R611.8(2) through R611.8(10).

Openings equal to or greater than 2 feet (610 mm) in width shall have a minimum of one No. 4 bar placed within 12 inches (305 mm) of the bottom of the opening. See Figure R611.8(1).

Horizontal reinforcement placed above and below an opening shall extend beyond the edges of the opening the dimension required to develop the bar in tension in accordance with Section R611.5.4.4.

R611.8.1.2 Vertical reinforcement. Not less than one No. 4 bar [Grade 40 (280 MPa)] shall be provided on each side of openings equal to or greater than 2 feet (610 mm) in width. The vertical reinforcement required by this section shall extend the full height of the wall story and shall be located within 12 inches (305 mm) of each side of the opening. The vertical reinforcement required on each side of an opening by this section is permitted to serve as reinforcement at the ends of solid wall segments in accordance with Section R611.7.2.2.2, provided it is located as required by the applicable detail in Figure R611.7(2). Where the vertical reinforcement required by this section is used to satisfy the requirements of Section R611.7.2.2.2 in waffle- and screen-grid walls, a concrete flange shall be created at the ends of the solid wall segments in accordance with Table R611.7(4), note e. In the top-most story, the reinforcement shall terminate in accordance with Section R611.6.4.

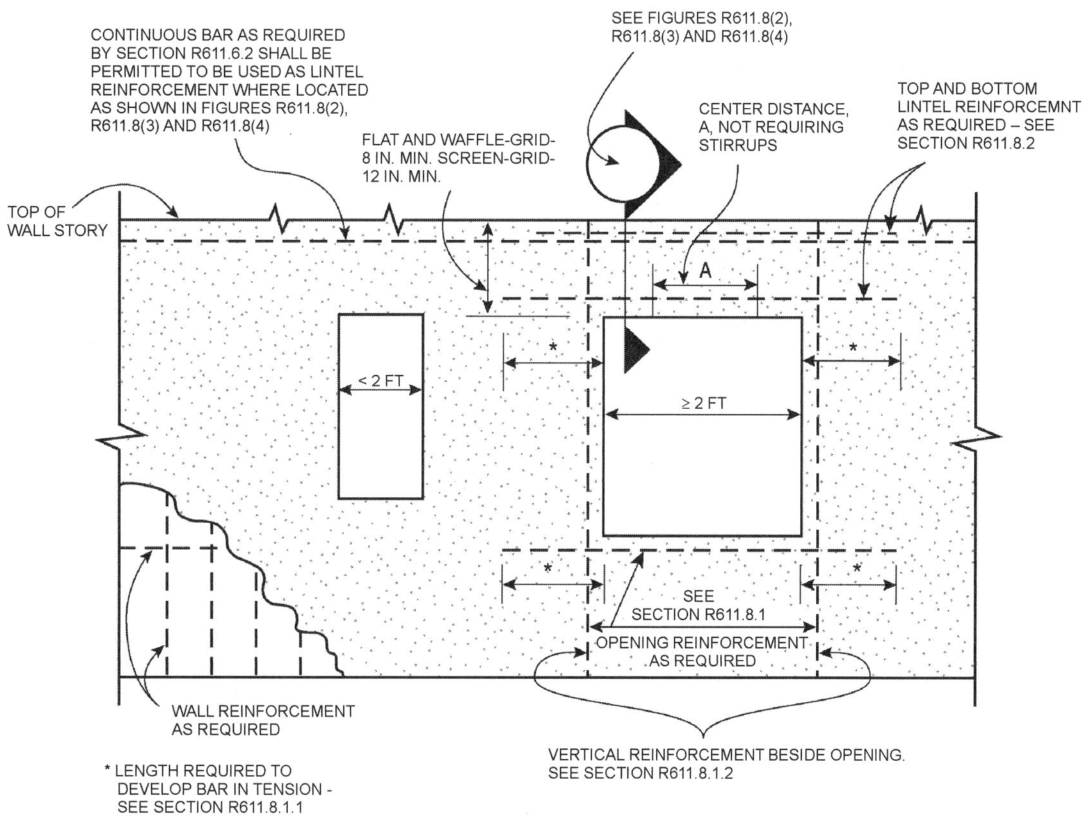

For SI: 1 inch = 25.4 mm, 1 foot = 304.8 mm.

**FIGURE R611.8(1)
REINFORCEMENT OF OPENINGS**

R611.8.2 Lintels. Lintels shall be provided over all openings equal to or greater than 2 feet (610 mm) in width. Lintels with uniform loading shall conform to Sections R611.8.2.1 and R611.8.2.2, or Section R611.8.2.3. Lintels supporting concentrated loads, such as from roof or floor beams or girders, shall be designed in accordance with ACI 318.

R611.8.2.1 Lintels designed for gravity load-bearing conditions. Where a lintel will be subjected to gravity load condition 1 through 5 of Table R611.8(1), the clear span of the lintel shall not exceed that permitted by Tables R611.8(2) through R611.8(8). The maximum clear span of lintels with and without stirrups in flat walls shall be determined in accordance with Tables R611.8(2) through R611.8(5), and constructed in accordance with Figure R611.8(2). The maximum clear span of lintels with and without stirrups in waffle-grid walls shall be determined in accordance with Tables R611.8(6) and R611.8(7), and constructed in accordance with Figure R611.8(3). The maximum clear span of lintels with and without stirrups in screen-grid walls shall be determined in accordance with Table R611.8(8), and constructed in accordance with Figure R611.8(4).

Where required by the applicable table, No. 3 stirrups shall be installed in lintels at a maximum spacing of *d/2* where d equals the depth of the lintel, *D*, less the cover of the concrete as shown in Figures R611.8(2) through R611.8(4). The smaller value of *d* computed for the top and bottom bar shall be used to determine the maximum stirrup spacing. Where stirrups are required in a lintel with a single bar or two bundled bars in the top and bottom, they shall be fabricated like the letter "c" or "s" with 135-degree (2.36 rad) standard hooks at each end that comply with Section R611.5.4.5 and Figure R611.5.4(3) and installed as shown in Figures R611.8(2) through R611.8(4). Where two bars are required in the top and bottom of the lintel and the bars are not bundled, the bars shall be separated by a minimum of 1 inch (25 mm). The free end of the stirrups shall be fabricated with 90- or 135-degree (1.57 or 2.36 rad) standard hooks that comply with Section R611.5.4.5 and Figure R611.5.4(3) and installed as shown in Figures R611.8(2) and R611.8(3). For flat, waffle-grid and screen-grid lintels, stirrups are not required in the center distance, A, portion of spans in accordance with Figure R611.8(1) and Tables R611.8(2) through R611.8(8). See Section R611.8.2.2, Item 5, for requirement for stirrups through out lintels with bundled bars.

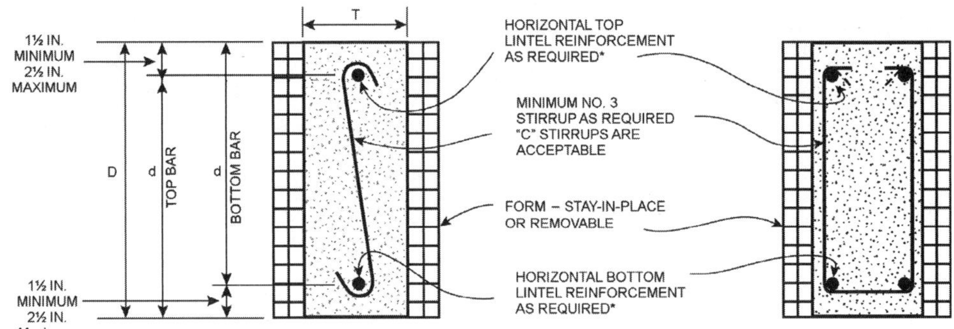

For SI: 1 inch = 25.4 mm.

FIGURE R611.8(2)
LINTEL FOR FLAT WALLS

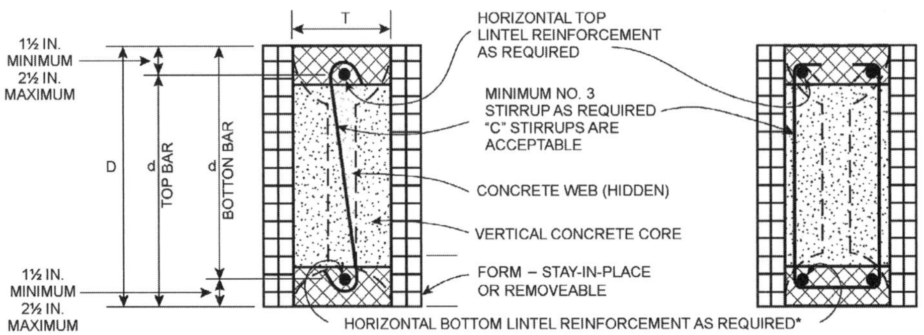

(a) SINGLE FORM HEIGHT SECTION CUT THROUGH VERTICAL CORE OF A WAFFLE-GRID LINTEL

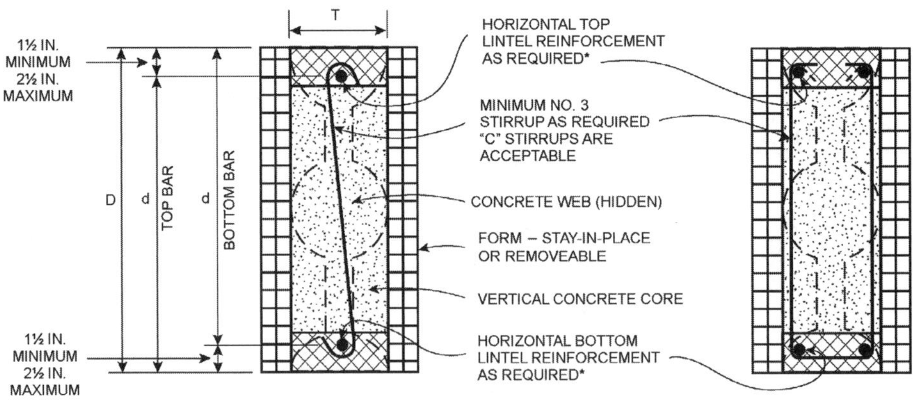

(b) DOUBLE FORM HEIGHT SECTION CUT THROUGH VERTICAL CORE OF A WAFFLE-GRID LINTEL

*FOR BUNDLED BARS, SEE SECTION R611.8.2.2.

NOTE: CROSS-HATCHING REPRESENTS THE AREA IN WHICH FORM MATERIAL SHALL BE REMOVED, IF NECESSARY, TO CREATE FLANGES CONTINUOUS THE LENGTH OF THE LINTEL. FLANGES SHALL HAVE A MINIMUM THICKNESS OF 3 IN., AND A MINIMUM WIDTH OF 5 IN. AND 7 IN. IN 6 IN. NOMINAL AND 8 IN. NOMINAL WAFFLE-GRID WALLS, RESPECTIVELY. SEE NOTE a TO TABLES R611.8(6) AND R611.8(10).

For SI: 1 inch = 25.4 mm.

FIGURE R611.8(3)
LINTELS FOR WAFFLE-GRID WALLS

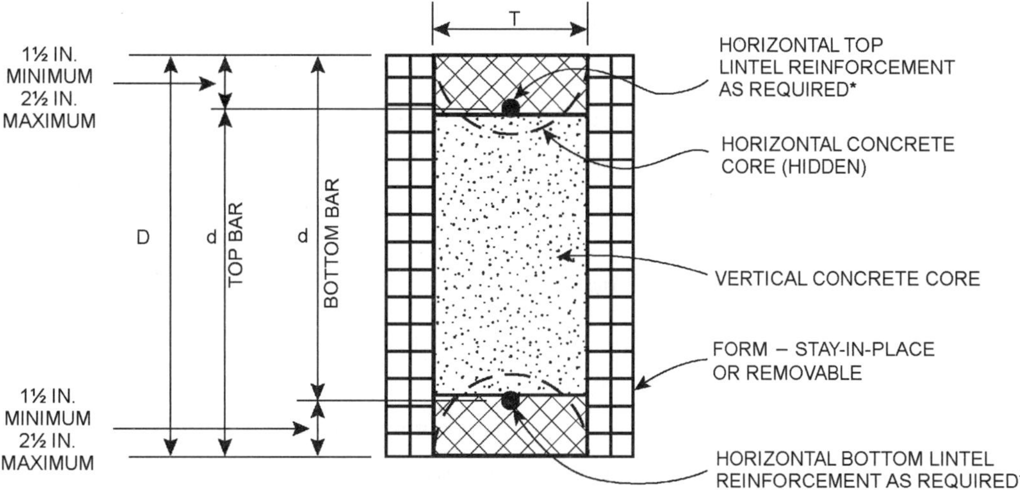

(a) SINGLE FORM HEIGHT SECTION CUT THROUGH VERTICAL CORE OF A SCREEN-GRID LINTEL

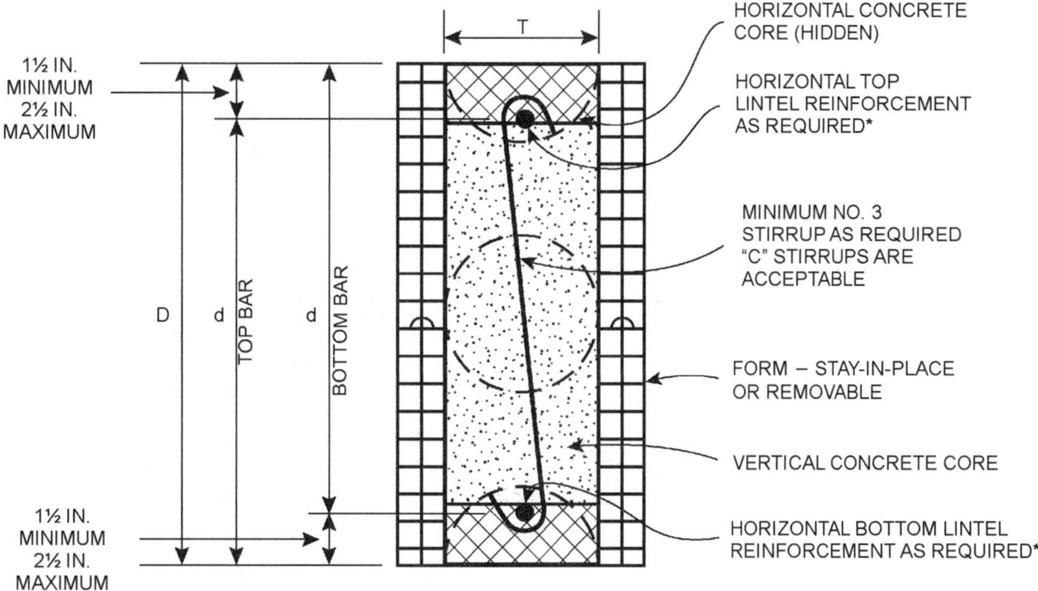

(b) DOUBLE FORM HEIGHT SECTION CUT THROUGH VERTICAL CORE OF A SCREEN-GRID LINTEL

*FOR BUNDLED BARS, SEE SECTION R611.8.2.2.

NOTE: CROSS-HATCHING REPRESENTS THE AREA IN WHICH FORM MATERIAL SHALL BE REMOVED, IF NECESSARY, TO CREATE FLANGES CONTINUOUS THE LENGTH OF THE LINTEL. FLANGES SHALL HAVE A MINIMUM THICKNESS OF 2.5 IN. AND A MINIMUM WIDTH OF 5 IN. SEE NOTE a TO TABLES R611.8(8) AND R611.8(10).

For SI: 1 inch = 25.4 mm.

FIGURE R611.8(4)
LINTELS FOR SCREEN-GRID WALLS

R611.8.2.2 Bundled bars in lintels. It is permitted to bundle two bars in contact with each other in lintels if all of the following are observed:

1. Bars no larger than No. 6 are bundled.
2. Where the wall thickness is not sufficient to provide not less than 3 inches (76 mm) of clear space beside bars (total on both sides) oriented horizontally in a bundle, the bundled bars shall be oriented in a vertical plane.
3. Where vertically oriented bundled bars terminate with standard hooks to develop the bars in tension beyond the support (see Section R611.5.4.4), the hook extensions shall be staggered to provide a minimum of 1 inch (25 mm) clear spacing between the extensions.
4. Bundled bars shall not be lap spliced within the lintel span and the length on each end of the lintel that is required to develop the bars in tension.
5. Bundled bars shall be enclosed within stirrups throughout the length of the lintel. Stirrups and the installation thereof shall comply with Section R611.8.2.1.

TABLE R611.8(1)
LINTEL DESIGN LOADING CONDITIONS[a, b, d]

DESCRIPTION OF LOADS AND OPENINGS ABOVE INFLUENCING DESIGN OF LINTEL			DESIGN LOAD CONDITION[c]
Opening in wall of top story of two-story building, or first story of one-story building			
Wall supporting loads from roof, including attic floor, if applicable, and	Top of lintel equal to or less than W/2 below top of wall		2
	Top of lintel greater than W/2 below top of wall		NLB
Wall not supporting loads from roof or attic floor			NLB
Opening in wall of first story of two-story building where wall immediately above is of concrete construction, or opening in basement wall of one-story building where wall immediately above is of concrete construction			
LB ledger board mounted to side of wall with bottom of ledger less than or equal to W/2 above top of lintel, and	Top of lintel greater than W/2 below bottom of opening in story above		1
	Top of lintel less than or equal to W/2 below bottom of opening in story above, and	Opening is entirely within the footprint of the opening in the story above	1
		Opening is partially within the footprint of the opening in the story above	4
LB ledger board mounted to side of wall with bottom of ledger more than W/2 above top of lintel			NLB
NLB ledger board mounted to side of wall with bottom of ledger less than or equal to W/2 above top of lintel, or no ledger board, and	Top of lintel greater than W/2 below bottom of opening in story above		NLB
	Top of lintel less than or equal to W/2 below bottom of opening in story above, and	Opening is entirely within the footprint of the opening in the story above	NLB
		Opening is partially within the footprint of the opening in the story above	1
Opening in basement wall of two-story building where walls of two stories above are of concrete construction			
LB ledger board mounted to side of wall with bottom of ledger less than or equal to W/2 above top of lintel, and	Top of lintel greater than W/2 below bottom of opening in story above		1
	Top of lintel less than or equal to W/2 below bottom of opening in story above, and	Opening is entirely within the footprint of the opening in the story above	1
		Opening is partially within the footprint of the opening in the story above	5
LB ledger board mounted to side of wall with bottom of ledger more than W/2 above top of lintel			NLB
NLB ledger board mounted to side of wall with bottom of ledger less than or equal to W/2 above top of lintel, or no ledger board, and	Top of lintel greater than W/2 below bottom of opening in story above		NLB
	Top of lintel less than or equal to W/2 below bottom of opening in story above, and	Opening is entirely within the footprint of the opening in the story above	NLB
		Opening is partially within the footprint of the opening in the story above	1
Opening in wall of first story of two-story building where wall immediately above is of light-framed construction, or opening in basement wall of one-story building, where wall immediately above is of light-framed construction			
Wall supporting loads from roof, second floor and top-story wall of light-framed construction, and	Top of lintel equal to or less than W/2 below top of wall		3
	Top of lintel greater than W/2 below top of wall		NLB
Wall not supporting loads from roof or second floor			NLB

a. LB means load bearing, NLB means nonload bearing, and W means width of opening.

b. Footprint is the area of the wall below an opening in the story above, bounded by the bottom of the opening and vertical lines extending downward from the edges of the opening.

c. For design loading condition "NLB" see Tables R611.8(9) and R611.8(10). For all other design loading conditions see Tables R611.8(2) through R611.8(8).

d. A NLB ledger board is a ledger attached to a wall that is parallel to the span of the floor, roof or ceiling framing that supports the edge of the floor, ceiling or roof.

TABLE R611.8(2)
MAXIMUM ALLOWABLE CLEAR SPANS FOR 4-INCH-NOMINAL THICK FLAT LINTELS IN LOAD-BEARING WALLS[a, b, c, d, e, f, m]
ROOF CLEAR SPAN 40 FEET AND FLOOR CLEAR SPAN 32 FEET

LINTEL DEPTH, D[g] (inches)	NUMBER OF BARS AND BAR SIZE IN TOP AND BOTTOM OF LINTEL	STEEL YIELD STRENGTH[h], f_y (psi)	DESIGN LOADING CONDITION DETERMINED FROM TABLE R611.8(1)								
			1	2		3		4		5	
			Maximum ground snow load (psf)								
			—	30	70	30	70	30	70	30	70
			Maximum clear span of lintel (feet - inches)								
8	Span without stirrups[i,j]		3-2	3-4	2-4	2-6	2-2	2-1	2-0	2-0	2-0
	1-#4	40,000	5-2	5-5	4-1	4-3	3-10	3-7	3-4	2-9	2-9
		60,000	6-2	6-5	4-11	5-1	4-6	4-2	3-8	2-11	2-10
	1-#5	40,000	6-3	6-7	5-0	5-2	4-6	4-2	3-8	2-11	2-10
		60,000	DR	DR	DR	DR	DR	DR	DR	DR	DR
	Center distance A[k, l]		1-1	1-2	0-8	0-9	0-7	0-6	0-5	0-4	0-4
12	Span without stirrups[i,j]		3-4	3-7	2-9	2-11	2-8	2-6	2-5	2-2	2-2
	1-#4	40,000	6-7	7-0	5-4	5-7	5-0	4-9	4-4	3-8	3-7
		60,000	7-11	8-6	6-6	6-9	6-0	5-9	5-3	4-5	4-4
	1-#5	40,000	8-1	8-8	6-7	6-10	6-2	5-10	5-4	4-6	4-5
		60,000	9-8	10-4	7-11	8-2	7-4	6-11	6-2	4-10	4-8
	2-#4 1-#6	40,000	9-1	9-8	7-4	7-8	6-10	6-6	6-0	4-10	4-8
		60,000	DR	DR	DR	DR	DR	DR	DR	DR	DR
	Center distance A[k, l]		1-8	1-11	1-1	1-3	1-0	0-11	0-9	0-6	0-6
16	Span without stirrups[i,j]		4-7	5-0	3-11	4-0	3-8	3-7	3-4	3-1	3-0
	1-#4	40,000	6-8	7-3	5-6	5-9	5-2	4-11	4-6	3-10	3-8
		60,000	9-3	10-1	7-9	8-0	7-2	6-10	6-3	5-4	5-2
	1-#4	40,000	9-6	10-4	7-10	8-2	7-4	6-11	6-5	5-5	5-3
		60,000	11-5	12-5	9-6	9-10	8-10	8-4	7-9	6-6	6-4
	2-#4 1-#6	40,000	10-7	11-7	8-10	9-2	8-3	7-9	7-2	6-1	5-11
		60,000	12-9	13-10	10-7	11-0	9-10	9-4	8-7	6-9	6-6
	2-#5	40,000	13-0	14-1	10-9	11-2	9-11	9-2	8-2	6-6	6-3
		60,000	DR	DR	DR	DR	DR	DR	DR	DR	DR
	Center distance[k, l]		2-3	2-8	1-7	1-8	1-4	1-3	1-0	0-9	0-8
20	Span without stirrups[i,j]		5-9	6-5	5-0	5-2	4-9	4-7	4-4	3-11	3-11
	1-#4	40,000	7-5	8-2	6-3	6-6	5-10	5-7	5-1	4-4	4-2
		60,000	9-0	10-0	7-8	7-11	7-1	6-9	6-3	5-3	5-1
	1-#5	40,000	9-2	10-2	7-9	8-1	7-3	6-11	6-4	5-4	5-2
		60,000	12-9	14-2	10-10	11-3	10-1	9-7	8-10	7-5	7-3
	2-#4 1-#6	40,000	11-10	13-2	10-1	10-5	9-4	8-11	8-2	6-11	6-9
		60,000	14-4	15-10	12-1	12-7	11-3	10-9	9-11	8-4	8-1
	2-#5	40,000	14-7	16-2	12-4	12-9	11-4	10-6	9-5	7-7	7-3
		60,000	17-5	19-2	14-9	15-3	13-5	12-4	11-0	8-8	8-4
	2-#6	40,000	16-4	18-11	12-7	13-3	11-4	10-6	9-5	7-7	7-3
		60,000	DR	DR	DR	DR	DR	DR	DR	DR	DR
	Center distance A[k, l]		2-9	3-5	2-0	2-2	1-9	1-7	1-4	0-11	0-11

(continued)

TABLE R611.8(2)—continued
MAXIMUM ALLOWABLE CLEAR SPANS FOR 4-INCH-NOMINAL THICK FLAT LINTELS IN LOAD-BEARING WALLS[a, b, c, d, e, f, m]
ROOF CLEAR SPAN 40 FEET AND FLOOR CLEAR SPAN 32 FEET

LINTEL DEPTH, D[g] (inches)	NUMBER OF BARS AND BAR SIZE IN TOP AND BOTTOM OF LINTEL	STEEL YIELD STRENGTH[h], f_y (psi)	DESIGN LOADING CONDITION DETERMINED FROM TABLE R611.8(1)								
			1	2		3		4		5	
			Maximum ground snow load (psf)								
			—	30	70	30	70	30	70	30	70
			Maximum clear span of lintel (feet - inches)								
24	Span without stirrups[i, j]		6-11	7-9	6-1	6-3	5-9	5-7	5-3	4-9	4-8
	1-#4	40,000	8-0	9-0	6-11	7-2	6-5	6-2	5-8	4-9	4-8
		60,000	9-9	11-0	8-5	8-9	7-10	7-6	6-11	5-10	5-8
	1-#5	40,000	10-0	11-3	8-7	8-11	8-0	7-7	7-0	5-11	5-9
		60,000	13-11	15-8	12-0	12-5	11-2	10-7	9-10	8-3	8-0
	2-#4 1-#6	40,000	12-11	14-6	11-2	11-6	10-5	9-10	9-1	7-8	7-5
		60,000	15-7	17-7	13-6	13-11	12-7	11-11	11-0	9-3	9-0
	2-#5	40,000	15-11	17-11	13-7	14-3	12-8	11-9	10-8	8-7	8-4
		60,000	19-1	21-6	16-5	17-1	15-1	14-0	12-6	9-11	9-7
	2-#6	40,000	17-7	21-1	14-1	14-10	12-8	11-9	10-8	8-7	8-4
		60,000	DR	DR	DR	DR	DR	DR	DR	DR	DR
	Center distance A[k, l]		3-3	4-1	2-5	2-7	2-1	1-11	1-7	1-2	1-1

For SI: 1 inch = 25.4 mm, 1 foot = 304.8 mm, 1 pound per square foot = 0.0479 kPa, Grade 40 = 280 MPa, Grade 60 = 420 MPa.

a. See Table R611.3 for tolerances permitted from nominal thickness.

b. Table values are based on concrete with a minimum specified compressive strength of 2,500 psi. See Note j.

c. Table values are based on uniform loading. See Section R611.8.2 for lintels supporting concentrated loads.

d. Deflection criterion is $L/240$, where L is the clear span of the lintel in inches, or $^1/_2$-inch, whichever is less.

e. Linear interpolation is permitted between ground snow loads and between lintel depths.

f. DR indicates design required.

g. Lintel depth, D, is permitted to include the available height of wall located directly above the lintel, provided that the increased lintel depth spans the entire length of the lintel.

h. Stirrups shall be fabricated from reinforcing bars with the same yield strength as that used for the main longitudinal reinforcement.

i. Allowable clear span without stirrups applicable to all lintels of the same depth, D. Top and bottom reinforcement for lintels without stirrups shall not be less than the least amount of reinforcement required for a lintel of the same depth and loading condition with stirrups. All other spans require stirrups spaced at not more than $d/2$.

j. Where concrete with a minimum specified compressive strength of 3,000 psi is used, clear spans for lintels without stirrups shall be permitted to be multiplied by 1.05. If the increased span exceeds the allowable clear span for a lintel of the same depth and loading condition with stirrups, the top and bottom reinforcement shall be equal to or greater than that required for a lintel of the same depth and loading condition that has an allowable clear span that is equal to or greater than that of the lintel without stirrups that has been increased.

k. Center distance, A, is the center portion of the clear span where stirrups are not required. This is applicable to all longitudinal bar sizes and steel yield strengths.

l. Where concrete with a minimum specified compressive strength of 3,000 psi is used, center distance, A, shall be permitted to be multiplied by 1.10.

m. The maximum clear opening width between two solid wall segments shall be 18 feet. See Section R611.7.2.1. Lintel clear spans in the table greater than 18 feet are shown for interpolation and information only.

TABLE R611.8(3)
MAXIMUM ALLOWABLE CLEAR SPANS FOR 6-INCH-NOMINAL THICK FLAT LINTELS IN LOAD-BEARING WALLS[a, b, c, d, e, f, m]
ROOF CLEAR SPAN 40 FEET AND FLOOR CLEAR SPAN 32 FEET

LINTEL DEPTH, D^g (inches)	NUMBER OF BARS AND BAR SIZE IN TOP AND BOTTOM OF LINTEL	STEEL YIELD STRENGTH[h], f_y (psi)	DESIGN LOADING CONDITION DETERMINED FROM TABLE R611.8(1)								
			1	2		3		4		5	
			Maximum ground snow load (psf)								
			—	30	70	30	70	30	70	30	70
			Maximum clear span of lintel (feet - inches)								
8	Span without stirrups[i, j]		4-2	4-8	3-1	3-3	2-10	2-6	2-3	2-0	2-0
	1-#4	40,000	5-1	5-5	4-2	4-3	3-10	3-6	3-3	2-8	2-7
		60,000	6-2	6-7	5-0	5-2	4-8	4-2	3-11	3-3	3-2
	1-#5	40,000	6-3	6-8	5-1	5-3	4-9	4-3	4-0	3-3	3-2
		60,000	7-6	8-0	6-1	6-4	5-8	5-1	4-9	3-8	3-6
	2-#4 1-#6	40,000	7-0	7-6	5-8	5-11	5-3	4-9	4-5	3-8	3-6
		60,000	DR	DR	DR	DR	DR	DR	DR	DR	DR
	Center distance $A^{k, l}$		1-7	1-10	1-1	1-2	0-11	0-9	0-8	0-5	0-5
12	Span without stirrups[i, j]		4-2	4-8	3-5	3-6	3-2	2-11	2-9	2-5	2-4
	1-#4	40,000	5-7	6-1	4-8	4-10	4-4	3-11	3-8	3-0	2-11
		60,000	7-9	8-6	6-6	6-9	6-1	5-6	5-1	4-3	4-1
	1-#5	40,000	7-11	8-8	6-8	6-11	6-2	5-7	5-2	4-4	4-2
		60,000	9-7	10-6	8-0	8-4	7-6	6-9	6-3	5-2	5-1
	2-#4 1-#6	40,000	8-11	9-9	7-6	7-9	6-11	6-3	5-10	4-10	4-8
		60,000	10-8	11-9	8-12	9-4	8-4	7-6	7-0	5-10	5-8
	2-#5	40,000	10-11	12-0	9-2	9-6	8-6	7-8	7-2	5-6	5-3
		60,000	12-11	14-3	10-10	11-3	10-1	9-0	8-1	6-1	5-10
	2-#6	40,000	12-9	14-0	10-8	11-1	9-7	8-1	7-3	5-6	5-3
		60,000	DR	DR	DR	DR	DR	DR	DR	DR	DR
	Center distance $A^{k, l}$		2-6	3-0	1-9	1-10	1-6	1-3	1-1	0-9	0-8
16	Span without stirrups[i, j]		5-7	6-5	4-9	4-11	4-5	4-0	3-10	3-4	3-4
	1-#4	40,000	6-5	7-2	5-6	5-9	5-2	4-8	4-4	3-7	3-6
		60,000	7-10	8-9	6-9	7-0	6-3	5-8	5-3	4-4	4-3
	1-#5	40,000	7-11	8-11	6-10	7-1	6-5	5-9	5-4	4-5	4-4
		60,000	11-1	12-6	9-7	9-11	8-11	8-0	7-6	6-2	6-0
	2-#4 1-#6	40,000	10-3	11-7	8-10	9-2	8-3	7-6	6-11	5-9	5-7
		60,000	12-5	14-0	10-9	11-1	10-0	9-0	8-5	7-0	6-9
	2-#5	40,000	12-8	14-3	10-11	11-4	10-2	9-2	8-7	6-9	6-6
		60,000	15-2	17-1	13-1	13-7	12-3	11-0	10-3	7-11	7-7
	2-#6	40,000	14-11	16-9	12-8	13-4	11-4	9-8	8-8	6-9	6-6
		60,000	DR	DR	DR	DR	DR	DR	DR	DR	DR
	Center distance $A^{k, l}$		3-3	4-1	2-5	2-7	2-1	1-9	1-6	1-0	1-0

(continued)

TABLE R611.8(3)—continued
MAXIMUM ALLOWABLE CLEAR SPANS FOR 6-INCH-NOMINAL THICK FLAT LINTELS IN LOAD-BEARING WALLS[a, b, c, d, e, f, m]
ROOF CLEAR SPAN 40 FEET AND FLOOR CLEAR SPAN 32 FEET

LINTEL DEPTH, D[g] (inches)	NUMBER OF BARS AND BAR SIZE IN TOP AND BOTTOM OF LINTEL	STEEL YIELD STRENGTH[h], f_y (psi)	DESIGN LOADING CONDITION DETERMINED FROM TABLE R611.8(1)								
			1	2		3		4		5	
			Maximum ground snow load (psf)								
			—	30	70	30	70	30	70	30	70
			Maximum clear span of lintel (feet - inches)								
20	Span without stirrups[i, j]		6-11	8-2	6-1	6-3	5-8	5-2	4-11	4-4	4-3
	1-#5	40,000	8-9	10-1	7-9	8-0	7-3	6-6	6-1	5-1	4-11
		60,000	10-8	12-3	9-5	9-9	8-10	8-0	7-5	6-2	6-0
	2-#4 1-#6	40,000	9-11	11-4	8-9	9-1	8-2	7-4	6-10	5-8	5-7
		60,000	13-9	15-10	12-2	12-8	11-5	10-3	9-7	7-11	7-9
	2-#5	40,000	14-0	16-2	12-5	12-11	11-7	10-6	9-9	7-11	7-8
		60,000	16-11	19-6	15-0	15-6	14-0	12-7	11-9	9-1	8-9
	2-#6	40,000	16-7	19-1	14-7	15-3	13-1	11-3	10-2	7-11	7-8
		60,000	19-11	22-10	17-4	18-3	15-6	13-2	11-10	9-1	8-9
	Center distance A[k, l]		3-11	5-2	3-1	3-3	2-8	2-2	1-11	1-4	1-3
24	Span without stirrups[i, j]		8-2	9-10	7-4	7-8	6-11	6-4	5-11	5-3	5-2
	1-#5	40,000	9-5	11-1	8-7	8-10	8-0	7-3	6-9	5-7	5-5
		60,000	11-6	13-6	10-5	10-9	9-9	8-9	8-2	6-10	6-8
	2-#4 1-#6	40,000	10-8	12-6	9-8	10-0	9-0	8-2	7-7	6-4	6-2
		60,000	12-11	15-2	11-9	12-2	11-0	9-11	9-3	7-8	7-6
	2-#5	40,000	15-2	17-9	13-9	14-3	12-10	11-7	10-10	9-0	8-9
		60,000	18-4	21-6	16-7	17-3	15-6	14-0	13-1	10-4	10-0
	2-#6	40,000	18-0	21-1	16-4	16-11	14-10	12-9	11-8	9-2	8-11
		60,000	21-7	25-4	19-2	20-4	17-2	14-9	13-4	10-4	10-0
	Center distance A[k, l]		4-6	6-2	3-8	4-0	3-3	2-8	2-3	1-7	1-6

For SI: 1 inch = 25.4 mm, 1 foot = 304.8 mm, 1 pounds per square foot = 0.0479 kPa, Grade 40 = 280 MPa, Grade 60 = 420 MPa.

a. See Table R611.3 for tolerances permitted from nominal thickness.

b. Table values are based on concrete with a minimum specified compressive strength of 2,500 psi. See Note j.

c. Table values are based on uniform loading. See Section R611.8.2 for lintels supporting concentrated loads.

d. Deflection criterion is $L/240$, where L is the clear span of the lintel in inches, or $^1/_2$-inch, whichever is less.

e. Linear interpolation is permitted between ground snow loads and between lintel depths.

f. DR indicates design required.

g. Lintel depth, D, is permitted to include the available height of wall located directly above the lintel, provided that the increased lintel depth spans the entire length of the lintel.

h. Stirrups shall be fabricated from reinforcing bars with the same yield strength as that used for the main longitudinal reinforcement.

i. Allowable clear span without stirrups applicable to all lintels of the same depth, D. Top and bottom reinforcement for lintels without stirrups shall not be less than the least amount of reinforcement required for a lintel of the same depth and loading condition with stirrups. All other spans require stirrups spaced at not more than $d/2$.

j. Where concrete with a minimum specified compressive strength of 3,000 psi is used, clear spans for lintels without stirrups shall be permitted to be multiplied by 1.05. If the increased span exceeds the allowable clear span for a lintel of the same depth and loading condition with stirrups, the top and bottom reinforcement shall be equal to or greater than that required for a lintel of the same depth and loading condition that has an allowable clear span that is equal to or greater than that of the lintel without stirrups that has been increased.

k. Center distance, A, is the center portion of the clear span where stirrups are not required. This is applicable to all longitudinal bar sizes and steel yield strengths.

l. Where concrete with a minimum specified compressive strength of 3,000 psi is used, center distance, A, shall be permitted to be multiplied by 1.10.

m. The maximum clear opening width between two solid wall segments shall be 18 feet. See Section R611.7.2.1. Lintel clear spans in the table greater than 18 feet are shown for interpolation and information only.

TABLE R611.8(4)
MAXIMUM ALLOWABLE CLEAR SPANS FOR 8-INCH-NOMINAL THICK FLAT LINTELS IN LOAD-BEARING WALLS[a, b, c, d, e, f, m]
ROOF CLEAR SPAN 40 FEET AND FLOOR CLEAR SPAN 32 FEET

LINTEL DEPTH, D[g] (inches)	NUMBER OF BARS AND BAR SIZE IN TOP AND BOTTOM OF LINTEL	STEEL YIELD STRENGTH[h], f_y (psi)	DESIGN LOADING CONDITION DETERMINED FROM TABLE R611.8(1)								
			1	2		3		4		5	
			Maximum ground snow load (psf)								
			—	30	70	30	70	30	70	30	70
			Maximum clear span of lintel (feet - inches)								
8	Span without stirrups[i, j]		4-4	4-9	3-7	3-9	3-4	2-10	2-7	2-1	2-0
	1-#4	40,000	4-4	4-9	3-7	3-9	3-4	2-11	2-9	2-3	2-2
		60,000	6-1	6-7	5-0	5-3	4-8	4-0	3-9	3-1	3-0
	1-#5	40,000	6-2	6-9	5-2	5-4	4-9	4-1	3-10	3-2	3-1
		60,000	7-5	8-1	6-2	6-5	5-9	4-11	4-7	3-9	3-8
	2-#4 1-#6	40,000	6-11	7-6	5-9	6-0	5-4	4-7	4-4	3-6	3-5
		60,000	8-3	9-0	6-11	7-2	6-5	5-6	5-2	4-2	4-1
	2-#5	40,000	8-5	9-2	7-0	7-3	6-6	5-7	5-3	4-2	4-0
		60,000	DR	DR	DR	DR	DR	DR	DR	DR	DR
	Center distance A[k, l]		2-1	2-6	1-5	1-6	1-3	0-11	0-10	0-6	0-6
12	Span without stirrups[i, j]		4-10	5-8	4-0	4-2	3-9	3-2	3-0	2-7	2-6
	1-#4	40,000	5-5	6-1	4-8	4-10	4-4	3-9	3-6	2-10	2-10
		60,000	6-7	7-5	5-8	5-11	5-4	4-7	4-3	3-6	3-5
	1-#5	40,000	6-9	7-7	5-9	6-0	5-5	4-8	4-4	3-7	3-6
		60,000	9-4	10-6	8-1	8-4	7-6	6-6	6-1	5-0	4-10
	2-#4 1-#6	40,000	8-8	9-9	7-6	7-9	7-0	6-0	5-8	4-7	4-6
		60,000	10-6	11-9	9-1	9-5	8-5	7-3	6-10	5-7	5-5
	2-#5	40,000	10-8	12-0	9-3	9-7	8-7	7-5	6-11	5-6	5-4
		60,000	12-10	14-5	11-1	11-6	10-4	8-11	8-4	6-7	6-4
	2-#6	40,000	12-7	14-2	10-10	11-3	10-2	8-3	7-6	5-6	5-4
		60,000	DR	DR	DR	DR	DR	DR	DR	DR	DR
	Center distance A[k, l]		3-2	4-0	2-4	2-6	2-0	1-6	1-4	0-11	0-10
16	Span without stirrups[i, j]		6-5	7-9	5-7	5-10	5-2	4-5	4-2	3-7	3-6
	1-#4	40,000	6-2	7-1	5-6	5-8	5-1	4-5	4-2	3-5	3-4
		60,000	7-6	8-8	6-8	6-11	6-3	5-5	5-1	4-2	4-0
	1-#5	40,000	7-8	8-10	6-10	7-1	6-4	5-6	5-2	4-3	4-1
		60,000	9-4	10-9	8-4	8-7	7-9	6-8	6-3	5-2	5-0
	2-#4 1-#6	40,000	8-8	10-0	7-8	8-0	7-2	6-2	5-10	4-9	4-8
		60,000	12-0	13-11	10-9	11-2	10-0	8-8	8-1	6-8	6-6
	2-#5	40,000	12-3	14-2	11-0	11-4	10-3	8-10	8-3	6-9	6-7
		60,000	14-10	17-2	13-3	13-8	12-4	10-8	10-0	7-11	7-8
	2-#6	40,000	14-6	16-10	13-0	13-5	12-1	10-1	9-2	6-11	6-8
		60,000	17-5	20-2	15-7	16-1	14-6	11-10	10-8	7-11	7-8
	Center distance[k, l]		4-1	5-5	3-3	3-6	2-10	2-1	1-10	1-3	1-2

(continued)

TABLE R611.8(4)—continued
MAXIMUM ALLOWABLE CLEAR SPANS FOR 8-INCH-NOMINAL THICK FLAT LINTELS IN LOAD-BEARING WALLS[a, b, c, d, e, f, m]
ROOF CLEAR SPAN 40 FEET AND FLOOR CLEAR SPAN 32 FEET

LINTEL DEPTH, D[g] (inches)	NUMBER OF BARS AND BAR SIZE IN TOP AND BOTTOM OF LINTEL	STEEL YIELD STRENGTH[h], f_y (psi)	DESIGN LOADING CONDITION DETERMINED FROM TABLE R611.8(1)								
			1	2		3		4		5	
			Maximum ground snow load (psf)								
			—	30	70	30	70	30	70	30	70
			Maximum clear span of lintel (feet - inches)								
20	Span without stirrups[i, j]		7-10	9-10	7-1	7-5	6-7	5-8	5-4	4-7	4-6
	1-#5	40,000	8-4	9-11	7-8	8-0	7-2	6-3	5-10	4-9	4-8
		60,000	10-2	12-1	9-5	9-9	8-9	7-7	7-1	5-10	5-8
	2-#4 1-#6	40,000	9-5	11-3	8-8	9-0	8-1	7-0	6-7	5-5	5-3
		60,000	11-6	13-8	10-7	11-0	9-11	8-7	8-0	6-7	6-5
	2-#5	40,000	11-9	13-11	10-10	11-2	10-1	8-9	8-2	6-8	6-7
		60,000	16-4	19-5	15-0	15-7	14-0	12-2	11-4	9-3	9-0
	2-#6	40,000	16-0	19-0	14-9	15-3	13-9	11-10	10-10	8-3	8-0
		60,000	19-3	22-11	17-9	18-5	16-7	13-7	12-4	9-3	9-0
	Center distance A[k, l]		4-10	6-10	4-1	4-5	3-7	2-8	2-4	1-7	1-6
24	Span without stirrups[i, j]		9-2	11-9	8-7	8-11	8-0	6-11	6-6	5-7	5-6
	1-#5	40,000	8-11	10-10	8-6	8-9	7-11	6-10	6-5	5-3	5-2
		60,000	10-11	13-3	10-4	10-8	9-8	8-4	7-10	6-5	6-3
	2-#4 1-#6	40,000	10-1	12-3	9-7	9-11	8-11	7-9	7-3	6-0	5-10
		60,000	12-3	15-0	11-8	12-1	10-11	9-5	8-10	7-3	7-1
	2-#5	40,000	12-6	15-3	11-11	12-4	11-1	9-7	9-0	7-5	7-3
		60,000	17-6	21-3	16-7	17-2	15-6	13-5	12-7	10-4	10-1
	2-#6	40,000	17-2	20-11	16-3	16-10	15-3	13-2	12-4	9-7	9-4
		60,000	20-9	25-3	19-8	20-4	18-5	15-4	14-0	10-7	10-3
	Center distance A[k, l]		5-6	8-1	4-11	5-3	4-4	3-3	2-10	1-11	1-10

For SI: 1 inch = 25.4 mm, 1 foot = 304.8 mm, 1 pound per square foot = 0.0479 kPa, Grade 40 = 280 MPa; Grade 60 = 420 MPa.

Note: Top and bottom reinforcement for lintels without stirrups shown in shaded cells shall be equal to or greater than that required for lintel of the same depth and loading condition that has an allowable clear span that is equal to or greater than that of the lintel without stirrups.

a. See Table R611.3 for tolerances permitted from nominal thickness.

b. Table values are based on concrete with a minimum specified compressive strength of 2,500 psi. See Note j.

c. Table values are based on uniform loading. See Section R611.8.2 for lintels supporting concentrated loads.

d. Deflection criterion is $L/240$, where L is the clear span of the lintel in inches, or $^1/_2$-inch, whichever is less.

e. Linear interpolation is permitted between ground snow loads and between lintel depths.

f. DR indicates design required.

g. Lintel depth, D, is permitted to include the available height of wall located directly above the lintel, provided that the increased lintel depth spans the entire length of the lintel.

h. Stirrups shall be fabricated from reinforcing bars with the same yield strength as that used for the main longitudinal reinforcement.

i. Allowable clear span without stirrups applicable to all lintels of the same depth, D. Top and bottom reinforcement for lintels without stirrups shall not be less than the least amount of reinforcement required for a lintel of the same depth and loading condition with stirrups. All other spans require stirrups spaced at not more than $d/2$.

j. Where concrete with a minimum specified compressive strength of 3,000 psi is used, clear spans for lintels without stirrups shall be permitted to be multiplied by 1.05. If the increased span exceeds the allowable clear span for a lintel of the same depth and loading condition with stirrups, the top and bottom reinforcement shall be equal to or greater than that required for a lintel of the same depth and loading condition that has an allowable clear span that is equal to or greater than that of the lintel without stirrups that has been increased.

k. Center distance, A, is the center portion of the clear span where stirrups are not required. This is applicable to all longitudinal bar sizes and steel yield strengths.

l. Where concrete with a minimum specified compressive strength of 3,000 psi is used, center distance, A, shall be permitted to be multiplied by 1.10.

m. The maximum clear opening width between two solid wall segments shall be 18 feet. See Section R611.7.2.1. Lintel clear spans in the table greater than 18 feet are shown for interpolation and information only.

TABLE R611.8(5)
MAXIMUM ALLOWABLE CLEAR SPANS FOR 10-INCH-NOMINAL THICK FLAT LINTELS IN LOAD-BEARING WALLS[a, b, c, d, e, f, m]
ROOF CLEAR SPAN 40 FEET AND FLOOR CLEAR SPAN 32 FEET

LINTEL DEPTH, D[g] (inches)	NUMBER OF BARS AND BAR SIZE IN TOP AND BOTTOM OF LINTEL	STEEL YIELD STRENGTH[h], f_y (psi)	DESIGN LOADING CONDITION DETERMINED FROM TABLE R611.8(1)								
			1	2		3		4		5	
			Maximum ground snow load (psf)								
			—	30	70	30	70	30	70	30	70
			Maximum clear span of lintel (feet - inches)								
8	Span without stirrups[i, j]		6-0	7-2	4-7	4-10	4-1	3-1	2-11	2-3	2-2
	1-#4	40,000	4-3	4-9	3-7	3-9	3-4	2-9	2-7	2-1	2-1
		60,000	5-11	6-7	5-0	5-3	4-8	3-10	3-8	2-11	2-11
	1-#5	40,000	6-1	6-9	5-2	5-4	4-9	3-11	3-9	3-0	2-11
		60,000	7-4	8-1	6-3	6-5	5-9	4-9	4-6	3-7	3-7
	2-#4 1-#6	40,000	6-10	7-6	5-9	6-0	5-5	4-5	4-2	3-4	3-4
		60,000	8-2	9-1	6-11	7-2	6-6	5-4	5-0	4-1	4-0
	2-#5	40,000	8-4	9-3	7-1	7-4	6-7	5-5	5-1	4-1	4-0
		60,000	9-11	11-0	8-5	8-9	7-10	6-6	6-1	4-8	4-6
	2-#6	40,000	9-9	10-10	8-3	8-7	7-9	6-4	5-10	4-1	4-0
		60,000	DR	DR	DR	DR	DR	DR	DR	DR	DR
	Center distance A[k, l]		2-6	3-1	1-10	1-11	1-7	1-1	0-11	0-7	0-7
12	Span without stirrups[i, j]		5-5	6-7	4-7	4-10	4-3	3-5	3-3	2-8	2-8
	1-#4	40,000	5-3	6-0	4-8	4-10	4-4	3-7	3-4	2-9	2-8
		60,000	6-5	7-4	5-8	5-10	5-3	4-4	4-1	3-4	3-3
	1-#5	40,000	6-6	7-6	5-9	6-0	5-5	4-5	4-2	3-5	3-4
		60,000	7-11	9-1	7-0	7-3	6-7	5-5	5-1	4-2	4-0
	2-#4 1-#6	40,000	7-4	8-5	6-6	6-9	6-1	5-0	4-9	3-10	3-9
		60,000	10-3	11-9	9-1	9-5	8-6	7-0	6-7	5-4	5-3
	2-#5	40,000	10-5	12-0	9-3	9-7	8-8	7-2	6-9	5-5	5-4
		60,000	12-7	14-5	11-2	11-6	10-5	8-7	8-1	6-6	6-4
	2-#6	40,000	12-4	14-2	10-11	11-4	10-2	8-5	7-8	5-7	5-5
		60,000	14-9	17-0	13-1	13-6	12-2	10-0	9-1	6-6	6-4
	Center distance A[k, l]		3-9	4-11	2-11	3-2	2-7	1-9	1-7	1-0	1-0
16	Span without stirrups[i, j]		7-1	9-0	6-4	6-8	5-10	4-9	4-6	3-9	3-8
	1-#4	40,000	5-11	7-0	5-5	5-8	5-1	4-3	4-0	3-3	3-2
		60,000	7-3	8-7	6-8	6-11	6-3	5-2	4-10	3-11	3-10
	1-#5	40,000	7-4	8-9	6-9	7-0	6-4	5-3	4-11	4-0	3-11
		60,000	9-0	10-8	8-3	8-7	7-9	6-5	6-0	4-11	4-9
	2-#4 1-#6	40,000	8-4	9-11	7-8	7-11	7-2	5-11	5-7	4-6	4-5
		60,000	10-2	12-0	9-4	9-8	8-9	7-3	6-10	5-6	5-5
	2-#5	40,000	10-4	12-3	9-6	9-10	8-11	7-4	6-11	5-8	5-6
		60,000	14-4	17-1	13-3	13-8	12-4	10-3	9-8	7-10	7-8
	2-#6	40,000	14-1	16-9	13-0	13-5	12-2	10-1	9-6	7-0	6-10
		60,000	17-0	20-2	15-8	16-2	14-7	12-0	10-11	8-0	7-9
	Center distance[k, l]		4-9	6-8	4-0	4-4	3-6	2-5	2-2	1-5	1-4

(continued)

TABLE R611.8(5)—continued
MAXIMUM ALLOWABLE CLEAR SPANS FOR 10-INCH-NOMINAL THICK FLAT LINTELS IN LOAD-BEARING WALLS[a, b, c, d, e, f, m]
ROOF CLEAR SPAN 40 FEET AND FLOOR CLEAR SPAN 32 FEET

LINTEL DEPTH, D[g] (inches)	NUMBER OF BARS AND BAR SIZE IN TOP AND BOTTOM OF LINTEL	STEEL YIELD STRENGTH[h], f_y (psi)	DESIGN LOADING CONDITION DETERMINED FROM TABLE R611.8(1)								
			1	2		3		4		5	
			Maximum ground snow load (psf)								
			—	30	70	30	70	30	70	30	70
			Maximum clear span of lintel (feet - inches)								
20	Span without stirrups[i, j]		8-7	11-4	8-1	8-5	7-5	6-1	5-9	4-10	4-9
	1-#4	40,000	6-5	7-10	6-2	6-4	5-9	4-9	4-6	3-8	3-7
		60,000	7-10	9-7	7-6	7-9	7-0	5-10	5-6	4-5	4-4
	1-#5	40,000	8-0	9-9	7-8	7-11	7-2	5-11	5-7	4-6	4-5
		60,000	9-9	11-11	9-4	9-8	8-9	7-3	6-10	5-6	5-5
	2-#4 1-#6	40,000	9-0	11-1	8-8	8-11	8-1	6-9	6-4	5-2	5-0
		60,000	11-0	13-6	10-6	10-11	9-10	8-2	7-9	6-3	6-2
	2-#5	40,000	11-3	13-9	10-9	11-1	10-0	8-4	7-10	6-5	6-3
		60,000	15-8	19-2	15-0	15-6	14-0	11-8	11-0	8-11	8-9
	2-#6	40,000	15-5	18-10	14-8	15-2	13-9	11-5	10-9	8-6	8-3
		60,000	18-7	22-9	17-9	18-5	16-7	13-10	12-9	9-5	9-2
	Center distance A[k, l]		5-7	8-4	5-1	5-5	4-5	3-1	2-9	1-10	1-9
24	Span without stirrups[i, j]		9-11	13-7	9-9	10-2	9-0	7-5	7-0	5-10	5-9
	1-#5	40,000	8-6	10-8	8-5	8-8	7-10	6-6	6-2	5-0	4-11
		60,000	10-5	13-0	10-3	10-7	9-7	8-0	7-6	6-1	6-0
	2-#4 1-#6	40,000	9-7	12-1	9-6	9-9	8-10	7-5	7-0	5-8	5-6
		60,000	11-9	14-9	11-7	11-11	10-10	9-0	8-6	6-11	6-9
	2-#5	40,000	12-0	15-0	11-9	12-2	11-0	9-2	8-8	7-1	6-11
		60,000	14-7	18-3	14-4	14-10	13-5	11-2	10-7	8-7	8-5
	2-#6	40,000	14-3	17-11	14-1	14-7	13-2	11-0	10-4	8-5	8-3
		60,000	19-11	25-0	19-7	20-3	18-4	15-3	14-5	10-10	10-7
	Center distance A[k, l]		6-3	9-11	6-1	6-6	5-4	3-9	3-4	2-2	2-1

For SI: 1 inch = 25.4 mm, 1 foot = 304.8 mm, 1 pound per square foot = 0.0479 kPa, Grade 40 = 280 MPa, Grade 60 = 420 MPa.

Note: Top and bottom reinforcement for lintels without stirrups shown in shaded cells shall be equal to or greater than that required for lintel of the same depth and loading condition that has an allowable clear span that is equal to or greater than that of the lintel without stirrups.

a. See Table R611.3 for tolerances permitted from nominal thickness.
b. Table values are based on concrete with a minimum specified compressive strength of 2,500 psi. See Note j.
c. Table values are based on uniform loading. See Section R611.8.2 for lintels supporting concentrated loads.
d. Deflection criterion is *L*/240, where *L* is the clear span of the lintel in inches, or $^1/_2$-inch, whichever is less.
e. Linear interpolation is permitted between ground snow loads and between lintel depths.
f. DR indicates design required.
g. Lintel depth, *D*, is permitted to include the available height of wall located directly above the lintel, provided that the increased lintel depth spans the entire length of the lintel.
h. Stirrups shall be fabricated from reinforcing bars with the same yield strength as that used for the main longitudinal reinforcement.
i. Allowable clear span without stirrups applicable to all lintels of the same depth, *D*. Top and bottom reinforcement for lintels without stirrups shall not be less than the least amount of reinforcement required for a lintel of the same depth and loading condition with stirrups. All other spans require stirrups spaced at not more than *d*/2.
j. Where concrete with a minimum specified compressive strength of 3,000 psi is used, clear spans for lintels without stirrups shall be permitted to be multiplied by 1.05. If the increased span exceeds the allowable clear span for a lintel of the same depth and loading condition with stirrups, the top and bottom reinforcement shall be equal to or greater than that required for a lintel of the same depth and loading condition that has an allowable clear span that is equal to or greater than that of the lintel without stirrups that has been increased.
k. Center distance, *A*, is the center portion of the clear span where stirrups are not required. This is applicable to all longitudinal bar sizes and steel yield strengths.
l. Where concrete with a minimum specified compressive strength of 3,000 psi is used, center distance, *A*, shall be permitted to be multiplied by 1.10.
m. The maximum clear opening width between two solid wall segments shall be 18 feet. See Section R611.7.2.1. Lintel clear spans in the table greater than 18 feet are shown for interpolation and information only.

TABLE R611.8(6)
MAXIMUM ALLOWABLE CLEAR SPANS FOR 6-INCH-THICK WAFFLE-GRID LINTELS IN LOAD-BEARING WALLS[a, b, c, d, e, f, o]
MAXIMUM ROOF CLEAR SPAN 40 FEET AND MAXIMUM FLOOR SPAN 32 FEET

LINTEL DEPTH, D[g] (inches)	NUMBER OF BARS AND BAR SIZE IN TOP AND BOTTOM OF LINTEL	STEEL YIELD STRENGTH[h], f_y (psi)	DESIGN LOADING CONDITION DETERMINED FROM TABLE R611.8(1)								
			1	2		3		4		5	
			Maximum ground snow load (psf)								
			—	30	70	30	70	30	70	30	70
			Maximum clear span of lintel (feet - inches)								
8[i]	Span without stirrups[k, l]		2-7	2-9	2-0	2-1	2-0	2-0	2-0	2-0	2-0
8[i]	1-#4	40,000	5-2	5-5	4-0	4-3	3-7	3-3	2-11	2-4	2-3
8[i]	1-#4	60,000	5-9	6-3	4-0	4-3	3-7	3-3	2-11	2-4	2-3
8[i]	1-#5	40,000	5-9	6-3	4-0	4-3	3-7	3-3	2-11	2-4	2-3
8[i]	1-#5	60,000	5-9	6-3	4-0	4-3	3-7	3-3	2-11	2-4	2-3
8[i]	2-#4 1-#6	40,000	5-9	6-3	4-0	4-3	3-7	3-3	2-11	2-4	2-3
8[i]	2-#4 1-#6	60,000	DR	DR	DR	DR	DR	DR	DR	DR	DR
8[i]	Center distance A[m, n]		0-9	0-10	0-6	0-6	0-5	0-5	0-4	STL	STL
12[i]	Span without stirrups[k, l]		2-11	3-1	2-6	2-7	2-5	2-4	2-3	2-1	2-0
12[i]	1-#4	40,000	5-9	6-2	4-8	4-10	4-4	4-1	3-9	3-2	3-1
12[i]	1-#4	60,000	8-0	8-7	6-6	6-9	6-0	5-5	4-11	3-11	3-10
12[i]	1-#5	40,000	8-1	8-9	6-8	6-11	6-0	5-5	4-11	3-11	3-10
12[i]	1-#5	60,000	9-1	10-3	6-8	7-0	6-0	5-5	4-11	3-11	3-10
12[i]	2-#4 1-#6	40,000	9-1	9-9	6-8	7-0	6-0	5-5	4-11	3-11	3-10
12[i]	Center distance A[m, n]		1-3	1-5	0-10	0-11	0-9	0-8	0-6	STL	STL
16[i]	Span without stirrups[k, l]		4-0	4-4	3-6	3-7	3-4	3-3	3-1	2-10	2-10
16[i]	1-#4	40,000	6-7	7-3	5-6	5-9	5-2	4-10	4-6	3-9	3-8
16[i]	1-#4	60,000	8-0	8-10	6-9	7-0	6-3	5-11	5-5	4-7	4-5
16[i]	1-#5	40,000	8-2	9-0	6-11	7-2	6-5	6-0	5-7	4-8	4-6
16[i]	1-#5	60,000	11-5	12-6	9-3	9-9	8-4	7-7	6-10	5-6	5-4
16[i]	2-#4 1-#6	40,000	10-7	11-7	8-11	9-3	8-3	7-7	6-10	5-6	5-4
16[i]	2-#4 1-#6	60,000	12-2	14-0	9-3	9-9	8-4	7-7	6-10	5-6	5-4
16[i]	2-#5	40,000	12-2	14-2	9-3	9-9	8-4	7-7	6-10	5-6	5-4
16[i]	2-#5	60,000	DR	DR	DR	DR	DR	DR	DR	DR	DR
16[i]	Center distance A[m, n]		1-8	2-0	1-2	1-3	1-0	0-11	0-9	STL	STL
20[i]	Span without stirrups[k, l]		5-0	5-6	4-6	4-7	4-3	4-1	4-0	3-8	3-8
20[i]	1-#4	40,000	7-2	8-2	6-3	6-6	5-10	5-6	5-1	4-3	4-2
20[i]	1-#4	60,000	8-11	9-11	7-8	7-11	7-1	6-8	6-2	5-2	5-0
20[i]	1-#5	40,000	9-1	10-2	7-9	8-1	7-3	6-10	6-4	5-4	5-2
20[i]	1-#5	60,000	12-8	14-2	10-11	11-3	10-2	9-6	8-9	7-1	6-10
20[i]	2-#4 1-#6	40,000	10-3	11-5	8-9	9-1	8-2	7-8	7-1	6-0	5-10
20[i]	2-#4 1-#6	60,000	14-3	15-11	11-9	12-5	10-8	9-9	8-9	7-1	6-10
20[i]	2-#5	40,000	14-6	16-3	11-6	12-1	10-4	9-6	8-6	6-11	6-8
20[i]	2-#5	60,000	DR	DR	DR	DR	DR	DR	DR	DR	DR
20[i]	Center distance A[m, n]		2-0	2-6	1-6	1-7	1-3	1-1	1-0	STL	STL

(continued)

TABLE R611.8(6)—continued
MAXIMUM ALLOWABLE CLEAR SPANS FOR 6-INCH-THICK WAFFLE-GRID LINTELS IN LOAD-BEARING WALLS[a, b, c, d, e, f, o]
MAXIMUM ROOF CLEAR SPAN 40 FEET AND MAXIMUM FLOOR SPAN 32 FEET

LINTEL DEPTH, D[g] (inches)	NUMBER OF BARS AND BAR SIZE IN TOP AND BOTTOM OF LINTEL	STEEL YIELD STRENGTH[h], f_y (psi)	DESIGN LOADING CONDITION DETERMINED FROM TABLE R611.8(1)								
			1	2		3		4		5	
			Maximum ground snow load (psf)								
			—	30	70	30	70	30	70	30	70
			Maximum clear span of lintel (feet - inches)								
24w[j]	Span without stirrups[k, l]		6-0	6-8	5-5	5-7	5-3	5-0	4-10	4-6	4-5
	1-#4	40,000	7-11	9-0	6-11	7-2	6-5	6-0	5-7	4-8	4-7
		60,000	9-8	10-11	8-5	8-9	7-10	7-4	6-10	5-9	5-7
	1-#5	40,000	9-10	11-2	8-7	8-11	8-0	7-6	7-0	5-10	5-8
		60,000	12-0	13-7	10-6	10-10	9-9	9-2	8-6	7-2	6-11
	2-#4 1-#6	40,000	11-1	12-7	9-8	10-1	9-1	8-6	7-10	6-7	6-5
		60,000	15-6	17-7	13-6	14-0	12-8	11-10	10-8	8-7	8-4
	2-#5	40,000	15-6	17-11	12-8	13-4	11-6	10-7	9-7	7-10	7-7
		60,000	DR	DR	DR	DR	DR	DR	DR	DR	DR
	Center distance A[m, n]		2-4	3-0	1-9	1-11	1-6	1-4	1-2	STL	STL

For SI: 1 inch = 25.4 mm, 1 pound per square foot = 0.0479 kPa, 1 foot = 304.8 mm, Grade 40 = 280 MPa, Grade 60 = 420 MPa.

a. Where lintels are formed with waffle-grid forms, form material shall be removed, if necessary, to create top and bottom flanges of the lintel that are not less than 3 inches in depth (in the vertical direction), are not less than 5 inches in width for 6-inch-nominal waffle-grid forms and not less than 7 inches in width for 8-inch-nominal waffle-grid forms. See Figure R611.8(3). Flat form lintels shall be permitted in place of waffle-grid lintels. See Tables R611.8(2) through R611.8(5).

b. See Table R611.3 for tolerances permitted from nominal thicknesses and minimum dimensions and spacing of cores.

c. Table values are based on concrete with a minimum specified compressive strength of 2,500 psi. See Notes l and n. Table values are based on uniform loading. See Section R611.8.2 for lintels supporting concentrated loads.

d. Deflection criterion is $L/240$, where L is the clear span of the lintel in inches, or $^1/_2$-inch, whichever is less.

e. Linear interpolation is permitted between ground snow loads.

f. DR indicates design required. STL - stirrups required throughout lintel.

g. Lintel depth, D, is permitted to include the available height of wall located directly above the lintel, provided that the increased lintel depth spans the entire length of the lintel.

h. Stirrups shall be fabricated from reinforcing bars with the same yield strength as that used for the main longitudinal reinforcement.

i. Lintels less than 24 inches in depth with stirrups shall be formed from flat-walls forms [see Tables R611.8(2) through R611.8(5)], or, if necessary, form material shall be removed from waffle-grid forms so as to provide the required cover for stirrups. Allowable spans for lintels formed with flat-wall forms shall be determined from Tables R611.8(2) through R611.8(5).

j. Where stirrups are required for 24-inch deep lintels, the spacing shall not exceed 12 inches on center.

k. Allowable clear span without stirrups applicable to all lintels of the same depth, D. Top and bottom reinforcement for lintels without stirrups shall not be less than the least amount of reinforcement required for a lintel of the same depth and loading condition with stirrups. All other spans require stirrups spaced at not more than $d/2$.

l. Where concrete with a minimum specified compressive strength of 3,000 psi is used, clear spans for lintels without stirrups shall be permitted to be multiplied by 1.05. If the increased span exceeds the allowable clear span for a lintel of the same depth and loading condition with stirrups, the top and bottom reinforcement shall be equal to or greater than that required for a lintel of the same depth and loading condition that has an allowable clear span that is equal to or greater than that of the lintel without stirrups that has been increased.

m. Center distance, A, is the center portion of the span where stirrups are not required. This is applicable to all longitudinal bar sizes and steel yield strengths.

n. Where concrete with a minimum specified compressive strength of 3,000 psi is used, center distance, A, shall be permitted to be multiplied by 1.10.

o. The maximum clear opening width between two solid wall segments shall be 18 feet. See Section R611.7.2.1. Lintel spans in the table greater than 18 feet are shown for interpolation and information only.

TABLE R611.8(7)
MAXIMUM ALLOWABLE CLEAR SPANS FOR 8-INCH-THICK WAFFLE-GRID LINTELS IN LOAD-BEARING WALLS[a, b, c, d, e, f, o]
MAXIMUM ROOF CLEAR SPAN 40 FEET AND MAXIMUM FLOOR CLEAR SPAN 32 FEET

LINTEL DEPTH, D[g] (inches)	NUMBER OF BARS AND BAR SIZE IN TOP AND BOTTOM OF LINTEL	STEEL YIELD STRENGTH[h], f_y (psi)	DESIGN LOADING CONDITION DETERMINED FROM TABLE R611.8(1)								
			1	2		3		4		5	
				Maximum ground snow load (psf)							
				30	70	30	70	30	70	30	70
			Maximum clear span of lintel (feet - inches)								
8[i]	Span with stirrups[k, l]		2-6	2-9	2-0	2-1	2-0	2-0	2-0	2-0	2-0
	1-#4	40,000	4-5	4-9	3-7	3-9	3-4	3-0	2-10	2-3	2-2
		60,000	5-6	6-2	4-0	4-3	3-7	3-1	2-10	2-3	2-2
	1-#5	40,000	5-6	6-2	4-0	4-3	3-7	3-1	2-10	2-3	2-2
	Center distance A[m, n]		0-9	0-10	0-6	0-6	0-5	0-4	0-4	STL	STL
12[i]	Span without stirrups[k, l]		2-10	3-1	2-6	2-7	2-5	2-3	2-2	2-0	2-0
	1-#4	40,000	5-7	6-1	4-8	4-10	4-4	3-11	3-8	3-0	2-11
		60,000	6-9	7-5	5-8	5-11	5-4	4-9	4-5	3-8	3-7
	1-#5	40,000	6-11	7-7	5-10	6-0	5-5	4-10	4-6	3-9	3-7
		60,000	8-8	10-1	6-7	7-0	5-11	5-2	4-8	3-9	3-7
	2-#4 1-#6	40,000	8-8	9-10	6-7	7-0	5-11	5-2	4-8	3-9	3-7
		60,000	8-8	10-1	6-7	7-0	5-11	5-2	4-8	3-9	3-7
	Center distance A[m, n]		1-2	1-5	0-10	0-11	0-9	0-7	0-6	STL	STL
16[i]	Span without stirrups[k, l]		3-10	4-3	3-6	3-7	3-4	3-2	3-0	2-10	2-9
	1-#4	40,000	6-5	7-2	5-6	5-9	5-2	4-8	4-4	3-7	3-6
		60,000	7-9	8-9	6-9	7-0	6-3	5-8	5-3	4-4	4-3
	1-#5	40,000	7-11	8-11	6-10	7-1	6-5	5-9	5-4	4-5	4-4
		60,000	9-8	10-11	8-4	8-8	7-10	7-0	6-6	5-2	5-1
	2-#4 1-#6	40,000	9-0	10-1	7-9	8-0	7-3	6-6	6-1	5-0	4-11
		60,000	11-5	13-10	9-2	9-8	8-3	7-2	6-6	5-2	5-1
	Center distance A[m, n]		1-6	1-11	1-2	1-3	1-0	0-10	0-8	STL	STL
20[i]	Span without stirrups[k, l]		4-10	5-5	4-5	4-7	4-3	4-0	3-11	3-7	3-7
	1-#4	40,000	7-0	8-1	6-3	6-5	5-10	5-3	4-11	4-1	3-11
		60,000	8-7	9-10	7-7	7-10	7-1	6-5	6-0	4-11	4-10
	1-#5	40,000	8-9	10-1	7-9	8-0	7-3	6-6	6-1	5-1	4-11
		60,000	10-8	12-3	9-6	9-10	8-10	8-0	7-5	6-2	6-0
	2-#4 1-#6	40,000	9-10	11-4	8-9	9-1	8-2	7-4	6-10	5-8	5-7
		60,000	12-0	13-10	10-8	11-0	9-11	9-0	8-4	6-8	6-6
	2-#5	40,000	12-3	14-1	10-10	11-3	10-2	8-11	8-1	6-6	6-4
		60,000	14-0	17-6	11-8	12-3	10-6	9-1	8-4	6-8	6-6
	Center distance A[m, n]		1-10	2-5	1-5	1-7	1-3	1-0	0-11	STL	STL

(continued)

TABLE R611.8(7)—continued
MAXIMUM ALLOWABLE CLEAR SPANS FOR 8-INCH-THICK WAFFLE-GRID LINTELS IN LOAD-BEARING WALLS[a, b, c, d, e, f, o]
MAXIMUM ROOF CLEAR SPAN 40 FEET AND MAXIMUM FLOOR CLEAR SPAN 32 FEET

LINTEL DEPTH, D[g] (inches)	NUMBER OF BARS AND BAR SIZE IN TOP AND BOTTOM OF LINTEL	STEEL YIELD STRENGTH[h], f_y (psi)	DESIGN LOADING CONDITION DETERMINED FROM TABLE R611.8(1)								
			1	2		3		4		5	
				Maximum ground snow load (psf)							
				30	70	30	70	30	70	30	70
			Maximum clear span of lintel (feet - inches)								
24[j]	Span without stirrups[k, l]		5-9	6-7	5-5	5-6	5-2	4-11	4-9	4-5	4-4
	1-#4	40,000	7-6	8-10	6-10	7-1	6-5	5-9	5-5	4-6	4-4
		60,000	9-2	10-9	8-4	8-8	7-10	7-1	6-7	5-6	5-4
	1-#5	40,000	9-5	11-0	8-6	8-10	8-0	7-2	6-8	5-7	5-5
		60,000	11-5	13-5	10-5	10-9	9-9	8-9	8-2	6-10	6-8
	2-#4 1-#6	40,000	10-7	12-5	9-8	10-0	9-0	8-1	7-7	6-3	6-2
		60,000	12-11	15-2	11-9	12-2	11-0	9-11	9-3	7-8	7-6
	2-#5	40,000	13-2	15-6	12-0	12-5	11-2	9-11	9-2	7-5	7-3
		60,000	16-3	21-0	14-1	14-10	12-9	11-1	10-1	8-1	7-11
	2-#6	40,000	14-4	18-5	12-6	13-2	11-5	9-11	9-2	7-5	7-3
	Center distance A[m, n]		2-1	2-11	1-9	1-10	1-6	1-3	1-1	STL	STL

For SI: 1 inch = 25.4 mm, 1 pound per square foot = 0.0479 kPa, 1 foot = 304.8 mm, Grade 40 = 280 MPa, Grade 60 = 420 MPa.

a. Where lintels are formed with waffle-grid forms, form material shall be removed, if necessary, to create top and bottom flanges of the lintel that are not less than 3 inches in depth (in the vertical direction), are not less than 5 inches in width for 6-inch-nominal waffle-grid forms and not less than 7 inches in width for 8-inch-nominal waffle-grid forms. See Figure R611.8(3). Flat form lintels shall be permitted in lieu of waffle-grid lintels. See Tables R611.8(2) through R611.8(5).

b. See Table R611.3 for tolerances permitted from nominal thicknesses and minimum dimensions and spacing of cores.

c. Table values are based on concrete with a minimum specified compressive strength of 2,500 psi. See Notes l and n. Table values are based on uniform loading. See Section R611.8.2 for lintels supporting concentrated loads.

d. Deflection criterion is $L/240$, where L is the clear span of the lintel in inches, or $^1/_2$-inch, whichever is less.

e. Linear interpolation is permitted between ground snow loads.

f. DR indicates design required. STL - stirrups required throughout lintel.

g. Lintel depth, D, is permitted to include the available height of wall located directly above the lintel, provided that the increased lintel depth spans the entire length of the lintel.

h. Stirrups shall be fabricated from reinforcing bars with the same yield strength as that used for the main longitudinal reinforcement.

i. Lintels less than 24 inches in depth with stirrups shall be formed from flat-walls forms [see Tables R611.8(2) through R611.8(5)], or, if necessary, form material shall be removed from waffle-grid forms so as to provide the required cover for stirrups. Allowable spans for lintels formed with flat-wall forms shall be determined from Tables R611.8(2) through R611.8(5).

j. Where stirrups are required for 24-inch deep lintels, the spacing shall not exceed 12 inches on center.

k. Allowable clear span without stirrups applicable to all lintels of the same depth, D. Top and bottom reinforcement for lintels without stirrups shall not be less than the least amount of reinforcement required for a lintel of the same depth and loading condition with stirrups. All other spans require stirrups spaced at not more than $d/2$.

l. Where concrete with a minimum specified compressive strength of 3,000 psi is used, clear spans for lintels without stirrups shall be permitted to be multiplied by 1.05. If the increased span exceeds the allowable clear span for a lintel of the same depth and loading condition with stirrups, the top and bottom reinforcement shall be equal to or greater than that required for a lintel of the same depth and loading condition that has an allowable clear span that is equal to or greater than that of the lintel without stirrups that has been increased.

m. Center distance, A, is the center portion of the span where stirrups are not required. This is applicable to all longitudinal bar sizes and steel yield strengths.

n. Where concrete with a minimum specified compressive strength of 3,000 psi is used, center distance, A, shall be permitted to be multiplied by 1.10.

o. The maximum clear opening width between two solid wall segments shall be 18 feet. See Section R611.7.2.1. Lintel spans in the table greater than 18 feet are shown for interpolation and information only.

TABLE R611.8(8)
MAXIMUM ALLOWABLE CLEAR SPANS FOR 6-INCH-THICK SCREEN-GRID LINTELS IN LOAD-BEARING WALLS[a, b, c, d, e, f, p]
ROOF CLEAR SPAN 40 FEET AND FLOOR CLEAR SPAN 32 FEET

LINTEL DEPTH, D[g] (inches)	NUMBER OF BARS AND BAR SIZE IN TOP AND BOTTOM OF LINTEL	STEEL YIELD STRENGTH[h], f_y (psi)	DESIGN LOADING CONDITION DETERMINED FROM TABLE R611.8(1)								
			1	2		3		4		5	
				Maximum ground snow load (psf)							
				30	70	30	70	30	70	30	70
			Maximum clear span of lintel (feet - inches)								
12[i,j]	Span without stirrups		2-9	2-11	2-4	2-5	2-3	2-3	2-2	2-0	2-0
16[i,j]	Span without stirrups		3-9	4-0	3-4	3-5	3-2	3-1	3-0	2-9	2-9
20[i,j]	Span without stirrups		4-9	5-1	4-3	4-4	4-1	4-0	3-10	3-7	3-7
24[k]	Span without stirrups[l, m]		5-8	6-3	5-2	5-3	5-0	4-10	4-8	4-4	4-4
	1-#4	40,000	7-11	9-0	6-11	7-2	6-5	6-1	5-8	4-9	4-7
		60,000	9-9	11-0	8-5	8-9	7-10	7-5	6-10	5-9	5-7
	1-#5	40,000	9-11	11-2	8-7	8-11	8-0	7-7	7-0	5-11	5-9
		60,000	12-1	13-8	10-6	10-10	9-9	9-3	8-6	7-2	7-0
	2-#4 1-#6	40,000	11-2	12-8	9-9	10-1	9-1	8-7	7-11	6-8	6-6
		60,000	15-7	17-7	12-8	13-4	11-6	10-8	9-8	7-11	7-8
	2-#5	40,000	14-11	18-0	12-2	12-10	11-1	10-3	9-4	7-8	7-5
		60,000	DR	DR	DR	DR	DR	DR	DR	DR	DR
	Center distance A[n, o]		2-0	2-6	1-6	1-7	1-4	1-2	1-0	STL	STL

For SI: 1 inch = 25.4 mm, 1 pound per square foot = 0.0479 kPa, 1 foot = 304.8 mm, Grade 40 = 280 MPa, Grade 60 = 420 MPa.

a. Where lintels are formed with screen-grid forms, form material shall be removed if necessary to create top and bottom flanges of the lintel that are not less than 5 inches in width and not less than 2.5 inches in depth (in the vertical direction). See Figure R611.8(4). Flat form lintels shall be permitted in lieu of screen-grid lintels. See Tables R611.8(2) through R611.8(5).

b. See Table R611.3 for tolerances permitted from nominal thickness and minimum dimensions and spacings of cores.

c. Table values are based on concrete with a minimum specified compressive strength of 2,500 psi. See Notes m and o. Table values are based on uniform loading. See Section R611.7.2.1 for lintels supporting concentrated loads.

d. Deflection criterion is $L/240$, where L is the clear span of the lintel in inches, or $^1/_2$-inch, whichever is less.

e. Linear interpolation is permitted between ground snow loads.

f. DR indicates design required. STL indicates stirrups required throughout lintel.

g. Lintel depth, D, is permitted to include the available height of wall located directly above the lintel, provided that the increased lintel depth spans the entire length of the lintel.

h. Stirrups shall be fabricated from reinforcing bars with the same yield strength as that used for the main longitudinal reinforcement.

i. Stirrups are not required for lintels less than 24 inches in depth fabricated from screen-grid forms. Top and bottom reinforcement shall consist of a No. 4 bar having a yield strength of 40,000 psi or 60,000 psi.

j. Lintels between 12 and 24 inches in depth with stirrups shall be formed from flat-wall forms [see Tables R611.8(2) through R611.8(5)], or form material shall be removed from screen-grid forms to provide a concrete section comparable to that required for a flat wall. Allowable spans for flat lintels with stirrups shall be determined from Tables R611.8(2) through R6111.8(5).

k. Where stirrups are required for 24-inch deep lintels, the spacing shall not exceed 12 inches on center.

l. Allowable clear span without stirrups applicable to all lintels of the same depth, D. Top and bottom reinforcement for lintels without stirrups shall not be less than the least amount of reinforcement required for a lintel of the same depth and loading condition with stirrups. All other spans require stirrups spaced at not more than 12 inches.

m. Where concrete with a minimum specified compressive strength of 3,000 psi is used, clear spans for lintels without stirrups shall be permitted to be multiplied by 1.05. If the increased span exceeds the allowable clear span for a lintel of the same depth and loading condition with stirrups, the top and bottom reinforcement shall be equal to or greater than that required for a lintel of the same depth and loading condition that has an allowable clear span that is equal to or greater than that of the lintel without stirrups that has been increased.

n. Center distance, A, is the center portion of the span where stirrups are not required. This is applicable to all longitudinal bar sizes and steel yield strengths.

o. Where concrete with a minimum specified compressive strength of 3,000 psi is used, center distance, A, shall be permitted to be multiplied by 1.10.

p. The maximum clear opening width between two solid wall segments shall be 18 feet. See Section R611.7.2.1. Lintel spans in the table greater than 18 feet are shown for interpolation and information only.

TABLE R611.8(9)
MAXIMUM ALLOWABLE CLEAR SPANS FOR FLAT LINTELS WITHOUT STIRRUPS IN NONLOAD-BEARING WALLS[a, b, c, d, e, g, h]

LINTEL DEPTH, D^f (inches)	NUMBER OF BARS AND BAR SIZE	STEEL YIELD STRENGTH, f_y (psi)	NOMINAL WALL THICKNESS (inches)							
			4		6		8		10	
			Lintel Supporting							
			Concrete Wall	Light-framed Gable	Concrete Wall	Light-framed Gable	Concrete Wall	Light-framed Gable	Concrete Wall	Light-framed Gable
			Maximum Clear Span of Lintel (feet - inches)							
8	1-#4	40,000	10-11	11-5	9-7	11-2	7-10	9-5	7-3	9-2
		60,000	12-5	11-7	10-11	13-5	9-11	13-2	9-3	12-10
	1-#5	40,000	12-7	11-7	11-1	13-8	10-1	13-5	9-4	13-1
		60,000	DR	DR	12-7	16-4	11-6	14-7	10-9	14-6
	2-#4 1-#6	40,000	DR	DR	12-0	15-3	10-11	15-0	10-2	14-8
		60,000	DR	DR	DR	DR	12-2	15-3	11-7	15-3
	2-#5	40,000	DR	DR	DR	DR	12-7	16-7	11-9	16-7
		60,000	DR	DR	DR	DR	DR	DR	13-3	16-7
	2-#6	40,000	DR	DR	DR	DR	DR	DR	13-2	17-8
		60,000	DR	DR	DR	DR	DR	DR	DR	DR
12	1-#4	40,000	11-5	9-10	10-6	12-0	9-6	11-6	8-9	11-1
		60,000	11-5	9-10	11-8	13-3	10-11	14-0	10-1	13-6
	1-#5	40,000	11-5	9-10	11-8	13-3	11-1	14-4	10-3	13-9
		60,000	11-5	9-10	11-8	13-3	11-10	16-0	11-9	16-9
	2-#4 1-#6	40,000	DR	DR	11-8	13-3	11-10	16-0	11-2	15-6
		60,000	DR	DR	11-8	13-3	11-10	16-0	11-11	18-4
	2-#5	40,000	DR	DR	11-8	13-3	11-10	16-0	11-11	18-4
		60,000	DR	DR	11-8	13-3	11-10	16-0	11-11	18-4
16	1-#4	40,000	13-6	13-0	11-10	13-8	10-7	12-11	9-11	12-4
		60,000	13-6	13-0	13-8	16-7	12-4	15-9	11-5	15-0
	1-#5	40,000	13-6	13-0	13-10	17-0	12-6	16-1	11-7	15-4
		60,000	13-6	13-0	13-10	17-1	14-0	19-7	13-4	18-8
	2-#4 1-#6	40,000	13-6	13-0	13-10	17-1	13-8	18-2	12-8	17-4
		60,000	13-6	13-0	13-10	17-1	14-0	20-3	14-1	—
	2-#5	40,000	13-6	13-0	13-10	17-1	14-0	20-3	14-1	—
		60,000	DR	DR	13-10	17-1	14-0	20-3	14-1	—
20	1-#4	40,000	14-11	15-10	13-0	14-10	11-9	13-11	10-10	13-2
		60,000	15-3	15-10	14-11	18-1	13-6	17-0	12-6	16-2
	1-#5	40,000	15-3	15-10	15-2	18-6	13-9	17-5	12-8	16-6
		60,000	15-3	15-10	15-8	20-5	15-9	—	14-7	20-1
	2-#4 1-#6	40,000	15-3	15-10	15-8	20-5	14-11	—	13-10	—
		60,000	15-3	15-10	15-8	20-5	15-10	—	15-11	—
	2-#5	40,000	15-3	15-10	15-8	20-5	15-10	—	15-11	—
		60,000	15-3	15-10	15-8	20-5	15-10	—	15-11	—

(continued)

TABLE R611.8(9)—continued
MAXIMUM ALLOWABLE CLEAR SPANS FOR FLAT LINTELS WITHOUT STIRRUPS IN NONLOAD-BEARING WALLS[a, b, c, d, e, g, h]

LINTEL DEPTH, *D*[f] (inches)	NUMBER OF BARS AND BAR SIZE	STEEL YIELD STRENGTH, f_y (psi)	NOMINAL WALL THICKNESS (inches)							
			4		6		8		10	
			Lintel Supporting							
			Concrete Wall	Light-framed Gable	Concrete Wall	Light-framed Gable	Concrete Wall	Light-framed Gable	Concrete Wall	Light-framed Gable
			Maximum Clear Span of Lintel (feet - inches)							
24	1-#4	40,000	16-1	17-1	13-11	15-10	12-7	14-9	11-8	13-10
		60,000	16-11	18-5	16-1	19-3	14-6	18-0	13-5	17-0
	1-#5	40,000	16-11	18-5	16-3	19-8	14-9	18-5	13-8	17-4
		60,000	16-11	18-5	17-4	—	17-0	—	15-8	—
	2-#4 1-#6	40,000	16-11	18-5	17-4	—	16-1	—	14-10	—
		60,000	16-11	18-5	17-4	—	17-6	—	17-1	—
	2-#5	40,000	16-11	18-5	17-4	—	17-6	—	17-4	—
		60,000	16-11	18-5	17-4	—	17-6	—	17-8	—

For SI: 1 inch = 25.4 mm, 1 foot = 304.8 mm, Grade 40 = 280 MPa, Grade 60 = 420 MPa.

a. See Table R611.3 for tolerances permitted from nominal thickness.

b. Table values are based on concrete with a minimum specified compressive strength of 2,500 psi. See Note e.

c. Deflection criterion is *L*/240, where *L* is the clear span of the lintel in inches, or $^1/_2$ inch, whichever is less.

d. Linear interpolation between lintels depths, *D*, is permitted provided the two cells being used to interpolate are shaded.

e. Where concrete with a minimum specified compressive strength of 3,000 psi is used, spans in cells that are shaded shall be permitted to be multiplied by 1.05.

f. Lintel depth, *D*, is permitted to include the available height of wall located directly above the lintel, provided that the increased lintel depth spans the entire length of the lintel.

g. DR indicates design required.

h. The maximum clear opening width between two solid wall segments shall be 18 feet. See Section R611.7.2.1. Lintel spans in the table greater than 18 feet are shown for interpolation and information purposes only.

TABLE R611.8(10)
MAXIMUM ALLOWABLE CLEAR SPANS FOR WAFFLE-GRID AND SCREEN-GRID LINTELS WITHOUT STIRRUPS IN NONLOAD-BEARING WALLS[c, d, e, f, g]

LINTEL DEPTH[h], *D* (inches)	FORM TYPE AND NOMINAL WALL THICKNESS (inches)					
	6-inch Waffle-grid[a]		8-inch Waffle-grid[a]		6-inch Screen-grid[b]	
	Lintel supporting					
	Concrete Wall	Light-framed Gable	Concrete Wall	Light-framed Gable	Concrete Wall	Light-framed Gable
	Maximum Clear Span of Lintel (feet - inches)					
8	10-3	8-8	8-8	8-3	—	—
12	9-2	7-6	7-10	7-1	8-8	6-9
16	10-11	10-0	9-4	9-3	—	—
20	12-5	12-2	10-7	11-2	—	—
24	13-9	14-2	11-10	12-11	13-0	12-9

For SI: 1 inch = 25.4 mm, 1 foot = 304.8 mm, Grade 40 = 280 MPa, Grade 60 = 420 MPa.

a. Where lintels are formed with waffle-grid forms, form material shall be removed, if necessary, to create top and bottom flanges of the lintel that are not less than 3 inches in depth (in the vertical direction), are not less than 5 inches in width for 6-inch waffle-grid forms and not less than 7 inches in width for 8-inch waffle-grid forms. See Figure R611.8(3). Flat form lintels shall be permitted in lieu of waffle-grid lintels. See Tables R611.8(2) through R611.8(5).

b. Where lintels are formed with screen-grid forms, form material shall be removed if necessary to create top and bottom flanges of the lintel that are not less than 5 inches in width and not less than 2.5 inches in depth (in the vertical direction). See Figure R611.8(4). Flat form lintels shall be permitted in lieu of screen-grid lintels. See Tables R611.8(2) through R611.8(5).

c. See Table R611.3 for tolerances permitted from nominal thickness and minimum dimensions and spacing of cores.

d. Table values are based on concrete with a minimum specified compressive strength of 2,500 psi. See Note g.

e. Deflection criterion is *L*/240, where *L* is the clear span of the lintel in inches, or $^1/_2$-inch, whichever is less.

f. Top and bottom reinforcement shall consist of a No. 4 bar having a minimum yield strength of 40,000 psi.

g. Where concrete with a minimum specified compressive strength of 3,000 psi is used, spans in shaded cells shall be permitted to be multiplied by 1.05.

h. Lintel depth, *D*, is permitted to include the available height of wall located directly above the lintel, provided that the increased lintel depth spans the entire length of the lintel.

R611.8.2.3 Lintels without stirrups designed for nonload-bearing conditions. The maximum clear span of lintels without stirrups designed for nonload-bearing conditions of Table R611.8(1).1 shall be determined in accordance with this section. The maximum clear span of lintels without stirrups in flat walls shall be determined in accordance with Table R611.8(9), and the maximum clear span of lintels without stirrups in walls of waffle-grid or screen-grid construction shall be determined in accordance with Table R611.8(10).

R611.9 Requirements for connections–general. Concrete walls shall be connected to footings, floors, ceilings and roofs in accordance with this section.

R611.9.1 Connections between concrete walls and light-framed floor, ceiling and roof systems. Connections between concrete walls and light-framed floor, ceiling and roof systems using the prescriptive details of Figures R611.9(1) through R611.9(12) shall comply with this section and Sections R611.9.2 and R611.9.3.

R611.9.1.1 Anchor bolts. Anchor bolts used to connect light-framed floor, ceiling and roof systems to concrete walls in accordance with Figures R611.9(1) through R611.9(12) shall have heads, or shall be rods with threads on both ends with a hex or square nut on the end embedded in the concrete. Bolts and threaded rods shall comply with Section R611.5.2.2. Anchor bolts with J- or L-hooks shall not be used where the connection details in these figures are used.

R611.9.1.2 Removal of stay-in-place form material at bolts. Holes in stay-in-place forms for installing bolts for attaching face-mounted wood ledger boards to the wall shall be a minimum of 4 inches (102 mm) in diameter for forms not greater than $1^1/_2$ inches (38 mm) in thickness, and increased 1 inch (25 mm) in diameter for each $^1/_2$-inch (13 mm) increase in form thickness. Holes in stay-in-place forms for installing bolts for attaching face-mounted cold-formed steel tracks to the wall shall be a minimum of 4 inches (102 mm) square. The wood ledger board or steel track shall be in direct contact with the concrete at each bolt location.

Exception: A vapor retarder or other material less than or equal to $^1/_{16}$ inch (1.6 mm) in thickness is permitted to be installed between the wood ledger or cold-formed track and the concrete.

R611.9.2 Connections between concrete walls and light-framed floor systems. Connections between concrete walls and light-framed floor systems shall be in accordance with one of the following:

1. For floor systems of wood frame construction, the provisions of Section R611.9.1 and the prescriptive details of Figures R611.9(1) through R611.9(4), where permitted by the tables accompanying those figures. Portions of connections of wood-framed floor systems not noted in the figures shall be in accordance with Section R502, or AF&PA/WFCM, if applicable.
2. For floor systems of cold-formed steel construction, the provisions of Section R611.9.1 and the prescriptive details of Figures R611.9(5) through R611.9(8), where permitted by the tables accompanying those figures. Portions of connections of cold-formed-steel framed floor systems not noted in the figures shall be in accordance with Section R505, or AISI S230, if applicable.
3. Proprietary connectors selected to resist loads and load combinations in accordance with Appendix A (ASD) or Appendix B (LRFD) of PCA 100.
4. An engineered design using loads and load combinations in accordance with Appendix A (ASD) or Appendix B (LRFD) of PCA 100.
5. An engineered design using loads and material design provisions in accordance with this code, or in accordance with ASCE 7, ACI 318, and AF&PA/NDS for wood frame construction or AISI S100 for cold-formed steel frame construction.

R611.9.3 Connections between concrete walls and light-framed ceiling and roof systems. Connections between concrete walls and light-framed ceiling and roof systems shall be in accordance with one of the following:

1. For ceiling and roof systems of wood frame construction, the provisions of Section R611.9.1 and the prescriptive details of Figures R611.9(9) and R611.9(10), where permitted by the tables accompanying those figures. Portions of connections of wood-framed ceiling and roof systems not noted in the figures shall be in accordance with Section R802, or AF&PA/WFCM, if applicable.
2. For ceiling and roof systems of cold-formed-steel construction, the provisions of Section R611.9.1 and the prescriptive details of Figures R611.9(11) and R611.9(12), where permitted by the tables accompanying those figures. Portions of connections of cold-formed-steel framed ceiling and roof systems not noted in the figures shall be in accordance with Section R804, or AISI S230, if applicable.
3. Proprietary connectors selected to resist loads and load combinations in accordance with Appendix A (ASD) or Appendix B (LRFD) of PCA 100.
4. An engineered design using loads and load combinations in accordance with Appendix A (ASD) or Appendix B (LRFD) of PCA 100.
5. An engineered design using loads and material design provisions in accordance with this code, or in accordance with ASCE 7, ACI 318, and AF&PA/NDS for wood-frame construction or AISI S100 for cold-formed-steel frame construction.

R611.10 Floor, roof and ceiling diaphragms. Floors and roofs in all buildings with exterior walls of concrete shall be designed and constructed as *diaphragms*. Where gable-end walls occur, ceilings shall also be designed and constructed as *diaphragms*. The design and construction of floors, roofs and ceilings of wood framing or cold-formed-steel framing serving as *diaphragms* shall comply with the applicable requirements of this code, or AF&PA/WFCM or AISI S230, if applicable.

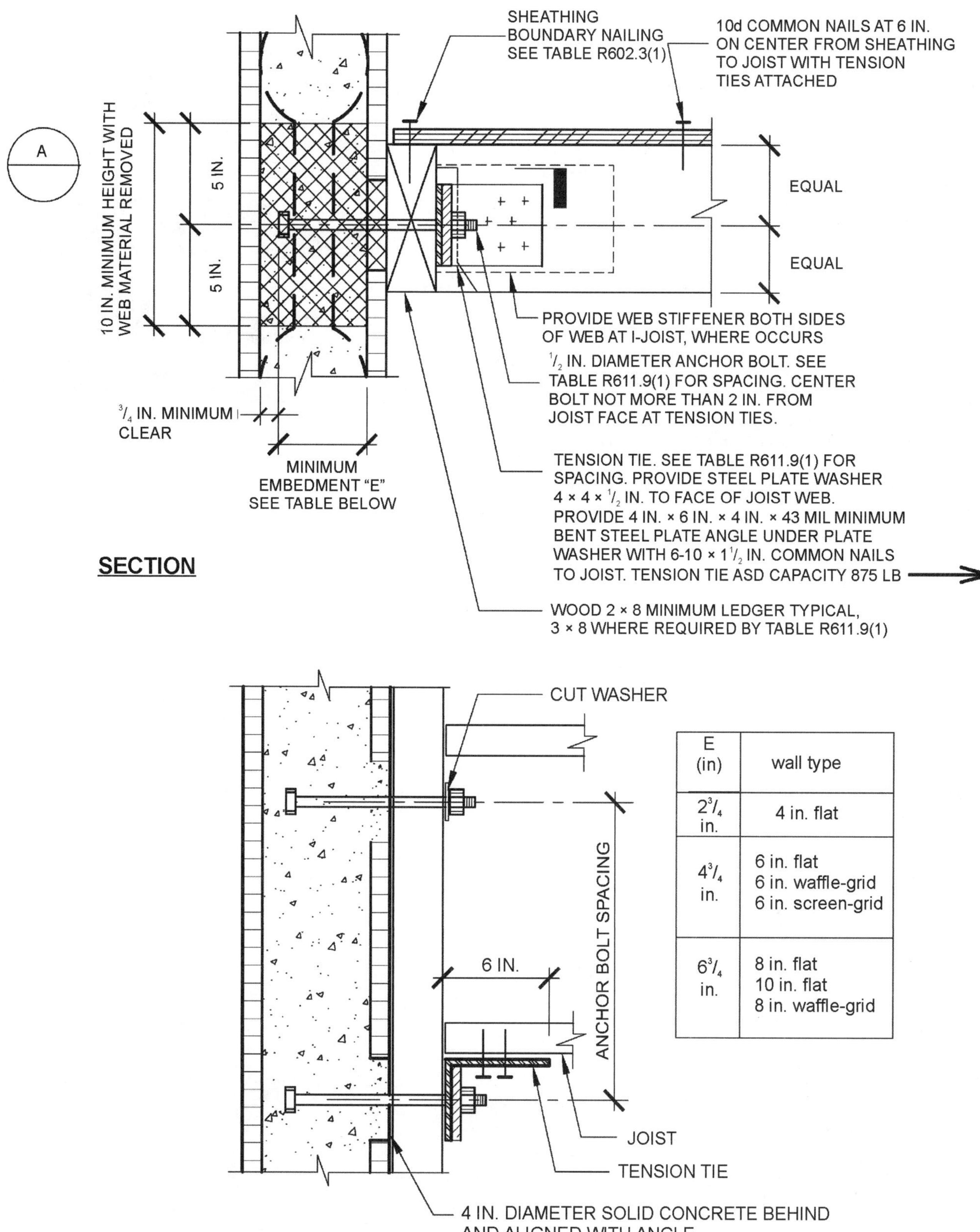

E (in)	wall type
2 3/4 in.	4 in. flat
4 3/4 in.	6 in. flat 6 in. waffle-grid 6 in. screen-grid
6 3/4 in.	8 in. flat 10 in. flat 8 in. waffle-grid

For SI: 1 mil = 0.0254 mm, 1 inch = 25.4 mm, 1 pound-force = 4.448 N.

FIGURE R611.9(1)
WOOD-FRAMED FLOOR TO SIDE OF CONCRETE WALL, FRAMING PERPENDICULAR

TABLE R611.9(1) WOOD-FRAMED FLOOR TO SIDE OF CONCRETE WALL, FRAMING PERPENDICULAR[a, b, c]

ANCHOR BOLT SPACING (inches)	TENSION TIE SPACING (inches)	BASIC WIND SPEED (mph)					
		85B	90B	100B	110B	120B	130B
				85C	90C	100C	110C
					85D	90D	100D
12	12						
12	24						
12	36						
12	48						
16	16					A	A
16	32						
16	48						
19.2	19.2	A	A	A	A	A	
19.2	38.4	A	A	A			

For SI: 1 inch = 25.4 mm, 1 mile per hour = 0.447 m/s.

a. This table is for use with the detail in Figure R611.9(1). Use of this detail is permitted where a cell is not shaded and prohibited where shaded.

b. Wall design per other provisions of Section R611 is required.

c. Letter "A" indicates that a minimum nominal 3 × 8 ledger is required.

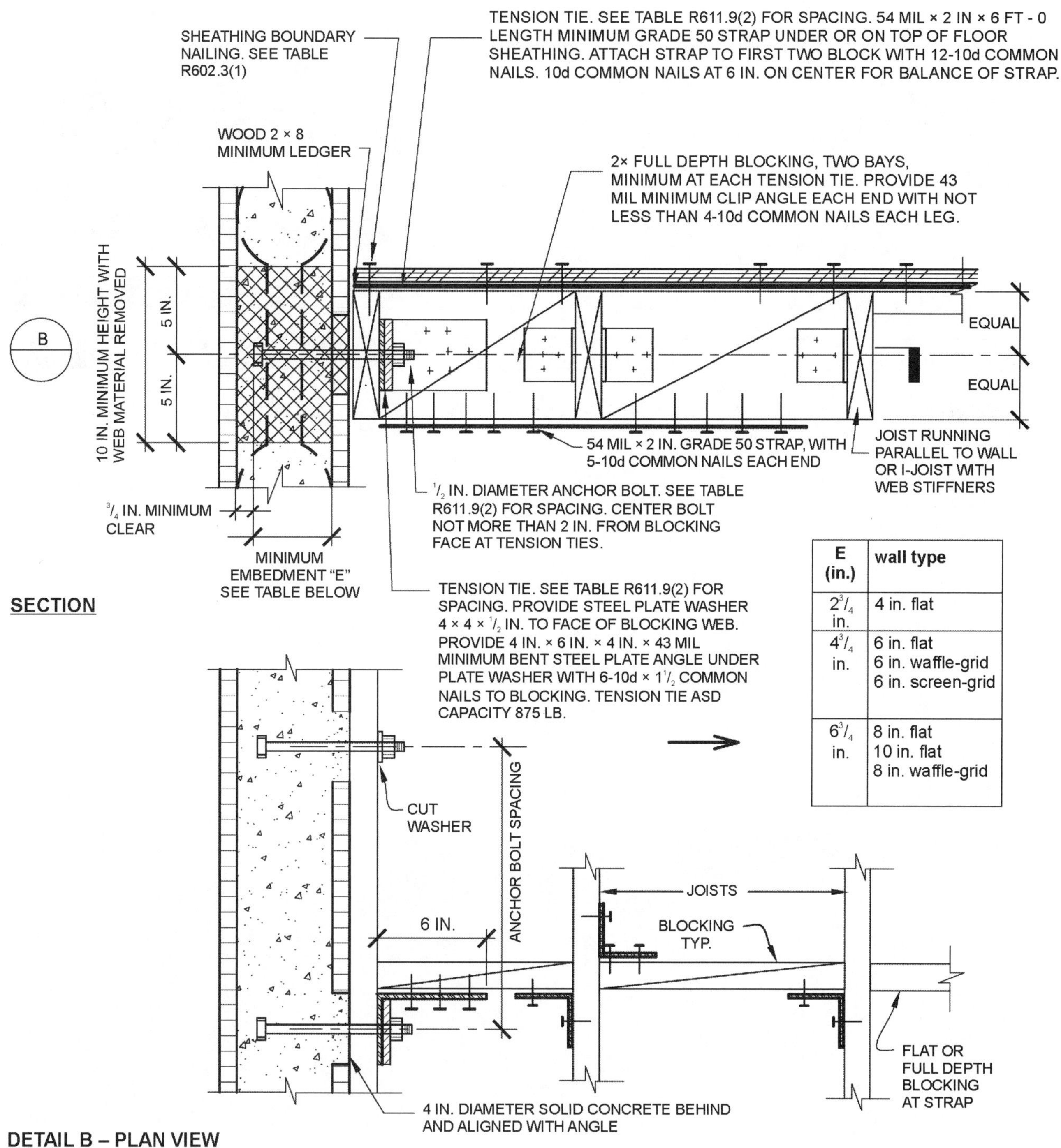

E (in.)	wall type
$2^3/_4$ in.	4 in. flat
$4^3/_4$ in.	6 in. flat 6 in. waffle-grid 6 in. screen-grid
$6^3/_4$ in.	8 in. flat 10 in. flat 8 in. waffle-grid

For SI: 1 mil = 0.0254 mm, 1 inch = 25.4 mm, 1 foot = 304.8 mm, 1 pound-force = 4.448 N.

FIGURE R611.9(2)
WOOD-FRAMED FLOOR TO SIDE OF CONCRETE WALL FRAMING PARALLEL

TABLE R611.9(2)
WOOD-FRAMED FLOOR TO SIDE OF CONCRETE WALL, FRAMING PARALLEL[a, b]

ANCHOR BOLT SPACING (inches)	TENSION TIE SPACING (inches)	BASIC WIND SPEED (mph) AND WIND EXPOSURE CATEGORY					
		85b	90B	100B	110B	120B	130B
				85C	90C	100C	110C
					85D	90D	100D
12	12						
12	24						
12	36						
12	48						
16	16						
16	32						
16	48						
19.2	19.2						
19.2	38.4						
24	24						
24	48						

For SI: 1 inch = 25.4 mm, 1 mile per hour = 0.447 m/s.

a. This table is for use with the detail in Figure R611.9(2). Use of this detail is permitted where a cell is not shaded and prohibited where shaded.

b. Wall design per other provisions of Section R611 is required.

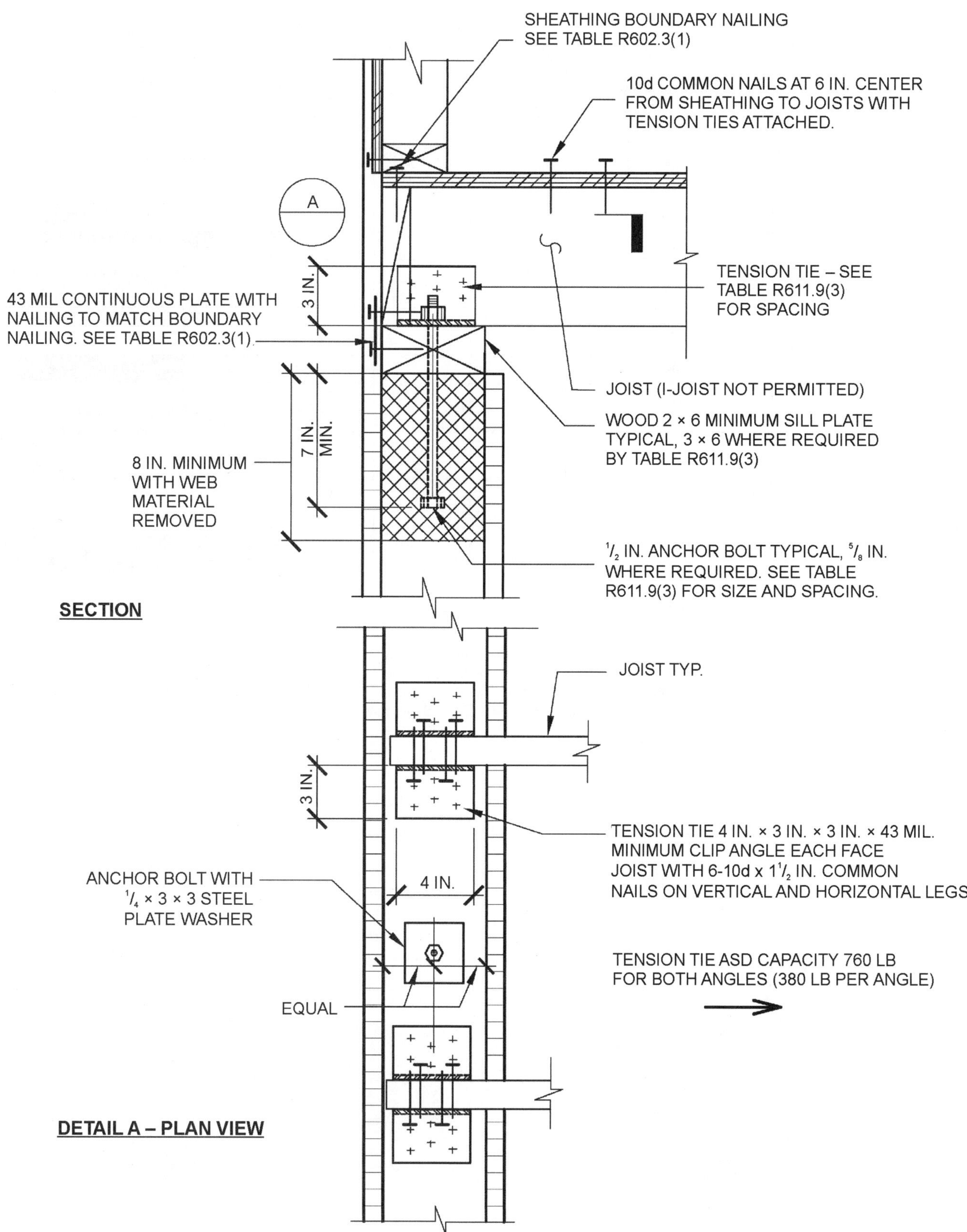

For SI: 1 mil = 0.0254 mm, 1 inch = 25.4 mm, 1 pound-force = 4.448 N.

FIGURE R611.9(3)
WOOD-FRAMED FLOOR TO TOP OF CONCRETE WALL FRAMING PERPENDICULAR

TABLE R611.9(3)
WOOD-FRAMED FLOOR TO TOP OF CONCRETE WALL, FRAMING PERPENDICULAR[a, b, c, d, e]

ANCHOR BOLT SPACING (inches)	TENSION TIE SPACING (inches)	BASIC WIND SPEED (mph) AND WIND EXPOSURE CATEGORY					
		85B	90B	100B	110B	120B	130B
				85C	90C	100C	110C
					85D	90D	100D
12	12						
12	24						
12	36						
12	48						
16	16					6 A	6 B
16	32					6 A	6 B
16	48						
19.2	19.2				6 A	6 A	6 B
19.2	38.4				6 A	6 A	
24	24			6 A	6 B	6 A	
24	48			6 A			

For SI: 1 inch = 25.4 mm, 1 mile per hour = 0.447 m/s.

a. This table is for use with the detail in Figure R611.9(3). Use of this detail is permitted where cell is not shaded, prohibited where shaded.

b. Wall design per other provisions in Section R611 is required.

c. For wind design, minimum 4-inch-nominal wall is permitted in unshaded cells with no number.

d. Number 6 indicates minimum permitted nominal wall thickness in inches necessary to develop required strength (capacity) of connection. As a minimum, this nominal thickness shall occur in the portion of the wall indicated by the cross-hatching in Figure R611.9(3). For the remainder of the wall, see Note b.

e. Letter "A" indicates that a minimum nominal 3 × 6 sill plate is required. Letter "B" indicates that a $^5/_8$-inch-diameter anchor bolt and a minimal nominal 3 × 6 sill plate are required.

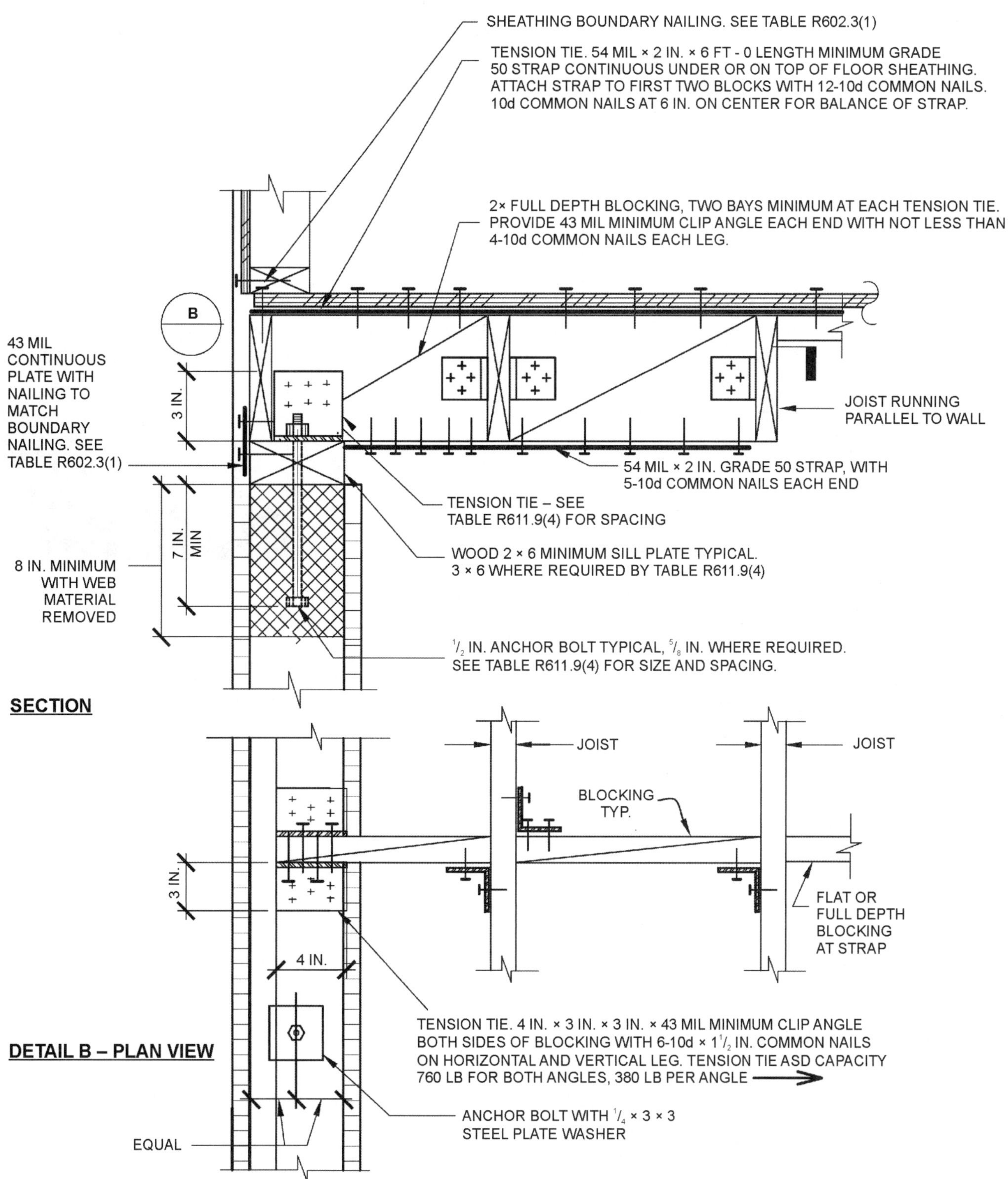

For SI: 1 mil = 0.0254 mm, 1 inch = 25.4 mm, 1 foot = 304.8 mm, 1 pound-force = 4.448 N.

FIGURE R611.9(4)
WOOD-FRAMED FLOOR TO TOP OF CONCRETE WALL FRAMING PARALLEL

TABLE R611.9(4)
WOOD-FRAMED FLOOR TO TOP OF CONCRETE WALL, FRAMING PARALLEL[a, b, c, d, e]

ANCHOR BOLT SPACING (inches)	TENSION TIE SPACING (inches)	BASIC WIND SPEED (mph) AND WIND EXPOSURE CATEGORY					
		85B	90B	100B	110B	120B	130B
				85C	90C	100C	110C
					85D	90D	100D
	12						
12	24						
12	36						
12	48						
16	16					6 A	6 B
16	32					6 A	6 B
16	48						
19.2	19.2				6 A	6 A	6 B
19.2	38.4				6 A	6 A	
24	24			6 A	6 B	6 B	
24	48			6 A			

For SI: 1 inch = 25.4 mm, 1 mile per hour = 0.447 m/s.

a. This table is for use with the detail in Figure R611.9(4). Use of this detail is permitted where a cell is not shaded, prohibited where shaded.

b. Wall design per other provisions of Section R611 is required.

c. For wind design, minimum 4-inch-nominal wall is permitted in unshaded cells with no number.

d. Number 6 indicates minimum permitted nominal wall thickness in inches necessary to develop required strength (capacity) of connection. As a minimum, this nominal thickness shall occur in the portion of the wall indicated by the cross-hatching in Figure R611.9(4). For the remainder of the wall, see Note b.

e. Letter "A" indicates that a minimum nominal 3 × 6 sill plate is required. Letter "B" indicates that a $^5/_8$-inch-diameter anchor bolt and a minimal nominal 3 × 6 sill plate are required.

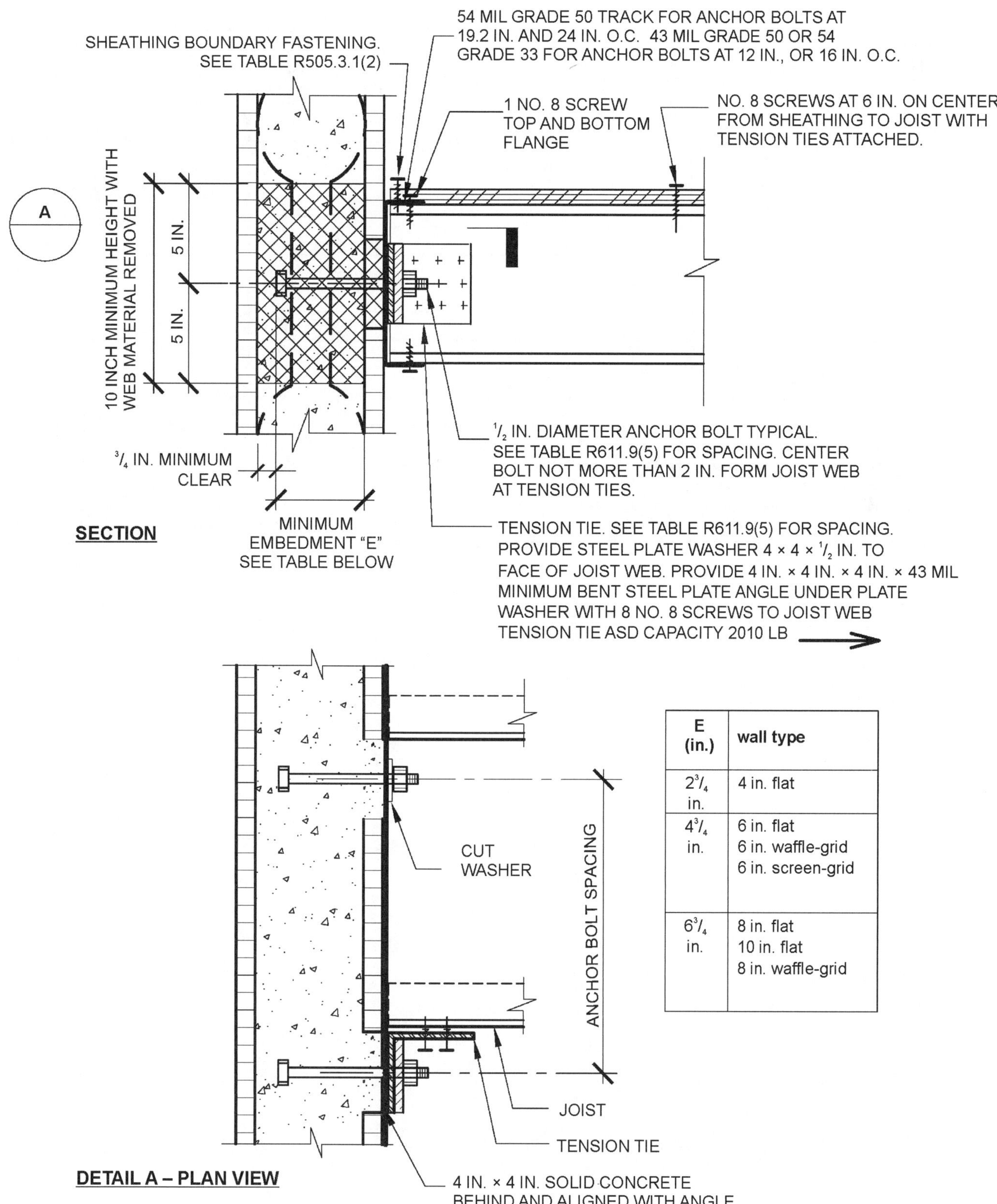

E (in.)	wall type
2 3/4 in.	4 in. flat
4 3/4 in.	6 in. flat 6 in. waffle-grid 6 in. screen-grid
6 3/4 in.	8 in. flat 10 in. flat 8 in. waffle-grid

For SI: 1 mil = 0.0254 mm, 1 inch = 25.4 mm, 1 pound-force = 4.448 N.

FIGURE R611.9(5)
COLD-FORMED STEEL FLOOR TO SIDE OF CONCRETE WALL, FRAMING PERPENDICULAR

TABLE R611.9(5)
COLD-FORMED STEEL-FRAMED FLOOR TO SIDE OF CONCRETE WALL, FRAMING PERPENDICULAR[a, b, c, d]

ANCHOR BOLT SPACING (inches)	TENSION TIE SPACING (inches)	BASIC WIND SPEED (mph) AND WIND EXPOSURE CATEGORY					
		85B	90B	100B	110B	120B	130B
				85C	90C	100C	110C
					85D	90D	100D
12	12						
12	24						
12	36						6
12	48					6	6
16	16						
16	32						
16	48					6	6
19.2	19.2						
19.2	38.4						6
24	24						
24	48					6	6

For SI: 1 inch = 25.4 mm, 1 mile per hour = 0.4470 m/s.

a. This table is for use with the detail in Figure R611.9(5). Use of this detail is permitted where a cell is not shaded.

b. Wall design per other provisions of Section R611 is required.

c. For wind design, minimum 4-inch-nominal wall is permitted in unshaded cells with no number.

d. Number 6 indicates minimum permitted nominal wall thickness in inches necessary to develop required strength (capacity) of connection. As a minimum, this nominal thickness shall occur in the portion of the wall indicated by the cross-hatching in Figure R611.9(5). For the remainder of the wall, see Note b.

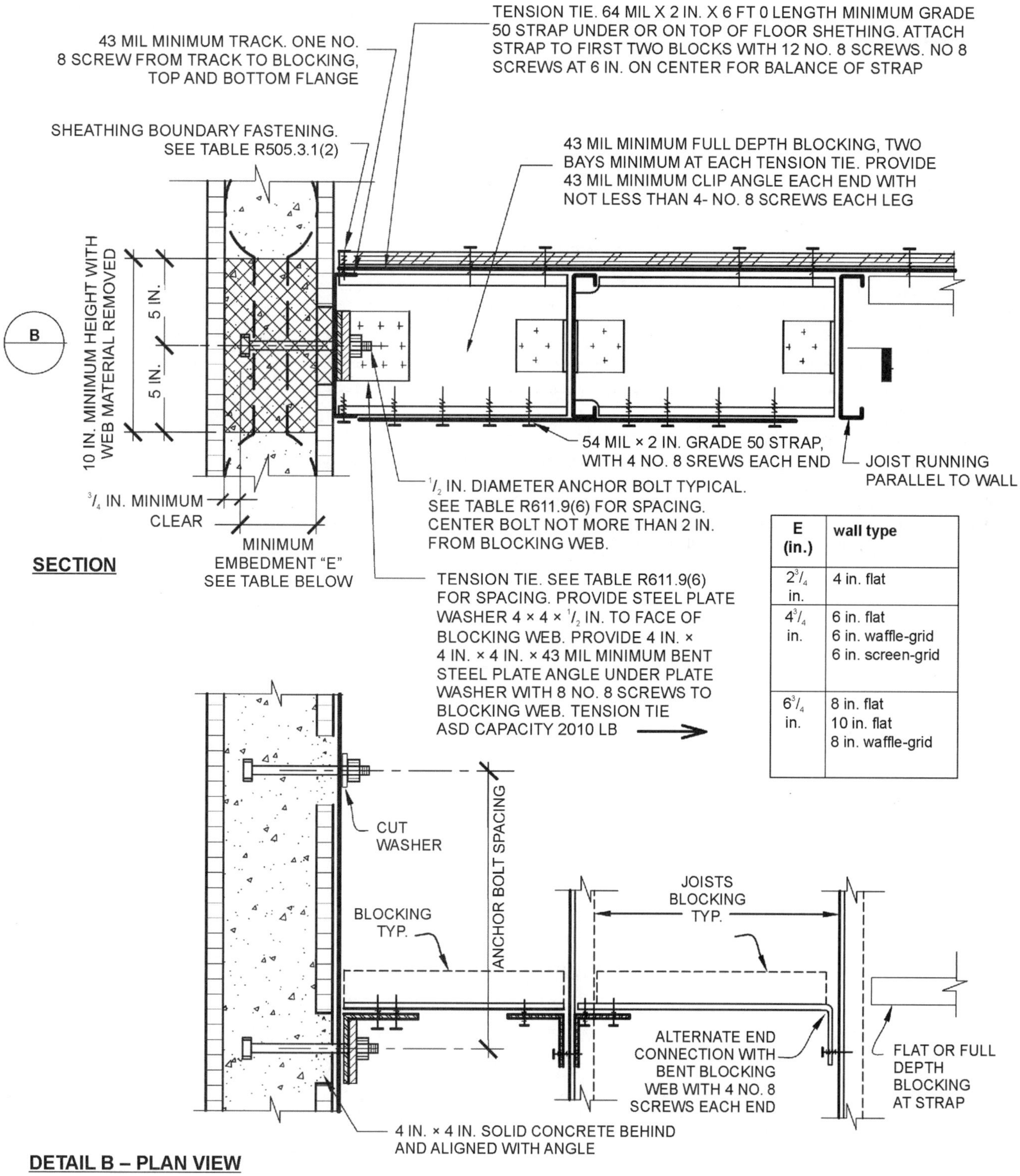

E (in.)	wall type
$2^{3}/_{4}$ in.	4 in. flat
$4^{3}/_{4}$ in.	6 in. flat 6 in. waffle-grid 6 in. screen-grid
$6^{3}/_{4}$ in.	8 in. flat 10 in. flat 8 in. waffle-grid

For SI: 1 mil = 0.0254 mm, 1 inch = 25.4 mm, 1 pound-force = 4.448 N.

FIGURE R611.9(6)
COLD-FORMED STEEL FLOOR TO SIDE OF CONCRETE WALL, FRAMING PARALLEL

TABLE R611.9(6)
COLD-FORMED STEEL-FRAMED FLOOR TO SIDE OF CONCRETE WALL, FRAMING PARALLEL[a, b, c, d]

ANCHOR BOLT SPACING (inches)	TENSION TIE SPACING (inches)	BASIC WIND SPEED (mph) AND WIND EXPOSURE CATEGORY					
		85B	90B	100B	110B	120B	130B
				85C	90C	100C	110C
					85D	90D	100D
12	12						
12	24						
12	36						6
12	48					6	6
16	16						
16	32						
16	48					6	6
19.2	19.2						
19.2	38.4						6
24	24						
24	48					6	6

For SI: 1 inch = 25.4 mm, 1 mile per hour = 0.447 m/s.

a. This table is for use with the detail in Figure R611.9(6). Use of this detail is permitted where a cell is not shaded.

b. Wall design per other provisions of Section R611 is required.

c. For wind design, minimum 4-inch-nominal wall is permitted in unshaded cells with no number.

d. Number 6 indicates minimum permitted nominal wall thick ness in inches necessary to develop required strength (capacity) of connection. As a minimum, this nominal thickness shall occur in the portion of the wall indicated by the cross-hatching in Figure R611.9(6). For the remainder of the wall, see Note b.

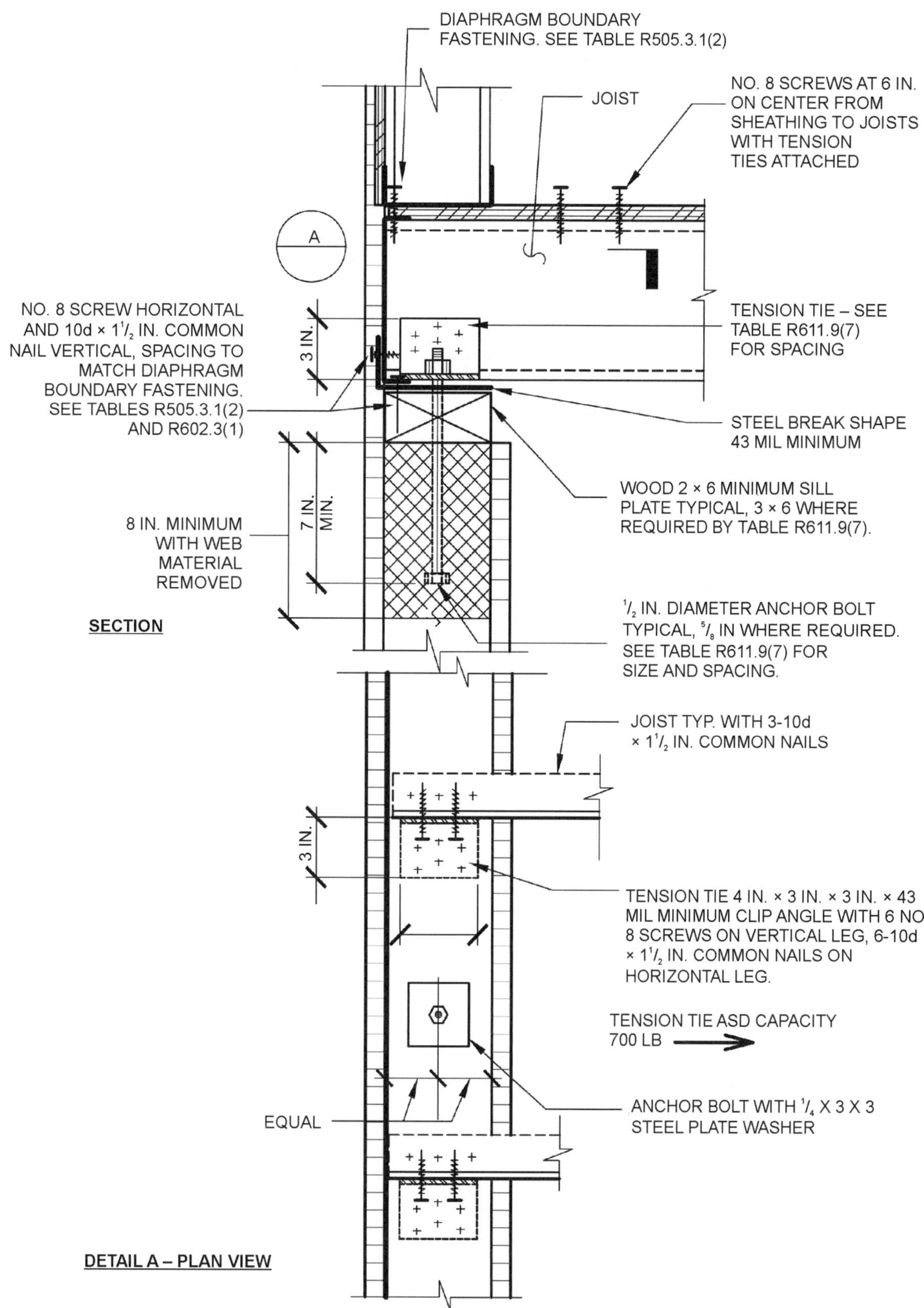

For SI: 1 mil = 0.0254 mm, 1 inch = 25.4 mm, 1 pound-force = 4.448 N.

FIGURE R611.9(7)
COLD-FORMED STEEL FLOOR TO TOP OF CONCRETE WALL FRAMING PERPENDICULAR

TABLE R611.9(7)
COLD-FORMED STEEL-FRAMED FLOOR TO TOP OF CONCRETE WALL, FRAMING PERPENDICULAR[a, b, c, d, e]

ANCHOR BOLT SPACING (inches)	TENSION TIE SPACING (inches)	BASIC WIND SPEED (mph) AND WIND EXPOSURE CATEGORY					
		85B	90B	100B	110B	120B	130B
				858C	90C	100C	110C
					85D	90D	100D
12	12						
12	24						
16	16					6 A	6 B
16	32					6 A	6 B
19.2	19.2				6 A	8 B	8 B
19.2	38.4				6 A	8 B	8 B
24	24			6 A	8 B	8 B	

For SI: 1 inch = 25.4 mm, 1 mile per hour = 0.447 m/s.

a. This table is for use with the detail in Figure R611.9(7). Use of this detail is permitted where a cell is not shaded, prohibited where shaded.

b. Wall design per other provisions of Section R611 is required.

c. For wind design, minimum 4-inch-nominal wall is permitted in unshaded cells with no number.

d. Numbers 6 and 8 indicate minimum permitted nominal wall thickness in inches necessary to develop required strength (capacity) of connection. As a minimum, this nominal thickness shall occur in the portion of the wall indicated by the cross-hatching in Figure R611.9(7). For the remainder of the wall, see Note b.

e. Letter "A" indicates that a minimum nominal 3 × 6 sill plate is required. Letter "B" indicates that a $^5/_8$-inch-diameter anchor bolt and a minimum nominal 3 × 6 sill plate are required.

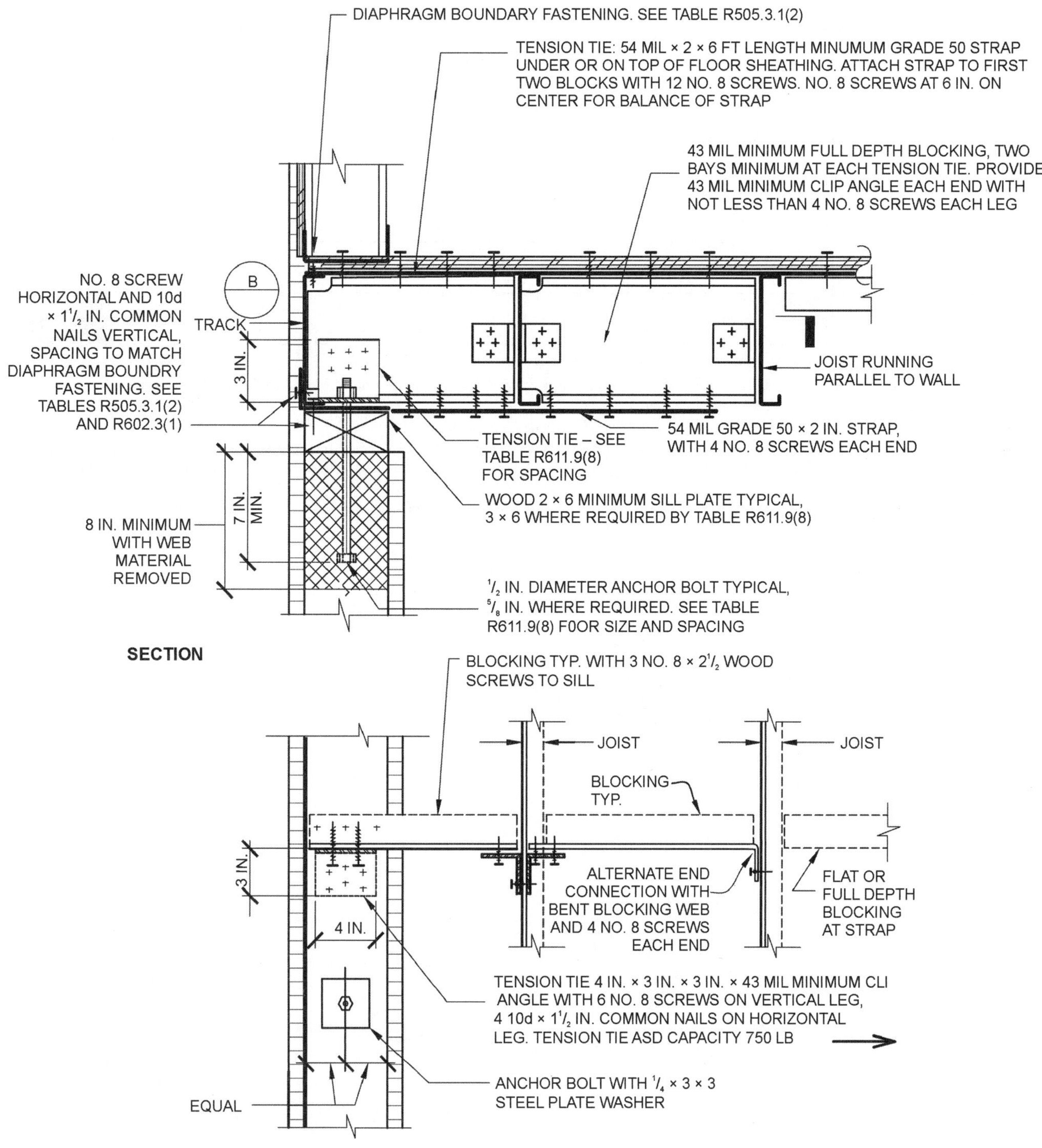

For SI: 1 mil = 0.0254 mm, 1 inch = 25.4 mm, 1 pound-force = 4.448 N.

FIGURE R611.9(8)
COLD-FORMED STEEL FLOOR TO TOP OF CONCRETE WALL, FRAMING PARALLEL

TABLE R611.9(8)
COLD-FORMED STEEL-FRAMED FLOOR TO TOP OF CONCRETE WALL, FRAMING PARALLEL[a, b, c, d, e]

ANCHOR BOLT SPACING (inches)	TENSION TIE SPACING (inches)	BASIC WIND SPEED (mph) AND WIND EXPOSURE CATEGORY					
		85B	90B	100B	110B	120B	130B
				85C	90C	100C	110C
					85D	90D	100D
12	12						
12	24						
16	16					6 A	6 B
16	32					6 A	6 B
19.2	19.2				6 A	8 B	8 B
19.2	38.4				6 A	8 B	8 B
24	24			6 A	8 B	8 B	

For SI: 1 inch = 25.4 mm, 1 mile per hour = 0.447 m/s.

a. This table is for use with the detail in Figure R611.9(8). Use of this detail is permitted where a cell is not shaded, prohibited where shaded.

b. Wall design per other provisions of Section R611 is required.

c. For wind design, minimum 4-inch-nominal wall is permitted in unshaded cells with no number.

d. Numbers 6 and 8 indicate minimum permitted nominal wall thickness in inches necessary to develop required strength (capacity) of connection. As a minimum, this nominal thickness shall occur in the portion of the wall indicated by the cross-hatching in Figure R611.9(8). For the remainder of the wall, see Note b.

e. Letter "A" indicates that a minimum nominal 3 × 6 sill plate is required. Letter "B" indicates that a $^5/_8$-inch-diameter anchor bolt and a minimum nominal 3 × 6 sill plate are required.

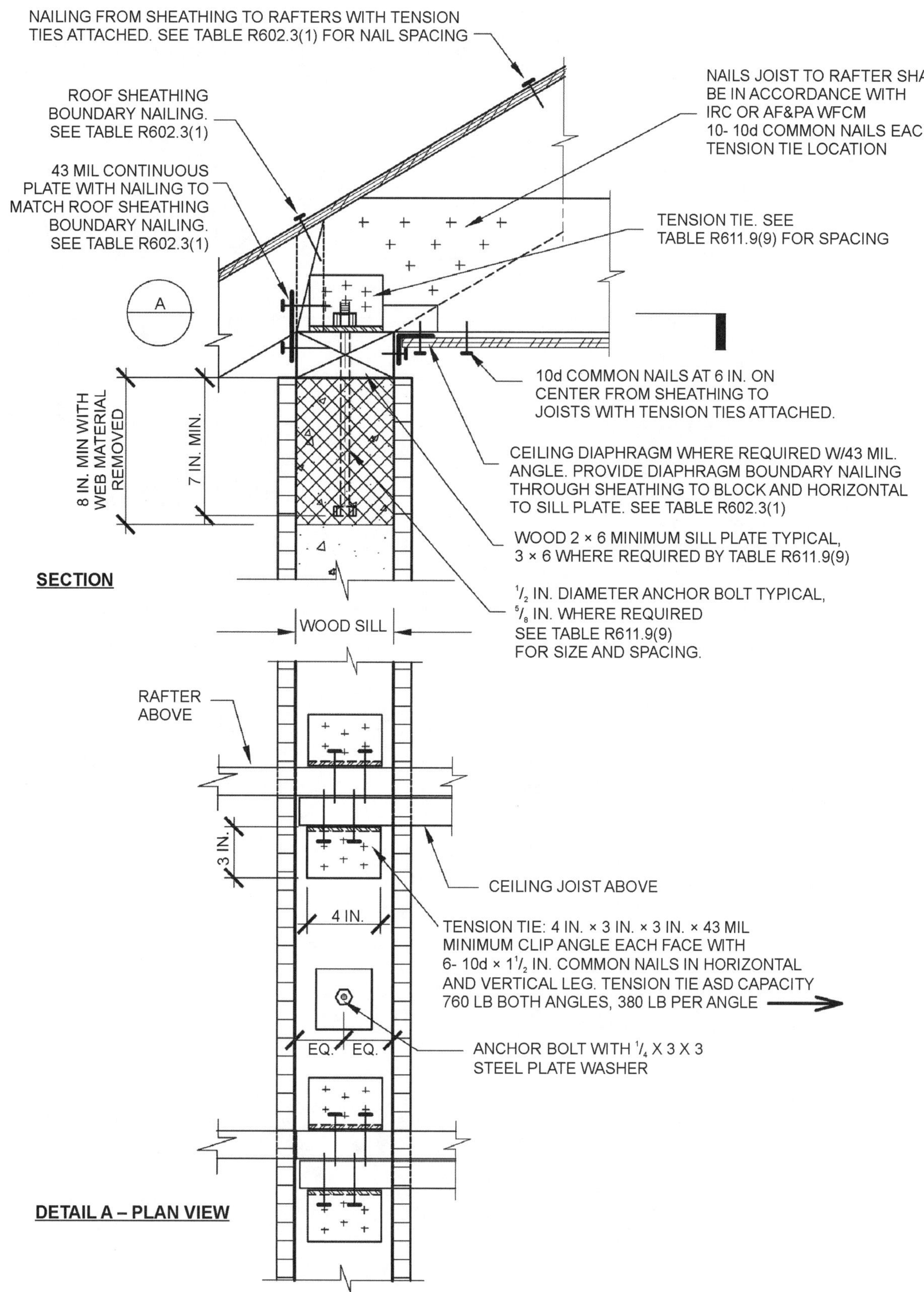

For SI: 1 mil = 0.0254 mm, 1 inch = 25.4 mm, 1 pound-force = 4.448 N.

FIGURE R611.9(9)
WOOD-FRAMED ROOF TO TOP OF CONCRETE WALL, FRAMING PERPENDICULAR

TABLE R611.9(9)
WOOD-FRAMED ROOF TO TOP OF CONCRETE WALL, FRAMING PERPENDICULAR[a, b, c, d, e]

ANCHOR BOLT SPACING (inches)	TENSION TIE SPACING (inches)	BASIC WIND SPEED (mph) AND WIND EXPOSURE CATEGORY					
		85B	90B	100B	110B	120B	130B
				85C	90C	100C	110C
					85D	90D	100D
12	12						
12	24						
12	36						
12	48						
16	16						6
16	32						6
16	48						
19.2	19.2					6	6 A
19.2	38.4					6	
24	24				6 A	6 A	6 B
24	48						

For SI: 1 inch = 25.4 mm, 1 mile per hour = 0.447 m/s.

a. This table is for use with the detail in Figure R611.9(9). Use of this detail is permitted where cell a is not shaded, prohibited where shaded.

b. Wall design per other provisions of Section R611 is required.

c. For wind design, minimum 4-inch-nominal wall is permitted in unshaded cells with no number.

d. Number 6 indicates minimum permitted nominal wall thickness in inches necessary to develop required strength (capacity) of connection. As a minimum, this nominal thickness shall occur in the portion of the wall indicated by the cross-hatching in Figure R611.9(9). For the remainder of the wall, see Note b.

e. Letter "A" indicates that a minimum nominal 3 × 6 sill plate is required. Letter "B" indicates that a $^5/_8$-inch-diameter anchor bolt and a minimum nominal 3 × 6 sill plate are required.

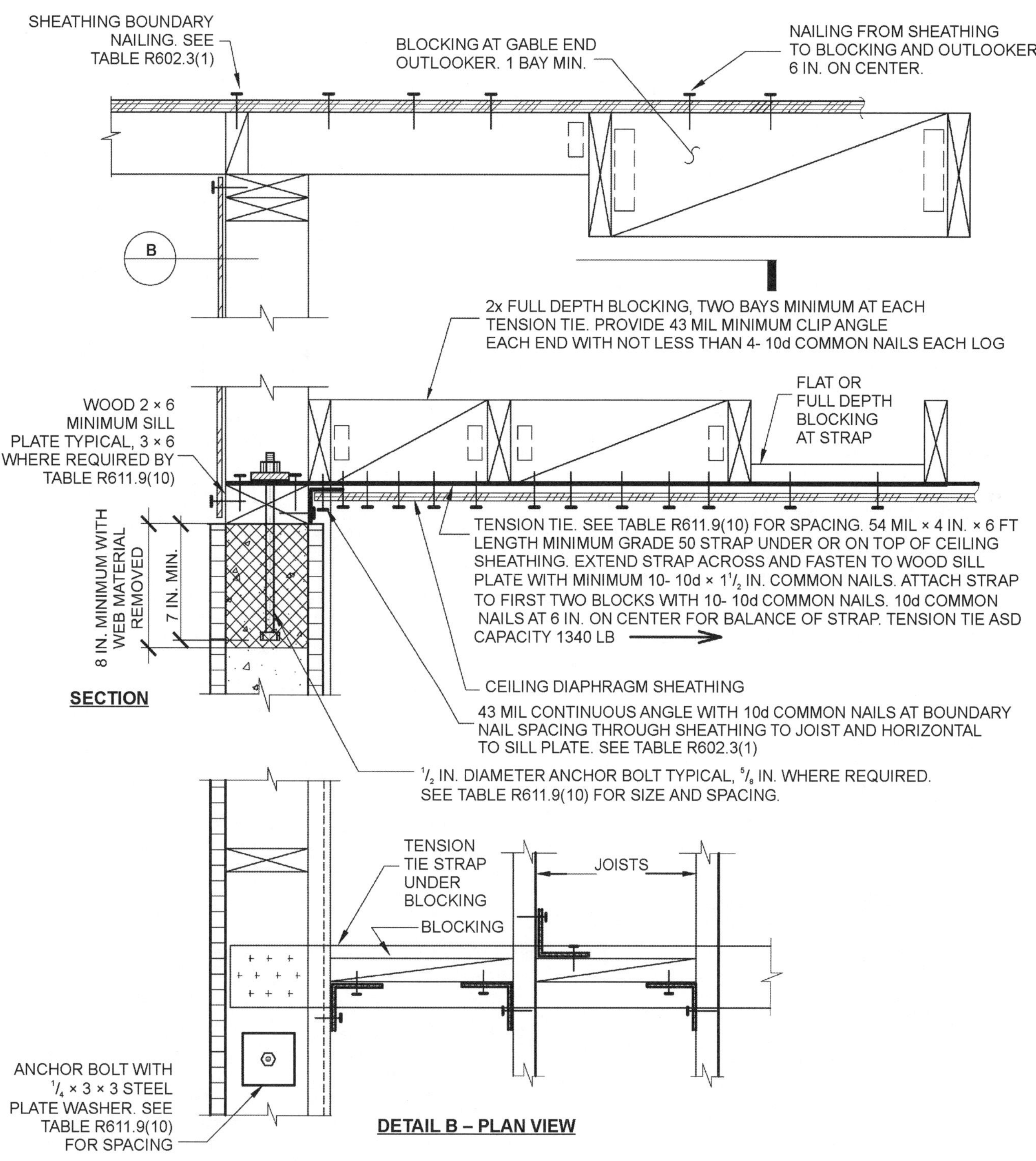

For SI: 1 mil = 0.0254 mm, 1 inch = 25.4 mm, 1 pound-force = 4.448 N.

FIGURE R611.9(10)
WOOD-FRAMED ROOF TO TOP OF CONCRETE WALL FRAMING PARALLEL

TABLE R611.9(10)
WOOD-FRAMED ROOF TO TOP OF CONCRETE WALL, FRAMING PARALLEL[a, b, c, d, e]

ANCHOR BOLT SPACING (inches)	TENSION TIE SPACING (inches)	BASIC WIND SPEED (mph) AND WIND EXPOSURE CATEGORY					
		85B	90B	100B	110B	120B	130B
				85C	90C	100C	110C
					85D	90D	100D
12	12						
12	24						
12	36						
12	48						
16	16					6	6
16	32					6	6
16	48					6	6
19.2	19.2				6	6	6 A
19.2	38.4				6	6	6 A
24	24			6	6 A	6 A	6 B
24	48			6	6 A	6 B	6 B

For SI: 1 inch = 25.4 mm, 1 mile per hour = 0.447 m/s.

a. This table is for use with the detail in Figure R611.9(10). Use of this detail is permitted where a cell is not shaded.

b. Wall design per other provisions of Section R611 is required.

c. For wind design, minimum 4-inch-nominal wall is permitted in cells with no number.

d. Number 6 indicates minimum permitted nominal wall thickness in inches necessary to develop required strength (capacity) of connection. As a minimum, this nominal thickness shall occur in the portion of the wall indicated by the cross-hatching in Figure R611.9(10). For the remainder of the wall, see Note b.

e. Letter "A" indicates that a minimum nominal 3 × 6 sill plate is required. Letter "B" indicates that a $^{5}/_{8}$-inch-diameter anchor bolt and a minimum nominal 3 × 6 sill plate are required.

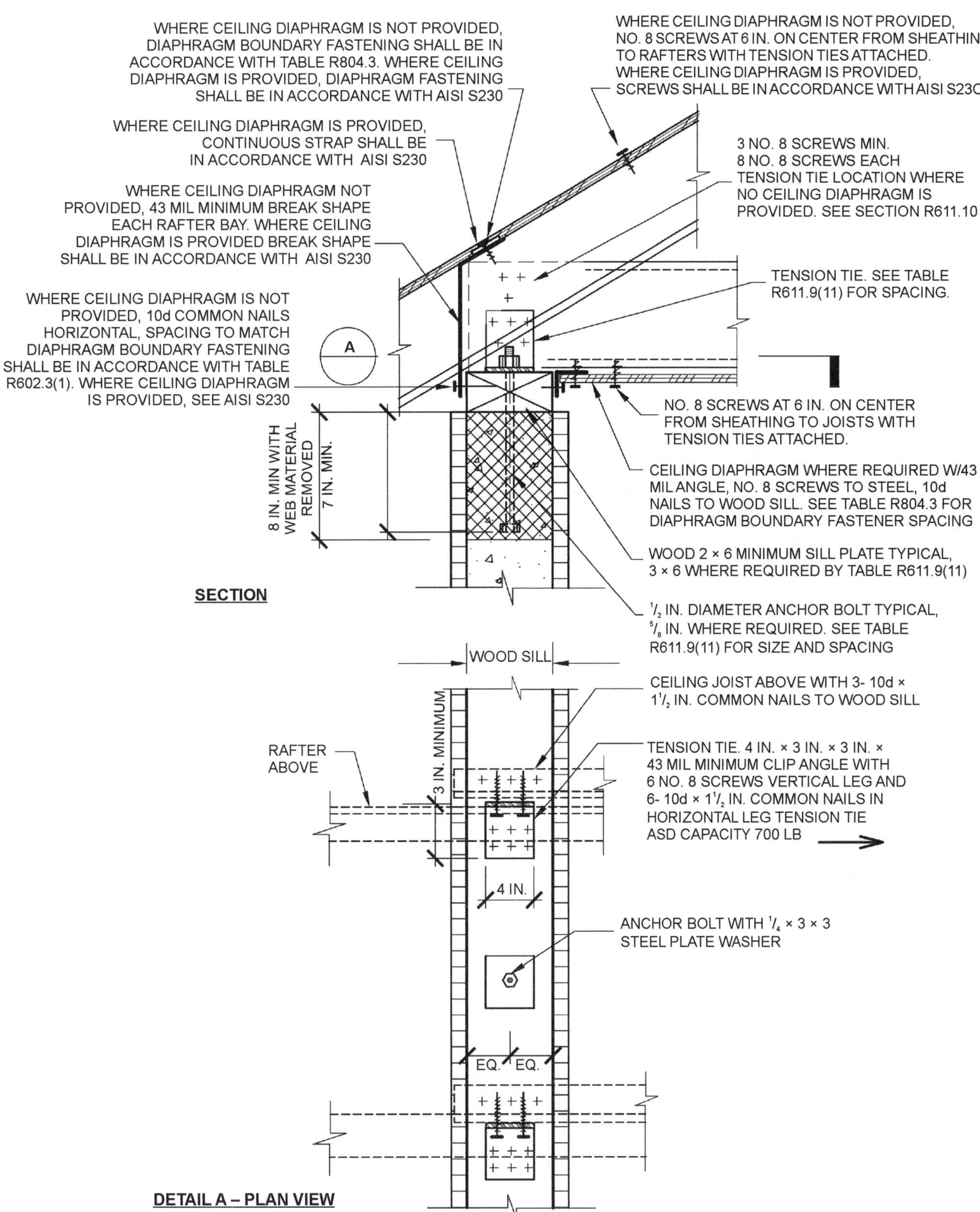

For SI: 1 mil = 0.0254 mm, 1 inch = 25.4 mm, 1 foot = 304.8 mm, 1 pound-force = 4.448 N.

FIGURE R611.9(11)
COLD-FORMED STEEL ROOF TO TOP OF CONCRETE WALL, FRAMING PERPENDICULAR

TABLE R611.9(11)
WOOD-FRAMED ROOF TO TOP OF CONCRETE WALL, FRAMING PERPENDICULAR[a, b, c, d, e]

ANCHOR BOLT SPACING (inches)	TENSION TIE SPACING (inches)	BASIC WIND SPEED (mph) AND WIND EXPOSURE CATEGORY					
		85B	90B	100B	110B	120B	130B
				85C	90C	100C	110C
					85D	90D	100D
12	12						
12	24						
16	16					6	6
16	32					6	6
19.2	19.2				6	6	8 B
19.2	38.4				6	6	8 B
24	24			6	6	8 B	

For SI: 1 inch = 25.4 mm, 1 mile per hour = 0.447 m/s.

a. This table is for use with the detail in Figure R611.9(11). Use of this detail is permitted where a cell is not shaded, prohibited where shaded.

b. Wall design per other provisions of Section R611 is required.

c. For wind design, minimum 4-inch-nominal wall is permitted in unshaded cells with no number.

d. Numbers 6 and 8 indicate minimum permitted nominal wall thickness in inches necessary to develop required strength (capacity) of connection. As a minimum, this nominal thick ness shall occur in the portion of the wall indicated by the cross-hatching in Figure R611.9(11). For the remainder of the wall, see Note b.

e. Letter "B" indicates that a $^5/_8$-inch-diameter anchor bolt and a minimum nominal 3 × 6 sill plate are required.

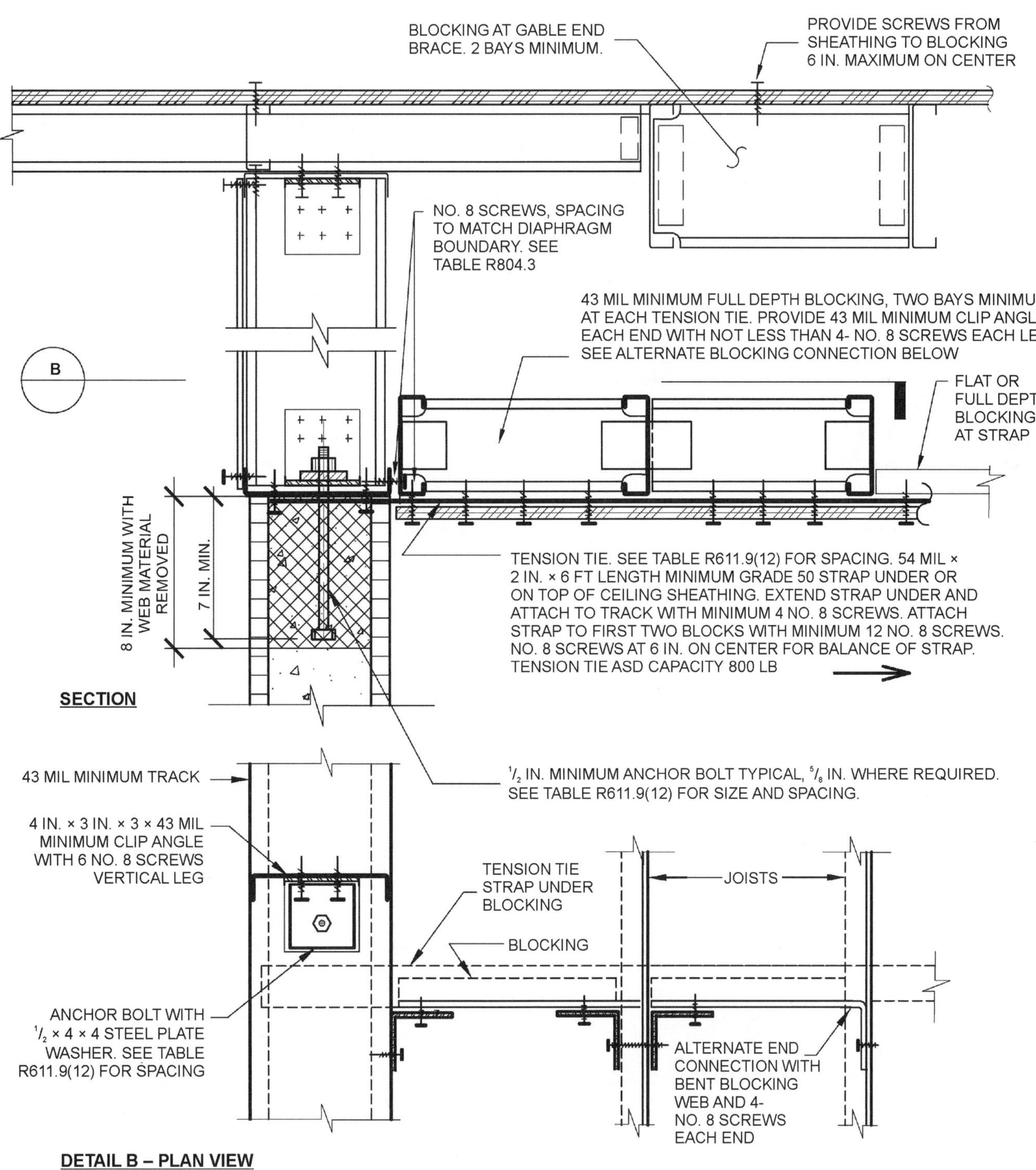

For SI: 1 mil = 0.0254 mm, 1 inch = 25.4 mm, 1 pound-force = 4.448 N.

FIGURE R611.9(12)
COLD-FORMED STEEL ROOF TO TOP OF CONCRETE WALL, FRAMING PARALLEL

TABLE R611.9(12)
COLD-FORMED STEEL ROOF TO TOP OF CONCRETE WALL, FRAMING PARALLEL[a, b, c, d, e]

ANCHOR BOLT SPACING (inches)	TENSION TIE SPACING (inches)	BASIC WIND SPEED (mph) AND WIND EXPOSURE CATEGORY					
		85B	90B	100B	110B	120B	130B
				85C	90C	100C	110C
					85D	90D	100D
12	12						
12	24						
16	16						
16	32						
19.2	19.2					6	6
19.2	38.4					6	6
24	24			6	6	8 B	8 B

For SI: 1 inch = 25.4 mm, 1 mile per hour = 0.447 m/s.

a. This table is for use with the detail in Figure R611.9(12). Use of this detail is permitted where a cell is not shaded.

b. Wall design per other provisions of Section R611 is required.

c. For wind design, minimum 4-inch-nominal wall is permitted in cells with no number.

d. Numbers 6 and 8 indicate minimum permitted nominal wall thickness in inches necessary to develop required strength (capacity) of connection. As a minimum, this nominal thickness shall occur in the portion of the wall indicated by the cross-hatching in Figure R611.9(12). For the remainder of the wall, see Note b.

e. Letter "B" indicates that a $^5/_8$-inch-diameter anchor bolt is required.

SECTION R612
EXTERIOR WINDOWS AND DOORS

R612.1 General. This section prescribes performance and construction requirements for exterior window and door installed in wall. Windows and doors shall be installed and flashed in accordance with the fenestration manufacturer's written installation instructions. Window and door openings shall be flashed in accordance with Section R703.8. Written installation instructions shall be provided by the fenestration manufacturer for each window or door.

R612.2 Performance. Exterior windows and doors shall be designed to resist the design wind loads specified in Table R301.2(2) adjusted for height and exposure in accordance with Table R301.2(3).

R612.3 Testing and labeling. Exterior windows and sliding doors shall be tested by an *approved* independent laboratory, and bear a *label* identifying manufacturer, performance characteristics and *approved* inspection agency to indicate compliance with AAMA/WDMA/CSA 101/I.S.2/A440. Exterior side-hinged doors shall be tested and *labeled* as conforming to AAMA/WDMA/CSA 101/I.S.2/A440 or comply with Section R612.5.

Exception: Decorative glazed openings.

R612.3.1 Comparative analysis. Structural wind load design pressures for window and door units smaller than the size tested in accordance with Section R612.3 shall be permitted to be higher than the design value of the tested unit provided such higher pressures are determined by accepted engineering analysis. All components of the small unit shall be the same as those of the tested unit. Where such calculated design pressures are used, they shall be validated by an additional test of the window or door unit having the highest allowable design pressure.

R612.4 Garage doors. Garage doors shall be tested in accordance with either ASTM E 330 or ANSI/ DASMA 108, and shall meet the acceptance criteria of ANSI/DASMA 108.

R612.5 Other exterior window and door assemblies. Exterior windows and door assemblies not included within the scope of Section R612.3 or Section R612.4 shall be tested in accordance with ASTM E 330. Glass in assemblies covered by this exception shall comply with Section R308.5

R612.6 Wind-borne debris protection. Protection of exterior windows and glass doors in buildings located in wind-borne debris regions shall be in accordance with Section R301.2.1.2.

R612.6.1 Fenestration testing and labeling. Fenestration shall be tested by an *approved* independent laboratory, listed by an *approved* entity, and bear a *label* identifying manufacturer, performance characteristics, and *approved* inspection agency to indicate compliance with the requirements of the following specification:

1. ASTM E 1886 and ASTM E 1996; or
2. AAMA 506.

R612.7 Anchorage methods. The methods cited in this section apply only to anchorage of window and glass door assemblies to the main force-resisting system.

R612.7.1 Anchoring requirements. Window and glass door assemblies shall be anchored in accordance with the published manufacturer's recommendations to achieve the design pressure specified. Substitute anchoring systems used for substrates not specified by the fenestration manufacturer shall provide equal or greater anchoring performance as demonstrated by accepted engineering practice.

R612.7.2 Anchorage details. Products shall be anchored in accordance with the minimum requirements illustrated in Figures R612.7.2(1), R612.7.2(2), R612.7.2(3), R612.7.2(4), R612.7.2(5), R612.7.2(6), R612.7.2(7) and R612.7.2(8).

R612.7.2.1 Masonry, concrete or other structural substrate. Where the wood shim or buck thickness is less than $1^1/_2$ inches (38 mm), window and glass door assemblies shall be anchored through the jamb, or by jamb clip and anchors shall be embedded directly into the masonry, concrete or other substantial substrate material. Anchors shall adequately transfer load from the window or door frame into the rough opening substrate [see Figures R612.7.2(1) and R612.7.2(2)].

Where the wood shim or buck thickness is $1^1/_2$ inches (38 mm) or more, the buck is securely fastened to the masonry, concrete or other substantial substrate, and the buck extends beyond the interior face of the window or door frame, window and glass door assemblies shall be anchored through the jamb, or by jamb clip, or through the flange to the secured wood buck. Anchors shall be embedded into the secured wood buck to adequately transfer load from the window or door frame assembly [Figures R612.7.2(3), R612.7.2(4) and R612.7.2(5)].

R612.7.2.2 Wood or other approved framing material. Where the framing material is wood or other *approved* framing material, window and glass door assemblies shall be anchored through the frame, or by frame clip, or through the flange. Anchors shall be embedded into the frame construction to adequately transfer load [Figures R612.7.2(6), R612.7.2(7) and R612.7.2(8)].

R612.8 Mullions. Mullions shall be tested by an *approved* testing laboratory in accordance with AAMA 450, or be engineered in accordance with accepted engineering practice. Mullions tested as stand-alone units or qualified by engineering shall use performance criteria cited in Sections R612.8.1, R612.8.2 and R612.8.3. Mullions qualified by an actual test of an entire assembly shall comply with Sections R612.8.1 and R612.8.3.

R612.8.1 Load transfer. Mullions shall be designed to transfer the design pressure loads applied by the window and door assemblies to the rough opening substrate.

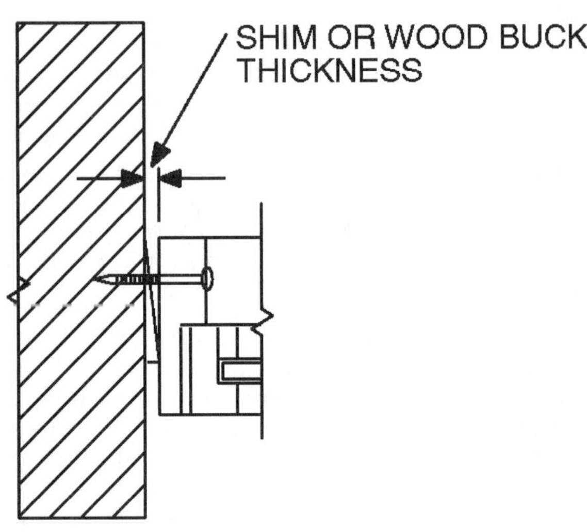

FIGURE R612.7.2(1)
THROUGH THE FRAME

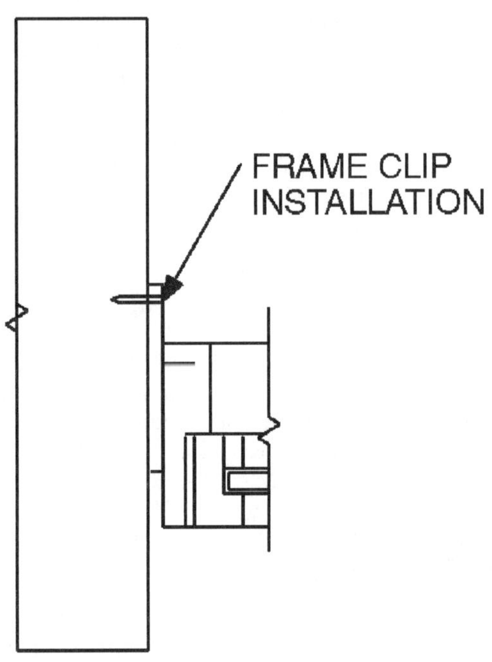

FIGURE R612.7.2(2)
FRAME CLIP

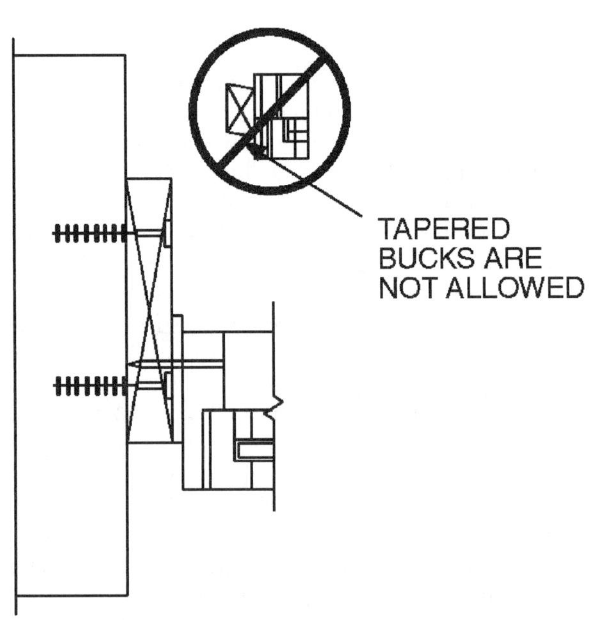

FIGURE R612.7.2(3)
THROUGH THE FRAME

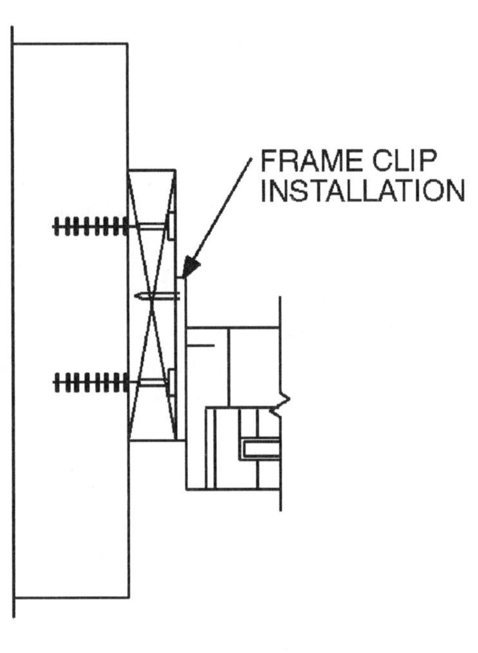

FIGURE R612.7.2(4)
FRAME CLIP

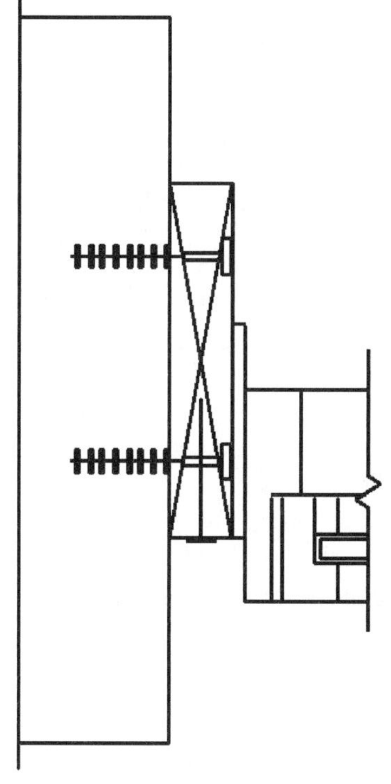

FIGURE R612.7.2(5)
THROUGH THE FLANGE

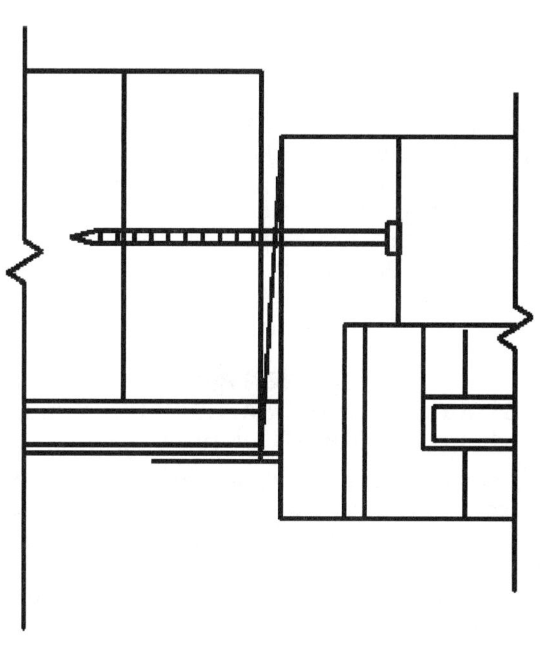

FIGURE R612.7.2(6)
THROUGH THE FLANGE

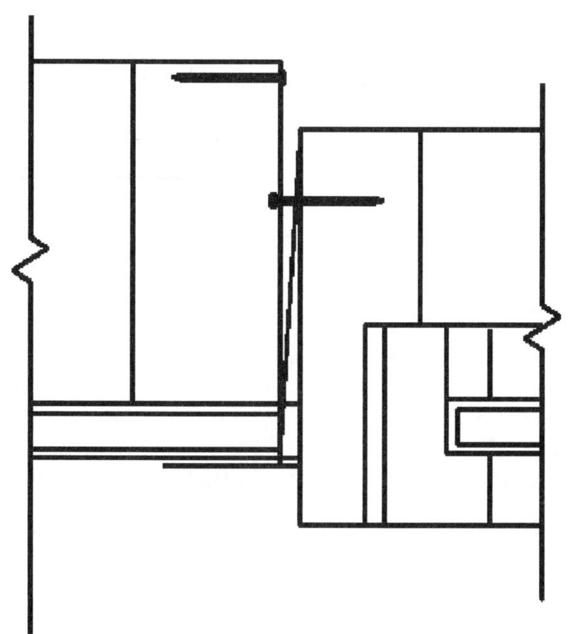

FIGURE R612.7.2(7)
FRAME CLIP

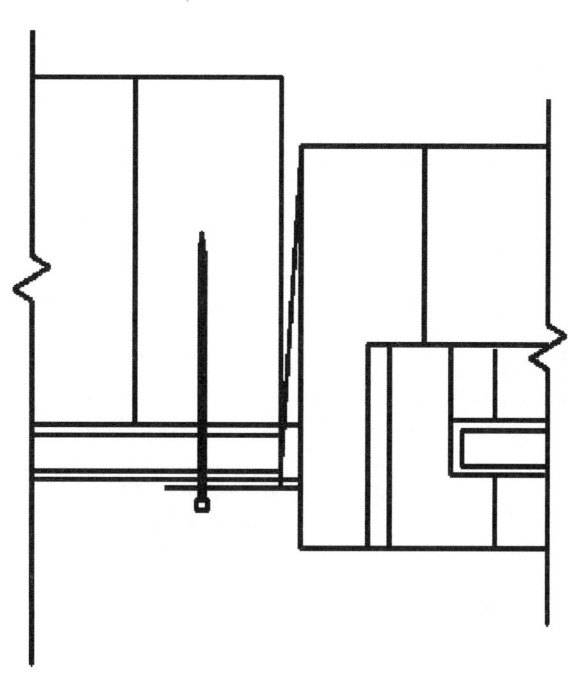

FIGURE R612.7.2(8)
THROUGH THE FLANGE

R612.8.2 Deflection. Mullions shall be capable of resisting the design pressure loads applied by the window and door assemblies to be supported without deflecting more than *L*/175, where *L* is the span of the mullion in inches.

R612.8.3 Structural safety factor. Mullions shall be capable of resisting a load of 1.5 times the design pressure loads applied by the window and door assemblies to be supported without exceeding the appropriate material stress levels. If tested by an *approved* laboratory, the 1.5 times the design pressure load shall be sustained for 10 seconds, and the permanent deformation shall not exceed 0.4 percent of the mullion span after the 1.5 times design pressure load is removed.

SECTION R613 STRUCTURAL INSULATED PANEL WALL CONSTRUCTION

R613.1 General. Structural insulated panel (SIP) walls shall be designed in accordance with the provisions of this section. When the provisions of this section are used to design structural insulated panel walls, project drawings, typical details and specifications are not required to bear the seal of the architect or engineer responsible for design, unless otherwise required by the state law of the *jurisdiction* having authority.

R613.2 Applicability limits. The provisions of this section shall control the construction of exterior structural insulated panel walls and interior load-bearing structural insulated panel walls for buildings not greater than 60 feet (18 288 mm) in length perpendicular to the joist or truss span, not greater than 40 feet (12 192 mm) in width parallel to the joist or truss span and not greater than two stories in height with each wall not greater than 10 feet (3048 mm) high. All exterior walls installed in accordance with the provisions of this section shall be considered as load-bearing walls. Structural insulated panel walls constructed in accordance with the provisions of this section shall be limited to sites subjected to a maximum design wind speed of 120 miles per hour (54 m/s), Exposure A or B or 110 miles per hour (49 m/s) Exposure C, and a maximum ground snow load of 70 pounds per foot (3.35 kPa), and Seismic Design Categories A, B and C.

R613.3 Materials. SIPs shall comply with the following criteria:

R613.3.1 Core. The core material shall be composed of foam plastic insulation meeting one of the following requirements:

1. ASTM C 578 and have a minimum density of 0.90 pounds per cubic feet (14.4 kg/m^3); or
2. Polyurethane meeting the physical properties shown in Table R613.3.1, or;
3. An *approved* alternative.

All cores shall meet the requirements of Section R316.

R613.3.2 Facing. Facing materials for SIPs shall be wood structural panels conforming to DOC PS 1 or DOC PS 2, each having a minimum nominal thickness of $^7/_{16}$ inch (11 mm) and shall meet the additional minimum properties specified in Table R613.3.2. Facing shall be identified by a grade mark or certificate of inspection issued by an *approved* agency.

TABLE R613.3.1
MINIMUM PROPERTIES FOR POLYURETHANE INSULATION USED AS SIPS CORE

PHYSICAL PROPERTY	POLYURETHANE
Density, core nominal (ASTM D 1622)	2.2 lb/ft^3
Compressive resistance at yield or 10% deformation, whichever occurs first (ASTM D 1621)	19 psi (perpendicular to rise)
Flexural strength, min. (ASTM C 203)	30 psi
Tensile strength, min. (ASTM D 1623)	35 psi
Shear strength, min. (ASTM C 273)	25 psi
Substrate adhesion, min. (ASTM D 1623)	22 psi
Water vapor permeance of 1.00-in. thickness, max. (ASTM E 96)	2.3 perm
Water absorption by total immersion, max. (ASTM C 272)	4.3% (volume)
Dimensional stability (change in dimensions), max. [ASTM D 2126 (7 days at 158°F/100% humidity and 7 days at -20°F)]	2%

For SI: 1 pound per cubic foot = 16.02 kg/m^3, 1 pound per square inch = 6.895 kPa, °C = [(°F) - 32]1.8.

TABLE R613.3.2
MINIMUM PROPERTIES[a] FOR ORIENTED STRAND BOARD FACER MATERIAL IN SIP WALLS

Thickness (in.)	Product	Flatwise Stiffness[b] (lbf-in^2/ft)		Flatwise Strength [c] (lbf-in/ft)		Tension [c] (lbf/ft)		Density[d] (pcf)
		Along	Across	Along	Across	Along	Across	
7/16	Sheathing	55,600	16,500	1,040	460	7,450	5,800	34

For SI: 1 inch = 25.4 mm, 1 lbf-in^2/ft = 9.415 × 10^{-6} kPa/m, 1 lbf-in/ft = 3.707 × 10^{-4} kN/m, 1 lbf/ft = 0.0146 N/mm, 1 pound per cubic foot = 16.018 kg/m^3.

a. Values listed in Table R613.3.2 are qualification test values and are not to be used for design purposes.

b. Mean test value shall be in accordance with Section 7.6 of DOC PS 2.

c. Characteristic test value (5th percent with 75% confidence).

d. Density shall be based on oven-dry weight and oven-dry volume.

R613.3.3 Adhesive. Adhesives used to structurally laminate the foam plastic insulation core material to the structural wood facers shall conform to ASTM D 2559 or *approved* alternative specifically intended for use as an adhesive used in the lamination of structural insulated panels. Each container of adhesive shall bear a *label* with the adhesive manufacturer's name, adhesive name and type and the name of the quality assurance agency.

R613.3.4 Lumber. The minimum lumber framing material used for SIPs prescribed in this document is NLGA graded No. 2 Spruce-pine-fir. Substitution of other wood species/grades that meet or exceed the mechanical properties and specific gravity of No. 2 Spruce-pine-fir shall be permitted.

R613.3.5 SIP screws. Screws used for the erection of SIPs as specified in Section R613.5 shall be fabricated from steel, shall be provided by the SIPs manufacturer and shall be sized to penetrate the wood member to which the assembly is being attached by a minimum of 1 inch (25 mm). The screws shall be corrosion resistant and have a minimum shank diameter of 0.188 inch (4.7 mm) and a minimum head diameter of 0.620 inch (15.5 mm).

R613.3.6 Nails. Nails specified in Section R613 shall be common or galvanized box unless otherwise stated.

R613.4 SIP wall panels. SIPs shall comply with Figure R613.4 and shall have minimum panel thickness in accordance with Tables R613.5(1) and R613.5(2) for above-grade walls. All SIPs shall be identified by grade mark or certificate of inspection issued by an *approved* agency.

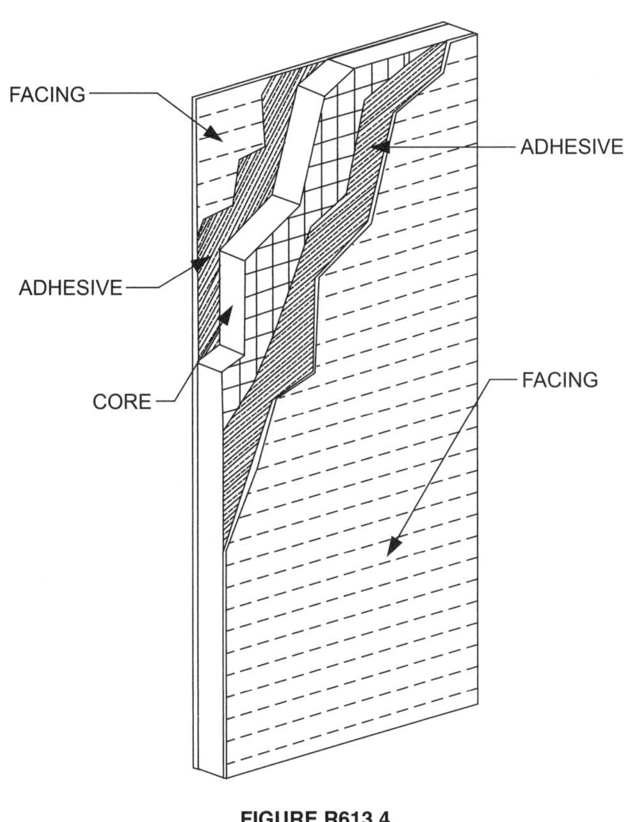

FIGURE R613.4
SIP WALL PANEL

R613.4.1 Labeling. All panels shall be identified by grade mark or certificate of inspection issued by an *approved* agency. Each (SIP) shall bear a stamp or *label* with the following minimum information:

1. Manufacturer name/logo.
2. Identification of the assembly.
3. Quality assurance agency.

R613.5 Wall construction. Exterior walls of SIP construction shall be designed and constructed in accordance with the provisions of this section and Tables R613.5(1) and R613.5(2) and Figures R613.5(1) through R613.5(5). SIP walls shall be fastened to other wood building components in accordance with Tables R602.3(1) through R602.3(4).

Framing shall be attached in accordance with Table R602.3(1) unless otherwise provided for in Section R613.

R613.5.1 Top plate connection. SIP walls shall be capped with a double top plate installed to provide overlapping at corner, intersections and splines in accordance with Figure R613.5.1. The double top plates shall be made up of a single 2 by top plate having a width equal to the width of the panel core, and shall be recessed into the SIP below. Over this top plate a cap plate shall be placed. The cap plate width shall match the SIP thickness and overlap the facers on both sides of the panel. End joints in top plates shall be offset at least 24 inches (610 mm).

R613.5.2 Bottom (sole) plate connection. SIP walls shall have full bearing on a sole plate having a width equal to the nominal width of the foam core. When SIP walls are supported directly on continuous foundations, the wall wood sill plate shall be anchored to the foundation in accordance with Figure R613.5.2 and Section R403.1.

R613.5.3 Wall bracing. SIP walls shall be braced in accordance with Section R602.10. SIP walls shall be considered continuous wood structural panel sheathing for purposes of computing required bracing. SIP walls shall meet the requirements of Section R602.10.4.2 except that SIPs corners shall be fabricated as shown in Figure R613.9. When SIP walls are used for wall bracing, the SIP bottom plate shall be attached to wood framing below in accordance with Table R602.3(1).

R613.6 Interior load-bearing walls. Interior load-bearing walls shall be constructed as specified for exterior walls.

R613.7 Drilling and notching. The maximum vertical chase penetration in SIPs shall have a maximum side dimension of 2 inches (51 mm) centered in the panel core. Vertical chases shall have a minimum spacing of 24-inches (610 mm) on center. Maximum of two horizontal chases shall be permitted in each wall panel, one at 14 inches (360 mm) from the bottom of the panel and one at mid-height of the wall panel. The maximum allowable penetration size in a wall panel shall be circular or rectangular with a maximum dimension of 12 inches (305 mm). Overcutting of holes in facing panels shall not be permitted.

R613.8 Connection. SIPs shall be connected at vertical in-plane joints in accordance with Figure R613.8 or by other *approved* methods.

R613.9 Corner framing. Corner framing of SIP walls shall be constructed in accordance with Figure R613.9.

R613.10 Headers. SIP headers shall be designed and constructed in accordance with Table R613.10 and Figure R613.5.1. SIPs headers shall be continuous sections without splines. Headers shall be at least $11^7/_8$ inches (302 mm) deep. Headers longer than 4 feet (1219 mm) shall be constructed in accordance with Section R602.7.

R613.10.1 Wood structural panel box headers. Wood structural panel box headers shall be allowed where SIP headers are not applicable. Wood structural panel box headers shall be constructed in accordance with Figure R602.7.2 and Table R602.7.2.

TABLE R613.5(1)
MINIMUM THICKNESS FOR SIP WALL SUPPORTING SIP OR LIGHT-FRAME ROOF ONLY (inches)[a]

Wind Speed (3-second gust)		Snow Load (psf)	Building Width (ft)														
			24			28			32			36			40		
			Wall Height (feet)			Wall Height (feet)			Wall Height (feet)			Wall Height (feet)			Wall Height (feet)		
Exp. A/B	Exp. C		8	9	10	8	9	10	8	9	10	8	9	10	8	9	10
85	—	20	4.5	4.5	4.5	4.5	4.5	4.5	4.5	4.5	4.5	4.5	4.5	4.5	4.5	4.5	4.5
		30	4.5	4.5	4.5	4.5	4.5	4.5	4.5	4.5	4.5	4.5	4.5	4.5	4.5	4.5	4.5
		50	4.5	4.5	4.5	4.5	4.5	4.5	4.5	4.5	4.5	4.5	4.5	4.5	4.5	4.5	4.5
		70	4.5	4.5	4.5	4.5	4.5	4.5	4.5	4.5	4.5	4.5	4.5	4.5	4.5	4.5	4.5
100	85	20	4.5	4.5	4.5	4.5	4.5	4.5	4.5	4.5	4.5	4.5	4.5	4.5	4.5	4.5	4.5
		30	4.5	4.5	4.5	4.5	4.5	4.5	4.5	4.5	4.5	4.5	4.5	4.5	4.5	4.5	4.5
		50	4.5	4.5	4.5	4.5	4.5	4.5	4.5	4.5	4.5	4.5	4.5	4.5	4.5	4.5	4.5
		70	4.5	4.5	4.5	4.5	4.5	4.5	4.5	4.5	4.5	4.5	4.5	6.5	4.5	4.5	N/A
110	100	20	4.5	4.5	4.5	4.5	4.5	4.5	4.5	4.5	4.5	4.5	4.5	4.5	4.5	4.5	4.5
		30	4.5	4.5	4.5	4.5	4.5	4.5	4.5	4.5	4.5	4.5	4.5	4.5	4.5	4.5	6.5
		50	4.5	4.5	4.5	4.5	4.5	6.5	4.5	4.5	6.5	4.5	4.5	N/A	4.5	4.5	N/A
		70	4.5	4.5	6.5	4.5	4.5	N/A	4.5	4.5	N/A	4.5	6.5	N/A	4.5	N/A	N/A
120	110	20	4.5	4.5	N/A	4.5	4.5	N/A	4.5	4.5	N/A	4.5	4.5	N/A	4.5	4.5	N/A
		30	4.5	4.5	N/A	4.5	4.5	N/A	4.5	4.5	N/A	4.5	4.5	N/A	4.5	6.5	N/A
		50	4.5	4.5	N/A	4.5	6.5	N/A	4.5	N/A	N/A	4.5	N/A	N/A	4.5	N/A	N/A
		70	4.5	N/A	N/A	4.5	N/A	N/A	4.5	N/A	N/A	N/A	N/A	N/A	N/A	N/A	N/A

For SI: 1 inch = 25.4 mm, 1 foot = 304.8 mm, 1 pound per square foot = 0.0479 kPa.

N/A = Not Applicable.

a. Design assumptions:

Deflection criteria: *L*/240.

Roof load: 7 psf.

Ceiling load: 5 psf.

Wind loads based on Table R301.2 (2).

Strength axis of facing materials applied vertically.

TABLE R613.5(2)
MINIMUM THICKNESS FOR SIP WALLS SUPPORTING SIP OR LIGHT-FRAME ONE STORY AND ROOF (inches)[a]

Wind Speed (3-second gust)		Snow Load (psf)	Building Width (ft)														
			24			28			32			36			40		
Exp. A/B	Exp. C		Wall Height (feet)			Wall Height (feet)			Wall Height (feet)			Wall Height (feet)			Wall Height (feet)		
			8	9	10	8	9	10	8	9	10	8	9	10	8	9	10
85	—	20	4.5	4.5	4.5	4.5	4.5	4.5	4.5	4.5	4.5	4.5	4.5	4.5	4.5	4.5	4.5
		30	4.5	4.5	4.5	4.5	4.5	4.5	4.5	4.5	4.5	4.5	4.5	4.5	4.5	4.5	4.5
		50	4.5	4.5	4.5	4.5	4.5	4.5	4.5	4.5	4.5	4.5	4.5	4.5	4.5	4.5	N/A
		70	4.5	4.5	4.5	4.5	4.5	4.5	4.5	4.5	6.5	4.5	4.5	N/A	4.5	N/A	N/A
100	85	20	4.5	4.5	4.5	4.5	4.5	4.5	4.5	4.5	6.5	4.5	4.5	N/A	4.5	4.5	N/A
		30	4.5	4.5	4.5	4.5	4.5	4.5	4.5	4.5	N/A	4.5	4.5	N/A	4.5	N/A	N/A
		50	4.5	4.5	6.5	4.5	4.5	N/A	4.5	4.5	N/A	4.5	N/A	N/A	N/A	N/A	N/A
		70	4.5	4.5	N/A	4.5	6.5	N/A	4.5	N/A	N/A	N/A	N/A	N/A	N/A	N/A	N/A
110	100	20	4.5	4.5	N/A	4.5	4.5	N/A	4.5	6.5	N/A	4.5	N/A	N/A	N/A	N/A	N/A
		30	4.5	4.5	N/A	4.5	4.5	N/A	4.5	N/A	N/A	4.5	N/A	N/A	N/A	N/A	N/A
		50	4.5	6.5	N/A	4.5	N/A	N/A	N/A	N/A	N/A	N/A	N/A	N/A	N/A	N/A	N/A
		70	4.5	N/A	N/A	N/A	N/A	N/A	N/A	N/A	N/A	N/A	N/A	N/A	N/A	N/A	N/A
120	110	20	4.5	N/A	N/A	4.5	N/A	N/A	N/A	N/A	N/A	N/A	N/A	N/A	N/A	N/A	N/A
		30	4.5	N/A	N/A	N/A	N/A	N/A	N/A	N/A	N/A	N/A	N/A	N/A	N/A	N/A	N/A
		50	N/A	N/A	N/A	N/A	N/A	N/A	N/A	N/A	N/A	N/A	N/A	N/A	N/A	N/A	N/A
		70	N/A	N/A	N/A	N/A	N/A	N/A	N/A	N/A	N/A	N/A	N/A	N/A	N/A	N/A	N/A

For SI: 1 inch = 25.4 mm, 1 foot = 304.8 mm, 1 pound per square foot = 0.0479 kPa.

N/A = Not Applicable.

a. Design assumptions:

Deflection criteria: *L*/240.

Roof load: 7 psf.

Ceiling load: 5 psf.

Second floor live load: 30 psf.

Second floor dead load: 10 psf.

Second floor dead load from walls: 10 psf.

Wind loads based on Table R301.2(2).

Strength axis of facing materials applied vertically.

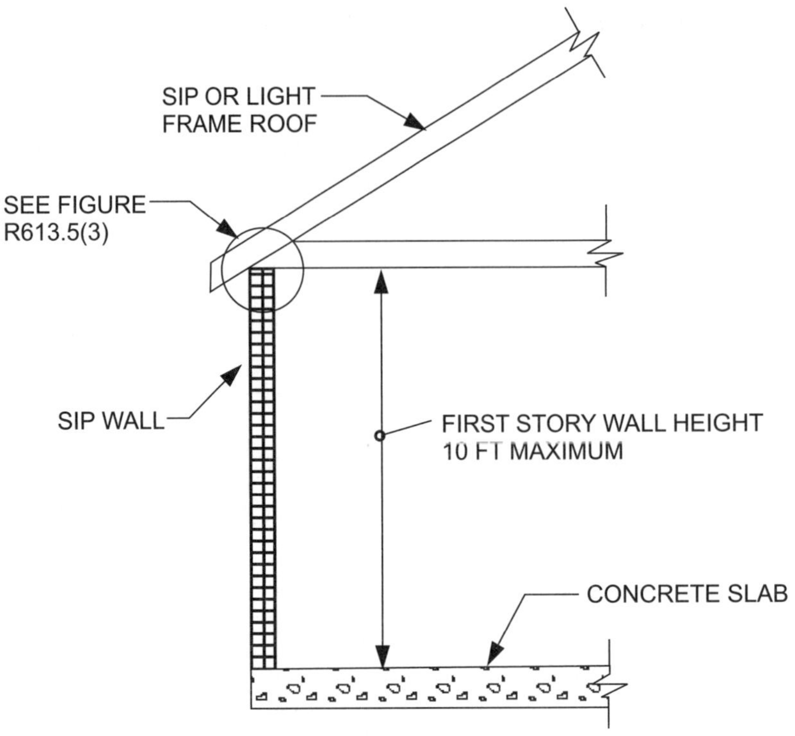

For SI: 1 foot = 304.8 mm.

FIGURE R613.5(1)
MAXIMUM ALLOWABLE HEIGHT OF SIP WALLS

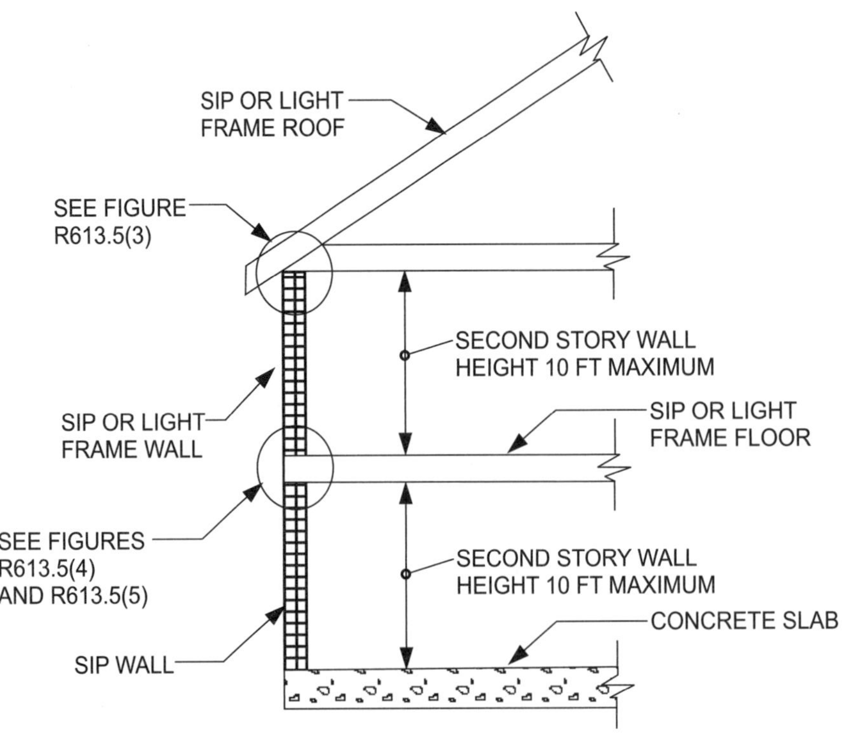

For SI: 1 foot = 304.8 mm.

FIGURE R613.5(2)
MAXIMUM ALLOWABLE HEIGHT OF SIP WALLS

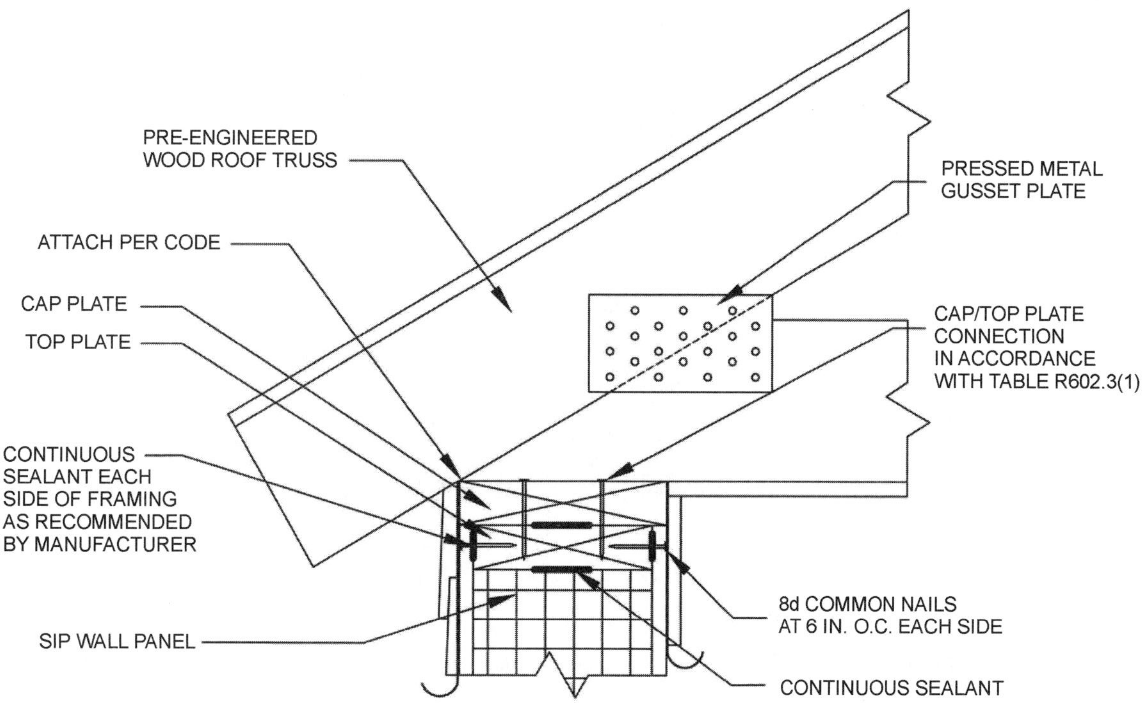

For SI: 1 inch = 25.4 mm.

FIGURE R613.5(3)
TRUSSED ROOF TO TOP PLATE CONNECTION

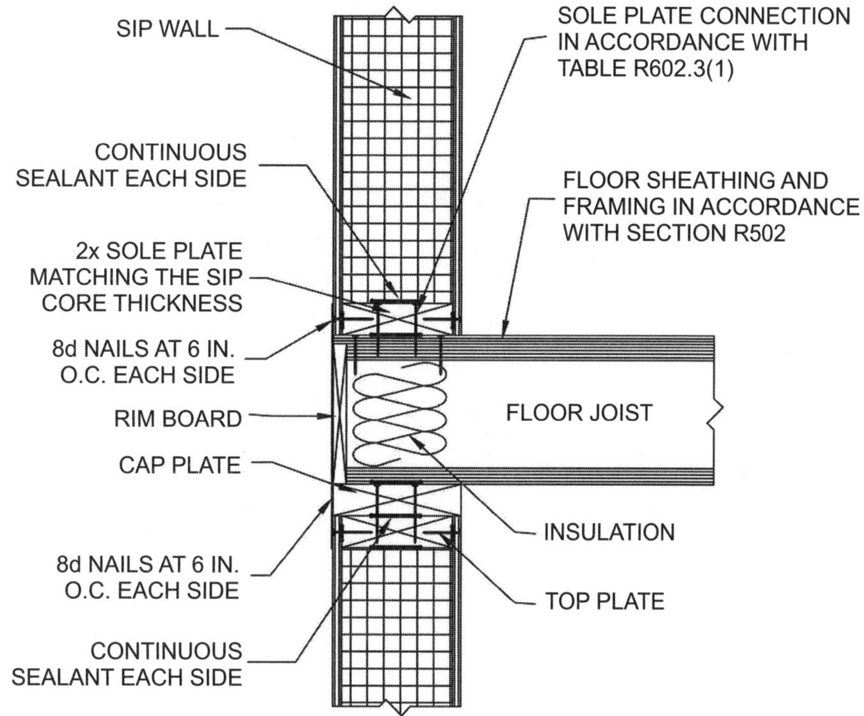

For SI: 1 inch = 25.4 mm.

Note: Figures illustrate SIP-specific attachment requirements. Other connections shall be made in accordance with Tables R602.3(1) and (2) as appropriate.

FIGURE R613.5(4)
SIP WALL-TO-WALL PLATFORM FRAME CONNECTION

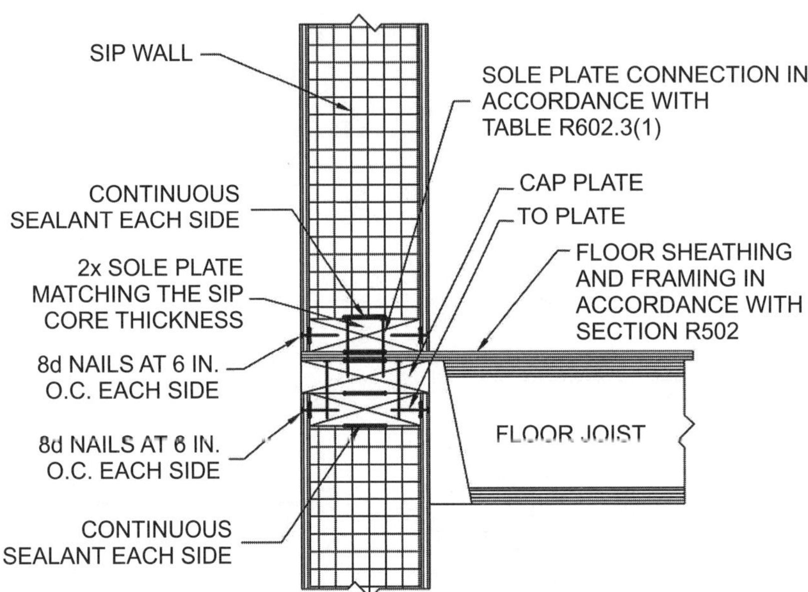

For SI: 1 inch = 25.4 mm.

Note: Figures illustrate SIP-specific attachment requirements. Other connections shall be made in accordance with Tables R602.3(1) and (2), as appropriate.

FIGURE R613.5(5)
SIP WALL-TO-WALL BALLOON FRAME CONNECTION (I-Joist floor shown for Illustration only)

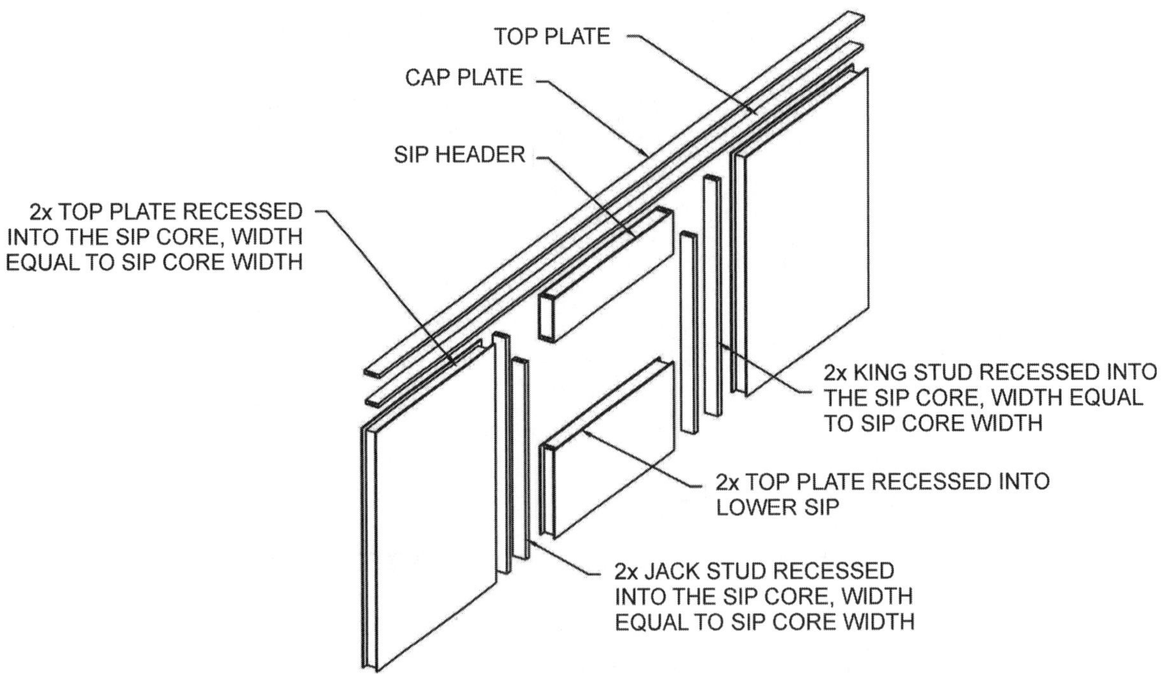

For SI: 1 inch = 25.4 mm.

Notes:

1. Top plates shall be continuous over header.
2. Lower 2x top plate shall have a width equal to the SIP core width and shall be recessed into the top edge of the panel. Cap plate shall be placed over the recessed top plate and shall have a width equal to the SIPs width.
3. SIP facing surfaces shall be nailed to framing and cripples with 8d common or galvanized box nails spaced 6 inches on center.
4. Galvanized nails shall be hot-dipped or tumbled. Framing shall be attached in accordance to Section R602.3(1) unless otherwise provide for in Section R613.

FIGURE R613.5.1
SIP WALL FRAMING CONFIGURATION

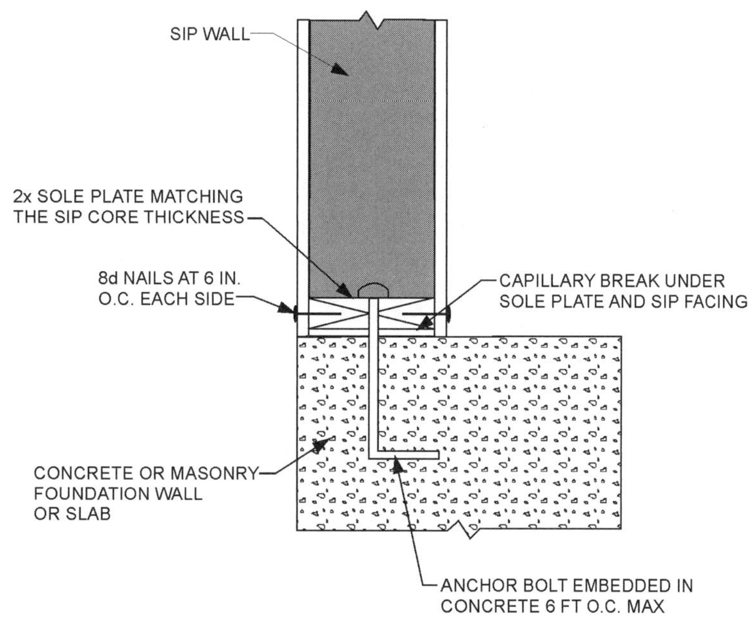

For SI: 1 inch = 25.4 mm, 1 foot = 304.8 mm.

FIGURE R613.5.2
SIP WALL TO CONCRETE SLAB FOR FOUNDATION WALL ATTACHMENT

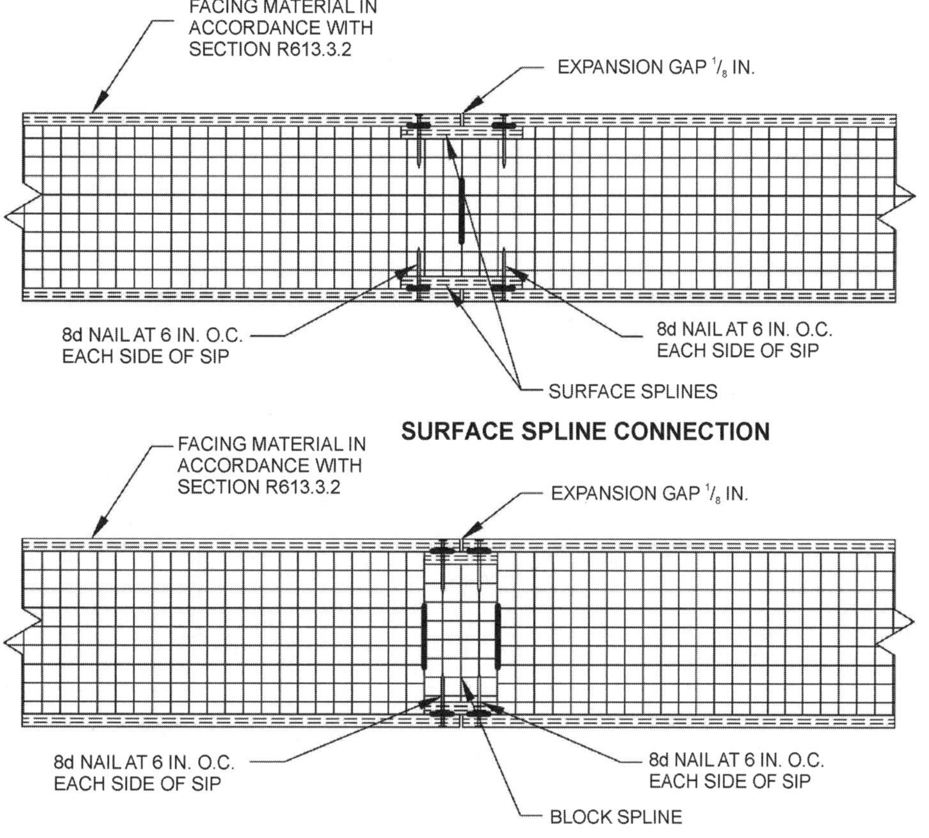

For SI: 1 inch = 25.4 mm.

FIGURE R613.8
TYPICAL SIP CONNECTION DETAILS FOR VERTICAL IN-PLANE JOINTS

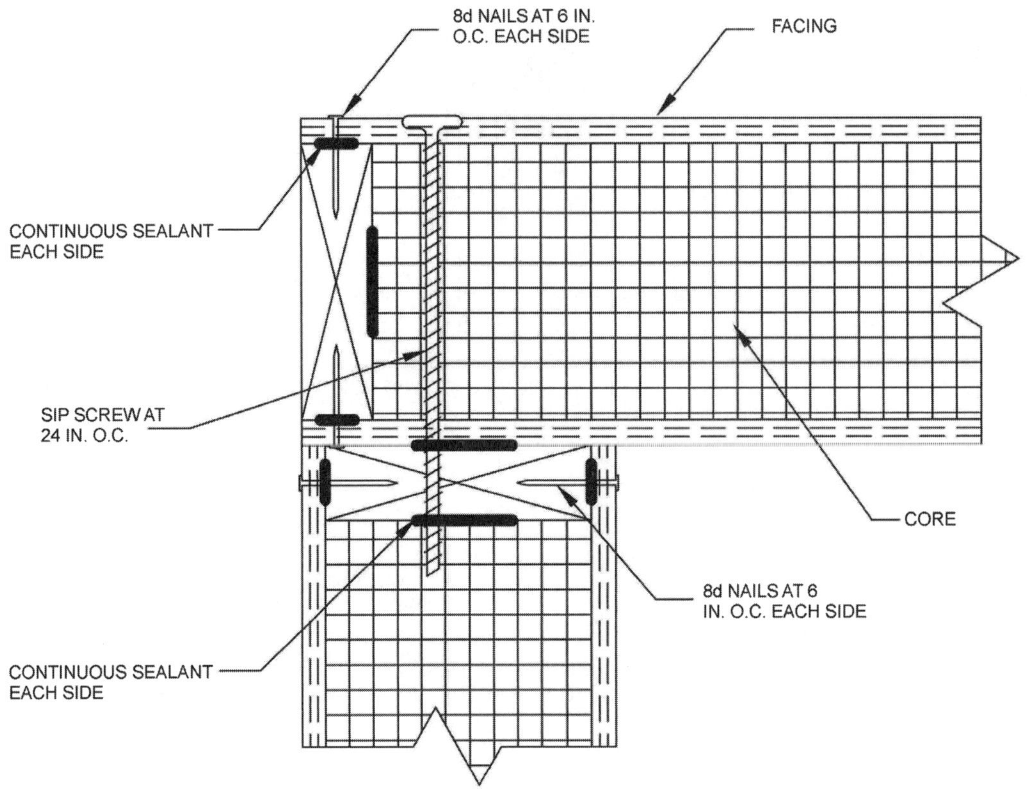

For SI: 1 inch = 25.4 mm.

FIGURE R613.9
SIP CORNER FRAMING DETAIL

TABLE R613.10
MAXIMUM SPANS FOR 11$^{7}/_{8}$-INCH-DEEP SIP HEADERS (feet)[a]

LOAD CONDITION	SNOW LOAD (psf)	BUILDING width (feet)				
		24	28	32	36	40
Supporting roof only	20	4	4	4	4	2
	30	4	4	4	2	2
	50	2	2	2	2	2
	70	2	2	2	N/A	N/A
Supporting roof and one-story	20	2	2	N/A	N/A	N/A
	30	2	2	N/A	N/A	N/A
	50	2	N/A	N/A	N/A	N/A
	70	N/A	N/A	N/A	N/A	N/A

For SI: 1 inch = 25.4 mm, 1 foot = 304.8 mm, 1 pound per square foot = 0.0479 kPa.

N/A = Not Applicable.

a. Design assumptions:

Maximum deflection criterion: *L*/360.

Maximum roof dead load: 10 psf.

Maximum ceiling load: 5 psf.

Maximum second floor live load: 30 psf.

Maximum second floor dead load: 10 psf.

Maximum second floor dead load from walls: 10 psf.

CHAPTER 7

WALL COVERING

SECTION R701 GENERAL

R701.1 Application. The provisions of this chapter shall control the design and construction of the interior and exterior wall covering for all buildings.

R701.2 Installation. Products sensitive to adverse weather shall not be installed until adequate weather protection for the installation is provided. Exterior sheathing shall be dry before applying exterior cover.

SECTION R702 INTERIOR COVERING

R702.1 General. Interior coverings or wall finishes shall be installed in accordance with this chapter and Table R702.1(1), Table R702.1(2), Table R702.1(3) and Table R702.3.5. Interior masonry veneer shall comply with the requirements of Section R703.7.1 for support and Section R703.7.4 for anchorage, except an air space is not required. Interior finishes and materials shall conform to the flame spread and smoke-development requirements of Section R302.9.

R702.2 Interior plaster.

R702.2.1 Gypsum plaster. Gypsum plaster materials shall conform to ASTM C 5, C 22, C 28, C 35, C 59, C 61, C 587, C 631, C 847, C 933, C 1032 and C 1047, and shall be installed or applied in compliance with ASTM C 843 and C 844. Gypsum lath or gypsum base for veneer plaster shall conform to ASTM C 1396. Plaster shall not be less than three coats when applied over metal lath and not less than two coats when applied over other bases permitted by this section, except that veneer plaster may be applied in one coat not to exceed $^{3}/_{16}$ inch (4.76 mm) thickness, provided the total thickness is in accordance with Table R702.1(1).

R702.2.2 Cement plaster. Cement plaster materials shall conform to ASTM C 91 (Type M, S or N), C 150 (Type I, II and III), C 595 [Type IP, I (PM), IS and I (SM), C 847, C 897, C 926, C 933, C 1032, C 1047 and C 1328, and shall be installed or applied in compliance with ASTM C 1063. Gypsum lath shall conform to ASTM C 1396. Plaster shall not be less than three coats when applied over metal lath and not less than two coats when applied over other bases permitted by this section, except that veneer plaster may be applied in one coat not to exceed $^{3}/_{16}$ inch (4.76 mm) thickness, provided the total thickness is in accordance with Table R702.1(1).

R702.2.2.1 Application. Each coat shall be kept in a moist condition for at least 24 hours prior to application of the next coat.

Exception: Applications installed in accordance with ASTM C 926.

R702.2.2.2 Curing. The finish coat for two-coat cement plaster shall not be applied sooner than 48 hours after application of the first coat. For three coat cement plaster the second coat shall not be applied sooner than 24 hours after application of the first coat. The finish coat for three-coat cement plaster shall not

**TABLE R702.1(1)
THICKNESS OF PLASTER**

PLASTER BASE	FINISHED THICKNESS OF PLASTER FROM FACE OF LATH, MASONRY, CONCRETE (inches)	
	Gypsum Plaster	Cement Plaster
Expanded metal lath	$^{5}/_{8}$, minimum[a]	$^{5}/_{8}$, minimum[a]
Wire lath	$^{5}/_{8}$, minimum[a]	$^{3}/_{4}$, minimum (interior)[b] $^{7}/_{8}$, minimum (exterior)[b]
Gypsum lath[g]	$^{1}/_{2}$, minimum	$^{3}/_{4}$, minimum (interior)[b]
Masonry walls[c]	$^{1}/_{2}$, minimum	$^{1}/_{2}$, minimum
Monolithic concrete walls[c, d]	$^{5}/_{8}$, maximum	$^{7}/_{8}$, maximum
Monolithic concrete ceilings[c, d]	$^{3}/_{8}$, maximum[e]	$^{1}/_{2}$, maximum
Gypsum veneer base[f, g]	$^{1}/_{16}$, minimum	$^{3}/_{4}$, minimum (interior)[b]
Gypsum sheathing[g]	—	$^{3}/_{4}$, minimum (interior)[b] $^{7}/_{8}$, minimum (exterior)[b]

For SI: 1 inch = 25.4 mm.

a. When measured from back plane of expanded metal lath, exclusive of ribs, or self-furring lath, plaster thickness shall be $^{3}/_{4}$ inch minimum.

b. When measured from face of support or backing.

c. Because masonry and concrete surfaces may vary in plane, thickness of plaster need not be uniform.

d. When applied over a liquid bonding agent, finish coat may be applied directly to concrete surface.

e. Approved acoustical plaster may be applied directly to concrete or over base coat plaster, beyond the maximum plaster thickness shown.

f. Attachment shall be in accordance with Table R702.3.5.

g. Where gypsum board is used as a base for cement plaster, a water-resistive barrier complying with Section R703.2 shall be provided.

be applied sooner than 48 hours after application of the second coat.

R702.2.3 Support. Support spacing for gypsum or metal lath on walls or ceilings shall not exceed 16 inches (406 mm) for $^3/_8$-inch-thick (9.5 mm) or 24 inches (610 mm) for $^1/_2$-inch-thick (12.7 mm) plain gypsum lath. Gypsum lath shall be installed at right angles to support framing with end joints in adjacent courses staggered by at least one framing space.

R702.3 Gypsum board.

R702.3.1 Materials. All gypsum board materials and accessories shall conform to ASTM C 22, C 475, C 514, C 1002, C 1047, C 1177, C 1178, C 1278, C 1396 or C 1658 and shall be installed in accordance with the provisions of this section. Adhesives for the installation of gypsum board shall conform to ASTM C 557.

R702.3.2 Wood framing. Wood framing supporting gypsum board shall not be less than 2 inches (51 mm) nominal thickness in the least dimension except that wood furring strips not less than 1-inch by 2-inch (25 mm by 51 mm) nominal dimension may be used over solid backing or framing spaced not more than 24 inches (610 mm) on center.

R702.3.3 Cold-formed steel framing. Cold-formed steel framing supporting gypsum board shall not be less than $1^1/_4$ inches (32 mm) wide in the least dimension. Nonload-bearing cold-formed steel framing shall comply with ASTM C 645. Load-bearing cold-formed steel framing and all cold-

TABLE R702.1(2)
GYPSUM PLASTER PROPORTIONS[a]

NUMBER	COAT	PLASTER BASE OR LATH	MAXIMUM VOLUME AGGREGATE PER 100 POUNDS NEAT PLASTER[b] (cubic feet)	
			Damp Loose Sand[a]	Perlite or Vermiculite[c]
Two-coat work	Base coat	Gypsum lath	2.5	2
	Base coat	Masonry	3	3
Three-coat work	First coat	Lath	2[d]	2
	Second coat	Lath	3[d]	2[e]
	First and second coats	Masonry	3	3

For SI: 1 inch = 25.4 mm, 1 cubic foot = 0.0283 m^3, 1 pound = 0.454 kg.

a. Wood-fibered gypsum plaster may be mixed in the proportions of 100 pounds of gypsum to not more than 1 cubic foot of sand where applied on masonry or concrete.

b. When determining the amount of aggregate in set plaster, a tolerance of 10 percent shall be allowed.

c. Combinations of sand and lightweight aggregate may be used, provided the volume and weight relationship of the combined aggregate to gypsum plaster is maintained.

d. If used for both first and second coats, the volume of aggregate may be 2.5 cubic feet.

e. Where plaster is 1 inch or more in total thickness, the proportions for the second coat may be increased to 3 cubic feet.

TABLE R702.1(3)
CEMENT PLASTER PROPORTIONS, PARTS BY VOLUME

COAT	CEMENT PLASTER TYPE	CEMENTITIOUS MATERIALS				VOLUME OF AGGREGATE PER SUM OF SEPARATE VOLUMES OF CEMENTITIOUS MATERIALS[b]
		Portland Cement Type I, II or III or Blended Cement Type IP, I (PM), IS or I (SM)	Plastic Cement	Masonry Cement Type M, S or N	Lime	
First	Portland or blended	1			$^3/_4$ - $1^1/_2$[a]	$2^1/_2$ - 4
	Masonry				1	$2^1/_2$ - 4
	Plastic		1			$2^1/_2$ - 4
Second	Portland or blended	1			$^3/_4$ - $1^1/_2$	3 - 5
	Masonry			1		3 - 5
	Plastic		1			3 - 5
Finish	Portland or blended	1			$^3/_4$ - 2	$1^1/_2$ - 3
	Masonry			1		$1^1/_2$ - 3
	Plastic		1			$1^1/_2$ - 3

For SI: 1 inch = 25.4 mm, 1 pound = 0.545 kg.

a. Lime by volume of 0 to 3/4 shall be used when the plaster will be placed over low-absorption surfaces such as dense clay tile or brick.

b. The same or greater sand proportion shall be used in the second coat than used in the first coat.

formed steel framing from 0.033 inch to 0.112 inch (1 mm to 3 mm) thick shall comply with ASTM C 955.

R702.3.4 Insulating concrete form walls. Foam plastics for insulating concrete form walls constructed in accordance with Sections R404.1.2 and R611 on the interior of *habitable spaces* shall be protected in accordance with Section R316.4. Use of adhesives in conjunction with mechanical fasteners is permitted. Adhesives used for interior and exterior finishes shall be compatible with the insulating form materials.

R702.3.5 Application. Maximum spacing of supports and the size and spacing of fasteners used to attach gypsum board shall comply with Table R702.3.5. Gypsum sheathing shall be attached to exterior walls in accordance with Table R602.3(1). Gypsum board shall be applied at right angles or parallel to framing members. All edges and ends of gypsum board shall occur on the framing members, except those edges and ends that are perpendicular to the framing members. Interior gypsum board shall not be installed where it is directly exposed to the weather or to water.

R702.3.6 Fastening. Screws for attaching gypsum board to wood framing shall be Type W or Type S in accordance with ASTM C 1002 and shall penetrate the wood not less than $^5/_8$ inch (16 mm). Gypsum board shall be attached to cold-formed steel framing with minimum No. 6 screws. Screws for attaching gypsum board to cold-formed steel framing less than 0.033 inch (1 mm) thick shall be Type S in accordance with ASTM C 1002 or bugle head style in accordance with ASTM C 1513 and shall penetrate the

TABLE R702.3.5
MINIMUM THICKNESS AND APPLICATION OF GYPSUM BOARD

THICKNESS OF GYPSUM BOARD (inches)	APPLICATION	ORIENTATION OF GYPSUM BOARD TO FRAMING	MAXIMUM SPACING OF FRAMING MEMBERS (inches o.c.)	MAXIMUM SPACING OF FASTENERS (inches)		SIZE OF NAILS FOR APPLICATION TO WOOD FRAMING[c]
				Nails[a]	Screws[b]	
Application without adhesive						
$^3/_8$	Ceiling[d]	Perpendicular	16	7	12	13 gage, $1^1/_4''$ long, $^{19}/_{64}''$ head; 0.098″ diameter, $1^1/_4''$ long, annular-ringed; or 4d cooler nail, 0.080″ diameter, $1^3/_8''$ long, $^7/_{32}''$ head.
	Wall	Either direction	16	8	16	
$^1/_2$	Ceiling	Either direction	16	7	12	13 gage, $1^3/_8''$ long, $^{19}/_{64}''$ head; 0.098″ diameter, $1^1/_4''$ long, annular-ringed; 5d cooler nail, 0.086″ diameter, $1^5/_8''$ long, $^{15}/_{64}''$ head; or gypsum board nail, 0.086? diameter, $1\,^5/_8''$ long, $^9/_{32}''$ head.
	Ceiling[d]	Perpendicular	24	7	12	
	Wall	Either direction	24	8	12	
	Wall	Either direction	16	8	16	
$^5/_8$	Ceiling	Either direction	16	7	12	13 gage, $1^5/_8''$ long, $^{19}/_{64}''$ head; 0.098″ diameter, $1^3/_8''$ long, annular-ringed; 6d cooler nail, 0.092″ diameter, $1^7/_8''$ long, $^1/_4''$ head; or gypsum board nail, 0.0915″ diameter, $1^7/_8''$ long, $^{19}/_{64}''$ head.
	Ceiling[e]	Perpendicular	24	7	12	
	Wall	Either direction	24	8	12	
	Wall	Either direction	16	8	16	
Application with adhesive						
$^3/_8$	Ceiling[d]	Perpendicular	16	16	16	Same as above for $^3/_8''$ gypsum board
	Wall	Either direction	16	16	24	
$^1/_2$ or $^5/_8$	Ceiling	Either direction	16	16	16	Same as above for $^1/_2''$ and $^5/_8''$ gypsum board, respectively
	Ceiling[d]	Perpendicular	24	12	16	
	Wall	Either direction	24	16	24	
Two $^3/_8$ layers	Ceiling	Perpendicular	16	16	16	Base ply nailed as above for $^1/_2''$ gypsum board; face ply installed with adhesive
	Wall	Either direction	24	24	24	

For SI: 1 inch = 25.4 mm.

a. For application without adhesive, a pair of nails spaced not less than 2 inches apart or more than $2^1/_2$ inches apart may be used with the pair of nails spaced 12 inches on center.

b. Screws shall be in accordance with Section R702.3.6. Screws for attaching gypsum board to structural insulated panels shall penetrate the wood structural panel facing not less than $^7/_{16}$ inch.

c. Where cold-formed steel framing is used with a clinching design to receive nails by two edges of metal, the nails shall be not less than $^5/_8$ inch longer than the gypsum board thickness and shall have ringed shanks. Where the cold-formed steel framing has a nailing groove formed to receive the nails, the nails shall have barbed shanks or be 5d, $13^1/_2$ gage, $^{15}/_8$ inches long, $^{15}/_{64}$-inch head for $^1/_2$-inch gypsum board; and 6d, 13 gage, $1^7/_8$ inches long, $^{15}/_{64}$-inch head for $^5/_8$-inch gypsum board.

d. Three-eighths-inch-thick single-ply gypsum board shall not be used on a ceiling where a water-based textured finish is to be applied, or where it will be required to support insulation above a ceiling. On ceiling applications to receive a water-based texture material, either hand or spray applied, the gypsum board shall be applied perpendicular to framing. When applying a water-based texture material, the minimum gypsum board thickness shall be increased from $^3/_8$ inch to $^1/_2$ inch for 16-inch on center framing, and from $^1/_2$ inch to $^5/_8$ inch for 24-inch on center framing or $^1/_2$-inch sag-resistant gypsum ceiling board shall be used.

e. Type X gypsum board for garage ceilings beneath habitable rooms shall be installed perpendicular to the ceiling framing and shall be fastened at maximum 6 inches o.c. by minimum $1^7/_8$ inches 6d coated nails or equivalent drywall screws.

steel not less than $^3/_8$ inch (9.5 mm). Screws for attaching gypsum board to cold-formed steel framing 0.033 inch to 0.112 inch (1 mm to 3 mm) thick shall be in accordance with ASTM C 954 or bugle head style in accordance with ASTM C 1513. Screws for attaching gypsum board to structural insulated panels shall penetrate the wood structural panel facing not less than $^7/_{16}$ inch (11 mm).

R702.3.7 Horizontal gypsum board diaphragm ceilings. Use of gypsum board shall be permitted on wood joists to create a horizontal *diaphragm* in accordance with Table R702.3.7. Gypsum board shall be installed perpendicular to ceiling framing members. End joints of adjacent courses of board shall not occur on the same joist. The maximum allowable *diaphragm* proportions shall be $1^1/_2$:1 between shear resisting elements. Rotation or cantilever conditions shall not be permitted. Gypsum board shall not be used in *diaphragm* ceilings to resist lateral forces imposed by masonry or concrete construction. All perimeter edges shall be blocked using wood members not less than 2-inch by 6-inch (51 mm by 152 mm) nominal dimension. Blocking material shall be installed flat over the top plate of the wall to provide a nailing surface not less than 2 inches (51 mm) in width for the attachment of the gypsum board.

R702.3.8 Water-resistant gypsum backing board. Gypsum board used as the base or backer for adhesive application of ceramic tile or other required nonabsorbent finish material shall conform to ASTM C 1396, C 1178 or C1278. Use of water-resistant gypsum backing board shall be permitted on ceilings where framing spacing does not exceed 12 inches (305 mm) on center for $^1/_2$-inch-thick (12.7 mm) or 16 inches (406 mm) for $^5/_8$-inch-thick (16 mm) gypsum board. Water-resistant gypsum board shall not be installed over a Class I or II vapor retarder in a shower or tub compartment. Cut or exposed edges, including those at wall intersections, shall be sealed as recommended by the manufacturer.

R702.3.8.1 Limitations. Water resistant gypsum backing board shall not be used where there will be direct exposure to water, or in areas subject to continuous high humidity.

R702.4 Ceramic tile.

R702.4.1 General. Ceramic tile surfaces shall be installed in accordance with ANSI A108.1, A108.4, A108.5, A108.6, A108.11, A118.1, A118.3, A136.1 and A137.1.

R702.4.2 Fiber-cement, fiber-mat reinforced cementitious backer units, glass mat gypsum backers and fiber-reinforced gypsum backers. Fiber-cement, fiber-mat reinforced cementitious backer units, glass mat gypsum backers or fiber-reinforced gypsum backers in compliance with ASTM C 1288, C 1325, C 1178 or C 1278, respectively, and installed in accordance with manufacturers' recommendations shall be used as backers for wall tile in tub and shower areas and wall panels in shower areas.

R702.5 Other finishes. Wood veneer paneling and hardboard paneling shall be placed on wood or cold-formed steel framing spaced not more than 16 inches (406 mm) on center. Wood veneer and hard board paneling less than $^1/_4$-inch (6 mm) nominal thickness shall not have less than a $^3/_8$-inch (10 mm) gypsum board backer. Wood veneer paneling not less than $^1/_4$-inch (6 mm) nominal thickness shall conform to ANSI/HPVA HP-1. Hardboard paneling shall conform to CPA/ANSI A135.5.

R702.6 Wood shakes and shingles. Wood shakes and shingles shall conform to CSSB *Grading Rules for Wood Shakes and Shingles* and shall be permitted to be installed directly to the studs with maximum 24 inches (610 mm) on-center spacing.

R702.6.1 Attachment. Nails, staples or glue are permitted for attaching shakes or shingles to the wall, and attachment of the shakes or shingles directly to the surface shall be permitted provided the fasteners are appropriate for the type of wall surface material. When nails or staples are used, two fasteners shall be provided and shall be placed so that they are covered by the course above.

R702.6.2 Furring strips. Where furring strips are used, they shall be 1 inch by 2 inches or 1 inch by 3 inches (25 mm by 51 mm or 25 mm by 76 mm), spaced a distance on center equal to the desired exposure, and shall be attached to the wall by nailing through other wall material into the studs.

TABLE R702.3.7
SHEAR CAPACITY FOR HORIZONTAL WOOD-FRAMED GYPSUM BOARD DIAPHRAGM CEILING ASSEMBLIES

MATERIAL	THICKNESS OF MATERIAL (min.) (inch)	SPACING OF FRAMING MEMBERS (max.) (inch)	SHEAR VALUE[a, b] (plf of ceiling)	MINIMUM FASTENER SIZE[c, d]
Gypsum board	$^1/_2$	16 o.c.	90	5d cooler or wallboard nail; $1^5/_8$-inch long; 0.086-inch shank; $^{15}/_{64}$-inch head
Gypsum board	$^1/_2$	24 o.c.	70	5d cooler or wallboard nail; $1^5/_8$-inch long; 0.086-inch shank; $^{15}/_{64}$-inch head

For SI: 1 inch = 25.4 mm, 1 pound per linear foot = 1.488 kg/m.

a. Values are not cumulative with other horizontal diaphragm values and are for short-term loading caused by wind or seismic loading. Values shall be reduced 25 percent for normal loading.

b. Values shall be reduced 50 percent in Seismic Design Categories D_0, D_1, D_2 and E.

c. $1^1/_4$-inch, #6 Type S or W screws may be substituted for the listed nails.

d. Fasteners shall be spaced not more than 7 inches on center at all supports, including perimeter blocking, and not less than $^3/_8$ inch from the edges and ends of the gypsum board.

** **R702.7 Vapor retarders.** Class I or II vapor retarders are required on the interior side of frame walls in Climate Zones 5, 6, 7, 8 and Marine 4.

Exceptions:

1. Basement walls.
2. Below grade portion of any wall.
3. Construction where moisture or its freezing will not damage the materials.

R702.7.1 Class III vapor retarders. Class III vapor retarders shall be permitted where any one of the conditions in Table R702.7.1 is met.

TABLE R702.7.1
CLASS III VAPOR RETARDERS

CLIMATE ZONE	CLASS III VAPOR RETARDERS PERMITTED FOR:[a]
Marine 4	Vented cladding over wood structural panels.
	Vented cladding over fiberboard.
	Vented cladding over gypsum.
	Insulated sheathing with *R*-value ≥ 2.5 over 2 × 4 wall.
	Insulated sheathing with *R*-value ≥ 3.75 over 2 × 6 wall.
5	Vented cladding over wood structural panels.
	Vented cladding over fiberboard.
	Vented cladding over gypsum.
	Insulated sheathing with *R*-value ≥ 5 over 2 × 4 wall.
	Insulated sheathing with *R*-value ≥ 7.5 over 2 × 6 wall.
6	Vented cladding over fiberboard.
	Vented cladding over gypsum.
	Insulated sheathing with *R*-value ≥ 7.5 over 2 × 4 wall.
	Insulated sheathing with *R*-value ≥ 11.25 over 2 × 6 wall.
7 and 8	Insulated sheathing with *R*-value ≥ 10 over 2 × 4 wall.
	Insulated sheathing with *R*-value ≥ 15 over 2 × 6 wall.

For SI: 1 pound per cubic foot = 16 kg/m^3.

a. Spray foam with a minimum density of 2 lb/ft^3 applied to the interior cavity side of wood structural panels, fiberboard, insulating sheathing or gypsum is deemed to meet the insulating sheathing requirement where the spray foam *R*-value meets or exceeds the specified insulating sheathing *R*-value.

R702.7.2 Material vapor retarder class. The vapor retarder class shall be based on the manufacturer's certified testing or a tested assembly.

The following shall be deemed to meet the class specified:

Class I: Sheet polyethylene, unperforated aluminum foil.

Class II: Kraft-faced fiberglass batts.

Class III: Latex or enamel paint.

R702.7.3 Minimum clear air spaces and vented openings for vented cladding. For the purposes of this section, vented cladding shall include the following minimum clear air spaces. Other openings with the equivalent vent area shall be permitted.

1. Vinyl lap or horizontal aluminum siding applied over a weather resistive barrier as specified in Table R703.4.
2. Brick veneer with a clear airspace as specified in Table R703.7.4.
3. Other approved vented claddings.

SECTION R703
EXTERIOR COVERING

R703.1 General. Exterior walls shall provide the building with a weather-resistant exterior wall envelope. The exterior wall envelope shall include flashing as described in Section R703.8.

R703.1.1 Water resistance. The exterior wall envelope shall be designed and constructed in a manner that prevents the accumulation of water within the wall assembly by providing a water-resistant barrier behind the exterior veneer as required by Section R703.2 and a means of draining to the exterior water that enters the assembly. Protection against condensation in the exterior wall assembly shall be provided in accordance with Section R702.7 of this code.

Exceptions:

1. A weather-resistant exterior wall envelope shall not be required over concrete or masonry walls designed in accordance with Chapter 6 and flashed according to Section R703.7 or R703.8.
2. Compliance with the requirements for a means of drainage, and the requirements of Sections R703.2 and R703.8, shall not be required for an exterior wall envelope that has been demonstrated to resist wind-driven rain through testing of the exterior wall envelope, including joints, penetrations and intersections with dissimilar materials, in accordance with ASTM E 331 under the following conditions:
   - 2.1. Exterior wall envelope test assemblies shall include at least one opening, one control joint, one wall/eave interface and one wall sill. All tested openings and penetrations shall be representative of the intended end-use configuration.
   - 2.2. Exterior wall envelope test assemblies shall be at least 4 feet by 8 feet (1219 mm by 2438 mm) in size.
   - 2.3. Exterior wall assemblies shall be tested at a minimum differential pressure of 6.24 pounds per square foot (299 Pa).
   - 2.4. Exterior wall envelope assemblies shall be subjected to the minimum test exposure for a minimum of 2 hours.

   The exterior wall envelope design shall be considered to resist wind-driven rain where the results of testing indicate that water did not penetrate control joints in the exterior wall envelope, joints at the perimeter of openings penetration or intersections of terminations with dissimilar materials.

R703.1.2 Wind resistance. Wall coverings, backing materials and their attachments shall be capable of resisting wind loads in accordance with Tables R301.2(2) and R301.2(3). Wind-pressure resistance of the siding and backing materials shall be determined by ASTM E 330 or other applicable standard test methods. Where wind-pressure resistance is determined by design analysis, data from approved design standards and analysis conforming to generally accepted engineering practice shall be used to evaluate the siding and backing material and its fastening. All applicable failure modes including bending rupture of siding, fastener withdrawal and fastener head pull-through shall be considered in the testing or design analysis. Where the wall covering and the backing material resist wind load as an assembly, use of the design capacity of the assembly shall be permitted.

R703.2 Water-resistive barrier. One layer of No. 15 asphalt felt, free from holes and breaks, complying with ASTM D 226 for Type 1 felt or other approved water-resistive barrier shall be applied over studs or sheathing of all exterior walls. Such felt or material shall be applied horizontally, with the upper layer lapped over the lower layer not less than 2 inches (51 mm). Where joints occur, felt shall be lapped not less than 6 inches (152 mm). The felt or other approved material shall be continuous to the top of walls and terminated at penetrations and building appendages in a manner to meet the requirements of the exterior wall envelope as described in Section R703.1.

Exception: Omission of the water-resistive barrier is permitted in the following situations:

1. In detached accessory buildings.
2. Under exterior wall finish materials as permitted in Table R703.4.
3. Under paperbacked stucco lath when the paper backing is an approved water-resistive barrier.

R703.3 Wood, hardboard and wood structural panel siding.

R703.3.1 Panel siding. Joints in wood, hardboard or wood structural panel siding shall be made as follows unless otherwise approved. Vertical joints in panel siding shall occur over framing members, unless wood or wood structural panel sheathing is used, and shall be shiplapped or covered with a batten. Horizontal joints in panel siding shall be lapped a minimum of 1 inch (25 mm) or shall be shiplapped or shall be flashed with Z-flashing and occur over solid blocking, wood or wood structural panel sheathing.

R703.3.2 Horizontal siding. Horizontal lap siding shall be installed in accordance with the manufacturer's recommendations. Where there are no recommendations the siding shall be lapped a minimum of 1 inch (25 mm), or $^1/_2$ inch (13 mm) if rabbeted, and shall have the ends caulked, covered with a batten or sealed and installed over a strip of flashing.

R703.4 Attachments. Unless specified otherwise, all wall coverings shall be securely fastened in accordance with Table R703.4 or with other *approved* aluminum, stainless steel, zinc-coated or other *approved* corrosion-resistive fasteners. Where the basic wind speed in accordance with Figure R301.2(4)A is 110 miles per hour (49 m/s) or higher, the attachment of wall coverings shall be designed to resist the component and cladding loads specified in Table R301.2(2), adjusted for height and exposure in accordance with Table R301.2(3).

R703.5 Wood shakes and shingles. Wood shakes and shingles shall conform to CSSB *Grading Rules for Wood Shakes and Shingles.*

R703.5.1 Application. Wood shakes or shingles shall be applied either single-course or double-course over nominal $^1/_2$-inch (13 mm) wood-based sheathing or to furring strips over $^1/_2$-inch (13 mm) nominal nonwood sheathing . A permeable water-resistive barrier shall be provided over all sheathing, with horizontal overlaps in the membrane of not less than 2 inches (51 mm) and vertical overlaps of not less than 6 inches (152 mm). Where furring strips are used, they shall be 1 inch by 3 inches or 1 inch by 4 inches (25 mm by 76 mm or 25 mm by 102 mm) and shall be fastened horizontally to the studs with 7d or 8d box nails and shall be spaced a distance on center equal to the actual weather exposure of the shakes or shingles, not to exceed the maximum exposure specified in Table R703.5.2. The spacing between adjacent shingles to allow for expansion shall not exceed $^1/_4$ inch (6 mm), and between adjacent shakes, it shall not exceed $^1/_2$ inch (13 mm). The offset spacing between joints in adjacent courses shall be a minimum of $1^1/_2$ inches (38 mm).

R703.5.2 Weather exposure. The maximum weather exposure for shakes and shingles shall not exceed that specified in Table R703.5.2.

TABLE R703.5.2
MAXIMUM WEATHER EXPOSURE FOR WOOD SHAKES AND SHINGLES ON EXTERIOR WALLS[a, b, c]
(Dimensions are in inches)

LENGTH	EXPOSURE FOR SINGLE COURSE	EXPOSURE FOR DOUBLE COURSE
Shingles[a]		
16	$7^1/_2$	12[b]
18	$8^1/_2$	14[c]
24	$11^1/_2$	16
Shakes[a]		
18	$8^1/_2$	14
24	$11^1/_2$	18

For SI: 1 inch = 25.4 mm.

a. Dimensions given are for No. 1 grade.

b. A maximum 10-inch exposure is permitted for No. 2 grade.

c. A maximum 11-inch exposure is permitted for No. 2 grade.

TABLE R703.4
WEATHER-RESISTANT SIDING ATTACHMENT AND MINIMUM THICKNESS

SIDING MATERIAL		NOMINAL THICKNESS[a] (inches)	JOINT TREATMENT	WATER-RESISTIVE BARRIER REQUIRED	TYPE OF SUPPORTS FOR THE SIDING MATERIAL AND FASTENERS[b, c, d]					
					Wood or wood structural panel sheathing into stud	Fiberboard sheathing into stud	Gypsum sheathing into stud	Foam plastic sheathing into stud	Direct to studs	Number or spacing of fasteners
Horizonal aluminum[e]	Without insulation	0.019[f]	Lap	Yes	0.120 nail 1 1/2″ long	0.120 nail 2″ long	0.120 nail 2″ long	0.120 nail[y]	Not allowed	Same as stud spacing
		0.024	Lap	Yes	0.120 nail 1 1/2″ long	0.120 nail 2″ long	0.120 nail 2″ long	0.120 nail[y]	Not allowed	
	With insulation	0.019	Lap	Yes	0.120 nail 1 1/2″ long	0.120 nail 2 1/2″ long	0.120 nail 2 1/2″ long	0.120 nail[y]	0.120 nail 1 1/2″ long	
Anchored veneer: brick, concrete, masonry or stone		2	Section R703	Yes	See Section R703 and Figure R703.7[g]					
Adhered veneer: concrete, stone or masonry[w]		—	Section R703	Yes Note w	See Section R703.6.1[g] or in accordance with the manufacturer's instructions.					
Hardboard[k] Panel siding-vertical		7/16	—	Yes	Note m	Note m	Note m	Note m	Note m	6″ panel edges 12″ inter. sup.[n]
Hardboard[k] Lap-siding-horizontal		7/16	Note p	Yes	Note o	Note o	Note o	Note o	Note o	Same as stud spacing 2 per bearing
Steel[h]		29 ga.	Lap	Yes	0.113 nail 1 3/4″ Staple-1 3/4″	0.113 nail 2 3/4″ Staple-2 1/2″	0.113 nail 2 1/2″ Staple-2 1/4″	0.113 nail[v] Staple[v]	Not allowed	Same as stud spacing
Particleboard panels		3/8 - 1/2	—	Yes	6d box nail (2″ × 0.099″)	6d box nail (2″ × 0.099″)	6d box nail (2″ × 0.099″)	box nail[v]	6d box nail (2″ × 0.099″), 3/8 not allowed	6″ panel edge, 12" inter. sup.
		5/8	—	Yes	6d box nail (2″ × 0.099″)	8d box nail (2 1/2″ × 0.113″)	8d box nail (2 1/2″ × 0.113″)	box nail[v]	6d box nail (2″" × 0.099″)	
Wood structural panel[i] ANSI/APA-PRP 210 siding[i] (exterior grade)		3/8 - 1/2	Note p	Yes	0.099 nail-2″	0.113 nail-2 1/2″	0.113 nail-2 1/2″	0.113 nail[v]	0.099 nail-2″	6″ panel edges, 12″ inter. sup.
Wood structural panel lapsiding		3/8 - 1/2	Note p Note x	Yes	0.099 nail-2″	0.113 nail-2 1/2″	0.113 nail-2 1/2″	0.113 nail[x]	0.099 nail-2″	8″ along bottom edge
Vinyl siding[l]		0.035	Lap	Yes	0.120 nail (shank) with a 0.313 head or 16-gage staple with 3/8 to 1/2-inch crown[y, z]	0.120 nail (shank) with a 0.313 head or 16-gage staple with 3/8 to 1/2-inch crown[y]	0.120 nail (shank) with a 0.313 head or 16-gage staple with 3/8 to 1/2-inch crown[y]	0.120 nail (shank) with a 0.313 head per Section R703.11.2	Not allowed	16 inches on center or specified by the manufacturer instructions or test report
Wood[j] rustic, drop	3/8 Min	Lap	Yes	Fastener penetration into stud-1″					0.113 nail-2 1/2″ Staple-2″	Face nailing up to 6″ widths, 1 nail per bearing; 8″ widths and over, 2 nails per bearing

(continued)

TABLE R703.4—continued
WEATHER-RESISTANT SIDING ATTACHMENT AND MINIMUM THICKNESS

SIDING MATERIAL	NOMINAL THICKNESS[a] (inches)	JOINT TREATMENT	WATER-RESISTIVE BARRIER REQUIRED	TYPE OF SUPPORTS FOR THE SIDING MATERIAL AND FASTENERS[b, c, d]					
				Wood or wood structural panel sheathing into stud	Fiberboard sheathing into stud	Gypsum sheathing into stud	Foam plastic sheathing into stud	Direct to studs	Number or spacing of fasteners
Shiplap	$^{19}/_{32}$ Average	Lap	Yes	Fastener penetration into stud-1″				0.113 nail-2$^{1}/_{2}$″ Staple-2″	Face nailing up to 6″ widths, 1 nail per bearing; 8″ widths and over, 2 nails per bearing
Bevel	$^{7}/_{16}$								
Butt tip	$^{3}/_{16}$	Lap	Yes						
Fiber cement panel siding[q]	$^{5}/_{16}$	Note q	Yes Note u	6d common corrosion-resistant nail[r]	6d common corrosion-resistant nail[r]	6d common corrosion-resistant nail[r]	6d common corrosion-resistant nail[r, v]	4d common corrosion-resistant nail[r]	6″ o.c. on edges, 12″ o.c. on intermed. studs
Fiber cement lap siding[s]	$^{5}/_{16}$	Note s	Yes Note u	6d common corrosion-resistant nail[r]	6d common corrosion-resistant nail[r]	6d common corrosion-resistant nail[r]	6d common corrosion-resistant nail[r, v]	6d common corrosion-resistant nail or 11-gage roofing nail[r]	Note t

For SI: 1 inch = 25.4 mm.

a. Based on stud spacing of 16 inches on center where studs are spaced 24 inches, siding shall be applied to sheathing approved for that spacing.
b. Nail is a general description and shall be T-head, modified round head, or round head with smooth or deformed shanks.
c. Staples shall have a minimum crown width of $^{7}/_{16}$-inch outside diameter and be manufactured of minimum 16-gage wire.
d. Nails or staples shall be aluminum, galvanized, or rust-preventative coated and shall be driven into the studs where fiberboard, gypsum, or foam plastic sheathing backing is used. Where wood or wood structural panel sheathing is used, fasteners shall be driven into studs unless otherwise permitted to be driven into sheathing in accordance with the siding manufacturer's installation instructions.
e. Aluminum nails shall be used to attach aluminum siding.
f. Aluminum (0.019 inch) shall be unbacked only when the maximum panel width is 10 inches and the maximum flat area is 8 inches. The tolerance for aluminum siding shall be +0.002 inch of the nominal dimension.
g. All attachments shall be coated with a corrosion-resistant coating.
h. Shall be of approved type.
i. Three-eighths-inch plywood shall not be applied directly to studs spaced more than 16 inches on center when long dimension is parallel to studs. Plywood $^{1}/_{2}$-inch or thinner shall not be applied directly to studs spaced more than 24 inches on center. The stud spacing shall not exceed the panel span rating provided by the manufacturer unless the panels are installed with the face grain perpendicular to the studs or over sheathing approved for that stud spacing.
j. Wood board sidings applied vertically shall be nailed to horizontal nailing strips or blocking set 24 inches on center. Nails shall penetrate 1$^{1}/_{2}$ inches into studs, studs and wood sheathing combined or blocking.
k. Hardboard siding shall comply with CPA/ANSI A135.6.
l. Vinyl siding shall comply with ASTM D 3679.
m. Minimum shank diameter of 0.092 inch, minimum head diameter of 0.225 inch, and nail length must accommodate sheathing and penetrate framing 1$^{1}/_{2}$ inches.
n. When used to resist shear forces, the spacing must be 4 inches at panel edges and 8 inches on interior supports.
o. Minimum shank diameter of 0.099 inch, minimum head diameter of 0.240 inch, and nail length must accommodate sheathing and penetrate framing 1$^{1}/_{2}$ inches.
p. Vertical end joints shall occur at studs and shall be covered with a joint cover or shall be caulked.
q. See Section R703.10.1.
r. Fasteners shall comply with the nominal dimensions in ASTM F 1667.
s. See Section R703.10.2.
t. Face nailing: one 6d common nail through the over lap ping planks at each stud. Concealed nailing: one 11 gage 1$^{1}/_{2}$ inch long galv. roofing nail through the top edge of each plank at each stud.
u. See Section R703.2 exceptions.
v. Minimum nail length must accommodate sheathing and penetrate framing 1$^{1}/_{2}$ inches.
w. Adhered masonry veneer shall comply with the requirements of Section R703.6.3 and shall comply with the requirements in Sections 6.1 and 6.3 of TMS-402 ACI 530/ASCE 5.
x. Vertical joints, if staggered shall be permitted to be away from studs if applied over wood structural panel sheathing.
y. Minimum fastener length must accommodate sheathing and penetrate framing 0.75 inches or in accordance with the manufacturer's installation instructions.
z. Where approved by the manufacturer's instructions or test report siding shall be permitted to be installed with fasteners penetrating not less than 0.75 inches through wood or wood structural sheathing with or without penetration into the framing.

R703.5.3 Attachment. Each shake or shingle shall be held in place by two hot-dipped zinc-coated, stainless steel, or aluminum nails or staples. The fasteners shall be long enough to penetrate the sheathing or furring strips by a minimum of $^1/_2$ inch (13 mm) and shall not be overdriven.

R703.5.3.1 Staple attachment. Staples shall not be less than 16 gage and shall have a crown width of not less than $^7/_{16}$ inch (11 mm), and the crown of the staples shall be parallel with the butt of the shake or shingle. In single-course application, the fasteners shall be concealed by the course above and shall be driven approximately 1 inch (25 mm) above the butt line of the succeeding course and $^3/_4$ inch (19 mm) from the edge. In double-course applications, the exposed shake or shingle shall be face-nailed with two casing nails, driven approximately 2 inches (51 mm) above the butt line and $^3/_4$ inch (19 mm) from each edge. In all applications, staples shall be concealed by the course above. With shingles wider than 8 inches (203 mm) two additional nails shall be required and shall be nailed approximately 1 inch (25 mm) apart near the center of the shingle.

R703.5.4 Bottom courses. The bottom courses shall be doubled.

R703.6 Exterior plaster. Installation of these materials shall be in compliance with ASTM C 926 and ASTM C 1063 and the provisions of this code.

R703.6.1 Lath. All lath and lath attachments shall be of corrosion-resistant materials. Expanded metal or woven wire lath shall be attached with $1^1/_2$-inch-long (38 mm), 11 gage nails having a $^7/_{16}$-inch (11.1 mm) head, or $^7/_8$-inch-long (22.2 mm), 16 gage staples, spaced at no more than 6 inches (152 mm), or as otherwise *approved.*

R703.6.2 Plaster. Plastering with portland cement plaster shall be not less than three coats when applied over metal lath or wire lath and shall be not less than two coats when applied over masonry, concrete, pressure-preservative treated wood or decay-resistant wood as specified in Section R317.1 or gypsum backing. If the plaster surface is completely covered by veneer or other facing material or is completely concealed, plaster application need be only two coats, provided the total thickness is as set forth in Table R702.1(1).

On wood-frame construction with an on-grade floor slab system, exterior plaster shall be applied to cover, but not extend below, lath, paper and screed.

The proportion of aggregate to cementitious materials shall be as set forth in Table R702.1(3).

R703.6.2.1 Weep screeds. A minimum 0.019-inch (0.5 mm) (No. 26 galvanized sheet gage), corrosion-resistant weep screed or plastic weep screed, with a minimum vertical attachment flange of $3^1/_2$ inches (89 mm) shall be provided at or below the foundation plate line on exterior stud walls in accordance with ASTM C 926. The weep screed shall be placed a minimum of 4 inches (102 mm) above the earth or 2 inches (51 mm) above paved areas and shall be of a type that will allow trapped water to drain to the exterior of the building. The weather-resistant barrier shall lap the attachment flange. The exterior lath shall cover and terminate on the attachment flange of the weep screed.

R703.6.3 Water-resistive barriers. Water-resistive barriers shall be installed as required in Section R703.2 and, where applied over wood-based sheathing, shall include a water-resistive vapor-permeable barrier with a performance at least equivalent to two layers of Grade D paper. The individual layers shall be installed independently such that each layer provides a separate continuous plane and any flashing (installed in accordance with Section R703.8) intended to drain to the water-resistive barrier is directed between the layers.

Exception: Where the water-resistive barrier that is applied over wood-based sheathing has a water resistance equal to or greater than that of 60-minute Grade D paper and is separated from the stucco by an intervening, substantially nonwater-absorbing layer or designed drainage space.

R703.6.4 Application. Each coat shall be kept in a moist condition for at least 48 hours prior to application of the next coat.

Exception: Applications installed in accordance with ASTM C 926.

R703.6.5 Curing. The finish coat for two-coat cement plaster shall not be applied sooner than seven days after application of the first coat. For three-coat cement plaster, the second coat shall not be applied sooner than 48 hours after application of the first coat. The finish coat for three-coat cement plaster shall not be applied sooner than seven days after application of the second coat.

R703.7 Stone and masonry veneer, general. Stone and masonry veneer shall be installed in accordance with this chapter, Table R703.4 and Figure R703.7. These veneers installed over a backing of wood or cold-formed steel shall be limited to the first *story* above-grade plane and shall not exceed 5 inches (127 mm) in thickness. See Section R602.10 for wall bracing requirements for masonry veneer for wood-framed construction and Section R603.9.5 for wall bracing requirements for masonry veneer for cold-formed steel construction.

Exceptions:

1. For all buildings in Seismic Design Categories A, B and C, exterior stone or masonry veneer, as specified in Table R703.7(1), with a backing of wood or steel framing shall be permitted to the height specified in Table R703.7(1) above a noncombustible foundation.
2. For detached one- or two-family *dwellings* in Seismic Design Categories D_0, D_1 and D_2, exterior stone or masonry veneer, as specified in Table R703.7(2), with a backing of wood framing shall be permitted to the height specified in Table R703.7(2) above a noncombustible foundation.

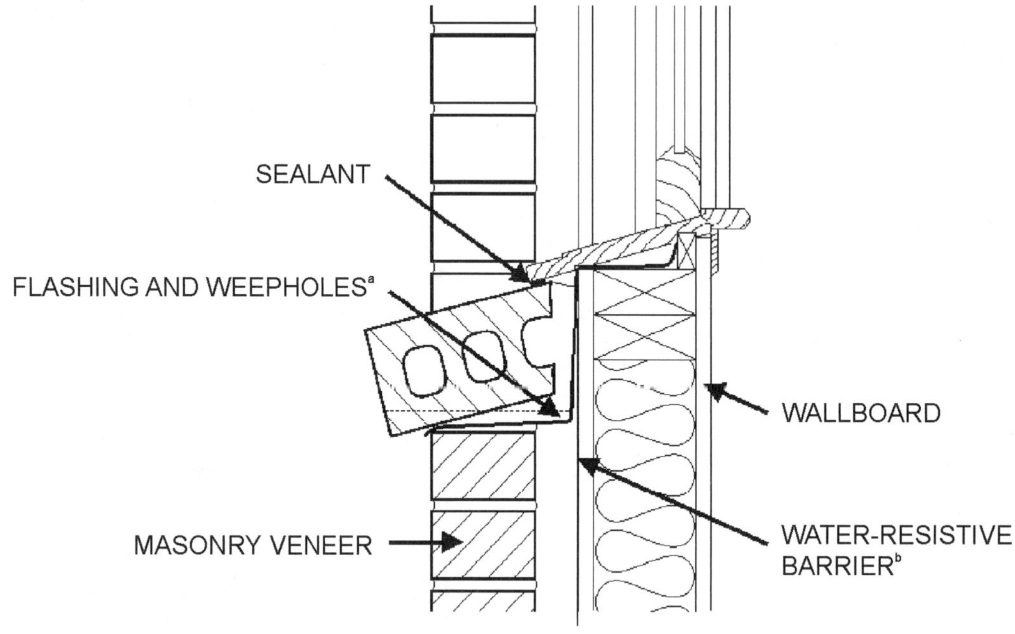

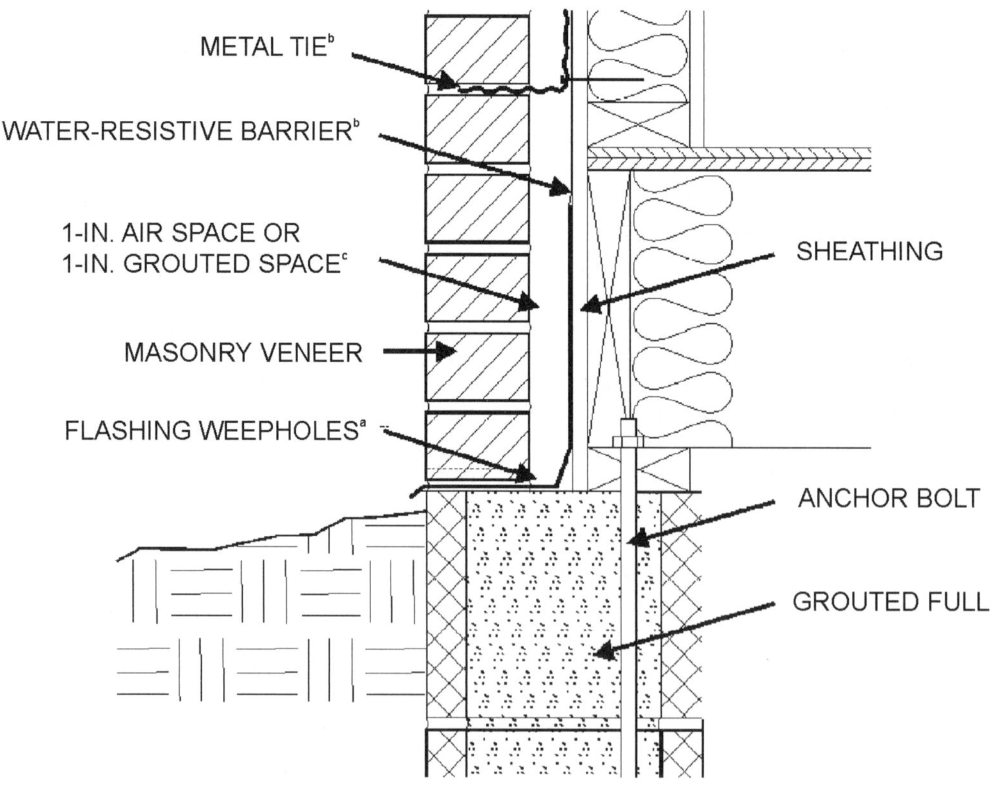

For SI: 1 inch = 24.5 mm.

FIGURE R703.7
MASONRY VENEER WALL DETAILS

(continued)

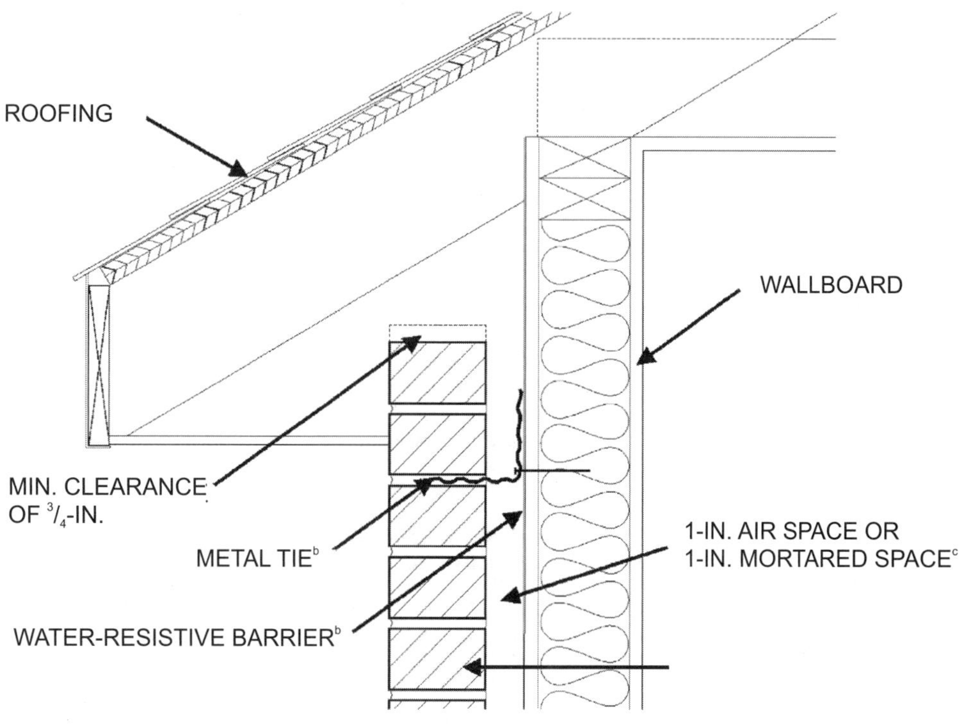

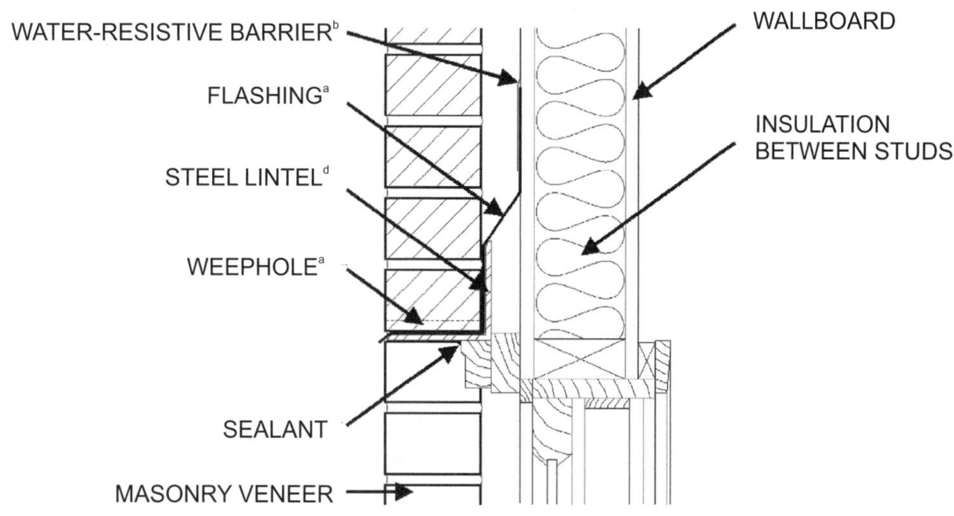

For SI: 1 inch = 25.4 mm.

a. See Sections R703.7.5, R703.7.6 and R703.8.

b. See Sections R703.2 and R703.7.4.

c. See Section R703.7.4.2 and Table R703.7.4.

d. See Section R703.7.3.

FIGURE R703.7—continued
MASONRY VENEER WALL DETAILS

TABLE R703.7(1)
STONE OR MASONRY VENEER LIMITATIONS AND REQUIREMENTS, WOOD OR STEEL FRAMING, SEISMIC DESIGN CATEGORIES A, B AND C

SEISMIC DESIGN CATEGORY	NUMBER OF WOOD OR STEEL-FRAMED STORIES	MAXIMUM HEIGHT OF VENEER ABOVE NONCOMBUSTIBLE FOUNDATION[a] (feet)	MAXIMUM NOMINAL THICKNESS OF VENEER (inches)	MAXIMUM WEIGHT OF VENEER (psf)[b]	WOOD OR STEEL-FRAMED STORY
A or B	Steel: 1 or 2 Wood: 1, 2 or 3	30	5	50	all
C	1	30	5	50	1 only
	2	30	5	50	top
					bottom
	Wood only: 3	30	5	50	top
					middle
					bottom

For SI: 1 inch = 25.4 mm, 1 foot = 304.8 mm, 1 pound per square foot = 0.479 kPa.

a. An additional 8 feet is permitted for gable end walls. See also story height limitations of Section R301.3.

b. Maximum weight is installed weight and includes weight of mortar, grout, lath and other materials used for installation. Where veneer is placed on both faces of a wall, the combined weight shall not exceed that specified in this table.

TABLE R703.7(2)
STONE OR MASONRY VENEER LIMITATIONS AND REQUIREMENTS, ONE- AND TWO-FAMILY DETACHED DWELLINGS, WOOD FRAMING, SEISMIC DESIGN CATEGORIES D_0, D_1 AND D_2

SEISMIC DESIGN CATEGORY	NUMBER OF WOOD FRAMED STORIES[a]	MAXIMUM HEIGHT OF VENEER ABOVE NONCOMBUSTIBLE FOUNDATION OR FOUNDATION WALL (feet)	MAXIMUM NOMINAL THICKNESS OF VENEER (inches)	MAXIMUM WEIGHT OF VENEER (psf)[b]
D_0	1	20[c]	4	40
	2	20[c]	4	40
	3	30[d]	4	40
D_1	1	20[c]	4	40
	2	20[c]	4	40
	3	20[c]	4	40
D_2	1	20[c]	3	30
	2	20[c]	3	30

For SI: 1 inch = 25.4 mm, 1 foot = 304.8 mm, 1 pound per square foot = 0.479 kPa, 1 pound-force = 4.448 N.

a. Cripple walls are not permitted in Seismic Design Categories D_0, D_1 and D_2.

b. Maximum weight is installed weight and includes weight of mortar, grout and lath, and other materials used for installation.

c. The veneer shall not exceed 20 feet in height above a noncombustible foundation, with an additional 8 feet permitted for gable end walls, or 30 feet in height with an additional 8 feet for gable end walls where the lower 10 feet has a backing of concrete or masonry wall. See also story height limitations of Section R301.3.

d. The veneer shall not exceed 30 feet in height above a noncombustible foundation, with an additional 8 feet permitted for gable end walls. See also story height limitations of Section R301.3.

R703.7.1 Interior veneer support. Veneers used as interior wall finishes shall be permitted to be supported on wood or cold-formed steel floors that are designed to support the loads imposed.

R703.7.2 Exterior veneer support. Except in Seismic Design Categories D_0, D_1 and D_2, exterior masonry veneers having an installed weight of 40 pounds per square foot (195 kg/m^2) or less shall be permitted to be supported on wood or cold-formed steel construction. When masonry veneer supported by wood or cold-formed steel construction adjoins masonry veneer supported by the foundation, there shall be a movement joint between the veneer supported by the wood or cold-formed steel construction and the veneer supported by the foundation. The wood or cold-formed steel construction supporting the masonry veneer shall be designed to limit the deflection to $^1/_{600}$ of the span for the supporting members. The design of the wood or cold-formed steel construction shall consider the weight of the veneer and any other loads.

R703.7.2.1 Support by steel angle. A minimum 6 inches by 4 inches by $^5/_{16}$ inch (152 mm by 102 mm by 8 mm) steel angle, with the long leg placed vertically, shall be anchored to double 2 inches by 4 inches (51 mm by 102 mm) wood studs at a maximum on-center spacing of 16 inches (406 mm). Anchorage of the steel angle at every double stud spacing shall be a minimum of two $^{7}/_{16}$ inch (11 mm) diameter by 4 inch (102 mm) lag screws. The steel angle shall have a minimum clearance to underlying construction of $^{1}/_{16}$ inch (2 mm). A minimum of two-thirds the width of the masonry veneer thickness shall bear on the steel angle. Flashing and weep holes shall be located in the masonry veneer wythe in accordance with Figure R703.7.2.1. The maximum height of masonry veneer above the steel angle support shall be 12 feet, 8 inches (3861 mm). The air space separating the masonry veneer from the wood backing shall be in accordance with Sections R703.7.4 and R703.7.4.2. The method of support for the masonry veneer on wood construction shall be constructed in accordance with Figure R703.7.2.1.

The maximum slope of the roof construction without stops shall be 7:12. Roof construction with slopes greater than 7:12 but not more than 12:12 shall have stops of a minimum 3 inch by 3 inch by $^{1}/_{4}$ inch (76 mm by 76 mm by 6 mm) steel plate welded to the angle at 24 inches (610 mm) on center along the angle or as *approved* by the *building official.*

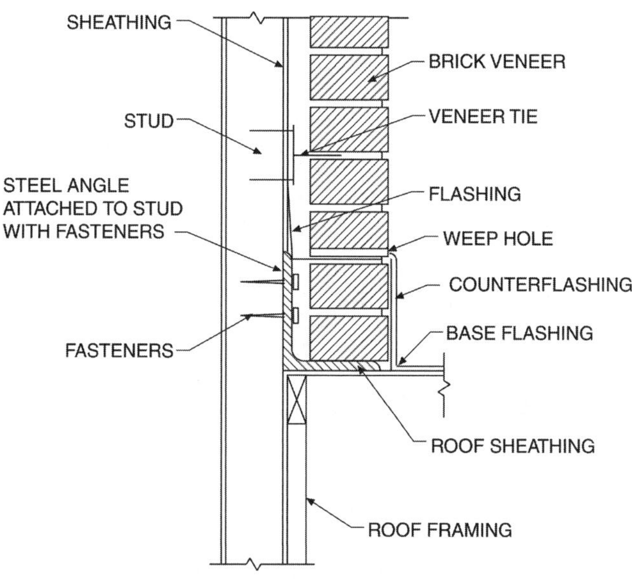

FIGURE R703.7.2.1
EXTERIOR MASONRY VENEER SUPPORT BY STEEL ANGLES

R703.7.2.2 Support by roof construction. A steel angle shall be placed directly on top of the roof construction. The roof supporting construction for the steel angle shall consist of a minimum of three 2 inch by 6 inch (51 mm by 152 mm) wood members. The wood member abutting the vertical wall stud construction shall be anchored with a minimum of three $^{5}/_{8}$-inch (16 mm) diameter by 5-inch (127 mm) lag screws to every wood stud spacing. Each additional roof member shall be anchored by the use of two 10d nails at every wood stud spacing. A minimum of two-thirds the width of the masonry veneer thickness shall bear on the steel angle. Flashing and weep holes shall be located in the masonry veneer wythe in accordance with Figure R703.7.2.2. The maximum height of the masonry veneer above the steel angle support shall be 12 feet, 8 inches (3861 mm). The air space separating the masonry veneer from the wood backing shall be in accordance with Sections R703.7.4 and R703.7.4.2. The support for the masonry veneer on wood construction shall be constructed in accordance with Figure R703.7.2.2.

The maximum slope of the roof construction without stops shall be 7:12. Roof construction with slopes greater than 7:12 but not more than 12:12 shall have stops of a minimum 3 inch by 3 inch by $^{1}/_{4}$ inch (76 mm by 76 mm by 6 mm) steel plate welded to the angle at 24 inches (610 mm) on center along the angle or as *approved* by the *building official.*

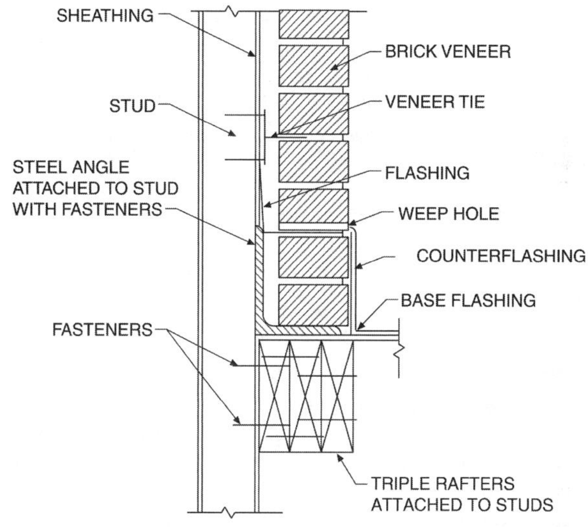

FIGURE R703.7.2.2
EXTERIOR MASONRY VENEER SUPPORT BY ROOF MEMBERS

R703.7.3 Lintels. Masonry veneer shall not support any vertical load other than the dead load of the veneer above. Veneer above openings shall be supported on lintels of noncombustible materials. The lintels shall have a length of bearing not less than 4 inches (102 mm). Steel lintels shall be shop coated with a rust-inhibitive paint, except for lintels made of corrosion-resistant steel or steel treated with coatings to provide corrosion resistance. Construction of openings shall comply with either Section R703.7.3.1 or 703.7.3.2.

R703.7.3.1 Allowable span. The allowable span shall not exceed the values set forth in Table R703.7.3.1.

R703.7.3.2 Maximum span. The allowable span shall not exceed 18 feet 3 inches (5562 mm) and shall be constructed to comply with Figure R703.7.3.2 and the following:

1. Provide a minimum length of 18 inches (457 mm) of masonry veneer on each side of opening as shown in Figure R703.7.3.2.

2. Provide a minimum 5-inch by $3^1/_2$-inch by $^5/_{16}$-inch (127 mm by 89 mm by 7.9 mm) steel angle above the opening and shore for a minimum of 7 days after installation.
3. Provide double-wire joint reinforcement extending 12 inches (305 mm) beyond each side of the opening. Lap splices of joint reinforcement a minimum of 12 inches (305 mm). Comply with one of the following:
   3.1. Double-wire joint reinforcement shall be $^3/_{16}$-inch (4.8 mm) diameter and shall be placed in the first two bed joints above the opening.
   3.2. Double-wire joint reinforcement shall be 9 gauge (0.144 inch or 3.66 mm diameter) and shall be placed in the first three bed joints above the opening.
4. Provide the height of masonry veneer above opening, in accordance with Table R703.7.3.2.

TABLE R703.7.3.2
HEIGHT OF MASONRY VENEER ABOVE OPENING

MINIMUM HEIGHT OF MASONRY VENEER ABOVE OPENING (INCH)	MAXIMUM HEIGHT OF MASONRY VENEER ABOVE OPENING (FEET)
13	< 5
24	5 to < 12
60	12 to height above support allowed by Section R703.7

For SI:1 inch = 25.4 mm, 1 foot = 304.8 mm.

R703.7.4 Anchorage. Masonry veneer shall be anchored to the supporting wall studs with corrosion-resistant metal ties embedded in mortar or grout and extending into the veneer a minimum of $1^1/_2$ inches (38 mm), with not less than $^5/_8$-inch (15.9 mm) mortar or grout cover to outside face. Masonry veneer shall conform to Table R703.7.4.

R703.7.4.1 Size and spacing. Veneer ties, if strand wire, shall not be less in thickness than No. 9 U.S. gage [(0.148 inch) (4 mm)] wire and shall have a hook embedded in the mortar joint, or if sheet metal, shall be not less than No. 22 U.S. gage by [(0.0299 inch) (0.76 mm)] $^7/_8$ inch (22 mm) corrugated. Each tie shall support not more than 2.67 square feet (0.25 m^2) of wall area and shall be spaced not more than 32 inches (813 mm) on center horizontally and 24 inches (635 mm) on center vertically.

Exception: In Seismic Design Category D_0, D_1 or D_2 or townhouses in Seismic Design Category C or in wind areas of more than 30 pounds per square foot pressure (1.44 kPa), each tie shall support not more than 2 square feet (0.2 m^2) of wall area.

R703.7.4.1.1 Veneer ties around wall openings. Additional metal ties shall be provided around all wall openings greater than 16 inches (406 mm) in either dimension. Metal ties around the perimeter of openings shall be spaced not more than 3 feet (9144 mm) on center and placed within 12 inches (305 mm) of the wall opening.

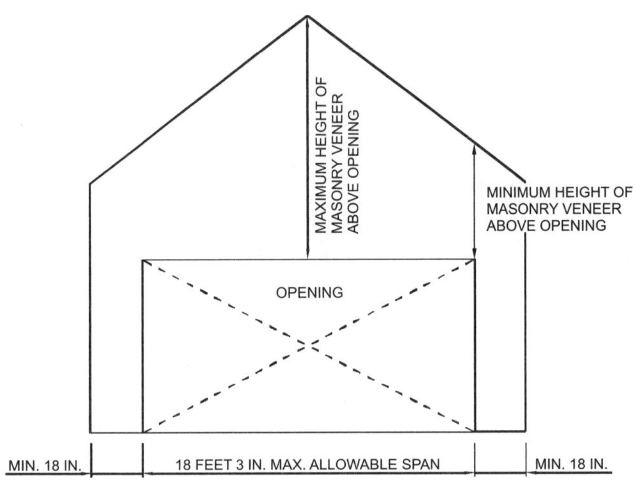

For SI: 1 inch = 25.4 mm, 1 foot = 304.8 mm.

FIGURE R703.7.3.2
MASONRY VENEER OPENING

TABLE R703.7.3.1
ALLOWABLE SPANS FOR LINTELS SUPPORTING MASONRY VENEER[a, b, c, d]

SIZE OF STEEL ANGLE[a, c, d] (inches)	NO STORY ABOVE	ONE STORY ABOVE	TWO STORIES ABOVE	NO. OF $^1/_2$-INCH OR EQUIVALENT REINFORCING BARS IN REINFORCED LINTEL[b, d]
$3 \times 3 \times ^1/_4$	6′-0″	4′-6″	3′-0″	1
$4 \times 3 \times ^1/_4$	8′-0″	6′-0″	4′-6″	1
$5 \times 3^1/_2 \times ^5/_{16}$	10′-0″	8′-0″	6′-0″	2
$6 \times 3^1/_2 \times ^5/_{16}$	14′-0″	9′-6″	7′-0″	2
$2\text{-}6 \times 3^1/_2 \times ^5/_{16}$	20′-0″	12′-0″	9′-6″	4

For SI: 1 inch = 25.4 mm, 1 foot = 304.8 mm.

a. Long leg of the angle shall be placed in a vertical position.

b. Depth of reinforced lintels shall not be less than 8 inches and all cells of hollow masonry lintels shall be grouted solid. Reinforcing bars shall extend not less than 8 inches into the support.

c. Steel members indicated are adequate typical examples; other steel members meeting structural design requirements may be used.

d. Either steel angle or reinforced lintel shall span opening.

**TABLE R703.7.4
TIE ATTACHMENT AND AIR SPACE REQUIREMENTS**

BACKING AND TIE	MINIMUM TIE	MINIMUM TIE FASTENER[a]	AIR SPACE	
Wood stud backing with corrugated sheet metal	22 U.S. gage (0.0299 in.) × $^7/_8$ in. wide	8d common nail [b] ($2^1/_2$ in. × 0.131 in.)	Nominal 1 in. between sheathing and veneer	
Wood stud backing with metal strand wire	W1.7 (No. 9 U.S. gage; 0.148 in.) with hook embedded in mortar joint	8d common nail [b] ($2^1/_2$ in. × 0.131 in.)	Minimum nominal 1 in. between sheathing and veneer	Maximum $4^1/_2$ in. between backing and veneer
Cold-formed steel stud backing with adjustable metal strand wire	W1.7 (No. 9 U.S. gage; 0.148 in.) with hook embedded in mortar joint	No. 10 screw extending through the steel framing a minimum of three exposed threads	Minimum nominal 1 in. between sheathing and veneer	Maximum $4^1/_2$ in. between backing and veneer

For SI: 1 inch = 25.4 mm.

a. In Seismic Design Category D_0, D_1 or D_2, the minimum tie fastener shall be an 8d ring-shank nail ($2^1/_2$ in. × 0.131 in.) or a No. 10 screw extending through the steel framing a minimum of three exposed threads.

b. All fasteners shall have rust-inhibitive coating suitable for the installation in which they are being used, or be manufactured from material not susceptible to corrosion.

R703.7.4.2 Grout fill. As an alternative to the air space required by Table R703.7.4, grout shall be permitted to fill the air space. When the air space is filled with grout, a water-resistive barrier is required over studs or sheathing. When filling the air space, replacing the sheathing and water-resistive barrier with a wire mesh and *approved* water-resistive barrier or an *approved* water-resistive barrier-backed reinforcement attached directly to the studs is permitted.

R703.7.5 Flashing. Flashing shall be located beneath the first course of masonry above finished ground level above the foundation wall or slab and at other points of support, including structural floors, shelf angles and lintels when masonry veneers are designed in accordance with Section R703.7. See Section R703.8 for additional requirements.

R703.7.6 Weepholes. Weepholes shall be provided in the outside wythe of masonry walls at a maximum spacing of 33 inches (838 mm) on center. Weepholes shall not be less than $^3/_{16}$ inch (5 mm) in diameter. Weepholes shall be located immediately above the flashing.

R703.8 Flashing. *Approved* corrosion-resistant flashing shall be applied shingle-fashion in a manner to prevent entry of water into the wall cavity or penetration of water to the building structural framing components. Self-adhered membranes used as flashing shall comply with AAMA 711. The flashing shall extend to the surface of the exterior wall finish. *Approved* corrosion-resistant flashings shall be installed at all of the following locations:

1. Exterior window and door openings. Flashing at exterior window and door openings shall extend to the surface of the exterior wall finish or to the water-resistive barrier for subsequent drainage. Flashing at exterior window and door openings shall be installed in accordance with one or more of the following:
   - 1.1. The fenestration manufacturer's installation and flashing instructions, or for applications not addressed in the fenestration manufacturer's instructions, in accordance with the flashing manufacturer's instructions. Where flashing instructions or details are not provided, pan flashing shall be installed at the sill of exterior window and door openings. Pan flashing shall be sealed or sloped in such a manner as to direct water to the surface of the exterior wall finish or to the water-resistive barrier for subsequent drainage. Openings using pan flashing shall also incorporate flashing or protection at the head and sides.
   - 1.2. In accordance with the flashing design or method of a registered design professional.
   - 1.3. In accordance with other approved methods.
2. At the intersection of chimneys or other masonry construction with frame or stucco walls, with projecting lips on both sides under stucco copings.
3. Under and at the ends of masonry, wood or metal copings and sills.
4. Continuously above all projecting wood trim.
5. Where exterior porches, decks or stairs attach to a wall or floor assembly of wood-frame construction.
6. At wall and roof intersections.
7. At built-in gutters.

R703.9 Exterior insulation and finish system (EIFS)/EIFS with drainage. Exterior Insulation and Finish System (EIFS) shall comply with this chapter and Sections R703.9.1 and R703.9.3. EIFS with drainage shall comply with this chapter and Sections R703.9.2, R703.9.3 and R703.9.4.

R703.9.1 Exterior insulation and finish system (EIFS). EIFS shall comply with ASTM E 2568.

R703.9.2 Exterior insulation and finish system (EIFS) with drainage. EIFS with drainage shall comply with ASTM E 2568 and shall have an average minimum drainage efficiency of 90 percent when tested in accordance with ASTM E 2273.

R703.9.2.1 Water-resistive barrier. The water-resistive barrier shall comply with Section R703.2 or ASTM E 2570.

R703.9.2.2 Installation. The water-resistive barrier shall be applied between the EIFS and the wall sheathing.

R703.9.3 Flashing, general. Flashing of EIFS shall be provided in accordance with the requirements of Section R703.8.

R703.9.4 EIFS/EIFS with drainage installation. All EIFS shall be installed in accordance with the manufacturer's installation instructions and the requirements of this section.

R703.9.4.1 Terminations. The EIFS shall terminate not less than 6 inches (152 mm) above the finished ground level.

R703.9.4.2 Decorative trim. Decorative trim shall not be face nailed though the EIFS.

R703.10 Fiber cement siding.

R703.10.1 Panel siding. Fiber-cement panels shall comply with the requirements of ASTM C 1186, Type A, minimum Grade II. Panels shall be installed with the long dimension either parallel or perpendicular to framing. Vertical and horizontal joints shall occur over framing members and shall be sealed with caulking, covered with battens or shall be designed to comply with Section R703.1. Panel siding shall be installed with fasteners according to Table R703.4 or *approved* manufacturer's installation instructions.

R703.10.2 Lap siding. Fiber-cement lap siding having a maximum width of 12 inches shall comply with the requirements of ASTM C 1186, Type A, minimum Grade II. Lap siding shall be lapped a minimum of $1^1/_4$ inches (32 mm) and lap siding not having tongue-and-groove end joints shall have the ends sealed with caulking, installed with an H-section joint cover, located over a strip of flashing or shall be designed to comply with Section R703.1. Lap siding courses may be installed with the fastener heads exposed or concealed, according to Table R703.4 or *approved* manufacturers' installation instructions.

R703.11 Vinyl siding. Vinyl siding shall be certified and *labeled* as conforming to the requirements of ASTM D 3679 by an *approved* quality control agency.

R703.11.1 Installation. Vinyl siding, soffit and accessories shall be installed in accordance with the manufacturer's installation instructions.

R703.11.1.1 Vinyl soffit panels. Soffit panels shall be individually fastened to a supporting component such as a nailing strip, fascia or subfascia component or as specified by the manufacturer's instructions.

R703.11.2 Foam plastic sheathing. Vinyl siding used with foam plastic sheathing shall be installed in accordance with Section R703.11.2.1, R703.11.2.2, or R703.11.2.3.

Exception: Where the foam plastic sheathing is applied directly over wood structural panels, fiberboard, gypsum sheathing or other *approved* backing capable of independently resisting the design wind pressure, the vinyl siding shall be installed in accordance with Section R703.11.1.

R703.11.2.1 Basic wind speed not exceeding 90 miles per hour and Exposure Category B. Where the basic wind speed does not exceed 90 miles per hour (40 m/s), the Exposure Category is B and gypsum wall board or equivalent is installed on the side of the wall opposite the foam plastic sheathing, the minimum siding fastener penetration into wood framing shall be $1^1/_4$ inches (32 mm) using minimum 0.120-inch diameter nail (shank) with a minimum 0.313-inch diameter head, 16 inches on center. The foam plastic sheathing shall be minimum $^1/_2$-inch-thick (12.7 mm) (nominal) extruded polystyrene per ASTM C 578, $^1/_2$-inch-thick (12.7 mm) (nominal) polyisocyanurate per ASTM C 1289, or 1-inch-thick (25 mm) (nominal) expanded polystyrene per ASTM C 578.

R703.11.2.2 Basic wind speed exceeding 90 miles per hour or Exposure Categories C and D. Where the basic wind speed exceeds 90 miles per hour (40 m/s) or the Exposure Category is C or D, or all conditions of Section R703.11.2.1 are not met, the adjusted design pressure rating for the assembly shall meet or exceed the loads listed in Tables R301.2(2) adjusted for height and exposure using Table R301.2(3). The design wind pressure rating of the vinyl siding for installation over solid sheathing as provided in the vinyl siding manufacturer's product specifications shall be adjusted for the following wall assembly conditions:

1. For wall assemblies with foam plastic sheathing on the exterior side and gypsum wall board or equivalent on the interior side of the wall, the vinyl siding's design wind pressure rating shall be multiplied by 0.39.
2. For wall assemblies with foam plastic sheathing on the exterior side and no gypsum wall board or equivalent on the interior side of wall, the vinyl siding's design wind pressure rating shall be multiplied by 0.27.

R703.11.2.3 Manufacturer specification. Where the vinyl siding manufacturer's product specifications provide an *approved* design wind pressure rating for installation over foam plastic sheathing, use of this design wind pressure rating shall be permitted and the siding shall be installed in accordance with the manufacturer's installation instructions.

R703.12 Adhered masonry veneer installation. Adhered masonry veneer shall be installed in accordance with the manufacturer's instructions.

R703.12.1 Clearances. On exterior stud walls, adhered masonry veneer shall be installed:

1. Minimum of 4 inches (102 mm) above the earth;
2. Minimum of 2 inches (51 mm) above paved areas; or
3. Minimum of $^1/_2$ inch (12 mm) above exterior walking surfaces which are supported by the same foundation that supports the exterior wall.

R703.12.2 Flashing at foundation. A corrosion-resistant screed or flashing of a minimum 0.019-inch (0.48 mm) or 26-gage galvanized or plastic with a minimum vertical attachment flange of $3^1/_2$ inches (89 mm) shall be installed to extend a minimum of 1 inch (25 mm) below the foundation plate line on exterior stud walls in accordance with Section R703.8. The water-resistive barrier, as required by Table R703.4, Footnote w, shall lap over the exterior of the attachment flange of the screed or flashing.

CHAPTER 8
ROOF-CEILING CONSTRUCTION

SECTION R801 GENERAL

R801.1 Application. The provisions of this chapter shall control the design and construction of the roof-ceiling system for all buildings.

R801.2 Requirements. Roof and ceiling construction shall be capable of accommodating all loads imposed according to Section R301 and of transmitting the resulting loads to the supporting structural elements.

R801.3 Roof drainage. In areas where expansive or collapsible soils are known to exist, all *dwellings* shall have a controlled method of water disposal from roofs that will collect and discharge roof drainage to the ground surface at least 5 feet (1524 mm) from foundation walls or to an *approved* drainage system.

SECTION R802 WOOD ROOF FRAMING

R802.1 Identification. Load-bearing dimension lumber for rafters, trusses and ceiling joists shall be identified by a grade mark of a lumber grading or inspection agency that has been approved by an accreditation body that complies with DOC PS 20. In lieu of a grade mark, a certificate of inspection issued by a lumber grading or inspection agency meeting the requirements of this section shall be accepted.

R802.1.1 Blocking. Blocking shall be a minimum of utility grade lumber.

R802.1.2 End-jointed lumber. *Approved* end-jointed lumber identified by a grade mark conforming to Section R802.1 may be used interchangeably with solid-sawn members of the same species and grade. End-jointed lumber used in an assembly required elsewhere in this code to have a fire-resistance rating shall have the designation "Heat-Resistant Adhesive" or "HRA" included in its grade mark.

R802.1.3 Fire-retardant-treated wood. Fire-retardant-treated wood (FRTW) is any wood product which, when impregnated with chemicals by a pressure process or other means during manufacture, shall have, when tested in accordance with ASTM E 84 or UL 723, a listed flame spread index of 25 or less and shows no evidence of significant progressive combustion when the test is continued for an additional 20-minute period. In addition, the flame front shall not progress more than 10.5 feet (3200 mm) beyond the center line of the burners at any time during the test.

R802.1.3.1 Pressure process. For wood products impregnated with chemicals by a pressure process, the process shall be performed in closed vessels under pressures not less than 50 pounds per square inch gauge (psig) (344.7 kPa).

R802.1.3.2 Other means during manufacture. For wood products produced by other means during manufacture the treatment shall be an integral part of the manufacturing process of the wood product. The treatment shall provide permanent protection to all surfaces of the wood product.

R802.1.3.3 Testing. For wood products produced by other means during manufacture, other than a pressure process, all sides of the wood product shall be tested in accordance with and produce the results required in Section R802.1.3. Testing of only the front and back faces of wood structural panels shall be permitted.

R802.1.3.4 Labeling. Fire-retardant-treated lumber and wood structural panels shall be *labeled.* The *label* shall contain:

1. The identification *mark* of an *approved agency* in accordance with Section 1703.5 of the *International Building Code.*
2. Identification of the treating manufacturer.
3. The name of the fire-retardant treatment.
4. The species of wood treated.
5. Flame spread index and smoke-developed index.
6. Method of drying after treatment.
7. Conformance to applicable standards in accordance with Sections R802.1.3.5 through R802.1.3.8.
8. For FRTW exposed to weather, or a damp or wet location, the words "No increase in the listed classification when subjected to the Standard Rain Test" (ASTM D 2898).

R802.1.3.5 Strength adjustments. Design values for untreated lumber and wood structural panels as specified in Section R802.1 shall be adjusted for fire-retardant-treated wood. Adjustments to design values shall be based upon an *approved* method of investigation which takes into consideration the effects of the anticipated temperature and humidity to which the fire-retardant-treated wood will be subjected, the type of treatment and redrying procedures.

R802.1.3.5.1 Wood structural panels. The effect of treatment and the method of redrying after treatment, and exposure to high temperatures and high humidities on the flexure properties of fire-retardant-treated softwood plywood shall be determined in accordance with ASTM D 5516. The test data developed by ASTM D 5516 shall be used to develop adjustment factors, maximum loads and spans, or both for untreated plywood design values in accordance with ASTM D 6305. Each manufacturer shall publish the allowable maximum loads and spans for service as floor and roof sheathing for their treatment.

R802.1.3.5.2 Lumber. For each species of wood treated, the effect of the treatment and the method of redrying after treatment and exposure to high temperatures and high humidities on the allowable design properties of fire-retardant-treated lumber shall be determined in accordance with ASTM D 5664. The test data developed by ASTM D 5664 shall be used to develop modification factors for use at or near room temperature and at elevated temperatures and humidity in accordance with ASTM D 6841. Each manufacturer shall publish the modification factors for service at temperatures of not less than 80°F (27°C) and for roof framing. The roof framing modification factors shall take into consideration the climatological location.

R802.1.3.6 Exposure to weather. Where fire-retardant-treated wood is exposed to weather or damp or wet locations, it shall be identified as "Exterior" to indicate there is no increase in the listed flame spread index as defined in Section R802.1.3 when subjected to ASTM D 2898.

R802.1.3.7 Interior applications. Interior fire-retardant-treated wood shall have a moisture content of not over 28 percent when tested in accordance with ASTM D 3201 procedures at 92 percent relative humidity. Interior fire-retardant-treated wood shall be tested in accordance with Section R802.1.3.5.1 or R802.1.3.5.2. Interior fire-retardant-treated wood designated as Type A shall be tested in accordance with the provisions of this section.

R802.1.3.8 Moisture content. Fire-retardant-treated wood shall be dried to a moisture content of 19 percent or less for lumber and 15 percent or less for wood structural panels before use. For wood kiln dried after treatment (KDAT) the kiln temperatures shall not exceed those used in kiln drying the lumber and plywood submitted for the tests described in Section R802.1.3.5.1 for plywood and R802.1.3.5.2 for lumber.

R802.1.4 Structural glued laminated timbers. Glued laminated timbers shall be manufactured and identified as required in ANSI/AITC A190.1 and ASTM D 3737.

R802.1.5 Structural log members. Stress grading of structural log members of nonrectangular shape, as typically used in log buildings, shall be in accordance with ASTM D 3957. Such structural log members shall be identified by the grade mark of an *approved* lumber grading or inspection agency. In lieu of a grade mark on the material, a certificate of inspection as to species and grade issued by a lumber-grading or inspection agency meeting the requirements of this section shall be permitted to be accepted.

R802.1.6 Structural composite lumber. Structural capacities for structural composite lumber shall be established and monitored in accordance with ASTM D 5456.

R802.2 Design and construction. The framing details required in Section R802 apply to roofs having a minimum slope of three units vertical in 12 units horizontal (25-percent slope) or greater. Roof-ceilings shall be designed and constructed in accordance with the provisions of this chapter and Figures R606.11(1), R606.11(2) and R606.11(3) or in accordance with AFPA/NDS. Components of roof-ceilings shall be fastened in accordance with Table R602.3(1).

R802.3 Framing details. Rafters shall be framed to ridge board or to each other with a gusset plate as a tie. Ridge board shall be at least 1-inch (25 mm) nominal thickness and not less in depth than the cut end of the rafter. At all valleys and hips there shall be a valley or hip rafter not less than 2-inch (51 mm) nominal thickness and not less in depth than the cut end of the rafter. Hip and valley rafters shall be supported at the ridge by a brace to a bearing partition or be designed to carry and distribute the specific load at that point. Where the roof pitch is less than three units vertical in 12 units horizontal (25-percent slope), structural members that support rafters and ceiling joists, such as ridge beams, hips and valleys, shall be designed as beams.

R802.3.1 Ceiling joist and rafter connections. Ceiling joists and rafters shall be nailed to each other in accordance with Table R802.5.1(9), and the rafter shall be nailed to the top wall plate in accordance with Table R602.3(1). Ceiling joists shall be continuous or securely joined in accordance with Table R802.5.1(9) where they meet over interior partitions and are nailed to adjacent rafters to provide a continuous tie across the building when such joists are parallel to the rafters.

Where ceiling joists are not connected to the rafters at the top wall plate, joists connected higher in the *attic* shall be installed as rafter ties, or rafter ties shall be installed to provide a continuous tie. Where ceiling joists are not parallel to rafters, rafter ties shall be installed. Rafter ties shall be a minimum of 2 inches by 4 inches (51 mm by 102 mm) (nominal), installed in accordance with the connection requirements in Table R802.5.1(9), or connections of equivalent capacities shall be provided. Where ceiling joists or rafter ties are not provided, the ridge formed by these rafters shall be supported by a wall or girder designed in accordance with accepted engineering practice.

Collar ties or ridge straps to resist wind uplift shall be connected in the upper third of the *attic* space in accordance with Table R602.3(1).

Collar ties shall be a minimum of 1 inch by 4 inches (25 mm by 102 mm) (nominal), spaced not more than 4 feet (1219 mm) on center.

R802.3.2 Ceiling joists lapped. Ends of ceiling joists shall be lapped a minimum of 3 inches (76 mm) or butted over bearing partitions or beams and toenailed to the bearing member. Where ceiling joists are used to provide resistance to rafter thrust, lapped joists shall be nailed together in accordance with Table R802.5.1(9) and butted joists shall be tied together in a manner to resist such thrust. Joists that do not resist thrust shall be permitted to be nailed in accordance with Table R602.3(1).

R802.4 Allowable ceiling joist spans. Spans for ceiling joists shall be in accordance with Tables R802.4(1) and R802.4(2). For other grades and species and for other loading conditions, refer to the AF&PA *Span Tables for Joists and Rafters.*

R802.5 Allowable rafter spans. Spans for rafters shall be in accordance with Tables R802.5.1(1) through R802.5.1(8). For other grades and species and for other loading conditions, refer to the AF&PA *Span Tables for Joists and Rafters*. The span of each rafter shall be measured along the horizontal projection of the rafter.

R802.5.1 Purlins. Installation of purlins to reduce the span of rafters is permitted as shown in Figure R802.5.1. Purlins shall be sized no less than the required size of the rafters that they support. Purlins shall be continuous and shall be supported by 2-inch by 4-inch (51 mm by 102 mm) braces installed to bearing walls at a slope not less than 45 degrees (0.785 rad) from the horizontal. The braces shall be spaced not more than 4 feet (1219 mm) on center and the unbraced length of braces shall not exceed 8 feet (2438 mm).

TABLE R802.4(1)
CEILING JOIST SPANS FOR COMMON LUMBER SPECIES
(Uninhabitable attics without storage, live load = 10 psf, L/Δ = 240)

CEILING JOIST SPACING (inches)	SPECIES AND GRADE		DEAD LOAD = 5 psf			
			2 × 4	2 × 6	2 × 8	2 × 10
			Maximum ceiling joist spans			
			(feet - inches)	(feet - inches)	(feet - inches)	(feet - inches)
12	Douglas fir-larch	SS	13-2	20-8	Note a	Note a
	Douglas fir-larch	#1	12-8	19-11	Note a	Note a
	Douglas fir-larch	#2	12-5	19-6	25-8	Note a
	Douglas fir-larch	#3	10-10	15-10	20-1	24-6
	Hem-fir	SS	12-5	19-6	25-8	Note a
	Hem-fir	#1	12-2	19-1	25-2	Note a
	Hem-fir	#2	11-7	18-2	24-0	Note a
	Hem-fir	#3	10-10	15-10	20-1	24-6
	Southern pine	SS	12-11	20-3	Note a	Note a
	Southern pine	#1	12-8	19-11	Note a	Note a
	Southern pine	#2	12-5	19-6	25-8	Note a
	Southern pine	#3	11-6	17-0	21-8	25-7
	Spruce-pine-fir	SS	12-2	19-1	25-2	Note a
	Spruce-pine-fir	#1	11-10	18-8	24-7	Note a
	Spruce-pine-fir	#2	11-10	18-8	24-7	Note a
	Spruce-pine-fir	#3	10-10	15-10	20-1	24-6
16	Douglas fir-larch	SS	11-11	18-9	24-8	Note a
	Douglas fir-larch	#1	11-6	18-1	23-10	Note a
	Douglas fir-larch	#2	11-3	17-8	23-0	Note a
	Douglas fir-larch	#3	9-5	13-9	17-5	21-3
	Hem-fir	SS	11-3	17-8	23-4	Note a
	Hem-fir	#1	11-0	17-4	22-10	Note a
	Hem-fir	#2	10-6	16-6	21-9	Note a
	Hem-fir	#3	9-5	13-9	17-5	21-3
	Southern pine	SS	11-9	18-5	24-3	Note a
	Southern pine	#1	11-6	18-1	23-1	Note a
	Southern pine	#2	11-3	17-8	23-4	Note a
	Southern pine	#3	10-0	14-9	18-9	22-2
	Spruce-pine-fir	SS	11-0	17-4	22-10	Note a
	Spruce-pine-fir	#1	10-9	16-11	22-4	Note a
	Spruce-pine-fir	#2	10-9	16-11	22-4	Note a
	Spruce-pine-fir	#3	9-5	13-9	17-5	21-3

(continued)

TABLE R802.4(1)—continued
CEILING JOIST SPANS FOR COMMON LUMBER SPECIES
(Uninhabitable attics without storage, live load = 10 psf, L/Δ = 240)

CEILING JOIST SPACING (inches)	SPECIES AND GRADE		DEAD LOAD = 5 psf			
			2 × 4	2 × 6	2 × 8	2 × 10
			Maximum ceiling joist spans			
			(feet - inches)	(feet - inches)	(feet - inches)	(feet - inches)
19.2	Douglas fir-larch	SS	11-3	17-8	23-3	Note a
	Douglas fir-larch	#1	10-10	17-0	22-5	Note a
	Douglas fir-larch	#2	10-7	16-7	21-0	25-8
	Douglas fir-larch	#3	8-7	12-6	15-10	19-5
	Hem-fir	SS	10-7	16-8	21-11	Note a
	Hem-fir	#1	10-4	16-4	21-6	Note a
	Hem-fir	#2	9-11	15-7	20-6	25-3
	Hem-fir	#3	8-7	12-6	15-10	19-5
	Southern -pine	SS	11-0	17-4	22-10	Note a
	Southern pine	#1	10-10	17-0	22-5	Note a
	Southern pine	#2	10-7	16-8	21-11	Note a
	Southern pine	#3	9-1	13-6	17-2	20-3
	Spruce-pine-fir	SS	10-4	16-4	21-6	Note a
	Spruce-pine-fir	#1	10-2	15-11	21-0	25-8
	Spruce-pine-fir	#2	10-2	15-11	21-0	25-8
	Spruce-pine-fir	#3	8-7	12-6	15-10	19-5
24	Douglas fir-larch	SS	10-5	16-4	21-7	Note a
	Douglas fir-larch	#1	10-0	15-9	20-1	24-6
	Douglas fir-larch	#2	9-10	14-10	18-9	22-11
	Douglas fir-larch	#3	7-8	11-2	14-2	17-4
	Hem-fir	SS	9-10	15-6	20-5	Note a
	Hem-fir	#1	9-8	15-2	19-7	23-11
	Hem-fir	#2	9-2	14-5	18-6	22-7
	Hem-fir	#3	7-8	11-2	14-2	17-4
	Southern pine	SS	10-3	16-1	21-2	Note a
	Southern pine	#1	10-0	15-9	20-10	Note a
	Southern pine	#2	9-10	15-6	20-1	23-11
	Southern pine	#3	8-2	12-0	15-4	18-1
	Spruce-pine-fir	SS	9-8	15-2	19-11	25-5
	Spruce-pine-fir	#1	9-5	14-9	18-9	22-11
	Spruce-pine-fir	#2	9-5	14-9	18-9	22-11
	Spruce-pine-fir	#3	7-8	11-2	14-2	17-4

Check sources for availability of lumber in lengths greater than 20 feet.

For SI: 1 inch = 25.4 mm, 1 foot = 304.8 mm, 1 pound per square foot = 0.0479kPa.

a. Span exceeds 26 feet in length.

TABLE R802.4(2)
CEILING JOIST SPANS FOR COMMON LUMBER SPECIES
(Uninhabitable attics with limited storage, live load = 20 psf, L/Δ = 240)

CEILING JOIST SPACING (inches)	SPECIES AND GRADE		DEAD LOAD = 10 psf			
			2 × 4	2 × 6	2 × 8	2 × 10
			Maximum ceiling joist spans			
			(feet - inches)	(feet - inches)	(feet - inches)	(feet - inches)
12	Douglas fir-larch	SS	10-5	16-4	21-7	Note a
	Douglas fir-larch	#1	10-0	15-9	20-1	24-6
	Douglas fir-larch	#2	9-10	14-10	18-9	22-11
	Douglas fir-larch	#3	7-8	11-2	14-2	17-4
	Hem-fir	SS	9-10	15-6	20-5	Note a
	Hem-fir	#1	9-8	15-2	19-7	23-11
	Hem-fir	#2	9-2	14-5	18-6	22-7
	Hem-fir	#3	7-8	11-2	14-2	17-4
	Southern pine	SS	10-3	16-1	21-2	Note a
	Southern pine	#1	10-0	15-9	20-10	Note a
	Southern pine	#2	9-10	15-6	20-1	23-11
	Southern pine	#3	8-2	12-0	15-4	18-1
	Spruce-pine-fir	SS	9-8	15-2	19-11	25-5
	Spruce-pine-fir	#1	9-5	14-9	18-9	22-11
	Spruce-pine-fir	#2	9-5	14-9	18-9	22-11
	Spruce-pine-fir	#3	7-8	11-2	14-2	17-4
16	Douglas fir-larch	SS	9-6	14-11	19-7	25-0
	Douglas fir-larch	#1	9-1	13-9	17-5	21-3
	Douglas fir-larch	#2	8-9	12-10	16-3	19-10
	Douglas fir-larch	#3	6-8	9-8	12-4	15-0
	Hem-fir	SS	8-11	14-1	18-6	23-8
	Hem-fir	#1	8-9	13-5	16-10	20-8
	Hem-fir	#2	8-4	12-8	16-0	19-7
	Hem-fir	#3	6-8	9-8	12-4	15-0
	Southern pine	SS	9-4	14-7	19-3	24-7
	Southern pine	#1	9-1	14-4	18-11	23-1
	Southern pine	#2	8-11	13-6	17-5	20-9
	Southern pine	#3	7-1	10-5	13-3	15-8
	Spruce-pine-fir	SS	8-9	13-9	18-1	23-1
	Spruce-pine-fir	#1	8-7	12-10	16-3	19-10
	Spruce-pine-fir	#2	8-7	12-10	16-3	19-10
	Spruce-pine-fir	#3	6-8	9-8	12-4	15-0

(continued)

TABLE R802.4(2)—continued
CEILING JOIST SPANS FOR COMMON LUMBER SPECIES
(Uninhabitable attics with limited storage, live load = 20 psf, L/Δ = 240)

CEILING JOIST SPACING (inches)	SPECIES AND GRADE		DEAD LOAD = 10 psf			
			2 × 4	2 × 6	2 × 8	2 × 10
			Maximum ceiling joist spans			
			(feet - inches)	(feet - inches)	(feet - inches)	(feet - inches)
19.2	Douglas fir-larch	SS	8-11	14-0	18-5	23-4
	Douglas fir-larch	#1	8-7	12-6	15-10	19-5
	Douglas fir-larch	#2	8-0	11-9	14-10	18-2
	Douglas fir-larch	#3	6-1	8-10	11-3	13-8
	Hem-fir	SS	8-5	13-3	17-5	22-3
	Hem-fir	#1	8-3	12-3	15-6	18-11
	Hem-fir	#2	7-10	11-7	14-8	17-10
	Hem-fir	#3	6-1	8-10	11-3	13-8
	Southern pine	SS	8-9	13-9	18-1	23-1
	Southern pine	#1	8-7	13-6	17-9	21-1
	Southern pine	#2	8-5	12-3	15-10	18-11
	Southern pine	#3	6-5	9-6	12-1	14-4
	Spruce-pine-fir	SS	8-3	12-11	17-1	21-8
	Spruce-pine-fir	#1	8-0	11-9	14-10	18-2
	Spruce-pine-fir	#2	8-0	11-9	14-10	18-2
	Spruce-pine-fir	#3	6-1	8-10	11-3	13-8
24	Douglas fir-larch	SS	8-3	13-0	17-1	20-11
	Douglas fir-larch	#1	7-8	11-2	14-2	17-4
	Douglas fir-larch	#2	7-2	10-6	13-3	16-3
	Douglas fir-larch	#3	5-5	7-11	10-0	12-3
	Hem-fir	SS	7-10	12-3	16-2	20-6
	Hem-fir	#1	7-6	10-11	13-10	16-11
	Hem-fir	#2	7-1	10-4	13-1	16-0
	Hem-fir	#3	5-5	7-11	10-0	12-3
	Southern pine	SS	8-1	12-9	16-10	21-6
	Southern pine	#1	8-0	12-6	15-10	18-10
	Southern pine	#2	7-8	11-0	14-2	16-11
	Southern pine	#3	5-9	8-6	10-10	12-10
	Spruce-pine-fir	SS	7-8	12-0	15-10	19-5
	Spruce-pine-fir	#1	7-2	10-6	13-3	16-3
	Spruce-pine-fir	#2	7-2	10-6	13-3	16-3
	Spruce-pine-fir	#3	5-5	7-11	10-0	12-3

Check sources for availability of lumber in lengths greater than 20 feet.

For SI: 1 inch = 25.4 mm, 1 foot = 304.8 mm, 1 pound per square foot = 0.0479kPa.

a. Span exceeds 26 feet in length.

TABLE R802.5.1(1)
RAFTER SPANS FOR COMMON LUMBER SPECIES
(Roof live load=20 psf, ceiling not attached to rafters, L/Δ = 180)

RAFTER SPACING (inches)	SPECIES AND GRADE		DEAD LOAD = 10 psf					DEAD LOAD = 20 psf				
			2 × 4	2 × 6	2 × 8	2 × 10	2 × 12	2 × 4	2 × 6	2 × 8	2 × 10	2 × 12
			Maximum rafter spans[a]									
			(feet - inches)	(feet - inches)	(feet - inches)	(feet - inches)	(feet - inches)	(feet - inches)	(feet - inches)	(feet - inches)	(feet - inches)	(feet - inches)
12	Douglas fir-larch	SS	11-6	18-0	23-9	Note b	Note b	11-6	18-0	23-5	Note b	Note b
	Douglas fir-larch	#1	11-1	17-4	22-5	Note b	Note b	10-6	15-4	19-5	23-9	Note b
	Douglas fir-larch	#2	10-10	16-7	21-0	25-8	Note b	9-10	14-4	18-2	22-3	25-9
	Douglas fir-larch	#3	8-7	12-6	15-10	19-5	22-6	7-5	10-10	13-9	16-9	19-6
	Hem-fir	SS	10-10	17-0	22-5	Note b	Note b	10-10	17-0	22-5	Note b	Note b
	Hem-fir	#1	10 -7	16-8	21-10	Note b	Note b	10-3	14-11	18-11	23-2	Note b
	Hem-fir	#2	10-1	15-11	20-8	25-3	Note b	9-8	14-2	17-11	21-11	25-5
	Hem-fir	#3	8-7	12-6	15-10	19-5	22-6	7-5	10-10	13-9	16-9	19-6
	Southern pine	SS	11-3	17-8	23-4	Note b	Note b	11-3	17-8	23-4	Note b	Note b
	Southern pine	#1	11-1	17-4	22-11	Note b	Note b	11-1	17-3	21-9	25-10	Note b
	Southern pine	#2	10-10	17-0	22-5	Note b	Note b	10-6	15-1	19-5	23-2	Note b
	Southern pine	#3	9-1	13-6	17-2	20-3	24-1	7-11	11-8	14-10	17-6	20-11
	Spruce-pine-fir	SS	10-7	16-8	21-11	Note b	Note b	10-7	16-8	21-9	Note b	Note b
	Spruce-pine-fir	#1	10-4	16-3	21-0	25-8	Note b	9-10	14-4	18-2	22-3	25-9
	Spruce-pine-fir	#2	10-4	16-3	21-0	25-8	Note b	9-10	14-4	18-2	22-3	25-9
	Spruce-pine-fir	#3	8-7	12-6	15-10	19-5	22-6	7-5	10-10	13-9	16-9	19-6
16	Douglas fir-larch	SS	10-5	16-4	21-7	Note b	Note b	10-5	16-0	20-3	24-9	Note b
	Douglas fir-larch	#1	10-0	15-4	19-5	23-9	Note b	9-1	13-3	16-10	20-7	23-10
	Douglas fir-larch	#2	9-10	14-4	18-2	22-3	25-9	8-6	12-5	15-9	19-3	22-4
	Douglas fir-larch	#3	7-5	10-10	13-9	16-9	19-6	6-5	9-5	11-11	14-6	16-10
	Hem-fir	SS	9-10	15-6	20-5	Note b	Note b	9-10	15-6	19-11	24-4	Note b
	Hem-fir	#1	9-8	14-11	18-11	23-2	Note b	8-10	12-11	16-5	20-0	23-3
	Hem-fir	#2	9-2	14-2	17-11	21-11	25-5	8-5	12-3	15-6	18-11	22-0
	Hem-fir	#3	7-5	10-10	13-9	16-9	19-6	6-5	9-5	11-11	14-6	16-10
	Southern pine	SS	10-3	16-1	21-2	Note b	Note b	10-3	16-1	21-2	Note b	Note b
	Southern pine	#1	10-0	15-9	20-10	25-10	Note b	10-0	15-0	18-10	22-4	Note b
	Southern pine	#2	9-10	15-1	19-5	23-2	Note b	9-1	13-0	16-10	20-1	23-7
	Southern pine	#3	7-11	11-8	14-10	17-6	20-11	6-10	10-1	12-10	15-2	18-1
	Spruce-pine-fir	SS	9-8	15-2	19-11	25-5	Note b	9-8	14-10	18-10	23-0	Note b
	Spruce-pine-fir	#1	9-5	14-4	18-2	22-3	25-9	8-6	12-5	15-9	19-3	22-4
	Spruce-pine-fir	#2	9-5	14-4	18-2	22-3	25-9	8-6	12-5	15-9	19-3	22-4
	Spruce-pine-fir	#3	7-5	10-10	13-9	16-9	19-6	6-5	9-5	11-11	14-6	16-10
19.2	Douglas fir-larch	SS	9-10	15-5	20-4	25-11	Note b	9-10	14-7	18-6	22-7	Note b
	Douglas fir-larch	#1	9-5	14-0	17-9	21-8	25-2	8-4	12-2	15-4	18-9	21-9
	Douglas fir-larch	#2	8-11	13-1	16-7	20-3	23-6	7-9	11-4	14-4	17-7	20-4
	Douglas fir-larch	#3	6-9	9-11	12-7	15-4	17-9	5-10	8-7	10-10	13-3	15-5
	Hem-fir	SS	9-3	14-7	19-2	24-6	Note b	9-3	14-4	18-2	22-3	25-9
	Hem-fir	#1	9-1	13-8	17-4	21-1	24-6	8-1	11-10	15-0	18-4	21-3
	Hem-fir	#2	8-8	12-11	16-4	20-0	23-2	7-8	11-2	14-2	17-4	20-1
	Hem-fir	#3	6-9	9-11	12-7	15-4	17-9	5-10	8-7	10-10	13-3	15-5
	Southern pine	SS	9-8	15-2	19-11	25-5	Note b	9-8	15-2	19-11	25-5	Note b
	Southern pine	#1	9-5	14-10	19-7	23-7	Note b	9-3	13-8	17-2	20-5	24-4
	Southern pine	#2	9-3	13-9	17-9	21-2	24-10	8-4	11-11	15-4	18-4	21-6
	Southern pine	#3	7-3	10-8	13-7	16-0	19-1	6-3	9-3	11-9	13-10	16-6
	Spruce-pine-fir	SS	9-1	14-3	18-9	23-11	Note b	9-1	13-7	17-2	21-0	24-4
	Spruce-pine-fir	#1	8-10	13-1	16-7	20-3	23-6	7-9	11-4	14-4	17-7	20-4
	Spruce-pine-fir	#2	8-10	13-1	16-7	20-3	23-6	7-9	11-4	14-4	17-7	20-4
	Spruce-pine-fir	#3	6-9	9-11	12-7	15-4	17-9	5-10	8-7	10-10	13-3	15-5

(continued)

TABLE R802.5.1(1)—continued
RAFTER SPANS FOR COMMON LUMBER SPECIES
(Roof live load=20 psf, ceiling not attached to rafters, L/Δ = 180)

RAFTER SPACING (inches)	SPECIES AND GRADE		DEAD LOAD = 10 psf					DEAD LOAD = 20 psf				
			2 × 4	2 × 6	2 × 8	2 × 10	2 × 12	2 × 4	2 × 6	2 × 8	2 × 10	2 × 12
			Maximum rafter spans[a]									
			(feet - inches)	(feet - inches)	(feet - inches)	(feet - inches)	(feet - inches)	(feet - inches)	(feet - inches)	(feet - inches)	(feet - inches)	(feet - inches)
24	Douglas fir-larch	SS	9-1	14-4	18-10	23-4	Note b	8-11	13-1	16-7	20-3	23-5
	Douglas fir-larch	#1	8-7	12-6	15-10	19-5	22-6	7-5	10-10	13-9	16-9	19-6
	Douglas fir-larch	#2	8-0	11-9	14-10	18-2	21-0	6-11	10-2	12-10	15-8	18-3
	Douglas fir-larch	#3	6-1	8-10	11-3	13-8	15-11	5-3	7-8	9-9	11-10	13-9
	Hem-fir	SS	8-7	13-6	17-10	22-9	Note b	8-7	12-10	16-3	19-10	23-0
	Hem-fir	#1	8-4	12-3	15-6	18-11	21-11	7-3	10-7	13-5	16-4	19-0
	Hem-fir	#2	7-11	11-7	14-8	17-10	20-9	6-10	10-0	12-8	15-6	17-11
	Hem-fir	#3	6-1	8-10	11-3	13-8	15-11	5-3	7-8	9-9	11-10	13-9
	Southern pine	SS	8-11	14-1	18-6	23-8	Note b	8-11	14-1	18-6	22-11	Note b
	Southern pine	#1	8-9	13-9	17-9	21-1	25-2	8-3	12-3	15-4	18-3	21-9
	Southern pine	#2	8-7	12-3	15-10	18-11	22-2	7-5	10-8	13-9	16-5	19-3
	Southern pine	#3	6-5	9-6	12-1	14-4	17-1	5-7	8-3	10-6	12-5	14-9
	Spruce-pine-fir	SS	8-5	13-3	17-5	21-8	25-2	8-4	12-2	15-4	18-9	21-9
	Spruce-pine-fir	#1	8-0	11-9	14-10	18-2	21-0	6-11	10-2	12-10	15-8	18-3
	Spruce-pine-fir	#2	8-0	11-9	14-10	18-2	21-0	6-11	10-2	12-10	15-8	18-3
	Spruce-pine-fir	#3	6-1	8-10	11-3	13-8	15-11	5-3	7-8	9-9	11-10	13-9

Check sources for availability of lumber in lengths greater than 20 feet.

For SI: 1 inch = 25.4 mm, 1 foot = 304.8 mm, 1 pound per square foot = 0.0479 kPa.

a. The tabulated rafter spans assume that ceiling joists are located at the bottom of the attic space or that some other method of resisting the outward push of the rafters on the bearing walls, such as rafter ties, is provided at that location. When ceiling joists or rafter ties are located higher in the attic space, the rafter spans shall be multiplied by the factors given below:

H_C/H_R	Rafter Span Adjustment Factor
1/3	0.67
1/4	0.76
1/5	0.83
1/6	0.90
1/7.5 or less	1.00

where:

H_C = Height of ceiling joists or rafter ties measured vertically above the top of the rafter support walls.

H_R = Height of roof ridge measured vertically above the top of the rafter support walls.

b. Span exceeds 26 feet in length.

TABLE R802.5.1(2)
RAFTER SPANS FOR COMMON LUMBER SPECIES
(Roof live load=20 psf, ceiling attached to rafters, L/Δ = 240)

RAFTER SPACING (inches)	SPECIES AND GRADE	DEAD LOAD = 10 psf					DEAD LOAD = 20 psf				
		2 × 4	2 × 6	2 × 8	2 × 10	2 × 12	2 × 4	2 × 6	2 × 8	2 × 10	2 × 12
		Maximum rafter spans[a]									
		(feet - inches)	(feet - inches)	(feet - inches)	(feet - inches)	(feet - inches)	(feet - inches)	(feet - inches)	(feet - inches)	(feet - inches)	(feet - inches)
12	Douglas fir-larch SS	10-5	16-4	21-7	Note b	Note b	10-5	16-4	21-7	Note b	Note b
	Douglas fir-larch #1	10-0	15-9	20-10	Note b	Note b	10-0	15-4	19-5	23-9	Note b
	Douglas fir-larch #2	9-10	15-6	20-5	25-8	Note b	9-10	14-4	18-2	22-3	25-9
	Douglas fir-larch #3	8-7	12-6	15-10	19-5	22-6	7-5	10-10	13-9	16-9	19-6
	Hem-fir SS	9-10	15-6	20-5	Note b	Note b	9-10	15-6	20-5	Note b	Note b
	Hem-fir #1	9-8	15-2	19-11	25-5	Note b	9-8	14-11	18-11	23-2	Note b
	Hem-fir #2	9-2	14-5	19-0	24-3	Note b	9-2	14-2	17-11	21-11	25-5
	Hem-fir #3	8-7	12-6	15-10	19-5	22-6	7-5	10-10	13-9	16-9	19-6
	Southern pine SS	10-3	16-1	21-2	Note b	Note b	10-3	16-1	21-2	Note b	Note b
	Southern pine #1	10-0	15-9	20-10	Note b	Note b	10-0	15-9	20-10	25-10	Note b
	Southern pine #2	9-10	15-6	20-5	Note b	Note b	9-10	15-1	19-5	23-2	Note b
	Southern pine #3	9-1	13-6	17-2	20-3	24-1	7-11	11-8	14-10	17-6	20-11
	Spruce-pine-fir SS	9-8	15-2	19-11	25-5	Note b	9-8	15-2	19-11	25-5	Note b
	Spruce-pine-fir #1	9-5	14-9	19-6	24-10	Note b	9-5	14-4	18-2	22-3	25-9
	Spruce-pine-fir #2	9-5	14-9	19-6	24-10	Note b	9-5	14-4	18-2	22-3	25-9
	Spruce-pine-fir #3	8-7	12-6	15-10	19-5	22-6	7-5	10-10	13-9	16-9	19-6
16	Douglas fir-larch SS	9-6	14-11	19-7	25-0	Note b	9-6	14-11	19-7	24-9	Note b
	Douglas fir-larch #1	9-1	14-4	18-11	23-9	Note b	9-1	13-3	16-10	20-7	23-10
	Douglas fir-larch #2	8-11	14-1	18-2	22-3	25-9	8-6	12-5	15-9	19-3	22-4
	Douglas fir-larch #3	7-5	10-10	13-9	16-9	19-6	6-5	9-5	11-11	14-6	16-10
	Hem-fir SS	8-11	14-1	18-6	23-8	Note b	8-11	14-1	18-6	23-8	Note b
	Hem-fir #1	8-9	13-9	18-1	23-1	Note b	8-9	12-11	16-5	20-0	23-3
	Hem-fir #2	8-4	13-1	17-3	21-11	25-5	8-4	12-3	15-6	18-11	22-0
	Hem-fir #3	7-5	10-10	13-9	16-9	19-6	6-5	9-5	11-11	14-6	16-10
	Southern pine SS	9-4	14-7	19-3	24-7	Note b	9-4	14-7	19-3	24-7	Note b
	Southern pine #1	9-1	14-4	18-11	24-1	Note b	9-1	14-4	18-10	22-4	Note b
	Southern pine #2	8-11	14-1	18-6	23-2	Note b	8-11	13-0	16-10	20-1	23-7
	Southern pine #3	7-11	11-8	14-10	17-6	20-11	6-10	10-1	12-10	15-2	18-1
	Spruce-pine-fir SS	8-9	13-9	18-1	23-1	Note b	8-9	13-9	18-1	23-0	Note b
	Spruce-pine-fir #1	8-7	13-5	17-9	22-3	25-9	8-6	12-5	15-9	19-3	22-4
	Spruce-pine-fir #2	8-7	13-5	17-9	22-3	25-9	8-6	12-5	15-9	19-3	22-4
	Spruce-pine-fir #3	7-5	10-10	13-9	16-9	19-6	6-5	9-5	11-11	14-6	16-10
19.2	Douglas fir-larch SS	8-11	14-0	18-5	23-7	Note b	8-11	14-0	18-5	22-7	Note b
	Douglas fir-larch #1	8-7	13-6	17-9	21-8	25-2	8-4	12-2	15-4	18-9	21-9
	Douglas fir-larch #2	8-5	13-1	16-7	20-3	23-6	7-9	11-4	14-4	17-7	20-4
	Douglas fir-larch #3	6-9	9-11	12-7	15-4	17-9	5-10	8-7	10-10	13-3	15-5
	Hem-fir SS	8-5	13-3	17-5	22-3	Note b	8-5	13-3	17-5	22-3	25-9
	Hem-fir #1	8-3	12-11	17-1	21-1	24-6	8-1	11-10	15-0	18-4	21-3
	Hem-fir #2	7-10	12-4	16-3	20-0	23-2	7-8	11-2	14-2	17-4	20-1
	Hem-fir #3	6-9	9-11	12-7	15-4	17-9	5-10	8-7	10-10	13-3	15-5

(continued)

TABLE R802.5.1(2)—continued
RAFTER SPANS FOR COMMON LUMBER SPECIES
(Roof live load=20 psf, ceiling attached to rafters, L/Δ = 240)

RAFTER SPACING (inches)	SPECIES AND GRADE		DEAD LOAD = 10 psf					DEAD LOAD = 20 psf				
			2 × 4	2 × 6	2 × 8	2 × 10	2 × 12	2 × 4	2 × 6	2 × 8	2 × 10	2 × 12
			Maximum rafter spans[a]									
			(feet - inches)	(feet - inches)	(feet - inches)	(feet - inches)	(feet - inches)	(feet - inches)	(feet - inches)	(feet - inches)	(feet - inches)	(feet - inches)
19.2	Southern pine	SS	8-9	13-9	18-1	23-1	Note b	8-9	13-9	18-1	23-1	Note b
	Southern pine	#1	8-7	13-6	17-9	22-8	Note b	8-7	13-6	17-2	20-5	24-4
	Southern pine	#2	8-5	13-3	17-5	21-2	24-10	8-4	11-11	15-4	18-4	21-6
	Southern pine	#3	7-3	10-8	13-7	16-0	19-1	6-3	9-3	11-9	13-10	16-6
	Spruce-pine-fir	SS	8-3	12-11	17-1	21-9	Note b	8-3	12-11	17-1	21-0	24-4
	Spruce-pine-fir	#1	8-1	12-8	16-7	20-3	23-6	7-9	11-4	14-4	17-7	20-4
	Spruce-pine-fir	#2	8-1	12-8	16-7	20-3	23-6	7-9	11-4	14-4	17-7	20-4
	Spruce-pine-fir	#3	6-9	9-11	12-7	15-4	17-9	5-10	8-7	10-10	13-3	15-5
24	Douglas fir-larch	SS	8-3	13-0	17-2	2 1-10	Note b	8-3	13-0	16-7	20-3	23-5
	Douglas fir-larch	#1	8-0	12-6	15-10	19-5	22-6	7-5	10-10	13-9	16-9	19-6
	Douglas fir-larch	#2	7-10	11-9	14-10	18-2	21-0	6-11	10-2	12-10	15-8	18-3
	Douglas fir-larch	#3	6-1	8-10	11-3	13-8	15-11	5-3	7-8	9-9	11-10	13-9
	Hem-fir	SS	7-10	12-3	16-2	20-8	25-1	7-10	12-3	16-2	19-10	23-0
	Hem-fir	#1	7-8	12-0	15-6	18-11	21-11	7-3	10-7	13-5	16-4	19-0
	Hem-fir	#2	7-3	11-5	14-8	17-10	20-9	6-10	10-0	12-8	15-6	17-11
	Hem-fir	#3	6-1	8-10	11-3	13-8	15-11	5-3	7-8	9-9	11-10	13-9
	Southern pine	SS	8-1	12-9	16-10	21-6	Note b	8-1	12-9	16-10	21-6	Note b
	Southern pine	#1	8-0	12-6	16-6	21-1	25-2	8-0	12-3	15-4	18-3	21-9
	Southern pine	#2	7-10	12-3	15-10	18-11	22-2	7-5	10-8	13-9	16-5	19-3
	Southern pine	#3	6-5	9-6	12-1	14-4	17-1	5-7	8-3	10-6	12-5	14-9
	Spruce-pine-fir	SS	7-8	12-0	15-10	20-2	24-7	7-8	12-0	15-4	18-9	21-9
	Spruce-pine-fir	#1	7-6	11-9	14-10	18-2	21-0	6-11	10-2	12-10	15-8	18-3
	Spruce-pine-fir	#2	7-6	11-9	14-10	18-2	21-0	6-11	10-2	12-10	15-8	18-3
	Spruce-pine-fir	#3	6-1	8-10	11-3	13-8	15-11	5-3	7-8	9-9	11-10	13-9

Check sources for availability of lumber in lengths greater than 20 feet.

For SI: 1 inch = 25.4 mm, 1 foot = 304.8 mm, 1 pound per square foot = 0.0479 kPa.

a. The tabulated rafter spans assume that ceiling joists are located at the bottom of the attic space or that some other method of resisting the outward push of the rafters on the bearing walls, such as rafter ties, is provided at that location. When ceiling joists or rafter ties are located higher in the attic space, the rafter spans shall be multiplied by the factors given below:

H_C/H_R	Rafter Span Adjustment Factor
1/3	0.67
1/4	0.76
1/5	0.83
1/6	0.90
1/7.5 or less	1.00

where:

H_C = Height of ceiling joists or rafter ties measured vertically above the top of the rafter support walls.

H_R = Height of roof ridge measured vertically above the top of the rafter support walls.

b. Span exceeds 26 feet in length.

TABLE R802.5.1(3)
RAFTER SPANS FOR COMMON LUMBER SPECIES
(Ground snow load=30 psf, ceiling not attached to rafters, L/Δ = 180)

RAFTER SPACING (inches)	SPECIES AND GRADE		DEAD LOAD = 10 psf					DEAD LOAD = 20 psf				
			2 × 4	2 × 6	2 × 8	2 × 10	2 × 12	2 × 4	2 × 6	2 × 8	2 × 10	2 × 12
			Maximum rafter spans[a]									
			(feet - inches)	(feet - inches)	(feet - inches)	(feet - inches)	(feet - inches)	(feet - inches)	(feet - inches)	(feet - inches)	(feet - inches)	(feet - inches)
12	Douglas fir-larch	SS	10-0	15-9	20-9	Note b	Note b	10-0	15-9	20-1	24-6	Note b
	Douglas fir-larch	#1	9-8	14-9	18-8	22-9	Note b	9-0	13-2	16-8	20-4	23-7
	Douglas fir-larch	#2	9-5	13-9	17-5	21-4	24-8	8-5	12-4	15-7	19-1	22-1
	Douglas fir-larch	#3	7-1	10-5	13-2	16-1	18-8	6-4	9-4	11-9	14-5	16-8
	Hem-fir	SS	9-6	14-10	19-7	25-0	Note b	9-6	14-10	19-7	24-1	Note b
	Hem-fir	#1	9-3	14-4	18-2	22-2	25-9	8-9	12-10	16-3	19-10	23-0
	Hem-fir	#2	8-10	13-7	17-2	21-0	24-4	8-4	12-2	15-4	18-9	21-9
	Hem-fir	#3	7-1	10-5	13-2	16-1	18-8	6-4	9-4	11-9	14-5	16-8
	Southern pine	SS	9-10	15-6	20-5	Note b	Note b	9-10	15-6	20-5	Note b	Note b
	Southern pine	#1	9-8	15-2	20-0	24-9	Note b	9-8	14-10	18-8	22-2	Note b
	Southern pine	#2	9-6	14-5	18-8	22-3	Note b	9-0	12-11	16-8	19-11	23-4
	Southern pine	#3	7-7	11-2	14-3	16-10	20-0	6-9	10-0	12-9	15-1	17-11
	Spruce-pine-fir	SS	9-3	14-7	19-2	24-6	Note b	9-3	14-7	18-8	22-9	Note b
	Spruce-pine-fir	#1	9-1	13-9	17-5	21-4	24-8	8-5	12-4	15-7	19-1	22-1
	Spruce-pine-fir	#2	9-1	13-9	17-5	21-4	24-8	8-5	12-4	15-7	19-1	22-1
	Spruce-pine-fir	#3	7-1	10-5	13-2	16-1	18-8	6-4	9-4	11-9	14-5	16-8
16	Douglas fir-larch	SS	9-1	14-4	18-10	23-9	Note b	9-1	13-9	17-5	21-3	24-8
	Douglas fir-larch	#1	8-9	12-9	16-2	19-9	22-10	7-10	11-5	14-5	17-8	20-5
	Douglas fir-larch	#2	8-2	11-11	15-1	18-5	21-5	7-3	10-8	13-6	16-6	19-2
	Douglas fir-larch	#3	6-2	9-0	11-5	13-11	16-2	5-6	8-1	10-3	12-6	14-6
	Hem-fir	SS	8-7	13-6	17-10	22-9	Note b	8-7	13-6	17-1	20-10	24-2
	Hem-fir	#1	8-5	12-5	15-9	19-3	22-3	7-7	11-1	14-1	17-2	19-11
	Hem-fir	#2	8-0	11-9	14-11	18-2	21-1	7-2	10-6	13-4	16-3	18-10
	Hem-fir	#3	6-2	9-0	11-5	13-11	16-2	5-6	8-1	10-3	12-6	14-6
	Southern pine	SS	8-11	14-1	18-6	23-8	Note b	8-11	14-1	18-6	23-8	Note b
	Southern pine	#1	8-9	13-9	18-1	21-5	25-7	8-8	12-10	16-2	19-2	22-10
	Southern pine	#2	8-7	12-6	16-2	19-3	22-7	7-10	11-2	14-5	17-3	20-2
	Southern pine	#3	6-7	9-8	12-4	14-7	17-4	5-10	8-8	11-0	13-0	15-6
	Spruce-pine-fir	SS	8-5	13-3	17-5	22-1	25-7	8-5	12-9	16-2	19-9	22-10
	Spruce-pine-fir	#1	8-2	11-11	15-1	18-5	21-5	7-3	10-8	13-6	16-6	19-2
	Spruce-pine-fir	#2	8-2	11-11	15-1	18-5	21-5	7-3	10-8	13-6	16-6	19-2
	Spruce-pine-fir	#3	6-2	9-0	11-5	13-11	16-2	5-6	8-1	10-3	12-6	14-6
19.2	Douglas fir-larch	SS	8-7	13-6	17-9	21-8	25-2	8-7	12-6	15-10	19-5	22-6
	Douglas fir-larch	#1	7-11	11-8	14-9	18-0	20-11	7-1	10-5	13-2	16-1	18-8
	Douglas fir-larch	#2	7-5	10-11	13-9	16-10	19-6	6-8	9-9	12-4	15-1	17-6
	Douglas fir-larch	#3	5-7	8-3	10-5	12-9	14-9	5-0	7-4	9-4	11-5	13-2
	Hem-fir	SS	8-1	12-9	16-9	21-4	24-8	8-1	12-4	15-7	19-1	22-1
	Hem-fir	#1	7-9	11-4	14-4	17-7	20-4	6-11	10-2	12-10	15-8	18-2
	Hem-fir	#2	7-4	10-9	13-7	16-7	19-3	6-7	9-7	12-2	14-10	17-3
	Hem-fir	#3	5-7	8-3	10-5	12-9	14-9	5-0	7-4	9-4	11-5	13-2

(continued)

TABLE R802.5.1(3)—continued
RAFTER SPANS FOR COMMON LUMBER SPECIES
(Ground snow load=30 psf, ceiling not attached to rafters, L/Δ = 180)

RAFTER SPACING (inches)	SPECIES AND GRADE		DEAD LOAD = 10 psf					DEAD LOAD = 20 psf				
			2 × 4	2 × 6	2 × 8	2 × 10	2 × 12	2 × 4	2 × 6	2 × 8	2 × 10	2 × 12
			Maximum rafter spans[a]									
			(feet - inches)	(feet - inches)	(feet - inches)	(feet - inches)	(feet - inches)	(feet - inches)	(feet - inches)	(feet - inches)	(feet - inches)	(feet - inches)
19.2	Southern pine	SS	8-5	13-3	17-5	22-3	Note b	8-5	13-3	17-5	22-0	25-9
	Southern pine	#1	8-3	13-0	16-6	19-7	23-4	7-11	11-9	14-9	17-6	20-11
	Southern pine	#2	7-11	11-5	14-9	17-7	20-7	7-1	10-2	13-2	15-9	18-5
	Southern pine	#3	6-0	8-10	11-3	13-4	15-10	5-4	7-11	10-1	11-11	14-2
	Spruce-pine-fir	SS	7-11	12-5	16-5	20-2	23-4	7-11	11-8	14-9	18-0	20-11
	Spruce-pine-fir	#1	7-5	10-11	13-9	16-10	19-6	6-8	9-9	12-4	15-1	17-6
	Spruce-pine-fir	#2	7-5	10-11	13-9	16-10	19-6	6-8	9-9	12-4	15-1	17-6
	Spruce-pine-fir	#3	5-7	8-3	10-5	12-9	14-9	5-0	7-4	9-4	11-5	13-2
24	Douglas fir-larch	SS	7-11	12-6	15-10	19-5	22-6	7-8	11-3	14-2	17-4	20-1
	Douglas fir-larch	#1	7-1	10-5	13-2	16-1	18-8	6-4	9-4	11-9	14-5	16-8
	Douglas fir-larch	#2	6-8	9-9	12-4	15-1	17-6	5-11	8-8	11-0	13-6	15-7
	Douglas fir-larch	#3	5-0	7-4	9-4	11-5	13-2	4-6	6-7	8-4	10-2	11-10
	Hem-fir	SS	7-6	11-10	15-7	19-1	22-1	7-6	11-0	13-11	17-0	19-9
	Hem-fir	#1	6-11	10-2	12-10	15-8	18-2	6-2	9-1	11-6	14-0	16-3
	Hem-fir	#2	6-7	9-7	12-2	14-10	17-3	5-10	8-7	10-10	13-3	15-5
	Hem-fir	#3	5-0	7-4	9-4	11-5	13-2	4-6	6-7	8-4	10-2	11-10
	Southern pine	SS	7-10	12-3	16-2	20-8	25-1	7-10	12-3	16-2	19-8	23-0
	Southern pine	#1	7-8	11-9	14-9	17-6	20-11	7-1	10-6	13-2	15-8	18-8
	Southern pine	#2	7-1	10-2	13-2	15-9	18-5	6-4	9-2	11-9	14-1	16-6
	Southern pine	#3	5-4	7-11	10-1	11-11	14-2	4-9	7-1	9-0	10-8	12-8
	Spruce-pine-fir	SS	7-4	11-7	14-9	18-0	20-11	7-1	10-5	13-2	16-1	18-8
	Spruce-pine-fir	#1	6-8	9-9	12-4	15-1	17-6	5-11	8-8	11-0	13-6	15-7
	Spruce-pine-fir	#2	6-8	9-9	12-4	15-1	17-6	5-11	8-8	11-0	13-6	15-7
	Spruce-pine-fir	#3	5-0	7-4	9-4	11-5	13-2	4-6	6-7	8-4	10-2	11-10

Check sources for availability of lumber in lengths greater than 20 feet.

For SI: 1 inch = 25.4 mm, 1 foot = 304.8 mm, 1 pound per square foot = 0.0479 kPa.

a. The tabulated rafter spans assume that ceiling joists are located at the bottom of the attic space or that some other method of resisting the outward push of the rafters on the bearing walls, such as rafter ties, is provided at that location. When ceiling joists or rafter ties are located higher in the attic space, the rafter spans shall be multiplied by the factors given below:

H_C/H_R	Rafter Span Adjustment Factor
1/3	0.67
1/4	0.76
1/5	0.83
1/6	0.90
1/7.5 or less	1.00

where:

H_C = Height of ceiling joists or rafter ties measured vertically above the top of the rafter support walls.

H_R = Height of roof ridge measured vertically above the top of the rafter support walls.

b. Span exceeds 26 feet in length.

TABLE R802.5.1(4)
RAFTER SPANS FOR COMMON LUMBER SPECIES
(Ground snow load=50 psf, ceiling not attached to rafters, L/Δ = 180)

RAFTER SPACING (inches)	SPECIES AND GRADE		DEAD LOAD = 10 psf					DEAD LOAD = 20 psf				
			2 × 4	2 × 6	2 × 8	2 × 10	2 × 12	2 × 4	2 × 6	2 × 8	2 × 10	2 × 12
			Maximum rafter spans[a]									
			(feet - inches)	(feet - inches)	(feet - inches)	(feet - inches)	(feet - inches)	(feet - inches)	(feet - inches)	(feet - inches)	(feet - inches)	(feet - inches)
12	Douglas fir-larch	SS	8-5	13-3	17-6	22-4	26-0	8-5	13-3	17-0	20-9	24-0
	Douglas fir-larch	#1	8-2	12-0	15-3	18-7	21-7	7-7	11-2	14-1	17-3	20-0
	Douglas fir-larch	#2	7-8	11-3	14-3	17-5	20-2	7-1	10-5	13-2	16-1	18-8
	Douglas fir-larch	#3	5-10	8-6	10-9	13-2	15-3	5-5	7-10	10-0	12-2	14-1
	Hem-fir	SS	8-0	12-6	16-6	21-1	25-6	8-0	12-6	16-6	20-4	23-7
	Hem-fir	#1	7-10	11-9	14-10	18-1	21-0	7-5	10-10	13-9	16-9	19-5
	Hem-fir	#2	7-5	11-1	14-0	17-2	19-11	7-0	10-3	13-0	15-10	18-5
	Hem-fir	#3	5-10	8-6	10-9	13-2	15-3	5-5	7-10	10-0	12-2	14-1
	Southern pine	SS	8-4	13-0	17-2	21-11	Note b	8-4	13-0	17-2	21-11	Note b
	Southern pine	#1	8-2	12-10	16-10	20-3	24-1	8-2	12-6	15-9	18-9	22-4
	Southern pine	#2	8-0	11-9	15-3	18-2	21-3	7-7	10-11	14-1	16-10	19-9
	Southern pine	#3	6-2	9-2	11-8	13-9	16-4	5-9	8-5	10-9	12-9	15-2
	Spruce-pine-fir	SS	7-10	12-3	16-2	20-8	24-1	7-10	12-3	15-9	19-3	22-4
	Spruce-pine-fir	#1	7-8	11-3	14-3	17-5	20-2	7-1	10-5	13-2	16-1	18-8
	Spruce-pine-fir	#2	7-8	11-3	14-3	17-5	20-2	7-1	10-5	13-2	16-1	18-8
	Spruce-pine-fir	#3	5-10	8-6	10-9	13-2	15-3	5-5	7-10	10-0	12-2	14-1
16	Douglas fir-larch	SS	7-8	12-1	15-10	19-5	22-6	7-8	11-7	14-8	17-11	20-10
	Douglas fir-larch	#1	7-1	10-5	13-2	16-1	18-8	6-7	9-8	12-2	14-11	17-3
	Douglas fir-larch	#2	6-8	9-9	12-4	15-1	17-6	6-2	9-0	11-5	13-11	16-2
	Douglas fir-larch	#3	5-0	7-4	9-4	11-5	13-2	4-8	6-10	8-8	10-6	12-3
	Hem-fir	SS	7-3	11-5	15-0	19-1	22-1	7-3	11-5	14-5	17-8	20-5
	Hem-fir	#1	6-11	10-2	12-10	15-8	18-2	6-5	9-5	11-11	14-6	16-10
	Hem-fir	#2	6-7	9-7	12-2	14-10	17-3	6-1	8-11	11-3	13-9	15-11
	Hem-fir	#3	5-0	7-4	9-4	11-5	13-2	4-8	6-10	8-8	10-6	12-3
	Southern pine	SS	7-6	11-10	15-7	19-11	24-3	7-6	11-10	15-7	19-11	23-10
	Southern pine	#1	7-5	11-7	14-9	17-6	20-11	7-4	10-10	13-8	16-2	19-4
	Southern pine	#2	7-1	10-2	13-2	15-9	18-5	6-7	9-5	12-2	14-7	17-1
	Southern pine	#3	5-4	7-11	10-1	11-11	14-2	4-11	7-4	9-4	11-0	13-1
	Spruce-pine-fir	SS	7-1	11-2	14-8	18-0	20-11	7-1	10-9	13-8	15-11	19-4
	Spruce-pine-fir	#1	6-8	9-9	12-4	15-1	17-6	6-2	9-0	11-5	13-11	16-2
	Spruce-pine-fir	#2	6-8	9-9	12-4	15-1	17-6	6-2	9-0	11-5	13-11	16-2
	Spruce-pine-fir	#3	5-0	7-4	9-4	11-5	13-2	4-8	6-10	8-8	10-6	12-3
19.2	Douglas fir-larch	SS	7-3	11-4	14-6	17-8	20-6	7-3	10-7	13-5	16-5	19-0
	Douglas fir-larch	#1	6-6	9-6	12-0	14-8	17-1	6-0	8-10	11-2	13-7	15-9
	Douglas fir-larch	#2	6-1	8-11	11-3	13-9	15-11	5-7	8-3	10-5	12-9	14-9
	Douglas fir-larch	#3	4-7	6-9	8-6	10-5	12-1	4-3	6-3	7-11	9-7	11-2
	Hem-fir	SS	6-10	10-9	14-2	17-5	20-2	6-10	10-5	13-2	16-1	18-8
	Hem-fir	#1	6-4	9-3	11-9	14-4	16-7	5-10	8-7	10-10	13-3	15-5
	Hem-fir	#2	6-0	8-9	11-1	13-7	15-9	5-7	8-1	10-3	12-7	14-7
	Hem-fir	#3	4-7	6-9	8-6	10-5	12-1	4-3	6-3	7-11	9-7	11-2

(continued)

TABLE R802.5.1(4)
RAFTER SPANS FOR COMMON LUMBER SPECIES
(Ground snow load=50 psf, ceiling not attached to rafters, L/Δ = 180)

RAFTER SPACING (inches)	SPECIES AND GRADE		DEAD LOAD = 10 psf					DEAD LOAD = 20 psf				
			2 × 4	2 × 6	2 × 8	2 × 10	2 × 12	2 × 4	2 × 6	2 × 8	2 × 10	2 × 12
			Maximum rafter spans[a]									
			(feet - inches)	(feet - inches)	(feet - inches)	(feet - inches)	(feet - inches)	(feet - inches)	(feet - inches)	(feet - inches)	(feet - inches)	(feet - inches)
19.2	Southern pine	SS	7-1	11-2	14-8	18-9	22-10	7-1	11-2	14-8	18 7	21-9
	Southern pine	#1	7-0	10-8	13-5	16-0	19-1	6-8	9-11	12-5	14-10	17-8
	Southern pine	#2	6-6	9-4	12-0	14-4	16-10	6-0	8-8	11-2	13-4	15-7
	Southern pine	#3	4-11	7-3	9-2	10-10	12-11	4-6	6-8	8-6	10-1	12-0
	Spruce-pine-fir	SS	6-8	10-6	13-5	16-5	19-1	6-8	9-10	12-5	15-3	17-8
	Spruce-pine-fir	#1	6-1	8-11	11-3	13-9	15-11	5-7	8-3	10-5	12-9	14-9
	Spruce-pine-fir	#2	6-1	8-11	11-3	13-9	15-11	5-7	8-3	10-5	12-9	14-9
	Spruce-pine-fir	#3	4-7	6-9	8-6	10-5	12-1	4-3	6-3	7-11	9-7	11-2
24	Douglas fir-larch	SS	6-8	10-	13-0	15-10	18-4	6-6	9-6	12-0	14-8	17-0
	Douglas fir-larch	#1	5-10	8-6	10-9	13-2	15-3	5-5	7-10	10-0	12-2	14-1
	Douglas fir-larch	#2	5-5	7-11	10-1	12-4	14-3	5-0	7-4	9-4	11-5	13-2
	Douglas fir-larch	#3	4-1	6-0	7-7	9-4	10-9	3-10	5-7	7-1	8-7	10-0
	Hem-fir	SS	6-4	9-11	12-9	15-7	18-0	6-4	9-4	11-9	14-5	16-8
	Hem-fir	#1	5-8	8-3	10-6	12-10	14-10	5-3	7-8	9-9	11-10	13-9
	Hem-fir	#2	5-4	7-10	9-11	12-1	14-1	4-11	7-3	9-2	11-3	13-0
	Hem-fir	#3	4-1	6-0	7-7	9-4	10-9	3-10	5-7	7-1	8-7	10-0
	Southern pine	SS	6-7	10-4	13-8	17-5	21-0	6-7	10-4	13-8	16-7	19-5
	Southern pine	#1	6-5	9-7	12-0	14-4	17-1	6-0	8-10	11-2	13-3	15-9
	Southern pine	#2	5-10	8-4	10-9	12-10	15-1	5-5	7-9	10-0	11-11	13-11
	Southern pine	#3	4-4	6-5	8-3	9-9	11-7	4-1	6-0	7-7	9-0	10-8
	Spruce-pine-fir	SS	6-2	9-6	12-0	14-8	17-1	6-0	8-10	11-2	13-7	15-9
	Spruce-pine-fir	#1	5-5	7-11	10-1	12-4	14-3	5-0	7-4	9-4	11-5	13-2
	Spruce-pine-fir	#2	5-5	7-11	10-1	12-4	14-3	5-0	7-4	9-4	11-5	13-2
	Spruce-pine-fir	#3	4-1	6-0	7-7	9-4	10-9	3-10	5-7	7-1	8-7	10-0

Check sources for availability of lumber in lengths greater than 20 feet.

For SI: 1 inch = 25.4 mm, 1 foot = 304.8 mm, 1 pound per square foot = 0.0479 kPa

a. The tabulated rafter spans assume that ceiling joists are located at the bottom of the attic space or that some other method of resisting the outward push of the rafters on the bearing walls, such as rafter ties, is provided at that location. When ceiling joists or rafter ties are located higher in the attic space, the rafter spans shall be multiplied by the factors given below:

H_C/H_R	Rafter Span Adjustment Factor
1/3	0.67
1/4	0.76
1/5	0.83
1/6	0.90
1/7.5 or less	1.00

where:

H_C = Height of ceiling joists or rafter ties measured vertically above the top of the rafter support walls.

H_R = Height of roof ridge measured vertically above the top of the rafter support walls.

b. Span exceeds 26 feet in length.

TABLE R802.5.1(5)
RAFTER SPANS FOR COMMON LUMBER SPECIES
(Ground snow load=30 psf, ceiling attached to rafters, L/Δ = 240)

RAFTER SPACING (inches)	SPECIES AND GRADE		DEAD LOAD = 10 psf					DEAD LOAD = 20 psf				
			2 × 4	2 × 6	2 × 8	2 × 10	2 × 12	2 × 4	2 × 6	2 × 8	2 × 10	2 × 12
			Maximum rafter spans[a]									
			(feet - inches)	(feet - inches)	(feet - inches)	(feet - inches)	(feet - inches)	(feet - inches)	(feet - inches)	(feet - inches)	(feet - inches)	(feet - inches)
12	Douglas fir-larch	SS	9-1	14-4	18-10	24-1	Note b	9-1	14-4	18-10	24-1	Note b
	Douglas fir-larch	#1	8-9	13-9	18-2	22-9	Note b	8-9	13-2	16-8	20-4	23-7
	Douglas fir-larch	#2	8-7	13-6	17-5	21-4	24-8	8-5	12-4	15-7	19-1	22-1
	Douglas fir-larch	#3	7-1	10-5	13-2	16-1	18-8	6-4	9-4	11-9	14-5	16-8
	Hem-fir	SS	8-7	13-6	17-10	22-9	Note b	8-7	13-6	17-10	22-9	Note b
	Hem-fir	#1	8-5	13-3	17-5	22-2	25-9	8-5	12-10	16-3	19-10	23-0
	Hem-fir	#2	8-0	12-7	16-7	21-0	24-4	8-0	12-2	15-4	18-9	21-9
	Hem-fir	#3	7-1	10-5	13-2	16-1	18-8	6-4	9-4	11-9	14-5	16-8
	Southern pine	SS	8-11	14-1	18-6	23-8	Note b	8-11	14-1	18-6	23-8	Note b
	Southern pine	#1	8-9	13-9	18-2	23-2	Note b	8-9	13-9	18-2	22-2	Note b
	Southern pine	#2	8-7	13-6	17-10	22-3	Note b	8-7	12-11	16-8	19-11	23-4
	Southern pine	#3	7-7	11-2	14-3	16-10	20-0	6-9	10-0	12-9	15-1	17-11
	Spruce-pine-fir	SS	8-5	13-3	17-5	22-3	Note b	8-5	13-3	17-5	22-3	Note b
	Spruce-pine-fir	#1	8-3	12-11	17-0	21-4	24-8	8-3	12-4	15-7	19-1	22-1
	Spruce-pine-fir	#2	8-3	12-11	17-0	21-4	24-8	8-3	12-4	15-7	19-1	22-1
	Spruce-pine-fir	#3	7-1	10-5	13-2	16-1	18-8	6-4	9-4	11-9	14-5	16-8
16	Douglas fir-larch	SS	8-3	13-0	17-2	21-10	Note b	8-3	13-0	17-2	21-3	24-8
	Douglas fir-larch	#1	8-0	12-6	16-2	19-9	22-10	7-10	11-5	14-5	17-8	20-5
	Douglas fir-larch	#2	7-10	11-11	15-1	18-5	21-5	7-3	10-8	13-6	16-6	19-2
	Douglas fir-larch	#3	6-2	9-0	11-5	13-11	16-2	5-6	8-1	10-3	12-6	14-6
	Hem-fir	SS	7-10	12-3	16-2	20-8	25-1	7-10	12-3	16-2	20-8	24-2
	Hem-fir	#1	7-8	12-0	15-9	19-3	22-3	7-7	11-1	14-1	17-2	19-11
	Hem-fir	#2	7-3	11-5	14-11	18-2	21-1	7-2	10-6	13-4	16-3	18-10
	Hem-fir	#3	6-2	9-0	11-5	13-11	16-2	5-6	8-1	10-3	12-6	14-6
	Southern pine	SS	8-1	12-9	16-10	21-6	Note b	8-1	12-9	16-10	21-6	Note b
	Southern pine	#1	8-0	12-6	16-6	21-1	25-7	8-0	12-6	16-2	19-2	22-10
	Southern pine	#2	7-10	12-3	16-2	19-3	22-7	7-10	11-2	14-5	17-3	20-2
	Southern pine	#3	6-7	9-8	12-4	14-7	17-4	5-10	8-8	11-0	13-0	15-6
	Spruce-pine-fir	SS	7-8	12-0	15-10	20-2	24-7	7-8	12-0	15-10	19-9	22-10
	Spruce-pine-fir	#1	7-6	11-9	15-1	18-5	21-5	7-3	10-8	13-6	16-6	19-2
	Spruce-pine-fir	#2	7-6	11-9	15-1	18-5	21-5	7-3	10-8	13-6	16-6	19-2
	Spruce-pine-fir	#3	6-2	9-0	11-5	13-11	16-2	5-6	8-1	10-3	12-6	14-6
19.2	Douglas fir-larch	SS	7-9	12-3	16-1	20-7	25-0	7-9	12-3	15-10	19-5	22-6
	Douglas fir-larch	#1	7-6	11-8	14-9	18-0	20-11	7-1	10-5	13-2	16-1	18-8
	Douglas fir-larch	#2	7-4	10-11	13-9	16-10	19-6	6-8	9-9	12-4	15-1	17-6
	Douglas fir-larch	#3	5-7	8-3	10-5	12-9	14-9	5-0	7-4	9-4	11-5	13-2
	Hem-fir	SS	7-4	11-7	15-3	19-5	23-7	7-4	11-7	15-3	19-1	22-1
	Hem-fir	#1	7-2	11-4	14-4	17-7	20-4	6-11	10-2	12-10	15-8	18-2
	Hem-fir	#2	6-10	10-9	13-7	16-7	19-3	6-7	9-7	12-2	14-10	17-3
	Hem-fir	#3	5-7	8-3	10-5	12-9	14-9	5-0	7-4	9-4	11-5	13-2

(continued)

TABLE R802.5.1(5)—continued
RAFTER SPANS FOR COMMON LUMBER SPECIES
(Ground snow load=30 psf, ceiling attached to rafters, L/Δ = 240)

RAFTER SPACING (inches)	SPECIES AND GRADE		DEAD LOAD = 10 psf					DEAD LOAD = 20 psf				
			2 × 4	2 × 6	2 × 8	2 × 10	2 × 12	2 × 4	2 × 6	2 × 8	2 × 10	2 × 12
			Maximum rafter spans[a]									
			(feet - inches)	(feet - inches)	(feet - inches)	(feet - inches)	(feet - inches)	(feet - inches)	(feet - inches)	(feet - inches)	(feet - inches)	(feet - inches)
19.2	Southern pine	SS	7-8	12-0	15-10	20-2	24-7	7-8	12-0	15-10	20-2	24-7
	Southern pine	#1	7-6	11-9	15-6	19-7	23-4	7-6	11-9	14-9	17-6	20-11
	Southern pine	#2	7-4	11-5	14-9	17-7	20-7	7-1	10-2	13-2	15-9	18-5
	Southern pine	#3	6-0	8-10	11-3	13-4	15-10	5-4	7-11	10-1	11-11	14-2
	Spruce-pine-fir	SS	7-2	11-4	14-11	19-0	23-1	7-2	11-4	14-9	18-0	20-11
	Spruce-pine-fir	#1	7-0	10-11	13-9	16-10	19-6	6-8	9-9	12-4	15-1	17-6
	Spruce-pine-fir	#2	7-0	10-11	13-9	16-10	19-6	6-8	9-9	12-4	15-1	17-6
	Spruce-pine-fir	#3	5-7	8-3	10-5	12-9	14-9	5-0	7-4	9-4	11-5	13-2
24	Douglas fir-larch	SS	7-3	11-4	15-0	19-1	22-6	7-3	11-3	14-2	17-4	20-1
	Douglas fir-larch	#1	7-0	10-5	13-2	16-1	18-8	6-4	9-4	11-9	14-5	16-8
	Douglas fir-larch	#2	6-8	9-9	12-4	15-1	17-6	5-11	8-8	11-0	13-6	15-7
	Douglas fir-larch	#3	5-0	7-4	9-4	11-5	13-2	4-6	6-7	8-4	10-2	11-10
	Hem-fir	SS	6-10	10-9	14-2	18-0	21-11	6-10	10-9	13-11	17-0	19-9
	Hem-fir	#1	6-8	10-2	12-10	15-8	18-2	6-2	9-1	11-6	14-0	16-3
	Hem-fir	#2	6-4	9-7	12-2	14-10	17-3	5-10	8-7	10-10	13-3	15-5
	Hem-fir	#3	5-0	7-4	9-4	11-5	13-2	4-6	6-7	8-4	10-2	11-10
	Southern pine	SS	7-1	11-2	14-8	18-9	22-10	7-1	11-2	14-8	18-9	22-10
	Southern pine	#1	7-0	10-11	14-5	17-6	20-11	7-0	10-6	13-2	15-8	18-8
	Southern pine	#2	6-10	10-2	13-2	15-9	18-5	6-4	9-2	11-9	14-1	16-6
	Southern pine	#3	5-4	7-11	10-1	11-11	14-2	4-9	7-1	9-0	10-8	12-8
	Spruce-pine-fir	SS	6-8	10-6	13-10	17-8	20-11	6-8	10-5	13-2	16-1	18-8
	Spruce-pine-fir	#1	6-6	9-9	12-4	15-1	17-6	5-11	8-8	11-0	13-6	15-7
	Spruce-pine-fir	#2	6-6	9-9	12-4	15-1	17-6	5-11	8-8	11-0	13-6	15-7
	Spruce-pine-fir	#3	5-0	7-4	9-4	11-5	13-2	4-6	6-7	8-4	10-2	11-10

Check sources for availability of lumber in lengths greater than 20 feet.

For SI: 1 inch = 25.4 mm, 1 foot = 304.8 mm, 1 pound per square foot = 0.0479 kPa.

a. The tabulated rafter spans assume that ceiling joists are located at the bottom of the attic space or that some other method of resisting the outward push of the rafters on the bearing walls, such as rafter ties, is provided at that location. When ceiling joists or rafter ties are located higher in the attic space, the rafter spans shall be multiplied by the factors given below:

H_C/H_R	Rafter Span Adjustment Factor
1/3	0.67
1/4	0.76
1/5	0.83
1/6	0.90
1/7.5 or less	1.00

where:

H_C = Height of ceiling joists or rafter ties measured vertically above the top of the rafter support walls.

H_R = Height of roof ridge measured vertically above the top of the rafter support walls.

b. Span exceeds 26 feet in length.

TABLE R802.5.1(6)
RAFTER SPANS FOR COMMON LUMBER SPECIES
(Ground snow load=50 psf, ceiling attached to rafters, L/Δ = 240)

RAFTER SPACING (inches)	SPECIES AND GRADE		DEAD LOAD = 10 psf					DEAD LOAD = 20 psf				
			2 × 4	2 × 6	2 × 8	2 × 10	2 × 12	2 × 4	2 × 6	2 × 8	2 × 10	2 × 12
			Maximum rafter spans[a]									
			(feet-inches)	(feet-inches)	(feet-inches)	(feet-inches)	(feet-inches)	(feet-inches)	(feet-inches)	(feet-inches)	(feet-inches)	(feet-inches)
12	Douglas fir-larch	SS	7-8	12-1	15-11	20-3	24-8	7-8	12-1	15-11	20-3	24-0
	Douglas fir-larch	#1	7-5	11-7	15-3	18-7	21-7	7-5	11-2	14-1	17-3	20-0
	Douglas fir-larch	#2	7-3	11-3	14-3	17-5	20-2	7-1	10-5	13-2	16-1	18-8
	Douglas fir-larch	#3	5-10	8-6	10-9	13-2	15-3	5-5	7-10	10-0	12-2	14-1
	Hem-fir	SS	7-3	11-5	15-0	19-2	23-4	7-3	11-5	15-0	19-2	23-4
	Hem-fir	#1	7-1	11-2	14-8	18-1	21-0	7-1	10-10	13-9	16-9	19-5
	Hem-fir	#2	6-9	10-8	14-0	17-2	19-11	6-9	10-3	13-0	15-10	18-5
	Hem-fir	#3	5-10	8-6	10-9	13-2	15-3	5-5	7-10	10-0	12-2	14-1
	Southern pine	SS	7-6	11-10	15-7	19-11	24-3	7-6	11-10	15-7	19-11	24-3
	Southern pine	#1	7-5	11-7	15-4	19-7	23-9	7-5	11-7	15-4	18-9	22-4
	Southern pine	#2	7-3	11-5	15-0	18-2	21-3	7-3	10-11	14-1	16-10	19-9
	Southern pine	#3	6-2	9-2	11-8	13-9	16-4	5-9	8-5	10-9	12-9	15-2
	Spruce-pine-fir	SS	7-1	11-2	14-8	18-9	22-10	7-1	11-2	14-8	18-9	22-4
	Spruce-pine-fir	#1	6-11	10-11	14-3	17-5	20-2	6-11	10-5	13-2	16-1	18-8
	Spruce-pine-fir	#2	6-11	10-11	14-3	17-5	20-2	6-11	10-5	13-2	16-1	18-8
	Spruce-pine-fir	#3	5-10	8-6	10-9	13-2	15-3	5-5	7-10	10-0	12-2	14-1
16	Douglas fir-larch	SS	7-0	11-0	14-5	18-5	22-5	7-0	11-0	14-5	17-11	20-10
	Douglas fir-larch	#1	6-9	10-5	13-2	16-1	18-8	6-7	9-8	12-2	14-11	17-3
	Douglas fir-larch	#2	6-7	9-9	12-4	15-1	17-6	6-2	9-0	11-5	13-11	16-2
	Douglas fir-larch	#3	5-0	7-4	9-4	11-5	13-2	4-8	6-10	8-8	10-6	12-3
	Hem-fir	SS	6-7	10-4	13-8	17-5	21-2	6-7	10-4	13-8	17-5	20-5
	Hem-fir	#1	6-5	10-2	12-10	15-8	18-2	6-5	9-5	11-11	14-6	16-10
	Hem-fir	#2	6-2	9-7	12-2	14-10	17-3	6-1	8-11	11-3	13-9	15-11
	Hem-fir	#3	5-0	7-4	9-4	11-5	13-2	4-8	6-10	8-8	10-6	12-3
	Southern pine	SS	6-10	10-9	14-2	18-1	22-0	6-10	10-9	14-2	18-1	22-0
	Southern pine	#1	6-9	10-7	13-11	17-6	20-11	6-9	10-7	13-8	16-2	19-4
	Southern pine	#2	6-7	10-2	13-2	15-9	18-5	6-7	9-5	12-2	14-7	17-1
	Southern pine	#3	5-4	7-11	10-1	11-11	14-2	4-11	7-4	9-4	11-0	13-1
	Spruce-pine-fir	SS	6-5	10-2	13-4	17-0	20-9	6-5	10-2	13-4	16-8	19-4
	Spruce-pine-fir	#1	6-4	9-9	12-4	15-1	17-6	6-2	9-0	11-5	13-11	16-2
	Spruce-pine-fir	#2	6-4	9-9	12-4	15-1	17-6	6-2	9-0	11-5	13-11	16-2
	Spruce-pine-fir	#3	5-0	7-4	9-4	11-5	13-2	4-8	6-10	8-8	10-6	12-3
19.2	Douglas fir-larch	SS	6-7	10-4	13-7	17-4	20-6	6-7	10-4	13-5	16-5	19-0
	Douglas fir-larch	#1	6-4	9-6	12-0	14-8	17-1	6-0	8-10	11-2	13-7	15-9
	Douglas fir-larch	#2	6-1	8-11	11-3	13-9	15-11	5-7	8-3	10-5	12-9	14-9
	Douglas fir-larch	#3	4-7	6-9	8-6	10-5	12-1	4-3	6-3	7-11	9-7	11-2
	Hem-fir	SS	6-2	9-9	12-10	16-5	19-11	6-2	9-9	12-10	16-1	18-8
	Hem-fir	#1	6-1	9-3	11-9	14-4	16-7	5-10	8-7	10-10	13-3	15-5
	Hem-fir	#2	5-9	8-9	11-1	13-7	15-9	5-7	8-1	10-3	12-7	14-7
	Hem-fir	#3	4-7	6-9	8-6	10-5	12-1	4-3	6-3	7-11	9-7	11-2

(continiued)

TABLE R802.5.1(6)—continued
RAFTER SPANS FOR COMMON LUMBER SPECIES
(Ground snow load=50 psf, ceiling attached to rafters, L/Δ = 240)

RAFTER SPACING (inches)	SPECIES AND GRADE		DEAD LOAD = 10 psf					DEAD LOAD = 20 psf				
			2 × 4	2 × 6	2 × 8	2 × 10	2 × 12	2 × 4	2 × 6	2 × 8	2 × 10	2 × 12
			Maximum rafter spans[a]									
			(feet-inches)	(feet-inches)	(feet-inches)	(feet-inches)	(feet-inches)	(feet-inches)	(feet-inches)	(feet-inches)	(feet-inches)	(feet-inches)
19.2	Southern pine	SS	6-5	10-2	13-4	17-0	20-9	6-5	10-2	13-4	17-0	20-9
	Southern pine	#1	6-4	9-11	13-1	16-0	19-1	6-4	9-11	12-5	14-10	17-8
	Southern pine	#2	6-2	9-4	12-0	14-4	16-10	6-0	8-8	11-2	13-4	15-7
	Southern pine	#3	4-11	7-3	9-2	10-10	12-11	4-6	6-8	8-6	10-1	12-0
	Spruce-pine-fir	SS	6-1	9-6	12-7	16-0	19-1	6-1	9-6	12-5	15-3	17-8
	Spruce-pine-fir	#1	5-11	8-11	11-3	13-9	15-11	5-7	8-3	10-5	12-9	14-9
	Spruce-pine-fir	#2	5-11	8-11	11-3	13-9	15-11	5-7	8-3	10-5	12-9	14-9
	Spruce-pine-fir	#3	4-7	6-9	8-6	10-5	12-1	4-3	6-3	7-11	9-7	11-2
24	Douglas fir-larch	SS	6-1	9-7	12-7	15-10	18-4	6-1	9-6	12-0	14-8	17-0
	Douglas fir-larch	#1	5-10	8-6	10-9	13-2	15-3	5-5	7-10	10-0	12-2	14-1
	Douglas fir-larch	#2	5-5	7-11	10-1	12-4	14-3	5-0	7-4	9-4	11-5	13-2
	Douglas fir-larch	#3	4-1	6-0	7-7	9-4	10-9	3-10	5-7	7-1	8-7	10-0
	Hem-fir	SS	5-9	9-1	11-11	15-2	18-0	5-9	9-1	11-9	14-5	15-11
	Hem-fir	#1	5-8	8-3	10-6	12-10	14-10	5-3	7-8	9-9	11-10	13-9
	Hem-fir	#2	5-4	7-10	9-11	12-1	14-1	4-11	7-3	9-2	11-3	13-0
	Hem-fir	#3	4-1	6-0	7-7	9-4	10-9	3-10	5-7	7-1	8-7	10-0
	Southern pine	SS	6-0	9-5	12-5	15-10	19-3	6-0	9-5	12-5	15-10	19-3
	Southern pine	#1	5-10	9-3	12-0	14-4	17-1	5-10	8-10	11-2	13-3	15-9
	Southern pine	#2	5-9	8-4	10-9	12-10	15-1	5-5	7-9	10-0	11-11	13-11
	Southern pine	#3	4-4	6-5	8-3	9-9	11-7	4-1	6-0	7-7	9-0	10-8
	Spruce-pine-fir	SS	5-8	8-10	11-8	14-8	17-1	5-8	8-10	11-2	13-7	15-9
	Spruce-pine-fir	#1	5-5	7-11	10-1	12-4	14-3	5-0	7-4	9-4	11-5	13-2
	Spruce-pine-fir	#2	5-5	7-11	10-1	12-4	14-3	5-0	7-4	9-4	11-5	13-2
	Spruce-pine-fir	#3	4-1	6-0	7-7	9-4	10-9	3-10	5-7	7-1	8-7	10-0

Check sources for availability of lumber in lengths greater than 20 feet.

For SI: 1 inch = 25.4 mm, 1 foot = 304.8 mm, 1 pound per square foot = 0.0479 kPa.

a. The tabulated rafter spans assume that ceiling joists are located at the bottom of the attic space or that some other method of resisting the outward push of the rafters on the bearing walls, such as rafter ties, is provided at that location. When ceiling joists or rafter ties are located higher in the attic space, the rafter spans shall be multiplied by the factors given below:

H_C/H_R	Rafter Span Adjustment Factor
1/3	0.67
1/4	0.76
1/5	0.83
1/6	0.90
1/7.5 or less	1.00

where:

H_C = Height of ceiling joists or rafter ties measured vertically above the top of the rafter support walls.

H_R = Height of roof ridge measured vertically above the top of the rafter support walls.

TABLE R802.5.1(7)
RAFTER SPANS FOR 70 PSF GROUND SNOW LOAD
(Ceiling not attached to rafters, L/Δ = 180)

RAFTER SPACING (inches)	SPECIES AND GRADE		DEAD LOAD = 10 psf					DEAD LOAD = 20 psf				
			2 × 4	2 × 6	2 × 8	2 × 10	2 × 12	2 × 4	2 × 6	2 × 8	2 × 10	2 × 12
			Maximum Rafter Spans[a]									
			(feet-inches)	(feet-inches)	(feet-inches)	(feet-inches)	(feet-inches)	(feet-inches)	(feet-inches)	(feet-inches)	(feet-inches)	(feet-inches)
12	Douglas fir-larch	SS	7-7	11-10	15-8	19-5	22-6	7-7	11-10	15-0	18-3	21-2
	Douglas fir-larch	#1	7-1	10-5	13-2	16-1	18-8	6-8	9-10	12-5	15-2	17-7
	Douglas fir-larch	#2	6-8	9-9	12-4	15-1	17-6	6-3	9-2	11-8	14-2	16-6
	Douglas fir-larch	#3	5-0	7-4	9-4	11-5	13-2	4-9	6-11	8-9	10-9	12-5
	Hem-fir	SS	7-2	11-3	14-9	18-10	22-1	7-2	11-3	14-8	18-0	20-10
	Hem-fir	#1	6-11	10-2	12-10	15-8	18-2	6-6	9-7	12-1	14-10	17-2
	Hem-fir	#2	6-7	9-7	12-2	14-10	17-3	6-2	9-1	11-5	14-0	16-3
	Hem-fir	#3	5-0	7-4	9-4	11-5	13-2	4-9	6-11	8-9	10-9	12-5
	Southern pine	SS	7-5	11-8	15-4	19-7	23-10	7-5	11-8	15-4	19-7	23-10
	Southern pine	#1	7-3	11-5	14-9	17-6	20-11	7-3	11-1	13-11	16-6	19-8
	Southern pine	#2	7-1	10-2	13-2	15-9	18-5	6-8	9-7	12-5	14-10	17-5
	Southern pine	#3	5-4	7-11	10-1	11-11	14-2	5-1	7-5	9-6	11-3	13-4
	Spruce-pine-fir	SS	7-0	11-0	14-6	18-0	20-11	7-0	11-0	13-11	17-0	19-8
	Spruce-pine-fir	#1	6-8	9-9	12-4	15-1	17-6	6-3	9-2	11-8	14-2	16-6
	Spruce-pine-fir	#2	6-8	9-9	12-4	15-1	17-6	6-3	9-2	11-8	14-2	16-6
	Spruce-pine-fir	#3	5-0	7-4	9-4	11-5	13-2	4-9	6-11	8-9	10-9	12-5
16	Douglas fir-larch	SS	6-10	10-9	13-9	16-10	19-6	6-10	10-3	13-0	15-10	18-4
	Douglas fir-larch	#1	6-2	9-0	11-5	13-11	16-2	5-10	8-6	10-9	13-2	15-3
	Douglas fir-larch	#2	5-9	8-5	10-8	13-1	15-2	5-5	7-11	10-1	12-4	14-3
	Douglas fir-larch	#3	4-4	6-4	8-1	9-10	11-5	4-1	6-0	7-7	9-4	10-9
	Hem-fir	SS	6-6	10-2	13-5	16-6	19-2	6-6	10-1	12-9	15-7	18-0
	Hem-fir	#1	6-0	8-9	11-2	13-7	15-9	5-8	8-3	10-6	12-10	14-10
	Hem-fir	#2	5-8	8-4	10-6	12-10	14-11	5-4	7-10	9-11	12-1	14-1
	Hem-fir	#3	4-4	6-4	8-1	9-10	11-5	4-1	6-0	7-7	9-4	10-9
	Southern pine	SS	6-9	10-7	14-0	17-10	21-8	6-9	10-7	14-0	17-10	21-0
	Southern pine	#1	6-7	10-2	12-9	15-2	18-1	6-5	9-7	12-0	14-4	17-1
	Southern pine	#2	6-2	8-10	11-5	13-7	16-0	5-10	8-4	10-9	12-10	15-1
	Southern pine	#3	4-8	6-10	8-9	10-4	12-3	4-4	6-5	8-3	9-9	11-7
	Spruce-pine-fir	SS	6-4	10-0	12-9	15-7	18-1	6-4	9-6	12-0	14-8	17-1
	Spruce-pine-fir	#1	5-9	8-5	10-8	13-1	15-2	5-5	7-11	10-1	12-4	14-3
	Spruce-pine-fir	#2	5-9	8-5	10-8	13-1	15-2	5-5	7-11	10-1	12-4	14-3
	Spruce-pine-fir	#3	4-4	6-4	8-1	9-10	11-5	4-1	6-0	7-7	9-4	10-9
19.2	Douglas fir-larch	SS	6-5	9-11	12-7	15-4	17-9	6-5	9-4	11-10	14-5	16-9
	Douglas fir-larch	#1	5-7	8-3	10-5	12-9	14-9	5-4	7-9	9-10	12-0	13-11
	Douglas fir-larch	#2	5-3	7-8	9-9	11-11	13-10	5-0	7-3	9-2	11-3	13-0
	Douglas fir-larch	#3	4-0	5-10	7-4	9-0	10-5	3-9	5-6	6-11	8-6	9-10
	Hem-fir	SS	6-1	9-7	12-4	15-1	17-4	6-1	9-2	11-8	14-2	15-5
	Hem-fir	#1	5-6	8-0	10-2	12-5	14-5	5-2	7-7	9-7	11-8	13-7
	Hem-fir	#2	5-2	7-7	9-7	11-9	13-7	4-11	7-2	9-1	11-1	12-10
	Hem-fir	#3	4-0	5-10	7-4	9-0	10-5	3-9	5-6	6-11	8-6	9-10

(continued)

TABLE R802.5.1(7)—continued
RAFTER SPANS FOR 70 PSF GROUND SNOW LOAD
(Ceiling not attached to rafters, L/Δ = 180)

RAFTER SPACING (inches)	SPECIES AND GRADE		DEAD LOAD = 10 psf					DEAD LOAD = 20 psf				
			2 × 4	2 × 6	2 × 8	2 × 10	2 × 12	2 × 4	2 × 6	2 × 8	2 × 10	2 × 12
			Maximum Rafter Spans[a]									
			(feet-inches)	(feet-inches)	(feet-inches)	(feet-inches)	(feet-inches)	(feet-inches)	(feet-inches)	(feet-inches)	(feet-inches)	(feet-inches)
19.2	Southern pine	SS	6-4	10-0	13-2	16-9	20-4	6-4	10-0	13-2	16-5	19-2
	Southern pine	#1	6-3	9-3	11-8	13-10	16-6	5-11	8-9	11-0	13-1	15-7
	Southern pine	#2	5-7	8-1	10-5	12-5	14-7	5-4	7-7	9-10	11-9	13-9
	Southern pine	#3	4-3	6-3	8-0	9-5	11-2	4-0	5-11	7-6	8-10	10-7
	Spruce-pine-fir	SS	6-0	9-2	11-8	14-3	16-6	5-11	8-8	11-0	13-5	15-7
	Spruce-pine-fir	#1	5-3	7-8	9-9	11-11	13-10	5-0	7-3	9-2	11-3	13-0
	Spruce-pine-fir	#2	5-3	7-8	9-9	11-11	13-10	5-0	7-3	9-2	11-3	13-0
	Spruce-pine-fir	#3	4-0	5-10	7-4	9-0	10-5	3-9	5-6	6-11	8-6	9-10
24	Douglas fir-larch	SS	6-0	8-10	11-3	13-9	15-11	5-9	8-4	10-7	12-11	15-0
	Douglas fir-larch	#1	5-0	7-4	9-4	11-5	13-2	4-9	6-11	8-9	10-9	12-5
	Douglas fir-larch	#2	4-8	6-11	8-9	10-8	12-4	4-5	6-6	8-3	10-0	11-8
	Douglas fir-larch	#3	3-7	5-2	6-7	8-1	9-4	3-4	4-11	6-3	7-7	8-10
	Hem-fir	SS	5-8	8-8	11-0	13-6	13-11	5-7	8-3	10-5	12-4	12-4
	Hem-fir	#1	4-11	7-2	9-1	11-1	12-10	4-7	6-9	8-7	10-6	12-2
	Hem-fir	#2	4-8	6-9	8-7	10-6	12-2	4-4	6-5	8-1	9-11	11-6
	Hem-fir	#3	3-7	5-2	6-7	8-1	9-4	3-4	4-11	6-3	7-7	8-10
	Southern pine	SS	5-11	9-3	12-2	15-7	18-2	5-11	9-3	12-2	14-8	17-2
	Southern pine	#1	5-7	8-3	10-5	12-5	14-9	5-3	7-10	9-10	11-8	13-11
	Southern pine	#2	5-0	7-3	9-4	11-1	13-0	4-9	6-10	8-9	10-6	12-4
	Southern pine	#3	3-9	5-7	7-1	8-5	10-0	3-7	5-3	6-9	7-11	9-5
	Spruce-pine-fir	SS	5-6	8-3	10-5	12-9	14-9	5-4	7-9	9-10	12-0	12-11
	Spruce-pine-fir	#1	4-8	6-11	8-9	10-8	12-4	4-5	6-6	8-3	10-0	11-8
	Spruce-pine-fir	#2	4-8	6-11	8-9	10-8	12-4	4-5	6-6	8-3	10-0	11-8
	Spruce-pine-fir	#3	3-7	5-2	6-7	8-1	9-4	3-4	4-11	6-3	7-7	8-10

Check sources for availability of lumber in lengths greater than 20 feet.

For SI: 1 inch = 25.4 mm, 1 foot = 304.8 mm, 1 pound per square foot = 0.0479 kPa.

a. The tabulated rafter spans assume that ceiling joists are located at the bottom of the attic space or that some other method of resisting the outward push of the rafters on the bearing walls, such as rafter ties, is provided at that location. When ceiling joists or rafter ties are located higher in the attic space, the rafter spans shall be multiplied by the factors given below:

H_C/H_R	Rafter Span Adjustment Factor
1/3	0.67
1/4	0.76
1/5	0.83
1/6	0.90
1/7.5 or less	1.00

where:

H_C = Height of ceiling joists or rafter ties measured vertically above the top of the rafter support walls.

H_R = Height of roof ridge measured vertically above the top of the rafter support walls.

TABLE R802.5.1(8)
RAFTER SPANS FOR 70 PSF GROUND SNOW LOAD
(Ceiling attached to rafters, L/Δ = 240)

RAFTER SPACING (inches)	SPECIES AND GRADE		DEAD LOAD = 10 psf					DEAD LOAD = 20 psf				
			2 × 4	2 × 6	2 × 8	2 × 10	2 × 12	2 × 4	2 × 6	2 × 8	2 × 10	2 × 12
			Maximum rafter spans[a]									
			(feet - inches)	(feet - inches)	(feet - inches)	(feet - inches)	(feet - inches)	(feet - inches)	(feet - inches)	(feet - inches)	(feet - inches)	(feet - inches)
12	Douglas fir-larch	SS	6-10	10-9	14-3	18-2	22-1	6-10	10-9	14-3	18-2	21-2
	Douglas fir-larch	#1	6-7	10-5	13-2	16-1	18-8	6-7	9-10	12-5	15-2	17-7
	Douglas fir-larch	#2	6-6	9-9	12-4	15-1	17-6	6-3	9-2	11-8	14-2	16-6
	Douglas fir-larch	#3	5-0	7-4	9-4	11-5	13-2	4-9	6-11	8-9	10-9	12-5
	Hem-fir	SS	6-6	10-2	13-5	17-2	20-10	6-6	10-2	13-5	17-2	20-10
	Hem-fir	#1	6-4	10-0	12-10	15-8	18-2	6-4	9-7	12-1	14-10	17-2
	Hem-fir	#2	6-1	9-6	12-2	14-10	17-3	6-1	9-1	11-5	14-0	16-3
	Hem-fir	#3	5-0	7-4	9-4	11-5	13-2	4-9	6-11	8-9	10-9	12-5
	Southern pine	SS	6-9	10-7	14-0	17-10	21-8	6-9	10-7	14-0	17-10	21-8
	Southern pine	#1	6-7	10-5	13-8	17-6	20-11	6-7	10-5	13-8	16-6	19-8
	Southern pine	#2	6-6	10-2	13-2	15-9	18-5	6-6	9-7	12-5	14-10	17-5
	Southern pine	#3	5-4	7-11	10-1	11-11	14-2	5-1	7-5	9-6	11-3	13-4
	Spruce-pine-fir	SS	6-4	10-0	13-2	16-9	20-5	6-4	10-0	13-2	16-9	19-8
	Spruce-pine-fir	#1	6-2	9-9	12-4	15-1	17-6	6-2	9-2	11-8	14-2	16-6
	Spruce-pine-fir	#2	6-2	9-9	12-4	15-1	17-6	6-2	9-2	11-8	14-2	16-6
	Spruce-pine-fir	#3	5-0	7-4	9-4	11-5	13-2	4-9	6-11	8-9	10-9	12-5
16	Douglas fir-larch	SS	6-3	9-10	12-11	16-6	19-6	6-3	9-10	12-11	15-10	18-4
	Douglas fir-larch	#1	6-0	9-0	11-5	13-11	16-2	5-10	8-6	10-9	13-2	15-3
	Douglas fir-larch	#2	5-9	8-5	10-8	13-1	15-2	5-5	7-11	10-1	12-4	14-3
	Douglas fir-larch	#3	4-4	6-4	8-1	9-10	11-5	4-1	6-0	7-7	9-4	10-9
	Hem-fir	SS	5-11	9-3	12-2	15-7	18-11	5-11	9-3	12-2	15-7	18-0
	Hem-fir	#1	5-9	8-9	11-2	13-7	15-9	5-8	8-3	10-6	12-10	14-10
	Hem-fir	#2	5-6	8-4	10-6	12-10	14-11	5-4	7-10	9-11	12-1	14-1
	Hem-fir	#3	4-4	6-4	8-1	9-10	11-5	4-1	6-0	7-7	9-4	10-9
	Southern pine	SS	6-1	9-7	12-8	16-2	19-8	6-1	9-7	12-8	16-2	19-8
	Southern pine	#1	6-0	9-5	12-5	15-2	18-1	6-0	9-5	12-0	14-4	17-1
	Southern pine	#2	5-11	8-10	11-5	13-7	16-0	5-10	8-4	10-9	12-10	15-1
	Southern pine	#3	4-8	6-10	8-9	10-4	12-3	4-4	6-5	8-3	9-9	11-7
	Spruce-pine-fir	SS	5-9	9-1	11-11	15-3	18-1	5-9	9-1	11-11	14-8	17-1
	Spruce-pine-fir	#1	5-8	8-5	10-8	13-1	15-2	5-5	7-11	10-1	12-4	14-3
	Spruce-pine-fir	#2	5-8	8-5	10-8	13-1	15-2	5-5	7-11	10-1	12-4	14-3
	Spruce-pine-fir	#3	4-4	6-4	8-1	9-10	11-5	4-1	6-0	7-7	9-4	10-9
19.2	Douglas fir-larch	SS	5-10	9-3	12-2	15-4	17-9	5-10	9-3	11-10	14-5	16-9
	Douglas fir-larch	#1	5-7	8-3	10-5	12-9	14-9	5-4	7-9	9-10	12-0	13-11
	Douglas fir-larch	#2	5-3	7-8	9-9	11-11	13-10	5-0	7-3	9-2	11-3	13-0
	Douglas fir-larch	#3	4-0	5-10	7-4	9-0	10-5	3-9	5-6	6-11	8-6	9-10
	Hem-fir	SS	5-6	8-8	11-6	14-8	17-4	5-6	8-8	11-6	14-2	15-5
	Hem-fir	#1	5-5	8-0	10-2	12-5	14-5	5-2	7-7	9-7	11-8	13-7
	Hem-fir	#2	5-2	7-7	9-7	11-9	13-7	4-11	7-2	9-1	11-1	12-10
	Hem-fir	#3	4-0	5-10	7-4	9-0	10-5	3-9	5-6	6-11	8-6	9-10

(continued)

TABLE R802.5.1(8)—continued
RAFTER SPANS FOR 70 PSF GROUND SNOW LOAD
(Ceiling attached to rafters, L/Δ = 240)

RAFTER SPACING (inches)	SPECIES AND GRADE		DEAD LOAD = 10 psf					DEAD LOAD = 20 psf				
			2 × 4	2 × 6	2 × 8	2 × 10	2 × 12	2 × 4	2 × 6	2 × 8	2 × 10	2 × 12
			Maximum rafter spans[a]									
			(feet - inches)	(feet - inches)	(feet - inches)	(feet - inches)	(feet - inches)	(feet - inches)	(feet - inches)	(feet - inches)	(feet - inches)	(feet - inches)
19.2	Southern pine	SS	5-9	9-1	11-11	15-3	18-6	5-9	9-1	11-11	15-3	18-6
	Southern pine	#1	5-8	8-11	11-8	13-10	16-6	5-8	8-9	11-0	13-1	15-7
	Southern pine	#2	5-6	8-1	10-5	12-5	14-7	5-4	7-7	9-10	11-9	13-9
	Southern pine	#3	4-3	6-3	8-0	9-5	11-2	4-0	5-11	7-6	8-10	10-7
	Spruce-pine-fir	SS	5-5	8-6	11-3	14-3	16-6	5-5	8-6	11-0	13-5	15-7
	Spruce-pine-fir	#1	5-3	7-8	9-9	11-11	13-10	5-0	7-3	9-2	11-3	13-0
	Spruce-pine-fir	#2	5-3	7-8	9-9	11-11	13-10	5-0	7-3	9-2	11-3	13-0
	Spruce-pine-fir	#3	4-0	5-10	7-4	9-0	10-5	3-9	5-6	6-11	8-6	9-10
24	Douglas fir-larch	SS	5-5	8-7	11-3	13-9	15-11	5-5	8-4	10-7	12-11	15-0
	Douglas fir-larch	#1	5-0	7-4	9-4	11-5	13-2	4-9	6-11	8-9	10-9	12-5
	Douglas fir-larch	#2	4-8	6-11	8-9	10-8	12-4	4-5	6-6	8-3	10-0	11-8
	Douglas fir-larch	#3	3-7	5-2	6-7	8-1	9-4	3-4	4-11	6-3	7-7	8-10
	Hem-fir	SS	5-2	8-1	10-8	13-6	13-11	5-2	8-1	10-5	12-4	12-4
	Hem-fir	#1	4-11	7-2	9-1	11-1	12-10	4-7	6-9	8-7	10-6	12-2
	Hem-fir	#2	4-8	6-9	8-7	10-6	12-2	4-4	6-5	8-1	9-11	11-6
	Hem-fir	#3	3-7	5-2	6-7	8-1	9-4	3-4	4-11	6-3	7-7	8-10
	Southern pine	SS	5-4	8-5	11-1	14-2	17-2	5-4	8-5	11-1	14-2	17-2
	Southern pine	#1	5-3	8-3	10-5	12-5	14-9	5-3	7-10	9-10	11-8	13-11
	Southern pine	#2	5-0	7-3	9-4	11-1	13-0	4-9	6-10	8-9	10-6	12-4
	Southern pine	#3	3-9	5-7	7-1	8-5	10-0	3-7	5-3	6-9	7-11	9-5
	Spruce-pine-fir	SS	5-0	7-11	10-5	12-9	14-9	5-0	7-9	9-10	12-0	12-11
	Spruce-pine-fir	#1	4-8	6-11	8-9	10-8	12-4	4-5	6-6	8-3	10-0	11-8
	Spruce-pine-fir	#2	4-8	6-11	8-9	10-8	12-4	4-5	6-6	8-3	10-0	11-8
	Spruce-pine-fir	#3	3-7	5-2	6-7	8-1	9-4	3-4	4-11	6-3	7-7	8-10

Check sources for availability of lumber in lengths greater than 20 feet.

For SI: 1 inch = 25.4 mm, 1 foot = 304.8 mm, 1 pound per square foot = 0.0479 kPa.

a. The tabulated rafter spans assume that ceiling joists are located at the bottom of the attic space or that some other method of resisting the outward push of the rafters on the bearing walls, such as rafter ties, is provided at that location. When ceiling joists or rafter ties are located higher in the attic space, the rafter spans shall be multiplied by the factors given below:

H_C/H_R	Rafter Span Adjustment Factor
1/3	0.67
1/4	0.76
1/5	0.83
1/6	0.90
1/7.5 or less	1.00

where:

H_C = Height of ceiling joists or rafter ties measured vertically above the top of the rafter support walls.

H_R = Height of roof ridge measured vertically above the top of the rafter support walls.

TABLE R802.5.1(9)
RAFTER/CEILING JOIST HEEL JOINT CONNECTIONS[a, b, c, d, e, f, h]

RAFTER SLOPE	RAFTER SPACING (inches)	GROUND SNOW LOAD (psf)															
		20[g]				30				50				70			
		Roof span (feet)															
		12	20	28	36	12	20	28	36	12	20	28	36	12	20	28	36
		Required number of 16d common nails[a, b] per heel joint splices[c, d, e, f]															
3:12	12	4	6	8	10	4	6	8	11	5	8	12	15	6	11	15	20
	16	5	8	10	13	5	8	11	14	6	11	15	20	8	14	20	26
	24	7	11	15	19	7	11	16	21	9	16	23	30	12	21	30	39
4:12	12	3	5	6	8	3	5	6	8	4	6	9	11	5	8	12	15
	16	4	6	8	10	4	6	8	11	5	8	12	15	6	11	15	20
	24	5	8	12	15	5	9	12	16	7	12	17	22	9	16	23	29
5:12	12	3	4	5	6	3	4	5	7	3	5	7	9	4	7	9	12
	16	3	5	6	8	3	5	7	9	4	7	9	12	5	9	12	16
	24	4	7	9	12	4	7	10	13	6	10	14	18	7	13	18	23
7:12	12	3	4	4	5	3	3	4	5	3	4	5	7	3	5	7	9
	16	3	4	5	6	3	4	5	6	3	5	7	9	4	6	9	11
	24	3	5	7	9	3	5	7	9	4	7	10	13	5	9	13	17
9:12	12	3	3	4	4	3	3	3	4	3	3	4	5	3	4	5	7
	16	3	4	4	5	3	3	4	5	3	4	5	7	3	5	7	9
	24	3	4	6	7	3	4	6	7	3	6	8	10	4	7	10	13
12:12	12	3	3	3	3	3	3	3	3	3	3	3	4	3	3	4	5
	16	3	3	4	4	3	3	3	4	3	3	4	5	3	4	5	7
	24	3	4	4	5	3	3	4	6	3	4	6	8	3	6	8	10

For SI: 1 inch = 25.4 mm, 1 foot = 304.8 mm, 1 pound per square foot = 0.0479 kPa.

a. 40d box nails shall be permitted to be substituted for 16d common nails.

b. Nailing requirements shall be permitted to be reduced 25 percent if nails are clinched.

c. Heel joint connections are not required when the ridge is supported by a load-bearing wall, header or ridge beam.

d. When intermediate support of the rafter is provided by vertical struts or purlins to a load-bearing wall, the tabulated heel joint connection requirements shall be permitted to be reduced proportionally to the reduction in span.

e. Equivalent nailing patterns are required for ceiling joist to ceiling joist lap splices.

f. When rafter ties are substituted for ceiling joists, the heel joint connection requirement shall be taken as the tabulated heel joint connection requirement for two-thirds of the actual rafter slope.

g. Applies to roof live load of 20 psf or less.

h. Tabulated heel joint connection requirements assume that ceiling joists or rafter ties are located at the bottom of the attic space. When ceiling joists or rafter ties are located higher in the attic, heel joint connection requirements shall be increased by the following factors:

H_C/H_R	Heel Joint Connection Adjustment Factor
1/3	1.5
1/4	1.33
1/5	1.25
1/6	1.2
1/10 or less	1.11

where:

H_C = Height of ceiling joists or rafter ties measured vertically above the top of the rafter support walls.

H_R = Height of roof ridge measured vertically above the top of the rafter support walls.

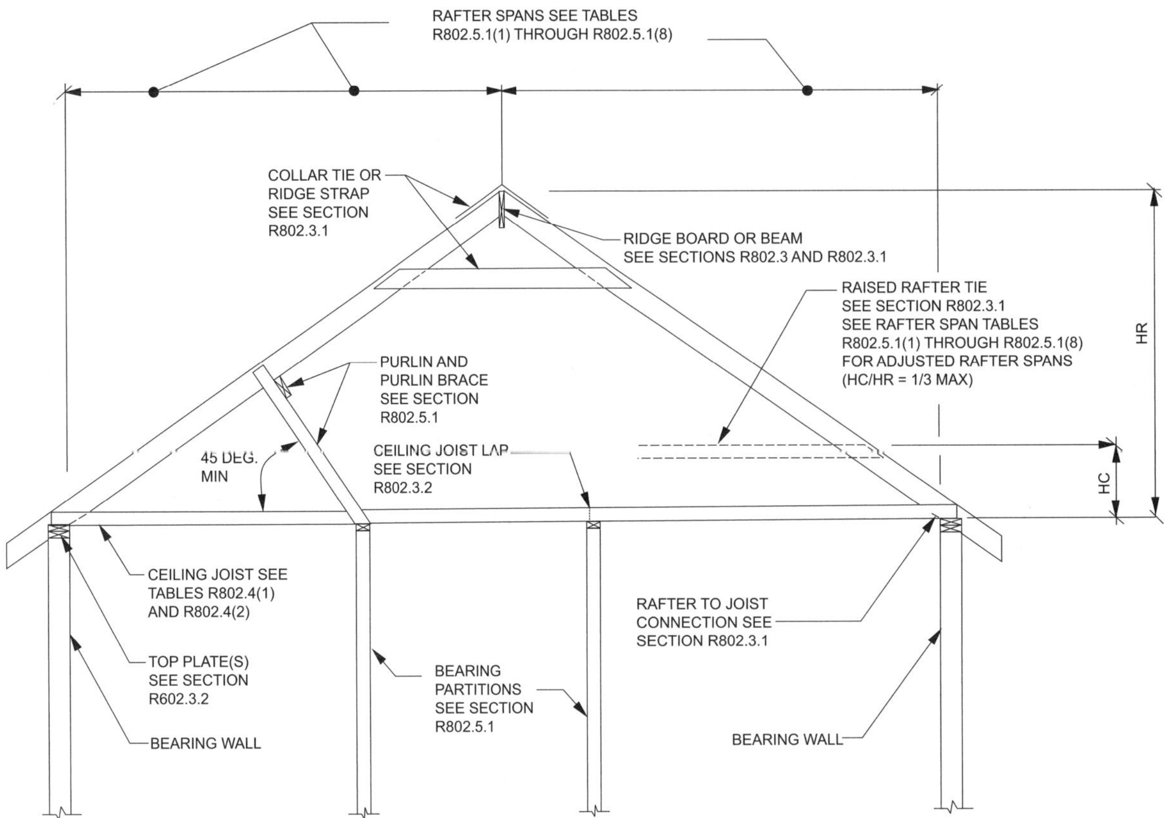

For SI: 1 inch = 25.4 mm, 1 foot = 305 mm, 1 degree = 0.018 rad.

Note: Where ceiling joists run perpendicular to the rafter, rafter ties shall be installed in accordance with Section R802.3.1.

H_C = Height of ceiling joists or rafter ties measured vertically above the top of rafter support walls.

H_R = Height of roof ridge measured vertically above the top of the rafter support walls.

FIGURE R802.5.1
BRACED RAFTER CONSTRUCTION

R802.6 Bearing. The ends of each rafter or ceiling joist shall have not less than $1^1/_2$ inches (38 mm) of bearing on wood or metal and not less than 3 inches (76 mm) on masonry or concrete. The bearing on masonry or concrete shall be direct, or a sill plate of 2-inch (51 mm) minimum nominal thickness shall be provided under the rafter or ceiling joist. The sill plate shall provide a minimum nominal bearing area of 48 square inches (30 865 mm^2).

R802.6.1 Finished ceiling material. If the finished ceiling material is installed on the ceiling prior to the attachment of the ceiling to the walls, such as in construction at a factory, a compression strip of the same thickness as the finish ceiling material shall be installed directly above the top plate of bearing walls if the compressive strength of the finish ceiling material is less than the loads it will be required to withstand. The compression strip shall cover the entire length of such top plate and shall be at least one-half the width of the top plate. It shall be of material capable of transmitting the loads transferred through it.

R802.7 Cutting, drilling and notching. Structural roof members shall not be cut, bored or notched in excess of the limitations specified in this section.

R802.7.1 Sawn lumber. Cuts, notches, and holes in solid lumber joists, rafters, blocking and beams shall comply with the provisions of R502.8.1 except that cantilevered portions of rafters shall be permitted in accordance with Section R802.7.1.1.

R802.7.1.1 Cantilevered portions of rafters. Notches on cantilevered portions of rafters are permitted provided the dimension of the remaining portion of the rafter is not less than $3^1/_2$ inches (89 mm) and the length of the cantilever does not exceed 24 inches (610 mm) in accordance with Figure R802.7.1.1.

R802.7.1.2 Ceiling joist taper cut. Taper cuts at the ends of the ceiling joist shall not exceed one-fourth the depth of the member in accordance with Figure R802.7.1.2.

R802.7.2 Engineered wood products. Cuts, notches and holes bored in trusses, structural composite lumber, structural glue-laminated members or I-joists are prohibited except where permitted by the manufacturer's recommendations or where the effects of such *alterations* are specifically considered in the design of the member by a registered *design professional.*

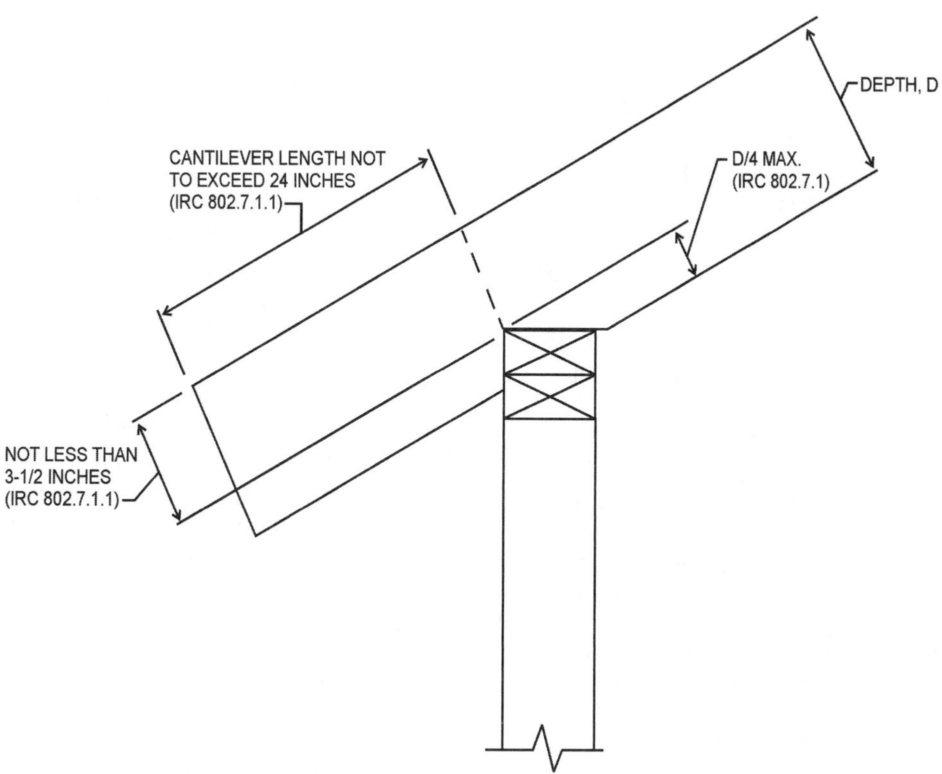

For SI: 1 inch = 25.4 mm.

FIGURE R802.7.1.1
RAFTER NOTCH

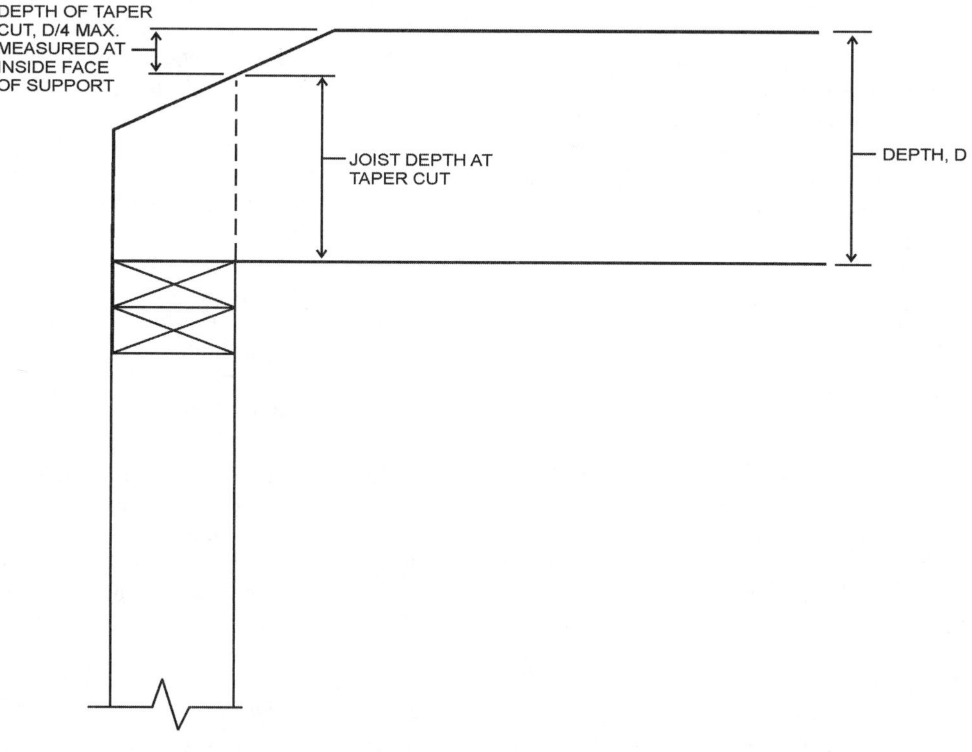

FIGURE R802.7.1.2
CEILING JOIST TAPER CUT

R802.8 Lateral support. Roof framing members and ceiling joists having a depth-to-thickness ratio exceeding 5 to 1 based on nominal dimensions shall be provided with lateral support at points of bearing to prevent rotation. For roof rafters with ceiling joists attached per Table R602.3(1), the depth-to-thickness ratio for the total assembly shall be determined using the combined thickness of the rafter plus the attached ceiling joist.

Exception: Roof trusses shall be braced in accordance with Section R802.10.3.

R802.8.1 Bridging. Rafters and ceiling joists having a depth-to-thickness ratio exceeding 6 to 1 based on nominal dimensions shall be supported laterally by solid blocking, diagonal bridging (wood or metal) or a continuous 1-inch by 3-inch (25 mm by 76 mm) wood strip nailed across the rafters or ceiling joists at intervals not exceeding 8 feet (2438 mm).

R802.9 Framing of openings. Openings in roof and ceiling framing shall be framed with header and trimmer joists. When the header joist span does not exceed 4 feet (1219 mm), the header joist may be a single member the same size as the ceiling joist or rafter. Single trimmer joists may be used to carry a single header joist that is located within 3 feet (914 mm) of the trimmer joist bearing. When the header joist span exceeds 4 feet (1219 mm), the trimmer joists and the header joist shall be doubled and of sufficient cross section to support the ceiling joists or rafter framing into the header. *Approved* hangers shall be used for the header joist to trimmer joist connections when the header joist span exceeds 6 feet (1829 mm). Tail joists over 12 feet (3658 mm) long shall be supported at the header by framing anchors or on ledger strips not less than 2 inches by 2 inches (51 mm by 51 mm).

R802.10 Wood trusses.

R802.10.1 Truss design drawings. Truss design drawings, prepared in conformance to Section R802.10.1, shall be provided to the *building official* and *approved* prior to installation. Truss design drawings shall include, at a minimum, the information specified below. Truss design drawings shall be provided with the shipment of trusses delivered to the jobsite.

1. Slope or depth, span and spacing.
2. Location of all joints.
3. Required bearing widths.
4. Design loads as applicable.
   - 4.1. Top chord live load (as determined from Section R301.6).
   - 4.2. Top chord dead load.
   - 4.3. Bottom chord live load.
   - 4.4. Bottom chord dead load.
   - 4.5. Concentrated loads and their points of application.
   - 4.6. Controlling wind and earthquake loads.
5. Adjustments to lumber and joint connector design values for conditions of use.
6. Each reaction force and direction.
7. Joint connector type and description (e.g., size, thickness or gage) and the dimensioned location of each joint connector except where symmetrically located relative to the joint interface.
8. Lumber size, species and *grade for each member.*
9. Connection requirements for:
   - 9.1. Truss to girder-truss.
   - 9.2. Truss ply to ply.
   - 9.3. Field splices.
10. Calculated deflection ratio and/or maximum description for live and total load.
11. Maximum axial compression forces in the truss members to enable the building designer to design the size, connections and anchorage of the permanent continuous lateral bracing. Forces shall be shown on the truss design drawing or on supplemental documents.
12. Required permanent truss member bracing location.

R802.10.2 Design. Wood trusses shall be designed in accordance with accepted engineering practice. The design and manufacture of metal-plate-connected wood trusses shall comply with ANSI/TPI 1. The truss design drawings shall be prepared by a registered professional where required by the statutes of the *jurisdiction* in which the project is to be constructed in accordance with Section R106.1.

R802.10.2.1 Applicability limits. The provisions of this section shall control the design of truss roof framing when snow controls for buildings not greater than 60 feet (18 288 mm) in length perpendicular to the joist, rafter or truss span, not greater than 36 feet (10 973 mm) in width parallel to the joist, rafter or truss span, not more than three stories above *grade plane* in height, and roof slopes not smaller than 3:12 (25 percent slope) or greater than 12:12 (100 percent slope). Truss roof framing constructed in accordance with the provisions of this section shall be limited to sites subjected to a maximum design wind speed of 110 miles per hour (49 m/s), Exposure A, B or C, and a maximum ground snow load of 70 psf (3352 Pa). For consistent loading of all truss types, roof snow load is to be computed as: 0.7 p_g.

R802.10.3 Bracing. Trusses shall be braced to prevent rotation and provide lateral stability in accordance with the requirements specified in the *construction documents* for the building and on the individual truss design drawings. In the absence of specific bracing requirements, trusses shall be braced in accordance with accepted industry practice such as the SBCA *Building Component Safety Information (BCSI) Guide to Good Practice for Handling, Installing & Bracing of Metal Plate Connected Wood Trusses.*

R802.10.4 Alterations to trusses. Truss members shall not be cut, notched, drilled, spliced or otherwise altered in any way without the approval of a registered *design pro-*

fessional. Alterations resulting in the addition of load (e.g., HVAC equipment, water heater) that exceeds the design load for the truss shall not be permitted without verification that the truss is capable of supporting such additional loading.

R802.11 Roof tie-down.

R802.11.1 Uplift resistance. Roof assemblies shall have uplift resistance in accordance with Sections R802.11.1.2 and R802.11.1.3.

Where the uplift force does not exceed 200 pounds, rafters and trusses spaced not more than 24 inches (610 mm) on center shall be permitted to be attached to their supporting wall assemblies in accordance with Table R602.3(1).

Where the basic wind speed does not exceed 90 mph, the wind exposure category is B, the roof pitch is 5:12 or greater, and the roof span is 32 feet (9754 mm) or less, rafters and trusses spaced not more than 24 inches (610 mm) on center shall be permitted to be attached to their supporting wall assemblies in accordance with Table R602.3(1).

R802.11.1.2 Truss uplift resistance. Trusses shall be attached to supporting wall assemblies by connections capable of resisting uplift forces as specified on the truss design drawings. Uplift forces shall be permitted to be determined as specified by Table R802.11, if applicable, or as determined by accepted engineering practice.

R802.11.1.3 Rafter uplift resistance. Individual rafters shall be attached to supporting wall assemblies by connections capable of resisting uplift forces as determined by Table R802.11 or as determined by accepted engineering practice. Connections for beams used in a roof system shall be designed in accordance with accepted engineering practice.

SECTION R803 ROOF SHEATHING

R803.1 Lumber sheathing. Allowable spans for lumber used as roof sheathing shall conform to Table R803.1. Spaced lumber sheathing for wood shingle and shake roofing shall conform to the requirements of Sections R905.7 and R905.8. Spaced lumber sheathing is not allowed in Seismic Design Category D_2.

TABLE R803.1 MINIMUM THICKNESS OF LUMBER ROOF SHEATHING

RAFTER OR BEAM SPACING (inches)	MINIMUM NET THICKNESS (inches)
24	$^5/_8$
48[a]	$1^1/_2$ T & G
60[b]	
72[c]	

For SI: 1 inch = 25.4 mm.

a. Minimum 270 F_b, 340,000 *E*.

b. Minimum 420 F_b, 660,000 *E*.

c. Minimum 600F_b, 1,150,000 *E*.

R803.2 Wood structural panel sheathing.

R803.2.1 Identification and grade. Wood structural panels shall conform to DOC PS 1, DOC PS 2 or, when manufactured in Canada, CSA 0437 or CSA 0325, and shall be identified for grade, bond classification, and Performance Category by a grade mark or certificate of inspection issued by an *approved* agency. Wood structural panels shall comply with the grades specified in Table R503.2.1.1(1).

R803.2.1.1 Exposure durability. All wood structural panels, when designed to be permanently exposed in outdoor applications, shall be of an exterior exposure durability. Wood structural panel roof sheathing exposed to the underside may be of interior type bonded with exterior glue, identified as Exposure 1.

R803.2.1.2 Fire-retardant-treated plywood. The allowable unit stresses for fire-retardant-treated plywood, including fastener values, shall be developed from an *approved* method of investigation that considers the effects of anticipated temperature and humidity to which the fire-retardant-treated plywood will be subjected, the type of treatment and redrying process. The fire-retardant-treated plywood shall be graded by an *approved agency.*

R803.2.2 Allowable spans. The maximum allowable spans for wood structural panel roof sheathing shall not exceed the values set forth in Table R503.2.1.1(1), or APA E30.

R803.2.3 Installation. Wood structural panel used as roof sheathing shall be installed with joints staggered or not staggered in accordance with Table R602.3(1), or APA E30 for wood roof framing or with Table R804.3 for steel roof framing.

SECTION R804 STEEL ROOF FRAMING

R804.1 General. Elements shall be straight and free of any defects that would significantly affect their structural performance. Cold-formed steel roof framing members shall comply with the requirements of this section.

R804.1.1 Applicability limits. The provisions of this section shall control the construction of cold-formed steel roof framing for buildings not greater than 60 feet (18 288 mm) perpendicular to the joist, rafter or truss span, not greater than 40 feet (12 192 mm) in width parallel to the joist span or truss, less than or equal to three stories above *grade* plane and with roof slopes not less than 3:12 (25-percent slope) or greater than 12:12 (100-percent slope). Cold-formed steel roof framing constructed in accordance with the provisions of this section shall be limited to sites subjected to a maximum design wind speed of 110 miles per hour (49 m/s), Exposure B or C, and a maximum ground snow load of 70 pounds per square foot (3350 Pa).

TABLE R802.11
RAFTER OR TRUSS UPLIFT CONNECTION FORCES FROM WIND (POUNDS PER CONNECTION)[a, b, c, d, e, f, g, h]

RAFTER OR TRUSS SPACING	ROOF SPAN (feet)	EXPOSURE B							
		Basic Wind Speed (mph)							
		85		90		100		110	
		Roof Pitch		Roof Pitch		Roof Pitch		Roof Pitch	
		< 5:12	≥ 5:12	< 5:12	≥ 5:12	< 5:12	≥ 5:12	< 5:12	≥ 5:12
12″ o.c.	12	47	41	62	54	93	81	127	110
	18	59	51	78	68	119	104	165	144
	24	70	61	93	81	145	126	202	176
	28	77	67	104	90	163	142	227	197
	32	85	74	115	100	180	157	252	219
	36	93	81	126	110	198	172	277	241
	42	105	91	143	124	225	196	315	274
	48	116	101	159	138	251	218	353	307
16″ o.c.	12	63	55	83	72	124	108	169	147
	18	78	68	103	90	159	138	219	191
	24	93	81	124	108	193	168	269	234
	28	102	89	138	120	217	189	302	263
	32	113	98	153	133	239	208	335	291
	36	124	108	168	146	264	230	369	321
	42	139	121	190	165	299	260	420	365
	48	155	135	212	184	335	291	471	410
24″ o.c.	12	94	82	124	108	186	162	254	221
	18	117	102	155	135	238	207	329	286
	24	140	122	186	162	290	252	404	351
	28	154	134	208	181	326	284	454	395
	32	170	148	230	200	360	313	504	438
	36	186	162	252	219	396	345	554	482
	42	209	182	285	248	449	391	630	548
	48	232	202	318	277	502	437	706	614
RAFTER OR TRUSS SPACING	ROOF SPAN (feet)	EXPOSURE C							
		Basic Wind Speed (mph)							
		85		90		100		110	
		Roof Pitch		Roof Pitch		Roof Pitch		Roof Pitch	
		< 5:12	≥ 5:12	< 5:12	≥ 5:12	< 5:12	≥ 5:12	< 5:12	≥ 5:12
12″ o.c.	12	94	82	114	99	157	137	206	179
	18	120	104	146	127	204	177	268	233
	24	146	127	179	156	251	218	330	287
	28	164	143	201	175	283	246	372	324
	32	182	158	224	195	314	273	414	360
	36	200	174	246	214	346	301	456	397
	42	227	197	279	243	394	343	520	452
	48	254	221	313	272	441	384	583	507

(continued)

TABLE R802.11—continued
RAFTER OR TRUSS UPLIFT CONNECTION FORCES FROM WIND (POUNDS PER CONNECTION)[a, b, c, d, e, f, g, h]

RAFTER OR TRUSS SPACING	ROOF SPAN (feet)	EXPOSURE C							
		Basic Wind Speed (mph)							
		85		90		100		110	
		Roof Pitch		Roof Pitch		Roof Pitch		Roof Pitch	
		< 5:12	≥ 5:12	< 5:12	≥ 5:12	< 5:12	≥ 5:12	< 5:12	≥ 5:12
16″ o.c.	12	125	109	152	132	209	182	274	238
	18	160	139	194	169	271	236	356	310
	24	194	169	238	207	334	291	439	382
	28	218	190	267	232	376	327	495	431
	32	242	211	298	259	418	364	551	479
	36	266	231	327	284	460	400	606	527
	42	302	263	372	324	524	456	691	601
	48	338	294	416	362	587	511	775	674
24″ o.c.	12	188	164	228	198	314	273	412	358
	18	240	209	292	254	408	355	536	466
	24	292	254	358	311	502	437	660	574
	28	328	285	402	350	566	492	744	647
	32	364	317	448	390	628	546	828	720
	36	400	348	492	428	692	602	912	793
	42	454	395	558	485	786	684	1040	905
	48	508	442	626	545	882	767	1166	1014

For SI: 1 inch = 25.4 mm, 1 foot = 304.8 mm, 1 mile per hour = 0.447 m/s, 1 pound = 0.454 kg, 1 pound per linear foot = 14.5 N/m.

a. The uplift connection forces are based on a maximum 33-foot mean roof height and Wind Exposure Category B or C. For Exposure D, the uplift connection force shall be selected from the Exposure C portion of the table using the next highest tabulated basic wind speed. The Adjustment Coefficients in Table R301.2(3) shall not be used to multiply the above forces for Exposures C and D or for other mean roof heights.

b. The uplift connection forces include an allowance for roof and ceiling assembly dead load of 15 psf.

c. The tabulated uplift connection forces are limited to a maximum roof overhang of 24 inches.

d. The tabulated uplift connection forces shall be permitted to be multiplied by 0.75 for connections not located within 8 feet of building corners.

e. For buildings with hip roofs with 5:12 and greater pitch, the tabulated uplift connection forces shall be permitted to be multiplied by 0.70. This reduction shall not be combined with any other reduction in tabulated forces.

f. For wall-to-wall and wall-to-foundation connections, the uplift connection force shall be permitted to be reduced by 60 plf for each full wall above.

g. Linear interpolation between tabulated roof spans and wind speeds shall be permitted.

h. The tabulated forces for a 12-inch on-center spacing shall be permitted to be used to determine the uplift load in pounds per linear foot.

R804.1.2 In-line framing. Cold-formed steel roof framing constructed in accordance with Section R804 shall be located in line with load-bearing studs in accordance with Figure R804.1.2 and the tolerances specified as follows:

1. The maximum tolerance shall be $^3/_4$ inch (19.1 mm) between the centerline of the horizontal framing member and the centerline of the vertical framing member.
2. Where the centerline of the horizontal framing member and bearing stiffener are located to one side of the center line of the vertical framing member, the maximum tolerance shall be $^1/_8$ inch (3 mm) between the web of the horizontal framing member and the edge of the vertical framing member.

R804.2 Structural framing. Load-bearing, cold-formed steel roof framing members shall comply with Figure R804.2(1) and with the dimensional and minimum thickness requirements specified in Tables R804.2(1) and R804.2(2). Tracks shall comply with Figure R804.2(2) and shall have a minimum flange width of $1^1/_4$ inches (32 mm).

R804.2.1 Material. Load-bearing, cold-formed steel framing members shall be cold-formed to shape from structural quality sheet steel complying with the requirements of one of the following:

1. ASTM A 653: *Grades* 33 and 50 (Class 1 and 3).
2. ASTM A 792: *Grades* 33 and 50A.
3. ASTM A 1003: Structural *Grades* 33 Type H and 50 Type H.

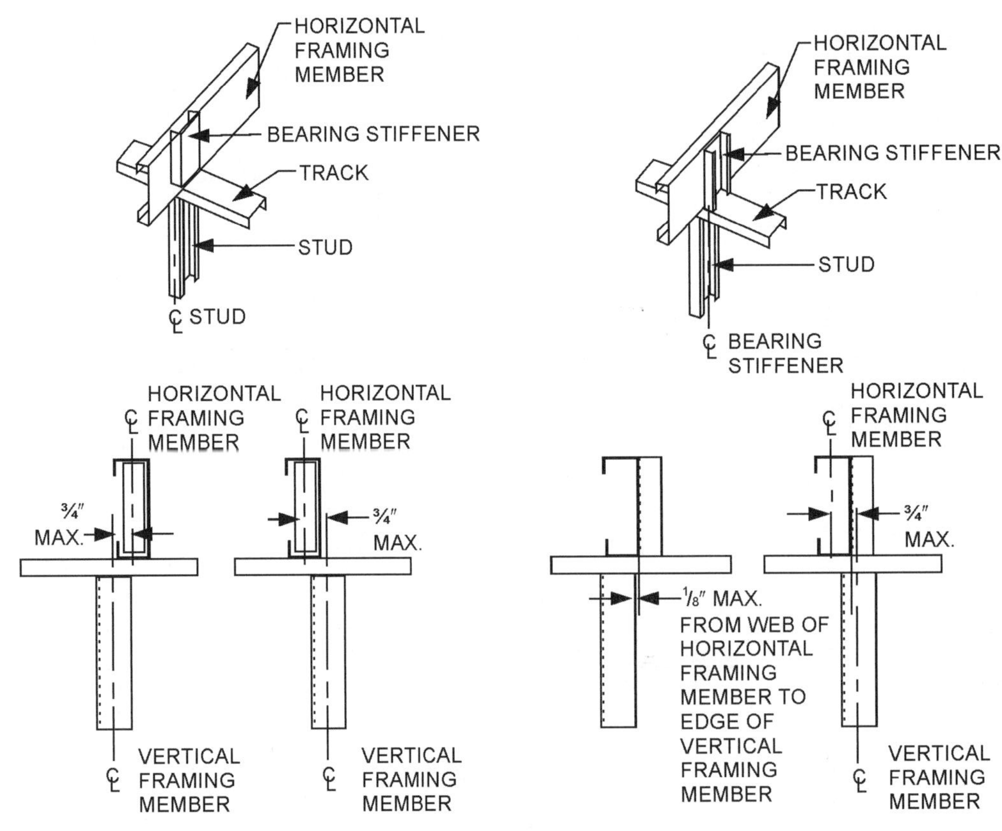

For SI: 1 inch = 25.4 mm.

FIGURE R804.1.2
IN-LINE FRAMING

TABLE R804.2(1)
LOAD-BEARING COLD-FORMED STEEL MEMBER SIZES

NOMINAL MEMBER SIZE MEMBER DESIGNATION[a]	WEB DEPTH (inches)	MINIMUM FLANGE WIDTH (inches)	MAXIMUM FLANGE WIDTH (inches)	MINIMUM LIP SIZE (inches)
350S162-t	3.5	1.625	2	0.5
550S162-t	5.5	1.625	2	0.5
800S162-t	8	1.625	2	0.5
1000S162-t	10	1.625	2	0.5
1200S162-t	12	1.625	2	0.5

For SI: 1 inch = 25.4 mm.

a. The member designation is defined by the first number representing the member depth in hundredths of an inch, the letter "S" representing a stud or joist member, the second number representing the flange width in hundredths of an inch, and the letter "t" shall be a number representing the minimum base metal thickness in mils [see Table R804.2(2)].

TABLE R804.2(2)
MINIMUM THICKNESS OF COLD-FORMED STEEL MEMBERS

DESIGNATION THICKNESS (mils)	MINIMUM BASE STEEL THICKNESS (inch)
33	0.0329
43	0.0428
54	0.0538
68	0.0677
97	0.0966

For SI: 1 inch = 25.4 mm, 1 mil = 0.0254 mm.

R804.2.2 Identification. Load-bearing, cold-formed steel framing members shall have a legible *label,* stencil, stamp or embossment with the following information as a minimum:

1. Manufacturer's identification.
2. Minimum base steel thickness in inches (mm).
3. Minimum coating designation.
4. Minimum yield strength, in kips per square inch (ksi) (MPa).

R804.2.3 Corrosion protection. Load-bearing, cold-formed steel framing shall have a metallic coating complying with ASTM A 1003 and one of the following:

1. A minimum of G 60 in accordance with ASTM A 653.
2. A minimum of AZ 50 in accordance with ASTM A 792.

R804.2.4 Fastening requirements. Screws for steel-to-steel connections shall be installed with a minimum edge distance and center-to-center spacing of $^1/_2$ inch (13 mm), shall be self-drilling tapping, and shall conform to ASTM C 1513. Structural sheathing shall be attached to cold-formed steel roof rafters with minimum No. 8 self-drilling tapping screws that conform to ASTM C 1513. Screws for attaching structural sheathing to cold-formed steel roof framing shall have a minimum head diameter of 0.292 inch (7.4 mm) with countersunk heads and shall be installed with a minimum edge distance of $^3/_8$ inch (10 mm). Gypsum board ceilings shall be attached to cold-formed steel joists with minimum No. 6 screws conforming to ASTM C 954 or ASTM C 1513 with a bugle-head style and shall be installed in accordance with Section R805. For all connections, screws shall extend through the steel a minimum of three exposed threads. All fasteners shall have rust-inhibitive coating suitable for the installation in which they are being used, or be manufactured from material not susceptible to corrosion.

Where No. 8 screws are specified in a steel-to-steel connection, reduction of the required number of screws in the connection is permitted in accordance with the reduction factors in Table R804.2.4 when larger screws are used or when one of the sheets of steel being connected is thicker than 33 mils (0.84 mm). When applying the reduction factor, the resulting number of screws shall be rounded up.

TABLE R804.2.4
SCREW SUBSTITUTION FACTOR

SCREW SIZE	THINNEST CONNECTED STEEL SHEET (mils)	
	33	43
#8	1.0	0.67
#10	0.93	0.62
#12	0.86	0.56

For SI: 1 mil = 0.0254 mm.

R804.2.5 Web holes, web hole reinforcing and web hole patching. Web holes, web hole reinforcing, and web hole patching shall be in accordance with this section.

R804.2.5.1 Web holes. Web holes in roof framing members shall comply with all of the following conditions:

1. Holes shall conform to Figure R804.2.5.1;
2. Holes shall be permitted only along the centerline of the web of the framing member;
3. Center-to-center spacing of holes shall not be less than 24 inches (610 mm);

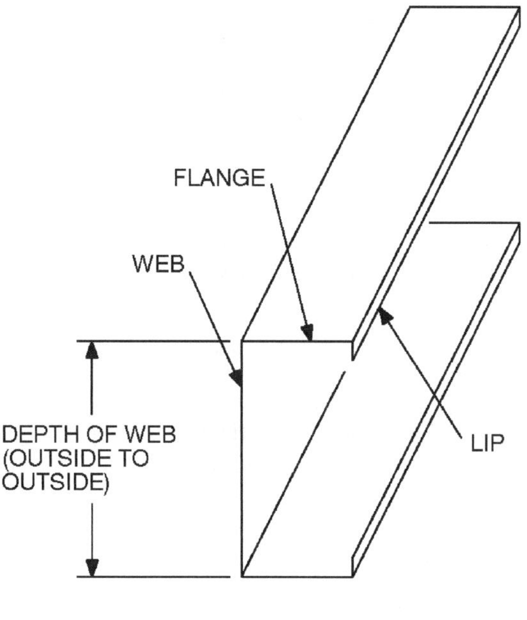

FIGURE R804.2(1)
C-SHAPED SECTION

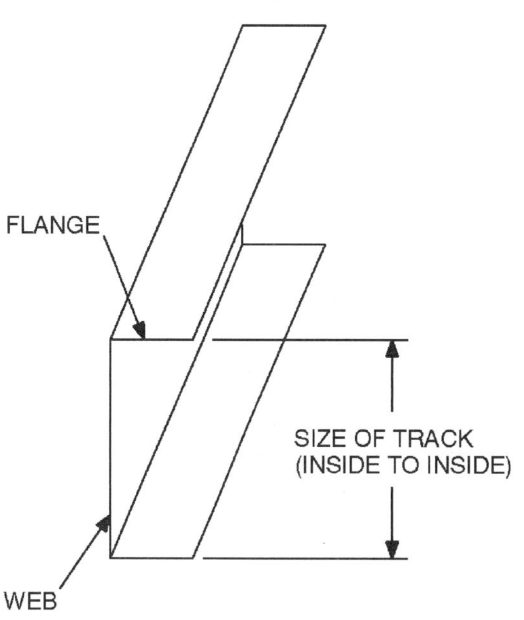

FIGURE R804.2(2)
TRACK SECTION

4. The web hole width shall not be greater than one-half the member depth, or $2^1/_2$ inches (64.5 mm);
5. Holes shall have a web hole length not exceeding $4^1/_2$ inches (114 mm); and
6. The minimum distance between the edge of the bearing surface and the edge of the web hole shall not be less than 10 inches (254 mm).

Framing members with web holes not conforming to the above requirements shall be reinforced in accordance with Section R804.2.5.2, patched in accordance with Section R804.2.5.3 or designed in accordance with accepted engineering practices.

R804.2.5.2 Web hole reinforcing. Reinforcement of web holes in ceiling joists not conforming to the requirements of Section R804.2.5.1 shall be permitted if the hole is located fully within the center 40 percent of the span and the depth and length of the hole does not exceed 65 percent of the flat width of the web. The reinforcing shall be a steel plate or C-shape section with a hole that does not exceed the web hole size limitations of Section R804.2.5.1 for the member being reinforced. The steel reinforcing shall be the same thickness as the receiving member and shall extend at least 1 inch (25.4 mm) beyond all edges of the hole. The steel reinforcing shall be fastened to the web of the receiving member with No. 8 screws spaced no greater than 1 inch (25.4 mm) center-to-center along the edges of the patch with minimum edge distance of $^1/_2$ inch (13 mm).

R804.2.5.3 Hole patching. Patching of web holes in roof framing members not conforming to the requirements in Section R804.2.5.1 shall be permitted in accordance with either of the following methods:

1. Framing members shall be replaced or designed in accordance with accepted engineering practices where web holes exceed the following size limits:
   1.1. The depth of the hole, measured across the web, exceeds 70 percent of the flat width of the web; or
   1.2. The length of the hole measured along the web, exceeds 10 inches (254 mm) or the depth of the web, whichever is greater.
2. Web holes not exceeding the dimensional requirements in Section R804.2.5.3, Item 1, shall be patched with a solid steel plate, stud section or track section in accordance with Figure R804.2.5.3. The steel patch shall, as a minimum, be the same thickness as the receiving member and shall extend at least 1 inch (25 mm) beyond all edges of the hole. The steel patch shall be fastened to the web of the receiving member with No.8 screws spaced no greater than 1 inch (25 mm) center-to-center along the edges of the patch with minimum edge distance of $^1/_2$ inch (13 mm).

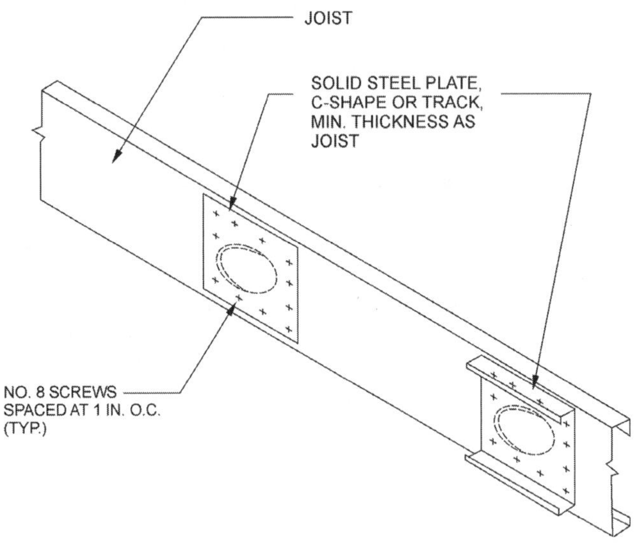

For SI: 1 inch = 25.4 mm.

**FIGURE R804.2.5.3
WEB HOLE PATCH**

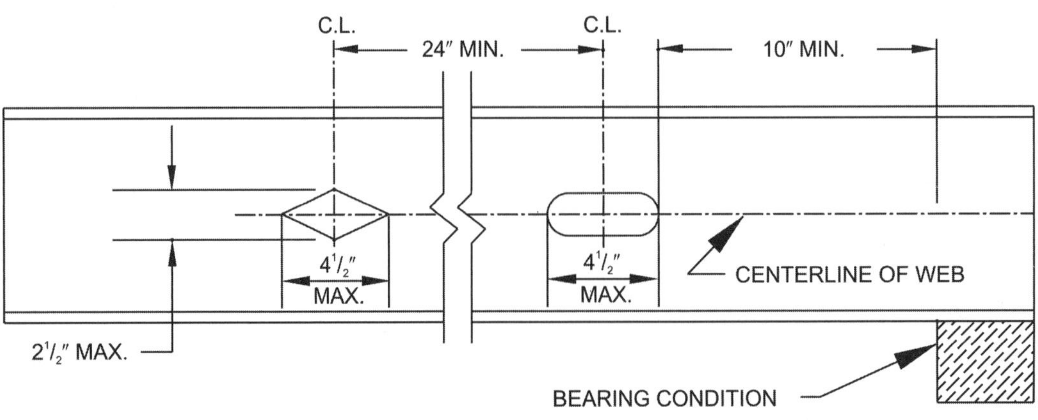

For SI: 1 inch = 25.4 mm.

**FIGURE R804.2.5.1
WEB HOLES**

R804.3 Roof construction. Cold-formed steel roof systems constructed in accordance with the provisions of this section shall consist of both ceiling joists and rafters in accordance with Figure R804.3 and fastened in accordance with Table R804.3, and hip framing in accordance with Section R804.3.3.

R804.3.1 Ceiling joists. Cold-formed steel ceiling joists shall be in accordance with this section.

R804.3.1.1 Minimum ceiling joist size. Ceiling joist size and thickness shall be determined in accordance with the limits set forth in Tables R804.3.1.1(1) through R804.3.1.1(8). When determining the size of ceiling joists, the lateral support of the top flange shall be classified as unbraced, braced at mid-span or braced at third points in accordance with Section R804.3.1.4. Where sheathing material is attached to the top flange of ceiling joists or where the bracing is spaced closer than third point of the joists, the "third point" values from Tables R804.3.1.1(1) through R804.3.1.1(8) shall be used.

Ceiling joists shall have a bearing support length of not less than $1^1/_2$ inches (38 mm) and shall be connected to roof rafters (heel joint) with No. 10 screws in accordance with Figures R804.3.1.1(1) and R804.3.1.1(2) and Table 804.3.1.1(9).

When continuous joists are framed across interior bearing supports, the interior bearing supports shall be located within 24 inches (610 mm) of midspan of the ceiling joist, and the individual spans shall not exceed the applicable spans in Tables R804.3.1.1(2), R804.3.1.1(4), R804.3.1.1(6) and R804.3.1.1(8).

When the *attic* is to be used as an *occupied space,* the ceiling joists shall be designed in accordance with Section R505.

R804.3.1.2 Ceiling joist bearing stiffeners. Where required in Tables R804.3.1.1(1) through R804.3.1.1(8), bearing stiffeners shall be installed at each bearing support in accordance with Figure R804.3.1.1(2). Bearing stiffeners shall be fabricated from a C-shaped or track member in accordance with the one of following:

1. C-shaped bearing stiffeners shall be a minimum 33 mils (0.84 mm) thick.
2. Track bearing stiffener shall be a minimum 43 mils (1.09 mm) thick.

The minimum length of a bearing stiffener shall be the depth of member being stiffened minus $^3/_8$ inch (9.5 mm). Each stiffener shall be fastened to the web of the ceiling joist with a minimum of four No. 8 screws equally spaced as shown in Figure R804.3.1.1(2). Installation of stiffeners shall be permitted on either side of the web.

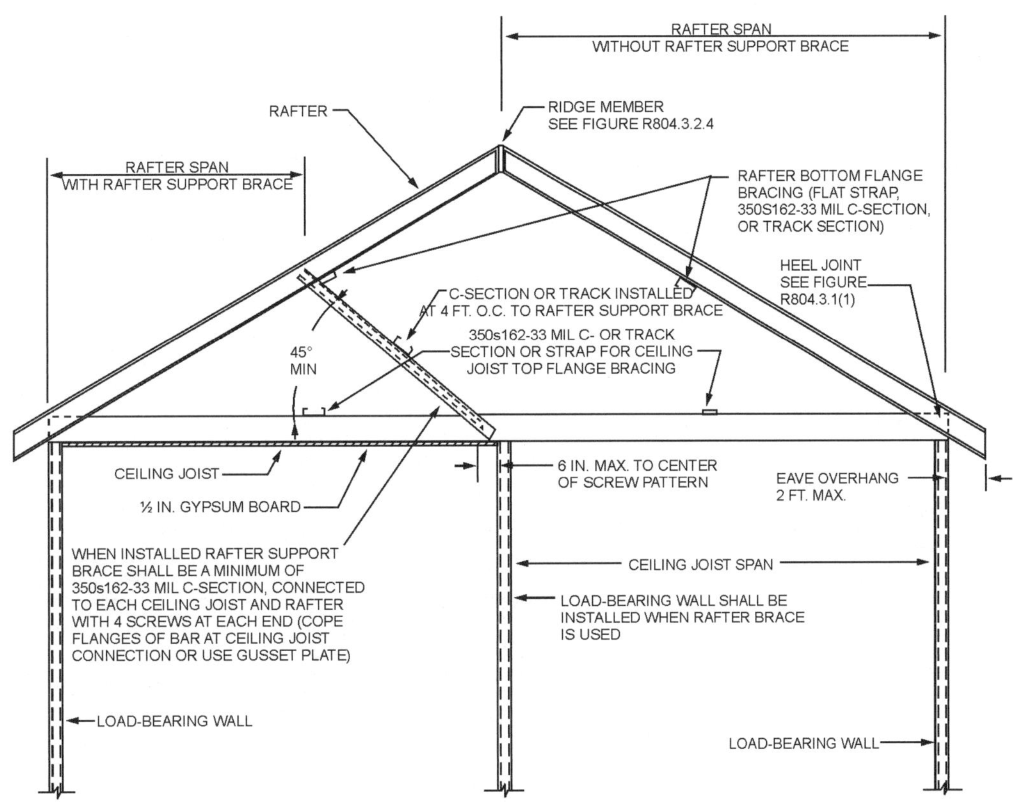

For SI: 1 inch = 25.4 mm, 1 foot = 304.8 mm, 1 mil = 0.0254 mm.

FIGURE R804.3
STEEL ROOF CONSTRUCTION

TABLE R804.3
ROOF FRAMING FASTENING SCHEDULE[a, b]

DESCRIPTION OF BUILDING ELEMENTS	NUMBER AND SIZE OF FASTENERS	SPACING OF FASTENERS
Ceiling joist to top track of load-bearing wall	2 No. 10 screws	Each joist
Roof sheathing (oriented strand board or plywood) to rafter	No. 8 screws	6″ o.c. on edges and 12″ o.c. at interior supports. 6″ o.c. at gable end truss
Truss to bearing wall[a]	2 No. 10 screws	Each truss
Gable end truss to end wall top track	No. 10 screws	12″ o.c.
Rafter to ceiling joist	Minimum No. 10 screws, per Table R804.3.1.1(9)	Evenly spaced, not less than $^1/_2$″ from all edges

For SI: 1 inch = 25.4 mm, 1 foot = 304.8 mm, 1 pound per square foot = 0.0479 kPa, 1 mil = 0.0254 mm.

a. Screws shall be applied through the flanges of the truss or ceiling joist or a 54-mil clip angle shall be used with two No. 10 screws in each leg. See Section R804.3.9 for additional requirements to resist uplift forces.

b. Spacing of fasteners on roof sheathing panel edges applies to panel edges supported by framing members and at all roof plane perimeters. Blocking of roof sheathing panel edges perpendicular to the framing members shall not be required except at the intersection of adjacent roof planes. Roof perimeter shall be supported by framing members or cold-formed blocking of the same depth and gage as the floor members.

TABLE R804.3.1.1(1)
CEILING JOIST SPANS
SINGLE SPANS WITH BEARING STIFFENERS
10 PSF LIVE LOAD (NO ATTIC STORAGE)[a, b, c] 33 KSI STEEL

MEMBER DESIGNATION	ALLOWABLE SPAN (feet-inches)					
	Lateral Support of Top (Compression) Flange					
	Unbraced		Mid-span Bracing		Third-point Bracing	
	Ceiling Joist Spacing (inches)					
	16	24	16	24	16	24
350S162-33	9′-5″	8′-6″	12′-2″	10′-4″	12′-2″	10′-7″
350S162-43	10′-3″	9′-2″	12′-10″	11′-2″	12′-10″	11′-2″
350S162-54	11′-1″	9′-11″	13′-9″	12′-0″	13′-9″	12′-0″
350S162-68	12′-1″	10′-9″	14′-8″	12′-10″	14′-8″	12′-10″
350S162-97	14′-4″	12′-7″	16′-4″	14′-3″	16′-4″	14′-3″
550S162-33	10′-7″	9′-6″	14′-10″	12′-10″	15′-11″	13′-4″
550S162-43	11′-8″	10′-6″	16′-4″	14′-3″	17′-10″	15′-3″
550S162-54	12′-6″	11′-2″	17′-7″	15′-7″	19′-5″	16′-10″
550S162-68	13′-6″	12′-1″	19′-2″	17′-1″	21′-0″	18′-4″
550S162-97	15′-9″	13′-11″	21′-8″	19′-3″	23′-5″	20′-5″
800S162-33	12′-2″	10′-11″	17′-8″	15′-10″	19′-10″	17′-1″
800S162-43	13′-0″	11′-9″	18′-10″	17′-0″	21′-6″	19′-1″
800S162-54	13′-10″	12′-5″	20′-0″	18′-0″	22′-9″	20′-4″
800S162-68	14′-11″	13′-4″	21′-3″	19′-1″	24′-1″	21′-8″
800S162-97	17′-1″	15′-2″	23′-10″	21′-3″	26′-7″	23′-10″
1000S162-43	13′-11″	12′-6″	20′-2″	18′-3″	23′-1″	20′-9″
1000S162-54	14′-9″	13′-3″	21′-4″	19′-3″	24′-4″	22′-0″
1000S162-68	15′-10″	14′-2″	22′-8″	20′-5″	25′-9″	23′-2″
1000S162-97	18′-0″	16′-0″	25′-3″	22′-7″	28′-3″	25′-4″
1200S162-43	14′-8″	13′-3″	21′-4″	19′-3″	24′-5″	21′-8″
1200S162-54	15′-7″	14′-0″	22′-6″	20′-4″	25′-9″	23′-2″
1200S162-68	16′-8″	14′-11″	23′-11″	21′-6″	27′-2″	24′-6″
1200S162-97	18′-9″	16′-9″	26′-6″	23′-8″	29′-9″	26′-9″

For SI: 1 inch = 25.4 mm, 1 foot = 304.8 mm, 1 pound per square foot = 0.0479 kPa.

a. Deflection criterion: *L*/240 for total loads.

b. Ceiling dead load = 5 psf.

c. Bearing stiffeners are required at all bearing points and concentrated load locations.

TABLE R804.3.1.1(2) CEILING JOIST SPANS TWO EQUAL SPANS WITH BEARING STIFFENERS 10 PSF LIVE LOAD (NO ATTIC STORAGE)[a, b, c] 33 KSI STEEL

MEMBER DESIGNATION	ALLOWABLE SPAN (feet-inches)					
	Lateral Support of Top (Compression) Flange					
	Unbraced		Mid-span Bracing		Third-point Bracing	
	Ceiling Joist Spacing (inches)					
	16	24	16	24	16	24
350S162-33	12′-11″	10′-11″	13′-5″	10′-11″	13′-5″	10′-11″
350S162-43	14′-2″	12′-8″	15′-10″	12′-11″	15′-10″	12′-11″
350S162-54	15′-6″	13′-10″	17′-1″	14′-6″	17′-9″	14′-6″
350S162-68	17′-3″	15′-3″	18′-6″	16′-1″	19′-8″	16′-1″
350S162-97	20′-10″	18′-4″	21′-5″	18′-10″	21′-11″	18′-10″
550S162-33	14′-4″	12′-11″	16′-7″	14′-1″	17′-3″	14′-1″
550S162-43	16′-0″	14′-1″	17′-11″	16′-1″	20′-7″	16′-10″
550S162-54	17′-4″	15′-6″	19′-5″	17′-6″	23′-2″	19′-0″
550S162-68	19′-1″	16′-11″	20′-10″	18′-8″	25′-2″	21′-5″
550S162-97	22′-8″	19′-9″	23′-6″	20′-11″	27′-11″	25′-1″
800S162-33	16′-5″	14′-10″	19′-2″	17′-3″	23′-1″	18′-3″
800S162-43	17′-9″	15′-11″	20′-6″	18′-5″	25′-0″	22′-6″
800S162-54	19′-1″	17′-1″	21′-8″	19′-6″	26′-4″	23′-9″
800S162-68	20′-9″	18′-6″	23′-1″	20′-9″	28′-0″	25′-2″
800S162-97	24′-5″	21′-6″	26′-0″	23′-2″	31′-1″	27′-9″
1000S162-43	18′-11″	17′-0″	21′-11″	19′-9″	26′-8″	24′-1″
1000S162-54	20′-3″	18′-2″	23′-2″	20′-10″	28′-2″	25′-5″
1000S162-68	21′-11″	19′-7″	24′-7″	22′-2″	29′-10″	26′-11″
1000S162-97	25′-7″	22′-7″	27′-6″	24′-6″	33′-0″	29′-7″
1200S162-43	19′-11″	17′-11″	23′-1″	20′-10″	28′-3″	25′-6″
1200S162-54	21′-3″	19′-1″	24′-5″	22′-0″	29′-9″	26′-10″
1200S162-68	23′-0″	20′-7″	25′-11″	23′-4″	31′-6″	28′-4″
1200S162-97	26′-7″	23′-6″	28′-9″	25′-10″	34′-8″	31′-1″

For SI: 1 inch = 25.4 mm, 1 foot = 304.8 mm, 1 pound per square foot = 0.0479 kPa.

a. Deflection criterion: *L*/240 for total loads.

b. Ceiling dead load = 5 psf.

c. Bearing stiffeners are required at all bearing points and concentrated load locations.

**TABLE R804.3.1.1(3)
CEILING JOIST SPANS
SINGLE SPANS WITH BEARING STIFFENERS
20 PSF LIVE LOAD (LIMITED ATTIC STORAGE)[a, b, c] 33 KSI STEEL**

MEMBER DESIGNATION	ALLOWABLE SPAN (feet-inches)					
	Lateral Support of Top (Compression) Flange					
	Unbraced		Mid-span Bracing		Third-point Bracing	
	Ceiling Joist Spacing (inches)					
	16	24	16	24	16	24
350S162-33	8′-2″	7′-2″	9′-9″	8′-1″	9′-11″	8′-1″
350S162-43	8′-10″	7′-10″	11′-0″	9′-5″	11′-0″	9′-7″
350S162-54	9′-6″	8′-6″	11′-9″	10′-3″	11′-9″	10′-3″
350S162-68	10′-4″	9′-2″	12′-7″	11′-0″	12′-7″	11′-0″
350S162-97	12′-1″	10′-8″	14′-0″	12′-0″	14′-0″	12′-0″
550S162-33	9′-2″	8′-3″	12′-2″	10′-2″	12′-6″	10′-5″
550S162-43	10′-1″	9′-1″	13′-7″	11′-7″	14′-5″	12′-2″
550S162-54	10′-9″	9′-8″	14′-10″	12′-10″	15′-11″	13′-6″
550S162-68	11′-7″	10′-4″	16′-4″	14′-0″	17′-5″	14′-11″
550S162-97	13′-4″	11′-10″	18′-5″	16′-2″	20′-1″	17′-1″
800S162-33	10′-7″	9′-6″	15′-1″	13′-0″	16′-2″	13′-7″
800S162-43	11′-4″	10′-2″	16′-5″	14′-6″	18′-2″	15′-9″
800S162-54	12′-0″	10′-9″	17′-4″	15′-6″	19′-6″	17′-0″
800S162-68	12′-10″	11′-6″	18′-5″	16′-6″	20′-10″	18′-3″
800S162-97	14′-7″	12′-11″	20′-5″	18′-3″	22′-11″	20′-5″
1000S162-43	12′-1″	10′-11″	17′-7″	15′-10″	19′-11″	17′-3″
1000S162-54	12′-10″	11′-6″	18′-7″	16′-9″	21′-2″	18′-10″
1000S162-68	13′-8″	12′-3″	19′-8″	17′-8″	22′-4″	20′-1″
1000S162-97	15′-4″	13′-8″	21′-8″	19′-5″	24′-5″	21′-11″
1200S162-43	12′-9″	11′-6″	18′-7″	16′-6″	20′-9″	18′-2″
1200S162-54	13′-6″	12′-2″	19′-7″	17′-8″	22′-5″	20′-2″
1200S162-68	14′-4″	12′-11″	20′-9″	18′-8″	23′-7″	21′-3″
1200S162-97	16′-1″	14′-4″	22′-10″	20′-6″	25′-9″	23′-2″

For SI: 1 inch = 25.4 mm, 1 foot = 304.8 mm, 1 pound per square foot = 0.0479 kPa.

a. Deflection criterion: *L*/240 for total loads.

b. Ceiling dead load = 5 psf.

c. Bearing stiffeners are required at all bearing points and concentrated load locations.

TABLE R804.3.1.1(4)
CEILING JOIST SPANS
TWO EQUAL SPANS WITH BEARING STIFFENERS
20 PSF LIVE LOAD (LIMITED ATTIC STORAGE)[a, b, c] 33 KSI STEEL

MEMBER DESIGNATION	ALLOWABLE SPAN (feet-inches)					
	Lateral Support of Top (Compression) Flange					
	Unbraced		Mid-span Bracing		Third-point Bracing	
	Ceiling Joist Spacing (inches)					
	16	24	16	24	16	24
350S162-33	10′-2″	8′-4″	10′-2″	8′-4″	10′-2″	8′-4″
350S162-43	12′-1″	9′-10″	12′-1″	9′-10″	12′-1″	9′-10″
350S162-54	13′-3″	11′-0″	13′-6″	11′-0″	13′-6″	11′-0″
350S162-68	14′-7″	12′-3″	15′-0″	12′-3″	15′-0″	12′-3″
350S162-97	17′-6″	14′-3″	17′-6″	14′-3″	17′-6″	14′-3″
550S162-33	12′-5″	10′-9″	13′-2″	10′-9″	13′-2″	10′-9″
550S162-43	13′-7″	12′-1″	15′-6″	12′-9″	15′-8″	12′-9″
550S162-54	14′-11″	13′-4″	16′-10″	14′-5″	17′-9″	14′-5″
550S162-68	16′-3″	14′-5″	18′-0″	16′-1″	20′-0″	16′-4″
550S162-97	19′-1″	16′-10″	20′-3″	18′-0″	23′-10″	19′-5″
800S162-33	14′-3″	12′-4″	16′-7″	12′-4″	16′-7″	12′-4″
800S162-43	15′-4″	13′-10″	17′-9″	16′-0″	21′-8″	17′-9″
800S162-54	16′-5″	14′-9″	18′-10″	16′-11″	22′-11″	20′-6″
800S162-68	17′-9″	15′-11″	20′-0″	18′-0″	24′-3″	21′-10″
800S162-97	20′-8″	18′-3″	22′-3″	19′-11″	26′-9″	24′-0″
1000S162-43	16′-5″	14′-9″	19′-0″	17′-2″	23′-3″	18′-11″
1000S162-54	17′-6″	15′-8″	20′-1″	18′-1″	24′-6″	22′-1″
1000S162-68	18′-10″	16′-10″	21′-4″	19′-2″	25′-11″	23′-4″
1000S162-97	21′-8″	19′-3″	23′-7″	21′-2″	28′-5″	25′-6″
1200S162-43	17′-3″	15′-7″	20′-1″	18′-2″	24′-6″	18′-3″
1200S162-54	18′-5″	16′-6″	21′-3″	19′-2″	25′-11″	23′-5″
1200S162-68	19′-9″	17′-8″	22′-6″	20′-3″	27′-4″	24′-8″
1200S162-97	22′-7″	20′-1″	24′-10″	22′-3″	29′-11″	26′-11″

For SI: 1 inch = 25.4 mm, 1 foot = 304.8 mm, 1 pound per square foot = 0.0479 kPa.

a. Deflection criterion: *L*/240 for total loads.

b. Ceiling dead load = 5 psf.

c. Bearing stiffeners are required at all bearing points and concentrated load locations.

**TABLE R804.3.1.1(5)
CEILING JOIST SPANS
SINGLE SPANS WITHOUT BEARING STIFFENERS
10 PSF LIVE LOAD (NO ATTIC STORAGE)[a, b] 33 KSI STEEL**

MEMBER DESIGNATION	ALLOWABLE SPAN (feet-inches)					
	Lateral Support of Top (Compression) Flange					
	Unbraced		Mid-span Bracing		Third-point Bracing	
	Ceiling Joist Spacing (inches)					
	16	24	16	24	16	24
350S162-33	9′-5″	8′-6″	12′-2″	10′-4″	12′-2″	10′-7″
350S162-43	10′-3″	9′-12″	13′-2″	11′-6″	13′-2″	11′-6″
350S162-54	11′-1″	9′-11″	13′-9″	12′-0″	13′-9″	12′-0″
350S162-68	12′-1″	10′-9″	14′-8″	12′-10″	14′-8″	12′-10″
350S162-97	14′-4″	12′-7″	16′-10″	14′-3″	16′-4″	14′-3″
550S162-33	10′-7″	9′-6″	14′-10″	12′-10″	15′-11″	13′-4″
550S162-43	11′-8″	10′-6″	16′-4″	14′-3″	17′-10″	15′-3″
550S162-54	12′-6″	11′-2″	17′-7″	15′-7″	19′-5″	16′-10″
550S162-68	13′-6″	12′-1″	19′-2″	17′-0″	21′-0″	18′-4″
550S162-97	15′-9″	13′-11″	21′-8″	19′-3″	23′-5″	20′-5″
800S162-33	—	—	—	—	—	—
800S162-43	13′-0″	11′-9″	18′-10″	17′-0″	21′-6″	19′-0″
800S162-54	13′-10″	12′-5″	20′-0″	18′-0″	22′-9″	20′-4″
800S162-68	14′-11″	13′-4″	21′-3″	19′-1″	24′-1″	21′-8″
800S162-97	17′-1″	15′-2″	23′-10″	21′-3″	26′-7″	23′-10″
1000S162-43	—	—	—	—	—	—
1000S162-54	14′-9″	13′-3″	21′-4″	19′-3″	24′-4″	22′-0″
1000S162-68	15′-10″	14′-2″	22′-8″	20′-5″	25′-9″	23′-2″
1000S162-97	18′-0″	16′-0″	25′-3″	22′-7″	28′-3″	25′-4″
1200S162-43	—	—	—	—	—	—
1200S162-54	—	—	—	—	—	—
1200S162-68	16′-8″	14′-11″	23′-11″	21′-6″	27′-2″	24′-6″
1200S162-97	18′-9″	16′-9″	26′-6″	23′-8″	29′-9″	26′-9″

For SI: 1 inch = 25.4 mm, 1 foot = 304.8 mm, 1 pound per square foot = 0.0479 kPa.

a. Deflection criterion: *L*/240 for total loads.

b. Ceiling dead load = 5 psf.

TABLE R804.3.1.1(6)
CEILING JOIST SPANS
TWO EQUAL SPANS WITHOUT BEARING STIFFENERS
10 PSF LIVE LOAD (NO ATTIC STORAGE)[a,b] 33 KSI STEEL

MEMBER DESIGNATION	ALLOWABLE SPAN (feet-inches)					
	Lateral Support of Top (Compression) Flange					
	Unbraced		Mid-span Bracing		Third-point Bracing	
	Ceiling Joist Spacing (inches)					
	16	24	16	24	16	24
350S162-33	11′-9″	8′-11″	11′-9″	8′-11″	11′-9″	8′-11″
350S162-43	14′-2″	11′-7″	14′-11″	11′-7″	14′-11″	11′-7″
350S162-54	15′-6″	13′-10″	17′-1″	13′-10″	17′-7″	13′-10″
350S162-68	17′-3″	15′-3″	18′-6″	16′-1″	19′-8″	16′-1″
350S162-97	20′-10″	18′-4″	21′-5″	18′-9″	21′-11″	18′-9″
550S162-33	13′-4″	9′-11″	13′-4″	9′-11″	13′-4″	9′-11″
550S162-43	16′-0″	13′-6″	17′-9″	13′-6″	17′-9″	13′-6″
550S162-54	17′-4″	15′-6″	19′-5″	16′-10″	21′-9″	16′-10″
550S162-68	19′-1″	16′-11″	20′-10″	18′-8″	24′-11″	20′-6″
550S162-97	22′-8″	20′-0″	23′-9″	21′-1″	28′-2″	25′-1″
800S162-33	—	—	—	—	—	—
800S162-43	17′-9″	15′-7″	20′-6″	15′-7″	21′-0″	15′-7″
800S162-54	19′-1″	17′-1″	21′-8″	19′-6″	26′-4″	23′-10″
800S162-68	20′-9″	18′-6″	23′-1″	20′-9″	28′-0″	25′-2″
800S162-97	24′-5″	21′-6″	26′-0″	23′-2″	31′-1″	27′-9″
1000S162-43	—	—	—	—	—	—
1000S162-54	20′-3″	18′-2″	23′-2″	20′-10″	28′-2″	21′-2″
1000S162-68	21′-11″	19′-7″	24′-7″	22′-2″	29′-10″	26′-11″
1000S162-97	25′-7″	22′-7″	27′-6″	24′-6″	33′-0″	29′-7″
1200S162-43	—	—	—	—	—	—
1200S162-54	—	—	—	—	—	—
1200S162-68	23′-0″	20′-7″	25′-11″	23′-4″	31′-6″	28′-4″
1200S162-97	26′-7″	23′-6″	28′-9″	25′-10″	34′-8″	31′-1″

For SI: 1 inch = 25.4 mm, 1 foot = 304.8 mm, 1 pound per square foot = 0.0479 kPa.

a. Deflection criterion: *L*/240 for total loads.

b. Ceiling dead load = 5 psf.

TABLE R804.3.1.1(7)
CEILING JOIST SPANS
SINGLE SPANS WITHOUT BEARING STIFFENERS
20 PSF LIVE LOAD (LIMITED ATTIC STORAGE)[a, b] 33 KSI STEEL

MEMBER DESIGNATION	ALLOWABLE SPAN (feet-inches)					
	Lateral Support of Top (Compression) Flange					
	Unbraced		Mid-span Bracing		Third-point Bracing	
	Ceiling Joist Spacing (inches)					
	16	24	16	24	16	24
350S162-33	8′-2″	6′-10″	9′-9″	6′-10″	9′-11″	6′-10″
350S162-43	8′-10″	7′-10″	11′-0″	9′-5″	11′-0″	9′-7″
350S162-54	9′-6″	8′-6″	11′-9″	10′-3″	11′-9″	10′-3″
350S162-68	10′-4″	9′-2″	12′-7″	11′-0″	12′-7″	11′-0″
350S162-97	12′-10″	10′-8″	13′-9″	12′-0″	13′-9″	12′-0″
550S162-33	9′-2″	8′-3″	12′-2″	8′-5″	12′-6″	8′-5″
550S162-43	10′-1″	9′-1″	13′-7″	11′-8″	14′-5″	12′-2″
550S162-54	10′-9″	9′-8″	14′-10″	12′-10″	15′-11″	13′-6″
550S162-68	11′-7″	10′-4″	16′-4″	14′-0″	17′-5″	14′-11″
550S162-97	13′-4″	11′-10″	18′-5″	16′-2″	20′-1″	17′-4″
800S162-33	—	—	—	—	—	—
800S162-43	11′-4″	10′-1″	16′-5″	13′-6″	18′-1″	13′-6″
800S162-54	20′-0″	10′-9″	17′-4″	15′-6″	19′-6″	27′-0″
800S162-68	12′-10″	11′-6″	18′-5″	16′-6″	20′-10″	18′-3″
800S162-97	14′-7″	12′-11″	20′-5″	18′-3″	22′-11″	20′-5″
1000S162-43	—	—	—	—	—	—
1000S162-54	12′-10″	11′-6″	18′-7″	16′-9″	21′-2″	15′-5″
1000S162-68	13′-8″	12′-3″	19′-8″	17′-8″	22′-4″	20′-1″
1000S162-97	15′-4″	13′-8″	21′-8″	19′-5″	24′-5″	21′-11″
1200S162-43	—	—	—	—	—	—
1200S162-54	—	—	—	—	—	—
1200S162-68	14′-4″	12′-11″	20′-9″	18′-8″	23′-7″	21′-3″
1200S162-97	16′-1″	14′-4″	22′-10″	20′-6″	25′-9″	23′-2″

For SI: 1 inch = 25.4 mm, 1 foot = 304.8 mm, 1 pound per square foot = 0.0479 kPa.

a. Deflection criterion: *L*/240 for total loads.

b. Ceiling dead load = 5 psf.

**TABLE R804.3.1.1(8)
CEILING JOIST SPANS
TWO EQUAL SPANS WITHOUT BEARING STIFFENERS
20 PSF LIVE LOAD (LIMITED ATTIC STORAGE)[a, b] 33 KSI STEEL**

MEMBER DESIGNATION	ALLOWABLE SPAN (feet-inches)					
	Lateral Support of Top (Compression) Flange					
	Unbraced		Mid-span Bracing		Third-point Bracing	
	Ceiling Joist Spacing (inches)					
	16	24	16	24	16	24
350S162-33	8′-1″	6′-1″	8′-1″	6′-1″	8′-1″	6′-1″
350S162-43	10′-7″	8′-1″	10′-7″	8′-1″	10′-7″	8′-1″
350S162-54	12′-8″	9′-10″	12′-8″	9′-10″	12′-8″	9′-10″
350S162-68	14′-7″	11′-10″	14′-11″	11′-10″	14′-11″	11′-10″
350S162-97	17′-6″	14′-3″	17′-6″	14′-3″	17′-6″	14′-3″
550S162-33	8′-11″	6′-8″	8′-11″	6′-8″	8′-11″	6′-8″
550S162-43	12′-3″	9′-2″	12′-3″	9′-2″	12′-3″	9′-2″
550S162-54	14′-11″	11′-8″	15′-4″	11′-8″	15′-4″	11′-8″
550S162-68	16′-3″	14′-5″	18′-0″	15′-8″	18′-10″	14′-7″
550S162-97	19′-1″	16′-10″	20′-3″	18′-0″	23′-9″	19′-5″
800S162-33	—	—	—	—	—	—
800S162-43	13′-11″	9′-10″	13′-11″	9′-10″	13′-11″	9′-10″
800S162-54	16′-5″	13′-9″	18′-8″	13′-9″	18′-8″	13′-9″
800S162-68	17′-9″	15′-11″	20′-0″	18′-0″	24′-1″	18′-3″
800S162-97	20′-8″	18′-3″	22′-3″	19′-11″	26′-9″	24′-0″
1000S162-43	—	—	—	—	—	—
1000S162-54	17′-6″	13′-11″	19′-1″	13′-11″	19′-1″	13′-11″
1000S162-68	18′-10″	16′-10″	21′-4″	19′-2″	25′-11″	19′-7″
1000S162-97	21′-8″	19′-3″	23′-7″	21′-2″	28′-5″	25′-6″
1200S162-43	—	—	—	—	—	—
1200S162-54	—	—	—	—	—	—
1200S162-68	19′-9″	17′-8″	22′-6″	19′-8″	26′-8″	19′-8″
1200S162-97	22′-7″	20′-1″	24′-10″	22′-3″	29′-11″	26′-11″

For SI: 1 inch = 25.4 mm, 1 foot = 304.8 mm, 1 pound per square foot = 0.0479 kPa.

a. Deflection criterion: *L*/240 for total loads.

b. Ceiling dead load = 5 psf.

TABLE R804.3.1.1(9)
NUMBER OF SCREWS REQUIRED FOR CEILING JOIST TO ROOF RAFTER CONNECTION[a]

ROOF SLOPE	NUMBER OF SCREWS																			
	Building width (feet)																			
	24				28				32				36				40			
	Ground snow load (psf)																			
	20	30	50	70	20	30	50	70	20	30	50	70	20	30	50	70	20	30	50	70
3/12	5	6	9	11	5	7	10	13	6	8	11	15	7	8	13	17	8	9	14	19
4/12	4	5	7	9	4	5	8	10	5	6	9	12	5	7	10	13	6	7	11	14
5/12	3	4	6	7	4	4	6	8	4	5	7	10	5	5	8	11	5	6	9	12
6/12	3	3	5	6	3	4	6	7	4	4	6	8	4	5	7	9	4	5	8	10
7/12	3	3	4	6	3	3	5	7	3	4	6	7	4	4	6	8	4	5	7	9
8/12	2	3	4	5	3	3	5	6	3	4	5	7	3	4	6	8	4	4	6	8
9/12	2	3	4	5	3	3	4	6	3	3	5	6	3	4	5	7	3	4	6	8
10/12	2	2	4	5	2	3	4	5	3	3	5	6	3	3	5	7	3	4	6	7
11/12	2	2	3	4	2	3	4	5	3	3	4	6	3	3	5	6	3	4	5	7
12/12	2	2	3	4	2	3	4	5	2	3	4	5	3	3	5	6	3	4	5	7

For SI: 1 inch = 25.4 mm, 1 foot = 304.8 mm, 1 pound per square foot = 0.0479kPa.

a. Screws shall be No. 10.

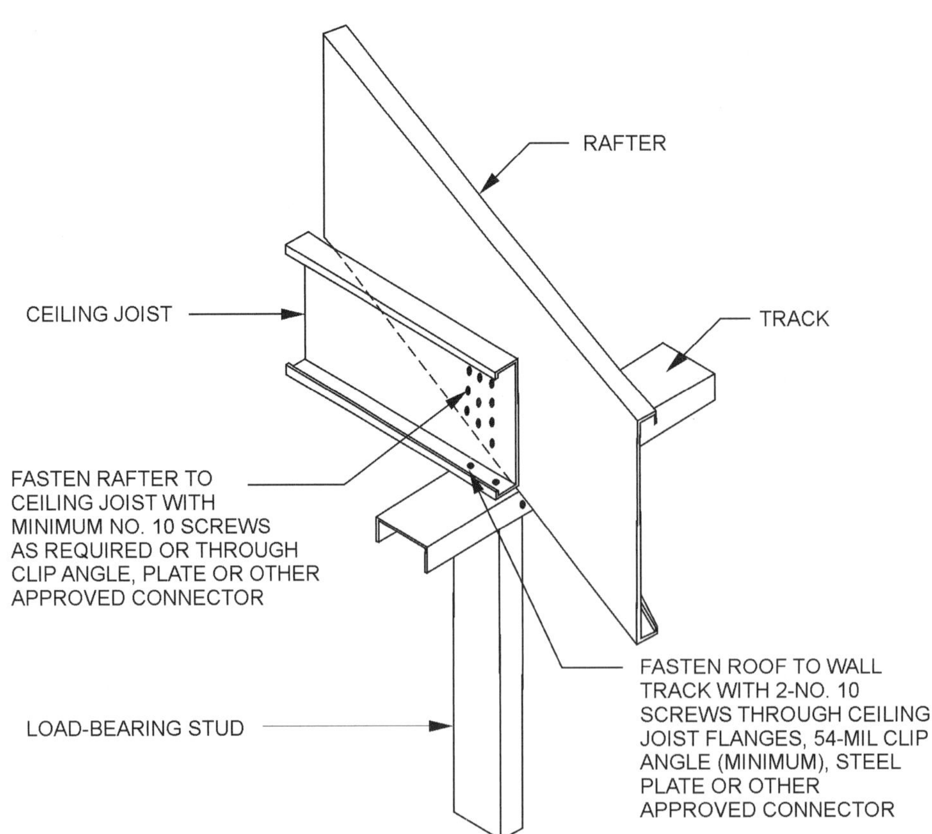

For SI: 1 mil = 0.0254 mm.

FIGURE R804.3.1.1(1)
JOIST TO RAFTER CONNECTION

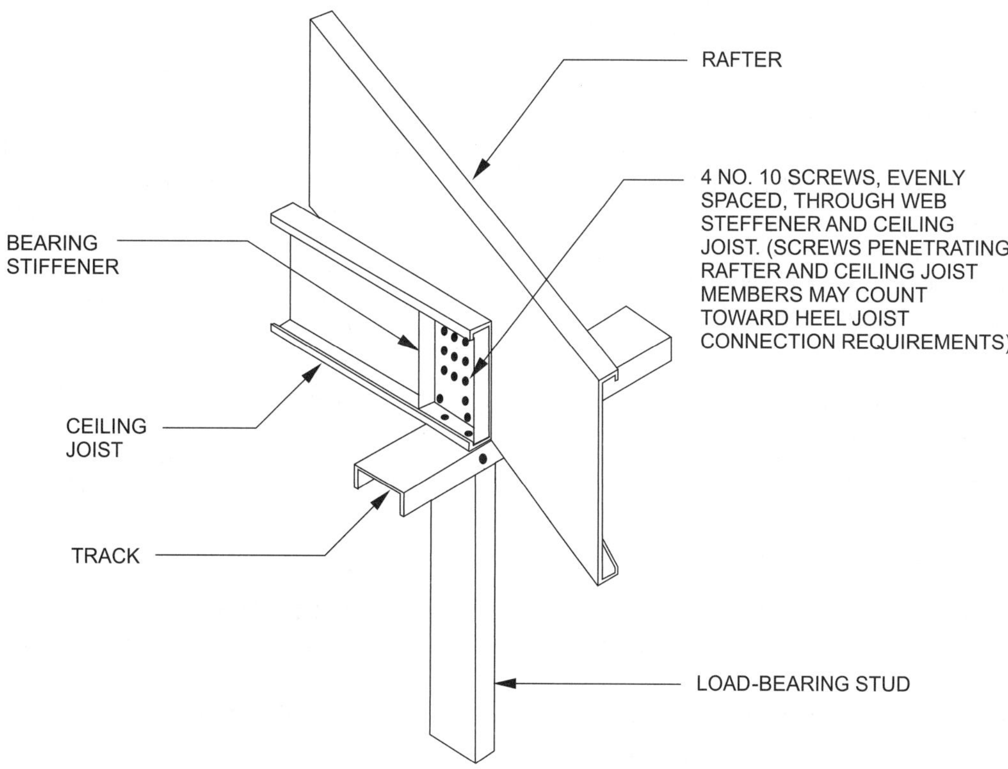

FIGURE R804.3.1.1(2)
BEARING STIFFENER

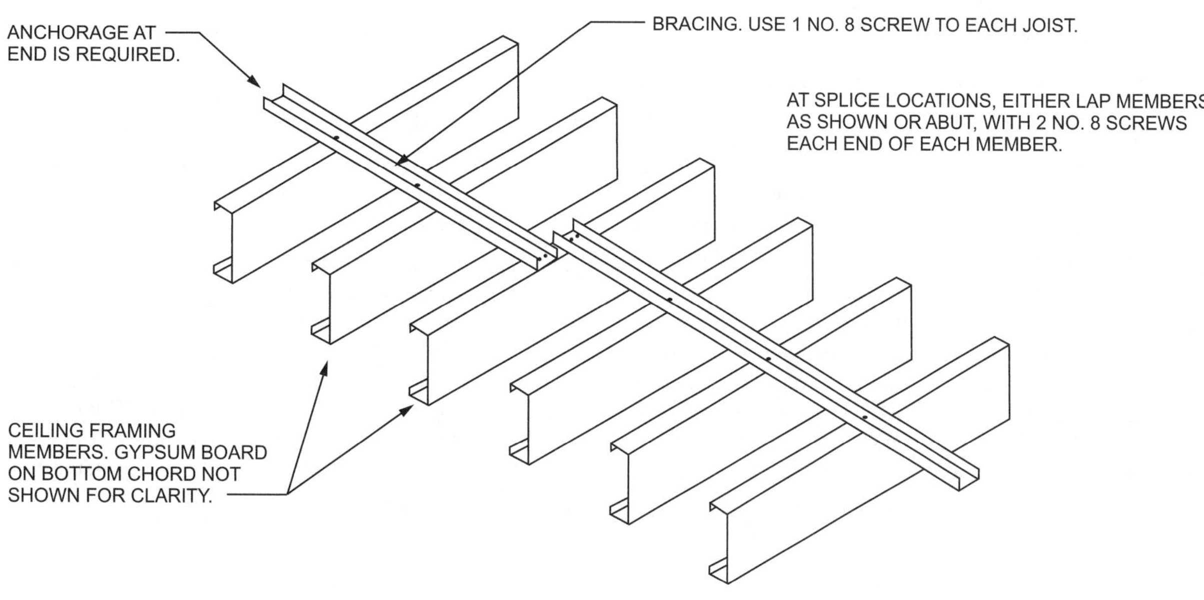

FIGURE R804.3.1.4(1)
CEILING JOIST TOP FLANGE BRACING WITH C-SHAPE, TRACK OR COLD-ROLLED CHANNEL

R804.3.1.3 Ceiling joist bottom flange bracing. The bottom flanges of ceiling joists shall be laterally braced by the application of gypsum board or continuous steel straps installed perpendicular to the joist run in accordance with one of the following:

1. Gypsum board shall be fastened with No. 6 screws in accordance with Section R702.
2. Steel straps with a minimum size of $1^1/_2$ inches by 33 mils (38 mm by 0.84 mm) shall be installed at a maximum spacing of 4 feet (1219 mm). Straps shall be fastened to the bottom flange at each joist with one No. 8 screw and shall be fastened to blocking with two No. 8 screws. Blocking shall be installed between joists at a maximum spacing of 12 feet (3658 mm) measured along a line of continuous strapping (perpendicular to the joist run). Blocking shall also be located at the termination of all straps.

R804.3.1.4 Ceiling joist top flange bracing. The top flanges of ceiling joists shall be laterally braced as required by Tables R804.3.1.1(1) through R804.3.1.1(8), in accordance with one of the following:

1. Minimum 33-mil (0.84 mm) C-shaped member in accordance with Figure R804.3.1.4(1).
2. Minimum 33-mil (0.84 mm) track section in accordance with Figure R804.3.1.4(1).
3. Minimum 33-mil (0.84 mm) hat section in accordance with Figure R804.3.1.4(1).
4. Minimum 54-mil (1.37 mm) $1^1/_2$-inch cold-rolled channel section in accordance with Figure R804.3.1.4(1).
5. Minimum $1^1/_2$-inch by 33-mil (38 mm by 0.84 mm) continuous steel strap in accordance with Figure R804.3.1.4(2).

Lateral bracing shall be installed perpendicular to the ceiling joists and shall be fastened to the top flange of each joist with one No. 8 screw. Blocking shall be installed between joists in line with bracing at a maximum spacing of 12 feet (3658 mm) measured perpendicular to the joists. Ends of lateral bracing shall be attached to blocking or anchored to a stable building component with two No. 8 screws.

R804.3.1.5 Ceiling joist splicing. Splices in ceiling joists shall be permitted, if ceiling joist splices are supported at interior bearing points and are constructed in accordance with Figure R804.3.1.5. The number of screws on each side of the splice shall be the same as required for the heel joint connection in Table R804.3.1.1(9).

R804.3.2 Roof rafters. Cold-formed steel roof rafters shall be in accordance with this section.

R804.3.2.1 Minimum roof rafter sizes. Roof rafter size and thickness shall be determined in accordance with the limits set forth in Tables R804.3.2.1(1) and R804.3.2.1(2) based on the horizontal projection of the roof rafter span. For determination of roof rafter sizes, reduction of roof spans shall be permitted when a roof rafter support brace is installed in accordance with Section R804.3.2.2. The reduced roof rafter span shall be taken as the larger of the distance from the roof rafter support brace to the ridge or to the heel measured horizontally.

For the purpose of determining roof rafter sizes in Tables R804.3.2.1(1) and R804.3.2.1(2), wind speeds shall be converted to equivalent ground snow loads in accordance with Table R804.3.2.1(3). Roof rafter sizes shall be based on the higher of the ground snow load or the equivalent snow load converted from the wind speed.

R804.3.2.1.1 Eave overhang. Eave overhangs shall not exceed 24 inches (610 mm) measured horizontally.

R804.3.2.1.2 Rake overhangs. Rake overhangs shall not exceed 12 inches (305 mm) measured horizontally. Outlookers at gable endwalls shall be installed in accordance with Figure R804.3.2.1.2.

R804.3.2.2 Roof rafter support brace. When used to reduce roof rafter spans in determining roof rafter sizes, a roof rafter support brace shall meet all of the following conditions:

1. Minimum 350S162-33 C-shaped brace member with maximum length of 8 feet (2438 mm).
2. Minimum brace member slope of 45 degrees (0.785 rad) to the horizontal.
3. Minimum connection of brace to a roof rafter and ceiling joist with four No.10 screws at each end.
4. Maximum 6 inches (152 mm) between brace/ceiling joist connection and load-bearing wall below.
5. Each roof rafter support brace greater than 4 feet (1219 mm) in length, shall be braced with a supplemental brace having a minimum size of 350S162-33 or 350T162-33 such that the maximum unsupported length of the roof rafter support brace is 4 feet (1219 mm). The supplemental brace shall be continuous and shall be connected to each roof rafter support brace using two No.8 screws.

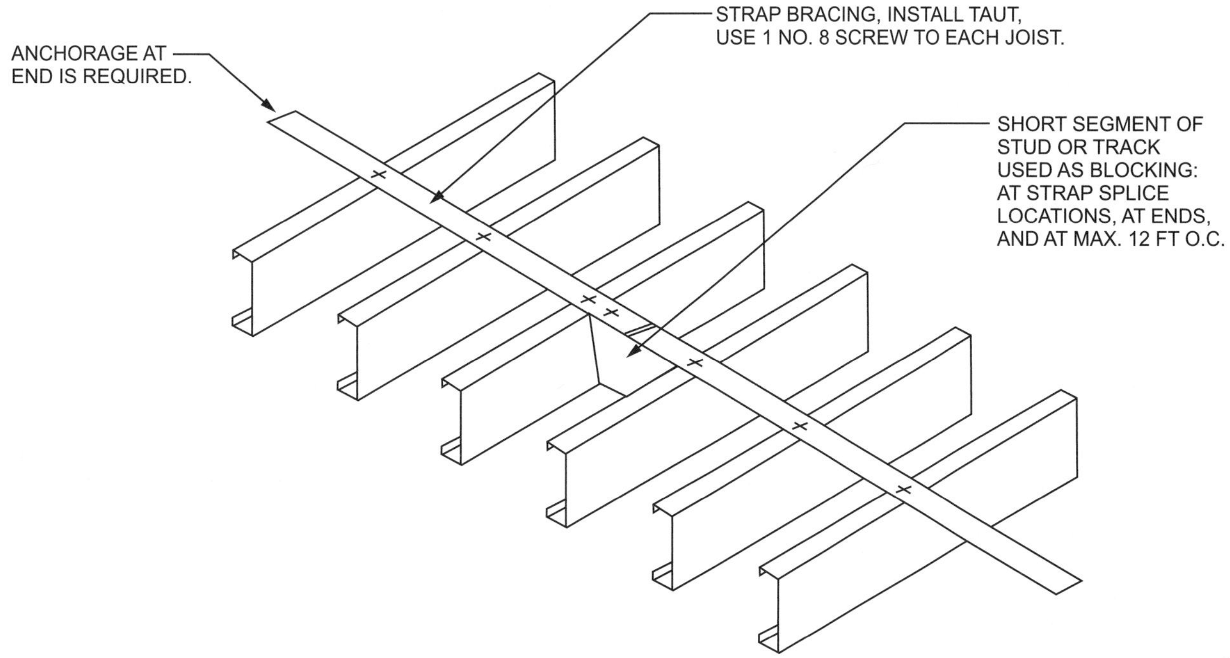

For SI: 1 foot = 304.8 mm.

FIGURE R804.3.1.4(2)
CEILING JOIST TOP FLANGE BRACING WITH CONTINUOUS STEEL STRAP AND BLOCKING

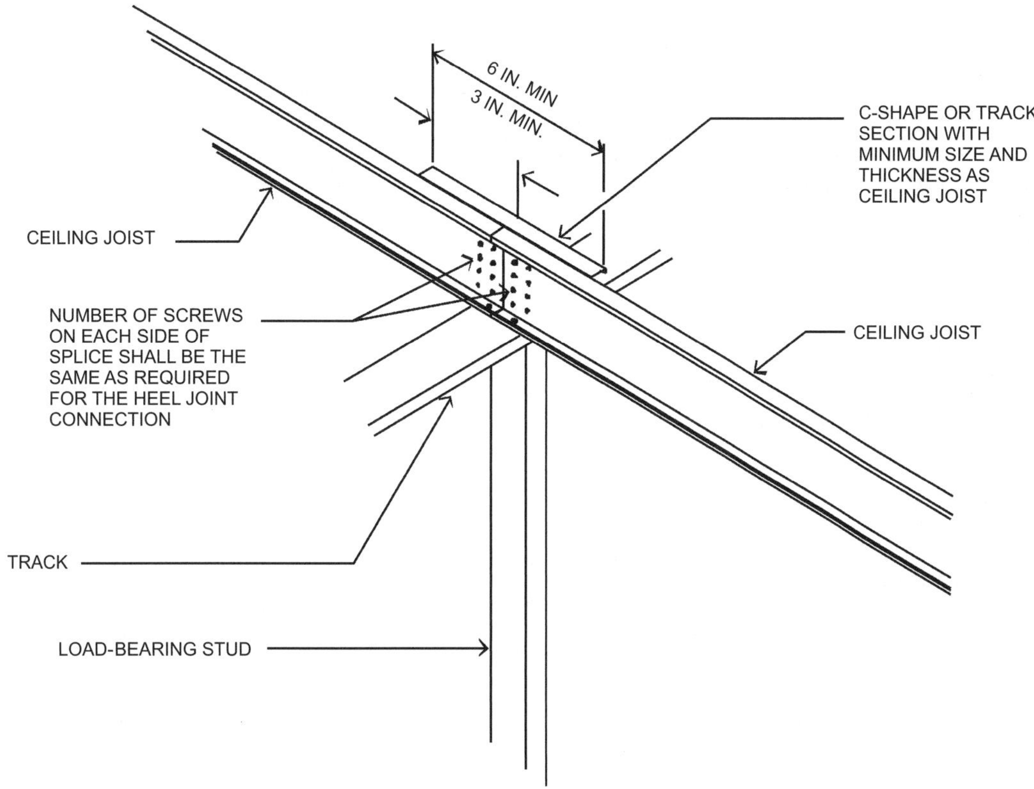

For SI: 1 inch = 25.4 mm.

FIGURE R804.3.1.5
SPLICED CEILING JOISTS

TABLE R804.3.2.1(1)
ROOF RAFTER SPANS[a, b, c]
33 KSI STEEL

MEMBER DESIGNATION	ALLOWABLE SPAN MEASURED HORIZONTALLY (feet-inches)							
	Ground snow load (psf)							
	20		30		50		70	
	Rafter spacing (inches)							
	16	24	16	24	16	24	16	24
550S162-33	14′-0″	11′-6″	11′-11″	9′-7″	9′-6″	7′-9″	8′-2″	6′-8″
550S162-43	16′-8″	13′-11″	14′-5″	11′-9″	11′-6″	9′-5″	9′-10″	8′-0″
550S162-54	17′-11″	15′-7″	15′-7″	13′-3″	12′-11″	10′-7″	11′-1″	9′-1″
550S162-68	19′-2″	16′-9″	16′-9″	14′-7″	14′-1″	11′-10″	12′-6″	10′-2″
550S162-97	21′-3″	18′-6″	18′-6″	16′-2″	15′-8″	13′-8″	14′-0″	12′-2″
800S162-33	16′-5″	13′-5″	13′-11″	11′-4″	11′-1″	8′-2″	9′-0″	6′-0″
800S162-43	19′-9″	16′-1″	16′-8″	13′-7″	13′-4″	10′-10″	11′-5″	9′-4″
800S162-54	22′-8″	18′-6″	19′-2″	15′-8″	15′-4″	12′-6″	13′-1″	10′-8″
800S162-68	25′-10″	21′-2″	21′-11″	17′-10″	17′-6″	14′-4″	15′-0″	12′-3″
800S162-97	21′-3″	18′-6″	18′-6″	16′-2″	15′-8″	13′-8″	14′-0″	12′-2″
1000S162-43	22′-3″	18′-2″	18′-9″	15′-8″	15′-0″	12′-3″	12′-10″	10′-6″
1000S162-54	25′-8″	20′-11″	21′-8″	17′-9″	17′-4″	14′-2″	14′-10″	12′-1″
1000S162-68	29′-7″	24′-2″	25′-0″	20′-5″	20′-0″	16′-4″	17′-2″	14′-0″
1000S162-97	34′-8″	30′-4″	30′-4″	25′-10″	25′-3″	20′-8″	21′-8″	17′-8″
1200S162-54	28′-3″	23′-1″	23′-11″	19′-7″	19′-2″	15′-7″	16′-5″	13′-5″
1200S162-68	32′-10″	26′-10″	27′-9″	22′-8″	22′-2″	18′-1″	19′-0″	15′-6″
1200S162-97	40′-6″	33′-5″	34′-6″	28′-3″	27′-7″	22′-7″	23′-8″	19′-4″

For SI: 1 inch = 25.4 mm, 1 foot = 304.8 mm, 1 pound per square foot = 0.0479 kPa.

a. Table provides maximum horizontal rafter spans in feet and inches for slopes between 3:12 and 12:12.

b. Deflection criterion: *L*/240 for live loads and *L*/180 for total loads.

c. Roof dead load = 12 psf.

TABLE R804.3.2.1(2)
ROOF RAFTER SPANS[a, b, c]
50 KSI STEEL

MEMBER DESIGNATION	ALLOWABLE SPAN MEASURED HORIZONTALLY (feet-inches)							
	Equivalent ground snow load (psf)							
	20		30		50		70	
	Rafter spacing (inches)							
	16	24	16	24	16	24	16	24
550S162-33	15′-4″	12′-11″	13′-4″	10′-11″	10′-9″	8′-9″	9′-2″	7′-6″
550S162-43	16′-8″	14′-7″	14′-7″	12′-9″	12′-3″	10′-6″	11′-0″	9′-0″
550S162-54	17′-11″	15′-7″	15′-7″	13′-8″	13′-2″	11′-6″	11′-9″	10′-3″
550S162-68	19′-2″	16′-9″	16′-9″	14′-7″	14′-1″	12′-4″	12′-7″	11′-0″
550S162-97	21′-3″	18′-6″	18′-6″	16′-2″	15′-8″	13′-8″	14′-0″	12′-3″
800S162-33	18′-10″	15′-5″	15′-11″	12′-9″	12′-3″	8′-2″	9′-0″	6′-0″
800S162-43	22′-3″	18′-2″	18′-10″	15′-5″	15′-1″	12′-3″	12′-11″	10′-6″
800S162-54	24′-2″	21′-2″	21′-1″	18′-5″	17′-10″	14′-8″	15′-5″	12′-7″
800S162-68	25′-11″	22′-8″	22′-8″	19′-9″	19′-1″	16′-8″	17′-1″	14′-9″
800S162-97	28′-10″	25′-2″	25′-2″	22′-0″	21′-2″	18′-6″	19′-0″	16′-7″
1000S162-43	25′-2″	20′-7″	21′-4″	17′-5″	17′-0″	13′-11″	14′-7″	10′-7″
1000S162-54	29′-0″	24′-6″	25′-4″	20′-9″	20′-3″	16′-7″	17′-5″	14′-2″
1000S162-68	31′-2″	27′-3″	27′-3″	23′-9″	20′-0″	19′-6″	20′-6″	16′-8″
1000S162-97	34′-8″	30′-4″	30′-4″	26′-5″	25′-7″	22′-4″	22′-10″	20′-0″
1200S162-54	33′-2″	27′-1″	28′-1″	22′-11″	22′-5″	18′-4″	19′-3″	15′-8″
1200S162-68	36′-4″	31′-9″	31′-9″	27′-0″	26′-5″	21′-6″	22′-6″	18′-6″
1200S162-97	40′-6″	35′-4″	35′-4″	30′-11″	29′-10″	26′-1″	26′-8″	23′-1″

For SI: 1 inch = 25.4 mm, 1 foot = 304.8 mm, 1 pound per square foot = 0.0479 kPa.

a. Table provides maximum horizontal rafter spans in feet and inches for slopes between 3:12 and 12:12.

b. Deflection criterion: *L*/240 for live loads and *L*/180 for total loads.

c. Roof dead load = 12 psf.

TABLE R804.3.2.1(3)
BASIC WIND SPEED TO EQUIVALENT SNOW LOAD CONVERSION

BASIC WIND SPEED AND EXPOSURE		EQUIVALENT GROUND SNOW LOAD (psf)									
		Roof slope									
Exp. B	Exp. C	3:12	4:12	5:12	6:12	7:12	8:12	9:12	10:12	11:12	12:12
85 mph	—	20	20	20	20	20	20	30	30	30	30
100 mph	85 mph	20	20	20	20	30	30	30	30	50	50
110 mph	100 mph	20	20	20	20	30	50	50	50	50	50
—	110 mph	30	30	30	50	50	50	70	70	70	—

For SI: 1 mile per hour = 0.447 m/s, 1 pound per square foot = 0.0479 kPa.

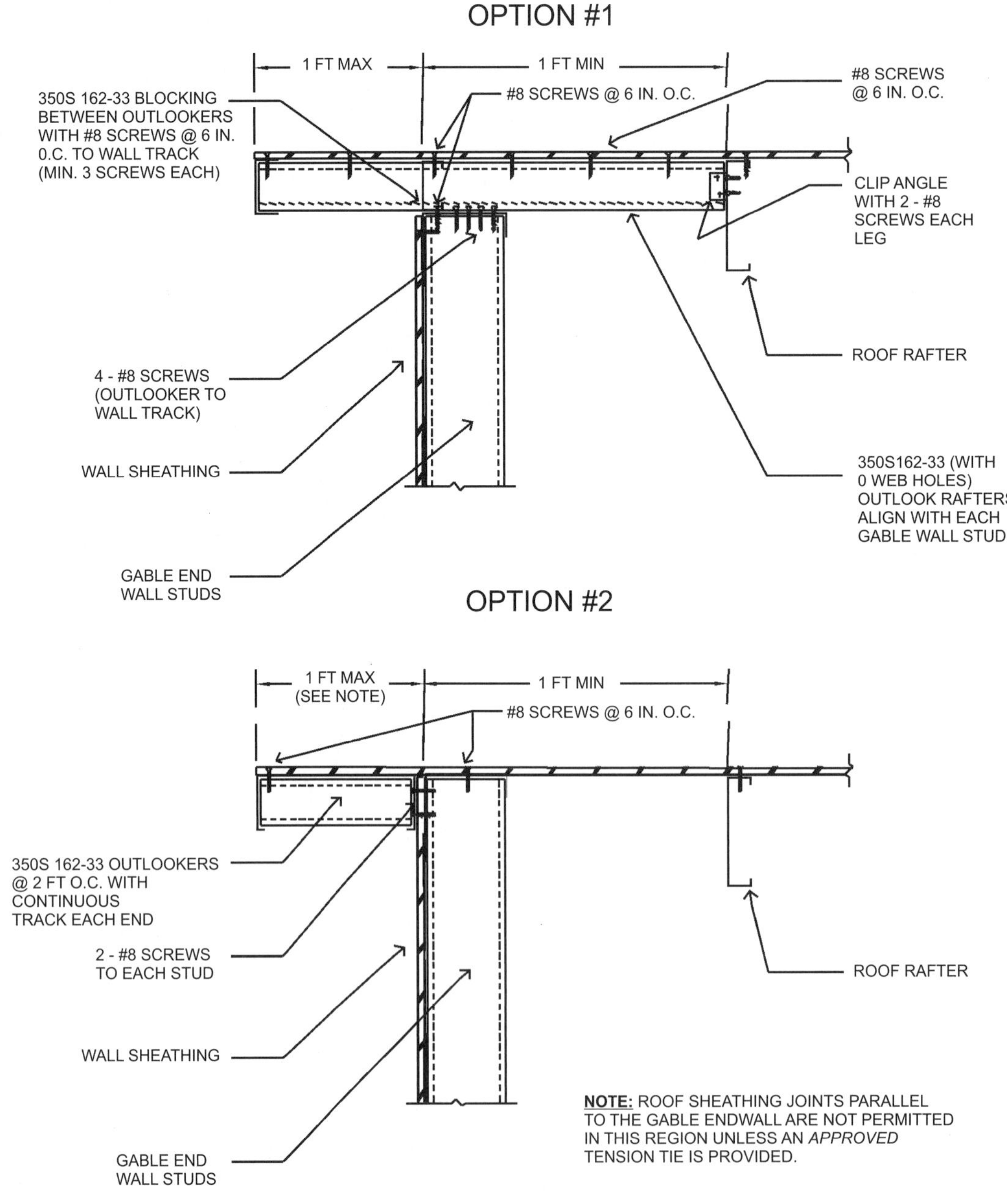

For SI: 1 inch = 25.4 mm, 1 foot = 304.8 mm.

FIGURE R804.3.2.1.2
GABLE ENDWALL OVERHANG DETAILS

R804.3.2.3 Roof rafter splice. Roof rafters shall not be spliced.

R804.3.2.4 Roof rafter to ceiling joist and ridge member connection. Roof rafters shall be connected to a parallel ceiling joist to form a continuous tie between exterior walls in accordance with Figure R804.3.1.1(1) or R804.3.1.1(2) and Table R804.3.1.1(9). Ceiling joists shall be connected to the top track of the load-bearing wall in accordance with Table R804.3, either with two No. 10 screws applied through the flange of the ceiling joist or by using a 54-mil (1.37 mm) clip angle with two No.10 screws in each leg. Roof rafters shall be connected to a ridge member with a minimum 2-inch by 2-inch (51 mm by 51 mm) clip angle fastened with No. 10 screws to the ridge member in accordance with Figure R804.3.2.4 and Table R804.3.2.4. The clip angle shall have a steel thickness equivalent to or greater than the roof rafter thickness and shall extend the depth of the roof rafter member to the extent possible. The ridge member shall be fabricated from a C-shaped member and a track section, which shall have a minimum size and steel thickness equivalent to or greater than that of adjacent roof rafters and shall be installed in accordance with Figure R804.3.2.4. The ridge member shall extend the full depth of the sloped roof rafter cut.

R804.3.2.5 Roof rafter bottom flange bracing. The bottom flanges of roof rafters shall be continuously braced, at a maximum spacing of 8 feet (2440 mm) as measured parallel to the roof rafters, with one of the following members:

1. Minimum 33-mil (0.84 mm) C-shaped member.
2. Minimum 33-mil (0.84 mm) track section.
3. Minimum $1^1/_2$-inch by 33-mil (38 mm by 0.84 mm) steel strap.

The bracing element shall be fastened to the bottom flange of each roof rafter with one No. 8 screw and

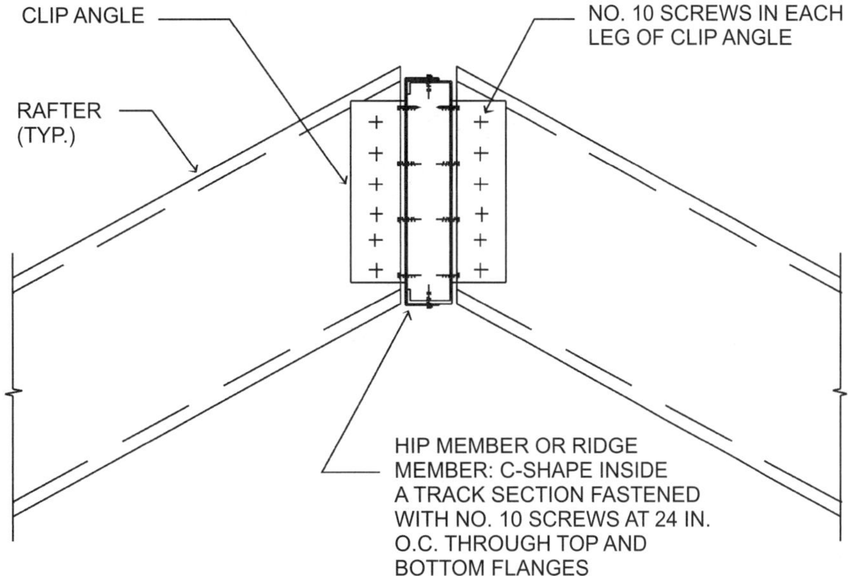

For SI: 1 inch = 25.4 mm.

FIGURE R804.3.2.4
HIP MEMBER OR RIDGE MEMBER CONNECTION

TABLE R804.3.2.4
SCREWS REQUIRED AT EACH LEG OF CLIP ANGLE FOR HIP RAFTER TO HIP MEMBER OR ROOF RAFTER TO RIDGE MEMBER CONNECTION[a]

BUILDING WIDTH (feet)	NUMBER OF SCREWS			
	Ground snow load (psf)			
	0 to 20	21 to 30	31 to 50	51 to 70
24	2	2	3	4
28	2	3	4	5
32	2	3	4	5
36	3	3	5	6
40	3	4	5	7

For SI: 1 inch = 25.4 mm, 1 foot = 304.8 mm, 1 pound per square foot = 0.0479 kPa.

a. Screws shall be No. 10 minimum.

shall be fastened to blocking with two No. 8 screws. Blocking shall be installed between roof rafters in-line with the continuous bracing at a maximum spacing of 12 feet (3658 mm) measured perpendicular to the roof rafters. The ends of continuous bracing shall be fastened to blocking or anchored to a stable building component with two No. 8 screws.

R804.3.3 Hip framing. Hip framing shall consist of jack-rafters, hip members, hip support columns and connections in accordance with this section, or shall be in accordance with an *approved* design. The provisions of this section for hip members and hip support columns shall apply only where the jack rafter slope is greater than or equal to the roof slope. For the purposes of determining member sizes in this section, wind speeds shall be converted to equivalent ground snow load in accordance with Table R804.3.2.1(3).

R804.3.3.1 Jack rafters. Jack rafters shall meet the requirements for roof rafters in accordance with Section R804.3.2, except that the requirements in Section R804.3.2.4 shall not apply.

R804.3.3.2 Hip members. Hip members shall be fabricated from C-shape members and track section, which shall have minimum sizes determined in accordance with Table R804.3.3.2. The C-shape member and track section shall be connected at a maximum spacing of 24 inches (610 mm) using No. 10 screws through top and bottom flanges in accordance with Figure R804.3.2.4. The depth of the hip member shall match that of the roof rafters and jack rafters, or shall be based on an *approved* design for a beam pocket at the corner of the supporting wall.

R804.3.3.3 Hip support columns. Hip support columns shall be used to support hip members at the ridge. A hip support column shall consist of a pair of C-shape members, with a minimum size determined in accordance with Table R804.3.3.3. The C-shape members shall be connected at a maximum spacing of 24 inches (610 mm) on center to form a box using minimum 3-inch by 33-mil (76 mm by 0.84 mm) strap connected to each of the flanges of the C-shape members with three-No. 10 screws. Hip support columns shall have a continuous load path to the foundation and shall be supported at the ceiling line by an interior wall or by an *approved* design for a supporting element.

TABLE R804.3.3.2
HIP MEMBER SIZES, 33 ksi STEEL

BUILDING WIDTH (feet)	HIP MEMBER DESIGNATION[a]			
	Equivalent ground snow load (psf)			
	0 to 20	21 to 30	31 to 50	51 to 70
24	800S162-68 800T150-68	800S162-68 800T150-68	800S162-97 800T150-97	1000S162-97 1000T150-97
28	1000S162-68 1000T150-68	1000S162-68 1000T150-68	1000S162-97 1000T150-97	1200S162-97 1200T150-97
32	1000S162-97 1000T150-97	1000S162-97 1000T150-97	1200S162-97 1200T150-97	—
36	1200S162-97 1200T150-97	—	—	—
40	—	—	—	—

For SI: 1 foot = 304.8 mm, 1 pound per square foot = 0.0479 kPa.

a. The web depth of the roof rafters and jack rafters is to match at the hip or they shall be installed in accordance with an approved design.

TABLE R804.3.3.3
HIP SUPPORT COLUMN SIZES

BUILDING WIDTH (feet)	HIP SUPPORT COLUMN DESIGNATION[a, b]			
	Equivalent ground snow load (psf)			
	0 to 20	21 to 30	31 to 50	51 to 70
24	2-350S162-33	2-350S162-33	2-350S162-43	2-350S162-54
28	2-350S162-54	2-550S162-54	2-550S162-68	2-550S162-68
32	2-550S162-68	2-550S162-68	2-550S162-97	—
36	2-550S162-97	—	—	—
40	—	—	—	—

For SI: 1 foot = 304.8 mm, 1 pound per square foot = 0.0479 kPa.

a. Box shape column only in accordance with Figure R804.3.3.4(2).

b. 33-ksi steel for 33- and 43-mil material; 50-ksi steel for thicker material.

R804.3.3.4 Hip framing connections. Hip rafter framing connections shall be installed in accordance with the following:

1. Jack rafters shall be connected at the eave to a parallel C-shape blocking member in accordance with Figure R804.3.3.4(1). The C-shape blocking member shall be attached to the supporting wall track with minimum two No. 10 screws.
2. Jack rafters shall be connected to a hip member with a minimum 2-inch by 2-inch (51 mm by 51 mm) clip angle fastened with No.10 screws to the hip member in accordance with Figure R804.3.2.4 and Table R804.3.2.4. The clip angle shall have a steel thickness equivalent to or greater than the jack rafter thickness and shall extend the depth of the jack rafter member to the extent possible.
3. The connection of the hip support columns at the ceiling line shall be in accordance with Figure R804.3.3.4(2), with an uplift strap sized in accordance with Table R804.3.3.4(1).
4. The connection of hip support members, ridge members and hip support columns at the ridge shall be in accordance with Figures R804.3.3.4(3) and R804.3.3.4(4) and Table R804.3.3.4(2).
5. The connection of hip members to the wall corner shall be in accordance with Figure R804.3.3.4(5) and Table R804.3.3.4(3).

R804.3.4 Cutting and notching. Flanges and lips of load-bearing, cold-formed steel roof framing members shall not be cut or notched.

R804.3.5 Headers. Roof-ceiling framing above wall openings shall be supported on headers. The allowable spans for headers in load-bearing walls shall not exceed the values set forth in Section R603.6 and Tables R603.6(1) through R603.6(24).

R804.3.6 Framing of openings in roofs and ceilings. Openings in roofs and ceilings shall be framed with header and trimmer joists. Header joist spans shall not exceed 4 feet (1219 mm) in length. Header and trimmer joists shall be fabricated from joist and track members having a minimum size and thickness at least equivalent to the adjacent ceiling joists or roof rafters and shall be installed in accordance with Figures R804.3.6(1) and R804.3.6(2). Each header joist shall be connected to trimmer joists with a minimum of four 2-inch by 2-inch (51 by 51 mm) clip angles. Each clip angle shall be fastened to both the header and trimmer joists with four No. 8 screws, evenly spaced, through each leg of the clip angle. The steel thickness of the clip angles shall be not less than that of the ceiling joist or roof rafter. Each track section for a built-up header or trimmer joist shall extend the full length of the joist (continuous).

R804.3.7 Roof trusses. Cold-formed steel trusses shall be designed and installed in accordance with AISI S100, Section D4. In the absence of specific bracing requirements, trusses shall be braced in accordance with accepted industry practices, such as the SBCA *Cold-Formed Steel Building Component Safety Information (CFSBCSI) Guide to Good Practice for Handling, Installing & Bracing of Cold-Formed Steel Trusses.* Trusses shall be connected to the top track of the load-bearing wall in accordance with Table R804.3, either with two No. 10 screws applied through the flange of the truss or by using a 54-mil (1.37 mm) clip angle with two No. 10 screws in each leg.

R804.3.8 Ceiling and roof diaphragms. Ceiling and roof diaphragms shall be in accordance with this section.

R804.3.8.1 Ceiling diaphragms. At gable endwalls a ceiling *diaphragm* shall be provided by attaching a minimum $^1/_2$-inch (12.7 mm) gypsum board in accordance with Tables R804.3.8(1) and R804.3.8(2) or a minimum $^3/_8$-inch (9.5 mm) wood structural panel sheathing, which complies with Section R803, in accordance with Table R804.3.8(3) to the bottom of ceiling joists or roof trusses and connected to wall framing in accordance with Figures R804.3.8(1) and R804.3.8(2), unless studs are designed as full height without bracing at the ceiling. Flat blocking shall consist of C-shape or track section with a minimum thickness of 33 mils (0.84 mm).

The ceiling *diaphragm* shall be secured with screws spaced at a maximum 6 inches (152 mm) o.c. at panel edges and a maximum 12 inches (305 mm) o.c. in the field. Multiplying the required lengths in Tables R804.3.8(1) and R804.3.8(2) for gypsum board sheathed ceiling diaphragms shall be permitted to be multiplied by 0.35 shall be permitted if all panel edges are blocked. Multiplying the required lengths in Tables R804.3.8(1) and R804.3.8(2) for gypsum board sheathed ceiling diaphragms by 0.9 shall be permitted if all panel edges are secured with screws spaced at 4 inches (102 mm) o.c.

R804.3.8.2 Roof diaphragm. A roof *diaphragm* shall be provided by attaching a minimum of $^3/_8$-inch (9.5 mm) wood structural panel which complies with Section R803 to roof rafters or truss top chords in accordance with Table R804.3. Buildings with 3:1 or larger plan *aspect ratio* and with roof rafter slope (pitch) of 9:12 or larger shall have the roof rafters and ceiling joists blocked in accordance with Figure R804.3.8(3).

R804.3.9 Roof tie-down. Roof assemblies subject to wind uplift pressures of 20 pounds per square foot (0.96 kPa) or greater, as established in Table R301.2(2), shall have rafter-to-bearing wall ties provided in accordance with Table R802.11.

TABLE R804.3.3.4(1)
UPLIFT STRAP CONNECTION REQUIREMENTS HIP SUPPORT COLUMN AT CEILING LINE

BUILDING WIDTH (feet)	BASIC WIND SPEED (mph) EXPOSURE B				
	85	100	110	—	—
	BASIC WIND SPEED (mph) EXPOSURE C				
	—	85	—	100	110
	Number of No. 10 screws in each end of each 3-inch by 54-mil steel strap[a, b, c]				
24	3	4	4	6	7
28	4	6	6	8	10
32	5	8	8	11	13
36	7	10	11	14	17
40	—	—	—	—	—

For SI: 1 foot = 304.8 mm, 1 pound per square foot = 0.0479 kPa, 1 mil = 0.0254 mm.

a. Two straps are required, one each side of the column.

b. Space screws at $^3/_4$ inch on center and provide $^3/_4$-inch end distance.

c. 50-ksi steel strap.

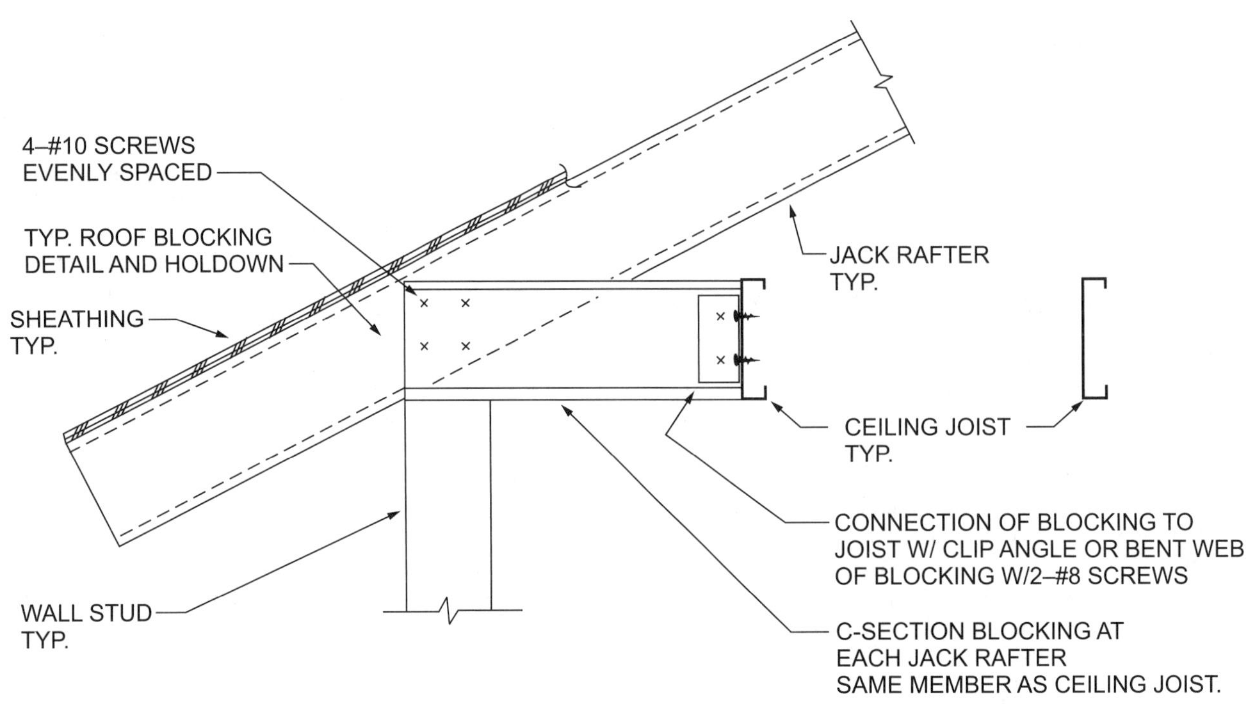

FIGURE R804.3.3.4(1)
JACK RAFTER CONNECTION AT EAVE

TABLE R804.3.3.4(2)
CONNECTION REQUIREMENTS HIP MEMBER TO HIP SUPPORT COLUMN

BUILDING WIDTH (feet)	NUMBER OF NO. 10 SCREWS IN EACH FRAMING ANGLE[a, b, c]			
	Equivalent ground snow load (psf)			
	0 to 20	21 to 30	31 to 50	51 to 70
24	10	10	10	12
28	10	10	14	18
32	10	12	—	—
36	14	—	—	—
40	—	—	—	—

For SI: 1 foot = 304.8 mm, 1 pound per square foot = 0.0479 kPa.

a. Screws to be divided equally between the connection to the hip member and the column. Refer to Figures R804.3.3.4(3) and R804.3.3.4(4).

b. The number of screws required in each framing angle is not to be less than shown in Table R804.3.3.4(1).

c. 50-ksi steel from the framing angle.

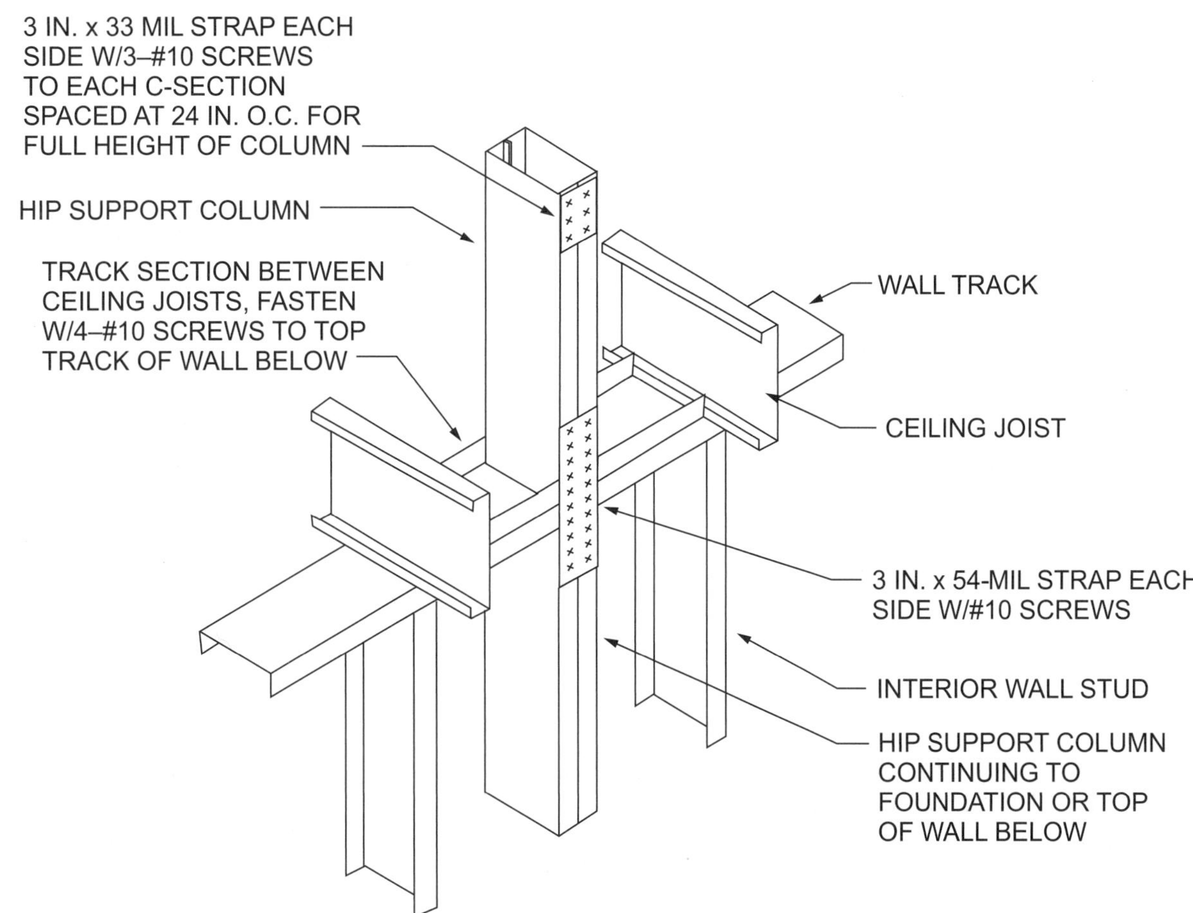

For SI: 1 inch = 25.4 mm, 1 mil = 0.0254 mm.

FIGURE R804.3.3.4(2)
HIP SUPPORT COLUMN

TABLE R804.3.3.4(3)
UPLIFT STRAP CONNECTION REQUIREMENTS HIP MEMBER TO WALL

BUILDING WIDTH (feet)	BASIC WIND SPEED (mph) EXPOSURE B				
	85	100	110	—	—
	BASIC WIND SPEED (mph) EXPOSURE C				
	—	85	—	100	110
	Number of No. 10 screws in each end of each 3-inch by 54-mil steel strap[a, b, c]				
24	2	2	3	3	4
28	2	3	3	4	5
32	3	4	4	6	7
36	3	5	5	7	8
40	—	—	—	—	—

For SI: 1 foot = 304.8 mm,1 pound per square foot = 0.0479 kPa.

a. Two straps are required, one each side of the column.

b. Space screws at $^3/_4$ inches on center and provide $^3/_4$-inch end distance.

c. 50-ksi steel strap.

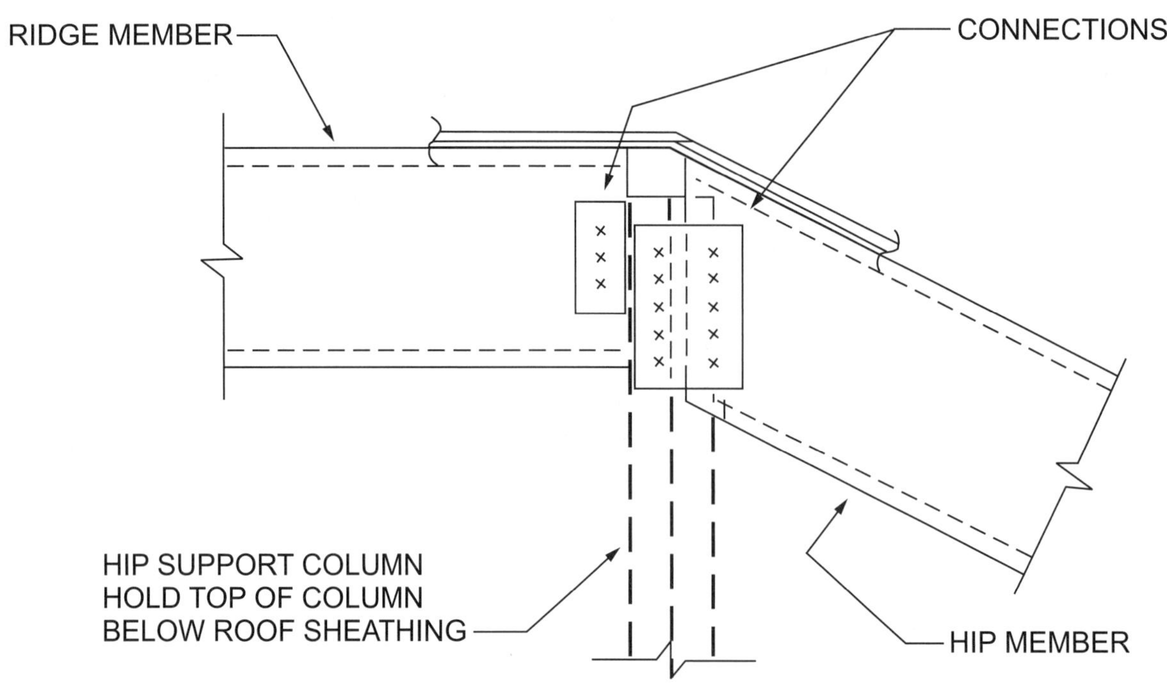

FIGURE R804.3.3.4(3)
HIP CONNECTIONS AT RIDGE

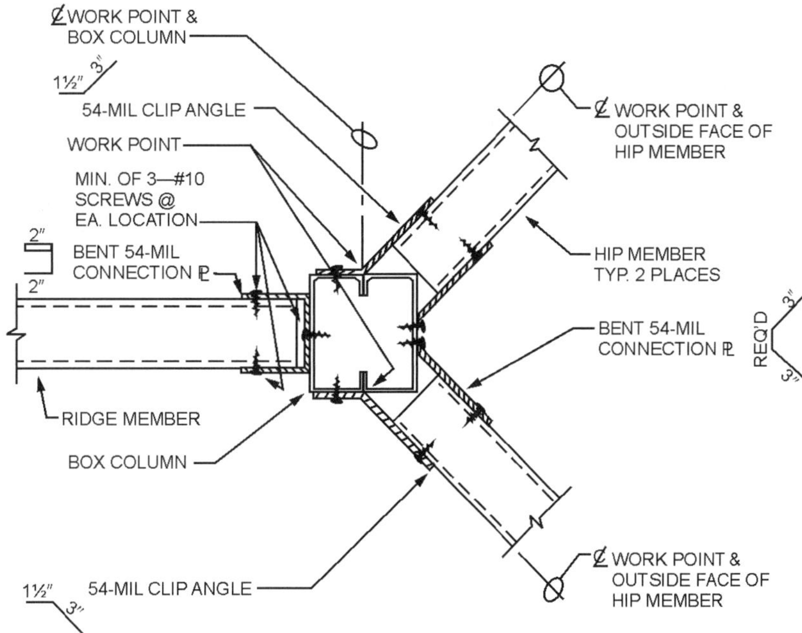

CONNECTION @ 3½″ BOX COLUMN

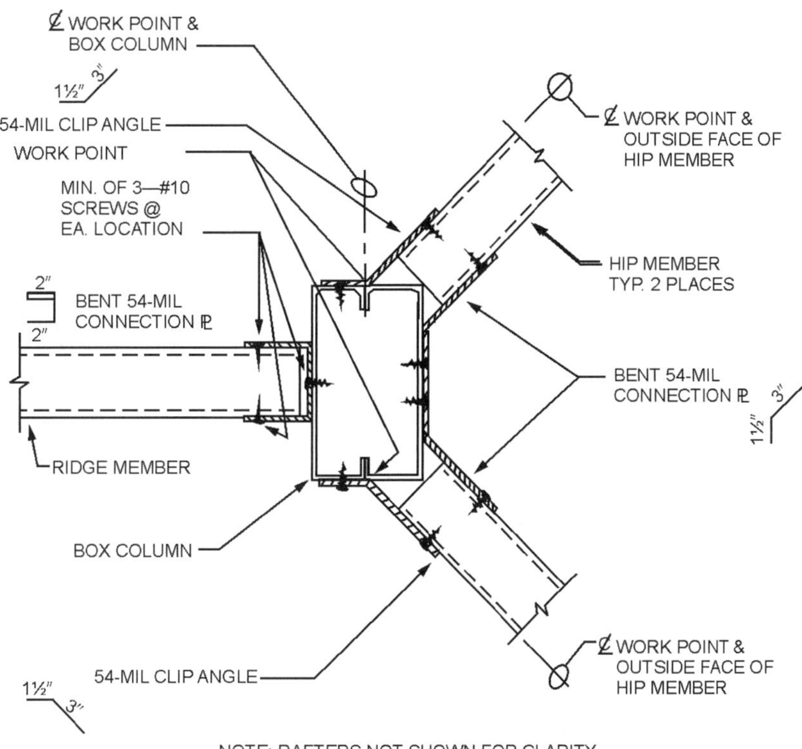

CONNECTION @ 5½″ BOX COLUMN

For SI: 1 inch = 25.4 mm, 1 mil = 0.0254 mm.

FIGURE R804.3.3.4(4)
HIP CONNECTIONS AT RIDGE AND BOX COLUMN

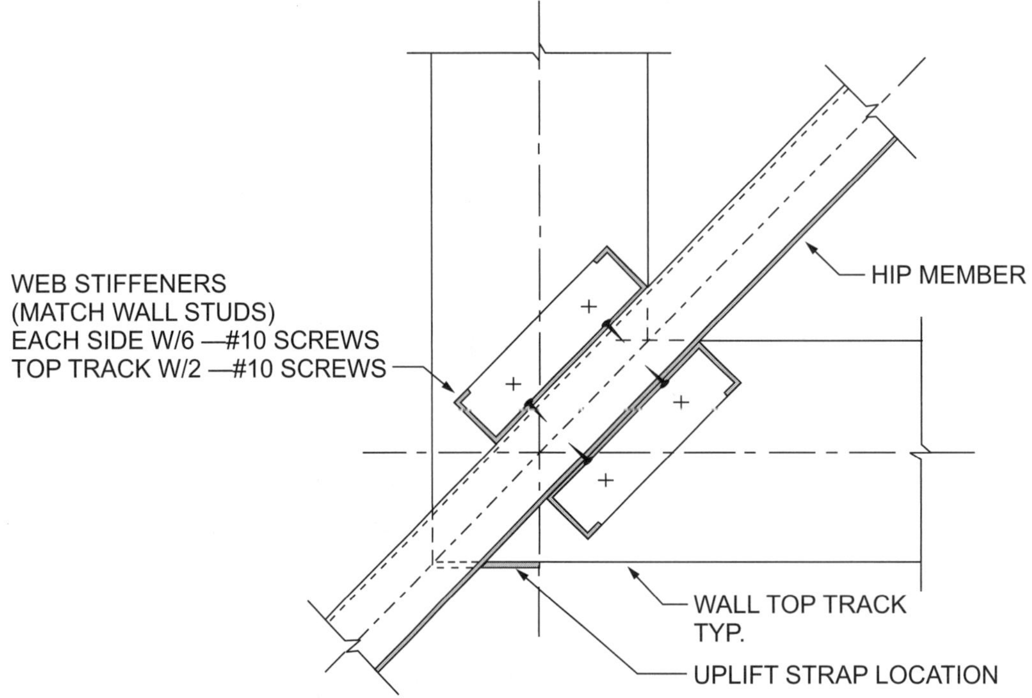

FIGURE R804.3.3.4(5)
HIP MEMBER CONNECTION AT WALL CORNER

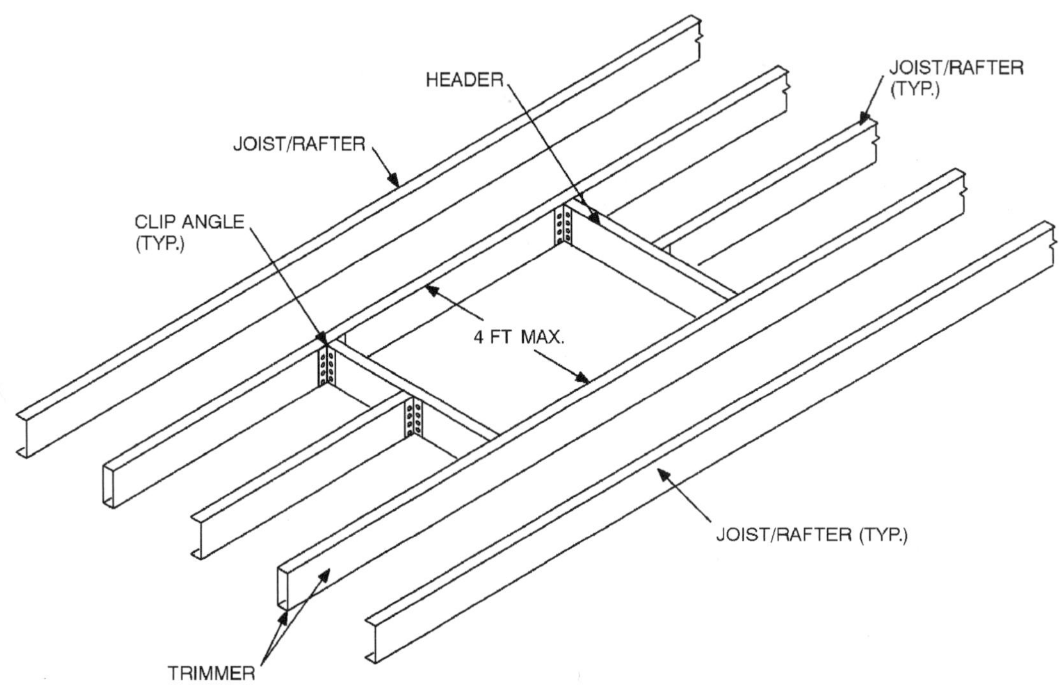

For SI: 1 foot = 304.8 mm.

FIGURE R804.3.6(1)
ROOF OR CEILING OPENING

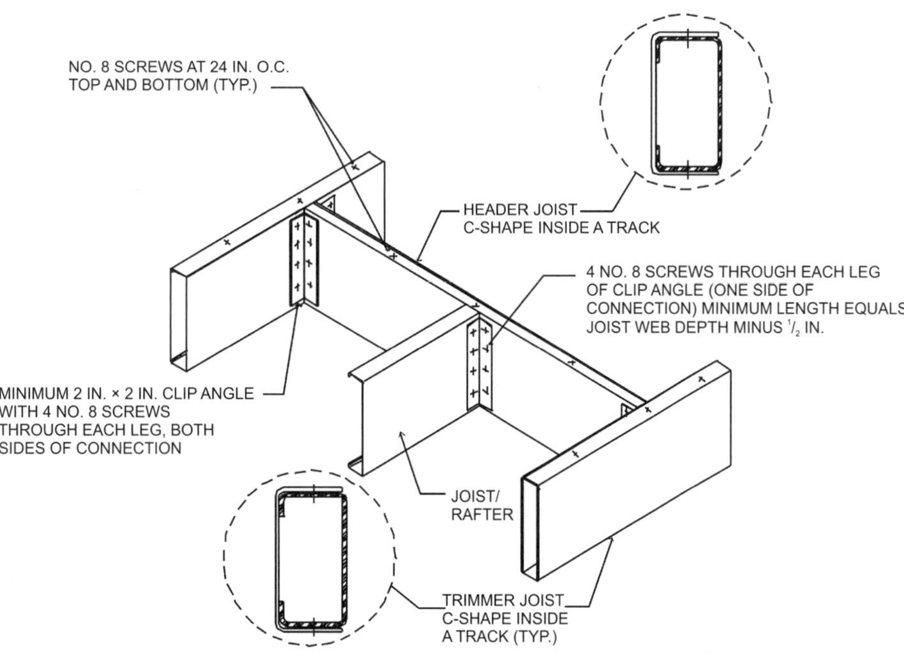

For SI: 1 inch = 25.4 mm.

FIGURE R804.3.6(2)
HEADER TO TRIMMER CONNECTION

TABLE R804.3.8(1)
REQUIRED LENGTHS FOR CEILING DIAPHRAGMS AT GABLE ENDWALLS GYPSUM BOARD SHEATHED, CEILING HEIGHT = 8 FEET[a, b, c, d, e, f]

Exposure B		BASIC WIND SPEED (mph)				
		85	100	110	—	—
Exposure C		—	85	—	100	110
Roof pitch	Building endwall width (feet)	Minimum diaphragm length (feet)				
3:12 to 6:12	24 - 28	14	20	22	28	32
	28 - 32	16	22	28	32	38
	32 - 36	20	26	32	38	44
	36 - 40	22	30	36	44	50
6:12 to 9:12	24 - 28	16	22	26	32	36
	28 - 32	20	26	32	38	44
	32 - 36	22	32	38	44	52
	36 - 40	26	36	44	52	60
9:12 to 12:12	24 - 28	18	26	30	36	42
	28 - 32	22	30	36	42	50
	32 - 36	26	36	42	50	60
	36 - 40	30	42	50	60	70

For SI: 1 inch = 25.4 mm, 1 pound per square foot = 0.0479 kPa, 1 mile per hour = 0.447 m/s, 1 foot = 304.8 mm, 1 mil = 0.0254 mm.

a. Ceiling diaphragm is composed of $^1/_2$-inch gypsum board (min. thickness) secured with screws spaced at 6 inches o.c. at panel edges and 12 inches o.c. infield. Use No. 8 screws (min.) when framing members have a designation thickness of 54 mils or less and No. 10 screws (min.) when framing members have a designation thickness greater than 54 mils.

b. Maximum aspect ratio (length/width) of diaphragms is 2:1.

c. Building width is in the direction of horizontal framing members supported by the wall studs.

d. Required diaphragm lengths are to be provided at each end of the structure.

e. Multiplying required diaphragm lengths by 0.35 is permitted if all panel edges are blocked.

f. Multiplying required diaphragm lengths by 0.9 is permitted if all panel edges are secured with screws spaced at 4 inches o.c.

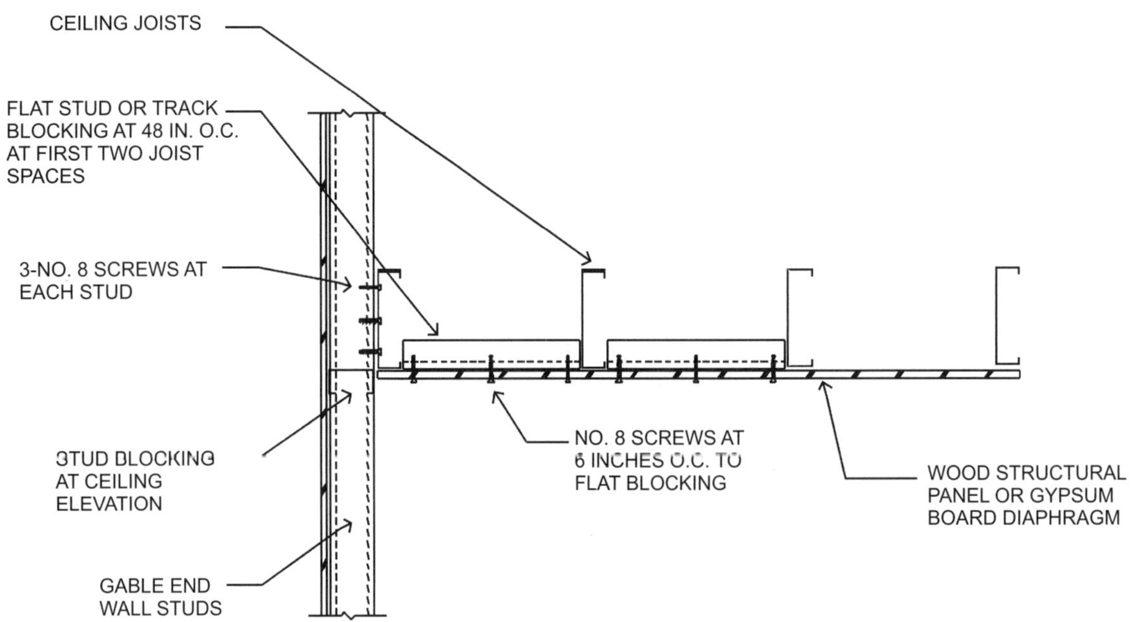

For SI: 1 inch = 25.4 mm.

**FIGURE R804.3.8(1)
CEILING DIAPHRAGM TO GABLE ENDWALL DETAIL**

**TABLE R804.3.8(2)
REQUIRED LENGTHS FOR CEILING DIAPHRAGMS AT GABLE ENDWALLS
GYPSUM BOARD SHEATHED CEILING HEIGHT = 9 OR 10 FEET[a, b, c, d, e, f]**

Exposure B		BASIC WIND SPEED (mph)				
		85	100	110	—	—
Exposure C		—	85	—	100	110
Roof pitch	Building endwall width (feet)	Minimum diaphragm length (feet)				
3:12 to 6:12	24 - 28	16	22	26	32	38
	28 - 32	20	26	32	38	44
	32 - 36	22	30	36	44	50
	36 - 40	26	36	42	50	58
6:12 to 9:12	24 - 28	18	26	30	36	42
	28 - 32	22	30	36	42	50
	32 - 36	26	36	42	50	58
	36 - 40	30	42	48	58	68
9:12 to 12:12	24 - 28	20	28	34	40	46
	28 - 32	24	34	40	48	56
	32 - 36	28	40	48	56	66
	36 - 40	34	46	56	66	78

For SI: 1 inch = 25.4 mm, 1 pound per square foot = 0.0479 kPa, 1 mile per hour = 0.447 m/s, 1 foot = 304.8 mm, 1 mil = 0.0254 mm.

a. Ceiling diaphragm is composed of $^1/_2$-inch gypsum board (min. thickness) secured with screws spaced at 6 inches o.c. at panel edges and 12 inches o.c. infield. Use No. 8 screws (min.) when framing members have a designation thickness of 54 mils or less and No. 10 screws (min.) when framing members have a designation thickness greater than 54 mils.

b. Maximum aspect ratio (length/width) of diaphragms is 2:1.

c. Building width is in the direction of horizontal framing members supported by the wall studs.

d. Required diaphragm lengths are to be provided at each end of the structure.

e. Required diaphragm lengths are permitted to be multiplied by 0.35 if all panel edges are blocked.

f. Required diaphragm lengths are permitted to be multiplied by 0.9 if all panel edges are secured with screws spaced at 4 inches o.c.

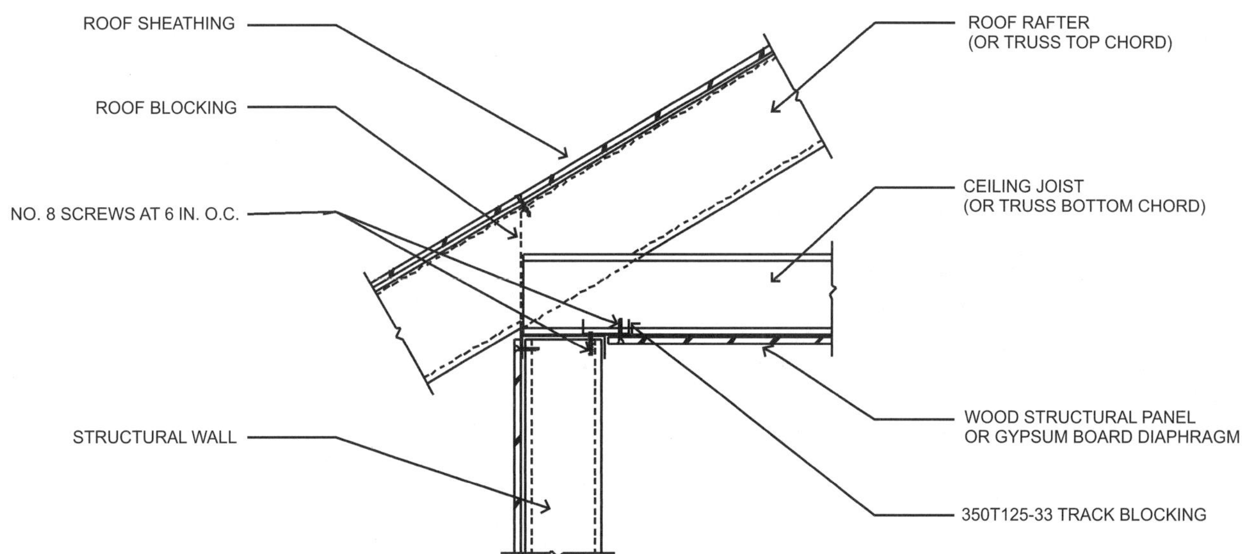

For SI: 1 inch = 25.4 mm.

FIGURE R804.3.8(2)
CEILING DIAPHRAGM TO SIDEWALL DETAIL

TABLE R804.3.8(3)
REQUIRED LENGTHS FOR CEILING DIAPHRAGMS AT GABLE ENDWALLS
WOOD STRUCTURAL PANEL SHEATHED CEILING HEIGHT = 8, 9 OR 10 FEET[a, b, c, d]

Exposure B		BASIC WIND SPEED (mph)				
		85	100	110	—	—
Exposure C		—	85	—	100	110
Roof pitch	Building endwall width (feet)	Minimum diaphragm length (feet)				
3:12 to 6:12	24 - 28	10	10	10	10	10
	28 - 32	12	12	12	12	12
	32 - 36	12	12	12	12	12
	36 - 40	14	14	14	14	14
6:12 to 9:12	24 - 28	10	10	10	10	10
	28 - 32	12	12	12	12	12
	32 - 36	12	12	12	12	12
	36 - 40	14	14	14	14	14
9:12 to 12:12	24 - 28	10	10	10	10	10
	28 - 32	12	12	12	12	12
	32 - 36	12	12	12	12	12
	36 - 40	14	14	14	14	14

For SI: 1 inch = 25.4 mm, 1 pound per square foot = 0.0479 kPa, 1 mile per hour = 0.447 m/s, 1 foot = 304.8 mm, 1 mil = 0.0254 mm.

a. Ceiling diaphragm is composed of $^{3}/_{8}$-inch wood structural panel sheathing (min. thickness) secured with screws spaced at 6 inches o.c. at panel edges and in field.
Use No. 8 screws (min.) when framing members have a designation thickness of 54 mils or less and No. 10 screws (min.) when framing members have a designation thickness greater than 54 mils.

b. Maximum aspect ratio (length/width) of diaphragms is 3:1.

c. Building width is in the direction of horizontal framing members supported by the wall studs.

d. Required diaphragm lengths are to be provided at each end of the structure.

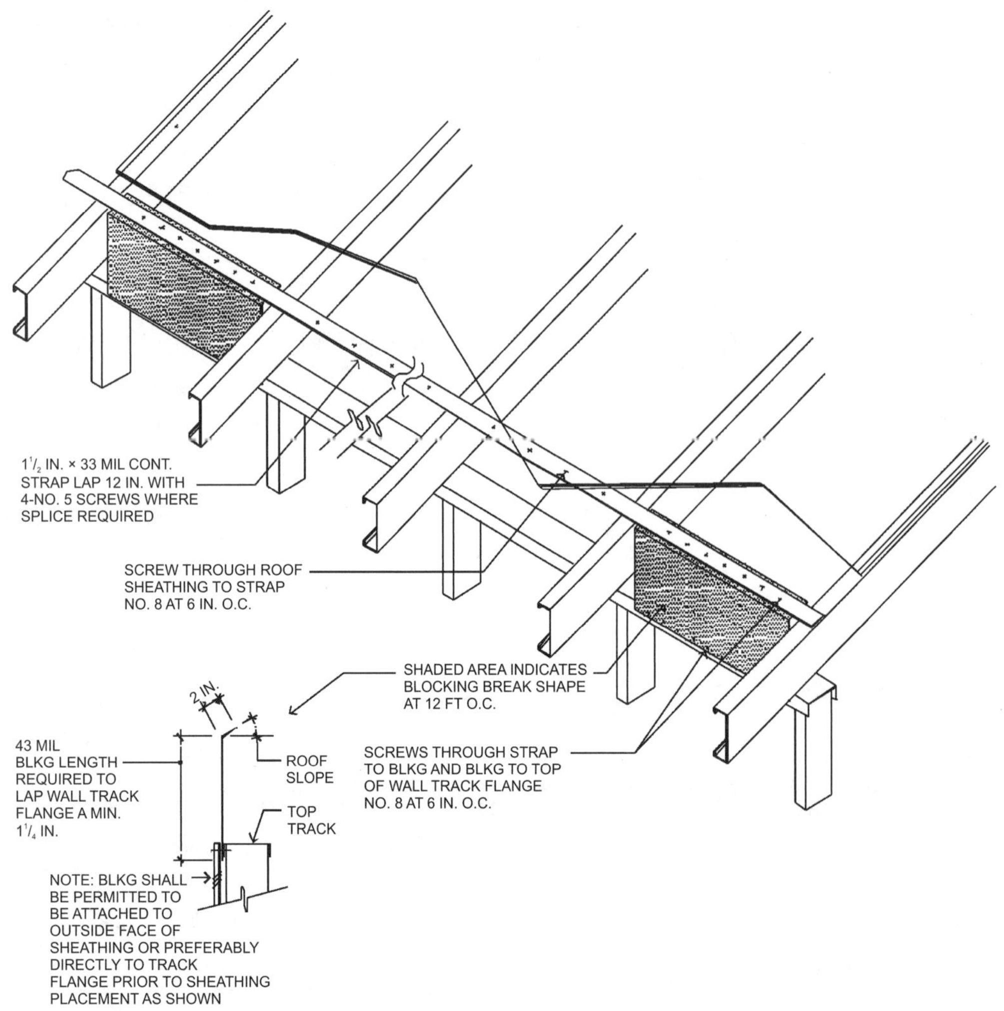

For SI: 1 mil = 0.0254 mm, 1 inch = 25.4 mm.

**FIGURE R804.3.8(3)
ROOF BLOCKING DETAIL**

SECTION R805
CEILING FINISHES

R805.1 Ceiling installation. Ceilings shall be installed in accordance with the requirements for interior wall finishes as provided in Section R702.

SECTION R806
ROOF VENTILATION

R806.1 Ventilation required. Enclosed *attics* and enclosed rafter spaces formed where ceilings are applied directly to the underside of roof rafters shall have cross ventilation for each separate space by ventilating openings protected against the entrance of rain or snow. Ventilation openings shall have a least dimension of $^1/_{16}$ inch (1.6 mm) minimum and $^1/_4$ inch (6.4 mm) maximum. Ventilation openings having a least dimension larger than $^1/_4$ inch (6.4 mm) shall be provided with corrosion-resistant wire cloth screening, hardware cloth, or similar material with openings having a least dimension of $^1/_{16}$ inch (1.6 mm) minimum and $^1/_4$ inch (6.4 mm) maximum. Openings in roof framing members shall conform to the requirements of Section R802.7. Required ventilation openings shall open directly to the outside air.

Exception: Attic ventilation shall not be required when determined not necessary by the code official due to atmospheric or climatic conditions.

R806.2 Minimum vent area. The minimum net free ventilating area shall be $^1/_{150}$ of the area of the vented space.

Exception: The minimum net free ventilation area shall be $^1/_{300}$ of the vented space provided one or more of the following conditions are met:

1. In Climate Zones 6, 7 and 8, a Class I or II vapor retarder is installed on the warm-in-winter side of the ceiling.
2. At least 40 percent and not more than 50 percent of the required ventilating area is provided by ventilators located in the upper portion of the attic or rafter space. Upper ventilators shall be located no more

than 3 feet (914 mm) below the ridge or highest point of the space, measured vertically, with the balance of the required ventilation provided by eave or cornice vents. Where the location of wall or roof framing members conflicts with the installation of upper ventilators, installation more than 3 feet (914 mm) below the ridge or highest point of the space shall be permitted.

R806.3 Vent and insulation clearance. Where eave or cornice vents are installed, insulation shall not block the free flow of air. A minimum of a 1-inch (25 mm) space shall be provided between the insulation and the roof sheathing and at the location of the vent.

R806.4 Installation and weather protection. Ventilators shall be installed in accordance with manufacturer's installation instructions. Installation of ventilators in roof systems shall be in accordance with the requirements of Section R903. Installation of ventilators in wall systems shall be in accordance with the requirements of Section R703.1.

R806.5 Unvented attic and unvented enclosed rafter assemblies. Unvented *attic* assemblies (spaces between the ceiling joists of the top *story* and the roof rafters) and unvented enclosed rafter assemblies (spaces between ceilings that are applied directly to the underside of roof framing members/rafters and the structural roof sheathing at the top of the roof framing members/rafters) shall be permitted if all the following conditions are met:

1. The unvented *attic* space is completely contained within the *building thermal envelope.*
2. No interior Class I vapor retarders are installed on the ceiling side (*attic* floor) of the unvented *attic* assembly or on the ceiling side of the unvented enclosed rafter assembly.
3. Where wood shingles or shakes are used, a minimum $^1/_4$-inch (6 mm) vented air space separates the shingles or shakes and the roofing underlayment above the structural sheathing.
4. In Climate Zones 5, 6, 7 and 8, any *air-impermeable insulation* shall be a Class II vapor retarder, or shall have a Class III vapor retarder coating or covering in direct contact with the underside of the insulation.
5. Either Items 5.1, 5.2 or 5.3 shall be met, depending on the air permeability of the insulation directly under the structural roof sheathing.
   - 5.1. *Air-impermeable insulation* only. Insulation shall be applied in direct contact with the underside of the structural roof sheathing.
   - 5.2. Air-permeable insulation only. In addition to the air-permeable insulation installed directly below the structural sheathing, rigid board or sheet insulation shall be installed directly above the structural roof sheathing as specified in Table R806.5 for condensation control.
   - 5.3. Air-impermeable and air-permeable insulation. The *air-impermeable insulation* shall be applied in direct contact with the underside of the structural roof sheathing as specified in Table R806.5 for condensation control. The air-permeable insulation shall be installed directly under the *air-impermeable insulation.*
   - 5.4. Where preformed insulation board is used as the air-impermeable insulation layer, it shall be sealed at the perimeter of each individual sheet interior surface to form a continuous layer.

TABLE R806.5
INSULATION FOR CONDENSATION CONTROL

CLIMATE ZONE	MINIMUM RIGID BOARD ON AIR-IMPERMEABLE INSULATION *R*-VALUE[a]
2B and 3B tile roof only	0 (none required)
1, 2A, 2B, 3A, 3B, 3C	R-5
4C	R-10
4A, 4B	R-15
5	R-20
6	R-25
7	R-30
8	R-35

a. Contributes to but does not supersede the requirements in Section N1103.2.1.

SECTION R807
ATTIC ACCESS

R807.1 Attic access. Buildings with combustible ceiling or roof construction shall have an *attic* access opening to *attic* areas that exceed 30 square feet (2.8 m^2) and have a vertical height of 30 inches (762 mm) or greater. The vertical height shall be measured from the top of the ceiling framing members to the underside of the roof framing members.

The rough-framed opening shall not be less than 22 inches by 30 inches (559 mm by 762 mm) and shall be located in a hallway or other readily accessible location. When located in a wall, the opening shall be a minimum of 22 inches wide by 30 inches high (559 mm wide by 762 mm high). When the access is located in a ceiling, minimum unobstructed headroom in the *attic* space shall be 30 inches (762 mm) at some point above the access measured vertically from the bottom of ceiling framing members. See Section M1305.1.3 for access requirements where mechanical *equipment* is located in *attics.*

CHAPTER 9

ROOF ASSEMBLIES

SECTION R901 GENERAL

R901.1 Scope. The provisions of this chapter shall govern the design, materials, construction and quality of roof assemblies.

SECTION R902 ROOF CLASSIFICATION

R902.1 Roofing covering materials. Roofs shall be covered with materials as set forth in Sections R904 and R905. Class A, B or C roofing shall be installed in areas designated by law as requiring their use or when the edge of the roof is less than 3 feet (914 mm) from a lot line. Classes A, B and C roofing required by this section to be listed shall be tested in accordance with UL 790 or ASTM E 108.

Exceptions:

1. Class A roof assemblies include those with coverings of brick, masonry and exposed concrete roof deck.
2. Class A roof assemblies also include ferrous or copper shingles or sheets, metal sheets and shingles, clay or concrete roof tile, or slate installed on noncombustible decks.
3. Class A roof assemblies include minimum 16 oz/ft^2 copper sheets installed over combustible decks.

R902.2 Fire-retardant-treated shingles and shakes. Fire-retardant-treated wood shakes and shingles shall be treated by impregnation with chemicals by the full-cell vacuum-pressure process, in accordance with AWPA C1. Each bundle shall be marked to identify the manufactured unit and the manufacturer, and shall also be *labeled* to identify the classification of the material in accordance with the testing required in Section R902.1, the treating company and the quality control agency.

SECTION R903 WEATHER PROTECTION

R903.1 General. Roof decks shall be covered with *approved* roof coverings secured to the building or structure in accordance with the provisions of this chapter. Roof assemblies shall be designed and installed in accordance with this code and the *approved* manufacturer's installation instructions such that the roof assembly shall serve to protect the building or structure.

R903.2 Flashing. Flashings shall be installed in a manner that prevents moisture from entering the wall and roof through joints in copings, through moisture permeable materials and at intersections with parapet walls and other penetrations through the roof plane.

R903.2.1 Locations. Flashings shall be installed at wall and roof intersections, wherever there is a change in roof slope or direction and around roof openings. A flashing shall be installed to divert the water away from where the eave of a sloped roof intersects a vertical sidewall. Where flashing is of metal, the metal shall be corrosion resistant with a thickness of not less than 0.019 inch (0.5 mm) (No. 26 galvanized sheet).

R903.2.2 Crickets and saddles. A cricket or saddle shall be installed on the ridge side of any chimney or penetration more than 30 inches (762 mm) wide as measured perpendicular to the slope. Cricket or saddle coverings shall be sheet metal or of the same material as the roof covering.

Exception: Unit skylights installed in accordance with Section R308.6 and flashed in accordance with the manufacturer's instructions shall be permitted to be installed without a cricket or saddle.

R903.3 Coping. Parapet walls shall be properly coped with noncombustible, weatherproof materials of a width no less than the thickness of the parapet wall.

R903.4 Roof drainage. Unless roofs are sloped to drain over roof edges, roof drains shall be installed at each low point of the roof.

R903.4.1 Secondary (emergency overflow) drains or scuppers. Where roof drains are required, secondary emergency overflow roof drains or scuppers shall be provided where the roof perimeter construction extends above the roof in such a manner that water will be entrapped if the primary drains allow buildup for any reason. Overflow drains having the same size as the roof drains shall be installed with the inlet flow line located 2 inches (51 mm) above the low point of the roof, or overflow scuppers having three times the size of the roof drains and having a minimum opening height of 4 inches (102 mm) shall be installed in the adjacent parapet walls with the inlet flow located 2 inches (51 mm) above the low point of the roof served. The installation and sizing of overflow drains, leaders and conductors shall comply with Sections 1106 and 1108 as applicable of the *International Plumbing Code.*

Overflow drains shall discharge to an *approved* location and shall not be connected to roof drain lines.

SECTION R904 MATERIALS

R904.1 Scope. The requirements set forth in this section shall apply to the application of roof covering materials specified herein. Roof assemblies shall be applied in accordance with this chapter and the manufacturer's installation instructions. Installation of roof assemblies shall comply with the applicable provisions of Section R905.

R904.2 Compatibility of materials. Roof assemblies shall be of materials that are compatible with each other and with the building or structure to which the materials are applied.

R904.3 Material specifications and physical characteristics. Roof covering materials shall conform to the applicable standards listed in this chapter. In the absence of applicable standards or where materials are of questionable suitability, testing by an *approved* testing agency shall be required by the *building official* to determine the character, quality and limitations of application of the materials.

R904.4 Product identification. Roof covering materials shall be delivered in packages bearing the manufacturer's identifying marks and *approved* testing agency *labels* when required. Bulk shipments of materials shall be accompanied by the same information issued in the form of a certificate or on a bill of lading by the manufacturer.

SECTION R905
REQUIREMENTS FOR ROOF COVERINGS

R905.1 Roof covering application. Roof coverings shall be applied in accordance with the applicable provisions of this section and the manufacturer's installation instructions. Unless otherwise specified in this section, roof coverings shall be installed to resist the component and cladding loads specified in Table R301.2(2), adjusted for height and exposure in accordance with Table R301.2(3).

R905.2 Asphalt shingles. The installation of asphalt shingles shall comply with the provisions of this section.

R905.2.1 Sheathing requirements. Asphalt shingles shall be fastened to solidly sheathed decks.

R905.2.2 Slope. Asphalt shingles shall be used only on roof slopes of two units vertical in 12 units horizontal (2:12) or greater. For roof slopes from two units vertical in 12 units horizontal (2:12) up to four units vertical in 12 units horizontal (4:12), double underlayment application is required in accordance with Section R905.2.7.

R905.2.3 Underlayment. Unless otherwise noted, required underlayment shall conform to ASTM D 226 Type I, ASTM D 4869 Type I, or ASTM D 6757.

Self-adhering polymer modified bitumen sheet shall comply with ASTM D 1970.

R905.2.4 Asphalt shingles. Asphalt shingles shall comply with ASTM D 225 or D 3462.

R905.2.4.1 Wind resistance of asphalt shingles. Asphalt shingles shall be tested in accordance with ASTM D 7158. Asphalt shingles shall meet the classification requirements of Table R905.2.4.1(1) for the appropriate maximum basic wind speed. Asphalt shingle packaging shall bear a *label* to indicate compliance with ASTM D 7158 and the required classification in Table R905.2.4.1(1).

Exception: Asphalt shingles not included in the scope of ASTM D 7158 shall be tested and *labeled* to indicate compliance with ASTM D 3161 and the required classification in Table R905.2.4.1(2).

TABLE R905.2.4.1(1)
CLASSIFICATION OF ASPHALT ROOF SHINGLES PER ASTM D 7158

MAXIMUM BASIC WIND SPEED FROM FIGURE 301.2(4)A (mph)	CLASSIFICATION REQUIREMENT
85	D, G or H
90	D, G or H
100	G or H
110	G or H
120	G or H
130	H
140	H
150	H

For SI:1 mile per hour = 0.447 m/s.

TABLE R905.2.4.1(2)
CLASSIFICATION OF ASPHALT SHINGLES PER ASTM D 3161

MAXIMUM BASIC WIND SPEED FROM FIGURE 301.2(4)A (mph)	CLASSIFICATION REQUIREMENT
85	A, D or F
90	A, D or F
100	A, D or F
110	F
120	F
130	F
140	F
150	F

For SI: 1 mile per hour = 0.447 m/s.

R905.2.5 Fasteners. Fasteners for asphalt shingles shall be galvanized steel, stainless steel, aluminum or copper roofing nails, minimum 12 gage [0.105 inch (3 mm)] shank with a minimum $^3/_8$-inch-diameter (10 mm) head, ASTM F 1667, of a length to penetrate through the roofing materials and a minimum of $^3/_4$ inch (19 mm) into the roof sheathing. Where the roof sheathing is less than $^3/_4$ inch (19 mm) thick, the fasteners shall penetrate through the sheathing. Fasteners shall comply with ASTM F 1667.

R905.2.6 Attachment. Asphalt shingles shall have the minimum number of fasteners required by the manufacturer, but not less than four fasteners per strip shingle or two fasteners per individual shingle. Where the roof slope exceeds 21 units vertical in 12 units horizontal (21:12, 175-percent slope), shingles shall be installed as required by the manufacturer.

R905.2.7 Underlayment application. For roof slopes from two units vertical in 12 units horizontal (17-percent slope), up to four units vertical in 12 units horizontal (33-percent slope), underlayment shall be two layers applied in the following manner. Apply a 19-inch (483 mm) strip of underlayment felt parallel to and starting at the eaves, fastened sufficiently to hold in place. Starting at the eave, apply 36-inch-wide (914 mm) sheets of underlayment, overlapping successive sheets 19 inches (483 mm), and fastened sufficiently to hold in place. Distortions in the underlayment shall not interfere with the ability of the shingles to seal. For roof slopes of four units vertical in 12 units horizontal (33-percent slope) or greater, underlayment shall be one layer applied in the following manner. Underlayment shall be applied shingle fashion, parallel to and starting from the eave and lapped 2 inches (51 mm), fastened sufficiently to hold in place. Distortions in the underlayment shall not interfere with the ability of the shingles to seal. End laps shall be offset by 6 feet (1829 mm).

R905.2.7.1 Ice barrier. In areas where there has been a history of ice forming along the eaves causing a backup of water as designated in Table R301.2(1), an ice barrier that consists of a least two layers of underlayment cemented together or of a self-adhering polymer modified bitumen sheet, shall be used in lieu of normal underlayment and extend from the lowest edges of all roof surfaces to a point at least 24 inches (610 mm) inside the exterior wall line of the building.

Exception: Detached *accessory structures* that contain no *conditioned floor area*.

R905.2.7.2 Underlayment and high winds. Underlayment applied in areas subject to high winds [above 110 mph (49 m/s) in accordance with Figure R301.2(4)A] shall be applied with corrosion-resistant fasteners in accordance with manufacturer's installation instructions. Fasteners are to be applied along the overlap not farther apart than 36 inches (914 mm) on center.

Underlayment installed where the basic wind speed equals or exceeds 120 mph (54 m/s) shall comply with ASTM D 226 Type II, ASTM D 4869 Type IV, or ASTM D 6757. The underlayment shall be attached in a grid pattern of 12 inches (305 mm) between side laps with a 6-inch (152 mm) spacing at the side laps. Underlayment shall be applied in accordance with Section R905.2.7 except all laps shall be a minimum of 4 inches (102 mm). Underlayment shall be attached using metal or plastic cap nails with a head diameter of not less than 1 inch (25.4 mm) with a thickness of at least 32-gauge sheet metal. The cap-nail shank shall be a minimum of 12 gauge (0.105 inches) with a length to penetrate through the roof sheathing or a minimum of $^3/_4$ inch (19 mm) into the roof sheathing.

Exception: As an alternative, adhered underlayment complying with ASTM D 1970 shall be permitted.

R905.2.8 Flashing. Flashing for asphalt shingles shall comply with this section.

R905.2.8.1 Base and cap flashing. Base and cap flashing shall be installed in accordance with manufacturer's installation instructions. Base flashing shall be of either corrosion-resistant metal of minimum nominal 0.019-inch (0.5 mm) thickness or mineral surface roll roofing weighing a minimum of 77 pounds per 100 square feet (4 kg/m^2). Cap flashing shall be corrosion-resistant metal of minimum nominal 0.019-inch (0.5 mm) thickness.

R905.2.8.2 Valleys. Valley linings shall be installed in accordance with the manufacturer's installation instructions before applying shingles. Valley linings of the following types shall be permitted:

1. For open valleys (valley lining exposed) lined with metal, the valley lining shall be at least 24 inches (610 mm) wide and of any of the corrosion-resistant metals in Table R905.2.8.2.
2. For open valleys, valley lining of two plies of mineral surfaced roll roofing, complying with ASTM D 3909 or ASTM D 6380 Class M, shall be permitted. The bottom layer shall be 18 inches (457 mm) and the top layer a minimum of 36 inches (914 mm) wide.
3. For closed valleys (valley covered with shingles), valley lining of one ply of smooth roll roofing complying with ASTM D 6380 and at least 36 inches wide (914 mm) or valley lining as described in Item 1 or 2 above shall be permitted. Self-adhering polymer modified bitumen underlayment complying with ASTM D 1970 shall be permitted in lieu of the lining material.

R905.2.8.3 Sidewall flashing. Base flashing against a vertical sidewall shall be continuous or step flashing and shall be a minimum of 4 inches (102 mm) in height and 4 inches (102 mm) in width and shall direct water away from the vertical sidewall onto the roof and/or into the gutter. Where siding is provided on the vertical sidewall, the vertical leg of the flashing shall be continuous under the siding. Where anchored masonry veneer is provided on the vertical sidewall, the base flashing shall be provided in accordance with this section and counterflashing shall be provided in accordance with

Section R703.7.2.2. Where exterior plaster or adhered masonry veneer is provided on the vertical sidewall, the base flashing shall be provided in accordance with this section and Section R703.6.3.

R905.2.8.4 Other flashing. Flashing against a vertical front wall, as well as soil stack, vent pipe and chimney flashing, shall be applied according to the asphalt shingle manufacturer's printed instructions.

R905.2.8.5 Drip edge. A drip edge shall be provided at eaves and gables of shingle roofs. Adjacent pieces of drip edge shall be overlapped a minimum of 2 inches (51 mm). Drip edges shall extend a minimum of 0.25 inch (6.4 mm) below the roof sheathing and extend up the roof deck a minimum of 2 inches (51 mm). Drip edges shall be mechanically fastened to the roof deck at a maximum of 12 inches (305 mm) o.c. with fasteners as specified in Section R905.2.5. Underlayment shall be installed over the drip edge along eaves and under the underlayment on gables. Unless specified differently by the shingle manufacturer, shingles are permitted to be flush with the drip edge.

R905.3 Clay and concrete tile. The installation of clay and concrete tile shall comply with the provisions of this section.

R905.3.1 Deck requirements. Concrete and clay tile shall be installed only over solid sheathing or spaced structural sheathing boards.

R905.3.2 Deck slope. Clay and concrete roof tile shall be installed on roof slopes of two and one-half units vertical in 12 units horizontal ($2^1/_2$:12) or greater. For roof slopes from two and one-half units vertical in 12 units horizontal ($2^1/_2$:12) to four units vertical in 12 units horizontal (4:12), double underlayment application is required in accordance with Section R905.3.3.

R905.3.3 Underlayment. Unless otherwise noted, required underlayment shall conform to ASTM D 226 Type II; ASTM D 2626 Type I; or ASTM D 6380 Class M mineral surfaced roll roofing.

R905.3.3.1 Low slope roofs. For roof slopes from two and one-half units vertical in 12 units horizontal ($2^1/_2$:12), up to four units vertical in 12 units horizontal (4:12), underlayment shall be a minimum of two layers underlayment applied as follows:

1. Starting at the eave, a 19-inch (483 mm) strip of underlayment shall be applied parallel with the eave and fastened sufficiently in place.
2. Starting at the eave, 36-inch-wide (914 mm) strips of underlayment felt shall be applied, overlapping successive sheets 19 inches (483 mm), and fastened sufficiently in place.

R905.3.3.2 High slope roofs. For roof slopes of four units vertical in 12 units horizontal (4:12) or greater, underlayment shall be a minimum of one layer of underlayment felt applied shingle fashion, parallel to and starting from the eaves and lapped 2 inches (51 mm), fastened sufficiently in place.

R905.3.3.3 Underlayment and high winds. Underlayment applied in areas subject to high wind [above 110 miles per hour (49 m/s) in accordance with Figure R301.2(4)A] shall be applied with corrosion-resistant fasteners in accordance with manufacturer's installation instructions. Fasteners are to be applied along the overlap not farther apart than 36 inches (914 mm) on center.

Underlayment installed where the basic wind speed equals or exceeds 120 mph (54 m/s) shall be attached in a grid pattern of 12 inches (305 mm) between side laps with a 6-inch (152 mm) spacing at the side laps. Underlayment shall be applied in accordance with Section R905.2.7 except all laps shall be a minimum of 4 inches (102 mm). Underlayment shall be attached using metal or plastic cap nails with a head diameter of not less than 1 inch (25.4 mm) with a thickness of at least 32-gauge sheet metal. The cap-nail shank shall be a minimum of 12 gauge (0.105 inches) with a length to penetrate through the roof sheathing or a minimum of $^3/_4$-inch (19 mm) into the roof sheathing.

Exception: As an alternative, adhered underlayment complying with ASTM D 1970 shall be permitted.

TABLE R905.2.8.2
VALLEY LINING MATERIAL

MATERIAL	MINIMUM THICKNESS (inches)	GAGE	WEIGHT (pounds)
Cold-rolled copper	0.0216 nominal	—	ASTM B 370, 16 oz. per square foot
Lead-coated copper	0.0216 nominal	—	ASTM B 101, 16 oz. per square foot
High-yield copper	0.0162 nominal	—	ASTM B 370, 12 oz. per square foot
Lead-coated high-yield copper	0.0162 nominal	—	ASTM B 101, 12 oz. per square foot
Aluminum	0.024	—	—
Stainless steel	—	28	—
Galvanized steel	0.0179	26 (zinc coated G90)	—
Zinc alloy	0.027	—	—
Lead	—	—	$2^1/_2$
Painted terne	—	—	20

For SI: 1 inch = 25.4 mm, 1 pound = 0.454 kg.

R905.3.4 Clay tile. Clay roof tile shall comply with ASTM C 1167.

R905.3.5 Concrete tile. Concrete roof tile shall comply with ASTM C 1492.

R905.3.6 Fasteners. Nails shall be corrosion resistant and not less than 11 gage, $^{5}/_{16}$-inch (11 mm) head, and of sufficient length to penetrate the deck a minimum of $^{3}/_{4}$ inch (19 mm) or through the thickness of the deck, whichever is less. Attaching wire for clay or concrete tile shall not be smaller than 0.083 inch (2 mm). Perimeter fastening areas include three tile courses but not less than 36 inches (914 mm) from either side of hips or ridges and edges of eaves and gable rakes.

R905.3.7 Application. Tile shall be applied in accordance with this chapter and the manufacturer's installation instructions, based on the following:

1. Climatic conditions.
2. Roof slope.
3. Underlayment system.
4. Type of tile being installed.

Clay and concrete roof tiles shall be fastened in accordance with this section and the manufacturer's installation instructions. Perimeter tiles shall be fastened with a minimum of one fastener per tile. Tiles with installed weight less than 9 pounds per square foot (0.4 kg/m^2) require a minimum of one fastener per tile regardless of roof slope. Clay and concrete roof tile attachment shall be in accordance with the manufacturer's installation instructions where applied in areas where the wind speed exceeds 100 miles per hour (45 m/s) and on buildings where the roof is located more than 40 feet (12 192 mm) above *grade*. In areas subject to snow, a minimum of two fasteners per tile is required. In all other areas, clay and concrete roof tiles shall be attached in accordance with Table R905.3.7.

TABLE R905.3.7
CLAY AND CONCRETE TILE ATTACHMENT

SHEATHING	ROOF SLOPE	NUMBER OF FASTENERS
Solid without battens	All	One per tile
Spaced or solid with battens and slope < 5:12	Fasteners not required	—
Spaced sheathing without battens	5:12 ≤ slope < 12:12	One per tile/every other row
	12:12 ≤ slope < 24:12	One per tile

R905.3.8 Flashing. At the juncture of roof vertical surfaces, flashing and counterflashing shall be provided in accordance with this chapter and the manufacturer's installation instructions and, where of metal, shall not be less than 0.019 inch (0.5 mm) (No. 26 galvanized sheet gage) corrosion-resistant metal. The valley flashing shall extend at least 11 inches (279 mm) from the centerline each way and have a splash diverter rib not less than 1 inch (25 mm) high at the flow line formed as part of the flashing. Sections of flashing shall have an end lap of not less than 4 inches (102 mm). For roof slopes of three units vertical in 12 units horizontal (25-percent slope) and greater, valley flashing shall have a 36-inch-wide (914 mm) underlayment of one layer of Type I underlayment running the full length of the valley, in addition to other required underlayment. In areas where the average daily temperature in January is 25°F (-4°C) or less, metal valley flashing underlayment shall be solid-cemented to the roofing underlayment for slopes less than seven units vertical in 12 units horizontal (58-percent slope) or be of self-adhering polymer modified bitumen sheet.

R905.4 Metal roof shingles. The installation of metal roof shingles shall comply with the provisions of this section.

R905.4.1 Deck requirements. Metal roof shingles shall be applied to a solid or closely fitted deck, except where the roof covering is specifically designed to be applied to spaced sheathing.

R905.4.2 Deck slope. Metal roof shingles shall not be installed on roof slopes below three units vertical in 12 units horizontal (25-percent slope).

R905.4.3 Underlayment. Underlayment shall comply with ASTM D 226, Type I or Type II, ASTM D 4869, Type I or Type II, or ASTM D 1970. Underlayment shall be installed in accordance with the manufacturer's installation instructions.

R905.4.3.1 Ice barrier. In areas where there has been a history of ice forming along the eaves causing a backup of water as designated in Table R301.2(1), an ice barrier that consists of at least two layers of underlayment cemented together or a self-adhering polymer modified bitumen sheet shall be used in place of normal underlayment and extend from the lowest edges of all roof surfaces to a point at least 24 inches (610 mm) inside the exterior wall line of the building.

Exception: Detached *accessory structures* that contain no *conditioned floor area*.

R905.4.3.2 Underlayment and high winds. Underlayment applied in areas subject to high winds [above 110 mph (49 m/s) in accordance with Figure R301.2(4)A] shall be applied with corrosion-resistant fasteners in accordance with manufacturer's installation instructions. Fasteners are to be applied along the overlap not farther apart than 36 inches (914 mm) on center.

Underlayment installed where the basic wind speed equals or exceeds 120 mph (54 m/s) shall comply with ASTM D 226 Type II, ASTM D 4869 Type IV, or ASTM D 1970. The underlayment shall be attached in a grid pattern of 12 inches (305 mm) between side laps with a 6-inch (152 mm) spacing at the side laps. Underlayment shall be applied in accordance with Section R905.2.7 except all laps shall be a minimum of 4 inches (102 mm). Underlayment shall be attached using metal or plastic cap nails with a head diameter of not less than 1 inch (25.4 mm) with a thickness of at least 32 gauge sheet metal. The cap-nail shank shall be a minimum of

12 gauge (0.105 inches) with a length to penetrate through the roof sheathing or a minimum of $^3/_4$ inch (19 mm) into the roof sheathing.

Exception: As an alternative, adhered underlayment complying with ASTM D 1970 shall be permitted.

R905.4.4 Material standards. Metal roof shingle roof coverings shall comply with Table R905.10.3(1). The materials used for metal roof shingle roof coverings shall be naturally corrosion resistant or be made corrosion resistant in accordance with the standards and minimum thicknesses listed in Table R905.10.3(2).

R905.4.5 Application. Metal roof shingles shall be secured to the roof in accordance with this chapter and the *approved* manufacturer's installation instructions.

R905.4.6 Flashing. Roof valley flashing shall be of corrosion-resistant metal of the same material as the roof covering or shall comply with the standards in Table R905.10.3(1). The valley flashing shall extend at least 8 inches (203 mm) from the centerline each way and shall have a splash diverter rib not less than $^3/_4$ inch (19 mm) high at the flow line formed as part of the flashing. Sections of flashing shall have an end lap of not less than 4 inches (102 mm). The metal valley flashing shall have a 36-inch-wide (914 mm) underlayment directly under it consisting of one layer of underlayment running the full length of the valley, in addition to underlayment required for metal roof shingles. In areas where the average daily temperature in January is 25°F (-4°C) or less, the metal valley flashing underlayment shall be solid cemented to the roofing underlayment for roof slopes under seven units vertical in 12 units horizontal (58-percent slope) or self-adhering polymer modified bitumen sheet.

R905.5 Mineral-surfaced roll roofing. The installation of mineral-surfaced roll roofing shall comply with this section.

R905.5.1 Deck requirements. Mineral-surfaced roll roofing shall be fastened to solidly sheathed roofs.

R905.5.2 Deck slope. Mineral-surfaced roll roofing shall not be applied on roof slopes below one unit vertical in 12 units horizontal (8-percent slope).

R905.5.3 Underlayment. Underlayment shall comply with ASTM D 226, Type I or ASTM D 4869, Type I or II.

R905.5.3.1 Ice barrier. In areas where there has been a history of ice forming along the eaves causing a backup of water as designated in Table R301.2(1), an ice barrier that consists of at least two layers of underlayment cemented together or a self-adhering polymer modified bitumen sheet shall be used in place of normal underlayment and extend from the lowest edges of all roof surfaces to a point at least 24 inches (610 mm) inside the exterior wall line of the building.

Exception: Detached *accessory structures* that contain no *conditioned floor area.*

R905.5.3.2 Underlayment and high winds. Underlayment applied in areas subject to high winds [above 110 mph (49 m/s) in accordance with Figure R301.2(4)A] shall be applied with corrosion-resistant fasteners in accordance with manufacturer's installation instructions. Fasteners are to be applied along the overlap not farther apart than 36 inches (914 mm) on center.

Underlayment installed where the basic wind speed equals or exceeds 120 mph (54 m/s) shall comply with ASTM D 226 Type II or ASTM D 4869 Type IV. The underlayment shall be attached in a grid pattern of 12 inches (305 mm) between side laps with a 6-inch (152 mm) spacing at the side laps. Underlayment shall be applied in accordance with Section R905.2.7 except all laps shall be a minimum of 4 inches (102 mm). Underlayment shall be attached using metal or plastic cap nails with a head diameter of not less than 1 inch (25.4 mm) with a thickness of at least 32-gauge sheet metal. The cap-nail shank shall be a minimum of 12 gauge (0.105 inches) with a length to penetrate through the roof sheathing or a minimum of $^3/_4$ inch (19 mm) into the roof sheathing.

Exception: As an alternative, adhered underlayment complying with ASTM D 1970 shall be permitted.

R905.5.4 Material standards. Mineral-surfaced roll roofing shall conform to ASTM D 3909 or ASTM D 6380, Class M.

R905.5.5 Application. Mineral-surfaced roll roofing shall be installed in accordance with this chapter and the manufacturer's installation instructions.

R905.6 Slate and slate-type shingles. The installation of slate and slate-type shingles shall comply with the provisions of this section.

R905.6.1 Deck requirements. Slate shingles shall be fastened to solidly sheathed roofs.

R905.6.2 Deck slope. Slate shingles shall be used only on slopes of four units vertical in 12 units horizontal (33-percent slope) or greater.

R905.6.3 Underlayment. Underlayment shall comply with ASTM D 226, Type I, or ASTM D 4869, Type I or II. Underlayment shall be installed in accordance with the manufacturer's installation instructions.

R905.6.3.1 Ice barrier. In areas where there has been a history of ice forming along the eaves causing a backup of water as designated in Table R301.2(1), an ice barrier that consists of at least two layers of underlayment cemented together or a self-adhering polymer modified bitumen sheet shall be used in lieu of normal underlayment and extend from the lowest edges of all roof surfaces to a point at least 24 inches (610 mm) inside the exterior wall line of the building.

Exception: Detached *accessory structures* that contain no *conditioned floor area.*

R905.6.3.2 Underlayment and high winds. Underlayment applied in areas subject to high winds [above 110 mph (49 m/s) in accordance with Figure R301.2(4)A] shall be applied with corrosion-resistant fasteners in accordance with manufacturer's installation instructions. Fasteners are to be applied along the overlap not farther apart than 36 inches (914 mm) on center.

Underlayment installed where the basic wind speed equals or exceeds 120 mph (54 m/s) shall comply with ASTM D 226 Type II or ASTM D 4869 Type IV. The underlayment shall be attached in a grid pattern of 12 inches (305 mm) between side laps with a 6-inch (152 mm) spacing at the side laps. Underlayment shall be applied in accordance with Section R905.2.7 except all laps shall be a minimum of 4 inches (102 mm). Underlayment shall be attached using metal or plastic cap nails with a head diameter of not less than 1 inch (25.4 mm) with a thickness of at least 32-gauge sheet metal. The cap-nail shank shall be a minimum of 12 gauge (0.105 inches) with a length to penetrate through the roof sheathing or a minimum of $^3/_4$ inch (19 mm) into the roof sheathing.

Exception: As an alternative, adhered underlayment complying with ASTM D 1970 shall be permitted.

R905.6.4 Material standards. Slate shingles shall comply with ASTM C 406.

R905.6.5 Application. Minimum headlap for slate shingles shall be in accordance with Table R905.6.5. Slate shingles shall be secured to the roof with two fasteners per slate. Slate shingles shall be installed in accordance with this chapter and the manufacturer's installation instructions.

TABLE R905.6.5
SLATE SHINGLE HEADLAP

SLOPE	HEADLAP (inches)
4:12 ≤ slope < 8:12	4
8:12 ≤ slope < 20:12	3
Slope ≤ 20:12	2

For SI: 1 inch = 25.4 mm.

R905.6.6 Flashing. Flashing and counterflashing shall be made with sheet metal. Valley flashing shall be a minimum of 15 inches (381 mm) wide. Valley and flashing metal shall be a minimum uncoated thickness of 0.0179-inch (0.5 mm) zinc coated G90. Chimneys, stucco or brick walls shall have a minimum of two plies of felt for a cap flashing consisting of a 4-inch-wide (102 mm) strip of felt set in plastic cement and extending 1 inch (25 mm) above the first felt and a top coating of plastic cement. The felt shall extend over the base flashing 2 inches (51 mm).

R905.7 Wood shingles. The installation of wood shingles shall comply with the provisions of this section.

R905.7.1 Deck requirements. Wood shingles shall be installed on solid or spaced sheathing. Where spaced sheathing is used, sheathing boards shall not be less than 1-inch by 4-inch (25.4 mm by 102 mm) nominal dimensions and shall be spaced on centers equal to the weather exposure to coincide with the placement of fasteners.

R905.7.1.1 Solid sheathing required. In areas where the average daily temperature in January is 25°F (-4°C) or less, solid sheathing is required on that portion of the roof requiring the application of an ice barrier.

R905.7.2 Deck slope. Wood shingles shall be installed on slopes of three units vertical in 12 units horizontal (25-percent slope) or greater.

R905.7.3 Underlayment. Underlayment shall comply with ASTM D 226, Type I or ASTM D 4869, Type I or II.

R905.7.3.1 Ice barrier. In areas where there has been a history of ice forming along the eaves causing a backup of water as designated in Table R301.2(1), an ice barrier that consists of at least two layers of underlayment cemented together or a self-adhering polymer modified bitumen sheet shall be used in lieu of normal underlayment and extend from the lowest edges of all roof surfaces to a point at least 24 inches (610 mm) inside the exterior wall line of the building.

Exception: Detached *accessory structures* that contain no *conditioned floor area.*

R905.7.3.2 Underlayment and high winds. Underlayment applied in areas subject to high winds [above 110 mph (49 m/s) in accordance with Figure R301.2(4)A] shall be applied with corrosion-resistant fasteners in accordance with manufacturer's installation instructions. Fasteners are to be applied along the overlap not farther apart than 36 inches (914mm) on center.

Underlayment installed where the basic wind speed equals or exceeds 120 mph (54 m/s) shall comply with ASTM D 226 Type II or ASTM D 4869 Type IV. The underlayment shall be attached in a grid pattern of 12 inches (305 mm) between side laps with a 6-inch (152 mm) spacing at the side laps. Underlayment shall be applied in accordance with Section R905.2.7 except all Head laps shall be a minimum of 4 inches (102 mm). Underlayment shall be attached using metal or plastic cap nails with a head diameter of not less than 1 inch (25.4 mm) with a thickness of at least 32-gauge sheet metal. The cap-nail shank shall be a minimum of 12 gauge (0.105 inches) with a length to penetrate through the roof sheathing or a minimum of $^3/_4$ inch (19 mm) into the roof sheathing.

Exception: As an alternative, adhered underlayment complying with ASTM D 1970 shall be permitted.

R905.7.4 Material standards. Wood shingles shall be of naturally durable wood and comply with the requirements of Table R905.7.4.

TABLE R905.7.4
WOOD SHINGLE MATERIAL REQUIREMENTS

MATERIAL	MINIMUM GRADES	APPLICABLE GRADING RULES
Wood shingles of naturally durable wood	1, 2 or 3	Cedar Shake and Shingle Bureau

R905.7.5 Application. Wood shingles shall be installed according to this chapter and the manufacturer's installation instructions. Wood shingles shall be laid with a side lap not less than $1^1/_2$ inches (38 mm) between joints in courses, and no two joints in any three adjacent courses shall be in direct alignment. Spacing between shingles shall not be less than $^1/_4$ inch to $^3/_8$ inch (6 mm to 10 mm). Weather exposure for wood shingles shall not exceed those set in Table R905.7.5. Fasteners for wood shingles shall be corrosion resistant with a minimum penetration of $^1/_2$ inch (13 mm) into the sheathing. For sheathing less than $^1/_2$ inch (13 mm) in thickness, the fasteners shall extend through the sheathing. Wood shingles shall be attached to the roof with two fasteners per shingle, positioned no more than $^3/_4$ inch (19 mm) from each edge and no more than 1 inch (25 mm) above the exposure line.

TABLE R905.7.5
WOOD SHINGLE WEATHER EXPOSURE AND ROOF SLOPE

ROOFING MATERIAL	LENGTH (inches)	GRADE	EXPOSURE (inches)	
			3:12 pitch to < 4:12	4:12 pitch or steeper
Shingles of naturally durable wood	16	No. 1	$3^3/_4$	5
		No. 2	$3^1/_2$	4
		No. 3	3	$3^1/_2$
	18	No. 1	$4^1/_4$	$5^1/_2$
		No. 2	4	$4^1/_2$
		No. 3	$3^1/_2$	4
	24	No. 1	$5^3/_4$	$7^1/_2$
		No. 2	$5^1/_2$	$6^1/_2$
		No. 3	5	$5^1/_2$

For SI: 1 inch = 25.4 mm.

R905.7.6 Valley flashing. Roof flashing shall be not less than No. 26 gage [0.019 inches (0.5 mm)] corrosion-resistant sheet metal and shall extend 10 inches (254 mm) from the centerline each way for roofs having slopes less than 12 units vertical in 12 units horizontal (100-percent slope), and 7 inches (178 mm) from the centerline each way for slopes of 12 units vertical in 12 units horizontal and greater. Sections of flashing shall have an end lap of not less than 4 inches (102 mm).

R905.7.7 Label required. Each bundle of shingles shall be identified by a *label* of an *approved* grading or inspection bureau or agency.

R905.8 Wood shakes. The installation of wood shakes shall comply with the provisions of this section.

R905.8.1 Deck requirements. Wood shakes shall be used only on solid or spaced sheathing. Where spaced sheathing is used, sheathing boards shall not be less than 1-inch by 4-inch (25 mm by 102 mm) nominal dimensions and shall be spaced on centers equal to the weather exposure to coincide with the placement of fasteners. Where 1-inch by 4-inch (25 mm by 102 mm) spaced sheathing is installed at 10 inches (254 mm) on center, additional 1-inch by 4-inch (25 mm by 102 mm) boards shall be installed between the sheathing boards.

R905.8.1.1 Solid sheathing required. In areas where the average daily temperature in January is 25°F (-4°C) or less, solid sheathing is required on that portion of the roof requiring an ice barrier.

R905.8.2 Deck slope. Wood shakes shall only be used on slopes of three units vertical in 12 units horizontal (25-percent slope) or greater.

R905.8.3 Underlayment. Underlayment shall comply with ASTM D 226, Type I or ASTM D 4869, Type I or II.

R905.8.3.1 Ice barrier. In areas where there has been a history of ice forming along the eaves causing a backup of water as designated in Table R301.2(1), an ice barrier that consists of at least two layers of underlayment cemented together or a self-adhering polymer modified bitumen sheet shall be used in place of normal underlayment and extend from the lowest edges of all roof surfaces to a point at least 24 inches (610 mm) inside the exterior wall line of the building.

Exception: Detached *accessory structures* that contain no *conditioned floor area.*

R905.8.3.2 Underlayment and high winds. Underlayment applied in areas subject to high winds [above 110 mph (49 m/s) in accordance with Figure R301.2(4)A] shall be applied with corrosion-resistant fasteners in accordance with manufacturer's installation instructions. Fasteners are to be applied along the overlap not farther apart than 36 inches (914 mm) on center.

Underlayment installed where the basic wind speed equals or exceeds 120 mph (54 m/s) shall comply with ASTM D 226 Type II or ASTM D 4869 Type IV. The underlayment shall be attached in a grid pattern of 12 inches (305 mm) between side laps with a 6-inch (152 mm) spacing at the side laps. Underlayment shall be applied in accordance with Section R905.2.7 except all laps shall be a minimum of 4 inches (102 mm). Underlayment shall be attached using metal or plastic cap nails with a head diameter of not less than 1 inch (25.4 mm) with a thickness of at least 32-gauge sheet metal. The cap-nail shank shall be a minimum of 12 gauge (0.105 inches) with a length to penetrate through the roof sheathing or a minimum of $^3/_4$ inch (19 mm) into the roof sheathing.

Exception: As an alternative, adhered underlayment complying with ASTM D 1970 shall be permitted.

R905.8.4 Interlayment. Interlayment shall comply with ASTM D 226, Type I.

R905.8.5 Material standards. Wood shakes shall comply with the requirements of Table R905.8.5.

TABLE R905.8.5
WOOD SHAKE MATERIAL REQUIREMENTS

MATERIAL	MINIMUM GRADES	APPLICABLE GRADING RULES
Wood shakes of naturally durable wood	1	Cedar Shake and Shingle Bureau
Taper sawn shakes of naturally durable wood	1 or 2	Cedar Shake and Shingle Bureau
Preservative-treated shakes and shingles of naturally durable wood	1	Cedar Shake and Shingle Bureau
Fire-retardant-treated shakes and shingles of naturally durable wood	1	Cedar Shake and Shingle Bureau
Preservative-treated taper sawn shakes of Southern pine treated in accordance with AWPA Standard U1 (Commodity Specification A, Use Category 3B and Section 5.6)	1 or 2	Forest Products Laboratory of the Texas Forest Services

R905.8.6 Application. Wood shakes shall be installed according to this chapter and the manufacturer's installation instructions. Wood shakes shall be laid with a side lap not less than $1^1/_2$ inches (38 mm) between joints in adjacent courses. Spacing between shakes in the same course shall be $^3/_8$ inch to $^5/_8$ inch (9.5 mm to 15.9 mm) for shakes and tapersawn shakes of naturally durable wood and shall be $^3/_8$ inch to $^5/_8$ inch (9.5 mm to 15.9 mm) for preservative-treated taper sawn shakes. Weather exposure for wood shakes shall not exceed those set forth in Table R905.8.6. Fasteners for wood shakes shall be corrosion-resistant, with a minimum penetration of $^1/_2$ inch (12.7 mm) into the sheathing. For sheathing less than $^1/_2$ inch (12.7 mm) thick, the fasteners shall extend through the sheathing. Wood shakes shall be attached to the roof with two fasteners per shake, positioned no more than 1 inch (25 mm) from each edge and no more than 2 inches (51 mm) above the exposure line.

TABLE R905.8.6
WOOD SHAKE WEATHER EXPOSURE AND ROOF SLOPE

ROOFING MATERIAL	LENGTH (inches)	GRADE	EXPOSURE (inches) 4:12 pitch or steeper
Shakes of naturally durable wood	18	No. 1	$7^1/_2$
	24	No. 1	10[a]
Preservative-treated taper sawn shakes of Southern Yellow Pine	18	No. 1	$7^1/_2$
	24	No. 1	10
	18	No. 2	$5^1/_2$
	24	No. 2	$7^1/_2$
Taper-sawn shakes of naturally durable wood	18	No. 1	$7^1/_2$
	24	No. 1	10
	18	No. 2	$5^1/_2$
	24	No. 2	$7^1/_2$

For SI: 1 inch = 25.4 mm.

a. For 24-inch by $^3/_8$-inch handsplit shakes, the maximum exposure is $7^1/_2$ inches.

R905.8.7 Shake placement. The starter course at the eaves shall be doubled and the bottom layer shall be either 15-inch (381 mm), 18-inch (457 mm) or 24-inch (610 mm) wood shakes or wood shingles. Fifteen-inch (381 mm) or 18-inch (457 mm) wood shakes may be used for the final course at the ridge. Shakes shall be interlaid with 18-inch-wide (457 mm) strips of not less than No. 30 felt shingled between each course in such a manner that no felt is exposed to the weather by positioning the lower edge of each felt strip above the butt end of the shake it covers a distance equal to twice the weather exposure.

R905.8.8 Valley flashing. Roof valley flashing shall not be less than No. 26 gage [0.019 inch (0.5 mm)] corrosion-resistant sheet metal and shall extend at least 11 inches (279 mm) from the centerline each way. Sections of flashing shall have an end lap of not less than 4 inches (102 mm).

R905.8.9 Label required. Each bundle of shakes shall be identified by a *label* of an *approved* grading or inspection bureau or agency.

R905.9 Built-up roofs. The installation of built-up roofs shall comply with the provisions of this section.

R905.9.1 Slope. Built-up roofs shall have a design slope of a minimum of one-fourth unit vertical in 12 units horizontal (2-percent slope) for drainage, except for coal-tar built-up roofs, which shall have a design slope of a minimum one-eighth unit vertical in 12 units horizontal (1-percent slope).

R905.9.2 Material standards. Built-up roof covering materials shall comply with the standards in Table R905.9.2 or UL 55A.

R905.9.3 Application. Built-up roofs shall be installed according to this chapter and the manufacturer's installation instructions.

R905.10 Metal roof panels. The installation of metal roof panels shall comply with the provisions of this section.

R905.10.1 Deck requirements. Metal roof panel roof coverings shall be applied to solid or spaced sheathing, except where the roof covering is specifically designed to be applied to spaced supports.

R905.10.2 Slope. Minimum slopes for metal roof panels shall comply with the following:

1. The minimum slope for lapped, nonsoldered-seam metal roofs without applied lap sealant shall be three units vertical in 12 units horizontal (25-percent slope).
2. The minimum slope for lapped, nonsoldered-seam metal roofs with applied lap sealant shall be one-half vertical unit in 12 units horizontal (4-percent slope). Lap sealants shall be applied in accordance with the *approved* manufacturer's installation instructions.
3. The minimum slope for standing-seam roof systems shall be one-quarter unit vertical in 12 units horizontal (2-percent slope).

R905.10.3 Material standards. Metal-sheet roof covering systems that incorporate supporting structural members shall be designed in accordance with the *International Building Code*. Metal-sheet roof coverings installed over structural decking shall comply with Table R905.10.3(1). The materials used for metal-sheet roof coverings shall be naturally corrosion resistant or provided with corrosion resistance in accordance with the standards and minimum thicknesses shown in Table R905.10.3(2).

TABLE R905.9.2
BUILT-UP ROOFING MATERIAL STANDARDS

MATERIAL STANDARD	STANDARD
Acrylic coatings used in roofing	ASTM D 6083
Aggregate surfacing	ASTM D 1863
Asphalt adhesive used in roofing	ASTM D 3747
Asphalt cements used in roofing	ASTM D 2822; D 3019; D 4586
Asphalt coated glass fiber base sheet	ASTM D 4601
Asphalt coatings used in roofing	ASTM D 1227; D 2823; D 2824; D 4479
Asphalt glass felt	ASTM D 2178
Asphalt primer used in roofing	ASTM D 41
Asphalt-saturated and asphalt-coated organic felt base sheet	ASTM D 2626
Asphalt-saturated organic felt (perforated)	ASTM D 226
Asphalt used in roofing	ASTM D 312
Coal-tar cements used in roofing	ASTM D 4022; D 5643
Coal-tar primer used in roofing, dampproofing and waterproofing	ASTM D 43
Coal-tar saturated organic felt	ASTM D 227
Coal-tar used in roofing	ASTM D 450, Type I or II
Glass mat, coal tar	ASTM D 4990
Glass mat, venting type	ASTM D 4897
Mineral-surfaced inorganic cap sheet	ASTM D 3909
Thermoplastic fabrics used in roofing	ASTM D 5665; D 5726

TABLE R905.10.3(1)
METAL ROOF COVERING STANDARDS

ROOF COVERING TYPE	STANDARD APPLICATION RATE/THICKNESS
Galvanized steel	ASTM A 653 G90 Zinc coated
Stainless steel	ASTM A 240, 300 Series alloys
Steel	ASTM A 924
Lead-coated copper	ASTM B 101
Cold-rolled copper	ASTM B 370 minimum 16 oz/sq ft and 12 oz/sq ft high-yield copper for metal-sheet roof-covering systems; 12 oz/sq ft for preformed metal shingle systems.
Hard lead	2 lb/sq ft
Soft lead	3 lb/sq ft
Aluminum	ASTM B 209, 0.024 minimum thickness for roll-formed panels and 0.019-inch minimum thickness for pressformed shingles.
Terne (tin) and terne-coated stainless	Terne coating of 40 lb per double base box, field painted where applicable in accordance with manufacturer's installation instructions.
Zinc	0.027 inch minimum thickness: 99.995% electrolytic high-grade zinc with alloy additives of copper (0.08 - 0.20%), titanium (0.07% - 0.12%) and aluminum (0.015%).

For SI: 1 ounce per square foot = 0.305 kg/m^2, 1 pound per square foot = 4.214 kg/m^2, 1 inch = 25.4 mm, 1 pound = 0.454 kg.

**TABLE R905.10.3(2)
MINIMUM CORROSION RESISTANCE**

55% aluminum-zinc alloy coated steel	ASTM A 792 AZ 50
5% aluminum alloy-coated steel	ASTM A 875 GF60
Aluminum-coated steel	ASTM A 463 T2 65
Galvanized steel	ASTM A 653 G-90
Prepainted steel	ASTM A 755[a]

a. Paint systems in accordance with ASTM A 755 shall be applied over steel products with corrosion-resistant coatings complying with ASTM A 792, ASTM A 875, ASTM A 463, or ASTM A 653.

R905.10.4 Attachment. Metal roof panels shall be secured to the supports in accordance with this chapter and the manufacturer's installation instructions. In the absence of manufacturer's installation instructions, the following fasteners shall be used:

1. Galvanized fasteners shall be used for steel roofs.
2. Copper, brass, bronze, copper alloy and 300-series stainless steel fasteners shall be used for copper roofs.
3. Stainless steel fasteners are acceptable for metal roofs.

R905.10.5 Underlayment. Underlayment shall be installed in accordance with the manufacturer's installation instructions.

R905.10.5.1 Underlayment and high winds. Underlayment applied in areas subject to high winds [above 110 mph (49 m/s) in accordance with Figure R301.2(4)A] shall be applied with corrosion-resistant fasteners in accordance with manufacturer's installation instructions. Fasteners are to be applied along the overlap not farther apart than 36 inches (914 mm) on center.

Underlayment installed where the basic wind speed equals or exceeds 120 mph (54 m/s) shall comply with ASTM D 226 Type II. The underlayment shall be attached in a grid pattern of 12 inches (305 mm) between side laps with a 6-inch (152 mm) spacing at the side laps. Underlayment shall be applied in accordance with Section R905.2.7 except all laps shall be a minimum of 4 inches (102 mm). Underlayment shall be attached using metal or plastic cap nails with a head diameter of not less than 1 inch (25.4 mm) with a thickness of at least 32-gauge sheet metal. The cap-nail shank shall be a minimum of 12 gauge (0.105 inches) with a length to penetrate through the roof sheathing or a minimum of $^3/_4$ inch (19 mm) into the roof sheathing.

Exception: As an alternative, adhered underlayment complying with ASTM D 1970 shall be permitted.

R905.11 Modified bitumen roofing. The installation of modified bitumen roofing shall comply with the provisions of this section.

R905.11.1 Slope. Modified bitumen membrane roofs shall have a design slope of a minimum of one-fourth unit vertical in 12 units horizontal (2-percent slope) for drainage.

R905.11.2 Material standards. Modified bitumen roof coverings shall comply with the standards in Table R905.11.2.

**TABLE R905.11.2
MODIFIED BITUMEN ROOFING MATERIAL STANDARDS**

MATERIAL	STANDARD
Acrylic coating	ASTM D 6083
Asphalt adhesive	ASTM D 3747
Asphalt cement	ASTM D 3019
Asphalt coating	ASTM D 1227; D 2824
Asphalt primer	ASTM D 41
Modified bitumen roof membrane	ASTM D 6162; D 6163; D 6164; D 6222; D 6223; D 6298; CGSB 37-GP-56M

R905.11.3 Application. Modified bitumen roofs shall be installed according to this chapter and the manufacturer's installation instructions.

R905.12 Thermoset single-ply roofing. The installation of thermoset single-ply roofing shall comply with the provisions of this section.

R905.12.1 Slope. Thermoset single-ply membrane roofs shall have a design slope of a minimum of one-fourth unit vertical in 12 units horizontal (2-percent slope) for drainage.

R905.12.2 Material standards. Thermoset single-ply roof coverings shall comply with ASTM D 4637, ASTM D 5019 or CGSB 37-GP-52M.

R905.12.3 Application. Thermoset single-ply roofs shall be installed according to this chapter and the manufacturer's installation instructions.

R905.13 Thermoplastic single-ply roofing. The installation of thermoplastic single-ply roofing shall comply with the provisions of this section.

R905.13.1 Slope. Thermoplastic single-ply membrane roofs shall have a design slope of a minimum of one-fourth unit vertical in 12 units horizontal (2-percent slope).

R905.13.2 Material standards. Thermoplastic single-ply roof coverings shall comply with ASTM D 4434, ASTM D 6754, ASTM D 6878 or CGSB CAN/CGSB 37.54.

R905.13.3 Application. Thermoplastic single-ply roofs shall be installed according to this chapter and the manufacturer's installation instructions.

R905.14 Sprayed polyurethane foam roofing. The installation of sprayed polyurethane foam roofing shall comply with the provisions of this section.

R905.14.1 Slope. Sprayed polyurethane foam roofs shall have a design slope of a minimum of one-fourth unit vertical in 12 units horizontal (2-percent slope) for drainage.

R905.14.2 Material standards. Spray-applied polyurethane foam insulation shall comply with ASTM C 1029, Type III or IV.

R905.14.3 Application. Foamed-in-place roof insulation shall be installed in accordance with this chapter and the manufacturer's installation instructions. A liquid-applied protective coating that complies with Table R905.14.3 shall be applied no less than 2 hours nor more than 72 hours following the application of the foam.

TABLE R905.14.3
PROTECTIVE COATING MATERIAL STANDARDS

MATERIAL	STANDARD
Acrylic coating	ASTM D 6083
Silicone coating	ASTM D 6694
Moisture-cured polyurethane coating	ASTM D 6947

R905.14.4 Foam plastics. Foam plastic materials and installation shall comply with Section R316.

R905.15 Liquid-applied roofing. The installation of liquid-applied roofing shall comply with the provisions of this section.

R905.15.1 Slope. Liquid-applied roofing shall have a design slope of a minimum of one-fourth unit vertical in 12 units horizontal (2-percent slope).

R905.15.2 Material standards. Liquid-applied roofing shall comply with ASTM C 836, C 957, D 1227, D 3468, D 6083, D 6694 or D 6947.

R905.15.3 Application. Liquid-applied roofing shall be installed according to this chapter and the manufacturer's installation instructions.

R905.16 Photovoltaic modules/shingles. The installation of photovoltaic modules/shingles shall comply with the provisions of this section.

R905.16.1 Material standards. Photovoltaic modules/shingles shall be listed and labeled in accordance with UL 1703.

R905.16.2 Attachment. Photovoltaic modules/shingles shall be attached in accordance with the manufacturer's installation instructions.

R905.16.3 Wind resistance. Photovoltaic modules/shingles shall be tested in accordance with procedures and acceptance criteria in ASTM D 3161. Photovoltaic modules/shingles shall comply with the classification requirements of Table R905.2.4.1(2) for the appropriate maximum basic wind speed. Photovoltaic modules/shingle packaging shall bear a label to indicate compliance with the procedures in ASTM D 3161 and the required classification from Table R905.2.4.1(2).

SECTION R906
ROOF INSULATION

R906.1 General. The use of above-deck thermal insulation shall be permitted provided such insulation is covered with an *approved* roof covering and complies with FM 4450 or UL 1256.

R906.2 Material standards. Above-deck thermal insulation board shall comply with the standards in Table R906.2.

TABLE R906.2
MATERIAL STANDARDS FOR ROOF INSULATION

Cellular glass board	ASTM C 552
Composite boards	ASTM C 1289, Type III, IV, V or VI
Expanded polystyrene	ASTM C 578
Extruded polystyrene board	ASTM C 578
Perlite board	ASTM C 728
Polyisocyanurate board	ASTM C 1289, Type I or II
Wood fiberboard	ASTM C 208

SECTION R907
REROOFING

R907.1 General. Materials and methods of application used for re-covering or replacing an existing roof covering shall comply with the requirements of Chapter 9.

Exception: Reroofing shall not be required to meet the minimum design slope requirement of one-quarter unit vertical in 12 units horizontal (2-percent slope) in Section R905 for roofs that provide positive roof drainage.

R907.2 Structural and construction loads. The structural roof components shall be capable of supporting the roof covering system and the material and equipment loads that will be encountered during installation of the roof covering system.

R907.3 Recovering versus replacement. New roof coverings shall not be installed without first removing all existing layers of roof coverings where any of the following conditions exist:

1. Where the existing roof or roof covering is water-soaked or has deteriorated to the point that the existing roof or roof covering is not adequate as a base for additional roofing.
2. Where the existing roof covering is wood shake, slate, clay, cement or asbestos-cement tile.
3. Where the existing roof has two or more applications of any type of roof covering.

Exceptions:

1. Complete and separate roofing systems, such as standing-seam metal roof systems, that are designed to transmit the roof loads directly to the building's structural system and that do not rely on existing roofs and roof coverings for support, shall not require the removal of existing roof coverings.
2. Installation of metal panel, metal shingle and concrete and clay tile roof coverings over existing wood shake roofs shall be permitted when the application is in accordance with Section R907.4.

3. The application of new protective coating over existing spray polyurethane foam roofing systems shall be permitted without tear-off of existing roof coverings.
4. Where the existing roof assembly includes an ice barrier membrane that is adhered to the roof deck, the existing ice barrier membrane shall be permitted to remain in place and covered with an additional layer of ice barrier membrane in accordance with Section R905.

R907.4 Roof recovering. Where the application of a new roof covering over wood shingle or shake roofs creates a combustible concealed space, the entire existing surface shall be covered with gypsum board, mineral fiber, glass fiber or other *approved* materials securely fastened in place.

R907.5 Reinstallation of materials. Existing slate, clay or cement tile shall be permitted for reinstallation, except that damaged, cracked or broken slate or tile shall not be reinstalled. Any existing flashings, edgings, outlets, vents or similar devices that are a part of the assembly shall be replaced when rusted, damaged or deteriorated. Aggregate surfacing materials shall not be reinstalled.

R907.6 Flashings. Flashings shall be reconstructed in accordance with *approved* manufacturer's installation instructions. Metal flashing to which bituminous materials are to be adhered shall be primed prior to installation.

CHAPTER 10

CHIMNEYS AND FIREPLACES

SECTION R1001
MASONRY FIREPLACES

R1001.1 General. Masonry fireplaces shall be constructed in accordance with this section and the applicable provisions of Chapters 3 and 4.

R1001.2 Footings and foundations. Footings for masonry fireplaces and their chimneys shall be constructed of concrete or *solid masonry* at least 12 inches (305 mm) thick and shall extend at least 6 inches (152 mm) beyond the face of the fireplace or foundation wall on all sides. Footings shall be founded on natural, undisturbed earth or engineered fill below frost depth. In areas not subjected to freezing, footings shall be at least 12 inches (305 mm) below finished *grade*.

R1001.2.1 Ash dump cleanout. Cleanout openings located within foundation walls below fireboxes, when provided, shall be equipped with ferrous metal or masonry doors and frames constructed to remain tightly closed except when in use. Cleanouts shall be accessible and located so that ash removal will not create a hazard to combustible materials.

R1001.3 Seismic reinforcing. Masonry or concrete chimneys in Seismic Design Category D_0, D_1 or D_2 shall be reinforced. Reinforcing shall conform to the requirements set forth in Table R1001.1 and Section R609, Grouted Masonry.

R1001.3.1 Vertical reinforcing. For chimneys up to 40 inches (1016 mm) wide, four No. 4 continuous vertical bars shall be placed between wythes of *solid masonry* or within the cells of hollow unit masonry and grouted in accordance with Section R609. Grout shall be prevented from bonding with the flue liner so that the flue liner is free to move with thermal expansion. For chimneys more than 40 inches (1016 mm) wide, two additional No. 4 vertical bars shall be provided for each additional flue incorporated into the chimney or for each additional 40 inches (1016 mm) in width or fraction thereof.

R1001.3.2 Horizontal reinforcing. Vertical reinforcement shall be placed within $^1/_4$-inch (6 mm) ties, or other reinforcing of equivalent net cross-sectional area, placed in the bed joints according to Section R607 at a minimum of every 18 inches (457 mm) of vertical height. Two such ties shall be installed at each bend in the vertical bars.

R1001.4 Seismic anchorage. Masonry or concrete chimneys in Seismic Design Category D_0, D_1 or D_2 shall be anchored at each floor, ceiling or roof line more than 6 feet (1829 mm) above *grade*, except where constructed completely within the exterior walls. Anchorage shall conform to the requirements of Section R1001.4.1.

R1001.4.1 Anchorage. Two $^3/_{16}$-inch by 1-inch (5 mm by 25 mm) straps shall be embedded a minimum of 12 inches (305 mm) into the chimney. Straps shall be hooked around the outer bars and extend 6 inches (152 mm) beyond the bend. Each strap shall be fastened to a minimum of four floor ceiling or floor joists or rafters with two $^1/_2$-inch (13 mm) bolts.

R1001.5 Firebox walls. Masonry fireboxes shall be constructed of *solid masonry* units, hollow masonry units grouted solid, stone or concrete. When a lining of firebrick at least 2 inches (51 mm) thick or other *approved* lining is provided, the minimum thickness of back and side walls shall each be 8 inches (203 mm) of *solid masonry*, including the lining. The width of joints between firebricks shall not be greater than $^1/_4$ inch (6 mm). When no lining is provided, the total minimum thickness of back and side walls shall be 10 inches (254 mm) of *solid masonry*. Firebrick shall conform to ASTM C 27 or C 1261 and shall be laid with medium duty refractory mortar conforming to ASTM C 199.

R1001.5.1 Steel fireplace units. Installation of steel fireplace units with *solid masonry* to form a masonry fireplace is permitted when installed either according to the requirements of their listing or according to the requirements of this section. Steel fireplace units incorporating a steel firebox lining, shall be constructed with steel not less than $^1/_4$ inch (6 mm) thick, and an air-circulating chamber which is ducted to the interior of the building. The firebox lining shall be encased with *solid masonry* to provide a total thickness at the back and sides of not less than 8 inches (203 mm), of which not less than 4 inches (102 mm) shall be of *solid masonry* or concrete. Circulating air ducts used with steel fireplace units shall be constructed of metal or masonry.

R1001.6 Firebox dimensions. The firebox of a concrete or masonry fireplace shall have a minimum depth of 20 inches (508 mm). The throat shall not be less than 8 inches (203 mm) above the fireplace opening. The throat opening shall not be less than 4 inches (102 mm) deep. The cross-sectional area of the passageway above the firebox, including the throat, damper and smoke chamber, shall not be less than the cross-sectional area of the flue.

Exception: Rumford fireplaces shall be permitted provided that the depth of the fireplace is at least 12 inches (305 mm) and at least one-third of the width of the fireplace opening, that the throat is at least 12 inches (305 mm) above the lintel and is at least $^1/_{20}$ the cross-sectional area of the fireplace opening.

R1001.7 Lintel and throat. Masonry over a fireplace opening shall be supported by a lintel of noncombustible material. The minimum required bearing length on each end of the fireplace opening shall be 4 inches (102 mm). The fireplace throat or damper shall be located a minimum of 8 inches (203 mm) above the lintel.

R1001.7.1 Damper. Masonry fireplaces shall be equipped with a ferrous metal damper located at least 8 inches (203 mm) above the top of the fireplace opening. Dampers shall

TABLE R1001.1
SUMMARY OF REQUIREMENTS FOR MASONRY FIREPLACES AND CHIMNEYS

ITEM	LETTER[a]	REQUIREMENTS
Hearth slab thickness	A	4″
Hearth extension (each side of opening)	B	8″ fireplace opening < 6 square foot. 12″ fireplace opening ≥ 6 square foot.
Hearth extension (front of opening)	C	16″ fireplace opening < 6 square foot. 20″ fireplace opening ≥ 6 square foot.
Hearth slab reinforcing	D	Reinforced to carry its own weight and all imposed loads.
Thickness of wall of firebox	E	10″ solid brick or 8″ where a firebrick lining is used. Joints in firebrick $^1/_4$″ maximum.
Distance from top of opening to throat	F	8″
Smoke chamber wall thickness Unlined walls	G	6″ 8″
Chimney Vertical reinforcing[b]	H	Four No. 4 full-length bars for chimney up to 40″ wide. Add two No. 4 bars for each additional 40″ or fraction of width or each additional flue.
Horizontal reinforcing	J	$^1/_4$″ ties at 18″ and two ties at each bend in vertical steel.
Bond beams	K	No specified requirements.
Fireplace lintel	L	Noncombustible material.
Chimney walls with flue lining	M	Solid masonry units or hollow masonry units grouted solid with at least 4-inch nominal thickness.
Distances between adjacent flues	—	See Section R1003.13.
Effective flue area (based on area of fireplace opening)	P	See Section R1003.15.
Clearances Combustible material Mantel and trim Above roof	R	 See Sections R1001.11 and R1003.18. See Section R1001.11, Exception 4. 3′ at roofline and 2′ at 10′.
Anchorage[b] Strap Number Embedment into chimney Fasten to Bolts	S	 $^3/_{16}$″ × 1″ Two 12″ hooked around outer bar with 6″ extension. 4 joists Two $^1/_2$″ diameter.
Footing Thickness Width	T	 12″ min. 6″ each side of fireplace wall.

For SI: 1 inch = 25.4 mm, 1 foot = 304.8 mm, 1 square foot = 0.0929 m^2.

Note: This table provides a summary of major requirements for the construction of masonry chimneys and fireplaces. Letter references are to Figure R1001.1, which shows examples of typical construction. This table does not cover all requirements, nor does it cover all aspects of the indicated requirements. For the actual mandatory requirements of the code, see the indicated section of text.

a. The letters refer to Figure R1001.1.

b. Not required in Seismic Design Category A, B or C.

be installed in the fireplace or the chimney venting the fireplace, and shall be operable from the room containing the fireplace.

R1001.8 Smoke chamber. Smoke chamber walls shall be constructed of *solid masonry* units, hollow masonry units grouted solid, stone or concrete. The total minimum thickness of front, back and side walls shall be 8 inches (203 mm) of *solid masonry*. The inside surface shall be parged smooth with refractory mortar conforming to ASTM C 199. When a lining of firebrick at least 2 inches (51 mm) thick, or a lining of vitrified clay at least $^5/_8$ inch (16 mm) thick, is provided, the total minimum thickness of front, back and side walls shall be 6 inches (152 mm) of *solid masonry*, including the lining. Firebrick shall conform to ASTM C 1261 and shall be laid with medium duty refractory mortar conforming to ASTM C 199. Vitrified clay linings shall conform to ASTM C 315.

R1001.8.1 Smoke chamber dimensions. The inside height of the smoke chamber from the fireplace throat to the beginning of the flue shall not be greater than the inside width of the fireplace opening. The inside surface of the smoke chamber shall not be inclined more than 45 degrees (0.79 rad) from vertical when prefabricated smoke chamber linings are used or when the smoke chamber walls are rolled or sloped rather than corbeled. When the inside surface of the smoke chamber is formed by corbeled masonry, the walls shall not be corbeled more than 30 degrees (0.52 rad) from vertical.

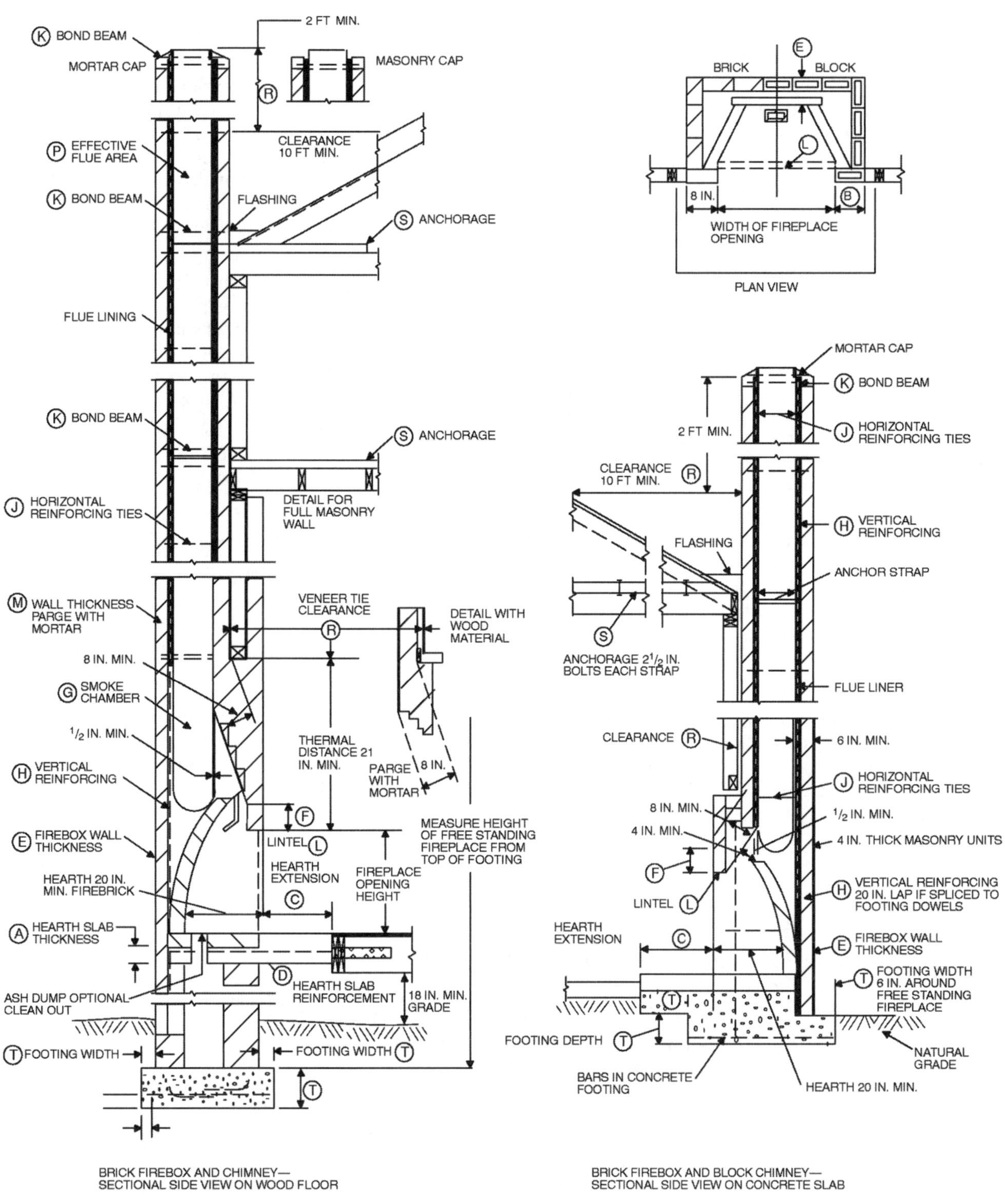

For SI: 1 inch = 25.4 mm, 1 foot = 304.8 mm.

FIGURE R1001.1
FIREPLACE AND CHIMNEY DETAILS

R1001.9 Hearth and hearth extension. Masonry fireplace hearths and hearth extensions shall be constructed of concrete or masonry, supported by noncombustible materials, and reinforced to carry their own weight and all imposed loads. No combustible material shall remain against the underside of hearths and hearth extensions after construction.

R1001.9.1 Hearth thickness. The minimum thickness of fireplace hearths shall be 4 inches (102 mm).

R1001.9.2 Hearth extension thickness. The minimum thickness of hearth extensions shall be 2 inches (51 mm).

Exception: When the bottom of the firebox opening is raised at least 8 inches (203 mm) above the top of the hearth extension, a hearth extension of not less than $^3/_8$-inch-thick (10 mm) brick, concrete, stone, tile or other *approved* noncombustible material is permitted.

R1001.10 Hearth extension dimensions. Hearth extensions shall extend at least 16 inches (406 mm) in front of and at least 8 inches (203 mm) beyond each side of the fireplace opening. Where the fireplace opening is 6 square feet (0.6 m^2) or larger, the hearth extension shall extend at least 20 inches (508 mm) in front of and at least 12 inches (305 mm) beyond each side of the fireplace opening.

R1001.11 Fireplace clearance. All wood beams, joists, studs and other combustible material shall have a clearance of not less than 2 inches (51 mm) from the front faces and sides of masonry fireplaces and not less than 4 inches (102 mm) from the back faces of masonry fireplaces. The air space shall not be filled, except to provide fire blocking in accordance with Section R1001.12.

Exceptions:

1. Masonry fireplaces *listed* and *labeled* for use in contact with combustibles in accordance with UL 127 and installed in accordance with the manufacturer's installation instructions are permitted to have combustible material in contact with their exterior surfaces.
2. When masonry fireplaces are part of masonry or concrete walls, combustible materials shall not be in contact with the masonry or concrete walls less than 12 inches (306 mm) from the inside surface of the nearest firebox lining.
3. Exposed combustible trim and the edges of sheathing materials such as wood siding, flooring and drywall shall be permitted to abut the masonry fireplace side walls and hearth extension in accordance with Figure R1001.11, provided such combustible trim or sheathing is a minimum of 12 inches (305 mm) from the inside surface of the nearest firebox lining.
4. Exposed combustible mantels or trim may be placed directly on the masonry fireplace front surrounding the fireplace opening providing such combustible materials are not placed within 6 inches (152 mm) of a fireplace opening. Combustible material within 12 inches (306 mm) of the fireplace opening shall not project more than $^1/_8$ inch (3 mm) for each 1-inch (25 mm) distance from such an opening.

R1001.12 Fireplace fireblocking. Fireplace fireblocking shall comply with the provisions of Section R602.8.

SECTION R1002
MASONRY HEATERS

R1002.1 Definition. A masonry heater is a heating *appliance* constructed of concrete or *solid masonry*, hereinafter referred to as masonry, which is designed to absorb and store heat from a solid-fuel fire built in the firebox by routing the exhaust gases through internal heat exchange channels in which the flow path downstream of the firebox may include flow in a horizontal or downward direction before entering the chimney and which delivers heat by radiation from the masonry surface of the heater.

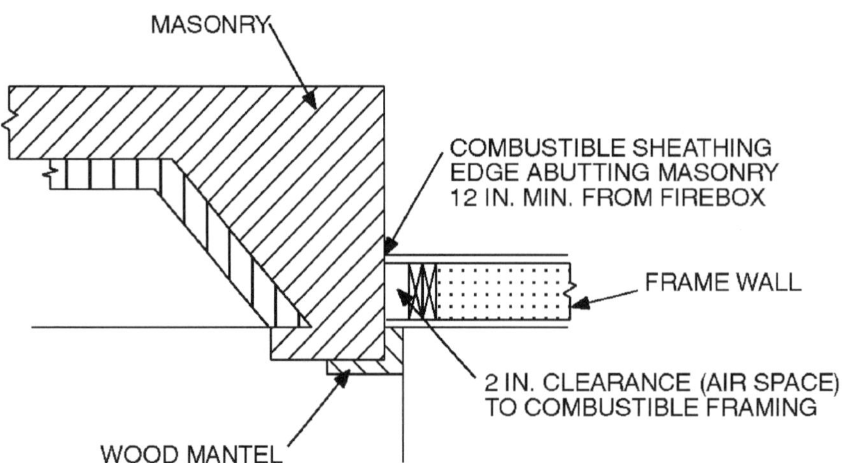

For SI: 1 inch = 25.4 mm.

FIGURE R1001.11
CLEARANCE FROM COMBUSTIBLES

R1002.2 Installation. Masonry heaters shall be installed in accordance with this section and comply with one of the following:

1. Masonry heaters shall comply with the requirements of ASTM E 1602; or
2. Masonry heaters shall be *listed* and *labeled* in accordance with UL 1482 and installed in accordance with the manufacturer's installation instructions.

R1002.3 Footings and foundation. The firebox floor of a masonry heater shall be a minimum thickness of 4 inches (102 mm) of noncombustible material and be supported on a noncombustible footing and foundation in accordance with Section R1003.2.

R1002.4 Seismic reinforcing. In Seismic Design Categories D_0, D_1 and D_2, masonry heaters shall be anchored to the masonry foundation in accordance with Section R1003.3. Seismic reinforcing shall not be required within the body of a masonry heater whose height is equal to or less than 3.5 times its body width and where the masonry chimney serving the heater is not supported by the body of the heater. Where the masonry chimney shares a common wall with the facing of the masonry heater, the chimney portion of the structure shall be reinforced in accordance with Section R1003.

R1002.5 Masonry heater clearance. Combustible materials shall not be placed within 36 inches (914 mm) of the outside surface of a masonry heater in accordance with NFPA 211 Section 8-7 (clearances for solid-fuel-burning *appliances*), and the required space between the heater and combustible material shall be fully vented to permit the free flow of air around all heater surfaces.

Exceptions:

1. When the masonry heater wall is at least 8 inches (203 mm) thick of *solid masonry* and the wall of the heat exchange channels is at least 5 inches (127 mm) thick of *solid masonry*, combustible materials shall not be placed within 4 inches (102 mm) of the outside surface of a masonry heater. A clearance of at least 8 inches (203 mm) shall be provided between the gas-tight capping slab of the heater and a combustible ceiling.
2. Masonry heaters listed and labeled in accordance with UL 1482 may be installed in accordance with the listing specifications and the manufacturer's written instructions.

SECTION R1003 MASONRY CHIMNEYS

R1003.1 Definition. A masonry chimney is a chimney constructed of *solid masonry* units, hollow masonry units grouted solid, stone or concrete, hereinafter referred to as masonry. Masonry chimneys shall be constructed, anchored, supported and reinforced as required in this chapter.

R1003.2 Footings and foundations. Footings for masonry chimneys shall be constructed of concrete or *solid masonry* at least 12 inches (305 mm) thick and shall extend at least 6 inches (152 mm) beyond the face of the foundation or support wall on all sides. Footings shall be founded on natural undisturbed earth or engineered fill below frost depth. In areas not subjected to freezing, footings shall be at least 12 inches (305 mm) below finished *grade*.

R1003.3 Seismic reinforcing. Masonry or concrete chimneys shall be constructed, anchored, supported and reinforced as required in this chapter. In Seismic Design Category D_0, D_1 or D_2 masonry and concrete chimneys shall be reinforced and anchored as detailed in Section R1003.3.1, R1003.3.2 and R1003.4. In Seismic Design Category A, B or C, reinforcement and seismic anchorage is not required.

R1003.3.1 Vertical reinforcing. For chimneys up to 40 inches (1016 mm) wide, four No. 4 continuous vertical bars, anchored in the foundation, shall be placed in the concrete, or between wythes of *solid masonry*, or within the cells of hollow unit masonry, and grouted in accordance with Section R609.1.1. Grout shall be prevented from bonding with the flue liner so that the flue liner is free to move with thermal expansion. For chimneys more than 40 inches (1016 mm) wide, two additional No. 4 vertical bars shall be installed for each additional 40 inches (1016 mm) in width or fraction thereof.

R1003.3.2 Horizontal reinforcing. Vertical reinforcement shall be placed enclosed within $^1/_4$-inch (6 mm) ties, or other reinforcing of equivalent net cross-sectional area, spaced not to exceed 18 inches (457 mm) on center in concrete, or placed in the bed joints of unit masonry, at a minimum of every 18 inches (457 mm) of vertical height. Two such ties shall be installed at each bend in the vertical bars.

R1003.4 Seismic anchorage. Masonry and concrete chimneys and foundations in Seismic Design Category D_0, D_1 or D_2 shall be anchored at each floor, ceiling or roof line more than 6 feet (1829 mm) above *grade*, except where constructed completely within the exterior walls. Anchorage shall conform to the requirements in Section R1003.4.1.

R1003.4.1 Anchorage. Two $^3/_{16}$-inch by 1-inch (5 mm by 25 mm) straps shall be embedded a minimum of 12 inches (305 mm) into the chimney. Straps shall be hooked around the outer bars and extend 6 inches (152 mm) beyond the bend. Each strap shall be fastened to a minimum of four floor joists with two $^1/_2$-inch (13 mm) bolts.

R1003.5 Corbeling. Masonry chimneys shall not be corbeled more than one-half of the chimney's wall thickness from a wall or foundation, nor shall a chimney be corbeled from a wall or foundation that is less than 12 inches (305 mm) thick unless it projects equally on each side of the wall, except that on the second *story* of a two-story *dwelling*, corbeling of chimneys on the exterior of the enclosing walls may equal the wall thickness. The projection of a single course shall not exceed one-half the unit height or one-third of the unit bed depth, whichever is less.

R1003.6 Changes in dimension. The chimney wall or chimney flue lining shall not change in size or shape within 6 inches (152 mm) above or below where the chimney passes through floor components, ceiling components or roof components.

R1003.7 Offsets. Where a masonry chimney is constructed with a fireclay flue liner surrounded by one wythe of masonry, the maximum offset shall be such that the centerline of the flue above the offset does not extend beyond the center of the chimney wall below the offset. Where the chimney offset is supported by masonry below the offset in an *approved* manner, the maximum offset limitations shall not apply. Each individual corbeled masonry course of the offset shall not exceed the projection limitations specified in Section R1003.5.

R1003.8 Additional load. Chimneys shall not support loads other than their own weight unless they are designed and constructed to support the additional load. Construction of masonry chimneys as part of the masonry walls or reinforced concrete walls of the building shall be permitted.

R1003.9 Termination. Chimneys shall extend at least 2 feet (610 mm) higher than any portion of a building within 10 feet (3048 mm), but shall not be less than 3 feet (914 mm) above the highest point where the chimney passes through the roof.

R1003.9.1 Chimney caps. Masonry chimneys shall have a concrete, metal or stone cap, sloped to shed water, a drip edge and a caulked bond break around any flue liners in accordance with ASTM C 1283.

R1003.9.2 Spark arrestors. Where a spark arrestor is installed on a masonry chimney, the spark arrestor shall meet all of the following requirements:

1. The net free area of the arrestor shall not be less than four times the net free area of the outlet of the chimney flue it serves.
2. The arrestor screen shall have heat and corrosion resistance equivalent to 19-gage galvanized steel or 24-gage stainless steel.
3. Openings shall not permit the passage of spheres having a diameter greater than $^1/_2$ inch (13 mm) nor block the passage of spheres having a diameter less than $^3/_8$ inch (10 mm).
4. The spark arrestor shall be accessible for cleaning and the screen or chimney cap shall be removable to allow for cleaning of the chimney flue.

R1003.9.3 Rain caps. Where a masonry or metal rain cap is installed on a masonry chimney, the net free area under the cap shall not be less than four times the net free area of the outlet of the chimney flue it serves.

R1003.10 Wall thickness. Masonry chimney walls shall be constructed of *solid masonry* units or hollow masonry units grouted solid with not less than a 4-inch (102 mm) nominal thickness.

R1003.10.1 Masonry veneer chimneys. Where masonry is used to veneer a frame chimney, through-flashing and weep holes shall be installed as required by Section R703.

R1003.11 Flue lining (material). Masonry chimneys shall be lined. The lining material shall be appropriate for the type of *appliance* connected, according to the terms of the *appliance* listing and manufacturer's instructions.

R1003.11.1 Residential-type appliances (general). Flue lining systems shall comply with one of the following:

1. Clay flue lining complying with the requirements of ASTM C 315.
2. Listed and labeled chimney lining systems complying with UL 1777.
3. Factory-built chimneys or chimney units listed for installation within masonry chimneys.
4. Other *approved* materials that will resist corrosion, erosion, softening or cracking from flue gases and condensate at temperatures up to 1,800°F (982°C).

R1003.11.2 Flue linings for specific appliances. Flue linings other than these covered in Section R1003.11.1, intended for use with specific types of *appliances*, shall comply with Sections R1003.11.3 through R1003.11.6.

R1003.11.3 Gas appliances. Flue lining systems for gas *appliances* shall be in accordance with Chapter 24.

R1003.11.4 Pellet fuel-burning appliances. Flue lining and vent systems for use in masonry chimneys with pellet fuel-burning *appliances* shall be limited to the following:

1. Flue lining systems complying with Section R1003.11.1.
2. Pellet vents listed for installation within masonry chimneys. (See Section R1003.11.6 for marking.)

R1003.11.5 Oil-fired appliances approved for use with Type L vent. Flue lining and vent systems for use in masonry chimneys with oil-fired *appliances approved* for use with Type L vent shall be limited to the following:

1. Flue lining systems complying with Section R1003.11.1.
2. Listed chimney liners complying with UL 641. (See Section R1003.11.6 for marking.)

R1003.11.6 Notice of usage. When a flue is relined with a material not complying with Section R1003.11.1, the chimney shall be plainly and permanently identified by a *label* attached to a wall, ceiling or other conspicuous location adjacent to where the connector enters the chimney. The *label* shall include the following message or equivalent language:

THIS CHIMNEY FLUE IS FOR USE ONLY WITH [TYPE OR CATEGORY OF *APPLIANCE*] *APPLIANCES* THAT BURN [TYPE OF FUEL]. DO NOT CONNECT OTHER TYPES OF *APPLIANCES*.

R1003.12 Clay flue lining (installation). Clay flue liners shall be installed in accordance with ASTM C 1283 and extend from a point not less than 8 inches (203 mm) below the lowest inlet or, in the case of fireplaces, from the top of the smoke chamber to a point above the enclosing walls. The lining shall be carried up vertically, with a maximum slope no greater than 30 degrees (0.52 rad) from the vertical.

Clay flue liners shall be laid in medium-duty water insoluble refractory mortar conforming to ASTM C 199 with tight mortar joints left smooth on the inside and installed to maintain an air space or insulation not to exceed the thickness of the flue liner separating the flue liners from the interior face of the chimney masonry walls. Flue liners shall be supported on all sides. Only enough mortar shall be placed to make the joint and hold the liners in position.

R1003.12.1 Listed materials. *Listed* materials used as flue linings shall be installed in accordance with the terms of their listings and manufacturer's instructions.

R1003.12.2 Space around lining. The space surrounding a chimney lining system or vent installed within a masonry chimney shall not be used to vent any other *appliance*.

Exception: This shall not prevent the installation of a separate flue lining in accordance with the manufacturer's installation instructions.

R1003.13 Multiple flues. When two or more flues are located in the same chimney, masonry wythes shall be built between adjacent flue linings. The masonry wythes shall be at least 4 inches (102 mm) thick and bonded into the walls of the chimney.

Exception: When venting only one *appliance*, two flues may adjoin each other in the same chimney with only the flue lining separation between them. The joints of the adjacent flue linings shall be staggered at least 4 inches (102 mm).

R1003.14 Flue area (appliance). Chimney flues shall not be smaller in area than that of the area of the connector from the *appliance* [see Tables R1003.14(1) and R1003.14(2)]. The sizing of a chimney flue to which multiple *appliance* venting systems are connected shall be in accordance with Section M1805.3.

R1003.15 Flue area (masonry fireplace). Flue sizing for chimneys serving fireplaces shall be in accordance with Section R1003.15.1 or Section R1003.15.2.

R1003.15.1 Option 1. Round chimney flues shall have a minimum net cross-sectional area of at least $^1/_{12}$ of the fireplace opening. Square chimney flues shall have a minimum net cross-sectional area of $^1/_{10}$ of the fireplace opening. Rectangular chimney flues with an *aspect ratio* less than 2 to 1 shall have a minimum net cross-sectional area of $^1/_{10}$ of the fireplace opening. Rectangular chimney flues with an *aspect ratio* of 2 to 1 or more shall have a minimum net cross-sectional area of $^1/_8$ of the fireplace opening. Cross-sectional areas of clay flue linings are shown in Tables R1003.14(1) and R1003.14(2) or as provided by the manufacturer or as measured in the field.

R1003.15.2 Option 2. The minimum net cross-sectional area of the chimney flue shall be determined in accordance with Figure R1003.15.2. A flue size providing at least the equivalent net cross-sectional area shall be used. Cross-sectional areas of clay flue linings are shown in Tables R1003.14(1) and R1003.14(2) or as provided by the manufacturer or as measured in the field. The height of the chimney shall be measured from the firebox floor to the top of the chimney flue.

R1003.16 Inlet. Inlets to masonry chimneys shall enter from the side. Inlets shall have a thimble of fireclay, rigid refractory material or metal that will prevent the connector from pulling out of the inlet or from extending beyond the wall of the liner.

R1003.17 Masonry chimney cleanout openings. Cleanout openings shall be provided within 6 inches (152 mm) of the base of each flue within every masonry chimney. The upper edge of the cleanout shall be located at least 6 inches (152 mm) below the lowest chimney inlet opening. The height of the opening shall be at least 6 inches (152 mm). The cleanout shall be provided with a noncombustible cover.

Exception: Chimney flues serving masonry fireplaces where cleaning is possible through the fireplace opening.

TABLE R1003.14(1)
NET CROSS-SECTIONAL AREA OF ROUND FLUE SIZES[a]

FLUE SIZE, INSIDE DIAMETER (inches)	CROSS-SECTIONAL AREA (square inches)
6	28
7	38
8	50
10	78
$10^3/_4$	90
12	113
15	176
18	254

For SI: 1 inch = 25.4 mm, 1 square inch = 645.16 mm^2.
a. Flue sizes are based on ASTM C 315.

TABLE R1003.14(2)
NET CROSS-SECTIONAL AREA OF SQUARE AND RECTANGULAR FLUE SIZES

FLUE SIZE, OUTSIDE NOMINAL DIMENSIONS (inches)	CROSS-SECTIONAL AREA (square inches)
4.5 × 8.5	23
4.5 × 13	34
8 × 8	42
8.5 × 8.5	49
8 × 12	67
8.5 × 13	76
12 × 12	102
8.5 × 18	101
13 × 13	127
12 × 16	131
13 × 18	173
16 × 16	181
16 × 20	222
18 × 18	233
20 × 20	298
20 × 24	335
24 × 24	431

For SI: 1 inch = 25.4 mm, 1 square inch = 645.16 mm^2.

R1003.18 Chimney clearances. Any portion of a masonry chimney located in the interior of the building or within the exterior wall of the building shall have a minimum air space clearance to combustibles of 2 inches (51 mm). Chimneys located entirely outside the exterior walls of the building, including chimneys that pass through the soffit or cornice, shall have a minimum air space clearance of 1 inch (25 mm). The air space shall not be filled, except to provide fire blocking in accordance with Section R1003.19.

Exceptions:

1. Masonry chimneys equipped with a chimney lining system listed and *labeled* for use in chimneys in contact with combustibles in accordance with UL 1777 and installed in accordance with the manufacturer's installation instructions are permitted to have combustible material in contact with their exterior surfaces.
2. When masonry chimneys are constructed as part of masonry or concrete walls, combustible materials shall not be in contact with the masonry or concrete wall less than 12 inches (305 mm) from the inside surface of the nearest flue lining.
3. Exposed combustible trim and the edges of sheathing materials, such as wood siding and flooring, shall be permitted to abut the masonry chimney side

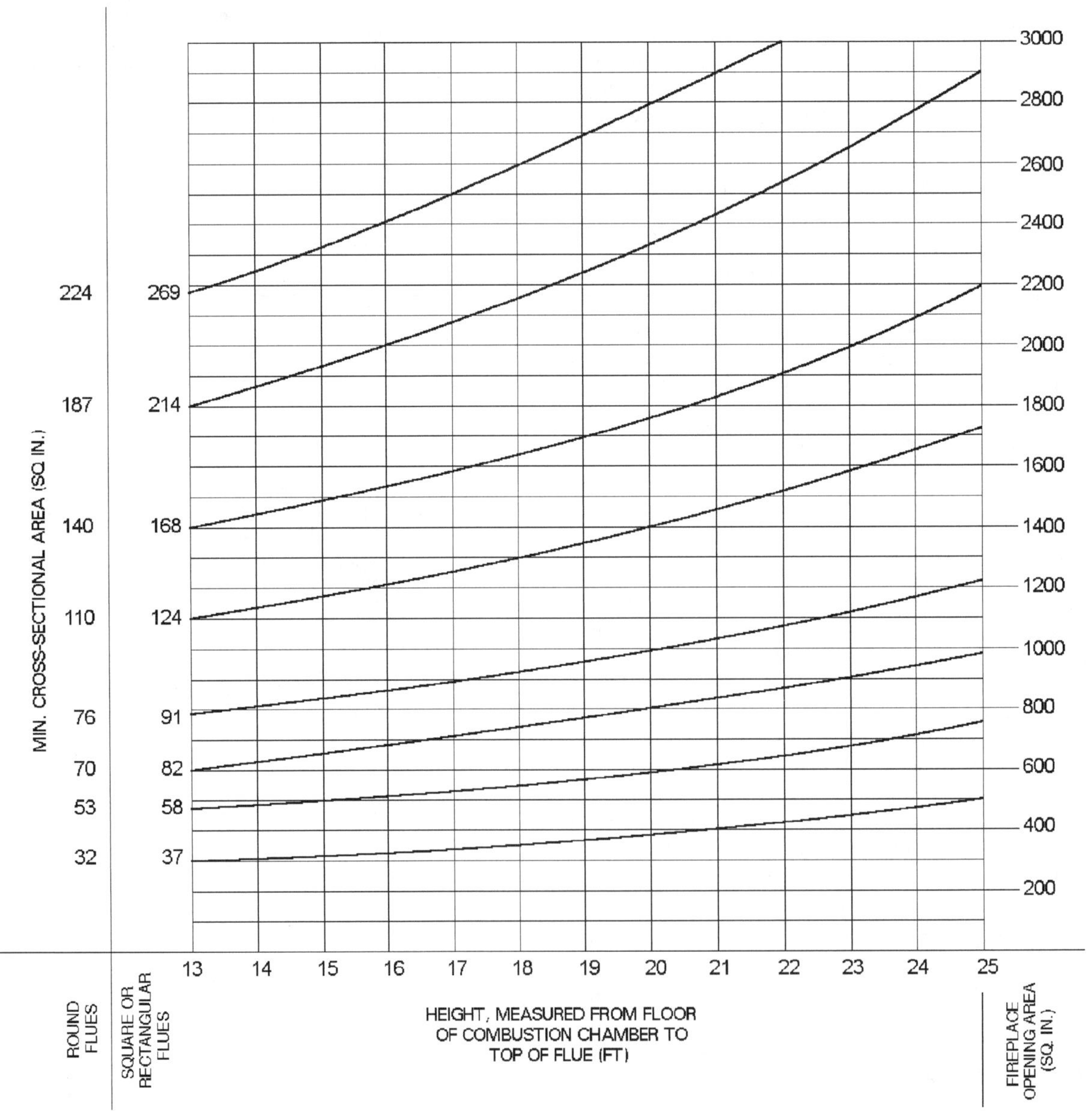

For SI: 1 foot = 304.8 mm, 1 square inch = 645.16 mm^2.

FIGURE R1003.15.2
FLUE SIZES FOR MASONRY CHIMNEYS

walls, in accordance with Figure R1003.18, provided such combustible trim or sheathing is a minimum of 12 inches (305 mm) from the inside surface of the nearest flue lining. Combustible material and trim shall not overlap the corners of the chimney by more than 1 inch (25 mm).

R1003.19 Chimney fireblocking. All spaces between chimneys and floors and ceilings through which chimneys pass shall be fireblocked with noncombustible material securely fastened in place. The fireblocking of spaces between chimneys and wood joists, beams or headers shall be self-supporting or be placed on strips of metal or metal lath laid across the spaces between combustible material and the chimney.

R1003.20 Chimney crickets. Chimneys shall be provided with crickets when the dimension parallel to the ridgeline is greater than 30 inches (762 mm) and does not intersect the ridgeline. The intersection of the cricket and the chimney shall be flashed and counterflashed in the same manner as normal roof-chimney intersections. Crickets shall be constructed in compliance with Figure R1003.20 and Table R1003.20.

**TABLE R1003.20
CRICKET DIMENSIONS**

ROOF SLOPE	H
12 - 12	$^1/_2$ of W
8 - 12	$^1/_3$ of W
6 - 12	$^1/_4$ of W
4 - 12	$^1/_6$ of W
3 - 12	$^1/_8$ of W

SECTION R1004
FACTORY-BUILT FIREPLACES

R1004.1 General. Factory-built fireplaces shall be *listed* and *labeled* and shall be installed in accordance with the conditions of the *listing*. Factory-built fireplaces shall be tested in accordance with UL 127.

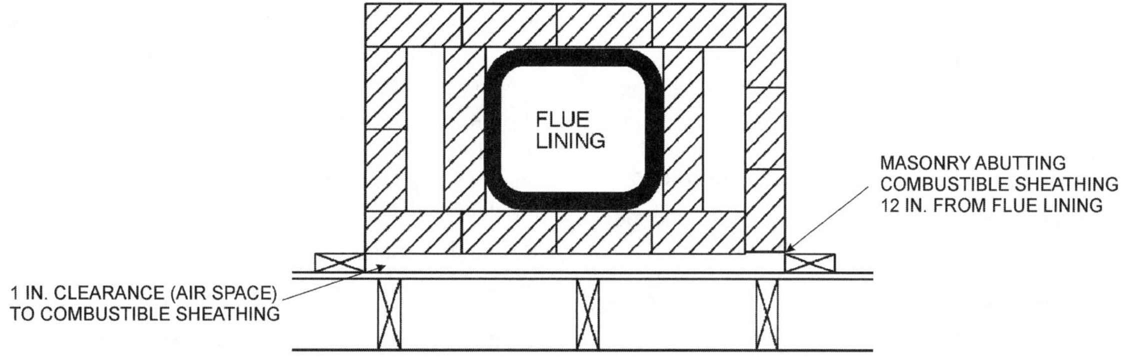

For SI: 1 inch = 25.4 mm.

**FIGURE R1003.18
CLEARANCE FROM COMBUSTIBLES**

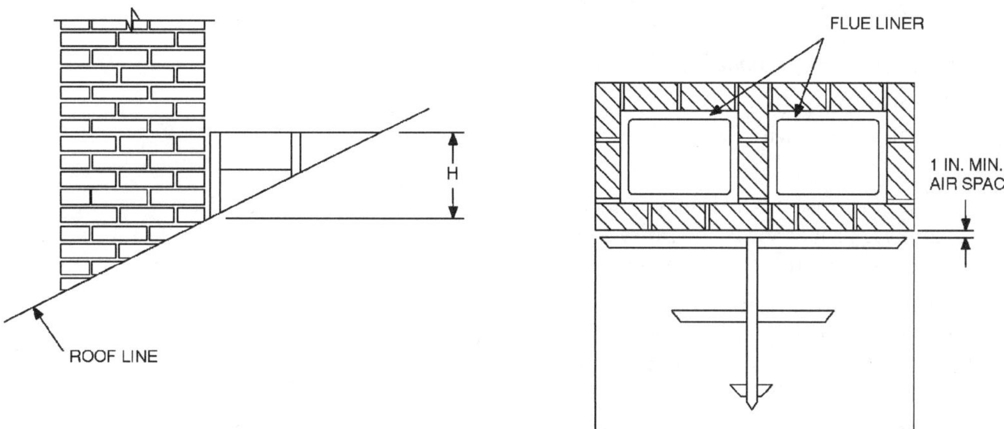

For SI: 1 inch = 25.4 mm.

**FIGURE R1003.20
CHIMNEY CRICKET**

R1004.2 Hearth extensions. Hearth extensions of *approved* factory-built fireplaces shall be installed in accordance with the *listing* of the fireplace. The hearth extension shall be readily distinguishable from the surrounding floor area. Listed and labeled hearth extensions shall comply with UL 1618.

R1004.3 Decorative shrouds. Decorative shrouds shall not be installed at the termination of chimneys for factory-built fireplaces except where the shrouds are listed and *labeled* for use with the specific factory-built fireplace system and installed in accordance with the manufacturer's installation instructions.

R1004.4 Unvented gas log heaters. An unvented gas log heater shall not be installed in a factory-built fireplace unless the fireplace system has been specifically tested, *listed* and *labeled* for such use in accordance with UL 127.

SECTION R1005 FACTORY-BUILT CHIMNEYS

R1005.1 Listing. Factory-built chimneys shall be *listed* and *labeled* and shall be installed and terminated in accordance with the manufacturer's installation instructions.

R1005.2 Decorative shrouds. Decorative shrouds shall not be installed at the termination of factory-built chimneys except where the shrouds are *listed* and *labeled* for use with the specific factory-built chimney system and installed in accordance with the manufacturer's installation instructions.

R1005.3 Solid-fuel appliances. Factory-built chimneys installed in *dwelling units* with solid-fuel-burning *appliances* shall comply with the Type HT requirements of UL 103 and shall be marked "Type HT and "Residential Type and Building Heating *Appliance* Chimney."

Exception: Chimneys for use with open combustion chamber fireplaces shall comply with the requirements of UL 103 and shall be marked "Residential Type and Building Heating *Appliance* Chimney."

Chimneys for use with open combustion chamber *appliances* installed in buildings other than *dwelling units* shall comply with the requirements of UL 103 and shall be marked "Building Heating *Appliance* Chimney" or "Residential Type and Building Heating *Appliance* Chimney."

R1005.4 Factory-built fireplaces. Chimneys for use with factory-built fireplaces shall comply with the requirements of UL 127.

R1005.5 Support. Where factory-built chimneys are supported by structural members, such as joists and rafters, those members shall be designed to support the additional load.

R1005.6 Medium-heat appliances. Factory-built chimneys for medium-heat *appliances* producing flue gases having a temperature above 1,000°F (538°C), measured at the entrance to the chimney shall comply with UL 959.

R1005.7 Factory-built chimney offsets. Where a factory-built chimney assembly incorporates offsets, no part of the chimney shall be at an angle of more than 30 degrees from vertical at any point in the assembly and the chimney assembly shall not include more than four elbows.

SECTION R1006 EXTERIOR AIR SUPPLY

R1006.1 Exterior air. Factory-built or masonry fireplaces covered in this chapter shall be equipped with an exterior air supply to assure proper fuel combustion unless the room is mechanically ventilated and controlled so that the indoor pressure is neutral or positive.

R1006.1.1 Factory-built fireplaces. Exterior *combustion air* ducts for factory-built fireplaces shall be a *listed* component of the fireplace and shall be installed according to the fireplace manufacturer's instructions.

R1006.1.2 Masonry fireplaces. *Listed combustion air* ducts for masonry fireplaces shall be installed according to the terms of their *listing* and the manufacturer's instructions.

R1006.2 Exterior air intake. The exterior air intake shall be capable of supplying all *combustion air* from the exterior of the *dwelling* or from spaces within the *dwelling* ventilated with outside air such as nonmechanically ventilated crawl or *attic* spaces. The exterior air intake shall not be located within the garage or *basement* of the *dwelling* nor shall the air intake be located at an elevation higher than the firebox. The exterior air intake shall be covered with a corrosion-resistant screen of $^1/_4$-inch (6 mm) mesh.

R1006.3 Clearance. Unlisted *combustion air* ducts shall be installed with a minimum 1-inch (25 mm) clearance to combustibles for all parts of the duct within 5 feet (1524 mm) of the duct outlet.

R1006.4 Passageway. The *combustion air* passageway shall be a minimum of 6 square inches (3870 mm^2) and not more than 55 square inches (0.035 m^2), except that *combustion air* systems for listed fireplaces shall be constructed according to the fireplace manufacturer's instructions.

R1006.5 Outlet. Locating the exterior air outlet in the back or sides of the firebox chamber or within 24 inches (610 mm) of the firebox opening on or near the floor is permitted. The outlet shall be closable and designed to prevent burning material from dropping into concealed combustible spaces.

Part IV—Energy Conservation

CHAPTER 11 [RE]

ENERGY EFFICIENCY

SECTION N1101 GENERAL

N1101.1 Scope. This chapter regulates the energy efficiency for the design and construction of buildings regulated by this code.

Note: *The text of the following Sections N1101.2 through N1105 is extracted from the 2012 edition of the* International Energy Conservation Code—Residential Provisions *and has been editorially revised to conform to the scope and application of this code. The section numbers appearing in parenthesis after each section number are the section numbers of the corresponding text in the* International Energy Conservation Code—Residential Provisions.

N1101.2 (R101.3) Intent. This code shall regulate the design and construction of buildings for the effective use and conservation of energy over the useful life of each building. This code is intended to provide flexibility to permit the use of innovative approaches and techniques to achieve this objective. This code is not intended to abridge safety, health or environmental requirements contained in other applicable codes or ordinances.

N1101.3 (R101.4.3) Additions, alterations, renovations or repairs. Additions, alterations, renovations or repairs to an existing building, building system or portion thereof shall conform to the provisions of this code as they relate to new construction without requiring the unaltered portion(s) of the existing building or building system to comply with this code. Additions, alterations, renovations or repairs shall not create an unsafe or hazardous condition or overload existing building systems. An addition shall be deemed to comply with this code if the addition alone complies or if the existing building and addition comply with this code as a single building.

Exception: The following need not comply provided the energy use of the building is not increased:

1. Storm windows installed over existing fenestration.
2. Glass only replacements in an existing sash and frame.
3. Existing ceiling, wall or floor cavities exposed during construction provided that these cavities are filled with insulation.
4. Construction where the existing roof, wall or floor cavity is not exposed.
5. Reroofing for roofs where neither the sheathing nor the insulation is exposed. Roofs without insulation in the cavity and where the sheathing or insulation is exposed during reroofing shall be insulated either above or below the sheathing.
6. Replacement of existing doors that separate *conditioned space* from the exterior shall not require the installation of a vestibule or revolving door, provided, however, that an existing vestibule that separates a *conditioned space* from the exterior shall not be removed.
7. Alterations that replace less than 50 percent of the luminaires in a space, provided that such alterations do not increase the installed interior lighting power.
8. Alterations that replace only the bulb and ballast within the existing luminaires in a space provided that the *alteration* does not increase the installed interior lighting power.

N1101.4 (R101.4.5) Change in space conditioning. Any nonconditioned space that is altered to become *conditioned space* shall be required to be brought into full compliance with this chapter.

N1101.5 (R101.5.1) Compliance materials. The *building official* shall be permitted to approve specific computer software, worksheets, compliance manuals and other similar materials that meet the intent of this code.

N1101.6 (R101.5.2) Low-energy buildings. The following buildings, or portions thereof, separated from the remainder of the building by *building thermal envelope* assemblies complying with this code shall be exempt from the *building thermal envelope* provisions of this code:

1. Those with a peak design rate of energy usage less than 3.4 Btu/h·ft^2 (10.7 W/m^2) or 1.0 watt/ft^2 (10.7 W/m^2) of floor area for space conditioning purposes.
2. Those that do not contain *conditioned space*.

N1101.7 (R102.1.1) Above code programs. The *building official* or other authority having jurisdiction shall be permitted to deem a national, state or local energy-efficiency program to exceed the energy efficiency required by this code. Buildings *approved* in writing by such an energy-efficiency program shall be considered in compliance with this code. The requirements identified as "mandatory" in Chapters 4 and 5 of this code, as applicable, shall be met.

N1101.8 (R103.2) Information on construction documents. Construction documents shall be drawn to scale upon suitable material. Electronic media documents are permitted to be submitted when *approved* by the *building official*. Construction documents shall be of sufficient clarity to indicate the location, nature and extent of the work proposed, and show in sufficient detail pertinent data and features of the building, systems and equipment as herein governed. Details shall include, but are not limited to, as applicable, insulation materials and their *R*-values; fenestration *U*-factors and SHGCs; area-weighted *U*-factor and SHGC calculations; mechanical system design criteria; mechanical and service water heating system and equipment types, sizes and efficiencies; economizer description; equipment and systems controls; fan motor horsepower (hp) and controls; duct sealing, duct and pipe insulation and location; lighting fixture schedule with wattage and control narrative; and air sealing details.

N1101.9 (R202) Defined terms. The following words and terms shall, for the purposes of this chapter, have the meanings shown herein.

ABOVE-GRADE WALL. A wall more than 50 percent above grade and enclosing *conditioned space*. This includes between-floor spandrels, peripheral edges of floors, roof and basement knee walls, dormer walls, gable end walls, walls enclosing a mansard roof and skylight shafts.

ACCESSIBLE. Admitting close approach as a result of not being guarded by locked doors, elevation or other effective means (see "Readily *accessible*").

ADDITION. An extension or increase in the *conditioned space* floor area or height of a building or structure.

AIR BARRIER. Material(s) assembled and joined together to provide a barrier to air leakage through the building envelope. An air barrier may be a single material or a combination of materials.

AUTOMATIC. Self-acting, operating by its own mechanism when actuated by some impersonal influence, as, for example, a change in current strength, pressure, temperature or mechanical configuration (see "Manual").

BASEMENT WALL. A wall 50 percent or more below grade and enclosing *conditioned space*.

BUILDING. Any structure used or intended for supporting or sheltering any use or occupancy, including any mechanical systems, service water heating systems and electric power and lighting systems located on the building site and supporting the building.

BUILDING SITE. A continguous area of land that is under the ownership or control of one entity.

BUILDING THERMAL ENVELOPE. The basement walls, exterior walls, floor, roof, and any other building elements that enclose *conditioned space* or provides a boundary between *conditioned space* and exempt or unconditioned space.

***C*-FACTOR (THERMAL CONDUCTANCE).** The coefficient of heat transmission (surface to surface) through a building component or assembly, equal to the time rate of heat flow per unit area and the unit temperature difference between the warm side and cold side surfaces (Btu/h · ft^2 · °F) [W/(m^2 · K)].

CONDITIONED FLOOR AREA. The horizontal projection of the floors associated with the *conditioned space*.

CONDITIONED SPACE. An area or room within a building being heated or cooled, containing uninsulated ducts, or with a fixed opening directly into an adjacent *conditioned space*.

CONTINUOUS AIR BARRIER. A combination of materials and assemblies that restrict or prevent the passage of air through the building thermal envelope.

CRAWL SPACE WALL. The opaque portion of a wall that encloses a crawl space and is partially or totally below grade.

DEMAND RECIRCULATION WATER SYSTEM. A water distribution system where pump(s) prime the service hot water piping with heated water upon demand for hot water.

DUCT. A tube or conduit utilized for conveying air. The air passages of self-contained systems are not to be construed as air ducts.

DUCT SYSTEM. A continuous passageway for the transmission of air that, in addition to ducts, includes duct fittings, dampers, plenums, fans and accessory air-handling equipment and appliances.

ENCLOSED SPACE. A volume surrounded by solid surfaces such as walls, floors, roofs, and openable devices such as doors and operable windows.

ENERGY ANALYSIS. A method for estimating the annual energy use of the *proposed design* and *standard reference design* based on estimates of energy use.

ENERGY COST. The total estimated annual cost for purchased energy for the building functions regulated by this code, including applicable demand charges.

ENERGY SIMULATION TOOL. An *approved* software program or calculation-based methodology that projects the annual energy use of a building.

ENTRANCE DOOR. Fenestration products used for ingress, egress and access in nonresidential buildings, including, but not limited to, exterior entrances that utilize latching hardware and automatic closers and contain over 50-percent glass specifically designed to withstand heavy use and possibly abuse.

EXTERIOR WALL. Walls including both above-grade walls and basement walls.

FENESTRATION. Skylights, roof windows, vertical windows (fixed or moveable), opaque doors, glazed doors, glazed block and combination opaque/glazed doors. Fenestration includes products with glass and nonglass glazing materials.

FENESTRATION PRODUCT, SITE-BUILT. A fenestration designed to be made up of field-glazed or field-assembled units using specific factory cut or otherwise factory-formed framing and glazing units. Examples of site-built fenestration include storefront systems, curtain walls, and atrium roof systems.

HEATED SLAB. Slab-on-grade construction in which the heating elements, hydronic tubing, or hot air distribution system is in contact with, or placed within or under, the slab.

HIGH-EFFICACY LAMPS. Compact fluorescent lamps, T-8 or smaller diameter linear fluorescent lamps, or lamps with a minimum efficacy of:

1. 60 lumens per watt for lamps over 40 watts;
2. 50 lumens per watt for lamps over 15 watts to 40 watts; and
3. 40 lumens per watt for lamps 15 watts or less.

INFILTRATION. The uncontrolled inward air leakage into a building caused by the pressure effects of wind or the effect of differences in the indoor and outdoor air density or both.

INSULATING SHEATHING. An insulating board with a core material having a minimum *R*-value of R-2.

LOW-VOLTAGE LIGHTING. Lighting equipment powered through a transformer such as a cable conductor, a rail conductor and track lighting.

MANUAL. Capable of being operated by personal intervention (see "Automatic").

PROPOSED DESIGN. A description of the proposed building used to estimate annual energy use for determining compliance based on total building performance.

READILY ACCESSIBLE. Capable of being reached quickly for operation, renewal or inspection without requiring those to whom ready access is requisite to climb over or remove obstacles or to resort to portable ladders or access equipment (see "*Accessible*").

REPAIR. The reconstruction or renewal of any part of an existing building.

RESIDENTIAL BUILDING. For this code, includes detached one- and two-family dwellings and multiple single-family dwellings (townhouses) as well as Group R-2, R-3 and R-4 buildings three stories or less in height above grade plane.

R-VALUE (THERMAL RESISTANCE). The inverse of the time rate of heat flow through a body from one of its bounding surfaces to the other surface for a unit temperature difference between the two surfaces, under steady state conditions, per unit area ($h \cdot ft^2 \cdot °F/Btu$) [$(m^2 \cdot K)/W$].

SERVICE WATER HEATING. Supply of hot water for purposes other than comfort heating.

SKYLIGHT. Glass or other transparent or translucent glazing material installed at a slope of less than 60 degrees (1.05 rad) from horizontal. Glazing material in skylights, including unit skylights, solariums, sunrooms, roofs and sloped walls is included in this definition.

SOLAR HEAT GAIN COEFFICIENT (SHGC). The ratio of the solar heat gain entering the space through the fenestration assembly to the incident solar radiation. Solar heat gain includes directly transmitted solar heat and absorbed solar radiation which is then reradiated, conducted or convected into the space.

STANDARD REFERENCE DESIGN. A version of the *proposed design* that meets the minimum requirements of this code and is used to determine the maximum annual energy use requirement for compliance based on total building performance.

SUNROOM. A one-story structure attached to a dwelling with a glazing area in excess of 40 percent of the gross area of the structure's exterior walls and roof.

THERMAL ISOLATION. Physical and space conditioning separation from *conditioned space(s)*. The *conditioned space*(s) shall be controlled as separate zones for heating and cooling or conditioned by separate equipment.

THERMOSTAT. An automatic control device used to maintain temperature at a fixed or adjustable set point.

***U*-FACTOR (THERMAL TRANSMITTANCE).** The coefficient of heat transmission (air to air) through a building component or assembly, equal to the time rate of heat flow per unit area and unit temperature difference between the warm side and cold side air films ($Btu/h \cdot ft^2 \cdot °F$) [$W/(m^2 \cdot K)$].

VENTILATION AIR. That portion of supply air that comes from outside (outdoors) plus any recirculated air that has been treated to maintain the desired quality of air within a designated space.

VISIBLE TRANSMITTANCE [VT]. The ratio of visible light entering the space through the fenestration product assembly to the incident visible light, visible transmittance, includes the effects of glazing material and frame and is expressed as a number between 0 and 1.

WHOLE HOUSE MECHANICAL VENTILATION SYSTEM. An exhaust system, supply system, or combination thereof that is designed to mechanically exchange indoor air with outdoor air when operating continuously or through a programmed intermittent schedule to satisfy the whole house ventilation rates.

ZONE. A space or group of spaces within a building with heating or cooling requirements that are sufficiently similar so that desired conditions can be maintained throughout using a single controlling device.

N1101.10 (R301.1) Climate zones. Climate zones from Figure N1101.10 or Table N1101.10 shall be used in determining the applicable requirements in Sections N1101 through N1105. Locations not in Table N1101.10 (outside the United States) shall be assigned a climate zone based on Section N101.10.2.

N1101.10.1 (R301.2) Warm humid counties. Warm humid counties are identified in Table N1101.10 by an asterisk.

N1101.10.2 (R301.3) International climate zones. The climate zone for any location outside the United States shall be determined by applying Table N1101.10.2(1) and then Table N1101.10.2(2).

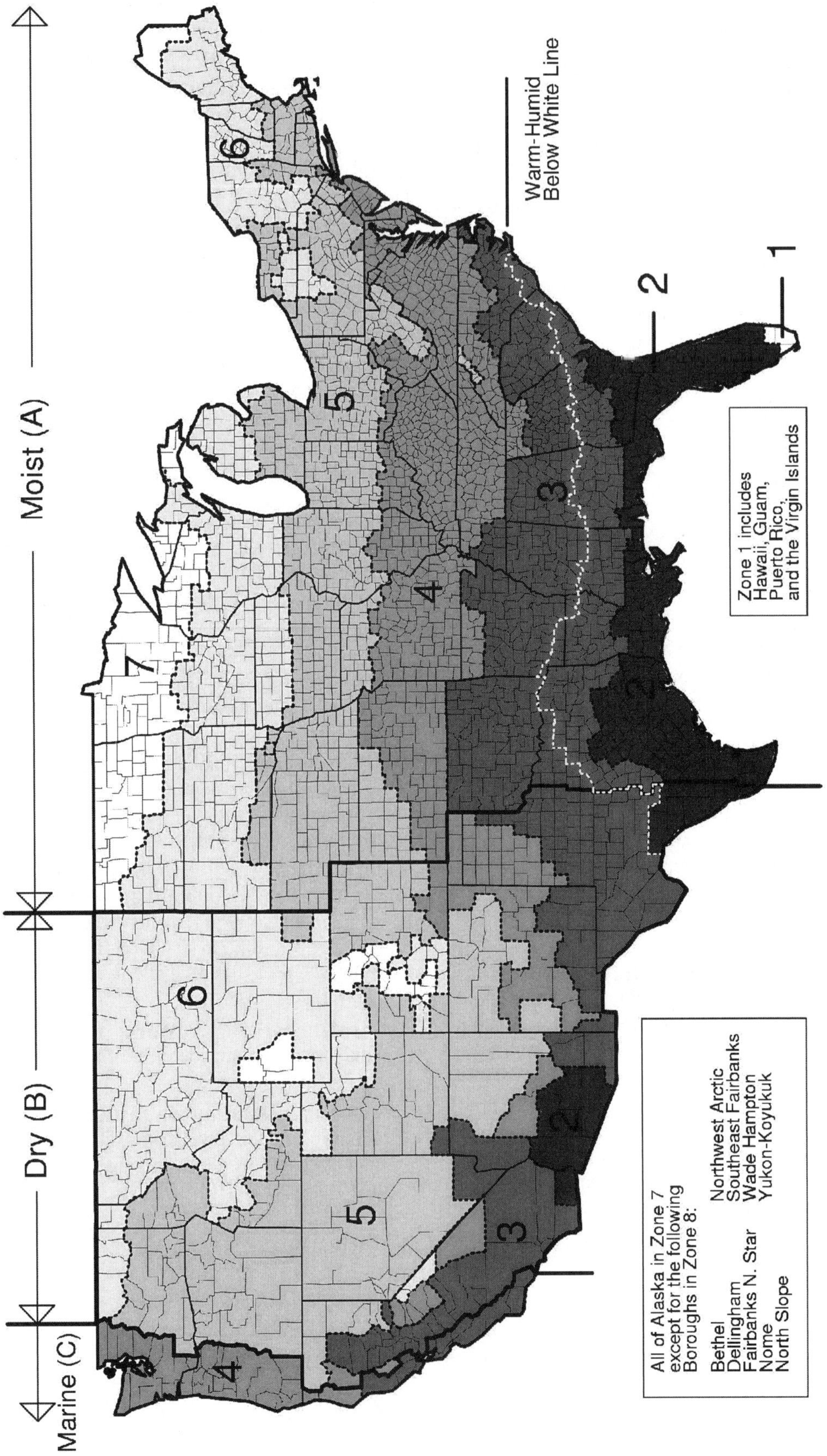

FIGURE N1101.10 (R301.1)
CLIMATE ZONES

TABLE N1101.10 (R301.1)
CLIMATE ZONES, MOISTURE REGIMES, AND WARM-HUMID DESIGNATIONS BY STATE, COUNTY AND TERRITORY

Key: A – Moist, B – Dry, C – Marine. Absence of moisture designation indicates moisture regime is irrelevant. Asterisk (*) indicates a warm-humid location.

US STATES

ALABAMA

3A Autauga*
2A Baldwin*
3A Barbour*
3A Bibb
3A Blount
3A Bullock*
3A Butler*
3A Calhoun
3A Chambers
3A Cherokee
3A Chilton
3A Choctaw*
3A Clarke*
3A Clay
3A Cleburne
3A Coffee*
3A Colbert
3A Conecuh*
3A Coosa
3A Covington*
3A Crenshaw*
3A Cullman
3A Dale*
3A Dallas*
3A DeKalb
3A Elmore*
3A Escambia*
3A Etowah
3A Fayette
3A Franklin
3A Geneva*
3A Greene
3A Hale
3A Henry*
3A Houston*
3A Jackson
3A Jefferson
3A Lamar
3A Lauderdale
3A Lawrence
3A Lee
3A Limestone
3A Lowndes*
3A Macon*
3A Madison
3A Marengo*
3A Marion
3A Marshall
2A Mobile*
3A Monroe*
3A Montgomery*
3A Morgan
3A Perry*
3A Pickens
3A Pike*
3A Randolph
3A Russell*
3A Shelby
3A St. Clair
3A Sumter
3A Talladega
3A Tallapoosa
3A Tuscaloosa
3A Walker
3A Washington*
3A Wilcox*
3A Winston

ALASKA

7 Aleutians East
7 Aleutians West
7 Anchorage
8 Bethel
7 Bristol Bay
7 Denali
8 Dillingham
8 Fairbanks North Star
7 Haines
7 Juneau
7 Kenai Peninsula
7 Ketchikan Gateway
7 Kodiak Island
7 Lake and Peninsula
7 Matanuska-Susitna
8 Nome
8 North Slope
8 Northwest Arctic
7 Prince of Wales Outer Ketchikan
7 Sitka
7 Skagway-Hoonah-Angoon
8 Southeast Fairbanks
7 Valdez-Cordova
8 Wade Hampton
7 Wrangell-Petersburg
7 Yakutat
8 Yukon-Koyukuk

ARIZONA

5B Apache
3B Cochise
5B Coconino
4B Gila
3B Graham
3B Greenlee
2B La Paz
2B Maricopa
3B Mohave
5B Navajo
2B Pima
2B Pinal
3B Santa Cruz
4B Yavapai
2B Yuma

ARKANSAS

3A Arkansas
3A Ashley
4A Baxter
4A Benton
4A Boone
3A Bradley
3A Calhoun
4A Carroll
3A Chicot
3A Clark
3A Clay
3A Cleburne
3A Cleveland
3A Columbia*
3A Conway
3A Craighead
3A Crawford
3A Crittenden
3A Cross
3A Dallas
3A Desha
3A Drew
3A Faulkner
3A Franklin
4A Fulton
3A Garland
3A Grant
3A Greene
3A Hempstead*
3A Hot Spring
3A Howard
3A Independence
4A Izard
3A Jackson
3A Jefferson
3A Johnson
3A Lafayette*
3A Lawrence
3A Lee
3A Lincoln
3A Little River*
3A Logan
3A Lonoke
4A Madison
4A Marion
3A Miller*
3A Mississippi
3A Monroe
3A Montgomery
3A Nevada
4A Newton
3A Ouachita
3A Perry
3A Phillips
3A Pike
3A Poinsett
3A Polk
3A Pope
3A Prairie
3A Pulaski
3A Randolph
3A Saline
3A Scott
4A Searcy
3A Sebastian
3A Sevier*
3A Sharp
3A St. Francis
4A Stone
3A Union*
3A Van Buren
4A Washington
3A White
3A Woodruff
3A Yell

CALIFORNIA

3C Alameda
6B Alpine
4B Amador
3B Butte
4B Calaveras
3B Colusa
3B Contra Costa
4C Del Norte
4B El Dorado
3B Fresno
3B Glenn

(continued)

**TABLE N1101.10 (R301.1)—continued
CLIMATE ZONES, MOISTURE REGIMES, AND WARM-HUMID
DESIGNATIONS BY STATE, COUNTY AND TERRITORY**

4C Humboldt
2B Imperial
4B Inyo
3B Kern
3B Kings
4B Lake
5B Lassen
3B Los Angeles
3B Madera
3C Marin
4B Mariposa
3C Mendocino
3B Merced
5B Modoc
6B Mono
3C Monterey
3C Napa
5B Nevada
3B Orange
3B Placer
5B Plumas
3B Riverside
3B Sacramento
3C San Benito
3B San Bernardino
3B San Diego
3C San Francisco
3B San Joaquin
3C San Luis Obispo
3C San Mateo
3C Santa Barbara
3C Santa Clara
3C Santa Cruz
3B Shasta
5B Sierra
5B Siskiyou
3B Solano
3C Sonoma
3B Stanislaus
3B Sutter
3B Tehama
4B Trinity
3B Tulare
4B Tuolumne
3C Ventura
3B Yolo
3B Yuba

COLORADO

5B Adams
6B Alamosa
5B Arapahoe
6B Archuleta
4B Baca
5B Bent
5B Boulder
6B Chaffee
5B Cheyenne
7 Clear Creek
6B Conejos
6B Costilla
5B Crowley
6B Custer
5B Delta
5B Denver
6B Dolores
5B Douglas
6B Eagle
5B Elbert
5B El Paso
5B Fremont
5B Garfield
5B Gilpin
7 Grand
7 Gunnison
7 Hinsdale
5B Huerfano
7 Jackson
5B Jefferson
5B Kiowa
5B Kit Carson
7 Lake
5B La Plata
5B Larimer
4B Las Animas
5B Lincoln
5B Logan
5B Mesa
7 Mineral
6B Moffat
5B Montezuma
5B Montrose
5B Morgan
4B Otero
6B Ouray
7 Park
5B Phillips
7 Pitkin
5B Prowers
5B Pueblo
6B Rio Blanco
7 Rio Grande
7 Routt
6B Saguache
7 San Juan
6B San Miguel
5B Sedgwick
7 Summit
5B Teller
5B Washington
5B Weld
5B Yuma

CONNECTICUT

5A (all)

DELAWARE

4A (all)

DISTRICT OF COLUMBIA

4A (all)

FLORIDA

2A Alachua*
2A Baker*
2A Bay*
2A Bradford*
2A Brevard*
1A Broward*
2A Calhoun*
2A Charlotte*
2A Citrus*
2A Clay*
2A Collier*
2A Columbia*
2A DeSoto*
2A Dixie*
2A Duval*
2A Escambia*
2A Flagler*
2A Franklin*
2A Gadsden*
2A Gilchrist*
2A Glades*
2A Gulf*
2A Hamilton*
2A Hardee*
2A Hendry*
2A Hernando*
2A Highlands*
2A Hillsborough*
2A Holmes*
2A Indian River*
2A Jackson*
2A Jefferson*
2A Lafayette*
2A Lake*
2A Lee*
2A Leon*
2A Levy*
2A Liberty*
2A Madison*
2A Manatee*
2A Marion*
2A Martin*
1A Miami-Dade*
1A Monroe*
2A Nassau*
2A Okaloosa*
2A Okeechobee*
2A Orange*
2A Osceola*
2A Palm Beach*
2A Pasco*
2A Pinellas*
2A Polk*
2A Putnam*
2A Santa Rosa*
2A Sarasota*
2A Seminole*
2A St. Johns*
2A St. Lucie*
2A Sumter*
2A Suwannee*
2A Taylor*
2A Union*
2A Volusia*
2A Wakulla*
2A Walton*
2A Washington*

GEORGIA

2A Appling*
2A Atkinson*
2A Bacon*
2A Baker*
3A Baldwin
4A Banks
3A Barrow
3A Bartow
3A Ben Hill*
2A Berrien*
3A Bibb
3A Bleckley*
2A Brantley*
2A Brooks*
2A Bryan*
3A Bulloch*
3A Burke
3A Butts
3A Calhoun*
2A Camden*
3A Candler*
3A Carroll
4A Catoosa
2A Charlton*
2A Chatham*
3A Chattahoochee*
4A Chattooga
3A Cherokee
3A Clarke
3A Clay*
3A Clayton
2A Clinch*
3A Cobb
3A Coffee*
2A Colquitt*
3A Columbia
2A Cook*
3A Coweta

(continued)

TABLE N1101.10 (R301.1)—continued
CLIMATE ZONES, MOISTURE REGIMES, AND WARM-HUMID DESIGNATIONS BY STATE, COUNTY AND TERRITORY

3A Crawford
3A Crisp*
4A Dade
4A Dawson
2A Decatur*
3A DeKalb
3A Dodge*
3A Dooly*
3A Dougherty*
3A Douglas
3A Early*
2A Echols*
2A Effingham*
3A Elbert
3A Emanuel*
2A Evans*
4A Fannin
3A Fayette
4A Floyd
3A Forsyth
4A Franklin
3A Fulton
4A Gilmer
3A Glascock
2A Glynn*
4A Gordon
2A Grady*
3A Greene
3A Gwinnett
4A Habersham
4A Hall
3A Hancock
3A Haralson
3A Harris
3A Hart
3A Heard
3A Henry
3A Houston*
3A Irwin*
3A Jackson
3A Jasper
2A Jeff Davis*
3A Jefferson
3A Jenkins*
3A Johnson*
3A Jones
3A Lamar
2A Lanier*
3A Laurens*
3A Lee*
2A Liberty*
3A Lincoln
2A Long*
2A Lowndes*
4A Lumpkin
3A Macon*
3A Madison
3A Marion*
3A McDuffie
2A McIntosh*
3A Meriwether
2A Miller*
2A Mitchell*
3A Monroe
3A Montgomery*
3A Morgan
4A Murray
3A Muscogee
3A Newton
3A Oconee
3A Oglethorpe
3A Paulding
3A Peach*
4A Pickens
2A Pierce*
3A Pike
3A Polk
3A Pulaski*
3A Putnam
3A Quitman*
4A Rabun
3A Randolph*
3A Richmond
3A Rockdale
3A Schley*
3A Screven*
2A Seminole*
3A Spalding
4A Stephens
3A Stewart*
3A Sumter*
3A Talbot
3A Taliaferro
2A Tattnall*
3A Taylor*
3A Telfair*
3A Terrell*
2A Thomas*
3A Tift*
2A Toombs*
4A Towns
3A Treutlen*
3A Troup
3A Turner*
3A Twiggs*
4A Union
3A Upson
4A Walker
3A Walton
2A Ware*
3A Warren
3A Washington
2A Wayne*
3A Webster*
3A Wheeler*
4A White
4A Whitfield
3A Wilcox*
3A Wilkes
3A Wilkinson
3A Worth*

HAWAII

1A (all)*

IDAHO

5B Ada
6B Adams
6B Bannock
6B Bear Lake
5B Benewah
6B Bingham
6B Blaine
6B Boise
6B Bonner
6B Bonneville
6B Boundary
6B Butte
6B Camas
5B Canyon
6B Caribou
5B Cassia
6B Clark
5B Clearwater
6B Custer
5B Elmore
6B Franklin
6B Fremont
5B Gem
5B Gooding
5B Idaho
6B Jefferson
5B Jerome
5B Kootenai
5B Latah
6B Lemhi
5B Lewis
5B Lincoln
6B Madison
5B Minidoka
5B Nez Perce
6B Oneida
5B Owyhee
5B Payette
5B Power
5B Shoshone
6B Teton
5B Twin Falls
6B Valley
5B Washington

ILLINOIS

5A Adams
4A Alexander
4A Bond
5A Boone
5A Brown
5A Bureau
5A Calhoun
5A Carroll
5A Cass
5A Champaign
4A Christian
5A Clark
4A Clay
4A Clinton
5A Coles
5A Cook
4A Crawford
5A Cumberland
5A DeKalb
5A De Witt
5A Douglas
5A DuPage
5A Edgar
4A Edwards
4A Effingham
4A Fayette
5A Ford
4A Franklin
5A Fulton
4A Gallatin
5A Greene
5A Grundy
4A Hamilton
5A Hancock
4A Hardin
5A Henderson
5A Henry
5A Iroquois
4A Jackson
4A Jasper
4A Jefferson
5A Jersey
5A Jo Daviess
4A Johnson
5A Kane
5A Kankakee
5A Kendall
5A Knox
5A Lake
5A La Salle
4A Lawrence
5A Lee
5A Livingston
5A Logan
5A Macon
4A Macoupin
4A Madison
4A Marion
5A Marshall
5A Mason
4A Massac
5A McDonough
5A McHenry

(continued)

TABLE N1101.10 (R301.1)—continued
CLIMATE ZONES, MOISTURE REGIMES, AND WARM-HUMID DESIGNATIONS BY STATE, COUNTY AND TERRITORY

5A McLean
5A Menard
5A Mercer
4A Monroe
4A Montgomery
5A Morgan
5A Moultrie
5A Ogle
5A Peoria
4A Perry
5A Piatt
5A Pike
4A Pope
4A Pulaski
5A Putnam
4A Randolph
4A Richland
5A Rock Island
4A Saline
5A Sangamon
5A Schuyler
5A Scott
4A Shelby
5A Stark
4A St. Clair
5A Stephenson
5A Tazewell
4A Union
5A Vermilion
4A Wabash
5A Warren
4A Washington
4A Wayne
4A White
5A Whiteside
5A Will
4A Williamson
5A Winnebago
5A Woodford

INDIANA

5A Adams
5A Allen
5A Bartholomew
5A Benton
5A Blackford
5A Boone
4A Brown
5A Carroll
5A Cass
4A Clark
5A Clay
5A Clinton
4A Crawford
4A Daviess
4A Dearborn
5A Decatur
5A De Kalb
5A Delaware
4A Dubois
5A Elkhart
5A Fayette
4A Floyd
5A Fountain
5A Franklin
5A Fulton
4A Gibson
5A Grant
4A Greene
5A Hamilton
5A Hancock
4A Harrison
5A Hendricks
5A Henry
5A Howard
5A Huntington
4A Jackson
5A Jasper
5A Jay
4A Jefferson
4A Jennings
5A Johnson
4A Knox
5A Kosciusko
5A Lagrange
5A Lake
5A La Porte
4A Lawrence
5A Madison
5A Marion
5A Marshall
4A Martin
5A Miami
4A Monroe
5A Montgomery
5A Morgan
5A Newton
5A Noble
4A Ohio
4A Orange
5A Owen
5A Parke
4A Perry
4A Pike
5A Porter
4A Posey
5A Pulaski
5A Putnam
5A Randolph
4A Ripley
5A Rush
4A Scott
5A Shelby
4A Spencer
5A Starke
5A Steuben
5A St. Joseph
4A Sullivan
4A Switzerland
5A Tippecanoe
5A Tipton
5A Union
4A Vanderburgh
5A Vermillion
5A Vigo
5A Wabash
5A Warren
4A Warrick
4A Washington
5A Wayne
5A Wells
5A White
5A Whitley

IOWA

5A Adair
5A Adams
6A Allamakee
5A Appanoose
5A Audubon
5A Benton
6A Black Hawk
5A Boone
6A Bremer
6A Buchanan
6A Buena Vista
6A Butler
6A Calhoun
5A Carroll
5A Cass
5A Cedar
6A Cerro Gordo
6A Cherokee
6A Chickasaw
5A Clarke
6A Clay
6A Clayton
5A Clinton
5A Crawford
5A Dallas
5A Davis
5A Decatur
6A Delaware
5A Des Moines
6A Dickinson
5A Dubuque
6A Emmet
6A Fayette
6A Floyd
6A Franklin
5A Fremont
5A Greene
6A Grundy
5A Guthrie
6A Hamilton
6A Hancock
6A Hardin
5A Harrison
5A Henry
6A Howard
6A Humboldt
6A Ida
5A Iowa
5A Jackson
5A Jasper
5A Jefferson
5A Johnson
5A Jones
5A Keokuk
6A Kossuth
5A Lee
5A Linn
5A Louisa
5A Lucas
6A Lyon
5A Madison
5A Mahaska
5A Marion
5A Marshall
5A Mills
6A Mitchell
5A Monona
5A Monroe
5A Montgomery
5A Muscatine
6A O'Brien
6A Osceola
5A Page
6A Palo Alto
6A Plymouth
6A Pocahontas
5A Polk
5A Pottawattamie
5A Poweshiek
5A Ringgold
6A Sac
5A Scott
5A Shelby
6A Sioux
5A Story
5A Tama
5A Taylor
5A Union
5A Van Buren
5A Wapello
5A Warren
5A Washington
5A Wayne
6A Webster
6A Winnebago

(continued)

TABLE N1101.10 (R301.1)—continued
CLIMATE ZONES, MOISTURE REGIMES, AND WARM-HUMID DESIGNATIONS BY STATE, COUNTY AND TERRITORY

6A Winneshiek
5A Woodbury
6A Worth
6A Wright

KANSAS

4A Allen
4A Anderson
4A Atchison
4A Barber
4A Barton
4A Bourbon
4A Brown
4A Butler
4A Chase
4A Chautauqua
4A Cherokee
5A Cheyenne
4A Clark
4A Clay
5A Cloud
4A Coffey
4A Comanche
4A Cowley
4A Crawford
5A Decatur
4A Dickinson
4A Doniphan
4A Douglas
4A Edwards
4A Elk
5A Ellis
4A Ellsworth
4A Finney
4A Ford
4A Franklin
4A Geary
5A Gove
5A Graham
4A Grant
4A Gray
5A Greeley
4A Greenwood
5A Hamilton
4A Harper
4A Harvey
4A Haskell
4A Hodgeman
4A Jackson
4A Jefferson
5A Jewell
4A Johnson
4A Kearny
4A Kingman
4A Kiowa
4A Labette
5A Lane
4A Leavenworth
4A Lincoln
4A Linn
5A Logan
4A Lyon
4A Marion
4A Marshall
4A McPherson
4A Meade
4A Miami
5A Mitchell
4A Montgomery
4A Morris
4A Morton
4A Nemaha
4A Neosho
5A Ness
5A Norton
4A Osage
5A Osborne
4A Ottawa
4A Pawnee
5A Phillips
4A Pottawatomie
4A Pratt
5A Rawlins
4A Reno
5A Republic
4A Rice
4A Riley
5A Rooks
4A Rush
4A Russell
4A Saline
5A Scott
4A Sedgwick
4A Seward
4A Shawnee
5A Sheridan
5A Sherman
5A Smith
4A Stafford
4A Stanton
4A Stevens
4A Sumner
5A Thomas
5A Trego
4A Wabaunsee
5A Wallace
4A Washington
5A Wichita
4A Wilson
4A Woodson
4A Wyandotte

KENTUCKY

4A (all)

LOUISIANA

2A Acadia*
2A Allen*
2A Ascension*
2A Assumption*
2A Avoyelles*
2A Beauregard*
3A Bienville*
3A Bossier*
3A Caddo*
2A Calcasieu*
3A Caldwell*
2A Cameron*
3A Catahoula*
3A Claiborne*
3A Concordia*
3A De Soto*
2A East Baton Rouge*
3A East Carroll
2A East Feliciana*
2A Evangeline*
3A Franklin*
3A Grant*
2A Iberia*
2A Iberville*
3A Jackson*
2A Jefferson*
2A Jefferson Davis*
2A Lafayette*
2A Lafourche*
3A La Salle*
3A Lincoln*
2A Livingston*
3A Madison*
3A Morehouse
3A Natchitoches*
2A Orleans*
3A Ouachita*
2A Plaquemines*
2A Pointe Coupee*
2A Rapides*
3A Red River*
3A Richland*
3A Sabine*
2A St. Bernard*
2A St. Charles*
2A St. Helena*
2A St. James*
2A St. John the Baptist*
2A St. Landry*
2A St. Martin*
2A St. Mary*
2A St. Tammany*
2A Tangipahoa*
3A Tensas*
2A Terrebonne*
3A Union*
2A Vermilion*
3A Vernon*
2A Washington*
3A Webster*
2A West Baton Rouge*
3A West Carroll
2A West Feliciana*
3A Winn*

MAINE

6A Androscoggin
7 Aroostook
6A Cumberland
6A Franklin
6A Hancock
6A Kennebec
6A Knox
6A Lincoln
6A Oxford
6A Penobscot
6A Piscataquis
6A Sagadahoc
6A Somerset
6A Waldo
6A Washington
6A York

MARYLAND

4A Allegany
4A Anne Arundel
4A Baltimore
4A Baltimore (city)
4A Calvert
4A Caroline
4A Carroll
4A Cecil
4A Charles
4A Dorchester
4A Frederick
5A Garrett
4A Harford
4A Howard
4A Kent
4A Montgomery
4A Prince George's
4A Queen Anne's
4A Somerset
4A St. Mary's
4A Talbot
4A Washington
4A Wicomico
4A Worcester

MASSACHSETTS

5A (all)

MICHIGAN

6A Alcona
6A Alger

(continued)

**TABLE N1101.10 (R301.1)—continued
CLIMATE ZONES, MOISTURE REGIMES, AND WARM-HUMID DESIGNATIONS BY STATE, COUNTY AND TERRITORY**

5A Allegan
6A Alpena
6A Antrim
6A Arenac
7 Baraga
5A Barry
5A Bay
6A Benzie
5A Berrien
5A Branch
5A Calhoun
5A Cass
6A Charlevoix
6A Cheboygan
7 Chippewa
6A Clare
5A Clinton
6A Crawford
6A Delta
6A Dickinson
5A Eaton
6A Emmet
5A Genesee
6A Gladwin
7 Gogebic
6A Grand Traverse
5A Gratiot
5A Hillsdale
7 Houghton
6A Huron
5A Ingham
5A Ionia
6A Iosco
7 Iron
6A Isabella
5A Jackson
5A Kalamazoo
6A Kalkaska
5A Kent
7 Keweenaw
6A Lake
5A Lapeer
6A Leelanau
5A Lenawee
5A Livingston
7 Luce
7 Mackinac
5A Macomb
6A Manistee
6A Marquette
6A Mason
6A Mecosta
6A Menominee
5A Midland
6A Missaukee
5A Monroe
5A Montcalm
6A Montmorency
5A Muskegon
6A Newaygo
5A Oakland
6A Oceana
6A Ogemaw
7 Ontonagon
6A Osceola
6A Oscoda
6A Otsego
5A Ottawa
6A Presque Isle
6A Roscommon
5A Saginaw
6A Sanilac
7 Schoolcraft
5A Shiawassee
5A St. Clair
5A St. Joseph
5A Tuscola
5A Van Buren
5A Washtenaw
5A Wayne
6A Wexford

MINNESOTA

7 Aitkin
6A Anoka
7 Becker
7 Beltrami
6A Benton
6A Big Stone
6A Blue Earth
6A Brown
7 Carlton
6A Carver
7 Cass
6A Chippewa
6A Chisago
7 Clay
7 Clearwater
7 Cook
6A Cottonwood
7 Crow Wing
6A Dakota
6A Dodge
6A Douglas
6A Faribault
6A Fillmore
6A Freeborn
6A Goodhue
7 Grant
6A Hennepin
6A Houston
7 Hubbard
6A Isanti
7 Itasca
6A Jackson
7 Kanabec
6A Kandiyohi
7 Kittson
7 Koochiching
6A Lac qui Parle
7 Lake
7 Lake of the Woods
6A Le Sueur
6A Lincoln
6A Lyon
7 Mahnomen
7 Marshall
6A Martin
6A McLeod
6A Meeker
7 Mille Lacs
6A Morrison
6A Mower
6A Murray
6A Nicollet
6A Nobles
7 Norman
6A Olmsted
7 Otter Tail
7 Pennington
7 Pine
6A Pipestone
7 Polk
6A Pope
6A Ramsey
7 Red Lake
6A Redwood
6A Renville
6A Rice
6A Rock
7 Roseau
6A Scott
6A Sherburne
6A Sibley
6A Stearns
6A Steele
6A Stevens
7 St. Louis
6A Swift
6A Todd
6A Traverse
6A Wabasha
7 Wadena
6A Waseca
6A Washington
6A Watonwan
7 Wilkin
6A Winona
6A Wright
6A Yellow Medicine

MISSISSIPPI

3A Adams*
3A Alcorn
3A Amite*
3A Attala
3A Benton
3A Bolivar
3A Calhoun
3A Carroll
3A Chickasaw
3A Choctaw
3A Claiborne*
3A Clarke
3A Clay
3A Coahoma
3A Copiah*
3A Covington*
3A DeSoto
3A Forrest*
3A Franklin*
3A George*
3A Greene*
3A Grenada
2A Hancock*
2A Harrison*
3A Hinds*
3A Holmes
3A Humphreys
3A Issaquena
3A Itawamba
2A Jackson*
3A Jasper
3A Jefferson*
3A Jefferson Davis*
3A Jones*
3A Kemper
3A Lafayette
3A Lamar*
3A Lauderdale
3A Lawrence*
3A Leake
3A Lee
3A Leflore
3A Lincoln*
3A Lowndes
3A Madison
3A Marion*
3A Marshall
3A Monroe
3A Montgomery
3A Neshoba
3A Newton
3A Noxubee
3A Oktibbeha
3A Panola
2A Pearl River*
3A Perry*
3A Pike*

(continued)

TABLE N1101.10 (R301.1)—continued
CLIMATE ZONES, MOISTURE REGIMES, AND WARM-HUMID DESIGNATIONS BY STATE, COUNTY AND TERRITORY

3A Pontotoc
3A Prentiss
3A Quitman
3A Rankin*
3A Scott
3A Sharkey
3A Simpson*
3A Smith*
2A Stone*
3A Sunflower
3A Tallahatchie
3A Tate
3A Tippah
3A Tishomingo
3A Tunica
3A Union
3A Walthall*
3A Warren*
3A Washington
3A Wayne*
3A Webster
3A Wilkinson*
3A Winston
3A Yalobusha
3A Yazoo

MISSOURI

5A Adair
5A Andrew
5A Atchison
4A Audrain
4A Barry
4A Barton
4A Bates
4A Benton
4A Bollinger
4A Boone
5A Buchanan
4A Butler
5A Caldwell
4A Callaway
4A Camden
4A Cape Girardeau
4A Carroll
4A Carter
4A Cass
4A Cedar
5A Chariton
4A Christian
5A Clark
4A Clay
5A Clinton
4A Cole
4A Cooper
4A Crawford
4A Dade
4A Dallas
5A Daviess
5A DeKalb
4A Dent
4A Douglas
4A Dunklin
4A Franklin
4A Gasconade
5A Gentry
4A Greene
5A Grundy
5A Harrison
4A Henry
4A Hickory
5A Holt
4A Howard
4A Howell
4A Iron
4A Jackson
4A Jasper
4A Jefferson
4A Johnson
5A Knox
4A Laclede
4A Lafayette
4A Lawrence
5A Lewis
4A Lincoln
5A Linn
5A Livingston
5A Macon
4A Madison
4A Maries
5A Marion
4A McDonald
5A Mercer
4A Miller
4A Mississippi
4A Moniteau
4A Monroe
4A Montgomery
4A Morgan
4A New Madrid
4A Newton
5A Nodaway
4A Oregon
4A Osage
4A Ozark
4A Pemiscot
4A Perry
4A Pettis
4A Phelps
5A Pike
4A Platte
4A Polk
4A Pulaski
5A Putnam
5A Ralls
4A Randolph
4A Ray
4A Reynolds
4A Ripley
4A Saline
5A Schuyler
5A Scotland
4A Scott
4A Shannon
5A Shelby
4A St. Charles
4A St. Clair
4A Ste. Genevieve
4A St. Francois
4A St. Louis
4A St. Louis (city)
4A Stoddard
4A Stone
5A Sullivan
4A Taney
4A Texas
4A Vernon
4A Warren
4A Washington
4A Wayne
4A Webster
5A Worth
4A Wright

MONTANA

6B (all)

NEBRASKA

5A (all)

NEVADA

5B Carson City (city)
5B Churchill
3B Clark
5B Douglas
5B Elko
5B Esmeralda
5B Eureka
5B Humboldt
5B Lander
5B Lincoln
5B Lyon
5B Mineral
5B Nye
5B Pershing
5B Storey
5B Washoe
5B White Pine

NEW HAMPSHIRE

6A Belknap
6A Carroll
5A Cheshire
6A Coos
6A Grafton
5A Hillsborough
6A Merrimack
5A Rockingham
5A Strafford
6A Sullivan

NEW JERSEY

4A Atlantic
5A Bergen
4A Burlington
4A Camden
4A Cape May
4A Cumberland
4A Essex
4A Gloucester
4A Hudson
5A Hunterdon
5A Mercer
4A Middlesex
4A Monmouth
5A Morris
4A Ocean
5A Passaic
4A Salem
5A Somerset
5A Sussex
4A Union
5A Warren

NEW MEXICO

4B Bernalillo
5B Catron
3B Chaves
4B Cibola
5B Colfax
4B Curry
4B DeBaca
3B Dona Ana
3B Eddy
4B Grant
4B Guadalupe
5B Harding
3B Hidalgo
3B Lea
4B Lincoln
5B Los Alamos
3B Luna
5B McKinley
5B Mora
3B Otero
4B Quay
5B Rio Arriba
4B Roosevelt
5B Sandoval
5B San Juan
5B San Miguel
5B Santa Fe
4B Sierra
4B Socorro

(continued)

**TABLE N1101.10 (R301.1)—continued
CLIMATE ZONES, MOISTURE REGIMES, AND WARM-HUMID DESIGNATIONS BY STATE, COUNTY AND TERRITORY**

5B Taos
5B Torrance
4B Union
4B Valencia

NEW YORK

5A Albany
6A Allegany
4A Bronx
6A Broome
6A Cattaraugus
5A Cayuga
5A Chautauqua
5A Chemung
6A Chenango
6A Clinton
5A Columbia
5A Cortland
6A Delaware
5A Dutchess
5A Erie
6A Essex
6A Franklin
6A Fulton
5A Genesee
5A Greene
6A Hamilton
6A Herkimer
6A Jefferson
4A Kings
6A Lewis
5A Livingston
6A Madison
5A Monroe
6A Montgomery
4A Nassau
4A New York
5A Niagara
6A Oneida
5A Onondaga
5A Ontario
5A Orange
5A Orleans
5A Oswego
6A Otsego
5A Putnam
4A Queens
5A Rensselaer
4A Richmond
5A Rockland
5A Saratoga
5A Schenectady
6A Schoharie
6A Schuyler
5A Seneca
6A Steuben
6A St. Lawrence
4A Suffolk
6A Sullivan
5A Tioga
6A Tompkins
6A Ulster
6A Warren
5A Washington
5A Wayne
4A Westchester
6A Wyoming
5A Yates

NORTH CAROLINA

4A Alamance
4A Alexander
5A Alleghany
3A Anson
5A Ashe
5A Avery
3A Beaufort
4A Bertie
3A Bladen
3A Brunswick*
4A Buncombe
4A Burke
3A Cabarrus
4A Caldwell
3A Camden
3A Carteret*
4A Caswell
4A Catawba
4A Chatham
4A Cherokee
3A Chowan
4A Clay
4A Cleveland
3A Columbus*
3A Craven
3A Cumberland
3A Currituck
3A Dare
3A Davidson
4A Davie
3A Duplin
4A Durham
3A Edgecombe
4A Forsyth
4A Franklin
3A Gaston
4A Gates
4A Graham
4A Granville
3A Greene
4A Guilford
4A Halifax
4A Harnett
4A Haywood
4A Henderson
4A Hertford
3A Hoke
3A Hyde
4A Iredell
4A Jackson
3A Johnston
3A Jones
4A Lee
3A Lenoir
4A Lincoln
4A Macon
4A Madison
3A Martin
4A McDowell
3A Mecklenburg
5A Mitchell
3A Montgomery
3A Moore
4A Nash
3A New Hanover*
4A Northampton
3A Onslow*
4A Orange
3A Pamlico
3A Pasquotank
3A Pender*
3A Perquimans
4A Person
3A Pitt
4A Polk
3A Randolph
3A Richmond
3A Robeson
4A Rockingham
3A Rowan
4A Rutherford
3A Sampson
3A Scotland
3A Stanly
4A Stokes
4A Surry
4A Swain
4A Transylvania
3A Tyrrell
3A Union
4A Vance
4A Wake
4A Warren
3A Washington
5A Watauga
3A Wayne
4A Wilkes
3A Wilson
4A Yadkin
5A Yancey

NORTH DAKOTA

6A Adams
7 Barnes
7 Benson
6A Billings
7 Bottineau
6A Bowman
7 Burke
6A Burleigh
7 Cass
7 Cavalier
6A Dickey
7 Divide
6A Dunn
7 Eddy
6A Emmons
7 Foster
6A Golden Valley
7 Grand Forks
6A Grant
7 Griggs
6A Hettinger
7 Kidder
6A LaMoure
6A Logan
7 McHenry
6A McIntosh
6A McKenzie
7 McLean
6A Mercer
6A Morton
7 Mountrail
7 Nelson
6A Oliver
7 Pembina
7 Pierce
7 Ramsey
6A Ransom
7 Renville
6A Richland
7 Rolette
6A Sargent
7 Sheridan
6A Sioux
6A Slope
6A Stark
7 Steele
7 Stutsman
7 Towner
7 Traill
7 Walsh
7 Ward
7 Wells
7 Williams

OHIO

4A Adams
5A Allen

(continued)

TABLE N1101.10 (R301.1)—continued
CLIMATE ZONES, MOISTURE REGIMES, AND WARM-HUMID DESIGNATIONS BY STATE, COUNTY AND TERRITORY

5A Ashland
5A Ashtabula
5A Athens
5A Auglaize
5A Belmont
4A Brown
5A Butler
5A Carroll
5A Champaign
5A Clark
4A Clermont
5A Clinton
5A Columbiana
5A Coshocton
5A Crawford
5A Cuyahoga
5A Darke
5A Defiance
5A Delaware
5A Erie
5A Fairfield
5A Fayette
5A Franklin
5A Fulton
4A Gallia
5A Geauga
5A Greene
5A Guernsey
4A Hamilton
5A Hancock
5A Hardin
5A Harrison
5A Henry
5A Highland
5A Hocking
5A Holmes
5A Huron
5A Jackson
5A Jefferson
5A Knox
5A Lake
4A Lawrence
5A Licking
5A Logan
5A Lorain
5A Lucas
5A Madison
5A Mahoning
5A Marion
5A Medina
5A Meigs
5A Mercer
5A Miami
5A Monroe
5A Montgomery
5A Morgan
5A Morrow
5A Muskingum
5A Noble
5A Ottawa
5A Paulding
5A Perry
5A Pickaway
4A Pike
5A Portage
5A Preble
5A Putnam
5A Richland
5A Ross
5A Sandusky
4A Scioto
5A Seneca
5A Shelby
5A Stark
5A Summit
5A Trumbull
5A Tuscarawas
5A Union
5A Van Wert
5A Vinton
5A Warren
4A Washington
5A Wayne
5A Williams
5A Wood
5A Wyandot

OKLAHOMA

3A Adair
3A Alfalfa
3A Atoka
4B Beaver
3A Beckham
3A Blaine
3A Bryan
3A Caddo
3A Canadian
3A Carter
3A Cherokee
3A Choctaw
4B Cimarron
3A Cleveland
3A Coal
3A Comanche
3A Cotton
3A Craig
3A Creek
3A Custer
3A Delaware
3A Dewey
3A Ellis
3A Garfield
3A Garvin
3A Grady
3A Grant
3A Greer
3A Harmon
3A Harper
3A Haskell
3A Hughes
3A Jackson
3A Jefferson
3A Johnston
3A Kay
3A Kingfisher
3A Kiowa
3A Latimer
3A Le Flore
3A Lincoln
3A Logan
3A Love
3A Major
3A Marshall
3A Mayes
3A McClain
3A McCurtain
3A McIntosh
3A Murray
3A Muskogee
3A Noble
3A Nowata
3A Okfuskee
3A Oklahoma
3A Okmulgee
3A Osage
3A Ottawa
3A Pawnee
3A Payne
3A Pittsburg
3A Pontotoc
3A Pottawatomie
3A Pushmataha
3A Roger Mills
3A Rogers
3A Seminole
3A Sequoyah
3A Stephens
4B Texas
3A Tillman
3A Tulsa
3A Wagoner
3A Washington
3A Washita
3A Woods
3A Woodward

OREGON

5B Baker
4C Benton
4C Clackamas
4C Clatsop
4C Columbia
4C Coos
5B Crook
4C Curry
5B Deschutes
4C Douglas
5B Gilliam
5B Grant
5B Harney
5B Hood River
4C Jackson
5B Jefferson
4C Josephine
5B Klamath
5B Lake
4C Lane
4C Lincoln
4C Linn
5B Malheur
4C Marion
5B Morrow
4C Multnomah
4C Polk
5B Sherman
4C Tillamook
5B Umatilla
5B Union
5B Wallowa
5B Wasco
4C Washington
5B Wheeler
4C Yamhill

PENNSYLVANIA

5A Adams
5A Allegheny
5A Armstrong
5A Beaver
5A Bedford
5A Berks
5A Blair
5A Bradford
4A Bucks
5A Butler
5A Cambria
6A Cameron
5A Carbon
5A Centre
4A Chester
5A Clarion
6A Clearfield
5A Clinton
5A Columbia
5A Crawford
5A Cumberland
5A Dauphin
4A Delaware
6A Elk
5A Erie
5A Fayette
5A Forest
5A Franklin
5A Fulton
5A Greene

(continued)

TABLE N1101.10 (R301.1)—continued
CLIMATE ZONES, MOISTURE REGIMES, AND WARM-HUMID DESIGNATIONS BY STATE, COUNTY AND TERRITORY

5A Huntingdon
5A Indiana
5A Jefferson
5A Juniata
5A Lackawanna
5A Lancaster
5A Lawrence
5A Lebanon
5A Lehigh
5A Luzerne
5A Lycoming
6A McKean
5A Mercer
5A Mifflin
5A Monroe
4A Montgomery
5A Montour
5A Northampton
5A Northumberland
5A Perry
4A Philadelphia
5A Pike
6A Potter
5A Schuylkill
5A Snyder
5A Somerset
5A Sullivan
6A Susquehanna
6A Tioga
5A Union
5A Venango
5A Warren
5A Washington
6A Wayne
5A Westmoreland
5A Wyoming
4A York

RHODE ISLAND

5A (all)

SOUTH CAROLINA

3A Abbeville
3A Aiken
3A Allendale*
3A Anderson
3A Bamberg*
3A Barnwell*
3A Beaufort*
3A Berkeley*
3A Calhoun
3A Charleston*
3A Cherokee
3A Chester
3A Chesterfield
3A Clarendon
3A Colleton*
3A Darlington
3A Dillon
3A Dorchester*
3A Edgefield
3A Fairfield
3A Florence
3A Georgetown*
3A Greenville
3A Greenwood
3A Hampton*
3A Horry*
3A Jasper*
3A Kershaw
3A Lancaster
3A Laurens
3A Lee
3A Lexington
3A Marion
3A Marlboro
3A McCormick
3A Newberry
3A Oconee
3A Orangeburg
3A Pickens
3A Richland
3A Saluda
3A Spartanburg
3A Sumter
3A Union
3A Williamsburg
3A York

SOUTH DAKOTA

6A Aurora
6A Beadle
5A Bennett
5A Bon Homme
6A Brookings
6A Brown
6A Brule
6A Buffalo
6A Butte
6A Campbell
5A Charles Mix
6A Clark
5A Clay
6A Codington
6A Corson
6A Custer
6A Davison
6A Day
6A Deuel
6A Dewey
5A Douglas
6A Edmunds
6A Fall River
6A Faulk
6A Grant
5A Gregory
6A Haakon
6A Hamlin
6A Hand
6A Hanson
6A Harding
6A Hughes
5A Hutchinson
6A Hyde
5A Jackson
6A Jerauld
6A Jones
6A Kingsbury
6A Lake
6A Lawrence
6A Lincoln
6A Lyman
6A Marshall
6A McCook
6A McPherson
6A Meade
5A Mellette
6A Miner
6A Minnehaha
6A Moody
6A Pennington
6A Perkins
6A Potter
6A Roberts
6A Sanborn
6A Shannon
6A Spink
6A Stanley
6A Sully
5A Todd
5A Tripp
6A Turner
5A Union
6A Walworth
5A Yankton
6A Ziebach

TENNESSEE

4A Anderson
4A Bedford
4A Benton
4A Bledsoe
4A Blount
4A Bradley
4A Campbell
4A Cannon
4A Carroll
4A Carter
4A Cheatham
3A Chester
4A Claiborne
4A Clay
4A Cocke
4A Coffee
3A Crockett
4A Cumberland
4A Davidson
4A Decatur
4A DeKalb
4A Dickson
3A Dyer
3A Fayette
4A Fentress
4A Franklin
4A Gibson
4A Giles
4A Grainger
4A Greene
4A Grundy
4A Hamblen
4A Hamilton
4A Hancock
3A Hardeman
3A Hardin
4A Hawkins
3A Haywood
3A Henderson
4A Henry
4A Hickman
4A Houston
4A Humphreys
4A Jackson
4A Jefferson
4A Johnson
4A Knox
3A Lake
3A Lauderdale
4A Lawrence
4A Lewis
4A Lincoln
4A Loudon
4A Macon
3A Madison
4A Marion
4A Marshall
4A Maury
4A McMinn
3A McNairy
4A Meigs
4A Monroe
4A Montgomery
4A Moore
4A Morgan
4A Obion
4A Overton
4A Perry
4A Pickett
4A Polk
4A Putnam
4A Rhea

(continued)

TABLE N1101.10 (R301.1)—continued
CLIMATE ZONES, MOISTURE REGIMES, AND WARM-HUMID DESIGNATIONS BY STATE, COUNTY AND TERRITORY

4A Roane
4A Robertson
4A Rutherford
4A Scott
4A Sequatchie
4A Sevier
3A Shelby
4A Smith
4A Stewart
4A Sullivan
4A Sumner
3A Tipton
4A Trousdale
4A Unicoi
4A Union
4A Van Buren
4A Warren
4A Washington
4A Wayne
4A Weakley
4A White
4A Williamson
4A Wilson

TEXAS

2A Anderson*
3B Andrews
2A Angelina*
2A Aransas*
3A Archer
4B Armstrong
2A Atascosa*
2A Austin*
4B Bailey
2B Bandera*
2A Bastrop*
3B Baylor
2A Bee*
2A Bell*
2A Bexar*
3A Blanco*
3B Borden
2A Bosque*
3A Bowie*
2A Brazoria*
2A Brazos*
3B Brewster
4B Briscoe
2A Brooks*
3A Brown*
2A Burleson*
3A Burnet*
2A Caldwell*
2A Calhoun*
3B Callahan
2A Cameron*
3A Camp*
4B Carson
3A Cass*
4B Castro
2A Chambers*
2A Cherokee*
3B Childress
3A Clay
4B Cochran
3B Coke
3B Coleman
3A Collin*
3B Collingsworth
2A Colorado*
2A Comal*
3A Comanche*
3B Concho
3A Cooke
2A Coryell*
3B Cottle
3B Crane
3B Crockett
3B Crosby
3B Culberson
4B Dallam
3A Dallas*
3B Dawson
4B Deaf Smith
3A Delta
3A Denton*
2A DeWitt*
3B Dickens
2B Dimmit*
4B Donley
2A Duval*
3A Eastland
3B Ector
2B Edwards*
3A Ellis*
3B El Paso
3A Erath*
2A Falls*
3A Fannin
2A Fayette*
3B Fisher
4B Floyd
3B Foard
2A Fort Bend*
3A Franklin*
2A Freestone*
2B Frio*
3B Gaines
2A Galveston*
3B Garza
3A Gillespie*
3B Glasscock
2A Goliad*
2A Gonzales*
4B Gray
3A Grayson
3A Gregg*
2A Grimes*
2A Guadalupe*
4B Hale
3B Hall
3A Hamilton*
4B Hansford
3B Hardeman
2A Hardin*
2A Harris*
3A Harrison*
4B Hartley
3B Haskell
2A Hays*
3B Hemphill
3A Henderson*
2A Hidalgo*
2A Hill*
4B Hockley
3A Hood*
3A Hopkins*
2A Houston*
3B Howard
3B Hudspeth
3A Hunt*
4B Hutchinson
3B Irion
3A Jack
2A Jackson*
2A Jasper*
3B Jeff Davis
2A Jefferson*
2A Jim Hogg*
2A Jim Wells*
3A Johnson*
3B Jones
2A Karnes*
3A Kaufman*
3A Kendall*
2A Kenedy*
3B Kent
3B Kerr
3B Kimble
3B King
2B Kinney*
2A Kleberg*
3B Knox
3A Lamar*
4B Lamb
3A Lampasas*
2B La Salle*
2A Lavaca*
2A Lee*
2A Leon*
2A Liberty*
2A Limestone*
4B Lipscomb
2A Live Oak*
3A Llano*
3B Loving
3B Lubbock
3B Lynn
2A Madison*
3A Marion*
3B Martin
3B Mason
2A Matagorda*
2B Maverick*
3B McCulloch
2A McLennan*
2A McMullen*
2B Medina*
3B Menard
3B Midland
2A Milam*
3A Mills*
3B Mitchell
3A Montague
2A Montgomery*
4B Moore
3A Morris*
3B Motley
3A Nacogdoches*
3A Navarro*
2A Newton*
3B Nolan
2A Nueces*
4B Ochiltree
4B Oldham
2A Orange*
3A Palo Pinto*
3A Panola*
3A Parker*
4B Parmer
3B Pecos
2A Polk*
4B Potter
3B Presidio
3A Rains*
4B Randall
3B Reagan
2B Real*
3A Red River*
3B Reeves
2A Refugio*
4B Roberts
2A Robertson*
3A Rockwall*
3B Runnels
3A Rusk*
3A Sabine*
3A San Augustine*
2A San Jacinto*
2A San Patricio*

(continued)

TABLE N1101.10 (R301.1)—continued
CLIMATE ZONES, MOISTURE REGIMES, AND WARM-HUMID DESIGNATIONS BY STATE, COUNTY AND TERRITORY

3A San Saba*
3B Schleicher
3B Scurry
3B Shackelford
3A Shelby*
4B Sherman
3A Smith*
3A Somervell*
2A Starr*
3A Stephens
3B Sterling
3B Stonewall
3B Sutton
4B Swisher
3A Tarrant*
3B Taylor
3B Terrell
3B Terry
3B Throckmorton
3A Titus*
3B Tom Green
2A Travis*
2A Trinity*
2A Tyler*
3A Upshur*
3B Upton
2B Uvalde*
2B Val Verde*
3A Van Zandt*
2A Victoria*
2A Walker*
2A Waller*
3B Ward
2A Washington*
2B Webb*
2A Wharton*
3B Wheeler
3A Wichita
3B Wilbarger
2A Willacy*
2A Williamson*
2A Wilson*
3B Winkler
3A Wise
3A Wood*
4B Yoakum
3A Young
2B Zapata*
2B Zavala*

UTAH

5B Beaver
6B Box Elder
6B Cache
6B Carbon
6B Daggett
5B Davis
6B Duchesne
5B Emery
5B Garfield
5B Grand
5B Iron
5B Juab
5B Kane
5B Millard
6B Morgan
5B Piute
6B Rich
5B Salt Lake
5B San Juan
5B Sanpete
5B Sevier
6B Summit
5B Tooele
6B Uintah
5B Utah
6B Wasatch
3B Washington
5B Wayne
5B Weber

VERMONT

6A (all)

VIRGINIA

4A (all)

WASHINGTON

5B Adams
5B Asotin
5B Benton
5B Chelan
4C Clallam
4C Clark
5B Columbia
4C Cowlitz
5B Douglas
6B Ferry
5B Franklin
5B Garfield
5B Grant
4C Grays Harbor
4C Island
4C Jefferson
4C King
4C Kitsap
5B Kittitas
5B Klickitat
4C Lewis
5B Lincoln
4C Mason
6B Okanogan
4C Pacific
6B Pend Oreille
4C Pierce
4C San Juan
4C Skagit
5B Skamania
4C Snohomish
5B Spokane
6B Stevens
4C Thurston
4C Wahkiakum
5B Walla Walla
4C Whatcom
5B Whitman
5B Yakima

WEST VIRGINIA

5A Barbour
4A Berkeley
4A Boone
4A Braxton
5A Brooke
4A Cabell
4A Calhoun
4A Clay
5A Doddridge
5A Fayette
4A Gilmer
5A Grant
5A Greenbrier
5A Hampshire
5A Hancock
5A Hardy
5A Harrison
4A Jackson
4A Jefferson
4A Kanawha
5A Lewis
4A Lincoln
4A Logan
5A Marion
5A Marshall
4A Mason
4A McDowell
4A Mercer
5A Mineral
4A Mingo
5A Monongalia
4A Monroe
4A Morgan
5A Nicholas
5A Ohio
5A Pendleton
4A Pleasants
5A Pocahontas
5A Preston
4A Putnam
5A Raleigh
5A Randolph
4A Ritchie
4A Roane
5A Summers
5A Taylor
5A Tucker
4A Tyler
5A Upshur
4A Wayne
5A Webster
5A Wetzel
4A Wirt
4A Wood
4A Wyoming

WISCONSIN

6A Adams
7 Ashland
6A Barron
7 Bayfield
6A Brown
6A Buffalo
7 Burnett
6A Calumet
6A Chippewa
6A Clark
6A Columbia
6A Crawford
6A Dane
6A Dodge
6A Door
7 Douglas
6A Dunn
6A Eau Claire
7 Florence
6A Fond du Lac
7 Forest
6A Grant
6A Green
6A Green Lake
6A Iowa
7 Iron
6A Jackson
6A Jefferson
6A Juneau
6A Kenosha
6A Kewaunee
6A La Crosse
6A Lafayette
7 Langlade
7 Lincoln
6A Manitowoc
6A Marathon
6A Marinette
6A Marquette
6A Menominee
6A Milwaukee
6A Monroe
6A Oconto
7 Oneida
6A Outagamie

(continued)

TABLE N1101.10 (R301.1)—continued
CLIMATE ZONES, MOISTURE REGIMES, AND WARM-HUMID DESIGNATIONS BY STATE, COUNTY AND TERRITORY

6A Ozaukee
6A Pepin
6A Pierce
6A Polk
6A Portage
7 Price
6A Racine
6A Richland
6A Rock
6A Rusk
6A Sauk
7 Sawyer
6A Shawano
6A Sheboygan
6A St. Croix
7 Taylor
6A Trempealeau
6A Vernon
7 Vilas
6A Walworth
7 Washburn
6A Washington
6A Waukesha
6A Waupaca
6A Waushara
6A Winnebago
6A Wood

WYOMING

6B Albany
6B Big Horn
6B Campbell
6B Carbon
6B Converse
6B Crook
6B Fremont
5B Goshen
6B Hot Springs
6B Johnson
6B Laramie
7 Lincoln
6B Natrona
6B Niobrara
6B Park
5B Platte
6B Sheridan
7 Sublette
6B Sweetwater
7 Teton
6B Uinta
6B Washakie
6B Weston

US TERRITORIES

AMERICAN SAMOA

1A (all)*

GUAM

1A (all)*

NORTHERN MARIANA ISLANDS

1A (all)*

PUERTO RICO

1A (all)*

VIRGIN ISLANDS

1A (all)*

TABLE N1101.10.2(1) [R302.3(1)]
INTERNATIONAL CLIMATE ZONE DEFINITIONS

MAJOR CLIMATE TYPE DEFINITIONS
Marine (C) Definition—Locations meeting all four criteria: 1. Mean temperature of coldest month between -3°C (27°F) and 18°C (65°F). 2. Warmest month mean < 22°C (72°F). 3. At least four months with mean temperatures over 10°C (50°F). 4. Dry season in summer. The month with the heaviest precipitation in the cold season has at least three times as much precipitation as the month with the least precipitation in the rest of the year. The cold season is October through March in the Northern Hemisphere and April through September in the Southern Hemisphere.
Dry (B) Definition—Locations meeting the following criteria: Not marine and $P_{in} < 0.44 \times (TF - 19.5)$ [$P_{cm} < 2.0 \times (TC + 7)$ in SI units] where: P_{in} = Annual precipitation in inches (cm) T = Annual mean temperature in °F (°C)
Moist (A) Definition—Locations that are not marine and not dry.
Warm-humid Definition—Moist (A) locations where either of the following wet-bulb temperature conditions shall occur during the warmest six consecutive months of the year: 1. 67°F (19.4°C) or higher for 3,000 or more hours; or 2. 73°F (22.8°C) or higher for 1,500 or more hours.

For SI: °C = [(°F)-32]/1.8, 1 inch = 2.54 cm.

TABLE N1101.10.2(2) [R301.3(2)]
INTERNATIONAL CLIMATE ZONE DEFINITIONS

ZONE NUMBER	THERMAL CRITERIA	
	IP Units	SI Units
1	9000 < CDD50°F	5000 < CDD10°C
2	6300 < CDD50°F ≤ 9000	3500 < CDD10°C ≤ 5000
3A and 3B	4500 < CDD50°F ≤ 6300 AND HDD65°F ≤ 5400	2500 < CDD10°C ≤ 3500 AND HDD18°C ≤ 3000
4A and 4B	CDD50°F ≤ 4500 AND HDD65°F ≤ 5400	CDD10°C ≤ 2500 AND HDD18°C ≤ 3000
3C	HDD65°F ≤ 3600	HDD18°C ≤ 2000
4C	3600 < HDD65°F ≤ 5400	2000 < HDD18°C ≤ 3000
5	5400 < HDD65°F ≤ 7200	3000 < HDD18°C ≤ 4000
6	7200 < HDD65°F ≤ 9000	4000 < HDD18°C ≤ 5000
7	9000 < HDD65°F ≤ 12600	5000 < HDD18°C ≤ 7000
8	12600 < HDD65°F	7000 < HDD18°C

For SI: °C = [(°F)-32]/1.8.

N1101.11 (R302.1) Interior design conditions. The interior design temperatures used for heating and cooling load calculations shall be a maximum of 72°F (22°C) for heating and minimum of 75°F (24°C) for cooling.

N1101.12 (R303.1) Identification. Materials, systems and equipment shall be identified in a manner that will allow a determination of compliance with the applicable provisions of this code.

N1101.12.1 (R303.1.1) Building thermal envelope insulation. An *R*-value identification mark shall be applied by the manufacturer to each piece of *building thermal envelope* insulation 12 inches (305 mm) or greater in width. Alternately, the insulation installers shall provide a certification listing the type, manufacturer and *R*-value of insulation installed in each element of the *building thermal envelope*. For blown or sprayed insulation (fiberglass and cellulose), the initial installed thickness, settled thickness, settled *R*-value, installed density, coverage area and number of bags installed shall be *listed* on the certification. For sprayed polyurethane foam (SPF) insulation, the installed thickness of the areas covered and *R*-value of installed thickness shall be *listed* on the certification. The insulation installer shall sign, date and post the certification in a conspicuous location on the job site.

N1101.12.1.1 (R303.1.1.1) Blown or sprayed roof/ceiling insulation. The thickness of blown-in or sprayed roof/ceiling insulation (fiberglass or cellulose) shall be written in inches (mm) on markers that are installed at least one for every 300 square feet (28 m^2) throughout the attic space. The markers shall be affixed to the trusses or joists and marked with the minimum initial installed thickness with numbers a minimum of 1 inch (25 mm) in height. Each marker shall face the attic access opening. Spray polyurethane foam thickness and installed *R*-value shall be *listed* on certification provided by the insulation installer.

N1101.12.2 (R303.1.2) Insulation mark installation. Insulating materials shall be installed such that the manufacturer's *R*-value mark is readily observable upon inspection.

N1101.12.3 (R303.1.3) Fenestration product rating. *U*-factors of fenestration products (windows, doors and skylights) shall be determined in accordance with NFRC 100 by an accredited, independent laboratory, and labeled and certified by the manufacturer. Products lacking such a labeled *U*-factor shall be assigned a default *U*-factor from Table N1101.12.3(1) or N1101.12.3(2). The solar heat gain coefficient (SHGC) and *visible transmittance* (VT) of glazed fenestration products (windows, glazed doors and skylights) shall be determined in accordance with NFRC 200 by an accredited, independent laboratory, and labeled and certified by the manufacturer. Products lacking such a labeled SHGC or VT shall be assigned a default SHGC or VT from Table N1101.12.3(3).

TABLE N1101.12.3(1) [R303.1.3(1)]
DEFAULT GLAZED FENESTRATION *U*-FACTOR

FRAME TYPE	SINGLE PANE	DOUBLE PANE	SKYLIGHT	
			Single	Double
Metal	1.20	0.80	2.00	1.30
Metal with Thermal Break	1.10	0.65	1.90	1.10
Nonmetal or Metal Clad	0.95	0.55	1.75	1.05
Glazed Block	0.60			

TABLE N1101.12.3(2) [R303.1.3(2)]
DEFAULT DOOR *U*-FACTORS

DOOR TYPE	*U*-FACTOR
Uninsulated Metal	1.20
Insulated Metal	0.60
Wood	0.50
Insulated, nonmetal edge, max 45% glazing, any glazing double pane	0.35

TABLE N1101.12.3(3) [R303.1.3(3)]
DEFAULT GLAZED FENESTRATION SHGC AND VT

	SINGLE GLAZED		DOUBLE GLAZED		GLAZED BLOCK
	Clear	Tinted	Clear	Tinted	
SHGC	0.8	0.7	0.7	0.6	0.6
VT	0.6	0.3	0.6	0.3	0.6

N1101.12.4 (R303.1.4) Insulation product rating. The thermal resistance (*R*-value) of insulation shall be determined in accordance with the U.S. Federal Trade Commission *R*-value rule (CFR Title 16, Part 460) in units of h × ft^2 × °F/Btu at a mean temperature of 75°F (24°C).

N1101.13 (R303.2) Installation. All materials, systems and equipment shall be installed in accordance with the manufacturer's installation instructions and this code.

N1101.13.1 (R303.2.1) Protection of exposed foundation insulation. Insulation applied to the exterior of basement walls, crawlspace walls and the perimeter of slab-on-grade floors shall have a rigid, opaque and weather-resistant protective covering to prevent the degradation of the insulation's thermal performance. The protective covering shall cover the exposed exterior insulation and extend a minimum of 6 inches (153 mm) below grade.

N1101.14 (R303.3) Maintenance information. Maintenance instructions shall be furnished for equipment and systems that require preventive maintenance. Required regular maintenance actions shall be clearly stated and incorporated on a readily accessible label. The label shall include the title or publication number for the operation and maintenance manual for that particular model and type of product.

N1101.15 (R401.2) Compliance. Projects shall comply with Sections identified as "mandatory" and with either sections identified as "prescriptive" or the performance approach in Section N1105.

N1101.16 (R401.3) Certificate (Mandatory). A permanent certificate shall be completed and posted on or in the electrical distribution panel by the builder or registered design professional. The certificate shall not cover or obstruct the visibility of the circuit directory label, service disconnect label or other required labels. The certificate shall list the predominant *R*-values of insulation installed in or on ceiling/roof, walls, foundation (slab, *basement wall,* crawl space wall and/or floor) and ducts outside conditioned spaces; *U*-factors for fenestration and the solar heat gain coefficient (SHGC) of fenestration, and the results from any required duct system and building envelope air leakage testing done on the building. Where there is more than one value for each component, the certificate shall list the value covering the largest area. The certificate shall list the types and efficiencies of heating, cooling and service water heating equipment. Where a gas-fired unvented room heater, electric furnace, or baseboard electric heater is installed in the residence, the certificate shall list "gas-fired unvented room heater," "electric furnace" or "baseboard electric heater," as appropriate. An efficiency shall not be *listed* for gas-fired unvented room heaters, electric furnaces or electric baseboard heaters.

TABLE N1102.1.1 (R402.1.1)
INSULATION AND FENESTRATION REQUIREMENTS BY COMPONENT[a]

CLIMATE ZONE	FENESTRATION *U*-FACTOR[b]	SKYLIGHT[b] *U*-FACTOR	GLAZED FENESTRATION SHGC[b, e]	CEILING *R*-VALUE	WOOD FRAME WALL *R*-VALUE	MASS WALL *R*-VALUE[i]	FLOOR *R*-VALUE	BASEMENT[c] WALL *R*-VALUE	SLAB[d] *R*-VALUE & DEPTH	CRAWL SPACE[c] WALL *R*-VALUE
1	NR	0.75	0.25	30	13	3/4	13	0	0	0
2	0.40	0.65	0.25	38	13	4/6	13	0	0	0
3	0.35	0.55	0.25	38	20 or 13 + 5[h]	8/13	19	5/13[f]	0	5/13
4 except Marine	0.35	0.55	0.40	49	20 or 13 + 5[h]	8/13	19	10 /13	10, 2 ft	10/13
5 and Marine 4	0.32	0.55	NR	49	20 or 13 + 5[h]	13/17	30[g]	15/19	10, 2 ft	15/19
6	0.32	0.55	NR	49	20 + 5 or 13 + 10[h]	15/20	30[g]	15/19	10, 4 ft	15/19
7 and 8	0.32	0.55	NR	49	20 + 5 or 13 + 10[h]	19/21	38[g]	15/19	10, 4 ft	15/19

For SI: 1 foot = 304.8 mm.

a. *R*-values are minimums. *U*-factors and SHGC are maximums. When insulation is installed in a cavity which is less than the label or design thickness of the insulation, the installed *R*-value of the insulation shall not be less than the *R*-value specified in the table.

b. The fenestration *U*-factor column excludes skylights. The SHGC column applies to all glazed fenestration.
Exception: Skylights may be excluded from glazed fenestration SHGC requirements in Climate Zones 1 through 3 where the SHGC for such skylights does not exceed 0.30.

c. "15/19" means R-15 continuous insulation on the interior or exterior of the home or R-19 cavity insulation at the interior of the basement wall. "15/19" shall be permitted to be met with R-13 cavity insulation on the interior of the basement wall plus R-5 continuous insulation on the interior or exterior of the home. "10/13" means R-10 continuous insulation on the interior or exterior of the home or R-13 cavity insulation at the interior of the basement wall.

d. R-5 shall be added to the required slab edge *R*-values for heated slabs. Insulation depth shall be the depth of the footing or 2 feet, whichever is less in Zones 1 through 3 for heated slabs.

e. There are no SHGC requirements in the Marine Zone.

f. Basement wall insulation is not required in warm-humid locations as defined by Figure N1101.10 and Table N1101.10.

g. Or insulation sufficient to fill the framing cavity, R-19 minimum.

h. First value is cavity insulation, second is continuous insulation or insulated siding, so "13 + 5" means R-13 cavity insulation plus R-5 continuous insulation or insulated siding. If structural sheathing covers 40 percent or less of the exterior, continuous insulation *R*-value shall be permitted to be reduced by no more than R-3 in the locations where structural sheathing is used – to maintain a consistent total sheathing thickness.

i. The second *R*-value applies when more than half the insulation is on the interior of the mass wall.

SECTION N1102 BUILDING THERMAL ENVELOPE

N1102.1 (R402.1) General (Prescriptive). The *building thermal envelope* shall meet the requirements of Sections N1102.1.1 through N1102.1.4.

N1102.1.1 (R402.1.1) Insulation and fenestration criteria. The *building thermal envelope* shall meet the requirements of Table N1102.1.1 based on the climate zone specified in Section N1101.10.

N1102.1.2 (R402.1.2) *R*-value computation. Insulation material used in layers, such as framing cavity insulation and insulating sheathing, shall be summed to compute the component *R*-value. The manufacturer's settled *R*-value shall be used for blown insulation. Computed *R*-values shall not include an *R*-value for other building materials or air films.

N1102.1.3 (R402.1.3) *U*-factor alternative. An assembly with a *U*-factor equal to or less than that specified in Table N1102.1.3 shall be permitted as an alternative to the *R*-value in Table N1102.1.1.

N1102.1.4 (R402.1.4) Total UA alternative. If the total *building thermal envelope* UA (sum of *U*-factor times assembly area) is less than or equal to the total UA resulting from using the *U*-factors in Table N1102.1.3 (multiplied by the same assembly area as in the proposed building), the building shall be considered in compliance with Table N1102.1.1. The UA calculation shall be done using a method consistent with the ASHRAE *Handbook of Fundamentals* and shall include the thermal bridging effects of framing materials. The SHGC requirements shall be met in addition to UA compliance.

N1102.2 (R402.2) Specific insulation requirements (Prescriptive). In addition to the requirements of Section N1102.1, insulation shall meet the specific requirements of Sections N1102.2.1 through N1102.2.12.

N1102.2.1 (R402.2.1) Ceilings with attic spaces. When Secton N1102.1.1 would require R-38 in the ceiling, R-30 shall be deemed to satisfy the requirement for R-38 wherever the full height of uncompressed R-30 insulation extends over the wall top plate at the eaves. Similarly, R-38 shall be deemed to satisfy the requirement for R-49 wherever the full height of uncompressed R-38 insulation extends over the wall top plate at the eaves. This reduction shall not apply to the *U*-factor alternative approach in Secton N1102.1.3 and the total UA alternative in Secton N1102.1.4.

N1102.2.2 (R402.2.2) Ceilings without attic spaces. Where Secton N1102.1.1 would require insulation levels above R-30 and the design of the roof/ceiling assembly does not allow sufficient space for the required insulation, the minimum required insulation for such roof/ceiling assemblies shall be R-30. This reduction of insulation from the requirements of Secton N1102.1.1 shall be limited to 500 square feet (46 m^2) or 20 percent of the total insulated ceiling area, whichever is less. This reduction shall not apply to the *U*-factor alternative approach in Secton N1102.1.3 and the total UA alternative in Secton N1102.1.4.

N1102.2.3 (R402.2.3) Eave baffle. For air permeable insulations in vented attics, a baffle shall be installed adjacent to soffit and eave vents. Baffles shall maintain an opening equal or greater than the size of the vent. The baffle shall extend over the top of the attic insulation. The baffle shall be permitted to be any solid material.

N1102.2.4 (R402.2.4) Access hatches and doors. Access doors from conditioned spaces to unconditioned spaces (e.g., attics and crawl spaces) shall be weatherstripped and insulated to a level equivalent to the insulation on the surrounding surfaces. Access shall be provided to all equipment that prevents damaging or compressing the insulation. A wood framed or equivalent baffle or retainer is required to be provided when loose fill insulation is installed, the purpose of which is to prevent the loose fill insulation from spilling into the living space when the attic access is opened, and to provide a permanent means of maintaining the installed *R*-value of the loose fill insulation.

N1102.2.5 (R402.2.5) Mass walls. Mass walls for the purposes of this chapter shall be considered above-grade walls of concrete block, concrete, insulated concrete form (ICF), masonry cavity, brick (other than brick veneer), earth (adobe, compressed earth block, rammed earth) and solid timber/logs.

TABLE N1102.1.3 (R402.1.3) EQUIVALENT *U*-FACTORS[a]

CLIMATE ZONE	FENESTRATION *U*-FACTOR	SKYLIGHT *U*-FACTOR	CEILING *U*-FACTOR	FRAME WALL *U*-FACTOR	MASS WALL *U*-FACTOR[b]	FLOOR *U*-FACTOR	BASEMENT WALL *U*-FACTOR	CRAWL SPACE WALL *U*-FACTOR
1	0.50	0.75	0.035	0.082	0.197	0.064	0.360	0.477
2	0.40	0.65	0.030	0.082	0.165	0.064	0.360	0.477
3	0.35	0.55	0.030	0.057	0.098	0.047	0.091[c]	0.136
4 except Marine	0.35	0.55	0.026	0.057	0.098	0.047	0.059	0.065
5 and Marine 4	0.32	0.55	0.026	0.057	0.082	0.033	0.050	0.055
6	0.32	0.55	0.026	0.048	0.060	0.033	0.050	0.055
7 and 8	0.32	0.55	0.026	0.048	0.057	0.028	0.050	0.055

a. Nonfenestration *U*-factors shall be obtained from measurement, calculation or an approved source.

b. When more than half the insulation is on the interior, the mass wall *U*-factors shall be a maximum of 0.17 in Zone 1, 0.14 in Zone 2, 0.12 in Zone 3, 0.087 in Zone 4 except Marine, 0.065 in Zone 5 and Marine 4, and 0.057 in Zones 6 through 8.

c. Basement wall *U*-factor of 0.360 in warm-humid locations as defined by Figure 301.1 and Table 301.1.

N1102.2.6 (R402.2.6) Steel-frame ceilings, walls, and floors. Steel-frame ceilings, walls, and floors shall meet the insulation requirements of Table N1102.2.6 or shall meet the *U*-factor requirements of Table N1102.1.3. The calculation of the *U*-factor for a steel-frame envelope assembly shall use a series-parallel path calculation method.

**TABLE N1102.2.6 (R402.2.6)
STEEL-FRAME CEILING, WALL AND FLOOR INSULATION
(*R*-VALUE)**

WOOD FRAME *R*-VALUE REQUIREMENT	COLD-FORMED STEEL EQUIVALENT *R*-VALUE[a]
Steel Truss Ceilings[b]	
R-30	R-38 or R-30 + 3 or R-26 + 5
R-38	R-49 or R-38 + 3
R-49	R-38 + 5
Steel Joist Ceilings[b]	
R-30	R-38 in 2 × 4 or 2 × 6 or 2 × 8 R-49 in any framing
R-38	R-49 in 2 × 4 or 2 × 6 or 2 × 8 or 2 × 10
Steel-Framed Wall, 16″ o.c.	
R-13	R-13 + 4.2 or R-19 + 2.1 or R-21 + 2.8 or R-0 + 9.3 or R-15 + 3.8 or R-21 + 3.1
R-13 + 3	R-0 + 11.2 or R-13 + 6.1 or R-15 + 5.7 or R-19 + 5.0 or R-21 + 4.7
R-20	R-0 + 14.0 or R-13 + 8.9 or R-15 + 8.5 or R-19 + 7.8 or R-19 + 6.2 or R-21 + 7.5
R-20 + 5	R-13 + 12.7 or R-15 + 12.3 or R-19 + 11.6 or R-21 + 11.3 or R-25 + 10.9
R-21	R-0 + 14.6 or R-13 + 9.5 or R-15 + 9.1 or R-19 + 8.4 or R-21 + 8.1 or R-25 + 7.7
Steel-Framed Wall, 24″ o.c.	
R-13	R-0 + 9.3 or R-13 + 3.0 or R-15 + 2.4
R-13 + 3	R-0 + 11.2 or R-13 + 4.9 or R-15 + 4.3 or R-19 + 3.5 or R-21 + 3.1
R-20	R-0 + 14.0 or R-13 + 7.7 or R-15 + 7.1 or R-19 + 6.3 or R-21 + 5.9
R-20 + 5	R-13 + 11.5 or R-15 + 10.9 or R-19 + 10.1 or R-21 + 9.7 or R-25 + 9.1
R-21	R-0 + 14.6 or R-13 + 8.3 or R-15 + 7.7 or R-19 + 6.9 or R-21 + 6.5 or R-25 + 5.9
Steel Joist Floor	
R-13	R-19 in 2 × 6, or R-19 + 6 in 2 × 8 or 2 × 10
R-19	R-19 + 6 in 2 × 6, or R-19 + 12 in 2 × 8 or 2 × 10

a. Cavity insulation *R*-value is listed first, followed by continuous insulation *R*-value.

b. Insulation exceeding the height of the framing shall cover the framing.

N1102.2.7 (R402.2.7) Floors. Floor insulation shall be installed to maintain permanent contact with the underside of the subfloor decking.

N1102.2.8 (R402.2.8) Basement walls. Walls associated with conditioned basements shall be insulated from the top of the *basement wall* down to 10 feet (3048 mm) below grade or to the basement floor, whichever is less. Walls associated with unconditioned basements shall meet this requirement unless the floor overhead is insulated in accordance with Sections N1102.1.1 and N1102.2.7.

N1102.2.9 (R402.2.9) Slab-on-grade floors. Slab-on-grade floors with a floor surface less than 12 inches (305 mm) below grade shall be insulated in accordance with Table N1102.1.1. The insulation shall extend downward from the top of the slab on the outside or inside of the foundation wall. Insulation located below grade shall be extended the distance provided in Table N1102.1.1 by any combination of vertical insulation, insulation extending under the slab or insulation extending out from the building. Insulation extending away from the building shall be protected by pavement or by a minimum of 10 inches (254 mm) of soil. The top edge of the insulation installed between the *exterior wall* and the edge of the interior slab shall be permitted to be cut at a 45-degree (0.79 rad) angle away from the *exterior wall.* Slab-edge insulation is not required in jurisdictions designated by the *building official* as having a very heavy termite infestation.

N1102.2.10 (R402.2.10) Crawl space walls. As an alternative to insulating floors over crawl spaces, crawl space walls shall be permitted to be insulated when the crawl space is not vented to the outside. Crawl space wall insulation shall be permanently fastened to the wall and extend downward from the floor to the finished grade level and then vertically and/or horizontally for at least an additional 24 inches (610 mm). Exposed earth in unvented crawl space foundations shall be covered with a continuous Class I vapor retarder in accordance with this code. All joints of the vapor retarder shall overlap by 6 inches (153 mm) and be sealed or taped. The edges of the vapor retarder shall extend at least 6 inches (153 mm) up the stem wall and shall be attached to the stem wall.

N1102.2.11 (R402.2.11) Masonry veneer. Insulation shall not be required on the horizontal portion of the foundation that supports a masonry veneer.

N1102.2.12 (R402.2.12) Sunroom insulation. All *sunrooms* enclosing conditioned spaces shall meet the insulation requirements of this code.

Exception: For *sunrooms* with *thermal isolation*, and enclosing conditioned spaces, the following exceptions to the insulation *requirements* of this code shall apply:

1. The minimum ceiling insulation *R*-values shall be R-19 in Zones 1 through 4 and R-24 in Zones 5 through 8; and
2. The minimum wall *R*-value shall be R-13 in all zones. Wall(s) separating a *sunroom* with a *thermal isolation* from *conditioned space* shall meet the *building thermal envelope* requirements of this code.

N1102.3 (R402.3) Fenestration (Prescriptive). In addition to the requirements of Secton N1102, fenestration shall comply with Sections N1102.3.1 through N1102.3.6.

N1102.3.1 (R402.3.1) *U*-factor. An area-weighted average of fenestration products shall be permitted to satisfy the *U*-factor requirements.

N1102.3.2 (R402.3.2) Glazed fenestration SHGC. An area-weighted average of fenestration products more than 50-percent glazed shall be permitted to satisfy the SHGC requirements.

N1102.3.3 (R402.3.3) Glazed fenestration exemption. Up to 15 square feet (1.4 m^2) of glazed fenestration per dwelling unit shall be permitted to be exempt from *U*-factor and SHGC requirements in Secton N1102.1.1. This exemption shall not apply to the *U*-factor alternative approach in Secton N1102.1.3 and the Total UA alternative in Secton N1102.1.4.

N1102.3.4 (R402.3.4) Opaque door exemption. Onc side-hinged opaque door assembly up to 24 square feet (2.22 m^2) in area is exempted from the *U*-factor requirement in Secton N1102.1.1. This exemption shall not apply to the *U*-factor alternative approach in Secton N1102.1.3 and the total UA alternative in Secton N1102.1.4.

N1102.3.5 (R402.3.5) Sunroom *U*-factor. All *sunrooms* enclosing conditioned spaces shall meet the fenestration requirements of this code.

Exception: For *sunrooms* with *thermal isolation* and enclosing conditioned spaces, in Zones 4 through 8, the following exceptions to the fenestration requirements of this code shall apply:

1. The maximum fenestration *U*-factor shall be 0.45; and
2. The maximum skylight *U*-factor shall be 0.70. New fenestration separating the *sunroom* with *thermal isolation* from *conditioned space* shall meet the *building thermal envelope* requirements of this code.

N1102.3.6 (R402.3.6) Replacement fenestration. Where some or all of an existing fenestration unit is replaced with a new fenestration product, including sash and glazing, the replacement fenestration unit shall meet the applicable requirements for *U*-factor and SHGC in Table N1102.1.1.

N1102.4 (R402.4) Air leakage (Mandatory). The building thermal envelope shall be constructed to limit air leakage in accordance with the requirements of Sections N1102.4.1 through N1102.4.4.

N1102.4.1 (R402.4.1) Building thermal envelope. The *building thermal envelope* shall comply with Sections N1102.4.1.1 and N1102.4.1.2. The sealing methods between dissimilar materials shall allow for differential expansion and contraction.

N1102.4.1.1 (R402.4.1.1) Installation. The components of the *building thermal envelope* as listed in Table N1102.4.1.1 shall be installed in accordance with the manufacturer's instructions and the criteria listed in Table N1102.4.1.1, as applicable to the method of construction. Where required by the *building official*, an *approved* third party shall inspect all components and verify compliance.

N1102.4.1.2 (R402.4.1.2) Testing. The building or dwelling unit shall be tested and verified as having an air leakage rate of not exceeding 5 air changes per hour in Zones 1 and 2, and 3 air changes per hour in Zones 3 through 8. Testing shall be conducted with a blower door at a pressure of 0.2 inches w.g. (50 Pascals). Where required by the *building official*, testing shall be conducted by an *approved* third party. A written report of the results of the test shall be signed by the party conducting the test and provided to the *building official*. Testing shall be performed at any time after creation of all penetrations of the *building thermal envelope*.

During testing:

1. Exterior windows and doors, fireplace and stove doors shall be closed, but not sealed, beyond the intended weatherstripping or other infiltration control measures;
2. Dampers including exhaust, intake, makeup air, backdraft and flue dampers shall be closed, but not sealed beyond intended infiltration control measures;
3. Interior doors, if installed at the time of the test, shall be open;
4. Exterior doors for continuous ventilation systems and heat recovery ventilators shall be closed and sealed;
5. Heating and cooling systems, if installed at the time of the test, shall be turned off; and
6. Supply and return registers, if installed at the time of the test, shall be fully open.

N1102.4.2 (R402.4.2) Fireplaces. New wood-burning fireplaces shall have tight-fitting flue dampers and outdoor combustion air.

N1102.4.3 (R402.4.3) Fenestration air leakage. Windows, skylights and sliding glass doors shall have an air infiltration rate of no more than 0.3 cfm per square foot (1.5 L/s/m^2), and swinging doors no more than 0.5 cfm per square foot (2.6 L/s/m^2), when tested according to NFRC 400 or AAMA/WDMA/CSA 101/I.S.2/A440 by an accredited, independent laboratory and *listed* and *labeled* by the manufacturer.

Exception: Site-built windows, skylights and doors.

N1102.4.4 (R402.4.4) Recessed lighting. Recessed luminaires installed in the *building thermal envelope* shall be sealed to limit air leakage between conditioned and unconditioned spaces. All recessed luminaires shall be IC-rated and *labeled* as having an air leakage rate not more than 2.0 cfm (0.944 L/s) when tested in accordance with ASTM E 283 at a 1.57 psf (75 Pa) pressure differential. All recessed luminaires shall be sealed with a gasket or caulk between the housing and the interior wall or ceiling covering.

N1102.5 (R402.5) Maximum fenestration *U*-factor and SHGC (Mandatory). The area-weighted average maximum fenestration *U*-factor permitted using tradeoffs from Secton N1102.1.4 or N1105 shall be 0.48 in Zones 4 and 5 and 0.40

in Zones 6 through 8 for vertical fenestration, and 0.75 in Zones 4 through 8 for skylights. The area-weighted average maximum fenestration SHGC permitted using tradeoffs from Secton N1105 in Zones 1 through 3 shall be 0.50.

SECTION N1103 SYSTEMS

N1103.1 (R403.1) Controls (Mandatory). At least one thermostat shall be provided for each separate heating and cooling system.

N1103.1.1 (R403.1.1) Programmable thermostat. Where the primary heating system is a forced-air furnace, at least one thermostat per dwelling unit shall be capable of controlling the heating and cooling system on a daily schedule to maintain different temperature set points at different times of the day. This thermostat shall include the capability to set back or temporarily operate the system to maintain zone temperatures down to 55°F (13°C) or up to 85°F (29°C). The thermostat shall initially be programmed with a heating temperature set point no higher

TABLE N1102.4.1.1 (R402.4.1.1)
AIR BARRIER AND INSULATION INSTALLATION

COMPONENT	CRITERIA[a]
Air barrier and thermal barrier	A continuous air barrier shall be installed in the building envelope. Exterior thermal envelope contains a continuous air barrier. Breaks or joints in the air barrier shall be sealed. Air-permeable insulation shall not be used as a sealing material.
Ceiling/attic	The air barrier in any dropped ceiling/soffit shall be aligned with the insulation and any gaps in the air barrier sealed. Access openings, drop down stair or knee wall doors to unconditioned attic spaces shall be sealed.
Walls	Corners and headers shall be insulated and the junction of the foundation and sill plate shall be sealed. The junction of the top plate and top of exterior walls shall be sealed. Exterior thermal envelope insulation for framed walls shall be installed in substantial contact and continuous alignment with the air barrier. Knee walls shall be sealed.
Windows, skylights and doors	The space between window/door jambs and framing and skylights and framing shall be sealed.
Rim joists	Rim joists shall be insulated and include the air barrier.
Floors (including above-garage and cantilevered floors)	Insulation shall be installed to maintain permanent contact with underside of subfloor decking. The air barrier shall be installed at any exposed edge of insulation.
Crawl space walls	Where provided in lieu of floor insulation, insulation shall be permanently attached to the crawlspace walls. Exposed earth in unvented crawl spaces shall be covered with a Class I vapor retarder with overlapping joints taped.
Shafts, penetrations	Duct shafts, utility penetrations, and flue shafts opening to exterior or unconditioned space shall be sealed.
Narrow cavities	Batts in narrow cavities shall be cut to fit, or narrow cavities shall be filled by insulation that on installation readily conforms to the available cavity space.
Garage separation	Air sealing shall be provided between the garage and conditioned spaces.
Recessed lighting	Recessed light fixtures installed in the building thermal envelope shall be air tight, IC rated, and sealed to the drywall.
Plumbing and wiring	Batt insulation shall be cut neatly to fit around wiring and plumbing in exterior walls, or insulation that on installation readily conforms to available space shall extend behind piping and wiring.
Shower/tub on exterior wall	Exterior walls adjacent to showers and tubs shall be insulated and the air barrier installed separating them from the showers and tubs.
Electrical/phone box on exterior walls	The air barrier shall be installed behind electrical or communication boxes or air-sealed boxes shall be installed.
HVAC register boots	HVAC register boots that penetrate building thermal envelope shall be sealed to the subfloor or drywall.
Fireplace	An air barrier shall be installed on fireplace walls. Fireplaces shall have gasketed doors.

a. In addition, inspection of log walls shall be in accordance with the provisions of ICC-400.

than 70°F (21°C) and a cooling temperature set point no lower than 78°F (26°C).

N1103.1.2 (R403.1.2) Heat pump supplementary heat (Mandatory). Heat pumps having supplementary electric-resistance heat shall have controls that, except during defrost, prevent supplemental heat operation when the heat pump compressor can meet the heating load.

N1103.2 (R403.2) Ducts. Ducts and air handlers shall be in accordance with Sections N1103.2.1 through N1103.2.3.

N1103.2.1 (R403.2.1) Insulation (Prescriptive). Supply ducts in attics shall be insulated to a minimum of R-8. All other ducts shall be insulated to a minimum of R-6.

Exception: Ducts or portions thereof located completely inside the *building thermal envelope.*

N1103.2.2 (R403.2.2) Sealing (Mandatory). Ducts, air handlers, and filter boxes shall be sealed. Joints and seams shall comply with Section M1601.4.1 of this code.

Exceptions:

1. Air-impermeable spray foam products shall be permitted to be applied without additional joint seals.
2. Where a duct connection is made that is partially inaccessible, three screws or rivets shall be equally spaced on the exposed portion of the joint so as to prevent a hinge effect.
3. Continuously welded and locking-type longitudinal joints and seams in ducts operating at static pressures less than 2 inches of water column (500 Pa) pressure classification shall not require additional closure systems.

Duct tightness shall be verified by either of the following:

1. Postconstruction test: Total leakage shall be less than or equal to 4 cfm (113.3 L/min) per 100 square feet (9.29 m^2) of conditioned floor area when tested at a pressure differential of 0.1 inches w.g. (25 Pa) across the entire system, including the manufacturer's air handler enclosure. All register boots shall be taped or otherwise sealed during the test.
2. Rough-in test: Total leakage shall be less than or equal to 4 cfm (113.3 L/min) per 100 ft^2 (9.29 m^2) of conditioned floor area when tested at a pressure differential of 0.1 inches w.g. (25 Pa) across the system, including the manufacturer's air handler enclosure. All registers shall be taped or otherwise sealed during the test. If the air handler is not installed at the time of the test, total leakage shall be less than or equal to 3 cfm (85 L/min) per 100 square feet (9.29 m^2) of conditioned floor area.

   Exception: The total leakage test is not required for ducts and air handlers located entirely within the building thermal envelope.

N1103.2.2.1 (R403.2.2.1) Sealed air handler. Air handlers shall have a manufacturer's designation for an air leakage of no more than 2 percent of the design air flow rate when tested in accordance with ASHRAE 193.

N1103.2.3 (R403.2.3) Building cavities (Mandatory). Building framing cavities shall not be used as ducts or plenums.

N1103.3 (R403.3) Mechanical system piping insulation (Mandatory). Mechanical system piping capable of carrying fluids above 105°F (41°C) or below 55°F (13°C) shall be insulated to a minimum of R-3.

N1103.3.1 (R403.3.1) Protection of piping insulation. Piping insulation exposed to weather shall be protected from damage, including that caused by sunlight, moisture, equipment maintenance, and wind, and shall provide shielding from solar radiation that can cause degradation of the material. Adhesive tape shall not be permitted.

N1103.4 (R403.4) Service hot water systems. Energy conservation measures for service hot water systems shall be in accordance with Sections N1103.4.1 and N1103.4.2.

N1103.4.1 (R403.4.1) Circulating hot water systems (Mandatory). Circulating hot water systems shall be provided with an automatic or readily *accessible* manual switch that can turn off the hot-water circulating pump when the system is not in use.

N1103.4.2 (R403.4.2) Hot water pipe insulation (Prescriptive). Insulation for hot water pipe with a minimum thermal resistance (*R*-value) of R-3 shall be applied to the following:

1. Piping larger than $^3/_4$-inch nominal diameter.
2. Piping serving more than one dwelling unit.
3. Piping from the water heater to kitchen outlets.
4. Piping located outside the conditioned space.
5. Piping from the water heater to a distribution manifold.
6. Piping located under a floor slab.
7. Buried piping.
8. Supply and return piping in recirculation systems other than demand recirculation systems.
9. Piping with run lengths greater than the maximum run lengths for the nominal pipe diameter given in Table N1103.4.2.

All remaining piping shall be insulated to at least R-3 or meet the run length requirements of Table N1103.4.2.

TABLE N1103.4.2 (R403.4.2)
MAXIMUM RUN LENGTH (feet)[a]

Nominal pipe diameter of largest diameter pipe in the run (inch)	$^3/_8$	$^1/_2$	$^3/_4$	> $^3/_4$
Maximum run length	30	20	10	5

For SI: 1 inch = 25.4 mm, 1 foot = 304.8 mm.

a. Total length of all piping from the distribution manifold or the recirculation loop to a point of use.

N1103.5 (R403.5) Mechanical ventilation (Mandatory). The building shall be provided with ventilation that meets the requirements of Section M1507 of this code or with other approved means of ventilation. Outdoor air intakes and exhausts shall have automatic or gravity dampers that close when the ventilation system is not operating.

N1103.5.1 (R403.5.1) Whole-house mechanical ventilation system fan efficacy. Mechanical ventilation system fans shall meet the efficacy requirements of Table N1103.5.1.

Exception: Where mechanical ventilation fans are integral to tested and listed HVAC equipment, they shall be powered by an electronically commutated motor.

N1103.6 (R403.6) Equipment sizing (Mandatory). Heating and cooling equipment shall be sized in accordance with ACCA Manual S based on building loads calculated in accordance with ACCA Manual J or other *approved* heating and cooling calculation methodologies.

N1103.7 (R403.7) Systems serving multiple dwelling units (Mandatory). Systems serving multiple dwelling units shall comply with Sections C403 and C404 of the IECC—Commercial Provisions in lieu of Secton N1103.

N1103.8 (R403.8) Snow melt system controls (Mandatory). Snow- and ice-melting systems, supplied through energy service to the building, shall include automatic controls capable of shutting off the system when the pavement temperature is above 50°F (10°C), and no precipitation is falling and an automatic or manual control that will allow shutoff when the outdoor temperature is above 40°F (4.8°C).

N1103.9 (R403.9) Pools and inground permanently installed spas (Mandatory). Pools and inground permanently installed spas shall comply with Sections N1103.9.1 through N1103.9.3.

N1103.9.1 (R403.9.1) Heaters. All heaters shall be equipped with a readily *accessible* on-off switch that is mounted outside of the heater to allow shutting off the heater without adjusting the thermostat setting. Gas-fired heaters shall not be equipped with constant burning pilot lights.

N1103.9.2 (R403.9.2) Time switches. Time switches or other control method that can automatically turn off and on heaters and pumps according to a preset schedule shall be installed on all heaters and pumps. Heaters, pumps and motors that have built in timers shall be deemed in compliance with this requirement.

Exceptions:

1. Where public health standards require 24-hour pump operation.
2. Where pumps are required to operate solar-and waste-heat-recovery pool heating systems.

N1103.9.3 (R403.9.3) Covers. Heated pools and inground permanently installed spas shall be provided with a vapor-retardant cover.

Exception: Pools deriving over 70 percent of the energy for heating from site-recovered energy, such as a heat pump or solar energy source computed over an operating season.

SECTION N1104
ELECTRICAL POWER AND LIGHTING SYSTEMS (MANDATORY)

N1104.1 (R404.1) Lighting equipment (Mandatory). A minimum of 75 percent of the lamps in permanently installed lighting fixtures shall be high-efficacy lamps or a minimum of 75 percent of the permanently installed lighting fixtures shall contain only high-efficacy lamps.

Exception: Low-voltage lighting shall not be required to utilize high-efficiency lamps.

N1104.1.1 (R404.1.1) Lighting equipment (Mandatory). Fuel gas lighting systems shall not have continuously burning pilot lights.

SECTION N1105
SIMULATED PERFORMANCE ALTERNATIVE (PERFORMANCE)

N1105.1 (R405.1) Scope. This section establishes criteria for compliance using simulated energy performance analysis. Such analysis shall include heating, cooling, and service water heating energy only.

N1105.2 (R405.2) Mandatory requirements. Compliance with this section requires that the mandatory provisions identified in Section N1101.15 be met. All supply and return ducts not completely inside the *building thermal envelope* shall be insulated to a minimum of R-6.

N1105.3 (R405.3) Performance-based compliance. Compliance based on simulated energy performance requires that a proposed residence (*proposed design)* be shown to have an annual energy cost that is less than or equal to the annual energy cost of the *standard reference design.* Energy prices shall be taken from a source *approved* by the *building official,* such as the Department of Energy, Energy Information Administration's *State Energy Price and Expenditure Report. Building officials* shall be permitted to require time-of-use pricing in energy cost calculations.

Exception: The energy use based on source energy expressed in Btu (J) or Btu per square foot (J/m^2) of *conditioned floor area* shall be permitted to be substituted for

TABLE N1103.5.1 (R403.5.1)
MECHANICAL VENTILATION SYSTEM FAN EFFICACY

FAN LOCATION	AIR FLOW RATE MINIMUM (CFM)	MINIMUM EFFICACY (CFM/WATT)	AIR FLOW RATE MAXIMUM (CFM)
Range hoods	Any	2.8 cfm/watt	Any
In-line fan	Any	2.8 cfm/watt	Any
Bathroom, utility room	10	1.4 cfm/watt	< 90
Bathroom, utility room	90	2.8 cfm/watt	Any

For SI: 1 cubic foot per minute = 28.3 L/min.

the energy cost. The source energy multiplier for electricity shall be 3.16. The source energy multiplier for fuels other than electricity shall be 1.1.

N1105.4 (R405.4) Documentation. Documentation of the software used for the performance design and the parameters for the building shall be in accordance with Sections N1105.4.1 through N1105.4.3.

N1105.4.1 (R405.4.1) Compliance software tools. Documentation verifying that the methods and accuracy of the compliance software tools conform to the provisions of this section shall be provided to the *building official.*

N1105.4.2 (R405.4.2) Compliance report. Compliance software tools shall generate a report that documents that the *proposed design* complies with Secton N1105.3. The compliance documentation shall include the following information:

1. Address or other identification of the residence;
2. An inspection checklist documenting the building component characteristics of the *proposed design* as listed in Table N1105.5.2(1). The inspection checklist shall show results for both the *standard reference design* and the *proposed design,* and shall document all inputs entered by the user necessary to reproduce the results;
3. Name of individual completing the compliance report; and
4. Name and version of the compliance software tool.

Exception: Multiple orientations. When an otherwise identical building model is offered in multiple orientations, compliance for any orientation shall be permitted by documenting that the building meets the performance requirements in each of the four cardinal (north, east, south and west) orientations.

N1105.4.3 (R405.4.3) Additional documentation. The *building official* shall be permitted to require the following documents:

1. Documentation of the building component characteristics of the *standard reference design.*
2. A certification signed by the builder providing the building component characteristics of the *proposed design* as given in Table N1105.5.2(1).
3. Documentation of the actual values used in the software calculations for the *proposed design.*

N1105.5 (R405.5) Calculation procedure. Calculations of the performance design shall be in accordance with Sections N1105.5.1 and N1105.5.2.

N1105.5.1 (R405.5.1) General. Except as specified by this section, the *standard reference design* and *proposed design* shall be configured and analyzed using identical methods and techniques.

N1105.5.2 (R405.5.2) Residence specifications. The *standard reference design* and *proposed design* shall be configured and analyzed as specified by Table N1105.5.2(1). Table N1105.5.2(1) shall include by reference all notes contained in Table N1102.1.1.

N1105.6 (R405.6) Calculation software tools. Calculation software, where used, shall be in accordance with Sections N1105.6.1 through N1105.6.3.

N1105.6.1 (R405.6.1) Minimum capabilities. Calculation procedures used to comply with this section shall be software tools capable of calculating the annual energy consumption of all building elements that differ between the *standard reference design* and the *proposed design* and shall include the following capabilities:

1. Computer generation of the *standard reference design* using only the input for the *proposed design.* The calculation procedure shall not allow the user to directly modify the building component characteristics of the *standard reference design.*
2. Calculation of whole-building (as a single *zone)* sizing for the heating and cooling equipment in the *standard reference design* residence in accordance with Secton N1103.6.
3. Calculations that account for the effects of indoor and outdoor temperatures and part-load ratios on the performance of heating, ventilating and air-conditioning equipment based on climate and equipment sizing.
4. Printed *building official* inspection checklist listing each of the *proposed design* component characteristics from Table N1105.5.2(1) determined by the analysis to provide compliance, along with their respective performance ratings (e.g., *R*-value, *U*-factor, SHGC, HSPF, AFUE, SEER, EF, etc.).

N1105.6.2 (R405.6.2) Specific approval. Performance analysis tools meeting the applicable sections of Secton N1105 shall be permitted to be *approved.* Tools are permitted to be *approved* based on meeting a specified threshold for a jurisdiction. The *building official* shall be permitted to approve tools for a specified application or limited scope.

N1105.6.3 (R405.6.3) Input values. When calculations require input values not specified by Sections N1102, N1103, N1104 and N1105, those input values shall be taken from an *approved* source.

TABLE N1105.5.2(1) [R405.5.2(1)]
SPECIFICATIONS FOR THE STANDARD REFERENCE AND PROPOSED DESIGNS

BUILDING COMPONENT	STANDARD REFERENCE DESIGN	PROPOSED DESIGN
Above-grade walls	Type: mass wall if proposed wall is mass; otherwise wood frame. Gross area: same as proposed *U*-factor: from Table N1102.1.3 Solar absorptance = 0.75 Remittance = 0.90	As proposed As proposed As proposed As proposed As proposed
Basement and crawl space walls	Type: same as proposed Gross area: same as proposed *U*-factor: from Table N1102.1.3, with insulation layer on interior side of walls.	As proposed As proposed As proposed
Above-grade floors	Type: wood frame Gross area: same as proposed *U*-factor: from Table N1102.1.3	As proposed As proposed As proposed
Ceilings	Type: wood frame Gross area: same as proposed *U*-factor: from Table N1102.1.3	As proposed As proposed As proposed
Roofs	Type: composition shingle on wood sheathing Gross area: same as proposed Solar absorptance = 0.75 Emittance = 0.90	As proposed As proposed As proposed As proposed
Attics	Type: vented with aperture = 1 ft^2 per 300 ft^2 ceiling area	As proposed
Foundations	Type: same as proposed foundation wall area above and below grade and soil characteristics: same as proposed.	As proposed As proposed
Doors	Area: 40 ft^2 Orientation: North *U*-factor: same as fenestration from Table N1102.1.3.	As proposed As proposed As proposed
Glazing[a]	Total area[b] = (a) The proposed glazing area; where proposed glazing area is less than 15% of the conditioned floor area. (b) 15% of the conditioned floor area; where the proposed glazing area is 15% or more of the conditioned floor area. Orientation: equally distributed to four cardinal compass orientations (N, E, S & W). *U*-factor: from Table N1102.1.3 SHGC: From Table N1102.1.1 except that for climates with no requirement (NR) SHGC = 0.40 shall be used. Interior shade fraction: 0.92-(0.21 × SHGC for the standard reference design) External shading: none	As proposed As proposed As proposed As proposed 0.92-(0.21 × SHGC as proposed) As proposed
Skylights	None	As proposed
Thermally isolated sunrooms	None	As proposed

(continued)

TABLE N1105.5.2(1) [R405.5.2(1)]—continued
SPECIFICATIONS FOR THE STANDARD REFERENCE AND PROPOSED DESIGNS

BUILDING COMPONENT	STANDARD REFERENCE DESIGN	PROPOSED DESIGN
Air exchange rate	Air leakage rate of 5 air changes per hour in Zones 1 and 2, and 3 air changes per hour in Zones 3 through 8 at a pressure of 0.2 inches w.g (50 Pa). The mechanical ventilation rate shall be in addition to the air leakage rate and the same as in the proposed design, but no greater than $0.01 \times CFA + 7.5 \times (N_{br} + 1)$ where: CFA = conditioned floor area N_{br} = number of bedrooms Energy recovery shall not be assumed for mechanical ventilation.	For residences that are not tested, the same air leakage rate as the standard reference design. For tested residences, the measured air exchange rate[c]. The mechanical ventilation rate[d] shall be in addition to the air leakage rate and shall be as proposed.
Mechanical ventilation	None, except where mechanical ventilation is specified by the proposed design, in which case: Annual vent fan energy use: $\text{kWh/yr} = 0.03942 \times CFA + 29.565 \times (N_{br} + 1)$ where: CFA = conditioned floor area N_{br} = number of bedrooms	As proposed
Internal gains	$\text{IGain} = 17{,}900 + 23.8 \times CFA + 4104 \times N_{br}$ (Btu/day per dwelling unit)	Same as standard reference design.
Internal mass	An internal mass for furniture and contents of 8 pounds per square foot of floor area.	Same as standard reference design, plus any additional mass specifically designed as a thermal storage element[e] but not integral to the building envelope or structure.
Structural mass	For masonry floor slabs, 80% of floor area covered by R-2 carpet and pad, and 20% of floor directly exposed to room air. For masonry basement walls, as proposed, but with insulation required by Table 402.1.3 located on the interior side of the walls. For other walls, for ceilings, floors, and interior walls, wood frame construction.	As proposed As proposed As proposed
Heating systems[f, g]	As proposed for other than electric heating without a heat pump. Where the proposed design utilizes electric heating without a heat pump the standard reference design shall be an air source heat pump meeting the requirements of Section C403 of the International Energy Conservation Code—Commercial Provisions. Capacity: sized in accordance with Secton N1103.6	As proposed
Cooling systems[f, h]	As proposed Capacity: sized in accordance with Secton N1103.6.	As proposed
Service water Heating[f, g, h, i]	As proposed Use: same as proposed design	As proposed $\text{gal/day} = 30 + (10 \times N_{br})$
Thermal distribution systems	None	Thermal distribution system efficiency shall be as tested or as specified in Table N1105.5.2(2) if not tested. Duct insulation shall be as proposed.
Thermostat	Type: Manual, cooling temperature setpoint = 75°F; Heating temperature setpoint = 72°F	Same as standard reference

(continued)

TABLE N1105.5.2(1) [R405.5.2(1)]—continued
SPECIFICATIONS FOR THE STANDARD REFERENCE AND PROPOSED DESIGNS

For SI: 1 square foot = 0.93 m^2, 1 British thermal unit = 1055 J, 1 pound per square foot = 4.88 kg/m^2, 1 gallon (U.S.) = 3.785 L, °C = (°F-3)/1.8, 1 degree = 0.79 rad, 1 inch water gauge = 1250 Pa.

a. Glazing shall be defined as sunlight-transmitting fenestration, including the area of sash, curbing or other framing elements, that enclose conditioned space. Glazing includes the area of sunlight-transmitting fenestration assemblies in walls bounding conditioned basements. For doors where the sunlight-transmitting opening is less than 50 percent of the door area, the glazing area is the sunlight transmitting opening area. For all other doors, the glazing area is the rough frame opening area for the door including the door and the frame.

b. For residences with conditioned basements, R-2 and R-4 residences and townhouses, the following formula shall be used to determine glazing area:

$AF = A_s \times FA \times F$

where:

AF = Total glazing area.

A_s = Standard reference design total glazing area.

FA = (Above-grade thermal boundary gross wall area)/(above-grade boundary wall area + 0.5 × below-grade boundary wall area).

F = (Above-grade thermal boundary wall area)/(above-grade thermal boundary wall area + common wall area) or 0.56, whichever is greater.

and where:

Thermal boundary wall is any wall that separates conditioned space from unconditioned space or ambient conditions.

Above-grade thermal boundary wall is any thermal boundary wall component not in contact with soil.

Below-grade boundary wall is any thermal boundary wall in soil contact.

Common wall area is the area of walls shared with an adjoining dwelling unit.

L and *CFA* are in the same units.

c. Where required by the *building official,* testing shall be conducted by an *approved* party. Hourly calculations as specified in the ASHRAE *Handbook of Fundamentals,* or the equivalent shall be used to determine the energy loads resulting from infiltration.

d. The combined air exchange rate for infiltration and mechanical ventilation shall be determined in accordance with Equation 43 of 2001 ASHRAE *Handbook of Fundamentals,* page 26.24 and the "Whole-house Ventilation" provisions of 2001 ASHRAE *Handbook of Fundamentals*, page 26.19 for intermittent mechanical ventilation.

e. Thermal storage element shall mean a component not part of the floors, walls or ceilings that is part of a passive solar system, and that provides thermal storage such as enclosed water columns, rock beds, or phase-change containers. A thermal storage element must be in the same room as fenestration that faces within 15 degrees (0.26 rad) of true south, or must be connected to such a room with pipes or ducts that allow the element to be actively charged.

f. For a proposed design with multiple heating, cooling or water heating systems using different fuel types, the applicable standard reference design system capacities and fuel types shall be weighted in accordance with their respective loads as calculated by accepted engineering practice for each equipment and fuel type present.

g. For a proposed design without a proposed heating system, a heating system with the prevailing federal minimum efficiency shall be assumed for both the standard reference design and proposed design.

h. For a proposed design home without a proposed cooling system, an electric air conditioner with the prevailing federal minimum efficiency shall be assumed for both the standard reference design and the proposed design.

i. For a proposed design with a nonstorage-type water heater, a 40-gallon storage-type water heater with the prevailing federal minimum energy factor for the same fuel as the predominant heating fuel type shall be assumed. For the case of a proposed design without a proposed water heater, a 40-gallon storage-type water heater with the prevailing federal minimum efficiency for the same fuel as the predominant heating fuel type shall be assumed for both the proposed design and standard reference design.

TABLE N1105.5.2(2) [R405.5.2(2)]
DEFAULT DISTRIBUTION SYSTEM EFFICIENCIES FOR PROPOSED DESIGNS[a]

DISTRIBUTION SYSTEM CONFIGURATION AND CONDITION	FORCED AIR SYSTEMS	HYDRONIC SYSTEMS[b]
Distribution system components located in unconditioned space	—	0.95
Untested distribution systems entirely located in conditioned space[c]	0.88	1
"Ductless" systems[d]	1	—

For SI: 1 cubic foot per minute = 0.47 L/s, 1 square foot = 0.093m^2, 1 pound per square inch = 6895 Pa, 1 inch water gauge = 1250 Pa.

a. Default values given by this table are for untested distribution systems, which must still meet minimum requirements for duct system insulation.

b. Hydronic systems shall mean those systems that distribute heating and cooling energy directly to individual spaces using liquids pumped through closed-loop piping and that do not depend on ducted, forced airflow to maintain space temperatures.

c. Entire system in conditioned space shall mean that no component of the distribution system, including the air handler unit, is located outside of the conditioned space.

d. Ductless systems shall be allowed to have forced airflow across a coil but shall not have any ducted airflow external to the manufacturer's air handler enclosure.

Part V—Mechanical

CHAPTER 12
MECHANICAL ADMINISTRATION

SECTION M1201
GENERAL

M1201.1 Scope. The provisions of Chapters 12 through 24 shall regulate the design, installation, maintenance, *alteration* and inspection of mechanical systems that are permanently installed and used to control environmental conditions within buildings. These chapters shall also regulate those mechanical systems, system components, *equipment* and *appliances* specifically addressed in this code.

M1201.2 Application. In addition to the general administration requirements of Chapter 1, the administrative provisions of this chapter shall also apply to the mechanical requirements of Chapters 13 through 24.

SECTION M1202
EXISTING MECHANICAL SYSTEMS

M1202.1 Additions, alterations or repairs. *Additions*, *alterations*, renovations or repairs to a mechanical system shall conform to the requirements for a new mechanical system without requiring the existing mechanical system to comply with all of the requirements of this code. *Additions, alterations* or repairs shall not cause an existing mechanical system to become unsafe, hazardous or overloaded. Minor *additions*, *alterations* or repairs to existing mechanical systems shall meet the provisions for new construction, unless such work is done in the same manner and arrangement as was in the existing system, is not hazardous, and is *approved*.

M1202.2 Existing installations. Except as otherwise provided for in this code, a provision in this code shall not require the removal, *alteration* or abandonment of, nor prevent the continued use and maintenance of, an existing mechanical system lawfully in existence at the time of the adoption of this code.

M1202.3 Maintenance. Mechanical systems, both existing and new, and parts thereof shall be maintained in proper operating condition in accordance with the original design and in a safe and sanitary condition. Devices or safeguards that are required by this code shall be maintained in compliance with the code edition under which installed. The owner or the owner's designated agent shall be responsible for maintenance of the mechanical systems. To determine compliance with this provision, the *building official* shall have the authority to require a mechanical system to be reinspected.

CHAPTER 13

GENERAL MECHANICAL SYSTEM REQUIREMENTS

SECTION M1301 GENERAL

M1301.1 Scope. The provisions of this chapter shall govern the installation of mechanical systems not specifically covered in other chapters applicable to mechanical systems. Installations of mechanical *appliances, equipment* and systems not addressed by this code shall comply with the applicable provisions of the *International Mechanical Code* and the *International Fuel Gas Code.*

M1301.1.1 Flood-resistant installation. In flood hazard areas as established by Table R301.2(1), mechanical *appliances, equipment* and systems shall be located or installed in accordance with Section R322.1.6.

M1301.2 Identification. Each length of pipe and tubing and each pipe fitting utilized in a mechanical system shall bear the identification of the manufacturer.

M1301.3 Installation of materials. All materials used shall be installed in strict accordance with the standards under which the materials are accepted and approved. In the absence of such installation procedures, the manufacturer's instructions shall be followed. Where the requirements of referenced standards or manufacturer's instructions do not conform to minimum provisions of this code, the provisions of this code shall apply.

M1301.4 Plastic pipe, fittings and components. Plastic pipe, fittings and components shall be third-party certified as conforming to NSF 14.

M1301.5 Third-party testing and certification. Piping, tubing and fittings shall comply with the applicable referenced standards, specifications and performance criteria of this code and shall be identified in accordance with Section M1301.2. Piping, tubing and fittings shall either be tested by an approved third-party testing agency or certified by an approved third-party certification agency.

SECTION M1302 APPROVAL

M1302.1 Listed and labeled. *Appliances* regulated by this code shall be *listed* and *labeled* for the application in which they are installed and used, unless otherwise *approved* in accordance with Section R104.11.

SECTION M1303 LABELING OF APPLIANCES

M1303.1 Label information. A permanent factory-applied nameplate(s) shall be affixed to *appliances* on which shall appear, in legible lettering, the manufacturer's name or trademark, the model number, a serial number and the seal or *mark* of the testing agency. A *label* shall also include the following:

1. Electrical *appliances*. Electrical rating in volts, amperes and motor phase; identification of individual electrical components in volts, amperes or watts and motor phase; and in Btu/h (W) output and required clearances.
2. Absorption units. Hourly rating in Btu/h (W), minimum hourly rating for units having step or automatic modulating controls, type of fuel, type of refrigerant, cooling capacity in Btu/h (W) and required clearances.
3. Fuel-burning units. Hourly rating in Btu/h (W), type of fuel *approved* for use with the *appliance* and required clearances.
4. Electric comfort-heating appliances. The electric rating in volts, amperes and phase; Btu/h (W) output rating; individual marking for each electrical component in amperes or watts, volts and phase; and required clearances from combustibles.
5. Maintenance instructions. Required regular maintenance actions and title or publication number for the operation and maintenance manual for that particular model and type of product.

SECTION M1304 TYPE OF FUEL

M1304.1 Fuel types. Fuel-fired *appliances* shall be designed for use with the type of fuel to which they will be connected and the altitude at which they are installed. *Appliances* that comprise parts of the building mechanical system shall not be converted for the use of a different fuel, except where *approved* and converted in accordance with the manufacturer's instructions. The fuel input rate shall not be increased or decreased beyond the limit rating for the altitude at which the *appliance* is installed.

SECTION M1305 APPLIANCE ACCESS

M1305.1 Appliance access for inspection service, repair and replacement. *Appliances* shall be accessible for inspection, service, repair and replacement without removing permanent construction, other *appliances*, or any other piping or ducts not connected to the *appliance* being inspected, serviced, repaired or replaced. A level working space at least 30 inches deep and 30 inches wide (762 mm by 762 mm) shall be provided in front of the control side to service an *appliance*. Installation of room heaters shall be permitted with at least an 18-inch (457 mm) working space. A platform shall not be required for room heaters.

M1305.1.1 Furnaces and air handlers. Furnaces and air handlers within compartments or alcoves shall have a minimum working space clearance of 3 inches (76 mm) along the sides, back and top with a total width of the enclosing space being at least 12 inches (305 mm) wider than the furnace or air handler. Furnaces having a firebox open to the atmosphere shall have at least a 6-inch (152 mm) working space along the front combustion chamber side. Combustion air openings at the rear or side of the compartment shall comply with the requirements of Chapter 17.

Exception: This section shall not apply to replacement *appliances* installed in existing compartments and alcoves where the working space clearances are in accordance with the *equipment* or *appliance* manufacturer's installation instructions.

M1305.1.2 Appliances in rooms. *Appliances* installed in a compartment, alcove, *basement* or similar space shall be accessed by an opening or door and an unobstructed passageway measuring not less than 24 inches (610 mm) wide and large enough to allow removal of the largest *appliance* in the space, provided there is a level service space of not less than 30 inches (762 mm) deep and the height of the *appliance*, but not less than 30 inches (762 mm), at the front or service side of the *appliance* with the door open.

M1305.1.3 Appliances in attics. *Attics* containing *appliances* shall be provided with an opening and a clear and unobstructed passageway large enough to allow removal of the largest *appliance*, but not less than 30 inches (762 mm) high and 22 inches (559 mm) wide and not more than 20 feet (6096 mm) long measured along the centerline of the passageway from the opening to the *appliance*. The passageway shall have continuous solid flooring in accordance with Chapter 5 not less than 24 inches (610 mm) wide. A level service space at least 30 inches (762 mm) deep and 30 inches (762 mm) wide shall be present along all sides of the *appliance* where access is required. The clear access opening dimensions shall be a minimum of 20 inches by 30 inches (508 mm by 762 mm), and large enough to allow removal of the largest appliance.

Exceptions:

1. The passageway and level service space are not required where the *appliance* can be serviced and removed through the required opening.
2. Where the passageway is unobstructed and not less than 6 feet (1829 mm) high and 22 inches (559 mm) wide for its entire length, the passageway shall be not more than 50 feet (15 250 mm) long.

M1305.1.3.1 Electrical requirements. A luminaire controlled by a switch located at the required passageway opening and a receptacle outlet shall be installed at or near the *appliance* location in accordance with Chapter 39.

M1305.1.4 Appliances under floors. Underfloor spaces containing *appliances* shall be provided with an unobstructed passageway large enough to remove the largest *appliance*, but not less than 30 inches (762 mm) high and 22 inches (559 mm) wide, nor more than 20 feet (6096 mm) long measured along the centerline of the passageway from the opening to the *appliance*. A level service space at least 30 inches (762 mm) deep and 30 inches (762 mm) wide shall be present at the front or service side of the *appliance*. If the depth of the passageway or the service space exceeds 12 inches (305 mm) below the adjoining grade, the walls of the passageway shall be lined with concrete or masonry extending 4 inches (102 mm) above the adjoining grade in accordance with Chapter 4. The rough-framed access opening dimensions shall be a minimum of 22 inches by 30 inches (559 mm by 762 mm), and large enough to remove the largest *appliance*.

Exceptions:

1. The passageway is not required where the level service space is present when the access is open, and the *appliance* can be serviced and removed through the required opening.
2. Where the passageway is unobstructed and not less than 6 feet high (1929 mm) and 22 inches (559 mm) wide for its entire length, the passageway shall not be limited in length.

M1305.1.4.1 Ground clearance. *Equipment* and *appliances* supported from the ground shall be level and firmly supported on a concrete slab or other *approved* material extending not less than 3 inches (76 mm) above the adjoining ground. Such support shall be in accordance with the manufacturer's installation instructions. *Appliances* suspended from the floor shall have a clearance of not less than 6 inches (152 mm) from the ground.

M1305.1.4.2 Excavations. Excavations for *appliance* installations shall extend to a depth of 6 inches (152 mm) below the *appliance* and 12 inches (305 mm) on all sides, except that the control side shall have a clearance of 30 inches (762 mm).

M1305.1.4.3 Electrical requirements. A luminaire controlled by a switch located at the required passageway opening and a receptacle outlet shall be installed at or near the *appliance* location in accordance with Chapter 39.

SECTION M1306 CLEARANCES FROM COMBUSTIBLE CONSTRUCTION

M1306.1 Appliance clearance. *Appliances* shall be installed with the clearances from unprotected combustible materials as indicated on the *appliance label* and in the manufacturer's installation instructions.

M1306.2 Clearance reduction. Reduction of clearances shall be in accordance with the *appliance* manufacturer's instructions and Table M1306.2. Forms of protection with ventilated air space shall conform to the following requirements:

1. Not less than 1-inch (25 mm) air space shall be provided between the protection and combustible wall surface.
2. Air circulation shall be provided by having edges of the wall protection open at least 1 inch (25 mm).
3. If the wall protection is mounted on a single flat wall away from corners, air circulation shall be provided by having the bottom and top edges, or the side and top edges open at least 1 inch (25 mm).
4. Wall protection covering two walls in a corner shall be open at the bottom and top edges at least 1 inch (25 mm).

M1306.2.1 Solid-fuel appliances. Table M1306.2 shall not be used to reduce the clearance required for solid-fuel *appliances* listed for installation with minimum clearances of 12 inches (305 mm) or less. For *appliances listed* for installation with minimum clearances greater than 12 inches (305 mm), Table M1306.2 shall not be used to reduce the clearance to less than 12 inches (305 mm).

SECTION M1307 APPLIANCE INSTALLATION

M1307.1 General. Installation of *appliances* shall conform to the conditions of their *listing* and *label* and the manufacturer's installation instructions. The manufacturer's operating and installation instructions shall remain attached to the *appliance*.

M1307.2 Anchorage of appliances. *Appliances* designed to be fixed in position shall be fastened or anchored in an *approved* manner. In Seismic Design Categories D_1 and D_2, water heaters shall be anchored or strapped to resist horizontal displacement caused by earthquake motion. Strapping shall be at points within the upper one-third and lower one-third of the *appliance's* vertical dimensions. At the lower point, the strapping shall maintain a minimum distance of 4 inches (102 mm) above the controls.

M1307.3 Elevation of ignition source. *Appliances* having an *ignition source* shall be elevated such that the source of ignition is not less than 18 inches (457 mm) above the floor in garages. For the purpose of this section, rooms or spaces that are not part of the living space of a *dwelling unit* and that communicate with a private garage through openings shall be considered to be part of the garage.

Exception: Elevation of the ignition source is not required for appliances that are listed as flammable vapor ignition resistant.

M1307.3.1 Protection from impact. *Appliances* shall not be installed in a location subject to vehicle damage except where protected by *approved* barriers.

M1307.4 Hydrogen generating and refueling operations. *Ventilation* shall be required in accordance with Section M1307.4.1, M1307.4.2 or M1307.4.3 in private garages that contain hydrogen-generating *appliances* or refueling systems. For the purpose of this section, rooms or spaces that are not part of the living space of a *dwelling unit* and that communicate directly with a private garage through openings shall be considered to be part of the private garage.

M1307.4.1 Natural ventilation. Indoor locations intended for hydrogen-generating or refueling operations shall be limited to a maximum floor area of 850 square feet (79 m^2) and shall communicate with the outdoors in accordance with Sections M1307.4.1.1 and M1307.4.1.2. The maximum rated output capacity of hydrogen-generating *appliances* shall not exceed 4 standard cubic feet per minute (1.9 L/s) of hydrogen for each 250 square feet (23 m^2) of floor area in such spaces. The minimum cross-sectional dimension of air openings shall be 3 inches (76 mm). Where ducts are used, they shall be of the same cross-sectional area as the free area of the openings to which they connect. In those locations, *equipment* and *appliances* having an *ignition source* shall be located so that the source of ignition is not within 12 inches (305 mm) of the ceiling.

M1307.4.1.1 Two openings. Two permanent openings shall be constructed within the garage. The upper opening shall be located entirely within 12 inches (305 mm) of the ceiling of the garage. The lower opening shall be located entirely within 12 inches (305 mm) of the floor of the garage. Both openings shall be constructed in the same exterior wall. The openings shall communicate directly with the outdoors and shall have a minimum free area of $^1/_2$ square foot per 1,000 cubic feet (1.7 m^2/1000 m^3) of garage volume.

M1307.4.1.2 Louvers and grilles. In calculating free area required by Section M1307.4.1, the required size of openings shall be based on the net free area of each opening. If the free area through a design of louver or grille is known, it shall be used in calculating the size opening required to provide the free area specified. If the design and free area are not known, it shall be assumed that wood louvers will have a 25-percent free area and metal louvers and grilles will have a 75-percent free area. Louvers and grilles shall be fixed in the open position.

M1307.4.2 Mechanical ventilation. Indoor locations intended for hydrogen-generating or refueling operations shall be ventilated in accordance with Section 502.16 of the *International Mechanical Code*. In these locations, *equipment* and *appliances* having an *ignition source* shall be located so that the source of ignition is below the mechanical *ventilation* outlet(s).

M1307.4.3 Specially engineered installations. As an alternative to the provisions of Sections M1307.4.1 and M1307.4.2, the necessary supply of air for *ventilation* and dilution of flammable gases shall be provided by an *approved* engineered system.

M1307.5 Electrical appliances. Electrical *appliances* shall be installed in accordance with Chapters 14, 15, 19, 20 and 34 through 43 of this code.

M1307.6 Plumbing connections. Potable water and drainage system connections to *equipment* and *appliances* regulated by this code shall be in accordance with Chapters 29 and 30.

TABLE M1306.2
REDUCTION OF CLEARANCES WITH SPECIFIED FORMS OF PROTECTION[a, c, d, e, f, g, h, i, j, k, l]

TYPE OF PROTECTION APPLIED TO AND COVERING ALL SURFACES OF COMBUSTIBLE MATERIAL WITHIN THE DISTANCE SPECIFIED AS THE REQUIRED CLEARANCE WITH NO PROTECTION (See Figures M1306.1 and M1306.2)	WHERE THE REQUIRED CLEARANCE WITH NO PROTECTION FROM APPLIANCE, VENT CONNECTOR, OR SINGLE WALL METAL PIPE IS:									
	36 inches		18 inches		12 inches		9 inches		6 inches	
	Allowable clearances with specified protection (Inches)[b]									
	Use column 1 for clearances above an appliance or horizontal connector. Use column 2 for clearances from an appliance, vertical connector and single-wall metal pipe.									
	Above column 1	Sides and rear column 2	Above column 1	Sides and rear column 2	Above column 1	Sides and rear column 2	Above column 1	Sides and rear column 2	Above column 1	Sides and rear column 2
$3^1/_2$-inch-thick masonry wall without ventilated air space	—	24	—	12	—	9	—	6	—	5
$^1/_2$-inch insulation board over 1-inch glass fiber or mineral wool batts	24	18	12	9	9	6	6	5	4	3
Galvanized sheet steel having a minimum thickness of 0.0236-inch (No. 24 gage) over 1-inch glass fiber or mineral wool batts reinforced with wire or rear face with a ventilated air space	18	12	9	6	6	4	5	3	3	3
$3^1/_2$-inch-thick masonry wall with ventilated air space	—	12	—	6	—	6	—	6	—	6
Galvanized sheet steel having a minimum thickness of 0.0236-inch (No. 24 gage) with a ventilated air space 1-inch off the combustible assembly	18	12	9	6	6	4	5	3	3	2
$^1/_2$-inch-thick insulation board with ventilated air space	18	12	9	6	6	4	5	3	3	3
Galvanized sheet steel having a minimum thickness of 0.0236-inch (No. 24 gage) with ventilated air space over 24 gage sheet steel with a ventilated space	18	12	9	6	6	4	5	3	3	3
1-inch glass fiber or mineral wool batts sandwiched between two sheets of galvanized sheet steel having a minimum thickness of 0.0236-inch (No. 24 gage) with a ventilated air space	18	12	9	6	6	4	5	3	3	3

For SI: 1 inch = 25.4 mm, 1 pound per cubic foot = 16.019 kg/m^3, °C = [(°F)-32/1.8], 1 Btu/(h × ft^2 × °F/in.) = 0.001442299 (W/cm^2 × °C/cm).

a. Reduction of clearances from combustible materials shall not interfere with combustion air, draft hood clearance and relief, and accessibility of servicing.

b. Clearances shall be measured from the surface of the heat producing appliance or equipment to the outer surface of the combustible material or combustible assembly.

c. Spacers and ties shall be of noncombustible material. No spacer or tie shall be used directly opposite appliance or connector.

d. Where all clearance reduction systems use a ventilated air space, adequate provision for air circulation shall be provided as described. (See Figures M1306.1 and M1306.2.)

e. There shall be at least 1 inch between clearance reduction systems and combustible walls and ceilings for reduction systems using ventilated air space.

f. If a wall protector is mounted on a single flat wall away from corners, adequate air circulation shall be permitted to be provided by leaving only the bottom and top edges or only the side and top edges open with at least a 1-inch air gap.

g. Mineral wool and glass fiber batts (blanket or board) shall have a minimum density of 8 pounds per cubic foot and a minimum melting point of 1,500°F.

h. Insulation material used as part of a clearance reduction system shall have a thermal conductivity of 1.0 Btu inch per square foot per hour °F or less. Insulation board shall be formed of noncombustible material.

i. There shall be at least 1 inch between the appliance and the protector. In no case shall the clearance between the appliance and the combustible surface be reduced below that allowed in this table.

j. All clearances and thicknesses are minimum; larger clearances and thicknesses are acceptable.

k. Listed single-wall connectors shall be permitted to be installed in accordance with the terms of their listing and the manufacturer's instructions.

l. For limitations on clearance reduction for solid-fuel-burning appliances see Section M1306.2.1.

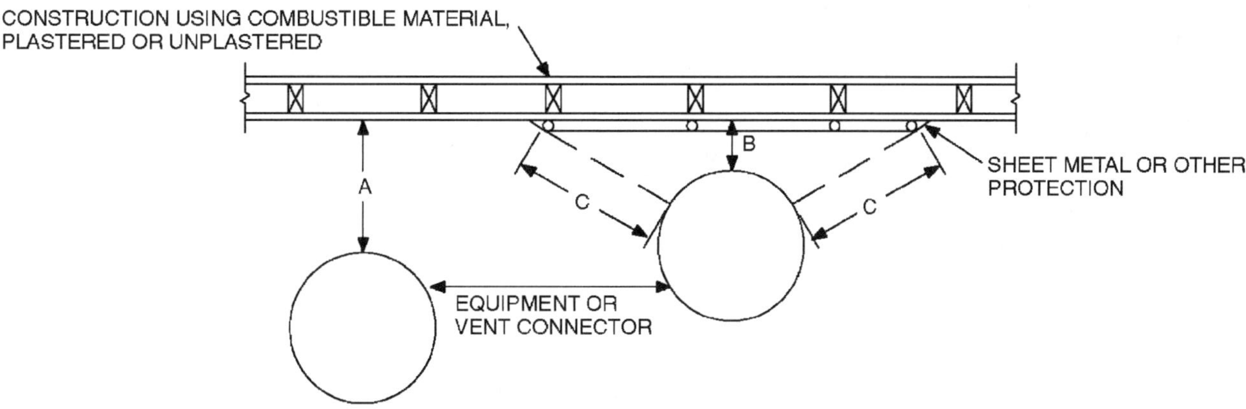

Note: "A" equals the required clearance with no protection. "B" equals the reduced clearance permitted in accordance with Table M1306.2. The protection applied to the construction using combustible material shall extend far enough in each direction to make "C" equal to "A."

FIGURE M1306.1
REDUCED CLEARANCE DIAGRAM

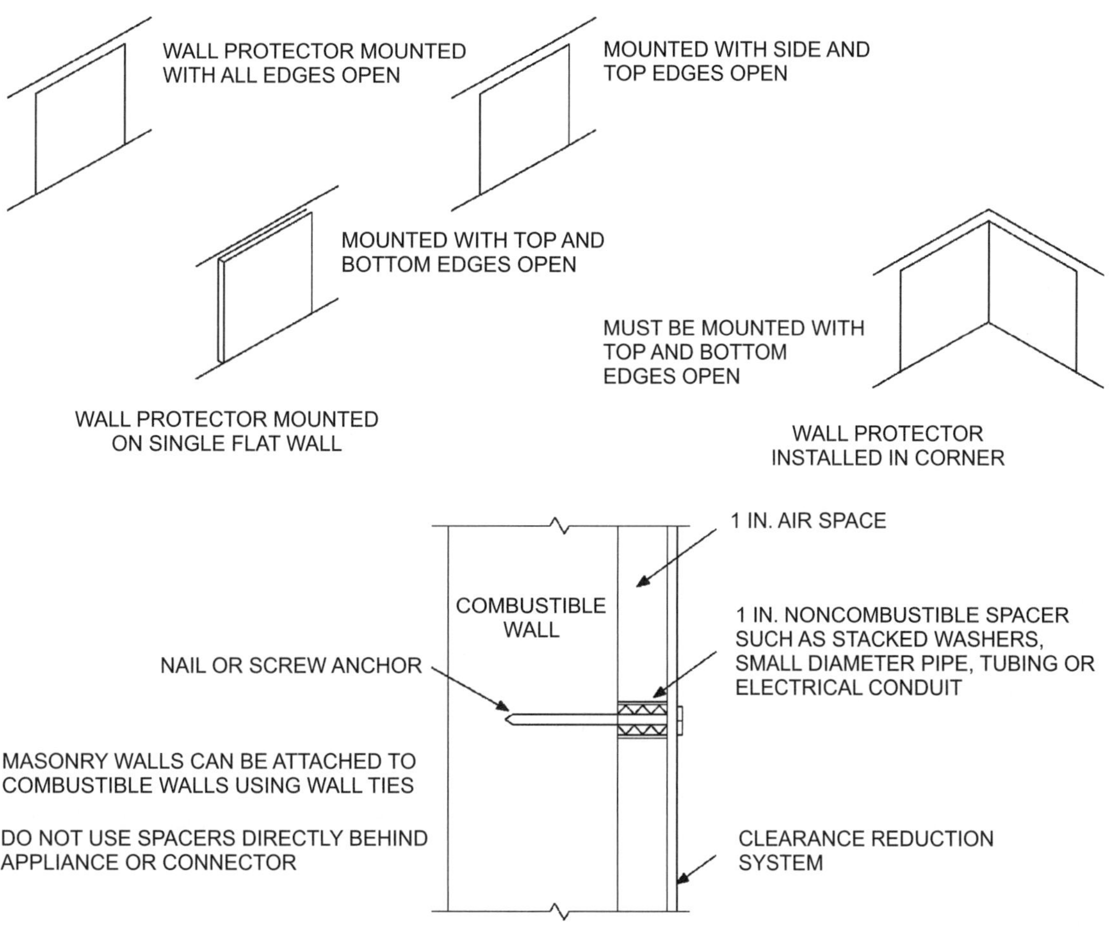

For SI: 1 inch = 25.4 mm.

FIGURE M1306.2
WALL PROTECTOR CLEARANCE REDUCTION SYSTEM

SECTION M1308
MECHANICAL SYSTEMS INSTALLATION

M1308.1 Drilling and notching. Wood-framed structural members shall be drilled, notched or altered in accordance with the provisions of Sections R502.8, R602.6, R602.6.1 and R802.7. Holes in load-bearing members of cold-formed steel light-frame construction shall be permitted only in accordance with Sections R505.2.5, R603.2.5 and R804.2.5. In accordance with the provisions of Sections R505.3.5, R603.3.4 and R804.3.4, cutting and notching of flanges and lips of load-bearing members of cold-formed steel light frame construction shall not be permitted. Structural insulated panels (SIPs) shall be drilled and notched or altered in accordance with the provisions of Section R613.7.

M1308.2 Protection against physical damage. In concealed locations where piping, other than cast-iron or galvanized steel, is installed through holes or notches in studs, joists, rafters or similar members less than 1.5 inches (38 mm) from the nearest edge of the member, the pipe shall be protected by shield plates. Protective steel shield plates having a minimum thickness of 0.0575-inch (1.463 mm) (No. 16 gage), shall cover the area of the pipe where the member is notched or bored, and shall extend a minimum of 2 inches (51 mm) above sole plates and below top plates.

CHAPTER 14

HEATING AND COOLING EQUIPMENT AND APPLIANCES

SECTION M1401 GENERAL

M1401.1 Installation. Heating and cooling *equipment* and *appliances* shall be installed in accordance with the manufacturer's installation instructions and the requirements of this code.

M1401.2 Access. Heating and cooling *equipment* and appliances shall be located with respect to building construction and other *equipment* and appliances to permit maintenance, servicing and replacement. Clearances shall be maintained to permit cleaning of heating and cooling surfaces; replacement of filters, blowers, motors, controls and vent connections; lubrication of moving parts; and adjustments.

> **Exception:** Access shall not be required for ducts, piping, or other components approved for concealment.

M1401.3 Sizing. Heating and cooling *equipment* and *appliances* shall be sized in accordance with ACCA Manual S based on building loads calculated in accordance with ACCA Manual J or other *approved* heating and cooling calculation methodologies.

M1401.4 Exterior installations. *Equipment* and *appliances* installed outdoors shall be *listed* and *labeled* for outdoor installation. Supports and foundations shall prevent excessive vibration, settlement or movement of the *equipment*. Supports and foundations shall be in accordance with Section M1305.1.4.1.

M1401.5 Flood hazard. In flood hazard areas as established by Table R301.2(1), heating and cooling *equipment* and *appliances* shall be located or installed in accordance with Section R322.1.6.

SECTION M1402 CENTRAL FURNACES

M1402.1 General. Oil-fired central furnaces shall conform to ANSI/UL 727. Electric furnaces shall conform to UL 1995.

M1402.2 Clearances. Clearances shall be provided in accordance with the *listing* and the manufacturer's installation instructions.

M1402.3 Combustion air. *Combustion air* shall be supplied in accordance with Chapter 17. *Combustion air* openings shall be unobstructed for a distance of not less than 6 inches (152 mm) in front of the openings.

SECTION M1403 HEAT PUMP EQUIPMENT

M1403.1 Heat pumps. The minimum unobstructed total area of the outside and return air ducts or openings to a heat pump shall be not less than 6 square inches per 1,000 Btu/h (13 208 mm^2/kW) output rating or as indicated by the conditions of the listing of the heat pump. Electric heat pumps shall conform to UL 1995.

M1403.2 Foundations and supports. Supports and foundations for the outdoor unit of a heat pump shall be raised at least 3 inches (76 mm) above the ground to permit free drainage of defrost water, and shall conform to the manufacturer's installation instructions.

SECTION M1404 REFRIGERATION COOLING EQUIPMENT

M1404.1 Compliance. Refrigeration cooling *equipment* shall comply with Section M1411.

SECTION M1405 BASEBOARD CONVECTORS

M1405.1 General. Electric baseboard convectors shall be installed in accordance with the manufacturer's installation instructions and Chapters 34 through 43 of this code. Electric baseboard heaters shall be listed and labeled in accordance with UL 1042.

SECTION M1406 RADIANT HEATING SYSTEMS

M1406.1 General. Electric radiant heating systems shall be installed in accordance with the manufacturer's installation instructions and Chapters 34 through 43 of this code and shall be listed for the application.

M1406.2 Clearances. Clearances for radiant heating panels or elements to any wiring, outlet boxes and junction boxes used for installing electrical devices or mounting luminaires shall comply with Chapters 34 through 43 of this code.

M1406.3 Installation of radiant panels. Radiant panels installed on wood framing shall conform to the following requirements:

1. Heating panels shall be installed parallel to framing members and secured to the surface of framing members or mounted between framing members.
2. Mechanical fasteners shall penetrate only the unheated portions provided for this purpose. Panels shall not be fastened at any point closer than $^1/_4$ inch (6.4 mm) to an element. Other methods of attachment of the panels shall be in accordance with the panel manufacturer's instructions.
3. Unless *listed* and *labeled* for field cutting, heating panels shall be installed as complete units.

M1406.4 Installation in concrete or masonry. Radiant heating systems installed in concrete or masonry shall conform to the following requirements:

1. Radiant heating systems shall be identified as being suitable for the installation, and shall be secured in place as specified in the manufacturer's installation instructions.
2. Radiant heating panels or radiant heating panel sets shall not be installed where they bridge expansion joints unless protected from expansion and contraction.

M1406.5 Finish surfaces. Finish materials installed over radiant heating panels or systems shall be installed in accordance with the manufacturer's installation instructions. Surfaces shall be secured so that nails or other fastenings do not pierce the radiant heating elements.

SECTION M1407 DUCT HEATERS

M1407.1 General. Electric duct heaters shall be installed in accordance with the manufacturer's instructions and Chapters 34 through 43 of this code. Electric duct heaters shall comply with UL 1996.

M1407.2 Installation. Electric duct heaters shall be installed so that they will not create a fire hazard. Class 1 ducts, duct coverings and linings shall be interrupted at each heater to provide the clearances specified in the manufacturer's installation instructions. Such interruptions are not required for duct heaters *listed* and *labeled* for zero clearance to combustible materials. Insulation installed in the immediate area of each heater shall be classified for the maximum temperature produced on the duct surface.

M1407.3 Installation with heat pumps and air conditioners. Duct heaters located within 4 feet (1219 mm) of a heat pump or air conditioner shall be *listed* and *labeled* for such installations. The heat pump or air conditioner shall additionally be *listed* and *labeled* for such duct heater installations.

M1407.4 Access. Duct heaters shall be accessible for servicing, and clearance shall be maintained to permit adjustment, servicing and replacement of controls and heating elements.

M1407.5 Fan interlock. The fan circuit shall be provided with an interlock to prevent heater operation when the fan is not operating.

SECTION M1408 VENTED FLOOR FURNACES

M1408.1 General. Oil-fired vented floor furnaces shall comply with UL 729 and shall be installed in accordance with their *listing*, the manufacturer's instructions and the requirements of this code.

M1408.2 Clearances. Vented floor furnaces shall be installed in accordance with their listing and the manufacturer's installation instructions.

M1408.3 Location. Location of floor furnaces shall conform to the following requirements:

1. Floor registers of floor furnaces shall be installed not less than 6 inches (152 mm) from a wall.
2. Wall registers of floor furnaces shall be installed not less than 6 inches (152 mm) from the adjoining wall at inside corners.
3. The furnace register shall be located not less than 12 inches (305 mm) from doors in any position, draperies or similar combustible objects.
4. The furnace register shall be located at least 5 feet (1524 mm) below any projecting combustible materials.
5. The floor furnace burner assembly shall not project into an occupied under-floor area.
6. The floor furnace shall not be installed in concrete floor construction built on grade.
7. The floor furnace shall not be installed where a door can swing within 12 inches (305 mm) of the grille opening.

M1408.4 Access. An opening in the foundation not less than 18 inches by 24 inches (457 mm by 610 mm), or a trap door not less than 22 inches by 30 inches (559 mm by 762 mm) shall be provided for access to a floor furnace. The opening and passageway shall be large enough to allow replacement of any part of the *equipment*.

M1408.5 Installation. Floor furnace installations shall conform to the following requirements:

1. Thermostats controlling floor furnaces shall be located in the room in which the register of the floor furnace is located.
2. Floor furnaces shall be supported independently of the furnace floor register.
3. Floor furnaces shall be installed not closer than 6 inches (152 mm) to the ground. Clearance may be reduced to 2 inches (51 mm), provided that the lower 6 inches (152 mm) of the furnace is sealed to prevent water entry.
4. Where excavation is required for a floor furnace installation, the excavation shall extend 30 inches (762 mm) beyond the control side of the floor furnace and 12 inches (305 mm) beyond the remaining sides. Excavations shall slope outward from the perimeter of the base of the excavation to the surrounding *grade* at an angle not exceeding 45 degrees (0.79 rad) from horizontal.
5. Floor furnaces shall not be supported from the ground.

SECTION M1409 VENTED WALL FURNACES

M1409.1 General. Oil-fired vented wall furnaces shall comply with UL 730 and shall be installed in accordance with their *listing*, the manufacturer's instructions and the requirements of this code.

M1409.2 Location. The location of vented wall furnaces shall conform to the following requirements:

1. Vented wall furnaces shall be located where they will not cause a fire hazard to walls, floors, combustible furnishings or doors. Vented wall furnaces installed between bathrooms and adjoining rooms shall not circulate air from bathrooms to other parts of the building.
2. Vented wall furnaces shall not be located where a door can swing within 12 inches (305 mm) of the furnace air inlet or outlet measured at right angles to the opening. Doorstops or door closers shall not be installed to obtain this clearance.

M1409.3 Installation. Vented wall furnace installations shall conform to the following requirements:

1. Required wall thicknesses shall be in accordance with the manufacturer's installation instructions.
2. Ducts shall not be attached to a wall furnace. Casing extensions or boots shall be installed only when listed as part of a *listed* and *labeled appliance*.
3. A manual shut off valve shall be installed ahead of all controls.

M1409.4 Access. Vented wall furnaces shall be provided with access for cleaning of heating surfaces; removal of burners; replacement of sections, motors, controls, filters and other working parts; and for adjustments and lubrication of parts requiring such attention. Panels, grilles and access doors that must be removed for normal servicing operations shall not be attached to the building construction.

SECTION M1410 VENTED ROOM HEATERS

M1410.1 General. Vented room heaters shall be tested in accordance with ASTM E 1509, UL 896 for oil-fired or UL 1482 for solid fuel-fired and installed in accordance with their *listing*, the manufacturer's installation instructions and the requirements of this code.

M1410.2 Floor mounting. Room heaters shall be installed on noncombustible floors or *approved* assemblies constructed of noncombustible materials that extend at least 18 inches (457 mm) beyond the *appliance* on all sides.

Exceptions:

1. *Listed* room heaters shall be installed on noncombustible floors, assemblies constructed of noncombustible materials or *listed* floor protectors with materials and dimensions in accordance with the *appliance* manufacturer's instructions.
2. Room heaters *listed* for installation on combustible floors without floor protection shall be installed in accordance with the *appliance* manufacturer's instructions.

SECTION M1411 HEATING AND COOLING EQUIPMENT

M1411.1 Approved refrigerants. Refrigerants used in direct refrigerating systems shall conform to the applicable provisions of ANSI/ASHRAE 34.

M1411.2 Refrigeration coils in warm-air furnaces. Where a cooling coil is located in the supply plenum of a warm-air furnace, the furnace blower shall be rated at not less than 0.5-inch water column (124 Pa) static pressure unless the furnace is *listed* and *labeled* for use with a cooling coil. Cooling coils shall not be located upstream from heat exchangers unless *listed* and *labeled* for such use. Conversion of existing furnaces for use with cooling coils shall be permitted provided the furnace will operate within the temperature rise specified for the furnace.

M1411.3 Condensate disposal. Condensate from all cooling coils or evaporators shall be conveyed from the drain pan outlet to an *approved* place of disposal. Such piping shall maintain a minimum horizontal slope in the direction of discharge of not less than $^1/_8$ unit vertical in 12 units horizontal (1-percent slope). Condensate shall not discharge into a street, alley or other areas where it would cause a nuisance.

M1411.3.1 Auxiliary and secondary drain systems. In addition to the requirements of Section M1411.3, a secondary drain or auxiliary drain pan shall be required for each cooling or evaporator coil where damage to any building components will occur as a result of overflow from the *equipment* drain pan or stoppage in the condensate drain piping. Such piping shall maintain a minimum horizontal slope in the direction of discharge of not less than $^1/_8$ unit vertical in 12 units horizontal (1-percent slope). Drain piping shall be a minimum of $^3/_4$-inch (19 mm) nominal pipe size. One of the following methods shall be used:

1. An auxiliary drain pan with a separate drain shall be installed under the coils on which condensation will occur. The auxiliary pan drain shall discharge to a conspicuous point of disposal to alert occupants in the event of a stoppage of the primary drain. The pan shall have a minimum depth of 1.5 inches (38 mm), shall not be less than 3 inches (76 mm) larger than the unit or the coil dimensions in width and length and shall be constructed of corrosion-resistant material. Galvanized sheet steel pans shall have a minimum thickness of not less than 0.0236-inch (0.6010 mm) (No. 24 Gage). Nonmetallic pans shall have a minimum thickness of not less than 0.0625 inch (1.6 mm).
2. A separate overflow drain line shall be connected to the drain pan installed with the *equipment*. This overflow drain shall discharge to a conspicuous point of disposal to alert occupants in the event of a stoppage of the primary drain. The overflow drain line shall connect to the drain pan at a higher level than the primary drain connection.

3. An auxiliary drain pan without a separate drain line shall be installed under the coils on which condensation will occur. This pan shall be equipped with a water level detection device conforming to UL 508 that will shut off the *equipment* served prior to overflow of the pan. The pan shall be equipped with a fitting to allow for drainage. The auxiliary drain pan shall be constructed in accordance with Item 1 of this section.
4. A water level detection device conforming to UL 508 shall be installed that will shut off the *equipment* served in the event that the primary drain is blocked. The device shall be installed in the primary drain line, the overflow drain line or the *equipment*-supplied drain pan, located at a point higher than the primary drain line connection and below the overflow rim of such pan.

M1411.3.1.1 Water-level monitoring devices. On down-flow units and all other coils that have no secondary drain or provisions to install a secondary or auxiliary drain pan, a water-level monitoring device shall be installed inside the primary drain pan. This device shall shut off the equipment served in the event that the primary drain becomes restricted. Devices shall not be installed in the drain line.

M1411.3.2 Drain pipe materials and sizes. Components of the condensate disposal system shall be cast iron, galvanized steel, copper, polybutylene, polyethylene, ABS, CPVC or PVC pipe or tubing. All components shall be selected for the pressure and temperature rating of the installation. Joints and connections shall be made in accordance with the materials specified in Chapter 30. Condensate waste and drain line size shall be not less than $^3/_4$-inch (19 mm) internal diameter and shall not decrease in size from the drain pan connection to the place of condensate disposal. Where the drain pipes from more than one unit are manifolded together for condensate drainage, the pipe or tubing shall be sized in accordance with an *approved* method.

M1411.3.3 Appliances, equipment and insulation in pans. Where *appliances, equipment* or insulation are subject to water damage when auxiliary drain pans fill, those portions of the *appliances, equipment* and insulation shall be installed above the flood level rim of the pan. Supports located inside of the pan to support the *appliance* or *equipment* shall be water resistant and *approved.*

M1411.4 Auxiliary drain pan. Category IV condensing *appliances* shall have an auxiliary drain pan where damage to any building component will occur as a result of stoppage in the condensate drainage system. These pans shall be installed in accordance with the applicable provisions of Section M1411.3.

Exception: Fuel-fired *appliances* that automatically shut down operation in the event of a stoppage in the condensate drainage system.

M1411.5 Insulation of refrigerant piping. Piping and fittings for refrigerant vapor (suction) lines shall be insulated with insulation having a thermal resistivity of at least R-4 and having external surface permeance not exceeding 0.05 perm [2.87 ng/(s · m^2 · Pa)] when tested in accordance with ASTM E 96.

M1411.6 Locking access port caps. Refrigerant circuit access ports located outdoors shall be fitted with locking-type tamper-resistant caps or shall be otherwise secured to prevent unauthorized access.

SECTION M1412 ABSORPTION COOLING EQUIPMENT

M1412.1 Approval of equipment. Absorption systems shall be installed in accordance with the manufacturer's installation instructions. Absorption equipment shall comply with UL 1995.

M1412.2 Condensate disposal. Condensate from the cooling coil shall be disposed of as provided in Section M1411.3.

M1412.3 Insulation of piping. Refrigerant piping, brine piping and fittings within a building shall be insulated to prevent condensation from forming on piping.

M1412.4 Pressure-relief protection. Absorption systems shall be protected by a pressure-relief device. Discharge from the pressure-relief device shall be located where it will not create a hazard to persons or property.

SECTION M1413 EVAPORATIVE COOLING EQUIPMENT

M1413.1 General. Evaporative cooling equipment and appliances shall comply with UL 1995 and shall be installed:

1. According to the manufacturer's instructions.
2. On level platforms in accordance with Section M1305.1.4.1.
3. So that openings in exterior walls are flashed in accordance with Section R703.8.
4. So as to protect the potable water supply in accordance with Section P2902.
5. So that air intake opening locations are in accordance with Section R303.5.1.

SECTION M1414 FIREPLACE STOVES

M1414.1 General. Fireplace stoves shall be *listed, labeled* and installed in accordance with the terms of the listing. Fireplace stoves shall be tested in accordance with UL 737.

M1414.2 Hearth extensions. Hearth extensions for fireplace stoves shall be installed in accordance with the *listing* of the fireplace stove. The supporting structure for a hearth extension for a fireplace stove shall be at the same level as the supporting structure for the fireplace unit. The hearth extension shall be readily distinguishable from the surrounding floor area.

SECTION M1415
MASONRY HEATERS

M1415.1 General. Masonry heaters shall be constructed in accordance with Section R1002.

CHAPTER 15
EXHAUST SYSTEMS

SECTION M1501 GENERAL

M1501.1 Outdoor discharge. The air removed by every mechanical exhaust system shall be discharged to the outdoors in accordance with Section M1506.2. Air shall not be exhausted into an attic, soffit, ridge vent or crawl space.

Exception: Whole-house *ventilation*-type *attic* fans that discharge into the *attic* space of *dwelling units* having private *attics* shall be permitted.

SECTION M1502 CLOTHES DRYER EXHAUST

M1502.1 General. Clothes dryers shall be exhausted in accordance with the manufacturer's instructions.

M1502.2 Independent exhaust systems. Dryer exhaust systems shall be independent of all other systems and shall convey the moisture to the outdoors.

Exception: This section shall not apply to *listed* and *labeled* condensing (ductless) clothes dryers.

M1502.3 Duct termination. Exhaust ducts shall terminate on the outside of the building. Exhaust duct terminations shall be in accordance with the dryer manufacturer's installation instructions. If the manufacturer's instructions do not specify a termination location, the exhaust duct shall terminate not less than 3 feet (914 mm) in any direction from openings into buildings. Exhaust duct terminations shall be equipped with a backdraft damper. Screens shall not be installed at the duct termination.

M1502.4 Dryer exhaust ducts. Dryer exhaust ducts shall conform to the requirements of Sections M1502.4.1 through M1502.4.6.

M1502.4.1 Material and size. Exhaust ducts shall have a smooth interior finish and be constructed of metal having a minimum thickness of 0.0157 inches (0.3950 mm) (No. 28 gage). The duct shall be 4 inches (102 mm) nominal in diameter.

M1502.4.2 Duct installation. Exhaust ducts shall be supported at intervals not to exceed 12 feet (3658 mm) and shall be secured in place. The insert end of the duct shall extend into the adjoining duct or fitting in the direction of airflow. Exhaust duct joints shall be sealed in accordance with Section M1601.4.1 and shall be mechanically fastened. Ducts shall not be joined with screws or similar fasteners that protrude more than $^1/_8$ inch (3.2 mm) into the inside of the duct.

M1502.4.3 Transition duct. Transition ducts used to connect the dryer to the exhaust *duct system* shall be a single length that is *listed* and *labeled* in accordance with UL 2158A. Transition ducts shall be a maximum of 8 feet (2438 mm) in length. Transition ducts shall not be concealed within construction.

M1502.4.4 Duct length. The maximum allowable exhaust duct length shall be determined by one of the methods specified in Section M1502.4.4.1 or M1502.4.4.2.

M1502.4.4.1 Specified length. The maximum length of the exhaust duct shall be 35 feet (10 668 mm) from the connection to the transition duct from the dryer to the outlet terminal. Where fittings are used, the maximum length of the exhaust duct shall be reduced in accordance with Table M1502.4.4.1. The maximum length of the exhaust duct does not include the transition duct.

M1502.4.4.2 Manufacturer's instructions. The size and maximum length of the exhaust duct shall be determined by the dryer manufacturer's installation instructions. The code official shall be provided with a copy of the installation instructions for the make and model of the dryer at the concealment inspection. In the absence of fitting equivalent length calculations from the clothes dryer manufacturer, Table M1502.4.4.1 shall be used.

M1502.4.5 Length identification. Where the exhaust duct is concealed within the building construction, the

TABLE M1502.4.4.1
DRYER EXHAUST DUCT FITTING EQUIVALENT LENGTH

DRYER EXHAUST DUCT FITTING TYPE	EQUIVALENT LENGTH
4 inch radius mitered 45 degree elbow	2 feet 6 inches
4 inch radius mitered 90 degree elbow	5 feet
6 inch radius smooth 45 degree elbow	1 foot
6 inch radius smooth 90 degree elbow	1 foot 9 inches
8 inch radius smooth 45 degree elbow	1 foot
8 inch radius smooth 90 degree elbow	1 foot 7 inches
10 inch radius smooth 45 degree elbow	9 inches
10 inch radius smooth 90 degree elbow	1 foot 6 inches

For SI: 1 inch = 25.4 mm, 1 foot = 304.8 mm, 1 degree = 0.0175 rad.

equivalent length of the exhaust duct shall be identified on a permanent label or tag. The label or tag shall be located within 6 feet (1829 mm) of the exhaust duct connection.

M1502.4.6 Exhaust duct required. Where space for a clothes dryer is provided, an exhaust *duct system* shall be installed. Where the clothes dryer is not installed at the time of occupancy the exhaust duct shall be capped or plugged in the space in which it originates and identified and marked "future use."

Exception: Where a *listed* condensing clothes dryer is installed prior to occupancy of the structure.

M1502.5 Protection required. Protective shield plates shall be placed where nails or screws from finish or other work are likely to penetrate the clothes dryer exhaust duct. Shield plates shall be placed on the finished face of all framing members where there is less than $1^{1}/_{4}$ inches (32 mm) between the duct and the finished face of the framing member. Protective shield plates shall be constructed of steel, shall have a minimum thickness of 0.062-inch (1.6 mm) and shall extend a minimum of 2 inches (51 mm) above sole plates and below top plates.

SECTION M1503
RANGE HOODS

M1503.1 General. Range hoods shall discharge to the outdoors through a single-wall duct. The duct serving the hood shall have a smooth interior surface, shall be air tight, shall be equipped with a back-draft damper, and shall be independent of all other exhaust systems. Ducts serving range hoods shall not terminate in an attic or crawl space or areas inside the building.

Exception: Where installed in accordance with the manufacturer's installation instructions, and where mechanical or natural *ventilation* is otherwise provided, *listed* and *labeled* ductless range hoods shall not be required to discharge to the outdoors.

M1503.2 Duct material. Single-wall ducts serving range hoods shall be constructed of galvanized steel, stainless steel or copper.

Exception: Ducts for domestic kitchen cooking *appliances* equipped with down-draft exhaust systems shall be permitted to be constructed of schedule 40 PVC pipe and fittings provided that the installation complies with all of the following:

1. The duct is installed under a concrete slab poured on grade;
2. The underfloor trench in which the duct is installed is completely backfilled with sand or gravel;
3. The PVC duct extends not more than 1 inch (25 mm) above the indoor concrete floor surface;
4. The PVC duct extends not more than 1 inch (25 mm) above grade *outside of the building;* and
5. The PVC ducts are solvent cemented.

M1503.3 Kitchen exhaust rates. Where domestic kitchen cooking *appliances* are equipped with ducted range hoods or down-draft exhaust systems, the fans shall be sized in accordance with Section M1507.4.

M1503.4 Makeup air required. Exhaust hood systems capable of exhausting in excess of 400 cubic feet per minute (0.19 m^3/s) shall be provided with makeup air at a rate approximately equal to the exhaust air rate. Such makeup air systems shall be equipped with a means of closure and shall be automatically controlled to start and operate simultaneously with the exhaust system.

SECTION M1504
INSTALLATION OF MICROWAVE OVENS

M1504.1 Installation of a microwave oven over a cooking appliance. The installation of a *listed* and *labeled* cooking *appliance* or microwave oven over a *listed* and *labeled* cooking *appliance* shall conform to the terms of the upper *appliance's listing* and *label* and the manufacturer's installation instructions. The microwave oven shall conform to UL 923.

SECTION M1505
OVERHEAD EXHAUST HOODS

M1505.1 General. Domestic open-top broiler units shall have a metal exhaust hood, having a minimum thickness of 0.0157-inch (0.3950 mm) (No. 28 gage) with $^{1}/_{4}$ inch (6.4 mm) clearance between the hood and the underside of combustible material or cabinets. A clearance of at least 24 inches (610 mm) shall be maintained between the cooking surface and the combustible material or cabinet. The hood shall be at least as wide as the broiler unit, extend over the entire unit, discharge to the outdoors and be equipped with a backdraft damper or other means to control infiltration/exfiltration when not in operation. Broiler units incorporating an integral exhaust system, and *listed* and *labeled* for use without an exhaust hood, need not have an exhaust hood.

SECTION M1506
EXHAUST DUCTS AND EXHAUST OPENINGS

M1506.1 Ducts. Where exhaust duct construction is not specified in this chapter, construction shall comply with Chapter 16.

M1506.2 Exhaust openings. Air exhaust openings shall terminate not less than 3 feet (914 mm) from property lines; 3 feet (914 mm) from operable and nonoperable openings into the building and 10 feet (3048 mm) from mechanical air intakes except where the opening is located 3 feet (914 mm) above the air intake. Openings shall comply with Sections R303.5.2 and R303.6.

SECTION M1507
MECHANICAL VENTILATION

M1507.1 General. Where local exhaust or whole-house mechanical ventilation is provided, the equipment shall be designed in accordance with this section.

M1507.2 Recirculation of air. Exhaust air from bathrooms and toilet rooms shall not be recirculated within a residence or to another *dwelling unit* and shall be exhausted directly to the outdoors. Exhaust air from bathrooms and toilet rooms shall not discharge into an *attic*, crawl space or other areas inside the building.

M1507.3 Whole-house mechanical ventilation system. Whole-house mechanical ventilation systems shall be designed in accordance with Sections M1507.3.1 through M1507.3.3.

M1507.3.1 System design. The whole-house ventilation system shall consist of one or more supply or exhaust fans, or a combination of such, and associated ducts and controls. Local exhaust or supply fans are permitted to serve as such a system. Outdoor air ducts connected to the return side of an air handler shall be considered to provide supply ventilation.

M1507.3.2 System controls. The whole-house mechanical ventilation system shall be provided with controls that enable manual override.

M1507.3.3 Mechanical ventilation rate. The whole-house mechanical ventilation system shall provide outdoor air at a continuous rate of not less than that determined in accordance with Table M1507.3.3(1).

Exception: The whole-house mechanical ventilation system is permitted to operate intermittently where the system has controls that enable operation for not less than 25-percent of each 4-hour segment and the ventilation rate prescribed in Table M1507.3.3(1) is multiplied by the factor determined in accordance with Table M1507.3.3(2).

M1507.4 Local exhaust rates. *Local exhaust* systems shall be designed to have the capacity to exhaust the minimum air flow rate determined in accordance with Table M1507.4.

TABLE M1507.4
MINIMUM REQUIRED LOCAL EXHAUST RATES FOR ONE- AND TWO-FAMILY DWELLINGS

AREA TO BE EXHAUSTED	EXHAUST RATES
Kitchens	100 cfm intermittent or 25 cfm continuous
Bathrooms-Toilet Rooms	Mechanical exhaust capacity of 50 cfm intermittent or 20 cfm continuous

For SI: 1 cubic foot per minute = 0.0004719 m^3/s.

TABLE M1507.3.3(1)
CONTINUOUS WHOLE-HOUSE MECHANICAL VENTILATION SYSTEM AIRFLOW RATE REQUIREMENTS

DWELLING UNIT FLOOR AREA (square feet)	NUMBER OF BEDROOMS				
	0 – 1	2 – 3	4 – 5	6 – 7	> 7
	Airflow in CFM				
< 1,500	30	45	60	75	90
1,501 - 3,000	45	60	75	90	105
3,001 - 4,500	60	75	90	105	120
4,501 - 6,000	75	90	105	120	135
6,001 - 7,500	90	105	120	135	150
> 7,500	105	120	135	150	165

For SI: 1 square foot = 0.0929 m^2, 1 cubic foot per minute = 0.0004719 m^3/s.

TABLE M1507.3.3(2)
INTERMITTENT WHOLE-HOUSE MECHANICAL VENTILATION RATE FACTORS[a, b]

RUN-TIME PERCENTAGE IN EACH 4-HOUR SEGMENT	25%	33%	50%	66%	75%	100%
Factor[a]	4	3	2	1.5	1.3	1.0

a. For ventilation system run time values between those given, the factors are permitted to be determined by interpolation.

b. Extrapolation beyond the table is prohibited.

CHAPTER 16
DUCT SYSTEMS

SECTION M1601
DUCT CONSTRUCTION

M1601.1 Duct design. *Duct systems* serving heating, cooling and *ventilation equipment* shall be installed in accordance with the provisions of this section and ACCA Manual D or other *approved* methods.

M1601.1.1 Above-ground duct systems. Above-ground *duct systems* shall conform to the following:

1. *Equipment* connected to *duct systems* shall be designed to limit discharge air temperature to a maximum of 250°F (121°C).
2. Factory-made air ducts shall be constructed of Class 0 or Class 1 materials as designated in Table M1601.1.1(1).
3. Fibrous duct construction shall conform to the SMACNA *Fibrous Glass Duct Construction Standards* or NAIMA *Fibrous Glass Duct Construction Standards*.
4. Minimum thickness of metal duct material shall be as listed in Table M1601.1.1(2). Galvanized steel shall conform to ASTM A 653. Metallic ducts shall be fabricated in accordance with SMACNA Duct Construction Standards Metal and Flexible.
5. Use of gypsum products to construct return air ducts or plenums is permitted, provided that the air temperature does not exceed 125°F (52°C) and exposed surfaces are not subject to condensation.
6. *Duct systems* shall be constructed of materials having a flame spread index not greater than 200.
7. Stud wall cavities and the spaces between solid floor joists to be used as air plenums shall comply with the following conditions:
   - 7.1. These cavities or spaces shall not be used as a plenum for supply air.
   - 7.2. These cavities or spaces shall not be part of a required fire-resistance-rated assembly.
   - 7.3. Stud wall cavities shall not convey air from more than one floor level.
   - 7.4. Stud wall cavities and joist-space plenums shall be isolated from adjacent concealed spaces by tight-fitting fireblocking in accordance with Section R602.8.
   - 7.5. Stud wall cavities in the outside walls of building envelope assemblies shall not be utilized as air plenums.

TABLE M1601.1.1(1)
CLASSIFICATION OF FACTORY-MADE AIR DUCTS

DUCT CLASS	MAXIMUM FLAME SPREAD INDEX
0	0
1	25

M1601.1.2 Underground duct systems. Underground *duct systems* shall be constructed of *approved* concrete, clay, metal or plastic. The maximum duct temperature for plastic ducts shall not be greater than 150°F (66°C). Metal ducts shall be protected from corrosion in an *approved* manner or shall be completely encased in concrete not less than 2 inches (51 mm) thick. Nonmetallic ducts shall be installed in accordance with the manufacturer's installation instructions. Plastic pipe and fitting materials shall conform to cell classification 12454-B of ASTM D 1248 or ASTM D 1784 and external loading properties of ASTM D 2412. All ducts shall slope to an accessible point for drainage. Where encased in concrete, ducts shall be sealed and secured prior to any concrete being poured. Metallic ducts having an *approved* protective coating and nonmetallic ducts shall be installed in accordance with the manufacturer's installation instructions.

M1601.2 Factory-made ducts. Factory-made air ducts or duct material shall be *approved* for the use intended, and shall be installed in accordance with the manufacturer's installa-

TABLE M1601.1.1(2)
GAGES OF METAL DUCTS AND PLENUMS USED FOR HEATING OR COOLING

DUCT SIZE	GALVANIZED		ALUMINUM
	Minimum Thickness (inches)	Equivalent Galvanized Gage No.	Minimum Thickness (inches)
Round ducts and enclosed rectangular ducts			
14 inches or less	0.0157	28	0.0145
16 and 18 inches	0.0187	26	0.018
20 inches and over	0.0236	24	0.023
Exposed rectangular ducts			
14 inches or	0.0157	28	0.0145
Over 14[a] inches	0.0187	26	0.018

For SI: 1 inch = 25.4 mm.

a. For duct gages and reinforcement requirements at static pressures of $^1/_2$ inch, 1 inch and 2 inches w.g., SMACNA *Duct Construction Standard*, Tables 2-1; 2-2 and 2-3 shall apply.

tion instructions. Each portion of a factory-made air *duct system* shall bear a *listing* and *label* indicating compliance with UL 181 and UL 181A or UL 181B.

M1601.2.1 Vibration isolators. Vibration isolators installed between mechanical *equipment* and metal ducts shall be fabricated from *approved* materials and shall not exceed 10 inches (254 mm) in length.

M1601.3 Duct insulation materials. Duct insulation materials shall conform to the following requirements:

1. Duct coverings and linings, including adhesives where used, shall have a flame spread index not higher than 25, and a smoke-developed index not over 50 when tested in accordance with ASTM E 84 or UL 723, using the specimen preparation and mounting procedures of ASTM E 2231.

   Exception: Spray application of polyurethane foam to the exterior of ducts in *attics* and crawl spaces shall be permitted subject to all of the following:

   1. The flame spread index is not greater than 25 and the smoke-developed index is not greater than 450 at the specified installed thickness.
   2. The foam plastic is protected in accordance with the ignition barrier requirements of Sections R316.5.3 and R316.5.4.
   3. The foam plastic complies with the requirements of Section R316.

2. Duct coverings and linings shall not flame, glow, smolder or smoke when tested in accordance with ASTM C 411 at the temperature to which they are exposed in service. The test temperature shall not fall below 250°F (121°C). Coverings and linings shall be listed and labeled.

3. External duct insulation and factory-insulated flexible ducts shall be legibly printed or identified at intervals not longer than 36 inches (914 mm) with the name of the manufacturer, the thermal resistance *R*-value at the specified installed thickness and the flame spread and smoke-developed indexes of the composite materials. Spray polyurethane foam manufacturers shall provide the same product information and properties, at the nominal installed thickness, to the customer in writing at the time of foam application. All duct insulation product *R*-values shall be based on insulation only, excluding air films, vapor retarders or other duct components, and shall be based on tested C-values at 75°F (24°C) mean temperature at the installed thickness, in accordance with recognized industry procedures. The installed thickness of duct insulation used to determine its *R*-value shall be determined as follows:

   3.1. For duct board, duct liner and factory-made rigid ducts not normally subjected to compression, the nominal insulation thickness shall be used.

   3.2. For ductwrap, the installed thickness shall be assumed to be 75 percent (25-percent compression) of nominal thickness.

   3.3. For factory-made flexible air ducts, The installed thickness shall be determined by dividing the difference between the actual outside diameter and nominal inside diameter by two.

   3.4. For spray polyurethane foam, the aged *R*-value per inch measured in accordance with recognized industry standards shall be provided to the customer in writing at the time of foam application. In addition, the total *R*-value for the nominal application thickness shall be provided.

M1601.4 Installation. Duct installation shall comply with Sections M1601.4.1 through M1601.4.7.

M1601.4.1 Joints, seams and connections. All longitudal and transverse joints, seams and connections in metallic and nonmetallic ducts shall be constructed as specified in SMACNA *HVAC Duct Construction Standards—Metal and Flexible* and NAIMA *Fibrous Glass Duct Construction Standards*. All joints, longitudinal and transverse seams, and connections in ductwork shall be securely fastened and sealed with welds, gaskets, mastics (adhesives), mastic-plus-embedded-fabric systems or tapes.

Closure systems used to seal flexible air ducts and flexible air connectors shall comply with UL 181B and shall be marked "181 B-FX" for pressure-sensitive tape or "181 B-M" for mastic. Duct connections to flanges of air distribution system equipment shall be sealed and mechanically fastened. Mechanical fasteners for use with flexible nonmetallic air ducts shall comply with UL 181B and shall be marked 181B-C. Crimp joints for round metallic ducts shall have a contact lap of not less than 1 inch (25.4 mm) and shall be mechanically fastened by means of not less than three sheet-metal screws or rivets equally spaced around the joint.

Closure systems used to seal metal ductwork shall be installed in accordance with the manufacturer's instructions. Round metallic ducts shall be mechanically fastened by means of at least three sheet metal screws or rivets spaced equally around the joint. Unlisted duct tape shall not be permitted as a sealant on any duct.

Exceptions:

1. Spray polyurethane foam shall be permitted to be applied without additional joint seals.
2. Where a duct connection is made that is partially inaccessible, three screws or rivets shall be equally spaced on the exposed portion of the joint so as to prevent a hinge effect.
3. Continuously welded and locking-type longitudinal joints and seams in ducts operating at static pressures less than 2 inches of water column (500 Pa) pressure classification shall not require additional closure systems.

M1601.4.2 Plastic duct joints. Joints between plastic ducts and plastic fittings shall be made in accordance with the manufacturer's installation instructions.

M1601.4.3 Support. Metal ducts shall be supported by $^1/_2$-inch-wide (13 mm) 18-gage metal straps or 12-gage galvanized wire at intervals not exceeding 10 feet (3048 mm) or other *approved* means. Nonmetallic ducts shall be supported in accordance with the manufacturer's installation instructions.

M1601.4.4 Fireblocking. Duct installations shall be fireblocked in accordance with Section R602.8.

M1601.4.5 Duct insulation. Duct insulation shall be installed in accordance with the following requirements:

1. A vapor retarder having a maximum permeance of 0.05 perm [2.87 ng/(s · m^2 · Pa)] in accordance with ASTM E 96, or aluminum foil with a minimum thickness of 2 mils (0.05 mm), shall be installed on the exterior of insulation on cooling supply ducts that pass through unconditioned spaces conducive to condensation except where the insulation is spray polyurethane foam with a maximum water vapor permeance of 3 perm per inch [1722 ng/(s · m^2 · Pa)] at the installed thickness.
2. Exterior *duct systems* shall be protected against the elements.
3. Duct coverings shall not penetrate a fireblocked wall or floor.

M1601.4.6 Factory-made air ducts. Factory-made air ducts shall not be installed in or on the ground, in tile or metal pipe, or within masonry or concrete.

M1601.4.7 Duct separation. Ducts shall be installed with at least 4 inches (102 mm) separation from earth except where they meet the requirements of Section M1601.1.2.

M1601.4.8 Ducts located in garages. Ducts in garages shall comply with the requirements of Section R302.5.2.

M1601.4.9 Flood hazard areas. In flood hazard areas as established by Table R301.2(1), *duct systems* shall be located or installed in accordance with Section R322.1.6.

M1601.5 Under-floor plenums. Under-floor plenums shall be prohibited in new structures. Modification or repairs to under-floor plenums in existing structures shall conform to the requirements of this section.

M1601.5.1 General. The space shall be cleaned of loose combustible materials and scrap, and shall be tightly enclosed. The ground surface of the space shall be covered with a moisture barrier having a minimum thickness of 4 mils (0.1 mm). Plumbing waste cleanouts shall not be located within the space.

Exception: Plumbing waste cleanouts shall be permitted to be located in unvented crawl spaces that receive *conditioned air* in accordance with Section R408.3.

M1601.5.2 Materials. The under-floor space, including the sidewall insulation, shall be formed by materials having flame spread index values not greater than 200 when tested in accordance with ASTM E 84 or UL 723.

M1601.5.3 Furnace connections. A duct shall extend from the furnace supply outlet to not less than 6 inches (152 mm) below the combustible framing. This duct shall comply with the provisions of Section M1601.1. A noncombustible receptacle shall be installed below any floor opening into the plenum in accordance with the following requirements:

1. The receptacle shall be securely suspended from the floor members and shall not be more than 18 inches (457 mm) below the floor opening.
2. The area of the receptacle shall extend 3 inches (76 mm) beyond the opening on all sides.
3. The perimeter of the receptacle shall have a vertical lip at least 1 inch (25 mm) high at the open sides.

M1601.5.4 Access. Access to an under-floor plenum shall be provided through an opening in the floor with minimum dimensions of 18 inches by 24 inches (457 mm by 610 mm).

M1601.5.5 Furnace controls. The furnace shall be equipped with an automatic control that will start the air-circulating fan when the air in the furnace bonnet reaches a temperature not higher than 150°F (66°C). The furnace shall additionally be equipped with an *approved* automatic control that limits the outlet air temperature to 200°F (93°C).

M1601.6 Independent garage HVAC systems. Furnaces and air-handling systems that supply air to living spaces shall not supply air to or return air from a garage.

SECTION M1602
RETURN AIR

M1602.1 Return air. Return air shall be taken from inside the *dwelling*. Dilution of return air with outdoor air shall be permitted.

M1602.2 Prohibited sources. Outdoor and return air for a forced-air heating or cooling system shall not be taken from the following locations:

1. Closer than 10 feet (3048 mm) to an *appliance* vent outlet, a vent opening from a plumbing drainage system or the discharge outlet of an exhaust fan, unless the outlet is 3 feet (914 mm) above the outside air inlet.
2. Where flammable vapors are present; or where located less than 10 feet (3048 mm) above the surface of any abutting public way or driveway; or where located at grade level by a sidewalk, street, alley or driveway.
3. A room or space, the volume of which is less than 25 percent of the entire volume served by the system. Where connected by a permanent opening having an area sized in accordance with ACCA Manual D, adjoining rooms or spaces shall be considered as a single room or space for the purpose of determining the volume of the rooms or spaces.

Exception: The minimum volume requirement shall not apply where the amount of return air taken from a room or space is less than or equal to the amount of supply air delivered to the room or space.

4. A closet, bathroom, toilet room, kitchen, garage, boiler room, furnace room, unconditioned *attic* or other *dwelling unit.*

   Exception: Dedicated forced-air systems serving only a garage shall not be prohibited from obtaining return air from the garage.

5. A room or space containing a fuel-burning *appliance* where such room or space serves as the sole source of return air.

   Exceptions:

   1. The fuel-burning *appliance* is a direct-vent *appliance* or an *appliance* not requiring a vent in accordance with Section M1801.1 or Chapter 24.
   2. The room or space complies with the following requirements:
      2.1. The return air shall be taken from a room or space having a volume exceeding 1 cubic foot for each 10 Btu/h (9.6 L/W) of combined input rating of all fuel-burning *appliances* therein.

      2.2. The volume of supply air discharged back into the same space shall be approximately equal to the volume of return air taken from the space.

      2.3. Return-air inlets shall not be located within 10 feet (3048 mm) of a draft hood in the same room or space or the combustion chamber of any atmospheric-burner *appliance* in the same room or space.
   3. Rooms or spaces containing solid-fuel burning *appliances*, if return-air inlets are located not less than 10 feet (3048 mm) from the firebox of those *appliances.*

6. An unconditioned crawl space by means of direct connection to the return side of a forced air system. Transfer openings in the crawl space enclosure shall not be prohibited.

M1602.3 Inlet opening protection. Outdoor air inlets shall be covered with screens having openings that are not less than $^1/_4$ inch (6.4 mm) and not greater than $^1/_2$ inch (12.7 mm).

CHAPTER 17

COMBUSTION AIR

SECTION M1701
GENERAL

M1701.1 Scope. Solid-fuel-burning *appliances* shall be provided with *combustion air* in accordance with the *appliance* manufacturer's installation instructions. Oil-fired *appliances* shall be provided with *combustion air* in accordance with NFPA 31. The methods of providing *combustion air* in this chapter do not apply to fireplaces, fireplace stoves and direct-vent *appliances*. The requirements for combustion and dilution air for gas-fired *appliances* shall be in accordance with Chapter 24.

M1701.2 Opening location. In flood hazard areas as established in Table R301.2(1), *combustion air* openings shall be located at or above the elevation required in Section R322.2.1 or R322.3.2.

CHAPTER 18

CHIMNEYS AND VENTS

SECTION M1801 GENERAL

M1801.1 Venting required. Fuel-burning *appliances* shall be vented to the outdoors in accordance with their *listing* and *label* and manufacturer's installation instructions except *appliances* listed and *labeled* for unvented use. Venting systems shall consist of *approved* chimneys or vents, or venting assemblies that are integral parts of *labeled appliances*. Gas-fired *appliances* shall be vented in accordance with Chapter 24.

M1801.2 Draft requirements. A venting system shall satisfy the draft requirements of the *appliance* in accordance with the manufacturer's installation instructions, and shall be constructed and installed to develop a positive flow to convey combustion products to the outside atmosphere.

M1801.3 Existing chimneys and vents. Where an *appliance* is permanently disconnected from an existing chimney or vent, or where an *appliance* is connected to an existing chimney or vent during the process of a new installation, the chimney or vent shall comply with Sections M1801.3.1 through M1801.3.4.

M1801.3.1 Size. The chimney or vent shall be resized as necessary to control flue gas condensation in the interior of the chimney or vent and to provide the *appliance*, or *appliances* served, with the required draft. For the venting of oil-fired *appliances* to masonry chimneys, the resizing shall be done in accordance with NFPA 31.

M1801.3.2 Flue passageways. The flue gas passageway shall be free of obstructions and combustible deposits and shall be cleaned if previously used for venting a solid- or liquid-fuel-burning *appliance* or fireplace. The flue liner, chimney inner wall or vent inner wall shall be continuous and free of cracks, gaps, perforations, or other damage or deterioration that would allow the escape of combustion products, including gases, moisture and creosote.

M1801.3.3 Cleanout. Masonry chimneys shall be provided with a cleanout opening complying with Section R1003.17.

M1801.3.4 Clearances. Chimneys and vents shall have airspace clearance to combustibles in accordance with this code and the chimney or vent manufacturer's installation instructions.

Exception: Masonry chimneys equipped with a chimney lining system tested and *listed* for installation in chimneys in contact with combustibles in accordance with UL 1777, and installed in accordance with the manufacturer's instruction, shall not be required to have a clearance between combustible materials and exterior surfaces of the masonry chimney. Noncombustible firestopping shall be provided in accordance with this code.

M1801.4 Space around lining. The space surrounding a flue lining system or other vent installed within a masonry chimney shall not be used to vent any other *appliance*. This shall not prevent the installation of a separate flue lining in accordance with the manufacturer's installation instructions and this code.

M1801.5 Mechanical draft systems. A mechanical draft system shall be used only with *appliances listed* and *labeled* for such use. Provisions shall be made to prevent the flow of fuel to the *equipment* when the draft system is not operating. Forced draft systems and all portions of induced draft systems under positive pressure during operation shall be designed and installed to prevent leakage of flue gases into a building.

M1801.6 Direct-vent appliances. Direct-vent *appliances* shall be installed in accordance with the manufacturer's installation instructions.

M1801.7 Support. Venting systems shall be adequately supported for the weight of the material used.

M1801.8 Duct penetrations. Chimneys, vents and vent connectors shall not extend into or through supply and return air ducts or plenums.

M1801.9 Fireblocking. Vent and chimney installations shall be fireblocked in accordance with Section R602.8.

M1801.10 Unused openings. Unused openings in any venting system shall be closed or capped.

M1801.11 Multiple-appliance venting systems. Two or more *listed* and *labeled appliances* connected to a common natural draft venting system shall comply with the following requirements:

1. *Appliances* that are connected to common venting systems shall be located on the same floor of the *dwelling*.

   Exception: Engineered systems as provided for in Section G2427.

2. Inlets to common venting systems shall be offset such that no portion of an inlet is opposite another inlet.
3. Connectors serving *appliances* operating under a natural draft shall not be connected to any portion of a mechanical draft system operating under positive pressure.

M1801.12 Multiple solid fuel prohibited. A solid-fuel-burning *appliance* or fireplace shall not connect to a chimney passageway venting another *appliance*.

SECTION M1802 VENT COMPONENTS

M1802.1 Draft hoods. Draft hoods shall be located in the same room or space as the *combustion air* openings for the *appliances*.

M1802.2 Vent dampers. Vent dampers shall comply with Sections M1802.2.1 and M1802.2.2.

M1802.2.1 Manually operated. Manually operated dampers shall not be installed except in connectors or chimneys serving solid-fuel-burning *appliances*.

M1802.2.2 Automatically operated. Automatically operated dampers shall conform to UL 17 and be installed in accordance with the terms of their *listing* and *label*. The installation shall prevent firing of the burner when the damper is not opened to a safe position.

M1802.3 Draft regulators. Draft regulators shall be provided for oil-fired *appliances* that must be connected to a chimney. Draft regulators provided for solid-fuel-burning *appliances* to reduce draft intensity shall be installed and set in accordance with the manufacturer's installation instructions.

M1802.3.1 Location. Where required, draft regulators shall be installed in the same room or enclosure as the *appliance* so that no difference in pressure will exist between the air at the regulator and the *combustion air* supply.

SECTION M1803
CHIMNEY AND VENT CONNECTORS

M1803.1 General. Connectors shall be used to connect fuel-burning *appliances* to a vertical chimney or vent except where the chimney or vent is attached directly to the *appliance*.

M1803.2 Connectors for oil and solid fuel appliances. Connectors for oil and solid-fuel-burning *appliances* shall be constructed of factory-built chimney material, Type L vent material or single-wall metal pipe having resistance to corrosion and heat and thickness not less than that of galvanized steel as specified in Table M1803.2.

TABLE M1803.2
THICKNESS FOR SINGLE-WALL METAL PIPE CONNECTORS

DIAMETER OF CONNECTOR (inches)	GALVANIZED SHEET METAL GAGE NUMBER	MINIMUM THICKNESS (inch)
Less than 6	26	0.019
6 to 10	24	0.024
Over 10 through 16	22	0.029

For SI: 1 inch = 25.4 mm.

M1803.3 Installation. Vent and chimney connectors shall be installed in accordance with the manufacturer's installation instructions and within the space where the *appliance* is located. *Appliances* shall be located as close as practical to the vent or chimney. Connectors shall be as short and straight as possible and installed with a slope of not less than $^1/_4$ inch (6 mm) rise per foot of run. Connectors shall be securely supported and joints shall be fastened with sheet metal screws or rivets. Devices that obstruct the flow of flue gases shall not be installed in a connector unless *listed* and *labeled* or *approved* for such installations.

M1803.3.1 Floor, ceiling and wall penetrations. A chimney connector or vent connector shall not pass through any floor or ceiling. A chimney connector or vent connector shall not pass through a wall or partition unless the connector is *listed* and *labeled* for wall pass-through, or is routed through a device *listed* and *labeled* for wall pass-through and is installed in accordance with the conditions of its *listing* and *label*. Connectors for oil-fired *appliances listed* and *labeled* for Type L vents, passing through walls or partitions shall be in accordance with the following:

1. Type L vent material for oil *appliances* shall be installed with not less than *listed* and *labeled* clearances to combustible material.
2. Single-wall metal pipe shall be *guarded* by a ventilated metal thimble not less than 4 inches (102 mm) larger in diameter than the vent connector. A minimum 6 inches (152 mm) of clearance shall be maintained between the thimble and combustibles.

M1803.3.2 Length. The horizontal run of an uninsulated connector to a natural draft chimney shall not exceed 75 percent of the height of the vertical portion of the chimney above the connector. The horizontal run of a *listed* connector to a natural draft chimney shall not exceed 100 percent of the height of the vertical portion of the chimney above the connector.

M1803.3.3 Size. A connector shall not be smaller than the flue collar of the *appliance*.

Exception: Where installed in accordance with the *appliance* manufacturer's installation instructions.

M1803.3.4 Clearance. Connectors shall be installed with clearance to combustibles as set forth in Table M1803.3.4. Reduced clearances to combustible materials shall be in accordance with Table M1306.2 and Figure M1306.1.

TABLE M1803.3.4
CHIMNEY AND VENT CONNECTOR CLEARANCES TO COMBUSTIBLE MATERIALS[a]

TYPE OF CONNECTOR	MINIMUM CLEARANCE (inches)
Single-wall metal pipe connectors:	
Oil and solid-fuel appliances	18
Oil appliances listed for use with Type L vents	9
Type L vent piping connectors:	
Oil and solid-fuel appliances	9
Oil appliances listed for use with Type L vents	3[b]

For SI: 1 inch = 25.4 mm.

a. These minimum clearances apply to unlisted single-wall chimney and vent connectors. Reduction of required clearances is permitted as in Table M1306.2.

b. When listed Type L vent piping is used, the clearance shall be in accordance with the vent listing.

M1803.3.5 Access. The entire length of a connector shall be accessible for inspection, cleaning and replacement.

M1803.4 Connection to fireplace flue. Connection of *appliances* to chimney flues serving fireplaces shall comply with Sections M1803.4.1 through M1803.4.4.

M1803.4.1 Closure and accessibility. A noncombustible seal shall be provided below the point of connection to prevent entry of room air into the flue. Means shall be provided for access to the flue for inspection and cleaning.

M1803.4.2 Connection to factory-built fireplace flue. A different *appliance* shall not be connected to a flue serving a factory-built fireplace unless the *appliance* is specifically *listed* for such an installation. The connection shall be made in conformance with the *appliance* manufacturer's instructions.

M1803.4.3 Connection to masonry fireplace flue. A connector shall extend from the *appliance* to the flue serving a masonry fireplace to convey the flue gases directly into the flue. The connector shall be accessible or removable for inspection and cleaning of both the connector and the flue. *Listed* direct-connection devices shall be installed in accordance with their *listing*.

M1803.4.4 Size of flue. The size of the fireplace flue shall be in accordance with Section M1805.3.1.

SECTION M1804 VENTS

M1804.1 Type of vent required. *Appliances* shall be provided with a *listed* and *labeled* venting system as set forth in Table M1804.1.

TABLE M1804.1 VENT SELECTION CHART

VENT TYPES	APPLIANCE TYPES
Type L oil vents	Oil-burning appliances listed and labeled for venting with Type L vents
Pellet vents	Pellet fuel-burning appliances listed and labeled for use with pellet vents

M1804.2 Termination. Vent termination shall comply with Sections M1804.2.1 through M1804.2.6.

M1804.2.1 Through the roof. Vents passing through a roof shall extend through flashing and terminate in accordance with the manufacturer's installation requirements.

M1804.2.2 Decorative shrouds. Decorative shrouds shall not be installed at the termination of vents except where the shrouds are *listed* and *labeled* for use with the specific venting system and are installed in accordance with the manufacturer's installation instructions.

M1804.2.3 Natural draft appliances. Vents for natural draft *appliances* shall terminate at least 5 feet (1524 mm) above the highest connected *appliance* outlet, and natural draft gas vents serving wall furnaces shall terminate at an elevation at least 12 feet (3658 mm) above the bottom of the furnace.

M1804.2.4 Type L vent. Type L venting systems shall conform to UL 641 and shall terminate with a *listed* and *labeled* cap in accordance with the vent manufacturer's installation instructions not less than 2 feet (610 mm) above the roof and not less than 2 feet (610 mm) above any portion of the building within 10 feet (3048 mm).

M1804.2.5 Direct vent terminations. Vent terminals for direct-vent *appliances* shall be installed in accordance with the manufacturer's installation instructions.

M1804.2.6 Mechanical draft systems. Mechanical draft systems shall comply with UL 378 and shall be installed in accordance with their *listing*, the manufacturer's instructions and, except for direct-vent *appliances*, the following requirements:

1. The vent terminal shall be located not less than 3 feet (914 mm) above a forced air inlet located within 10 feet (3048 mm).
2. The vent terminal shall be located not less than 4 feet (1219 mm) below, 4 feet (1219 mm) horizontally from, or 1 foot (305 mm) above any door, window or gravity air inlet into a *dwelling*.
3. The vent termination point shall not be located closer than 3 feet (914 mm) to an interior corner formed by two walls perpendicular to each other.
4. The bottom of the vent terminal shall be located at least 12 inches (305 mm) above finished ground level.
5. The vent termination shall not be mounted directly above or within 3 feet (914 mm) horizontally of an oil tank vent or gas meter.
6. Power exhauster terminations shall be located not less than 10 feet (3048 mm) from *lot lines* and adjacent buildings.
7. The discharge shall be directed away from the building.

M1804.3 Installation. Type L and pellet vents shall be installed in accordance with the terms of their *listing* and *label* and the manufacturer's installation instructions.

M1804.3.1 Size of single-appliance venting systems. An individual vent for a single *appliance* shall have a cross-sectional area equal to or greater than the area of the connector to the *appliance*, but not less than 7 square inches (4515 mm^2) except where the vent is an integral part of a *listed* and *labeled appliance*.

SECTION M1805 MASONRY AND FACTORY-BUILT CHIMNEYS

M1805.1 General. Masonry and factory-built chimneys shall be built and installed in accordance with Sections R1003 and R1005, respectively. Flue lining for masonry chimneys shall comply with Section R1003.11.

M1805.2 Masonry chimney connection. A chimney connector shall enter a masonry chimney not less than 6 inches (152 mm) above the bottom of the chimney. Where it is not possible to locate the connector entry at least 6 inches (152 mm) above the bottom of the chimney flue, a cleanout shall be provided by installing a capped tee in the connector next to the chimney. A connector entering a masonry chimney shall extend through, but not beyond, the wall and shall be flush with the inner face of the liner. Connectors, or thimbles where used, shall be firmly cemented into the masonry.

M1805.3 Size of chimney flues. The effective area of a natural draft chimney flue for one *appliance* shall be not less than the area of the connector to the *appliance*. The area of chimney flues connected to more than one *appliance* shall be not less than the area of the largest connector plus 50 percent of the areas of additional chimney connectors.

Exception: Chimney flues serving oil-fired *appliances* sized in accordance with NFPA 31.

M1805.3.1 Size of chimney flue for solid-fuel appliance. Except where otherwise specified in the manufacturer's installation instructions, the cross-sectional area of a flue connected to a solid-fuel-burning *appliance* shall be not less than the area of the flue collar or connector, and not larger than three times the area of the flue collar.

CHAPTER 19

SPECIAL APPLIANCES, EQUIPMENT AND SYSTEMS

SECTION M1901 RANGES AND OVENS

M1901.1 Clearances. Freestanding or built-in ranges shall have a vertical clearance above the cooking top of not less than 30 inches (762 mm) to unprotected combustible material. Reduced clearances are permitted in accordance with the *listing* and *labeling* of the range hoods or *appliances*. The installation of a listed and labeled cooking appliance or microwave oven over a listed and labeled cooking appliance shall be in accordance with Section M1504.1. The clearances for a domestic open-top broiler unit shall be in accordance with Section M1505.1.

M1901.2 Cooking appliances. Cooking *appliances* shall be *listed* and *labeled* for household use and shall be installed in accordance with the manufacturer's instructions. The installation shall not interfere with *combustion air* or access for operation and servicing. Electric cooking appliances shall comply with UL 1026 or UL 858. Solid-fuel-fired fireplace stoves shall comply with UL 737.

M1901.3 Prohibited location. Cooking appliances designed, tested, listed and labeled for use in commercial occupancies shall not be installed within dwelling units or within any area where domestic cooking operations occur.

SECTION M1902 SAUNA HEATERS

M1902.1 Locations and protection. Sauna heaters shall be protected from accidental contact by persons with a guard of material having a low thermal conductivity, such as wood. The guard shall have no substantial effect on the transfer of heat from the heater to the room.

M1902.2 Installation. Sauna heaters shall be installed in accordance with the manufacturer's installation instructions. Sauna heaters shall comply with UL 875.

M1902.3 Combustion air. *Combustion air* and venting for a nondirect vent-type heater shall be provided in accordance with Chapters 17 and 18, respectively.

M1902.4 Controls. Sauna heaters shall be equipped with a thermostat that will limit room temperature to not greater than 194°F (90°C). Where the thermostat is not an integral part of the heater, the heat-sensing element shall be located within 6 inches (152 mm) of the ceiling.

SECTION M1903 STATIONARY FUEL CELL POWER PLANTS

M1903.1 General. Stationary fuel cell power plants having a power output not exceeding 1,000 kW, shall comply with ANSI/CSA America FC 1 and shall be installed in accordance with the manufacturer's instructions and NFPA 853.

SECTION M1904 GASEOUS HYDROGEN SYSTEMS

M1904.1 Installation. Gaseous hydrogen systems shall be installed in accordance with the applicable requirements of Sections M1307.4 and M1903.1 and the *International Fuel Gas Code*, the *International Fire Code* and the *International Building Code*.

CHAPTER 20

BOILERS AND WATER HEATERS

SECTION M2001 BOILERS

M2001.1 Installation. In addition to the requirements of this code, the installation of boilers shall conform to the manufacturer's instructions. The manufacturer's rating data, the nameplate and operating instructions of a permanent type shall be attached to the boiler. Boilers shall have all controls set, adjusted and tested by the installer. A complete control diagram together with complete boiler operating instructions shall be furnished by the installer. Solid- and liquid-fuel-burning boilers shall be provided with *combustion air* as required by Chapter 17.

M2001.1.1 Standards. Oil-fired boilers and their control systems shall be listed and *labeled* in accordance with UL 726. Electric boilers and their control systems shall be *listed* in accordance with UL 834. Solid-fuel-fired boilers shall be listed and labeled in accordance with UL 2523. Boilers shall be designed and constructed in accordance with the requirements of ASME CSD-1 and as applicable, the ASME *Boiler and Pressure Vessel Code*, Sections I and IV. Gas-fired boilers shall conform to the requirements *listed* in Chapter 24.

M2001.2 Clearance. Boilers shall be installed in accordance with their *listing* and *label*.

M2001.3 Valves. Every boiler or modular boiler shall have a shutoff valve in the supply and return piping. For multiple boiler or multiple modular boiler installations, each boiler or modular boiler shall have individual shutoff valves in the supply and return piping.

Exception: Shutoff valves are not required in a system having a single low-pressure steam boiler.

M2001.4 Flood-resistant installation. In flood hazard areas established in Table R301.2(1), boilers, water heaters and their control systems shall be located or installed in accordance with Section R322.1.6.

SECTION M2002 OPERATING AND SAFETY CONTROLS

M2002.1 Safety controls. Electrical and mechanical operating and safety controls for boilers shall be *listed* and *labeled.*

M2002.2 Hot water boiler gauges. Every hot water boiler shall have a pressure gauge and a temperature gauge, or combination pressure and temperature gauge. The gauges shall indicate the temperature and pressure within the normal range of the system's operation.

M2002.3 Steam boiler gauges. Every steam boiler shall have a water-gauge glass and a pressure gauge. The pressure gauge shall indicate the pressure within the normal range of the system's operation. The gauge glass shall be installed so that the midpoint is at the normal water level.

M2002.4 Pressure-relief valve. Boilers shall be equipped with pressure-relief valves with minimum rated capacities for the *equipment* served. Pressure-relief valves shall be set at the maximum rating of the boiler. Discharge shall be piped to drains by gravity to within 18 inches (457 mm) of the floor or to an open receptor.

M2002.5 Boiler low-water cutoff. All steam and hot water boilers shall be protected with a low-water cutoff control. The low-water cutoff shall automatically stop the combustion operation of the *appliance* when the water level drops below the lowest safe water level as established by the manufacturer.

SECTION M2003 EXPANSION TANKS

M2003.1 General. Hot water boilers shall be provided with expansion tanks. Nonpressurized expansion tanks shall be securely fastened to the structure or boiler and supported to carry twice the weight of the tank filled with water. Provisions shall be made for draining nonpressurized tanks without emptying the system.

M2003.1.1 Pressurized expansion tanks. Pressurized expansion tanks shall be consistent with the volume and capacity of the system. Tanks shall be capable of withstanding a hydrostatic test pressure of two and one-half times the allowable working pressure of the system.

M2003.2 Minimum capacity. The minimum capacity of expansion tanks shall be determined from Table M2003.2.

TABLE M2003.2 EXPANSION TANK MINIMUM CAPACITY[a] FOR FORCED HOT-WATER SYSTEMS

SYSTEM VOLUME[b] (gallons)	PRESSURIZED DIAPHRAGM TYPE	NONPRESSURIZED TYPE
10	1.0	1.5
20	1.5	3.0
30	2.5	4.5
40	3.0	6.0
50	4.0	7.5
60	5.0	9.0
70	6.0	10.5
80	6.5	12.0
90	7.5	13.5
100	8.0	15.0

For SI: 1 gallon = 3.785 L, 1 pound per square inch gauge = 6.895 kPa, °C = [(°F)-32]/1.8.

a. Based on average water temperature of 195°F (91°C), fill pressure of 12 psig and a maximum operating pressure of 30 psig.

b. System volume includes volume of water in boiler, convectors and piping, not including the expansion tank.

SECTION M2004
WATER HEATERS USED FOR SPACE HEATING

M2004.1 General. Water heaters used to supply both potable hot water and hot water for space heating shall be installed in accordance with this chapter, Chapter 24, Chapter 28 and the manufacturer's installation instructions.

SECTION M2005
WATER HEATERS

M2005.1 General. Water heaters shall be installed in accordance with the manufacturer's instructions and the requirements of this code. Water heaters installed in an *attic* shall comply with the requirements of Section M1305.1.3. Gas-fired water heaters shall comply with the requirements in Chapter 24. Domestic electric water heaters shall comply with UL 174. Oiled-fired water heaters shall comply with UL 732. Thermal solar water heaters shall comply with Chapter 23 and UL 174. Solid-fuel-fired water heaters shall comply with UL 2523.

M2005.2 Prohibited locations. Fuel-fired water heaters shall not be installed in a room used as a storage closet. Water heaters located in a bedroom or bathroom shall be installed in a sealed enclosure so that *combustion air* will not be taken from the living space. Installation of direct-vent water heaters within an enclosure is not required.

M2005.2.1 Water heater access. Access to water heaters that are located in an *attic* or underfloor crawl space is permitted to be through a closet located in a sleeping room or bathroom where *ventilation* of those spaces is in accordance with this code.

M2005.3 Electric water heaters. Electric water heaters shall also be installed in accordance with the applicable provisions of Chapters 34 through 43.

M2005.4 Supplemental water-heating devices. Potable water heating devices that use refrigerant-to-water heat exchangers shall be *approved* and installed in accordance with the manufacturer's installation instructions.

SECTION M2006
POOL HEATERS

M2006.1 General. Pool and spa heaters shall be installed in accordance with the manufacturer's installation instructions. Oil-fired pool heaters shall comply with UL 726. Electric pool and spa heaters shall comply with UL 1261.

M2006.2 Clearances. In no case shall the clearances interfere with *combustion air*, draft hood or flue terminal relief, or accessibility for servicing.

M2006.3 Temperature-limiting devices. Pool heaters shall have temperature-relief valves.

M2006.4 Bypass valves. Where an integral bypass system is not provided as a part of the pool heater, a bypass line and valve shall be installed between the inlet and outlet piping for use in adjusting the flow of water through the heater.

CHAPTER 21

HYDRONIC PIPING

SECTION M2101
HYDRONIC PIPING SYSTEMS INSTALLATION

M2101.1 General. Hydronic piping shall conform to Table M2101.1. *Approved* piping, valves, fittings and connections shall be installed in accordance with the manufacturer's installation instructions. Pipe and fittings shall be rated for use at the operating temperature and pressure of the hydronic system. Used pipe, fittings, valves or other materials shall be free of foreign materials.

M2101.2 System drain down. Hydronic piping systems shall be installed to permit draining of the system. Where the system drains to the plumbing drainage system, the installation shall conform to the requirements of Chapters 25 through 32 of this code.

> **Exception:** The buried portions of systems embedded underground or under floors.

M2101.3 Protection of potable water. The potable water system shall be protected from backflow in accordance with the provisions listed in Section P2902.

M2101.4 Pipe penetrations. Openings through concrete or masonry building elements shall be sleeved.

M2101.5 Contact with building material. A hydronic piping system shall not be in direct contact with any building material that causes the piping material to degrade or corrode.

M2101.6 Drilling and notching. Wood-framed structural members shall be drilled, notched or altered in accordance with the provisions of Sections R502.8, R602.6, R602.6.1 and R802.7. Holes in load bearing members of cold-formed steel light-frame construction shall be permitted only in accordance with Sections R505.2.5, R603.2.5 and R804.2.5. In accordance with the provisions of Sections R505.3.5, R603.3.4 and R804.3.4, cutting and notching of flanges and lips of load-bearing members of cold-formed steel light-frame construction shall not be permitted. Structural insulated panels (SIPs) shall be drilled and notched or altered in accordance with the provisions of Section R613.

M2101.7 Prohibited tee applications. Fluid in the supply side of a hydronic system shall not enter a tee fitting through the branch opening.

M2101.8 Expansion, contraction and settlement. Piping shall be installed so that piping, connections and *equipment* shall not be subjected to excessive strains or stresses. Provisions shall be made to compensate for expansion, contraction, shrinkage and structural settlement.

M2101.9 Piping support. Hangers and supports shall be of material of sufficient strength to support the piping, and shall be fabricated from materials compatible with the piping material. Piping shall be supported at intervals not exceeding the spacing specified in Table M2101.9.

TABLE M2101.9
HANGER SPACING INTERVALS

PIPING MATERIAL	MAXIMUM HORIZONTAL SPACING (feet)	MAXIMUM VERTICAL SPACING (feet)
ABS	4	10
CPVC ≤ 1-inch pipe or tubing	3	5
CPVC ≥ $1^1/_4$ inches	4	10
Copper or copper alloy pipe	12	10
Copper or copper alloy tubing	6	10
PB pipe or tubing	2.67	4
PE pipe or tubing	2.67	4
PEX tubing	2.67	4
PP < 1-inch pipe or tubing	2.67	4
PP > $1^1/_4$ inches	4	10
PVC	4	10
Steel pipe	12	15
Steel tubing	8	10

For SI: 1 inch = 25.4 mm, 1 foot = 304.8 mm.

M2101.10 Tests. Hydronic piping shall be tested hydrostatically at a pressure of not less than 100 pounds per square inch (690 kPa) for a duration of not less than 15 minutes.

SECTION M2102
BASEBOARD CONVECTORS

M2102.1 General. Baseboard convectors shall be installed in accordance with the manufacturer's installation instructions. Convectors shall be supported independently of the hydronic piping.

SECTION M2103
FLOOR HEATING SYSTEMS

M2103.1 Piping materials. Piping for embedment in concrete or gypsum materials shall be standard-weight steel pipe, copper tubing, cross-linked polyethylene/aluminum/cross-linked polyethylene (PEX-AL-PEX) pressure pipe, chlorinated polyvinyl chloride (CPVC), polybutylene, cross-linked polyethylene (PEX) tubing or polypropylene (PP) with a minimum rating of 100 psi at 180°F (690 kPa at 82°C).

M2103.2 Thermal barrier required. Radiant floor heating systems shall have a thermal barrier in accordance with Sections M2103.2.1 through M2103.2.4.

> **Exception:** Insulation shall not be required in engineered systems where it can be demonstrated that the insulation will decrease the efficiency or have a negative effect on the installation.

TABLE M2101.1
HYDRONIC PIPING MATERIALS

MATERIAL	USE CODE[a]	STANDARD[b]	JOINTS	NOTES
Brass pipe	1	ASTM B 43	Brazed, welded, threaded, mechanical and flanged fittings	
Brass tubing	1	ASTM B 135	Brazed, soldered and mechanical fittings	
Chlorinated poly (vinyl chloride) (CPVC) pipe and tubing	1, 2, 3	ASTM D 2846	Solvent cement joints, compression joints and threaded adapters	
Copper pipe	1	ASTM B 42, B 302	Brazed, soldered and mechanical fittings threaded, welded and flanged	
Copper tubing (type K, L or M)	1, 2	ASTM B 75, B 88, B 251, B 306	Brazed, soldered and flared mechanical fittings	Joints embedded in concrete
Cross-linked polyethylene (PEX)	1, 2, 3	ASTM F 876, F 877	(See PEX fittings)	Install in accordance with manufacturer's instructions
Cross-linked polyethylene/aluminum/cross-linked polyethylene-(PEX-AL-PEX) pressure pipe	1, 2	ASTM F 1281 or CAN/ CSA B137.10	Mechanical, crimp/insert	Install in accordance with manufacturer's instructions
PEX fittings		ASTM F 877 ASTM F 1807 ASTM F 1960 ASTM F 2098 ASTM F 2159 ASTM F 2735	Copper-crimp/insert fittings, cold expansion fittings, stainless steel clamp, insert fittings	Install in accordance with manufacturer's instructions
Polybutylene (PB) pipe and tubing	1, 2, 3	ASTM D 3309	Heat-fusion, crimp/insert and compression	Joints in concrete shall be heat-fused
Polyethylene (PE) pipe, tubing and fittings (for ground source heat pump loop systems)	1, 2, 4	ASTM D 2513 ASTM D 3350 ASTM D 2513 ASTM D 3035 ASTM D 2447 ASTM D 2683 ASTM F 1055 ASTM D 2837 ASTM D 3350 ASTM D 1693	Heat-fusion	
Polyethylene/aluminum/polyethylene (PE-AL-PE) pressure pipe	1, 2, 3	ASTM F 1282 CSA B 137.9	Mechanical, crimp/insert	
Polypropylene (PP)	1, 2, 3	ISO 15874 ASTM F 2389	Heat-fusion joints, mechanical fittings, threaded adapters, compression joints	
Raised temperature polyethylene (PE-RT)	1, 2, 3	ASTM F 2623 ASTM F 2769	Copper crimp/insert fitting stainless steel clamp, insert fittings	
Soldering fluxes	1	ASTM B 813	Copper tube joints	
Steel pipe	1, 2	ASTM A 53, A 106	Brazed, welded, threaded, flanged and mechanical fittings	Joints in concrete shall be welded. Galvanized pipe shall not be welded or brazed.
Steel tubing	1	ASTM A 254	Mechanical fittings, welded	

For SI: °C = [(°F)-32]/1.8.

a. Use code:
1. Above ground.
2. Embedded in radiant systems.
3. Temperatures below 180°F only.
4. Low temperature (below 130°F) applications only.

b. Standards as listed in Chapter 44.

M2103.2.1 Slab-on-grade installation. Radiant piping used in slab-on-grade applications shall have insulating materials having a minimum *R*-value of 5 installed beneath the piping.

M2103.2.2 Suspended floor installation. In suspended floor applications, insulation shall be installed in the joist bay cavity serving the heating space above and shall consist of materials having a minimum *R*-value of 11.

M2103.2.3 Thermal break required. A thermal break consisting of asphalt expansion joint materials or similar insulating materials shall be provided at a point where a heated slab meets a foundation wall or other conductive slab.

M2103.2.4 Thermal barrier material marking. Insulating materials used in thermal barriers shall be installed so that the manufacturer's *R*-value mark *is readily observable upon inspection.*

M2103.3 Piping joints. Piping joints that are embedded shall be installed in accordance with the following requirements:

1. Steel pipe joints shall be welded.
2. Copper tubing shall be joined with brazing material having a melting point exceeding 1,000°F (538°C).
3. Polybutylene pipe and tubing joints shall be installed with socket-type heat-fused polybutylene fittings.
4. CPVC tubing shall be joined using solvent cement joints.
5. Polypropylene pipe and tubing joints shall be installed with socket-type heat-fused polypropylene fittings.
6. Cross-linked polyethylene (PEX) tubing shall be joined using cold expansion, insert or compression fittings.

M2103.4 Testing. Piping or tubing to be embedded shall be tested by applying a hydrostatic pressure of not less than 100 psi (690 kPa). The pressure shall be maintained for 30 minutes, during which all joints shall be visually inspected for leaks.

SECTION M2104
LOW TEMPERATURE PIPING

M2104.1 Piping materials. Low temperature piping for embedment in concrete or gypsum materials shall be as indicated in Table M2101.1.

M2104.2 Piping joints. Piping joints, other than those in Section M2103.3, that are embedded shall comply with the following requirements:

1. Cross-linked polyethylene (PEX) tubing shall be installed in accordance with the manufacturer's instructions.
2. Polyethylene tubing shall be installed with heat fusion joints.
3. Polypropylene (PP) tubing shall be installed in accordance with the manufacturer's instructions.

M2104.2.1 Polyethylene plastic pipe and tubing for ground source heat pump loop systems. Joints between polyethylene plastic pipe and tubing or fittings for ground source heat pump loop systems shall be heat fusion joints conforming to Section M2104.2.1.1, electrofusion joints conforming to Section M2104.2.1.2 or stab-type insertion joints conforming to Section M2104.2.1.3.

M2104.2.1.1 Heat-fusion joints. Joints shall be of the socket-fusion, saddle-fusion or butt-fusion type, fabricated in accordance with the piping manufacturer's instructions. Joint surfaces shall be clean and free of moisture. Joint surfaces shall be heated to melt temperatures and joined. The joint shall be undisturbed until cool. Fittings shall be manufactured in accordance with ASTM D 2683.

M2104.2.1.2 Electrofusion joints. Joint surfaces shall be clean and free of moisture, and scoured to expose virgin resin. Joint surfaces shall be heated to melt temperatures for the period of time specified by the manufacturer. The joint shall be undisturbed until cool. Fittings shall be manufactured in accordance with ASTM F 1055.

M2104.2.1.3 Stab-type insert fittings. Joint surfaces shall be clean and free of moisture. Pipe ends shall be chamfered and inserted into the fitting to full depth. Fittings shall be manufactured in accordance with ASTM D 2513.

M2104.3 Raised temperature polyethylene (PE-RT) plastic tubing. Joints between raised temperature polyethylene tubing and fittings shall conform to Sections M2104.3.1 and M2104.3.2. Mechanical joints shall be installed in accordance with the manufacturer's instructions.

M2104.3.1 Compression-type fittings. Where compression-type fittings include inserts and ferrules or O-rings, the fittings shall be installed without omitting such inserts and ferrules or O-rings.

M2104.3.2 PE-RT-to-metal connections. Solder joints in a metal pipe shall not occur within 18 inches (457 mm) of a transition from such metal pipe to PE-RT pipe.

M2104.4 Polyethylene/Aluminum/Polyethylene (PE-AL-PE) pressure pipe. Joints between polyethylene/aluminum/polyethylene pressure pipe and fittings shall conform to Sections M2104.4.1 and M2104.4.2. Mechanical joints shall be installed in accordance with the manufacturer's instructions.

M2104.4.1 Compression-type fittings. Where compression-type fittings include inserts and ferrules or O-rings, the fittings shall be installed without omitting such inserts and ferrules or O-rings.

M2104.4.2 PE-AL-PE to metal connections. Solder joints in a metal pipe shall not occur within 18 inches (457 mm) of a transition from such metal pipe to PE-AL-PE pipe.

SECTION M2105
GROUND SOURCE HEAT PUMP SYSTEM LOOP PIPING

M2105.1 Testing. The assembled loop system shall be pressure tested with water at 100 psi (690 kPa) for 30 minutes with no observed leaks before connection (header) trenches are backfilled. Flow rates and pressure drops shall be compared to calculated values. If actual flow rate or pressure drop figures differ from calculated values by more than 10 percent, the problem shall be identified and corrected.

CHAPTER 22

SPECIAL PIPING AND STORAGE SYSTEMS

SECTION M2201 OIL TANKS

M2201.1 Materials. Supply tanks shall be *listed* and *labeled* and shall conform to UL 58 for underground tanks and UL 80 for indoor tanks.

M2201.2 Above-ground tanks. The maximum amount of fuel oil stored above ground or inside of a building shall be 660 gallons (2498 L). The supply tank shall be supported on rigid noncombustible supports to prevent settling or shifting.

> **Exception:** The storage of fuel oil, used for space or water heating, above ground or inside buildings in quantities exceeding 660 gallons (2498 L) shall comply with NFPA 31.

M2201.2.1 Tanks within buildings. Supply tanks for use inside of buildings shall be of such size and shape to permit installation and removal from *dwellings* as whole units. Supply tanks larger than 10 gallons (38 L) shall be placed not less than 5 feet (1524 mm) from any fire or flame either within or external to any fuel-burning *appliance*.

M2201.2.2 Outside above-ground tanks. Tanks installed outside above ground shall be a minimum of 5 feet (1524 mm) from an adjoining property line. Such tanks shall be suitably protected from the weather and from physical damage.

M2201.3 Underground tanks. Excavations for underground tanks shall not undermine the foundations of existing structures. The clearance from the tank to the nearest wall of a *basement*, pit or property line shall not be less than 1 foot (305 mm). Tanks shall be set on and surrounded with noncorrosive inert materials such as clean earth, sand or gravel well tamped in place. Tanks shall be covered with not less than 1 foot (305 mm) of earth. Corrosion protection shall be provided in accordance with Section M2203.7.

M2201.4 Multiple tanks. Cross connection of two supply tanks shall be permitted in accordance with Section M2203.6.

M2201.5 Oil gauges. Inside tanks shall be provided with a device to indicate when the oil in the tank has reached a predetermined safe level. Glass gauges or a gauge subject to breakage that could result in the escape of oil from the tank shall not be used. Liquid-level indicating gauges shall comply with UL 180.

M2201.6 Flood-resistant installation. In flood hazard areas as established by Table R301.2(1), tanks shall be installed at or above the elevation required in Section R322.2.1 or R322.3.2 or shall be anchored to prevent flotation, collapse and lateral movement under conditions of the design flood.

M2201.7 Tanks abandoned or removed. Exterior above-grade fill piping shall be removed when tanks are abandoned or removed. Tank abandonment and removal shall be in accordance with the *International Fire Code.*

SECTION M2202 OIL PIPING, FITTING AND CONNECTIONS

M2202.1 Materials. Piping shall consist of steel pipe, copper tubing or steel tubing conforming to ASTM A 539. Aluminum tubing shall not be used between the fuel-oil tank and the burner units.

M2202.2 Joints and fittings. Piping shall be connected with standard fittings compatible with the piping material. Cast iron fittings shall not be used for oil piping. Unions requiring gaskets or packings, right or left couplings, and sweat fittings employing solder having a melting point less than 1,000°F (538°C) shall not be used for oil piping. Threaded joints and connections shall be made tight with a lubricant or pipe thread compound.

M2202.3 Flexible connectors. Flexible metallic hoses shall be *listed* and *labeled* in accordance with UL 536 and shall be installed in accordance with their *listing* and *labeling* and the manufacturer's installation instructions. Connectors made from combustible materials shall not be used inside of buildings or above ground outside of buildings.

SECTION M2203 INSTALLATION

M2203.1 General. Piping shall be installed in a manner to avoid placing stresses on the piping, and to accommodate expansion and contraction of the piping system.

M2203.2 Supply piping. Supply piping used in the installation of oil burners and *appliances* shall be not smaller than $^3/_8$-inch (9 mm) pipe or $^3/_8$-inch (9 mm) outside diameter tubing. Copper tubing and fittings shall be a minimum of Type L.

M2203.3 Fill piping. Fill piping shall terminate outside of buildings at a point at least 2 feet (610 mm) from any building opening at the same or lower level. Fill openings shall be equipped with a tight metal cover.

M2203.4 Vent piping. Vent piping shall be not smaller than $1^1/_4$-inch (32 mm) pipe. Vent piping shall be laid to drain toward the tank without sags or traps in which the liquid can collect. Vent pipes shall not be cross connected with fill pipes, lines from burners or overflow lines from auxiliary tanks. The lower end of a vent pipe shall enter the tank through the top and shall extend into the tank not more than 1 inch (25 mm).

M2203.5 Vent termination. Vent piping shall terminate outside of buildings at a point not less than 2 feet (610 mm), measured vertically or horizontally, from any building opening. Outer ends of vent piping shall terminate in a weatherproof cap or fitting having an unobstructed area at least equal to the cross-sectional area of the vent pipe, and shall be located sufficiently above the ground to avoid being obstructed by snow and ice.

M2203.6 Cross connection of tanks. Cross connection of two supply tanks, not exceeding 660 gallons (2498 L) aggregate capacity, with gravity flow from one tank to another, shall be acceptable providing that the two tanks are on the same horizontal plane.

M2203.7 Corrosion protection. Underground tanks and buried piping shall be protected by corrosion-resistant coatings or special alloys or fiberglass-reinforced plastic.

SECTION M2204 OIL PUMPS AND VALVES

M2204.1 Pumps. Oil pumps shall be positive displacement types that automatically shut off the oil supply when stopped. Automatic pumps shall be *listed* and *labeled* in accordance with UL 343 and shall be installed in accordance with their *listing*.

M2204.2 Shutoff valves. A *readily accessible* manual shutoff valve shall be installed between the oil supply tank and the burner. Where the shutoff valve is installed in the discharge line of an oil pump, a pressure-relief valve shall be incorporated to bypass or return surplus oil. Valves shall comply with UL 842.

M2204.3 Maximum pressure. Pressure at the oil supply inlet to an *appliance* shall be not greater than 3 pounds per square inch (20.7 kPa).

M2204.4 Relief valves. Fuel-oil lines incorporating heaters shall be provided with relief valves that will discharge to a return line when excess pressure exists.

CHAPTER 23
SOLAR ENERGY SYSTEMS

SECTION M2301 THERMAL SOLAR ENERGY SYSTEMS

M2301.1 General. This section provides for the design, construction, installation, *alteration* and repair of *equipment* and systems using thermal solar energy to provide space heating or cooling, hot water heating and swimming pool heating.

M2301.2 Installation. Installation of solar energy systems shall comply with Sections M2301.2.1 through M2301.2.9.

M2301.2.1 Access. Solar energy collectors, controls, dampers, fans, blowers and pumps shall be accessible for inspection, maintenance, repair and replacement.

M2301.2.2 Roof-mounted collectors. The roof shall be constructed to support the loads imposed by roof-mounted solar collectors. Roof-mounted solar collectors that serve as a roof covering shall conform to the requirements for roof coverings in Chapter 9 of this code. Where mounted on or above the roof coverings, the collectors and supporting structure shall be constructed of noncombustible materials or fire-retardant-treated wood equivalent to that required for the roof construction.

M2301.2.3 Pressure and temperature relief. System components containing fluids shall be protected with pressure- and temperature-relief valves. Relief devices shall be installed in sections of the system so that a section cannot be valved off or isolated from a relief device.

M2301.2.4 Vacuum relief. System components that might be subjected to pressure drops below atmospheric pressure during operation or shutdown shall be protected by a vacuum-relief valve.

M2301.2.5 Protection from freezing. System components shall be protected from damage resulting from freezing of heat-transfer liquids at the winter design temperature provided in Table R301.2(1). Freeze protection shall be provided by heating, insulation, thermal mass and heat transfer fluids with freeze points lower than the winter design temperature, heat tape or other *approved* methods, or combinations thereof.

Exception: Where the winter design temperature is greater than 32°F (0°C).

M2301.2.6 Expansion tanks. Expansion tanks in solar energy systems shall be installed in accordance with Section M2003 in closed fluid loops that contain heat transfer fluid.

M2301.2.7 Roof and wall penetrations. Roof and wall penetrations shall be flashed and sealed in accordance with Chapter 9 of this code to prevent entry of water, rodents and insects.

M2301.2.8 Solar loop isolation. Valves shall be installed to allow the solar collectors to be isolated from the remainder of the system. Each isolation valve shall be labeled with the open and closed position.

M2301.2.9 Maximum temperature limitation. Systems shall be equipped with means to limit the maximum water temperature of the system fluid entering or exchanging heat with any pressurized vessel inside the *dwelling* to 180°F (82°C). This protection is in addition to the required temperature- and pressure-relief valves required by Section M2301.2.3.

M2301.3 Labeling. *Labeling* shall comply with Sections M2301.3.1 and M2301.3.2.

M2301.3.1 Collectors. Collectors shall be *listed* and *labeled* to show the manufacturer's name, model number, serial number, collector weight, collector maximum allowable temperatures and pressures, and the type of heat transfer fluids that are compatible with the collector. The *label* shall clarify that these specifications apply only to the collector.

M2301.3.2 Thermal storage units. Pressurized thermal storage units shall be *listed* and *labeled* to show the manufacturer's name, model number, serial number, storage unit maximum and minimum allowable operating temperatures and pressures, and the type of heat transfer fluids that are compatible with the storage unit. The *label* shall clarify that these specifications apply only to the thermal storage unit.

M2301.4 Prohibited heat transfer fluids. Flammable gases and liquids shall not be used as heat transfer fluids.

M2301.5 Backflow protection. Connections from the potable water supply to solar systems shall comply with Section P2902.5.5.

SECTION M2302 PHOTOVOLTAIC SOLAR ENERGY SYSTEMS

M2302.1 General. This section provides for the design, construction, installation, alteration, and repair of photovoltaic equipment and systems.

M2302.2 Requirements. The installation, inspection, maintenance, repair and replacement of photovoltaic systems and all system components shall comply with the manufacturer's instructions, Sections M2302.2.1 through M2302.2.3 and NFPA 70.

M2302.2.1 Roof-mounted panels and modules. Where photovoltaic panels and modules are installed on roofs, the roof shall be constructed to support the loads imposed by such modules. Roof-mounted photovoltaic panels and modules that serve as roof covering shall conform to the requirements for roof coverings in Chapter 9. Where mounted on or above the roof coverings, the photovoltaic panels and modules and supporting structure shall be constructed of noncombustible materials or fire-retardant-treated wood equivalent to that required for the roof construction.

M2302.2.2 Roof and wall penetrations. Roof and wall penetrations shall be flashed and sealed in accordance with Chapter 9 to prevent entry of water, rodents, and insects.

M2302.2.3 Ground-mounted panels and modules. Ground-mounted panels and modules shall be installed in accordance with the manufacturer's instructions.

M2302.3 Photovoltaic panels and modules. Photovoltaic panels and modules shall be listed and labeled in accordance with UL 1703.

M2302.4 Inverters. Inverters shall be listed and labeled in accordance with UL 1741. Systems connected to the utility grid shall use inverters listed for utility interaction.

Part VI—Fuel Gas

CHAPTER 24

FUEL GAS

The text of this chapter is extracted from the 2012 edition of the *International Fuel Gas Code* and has been modified where necessary to conform to the scope of application of the *International Residential Code for One- and Two-Family Dwellings*. The section numbers appearing in parentheses after each section number are the section numbers of the corresponding text in the *International Fuel Gas Code.*

SECTION G2401 (101) GENERAL

G2401.1 (101.2) Application. This chapter covers those *fuel gas piping systems*, fuel-gas *appliances* and related accessories, *venting systems* and *combustion air* configurations most commonly encountered in the construction of one- and two-family dwellings and structures regulated by this *code*.

Coverage of *piping systems* shall extend from the *point of delivery* to the outlet of the *appliance* shutoff *valves* (see definition of *"Point of delivery"*). *Piping systems* requirements shall include design, materials, components, fabrication, assembly, installation, testing, inspection, operation and maintenance. Requirements for gas *appliances* and related accessories shall include installation, *combustion* and ventilation air and venting and connections to *piping systems*.

The omission from this chapter of any material or method of installation provided for in the *International Fuel Gas Code* shall not be construed as prohibiting the use of such material or method of installation. *Fuel-gas piping systems*, fuel-gas *appliances* and related accessories, *venting systems* and *combustion air* configurations not specifically covered in these chapters shall comply with the applicable provisions of the *International Fuel Gas Code*.

Gaseous hydrogen systems shall be regulated by Chapter 7 of the *International Fuel Gas Code*.

This chapter shall not apply to the following:

1. Liquified natural gas (LNG) installations.
2. Temporary LP-*gas piping* for buildings under construction or renovation that is not to become part of the permanent *piping system*.
3. Except as provided in Section G2412.1.1, *gas piping, meters*, gas *pressure regulators*, and other appurtenances used by the serving gas supplier in the distribution of gas, other than undiluted LP-gas.
4. Portable LP-gas *appliances* and *equipment* of all types that is not connected to a fixed fuel *piping system*.
5. Portable fuel cell *appliances* that are neither connected to a fixed *piping system* nor interconnected to a power grid.
6. Installation of hydrogen gas, LP-gas and compressed natural gas (CNG) systems on vehicles.

SECTION G2402 (201) GENERAL

G2402.1 (201.1) Scope. Unless otherwise expressly stated, the following words and terms shall, for the purposes of this chapter, have the meanings indicated in this chapter.

G2402.2 (201.2) Interchangeability. Words used in the present tense include the future; words in the masculine gender include the feminine and neuter; the singular number includes the plural and the plural, the singular.

G2402.3 (201.3) Terms defined in other codes. Where terms are not defined in this *code* and are defined in the *International Building Code*, *International Fire Code*, *International Mechanical Code* or *International Plumbing Code*, such terms shall have meanings ascribed to them as in those *codes*.

SECTION G2403 (202) GENERAL DEFINITIONS

AIR CONDITIONING, GAS FIRED. A gas-burning, automatically operated *appliance* for supplying cooled and/or dehumidified air or chilled liquid.

AIR, EXHAUST. Air being removed from any space or piece of *equipment* or *appliance* and conveyed directly to the atmosphere by means of openings or ducts.

AIR-HANDLING UNIT. A blower or fan used for the purpose of distributing supply air to a room, space or area.

AIR, MAKEUP. Air that is provided to replace air being exhausted.

ALTERATION. A change in a system that involves an extension, addition or change to the arrangement, type or purpose of the original installation.

ANODELESS RISER. A transition assembly in which plastic *piping* is installed and terminated above ground outside of a building.

APPLIANCE. Any apparatus or device that utilizes a fuel or raw material to produce light, heat, power, refrigeration or air conditioning.

APPLIANCE, AUTOMATICALLY CONTROLLED. Appliances equipped with an automatic *burner* ignition and safety shut-off device and other automatic devices, which accomplish complete turn-on and shut-off of the gas to the *main burner* or *burners*, and graduate the gas supply to the *burner* or *burners*, but do not affect complete shut-off of the gas.

APPLIANCE, FAN-ASSISTED COMBUSTION. An *appliance* equipped with an integral mechanical means to either draw or force products of *combustion* through the *combustion* chamber or heat exchanger.

APPLIANCE, UNVENTED. An *appliance* designed or installed in such a manner that the products of *combustion* are not conveyed by a vent or *chimney* directly to the outside atmosphere.

APPLIANCE, VENTED. An *appliance* designed and installed in such a manner that all of the products of *combustion* are conveyed directly from the *appliance* to the outside atmosphere through an *approved chimney* or vent system.

APPROVED. Acceptable to the *code official* or other authority having jurisdiction.

ATMOSPHERIC PRESSURE. The pressure of the weight of air and water vapor on the surface of the earth, approximately 14.7 pounds per square inch (psia) (101 kPa absolute) at sea level.

AUTOMATIC IGNITION. Ignition of gas at the *burner(s)* when the gas controlling device is turned on, including reignition if the flames on the *burner(s)* have been extinguished by means other than by the closing of the gas controlling device.

BAROMETRIC DRAFT REGULATOR. A balanced *damper* device attached to a *chimney,* vent *connector,* breeching or flue gas manifold to protect *combustion appliances* by controlling *chimney draft*. A double-acting *barometric draft regulator* is one whose balancing *damper* is free to move in either direction to protect *combustion appliances* from both excessive *draft* and backdraft.

BOILER, LOW-PRESSURE. A self-contained gas-fired *appliance* for supplying steam or hot water.

Hot water heating boiler. A boiler in which no steam is generated, from which hot water is circulated for heating purposes and then returned to the boiler, and that operates at water pressures not exceeding 160 psig (1100 kPa gauge) and at water temperatures not exceeding 250°F (121°C) at or near the boiler outlet.

Hot water supply boiler. A boiler, completely filled with water, which furnishes hot water to be used externally to itself, and that operates at water pressures not exceeding 160 psig (1100 kPa gauge) and at water temperatures not exceeding 250°F (121°C) at or near the boiler outlet.

Steam heating boiler. A boiler in which steam is generated and that operates at a steam pressure not exceeding 15 psig (100 kPa gauge).

BONDING JUMPER. A conductor installed to electrically connect metallic *gas piping* to the grounding electrode system.

BRAZING. A metal joining process wherein coalescence is produced by the use of a nonferrous filler metal having a melting point above 1,000°F (538°C), but lower than that of the base metal being joined. The filler material is distributed between the closely fitted surfaces of the joint by capillary action.

BTU. Abbreviation for British thermal unit, which is the quantity of heat required to raise the temperature of 1 pound (454 g) of water 1°F (0.56°C) (1 *Btu* = 1055 J).

BURNER. A device for the final conveyance of the gas, or a mixture of gas and air, to the *combustion* zone.

Induced-draft. A *burner* that depends on *draft* induced by a fan that is an integral part of the *appliance* and is located downstream from the *burner.*

Power. A *burner* in which gas, air or both are supplied at pressures exceeding, for gas, the line pressure, and for air, *atmospheric pressure*, with this added pressure being applied at the *burner.*

CHIMNEY. A primarily vertical structure containing one or more flues, for the purpose of carrying gaseous products of *combustion* and air from an *appliance* to the outside atmosphere.

Factory-built chimney. A listed and labeled *chimney* composed of factory-made components, assembled in the field in accordance with manufacturer's instructions and the conditions of the listing.

Masonry chimney. A field-constructed *chimney* composed of solid masonry units, bricks, stones or concrete.

CLEARANCE. The minimum distance through air measured between the heat-producing surface of the mechanical *appliance*, device or *equipment* and the surface of the combustible material or assembly.

CLOTHES DRYER. An *appliance* used to dry wet laundry by means of heated air.

Type 1. Factory-built package, multiple production. Primarily used in the family living environment. Usually the smallest unit physically and in function output.

CODE. These regulations, subsequent amendments thereto, or any emergency rule or regulation that the administrative authority having jurisdiction has lawfully adopted.

CODE OFFICIAL. The officer or other designated authority charged with the administration and enforcement of this *code*, or a duly authorized representative.

COMBUSTIBLE ASSEMBLY. Wall, floor, ceiling or other assembly constructed of one or more component materials that are not defined as noncombustible.

COMBUSTIBLE MATERIAL. Any material not defined as noncombustible.

COMBUSTION. In the context of this *code*, refers to the rapid oxidation of fuel accompanied by the production of heat or heat and light.

COMBUSTION AIR. Air necessary for complete *combustion* of a fuel, including theoretical air and excess air.

COMBUSTION CHAMBER. The portion of an *appliance* within which *combustion* occurs.

COMBUSTION PRODUCTS. Constituents resulting from the *combustion* of a fuel with the oxygen of the air, including the inert gases, but excluding excess air.

CONCEALED LOCATION. A location that cannot be accessed without damaging permanent parts of the building structure or finish surface. Spaces above, below or behind readily removable panels or doors shall not be considered as concealed.

CONCEALED PIPING. *Piping* that is located in a *concealed location* (see *"Concealed location"*).

CONDENSATE. The liquid that condenses from a gas (including flue gas) caused by a reduction in temperature or increase in pressure.

CONNECTOR, APPLIANCE (Fuel). Rigid metallic *pipe* and fittings, semirigid metallic *tubing* and fittings or a listed and labeled device that connects an *appliance* to the *gas piping system*.

CONNECTOR, CHIMNEY OR VENT. The *pipe* that connects an *appliance* to a *chimney* or vent.

CONTROL. A manual or automatic device designed to regulate the gas, air, water or electrical supply to, or operation of, a mechanical system.

CONVERSION BURNER. A unit consisting of a *burner* and its *controls* for installation in an *appliance* originally utilizing another fuel.

CUBIC FOOT. The amount of gas that occupies 1 *cubic foot* (0.02832 m^3) when at a temperature of 60°F (16°C), saturated with water vapor and under a pressure equivalent to that of 30 inches of mercury (101 kPa).

DAMPER. A manually or automatically controlled device to regulate *draft* or the rate of flow of air or *combustion* gases.

DECORATIVE GAS APPLIANCE, VENTED. A *vented appliance* wherein the primary function lies in the aesthetic effect of the flames.

DECORATIVE GAS APPLIANCES FOR INSTALLATION IN VENTED FIREPLACES. A *vented appliance* designed for installation within the fire chamber of a vented *fireplace*, wherein the primary function lies in the aesthetic effect of the flames.

DEMAND. The maximum amount of gas input required per unit of time, usually expressed in cubic feet per hour, or *Btu*/h (1 *Btu*/h = 0.2931 W).

DESIGN FLOOD ELEVATION. The elevation of the "design flood," including wave height, relative to the datum specified on the community's legally designated flood hazard map.

DILUTION AIR. Air that is introduced into a *draft hood* and is mixed with the *flue gases*.

DIRECT-VENT APPLIANCES. *Appliances* that are constructed and installed so that all air for *combustion* is derived directly from the outside atmosphere and all *flue gases* are discharged directly to the outside atmosphere.

DRAFT. The pressure difference existing between the *appliance* or any component part and the atmosphere, that causes a continuous flow of air and products of *combustion* through the gas passages of the *appliance* to the atmosphere.

Mechanical or induced draft. The pressure difference created by the action of a fan, blower or ejector that is located between the *appliance* and the *chimney* or vent termination.

Natural draft. The pressure difference created by a vent or *chimney* because of its height, and the temperature difference between the *flue gases* and the atmosphere.

DRAFT HOOD. A nonadjustable device built into an *appliance*, or made as part of the vent *connector* from an *appliance*, that is designed to (1) provide for ready escape of the *flue gases* from the *appliance* in the event of no *draft,* backdraft, or stoppage beyond the *draft hood*, (2) prevent a backdraft from entering the *appliance*, and (3) neutralize the effect of stack action of the *chimney* or gas vent upon operation of the *appliance*.

DRAFT REGULATOR. A device that functions to maintain a desired *draft* in the *appliance* by automatically reducing the *draft* to the desired value.

DRIP. The container placed at a low point in a system of *piping* to collect *condensate* and from which the *condensate* is removable.

DUCT FURNACE. A warm-air *furnace* normally installed in an air-distribution duct to supply warm air for heating. This definition shall apply only to a warm-air heating *appliance* that depends for air circulation on a blower not furnished as part of the *furnace*.

DWELLING UNIT. A *single* unit providing complete, independent living facilities for one or more persons, including permanent provisions for living, sleeping, eating, cooking and sanitation.

EQUIPMENT. Apparatus and devices other than *appliances*.

EXCESS FLOW VALVE (EFV). A valve designed to activate when the fuel gas passing through it exceeds a prescribed flow rate.

EXTERIOR MASONRY CHIMNEYS. *Masonry chimneys* exposed to the outdoors on one or more sides below the roof line.

FIREPLACE. A fire chamber and hearth constructed of noncombustible material for use with solid fuels and provided with a *chimney*.

Masonry fireplace. A hearth and fire chamber of solid masonry units such as bricks, stones, listed masonry units or reinforced concrete, provided with a suitable *chimney*.

Factory-built fireplace. A *fireplace* composed of listed factory-built components assembled in accordance with the terms of listing to form the completed *fireplace*.

FLAME SAFEGUARD. A device that will automatically shut off the fuel supply to a *main burner* or group of *burners* when the means of ignition of such *burners* becomes inoperative, and when flame failure occurs on the *burner* or group of *burners*.

FLASHBACK ARRESTOR CHECK VALVE. A device that will prevent the backflow of one gas into the supply system of another gas and prevent the passage of flame into the gas supply system.

FLOOD HAZARD AREA. The greater of the following two areas:

1. The area within a floodplain subject to a 1 percent or greater chance of flooding in any given year.
2. This area designated as a *flood hazard area* on a community's flood hazard map, or otherwise legally designated.

FLOOR FURNACE. A completely self-contained *furnace* suspended from the floor of the space being heated, taking air for *combustion* from outside such space and with means for observing flames and lighting the *appliance* from such space.

FLUE, APPLIANCE. The passage(s) within an *appliance* through which *combustion products* pass from the *combustion chamber* of the *appliance* to the *draft hood* inlet opening on an *appliance* equipped with a *draft hood* or to the outlet of the *appliance* on an *appliance* not equipped with a *draft hood.*

FLUE COLLAR. That portion of an *appliance* designed for the attachment of a *draft hood, vent connector* or venting system.

FLUE GASES. Products of *combustion* plus excess air in *appliance flues* or heat exchangers.

FLUE LINER (LINING). A system or material used to form the inside surface of a flue in a *chimney* or vent, for the purpose of protecting the surrounding structure from the effects of *combustion products* and for conveying *combustion products* without leakage to the atmosphere.

FUEL GAS. A natural gas, manufactured gas, *liquefied petroleum gas* or mixtures of these gases.

FUEL GAS UTILIZATION EQUIPMENT. See *"Appliance."*

FURNACE. A completely self-contained heating unit that is designed to supply heated air to spaces remote from or adjacent to the *appliance* location.

FURNACE, CENTRAL FURNACE. A self-contained *appliance* for heating air by transfer of heat of *combustion* through metal to the air, and designed to supply heated air through ducts to spaces remote from or adjacent to the *appliance* location.

FURNACE PLENUM. An air compartment or chamber to which one or more ducts are connected and which forms part of an air distribution system.

GAS CONVENIENCE OUTLET. A permanently mounted, manually operated device that provides the means for connecting an *appliance* to, and disconnecting an *appliance* from, the gas supply *piping*. The device includes an integral, manually operated *valve* with a nondisplaceable *valve* member and is designed so that disconnection of an *appliance* only occurs when the manually operated *valve* is in the closed position.

GAS PIPING. An installation of *pipe*, *valves* or fittings installed on a premises or in a building and utilized to convey *fuel gas.*

HAZARDOUS LOCATION. Any location considered to be a fire hazard for flammable vapors, dust, combustible fibers or other highly combustible substances. The location is not necessarily categorized in the *International Building Code* as a high-hazard use group classification.

HOUSE PIPING. See *"Piping system."*

IGNITION PILOT. A *pilot* that operates during the lighting cycle and discontinues during *main burner* operation.

IGNITION SOURCE. A flame spark or hot surface capable of igniting flammable vapors or fumes. Such sources include *appliance burners, burner* ignitors and electrical switching devices.

INFRARED RADIANT HEATER. A heater which directs a substantial amount of its energy output in the form of infrared radiant energy into the area to be heated. Such heaters are of either the vented or unvented type.

JOINT, FLARED. A metal-to-metal compression joint in which a conical spread is made on the end of a tube that is compressed by a flare nut against a mating flare.

JOINT, MECHANICAL. A general form of gas-tight joints obtained by the joining of metal parts through a positive-holding mechanical construction, such as press joint, flanged joint, threaded joint, flared joint or compression joint.

JOINT, PLASTIC ADHESIVE. A joint made in thermoset *plastic piping* by the use of an adhesive substance which forms a continuous bond between the mating surfaces without dissolving either one of them.

LEAK CHECK. An operation performed on a *gas piping system* to verify that the system does not leak.

LIQUEFIED PETROLEUM GAS or LPG (LP-GAS). *Liquefied petroleum gas* composed predominately of propane, propylene, butanes or butylenes, or mixtures thereof that is gaseous under normal atmospheric conditions, but is capable of being liquefied under moderate pressure at normal temperatures.

LIVING SPACE. Space within a *dwelling unit* utilized for living, sleeping, eating, cooking, bathing, washing and sanitation purposes.

LOG LIGHTER, GAS-FIRED. A manually operated solid-fuel ignition *appliance* for installation in a vented solid-fuel-burning *fireplace.*

MAIN BURNER. A device or group of devices essentially forming an integral unit for the final conveyance of gas or a mixture of gas and air to the *combustion* zone, and on which *combustion* takes place to accomplish the function for which the *appliance* is designed.

METER. The instrument installed to measure the volume of gas delivered through it.

MODULATING. Modulating or throttling is the action of a *control* from its maximum to minimum position in either predetermined steps or increments of movement as caused by its actuating medium.

NONCOMBUSTIBLE MATERIALS. Materials that, when tested in accordance with ASTM E 136, have at least three of four specimens tested meeting all of the following criteria:

1. The recorded temperature of the surface and interior thermocouples shall not at any time during the test rise more than 54°F (30°C) above the furnace temperature at the beginning of the test.
2. There shall not be flaming from the specimen after the first 30 seconds.
3. If the weight loss of the specimen during testing exceeds 50 percent, the recorded temperature of the surface and interior thermocouples shall not at any time during the test rise above the furnace air temperature at the beginning of the test, and there shall not be flaming of the specimen.

OFFSET (VENT). A combination of *approved* bends that make two changes in direction bringing one section of the vent out of line, but into a line parallel with the other section.

OUTLET. The point at which a gas-fired *appliance* connects to the *gas piping system.*

OXYGEN DEPLETION SAFETY SHUTOFF SYSTEM (ODS). A system designed to act to shut off the gas supply to the main and *pilot burners* if the oxygen in the surrounding atmosphere is reduced below a predetermined level.

PILOT. A small flame that is utilized to ignite the gas at the *main burner* or *burners.*

PIPING. Where used in this *code, "piping"* refers to either *pipe* or *tubing*, or both.

Pipe. A rigid conduit of iron, steel, copper, brass or plastic.

Tubing. Semirigid conduit of copper, aluminum, plastic or steel.

PIPING SYSTEM. All fuel *piping, valves* and fittings from the outlet of the *point of delivery* to the outlets of the *appliance* shutoff valves.

PLASTIC, THERMOPLASTIC. A plastic that is capable of being repeatedly softened by increase of temperature and hardened by decrease of temperature.

POINT OF DELIVERY. For natural gas systems, the *point of delivery* is the outlet of the service meter assembly or the outlet of the service regulator or service shutoff valve where a meter is not provided. Where a valve is provided at the outlet of the service meter assembly, such valve shall be considered to be downstream of the *point of delivery*. For undiluted liquefied petroleum gas systems, the point of delivery shall be considered to be the outlet of the service pressure regulator, exclusive of line gas regulators, in the system.

PRESSURE DROP. The loss in pressure due to friction or obstruction in pipes, *valves*, fittings, *regulators* and *burners.*

PRESSURE TEST. An operation performed to verify the gas-tight integrity of *gas piping* following its installation or modification.

READY ACCESS (TO). That which enables a device, *appliance* or *equipment* to be directly reached, without requiring the removal or movement of any panel, door or similar obstruction. (See "Access.")

REGULATOR. A device for controlling and maintaining a uniform gas supply pressure, either pounds-to-inches water column (MP *regulator*) or inches-to-inches water column (*appliance regulator*).

REGULATOR, GAS APPLIANCE. A *pressure regulator* for controlling pressure to the manifold of the gas *appliance.*

REGULATOR, LINE GAS PRESSURE. A device placed in a gas line between the *service pressure regulator* and the *appliance* for controlling, maintaining or reducing the pressure in that portion of the *piping system* downstream of the device.

REGULATOR, MEDIUM-PRESSURE (MP Regulator). A line *pressure regulator* that reduces gas pressure from the range of greater than 0.5 psig (3.4 kPa) and less than or equal to 5 psig (34.5 kPa) to a lower pressure.

REGULATOR, PRESSURE. A device placed in a gas line for reducing, controlling and maintaining the pressure in that portion of the *piping system* downstream of the device.

REGULATOR, SERVICE PRESSURE. For natural gas systems, a device installed by the serving gas supplier to reduce and limit the service line pressure to delivery pressure. For undiluted liquefied petroleum gas systems, the regulator located upstream from all line gas pressure regulators, where installed, and downstream from any first stage or a high pressure regulator in the system.

RELIEF OPENING. The opening provided in a *draft hood* to permit the ready escape to the atmosphere of the flue products from the *draft hood* in the event of no *draft*, backdraft or stoppage beyond the *draft hood*, and to permit air into the *draft hood* in the event of a strong *chimney* updraft.

RELIEF VALVE (DEVICE). A safety *valve* designed to forestall the development of a dangerous condition by relieving either pressure, temperature or vacuum in the hot water supply system.

RELIEF VALVE, PRESSURE. An *automatic valve* which opens and closes a *relief vent*, depending on whether the pressure is above or below a predetermined value.

RELIEF VALVE, TEMPERATURE.

Manual reset type. A *valve* which automatically opens a *relief* vent at a predetermined temperature and which must be manually returned to the closed position.

Reseating or self-closing type. An *automatic valve* which opens and closes a *relief* vent, depending on whether the temperature is above or below a predetermined value.

RELIEF VALVE, VACUUM. A *valve* that automatically opens and closes a vent for relieving a vacuum within the hot water supply system, depending on whether the vacuum is above or below a predetermined value.

RISER, GAS. A vertical *pipe* supplying *fuel gas.*

ROOM HEATER, UNVENTED. See *"Unvented room heater."*

ROOM HEATER, VENTED. A free-standing gas-fired heating unit used for direct heating of the space in and adjacent to that in which the unit is located. (See also *"Vented room heater."*)

SAFETY SHUTOFF DEVICE. See *"Flame safeguard."*

SHAFT. An enclosed space extending through one or more stories of a building, connecting vertical openings in successive floors, or floors and the roof.

SPECIFIC GRAVITY. As applied to gas, *specific gravity* is the ratio of the weight of a given volume to that of the same volume of air, both measured under the same condition.

THERMOSTAT.

Electric switch type. A device that senses changes in temperature and controls electrically, by means of separate components, the flow of gas to the *burner(s)* to maintain selected temperatures.

Integral gas valve type. An automatic device, actuated by temperature changes, designed to control the gas supply to the *burner(s)* in order to maintain temperatures between predetermined limits, and in which the thermal actuating element is an integral part of the device.

1. Graduating thermostat. A *thermostat* in which the motion of the *valve* is approximately in direct proportion to the effective motion of the thermal element induced by temperature change.
2. Snap-acting thermostat. A *thermostat* in which the thermostatic valve travels instantly from the closed to the open position, and vice versa.

THIRD-PARTY CERTIFICATION AGENCY. An approved agency operating a product or material certification system that incorporates initial product testing, assessment and surveillance of a manufacturer's quality control system.

THIRD-PARTY CERTIFIED. Certification obtained by the manufacturer indicating that the function and performance characteristics of a product or material have been determined by testing and ongoing surveillance by an approved third-party certification agency. Assertion of certification is in the form of identification in accordance with the requirements of the third-party certification agency.

THIRD-PARTY TESTED. Procedure by which an approved testing laboratory provides documentation that a product, material or system conforms to specified requirements.

TRANSITION FITTINGS, PLASTIC TO STEEL. An adapter for joining plastic *pipe* to steel *pipe*. The purpose of this fitting is to provide a permanent, pressure-tight connection between two materials that cannot be joined directly one to another.

UNIT HEATER.

High-static pressure type. A self-contained, automatically controlled, *vented appliance* having integral means for circulation of air against 0.2 inch w.c. (50 Pa) or greater static pressure. Such *appliance* is equipped with provisions for attaching an outlet air duct and, where the *appliance* is for indoor installation remote from the space to be heated, is also equipped with provisions for attaching an inlet air duct.

Low-static pressure type. A self-contained, automatically controlled, *vented appliance*, intended for installation in the space to be heated without the use of ducts, having integral means for circulation of air. Such units are allowed to be equipped with louvers or face extensions made in accordance with the manufacturer's specifications.

UNVENTED ROOM HEATER. An unvented heating *appliance* designed for stationary installation and utilized to provide comfort heating. Such *appliances* provide radiant heat or convection heat by gravity or fan circulation directly from the heater and do not utilize ducts.

VALVE. A device used in *piping* to control the gas supply to any section of a system of *piping* or to an *appliance.*

Appliance shutoff. A *valve* located in the *piping system*, used to isolate individual *appliances* for purposes such as service or replacement.

Automatic. An automatic or semiautomatic device consisting essentially of a *valve* and an operator that control the gas supply to the *burner(s)* during operation of an *appliance*. The operator shall be actuated by application of gas pressure on a flexible diaphragm, by electrical means, by mechanical means or by other *approved* means.

Automatic gas shutoff. A *valve* used in conjunction with an automatic gas shutoff device to shut off the gas supply to a water heating system. It shall be constructed integrally with the gas shutoff device or shall be a separate assembly.

Individual main burner. A *valve* that controls the gas supply to an individual *main burner.*

Main burner control. A *valve* that controls the gas supply to the *main burner* manifold.

Manual main gas-control. A manually operated *valve* in the gas line for the purpose of completely turning on or shutting off the gas supply to the *appliance*, except to a *pilot* or pilots that have independent shutoff.

Manual reset. An automatic shutoff *valve* installed in the gas supply *piping* and set to shut off when unsafe conditions occur. The device remains closed until manually reopened.

Service shutoff. A *valve*, installed by the serving gas supplier between the service *meter* or source of supply and the customer *piping system,* to shut off the entire *piping system.*

VENT. A *pipe* or other conduit composed of factory-made components, containing a passageway for conveying *combustion products* and air to the atmosphere, listed and labeled for use with a specific type or class of *appliance.*

Special gas vent. A vent listed and labeled for use with listed Category II, III and IV gas *appliances.*

Type B vent. A vent listed and labeled for use with *appliances* with *draft hoods* and other Category I *appliances* that are listed for use with Type B vents.

Type BW vent. A vent listed and labeled for use with wall *furnaces*.

Type L vent. A vent listed and labeled for use with *appliances* that are listed for use with Type L or Type B vents.

VENT CONNECTOR. See "Connector."

VENT PIPING.

Breather. *Piping* run from a pressure-regulating device to the outdoors, designed to provide a reference to *atmospheric pressure*. If the device incorporates an integral pressure *relief* mechanism, a breather vent can also serve as a *relief* vent.

Relief. *Piping* run from a pressure-regulating or pressure-limiting device to the outdoors, designed to provide for the safe venting of gas in the event of excessive pressure in the *gas piping system*.

VENTED GAS APPLIANCE CATEGORIES. *Appliances* that are categorized for the purpose of vent selection are classified into the following four categories:

Category I. An *appliance* that operates with a nonpositive vent static pressure and with a vent gas temperature that avoids excessive *condensate* production in the vent.

Category II. An *appliance* that operates with a nonpositive *vent* static pressure and with a vent gas temperature that is capable of causing excessive *condensate* production in the vent.

Category III. An *appliance* that operates with a positive vent static pressure and with a vent gas temperature that avoids excessive *condensate* production in the vent.

Category IV. An *appliance* that operates with a positive vent static pressure and with a vent gas temperature that is capable of causing excessive *condensate* production in the vent.

VENTED ROOM HEATER. A vented self-contained, free-standing, nonrecessed *appliance* for furnishing warm air to the space in which it is installed, directly from the heater without duct connections.

VENTED WALL FURNACE. A self-contained *vented appliance* complete with grilles or equivalent, designed for incorporation in or permanent attachment to the structure of a building, mobile home or travel trailer, and furnishing heated air circulated by gravity or by a fan directly into the space to be heated through openings in the casing. This definition shall exclude *floor furnaces, unit heaters* and *central furnaces* as herein defined.

VENTING SYSTEM. A continuous open passageway from the *flue collar* or *draft hood* of an *appliance* to the outside atmosphere for the purpose of removing flue or vent gases. A venting system is usually composed of a vent or a *chimney* and *vent connector*, if used, assembled to form the open passageway.

WATER HEATER. Any heating *appliance* or *equipment* that heats potable water and supplies such water to the potable hot water distribution system.

WALL HEATER, UNVENTED TYPE. A room heater of the type designed for insertion in or attachment to a wall or partition. Such heater does not incorporate concealed venting arrangements in its construction and discharges all products of *combustion* through the front into the room being heated.

SECTION G2404 (301) GENERAL

G2404.1 (301.1) Scope. This section shall govern the approval and installation of all *equipment* and *appliances* that comprise parts of the installations regulated by this *code* in accordance with Section G2401.

G2404.2 (301.1.1) Other fuels. The requirements for *combustion* and *dilution air* for gas-fired *appliances* shall be governed by Section G2407. The requirements for *combustion* and *dilution air* for *appliances* operating with fuels other than fuel gas shall be regulated by Chapter 17.

G2404.3 (301.3) Listed and labeled. *Appliances* regulated by this *code* shall be listed and labeled for the application in which they are used unless otherwise *approved* in accordance with Section R104.11. The approval of unlisted *appliances* in accordance with Section R104.11 shall be based upon *approved* engineering evaluation.

G2404.4 (301.8) Vibration isolation. Where means for isolation of vibration of an *appliance* is installed, an *approved* means for support and restraint of that *appliance* shall be provided.

G2404.5 (301.9) Repair. Defective material or parts shall be replaced or repaired in such a manner so as to preserve the original approval or listing.

G2404.6 (301.10) Wind resistance. *Appliances* and supports that are exposed to wind shall be designed and installed to resist the wind pressures determined in accordance with this *code*.

G2404.7 (301.11) Flood hazard. For structures located in flood hazard areas, the appliance, equipment and system installations regulated by this code shall be located at or above the elevation required by Section R322 for utilities and attendant equipment.

Exception: The appliance, equipment and system installations regulated by this code are permitted to be located below the elevation required by Section R322 for utilities and attendant equipment provided that they are designed and installed to prevent water from entering or accumulating within the components and to resist hydrostatic and hydrodynamic loads and stresses, including the effects of buoyancy, during the occurrence of flooding to such elevation.

G2404.8 (301.12) Seismic resistance. When earthquake loads are applicable in accordance with this *code*, the supports shall be designed and installed for the seismic forces in accordance with this *code*.

G2404.9 (301.14) Rodentproofing. Buildings or structures and the walls enclosing habitable or occupiable rooms and spaces in which persons live, sleep or work, or in which feed, food or foodstuffs are stored, prepared, processed, served or sold, shall be constructed to protect against the entry of rodents.

G2404.10 (307.5) Auxiliary drain pan. Category IV condensing *appliances* shall be provided with an auxiliary drain pan where damage to any building component will occur as a result of stoppage in the *condensate* drainage system. Such pan shall be installed in accordance with the applicable provisions of Section M1411.

Exception: An auxiliary drain pan shall not be required for *appliances* that automatically shut down operation in the event of a stoppage in the *condensate* drainage system.

SECTION G2405 (302) STRUCTURAL SAFETY

G2405.1 (302.1) Structural safety. The building shall not be weakened by the installation of any *gas piping*. In the process of installing or repairing any *gas piping*, the finished floors, walls, ceilings, tile work or any other part of the building or premises which are required to be changed or replaced shall be left in a safe structural condition in accordance with the requirements of this *code*.

G2405.2 (302.4) Alterations to trusses. Truss members and components shall not be cut, drilled, notched, spliced or otherwise altered in any way without the written concurrence and approval of a registered design professional. *Alterations* resulting in the addition of loads to any member (e.g., HVAC *equipment, water heaters*) shall not be permitted without verification that the truss is capable of supporting such additional loading.

G2405.3 (302.3.1) Engineered wood products. Cuts, notches and holes bored in trusses, structural composite lumber, structural glued-laminated members and I-joists are prohibited except where permitted by the manufacturer's recommendations or where the effects of such *alterations* are specifically considered in the design of the member by a registered design professional.

SECTION G2406 (303) APPLIANCE LOCATION

G2406.1 (303.1) General. *Appliances* shall be located as required by this section, specific requirements elsewhere in this *code* and the conditions of the *equipment* and *appliance* listing.

G2406.2 (303.3) Prohibited locations. *Appliances* shall not be located in sleeping rooms, bathrooms, toilet rooms, storage closets or surgical rooms, or in a space that opens only into such rooms or spaces, except where the installation complies with one of the following:

1. The *appliance* is a direct-vent *appliance* installed in accordance with the conditions of the listing and the manufacturer's instructions.
2. *Vented room heaters*, wall *furnaces*, vented decorative *appliances*, vented gas *fireplaces*, vented gas *fireplace* heaters and decorative *appliances* for installation in vented solid fuel-burning *fireplaces* are installed in rooms that meet the required volume criteria of Section G2407.5.
3. A single wall-mounted *unvented room heater* is installed in a bathroom and such *unvented room heater* is equipped as specified in Section G2445.6 and has an input rating not greater than 6,000 *Btu*/h (1.76 kW). The bathroom shall meet the required volume criteria of Section G2407.5.
4. A single wall-mounted *unvented room heater* is installed in a bedroom and such *unvented room heater* is equipped as specified in Section G2445.6 and has an input rating not greater than 10,000 *Btu*/h (2.93 kW). The bedroom shall meet the required volume criteria of Section G2407.5.
5. The *appliance* is installed in a room or space that opens only into a bedroom or bathroom, and such room or space is used for no other purpose and is provided with a solid weather-stripped door equipped with an *approved* self-closing device. All *combustion air* shall be taken directly from the outdoors in accordance with Section G2407.6.

G2406.3 (303.6) Outdoor locations. *Appliances* installed in outdoor locations shall be either listed for outdoor installation or provided with protection from outdoor environmental factors that influence the operability, durability and safety of the *appliance*.

SECTION G2407 (304) COMBUSTION, VENTILATION AND DILUTION AIR

G2407.1 (304.1) General. Air for *combustion*, ventilation and dilution of *flue gases* for *appliances* installed in buildings shall be provided by application of one of the methods prescribed in Sections G2407.5 through G2407.9. Where the requirements of Section G2407.5 are not met, outdoor air shall be introduced in accordance with one of the methods prescribed in Sections G2407.6 through G2407.9. *Direct-vent appliances*, gas *appliances* of other than *natural draft* design and vented gas *appliances* other than Category I shall be provided with *combustion*, ventilation and *dilution air* in accordance with the *appliance* manufacturer's instructions.

Exception: *Type 1 clothes dryers* that are provided with *makeup air* in accordance with Section G2439.4.

G2407.2 (304.2) Appliance location. *Appliances* shall be located so as not to interfere with proper circulation of *combustion*, ventilation and *dilution air*.

G2407.3 (304.3) Draft hood/regulator location. Where used, a *draft hood* or a *barometric draft regulator* shall be installed in the same room or enclosure as the *appliance* served so as to prevent any difference in pressure between the hood or *regulator* and the *combustion air* supply.

G2407.4 (304.4) Makeup air provisions. Where exhaust fans, *clothes dryers* and kitchen ventilation systems interfere with the operation of *appliances, makeup air* shall be provided.

G2407.5 (304.5) Indoor combustion air. The required volume of indoor air shall be determined in accordance with Section G2407.5.1 or G2407.5.2, except that where the air infiltration rate is known to be less than 0.40 air changes per hour (ACH), Section G2407.5.2 shall be used. The total required volume shall be the sum of the required volume calculated for all *appliances* located within the space. Rooms communicating directly with the space in which the *appliances* are installed through openings not furnished with doors, and through *combustion air* openings sized and located in accordance with Section G2407.5.3, are considered to be part of the required volume.

G2407.5.1 (304.5.1) Standard method. The minimum required volume shall be 50 cubic feet per 1,000 *Btu*/h (4.8 m^3/kW).

G2407.5.2 (304.5.2) Known air-infiltration-rate method. Where the air infiltration rate of a structure is known, the minimum required volume shall be determined as follows:

For *appliances* other than fan assisted, calculate volume using Equation 24-1.

$$\text{Required Volume}_{\text{other}} \geq \frac{21\text{ft}^3}{ACH}\left(\frac{I_{other}}{1{,}000\text{ BTU/hr}}\right)$$

(Equation 24-1)

For fan-assisted *appliances*, calculate volume using Equation 24-2.

$$\text{Required Volume}_{\text{fan}} \geq \frac{15\text{ft}^3}{ACH}\left(\frac{I_{fan}}{1{,}000\text{ BTU/hr}}\right)$$

(Equation 24-2

where:

I_{other} = All *appliances* other than fan assisted (input in *Btu*/h).

I_{fan} = Fan-assisted *appliance* (input in *Btu*/h).

ACH = Air change per hour (percent of volume of space exchanged per hour, expressed as a decimal).

For purposes of this calculation, an infiltration rate greater than 0.60 ACH shall not be used in Equations 24-1 and 24-2.

G2407.5.3 (304.5.3) Indoor opening size and location. Openings used to connect indoor spaces shall be sized and located in accordance with Sections G2407.5.3.1 and G2407.5.3.2 (see Figure G2407.5.3).

G2407.5.3.1 (304.5.3.1) Combining spaces on the same story. Each opening shall have a minimum free area of 1 square inch per 1,000 *Btu*/h (2,200 mm^2/kW) of the total input rating of all *appliances* in the space, but not less than 100 square inches (0.06 m^2). One opening shall commence within 12 inches (305 mm) of the top and one opening shall commence within 12 inches (305 mm) of the bottom of the enclosure. The minimum dimension of air openings shall be not less than 3 inches (76 mm).

G2407.5.3.2 (304.5.3.2) Combining spaces in different stories. The volumes of spaces in different stories shall be considered as communicating spaces where such spaces are connected by one or more openings in doors or floors having a total minimum free area of 2 square inches per 1,000 *Btu*/h (4402 mm^2/kW) of total input rating of all *appliances*.

G2407.6 (304.6) Outdoor combustion air. Outdoor *combustion* air shall be provided through opening(s) to the outdoors in accordance with Section G2407.6.1 or G2407.6.2. The minimum dimension of air openings shall be not less than 3 inches (76 mm).

G2407.6.1 (304.6.1) Two-permanent-openings method. Two permanent openings, one commencing within 12 inches (305 mm) of the top and one commencing within 12 inches (305 mm) of the bottom of the enclosure, shall be provided. The openings shall communicate directly, or by ducts, with the outdoors or spaces that freely communicate with the outdoors.

Where directly communicating with the outdoors, or where communicating with the outdoors through vertical ducts, each opening shall have a minimum free area of 1 square inch per 4,000 *Btu*/h (550 mm^2/kW) of total input rating of all *appliances* in the enclosure [see Figures G2407.6.1(1) and G2407.6.1(2)].

Where communicating with the outdoors through horizontal ducts, each opening shall have a minimum free area of not less than 1 square inch per 2,000 *Btu*/h (1,100 mm^2/kW) of total input rating of all *appliances* in the enclosure [see Figure G2407.6.1(3)].

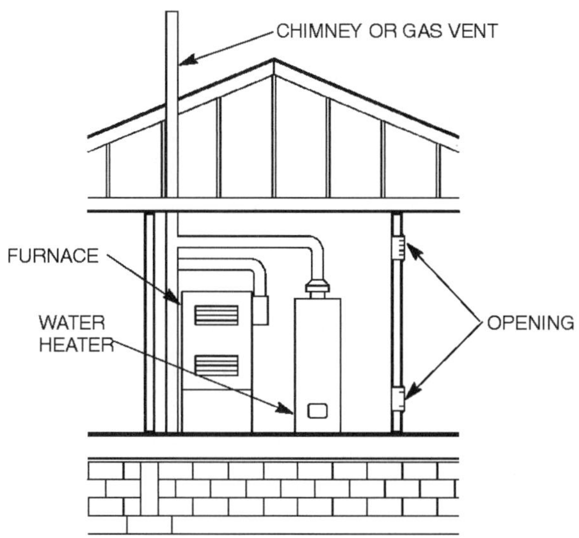

FIGURE G2407.5.3 (304.5.3)
ALL AIR FROM INSIDE THE BUILDING
(see Section G2407.5.3)

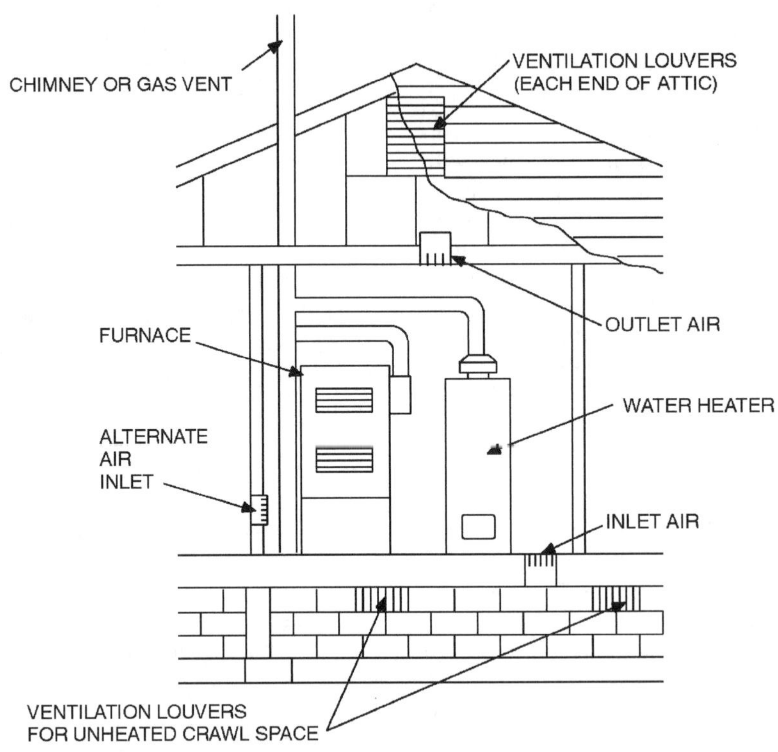

FIGURE G2407.6.1(1) [304.6.1(1)]
ALL AIR FROM OUTDOOR-INLET AIR FROM VENTILATED CRAWL SPACE AND OUTLET AIR TO VENTILATED ATTIC
(see Section G2407.6.1)

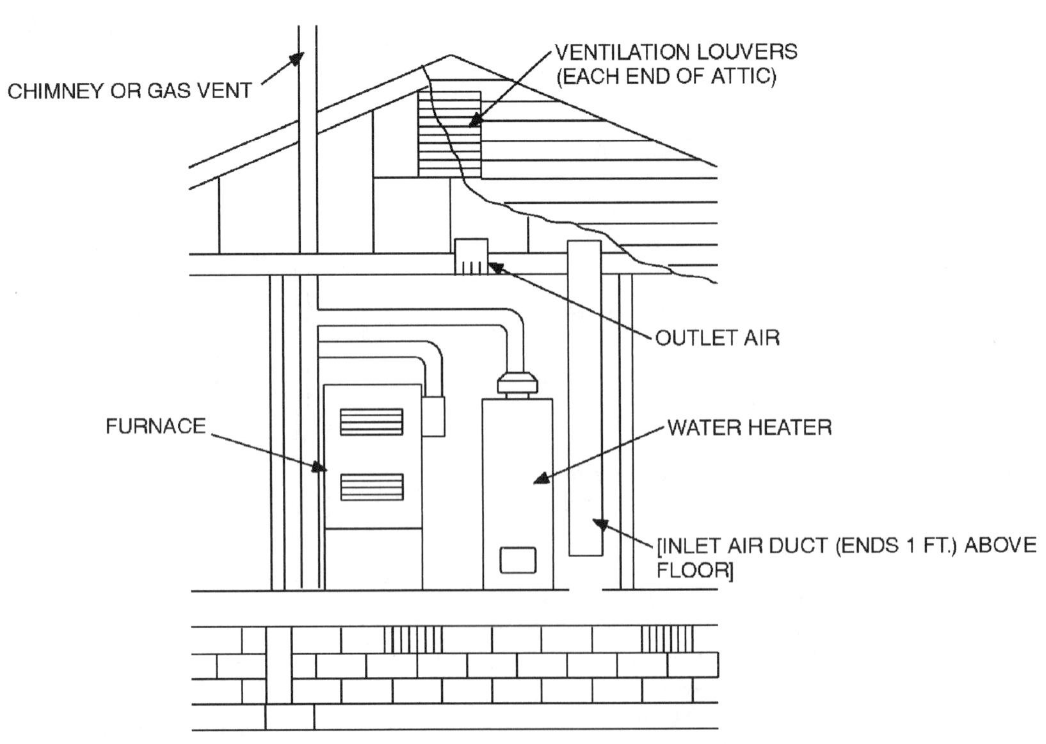

For SI: 1 foot = 304.8 mm.

FIGURE G2407.6.1(2) [304.6.1(2)]
ALL AIR FROM OUTDOORS THROUGH VENTILATED ATTIC (see Section G2407.6.1)

G2407.6.2 (304.6.2) One-permanent-opening method. One permanent opening, commencing within 12 inches (305 mm) of the top of the enclosure, shall be provided. The *appliance* shall have *clearances* of at least 1 inch (25 mm) from the sides and back and 6 inches (152 mm) from the front of the *appliance*. The opening shall directly communicate with the outdoors or through a vertical or horizontal duct to the outdoors, or spaces that freely communicate with the outdoors (see Figure G2407.6.2) and shall have a minimum free area of 1 square inch per 3,000 *Btu*/h (734 mm^2/kW) of the total input rating of all *appliances* located in the enclosure and not less than the sum of the areas of all *vent connectors* in the space.

G2407.7 (304.7) Combination indoor and outdoor combustion air. The use of a combination of indoor and outdoor *combustion air* shall be in accordance with Sections G2407.7.1 through G2407.7.3.

G2407.7.1 (304.7.1) Indoor openings. Where used, openings connecting the interior spaces shall comply with Section G2407.5.3.

G2407.7.2 (304.7.2) Outdoor opening location. Outdoor opening(s) shall be located in accordance with Section G2407.6.

G2407.7.3 (304.7.3) Outdoor opening(s) size. The outdoor opening(s) size shall be calculated in accordance with the following:

1. The ratio of interior spaces shall be the available volume of all communicating spaces divided by the required volume.
2. The outdoor size reduction factor shall be one minus the ratio of interior spaces.
3. The minimum size of outdoor opening(s) shall be the full size of outdoor opening(s) calculated in accordance with Section G2407.6, multiplied by the reduction factor. The minimum dimension of air openings shall be not less than 3 inches (76 mm).

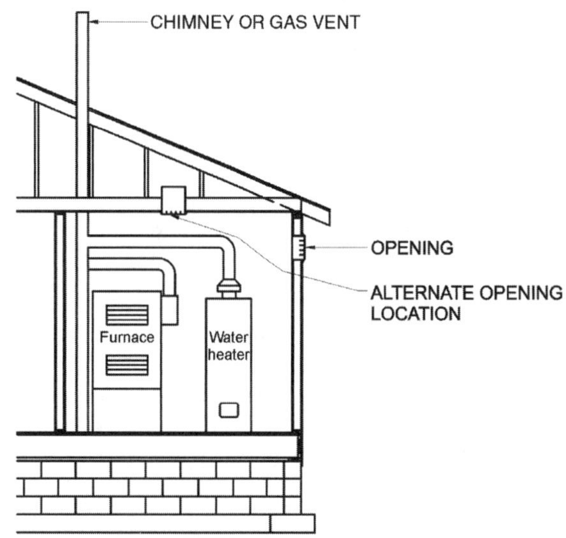

**FIGURE G2407.6.2 (304.6.2)
SINGLE COMBUSTION AIR OPENING,
ALL AIR FROM OUTDOORS
(see Section G2407.6.2)**

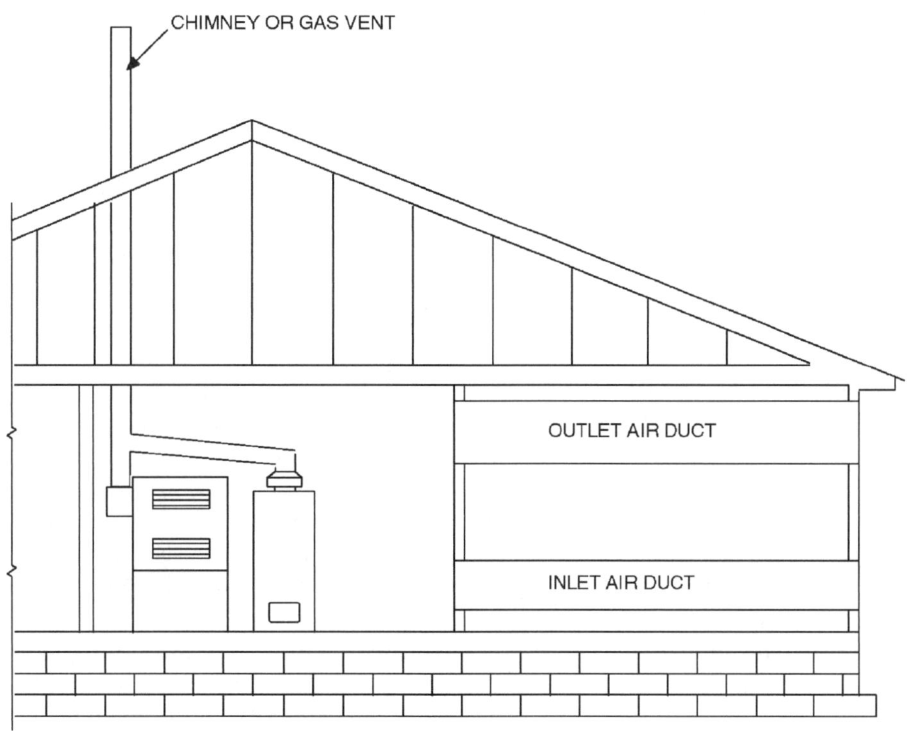

**FIGURE G2407.6.1(3) [304.6.1(3)]
ALL AIR FROM OUTDOORS (see Section G2407.6.1)**

G2407.8 (304.8) Engineered installations. Engineered *combustion air* installations shall provide an adequate supply of *combustion*, ventilation and *dilution air* and shall be *approved.*

G2407.9 (304.9) Mechanical combustion air supply. Where all *combustion air* is provided by a mechanical air supply system, the *combustion air* shall be supplied from the outdoors at a rate not less than 0.35 cubic feet per minute per 1,000 *Btu*/h (0.034 m^3/min per kW) of total input rating of all *appliances* located within the space.

G2407.9.1 (304.9.1) Makeup air. Where exhaust fans are installed, *makeup air* shall be provided to replace the exhausted air.

G2407.9.2 (304.9.2) Appliance interlock. Each of the *appliances* served shall be interlocked with the mechanical air supply system to prevent *main burner* operation when the mechanical air supply system is not in operation.

G2407.9.3 (304.9.3) Combined combustion air and ventilation air system. Where *combustion air* is provided by the building's mechanical ventilation system, the system shall provide the specified *combustion air* rate in addition to the required ventilation air.

G2407.10 (304.10) Louvers and grilles. The required size of openings for *combustion*, ventilation and *dilution air* shall be based on the net free area of each opening. Where the free area through a design of louver, grille or screen is known, it shall be used in calculating the size opening required to provide the free area specified. Where the design and free area of louvers and grilles are not known, it shall be assumed that wood louvers will have 25-percent free area and metal louvers and grilles will have 75-percent free area. Screens shall have a mesh size not smaller than $^1/_4$ inch (6.4 mm). Nonmotorized louvers and grilles shall be fixed in the open position. Motorized louvers shall be interlocked with the *appliance* so that they are proven to be in the full open position prior to *main burner* ignition and during *main burner* operation. Means shall be provided to prevent the *main burner* from igniting if the louvers fail to open during *burner* start-up and to shut down the *main burner* if the louvers close during operation.

G2407.11 (304.11) Combustion air ducts. *Combustion air* ducts shall comply with all of the following:

1. Ducts shall be constructed of galvanized steel complying with Chapter 16 or of a material having equivalent corrosion resistance, strength and rigidity.

   Exception: Within dwellings units, unobstructed stud and joist spaces shall not be prohibited from conveying *combustion air*, provided that not more than one required fireblock is removed.

2. Ducts shall terminate in an unobstructed space allowing free movement of *combustion air* to the *appliances*.
3. Ducts shall serve a single enclosure.
4. Ducts shall not serve both upper and lower *combustion air* openings where both such openings are used. The separation between ducts serving upper and lower *combustion air* openings shall be maintained to the source of *combustion air.*
5. Ducts shall not be screened where terminating in an attic space.
6. Horizontal upper *combustion air* ducts shall not slope downward toward the source of *combustion air.*
7. The remaining space surrounding a *chimney* liner, gas vent, special gas vent or plastic *piping* installed within a masonry, metal or factory-built *chimney* shall not be used to supply *combustion air.*

   Exception: Direct-vent gas-fired *appliances* designed for installation in a solid fuel-burning *fireplace* where installed in accordance with the manufacturer's instructions.

8. *Combustion air* intake openings located on the exterior of a building shall have the lowest side of such openings located not less than 12 inches (305 mm) vertically from the adjoining finished ground level.

G2407.12 (304.12) Protection from fumes and gases. Where corrosive or flammable process fumes or gases, other than products of *combustion*, are present, means for the disposal of such fumes or gases shall be provided. Such fumes or gases include carbon monoxide, hydrogen sulfide, ammonia, chlorine and halogenated hydrocarbons.

In barbershops, beauty shops and other facilities where chemicals that generate corrosive or flammable products, such as aerosol sprays, are routinely used, nondirect vent-type *appliances* shall be located in a mechanical room separated or partitioned off from other areas with provisions for *combustion air* and *dilution air* from the outdoors. *Direct-vent appliances* shall be installed in accordance with the *appliance* manufacturer's installation instructions.

SECTION G2408 (305)
INSTALLATION

G2408.1 (305.1) General. *Equipment* and *appliances* shall be installed as required by the terms of their approval, in accordance with the conditions of listing, the manufacturer's instructions and this *code*. Manufacturer's installation instructions shall be available on the job site at the time of inspection. Where a *code* provision is less restrictive than the conditions of the listing of the *equipment* or *appliance* or the manufacturer's installation instructions, the conditions of the listing and the manufacturer's installation instructions shall apply.

Unlisted *appliances approved* in accordance with Section G2404.3 shall be limited to uses recommended by the manufacturer and shall be installed in accordance with the manufacturer's instructions, the provisions of this *code* and the requirements determined by the *code official.*

G2408.2 (305.3) Elevation of ignition source. *Equipment* and *appliances* having an *ignition source* shall be elevated such that the source of ignition is not less than 18 inches (457 mm) above the floor in *hazardous locations* and public garages, private garages, repair garages, motor fuel-dispensing facilities and parking garages. For the purpose of this section, rooms or spaces that are not part of the *living space* of a *dwelling unit* and that communicate directly with a private garage through openings shall be considered to be part of the private garage.

Exception: Elevation of the *ignition source* is not required for *appliances* that are listed as flammable vapor ignition resistant.

G2408.2.1 (305.3.1) Installation in residential garages. In residential garages where *appliances* are installed in a separate, enclosed space having access only from outside of the garage, such *appliances* shall be permitted to be installed at floor level, provided that the required *combustion air* is taken from the exterior of the garage.

G2408.3 (305.5) Private garages. *Appliances* located in private garages shall be installed with a minimum *clearance* of 6 feet (1829 mm) above the floor.

Exception: The requirements of this section shall not apply where the *appliances* are protected from motor vehicle impact and installed in accordance with Section G2408.2.

G2408.4 (305.7) Clearances from grade. *Equipment* and *appliances* installed at grade level shall be supported on a level concrete slab or other *approved* material extending not less than 3 inches (76 mm) above adjoining grade or shall be suspended not less than 6 inches (152 mm) above adjoining grade. Such supports shall be installed in accordance with the manufacturer's installation instructions.

G2408.5 (305.8) Clearances to combustible construction. Heat-producing *equipment* and *appliances* shall be installed to maintain the required clearances to combustible construction as specified in the listing and manufacturer's instructions. Such *clearances* shall be reduced only in accordance with Section G2409. *Clearances* to combustibles shall include such considerations as door swing, drawer pull, overhead projections or shelving and window swing. Devices, such as door stops or limits and closers, shall not be used to provide the required *clearances*.

G2408.6 (305.12) Avoid strain on gas piping. *Appliances* shall be supported and connected to the *piping* so as not to exert undue strain on the connections.

SECTION G2409 (308) CLEARANCE REDUCTION

G2409.1 (308.1) Scope. This section shall govern the reduction in required clearances to *combustible materials*, including gypsum board, and *combustible assemblies* for chimneys, vents, appliances, devices and equipment. Clearance requirements for air-conditioning equipment and central heating boilers and furnaces shall comply with Sections G2409.3 and G2409.4.

G2409.2 (308.2) Reduction table. The allowable *clearance* reduction shall be based on one of the methods specified in Table G2409.2 or shall utilize an assembly listed for such application. Where required *clearances* are not listed in Table G2409.2, the reduced *clearances* shall be determined by linear interpolation between the distances listed in the table. Reduced *clearances* shall not be derived by extrapolation below the range of the table. The reduction of the required *clearances* to combustibles for listed and labeled *appliances* and *equipment* shall be in accordance with the requirements of this section except that such *clearances* shall not be reduced where reduction is specifically prohibited by the terms of the *appliance* or *equipment* listing [see Figures G2409.2(1), G2409.2(2) and G2409.2(3)].

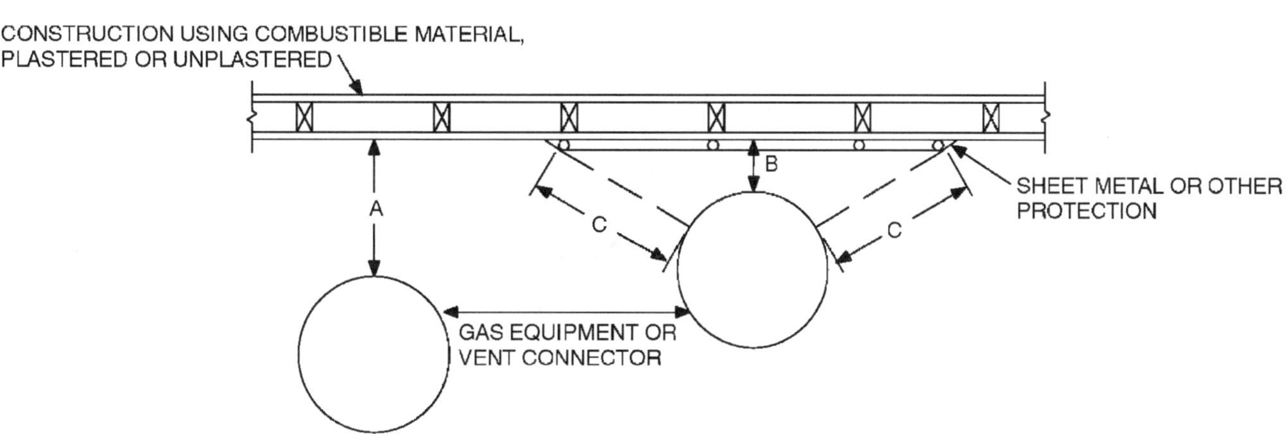

NOTES:
"A" equals the clearance with no protection.
"B" equals the reduced clearance permitted in accordance with Table G2409.2. The protection applied to the construction using combustible material shall extend far enough in each direction to make "C" equal to "A."

FIGURE G2409.2(1) [308.2(1)]
EXTENT OF PROTECTION NECESSARY TO REDUCE CLEARANCES FROM GAS EQUIPMENT OR VENT CONNECTORS

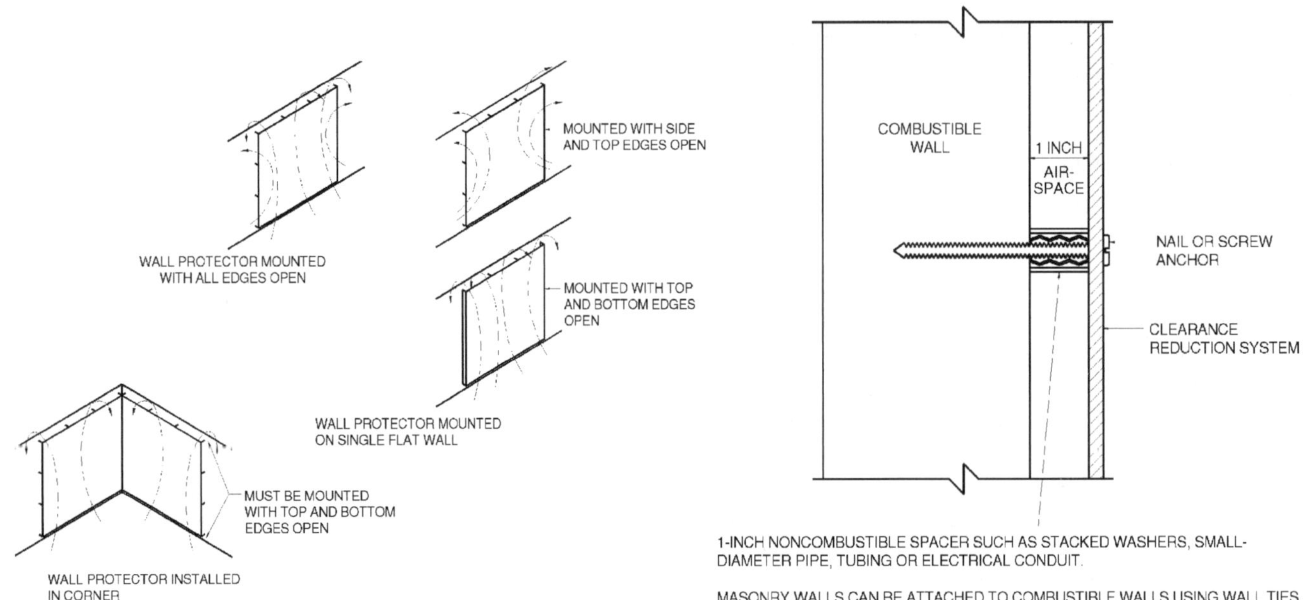

For SI: 1 inch = 25.4 mm.

FIGURE G2409.2(2) [308.2(2)]
WALL PROTECTOR CLEARANCE REDUCTION SYSTEM

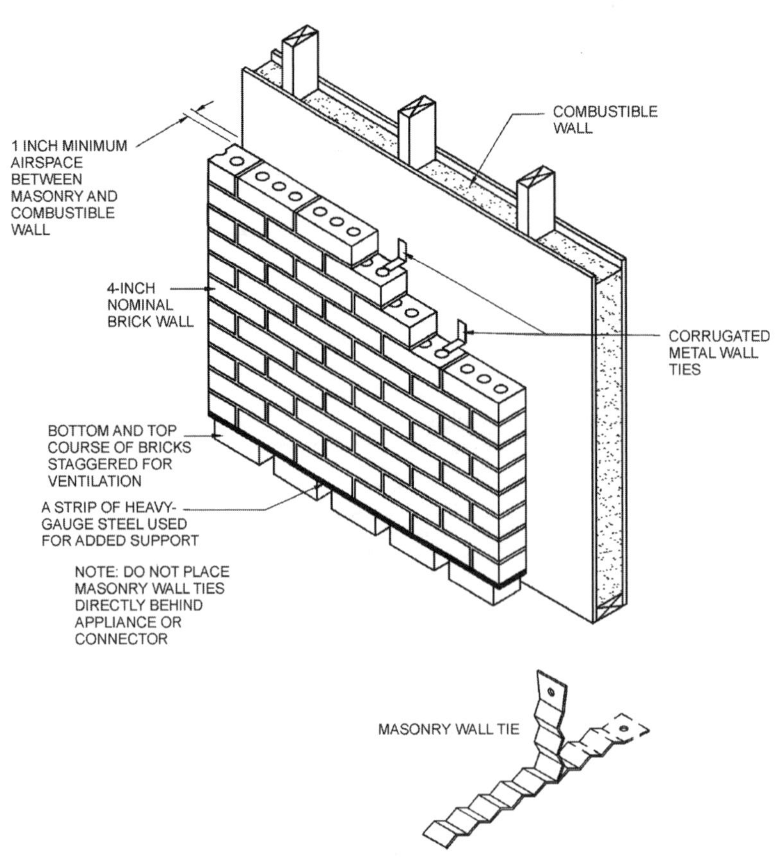

For SI: 1 inch = 25.4 mm.

FIGURE G2409.2(3) [308.2(3)]
MASONRY CLEARANCE REDUCTION SYSTEM

TABLE G2409.2 (308.2)[a through k]
REDUCTION OF CLEARANCES WITH SPECIFIED FORMS OF PROTECTION

TYPE OF PROTECTION APPLIED TO AND COVERING ALL SURFACES OF COMBUSTIBLE MATERIAL WITHIN THE DISTANCE SPECIFIED AS THE REQUIRED CLEARANCE WITH NO PROTECTION [see Figures G2409.2(1), G2409.2(2), and G2409.2(3)]	WHERE THE REQUIRED CLEARANCE WITH NO PROTECTION FROM APPLIANCE, VENT CONNECTOR, OR SINGLE-WALL METAL PIPE IS: (inches)									
	36		18		12		9		6	
	Allowable clearances with specified protection (inches)									
	Use Column 1 for clearances above appliance or horizontal connector. Use Column 2 for clearances from appliance, vertical connector and single-wall metal pipe.									
	Above Col. 1	Sides and rear Col. 2	Above Col. 1	Sides and rear Col. 2	Above Col. 1	Sides and rear Col. 2	Above Col. 1	Sides and rear Col. 2	Above Col. 1	Sides and rear Col. 2
1. 3$^1/_2$-inch-thick masonry wall without ventilated airspace	—	24	—	12	—	9	—	6	—	5
2. $^1/_2$-inch insulation board over 1-inch glass fiber or mineral wool batts	24	18	12	9	9	6	6	5	4	3
3. 0.024-inch (nominal 24 gage) sheet metal over 1-inch glass fiber or mineral wool batts reinforced with wire on rear face with ventilated airspace	18	12	9	6	6	4	5	3	3	3
4. 3$^1/_2$-inch-thick masonry wall with ventilated airspace	—	12	—	6	—	6	—	6	—	6
5. 0.024-inch (nominal 24 gage) sheet metal with ventilated airspace	18	12	9	6	6	4	5	3	3	2
6. $^1/_2$-inch-thick insulation board with ventilated airspace	18	12	9	6	6	4	5	3	3	3
7. 0.024-inch (nominal 24 gage) sheet metal with ventilated airspace over 0.024-inch (nominal 24 gage) sheet metal with ventilated airspace	18	12	9	6	6	4	5	3	3	3
8. 1-inch glass fiber or mineral wool batts sandwiched between two sheets 0.024-inch (nominal 24 gage) sheet metal with ventilated airspace	18	12	9	6	6	4	5	3	3	3

For SI: 1 inch = 25.4 mm, °C = [(°F - 32)/1.8], 1 pound per cubic foot = 16.02 kg/m^3, 1 Btu per inch per square foot per hour per °F = 0.144 W/m^2 · K.

a. Reduction of clearances from combustible materials shall not interfere with combustion air, draft hood clearance and relief, and accessibility of servicing.

b. All clearances shall be measured from the outer surface of the combustible material to the nearest point on the surface of the appliance, disregarding any intervening protection applied to the combustible material.

c. Spacers and ties shall be of noncombustible material. No spacer or tie shall be used directly opposite an appliance or connector.

d. For all clearance reduction systems using a ventilated airspace, adequate provision for air circulation shall be provided as described [see Figures G2409.2(2) and G2409.2(3)].

e. There shall be at least 1 inch between clearance reduction systems and combustible walls and ceilings for reduction systems using ventilated airspace.

f. Where a wall protector is mounted on a single flat wall away from corners, it shall have a minimum 1-inch air gap. To provide air circulation, the bottom and top edges, or only the side and top edges, or all edges shall be left open.

g. Mineral wool batts (blanket or board) shall have a minimum density of 8 pounds per cubic foot and a minimum melting point of 1,500°F.

h. Insulation material used as part of a clearance reduction system shall have a thermal conductivity of 1.0 Btu per inch per square foot per hour per °F or less.

i. There shall be at least 1 inch between the appliance and the protector. In no case shall the clearance between the appliance and the combustible surface be reduced below that allowed in this table.

j. All clearances and thicknesses are minimum; larger clearances and thicknesses are acceptable.

k. Listed single-wall connectors shall be installed in accordance with the manufacturer's installation instructions.

G2409.3 (308.3) Clearances for indoor air-conditioning appliances. *Clearance* requirements for indoor air-conditioning *appliances* shall comply with Sections G2409.3.1 through G2409.3.4.

G2409.3.1 (308.3.1) Appliances clearances. Air-conditioning appliances shall be installed with clearances in accordance with the manufacturer's instructions.

G2409.3.2 (308.3.2) Clearance reduction. Air-conditioning appliances shall be permitted to be installed with reduced clearances to *combustible material*, provided that the *combustible material* or *appliance* is protected as described in Table G2409.2 and such reduction is allowed by the manufacturer's instructions.

G2409.3.3 (308.3.3) Plenum clearances. Where the *furnace plenum* is adjacent to plaster on metal lath or noncombustible material attached to combustible material, the *clearance* shall be measured to the surface of the plaster or other noncombustible finish where the *clearance* specified is 2 inches (51 mm) or less.

G2409.3.4 (308.3.4) Clearance from supply ducts. Supply air ducts connecting to listed central heating furnaces shall have the same minimum clearance to combustibles as required for the furnace supply plenum for a distance of not less than 3 feet (914 mm) from the supply plenum. Clearance is not required beyond the 3-foot (914 mm) distance.

G2409.4 (308.4) Central heating boilers and furnaces. *Clearance* requirements for central-heating boilers and *furnaces* shall comply with Sections G2409.4.1 through G2409.4.5. The *clearance* to these *appliances* shall not interfere with *combustion air*; *draft hood clearance* and *relief*; and accessibility for servicing.

G2409.4.1 (308.4.1) Appliances clearances. Central-heating furnaces and low-pressure boilers shall be installed with clearances in accordance with the manufacturer's instructions

G2409.4.2 (308.4.2) Clearance reduction. Central-heating furnaces and low-pressure boilers shall be permitted to be installed with reduced clearances to *combustible material* provided that the *combustible material* or *appliance* is protected as described in Table G2409.2 and such reduction is allowed by the manufacturer's instructions.

G2409.4.3 (308.4.4) Plenum clearances. Where the *furnace plenum* is adjacent to plaster on metal lath or noncombustible material attached to combustible material, the *clearance* shall be measured to the surface of the plaster or other noncombustible finish where the *clearance* specified is 2 inches (51 mm) or less.

G2409.4.4 (308.4.5) Clearance from supply ducts. Supply air ducts connecting to listed central heating furnaces shall have the same minimum clearance to combustibles as required for the furnace supply plenum for a distance of not less than 3 feet (914 mm) from the supply plenum. Clearance is not required beyond the 3-foot (914 mm) distance.

G2409.4.5 (308.4.3) Clearance for servicing appliances. Front *clearance* shall be sufficient for servicing the *burner* and the *furnace* or boiler.

SECTION G2410 (309) ELECTRICAL

G2410.1 (309.1) Grounding. *Gas piping* shall not be used as a *grounding electrode*.

G2410.2 (309.2) Connections. Electrical connections between *appliances* and the building wiring, including the grounding of the *appliances,* shall conform to Chapters 34 through 43.

SECTION G2411 (310) ELECTRICAL BONDING

G2411.1 (310.1) Pipe and *tubing* other than CSST. Each above-ground portion of a *gas piping system* other than corrugated stainless steel tubing (CSST), that is likely to become energized shall be electrically continuous and bonded to an effective ground-fault current path. *Gas piping*, other than CSST, shall be considered to be bonded where it is connected to *appliances* that are connected to the equipment grounding conductor of the circuit supplying that *appliance*.

G2411.1.1 (310.1.1) CSST. Corrugated stainless steel tubing (CSST) gas *piping* systems shall be bonded to the electrical service grounding electrode system. The bonding jumper shall connect to a metallic pipe or fitting between the *point of delivery* and the first downstream CSST fitting. The bonding jumper shall be not smaller than 6 AWG copper wire or equivalent. Gas piping systems that contain one or more segments of CSST shall be bonded in accordance with this section.

SECTION G2412 (401) GENERAL

G2412.1 (401.1) Scope. This section shall govern the design, installation, modification and maintenance of *piping systems*. The applicability of this *code* to *piping systems* extends from the *point of delivery* to the connections with the *appliances* and includes the design, materials, components, fabrication, assembly, installation, testing, inspection, operation and maintenance of such *piping systems*.

G2412.1.1 (401.1.1) Utility piping systems located within buildings. Utility service *piping* located within buildings shall be installed in accordance with the structural safety and fire protection provisions of this *code*.

G2412.2 (401.2) Liquefied petroleum gas storage. The storage system for *liquefied petroleum gas* shall be designed and installed in accordance with the *International Fire Code* and NFPA 58.

G2412.3 (401.3) Modifications to existing systems. In modifying or adding to existing *piping systems*, sizes shall be maintained in accordance with this chapter.

G2412.4 (401.4) Additional appliances. Where an additional *appliance* is to be served, the existing *piping* shall be checked to determine if it has adequate capacity for all *appliances* served. If inadequate, the existing system shall be enlarged as required or separate *piping* of adequate capacity shall be provided.

G2412.5 (401.5) Identification. For other than steel *pipe*, exposed *piping* shall be identified by a yellow label marked "Gas" in black letters. The marking shall be spaced at intervals not exceeding 5 feet (1524 mm). The marking shall not be required on *pipe* located in the same room as the *appliance* served.

G2412.6 (401.6) Interconnections. Where two or more *meters* are installed on the same premises, but supply separate consumers, the *piping systems* shall not be interconnected on the outlet side of the *meters*.

G2412.7 (401.7) Piping meter identification. *Piping* from multiple *meter* installations shall be marked with an *approved* permanent identification by the installer so that the *piping system* supplied by each *meter* is readily identifiable.

G2412.8 (401.8) Minimum sizes. All *pipe* utilized for the installation, extension and *alteration* of any *piping system* shall be sized to supply the full number of outlets for the intended purpose and shall be sized in accordance with Section G2413.

G2412.9 (401.9) Identification. Each length of pipe and tubing and each pipe fitting, utilized in a fuel gas system, shall bear the identification of the manufacturer.

G2412.10 (401.10) Third-party testing and certification. All piping, tubing and fittings shall comply with the applicable referenced standards, specifications and performance criteria of this code and shall be identified in accordance with Section G2412.9. Piping, tubing and fittings shall either be tested by an approved third-party testing agency or certified by an approved *third-party certification agency*.

SECTION G2413 (402)
PIPE SIZING

G2413.1 (402.1) General considerations. *Piping systems* shall be of such size and so installed as to provide a supply of gas sufficient to meet the maximum *demand* and supply gas to each *appliance* inlet at not less than the minimum supply pressure required by the *appliance*.

G2413.2 (402.2) Maximum gas demand. The volumetric flow rate of gas to be provided, in cubic feet per hour, shall be calculated using the manufacturer's input ratings of the *appliances* served adjusted for altitude. Where an input rating is not indicated, the gas supplier, *appliance* manufacturer or a qualified agency shall be contacted, or the rating from Table G2413.2 shall be used for estimating the volumetric flow rate of gas to be supplied.

The total connected hourly load shall be used as the basis for pipe sizing, assuming that all appliances could be operating at full capacity simultaneously. Where a diversity of load can be established, pipe sizing shall be permitted to be based on such loads.

G2413.3 (402.3) Sizing. *Gas piping* shall be sized in accordance with one of the following:

1. *Pipe* sizing tables or sizing equations in accordance with Section G2413.4.
2. The sizing tables included in a listed *piping* system's manufacturer's installation instructions.
3. Other *approved* engineering methods.

TABLE G2413.2 (402.2)
APPROXIMATE GAS INPUT FOR TYPICAL APPLIANCES

APPLIANCE	INPUT BTU/H (Approx.)
Space Heating Units	
Hydronic boiler	
Single family	100,000
Multifamily, per unit	60,000
Warm-air furnace	
Single family	100,000
Multifamily, per unit	60,000
Space and Water Heating Units	
Hydronic boiler	
Single family	120,000
Multifamily, per unit	75,000
Water Heating Appliances	
Water heater, automatic instantaneous	
Capacity at 2 gal./minute	142,800
Capacity at 4 gal./minute	285,000
Capacity at 6 gal./minute	428,400
Water heater, automatic storage, 30- to 40-gal. tank	35,000
Water heater, automatic storage, 50-gal. tank	50,000
Water heater, domestic, circulating or side-arm	35,000
Cooking Appliances	
Built-in oven or broiler unit, domestic	25,000
Built-in top unit, domestic	40,000
Range, free-standing, domestic	65,000
Other Appliances	
Barbecue	40,000
Clothes dryer, Type 1 (domestic)	35,000
Gas fireplace, direct-vent	40,000
Gas light	2,500
Gas log	80,000
Refrigerator	3,000

For SI: 1 British thermal unit per hour = 0.293 W, 1 gallon = 3.785 L, 1 gallon per minute = 3.785 L/m.

G2413.4 (402.4) Sizing tables and equations. Where Tables G2413.4(1) through G2413.4(21) are used to size *piping* or *tubing*, the *pipe* length shall be determined in accordance with Section G2413.4.1, G2413.4.2 or G2413.4.3.

Where Equations 24-3 and 24-4 are used to size *piping* or *tubing*, the *pipe* or *tubing* shall have smooth inside walls and the pipe length shall be determined in accordance with Section G2413.4.1, G2413.4.2 or G2413.4.3.

1. Low-pressure gas equation [Less than 1.5 pounds per square inch (psi) (10.3 kPa)]:

$$D = \frac{Q^{0.381}}{19.17\left(\frac{\Delta H}{C_r \times L}\right)^{0.206}} \quad \text{(Equation 24-3)}$$

2. High-pressure gas equation [1.5 psi (10.3 kPa) and above]:

$$D = \frac{Q^{0.381}}{18.93\left[\frac{(P_1^2 - P_2^2) \times Y}{C_r \times L}\right]^{0.206}} \quad \text{(Equation 24-4)}$$

where:

D = Inside diameter of *pipe*, inches (mm).

Q = Input rate *appliance(s)*, cubic feet per hour at 60°F (16°C) and 30-inch mercury column.

P_1 = Upstream pressure, psia (P_1 + 14.7).

P_2 = Downstream pressure, psia (P_2 + 14.7).

L = Equivalent length of *pipe*, feet.

ΔH = *Pressure drop*, inch water column (27.7 inch water column = 1 psi).

TABLE G2413.4 (402.4)
C_r AND Y VALUES FOR NATURAL GAS AND UNDILUTED PROPANE AT STANDARD CONDITIONS

GAS	EQUATION FACTORS	
	C_r	Y
Natural gas	0.6094	0.9992
Undiluted propane	1.2462	0.9910

For SI: 1 cubic foot = 0.028 m^3, 1 foot = 305 mm,
1 inch water column = 0.249 kPa,
1 pound per square inch = 6.895 kPa,
1 British thermal unit per hour = 0.293 W.

G2413.4.1 (402.4.1) Longest length method. The *pipe* size of each section of *gas piping* shall be determined using the longest length of *piping* from the *point of delivery* to the most remote *outlet* and the load of the section.

G2413.4.2 (402.4.2) Branch length method. *Pipe* shall be sized as follows:

1. *Pipe* size of each section of the longest *pipe* run from the *point of delivery* to the most remote *outlet* shall be determined using the longest run of *piping* and the load of the section.
2. The *pipe* size of each section of branch *piping* not previously sized shall be determined using the length of *piping* from the *point of delivery* to the most remote *outlet* in each branch and the load of the section.

G2413.4.3 (402.4.3) Hybrid pressure. The *pipe* size for each section of higher pressure *gas piping* shall be determined using the longest length of *piping* from the *point of delivery* to the most remote line *pressure regulator*. The *pipe* size from the line *pressure regulator* to each *outlet* shall be determined using the length of *piping* from the *regulator* to the most remote outlet served by the *regulator*.

G2413.5 (402.5) Allowable pressure drop. The design pressure loss in any *piping system* under maximum probable flow conditions, from the *point of delivery* to the inlet connection of the *appliance*, shall be such that the supply pressure at the *appliance* is greater than or equal to the minimum pressure required by the *appliance*.

G2413.6 (402.6) Maximum design operating pressure. The maximum design operating pressure for *piping systems* located inside buildings shall not exceed 5 pounds per square inch gauge (psig) (34 kPa gauge) except where one or more of the following conditions are met:

1. The *piping system* is welded.
2. The *piping* is located in a ventilated chase or otherwise enclosed for protection against accidental gas accumulation.
3. The *piping* is a temporary installation for buildings under construction.

G2413.6.1 (402.6.1) Liquefied petroleum gas systems. LP-gas systems designed to operate below -5°F (-21°C) or with butane or a propane-butane mix shall be designed to either accommodate liquid LP-gas or prevent LP-gas vapor from condensing into a liquid.

SECTION G2414 (403) PIPING MATERIALS

G2414.1 (403.1) General. Materials used for piping systems shall comply with the requirements of this chapter or shall be *approved*.

G2414.2 (403.2) Used materials. *Pipe*, fittings, *valves* or other materials shall not be used again unless they are free of foreign materials and have been ascertained to be adequate for the service intended.

G2414.3 (403.3) Other materials. Material not covered by the standards specifications listed herein shall be investigated and tested to determine that it is safe and suitable for the proposed service, and, in addition, shall be recommended for that service by the manufacturer and shall be *approved* by the *code official*.

G2414.4 (403.4) Metallic pipe. Metallic *pipe* shall comply with Sections G2414.4.1 and G2414.4.2.

G2414.4.1 (403.4.1) Cast iron. Cast-iron *pipe* shall not be used.

TABLE G2413.4(1) [402.4(2)]
SCHEDULE 40 METALLIC PIPE

Gas	Natural
Inlet Pressure	Less than 2 psi
Pressure Drop	0.5 in. w.c.
Specific Gravity	0.60

	PIPE SIZE (inch)													
Nominal	1/2	3/4	1	1 1/4	1 1/2	2	2 1/2	3	4	5	6	8	10	12
Actual ID	0.622	0.824	1.049	1.380	1.610	2.067	2.469	3.068	4.026	5.047	6.065	7.981	10.020	11.938
Length (ft)	**Capacity in Cubic Feet of Gas per Hour**													
10	172	360	678	1,390	2,090	4,020	6,400	11,300	23,100	41,800	67,600	139,000	252,000	399,000
20	118	247	466	957	1,430	2,760	4,400	7,780	15,900	28,700	46,500	95,500	173,000	275,000
30	95	199	374	768	1,150	2,220	3,530	6,250	12,700	23,000	37,300	76,700	139,000	220,000
40	81	170	320	657	985	1,900	3,020	5,350	10,900	19,700	31,900	65,600	119,000	189,000
50	72	151	284	583	873	1,680	2,680	4,740	9,660	17,500	28,300	58,200	106,000	167,000
60	65	137	257	528	791	1,520	2,430	4,290	8,760	15,800	25,600	52,700	95,700	152,000
70	60	126	237	486	728	1,400	2,230	3,950	8,050	14,600	23,600	48,500	88,100	139,000
80	56	117	220	452	677	1,300	2,080	3,670	7,490	13,600	22,000	45,100	81,900	130,000
90	52	110	207	424	635	1,220	1,950	3,450	7,030	12,700	20,600	42,300	76,900	122,000
100	50	104	195	400	600	1,160	1,840	3,260	6,640	12,000	19,500	40,000	72,600	115,000
125	44	92	173	355	532	1,020	1,630	2,890	5,890	10,600	17,200	35,400	64,300	102,000
150	40	83	157	322	482	928	1,480	2,610	5,330	9,650	15,600	32,100	58,300	92,300
175	37	77	144	296	443	854	1,360	2,410	4,910	8,880	14,400	29,500	53,600	84,900
200	34	71	134	275	412	794	1,270	2,240	4,560	8,260	13,400	27,500	49,900	79,000
250	30	63	119	244	366	704	1,120	1,980	4,050	7,320	11,900	24,300	44,200	70,000
300	27	57	108	221	331	638	1,020	1,800	3,670	6,630	10,700	22,100	40,100	63,400
350	25	53	99	203	305	587	935	1,650	3,370	6,100	9,880	20,300	36,900	58,400
400	23	49	92	189	283	546	870	1,540	3,140	5,680	9,190	18,900	34,300	54,300
450	22	46	86	177	266	512	816	1,440	2,940	5,330	8,620	17,700	32,200	50,900
500	21	43	82	168	251	484	771	1,360	2,780	5,030	8,150	16,700	30,400	48,100
550	20	41	78	159	239	459	732	1,290	2,640	4,780	7,740	15,900	28,900	45,700
600	19	39	74	152	228	438	699	1,240	2,520	4,560	7,380	15,200	27,500	43,600
650	18	38	71	145	218	420	669	1,180	2,410	4,360	7,070	14,500	26,400	41,800
700	17	36	68	140	209	403	643	1,140	2,320	4,190	6,790	14,000	25,300	40,100
750	17	35	66	135	202	389	619	1,090	2,230	4,040	6,540	13,400	24,400	38,600
800	16	34	63	130	195	375	598	1,060	2,160	3,900	6,320	13,000	23,600	37,300
850	16	33	61	126	189	363	579	1,020	2,090	3,780	6,110	12,600	22,800	36,100
900	15	32	59	122	183	352	561	992	2,020	3,660	5,930	12,200	22,100	35,000
950	15	31	58	118	178	342	545	963	1,960	3,550	5,760	11,800	21,500	34,000
1,000	14	30	56	115	173	333	530	937	1,910	3,460	5,600	11,500	20,900	33,100
1,100	14	28	53	109	164	316	503	890	1,810	3,280	5,320	10,900	19,800	31,400
1,200	13	27	51	104	156	301	480	849	1,730	3,130	5,070	10,400	18,900	30,000
1,300	12	26	49	100	150	289	460	813	1,660	3,000	4,860	9,980	18,100	28,700
1,400	12	25	47	96	144	277	442	781	1,590	2,880	4,670	9,590	17,400	27,600
1,500	11	24	45	93	139	267	426	752	1,530	2,780	4,500	9,240	16,800	26,600
1,600	11	23	44	89	134	258	411	727	1,480	2,680	4,340	8,920	16,200	25,600
1,700	11	22	42	86	130	250	398	703	1,430	2,590	4,200	8,630	15,700	24,800
1,800	10	22	41	84	126	242	386	682	1,390	2,520	4,070	8,370	15,200	24,100
1,900	10	21	40	81	122	235	375	662	1,350	2,440	3,960	8,130	14,800	23,400
2,000	NA	20	39	79	119	229	364	644	1,310	2,380	3,850	7,910	14,400	22,700

For SI: 1 inch = 25.4 mm, 1 foot = 304.8 mm, 1 pound per square inch = 6.895 kPa, 1-inch water column = 0.2488 kPa, 1 British thermal unit per hour = 0.2931 W, 1 cubic foot per hour = 0.0283 m^3/h, 1 degree = 0.01745 rad.

Notes:

1. NA means a flow of less than 10 cfh.
2. All table entries have been rounded to three significant digits.

Gas	Natural
Inlet Pressure	2.0 psi
Pressure Drop	1.0 psi
Specific Gravity	0.60

TABLE G2413.4(2) [402.4(5)] SCHEDULE 40 METALLIC PIPE

	PIPE SIZE (inch)								
Nominal	$^1/_2$	$^3/_4$	1	$1^1/_4$	$1^1/_2$	2	$2^1/_2$	3	4
Actual ID	0.622	0.824	1.049	1.380	1.610	2.067	2.469	3.068	4.026
Length (ft)	Capacity in Cubic Feet of Gas per Hour								
10	1,510	3,040	5,560	11,400	17,100	32,900	52,500	92,800	189,000
20	1,070	2,150	3,930	8,070	12,100	23,300	37,100	65,600	134,000
30	869	1,760	3,210	6,590	9,880	19,000	30,300	53,600	109,000
40	753	1,520	2,780	5,710	8,550	16,500	26,300	46,400	94,700
50	673	1,360	2,490	5,110	7,650	14,700	23,500	41,500	84,700
60	615	1,240	2,270	4,660	6,980	13,500	21,400	37,900	77,300
70	569	1,150	2,100	4,320	6,470	12,500	19,900	35,100	71,600
80	532	1,080	1,970	4,040	6,050	11,700	18,600	32,800	67,000
90	502	1,010	1,850	3,810	5,700	11,000	17,500	30,900	63,100
100	462	934	1,710	3,510	5,260	10,100	16,100	28,500	58,200
125	414	836	1,530	3,140	4,700	9,060	14,400	25,500	52,100
150	372	751	1,370	2,820	4,220	8,130	13,000	22,900	46,700
175	344	695	1,270	2,601	3,910	7,530	12,000	21,200	43,300
200	318	642	1,170	2,410	3,610	6,960	11,100	19,600	40,000
250	279	583	1,040	2,140	3,210	6,180	9,850	17,400	35,500
300	253	528	945	1,940	2,910	5,600	8,920	15,800	32,200
350	232	486	869	1,790	2,670	5,150	8,210	14,500	29,600
400	216	452	809	1,660	2,490	4,790	7,640	13,500	27,500
450	203	424	759	1,560	2,330	4,500	7,170	12,700	25,800
500	192	401	717	1,470	2,210	4,250	6,770	12,000	24,400
550	182	381	681	1,400	2,090	4,030	6,430	11,400	23,200
600	174	363	650	1,330	2,000	3,850	6,130	10,800	22,100
650	166	348	622	1,280	1,910	3,680	5,870	10,400	21,200
700	160	334	598	1,230	1,840	3,540	5,640	9,970	20,300
750	154	322	576	1,180	1,770	3,410	5,440	9,610	19,600
800	149	311	556	1,140	1,710	3,290	5,250	9,280	18,900
850	144	301	538	1,100	1,650	3,190	5,080	8,980	18,300
900	139	292	522	1,070	1,600	3,090	4,930	8,710	17,800
950	135	283	507	1,040	1,560	3,000	4,780	8,460	17,200
1,000	132	275	493	1,010	1,520	2,920	4,650	8,220	16,800
1,100	125	262	468	960	1,440	2,770	4,420	7,810	15,900
1,200	119	250	446	917	1,370	2,640	4,220	7,450	15,200
1,300	114	239	427	878	1,320	2,530	4,040	7,140	14,600
1,400	110	230	411	843	1,260	2,430	3,880	6,860	14,000
1,500	106	221	396	812	1,220	2,340	3,740	6,600	13,500
1,600	102	214	382	784	1,180	2,260	3,610	6,380	13,000
1,700	99	207	370	759	1,140	2,190	3,490	6,170	12,600
1,800	96	200	358	736	1,100	2,120	3,390	5,980	12,200
1,900	93	195	348	715	1,070	2,060	3,290	5,810	11,900
2,000	91	189	339	695	1,040	2,010	3,200	5,650	11,500

For SI: 1 inch = 25.4 mm, 1 foot = 304.8 mm, 1 pound per square inch = 6.895 kPa, 1-inch water column = 0.2488 kPa, 1 British thermal unit per hour = 0.2931 W, 1 cubic foot per hour = 0.0283 m^3/h, 1 degree = 0.01745 rad.

Note: All table entries have been rounded to three significant digits.

Gas	Natural
Inlet Pressure	Less than 2 psi
Pressure Drop	0.5 in. w.c.
Specific Gravity	0.60

TABLE G2413.4(3) [402.4(9)]
SEMIRIGID COPPER TUBING

		TUBE SIZE (inch)								
Nominal	K & L	1/4	3/8	1/2	5/8	3/4	1	1 1/4	1 1/2	2
	ACR	3/8	1/2	5/8	3/4	7/8	1 1/8	1 3/8	—	—
Outside		0.375	0.500	0.625	0.750	0.875	1.125	1.375	1.625	2.125
Inside		0.305	0.402	0.527	0.652	0.745	0.995	1.245	1.481	1.959
Length (ft)		Capacity in Cubic Feet of Gas per Hour								
10		27	55	111	195	276	590	1,060	1,680	3,490
20		18	38	77	134	190	406	730	1,150	2,400
30		15	30	61	107	152	326	586	925	1,930
40		13	26	53	92	131	279	502	791	1,650
50		11	23	47	82	116	247	445	701	1,460
60		10	21	42	74	105	224	403	635	1,320
70		NA	19	39	68	96	206	371	585	1,220
80		NA	18	36	63	90	192	345	544	1,130
90		NA	17	34	59	84	180	324	510	1,060
100		NA	16	32	56	79	170	306	482	1,000
125		NA	14	28	50	70	151	271	427	890
150		NA	13	26	45	64	136	245	387	806
175		NA	12	24	41	59	125	226	356	742
200		NA	11	22	39	55	117	210	331	690
250		NA	NA	20	34	48	103	186	294	612
300		NA	NA	18	31	44	94	169	266	554
350		NA	NA	16	28	40	86	155	245	510
400		NA	NA	15	26	38	80	144	228	474
450		NA	NA	14	25	35	75	135	214	445
500		NA	NA	13	23	33	71	128	202	420
550		NA	NA	13	22	32	68	122	192	399
600		NA	NA	12	21	30	64	116	183	381
650		NA	NA	12	20	29	62	111	175	365
700		NA	NA	11	20	28	59	107	168	350
750		NA	NA	11	19	27	57	103	162	338
800		NA	NA	10	18	26	55	99	156	326
850		NA	NA	10	18	25	53	96	151	315
900		NA	NA	NA	17	24	52	93	147	306
950		NA	NA	NA	17	24	50	90	143	297
1,000		NA	NA	NA	16	23	49	88	139	289
1,100		NA	NA	NA	15	22	46	84	132	274
1,200		NA	NA	NA	15	21	44	80	126	262
1,300		NA	NA	NA	14	20	42	76	120	251
1,400		NA	NA	NA	13	19	41	73	116	241
1,500		NA	NA	NA	13	18	39	71	111	232
1,600		NA	NA	NA	13	18	38	68	108	224
1,700		NA	NA	NA	12	17	37	66	104	217
1,800		NA	NA	NA	12	17	36	64	101	210
1,900		NA	NA	NA	11	16	35	62	98	204
2,000		NA	NA	NA	11	16	34	60	95	199

For SI: 1 inch = 25.4 mm, 1 foot = 304.8 mm, 1 pound per square inch = 6.895 kPa, 1-inch water column = 0.2488 kPa, 1 British thermal unit per hour = 0.2931 W, 1 cubic foot per hour = 0.0283 m^3/h, 1 degree = 0.01745 rad.

Notes:

1. Table capacities are based on Type K copper tubing inside diameter (shown), which has the smallest inside diameter of the copper tubing products.
2. NA means a flow of less than 10 cfh.
3. All table entries have been rounded to three significant digits.

Gas	Natural
Inlet Pressure	2.0 psi
Pressure Drop	1.0 psi
Specific Gravity	0.60

TABLE G2413.4(4) [402.4(12)]
SEMIRIGID COPPER TUBING

TUBE SIZE (inch)										
Nominal	K & L	$^1/_4$	$^3/_8$	$^1/_2$	$^5/_8$	$^3/_4$	1	$1^1/_4$	$1^1/_2$	2
	ACR	$^3/_8$	$^1/_2$	$^5/_8$	$^3/_4$	$^7/_8$	$1^1/_8$	$1^3/_8$	—	—
Outside		0.375	0.500	0.625	0.750	0.875	1.125	1.375	1.625	2.125
Inside		0.305	0.402	0.527	0.652	0.745	0.995	1.245	1.481	1.959
Length (ft)		Capacity in Cubic Feet of Gas per Hour								
10		245	506	1,030	1,800	2,550	5,450	9,820	15,500	32,200
20		169	348	708	1,240	1,760	3,750	6,750	10,600	22,200
30		135	279	568	993	1,410	3,010	5,420	8,550	17,800
40		116	239	486	850	1,210	2,580	4,640	7,310	15,200
50		103	212	431	754	1,070	2,280	4,110	6,480	13,500
60		93	192	391	683	969	2,070	3,730	5,870	12,200
70		86	177	359	628	891	1,900	3,430	5,400	11,300
80		80	164	334	584	829	1,770	3,190	5,030	10,500
90		75	154	314	548	778	1,660	2,990	4,720	9,820
100		71	146	296	518	735	1,570	2,830	4,450	9,280
125		63	129	263	459	651	1,390	2,500	3,950	8,220
150		57	117	238	416	590	1,260	2,270	3,580	7,450
175		52	108	219	383	543	1,160	2,090	3,290	6,850
200		49	100	204	356	505	1,080	1,940	3,060	6,380
250		43	89	181	315	448	956	1,720	2,710	5,650
300		39	80	164	286	406	866	1,560	2,460	5,120
350		36	74	150	263	373	797	1,430	2,260	4,710
400		33	69	140	245	347	741	1,330	2,100	4,380
450		31	65	131	230	326	696	1,250	1,970	4,110
500		30	61	124	217	308	657	1,180	1,870	3,880
550		28	58	118	206	292	624	1,120	1,770	3,690
600		27	55	112	196	279	595	1,070	1,690	3,520
650		26	53	108	188	267	570	1,030	1,620	3,370
700		25	51	103	181	256	548	986	1,550	3,240
750		24	49	100	174	247	528	950	1,500	3,120
800		23	47	96	168	239	510	917	1,450	3,010
850		22	46	93	163	231	493	888	1,400	2,920
900		22	44	90	158	224	478	861	1,360	2,830
950		21	43	88	153	217	464	836	1,320	2,740
1,000		20	42	85	149	211	452	813	1,280	2,670
1,100		19	40	81	142	201	429	772	1,220	2,540
1,200		18	38	77	135	192	409	737	1,160	2,420
1,300		18	36	74	129	183	392	705	1,110	2,320
1,400		17	35	71	124	176	376	678	1,070	2,230
1,500		16	34	68	120	170	363	653	1,030	2,140
1,600		16	33	66	116	164	350	630	994	2,070
1,700		15	31	64	112	159	339	610	962	2,000
1,800		15	30	62	108	154	329	592	933	1,940
1,900		14	30	60	105	149	319	575	906	1,890
2,000		14	29	59	102	145	310	559	881	1,830

For SI: 1 inch = 25.4 mm, 1 foot = 304.8 mm, 1 pound per square inch = 6.895 kPa, 1-inch water column = 0.2488 kPa, 1 British thermal unit per hour = 0.2931 W, 1 cubic foot per hour = 0.0283 m^3/h, 1 degree = 0.01745 rad.

Notes:

1. Table capacities are based on Type K copper tubing inside diameter (shown), which has the smallest inside diameter of the copper tubing products.
2. All table entries have been rounded to three significant digits.

TABLE G2413.4(5) [402.4(15)] CORRUGATED STAINLESS STEEL TUBING (CSST)

Gas	Natural
Inlet Pressure	Less than 2 psi
Pressure Drop	0.5 in. w.c.
Specific Gravity	0.60

	TUBE SIZE (EHD)													
Flow Designation	13	15	18	19	23	25	30	31	37	39	46	48	60	62
Length (ft)	Capacity in Cubic Feet of Gas per Hour													
5	46	63	115	134	225	270	471	546	895	1,037	1,790	2,070	3,660	4,140
10	32	44	82	95	161	192	330	383	639	746	1,260	1,470	2,600	2,930
15	25	35	66	77	132	157	267	310	524	615	1,030	1,200	2,140	2,400
20	22	31	58	67	116	137	231	269	456	536	888	1,050	1,850	2,080
25	19	27	52	60	104	122	206	240	409	482	793	936	1,660	1,860
30	18	25	47	55	96	112	188	218	374	442	723	856	1,520	1,700
40	15	21	41	47	83	97	162	188	325	386	625	742	1,320	1,470
50	13	19	37	42	75	87	144	168	292	347	559	665	1,180	1,320
60	12	17	34	38	68	80	131	153	267	318	509	608	1,080	1,200
70	11	16	31	36	63	74	121	141	248	295	471	563	1,000	1,110
80	10	15	29	33	60	69	113	132	232	277	440	527	940	1,040
90	10	14	28	32	57	65	107	125	219	262	415	498	887	983
100	9	13	26	30	54	62	101	118	208	249	393	472	843	933
150	7	10	20	23	42	48	78	91	171	205	320	387	691	762
200	6	9	18	21	38	44	71	82	148	179	277	336	600	661
250	5	8	16	19	34	39	63	74	133	161	247	301	538	591
300	5	7	15	17	32	36	57	67	95	148	226	275	492	540

For SI: 1 inch = 25.4 mm, 1 foot = 304.8 mm, 1 pound per square inch = 6.895 kPa, 1-inch water column = 0.2488 kPa, 1 British thermal unit per hour = 0.2931 W, 1 cubic foot per hour = 0.0283 m^3/h, 1 degree = 0.01745 rad.

Notes:

1. Table includes losses for four 90-degree bends and two end fittings. Tubing runs with larger numbers of bends and/or fittings shall be increased by an equivalent length of tubing to the following equation: $L = 1.3n$, where L is additional length (feet) of tubing and n is the number of additional fittings and/or bends.
2. EHD—Equivalent Hydraulic Diameter, which is a measure of the relative hydraulic efficiency between different tubing sizes. The greater the value of EHD, the greater the gas capacity of the tubing.
3. All table entries have been rounded to three significant digits.

TABLE G2413.4(6) [402.4(18)]
CORRUGATED STAINLESS STEEL TUBING (CSST)

Gas	Natural
Inlet Pressure	2.0 psi
Pressure Drop	1.0 psi
Specific Gravity	0.60

	TUBE SIZE (EHD)													
Flow Designation	13	15	18	19	23	25	30	31	37	39	46	48	60	62
Length (ft)	Capacity in Cubic Feet of Gas Per Hour													
10	270	353	587	700	1,100	1,370	2,590	2,990	4,510	5,037	9,600	10,700	18,600	21,600
25	166	220	374	444	709	876	1,620	1,870	2,890	3,258	6,040	6,780	11,900	13,700
30	151	200	342	405	650	801	1,480	1,700	2,640	2,987	5,510	6,200	10,900	12,500
40	129	172	297	351	567	696	1,270	1,470	2,300	2,605	4,760	5,380	9,440	10,900
50	115	154	266	314	510	624	1,140	1,310	2,060	2,343	4,260	4,820	8,470	9,720
75	93	124	218	257	420	512	922	1,070	1,690	1,932	3,470	3,950	6,940	7,940
80	89	120	211	249	407	496	892	1,030	1,640	1,874	3,360	3,820	6,730	7,690
100	79	107	189	222	366	445	795	920	1,470	1,685	3,000	3,420	6,030	6,880
150	64	87	155	182	302	364	646	748	1,210	1,389	2,440	2,800	4,940	5,620
200	55	75	135	157	263	317	557	645	1,050	1,212	2,110	2,430	4,290	4,870
250	49	67	121	141	236	284	497	576	941	1,090	1,890	2,180	3,850	4,360
300	44	61	110	129	217	260	453	525	862	999	1,720	1,990	3,520	3,980
400	38	52	96	111	189	225	390	453	749	871	1,490	1,730	3,060	3,450
500	34	46	86	100	170	202	348	404	552	783	1,330	1,550	2,740	3,090

For SI: 1 inch = 25.4 mm, 1 foot = 304.8 mm, 1 pound per square inch = 6.895 kPa, 1-inch water column = 0.2488 kPa, 1 British thermal unit per hour = 0.2931 W, 1 cubic foot per hour = 0.0283 m^3/h, 1 degree = 0.01745 rad.

Notes:

1. Table does not include effect of pressure drop across the line regulator. Where regulator loss exceeds $^3/_4$ psi, DO NOT USE THIS TABLE. Consult with the regulator manufacturer for pressure drops and capacity factors. Pressure drops across a regulator can vary with flow rate.
2. CAUTION: Capacities shown in the table might exceed maximum capacity for a selected regulator. Consult with the regulator or tubing manufacturer for guidance.
3. Table includes losses for four 90-degree bends and two end fittings. Tubing runs with larger numbers of bends and/or fittings shall be increased by an equivalent length of tubing to the following equation: $L = 1.3n$ where L is additional length (feet) of tubing and n is the number of additional fittings and/or bends.
4. EHD—Equivalent Hydraulic Diameter, which is a measure of the relative hydraulic efficiency between different tubing sizes. The greater the value of EHD, the greater the gas capacity of the tubing.
5. All table entries have been rounded to three significant digits.

TABLE G2413.4(7) [402.4(21)]
POLYETHYLENE PLASTIC PIPE

Gas	Natural
Inlet Pressure	Less than 2 psi
Pressure Drop	0.5 in. w.c.
Specific Gravity	0.60

PIPE SIZE (in.)						
Nominal OD	1/2	3/4	1	1 1/4	1 1/2	2
Designation	SDR 9.33	SDR 11.0	SDR 11.00	SDR 10.00	SDR 11.00	SDR 11.00
Actual ID	0.660	0.860	1.077	1.328	1.554	1.943
Length (ft)	**Capacity in Cubic Feet of Gas per Hour**					
10	201	403	726	1,260	1,900	3,410
20	138	277	499	865	1,310	2,350
30	111	222	401	695	1,050	1,880
40	95	190	343	594	898	1,610
50	84	169	304	527	796	1,430
60	76	153	276	477	721	1,300
70	70	140	254	439	663	1,190
80	65	131	236	409	617	1,110
90	61	123	221	383	579	1,040
100	58	116	209	362	547	983
125	51	103	185	321	485	871
150	46	93	168	291	439	789
175	43	86	154	268	404	726
200	40	80	144	249	376	675
250	35	71	127	221	333	598
300	32	64	115	200	302	542
350	29	59	106	184	278	499
400	27	55	99	171	258	464
450	26	51	93	160	242	435
500	24	48	88	152	229	411

For SI: 1 inch = 25.4 mm, 1 foot = 304.8 mm, 1 pound per square inch = 6.895 kPa, 1-inch water column = 0.2488 kPa, 1 British thermal unit per hour = 0.2931 W, 1 cubic foot per hour = 0.0283 m^3/h, 1 degree = 0.01745 rad.

Note: All table entries have been rounded to three significant digits.

Gas	Natural
Inlet Pressure	2.0 psi
Pressure Drop	1.0 psi
Specific Gravity	0.60

TABLE G2413.4(8) [402.4(22)] POLYETHYLENE PLASTIC PIPE

	PIPE SIZE (in.)					
Nominal OD	$^1/_2$	$^3/_4$	1	$1^1/_4$	$1^1/_2$	2
Designation	SDR 9.33	SDR 11.0	SDR 11.00	SDR 10.00	SDR 11.00	SDR 11.00
Actual ID	0.660	0.860	1.077	1.328	1.554	1.943
Length (ft)	Capacity in Cubic Feet of Gas per Hour					
10	1,860	3,720	6,710	11,600	17,600	31,600
20	1,280	2,560	4,610	7,990	12,100	21,700
30	1,030	2,050	3,710	6,420	9,690	17,400
40	878	1,760	3,170	5,490	8,300	14,900
50	778	1,560	2,810	4,870	7,350	13,200
60	705	1,410	2,550	4,410	6,660	12,000
70	649	1,300	2,340	4,060	6,130	11,000
80	603	1,210	2,180	3,780	5,700	10,200
90	566	1,130	2,050	3,540	5,350	9,610
100	535	1,070	1,930	3,350	5,050	9,080
125	474	949	1,710	2,970	4,480	8,050
150	429	860	1,550	2,690	4,060	7,290
175	395	791	1,430	2,470	3,730	6,710
200	368	736	1,330	2,300	3,470	6,240
250	326	652	1,180	2,040	3,080	5,530
300	295	591	1,070	1,850	2,790	5,010
350	272	544	981	1,700	2,570	4,610
400	253	506	913	1,580	2,390	4,290
450	237	475	856	1,480	2,240	4,020
500	224	448	809	1,400	2,120	3,800
550	213	426	768	1,330	2,010	3,610
600	203	406	733	1,270	1,920	3,440
650	194	389	702	1,220	1,840	3,300
700	187	374	674	1,170	1,760	3,170
750	180	360	649	1,130	1,700	3,050
800	174	348	627	1,090	1,640	2,950
850	168	336	607	1,050	1,590	2,850
900	163	326	588	1,020	1,540	2,770
950	158	317	572	990	1,500	2,690
1,000	154	308	556	963	1,450	2,610
1,100	146	293	528	915	1,380	2,480
1,200	139	279	504	873	1,320	2,370
1,300	134	267	482	836	1,260	2,270
1,400	128	257	463	803	1,210	2,180
1,500	124	247	446	773	1,170	2,100
1,600	119	239	431	747	1,130	2,030
1,700	115	231	417	723	1,090	1,960
1,800	112	224	404	701	1,060	1,900
1,900	109	218	393	680	1,030	1,850
2,000	106	212	382	662	1,000	1,800

For SI: 1 inch = 25.4 mm, 1 foot = 304.8 mm, 1 pound per square inch = 6.895 kPa, 1-inch water column = 0.2488 kPa, 1 British thermal unit per hour = 0.2931 W, 1 cubic foot per hour = 0.0283 m^3/h, 1 degree = 0.01745 rad.

Note: All table entries have been rounded to three significant digits.

Gas	Undiluted Propane
Inlet Pressure	10.0 psi
Pressure Drop	1.0 psi
Specific Gravity	1.50

TABLE G2413.4(9) [402.4(25]
SCHEDULE 40 METALLIC PIPE

INTENDED USE	Pipe sizing between first stage (high-pressure regulator) and second stage (low-pressure regulator).								
	PIPE SIZE (in.)								
Nominal	$^{1}/_{2}$	$^{3}/_{4}$	1	$1^{1}/_{4}$	$1^{1}/_{2}$	2	$2^{1}/_{2}$	3	4
Actual ID	0.622	0.824	1.049	1.380	1.610	2.067	2.469	3.068	4.026
Length (ft)	Capacity in Thousands of Btu per Hour								
10	3,320	6,950	13,100	26,900	40,300	77,600	124,000	219,000	446,000
20	2,280	4,780	9,000	18,500	27,700	53,300	85,000	150,000	306,000
30	1,830	3,840	7,220	14,800	22,200	42,800	68,200	121,000	246,000
40	1,570	3,280	6,180	12,700	19,000	36,600	58,400	103,000	211,000
50	1,390	2,910	5,480	11,300	16,900	32,500	51,700	91,500	187,000
60	1,260	2,640	4,970	10,200	15,300	29,400	46,900	82,900	169,000
70	1,160	2,430	4,570	9,380	14,100	27,100	43,100	76,300	156,000
80	1,080	2,260	4,250	8,730	13,100	25,200	40,100	70,900	145,000
90	1,010	2,120	3,990	8,190	12,300	23,600	37,700	66,600	136,000
100	956	2,000	3,770	7,730	11,600	22,300	35,600	62,900	128,000
125	848	1,770	3,340	6,850	10,300	19,800	31,500	55,700	114,000
150	768	1,610	3,020	6,210	9,300	17,900	28,600	50,500	103,000
175	706	1,480	2,780	5,710	8,560	16,500	26,300	46,500	94,700
200	657	1,370	2,590	5,320	7,960	15,300	24,400	43,200	88,100
250	582	1,220	2,290	4,710	7,060	13,600	21,700	38,300	78,100
300	528	1,100	2,080	4,270	6,400	12,300	19,600	34,700	70,800
350	486	1,020	1,910	3,930	5,880	11,300	18,100	31,900	65,100
400	452	945	1,780	3,650	5,470	10,500	16,800	29,700	60,600
450	424	886	1,670	3,430	5,140	9,890	15,800	27,900	56,800
500	400	837	1,580	3,240	4,850	9,340	14,900	26,300	53,700
550	380	795	1,500	3,070	4,610	8,870	14,100	25,000	51,000
600	363	759	1,430	2,930	4,400	8,460	13,500	23,900	48,600
650	347	726	1,370	2,810	4,210	8,110	12,900	22,800	46,600
700	334	698	1,310	2,700	4,040	7,790	12,400	21,900	44,800
750	321	672	1,270	2,600	3,900	7,500	12,000	21,100	43,100
800	310	649	1,220	2,510	3,760	7,240	11,500	20,400	41,600
850	300	628	1,180	2,430	3,640	7,010	11,200	19,800	40,300
900	291	609	1,150	2,360	3,530	6,800	10,800	19,200	39,100
950	283	592	1,110	2,290	3,430	6,600	10,500	18,600	37,900
1,000	275	575	1,080	2,230	3,330	6,420	10,200	18,100	36,900
1,100	261	546	1,030	2,110	3,170	6,100	9,720	17,200	35,000
1,200	249	521	982	2,020	3,020	5,820	9,270	16,400	33,400
1,300	239	499	940	1,930	2,890	5,570	8,880	15,700	32,000
1,400	229	480	903	1,850	2,780	5,350	8,530	15,100	30,800
1,500	221	462	870	1,790	2,680	5,160	8,220	14,500	29,600
1,600	213	446	840	1,730	2,590	4,980	7,940	14,000	28,600
1,700	206	432	813	1,670	2,500	4,820	7,680	13,600	27,700
1,800	200	419	789	1,620	2,430	4,670	7,450	13,200	26,900
1,900	194	407	766	1,570	2,360	4,540	7,230	12,800	26,100
2,000	189	395	745	1,530	2,290	4,410	7,030	12,400	25,400

For SI: 1 inch = 25.4 mm, 1 foot = 304.8 mm, 1 pound per square inch = 6.895 kPa, 1-inch water column = 0.2488 kPa, 1 British thermal unit per hour = 0.2931 W, 1 cubic foot per hour = 0.0283 m^3/h, 1 degree = 0.01745 rad.

Note: All table entries have been rounded to three significant digits.

Gas	Undiluted Propane
Inlet Pressure	10.0 psi
Pressure Drop	3.0 psi
Specific Gravity	1.50

TABLE G2413.4(10) [402.4(26)]
SCHEDULE 40 METALLIC PIPE

INTENDED USE	Pipe sizing between first stage (high-pressure regulator) and second stage (low-pressure regulator).								
	PIPE SIZE (in)								
Nominal	$^{1}/_{2}$	$^{3}/_{4}$	1	$1^{1}/_{4}$	$1^{1}/_{2}$	2	$2^{1}/_{2}$	3	4
Actual ID	0.622	0.824	1.049	1.380	1.610	2.067	2.469	3.068	4.026
Length (ft)	Capacity in Thousands of Btu per Hour								
10	5,890	12,300	23,200	47,600	71,300	137,000	219,000	387,000	789,000
20	4,050	8,460	15,900	32,700	49,000	94,400	150,000	266,000	543,000
30	3,250	6,790	12,800	26,300	39,400	75,800	121,000	214,000	436,000
40	2,780	5,810	11,000	22,500	33,700	64,900	103,000	183,000	373,000
50	2,460	5,150	9,710	19,900	29,900	57,500	91,600	162,000	330,000
60	2,230	4,670	8,790	18,100	27,100	52,100	83,000	147,000	299,000
70	2,050	4,300	8,090	16,600	24,900	47,900	76,400	135,000	275,000
80	1,910	4,000	7,530	15,500	23,200	44,600	71,100	126,000	256,000
90	1,790	3,750	7,060	14,500	21,700	41,800	66,700	118,000	240,000
100	1,690	3,540	6,670	13,700	20,500	39,500	63,000	111,000	227,000
125	1,500	3,140	5,910	12,100	18,200	35,000	55,800	98,700	201,000
150	1,360	2,840	5,360	11,000	16,500	31,700	50,600	89,400	182,000
175	1,250	2,620	4,930	10,100	15,200	29,200	46,500	82,300	167,800
200	1,160	2,430	4,580	9,410	14,100	27,200	43,300	76,500	156,100
250	1,030	2,160	4,060	8,340	12,500	24,100	38,400	67,800	138,400
300	935	1,950	3,680	7,560	11,300	21,800	34,800	61,500	125,400
350	860	1,800	3,390	6,950	10,400	20,100	32,000	56,500	115,300
400	800	1,670	3,150	6,470	9,690	18,700	29,800	52,600	107,300
450	751	1,570	2,960	6,070	9,090	17,500	27,900	49,400	100,700
500	709	1,480	2,790	5,730	8,590	16,500	26,400	46,600	95,100
550	673	1,410	2,650	5,450	8,160	15,700	25,000	44,300	90,300
600	642	1,340	2,530	5,200	7,780	15,000	23,900	42,200	86,200
650	615	1,290	2,420	4,980	7,450	14,400	22,900	40,500	82,500
700	591	1,240	2,330	4,780	7,160	13,800	22,000	38,900	79,300
750	569	1,190	2,240	4,600	6,900	13,300	21,200	37,400	76,400
800	550	1,150	2,170	4,450	6,660	12,800	20,500	36,200	73,700
850	532	1,110	2,100	4,300	6,450	12,400	19,800	35,000	71,400
900	516	1,080	2,030	4,170	6,250	12,000	19,200	33,900	69,200
950	501	1,050	1,970	4,050	6,070	11,700	18,600	32,900	67,200
1,000	487	1,020	1,920	3,940	5,900	11,400	18,100	32,000	65,400
1,100	463	968	1,820	3,740	5,610	10,800	17,200	30,400	62,100
1,200	442	923	1,740	3,570	5,350	10,300	16,400	29,000	59,200
1,300	423	884	1,670	3,420	5,120	9,870	15,700	27,800	56,700
1,400	406	849	1,600	3,280	4,920	9,480	15,100	26,700	54,500
1,500	391	818	1,540	3,160	4,740	9,130	14,600	25,700	52,500
1,600	378	790	1,490	3,060	4,580	8,820	14,100	24,800	50,700
1,700	366	765	1,440	2,960	4,430	8,530	13,600	24,000	49,000
1,800	355	741	1,400	2,870	4,300	8,270	13,200	23,300	47,600
1,900	344	720	1,360	2,780	4,170	8,040	12,800	22,600	46,200
2,000	335	700	1,320	2,710	4,060	7,820	12,500	22,000	44,900

For SI: 1 inch = 25.4 mm, 1 foot = 304.8 mm, 1 pound per square inch = 6.895 kPa, 1-inch water column = 0.2488 kPa, 1 British thermal unit per hour = 0.2931 W, 1 cubic foot per hour = 0.0283 m^3/h, 1 degree = 0.01745 rad.

Note: All table entries have been rounded to three significant digits.

Gas	Undiluted Propane
Inlet Pressure	2.0 psi
Pressure Drop	1.0 psi
Specific Gravity	1.50

TABLE G2413.4(11) [402.4(27)]
SCHEDULE 40 METALLIC PIPE

INTENDED USE	Pipe sizing between 2 psig service and line pressure regulator.								
	PIPE SIZE (in.)								
Nominal	$^1/_2$	$^3/_4$	1	$1^1/_4$	$1^1/_2$	2	$2^1/_2$	3	4
Actual ID	0.622	0.824	1.049	1.380	1.610	2.067	2.469	3.068	4.026
Length (ft)	**Capacity in Thousands of Btu per Hour**								
10	2,680	5,590	10,500	21,600	32,400	62,400	99,500	176,000	359,000
20	1,840	3,850	7,240	14,900	22,300	42,900	68,400	121,000	247,000
30	1,480	3,090	5,820	11,900	17,900	34,500	54,900	97,100	198,000
40	1,260	2,640	4,980	10,200	15,300	29,500	47,000	83,100	170,000
50	1,120	2,340	4,410	9,060	13,600	26,100	41,700	73,700	150,000
60	1,010	2,120	4,000	8,210	12,300	23,700	37,700	66,700	136,000
70	934	1,950	3,680	7,550	11,300	21,800	34,700	61,400	125,000
80	869	1,820	3,420	7,020	10,500	20,300	32,300	57,100	116,000
90	815	1,700	3,210	6,590	9,880	19,000	30,300	53,600	109,000
100	770	1,610	3,030	6,230	9,330	18,000	28,600	50,600	103,000
125	682	1,430	2,690	5,520	8,270	15,900	25,400	44,900	91,500
150	618	1,290	2,440	5,000	7,490	14,400	23,000	40,700	82,900
175	569	1,190	2,240	4,600	6,890	13,300	21,200	37,400	76,300
200	529	1,110	2,080	4,280	6,410	12,300	19,700	34,800	71,000
250	469	981	1,850	3,790	5,680	10,900	17,400	30,800	62,900
300	425	889	1,670	3,440	5,150	9,920	15,800	27,900	57,000
350	391	817	1,540	3,160	4,740	9,120	14,500	25,700	52,400
400	364	760	1,430	2,940	4,410	8,490	13,500	23,900	48,800
450	341	714	1,340	2,760	4,130	7,960	12,700	22,400	45,800
500	322	674	1,270	2,610	3,910	7,520	12,000	21,200	43,200
550	306	640	1,210	2,480	3,710	7,140	11,400	20,100	41,100
600	292	611	1,150	2,360	3,540	6,820	10,900	19,200	39,200
650	280	585	1,100	2,260	3,390	6,530	10,400	18,400	37,500
700	269	562	1,060	2,170	3,260	6,270	9,990	17,700	36,000
750	259	541	1,020	2,090	3,140	6,040	9,630	17,000	34,700
800	250	523	985	2,020	3,030	5,830	9,300	16,400	33,500
850	242	506	953	1,960	2,930	5,640	9,000	15,900	32,400
900	235	490	924	1,900	2,840	5,470	8,720	15,400	31,500
950	228	476	897	1,840	2,760	5,310	8,470	15,000	30,500
1,000	222	463	873	1,790	2,680	5,170	8,240	14,600	29,700
1,100	210	440	829	1,700	2,550	4,910	7,830	13,800	28,200
1,200	201	420	791	1,620	2,430	4,680	7,470	13,200	26,900
1,300	192	402	757	1,550	2,330	4,490	7,150	12,600	25,800
1,400	185	386	727	1,490	2,240	4,310	6,870	12,100	24,800
1,500	178	372	701	1,440	2,160	4,150	6,620	11,700	23,900
1,600	172	359	677	1,390	2,080	4,010	6,390	11,300	23,000
1,700	166	348	655	1,340	2,010	3,880	6,180	10,900	22,300
1,800	161	337	635	1,300	1,950	3,760	6,000	10,600	21,600
1,900	157	327	617	1,270	1,900	3,650	5,820	10,300	21,000
2,000	152	318	600	1,230	1,840	3,550	5,660	10,000	20,400

For SI: 1 inch = 25.4 mm, 1 foot = 304.8 mm, 1 pound per square inch = 6.895 kPa, 1-inch water column = 0.2488 kPa,
1 British thermal unit per hour = 0.2931 W, 1 cubic foot per hour = 0.0283 m^3/h, 1 degree = 0.01745 rad.

Note: All table entries have been rounded to three significant digits.

Gas	Undiluted Propane
Inlet Pressure	11.0 in. w.c.
Pressure Drop	0.5 in. w.c.
Specific Gravity	1.50

TABLE G2413.4(12) [402.4(28)]
SCHEDULE 40 METALLIC PIPE

INTENDED USE	Pipe sizing between single- or second-stage (low pressure) regulator and appliance.							
PIPE SIZE (in.)								
Nominal	$^1/_2$	$^3/_4$	1	$1^1/_4$	$1^1/_2$	2	$2^1/_2$	3	4
Actual ID	0.622	0.824	1.049	1.380	1.610	2.067	2.469	3.068	4.026
Length (ft)	Capacity in Thousands of Btu per Hour								
10	291	608	1,150	2,350	3,520	6,790	10,800	19,100	39,000
20	200	418	787	1,620	2,420	4,660	7,430	13,100	26,800
30	160	336	632	1,300	1,940	3,750	5,970	10,600	21,500
40	137	287	541	1,110	1,660	3,210	5,110	9,030	18,400
50	122	255	480	985	1,480	2,840	4,530	8,000	16,300
60	110	231	434	892	1,340	2,570	4,100	7,250	14,800
80	101	212	400	821	1,230	2,370	3,770	6,670	13,600
100	94	197	372	763	1,140	2,200	3,510	6,210	12,700
125	89	185	349	716	1,070	2,070	3,290	5,820	11,900
150	84	175	330	677	1,010	1,950	3,110	5,500	11,200
175	74	155	292	600	899	1,730	2,760	4,880	9,950
200	67	140	265	543	814	1,570	2,500	4,420	9,010
250	62	129	243	500	749	1,440	2,300	4,060	8,290
300	58	120	227	465	697	1,340	2,140	3,780	7,710
350	51	107	201	412	618	1,190	1,900	3,350	6,840
400	46	97	182	373	560	1,080	1,720	3,040	6,190
450	42	89	167	344	515	991	1,580	2,790	5,700
500	40	83	156	320	479	922	1,470	2,600	5,300
550	37	78	146	300	449	865	1,380	2,440	4,970
600	35	73	138	283	424	817	1,300	2,300	4,700
650	33	70	131	269	403	776	1,240	2,190	4,460
700	32	66	125	257	385	741	1,180	2,090	4,260
750	30	64	120	246	368	709	1,130	2,000	4,080
800	29	61	115	236	354	681	1,090	1,920	3,920
850	28	59	111	227	341	656	1,050	1,850	3,770
900	27	57	107	220	329	634	1,010	1,790	3,640
950	26	55	104	213	319	613	978	1,730	3,530
1,000	25	53	100	206	309	595	948	1,680	3,420
1,100	25	52	97	200	300	578	921	1,630	3,320
1,200	24	50	95	195	292	562	895	1,580	3,230
1,300	23	48	90	185	277	534	850	1,500	3,070
1,400	22	46	86	176	264	509	811	1,430	2,930
1,500	21	44	82	169	253	487	777	1,370	2,800
1,200	24	50	95	195	292	562	895	1,580	3,230
1,300	23	48	90	185	277	534	850	1,500	3,070
1,400	22	46	86	176	264	509	811	1,430	2,930
1,500	21	44	82	169	253	487	777	1,370	2,800
1,600	20	42	79	162	243	468	746	1,320	2,690
1,700	19	40	76	156	234	451	719	1,270	2,590
1,800	19	39	74	151	226	436	694	1,230	2,500
1,900	18	38	71	146	219	422	672	1,190	2,420
2,000	18	37	69	142	212	409	652	1,150	2,350

For SI: 1 inch = 25.4 mm, 1 foot = 304.8 mm, 1 pound per square inch = 6.895 kPa, 1-inch water column = 0.2488 kPa, 1 British thermal unit per hour = 0.2931 W, 1 cubic foot per hour = 0.0283 m^3/h, 1 degree = 0.01745 rad.

Note: All table entries have been rounded to three significant digits.

Gas	Undiluted Propane
Inlet Pressure	10.0 psi
Pressure Drop	1.0 psi
Specific Gravity	1.50

TABLE G2413.4(13) [402.4(29)]
SEMIRIGID COPPER TUBING

INTENDED USE		Sizing between first stage (high-pressure regulator) and second stage (low-pressure regulator).								
TUBE SIZE (in.)										
Nominal	K & L	1/4	3/8	1/2	5/8	3/4	1	1 1/4	1 1/2	2
	ACR	3/8	1/2	5/8	3/4	7/8	1 1/8	1 3/8	—	—
Outside		0.375	0.500	0.625	0.750	0.875	1.125	1.375	1.625	2.125
Inside		0.305	0.402	0.527	0.652	0.745	0.995	1.245	1.481	1.959
Length (ft)		Capacity in Thousands of Btu per Hour								
10		513	1,060	2,150	3,760	5,330	11,400	20,500	32,300	67,400
20		352	727	1,480	2,580	3,670	7,830	14,100	22,200	46,300
30		283	584	1,190	2,080	2,940	6,290	11,300	17,900	37,200
40		242	500	1,020	1,780	2,520	5,380	9,690	15,300	31,800
50		215	443	901	1,570	2,230	4,770	8,590	13,500	28,200
60		194	401	816	1,430	2,020	4,320	7,780	12,300	25,600
70		179	369	751	1,310	1,860	3,980	7,160	11,300	23,500
80		166	343	699	1,220	1,730	3,700	6,660	10,500	21,900
90		156	322	655	1,150	1,630	3,470	6,250	9,850	20,500
100		147	304	619	1,080	1,540	3,280	5,900	9,310	19,400
125		131	270	549	959	1,360	2,910	5,230	8,250	17,200
150		118	244	497	869	1,230	2,630	4,740	7,470	15,600
175		109	225	457	799	1,130	2,420	4,360	6,880	14,300
200		101	209	426	744	1,060	2,250	4,060	6,400	13,300
250		90	185	377	659	935	2,000	3,600	5,670	11,800
300		81	168	342	597	847	1,810	3,260	5,140	10,700
350		75	155	314	549	779	1,660	3,000	4,730	9,840
400		70	144	292	511	725	1,550	2,790	4,400	9,160
450		65	135	274	480	680	1,450	2,620	4,130	8,590
500		62	127	259	453	643	1,370	2,470	3,900	8,120
550		59	121	246	430	610	1,300	2,350	3,700	7,710
600		56	115	235	410	582	1,240	2,240	3,530	7,350
650		54	111	225	393	558	1,190	2,140	3,380	7,040
700		51	106	216	378	536	1,140	2,060	3,250	6,770
750		50	102	208	364	516	1,100	1,980	3,130	6,520
800		48	99	201	351	498	1,060	1,920	3,020	6,290
850		46	96	195	340	482	1,030	1,850	2,920	6,090
900		45	93	189	330	468	1,000	1,800	2,840	5,910
950		44	90	183	320	454	970	1,750	2,750	5,730
1,000		42	88	178	311	442	944	1,700	2,680	5,580
1,100		40	83	169	296	420	896	1,610	2,540	5,300
1,200		38	79	161	282	400	855	1,540	2,430	5,050
1,300		37	76	155	270	383	819	1,470	2,320	4,840
1,400		35	73	148	260	368	787	1,420	2,230	4,650
1,500		34	70	143	250	355	758	1,360	2,150	4,480
1,600		33	68	138	241	343	732	1,320	2,080	4,330
1,700		32	66	134	234	331	708	1,270	2,010	4,190
1,800		31	64	130	227	321	687	1,240	1,950	4,060
1,900		30	62	126	220	312	667	1,200	1,890	3,940
2,000		29	60	122	214	304	648	1,170	1,840	3,830

For SI: 1 inch = 25.4 mm, 1 foot = 304.8 mm, 1 pound per square inch = 6.895 kPa, 1-inch water column = 0.2488 kPa, 1 British thermal unit per hour = 0.2931 W, 1 cubic foot per hour = 0.0283 m^3/h, 1 degree = 0.01745 rad.

Notes:

1. Table capacities are based on Type K copper tubing inside diameter (shown), which has the smallest inside diameter of the copper tubing products.
2. All table entries have been rounded to three significant digits.

Gas	Undiluted Propane
Inlet Pressure	11.0 in. w.c.
Pressure Drop	0.5 in. w.c.
Specific Gravity	1.50

TABLE G2413.4(14) [402.4(30)] SEMIRIGID COPPER TUBING

INTENDED USE		Sizing between single- or second-stage (low-pressure regulator) and appliance.								
		TUBE SIZE (in.)								
Nominal	**K & L**	1/4	3/8	1/2	5/8	3/4	1	1 1/4	1 1/2	2
	ACR	3/8	1/2	5/8	3/4	7/8	1 1/8	1 3/8	—	—
Outside		0.375	0.500	0.625	0.750	0.875	1.125	1.375	1.625	2.125
Inside		0.305	0.402	0.527	0.652	0.745	0.995	1.245	1.481	1.959
Length (ft)		**Capacity in Thousands of Btu per Hour**								
10		45	93	188	329	467	997	1,800	2,830	5,890
20		31	64	129	226	321	685	1,230	1,950	4,050
30		25	51	104	182	258	550	991	1,560	3,250
40		21	44	89	155	220	471	848	1,340	2,780
50		19	39	79	138	195	417	752	1,180	2,470
60		17	35	71	125	177	378	681	1,070	2,240
70		16	32	66	115	163	348	626	988	2,060
80		15	30	61	107	152	324	583	919	1,910
90		14	28	57	100	142	304	547	862	1,800
100		13	27	54	95	134	287	517	814	1,700
125		11	24	48	84	119	254	458	722	1,500
150		10	21	44	76	108	230	415	654	1,360
175		NA	20	40	70	99	212	382	602	1,250
200		NA	18	37	65	92	197	355	560	1,170
250		NA	16	33	58	82	175	315	496	1,030
300		NA	15	30	52	74	158	285	449	936
350		NA	14	28	48	68	146	262	414	861
400		NA	13	26	45	63	136	244	385	801
450		NA	12	24	42	60	127	229	361	752
500		NA	11	23	40	56	120	216	341	710
550		NA	11	22	38	53	114	205	324	674
600		NA	10	21	36	51	109	196	309	643
650		NA	NA	20	34	49	104	188	296	616
700		NA	NA	19	33	47	100	180	284	592
750		NA	NA	18	32	45	96	174	274	570
800		NA	NA	18	31	44	93	168	264	551
850		NA	NA	17	30	42	90	162	256	533
900		NA	NA	17	29	41	87	157	248	517
950		NA	NA	16	28	40	85	153	241	502
1,000		NA	NA	16	27	39	83	149	234	488
1,100		NA	NA	15	26	37	78	141	223	464
1,200		NA	NA	14	25	35	75	135	212	442
1,300		NA	NA	14	24	34	72	129	203	423
1,400		NA	NA	13	23	32	69	124	195	407
1,500		NA	NA	13	22	31	66	119	188	392
1,600		NA	NA	12	21	30	64	115	182	378
1,700		NA	NA	12	20	29	62	112	176	366
1,800		NA	NA	11	20	28	60	108	170	355
1,900		NA	NA	11	19	27	58	105	166	345
2,000		NA	NA	11	19	27	57	102	161	335

For SI: 1 inch = 25.4 mm, 1 foot = 304.8 mm, 1 pound per square inch = 6.895 kPa, 1-inch water column = 0.2488 kPa, 1 British thermal unit per hour = 0.2931 W, 1 cubic foot per hour = 0.0283 m^3/h, 1 degree = 0.01745 rad.

Notes:

1. Table capacities are based on Type K copper tubing inside diameter (shown), which has the smallest inside diameter of the copper tubing products.
2. NA means a flow of less than 10,000 Btu/hr.
3. All table entries have been rounded to three significant digits.

Gas	Undiluted Propane
Inlet Pressure	2.0 psi
Pressure Drop	1.0 psi
Specific Gravity	1.50

TABLE G2413.4(15) [402.4(31)] SEMIRIGID COPPER TUBING

INTENDED USE		Tube sizing between 2 psig service and line pressure regulator.								
		TUBE SIZE (in.)								
Nominal	K & L	$^1/_4$	$^3/_8$	$^1/_2$	$^5/_8$	$^3/_4$	1	$1^1/_4$	$1^1/_2$	2
	ACR	$^3/_8$	$^1/_2$	$^5/_8$	$^3/_4$	$^7/_8$	$1^1/_8$	$1^3/_8$	—	—
Outside		0.375	0.500	0.625	0.750	0.875	1.125	1.375	1.625	2.125
Inside		0.305	0.402	0.527	0.652	0.745	0.995	1.245	1.481	1.959
Length (ft)		Capacity in Thousands of Btu per Hour								
10		413	852	1,730	3,030	4,300	9,170	16,500	26,000	54,200
20		284	585	1,190	2,080	2,950	6,310	11,400	17,900	37,300
30		228	470	956	1,670	2,370	5,060	9,120	14,400	29,900
40		195	402	818	1,430	2,030	4,330	7,800	12,300	25,600
50		173	356	725	1,270	1,800	3,840	6,920	10,900	22,700
60		157	323	657	1,150	1,630	3,480	6,270	9,880	20,600
70		144	297	605	1,060	1,500	3,200	5,760	9,090	18,900
80		134	276	562	983	1,390	2,980	5,360	8,450	17,600
90		126	259	528	922	1,310	2,790	5,030	7,930	16,500
100		119	245	498	871	1,240	2,640	4,750	7,490	15,600
125		105	217	442	772	1,100	2,340	4,210	6,640	13,800
150		95	197	400	700	992	2,120	3,820	6,020	12,500
175		88	181	368	644	913	1,950	3,510	5,540	11,500
200		82	168	343	599	849	1,810	3,270	5,150	10,700
250		72	149	304	531	753	1,610	2,900	4,560	9,510
300		66	135	275	481	682	1,460	2,620	4,140	8,610
350		60	124	253	442	628	1,340	2,410	3,800	7,920
400		56	116	235	411	584	1,250	2,250	3,540	7,370
450		53	109	221	386	548	1,170	2,110	3,320	6,920
500		50	103	209	365	517	1,110	1,990	3,140	6,530
550		47	97	198	346	491	1,050	1,890	2,980	6,210
600		45	93	189	330	469	1,000	1,800	2,840	5,920
650		43	89	181	316	449	959	1,730	2,720	5,670
700		41	86	174	304	431	921	1,660	2,620	5,450
750		40	82	168	293	415	888	1,600	2,520	5,250
800		39	80	162	283	401	857	1,540	2,430	5,070
850		37	77	157	274	388	829	1,490	2,350	4,900
900		36	75	152	265	376	804	1,450	2,280	4,750
950		35	72	147	258	366	781	1,410	2,220	4,620
1,000		34	71	143	251	356	760	1,370	2,160	4,490
1,100		32	67	136	238	338	721	1,300	2,050	4,270
1,200		31	64	130	227	322	688	1,240	1,950	4,070
1,300		30	61	124	217	309	659	1,190	1,870	3,900
1,400		28	59	120	209	296	633	1,140	1,800	3,740
1,500		27	57	115	201	286	610	1,100	1,730	3,610
1,600		26	55	111	194	276	589	1,060	1,670	3,480
1,700		26	53	108	188	267	570	1,030	1,620	3,370
1,800		25	51	104	182	259	553	1,000	1,570	3,270
1,900		24	50	101	177	251	537	966	1,520	3,170
2,000		23	48	99	172	244	522	940	1,480	3,090

For SI: 1 inch = 25.4 mm, 1 foot = 304.8 mm, 1 pound per square inch = 6.895 kPa, 1-inch water column = 0.2488 kPa,
1 British thermal unit per hour = 0.2931 W, 1 cubic foot per hour = 0.0283 m^3/h, 1 degree = 0.01745 rad.

Notes:

1. Table capacities are based on Type K copper tubing inside diameter (shown), which has the smallest inside diameter of the copper tubing products.
2. All table entries have been rounded to three significant digits.

Gas	Undiluted Propane
Inlet Pressure	11.0 in. w.c.
Pressure Drop	0.5 in. w.c.
Specific Gravity	1.50

TABLE G2413.4(16) [402.4(32)]
CORRUGATED STAINLESS STEEL TUBING (CSST)

INTENDED USE	Sizing between single or second stage (low pressure) regulator and the appliance shutoff valve.													
	TUBE SIZE (EHD)													
Flow Designation	13	15	18	19	23	25	30	31	37	39	46	48	60	62
Length (ft)	Capacity in Thousands of Btu per Hour													
5	72	99	181	211	355	426	744	863	1,420	1,638	2,830	3,270	5,780	6,550
10	50	69	129	150	254	303	521	605	971	1,179	1,990	2,320	4,110	4,640
15	39	55	104	121	208	248	422	490	775	972	1,620	1,900	3,370	3,790
20	34	49	91	106	183	216	365	425	661	847	1,400	1,650	2,930	3,290
25	30	42	82	94	164	192	325	379	583	762	1,250	1,480	2,630	2,940
30	28	39	74	87	151	177	297	344	528	698	1,140	1,350	2,400	2,680
40	23	33	64	74	131	153	256	297	449	610	988	1,170	2,090	2,330
50	20	30	58	66	118	137	227	265	397	548	884	1,050	1,870	2,080
60	19	26	53	60	107	126	207	241	359	502	805	961	1,710	1,900
70	17	25	49	57	99	117	191	222	330	466	745	890	1,590	1,760
80	15	23	45	52	94	109	178	208	307	438	696	833	1,490	1,650
90	15	22	44	50	90	102	169	197	286	414	656	787	1,400	1,550
100	14	20	41	47	85	98	159	186	270	393	621	746	1,330	1,480
150	11	15	31	36	66	75	123	143	217	324	506	611	1,090	1,210
200	9	14	28	33	60	69	112	129	183	283	438	531	948	1,050
250	8	12	25	30	53	61	99	117	163	254	390	476	850	934
300	8	11	23	26	50	57	90	107	147	234	357	434	777	854

For SI: 1 inch = 25.4 mm, 1 foot = 304.8 mm, 1 pound per square inch = 6.895 kPa, 1-inch water column = 0.2488 kPa, 1 British thermal unit per hour = 0.2931 W, 1 cubic foot per hour = 0.0283 m^3/h, 1 degree = 0.01745 rad.

Notes:

1. Table includes losses for four 90-degree bends and two end fittings. Tubing runs with larger numbers of bends and/or fittings shall be increased by an equivalent length of tubing to the following equation: $L = 1.3n$ where L is additional length (feet) of tubing and n is the number of additional fittings and/or bends.
2. EHD—Equivalent Hydraulic Diameter, which is a measure of the relative hydraulic efficiency between different tubing sizes. The greater the value of EHD, the greater the gas capacity of the tubing.
3. All table entries have been rounded to three significant digits.

Gas	Undiluted Propane
Inlet Pressure	2.0 psi
Pressure Drop	1.0 psi
Specific Gravity	1.50

TABLE G2413.4(17) [402.4(33)]
CORRUGATED STAINLESS STEEL TUBING (CSST)

INTENDED USE	Sizing between 2 psi service and the line pressure regulator.													
	TUBE SIZE (EHD)													
Flow Designation	13	15	18	19	23	25	30	31	37	39	46	48	60	62
Length (ft)	Capacity in Thousands of Btu per Hour													
10	426	558	927	1,110	1,740	2,170	4,100	4,720	7,130	7,958	15,200	16,800	29,400	34,200
25	262	347	591	701	1,120	1,380	2,560	2,950	4,560	5,147	9,550	10,700	18,800	21,700
30	238	316	540	640	1,030	1,270	2,330	2,690	4,180	4,719	8,710	9,790	17,200	19,800
40	203	271	469	554	896	1,100	2,010	2,320	3,630	4,116	7,530	8,500	14,900	17,200
50	181	243	420	496	806	986	1,790	2,070	3,260	3,702	6,730	7,610	13,400	15,400
75	147	196	344	406	663	809	1,460	1,690	2,680	3,053	5,480	6,230	11,000	12,600
80	140	189	333	393	643	768	1,410	1,630	2,590	2,961	5,300	6,040	10,600	12,200
100	124	169	298	350	578	703	1,260	1,450	2,330	2,662	4,740	5,410	9,530	10,900
150	101	137	245	287	477	575	1,020	1,180	1,910	2,195	3,860	4,430	7,810	8,890
200	86	118	213	248	415	501	880	1,020	1,660	1,915	3,340	3,840	6,780	7,710
250	77	105	191	222	373	448	785	910	1,490	1,722	2,980	3,440	6,080	6,900
300	69	96	173	203	343	411	716	829	1,360	1,578	2,720	3,150	5,560	6,300
400	60	82	151	175	298	355	616	716	1,160	1,376	2,350	2,730	4,830	5,460
500	53	72	135	158	268	319	550	638	1,030	1,237	2,100	2,450	4,330	4,880

For SI: 1 inch = 25.4 mm, 1 foot = 304.8 mm, 1 pound per square inch = 6.895 kPa, 1-inch water column = 0.2488 kPa,
1 British thermal unit per hour = 0.2931 W, 1 cubic foot per hour = 0.0283 m^3/h, 1 degree = 0.01745 rad.

Notes:

1. Table does not include effect of pressure drop across the line regulator. Where regulator loss exceeds $^1/_2$ psi (based on 13 in. w.c. outlet pressure), DO NOT USE THIS TABLE. Consult with the regulator manufacturer for pressure drops and capacity factors. Pressure drops across a regulator can vary with flow rate.
2. CAUTION: Capacities shown in the table might exceed maximum capacity for a selected regulator. Consult with the regulator or tubing manufacturer for guidance.
3. Table includes losses for four 90-degree bends and two end fittings. Tubing runs with larger numbers of bends and/or fittings shall be increased by an equivalent length of tubing to the following equation: $L = 1.3n$ where L is additional length (feet) of tubing and n is the number of additional fittings and/or bends.
4. EHD—Equivalent Hydraulic Diameter, which is a measure of the relative hydraulic efficiency between different tubing sizes. The greater the value of EHD, the greater the gas capacity of the tubing.
5. All table entries have been rounded to three significant digits.

Gas	Undiluted Propane
Inlet Pressure	5.0 psi
Pressure Drop	3.5 psi
Specific Gravity	1.50

TABLE G2413.4(18) [402.4(34)]
CORRUGATED STAINLESS STEEL TUBING (CSST)

	TUBE SIZE (EHD)													
Flow Designation	13	15	18	19	23	25	30	31	37	39	46	48	60	62
Length (ft)	Capacity in Thousands of Btu per Hour													
10	826	1,070	1,710	2,060	3,150	4,000	7,830	8,950	13,100	14,441	28,600	31,200	54,400	63,800
25	509	664	1,090	1,310	2,040	2,550	4,860	5,600	8,400	9,339	18,000	19,900	34,700	40,400
30	461	603	999	1,190	1,870	2,340	4,430	5,100	7,680	8,564	16,400	18,200	31,700	36,900
40	396	520	867	1,030	1,630	2,030	3,820	4,400	6,680	7,469	14,200	15,800	27,600	32,000
50	352	463	777	926	1,460	1,820	3,410	3,930	5,990	6,717	12,700	14,100	24,700	28,600
75	284	376	637	757	1,210	1,490	2,770	3,190	4,920	5,539	10,300	11,600	20,300	23,400
80	275	363	618	731	1,170	1,450	2,680	3,090	4,770	5,372	9,990	11,200	19,600	22,700
100	243	324	553	656	1,050	1,300	2,390	2,760	4,280	4,830	8,930	10,000	17,600	20,300
150	196	262	453	535	866	1,060	1,940	2,240	3,510	3,983	7,270	8,210	14,400	16,600
200	169	226	393	464	755	923	1,680	1,930	3,050	3,474	6,290	7,130	12,500	14,400
250	150	202	352	415	679	828	1,490	1,730	2,740	3,124	5,620	6,390	11,200	12,900
300	136	183	322	379	622	757	1,360	1,570	2,510	2,865	5,120	5,840	10,300	11,700
400	117	158	279	328	542	657	1,170	1,360	2,180	2,498	4,430	5,070	8,920	10,200
500	104	140	251	294	488	589	1,050	1,210	1,950	2,247	3,960	4,540	8,000	9,110

For SI: 1 inch = 25.4 mm, 1 foot = 304.8 mm, 1 pound per square inch = 6.895 kPa, 1-inch water column = 0.2488 kPa, 1 British thermal unit per hour = 0.2931 W, 1 cubic foot per hour = 0.0283 m^3/h, 1 degree = 0.01745 rad.

Notes:

1. Table does not include effect of pressure drop across line regulator. Where regulator loss exceeds1 psi, DO NOT USE THIS TABLE. Consult with the regulator manufacturer for pressure drops and capacity factors. Pressure drop across regulator can vary with the flow rate.
2. CAUTION: Capacities shown in the table might exceed maximum capacity of selected regulator. Consult with the tubing manufacturer for guidance.
3. Table includes losses for four 90-degree bends and two end fittings. Tubing runs with larger numbers of bends and/or fittings shall be increased by an equivalent length of tubing to the following equation: $L = 1.3n$ where L is additional length (feet) of tubing and n is the number of additional fittings and/or bends.
4. EHD—Equivalent Hydraulic Diameter, which is a measure of the relative hydraulic efficiency between different tubing sizes. The greater the value of EHD, the greater the gas capacity of the tubing.
5. All table entries have been rounded to three significant digits.

TABLE G2413.4(19) [402.4(35)] POLYETHYLENE PLASTIC PIPE

Gas	Undiluted Propane
Inlet Pressure	11.0 in. w.c.
Pressure Drop	0.5 in. w.c.
Specific Gravity	1.50

INTENDED USE	**PE pipe sizing between integral 2-stage regulator at tank or second stage (low pressure regulator) and building.**					
PIPE SIZE (in.)						
Nominal OD	$^1/_2$	$^3/_4$	1	$1^1/_4$	$1^1/_2$	2
Designation	SDR 9.33	SDR 11.0	SDR 11.00	SDR 10.00	SDR 11.00	SDR 11.00
Actual ID	0.660	0.860	1.077	1.328	1.554	1.943
Length (ft)	**Capacity in Thousands of Btu per Hour**					
10	340	680	1,230	2,130	3,210	5,770
20	233	468	844	1,460	2,210	3,970
30	187	375	677	1,170	1,770	3,180
40	160	321	580	1,000	1,520	2,730
50	142	285	514	890	1,340	2,420
60	129	258	466	807	1,220	2,190
70	119	237	428	742	1,120	2,010
80	110	221	398	690	1,040	1,870
90	103	207	374	648	978	1,760
100	98	196	353	612	924	1,660
125	87	173	313	542	819	1,470
150	78	157	284	491	742	1,330
175	72	145	261	452	683	1,230
200	67	135	243	420	635	1,140
250	60	119	215	373	563	1,010
300	54	108	195	338	510	916
350	50	99	179	311	469	843
400	46	92	167	289	436	784
450	43	87	157	271	409	736
500	41	82	148	256	387	695

For SI: 1 inch = 25.4 mm, 1 foot = 304.8 mm, 1 pound per square inch = 6.895 kPa, 1-inch water column = 0.2488 kPa, 1 British thermal unit per hour = 0.2931 W, 1 cubic foot per hour = 0.0283 m^3/h, 1 degree = 0.01745 rad.

Note: All table entries have been rounded to three significant digits.

Gas	Undiluted Propane
Inlet Pressure	2.0 psi
Pressure Drop	1.0 psi
Specific Gravity	1.50

TABLE G2413.4(20) [402.4(36)]
POLYETHYLENE PLASTIC PIPE

INTENDED USE	PE pipe sizing between 2 psig service regulator and line pressure regulator.					
	PIPE SIZE (in.)					
Nominal OD	1/2	3/4	1	1 1/4	1 1/2	2
Designation	SDR 9.33	SDR 11.0	SDR 11.00	SDR 10.00	SDR 11.00	SDR 11.00
Actual ID	0.660	0.860	1.077	1.328	1.554	1.943
Length (ft)	Capacity in Thousands of Btu per Hour					
10	3,130	6,260	11,300	19,600	29,500	53,100
20	2,150	4,300	7,760	13,400	20,300	36,500
30	1,730	3,450	6,230	10,800	16,300	29,300
40	1,480	2,960	5,330	9,240	14,000	25,100
50	1,310	2,620	4,730	8,190	12,400	22,200
60	1,190	2,370	4,280	7,420	11,200	20,100
70	1,090	2,180	3,940	6,830	10,300	18,500
80	1,010	2,030	3,670	6,350	9,590	17,200
90	952	1,910	3,440	5,960	9,000	16,200
100	899	1,800	3,250	5,630	8,500	15,300
125	797	1,600	2,880	4,990	7,530	13,500
150	722	1,450	2,610	4,520	6,830	12,300
175	664	1,330	2,400	4,160	6,280	11,300
200	618	1,240	2,230	3,870	5,840	10,500
250	548	1,100	1,980	3,430	5,180	9,300
300	496	994	1,790	3,110	4,690	8,430
350	457	914	1,650	2,860	4,320	7,760
400	425	851	1,530	2,660	4,020	7,220
450	399	798	1,440	2,500	3,770	6,770
500	377	754	1,360	2,360	3,560	6,390
550	358	716	1,290	2,240	3,380	6,070
600	341	683	1,230	2,140	3,220	5,790
650	327	654	1,180	2,040	3,090	5,550
700	314	628	1,130	1,960	2,970	5,330
750	302	605	1,090	1,890	2,860	5,140
800	292	585	1,050	1,830	2,760	4,960
850	283	566	1,020	1,770	2,670	4,800
900	274	549	990	1,710	2,590	4,650
950	266	533	961	1,670	2,520	4,520
1,000	259	518	935	1,620	2,450	4,400
1,100	246	492	888	1,540	2,320	4,170
1,200	234	470	847	1,470	2,220	3,980
1,300	225	450	811	1,410	2,120	3,810
1,400	216	432	779	1,350	2,040	3,660
1,500	208	416	751	1,300	1,960	3,530
1,600	201	402	725	1,260	1,900	3,410
1,700	194	389	702	1,220	1,840	3,300
1,800	188	377	680	1,180	1,780	3,200
1,900	183	366	661	1,140	1,730	3,110
2,000	178	356	643	1,110	1,680	3,020

For SI: 1 inch = 25.4 mm, 1 foot = 304.8 mm, 1 pound per square inch = 6.895 kPa, 1-inch water column = 0.2488 kPa, 1 British thermal unit per hour = 0.2931 W, 1 cubic foot per hour = 0.0283 m^3/h, 1 degree = 0.01745 rad.

Note: All table entries have been rounded to three significant digits.

Gas	Undiluted Propane
Inlet Pressure	11.0 in. w.c.
Pressure Drop	0.5 in. w.c.
Specific Gravity	1.50

TABLE G2413.4(21) [402.4(37)] POLYETHYLENE PLASTIC TUBING

INTENDED USE	PE pipe sizing between integral 2-stage regulator at tank or second stage (low pressure regulator) and building.	
	Plastic Tubing Size (CTS) (in.)	
Nominal OD	$^{1}/_{2}$	1
Designation	SDR 7.00	SDR 11.00
Actual ID	0.445	0.927
Length (ft)	Capacity in Cubic Feet of Gas per Hour	
10	121	828
20	83	569
30	67	457
40	57	391
50	51	347
60	46	314
70	42	289
80	39	269
90	37	252
100	35	238
125	31	211
150	28	191
175	26	176
200	24	164
225	22	154
250	21	145
275	20	138
300	19	132
350	18	121
400	16	113
450	15	106
500	15	100

For SI: 1 inch = 25.4 mm, 1 foot = 304.8 mm, 1 pound per square inch = 6.895 kPa, 1-inch water column = 0.2488 kPa, 1 British thermal unit per hour = 0.2931 W, 1 cubic foot per hour = 0.0283 m^3/h, 1 degree = 0.01745 rad.

Note: All table entries have been rounded to three significant digits.

G2414.4.2 (403.4.2) Steel. Steel and wrought-iron *pipe* shall be at least of standard weight (Schedule 40) and shall comply with one of the following:

1. ASME B 36.10, 10M;
2. ASTM A 53/A 53M; or
3. ASTM A 106.

G2414.5 (403.5) Metallic tubing. Seamless copper, aluminum alloy or steel *tubing* shall be permitted to be used with gases not corrosive to such material.

G2414.5.1 (403.5.1) Steel tubing. Steel *tubing* shall comply with ASTM A 254.

G2414.5.2 (403.5.2) Copper tubing. Copper *tubing* shall comply with standard Type K or L of ASTM B 88 or ASTM B 280.

Copper and brass *tubing* shall not be used if the gas contains more than an average of 0.3 grains of hydrogen sulfide per 100 standard cubic feet of gas (0.7 milligrams per 100 liters).

G2414.5.3 (403.5.4) Corrugated stainless steel tubing. Corrugated stainless steel *tubing* shall be listed in accordance with ANSI LC 1/CSA 6.26.

G2414.6 (403.6) Plastic pipe, tubing and fittings. Polyethylene plastic pipe, tubing and fittings used to supply fuel gas shall conform to the 2009 edition of ASTM D 2513. Such pipe shall be marked "Gas" and "ASTM D 2513."

Plastic pipe, tubing and fittings, other than polyethylene, shall be identified and conform to the 2008 edition of ASTM D 2513. Such pipe shall be marked "Gas" and "ASTM D 2513."

G2414.6.1 (403.6.1) Anodeless risers. Plastic pipe, tubing and anodeless risers shall comply with the following:

1. Factory-assembled anodeless risers shall be recommended by the manufacturer for the gas used and shall be leak tested by the manufacturer in accordance with written procedures.
2. Service head adapters and field-assembled anodeless risers incorporating service head adapters shall be recommended by the manufacturer for the gas used, and shall be designed and certified to meet the requirements of Category I of the 2009 edition of ASTM D 2513, and U.S. Department of Transportation, Code of Federal Regulations, Title 49, Part 192.281(e). The manufacturer shall provide the user with qualified installation instructions as prescribed by the U.S. Department of Transportation, Code of Federal Regulations, Title 49, Part 192.283(b).

G2414.6.2 (403.6.2) LP-gas systems. The use of plastic pipe, tubing and fittings in undiluted liquefied petroleum gas *piping* systems shall be in accordance with NFPA 58.

G2414.6.3 (403.6.3) Regulator vent piping. Plastic *pipe*, *tubing* and fittings used to connect *regulator* vents to remote vent terminations shall be of PVC conforming to ANSI/UL 651. PVC vent *piping* shall not be installed indoors.

G2414.7 (403.7) Workmanship and defects. *Pipe* or *tubing* and fittings shall be clear and free from cutting burrs and defects in structure or threading, and shall be thoroughly brushed, and chip and scale blown.

Defects in *pipe* or *tubing* or fittings shall not be repaired. Defective *pipe*, *tubing* or fittings shall be replaced. (See Section G2417.1.2.)

G2414.8 (403.8) Protective coating. Where in contact with material or atmosphere exerting a corrosive action, metallic *piping* and fittings coated with a corrosion-resistant material shall be used. External or internal coatings or linings used on *piping* or components shall not be considered as adding strength.

G2414.9 (403.9) Metallic pipe threads. Metallic *pipe* and fitting threads shall be taper *pipe* threads and shall comply with ASME B 1.20.1.

G2414.9.1 (403.9.1) Damaged threads. *Pipe* with threads that are stripped, chipped, corroded or otherwise damaged shall not be used. If a weld opens during the operation of cutting or threading, that portion of the *pipe* shall not be used.

G2414.9.2 (403.9.2) Number of threads. Field threading of metallic *pipe* shall be in accordance with Table G2414.9.2.

TABLE G2414.9.2 (403.9.2)
SPECIFICATIONS FOR THREADING METALLIC PIPE

IRON PIPE SIZE (inches)	APPROXIMATE LENGTH OF THREADED PORTION (inches)	APPROXIMATE NO. OF THREADS TO BE CUT
$^1/_2$	$^3/_4$	10
$^3/_4$	$^3/_4$	10
1	$^7/_8$	10
$1^1/_4$	1	11
$1^1/_2$	1	11

For SI: 1 inch = 25.4 mm.

G2414.9.3 (403.9.3) Thread joint compounds. Thread joint compounds shall be resistant to the action of liquefied petroleum gas or to any other chemical constituents of the gases to be conducted through the *piping*.

G2414.10 (403.10) Metallic piping joints and fittings. The type of *piping* joint used shall be suitable for the pressure-temperature conditions and shall be selected giving consideration to joint tightness and mechanical strength under the service conditions. The joint shall be able to sustain the maximum end force due to the internal pressure and any additional forces due to temperature expansion or contraction, vibration, fatigue, or to the weight of the *pipe* and its contents.

G2414.10.1 (403.10.1) Pipe joints. *Pipe* joints shall be threaded, flanged, brazed or welded. Where nonferrous *pipe* is brazed, the *brazing* materials shall have a melting point in excess of 1,000°F (538°C). *Brazing* alloys shall not contain more than 0.05-percent phosphorus.

G2414.10.2 (403.10.2) Tubing joints. *Tubing* joints shall be made with *approved gas tubing* fittings or be brazed with a material having a melting point in excess of 1,000°F (538°C) or made with press-connect fittings complying with ANSI LC-4. *Brazing alloys* shall not contain more than 0.05-percent phosphorus.

G2414.10.3 (403.10.3) Flared joints. *Flared joints* shall be used only in systems constructed from nonferrous *pipe* and *tubing* where experience or tests have demonstrated that the joint is suitable for the conditions and where provisions are made in the design to prevent separation of the joints.

G2414.10.4 (403.10.4) Metallic fittings. Metallic fittings, including *valves*, strainers and filters shall comply with the following:

1. Fittings used with steel or wrought-iron *pipe* shall be steel, brass, bronze, malleable iron, ductile iron or cast iron.
2. Fittings used with copper or brass *pipe* shall be copper, brass or bronze.
3. Cast-iron bushings shall be prohibited.
4. Special fittings. Fittings such as couplings, proprietary-type joints, saddle tees, gland-type compression fittings, and flared, flareless or compression-type *tubing* fittings shall be: used within the fitting manufacturer's pressure-temperature recommendations; used within the service conditions anticipated with respect to vibration, fatigue, thermal expansion or contraction; installed or braced to prevent separation of the joint by gas pressure or external physical damage; and shall be *approved*.

G2414.11 (403.11) Plastic piping, joints and fittings. Plastic *pipe, tubing* and fittings shall be joined in accordance with the manufacturers' instructions. Such joints shall comply with the following:

1. The joints shall be designed and installed so that the longitudinal pull-out resistance of the joints will be at least equal to the tensile strength of the plastic *piping* material.
2. Heat-fusion joints shall be made in accordance with qualified procedures that have been established and proven by test to produce gas-tight joints at least as strong as the *pipe* or *tubing* being joined. Joints shall be made with the joining method recommended by the *pipe* manufacturer. Heat fusion fittings shall be marked "ASTM D 2513."
3. Where compression-type *mechanical joints* are used, the gasket material in the fitting shall be compatible with the plastic *piping* and with the gas distributed by the system. An internal tubular rigid stiffener shall be used in conjunction with the fitting. The stiffener shall be flush with the end of the *pipe* or *tubing* and shall extend at least to the outside end of the compression fitting when installed. The stiffener shall be free of rough or sharp edges and shall not be a force fit in the plastic. Split tubular stiffeners shall not be used.
4. Plastic *piping* joints and fittings for use in *liquefied petroleum gas piping systems* shall be in accordance with NFPA 58.

G2414.12 (403.13) Flange gaskets. Material for gaskets shall be capable of withstanding the design temperature and pressure of the *piping* system, and the chemical constituents of the gas being conducted, without change to its chemical and physical properties. The effects of fire exposure to the joint shall be considered in choosing material. Acceptable materials include metal (plain or corrugated), composition, and aluminum "O" rings and spiral wound metal gaskets. When a flanged joint is opened, the gasket shall be replaced. Full-face gaskets shall be used with all bronze and cast-iron flanges.

SECTION G2415 (404) PIPING SYSTEM INSTALLATION

G2415.1 (404.1) Installation of materials. All materials used shall be installed in strict accordance with the standards under which the materials are accepted and approved. In the absence of such installation procedures, the manufacturer's instructions shall be followed. Where the requirements of referenced standards or manufacturer's instructions do not conform to minimum provisions of this code, the provisions of this code shall apply.

G2415.2 (404.2) CSST. CSST piping systems shall be installed in accordance with the terms of their approval, the conditions of listing, the manufacturer's instructions and this code.

G2415.3 (404.3) Prohibited locations. *Piping* shall not be installed in or through a ducted supply, return or exhaust, or a clothes chute, *chimney* or gas vent, dumbwaiter or elevator *shaft*. *Piping* installed downstream of the *point of delivery* shall not extend through any townhouse unit other than the unit served by such *piping*.

G2415.4 (404.4) Piping in solid partitions and walls. *Concealed piping* shall not be located in solid partitions and solid walls, unless installed in a chase or casing.

G2415.5 (404.5) Piping in concealed locations. Portions of a *piping system* installed in *concealed locations* shall not have unions, *tubing* fittings, right and left couplings, bushings, compression couplings, and swing joints made by combinations of fittings.

Exceptions:

1. *Tubing* joined by *brazing*.
2. Fittings listed for use in *concealed locations*.

G2415.6 (404.6) Underground penetrations prohibited. *Gas piping* shall not penetrate building foundation walls at any point below grade. *Gas piping* shall enter and exit a building at a point above grade and the annular space between the *pipe* and the wall shall be sealed.

G2415.7 (404.7) Protection against physical damage. In *concealed locations*, where *piping* other than black or galvanized steel is installed through holes or notches in wood studs, joists, rafters or similar members less than $1^1/_2$ inches (38 mm) from the nearest edge of the member, the *pipe* shall be protected by shield plates. Protective steel shield plates having a minimum thickness of 0.0575-inch (1.463 mm) (No. 16 Gage) shall cover the area of the *pipe* where the member is notched or bored and shall extend a minimum of 4 inches (102 mm) above sole plates, below top plates and to each side of a stud, joist or rafter.

G2415.8 (404.8) Piping in solid floors. *Piping* in solid floors shall be laid in channels in the floor and covered in a manner that will allow access to the *piping* with a minimum amount of damage to the building. Where such *piping* is subject to exposure to excessive moisture or corrosive substances, the *piping* shall be protected in an *approved* manner. As an alternative to installation in channels, the *piping* shall be installed in a conduit of Schedule 40 steel, wrought iron, PVC or ABS *pipe* in accordance with Section G2415.6.1 or G2415.6.2.

G2415.8.1 (404.8.1) Conduit with one end terminating outdoors. The conduit shall extend into an occupiable portion of the building and, at the point where the conduit terminates in the building, the space between the conduit and the *gas piping* shall be sealed to prevent the possible entrance of any gas leakage. The conduit shall extend not less than 2 inches (51 mm) beyond the point where the *pipe* emerges from the floor. If the end sealing is capable of withstanding the full pressure of the gas *pipe,* the conduit shall be designed for the same pressure as the *pipe*. Such conduit shall extend not less than 4 inches (102 mm) outside of the building, shall be vented above grade to the outdoors and shall be installed to prevent the entrance of water and insects.

G2415.8.2 (404.8.2) Conduit with both ends terminating indoors. Where the conduit originates and terminates within the same building, the conduit shall originate and terminate in an accessible portion of the building and shall not be sealed. The conduit shall extend not less than 2 inches (51 mm) beyond the point where the *pipe* emerges from the floor.

G2415.9 (404.9) Above-ground piping outdoors. All *piping* installed outdoors shall be elevated not less than $3^1/_2$ inches (152 mm) above ground and where installed across roof surfaces, shall be elevated not less than $3^1/_2$ inches (152 mm) above the roof surface. *Piping* installed above ground, outdoors, and installed across the surface of roofs shall be securely supported and located where it will be protected from physical damage. Where passing through an outside wall, the *piping* shall also be protected against corrosion by coating or wrapping with an inert material. Where *piping* is encased in a protective *pipe* sleeve, the annular space between the *piping* and the sleeve shall be sealed.

G2415.10 (404.10) Isolation. Metallic *piping* and metallic *tubing* that conveys *fuel gas* from an LP-gas storage container shall be provided with an *approved* dielectric fitting to electrically isolate the underground portion of the *pipe* or tube from the above ground portion that enters a building. Such dielectric fitting shall be installed aboveground outdoors.

G2415.11 (404.11) Protection against corrosion. Metallic *pipe* or *tubing* exposed to corrosive action, such as soil condition or moisture, shall be protected in an *approved* manner. Zinc coatings (galvanizing) shall not be deemed adequate protection for *gas piping* underground. Where dissimilar metals are joined underground, an insulating coupling or fitting shall be used. *Piping* shall not be laid in contact with cinders.

G2415.11.1 (404.11.1) Prohibited use. Uncoated threaded or socket welded joints shall not be used in *piping* in contact with soil or where internal or external crevice corrosion is known to occur.

G2415.11.2 (404.11.2) Protective coatings and wrapping. *Pipe* protective coatings and wrappings shall be *approved* for the application and shall be factory applied.

Exception: Where installed in accordance with the manufacturer's installation instructions, field application of coatings and wrappings shall be permitted for *pipe* nipples, fittings and locations where the factory coating or wrapping has been damaged or necessarily removed at joints.

G2415.12 (404.12) Minimum burial depth. Underground *piping systems* shall be installed a minimum depth of 12 inches (305 mm) below grade, except as provided for in Section G2415.10.1.

G2415.12.1 (404.12.1) Individual outside appliances. Individual lines to outside lights, grills or other *appliances* shall be installed a minimum of 8 inches (203 mm) below finished grade, provided that such installation is *approved* and is installed in locations not susceptible to physical damage.

G2415.13 (404.13) Trenches. The trench shall be graded so that the *pipe* has a firm, substantially continuous bearing on the bottom of the trench.

G2415.14 (404.14) Piping underground beneath buildings. *Piping* installed underground beneath buildings is prohibited except where the *piping* is encased in a conduit of wrought iron, plastic *pipe*, steel *pipe* or other *approved* conduit material designed to withstand the superimposed loads. The conduit shall be protected from corrosion in accordance with Section G2415.9 and shall be installed in accordance with Section G2415.12.1 or G2415.12.2.

G2415.14.1 (404.14.1) Conduit with one end terminating outdoors. The conduit shall extend into an occupiable portion of the building and, at the point where the conduit terminates in the building, the space between the conduit and the *gas piping* shall be sealed to prevent the possible entrance of any gas leakage. The conduit shall extend not less than 2 inches (51 mm) beyond the point where the *pipe* emerges from the floor. Where the end sealing is capable of withstanding the full pressure of the gas *pipe*, the conduit shall be designed for the same pressure as the *pipe*. Such conduit shall extend not less than 4 inches (102 mm) outside the building, shall be vented above grade to the outdoors and shall be installed so as to prevent the entrance of water and insects.

G2415.14.2 (404.14.2) Conduit with both ends terminating indoors. Where the conduit originates and terminates within the same building, the conduit shall originate and terminate in an accessible portion of the building and shall not be sealed. The conduit shall extend not less than 2 inches (51 mm) beyond the point where the *pipe* emerges from the floor.

G2415.15 (404.15) Outlet closures. Gas *outlets* that do not connect to *appliances* shall be capped gas tight.

Exception: Listed and labeled flush-mounted-type quick-disconnect devices and listed and labeled *gas convenience outlets* shall be installed in accordance with the manufacturer's installation instructions.

G2415.16 (404.16) Location of outlets. The unthreaded portion of *piping outlets* shall extend not less than 1 inch (25 mm) through finished ceilings and walls and where extending through floors, outdoor patios and slabs, shall not be less than 2 inches (51 mm) above them. The *outlet* fitting or *piping* shall be securely supported. *Outlets* shall not be placed behind doors. *Outlets* shall be located in the room or space where the *appliance* is installed.

Exception: Listed and labeled flush-mounted-type quick-disconnect devices and listed and labeled *gas convenience outlets* shall be installed in accordance with the manufacturer's installation instructions.

G2415.17 (404.17) Plastic pipe. The installation of plastic *pipe* shall comply with Sections G2415.15.1 through G2415.15.3.

G2415.17.1 (404.17.1) Limitations. Plastic *pipe* shall be installed outdoors underground only. Plastic *pipe* shall not be used within or under any building or slab or be operated at pressures greater than 100 psig (689 kPa) for natural gas or 30 psig (207 kPa) for LP-gas.

Exceptions:

1. Plastic *pipe* shall be permitted to terminate above ground outside of buildings where installed in premanufactured *anodeless risers* or service head adapter risers that are installed in accordance with the manufacturer's installation instructions.
2. Plastic *pipe* shall be permitted to terminate with a wall head adapter within buildings where the plastic *pipe* is inserted in a *piping* material for *fuel gas* use in buildings.
3. Plastic pipe shall be permitted under outdoor patio, walkway and driveway slabs provided that the burial depth complies with Section G2415.10.

G2415.17.2 (404.17.2) Connections. Connections outdoors and underground between metallic and plastic *piping* shall be made only with transition fittings conforming to ASTM D 2513 Category I or ASTM F 1973.

G2415.17.3 (404.17.3) Tracer. A yellow insulated copper tracer wire or other *approved* conductor shall be installed adjacent to underground nonmetallic *piping*. Access shall be provided to the tracer wire or the tracer wire shall terminate above ground at each end of the nonmetallic *piping*. The tracer wire size shall not be less than 18 AWG and the insulation type shall be suitable for direct burial.

G2415.18 (404.18) Prohibited devices. A device shall not be placed inside the *piping* or fittings that will reduce the cross-sectional area or otherwise obstruct the free flow of gas.

Exceptions:

1. Approved gas filters.
2. An approved fitting or device where the gas piping system has been sized to accommodate the pressure drop of the fitting or device.

G2415.19 (404.19) Testing of piping. Before any system of *piping* is put in service or concealed, it shall be tested to ensure that it is gas tight. Testing, inspection and purging of *piping systems* shall comply with Section G2417.

SECTION G2416 (405) PIPING BENDS AND CHANGES IN DIRECTION

G2416.1 (405.1) General. Changes in direction of *pipe* shall be permitted to be made by the use of fittings, factory bends or field bends.

G2416.2 (405.2) Metallic pipe. Metallic *pipe* bends shall comply with the following:

1. Bends shall be made only with bending tools and procedures intended for that purpose.
2. All bends shall be smooth and free from buckling, cracks or other evidence of mechanical damage.
3. The longitudinal weld of the *pipe* shall be near the neutral axis of the bend.
4. *Pipe* shall not be bent through an arc of more than 90 degrees (1.6 rad).
5. The inside radius of a bend shall be not less than six times the outside diameter of the *pipe*.

G2416.3 (405.3) Plastic pipe. Plastic *pipe* bends shall comply with the following:

1. The *pipe* shall not be damaged and the internal diameter of the *pipe* shall not be effectively reduced.
2. Joints shall not be located in *pipe* bends.
3. The radius of the inner curve of such bends shall not be less than 25 times the inside diameter of the *pipe*.
4. Where the *piping* manufacturer specifies the use of special bending tools or procedures, such tools or procedures shall be used.

SECTION G2417 (406) INSPECTION, TESTING AND PURGING

G2417.1 (406.1) General. Prior to acceptance and initial operation, all *piping* installations shall be visually inspected and pressure tested to determine that the materials, design, fabrication and installation practices comply with the requirements of this code.

G2417.1.1 (406.1.1) Inspections. Inspection shall consist of visual examination, during or after manufacture, fabrication, assembly or *pressure tests* as appropriate.

G2417.1.2 (406.1.2) Repairs and additions. In the event repairs or additions are made after the *pressure test*, the affected *piping* shall be tested.

Minor repairs and additions are not required to be *pressure tested* provided that the work is inspected and connections are tested with a noncorrosive leak-detecting fluid or other *approved* leak-detecting methods.

G2417.1.3 (406.1.3) New branches. Where new branches are installed to new *appliances*, only the newly installed branches shall be required to be *pressure tested*. Connections between the new *piping* and the existing *piping* shall be tested with a noncorrosive leak-detecting fluid or other *approved* leak-detecting methods.

G2417.1.4 (406.1.4) Section testing. A *piping system* shall be permitted to be tested as a complete unit or in sections. Under no circumstances shall a *valve* in a line be used as a bulkhead between gas in one section of the *piping system* and test medium in an adjacent section, unless two *valves* are installed in series with a valved "tell-tale" located between these *valves*. A valve shall not be subjected to the test pressure unless it can be determined that the valve, including the valve closing mechanism, is designed to safely withstand the test pressure.

G2417.1.5 (406.1.5) Regulators and valve assemblies. *Regulator* and valve assemblies fabricated independently of the *piping system* in which they are to be installed shall be permitted to be tested with inert gas or air at the time of fabrication.

G2417.1.6 (406.1.6) Pipe clearing. Prior to testing, the interior of the pipe shall be cleared of all foreign material.

G2417.2 (406.2) Test medium. The test medium shall be air, nitrogen, carbon dioxide or an inert gas. Oxygen shall not be used.

G2417.3 (406.3) Test preparation. *Pipe* joints, including welds, shall be left exposed for examination during the test.

Exception: Covered or *concealed pipe* end joints that have been previously tested in accordance with this *code*.

G2417.3.1 (406.3.1) Expansion joints. Expansion joints shall be provided with temporary restraints, if required, for the additional thrust load under test.

G2417.3.2 (406.3.2) Equipment isolation. *Equipment* that is not to be included in the test shall be either disconnected from the *piping* or isolated by blanks, blind flanges or caps.

G2417.3.3 (406.3.3) Appliance and equipment disconnection. Where the *piping system* is connected to *appliances* or *equipment* designed for operating pressures of less than the test pressure, such *appliances* or *equipment* shall be isolated from the *piping system* by disconnecting them and capping the *outlet(s)*.

G2417.3.4 (406.3.4) Valve isolation. Where the *piping system* is connected to *appliances* or *equipment* designed for operating pressures equal to or greater than the test pressure, such *appliances* or *equipment* shall be isolated from the *piping system* by closing the individual *appliance* or *equipment* shutoff valve(s).

G2417.3.5 (406.3.5) Testing precautions. All testing of *piping* systems shall be performed in a manner that protects the safety of employees and the public during the test.

G2417.4 (406.4) Test pressure measurement. Test pressure shall be measured with a manometer or with a pressure-measuring device designed and calibrated to read, record, or indicate a pressure loss caused by leakage during the *pressure test* period. The source of pressure shall be isolated before the *pressure tests* are made. Mechanical gauges used to measure test pressures shall have a range such that the highest end of the scale is not greater than five times the test pressure.

G2417.4.1 (406.4.1) Test pressure. The test pressure to be used shall be not less than one and one-half times the proposed maximum working pressure, but not less than 3 psig (20 kPa gauge), irrespective of design pressure. Where the test pressure exceeds 125 psig (862 kPa gauge), the test pressure shall not exceed a value that produces a hoop stress in the *piping* greater than 50 percent of the specified minimum yield strength of the *pipe*.

G2417.4.2 (406.4.2) Test duration. The test duration shall be not less than 10 minutes.

G2417.5 (406.5) Detection of leaks and defects. The *piping system* shall withstand the test pressure specified without showing any evidence of leakage or other defects. Any reduction of test pressures as indicated by pressure gauges shall be deemed to indicate the presence of a leak unless such reduction can be readily attributed to some other cause.

G2417.5.1 (406.5.1) Detection methods. The leakage shall be located by means of an *approved* combustible gas detector, a noncorrosive leak detection fluid or an equivalent nonflammable solution. Matches, candles, open flames or other methods that could provide a source of ignition shall not be used.

G2417.5.2 (406.5.2) Corrections. Where leakage or other defects are located, the affected portion of the *piping system* shall be repaired or replaced and retested.

G2417.6 (406.6) Piping system and equipment leakage check. Leakage checking of systems and *equipment* shall be in accordance with Sections G2417.6.1 through G2417.6.4.

G2417.6.1 (406.6.1) Test gases. *Fuel gas* shall be permitted to be used for *leak checks* in *piping systems* that have been tested in accordance with Section G2417.

G2417.6.2 (406.6.2) Turning gas on. During the process of turning gas on into a system of new *gas piping*, the entire system shall be inspected to determine that there are no open fittings or ends and that all *valves* at unused outlets are closed and plugged or capped.

G2417.6.3 (406.6.3) Leak check. Immediately after the gas is turned on into a new system or into a system that has been initially restored after an interruption of service, the *piping system* shall be checked for leakage. Where leakage is indicated, the gas supply shall be shut off until the necessary repairs have been made.

G2417.6.4 (406.6.4) Placing appliances and equipment in operation. *Appliance*s and *equipment* shall not be placed in operation until after the *piping* system has been checked for leakage in accordance with Section G2417.6.3, the piping system has been purged in accordance with Section G2417.7 and the connections to the *appliances* have been checked for leakage.

G2417.7 (406.7) Purging. The purging of piping shall be in accordance with Sections G2417.7.1 through 2417.7.3.

G2417.7.1 (406.7.1) Piping systems required to be purged outdoors. The purging of piping systems shall be in accordance with the provisions of Sections G2417.7.1.1 through G2417.7.1.4 where the *piping system* meets either of the following:

1. The design operating gas pressure is greater than 2 psig (13.79 kPa).
2. The piping being purged contains one or more sections of pipe or tubing meeting the size and length criteria of Table G2417.7.1.1.

G2417.7.1.1 (406.7.1.1) Removal from service. Where existing gas piping is opened, the section that is opened shall be isolated from the gas supply and the line pressure vented in accordance with Section G2417.7.1.3. Where gas piping meeting the criteria of Table G2417.7.1.1 is removed from service, the residual fuel gas in the piping shall be displaced with an inert gas.

TABLE G2417.7.1.1
SIZE AND LENGTH OF PIPING

NOMINAL PIPE SIZE (inches)[a]	LENGTH OF PIPING (feet)
$\geq 2^1/_2 < 3$	> 50
≥ 3 < 4	> 30
≥ 4 < 6	> 15
≥ 6 < 8	> 10
≥ 8	Any length

For SI: 1 inch = 25.4 mm, 1 foot = 304.8 mm.
a. CSST EHD size of 62 is equivalent to nominal 2-inch pipe or tubing size.

G2417.7.1.2 (406.7.1.2) Placing in operation. Where gas piping containing air and meeting the criteria of Table G2417.7.1.1 is placed in operation, the air in the piping shall first be displaced with an inert gas. The inert gas shall then be displaced with fuel gas in accordance with Section G2417.7.1.3.

G2417.7.1.3 (406.7.1.3) Outdoor discharge of purged gases. The open end of a piping system being pressure vented or purged shall discharge directly to an outdoor location. Purging operations shall comply will all of the following requirements:

1. The point of discharge shall be controlled with a shutoff valve.
2. The point of discharge shall be located at least 10 feet (3048 mm) from sources of ignition, at least 10 feet (3048 mm) from building openings and at least 25 feet (7620 mm) from mechanical air intake openings.
3. During discharge, the open point of discharge shall be continuously attended and monitored with a combustible gas indicator that complies with Section G2417.7.1.4.
4. Purging operations introducing fuel gas shall be stopped when 90 percent fuel gas by volume is detected within the pipe.
5. Persons not involved in the purging operations shall be evacuated from all areas within 10 feet (3048 mm) of the point of discharge.

G2417.7.1.4 (406.7.1.4) Combustible gas indicator. Combustible gas indicators shall be listed and shall be calibrated in accordance with the manufacturer's instructions. Combustible gas indicators shall numerically display a volume scale from zero percent to 100 percent in 1 percent or smaller increments.

G2417.7.2 (406.7.2) Piping systems allowed to be purged indoors or outdoors. The purging of piping systems shall be in accordance with the provisions of Section G2417.7.2.1 where the piping system meets both of the following:

1. The design operating gas pressure is 2 psig (13.79 kPa) or less.
2. The piping being purged is constructed entirely from pipe or tubing not meeting the size and length criteria of Table G2417.7.1.1.

G2417.7.2.1 (406.7.2.1) Purging procedure. The piping system shall be purged in accordance with one or more of the following:

1. The piping shall be purged with fuel gas and shall discharge to the outdoors.
2. The piping shall be purged with fuel gas and shall discharge to the indoors or outdoors through an *appliance* burner not located in a combustion chamber. Such burner shall be provided with a continuous source of ignition.
3. The piping shall be purged with fuel gas and shall discharge to the indoors or outdoors through a burner that has a continuous source of ignition and that is designed for such purpose.
4. The piping shall be purged with fuel gas that is discharged to the indoors or outdoors, and the point of discharge shall be monitored with a listed combustible gas detector in accordance with Section G2417.7.2.2. Purging shall be stopped when fuel gas is detected.
5. The piping shall be purged by the gas supplier in accordance with written procedures.

G2417.7.2.2 (406.7.2.2) Combustible gas detector. Combustible gas detectors shall be listed and shall be calibrated or tested in accordance with the manufacturer's instructions. Combustible gas detectors shall be capable of indicating the presence of fuel gas.

G2417.7.3 (406.7.3) Purging appliances and equipment. After the *piping system* has been placed in operation, *appliances* and *equipment* shall be purged before being placed into operation.

SECTION G2418 (407) PIPING SUPPORT

G2418.1 (407.1) General. *Piping* shall be provided with support in accordance with Section G2418.2.

G2418.2 (407.2) Design and installation. *Piping* shall be supported with metal *pipe* hooks, metal *pipe* straps, metal bands, metal brackets, metal hangers or building structural components suitable for the size of *piping*, of adequate strength and quality, and located at intervals so as to prevent or damp out excessive vibration. *Piping* shall be anchored to prevent undue strains on connected *appliances* and shall not be supported by other *piping*. *Pipe* hangers and supports shall conform to the requirements of MSS SP-58 and shall be spaced in accordance with Section G2424. Supports, hangers and anchors shall be installed so as not to interfere with the free expansion and contraction of the *piping* between anchors. All parts of the supporting *equipment* shall be designed and installed so that they will not be disengaged by movement of the supported *piping*.

SECTION G2419 (408) DRIPS AND SLOPED PIPING

G2419.1 (408.1) Slopes. *Piping* for other than dry gas conditions shall be sloped not less than 0.25 inch in 15 feet (6.4 mm in 4572 mm) to prevent traps.

G2419.2 (408.2) Drips. Where wet gas exists, a *drip* shall be provided at any point in the line of *pipe* where *condensate* could collect. A *drip* shall also be provided at the outlet of the *meter* and shall be installed so as to constitute a trap wherein an accumulation of *condensate* will shut off the flow of gas before the *condensate* will run back into the *meter*.

G2419.3 (408.3) Location of drips. *Drips* shall be provided with *ready access* to permit cleaning or emptying. A *drip* shall not be located where the *condensate* is subject to freezing.

G2419.4 (408.4) Sediment trap. Where a sediment trap is not incorporated as part of the appliance, a sediment trap shall be installed downstream of the appliance shutoff valve as close to the inlet of the appliance as practical. The sediment trap shall be either a tee fitting having a capped nipple of any length installed vertically in the bottommost opening of the tee as illustrated in Figure G2419.4 or other device approved as an effective sediment trap. Illuminating appliances, ranges, clothes dryers, decorative vented appliances for installation in vented fireplaces, gas fireplaces, and outdoor grills need not be so equipped.

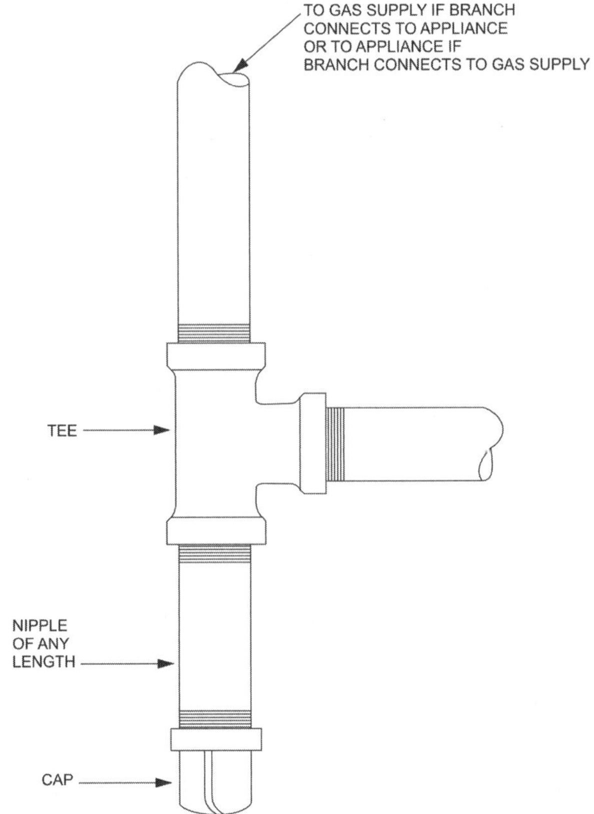

FIGURE G2419.4 METHOD OF INSTALLING A TEE FITTING SEDIMENT TRAP

SECTION G2420 (409) GAS SHUTOFF VALVES

G2420.1 (409.1) General. *Piping systems* shall be provided with shutoff *valves* in accordance with this section.

G2420.1.1 (409.1.1) Valve approval. Shutoff *valves* shall be of an *approved* type; shall be constructed of materials compatible with the *piping*; and shall comply with the standard that is applicable for the pressure and application, in accordance with Table G2420.1.1.

G2420.1.2 (409.1.2) Prohibited locations. Shutoff *valves* shall be prohibited in *concealed locations* and *furnace plenums*.

G2420.1.3 (409.1.3) Access to shutoff valves. Shutoff *valves* shall be located in places so as to provide access for operation and shall be installed so as to be protected from damage.

G2420.2 (409.2) Meter valve. Every *meter* shall be equipped with a shutoff *valve* located on the supply side of the *meter*.

G2420.3 (409.3.2) Individual buildings. In a common system serving more than one building, shutoff *valves* shall be installed outdoors at each building.

G2420.4 (409.4) MP regulator valves. A listed shutoff valve shall be installed immediately ahead of each MP *regulator*.

TABLE G2420.1.1
MANUAL GAS VALVE STANDARDS

VALVE STANDARDS	APPLIANCE SHUTOFF VALVE APPLICATION UP TO $^1/_2$ psig PRESSURE	OTHER VALVE APPLICATIONS			
		UP TO $^1/_2$ psig PRESSURE	UP TO 2 psig PRESSURE	UP TO 5 psig PRESSURE	UP TO 125 psig PRESSURE
ANSI Z21.15	X	—	—	—	—
ASME B 16.44	X	X	X[a]	X[b]	—
ASME B 16.33	X	X	X	X	X

For SI: 1 pound per square inch gauge = 6.895 kPa.
a. If labeled 2G.
b. If labeled 5G.

G2420.5 (409.5) Appliance shutoff valve. Each *appliance* shall be provided with a shutoff *valve* in accordance with Section G2420.5.1, G2420.5.2 or G2420.5.3.

G2420.5.1 (409.5.1) Located within same room. The shutoff *valve* shall be located in the same room as the *appliance*. The shutoff *valve* shall be within 6 feet (1829 mm) of the *appliance*, and shall be installed upstream of the union, connector or quick disconnect device it serves. Such shutoff *valves* shall be provided with access. *Appliance shutoff valves* located in the firebox of a *fireplace* shall be installed in accordance with the *appliance* manufacturer's instructions.

G2420.5.2 (409.5.2) Vented decorative appliances and room heaters. Shutoff *valves* for vented decorative *appliances*, room heaters and decorative *appliances* for installation in vented fireplaces shall be permitted to be installed in an area remote from the *appliances* where such *valves* are provided with *ready access*. Such *valves* shall be permanently identified and shall serve no other *appliance*. The *piping* from the shutoff *valve* to within 6 feet (1829 mm) of the *appliance* shall be designed, sized and installed in accordance with Sections G2412 through G2419.

G2420.5.3 (409.5.3) Located at manifold. Where the *appliance shutoff valve* is installed at a manifold, such shutoff valve shall be located within 50 feet (15 240 mm) of the *appliance* served and shall be readily accessible and permanently identified. The *piping* from the manifold to within 6 feet (1829 mm) of the *appliance* shall be designed, sized and installed in accordance with Sections G2412 through G2419.

SECTION G2421 (410)
FLOW CONTROLS

G2421.1 (410.1) Pressure regulators. A line *pressure regulator* shall be installed where the *appliance* is designed to operate at a lower pressure than the supply pressure. *Line gas pressure regulators* shall be listed as complying with ANSI Z21.80. Access shall be provided to *pressure regulators*. *Pressure regulators* shall be protected from physical damage. *Regulators* installed on the exterior of the building shall be *approved* for outdoor installation.

G2421.2 (410.2) MP regulators. MP *pressure regulators* shall comply with the following:

1. The MP *regulator* shall be *approved* and shall be suitable for the inlet and *outlet* gas pressures for the application.
2. The MP *regulator* shall maintain a reduced *outlet* pressure under lockup (no-flow) conditions.
3. The capacity of the MP *regulator*, determined by published ratings of its manufacturer, shall be adequate to supply the *appliances* served.
4. The MP *pressure regulator* shall be provided with access. Where located indoors, the *regulator* shall be vented to the outdoors or shall be equipped with a leak-limiting device, in either case complying with Section G2421.3.
5. A tee fitting with one opening capped or plugged shall be installed between the MP *regulator* and its upstream shutoff *valve*. Such tee fitting shall be positioned to allow connection of a pressure measuring instrument and to serve as a sediment trap.
6. A tee fitting with one opening capped or plugged shall be installed not less than 10 *pipe* diameters downstream of the MP *regulator* outlet. Such tee fitting shall be positioned to allow connection of a pressure measuring instrument.

G2421.3 (410.3) Venting of regulators. *Pressure regulators* that require a vent shall be vented directly to the outdoors. The vent shall be designed to prevent the entry of insects, water and foreign objects.

Exception: A vent to the outdoors is not required for *regulators* equipped with and labeled for utilization with an *approved* vent-limiting device installed in accordance with the manufacturer's instructions.

G2421.3.1 (410.3.1) Vent piping. Vent *piping* for *relief* vents and *breather* vents shall be constructed of materials allowed for *gas piping* in accordance with Section G2414. Vent *piping* shall be not smaller than the vent connection on the pressure regulating device. Vent *piping* serving *relief* vents and combination relief and *breather* vents shall be run independently to the outdoors and shall serve only a single device vent. Vent *piping* serving only *breather* vents is permitted to be connected in a manifold arrangement where sized in accordance with an *approved* design that minimizes back pressure in the event of diaphragm rupture. *Regulator* vent *piping* shall not exceed the length specified in the *regulator* manufacturer's installation instructions.

G2421.4 (410.4) Excess flow valves. Where automatic *excess flow valves* are installed, they shall be listed for the application and shall be sized and installed in accordance with the manufacturer's instructions.

G2421.5 (410.5) Flashback arrestor check valve. Where fuel gas is used with oxygen in any hot work operation, a listed protective device that serves as a combination flashback arrestor and backflow check valve shall be installed at an approved location on both the fuel gas and oxygen supply lines. Where the pressure of the piped fuel gas supply is insufficient to ensure such safe operation, approved equipment shall be installed between the gas meter and the appliance that increases pressure to the level required for such safe operation.

SECTION G2422 (411) APPLIANCE CONNECTIONS

G2422.1 (411.1) Connecting appliances. *Appliances* shall be connected to the *piping system* by one of the following:

1. Rigid metallic *pipe* and fittings.
2. Corrugated stainless steel *tubing* (CSST) where installed in accordance with the manufacturer's instructions.
3. Listed and labeled *appliance connectors* in compliance with ANSI Z21.24 and installed in accordance with the manufacturer's installation instructions and located entirely in the same room as the *appliance*.
4. Listed and labeled quick-disconnect devices used in conjunction with listed and labeled *appliance connectors*.
5. Listed and labeled convenience outlets used in conjunction with listed and labeled *appliance connectors*.
6. Listed and labeled outdoor *appliance connectors* in compliance with ANSI Z21.75/CSA 6.27 and installed in accordance with the manufacturer's installation instructions.

G2422.1.1 (411.1.2) Protection from damage. Connectors and *tubing* shall be installed so as to be protected against physical damage.

G2422.1.2 (411.1.3) Connector installation. *Appliance* fuel connectors shall be installed in accordance with the manufacturer's instructions and Sections G2422.1.2.1 through G2422.1.2.4.

G2422.1.2.1 (411.1.3.1) Maximum length. Connectors shall not exceed 6 feet (1829 mm) in overall length. Measurement shall be made along the centerline of the connector. Only one connector shall be used for each *appliance*.

Exception: Rigid metallic *piping* used to connect an *appliance* to the *piping system* shall be permitted to have a total length greater than 6 feet (1829 mm) provided that the connecting *pipe* is sized as part of the *piping system* in accordance with Section G2413 and the location of the *appliance shutoff valve* complies with Section G2420.5.

G2422.1.2.2 (411.1.3.2) Minimum size. Connectors shall have the capacity for the total *demand* of the connected *appliance*.

G2422.1.2.3 (411.1.3.3) Prohibited locations and penetrations. Connectors shall not be concealed within, or extended through, walls, floors, partitions, ceilings or *appliance* housings.

Exceptions:

1. Connectors constructed of materials allowed for *piping systems* in accordance with Section G2414 shall be permitted to pass through walls, floors, partitions and ceilings where installed in accordance with Section G2420.5.2 or G2420.5.3.
2. Rigid steel *pipe* connectors shall be permitted to extend through openings in *appliance* housings.
3. *Fireplace* inserts that are factory equipped with grommets, sleeves or other means of protection in accordance with the listing of the *appliance*.
4. Semirigid *tubing* and listed connectors shall be permitted to extend through an opening in an *appliance* housing, cabinet or casing where the tubing or connector is protected against damage.

G2422.1.2.4 (411.1.3.4) Shutoff valve. A shutoff *valve* not less than the nominal size of the connector shall be installed ahead of the connector in accordance with Section G2420.5.

G2422.1.3 (411.1.5) Connection of gas engine-powered air conditioners. Internal *combustion* engines shall not be rigidly connected to the gas supply *piping*.

G2422.1.4 (411.1.6) Unions. A union fitting shall be provided for *appliances* connected by rigid metallic *pipe*. Such unions shall be accessible and located within 6 feet (1829 mm) of the *appliance*.

G2422.1.5 (411.1.4) Movable appliances. Where *appliances* are equipped with casters or are otherwise subject to periodic movement or relocation for purposes such as routine cleaning and maintenance, such *appliances* shall be connected to the supply system *piping* by means of an *approved* flexible connector designed and labeled for the application. Such flexible connectors shall be installed and protected against physical damage in accordance with the manufacturer's installation instructions.

G2422.2 (411.3) Suspended low-intensity infrared tube heaters. Suspended low-intensity infrared tube heaters shall be connected to the building *piping system* with a connector listed for the application complying with ANSI Z21.24/CGA 6.10. The connector shall be installed as specified by the tube heater manufacturer's instructions.

SECTION G2423 (413) CNG GAS-DISPENSING SYSTEMS

G2423.1 (413.1) General. Motor fuel-dispensing facilities for CNG fuel shall be in accordance with Section 413 of the *International Fuel Gas Code.*

SECTION G2424 (415) PIPING SUPPORT INTERVALS

G2424.1 (415.1) Interval of support. *Piping* shall be supported at intervals not exceeding the spacing specified in Table G2424.1. Spacing of supports for CSST shall be in accordance with the CSST manufacturer's instructions.

TABLE G2424.1 SUPPORT OF PIPING

STEEL PIPE, NOMINAL SIZE OF PIPE (inches)	SPACING OF SUPPORTS (feet)	NOMINAL SIZE OF TUBING SMOOTH-WALL (inch O.D.)	SPACING OF SUPPORTS (feet)
$^1/_2$	6	$^1/_2$	4
$^3/_4$ or 1	8	$^5/_8$ or $^3/_4$	6
$1^1/_4$ or larger (horizontal)	10	$^7/_8$ or 1 (horizontal)	8
$1^1/_4$ or larger (vertical)	Every floor level	1 or larger (vertical)	Every floor level

For SI: 1 inch = 25.4 mm, 1 foot = 304.8 mm.

SECTION G2425 (501) GENERAL

G2425.1 (501.1) Scope. This section shall govern the installation, maintenance, repair and approval of factory-built and *masonry chimneys, chimney* liners, vents and connectors serving gas-fired *appliances.*

G2425.2 (501.2) General. Every *appliance* shall discharge the products of *combustion* to the outdoors, except for *appliances* exempted by Section G2425.8.

G2425.3 (501.3) Masonry chimneys. *Masonry chimneys* shall be constructed in accordance with Section G2427.5 and Chapter 10.

G2425.4 (501.4) Minimum size of chimney or vent. *Chimneys* and vents shall be sized in accordance with Sections G2427 and G2428.

G2425.5 (501.5) Abandoned inlet openings. Abandoned inlet openings in *chimneys* and vents shall be closed by an *approved* method.

G2425.6 (501.6) Positive pressure. Where an *appliance* equipped with a mechanical forced *draft* system creates a positive pressure in the venting system, the venting system shall be designed for positive pressure applications.

G2425.7 (501.7) Connection to fireplace. Connection of *appliances* to *chimney* flues serving *fireplaces* shall be in accordance with Sections G2425.7.1 through G2425.7.3.

G2425.7.1 (501.7.1) Closure and access. A noncombustible seal shall be provided below the point of connection to prevent entry of room air into the flue. Means shall be provided for access to the flue for inspection and cleaning.

G2425.7.2 (501.7.2) Connection to factory-built fireplace flue. An *appliance* shall not be connected to a flue serving a *factory-built fireplace* unless the *appliance* is specifically listed for such installation. The connection shall be made in accordance with the *appliance* manufacturer's installation instructions.

G2425.7.3 (501.7.3) Connection to masonry fireplace flue. A connector shall extend from the *appliance* to the flue serving a *masonry fireplace* such that the *flue gases* are exhausted directly into the flue. The connector shall be accessible or removable for inspection and cleaning of both the connector and the flue. Listed direct connection devices shall be installed in accordance with their listing.

G2425.8 (501.8) Appliances not required to be vented. The following *appliances* shall not be required to be vented:

1. Ranges.
2. Built-in domestic cooking units listed and marked for optional venting.
3. Hot plates and laundry stoves.
4. *Type 1 clothes dryers* (*Type 1 clothes dryers* shall be exhausted in accordance with the requirements of Section G2439).
5. Refrigerators.
6. Counter *appliances.*
7. Room heaters listed for unvented use.

Where the *appliances* listed in Items 5 through 7 above are installed so that the aggregate input rating exceeds 20 *Btu* per hour per *cubic foot* (207 W/m^3) of volume of the room or space in which such *appliances* are installed, one or more shall be provided with venting *systems* or other *approved* means for conveying the *vent gases* to the outdoor atmosphere so that the aggregate input rating of the remaining *unvented appliances* does not exceed 20 *Btu* per hour per *cubic foot* (207 W/m^3). Where the room or space in which the *appliance* is installed is directly connected to another room or space by a doorway, archway or other opening of comparable size that cannot be closed, the volume of such adjacent room or space shall be permitted to be included in the calculations.

G2425.9 (501.9) Chimney entrance. Connectors shall connect to a *masonry chimney* flue at a point not less than 12 inches (305 mm) above the lowest portion of the interior of the *chimney* flue.

G2425.10 (501.10) Connections to exhauster. *Appliance* connections to a *chimney* or vent equipped with a power exhauster shall be made on the inlet side of the exhauster. Joints on the positive pressure side of the exhauster shall be sealed to prevent flue-gas leakage as specified by the manufacturer's installation instructions for the exhauster.

G2425.11 (501.11) Masonry chimneys. *Masonry chimneys* utilized to vent *appliances* shall be located, constructed and sized as specified in the manufacturer's installation instructions for the *appliances* being vented and Section G2427.

G2425.12 (501.12) Residential and low-heat appliances flue lining systems. *Flue lining* systems for use with residential-type and low-heat *appliances* shall be limited to the following:

1. Clay *flue lining* complying with the requirements of ASTM C 315 or equivalent. Clay *flue lining* shall be installed in accordance with Chapter 10.
2. Listed *chimney* lining systems complying with UL 1777.
3. Other *approved* materials that will resist, without cracking, softening or corrosion, *flue gases* and *condensate* at temperatures up to 1,800°F (982°C).

G2425.13 (501.13) Category I appliance flue lining systems. *Flue lining* systems for use with Category I *appliances* shall be limited to the following:

1. *Flue lining* systems complying with Section G2425.12.
2. *Chimney* lining systems listed and labeled for use with *appliances* with *draft hoods* and other Category I gas *appliances* listed and labeled for use with Type B vents.

G2425.14 (501.14) Category II, III and IV appliance venting systems. The design, sizing and installation of vents for Category II, III and IV *appliances* shall be in accordance with the *appliance* manufacturer's installation instructions.

G2425.15 (501.15) Existing chimneys and vents. Where an *appliance* is permanently disconnected from an existing *chimney* or vent, or where an *appliance* is connected to an existing *chimney* or vent during the process of a new installation, the *chimney* or vent shall comply with Sections G2425.15.1 through G2425.15.4.

G2425.15.1 (501.15.1) Size. The *chimney* or vent shall be resized as necessary to control flue gas condensation in the interior of the *chimney* or vent and to provide the *appliance* or *appliances* served with the required *draft*. For Category I *appliances*, the resizing shall be in accordance with Section G2426.

G2425.15.2 (501.15.2) Flue passageways. The flue gas passageway shall be free of obstructions and combustible deposits and shall be cleaned if previously used for venting a solid or liquid fuel-burning appliance or *fireplace*. The *flue liner*, *chimney* inner wall or vent inner wall shall be continuous and shall be free of cracks, gaps, perforations, or other damage or deterioration that would allow the escape of *combustion products*, including gases, moisture and creosote.

G2425.15.3 (501.15.3) Cleanout. *Masonry chimney* flues shall be provided with a cleanout opening having a minimum height of 6 inches (152 mm). The upper edge of the opening shall be located not less than 6 inches (152 mm) below the lowest *chimney* inlet opening. The cleanout shall be provided with a tight-fitting, noncombustible cover.

G2425.15.4 (501.15.4) Clearances. *Chimneys* and vents shall have airspace *clearance* to combustibles in accordance with Chapter 10 and the *chimney* or vent manufacturer's installation instructions.

Exception: *Masonry chimneys* without the required air-space *clearances* shall be permitted to be used if lined or relined with a *chimney* lining system listed for use in *chimneys* with reduced *clearances* in accordance with UL 1777. The *chimney clearance* shall be not less than that permitted by the terms of the *chimney* liner listing and the manufacturer's instructions.

G2425.15.4.1 (501.15.4.1) Fireblocking. Noncombustible fireblocking shall be provided in accordance with Chapter 10.

SECTION G2426 (502)
VENTS

G2426.1 (502.1) General. All vents, except as provided in Section G2427.7, shall be listed and labeled. Type B and BW vents shall be tested in accordance with UL 441. Type L vents shall be tested in accordance with UL 641. Vents for Category II and III *appliances* shall be tested in accordance with UL 1738. Plastic vents for Category IV *appliances* shall not be required to be listed and labeled where such vents are as specified by the *appliance* manufacturer and are installed in accordance with the *appliance* manufacturer's installation instructions.

G2426.2 (502.2) Connectors required. Connectors shall be used to connect *appliances* to the vertical *chimney* or vent, except where the *chimney* or vent is attached directly to the *appliance*. Vent *connector* size, material, construction and installation shall be in accordance with Section G2427.

G2426.3 (502.3) Vent application. The application of vents shall be in accordance with Table G2427.4.

G2426.4 (502.4) Insulation shield. Where vents pass through insulated assemblies, an insulation shield constructed of steel having a minimum thickness of 0.0187 inch (0.4712 mm) (26 gage) shall be installed to provide *clearance* between the vent and the insulation material. The *clearance* shall not be less than the *clearance* to combustibles specified by the vent manufacturer's installation instructions. Where vents pass through attic space, the shield shall terminate not less than 2 inches (51 mm) above the insulation materials and shall be secured in place to prevent displacement. Insulation shields provided as part of a listed vent system shall be installed in accordance with the manufacturer's installation instructions.

G2426.5 (502.5) Installation. Vent systems shall be sized, installed and terminated in accordance with the vent and *appliance* manufacturer's installation instructions and Section G2427.

G2426.6 (502.6) Support of vents. All portions of vents shall be adequately supported for the design and weight of the materials employed.

G2426.7 (502.7) Protection against physical damage. In *concealed locations*, where a vent is installed through holes or notches in studs, joists, rafters or similar members less than $1^1/_2$ inches (38 mm) from the nearest edge of the member, the vent shall be protected by shield plates. Protective steel shield plates having a minimum thickness of 0.0575-inch (1.463 mm) (16 gage) shall cover the area of the vent where the member is notched or bored and shall extend a minimum of 4 inches (102 mm) above sole plates, below top plates and to each side of a stud, joist or rafter.

SECTION G2427 (503) VENTING OF APPLIANCES

G2427.1 (503.1) General. The venting of appliances shall be in accordance with Sections G2427.2 through G2427.16.

G2427.2 (503.2) Venting systems required. Except as permitted in Sections G2427.2.1, G2427.2.2 and G2425.8, all *appliances* shall be connected to *venting systems*.

G2427.2.1 (503.2.3) Direct-vent appliances. Listed *direct-*vent *appliances* shall be installed in accordance with the manufacturer's instructions and Section G2427.8, Item 3.

G2427.2.2 (503.2.4) Appliances with integral vents. *Appliances* incorporating integral venting means shall be considered properly vented where installed in accordance with the manufacturer's instructions and Section G2427.8, Items 1 and 2.

G2427.3 (503.3) Design and construction. A venting system shall be designed and constructed so as to develop a positive flow adequate to convey flue or *vent gases* to the outdoors.

G2427.3.1 (503.3.1) Appliance draft requirements. A venting system shall satisfy the *draft* requirements of the *appliance* in accordance with the manufacturer's instructions.

G2427.3.2 (503.3.2) Design and construction. *Appliances* required to be vented shall be connected to a venting system designed and installed in accordance with the provisions of Sections G2427.4 through G2427.16.

G2427.3.3 (503.3.3) Mechanical draft systems. *Mechanical draft* systems shall comply with the following:

1. *Mechanical draft* systems shall be listed and shall be installed in accordance with the manufacturer's installation instructions for both the *appliance* and the *mechanical draft* system.
2. *Appliances*, except incinerators, requiring venting shall be permitted to be vented by means of *mechanical draft* systems of either forced or *induced draft* design.
3. Forced *draft* systems and all portions of *induced draft* systems under positive pressure during operation shall be designed and installed so as to prevent leakage of flue or *vent gases* into a building.
4. *Vent connectors* serving *appliances* vented by *natural draft* shall not be connected into any portion of *mechanical draft* systems operating under positive pressure.
5. Where a *mechanical draft* system is employed, provisions shall be made to prevent the flow of gas to the *main burners* when the *draft* system is not performing so as to satisfy the operating requirements of the *appliance* for safe performance.
6. The exit terminals of *mechanical draft* systems shall be not less than 7 feet (2134 mm) above finished ground level where located adjacent to public walkways and shall be located as specified in Section G2427.8, Items 1 and 2.

G2427.3.4 (503.3.5) Air ducts and furnace plenums. *Venting systems* shall not extend into or pass through any fabricated air duct or *furnace plenum*.

G2427.3.5 (503.3.6) Above-ceiling air-handling spaces. Where a venting system passes through an above-ceiling air-handling space or other nonducted portion of an air-handling system, the venting system shall conform to one of the following requirements:

1. The venting system shall be a listed special gas vent; other venting system serving a Category III or Category IV *appliance*; or other positive pressure vent, with joints sealed in accordance with the *appliance* or vent manufacturer's instructions.
2. The venting system shall be installed such that fittings and joints between sections are not installed in the above-ceiling space.
3. The venting system shall be installed in a conduit or enclosure with sealed joints separating the interior of the conduit or enclosure from the ceiling space.

G2427.4 (503.4) Type of venting system to be used. The type of venting system to be used shall be in accordance with Table G2427.4.

G2427.4.1 (503.4.1) Plastic piping. Plastic *piping* used for venting *appliances* listed for use with such venting materials shall be *approved*.

G2427.4.1.1 (503.4.1.1) (IFGS) Plastic vent joints. Plastic *pipe* and fittings used to vent *appliances* shall be installed in accordance with the *appliance* manufacturer's installation instructions. Where a primer is required, it shall be of a contrasting color.

G2427.4.2 (503.4.2) Special gas vent. *Special gas vent* shall be listed and installed in accordance with the *special gas vent* manufacturer's installation instructions.

G2427.5 (503.5) Masonry, metal and factory-built chimneys. Masonry, metal and factory-built *chimneys* shall comply with Sections G2427.5.1 through G2427.5.9.

G2427.5.1 (503.5.1) Factory-built chimneys. Factory-built *chimneys* shall be installed in accordance with the manufacturer's installation instructions. Factory-built *chimneys* used to vent *appliances* that operate at a positive vent pressure shall be listed for such application.

TABLE G2427.4
TYPE OF VENTING SYSTEM TO BE USED

APPLIANCES	TYPE OF VENTING SYSTEM
Listed Category I appliances Listed appliances equipped with draft hood Appliances listed for use with Type B gas vent	Type B gas vent (Section G2427.6) Chimney (Section G2427.5) Single-wall metal pipe (Section G2427.7) Listed chimney lining system for gas venting (Section G2427.5.2) Special gas vent listed for these appliances (Section G2427.4.2)
Listed vented wall furnaces	Type B-W gas vent (Sections G2427.6, G2436)
Category II appliances	As specified or furnished by manufacturers of listed appliances (Sections G2427.4.1, G2427.4.2)
Category III appliances	As specified or furnished by manufacturers of listed appliances (Sections G2427.4.1, G2427.4.2)
Category IV appliances	As specified or furnished by manufacturers of listed appliances (Sections G2427.4.1, G2427.4.2)
Unlisted appliances	Chimney (Section G2427.5)
Decorative appliances in vented fireplaces	Chimney
Direct-vent appliances	See Section G2427.2.1
Appliances with integral vent	See Section G2427.2.2

G2427.5.2 (503.5.3) Masonry chimneys. Masonry *chimneys* shall be built and installed in accordance with NFPA 211 and shall be lined with *approved* clay *flue lining*, a listed *chimney* lining system or other *approved* material that will resist corrosion, erosion, softening or cracking from vent gases at temperatures up to 1,800°F (982°C).

Exception: Masonry *chimney* flues serving listed gas *appliances* with *draft hoods*, Category I *appliances* and other gas *appliances* listed for use with Type B vents shall be permitted to be lined with a *chimney* lining system specifically listed for use only with such *appliances*. The liner shall be installed in accordance with the liner manufacturer's installation instructions. A permanent identifying label shall be attached at the point where the connection is to be made to the liner. The label shall read: "This *chimney* liner is for *appliances* that burn gas only. Do not connect to solid or liquid fuel-burning appliances or incinerators."

G2427.5.3 (503.5.4) Chimney termination. *Chimneys* for residential-type or low-heat *appliances* shall extend at least 3 feet (914 mm) above the highest point where they pass through a roof of a building and at least 2 feet (610 mm) higher than any portion of a building within a horizontal distance of 10 feet (3048 mm) (see Figure G2427.5.3). *Chimneys* for medium-heat *appliances* shall extend at least 10 feet (3048 mm) higher than any portion of any building within 25 feet (7620 mm). *Chimneys* shall extend at least 5 feet (1524 mm) above the highest connected *appliance draft hood* outlet or *flue collar*. Decorative shrouds shall not be installed at the termination of factory-built *chimneys* except where such shrouds are listed and labeled for use with the specific factory-built *chimney* system and are installed in accordance with the manufacturer's installation instructions.

G2427.5.4 (503.5.5) Size of chimneys. The effective area of a *chimney* venting system serving listed *appliances* with *draft hoods*, Category I *appliances*, and other *appliances* listed for use with Type B vents shall be determined in accordance with one of the following methods:

1. The provisions of Section G2428.
2. For sizing an individual *chimney* venting system for a single *appliance* with a *draft hood*, the effective areas of the *vent connector* and *chimney* flue shall be not less than the area of the *appliance flue collar* or *draft hood* outlet, nor greater than seven times the *draft hood* outlet area.
3. For sizing a *chimney* venting system connected to two *appliances* with *draft hoods*, the effective area of the *chimney* flue shall be not less than the area of the larger *draft hood* outlet plus 50 percent of the area of the smaller *draft hood* outlet, nor greater than seven times the smallest *draft hood* outlet area.
4. *Chimney venting systems* using *mechanical draft* shall be sized in accordance with *approved* engineering methods.
5. Other *approved* engineering methods.

G2427.5.5 (503.5.6) Inspection of chimneys. Before replacing an existing *appliance* or connecting a vent *connector* to a *chimney*, the *chimney* passageway shall be examined to ascertain that it is clear and free of obstructions and it shall be cleaned if previously used for venting solid or liquid fuel-burning appliances or *fireplaces*.

G2427.5.5.1 (503.5.6.1) Chimney lining. *Chimneys* shall be lined in accordance with NFPA 211.

Exception: Where an existing chimney complies with Sections G2427.5.5 through G2427.5.5.3 and its sizing is in accordance with Section G2427.5.4, its continued use shall be allowed where the *appliance* vented by that *chimney* is replaced by an *appliance* of similar type, input rating and efficiency.

G2427.5.5.2 (503.5.6.2) Cleanouts. Cleanouts shall be examined to determine that they will remain tightly closed when not in use.

G2427.5.5.3 (503.5.6.3) Unsafe chimneys. Where inspection reveals that an existing *chimney* is not safe for the intended application, it shall be repaired, rebuilt, lined, relined or replaced with a vent or *chimney* to conform to NFPA 211 and it shall be suitable for the *appliances* to be vented.

G2427.5.6 (503.5.7) Chimneys serving appliances burning other fuels. *Chimneys* serving *appliances* burning other fuels shall comply with Sections G2427.5.6.1 through G2427.5.6.4.

G2427.5.6.1 (503.5.7.1) Solid fuel-burning appliances. An *appliance* shall not be connected to a *chimney* flue serving a separate appliance designed to burn solid fuel.

G2427.5.6.2 (503.5.7.2) Liquid fuel-burning appliances. Where one *chimney* flue serves gas *appliances* and liquid fuel-burning appliances, the appliances shall be connected through separate openings or shall be connected through a single opening where joined by a suitable fitting located as close as practical to the *chimney*. Where two or more openings are provided into one *chimney* flue, they shall be at different levels. Where the appliances are automatically controlled, they shall be equipped with *safety shutoff devices*.

G2427.5.6.3 (503.5.7.3) Combination gas- and solid fuel-burning appliances. A combination gas- and solid fuel-burning *appliance* equipped with a manual reset device to shut off gas to the *main burner* in the

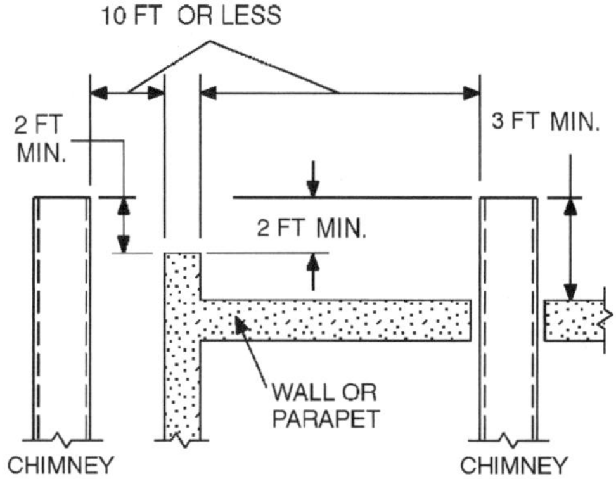

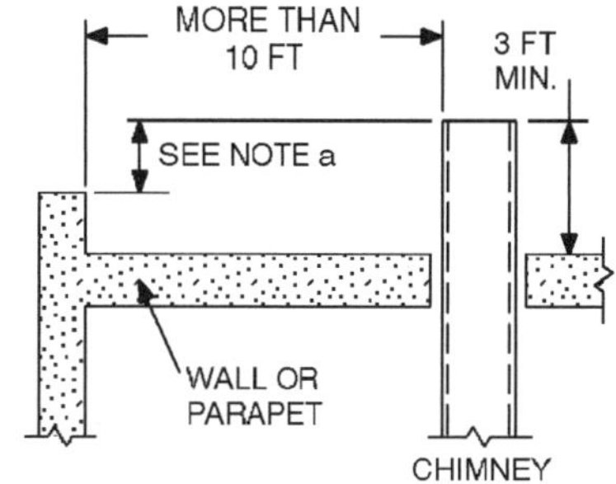

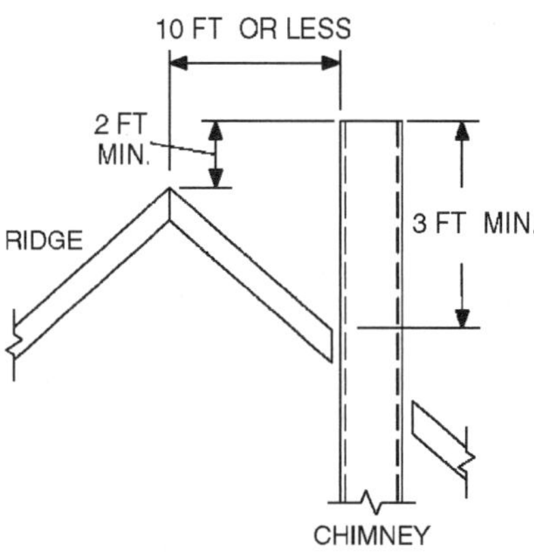

A. TERMINATION 10 FT OR LESS FROM RIDGE, WALL OR PARAPET

MORE THAN 10 FT
10 FT
RIDGE
SEE NOTE b
2 FT MIN.
3 FT MIN.
CHIMNEY

B. TERMINATION MORE THAN 10 FT FROM RIDGE, WALL OR PARAPET

For SI: 1 inch = 25.4 mm, 1 foot = 304.8 mm.

NOTES:

a. No height above parapet required when distance from walls or parapet is more than 10 feet.

b. Height above any roof surface within 10 feet horizontally.

FIGURE G2427.5.3 (503.5.4)
TYPICAL TERMINATION LOCATIONS FOR CHIMNEYS AND SINGLE-WALL METAL PIPES SERVING RESIDENTIAL-TYPE AND LOW-HEAT APPLIANCES

event of sustained backdraft or flue gas spillage shall be permitted to be connected to a single *chimney* flue. The *chimney* flue shall be sized to properly vent the *appliance*.

G2427.5.6.4 (503.5.7.4) Combination gas- and oil fuel-burning appliances. A listed combination gas- and oil fuel-burning *appliance* shall be permitted to be connected to a single *chimney* flue. The *chimney* flue shall be sized to properly vent the *appliance*.

G2427.5.7 (503.5.8) Support of chimneys. All portions of *chimneys* shall be supported for the design and weight of the materials employed. Factory-built *chimneys* shall be supported and spaced in accordance with the manufacturer's installation instructions.

G2427.5.8 (503.5.9) Cleanouts. Where a *chimney* that formerly carried flue products from liquid or solid fuel-burning appliances is used with an *appliance* using *fuel gas*, an accessible cleanout shall be provided. The cleanout shall have a tight-fitting cover and be installed so its upper edge is at least 6 inches (152 mm) below the lower edge of the lowest *chimney* inlet opening.

G2427.5.9 (503.5.10) Space surrounding lining or vent. The remaining space surrounding a *chimney* liner, gas vent, *special gas vent* or plastic *piping* installed within a *masonry chimney* flue shall not be used to vent another *appliance*. The insertion of another liner or vent within the *chimney* as provided in this *code* and the liner or vent manufacturer's instructions shall not be prohibited.

The remaining space surrounding a *chimney* liner, gas vent, special gas vent or plastic *piping* installed within a masonry, metal or factory-built *chimney* shall not be used to supply *combustion air*. Such space shall not be prohibited from supplying *combustion air* to *direct-vent appliances* designed for installation in a solid fuel-burning *fireplace* and installed in accordance with the manufacturer's installation instructions.

G2427.6 (503.6) Gas vents. Gas vents shall comply with Sections G2427.6.1 through G2427.6.11. (See Section G2403, Definitions.)

G2427.6.1 (503.6.1) Installation, general. Gas vents shall be installed in accordance with the terms of their listings and the manufacturer's instructions.

G2427.6.2 (503.6.2) Type B-W vent capacity. A Type B-W gas vent shall have a listed capacity not less than that of the listed *vented wall furnace* to which it is connected.

G2427.6.3 (503.6.4) Gas vent termination. A gas vent shall terminate in accordance with one of the following:

1. Gas vents that are 12 inches (305 mm) or less in size and located not less than 8 feet (2438 mm) from a vertical wall or similar obstruction shall terminate above the roof in accordance with Figure G2427.6.3.
2. Gas vents that are over 12 inches (305 mm) in size or are located less than 8 feet (2438 mm) from a vertical wall or similar obstruction shall terminate not less than 2 feet (610 mm) above the highest point where they pass through the roof and not less than 2 feet (610 mm) above any portion of a building within 10 feet (3048 mm) horizontally.
3. As provided for direct-vent systems in Section G2427.2.1.
4. As provided for *appliances* with integral vents in Section G2427.2.2.
5. As provided for *mechanical draft* systems in Section G2427.3.3.

G2427.6.3.1 (503.6.4.1) Decorative shrouds. Decorative shrouds shall not be installed at the termination of gas vents except where such shrouds are listed for use with the specific gas venting system and are installed in accordance with manufacturer's installation instructions.

G2427.6.4 (503.6.5) Minimum height. A Type B or L gas vent shall terminate at least 5 feet (1524 mm) in vertical height above the highest connected *appliance draft hood* or *flue collar*. A Type B-W gas vent shall terminate at least 12 feet (3658 mm) in vertical height above the bottom of the wall *furnace*.

G2427.6.5 (503.6.6) Roof terminations. Gas vents shall extend through the roof flashing, roof jack or roof thimble and terminate with a listed cap or listed roof assembly.

G2427.6.6 (503.6.7) Forced air inlets. Gas vents shall terminate not less than 3 feet (914 mm) above any forced air inlet located within 10 feet (3048 mm).

G2427.6.7 (503.6.8) Exterior wall penetrations. A gas *vent* extending through an exterior wall shall not terminate adjacent to the wall or below eaves or parapets, except as provided in Sections G2427.2.1 and G2427.3.3.

G2427.6.8 (503.6.9) Size of gas vents. *Venting systems* shall be sized and constructed in accordance with Section G2428 or other *approved* engineering methods and the gas vent and *appliance* manufacturer's installation instructions.

G2427.6.8.1 (503.6.9.1) Category I appliances. The sizing of *natural draft venting systems* serving one or more listed *appliances* equipped with a *draft hood* or *appliances* listed for use with Type B gas vent, installed in a single story of a building, shall be in accordance with one of the following methods:

1. The provisions of Section G2428.
2. For sizing an individual gas vent for a single, draft-hood-equipped *appliance*, the effective area of the vent *connector* and the gas vent shall be not less than the area of the *appliance draft hood* outlet, nor greater than seven times the *draft hood* outlet area.
3. For sizing a gas vent connected to two *appliances* with *draft hoods*, the effective area of the vent shall be not less than the area of the larger *draft hood* outlet plus 50 percent of the area of the smaller *draft hood* outlet, nor greater than seven times the smaller *draft hood* outlet area.
4. *Approved* engineering practices.

G2427.6.8.2 (503.6.9.2) Vent offsets. Type B and L vents sized in accordance with Item 2 or 3 of Section G2427.6.8.1 shall extend in a generally vertical direction with offsets not exceeding 45 degrees (0.79 rad), except that a vent system having not more than one 60-degree (1.04 rad) *offset* shall be permitted. Any angle greater than 45 degrees (0.79 rad) from the vertical is considered horizontal. The total horizontal distance of a vent plus the horizontal vent *connector* serving *draft hood*-equipped *appliances* shall be not greater than 75 percent of the vertical height of the vent.

G2427.6.8.3 (503.6.9.3) Category II, III and IV appliances. The sizing of gas vents for Category II, III and IV appliances shall be in accordance with the appliance manufacturer's instructions.

G2427.6.8.4 (503.6.9.4) Mechanical draft. *Chimney venting systems* using *mechanical draft* shall be sized in accordance with *approved* engineering methods.

G2427.6.9 (503.6.11) Support of gas vents. Gas vents shall be supported and spaced in accordance with the manufacturer's installation instructions.

G2427.6.10 (503.6.12) Marking. In those localities where solid and liquid fuels are used extensively, gas vents shall be permanently identified by a label attached to the wall or ceiling at a point where the *vent connector* enters the gas vent. The determination of where such localities exist shall be made by the *code official*. The label shall read:

"This gas vent is for *appliances* that burn gas. Do not connect to solid or liquid fuel-burning appliances or incinerators."

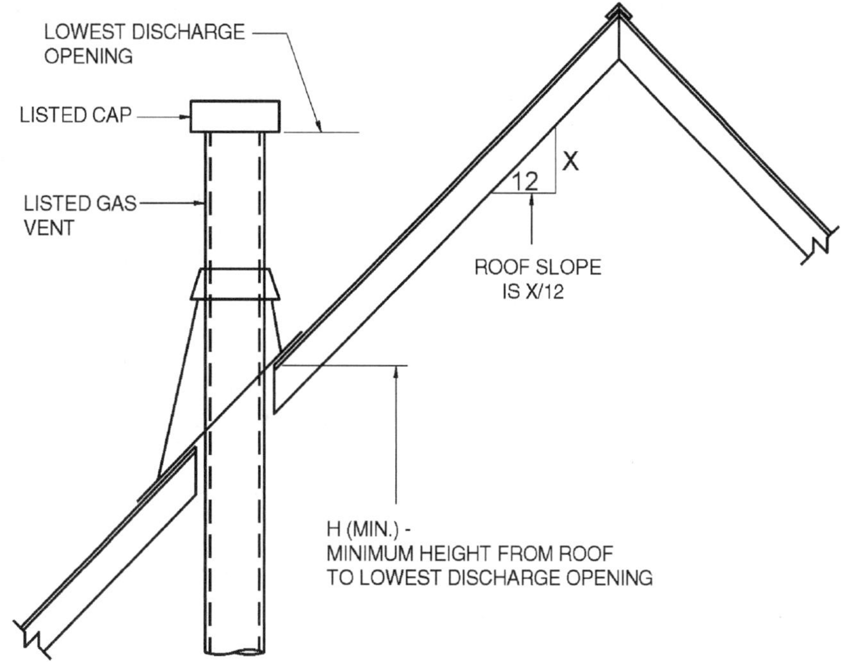

ROOF SLOPE	H (minimum) ft
Flat to $^{6}/_{12}$	1.0
Over $^{6}/_{12}$ to $^{7}/_{12}$	1.25
Over $^{7}/_{12}$ to $^{8}/_{12}$	1.5
Over $^{8}/_{12}$ to $^{9}/_{12}$	2.0
Over $^{9}/_{12}$ to $^{10}/_{12}$	2.5
Over $^{10}/_{12}$ to $^{11}/_{12}$	3.25
Over $^{11}/_{12}$ to $^{12}/_{12}$	4.0
Over $^{12}/_{12}$ to $^{14}/_{12}$	5.0
Over $^{14}/_{12}$ to $^{16}/_{12}$	6.0
Over $^{16}/_{12}$ to $^{18}/_{12}$	7.0
Over $^{18}/_{12}$ to $^{20}/_{12}$	7.5
Over $^{20}/_{12}$ to $^{21}/_{12}$	8.0

For SI: 1 foot = 304.8 mm.

FIGURE G2427.6.3 (503.6.4)
GAS VENT TERMINATION LOCATIONS FOR LISTED CAPS 12 INCHES OR LESS IN SIZE AT LEAST 8 FEET FROM A VERTICAL WALL

G2427.6.11 (503.6.13) Fastener penetrations. Screws, rivets and other fasteners shall not penetrate the inner wall of double-wall gas vents, except at the transition from an *appliance draft hood* outlet, a *flue collar* or a single-wall metal connector to a double-wall vent.

G2427.7 (503.7) Single-wall metal pipe. Single-wall metal *pipe* vents shall comply with Sections G2427.7.1 through G2427.7.13.

G2427.7.1 (503.7.1) Construction. Single-wall metal pipe shall be constructed of galvanized sheet steel not less than 0.0304 inch (0.7 mm) thick, or other *approved*, noncombustible, corrosion-resistant material.

G2427.7.2 (503.7.2) Cold climate. Uninsulated single-wall metal pipe shall not be used outdoors for venting *appliances* in regions where the 99-percent winter design temperature is below 32°F (0°C).

G2427.7.3 (503.7.3) Termination. Single-wall metal pipe shall terminate at least 5 feet (1524 mm) in vertical height above the highest connected *appliance draft hood* outlet or *flue collar*. Single-wall metal pipe shall extend at least 2 feet (610 mm) above the highest point where it passes through a roof of a building and at least 2 feet (610 mm) higher than any portion of a building within a horizontal distance of 10 feet (3048 mm) (see Figure G2427.5.3). An *approved* cap or roof assembly shall be attached to the terminus of a single-wall metal pipe (see also Section G2427.7.9, Item 3).

G2427.7.4 (503.7.4) Limitations of use. Single-wall metal pipe shall be used only for runs directly from the space in which the *appliance* is located through the roof or exterior wall to the outdoor atmosphere.

G2427.7.5 (503.7.5) Roof penetrations. A pipe passing through a roof shall extend without interruption through the roof flashing, roof jack, or roof thimble. Where a single-wall metal pipe passes through a roof constructed of combustible material, a noncombustible, nonventilating thimble shall be used at the point of passage. The thimble shall extend at least 18 inches (457 mm) above and 6 inches (152 mm) below the roof with the annular space open at the bottom and closed only at the top. The thimble shall be sized in accordance with Section G2427.7.7.

G2427.7.6 (503.7.6) Installation. Single-wall metal pipe shall not originate in any unoccupied attic or concealed space and shall not pass through any attic, inside wall, concealed space, or floor. The installation of a single-wall metal pipe through an exterior combustible wall shall comply with Section G2427.7.7. Single-wall metal pipe used for venting an incinerator shall be exposed and readily examinable for its full length and shall have suitable *clearances* maintained.

G2427.7.7 (503.7.7) Single-wall penetrations of combustible walls. Single-wall metal pipe shall not pass through a combustible exterior wall unless guarded at the point of passage by a ventilated metal thimble not smaller than the following:

1. For listed *appliances* equipped with *draft hoods* and *appliances* listed for use with Type B gas vents, the thimble shall be not less than 4 inches (102 mm) larger in diameter than the metal pipe. Where there is a run of not less than 6 feet (1829 mm) of metal pipe in the open between the *draft hood* outlet and the thimble, the thimble shall be permitted to be not less than 2 inches (51 mm) larger in diameter than the metal pipe.
2. For unlisted *appliances* having *draft hoods*, the thimble shall be not less than 6 inches (152 mm) larger in diameter than the metal pipe.
3. For residential and low-heat *appliances*, the thimble shall be not less than 12 inches (305 mm) larger in diameter than the metal pipe.

Exception: In lieu of thimble protection, all combustible material in the wall shall be removed a sufficient distance from the metal pipe to provide the specified *clearance* from such metal pipe to combustible material. Any material used to close up such opening shall be noncombustible.

G2427.7.8 (503.7.8) Clearances. Minimum *clearances* from single-wall metal pipe to combustible material shall be in accordance with Table G2427.10.5. The *clearance* from single-wall metal pipe to combustible material shall be permitted to be reduced where the combustible material is protected as specified for *vent connectors* in Table G2409.2.

G2427.7.9 (503.7.9) Size of single-wall metal pipe. A venting system constructed of single-wall metal pipe shall be sized in accordance with one of the following methods and the *appliance* manufacturer's instructions:

1. For a draft-hood-equipped *appliance*, in accordance with Section G2428.
2. For a venting system for a single *appliance* with a *draft hood*, the areas of the connector and the pipe each shall be not less than the area of the *appliance flue collar* or *draft hood* outlet, whichever is smaller. The vent area shall not be greater than seven times the *draft hood* outlet area.
3. Other *approved* engineering methods.

G2427.7.10 (503.7.10) Pipe geometry. Any shaped single-wall metal pipe shall be permitted to be used, provided that its equivalent effective area is equal to the effective area of the round pipe for which it is substituted, and provided that the minimum internal dimension of the pipe is not less than 2 inches (51 mm).

G2427.7.11 (503.7.11) Termination capacity. The vent cap or a roof assembly shall have a venting capacity not less than that of the pipe to which it is attached.

G2427.7.12 (503.7.12) Support of single-wall metal pipe. All portions of single-wall metal pipe shall be supported for the design and weight of the material employed.

G2427.7.13 (503.7.13) Marking. Single-wall metal pipe shall comply with the marking provisions of Section G2427.6.10.

G2427.8 (503.8) Venting system termination location. The location of venting system terminations shall comply with the following (see Appendix C):

1. A *mechanical draft* venting system shall terminate at least 3 feet (914 mm) above any forced-air inlet located within 10 feet (3048 mm).

   Exceptions:

   1. This provision shall not apply to the *combustion air* intake of a direct-vent *appliance*.
   2. This provision shall not apply to the separation of the integral outdoor air inlet and flue gas discharge of listed outdoor *appliances*.

2. A *mechanical draft* venting system, excluding *direct-vent appliances*, shall terminate at least 4 feet (1219 mm) below, 4 feet (1219 mm) horizontally from, or 1 foot (305 mm) above any door, operable window, or gravity air inlet into any building. The bottom of the vent terminal shall be located at least 12 inches (305 mm) above finished ground level.
3. The vent terminal of a *direct*-vent *appliance* with an input of 10,000 *Btu* per hour (3 kW) or less shall be located at least 6 inches (152 mm) from any air opening into a building, and such an *appliance* with an input over 10,000 *Btu* per hour (3 kW) but not over 50,000 *Btu* per hour (14.7 kW) shall be installed with a 9-inch (230 mm) vent termination *clearance*, and an *appliance* with an input over 50,000 *Btu*/h (14.7 kW) shall have at least a 12-inch (305 mm) vent termination *clearance*. The bottom of the vent terminal and the air intake shall be located at least 12 inches (305 mm) above grade finished ground level.
4. Through-the-wall vents for Category II and IV *appliances* and noncategorized condensing *appliances* shall not terminate over public walkways or over an area where *condensate* or vapor could create a nuisance or hazard or could be detrimental to the operation of *regulators, relief valves*, or other *equipment*. Where local experience indicates that *condensate* is a problem with Category I and III *appliances*, this provision shall also apply. Drains for *condensate* shall be installed in accordance with the appliance and vent manufacturer's installation instructions.

G2427.9 (503.9) Condensation drainage. Provisions shall be made to collect and dispose of *condensate* from *venting systems* serving Category II and IV *appliances* and noncategorized condensing *appliances* in accordance with Section G2427.8, Item 4. Where local experience indicates that condensation is a problem, provision shall be made to drain off and dispose of *condensate* from *venting systems* serving Category I and III *appliances* in accordance with Section G2427.8, Item 4.

G2427.10 (503.10) Vent connectors for Category I appliances. Vent *connectors* for Category I *appliances* shall comply with Sections G2427.10.1 through G2427.10.14.

G2427.10.1 (503.10.1) Where required. A vent *connector* shall be used to connect an *appliance* to a gas vent, *chimney* or single-wall metal pipe, except where the gas vent, *chimney* or single-wall metal pipe is directly connected to the *appliance*.

G2427.10.2 (503.10.2) Materials. *Vent connectors* shall be constructed in accordance with Sections G2427.10.2.1 through G2427.10.2.4.

G2427.10.2.1 (503.10.2.1) General. A *vent connector* shall be made of noncombustible corrosion-resistant material capable of withstanding the vent gas temperature produced by the *appliance* and of sufficient thickness to withstand physical damage.

G2427.10.2.2 (503.10.2.2) Vent connectors located in unconditioned areas. Where the *vent connector* used for an *appliance* having a *draft hood* or a Category I appliance is located in or passes through attics, crawl spaces or other unconditioned spaces, that portion of the *vent connector* shall be listed Type B, Type L or listed vent material having equivalent insulation properties.

Exception: Single-wall metal pipe located within the exterior walls of the building in areas having a local 99-percent winter design temperature of 5°F (-15°C) or higher shall be permitted to be used in unconditioned spaces other than attics and crawl spaces.

G2427.10.2.3 (503.10.2.3) Residential-type appliance connectors. Where *vent connectors* for residential-type *appliances* are not installed in attics or other unconditioned spaces, connectors for listed *appliances* having *draft hoods, appliances* having *draft hoods* and equipped with listed *conversion burners* and Category I *appliances* shall be one of the following:

1. Type B or L vent material;
2. Galvanized sheet steel not less than 0.018 inch (0.46 mm) thick;
3. Aluminum (1100 or 3003 alloy or equivalent) sheet not less than 0.027 inch (0.69 mm) thick;
4. Stainless steel sheet not less than 0.012 inch (0.31 mm) thick;
5. Smooth interior wall metal pipe having resistance to heat and corrosion equal to or greater than that of Item 2, 3 or 4 above; or
6. A listed vent *connector*.

Vent connectors shall not be covered with insulation.

Exception: Listed insulated *vent connectors* shall be installed in accordance with the manufacturer's installation instructions.

G2427.10.2.4 (503.10.2.4) Low-heat appliance. A *vent connector* for a nonresidential, low-heat *appliance* shall be a factory-built *chimney* section or steel *pipe* having resistance to heat and corrosion equivalent to that for the appropriate galvanized pipe as specified in Table G2427.10.2.4. Factory-built *chimney* sections shall be joined together in accordance with the *chimney* manufacturer's instructions.

TABLE G2427.10.2.4 (503.10.2.4)
MINIMUM THICKNESS FOR GALVANIZED STEEL VENT CONNECTORS FOR LOW-HEAT APPLIANCES

DIAMETER OF CONNECTOR (inches)	MINIMUM THICKNESS (inch)
Less than 6	0.019
6 to less than 10	0.023
10 to 12 inclusive	0.029
14 to 16 inclusive	0.034
Over 16	0.056

For SI: 1 inch = 25.4 mm.

G2427.10.3 (503.10.3) Size of vent connector. *Vent connectors* shall be sized in accordance with Sections G2427.10.3.1 through G2427.3.5.

G2427.10.3.1 (503.10.3.1) Single draft hood and fan-assisted. A *vent connector* for an *appliance* with a single *draft hood* or for a Category I fan-assisted *combustion* system *appliance* shall be sized and installed in accordance with Section G2428 or other *approved* engineering methods.

G2427.10.3.2 (503.10.3.2) Multiple draft hood. For a single *appliance* having more than one *draft hood* outlet or *flue collar*, the manifold shall be constructed according to the instructions of the *appliance* manufacturer. Where there are no instructions, the manifold shall be designed and constructed in accordance with *approved* engineering practices. As an alternate method, the effective area of the manifold shall equal the combined area of the *flue collars* or *draft hood* outlets and the *vent connectors* shall have a minimum 1-foot (305 mm) rise.

G2427.10.3.3 (503.10.3.3) Multiple appliances. Where two or more *appliances* are connected to a common *vent* or *chimney*, each *vent connector* shall be sized in accordance with Section G2428 or other *approved* engineering methods.

As an alternative method applicable only when all of the *appliances* are *draft hood* equipped, each *vent connector* shall have an effective area not less than the area of the *draft hood* outlet of the *appliance* to which it is connected.

G2427.10.3.4 (503.10.3.4) Common connector/manifold. Where two or more *appliances* are vented through a common *vent connector* or vent manifold, the common *vent connector* or vent manifold shall be located at the highest level consistent with available headroom and the required *clearance* to combustible materials and shall be sized in accordance with Section G2428 or other *approved* engineering methods.

As an alternate method applicable only where there are two *draft hood*-equipped *appliances*, the effective area of the common *vent connector* or vent manifold and all junction fittings shall be not less than the area of the larger *vent connector* plus 50 percent of the area of the smaller *flue collar* outlet.

G2427.10.3.5 (503.10.3.5) Size increase. Where the size of a *vent connector* is increased to overcome installation limitations and obtain connector capacity equal to the *appliance* input, the size increase shall be made at the *appliance draft hood* outlet.

G2427.10.4 (503.10.4) Two or more appliances connected to a single vent or chimney. Where two or more *vent connectors* enter a common gas vent, *chimney* flue, or single-wall metal pipe, the smaller connector shall enter at the highest level consistent with the available headroom or *clearance* to combustible material. *Vent connectors* serving Category I *appliances* shall not be connected to any portion of a *mechanical draft* system operating under positive static pressure, such as those serving Category III or IV *appliances*.

G2427.10.4.1 (503.10.4.1) Two or more openings. Where two or more openings are provided into one *chimney* flue or vent, the openings shall be at different levels, or the connectors shall be attached to the vertical portion of the *chimney* or vent at an angle of 45 degrees (0.79 rad) or less relative to the vertical.

G2427.10.5 (503.10.5) Clearance. Minimum *clearances* from *vent connectors* to combustible material shall be in accordance with Table G2427.10.5.

Exception: The *clearance* between a *vent connector* and combustible material shall be permitted to be reduced where the combustible material is protected as specified for *vent connectors* in Table G2409.2.

G2427.10.6 (503.10.6) Joints. Joints between sections of connector piping and connections to *flue collars* and *draft hood* outlets shall be fastened by one of the following methods:

1. Sheet metal screws.
2. *Vent connectors* of listed vent material assembled and connected to *flue collars* or *draft hood* outlets in accordance with the manufacturers' instructions.
3. Other *approved* means.

G2427.10.7 (503.10.7) Slope. A *vent connector* shall be installed without dips or sags and shall slope upward toward the vent or *chimney* at least $^1/_4$ inch per foot (21 mm/m).

Exception: *Vent connectors* attached to a *mechanical draft* system installed in accordance with the *appliance* and *draft* system manufacturers' instructions.

G2427.10.8 (503.10.8) Length of vent connector. A *vent connector* shall be as short as practical and the *appliance* located as close as practical to the *chimney* or vent. The maximum horizontal length of a single-wall connector shall be 75 percent of the height of the *chimney* or vent except for engineered systems. The maximum horizontal length of a Type B double-wall connector shall be 100 percent of the height of the *chimney* or vent except for engineered systems.

G2427.10.9 (503.10.9) Support. A *vent connector* shall be supported for the design and weight of the material employed to maintain *clearances* and prevent physical damage and separation of joints.

G2427.10.10 (503.10.10) Chimney connection. Where entering a flue in a masonry or metal *chimney*, the *vent connector* shall be installed above the extreme bottom to avoid stoppage. Where a thimble or slip joint is used to facilitate removal of the connector, the connector shall be firmly attached to or inserted into the thimble or slip joint to prevent the connector from falling out. Means shall be employed to prevent the connector from entering so far as to restrict the space between its end and the opposite wall of the *chimney* flue (see Section G2425.9).

G2427.10.11 (503.10.11) Inspection. The entire length of a *vent connector* shall be provided with *ready access* for inspection, cleaning, and replacement.

G2427.10.12 (503.10.12) Fireplaces. A *vent connector* shall not be connected to a *chimney* flue serving a *fireplace* unless the *fireplace* flue opening is permanently sealed.

G2427.10.13 (503.10.13) Passage through ceilings, floors or walls. Single-wall metal pipe connectors shall not pass through any wall, floor or ceiling except as permitted by Section G2427.7.4.

G2427.11 (503.11) Vent connectors for Category II, III and IV appliances. *Vent connectors* for Category II, III and IV *appliances* shall be as specified for the *venting systems* in accordance with Section G2427.4.

G2427.12 (503.12) Draft hoods and draft controls. The installation of *draft hoods* and *draft controls* shall comply with Sections G2427.12.1 through G2427.12.7.

G2427.12.1 (503.12.1) Appliances requiring draft hoods. *Vented appliances* shall be installed with *draft hoods*.

Exception: Dual oven-type combination ranges; incinerators; *direct*-vent *appliances*; fan-assisted *combustion* system *appliances*; *appliances* requiring *chimney draft* for operation; single firebox boilers equipped with *conversion burners* with inputs greater than 400,000 *Btu* per hour (117 kW); *appliances* equipped with blast, power or pressure *burners* that are not listed for use with *draft hoods*; and *appliances* designed for forced venting.

G2427.12.2 (503.12.2) Installation. A *draft hood* supplied with or forming a part of a listed *vented appliance* shall be installed without *alteration*, exactly as furnished and specified by the *appliance* manufacturer.

G2427.12.2.1 (503.12.2.1) Draft hood required. If a *draft hood* is not supplied by the *appliance* manufacturer where one is required, a *draft hood* shall be installed, shall be of a listed or *approved* type and, in the absence of other instructions, shall be of the same size as the *appliance flue* collar. Where a *draft hood* is required with a *conversion burner*, it shall be of a listed or *approved* type.

G2427.12.2.2 (503.12.2.2) Special design draft hood. Where it is determined that a *draft hood* of special design is needed or preferable for a particular installation, the installation shall be in accordance with the recommendations of the *appliance* manufacturer and shall be *approved*.

G2427.12.3 (503.12.3) Draft control devices. Where a *draft control* device is part of the *appliance* or is supplied by the *appliance* manufacturer, it shall be installed in accordance with the manufacturer's instructions. In the absence of manufacturer's instructions, the device shall be attached to the *flue collar* of the *appliance* or as near to the *appliance* as practical.

G2427.12.4 (503.12.4) Additional devices. *Appliances* (except incinerators) requiring a controlled *chimney draft* shall be permitted to be equipped with a listed double-acting barometric-*draft regulator* installed and adjusted in accordance with the manufacturer's instructions.

G2427.12.5 (503.12.5) Location. *Draft hoods* and *barometric draft regulators* shall be installed in the same room or enclosure as the *appliance* in such a manner as to prevent any difference in

TABLE G2427.10.5 (503.10.5)[a]
CLEARANCES FOR CONNECTORS

APPLIANCE	MINIMUM DISTANCE FROM COMBUSTIBLE MATERIAL			
	Listed Type B gas vent material	Listed Type L vent material	Single-wall metal pipe	Factory-built chimney sections
Listed appliances with draft hoods and appliances listed for use with Type B gas vents	As listed	As listed	6 inches	As listed
Residential boilers and furnaces with listed gas conversion burner and with draft hood	6 inches	6 inches	9 inches	As listed
Residential appliances listed for use with Type L vents	Not permitted	As listed	9 inches	As listed
Listed gas-fired toilets	Not permitted	As listed	As listed	As listed
Unlisted residential appliances with draft hood	Not permitted	6 inches	9 inches	As listed
Residential and low-heat appliances other than above	Not permitted	9 inches	18 inches	As listed
Medium-heat appliances	Not permitted	Not permitted	36 inches	As listed

For SI: 1 inch = 25.4 mm.

a. These clearances shall apply unless the manufacturer's installation instructions for a listed appliance or connector specify different clearances, in which case the listed clearances shall apply.

pressure between the hood or *regulator* and the *combustion air* supply.

G2427.12.6 (503.12.6) Positioning. *Draft hoods* and *draft regulators* shall be installed in the position for which they were designed with reference to the horizontal and vertical planes and shall be located so that the *relief opening* is not obstructed by any part of the *appliance* or adjacent construction. The *appliance* and its *draft hood* shall be located so that the *relief opening* is accessible for checking *vent* operation.

G2427.12.7 (503.12.7) Clearance. A *draft hood* shall be located so its *relief opening* is not less than 6 inches (152 mm) from any surface except that of the *appliance* it serves and the venting system to which the *draft hood* is connected. Where a greater or lesser *clearance* is indicated on the *appliance* label, the *clearance* shall be not less than that specified on the label. Such *clearances* shall not be reduced.

G2427.13 (503.13) Manually operated dampers. A manually operated *damper* shall not be placed in the vent *connector* for any *appliance*. Fixed baffles shall not be classified as manually operated *dampers*.

G2427.14 (503.14) Automatically operated vent dampers. An automatically operated vent damper shall be of a listed type.

G2427.15 (503.15) Obstructions. Devices that retard the flow of *vent gases* shall not be installed in a *vent connector*, *chimney*, or vent. The following shall not be considered as obstructions:

1. *Draft regulators* and safety *controls* specifically listed for installation in *venting systems* and installed in accordance with the manufacturer's installation instructions.
2. *Approved draft regulators* and safety *controls* that are designed and installed in accordance with *approved* engineering methods.
3. Listed heat reclaimers and automatically operated vent dampers installed in accordance with the manufacturer's installation instructions.
4. *Approved* economizers, heat reclaimers, and recuperators installed in *venting systems* of *appliances* not required to be equipped with *draft hoods*, provided that the *appliance* manufacturer's instructions cover the installation of such a device in the venting system and performance in accordance with Sections G2427.3 and G2427.3.1 is obtained.
5. Vent dampers serving listed *appliances* installed in accordance with Sections G2428.2.1 and G2428.3.1 or other *approved* engineering methods.

G2427.16 (503.16) (IFGS) Outside wall penetrations. Where vents, including those for *direct-vent appliances*, penetrate outside walls of buildings, the annular spaces around such penetrations shall be permanently sealed using *approved* materials to prevent entry of *combustion products* into the building.

SECTION G2428 (504) SIZING OF CATEGORY I APPLIANCE VENTING SYSTEMS

G2428.1 (504.1) Definitions. The following definitions apply to tables in this section.

APPLIANCE CATEGORIZED VENT DIAMETER/AREA. The minimum vent area/diameter permissible for Category I *appliances* to maintain a nonpositive vent static pressure when tested in accordance with nationally recognized standards.

FAN-ASSISTED COMBUSTION SYSTEM. An *appliance* equipped with an integral mechanical means to either draw or force products of *combustion* through the *combustion chamber* or heat exchanger.

FAN MIN. The minimum input rating of a Category I fan-assisted *appliance* attached to a vent or connector.

FAN MAX. The maximum input rating of a Category I fan-assisted *appliance* attached to a vent or connector.

NAT MAX. The maximum input rating of a Category I draft-hood-equipped *appliance* attached to a vent or connector.

FAN + FAN. The maximum combined *appliance* input rating of two or more Category I fan-assisted *appliances* attached to the common vent.

FAN + NAT. The maximum combined *appliance* input rating of one or more Category I fan-assisted *appliances* and one or more Category I draft-hood-equipped *appliances* attached to the common vent.

NA. Vent configuration is not permitted due to potential for *condensate* formation or pressurization of the venting system, or not applicable due to physical or geometric restraints.

NAT + NAT. The maximum combined *appliance* input rating of two or more Category I draft-hood-equipped *appliances* attached to the common vent.

G2428.2 (504.2) Application of single appliance vent Tables G2428.2(1) and G2428.2(2). The application of Tables G2428.2(1) and G2428.2(2) shall be subject to the requirements of Sections G2428.2.1 through G2428.2.16.

G2428.2.1 (504.2.1) Vent obstructions. These venting tables shall not be used where obstructions, as described in Section G2427.15, are installed in the venting system. The installation of vents serving listed *appliances* with vent dampers shall be in accordance with the *appliance* manufacturer's instructions or in accordance with the following:

1. The maximum capacity of the vent system shall be determined using the "NAT Max" column.
2. The minimum capacity shall be determined as if the *appliance* were a fan-assisted *appliance*, using the "FAN Min" column to determine the minimum capacity of the vent system. Where the corresponding "FAN Min" is "NA," the vent configuration shall not be permitted and an alternative venting configuration shall be utilized.

G2428.2.2 (504.2.2) Minimum size. Where the vent size determined from the tables is smaller than the *appliance draft hood* outlet or *flue collar*, the smaller size shall be permitted to be used provided all of the following are met:

1. The total vent height (H) is at least 10 feet (3048 mm).
2. Vents for *appliance draft hood* outlets or *flue collars* 12 inches (305 mm) in diameter or smaller are not reduced more than one table size.
3. Vents for *appliance draft hood* outlets or *flue collars* larger than 12 inches (305 mm) in diameter are not reduced more than two table sizes.
4. The maximum capacity listed in the tables for a fan-assisted *appliance* is reduced by 10 percent (0.90 by maximum table capacity).
5. The *draft hood* outlet is greater than 4 inches (102 mm) in diameter. Do not connect a 3-inch-diameter (76 mm) vent to a 4-inch-diameter (102 mm) *draft hood* outlet. This provision shall not apply to fan-assisted *appliances*.

G2428.2.3 (504.2.3) Vent offsets. Single-*appliance* venting configurations with zero (0) lateral lengths in Tables G2428.2(1) and G2428.2(2) shall not have elbows in the *venting system*. Single-*appliance* venting configurations with lateral lengths include two 90-degree (1.57 rad) elbows. For each additional elbow up to and including 45 degrees (0.79 rad), the maximum capacity listed in the venting tables shall be reduced by 5 percent. For each additional elbow greater than 45 degrees (0.79 rad) up to and including 90 degrees (1.57 rad), the maximum capacity listed in the venting tables shall be reduced by 10 percent. Where multiple *offsets* occur in a vent, the total lateral length of all *offsets* combined shall not exceed that specified in Tables G2428.2(1) and G2428.2(2).

G2428.2.4 (504.2.4) Zero lateral. Zero (0) lateral (L) shall apply only to a straight vertical vent attached to a top outlet *draft hood* or *flue collar*.

G2428.2.5 (504.2.5) High altitude installations. Sea level input ratings shall be used when determining maximum capacity for high altitude installation. Actual input, derated for altitude, shall be used for determining minimum capacity for high altitude installation.

G2428.2.6 (504.2.6) Multiple input rate appliances. For *appliances* with more than one input rate, the minimum vent capacity (FAN Min) determined from the tables shall be less than the lowest *appliance* input rating, and the maximum vent capacity (FAN Max/NAT Max) determined from the tables shall be greater than the highest *appliance* rating input.

G2428.2.7 (504.2.7) Liner system sizing and connections. Listed corrugated metallic *chimney* liner systems in *masonry chimneys* shall be sized by using Table G2428.2(1) or G2428.2(2) for Type B vents with the maximum capacity reduced by 20 percent (0.80 × maximum capacity) and the minimum capacity as shown in Table G2428.2(1) or G2428.2(2). Corrugated metallic liner systems installed with bends or offsets shall have their maximum capacity further reduced in accordance with Section G2428.2.3. The 20-percent reduction for corrugated metallic *chimney* liner systems includes an allowance for one long-radius 90-degree (1.57 rad) turn at the bottom of the liner.

Connections between *chimney* liners and listed double-wall connectors shall be made with listed adapters designed for such purpose.

G2428.2.8 (504.2.8) Vent area and diameter. Where the vertical vent has a larger diameter than the *vent connector*, the vertical vent diameter shall be used to determine the minimum vent capacity, and the connector diameter shall be used to determine the maximum vent capacity. The flow area of the vertical vent shall not exceed seven times the flow area of the listed *appliance* categorized vent area, *flue collar* area, or *draft hood* outlet area unless designed in accordance with *approved* engineering methods.

G2428.2.9 (504.2.9) Chimney and vent locations. Tables G2428.2(1) and G2428.2(2) shall be used only for chimneys and vents not exposed to the outdoors below the roof line. A Type B vent or listed chimney lining system passing through an unused masonry chimney flue shall not be considered to be exposed to the outdoors. Where vents extend outdoors above the roof more than 5 feet (1524 mm) higher than required by Figure G2427.6.3 and where vents terminate in accordance with Section G2427.6.3, Item 2, the outdoor portion of the vent shall be enclosed as required by this section for vents not considered to be exposed to the outdoors or such venting system shall be engineered. A Type B vent shall not be considered to be exposed to the outdoors where it passes through an unventilated enclosure or chase insulated to a value of not less than R8.

G2428.2.10 (504.2.10) Corrugated vent connector size. Corrugated *vent connectors* shall be not smaller than the listed *appliance* categorized *vent* diameter, *flue collar* diameter, or *draft hood* outlet diameter.

G2428.2.11 (504.2.11) Vent connector size limitation. *Vent connectors* shall not be increased in size more than two sizes greater than the listed *appliance* categorized vent diameter, *flue collar* diameter or *draft hood* outlet diameter.

G2428.2.12 (504.2.12) Component commingling. In a single run of vent or *vent connector*, different diameters and types of vent and connector components shall be permitted to be used, provided that all such sizes and types are permitted by the tables.

G2428.2.13 (504.2.13) Draft hood conversion accessories. *Draft hood* conversion accessories for use with *masonry chimneys* venting listed Category I fan-assisted *appliances* shall be listed and installed in accordance with the manufacturer's installation instructions for such listed accessories.

G2428.2.14 (504.2.14) Table interpolation. Interpolation shall be permitted in calculating capacities for vent dimensions that fall between the table entries (see Example 3, Appendix B).

G2428.2.15 (504.2.15) Extrapolation prohibited. Extrapolation beyond the table entries shall not be permitted.

G2428.2.16 (504.2.16) Engineering calculations. For *vent* heights less than 6 feet (1829 mm) and greater than shown in the tables, engineering methods shall be used to calculate *vent* capacities.

G2428.2.17 (504.2.17) Height entries. Where the actual height of a vent falls between entries in the height column of the applicable table in Tables G2428.2(1) and G2428.2(2), either interpolation shall be used or the lower appliance input rating shown in the table entries shall be used for FAN MAX and NAT MAX column values and the higher appliance input rating shall be used for the FAN MIN column values.

G2428.3 (504.3) Application of multiple appliance vent Tables G2428.3(1) through G2428.3(4). The application of Tables G2428.3(1) through G2428.3(4) shall be subject to the requirements of Sections G2428.3.1 through G2428.3.23.

G2428.3.1 (504.3.1) Vent obstructions. These venting tables shall not be used where obstructions, as described in Section G2427.15, are installed in the venting system. The installation of vents serving listed *appliances* with vent dampers shall be in accordance with the *appliance* manufacturer's instructions or in accordance with the following:

1. The maximum capacity of the *vent connector* shall be determined using the NAT Max column.
2. The maximum capacity of the vertical vent or *chimney* shall be determined using the FAN+NAT column when the second *appliance* is a fan-assisted *appliance*, or the NAT+NAT column when the second *appliance* is equipped with a *draft hood*.
3. The minimum capacity shall be determined as if the *appliance* were a fan-assisted *appliance*.
   - 3.1. The minimum capacity of the *vent connector* shall be determined using the FAN Min column.
   - 3.2. The FAN+FAN column shall be used when the second *appliance* is a fan-assisted *appliance*, and the FAN+NAT column shall be used when the second *appliance* is equipped with a *draft hood*, to determine whether the vertical vent or *chimney* configuration is not permitted (NA). Where the vent configuration is NA, the vent configuration shall not be permitted and an alternative venting configuration shall be utilized.

G2428.3.2 (504.3.2) Connector length limit. The *vent connector* shall be routed to the vent utilizing the shortest possible route. Except as provided in Section G2428.3.3, the maximum *vent connector* horizontal length shall be 1.5 feet (457 mm) for each inch (18 mm per mm) of connector diameter as shown in Table G2428.3.2.

TABLE G2428.3.2 (504.3.2)
MAXIMUM VENT CONNECTOR LENGTH

CONNECTOR DIAMETER	CONNECTOR HORIZONTAL
Maximum (inches)	Length (feet)
3	4.5
4	6
5	7.5
6	9
7	10.5
8	12
9	13.5

For SI: 1 inch = 25.4 mm, 1 foot = 304.8 mm.

G2428.3.3 (504.3.3) Connectors with longer lengths. Connectors with longer horizontal lengths than those listed in Section G2428.3.2 are permitted under the following conditions:

1. The maximum capacity (FAN Max or NAT Max) of the *vent connector* shall be reduced 10 percent for each additional multiple of the length listed above. For example, the maximum length listed above for a 4-inch (102 mm) connector is 6 feet (1829 mm). With a connector length greater than 6 feet (1829 mm), but not exceeding 12 feet (3658 mm), the maximum capacity must be reduced by 10 percent (0.90 × maximum vent *connector* capacity). With a connector length greater than 12 feet (3658 mm), but not exceeding 18 feet (5486 mm), the maximum capacity must be reduced by 20 percent (0.80 × maximum vent capacity).
2. For a connector serving a fan-assisted *appliance*, the minimum capacity (FAN Min) of the connector shall be determined by referring to the corresponding single *appliance* table. For Type B double-wall connectors, Table G2428.2(1) shall be used. For single-wall connectors, Table G2428.2(2) shall be used. The height (H) and lateral (L) shall be measured according to the procedures for a single *appliance* vent, as if the other *appliances* were not present.

G2428.3.4 (504.3.4) Vent connector manifold. Where the *vent connectors* are combined prior to entering the vertical portion of the common vent to form a common vent manifold, the size of the common vent manifold and the common vent shall be determined by applying a 10-percent reduction (0.90 × maximum common vent capacity) to the common vent capacity part of the common vent tables. The length of the common *vent connector* manifold (L_M) shall not exceed $1^1/_2$ feet for each inch (18 mm per mm) of common *vent connector* manifold diameter (D) (see Appendix B Figure B-11).

G2428.3.5 (504.3.5) Common vertical vent offset. Where the common vertical vent is *offset*, the maximum capacity of the common vent shall be reduced in accordance with Section G2428.3.6. The horizontal length of the common vent *offset* (L_o) shall not exceed $1^1/_2$ feet for each inch (18 mm per mm) of common vent diameter (D). Where multiple *offsets* occur in a common vent, the total horizontal length of all *offsets* combined shall not exceed $1^1/_2$ feet for each inch (18 mm/mm per) of the common vent diameter (D).

G2428.3.6 (504.3.6) Elbows in vents. For each elbow up to and including 45 degrees (0.79 rad) in the common vent, the maximum common vent capacity listed in the venting tables shall be reduced by 5 percent. For each elbow greater than 45 degrees (0.79 rad) up to and including 90 degrees (1.57 rad), the maximum common vent capacity listed in the venting tables shall be reduced by 10 percent.

G2428.3.7 (504.3.7) Elbows in connectors. The *vent connector* capacities listed in the common vent sizing tables include allowance for two 90-degree (1.57 rad) elbows. For each additional elbow up to and including 45 degrees (0.79 rad), the maximum *vent connector* capacity listed in the venting tables shall be reduced by 5 percent. For each elbow greater than 45 degrees (0.79 rad) up to and including 90 degrees (1.57 rad), the maximum *vent connector* capacity listed in the venting tables shall be reduced by 10 percent.

G2428.3.8 (504.3.8) Common vent minimum size. The cross-sectional area of the common vent shall be equal to or greater than the cross-sectional area of the largest connector.

G2428.3.9 (504.3.9) Common vent fittings. At the point where tee or wye fittings connect to a common vent, the opening size of the fitting shall be equal to the size of the common vent. Such fittings shall not be prohibited from having reduced-size openings at the point of connection of *appliance vent connectors*.

G2428.3.9.1 (504.3.9.1) Tee and wye fittings. Tee and wye fittings connected to a common gas vent shall be considered as part of the common gas vent and shall be constructed of materials consistent with that of the common gas vent.

G2428.3.10 (504.3.10) High altitude installations. Sea-level input ratings shall be used when determining maximum capacity for high altitude installation. Actual input, derated for altitude, shall be used for determining minimum capacity for high altitude installation.

G2428.3.11 (504.3.11) Connector rise measurement. Connector rise (*R*) for each *appliance connector* shall be measured from the *draft hood* outlet or *flue collar* to the centerline where the vent gas streams come together.

G2428.3.12 (504.3.12) Vent height measurement. For multiple *appliances* all located on one floor, available total height (H) shall be measured from the highest *draft hood* outlet or *flue collar* up to the level of the outlet of the common vent.

G2428.3.13 (504.3.17) Vertical vent maximum size. Where two or more *appliances* are connected to a vertical vent or *chimney*, the flow area of the largest section of vertical vent or *chimney* shall not exceed seven times the smallest listed *appliance* categorized vent areas, *flue collar* area, or *draft hood* outlet area unless designed in accordance with *approved* engineering methods.

G2428.3.14 (504.3.18) Multiple input rate appliances. For *appliances* with more than one input rate, the minimum *vent connector* capacity (FAN Min) determined from the tables shall be less than the lowest *appliance* input rating, and the maximum *vent connector* capacity (FAN Max or NAT Max) determined from the tables shall be greater than the highest *appliance* input rating.

G2428.3.15 (504.3.19) Liner system sizing and connections. Listed, corrugated metallic *chimney* liner systems in *masonry chimneys* shall be sized by using Table G2428.3(1) or G2428.3(2) for Type B vents, with the maximum capacity reduced by 20 percent (0.80 × maximum capacity) and the minimum capacity as shown in Table G2428.3(1) or G2428.3(2). Corrugated metallic liner systems installed with bends or offsets shall have their maximum capacity further reduced in accordance with Sections G2428.3.5 and G2428.3.6. The 20-percent reduction for corrugated metallic *chimney* liner systems includes an allowance for one long-radius 90-degree (1.57 rad) turn at the bottom of the liner. Where double-wall connectors are required, tee and wye fittings used to connect to the common vent *chimney* liner shall be listed double-wall fittings. Connections between *chimney* liners and listed double-wall fittings shall be made with listed adapter fittings designed for such purpose.

G2428.3.16 (504.3.20) Chimney and vent location. Tables G2428.3(1), G2428.3(2), G2428.3(3) and G2428.3(4) shall be used only for chimneys and vents not exposed to the outdoors below the roof line. A Type B vent or listed chimney lining system passing through an unused masonry chimney flue shall not be considered to be exposed to the outdoors. Where vents extend outdoors above the roof more than 5 feet (1524 mm) higher than required by Figure G2427.6.3 and where vents terminate in accordance with Section G2427.6.3, Item 2, the outdoor portion of the vent shall be enclosed as required by this section for vents not considered to be exposed to the outdoors or such venting system shall be engineered. A Type B vent shall not be considered to be exposed to the outdoors where it passes through an unventilated enclosure or chase insulated to a value of not less than R8.

G2428.3.17 (504.3.21) Connector maximum and minimum size. *Vent connectors* shall not be increased in size more than two sizes greater than the listed *appliance* categorized vent diameter, *flue collar* diameter, or *draft hood* outlet diameter. *Vent connectors* for draft-hood-equipped *appliances* shall not be smaller than the *draft hood* outlet diameter. Where a *vent connector* size(s) determined from the tables for a fan-assisted *appliance(s)* is smaller than the *flue collar* diameter, the use of the smaller size(s) shall be permitted provided that the installation complies with all of the following conditions:

1. *Vent connectors* for fan-assisted *appliance flue collars* 12 inches (305 mm) in diameter or smaller are not reduced by more than one table size [e.g., 12 inches to 10 inches (305 mm to 254 mm) is a one-size reduction] and those larger than 12 inches (305

mm) in diameter are not reduced more than two table sizes [e.g., 24 inches to 20 inches (610 mm to 508 mm) is a two-size reduction].

2. The fan-assisted *appliance(s)* is common vented with a draft-hood-equipped *appliance(s)*.

3. The vent *connector* has a smooth interior wall.

G2428.3.18 (504.3.22) Component commingling. All combinations of pipe sizes, single-wall, and double-wall metal pipe shall be allowed within any connector run(s) or within the common vent, provided all of the appropriate tables permit all of the desired sizes and types of pipe, as if they were used for the entire length of the subject connector or vent. Where single-wall and Type B double-wall metal pipes are used for *vent connectors* within the same venting system, the common vent must be sized using Table G2428.3(2) or G2428.3(4), as appropriate.

G2428.3.19 (504.3.23) Draft hood conversion accessories. *Draft hood* conversion accessories for use with *masonry chimneys* venting listed Category I fan-assisted *appliances* shall be listed and installed in accordance with the manufacturer's installation instructions for such listed accessories.

G2428.3.20 (504.3.24) Multiple sizes permitted. Where a table permits more than one diameter of pipe to be used for a connector or vent, all the permitted sizes shall be permitted to be used.

G2428.3.21 (504.3.25) Table interpolation. Interpolation shall be permitted in calculating capacities for vent dimensions that fall between table entries. (See Example 3, Appendix B.)

G2428.3.22 (504.3.26) Extrapolation prohibited. Extrapolation beyond the table entries shall not be permitted.

G2428.3.23 (504.3.27) Engineering calculations. For vent heights less than 6 feet (1829 mm) and greater than shown in the tables, engineering methods shall be used to calculate vent capacities.

G2428.3.24 (504.3.28) Height entries. Where the actual height of a vent falls between entries in the height column of the applicable table in Tables G2428.3(1) through G2428.3(4), either interpolation shall be used or the lower appliance input rating shown in the table shall be used for FAN MAX and NAT MAX column values and the higher appliance input rating shall be used for the FAN MIN column values.

SECTION G2429 (505) DIRECT-VENT, INTEGRAL VENT, MECHANICAL VENT AND VENTILATION/EXHAUST HOOD VENTING

G2429.1 (505.1) General. The installation of direct-vent and integral vent *appliances* shall be in accordance with Section G2427. Mechanical *venting systems* shall be designed and installed in accordance with Section G2427.

SECTION G2430 (506) FACTORY-BUILT CHIMNEYS

G2430.1 (506.1) Listing. Factory-built *chimneys* for building heating *appliances* producing *flue gases* having a temperature not greater than 1,000°F (538°C), measured at the entrance to the *chimney*, shall be listed and labeled in accordance with UL 103 and shall be installed and terminated in accordance with the manufacturer's installation instructions.

G2430.2 (506.2) Support. Where factory-built *chimneys* are supported by structural members, such as joists and rafters, such members shall be designed to support the additional load.

SECTION G2431 (601) GENERAL

G2431.1 (601.1) Scope. Sections G2432 through G2453 shall govern the approval, design, installation, construction, maintenance, *alteration* and repair of the *appliances* and *equipment* specifically identified herein.

SECTION G2432 (602) DECORATIVE APPLIANCES FOR INSTALLATION IN FIREPLACES

G2432.1 (602.1) General. Decorative *appliances* for installation in *approved* solid fuel burning *fireplaces* shall be tested in accordance with ANSI Z21.60 and shall be installed in accordance with the manufacturer's installation instructions. Manually lighted natural gas decorative *appliances* shall be tested in accordance with ANSI Z21.84.

G2432.2 (602.2) Flame safeguard device. Decorative *appliances* for installation in *approved* solid fuel-burning *fireplaces*, with the exception of those tested in accordance with ANSI Z21.84, shall utilize a direct ignition device, an ignitor or a *pilot* flame to ignite the fuel at the *main burner*, and shall be equipped with a *flame safeguard* device. The *flame safeguard* device shall automatically shut off the fuel supply to a *main burner* or group of *burners* when the means of ignition of such *burners* becomes inoperative.

G2432.3 (602.3) Prohibited installations. Decorative *appliances* for installation in *fireplaces* shall not be installed where prohibited by Section G2406.2.

SECTION G2433 (603) LOG LIGHTERS

G2433.1 (603.1) General. Log lighters shall be tested in accordance with CSA 8 and shall be installed in accordance with the manufacturer's installation instructions.

TABLE G2428.2(1) [504.2(1)] TYPE B DOUBLE-WALL GAS VENT

Number of Appliances	Single
Appliance Type	Category I
Appliance Vent Connection	Connected directly to vent

HEIGHT (H) (feet)	LATERAL (L) (feet)	VENT DIAMETER—(D) inches																				
		3			4			5			6			7			8			9		
		APPLIANCE INPUT RATING IN THOUSANDS OF BTU/H																				
		FAN		NAT	FAN		NAT	FAN		NAT	FAN		NAT	FAN		NAT	FAN		NAT	FAN		NAT
		Min	Max	Max	Min	Max	Max	Min	Max	Max	Min	Max	Max	Min	Max	Max	Min	Max	Max	Min	Max	Max
6	0	0	78	46	0	152	86	0	251	141	0	375	205	0	524	285	0	698	370	0	897	470
	2	13	51	36	18	97	67	27	157	105	32	232	157	44	321	217	53	425	285	63	543	370
	4	21	49	34	30	94	64	39	153	103	50	227	153	66	316	211	79	419	279	93	536	362
	6	25	46	32	36	91	61	47	149	100	59	223	149	78	310	205	93	413	273	110	530	354
8	0	0	84	50	0	165	94	0	276	155	0	415	235	0	583	320	0	780	415	0	1,006	537
	2	12	57	40	16	109	75	25	178	120	28	263	180	42	365	247	50	483	322	60	619	418
	5	23	53	38	32	103	71	42	171	115	53	255	173	70	356	237	83	473	313	99	607	407
	8	28	49	35	39	98	66	51	164	109	64	247	165	84	347	227	99	463	303	117	596	396
10	0	0	88	53	0	175	100	0	295	166	0	447	255	0	631	345	0	847	450	0	1,096	585
	2	12	61	42	17	118	81	23	194	129	26	289	195	40	402	273	48	533	355	57	684	457
	5	23	57	40	32	113	77	41	187	124	52	280	188	68	392	263	81	522	346	95	671	446
	10	30	51	36	41	104	70	54	176	115	67	267	175	88	376	245	104	504	330	122	651	427
15	0	0	94	58	0	191	112	0	327	187	0	502	285	0	716	390	0	970	525	0	1,263	682
	2	11	69	48	15	136	93	20	226	150	22	339	225	38	475	316	45	633	414	53	815	544
	5	22	65	45	30	130	87	39	219	142	49	330	217	64	463	300	76	620	403	90	800	529
	10	29	59	41	40	121	82	51	206	135	64	315	208	84	445	288	99	600	386	116	777	507
	15	35	53	37	48	112	76	61	195	128	76	301	198	98	429	275	115	580	373	134	755	491
20	0	0	97	61	0	202	119	0	349	202	0	540	307	0	776	430	0	1,057	575	0	1,384	752
	2	10	75	51	14	149	100	18	250	166	20	377	249	33	531	346	41	711	470	50	917	612
	5	21	71	48	29	143	96	38	242	160	47	367	241	62	519	337	73	697	460	86	902	599
	10	28	64	44	38	133	89	50	229	150	62	351	228	81	499	321	95	675	443	112	877	576
	15	34	58	40	46	124	84	59	217	142	73	337	217	94	481	308	111	654	427	129	853	557
	20	48	52	35	55	116	78	69	206	134	84	322	206	107	464	295	125	634	410	145	830	537

(continued)

TABLE G2428.2(1) [504.2(1)]—continued
TYPE B DOUBLE-WALL GAS VENT

Number of Appliances	Single
Appliance Type	Category I
Appliance Vent Connection	Connected directly to vent

HEIGHT (H) (feet)	LATERAL (L) (feet)	VENT DIAMETER—(D) inches																				
		3			4			5			6			7			8			9		
		APPLIANCE INPUT RATING IN THOUSANDS OF BTU/H																				
		FAN		NAT	FAN		NAT	FAN		NAT	FAN		NAT	FAN		NAT	FAN		NAT	FAN		NAT
		Min	Max	Max	Min	Max	Max	Min	Max	Max	Min	Max	Max	Min	Max	Max	Min	Max	Max	Min	Max	Max
30	0	0	100	64	0	213	128	0	374	220	0	587	336	0	853	475	0	1,173	650	0	1,548	855
	2	9	81	56	13	166	112	14	283	185	18	432	280	27	613	394	33	826	535	42	1,072	700
	5	21	77	54	28	160	108	36	275	176	45	421	273	58	600	385	69	811	524	82	1,055	688
	10	27	70	50	37	150	102	48	262	171	59	405	261	77	580	371	91	788	507	107	1,028	668
	15	33	64	NA	44	141	96	57	249	163	70	389	249	90	560	357	105	765	490	124	1,002	648
	20	56	58	NA	53	132	90	66	237	154	80	374	237	102	542	343	119	743	473	139	977	628
	30	NA	NA	NA	73	113	NA	88	214	NA	104	346	219	131	507	321	149	702	444	171	929	594
50	0	0	101	67	0	216	134	0	397	232	0	633	363	0	932	518	0	1,297	708	0	1,730	952
	2	8	86	61	11	183	122	14	320	206	15	497	314	22	715	445	26	975	615	33	1,276	813
	5	20	82	NA	27	177	119	35	312	200	43	487	308	55	702	438	65	960	605	77	1,259	798
	10	26	76	NA	35	168	114	45	299	190	56	471	298	73	681	426	86	935	589	101	1,230	773
	15	59	70	NA	42	158	NA	54	287	180	66	455	288	85	662	413	100	911	572	117	1,203	747
	20	NA	NA	NA	50	149	NA	63	275	169	76	440	278	97	642	401	113	888	556	131	1,176	722
	30	NA	NA	NA	69	131	NA	84	250	NA	99	410	259	123	605	376	141	844	522	161	1,125	670

For SI: 1 inch = 25.4 mm, 1 foot = 304.8 mm, 1 British thermal unit per hour = 0.2931 W.

TABLE G2428.2(2) [504.2(2)]
TYPE B DOUBLE-WALL GAS VENT

Number of Appliances	Single
Appliance Type	Category I
Appliance Vent Connection	Single-wall metal connector

HEIGHT (H) (feet)	LATERAL (L) (feet)	VENT DIAMETER—(D) inches																										
		3			4			5			6			7			8			9			10			12		
		APPLIANCE INPUT RATING IN THOUSANDS OF BTU/H																										
		FAN		NAT	FAN		NAT	FAN		NAT	FAN		NAT	FAN		NAT	FAN		NAT	FAN		NAT	FAN		NAT	FAN		NAT
		Min	Max	Max	Min	Max	Max	Min	Max	Max	Min	Max	Max	Min	Max	Max	Min	Max	Max	Min	Max	Max	Min	Max	Max	Min	Max	Max
6	0	38	77	45	59	151	85	85	249	140	126	373	204	165	522	284	211	695	369	267	894	469	371	1,118	569	537	1,639	849
	2	39	51	36	60	96	66	85	156	104	123	231	156	159	320	213	201	423	284	251	541	368	347	673	453	498	979	648
	4	NA	NA	33	74	92	63	102	152	102	146	225	152	187	313	208	237	416	277	295	533	360	409	664	443	584	971	638
	6	NA	NA	31	83	89	60	114	147	99	163	220	148	207	307	203	263	409	271	327	526	352	449	656	433	638	962	627
8	0	37	83	50	58	164	93	83	273	154	123	412	234	161	580	319	206	777	414	258	1,002	536	360	1,257	658	521	1,852	967
	2	39	56	39	59	108	75	83	176	119	121	261	179	155	363	246	197	482	321	246	617	417	339	768	513	486	1,120	743
	5	NA	NA	37	77	102	69	107	168	114	151	252	171	193	352	235	245	470	311	305	604	404	418	754	500	598	1,104	730
	8	NA	NA	33	90	95	64	122	161	107	175	243	163	223	342	225	280	458	300	344	591	392	470	740	486	665	1,089	715
10	0	37	87	53	57	174	99	82	293	165	120	444	254	158	628	344	202	844	449	253	1,093	584	351	1,373	718	507	2,031	1,057
	2	39	61	41	59	117	80	82	193	128	119	287	194	153	400	272	193	531	354	242	681	456	332	849	559	475	1,242	848
	5	52	56	39	76	111	76	105	185	122	148	277	186	190	388	261	241	518	344	299	667	443	409	834	544	584	1,224	825
	10	NA	NA	34	97	100	68	132	171	112	188	261	171	237	369	241	296	497	325	363	643	423	492	808	520	688	1,194	788
15	0	36	93	57	56	190	111	80	325	186	116	499	283	153	713	388	195	966	523	244	1,259	681	336	1,591	838	488	2,374	1,237
	2	38	69	47	57	136	93	80	225	149	115	337	224	148	473	314	187	631	413	232	812	543	319	1,015	673	457	1,491	983
	5	51	63	44	75	128	86	102	216	140	144	326	217	182	459	298	231	616	400	287	795	526	392	997	657	562	1,469	963
	10	NA	NA	39	95	116	79	128	201	131	182	308	203	228	438	284	284	592	381	349	768	501	470	966	628	664	1,433	928
	15	NA	NA	NA	NA	NA	72	158	186	124	220	290	192	272	418	269	334	568	367	404	742	484	540	937	601	750	1,399	894
20	0	35	96	60	54	200	118	78	346	201	114	537	306	149	772	428	190	1,053	573	238	1,379	750	326	1,751	927	473	2,631	1,346
	2	37	74	50	56	148	99	78	248	165	113	375	248	144	528	344	182	708	468	227	914	611	309	1,146	754	443	1,689	1,098
	5	50	68	47	73	140	94	100	239	158	141	363	239	178	514	334	224	692	457	279	896	596	381	1,126	734	547	1,665	1,074
	10	NA	NA	41	93	129	86	125	223	146	177	344	224	222	491	316	277	666	437	339	866	570	457	1,092	702	646	1,626	1,037
	15	NA	NA	NA	NA	NA	80	155	208	136	216	325	210	264	469	301	325	640	419	393	838	549	526	1,060	677	730	1,587	1,005
	20	NA	NA	NA	NA	NA	NA	186	192	126	254	306	196	309	448	285	374	616	400	448	810	526	592	1,028	651	808	1,550	973

(continued)

TABLE G2428.2(2) [504.2(2)]—continued
TYPE B DOUBLE-WALL GAS VENT

Number of Appliances	Single
Appliance Type	Category I
Appliance Vent Connection	Single-wall metal connector

HEIGHT (H) (feet)	LATERAL (L) (feet)	VENT DIAMETER—(D) inches																										
		3			4			5			6			7			8			9			10			12		
		APPLIANCE INPUT RATING IN THOUSANDS OF BTU/H																										
		FAN		NAT	FAN		NAT	FAN		NAT	FAN		NAT	FAN		NAT	FAN		NAT	FAN		NAT	FAN		NAT	FAN		NAT
		Min	Max	Max	Min	Max	Max	Min	Max	Max	Min	Max	Max	Min	Max	Max	Min	Max	Max	Min	Max	Max	Min	Max	Max	Min	Max	Max
30	0	34	99	63	53	211	127	76	372	219	110	584	334	144	849	472	184	1,168	647	229	1,542	852	312	1,971	1,056	454	2,996	1,545
	2	37	80	56	55	164	111	76	281	183	109	429	279	139	610	392	175	823	533	219	1,069	698	296	1,346	863	424	1,999	1,308
	5	49	74	52	72	157	106	98	271	173	136	417	271	171	595	382	215	806	521	269	1,049	684	366	1,324	846	524	1,971	1,283
	10	NA	NA	NA	91	144	98	122	255	168	171	397	257	213	570	367	265	777	501	327	1,017	662	440	1,287	821	620	1,927	1,234
	15	NA	NA	NA	115	131	NA	151	239	157	208	377	242	255	547	349	312	750	481	379	985	638	507	1,251	794	702	1,884	1,205
	20	NA	NA	NA	NA	NA	NA	181	223	NA	246	357	228	298	524	333	360	723	461	433	955	615	570	1,216	768	780	1,841	1,166
	30	NA	NA	NA	NA	NA	NA	NA	NA	NA	NA	NA	NA	389	477	305	461	670	426	541	895	574	704	1,147	720	937	1,759	1,101
50	0	33	99	66	51	213	133	73	394	230	105	629	361	138	928	515	176	1,292	704	220	1,724	948	295	2,223	1,189	428	3,432	1,818
	2	36	84	61	53	181	121	73	318	205	104	495	312	133	712	443	168	971	613	209	1,273	311	280	1,615	1,007	401	2,426	1,509
	5	48	80	NA	70	174	117	94	308	198	131	482	305	164	696	435	204	953	602	257	1,252	795	347	1,591	991	496	2,396	1,490
	10	NA	NA	NA	89	160	NA	118	292	186	162	461	292	203	671	420	253	923	583	313	1,217	765	418	1,551	963	589	2,347	1,455
	15	NA	NA	NA	112	148	NA	145	275	174	199	441	280	244	646	405	299	894	562	363	1,183	736	481	1,512	934	668	2,299	1,421
	20	NA	NA	NA	NA	NA	NA	176	257	NA	236	420	267	285	622	389	345	866	543	415	1,150	708	544	1,473	906	741	2,251	1,387
	30	NA	NA	NA	NA	NA	NA	NA	NA	NA	315	376	NA	373	573	NA	442	809	502	521	1,086	549	674	1,399	848	892	2,159	1,318

For SI:1 inch = 25.4 mm, 1 foot = 304.8 mm, 1 British thermal unit per hour = 0.2931 W.

TABLE G2428.3(1) [504.3(1)]
TYPE B DOUBLE-WALL VENT

Number of Appliances	Two or more
Appliances Type	Category I
Appliances Vent Connection	Type B double-wall connector

VENT CONNECTOR CAPACITY

VENT HEIGHT (H) (feet)	CONNECTOR RISE (R) (feet)	TYPE B DOUBLE-WALL VENT AND CONNECTOR DIAMETER—(*D*) inches																							
		3			4			5			6			7			8			9			10		
		APPLIANCE INPUT RATING LIMITS IN THOUSANDS OF BTU/H																							
		FAN		NAT	FAN		NAT	FAN		NAT	FAN		NAT	FAN		NAT	FAN		NAT	FAN		NAT	FAN		NAT
		Min	Max	Max	Min	Max	Max	Min	Max	Max	Min	Max	Max	Min	Max	Max	Min	Max	Max	Min	Max	Max	Min	Max	Max
6	1	22	37	26	35	66	46	46	106	72	58	164	104	77	225	142	92	296	185	109	376	237	128	466	289
	2	23	41	31	37	75	55	48	121	86	60	183	124	79	253	168	95	333	220	112	424	282	131	526	345
	3	24	44	35	38	81	62	49	132	96	62	199	139	82	275	189	97	363	248	114	463	317	134	575	386
8	1	22	40	27	35	72	48	49	114	76	64	176	109	84	243	148	100	320	194	118	408	248	138	507	303
	2	23	44	32	36	80	57	51	128	90	66	195	129	86	269	175	103	356	230	121	454	294	141	564	358
	3	24	47	36	37	87	64	53	139	101	67	210	145	88	290	198	105	384	258	123	492	330	143	612	402
10	1	22	43	28	34	78	50	49	123	78	65	189	113	89	257	154	106	341	200	125	436	257	146	542	314
	2	23	47	33	36	86	59	51	136	93	67	206	134	91	282	182	109	374	238	128	479	305	149	596	372
	3	24	50	37	37	92	67	52	146	104	69	220	150	94	303	205	111	402	268	131	515	342	152	642	417
15	1	21	50	30	33	89	53	47	142	83	64	220	120	88	298	163	110	389	214	134	493	273	162	609	333
	2	22	53	35	35	96	63	49	153	99	66	235	142	91	320	193	112	419	253	137	532	323	165	658	394
	3	24	55	40	36	102	71	51	163	111	68	248	160	93	339	218	115	445	286	140	565	365	167	700	444
20	1	21	54	31	33	99	56	46	157	87	62	246	125	86	334	171	107	436	224	131	552	285	158	681	347
	2	22	57	37	34	105	66	48	167	104	64	259	149	89	354	202	110	463	265	134	587	339	161	725	414
	3	23	60	42	35	110	74	50	176	116	66	271	168	91	371	228	113	486	300	137	618	383	164	764	466
30	1	20	62	33	31	113	59	45	181	93	60	288	134	83	391	182	103	512	238	125	649	305	151	802	372
	2	21	64	39	33	118	70	47	190	110	62	299	158	85	408	215	105	535	282	129	679	360	155	840	439
	3	22	66	44	34	123	79	48	198	124	64	309	178	88	423	242	108	555	317	132	706	405	158	874	494

COMMON VENT CAPACITY

VENT HEIGHT (H) (feet)	TYPE B DOUBLE-WALL COMMON VENT DIAMETER (*D*)—inches																				
	4			5			6			7			8			9			10		
	COMBINED APPLIANCE INPUT RATING IN THOUSANDS OF BTU/H																				
	FAN +FAN	FAN +NAT	NAT +NAT	FAN +FAN	FAN +NAT	NAT +NAT	FAN +FAN	FAN +NAT	NAT +NAT	FAN +FAN	FAN +NAT	NAT +NAT	FAN +FAN	FAN +NAT	NAT +NAT	FAN +FAN	FAN +NAT	NAT +NAT	FAN +FAN	FAN +NAT	NAT +NAT
6	92	81	65	140	116	103	204	161	147	309	248	200	404	314	260	547	434	335	672	520	410
8	101	90	73	155	129	114	224	178	163	339	275	223	444	348	290	602	480	378	740	577	465
10	110	97	79	169	141	124	243	194	178	367	299	242	477	377	315	649	522	405	800	627	495
15	125	112	91	195	164	144	283	228	206	427	352	280	556	444	365	753	612	465	924	733	565
20	136	123	102	215	183	160	314	255	229	475	394	310	621	499	405	842	688	523	1,035	826	640
30	152	138	118	244	210	185	361	297	266	547	459	360	720	585	470	979	808	605	1,209	975	740
50	167	153	134	279	244	214	421	353	310	641	547	423	854	706	550	1,164	977	705	1,451	1,188	860

For SI: 1 inch = 25.4 mm, 1 foot = 304.8 mm, 1 British thermal unit per hour = 0.2931 W.

TABLE G2428.3(2) [504.3(2)]
TYPE B DOUBLE-WALL VENT

Number of Appliances	Two or more
Appliances Type	Category I
Appliances Vent Connection	Type B double-wall connector

VENT CONNECTOR CAPACITY

VENT HEIGHT (H) (feet)	CONNECTOR RISE (R) (feet)	SINGLE-WALL METAL VENT CONNECTOR DIAMETER—(D) inches																							
		3			4			5			6			7			8			9			10		
		APPLIANCE INPUT RATING LIMITS IN THOUSANDS OF BTU/H																							
		FAN		NAT	FAN		NAT	FAN		NAT	FAN		NAT	FAN		NAT	FAN		NAT	FAN		NAT	FAN		NAT
		Min	Max	Max	Min	Max	Max	Min	Max	Max	Min	Max	Max	Min	Max	Max	Min	Max	Max	Min	Max	Max	Min	Max	Max
6	1	NA	NA	26	NA	NA	46	NA	NA	71	NA	NA	102	207	223	140	262	293	183	325	373	234	447	463	286
	2	NA	NA	31	NA	NA	55	NA	NA	85	168	182	123	215	251	167	271	331	219	334	422	281	458	524	344
	3	NA	NA	34	NA	NA	62	121	131	95	175	198	138	222	273	188	279	361	247	344	462	316	468	574	385
8	1	NA	NA	27	NA	NA	48	NA	NA	75	NA	NA	106	226	240	145	285	316	191	352	403	244	481	502	299
	2	NA	NA	32	NA	NA	57	125	126	89	184	193	127	234	266	173	293	353	228	360	450	292	492	560	355
	3	NA	NA	35	NA	NA	64	130	138	100	191	208	144	241	287	197	302	381	256	370	489	328	501	609	400
10	1	NA	NA	28	NA	NA	50	119	121	77	182	186	110	240	253	150	302	335	196	372	429	252	506	534	308
	2	NA	NA	33	84	85	59	124	134	91	189	203	132	248	278	183	311	369	235	381	473	302	517	589	368
	3	NA	NA	36	89	91	67	129	144	102	197	217	148	257	299	203	320	398	265	391	511	339	528	637	413
15	1	NA	NA	29	79	87	52	116	138	81	177	214	116	238	291	158	312	380	208	397	482	266	556	596	324
	2	NA	NA	34	83	94	62	121	150	97	185	230	138	246	314	189	321	411	248	407	522	317	568	646	387
	3	NA	NA	39	87	100	70	127	160	109	193	243	157	255	333	215	331	438	281	418	557	360	579	690	437
20	1	49	56	30	78	97	54	115	152	84	175	238	120	233	325	165	306	425	217	390	538	276	546	664	336
	2	52	59	36	82	103	64	120	163	101	182	252	144	243	346	197	317	453	259	400	574	331	558	709	403
	3	55	62	40	87	107	72	125	172	113	190	264	164	252	363	223	326	476	294	412	607	375	570	750	457
30	1	47	60	31	77	110	57	112	175	89	169	278	129	226	380	175	296	497	230	378	630	294	528	779	358
	2	51	62	37	81	115	67	117	185	106	177	290	152	236	397	208	307	521	274	389	662	349	541	819	425
	3	54	64	42	85	119	76	122	193	120	185	300	172	244	412	235	316	542	309	400	690	394	555	855	482

COMMON VENT CAPACITY

VENT HEIGHT (H) (feet)	TYPE B DOUBLE-WALL COMMON VENT DIAMETER—(D) inches																				
	4			5			6			7			8			9			10		
	COMBINED APPLIANCE INPUT RATING IN THOUSANDS OF BTU/H																				
	FAN +FAN	FAN +NAT	NAT +NAT	FAN +FAN	FAN +NAT	NAT +NAT	FAN +FAN	FAN +NAT	NAT +NAT	FAN +FAN	FAN +NAT	NAT +NAT	FAN +FAN	FAN +NAT	NAT +NAT	FAN +FAN	FAN +NAT	NAT +NAT	FAN +FAN	FAN +NAT	NAT +NAT
6	NA	78	64	NA	113	99	200	158	144	304	244	196	398	310	257	541	429	332	665	515	407
8	NA	87	71	NA	126	111	218	173	159	331	269	218	436	342	285	592	473	373	730	569	460
10	NA	94	76	163	137	120	237	189	174	357	292	236	467	369	309	638	512	398	787	617	487
15	121	108	88	189	159	140	275	221	200	416	343	274	544	434	357	738	599	456	905	718	553
20	131	118	98	208	177	156	305	247	223	463	383	302	606	487	395	824	673	512	1,013	808	626
30	145	132	113	236	202	180	350	286	257	533	446	349	703	570	459	958	790	593	1,183	952	723
50	159	145	128	268	233	208	406	337	296	622	529	410	833	686	535	1,139	954	689	1,418	1,157	838

For SI: 1 inch = 25.4 mm, 1 foot = 304.8 mm, 1 British thermal unit per hour = 0.2931 W.

Number of Appliances	Two or more
Appliances Type	Category I
Appliances Vent Connection	Type B double-wall connector

TABLE G2428.3(3) [504.3(3)] MASONRY CHIMNEY

VENT CONNECTOR CAPACITY

VENT HEIGHT (H) (feet)	CONNECTOR RISE (R) (feet)	TYPE B DOUBLE-WALL VENT CONNECTOR DIAMETER—(*D*) inches																							
		3			4			5			6			7			8			9			10		
		APPLIANCE INPUT RATING LIMITS IN THOUSANDS OF BTU/H																							
		FAN		NAT	FAN		NAT	FAN		NAT	FAN		NAT	FAN		NAT	FAN		NAT	FAN		NAT	FAN		NAT
		Min	Max	Max	Min	Max	Max	Min	Max	Max	Min	Max	Max	Min	Max	Max	Min	Max	Max	Min	Max	Max	Min	Max	Max
6	1	24	33	21	39	62	40	52	106	67	65	194	101	87	274	141	104	370	201	124	479	253	145	599	319
	2	26	43	28	41	79	52	53	133	85	67	230	124	89	324	173	107	436	232	127	562	300	148	694	378
	3	27	49	34	42	92	61	55	155	97	69	262	143	91	369	203	109	491	270	129	633	349	151	795	439
8	1	24	39	22	39	72	41	55	117	69	71	213	105	94	304	148	113	414	210	134	539	267	156	682	335
	2	26	47	29	40	87	53	57	140	86	73	246	127	97	350	179	116	473	240	137	615	311	160	776	394
	3	27	52	34	42	97	62	59	159	98	75	269	145	99	383	206	119	517	276	139	672	358	163	848	452
10	1	24	42	22	38	80	42	55	130	71	74	232	108	101	324	153	120	444	216	142	582	277	165	739	348
	2	26	50	29	40	93	54	57	153	87	76	261	129	103	366	184	123	498	247	145	652	321	168	825	407
	3	27	55	35	41	105	63	58	170	100	78	284	148	106	397	209	126	540	281	147	705	366	171	893	463
15	1	24	48	23	38	93	44	54	154	74	72	277	114	100	384	164	125	511	229	153	658	297	184	824	375
	2	25	55	31	39	105	55	56	174	89	74	299	134	103	419	192	128	558	260	156	718	339	187	900	432
	3	26	59	35	41	115	64	57	189	102	76	319	153	105	448	215	131	597	292	159	760	382	190	960	486
20	1	24	52	24	37	102	46	53	172	77	71	313	119	98	437	173	123	584	239	150	752	312	180	943	397
	2	25	58	31	39	114	56	55	190	91	73	335	138	101	467	199	126	625	270	153	805	354	184	1,011	452
	3	26	63	35	40	123	65	57	204	104	75	353	157	104	493	222	129	661	301	156	851	396	187	1,067	505

COMMON VENT CAPACITY

VENT HEIGHT (H) (feet)	MINIMUM INTERNAL AREA OF MASONRY CHIMNEY FLUE (square inches)																							
	12			19			28			38			50			63			78			113		
	COMBINED APPLIANCE INPUT RATING IN THOUSANDS OF BTU/H																							
	FAN +FAN	FAN +NAT	NAT +NAT	FAN +FAN	FAN +NAT	NAT +NAT	FAN +FAN	FAN +NAT	NAT +NAT	FAN +FAN	FAN +NAT	NAT +NAT	FAN +FAN	FAN +NAT	NAT +NAT	FAN +FAN	FAN +NAT	NAT +NAT	FAN +FAN	FAN +NAT	NAT +NAT	FAN +FAN	FAN +NAT	NAT +NAT
6	NA	74	25	NA	119	46	NA	178	71	NA	257	103	NA	351	143	NA	458	188	NA	582	246	1,041	853	NA
8	NA	80	28	NA	130	53	NA	193	82	NA	279	119	NA	384	163	NA	501	218	724	636	278	1,144	937	408
10	NA	84	31	NA	138	56	NA	207	90	NA	299	131	NA	409	177	606	538	236	776	686	302	1,226	1,010	454
15	NA	NA	36	NA	152	67	NA	233	106	NA	334	152	523	467	212	682	611	283	874	781	365	1,374	1,156	546
20	NA	NA	41	NA	NA	75	NA	250	122	NA	368	172	565	508	243	742	668	325	955	858	419	1,513	1,286	648
30	NA	NA	NA	NA	NA	NA	NA	270	137	NA	404	198	615	564	278	816	747	381	1,062	969	496	1,702	1,473	749
50	NA	NA	NA	NA	NA	NA	NA	NA	NA	NA	NA	NA	NA	620	328	879	831	461	1,165	1,089	606	1,905	1,692	922

For SI: 1 inch = 25.4 mm, 1 square inch = 645.16 mm^2, 1 foot = 304.8 mm, 1 British thermal unit per hour = 0.2931 W.

TABLE G2428.3(4) [504.3(4)] MASONRY CHIMNEY

Number of Appliances	Two or more
Appliances Type	Category I
Appliances Vent Connection	Single-wall connector

VENT CONNECTOR CAPACITY

VENT HEIGHT (H) (feet)	CONNECTOR RISE (R) (feet)	SINGLE-WALL METAL VENT CONNECTOR DIAMETER (D)—inches																							
		3			4			5			6			7			8			9			10		
		APPLIANCE INPUT RATING LIMITS IN THOUSANDS OF BTU/H																							
		FAN		NAT	FAN		NAT	FAN		NAT	FAN		NAT	FAN		NAT	FAN		NAT	FAN		NAT	FAN		NAT
		Min	Max	Max	Min	Max	Max	Min	Max	Max	Min	Max	Max	Min	Max	Max	Min	Max	Max	Min	Max	Max	Min	Max	Max
6	1	NA	NA	21	NA	NA	39	NA	NA	66	179	191	100	231	271	140	292	366	200	362	474	252	499	594	316
	2	NA	NA	28	NA	NA	52	NA	NA	84	186	227	123	239	321	172	301	432	231	373	557	299	509	696	376
	3	NA	NA	34	NA	NA	61	134	153	97	193	258	142	247	365	202	309	491	269	381	634	348	519	793	437
8	1	NA	NA	21	NA	NA	40	NA	NA	68	195	208	103	250	298	146	313	407	207	387	530	263	529	672	331
	2	NA	NA	28	NA	NA	52	137	139	85	202	240	125	258	343	177	323	465	238	397	607	309	540	766	391
	3	NA	NA	34	NA	NA	62	143	156	98	210	264	145	266	376	205	332	509	274	407	663	356	551	838	450
10	1	NA	NA	22	NA	NA	41	130	151	70	202	225	106	267	316	151	333	434	213	410	571	273	558	727	343
	2	NA	NA	29	NA	NA	53	136	150	86	210	255	128	276	358	181	343	489	244	420	640	317	569	813	403
	3	NA	NA	34	97	102	62	143	166	99	217	277	147	284	389	207	352	530	279	430	694	363	580	880	459
15	1	NA	NA	23	NA	NA	43	129	151	73	199	271	112	268	376	161	349	502	225	445	646	291	623	808	366
	2	NA	NA	30	92	103	54	135	170	88	207	295	132	277	411	189	359	548	256	456	706	334	634	884	424
	3	NA	NA	34	96	112	63	141	185	101	215	315	151	286	439	213	368	586	289	466	755	378	646	945	479
20	1	NA	NA	23	87	99	45	128	167	76	197	303	117	265	425	169	345	569	235	439	734	306	614	921	347
	2	NA	NA	30	91	111	55	134	185	90	205	325	136	274	455	195	355	610	266	450	787	348	627	986	443
	3	NA	NA	35	96	119	64	140	199	103	213	343	154	282	481	219	365	644	298	461	831	391	639	1,042	496

COMMON VENT CAPACITY

VENT HEIGHT (H) (feet)	MINIMUM INTERNAL AREA OF MASONRY CHIMNEY FLUE (square inches)																							
	12			19			28			38			50			63			78			113		
	COMBINED APPLIANCE INPUT RATING IN THOUSANDS OF BTU/H																							
	FAN +FAN	FAN +NAT	NAT +NAT	FAN +FAN	FAN +NAT	NAT +NAT	FAN +FAN	FAN +NAT	NAT +NAT	FAN +FAN	FAN +NAT	NAT +NAT	FAN +FAN	FAN +NAT	NAT +NAT	FAN +FAN	FAN +NAT	NAT +NAT	FAN +FAN	FAN +NAT	NAT +NAT	FAN +FAN	FAN +NAT	NAT +NAT
6	NA	NA	25	NA	118	45	NA	176	71	NA	255	102	NA	348	142	NA	455	187	NA	579	245	NA	846	NA
8	NA	NA	28	NA	128	52	NA	190	81	NA	276	118	NA	380	162	NA	497	217	NA	633	277	1,136	928	405
10	NA	NA	31	NA	136	56	NA	205	89	NA	295	129	NA	405	175	NA	532	234	171	680	300	1,216	1,000	450
15	NA	NA	36	NA	NA	66	NA	230	105	NA	335	150	NA	400	210	677	602	280	866	772	360	1,359	1,139	540
20	NA	NA	NA	NA	NA	74	NA	247	120	NA	362	170	NA	503	240	765	661	321	947	849	415	1,495	1,264	640
30	NA	NA	NA	NA	NA	NA	NA	NA	135	NA	398	195	NA	558	275	808	739	377	1,052	957	490	1,682	1,447	740
50	NA	NA	NA	NA	NA	NA	NA	NA	NA	NA	NA	NA	NA	612	325	NA	821	456	1,152	1,076	600	1,879	1,672	910

For SI: 1 inch = 25.4 mm, 1 square inch = 645.16 mm^2, 1 foot = 304.8 mm, 1 British thermal unit per hour = 0.2931 W.

SECTION G2434 (604) VENTED GAS FIREPLACES (DECORATIVE APPLIANCES)

G2434.1 (604.1) General. Vented gas *fireplaces* shall be tested in accordance with ANSI Z21.50, shall be installed in accordance with the manufacturer's installation instructions and shall be designed and equipped as specified in Section G2432.2.

G2434.2 (604.2) Access. Panels, grilles, and access doors that are required to be removed for normal servicing operations shall not be attached to the building.

SECTION G2435 (605) VENTED GAS FIREPLACE HEATERS

G2435.1 (605.1) General. Vented gas *fireplace* heaters shall be installed in accordance with the manufacturer's installation instructions, shall be tested in accordance with ANSI Z21.88 and shall be designed and equipped as specified in Section G2432.2.

SECTION G2436 (608) VENTED WALL FURNACES

G2436.1 (608.1) General. *Vented wall furnaces* shall be tested in accordance with ANSI Z21.86/CSA 2.32 and shall be installed in accordance with the manufacturer's installation instructions.

G2436.2 (608.2) Venting. *Vented wall furnaces* shall be vented in accordance with Section G2427.

G2436.3 (608.3) Location. *Vented wall furnaces* shall be located so as not to cause a fire hazard to walls, floors, combustible furnishings or doors. *Vented wall furnaces* installed between bathrooms and adjoining rooms shall not circulate air from bathrooms to other parts of the building.

G2436.4 (608.4) Door swing. *Vented wall furnaces* shall be located so that a door cannot swing within 12 inches (305 mm) of an air inlet or air outlet of such *furnace* measured at right angles to the opening. Doorstops or door closers shall not be installed to obtain this *clearance*.

G2436.5 (608.5) Ducts prohibited. Ducts shall not be attached to wall *furnaces*. Casing extension boots shall not be installed unless listed as part of the *appliance*.

G2436.6 (608.6) Access. *Vented wall furnaces* shall be provided with access for cleaning of heating surfaces, removal of *burners*, replacement of sections, motors, *controls*, filters and other working parts, and for adjustments and lubrication of parts requiring such attention. Panels, grilles and access doors that are required to be removed for normal servicing operations shall not be attached to the building construction.

SECTION G2437 (609) FLOOR FURNACES

G2437.1 (609.1) General. *Floor furnaces* shall be tested in accordance with ANSI Z21.86/CSA 2.32 and shall be installed in accordance with the manufacturer's installation instructions.

G2437.2 (609.2) Placement. The following provisions apply to *floor furnaces*:

1. Floors. *Floor furnaces* shall not be installed in the floor of any doorway, stairway landing, aisle or passageway of any enclosure, public or private, or in an exitway from any such room or space.
2. Walls and corners. The register of a *floor furnace* with a horizontal warm air outlet shall not be placed closer than 6 inches (152 mm) to the nearest wall. A distance of at least 18 inches (457 mm) from two adjoining sides of the *floor furnace* register to walls shall be provided to eliminate the necessity of occupants walking over the warm air discharge. The remaining sides shall be permitted to be placed not closer than 6 inches (152 mm) to a wall. Wall-register models shall not be placed closer than 6 inches (152 mm) to a corner.
3. Draperies. The *furnace* shall be placed so that a door, drapery, or similar object cannot be nearer than 12 inches (305 mm) to any portion of the register of the *furnace*.
4. Floor construction. *Floor furnaces* shall not be installed in concrete floor construction built on grade.
5. *Thermostat.* The controlling *thermostat* for a *floor furnace* shall be located within the same room or space as the *floor furnace* or shall be located in an adjacent room or space that is permanently open to the room or space containing the *floor furnace*.

G2437.3 (609.3) Bracing. The floor around the *furnace* shall be braced and headed with a support framework designed in accordance with Chapter 5.

G2437.4 (609.4) Clearance. The lowest portion of the *floor furnace* shall have not less than a 6-inch (152 mm) *clearance* from the grade level; except where the lower 6-inch (152 mm) portion of the *floor furnace* is sealed by the manufacturer to prevent entrance of water, the minimum *clearance* shall be reduced to not less than 2 inches (51 mm). Where these *clearances* cannot be provided, the ground below and to the sides shall be excavated to form a pit under the *furnace* so that the required *clearance* is provided beneath the lowest portion of the *furnace*. A 12-inch (305 mm) minimum clearance shall be provided on all sides except the *control* side, which shall have an 18-inch (457 mm) minimum *clearance*.

G2437.5 (609.5) First floor installation. Where the basement story level below the floor in which a *floor furnace* is installed is utilized as habitable space, such *floor furnaces* shall be enclosed as specified in Section G2437.6 and shall project into a nonhabitable space.

G2437.6 (609.6) Upper floor installations. *Floor furnaces* installed in upper stories of buildings shall project below into nonhabitable space and shall be separated from the nonhabitable space by an enclosure constructed of noncombustible materials. The *floor furnace* shall be provided with access, *clearance* to all sides and bottom of not less than 6 inches (152 mm) and *combustion air* in accordance with Section G2407.

SECTION G2438 (613)
CLOTHES DRYERS

G2438.1 (613.1) General. *Clothes dryers* shall be tested in accordance with ANSI Z21.5.1 and shall be installed in accordance with the manufacturer's installation instructions.

SECTION G2439 (614)
CLOTHES DRYER EXHAUST

G2439.1 (614.1) Installation. *Clothes dryers* shall be exhausted in accordance with the manufacturer's instructions. Dryer exhaust systems shall be independent of all other systems and shall convey the moisture and any products of *combustion* to the outside of the building.

G2439.2 (614.2) Duct penetrations. Ducts that exhaust *clothes dryers* shall not penetrate or be located within any fireblocking, draftstopping or any wall, floor/ceiling or other assembly required by this *code* to be fire-resistance rated, unless such duct is constructed of galvanized steel or aluminum of the thickness specified in the mechanical provisions of this *code* and the fire-resistance rating is maintained in accordance with this *code*. Fire dampers shall not be installed in *clothes dryer* exhaust duct systems.

G2439.3 (614.4) Exhaust installation. Dryer exhaust ducts for *clothes dryers* shall terminate on the outside of the building and shall be equipped with a backdraft *damper*. Screens shall not be installed at the duct termination. Ducts shall not be connected or installed with sheet metal screws or other fasteners that will obstruct the flow. *Clothes dryer* exhaust ducts shall not be connected to a *vent connector*, vent or *chimney*. *Clothes dryer* exhaust ducts shall not extend into or through ducts or plenums.

G2439.4 (614.5) Makeup air. Installations exhausting more than 200 cfm (0.09 m^3/s) shall be provided with *makeup air*. Where a closet is designed for the installation of a *clothes dryer*, an opening having an area of not less than 100 square inches (0.0645 m^2) for *makeup air* shall be provided in the closet enclosure, or *makeup air* shall be provided by other *approved* means.

G2439.5 (614.6) Domestic clothes dryer exhaust ducts. Exhaust ducts for domestic *clothes dryers* shall conform to the requirements of Sections G2439.5.1 through G2439.5.7.

G2439.5.1 (614.6.1) Material and size. Exhaust ducts shall have a smooth interior finish and shall be constructed of metal a minimum 0.016-inch (0.4 mm) thick. The exhaust duct size shall be 4 inches (102 mm) nominal in diameter.

G2439.5.2 (614.6.2) Duct installation. Exhaust ducts shall be supported at 4 foot (1219 mm) intervals and secured in place. The insert end of the duct shall extend into the adjoining duct or fitting in the direction of airflow. Ducts shall not be joined with screws or similar fasteners that protrude into the inside of the duct.

G2439.5.3 (614.6.3) Protection required. Protective shield plates shall be placed where nails or screws from finish or other work are likely to penetrate the *clothes dryer* exhaust duct. Shield plates shall be placed on the finished face of all framing members where there is less than $1^1/_4$ inches (32 mm) between the duct and the finished face of the framing member. Protective shield plates shall be constructed of steel, shall have a minimum thickness of 0.062 inch (1.6 mm) and shall extend a minimum of 2 inches (51 mm) above sole plates and below top plates.

G2439.5.4 (614.6.4) Transition ducts. Transition ducts used to connect the dryer to the exhaust duct system shall be a single length that is listed and labeled in accordance with UL 2158A. Transition ducts shall be a maximum of 8 feet (2438 mm) in length and shall not be concealed within construction.

G2439.5.5 (614.6.5) Duct length. The maximum allowable exhaust duct length shall be determined by one of the methods specified in Section G2439.5.5.1 or G2439.5.5.2.

G2439.5.5.1 (614.6.5.1) Specified length. The maximum length of the exhaust duct shall be 35 feet (10 668 mm) from the connection to the transition duct from the dryer to the outlet terminal. Where fittings are used, the maximum length of the exhaust duct shall be reduced in accordance with Table G2439.5.5.1.

TABLE G2439.5.5.1 (TABLE 614.6.5.1)
DRYER EXHAUST DUCT FITTING EQUIVALENT LENGTH

DRYER EXHAUST DUCT FITTING TYPE	EQUIVALENT LENGTH
4 inch radius mitered 45 degree elbow	2 feet 6 inches
4 inch radius mitered 90 degree elbow	5 feet
6 inch radius smooth 45 degree elbow	1 foot
6 inch radius smooth 90 degree elbow	1 foot 9 inches
8 inch radius smooth 45 degree elbow	1 foot
8 inch radius smooth 90 degree elbow	1 foot 7 inches
10 inch radius smooth 45 degree elbow	9 inches
10 inch radius smooth 90 degree elbow	1 foot 6 inches

For SI: 1 inch = 25.4 mm, 1 foot = 304.8 mm, 1 degree = 0.0175 rad.

G2439.5.5.2 (614.6.5.2) Manufacturer's instructions. The maximum length of the exhaust duct shall be determined by the dryer manufacturer's installation instructions. The *code official* shall be provided with a copy of the installation instructions for the make and model of the dryer. Where the exhaust duct is to be concealed, the installation instructions shall be provided to the *code official* prior to the concealment inspection. In the absence of fitting equivalent length calculations from the clothes dryer manufacturer, Table G2439.5.5.1 shall be used.

G2439.5.6 (614.6.5) Length identification. Where the exhaust duct is concealed within the building construction, the equivalent length of the exhaust duct shall be identified on a permanent label or tag. The label or tag shall be located within 6 feet (1829 mm) of the exhaust duct connection.

G2439.5.7 (614.6.6) Exhaust duct required. Where space for a *clothes dryer* is provided, an exhaust duct system shall be installed. Where the *clothes dryer* is not installed at the time of occupancy, the exhaust duct shall be capped at location of the future dryer.

> **Exception:** Where a listed condensing *clothes dryer* is installed prior to occupancy of the structure.

SECTION G2440 (615) SAUNA HEATERS

G2440.1 (615.1) General. Sauna heaters shall be installed in accordance with the manufacturer's installation instructions.

G2440.2 (615.2) Location and protection. Sauna heaters shall be located so as to minimize the possibility of accidental contact by a person in the room.

G2440.2.1 (615.2.1) Guards. Sauna heaters shall be protected from accidental contact by an *approved* guard or barrier of material having a low coefficient of thermal conductivity. The guard shall not substantially affect the transfer of heat from the heater to the room.

G2440.3 (615.3) Access. Panels, grilles and access doors that are required to be removed for normal servicing operations, shall not be attached to the building.

G2440.4 (615.4) Combustion and dilution air intakes. Sauna heaters of other than the direct-vent type shall be installed with the *draft hood* and *combustion air* intake located outside the sauna room. Where the *combustion air* inlet and the *draft hood* are in a dressing room adjacent to the sauna room, there shall be provisions to prevent physically blocking the *combustion air* inlet and the *draft hood* inlet, and to prevent physical contact with the *draft hood* and vent assembly, or warning notices shall be posted to avoid such contact. Any warning notice shall be easily readable, shall contrast with its background, and the wording shall be in letters not less than 0.25 inch (6.4 mm) high.

G2440.5 (615.5) Combustion and ventilation air. Combustion air shall not be taken from inside the sauna room. *Combustion* and ventilation air for a sauna heater not of the direct-vent type shall be provided to the area in which the *combustion air* inlet and *draft hood* are located in accordance with Section G2407.

G2440.6 (615.6) Heat and time controls. Sauna heaters shall be equipped with a *thermostat* which will limit room temperature to 194°F (90°C). If the *thermostat* is not an integral part of the sauna heater, the heat-sensing element shall be located within 6 inches (152 mm) of the ceiling. If the heat-sensing element is a capillary tube and bulb, the assembly shall be attached to the wall or other support, and shall be protected against physical damage.

G2440.6.1 (615.6.1) Timers. A timer, if provided to *control main burner* operation, shall have a maximum operating time of 1 hour. The *control* for the timer shall be located outside the sauna room.

G2440.7 (615.7) Sauna room. A ventilation opening into the sauna room shall be provided. The opening shall be not less than 4 inches by 8 inches (102 mm by 203 mm) located near the top of the door into the sauna room.

SECTION G2441 (617) POOL AND SPA HEATERS

G2441.1 (617.1) General. Pool and spa heaters shall be tested in accordance with ANSI Z21.56 and shall be installed in accordance with the manufacturer's installation instructions.

SECTION G2442 (618) FORCED-AIR WARM-AIR FURNACES

G2442.1 (618.1) General. Forced-air warm-air *furnaces* shall be tested in accordance with ANSI Z21.47 or UL 795 and shall be installed in accordance with the manufacturer's installation instructions.

G2442.2 (618.2) Forced-air furnaces. The minimum unobstructed total area of the outside and return air ducts or openings to a forced-air warm-air *furnace* shall be not less than 2 square inches for each 1,000 *Btu*/h (4402 mm^2/W) output rating capacity of the *furnace* and not less than that specified in the *furnace* manufacturer's installation instructions. The minimum unobstructed total area of supply ducts from a forced-air warm-air *furnace* shall be not less than 2 square inches for each 1,000 *Btu*/h (4402 mm^2/W) output rating capacity of the *furnace* and not less than that specified in the *furnace* manufacturer's installation instructions.

> **Exception:** The total area of the supply air ducts and outside and return air ducts shall not be required to be larger than the minimum size required by the *furnace* manufacturer's installation instructions.

G2442.3 (618.3) Dampers. Volume dampers shall not be placed in the air inlet to a *furnace* in a manner that will reduce the required air to the *furnace*.

G2442.4 (618.4) Prohibited sources. Outdoor or return air for forced-air heating and cooling systems shall not be taken from the following locations:

1. Closer than 10 feet (3048 mm) from an *appliance* vent outlet, a vent opening from a plumbing drainage system or the discharge outlet of an exhaust fan, unless the outlet is 3 feet (914 mm) above the outside air inlet.
2. Where there is the presence of objectionable odors, fumes or flammable vapors; or where located less than 10 feet (3048 mm) above the surface of any abutting public way or driveway; or where located at grade level by a sidewalk, street, alley or driveway.
3. A hazardous or insanitary location or a refrigeration machinery room as defined in the *International Mechanical Code*.
4. A room or space, the volume of which is less than 25 percent of the entire volume served by such system. Where connected by a permanent opening having an area sized in accordance with Section G2442.2, adjoining rooms or spaces shall be considered as a single room or space for the purpose of determining the volume of such rooms or spaces.

   Exception: The minimum volume requirement shall not apply where the amount of return air taken from a room or space is less than or equal to the amount of supply air delivered to such room or space.

5. A room or space containing an *appliance* where such a room or space serves as the sole source of return air.

   Exception: This shall not apply where:

   1. The *appliance* is a direct-vent *appliance* or an *appliance* not requiring a vent in accordance with Section G2425.8.

   2. The room or space complies with the following requirements:

      2.1. The return air shall be taken from a room or space having a volume exceeding 1 cubic foot for each 10 Btu/h (9.6L/W) of combined input rating of all fuel-burning appliances therein.

      2.2. The volume of supply air discharged back into the same space shall be approximately equal to the volume of return air taken from the space.

      2.3. Return-air inlets shall not be located within 10 feet (3048 mm) of a draft hood in the same room or space or the combustion chamber of any atmospheric burner *appliance* in the same room or space.

   3. Rooms or spaces containing solid fuel-burning appliances, provided that return-air inlets are located not less than 10 feet (3048 mm) from the firebox of such appliances.

6. A closet, bathroom, toilet room, kitchen, garage, boiler room, furnace room or unconditioned attic.

   Exceptions:

   1. Where return air intakes are located not less than 10 feet (3048 mm) from cooking appliances and serve only the kitchen area, taking return air from a kitchen area shall not be prohibited.

   2. Dedicated forced air systems serving only a garage shall not be prohibited from obtaining return air from the garage.

7. A crawl space by means of direct connection to the return side of a forced-air system. Transfer openings in the crawl space enclosure shall not be prohibited.

G2442.5 (618.5) Screen. Required outdoor air inlets shall be covered with a screen having $^1/_4$-inch (6.4 mm) openings. Required outdoor air inlets serving a nonresidential portion of a building shall be covered with screen having openings larger than $^1/_4$ inch (6.4 mm) and not larger than 1 inch (25 mm).

G2442.6 (618.6) Return-air limitation. Return air from one *dwelling unit* shall not be discharged into another *dwelling unit.*

G2442.7 (618.7) Furnace plenums and air ducts. Where a *furnace* is installed so that supply ducts carry air circulated by the *furnace* to areas outside of the space containing the *furnace*, the return air shall also be handled by a duct(s) sealed to the *furnace* casing and terminating outside of the space containing the *furnace*.

SECTION G2443 (619) CONVERSION BURNERS

G2443.1 (619.1) Conversion burners. The installation of *conversion burners* shall conform to ANSI Z21.8.

SECTION G2444 (620) UNIT HEATERS

G2444.1 (620.1) General. *Unit heaters* shall be tested in accordance with ANSI Z83.8 and shall be installed in accordance with the manufacturer's installation instructions.

G2444.2 (620.2) Support. Suspended-type *unit heaters* shall be supported by elements that are designed and constructed to accommodate the weight and dynamic loads. Hangers and brackets shall be of noncombustible material.

G2444.3 (620.3) Ductwork. Ducts shall not be connected to a unit heater unless the heater is listed for such installation.

G2444.4 (620.4) Clearance. Suspended-type *unit heaters* shall be installed with *clearances* to combustible materials of not less than 18 inches (457 mm) at the sides, 12 inches (305 mm) at the bottom and 6 inches (152 mm) above the top where the unit heater has an internal *draft hood* or 1 inch (25 mm) above the top of the sloping side of the vertical *draft hood.*

Floor-mounted-type *unit heaters* shall be installed with *clearances* to combustible materials at the back and one side only of not less than 6 inches (152 mm). Where the *flue gases* are vented horizontally, the 6-inch (152 mm) *clearance* shall be measured from the *draft hood* or *vent* instead of the rear wall of the unit heater. Floor-mounted-type *unit heaters* shall not be installed on combustible floors unless listed for such installation.

Clearance for servicing all *unit heaters* shall be in accordance with the manufacturer's installation instructions.

Exception: *Unit heaters* listed for reduced *clearance* shall be permitted to be installed with such *clearances* in accordance with their listing and the manufacturer's instructions.

SECTION G2445 (621) UNVENTED ROOM HEATERS

G2445.1 (621.1) General. *Unvented room heaters* shall be tested in accordance with ANSI Z 21.11.2 and shall be installed in accordance with the conditions of the listing and the manufacturer's installation instructions.

G2445.2 (621.2) Prohibited use. One or more *unvented room heaters* shall not be used as the sole source of comfort heating in a *dwelling unit.*

G2445.3 (621.3) Input rating. *Unvented room heaters* shall not have an input rating in excess of 40,000 *Btu*/h (11.7 kW).

G2445.4 (621.4) Prohibited locations. The location of *unvented room heaters* shall comply with Section G2406.2.

G2445.5 (621.5) Room or space volume. The aggregate input rating of all *unvented appliances* installed in a room or space shall not exceed 20 *Btu*/h per *cubic foot* (0.21 kW/m^3) of volume of such room or space. Where the room or space in which the *appliance* is installed is directly connected to another room or space by a doorway, archway or other opening of comparable size that cannot be closed, the volume of such adjacent room or space shall be permitted to be included in the calculations.

G2445.6 (621.6) Oxygen-depletion safety system. *Unvented room heaters* shall be equipped with an oxygen-depletion-sensitive safety shutoff system. The system shall shut off the gas supply to the main and *pilot burners* when the oxygen in the surrounding atmosphere is depleted to the percent concentration specified by the manufacturer, but not lower than 18 percent. The system shall not incorporate field adjustment means capable of changing the set point at which the system acts to shut off the gas supply to the room heater.

G2445.7 (621.7) Unvented decorative room heaters. An unvented decorative room heater shall not be installed in a *factory-built fireplace* unless the *fireplace* system has been specifically tested, listed and labeled for such use in accordance with UL 127.

G2445.7.1 (621.7.1) Ventless firebox enclosures. Ventless firebox enclosures used with unvented decorative room heaters shall be listed as complying with ANSI Z21.91.

SECTION G2446 (622) VENTED ROOM HEATERS

G2446.1 (622.1) General. *Vented room heaters* shall be tested in accordance with ANSI Z21.86/CSA 2.32, shall be designed and equipped as specified in Section G2432.2 and shall be installed in accordance with the manufacturer's installation instructions.

SECTION G2447 (623) COOKING APPLIANCES

G2447.1 (623.1) Cooking appliances. Cooking *appliances* that are designed for permanent installation, including ranges, ovens, stoves, broilers, grills, fryers, griddles, hot plates and barbecues, shall be tested in accordance with ANSI Z21.1 or ANSI Z21.58 and shall be installed in accordance with the manufacturer's installation instructions.

G2447.2 (623.2) Prohibited location. Cooking *appliances* designed, tested, listed and labeled for use in commercial occupancies shall not be installed within *dwelling units* or within any area where domestic cooking operations occur.

G2447.3 (623.3) Domestic appliances. Cooking *appliances* installed within *dwelling units* and within areas where domestic cooking operations occur shall be listed and labeled as household-type *appliances* for domestic use.

G2447.4 (623.4) Range installation. Ranges installed on combustible floors shall be set on their own bases or legs and shall be installed with *clearances* of not less than that shown on the label.

G2447.5 (623.7) Vertical clearance above cooking top. Household cooking *appliances* shall have a vertical *clearance* above the cooking top of not less than 30 inches (760 mm) to combustible material and metal cabinets. A minimum *clearance* of 24 inches (610 mm) is permitted where one of the following is installed:

1. The underside of the combustible material or metal cabinet above the cooking top is protected with not less than $^1/_4$ inch (6 mm) thick insulating millboard covered with sheet metal not less than 0.0122 inch (0.3 mm) thick.
2. A metal ventilating hood constructed of sheet metal not less than 0.0122 inch (0.3 mm) thick is installed above the cooking top with a *clearance* of not less than $^1/_4$ inch (6 mm) between the hood and the underside of the combustible material or metal cabinet. The hood shall have a width not less than the width of the *appliance* and shall be centered over the *appliance*.
3. A listed cooking *appliance* or microwave oven is installed over a listed cooking *appliance* and in compliance with the terms of the manufacturer's installation instructions for the upper *appliance*.

SECTION G2448 (624) WATER HEATERS

G2448.1 (624.1) General. Water heaters shall be tested in accordance with ANSI Z 21.10.1 and ANSI Z 21.10.3 and shall be installed in accordance with the manufacturer's installation instructions.

G2448.1.1 (624.1.1) Installation requirements. The requirements for *water heaters* relative to sizing, *relief valves*, drain pans and scald protection shall be in accordance with this *code*.

G2448.2 (624.2) Water heaters utilized for space heating. *Water heaters* utilized both to supply potable hot water and provide hot water for space-heating applications shall be listed and labeled for such applications by the manufacturer and shall be installed in accordance with the manufacturer's installation instructions and this *code*.

SECTION G2449 (627) AIR CONDITIONING APPLIANCES

G2449.1 (627.1) General. Air conditioning *appliances* shall be tested in accordance with ANSI Z21.40.1 or ANSI Z21.40.2 and shall be installed in accordance with the manufacturer's installation instructions.

G2449.2 (627.2) Independent piping. *Gas piping* serving heating *appliances* shall be permitted to also serve cooling *appliances* where such heating and cooling *appliances* cannot be operated simultaneously. (See Section G2413.)

G2449.3 (627.3) Connection of gas engine-powered air conditioners. To protect against the effects of normal vibration in service, gas engines shall not be rigidly connected to the gas supply *piping*.

G2449.4 (627.6) Installation. Air conditioning *appliances* shall be installed in accordance with the manufacturer's instructions. Unless the *appliance* is listed for installation on a combustible surface such as a floor or roof, or unless the surface is protected in an *approved* manner, the *appliance* shall be installed on a surface of noncombustible construction with noncombustible material and surface finish and with no combustible material against the underside thereof.

SECTION G2450 (628) ILLUMINATING APPLIANCES

G2450.1 (628.1) General. Illuminating *appliances* shall be tested in accordance with ANSI Z21.42 and shall be installed in accordance with the manufacturer's installation instructions.

G2450.2 (628.2) Mounting on buildings. Illuminating *appliances* designed for wall or ceiling mounting shall be securely attached to substantial structures in such a manner that they are not dependent on the *gas piping* for support.

G2450.3 (628.3) Mounting on posts. Illuminating *appliances* designed for post mounting shall be securely and rigidly attached to a post. Posts shall be rigidly mounted. The strength and rigidity of posts greater than 3 feet (914 mm) in height shall be at least equivalent to that of a 2.5-inch-diameter (64 mm) post constructed of 0.064-inch-thick (1.6 mm) steel or a 1-inch (25 mm) Schedule 40 steel *pipe*. Posts 3 feet (914 mm) or less in height shall not be smaller than $^3/_4$-inch (19.1 mm) Schedule 40 steel *pipe*. Drain openings shall be provided near the base of posts where there is a possibility of water collecting inside them.

G2450.4 (628.4) Appliance pressure regulators. Where an *appliance pressure regulator* is not supplied with an illuminating *appliance* and the service line is not equipped with a *service pressure regulator*, an *appliance pressure regulator* shall be installed in the line to the illuminating *appliance*. For multiple installations, one *regulator* of adequate capacity shall be permitted to serve more than one illuminating *appliance*.

SECTION G2451 (630) INFRARED RADIANT HEATERS

G2451.1 (630.1) General. Infrared radiant heaters shall be tested in accordance with ANSI Z83.19 or Z83.20 and shall be installed in accordance with the manufacturer's instructions.

G2451.2 (630.2) Support. *Infrared radiant heaters* shall be fixed in a position independent of gas and electric supply lines. Hangers and brackets shall be of noncombustible material.

SECTION G2452 (631) BOILERS

G2452.1 (631.1) Standards. Boilers shall be listed in accordance with the requirements of ANSI Z21.13 or UL 795. If applicable, the boiler shall be designed and constructed in accordance with the requirements of ASME CSD-1 and as applicable, the ASME *Boiler and Pressure Vessel Code*, Sections I, II, IV, V and IX and NFPA 85.

G2452.2 (631.2) Installation. In addition to the requirements of this *code*, the installation of boilers shall be in accordance with the manufacturer's instructions and this *code*. Operating instructions of a permanent type shall be attached to the boiler. Boilers shall have all *controls* set, adjusted and tested by the installer. A complete *control* diagram together with complete boiler operating instructions shall be furnished by the installer. The manufacturer's rating data and the nameplate shall be attached to the boiler.

G2452.3 (631.3) Clearance to combustible material. *Clearances* to combustible materials shall be in accordance with Section G2409.4.

SECTION G2453 (634) CHIMNEY DAMPER OPENING AREA

G2453.1 (634.1) Free opening area of chimney dampers. Where an unlisted decorative *appliance* for installation in a vented *fireplace* is installed, the *fireplace damper* shall have a permanent free opening equal to or greater than specified in Table G2453.1.

SECTION G2454 (636) OUTDOOR DECORATIVE APPLIANCES

G2454.1 (636.1) General. Permanently fixed-in-place outdoor decorative appliances shall be tested in accordance with ANSI Z21.97 and shall be installed in accordance with the manufacturer's instructions.

TABLE G2453.1 (634.1)
FREE OPENING AREA OF CHIMNEY DAMPER FOR VENTING FLUE GASES FROM UNLISTED DECORATIVE APPLIANCES FOR INSTALLATION IN VENTED FIREPLACES

CHIMNEY HEIGHT(feet)	MINIMUM PERMANENT FREE OPENING (square inches)[a]						
	8	13	20	29	39	51	64
	Appliance input rating (Btu per hour)						
6	7,800	14,000	23,200	34,000	46,400	62,400	80,000
8	8,400	15,200	25,200	37,000	50,400	68,000	86,000
10	9,000	16,800	27,600	40,400	55,800	74,400	96,400
15	9,800	18,200	30,200	44,600	62,400	84,000	108,800
20	10,600	20,200	32,600	50,400	68,400	94,000	122,200
30	11,200	21,600	36,600	55,200	76,800	105,800	138,600

For SI: 1 inch = 25.4 mm, 1 foot = 304.8 mm, 1 square inch = 645.16 mm^2, 1,000 Btu per hour = 0.293 kW.

a. The first six minimum permanent free openings (8 square inches to 51 square inches) correspond approximately to the cross-sectional areas of chimneys having diameters of 3 inches through 8 inches, respectively. The 64-square inch opening corresponds to the cross-sectional area of standard 8-inch by 8-inch chimney tile.

Part VII—Plumbing

CHAPTER 25

PLUMBING ADMINISTRATION

SECTION P2501 GENERAL

P2501.1 Scope. The provisions of this chapter shall establish the general administrative requirements applicable to plumbing systems and inspection requirements of this code.

P2501.2 Application. In addition to the general administration requirements of Chapter 1, the administrative provisions of this chapter shall also apply to the plumbing requirements of Chapters 25 through 32.

SECTION P2502 EXISTING PLUMBING SYSTEMS

P2502.1 Existing building sewers and drains. Existing *building sewers* and drains shall be used in connection with new systems when found by examination and/or test to conform to the requirements prescribed by this document.

P2502.2 Additions, alterations or repairs. Additions, *alterations*, renovations or repairs to any plumbing system shall conform to that required for a new plumbing system without requiring the existing plumbing system to comply with all the requirements of this code. Additions, *alterations* or repairs shall not cause an existing system to become unsafe, insanitary or overloaded.

Minor additions, *alterations*, renovations and repairs to existing plumbing systems shall be permitted in the same manner and arrangement as in the existing system, provided that such repairs or replacement are not hazardous and are *approved*.

SECTION P2503 INSPECTION AND TESTS

P2503.1 Inspection required. New plumbing work and parts of existing systems affected by new work or *alterations* shall be inspected by the *building official* to ensure compliance with the requirements of this code.

P2503.2 Concealment. A plumbing or drainage system, or part thereof, shall not be covered, concealed or put into use until it has been tested, inspected and *approved* by the *building official*.

P2503.3 Responsibility of permittee. Test equipment, materials and labor shall be furnished by the permittee.

P2503.4 Building sewer testing. The *building sewer* shall be tested by insertion of a test plug at the point of connection with the public sewer and filling the *building sewer* with water, testing with not less than a 10-foot (3048 mm) head of water and be able to maintain such pressure for 15 minutes.

P2503.5 DWV systems testing. Rough and finished plumbing installations shall be tested in accordance with Sections P2503.5.1 and P2503.5.2.

P2503.5.1 Rough plumbing. DWV systems shall be tested on completion of the rough piping installation by water or for piping systems other than plastic, by air with no evidence of leakage. Either test shall be applied to the drainage system in its entirety or in sections after rough piping has been installed, as follows:

1. Water test. Each section shall be filled with water to a point not less than 10 feet (3048 mm) above the highest fitting connection in that section, or to the highest point in the completed system. Water shall be held in the section under test for a period of 15 minutes. The system shall prove leak free by visual inspection.
2. Air test. The portion under test shall be maintained at a gauge pressure of 5 pounds per square inch (psi) (34 kPa) or 10 inches of mercury column (34 kPa). This pressure shall be held without introduction of additional air for a period of 15 minutes.

P2503.5.2 Finished plumbing. After the plumbing fixtures have been set and their traps filled with water, their connections shall be tested and proved gas tight and/or water tight as follows:

1. Water tightness. Each fixture shall be filled and then drained. Traps and fixture connections shall be proven water tight by visual inspection.
2. Gas tightness. When required by the local administrative authority, a final test for gas tightness of the DWV system shall be made by the smoke or peppermint test as follows:
    - 2.1. Smoke test. Introduce a pungent, thick smoke into the system. When the smoke appears at vent terminals, such terminals shall be sealed and a pressure equivalent to a 1-inch water column (249 Pa) shall be applied and maintained for a test period of not less than 15 minutes.
    - 2.2. Peppermint test. Introduce 2 ounces (59 mL) of oil of peppermint into the system. Add 10 quarts (9464 mL) of hot water and seal all vent terminals. The odor of peppermint shall not be detected at any trap or other point in the system.

P2503.6 Shower liner test. Where shower floors and receptors are made water tight by the application of materials required by Section P2709.2, the completed liner installation shall be tested. The pipe from the shower drain shall be plugged water tight for the test. The floor and receptor area shall be filled with potable water to a depth of not less than 2 inches (51 mm) measured at the threshold. Where a threshold of at least 2 inches high does not exist, a temporary threshold shall be constructed to retain the test water in the lined floor or receptor area to a level not less than 2 inches deep measured at the threshold. The water shall be retained for a test period of not less than 15 minutes and there shall be no evidence of leakage.

P2503.7 Water-supply system testing. Upon completion of the water-supply system or a section of it, the system or portion completed shall be tested and proved tight under a water pressure of not less than the working pressure of the system or, for piping systems other than plastic, by an air test of not less than 50 psi (345 kPa). This pressure shall be held for not less than 15 minutes. The water used for tests shall be obtained from a potable water source.

P2503.8 Inspection and testing of backflow prevention devices. Inspection and testing of backflow prevention devices shall comply with Sections P2503.8.1 and P2503.8.2.

P2503.8.1 Inspections. Inspections shall be made of all backflow prevention assemblies to determine whether they are operable.

P2503.8.2 Testing. Reduced pressure principle, double check, double check detector and pressure vacuum breaker backflow preventer assemblies shall be tested at the time of installation, immediately after repairs or relocation and at least annually.

P2503.9 Test gauges. Gauges used for testing shall be as follows:

1. Tests requiring a pressure of 10 psi or less shall utilize a testing gauge having increments of 0.10 psi (0.69 kPa) or less.
2. Tests requiring a pressure higher than 10 psi (0.69 kPa) but less than or equal to 100 psi (690 kPa) shall use a testing gauge having increments of 1 psi (6.9 kPa) or less.
3. Tests requiring a pressure higher than 100 psi (690 kPa) shall use a testing gauge having increments of 2 psi (14 kPa) or less.

CHAPTER 26

GENERAL PLUMBING REQUIREMENTS

SECTION P2601 GENERAL

P2601.1 Scope. The provisions of this chapter shall govern the installation of plumbing not specifically covered in other chapters applicable to plumbing systems. The installation of plumbing, *appliances, equipment* and systems not addressed by this code shall comply with the applicable provisions of the *International Plumbing Code*.

P2601.2 Connections to drainage system. Plumbing fixtures, drains, appurtenances and appliances used to receive or discharge liquid wastes or sewage shall be directly connected to the sanitary drainage system of the building or premises, in accordance with the requirements of this code. This section shall not be construed to prevent indirect waste systems.

Exception: Bathtubs, showers, lavatories, clothes washers and laundry trays shall not be required to discharge to the sanitary drainage system where such fixtures discharge to an approved gray water system for flushing of water closets and urinals or for subsurface landscape irrigation.

P2601.3 Flood hazard areas. In flood hazard areas as established by Table R301.2(1), plumbing fixtures, drains, and appliances shall be located or installed in accordance with Section R322.1.6.

SECTION P2602 INDIVIDUAL WATER SUPPLY AND SEWAGE DISPOSAL

P2602.1 General. The water-distribution and drainage system of any building or premises where plumbing fixtures are installed shall be connected to a public water supply or sewer system, respectively, if available. When either a public water-supply or sewer system, or both, are not available, or connection to them is not feasible, an individual water supply or individual (private) sewage-disposal system, or both, shall be provided.

P2602.2 Flood-resistant installation. In flood hazard areas as established by Table R301.2(1):

1. Water supply systems shall be designed and constructed to prevent infiltration of floodwaters.
2. Pipes for sewage disposal systems shall be designed and constructed to prevent infiltration of floodwaters into the systems and discharges from the systems into floodwaters.

SECTION P2603 STRUCTURAL AND PIPING PROTECTION

P2603.1 General. In the process of installing or repairing any part of a plumbing and drainage installation, the finished floors, walls, ceilings, tile work or any other part of the building or premises that must be changed or replaced shall be left in a safe structural condition in accordance with the requirements of the building portion of this code.

P2603.2 Drilling and notching. Wood-framed structural members shall not be drilled, notched or altered in any manner except as provided in Sections R502.8, R602.6, R802.7 and R802.7.1. Holes in load-bearing members of cold-formed steel light-frame construction shall be permitted only in accordance with Sections R505.2.5, R603.2.5 and R804.2.5. In accordance with the provisions in Sections R505.3.5, R603.3.4 and R804.3.4, cutting and notching of flanges and lips of load-bearing members of cold-formed steel light-frame construction shall not be permitted. Structural insulated panels (SIPs) shall be drilled and notched or altered in accordance with the provisions of Section R613.7.

P2603.2.1 Protection against physical damage. In concealed locations, where piping, other than cast-iron or galvanized steel, is installed through holes or notches in studs, joists, rafters or similar members less than $1^1/_2$ inches (38 mm) from the nearest edge of the member, the pipe shall be protected by steel shield plates. Such shield plates shall have a thickness of not less than 0.0575 inch (1.463 mm) (No. 16 Gage). Such plates shall cover the area of the pipe where the member is notched or bored, and shall extend not less than 2 inches (51 mm) above sole plates and below top plates.

P2603.3 Breakage and corrosion. Pipes passing through concrete or cinder walls and floors, cold-formed steel framing or other corrosive material shall be protected against external corrosion by a protective sheathing or wrapping or other means that will withstand any reaction from lime and acid of concrete, cinder or other corrosive material. Sheathing or wrapping shall allow for movement including expansion and contraction of piping. The wall thickness of material shall be not less than 0.025 inch (0.64 mm).

P2603.4 Pipes through foundation walls. A pipe that passes through a foundation wall shall be provided with a relieving arch, or a pipe sleeve shall be built into the foundation wall. The sleeve shall be two pipe sizes greater than the pipe passing through the wall.

P2603.5 Freezing. In localities having a winter design temperature of 32°F (0°C) or lower as shown in Table R301.2(1) of this code, a water, soil or waste pipe shall not be installed outside of a building, in exterior walls, in *attics* or crawl spaces, or in any other place subjected to freezing temperature unless adequate provision is made to protect it from freezing by insulation or heat or both. Water service pipe shall be installed not less than 12 inches (305 mm) deep and not less than 6 inches (152 mm) below the frost line.

P2603.5.1 Sewer depth. *Building sewers* that connect to private sewage disposal systems shall be a not less than **[NUMBER]** inches (mm) below finished *grade* at the point of septic tank connection. *Building sewers* shall be not less than **[NUMBER]** inches (mm) below *grade*.

SECTIONP2604 TRENCHING AND BACKFILLING

P2604.1 Trenching and bedding. Where trenches are excavated such that the bottom of the trench forms the bed for the pipe, solid and continuous load-bearing support shall be provided between joints. Where over-excavated, the trench shall be backfilled to the proper grade with compacted earth, sand, fine gravel or similar granular material. Piping shall not be supported on rocks or blocks at any point. Rocky or unstable soil shall be over-excavated by two or more pipe diameters and brought to the proper grade with suitable compacted granular material.

P2604.2 Common trench. See Section P2905.4.2.

P2604.3 Backfilling. Backfill shall be free from discarded construction material and debris. Backfill shall be free from rocks, broken concrete and frozen chunks until the pipe is covered by not less than 12 inches (305 mm) of tamped earth. Backfill shall be placed evenly on both sides of the pipe and tamped to retain proper alignment. Loose earth shall be carefully placed in the trench in 6-inch (152 mm) layers and tamped in place.

P2604.4 Protection of footings. Trenching installed parallel to footings shall not extend below the 45-degree (0.79 rad) bearing plane of the bottom edge of a wall or footing (see Figure P2604.4).

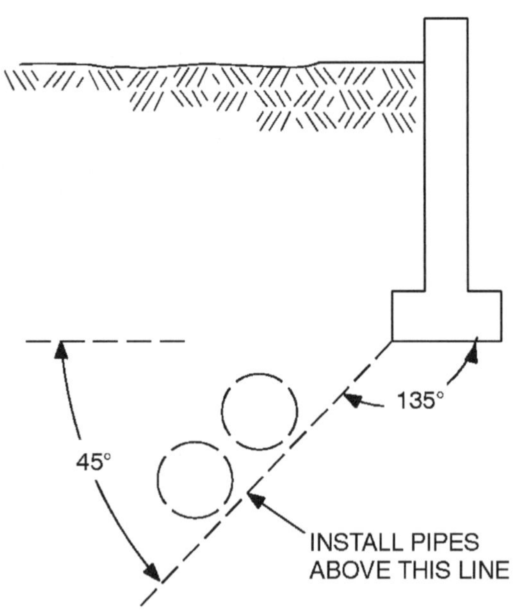

For SI: 1 degree = 0.018 rad.

FIGURE P2604.4
PIPE LOCATION WITH RESPECT TO FOOTINGS

SECTION P2605 SUPPORT

P2605.1 General. Piping shall be supported in accordance with the following:

1. Piping shall be supported to ensure alignment and prevent sagging, and allow movement associated with the expansion and contraction of the piping system.
2. Piping in the ground shall be laid on a firm bed for its entire length, except where support is otherwise provided.
3. Hangers and anchors shall be of sufficient strength to maintain their proportional share of the weight of pipe and contents and of sufficient width to prevent distortion to the pipe. Hangers and strapping shall be of *approved* material that will not promote galvanic action. Rigid support sway bracing shall be provided at changes in direction greater than 45 degrees (0.79 rad) for pipe sizes 4 inches (102 mm) and larger.
4. Piping shall be supported at distances not to exceed those indicated in Table P2605.1.

SECTION P2606 PENETRATIONS

P2606.1 Sealing of annular spaces. The annular space between the outside of a pipe and the inside of a pipe sleeve or between the outside of a pipe and an opening in a building envelope wall, floor, or ceiling assembly penetrated by a pipe shall be sealed with caulking material or foam sealant or closed with a gasketing system. The caulking material, foam sealant or gasketing system shall be designed for the conditions at the penetration location and shall be compatible with the pipe, sleeve and building materials in contact with the sealing materials. Annular spaces created by pipes penetrating fire-resistance-rated assemblies or membranes of such assemblies shall be sealed or closed in accordance with the building portion of this code.

SECTION P2607 WATERPROOFING OF OPENINGS

P2607.1 General. Roof and exterior wall penetrations shall be made water tight. Joints at the roof, around vent pipes, shall be made water tight by the use of lead, copper or galvanized iron flashings or an *approved* elastomeric material. Counterflashing shall not restrict the required internal cross-sectional area of any vent.

SECTION P2608 WORKMANSHIP

P2608.1 General. Valves, pipes and fittings shall be installed in correct relationship to the direction of the flow. Burred ends shall be reamed to the full bore of the pipe.

SECTION P2609 MATERIALS EVALUATION AND LISTING

P2609.1 Identification. Each length of pipe and each pipe fitting, trap, fixture, material and device utilized in a plumbing system shall bear the identification of the manufacturer and any markings required by the applicable referenced standards.

P2609.2 Installation of materials. All materials used shall be installed in strict accordance with the standards under which the materials are accepted and *approved*. In the absence of such installation procedures, the manufacturer's instructions shall be followed. Where the requirements of referenced standards or manufacturer's instructions do not conform to the minimum provisions of this code, the provisions of this code shall apply.

P2609.3 Plastic pipe, fittings and components. All plastic pipe, fittings and components shall be third-party certified as conforming to NSF 14.

P2609.4 Third-party certification. All plumbing products and materials shall be listed by a third-party certification agency as complying with the referenced standards. Products and materials shall be identified in accordance with Section P2609.1.

P2609.5 Water supply systems. Water service pipes, water distribution pipes and the necessary connecting pipes, fittings, control valves, faucets and appurtenances used to dispense water intended for human ingestion shall be evaluated and listed as conforming to the requirements of NSF 61.

TABLE P2605.1 PIPING SUPPORT

PIPING MATERIAL	MAXIMUM HORIZONTAL SPACING (feet)	MAXIMUM VERTICAL SPACING
ABS pipe	4	10[b]
Aluminum tubing	10	15
Brass pipe	10	10
Cast-iron pipe	5[a]	15
Copper or copper alloy pipe	12	10
Copper or copper alloy tubing ($1^1/_4$ inches in diameter and smaller)	6	10
Copper or copper alloy tubing ($1^1/_2$ inches in diameter and larger)	10	10
Cross-linked polyethylene (PEX) pipe	2.67 (32 inches)	10[b]
Cross-linked polyethylene/aluminum/cross-linked polyethylene (PEX-AL-PEX) pipe	2.67 (32 inches)	4[b]
CPVC pipe or tubing (1 inch in diameter and smaller)	3	10[b]
CPVC pipe or tubing ($1^1/_4$ inches in diameter and larger)	4	10[b]
Lead pipe	Continuous	4
PB pipe or tubing	2.67 (32 inches)	4
Polyethylene/aluminum/polyethylene (PE-AL-PE) pipe	2.67 (32 inches)	4[b]
Polyethylene of raised temperature (PE-RT) pipe	2.67 (32 inches)	10[b]
Polypropylene (PP) pipe or tubing (1 inch and smaller)	2.67 (32 inches)	10[b]
Polypropylene (PP) pipe or tubing ($1^1/_4$ inches and larger)	4	10[b]
PVC pipe	4	10[b]
Stainless steel drainage systems	10	10[b]
Steel pipe	12	15

For SI: 1 inch = 25.4 mm, 1 foot = 304.8 mm.

a. The maximum horizontal spacing of cast-iron pipe hangers shall be increased to 10 feet where 10-foot lengths of pipe are installed.

b. Midstory guide for sizes 2 inches and smaller.

CHAPTER 27

PLUMBING FIXTURES

SECTION P2701
FIXTURES, FAUCETS AND FIXTURE FITTINGS

P2701.1 Quality of fixtures. Plumbing fixtures, faucets and fixture fittings shall be constructed of *approved* materials, shall have smooth impervious surfaces, shall be free from defects and concealed fouling surfaces, and shall conform to the standards cited in this code. Plumbing fixtures shall be provided with an adequate supply of potable water to flush and keep the fixtures in a clean and sanitary condition without danger of backflow or cross connection.

SECTION P2702
FIXTURE ACCESSORIES

P2702.1 Plumbing fixtures. Plumbing fixtures, other than water closets, shall be provided with *approved* strainers.

Exception: Hub drains and standpipes.

P2702.2 Waste fittings. Waste fittings shall conform to ASME A112.18.2/CSA B125.2, ASTM F 409 or to one of the standards listed in Table P3002.1(1) for above-ground drainage and vent pipe and fittings.

P2702.3 Plastic tubular fittings. Plastic tubular fittings shall conform to ASTM F 409 listed in Table P2701.1.

P2702.4 Carriers for wall-hung water closets. Carriers for wall-hung water closets shall conform to ASME A112.6.1 or ASME A112.6.2.

SECTION P2703
TAIL PIECES

P2703.1 Minimum size. Fixture tail pieces shall be not less than $1^1/_2$ inches (38 mm) in diameter for sinks, dishwashers, laundry tubs, bathtubs and similar fixtures, and not less than $1^1/_4$ inches (32 mm) in diameter for bidets, lavatories and similar fixtures.

SECTION P2704
ACCESS TO CONNECTIONS

P2704.1 General. Slip joints shall be made with an *approved* elastomeric gasket and shall be installed only on the trap outlet, trap inlet and within the trap seal. Fixtures with concealed slip-joint connections shall be provided with an access panel or utility space not less than 12 inches (305 mm) in its smallest dimension or other *approved* arrangement so as to provide access to the slip connections for inspection and repair.

SECTION P2705
INSTALLATION

P2705.1 General. The installation of fixtures shall conform to the following:

1. Floor-outlet or floor-mounted fixtures shall be secured to the drainage connection and to the floor, where so designed, by screws, bolts, washers, nuts and similar fasteners of copper, brass or other corrosion-resistant material.
2. Wall-hung fixtures shall be rigidly supported so that strain is not transmitted to the plumbing system.
3. Where fixtures come in contact with walls and floors, the contact area shall be water tight.
4. Plumbing fixtures shall be usable.
5. Water closets, lavatories and bidets. A water closet, lavatory or bidet shall not be set closer than 15 inches (381 mm) from its center to any side wall, partition or vanity or closer than 30 inches (762 mm) center-to-center between adjacent fixtures. There shall be a clearance of not less than 21 inches (533 mm) in front of a water closet, lavatory or bidet to any wall, fixture or door.
6. The location of piping, fixtures or equipment shall not interfere with the operation of windows or doors.
7. In flood hazard areas as established by Table R301.2(1), plumbing fixtures shall be located or installed in accordance with Section R322.1.7.
8. Integral fixture-fitting mounting surfaces on manufactured plumbing fixtures or plumbing fixtures constructed on site, shall meet the design requirements of ASME A112.19.2/CSA B45.1 or ASME A112.19.3/CSA B45.1.

SECTION P2706
WASTE RECEPTORS

P2706.1 General. Waste receptors shall be of an *approved* type. Plumbing fixtures or other receptors receiving the discharge of indirect waste pipes shall be shaped and have a capacity to prevent splashing or flooding and shall be readily accessible for inspection and cleaning. Waste receptors and standpipes shall be trapped and vented and shall connect to the building drainage system. A removable strainer or basket shall cover the waste outlet of waste receptors. Waste receptors shall be installed in ventilated spaces. Waste receptors shall not be installed in bathrooms, attics, crawl spaces, interstitial spaces above ceilings and below floors or in any inaccessible or unventilated space such as a closet. Ready access shall be provided to waste receptors.

Exceptions:

1. Open hub waste receptors shall be permitted in the form of a hub or pipe extending not less than 1 inch (25 mm) above a water-impervious floor, and are not required to have a strainer.
2. Clothes washer standpipes shall not be prohibited in bathrooms.

P2706.2 Standpipes. Standpipes shall extend not less than of 18 inches (457 mm) but not greater than 42 inches (1067 mm)

TABLE P2701.1
PLUMBING FIXTURES, FAUCETS AND FIXTURE FITTINGS

MATERIAL	STANDARD
Air gap fittings for use with plumbing fixtures, appliances and appurtenances	ASME A 112.1.3
Bathtub/whirlpool pressure-sealed doors	ASME A 112.19.15
Diverters for faucets with hose spray, anti-syphon type, residential application	ASTM A 112.18.1/CSA B125.1
Enameled cast-iron plumbing fixtures	ASME A 112.19.1M, CSA B45.2
Floor drains	ASME A 112.6.3
Floor-affixed supports for off-the-floor plumbing fixtures for public use	ASME A 112.6.1M
Framing-affixed supports for off-the-floor water closets with concealed tanks	ASME A 112.6.2
Hose connection vacuum breaker	ASSE 1052
Hot water dispensers, household storage type, electrical	ASSE 1023
Household disposers	ASSE 1008
Hydraulic performance for water closets and urinals	ASME A 112.19.2/CSA B45.1
Individual automatic compensating valves for individual fixture fittings	ASTM A 112.18.1/CSA B125.1
Individual shower control valves anti-scald	ASSE 1016, CSA B125
Macerating toilet systems and related components	ASME A 112.3.4 , CSA B54.9
Nonvitreous ceramic plumbing fixtures	ASME A 112.19.2/CSA B45.1
Plastic bathtub units	ANSI Z124.1.2, ASME A112.19.2/CSA B45.1
Plastic lavatories	ANSI Z124.3, CSA B45.5
Plastic shower receptors and shower stall	ANSI Z124.2, CSA B45.5
Plastic sinks	ANSI Z124.6, CSA B45.5
Plastic water closet bowls and tanks	ANSI Z124.4, CSA B45.5
Plumbing fixture fittings	ASME A 112.18.1/CSA B125.1
Plumbing fixture waste fittings	ASME A 112.18.2/CSA B125.2, ASTM F 409
Porcelain-enameled formed steel plumbing fixtures	ASME A 112.19.1/CSA B45.2
Pressurized flushing devices for plumbing fixtures	ASSE 1037, CSA B125.3
Specification for copper sheet and strip for building construction	ASTM B 370
Stainless steel plumbing fixtures	ASME A 112.19.3/CSA B45.4
Suction fittings for use in whirlpool bathtub appliances	ASME A 112.19.7 /CSA B45.10
Temperature-actuated, flow reduction valves to individual fixture fittings	ASSE 1062
Thermoplastic accessible and replaceable plastic tube and tubular fittings	ASTM F 409
Trench drains	ASME A 112.6.3
Trim for water closet bowls, tanks and urinals	ASME A 112.19.5/CSA B45.15
Vacuum breaker wall hydrant-frost-resistant, automatic-draining type	ASSE 1019
Vitreous china plumbing fixtures	ASME A 112.19.2/CSA B45.1
Wall-mounted and pedestal-mounted, adjustable and pivoting lavatory and sink carrier systems	ASME A 112.19.12
Water closet flush tank fill valves	ASSE 1002, CSA B125.3
Whirlpool bathtub appliances	ASME A 112.19.7 /CSA B45.10

above the trap weir. Access shall be provided to standpipe traps and drains for rodding.

P2706.2.1 Laundry tray connection. A laundry tray waste line is permitted to connect into a standpipe for the automatic clothes washer drain. The standpipe shall extend not less than 30 inches (762 mm) above the trap weir and shall extend above the flood level rim of the laundry tray. The outlet of the laundry tray shall not be greater than 30 inches (762 mm) horizontal distance from the standpipe trap.

P2706.3 Prohibited waste receptors. Plumbing fixtures that are used for washing or bathing shall not be used to receive the discharge of indirect waste piping.

Exceptions:

1. A kitchen sink trap is acceptable for use as a receptor for a dishwasher.
2. A laundry tray is acceptable for use as a receptor for a clothes washing machine.

SECTION P2707 DIRECTIONAL FITTINGS

P2707.1 Directional fitting required. *Approved* directional-type branch fittings shall be installed in fixture tailpieces receiving the discharge from food waste disposal units or dishwashers.

SECTION P2708 SHOWERS

P2708.1 General. Shower compartments shall have not less than 900 square inches (0.6 m^2) of interior cross-sectional area. Shower compartments shall be not less than 30 inches (762 mm) in minimum dimension measured from the finished interior dimension of the shower compartment, exclusive of fixture valves, shower heads, soap dishes, and safety grab bars or rails. The minimum required area and dimension shall be measured from the finished interior dimension at a height equal to the top of the threshold and at a point tangent to its centerline and shall be continued to a height of not less than 70 inches (1778 mm) above the shower drain outlet. Hinged shower doors shall open outward. The wall area above built-in tubs having installed shower heads and in shower compartments shall be constructed in accordance with Section R702.4. Such walls shall form a water-tight joint with each other and with either the tub, receptor or shower floor.

Exceptions:

1. Fold-down seats shall be permitted in the shower, provided the required 900-square-inch (0.6 m^2) dimension is maintained when the seat is in the folded-up position.
2. Shower compartments having not less than 25 inches (635 mm) in minimum dimension measured from the finished interior dimension of the compartment provided that the shower compartment has a cross-sectional area of not less than 1,300 square inches (0.838 m^2).

P2708.1.1 Access. The shower compartment access and egress opening shall have a clear and unobstructed finished width of not less than 22 inches (559 mm).

P2708.2 Water supply riser. Water supply risers from the shower valve to the shower head outlet, whether exposed or concealed, shall be attached to the structure using support devices designed for use with the specific piping material or fittings anchored with screws.

P2708.3 Shower control valves. Individual shower and tub/shower combination valves shall be equipped with control valves of the pressure-balance, thermostatic-mixing or combination pressure-balance/thermostatic-mixing valve types with a high limit stop in accordance with ASSE 1016 or ASME A112.18.1/CSA B125.1. The high limit stop shall be set to limit the water temperature to not greater than 120°F (49°C). In-line thermostatic valves shall not be used for compliance with this section.

P2708.4 Hand showers. Hand-held showers shall conform to ASME A112.18.1/CSA B125.1. Hand-held showers shall provide backflow protection in accordance with ASME A112.18.1/CSA B125.1 or shall be protected against backflow by a device complying with ASME A112.18.3.

SECTION P2709 SHOWER RECEPTORS

P2709.1 Construction. Where a shower receptor has a finished curb threshold, it shall be not less than 1 inch (25 mm) below the sides and back of the receptor. The curb shall be not less than 2 inches (51 mm) and not more than 9 inches (229 mm) deep when measured from the top of the curb to the top of the drain. The finished floor shall slope uniformly toward the drain not less than $^1/_4$ unit vertical in 12 units horizontal (2-percent slope) nor more than $^1/_2$ unit vertical per 12 units horizontal (4-percent slope) and floor drains shall be flanged to provide a water-tight joint in the floor.

P2709.2 Lining required. The adjoining walls and floor framing enclosing on-site built-up shower receptors shall be lined with one of the following materials:

1. Sheet lead;
2. Sheet copper;
3. Plastic liner material that complies with ASTM D 4068 or ASTM D 4551;
4. Hot mopping in accordance with Section P2709.2.3; or
5. Sheet-applied load-bearing, bonded waterproof membranes that comply with ANSI A118.10.

The lining material shall extend not less than 2 inches (51 mm) beyond or around the rough jambs and not less than 2 inches (51 mm) above finished thresholds. Sheet-applied load bearing, bonded waterproof membranes shall be applied in accordance with the manufacturer's instructions.

P2709.2.1 PVC sheets. Plasticized polyvinyl chloride (PVC) sheet shall meet the requirements of ASTM D 4551. Sheets shall be joined by solvent welding in accordance with the manufacturer's instructions.

P2709.2.2 Chlorinated polyethylene (CPE) sheets. Nonplasticized chlorinated polyethylene sheet shall meet the requirements of ASTM D 4068. The liner shall be joined in accordance with the manufacturer's instructions.

P2709.2.3 Hot-mopping. Shower receptors lined by hot mopping shall be built-up with not less than three layers of standard grade Type 15 asphalt-impregnated roofing felt. The bottom layer shall be fitted to the formed subbase and each succeeding layer thoroughly hot-mopped to that below. All corners shall be carefully fitted and shall be made strong and water tight by folding or lapping, and each corner shall be reinforced with suitable webbing hot-mopped in place. All folds, laps and reinforcing webbing shall extend not less than 4 inches (102 mm) in all directions from the corner and all webbing shall be of *approved* type and mesh, producing a tensile strength of not less than 50 pounds per inch (893 kg/m) in either direction.

P2709.2.4 Liquid-type, trowel-applied, load-bearing, bonded waterproof materials. Liquid-type, trowel-applied, load-bearing, bonded waterproof materials shall meet the requirements of ANSI A118.10 and shall be applied in accordance with the manufacturer's instructions.

P2709.3 Installation. Lining materials shall be sloped one-fourth unit vertical in 12 units horizontal (2-percent slope) to weep holes in the subdrain by means of a smooth, solidly formed subbase, shall be properly recessed and fastened to *approved* backing so as not to occupy the space required for the wall covering, and shall not be nailed or perforated at any point less than 1 inch (25.4 mm) above the finished threshold.

P2709.3.1 Materials. Lead and copper linings shall be insulated from conducting substances other than the connecting drain by 15-pound (6.80 kg) asphalt felt or its equivalent. Sheet lead liners shall weigh not less than 4 pounds per square foot (19.5 kg/m^2). Sheet copper liners shall weigh not less than 12 ounces per square foot (3.7 kg/m^2). Joints in lead and copper pans or liners shall be burned or silver brazed, respectively. Joints in plastic liner materials shall be jointed per the manufacturer's instructions.

P2709.4 Receptor drains. An *approved* flanged drain shall be installed with shower subpans or linings. The flange shall be placed flush with the subbase and be equipped with a clamping ring or other device to make a water-tight connection between the lining and the drain. The flange shall have weep holes into the drain.

SECTION P2710
SHOWER WALLS

P2710.1 Bathtub and shower spaces. Shower walls shall be finished in accordance with Section R307.2.

SECTION P2711
LAVATORIES

P2711.1 Approval. Lavatories shall conform to ANSI Z124.3, ASME A112.19.1/CSA B45.2, ASME A112.19.2/CSA B45.1 or ASME A112.19.3/CSA B45.4.

P2711.2 Cultured marble lavatories. Cultured marble vanity tops with an integral lavatory shall conform to ANSI Z124.3 or CSA B45.5.

P2711.3 Lavatory waste outlets. Lavatories shall have waste outlets not less than $1^1/_4$ inch (32 mm) in diameter. A strainer, pop-up stopper, crossbar or other device shall be provided to restrict the clear opening of the waste outlet.

P2711.4 Movable lavatory systems. Movable lavatory systems shall comply with ASME A112.19.12.

SECTION P2712
WATER CLOSETS

P2712.1 Approval. Water closets shall conform to the water consumption requirements of Section P2903.2 and shall conform to ANSI Z124.4, ASME A112.19.2/CSA B45.1, ASME A112.19.3/CSA B45.4 or CSA B45.5. Water closets shall conform to the hydraulic performance requirements of ASME A112.19.2/CSA B45.1. Water closet tanks shall conform to ANSI Z124.4, ASME A112.19.2/CSA B45.1, ASME A112.19.3/CSA B45.4 or CSA B45.5. Water closets that have an invisible seal and unventilated space or walls that are not thoroughly washed at each discharge shall be prohibited. Water closets that permit backflow of the contents of the bowl into the flush tank shall be prohibited.

P2712.2 Flushing devices required. Water closets shall be provided with a flush tank, flushometer tank or flushometer valve designed and installed to supply water in sufficient quantity and flow to flush the contents of the fixture, to cleanse the fixture and refill the fixture trap in accordance with ASME A112.19.2/CSA B45.1.

P2712.3 Water supply for flushing devices. An adequate quantity of water shall be provided to flush and clean the fixture served. The water supply to flushing devices equipped for manual flushing shall be controlled by a float valve or other automatic device designed to refill the tank after each discharge and to completely shut off the water flow to the tank when the tank is filled to operational capacity. Provision shall be made to automatically supply water to the fixture so as to refill the trap after each flushing.

P2712.4 Flush valves in flush tanks. Flush valve seats in tanks for flushing water closets shall be not less than 1 inch (25 mm) above the flood-level rim of the bowl connected thereto, except an *approved* water closet and flush tank combination designed so that when the tank is flushed and the fixture is clogged or partially clogged, the flush valve will close tightly so that water will not spill continuously over the rim of the bowl or backflow from the bowl to the tank.

P2712.5 Overflows in flush tanks. Flush tanks shall be provided with overflows discharging to the water closet connected thereto and such overflow shall be of sufficient size to prevent flooding the tank at the maximum rate at which the tanks are supplied with water according to the manufacturer's design conditions.

P2712.6 Access. All parts in a flush tank shall be accessible for repair and replacement.

P2712.7 Water closet seats. Water closets shall be equipped with seats of smooth, nonabsorbent material and shall be properly sized for the water closet bowl type.

P2712.8 Flush tank lining. Sheet copper used for flush tank linings shall have a weight of not less than 10 ounces per square foot (3 kg/m^2).

P2712.9 Electro-hydraulic water closets. Electro-hydraulic water closets shall conform to ASME A112.19.2/CSA B45.1.

SECTION P2713 BATHTUBS

P2713.1 Bathtub waste outlets and overflows. Bathtubs shall be equipped with a waste outlet and an overflow outlet. The outlets shall be connected to waste tubing or piping not less than $1^1/_2$ inches (38 mm) in diameter. The waste outlet shall be equipped with a water-tight stopper.

P2713.2 Bathtub enclosures. Doors within a bathtub enclosure shall conform to ASME A112.19.15.

P2713.3 Bathtub and whirlpool bathtub valves. Hot water supplied to bathtubs and whirlpool bathtubs shall be limited to a temperature of not greater than 120°F (49°C) by a water-temperature limiting device that conforms to ASSE 1070 or CSA B125.3, except where such protection is otherwise provided by a combination tub/shower valve in accordance with Section P2708.3.

SECTION P2714 SINKS

P2714.1 Sink waste outlets. Sinks shall be provided with waste outlets not less than $1^1/_2$ inches (38 mm) in diameter. A strainer, crossbar or other device shall be provided to restrict the clear opening of the waste outlet.

P2714.2 Movable sink systems. Movable sink systems shall comply with ASME A112.19.12.

SECTION P2715 LAUNDRY TUBS

P2715.1 Laundry tub waste outlet. Each compartment of a laundry tub shall be provided with a waste outlet not less than $1^1/_2$ inches (38 mm) in diameter and a strainer or crossbar to restrict the clear opening of the waste outlet.

SECTION P2716 FOOD WASTE GRINDER

P2716.1 Food waste grinder waste outlets. Food waste grinders shall be connected to a drain of not less than $1^1/_2$ inches (38 mm) in diameter.

P2716.2 Water supply required. Food waste grinders shall be provided with an adequate supply of water at a sufficient flow rate to ensure proper functioning of the unit.

SECTION P2717 DISHWASHING MACHINES

P2717.1 Protection of water supply. The water supply for dishwashers shall be protected by an air gap or integral backflow preventer.

P2717.2 Sink and dishwasher. A sink and dishwasher are permitted to discharge through a single $1^1/_2$-inch (38 mm) trap. The discharge pipe from the dishwasher shall be increased to not less than $^3/_4$ inch (19 mm) in diameter and shall be connected with a wye fitting to the sink tailpiece. The dishwasher waste line shall rise and be securely fastened to the underside of the counter before connecting to the sink tailpiece.

P2717.3 Sink, dishwasher and food grinder. The combined discharge from a sink, dishwasher, and waste grinder is permitted to discharge through a single $1^1/_2$-inch (38 mm) trap. The discharge pipe from the dishwasher shall be increased to not less than $^3/_4$ inch (19 mm) in diameter and shall connect with a wye fitting between the discharge of the food-waste grinder and the trap inlet or to the head of the food grinder. The dishwasher waste line shall rise and be securely fastened to the underside of the counter before connecting to the sink tail piece or the food grinder.

SECTION P2718 CLOTHES WASHING MACHINE

P2718.1 Waste connection. The discharge from a clothes washing machine shall be through an *air break*.

SECTION P2719 FLOOR DRAINS

P2719.1 Floor drains. Floor drains shall have waste outlets not less than 2 inches (51 mm) in diameter and a removable strainer. The floor drain shall be constructed so that the drain can be cleaned. Access shall be provided to the drain inlet. Floor drains shall not be located under or have their access restricted by permanently installed appliances.

SECTION P2720 WHIRLPOOL BATHTUBS

P2720.1 Access to pump. Access shall be provided to circulation pumps in accordance with the fixture or pump manufacturer's installation instructions. Where the manufacturer's instructions do not specify the location and minimum size of field-fabricated access openings, an opening of not less than 12-inches by 12-inches (305 mm by 305 mm) shall be installed for access to the circulation pump. Where pumps are located more than 2 feet (610 mm) from the access opening, an opening of not less than 18 inches by 18 inches (457 mm by 457 mm) shall be installed. A door or panel shall be permitted to close the opening. In all cases, the access opening shall be unobstructed and be of the size necessary to permit the removal and replacement of the circulation pump.

P2720.2 Piping drainage. The circulation pump shall be accessibly located above the crown weir of the trap. The pump drain line shall be properly graded to ensure minimum water retention in the volute after fixture use. The circulation piping shall be installed to be self-draining.

P2720.3 Leak testing. Leak testing and pump operation shall be performed in accordance with the manufacturer's instructions.

P2720.4 Manufacturer's instructions. The product shall be installed in accordance with the manufacturer's instructions.

SECTION P2721 BIDET INSTALLATIONS

P2721.1 Water supply. The bidet shall be equipped with either an air-gap-type or vacuum-breaker-type fixture supply fitting.

P2721.2 Bidet water temperature. The discharge water temperature from a bidet fitting shall be limited to not greater than 110°F (43°C) by a water-temperature-limiting device conforming to ASSE 1070 or CSA B125.3.

SECTION P2722 FIXTURE FITTING

P2722.1 General. Fixture supply valves and faucets shall comply with ASME A112.18.1/CSA B125.1 as listed in Table P2701.1. Faucets and fixture fittings that supply drinking water for human ingestion shall conform to the requirements of NSF 61, Section 9. Flexible water connectors shall conform to the requirements of Section P2905.7.

P2722.2 Hot water. Fixture fittings and faucets that are supplied with both hot and cold water shall be installed and adjusted so that the left-hand side of the water temperature control represents the flow of hot water when facing the outlet.

> **Exception:** Shower and tub/shower mixing valves conforming to ASSE 1016 or ASME A112.18.1/CSA B125.1, where the water temperature control corresponds to the markings on the device.

P2722.3 Hose-connected outlets. Faucets and fixture fittings with hose-connected outlets shall conform to ASME A112.18.3 or ASME A112.18.1/CSA B125.1.

P2722.4 Individual pressure-balancing in-line valves for individual fixture fittings. Where individual pressure-balancing in-line valves for individual fixture fittings are installed, the valves shall comply with ASSE 1066. Such valves shall be installed in an accessible location and shall not be used alone as a substitute for the balanced pressure, thermostatic or combination shower valves required in Section P2708.3.

P2722.5 Water closet personal hygiene devices. Personal hygiene devices integral to water closets or water closet seats shall conform to the requirements of ASME A112.4.2.

SECTION P2723 MACERATING TOILET SYSTEMS

P2723.1 General. Macerating toilet systems shall be installed in accordance with manufacturer's instructions.

P2723.2 Drain. The size of the drain from the macerating toilet system shall be not less than $^3/_4$ inch (19 mm) in diameter.

SECTION P2724 SPECIALTY TEMPERATURE CONTROL DEVICES AND VALVES

P2724.1 Temperature-actuated mixing valves. Temperature-actuated mixing valves, which are installed to reduce water temperatures to defined limits, shall comply with ASSE 1017. Such valves shall be installed at the hot water source.

P2724.2 Temperature-actuated, flow-reduction devices for individual fixtures. Temperature-actuated, flow-reduction devices, where installed for individual fixture fittings, shall conform to ASSE 1062. Such valves shall not be used alone as a substitute for the balanced pressure, thermostatic or combination shower valves required for showers in Section P2708.3.

CHAPTER 28

WATER HEATERS

SECTION P2801 GENERAL

P2801.1 Required. Each *dwelling* shall have an *approved* automatic water heater or other type of domestic water-heating system sufficient to supply hot water to plumbing fixtures and appliances intended for bathing, washing or culinary purposes. Storage tanks shall be constructed of noncorrosive metal or shall be lined with noncorrosive material.

P2801.2 Installation. Water heaters shall be installed in accordance with this chapter and Chapters 20 and 24.

P2801.3 Location. Water heaters and storage tanks shall be installed in accordance with Section M1305 and shall be located and connected to provide access for observation, maintenance, servicing and replacement.

P2801.4 Prohibited locations. Water heaters shall be located in accordance with Chapter 20.

P2801.5 Required pan. Where a storage tank-type water heater or a hot water storage tank is installed in a location where water leakage from the tank will cause damage, the tank shall be installed in a galvanized steel pan having a material thickness of not less than 0.0236 inch (0.6010 mm) (No. 24 gage), or other pans approved for such use. Listed pans shall comply with CSA LC3.

P2801.5.1 Pan size and drain. The pan shall be not less than $1^1/_2$ inches (38 mm) deep and shall be of sufficient size and shape to receive all dripping or condensate from the tank or water heater. The pan shall be drained by an indirect waste pipe of not less than $^3/_4$ inch (19 mm) diameter. Piping for safety pan drains shall be of those materials listed in Table P2905.5.

P2801.5.2 Pan drain termination. The pan drain shall extend full-size and terminate over a suitably located indirect waste receptor or shall extend to the exterior of the building and terminate not less than 6 inches (152 mm) and not more than 24 inches (610 mm) above the adjacent ground surface.

P2801.6 Water heaters installed in garages. Water heaters having an *ignition source* shall be elevated such that the source of ignition is not less than 18 inches (457 mm) above the garage floor.

Exception: Elevation of the ignition source is not required for appliances that are listed as flammable vapor ignition-resistant.

P2801.7 Water heater seismic bracing. In Seismic Design Categories D_0, D_1 and D_2 and townhouses in Seismic Design Category C, water heaters shall be anchored or strapped in the upper one-third and in the lower one-third of the appliance to resist a horizontal force equal to one-third of the operating weight of the water heater, acting in any horizontal direction, or in accordance with the appliance manufacturer's recommendations.

SECTION P2802 WATER HEATERS USED FOR SPACE HEATING

P2802.1 Protection of potable water. Piping and components connected to a water heater for space heating applications shall be suitable for use with potable water in accordance with Chapter 29. Water heaters that will be used to supply potable water shall not be connected to a heating system or components previously used with nonpotable-water heating *appliances*. Chemicals for boiler treatment shall not be introduced into the water heater.

P2802.2 Temperature control. Where a combination water heater-space heating system requires water for space heating at temperatures exceeding 140°F (60°C), a master thermostatic mixing valve complying with ASSE 1017 shall be installed to temper the water to a temperature of not greater than 140°F (60°C) for domestic uses.

SECTION P2803 RELIEF VALVES

P2803.1 Relief valves required. Appliances and equipment used for heating water or storing hot water shall be protected by:

1. A separate pressure-relief valve and a separate temperature-relief valve; or
2. A combination pressure- and temperature-relief valve.

P2803.2 Rating. Relief valves shall have a minimum rated capacity for the equipment served and shall conform to ANSI Z21.22.

P2803.3 Pressure-relief valves. Pressure-relief valves shall have a relief rating adequate to meet the pressure conditions for the appliances or equipment protected. In tanks, they shall be installed directly into a tank tapping or in a water line close to the tank. They shall be set to open at not less than 25 psi (172 kPa) above the system pressure but not over 150 psi (1034 kPa). The relief-valve setting shall not exceed the tanks rated working pressure.

P2803.4 Temperature-relief valves. Temperature-relief valves shall have a relief rating compatible with the temperature conditions of the appliances or equipment protected. The valves shall be installed such that the temperature-sensing element monitors the water within the top 6 inches (152 mm) of the tank. The valve shall be set to open at a temperature of not greater than 210°F (99°C).

P2803.5 Combination pressure-/temperature-relief valves. Combination pressure-/temperature-relief valves shall comply with all the requirements for separate pressure- and temperature-relief valves.

P2803.6 Installation of relief valves. A check or shutoff valve shall not be installed in the following locations:

1. Between a relief valve and the termination point of the relief valve discharge pipe;
2. Between a relief valve and a tank; or
3. Between a relief valve and heating appliances or equipment.

P2803.6.1 Requirements for discharge pipe. The discharge piping serving a pressure-relief valve, temperature-relief valve or combination valve shall:

1. Not be directly connected to the drainage system.
2. Discharge through an air gap located in the same room as the water heater.
3. Not be smaller than the diameter of the outlet of the valve served and shall discharge full size to the air gap.
4. Serve a single relief device and shall not connect to piping serving any other relief device or equipment.
5. Discharge to the floor, to the pan serving the water heater or storage tank, to a waste receptor or to the outdoors.
6. Discharge in a manner that does not cause personal injury or structural damage.
7. Discharge to a termination point that is readily observable by the building occupants.
8. Not be trapped.
9. Be installed to flow by gravity.
10. Not terminate more than 6 inches (152 mm) above the floor or waste receptor.
11. Not have a threaded connection at the end of the piping.
12. Not have valves or tee fittings.
13. Be constructed of those materials listed in Section P2904.5 or materials tested, rated and *approved* for such use in accordance with ASME A112.4.1.

P2803.7 Vacuum-relief valve. Bottom fed tank-type water heaters and bottom fed tanks connected to water heaters shall have a vacuum-relief valve installed that complies with ANSI Z21.22.

CHAPTER 29
WATER SUPPLY AND DISTRIBUTION

SECTION P2901
GENERAL

P2901.1 Potable water required. *Dwelling units* shall be supplied with potable water in the amounts and pressures specified in this chapter. Where a nonpotable water-distribution system is installed, the nonpotable system shall be identified by color marking, metal tags or other appropriate method. Where color is used for marking, purple shall be used to identify municipally reclaimed water, rainwater and graywater distribution systems. Nonpotable outlets that could inadvertently be used for drinking or domestic purposes shall be posted.

SECTION P2902
PROTECTION OF POTABLE WATER SUPPLY

P2902.1 General. A potable water supply system shall be designed and installed as to prevent contamination from nonpotable liquids, solids or gases being introduced into the potable water supply. Connections shall not be made to a potable water supply in a manner that could contaminate the water supply or provide a cross-connection between the supply and a source of contamination except where approved methods are installed to protect the potable water supply. Cross-connections between an individual water supply and a potable public water supply shall be prohibited.

P2902.2 Plumbing fixtures. The supply lines and fittings for every plumbing fixture shall be installed so as to prevent backflow. Plumbing fixture fittings shall provide backflow protection in accordance with ASME A112.18.1/CSA B125.1.

P2902.3 Backflow protection. A means of protection against backflow shall be provided in accordance with Sections P2902.3.1 through P2902.3.6. Backflow prevention applications shall conform to Table P2902.3, except as specifically stated in Sections P2902.4 through P2902.5.5.

P2902.3.1 Air gaps. Air gaps shall comply with ASME A112.1.2 and air gap fittings shall comply with ASME A112.1.3. The minimum air gap shall be measured vertically from the lowest end of a water supply outlet to the flood level rim of the fixture or receptor into which such potable water outlets discharge. The minimum required air gap shall be twice the diameter of the effective opening of the outlet, but in no case less than the values specified in Table P2902.3.1. An air gap is required at the discharge point of a relief valve or piping. Air gap devices shall be incorporated in dishwashing and clothes washing *appliances*.

P2902.3.2 Atmospheric-type vacuum breakers. Pipe-applied atmospheric-type vacuum breakers shall conform to ASSE 1001 or CSA B64.1.1. Hose-connection vacuum breakers shall conform to ASSE 1011, ASSE 1019, ASSE 1035, ASSE 1052, CSA B64.2, CSA B64.2.1, CSA B64.2.1.1, CSA B64.2.2 or CSA B64.7. These devices shall operate under normal atmospheric pressure when the critical level is installed at the required height.

P2902.3.3 Backflow preventer with intermediate atmospheric vent. Backflow preventers with intermediate atmospheric vents shall conform to ASSE 1012 or CAN/CSA B64.3. These devices shall be permitted to be installed where subject to continuous pressure conditions. The relief opening shall discharge by air gap and shall be prevented from being submerged.

P2902.3.4 Pressure vacuum breaker assemblies. Pressure vacuum breaker assemblies shall conform to ASSE 1020 or CSA B64.1.2. Spill-resistant vacuum breaker assemblies shall comply with ASSE 1056. These assemblies are designed for installation under continuous pressure conditions where the critical level is installed at the required height. Pressure vacuum breaker assemblies shall not be installed in locations where spillage could cause damage to the structure.

TABLE P2902.3.1
MINIMUM AIR GAPS

FIXTURE	MINIMUM AIR GAP	
	Away from a wall[a] (inches)	Close to a wall (inches)
Effective openings greater than 1 inch	Two times the diameter of the effective opening	Three times the diameter of the effective opening
Lavatories and other fixtures with effective opening not greater than $^1/_2$ inch in diameter	1	1.5
Over-rim bath fillers and other fixtures with effective openings not greater than 1 inch in diameter	2	3
Sink, laundry trays, gooseneck back faucets and other fixtures with effective openings not greater than $^3/_4$ inch in diameter	1.5	2.5

For SI: 1 inch = 25.4 mm.

a. Applicable where walls or obstructions are spaced from the nearest inside edge of the spout opening a distance greater than three times the diameter of the effective opening for a single wall, or a distance greater than four times the diameter of the effective opening for two intersecting walls.

TABLE P2902.3
APPLICATION FOR BACKFLOW PREVENTERS

DEVICE	DEGREE OF HAZARD[a]	APPLICATION[b]	APPLICABLE STANDARDS
Air gap	High or low hazard	Backsiphonage or backpressure	ASME A112.1.2
Air gap fittings for use with plumbing fixtures, appliances and appurtenances	High or low hazard	Backsiphonage or backpressure	ASME A112.1.3
Antisiphon-type fill valves for gravity water closet flush tanks	High hazard	Backsiphonage only	ASSE 1002, CSA B125.3
Backflow preventer with intermediate atmospheric vents	Low hazard	Backpressure or backsiphonage Sizes $^1/_4$″ – $^3/_4$″	ASSE 1012, CSA B64.3
Double check backflow prevention assembly and double check fire protection backflow prevention assembly	Low hazard	Backpressure or backsiphonage Sizes $^3/_8$″ – 16″	ASSE 1015, AWWA C510, CSA B64.5, CSA B64.5.1
Double check detector fire protection backflow prevention assemblies	Low hazard	Backpressure or backsiphonage (Fire sprinkler systems) Sizes 2″ – 16″	ASSE 1048
Dual-check-valve-type backflow preventer	Low hazard	Backpressure or backsiphonage Sizes $^1/_4$″ – 1″	ASSE 1024, CSA B64.6
Hose-connection backflow preventer	High or low hazard	Low head backpressure, rated working pressure backpressure or backsiphonage Sizes $^1/_2$″ – 1″	ASSE 1052, CSA B64.2.1.1
Hose-connection vacuum breaker	High or low hazard	Low head backpressure or backsiphonage Sizes $^1/_2$″, $^3/_4$″, 1″	ASSE 1011, CSA B64.2, CSA B64.2.1
Laboratory faucet backflow preventer	High or low hazard	Low head backpressure and backsiphonage	ASSE 1035, CSA B64.7
Pipe-applied atmospheric-type vacuum breaker	High or low hazard	Backsiphonage only Sizes $^1/_4$″ – 4″	ASSE 1001, CSA B64.1.1
Pressure vacuum breaker assembly	High or low hazard	Backsiphonage only Sizes $^1/_2$″ – 2″	ASSE 1020, CSA B64.1.2
Reduced pressure detector fire protection backflow prevention assemblies	High or low hazard	Backsiphonage or backpressure (Fire sprinkler systems)	ASSE 1047
Reduced pressure principle backflow prevention assembly and reduced pressure principle fire protection backflow prevention assembly	High or low hazard	Backpressure or backsiphonage Sizes $^3/_8$″ – 16″	ASSE 1013, AWWA C511, CSA B64.4, CSA B64.4.1
Spill-resistant pressure vacuum breaker	High or low hazard	Backsiphonage only Sizes $^1/_4$″ – 2″	ASSE 1056, CSA B64.1.3
Vacuum breaker wall hydrants, frost-resistant, automatic-draining type	High or low hazard	Low head backpressure or backsiphonage Sizes $^3/_4$″ – 1″	ASSE 1019, CSA B64.2.2

For SI: 1 inch = 25.4 mm.

a. Low hazard—See Pollution (Section R202). High hazard—See Contamination (Section R202).

b. See Backpressure (Section R202). See Backpressure, Low Head (Section R202). See Backsiphonage (Section R202).

P2902.3.5 Reduced pressure principle backflow prevention assemblies. Reduced pressure principle backflow prevention assemblies and reduced pressure principle fire protection backflow prevention assemblies shall conform to ASSE 1013, AWWA C511, CSA B64.4 or CSA B64.4.1. Reduced pressure detector fire protection backflow prevention assemblies shall conform to ASSE 1047. These devices shall be permitted to be installed where subject to continuous pressure conditions. The relief opening shall discharge by air gap and shall be prevented from being submerged.

P2902.3.6 Double check-valve assemblies. Double check-valve assemblies shall conform to ASSE 1015, CSA B64.5, CSA B64.5.1 or AWWA C510. Double-detector check-valve assemblies shall conform to ASSE 1048. These devices shall be capable of operating under continuous pressure conditions.

P2902.4 Protection of potable water outlets. Potable water openings and outlets shall be protected by an air gap, reduced pressure principle backflow preventer with atmospheric vent, atmospheric-type vacuum breaker, pressure-type vacuum breaker or hose connection backflow preventer.

P2902.4.1 Fill valves. Flush tanks shall be equipped with an antisiphon fill valve conforming to ASSE 1002 or CSA B125.3. The fill valve backflow preventer shall be located not less than 1 inch (25 mm) above the full opening of the overflow pipe.

P2902.4.2 Deck-mounted and integral vacuum breakers. *Approved* deck-mounted or equipment-mounted vacuum breakers and faucets with integral atmospheric vacuum breakers or spill–resistant vacuum breaker assemblies shall be installed in accordance with the manufacturer's instructions and the requirements for labeling. The critical level of the breakers and assemblies shall be located at not less than 1 inch (25 mm) above the *flood level rim*.

P2902.4.3 Hose connection. Sillcocks, hose bibbs, wall hydrants and other openings with a hose connection shall be protected by an atmospheric-type or pressure-type vacuum breaker or a permanently attached hose connection vacuum breaker.

Exceptions:

1. This section shall not apply to water heater and boiler drain valves that are provided with hose connection threads and that are intended only for tank or vessel draining.
2. This section shall not apply to water supply valves intended for connection of clothes washing machines where backflow prevention is otherwise provided or is integral with the machine.

P2902.5 Protection of potable water connections. Connections to the potable water shall conform to Sections P2902.5.1 through P2902.5.5.

P2902.5.1 Connections to boilers. The potable supply to the boiler shall be equipped with a backflow preventer with an intermediate atmospheric vent complying with ASSE 1012 or CSA B64.3. Where conditioning chemicals are introduced into the system, the potable water connection shall be protected by an air gap or a reduced pressure principle backflow preventer complying with ASSE 1013, CSA B64.4 or AWWA C511.

P2902.5.2 Heat exchangers. Heat exchangers using an essentially toxic transfer fluid shall be separated from the potable water by double-wall construction. An air gap open to the atmosphere shall be provided between the two walls. Heat exchangers utilizing an essentially nontoxic transfer fluid shall be permitted to be of single-wall construction.

P2902.5.3 Lawn irrigation systems. The potable water supply to lawn irrigation systems shall be protected against backflow by an atmospheric vacuum breaker, a pressure vacuum breaker assembly or a reduced pressure principle backflow prevention assembly. Valves shall not be installed downstream from an atmospheric vacuum breaker. Where chemicals are introduced into the system, the potable water supply shall be protected against backflow by a reduced pressure principle backflow prevention assembly.

P2902.5.4 Connections to automatic fire sprinkler systems. The potable water supply to automatic fire sprinkler shall be protected against backflow by a double check backflow prevention assembly, a double check fire protection backflow prevention assembly, a reduced pressure principle backflow prevention assembly or a reduced pressure principle fire protection backflow prevention assembly.

Exception: Where systems are installed as a portion of the water distribution system in accordance with the requirements of this code and are not provided with a fire department connection, backflow protection for the water supply system shall not be required.

P2902.5.4.1 Additives or nonpotable source. Where systems contain chemical additives or antifreeze, or where systems are connected to a nonpotable secondary water supply, the potable water supply shall be protected against backflow by a reduced pressure principle backflow prevention assembly or a reduced pressure principle fire protection backflow prevention assembly. Where chemical additives or antifreeze is added to only a portion of an automatic fire sprinkler or standpipe system, the reduced pressure principle fire protection backflow preventer shall be permitted to be located so as to isolate that portion of the system.

P2902.5.5 Solar systems. The potable water supply to a solar system shall be equipped with a backflow preventer with intermediate atmospheric vent complying with ASSE 1012 or a reduced pressure principle backflow preventer complying with ASSE 1013. Where chemicals are used, the potable water supply shall be protected by a reduced pressure principle backflow preventer.

Exception: Where all solar system piping is a part of the potable water distribution system, in accordance with the requirements of the *International Plumbing Code*, and all components of the piping system are listed for potable water use, cross-connection protection measures shall not be required.

P2902.6 Location of backflow preventers. Access shall be provided to backflow preventers as specified by the manufacturer's installation instructions.

P2902.6.1 Outdoor enclosures for backflow prevention devices. Outdoor enclosures for backflow prevention devices shall comply with ASSE 1060.

P2902.6.2 Protection of backflow preventers. Backflow preventers shall not be located in areas subject to freezing except where they can be removed by means of unions, or are protected by heat, insulation or both.

P2902.6.3 Relief port piping. The termination of the piping from the relief port or air gap fitting of the backflow preventer shall discharge to an *approved* indirect waste receptor or to the outdoors where it will not cause damage or create a nuisance.

SECTION P2903 WATER-SUPPLY SYSTEM

P2903.1 Water supply system design criteria. The water service and water distribution systems shall be designed and pipe sizes shall be selected such that under conditions of peak demand, the capacities at the point of outlet discharge shall not be less than shown in Table P2903.1.

P2903.2 Maximum flow and water consumption. The maximum water consumption flow rates and quantities for all plumbing fixtures and fixture fittings shall be in accordance with Table P2903.2.

TABLE P2903.1
REQUIRED CAPACITIES AT POINT OF OUTLET DISCHARGE

FIXTURE AT POINT OF OUTLET	FLOW RATE (gpm)	FLOW PRESSURE (psi)
Bathtub, pressure-balanced or thermostatic mixing valve	4	20
Bidet, thermostatic mixing	2	20
Dishwasher	2.75	8
Laundry tub	4	8
Lavatory	2	8
Shower, pressure-balancing or thermostatic mixing valve	3	20
Sillcock, hose bibb	5	8
Sink	2.5	8
Water closet, flushometer tank	1.6	20
Water closet, tank, close coupled	3	20
Water closet, tank, one-piece	6	20

For SI: 1 gallon per minute = 3.785 L/m,
1 pound per square inch = 6.895 kPa.

TABLE P2903.2
MAXIMUM FLOW RATES AND CONSUMPTION FOR PLUMBING FIXTURES AND FIXTURE FITTINGS[b]

PLUMBING FIXTURE OR FIXTURE FITTING	PLUMBING FIXTURE OR FIXTURE FITTING
Lavatory faucet	2.2 gpm at 60 psi
Shower head[a]	2.5 gpm at 80 psi
Sink faucet	2.2 gpm at 60 psi
Water closet	1.6 gallons per flushing cycle

For SI: 1 gallon per minute = 3.785 L/m,
1 pound per square inch = 6.895 kPa.

a. A handheld shower spray is also a shower head.

b. Consumption tolerances shall be determined from referenced standards.

P2903.3 Minimum pressure. The static water pressure (as determined by the local water authority) at the building entrance for either public or private water service shall be not less than 40 psi (276 kPa).

P2903.3.1 Maximum pressure. The static water pressure shall be not greater than 80 psi (551 kPa). When main pressure exceeds 80 psi (551 kPa), an approved pressure-reducing valve conforming to ASSE 1003 or CSA B356 shall be installed on the domestic water branch main or riser at the connection to the water-service pipe.

P2903.4 Thermal expansion control. A means for controlling increased pressure caused by thermal expansion shall be installed where required in accordance with Sections P2903.4.1 and P2903.4.2.

P2903.4.1 Pressure-reducing valve. For water service system sizes up to and including 2 inches (51 mm), a device for controlling pressure shall be installed where, because of thermal expansion, the pressure on the downstream side of a pressure-reducing valve exceeds the pressure-reducing valve setting.

P2903.4.2 Backflow prevention device or check valve. Where a backflow prevention device, check valve or other device is installed on a water supply system using storage water heating equipment such that thermal expansion causes an increase in pressure, a device for controlling pressure shall be installed.

P2903.5 Water hammer. The flow velocity of the water distribution system shall be controlled to reduce the possibility of water hammer. Water-hammer arrestors shall be installed in accordance with the manufacturer's installation instructions. Water hammer arrestors shall conform to ASSE 1010.

P2903.6 Determining water-supply fixture units. Supply loads in the building water-distribution system shall be determined by total load on the pipe being sized, in terms of water-supply fixture units (w.s.f.u.), as shown in Table P2903.6, and gallon per minute (gpm) flow rates [see Table P2903.6(1)]. For fixtures not listed, choose a w.s.f.u. value of a fixture with similar flow characteristics.

P2903.7 Size of water-service mains, branch mains and risers. The size of the water service pipe shall be not less than $^3/_4$ inch (19 mm) diameter. The size of water service mains, branch mains and risers shall be determined according to water supply demand [gpm (L/m)], available water pressure [psi (kPa)] and friction loss caused by the water meter and *developed length* of pipe [feet (m)], including *equivalent length* of fittings. The size of each water distribution system shall be determined according to design methods conforming to acceptable engineering practice, such as those methods in Appendix P and shall be *approved* by the code official.

P2903.8 Gridded and parallel water distribution system manifolds. Hot water and cold water manifolds installed with gridded or parallel-connected individual distribution lines to each fixture or fixture fittings shall be designed in accordance with Sections P2903.8.1 through P2903.8.6.

P2903.8.1 Sizing of manifolds. Manifolds shall be sized in accordance with Table P2903.8.1. Total gallons per minute is the demand for all outlets.

TABLE P2903.8.1
MANIFOLD SIZING

PLASTIC		METALLIC	
Nominal Size ID (inches)	Maximum[a] gpm	Nominal Size ID (inches)	Maximum[a] gpm
$^3/_4$	17	$^3/_4$	11
1	29	1	20
$1^1/_4$	46	$1^1/_4$	31
$1^1/_2$	66	$1^1/_2$	44

For SI: 1 inch = 25.4 mm, 1 gallon per minute = 3.785 L/m,
1 foot per second = 0.3048 m/s.

Note: See Table P2903.6(1) for w.s.f.u and Table 2903.6(1) for gallon-per-minute (gpm) flow rates.

a. Based on velocity limitation: plastic-12 fps; metal-8 fps.

TABLE P2903.6
WATER-SUPPLY FIXTURE-UNIT VALUES FOR VARIOUS PLUMBING FIXTURES AND FIXTURE GROUPS

TYPE OF FIXTURES OR GROUP OF FIXTURES	WATER-SUPPLY FIXTURE-UNIT VALUE (w.s.f.u.)		
	Hot	Cold	Combined
Bathtub (with/without overhead shower head)	1.0	1.0	1.4
Clothes washer	1.0	1.0	1.4
Dishwasher	1.4	—	1.4
Full-bath group with bathtub (with/without shower head) or shower stall	1.5	2.7	3.6
Half-bath group (water closet and lavatory)	0.5	2.5	2.6
Hose bibb (sillcock)[a]	—	2.5	2.5
Kitchen group (dishwasher and sink with/without garbage grinder)	1.9	1.0	2.5
Kitchen sink	1.0	1.0	1.4
Laundry group (clothes washer standpipe and laundry tub)	1.8	1.8	2.5
Laundry tub	1.0	1.0	1.4
Lavatory	0.5	0.5	0.7
Shower stall	1.0	1.0	1.4
Water closet (tank type)	—	2.2	2.2

For SI: 1 gallon per minute = 3.785 L/m.

a. The fixture unit value 2.5 assumes a flow demand of 2.5 gpm, such as for an individual lawn sprinkler device. If a hose bibb/sill cock will be required to furnish a greater flow, the equivalent fixture-unit value may be obtained from this table or Table P2903.6(1).

TABLE P2903.6(1)
CONVERSIONS FROM WATER SUPPLY FIXTURE UNIT TO GALLON PER MINUTE FLOW RATES

SUPPLY SYSTEMS PREDOMINANTLY FOR FLUSH TANKS			SUPPLY SYSTEMS PREDOMINANTLY FOR FLUSHOMETER VALVES		
Load	Demand		Load	Demand	
(Water supply fixture units)	(Gallons per minute)	(Cubic feet per minute)	(Water supply fixture units)	(Gallons per minute)	(Cubic feet per minute)
1	3.0	0.04104	—	—	—
2	5.0	0.0684	—	—	—
3	6.5	0.86892	—	—	—
4	8.0	1.06944	—	—	—
5	9.4	1.256592	5	15.0	2.0052
6	10.7	1.430376	6	17.4	2.326032
7	11.8	1.577424	7	19.8	2.646364
8	12.8	1.711104	8	22.2	2.967696
9	13.7	1.831416	9	24.6	3.288528
10	14.6	1.951728	10	27.0	3.60936
11	15.4	2.058672	11	27.8	3.716304
12	16.0	2.13888	12	28.6	3.823248
13	16.5	2.20572	13	29.4	3.930192
14	17.0	2.27256	14	30.2	4.037136
15	17.5	2.3394	15	31.0	4.14408
16	18.0	2.90624	16	31.8	4.241024
17	18.4	2.459712	17	32.6	4.357968
18	18.8	2.513184	18	33.4	4.464912
19	19.2	2.566656	19	34.2	4.571856
20	19.6	2.620128	20	35.0	4.6788

(continued)

TABLE P2903.6(1)—continued
CONVERSIONS FROM WATER SUPPLY FIXTURE UNIT TO GALLON PER MINUTE FLOW RATES

SUPPLY SYSTEMS PREDOMINANTLY FOR FLUSH TANKS			SUPPLY SYSTEMS PREDOMINANTLY FOR FLUSHOMETER VALVES		
Load	Demand		Load	Demand	
(Water supply fixture units)	(Gallons per minute)	(Cubic feet per minute)	(Water supply fixture units)	(Gallons per minute)	(Cubic feet per minute)
25	21.5	2.87412	25	38.0	5.07984
30	23.3	3.114744	30	42.0	5.61356
35	24.9	3.328632	35	44.0	5.88192
40	26.3	3.515784	40	46.0	6.14928
45	27.7	3.702936	45	48.0	6.41664
50	29.1	3.890088	50	50.0	6.684

For SI: 1 gallon per minute = 3.785 L/m, 1 cubic foot per minute = 0.4719 L/s.

P2903.8.2 Minimum size. Where the *developed length* of the distribution line is 60 feet (18 288 mm) or less, and the available pressure at the meter is not less than 40 pounds per square inch (276 kPa), the size of individual distribution lines shall be not less than $^3/_8$ inch (10 mm) diameter. Certain fixtures such as one-piece water closets and whirlpool bathtubs shall require a larger size where specified by the manufacturer. If a water heater is fed from the end of a cold water manifold, the manifold shall be one size larger than the water heater feed.

P2903.8.3 Orientation. Manifolds shall be permitted to be installed in a horizontal or vertical position.

P2903.8.4 Support and protection. Plastic piping bundles shall be secured in accordance with the manufacturer's instructions and supported in accordance with Section P2605. Bundles that have a change in direction equal to or greater than 45 degrees (0.79 rad) shall be protected from chafing at the point of contact with framing members by sleeving or wrapping.

P2903.8.5 Valving. Fixture valves, when installed, shall be located either at the fixture or at the manifold. If valves are installed at the manifold, they shall be labeled indicating the fixture served.

P2903.8.6 Hose bibb bleed. A *readily accessible* air bleed shall be installed in hose bibb supplies at the manifold or at the hose bibb exit point.

P2903.9 Valves. Valves shall be installed in accordance with Sections P2903.9.1 through P2903.9.5.

P2903.9.1 Service valve. Each *dwelling unit* shall be provided with an accessible main shutoff valve near the entrance of the water service. The valve shall be of a full-open type having nominal restriction to flow, with provision for drainage such as a bleed orifice or installation of a separate drain valve. Additionally, the water service shall be valved at the curb or lot line in accordance with local requirements.

P2903.9.2 Water heater valve. A *readily accessible* full-open valve shall be installed in the cold-water supply pipe to each water heater at or near the water heater.

P2903.9.3 Fixture valves and access. Valves serving individual fixtures, *appliances*, risers and branches shall be provided with access. An individual shutoff valve shall be required on the fixture supply pipe to each plumbing fixture other than bathtubs and showers.

P2903.9.4 Valve requirements. Valves shall be of an *approved* type and compatible with the type of piping material installed in the system. Ball valves, gate valves, globe valves and plug valves intended to supply drinking water shall meet the requirements of NSF 61.

P2903.9.5 Valves and outlets prohibited below grade. Potable water outlets and combination stop-and-waste valves shall not be installed underground or below grade. Freezeproof yard hydrants that drain the riser into the ground are considered to be stop-and-waste valves.

Exception: Installation of freezeproof yard hydrants that drain the riser into the ground shall be permitted if the potable water supply to such hydrants is protected upstream of the hydrants in accordance with Section P2902 and the hydrants are permanently identified as nonpotable outlets by *approved* signage that reads as follows: "Caution, Nonpotable Water. Do Not Drink."

P2903.10 Hose bibb. Hose bibbs subject to freezing, including the "frostproof" type, shall be equipped with an accessible stop-and-waste-type valve inside the building so that they can be controlled and drained during cold periods.

Exception: Frostproof hose bibbs installed such that the stem extends through the building insulation into an open heated or semi*conditioned space* need not be separately valved (see Figure P2903.10).

SECTION P2904
DWELLING UNIT FIRE SPRINKLER SYSTEMS

P2904.1 General. The design and installation of residential fire sprinkler systems shall be in accordance with NFPA 13D or Section P2904, which shall be considered equivalent to NFPA 13D. Partial residential sprinkler systems shall be permitted to be installed only in buildings not required to be equipped with a residential sprinkler system. Section P2904 shall apply to stand-alone and multipurpose wet-pipe sprinkler systems that do not include the use of antifreeze. A multipurpose fire sprinkler system shall provide domestic water to both fire sprinklers and plumbing fixtures. A stand-alone sprinkler system shall be separate and independent from the water distribution system. A backflow preventer shall not be required to separate a stand-alone sprinkler system from the water distribution system.

P2904.1.1 Required sprinkler locations. Sprinklers shall be installed to protect all areas of a *dwelling unit.*

Exceptions:

1. Attics, crawl spaces and normally unoccupied concealed spaces that do not contain fuel-fired appliances do not require sprinklers. In *attics*, crawl spaces and normally unoccupied concealed spaces that contain fuel-fired equipment, a sprinkler shall be installed above the equipment; however, sprinklers shall not be required in the remainder of the space.
2. Clothes closets, linen closets and pantries not exceeding 24 square feet (2.2 m^2) in area, with the smallest dimension not greater than 3 feet (915 mm) and having wall and ceiling surfaces of gypsum board.
3. Bathrooms not more than 55 square feet (5.1 m^2) in area.
4. Garages; carports; exterior porches; unheated entry areas, such as mud rooms, that are adjacent to an exterior door; and similar areas.

P2904.2 Sprinklers. Sprinklers shall be new listed residential sprinklers and shall be installed in accordance with the sprinkler manufacturer's installation instructions.

P2904.2.1 Temperature rating and separation from heat sources. Except as provided for in Section P2904.2.2, sprinklers shall have a temperature rating of not less than 135°F (57°C) and not more than 170°F (77°C). Sprinklers shall be separated from heat sources as required by the sprinkler manufacturer's installation instructions.

P2904.2.2 Intermediate temperature sprinklers. Sprinklers shall have an intermediate temperature rating not less than 175°F (79°C) and not more than 225°F (107°C) where installed in the following locations:

1. Directly under skylights, where the sprinkler is exposed to direct sunlight.
2. In *attics.*
3. In concealed spaces located directly beneath a roof.
4. Within the distance to a heat source as specified in Table P2904.2.2

P2904.2.3 Freezing areas. Piping shall be protected from freezing as required by Section P2603.6. Where sprinklers are required in areas that are subject to freezing, dry-sidewall or dry-pendent sprinklers extending from a nonfreezing area into a freezing area shall be installed.

P2904.2.4 Sprinkler coverage. Sprinkler coverage requirements and sprinkler obstruction requirements shall be in accordance with Sections P2904.2.4.1 and P2904.2.4.2.

P2904.2.4.1 Coverage area limit. The area of coverage of a single sprinkler shall not exceed 400 square feet (37 m^2) and shall be based on the sprinkler listing and the sprinkler manufacturer's installation instructions.

P2904.2.4.2 Obstructions to coverage. Sprinkler discharge shall not be blocked by obstructions unless additional sprinklers are installed to protect the obstructed area. Additional sprinklers shall not be required where the sprinkler separation from obstructions complies with either the minimum distance indicated in Figure P2904.2.4.2 or the minimum distances specified in the sprinkler manufacturer's instructions where the manufacturer's instructions permit a lesser distance.

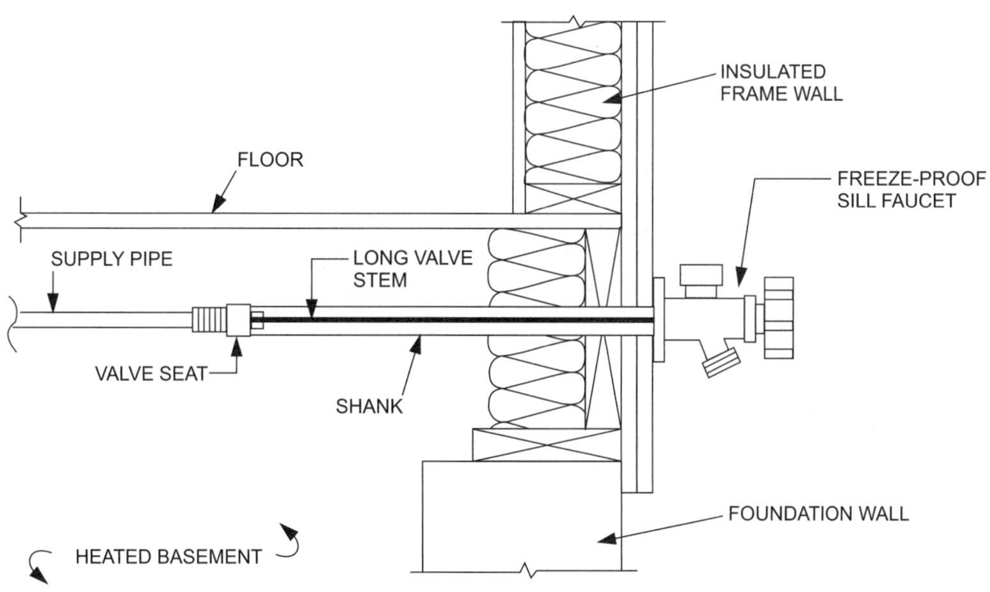

FIGURE P2903.10
TYPICAL FROSTPROOF HOSE BIBB INSTALLATION NOT REQUIRING SEPARATE VALUE

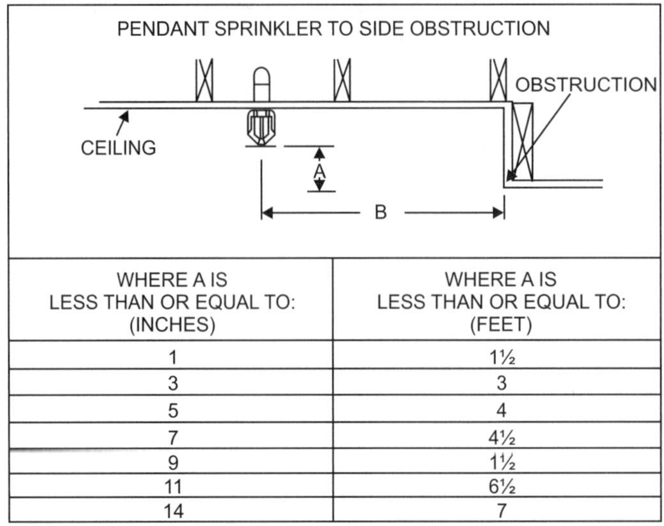

WHERE A IS LESS THAN OR EQUAL TO: (INCHES)	WHERE A IS LESS THAN OR EQUAL TO: (FEET)
1	1½
3	3
5	4
7	4½
9	1½
11	6½
14	7

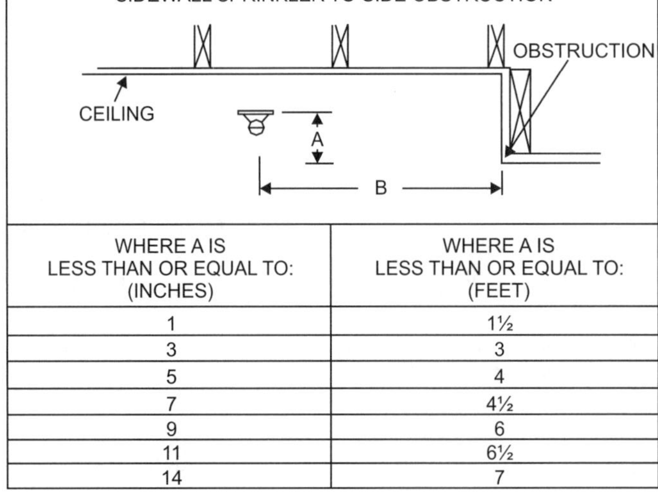

WHERE A IS LESS THAN OR EQUAL TO: (INCHES)	WHERE A IS LESS THAN OR EQUAL TO: (FEET)
1	1½
3	3
5	4
7	4½
9	6
11	6½
14	7

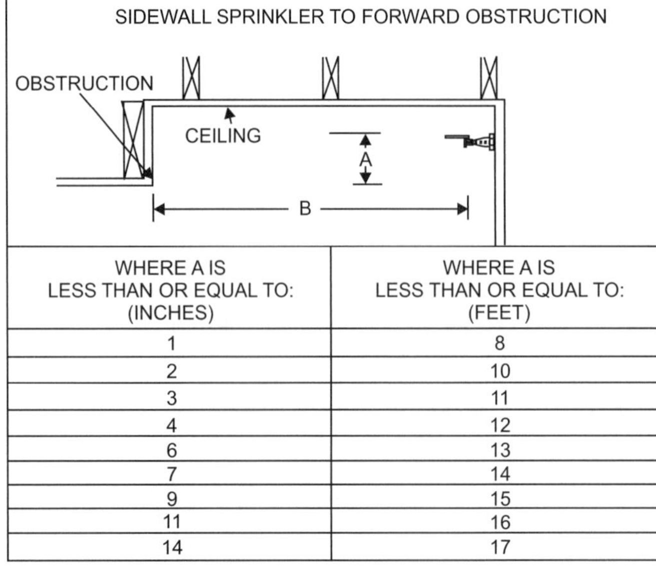

WHERE A IS LESS THAN OR EQUAL TO: (INCHES)	WHERE A IS LESS THAN OR EQUAL TO: (FEET)
1	8
2	10
3	11
4	12
6	13
7	14
9	15
11	16
14	17

For SI: 1 inch = 25.4 mm, 1 foot = 304.8 mm.

FIGURE P2904.2.4.2
MINIMUM ALLOWABLE DISTANCE BETWEEN SPRINKLER AND OBSTRUCTION

TABLE P2904.2.2
LOCATIONS WHERE INTERMEDIATE TEMPERATURE SPRINKLERS ARE REQUIRED

HEAT SOURCE	RANGE OF DISTANCE FROM HEAT SOURCE WITHIN WHICH INTERMEDIATE TEMPERATURE SPRINKLERS ARE REQUIRED[a, b] (inches)
Fireplace, side of open or recessed fireplace	12 to 36
Fireplace, front of recessed fireplace	36 to 60
Coal and wood burning stove	12 to 42
Kitchen range top	9 to 18
Oven	9 to 18
Vent connector or chimney connector	9 to 18
Heating duct, not insulated	9 to 18
Hot water pipe, not insulated	6 to 12
Side of ceiling or wall warm air register	12 to 24
Front of wall mounted warm air register	18 to 36
Water heater, furnace or boiler	3 to 6
Luminaire up to 250 watts	3 to 6
Luminaire 250 watts up to 499 watts	6 to 12

For SI: 1 inch = 25.4 mm.

a. Sprinklers shall not be located at distances less than the minimum table distance unless the sprinkler listing allows a lesser distance.

b. Distances shall be measured in a straight line from the nearest edge of the heat source to the nearest edge of the sprinkler.

P2904.2.4.2.1 Additional requirements for pendent sprinklers. Pendent sprinklers within 3 feet (915 mm) of the center of a ceiling fan, surface-mounted ceiling luminaire or similar object shall be considered to be obstructed, and additional sprinklers shall be installed.

P2904.2.4.2.2 Additional requirements for sidewall sprinklers. Sidewall sprinklers within 5 feet (1524 mm) of the center of a ceiling fan, surface-mounted ceiling luminaire or similar object shall be considered to be obstructed, and additional sprinklers shall be installed.

P2904.2.5 Sprinkler installation on systems assembled with solvent cement. The solvent cementing of threaded adapter fittings shall be completed and threaded adapters for sprinklers shall be verified as being clear of excess cement prior to the installation of sprinklers on systems assembled with solvent cement.

P2904.2.6 Sprinkler modifications prohibited. Painting, caulking or modifying of sprinklers shall be prohibited. Sprinklers that have been painted, caulked, modified or damaged shall be replaced with new sprinklers.

P2904.3 Sprinkler piping system. Sprinkler piping shall be supported in accordance with the requirements for cold water distribution piping. Sprinkler piping shall comply with all requirements for cold water distribution piping. For multipurpose piping systems, the sprinkler piping shall connect to and be a part of the cold water distribution piping system.

P2904.3.1 Nonmetallic pipe and tubing. Nonmetallic pipe and tubing, such as CPVC, PEX, and PE-RT shall be listed for use in residential fire sprinkler systems.

P2904.3.1.1 Nonmetallic pipe protection. Nonmetallic pipe and tubing systems shall be protected from exposure to the living space by a layer of not less than $^{3}/_{8}$-inch-thick (9.5 mm) gypsum wallboard, $^{1}/_{2}$-inch-thick (13 mm) plywood, or other material having a 15-minute fire rating.

Exceptions:

1. Pipe protection shall not be required in areas that do not require protection with sprinklers as specified in Section P2904.1.1.
2. Pipe protection shall not be required where exposed piping is permitted by the pipe listing.

P2904.3.2 Shutoff valves prohibited. With the exception of shutoff valves for the entire water distribution system, valves shall not be installed in any location where the valve would isolate piping serving one or more sprinklers.

P2904.3.3 Single dwelling limit. Piping beyond the service valve located at the beginning of the water distribution system shall not serve more than one *dwelling*.

P2904.3.4 Drain. A means to drain the sprinkler system shall be provided on the system side of the water distribution shutoff valve.

P2904.4 Determining system design flow. The flow for sizing the sprinkler piping system shall be based on the flow rating of each sprinkler in accordance with Section P2904.4.1 and the calculation in accordance with Section P2904.4.2.

P2904.4.1 Determining required flow rate for each sprinkler. The minimum required flow for each sprinkler shall be determined using the sprinkler manufacturer's published data for the specific sprinkler model based on all of the following:

1. The area of coverage.
2. The ceiling configuration.

3. The temperature rating.
4. Any additional conditions specified by the sprinkler manufacturer.

P2904.4.2 System design flow rate. The design flow rate for the system shall be based on the following:

1. The design flow rate for a room having only one sprinkler shall be the flow rate required for that sprinkler, as determined by Section P2904.4.1.
2. The design flow rate for a room having two or more sprinklers a shall be determined by identifying the sprinkler in that room with the highest required flow rate, based on Section P2904.4.1, and multiplying that flow rate by 2.
3. Where the sprinkler manufacturer specifies different criteria for ceiling configurations that are not smooth, flat and horizontal, the required flow rate for that room shall comply with the sprinkler manufacturer's instructions.
4. The design flow rate for the sprinkler system shall be the flow required by the room with the largest flow rate, based on Items 1, 2 and 3.
5. For the purpose of this section, it shall be permissible to reduce the design flow rate for a room by subdividing the space into two or more rooms, where each room is evaluated separately with respect to the required design flow rate. Each room shall be bounded by walls and a ceiling. Openings in walls shall have a lintel not less than 8 inches (203 mm) in depth and each lintel shall form a solid barrier between the ceiling and the top of the opening.

P2904.5 Water supply. The water supply shall provide not less than the required design flow rate for sprinklers in accordance with Section P2904.4.2 at a pressure not less than that used to comply with Section P2904.6.

P2904.5.1 Water supply from individual sources. Where a *dwelling unit* water supply is from a tank system, a private well system or a combination of these, the available water supply shall be based on the minimum pressure control setting for the pump.

P2904.5.2 Required capacity. The water supply shall have the capacity to provide the required design flow rate for sprinklers for a period of time as follows:

1. Seven minutes for *dwelling units* one *story* in height and less than 2,000 square feet (186 m^2) in area.
2. Ten minutes for *dwelling units* two or more stories in height or equal to or greater than 2,000 square feet (186 m^2) in area.

Where a well system, a water supply tank system or a combination thereof is used, any combination of well capacity and tank storage shall be permitted to meet the capacity requirement.

P2904.6 Pipe sizing. The piping to sprinklers shall be sized for the flow required by Section P2904.4.2. The flow required to supply the plumbing fixtures shall not be required to be added to the sprinkler design flow.

P2904.6.1 Method of sizing pipe. Piping supplying sprinklers shall be sized using the prescriptive method in Section P2904.6.2 or by hydraulic calculation in accordance with NFPA 13D. The minimum pipe size from the water supply source to any sprinkler shall be $^3/_4$ inch (19 mm) nominal. Threaded adapter fittings at the point where sprinklers are attached to the piping shall be a minimum of $^1/_2$ inch (13 mm) nominal.

P2904.6.2 Prescriptive pipe sizing method. Pipe shall be sized by determining the available pressure to offset friction loss in piping and identifying a piping material, diameter and length using the equation in Section P2904.6.2.1 and the procedure in Section P2904.6.2.2.

P2904.6.2.1 Available pressure equation. The pressure available to offset friction loss in the interior piping system (P_t) shall be determined in accordance with the Equation 29-1.

$$P_t = P_{sup} - PL_{svc} - PL_m - PL_d - PL_e - P_{sp} \quad \textbf{(Equation 29-1)}$$

where:

P_t = Pressure used in applying Tables P2904.6.2(4) through P2904.6.2(9).

P_{sup} = Pressure available from the water supply source.

PL_{svc} = Pressure loss in the water-service pipe.

PL_m = Pressure loss in the water meter.

PL_d = Pressure loss from devices other than the water meter.

PL_e = Pressure loss associated with changes in elevation.

P_{sp} = Maximum pressure required by a sprinkler.

P2904.6.2.2 Calculation procedure. Determination of the required size for water distribution piping shall be in accordance with the following procedure:

Step 1—Determine P_{sup}

Obtain the static supply pressure that will be available from the water main from the water purveyor, or for an individual source, the available supply pressure shall be in accordance with Section P2904.5.1.

Step 2—Determine PL_{svc}

Use Table P2904.6.2(1) to determine the pressure loss in the water service pipe based on the selected size of the water service.

Step 3—Determine PL_m

Use Table P2904.6.2(2) to determine the pressure loss from the water meter, based on the selected water meter size.

Step 4—Determine PL_d

Determine the pressure loss from devices other than the water meter installed in the piping system supplying sprinklers, such as pressure-reducing valves, backflow preventers, water softeners or water filters. Device pressure losses shall be based on the device manufacturer's specifications. The flow rate used to determine pressure loss shall be the rate from Section P2904.4.2, except that 5 gpm (0.3 L/s) shall be added where the device is installed in a water-service pipe that supplies more than one *dwelling*. As alternative to deducting pressure loss for a device, an automatic bypass valve shall be installed to divert flow around the device when a sprinkler activates.

Step 5—Determine PL_e

Use Table P2904.6.2(3) to determine the pressure loss associated with changes in elevation. The elevation used in applying the table shall be the difference between the elevation where the water source pressure was measured and the elevation of the highest sprinkler.

Step 6—Determine P_{sp}

Determine the maximum pressure required by any individual sprinkler based on the flow rate from Section P2904.4.1. The required pressure is provided in the sprinkler manufacturer's published data for the specific sprinkler model based on the selected flow rate.

Step 7—Calculate P_t

Using Equation 29-1, calculate the pressure available to offset friction loss in water-distribution piping between the service valve and the sprinklers.

Step 8—Determine the maximum allowable pipe length

Use Tables P2904.6.2(4) through P2904.6.2(9) to select a material and size for water distribution piping. The piping material and size shall be acceptable if the *developed length* of pipe between the service valve and the most remote sprinkler does not exceed the maximum allowable length specified by the applicable table. Interpolation of P_t between the tabular values shall be permitted.

The maximum allowable length of piping in Tables P2904.6.2(4) through P2904.6.2(9) incorporates an adjustment for pipe fittings, and no additional consideration of friction losses associated with pipe fittings shall be required.

P2904.7 Instructions and signs. An owner's manual for the fire sprinkler system shall be provided to the owner. A sign or valve tag shall be installed at the main shutoff valve to the water distribution system stating the following: "Warning, the water system for this home supplies fire sprinklers that require certain flows and pressures to fight a fire. Devices that restrict the flow or decrease the pressure or automatically shut off the water to the fire sprinkler system, such as water softeners, filtration systems and automatic shutoff valves, shall not be added to this system without a review of the fire sprinkler system by a fire protection specialist. Do not remove this sign."

TABLE P2904.6.2(1)
WATER SERVICE PRESSURE LOSS (PL_{svc})[a, b]

FLOW RATE[c] (gpm)	$^3/_4$-INCH WATER SERVICE PRESSURE LOSS (psi)				1-INCH WATER SERVICE PRESSURE LOSS (psi)				$1^1/_4$-INCH WATER SERVICE PRESSURE LOSS (psi)			
	Length of water service pipe (feet)				Length of water service pipe (feet)				Length of water service pipe (feet)			
	40 or less	41 to 75	76 to 100	101 to 150	40 or less	41 to 75	76 to 100	101 to 150	40 or less	41 to 75	76 to 100	101 to 150
8	5.1	8.7	11.8	17.4	1.5	2.5	3.4	5.1	0.6	1.0	1.3	1.9
10	7.7	13.1	17.8	26.3	2.3	3.8	5.2	7.7	0.8	1.4	2.0	2.9
12	10.8	18.4	24.9	NP	3.2	5.4	7.3	10.7	1.2	2.0	2.7	4.0
14	14.4	24.5	NP	NP	4.2	7.1	9.6	14.3	1.6	2.7	3.6	5.4
16	18.4	NP	NP	NP	5.4	9.1	12.4	18.3	2.0	3.4	4.7	6.9
18	22.9	NP	NP	NP	6.7	11.4	15.4	22.7	2.5	4.3	5.8	8.6
20	27.8	NP	NP	NP	8.1	13.8	18.7	27.6	3.1	5.2	7.0	10.4
22	NP	NP	NP	NP	9.7	16.5	22.3	NP	3.7	6.2	8.4	12.4
24	NP	NP	NP	NP	11.4	19.3	26.2	NP	4.3	7.3	9.9	14.6
26	NP	NP	NP	NP	13.2	22.4	NP	NP	5.0	8.5	11.4	16.9
28	NP	NP	NP	NP	15.1	25.7	NP	NP	5.7	9.7	13.1	19.4
30	NP	NP	NP	NP	17.2	NP	NP	NP	6.5	11.0	14.9	22.0
32	NP	NP	NP	NP	19.4	NP	NP	NP	7.3	12.4	16.8	24.8
34	NP	NP	NP	NP	21.7	NP	NP	NP	8.2	13.9	18.8	NP
36	NP	NP	NP	NP	24.1	NP	NP	NP	9.1	15.4	20.9	NP

For SI: 1 inch = 25.4 mm, 1 foot = 304.8 mm, 1 gallon per minute = 0.063 L/s, 1 pound per square inch = 6.895 kPa.

NP = Not permitted. Pressure loss exceeds reasonable limits.

a. Values are applicable for underground piping materials listed in Table P2905.4 and are based on an SDR of 11 and a Hazen Williams C Factor of 150.

b. Values include the following length allowances for fittings: 25% length increase for actual lengths up to 100 feet and 15% length increase for actual lengths over 100 feet.

c. Flow rate from Section P2904.4.2. Add 5 gpm to the flow rate required by Section P2904.4.2 where the water-service pipe supplies more than one dwelling.

TABLE P2904.6.2(2)
MINIMUM WATER METER PRESSURE LOSS (PL_m)[a]

FLOW RATE (gallons per minute, gpm)[b]	5/8-INCH METER PRESSURE LOSS (pounds per square inch, psi)	3/4-INCH METER PRESSURE LESS (pounds per square inch, psi)	1-INCH METER PRESSURE LOSS (pounds per square inch, psi)
8	2	1	1
10	3	1	1
12	4	1	1
14	5	2	1
16	7	3	1
18	9	4	1
20	11	4	2
22	NP	5	2
24	NP	5	2
26	NP	6	2
28	NP	6	2
30	NP	7	2
32	NP	7	3
34	NP	8	3
36	NP	8	3

For SI: 1 inch = 25.4 mm, 1 pound per square inch = 6.895 kPa, 1 gallon per minute = 0.063 L/s.

NP—Not permitted unless the actual water meter pressure loss is known.

a. Table P2904.6.2(2) establishes conservative values for water meter pressure loss or installations where the water meter loss is unknown. Where the actual water meter pressure loss is known, P_m shall be the actual loss.

b. Flow rate from Section P2904.4.2. Add 5 gpm to the flow rate required by Section P2904.4.2 where the water-service pipe supplies more than one dwelling.

TABLE P2904.6.2(3)
ELEVATION LOSS (PL_e)

ELEVATION (feet)	PRESSURE LOSS (psi)
5	2.2
10	4.4
15	6.5
20	8.7
25	10.9
30	13
35	15.2
40	17.4

For SI: 1 foot = 304.8 mm, 1 pound per square inch = 6.895 kPa.

TABLE P2904.6.2(4)
ALLOWABLE PIPE LENGTH FOR $^{3}/_{4}$-INCH TYPE M COPPER WATER TUBING

SPRINKLER FLOW RATE[a] (gpm)	WATER DISTRIBUTION SIZE (inch)	AVAILABLE PRESSURE—P_t (psi)									
		15	20	25	30	35	40	45	50	55	60
		Allowable length of pipe from service valve to farthest sprinkler (feet)									
8	$^{3}/_{4}$	217	289	361	434	506	578	650	723	795	867
9	$^{3}/_{4}$	174	232	291	349	407	465	523	581	639	697
10	$^{3}/_{4}$	143	191	239	287	335	383	430	478	526	574
11	$^{3}/_{4}$	120	160	200	241	281	321	361	401	441	481
12	$^{3}/_{4}$	102	137	171	205	239	273	307	341	375	410
13	$^{3}/_{4}$	88	118	147	177	206	235	265	294	324	353
14	$^{3}/_{4}$	77	103	128	154	180	205	231	257	282	308
15	$^{3}/_{4}$	68	90	113	136	158	181	203	226	248	271
16	$^{3}/_{4}$	60	80	100	120	140	160	180	200	220	241
17	$^{3}/_{4}$	54	72	90	108	125	143	161	179	197	215
18	$^{3}/_{4}$	48	64	81	97	113	129	145	161	177	193
19	$^{3}/_{4}$	44	58	73	88	102	117	131	146	160	175
20	$^{3}/_{4}$	40	53	66	80	93	106	119	133	146	159
21	$^{3}/_{4}$	36	48	61	73	85	97	109	121	133	145
22	$^{3}/_{4}$	33	44	56	67	78	89	100	111	122	133
23	$^{3}/_{4}$	31	41	51	61	72	82	92	102	113	123
24	$^{3}/_{4}$	28	38	47	57	66	76	85	95	104	114
25	$^{3}/_{4}$	26	35	44	53	61	70	79	88	97	105
26	$^{3}/_{4}$	24	33	41	49	57	65	73	82	90	98
27	$^{3}/_{4}$	23	30	38	46	53	61	69	76	84	91
28	$^{3}/_{4}$	21	28	36	43	50	57	64	71	78	85
29	$^{3}/_{4}$	20	27	33	40	47	53	60	67	73	80
30	$^{3}/_{4}$	19	25	31	38	44	50	56	63	69	75
31	$^{3}/_{4}$	18	24	29	35	41	47	53	59	65	71
32	$^{3}/_{4}$	17	22	28	33	39	44	50	56	61	67
33	$^{3}/_{4}$	16	21	26	32	37	42	47	53	58	63
34	$^{3}/_{4}$	NP	20	25	30	35	40	45	50	55	60
35	$^{3}/_{4}$	NP	19	24	28	33	38	42	47	52	57
36	$^{3}/_{4}$	NP	18	22	27	31	36	40	45	49	54
37	$^{3}/_{4}$	NP	17	21	26	30	34	38	43	47	51
38	$^{3}/_{4}$	NP	16	20	24	28	32	36	40	45	49
39	$^{3}/_{4}$	NP	15	19	23	27	31	35	39	42	46
40	$^{3}/_{4}$	NP	NP	18	22	26	29	33	37	40	44

For SI: 1 inch = 25.4 mm, 1 foot = 304.8 mm, 1 pound per square inch = 6.895 kPa, 1 gallon per minute = 0.963 L/s.

NP—Not permitted

a. Flow rate from Section P2904.4.2.

TABLE P2904.6.2(5)
ALLOWABLE PIPE LENGTH FOR 1-INCH TYPE M COPPER WATER TUBING

SPRINKLER FLOW RATE[a] (gpm)	WATER DISTRIBUTION SIZE (inch)	AVAILABLE PRESSURE—P_t (psi)									
		15	20	25	30	35	40	45	50	55	60
		Allowable length of pipe from service valve to farthest sprinkler (feet)									
8	1	806	1075	1343	1612	1881	2149	2418	2687	2955	3224
9	1	648	864	1080	1296	1512	1728	1945	2161	2377	2593
10	1	533	711	889	1067	1245	1422	1600	1778	1956	2134
11	1	447	586	745	894	1043	1192	1341	1491	1640	1789
12	1	381	508	634	761	888	1015	1142	1269	1396	1523
13	1	328	438	547	657	766	875	985	1094	1204	1313
14	1	286	382	477	572	668	763	859	954	1049	1145
15	1	252	336	420	504	588	672	756	840	924	1008
16	1	224	298	373	447	522	596	671	745	820	894
17	1	200	266	333	400	466	533	600	666	733	799
18	1	180	240	300	360	420	479	539	599	659	719
19	1	163	217	271	325	380	434	488	542	597	651
20	1	148	197	247	296	345	395	444	493	543	592
21	1	135	180	225	270	315	360	406	451	496	541
22	1	124	165	207	248	289	331	372	413	455	496
23	1	114	152	190	228	267	305	343	381	419	457
24	1	106	141	176	211	246	282	317	352	387	422
25	1	98	131	163	196	228	261	294	326	359	392
26	1	91	121	152	182	212	243	273	304	334	364
27	1	85	113	142	170	198	226	255	283	311	340
28	1	79	106	132	159	185	212	238	265	291	318
29	1	74	99	124	149	174	198	223	248	273	298
30	1	70	93	116	140	163	186	210	233	256	280
31	1	66	88	110	132	153	175	197	219	241	263
32	1	62	83	103	124	145	165	186	207	227	248
33	1	59	78	98	117	137	156	176	195	215	234
34	1	55	74	92	111	129	148	166	185	203	222
35	1	53	70	88	105	123	140	158	175	193	210
36	1	50	66	83	100	116	133	150	166	183	199
37	1	47	63	79	95	111	126	142	158	174	190
38	1	45	60	75	90	105	120	135	150	165	181
39	1	43	57	72	86	100	115	129	143	158	172
40	1	41	55	68	82	96	109	123	137	150	164

For SI: 1 inch = 25.4 mm, 1 foot = 304.8 mm, 1 pound per square inch = 6.895 kPa, 1 gallon per minute = 0.963 L/s.

a. Flow rate from Section P2904.4.2.

TABLE P2904.6.2(6)
ALLOWABLE PIPE LENGTH FOR $^3/_4$-INCH CPVC PIPE

SPRINKLER FLOW RATE[a] (gpm)	WATER DISTRIBUTION SIZE (inch)	AVAILABLE PRESSURE—P_t (psi)									
		15	20	25	30	35	40	45	50	55	60
		Allowable length of pipe from service valve to farthest sprinkler (feet)									
8	$^3/_4$	348	465	581	697	813	929	1045	1161	1278	1394
9	$^3/_4$	280	374	467	560	654	747	841	934	1027	1121
10	$^3/_4$	231	307	384	461	538	615	692	769	845	922
11	$^3/_4$	193	258	322	387	451	515	580	644	709	773
12	$^3/_4$	165	219	274	329	384	439	494	549	603	658
13	$^3/_4$	142	189	237	284	331	378	426	473	520	568
14	$^3/_4$	124	165	206	247	289	330	371	412	454	495
15	$^3/_4$	109	145	182	218	254	290	327	363	399	436
16	$^3/_4$	97	129	161	193	226	258	290	322	354	387
17	$^3/_4$	86	115	144	173	202	230	259	288	317	346
18	$^3/_4$	78	104	130	155	181	207	233	259	285	311
19	$^3/_4$	70	94	117	141	164	188	211	234	258	281
20	$^3/_4$	64	85	107	128	149	171	192	213	235	256
21	$^3/_4$	58	78	97	117	136	156	175	195	214	234
22	$^3/_4$	54	71	89	107	125	143	161	179	197	214
23	$^3/_4$	49	66	82	99	115	132	148	165	181	198
24	$^3/_4$	46	61	76	91	107	122	137	152	167	183
25	$^3/_4$	42	56	71	85	99	113	127	141	155	169
26	$^3/_4$	39	52	66	79	92	105	118	131	144	157
27	$^3/_4$	37	49	61	73	86	98	110	122	135	147
28	$^3/_4$	34	46	57	69	80	92	103	114	126	137
29	$^3/_4$	32	43	54	64	75	86	96	107	118	129
30	$^3/_4$	30	40	50	60	70	81	91	101	111	121
31	$^3/_4$	28	38	47	57	66	76	85	95	104	114
32	$^3/_4$	27	36	45	54	63	71	80	89	98	107
33	$^3/_4$	25	34	42	51	59	68	76	84	93	101
34	$^3/_4$	24	32	40	48	56	64	72	80	88	96
35	$^3/_4$	23	30	38	45	53	61	68	76	83	91
36	$^3/_4$	22	29	36	43	50	57	65	72	79	86
37	$^3/_4$	20	27	34	41	48	55	61	68	75	82
38	$^3/_4$	20	26	33	39	46	52	59	65	72	78
39	$^3/_4$	19	25	31	37	43	50	56	62	68	74
40	$^3/_4$	18	24	30	35	41	47	53	59	65	71

For SI: 1 inch = 25.4 mm, 1 foot = 304.8 mm, 1 pound per square inch = 6.895 kPa, 1 gallon per minute = 0.963 L/s.

a. Flow rate from Section P2904.4.2.

TABLE P2904.6.2(7)
ALLOWABLE PIPE LENGTH FOR 1-INCH CPVC PIPE

SPRINKLER FLOW RATE[a] (gpm)	WATER DISTRIBUTION SIZE (inch)	AVAILABLE PRESSURE—P_t (psi)									
		15	20	25	30	35	40	45	50	55	60
		Allowable length of pipe from service valve to farthest sprinkler (feet)									
8	1	1049	1398	1748	2098	2447	2797	3146	3496	3845	4195
9	1	843	1125	1406	1687	1968	2249	2530	2811	3093	3374
10	1	694	925	1157	1388	1619	1851	2082	2314	2545	2776
11	1	582	776	970	1164	1358	1552	1746	1940	2133	2327
12	1	495	660	826	991	1156	1321	1486	1651	1816	1981
13	1	427	570	712	854	997	1139	1281	1424	1566	1709
14	1	372	497	621	745	869	993	1117	1241	1366	1490
15	1	328	437	546	656	765	874	983	1093	1202	1311
16	1	291	388	485	582	679	776	873	970	1067	1164
17	1	260	347	433	520	607	693	780	867	954	1040
18	1	234	312	390	468	546	624	702	780	858	936
19	1	212	282	353	423	494	565	635	706	776	847
20	1	193	257	321	385	449	513	578	642	706	770
21	1	176	235	293	352	410	469	528	586	645	704
22	1	161	215	269	323	377	430	484	538	592	646
23	1	149	198	248	297	347	396	446	496	545	595
24	1	137	183	229	275	321	366	412	458	504	550
25	1	127	170	212	255	297	340	382	425	467	510
26	1	118	158	197	237	276	316	355	395	434	474
27	1	111	147	184	221	258	295	332	368	405	442
28	1	103	138	172	207	241	275	310	344	379	413
29	1	97	129	161	194	226	258	290	323	355	387
30	1	91	121	152	182	212	242	273	303	333	364
31	1	86	114	143	171	200	228	257	285	314	342
32	1	81	108	134	161	188	215	242	269	296	323
33	1	76	102	127	152	178	203	229	254	280	305
34	1	72	96	120	144	168	192	216	240	265	289
35	1	68	91	114	137	160	182	205	228	251	273
36	1	65	87	108	130	151	173	195	216	238	260
37	1	62	82	103	123	144	165	185	206	226	247
38	1	59	78	98	117	137	157	176	196	215	235
39	1	56	75	93	112	131	149	168	187	205	224
40	1	53	71	89	107	125	142	160	178	196	214

For SI: 1 inch = 25.4 mm, 1 foot = 304.8 mm, 1 pound per square inch = 6.895 kPa, 1 gallon per minute = 0.963 L/s.

a. Flow rate from Section P2904.4.2.

TABLE P2904.6.2(8)
ALLOWABLE PIPE LENGTH FOR $^3/_4$-INCH PEX AND PE-RT TUBING

SPRINKLER FLOW RATE[a] (gpm)	WATER DISTRIBUTION SIZE (inch)	AVAILABLE PRESSURE—P_t (psi)									
		15	20	25	30	35	40	45	50	55	60
		Allowable length of pipe from service valve to farthest sprinkler (feet)									
8	$^3/_4$	93	123	154	185	216	247	278	309	339	370
9	$^3/_4$	74	99	124	149	174	199	223	248	273	298
10	$^3/_4$	61	82	102	123	143	163	184	204	225	245
11	$^3/_4$	51	68	86	103	120	137	154	171	188	205
12	$^3/_4$	44	58	73	87	102	117	131	146	160	175
13	$^3/_4$	38	50	63	75	88	101	113	126	138	151
14	$^3/_4$	33	44	55	66	77	88	99	110	121	132
15	$^3/_4$	29	39	48	58	68	77	87	96	106	116
16	$^3/_4$	26	34	43	51	60	68	77	86	94	103
17	$^3/_4$	23	31	38	46	54	61	69	77	84	92
18	$^3/_4$	21	28	34	41	48	55	62	69	76	83
19	$^3/_4$	19	25	31	37	44	50	56	62	69	75
20	$^3/_4$	17	23	28	34	40	45	51	57	62	68
21	$^3/_4$	16	21	26	31	36	41	47	52	57	62
22	$^3/_4$	NP	19	24	28	33	38	43	47	52	57
23	$^3/_4$	NP	17	22	26	31	35	39	44	48	52
24	$^3/_4$	NP	16	20	24	28	32	36	40	44	49
25	$^3/_4$	NP	NP	19	22	26	30	34	37	41	45
26	$^3/_4$	NP	NP	17	21	24	28	31	35	38	42
27	$^3/_4$	NP	NP	16	20	23	26	29	33	36	39
28	$^3/_4$	NP	NP	15	18	21	24	27	30	33	36
29	$^3/_4$	NP	NP	NP	17	20	23	26	28	31	34
30	$^3/_4$	NP	NP	NP	16	19	21	24	27	29	32
31	$^3/_4$	NP	NP	NP	15	18	20	23	25	28	30
32	$^3/_4$	NP	NP	NP	NP	17	19	21	24	26	28
33	$^3/_4$	NP	NP	NP	NP	16	18	20	22	25	27
34	$^3/_4$	NP	NP	NP	NP	NP	17	19	21	23	25
35	$^3/_4$	NP	NP	NP	NP	NP	16	18	20	22	24
36	$^3/_4$	NP	NP	NP	NP	NP	15	17	19	21	23
37	$^3/_4$	NP	NP	NP	NP	NP	NP	16	18	20	22
38	$^3/_4$	NP	NP	NP	NP	NP	NP	16	17	19	21
39	$^3/_4$	NP	NP	NP	NP	NP	NP	NP	16	18	20
40	$^3/_4$	NP	NP	NP	NP	NP	NP	NP	16	17	19

For SI: 1 inch = 25.4 mm, 1 foot = 304.8 mm, 1 pound per square inch = 6.895 kPa, 1 gallon per minute = 0.963 L/s.

NP— Not permitted.

a. Flow rate from Section P2904.4.2.

TABLE P2904.6.2(9)
ALLOWABLE PIPE LENGTH FOR 1-INCH PEX AND PE-RT TUBING

SPRINKLER FLOW RATE[a] (gpm)	WATER DISTRIBUTION SIZE (inch)	AVAILABLE PRESSURE—P_t (psi)									
		15	20	25	30	35	40	45	50	55	60
		Allowable length of pipe from service valve to farthest sprinkler (feet)									
8	1	314	418	523	628	732	837	941	1046	1151	1255
9	1	252	336	421	505	589	673	757	841	925	1009
10	1	208	277	346	415	485	554	623	692	761	831
11	1	174	232	290	348	406	464	522	580	638	696
12	1	148	198	247	296	346	395	445	494	543	593
13	1	128	170	213	256	298	341	383	426	469	511
14	1	111	149	186	223	260	297	334	371	409	446
15	1	98	131	163	196	229	262	294	327	360	392
16	1	87	116	145	174	203	232	261	290	319	348
17	1	78	104	130	156	182	208	233	259	285	311
18	1	70	93	117	140	163	187	210	233	257	280
19	1	63	84	106	127	148	169	190	211	232	253
20	1	58	77	96	115	134	154	173	192	211	230
21	1	53	70	88	105	123	140	158	175	193	211
22	1	48	64	80	97	113	129	145	161	177	193
23	1	44	59	74	89	104	119	133	148	163	178
24	1	41	55	69	82	96	110	123	137	151	164
25	1	38	51	64	76	89	102	114	127	140	152
26	1	35	47	59	71	83	95	106	118	130	142
27	1	33	44	55	66	77	88	99	110	121	132
28	1	31	41	52	62	72	82	93	103	113	124
29	1	29	39	48	58	68	77	87	97	106	116
30	1	27	36	45	54	63	73	82	91	100	109
31	1	26	34	43	51	60	68	77	85	94	102
32	1	24	32	40	48	56	64	72	80	89	97
33	1	23	30	38	46	53	61	68	76	84	91
34	1	22	29	36	43	50	58	65	72	79	86
35	1	20	27	34	41	48	55	61	68	75	82
36	1	19	26	32	39	45	52	58	65	71	78
37	1	18	25	31	37	43	49	55	62	68	74
38	1	18	23	29	35	41	47	53	59	64	70
39	1	17	22	28	33	39	45	50	56	61	67
40	1	16	21	27	32	37	43	48	53	59	64

For SI: 1 inch = 25.4 mm, 1 foot = 304.8 mm, 1 pound per square inch = 6.895 kPa, 1 gallon per minute = 0.963 L/s.

a. Flow rate from Section P2904.4.2.

P2904.8 Inspections. The water distribution system shall be inspected in accordance with Sections P2904.8.1 and P2904.8.2.

P2904.8.1 Preconcealment inspection. The following items shall be verified prior to the concealment of any sprinkler system piping:

1. Sprinklers are installed in all areas as required by Section P2904.1.1.
2. Where sprinkler water spray patterns are obstructed by construction features, luminaires or ceiling fans, additional sprinklers are installed as required by Section P2904.2.4.2.
3. Sprinklers are the correct temperature rating and are installed at or beyond the required separation distances from heat sources as required by Sections P2904.2.1 and P2904.2.2.
4. The pipe size equals or exceeds the size used in applying Tables P2904.6.2(4) through P2904.6.2(9) or, if the piping system was hydraulically calculated in accordance with Section P2904.6.1, the size used in the hydraulic calculation.
5. The pipe length does not exceed the length permitted by Tables P2904.6.2(4) through P2904.6.2(9) or, if the piping system was hydraulically calculated in accordance with Section P2904.6.1, pipe lengths and fittings do not exceed those used in the hydraulic calculation.
6. Nonmetallic piping that conveys water to sprinklers is listed for use with fire sprinklers.
7. Piping is supported in accordance with the pipe manufacturer's and sprinkler manufacturer's installation instructions.
8. The piping system is tested in accordance with Section P2503.7.

P2904.8.2 Final inspection. The following items shall be verified upon completion of the system:

1. Sprinkler are not painted, damaged or otherwise hindered from operation.
2. Where a pump is required to provide water to the system, the pump starts automatically upon system water demand.
3. Pressure-reducing valves, water softeners, water filters or other impairments to water flow that were not part of the original design have not been installed.
4. The sign or valve tag required by Section P2904.7 is installed and the owner's manual for the system is present.

SECTION P2905
MATERIALS, JOINTS AND CONNECTIONS

P2905.1 Soil and groundwater. The installation of water service pipe, water distribution pipe, fittings, valves, appurtenances and gaskets shall be prohibited in soil and groundwater that is contaminated with solvents, fuels, organic compounds or other detrimental materials that cause permeation, corrosion, degradation or structural failure of the water service or water distribution piping material.

P2905.1.1 Investigation required. Where detrimental conditions are suspected by or brought to the attention of the *building official*, a chemical analysis of the soil and groundwater conditions shall be required to ascertain the acceptability of the water service material for the specific installation.

P2905.1.2 Detrimental condition. When a detrimental condition exists, *approved* alternate materials or alternate routing shall be required.

P2905.2 Lead content. Pipe and fittings used in the water-supply system shall have lead content of not greater than 8 percent lead.

P2905.3 Polyethylene plastic piping installation. Polyethylene pipe shall be cut square using a cutter designed for plastic pipe. Except where joined by heat fusion, pipe ends shall be chamfered to remove sharp edges. Pipe that has been kinked shall not be installed. For bends, the installed radius of pipe curvature shall be greater than 30 pipe diameters or the coil radius when bending with the coil. Coiled pipe shall not be bent beyond straight. Bends shall not be permitted within 10 pipe diameters of any fitting or valve. Joints between polyethylene plastic pipe and fittings shall comply with Sections P2905.3.1 and P2905.3.2.

P2905.3.1 Heat-fusion joints. Joint surfaces shall be clean and free from moisture. Joint surfaces shall be heated to melting temperature and joined. The joint shall be undisturbed until cool. Joints shall be made in accordance with ASTM D 2657.

P2905.3.2 Mechanical joints. Mechanical joints shall be installed in accordance with the manufacturer's instructions.

P2905.4 Water service pipe. Water service pipe shall conform to NSF 61 and shall conform to one of the standards listed in Table P2905.4. Water service pipe or tubing, installed underground and outside of the structure, shall have a working pressure rating of not less than 160 pounds per square inch at 73°F (1103 kPa at 23°C). Where the water pressure exceeds 160 pounds per square inch (1103 kPa), piping material shall have a rated working pressure equal to or greater than the highest available pressure. Water service piping materials not third-party certified for water distribution shall terminate at or before the full open valve located at the entrance to the structure. Ductile iron water service piping shall be cement mortar lined in accordance with AWWA C104.

P2905.4.1 Dual check-valve-type backflow preventer. Where a dual check-valve backflow preventer is installed on the water supply system, it shall comply with ASSE 1024 or CSA B64.6.

P2905.4.2 Water service installation. Trenching, pipe installation and backfilling shall be in accordance with Section P2604. Water-service pipe is permitted to be located in the same trench with a *building sewer* provided such sewer is constructed of materials listed for under-

TABLE P2905.4
WATER SERVICE PIPE

MATERIAL	STANDARD
Acrylonitrile butadiene styrene (ABS) plastic pipe	ASTM D 1527; ASTM D 2282
Asbestos-cement pipe	ASTM C 296
Brass pipe	ASTM B 43
Chlorinated polyvinyl chloride (CPVC) plastic pipe	ASTM D 2846; ASTM F 441; ASTM F 442; CSA B137.6
Copper or copper-alloy pipe	ASTM B 42; ASTM B 302
Copper or copper-alloy tubing (Type K, WK, L, WL, M or WM)	ASTM B 75; ASTM B 88; ASTM B 251; ASTM B 447
Cross-linked polyethylene/aluminum/cross-linked polyethylene (PEX-AL-PEX) pipe	ASTM F 1281; ASTM F 2262; CSA B137.10M
Cross-linked polyethylene/aluminum/high-density polyethylene (PEX-AL-HDPE)	ASTM F 1986
Cross-linked polyethylene (PEX) plastic tubing	ASTM F 876; ASTM F 877; CSA B137.5
Ductile iron water pipe	AWWA C151; AWWA C115
Galvanized steel pipe	ASTM A 53
Polyethylene/aluminum/polyethylene (PE-AL-PE) pipe	ASTM F 1282; CSA CAN/CSA-B137.9M
Polyethylene (PE) plastic pipe	ASTM D 2104; ASTM D 2239; AWWA C901; CSA B137.1
Polyethylene (PE) plastic tubing	ASTM D 2737; AWWA C901; CSA B137.1
Polyethylene of raised temperature (PE-RT) plastic tubing	ASTM F 2769
Polyvinyl chloride (PVC) plastic pipe	ASTM D 1785; ASTM D 2241; ASTM D 2672; CSA B137.3
Stainless steel (Type 304/304L) pipe	ASTM A 312; ASTM A 778
Stainless steel (Type 316/316L) pipe	ASTM A 312; ASTM A 778

ground use within a building in Section P3002.1. If the *building sewer* is not constructed of materials listed in Section P3002.1, the water-service pipe shall be separated from the *building sewer* by not less than 5 feet (1524 mm), measured horizontally, of undisturbed or compacted earth or placed on a solid ledge not less than 12 inches (305 mm) above and to one side of the highest point in the sewer line.

Exception: The required separation distance shall not apply where a water service pipe crosses a sewer pipe, provided that the water service pipe is sleeved not less than 5 feet (1524 mm), horizontally from the sewer pipe centerline, on both sides of the crossing with pipe materials listed in Table P2905.4, P3002.1(1), P3002.1(2) or P3002.2.

P2905.5 Water-distribution pipe. Water-distribution piping within *dwelling units* shall conform to NSF 61 and shall conform to one of the standards listed in Table P2905.5. All hot-water-distribution pipe and tubing shall have a pressure rating of not less than 100 psi at 180°F (689 kPa at 82°C).

P2905.6 Fittings. Pipe fittings shall be *approved* for installation with the piping material installed and shall comply with the applicable standards listed in Table P2905.6. All pipe fittings used in water supply systems shall also comply with NSF 61.

P2905.7 Flexible water connectors. Flexible water connectors, exposed to continuous pressure, shall conform to ASME A112.18.6/CSA B125.6. Access shall be provided to all flexible water connectors.

P2905.8 Joint and connection tightness. Joints and connections in the plumbing system shall be gas tight and water tight for the intended use or required test pressure.

P2905.9 Plastic pipe joints. Joints in plastic piping shall be made with *approved* fittings by solvent cementing, heat fusion, corrosion-resistant metal clamps with insert fittings or compression connections. Flared joints for polyethylene pipe shall be permitted in accordance with Section P2905.3.

P2905.9.1 Solvent cementing. Solvent-cemented joints shall comply with Sections P2905.9.1.1 through P2905.9.1.3.

P2905.9.1.1 ABS plastic pipe. Solvent cement for ABS plastic pipe conforming to ASTM D 2235 shall be applied to all joint surfaces.

P2905.9.1.2 CPVC plastic pipe. Joint surfaces shall be clean and free from moisture and an *approved* primer shall be applied. Solvent cement for CPVC plastic pipe, orange in color and conforming to ASTM F 493, shall be applied to all joint surfaces. The parts shall be joined while the cement is wet and in accordance with ASTM D 2846 or ASTM F 493. Solvent-cement joints shall be permitted above or below ground.

Exception: A primer shall not be required where all of the following conditions apply:

1. The solvent cement used is third-party certified as conforming to ASTM F 493.
2. The solvent cement used is yellow in color.

3. The solvent cement is used only for joining $^1/_2$ inch (13 mm) through 2 inch (51 mm) diameter CPVC pipe and fittings.
4. The CPVC pipe and fittings are manufactured in accordance with ASTM D 2846.

P2905.9.1.3 PVC plastic pipe. A purple primer that conforms to ASTM F 656 shall be applied to PVC solvent-cemented joints. Solvent cement for PVC plastic pipe conforming to ASTM D 2564 shall be applied to all joint surfaces.

TABLE P2905.5
WATER DISTRIBUTION PIPE

MATERIAL	STANDARD
Brass pipe	ASTM B 43
Chlorinated polyvinyl chloride (CPVC) plastic pipe and tubing	ASTM D 2846; ASTM F 441; ASTM F 442; CSA B137.6
Copper or copper-alloy pipe	ASTM B 42; ASTM B 302
Copper or copper-alloy tubing (Type K, WK, L, WL, M or WM)	ASTM B 75; ASTM B 88; ASTM B 251; ASTM B 447
Cross-linked polyethylene (PEX) plastic tubing	ASTM F 876; ASTM F 877; CSA B137.5
Cross-linked polyethylene/aluminum/cross-linked polyethylene (PEX-AL-PEX) pipe	ASTM F 1281; ASTM F 2262; CSA B137.10M
Cross-linked polyethylene/aluminum/high-density polyethylene (PEX-AL-HDPE)	ASTM F 1986
Galvanized steel pipe	ASTM A 53
Polyethylene/aluminum/polyethylene (PE-AL-PE) composite pipe	ASTM F 1282
Polyethylene of raised temperature (PE-RT) plastic tubing	ASTM F 2769
Polypropylene (PP) plastic pipe or tubing	ASTM F 2389; CSA B137.11
Stainless steel (Type 304/304L) pipe	ASTM A 312; ASTM A 778

TABLE P2905.6
PIPE FITTINGS

MATERIAL	STANDARD
Acrylonitrile butadiene styrene (ABS) plastic	ASTM D 2468
Brass	ASTM F1974
Cast-iron	ASME B16.4; ASME B16.12
Chlorinated polyvinyl chloride (CPVC) plastic	ASSE 1061; ASTM D 2846; ASTM F 437; ASTM F 438; ASTM F 439; CSA B137.6
Copper or copper alloy	ASSE 1061; ASME B16.15; ASME B16.18; ASME B16.22; ASME B16.23; ASME B16.26; ASME B16.29
Cross-linked polyethylene/aluminum/high-density polyethylene (PEX-AL-HDPE)	ASTM F 1986
Fittings for cross-linked polyethylene (PEX) plastic tubing	ASSE 1061; ASTM F 877; ASTM F 1807; ASTM F 1960; ASTM F 2080; ASTM F 2098; ASTM F 2159; ASTM F 2434; ASTM F 2735; CSA B137.5
Gray iron and ductile iron	AWWA C110; AWWA C153
Malleable iron	ASME B16.3
Insert fittings for Polyethylene/aluminum/polyethylene (PE-AL-PE) and cross-linked polyethylene/aluminum/polyethylene (PEX-AL-PEX)	ASTM F 1974; ASTM F 1281; ASTM F 1282; CSA B137.9; CSA B137.10
Polyethylene (PE) plastic	ASTM D 2609; CSA B137.1
Fittings for polyethylene of raised temperature (PE-RT) plastic tubing	ASTM F 1807; ASTM F2098; ASTM F 2159; ASTM F 2735
Polypropylene (PP) plastic pipe or tubing	ASTM F 2389; CSA B137.11
Polyvinyl chloride (PVC) plastic	ASTM D 2464; ASTM D 2466; ASTM D 2467; CSA B137.2; CSA B137.3
Stainless steel (Type 304/304L) pipe	ASTM A 312; ASTM A 778
Stainless steel (Type 316/316L) pipe	ASTM A 312; ASTM A 778
Steel	ASME B16.9; ASME B16.11; ASME B16.28

P2905.9.1.4 Cross-linked polyethylene plastic (PEX). Joints between cross-linked polyethylene plastic tubing or fittings shall comply with Section P2905.9.1.4.1 or Section P2905.9.1.4.2.

P2905.9.1.4.1 Flared joints. Flared pipe ends shall be made by a tool designed for that operation.

P2905.9.1.4.2 Mechanical joints. Mechanical joints shall be installed in accordance with the manufacturer's instructions. Fittings for cross-linked polyethylene (PEX) plastic tubing shall comply with the applicable standards listed in Table P2905.6 and shall be installed in accordance with the manufacturer's instructions. PEX tubing shall be factory marked with the applicable standards for the fittings that the PEX manufacturer specifies for use with the tubing.

P2905.10 Polypropylene (PP) plastic. Joints between polypropylene plastic pipe and fittings shall comply with Section P2905.10.1 or P2905.10.2.

P2905.10.1 Heat-fusion joints. Heat fusion joints for polypropylene pipe and tubing joints shall be installed with socket-type heat-fused polypropylene fittings, butt-fusion polypropylene fittings or electrofusion polypropylene fittings. Joint surfaces shall be clean and free from moisture. The joint shall be undisturbed until cool. Joints shall be made in accordance with ASTM F 2389.

P2905.10.2 Mechanical and compression sleeve joints. Mechanical and compression sleeve joints shall be installed in accordance with the manufacturer's instructions.

P2905.11 Cross-linked polyethylene/aluminum/cross-linked polyethylene. Joints between polyethylene/aluminum/polyethylene (PE-AL-PE) and cross-linked polyethylene/aluminum/cross-linked polyethylene (PEX-AL-PEX) pipe and fittings shall comply with Section P2905.11.1.

P2905.11.1 Mechanical joints. Mechanical joints shall be installed in accordance with the manufacturer's instructions. Fittings for PE-AL-PE and PEX-AL-PEX as described in ASTM F 1974, ASTM F 1281, ASTM F 1282, CSA B137.9 and CSA B137.10 shall be installed in accordance with the manufacturer's instructions.

P2905.12 Stainless steel. Joints between stainless steel pipe and fittings shall comply with Sections P2905.12.1 and P2905.12.2.

P2905.12.1 Mechanical joints. Mechanical joints shall be installed in accordance with the manufacturer's instructions.

P2905.12.2 Welded joints. Joint surfaces shall be cleaned. The joint shall be welded autogenously or with an *approved* filler metal in accordance with ASTM A 312.

P2905.13 Threaded pipe joints. Threaded joints shall conform to American National Taper Pipe Thread specifications. Pipe ends shall be deburred and chips removed. Pipe joint compound shall be used only on male threads.

P2905.14 Soldered joints. Soldered joints in tubing shall be made with fittings *approved* for water piping and shall conform to ASTM B 828. Surfaces to be soldered shall be cleaned bright. The joints shall be properly fluxed and made with *approved* solder. Solders and fluxes used in potable water-supply systems shall have a lead content of not greater than 0.2 percent. Fluxes shall conform to ASTM B 813.

P2905.15 Flared joints. Flared joints in water tubing shall be made with *approved* fittings. The tubing shall be reamed and then expanded with a flaring tool.

P2905.16 Above-ground joints. Joints within the building between copper pipe or CPVC tubing, in any combination with compatible outside diameters, shall be permitted to be made with the use of *approved* push-in mechanical fittings of a pressure-lock design.

P2905.17 Joints between different materials. Joints between different piping materials shall be made in accordance with Sections P2905.17.1, P2905.17.2 and P2905.17.3 or with a mechanical joint of the compression or mechanical sealing type having an elastomeric seal conforming to ASTM D 1869 or ASTM F 477. Joints shall be installed in accordance with the manufacturer's instructions.

P2905.17.1 Copper or copper-alloy tubing to galvanized steel pipe. Joints between copper or copper-alloy tubing and galvanized steel pipe shall be made with a brass fitting or dielectric fitting. The copper tubing shall be joined to the fitting in an *approved* manner, and the fitting shall be screwed to the threaded pipe.

P2904.17.2 Plastic pipe or tubing to other piping material. Joints between different types of plastic pipe or between plastic pipe and other piping material shall be made with an approved adapter fitting.

P2905.17.3 Stainless steel. Joints between stainless steel and different piping materials shall be made with a mechanical joint of the compression or mechanical-sealing type or a dielectric fitting.

P2905.18 Press joints. Press-type mechanical joints in copper tubing shall be made in accordance with the manufacturer's instructions using *approved* tools which affix the copper fitting with integral O-ring to the tubing.

P2905.19 Polyethylene of raised temperature plastic. Joints between polyethylene of raised temperature plastic tubing and fittings shall be in accordance with Section P2905.19.1 and Section P2905.19.2

P2905.19.1 Flared joints. Flared pipe ends shall be made by a tool designed for that operation.

P2905.19.2 Mechanical joints. Mechanical joints shall be installed in accordance with the manufacturer's instructions. Fittings for polyethylene of raised temperature plastic tubing shall comply with the applicable standards listed in Table P2905.6 and shall be installed in accordance with the manufacturer's instructions. Polyethylene of raised temperature plastic tubing shall be factory marked with the applicable standards for the fittings that the manufacturer of the tubing specifies for use with the tubing.

SECTION P2906
CHANGES IN DIRECTION

P2906.1 Bends. Changes in direction in copper tubing are permitted to be made with bends having a radius of not less than four diameters of the tube, providing such bends are made by use of forming equipment that does not deform or create loss in cross-sectional area of the tube.

SECTION P2907
SUPPORT

P2907.1 General. Pipe and tubing support shall conform to Section P2605.

SECTION P2908
DRINKING WATER TREATMENT UNITS

P2908.1 Design. Drinking water treatment units shall meet the requirements of NSF42, NSF 44, NSF 53, NSF 60 or CSA B483.1.

P2908.2 Reverse osmosis drinking water treatment units. Point-of-use reverse osmosis drinking water treatment units, designed for residential use, shall meet the requirements of NSF 58 or CSA B483.1. Waste or discharge from reverse osmosis drinking water treatment units shall enter the drainage system through an air gap or an air gap device that meets the requirements of NSF 58.

P2908.3 Connection tubing. The tubing to and from drinking water treatment units shall be of a size and material as recommended by the manufacturer. The tubing shall comply with NSF 14, NSF 42, NSF 44, NSF 53, NSF 58 or NSF 61.

CHAPTER 30

SANITARY DRAINAGE

SECTION P3001 GENERAL

P3001.1 Scope. The provisions of this chapter shall govern the materials, design, construction and installation of sanitary drainage systems. Plumbing materials shall conform to the requirements of this chapter. The drainage, waste and vent (DWV) system shall consist of all piping for conveying wastes from plumbing fixtures, appliances and appurtenances, including fixture traps; above-grade drainage piping; below-grade drains within the building (*building drain*); below- and above-grade venting systems; and piping to the public sewer or private septic system.

P3001.2 Protection from freezing. No portion of the above grade DWV system other than vent terminals shall be located outside of a building, in *attics* or crawl spaces, concealed in outside walls, or in any other place subjected to freezing temperatures unless adequate provision is made to protect them from freezing by insulation or heat or both, except in localities having a winter design temperature above 32°F (0°C) (ASHRAE 97.5 percent column, winter, see Chapter 3).

P3001.3 Flood-resistant installation. In flood hazard areas as established by Table R301.2(1), drainage, waste and vent systems shall be located and installed to prevent infiltration of floodwaters into the systems and discharges from the systems into floodwaters.

SECTION P3002 MATERIALS

P3002.1 Piping within buildings. Drain, waste and vent (DWV) piping in buildings shall be as shown in Tables P3002.1(1) and P3002.1(2) except that galvanized wrought-iron or galvanized steel pipe shall not be used underground and shall be maintained not less than 6 inches (152 mm) above ground. Allowance shall be made for the thermal expansion and contraction of plastic piping.

P3002.2 Building sewer. *Building sewer* piping shall be as shown in Table P3002.2. Forced main sewer piping shall conform to one of the standards for ABS plastic pipe, copper or copper-alloy tubing, PVC plastic pipe or pressure-rated pipe listed in Table P3002.2.

P3002.3 Fittings. Pipe fittings shall be *approved* for installation with the piping material installed and shall comply with the applicable standards listed in Table P3002.3.

P3002.3.1 Drainage. Drainage fittings shall have a smooth interior waterway of the same diameter as the piping served. All fittings shall conform to the type of pipe used. Drainage fittings shall have no ledges, shoulders or reductions which can retard or obstruct drainage flow in the piping. Threaded drainage pipe fittings shall be of the recessed drainage type, black or galvanized. Drainage fittings shall be designed to maintain one-fourth unit vertical in 12 units horizontal (2-percent slope) grade. This section shall not be applicable to tubular waste fittings used to convey vertical flow upstream of the trap seal liquid level of a fixture trap.

P3002.4 Other materials. Sheet lead, lead bends, lead traps and sheet copper shall comply with Sections P3002.4.1 through P3002.4.3.

P3002.4.1 Sheet lead. Sheet lead for the following uses shall weigh not less than indicated below:

1. Flashing of vent terminals, 3 psf (15 kg/m^2).
2. Prefabricated flashing for vent pipes, $2^1/_2$ psf (12 kg/m^2).

TABLE P3002.1(1)
ABOVE-GROUND DRAINAGE AND VENT PIPE

PIPE	STANDARD
Acrylonitrile butadiene styrene (ABS) plastic pipe in IPS diameters, including schedule 40, DR 22 (PS 200) and DR 24 (PS 140); with a solid, cellular core or composite wall	ASTM D 2661; ASTM F 628; ASTM F 1488; CSA B181.1
Brass pipe	ASTM B 43
Cast-iron pipe	ASTM A 74; CISPI 301; ASTM A 888
Copper or copper-alloy pipe	ASTM B 42; ASTM B 302
Copper or copper-alloy tubing (Type K, L, M or DWV)	ASTM B 75; ASTM B 88; ASTM B 251; ASTM B 306
Galvanized steel pipe	ASTM A 53
Polyolefin pipe	CSA B181.3
Polyvinyl chloride (PVC) plastic pipe in IPS diameters, including schedule 40, DR 22 (PS 200) and DR 24 (PS 140); with a solid, cellular core or composite wall	ASTM D 2665; ASTM F 891; CSA B181.2; ASTM F 1488
Polyvinyl chloride (PVC) plastic pipe with a 3.25 inch O.D. and a solid, cellular core or composite wall	ASTM D 2949; ASTM F 1488
Stainless steel drainage systems, Types 304 and 316L	ASME A 112.3.1

For SI: 1 inch = 25.4 mm.

**TABLE P3002.1(2)
UNDERGROUND BUILDING DRAINAGE AND VENT PIPE**

PIPE	STANDARD
Acrylonitrile butadiene styrene (ABS) plastic pipe in IPS diameters, including schedule 40, DR 22 (PS 200) and DR 24 (PS 140); with a solid, cellular core or composite wall	ASTM D 2661; ASTM F 628; ASTM F 1488; CSA B181.1
Asbestos-cement pipe	ASTM C 428
Cast-iron pipe	ASTM A 74; CISPI 301; ASTM A 888
Copper or copper alloy tubing (Type K, L, M or DWV)	ASTM B 75; ASTM B 88; ASTM B 251; ASTM B 306
Polyolefin pipe	ASTM F 1412; CSA B181.3
Polyvinyl chloride (PVC) plastic pipe in IPS diameters, including schedule 40, DR 22 (PS 200) and DR 24 (PS 140); with a solid, cellular core or composite wall	ASTM D 2665; ASTM F 891; ASTM F 1488; CSA B181.2
Polyvinyl chloride (PVC) plastic pipe with a 3.25 inch O.D. and a solid, cellular core or composite wall	ASTM D 2949; ASTM F 1488
Stainless steel drainage systems, Type 316L	ASME A 112.3.1

For SI: 1 inch = 25.4 mm.

**TABLE P3002.2
BUILDING SEWER PIPE**

MATERIAL	STANDARD
Acrylonitrile butadiene styrene (ABS) plastic pipe in IPS diameters, including schedule 40, DR 22 (PS 200) and DR 24 (PS 140); with a solid, cellular core or composite wall	ASTM D 2661; ASTM F 628; ASTM F 1488
Asbestos-cement pipe	ASTM C 428
Cast-iron pipe	ASTM A 74; ASTM A 888; CISPI 301
Acrylonitrile butadiene styrene (ABS) plastic pipe in sewer and drain diameters, including SDR 42 (PS 20), PS35, SDR 35 (PS 45), PS50, PS100, PS140, SDR 23.5 (PS 150) and PS200; with a solid, cellular core or composite wall	ASTM F 1488; ASTM D 2751
Polyvinyl chloride (PVC) plastic pipe in sewer and drain diameters, including PS 25, SDR 41 (PS 28), PS 35, SDR 35 (PS 46), PS 50, PS 100, SDR 26 (PS 115), PS140 and PS 200; with a solid, cellular core or composite wall	ASTM F 891; ASTM F 1488; ASTM D 3034; CSA B182.2; CSA B182.4
Concrete pipe	ASTM C 14; ASTM C 76; CSA A257.1M; CSA A257.2M
Copper or copper-alloy tubing (Type K or L)	ASTM B 75; ASTM B 88; ASTM B 251
Polyethylene (PE) plastic pipe (SDR-PR)	ASTM F 714
Polyolefin pipe	ASTM F 1412; CSA B181.3
Polyvinyl chloride (PVC) plastic pipe in IPS diameters, including schedule 40, DR 22 (PS 200) and DR 24 (PS 140); with solid, cellular core or composite wall	ASTM D 2665; ASTM D 2949; ASTM D 3034; ASTM F 1412; CSA B182.2; CSA B182.4
Polyvinyl chloride (PVC) plastic pipe with a 3.25 inch O.D. and a solid, cellular core or composite wall	ASTM D 2949, ASTM F 1488
Stainless steel drainage systems, Types 304 and 316L	ASME A 112.3.1
Vitrified clay pipe	ASTM C 425; ASTM C 700

For SI: 1 inch = 25.4 mm.

**TABLE P3002.3
PIPE FITTINGS**

PIPE MATERIAL	FITTING STANDARD
Acrylonitrile butadiene styrene (ABS) plastic pipe in IPS diameters	ASTM D 2661; ASTM D 3311; ASTM F 628; CSA B181.1
Asbestos cement	ASTM C 428
Cast-iron	ASME B 16.4; ASME B 16.12; ASTM A 74; ASTM A 888; CISPI 301
Acrylonotrile butadiene styrene (ABS) plastic pipe in sewer and drain diameters	ASTM D 2751
Polyvinyl chloride (PVC) plastic pipe in sewer and drain diameters	ASTM D 3034
Copper or copper alloy	ASME B 16.15; ASME B 16.18; ASME B 16.22; ASME B 16.23; ASME B 16.26; ASME B 16.29
Gray iron and ductile iron	AWWA C 110
Polyolefin	ASTM F 1412; CSA B181.3
Polyvinyl chloride (PVC) plastic in IPS diameters	ASTM D 2665; ASTM D 3311; ASTM F 1866
Polyvinyl chloride (PVC) plastic pipe with a 3.25 inch O.D.	ASTM D 2949
PVC fabricated fittings	ASTM F 1866
Stainless steel drainage systems, Types 304 and 316L	ASME A 112.3.1
Vitrified clay	ASTM C 700

For SI: 1 inch = 25.4 mm.

P3002.4.2 Lead bends and traps. Lead bends and lead traps shall not be less than $^1/_8$-inch (3 mm) wall thickness.

P3002.4.3 Sheet copper. Sheet copper for the following uses shall weigh not less than indicated below:

1. General use, 12 ounces per square feet (4 kg/m^2).
2. Flashing for vent pipes, 8 ounces per square feet (2.5 kg/m^2).

SECTION P3003 JOINTS AND CONNECTIONS

P3003.1 Tightness. Joints and connections in the DWV system shall be gas tight and water tight for the intended use or pressure required by test.

P3003.1.1 Threaded joints, general. Pipe and fitting threads shall be tapered.

P3003.2 Prohibited joints. Running threads and bands shall not be used in the drainage system. Drainage and vent piping shall not be drilled, tapped, burned or welded.

The following types of joints and connections shall be prohibited:

1. Cement or concrete.
2. Mastic or hot-pour bituminous joints.
3. Joints made with fittings not *approved* for the specific installation.
4. Joints between different diameter pipes made with elastomeric rolling O-rings.
5. Solvent-cement joints between different types of plastic pipe.
6. Saddle-type fittings.

P3003.3 ABS plastic. Joints between ABS plastic pipe or fittings shall comply with Sections P3003.3.1 through P3003.3.3.

P3003.3.1 Mechanical joints. Mechanical joints on drainage pipes shall be made with an elastomeric seal conforming to ASTM C 1173, ASTM D 3212 or CSA B602. Mechanical joints shall be installed only in underground systems unless otherwise *approved.* Joints shall be installed in accordance with the manufacturer's instructions.

P3003.3.2 Solvent cementing. Joint surfaces shall be clean and free from moisture. Solvent cement that conforms to ASTM D 2235 or CSA B181.1 shall be applied to all joint surfaces. The joint shall be made while the cement is wet. Joints shall be made in accordance with ASTM D 2235, ASTM D 2661, ASTM F 628 or CSA B181.1. Solvent-cement joints shall be permitted above or below ground.

P3003.3.3 Threaded joints. Threads shall conform to ASME B1.20.1. Schedule 80 or heavier pipe shall be permitted to be threaded with dies specifically designed for plastic pipe. *Approved* thread lubricant or tape shall be applied on the male threads only.

P3003.4 Asbestos-cement. Joints between asbestos-cement pipe or fittings shall be made with a sleeve coupling of the same composition as the pipe, sealed with an elastomeric ring conforming to ASTM D 1869.

P3003.5 Brass. Joints between brass pipe or fittings shall comply with Sections P3003.5.1 through P3003.5.3.

P3003.5.1 Brazed joints. All joint surfaces shall be cleaned. An *approved* flux shall be applied where required. The joint shall be brazed with a filler metal conforming to AWS A5.8.

P3003.5.2 Mechanical joints. Mechanical joints shall be installed in accordance with the manufacturer's instructions.

P3003.5.3 Threaded joints. Threads shall conform to ASME B1.20.1. Pipe-joint compound or tape shall be applied on the male threads only.

P3003.6 Cast iron. Joints between cast-iron pipe or fittings shall comply with Sections P3003.6.1 through P3003.6.3.

P3003.6.1 Caulked joints. Joints for hub and spigot pipe shall be firmly packed with oakum or hemp. Molten lead shall be poured in one operation to a depth of not less than 1 inch (25 mm). The lead shall not recede more than $^{1}/_{8}$ inch (3 mm) below the rim of the hub and shall be caulked tight. Paint, varnish or other coatings shall not be permitted on the jointing material until after the joint has been tested and *approved*. Lead shall be run in one pouring and shall be caulked tight. Acid-resistant rope and acidproof cement shall be permitted.

P3003.6.2 Compression gasket joints. Compression gaskets for hub and spigot pipe and fittings shall conform to ASTM C 564. Gaskets shall be compressed when the pipe is fully inserted.

P3003.6.3 Mechanical joint coupling. Mechanical joint couplings for hubless pipe and fittings shall comply with CISPI 310 or ASTM C 1277. The elastomeric sealing sleeve shall conform to ASTM C 564 or CSA B602 and shall have a center stop. Mechanical joint couplings shall be installed in accordance with the manufacturer's installation instructions.

P3003.7 Concrete joints. Joints between concrete pipe and fittings shall be made with an elastomeric seal conforming to ASTM C 443, ASTM C 1173, CSA A257.3M or CSA B602.

P3003.8 Coextruded composite ABS pipe. Joints between coextruded composite pipe with an ABS outer layer or ABS fittings shall comply with Sections P3003.8.1 and P3003.8.2.

P3003.8.1 Mechanical joints. Mechanical joints on drainage pipe shall be made with an elastomeric seal conforming to ASTM C 1173, ASTM D 3212 or CSA B602. Mechanical joints shall not be installed in above-ground systems, unless otherwise *approved*. Joints shall be installed in accordance with the manufacturer's instructions.

P3003.8.2 Solvent cementing. Joint surfaces shall be clean and free from moisture. Solvent cement that conforms to ASTM D 2235 or CSA B181.1 shall be applied to all joint surfaces. The joint shall be made while the cement is wet. Joints shall be made in accordance with ASTM D 2235, ASTM D 2661, ASTM F 628 or CSA B181.1. Solvent-cement joints shall be permitted above or below ground.

P3003.9 Coextruded composite PVC pipe. Joints between coextruded composite pipe with a PVC outer layer or PVC fittings shall comply with Sections P3003.9.1 and P3003.9.2.

P3003.9.1 Mechanical joints. Mechanical joints on drainage pipe shall be made with an elastomeric seal conforming to ASTM D 3212. Mechanical joints shall not be installed in above-ground systems, unless otherwise *approved*. Joints shall be installed in accordance with the manufacturer's instructions.

P3003.9.2 Solvent cementing. Joint surfaces shall be clean and free from moisture. A purple primer that conforms to ASTM F 656 shall be applied. Solvent cement not purple in color and conforming to ASTM D 2564, CSA B137.3 or CSA B181.2 shall be applied to all joint surfaces. The joint shall be made while the cement is wet, and shall be in accordance with ASTM D 2855. Solvent-cement joints shall be permitted above or below ground.

P3003.10 Copper pipe. Joints between copper or copper-alloy pipe or fittings shall comply with Sections P3003.10.1 through P3003.10.4.

P3003.10.1 Brazed joints. All joint surfaces shall be cleaned. An *approved* flux shall be applied where required. The joint shall be brazed with a filler metal conforming to AWS A5.8.

P3003.10.2 Mechanical joints. Mechanical joints shall be installed in accordance with the manufacturer's instructions.

P3003.10.3 Soldered joints. Solder joints shall be made in accordance with the methods of ASTM B 828. All cut tube ends shall be reamed to the full inside diameter of the tube end. All joint surfaces shall be cleaned. A flux conforming to ASTM B 813 shall be applied. The joint shall be soldered with a solder conforming to ASTM B 32.

P3003.10.4 Threaded joints. Threads shall conform to ASME B1.20.1. Pipe-joint compound or tape shall be applied on the male threads only.

P3003.11 Copper tubing. Joints between copper or copper-alloy tubing or fittings shall comply with Sections P3003.11.1 through P3003.11.3.

P3003.11.1 Brazed joints. All joint surfaces shall be cleaned. An *approved* flux shall be applied where required. The joint shall be brazed with a filler metal conforming to AWS A5.8.

P3003.11.2 Mechanical joints. Mechanical joints shall be installed in accordance with the manufacturer's instructions.

P3003.11.3 Soldered joints. Solder joints shall be made in accordance with the methods of ASTM B 828. Cut tube ends shall be reamed to the full inside diameter of the tube end. All joint surfaces shall be cleaned. A flux conforming to ASTM B 813 shall be applied. The joint shall be soldered with a solder conforming to ASTM B 32.

P3003.12 Steel. Joints between galvanized steel pipe or fittings shall comply with Sections P3003.12.1 and P3003.12.2.

P3003.12.1 Threaded joints. Threads shall conform to ASME B1.20.1. Pipe-joint compound or tape shall be applied on the male threads only.

P3003.12.2 Mechanical joints. Joints shall be made with an *approved* elastomeric seal. Mechanical joints shall be installed in accordance with the manufacturer's instructions.

P3003.13 Lead. Joints between lead pipe or fittings shall comply with Sections P3003.13.1 and P3003.13.2.

P3003.13.1 Burned. Burned joints shall be uniformly fused together into one continuous piece. The thickness of the joint shall be at least as thick as the lead being joined. The filler metal shall be of the same material as the pipe.

P3003.13.2 Wiped. Joints shall be fully wiped, with an exposed surface on each side of the joint not less than $^3/_4$ inch (19 mm). The joint shall be at least $^3/_8$ inch (9.5 mm) thick at the thickest point.

P3003.14 PVC plastic. Joints between PVC plastic pipe or fittings shall comply with Sections P3003.14.1 through P3003.14.3.

P3003.14.1 Mechanical joints. Mechanical joints on drainage pipe shall be made with an elastomeric seal conforming to ASTM C 1173, ASTM D 3212 or CSA B602. Mechanical joints shall not be installed in above-ground systems, unless otherwise *approved.* Joints shall be installed in accordance with the manufacturer's instructions.

P3003.14.2 Solvent cementing. Joint surfaces shall be clean and free from moisture. A purple primer that conforms to ASTM F 656 shall be applied. Solvent cement not purple in color and conforming to ASTM D 2564, CSA B137.3 or CSA B181.2 shall be applied to all joint surfaces. The joint shall be made while the cement is wet, and shall be in accordance with ASTM D 2855. Solvent-cement joints shall be permitted above or below ground.

P3003.14.3 Threaded joints. Threads shall conform to ASME B1.20.1. Schedule 80 or heavier pipe shall be permitted to be threaded with dies specifically designed for plastic pipe. *Approved* thread lubricant or tape shall be applied on the male threads only.

P3003.15 Vitrified clay. Joints between vitrified clay pipe or fittings shall be made with an elastomeric seal conforming to ASTM C 425, ASTM C 1173 or CSA B602.

P3003.16 Polyolefin plastic. Joints between polyolefin plastic pipe and fittings shall comply with Sections P3003.16.1 and P3003.16.2.

P3003.16.1 Heat-fusion joints. Heat-fusion joints for polyolefin pipe and tubing joints shall be installed with socket-type heat-fused polyolefin fittings or electrofusion polyolefin fittings. Joint surfaces shall be clean and free from moisture. The joint shall be undisturbed until cool. Joints shall be made in accordance with ASTM F 1412 or CSA B181.3.

P3003.16.2 Mechanical and compression sleeve joints. Mechanical and compression sleeve joints shall be installed in accordance with the manufacturer's instructions.

P3003.17 Polyethylene plastic pipe. Joints between polyethylene plastic pipe and fittings shall be underground and shall comply with Section P3003.17.1 or P3003.17.2.

P3003.17.1 Heat fusion joints. Joint surfaces shall be clean and free from moisture. All joint surfaces shall be cut, heated to melting temperature and joined using tools specifically designed for the operation. Joints shall be undisturbed until cool. Joints shall be made in accordance with ASTM D 2657 and the manufacturer's instructions.

P3003.17.2 Mechanical joints. Mechanical joints in drainage piping shall be made with an elastomeric seal conforming to ASTM C 1173, ASTM D 3212 or CSA B602. Mechanical joints shall be installed in accordance with the manufacturer's instructions.

P3003.18 Joints between different materials. Joints between different piping materials shall be made with a mechanical joint of the compression or mechanical-sealing type conforming to ASTM C 1173, ASTM C 1460 or ASTM C 1461. Connectors and adapters shall be *approved* for the application and such joints shall have an elastomeric seal conforming to ASTM C 425, ASTM C 443, ASTM C 564, ASTM C 1440, ASTM D 1869, ASTM F 477, CSA A257.3M or CSA B602, or as required in Sections P3003.18.1 through P3003.18.6. Joints between glass pipe and other types of materials shall be made with adapters having a TFE seal. Joints shall be installed in accordance with the manufacturer's instructions.

P3003.18.1 Copper or copper-alloy tubing to cast-iron hub pipe. Joints between copper or copper-alloy tubing and cast-iron hub pipe shall be made with a brass ferrule or compression joint. The copper or copper-alloy tubing shall be soldered to the ferrule in an *approved* manner, and the ferrule shall be joined to the cast-iron hub by a caulked joint or a mechanical compression joint.

P3003.18.2 Copper or copper-alloy tubing to galvanized steel pipe. Joints between copper or copper-alloy tubing and galvanized steel pipe shall be made with a brass converter fitting or dielectric fitting. The copper tubing shall be soldered to the fitting in an *approved* manner, and the fitting shall be screwed to the threaded pipe.

P3003.18.3 Cast-iron pipe to galvanized steel or brass pipe. Joints between cast-iron and galvanized steel or brass pipe shall be made by either caulked or threaded joints or with an *approved* adapter fitting.

P3003.18.4 Plastic pipe or tubing to other piping material. Joints between different types of plastic pipe or between plastic pipe and other piping material shall be made with an *approved* adapter fitting. Joints between plastic pipe and cast-iron hub pipe shall be made by a caulked joint or a mechanical compression joint.

P3003.18.5 Lead pipe to other piping material. Joints between lead pipe and other piping material shall be made by a wiped joint to a caulking ferrule, soldering nipple, or bushing or shall be made with an *approved* adapter fitting.

P3003.18.6 Stainless steel drainage systems to other materials. Joints between stainless steel drainage systems and other piping materials shall be made with *approved* mechanical couplings.

P3003.19 Joints between drainage piping and water closets. Joints between drainage piping and water closets or similar fixtures shall be made by means of a closet flange or a waste connector and sealing gasket compatible with the drainage system material, securely fastened to a structurally firm base. The inside diameter of the drainage pipe shall not be used as a socket fitting for a 4-inch by 3-inch (102 mm by 76 mm) closet flange. The joint shall be bolted, with an approved gasket flange to fixture connection complying with ASME A112.4.3 or setting compound between the fixture and the closet flange or waste connector and sealing gasket. The waste connector and sealing gasket joint shall comply with the joint-tightness test of ASME A112.4.3 and shall be installed in accordance with the manufacturer's installation instructions.

SECTION P3004 DETERMINING DRAINAGE FIXTURE UNITS

P3004.1 DWV system load. The load on DWV-system piping shall be computed in terms of drainage fixture unit (d.f.u.) values in accordance with Table P3004.1.

SECTION P3005 DRAINAGE SYSTEM

P3005.1 Drainage fittings and connections. Changes in direction in drainage piping shall be made by the appropriate use of sanitary tees, wyes, sweeps, bends or by a combination of these drainage fittings in accordance with Table P3005.1. Change in direction by combination fittings, heel or side inlets or increasers shall be installed in accordance with Table P3005.1 and Sections P3005.1.1 through P3005.1.4. based on the pattern of flow created by the fitting.

P3005.1.1 Horizontal to vertical (multiple connection fittings). Double fittings such as double sanitary tees and tee-wyes or *approved* multiple connection fittings and back-to-back fixture arrangements that connect two or more branches at the same level shall be permitted as long as directly opposing connections are the same size and the

TABLE P3004.1
DRAINAGE FIXTURE UNIT (d.f.u.) VALUES FOR VARIOUS PLUMBING FIXTURES

TYPE OF FIXTURE OR GROUP OF FIXTURES	DRAINAGE FIXTURE UNIT VALUE (d.f.u.)[a]
Bar sink	1
Bathtub (with or without shower head and/or whirlpool attachments)	2
Bidet	1
Clothes washer standpipe	2
Dishwasher	2
Floor drain[b]	0
Kitchen sink	2
Lavatory	1
Laundry tub	2
Shower stall	2
Water closet (1.6 gallons per flush)	3
Water closet (greater than 1.6 gallons per flush)	4
Full-bath group with bathtub (with 1.6 gallon per flush water closet, and with or without shower head and/or whirlpool attachment on the bathtub or shower stall)	5
Full-bath group with bathtub (water closet greater than 1.6 gallon per flush, and with or without shower head and/or whirlpool attachment on the bathtub or shower stall)	6
Half-bath group (1.6 gallon per flush water closet plus lavatory)	4
Half-bath group (water closet greater than 1.6 gallon per flush plus lavatory)	5
Kitchen group (dishwasher and sink with or without garbage grinder)	2
Laundry group (clothes washer standpipe and laundry tub)	3
Multiple-bath groups[c]: 1.5 baths 2 baths 2.5 baths 3 baths 3.5 baths	 7 8 9 10 11

For SI: 1 gallon = 3.785 L.

a. For a continuous or semicontinuous flow into a drainage system, such as from a pump or similar device, 1.5 fixture units shall be allowed per gpm of flow. For a fixture not listed, use the highest d.f.u. value for a similar listed fixture.

b. A floor drain itself adds no hydraulic load. However, where used as a receptor, the fixture unit value of the fixture discharging into the receptor shall be applicable.

c. Add 2 d.f.u. for each additional full bath.

discharge into directly opposing connections is from similar fixture types or fixture groups. Double sanitary tee patterns shall not receive the discharge of back-to-back water closets and fixtures or appliances with pumping action discharge.

Exception: Back-to-back water closet connections to double sanitary tee patterns shall be permitted where the horizontal *developed length* between the outlet of the water closet and the connection to the double sanitary tee is 18 inches (457 mm) or greater.

P3005.1.2 Heel- or side-inlet quarter bends, drainage. Heel-inlet quarter bends shall be an acceptable means of connection, except where the quarter bends serves a water closet. A low-heel inlet shall not be used as a wet-vented connection. Side-inlet quarter bends shall be an acceptable means of connection for both drainage, wet venting and stack venting arrangements.

P3005.1.3 Heel- or side-inlet quarter bends, venting. Heel-inlet or side-inlet quarter bends, or any arrangement of pipe and fittings producing a similar effect, shall be acceptable as a dry vent where the inlet is placed in a vertical position. The inlet is permitted to be placed in a horizontal position only where the entire fitting is part of a dry vent arrangement.

P3005.1.4 Water closet connection between flange and pipe. One-quarter bends 3 inches (76 mm) in diameter shall be acceptable for water closet or similar connections, provided a 4-inch by 3-inch (102 mm by 76 mm) flange is installed to receive the closet fixture horn. Alternately, a 4-inch by 3-inch (102 mm by 76 mm) elbow shall be acceptable with a 4-inch (102 mm) flange.

P3005.1.5 Dead ends. Dead ends shall be prohibited except where necessary to extend a cleanout or as an *approved* part of a rough-in more than 2 feet (610 mm) in length.

P3005.1.6 Provisions for future fixtures. Where drainage has been roughed-in for future fixtures, the drainage unit values of the future fixtures shall be considered in determining the required drain sizes. Such future installations shall be terminated with an accessible permanent plug or cap fitting.

P3005.1.7 Change in size. The size of the drainage piping shall not be reduced in size in the direction of the flow. A 4-inch by 3-inch (102 mm by 76 mm) water closet connection shall not be considered as a reduction in size.

P3005.2 Drainage pipe cleanouts. Drainage pipe cleanouts shall comply with Sections P3005.2.1 through P3005.2.11.

Exception: These provisions shall not apply to pressurized *building drains* and *building sewers* that convey the discharge of automatic pumping equipment to a gravity drainage system.

P3005.2.1 Materials. Cleanouts shall be liquid and gas tight. Cleanout plugs shall be brass or plastic.

P3005.2.2 Spacing. Cleanouts shall be installed not more than 100 feet (30 480 mm) apart in horizontal drainage lines measured from the upstream entrance of the cleanout.

P3005.2.3 Underground drainage cleanouts. When installed in underground drains, cleanouts shall be extended vertically to or above finished grade either inside or outside the building.

P3005.2.4 Change of direction. Cleanouts shall be installed at each fitting with a change of direction more than 45 degrees (0.79 rad) in the *building sewer, building drain* and horizontal waste or soil lines. Where more than one change of direction occurs in a run of piping, only one cleanout shall be required in each 40 feet (12 192 mm) of *developed length* of the drainage piping.

P3005.2.5 Accessibility. Cleanouts shall be accessible. The clearance in front of cleanouts shall be not less than 18 inches (457 mm) on 3-inch (76 mm) and larger pipes, and not less than 12 inches (305 mm) on smaller pipes. Concealed cleanouts shall be provided with access of sufficient size to permit removal of the cleanout plug and rodding of the system. Cleanout plugs shall not be concealed by permanent finishing material.

TABLE P3005.1
FITTINGS FOR CHANGE IN DIRECTION

TYPE OF FITTING PATTERN	CHANGE IN DIRECTION		
	Horizontal to vertical[c]	Vertical to horizontal	Horizontal to horizontal
Sixteenth bend	X	X	X
Eighth bend	X	X	X
Sixth bend	X	X	X
Quarter bend	X	X[a]	X[a]
Short sweep	X	X[a,b]	X[a]
Long sweep	X	X	X
Sanitary tee	X[c]	—	—
Wye	X	X	X
Combination wye and eighth bend	X	X	X

For SI: 1 inch = 25.4 mm.

a. The fittings shall only be permitted for a 2-inch or smaller fixture drain.

b. Three inches and larger.

c. For a limitation on multiple connection fittings, see Section P3005.1.1.

P3005.2.6 Base of stacks. A cleanout shall be provided at the base of each waste or soil stack.

P3005.2.7 Building drain and building sewer junction. There shall be a cleanout near the junction of the *building drain* and *building sewer*. This cleanout shall be either inside or outside the building wall, provided that it is brought up to finish grade or to the lowest floor level. An *approved* two-way cleanout shall be permitted to serve as the required cleanout for both the *building drain* and the *building sewer*. The cleanout at the junction of the *building drain* and *building sewer* shall not be required where a cleanout on a 3-inch (76 mm) or larger diameter soil stack is located within a *developed length* of 10 feet (3048 mm) of the *building drain* and *building sewer* junction.

P3005.2.8 Direction of flow. Cleanouts shall be installed so that the cleanout opens to allow cleaning in the direction of the flow of the drainage line.

P3005.2.9 Cleanout size. Cleanouts shall be the same nominal size as the pipe they serve up to 4 inches (102 mm). For pipes larger than 4 inches (102 mm) nominal size, the size of the cleanout shall be not less than 4 inches (102 mm).

Exceptions:

1. "P" trap connections with slip joints or ground joint connections, or stack cleanouts that are not more than one pipe diameter smaller than the drain served, shall be permitted.
2. Cast-iron cleanouts sized in accordance with the referenced standards in Table P3002.3, ASTM A 74 for hub and spigot fittings or ASTM A 888 or CISPI 301 for hubless fittings.

P3005.2.10 Cleanout equivalent. A fixture trap or a fixture with integral trap, readily removable without disturbing concealed piping shall be acceptable as a cleanout equivalent.

P3005.2.11 Connections to cleanouts prohibited. Cleanout openings shall not be used for the installation of new fixtures except where *approved* and an acceptable alternate cleanout is provided.

P3005.3 Horizontal drainage piping slope. Horizontal drainage piping shall be installed in uniform alignment at uniform slopes not less than $^1/_4$ unit vertical in 12 units horizontal (2-percent slope) for $2^1/_2$ inch (64 mm) diameter and less, and not less than $^1/_8$ unit vertical in 12 units horizontal (1-percent slope) for diameters of 3 inches (76 mm) or more.

P3005.4 Drain pipe sizing. Drain pipes shall be sized according to drainage fixture unit (d.f.u.) loads. The size of the drainage piping shall not be reduced in size in the direction of flow. The following general procedure is permitted to be used:

1. Draw an isometric layout or riser diagram denoting fixtures on the layout.
2. Assign d.f.u. values to each fixture group plus individual fixtures using Table P3004.1.
3. Starting with the top floor or most remote fixtures, work downstream toward the *building drain* accumulating d.f.u. values for fixture groups plus individual fixtures for each branch. Where multiple bath groups are being added, use the reduced d.f.u. values in Table P3004.1, which take into account probability factors of simultaneous use.
4. Size branches and stacks by equating the assigned d.f.u. values to pipe sizes shown in Table P3005.4.1.
5. Determine the pipe diameter and slope of the *building drain* and *building sewer* based on the accumulated d.f.u. values, using Table P3005.4.2.

P3005.4.1 Branch and stack sizing. Branches and stacks shall be sized in accordance with Table P3005.4.1. Below grade drain pipes shall be not less than $1^1/_2$ inches (38 mm) in diameter. Drain stacks shall be not smaller than the largest horizontal branch connected.

Exceptions:

1. A 4-inch by 3-inch (102 mm by 76 mm) closet bend or flange.
2. A 4-inch (102 mm) closet bend connected to a 3-inch (76 mm) stack tee shall not be prohibited.

**TABLE P3005.4.1
MAXIMUM FIXTURE UNITS ALLOWED TO BE CONNECTED TO BRANCHES AND STACKS**

NOMINAL PIPE SIZE (inches)	ANY HORIZONTAL FIXTURE BRANCH	ANY ONE VERTICAL STACK OR DRAIN
$1^1/_4$[a]	—	—
$1^1/_2$[b]	3	4
2[b]	6	10
$2^1/_2$[b]	12	20
3	20	48
4	160	240

For SI: 1 inch = 25.4 mm.

a. $1^1/_4$-inch pipe size limited to a single-fixture drain or trap arm. See Table P3201.7.

b. No water closets.

P3005.4.2 Building drain and sewer size and slope. Pipe sizes and slope shall be determined from Table P3005.4.2 on the basis of drainage load in fixture units (d.f.u.) computed from Table P3004.1.

**TABLE P3005.4.2
MAXIMUM NUMBER OF FIXTURE UNITS ALLOWED TO BE CONNECTED TO THE BUILDING DRAIN, BUILDING DRAIN BRANCHES OR THE BUILDING SEWER**

DIAMETER OF PIPE (inches)	SLOPE PER FOOT		
	$^1/_8$ inch	$^1/_4$ inch	$^1/_2$ inch
$1^1/_2$[a, b]	—	Note a	Note a
2[b]	—	21	27
$2^1/_2$[b]	—	24	31
3	36	42	50
4	180	216	250

For SI: 1 inch = 25.4 mm, 1 foot = 304.8 mm.

a. $1^1/_2$-inch pipe size limited to a building drain branch serving not more than two waste fixtures, or not more than one waste fixture if serving a pumped discharge fixture or garbage grinder discharge.

b. No water closets.

P3005.5 Connections to offsets and bases of stacks. Horizontal branches shall connect to the bases of stacks at a point located not less than 10 times the diameter of the drainage stack downstream from the stack. Horizontal branches shall connect to horizontal stack offsets at a point located not less than 10 times the diameter of the drainage stack downstream from the upper stack.

SECTION P3006 SIZING OF DRAIN PIPE OFFSETS

P3006.1 Vertical offsets. An offset in a vertical drain, with a change of direction of 45 degrees (0.79 rad) or less from the vertical, shall be sized as a straight vertical drain.

P3006.2 Horizontal offsets above the lowest branch. A stack with an offset of more than 45 degrees (0.79 rad) from the vertical shall be sized as follows:

1. The portion of the stack above the offset shall be sized as for a regular stack based on the total number of fixture units above the offset.
2. The offset shall be sized as for a *building drain* in accordance with Table P3005.4.2.
3. The portion of the stack below the offset shall be sized as for the offset or based on the total number of fixture units on the entire stack, whichever is larger.

P3006.3 Horizontal offsets below the lowest branch. In soil or waste stacks below the lowest horizontal branch, there shall be no change in diameter required if the offset is made at an angle not greater than 45 degrees (0.79 rad) from the vertical. If an offset greater than 45 degrees (0.79 rad) from the vertical is made, the offset and stack below it shall be sized as a *building drain* (see Table P3005.4.2).

SECTION P3007 SUMPS AND EJECTORS

P3007.1 Building subdrains. Building subdrains that cannot be discharged to the sewer by gravity flow shall be discharged into a tightly covered and vented sump from which the liquid shall be lifted and discharged into the building gravity drainage system by automatic pumping equipment or other *approved* method. In other than existing structures, the sump shall not receive drainage from any piping within the building capable of being discharged by gravity to the *building sewer.*

P3007.2 Valves required. A check valve and a full open valve located on the discharge side of the check valve shall be installed in the pump or ejector discharge piping between the pump or ejector and the gravity drainage system. Access shall be provided to such valves. Such valves shall be located above the sump cover required by Section P3007.3.2 or, where the discharge pipe from the ejector is below grade, the valves shall be accessibly located outside the sump below grade in an access pit with a removable access cover.

P3007.3 Sump design. The sump pump, pit and discharge piping shall conform to the requirements of Sections P3007.3.1 through P3007.3.5.

P3007.3.1 Sump pump. The sump pump capacity and head shall be appropriate to anticipated use requirements.

P3007.3.2 Sump pit. The sump pit shall be not less than 18 inches (457 mm) in diameter and 24 inches (610 mm) deep, unless otherwise *approved.* The pit shall be accessible and located so that all drainage flows into the pit by gravity. The sump pit shall be constructed of tile, concrete, steel, plastic or other *approved* materials. The pit bottom shall be solid and provide permanent support for the pump. The sump pit shall be fitted with a gastight removable cover adequate to support anticipated loads in the area of use. The sump pit shall be vented in accordance with Chapter 31.

P3007.3.3 Discharge pipe and fittings. Discharge pipe and fittings serving sump pumps and ejectors shall be constructed of materials in accordance with Sections P3007.3.3.1 and P3007.3.3.2 and shall be *approved.*

P3007.3.3.1 Materials. Pipe and fitting materials shall be constructed of brass, copper, CPVC, ductile iron, PE, or PVC.

P3007.3.3.2 Ratings. Pipe and fittings shall be rated for the maximum system operating pressure and temperature. Pipe fitting materials shall be compatible with the pipe material. Where pipe and fittings are buried in the earth, they shall be suitable for burial.

P3007.3.4 Maximum effluent level. The effluent level control shall be adjusted and maintained to at all times prevent the effluent in the sump from rising to within 2 inches (51 mm) of the invert of the gravity drain inlet into the sump.

P3007.3.5 Ejector connection to the drainage system. Pumps connected to the drainage system shall connect to a *building sewer, building drain,* soil stack, waste stack or horizontal branch drain. Where the discharge line connects into horizontal drainage piping, the connection shall be made through a wye fitting into the top of the drainage piping and such wye fitting shall be located not less than 10 pipe diameters from the base of any soil stack, waste stack or fixture drain.

P3007.4 Sewage pumps and sewage ejectors. A sewage pump or sewage ejector shall automatically discharge the contents of the sump to the building drainage system.

P3007.5 Macerating toilet systems. Macerating toilet systems shall comply with CSA B45.9 or ASME A112.3.4 and shall be installed in accordance with the manufacturer's installation instructions.

P3007.6 Capacity. A sewage pump or sewage ejector shall have the capacity and head for the application requirements. Pumps or ejectors that receive the discharge of water closets shall be capable of handling spherical solids with a diameter of up to and including 2 inches (51 mm). Other pumps or ejectors shall be capable of handling spherical solids with a diameter of up to and including 1 inch (25.4 mm). The minimum capacity of a pump or ejector based on the diameter of the discharge pipe shall be in accordance with Table 3007.6.

Exceptions:

1. Grinder pumps or grinder ejectors that receive the discharge of water closets shall have a minimum discharge opening of $1^1/_4$ inches (32 mm).
2. Macerating toilet assemblies that serve single water closets shall have a minimum discharge opening of $^3/_4$ inch (19 mm).

TABLE 3007.6
MINIMUM CAPACITY OF SEWAGE PUMP OR SEWAGE EJECTOR

DIAMETER OF THE DISCHARGE PIPE (inches)	CAPACITY OF PUMP OR EJECTOR (gpm)
2	21
$2^1/_2$	30
3	46

For SI: 1 inch = 25.4 mm, 1 gallon per minute = 3.785 L/m.

SECTION P3008
BACKWATER VALVES

P3008.1 Sewage backflow. Where the flood level rims of plumbing fixtures are below the elevation of the manhole cover of the next upstream manhole in the public sewer, the fixtures shall be protected by a backwater valve installed in the *building drain*, branch of the *building drain* or horizontal branch serving such fixtures. Plumbing fixtures having flood level rims above the elevation of the manhole cover of the next upstream manhole in the public sewer shall not discharge through a backwater valve.

P3008.2 Material. All bearing parts of backwater valves shall be of corrosion-resistant material. Backwater valves shall comply with ASME A112.14.1, CSA B181.1 or CSA B181.2.

P3008.3 Seal. Backwater valves shall be constructed to provide a mechanical seal against backflow.

P3008.4 Diameter. Backwater valves, when fully opened, shall have a capacity not less than that of the pipes in which they are installed.

P3008.5 Location. Backwater valves shall be installed so that access is provided to the working parts for service and repair.

SECTION P3009
GRAY WATER RECYCLING SYSTEMS

P3009.1 Scope. The provisions of Section P3009 shall govern the materials, design, construction and installation of gray water systems for flushing of water closets and urinals and for subsurface landscape irrigation. See Figures P3009.1(1) and P3009.1(2).

P3009.2 Installation. In addition to the provisions of Section P3009, systems for flushing of water closets and urinals shall comply with Section P3009.13 and systems for subsurface landscape irrigation shall comply with Section P3009.14. Except as provided for in Section P3009, all systems shall comply with the provisions of the other sections of this code.

P3009.3 Materials. Above-ground drain, waste and vent piping for gray water systems shall conform to one of the standards listed in Table P3002.1(1). Gray water underground building drainage and vent pipe shall conform to one of the standards listed in Table P3002.1(2).

P3009.4 Tests. Drain, waste and vent piping for gray water systems shall be tested in accordance with Section P2503.

P3009.5 Inspections. Gray water systems shall be inspected in accordance with Section P2503.

P3009.6 Potable water connections. Only connections in accordance with Section 3009.13.1 shall be made between a gray water recycling system and a potable water system.

P3009.7 Waste water connections. Gray water recycling systems shall receive only the waste discharge of bathtubs, showers, lavatories, clothes washers or laundry trays.

P3009.8 Collection reservoir. Gray water shall be collected in an approved reservoir constructed of durable, nonabsorbent and corrosion-resistant materials. The reservoir shall be a closed and gas-tight vessel. Access openings shall be provided to allow inspection and cleaning of the reservoir interior.

P3009.9 Filtration. Gray water entering the reservoir shall pass through an approved filter such as a media, sand or diatomaceous earth filter.

P3009.9.1 Required valve. A full-open valve shall be installed downstream of the last fixture connection to the gray water discharge pipe before entering the required filter.

P3009.10 Overflow. The collection reservoir shall be equipped with an overflow pipe having the same or larger diameter as the influent pipe for the gray water. The overflow pipe shall be trapped and shall be indirectly connected to the sanitary drainage system.

P3009.11 Drain. A drain shall be located at the lowest point of the collection reservoir and shall be indirectly connected to the sanitary drainage system. The drain shall be the same diameter as the overflow pipe required in Section P3009.10.

P3009.12 Vent required. The reservoir shall be provided with a vent sized in accordance with Chapter 31 and based on the diameter of the reservoir influent pipe.

P3009.13 Flushing water systems. Systems for flushing water closets and urinals shall comply with Sections P3009.13.1 through P3009.13.6

P3009.13.1 Collection reservoir. The holding capacity of the reservoir shall be a minimum of twice the volume of water required to meet the daily flushing requirements of the fixtures supplied with gray water, but not less than 50 gallons (189 L). The reservoir shall be sized to limit the retention time of gray water to a maximum of 72 hours.

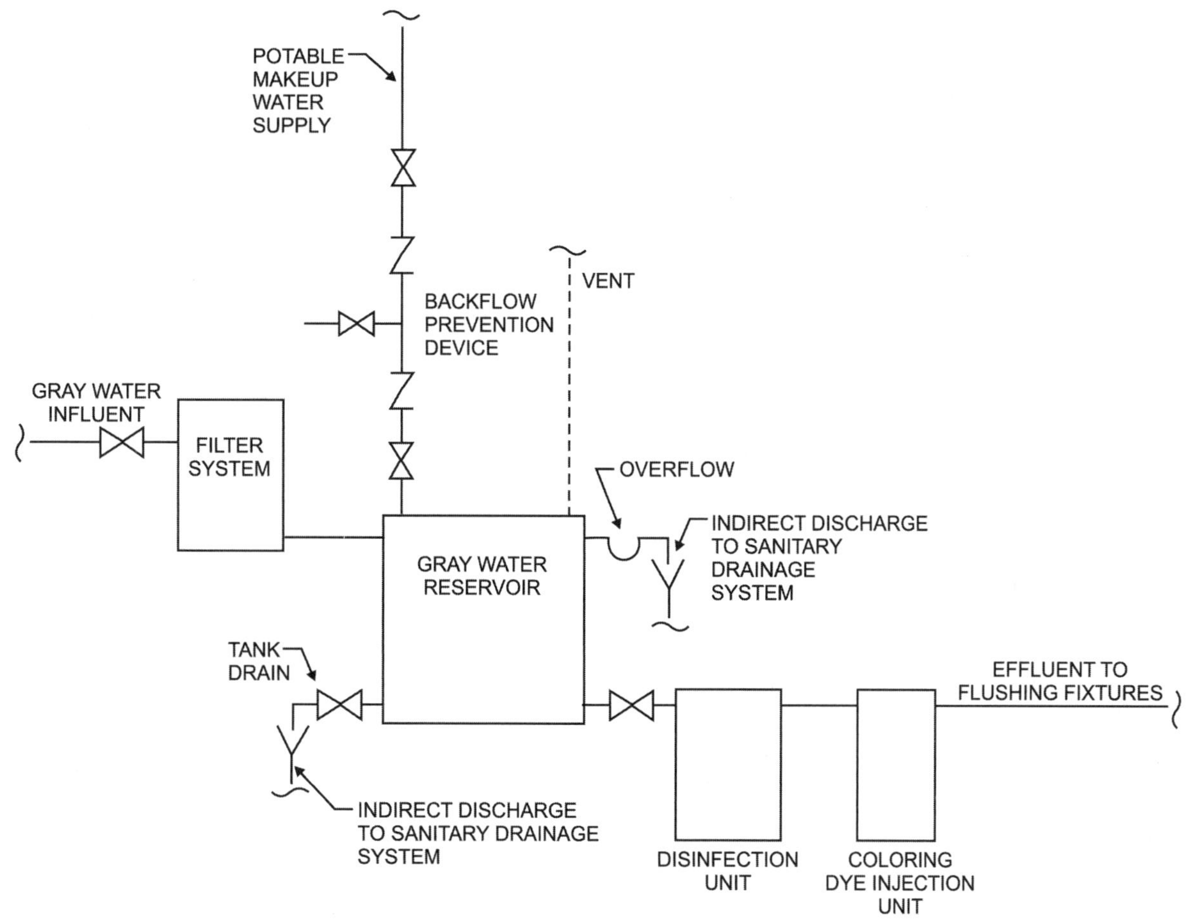

FIGURE P3009.1(1)
GRAY WATER RECYCLING SYSTEM FOR FLUSHING WATER CLOSETS AND URINALS

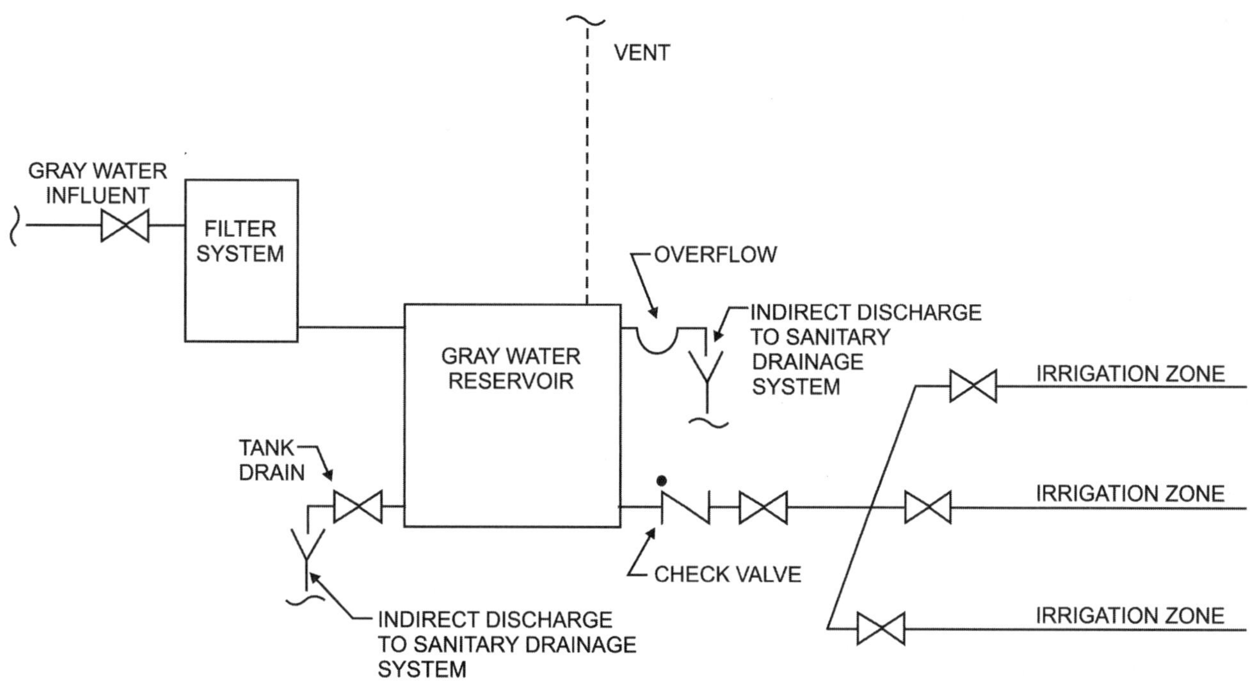

FIGURE P3009.1(2)
GRAY WATER RECYCLING SYSTEM FOR SUBSURFACE LANDSCAPE IRRIGATION

P3009.13.2 Disinfection. Gray water shall be disinfected by an approved method that employs one or more disinfectants such as chlorine, iodine or ozone that are recommended for use with the pipes, fittings and equipment by the manufacturer of the pipes, fittings and equipment.

P3009.13.3 Makeup water. Potable water shall be supplied as a source of makeup water for the gray water system. The potable water supply shall be protected against backflow in accordance with Section P2902. There shall be a full-open valve located on the makeup water supply line to the collection reservoir.

P3009.13.4 Coloring. The gray water shall be dyed blue or green with a food-grade vegetable dye before such water is supplied to the fixtures.

P3009.13.5 Materials. Distribution piping shall conform to one of the standards listed in Table P2905.4.

P3009.13.6 Identification. Distribution piping and reservoirs shall be identified as containing nonpotable water. Piping identification shall be in accordance with Section P2901.1.

P3009.14 Landscape irrigation systems. Subsurface landscape irrigation systems shall comply with Sections P3009.14.1 through P3009.14.11

P3009.14.1 Collection reservoir. Reservoirs shall be sized to limit the retention time of gray water to a maximum of 24 hours.

P3009.14.1.1 Identification. The reservoir shall be identified as containing nonpotable water.

P3009.14.2 Valves required. A check valve and a full-open valve located on the discharge side of the check valve shall be installed on the effluent pipe of the collection reservoir.

P3009.14.3 Makeup water. Makeup water shall not be required for subsurface landscape irrigation systems. Where makeup water is provided, the installation shall be in accordance with Section 3009.13.3.

P3009.14.4 Disinfection. Disinfection shall not be required for gray water used or subsurface landscape irrigation systems.

P3009.14.5 Coloring. Gray water used for subsurface landscape irrigation systems shall not be required to be dyed.

P3009.14.6 Estimating gray water discharge. The system shall be sized in accordance with the gallons-per-day-per-occupant number based on the type of fixtures connected to the gray water system. The discharge shall be calculated by the following equation:

(Equation 30-1)

where:

$C = A \times B$

A = Number of occupants:

Number of occupants shall be determined by the actual number of occupants, but not less than two occupants for one bedroom and one occupant for each additional bedroom.

B = Estimated flow demands for each occupant:

Residential—25 gallons per day (94.6 lpd) per occupant for showers, bathtubs and lavatories and 15 gallons per day (56.7 lpd) per occupant for clothes washers or laundry trays.

C = Estimated gray water discharge based on the total number of occupants.

P3009.14.7 Percolation tests. The permeability of the soil in the proposed absorption system shall be determined by percolation tests or permeability evaluation.

P3009.14.7.1 Percolation tests and procedures. At least three percolation tests in each system area shall be conducted. The holes shall be spaced uniformly in relation to the bottom depth of the proposed absorption system. More percolation tests shall be made where necessary, depending on system design.

P3009.14.7.1.1 Percolation test hole. The test hole shall be dug or bored. The test hole shall have vertical sides and a horizontal dimension of 4 inches to 8 inches (102 mm to 203 mm). The bottom and sides of the hole shall be scratched with a sharp-pointed instrument to expose the natural soil. All loose material shall be removed from the hole and the bottom shall be covered with 2 inches (51 mm) of gravel or coarse sand.

P3009.14.7.1.2 Test procedure, sandy soils. The hole shall be filled with clear water to a minimum of 12 inches (305 mm) above the bottom of the hole for tests in sandy soils. The time for this amount of water to seep away shall be determined, and this procedure shall be repeated if the water from the second filling of the hole seeps away in 10 minutes or less. The test shall proceed as follows: Water shall be added to a point not more than 6 inches (152 mm) above the gravel or coarse sand. Thereupon, from a fixed reference point, water levels shall be measured at 10-minute intervals for a period of 1 hour. Where 6 inches (152 mm) of water seeps away in less than 10 minutes, a shorter interval between measurements shall be used, but in no case shall the water depth exceed 6 inches (152 mm). Where 6 inches (152 mm) of water seeps away in less than 2 minutes, the test shall be stopped and a rate of less than 3 minutes per inch (7.2 s/mm) shall be reported. The final water level drop shall be used to calculate the percolation rate. Soils not meeting the above requirements shall be tested in accordance with Section 3009.14.7.1.3.

P3009.14.7.1.3 Test procedure, other soils. The hole shall be filled with clear water, and a minimum water depth of 12 inches (305 mm) shall be maintained above the bottom of the hole for a 4-hour period by refilling whenever necessary or by use of an automatic siphon. Water remaining in the hole after 4 hours shall not be removed. Thereafter, the soil shall be allowed to swell not less than 16 hours or more than 30 hours. Immediately after the soil swelling period, the measurements for determining the percolation rate shall be made as follows: Any soil sloughed into the hole shall be removed and the water level shall be adjusted to 6 inches (152 mm) above the gravel or coarse sand. Thereupon, from a fixed reference point, the water level shall be measured at 30-minute intervals for a period of 4 hours, unless two successive water level drops do not vary by more than $^{1}/_{16}$ inch (1.59 mm). At least three water level drops shall be observed and recorded. The hole shall be filled with clear water to a point not more than 6 inches (152 mm) above the gravel or coarse sand whenever it becomes nearly empty. Adjustments of the water level shall not be made during the three measurement periods except to the limits of the last measured water level drop. When the first 6 inches (152 mm) of water seeps away in less than 30 minutes, the time interval between measurements shall be 10 minutes and the test run for 1 hour. The water depth shall not exceed 5 inches (127 mm) at any time during the measurement period. The drop that occurs during the final measurement period shall be used in calculating the percolation rate.

P3009.14.7.1.4 Mechanical test equipment. Mechanical percolation test equipment shall be of an approved type.

P3009.14.7.2 Permeability evaluation. Soil shall be evaluated for estimated percolation based on structure and texture in accordance with accepted soil evaluation practices. Borings shall be made in accordance with Section P3009.14.7.1 for evaluating the soil.

P3009.14.8 Subsurface landscape irrigation site location. The surface grade of all soil absorption systems shall be located at a point lower than the surface grade of any water well or reservoir on the same or adjoining lot. Where this is not possible, the site shall be located so that surface water drainage from the site is not directed toward a well or reservoir. The soil absorption system shall be located with a minimum horizontal distance between various elements as indicated in Table P3009.14.8. Private sewage disposal systems in compacted areas, such as parking lots and driveways, are prohibited. Surface water shall be diverted away from any soil absorption site on the same or neighboring lots.

TABLE P3009.14.8
LOCATION OF GRAY WATER SYSTEM

ELEMENT	MINIMUM HORIZONTAL DISTANCE	
	HOLDING TANK (feet)	IRRIGATION DISPOSAL FIELD (feet)
Buildings	5	2
Property line adjoining private property	5	5
Public water main	10	10
Seepage pits	5	5
Septic tanks	0	5
Streams and lakes	50	50
Water service	5	5
Water wells	50	100

For SI: 1 foot = 304.8 mm.

P3009.14.9 Installation. Absorption systems shall be installed in accordance with Sections P3009.14.9.1 through P3009.14.9.5 to provide landscape irrigation without surfacing of gray water.

P3009.14.9.1 Absorption area. The total absorption area required shall be computed from the estimated daily gray water discharge and the design-loading rate based on the percolation rate for the site. The required absorption area equals the estimated gray water discharge divided by the design-loading rate from Table P3009.14.9.1.

TABLE P3009.14.9.1
DESIGN LOADING RATE

PERCOLATION RATE (minutes per inch)	DESIGN LOADING FACTOR (gallons per square foot per day)
0 to less than 10	1.2
10 to less than 30	0.8
30 to less than 45	0.72
45 to 60	0.4

For SI: 1 minute per inch = min/25.4 mm,
1 gallon per square foot = 40.7 L/m^2.

P3009.14.9.2 Seepage trench excavations. Seepage trench excavations shall be a minimum of 1 foot (304 mm) to a maximum of 5 feet (1524 mm) wide. Trench excavations shall be spaced a minimum of 2 feet (610 mm) apart. The soil absorption area of a seepage trench shall be computed by using the bottom of the trench area (width) multiplied by the length of pipe. Individual seepage trenches shall be a maximum of 100 feet (30 480 mm) in developed length.

P3009.14.9.3 Seepage bed excavations. Seepage bed excavations shall be a minimum of 5 feet (1524 mm) wide and have more than one distribution pipe. The absorption area of a seepage bed shall be computed by using the bottom of the trench area. Distribution piping in a seepage bed shall be uniformly spaced a maximum of 5 feet (1524 mm) and a minimum of 3 feet (914 mm) apart, and a maximum of 3 feet (914 mm) and a minimum of 1 foot (305 mm) from the sidewall or headwall.

P3009.14.9.4 Excavation and construction. The bottom of a trench or bed excavation shall be level. Seepage trenches or beds shall not be excavated where the soil is so wet that such material rolled between the hands forms a soil wire. All smeared or compacted soil surfaces in the sidewalls or bottom of seepage trench or bed excavations shall be scarified to the depth of smearing or compaction and the loose material removed. Where rain falls on an open excavation, the soil shall be left until sufficiently dry so a soil wire will not form when soil from the excavation bottom is rolled between the hands. The bottom area shall then be scarified and loose material removed.

P3009.14.9.5 Aggregate and backfill. A minimum of 6 inches (152 mm) of aggregate ranging in size from $^{1}/_{2}$ inch to $2^{1}/_{2}$ inches (12.7 mm to 64 mm) shall be laid into the trench below the distribution piping elevation. The aggregate shall be evenly distributed a minimum of 2 inches (51 mm) over the top of the distribution pipe. The aggregate shall be covered with approved synthetic materials or 9 inches (229 mm) of uncompacted marsh hay or straw. Building paper shall not be used to cover the aggregate. A minimum of 9 inches (229 mm) of soil backfill shall be provided above the covering.

P3009.14.10 Distribution piping. Distribution piping shall be not less than 3 inches (76 mm) in diameter. Materials shall comply with Table P3009.14.10. The top of the distribution pipe shall be not less than 8 inches (203 mm) below the original surface. The slope of the distribution pipes shall be a minimum of 2 inches (51 mm) and a maximum of 4 inches (102 mm) per 100 feet (30 480 mm).

TABLE P3009.14.10
DISTRIBUTION PIPE

MATERIAL	STANDARD
Polyethylene (PE) plastic pipe	ASTM F 405
Polyvinyl chloride (PVC) plastic pipe	ASTM D 2729
Polyvinyl chloride (PVC) plastic pipe with a 3.5-inch O.D. and solid cellular core or composite wall	ASTM F 1488

P3009.14.11 Joints. Joints in distribution pipe shall be made in accordance with Section P3003.

CHAPTER 31

VENTS

SECTION P3101 VENT SYSTEMS

P3101.1 General. This chapter shall govern the selection and installation of piping, tubing and fittings for vent systems. This chapter shall control the minimum diameter of vent pipes, circuit vents, branch vents and individual vents, and the size and length of vents and various aspects of vent stacks and stack vents. Additionally, this chapter regulates vent grades and connections, height above fixtures and relief vents for stacks and fixture traps, and the venting of sumps and sewers.

P3101.2 Trap seal protection. The plumbing system shall be provided with a system of vent piping that will permit the admission or emission of air so that the seal of any fixture trap shall not be subjected to a pneumatic pressure differential of more than 1 inch of water column (249 Pa).

P3101.2.1 Venting required. Every trap and trapped fixture shall be vented in accordance with one of the venting methods specified in this chapter.

P3101.3 Use limitations. The plumbing vent system shall not be used for purposes other than the venting of the plumbing system.

P3101.4 Extension outside a structure. In climates where the 97.5-percent value for outside design temperature is 0°F (-18°C) or less (ASHRAE 97.5-percent column, winter, see Chapter 3), vent pipes installed on the exterior of the structure shall be protected against freezing by insulation, heat or both. Vent terminals shall be protected from frost closure in accordance with Section P3103.2.

P3101.5 Flood resistance. In flood hazard areas as established by Table R301.2(1), vents shall be located at or above the elevation required in Section R322.1 (flood hazard areas including A Zones) or R322.2 (coastal high-hazard areas including V Zones).

SECTION P3102 VENT STACKS AND STACK VENTS

P3102.1 Required vent extension. The vent system serving each *building drain* shall have at least one vent pipe that extends to the outdoors.

P3102.2 Installation. The required vent shall be a dry vent that connects to the *building drain* or an extension of a drain that connects to the *building drain*. Such vent shall not be an island fixture vent as permitted by Section P3112.

P3102.3 Size. The required vent shall be sized in accordance with Section P3113.1 based on the required size of the *building drain*.

SECTION P3103 VENT TERMINALS

P3103.1 Roof extension. Open vent pipes that extend through a roof shall be terminated not less than 6 inches (152 mm) above the roof or 6 inches (152 mm) above the anticipated snow accumulation, whichever is greater, except that where a roof is to be used for any purpose other than weather protection, the vent extension shall be run not less than 7 feet (2134 mm) above the roof.

P3103.2 Frost closure. Where the 97.5-percent value for outside design temperature is 0°F (-18°C) or less, every vent extension through a roof or wall shall be not less than 3 inches (76 mm) in diameter. Any increase in the size of the vent shall be made inside the structure not less than 1 foot (305 mm) below the roof or inside the wall.

P3103.3 Flashings and sealing. The juncture of each vent pipe with the roof line shall be made water tight by an *approved* flashing. Vent extensions in walls and soffits shall be made weather tight by caulking.

P3103.4 Prohibited use. A vent terminal shall not be used for any purpose other than a vent terminal.

P3103.5 Location of vent terminal. An open vent terminal from a drainage system shall not be located less than 4 feet (1219 mm) directly beneath any door, openable window, or other air intake opening of the building or of an adjacent building, nor shall any such vent terminal be within 10 feet (3048 mm) horizontally of such an opening unless it is not less than 3 feet (914 mm) above the top of such opening

P3103.6 Extension through the wall. Vent terminals extending through the wall shall terminate not less than 10 feet (3048 mm) from the *lot line* and 10 feet (3048 mm) above the highest adjacent *grade* within 10 feet (3048 mm) horizontally of the vent terminal. Vent terminals shall not terminate under the overhang of a structure with soffit vents. Side wall vent terminals shall be protected to prevent birds or rodents from entering or blocking the vent opening.

SECTION P3104 VENT CONNECTIONS AND GRADES

P3104.1 Connection. All individual branch and circuit vents shall connect to a vent stack, stack vent or extend to the open air.

Exception: Individual, branch and circuit vents shall be permitted to terminate at an *air admittance valve* in accordance with Section P3114.

P3104.2 Grade. Vent and branch vent pipes shall be graded, connected and supported to allow moisture and condensate to drain back to the soil or waste pipe by gravity.

P3104.3 Vent connection to drainage system. A dry vent connecting to a horizontal drain shall connect above the centerline of the horizontal drain pipe.

P3104.4 Vertical rise of vent. A dry vent shall rise vertically to not less than 6 inches (152 mm) above the flood level rim of the highest trap or trapped fixture being vented.

P3104.5 Height above fixtures. A connection between a vent pipe and a vent stack or stack vent shall be made not less than 6 inches (152 mm) above the flood level rim of the highest fixture served by the vent. Horizontal vent pipes forming branch vents shall be not less than 6 inches (152 mm) above the flood level rim of the highest fixture served.

P3104.6 Vent for future fixtures. Where the drainage piping has been roughed-in for future fixtures, a rough-in connection for a vent, not less than one-half the diameter of the drain, shall be installed. The vent rough-in shall connect to the vent system or shall be vented by other means as provided in this chapter. The connection shall be identified to indicate that the connection is a vent.

SECTION P3105 FIXTURE VENTS

P3105.1 Distance of trap from vent. Each fixture trap shall have a protecting vent located so that the slope and the *developed length* in the *fixture drain* from the trap weir to the vent fitting are within the requirements set forth in Table P3105.1.

> **Exception:** The *developed length* of the *fixture drain* from the trap weir to the vent fitting for self-siphoning fixtures, such as water closets, shall not be limited.

TABLE P3105.1 MAXIMUM DISTANCE OF FIXTURE TRAP FROM VENT

SIZE OF TRAP (inches)	SLOPE (inch per foot)	DISTANCE FROM TRAP (feet)
$1^1/_4$	$^1/_4$	5
$1^1/_2$	$^1/_4$	6
2	$^1/_4$	8
3	$^1/_8$	12
4	$^1/_8$	16

For SI: 1 inch = 25.4 mm, 1 foot = 304.8 mm, 1 inch per foot = 83.3 mm/m.

P3105.2 Fixture drains. The total fall in a *fixture drain* resulting from pipe slope shall not exceed one pipe diameter, nor shall the vent pipe connection to a *fixture drain*, except for water closets, be below the weir of the trap.

P3105.3 Crown vent prohibited. A vent shall not be installed within two pipe diameters of the trap weir.

SECTION P3106 INDIVIDUAL VENT

P3106.1 Individual vent permitted. Each trap and trapped fixture shall be permitted to be provided with an individual vent. The individual vent shall connect to the *fixture drain* of the trap or trapped fixture being vented.

SECTION P3107 COMMON VENT

P3107.1 Individual vent as common vent. An individual vent shall be permitted to vent two traps or trapped fixtures as a common vent. The traps or trapped fixtures being common vented shall be located on the same floor level.

P3107.2 Connection at the same level. Where the *fixture drains* being common vented connect at the same level, the vent connection shall be at the interconnection of the *fixture drains* or downstream of the interconnection.

P3107.3 Connection at different levels. Where the *fixture drains* connect at different levels, the vent shall connect as a vertical extension of the vertical drain. The vertical drain pipe connecting the two *fixture drains* shall be considered the vent for the lower *fixture drain*, and shall be sized in accordance with Table P3107.3. The upper fixture shall not be a water closet.

TABLE P3107.3 COMMON VENT SIZES

PIPE SIZE (inches)	MAXIMUM DISCHARGE FROM UPPER FIXTURE DRAIN (d.f.u.)
$1^1/_2$	1
2	4
$2^1/_2$ to 3	6

For SI: 1 inch = 25.4 mm.

SECTION P3108 WET VENTING

P3108.1 Horizontal wet vent permitted. Any combination of fixtures within two *bathroom groups* located on the same floor level shall be permitted to be vented by a horizontal wet vent. The wet vent shall be considered the vent for the fixtures and shall extend from the connection of the dry vent along the direction of the flow in the drain pipe to the most downstream *fixture drain* connection. Each *fixture drain* shall connect horizontally to the horizontal branch being wet vented or shall have a dry vent. Each wet-vented *fixture drain* shall connect independently to the horizontal wet vent. Only the fixtures within the *bathroom groups* shall connect to the wet-vented horizontal branch drain. Any additional fixtures shall discharge downstream of the horizontal wet vent.

P3108.2 Dry vent connection. The required dry-vent connection for wet-vented systems shall comply with Sections P3108.2.1 and P3108.2.2.

P3108.2.1 Horizontal wet vent. The dry-vent connection for a horizontal wet-vent system shall be an individual vent or a common vent for any *bathroom group* fixture, except an emergency floor drain. Where the dry vent connects to a water closet *fixture drain*, the drain shall connect horizontally to the horizontal wet vent system. Not more than one wet-vented *fixture drain* shall discharge upstream of the dry-vented *fixture drain* connection.

P3108.2.2 Vertical wet vent. The dry-vent connection for a vertical wet-vent system shall be an individual vent or common vent for the most upstream *fixture drain*.

P3108.3 Size. Horizontal and vertical wet vents shall be not less than the size as specified in Table P3108.3, based on the fixture unit discharge to the wet vent. The dry vent serving the wet vent shall be sized based on the largest required diameter of pipe within the wet-vent system served by the dry vent.

TABLE P3108.3
WET VENT SIZE

WET VENT PIPE SIZE (inches)	FIXTURE UNIT LOAD (d.f.u.)
$1^1/_2$	1
2	4
$2^1/_2$	6
3	12
4	32

For SI: 1 inch = 25.4 mm.

P3108.4 Vertical wet vent permitted. A combination of fixtures located on the same floor level shall be permitted to be vented by a vertical wet vent. The vertical wet vent shall be considered the vent for the fixtures and shall extend from the connection of the dry vent down to the lowest *fixture drain* connection. Each wet-vented fixture shall connect independently to the vertical wet vent. All water closet drains shall connect at the same elevation. Other *fixture drains* shall connect above or at the same elevation as the water closet *fixture drains*. The dry vent connection to the vertical wet vent shall be an individual or common vent serving one or two fixtures.

P3108.5 Trap weir to wet vent distances. The maximum *developed length* of wet-vented *fixture drains* shall comply with Table P3105.1.

SECTION P3109
WASTE STACK VENT

P3109.1 Waste stack vent permitted. A waste stack shall be considered a vent for all of the fixtures discharging to the stack where installed in accordance with the requirements of this section.

P3109.2 Stack installation. The waste stack shall be vertical, and both horizontal and vertical offsets shall be prohibited between the lowest *fixture drain* connection and the highest *fixture drain* connection to the stack. Every *fixture drain* shall connect separately to the waste stack. The stack shall not receive the discharge of water closets or urinals.

P3109.3 Stack vent. A stack vent shall be installed for the waste stack. The size of the stack vent shall be not less than the size of the waste stack. Offsets shall be permitted in the stack vent and shall be located not less than 6 inches (152 mm) above the flood level of the highest fixture, and shall be in accordance with Section P3104.5. The stack vent shall be permitted to connect with other stack vents and vent stacks in accordance with Section P3113.3.

P3109.4 Waste stack size. The waste stack shall be sized based on the total discharge to the stack and the discharge within a *branch interval* in accordance with Table P3109.4. The waste stack shall be the same size throughout the length of the waste stack.

TABLE P3109.4
WASTE STACK VENT SIZE

STACK SIZE (inches)	MAXIMUM NUMBER OF FIXTURE UNITS (d.f.u.)	
	Total discharge into one branch interval	Total discharge for stack
$1^1/_2$	1	2
2	2	4
$2^1/_2$	No limit	8
3	No limit	24
4	No limit	50

For SI: 1 inch = 25.4 mm.

SECTION P3110
CIRCUIT VENTING

P3110.1 Circuit vent permitted. A maximum of eight fixtures connected to a horizontal branch drain shall be permitted to be circuit vented. Each *fixture drain* shall connect horizontally to the horizontal branch being circuit vented. The horizontal branch drain shall be classified as a vent from the most downstream *fixture drain* connection to the most upstream *fixture drain* connection to the horizontal branch.

P3110.2 Vent connection. The circuit vent connection shall be located between the two most upstream *fixture drains*. The vent shall connect to the horizontal branch and shall be installed in accordance with Section P3104. The circuit vent pipe shall not receive the discharge of any soil or waste.

P3110.3 Slope and size of horizontal branch. The slope of the vent section of the horizontal branch drain shall be not greater than one unit vertical in 12 units horizontal (8-percent slope). The entire length of the vent section of the horizontal branch drain shall be sized for the total drainage discharge to the branch in accordance with Table P3005.4.1.

P3110.4 Additional fixtures. Fixtures, other than the circuit vented fixtures shall be permitted to discharge, to the horizontal branch drain. Such fixtures shall be located on the same floor as the circuit vented fixtures and shall be either individually or common vented.

SECTION P3111
COMBINATION WASTE AND VENT SYSTEM

P3111.1 Type of fixtures. A combination waste and vent system shall not serve fixtures other than floor drains, sinks and lavatories. A combination waste and vent system shall not receive the discharge of a food waste grinder.

P3111.2 Installation. The only vertical pipe of a combination waste and vent system shall be the connection between the fixture drain and the horizontal combination waste and vent pipe. The vertical distance shall be not greater than 8 feet (2438 mm).

P3111.2.1 Slope. The horizontal combination waste and vent pipe shall have a slope of not greater than $^1/_2$ unit vertical in 12 units horizontal (4-percent slope). The minimum slope shall be in accordance with Section P3005.3.

P3111.2.2 Connection. The combination waste and vent pipe shall connect to a horizontal drain that is vented or a vent shall connect to the combination waste and vent. The vent connecting to the combination waste and vent pipe shall extend vertically not less than 6 inches (152 mm) above the flood level rim of the highest fixture being vented before offsetting horizontally.

P3111.2.3 Vent size. The vent shall be sized for the total fixture unit load in accordance with Section P3113.1.

P3111.2.4 Fixture branch or drain. The fixture branch or *fixture drain* shall connect to the combination waste and vent within a distance specified in Table P3105.1. The combination waste and vent pipe shall be considered the vent for the fixture.

P3111.3 Size. The size of a combination drain and vent pipe shall be not less than that specified in Table 3111.3. The horizontal length of a combination drain and vent system shall be unlimited.

TABLE P3111.3
SIZE OF COMBINATION WASTE AND VENT PIPE

DIAMETER PIPE (inches)	MAXIMUM NUMBER OF FIXTURE UNITS (d.f.u.)	
	Connecting to a horizontal branch or stack	Connecting to a building drain or building subdrain
2	3	4
$2^1/_2$	6	26
3	12	31
4	20	50

For SI: 1 inch = 25.4 mm.

SECTION P3112
ISLAND FIXTURE VENTING

P3112.1 Limitation. Island fixture venting shall not be permitted for fixtures other than sinks and lavatories. Kitchen sinks with a dishwasher waste connection, a food waste grinder, or both, in combination with the kitchen sink waste, shall be permitted to be vented in accordance with this section.

P3112.2 Vent connection. The island fixture vent shall connect to the *fixture drain* as required for an individual or common vent. The vent shall rise vertically to above the drainage outlet of the fixture being vented before offsetting horizontally or vertically downward. The vent or branch vent for multiple island fixture vents shall extend not less than 6 inches (152 mm) above the highest island fixture being vented before connecting to the outside vent terminal.

P3112.3 Vent installation below the fixture flood level rim. The vent located below the flood level rim of the fixture being vented shall be installed as required for drainage piping in accordance with Chapter 30, except for sizing. The vent shall be sized in accordance with Section P3113.1. The lowest point of the island fixture vent shall connect full size to the drainage system. The connection shall be to a vertical drain pipe or to the top half of a horizontal drain pipe. Cleanouts shall be provided in the island fixture vent to permit rodding of all vent piping located below the flood level rim of the fixtures. Rodding in both directions shall be permitted through a cleanout.

SECTION P3113
VENT PIPE SIZING

P3113.1 Size of vents. The required diameter of individual vents, branch vents, circuit vents, vent stacks and stack vents shall be not less than one-half the required diameter of the drain served. The required size of the drain shall be determined in accordance with Chapter 30. Vent pipes shall be not less than $1^1/_4$ inches (32 mm) in diameter. Vents exceeding 40 feet (12 192 mm) in *developed length* shall be increased by one nominal pipe size for the entire *developed length* of the vent pipe.

P3113.2 Developed length. The *developed length* of individual, branch, and circuit vents shall be measured from the farthest point of vent connection to the drainage system, to the point of connection to the vent stack, stack vent or termination outside of the building.

P3113.3 Branch vents. Where branch vents are connected to a common branch vent, the common branch vent shall be sized in accordance with this section, based on the size of the common horizontal drainage branch that is or would be required to serve the total drainage fixture unit (d.f.u.) load being vented.

P3113.4 Sump vents. Sump vent sizes shall be determined in accordance with Sections P3113.4.1 and P3113.4.2.

P3113.4.1 Sewage pumps and sewage ejectors other than pneumatic. Drainage piping below sewer level shall be vented in the same manner as that of a gravity system. Building sump vent sizes for sumps with sewage pumps or sewage ejectors, other than pneumatic, shall be determined in accordance with Table P3113.4.1.

P3113.4.2 Pneumatic sewage ejectors. The air pressure relief pipe from a pneumatic sewage ejector shall be connected to an independent vent stack terminating as required for vent extensions through the roof. The relief pipe shall be sized to relieve air pressure inside the ejector to atmospheric pressure, but shall not be less than $1^1/_4$ inches (32 mm) in size.

SECTION P3114
AIR ADMITTANCE VALVES

P3114.1 General. Vent systems using *air admittance valves* shall comply with this section. Individual and branch-type air admittance valves shall conform to ASSE 1051. Stack-type air admittance valves shall conform to ASSE 1050.

P3114.2 Installation. The valves shall be installed in accordance with the requirements of this section and the manufacturer's instructions. *Air admittance valves* shall be installed after the DWV testing required by Section P2503.5.1 or P2503.5.2 has been performed.

P3114.3 Where permitted. Individual vents, branch vents, circuit vents and stack vents shall be permitted to terminate with a connection to an *air admittance valve*. Individual and branch type air admittance valves shall vent only fixtures that are on the same floor level and connect to a horizontal branch drain.

P3114.4 Location. Individual and branch *air admittance valves* shall be located not less than 4 inches (102 mm) above the horizontal branch drain or *fixture drain* being vented. Stack-type air admittance valves shall be located not less than 6 inches (152 mm) above the flood level rim of the highest fixture being vented. The *air admittance valve* shall be located within the maximum *developed length* permitted for the vent. The *air admittance valve* shall be installed not less than 6 inches (152 mm) above insulation materials where installed in *attics*.

P3114.5 Access and ventilation. Access shall be provided to all *air admittance valves*. The valve shall be located within a ventilated space that allows air to enter the valve.

P3114.6 Size. The *air admittance valve* shall be rated for the size of the vent to which the valve is connected.

P3114.7 Vent required. Within each plumbing system, not less than one stack vent or a vent stack shall extend outdoors to the open air.

P3114.8 Prohibited installations. *Air admittance valves* without an engineered design shall not be used to vent sumps or tanks of any type.

TABLE P3113.4.1
SIZE AND LENGTH OF SUMP VENTS

DISCHARGE CAPACITY OF PUMP (gpm)	MAXIMUM DEVELOPED LENGTH OF VENT (feet)[a]				
	Diameter of vent (inches)				
	$1^1/_4$	$1^1/_2$	2	$2^1/_2$	3
10	No limit[b]	No limit	No limit	No limit	No limit
20	270	No limit	No limit	No limit	No limit
40	72	160	No limit	No limit	No limit
60	31	75	270	No limit	No limit

For SI: 1 inch = 25.4 mm, 1 foot = 304.8 mm, 1 gallon per minute (gpm) = 3.785 L/m.

a. Developed length plus an appropriate allowance for entrance losses and friction caused by fittings, changes in direction and diameter. Suggested allowances shall be obtained from NBS Monograph 31 or other approved sources. An allowance of 50 percent of the developed length shall be assumed if a more precise value is not available.

b. Actual values greater than 500 feet.

CHAPTER 32

TRAPS

SECTION P3201 FIXTURE TRAPS

P3201.1 Design of traps. Traps shall be of standard design, shall have smooth uniform internal waterways, shall be self-cleaning and shall not have interior partitions except where integral with the fixture. Traps shall be constructed of lead, cast iron, cast or drawn brass or *approved* plastic. Tubular brass traps shall be not less than No. 20 gage (0.8 mm) thickness. Solid connections, slip joints and couplings shall be permitted to be used on the trap inlet, trap outlet, or within the trap seal. Slip joints shall be accessible.

P3201.2 Trap seals and trap seal protection. Traps shall have a liquid seal not less than 2 inches (51 mm) and not more than 4 inches (102 mm). Traps for floor drains shall be fitted with a trap primer or shall be of the deep seal design. Trap seal primer valves shall connect to the trap at a point above the level of the trap seal.

P3201.3 Trap setting and protection. Traps shall be set level with respect to their water seals and shall be protected from freezing. Trap seals shall be protected from siphonage, aspiration or back pressure by an *approved* system of venting (see Section P3101).

P3201.4 Building traps. Building traps shall not be installed, except in special cases where sewer gases are extremely corrosive or noxious, as directed by the *building official*.

P3201.5 Prohibited trap designs. The following types of traps are prohibited:

1. Bell traps.
2. Separate fixture traps with interior partitions, except those lavatory traps made of plastic, stainless steel or other corrosion-resistant material.
3. "S" traps.
4. Drum traps.
5. Trap designs with moving parts.

P3201.6 Number of fixtures per trap. Each plumbing fixture shall be separately trapped by a water seal trap. The vertical distance from the fixture outlet to the trap weir shall not exceed 24 inches (610 mm) and the horizontal distance shall not exceed 30 inches (762 mm) measured from the center line of the fixture outlet to the centerline of the inlet of the trap. The height of a clothes washer standpipe above a trap shall conform to Section P2706.2. Fixtures shall not be double trapped.

Exceptions:

1. Fixtures that have integral traps.
2. A single trap shall be permitted to serve two or three like fixtures limited to kitchen sinks, laundry tubs and lavatories. Such fixtures shall be adjacent to each other and located in the same room with a continuous waste arrangement. The trap shall be installed at the center fixture where three fixtures are installed. Common trapped fixture outlets shall be not more than 30 inches (762 mm) apart.
3. Connection of a laundry tray waste line into a standpipe for the automatic clothes-washer drain shall be permitted in accordance with Section P2706.2.1.

P3201.7 Size of fixture traps. Fixture trap size shall be sufficient to drain the fixture rapidly and not less than the size indicated in Table P3201.7. A trap shall not be larger than the drainage pipe into which the trap discharges.

TABLE P3201.7
SIZE OF TRAPS AND TRAP ARMS FOR PLUMBING FIXTURES

PLUMBING FIXTURE	TRAP SIZE MINIMUM (inches)
Bathtub (with or without shower head and/or whirlpool attachments)	$1^1/_2$
Bidet	$1^1/_4$
Clothes washer standpipe	2
Dishwasher (on separate trap)	$1^1/_2$
Floor drain	2
Kitchen sink (one or two traps, with or without dishwasher and garbage grinder)	$1^1/_2$
Laundry tub (one or more compartments)	$1^1/_2$
Lavatory	$1^1/_4$
Shower (based on the total flow rate through showerheads and bodysprays) Flow rate: 5.7 gpm and less More than 5.7 gpm up to 12.3 gpm More than 12.3 gpm up to 25.8 gpm More than 25.8 gpm up to 55.6 gpm	 $1^1/_2$ 2 3 4
Water closet	Note a

For SI: 1 inch = 25.4 mm.

a. Consult fixture standards for trap dimensions of specific bowls.

CHAPTER 33

STORM DRAINAGE

SECTION P3301 GENERAL

P3301.1 Scope. The provisions of this chapter shall govern the materials, design, construction and installation of storm drainage.

SECTION P3302 SUBSOIL DRAINS

P3302.1 Subsoil drains. Subsoil drains shall be open-jointed, horizontally split or perforated pipe conforming to one of the standards listed in Table P3302.1. Such drains shall be not less than 4 inches (102 mm) in diameter. Where the building is subject to backwater, the subsoil drain shall be protected by an accessibly located backwater valve. Subsoil drains shall discharge to a trapped area drain, sump, dry well or *approved* location above ground. The subsoil sump shall not be required to have either a gas-tight cover or a vent. The sump and pumping system shall comply with Section P3303.

SECTION P3303 SUMPS AND PUMPING SYSTEMS

P3303.1 Pumping system. The sump pump, pit and discharge piping shall conform to Sections P3303.1.1 through P3303.1.4.

P3303.1.1 Pump capacity and head. The sump pump shall be of a capacity and head appropriate to anticipated use requirements.

P3303.1.2 Sump pit. The sump pit shall be not less than 18 inches (457 mm) in diameter and 24 inches (610 mm) deep, unless otherwise *approved*. The pit shall be accessible and located so that all drainage flows into the pit by gravity. The sump pit shall be constructed of tile, steel, plastic, cast-iron, concrete or other *approved* material, with a removable cover adequate to support anticipated loads in the area of use. The pit floor shall be solid and provide permanent support for the pump.

P3303.1.3 Electrical. Electrical outlets shall meet the requirements of Chapters 34 through 43.

P3303.1.4 Piping. Discharge piping shall meet the requirements of Sections P3002.1, P3002.2, P3002.3 and P3003. Discharge piping shall include an accessible full flow check valve. Pipe and fittings shall be the same size as, or larger than, pump discharge tapping.

**TABLE P3302.1
SUBSOIL DRAIN PIPE**

MATERIAL	STANDARD
Asbestos-cement pipe	ASTM C 508
Cast-iron pipe	ASTM A 74; ASTM A 888; CISPI 301
Polyethylene (PE) plastic pipe	ASTM F 405; CSA B182.1; CSA B182.6; CSA B182.8
Polyvinyl chloride (PVC) Plastic pipe (type sewer pipe, PS25, PS50 or PS100)	ASTM D 2729; ASTM F 891; CSA B182.2; CSA B182.4
Stainless steel drainage systems, Type 316L	ASME A112.3.1
Vitrified clay pipe	ASTM C 4; ASTM C 700

Part VIII—Electrical

CHAPTER 34
GENERAL REQUIREMENTS

SECTION E3401
GENERAL

E3401.1 Applicability. The provisions of Chapters 34 through 43 shall establish the general scope of the electrical system and equipment requirements of this code. Chapters 34 through 43 cover those wiring methods and materials most commonly encountered in the construction of one- and two-family dwellings and structures regulated by this code. Other wiring methods, materials and subject matter covered in NFPA 70 are also allowed by this code.

E3401.2 Scope. Chapters 34 through 43 shall cover the installation of electrical systems, equipment and components indoors and outdoors that are within the scope of this code, including services, power distribution systems, fixtures, appliances, devices and appurtenances. Services within the scope of this code shall be limited to 120/240-volt, 0- to 400-ampere, single-phase systems. These chapters specifically cover the equipment, fixtures, appliances, wiring methods and materials that are most commonly used in the construction or alteration of one- and two-family dwellings and accessory structures regulated by this code. The omission from these chapters of any material or method of construction provided for in the referenced standard NFPA 70 shall not be construed as prohibiting the use of such material or method of construction. Electrical systems, equipment or components not specifically covered in these chapters shall comply with the applicable provisions of NFPA 70.

E3401.3 Not covered. Chapters 34 through 43 do not cover the following:

1. Installations, including associated lighting, under the exclusive control of communications utilities and electric utilities.
2. Services over 400 amperes.

E3401.4 Additions and alterations. Any addition or alteration to an existing electrical system shall be made in conformity to the provisions of Chapters 34 through 43. Where additions subject portions of existing systems to loads exceeding those permitted herein, such portions shall be made to comply with Chapters 34 through 43.

SECTION E3402 BUILDING STRUCTURE PROTECTION

E3402.1 Drilling and notching. Wood-framed structural members shall not be drilled, notched or altered in any manner except as provided for in this code.

E3402.2 Penetrations of fire-resistance-rated assemblies. Electrical installations in hollow spaces, vertical shafts and ventilation or air-handling ducts shall be made so that the possible spread of fire or products of combustion will not be substantially increased. Electrical penetrations into or through fire-resistance-rated walls, partitions, floors or ceilings shall be protected by approved methods to maintain the fire-resistance rating of the element penetrated. Penetrations of fire-resistance-rated walls shall be limited as specified in Section R317.3.

E3402.3 Penetrations of firestops and draftstops. Penetrations through fire blocking and draftstopping shall be protected in an approved manner to maintain the integrity of the element penetrated.

SECTION E3403 INSPECTION AND APPROVAL

E3403.1 Approval. Electrical materials, components and equipment shall be approved.

E3403.2 Inspection required. New electrical work and parts of existing systems affected by new work or alterations shall be inspected by the building official to ensure compliance with the requirements of Chapters 34 through 43.

E3403.3 Listing and labeling. Electrical materials, components, devices, fixtures and equipment shall be listed for the application, shall bear the label of an approved agency and shall be installed, and used, or both, in accordance with the manufacturer's installation instructions.

SECTION E3404 GENERAL EQUIPMENT REQUIREMENTS

E3404.1 Voltages. Throughout Chapters 34 through 43, the voltage considered shall be that at which the circuit operates.

E3404.2 Interrupting rating. Equipment intended to interrupt current at fault levels shall have a minimum interrupting rating of 10,000 amperes. Equipment intended to interrupt current at levels other than fault levels shall have an interrupting rating at nominal circuit voltage of not less than the current that must be interrupted.

E3404.3 Circuit characteristics. The overcurrent protective devices, total impedance, equipment short-circuit current ratings and other characteristics of the circuit to be protected shall be so selected and coordinated as to permit the circuit protective devices that are used to clear a fault to do so without extensive damage to the electrical equipment of the circuit. This fault shall be assumed to be either between two or more of the circuit conductors or between any circuit conductor and the equipment grounding conductors permitted in Section E3908.8. Listed equipment applied in accordance with its listing shall be considered to meet the requirements of this section.

E3404.4 Enclosure types. Enclosures, other than surrounding fences or walls, of panelboards, meter sockets, enclosed switches, transfer switches, circuit breakers, pullout switches and motor controllers, rated not over 600 volts nominal and intended for such locations, shall be marked with an enclosure-type number as shown in Table E3404.4.

Table E3404.4 shall be used for selecting these enclosures for use in specific locations other than hazardous (classified) locations. The enclosures are not intended to protect against conditions such as condensation, icing, corrosion, or contamination that might occur within the enclosure or enter through the conduit or unsealed openings.

E3404.5 Protection of equipment. Equipment not identified for outdoor use and equipment identified only for indoor use, such as "dry locations," "indoor use only" "damp locations," or enclosure Type 1, 2, 5, 12, 12K and/or 13, shall be protected against damage from the weather during construction.

E3404.6 Unused openings. Unused openings, other than those intended for the operation of equipment, those intended for mounting purposes, and those permitted as part of the design for listed equipment, shall be closed to afford protection substantially equivalent to the wall of the equipment. Where metallic plugs or plates are used with nonmetallic enclosures they shall be recessed at least $^1/_4$ inch (6.4 mm) from the outer surface of the enclosure.

E3404.7 Integrity of electrical equipment. Internal parts of electrical equipment, including busbars, wiring terminals, insulators and other surfaces, shall not be damaged or contaminated by foreign materials such as paint, plaster, cleaners or abrasives, and corrosive residues. There shall not be any damaged parts that might adversely affect safe operation or mechanical strength of the equipment such as parts that are broken; bent; cut; deteriorated by corrosion, chemical action, or overheating. Foreign debris shall be removed from equipment.

E3404.8 Mounting. Electrical equipment shall be firmly secured to the surface on which it is mounted. Wooden plugs driven into masonry, concrete, plaster, or similar materials shall not be used.

E3404.9 Energized parts guarded against accidental contact. Approved enclosures shall guard energized parts that are operating at 50 volts or more against accidental contact.

E3404.10 Prevent physical damage. In locations where electrical equipment is likely to be exposed to physical damage, enclosures or guards shall be so arranged and of such strength as to prevent such damage.

E3404.11 Equipment identification. The manufacturer's name, trademark or other descriptive marking by which the organization responsible for the product can be identified shall be placed on all electric equipment. Other markings shall be provided that indicate voltage, current, wattage or other ratings as specified elsewhere in Chapters 34 through 43. The marking shall have the durability to withstand the environment involved.

E3404.12 Identification of disconnecting means. Each disconnecting means shall be legibly marked to indicate its purpose, except where located and arranged so that the purpose is evident. The marking shall have the durability to withstand the environment involved.

SECTION E3405 EQUIPMENT LOCATION AND CLEARANCES

E3405.1 Working space and clearances. Sufficient access and working space shall be provided and maintained around all electrical equipment to permit ready and safe operation and maintenance of such equipment in accordance with this section and Figure E3405.1.

E3405.2 Working clearances for energized equipment and panelboards. Except as otherwise specified in Chapters 34 through 43, the dimension of the working space in the direction of access to panelboards and live parts likely to require examination, adjustment, servicing or maintenance while energized shall be not less than 36 inches (914 mm) in depth. Distances shall be measured from the energized parts where such parts are exposed or from the enclosure front or opening where such parts are enclosed. In addition to the 36-inch dimension (914 mm), the work space shall not be less than 30 inches (762 mm) wide in front of the electrical equipment and not less than the width of such equipment. The work space shall be clear and shall extend from the floor or platform to a height of 6.5 feet (1981 mm) or the height of the equipment, whichever is greater. In all cases, the work space shall allow at least a 90-degree (1.57 rad) opening of equipment doors or hinged panels. Equipment associated with the electrical installation located above or below the electrical equipment shall be permitted to extend not more than 6 inches (152 mm) beyond the front of the electrical equipment.

Exceptions:

1. In existing dwelling units, service equipment and panelboards that are not rated in excess of 200 amperes shall be permitted in spaces where the height of the working space is less than 6.5 feet (1981 mm).
2. Meters that are installed in meter sockets shall be permitted to extend beyond the other equipment. Meter sockets shall not be exempt from the requirements of this section.

TABLE E3404.4 ENCLOSURE SELECTION

PROVIDES A DEGREE OF PROTECTION AGAINST THE FOLLOWING ENVIRONMENTAL CONDITIONS	FOR OUTDOOR USE									
	Enclosure-type Number									
	3	3R	3S	3X	3RX	3SX	4	4X	6	6P
Incidental contact with the enclosed equipment	X	X	X	X	X	X	X	X	X	X
Rain, snow and sleet	X	X	X	X	X	X	X	X	X	X
Sleet[a]	—	—	X	—	—	X	—	—	—	—
Windblown dust	X	—	X	X	—	X	X	X	X	X
Hosedown	—	—	—	—	—	—	X	X	X	X
Corrosive agents	—	—	—	X	X	X	—	X	—	X
Temporary submersion	—	—	—	—	—	—	—	—	X	X
Prolonged submersion	—	—	—	—	—	—	—	—	—	X
PROVIDES A DEGREE OF PROTECTION AGAINST THE FOLLOWING ENVIRONMENTAL CONDITIONS	FOR INDOOR USE									
	Enclosure-type Number									
	1	2	4	4X	5	6	6P	12	12K	13
Incidental contact with the enclosed equipment	X	X	X	X	X	X	X	X	X	X
Falling dirt	X	X	X	X	X	X	X	X	X	X
Falling liquids and light splashing	—	X	X	X	X	X	X	X	X	X
Circulating dust, lint, fibers and flyings	—	—	X	X	—	X	X	X	X	X
Settling airborne dust, lint, fibers and flings	—	—	X	X	X	X	X	X	X	X
Hosedown and splashing water	—	—	X	X	—	X	X	—	—	—
Oil and coolant seepage	—	—	—	—	—	—	—	X	X	X
Oil or coolant spraying and splashing	—	—	—	—	—	—	—	—	—	X
Corrosive agents	—	—	—	X	—	—	X	—	—	—
Temporary submersion	—	—	—	—	—	X	X	—	—	—
Prolonged submersion	—	—	—	—	—	—	X	—	—	—

a. Mechanism shall be operable when ice covered.

Note 1: The term raintight is typically used in conjunction with Enclosure Types 3, 3S, 3SX, 3X, 4, 4X, 6 and 6P. The term rainproof is typically used in conjunction with Enclosure Types 3R and 3RX. The term watertight is typically used in conjunction with Enclosure Types 4, 4X, 6 and 6P. The term driptight is typically used in conjunction with Enclosure Types 2, 5, 12, 12K and 13. The term dusttight is typically used in conjunction with Enclosure Types 3, 3S, 3SX, 3X, 5, 12, 12K and 13.

Note 2: Ingress protection (IP) ratings are found in ANSI/NEMA 60529, *Degrees of Protection Provided by Enclosures*. IP ratings are not a substitute for enclosure-type ratings.

E3405.3 Dedicated panelboard space. The space equal to the width and depth of the panelboard and extending from the floor to a height of 6 feet (1829 mm) above the panelboard, or to the structural ceiling, whichever is lower, shall be dedicated to the electrical installation. Piping, ducts, leak protection apparatus and other equipment foreign to the electrical installation shall not be installed in such dedicated space. The area above the dedicated space shall be permitted to contain foreign systems, provided that protection is installed to avoid damage to the electrical equipment from condensation, leaks and breaks in such foreign systems (see Figure E3405.1).

Exception: Suspended ceilings with removable panels shall be permitted within the 6-foot (1829 mm) dedicated space.

E3405.4 Location of working spaces and equipment. Required working space shall not be designated for storage. Panelboards and overcurrent protection devices shall not be located in clothes closets, in bathrooms, or over the steps of a stairway.

E3405.5 Access and entrance to working space. Access shall be provided to the required working space.

E3405.6 Illumination. Artificial illumination shall be provided for all working spaces for service equipment and panelboards installed indoors and shall not be controlled by automatic means only. Additional lighting outlets shall not be required where the work space is illuminated by an adjacent light source or as permitted by Exception 1 of Section E3903.2 for switched receptacles.

SECTION E3406 ELECTRICAL CONDUCTORS AND CONNECTIONS

E3406.1 General. This section provides general requirements for conductors, connections and splices. These requirements do not apply to conductors that form an integral part of equipment, such as motors, appliances and similar equipment, or to conductors specifically provided for elsewhere in Chapters 34 through 43.

E3406.2 Conductor material. Conductors used to conduct current shall be of copper except as otherwise provided in Chapters 34 through 43. Where the conductor material is not specified, the material and the sizes given in these chapters shall apply to copper conductors. Where other materials are used, the conductor sizes shall be changed accordingly.

E3406.3 Minimum size of conductors. The minimum size of conductors for feeders and branch circuits shall be 14 AWG copper and 12 AWG aluminum. The minimum size of service conductors shall be as specified in Chapter 36. The minimum size of Class 2 remote control, signaling and power-limited circuits conductors shall be as specified in Chapter 43.

E3406.4 Stranded conductors. Where installed in raceways, conductors 8 AWG and larger shall be stranded. A solid 8 AWG conductor shall be permitted to be installed in a raceway only to meet the requirements of Sections E3610.2 and E4204.

E3406.5 Individual conductor insulation. Except where otherwise permitted in Sections E3605.1 and E3908.9, and E4303, current-carrying conductors shall be insulated. Insulated conductors shall have insulation types identified as RHH, RHW, RHW-2, THHN, THHW, THW, THW-2, THWN, THWN-2, TW, UF, USE, USE-2, XHHW or XHHW-2. Insulation types shall be approved for the application.

E3406.6 Conductors in parallel. Circuit conductors that are connected in parallel shall be limited to sizes 1/0 AWG and larger. Conductors in parallel shall: be of the same length; consist of the same conductor material; be the same circular mil area and have the same insulation type. Conductors in parallel shall be terminated in the same manner. Where run in separate raceways or cables, the raceway or cables shall have the same physical characteristics. Where conductors are in separate raceways or cables, the same number of conductors shall be used in each raceway or cable.

E3406.7 Conductors of the same circuit. All conductors of the same circuit and, where used, the grounded conductor and all equipment grounding conductors and bonding conductors shall be contained within the same raceway, cable or cord.

E3406.8 Aluminum and copper connections. Terminals and splicing connectors shall be identified for the material of the conductors joined. Conductors of dissimilar metals shall not be joined in a terminal or splicing connector where physical contact occurs between dissimilar conductors such as copper and aluminum, copper and copper-clad aluminum, or aluminum and copper-clad aluminum, except where the device is listed for the purpose and conditions of application. Materials such as inhibitors and compounds shall be suitable for the application and shall be of a type that will not adversely affect the conductors, installation or equipment.

E3406.9 Fine stranded conductors. Connectors and terminals for conductors that are more finely stranded than Class B and Class C stranding as shown in Table E3406.9, shall be identified for the specific conductor class or classes.

E3406.10 Terminals. Connection of conductors to terminal parts shall be made without damaging the conductors and shall be made by means of pressure connectors, including setscrew type, by means of splices to flexible leads, or for conductor sizes of 10 AWG and smaller, by means of wire binding screws or studs and nuts having upturned lugs or the equivalent. Terminals for more than one conductor and terminals for connecting aluminum conductors shall be identified for the application.

E3406.11 Splices. Conductors shall be spliced or joined with splicing devices listed for the purpose. Splices and joints and the free ends of conductors shall be covered with an insulation equivalent to that of the conductors or with an insulating device listed for the purpose. Wire connectors or splicing means installed on conductors for direct burial shall be listed for such use.

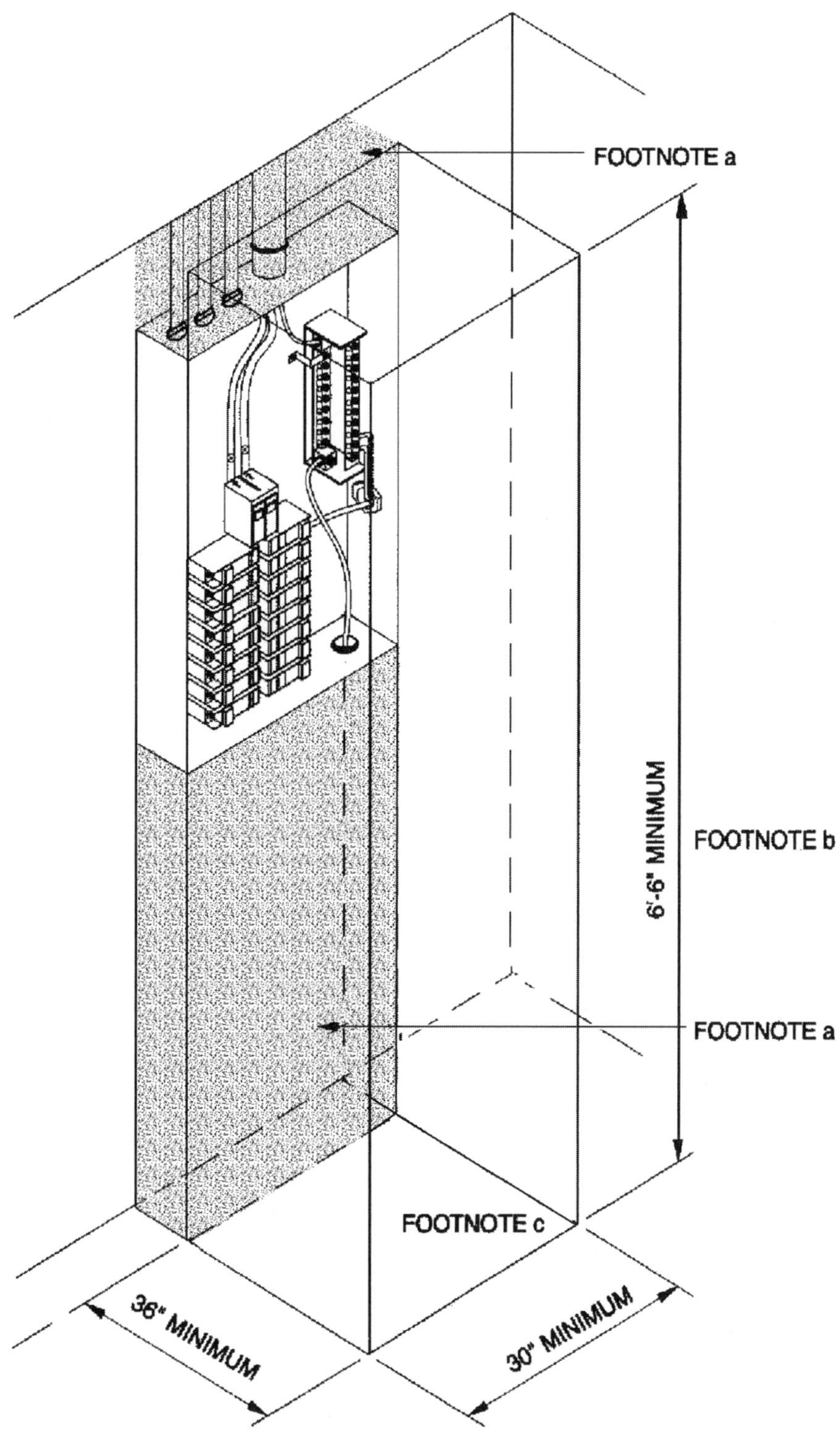

For SI: 1 inch = 25.4 mm, 1 foot = 304.8 mm.

a. Equipment, piping and ducts foreign to the electrical installation shall not be placed in the shaded areas extending from the floor to a height of 6 feet above the panelboard enclosure, or to the structural ceiling, whichever is lower.

b. The working space shall be clear and unobstructed from the floor to a height of 6.5 feet or the height of the equipment, whichever is greater.

c. The working space shall not be designated for storage.

d. Panelboards, service equipment and similar enclosures shall not be located in bathrooms, toilet rooms, clothes closets or over the steps of a stairway.

e. Such work spaces shall be provided with artificial lighting where located indoors and shall not be controlled by automatic means only.

FIGURE E3405.1[a, b, c, d, e]
WORKING SPACE AND CLEARANCES

TABLE E3406.9
CONDUCTOR STRANDING[c]

CONDUCTOR SIZE		NUMBER OF STRANDS		
		Copper		Aluminum
AWG or kcmil	mm^2	Class B	Class C	Class B
24-30	0.20-0.05	a	—	—
22	0.32	7	—	—
20	0.52	10	—	—
18	0.82	16	—	—
16	1.3	26	—	—
14-2	2.1-33.6	7	19	7[b]
1-4/0	42.4-107	19	37	19
250-500	127-253	37	61	37
600-1000	304-508	61	91	61
1250-1500	635-759	91	127	91
1750-2000	886-1016	127	271	127

a. Number of strands vary.

b. Aluminum 14 AWG (2.1 mm^2) is not available.

c. With the permission of Underwriters Laboratories, Inc., this material is reproduced from UL Standard 486A-B, Wire Connectors, which is copyrighted by Underwriters Laboratories, Inc., Northbrook, Illinois. While use of this material has been authorized, UL shall not be responsible for the manner in which the information is presented, nor for any interpretations thereof.

E3406.11.1 Continuity. Conductors in raceways shall be continuous between outlets, boxes, and devices and shall be without splices or taps in the raceway.

Exception: Splices shall be permitted within surface-mounted raceways that have a removable cover.

E3406.11.2 Device connections. The continuity of a grounded conductor in multiwire branch circuits shall not be dependent on connection to devices such as receptacles and lampholders. The arrangement of grounding connections shall be such that the disconnection or the removal of a receptacle, luminaire or other device fed from the box does not interfere with or interrupt the grounding continuity.

E3406.11.3 Length of conductor for splice or termination. Where conductors are to be spliced, terminated or connected to fixtures or devices, a minimum length of 6 inches (152 mm) of free conductor shall be provided at each outlet, junction or switch point. The required length shall be measured from the point in the box where the conductor emerges from its raceway or cable sheath. Where the opening to an outlet, junction or switch point is less than 8 inches (200 mm) in any dimension, each conductor shall be long enough to extend at least 3 inches (75 mm) outside of such opening.

E3406.12 Grounded conductor continuity. The continuity of a grounded conductor shall not depend on connection to a metallic enclosure, raceway or cable armor.

SECTION E3407
CONDUCTOR AND TERMINAL IDENTIFICATION

E3407.1 Grounded conductors. Insulated grounded conductors of sizes 6 AWG or smaller shall be identified by a continuous white or gray outer finish or by three continuous white stripes on other than green insulation along the entire length of the conductors. Conductors of sizes 4 AWG or larger shall be identified either by a continuous white or gray outer finish or by three continuous white stripes on other than green insulation along its entire length or at the time of installation by a distinctive white or gray marking at its terminations. This marking shall encircle the conductor or insulation.

E3407.2 Equipment grounding conductors. Equipment grounding conductors of sizes 6 AWG and smaller shall be identified by a continuous green color or a continuous green color with one or more yellow stripes on the insulation or covering, except where bare. Conductors with insulation or individual covering that is green, green with one or more yellow stripes, or otherwise identified as permitted by this section shall not be used for ungrounded or grounded circuit conductors.

Equipment grounding conductors larger than 6 AWG that are not identified as required for conductors of sizes 6 AWG and smaller shall, at the time of installation, be permanently identified as an equipment grounding conductor at each end and at every point where the conductor is accessible, except where such conductors are bare.

The required identification for conductors larger than 6 AWG shall encircle the conductor and shall be accomplished by one of the following:

1. Stripping the insulation or covering from the entire exposed length.
2. Coloring the exposed insulation or covering green at the termination.
3. Marking the exposed insulation or covering with green tape or green adhesive labels at the termination.

Exceptions:

1. Conductors larger than 6 AWG shall not be required to be identified in conduit bodies that do not contain splices or unused hubs.
2. Power-limited, Class 2 or Class 3 circuit cables containing only circuits operating at less than 50 volts shall be permitted to use a conductor with green insulation for other than equipment grounding purposes.

E3407.3 Ungrounded conductors. Insulation on the ungrounded conductors shall be a continuous color other than white, gray and green.

Exceptions:

1. An insulated conductor that is part of a cable or flexible cord assembly and that has a white or gray finish or a finish marking with three continuous white stripes shall be permitted to be used as an ungrounded conductor where it is permanently reidentified to indicate its use as an ungrounded conductor at all terminations and at each location where the conductor is visible and accessible. Identification shall encircle the insulation and shall be a color other than white, gray, and green. Where used for single-pole, 3-way or 4-way switch loops, the reidentified conductor with white or gray insulation or three continuous white stripes shall be used only as a return conductor from the switch to the outlet.

E3407.4 Identification of terminals. Terminals for attachment to conductors shall be identified in accordance with Sections E3407.4.1 and E3407.4.2.

E3407.4.1 Device terminals. All devices excluding panelboards, provided with terminals for the attachment of conductors and intended for connection to more than one side of the circuit shall have terminals properly marked for identification, except where the terminal intended to be connected to the grounded conductor is clearly evident.

Exception: Terminal identification shall not be required for devices that have a normal current rating of over 30 amperes, other than polarized attachment caps and polarized receptacles for attachment caps as required in Section E3407.4.2.

E3407.4.2 Receptacles, plugs and connectors. Receptacles, polarized attachment plugs and cord connectors for plugs and polarized plugs shall have the terminal intended for connection to the grounded (white) conductor identified. Identification shall be by a metal or metal coating substantially white in color or by the word "white" or the letter "W" located adjacent to the identified terminal. Where the terminal is not visible, the conductor entrance hole for the connection shall be colored white or marked with the word "white" or the letter "W."

CHAPTER 35

ELECTRICAL DEFINITIONS

SECTION E3501 GENERAL

E3501.1 Scope. This chapter contains definitions that shall apply only to the electrical requirements of Chapters 34 through 43. Unless otherwise expressly stated, the following terms shall, for the purpose of this code, have the meanings indicated in this chapter. Words used in the present tense include the future; the singular number includes the plural and the plural the singular. Where terms are not defined in this section and are defined in Section R202 of this code, such terms shall have the meanings ascribed to them in that section. Where terms are not defined in these sections, they shall have their ordinarily accepted meanings or such as the context implies.

ACCESSIBLE. (As applied to equipment.) Admitting close approach; not guarded by locked doors, elevation or other effective means.

ACCESSIBLE. (As applied to wiring methods.) Capable of being removed or exposed without damaging the building structure or finish, or not permanently closed in by the structure or finish of the building.

ACCESSIBLE, READILY. Capable of being reached quickly for operation, renewal or inspections, without requiring those to whom ready access is requisite to climb over or remove obstacles or to resort to portable ladders, etc.

AMPACITY. The maximum current in amperes that a conductor can carry continuously under the conditions of use without exceeding its temperature rating.

APPLIANCE. Utilization equipment, normally built in standardized sizes or types, that is installed or connected as a unit to perform one or more functions such as clothes washing, air conditioning, food mixing, deep frying, etc.

APPROVED. Acceptable to the authority having jurisdiction.

ARC-FAULT CIRCUIT INTERRUPTER. A device intended to provide protection from the effects of arc-faults by recognizing characteristics unique to arcing and by functioning to de-energize the circuit when an arc-fault is detected.

ATTACHMENT PLUG (PLUG CAP) (PLUG). A device that, by insertion into a receptacle, establishes connection between the conductors of the attached flexible cord and the conductors connected permanently to the receptacle.

AUTOMATIC. Performing a function without the necessity of human intervention.

BATHROOM. An area, including a basin, with one or more of the following: a toilet, a urinal, a tub, a shower, a bidet, or similar plumbing fixture.

BONDED (BONDING). Connected to establish electrical continuity and conductivity.

BONDING CONDUCTOR OR JUMPER. A reliable conductor to ensure the required electrical conductivity between metal parts required to be electrically connected.

BONDING JUMPER (EQUIPMENT). The connection between two or more portions of the equipment grounding conductor.

BONDING JUMPER, MAIN. The connection between the grounded circuit conductor and the equipment grounding conductor at the service.

BONDING JUMPER, SUPPLY-SIDE. A conductor installed on the supply side of a service or within a service equipment enclosure(s) that ensures the required electrical conductivity between metal parts required to be electrically connected.

BRANCH CIRCUIT. The circuit conductors between the final overcurrent device protecting the circuit and the outlet(s).

BRANCH CIRCUIT, APPLIANCE. A branch circuit that supplies energy to one or more outlets to which appliances are to be connected, and that has no permanently connected luminaires that are not a part of an appliance.

BRANCH CIRCUIT, GENERAL PURPOSE. A branch circuit that supplies two or more receptacle outlets or outlets for lighting and appliances.

BRANCH CIRCUIT, INDIVIDUAL. A branch circuit that supplies only one utilization equipment.

BRANCH CIRCUIT, MULTIWIRE. A branch circuit consisting of two or more ungrounded conductors having voltage difference between them, and a grounded conductor having equal voltage difference between it and each ungrounded conductor of the circuit, and that is connected to the neutral or grounded conductor of the system.

CABINET. An enclosure designed either for surface or flush mounting and provided with a frame, mat or trim in which a swinging door or doors are or may be hung.

CIRCUIT BREAKER. A device designed to open and close a circuit by nonautomatic means and to open the circuit automatically on a predetermined overcurrent without damage to itself when properly applied within its rating.

CLOTHES CLOSET. A nonhabitable room or space intended primarily for storage of garments and apparel.

CONCEALED. Rendered inaccessible by the structure or finish of the building. Wires in concealed raceways are considered to be concealed, even though they become accessible upon withdrawing them [see "Accessible (As applied to wiring methods)"].

CONDUCTOR

Bare. A conductor having no covering or electrical insulation whatsoever.

Covered. A conductor encased within material of composition or thickness that is not recognized by this code as electrical insulation.

Insulated. A conductor encased within material of composition and thickness that is recognized by this code as electrical insulation.

CONDUIT BODY. A separate portion of a conduit or tubing system that provides access through a removable cover(s) to the interior of the system at a junction of two or more sections of the system or at a terminal point of the system. Boxes such as FS and FD or larger cast or sheet metal boxes are not classified as conduit bodies.

CONNECTOR, PRESSURE (SOLDERLESS). A device that establishes a connection between two or more conductors or between one or more conductors and a terminal by means of mechanical pressure and without the use of solder.

CONTINUOUS LOAD. A load where the maximum current is expected to continue for 3 hours or more.

COOKING UNIT, COUNTER-MOUNTED. A cooking appliance designed for mounting in or on a counter and consisting of one or more heating elements, internal wiring and built-in or separately mountable controls.

COPPER-CLAD ALUMINUM CONDUCTORS. Conductors drawn from a copper-clad aluminum rod with the copper metallurgically bonded to an aluminum core. The copper forms a minimum of 10 percent of the cross-sectional area of a solid conductor or each strand of a stranded conductor.

CUTOUT BOX. An enclosure designed for surface mounting and having swinging doors or covers secured directly to and telescoping with the walls of the box proper (see "Cabinet").

DEAD FRONT. Without live parts exposed to a person on the operating side of the equipment.

DEMAND FACTOR. The ratio of the maximum demand of a system, or part of a system, to the total connected load of a system or the part of the system under consideration.

DEVICE. A unit of an electrical system that carries or controls electrical energy as it principal function.

DISCONNECTING MEANS. A device, or group of devices, or other means by which the conductors of a circuit can be disconnected from their source of supply.

DWELLING

Dwelling unit. A single unit, providing complete and independent living facilities for one or more persons, including permanent provisions for living, sleeping, cooking and sanitation.

One-family dwelling. A building consisting solely of one dwelling unit.

Two-family dwelling. A building consisting solely of two dwelling units.

ENCLOSED. Surrounded by a case, housing, fence or walls that will prevent persons from accidentally contacting energized parts.

ENCLOSURE. The case or housing of apparatus, or the fence or walls surrounding an installation, to prevent personnel from accidentally contacting energized parts or to protect the equipment from physical damage.

ENERGIZED. Electrically connected to, or is, a source of voltage.

EQUIPMENT. A general term including material, fittings, devices, appliances, luminaires, apparatus, machinery and the like used as a part of, or in connection with, an electrical installation.

EXPOSED. (As applied to live parts.) Capable of being inadvertently touched or approached nearer than a safe distance by a person. It is applied to parts not suitably guarded, isolated or insulated.

EXPOSED. (As applied to wiring methods.) On or attached to the surface or behind panels designed to allow access.

EXTERNALLY OPERABLE. Capable of being operated without exposing the operator to contact with live parts.

FEEDER. All circuit conductors between the service equipment, or the source of a separately derived system, or other power supply source and the final branch-circuit overcurrent device.

FITTING. An accessory such as a locknut, bushing or other part of a wiring system that is intended primarily to perform a mechanical rather than an electrical function.

GROUND. The earth.

GROUNDED (GROUNDING). Connected (connecting) to ground or to a conductive body that extends the ground connection.

GROUNDED, EFFECTIVELY. Intentionally connected to earth through a ground connection or connections of sufficiently low impedance and having sufficient current-carrying capacity to prevent the buildup of voltages that may result in undue hazards to connected equipment or to persons.

GROUNDED CONDUCTOR. A system or circuit conductor that is intentionally grounded.

GROUNDING CONDUCTOR, EQUIPMENT (EGC). The conductive path(s) installed to connect normally noncurrent-carrying metal parts of equipment together and, to the system grounded conductor, the grounding electrode conductor or both.

GROUNDING ELECTRODE. A conducting object through which a direct connection to earth is established.

GROUNDING ELECTRODE CONDUCTOR. A conductor used to connect the system grounded conductor or the equipment to a grounding electrode or to a point on the grounding electrode system.

GROUND-FAULT CIRCUIT-INTERRUPTER. A device intended for the protection of personnel that functions to de-energize a circuit or portion thereof within an established period of time when a current to ground exceeds the value for a Class A device.

GUARDED. Covered, shielded, fenced, enclosed or otherwise protected by means of suitable covers, casings, barriers, rails, screens, mats or platforms to remove the likelihood of approach or contact by persons or objects to a point of danger.

IDENTIFIED. (As applied to equipment.) Recognizable as suitable for the specific purpose, function, use, environment, application, etc., where described in a particular code requirement.

INTERRUPTING RATING. The highest current at rated voltage that a device is identified to interrupt under standard test conditions.

INTERSYSTEM BONDING TERMINATION. A device that provides a means for connecting bonding conductors for communications systems to the grounding electrode system.

ISOLATED. (As applied to location.) Not readily accessible to persons unless special means for access are used.

KITCHEN. An area with a sink and permanent provisions for food preparation and cooking.

LABELED. Equipment or materials to which has been attached a label, symbol or other identifying mark of an organization acceptable to the authority having jurisdiction and concerned with product evaluation that maintains periodic inspection of production of labeled equipment or materials and by whose labeling the manufacturer indicates compliance with appropriate standards or performance in a specified manner.

LIGHTING OUTLET. An outlet intended for the direct connection of a lampholder or luminaire.

LISTED. Equipment, materials or services included in a list published by an organization that is acceptable to the authority having jurisdiction and concerned with evaluation of products or services, that maintains periodic inspection of production of listed equipment or materials or periodic evaluation of services, and whose listing states either that the equipment, material or services meets identified standards or has been tested and found suitable for a specified purpose.

LIVE PARTS. Energized conductive components.

LOCATION, DAMP. Location protected from weather and not subject to saturation with water or other liquids but subject to moderate degrees of moisture. Examples of such locations include partially protected locations under canopies, marquees, roofed open porches and like locations, and interior locations subject to moderate degrees of moisture, such as some basements, some barns and some cold-storage warehouses.

LOCATION, DRY. A location not normally subject to dampness or wetness. A location classified as dry may be temporarily subject to dampness or wetness, as in the case of a building under construction.

LOCATION, WET. Installations underground or in concrete slabs or masonry in direct contact with the earth and locations subject to saturation with water or other liquids, such as vehicle-washing areas, and locations exposed to weather.

LUMINAIRE. A complete lighting unit consisting of a light source such as a lamp or lamps together with the parts designed to position the light source and connect it to the power supply. A luminaire can include parts to protect the light source or the ballast or to distribute the light. A lampholder itself is not a luminaire.

MULTIOUTLET ASSEMBLY. A type of surface, or flush, or freestanding raceway; designed to hold conductors and receptacles, assembled in the field or at the factory.

NEUTRAL CONDUCTOR. The conductor connected to the neutral point of a system that is intended to carry current under normal conditions.

NEUTRAL POINT. The common point on a wye-connection in a polyphase system or midpoint on a single-phase, 3-wire system, or midpoint of a single-phase portion of a 3-phase delta system, or a midpoint of a 3-wire, direct-current system.

OUTLET. A point on the wiring system at which current is taken to supply utilization equipment.

OVERCURRENT. Any current in excess of the rated current of equipment or the ampacity of a conductor. Such current might result from overload, short circuit or ground fault.

OVERLOAD. Operation of equipment in excess of normal, full-load rating, or of a conductor in excess of rated ampacity that, when it persists for a sufficient length of time, would cause damage or dangerous overheating. A fault, such as a short circuit or ground fault, is not an overload.

PANELBOARD. A single panel or group of panel units designed for assembly in the form of a single panel, including buses and automatic overcurrent devices, and equipped with or without switches for the control of light, heat or power circuits, designed to be placed in a cabinet or cutout box placed in or against a wall, partition or other support and accessible only from the front.

PLENUM. A compartment or chamber to which one or more air ducts are connected and that forms part of the air distribution system.

POWER OUTLET. An enclosed assembly that may include receptacles, circuit breakers, fuseholders, fused switches, buses and watt-hour meter mounting means, intended to supply and control power to mobile homes, recreational vehicles or boats, or to serve as a means for distributing power required to operate mobile or temporarily installed equipment.

PREMISES WIRING (SYSTEM). Interior and exterior wiring, including power, lighting, control and signal circuit wiring together with all of their associated hardware, fittings and wiring devices, both permanently and temporarily installed. This includes wiring from the service point or power source to the outlets and wiring from and including the power source to the outlets where there is no service point. Such wiring does not include wiring internal to appliances, luminaires, motors, controllers, and similar equipment.

QUALIFIED PERSON. One who has the skills and knowledge related to the construction and operation of the electrical equipment and installations and has received safety training to recognize and avoid the hazards involved.

RACEWAY. An enclosed channel of metal or nonmetallic materials designed expressly for holding wires, cables, or busbars, with additional functions as permitted in this code. Raceways include, but are not limited to, rigid metal conduit, rigid nonmetallic conduit, intermediate metal conduit, liquid-tight flexible conduit, flexible metallic tubing, flexible metal conduit, electrical nonmetallic tubing, electrical metallic tubing, underfloor raceways, cellular concrete floor raceways, cellular metal floor raceways, surface raceways, wireways and busways.

RAINPROOF. Constructed, protected or treated so as to prevent rain from interfering with the successful operation of the apparatus under specified test conditions.

RAIN TIGHT. Constructed or protected so that exposure to a beating rain will not result in the entrance of water under specified test conditions.

RECEPTACLE. A receptacle is a contact device installed at the outlet for the connection of an attachment plug. A single receptacle is a single contact device with no other contact device on the same yoke. A multiple receptacle is two or more contact devices on the same yoke.

RECEPTACLE OUTLET. An outlet where one or more receptacles are installed.

SERVICE. The conductors and equipment for delivering energy from the serving utility to the wiring system of the premises served.

SERVICE CABLE. Service conductors made up in the form of a cable.

SERVICE CONDUCTORS. The conductors from the service point to the service disconnecting means.

SERVICE CONDUCTORS, OVERHEAD. The overhead conductors between the service point and the first point of connection to the service-entrance conductors at the building or other structure.

SERVICE CONDUCTORS, UNDERGROUND. The underground conductors between the service point and the first point of connection to the service-entrance conductors in a terminal box, meter, or other enclosure, inside or outside of the building wall.

SERVICE DROP. The overhead service conductors between the utility electric supply system and the service point.

SERVICE-ENTRANCE CONDUCTORS, OVERHEAD SYSTEM. The service conductors between the terminals of the service equipment and a point usually outside of the building, clear of building walls, where joined by tap or splice to the service drop or overhead service conductors.

SERVICE-ENTRANCE CONDUCTORS, UNDERGROUND SYSTEM. The service conductors between the terminals of the service equipment and the point of connection to the service lateral or underground service conductors.

SERVICE EQUIPMENT. The necessary equipment, usually consisting of a circuit breaker(s) or switch(es) and fuse(s), and their accessories, connected to the load end of the service conductors to a building or other structure, or an otherwise designated area, and intended to constitute the main control and cutoff of the supply.

SERVICE LATERAL. The underground service conductors between the electric utility supply system and the service point.

SERVICE POINT. The point of connection between the facilities of the serving utility and the premises wiring.

STRUCTURE. That which is built or constructed.

SWITCHES

General-use switch. A switch intended for use in general distribution and branch circuits. It is rated in amperes and is capable of interrupting its rated current at its rated voltage.

General-use snap switch. A form of general-use switch constructed so that it can be installed in device boxes or on box covers or otherwise used in conjunction with wiring systems recognized by this code.

Isolating switch. A switch intended for isolating an electric circuit from the source of power. It has no interrupting rating and is intended to be operated only after the circuit has been opened by some other means.

Motor-circuit switch. A switch, rated in horsepower that is capable of interrupting the maximum operating overload current of a motor of the same horsepower rating as the switch at the rated voltage.

UNGROUNDED. Not connected to ground or to a conductive body that extends the ground connection.

UTILIZATION EQUIPMENT. Equipment that utilizes electric energy for electronic, electromechanical, chemical, heating, lighting or similar purposes.

VENTILATED. Provided with a means to permit circulation of air sufficient to remove an excess of heat, fumes or vapors.

VOLTAGE (OF A CIRCUIT). The greatest root-mean-square (rms) (effective) difference of potential between any two conductors of the circuit concerned.

VOLTAGE, NOMINAL. A nominal value assigned to a circuit or system for the purpose of conveniently designating its voltage class (e.g., 120/240). The actual voltage at which a circuit operates can vary from the nominal within a range that permits satisfactory operation of equipment.

VOLTAGE TO GROUND. For grounded circuits, the voltage between the given conductor and that point or conductor of the circuit that is grounded. For ungrounded circuits, the greatest voltage between the given conductor and any other conductor of the circuit.

WATERTIGHT. Constructed so that moisture will not enter the enclosure under specified test conditions.

WEATHERPROOF. Constructed or protected so that exposure to the weather will not interfere with successful operation.

CHAPTER 36

SERVICES

SECTION E3601 GENERAL SERVICES

E3601.1 Scope. This chapter covers service conductors and equipment for the control and protection of services and their installation requirements.

E3601.2 Number of services. One- and two-family dwellings shall be supplied by only one service.

E3601.3 One building or other structure not to be supplied through another. Service conductors supplying a building or other structure shall not pass through the interior of another building or other structure.

E3601.4 Other conductors in raceway or cable. Conductors other than service conductors shall not be installed in the same service raceway or service cable.

Exceptions:

1. Grounding conductors and bonding jumpers.
2. Load management control conductors having overcurrent protection.

E3601.5 Raceway seal. Where a service raceway enters from an underground distribution system, it shall be sealed in accordance with Section E3803.6.

E3601.6 Service disconnect required. Means shall be provided to disconnect all conductors in a building or other structure from the service entrance conductors.

E3601.6.1 Marking of service equipment and disconnects. Service disconnects shall be permanently marked as a service disconnect.

E3601.6.2 Service disconnect location. The service disconnecting means shall be installed at a readily accessible location either outside of a building or inside nearest the point of entrance of the service conductors. Service disconnecting means shall not be installed in bathrooms. Each occupant shall have access to the disconnect serving the dwelling unit in which they reside.

E3601.7 Maximum number of disconnects. The service disconnecting means shall consist of not more than six switches or six circuit breakers mounted in a single enclosure or in a group of separate enclosures.

SECTION E3602 SERVICE SIZE AND RATING

E3602.1 Ampacity of ungrounded conductors. Ungrounded service conductors shall have an ampacity of not less than the load served. For one-family dwellings, the ampacity of the ungrounded conductors shall be not less than 100 amperes, 3 wire. For all other installations, the ampacity of the ungrounded conductors shall be not less than 60 amperes.

E3602.2 Service load. The minimum load for ungrounded service conductors and service devices that serve 100 percent of the dwelling unit load shall be computed in accordance with Table E3602.2. Ungrounded service conductors and service devices that serve less than 100 percent of the dwelling unit load shall be computed as required for feeders in accordance with Chapter 37.

TABLE E3602.2 MINIMUM SERVICE LOAD CALCULATION

LOADS AND PROCEDURE
3 volt-amperes per square foot of floor area for general lighting and general use receptacle outlets.
Plus
1,500 volt-amperes multiplied by total number of 20-ampere-rated small appliance and laundry circuits.
Plus
The nameplate volt-ampere rating of all fastened-in-place, permanently connected or dedicated circuit-supplied appliances such as ranges, ovens, cooking units, clothes dryers not connected to the laundry branch circuit and water heaters.
Apply the following demand factors to the above subtotal:
The minimum subtotal for the loads above shall be 100 percent of the first 10,000 volt-amperes of the sum of the above loads plus 40 percent of any portion of the sum that is in excess of 10,000 volt-amperes.
Plus the largest of the following:
One-hundred percent of the nameplate rating(s) of the air-conditioning and cooling equipment.
One hundred percent of the nameplate rating(s) of the heat pump where a heat pump is used without any supplemental electric heating.
One-hundred percent of the nameplate rating of the electric thermal storage and other heating systems where the usual load is expected to be continuous at the full nameplate value. Systems qualifying under this selection shall not be figured under any other category in this table.
One-hundred percent of nameplate rating of the heat pump compressor and sixty-five percent of the supplemental electric heating load for central electric space-heating systems. If the heat pump compressor is prevented from operating at the same time as the supplementary heat, the compressor load does not need to be added to the supplementary heat load for the total central electric space-heating load.
Sixty-five percent of nameplate rating(s) of electric space-heating units if less than four separately controlled units.
Forty percent of nameplate rating(s) of electric space-heating units of four or more separately controlled units.
The minimum total load in amperes shall be the volt-ampere sum calculated above divided by 240 volts.

E3602.2.1 Services under 100 amperes. Services that are not required to be 100 amperes shall be sized in accordance with Chapter 37.

E3602.3 Rating of service disconnect. The combined rating of all individual service disconnects serving a single dwelling unit shall not be less than the load determined from Table E3602.2 and shall not be less than as specified in Section E3602.1.

E3602.4 Voltage rating. Systems shall be three-wire, 120/240-volt, single-phase with a grounded neutral.

SECTION E3603 SERVICE, FEEDER AND GROUNDING ELECTRODE CONDUCTOR SIZING

E3603.1 Grounded and ungrounded service conductor size. Conductors used as ungrounded service entrance conductors, service lateral conductors, and feeder conductors that serve as the main power feeder to a dwelling unit shall be those listed in Table E3603.1. The main power feeder shall be the feeder(s) between the main disconnect and the panelboard that supplies, either by branch circuits or by feeders, or both, all loads that are part of or are associated with the dwelling unit. The feeder conductors to a dwelling unit shall not be required to have an allowable ampacity greater than that of the service-entrance conductors that supply them. Ungrounded service conductors shall have a minimum size in accordance with Table E3603.1. The grounded conductor ampacity shall be not less than the maximum unbalance of the load and its size shall be not smaller than the required minimum grounding electrode conductor size specified in Table E3603.1.

E3603.2 Ungrounded service conductors for accessory buildings and structures. Ungrounded conductors for other than dwelling units shall have an ampacity of not less than 60 amperes and shall be sized as required for feeders in Chapter 37.

Exceptions:

1. For limited loads of a single branch circuit, the service conductors shall have an ampacity of not less than 15 amperes.
2. For loads consisting of not more than two two-wire branch circuits, the service conductors shall have an ampacity of not less than 30 amperes.

E3603.3 Overload protection. Each ungrounded service conductor shall have overload protection.

E3603.3.1 Ungrounded conductor. Overload protection shall be provided by an overcurrent device installed in series with each ungrounded service conductor. The overcurrent device shall have a rating or setting not higher than the allowable service or feeder rating specified in Table E3603.1. A set of fuses shall be considered all the fuses required to protect all of the ungrounded conductors of a circuit. Single pole circuit breakers, grouped in accordance with Section E3601.7, shall be considered as one protective device.

Exception: Two to six circuit breakers or sets of fuses shall be permitted as the overcurrent device to provide

TABLE E3603.1
SERVICE CONDUCTOR AND GROUNDING ELECTRODE CONDUCTOR SIZING

CONDUCTOR TYPES AND SIZES-THHN, THHW, THW, THWN, USE, RHH, RHW, XHHW, RHW-2, THW-2, THWN-2, XHHW-2, SE, USE-2 (Parallel sets of 1/0 and larger conductors are permitted in either a single raceway or in separate raceways)		SERVICE OR FEEDER RATING (AMPERES)	MINIMUM GROUNDING ELECTRODE CONDUCTOR SIZE[a]	
Copper (AWG)	Aluminum and copper-clad aluminum (AWG)	Maximum load (amps)	Copper (AWG)	Aluminum (AWG)
4	2	100	8[b]	6[c]
3	1	110	8[b]	6[c]
2	1/0	125	8[b]	6[c]
1	2/0	150	6[c]	4
1/0	3/0	175	6[c]	4
2/0	4/0 or two sets of 1/0	200	4[d]	2[d]
3/0	250 kcmil or two sets of 2/0	225	4[d]	2[d]
4/0 or two sets of 1/0	300 kcmil ortwo sets of 3/0	250	2[d]	1/0[d]
250 kcmil or two sets of 2/0	350 kcmil or two sets of 4/0	300	2[d]	1/0[d]
350 kcmil or two sets of 3/0	500 kcmil or two sets of 250 kcmil	350	2[d]	1/0[d]
400 kcmil or two sets of 4/0	600 kcmil or two sets of 300 kcmil	400	1/0[d]	3/0[d]

For SI: 1 inch = 25.4 mm.

a. Where protected by a ferrous metal raceway, grounding electrode conductors shall be electrically bonded to the ferrous metal raceway at both ends.

b. An 8 AWG grounding electrode conductor shall be protected with rigid metal conduit, intermediate metal conduit, rigid polyvinyl chloride (Type PVC) nonmetallic conduit, rigid thermosetting resin (Type RTRC) nonmetallic conduit, electrical metallic tubing or cable armor.

c. Where not protected, 6 AWG grounding electrode conductor shall closely follow a structural surface for physical protection. The supports shall be spaced not more than 24 inches on center and shall be within 12 inches of any enclosure or termination.

d. Where the sole grounding electrode system is a ground rod or pipe as covered in Section E3608.2, the grounding electrode conductor shall not be required to be larger than 6 AWG copper or 4 AWG aluminum. Where the sole grounding electrode system is the footing steel as covered in Section E3608.1.2, the grounding electrode conductor shall not be required to be larger than 4 AWG copper conductor.

the overload protection. The sum of the ratings of the circuit breakers or fuses shall be permitted to exceed the ampacity of the service conductors, provided that the calculated load does not exceed the ampacity of the service conductors.

E3603.3.2 Not in grounded conductor. Overcurrent devices shall not be connected in series with a grounded service conductor except where a circuit breaker is used that simultaneously opens all conductors of the circuit.

E3603.3.3 Location. The service overcurrent device shall be an integral part of the service disconnecting means or shall be located immediately adjacent thereto.

E3603.4 Grounding electrode conductor size. The grounding electrode conductors shall be sized based on the size of the service entrance conductors as required in Table E3603.1.

E3603.5 Temperature limitations. Except where the equipment is marked otherwise, conductor ampacities used in determining equipment termination provisions shall be based on Table E3603.1.

SECTION E3604 OVERHEAD SERVICE AND SERVICE-ENTRANCE CONDUCTOR INSTALLATION

E3604.1 Clearances on buildings. Open conductors and multiconductor cables without an overall outer jacket shall have a clearance of not less than 3 feet (914 mm) from the sides of doors, porches, decks, stairs, ladders, fire escapes and balconies, and from the sides and bottom of windows that open. See Figure E3604.1.

E3604.2 Vertical clearances. Overhead service conductors shall not have ready access and shall comply with Sections E3604.2.1 and E3604.2.2.

E3604.2.1 Above roofs. Conductors shall have a vertical clearance of not less than 8 feet (2438 mm) above the roof surface. The vertical clearance above the roof level shall be maintained for a distance of not less than 3 feet (914 mm) in all directions from the edge of the roof. See Figure E3604.2.1.

Exceptions:

1. Conductors above a roof surface subject to pedestrian traffic shall have a vertical clearance from the roof surface in accordance with Section E3604.2.2.
2. Where the roof has a slope of 4 inches (102 mm) in 12 inches (305 mm), or greater, the minimum clearance shall be 3 feet (914 mm).
3. The minimum clearance above only the overhanging portion of the roof shall not be less than 18 inches (457 mm) where not more than 6 feet (1829 mm) of overhead service conductor length

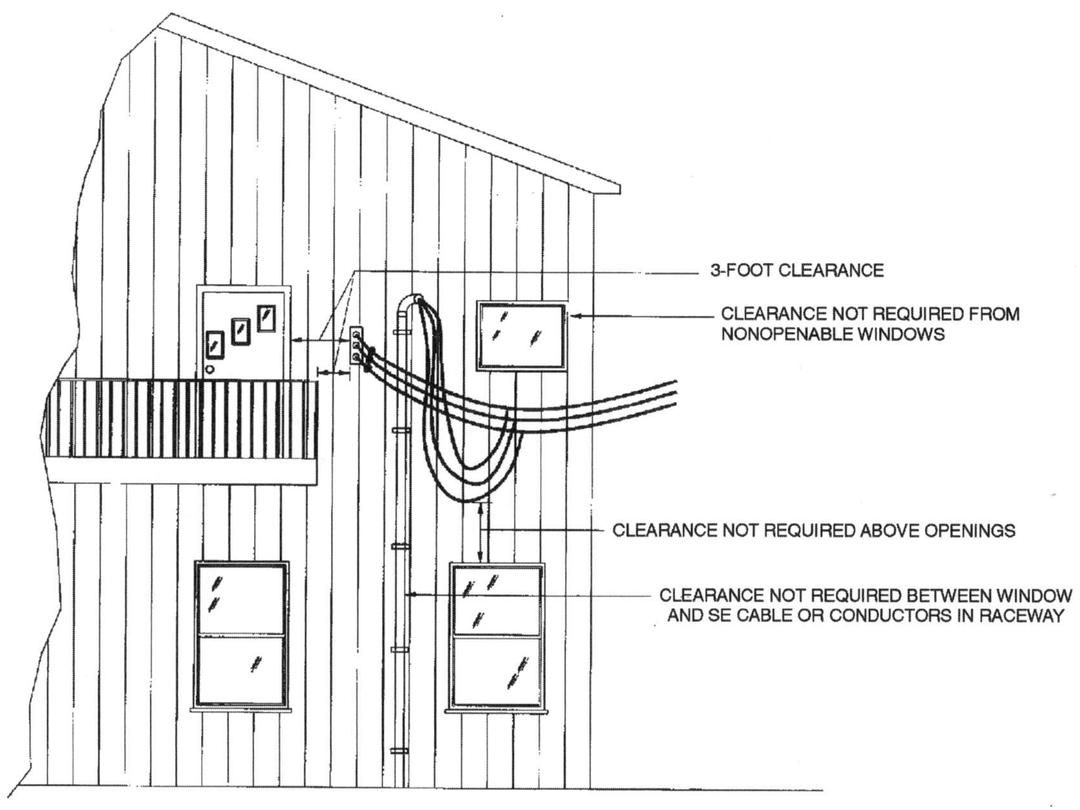

For SI: 1 foot = 304.8 mm.

FIGURE E3604.1 CLEARANCES FROM BUILDING OPENINGS

passes over 4 feet (1219 mm) or less of roof surface measured horizontally and such conductors are terminated at a through-the-roof raceway or approved support.

4. The requirement for maintaining the vertical clearance for a distance of 3 feet (914 mm) from the edge of the roof shall not apply to the final conductor span where the service drop is attached to the side of a building.
5. Where the voltage between conductors does not exceed 300 and the roof area is guarded or isolated, a reduction in clearance to 3 feet (914 mm) shall be permitted.

E3604.2.2 Vertical clearance from grade. Overhead service conductors shall have the following minimum clearances from final grade:

1. For conductors supported on and cabled together with a grounded bare messenger wire, the minimum vertical clearance shall be 10 feet (3048 mm) at the electric service entrance to buildings, at the lowest point of the drip loop of the building electric entrance, and above areas or sidewalks accessed by pedestrians only. Such clearance shall be measured from final grade or other accessible surfaces.
2. Twelve feet (3658 mm)—over residential property and driveways.
3. Eighteen feet (5486 mm)—over public streets, alleys, roads or parking areas subject to truck traffic.

E3604.3 Point of attachment. The point of attachment of the service-drop conductors to a building or other structure shall provide the minimum clearances as specified in Sections E3604.1 through E3604.2.2. In no case shall the point of attachment be less than 10 feet (3048 mm) above finished grade.

E3604.4 Means of attachment. Multiconductor cables used for overhead service conductors shall be attached to buildings or other structures by fittings approved for the purpose.

E3604.5 Service masts as supports. Where a service mast is used for the support of service-drop conductors, it shall be of adequate strength or be supported by braces or guys to withstand the strain imposed by the service drop. Where raceway-type service masts are used, all equipment shall be approved. Only power service drop conductors shall be permitted to be attached to a service mast.

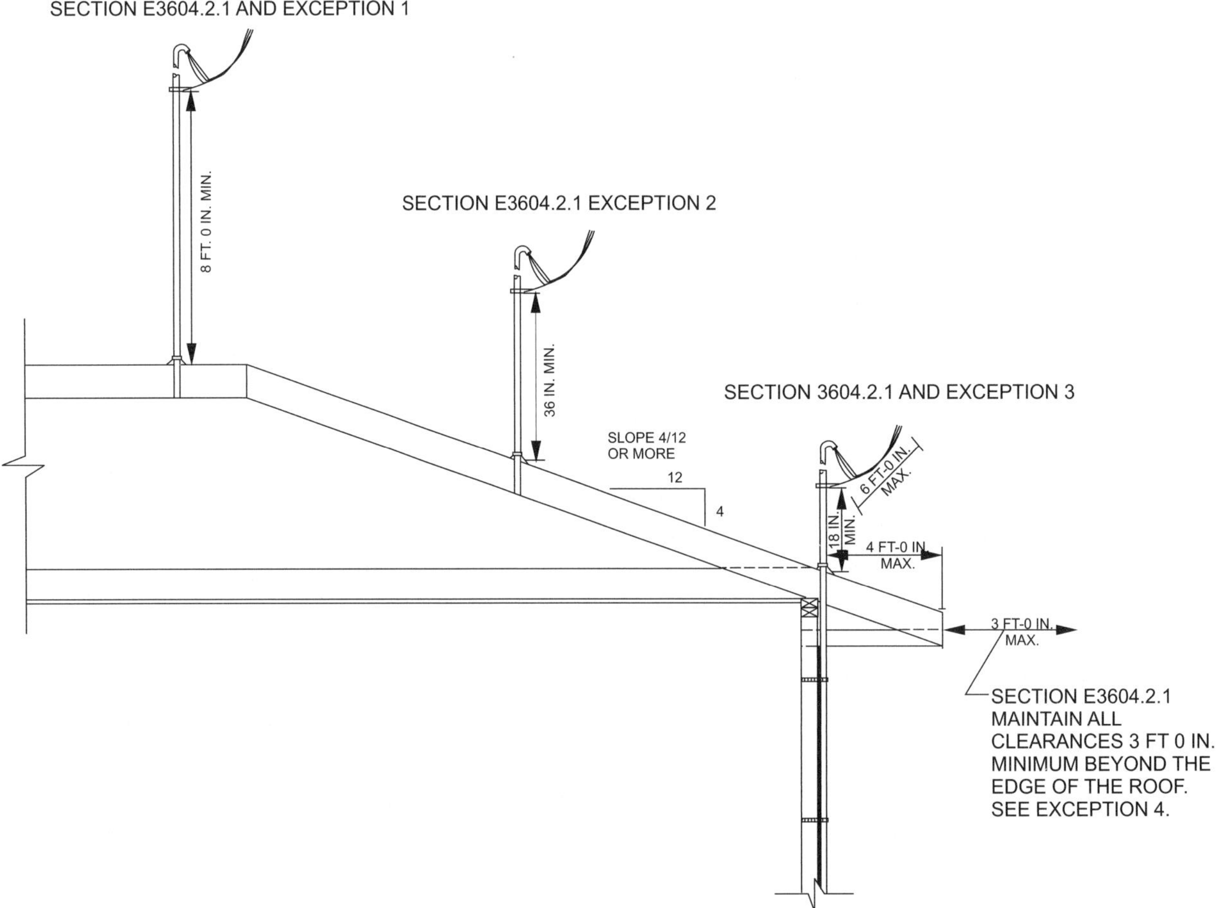

For SI: 1 inch = 25.4 mm, 1 foot = 304.8 mm.

FIGURE E3604.2.1
CLEARANCES FROM ROOFS

E3604.6 Supports over buildings. Service conductors passing over a roof shall be securely supported. Where practicable, such supports shall be independent of the building.

SECTION E3605 SERVICE-ENTRANCE CONDUCTORS

E3605.1 Insulation of service-entrance conductors. Service-entrance conductors entering or on the exterior of buildings or other structures shall be insulated in accordance with Section E3406.5.

Exceptions:

1. A copper grounded conductor shall not be required to be insulated where it is:
   1.1. In a raceway or part of a service cable assembly,
   1.2. Directly buried in soil of suitable condition, or
   1.3. Part of a cable assembly listed for direct burial without regard to soil conditions.
2. An aluminum or copper-clad aluminum grounded conductor shall not be required to be insulated where part of a cable or where identified for direct burial or utilization in underground raceways.

E3605.2 Wiring methods for services. Service-entrance wiring methods shall be installed in accordance with the applicable requirements in Chapter 38.

E3605.3 Spliced conductors. Service-entrance conductors shall be permitted to be spliced or tapped. Splices shall be made in enclosures or, if directly buried, with listed underground splice kits. Conductor splices shall be made in accordance with Chapters 34, 37, 38 and 39.

E3605.4 Protection of underground service entrance conductors. Underground service-entrance conductors shall be protected against physical damage in accordance with Chapter 38.

E3605.5 Protection of all other service cables. Aboveground service-entrance cables, where subject to physical damage, shall be protected by one or more of the following: rigid metal conduit, intermediate metal conduit, Schedule 80 PVC conduit, electrical metallic tubing or other approved means.

E3605.6 Locations exposed to direct sunlight. Insulated conductors and cables used where exposed to direct rays of the sun shall comply with one of the following:

1. The conductors and cables shall be listed, or listed and marked, as being sunlight resistant.
2. The conductors and cables are covered with insulating material, such as tape or sleeving, that is listed, or listed and marked, as being sunlight resistant.

E3605.7 Mounting supports. Service-entrance cables shall be supported by straps or other approved means within 12 inches (305 mm) of every service head, gooseneck or connection to a raceway or enclosure and at intervals not exceeding 30 inches (762 mm).

E3605.8 Raceways to drain. Where exposed to the weather, raceways enclosing service-entrance conductors shall be suitable for use in wet locations and arranged to drain. Where embedded in masonry, raceways shall be arranged to drain.

E3605.9 Overhead service locations. Connections at service heads shall be in accordance with Sections E3605.9.1 through E3605.9.7.

E3605.9.1 Rain-tight service head. Service raceways shall be equipped with a service head at the point of connection to service-drop or overhead conductors. The service head shall be listed for use in wet locations.

E3605.9.2 Service cable, service head or gooseneck. Service-entrance cable shall be equipped with a service head or shall be formed into a gooseneck in an approved manner. The service head shall be listed for use in wet locations.

E3605.9.3 Service-head location. Service heads, and goosenecks in service-entrance cables, shall be located above the point of attachment of the service-drop or overhead service conductors to the building or other structure.

Exception: Where it is impracticable to locate the service head or gooseneck above the point of attachment, the service head or gooseneck location shall be not more than 24 inches (610 mm) from the point of attachment.

E3605.9.4 Separately bushed openings. Service heads shall have conductors of different potential brought out through separately bushed openings.

E3605.9.5 Drip loops. Drip loops shall be formed on individual conductors. To prevent the entrance of moisture, service-entrance conductors shall be connected to the service-drop or overhead conductors either below the level of the service head or below the level of the termination of the service-entrance cable sheath.

E3605.9.6 Conductor arrangement. Service-entrance and overhead service conductors shall be arranged so that water will not enter service raceways or equipment.

E3605.9.7 Secured. Service-entrance cables shall be held securely in place.

SECTION E3606 SERVICE EQUIPMENT—GENERAL

E3606.1 Service equipment enclosures. Energized parts of service equipment shall be enclosed.

E3606.2 Working space. In no case shall the working space in the vicinity of service equipment be less than that specified in Chapter 34.

E3606.3 Available short-circuit current. Service equipment shall be suitable for the maximum fault current available at its supply terminals, but not less than 10,000 amperes.

E3606.4 Marking. Service equipment shall be marked to identify it as being suitable for use as service equipment. Service equipment shall be listed. Individual meter socket enclosures shall not be considered as service equipment.

SECTION E3607
SYSTEM GROUNDING

E3607.1 System service ground. The premises wiring system shall be grounded at the service with a grounding electrode conductor connected to a grounding electrode system as required by this code. Grounding electrode conductors shall be sized in accordance with Table E3603.1.

E3607.2 Location of grounding electrode conductor connection. The grounding electrode conductor shall be connected to the grounded service conductor at any accessible point from the load end of the service drop or service lateral to and including the terminal or bus to which the grounded service conductor is connected at the service disconnecting means. A grounding connection shall not be made to any grounded circuit conductor on the load side of the service disconnecting means, except as provided in Section E3607.3.2.

E3607.3 Buildings or structures supplied by feeder(s) or branch circuit(s). Buildings or structures supplied by feeder(s) or branch circuit(s) shall have a grounding electrode or grounding electrode system installed in accordance with Section E3608. The grounding electrode conductor(s) shall be connected in a manner specified in Section E3607.3.1 or, for existing premises wiring systems only, Section E3607.3.2. Where there is no existing grounding electrode, the grounding electrode(s) required in Section E3608 shall be installed.

Exception: A grounding electrode shall not be required where only one branch circuit, including a multiwire branch circuit, supplies the building or structure and the branch circuit includes an equipment grounding conductor for grounding the noncurrent-carrying parts of all equipment. For the purposes of this section, a multiwire branch circuit shall be considered as a single branch circuit.

E3607.3.1 Equipment grounding conductor. An equipment grounding conductor as described in Section E3908 shall be run with the supply conductors and connected to the building or structure disconnecting means and to the grounding electrode(s). The equipment grounding conductor shall be used for grounding or bonding of equipment, structures or frames required to be grounded or bonded. The equipment grounding conductor shall be sized in accordance with Section E3908.12. Any installed grounded conductor shall not be connected to the equipment grounding conductor or to the grounding electrode(s).

E3607.3.2 Grounded conductor, existing premises. For installations made in compliance with previous editions of this code that permitted such connection and where an equipment grounding conductor is not run with the supply conductors to the building or structure, there are no continuous metallic paths bonded to the grounding system in both buildings or structures involved, and ground-fault protection of equipment has not been installed on the supply side of the feeder(s), the grounded conductor run with the supply to the buildings or structure shall be connected to the building or structure disconnecting means and to the grounding electrode(s) and shall be used for grounding or bonding of equipment, structures, or frames required to be grounded or bonded. Where used for grounding in accordance with this provision, the grounded conductor shall be not smaller than the larger of:

1. That required by Section E3704.3.
2. That required by Section E3908.12.

E3607.4 Grounding electrode conductor. A grounding electrode conductor shall be used to connect the equipment grounding conductors, the service equipment enclosures, and the grounded service conductor to the grounding electrode(s). This conductor shall be sized in accordance with Table E3603.1.

E3607.5 Main bonding jumper. An unspliced main bonding jumper shall be used to connect the equipment grounding conductor(s) and the service-disconnect enclosure to the grounded conductor of the system within the enclosure for each service disconnect.

E3607.6 Common grounding electrode. Where an ac system is connected to a grounding electrode in or at a building or structure, the same electrode shall be used to ground conductor enclosures and equipment in or on that building or structure. Where separate services, feeders or branch circuits supply a building and are required to be connected to a grounding electrode(s), the same grounding electrode(s) shall be used. Two or more grounding electrodes that are effectively bonded together shall be considered as a single grounding electrode system.

SECTION E3608
GROUNDING ELECTRODE SYSTEM

E3608.1 Grounding electrode system. All electrodes specified in Sections E3608.1.1, E3608.1.2, E3608.1.3, E3608.1.4 E3608.1.5 and E3608.1.6 that are present at each building or structure served shall be bonded together to form the grounding electrode system. Where none of these electrodes are present, one or more of the electrodes specified in Sections E3608.1.3, E3608.1.4, E3608.1.5 and E3608.1.6 shall be installed and used.

Exception: Concrete-encased electrodes of existing buildings or structures shall not be required to be part of the grounding electrode system where the steel reinforcing bars or rods are not accessible for use without disturbing the concrete.

E3608.1.1 Metal underground water pipe. A metal underground water pipe that is in direct contact with the earth for 10 feet (3048 mm) or more, including any well casing effectively bonded to the pipe and that is electrically continuous, or made electrically continuous by bonding around insulating joints or insulating pipe to the points of connection of the grounding electrode conductor and the bonding conductors, shall be considered as a grounding electrode (see Section E3608.1). Interior metal water piping located more than 5 feet (1524 mm) from the entrance to the building shall not be used as part of the grounding electrode system or as a conductor to interconnect electrodes that are part of the grounding electrode system.

E3608.1.1.1 Interior metal water piping. Interior metal water piping located more than 5 feet (1524 mm) from the entrance to the building shall not be used as a conductor to interconnect electrodes that are part of the grounding electrode system.

E3608.1.1.2 Installation. Continuity of the grounding path or the bonding connection to interior piping shall not rely on water meters, filtering devices and similar equipment. A metal underground water pipe shall be supplemented by an additional electrode of a type specified in Sections E3608.1.2 through E3608.1.6. The supplemental electrode shall be bonded to the grounding electrode conductor, the grounded service-entrance conductor, a nonflexible grounded service raceway, any grounded service enclosure or to the equipment grounding conductor provided in accordance with Section E3607.3.1. Where the supplemental electrode is a rod, pipe or plate electrode in accordance with Section E3608.1.4 or E3608.1.5, it shall comply with Section E3608.4.

Where the supplemental electrode is a rod, pipe or plate electrode in accordance with Section E3608.1.4 or E3608.1.5, that portion of the bonding jumper that is the sole connection to the supplemental grounding electrode shall not be required to be larger than 6 AWG copper or 4 AWG aluminum wire.

E3608.1.2 Concrete-encased electrode. A concrete-encased electrode consisting of at least 20 feet (6096 mm) of either of the following shall be considered as a grounding electrode:

1. One or more bare or zinc galvanized or other electrically conductive coated steel reinforcing bars or rods not less than $^1/_2$ inch (13 mm) in diameter, installed in one continuous 20-foot (6096 mm) length, or if in multiple pieces connected together by the usual steel tie wires, exothermic welding, welding, or other effective means to create a 20-foot (6096 mm) or greater length.
2. A bare copper conductor not smaller than 4 AWG.

Metallic components shall be encased by at least 2 inches (51 mm) of concrete and shall be located horizontally within that portion of a concrete foundation or footing that is in direct contact with the earth or within vertical foundations or structural components or members that are in direct contact with the earth.

Where multiple concrete-encased electrodes are present at a building or structure, only one shall be required to be bonded into the grounding electrode system.

E3608.1.3 Ground rings. A ground ring encircling the building or structure, in direct contact with the earth at a depth below the earth's surface of not less than 30 inches (762 mm), consisting of at least 20 feet (6096 mm) of bare copper conductor not smaller than 2 AWG shall be considered as a grounding electrode.

E3608.1.4 Rod and pipe electrodes. Rod and pipe electrodes not less than 8 feet (2438 mm) in length and consisting of the following materials shall be considered as a grounding electrode:

1. Grounding electrodes of pipe or conduit shall not be smaller than trade size $^3/_4$ (metric designator 21) and, where of iron or steel, shall have the outer surface galvanized or otherwise metal-coated for corrosion protection.
2. Rod-type grounding electrodes of stainless steel and copper or zinc-coated steel shall be at least $^5/_8$ inch (15.9 mm) in diameter unless listed.

E3608.1.4.1 Installation. The rod and pipe electrodes shall be installed such that at least 8 feet (2438 mm) of length is in contact with the soil. They shall be driven to a depth of not less than 8 feet (2438 mm) except that, where rock bottom is encountered, electrodes shall be driven at an oblique angle not to exceed 45 degrees (0.79 rad) from the vertical or shall be buried in a trench that is at least 30 inches (762 mm) deep. The upper end of the electrodes shall be flush with or below ground level except where the aboveground end and the grounding electrode conductor attachment are protected against physical damage.

E3608.1.5 Plate electrodes. A plate electrode that exposes not less than 2 square feet (0.186 m^2) of surface to exterior soil shall be considered as a grounding electrode. Electrodes of bare or conductively coated iron or steel plates shall be at least $^1/_4$ inch (6.4 mm) in thickness. Solid, uncoated electrodes of nonferrous metal shall be at least 0.06 inch (1.5 mm) in thickness. Plate electrodes shall be installed not less than 30 inches (762 mm) below the surface of the earth.

E3608.1.6 Other electrodes. In addition to the grounding electrodes specified in Sections E3608.1.1 through E3608.1.5, other listed grounding electrodes shall be permitted.

E3608.2 Bonding jumper. The bonding jumper(s) used to connect the grounding electrodes together to form the grounding electrode system shall be installed in accordance with Sections E3610.2, and E3610.3, shall be sized in accordance with Section E3603.4, and shall be connected in the manner specified in Section E3611.1.

E3608.3 Rod, pipe and plate electrode requirements. Where practicable, rod, pipe and plate electrodes shall be embedded below permanent moisture level. Such electrodes shall be free from nonconductive coatings such as paint or enamel. Where more than one such electrode is used, each electrode of one grounding system shall be not less than 6 feet (1829 mm) from any other electrode of another grounding system. Two or more grounding electrodes that are effectively bonded together shall be considered as a single grounding electrode system. That portion of a bonding jumper that is the sole connection to a rod, pipe or plate electrode shall not be required to be larger than 6 AWG copper or 4 AWG aluminum wire.

E3608.4 Supplemental electrode required. A single rod, pipe, or plate electrode shall be supplemented by an additional electrode of a type specified in Sections E3608.1.2 through E3608.1.6. The supplemental electrode shall be bonded to one of the following:

1. A rod, pipe, or plate electrode.
2. A grounding electrode conductor.
3. A grounded service-entrance conductor.
4. A nonflexible grounded service raceway.
5. A grounded service enclosure.

Where multiple rod, pipe, or plate electrodes are installed to meet the requirements of this section, they shall not be less than 6 feet (1829 mm) apart.

Exception: Where a single rod, pipe, or plate grounding electrode has a resistance to earth of 25 ohms or less, the supplemental electrode shall not be required.

E3608.5 Aluminum electrodes. Aluminum electrodes shall not be permitted.

E3608.6 Metal underground gas piping system. A metal underground gas piping system shall not be used as a grounding electrode.

SECTION E3609
BONDING

E3609.1 General. Bonding shall be provided where necessary to ensure electrical continuity and the capacity to conduct safely any fault current likely to be imposed.

E3609.2 Bonding of equipment for services. The noncurrent-carrying metal parts of the following equipment shall be effectively bonded together:

1. Raceways or service cable armor or sheath that enclose, contain, or support service conductors.
2. Service enclosures containing service conductors, including meter fittings, and boxes, interposed in the service raceway or armor.

E3609.3 Bonding for other systems. An intersystem bonding termination for connecting intersystem bonding conductors required for other systems shall be provided external to enclosures at the service equipment or metering equipment enclosure and at the disconnecting means for any additional buildings or structures. The intersystem bonding termination shall comply with all of the following:

1. It shall be accessible for connection and inspection.
2. It shall consist of a set of terminals with the capacity for connection of not less than three intersystem bonding conductors.
3. It shall not interfere with opening of the enclosure for a service, building or structure disconnecting means, or metering equipment.
4. Where located at the service equipment, it shall be securely mounted and electrically connected to an enclosure for the service equipment, to the meter enclosure, or to an exposed nonflexible metallic service raceway, or shall be mounted at one of these enclosures and connected to the enclosure or to the grounding electrode conductor with a 6 AWG or larger copper conductor.
5. Where located at the disconnecting means for a building or structure, it shall be securely mounted and electrically connected to the metallic enclosure for the building or structure disconnecting means, or shall be mounted at the disconnecting means and connected to the metallic enclosure or to the grounding electrode conductor with a 6 AWG or larger copper conductor.
6. It shall be listed as grounding and bonding equipment.

E3609.4 Method of bonding at the service. Bonding jumpers meeting the requirements of this chapter shall be used around impaired connections, such as reducing washers or oversized, concentric, or eccentric knockouts. Standard locknuts or bushings shall not be the only means for the bonding required by this section but shall be permitted to be installed to make mechanical connections of raceways. Electrical continuity at service equipment, service raceways and service conductor enclosures shall be ensured by one or more of the methods specified in Sections E3609.4.1 through E3609.4.4.

E3609.4.1 Grounded service conductor. Equipment shall be bonded to the grounded service conductor in a manner provided in this code.

E3609.4.2 Threaded connections. Equipment shall be bonded by connections using threaded couplings or threaded hubs on enclosures. Such connections shall be made wrench tight.

E3609.4.3 Threadless couplings and connectors. Equipment shall be bonded by threadless couplings and connectors for metal raceways and metal-clad cables. Such couplings and connectors shall be made wrench tight. Standard locknuts or bushings shall not be used for the bonding required by this section.

E3609.4.4 Other devices. Equipment shall be bonded by other listed devices, such as bonding-type locknuts, bushings and bushings with bonding jumpers.

E3609.5 Sizing supply-side bonding jumper and main bonding jumper. The bonding jumper shall not be smaller than the sizes shown in Table E3603.1 for grounding electrode conductors. Where the service-entrance conductors are paralleled in two or more raceways or cables, and an individual supply-side bonding jumper is used for bonding these raceways or cables, the supply-side bonding jumper for each raceway or cable shall be selected from Table E3603.1 based on the size of the ungrounded supply conductors in each raceway or cable. A single supply-side bonding jumper installed for bonding two or more raceways or cables shall be sized in accordance with Table E3603.1 based on the largest set of parallel ungrounded supply conductors.

E3609.6 Metal water piping bonding. The metal water piping system shall be bonded to the service equipment enclosure, the grounded conductor at the service, the grounding electrode conductor where of sufficient size, or to the one or more grounding electrodes used. The bonding jumper shall be sized in accordance with Table E3603.1. The points of attachment of the bonding jumper(s) shall be accessible.

E3609.7 Bonding other metal piping. Where installed in or attached to a building or structure, metal piping systems, including gas piping, capable of becoming energized shall be bonded to the service equipment enclosure, the grounded conductor at the service, the grounding electrode conductor where of sufficient size, or to the one or more grounding electrodes used. The bonding conductor(s) or jumper(s) shall be sized in accordance with Table E3908.12 using the rating of the circuit capable of energizing the piping. The equipment grounding conductor for the circuit that is capable of energizing the piping shall be permitted to serve as the bonding means. The points of attachment of the bonding jumper(s) shall be accessible.

SECTION E3610 GROUNDING ELECTRODE CONDUCTORS

E3610.1 Continuous. The grounding electrode conductor shall be installed in one continuous length without splices or joints and shall run to any convenient grounding electrode available in the grounding electrode system where the other electrode(s), if any, are connected by bonding jumpers in accordance with Section E3608.2, or to one or more grounding electrode(s) individually. The grounding electrode conductor shall be sized for the largest grounding electrode conductor required among all of the electrodes connected to it.

> **Exception:** Splicing of the grounding electrode conductor by irreversible compression-type connectors listed as grounding and bonding equipment or by the exothermic welding process shall not be prohibited.

E3610.2 Securing and protection against physical damage. Where exposed, a grounding electrode conductor or its enclosure shall be securely fastened to the surface on which it is carried. Grounding electrode conductors shall be permitted to be installed on or through framing members. A 4 AWG or larger conductor shall be protected where exposed to physical damage. A 6 AWG grounding conductor that is free from exposure to physical damage shall be permitted to be run along the surface of the building construction without metal covering or protection where it is and securely fastened to the construction; otherwise, it shall be in rigid metal conduit, intermediate metal conduit, rigid polyvinyl chloride (PVC), nonmetallic conduit, reinforced thermosetting resin (RTRC) nonmetallic conduit, electrical metallic tubing or cable armor. Grounding electrode conductors smaller than 6 AWG shall be in rigid metal conduit, intermediate metal conduit, rigid polyvinyl chloride (PVC) nonmetallic conduit, reinforced thermoseting resin (RTRC) nonmetallic conduit, electrical metallic tubing or cable armor.

Bare aluminum or copper-clad aluminum grounding electrode conductors shall not be used where in direct contact with masonry or the earth or where subject to corrosive conditions. Where used outside, aluminum or copper-clad aluminum grounding electrode conductors shall not be installed within 18 inches (457 mm) of the earth.

E3610.3 Enclosures for grounding electrode conductors. Ferrous metal enclosures for grounding electrode conductors shall be electrically continuous from the point of attachment to cabinets or equipment to the grounding electrode, and shall be securely fastened to the ground clamp or fitting. Nonferrous metal enclosures shall not be required to be electrically continuous. Ferrous metal enclosures that are not physically continuous from cabinet or equipment to the grounding electrode shall be made electrically continuous by bonding each end to the grounding electrode conductor. Bonding methods in compliance with Section E3609.4 for installations at service equipment locations and with E3609.4.2(B)(2) through E3609.4.4 for other than service equipment locations shall apply at each end and to all intervening ferrous raceways, boxes, and enclosures between the cabinets or equipment and the grounding electrode. The bonding jumper for a grounding electrode conductor raceway shall be the same size or larger than the required enclosed grounding electrode conductor.

Where a raceway is used as protection for a grounding conductor, the installation shall comply with the requirements of Chapter 38.

E3610.4 Prohibited use. An equipment grounding conductor shall not be used as a grounding electrode conductor.

SECTION E3611 GROUNDING ELECTRODE CONDUCTOR CONNECTION TO THE GROUNDING ELECTRODES

E3611.1 Methods of grounding conductor connection to electrodes. The grounding or bonding conductor shall be connected to the grounding electrode by exothermic welding, listed lugs, listed pressure connectors, listed clamps or other listed means. Connections depending on solder shall not be used. Ground clamps shall be listed for the materials of the grounding electrode and the grounding electrode conductor and, where used on pipe, rod or other buried electrodes, shall also be listed for direct soil burial or concrete encasement. Not more than one conductor shall be connected to the grounding electrode by a single clamp or fitting unless the clamp or fitting is listed for multiple conductors. One of the methods indicated in the following items shall be used:

1. A pipe fitting, pipe plug or other approved device screwed into a pipe or pipe fitting.
2. A listed bolted clamp of cast bronze or brass, or plain or malleable iron.
3. For indoor communications purposes only, a listed sheet metal strap-type ground clamp having a rigid metal base that seats on the electrode and having a strap of such material and dimensions that it is not likely to stretch during or after installation.
4. Other equally substantial approved means.

E3611.2 Accessibility. All mechanical elements used to terminate a grounding electrode conductor or bonding jumper to the grounding electrodes that are not buried or concrete encased shall be accessible.

E3611.3 Effective grounding path. The connection of the grounding electrode conductor or bonding jumper shall be made in a manner that will ensure a permanent and effective grounding path. Where necessary to ensure effective grounding for a metal piping system used as a grounding electrode, effective bonding shall be provided around insulated joints and sections and around any equipment that is likely to be disconnected for repairs or replacement. Bonding jumpers shall be of sufficient length to permit removal of such equipment while retaining the integrity of the grounding path.

E3611.4 Interior metal water piping. Where grounding electrode conductors and bonding jumpers are connected to interior metal water piping as a means to extend the grounding electrode conductor connection to an electrode(s), such piping shall be located not more than 5 feet (1524 mm) from the point of entry into the building.

Where interior metal water piping is used as a conductor to interconnect electrodes that are part of the grounding electrode system, such piping shall be located not more than 5 feet (1524 mm) from the point of entry into the building.

E3611.5 Protection of ground clamps and fittings. Ground clamps or other fittings shall be approved for applications without protection or shall be protected from physical damage by installing them where they are not likely to be damaged or by enclosing them in metal, wood or equivalent protective coverings.

E3611.6 Clean surfaces. Nonconductive coatings (such as paint, enamel and lacquer) on equipment to be grounded shall be removed from threads and other contact surfaces to ensure good electrical continuity or shall be connected by fittings that make such removal unnecessary.

CHAPTER 37

BRANCH CIRCUIT AND FEEDER REQUIREMENTS

SECTION E3701 GENERAL

E3701.1 Scope. This chapter covers branch circuits and feeders and specifies the minimum required branch circuits, the allowable loads and the required overcurrent protection for branch circuits and feeders that serve less than 100 percent of the total dwelling unit load. Feeder circuits that serve 100 percent of the dwelling unit load shall be sized in accordance with the procedures in Chapter 36.

E3701.2 Branch-circuit and feeder ampacity. Branch-circuit and feeder conductors shall have ampacities not less than the maximum load to be served. Where a branch circuit or a feeder supplies continuous loads or any combination of continuous and noncontinuous loads, the minimum branch-circuit or feeder conductor size, before the application of any adjustment or correction factors, shall have an allowable ampacity equal to or greater than the noncontinuous load plus 125 percent of the continuous load.

> **Exception:** The grounded conductors of feeders that are not connected to an overcurrent device shall be permitted to be sized at 100 percent of the continuous and noncontinuous load.

E3701.3 Selection of ampacity. Where more than one calculated or tabulated ampacity could apply for a given circuit length, the lowest value shall be used.

> **Exception:** Where two different ampacities apply to adjacent portions of a circuit, the higher ampacity shall be permitted to be used beyond the point of transition, a distance equal to 10 feet (3048 mm) or 10 percent of the circuit length figured at the higher ampacity, whichever is less.

E3701.4 Branch circuits with more than one receptacle. Conductors of branch circuits supplying more than one receptacle for cord-and-plug-connected portable loads shall have ampacities of not less than the rating of the branch circuit.

E3701.5 Multiwire branch circuits. All conductors for multiwire branch circuits shall originate from the same panelboard or similar distribution equipment. Except where all ungrounded conductors are opened simultaneously by the branch-circuit overcurrent device, multiwire branch circuits shall supply only line-to-neutral loads or only one appliance.

E3701.5.1 Disconnecting means. Each multiwire branch circuit shall be provided with a means that will simultaneously disconnect all ungrounded conductors at the point where the branch circuit originates.

E3701.5.2 Grouping. The ungrounded and grounded circuit conductors of each multiwire branch circuit shall be grouped by cable ties or similar means in at least one location within the panelboard or other point of origination.

> **Exception:** Grouping shall not be required where the circuit conductors enter from a cable or raceway unique to the circuit, thereby making the grouping obvious.

SECTION E3702 BRANCH CIRCUIT RATINGS

E3702.1 Branch-circuit voltage limitations. The voltage ratings of branch circuits that supply luminaires or receptacles for cord-and-plug-connected loads of up to 1,400 volt-amperes or of less than $^1/_4$ horsepower (0.186 kW) shall be limited to a maximum rating of 120 volts, nominal, between conductors.

Branch circuits that supply cord-and-plug-connected or permanently connected utilization equipment and appliances rated at over 1,440 volt-amperes or $^1/_4$ horsepower (0.186 kW) and greater shall be rated at 120 volts or 240 volts, nominal.

E3702.2 Branch-circuit ampere rating. Branch circuits shall be rated in accordance with the maximum allowable ampere rating or setting of the overcurrent protection device. The rating for other than individual branch circuits shall be 15, 20, 30, 40 and 50 amperes. Where conductors of higher ampacity are used, the ampere rating or setting of the specified over-current device shall determine the circuit rating.

E3702.3 Fifteen- and 20-ampere branch circuits. A 15- or 20-ampere branch circuit shall be permitted to supply lighting units, or other utilization equipment, or a combination of both. The rating of any one cord-and-plug-connected utilization equipment not fastened in place shall not exceed 80 percent of the branch-circuit ampere rating. The total rating of utilization equipment fastened in place, other than luminaires, shall not exceed 50 percent of the branch-circuit ampere rating where lighting units, cord-and-plug-connected utilization equipment not fastened in place, or both, are also supplied.

E3702.4 Thirty-ampere branch circuits. A 30-ampere branch circuit shall be permitted to supply fixed utilization equipment. A rating of any one cord-and-plug-connected utilization equipment shall not exceed 80 percent of the branch-circuit ampere rating.

E3702.5 Branch circuits serving multiple loads or outlets. General-purpose branch circuits shall supply lighting outlets, appliances, equipment or receptacle outlets, and combinations of such. Multi-outlet branch circuits serving lighting or receptacles shall be limited to a maximum branch-circuit rating of 20 amperes.

E3702.6 Branch circuits serving a single motor. Branch-circuit conductors supplying a single motor shall have an ampacity not less than 125 percent of the motor full-load current rating.

E3702.7 Branch circuits serving motor-operated and combination loads. For circuits supplying loads consisting of motor-operated utilization equipment that is fastened in place and that has a motor larger than $^1/_8$ horsepower (0.093 kW) in combination with other loads, the total calculated load shall be based on 125 percent of the largest motor load plus the sum of the other loads.

E3702.8 Branch-circuit inductive and LED lighting loads. For circuits supplying luminaires having ballasts or LED drivers, the calculated load shall be based on the total ampere ratings of such units and not on the total watts of the lamps.

E3702.9 Branch-circuit load for ranges and cooking appliances. It shall be permissible to calculate the branch-circuit load for one range in accordance with Table E3704.2(2). The branch-circuit load for one wall-mounted oven or one counter-mounted cooking unit shall be the nameplate rating of the appliance. The branch-circuit load for a counter-mounted cooking unit and not more than two wall-mounted ovens all supplied from a single branch circuit and located in the same room shall be calculated by adding the nameplate ratings of the individual appliances and treating the total as equivalent to one range.

E3702.9.1 Minimum branch circuit for ranges. Ranges with a rating of 8.75 kVA or more shall be supplied by a branch circuit having a minimum rating of 40 amperes.

E3702.10 Branch circuits serving heating loads. Electric space-heating and water-heating appliances shall be considered continuous loads. Branch circuits supplying two or more outlets for fixed electric space-heating equipment shall be rated 15, 20, 25 or 30 amperes.

E3702.11 Branch circuits for air-conditioning and heat pump equipment. The ampacity of the conductors supplying multimotor and combination load equipment shall not be less than the minimum circuit ampacity marked on the equipment. The branch-circuit overcurrent device rating shall be the size and type marked on the appliance.

E3702.12 Branch circuits serving room air conditioners. A room air conditioner shall be considered as a single motor unit in determining its branch-circuit requirements where all the following conditions are met:

1. It is cord- and attachment plug-connected.
2. The rating is not more than 40 amperes and 250 volts; single phase.
3. Total rated-load current is shown on the room air-conditioner nameplate rather than individual motor currents.
4. The rating of the branch-circuit short-circuit and ground-fault protective device does not exceed the ampacity of the branch-circuit conductors, or the rating of the branch-circuit conductors, or the rating of the receptacle, whichever is less.

E3702.12.1 Where no other loads are supplied. The total marked rating of a cord- and attachment plug-connected room air conditioner shall not exceed 80 percent of the rating of a branch circuit where no other appliances are also supplied.

E3702.12.2 Where lighting units or other appliances are also supplied. The total marked rating of a cord- and attachment plug-connected room air conditioner shall not exceed 50 percent of the rating of a branch circuit where lighting or other appliances are also supplied. Where the circuitry is interlocked to prevent simultaneous operation of the room air conditioner and energization of other outlets on the same branch circuit, a cord- and attachment-plug-connected room air conditioner shall not exceed 80 percent of the branch-circuit rating.

E3702.13 Branch-circuit requirement—summary. The requirements for circuits having two or more outlets, or receptacles, other than the receptacle circuits of Sections E3703.2, E3703.3 and E3703.4, are summarized in Table E3702.13. Branch circuits in dwelling units shall supply only loads within that dwelling unit or loads associated only with that dwelling unit. Branch circuits installed for the purpose of lighting, central alarm, signal, communications or other purposes for public or common areas of a two-family dwelling shall not be supplied from equipment that supplies an individual dwelling unit.

TABLE E3702.13
BRANCH-CIRCUIT REQUIREMENTS-SUMMARY[a, b]

	CIRCUIT RATING		
	15 amp	20 amp	30 amp
Conductors: Minimum size (AWG) circuit conductors	14	12	10
Maximum overcurrent-protection device rating Ampere rating	15	20	30
Outlet devices: Lampholders permitted Receptacle rating (amperes)	Any type 15 maximum	Any type 15 or 20	N/A 30
Maximum load (amperes)	15	20	30

a. These gages are for copper conductors.
b. N/A means not allowed.

SECTION E3703 REQUIRED BRANCH CIRCUITS

E3703.1 Branch circuits for heating. Central heating equipment other than fixed electric space heating shall be supplied by an individual branch circuit. Permanently connected air-conditioning equipment, and auxiliary equipment directly associated with the central heating equipment such as pumps, motorized valves, humidifiers and electrostatic air cleaners, shall not be prohibited from connecting to the same branch circuit as the central heating equipment.

E3703.2 Kitchen and dining area receptacles. A minimum of two 20-ampere-rated branch circuits shall be provided to serve all wall and floor receptacle outlets located in the kitchen, pantry, breakfast area, dining area or similar area of a dwelling. The kitchen countertop receptacles shall be served by a minimum of two 20-ampere-rated branch circuits, either or both of which shall also be permitted to supply other receptacle outlets in the same kitchen, pantry, breakfast and dining area including receptacle outlets for refrigeration appliances.

Exception: The receptacle outlet for refrigeration appliances shall be permitted to be supplied from an individual branch circuit rated 15 amperes or greater.

E3703.3 Laundry circuit. A minimum of one 20-ampere-rated branch circuit shall be provided for receptacles located in the laundry area and shall serve only receptacle outlets located in the laundry area.

E3703.4 Bathroom branch circuits. A minimum of one 20-ampere branch circuit shall be provided to supply bathroom receptacle outlet(s). Such circuits shall have no other outlets.

> **Exception:** Where the 20-ampere circuit supplies a single bathroom, outlets for other equipment within the same bathroom shall be permitted to be supplied in accordance with Section E3702.

E3703.5 Number of branch circuits. The minimum number of branch circuits shall be determined from the total calculated load and the size or rating of the circuits used. The number of circuits shall be sufficient to supply the load served. In no case shall the load on any circuit exceed the maximum specified by Section E3702.

E3703.6 Branch-circuit load proportioning. Where the branch-circuit load is calculated on a volt-amperes-per-square-foot (m^2) basis, the wiring system, up to and including the branch-circuit panelboard(s), shall have the capacity to serve not less than the calculated load. This load shall be evenly proportioned among multioutlet branch circuits within the panelboard(s). Branch-circuit overcurrent devices and circuits shall only be required to be installed to serve the connected load.

SECTION E3704 FEEDER REQUIREMENTS

E3704.1 Conductor size. Feeder conductors that do not serve 100 percent of the dwelling unit load and branch-circuit conductors shall be of a size sufficient to carry the load as determined by this chapter. Feeder conductors shall not be required to be larger than the service-entrance conductors that supply the dwelling unit. The load for feeder conductors that serve as the main power feeder to a dwelling unit shall be determined as specified in Chapter 36 for services.

E3704.2 Feeder loads. The minimum load in volt-amperes shall be calculated in accordance with the load calculation procedure prescribed in Table E3704.2(1). The associated table demand factors shall be applied to the actual load to determine the minimum load for feeders.

E3704.3 Feeder neutral load. The feeder neutral load shall be the maximum unbalance of the load determined in accordance with this chapter. The maximum unbalanced load shall be the maximum net calculated load between the neutral and any one ungrounded conductor. For a feeder or service supplying electric ranges, wall-mounted ovens, counter-mounted cooking units and electric dryers, the maximum unbalanced load shall be considered as 70 percent of the load on the ungrounded conductors.

E3704.4 Lighting and general use receptacle load. A unit load of not less than 3 volt-amperes shall constitute the minimum lighting and general use receptacle load for each square foot of floor area (33 VA for each square meter of floor area). The floor area for each floor shall be calculated from the outside dimensions of the building. The calculated floor area shall not include open porches, garages, or unused or unfinished spaces not adaptable for future use.

E3704.5 Ampacity and calculated loads. The calculated load of a feeder shall be not less than the sum of the loads on the branch circuits supplied, as determined by Section E3704, after any applicable demand factors permitted by Section E3704 have been applied.

TABLE E3704.2(1) FEEDER LOAD CALCULATION

LOAD CALCULATION PROCEDURE	APPLIED DEMAND FACTOR
Lighting and receptacles: A unit load of not less than 3 VA per square foot of total floor area shall constitute the lighting and 120-volt, 15- and 20-ampere general use receptacle load. 1,500 VA shall be added for each 20-ampere branch circuit serving receptacles in the kitchen, dining room, pantry, breakfast area and laundry area.	100 percent of first 3,000 VA or less and 35 percent of that in excess of 3,000 VA.
Plus	
Appliances and motors: The nameplate rating load of all fastened-in-place appliances other than dryers, ranges, air-conditioning and space-heating equipment.	100 percent of load for three or less appliances. 75 percent of load for four or more appliances.
Plus	
Fixed motors: Full-load current of motors plus 25 percent of the full load current of the largest motor.	
Plus	
Electric clothes dryer: The dryer load shall be 5,000 VA for each dryer circuit or the nameplate rating load of each dryer, whichever is greater.	
Plus	
Cooking appliances: The nameplate rating of ranges, wall-mounted ovens, counter-mounted cooking units and other cooking appliances rated in excess of 1.75 kVA shall be summed.	Demand factors shall be as allowed by Table E3704.2(2).
Plus the largest of either the heating or cooling load	
Largest of the following two selections: 1. 100 percent of the nameplate rating(s) of the air conditioning and cooling, including heat pump compressors. 2. 100 percent of the fixed electric space heating.	

For SI: 1 square foot = $0.0929\ m^2$.

TABLE E3704.2(2)
DEMAND LOADS FOR ELECTRIC RANGES, WALL-MOUNTED OVENS, COUNTER-MOUNTED COOKING UNITS AND OTHER COOKING APPLIANCES OVER $1^3/_4$ kVA RATING[a, b]

NUMBER OF APPLIANCES	MAXIMUM DEMAND[b, c]	DEMAND FACTORS (percent)[d]	
	Column A maximum 12 kVA rating	Column B less than $3^1/_2$ kVA rating	Column C $3^1/_2$ to $8^3/_4$ kVA rating
1	8 kVA	80	80
2	11 kVA	75	65

a. Column A shall be used in all cases except as provided for in Footnote d.

b. For ranges all having the same rating and individually rated more than 12 kVA but not more than 27 kVA, the maximum demand in Column A shall be increased 5 percent for each additional kVA of rating or major fraction thereof by which the rating of individual ranges exceeds 12 kVA.

c. For ranges of unequal ratings and individually rated more than 8.75 kVA, but none exceeding 27 kVA, an average value of rating shall be computed by adding together the ratings of all ranges to obtain the total connected load (using 12 kVA for any ranges rated less than 12 kVA) and dividing by the total number of ranges; and then the maximum demand in Column A shall be increased 5 percent for each kVA or major fraction thereof by which this average value exceeds 12 kVA.

d. Over 1.75 kVA through 8.75 kVA. As an alternative to the method provided in Column A, the nameplate ratings of all ranges rated more than 1.75 kVA but not more than 8.75 kVA shall be added and the sum shall be multiplied by the demand factor specified in Column B or C for the given number of appliances.

E3704.6 Equipment grounding conductor. Where a feeder supplies branch circuits in which equipment grounding conductors are required, the feeder shall include or provide an equipment grounding conductor that is one or more or a combination of the types specified in Section E3908.8, to which the equipment grounding conductors of the branch circuits shall be connected. Where the feeder supplies a separate building or structure, the requirements of Section E3607.3.1 shall apply.

SECTION E3705 CONDUCTOR SIZING AND OVERCURRENT PROTECTION

E3705.1 General. Ampacities for conductors shall be determined based in accordance with Table E3705.1 and Sections E3705.2 and E3705.3.

E3705.2 Correction factor for ambient temperatures. For ambient temperatures other than 30°C (86°F), multiply the allowable ampacities specified in Table E3705.1 by the appropriate correction factor shown in Table E3705.2.

E3705.3 Adjustment factor for conductor proximity. Where the number of current-carrying conductors in a raceway or cable exceeds three, or where single conductors or multiconductor cables are stacked or bundled for distances greater than 24 inches (610 mm) without maintaining spacing and are not installed in raceways, the allowable ampacity of each conductor shall be reduced as shown in Table E3705.3.

Exceptions:

1. Adjustment factors shall not apply to conductors in nipples having a length not exceeding 24 inches (610 mm).
2. Adjustment factors shall not apply to underground conductors entering or leaving an outdoor trench if those conductors have physical protection in the form of rigid metal conduit, intermediate metal conduit, or rigid nonmetallic conduit having a length not exceeding 10 feet (3048 mm) and the number of conductors does not exceed four.
3. Adjustment factors shall not apply to type AC cable or to type MC cable without an overall outer jacket meeting all of the following conditions:
   - 3.1. Each cable has not more than three current-carrying conductors.
   - 3.2. The conductors are 12 AWG copper.
   - 3.3. Not more than 20 current-carrying conductors are bundled, stacked or supported on bridle rings.
4. An adjustment factor of 60 percent shall be applied to Type AC cable and Type MC cable where all of the following conditions apply:
   - 4.1. The cables do not have an overall outer jacket.
   - 4.2. The number of current-carrying conductors exceeds 20.
   - 4.3. The cables are stacked or bundled longer than 24 inches (607 mm) without spacing being maintained.

TABLE E3705.3
CONDUCTOR PROXIMITY ADJUSTMENT FACTORS

NUMBER OF CURRENT-CARRYING CONDUCTORS IN CABLE OR RACEWAY	PERCENT OF VALUES IN TABLE E3705.1
4-6	80
7-9	70
10-20	50
21-30	45
31-40	40
41 and above	35

TABLE E3705.1
ALLOWABLE AMPACITIES

CONDUCTOR SIZE	CONDUCTOR TEMPERATURE RATING						CONDUCTOR SIZE
	60°C	75°C	90°C	60°C	75°C	90°C	
AWG kcmil	Types TW, UF	Types RHW, THHW, THW, THWN, USE, XHHW	Types RHW-2, THHN, THHW, THW-2, THWN-2, XHHW, XHHW-2, USE-2	Types TW, UF	Types RHW, THHW, THW, THWN, USE, XHHW	Types RHW-2, THHN, THHW, THW-2, THWN-2, XHHW, XHHW-2, USE-2	AWG kcmil
	Copper			Aluminum or copper-clad aluminum			
14[a]	15	20	25	—	—	—	—
12[a]	20	25	30	15	20	25	12[a]
10[a]	30	35	40	25	30	35	10[a]
8	40	50	55	35	40	45	8
6	55	65	75	40	50	55	6
4	70	85	95	55	65	75	4
3	85	100	115	65	75	85	3
2	95	115	130	75	90	100	2
1	110	130	145	85	100	115	1
1/0	125	150	170	100	120	135	1/0
2/0	145	175	195	115	135	150	2/0
3/0	165	200	225	130	155	175	3/0
4/0	195	230	260	150	180	205	4/0

For SI: °C = [(°F) - 32]/1.8.

a. See Table E3705.5.3 for conductor overcurrent protection limitations.

TABLE E3705.2
AMBIENT TEMPERATURE CORRECTION FACTORS

For ambient temperatures other than 30°C (86°F), multiply the allowable ampacities specified in the ampacity tables by the appropriate correction factor shown below.				
Ambient Temperature (°C)	Temperature Rating of Conductor			Ambient Temperature (°F)
	60°C	75°C	90°C	
10 or less	1.29	1.20	1.15	50 or less
11-15	1.22	1.15	1.12	51-59
16-20	1.15	1.11	1.08	60-68
21-25	1.08	1.05	1.04	69-77
26-30	1.00	1.00	1.00	78-86
31-35	0.91	0.94	0.96	87-95
36-40	0.82	0.88	0.91	96-104
41-45	0.71	0.82	0.87	105-113
46-50	0.58	0.75	0.82	114-122
51-55	0.41	0.67	0.76	123-131
56-60	—	0.58	0.71	132-140
61-65	—	0.47	0.65	141-149
66-70	—	0.33	0.58	150-158
71-75	—	—	0.50	159-167
76-80	—	—	0.41	168-176
81-85	—	—	0.29	177-185

For SI: 1 °C = [(°F) - 32]/1.8.

E3705.4 Temperature limitations. The temperature rating associated with the ampacity of a conductor shall be so selected and coordinated to not exceed the lowest temperature rating of any connected termination, conductor or device. Conductors with temperature ratings higher than specified for terminations shall be permitted to be used for ampacity adjustment, correction, or both. Except where the equipment is marked otherwise, conductor ampacities used in determining equipment termination provisions shall be based on Table E3705.1.

E3705.4.1 Conductors rated 60°C. Except where the equipment is marked otherwise, termination provisions of equipment for circuits rated 100 amperes or less, or marked for 14 AWG through 1 AWG conductors, shall be used only for one of the following:

1. Conductors rated 60°C (140°F);
2. Conductors with higher temperature ratings, provided that the ampacity of such conductors is determined based on the 60°C (140°F) ampacity of the conductor size used;
3. Conductors with higher temperature ratings where the equipment is listed and identified for use with such conductors; or
4. For motors marked with design letters B, C, or D conductors having an insulation rating of 75°C (167°F) or higher shall be permitted to be used provided that the ampacity of such conductors does not exceed the 75°C (167°F) ampacity.

E3705.4.2 Conductors rated 75°C. Termination provisions of equipment for circuits rated over 100 amperes, or marked for conductors larger than 1 AWG, shall be used only for:

1. Conductors rated 75°C (167°F).
2. Conductors with higher temperature ratings provided that the ampacity of such conductors does not exceed the 75°C (167°F) ampacity of the conductor size used, or provided that the equipment is listed and identified for use with such conductors.

E3705.4.3 Separately installed pressure connectors. Separately installed pressure connectors shall be used with conductors at the ampacities not exceeding the ampacity at the listed and identified temperature rating of the connector.

E3705.4.4 Conductors of Type NM cable. Conductors in NM cable assemblies shall be rated at 90°C (194°F). Types NM, NMC, and NMS cable identified by the markings NM-B, NMC-B, and NMS-B meet this requirement. The allowable ampacity of Types NM, NMC, and NMS cable shall not exceed that of 60°C (140°F) rated conductors and shall comply with Section E3705.1 and Table E3705.5.3. The 90°C (194°F) rating shall be permitted to be used for ampacity adjustment and calculations provided that the final corrected or adjusted ampacity does not exceed that for a 60°C (140°F) rated conductor. Where more than two NM cables containing two or more current-carrying conductors are installed, without maintaining spacing between the cables, through the same opening in wood framing that is to be sealed with thermal insulation, caulk or sealing foam, the allowable ampacity of each conductor shall be adjusted in accordance with Table E3705.3. Where more than two NM cables containing two or more current-carrying conductors are installed in contact with thermal insulation without maintaining spacing between cables, the allowable ampacity of each conductor shall be adjusted in accordance with Table E3705.3.

E3705.5 Overcurrent protection required. All ungrounded branch-circuit and feeder conductors shall be protected against overcurrent by an overcurrent device installed at the point where the conductors receive their supply. Overcurrent devices shall not be connected in series with a grounded conductor. Overcurrent protection and allowable loads for branch circuits and feeders that do not serve as the main power feeder to the dwelling unit load shall be in accordance with this chapter.

Branch-circuit conductors and equipment shall be protected by overcurrent protective devices having a rating or setting not exceeding the allowable ampacity specified in Table E3705.1 and Sections E3705.2, E3705.3 and E3705.4 except where otherwise permitted or required in Sections E3705.5.1 through E3705.5.3.

E3705.5.1 Cords. Cords shall be protected in accordance with Section E3909.2.

E3705.5.2 Overcurrent devices of the next higher rating. The next higher standard overcurrent device rating, above the ampacity of the conductors being protected, shall be permitted to be used, provided that all of the following conditions are met:

1. The conductors being protected are not part of a branch circuit supplying more than one receptacle for cord- and plug-connected portable loads.
2. The ampacity of conductors does not correspond with the standard ampere rating of a fuse or a circuit breaker without overload trip adjustments above its rating (but that shall be permitted to have other trip or rating adjustments).
3. The next higher standard device rating does not exceed 400 amperes.

E3705.5.3 Small conductors. Except as specifically permitted by Section E3705.5.4, the rating of overcurrent protection devices shall not exceed the ratings shown in Table E3705.5.3 for the conductors specified therein.

TABLE E3705.5.3
OVERCURRENT-PROTECTION RATING

COPPER		ALUMINUM OR COPPER-CLAD ALUMINUM	
Size (AWG)	Maximum overcurrent-protection-device rating[a] (amps)	Size (AWG)	Maximum overcurrent-protection-device rating[a] (amps)
14	15	12	15
12	20	10	25
10	30	8	30

a. The maximum overcurrent-protection-device rating shall not exceed the conductor allowable ampacity determined by the application of the correction and adjustment factors in accordance with Sections E3705.2 and E3705.3.

E3705.5.4 Air-conditioning and heat pump equipment. Air-conditioning and heat pump equipment circuit conductors shall be permitted to be protected against overcurrent in accordance with Section E3702.11.

E3705.6 Fuses and fixed trip circuit breakers. The standard ampere ratings for fuses and inverse time circuit breakers shall be considered 15, 20, 25, 30, 35, 40, 45, 50, 60, 70, 80, 90, 100, 110, 125, 150, 175, 200, 225, 250, 300, 350 and 400 amperes.

E3705.7 Location of overcurrent devices in or on premises. Overcurrent devices shall:

1. Be readily accessible.
2. Not be located where they will be exposed to physical damage.
3. Not be located where they will be in the vicinity of easily ignitible material such as in clothes closets.
4. Not be located in bathrooms.
5. Not be located over steps of a stairway.
6. Be installed so that the center of the grip of the operating handle of the switch or circuit breaker, when in its highest position, is not more than 6 feet 7 inches (2007 mm) above the floor or working platform.

Exceptions:

1. This section shall not apply to supplementary overcurrent protection that is integral to utilization equipment.
2. Overcurrent devices installed adjacent to the utilization equipment that they supply shall be permitted to be accessible by portable means.

E3705.8 Ready access for occupants. Each occupant shall have ready access to all overcurrent devices protecting the conductors supplying that occupancy.

E3705.9 Enclosures for overcurrent devices. Overcurrent devices shall be enclosed in cabinets, cutout boxes, or equipment assemblies. The operating handle of a circuit breaker shall be permitted to be accessible without opening a door or cover.

SECTION E3706 PANELBOARDS

E3706.1 Panelboard rating. All panelboards shall have a rating not less than that of the minimum service or feeder capacity required for the calculated load.

E3706.2 Panelboard circuit identification. All circuits and circuit modifications shall be legibly identified as to their clear, evident, and specific purpose or use. The identification shall include sufficient detail to allow each circuit to be distinguished from all others. Spare positions that contain unused overcurrent devices or switches shall be described accordingly. The identification shall be included in a circuit directory located on the face of the panelboard enclosure or inside the panel door. Circuits shall not be described in a manner that depends on transient conditions of occupancy.

E3706.3 Panelboard overcurrent protection. In addition to the requirement of Section E3706.1, a panelboard shall be protected by an overcurrent protective device having a rating not greater than that of the panelboard. Such overcurrent protective device shall be located within or at any point on the supply side of the panelboard.

E3706.4 Grounded conductor terminations. Each grounded conductor shall terminate within the panelboard on an individual terminal that is not also used for another conductor, except that grounded conductors of circuits with parallel conductors shall be permitted to terminate on a single terminal where the terminal is identified for connection of more than one conductor.

E3706.5 Back-fed devices. Plug-in-type overcurrent protection devices or plug-in-type main lug assemblies that are back-fed and used to terminate field-installed ungrounded supply conductors shall be secured in place by an additional fastener that requires other than a pull to release the device from the mounting means on the panel.

CHAPTER 38

WIRING METHODS

SECTION E3801 GENERAL REQUIREMENTS

E3801.1 Scope. This chapter covers the wiring methods for services, feeders and branch circuits for electrical power and distribution.

E3801.2 Allowable wiring methods. The allowable wiring methods for electrical installations shall be those listed in Table E3801.2. Single conductors shall be used only where part of one of the recognized wiring methods listed in Table E3801.2. As used in this code, abbreviations of the wiring-method types shall be as indicated in Table E3801.2.

TABLE E3801.2 ALLOWABLE WIRING METHODS

ALLOWABLE WIRING METHOD	DESIGNATED ABBREVIATION
Armored cable	AC
Electrical metallic tubing	EMT
Electrical nonmetallic tubing	ENT
Flexible metal conduit	FMC
Intermediate metal conduit	IMC
Liquidtight flexible conduit	LFC
Metal-clad cable	MC
Nonmetallic sheathed cable	NM
Rigid polyvinyl chloride conduit (Type PVC)	RNC
Rigid metallic conduit	RMC
Service entrance cable	SE
Surface raceways	SR
Underground feeder cable	UF
Underground service cable	USE

E3801.3 Circuit conductors. All conductors of a circuit, including equipment grounding conductors and bonding conductors, shall be contained in the same raceway, trench, cable or cord.

E3801.4 Wiring method applications. Wiring methods shall be applied in accordance with Table E3801.4.

SECTION E3802 ABOVE-GROUND INSTALLATION REQUIREMENTS

E3802.1 Installation and support requirements. Wiring methods shall be installed and supported in accordance with Table E3802.1.

E3802.2 Cables in accessible attics. Cables in attics or roof spaces provided with access shall be installed as specified in Sections E3802.2.1 and E3802.2.2.

E3802.2.1 Across structural members. Where run across the top of floor joists, or run within 7 feet (2134 mm) of floor or floor joists across the face of rafters or studding, in attics and roof spaces that are provided with access, the cable shall be protected by substantial guard strips that are at least as high as the cable. Where such spaces are not provided with access by permanent stairs or ladders, protection shall only be required within 6 feet (1829 mm) of the nearest edge of the attic entrance.

E3802.2.2 Cable installed through or parallel to framing members. Where cables are installed through or parallel to the sides of rafters, studs or floor joists, guard strips and running boards shall not be required, and the installation shall comply with Table E3802.1.

E3802.3 Exposed cable. In exposed work, except as provided for in Sections E3802.2 and E3802.4, cable assemblies shall be installed as specified in Sections E3802.3.1 and E3802.3.2.

E3802.3.1 Surface installation. Cables shall closely follow the surface of the building finish or running boards.

E3802.3.2 Protection from physical damage. Where subject to physical damage, cables shall be protected by rigid metal conduit, intermediate metal conduit, electrical metallic tubing, Schedule 80 PVC rigid nonmetallic conduit, or other approved means. Where passing through a floor, the cable shall be enclosed in rigid metal conduit, intermediate metal conduit, electrical metallic tubing, Schedule 80 PVC rigid nonmetallic conduit or other approved means extending not less than 6 inches (152 mm) above the floor.

E3802.3.3 Locations exposed to direct sunlight. Insulated conductors and cables used where exposed to direct rays of the sun shall be listed or listed and marked, as being "sunlight resistant," or shall be covered with insulating material, such as tape or sleeving, that is listed or listed and marked as being "sunlight resistant."

E3802.4 In unfinished basements and crawl spaces. Where type NM or SE cable is run at angles with joists in unfinished basements and crawl spaces, cable assemblies containing two or more conductors of sizes 6 AWG and larger and assemblies containing three or more conductors of sizes 8 AWG and larger shall not require additional protection where attached directly to the bottom of the joists. Smaller cables shall be run either through bored holes in joists or on running boards. Type NM or SE cable installed on the wall of an unfinished basement shall be permitted to be installed in a listed conduit or tubing or shall be protected in accordance with Table E3802.1. Conduit or tubing shall be provided with a suitable insulating bushing or adapter at the point where the cable enters the raceway. The sheath of the Type NM or SE cable shall extend through the conduit or tubing and into the outlet or device box not less than $^1/_4$ inch (6.4 mm). The cable shall be secured within 12 inches (305 mm) of the point where the cable enters the conduit or tubing. Metal conduit, tubing, and metal outlet boxes shall be connected to an equipment grounding conductor complying with Section E3908.13.

E3802.5 Bends. Bends shall be made so as not to damage the wiring method or reduce the internal diameter of raceways.

For types NM and SE cable, bends shall be so made, and other handling shall be such that the cable will not be damaged and the radius of the curve of the inner edge of any bend shall be not less than five times the diameter of the cable.

E3802.6 Raceways exposed to different temperatures. Where portions of a raceway or sleeve are known to be subjected to different temperatures and where condensation is known to be a problem, as in cold storage areas of buildings or where passing from the interior to the exterior of a building, the raceway or sleeve shall be filled with an approved material to prevent the circulation of warm air to a colder section of the raceway or sleeve.

E3802.7 Raceways in wet locations above grade. Where raceways are installed in wet locations above grade, the interior of such raceways shall be considered to be a wet location. Insulated conductors and cables installed in raceways in wet locations above grade shall be listed for use in wet locations.

SECTION E3803
UNDERGROUND INSTALLATION REQUIREMENTS

E3803.1 Minimum cover requirements. Direct buried cable or raceways shall be installed in accordance with the minimum cover requirements of Table E3803.1.

E3803.2 Warning ribbon. Underground service conductors that are not encased in concrete and that are buried 18 inches (457 mm) or more below grade shall have their location identified by a warning ribbon that is placed in the trench not less than 12 inches (305 mm) above the underground installation.

TABLE E3801.4
ALLOWABLE APPLICATIONS FOR WIRING METHODS[a, b, c, d, e, f, g, h, i, j, k]

ALLOWABLE APPLICATIONS (application allowed where marked with an "A")	AC	EMT	ENT	FMC	IMC RMC RNC	LFC[a, g]	MC	NM	SR	SE	UF	USE
Services	—	A	A[h]	A[i]	A	A[i]	A	—	—	A	—	A
Feeders	A	A	A	A	A	A	A	A	—	A[b]	A	A[b]
Branch circuits	A	A	A	A	A	A	A	A	A	A[c]	A	—
Inside a building	A	A	A	A	A	A	A	A	A	A	A	—
Wet locations exposed to sunlight	—	A	A[h]	—	A	A	A	—	—	A	A[e]	A[e]
Damp locations	—	A	A	A[d]	A	A	A	—	—	A	A	A
Embedded in noncinder concrete in dry location	—	A	A	—	A	A[j]	—	—	—	—	—	—
In noncinder concrete in contact with grade	—	A[f]	A	—	A[f]	A[j]	—	—	—	—	—	—
Embedded in plaster not exposed to dampness	A	A	A	A	A	A	A	—	—	A	A	—
Embedded in masonry	—	A	A	—	A[f]	A	A	—	—	—	—	—
In masonry voids and cells exposed to dampness or below grade line	—	A[f]	A	A[d]	A[f]	A	A	—	—	A	A	—
Fished in masonry voids	A	—	—	A	—	A	A	A	—	A	A	—
In masonry voids and cells not exposed to dampness	A	A	A	A	A	A	A	A	—	A	A	—
Run exposed	A	A	A	A	A	A	A	A	A	A	A	—
Run exposed and subject to physical damage	—	—	—	—	A[g]	—	—	—	—	—	—	—
For direct burial	—	A[f]	—	—	A[f]	A	A[f]	—	—	—	A	A

For SI: 1 foot = 304.8 mm.

a. Liquid-tight flexible nonmetallic conduit without integral reinforcement within the conduit wall shall not exceed 6 feet in length.

b. Type USE cable shall not be used inside buildings.

c. The grounded conductor shall be insulated.

d. Conductors shall be a type approved for wet locations and the installation shall prevent water from entering other raceways.

e. Shall be listed as "Sunlight Resistant."

f. Metal raceways shall be protected from corrosion and approved for the application. Aluminum RMC requires approved supplementary corrosion protection.

g. RNC shall be Schedule 80.

h. Shall be listed as "Sunlight Resistant" where exposed to the direct rays of the sun.

i. Conduit shall not exceed 6 feet in length.

j. Liquid-tight flexible nonmetallic conduit is permitted to be encased in concrete where listed for direct burial and only straight connectors listed for use with LFNC are used.

k. In wet locations under any of the following conditions:

1. The metallic covering is impervious to moisture.
2. A lead sheath or moisture-impervious jacket is provided under the metal covering.
3. The insulated conductors under the metallic covering are listed for use in wet locations and a corrosion-resistant jacket is provided over the metallic sheath.

TABLE E3802.1
GENERAL INSTALLATION AND SUPPORT REQUIREMENTS FOR WIRING METHODS[a, b, c, d, e, f, g, h, i, j, k]

INSTALLATION REQUIREMENTS (Requirement applicable only to wiring methods marked "A")	AC MC	EMT IMC RMC	ENT	FMC LFC	NM UF	RNC	SE	SR[a]	USE
Where run parallel with the framing member or furring strip, the wiring shall be not less than $1\frac{1}{4}$ inches from the edge of a furring strip or a framing member such as a joist, rafter or stud or shall be physically protected.	A	—	A	A	A	—	A	—	—
Bored holes in framing members for wiring shall be located not less than $1\frac{1}{4}$ inches from the edge of the framing member or shall be protected with a minimum 0.0625-inch steel plate or sleeve, a listed steel plate or other physical protection.	A[k]	—	A[k]	A[k]	A[k]	—	A[k]	—	—
Where installed in grooves, to be covered by wallboard, siding, paneling, carpeting, or similar finish, wiring methods shall be protected by 0.0625-inch-thick steel plate, sleeve, or equivalent, a listed steel plate or by not less than $1\frac{1}{4}$-inch free space for the full length of the groove in which the cable or raceway is installed.	A	—	A	A	A	—	A	A	A
Securely fastened bushings or grommets shall be provided to protect wiring run through openings in metal framing members.	—	—	A[j]	—	A[j]	—	A[j]	—	—
The maximum number of 90-degree bends shall not exceed four between junction boxes.	—	A	A	A	—	A	—	—	—
Bushings shall be provided where entering a box, fitting or enclosure unless the box or fitting is designed to afford equivalent protection.	A	A	A	A	—	A	—	A	—
Ends of raceways shall be reamed to remove rough edges.	—	A	A	A	—	A	—	A	—
Maximum allowable on center support spacing for the wiring method in feet.	4.5[b, c]	10[l]	3[b]	4.5[b]	4.5[i]	3[d, l]	2.5[e]	—	2.5
Maximum support distance in inches from box or other terminations.	12[b, f]	36	36	12[b, g]	12[h, i]	36	12	—	—

For SI: 1 inch = 25.4 mm, 1 foot = 304.8 mm, 1 degree = 0.0175 rad.

a. Installed in accordance with listing requirements.

b. Supports not required in accessible ceiling spaces between light fixtures where lengths do not exceed 6 feet.

c. Six feet for MC cable.

d. Five feet for trade sizes greater than 1 inch.

e. Two and one-half feet where used for service or outdoor feeder and 4.5 feet where used for branch circuit or indoor feeder.

f. Twenty-four inches where flexibility is necessary.

g. Where flexibility after installation is necessary, lengths of flexible metal conduit and liquidtight flexible metal conduit measured from the last point where the raceway is securely fastened shall not exceed: 36 inches for trade sizes $\frac{1}{2}$ through $1\frac{1}{4}$, 48 inches for trade sizes $1\frac{1}{2}$ through 2 and 5 feet for trade sizes $2\frac{1}{2}$ and larger.

h. Within 8 inches of boxes without cable clamps.

i. Flat cables shall not be stapled on edge.

j. Bushings and grommets shall remain in place and shall be listed for the purpose of cable protection.

k. See Sections R502.8 and R802.7 for additional limitations on the location of bored holes in horizontal framing members.

E3803.3 Protection from damage. Direct buried conductors and cables emerging from the ground shall be protected by enclosures or raceways extending from the minimum cover distance below grade required by Section E3803.1 to a point at least 8 feet (2438 mm) above finished grade. In no case shall the protection be required to exceed 18 inches (457 mm) below finished grade. Conductors entering a building shall be protected to the point of entrance. Where the enclosure or raceway is subject to physical damage, the conductors shall be installed in rigid metal conduit, intermediate metal conduit, Schedule 80 rigid nonmetallic conduit or the equivalent.

E3803.4 Splices and taps. Direct buried conductors or cables shall be permitted to be spliced or tapped without the use of splice boxes. The splices or taps shall be made by approved methods with materials listed for the application.

E3803.5 Backfill. Backfill containing large rock, paving materials, cinders, large or sharply angular substances, or corrosive material shall not be placed in an excavation where such materials cause damage to raceways, cables or other substructures or prevent adequate compaction of fill or contribute to corrosion of raceways, cables or other substructures. Where necessary to prevent physical damage to the raceway or cable, protection shall be provided in the form of granular or selected material, suitable boards, suitable sleeves or other approved means.

E3803.6 Raceway seals. Conduits or raceways shall be sealed or plugged at either or both ends where moisture will enter and contact live parts.

E3803.7 Bushing. A bushing, or terminal fitting, with an integral bushed opening shall be installed on the end of a conduit or other raceway that terminates underground where the conductors or cables emerge as a direct burial wiring method. A seal incorporating the physical protection characteristics of a bushing shall be considered equivalent to a bushing.

E3803.8 Single conductors. All conductors of the same circuit and, where present, the grounded conductor and all equipment grounding conductors shall be installed in the same raceway or shall be installed in close proximity in the same trench.

Exception: Conductors shall be permitted to be installed in parallel in raceways, multiconductor cables, and direct-buried single conductor cables. Each raceway or multiconductor cable shall contain all conductors of the same circuit, including equipment grounding conductors. Each direct-buried single conductor cable shall be located in close proximity in the trench to the other single conductor cables in the same parallel set of conductors in the circuit, including equipment grounding conductors.

E3803.9 Ground movement. Where direct buried conductors, raceways or cables are subject to movement by settlement or frost, direct buried conductors, raceways or cables shall be arranged to prevent damage to the enclosed conductors or to equipment connected to the raceways.

E3803.10 Wet locations. The interior of enclosures or raceways installed underground shall be considered to be a wet location. Insulated conductors and cables installed in such enclosures or raceways in underground installations shall be listed for use in wet locations. Connections or splices in an underground installation shall be approved for wet locations.

E3803.11 Under buildings. Underground cable installed under a building shall be in a raceway.

Exception: Type MC Cable shall be permitted under a building without installation in a raceway where the cable

TABLE E3803.1
MINIMUM COVER REQUIREMENTS, BURIAL IN INCHES[a, b, c, d, e]

LOCATION OF WIRING METHOD OR CIRCUIT	TYPE OF WIRING METHOD OR CIRCUIT				
	1 Direct burial cables or conductors	2 Rigid metal conduit or intermediate metal conduit	3 Nonmetallic raceways listed for direct burial without concrete encasement or other approved raceways	4 Residential branch circuits rated 120 volts or less with GFCI protection and maximum overcurrent protection of 20 amperes	5 Circuits for control of irrigation and landscape lighting limited to not more than 30 volts and installed with type UF or in other identified cable or raceway
All locations not specified below	24	6	18	12	6
In trench below 2-inch-thick concrete or equivalent	18	6	12	6	6
Under a building	0 (In raceway only or Type MC identified for direct burial)	0	0	0 (In raceway only or Type MC identified for direct burial)	0 (In raceway only or Type MC identified for direct burial)
Under minimum of 4-inch-thick concrete exterior slab with no vehicular traffic and the slab extending not less than 6 inches beyond the underground installation	18	4	4	6 (Direct burial) 4 (In raceway)	6 (Direct burial) 4 (In raceway)
Under streets, highways, roads, alleys, driveways and parking lots	24	24	24	24	24
One- and two-family dwelling driveways and outdoor parking areas, and used only for dwelling-related purposes	18	18	18	12	18
In solid rock where covered by minimum of 2 inches concrete extending down to rock	2 (In raceway only)	2	2	2 (In raceway only)	2 (In raceway only)

For SI: 1 inch = 25.4 mm.

a. Raceways approved for burial only where encased concrete shall require concrete envelope not less than 2 inches thick.

b. Lesser depths shall be permitted where cables and conductors rise for terminations or splices or where access is otherwise required.

c. Where one of the wiring method types listed in columns 1 to 3 is combined with one of the circuit types in columns 4 and 5, the shallower depth of burial shall be permitted.

d. Where solid rock prevents compliance with the cover depths specified in this table, the wiring shall be installed in metal or nonmetallic raceway permitted for direct burial. The raceways shall be covered by a minimum of 2 inches of concrete extending down to the rock.

e. Cover is defined as the shortest distance in inches (millimeters) measured between a point on the top surface of any direct-buried conductor, cable, conduit or other raceway and the top surface of finished grade, concrete, or similar cover.

is listed and identified for direct burial or concrete encasement and one or more of the following applies:

1. The metallic covering is impervious to moisture.
2. A moisture-impervious jacket is provided under the metal covering.
3. The insulated conductors under the metallic covering are listed for use in wet locations, and a corrosion-resistant jacket is provided over the metallic sheath.

CHAPTER 39

POWER AND LIGHTING DISTRIBUTION

SECTION E3901 RECEPTACLE OUTLETS

E3901.1 General. Outlets for receptacles rated at 125 volts, 15- and 20-amperes shall be provided in accordance with Sections E3901.2 through E3901.11. Receptacle outlets required by this section shall be in addition to any receptacle that is:

1. Part of a luminaire or appliance;
2. Located within cabinets or cupboards;
3. Controlled by a wall switch in accordance with Section E3903.2, Exception 1; or
4. Located over 5.5 feet (1676 mm) above the floor.

Permanently installed electric baseboard heaters equipped with factory-installed receptacle outlets, or outlets provided as a separate assembly by the baseboard manufacturer shall be permitted as the required outlet or outlets for the wall space utilized by such permanently installed heaters. Such receptacle outlets shall not be connected to the heater circuits.

E3901.2 General purpose receptacle distribution. In every kitchen, family room, dining room, living room, parlor, library, den, sun room, bedroom, recreation room, or similar room or area of dwelling units, receptacle outlets shall be installed in accordance with the general provisions specified in Sections E3901.2.1 through E3901.2.3 (see Figure E3901.2).

E3901.2.1 Spacing. Receptacles shall be installed so that no point measured horizontally along the floor line of any wall space is more than 6 feet (1829 mm), from a receptacle outlet.

E3901.2.2 Wall space. As used in this section, a wall space shall include the following:

1. Any space that is 2 feet (610 mm) or more in width, including space measured around corners, and that is unbroken along the floor line by doorways and similar openings, fireplaces, and fixed cabinets.
2. The space occupied by fixed panels in exterior walls, excluding sliding panels.
3. The space created by fixed room dividers such as railings and freestanding bar-type counters.

E3901.2.3 Floor receptacles. Receptacle outlets in floors shall not be counted as part of the required number of receptacle outlets except where located within 18 inches (457 mm) of the wall.

E3901.2.4 Countertop receptacles. Receptacles installed for countertop surfaces as specified in Section E3901.4 shall not be considered as the receptacles required by Section E3901.2.

E3901.3 Small appliance receptacles. In the kitchen, pantry, breakfast room, dining room, or similar area of a dwelling unit, the two or more 20-ampere small-appliance branch circuits required by Section E3703.2, shall serve all wall and floor receptacle outlets covered by Sections E3901.2 and E3901.4 and those receptacle outlets provided for refrigeration appliances.

Exceptions:

1. In addition to the required receptacles specified by Sections E3901.1 and E3901.2, switched receptacles supplied from a general-purpose branch circuit as defined in Section E3903.2, Exception 1 shall be permitted.
2. The receptacle outlet for refrigeration appliances shall be permitted to be supplied from an individual branch circuit rated at 15 amperes or greater.

E3901.3.1 Other outlets prohibited. The two or more small-appliance branch circuits specified in Section E3901.3 shall serve no other outlets.

Exceptions:

1. A receptacle installed solely for the electrical supply to and support of an electric clock in any of the rooms specified in Section E3901.3.

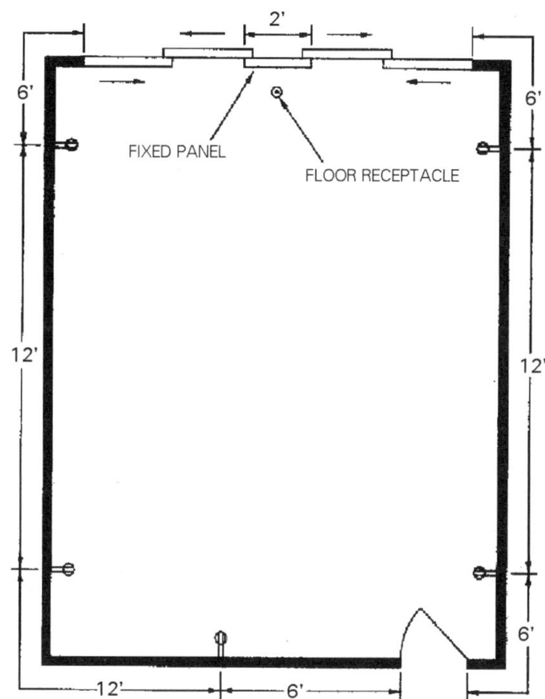

For SI: 1 foot = 304.8 mm.

FIGURE E3901.2
GENERAL USE RECEPTACLE DISTRIBUTION

2. Receptacles installed to provide power for supplemental equipment and lighting on gas-fired ranges, ovens, and counter-mounted cooking units.

E3901.3.2 Limitations. Receptacles installed in a kitchen to serve countertop surfaces shall be supplied by not less than two small-appliance branch circuits, either or both of which shall also be permitted to supply receptacle outlets in the same kitchen and in other rooms specified in Section E3901.3. Additional small-appliance branch circuits shall be permitted to supply receptacle outlets in the kitchen and other rooms specified in Section E3901.3. A small-appliance branch circuit shall not serve more than one kitchen.

E3901.4 Countertop receptacles. In kitchens pantries, breakfast rooms, dining rooms and similar areas of dwelling units, receptacle outlets for countertop spaces shall be installed in accordance with Sections E3901.4.1 through E3901.4.5 (see Figure E3901.4).

E3901.4.1 Wall countertop space. A receptacle outlet shall be installed at each wall countertop space 12 inches (305 mm) or wider. Receptacle outlets shall be installed so that no point along the wall line is more than 24 inches (610 mm), measured horizontally from a receptacle outlet in that space.

Exception: Receptacle outlets shall not be required on a wall directly behind a range, counter-mounted cooking unit or sink in the installation described in Figure E3901.4.1.

E3901.4.2 Island countertop spaces. At least one receptacle outlet shall be installed at each island countertop space with a long dimension of 24 inches (610 mm) or greater and a short dimension of 12 inches (305 mm) or greater.

E3901.4.3 Peninsular countertop space. At least one receptacle outlet shall be installed at each peninsular countertop space with a long dimension of 24 inches (610 mm) or greater and a short dimension of 12 inches (305 mm) or greater. A peninsular countertop is measured from the connecting edge.

E3901.4.4 Separate spaces. Countertop spaces separated by range tops, refrigerators, or sinks shall be considered as separate countertop spaces in applying the requirements of Sections E3901.4.1, E3901.4.2 and E3901.4.3. Where a range, counter-mounted cooking unit, or sink is installed in an island or peninsular countertop and the depth of the countertop behind the range, counter-mounted cooking unit, or sink is less than 12 inches (305 mm), the range, counter-mounted cooking unit, or sink has divided the countertop space into two separate countertop spaces as defined in Section E3901.4.4. Each separate countertop space shall comply with the applicable requirements of this section.

For SI: 1 foot = 304.8 mm.

FIGURE E3901.4
COUNTERTOP RECEPTACLES

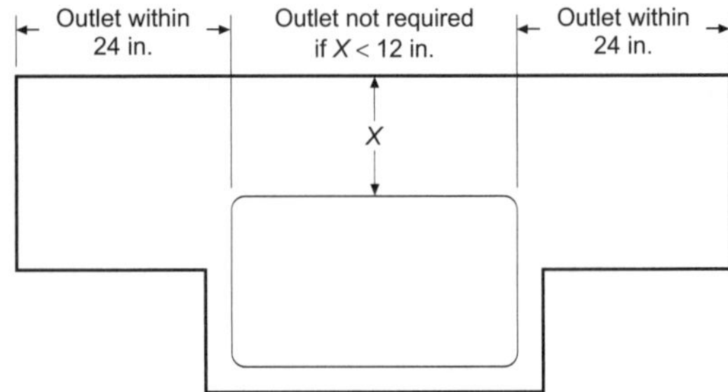

Sink, range or counter-mounted cooking unit extending from face of counter

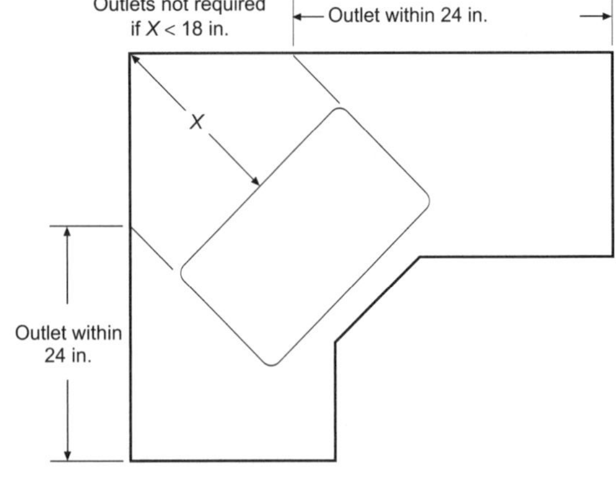

Sink, range or counter-mounted cooking unit mounted in corner

For SI: 1 inch = 25.4 mm.

FIGURE E3901.4.1
DETERMINATION OF AREA BEHIND SINK OR RANGE

E3901.4.5 Receptacle outlet location. Receptacle outlets shall be located not more than 20 inches (508 mm) above the countertop. Receptacle outlet assemblies installed in countertops shall be listed for the application. Receptacle outlets shall not be installed in a face-up position in the work surfaces or countertops. Receptacle outlets rendered not readily accessible by appliances fastened in place, appliance garages, sinks or rangetops as addressed in the exception to Section E3901.4.1, or appliances occupying dedicated space shall not be considered as these required outlets.

Exception: Receptacle outlets shall be permitted to be mounted not more than 12 inches (305 mm) below the countertop in construction designed for the physically impaired and for island and peninsular countertops where the countertop is flat across its entire surface and there are no means to mount a receptacle within 20 inches (508 mm) above the countertop, such as in an overhead cabinet. Receptacles mounted below the countertop in accordance with this exception shall not be located where the countertop extends more than 6 inches (152 mm) beyond its support base.

E3901.5 Appliance receptacle outlets. Appliance receptacle outlets installed for specific appliances, such as laundry equipment, shall be installed within 6 feet (1829 mm) of the intended location of the appliance.

E3901.6 Bathroom. At least one wall receptacle outlet shall be installed in bathrooms and such outlet shall be located within 36 inches (914 mm) of the outside edge of each lavatory basin. The receptacle outlet shall be located on a wall or partition that is adjacent to the lavatory basin location, located on the countertop, or installed on the side or face of the basin cabinet not more than 12 inches (305 mm) below the countertop.

Receptacle outlets shall not be installed in a face-up position in the work surfaces or countertops in a bathroom basin location. Receptacle outlet assemblies installed in countertops shall be listed for the application.

E3901.7 Outdoor outlets. At least one receptacle outlet that is accessible while standing at grade level and located not more than 6 feet, 6 inches (1981 mm) above grade, shall be installed outdoors at the front and back of each dwelling unit having direct access to grade. Balconies, decks, and porches that are accessible from inside of the dwelling unit shall have at least one receptacle outlet installed within the perimeter of the balcony, deck, or porch. The receptacle shall be located not more than 6 feet, 6 inches (1981 mm) above the balcony, deck, or porch surface.

E3901.8 Laundry areas. At least one receptacle outlet shall be installed to serve laundry appliances.

E3901.9 Basements, garages and accessory buildings. At least one receptacle outlet, in addition to any provided for specific equipment, shall be installed in each basement and in each attached garage, and in each detached garage or accessory building that is provided with electrical power. Where a portion of the basement is finished into one or more habitable room(s), each separate unfinished portion shall have a receptacle outlet installed in accordance with this section.

E3901.10 Hallways. Hallways of 10 feet (3048 mm) or more in length shall have at least one receptacle outlet. The hall length shall be considered the length measured along the centerline of the hall without passing through a doorway.

E3901.11 Foyers. Foyers that are not part of a hallway in accordance with Section E3901.10 and that have an area that is greater than 60 ft^2 (5.57 m^2) shall have a receptacle(s) located in each wall space that is 3 feet (914 mm) or more in width and unbroken by doorways, floor-to-ceiling windows, and similar openings.

E3901.12 HVAC outlet. A 125-volt, single-phase, 15- or 20-ampere-rated receptacle outlet shall be installed at an accessible location for the servicing of heating, air-conditioning and refrigeration equipment. The receptacle shall be located on the same level and within 25 feet (7620 mm) of the heating, air-conditioning and refrigeration equipment. The receptacle outlet shall not be connected to the load side of the HVAC equipment disconnecting means.

Exception: A receptacle outlet shall not be required for the servicing of evaporative coolers.

SECTION E3902 GROUND-FAULT AND ARC-FAULT CIRCUIT-INTERRUPTER PROTECTION

E3902.1 Bathroom receptacles. All 125-volt, single-phase, 15- and 20-ampere receptacles installed in bathrooms shall have ground-fault circuit-interrupter protection for personnel.

E3902.2 Garage and accessory building receptacles. All 125-volt, single-phase, 15- or 20-ampere receptacles installed in garages and grade-level portions of unfinished accessory buildings used for storage or work areas shall have ground-fault circuit-interrupter protection for personnel.

E3902.3 Outdoor receptacles. All 125-volt, single-phase, 15- and 20-ampere receptacles installed outdoors shall have ground-fault circuit-interrupter protection for personnel.

Exception: Receptacles as covered in Section E4101.7.

E3902.4 Crawl space receptacles. Where a crawl space is at or below grade level, all 125-volt, single-phase, 15- and 20-ampere receptacles installed in such spaces shall have ground-fault circuit-interrupter protection for personnel.

E3902.5 Unfinished basement receptacles. All 125-volt, single-phase, 15- and 20-ampere receptacles installed in unfinished basements shall have ground-fault circuit-interrupter protection for personnel. For purposes of this section, unfinished basements are defined as portions or areas of the basement not intended as habitable rooms and limited to storage areas, work areas, and the like.

Exception: A receptacle supplying only a permanently installed fire alarm or burglar alarm system.

E3902.6 Kitchen receptacles. All 125-volt, single-phase, 15- and 20-ampere receptacles that serve countertop surfaces shall have ground-fault circuit-interrupter protection for personnel.

E3902.7 Sink receptacles. All 125-volt, single-phase, 15- and 20-ampere receptacles that are located within 6 feet

(1829 mm) of the outside edge of a sink that is located in an area other than a kitchen, shall have ground-fault circuit-interrupter protection for personnel. Receptacle outlets shall not be installed in a face-up position in the work surfaces or countertops.

E3902.8 Boathouse receptacles. All 125-volt, single-phase, 15- or 20-ampere receptacles installed in boathouses shall have ground-fault circuit-interrupter protection for personnel.

E3902.9 Boat hoists. Ground-fault circuit-interrupter protection for personnel shall be provided for 240-volt and less outlets that supply boat hoists.

E3902.10 Electrically heated floors. Ground-fault circuit-interrupter protection for personnel shall be provided for electrically heated floors in bathrooms, kitchens and in hydromassage bathtub, spa and hot tub locations.

E3902.11 Location of ground-fault circuit interrupters. Ground-fault circuit interrupters shall be installed in a readily accessible location.

E3902.12 Arc-fault circuit-interrupter protection. All branch circuits that supply 120-volt, single-phase, 15- and 20-ampere outlets installed in family rooms, dining rooms, living rooms, parlors, libraries, dens, bedrooms, sunrooms, recreations rooms, closets, hallways and similar rooms or areas shall be protected by a combination type arc-fault circuit interrupter installed to provide protection of the branch circuit.

Exception:

1. Where an outlet branch-circuit type AFCI is installed at the first outlet to provide protection for the remaining portion of the branch circuit, the portion of the branch circuit between the branch-circuit overcurrent device and the first outlet shall be installed with metal outlet and junction boxes and RMC, IMC, EMT, type MC, or steel armored type AC cables meeting the requirements of Section E 3908.8.
2. Where an outlet branch-circuit type AFCI is installed at the first outlet to provide protection for the remaining portion of the branch circuit, the portion of the branch circuit between the branch-circuit overcurrent device and the first outlet shall be installed with metal or nonmetallic conduit or tubing that is encased in not less than 2 inches (51 mm) of concrete.
3. AFCI protection is not required for an individual branch circuit supplying only a fire alarm system where the branch circuit is wired with metal outlet and junction boxes and RMC, IMC, EMT or steel-sheathed armored cable Type AC, or Type MC meeting the requirements of Section E3908.8.

E3902.13 Arc-fault circuit interrupter protection for branch circuit extensions or modifications. Where branch-circuit wiring is modified, replaced, or extended in any of the areas specified in Section E3902.12, the branch circuit shall be protected by one of the following:

1. A combination-type AFCI located at the origin of the branch circuit
2. An outlet branch-circuit type AFCI located at the first receptacle outlet of the existing branch circuit.

SECTION E3903 LIGHTING OUTLETS

E3903.1 General. Lighting outlets shall be provided in accordance with Sections E3903.2 through E3903.4.

E3903.2 Habitable rooms. At least one wall switch-controlled lighting outlet shall be installed in every habitable room and bathroom.

Exceptions:

1. In other than kitchens and bathrooms, one or more receptacles controlled by a wall switch shall be considered equivalent to the required lighting outlet.
2. Lighting outlets shall be permitted to be controlled by occupancy sensors that are in addition to wall switches, or that are located at a customary wall switch location and equipped with a manual override that will allow the sensor to function as a wall switch.

E3903.3 Additional locations. At least one wall-switch-controlled lighting outlet shall be installed in hallways, stairways, attached garages, and detached garages with electric power. At least one wall-switch-controlled lighting outlet shall be installed to provide illumination on the exterior side of each outdoor egress door having grade level access, including outdoor egress doors for attached garages and detached garages with electric power. A vehicle door in a garage shall not be considered as an outdoor egress door. Where one or more lighting outlets are installed for interior stairways, there shall be a wall switch at each floor level and landing level that includes an entryway to control the lighting outlets where the stairway between floor levels has six or more risers.

Exception: In hallways, stairways, and at outdoor egress doors, remote, central, or automatic control of lighting shall be permitted.

E3903.4 Storage or equipment spaces. In attics, under-floor spaces, utility rooms and basements, at least one lighting outlet shall be installed where these spaces are used for storage or contain equipment requiring servicing. Such lighting outlet shall be controlled by a wall switch or shall have an integral switch. At least one point of control shall be at the usual point of entry to these spaces. The lighting outlet shall be provided at or near the equipment requiring servicing.

SECTION E3904
GENERAL INSTALLATION REQUIREMENTS

E3904.1 Electrical continuity of metal raceways and enclosures. Metal raceways, cable armor and other metal enclosures for conductors shall be mechanically joined together into a continuous electric conductor and shall be connected to all boxes, fittings and cabinets so as to provide effective electrical continuity. Raceways and cable assemblies shall be mechanically secured to boxes, fittings cabinets and other enclosures.

Exception: Short sections of raceway used to provide cable assemblies with support or protection against physical damage.

E3904.2 Mechanical continuity—raceways and cables. Metal or nonmetallic raceways, cable armors and cable sheaths shall be continuous between cabinets, boxes, fittings or other enclosures or outlets.

Exception: Short sections of raceway used to provide cable assemblies with support or protection against physical damage.

E3904.3 Securing and supporting. Raceways, cable assemblies, boxes, cabinets and fittings shall be securely fastened in place.

E3904.3.1 Prohibited means of support. Cable wiring methods shall not be used as a means of support for other cables, raceways and nonelectrical equipment.

E3904.4 Raceways as means of support. Raceways shall be used as a means of support for other raceways, cables or nonelectric equipment only under the following conditions:

1. Where the raceway or means of support is identified for the purpose; or
2. Where the raceway contains power supply conductors for electrically controlled equipment and is used to support Class 2 circuit conductors or cables that are solely for the purpose of connection to the control circuits of the equipment served by such raceway; or
3. Where the raceway is used to support boxes or conduit bodies in accordance with Sections E3906.8.4 and E3906.8.5.

E3904.5 Raceway installations. Raceways shall be installed complete between outlet, junction or splicing points prior to the installation of conductors.

Exception: Short sections of raceways used to contain conductors or cable assemblies for protection from physical damage shall not be required to be installed complete between outlet, junction, or splicing points.

E3904.6 Conduit and tubing fill. The maximum number of conductors installed in conduit or tubing shall be in accordance with Tables E3904.6(1) through E3904.6(10).

E3904.7 Air handling-stud cavity and joist spaces. Where wiring methods having a nonmetallic covering pass through stud cavities and joist spaces used for air handling, such wiring shall pass through such spaces perpendicular to the long dimension of the spaces.

SECTION E3905
BOXES, CONDUIT BODIES AND FITTINGS

E3905.1 Box, conduit body or fitting—where required. A box or conduit body shall be installed at each conductor splice point, outlet, switch point, junction point and pull point except as otherwise permitted in Sections E3905.1.1 through E3905.1.6.

Fittings and connectors shall be used only with the specific wiring methods for which they are designed and listed.

E3905.1.1 Equipment. An integral junction box or wiring compartment that is part of listed equipment shall be permitted to serve as a box or conduit body.

E3905.1.2 Protection. A box or conduit body shall not be required where cables enter or exit from conduit or tubing that is used to provide cable support or protection against physical damage. A fitting shall be provided on the end(s) of the conduit or tubing to protect the cable from abrasion.

E3905.1.3 Integral enclosure. A wiring device with integral enclosure identified for the use, having brackets that securely fasten the device to walls or ceilings of conventional on-site frame construction, for use with nonmetallic-sheathed cable, shall be permitted in lieu of a box or conduit body.

E3905.1.4 Fitting. A fitting identified for the use shall be permitted in lieu of a box or conduit body where such fitting is accessible after installation and does not contain spliced or terminated conductors.

E3905.1.5 Buried conductors. Splices and taps in buried conductors and cables shall not be required to be enclosed in a box or conduit body where installed in accordance with Section E3803.4.

E3905.1.6 Luminaires. Where a luminaire is listed to be used as a raceway, a box or conduit body shall not be required for wiring installed therein.

E3905.2 Metal boxes. All metal boxes shall be grounded.

E3905.3 Nonmetallic boxes. Nonmetallic boxes shall be used only with cabled wiring methods with entirely nonmetallic sheaths, flexible cords and nonmetallic raceways.

Exceptions:

1. Where internal bonding means are provided between all entries, nonmetallic boxes shall be permitted to be used with metal raceways and metal-armored cables.
2. Where integral bonding means with a provision for attaching an equipment grounding jumper inside the box are provided between all threaded entries in nonmetallic boxes listed for the purpose, nonmetallic boxes shall be permitted to be used with metal raceways and metal-armored cables.

E3905.3.1 Nonmetallic-sheathed cable and nonmetallic boxes. Where nonmetallic-sheathed cable is used, the cable assembly, including the sheath, shall extend into the box not less than $^1/_4$ inch (6.4 mm) through a nonmetallic-sheathed cable knockout opening.

TABLE E3904.6(1)
MAXIMUM NUMBER OF CONDUCTORS IN ELECTRICAL METALLIC TUBING (EMT)[a]

TYPE LETTERS	CONDUCTOR SIZE AWG/kcmil	TRADE SIZES (inches)					
		1/2	3/4	1	1 1/4	1 1/2	2
RHW, RHW-2	14	4	7	11	20	27	46
	12	3	6	9	17	23	38
	10	2	5	8	13	18	30
	8	1	2	4	7	9	16
	6	1	1	3	5	8	13
	4	1	1	2	4	6	10
	3	1	1	1	4	5	9
	2	1	1	1	3	4	7
	1	0	1	1	1	3	5
	1/0	0	1	1	1	2	4
	2/0	0	1	1	1	2	4
	3/0	0	0	1	1	1	3
	4/0	0	0	1	1	1	3
TW	14	8	15	25	43	58	96
	12	6	11	19	33	45	74
	10	5	8	14	24	33	55
	8	2	5	8	13	18	30
RHW[a], RHW-2[a], THHW, THW, THW-2	14	6	10	16	28	39	64
	12	4	8	13	23	31	51
	10	3	6	10	18	24	40
	8	1	4	6	10	14	24
RHW[a], RHW-2[a], TW, THW, THHW, THW-2	6	1	3	4	8	11	18
	4	1	1	3	6	8	13
	3	1	1	3	5	7	12
	2	1	1	2	4	6	10
	1	1	1	1	3	4	7
	1/0	0	1	1	2	3	6
	2/0	0	1	1	1	3	5
	3/0	0	1	1	1	2	4
	4/0	0	0	1	1	1	3
THHN, THWN, THWN-2	14	12	22	35	61	84	138
	12	9	16	26	45	61	101
	10	5	10	16	28	38	63
	8	3	6	9	16	22	36
	6	2	4	7	12	16	26
	4	1	2	4	7	10	16
	3	1	1	3	6	8	13
	2	1	1	3	5	7	11
	1	1	1	1	4	5	8
	1/0	1	1	1	3	4	7
	2/0	0	1	1	2	3	6
	3/0	0	1	1	1	3	5
	4/0	0	1	1	1	2	4
XHHW, XHHW-2	14	8	15	25	43	58	96
	12	6	11	19	33	45	74
	10	5	8	14	24	33	55
	8	2	5	8	13	18	30
	6	1	3	6	10	14	22
	4	1	2	4	7	10	16
	3	1	1	3	6	8	14
	2	1	1	3	5	7	11
	1	1	1	1	4	5	8
	1/0	1	1	1	3	4	7
	2/0	0	1	1	2	3	6
	3/0	0	1	1	1	3	5
	4/0	0	1	1	1	2	4

For SI: 1 inch = 25.4 mm.

a. Types RHW, and RHW-2 without outer covering.

TABLE E3904.6(2)
MAXIMUM NUMBER OF CONDUCTORS IN ELECTRICAL NONMETALLIC TUBING (ENT)[a]

TYPE LETTERS	CONDUCTOR SIZE AWG/kcmil	TRADE SIZES (inches)					
		1/2	3/4	1	1 1/4	1 1/2	2
RHW, RHW-2	14	3	6	10	19	26	43
	12	2	5	9	16	22	36
	10	1	4	7	13	17	29
	8	1	1	3	6	9	15
	6	1	1	3	5	7	12
	4	1	1	2	4	6	9
	3	1	1	1	3	5	8
	2	0	1	1	3	4	7
	1	0	1	1	1	3	5
	1/0	0	0	1	1	2	4
	2/0	0	0	1	1	1	3
RHW, RHW-2	3/0	0	0	1	1	1	3
	4/0	0	0	1	1	1	2
TW	14	7	13	22	40	55	92
	12	5	10	17	31	42	71
	10	4	7	13	23	32	52
	8	1	4	7	13	17	29
RHW[a], RHW-2[a], THHW, THW, THW-2	14	4	8	15	27	37	61
	12	3	7	12	21	29	49
	10	3	5	9	17	23	38
	8	1	3	5	10	14	23
RHW[a], RHW-2[a], TW, THW, THHW, THW-2	6	1	2	4	7	10	17
	4	1	1	3	5	8	13
	3	1	1	2	5	7	11
	2	1	1	2	4	6	9
	1	0	1	1	3	4	6
	1/0	0	1	1	2	3	5
	2/0	0	1	1	1	3	5
	3/0	0	0	1	1	2	4
	4/0	0	0	1	1	1	3
THHN, THWN, THWN-2	14	10	18	32	58	80	132
	12	7	13	23	42	58	96
	10	4	8	15	26	36	60
	8	2	5	8	15	21	35
	6	1	3	6	11	15	25
	4	1	1	4	7	9	15
	3	1	1	3	5	8	13
	2	1	1	2	5	6	11
THHN, THWN, THWN-2	1	1	1	1	3	5	8
	1/0	0	1	1	3	4	7
	2/0	0	1	1	2	3	5
	3/0	0	1	1	1	3	4
	4/0	0	0	1	1	2	4
XHHW, XHHW-2	14	7	13	22	40	55	92
	12	5	10	17	31	42	71
	10	4	7	13	23	32	52
	8	1	4	7	13	17	29
	6	1	3	5	9	13	21
	4	1	1	4	7	9	15
	3	1	1	3	6	8	13
	2	1	1	2	5	6	11
	1	1	1	1	3	5	8
	1/0	0	1	1	3	4	7
	2/0	0	1	1	2	3	6
	3/0	0	1	1	1	3	5
	4/0	0	0	1	1	2	4

For SI: 1 inch = 25.4 mm.

a. Types RHW, and RHW-2 without outer covering.

TABLE E3904.6(3)
MAXIMUM NUMBER OF CONDUCTORS IN FLEXIBLE METALLIC CONDUIT (FMC)[a]

TYPE LETTERS	CONDUCTOR SIZE AWG/kcmil	TRADE SIZES (inches)					
		1/2	3/4	1	1 1/4	1 1/2	2
RHW, RHW-2	14	4	7	11	17	25	44
	12	3	6	9	14	21	37
	10	3	5	7	11	17	30
	8	1	2	4	6	9	15
	6	1	1	3	5	7	12
	4	1	1	2	4	5	10
	3	1	1	1	3	5	7
RHW, RHW-2	2	1	1	1	3	4	7
	1	0	1	1	1	2	5
	1/0	0	1	1	1	2	4
	2/0	0	1	1	1	1	3
	3/0	0	0	1	1	1	3
TW	14	9	15	23	36	53	94
	12	7	11	18	28	41	72
	10	5	8	13	21	30	54
	8	3	5	7	11	17	30
RHW[a], RHW-2[a], THHW, THW, THW-2	14	6	10	15	24	35	62
	12	5	8	12	19	28	50
	10	4	6	10	15	22	39
	8	1	4	6	9	13	23
RHW[a], RHW-2[a], TW, THW, THHW, THW-2	6	1	3	4	7	10	18
	4	1	1	3	5	7	13
	3	1	1	3	4	6	11
	2	1	1	2	4	5	10
	1	1	1	1	2	4	7
	1/0	0	1	1	1	3	6
	2/0	0	1	1	1	3	5
	3/0	0	1	1	1	2	4
	4/0	0	0	1	1	1	3
	4/0	0	0	1	1	1	2
THHN, THWN, THWN-2	14	13	22	33	52	76	134
	12	9	16	24	38	56	98
	10	6	10	15	24	35	62
	8	3	6	9	14	20	35
THHN, THWN, THWN-2	6	2	4	6	10	14	25
	4	1	2	4	6	9	16
	3	1	1	3	5	7	13
	2	1	1	3	4	6	11
	1	1	1	1	3	4	8
	1/0	1	1	1	2	4	7
	2/0	0	1	1	1	3	6
	3/0	0	1	1	1	2	5
	4/0	0	1	1	1	1	4
XHHW, XHHW-2	14	9	15	23	36	53	94
	12	7	11	18	28	41	72
	10	5	8	13	21	30	54
	8	3	5	7	11	17	30
	6	1	3	5	8	12	22
	4	1	2	4	6	9	16
	3	1	1	3	5	7	13
	2	1	1	3	4	6	11
	1	1	1	1	3	5	8
	1/0	1	1	1	2	4	7
	2/0	0	1	1	2	3	6
	3/0	0	1	1	1	3	5
	4/0	0	1	1	1	2	4

For SI: 1 inch = 25.4 mm.

a. Types RHW, and RHW-2 without outer covering.

TABLE E3904.6(4) MAXIMUM NUMBER OF CONDUCTORS IN INTERMEDIATE METALLIC CONDUIT (IMC)[a]

TYPE LETTERS	CONDUCTOR SIZE AWG/kcmil	TRADE SIZES (inches)					
		1/2	3/4	1	1 1/4	1 1/2	2
RHW, RHW-2	14	4	8	13	22	30	49
	12	4	6	11	18	25	41
	10	3	5	8	15	20	33
	8	1	3	4	8	10	17
	6	1	1	3	6	8	14
	4	1	1	3	5	6	11
	3	1	1	2	4	6	9
	2	1	1	1	3	5	8
	1	0	1	1	2	3	5
	1/0	0	1	1	1	3	4
	2/0	0	1	1	1	2	4
	3/0	0	0	1	1	1	3
	4/0	0	0	1	1	1	3
TW	14	10	17	27	47	64	104
	12	7	13	21	36	49	80
	10	5	9	15	27	36	59
	8	3	5	8	15	20	33
RHW[a], RHW-2[a], THHW, THW, THW-2	14	6	11	18	31	42	69
	12	5	9	14	25	34	56
	10	4	7	11	19	26	43
	8	2	4	7	12	16	26
RHW[a], RHW-2[a], TW, THW, THHW, THW-2	6	1	3	5	9	12	20
	4	1	2	4	6	9	15
	3	1	1	3	6	8	13
	2	1	1	3	5	6	11
	1	1	1	1	3	4	7
	1/0	1	1	1	3	4	6
	2/0	0	1	1	2	3	5
	3/0	0	1	1	1	3	4
	4/0	0	1	1	1	2	4
THHN, THWN, THWN-2	14	14	24	39	68	91	149
	12	10	17	29	49	67	109
	10	6	11	18	31	42	68
	8	3	6	10	18	24	39
	6	2	4	7	13	17	28
	4	1	3	4	8	10	17
	3	1	2	4	6	9	15
	2	1	1	3	5	7	12
	1	1	1	2	4	5	9
	1/0	1	1	1	3	4	8
	2/0	1	1	1	3	4	6
THHN, THWN, THWN-2	3/0	0	1	1	2	3	5
	2/0	0	1	1	1	2	4
XHHW, XHHW-2	14	10	17	27	47	64	104
	12	7	13	21	36	49	80
	10	5	9	15	27	36	59
	8	3	5	8	15	20	33
	6	1	4	6	11	15	24
	4	1	3	4	8	11	18
	3	1	2	4	7	9	15
	2	1	1	3	5	7	12
	1	1	1	2	4	5	9
	1/0	1	1	1	3	5	8
	2/0	1	1	1	3	4	6
	3/0	0	1	1	2	3	5
	4/0	0	1	1	1	2	4

For SI: 1 inch = 25.4 mm.

a. Types RHW, and RHW-2 without outer covering.

TABLE E3904.6(5)
MAXIMUM NUMBER OF CONDUCTORS IN LIQUID-TIGHT FLEXIBLE NONMETALLIC CONDUIT (FNMC-B)[a]

TYPE LETTERS	CONDUCTOR SIZE AWG/kcmil	TRADE SIZES (inches)						
		$^3/_8$	$^1/_2$	$^3/_4$	1	$1^1/_4$	$1^1/_2$	2
RHW, RHW-2	14	2	4	7	12	21	27	44
	12	1	3	6	10	17	22	36
	10	1	3	5	8	14	18	29
	8	1	1	2	4	7	9	1
	6	1	1	1	3	6	7	12
	4	0	1	1	2	4	6	9
RHW, RHW-2	3	0	1	1	1	4	5	8
	2	0	1	1	1	3	4	7
	1	0	0	1	1	1	3	5
	1/0	0	0	1	1	1	2	4
	2/0	0	0	1	1	1	1	3
	3/0	0	0	0	1	1	1	3
	4/0	0	0	0	1	1	1	2
TW	14	5	9	15	25	44	57	93
	12	4	7	12	19	33	43	71
	10	3	5	9	14	25	32	53
	8	1	3	5	8	14	18	29
RHW[a], RGW-2[a], THHW, THW, THW-2	14	3	6	10	16	29	38	62
	12	3	5	8	13	23	30	50
	10	1	3	6	10	18	23	39
	8	1	1	4	6	11	14	23
RHW[a], RHW-2[a], TW, THW, THHW, THW-2	6	1	1	3	5	8	11	18
	4	1	1	1	3	6	8	13
	3	1	1	1	3	5	7	11
	2	0	1	1	2	4	6	9
	1	0	1	1	1	3	4	7
	1/0	0	0	1	1	2	3	6
	2/0	0	0	1	1	2	3	5
	3/0	0	0	1	1	1	2	4
	4/0	0	0	0	1	1	1	3
THHN, THWN, THWN-2	14	8	13	22	36	63	81	133
	12	5	9	16	26	46	59	97
	10	3	6	10	16	29	37	61
	8	1	3	6	9	16	21	35
THHN, THWN, THWN-2	6	1	2	4	7	12	15	25
	4	1	1	2	4	7	9	15
	3	1	1	1	3	6	8	13
	2	1	1	1	3	5	7	11
	1	0	1	1	1	4	5	8
	1/0	0	1	1	1	3	4	7
	2/0	0	0	1	1	2	3	6
	3/0	0	0	1	1	1	3	5
	4/0	0	0	1	1	1	2	4
XHHW, XHHW-2	14	5	9	15	25	44	57	93
	12	4	7	12	19	33	43	71
	10	3	5	9	14	25	32	53
	8	1	3	5	8	14	18	29
	6	1	1	3	6	10	13	22
	4	1	1	2	4	7	9	16
	3	1	1	1	3	6	8	13
	2	1	1	1	3	5	7	11
	1	0	1	1	1	4	5	8
	1/0	0	1	1	1	3	4	7
	2/0	0	0	1	1	2	3	6
	3/0	0	0	1	1	1	3	5
	4/0	0	0	1	1	1	2	4

For SI: 1 inch = 25.4 mm.

a. Types RHW, and RHW-2 without outer covering.

TABLE E3904.6(6)
MAXIMUM NUMBER OF CONDUCTORS IN LIQUID-TIGHT FLEXIBLE NONMETALLIC CONDUIT (FNMC-A)[a]

TYPE LETTERS	CONDUCTOR SIZE AWG/kcmil	TRADE SIZES (inches)						
		3/8	1/2	3/4	1	1 1/4	1 1/2	2
RHW, RHW-2	14	2	4	7	11	20	27	45
	12	1	3	6	9	17	23	38
	10	1	3	5	8	13	18	30
	8	1	1	2	4	7	9	16
	6	1	1	1	3	5	7	13
	4	0	1	1	2	4	6	10
	3	0	1	1	1	4	5	8
	2	0	1	1	1	3	4	7
	1	0	0	1	1	1	3	5
	1/0	0	0	1	1	1	2	4
	2/0	0	0	1	1	1	1	4
	3/0	0	0	0	1	1	1	3
	4/0	0	0	0	1	1	1	3
TW	14	5	9	15	24	43	58	96
	12	4	7	12	19	33	44	74
	10	3	5	9	14	24	33	55
	8	1	3	5	8	13	18	30
RHW[a], RHW-2[a], THHW, THW, THW-2	14	3	6	10	16	28	38	64
	12	3	4	8	13	23	31	51
	10	1	3	6	10	18	24	40
	8	1	1	4	6	10	14	24
RHW[a], RHW-2[a], TW, THW, THHW, THW-2	6	1	1	3	4	8	11	18
	4	1	1	1	3	6	8	13
	3	1	1	1	3	5	7	11
	2	0	1	1	2	4	6	10
	1	0	1	1	1	3	4	7
RHW[a], RHW-2[a], TW, THW, THHW, THW-2	1/0	0	0	1	1	2	3	6
	2/0	0	0	1	1	1	3	5
	3/0	0	0	1	1	1	2	4
	4/0	0	0	0	1	1	1	3
THHN, THWN, THWN-2	14	8	13	22	35	62	83	137
	12	5	9	16	25	45	60	100
	10	3	6	10	16	28	38	63
	8	1	3	6	9	16	22	36
	6	1	2	4	6	12	16	26
	4	1	1	2	4	7	9	16
	3	1	1	1	3	6	8	13
	2	1	1	1	3	5	7	11
	1	0	1	1	1	4	5	8
	1/0	0	1	1	1	3	4	7
	2/0	0	0	1	1	2	3	6
	3/0	0	0	1	1	1	3	5
	4/0	0	0	1	1	1	2	4
XHHW, XHHW-2	14	5	9	15	24	43	58	96
	12	4	7	12	19	33	44	74
	10	3	5	9	14	24	33	55
	8	1	3	5	8	13	18	30
	6	1	1	3	5	10	13	22
	4	1	1	2	4	7	10	16
	3	1	1	1	3	6	8	14
	2	1	1	1	3	5	7	11
	1	0	1	1	1	4	5	8
XHHW, XHHW-2	1/0	0	1	1	1	3	4	7
	2/0	0	0	1	1	2	3	6
	3/0	0	0	1	1	1	3	5
	4/0	0	0	1	1	1	2	4

For SI: 1 inch = 25.4 mm.

a. Types RHW, and RHW-2 without outer covering.

TABLE E3904.6(7)
MAXIMUM NUMBER OF CONDUCTORS IN LIQUID-TIGHT FLEXIBLE METAL CONDUIT (LFMC)[a]

TYPE LETTERS	CONDUCTOR SIZE AWG/kcmil	TRADE SIZES (inches)					
		1/2	3/4	1	1 1/4	1 1/2	2
RHW, RHW-2	14	4	7	12	21	27	44
	12	3	6	10	17	22	36
	10	3	5	8	14	18	29
	8	1	2	4	7	9	15
	6	1	1	3	6	7	12
	4	1	1	2	4	6	9
	3	1	1	1	4	5	8
	2	1	1	1	3	4	7
	1	0	1	1	1	3	5
	1/0	0	1	1	1	2	4
	2/0	0	1	1	1	1	3
	3/0	0	0	1	1	1	3
	4/0	0	0	1	1	1	2
TW	14	9	15	25	44	57	93
	12	7	12	19	33	43	71
	10	5	9	14	25	32	53
	8	3	5	8	14	18	29
RHW[a], RHW-2[a], THHW, THW, THW-2	14	6	10	16	29	38	62
	12	5	8	13	23	30	50
	10	3	6	10	18	23	39
	8	1	4	6	11	14	23
RHW[a], RHW-2[a], TW, THW, THHW, THW-2	6	1	3	5	8	11	18
	4	1	1	3	6	8	13
	3	1	1	3	5	7	11
	2	1	1	2	4	6	9
	1	1	1	1	3	4	7
	1/0	0	1	1	2	3	6
	2/0	0	1	1	2	3	5
	3/0	0	1	1	1	2	4
	4/0	0	0	1	1	1	3
THHN, THWN, THWN-2	14	13	22	36	63	81	133
	12	9	16	26	46	59	97
	10	6	10	16	29	37	61
	8	3	6	9	16	21	35
	6	2	4	7	12	15	25
	4	1	2	4	7	9	15
	3	1	1	3	6	8	13
	2	1	1	3	5	7	11
	1	1	1	1	4	5	8
	1/0	1	1	1	3	4	7
	2/0	0	1	1	2	3	6
	3/0	0	1	1	1	3	5
	4/0	0	1	1	1	2	4
XHHW, XHHW-2	14	9	15	25	44	57	93
	12	7	12	19	33	43	71
	10	5	9	14	25	32	53
	8	3	5	8	14	18	29
XHHW, XHHW-2	6	1	3	6	10	13	22
	4	1	2	4	7	9	16
	3	1	1	3	6	8	13
	2	1	1	3	5	7	11
	1	1	1	1	4	5	8
	1/0	1	1	1	3	4	7
	2/0	0	1	1	2	3	6
	3/0	0	1	1	1	3	5
	4/0	0	1	1	1	2	4

For SI: 1 inch = 25.4 mm.

a. Types RHW, and RHW-2 without outer covering.

TABLE E3904.6(8)
MAXIMUM NUMBER OF CONDUCTORS IN RIGID METAL CONDUIT (RMC)[a]

TYPE LETTERS	CONDUCTOR SIZE AWG/kcmil	TRADE SIZES (inches)					
		1/2	3/4	1	1 1/4	1 1/2	2
RHW, RHW-2	14	4	7	12	21	28	46
	12	3	6	10	17	23	38
	10	3	5	8	14	19	31
	8	1	2	4	7	10	16
	6	1	1	3	6	8	13
	4	1	1	2	4	6	10
	3	1	1	2	4	5	9
	2	1	1	1	3	4	7
	1	0	1	1	1	3	5
	1/0	0	1	1	1	2	4
	2/0	0	1	1	1	2	4
	3/0	0	0	1	1	1	3
	4/0	0	0	1	1	1	3
TW	14	9	15	25	44	59	98
	12	7	12	19	33	45	75
	10	5	9	14	25	34	56
	8	3	5	8	14	19	31
RHW[a], RHW-2[a], THHW, THW, THW-2	14	6	10	17	29	39	65
	12	5	8	13	23	32	52
	10	3	6	10	18	25	41
	8	1	4	6	11	15	24
RHW[a], RHW-2[a], TW, THW, THHW, THW-2	6	1	3	5	8	11	18
	4	1	1	3	6	8	14
	3	1	1	3	5	7	12
	2	1	1	2	4	6	10
	1	1	1	1	3	4	7
	1/0	0	1	1	2	3	6
	2/0	0	1	1	2	3	5
	3/0	0	1	1	1	2	4
	4/0	0	0	1	1	1	3
THHN, THWN, THWN-2	14	13	22	36	63	85	140
	12	9	16	26	46	62	102
	10	6	10	17	29	39	64
	8	3	6	9	16	22	37
	6	2	4	7	12	16	27
	4	1	2	4	7	10	16
	3	1	1	3	6	8	14
	2	1	1	3	5	7	11
	1	1	1	1	4	5	8
THHN, THWN, THWN-2	1/0	1	1	1	3	4	7
	2/0	0	1	1	2	3	6
	3/0	0	1	1	1	3	5
	4/0	0	1	1	1	2	4
XHHW, XHHW-2	14	9	15	25	44	59	98
	12	7	12	19	33	45	75
	10	5	9	14	25	34	56
	8	3	5	8	14	19	31
	6	1	3	6	10	14	23
	4	1	2	4	7	10	16
	3	1	1	3	6	8	14
	2	1	1	3	5	7	12
	1	1	1	1	4	5	9
	1/0	1	1	1	3	4	7
	2/0	0	1	1	2	3	6
	3/0	0	1	1	1	3	5
	4/0	0	1	1	1	2	4

For SI: 1 inch = 25.4 mm.

a. Types RHW, and RHW-2 without outer covering.

TABLE E3904.6(9)
MAXIMUM NUMBER OF CONDUCTORS IN RIGID PVC CONDUIT, SCHEDULE 80 (PVC-80)[a]

TYPE LETTERS	CONDUCTOR SIZE AWG/kcmil	TRADE SIZES (inches)					
		$^1/_2$	$^3/_4$	1	$1^1/_4$	$1^1/_2$	2
RHW, RHW-2	14	3	5	9	17	23	39
	12	2	4	7	14	19	32
	10	1	3	6	11	15	26
	8	1	1	3	6	8	13
	6	1	1	2	4	6	11
	4	1	1	1	3	5	8
RHW, RHW-2	3	0	1	1	3	4	7
	2	0	1	1	3	4	6
	1	0	1	1	1	2	4
	1/0	0	0	1	1	1	3
	2/0	0	0	1	1	1	3
	3/0	0	0	1	1	1	3
	4/0	0	0	0	1	1	2
TW	14	6	11	20	35	49	82
	12	5	9	15	27	38	63
	10	3	6	11	20	28	47
	8	1	3	6	11	15	26
RHW[a], RHW-2[a], THHW, THW, THW-2	14	4	8	13	23	32	55
	12	3	6	10	19	26	44
	10	2	5	8	15	20	34
	8	1	3	5	9	12	20
RHW[a], RHW-2[a], TW, THW, THHW, THW-2	6	1	1	3	7	9	16
	4	1	1	3	5	7	12
	3	1	1	2	4	6	10
	2	1	1	1	3	5	8
	1	0	1	1	2	3	6
	1/0	0	1	1	1	3	5
RHW[a], RHW-2[a], TW, THW, THHW, THW-2	2/0	0	1	1	1	2	4
	3/0	0	0	1	1	1	3
	4/0	0	0	1	1	1	3
THHN, THWN, THWN-2	14	9	17	28	51	70	118
	12	6	12	20	37	51	86
	10	4	7	13	23	32	54
	8	2	4	7	13	18	31
	6	1	3	5	9	13	22
	4	1	1	3	6	8	14
	3	1	1	3	5	7	12
	2	1	1	2	4	6	10
	1	0	1	1	3	4	7
	1/0	0	1	1	2	3	6
	2/0	0	1	1	1	3	5
	3/0	0	1	1	1	2	4
	4/0	0	0	1	1	1	3
XHHW, XHHW-2	14	6	11	20	35	49	82
	12	5	9	15	27	38	63
	10	3	6	11	20	28	47
	8	1	3	6	118	15	26
	6	1	2	4		11	19

(continued)

TABLE E3904.6(9)—continued
MAXIMUM NUMBER OF CONDUCTORS IN RIGID PVC CONDUIT, SCHEDULE 80 (PVC-80)[a]

TYPE LETTERS	CONDUCTOR SIZE AWG/kcmil	TRADE SIZES (inches)					
		1/2	3/4	1	1 1/4	1 1/2	2
XHHW, XHHW-2	14	6	11	20	35	49	82
	12	5	9	15	27	38	63
	10	3	6	11	20	28	47
	8	1	3	6	11	15	26
	6	1	2	4	8	11	19
	4	1	1	3	6	8	14
	3	1	1	3	5	7	12
	2	1	1	2	4	6	10
	1	0	1	1	3	4	7
	1/0	0	1	1	2	3	6
	2/0	0	1	1	1	3	5
	3/0	0	1	1	1	2	4
	4/0	0	0	1	1	1	3

For SI: 1 inch = 25.4 mm.

a. Types RHW, and RHW-2 without outer covering.

TABLE E3904.6(10)
MAXIMUM NUMBER OF CONDUCTORS IN RIGID PVC CONDUIT SCHEDULE 40 (PVC-40)[a]

TYPE LETTERS	CONDUCTOR SIZE AWG/kcmil	TRADE SIZES (inches)					
		1/2	3/4	1	1 1/4	1 1/2	2
RHW, RHW-2	14	4	7	11	20	27	45
	12	3	5	9	16	22	37
	10	2	4	7	13	18	30
	8	1	2	4	7	9	15
	6	1	1	3	5	7	12
	4	1	1	2	4	6	10
	3	1	1	1	4	5	8
	2	1	1	1	3	4	7
RHW, RHW-2	1	0	1	1	1	3	5
	1/0	0	1	1	1	2	4
	2/0	0	0	1	1	1	3
	3/0	0	0	1	1	1	3
	4/0	0	0	1	1	1	2
TW	14	8	14	24	42	57	94
	12	6	11	18	32	44	72
	10	4	8	13	24	32	54
	8	2	4	7	13	18	30
RHW[a], RHW-2[a], THHW, THW, THW-2	14	5	9	16	28	38	63
	12	4	8	12	22	30	50
	10	3	6	10	17	24	39
	8	1	3	6	10	14	23
RHW[a], RHW-2[a], TW, THW, THHW, THW-2	6	1	2	4	8	11	18
	4	1	1	3	6	8	13
	3	1	1	3	5	7	11
	2	1	1	2	4	6	10
	1	0	1	1	3	4	7
	1/0	0	1	1	2	3	6
	2/0	0	1	1	1	3	5
	3/0	0	1	1	1	2	4
	4/0	0	0	1	1	1	3
THHN, THWN, THWN-2	14	11	21	34	60	82	135
	12	8	15	25	43	59	99
	10	5	9	15	27	37	62
	8	3	5	9	16	21	36
THHN, THWN, THWN-2	6	1	4	6	11	15	26
	4	1	2	4	7	9	16
	3	1	1	3	6	8	13
	2	1	1	3	5	7	11
	1	1	1	1	3	5	8
	1/0	1	1	1	3	4	7
	2/0	0	1	1	2	3	6
	3/0	0	1	1	1	3	5
	4/0	0	1	1	1	2	4
XHHW, XHHW-2	14	8	14	24	42	57	94
	12	6	11	18	32	44	72
	10	4	8	13	24	32	54
	8	2	4	7	13	18	30
	6	1	3	5	10	13	22
	4	1	2	4	7	9	16
	3	1	1	3	6	8	13
	2	1	1	3	5	7	11
	1	1	1	1	3	5	8
	1/0	1	1	1	3	4	7
	2/0	0	1	1	2	3	6
	3/0	0	1	1	1	3	5
	4/0	0	1	1	1	2	4

For SI: 1 inch = 25.4 mm.

a. Types RHW, and RHW-2 without outer covering.

E3905.3.2 Securing to box. All permitted wiring methods shall be secured to the boxes.

Exception: Where nonmetallic-sheathed cable is used with boxes not larger than a nominal size of $2^1/_4$ inches by 4 inches (57 mm by 102 mm) mounted in walls or ceilings, and where the cable is fastened within 8 inches (203 mm) of the box measured along the sheath, and where the sheath extends through a cable knockout not less than $^1/_4$ inch (6.4 mm), securing the cable to the box shall not be required.

E3905.3.3 Conductor rating. Nonmetallic boxes shall be suitable for the lowest temperature-rated conductor entering the box.

E3905.4 Minimum depth of boxes for outlets, devices, and utilization equipment. Outlet and device boxes shall have sufficient depth to allow equipment installed within them to be mounted properly and without the likelihood of damage to conductors within the box.

E3905.4.1 Outlet boxes without enclosed devices or utilization equipment. Outlet boxes that do not enclose devices or utilization equipment shall have an internal depth of not less than $^1/_2$ inch (12.7 mm).

E3905.4.2 Utilization equipment. Outlet and device boxes that enclose devices or utilization equipment shall have a minimum internal depth that accommodates the rearward projection of the equipment and the size of the conductors that supply the equipment. The internal depth shall include that of any extension boxes, plaster rings, or raised covers. The internal depth shall comply with all of the applicable provisions that follow.

Exception: Utilization equipment that is listed to be installed with specified boxes.

1. Large equipment. Boxes that enclose devices or utilization equipment that projects more than $1^7/_8$ inches (48 mm) rearward from the mounting plane of the box shall have a depth that is not less than the depth of the equipment plus $^1/_4$ inch (6.4 mm).
2. Conductors larger than 4 AWG. Boxes that enclose devices or utilization equipment supplied by conductors larger than 4 AWG shall be identified for their specific function.
3. Conductors 8, 6, or 4 AWG. Boxes that enclose devices or utilization equipment supplied by 8, 6, or 4 AWG conductors shall have an internal depth that is not less than $2^1/_{16}$ inches (52.4 mm).
4. Conductors 12 or 10 AWG. Boxes that enclose devices or utilization equipment supplied by 12 or 10 AWG conductors shall have an internal depth that is not less than $1^3/_{16}$ inches (30.2 mm). Where the equipment projects rearward from the mounting plane of the box by more than 1 inch (25.4 mm), the box shall have a depth that is not less than that of the equipment plus $^1/_4$ inch (6.4 mm).
5. Conductors 14 AWG and smaller. Boxes that enclose devices or utilization equipment supplied by 14 AWG or smaller conductors shall have a depth that is not less than $^{15}/_{16}$ inch (23.8 mm).

E3905.5 Boxes enclosing flush-mounted devices. Boxes enclosing flush-mounted devices shall be of such design that the devices are completely enclosed at the back and all sides and shall provide support for the devices. Screws for supporting the box shall not be used for attachment of the device contained therein.

E3905.6 Boxes at luminaire outlets. Outlet boxes used at luminaire or lampholder outlets shall be designed for the support of luminaires and lampholders and shall be installed as required by Section E3904.3.

E3905.6.1 Wall Outlets. Boxes used at luminaire or lampholder outlets in a wall shall be marked on the interior of the box to indicate the maximum weight of the luminaire or lamp holder that is permitted to be supported by the box in the wall, if other than 50 pounds (22.7 kg).

Exception: A wall-mounted luminaire or lampholder weighing not more than 6 pounds (2.7 kg) shall be permitted to be supported on other boxes or plaster rings that are secured to other boxes, provided that the luminaire or its supporting yoke is secured to the box with not fewer than two No. 6 or larger screws.

E3905.6.2 Ceiling outlets. At every outlet used exclusively for lighting, the box shall be designed or installed so that a luminaire or lampholder can be attached. Such boxes shall be capable of supporting a luminaire weighing up to 50 pounds (22.7 kg). A luminaire that weighs more than 50 pounds (22.7 kg) shall be supported independently of the outlet box, unless the outlet box is listed and marked for the maximum weight to be supported.

E3905.7 Floor boxes. Where outlet boxes for receptacles are installed in the floor, such boxes shall be listed specifically for that application.

E3905.8 Boxes at fan outlets. Outlet boxes and outlet box systems used as the sole support of ceiling-suspended fans (paddle) shall be marked by their manufacturer as suitable for this purpose and shall not support ceiling-suspended fans (paddle) that weigh more than 70 pounds (31.8 kg). For outlet boxes and outlet box systems designed to support ceiling-suspended fans (paddle) that weigh more than 35 pounds (15.9 kg), the required marking shall include the maximum weight to be supported.

Where spare, separately switched, ungrounded conductors are provided to a ceiling-mounted outlet box and such box is in a location acceptable for a ceiling-suspended (paddle) fan, the outlet box or outlet box system shall be listed for sole support of a ceiling-suspended (paddle) fan.

E3905.9 Utilization equipment. Boxes used for the support of utilization equipment other than ceiling-suspended (paddle) fans shall meet the requirements of Sections E3905.6.1 and E3905.6.2 for the support of a luminaire that is the same size and weight.

Exception: Utilization equipment weighing not more than 6 pounds (2.7 kg) shall be permitted to be supported on other boxes or plaster rings that are secured to other boxes, provided that the equipment or its supporting yoke is secured to the box with not fewer than two No. 6 or larger screws.

E3905.10 Conduit bodies and junction, pull and outlet boxes to be accessible. Conduit bodies and junction, pull and outlet boxes shall be installed so that the wiring therein can be accessed without removing any part of the building or, in underground circuits, without excavating sidewalks, paving, earth or other substance used to establish the finished grade.

Exception: Boxes covered by gravel, light aggregate or noncohesive granulated soil shall be listed for the application, and the box locations shall be effectively identified and access shall be provided for excavation.

E3905.11 Damp or wet locations. In damp or wet locations, boxes, conduit bodies and fittings shall be placed or equipped so as to prevent moisture from entering or accumulating within the box, conduit body or fitting. Boxes, conduit bodies and fittings installed in wet locations shall be listed for use in wet locations.

E3905.12 Number of conductors in outlet, device, and junction boxes, and conduit bodies. Boxes and conduit bodies shall be of sufficient size to provide free space for all enclosed conductors. In no case shall the volume of the box, as calculated in Section E3905.12.1, be less than the box fill calculation as calculated in Section E3905.12.2. The minimum volume for conduit bodies shall be as calculated in Section E3905.12.3. The provisions of this section shall not apply to terminal housings supplied with motors or generators.

E3905.12.1 Box volume calculations. The volume of a wiring enclosure (box) shall be the total volume of the assembled sections, and, where used, the space provided by plaster rings, domed covers, extension rings, etc., that are marked with their volume in cubic inches or are made from boxes the dimensions of which are listed in Table E3905.12.1.

E3905.12.1.1 Standard boxes. The volumes of standard boxes that are not marked with a cubic-inch capacity shall be as given in Table E3905.12.1.

TABLE E3905.12.2.1
VOLUME ALLOWANCE REQUIRED PER CONDUCTOR

SIZE OF CONDUCTOR	FREE SPACE WITHIN BOX FOR EACH CONDUCTOR (cubic inches)
18 AWG	1.50
16 AWG	1.75
14 AWG	2.00
12 AWG	2.25
10 AWG	2.50
8 AWG	3.00
6 AWG	5.00

For SI: 1 cubic inch = 16.4 cm^3.

TABLE E3905.12.1
MAXIMUM NUMBER OF CONDUCTORS IN METAL BOXES[a]

BOX DIMENSIONS (inches trade size and type)	MAXIMUM CAPACITY (cubic inches)	MAXIMUM NUMBER OF CONDUCTORS[a]						
		18 Awg	16 Awg	14 Awg	12 Awg	10 Awg	8 Awg	6 Awg
$4 \times 1^{1}/_{4}$ round or octagonal	12.5	8	7	6	5	5	4	2
$4 \times 1^{1}/_{2}$ round or octagonal	15.5	10	8	7	6	6	5	3
$4 \times 2^{1}/_{8}$ round or octagonal	21.5	14	12	10	9	8	7	4
$4 \times 1^{1}/_{4}$ square	18.0	12	10	9	8	7	6	3
$4 \times 1^{1}/_{2}$ square	21.0	14	12	10	9	8	7	4
$4 \times 2^{1}/_{8}$ square	30.3	20	17	15	13	12	10	6
$4^{11}/_{16} \times {}^{11}/_{4}$ square	25.5	17	14	12	11	10	8	5
$4^{11}/_{16} \times {}^{11}/_{2}$ square	29.5	19	16	14	13	11	9	5
$4^{11}/_{16} \times 2^{1}/_{8}$ square	42.0	28	24	21	18	16	14	8
$3 \times 2 \times 1^{1}/_{2}$ device	7.5	5	4	3	3	3	2	1
$3 \times 2 \times 2$ device	10.0	6	5	5	4	4	3	2
$3 \times 2 \times 2^{1}/_{4}$ device	10.5	7	6	5	4	4	3	2
$3 \times 2 \times 2^{1}/_{2}$ device	12.5	8	7	6	5	5	4	2
$3 \times 2 \times 2^{3}/_{4}$ device	14.0	9	8	7	6	5	4	2
$3 \times 2 \times 3^{1}/_{2}$ device	18.0	12	10	9	8	7	6	3
$4 \times 2^{1}/_{8} \times 1^{1}/_{2}$ device	10.3	6	5	5	4	4	3	2
$4 \times 2^{1}/_{8} \times 1^{7}/_{8}$ device	13.0	8	7	6	5	5	4	2
$4 \times 2^{1}/_{8} \times 2^{1}/_{8}$ device	14.5	9	8	7	6	5	4	2
$3^{3}/_{4} \times 2 \times 2^{1}/_{2}$ masonry box/gang	14.0	9	8	7	6	5	4	2
$3^{3}/_{4} \times 2 \times 3^{1}/_{2}$ masonry box/gang	21.0	14	12	10	9	8	7	4

For SI: 1 inch = 25.4 mm, 1 cubic inch = 16.4 cm^3.

a. Where volume allowances are not required by Sections E3905.12.2.2 through E3905.12.2.5.

E3905.12.1.2 Other boxes. Boxes 100 cubic inches (1640 cm^3) or less, other than those described in Table E3905.12.1, and nonmetallic boxes shall be durably and legibly marked by the manufacturer with their cubic-inch capacity. Boxes described in Table E3905.12.1 that have a larger cubic inch capacity than is designated in the table shall be permitted to have their cubic-inch capacity marked as required by this section.

E3905.12.2 Box fill calculations. The volumes in Section E3905.12.2.1 through Section E3905.12.2.5, as applicable, shall be added together. No allowance shall be required for small fittings such as locknuts and bushings.

E3905.12.2.1 Conductor fill. Each conductor that originates outside the box and terminates or is spliced within the box shall be counted once, and each conductor that passes through the box without splice or termination shall be counted once. Each loop or coil of unbroken conductor having a length equal to or greater than twice that required for free conductors by Section E3406.10.3, shall be counted twice. The conductor fill, in cubic inches, shall be computed using Table E3905.12.2.1. A conductor, no part of which leaves the box, shall not be counted.

Exception: An equipment grounding conductor or not more than four fixture wires smaller than No. 14, or both, shall be permitted to be omitted from the calculations where such conductors enter a box from a domed fixture or similar canopy and terminate within that box.

E3905.12.2.2 Clamp fill. Where one or more internal cable clamps, whether factory or field supplied, are present in the box, a single volume allowance in accordance with Table E3905.12.2.1 shall be made based on the largest conductor present in the box. No allowance shall be required for a cable connector with its clamping mechanism outside the box.

E3905.12.2.3 Support fittings fill. Where one or more fixture studs or hickeys are present in the box, a single volume allowance in accordance with Table E3905.12.2.1 shall be made for each type of fitting based on the largest conductor present in the box.

E3905.12.2.4 Device or equipment fill. For each yoke or strap containing one or more devices or equipment, a double volume allowance in accordance with Table E3905.12.2.1 shall be made for each yoke or strap based on the largest conductor connected to a device(s) or equipment supported by that yoke or strap. For a device or utilization equipment that is wider than a single 2-inch (51 mm) device box as described in Table E3905.12.1, a double volume allowance shall be made for each ganged portion required for mounting of the device or equipment.

E3905.12.2.5 Equipment grounding conductor fill. Where one or more equipment grounding conductors or equipment bonding jumpers enters a box, a single volume allowance in accordance with Table E3905.12.2.1 shall be made based on the largest equipment grounding conductor or equipment bonding jumper present in the box.

E3905.12.3 Conduit bodies. Conduit bodies enclosing 6 AWG conductors or smaller, other than short-radius conduit bodies, shall have a cross-sectional area not less than twice the cross-sectional area of the largest conduit or tubing to which they can be attached. The maximum number of conductors permitted shall be the maximum number permitted by Section E3904.6 for the conduit to which it is attached.

E3905.12.3.1 Splices, taps or devices. Only those conduit bodies that are durably and legibly marked by the manufacturer with their cubic inch capacity shall be permitted to contain splices, taps or devices. The maximum number of conductors shall be calculated using the same procedure for similar conductors in other than standard boxes.

E3905.12.3.2 Short-radius conduit bodies. Conduit bodies such as capped elbows and service-entrance elbows that enclose conductors 6 AWG or smaller and that are only intended to enable the installation of the raceway and the contained conductors, shall not contain splices, taps, or devices and shall be of sufficient size to provide free space for all conductors enclosed in the conduit body.

SECTION E3906 INSTALLATION OF BOXES, CONDUIT BODIES AND FITTINGS

E3906.1 Conductors entering boxes, conduit bodies or fittings. Conductors entering boxes, conduit bodies or fittings shall be protected from abrasion.

E3906.1.1 Insulated fittings. Where raceways contain 4 AWG or larger insulated circuit conductors and these conductors enter a cabinet, box enclosure, or raceway, the conductors shall be protected by an identified fitting providing a smoothly rounded insulating surface, unless the conductors are separated from the fitting or raceway by identified insulating material securely fastened in place.

Exception: Where threaded hubs or bosses that are an integral part of a cabinet, box enclosure, or raceway provide a smoothly rounded or flared entry for conductors.

Conduit bushings constructed wholly of insulating material shall not be used to secure a fitting or raceway. The insulating fitting or insulating material shall have a temperature rating not less than the insulation temperature rating of the installed conductors.

E3906.2 Openings. Openings through which conductors enter shall be adequately closed.

E3906.3 Metal boxes and conduit bodies. Where raceway or cable is installed with metal boxes, or conduit bodies, the raceway or cable shall be secured to such boxes and conduit bodies.

E3906.4 Unused openings. Unused openings other than those intended for the operation of equipment, those intended for mounting purposes, or those permitted as part of the design for listed equipment, shall be closed to afford protection substantially equivalent to that of the wall of the equipment. Metal plugs or plates used with nonmetallic boxes or conduit bodies shall be recessed at least $^1/_4$ inch (6.4 mm) from the outer surface of the box or conduit body.

E3906.5 In wall or ceiling. In walls or ceilings of concrete, tile or other noncombustible material, boxes employing a flush-type cover or faceplate shall be installed so that the front edge of the box, plaster ring, extension ring, or listed extender will not be set back from the finished surface more than $^1/_4$ inch (6.4 mm). In walls and ceilings constructed of wood or other combustible material, boxes, plaster rings, extension rings and listed extenders shall be flush with the finished surface or project therefrom.

E3906.6 Noncombustible surfaces. Openings in noncombustible surfaces that accommodate boxes employing a flush-type cover or faceplate shall be made so that there are no gaps or open spaces greater than $^1/_8$ inch (3.2 mm) around the edge of the box.

E3906.7 Surface extensions. Surface extensions shall be made by mounting and mechanically securing an extension ring over the box.

Exception: A surface extension shall be permitted to be made from the cover of a flush-mounted box where the cover is designed so it is unlikely to fall off, or be removed if its securing means becomes loose. The wiring method shall be flexible for a length sufficient to permit removal of the cover and provide access to the box interior and arranged so that any bonding or grounding continuity is independent of the connection between the box and cover.

E3906.8 Supports. Boxes and enclosures shall be supported in accordance with one or more of the provisions in Sections E3906.8.1 through E3906.8.6.

E3906.8.1 Surface mounting. An enclosure mounted on a building or other surface shall be rigidly and securely fastened in place. If the surface does not provide rigid and secure support, additional support in accordance with other provisions of Section E3906.8 shall be provided.

E3906.8.2 Structural mounting. An enclosure supported from a structural member of a building or from grade shall be rigidly supported either directly, or by using a metal, polymeric or wood brace.

E3906.8.2.1 Nails and screws. Nails and screws, where used as a fastening means, shall be attached by using brackets on the outside of the enclosure, or they shall pass through the interior within $^1/_4$ inch (6.4 mm) of the back or ends of the enclosure. Screws shall not be permitted to pass through the box except where exposed threads in the box are protected by an approved means to avoid abrasion of conductor insulation.

E3906.8.2.2 Braces. Metal braces shall be protected against corrosion and formed from metal that is not less than 0.020 inch (0.508 mm) thick uncoated. Wood braces shall have a cross section not less than nominal 1 inch by 2 inches (25.4 mm by 51 mm). Wood braces in wet locations shall be treated for the conditions. Polymeric braces shall be identified as being suitable for the use.

E3906.8.3 Mounting in finished surfaces. An enclosure mounted in a finished surface shall be rigidly secured there to by clamps, anchors, or fittings identified for the application.

E3906.8.4 Raceway supported enclosures without devices or fixtures. An enclosure that does not contain a device(s), other than splicing devices, or support a luminaire, lampholder or other equipment, and that is supported by entering raceways shall not exceed 100 cubic inches (1640 cm^3) in size. The enclosure shall have threaded entries or have hubs identified for the purpose. The enclosure shall be supported by two or more conduits threaded wrenchtight into the enclosure or hubs. Each conduit shall be secured within 3 feet (914 mm) of the enclosure, or within 18 inches (457 mm) of the enclosure if all entries are on the same side of the enclosure.

Exception: Rigid metal, intermediate metal, or rigid polyvinyl chloride nonmetallic conduit or electrical metallic tubing shall be permitted to support a conduit body of any size, provided that the conduit body is not larger in trade size than the largest trade size of the supporting conduit or electrical metallic tubing.

E3906.8.5 Raceway supported enclosures, with devices or luminaire. An enclosure that contains a device(s), other than splicing devices, or supports a luminaire, lampholder or other equipment and is supported by entering raceways shall not exceed 100 cubic inches (1640 cm^3) in size. The enclosure shall have threaded entries or have hubs identified for the purpose. The enclosure shall be supported by two or more conduits threaded wrench-tight into the enclosure or hubs. Each conduit shall be secured within 18 inches (457 mm) of the enclosure.

Exceptions:

1. Rigid metal or intermediate metal conduit shall be permitted to support a conduit body of any size, provided that the conduit bodies are not larger in trade size than the largest trade size of the supporting conduit.
2. An unbroken length(s) of rigid or intermediate metal conduit shall be permitted to support a box used for luminaire or lampholder support, or to support a wiring enclosure that is an integral part of a luminaire and used in lieu of a box in accordance with Section E3905.1.1, where all of the following conditions are met:
   - 2.1. The conduit is securely fastened at a point so that the length of conduit beyond the last point of conduit support does not exceed 3 feet (914 mm).
   - 2.2. The unbroken conduit length before the last point of conduit support is 12 inches (305 mm) or greater, and that portion of the conduit is securely fastened at some

point not less than 12 inches (305 mm) from its last point of support.

2.3. Where accessible to unqualified persons, the luminaire or lampholder, measured to its lowest point, is not less than 8 feet (2438 mm) above grade or standing area and at least 3 feet (914 mm) measured horizontally to the 8-foot (2438 mm) elevation from windows, doors, porches, fire escapes, or similar locations.

2.4. A luminaire supported by a single conduit does not exceed 12 inches (305 mm) in any direction from the point of conduit entry.

2.5. The weight supported by any single conduit does not exceed 20 pounds (9.1 kg).

2.6. At the luminaire or lampholder end, the conduit(s) is threaded wrenchtight into the box, conduit body, or integral wiring enclosure, or into hubs identified for the purpose. Where a box or conduit body is used for support, the luminaire shall be secured directly to the box or conduit body, or through a threaded conduit nipple not over 3 inches (76 mm) long.

E3906.8.6 Enclosures in concrete or masonry. An enclosure supported by embedment shall be identified as being suitably protected from corrosion and shall be securely embedded in concrete or masonry.

E3906.9 Covers and canopies. Outlet boxes shall be effectively closed with a cover, faceplate or fixture canopy.

E3906.10 Metal covers and plates. Metal covers and plates shall be grounded.

E3906.11 Exposed combustible finish. Combustible wall or ceiling finish exposed between the edge of a fixture canopy or pan and the outlet box shall be covered with noncombustible material.

SECTION E3907 CABINETS AND PANELBOARDS

E3907.1 Switch and overcurrent device enclosures with splices, taps, and feed-through conductors. Where the wiring space of enclosures for switches or overcurrent devices contains conductors that are feeding through, spliced, or tapping off to other enclosures, switches, or overcurrent devices, all of the following conditions shall apply:

1. The total area of all conductors installed at any cross section of the wiring space shall not exceed 40 percent of the cross-sectional area of that space.
2. The total area of all conductors, splices, and taps installed at any cross section of the wiring space shall not exceed 75 percent of the cross-sectional area of that space.
3. A warning label shall be applied to the enclosure that identifies the closest disconnecting means for any feed-through conductors.

E3907.2 Damp and wet locations. In damp or wet locations, cabinets and panelboards of the surface type shall be placed or equipped so as to prevent moisture or water from entering and accumulating within the cabinet, and shall be mounted to provide an air-space not less than $^1/_4$ inch (6.4 mm) between the enclosure and the wall or other supporting surface. Cabinets installed in wet locations shall be weatherproof. For enclosures in wet locations, raceways and cables entering above the level of uninsulated live parts shall be installed with fittings listed for wet locations.

Exception: Nonmetallic enclosures installed on concrete, masonry, tile, or similar surfaces shall not be required to be installed with an air space between the enclosure and the wall or supporting surface.

E3907.3 Position in wall. In walls of concrete, tile or other noncombustible material, cabinets and panelboards shall be installed so that the front edge of the cabinet will not set back of the finished surface more than $^1/_4$ inch (6.4 mm). In walls constructed of wood or other combustible material, cabinets shall be flush with the finished surface or shall project therefrom.

E3907.4 Repairing noncombustible surfaces. Noncombustible surfaces that are broken or incomplete shall be repaired so that there will not be gaps or open spaces greater than $^1/_8$ inch (3.2 mm) at the edge of the cabinet or cutout box employing a flush-type cover.

E3907.5 Unused openings. Unused openings, other than those intended for the operation of equipment, those intended for mounting purposes, and those permitted as part of the design for listed equipment, shall be closed to afford protection substantially equivalent to that of the wall of the equipment. Metal plugs and plates used with nonmetallic cabinets shall be recessed at least $^1/_4$ inch (6.4 mm) from the outer surface. Unused openings for circuit breakers and switches shall be closed using identified closures, or other approved means that provide protection substantially equivalent to the wall of the enclosure.

E3907.6 Conductors entering cabinets. Conductors entering cabinets and panelboards shall be protected from abrasion and shall comply with Section E3906.1.1.

E3907.7 Openings to be closed. Openings through which conductors enter cabinets, panelboards and meter sockets shall be adequately closed.

E3907.8 Cables. Where cables are used, each cable shall be secured to the cabinet, panelboard, cutout box, or meter socket enclosure.

Exception: Cables with entirely nonmetallic sheaths shall be permitted to enter the top of a surface-mounted enclosure through one or more sections of rigid raceway not less than 18 inches (457 mm) nor more than 10 feet (3048 mm) in length, provided all the following conditions are met:

1. Each cable is fastened within 12 inches (305 mm), measured along the sheath, of the outer end of the raceway.
2. The raceway extends directly above the enclosure and does not penetrate a structural ceiling.
3. A fitting is provided on each end of the raceway to protect the cable(s) from abrasion and the fittings remain accessible after installation.
4. The raceway is sealed or plugged at the outer end using approved means so as to prevent access to the enclosure through the raceway.
5. The cable sheath is continuous through the raceway and extends into the enclosure beyond the fitting not less than $^1/_4$ inch (6.4 mm).
6. The raceway is fastened at its outer end and at other points in accordance with Section E3802.1.
7. The allowable cable fill shall not exceed that permitted by Table E3907.8.A multiconductor cable having two or more conductors shall be treated as a single conductor for calculating the percentage of conduit fill area. For cables that have elliptical cross sections, the cross-sectional area calculation shall be based on the major diameter of the ellipse as a circle diameter.

TABLE E3907.8
PERCENT OF CROSS SECTION OF CONDUIT AND TUBING FOR CONDUCTORS

NUMBER OF CONDUCTORS	MAXIMUM PERCENT OF CONDUIT AND TUBING AREA FILLED BY CONDUCTORS
1	53
2	31
Over 2	40

SECTION E3908 GROUNDING

E3908.1 Metal enclosures. Metal enclosures of conductors, devices and equipment shall be connected to the equipment grounding conductor.

Exceptions:

1. Short sections of metal enclosures or raceways used to provide cable assemblies with support or protection against physical damage.
2. A metal elbow that is installed in an underground installation of rigid nonmetallic conduit and is isolated from possible contact by a minimum cover of 18 inches (457 mm) to any part of the elbow or that is encased in not less than 2 inches (51 mm) of concrete.

E3908.2 Equipment fastened in place or connected by permanent wiring methods (fixed). Exposed, normally noncurrent-carrying metal parts of fixed equipment supplied by or enclosing conductors or components that are likely to become energized shall be connected to the equipment grounding conductor where any of the following conditions apply:

1. Where within 8 feet (2438 mm) vertically or 5 feet (1524 mm) horizontally of earth or grounded metal objects and subject to contact by persons;
2. Where located in a wet or damp location and not isolated; or
3. Where in electrical contact with metal.

E3908.3 Specific equipment fastened in place (fixed) or connected by permanent wiring methods. Exposed, normally noncurrent-carrying metal parts of the following equipment and enclosures shall be connected to an equipment grounding conductor:

1. Luminaires as provided in Chapter 40.
2. Motor-operated water pumps, including submersible types.Where a submersible pump is used in a metal well casing, the well casing shall be connected to the pump circuit equipment grounding conductor.

E3908.4 Effective ground-fault current path. Electrical equipment and wiring and other electrically conductive material likely to become energized shall be installed in a manner that creates a low-impedance circuit facilitating the operation of the overcurrent device or ground detector for high-impedance grounded systems. Such circuit shall be capable of safely carrying the maximum ground-fault current likely to be imposed on it from any point on the wiring system where a ground fault might occur to the electrical supply source.

E3908.5 Earth as a ground-fault current path. The earth shall not be considered as an effective ground-fault current path.

E3908.6 Load-side grounded conductor neutral. A grounded conductor shall not be connected to normally noncurrent-carrying metal parts of equipment, to equipment grounding conductor(s), or be reconnected to ground on the load side of the service disconnecting means.

E3908.7 Load-side equipment. A grounded circuit conductor shall not be used for grounding noncurrent-carrying metal parts of equipment on the load side of the service disconnecting means.

E3908.8 Types of equipment grounding conductors. The equipment grounding conductor run with or enclosing the circuit conductors shall be one or more or a combination of the following:

1. A copper, aluminum or copper-clad conductor.This conductor shall be solid or stranded; insulated, covered or bare; and in the form of a wire or a busbar of any shape.
2. Rigid metal conduit.
3. Intermediate metal conduit.
4. Electrical metallic tubing.
5. Armor of Type AC cable in accordance with Section E3908.4.

6. Type MC cable that provides an effective ground-fault current path in accordance with one or more of the following:
   6.1. It contains an insulated or uninsulated equipment grounding conductor in compliance with Item 1 of this section.
   6.2. The combined metallic sheath and uninsulated equipment grounding/bonding conductor of interlocked metal tape-type MC cable that is listed and identified as an equipment grounding conductor.
   6.3. The metallic sheath or the combined metallic sheath and equipment grounding conductors of the smooth or corrugated tube-type MC cable that is listed and identified as an equipment grounding conductor.
7. Other electrically continuous metal raceways and auxiliary gutters.
8. Surface metal raceways listed for grounding.

E3908.8.1 Flexible metal conduit. Flexible metal conduit shall be permitted as an equipment grounding conductor where all of the following conditions are met:

1. The conduit is terminated in listed fittings.
2. The circuit conductors contained in the conduit are protected by overcurrent devices rated at 20 amperes or less.
3. The combined length of flexible metal conduit and flexible metallic tubing and liquid-tight flexible metal conduit in the same ground return path does not exceed 6 feet (1829 mm).

If used to connect equipment where flexibility is necessary to minimize the transmission of vibration from equipment or to provide flexibility for equipment that requires movement after installation, an equipment grounding conductor shall be installed.

E3908.8.2 Liquid-tight flexible metal conduit. Liquid-tight flexible metal conduit shall be permitted as an equipment grounding conductor where all of the following conditions are met:

1. The conduit is terminated in listed fittings.
2. For trade sizes $^3/_8$ through $^1/_2$ (metric designator 12 through 16), the circuit conductors contained in the conduit are protected by overcurrent devices rated at 20 amperes or less.
3. For trade sizes $^3/_4$ through $1^1/_4$ (metric designator 21 through 35), the circuit conductors contained in the conduit are protected by overcurrent devices rated at not more than 60 amperes and there is no flexible metal conduit, flexible metallic tubing, or liquid-tight flexible metal conduit in trade sizes $^3/_8$ inch or $^1/_2$ inch (9.5 mm through 12.7 mm) in the ground fault current path.
4. The combined length of flexible metal conduit and flexible metallic tubing and liquid-tight flexible metal conduit in the same ground return path does not exceed 6 feet (1829 mm).

If used to connect equipment where flexibility is necessary to minimize the transmission of vibration from equipment or to provide flexibility for equipment that requires movement after installation, an equipment grounding conductor shall be installed.

E3908.8.3 Nonmetallic sheathed cable (Type NM). In addition to the insulated conductors, the cable shall have an insulated, covered, or bare equipment grounding conductor. Equipment grounding conductors shall be sized in accordance with Table E3908.12.

E3908.9 Equipment fastened in place or connected by permanent wiring methods. Noncurrent-carrying metal parts of equipment, raceways and other enclosures, where required to be grounded, shall be grounded by one of the following methods:

1. By any of the equipment grounding conductors permitted by Sections E3908.8 through E3908.8.3.
2. By an equipment grounding conductor contained within the same raceway, cable or cord, or otherwise run with the circuit conductors.Equipment grounding conductors shall be identified in accordance with Section E3407.2.

E3908.10 Methods of equipment grounding. Fixtures and equipment shall be considered grounded where mechanically connected to an equipment grounding conductor as specified in Sections E3908.8 through E3908.8.3. Wire type equipment grounding conductors shall be sized in accordance with Section E3908.12.

E3908.11 Equipment grounding conductor installation. Where an equipment grounding conductor consists of a raceway, cable armor or cable sheath or where such conductor is a wire within a raceway or cable, it shall be installed in accordance with the provisions of this chapter and Chapters 34 and 38 using fittings for joints and terminations approved for installation with the type of raceway or cable used. All connections, joints and fittings shall be made tight using suitable tools.

E3908.12 Equipment grounding conductor size. Copper, aluminum and copper-clad aluminum equipment grounding conductors of the wire type shall be not smaller than shown in Table E3908.12, but in no case shall they be required to be larger than the circuit conductors supplying the equipment. Where a raceway or a cable armor or sheath is used as the equipment grounding conductor, as provided in Section E3908.8, it shall comply with Section E3908.4. Where ungrounded connectors are increased in size, equipment grounding conductors shall be increased proportionally according to the circular mil area of the ungrounded conductors.

E3908.12.1 Multiple circuits. Where a single equipment grounding conductor is run with multiple circuits in the same raceway or cable, it shall be sized for the largest overcurrent device protecting conductors in the raceway or cable.

E3908.13 Continuity and attachment of equipment grounding conductors to boxes. Where circuit conductors are spliced within a box or terminated on equipment within or supported by a box, any equipment grounding conductors associated with the circuit conductors shall be connected within the box or to the box with devices suitable for the use. Connections depending solely on solder shall not be used. Splices shall be made in accordance with Section E3406.10 except that insulation shall not be required. The arrangement of grounding connections shall be such that the disconnection or removal of a receptacle, luminaire or other device fed from the box will not interfere with or interrupt the grounding continuity.

E3908.14 Connecting receptacle grounding terminal to box. An equipment bonding jumper, sized in accordance with Table E3908.12 based on the rating of the overcurrent device protecting the circuit conductors, shall be used to connect the grounding terminal of a grounding-type receptacle to a grounded box except where grounded in accordance with one of the following:

1. Surface mounted box.Where the box is mounted on the surface, direct metal-to-metal contact between the device yoke and the box shall be permitted to ground the receptacle to the box. At least one of the insulating washers shall be removed from receptacles that do not have a contact yoke or device designed and listed to be used in conjunction with the supporting screws to establish the grounding circuit between the device yoke and flush-type boxes. This provision shall not apply to cover-mounted receptacles except where the box and cover combination are listed as providing satisfactory ground continuity between the box and the receptacle. A listed exposed work cover shall be considered to be the grounding and bonding means where the device is attached to the cover with at least two fasteners that are permanent, such as a rivet or have a thread locking or screw locking means and where the cover mounting holes are located on a flat non-raised portion of the cover.
2. Contact devices or yokes.Contact devices or yokes designed and listed for the purpose shall be permitted in conjunction with the supporting screws to establish the grounding circuit between the device yoke and flush-type boxes.
3. Floor boxes.The receptacle is installed in a floor box designed for and listed as providing satisfactory ground continuity between the box and the device.

E3908.15 Metal boxes. A connection shall be made between the one or more equipment grounding conductors and a metal box by means of a grounding screw that shall be used for no other purpose, equipment listed for grounding or by means of a listed grounding device. Where screws are used to connect grounding conductors or connection devices to boxes, such screws shall be:

1. Machine screw-type fasteners that engage not less than two threads,
2. Secured with a nut, or
3. Thread-forming machine screws that engage not less than two threads in the enclosure.

E3908.16 Nonmetallic boxes. One or more equipment grounding conductors brought into a nonmetallic outlet box shall be arranged to allow connection to fittings or devices installed in that box.

E3908.17 Clean surfaces. Nonconductive coatings such as paint, lacquer and enamel on equipment to be grounded shall be removed from threads and other contact surfaces to ensure electrical continuity or the equipment shall be connected by means of fittings designed so as to make such removal unnecessary.

E3908.18 Bonding other enclosures. Metal raceways, cable armor, cable sheath, enclosures, frames, fittings and other metal noncurrent-carrying parts that serve as equipment grounding conductors, with or without the use of supplementary equipment grounding conductors, shall be effectively bonded where necessary to ensure electrical continuity and the capacity to conduct safely any fault current likely to be imposed on them. Any nonconductive paint, enamel and similar coating shall be removed at threads, contact points and contact surfaces, or connections shall be made by means of fittings designed so as to make such removal unnecessary.

E3908.19 Size of equipment bonding jumper on load side of an overcurrent device. The equipment bonding jumper on the load side of an overcurrent devices shall be sized, as a minimum, in accordance with Table E3908.12, but shall not be required to be larger than the circuit conductors supplying

TABLE E3908.12
EQUIPMENT GROUNDING CONDUCTOR SIZING

RATING OR SETTING OF AUTOMATIC OVERCURRENT DEVICE IN CIRCUIT AHEAD OF EQUIPMENT, CONDUIT, ETC., NOT EXCEEDING THE FOLLOWING RATINGS (amperes)	MINIMUM SIZE	
	Copper wire No. (AWG)	Aluminum or copper-clad aluminum wire No. (AWG)
15	14	12
20	12	10
60	10	8
100	8	6
200	6	4
300	4	2
400	3	1

the equipment. An equipment bonding conductor shall be not smaller than No. 14 AWG.

A single common continuous equipment bonding jumper shall be permitted to connect two or more raceways or cables where the bonding jumper is sized in accordance with Table E3908.12 for the largest overcurrent device supplying circuits therein.

E3908.20 Installation equipment bonding jumper. Bonding jumpers or conductors and equipment bonding jumpers shall be installed either inside or outside of a raceway or an enclosure in accordance with Sections E3908.20.1 and E3908.20.2.

E3908.20.1 Inside raceway or enclosure. Where installed inside a raceway or enclosure, equipment bonding jumpers and bonding jumpers or conductors shall comply with the requirements of Sections E3407.2 and E3908.13.

E3908.20.2 Outside raceway or enclosure. Where installed outside of a raceway or enclosure, the length of the bonding jumper or conductor or equipment bonding jumper shall not exceed 6 feet (1829 mm) and shall be routed with the raceway or enclosure.

Equipment bonding jumpers and supply-side bonding jumpers installed for bonding grounding electrodes and installed at outdoor pole locations for the purpose of bonding or grounding isolated sections of metal raceways or elbows installed in exposed risers of metal conduit or other metal raceway, shall not be limited in length and shall not be required to be routed with a raceway or enclosure.

E3908.20.3 Protection. Bonding jumpers or conductors and equipment bonding jumpers shall be installed in accordance with Section E3610.2.

SECTION E3909
FLEXIBLE CORDS

E3909.1 Where permitted. Flexible cords shall be used only for the connection of appliances where the fastening means and mechanical connections of such appliances are designed to permit ready removal for maintenance, repair or frequent interchange and the appliance is listed for flexible cord connection. Flexible cords shall not be installed as a substitute for the fixed wiring of a structure; shall not be run through holes in walls, structural ceilings, suspended ceilings, dropped ceilings or floors; shall not be concealed behind walls, floors, ceilings or located above suspended or dropped ceilings.

E3909.2 Loading and protection. The ampere load of flexible cords serving fixed appliances shall be in accordance with Table E3909.2. This table shall be used in conjunction with applicable end use product standards to ensure selection of the proper size and type. Where flexible cord is approved for and used with a specific listed appliance, it shall be considered to be protected where applied within the appliance listing requirements.

TABLE E3909.2
MAXIMUM AMPERE LOAD FOR FLEXIBLE CORDS

CORD SIZE (AWG)	CORD TYPES S, SE, SEO, SJ, SJE, SJEO, SJO, SJOO, SJT, SJTO, SJTOO, SO, SOO, SRD, SRDE, SRDT, ST, STD, SV, SVO, SVOO, SVTO, SVTOO	
	Maximum ampere load	
	Three current-carrying conductors	Two current-carrying conductors
18	7	10
16	10	13
14	15	18
12	20	25

E3909.3 Splices. Flexible cord shall be used only in continuous lengths without splices or taps.

E3909.4 Attachment plugs. Where used in accordance with Section E3909.1, each flexible cord shall be equipped with an attachment plug and shall be energized from a receptacle outlet.

CHAPTER 40

DEVICES AND LUMINAIRES

SECTION E4001 SWITCHES

E4001.1 Rating and application of snap switches. General-use snap switches shall be used within their ratings and shall control only the following loads:

1. Resistive and inductive loads not exceeding the ampere rating of the switch at the voltage involved.
2. Tungsten-filament lamp loads not exceeding the ampere rating of the switch at 120 volts.
3. Motor loads not exceeding 80 percent of the ampere rating of the switch at its rated voltage.

E4001.2 CO/ALR snap switches. Snap switches rated 20 amperes or less directly connected to aluminum conductors shall be marked CO/ALR.

E4001.3 Indicating. General-use and motor-circuit switches and circuit breakers shall clearly indicate whether they are in the open OFF or closed ON position. Where single-throw switches or circuit breaker handles are operated vertically rather than rotationally or horizontally, the up position of the handle shall be the ON position.

E4001.4 Time switches and similar devices. Time switches and similar devices shall be of the enclosed type or shall be mounted in cabinets or boxes or equipment enclosures. A barrier shall be used around energized parts to prevent operator exposure when making manual adjustments or switching.

E4001.5 Grounding of enclosures. Metal enclosures for switches or circuit breakers shall be connected to an equipment grounding conductor. Metal enclosures for switches or circuit breakers used as service equipment shall comply with the provisions of Section E3609.4. Where nonmetallic enclosures are used with metal raceways or metal-armored cables, provisions shall be made for connecting the equipment grounding conductor.

Nonmetallic boxes for switches shall be installed with a wiring method that provides or includes an equipment grounding conductor.

E4001.6 Access. All switches and circuit breakers used as switches shall be located to allow operation from a readily accessible location. Such devices shall be installed so that the center of the grip of the operating handle of the switch or circuit breaker, when in its highest position, will not be more than 6 feet 7 inches (2007 mm) above the floor or working platform.

E4001.7 Damp or wet locations. A surface mounted switch or circuit breaker located in a damp or wet location or outside of a building shall be enclosed in a weatherproof enclosure or cabinet. A flush-mounted switch or circuit breaker in a damp or wet location shall be equipped with a weatherproof cover. Switches shall not be installed within wet locations in tub or shower spaces unless installed as part of a listed tub or shower assembly.

E4001.8 Grounded conductors. Switches or circuit breakers shall not disconnect the grounded conductor of a circuit except where the switch or circuit breaker simultaneously disconnects all conductors of the circuit.

E4001.9 Switch connections. Three- and four-way switches shall be wired so that all switching occurs only in the ungrounded circuit conductor. Color coding of switch connection conductors shall comply with Section E3407.3. Where in metal raceways or metal-jacketed cables, wiring between switches and outlets shall be in accordance with Section E3406.7.

Exception: Switch loops do not require a grounded conductor.

E4001.10 Box mounted. Flush-type snap switches mounted in boxes that are recessed from the finished wall surfaces as covered in Section E3906.5 shall be installed so that the extension plaster ears are seated against the surface of the wall. Flush-type snap switches mounted in boxes that are flush with the finished wall surface or project therefrom shall be installed so that the mounting yoke or strap of the switch is seated against the box.

E4001.11 Snap switch faceplates. Faceplates provided for snap switches mounted in boxes and other enclosures shall be installed so as to completely cover the opening and, where the switch is flush mounted, seat against the finished surface.

E4001.11.1 Faceplate grounding. Snap switches, including dimmer and similar control switches, shall be connected to an equipment grounding conductor and shall provide a means to connect metal faceplates to the equipment grounding conductor, whether or not a metal faceplate is installed. Snap switches shall be considered to be part of an effective ground-fault current path if either of the following conditions is met:

1. The switch is mounted with metal screws to a metal box or metal cover that is connected to an equipment grounding conductor or to a nonmetallic box with integral means for connecting to an equipment grounding conductor.
2. An equipment grounding conductor or equipment bonding jumper is connected to an equipment grounding termination of the snap switch.

Exceptions:

1. Where a means to connect to an equipment grounding conductor does not exist within the snap-switch enclosure or where the wiring method does not include or provide an equipment grounding conductor, a snap switch without a grounding connection to an equipment grounding conductor shall be permitted for replacement purposes only. A snap switch wired under the provisions of this exception and located within 8 feet

(2438 mm) vertically or 5 feet (1524 mm) horizontally of ground or exposed grounded metal objects, shall be provided with a faceplate of nonconducting noncombustible material with nonmetallic attachment screws, except where the switch-mounting strap or yoke is nonmetallic or the circuit is protected by a ground-fault circuit interrupter.

2. Listed kits or listed assemblies shall not be required to be connected to an equipment grounding conductor if all of the following conditions apply:

   2.1. The device is provided with a nonmetallic faceplate that cannot be installed on any other type of device.

   2.2. The device does not have mounting means to accept other configurations of faceplates.

   2.3. The device is equipped with a nonmetallic yoke.

   2.4. All parts of the device that are accessible after installation of the faceplate are manufactured of nonmetallic materials.

3. Connection to an equipment grounding conductor shall not be required for snap switches that have an integral nonmetallic enclosure complying with Section E3905.13.

E4001.12 Dimmer switches. General-use dimmer switches shall be used only to control permanently installed incandescent luminaires (lighting fixtures) except where listed for the control of other loads and installed accordingly.

E4001.13 Multipole snap switches. A multipole, general-use snap switch shall not be fed from more than a single circuit unless it is listed and marked as a two-circuit or three-circuit switch, or unless its voltage rating is not less than the nominal line-to-line voltage of the system supplying the circuits.

E4001.14 Cord-and-plug-connected loads. Where snap switches are used to control cord-and-plug-connected equipment on a general-purpose branch circuit, each snap switch controlling receptacle outlets or cord connectors that are supplied by permanently connected cord pendants shall be rated at not less than the rating of the maximum permitted ampere rating or setting of the overcurrent device protecting the receptacles or cord connectors, as provided in Sections E4002.1.1 and E4002.1.2.

E4001.15 Switches controlling lighting loads. Where switches control lighting loads supplied by a grounded general purpose branch circuit, the grounded circuit conductor for the controlled lighting circuit shall be provided at the switch location.

Exception: The grounded circuit conductor is not required to be provided at the switch enclosure where either of the following conditions apply:

1. The conductors enter the box through a raceway. The raceway shall have sufficient cross-sectional area to accommodate the extension of the grounded circuit conductor of the lighting circuit to the switch location whether or not the conductors in the raceway are required to be increased in size to comply with Section E3705.3.

2. Cable assemblies enter the box through a framing cavity that is open at the top or bottom on the same floor level, or through a wall, floor, or ceiling that is unfinished on one side.

SECTION E4002 RECEPTACLES

E4002.1 Rating and type. Receptacles and cord connectors shall be rated at not less than 15 amperes, 125 volts, or 15 amperes, 250 volts, and shall not be a lampholder type. Receptacles shall be rated in accordance with this section.

E4002.1.1 Single receptacle. A single receptacle installed on an individual branch circuit shall have an ampere rating not less than that of the branch circuit.

E4002.1.2 Two or more receptacles. Where connected to a branch circuit supplying two or more receptacles or outlets, receptacles shall conform to the values listed in Table E4002.1.2.

TABLE E4002.1.2 RECEPTACLE RATINGS FOR VARIOUS SIZE MULTI-OUTLET CIRCUITS

CIRCUIT RATING (amperes)	RECEPTACLE RATING (amperes)
15	15
20	15 or 20
30	30
40	40 or 50
50	50

E4002.2 Grounding type. Receptacles installed on 15- and 20-ampere-rated branch circuits shall be of the grounding type.

E4002.3 CO/ALR receptacles. Receptacles rated at 20 amperes or less and directly connected to aluminum conductors shall be marked CO/ALR.

E4002.4 Faceplates. Metal face plates shall be grounded.

E4002.5 Position of receptacle faces. After installation, receptacle faces shall be flush with or project from face plates of insulating material and shall project a minimum of 0.015 inch (0.381 mm) from metal face plates. Faceplates shall be installed so as to completely cover the opening and seat against the mounting surface. Receptacle faceplates mounted inside of a box having a recess-mounted receptacle shall effectively close the opening and seat against the mounting surface.

Exception: Listed kits or assemblies encompassing receptacles and nonmetallic faceplates that cover the receptacle face, where the plate cannot be installed on any other receptacle, shall be permitted.

E4002.6 Receptacle mounted in boxes. Receptacles mounted in boxes that are set back from the finished wall surface as permitted by Section E3906.5 shall be installed so that the mounting yoke or strap of the receptacle is held rigidly at the finished surface of the wall. Receptacles mounted in boxes that are flush with the wall surface or project therefrom shall be so installed that the mounting yoke or strap is seated against the box or raised cover.

E4002.7 Receptacles mounted on covers. Receptacles mounted to and supported by a cover shall be held rigidly against the cover by more than one screw or shall be a device assembly or box cover listed and identified for securing by a single screw.

E4002.8 Damp locations. A receptacle installed outdoors in a location protected from the weather or in other damp locations shall have an enclosure for the receptacle that is weatherproof when the receptacle cover(s) is closed and an attachment plug cap is not inserted. An installation suitable for wet locations shall also be considered suitable for damp locations. A receptacle shall be considered to be in a location protected from the weather where located under roofed open porches, canopies and similar structures and not subject to rain or water runoff. Fifteen- and 20-ampere, 125- and 250-volt nonlocking receptacles installed in damp locations shall be listed a weather-resistant type.

E4002.9 Fifteen- and 20-ampere receptacles in wet locations. Where installed in a wet location, 15- and 20-ampere, 125- and 250-volt receptacles shall have an enclosure that is weatherproof whether or not the attachment plug cap is inserted. Fifteen- and 20-ampere, 125- and 250-volt nonlocking receptacles installed in wet locations shall be a listed weather-resistant type.

E4002.10 Other receptacles in wet locations. Where a receptacle other than a 15- or 20-amp, 125- or 250-volt receptacle is installed in a wet location and where the product intended to be plugged into it is not attended while in use, the receptacle shall have an enclosure that is weatherproof both when the attachment plug cap is inserted and when it is removed. Where such receptacle is installed in a wet location and where the product intended to be plugged into it will be attended while in use, the receptacle shall have an enclosure that is weatherproof when the attachment plug cap is removed.

E4002.11 Bathtub and shower space. A receptacle shall not be installed within or directly over a bathtub or shower stall.

E4002.12 Flush mounting with faceplate. In damp or wet locations, the enclosure for a receptacle installed in an outlet box flush-mounted in a finished surface shall be made weatherproof by means of a weatherproof faceplate assembly that provides a water-tight connection between the plate and the finished surface.

E4002.13 Exposed terminals. Receptacles shall be enclosed so that live wiring terminals are not exposed to contact.

E4002.14 Tamper-resistant receptacles. In areas specified in Section E3901.1, 125-volt, 15- and 20-ampere receptacles shall be listed tamper-resistant receptacles.

Exception: Receptacles in the following locations shall not be required to be tamper resistant:

1. Receptacles located more than 5.5 feet (1676 mm) above the floor.
2. Receptacles that are part of a luminaire or appliance.
3. A single receptacle for a single appliance or a duplex receptacle for two appliances where such receptacles are located in spaces dedicated for the appliances served and, under conditions of normal use, the appliances are not easily moved from one place to another. The appliances shall be cord-and-plug-connected to such receptacles in accordance with Section E3909.4.

SECTION E4003 FIXTURES

E4003.1 Energized parts. Luminaires, lampholders, and lamps shall not have energized parts normally exposed to contact.

E4003.2 Luminaires near combustible material. Luminaires shall be installed so that combustible material will not be subjected to temperatures in excess of 90°C (194°F).

E4003.3 Exposed conductive parts. The exposed metal parts of luminaires shall be connected to an equipment grounding conductor or shall be insulated from the equipment grounding conductor and other conducting surfaces. Lamp tie wires, mounting screws, clips and decorative bands on glass spaced at least $1^1/_2$ inches (38 mm) from lamp terminals shall not be required to be grounded.

E4003.4 Screw-shell type. Lampholders of the screw-shell type shall be installed for use as lampholders only.

E4003.5 Recessed incandescent luminaires. Recessed incandescent luminaires shall have thermal protection and shall be listed as thermally protected.

Exceptions:

1. Thermal protection shall not be required in recessed luminaires listed for the purpose and installed in poured concrete.
2. Thermal protection shall not be required in recessed luminaires having design, construction, and thermal performance characteristics equivalent to that of thermally protected luminaires, and such luminaires are identified as inherently protected.

E4003.6 Thermal protection. The ballast of a fluorescent luminaire installed indoors shall have integral thermal protection. Replacement ballasts shall also have thermal protection integral with the ballast. A simple reactance ballast in a fluorescent luminaire with straight tubular lamps shall not be required to be thermally protected.

E4003.7 High-intensity discharge luminaires. Recessed high-intensity luminaires designed to be installed in wall or ceiling cavities shall have thermal protection and be identified as thermally protected. Thermal protection shall not be required in recessed high-intensity luminaires having design, construction and thermal performance characteristics equivalent to that of thermally protected luminaires, and such luminaires are identified as inherently protected. Thermal protection shall not be required in recessed high-intensity discharge luminaires installed in and identified for use in poured concrete. A recessed remote ballast for a high-intensity discharge luminaire shall have thermal protection that is integral with the ballast and shall be identified as thermally protected.

E4003.8 Metal halide lamp containment. Luminaires that use a metal halide lamp other than a thick-glass parabolic reflector lamp (PAR) shall be provided with a containment barrier that encloses the lamp, or shall be provided with a physical means that allows the use of only a lamp that is Type O.

E4003.9 Wet or damp locations. Luminaires installed in wet or damp locations shall be installed so that water cannot enter or accumulate in wiring compartments, lampholders or other electrical parts. All luminaires installed in wet locations shall be marked SUITABLE FOR WET LOCATIONS. All luminaires installed in damp locations shall be marked SUITABLE FOR WET LOCATIONS or SUITABLE FOR DAMP LOCATIONS.

E4003.10 Lampholders in wet or damp locations. Lampholders installed in wet locations shall be listed for use in wet locations. Lampholders installed in damp locations shall be listed for damp locations or shall be listed for wet locations.

E4003.11 Bathtub and shower areas. Cord-connected luminaires, chain-, cable-, or cord-suspended-luminaires, lighting track, pendants, and ceiling-suspended (paddle) fans shall not have any parts located within a zone measured 3 feet (914 mm) horizontally and 8 feet (2438 mm) vertically from the top of a bathtub rim or shower stall threshold. This zone is all encompassing and includes the space directly over the tub or shower. Luminaires within the actual outside dimension of the bathtub or shower to a height of 8 feet (2438 mm) vertically from the top of the bathtub rim or shower threshold shall be marked for damp locations and where subject to shower spray, shall be marked for wet locations.

E4003.12 Luminaires in clothes closets. For the purposes of this section, storage space shall be defined as a volume bounded by the sides and back closet walls and planes extending from the closet floor vertically to a height of 6 feet (1829 mm) or the highest clothes-hanging rod and parallel to the walls at a horizontal distance of 24 inches (610 mm) from the sides and back of the closet walls respectively, and continuing vertically to the closet ceiling parallel to the walls at a horizontal distance of 12 inches (305 mm) or the width of the shelf, whichever is greater. For a closet that permits access to both sides of a hanging rod, the storage space shall include the volume below the highest rod extending 12 inches (305 mm) on either side of the rod on a plane horizontal to the floor extending the entire length of the rod (see Figure E4003.12).

The types of luminaires installed in clothes closets shall be limited to surface-mounted or recessed incandescent or LED luminaires with completely enclosed light sources, surface-mounted or recessed fluorescent luminaires, and surface-mounted fluorescent or LED luminaires identified as suitable for installation within the closet storage area. Incandescent luminaires with open or partially enclosed lamps and pendant luminaires or lamp-holders shall be prohibited. The minimum clearance between luminaires installed in clothes closets and the nearest point of a closet storage area shall be as follows:

1. Surface-mounted incandescent or LED luminaires with a completely enclosed light source shall be installed on the wall above the door or on the ceiling, provided that there is a minimum clearance of 12 inches (305 mm) between the fixture and the nearest point of a storage space.
2. Surface-mounted fluorescent luminaires shall be installed on the wall above the door or on the ceiling,

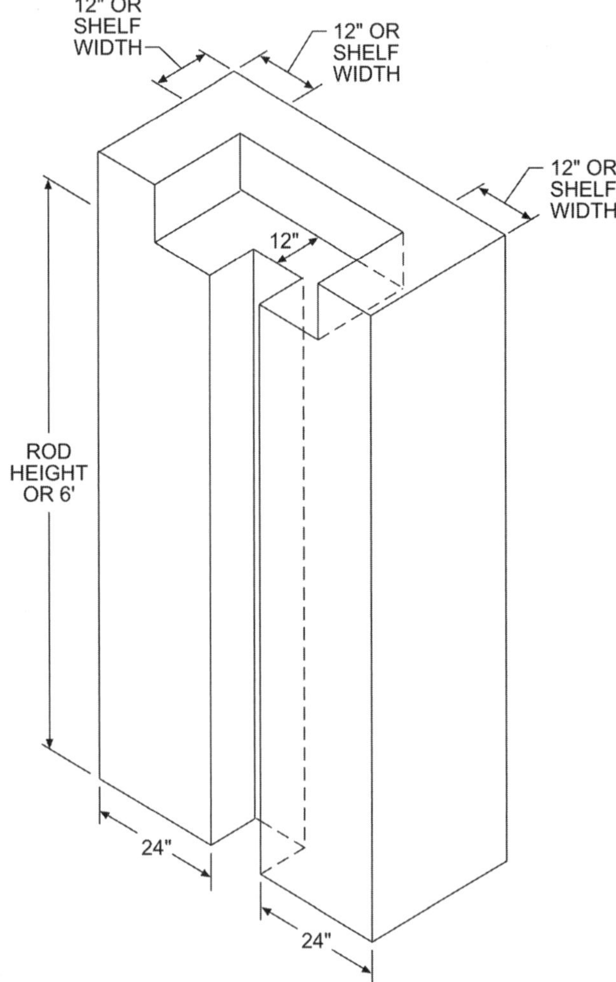

For SI: 1 inch = 25.4 mm, 1 foot = 304.8 mm.

FIGURE E4003.12
CLOSET STORAGE SPACE

provided that there is a minimum clearance of 6 inches (152 mm).

3. Recessed incandescent luminaires or LED luminaires with a completely enclosed light source shall be installed in the wall or the ceiling provided that there is a minimum clearance of 6 inches (152 mm).
4. Recessed fluorescent luminaires shall be installed in the wall or on the ceiling provided that there is a minimum clearance of 6 inches (152 mm) between the fixture and the nearest point of a storage space.
5. Surface-mounted fluorescent or LED luminaires shall be permitted to be installed within the closet storage space where identified for this use.

E4003.13 Luminaire wiring—general. Wiring on or within luminaires shall be neatly arranged and shall not be exposed to physical damage. Excess wiring shall be avoided. Conductors shall be arranged so that they are not subjected to temperatures above those for which the conductors are rated.

E4003.13.1 Polarization of luminaires. Luminaires shall be wired so that the screw shells of lampholders will be connected to the same luminaire or circuit conductor or terminal. The grounded conductor shall be connected to the screw shell.

E4003.13.2 Luminaires as raceways. Luminaires shall not be used as raceways for circuit conductors except where such luminaires are listed and marked for use as a raceway. Luminaires used as a raceway shall be identified for through-wiring. Luminaires designed for end-to-end connection to form a continuous assembly, and luminaires connected together by recognized wiring methods, shall not be required to be listed as a raceway where they contain the conductors of one 2-wire branch circuit or one multiwire branch circuit and such conductors supply the connected luminaires. One additional 2-wire branch circuit that separately supplies one or more of the connected luminaires shall also be permitted.

SECTION E4004 LUMINAIRE INSTALLATION

E4004.1 Outlet box covers. In a completed installation, each outlet box shall be provided with a cover except where covered by means of a luminaire canopy, lampholder or device with a faceplate.

E4004.2 Combustible material at outlet boxes. Combustible wall or ceiling finish exposed between the inside edge of a luminaire canopy or pan and the outlet box to which the luminaire connects shall be covered with a noncombustible material.

E4004.3 Access. Luminaires shall be installed so that the connections between the luminaire conductors and the circuit conductors can be accessed without requiring the disconnection of any part of the wiring. Luminaires that are connected by attachment plugs and receptacles meet the requirement of this section.

E4004.4 Supports. Luminaires and lampholders shall be securely supported. A luminaire that weighs more than 6 pounds (2.72 kg) or exceeds 16 inches (406 mm) in any dimension shall not be supported by the screw shell of a lampholder.

E4004.5 Means of support. Outlet boxes or fittings installed as required by Sections E3905 and E3906 shall be permitted to support luminaires.

E4004.6 Exposed components. Luminaires having exposed ballasts, transformers, LED drivers or power supplies shall be installed so that such ballasts, transformers, LED drivers or power supplies are not in contact with combustible material unless listed for such condition.

E4004.7 Combustible low-density cellulose fiberboard. Where a surface-mounted luminaire containing a ballast, transformer, LED driver or power supply is installed on combustible low-density cellulose fiberboard, the luminaire shall be marked for this purpose or it shall be spaced not less than $1^1/_2$ inches (38 mm) from the surface of the fiberboard. Where such luminaires are partially or wholly recessed, the provisions of Sections E4004.8 and E4004.9 shall apply.

E4004.8 Recessed luminaire clearance. A recessed luminaire that is not identified for contact with insulation shall have all recessed parts spaced at least $^1/_2$ inch (12.7 mm) from combustible materials. The points of support and the finish trim parts at the opening in the ceiling, wall or other finished surface shall be permitted to be in contact with combustible materials. A recessed luminaire that is identified for contact with insulation, Type IC, shall be permitted to be in contact with combustible materials at recessed parts, points of support, and portions passing through the building structure and at finish trim parts at the opening in the ceiling or wall.

E4004.9 Recessed luminaire installation. Thermal insulation shall not be installed above a recessed luminaire or within 3 inches (76 mm) of the recessed luminaire's enclosure, wiring compartment, ballast, transformer, LED driver or power supply except where such luminaire is identified for contact with insulation, Type IC.

SECTION E4005 TRACK LIGHTING

E4005.1 Installation. Lighting track shall be permanently installed and permanently connected to a branch circuit having a rating not more than that of the track.

E4005.2 Fittings. Fittings identified for use on lighting track shall be designed specifically for the track on which they are to be installed. Fittings shall be securely fastened to the track, shall maintain polarization and connection to the equipment grounding conductor, and shall be designed to be suspended directly from the track. Only lighting track fittings shall be installed on lighting track. Lighting track fittings shall not be equipped with general-purpose receptacles.

E4005.3 Connected load. The connected load on lighting track shall not exceed the rating of the track. The rating of the branch circuit that supplies the track shall not exceed the rating of the track.

E4005.4 Prohibited locations. Lighting track shall not be installed in the following locations:

1. Where likely to be subjected to physical damage.
2. In wet or damp locations.
3. Where subject to corrosive vapors.
4. In storage battery rooms.
5. In hazardous (classified) locations.
6. Where concealed.
7. Where extended through walls or partitions.
8. Less than 5 feet (1524 mm) above the finished floor except where protected from physical damage or the track operates at less than 30 volts rms open-circuit voltage.
9. Where prohibited by Section E4003.11.

E4005.5 Fastening. Lighting track shall be securely mounted so that each fastening will be suitable for supporting the maximum weight of luminaires that can be installed. Except where identified for supports at greater intervals, a single section 4 feet (1219 mm) or shorter in length shall have two supports and, where installed in a continuous row, each individual section of not more than 4 feet (1219 mm) in length shall have one additional support.

E4005.6 Grounding. Lighting track shall be grounded in accordance with Chapter 39, and the track sections shall be securely coupled to maintain continuity of the circuitry, polarization and grounding throughout.

CHAPTER 41

APPLIANCE INSTALLATION

SECTION E4101
GENERAL

E4101.1 Scope. This section covers installation requirements for appliances and fixed heating equipment.

E4101.2 Installation. Appliances and equipment shall be installed in accordance with the manufacturer's installation instructions. Electrically heated appliances and equipment shall be installed with the required clearances to combustible materials.

E4101.3 Flexible cords. Cord-and-plug-connected appliances shall use cords suitable for the environment and physical conditions likely to be encountered. Flexible cords shall be used only where the appliance is listed to be connected with a flexible cord. The cord shall be identified as suitable for the purpose in the installation instructions of the appliance manufacturer. Receptacles for cord-and-plug-connected appliances shall be accessible and shall be located to avoid physical damage to the flexible cord. Except for a listed appliance marked to indicate that it is protected by a system of double-insulation, the flexible cord supplying an appliance shall terminate in a grounding-type attachment plug. A receptacle for a cord-and-plug-connected range hood shall be supplied by an individual branch circuit. Specific appliances have additional requirements as specified in Table E4101.3 (see Section E3909).

TABLE E4101.3
FLEXIBLE CORD LENGTH

APPLIANCE	MINIMUM CORD LENGTH (inches)	MAXIMUM CORD LENGTH (inches)
Kitchen waste disposal	18	36
Built-in dishwasher	36	48
Trash compactor	36	48
Range hoods	18	36

For SI: 1 inch = 25.4 mm.

E4101.4 Overcurrent protection. Each appliance shall be protected against overcurrent in accordance with the rating of the appliance and its listing.

E4101.4.1 Single nonmotor-operated appliance. The overcurrent protection for a branch circuit that supplies a single nonmotor-operated appliance shall not exceed that marked on the appliance. Where the overcurrent protection rating is not marked and the appliance is rated at over 13.3 amperes, the overcurrent protection shall not exceed 150 percent of the appliance rated current. Where 150 percent of the appliance rating does not correspond to a standard overcurrent device ampere rating, the next higher standard rating shall be permitted. Where the overcurrent protection rating is not marked and the appliance is rated at 13.3 amperes or less, the overcurrent protection shall not exceed 20 amperes.

E4101.5 Disconnecting means. Each appliance shall be provided with a means to disconnect all ungrounded supply conductors. For fixed electric space-heating equipment, means shall be provided to disconnect the heater and any motor controller(s) and supplementary overcurrent-protective devices. Switches and circuit breakers used as a disconnecting means shall be of the indicating type. Disconnecting means shall be as set forth in Table E4101.5.

E4101.6 Support of ceiling-suspended paddle fans. Ceiling-suspended fans (paddle) shall be supported independently of an outlet box or by a listed outlet box or outlet box system identified for the use and installed in accordance with Section E3905.9.

E4101.7 Snow-melting and deicing equipment protection. Outdoor receptacles that are not readily accessible and are supplied from a dedicated branch circuit for electric snow-melting or deicing equipment shall be permitted to be installed without ground-fault circuit-interrupter protection for personnel. However, ground-fault protection of equipment shall be provided for fixed outdoor electric deicing and snow-melting equipment.

TABLE E4101.5
DISCONNECTING MEANS

DESCRIPTION	ALLOWED DISCONNECTING MEANS
Permanently connected appliance rated at not over 300 volt-amperes or $^1/_8$ horsepower.	Branch-circuit overcurrent device.
Permanently connected appliances rated in excess of 300 volt-amperes.	Branch circuit breaker or switch located within sight of appliance or such devices in any location that are capable of being locked in the open position. The provision for locking or adding a lock to the disconnecting means shall be installed on or at the switch or circuit breaker used as the disconnecting means and shall remain in place with or without the lock installed.
Motor-operated appliances rated over $^1/_8$ horsepower.	For permanently connected motor-operated appliances with motors rated over $^1/_8$ horsepower, the branch circuit switch or circuit breaker shall be permitted to serve as the disconnecting means where the switch or circuit breaker is within sight from the appliance. Where the branch circuit switch is not located within sight from the appliance, the disconnecting means shall be one of the following types: a listed motor-circuit switch rated in horsepower, a listed molded case circuit breaker, a listed molded case switch, a listed manual motor controller additionally marked "Suitable as Motor Disconnect" where installed between the final motor branch-circuit short-circuit protective device and the motor. For stationary motors rated at 2 hp or less and 300 volts or less, the disconnecting means shall be permitted to be one of the following devices: 1. A general-use switch having an ampere rating not less than twice the full-load current rating of the motor. 2. On AC circuits, a general-use snap switch suitable only for use on AC, not general-use AC–DC snap switches, where the motor full-load current rating is not more than 80 percent of the ampere rating of the switch. 3. A listed manual motor controller having a horsepower rating not less than the rating of the motor and marked "Suitable as Motor Disconnect". The disconnecting means for motor circuits rated 600 volts, nominal, or less shall have an ampere rating not less than 115 percent of the full-load current rating of the motor except that a listed unfused motor-circuit switch having a horsepower rating not less than the motor horsepower shall be permitted to have an ampere rating less than 115 percent of the full-load current rating of the motor. **Exception:** A unit switch with a marked-off position that is a part of an appliance and disconnects all ungrounded conductors shall be permitted as the disconnecting means and the switch or circuit breaker serving as the other disconnecting means shall be permitted to be out of sight from the appliance.
Appliances listed for cord-and-plug connection.	A separable connector or attachment plug and receptacle provided with access.

(continued)

TABLE E4101.5—continued
DISCONNECTING MEANS

DESCRIPTION	ALLOWED DISCONNECTING MEANS
Permanently installed heating equipment with motors rated at not over $^1/_8$ horsepower with supplementary overcurrent protection.	Disconnect, on the supply side of fuses, in sight from the supplementary overcurrent device, and in sight of the heating equipment or, in any location, if capable of being locked in the open position.
Heating equipment containing motors rated over $^1/_8$ horsepower with supplementary overcurrent protection.	Disconnect permitted to serve as required disconnect for both the heating equipment and the controller where, on the supply side of fuses, and in sight from the supplementary overcurrent devices, if the disconnecting means is also in sight from the controller, or is capable of being locked off and simultaneously disconnects the heater, motor controller(s) and supplementary overcurrent protective devices from all ungrounded conductors. The provision for locking or adding a lock to the disconnecting means shall be installed on or at the switch or circuit breaker used as the disconnecting means and shall remain in place with or without the lock installed. The disconnecting means shall have an ampere rating not less than 125 percent of the total load of the motors and the heaters.
Heating equipment containing no motor rated over $^1/_8$ horsepower without supplementary overcurrent protection.	Branch-circuit switch or circuit breaker where within sight from the heating equipment or capable of being locked off and simultaneously disconnects the heater, motor controller(s) and supplementary overcurrent protective devices from all ungrounded conductors. The provision for locking or adding a lock to the disconnecting means shall be installed on or at the switch or circuit breaker used as the disconnecting means and shall remain in place with or without the lock installed. The disconnecting means shall have an ampere rating not less than 125 percent of the total load of the motors and the heaters.
Heating equipment containing motors rated over $^1/_8$ horsepower without supplementary overcurrent protection.	Disconnecting means in sight from motor controller or as provided for heating equipment with motor rated over $^1/_8$ horsepower with supplementary overcurrent protection and simultaneously disconnects the heater, motor controller(s) and supplementary overcurrent protective devices from all ungrounded conductors. The provision for locking or adding a lock to the disconnecting means shall be installed on or at the switch or circuit breaker used as the disconnecting means and shall remain in place with or without the lock installed. The disconnecting means shall have an ampere rating not less than 125 percent of the total load of the motors and the heaters.
Air-conditioning condensing units and heat pump units.	A readily accessible disconnect within sight from unit as the only allowable means.[a]
Appliances and fixed heating equipment with unit switches having a marked OFF position.	Unit switch where an additional individual switch or circuit breaker serves as a redundant disconnecting means.
Thermostatically controlled fixed heating equipment.	Thermostats with a marked OFF position that directly open all ungrounded conductors, which when manually placed in the OFF position are designed so that the circuit cannot be energized automatically and that are located within sight of the equipment controlled.

For SI: 1 horsepower = 0.746 kW.

a. The disconnecting means shall be permitted to be installed on or within the unit. It shall not be located on panels designed to allow access to the unit or located so as to obscure the air-conditioning equipment nameplate(s).

CHAPTER 42

SWIMMING POOLS

SECTION E4201 GENERAL

E4201.1 Scope. The provisions of this chapter shall apply to the construction and installation of electric wiring and equipment associated with all swimming pools, wading pools, decorative pools, fountains, hot tubs and spas, and hydromassage bathtubs, whether permanently installed or storable, and shall apply to metallic auxiliary equipment, such as pumps, filters and similar equipment. Sections E4202 through E4206 provide general rules for permanent pools, spas and hot tubs. Section E4207 provides specific rules for storable pools. Section E4208 provides specific rules for spas and hot tubs. Section E4209 provides specific rules for hydromassage bathtubs.

E4201.2 Definitions.

CORD-AND-PLUG-CONNECTED LIGHTING ASSEMBLY. A lighting assembly consisting of a cord-and-plug-connected transformer and a luminaire intended for installation in the wall of a spa, hot tub, or storable pool.

DRY-NICHE LUMINAIRE. A luminaire intended for installation in the floor or wall of a pool, spa or fountain in a niche that is sealed against the entry of water.

FORMING SHELL. A structure designed to support a wet-niche luminaire assembly and intended for mounting in a pool or fountain structure.

FOUNTAIN. Fountains, ornamental pools, display pools, and reflection pools. The definition does not include drinking fountains.

HYDROMASSAGE BATHTUB. A permanently installed bathtub equipped with a recirculating piping system, pump, and associated equipment. It is designed so it can accept, circulate and discharge water upon each use.

LOW VOLTAGE CONTACT LIMIT. A voltage not exceeding the following values:

1. 15 volts (RMS) for sinusoidal AC
2. 21.2 volts peak for nonsinusoidal AC
3. 30 volts for continuous DC
4. 12.4 volts peak for DC that is interrupted at a rate of 10 to 200 Hz

MAXIMUM WATER LEVEL. The highest level that water can reach before it spills out.

NO-NICHE LUMINAIRE. A luminaire intended for installation above or below the water without a niche.

PACKAGED SPA OR HOT TUB EQUIPMENT ASSEMBLY. A factory-fabricated unit consisting of water-circulating, heating and control equipment mounted on a common base, intended to operate a spa or hot tub. Equipment may include pumps, air blowers, heaters, luminaires, controls and sanitizer generators.

PERMANENTLY INSTALLED SWIMMING, WADING, IMMERSION AND THERAPEUTIC POOLS. Those that are constructed in the ground or partially in the ground, and all others capable of holding water with a depth greater than 42 inches (1067 mm), and all pools installed inside of a building, regardless of water depth, whether or not served by electrical circuits of any nature.

POOL. Manufactured or field-constructed equipment designed to contain water on a permanent or semipermanent basis and used for swimming, wading, immersion, or therapeutic purposes.

POOL COVER, ELECTRICALLY OPERATED. Motor-driven equipment designed to cover and uncover the water surface of a pool by means of a flexible sheet or rigid frame.

SELF-CONTAINED SPA OR HOT TUB. A factory-fabricated unit consisting of a spa or hot tub vessel with all water-circulating, heating and control equipment integral to the unit. Equipment may include pumps, air blowers, heaters, luminaires, controls and sanitizer generators.

SPA OR HOT TUB. A hydromassage pool, or tub for recreational or therapeutic use, not located in health care facilities, designed for immersion of users, and usually having a filter, heater, and motor-driven blower. They are installed indoors or outdoors, on the ground or supporting structure, or in the ground or supporting structure. Generally, a spa or hot tub is not designed or intended to have its contents drained or discharged after each use.

STORABLE SWIMMING OR WADING POOL. Those that are constructed on or above the ground and are capable of holding water with a maximum depth of 42 inches (1067 mm), or a pool with nonmetallic, molded polymeric walls or inflatable fabric walls regardless of dimension.

THROUGH-WALL LIGHTING ASSEMBLY. A lighting assembly intended for installation above grade, on or through the wall of a pool, consisting of two interconnected groups of components separated by the pool wall.

WET-NICHE LUMINAIRE. A luminaire intended for installation in a forming shell mounted in a pool or fountain structure where the luminaire will be completely surrounded by water.

SECTION E4202 WIRING METHODS FOR POOLS, SPAS, HOT TUBS AND HYDROMASSAGE BATHTUBS

E4202.1 General. Wiring methods used in conjunction with permanently installed swimming pools, spas, hot tubs or hydromassage bathtubs shall be installed in accordance with Table E4202.1 and Chapter 38 except as otherwise stated in this section. Storable swimming pools shall comply with Section E4207.

E4202.2 Flexible cords. Flexible cords used in conjunction with a pool, spa, hot tub or hydromassage bathtub shall be installed in accordance with the following:

1. For other than underwater luminaires, fixed or stationary equipment shall be permitted to be connected with a flexible cord to facilitate removal or disconnection for maintenance or repair. For other than storable pools, the flexible cord shall not exceed 3 feet (914 mm) in length. Cords that supply swimming pool equipment shall have a copper equipment grounding conductor not smaller than 12 AWG and shall terminate in a grounding-type attachment plug.
2. Other than listed low-voltage lighting systems not requiring grounding, wet-niche luminaires that are supplied by a flexible cord or cable shall have all exposed noncurrent-carrying metal parts grounded by an insulated copper equipment grounding conductor that is an integral part of the cord or cable. Such grounding conductor shall be connected to a grounding terminal in the supply junction box, transformer enclosure, or other enclosure and shall be not smaller than the supply conductors and not smaller than 16 AWG.
3. A listed packaged spa or hot tub installed outdoors that is GFCI protected shall be permitted to be cord-and-plug-connected provided that such cord does not exceed 15 feet (4572 mm) in length.
4. A listed packaged spa or hot tub rated at 20 amperes or less and installed indoors shall be permitted to be cord-and-plug-connected to facilitate maintenance and repair.
5. For other than underwater and storable pool lighting luminaire, the requirements of Item 1 shall apply to any

TABLE E4202.1
ALLOWABLE APPLICATIONS FOR WIRING METHODS[a, b, c, d, e, f, g, h, l]

WIRING LOCATION OR PURPOSE (Application allowed where marked with an "A")	AC, FMC, NM, SR, SE	EMT	ENT	IMC[j], RMC[j], RNC[i]	LFMC	LFNMC	UF	MC[k]	FLEX CORD
Panelboard(s) that supply pool equipment: from service equipment to panelboard	A[b, e] SR not permitted	A[c]	A[b]	A	—	A	A[e]	A[e]	—
Wet-niche and no-niche luminaires: from branch circuit OCPD to deck or junction box	AC[b] only	A[c]	A[b]	A	—	A	—	A[b]	—
Wet-niche and no-niche luminaires: from deck or junction box to forming shell	—	—	—	A[d]	—	A	—	—	A[h]
Dry niche: from branch circuit OCPD to luminaires	AC[b] only	A[c]	A[b]	A	—	A	—	A[b]	—
Pool-associated motors: from branch circuit OCPD to motor	A[b]	A[c]	A[b]	A	A[f]	A[f]	A[b]	A	A[h]
Packaged or self-contained outdoor spas and hot tubs with underwater luminaire: from branch circuit OCPD to spa or hot tub	AC[b] only	A[c]	A[b]	A	A[g]	A[g]	—	A[b]	A[h]
Packaged or self-contained outdoor spas and hot tubs without underwater luminaire: from branch circuit OCPD to spa or hot tub	A[b]	A[c]	A[b]	A	A[g]	A[g]	A[b]	A	A[h]
Indoor spas and hot tubs, hydromassage bathtubs, and other pool, spa or hot tub associated equipment: from branch circuit OCPD to equipment	A[b]	A[c]	A[b]	A	A	A	A	A	A[h]
Connection at pool lighting transformers or power supplies	AC[b] only	A[c]	A[b]	A	A[m, g]	A[g]	—	A[b]	—

For SI: 1 foot = 304.8 mm.

a. For all wiring methods, see Section E4205 for equipment grounding conductor requirements.
b. Limited to use within buildings.
c. Limited to use on or within buildings.
d. Metal conduit shall be constructed of brass or other approved corrosion-resistant metal.
e. Permitted only for existing installations in accordance with the exception to Section E4205.6.
f. Limited to where necessary to employ flexible connections at or adjacent to a pool motor.
g. Sections installed external to spa or hot tub enclosure limited to individual lengths not to exceed 6 feet. Length not limited inside spa or hot tub enclosure.
h. Flexible cord shall be installed in accordance with Section E4202.2.
i. Nonmetallic conduit shall be rigid polyvinyl chloride conduit Type PVC or reinforced thermosetting resin conduit Type RTRC.
j. Aluminum conduits shall not be permitted in the pool area where subject to corrosion.
k. Where installed as direct burial cable or in wet locations, Type MC cable shall be listed and identified for the location.
l. See Section E4202.3 for listed, double-insulated pool pump motors.
m. Limited to use in individual lengths not to exceed 6 feet. The total length of all individual runs of LFMC shall not exceed 10 feet.

cord-equipped luminaire that is located within 16 feet (4877 mm) radially from any point on the water surface.

E4202.3 Double insulated pool pumps. A listed cord and plug-connected pool pump incorporating an approved system of double insulation that provides a means for grounding only the internal and nonaccessible, noncurrent-carrying metal parts of the pump shall be connected to any wiring method recognized in Chapter 38 that is suitable for the location. Where the bonding grid is connected to the equipment grounding conductor of the motor circuit in accordance with Section E4204.2, Item 6.1, the branch circuit wiring shall comply with Sections E4202.1 and E4205.5.

SECTION E4203
EQUIPMENT LOCATION AND CLEARANCES

E4203.1 Receptacle outlets. Receptacles outlets shall be installed and located in accordance with Sections E4203.1.1 through E4203.1.5. Distances shall be measured as the shortest path that an appliance supply cord connected to the receptacle would follow without penetrating a floor, wall, ceiling, doorway with hinged or sliding door, window opening, or other effective permanent barrier.

E4203.1.1 Location. Receptacles that provide power for water-pump motors or other loads directly related to the circulation and sanitation system shall be permitted to be located between 6 feet and 10 feet (1829 mm and 3048 mm) from the inside walls of pools and outdoor spas and hot tubs, where the receptacle is single and of the locking and grounding type and protected by ground-fault circuit interrupters.

Other receptacles on the property shall be located not less than 6 feet (1829 mm) from the inside walls of pools and outdoor spas and hot tubs.

E4203.1.2 Where required. At least one 125-volt, 15- or 20-ampere receptacle supplied by a general-purpose branch circuit shall be located a minimum of 6 feet (1829 mm) from and not more than 20 feet (6096 mm) from the inside wall of pools and outdoor spas and hot tubs. This receptacle shall be located not more than 6 feet, 6 inches (1981 mm) above the floor, platform or grade level serving the pool, spa or hot tub.

E4203.1.3 GFCI protection. All 15- and 20-ampere, single phase, 125-volt receptacles located within 20 feet (6096 mm) of the inside walls of pools and outdoor spas and hot tubs shall be protected by a ground-fault circuit-interrupter. Outlets supplying pool pump motors from branch circuits with short-circuit and ground-fault protection rated 15 or 20 amperes, 125 volts through 240 volts, single phase, whether by receptacle or direct connection, shall be provided with ground-fault circuit-interrupter protection for personnel.

E4203.1.4 Indoor locations. Receptacles shall be located not less than 6 feet (1829 mm) from the inside walls of indoor spas and hot tubs. A minimum of one 125-volt receptacle shall be located between 6 feet (1829 mm) and 10 feet (3048 mm) from the inside walls of indoor spas or hot tubs.

E4203.1.5 Indoor GFCI protection. All 125-volt receptacles rated 30 amperes or less and located within 10 feet (3048 mm) of the inside walls of spas and hot tubs installed indoors, shall be protected by ground-fault circuit-interrupters.

E4203.2 Switching devices. Switching devices shall be located not less than 5 feet (1524 mm) horizontally from the inside walls of pools, spas and hot tubs except where separated from the pool, spa or hot tub by a solid fence, wall, or other permanent barrier or the switches are listed for use within 5 feet (1524 mm). Switching devices located in a room or area containing a hydromassage bathtub shall be located in accordance with the general requirements of this code.

E4203.3 Disconnecting means. One or more means to simultaneously disconnect all ungrounded conductors for all utilization equipment, other than lighting, shall be provided. Each of such means shall be readily accessible and within sight from the equipment it serves and shall be located at least 5 feet (1524 mm) horizontally from the inside walls of a pool, spa, or hot tub unless separated from the open water by a permanently installed barrier that provides a 5-foot (1524 mm) or greater reach path. This horizontal distance shall be measured from the water's edge along the shortest path required to reach the disconnect.

E4203.4 Luminaires and ceiling fans. Lighting outlets, luminaires, and ceiling-suspended paddle fans shall be installed and located in accordance with Sections E4203.4.1 through E4203.4.5.

E4203.4.1 Outdoor location. In outdoor pool, outdoor spas and outdoor hot tubs areas, luminaires, lighting outlets, and ceiling-suspended paddle fans shall not be installed over the pool or over the area extending 5 feet (1524 mm) horizontally from the inside walls of a pool except where no part of the luminaire or ceiling-suspended paddle fan is less than 12 feet (3658 mm) above the maximum water level.

E4203.4.2 Indoor locations. In indoor pool areas, the limitations of Section E4203.4.1 shall apply except where the luminaires, lighting outlets and ceiling-suspended paddle fans comply with all of the following conditions:

1. The luminaires are of a totally enclosed type;
2. A ground-fault circuit interrupter is installed in the branch circuit supplying the luminaires or ceiling-suspended (paddle) fans; and
3. The distance from the bottom of the luminaire or ceiling-suspended (paddle) fan to the maximum water level is not less than 7 feet, 6 inches (2286 mm).

E4203.4.3 Existing lighting outlets and luminaires. Existing lighting outlets and luminaires that are located within 5 feet (1524 mm) horizontally from the inside walls of pools and outdoor spas and hot tubs shall be permitted to be located not less than 5 feet (1524 mm) vertically above the maximum water level, provided that such luminaires and outlets are rigidly attached to the existing structure and are protected by a ground-fault circuit-interrupter.

E4203.4.4 Indoor spas and hot tubs.

1. Luminaires, lighting outlets, and ceiling-suspended paddle fans located over the spa or hot tub or within 5 feet (1524 mm) from the inside walls of the spa or hot tub shall be a minimum of 7 feet, 6 inches (2286 mm) above the maximum water level and shall be protected by a ground-fault circuit interrupter.

   Luminaires, lighting outlets, and ceiling-suspended paddle fans that are located 12 feet (3658 mm) or more above the maximum water level shall not require ground-fault circuit interrupter protection.

2. Luminaires protected by a ground-fault circuit interrupter and complying with Item 2.1 or 2.2 shall be permitted to be installed less than 7 feet, 6 inches (2286 mm) over a spa or hot tub.

   2.1. Recessed luminaires shall have a glass or plastic lens and nonmetallic or electrically isolated metal trim, and shall be suitable for use in damp locations.

   2.2. Surface-mounted luminaires shall have a glass or plastic globe and a nonmetallic body or a metallic body isolated from contact. Such luminaires shall be suitable for use in damp locations.

E4203.4.5 GFCI protection in adjacent areas. Luminaires and outlets that are installed in the area extending between 5 feet (1524 mm) and 10 feet (3048 mm) from the inside walls of pools and outdoor spas and hot tubs shall be protected by ground-fault circuit-interrupters except where such fixtures and outlets are installed not less than 5 feet (1524 mm) above the maximum water level and are rigidly attached to the structure.

E4203.5 Other outlets. Other outlets such as for remote control, signaling, fire alarm and communications shall be not less than 10 feet (3048 mm) from the inside walls of the pool. Measurements shall be determined in accordance with Section E4203.1.

E4203.6 Overhead conductor clearances. Except where installed with the clearances specified in Table E4203.5, the following parts of pools and outdoor spas and hot tubs shall not be placed under existing service-drop conductors or any other open overhead wiring; nor shall such wiring be installed above the following:

1. Pools and the areas extending 10 feet (3048 mm) horizontally from the inside of the walls of the pool;
2. Diving structures; or
3. Observation stands, towers, and platforms.

Overhead conductors of network-powered broadband communications systems shall comply with the provisions in Table E4203.5 for conductors operating at 0 to 750 volts to ground.

Utility-owned, -operated and -maintained communications conductors, community antenna system coaxial cables and the supporting messengers shall be permitted at a height of not less than 10 feet (3048 mm) above swimming and wading pools, diving structures, and observation stands, towers, and platforms.

E4203.7 Underground wiring. Underground wiring shall not be installed under or within the area extending 5 feet (1524 mm) horizontally from the inside walls of pools and outdoor hot tubs and spas except where the wiring is installed to supply pool, spa or hot tub equipment or where space limitations prevent wiring from being routed 5 feet (1524 mm) or more horizontally from the inside walls. Where installed within 5 feet (1524 mm) of the inside walls, the wiring method shall be a complete raceway system of rigid metal conduit, intermediate metal conduit or a nonmetallic raceway system. Metal conduit shall be corrosion resistant and suitable for the location. The minimum cover depth shall be in accordance with Table E4203.7.

SECTION E4204 BONDING

E4204.1 Performance. The equipotential bonding required by this section shall be installed to reduce voltage gradients in the pool area as prescribed.

E4204.2 Bonded parts. The parts of pools, spas, and hot tubs specified in Items 1 through 7 shall be bonded together using insulated, covered or bare solid copper conductors not smaller than 8 AWG or using rigid metal conduit of brass or other identified corrosion-resistant metal. An 8 AWG or larger solid copper bonding conductor provided to reduce voltage gradients in the pool, spa, or hot tub area shall not be required to be extended or attached to remote panelboards,

TABLE E4203.5 OVERHEAD CONDUCTOR CLEARANCES

	INSULATED SUPPLY OR SERVICE DROP CABLES, 0-750 VOLTS TO GROUND, SUPPORTED ON AND CABLED TOGETHER WITH AN EFFECTIVELY GROUNDED BARE MESSENGER OR EFFECTIVELY GROUNDED NEUTRAL CONDUCTOR (feet)	ALL OTHER SUPPLY OR SERVICE DROP CONDUCTORS (feet)	
		Voltage to ground	
		0-15 kV	Greater than 15 to 50 kV
A. Clearance in any direction to the water level, edge of water surface, base of diving platform, or permanently anchored raft	22.5	25	27
B. Clearance in any direction to the diving platform	14.5	17	18

For SI: 1 foot = 304.8 mm.

service equipment, or electrodes. Connections shall be made by exothermic welding, by listed pressure connectors or clamps that are labeled as being suitable for the purpose and that are made of stainless steel, brass, copper or copper alloy, machine screw-type fasteners that engage not less than two threads or are secured with a nut, thread-forming machine screws that engage not less than two-threads, or terminal bars. Connection devices or fittings that depend solely on solder shall not be used. Sheet metal screws shall not be used to connect bonding conductors or connection devices:

1. Conductive pool shells. Bonding to conductive pool shells shall be provided as specified in Item 1.1 or 1.2. Poured concrete, pneumatically applied or sprayed concrete, and concrete block with painted or plastered coatings shall be considered to be conductive materials because of their water permeability and porosity. Vinyl liners and fiberglass composite shells shall be considered to be nonconductive materials.

    1.1. Structural reinforcing steel. Unencapsulated structural reinforcing steel shall be bonded together by steel tie wires or the equivalent. Where structural reinforcing steel is encapsulated in a nonconductive compound, a copper conductor grid shall be installed in accordance with Item 1.2.

    1.2. Copper conductor grid. A copper conductor grid shall be provided and shall comply with Items 1.2.1 through 1.2.4:

        1.2.1. It shall be constructed of minimum 8 AWG bare solid copper conductors bonded to each other at all points of crossing.

        1.2.2. It shall conform to the contour of the pool and the pool deck.

        1.2.3. It shall be arranged in a 12-inch (305 mm) by 12-inch (305 mm) network of conductors in a uniformly spaced perpendicular grid pattern with a tolerance of 4 inches (102 mm).

        1.2.4. It shall be secured within or under the pool not more than 6 inches (152 mm) from the outer contour of the pool shell.

2. Perimeter surfaces. The perimeter surface shall extend for 3 feet (914 mm) horizontally beyond the inside walls of the pool and shall include unpaved surfaces, poured concrete surfaces and other types of paving. Perimeter surfaces that extend less than 3 feet (914 mm) beyond the inside wall of the pool and that are separated from the pool by a permanent wall or building 5 feet (1524 mm) or more in height shall require equipotential bonding on the pool side of the permanent wall or building. Bonding to perimeter surfaces shall be provided as specified in Item 2.1 or 2.2 and shall be attached to the pool, spa, or hot tub reinforcing steel or copper conductor grid at a minimum of four points uniformly spaced around the perimeter of the pool, spa, or hot tub. For nonconductive pool shells, bonding at four points shall not be required.

    Exception: The equipotential bonding requirements for perimeter surfaces shall not apply to a listed self-contained spa or hot tub located indoors and installed above a finished floor.

    2.1. Structural reinforcing steel. Structural reinforcing steel shall be bonded in accordance with Item 1.1.

    2.2. Alternate means. Where structural reinforcing steel is not available or is encapsulated in a nonconductive compound, a copper conductor(s) shall be used in accordance with Items 2.2.1 through 2.2.5:

        2.2.1. At least one minimum 8 AWG bare solid copper conductor shall be provided.

        2.2.2. The conductors shall follow the contour of the perimeter surface.

        2.2.3. Splices shall be listed.

        2.2.4. The required conductor shall be 18 to 24 inches (457 to 610 mm) from the inside walls of the pool.

        2.2.5. The required conductor shall be secured within or under the perimeter surface 4 to 6 inches (102 mm to 152 mm) below the subgrade.

3. Metallic components. All metallic parts of the pool structure, including reinforcing metal not addressed in Item 1.1, shall be bonded. Where reinforcing steel is encapsulated with a nonconductive compound, the reinforcing steel shall not be required to be bonded.

TABLE E4203.7
MINIMUM BURIAL DEPTHS

WIRING METHOD	UNDERGROUND WIRING (inches)
Rigid metal conduit	6
Intermediate metal conduit	6
Nonmetallic raceways listed for direct burial and under concrete exterior slab not less than 4 inches in thickness and extending not less than 6 inches (162 mm) beyond the underground installation	6
Nonmetallic raceways listed for direct burial without concrete encasement	18
Other approved raceways[a]	18

For SI: 1 inch = 25.4 mm.

a. Raceways approved for burial only where concrete-encased shall require a concrete envelope not less than 2 inches in thickness.

4. Underwater lighting. All metal forming shells and mounting brackets of no-niche luminaires shall be bonded.

   Exception: Listed low-voltage lighting systems with nonmetallic forming shells shall not require bonding.

5. Metal fittings. All metal fittings within or attached to the pool structure shall be bonded. Isolated parts that are not over 4 inches (102 mm) in any dimension and do not penetrate into the pool structure more than 1 inch (25.4 mm) shall not require bonding.

6. Electrical equipment. Metal parts of electrical equipment associated with the pool water circulating system, including pump motors and metal parts of equipment associated with pool covers, including electric motors, shall be bonded.

   Exception: Metal parts of listed equipment incorporating an approved system of double insulation shall not be bonded.

   6.1. Double-insulated water pump motors. Where a double-insulated water pump motor is installed under the provisions of this item, a solid 8 AWG copper conductor of sufficient length to make a bonding connection to a replacement motor shall be extended from the bonding grid to an accessible point in the vicinity of the pool pump motor. Where there is no connection between the swimming pool bonding grid and the equipment grounding system for the premises, this bonding conductor shall be connected to the equipment grounding conductor of the motor circuit.

   6.2. Pool water heaters. For pool water heaters rated at more than 50 amperes and having specific instructions regarding bonding and grounding, only those parts designated to be bonded shall be bonded and only those parts designated to be grounded shall be grounded.

7. All fixed metal parts including, but not limited to, metal-sheathed cables and raceways, metal piping, metal awnings, metal fences and metal door and window frames.

   Exceptions:

   1. Those separated from the pool by a permanent barrier that prevents contact by a person shall not be required to be bonded.
   2. Those greater than 5 feet (1524 mm) horizontally from the inside walls of the pool shall not be required to be bonded.
   3. Those greater than 12 feet (3658 mm) measured vertically above the maximum water level of the pool, or as measured vertically above any observation stands, towers, or platforms, or any diving structures, shall not be required to be bonded.

E4204.3 Pool water. The pool water shall be intentionally bonded by means of a conductive surface area not less than 9 square inches (5806 mm^2) installed in contact with the pool water. This bond shall be permitted to consist of parts that are required to be bonded in Section E4204.2.

E4204.4 Bonding of outdoor hot tubs and spas. Outdoor hot tubs and spas shall comply with the bonding requirements of Sections E4204.1 through E4204.3. Bonding by metal-to-metal mounting on a common frame or base shall be permitted. The metal bands or hoops used to secure wooden staves shall not be required to be bonded as required in Section E4204.2.

E4204.5 Bonding of indoor hot tubs and spas. The following parts of indoor hot tubs and spas shall be bonded together:

1. All metal fittings within or attached to the hot tub or spa structure.
2. Metal parts of electrical equipment associated with the hot tub or spa water circulating system, including pump motors unless part of a listed self-contained spa or hot tub.
3. Metal raceway and metal piping that are within 5 feet (1524 mm) of the inside walls of the hot tub or spa and that are not separated from the spa or hot tub by a permanent barrier.
4. All metal surfaces that are within 5 feet (1524 mm) of the inside walls of the hot tub or spa and that are not separated from the hot tub or spa area by a permanent barrier.

   Exception: Small conductive surfaces not likely to become energized, such as air and water jets and drain fittings, where not connected to metallic piping, towel bars, mirror frames, and similar nonelectrical equipment, shall not be required to be bonded.

5. Electrical devices and controls that are not associated with the hot tubs or spas and that are located less than 5 feet (1524 mm) from such units.

E4204.5.1 Methods. All metal parts associated with the hot tub or spa shall be bonded by any of the following methods:

1. The interconnection of threaded metal piping and fittings.
2. Metal-to-metal mounting on a common frame or base
3. The provision of an insulated, covered or bare solid copper bonding jumper not smaller than 8 AWG.It shall not be the intent to require that the 8 AWG or larger solid copper bonding conductor be extended or attached to any remote panelboard, service equipment, or any electrode, but only that it shall be employed to eliminate voltage gradients in the hot tub or spa area as prescribed.

E4204.5.2 Connections. Connections shall be made by exothermic welding or by listed pressure connectors or clamps that are labeled as being suitable for the purpose and that are made of stainless steel, brass, copper or copper alloy. Connection devices or fittings that depend solely on solder shall not be used. Sheet metal screws shall not be used to connect bonding conductors or connection devices.

SECTION E4205 GROUNDING

E4205.1 Equipment to be grounded. The following equipment shall be grounded:

1. Through-wall lighting assemblies and underwater luminaires other than those low-voltage lighting products listed for the application without a grounding conductor.
2. All electrical equipment located within 5 feet (1524 mm) of the inside wall of the pool, spa or hot tub.
3. All electrical equipment associated with the recirculating system of the pool, spa or hot tub.
4. Junction boxes.
5. Transformer and power supply enclosures.
6. Ground-fault circuit-interrupters.
7. Panelboards that are not part of the service equipment and that supply any electrical equipment associated with the pool, spa or hot tub.

E4205.2 Luminaires and related equipment. Other than listed low-voltage luminaires not requiring grounding, all through-wall lighting assemblies, wet-niche, dry-niche, or no-niche luminaires shall be connected to an insulated copper equipment grounding conductor sized in accordance with Table E3908.12 but not smaller than 12 AWG. The equipment grounding conductor between the wiring chamber of the secondary winding of a transformer and a junction box shall be sized in accordance with the overcurrent device in such circuit. The junction box, transformer enclosure, or other enclosure in the supply circuit to a wet-niche or no-niche luminaire and the field-wiring chamber of a dry-niche luminaire shall be grounded to the equipment grounding terminal of the panelboard. The equipment grounding terminal shall be directly connected to the panelboard enclosure. The equipment grounding conductor shall be installed without joint or splice.

Exceptions:

1. Where more than one underwater luminaire is supplied by the same branch circuit, the equipment grounding conductor, installed between the junction boxes, transformer enclosures, or other enclosures in the supply circuit to wet-niche luminaires, or between the field-wiring compartments of dry-niche luminaires, shall be permitted to be terminated on grounding terminals.
2. Where an underwater luminaire is supplied from a transformer, ground-fault circuit-interrupter, clock-operated switch, or a manual snap switch that is located between the panelboard and a junction box connected to the conduit that extends directly to the underwater luminaire, the equipment grounding conductor shall be permitted to terminate on grounding terminals on the transformer, ground-fault circuit-interrupter, clock-operated switch enclosure, or an outlet box used to enclose a snap switch.

E4205.3 Nonmetallic conduit. Where a nonmetallic conduit is installed between a forming shell and a junction box, transformer enclosure, or other enclosure, a 8 AWG insulated copper bonding jumper shall be installed in this conduit except where a listed low-voltage lighting system not requiring grounding is used. The bonding jumper shall be terminated in the forming shell, junction box or transformer enclosure, or ground-fault circuit-interrupter enclosure. The termination of the 8 AWG bonding jumper in the forming shell shall be covered with, or encapsulated in, a listed potting compound to protect such connection from the possible deteriorating effect of pool water.

E4205.4 Flexible cords. Other than listed low-voltage lighting systems not requiring grounding, wet-niche luminaires that are supplied by a flexible cord or cable shall have all exposed noncurrent-carrying metal parts grounded by an insulated copper equipment grounding conductor that is an integral part of the cord or cable. This grounding conductor shall be connected to a grounding terminal in the supply junction box, transformer enclosure, or other enclosure. The grounding conductor shall not be smaller than the supply conductors and not smaller than 16 AWG.

E4205.5 Motors. Pool-associated motors shall be connected to an insulated copper equipment grounding conductor sized in accordance with Table E3908.12, but not smaller than 12 AWG. Where the branch circuit supplying the motor is installed in the interior of a one-family dwelling or in the interior of accessory buildings associated with a one-family dwelling, using a cable wiring method permitted by Table E4202.1, an uninsulated equipment grounding conductor shall be permitted provided that it is enclosed within the outer sheath of the cable assembly.

E4205.6 Feeders. An equipment grounding conductor shall be installed with the feeder conductors between the grounding terminal of the pool equipment panelboard and the grounding terminal of the applicable service equipment or source of a separately derived system. The equipment grounding conductor shall be insulated, shall be sized in accordance with Table E3908.12, and shall be not smaller than 12 AWG.

Exception: An existing feeder between an existing remote panelboard and service equipment shall be permitted to run in flexible metal conduit or an approved cable assembly that includes an equipment grounding conductor within its outer sheath. The equipment grounding conductor shall not be connected to the grounded conductor in the remote panelboard.

E4205.6.1 Separate buildings. A feeder to a separate building or structure shall be permitted to supply swimming pool equipment branch circuits, or feeders supplying swimming pool equipment branch circuits, provided that the grounding arrangements in the separate building meet the requirements of Section E3607.3. Where installed in other than existing feeders covered in the exception to

Section E4205.6, a separate equipment grounding conductor shall be an insulated conductor.

E4205.7 Cord-connected equipment. Where fixed or stationary equipment is connected with a flexible cord to facilitate removal or disconnection for maintenance, repair, or storage, as provided in Section E4202.2, the equipment grounding conductors shall be connected to a fixed metal part of the assembly. The removable part shall be mounted on or bonded to the fixed metal part.

E4205.8 Other equipment. Other electrical equipment shall be grounded in accordance with Section E3908.

SECTION E4206 EQUIPMENT INSTALLATION

E4206.1 Transformers and power supplies. Transformers and power supplies used for the supply of underwater luminaires, together with the transformer or power supply enclosure, shall be listed for swimming pool and spa use. The transformer or power supply shall incorporate either a transformer of the isolated-winding type with an ungrounded secondary that has a grounded metal barrier between the primary and secondary windings, or a transformer that incorporates an approved system of double insulation between the primary and secondary windings.

E4206.2 Ground-fault circuit-interrupters. Ground-fault circuit-interrupters shall be self-contained units, circuit-breaker types, receptacle types or other approved types.

E4206.3 Wiring on load side of ground-fault circuit-interrupters and transformers. For other than grounding conductors, conductors installed on the load side of a ground-fault circuit-interrupter or transformer used to comply with the provisions of Section E4206.4, shall not occupy raceways, boxes, or enclosures containing other conductors except where the other conductors are protected by ground-fault circuit interrupters or are grounding conductors. Supply conductors to a feed-through type ground-fault circuit interrupter shall be permitted in the same enclosure. Ground-fault circuit interrupters shall be permitted in a panelboard that contains circuits protected by other than ground-fault circuit interrupters.

E4206.4 Underwater luminaires. The design of an underwater luminaire supplied from a branch circuit either directly or by way of a transformer or power supply meeting the requirements of Section E4206.1, shall be such that, where the fixture is properly installed without a ground-fault circuit-interrupter, there is no shock hazard with any likely combination of fault conditions during normal use (not relamping). In addition, a ground-fault circuit-interrupter shall be installed in the branch circuit supplying luminaires operating at more than the low-voltage contact limit, such that there is no shock hazard during relamping. The installation of the ground-fault circuit-interrupter shall be such that there is no shock hazard with any likely fault-condition combination that involves a person in a conductive path from any ungrounded part of the branch circuit or the luminaire to ground. Compliance with this requirement shall be obtained by the use of a listed underwater luminaire and by installation of a listed ground-fault circuit-interrupter in the branch circuit or a listed transformer or power supply for luminaires operating at more than the low-voltage contact limit. Luminaires that depend on submersion for safe operation shall be inherently protected against the hazards of overheating when not submerged.

E4206.4.1 Maximum voltage. Luminaires shall not be installed for operation on supply circuits over 150 volts between conductors.

E4206.4.2 Luminaire location. Luminaires mounted in walls shall be installed with the top of the fixture lens not less than 18 inches (457 mm) below the normal water level of the pool, except where the luminaire is listed and identified for use at a depth of not less than 4 inches (102 mm) below the normal water level of the pool. A luminaire facing upward shall have the lens adequately guarded to prevent contact by any person or shall be listed for use without a guard.

E4206.5 Wet-niche luminaires. Forming shells shall be installed for the mounting of all wet-niche underwater luminaires and shall be equipped with provisions for conduit entries. Conduit shall extend from the forming shell to a suitable junction box or other enclosure located as provided in Section E4206.9. Metal parts of the luminaire and forming shell in contact with the pool water shall be of brass or other approved corrosion-resistant metal.

The end of flexible-cord jackets and flexible-cord conductor terminations within a luminaire shall be covered with, or encapsulated in, a suitable potting compound to prevent the entry of water into the luminaire through the cord or its conductors. If present, the grounding connection within a luminaire shall be similarly treated to protect such connection from the deteriorating effect of pool water in the event of water entry into the luminaire.

Luminaires shall be bonded to and secured to the forming shell by a positive locking device that ensures a low-resistance contact and requires a tool to remove the luminaire from the forming shell.

E4206.5.1 Servicing. All wet-niche luminaires shall be removable from the water for inspection, relamping, or other maintenance. The forming shell location and length of cord in the forming shell shall permit personnel to place the removed luminaire on the deck or other dry location for such maintenance. The luminaire maintenance location shall be accessible without entering or going into the pool water.

E4206.6 Dry-niche luminaires. Dry-niche luminaires shall have provisions for drainage of water. Other than listed low-voltage luminaires not requiring grounding, a dry-niche luminaire shall have means for accommodating one equipment grounding conductor for each conduit entry. Junction boxes shall not be required but, if used, shall not be required to be elevated or located as specified in Section E4206.9 if the luminaire is specifically identified for the purpose.

E4206.7 No-niche luminaires. No-niche luminaires shall be listed for the purpose and shall be installed in accordance with the requirements of Section E4206.5. Where connection to a forming shell is specified, the connection shall be to the mounting bracket.

E4206.8 Through-wall lighting assembly. A through-wall lighting assembly shall be equipped with a threaded entry or hub, or a nonmetallic hub, for the purpose of accommodating the termination of the supply conduit. A through-wall lighting assembly shall meet the construction requirements of Section E4205.4 and be installed in accordance with the requirements of Section E4206.5 Where connection to a forming shell is specified, the connection shall be to the conduit termination point.

E4206.9 Junction boxes and enclosures for transformers or ground-fault circuit interrupters. Junction boxes for underwater luminaires and enclosures for transformers and ground-fault circuit-interrupters that supply underwater luminaires shall comply with the following:

E4206.9.1 Junction boxes. A junction box connected to a conduit that extends directly to a forming shell or mounting bracket of a no-niche luminaire shall be:

1. Listed as a swimming pool junction box;
2. Equipped with threaded entries or hubs or a nonmetallic hub;
3. Constructed of copper, brass, suitable plastic, or other approved corrosion-resistant material;
4. Provided with electrical continuity between every connected metal conduit and the grounding terminals by means of copper, brass, or other approved corrosion-resistant metal that is integral with the box; and
5. Located not less than 4 inches (102 mm), measured from the inside of the bottom of the box, above the ground level, or pool deck, or not less than 8 inches (203 mm) above the maximum pool water level, whichever provides the greatest elevation, and shall be located not less than 4 feet (1219 mm) from the inside wall of the pool, unless separated from the pool by a solid fence, wall or other permanent barrier. Where used on a lighting system operating at the low-voltage contact limit or less, a flush deck box shall be permitted provided that an approved potting compound is used to fill the box to prevent the entrance of moisture; and the flush deck box is located not less than 4 feet (1219 mm) from the inside wall of the pool.

E4206.9.2 Other enclosures. An enclosure for a transformer, ground-fault circuit-interrupter or a similar device connected to a conduit that extends directly to a forming shell or mounting bracket of a no-niche luminaire shall be:

1. Listed and labeled for the purpose, comprised of copper, brass, suitable plastic, or other approved corrosion-resistant material;
2. Equipped with threaded entries or hubs or a nonmetallic hub;
3. Provided with an approved seal, such as duct seal at the conduit connection, that prevents circulation of air between the conduit and the enclosures;
4. Provided with electrical continuity between every connected metal conduit and the grounding terminals by means of copper, brass or other approved corrosion-resistant metal that is integral with the enclosures; and
5. Located not less than 4 inches (102 mm), measured from the inside bottom of the enclosure, above the ground level or pool deck, or not less than 8 inches (203 mm) above the maximum pool water level, whichever provides the greater elevation, and shall be located not less than 4 feet (1219 mm) from the inside wall of the pool, except where separated from the pool by a solid fence, wall or other permanent barrier.

E4206.9.3 Protection of junction boxes and enclosures. Junction boxes and enclosures mounted above the grade of the finished walkway around the pool shall not be located in the walkway unless afforded additional protection, such as by location under diving boards or adjacent to fixed structures.

E4206.9.4 Grounding terminals. Junction boxes, transformer and power supply enclosures, and ground-fault circuit-interrupter enclosures connected to a conduit that extends directly to a forming shell or mounting bracket of a no-niche luminaire shall be provided with grounding terminals in a quantity not less than the number of conduit entries plus one.

E4206.9.5 Strain relief. The termination of a flexible cord of an underwater luminaire within a junction box, transformer or power supply enclosure, ground-fault circuit-interrupter, or other enclosure shall be provided with a strain relief.

E4206.10 Underwater audio equipment. Underwater audio equipment shall be identified for the purpose.

E4206.10.1 Speakers. Each speaker shall be mounted in an approved metal forming shell, the front of which is enclosed by a captive metal screen, or equivalent, that is bonded to and secured to the forming shell by a positive locking device that ensures a low-resistance contact and requires a tool to open for installation or servicing of the speaker. The forming shell shall be installed in a recess in the wall or floor of the pool.

E4206.10.2 Wiring methods. Rigid metal conduit of brass or other identified corrosion-resistant metal, rigid polyvinyl chloride conduit, rigid thermosetting resin conduit or liquid-tight flexible nonmetallic conduit (LFNC-B) shall extend from the forming shell to a suitable junction box or other enclosure as provided in Section E4206.9. Where rigid nonmetallic conduit or liquid-tight flexible nonmetallic conduit is used, an 8 AWG solid or stranded insulated copper bonding jumper shall be installed in this conduit with provisions for terminating in the forming shell and the junction box. The termination of the 8 AWG bonding jumper in the forming shell shall be covered with, or encapsulated in, a suitable potting compound to protect such connection from the possible deteriorating effect of pool water.

E4206.10.3 Forming shell and metal screen. The forming shell and metal screen shall be of brass or other approved corrosion-resistant metal. All forming shells shall include provisions for terminating an 8 AWG copper conductor.

E4206.11 Electrically operated pool covers. The electric motors, controllers, and wiring for pool covers shall be located not less than 5 feet (1524 mm) from the inside wall of the pool except where separated from the pool by a wall, cover, or other permanent barrier. Electric motors installed below grade level shall be of the totally enclosed type. The electric motor and controller shall be connected to a circuit protected by a ground-fault circuit-interrupter. The device that controls the operation of the motor for an electrically operated pool cover shall be located so that the operator has full view of the pool.

E4206.12 Electric pool water heaters. All electric pool water heaters shall have the heating elements subdivided into loads not exceeding 48 amperes and protected at not more than 60 amperes. The ampacity of the branch-circuit conductors and the rating or setting of overcurrent protective devices shall be not less than 125 percent of the total nameplate load rating.

E4206.13 Pool area heating. The provisions of Sections E4206.13.1 through E4206.13.3 shall apply to all pool deck areas, including a covered pool, where electrically operated comfort heating units are installed within 20 feet (6096 mm) of the inside wall of the pool.

E4206.13.1 Unit heaters. Unit heaters shall be rigidly mounted to the structure and shall be of the totally enclosed or guarded types. Unit heaters shall not be mounted over the pool or within the area extending 5 feet (1524 mm) horizontally from the inside walls of a pool.

E4206.13.2 Permanently wired radiant heaters. Electric radiant heaters shall be suitably guarded and securely fastened to their mounting devices. Heaters shall not be installed over a pool or within the area extending 5 feet (1524 mm) horizontally from the inside walls of the pool and shall be mounted not less than 12 feet (3658 mm) vertically above the pool deck.

E4206.13.3 Radiant heating cables prohibited. Radiant heating cables embedded in or below the deck shall be prohibited.

SECTION E4207
STORABLE SWIMMING POOLS

E4207.1 Pumps. A cord and plug-connected pool filter pump for use with storable pools shall incorporate an approved system of double insulation or its equivalent and shall be provided with means for grounding only the internal and nonaccessible noncurrent-carrying metal parts of the appliance.

The means for grounding shall be an equipment grounding conductor run with the power-supply conductors in a flexible cord that is properly terminated in a grounding-type attachment plug having a fixed grounding contact. Cord and plug-connected pool filter pumps shall be provided with a ground-fault circuit interrupter that is an integral part of the attachment plug or located in the power supply cord within 12 inches (305 mm) of the attachment plug.

E4207.2 Ground-fault circuit-interrupters required. Electrical equipment, including power-supply cords, used with storable pools shall be protected by ground-fault circuit-interrupters. All 125-volt, 15- and 20-ampere receptacles located within 20 feet (6096 mm) of the inside walls of a storable pool shall be protected by a ground-fault circuit interrupter. In determining these dimensions, the distance to be measured shall be the shortest path that the supply cord of an appliance connected to the receptacle would follow without passing through a floor, wall, ceiling, doorway with hinged or sliding door, window opening, or other effective permanent barrier.

E4207.3 Luminaires. Luminaires for storable pools shall not have exposed metal parts and shall be listed for the purpose as an assembly. In addition, luminaires for storable pools shall comply with the requirements of Section E4207.3.1 or E4207.3.2.

E4207.3.1 Within the low-voltage contact limit. A luminaire installed in or on the wall of a storable pool shall be part of a cord and plug-connected lighting assembly. The assembly shall:

1. Have a luminaire lamp that is suitable for the use at the supplied voltage;
2. Have an impact-resistant polymeric lens, luminaire body, and transformer enclosure;
3. Have a transformer meeting the requirements of section E4206.1 with a primary rating not over 150 volts; and
4. Have no exposed metal parts.

E4207.3.2 Over the low-voltage contact limit but not over 150 volts. A lighting assembly without a transformer or power supply, and with the luminaire lamp(s) operating at over the low-voltage contact limit, but not over 150 volts, shall be permitted to be cord and plug-connected where the assembly is listed as an assembly for the purpose and complies with all of the following:

1. It has an impact-resistant polymeric lens and luminaire body.
2. A ground-fault circuit interrupter with open neutral conductor protection is provided as an integral part of the assembly.
3. The luminaire lamp is permanently connected to the ground-fault circuit interrupter with open-neutral protection.
4. It complies with the requirements of Section E4206.4.
5. It has no exposed metal parts.

E4207.4 Receptacle locations. Receptacles shall be located not less than 6 feet (1829 mm) from the inside walls of a pool. In determining these dimensions, the distance to be measured shall be the shortest path that the supply cord of an appliance connected to the receptacle would follow without passing through a floor, wall, ceiling, doorway with hinged or sliding door, window opening, or other effective permanent barrier.

SECTION E4208 SPAS AND HOT TUBS

E4208.1 Ground-fault circuit-interrupters. The outlet(s) that supplies a self-contained spa or hot tub, or a packaged spa or hot tub equipment assembly, or a field-assembled spa or hot tub with a heater load of 50 amperes or less, shall be protected by a ground-fault circuit-interrupter.

A listed self-contained unit or listed packaged equipment assembly marked to indicate that integral ground-fault circuit-interrupter protection is provided for all electrical parts within the unit or assembly, including pumps, air blowers, heaters, lights, controls, sanitizer generators and wiring, shall not require that the outlet supply be protected by a ground-fault circuit interrupter.

E4208.2 Electric water heaters. Electric spa and hot tub water heaters shall be listed and shall have the heating elements subdivided into loads not exceeding 48 amperes and protected at not more than 60 amperes. The ampacity of the branch-circuit conductors, and the rating or setting of overcurrent protective devices, shall be not less than 125 percent of the total nameplate load rating.

E4208.3 Underwater audio equipment. Underwater audio equipment used with spas and hot tubs shall comply with the provisions of Section E4206.10.

E4208.4 Emergency switch for spas and hot tubs. A clearly labeled emergency shutoff or control switch for the purpose of stopping the motor(s) that provides power to the recirculation system and jet system shall be installed at a point that is readily accessible to the users, adjacent to and within sight of the spa or hot tub and not less than 5 feet (1524 mm) away from the spa or hot tub. This requirement shall not apply to single-family dwellings.

SECTION E4209 HYDROMASSAGE BATHTUBS

E4209.1 Ground-fault circuit-interrupters. Hydromassage bathtubs and their associated electrical components shall be supplied by an individual branch circuit(s) and protected by a readily accessible ground-fault circuit-interrupter. All 125-volt, single-phase receptacles not exceeding 30 amperes and located within 6 feet (1829 mm) measured horizontally of the inside walls of a hydromassage tub shall be protected by a ground-fault circuit interrupter(s).

E4209.2 Other electric equipment. Luminaires, switches, receptacles, and other electrical equipment located in the same room, and not directly associated with a hydromassage bathtub, shall be installed in accordance with the requirements of this code relative to the installation of electrical equipment in bathrooms.

E4209.3 Accessibility. Hydromassage bathtub electrical equipment shall be accessible without damaging the building structure or building finish. Where the hydromassage bathtub is cord- and plug-connected with the supply receptacle accessible only through a service access opening, the receptacle shall be installed so that its face is within direct view and not more than 12 inches (305 mm) from the plane of the opening.

E4209.4 Bonding. All metal piping systems and all grounded metal parts in contact with the circulating water shall be bonded together using an insulated, covered or bare solid copper bonding jumper not smaller than 8 AWG. The bonding jumper shall be connected to the terminal on the circulating pump motor that is intended for this purpose. The bonding jumper shall not be required to be connected to a double insualted circulating pump motor. The 8 AWG or larger solid copper bonding jumper shall be required for equipotential bonding in the area of the hydromassage bathtub and shall not be required to be extended or attached to any remote panelboard, service equipment, or any electrode. Where a double-insulated circulating pump motor is used, the 8 AWG or larger solid copper bonding jumper shall be long enough to terminate on a replacement nondouble-insulated pump motor and shall be terminated to the equipment grounding conductor of the branch circuit for the motor.

CHAPTER 43

CLASS 2 REMOTE-CONTROL, SIGNALING AND POWER-LIMITED CIRCUITS

SECTION E4301 GENERAL

E4301.1 Scope. This chapter contains requirements for power supplies and wiring methods associated with Class 2 remote-control, signaling, and power-limited circuits that are not an integral part of a device or appliance. Other classes of remote-control, signaling and power-limited conductors shall comply with Article 725 of NFPA 70.

E4301.2 Definitions.

CLASS 2 CIRCUIT. That portion of the wiring system between the load side of a Class 2 power source and the connected equipment. Due to its power limitations, a Class 2 circuit considers safety from a fire initiation standpoint and provides acceptable protection from electric shock.

REMOTE-CONTROL CIRCUIT. Any electrical circuit that controls any other circuit through a relay or an equivalent device.

SIGNALING CIRCUIT. Any electrical circuit that energizes signaling equipment.

SECTION E4302 POWER SOURCES

E4302.1 Power sources for Class 2 circuits. The power source for a Class 2 circuit shall be one of the following:

1. A listed Class 2 transformer.
2. A listed Class 2 power supply.
3. Other listed equipment marked to identify the Class 2 power source.
4. Listed information technology (computer) equipment limited power circuits.
5. A dry-cell battery provided that the voltage is 30 volts or less and the capacity is equal to or less than that available from series connected No. 6 carbon zinc cells.

E4302.2 Interconnection of power sources. A Class 2 power source shall not have its output connections paralleled or otherwise interconnected with another Class 2 power source except where listed for such interconnection.

SECTION E4303 WIRING METHODS

E4303.1 Wiring methods on supply side of Class 2 power source. Conductors and equipment on the supply side of the power source shall be installed in accordance with the appropriate requirements of Chapters 34 through 41. Transformers or other devices supplied from electric light or power circuits shall be protected by an over-current device rated at not over 20 amperes. The input leads of a transformer or other power source supplying Class 2 circuits shall be permitted to be smaller than 14 AWG, if not over 12 inches (305 mm) long and if the conductor insulation is rated at not less than 600 volts. In no case shall such leads be smaller than 18 AWG.

E4303.2 Wiring methods and materials on load side of the Class 2 power source. Class 2 cables installed as wiring within buildings shall be listed as being resistant to the spread of fire and listed as meeting the criteria specified in Sections E4303.2.1 through E4303.2.3. Cables shall be marked in accordance with Section E4303.2.4. Cable substitutions as described in Table E4303.2 and wiring methods covered in Chapter 38 shall also be permitted.

TABLE E4303.2
CABLE USES AND PERMITTED SUBSTITUTIONS

CABLE TYPE	USE	PERMITTED SUBSTITUTIONS[a]
CL2P	Class 2 Plenum Cable	CMP, CL3P
CL2	Class 2 Cable	CMP, CL3P, CL2P, CMR, CL3R, CL2R CMG, CM, CL3
CL2X	Class 2 Cable, Limited Use	CMP, CL3P CL2P, CMR, CL3R, CL2R, CMG, CM, CL3, CL2, CMX, CL3X

a. For identification of cables other than Class 2 cables, see NFPA 70.

E4303.2.1 Type CL2P cables. Cables installed in ducts, plenums and other spaces used to convey environmental air shall be Type CL2P cables listed as being suitable for the use and listed as having adequate fire-resistant and low smoke-producing characteristics.

E4303.2.2 Type CL2 cables. Cables for general-purpose use, shall be listed as being resistant to the spread of fire and listed for the use.

E4303.2.3 Type CL2X cables. Type CL2X limited-use cable shall be listed as being suitable for use in dwellings and for the use and in raceways and shall also be listed as being flame retardant. Cables with a diameter of less than $^1/_4$ inch (6.4 mm) shall be permitted to be installed without a raceway.

E4303.2.4 Marking. Cables shall be marked in accordance with Table E4303.2. Voltage ratings shall not be marked on cables.

SECTION E4304 INSTALLATION REQUIREMENTS

E4304.1 Separation from other conductors. In cables, compartments, enclosures, outlet boxes, device boxes, and raceways, conductors of Class 2 circuits shall not be placed in any cable, compartment, enclosure, outlet box, device box, raceway, or similar fitting with conductors of electric light, power, Class 1 and nonpower-limited fire alarm circuits.

Exceptions:

1. Where the conductors of the electric light, power, Class 1 and nonpower-limited fire alarm circuits are separated by a barrier from the Class 2 circuits. In enclosures, Class 2 circuits shall be permitted to be installed in a raceway within the enclosure to separate them from Class 1, electric light, power and nonpower-limited fire alarm circuits.
2. Class 2 conductors in compartments, enclosures, device boxes, outlet boxes and similar fittings where electric light, power, Class 1 or nonpower-limited fire alarm circuit conductors are introduced solely to connect to the equipment connected to the Class 2 circuits. The electric light, power, Class 1 and nonpower-limited fire alarm circuit conductors shall be routed to maintain a minimum of $^1/_4$ inch (6.4 mm) separation from the conductors and cables of the Class 2 circuits; or the electric light power, Class 1 and nonpower-limited fire alarm circuit conductors operate at 150 volts or less to ground and the Class 2 circuits are installed using Types CL3, CL3R, or CL3P or permitted substitute cables, and provided that these Class 3 cable conductors extending beyond their jacket are separated by a minimum of $^1/_4$ inch (6.4 mm) or by a nonconductive sleeve or nonconductive barrier from all other conductors.

E4304.2 Other applications. Conductors of Class 2 circuits shall be separated by not less than 2 inches (51 mm) from conductors of any electric light, power, Class 1 or nonpower-limited fire alarm circuits except where one of the following conditions is met:

1. All of the electric light, power, Class 1 and nonpower-limited fire alarm circuit conductors are in raceways or in metal-sheathed, metal-clad, nonmetallic-sheathed or Type UF cables.
2. All of the Class 2 circuit conductors are in raceways or in metal-sheathed, metal-clad, nonmetallic-sheathed or Type UF cables.

E4304.3 Class 2 circuits with communications circuits. Where Class 2 circuit conductors are in the same cable as communications circuits, the Class 2 circuits shall be classified as communications circuits and shall meet the requirements of Article 800 of NFPA 70. The cables shall be listed as communications cables or multipurpose cables.

Cables constructed of individually listed Class 2 and communications cables under a common jacket shall be permitted to be classified as communications cables. The fire-resistance rating of the composite cable shall be determined by the performance of the composite cable.

E4304.4 Class 2 cables with other circuit cables. Jacketed cables of Class 2 circuits shall be permitted in the same enclosure or raceway with jacketed cables of any of the following:

1. Power-limited fire alarm systems in compliance with Article 760 of NFPA 70.
2. Nonconductive and conductive optical fiber cables in compliance with Article 770 of NFPA 70.
3. Communications circuits in compliance with Article 800 of NFPA 70.
4. Community antenna television and radio distribution systems in compliance with Article 820 of NFPA 70.
5. Low-power, network-powered broadband communications in compliance with Article 830 of NFPA 70.

E4304.5 Installation of conductors and cables. Cables and conductors installed exposed on the surface of ceilings and sidewalls shall be supported by the building structure in such a manner that they will not be damaged by normal building use. Such cables shall be supported by straps, staples, hangers, cable ties or similar fittings designed so as to not damage the cable. The installation shall comply with Table E3802.1 regarding cables run parallel with framing members and furring strips. The installation of wires and cables shall not prevent access to equipment nor prevent removal of panels, including suspended ceiling panels. Raceways shall not be used as a means of support for Class 2 circuit conductors, except where the supporting raceway contains conductors supplying power to the functionally associated equipment controlled by the Class 2 conductors.

Part IX—Referenced Standards

CHAPTER 44

REFERENCED STANDARDS

This chapter lists the standards that are referenced in various sections of this document. The standards are listed herein by the promulgating agency of the standard, the standard identification, the effective date and title, and the section or sections of this document that reference the standard. The application of the referenced standards shall be as specified in Section R102.4.

AAMA

American Architectural Manufacturers Association
1827 Walden Office Square, Suite 550
Schaumburg, IL 60173

Standard reference number	Title	Referenced in code section number
AAMA/WDMA/CSA 101/I.S.2/A440—11	North American Fenestration Standards/Specifications for Windows, Doors and Skylights	R308.6.9, R612.3, N1102.4.3
450—09	Voluntary Performance Rating Method for Mulled Fenestration Assemblies	R612.8
506—08	Voluntary Specifications for Hurricane Impact and Cycle Testing of Fenestration Products	R612.6.1
711—07	Voluntary Specification for Self-adhering Flashing Used for Installation of Exterior Wall Fenestration Products	R703.8

ACI

American Concrete Institute
38800 Country Club Drive
Farmington Hills, MI 48331

Standard reference number	Title	Referenced in code section number
318—11	Building Code Requirements for Structural Concrete	R301.2.2.2.4, R301.2.2.3.4, Table R402.1.2(2), R402.2, R404.1.2, Table R404.1.2(5), Table R404.1.2(7), Table R404.1.2(8), Table R404.1.2(9), R404.1.2.1, R404.1.2.3, R404.1.2.4, R404.1.4.2, R404.5.1, R611.1, R611.1.1, R611.1.2, R611.2, R611.5.1, R611.6.1, R611.8.2, R611.9.2, R611.9.3
332—10	Code Requirements for Residential Concrete Construction	R402.2, R403.1, R404.1.2, R404.1.2.4, R404.1.4.2, R506.1
530—11	Building Code Requirements for Masonry Structures	R404.1.1, R606.1, R606.1.1, R606.12.1, R606.12.2.2.2, R606.12.2.3.1, R606.12.3.1, Table R703.4
530.1—11	Specification for Masonry Structures	R404.1.1, R606.1, R606.1.1, R606.12.1, R606.12.2.2.1, R606.12.2.2.2, R606.12.3.1, Table R703.4

ACCA

Air Conditioning Contractors of America
2800 Shirlington Road, Suite 300
Arlington, VA 22206

Standard reference number	Title	Referenced in code section number
Manual D—-09	Residential Duct Systems	M1601.1, M1602.2
Manual J—-11	Residential Load Calculation—Eighth Edition	N1103.6, M1401.3
Manual S—-10	Residential Equipment Selection	N1103.6, M1401.3

AFPA

American Forest and Paper Association
1111 19th Street, NW, Suite 800
Washington, DC 20036

Standard reference number	Title	Referenced in code section number
AFPA—2012	Span Tables for Joists and Rafters	R502.3, R802.4, R802.5
ANSI/AF&PA WFCM—2012	Wood Frame Construction Manual for One- and Two-family Dwellings	R301.1.1, R301.2.1.1, R611.9.2, R611.9.3, R611.10
NDS—2012	National Design Specification (NDS) for Wood Construction—with 2005 Supplement	R404.2.2, R502.2, Table R503.1, R602.3, Table R602.3.1, R611.9.2, R611.9.3, R802.2
PWF—2007	Permanent Wood Foundation Design Specification	R401.1, R404.2.3

AISI

American Iron and Steel Institute
1140 Connecticut Ave, Suite 705
Washington, DC 20036

Standard reference number	Title	Referenced in code section number
AISI S100—07/S1—10	North American Specification for the Design of Cold-formed Steel Structural Members	R505.1.3, R603.6, R611.9.2, R611.9.3, R804.3.7
AISI S230—07	Standard for Cold-formed Steel Framing—Prescriptive Method for One- and Two-family Dwellings, with Supplement 2, dated 2008	R301.1.1, R301.2.1.1, R301.2.2.3.1, R301.2.2.3.5, R603.6, R603.9.4.1, R603.9.4.2, R611.9.2, R611.9.3, R611.10

AITC

American Institute of Timber Construction
7012 S. Revere Parkway, Suite 140
Centennial, CO 80112

Standard reference number	Title	Referenced in code section number
ANSI/AITC A 190.1—07	Structural Glued-laminated Timber	R502.1.5, R602.1.2, R802.1.4

ANSI

American National Standards Institute
25 West 43rd Street, Fourth Floor
New York, NY 10036

Standard reference number	Title	Referenced in code section number
A108.1A—99	Installation of Ceramic Tile in the Wet-set Method, with Portland Cement Mortar	R702.4.1
A108.1B—99	Installation of Ceramic Tile, Quarry Tile on a Cured Portland Cement Mortar Setting Bed with Dry-set or Latex Portland Mortar	R702.4.1
A108.4—99	Installation of Ceramic Tile with Organic Adhesives or Water-Cleanable Tile-setting Epoxy Adhesive	R702.4.1
A108.5—99	Installation of Ceramic Tile with Dry-set Portland Cement Mortar or Latex Portland Cement Mortar	R702.4.1
A108.6—99	Installation of Ceramic Tile with Chemical-resistant, Water-cleanable Tile-setting and -grouting Epoxy	R702.4.1
A108.11—99	Interior Installation of Cementitious Backer Units	R702.4.1
A118.1—99	American National Standard Specifications for Dry-set Portland Cement Mortar	R702.4.1
A118.3—99	American National Standard Specifications for Chemical-resistant, Water-cleanable Tile-setting and -grouting Epoxy, and Water-cleanable Tile-setting Epoxy Adhesive	R702.4.1
A118.10—99	Specification for Load-bearing, Bonded, Waterproof Membranes for Thin-set Ceramic Tile and Dimension Stone Installation	P2709.2 , P2709.2.4

ANSI—continued

A136.1—99	American National Standard Specifications for Organic Adhesives for Installation of Ceramic Tile	R702.4.1
A137.1—88	American National Standard Specifications for Ceramic Tile	R702.4.1
A208.1—2009	Particleboard	R503.3.1, R605.1
LC1/CSA 6.26—05	Fuel Gas Piping Systems Using Corrugated Stainless Steel Tubing (CSST)	G2414.5.3
LC4—07	Press-connect Copper and Copper-alloy Fittings for Use in Fuel Gas Distribution Systems	G2414.10.2
Z21.1—05	Household Cooking Gas Appliances	G2447.1
Z21.5.1/CSA 7.1—06	Gas Clothes Dryers—Volume I—Type I Clothes Dryers	G2438.1
Z21.8—94 (R2002)	Installation of Domestic Gas Conversion Burners	G2443.1
Z21.10.1/CSA 4.1—09	Gas Water Heaters—Volume I—Storage Water Heaters with Input Ratings of 75,000 Btu per hour or Less	G2448.1
Z21.10.3/CSA 4.3—04	Gas Water Heaters—Volume III—Storage Water Heaters with Input Ratings above 75,000 Btu per hour, Circulating and Instantaneous	G2448.1
Z21.11.2—07	Gas-fired Room Heaters—Volume II—Unvented Room Heaters	G2445.1
Z21.13/CSA 4.9—10	Gas-fired Low-pressure Steam and Hot Water Boilers	G2452.1
Z21.15/CSA 9.1—09	Manually Operated Gas Valves for Appliances, Appliance Connector Valves and Hose End Valves	Table G2420.1.1
Z21.22—99 (R2003)	Relief Valves for Hot Water Supply Systems—with Addenda Z21.22a—2000 (R2003) and 21.22b—2001 (R2003)	P2803.2, P2803.7
Z21.24—97	Connectors for Gas Appliances	G2422.1
Z21.40.1/ CSA 2.91—96 (R2002)	Gas-fired, Heat-activated Air-conditioning and Heat Pump Appliances	G2449.1
Z21.40.2/ CSA 2.92—96 (R2002)	Gas-fired, Work-activated Air-conditioning and Heat Pump Appliances (Internal Combustion)	G2449.1
Z21.42—93 (R2002)	Gas-fired Illuminating Appliances	G2450.1
Z21.47/CSA 2.3—06	Gas-fired Central Furnaces	G2442.1
Z21.50/CSA 2.22—07	Vented Gas Fireplaces	G2434.1
Z21.56/CSA 4.7—06	Gas-fired Pool Heaters	G2441.1
Z21.58—95/CSA 1.6—07	Outdoor Cooking Gas Appliances	G2447.1
Z21.60/CSA 2.26—03	Decorative Gas Appliances for Installation in Solid Fue-burning Fireplaces	G2432.1
Z21.75/CSA 6.27—07	Connectors for Outdoor Gas Appliances	G2422.1
Z21.80—03	Line Pressure Regulators	G2421.1
Z21.83—98	Fuel Cell Power Plants	M1903.1
Z21.84—02	Manually Listed, Natural Gas Decorative Gas Appliances for Installation in Solid Fuel-burning Fireplaces—with Addenda Z21.84a—2003	G2432.1, G2432.2
Z21.86—04	Gas-fired Vented Space Heating Appliances	G2436.1, G2437.1, G2446.1
Z21.88—02	Vented Gas Fireplace Heaters—with Addenda A21.88a—2003 and Z21.88b—2004	G2435.1
Z21.91—01	Ventless Firebox Enclosures for Gas-fired Unvented Decorative Room Heaters	G2445.7.1
Z21.97—09	Outdoor Decorative Appliances	G2454.1
Z83.6—90 (R1998)	Gas-fired Infrared Heaters	G2451.1
Z83.8—02	Gas-fired Unit Heaters and Gas-fired Duct Furnaces—with Addenda Z83.8a—2003	G2444.1
Z83.19—01 (R2005)	Gas-fuel High-intensity Infrared Heaters	G2451.1
Z83.20—08	Gas-fired Low-intensity Infared Heaters Outdoor Decorative Appliances	G2451.1
Z97.1—09	Safety Glazing Materials Used in Buildings—Safety Performance Specifications and Methods of Test	R308.1.1, R308.3.1
Z124.1—95	Plastic Bathtub Units	Table P2701.1
Z124.2—95	Plastic Shower Receptors and Shower Stalls	Table P2701.1
Z124.3—95	Plastic Lavatories	Table P2701.1, P2711.1, P2711.2
Z124.4—96	Plastic Water Closet Bowls and Tanks	Table P2701.1, P2712.1
Z124.6—97	Plastic Sinks	Table P2701.1

APA

APA—The Engineered Wood Association
7011 South 19th
Tacoma, WA 98466

Standard reference number	Title	Referenced in code section number
ANSI/APA PRP 210—08	Standard for Performance-rated Engineered Wood Siding	R604.1, Table R703.4
APA E30—03	Engineered Wood Construction Guide	Table R503.2.1.1(1), R503.2.2, R803.2.2, R803.2.3

APSP

The Association of Pool & Spa Professionals
2111 Eisenhower Avenue
Alexandria, VA 22314

Standard reference number	Title	Referenced in code section number
ANSI/APSP 7—06	Standard for Suction Entrapment Avoidance in Swimming Pools Wading Pools, Spas, Hot Tubs and Catch Basins	AG106.1
ANSI/NSPI 3—99	Standard for Permanently Installed Residential Spas	AG104.1
ANSI/NSPI 4—2007	Standard for Above-ground/On-ground Residential Swimming Pools	AG103.2
ANSI/NSPI 5—2003	Standard for Residential In-ground Swimming Pools	AG103.1
ANSI/NSPI 6—99	Standard for Residential Portable Spas	AG104.2

ASCE/SEI

American Society of Civil Engineers
Structural Engineering Institute
1801 Alexander Bell Drive
Reston, VA 20191

Standard reference number	Title	Referenced in code section number
7—10	Minimum Design Loads for Buildings and Other Structures	R301.2.1.1, R301.2.1.2, R301.2.1.5, R301.2.1.5.1, Table R611.6(1), Table R611.6(2), Table R611.6(3), Table R611.6(4), Table R611.7(1A), R611.9.2, R611.9.3, Table R802.11, AH107.4.3
24—05	Flood-resistant Design and Construction	R301.2.4, R301.2.4.1, R322.1, R322.1.1, R322.1.6, R322.1.9, R322.2.2, R3223.3, AG103.3
32—01	Design and Construction of Frost-protected Shallow Foundations	R403.1.4.1
402—11	Building Code Requirements for Masonry Structures	R404.1.1, R606.1, R606.1.1, R606.12.1, R606.12.2.2.1, R606.12.2.2.2, R606.12.3.1, Table R703.4
602—11	Specification for Masonry Structures	R606.1, R606.1.1, R606.12.1, R606.12.2.2.1, R606.12.2.2.2, R606.12.3.1, Table R703.4

ASHRAE

American Society of Heating, Refrigerating
and Air-Conditioning Engineers, Inc.
1791 Tullie Circle, NE
Atlanta, GA 30329

Standard reference number	Title	Referenced in code section number
ASHRAE—2009	ASHRAE Handbook of Fundamentals	N1102.1.4, Table N1105.5.2(1), P3001.2, P3101.4, P3103.2
ASHRAE 193—2010	Method of Test for Determining Air Tightness of HVAC Equipment	N1103.2.2.1
34—2010	Designation and Safety Classification of Refrigerants	M1411.1

ASME

American Society of Mechanical Engineers
Three Park Avenue
New York, NY 10016-5990

Standard reference number	Title	Referenced in code section number
ASME/A17.1—2007/ CSA B44—2007	Safety Code for Elevators and Escalators—with A17.1a/CSA B44a—08 Addenda	R321.1
A18.1—2008	Safety Standard for Platforms and Stairway Chair Lifts	R321.2
A112.1.2—2004	Air Gaps in Plumbing Systems	Table P2902.3, P2902.3.1
A112.1.3—2000 (Reaffirmed 2005)	Air Gap Fittings for Use with Plumbing Fixtures, Appliances and Appurtenances	Table P2701.1, P2902.3.1

ASME—continued

Standard	Title	Referenced in code section number
A112.3.1—2007	Stainless Steel Drainage Systems for Sanitary, DWV, Storm and Vacuum Applications Above and Below Ground	Table P3002.1(1), Table P3002.1(2), Table P3002.2, Table P3002.3, Table P3302.1
A112.3.4—2000 (R2004)	Macerating Toilet Systems and Related Components	Table P2701.1, P3007.5
A112.4.1—1993 (R2002)	Water Heater Relief Valve Drain Tubes	P2803.6.1
A112.4.2—2003 (R2008)	Water-closet Personal Hygiene Devices	P2722.5
A112.4.3—1999 (R2004)	Plastic Fittings for Connecting Water Closets to the Sanitary Drainage System	P3003.19
A112.6.1M—1997 (R2002)	Floor-affixed Supports for Off-the-floor Plumbing Fixtures for Public Use	Table P2701.1, P2702.4
A112.6.2—2000 (R2004)	Framing-affixed Supports for Off-the-floor Water Closets with Concealed Tanks	Table P2701.1, P2702.4
A112.6.3—2001 (R2007)	Floor and Trench Drains	Table P2701.1
A112.14.1—03 (Reaffirmed 2008)	Backwater Valves	P3008.2
A112.18.1—2005/ CSA B125.1—2010	Plumbing Supply Fittings	Table P2701.1, P2708.4, P2708.5, P2722.1, P2902.2
A112.18.2—2005/ CSA B125.2—2005	Plumbing Waste Fittings with 2007 and 2008 Supplements	Table P2701.1, P2702.2
A112.18.3—2002 (Reaffirmed 2008)	Performance Requirements for Backflow Protection Devices and Systems in Plumbing Fixture Fittings	P2708.4, P2722.3
A112.18.6/ CSA B125.6—2010	Flexible Water Connectors	P2905.7
A112.19.1M/ CSA B45.2—2008	Enameled Cast-iron and Enameled Steel Plumbing Fixtures	Table P2701.1, P2711.1
A112.19.2—2008/ CSA B45.1—2008	Ceramic Plumbing Fixtures	Table P2701.1, P2705.1, P2711.1, P2712.1, P2712.2, P2712.9
A112.19.3—2008/ CSA B45.4—2008	Stainless Steel Plumbing Fixtures	Table P2701.1, P2705.1, P2711.1, P2712.1
A112.19.4M—1994 (R2004)	Porcelain-enameled-formed Steel Plumbing Fixtures—with 1998 and 2000 Supplements	Table P2701.1, P2711.1
A112.19.5/ CSA B45.X—2009	Trim for Water-closet Bowls, Tanks and Urinals	Table P2701.1
A112.19.7/ CSA B45.10—2009	Hydromassage Bathtub Appliances	Table P2701.1
A112.19.9M—1991 (R2002)	Nonvitreous Ceramic Plumbing Fixtures—with 2002 Supplement	Table P2701.1, P2711.1, P2712.1
A112.19.12—2006	Wall-mounted and Pedestal-mounted, Adjustable and Pivoting Lavatory and Sink Carrier Systems	Table P2701.1, P2711.4, P2714.2
A112.19.15—2005	Bathtub/Whirlpool Bathtubs with Pressure-sealed Doors	Table P2701.1, P2713.2
B1.20.1—1983 (R2006)	Pipe Threads, General-purpose (Inch)	G2414.9, P3003.3.3, P3003.5.3, P3003.10.4, P3003.12.1, P3003.14.3
B16.3—2006	Malleable-iron-threaded Fittings, Classes 150 and 300	Table P2905.6
B16.4—2006	Gray-iron-threaded Fittings Classes, 125 and 250	Table P2905.6, Table P3002.3
B16.9—2007	Factory-made, Wrought-steel Buttwelding Fittings	Table P2905.6
B16.11—2005	Forged Fittings, Socket-welding and Threaded	Table P2905.6
B16.12—1998	Cast-iron-threaded Drainage Fittings	Table P2905.6, Table P3002.3(R2006)
B16.15—2006	Cast-bronze-threaded Fittings	Table P2905.6, Table P3002.3
B16.18—2001 (R2005)	Cast-copper-alloy Solder Joint Pressure Fittings	Table P2905.6, Table P3002.3
B16.22—2001(R2005)	Wrought-copper and Copper-alloy Solder Joint Pressure Fittings	Table P2905.6, Table P3002.3
B16.23—2002 (R2006)	Cast-copper-alloy Solder Joint Drainage Fittings (DWV)	Table P2905.6, Table P3002.3
B16.26—2006	Cast-copper-alloy Fittings for Flared Copper Tubes	Table P2905.6, Table P3002.3
B16.28—1994	Wrought-steel Buttwelding Short Radius Elbows and Returns	Table P2905.6
B16.29—2007	Wrought-copper and Wrought-copper-alloy Solder Joint Drainage Fittings (DWV)	Table P2905.6, Table P3002.3
B16.33—2002 (Reaffirmed 2007)	Manually Operated Metallic Gas Valves for Use in Gas Piping Systems up to 125 psig (Sizes $^1/_2$ through 2)	Table G2420.1.1
B16.44—2002 (Reaffirmed 2007)	Manually Operated Metallic Gas Valves for Use in Above-ground Piping Systems up to 5 psi	Table G2420.1.1
B36.10M—2004	Welded and Seamless Wrought-steel Pipe	G2414.4.2
BPVC—2007	ASME Boiler and Pressure Vessel Code (2007 Edition)	M2001.1.1, G2452.1
CSD-1—2009	Controls and Safety Devices for Automatically Fired Boilers	M2001.1.1, G2452.1

ASSE

American Society of Sanitary Engineering
901 Canterbury, Suite A
Westlake, OH 44145

Standard reference number	Title	Referenced in code section number
1001—2008	Performance Requirements for Atmospheric-type Vacuum Breakers	Table P2902.3, P2902.3.2
1002—2008	Performance Requirements for Anti-siphon Fill Valves for Water Closet Flush Tank	Table P2701.1, Table P2902.3, P2902.4.1
1003—2009	Performance Requirements for Water-pressure-reducing Valves for Domestic Water Distribution Systems	P2903.3.1
1008—2006	Performance Requirements for Plumbing Aspects of Residential Food Waste Disposer Units	Table P2701.1
1010—2004	Performance Requirements for Water Hammer Arresters	P2903.5
1011—2004	Performance Requirements for Hose Connection Vacuum Breakers	Table P2902.3, P2902.3.2
1012—2009	Performance Requirements for Backflow Preventers with Intermediate Atmospheric Vent	Table P2902.3, P2902.3.3, P2902.5.1, P2902.5.5
1013—2009	Performance Requirements for Reduced Pressure Principle Backflow Preventers and Reduced Pressure Principle Fire Protection Backflow Preventers	Table P2902.3, P2902.3.5, P2902.5.1, P2902.5.5
1015—2009	Performance Requirements for Double Check Backflow Prevention Assemblies and Double Check Fire Protection Backflow Prevention Assemblies	Table P2902.3, P2902.3.6
1016—2010	Performance Requirements for Automatic Compensating Valves for Individual Showers and Tub/Shower Combinations	Table P2701.1, P2708.3, P2722.2
1017—2010	Performance Requirements for Temperature-actuated Mixing Valves for Hot Water Distribution Systems	P2724.1, P2802.2
1019—2010	Performance Requirements for Freeze-resistant, Wall Hydrants, Vacuum Breaker, Draining Types	Table P2701.1, P2902.3
1020—2004	Performance Requirements for Pressure Vacuum Breaker Assembly	Table P2902.3, P2902.3.4
1023—2010	Performance Requirements for Hot Water Dispensers, Household-storage-type—Electrical	Table P2701.1
1024—2004	Performance Requirements for Dual Check Backflow Preventers, Anti-siphon-type, Residential Applications	Table P2902.3
1035—2008	Performance Requirements for Laboratory Faucet Backflow Preventers	Table P2902.3, P2902.3.2
1037—2010	Performance Requirements for Pressurized Flushing Devices (Flushometer) for Plumbing Fixtures	Table P2701.1
1047—2009	Performance Requirements for Reduced Pressure Detector Fire Protection Backflow Prevention Assemblies	Table P2902.3, P2902.3.5
1048—2009	Performance Requirements for Double Check Detector Fire Protection Backflow Prevention Assemblies	Table P2902.3, P2902.3.6
1050—2009	Performance Requirements for Stack Air Admittance Valves for Sanitary Drainage Systems	P3114.1
1051—2009	Performance Requirements for Individual and Branch-type Air Admittance Valves for Plumbing Drainage Systems	P3114.1
1052—2004	Performance Requirements for Hose Connection Backflow Preventers	Table P2701.1, Table P2902.3, P2902.3.2
1056—2010	Performance Requirements for Spill-resistant Vacuum Breakers	Table P2902.3, P2902.3.4
1060—2006	Performance Requirements for Outdoor Enclosures for Fluid-conveying Components	P2902.6.1
1061—2010	Performance Requirements for Removable and Nonremovable Push Fit Fittings	Table P2905.6
1062—2006	Performance Requirements for Temperature-actuated, Flow Reduction (TAFR) Valves for Individual Supply Fittings	Table P2701.1, P2724.2
1066—2009	Performance Requirements for Individual Pressure Balancing In-line Valves for Individual Fixture Fittings	P2722.4
1070—2004	Performance Requirements for Water-temperature-limiting Devices	P2713.3, P2721.2

ASTM

ASTM International
100 Barr Harbor Drive
West Conshohocken, PA 19428

Standard reference number	Title	Referenced in code section number
A 36/A 36M—08	Specification for Carbon Structural Steel	R606.15, R611.5.2.2
A 53/A 53M—07	Specification for Pipe, Steel, Black and Hot-dipped, Zinc-coated Welded and Seamless	R402.3, Table M2101.1, G2414.4.2, Table P2905.4, Table P2905.5, Table P3002.1(1)

ASTM—continued

Standard reference number	Title	Referenced in code section number
A 74—09	Specification for Cast-iron Soil Pipe and Fittings	Table P3002.1(1), Table P3002.1(2), Table P3002.2, Table P3002.3, P3005.2.9, Table P3302.1
A 82/A 82M—05a	Specification for Steel Wire, Plain, for Concrete Reinforcement	R606.15
A 106/A 106M—08	Specification for Seamless Carbon Steel Pipe for High-temperature Service	Table M2101.1, G2414.4.2
A 153/A 153M—05	Specification for Zinc Coating (Hot Dip) on Iron and Steel Hardware	R317.3, Table R606.15.1
A 167—99 (2009)	Specification for Stainless and Heat-resisting Chromium-nickel Steel Plate, Sheet and Strip	R606.15, Table R606.15.1
A 240/A 240M—09a	Standard Specification for Chromium and Chromium-nickel Stainless Steel Plate, Sheet and Strip for Pressure Vessels and for General Applications	Table R905.10.3(1)
A 254—97 (2007)	Specification for Copper Brazed Steel Tubing	Table M2101.1 , G2414.5.1
A 307—07b	Specification for Carbon Steel Bolts and Studs, 6000 psi Tensile Strength	R611.5.2.2
A 312/A 312M—08a	Specification for Seamless, Welded and Heavily Cold Worked Austenitic Stainless Steel Pipes	Table P2905.4, Table P2905.5, Table P2905.6, P2905.12.2
A 463/A 463M—06	Standard Specification for Steel Sheet, Aluminum-coated by the Hot-dip Process	Table R905.10.3(2)
A 510—08	Specification for General Requirements for Wire Rods and Coarse Round Wire, Carbon Steel	R606.15
A 539—99	Specification for Electric-resistance-welded Coiled Steel Tubing for Gas and Fuel Oil Lines	M2202.1
A 615/A 615M—09	Specification for Deformed and Plain Billet-steel Bars for Concrete Reinforcement	R402.3.1, R404.1.2.3.7.1, R611.5.2.1
A 641/A 641M—09a	Specification for Zinc-coated (Galvanized) Carbon Steel Wire	Table R606.15.1
A 653/A 653M—08	Specification for Steel Sheet, Zinc-coated (Galvanized) or Zinc-iron Alloy-coated Galvanized) by the Hot-dip Process	R317.3.1, R505.2.1, R505.2.3, R603.2.1, R603.2.3, Table R606.15.1, R611.5.2.3, R804.2.1, R804.2.3, Table R905.10.3(1), Table R905.10.3(2), M1601.1.1
A 706/A 706M—09	Specification for Low-alloy Steel Deformed and Plain Bars for Concrete Reinforcement	R402.3.1, R404.1.2.3.7.1, R611.5.2.1
A 755/A 755M—03 (2008)	Specification for Steel Sheet, Metallic Coated by the Hot-dip Process and Prepainted by the Coil-coating Process for Exterior Exposed Building Products	Table R905.10.3(2)
A 778—01	Specification for Welded Unannealed Austenitic Stainless Steel Tubular Products	Table P2905.4, Table P2905.5, Table P2905.6
A 792/A 792M—08	Specification for Steel Sheet, 55% Aluminum-zinc Alloy-coated by the Hot-dip Process	R505.2.1, R505.2.3, R603.2.1, R603.2.3, R611.5.2. 3, R804.2.1, R804.2.3, Table 905.10.3(2)
A 875/A 875M—06	Specification for Steel Sheet, Zinc-5%, Aluminum Alloy-coated by the Hot-dip Process	R611.5.2.3, Table R905.10.3(2)
A 888—09	Specification for Hubless Cast Iron Soil Pipe and Fittings for Sanitary and Storm Drain, Waste and Vent Piping Application	Table P3002.1(1), Table P3002.1(2), Table P3002.2, Table P3002.3, P3005.2.9, Table P3302.1
A 924/A 924M—08a	Standard Specification for General Requirements for Steel Sheet, Metallic-coated by the Hot-Dip Process	Table R905.10.3(1)
A 951—06	Specification for Steel Wire Masonry Joint Reinforcement	R606.15
A 996/A 996M—09	Specifications for Rail-steel and Axle-steel Deformed Bars for Concrete Reinforcement	Table R404.1.2(9), R404.1.2.3.7.1, R611.5.2.1, Table R611.5.4(2)
A 1003/A 1003M—08	Standard Specification for Steel Sheet, Carbon, Metallic and Nonmetallic-coated for Cold-formed Framing Members	R505.2.1, R505.2.3, R603.2.1, R603.2.3, R804.2.1, R804.2.3
B 32—08	Specification for Solder Metal	P3003.10.3, P3003.11.3
B 42—02e01	Specification for Seamless Copper Pipe, Standard Sizes	Table M2101.1, Table P2905.4, Table P2905.5, Table P3002.1(1)
B 43—98 (2004)	Specification for Seamless Red Brass Pipe, Standard Sizes	Table M2101.1, G2413.5.2, Table P2905.4, Table P3002.1(1)
B 75—02	Specification for Seamless Copper Tube	Table M2101.1, Table P2905.4, Table P2905.5, Table P3002.1(1), Table P3002.1(2), Table P3002.2
B 88—03	Specification for Seamless Copper Water Tube	Table M2101.1, G2414.5.2, Table, P2905.4, Table P2905.5, Table P3002.1(1), Table P3002.1(2), Table P3002.2
B 101—07	Specification for Lead-coated Copper Sheet and Strip for Building Construction	Table R905.2.8.2, Table R905.10.3(1)
B 135—08a	Specification for Seamless Brass Tube	Table M2101.1
B 209—07	Specification for Aluminum and Aluminum-alloy Sheet and Plate	Table 905.10.3(1)
B 227—04	Specification for Hard-drawn Copper-clad Steel Wire	R606.15

ASTM—continued

B 251—02e01	Specification for General Requirements for Wrought Seamless Copper and Copper-alloy Tube	Table M2101.1, Table P2905.4, Table P2905.5, Table P3002.1(1), Table P3002.2
B 302—07	Specification for Threadless Copper Pipe, Standard Sizes	Table M2101.1, Table P2905.4, Table P2905.5, Table P3002.1(1)
B 306—09	Specification for Copper Drainage Tube (DWV)	Table M2101.1, Table P3002.1(1), Table P3002.1(2), Table P3002.2
B 370—09	Specification for Copper Sheet and Strip for Building Construction	Table R905.2.8.2, Table R905.10.3(1), Table P2701.1
B 447—07	Specification for Welded Copper Tube	Table P2904.4, Table P2905.5
B 695—04	Standard Specification for Coatings of Zinc Mechanically Deposited on Iron and Steel	R317.3.1, R317.3.3
B 813—00 (2009)	Specification for Liquid and Paste Fluxes for Soldering Applications of Copper and Copper Alloy Tube	Table M2101.1, P2905.14, P3003.10.3, P3003.11.3
B 828—02	Practice for Making Capillary Joints by Soldering of Copper and Copper Alloy Tube and Fittings	P2905.14, P3003.10.3, P3003.11.3
C 4—04e01	Specification for Clay Drain Tile and Perforated Clay Drain Tile	Table P3302.1
C 5—03	Specification for Quicklime for Structural Purposes	R702.2.1
C 14—07	Specification for Non-reinforced Concrete Sewer, Storm Drain and Culvert Pipe	Table P3002.2
C 22/C 22M—00 (2005e01)	Specification for Gypsum	R702.2.1, R702.3.1
C 27—98 (2008)	Specification for Standard Classification of Fireclay and High-alumina Refractory Brick	R1001.5, R1001.8
C 28/C 28M—00 (2005)	Specification for Gypsum Plasters	R702.2.1
C 33/C 33M—08	Specification for Concrete Aggregates	R403.4.1
C 34—03	Specification for Structural Clay Load-bearing Wall Tile	Table R301.2(1)
C 35—01(2005)	Specification for Inorganic Aggregates for Use in Gypsum Plaster	R702.2.1
C 55—06e01	Specification for Concrete Building Brick	R202, Table R301.2(1)
C 59/C 59M—00 (2006)	Specification for Gypsum Casting and Molding Plaster	R702.2
C 61/C 61M—00 (2006)	Specification for Gypsum Keene's Cement	R702.2.1
C 62—08	Specification for Building Brick (Solid Masonry Units Made from Clay or Shale)	R202, Table R301.2(1)
C 73—05	Specification for Calcium Silicate Face Brick (Sand Lime Brick)	R202, Table R301.2(1)
C 76—08a	Specification for Reinforced Concrete Culvert, Storm Drain and Sewer Pipe	Table P3002.2
C 90—08	Specification for Load-bearing Concrete Masonry Units	Table R301.2(1)
C 91—05	Specification for Masonry Cement	R702.2.2
C 94/C 94M—09	Specification for Ready-mixed Concrete	R404.1.2.3.2, R611.5.1.1
C 129—06	Specification for Nonload-bearing Concrete Masonry Units	Table R301.2(1)
C 143/C 143M—08	Test Method for Slump or Hydraulic Cement Concrete	R404.1.2.3.4, R611.5.1.3, R611.6.1
C 145—85	Specification for Solid Load-bearing Concrete Masonry Units	R202, Table R301.2(1)
C 150—07	Specification for Portland Cement	R702.2.2
C 199—84 (2005)	Test Method for Pier Test for Refractory Mortar	R1001.5, R1001.8, R1003.12
C 203—05a	Standard Test Methods for Breaking Load and Flexural Properties of Block-type Thermal Insulation	Table R613.3.1
C 207—06	Specification for Hydrated Lime for Masonry Purposes	Table R607.1
C 208—08a	Specification for Cellulosic Fiber Insulating Board	Table R602.3(1), Table R906.2
C 216—07a	Specification for Facing Brick (Solid Masonry Units Made from Clay or Shale)	R202, Table R301.2(1)
C 270—08a	Specification for Mortar for Unit Masonry	R607.1, AE602
C 272—01(2007)	Standard Test Method for Water Absorption of Core Materials for Structural Sandwich Constructions	Table R613.3.1
C 273/C 273M—07a	Standard Test Method for Shear Properties of Sandwich Core Materials	Table R613.3.1

ASTM—continued

Standard	Title	Referenced in code section number
C 296—00 (2004)	Specification for Asbestos Cement Pressure Pipe	Table P2905.4
C 315—07	Specification for Clay Flue Liners and Chimney Pots	R1001.8, R1003.11.1, Table R1003.14(1), G2425.12
C 406—06e01	Specifications for Roofing Slate	R905.6.4
C 411—05	Test Method for Hot-surface Performance of High-temperature Thermal Insulation	M1601.3
C 425—04	Specification for Compression Joints for Vitrified Clay Pipe and Fittings	Table P3002.2, P3003.15, P3003.18
C 428—05 (2006)	Specification for Asbestos-cement Nonpressure Sewer Pipe	Table P3002.2
C 443—05a	Specification for Joints for Concrete Pipe and Manholes, Using Rubber Gaskets	P3003.7, P3003.18
C 475/C 475—02 (2007)	Specification for Joint Compound and Joint Tape for Finishing Gypsum Wallboard	R702.3.1
C 476—08	Specification for Grout for Masonry	R609.1.1
C 508—04	Specification for Asbestos-cement Underdrain Pipe	Table P3302.1
C 514—04	Specification for Nails for the Application of Gypsum Wallboard	R702.3.1
C 552—07	Standard Specification for Cellular Glass Thermal Insulation	Table R906.2
C 557—03e01	Specification for Adhesives for Fastening Gypsum Wallboard to Wood Framing	R702.3.1
C 564—08	Specification for Rubber Gaskets for Cast Iron Soil Pipe and Fittings	P3003.6.2, P3003.6.3, P3003.18
C 578—08b	Specification for Rigid, Cellular Polystyrene Thermal Insulation	R403.3, R613.3.1, R703.11.2.1, Table R906.2
C 587—04	Specification for Gypsum Veneer Plaster	R702.2.1
C 595—08a	Specification for Blended Hydraulic Cements	R702.2.2
C 631—09	Specification for Bonding Compounds for Interior Gypsum Plastering	R702.2.1
C 645—08a	Specification for Nonstructural Steel Framing Members	R702.3.3
C 652—09	Specification for Hollow Brick (Hollow Masonry Units Made from Clay or Shale)	R202, Table R301.2(1)
C 685/C 685M—07	Specification for Concrete Made by Volumetric Batching and Continuous Mixing	R404.1.2.3.2, R611.5.1.1
C 700—07a	Specification for Vitrified Clay Pipe, Extra Strength, Standard Strength and Perforated	Table P3002.2, Table P3002.3, Table P3302.1
C 728—05	Standard Specification for Perlite Thermal Insulation Board	Table R906.2
C 836—06	Specification for High Solids Content, Cold Liquid-applied Elastomeric Waterproofing Membrane for Use with Separate Wearing Course	R905.15.2
C 843—99 (2006)	Specification for Application of Gypsum Veneer Plaster	R702.2.1
C 844—04	Specification for Application of Gypsum Base to Receive Gypsum Veneer Plaster	R702.2.1
C 847—09	Specification for Metal Lath	R702.2.1, R702.2.2
C 887—05	Specification for Packaged, Dry, Combined Materials for Surface Bonding Mortar	R406.1
C 897—05	Specification for Aggregate for Job-mixed Portland Cement-based Plasters	R702.2.2
C 920—08	Standard Specification for Elastomeric Joint Sealants	R406.4.1
C 926—06	Specification for Application of Portland Cement-based Plaster	R702.2.2, R702.2.2.1, R703.6, R703.6.2.1, R703.6.4
C 933—07b	Specification for Welded Wire Lath	R702.2.1, R702.2.2
C 954—07	Specification for Steel Drill Screws for the Application of Gypsum Panel Products or Metal Plaster Bases to Steel Studs from 0.033 in (0.84 mm) or to 0.112 in. (2.84 mm) in Thickness	R505.2.4, R603.2.4, R702.3.6, R804.2.4
C 955—09	Specification for Load-bearing (Transverse and Axial) Steel Studs, Runners (Tracks), and Bracing or Bridging for Screw Application of Gypsum Panel Products and Metal Plaster Bases	R702.3.3
C 957—06	Specification for High-solids Content, Cold Liquid-applied Elastomeric Waterproofing Membrane for Use with Integral Wearing Surface	R905.15.2
C 1002—07	Specification for Steel Drill Screws for the Application of Gypsum Panel Products or Metal Plaster Bases	R702.3.1, R702.3.6
C 1029—08	Specification for Spray-applied Rigid Cellular Polyurethane Thermal Insulation	R905.14.2
C 1032—06	Specification for Woven Wire Plaster Base	R702.2.1, R702.2.2
C 1047—09	Specification for Accessories for Gypsum Wallboard and Gypsum Veneer Base	R702.2.1, R702.2.2, R702.3.1
C 1063—08	Specification for Installation of Lathing and Furring to Receive Interior and Exterior Portland Cement-based Plaster	R702.2.2, R703.6
C 1107/C 1107M—08	Standard Specification for Packaged Dry, Hydraulic-cement Grout (Nonshrink)	R402.3.1
C 1116/C 116M—08a	Standard Specification for Fiber-reinforced Concrete and Shotcrete	R402.3.1
C 1167—03	Specification for Clay Roof Tiles	R905.3.4
C 1173—08	Specification for Flexible Transition Couplings for Underground Piping Systems	P3003.3, P3003.7, P3003.8.1, P3003.14.1, P3003.15, P3003.17.2, P3003.18
C 1177/C 1177M—08	Specification for Glass Mat Gypsum Substrate for Use as Sheathing	R702.3.1

ASTM—continued

C 1178/C 1178M—08 Specification for Glass Mat Water-resistant Gypsum Backing Panel R702.3.1, R702.3.8, R702.4.2

C 1186—08 Specification for Flat Nonasbestos Fiber Cement Sheets R703.10.1, R703.10.2

C 1261—07 Specification for Firebox Brick for Residential Fireplaces R1001.5, R1001.8

C 1277—08 Specification for Shielded Couplings Joining Hubless Cast Iron Soil Pipe and Fittings P3003.6.3

C 1278/C 1278M—07a Specification for Fiber-reinforced Gypsum Panels R702.3.1, R702.3.8, R702.4.2

C 1283—07a Practice for Installing Clay Flue Lining R1003.9.1, R1003.12

C 1288—99 (2004)e1 Standard Specification for Discrete Nonasbestos Fiber-cement Interior Substrate Sheets R702.4.2

C 1289—08 Standard Specification for Faced Rigid Cellular Polyisocyanurate Thermal Insulation Board R703.11.2.1, Table R906.2

C 1325—08b Standard Specification for Nonasbestos Fiber-mat Reinforced Cement Interior Substrate Sheets R702.4.2

C 1328—05 Specification for Plastic (Stucco) Cement R702.2.2

C 1396/C 1396M—06a Specification for Gypsum Board Table R602.3(1), R702.2.1, R702.2.2, R702.3.1, R702.3.8

C 1440—08 Specification for Thermoplastic Elastomeric (TPE) Gasket Materials for Drain, Waste and Vent (DWV), Sewer, Sanitary and Storm Plumbing Systems P3003.18

C 1460—08 Specification for Shielded Transition Couplings for Use with Dissimilar DWV Pipe and Fittings Above Ground P3003.18

C 1461—08 Specification for Mechanical Couplings Using Thermoplastic Elastomeric (TPE) Gaskets for Joining Drain, Waste and Vent (DWV) Sewer, Sanitary and Storm Plumbing Systems for Above and Below Ground Use P3003.18

C 1492—03 Specification for Concrete Roof Tile R905.3.5

C 1513—04 Standard Specification for Steel Tapping Screws for Cold-formed Steel Framing Connections R505.2.4, R603.2.4, R702.3.6, R804.2.4

C 1658/C 1658M—06 Standard Specification for Glass Mat Gypsum Panels R702.3.1

D 41—05 Specification for Asphalt Primer Used in Roofing, Dampproofing and Waterproofing Table R905.9.2, Table R905.11.2

D 43—00(2006) Specification for Coal Tar Primer Used in Roofing, Dampproofing and Waterproofing Table R905.9.2

D 225—07 Specification for Asphalt Shingles (Organic Felt) Surfaced with Mineral Granules R905.2.4

D 226—06 Specification for Asphalt-saturated (Organic Felt) Used in Roofing and Waterproofing R703.2, R905.2.3, R905.3.3, R905.4.3, R905.4.3.2, R905.5.3, R905.5.3.2, R905.6.3, R905.6.3.2, R905.7.3, R905.8.3, R905.8.3.2, R905.8.4, Table R905.9.2, R905.10.5.1

D 227—03 Specification for Coal Tar Saturated (Organic Felt) Used in Roofing and Waterproofing Table R905.9.2

D 312—00 (2006) Specification for Asphalt Used in Roofing Table R905.9.2

D 422—63 (2007) Test Method for Particle-size Analysis of Soils R403.1.8.1

D 449—03 (2008) Specification for Asphalt Used in Dampproofing and Waterproofing R406.2

D 450—07 Specification for Coal-tar Pitch Used in Roofing, Dampproofing and Waterproofing Table R905.9.2

D 1227—95 (2007) Specification for Emulsified Asphalt Used as a Protective Coating for Roofing Table R905.9.2, Table R905.11.2, R905.15.2

D 1248—05 Specification for Polyethylene Plastics Extrusion Materials for Wire and Cable M1601.1.2

D 1527—99 (2005) Specification for Acrylonitrile-butadiene-styrene (ABS) Plastic Pipe, Schedules 40 and 80 Table P2905.4

D1621—04a Standard Test Method for Compressive Properties of Rigid Cellular Plastics Table R613.3.1

D 1622—08 Standard Test Method for Apparent Density of Rigid Cellular Plastics Table R613.3.1

D 1623—03 Standard Test Method for Tensile and Tensile Adhesion Properties of Rigid Cellular Plastics Table R613.3.1

D 1693—08 Test Method for Environmental Stress-cracking of Ethylene Plastics Table M2101.1

D 1784—08 Standard Specification for Rigid Poly (Vinyl Chloride) (PVC) Compounds and Chlorinated Poly (Vinyl Chloride) (CPVC) Compounds M1601.1.2

D 1785—06 Specification for Poly (Vinyl Chloride) (PVC) Plastic Pipe, Schedules 40, 80 and 120 Table P2905.4

D 1863—05 Specification for Mineral Aggregate Used in Built-up Roofs Table R905.9.2

D 1869—95 (2005)e1 Specification for Rubber Rings for Asbestos-cement Pipe P2904.17, P3003.4, P3003.18

D 1970—09 Specification for Self-adhering Polymer Modified Bitumen Sheet Materials Used as Steep Roofing Underlayment for Ice Dam Protection R905.2.3, R905.2.7.2, R905.2.8.2, R905.3.3.3, R905.4.3, R905.4.3.2, R905.5.3.2, R905.6.3.2, R905.7.3.2, R905.8.3.2, R905.10.5.1

D 2104—03 Specification for Polyethylene (PE) Plastic Pipe, Schedule 40 Table P2905.4

D 2126—04 Standard Test Method for Response of Rigid Cellular Plastics to Thermal and Humid Aging Table R613.3.1

D 2178—04 Specification for Asphalt Glass Felt Used in Roofing and Waterproofing Table R905.9.2

D 2235—04 Specification for Solvent Cement for Acrylonitrile-butadiene-styrene (ABS) Plastic Pipe and Fittings P2905.9.1.1, P3003.3.2, P3003.8.2

ASTM—continued

Standard	Title	Referenced in
D 2239—03	Specification for Polyethylene (PE) Plastic Pipe (SIDR-PR) Based on Controlled Inside Diameter	Table P2905.4
D 2241—05	Specification for Poly (Vinyl Chloride) (PVC) Pressure-rated Pipe (SDR-Series)	Table P2905.4
D 2282—05	Specification for Acrylonitrile-butadiene-styrene (ABS) Plastic Pipe (SDR-PR)	Table P2905.4
D 2412—02 (2008)	Test Method for Determination of External Loading Characteristics of Plastic Pipe by Parallel-plate Loading	M1601.1.2
D 2447—03	Specification for Polyethylene (PE) Plastic Pipe Schedules 40 and 80, Based on Outside Diameter	Table M2101.1
D 2464—06	Specification for Threaded Poly (Vinyl Chloride) (PVC) Plastic Pipe Fittings, Schedule 80	Table P2905.6
D 2466—06	Specification for Poly (Vinyl Chloride) (PVC) Plastic Pipe Fittings, Schedule 40	Table P2905.6
D 2467—06	Specification for Poly (Vinyl Chloride) (PVC) Plastic Pipe Fittings, Schedule 80	Table P2905.6
D 2468—96a	Specification for Acrylonitrile-butadiene-styrene (ABS) Plastic Pipe Fittings, Schedule 40	Table P2905.6
D 2513—08b	Specification for Thermoplastic Gas Pressure Pipe, Tubing and Fittings	Table M2101.1, M2104.2.1.3, G2414.6, G2414.6.1, G2414.11, G2415.15.2
D 2559—04	Standard Specification for Adhesives for Structural Laminated Wood Products for Use Under Exterior (West Use) Exposure Conditions	R613.3.3
D 2564—04e01	Specification for Solvent Cements for Poly (Vinyl Chloride) (PVC) Plastic Piping Systems	P2905.9.1.3, Table P3002.2, P3003.9.2, P3003.14.2
D 2609—02 (2008)	Specification for Plastic Insert Fittings for Polyethylene (PE) Plastic Pipe	Table P2905.6
D 2626—04	Specification for Asphalt-saturated and Coated Organic Felt Base Sheet Used in Roofing	R905.3.3, Table R905.9.2
D 2657—07	Standard Practice for Heat Fusion-joining of Polyolefin Pipe Fittings	P2905.3.1, P3003.17.1
D 2661—08	Specification for Acrylonitrile-butadiene-styrene (ABS) Schedule 40 Plastic Drain, Waste, and Vent Pipe and Fittings	Table P3002.1(1), Table P3002.1(2), Table P3002.2, Table P3002.3, P3003.3.2, P3003.8.2
D 2665—09	Specification for Poly (Vinyl Chloride) (PVC) Plastic Drain, Waste and Vent Pipe and Fittings	Table P3002.1(1), Table P3002.1(2), Table P3002.2, Table P3002.3
D 2672—96a (2003)	Specification for Joints for IPS PVC Pipe Using Solvent Cement	Table P2905.4
D 2683—04	Specification for Socket-type Polyethylene Fittings for Outside Diameter-controlled Polyethylene Pipe and Tubing	Table M2101.1, M2104.2.1.1
D 2729—03	Specification for Poly (Vinyl Chloride) (PVC) Sewer Pipe and Fittings	P3009.14.10, Table P3302.1, Table AO103.10
D 2737—03	Specification for Polyethylene (PE) Plastic Tubing	Table P2905.4
D 2751—05	Specification for Acrylonitrile-butadiene-styrene (ABS) Sewer Pipe and Fittings	Table P3002.2, Table P3002.3
D 2822—05	Specification for Asphalt Roof Cement, Asbestos Containing	Table R905.9.2
D 2823—05	Specification for Asphalt Roof Coatings, Asbestos Containing	Table R905.9.2
D 2824—06	Specification for Aluminum-pigmented Asphalt Roof Coatings, Nonfibered, Asbestos Fibered and Fibered without Asbestos	Table R905.9.2, Table R905.11.2
D 2837—08	Test Method for Obtaining Hydrostatic Design Basis for Thermoplastic Pipe Materials or Pressure Design Basis for Thermoplastic Pipe Products	Table M2101.1
D 2846/D 2846M—09	Specification for Chlorinated Poly (Vinyl Chloride) (CPVC) Plastic Hot- and Cold-water Distribution Systems	Table M2101.1, P2904.9.1.2, Table P2905.4, Table P2905.5, Table P2905.6
D 2855—96 (2002)	Standard Practice for Making Solvent-cemented Joints with Poly (Vinyl Chloride) (PVC) Pipe and Fittings	P3003.9.2, P3003.14.2
D 2898—04	Test Methods for Accelerated Weathering of Fire-retardant-treated Wood for Fire Testing	R802.1.3.4, R802.1.3.6
D 2949—01a (2008)	Specification for 3.25-in. Outside Diameter Poly (Vinyl Chloride) (PVC) Plastic Drain, Waste and Vent Pipe and Fittings	Table P3002.1(1), Table P3002.1(2), Table P3002.2, Table P3002.3
D 3019—08	Specification for Lap Cement Used with Asphalt Roll Roofing, Nonfibered, Asbestos Fibered and Nonasbestos Fibered	Table R905.9.2, Table R905.11.2
D 3034—08	Specification for Type PSM Poly (Vinyl Chloride) (PVC) Sewer Pipe and Fittings	Table P3002.2, Table P3002.3
D 3035—08	Specification for Polyethylene (PE) Plastic Pipe (DR-PR) Based On Controlled Outside Diameter	Table M2101.1

ASTM—continued

Standard	Title	Referenced in code section number
D 3161—09	Test Method for Wind Resistance of Asphalt Shingles (Fan Induced Method)	R905.2.4.1, Table R905.2.4.1(2) , R905.16.3
D 3201—08a	Test Method for Hygroscopic Properties of Fire-retardant Wood and Wood-base Products	R802.1.3.7
D 3212—07	Specification for Joints for Drain and Sewer Plastic Pipes Using Flexible Elastomeric Seals	P3003.3.1, P3003.8.1, P3003.9.1, P3003.14.1, P3003.17.2
D 3309—96a (2002)	Specification for Polybutylene (PB) Plastic Hot- and Cold-water Distribution System	Table M2101.1
D 3311—08	Specification for Drain, Waste and Vent (DWV) Plastic Fittings Patterns	P3002.3
D 3350—08	Specification for Polyethylene Plastic Pipe and Fitting Materials	Table M2101.1
D 3462—09	Specification for Asphalt Shingles Made From Glass Felt and Surfaced with Mineral Granules	R905.2.4
D 3468—99 (2006)e01	Specification for Liquid-applied Neoprene and Chlorosulfanated Polyethylene Used in Roofing and Waterproofing	R905.15.2
D 3679—09	Specification for Rigid Poly (Vinyl Chloride) (PVC) Siding	Table R703.4, R703.11
D 3737—08	Practice for Establishing Allowable Properties for Structural Glued Laminated Timber (Glulam)	R502.1.5, R602.1.2, R802.1.4
D 3747—79 (2007)	Specification for Emulsified Asphalt Adhesive for Adhering Roof Insulation	Table R905.9.2, Table R905.11.2
D 3909—97b (2004)e1	Specification for Asphalt Roll Roofing (Glass Felt) Surfaced with Mineral Granules	R905.2.8.2, R905.5.4, Table R905.9.2
D 3957—06	Standard Practices for Establishing Stress Grades for Structural Members Used in Log Buildings	R502.1.6, R602.1.3, R802.1.5
D 4022—07	Specification for Coal Tar Roof Cement, Asbestos Containing	Table R905.9.2
D 4068—01	Specification for Chlorinated Polyethylene (CPE) Sheeting for Concealed Water Containment Membrane	P2709.2, P2709.2.2
D 4318—05	Test Methods for Liquid Limit, Plastic Limit and Plasticity Index of Soils	R403.1.8.1
D 4434/D 4434M—09	Specification for Poly (Vinyl Chloride) Sheet Roofing	R905.13.2
D 4479—07	Specification for Asphalt Roof Coatings-asbestos-free	Table R905.9.2
D 4551—96 (2008)e1	Specification for Poly (Vinyl) Chloride (PVC) Plastic Flexible Concealed Water-containment Membrane	P2709.2, P2709.2.1
D 4586—07	Specification for Asphalt Roof Cement-asbestos-free	Table R905.9.2
D 4601—08	Specification for Asphalt-coated Glass Fiber Base Sheet Used in Roofing	Table R905.9.2
D 4637—08	Specification for EPDM Sheet Used in Single-ply Roof Membrane	R905.12.2
D 4829—08a	Test Method for Expansion Index of Soils	R403.1.8.1
D 4869—05e01	Specification for Asphalt-saturated (Organic Felt) Underlayment Used in Steep Slope Roofing	R905.2.3, R905.2.7.2, R905.4.3, R905.4.3.2, R905.5.3, R905.5.3.2, R905.6.3, R905.6.3.2, R905.7.3, R905.7.3.2, R905.8.3, R905.8.3.2
D 4897—01	Specification for Asphalt Coated Glass-fiber Venting Base Sheet Used in Roofing	Table R905.9.2
D 4990—97a (2005)e01	Specification for Coal Tar Glass Felt Used in Roofing and Waterproofing	Table R905.9.2
D 5019—07a	Specification for Reinforced Nonvulcanized Polymeric Sheet Used in Roofing Membrane	R905.12.2
D 5055—09	Specification for Establishing and Monitoring Structural Capacities of Prefabricated Wood I-joists	R502.1.4
D 5456—09	Standard Specification for Evaluation of Structural Composite Lumber Products	R502.1.8, R602.14, R802.1.6
D 5516—03	Test Method for Evaluating the Flexural Properties of Fire-retardant-treated Softwood Plywood Exposed to the Elevated Temperatures	R802.1.3.5.1
D 5643—06	Specification for Coal Tar Roof Cement Asbestos-free	Table R905.9.2
D 5664—08	Test Methods For Evaluating the Effects of Fire-retardant Treatments and Elevated Temperatures on Strength Properties of Fire-retardant-treated Lumber	R802.1.3.5.2
D 5665—99a (2006)	Specification for Thermoplastic Fabrics Used in Cold-applied Roofing and Waterproofing	Table R905.9.2
D 5726—98 (2005)	Specification for Thermoplastic Fabrics Used in Hot-applied Roofing and Waterproofing	Table R905.9.2
D 6083—05e01	Specification for Liquid-applied Acrylic Coating Used in Roofing	Table R905.9.2, Table R905.11.2, Table R905.14.3, R905.15.2
D 6162—00a (2008)	Specification for Styrene Butadiene Styrene (SBS) Modified Bituminous Sheet Materials Using a Combination of Polyester and Glass Fiber Reinforcements	Table R905.11.2
D 6163—00 (2008)	Specification for Styrene Butadiene Styrene (SBS) Modified Bituminous Sheet Materials Using Glass Fiber Reinforcements	Table R905.11.2
D 6164—05e1	Specification for Styrene Butadiene Styrene (SBS) Modified Bituminous Sheet Materials Using Polyester Reinforcements	Table R905.11.2
D 6222—08	Specification for Atactic Polypropylene (APP) Modified Bituminous Sheet Materials Using Polyester Reinforcements	Table R905.11.2

ASTM—continued

Standard	Title	Referenced in code section number
D 6223—02	Specification for Atactic Polypropylene (APP) Modified Bituminous Sheet Materials Using a Combination of Polyester and Glass Fiber Reinforcement	Table R905.11.2
D 6298—05e1	Specification for Fiberglass-reinforced Styrene Butadiene Styrene (SBS) Modified Bituminous Sheets with a Factory Applied Metal Surface	Table R905.11.2
D 6305—08	Practice for Calculating Bending Strength Design Adjustment Factors for Fire-retardant-treated Plywood Roof Sheathing	R802.1.3.5.1
D 6380—03 (2009)	Standard Specification for Asphalt Roll Roofing (Organic Felt)	R905.2.8.2, R905.3.3, R905.5.4
D 6694—08	Standard Specification for Liquid-applied Silicone Coating Used in Spray Polyurethane Foam Roofing Systems	Table R905.14.3, R905.15.2
D 6754—02	Standard Specification for Ketone-ethylene-ester-based Sheet Roofing	R905.13.2
D 6757—07	Standard Specification for Inorganic Underlayment for Use with Steep Slope Roofing Products	R905.2.3 , R905.2.7.2
D 6841—08	Standard Practice for Calculating Design Value Treatment Adjustment Factors for Fire-retardant-treated Lumber	R802.1.3.5.2
D 6878—08e1	Standard Specification for Thermoplastic-polyolefin-based Sheet Roofing	R905.13.2
D 6947—07	Standard Specification for Liquid Applied Moisture Cured Polyurethane Coating Used in Spray Polyurethane Foam Roofing System	Table R905.14.3, R905.15.2
D 7032—08	Standard Specification for Establishing Performance Ratings for Wood-plastic Composite Deck Boards and Guardrail Systems (Guards or Handrails)	R317.4, R317.4.1, R507.3
D 7158—08d	Standard Test Method for Wind Resistance of Sealed Asphalt Shingles (Uplift Force/Uplift Resistance Method)	R905.2.4.1, Table R905.2.4.1(1)
E 84—09	Test Method for Surface Burning Characteristics of Building Materials	R202, R302.9.3, R302.9.4, R302.10.1, R302.10.2, R316.3, R316.4, R316.5.9, R316.5.11, R802.1.3, M1601.3, M1601.5.2
E 90—04	Test Method for Laboratory Measurement of Airborne Sound Transmission Loss of Building Partitions and Elements	AK102, AK102.1.1
E 96/E 96M—05	Test Method for Water Vapor Transmission of Materials	, R202, Table R613.3.1, M1411.5, M1601.4.5
E 108—07a	Test Methods for Fire Tests of Roof Coverings	R902.1
E 119—08a	Test Methods for Fire Tests of Building Construction and Materials	R302.2, R302.3, R302.4.1, R316.4
E 136—09	Test Method for Behavior of Materials in a Vertical Tube Furnace at 750°C	R202, Table R302.1(1), Table R302.1(2), R302.11
E 283—04	Test Method for Determining the Rate of Air Leakage Through Exterior Windows, Curtain Walls and Doors Under Specified Pressure Differences Across the Specimen	N1102.4.4
E 330—02	Test Method for Structural Performance of Exterior Windows, Curtain Walls and Doors by Uniform Static Air Pressure Difference	R612.4, R612.5, R703.1.2
E 331—00 (2009)	Test Method for Water Penetration of Exterior Windows, Skylights, Doors and Curtain Walls by Uniform Static Air Pressure Difference	R703.1.1
E 492—09	Specification for Laboratory Measurement of Impact Sound Transmission through Floor-ceiling Assemblies Using the Tapping Machine	AK103
E 814—08b	Test Method for Fire Tests of Through-penetration Firestops	R302.4.1.2
E 970—08a	Test Method for Critical Radiant Flux of Exposed Attic Floor Insulation Using a Radiant Heat Energy Source	R302.10.5
E 1509—04	Standard Specification for Room Heaters, Pellet Fuel-burning Type	M1410.1
E 1602—03	Guide for Construction of Solid Fuel Burning Masonry Heaters	R1002.2
E 1886—05	Test Method for Performance of Exterior Windows, Curtain Walls, Doors and Storm Shutters Impacted by Missles and Exposed to Cyclic Pressure Differentials	R301.2.1.2, R612.6.1
E 1996—09	Standard Specification for Performance of Exterior Windows, Curtain Walls, Doors and Impact Protective Systems Impacted by Windborne Debris in Hurricanes	R301.2.1.2, R612.6.1
E 2178—03	Standard Test Method for Air Permeance of Building Materials	R202
E 2231—04	Standard Practice for Specimen Preparation and Mounting of Pipe and Duct Insulation Materials to Assess Surface Burning Characteristics	M1601.3
E 2273—03	Standard Test Method for Determining the Drainage Efficiency of Exterior Insulation and Finish Systems (EIFS) Clad Wall Assemblies	R703.9.2
E 2568—09e1	Standard Specification for PB Exterior Insulation and Finish Systems (EIFS)	R703.9.1, R703.9.2
E 2570—07	Standard Test Methods for Evaluating Water-resistive Barrier (WRB) Coatings Used Under Exterior Insulation and Finish Systems (EIFS) or EIFS with Drainage	R703.9.2.1
E 2634—08	Standard Specification for Flat Wall Insulating Concrete Form (ICF) Systems	R404.1.2.3.6.1, R611.4.4
F 405—05	Specification for Corrugated Polyethylene (PE) Pipe and Fittings	Table P3009.14.10, Table P3302.1, Table AO103.10

ASTM—continued

Standard	Title	Referenced in code section number
F 409—02(2008)	Specification for Thermoplastic Accessible and Replaceable Plastic Tube and Tubular Fittings	Table P2701.1, P2702.2, P2702.3
F 437—06	Specification for Threaded Chlorinated Poly (Vinyl Chloride) (CPVC) Plastic Pipe Fittings, Schedule 80	Table P2905.6
F 438—04	Specification for Socket-type Chlorinated Poly (Vinyl Chloride) (CPVC) Plastic Pipe Fittings, Schedule 40	Table P2905.6
F 439—06	Specification for Socket-type Chlorinated Poly (Vinyl Chloride) (CPVC) Plastic Pipe Fittings, Schedule 80	Table P2905.6
F 441/F 441M—02 (2008)	Specification for Chlorinated Poly (Vinyl Chloride) (CPVC) Plastic Pipe, Schedules 40 and 80	Table P2905.4, Table P2905.5
F 442/F 442M—99 (2005)e1	Specification for Chlorinated Poly (Vinyl Chloride) (CPVC) Plastic Pipe (SDR-PR)	Table P2905.4, Table P2905.5
F 477—08	Specification for Elastomeric Seals (Gaskets) for Joining Plastic Pipe	P2905.17, P3003.18
F 493—04	Specification for Solvent Cements for Chlorinated Poly (Vinyl Chloride) (CPVC) Plastic Pipe and Fittings	P2905.9.1.2
F 628—08	Specification for Acrylonitrile-butadiene-styrene (ABS) Schedule 40 Plastic Drain, Waste and Vent Pipe with a Cellular Core	Table P3002.1(1), Table P3002.1(2), Table P3002.2, Table P3002.3, P3003.3.2, P3003.8.2
F 656—08	Specification for Primers for Use in Solvent Cement Joints of Poly (Vinyl Chloride) (PVC) Plastic Pipe and Fittings	P2905.9.1.3, P3003.9.2, P3003.14.2
F 714—08	Specification for Polyethylene (PE) Plastic Pipe (SDR-PR) Based on Outside Diameter	Table P3002.2
F 876—08b	Specification for Cross-linked Polyethylene (PEX) Tubing	Table M2101.1, Table P2905.4, Table P2905.5
F 877—07	Specification for Cross-linked Polyethylene (PEX) Plastic Hot- and Cold-water Distribution Systems	Table M2101.1, Table P2905.4, Table P2905.5, Table P2905.6
F 891—07	Specification for Coextruded Poly (Vinyl Chloride) (PVC) Plastic Pipe with a Cellular Core	P2905.6, Table P3002.1(1), Table P3002.1(2), Table P3002.2, Table P3302.1
F 1055—98 (2006)	Specification for Electrofusion Type Polyethylene Fittings for Outside Diameter Controlled Polyethylene Pipe and Tubing	Table M2101.1, M2104.2.1.2
F 1281—07	Specification for Cross-linked Polyethylene/Aluminum/Cross-linked Polyethylene (PEX-AL-PEX) Pressure Pipe	Table M2101.1, Table P2905.4, Table P2905.5, Table P2905.6, P2505.11.1
F 1282—06	Specification for Polyethylene/Aluminum/Polyethylene (PE-AL-PE) Composite Pressure Pipe	Table M2101.1, Table P2905.4, Table P2905.5, Table P2905.6, P2905.11.1
F 1346—91 (2003)	Performance Specification for Safety Covers and Labeling Requirements for All Covers for Swimming Pools, Spas and Hot Tubs	AG105.2, AG105.5
F 1412—09	Specification for Polyolefin Pipe and Fittings for Corrosive Waste Drainage	Table P3002.1(2), Table P3002.2, Table P3002.3, P3003.16.1
F 1488—03	Specification for Coextruded Composite Pipe	Table P3002.1(1), Table P3002.1(2), Table P3002.2, Table P3009.14.10, Table AO103.10
F 1554—07a	Specification for Anchor Bolts, Steel, 36, 55 and 105-ksi Yield Strength	R611.5.2.2
F 1667—05	Specification for Driven Fasteners, Nails, Spikes and Staples	Table R703.4, R905.2.5
F 1807—08	Specification for Metal Insert Fittings Utilizing a Copper Crimp Ring for SDR9 Cross-linked Polyethylene (PEX) Tubing	Table M2101.1, Table P2905.6
F 1866—07	Specification for Poly (Vinyl Chloride) (PVC) Plastic Schedule 40 Drainage and DWV Fabricated Fittings	Table P3002.3
F 1960—09	Specification for Cold Expansion Fittings with PEX Reinforcing Rings for Use with Cross-linked Polyethylene (PEX) Tubing	Table M2101.1, Table P2905.6
F 1973—08	Standard Specification for Factory Assembled Anodeless Risers and Transition Fittings in Polyethylene (PE) and Polyamide 11 (PA 11) Fuel Gas Distribution Systems	G2415.15.2
F 1974—08	Specification for Metal Insert Fittings for Polyethylene/Aluminum/Polyethylene and Cross-linked Polyethylene/Aluminum/Cross-linked Polyethylene Composite Pressure Pipe	P2505.11.1, Table P2905.6
F 1986—01 (2006)	Multilayer Pipe Type 2, Compression Joints for Hot and Cold Drinking Water Systems	Table P2905.4, Table P2905.5, Table P2905.6
F 2080—08	Specification for Cold-expansion Fittings with Metal Compression-sleeves for Cross-linked Polyethylene (PEX) Pipe	P2905.6

ASTM—continued

F 2090—08	Specification for Window Fall Prevention Devices—with Emergency Escape (Egress) Release Mechanisms	R612.2, R612.3
F 2098—08	Standard Specification for Stainless Steel Clamps for SDR9 PEX Tubing to Metal Insert Fittings	Table M2101.1, Table P2905.6
F 2159—08	Standard Specification for Plastic Insert Fittings Utilizing a Copper Crimp Ring for SDR9 Cross-linked Polyethylene (PEX) Tubing	P2905.6
F 2200—05	Standard Specification for Automated Vehicular Gate Construction	AR103.1
F 2262—05	Standard Specification for Cross-linked Polyethylene/Aluminum/Cross-linked Polyethylene Tubing OD Controlled SDR9	Table P2905.4, Table P2905.5
F 2389—07e1	Standard for Pressure-rated Polypropylene (PP) Piping Systems	Table M2101.1, Table P2905.4, Table P2905.5, Table P2905.6, P2905.10.1
F 2434—08	Standard Specification for Metal Insert Fittings Utilizing a Copper Crimp Ring for Polyethylene/Aluminum/Cross-linked Polyethylene (PEX-AL-PEX) Tubing	Table P2905.6
F 2623—08	Standard Specification for Polyethylene of Raised Temperature (PE-RT) SDRG Tubing	Table M2101.1
F 2735—09	Standard Specification for SDR9 Cross-linked Polyethylene (PEX) and Raised Temperature (PE-RT) Tubing	Table 605.3, Table 605.4, Table M2101.1
F 2769—09	Polyethylene or Raised Temperature (PE-RT) Plastic Hot and Cold-Water Tubing and Distribution Systems	Table M2101.1, Table P2905.4

AWPA

American Wood Protection Association
P.O. Box 361784
Birmingham, AL 35236-1784

Standard reference number	Title	Referenced in code section number
C1—03	All Timber Products—Preservative Treatment by Pressure Processes	R902.2
M4—08	Standard for the Care of Preservative-treated Wood Products	R317.1.1, R318.1.2
U1—11	USE CATEGORY SYSTEM: User Specification for Treated Wood Except Section 6 Commodity Specification H	R317.1, R322.1.8, R402.1.2, R504.3, Table R905.8.5

AWS

American Welding Society
550 N. W. LeJeune Road
Miami, FL 33126

Standard reference number	Title	Referenced in code section number
A5.8—04	Specifications for Filler Metals for Brazing and Braze Welding	P3003.5.1, P3003.10.1, P3003.11.1

AWWA

American Water Works Association
6666 West Quincy Avenue
Denver, CO 80235

Standard reference number	Title	Referenced in code section number
C104—98	Standard for Cement-mortar Lining for Ductile-iron Pipe and Fittings for Water	P2905.4
C110/A21.10—03	Standard for Ductile-iron and Gray-iron Fittings, 3 Inches through 48 Inches, for Water	Table P2905.6, Table P3002.3
C115/A21.15—99	Standard for Flanged Ductile-iron Pipe with Ductile-iron or Gray-iron Threaded Flanges	Table P2905.4
C151/A21.51—02	Standard for Ductile-iron Pipe, Centrifugally Cast, for Water	Table P2905.4
C153/A21.53—00	Standard for Ductile-iron Compact Fittings for Water Service	Table P2905.6
C510—00	Double Check Valve Backflow Prevention Assembly	Table P2902.3, P2902.3.6
C511—00	Reduced-pressure Principle Backflow Prevention Assembly	Table P2902.3, P2902.3.5, P2902.5.1

AWWA—continued

C901—08	Polyethylene (PE) Pressure Pipe and Tubing $^1/_2$ in. (13 mm) through 3 in. (76 mm) for Water Service	P2905.4
C904—06	Cross-Linked Polyethylene (PEX) Pressure Pipe, $^1/_2$ in. (12 mm) through 3 in. (76 mm) for Water Service	P2905.4

CGSB

Canadian General Standards Board
Place du Portage 111, 6B1
11 Laurier Street
Gatineau, Quebec, Canada KIA 1G6

Standard reference number	Title	Referenced in code section number
CAN/CGSB-37.54—95	Polyvinyl Chloride Roofing and Waterproofing Membrane	R905.13.2
37 GP 52M (1984)	Roofing and Waterproofing Membrane, Sheet Applied, Elastomeric	R905.12.2
37-GP-56M—(1980)	Membrane, Modified Bituminous, Prefabricated and Reinforced for Roofing—with December 1985 Amendment	Table R905.11.2

CISPI

Cast Iron Soil Pipe Institute
5959 Shallowford Road, Suite 419
Chattanooga, TN 37421

Standard reference number	Title	Referenced in code section number
301—04a	Standard Specification for Hubless Cast Iron Soil Pipe and Fittings for Sanitary and Storm Drain, Waste and Vent Piping Applications	Table P3002.1(1), Table P3002.1(2), Table P3002.2, Table P3002.3, P3005.2.9, Table P3302.1
310—04	Standard Specification for Coupling for Use in Connection with Hubless Cast Iron Soil Pipe and Fittings for Sanitary and Storm Drain, Waste and Vent Piping Applications	P3003.6.3

CPA

Composite Panel Association
19465 Deerfield Avenue, Suite 306
Leesburg, VA 20176

Standard reference number	Title	Referenced in code section number
ANSI A135.4—04	Basic Hardboard	Table R602.3(2)
ANSI A135.5—04	Prefinished Hardboard Paneling	R702.5
ANSI A135.6—06	Hardboard Siding	Table R703.4

CPSC

Consumer Product Safety Commission
4330 East West Highway
Bethesda, MD 20814-4408

Standard reference number	Title	Referenced in code section number
16 CFR, Part 1201—(2002)	Safety Standard for Architectural Glazing	R308.1.1, R308.3.1, Table R308.3.1(1)
16 CFR, Part 1209—(2002)	Interim Safety Standard for Cellulose Insulation	R302.10.3
16 CFR, Part 1404—(2002)	Cellulose Insulation	R302.10.3

CSA

Canadian Standards Association
5060 Spectrum Way
Mississauga, Ontario, Canada L4N 5N6

Standard reference number	Title	Referenced in code section number
AAMA/WDMA/CSA 101/I.S.2/A440—11	North American Fenestration Standard/Specification for Windows, Doors and Unit Skylights	N1102.4.3
ANSI/CSA America FCI—03	Stationary Fuel Cell Power Systems	M1903.1
ASME A112.18.1/ CSA B125.1—2005	Plumbing Supply Fittings	P2708.4, P2722.1, P2722.2, P2722.3
ASME A112.18.2/ CSA B125.2—05	Plumbing Waste Fittings	Table P2701.1, P2702.2
ASME A112.9.2/ CSA B45.1—08	Ceramic Plumbing Fixtures	Table P2701.1, P2711.1, P2712.1
ASME A112.19.1/ CSA B45.2—02	Enameled Cast-Iron and Enameled Steel Plumbing Fixtures	Table 2701.1, P2711.1
ASME A112.19.3/ CSA Requirement 3—88	Manually Operated Gas Valves for Use in House Piping Systems	Table G2420.1.1
CSA 8—93	Requirements for Gas Fired Log Lighters for Wood Burning Fireplaces—with revisions through January 1999	G2433.1
0325—07	Construction Sheathing	R503.2.1, R602.3, R604.1, R903.2.1
0437-Series—93	Standards on OSB and Waferboard (Reaffirmed 2006)	R503.2.1, R602.3, R604.1, R803.2.1
CAN/CSA A257.1M—92	Circular Concrete Culvert, Storm Drain, Sewer Pipe and Fittings	Table P3002.2
CAN/CSA A257.2M—92	Reinforced Circular Concrete Culvert, Storm Drain, Sewer Pipe and Fittings	Table P3002.2
CAN/CSA A257.3M—92	Joints for Circular Concrete Sewer and Culvert Pipe, Manhole Sections and Fittings Using Rubber Gaskets	P3003.7, P3003.18
CAN/CSA B64.1.1—01	Vacuum Breakers, Atmospheric Type (AVB)	Table P2902.2, P2902.3.2
CAN/CSA B64.2—01	Vacuum Breakers, Hose Connection Type (HCVP)	Table P2902.2, P2902.3.2
CAN/CSA B64.2.2—01	Vacuum Breakers, Hose Connection Type (HCVP) with Automatic Draining Feature	Table P2902.2, P2902.3.2
CAN/CSA B64.3—01	Backflow Preventers, Dual Check Valve Type with Atmospheric Port (DCAP)	Table P2902.2, P2902.3.3
CAN/CSA B64.4—01	Backflow Preventers, Reduced Pressure Principle Type (RP)	Table P2902.3,
101/I.S.2/A440—08	Specifications for Windows, Doors and Unit Skylights	R308.6.9, R612.6
B45.4—02	Stainless Steel Plumbing Fixtures	Table P2701.1, P2711.1, P2712.1
B45.5—02 (R2008)	Plastic Plumbing Fixtures	Table P2701.1, P2711.2, P2712.1
B45.9—99 (R2008)	Macerating Systems and Related Components	P3007.1, P3007.2.1, P3007.5
B64.1.2—07	Pressure Vacuum Breakers (PVB)	Table P2902.2, P2902.3.4
B64.1.3—07	Spill Resistant Pressure Vacuum Breakers (SRPVB)	P2902.3
B64.2.1—07	Hose Connection Vacuum Breakers (HCVB) with Manual Draining Feature	Table P2902.2, P2902.3.2
B64.2.1.1—07	Hose Connection Dual Check Vacuum Breakers (HCDVB)	Table P2902.2, P2902.3.2
B64.3—07	Dual Check Backflow Preventers with Atmospheric Port (DCAP)	Table P2902.2
B64.4.1—07	Reduced Pressure Principle for Fire Sprinklers (RPF)	Table P2902.2
B64.5—07	Double Check Backflow Preventers (DCVA)	Table P2902.2, P2902.3.6
B64.5.1—07	Double Check Valve Backflow Preventers, Type for Fire Systems (DCVAF)	Table P2902.2, P2902.3.6
B64.6—07	Dual Check Valve Backflow Preventers (DuC)	Table P2902.3
B64.7—07	Laboratory Faucet Vacuum Breakers (LFVB)	Table P2902.2, P2902.3.2
B125.1—01	Plumbing Fittings	Table P2701.1, P2708.3, P2722.2, P2722.3
B125.3—05	Plumbing Fittings	Table 2701.1, P2713.3, P2721.2
B137.1—05	Polyethylene (PE) Pipe, Tubing and Fittings for Cold Water Pressure Services	Table P2905.4, Table P2905.6
B137.2—05	Polyvinylchloride PVC Injection-moulded Gasketed Fittings for Pressure Applications	Table P2905.6
B137.3—05	Rigid Poly (Vinyl Chloride) (PVC) Pipe for Pressure Applications	Table P2905.4, P3003.9.2, P3003.14.2
B137.5—05	Cross-linked Polyethylene (PEX) Tubing Systems for Pressure Applications	Table P2905.4, Table P2905.5, Table P2905.6
B137.6—05	Chlorinated polyvinylchloride CPVC Pipe, Tubing and Fittings For Hot- and Cold-water Distribution Systems	Table P2905.4, Table P2905.5, Table 2905.6
B137.9—05	Polyethylene/Aluminum/Polyethylene (PE-AL-PE) Composite Pressure Pipe Systems	P2505.11.1, Table P2905.4

CSA—continued

B137.10M—05	Cross-linked Polyethylene/Aluminum/Polyethylene (PE-AL-PE) Composite Pressure Pipe Systems	Table M2101.1, P2505.11.1, Table P2905.4, Table P2905.5
B137.11—05	Polypropylene (PP-R) Pipe and Fittings for Pressure Applications	Table P2905.4.1, Table 2905.4, Table P2905.6
B181.1—06	Acrylonitrile-butadiene-styrene (ABS) Drain, Waste and Vent Pipe and Pipe Fittings	Table P3002.1(1), Table P3002.1(2), Table P3002.2, Table P3002.3, P3003.3.2, P3003.8.2
B181.2—06	Polyvinylchloride (PVC) and chlorinated polyvinylchloride (CPVC) Drain, Waste and Vent Pipe and Pipe Fittings	Table P3002.1(1), Table P3002.1(2), Table P3002.2, Table P3002.3, P3003.9.2, P3003.14.2, P3008.2, Table P3302.1
B181.3—06	Polyolefin and polyvinylidene (PVDF) Laboratory Drainage Systems	Table P3002.1(1), Table P3002.1(2), Table P3002.2, Table P3002.3, P3003.16.1
B182.2—06	PSM Type polyvinylchloride (PVC) Sewer Pipe and Fittings	Table P3002.1(1), Table P3002.1(2), Table P3002.2, Table P3002.3, Table P3302.1
B182.4—06	Profile polyvinylchloride (PVC) Sewer Pipe & Fittings	Table P3002.2, Table P3002.3, Table P3302.1
B182.6—06	Profile Polyethylene Sewer Pipe and Fittings	Table P3302.1
B182.8—06	Profile Polyethylene (PE) Storm Sewer and Drainage Pipe and Fittings	Table P3302.1
B356—00	Water Pressure Reducing Valves for Domestic Water Supply Systems	P2903.3.1
B483.1—07	Drinking Water Treatment Systems	P2908.1, P2908.2
B602—05	Mechanical Couplings for Drain, Waste and Vent Pipe and Sewer Pipe	P3003.3.1, P3003.6.3, P3003.7, P3003.8.1, P3003.14.1, P3003.15, P3003.17.2
LC3—00	Appliance Stands and Drain Pans	P2801.5

P2902.3.5, P2902.5.1

CSSB

Cedar Shake & Shingle Bureau
P. O. Box 1178
Sumas, WA 98295-1178

Standard reference number	Title	Referenced in code section number
CSSB—97	Grading and Packing Rules for Western Red Cedar Shakes and Western Red Shingles of the Cedar Shake and Shingle Bureau	R702.6, R703.5, Table R905.7.4, Table R905.8.5

DASMA

Door and Access Systems Manufacturers Association International
1300 Summer Avenue
Cleveland, OH 44115-2851

Standard reference number	Title	Referenced in code section number
108—05	Standard Method for Testing Garage Doors: Determination of Structural Performance Under Uniform Static Air Pressure Difference	R612 .4
115—05	Standard Method for Testing Garage Doors: Determination of Structural Performance Under Missile Impact and Cyclic Wind Pressure	R301.2.1.2

DOC

United States Department of Commerce
1401 Constitution Avenue, NW
Washington, DC 20230

Standard reference number	Title	Referenced in code section number
PS 1—09	Structural Plywood	R404.2.1, Table R404.2.3, R503.2.1, R602.3, R604.1, R613.3.2, R803.2.1

DOC—continued

PS 2—10	Performance Standard for Wood-based Structural-use Panels	R404.2.1, Table R404.2.3, R503.2.1, R602.3, R604.1, R613.3.2, Table 613.3.2, R803.2.1
PS 20—05	American Softwood Lumber Standard	R404.2.1, R502.1, R602.1, R802.1

DOTn

Department of Transportation
1200 New Jersey Avenue SE
East Building, 2nd floor
Washington, DC 20590

Standard reference number	Title	Referenced in code section number
49 CFR, Parts 192.281(e) & 192.283 (b) (2009)	Transportation of Natural and Other Gas by Pipeline: Minimum Federal Safety Standards	G2414.6.1

FEMA

Federal Emergency Management Agency
500 C Street, SW
Washington, DC 20472

Standard reference number	Title	Referenced in code section number
FA/TB-2—08	Flood-damage Resistant Materials Requirements	R322.1.8
FEMA-TB-11—01	Crawlspace Construction for Buildings Located in Special Flood Hazard Area	R408.7

FM

Factory Mutual Global Research
Standards Laboratories Department
1301 Atwood Avenue, P. O. Box 7500
Johnson, RI 02919

Standard reference number	Title	Referenced in code section number
4450—(1989)	Approval Standard for Class 1 Insulated Steel Deck Roofs—with Supplements through July 1992	R906.1
4880—(2005)	American National Standard for Evaluating Insulated Wall or Wall and Roof/Ceiling Assemblies, Plastic Interior Finish Materials, Plastic Exterior Building Panels, Wall/Ceiling Coating Systems, Interior or Exterior Finish Systems	R316.4, R316.6

GA

Gypsum Association
810 First Street, Northeast, Suite 510
Washington, DC 20002-4268

Standard reference number	Title	Referenced in code section number
GA-253—07	Application of Gypsum Sheathing	Table R602.3(1)

HPVA

Hardwood Plywood & Veneer Association
1825 Michael Faraday Drive
Reston, Virginia 20190-5350

Standard reference number	Title	Referenced in code section number
ANSI/HP-1—2009	Standard for Hardwood and Decorative Plywood	R702.5

HVI

Home Ventilating Institute
1000 N. Rand Road, Ste 214
Wauconda, IL 60084

Standard reference number	Title	Referenced in code section number
HVI 916—09	Airflow Test Procedure	Table M1507.3

ICC

International Code Council, Inc.
500 New Jersey Avenue, NW
6th Floor
Washington, DC 20001

Standard reference number	Title	Referenced in code section number
IBC—12	International Building Code®	R101.2, R110.2, R301.1, R301.1.3, R301.2.2.1.1, R301.2.2.1.2, R301.2.2.4, R301.3, R308.5, R320.1, R321.3, R322.1, R403.1.8, R602.10.3(3), R606.12.2.1, R802.1.3.4, R905.10.3, N1102.2.10, G2402.3, Table AH107.4(1), AH107.4.3
ICC/ANSI A117.1—09	Accessible and Usable Buildings and Facilities	R321.3
ICC 400—12	Standard on the Design and Construction of Log Structures	R301.1.1
ICC 500—08	ICC/NSSA Standard on the Design and Construction of Storm Shelters	R323.1
ICC 600—08	Standard for Residential Construction in High-wind Regions	R301.2.1.1
IECC—12	International Energy Conservation Code®	N1101.2
IFC—12	International Fire Code®	R102.7, M2201.7, G2402.3, G2412.2, G2423.1
IFGC—12	International Fuel Gas Code®	G2401.1, G2423.1
IMC—12	International Mechanical Code®	N1103.2.2, G2402.3
IPC—12	International Plumbing Code®	Table R301.2(1), R903.4.1, G2402.3, R2601.1, AO102.6
IPMC—12	International Property Maintenance Code®	R102.7
IPSDC—12	International Private Sewage Disposal Code®	R322.1.7, AI101.1

ISO

International Organization for Standardization
1, ch. de la Voie - Creuse
Case postale 56
CH-1211 Geneva 20, Switzerland

Standard reference number	Title	Referenced in code section number
15874—2002	Polypropylene Plastic Piping Systems for Hot and Cold Water Installations	Table M2101.1

MSS

Manufacturers Standardization Society of the Valve and Fittings Industry
127 Park Street, Northeast
Vienna, VA 22180

Standard reference number	Title	Referenced in code section number
SP-58—93	Pipe Hangers and Supports—Materials, Design and Manufacture	G2418.2

NAIMA

North American Insulation Manufacturers Association
44 Canal Center Plaza, Suite 310
Alexandria, VA 22314

Standard reference number	Title	Referenced in code section number
AH 116—09	Fibrous Glass Duct Construction Standards, Fifth Edition	M1601.1.1

NCMA

National Concrete Masonry Association
13750 Sunrise Valley Drive
Herndon, VA 20171-4662

Standard reference number	Title	Referenced in code section number
TR 68-A—75	Design and Construction of Plain and Reinforced Concrete Masonry and Basement and Foundation Walls	R404.1.1

NFPA

National Fire Protection Association
1 Batterymarch Park
Quincy, MA 02269

Standard reference number	Title	Referenced in code section number
13—10	Installation of Sprinkler Systems	R302.3
13D—10	Standard for the Installation of Sprinkler Systems in One- and Two-family Dwellings and Manufactured Homes	R313.2.1, P2904.1, P2904.2, P2904.6.1
31—11	Installation of Oil-burning Equipment	M1801.3.1, M1805.3
58—11	Liquefied Petroleum Gas Code	G2412.2, G2414.6.2
70—11	National Electrical Code	E3401.1, E3401.2, E4301.1, Table E4303.2, E4304.3, E4304.4
72—11	National Fire Alarm Code	R314.1, R314.2
82—09	Standard on Incinerators and Waste Linen Handling Systems and Equipment	G2427.2.3
85—11	Boiler and Construction Systems Hazards Code	G2452.1
211—10	Chimneys, Fireplaces, Vents and Solid Fuel Burning Appliances	R1002.5, G2427.5.5.1
259—08	Test Method for Potential Heat of Building Materials	R316.5.7, R316.5.8
275—09	Standard Method of Fire Tests for the Evaluation of Thermal Barriers Used Over Foam Plastic Insulation	R316.4
286—11	Standard Methods of Fire Tests for Evaluating Contribution of Wall and Ceiling Interior Finish to Room Fire Growth	R302.9.4, R316.4, R316.5.8, R316.6
501—10	Standard on Manufactured Housing	R202, AE201
720—09	Standard for the Installation of Carbon Monoxide (CO) Detectors and Warning Equipment	R316.4
853—10	Standard for the Installation of Stationary Fuel Cell Power Systems	M1903.1

NSF

NSF International
789 N. Dixboro
Ann Arbor, MI 48105

Standard reference number	Title	Referenced in code section number
14—2008e	Plastics Piping System Components and Related Materials	M1301.4, P2608.3, P2908.3
42—2007ae	Drinking Water Treatment Units—Anesthetic Effects	P2908.1, P2908.3
44—2007	Residential Cation Exchange Water Softeners	P2908.1, P2908.3
53—2007a	Drinking Water Treatment Units—Health Effects	P2908.1, P2908.3
58—2007	Reverse Osmosis Drinking Water Treatment Systems	P2908.2, P2908.3
61—2008	Drinking Water System Components—Health Effects	P2608.5, P2722.1, P2903.9.4, P2905.4, P2905.5, P2905.6, P2907.3

PCA

Portland Cement Association
5420 Old Orchard Road
Skokie, IL 60077

Standard reference number	Title	Referenced in code section number
100—10	Prescriptive Design of Exterior Concrete Walls for One- and Two-family Dwellings (Pub. No. EB241)	R301.2.2.2.4, R301.2.2.3.4, R404.1.2, R404.1.2.2.1, R404.1.2.2.2, R404.1.2.4, R404.1.4.2, R611.1, R611.2, R611.9.2, R611.9.3

SBCA

Structural Building Components Association
6300 Enterprise Lane
Madison, WI 53719

Standard reference number	Title	Referenced in code section number
BCSI—2008	Building Component Safety Information Guide to Good Practice for Handling, Installing, Restraining & Bracing of Metal Plate Connected Wood Trusses	502.11.2, 802.10.3
CFS-BCSI—2008	Guide to Good Practice for Handling, Installing & Bracing of Cold-formed Steel Trusses	505.1.3, 804.3.7

SMACNA

Sheet Metal & Air Conditioning Contractors National Assoc. Inc.
4021 Lafayette Center Road
Chantilly, VA 22021

Standard reference number	Title	Referenced in code section number
SMACNA—10	Fibrous Glass Duct Construction Standards (2003)	M1601.1.1, M1604.4.1

TMS

The Masonry Society
3970 Broadway, Suite 201-D
Boulder, CO 80304

Standard reference number	Title	Referenced in code section number
302—07	Standard Method for Determining the Sound Transmission Class Rating for Masonry Walls	AK102.1.1
402—11	Building Code Requirements for Masonry Structures	R404.1.1, R606.1, R606.1.1, R606.12.1, R606.12.2.3.1, R606.12.3.2, Table R703.4
403—10	Direct Design Handbook for Masonry Structures	R606.1, R606.1.1, R606.12.1, R606.12.3.1
602—11	Specification for Masonry Structures	R404.1.1, R606.12.3.1, Table R703.4

TPI

Truss Plate Institute
583 D'Onofrio Drive, Suite 200
Madison, WI 53719

Standard reference number	Title	Referenced in code section number
TPI 1—2002	National Design Standard for Metal-plate-connected Wood Truss Construction	R502.11.1, R802.10.2

UL

Underwriters Laboratories, Inc.
333 Pfingsten Road
Northbrook, IL 60062

Standard reference number	Title	Referenced in code section number
17—2008	Vent or Chimney Connector Dampers for Oil-fired Appliances	M1802.2.2
55A—04	Materials for Built-up Roof Coverings	R905.9.2
58—96	Steel Underground Tanks for Flammable and Combustible Liquids—with revisions through July 1998	M2201.1
80—07	Steel Tanks for Oil-burner Fuel	M2201.1
103—01	Factory-built Chimneys for Residential Type and Building Heating Appliances—with revisions through March 2010	R202, R1005.3, G2430.1
127—08	Factory-built Fireplaces—with revisions through January 2010	R1001.11, R1004.1, R1004.4, R1005.4, G2445.7,
174—04	Household Electric Storage Tank Water Heaters—with revisions through November 2005	M2005.1
180—03	Liquid-level Indicating Gauges for Oil Burner Fuels—with revision through March 2007	M2201.5
181—05	Factory-made Air Ducts and Air Connectors—with revisions through May 2003	M1601.2, M1601.4.1
181A—05	Closure Systems for Use with Rigid Air Ducts and Air Connectors—with revisions through December 1998	M1601.2, M1601.4.1
181B—05	Closure Systems for Use with Flexible Air Ducts and Air Connectors—with revisions through August 2003	M1601.2, M1601.4.1
217—06	Single- and Multiple-station Smoke Alarms—with revisions through April 2010	R314.1
263—03	Standards for Fire Test of Building Construction and Materials—with revisions through October 2007	Table R302.1(2), R302.2, R302.3, R302.4.1, R316.4
325—02	Door, Drapery, Gate, Louver and Window Operations and Systems—with revisions through February 2010	R309.4
343—97	Pumps for Oil-burning Appliances—with revisions through May 2002	M2204.1
378—06	Draft Equipment	M1804.2.6
441—10	Gas Vents	G2426.1
508—99	Industrial Control Equipment—with revisions through April 2010	M1411.3.1
536—97	Flexible Metallic Hose—with revisions through June 2003	M2202.3
641—95	Type L, Low-temperature Venting Systems—with revisions through July 2009	R202, R1003.11.5, M1804.2.4, G2426.1
651—05	Schedule 40 and Schedule 80 Rigid PVC Conduit and Fittings—with revisions through March 2010	G2414.6.3
723—03	Standard for Test for Surface Burning Characteristics of Building Materials—with revisions through May 2005	R302.9.3, R302.9.4, R302.10.1, R302.10.2, R316.3, R316.5.9, R316.5.10, R316.6 , R802.1.3, M1601.3
726—95	Oil-fired Boiler Assemblies—with revisions through April 2010	M2001.1.1, M2006.1
727—06	Oil-fired Central Furnaces—with revisions through April 2010	M1402.1
729—03	Oil-fired Floor Furnaces—with revisions through April 2010	M1408.1
730—03	Oil-fired Wall Furnaces—with revisions through April 2010	M1409.1
732—95	Oil-fired Storage Tank Water Heaters—with revisions through April 2010	M2005.1
737—07	Fireplaces Stoves	M1414.1, M1901.2
790—04	Standard Test Methods for Fire Tests of Roof Coverings	R902.1
795—06	Commercial-industrial Gas Heating Equipment—with revisions through April 2010	G2442.1, G2452.1
834—04	Heating, Water Supply and Power Boilers—Electric—with revisions through December 2009	M2001.1.1
842—07	Valves for Flammable Fluids	M2204.2
858—05	Household Electric Ranges—with revisions through November 2007	M1901.2
875—09	Electric Dry-bath Heaters	M1902.2
896—93	Oil-burning Stoves—with revisions through May 2010	M1410.1
923—08	Microwave Cooking Appliances—with revisions through June 2010	M1504.1
959—01	Medium Heat Appliance Factory-built Chimneys—with revisions through June 2010	R1005.6
1026—07	Electric Household Cooking and Food Serving Appliances	M1901.2
1040—96	Fire Test of Insulated Wall Construction—with revisions through June 2001	R316.4, R316.6
1042—94	Electric Baseboard Heating Equipment—with revisions through February 2008	M1405.1
1256—02	Fire Test of Roof Deck Construction	R906.1
1261—01	Electric Water Heaters for Pools and Tubs—with revisions through June 2004	M2006.1
1453—04	Electronic Booster and Commercial Storage Tank Water Heaters—with revisions through December 2009	M2005.1
1479—03	Fire Tests of Through-Penetration Firestops—with revisions through March 2010	R302.4.1.2
1482—10	Solid-Fuel-type Room Heaters	R1002.2, R1002.5, M1410.1

UL—continued

1618—09	Wall Protectors, Floor Protectors, and Hearth Extensions	M1004.2
1673—96	Electric Space Heating Cables—with revisions through July 2003	M1406.1
1693—02	Electric Radiant Heating Panels and Heating Panel Sets	M1406.1
1703—02	Flat-plate Photovoltaic Modules and Panels—with revisions through April 2005	M2302.3
1741—99	Inverters, Converters, Controllers and Interconnection System Equipment with Distributed Energy Resources—with revisions through November 2005	M2302.4
1715—97	Fire Test of Interior Finish Material—with revisions through March 2004	R316.6
1738—06	Venting Systems for Gas-burning Appliances, Categories II, III and IV	G2426.1
1741—99	Inverters, Converters, Controllers and Interconnection System Equipment for use with Distributed Energy Resources—with revisions through November 2005	M2302.4
1777—04	Chimney Liners with revisions through June 2009	R1003.11.1, R1003.18, G2425.12, G2425.15.4, M1801.3.4,
1995—05	Heating and Cooling Equipment—with revisions through July 2009	M1402.1, M1403.1, M1407.1, M1412.1, M1413.1
1996—04	Electric Duct Heaters—with revisions through December 2006	M1402.1
2017—08	Standard for General-purpose Signaling Devices and Systems—with revisions through October 2009	AG105.2
2523—03	Outline of Investigation for Sealed Fuel-fired Water Heaters and Boilers	M2001.1.1
2034—08	Standard for Single- and Multiple-station Carbon Monoxide Alarms	R315.4
2075—04	Standard for Gas and Vapor Detectors and Sensors—with revisions through September 28, 2007	R315.2
2158A—06	Outline of Investigation for Clothes Dryer Transition Duct	M1502.4.3

ULC

Underwriters' Laboratories of Canada
7 Underwriters Road
Toronto, Ontario, Canada M1R 3B4

Standard reference number	Title	Referenced in code section number
CAN/ULC S 102—1988	Standard Methods for Test for Surface Burning Characteristics of Building Materials and Assemblies—with 2000 revisions	R302.10.2

US-FTC

United States-Federal Trade Commission
600 Pennsylvania Avenue NW
Washington, DC 20580

Standard reference number	Title	Referenced in code section number
CFR Title 16 (May 31, 2005)	R-value Rule	303.1.4

WDMA

Window & Door Manufacturers Association
1400 East Touhy Avenue, Suite 470
Des Plaines, IL 60018

Standard reference number	Title	Referenced in code section number
AAMA/WDMA/CSA 101/I.S2/A440—08	Specifications for Windows, Doors and Skylights	R308.6.9, R612.3, N1102.4.3

APPENDIX A

SIZING AND CAPACITIES OF GAS PIPING

(This appendix is informative and is not part of the *code*. This appendix is an excerpt from the 2012 *International Fuel Gas Code*, coordinated with the section numbering of the *International Residential Code*.)

A.1 General piping considerations. The first goal of determining the pipe sizing for a fuel gas *piping* system is to make sure that there is sufficient gas pressure at the inlet to each *appliance*. The majority of systems are residential and the appliances will all have the same, or nearly the same, requirement for minimum gas pressure at the *appliance* inlet. This pressure will be about 5-inch water column (w.c.) (1.25 kPa), which is enough for proper operation of the *appliance* regulator to deliver about 3.5-inches water column (w.c.) (875 kPa) to the burner itself. The pressure drop in the *piping* is subtracted from the source delivery pressure to verify that the minimum is available at the *appliance*.

There are other systems, however, where the required inlet pressure to the different appliances may be quite varied. In such cases, the greatest inlet pressure required must be satisfied, as well as the farthest *appliance*, which is almost always the critical *appliance* in small systems.

There is an additional requirement to be observed besides the capacity of the system at 100-percent flow. That requirement is that at minimum flow, the pressure at the inlet to any *appliance* does not exceed the pressure rating of the *appliance* regulator. This would seldom be of concern in small systems if the source pressure is $^1/_2$ psi (14-inch w.c.) (3.5 kPa) or less but it should be verified for systems with greater gas pressure at the point of supply.

To determine the size of *piping* used in a gas *piping* system, the following factors must be considered:

(1) Allowable loss in pressure from *point of delivery* to *appliance*.

(2) Maximum gas demand.

(3) Length of *piping* and number of fittings.

(4) Specific gravity of the gas.

(5) Diversity factor.

For any gas *piping* system, or special *appliance*, or for conditions other than those covered by the tables provided in this code, such as longer runs, greater gas demands or greater pressure drops, the size of each gas *piping* system should be determined by standard engineering practices acceptable to the code official.

A.2 Description of tables.

A.2.1 General. The quantity of gas to be provided at each *outlet* should be determined, whenever possible, directly from the manufacturer's gas input Btu/h rating of the *appliance* that will be installed. In case the ratings of the appliances to be installed are not known, Table 402.2 shows the approximate consumption (in Btu per hour) of certain types of typical household appliances.

To obtain the cubic feet per hour of gas required, divide the total Btu/h input of all appliances by the average Btu heating value per cubic feet of the gas. The average Btu per cubic feet of the gas in the area of the installation can be obtained from the serving gas supplier.

A.2.2 Low pressure natural gas tables. Capacities for gas at low pressure [less than 2.0 psig (13.8 kPa gauge)] in cubic feet per hour of 0.60 specific gravity gas for different sizes and lengths are shown in Tables 402.4(1) and 402.4(2) for iron pipe or equivalent rigid pipe; in Tables 402.4(8) through 402.4(11) for smooth wall semirigid tubing; and in Tables 402.4(15) through 402.4(17) for corrugated stainless steel tubing. Tables 402.4(1) and 402.4(6) are based upon a pressure drop of 0.3-inch w.c. (75 Pa), whereas Tables 402.4(2), 402.4(9) and 402.4(15) are based upon a pressure drop of 0.5-inch w.c. (125 Pa). Tables 402.4(3), 402.4(4), 402.4(10), 402.4(11), 402.4(16) and 402.4(17) are special low-pressure applications based upon pressure drops greater than 0.5-inch w.c. (125 Pa). In using these tables, an allowance (in equivalent length of pipe) should be considered for any *piping* run with four or more fittings (see Table A.2.2).

A.2.3 Undiluted liquefied petroleum tables. Capacities in thousands of Btu per hour of undiluted liquefied petroleum gases based on a pressure drop of 0.5-inch w.c. (125 Pa) for different sizes and lengths are shown in Table 402.4(28) for iron pipe or equivalent rigid pipe, in Table 402.4(30) for smooth wall semi-rigid tubing, in Table 402.4(32) for corrugated stainless steel tubing, and in Tables 402.4(35) and 402.4(37) for polyethylene plastic pipe and tubing. Tables 402.4(33) and 402.4(34) for corrugated stainless steel tubing and Table 402.4(36) for polyethylene plastic pipe are based on operating pressures greater than $1^1/_2$ pounds per square inch (psi) (3.5 kPa) and pressure drops greater than 0.5-inch w.c. (125 Pa). In using these tables, an allowance (in equivalent length of pipe) should be considered for any *piping* run with four or more fittings [see Table A.2.2].

A.2.4 Natural gas specific gravity. Gas *piping* systems that are to be supplied with gas of a specific gravity of 0.70 or less can be sized directly from the tables provided in this code, unless the code official specifies that a gravity factor be applied. Where the specific gravity of the gas is greater than 0.70, the gravity factor should be applied.

Application of the gravity factor converts the figures given in the tables provided in this code to capacities for another gas of different specific gravity. Such application is accomplished by multiplying the capacities given in the tables by the multipliers shown in Table A.2.4. In case the exact specific gravity does not appear in the table, choose the next higher value specific gravity shown.

TABLE A.2.2
EQUIVALENT LENGTHS OF PIPE FITTINGS AND VALVES

		SCREWED FITTINGS[1]				90° WELDING ELBOWS AND SMOOTH BENDS[2]					
		45°/Ell	90°/Ell	180°close return bends	Tee	R/d = 1	R/d = 1 1/3	R/d = 2	R/d = 4	R/d = 6	R/d = 8
k factor =		0.42	0.90	2.00	1.80	0.48	0.36	0.27	0.21	0.27	0.36
L/d' ratio[4] *n* =		1 4	30	67	60	16	12	9	7	9	12
Nominal pipe size, inches	Inside diameter d, inches, Schedule 40[6]	L = Equivalent Length In Feet of Schedule 40 (Standard-weight) Straight Pipe[6]									
1/2	0.622	0.73	1.55	3.47	3.10	0.83	0.62	0.47	0.36	0.47	0.62
3/4	0.824	0.96	2.06	4.60	4.12	1.10	0.82	0.62	0.48	0.62	0.82
1	1.049	1.22	2.62	5.82	5.24	1.40	1.05	0.79	0.61	0.79	1.05
1 1/4	1.380	1.61	3.45	7.66	6.90	1.84	1.38	1.03	0.81	1.03	1.38
1 1/2	1.610	1.88	4.02	8.95	8.04	2.14	1.61	1.21	0.94	1.21	1.61
2	2.067	2.41	5.17	11.5	10.3	2.76	2.07	1.55	1.21	1.55	2.07
2 1/2	2.469	2.88	6.16	13.7	12.3	3.29	2.47	1.85	1.44	1.85	2.47
3	3.068	3.58	7.67	17.1	15.3	4.09	3.07	2.30	1.79	2.30	3.07
4	4.026	4.70	10.1	22.4	20.2	5.37	4.03	3.02	2.35	3.02	4.03
5	5.047	5.88	12.6	28.0	25.2	6.72	5.05	3.78	2.94	3.78	5.05
6	6.065	7.07	15.2	33.8	30.4	8.09	6.07	4.55	3.54	4.55	6.07
8	7.981	9.31	20.0	44.6	40.0	10.6	7.98	5.98	4.65	5.98	7.98
10	10.02	11.7	25.0	55.7	50.0	13.3	10.0	7.51	5.85	7.51	10.0
12	11.94	13.9	29.8	66.3	59.6	15.9	11.9	8.95	6.96	8.95	11.9
14	13.13	15.3	32.8	73.0	65.6	17.5	13.1	9.85	7.65	9.85	13.1
16	15.00	17.5	37.5	83.5	75.0	20.0	15.0	11.2	8.75	11.2	15.0
18	16.88	19.7	42.1	93.8	84.2	22.5	16.9	12.7	9.85	12.7	16.9
20	18.81	22.0	47.0	105.0	94.0	25.1	18.8	14.1	11.0	14.1	18.8
24	22.63	26.4	56.6	126.0	113.0	30.2	22.6	17.0	13.2	17.0	22.6

(continued)

TABLE A.2.2—continued
EQUIVALENT LENGTHS OF PIPE FITTINGS AND VALVES

		MITER ELBOWS[3] (No. of miters)					WELDING TEES		VALVES (screwed, flanged, or welded)			
		1-45°	1-60°	1-90°	2-90°[5]	3-90°[5]	Forged	Miter[3]	Gate	Globe	Angle	Swing Check
k factor =		0.45	0.90	1.80	0.60	0.45	1.35	1.80	0.21	10	5.0	2.5
L/d' ratio[4] n =		15	30	60	20	15	45	60	7	333	167	83
Nominal pipe size, inches	Inside diameter d, inches, Schedule 40[6]	L = Equivalent Length In Feet of Schedule 40 (Standard-weight) Straight Pipe[6]										
$^{1}/_{2}$	0.622	0.78	1.55	3.10	1.04	0.78	2.33	3.10	0.36	17.3	8.65	4.32
$^{3}/_{4}$	0.824	1.03	2.06	4.12	1.37	1.03	3.09	4.12	0.48	22.9	11.4	5.72
1	1.049	1.31	2.62	5.24	1.75	1.31	3.93	5.24	0.61	29.1	14.6	7.27
$1^{1}/_{4}$	1.380	1.72	3.45	6.90	2.30	1.72	5.17	6.90	0.81	38.3	19.1	9.58
$1^{1}/_{2}$	1.610	2.01	4.02	8.04	2.68	2.01	6.04	8.04	0.94	44.7	22.4	11.2
2	2.067	2.58	5.17	10.3	3.45	2.58	7.75	10.3	1.21	57.4	28.7	14.4
$2^{1}/_{2}$	2.469	3.08	6.16	12.3	4.11	3.08	9.25	12.3	1.44	68.5	34.3	17.1
3	3.068	3.84	7.67	15.3	5.11	3.84	11.5	15.3	1.79	85.2	42.6	21.3
4	4.026	5.04	10.1	20.2	6.71	5.04	15.1	20.2	2.35	112.0	56.0	28.0
5	5.047	6.30	12.6	25.2	8.40	6.30	18.9	25.2	2.94	140.0	70.0	35.0
6	6.065	7.58	15.2	30.4	10.1	7.58	22.8	30.4	3.54	168.0	84.1	42.1
8	7.981	9.97	20.0	40.0	13.3	9.97	29.9	40.0	4.65	22.0	111.0	55.5
10	10.02	12.5	25.0	50.0	16.7	12.5	37.6	50.0	5.85	278.0	139.0	69.5
12	11.94	14.9	29.8	59.6	19.9	14.9	44.8	59.6	6.96	332.0	166.0	83.0
14	13.13	16.4	32.8	65.6	21.9	16.4	49.2	65.6	7.65	364.0	182.0	91.0
16	15.00	18.8	37.5	75.0	25.0	18.8	56.2	75.0	8.75	417.0	208.0	104.0
18	16.88	21.1	42.1	84.2	28.1	21.1	63.2	84.2	9.85	469.0	234.0	117.0
20	18.81	23.5	47.0	94.0	31.4	23.5	70.6	94.0	11.0	522.0	261.0	131.0
24	22.63	28.3	56.6	113.0	37.8	28.3	85.0	113.0	13.2	629.0	314.0	157.0

For SI: 1 foot = 305 mm, 1 degree = 0.01745 rad.

Note: Values for welded fittings are for conditions where bore is not obstructed by weld spatter or backing rings. If appreciably obstructed, use values for "Screwed Fittings."

1. Flanged fittings have three-fourths the resistance of screwed elbows and tees.
2. Tabular figures give the extra resistance due to curvature alone to which should be added the full length of travel.
3. Small size socket-welding fittings are equivalent to miter elbows and miter tees.
4. Equivalent resistance in number of diameters of straight pipe computed for a value of (f - 0.0075) from the *relation* (n - $k/4f$).
5. For condition of minimum resistance where the centerline length of each miter is between d and $2^{1}/_{2}d$.
6. For pipe having other inside diameters, the equivalent resistance may be computed from the above n values.

Source: Crocker, S. *Piping Handbook*, 4th ed., Table XIV, pp. 100-101. Copyright 1945 by McGraw-Hill, Inc. Used by permission of McGraw-Hill Book Company.

TABLE A.2.4
MULTIPLIERS TO BE USED WITH TABLES 402.4(1) THROUGH 402.4(22) WHERE THE SPECIFIC GRAVITY OF THE GAS IS OTHER THAN 0.60

SPECIFIC GRAVITY	MULTIPLIER	SPECIFIC GRAVITY	MULTIPLIER
0.35	1.31	1.00	0.78
0.40	1.23	1.10	0.74
0.45	1.16	1.20	0.71
0.50	1.10	1.30	0.68
0.55	1.04	1.40	0.66
0.60	1.00	1.50	0.63
0.65	0.96	1.60	0.61
0.70	0.93	1.70	0.59
0.75	0.90	1.80	0.58
0.80	0.87	1.90	0.56
0.85	0.84	2.00	0.55
0.90	0.82	2.10	0.54

A.2.5 Higher pressure natural gas tables. Capacities for gas at pressures 2.0 psig (13.8 kPa) or greater in cubic feet per hour of 0.60 specific gravity gas for different sizes and lengths are shown in Tables 402.4(5) through 402.4(7) for iron pipe or equivalent rigid pipe; Tables 402.4(12) to 402.4(14) for semirigid tubing; Tables 402.4(18) and 402.4(19) for corrugated stainless steel tubing; and Table 402.4(22) for polyethylene plastic pipe.

A.3 Use of capacity tables.

A.3.1 Longest length method. This sizing method is conservative in its approach by applying the maximum operating conditions in the system as the norm for the system and by setting the length of pipe used to size any given part of the *piping* system to the maximum value.

To determine the size of each section of gas *piping* in a system within the range of the capacity tables, proceed as follows (also see sample calculations included in this Appendix):

(1) Divide the *piping* system into appropriate segments consistent with the presence of tees, branch lines and main runs. For each segment, determine the gas load (assuming all appliances operate simultaneously) and its overall length. An allowance (in equivalent length of pipe) as determined from Table A.2.2 shall be considered for *piping* segments that include four or more fittings.

(2) Determine the gas demand of each *appliance* to be attached to the *piping* system. Where Tables 402.4(1) through 402.4(24) are to be used to select the *piping* size, calculate the gas demand in terms of cubic feet per hour for each *piping* system *outlet*. Where Tables 402.4(25) through 402.4(37) are to be used to select the *piping* size, calculate the gas demand in terms of thousands of Btu per hour for each *piping* system *outlet*.

(3) Where the *piping* system is for use with other than undiluted liquefied petroleum gases, determine the design system pressure, the allowable loss in pressure (pressure drop), and specific gravity of the gas to be used in the *piping* system.

(4) Determine the length of *piping* from the *point of delivery* to the most remote *outlet* in the building/ *piping* system.

(5) In the appropriate capacity table, select the row showing the measured length or the next longer length if the table does not give the exact length. This is the only length used in determining the size of any section of gas *piping*. If the gravity factor is to be applied, the values in the selected row of the table are multiplied by the appropriate multiplier from Table A.2.4.

(6) Use this horizontal row to locate ALL gas demand figures for this particular system of *piping*.

(7) Starting at the most remote *outlet*, find the gas demand for that *outlet* in the horizontal row just selected. If the exact figure of demand is not shown, choose the next larger figure left in the row.

(8) Opposite this demand figure, in the first row at the top, the correct size of gas *piping* will be found.

(9) Proceed in a similar manner for each *outlet* and each section of gas *piping*. For each section of *piping*, determine the total gas demand supplied by that section.

When a large number of *piping* components (such as elbows, tees and valves) are installed in a pipe run, additional pressure loss can be accounted for by the use of equivalent lengths. Pressure loss across any *piping* component can be equated to the pressure drop through a length of pipe. The equivalent length of a combination of only four elbows/tees can result in a jump to the next larger length row, resulting in a significant reduction in capacity. The equivalent lengths in feet shown in Table A.2.2 have been computed on a basis that the inside diameter corresponds to that of Schedule 40 (standard-weight) steel pipe, which is close enough for most purposes involving other schedules of pipe. Where a more specific solution for equivalent length is desired, this may be made by multiplying the actual inside diameter of the pipe in inches by $n/12$, or the actual inside diameter in feet by n (n can be read from the table heading). The equivalent length values can be used with reasonable accuracy for copper or brass fittings and bends although the resistance per foot of copper or brass pipe is less than that of steel. For copper or brass valves, however, the equivalent length of pipe should be taken as 45 percent longer than the values in the table, which are for steel pipe.

A.3.2 Branch length method. This sizing method reduces the amount of conservatism built into the traditional Longest Length Method. The longest length as measured from the meter to the furthest remote *appliance* is only used to size the initial parts of the overall *piping* system. The Branch Length Method is applied in the following manner:

(1) Determine the gas load for each of the connected appliances.

(2) Starting from the meter, divide the *piping* system into a number of connected segments, and determine the length and amount of gas that each segment would carry assuming that all appliances were operated simultaneously. An allowance (in equivalent length of pipe) as determined from Table A.2.2 should be considered for piping segments that include four or more fittings.

(3) Determine the distance from the *outlet* of the gas meter to the *appliance* furthest removed from the meter.

(4) Using the longest distance (found in Step 3), size each *piping* segment from the meter to the most remote *appliance outlet*.

(5) For each of these *piping* segments, use the longest length and the calculated gas load for all of the connected appliances for the segment and begin the sizing process in Steps 6 through 8.

(6) Referring to the appropriate sizing table (based on operating conditions and *piping* material), find the longest length distance in the first column or the next larger distance if the exact distance is not listed. The use of alternative operating pressures and/or pressure drops will require the use of a different sizing table, but will not alter the sizing methodology. In many cases, the use of alternative operating pressures and/or pressure drops will require the approval of both the code official and the local gas serving utility.

(7) Trace across this row until the gas load is found or the closest larger capacity if the exact capacity is not listed.

(8) Read up the table column and select the appropriate pipe size in the top row. Repeat Steps 6, 7 and 8 for each pipe segment in the longest run.

(9) Size each remaining section of branch *piping* not previously sized by measuring the distance from the gas meter location to the most remote *outlet* in that branch, using the gas load of attached appliances and following the procedures of Steps 2 through 8.

A.3.3 Hybrid pressure method. The sizing of a 2 psi (13.8 kPa) gas *piping* system is performed using the traditional Longest Length Method but with modifications. The 2 psi (13.8 kPa) system consists of two independent pressure zones, and each zone is sized separately. The Hybrid Pressure Method is applied as follows:

The sizing of the 2 psi (13.8 kPa) section (from the meter to the line regulator) is as follows:

(1) Calculate the gas load (by adding up the name plate ratings) from all connected appliances. (In certain circumstances the installed gas load may be increased up to 50 percent to accommodate future addition of appliances.) Ensure that the line regulator capacity is adequate for the calculated gas load and that the required pressure drop (across the regulator) for that capacity does not exceed $^3/_4$ psi (5.2 kPa) for a 2 psi (13.8 kPa) system. If the pressure drop across the regulator is too high (for the connected gas load), select a larger regulator.

(2) Measure the distance from the meter to the line regulator located inside the building.

(3) If there are multiple line regulators, measure the distance from the meter to the regulator furthest removed from the meter.

(4) The maximum allowable pressure drop for the 2 psi (13.8 kPa) section is 1 psi (6.9 kPa).

(5) Referring to the appropriate sizing table (based on *piping* material) for 2 psi (13.8 kPa) systems with a 1 psi (6.9 kPa) pressure drop, find this distance in the first column, or the closest larger distance if the exact distance is not listed.

(6) Trace across this row until the gas load is found or the closest larger capacity if the exact capacity is not listed.

(7) Read up the table column to the top row and select the appropriate pipe size.

(8) If there are multiple regulators in this portion of the *piping* system, each line segment must be sized for its actual gas load, but using the longest length previously determined above.

The low pressure section (all *piping* downstream of the line regulator) is sized as follows:

(1) Determine the gas load for each of the connected appliances.

(2) Starting from the line regulator, divide the piping system into a number of connected segments and/or independent parallel piping segments, and determine the amount of gas that each segment would carry assuming that all appliances were operated simultaneously. An allowance (in equivalent length of pipe) as determined from Table A.2.2 should be considered for piping segments that include four or more fittings.

(3) For each piping segment, use the actual length or longest length (if there are sub-branchlines) and the calculated gas load for that segment and begin the sizing process as follows:

(a) Referring to the appropriate sizing table (based on operating pressure and piping material), find the longest length distance in the first column or the closest larger distance if the exact distance is not listed. The use of alternative operating pressures and/or pressure drops will require the use of a different sizing table, but will not alter the sizing methodology. In many cases, the use of alternative operating pressures and/or pressure drops may require the approval of the code official.

(b) Trace across this row until the appliance gas load is found or the closest larger capacity if the exact capacity is not listed.

(c) Read up the table column to the top row and select the appropriate pipe size.

(d) Repeat this process for each segment of the piping system.

A.3.4 Pressure drop per 100 feet method. This sizing method is less conservative than the others, but it allows the designer to immediately see where the largest pressure drop occurs in the system. With this information, modifications can be made to bring the total drop to the critical *appliance* within the limitations that are presented to the designer.

Follow the procedures described in the Longest Length Method for Steps (1) through (4) and (9).

For each *piping* segment, calculate the pressure drop based on pipe size, length as a percentage of 100 feet (30 480 mm) and gas flow. Table A.3.4 shows pressure drop per 100 feet (30 480 mm) for pipe sizes from $^1/_2$ inch (12.7 mm) through 2 inches (51 mm). The sum of pressure drops to the critical *appliance* is subtracted from the supply pressure to verify that sufficient pressure will be available. If not, the layout can be examined to find the high drop section(s) and sizing selections modified.

Note: Other values can be obtained by using the following equation:

$$\text{Desired Value} = MBH \times \sqrt{\frac{\text{Desired Drop}}{\text{Table Drop}}}$$

For example, if it is desired to get flow through $^3/_4$-inch (19.1 mm) pipe at 2 inches/100 feet, multiply the capacity of $^3/_4$-inch pipe at 1 inch/100 feet by the square root of the pressure ratio:

$$147\ MBH \times \sqrt{\frac{2''\text{ w.c.}}{1''\text{ w.c.}}} = 147 \times 1.414 = 208\ MBH$$

$$(MBH = 1000\text{ Btu/h})$$

A.4 Use of sizing equations. Capacities of smooth wall pipe or tubing can also be determined by using the following formulae:

(1) High Pressure [1.5 psi (10.3 kPa) and above]:

$$Q = 181.6\sqrt{\frac{D^5 \cdot (P_1^2 - P_2^2) \cdot Y}{C_r \cdot fba \cdot L}}$$

$$= 2237\ D^{2.623}\left[\frac{(P_1^2 - P_2^2) \cdot Y}{C_r \cdot L}\right]^{0.541}$$

(2) Low Pressure [Less than 1.5 psi (10.3 kPa)]:

$$Q = 187.3\sqrt{\frac{D^5 \cdot \Delta H}{C_r \cdot fba \cdot L}}$$

$$= 2313 D^{2.623}\left(\frac{\Delta H}{C_r \cdot L}\right)^{0.541}$$

where:

Q = Rate, cubic feet per hour at 60°F and 30-inch mercury column

D = Inside diameter of pipe, in.

P_1 = Upstream pressure, psia

P_2 = Downstream pressure, psia

Y = Superexpansibility factor = 1/supercompressibility factor

C_r = Factor for viscosity, density and temperature*

$$= 0.00354\ ST\left(\frac{Z}{S}\right)^{0.152}$$

Note: See Table 402.4 for Y and C_r for natural gas and propane.

TABLE A.3.4
THOUSANDS OF BTU/H (MBH) OF NATURAL GAS PER 100 FEET OF PIPE AT VARIOUS PRESSURE DROPS AND PIPE DIAMETERS

PRESSURE DROP PER 100 FEET IN INCHES W.C.	PIPE SIZES (inch)					
	$^1/_2$	$^3/_4$	1	$1^1/_4$	$1^1/_2$	2
0.2	31	64	121	248	372	716
0.3	38	79	148	304	455	877
0.5	50	104	195	400	600	1160
1.0	71	147	276	566	848	1640

For SI: 1 inch = 25.4 mm, 1 foot = 304.8 mm.

S = Specific gravity of gas at 60°F and 30-inch mercury column (0.60 for natural gas, 1.50 for propane), or = 1488μ

T = Absolute temperature, °F or = $t + 460$

t = Temperature, °F

Z = Viscosity of gas, centipoise (0.012 for natural gas, 0.008 for propane), or = 1488μ

fba = Base friction factor for air at 60°F (CF = 1)

L = Length of pipe, ft

ΔH = Pressure drop, in. w.c. (27.7 in. H_2O = 1 psi)

(For SI, see Section 402.4)

A.5 Pipe and tube diameters. Where the internal diameter is determined by the formulas in Section 402.4, Tables A.5.1 and A.5.2 can be used to select the nominal or standard pipe size based on the calculated internal diameter.

TABLE A.5.1
SCHEDULE 40 STEEL PIPE STANDARD SIZES

NOMINAL SIZE (inch)	INTERNAL DIAMETER (inch)	NOMINAL SIZE (inch)	INTERNAL DIAMETER (inch)
$^1/_4$	0.364	$1^1/_2$	1.610
$^3/_8$	0.493	2	2.067
$^1/_2$	0.622	$2^1/_2$	2.469
$^3/_4$	0.824	3	3.068
1	1.049	$3^1/_2$	3.548
$1^1/_4$	1.380	4	4.026

For SI: 1 inch = 25.4 mm.

A.6 Use of sizing charts. A third method of sizing gas *piping* is detailed below as an option that is useful when large quantities of *piping* are involved in a job (e.g., an apartment house) and material costs are of concern. If the user is not completely familiar with this method, the resulting pipe sizing should be checked by a knowledgeable gas engineer. The sizing charts are applied as follows:

(1) With the layout developed according to Section 106.3.1 of the code, indicate in each section the design gas flow under maximum operation conditions. For many layouts, the maximum design flow will be the sum of all connected loads; however, in some cases, certain combinations of appliances will not occur simultaneously (e.g., gas heating and *air conditioning*). For these cases, the design flow is the greatest gas flow that can occur at any one time.

(2) Determine the inlet gas pressure for the system being designed. In most cases, the point of inlet will be the gas meter or service regulator, but in the case of a system addition, it could be the point of connection to the existing system.

(3) Determine the minimum pressure required at the inlet to the critical *appliance*. Usually, the critical item will be the *appliance* with the highest required pressure for satisfactory operation. If several items have the same required pressure, it will be the one with the greatest length of *piping* from the system inlet.

(4) The difference between the inlet pressure and critical item pressure is the allowable system pressure drop. Figures A.6(a) and A.6(b) show the relationship between gas flow, pipe size and pipe length for natural gas with 0.60 specific gravity.

(5) To use Figure A.6(a) (low pressure applications), calculate the *piping* length from the inlet to the critical

TABLE A.5.2
COPPER TUBE STANDARD SIZES

TUBE TYPE	NOMINAL OR STANDARD SIZE (inches)	INTERNAL DIAMETER (inches)
K	$^1/_4$	0.305
L	$^1/_4$	0.315
ACR (D)	$^3/_8$	0.315
ACR (A)	$^3/_8$	0.311
K	$^3/_8$	0.402
L	$^3/_8$	0.430
ACR (D)	$^1/_2$	0.430
ACR (A)	$^1/_2$	0.436
K	$^1/_2$	0.527
L	$^1/_2$	0.545
ACR (D)	$^5/_8$	0.545
ACR (A)	$^5/_8$	0.555
K	$^5/_8$	0.652
L	$^5/_8$	0.666
ACR (D)	$^3/_4$	0.666
ACR (A)	$^3/_4$	0.680
K	$^3/_4$	0.745
L	$^3/_4$	0.785
ACR	$^7/_8$	0.785
K	1	0.995
L	1	1.025
ACR	$1^1/_8$	1.025
K	$1^1/_4$	1.245
L	$1^1/_4$	1.265
ACR	$1^3/_8$	1.265
K	$1^1/_2$	1.481
L	$1^1/_2$	1.505
ACR	$1^5/_8$	1.505
K	2	1.959
L	2	1.985
ACR	$2^1/_8$	1.985
K	$2^1/_2$	2.435
L	$2^1/_2$	2.465
ACR	$2^5/_8$	2.465
K	3	2.907
L	3	2.945
ACR	$3^1/_8$	2.945

For SI: 1 inch = 25.4 mm.

appliance. Increase this length by 50 percent to allow for fittings. Divide the allowable pressure drop by the equivalent length (in hundreds of feet) to determine the allowable pressure drop per 100 feet (30 480 mm). Select the pipe size from Figure A.6(a) for the required volume of flow.

(6) To use Figure A.6(b) (high pressure applications), calculate the equivalent length as above. Calculate the index number for Figure A.6(b) by dividing the difference between the squares of the absolute values of inlet and *outlet* pressures by the equivalent length (in hundreds of feet). Select the pipe size from Figure A.6(b) for the gas volume required.

A.7 Examples of piping system design and sizing.

A.7.1 Example 1: Longest length method. Determine the required pipe size of each section and *outlet* of the *piping* system shown in Figure A.7.1, with a designated pressure drop of 0.5-inch w.c. (125 Pa) using the Longest Length Method. The gas to be used has 0.60 specific gravity and a heating value of 1,000 Btu/ft^3 (37.5 MJ/m^3).

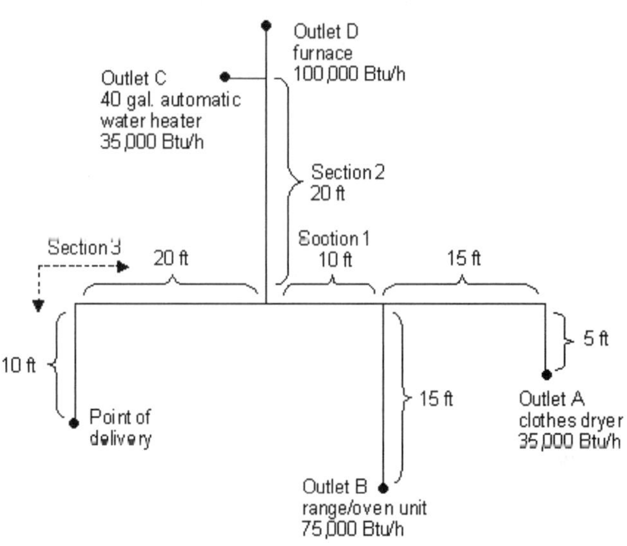

FIGURE A.7.1
PIPING PLAN SHOWING A STEEL PIPING SYSTEM

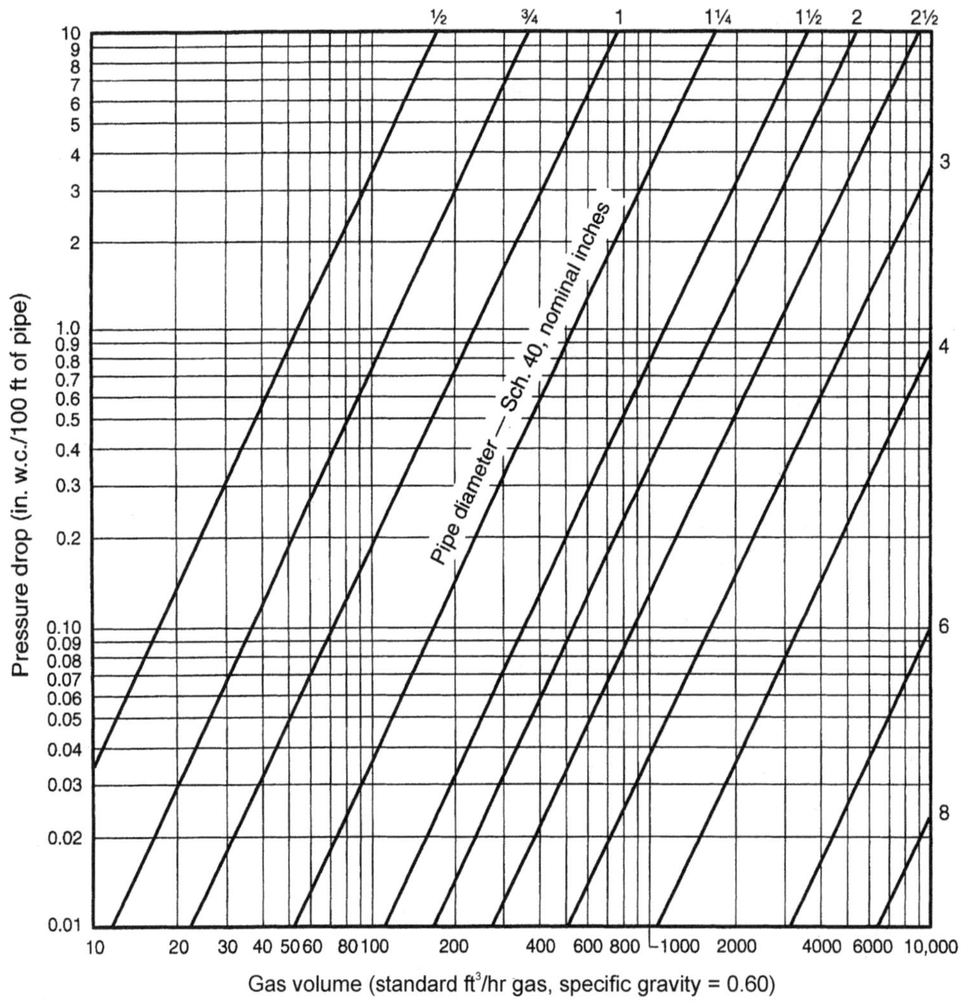

FIGURE A.6(a)
CAPACITY OF NATURAL GAS PIPING, LOW PRESSURE (0.60 WC)

Solution:

(1) Maximum gas demand for *Outlet* A:

$$\frac{\text{Consumption (rating plate input, or Table 402.2 if necessary}}{\text{Btu of gas}} =$$

$$\frac{\text{35,000 Btu per hour rating}}{\text{1,000 Btu per cubic foot}} = \text{35 cubic feet per hour} = \text{35 cfh}$$

Maximum gas demain for *Outlet* B:

$$\frac{\text{Consumption}}{\text{Btu of gas}} = \frac{75{,}000}{1{,}000} = \text{75 cfh}$$

Maximum gas demain for *Outlet* C:

$$\frac{\text{Consumption}}{\text{Btu of gas}} = \frac{35{,}000}{1{,}000} = \text{35 cfh}$$

Maximum gas demain for *Outlet* D:

$$\frac{\text{Consumption}}{\text{Btu of gas}} = \frac{100{,}000}{1{,}000} = \text{100 cfh}$$

(2) The length of pipe from the *point of delivery* to the most remote *outlet* (A) is 60 feet (18 288 mm). This is the only distance used.

(3) Using the row marked 60 feet (18 288 mm) in Table 402.4(2):

(a) *Outlet* A, supplying 35 cfh (0.99 m^3/hr), requires $^1/_2$-inch pipe.

(b) *Outlet* B, supplying 75 cfh (2.12 m^3/hr), requires $^3/_4$-inch pipe.

(c) Section 1, supplying *Outlets* A and B, or 110 cfh (3.11 m^3/hr), requires $^3/_4$-inch pipe.

(d) Section 2, supplying *Outlets* C and D, or 135 cfh (3.82 m^3/hr), requires $^3/_4$-inch pipe.

(e) Section 3, supplying *Outlets* A, B, C and D, or 245 cfh (6.94 m^3/hr), requires 1-inch pipe.

(4) If a different gravity factor is applied to this example, the values in the row marked 60 feet (18 288 mm) of Table 402.4(2) would be multiplied by the appropriate multiplier from Table A.2.4 and the resulting cubic feet per hour values would be used to size the *piping*.

A.7.2 Example 2: Hybrid or dual pressure systems. Determine the required CSST size of each section of the *piping* system shown in Figure A.7.2, with a designated pressure

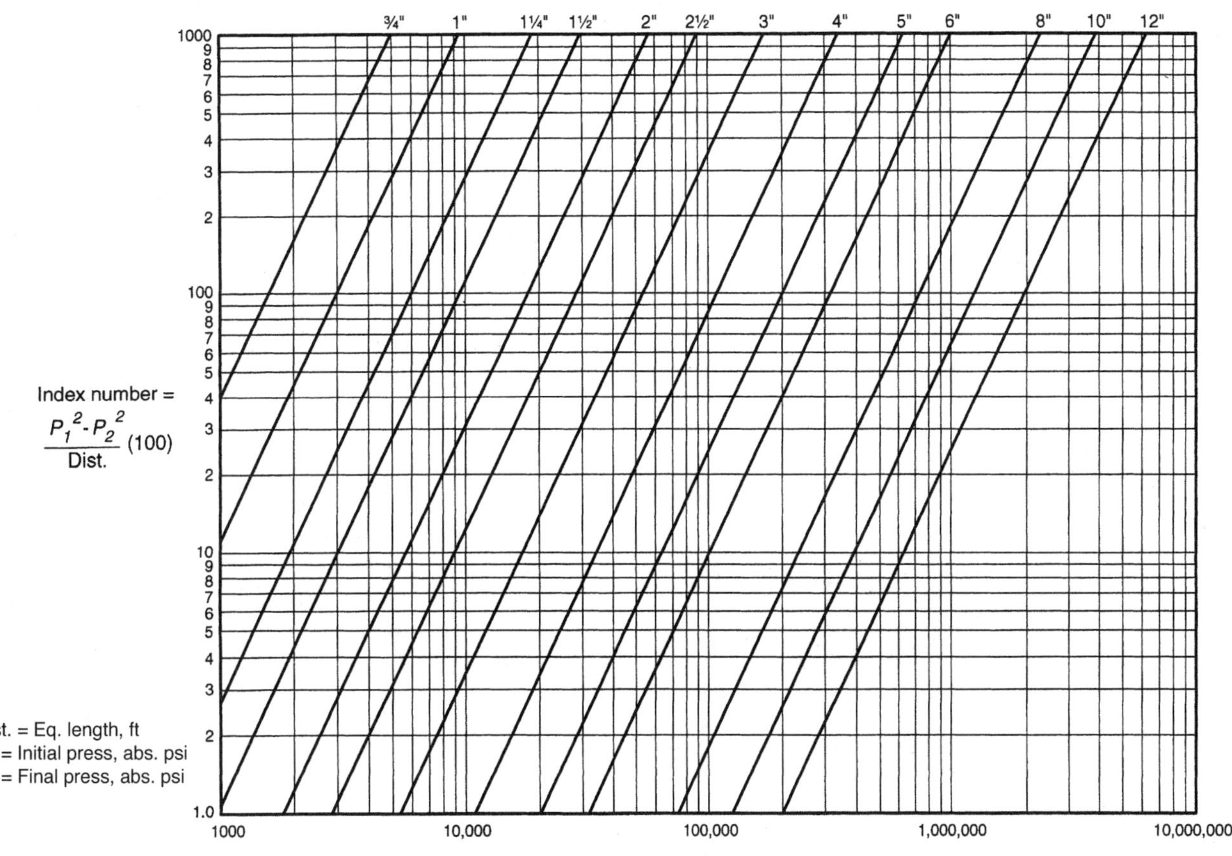

FIGURE A.6(b)
CAPACITY OF NATURAL GAS PIPING, HIGH PRESSURE (1.5 psi and above)

drop of 1 psi (6.9 kPa) for the 2 psi (13.8 kPa) section and 3-inch w.c. (0.75 kPa) pressure drop for the 13-inch w.c. (2.49 kPa) section. The gas to be used has 0.60 specific gravity and a heating value of 1,000 Btu/ft^3 (37.5 MJ/ m^3).

Solution:

(1) Size 2 psi (13.8 kPa) line using Table 402.4(18).

(2) Size 10-inch w.c. (2.5 kPa) lines using Table 402.4(16).

(3) Using the following, determine if sizing tables can be used.

(a) Total gas load shown in Figure A.7.2 equals 110 cfh (3.11 m^3/hr).

(b) Determine pressure drop across regulator [see notes in Table 402.4(18)].

(c) If pressure drop across regulator exceeds $^3/_4$ psig (5.2 kPa), Table 402.4(18) cannot be used. Note: If pressure drop exceeds $^3/_4$ psi (5.2 kPa), then a larger regulator must be selected or an alternative sizing method must be used.

(d) Pressure drop across the line regulator [for 110 cfh (3.11 m^3/hr)] is 4-inch w.c. (0.99 kPa) based on manufacturer's performance data.

(e) Assume the CSST manufacturer has tubing sizes or EHDs of 13, 18, 23 and 30.

(4) Section A [2 psi (13.8 kPa) zone]

(a) Distance from meter to regulator = 100 feet (30 480 mm).

(b) Total load supplied by A = 110 cfh (3.11 m^3/hr) (furnace + water heater + dryer).

(c) Table 402.4(18) shows that EHD size 18 should be used.

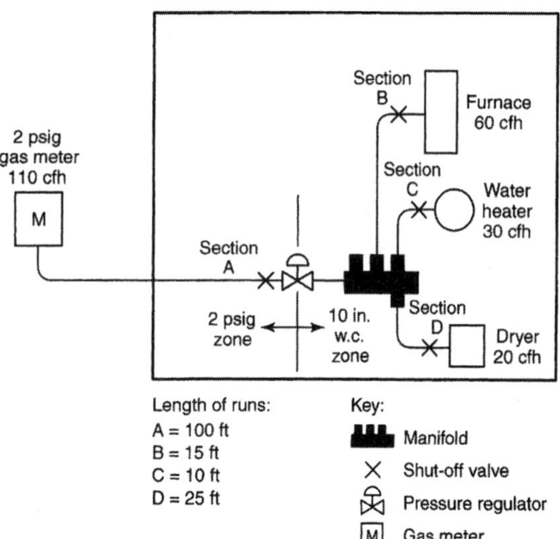

FIGURE A.7.2
PIPING PLAN SHOWING A CSST SYSTEM

Note: It is not unusual to oversize the supply line by 25 to 50 percent of the as-installed load. EHD size 18 has a capacity of 189 cfh (5.35 m^3/hr).

(5) Section B (low pressure zone)

(a) Distance from regulator to furnace is 15 feet (4572 mm).

(b) Load is 60 cfh (1.70 m^3/hr).

(c) Table 402.4(16) shows that EHD size 13 should be used.

(6) Section C (low pressure zone)

(a) Distance from regulator to water heater is 10 feet (3048 mm).

(b) Load is 30 cfh (0.85 m^3/hr).

(c) Table 402.4(16) shows that EHD size 13 should be used.

(7) Section D (low pressure zone)

(a) Distance from regulator to dryer is 25 feet (7620 mm).

(b) Load is 20 cfh (0.57 m^3/hr).

(c) Table 402.4(16) shows that EHD size 13 should be used.

A.7.3 Example 3: Branch length method. Determine the required semirigid copper tubing size of each section of the *piping* system shown in Figure A.7.3, with a designated pressure drop of 1-inch w.c. (250 Pa) (using the Branch Length Method). The gas to be used has 0.60 specific gravity and a heating value of 1,000 Btu/ft^3 (37.5 MJ/m^3).

Solution:

(1) Section A

(a) The length of tubing from the *point of delivery* to the most remote *appliance* is 50 feet (15 240 mm), A + C.

(b) Use this longest length to size Sections A and C.

(c) Using the row marked 50 feet (15 240 mm) in Table 402.4(10), Section A, supplying 220 cfh (6.2 m^3/hr) for four appliances requires 1-inch tubing.

(2) Section B

(a) The length of tubing from the *point of delivery* to the range/oven at the end of Section B is 30 feet (9144 mm), A + B.

(b) Use this branch length to size Section B only.

(c) Using the row marked 30 feet (9144 mm) in Table 402.4(10), Section B, supplying 75 cfh (2.12 m^3/hr) for the range/oven requires $^1/_2$-inch tubing.

(3) Section C

(a) The length of tubing from the *point of delivery* to the dryer at the end of Section C is 50 feet (15 240 mm), A + C.

(b) Use this branch length (which is also the longest length) to size Section C.

(c) Using the row marked 50 feet (15 240 mm) in Table 402.4(10), Section C, supplying 30 cfh (0.85 m^3/hr) for the dryer requires $^3/_8$-inch tubing.

(4) Section D

(a) The length of tubing from the *point of delivery* to the water heater at the end of Section D is 30 feet (9144 mm), A + D.

(b) Use this branch length to size Section D only.

(c) Using the row marked 30 feet (9144 mm) in Table 402.4(10), Section D, supplying 35 cfh (0.99 m^3/hr) for the water heater requires $^3/_8$-inch tubing.

(5) Section E

(a) The length of tubing from the *point of delivery* to the furnace at the end of Section E is 30 feet (9144 mm), A + E.

(b) Use this branch length to size Section E only.

(c) Using the row marked 30 feet (9144 mm) in Table 402.4(10), Section E, supplying 80 cfh (2.26 m^3/hr) for the furnace requires $^1/_2$-inch tubing.

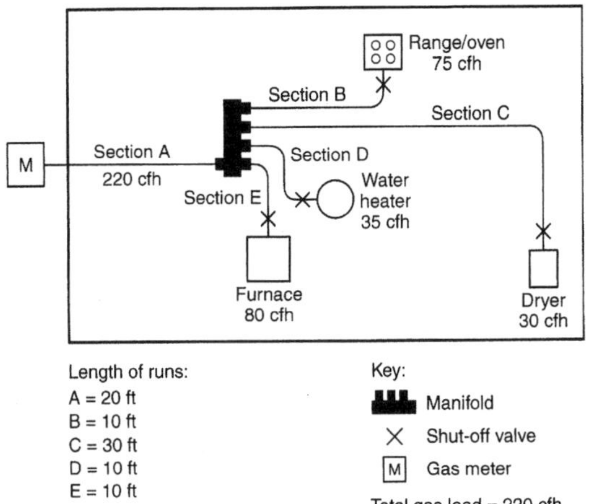

FIGURE A.7.3
PIPING PLAN SHOWING A COPPER TUBING SYSTEM

A.7.4 Example 4: Modification to existing piping system. Determine the required CSST size for Section G (retrofit application) of the *piping* system shown in Figure A.7.4, with a designated pressure drop of 0.5-inch w.c. (125 Pa) using the branch length method. The gas to be used has 0.60 specific gravity and a heating value of 1,000 Btu/ft^3 (37.5 MJ/m^3).

Solution:

(1) The length of pipe and CSST from the *point of delivery* to the retrofit *appliance* (barbecue) at the end of Section G is 40 feet (12 192 mm), A + B + G.

(2) Use this branch length to size Section G.

(3) Assume the CSST manufacturer has tubing sizes or EHDs of 13, 18, 23 and 30.

(4) Using the row marked 40 feet (12 192 mm) in Table 402.4(15), Section G, supplying 40 cfh (1.13 m^3/hr) for the barbecue requires EHD 18 CSST.

(5) The sizing of Sections A, B, F and E must be checked to ensure adequate gas carrying capacity since an *appliance* has been added to the *piping* system (see A.7.1 for details).

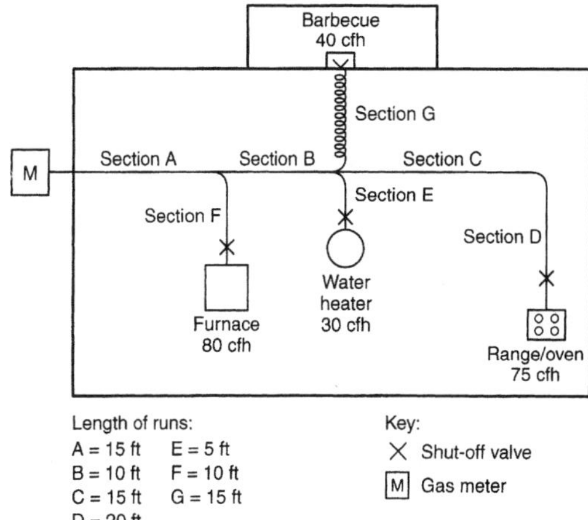

FIGURE A.7.4
PIPING PLAN SHOWING A MODIFICATION TO EXISTING PIPING SYSTEM

A.7.5 Example 5: Calculating pressure drops due to temperature changes. A test *piping* system is installed on a warm autumn afternoon when the temperature is 70°F (21°C). In accordance with local custom, the new *piping* system is subjected to an air pressure test at 20 psig (138 kPa). Overnight, the temperature drops and when the inspector shows up first thing in the morning the temperature is 40°F (4°C).

If the volume of the *piping* system is unchanged, then the formula based on Boyle's and Charles' law for determining the new pressure at a reduced temperature is as follows:

$$\frac{T_1}{T_2} = \frac{P_1}{P_2}$$

where:

T_1 = Initial temperature, absolute (T_1 + 459)

T_2 = Final temperature, absolute (T_2 + 459)

P_1 = Initial pressure, psia (P_1 + 14.7)

P_2 = Final pressure, psia (P_2 + 14.7)

$$\frac{(70 + 459)}{(40 + 459)} = \frac{(20 + 14.7)}{(P_2 + 14.7)}$$

$$\frac{529}{499} = \frac{34.7}{(P_2 + 14.7)}$$

$$(P_2 + 14.7) \times \frac{529}{499} = 34.7$$

$$(P_2 + 14.7) \times \frac{34.7}{1.060}$$

P_2 = 32.7 - 14.7

P_2 = 18 *psig*

Therefore, the gauge could be expected to register 18 psig (124 kPa) when the ambient temperature is 40°F (4°C).

A.7.6 Example 6: Pressure drop per 100 feet of pipe method. Using the layout shown in Figure A.7.1 and ΔH = pressure drop, in w.c. (27.7 in. H_2O = 1 psi), proceed as follows:

(1) Length to A = 20 feet, with 35,000 Btu/hr.

For $^1/_2$-inch pipe, $\Delta H = {}^{20\text{ feet}}/_{100\text{ feet}} \times 0.3$ inch w.c. = 0.06 in w.c.

(2) Length to B = 15 feet, with 75,000 Btu/hr.

For $^3/_4$-inch pipe, $\Delta H = {}^{15\text{ feet}}/_{100\text{ feet}} \times 0.3$ inch w.c. = 0.045 in w.c.

(3) Section 1 = 10 feet, with 110,000 Btu/hr. Here there is a choice:

For 1 inch pipe: $\Delta H = {}^{10\text{ feet}}/_{100\text{ feet}} \times 0.2$ inch w.c. = 0.02 in w.c.

For $^3/_4$-inch pipe: $\Delta H = {}^{10\text{ feet}}/_{100\text{ feet}} \times$ [0.5 inch w.c. + $^{(110{,}000\text{ Btu/hr}-104{,}000\text{ Btu/hr})}/_{(147{,}000\text{ Btu/hr}-104{,}000\text{ Btu/hr})} \times$ (1.0 inches w.c. - 0.5 inch w.c.)] = 0.1 × 0.57 inch w.c.≈ 0.06 inch w.c.

Note that the pressure drop between 104,000 Btu/hr and 147,000 Btu/hr has been interpolated as 110,000 Btu/hr.

(4) Section 2 = 20 feet, with 135,000 Btu/hr. Here there is a choice:

For 1-inch pipe: $\Delta H = {}^{20\text{ feet}}/_{100\text{ feet}} \times$ [0.2 inch w.c. + $^{(14{,}000\text{ Btu/hr})}/_{(27{,}000\text{ Btu/hr})} \times$ 0.1 inch w.c.] = 0.05 inch w.c.

For $^3/_4$-inch pipe: $\Delta H = {}^{20\text{ feet}}/_{100\text{ feet}} \times 1.0$ inch w.c. = 0.2 inch w.c.

Note that the pressure drop between 121,000 Btu/hr and 148,000 Btu/hr has been interpolated as 135,000 Btu/hr, but interpolation for the ¾-inch pipe (trivial for 104,000 Btu/hr to 147,000 Btu/hr) was not used.

(5) Section 3 = 30 feet, with 245,000 Btu/hr. Here there is a choice:

For 1-inch pipe: $\Delta H = {}^{30\text{ feet}}/_{100\text{ feet}} \times 1.0$ inches w.c. = 0.3 inch w.c.

For $1^1/_4$-inch pipe: $\Delta H = {}^{30\text{ feet}}/_{100\text{ feet}} \times 0.2$ inch w.c. = 0.06 inch w.c.

Note that interpolation for these options is ignored since the table values are close to the 245,000 Btu/hr carried by that section.

(6) The total pressure drop is the sum of the section approaching A, Sections 1 and 3, or either of the following, depending on whether an absolute minimum is needed or the larger drop can be accommodated.

Minimum pressure drop to farthest *appliance*:

ΔH = 0.06 inch w.c. + 0.02 inch w.c. + 0.06 inch w.c. = 0.14 inch w.c.

Larger pressure drop to the farthest *appliance*:

ΔH = 0.06 inch w.c. + 0.06 inch w.c. + 0.3 inch w.c. = 0.42 inch w.c.

Notice that Section 2 and the run to B do not enter into this calculation, provided that the appliances have similar input pressure requirements.

For SI units: 1 Btu/hr = 0.293 W, 1 cubic foot = 0.028 m^3, 1 foot = 0.305 m, 1 inch w.c. = 249 Pa.

APPENDIX B

SIZING OF VENTING SYSTEMS SERVING APPLIANCES EQUIPPED WITH DRAFT HOODS, CATEGORY I APPLIANCES, AND APPLIANCES LISTED FOR USE WITH TYPE B VENTS

(This appendix is informative and is not part of the *code*. This appendix is an excerpt from the 2012 *International Fuel Gas Code,* coordinated with the section numbering of the *International Residential Code.*)

EXAMPLES USING SINGLE APPLIANCE VENTING TABLES

Example 1: Single draft-hood-equipped appliance.

An installer has a 120,000 British thermal unit (Btu) per hour input *appliance* with a 5-inch-diameter draft hood outlet that needs to be vented into a 10-foot-high Type B vent system. What size vent should be used assuming (a) a 5-foot lateral single-wall metal vent connector is used with two 90-degree elbows, or (b) a 5-foot lateral single-wall metal vent connector is used with three 90-degree elbows in the vent system?

Solution:

Table 504.2(2) should be used to solve this problem, because single-wall metal vent connectors are being used with a Type B vent.

(a) Read down the first column in Table 504.2(2) until the row associated with a 10-foot height and 5-foot lateral is found. Read across this row until a vent capacity greater than 120,000 Btu per hour is located in the shaded columns *labeled* "NAT Max" for draft-hood-equipped appliances. In this case, a 5-inch-diameter vent has a capacity of 122,000 Btu per hour and may be used for this application.

(b) If three 90-degree elbows are used in the vent system, then the maximum vent capacity listed in the tables must be reduced by 10 percent (see Section 504.2.3 for single *appliance* vents). This implies that the 5-inch-diameter vent has an adjusted capacity of only 110,000 Btu per hour. In this case, the vent system must be increased to 6 inches in diameter (see calculations below).

122,000 (0.90) = 110,000 for 5-inch vent
From Table 504.2(2), Select 6-inch vent
186,000 (0.90) = 167,000; This is greater than the required 120,000. Therefore, use a 6-inch vent and connector where three elbows are used.

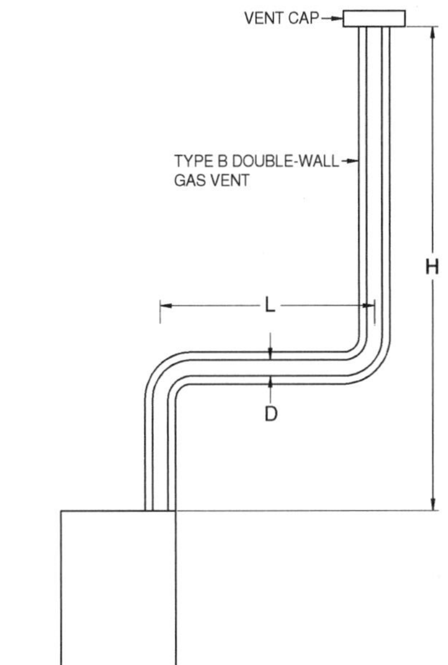

For SI: 1 foot = 304.8 mm, 1 British thermal unit per hour = 0.2931 W.

Table 504.2(1) is used when sizing Type B double-wall gas vent connected directly to the appliance.

Note: The appliance may be either Category I draft hood equipped or fan-assisted type.

FIGURE B-1
TYPE B DOUBLE-WALL VENT SYSTEM SERVING A SINGLE APPLIANCE WITH A TYPE B DOUBLE-WALL VENT

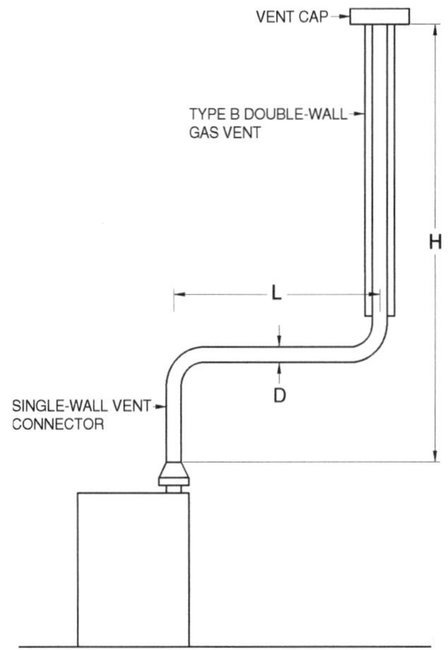

For SI: 1 foot = 304.8 mm, 1 British thermal unit per hour = 0.2931 W.

Table 504.2(2) is used when sizing a single-wall metal vent connector attached to a Type B double-wall gas vent.

Note: The appliance may be either Category I draft hood equipped or fan-assisted type.

FIGURE B-2
TYPE B DOUBLE-WALL VENT SYSTEM SERVING A SINGLE APPLIANCE WITH A SINGLE-WALL METAL VENT CONNECTOR

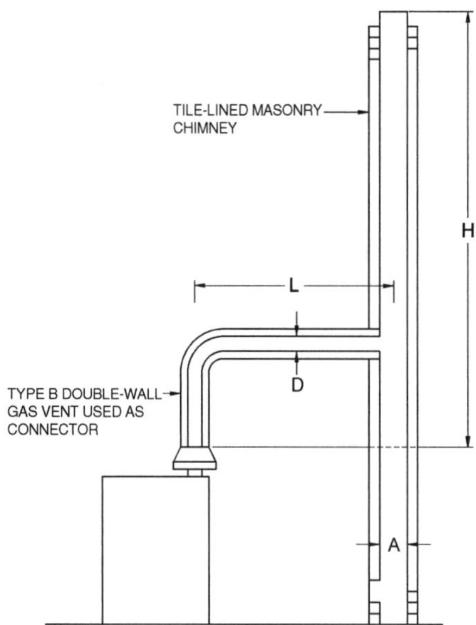

Table 504.2(3) is used when sizing a Type B double-wall gas vent connector attached to a tile-lined masonry chimney.

Note: "A" is the equivalent cross-sectional area of the tile liner.

Note: The appliance may be either Category I draft hood equipped or fan-assisted type.

FIGURE B-3
VENT SYSTEM SERVING A SINGLE APPLIANCE WITH A MASONRY CHIMNEY OF TYPE B DOUBLE-WALL VENT CONNECTOR

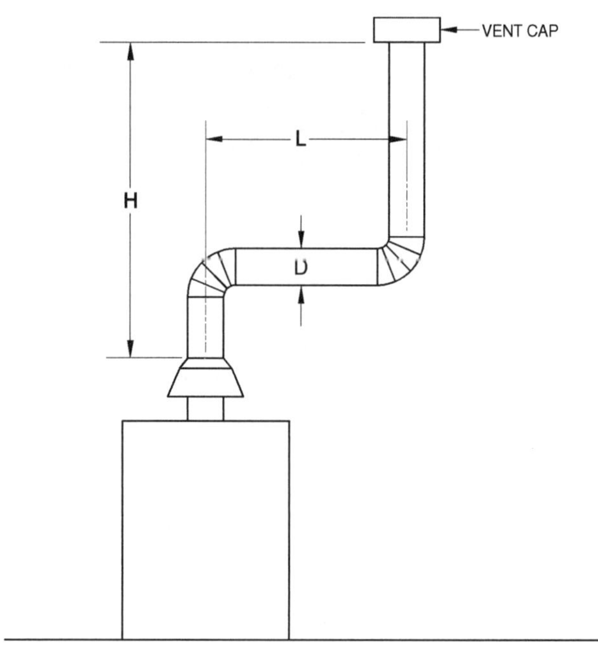

Asbestos cement Type B or single-wall metal vent serving a single draft-hood-equipped appliance [see Table 504.2(5)].

FIGURE B-5
ASBESTOS CEMENT TYPE B OR SINGLE-WALL METAL VENT SYSTEM SERVING A SINGLE DRAFT-HOOD-EQUIPPED APPLIANCE

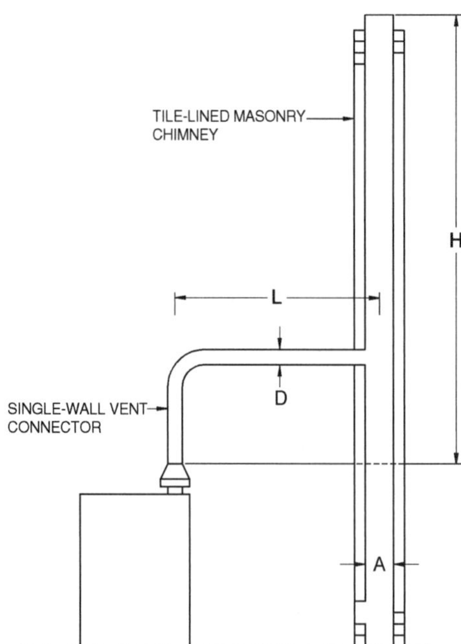

Table 504.2(4) is used when sizing a single-wall vent connector attached to a tile-lined masonry chimney.

Note: "A" is the equivalent cross-sectional area of the tile liner.

Note: The appliance may be either Category I draft hood equipped or fan-assisted type.

FIGURE B-4
VENT SYSTEM SERVING A SINGLE APPLIANCE USING A MASONRY CHIMNEY AND A SINGLE-WALL METAL VENT CONNECTOR

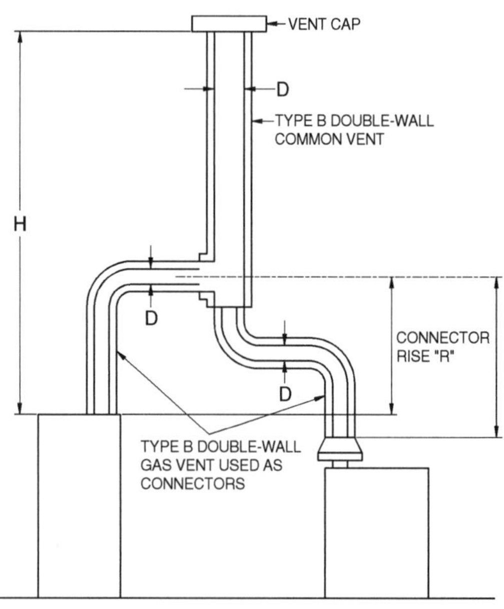

Table 504.3(1) is used when sizing Type B double-wall vent connectors attached to a Type B double-wall common vent.

Note: Each appliance may be either Category I draft hood equipped or fan-assisted type.

FIGURE B-6
VENT SYSTEM SERVING TWO OR MORE APPLIANCES WITH TYPE B DOUBLE-WALL VENT AND TYPE B DOUBLE-WALL VENT CONNECTOR

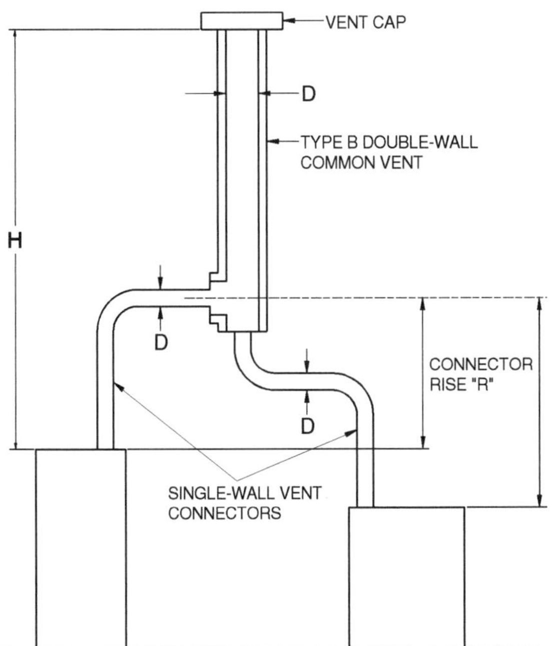

Table 504.3(2) is used when sizing single-wall vent connectors attached to a Type B double-wall common vent.

Note: Each appliance may be either Category I draft hood equipped or fan-assisted type.

FIGURE B-7
VENT SYSTEM SERVING TWO OR MORE APPLIANCES WITH TYPE B DOUBLE-WALL VENT AND SINGLE-WALL METAL VENT CONNECTORS

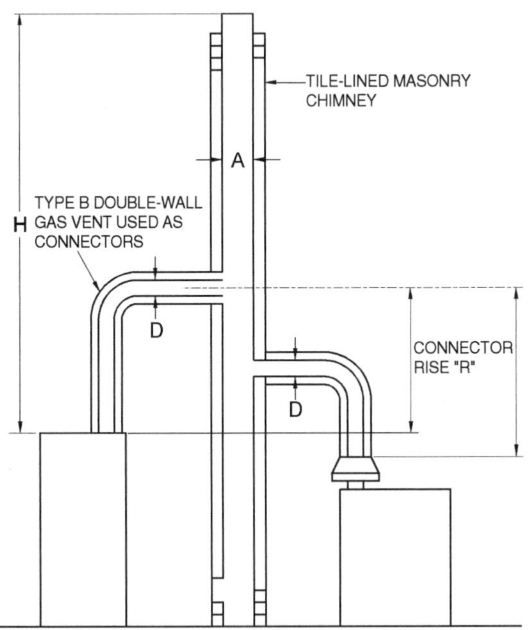

Table 504.3(3) is used when sizing Type B double-wall vent connectors attached to a tile-lined masonry chimney.

Note: "A" is the equivalent cross-sectional area of the tile liner.

Note: Each appliance may be either Category I draft hood equipped or fan-assisted type.

FIGURE B-8
MASONRY CHIMNEY SERVING TWO OR MORE APPLIANCES WITH TYPE B DOUBLE-WALL VENT CONNECTOR

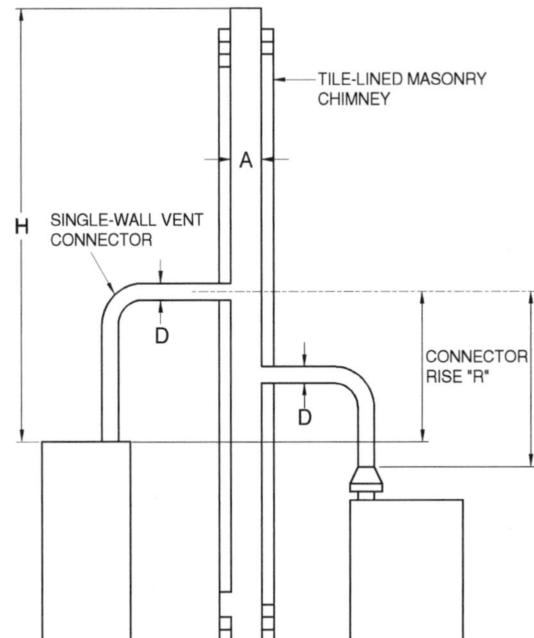

Table 504.3(4) is used when sizing single-wall metal vent connectors attached to a tile-lined masonry chimney.

Note: "A" is the equivalent cross-sectional area of the tile liner.

Note: Each appliance may be either Category I draft hood equipped or fan-assisted type.

FIGURE B-9
MASONRY CHIMNEY SERVING TWO OR MORE APPLIANCES WITH SINGLE-WALL METAL VENT CONNECTORS

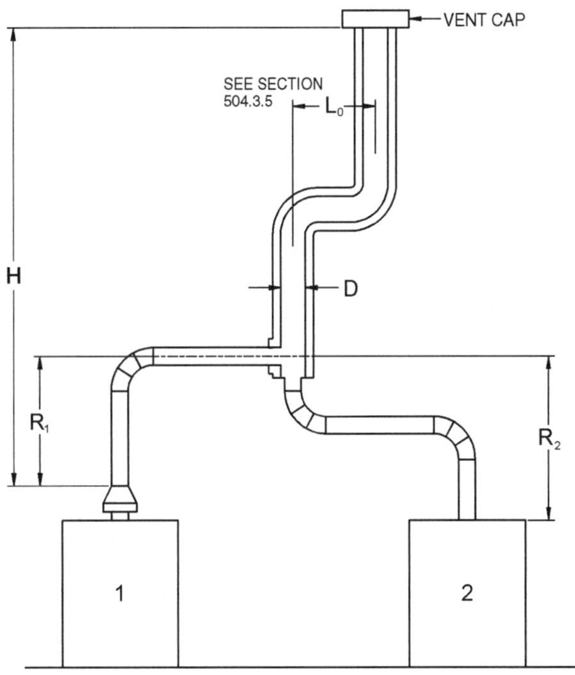

Asbestos cement Type B or single-wall metal pipe vent serving two or more draft-hood-equipped appliances [see Table 504.3(5)].

FIGURE B-10
ASBESTOS CEMENT TYPE B OR SINGLE-WALL METAL VENT SYSTEM SERVING TWO OR MORE DRAFT-HOOD-EQUIPPED APPLIANCES

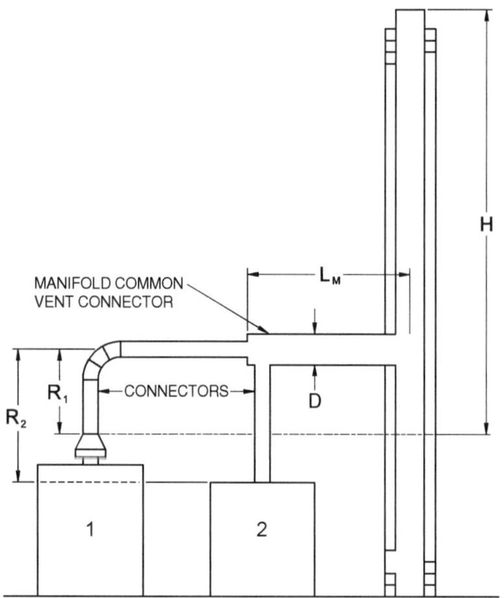

Example: Manifolded Common Vent Connector L_M shall be no greater than 18 times the common vent connector manifold inside diameter; i.e., a 4-inch (102 mm) inside diameter common vent connector manifold shall not exceed 72 inches (1829 mm) in length (see Section 504.3.4).

Note: This is an illustration of a typical manifolded vent connector. Different appliance, vent connector, or common vent types are possible. Consult Section 502.3.

FIGURE B-11
USE OF MANIFOLD COMMON VENT CONNECTOR

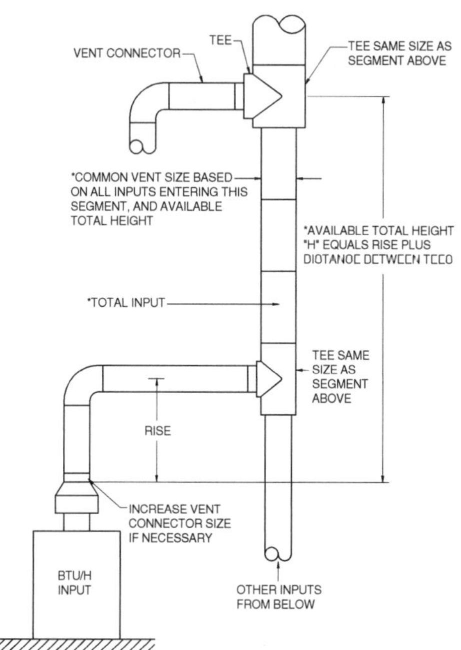

Vent connector size depends on:	Common vent size depends on:
• Input	• Combined inputs
• Rise	• Available total height "H"
• Available total height "H"	• Table 504.3(1) common vent
• Table 504.3(1) connectors	

FIGURE B-13
MULTISTORY GAS VENT DESIGN PROCEDURE FOR EACH SEGMENT OF SYSTEM

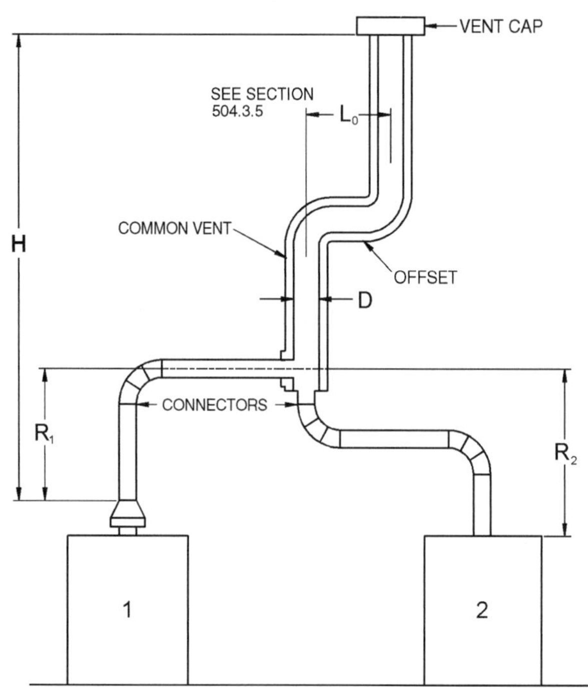

Example: Offset Common Vent

Note: This is an illustration of a typical offset vent. Different appliance, vent connector, or vent types are possible. Consult Sections 504.2 and 504.3.

FIGURE B-12
USE OF OFFSET COMMON VENT

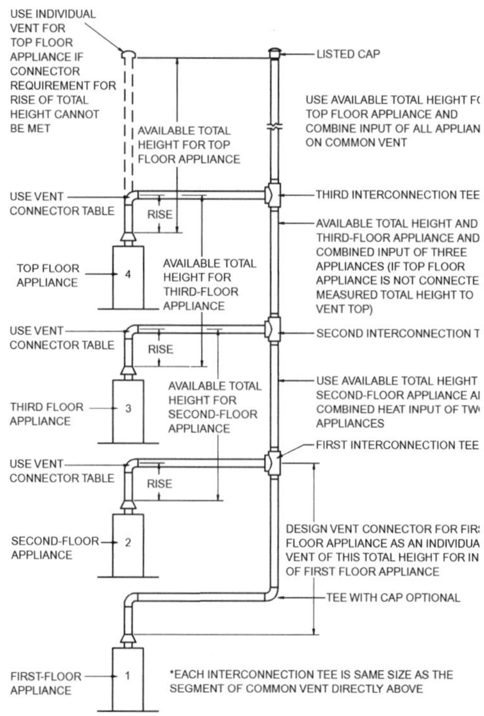

Principles of design of multistory vents using vent connector and common vent design tables (see Sections 504.3.11 through 504.3.17).

FIGURE B-14
MULTISTORY VENT SYSTEMS

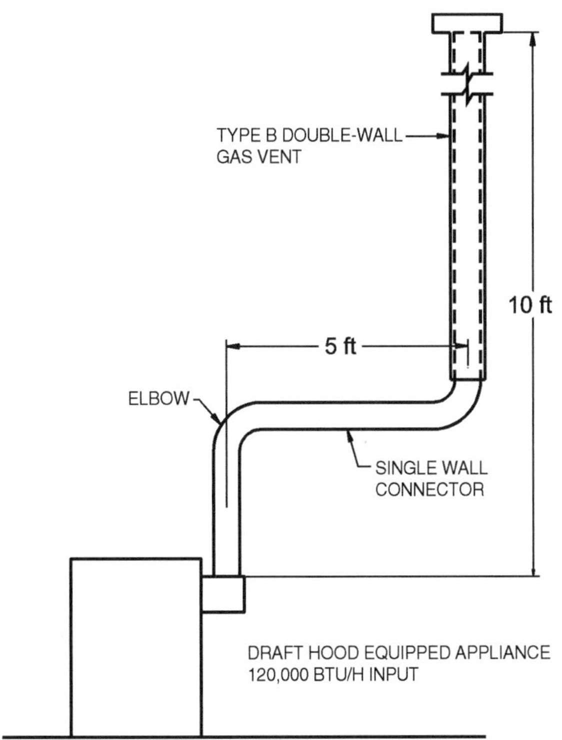

For SI: 1 foot = 304.8 mm, 1 British thermal unit per hour = 0.2931 W.

FIGURE B-15 (EXAMPLE 1)
SINGLE DRAFT-HOOD-EQUIPPED APPLIANCE

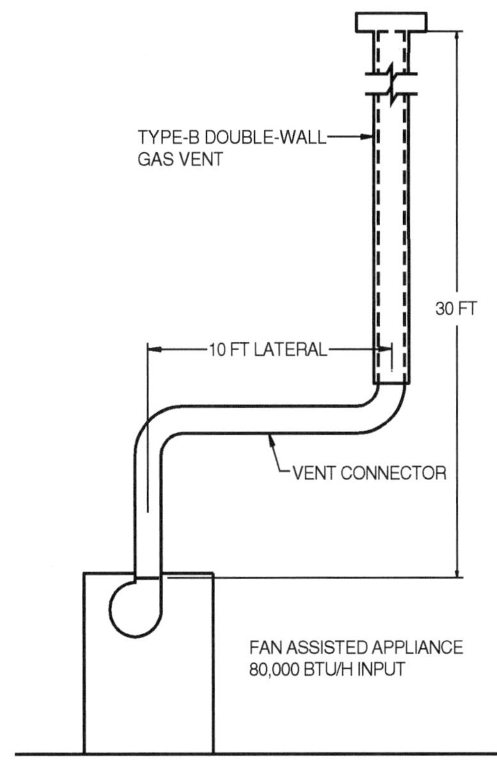

For SI: 1 foot = 304.8 mm, 1 British thermal unit per hour = 0.2931 W.

FIGURE B-16 (EXAMPLE 2)
SINGLE FAN-ASSISTED APPLIANCE

Example 2: Single fan-assisted appliance.

An installer has an 80,000 Btu per hour input fan-assisted *appliance* that must be installed using 10 feet of lateral connector attached to a 30-foot-high Type B vent. Two 90-degree elbows are needed for the installation. Can a single-wall metal vent connector be used for this application?

Solution:

Table 504.2(2) refers to the use of single-wall metal vent connectors with Type B vent. In the first column find the row associated with a 30-foot height and a 10-foot lateral. Read across this row, looking at the FAN Min and FAN Max columns, to find that a 3-inch-diameter single-wall metal vent connector is not recommended. Moving to the next larger size single wall connector (4 inches), note that a 4-inch-diameter single-wall metal connector has a recommended minimum vent capacity of 91,000 Btu per hour and a recommended maximum vent capacity of 144,000 Btu per hour. The 80,000 Btu per hour fan-assisted *appliance* is outside this range, so the conclusion is that a single-wall metal vent connector cannot be used to vent this *appliance* using 10 feet of lateral for the connector.

However, if the 80,000 Btu per hour input *appliance* could be moved to within 5 feet of the vertical vent, then a 4-inch single-wall metal connector could be used to vent the *appliance*. Table 504.2(2) shows the acceptable range of vent capacities for a 4-inch vent with 5 feet of lateral to be between 72,000 Btu per hour and 157,000 Btu per hour.

If the *appliance* cannot be moved closer to the vertical vent, then Type B vent could be used as the connector material. In this case, Table 504.2(1) shows that for a 30-foot-high vent with 10 feet of lateral, the acceptable range of vent capacities for a 4-inch-diameter vent attached to a fan-assisted *appliance* is between 37,000 Btu per hour and 150,000 Btu per hour.

Example 3: Interpolating between table values.

An installer has an 80,000 Btu per hour input *appliance* with a 4-inch-diameter draft hood outlet that needs to be vented into a 12-foot-high Type B vent. The vent connector has a 5-foot lateral length and is also Type B. Can this *appliance* be vented using a 4-inch-diameter vent?

Solution:

Table 504.2(1) is used in the case of an all Type B vent system. However, since there is no entry in Table 504.2(1) for a height of 12 feet, interpolation must be used. Read down the 4-inch diameter NAT Max column to the row associated with 10-foot height and 5-foot lateral to find the capacity value of 77,000 Btu per hour. Read further down to the 15-foot height, 5-foot lateral row to find the capacity value of 87,000 Btu per hour. The difference between the 15-foot height capacity value and the 10-foot height capacity value is 10,000 Btu per hour. The capacity for a vent system with a 12-foot height is equal to the capacity for a 10-foot height plus $^2/_5$ of the difference between the 10-foot and 15-foot height values, or $77{,}000 + {}^2/_5\,(10{,}000) = 81{,}000$ Btu per hour. Therefore, a 4-inch-diameter vent may be used in the installation.

EXAMPLES USING COMMON VENTING TABLES

Example 4: Common venting two draft-hood-equipped appliances.

A 35,000 Btu per hour water heater is to be common vented with a 150,000 Btu per hour furnace using a common vent with a total height of 30 feet. The connector rise is 2 feet for the water heater with a horizontal length of 4 feet. The connector rise for the furnace is 3 feet with a horizontal length of 8 feet. Assume single-wall metal connectors will be used with Type B vent. What size connectors and combined vent should be used in this installation?

Solution:

Table 504.3(2) should be used to size single-wall metal vent connectors attached to Type B vertical vents. In the vent connector capacity portion of Table 504.3(2), find the row associated with a 30-foot vent height. For a 2-foot rise on the vent connector for the water heater, read the shaded columns for draft-hood-equipped appliances to find that a 3-inch-diameter vent connector has a capacity of 37,000 Btu per hour. Therefore, a 3-inch single-wall metal vent connector may be used with the water heater. For a draft-hood-equipped furnace with a 3-foot rise, read across the appropriate row to find that a 5-inch-diameter vent connector has a maximum capacity of 120,000 Btu per hour (which is too small for the furnace) and a 6-inch-diameter vent connector has a maximum vent capacity of 172,000 Btu per hour. Therefore, a 6-inch-diameter vent connector should be used with the 150,000 Btu per hour furnace. Since both vent connector horizontal lengths are less than the maximum lengths *listed* in Section 504.3.2, the table values may be used without adjustments.

In the common vent capacity portion of Table 504.3(2), find the row associated with a 30-foot vent height and read over to the NAT + NAT portion of the 6-inch-diameter column to find a maximum combined capacity of 257,000 Btu per hour. Since the two appliances total only 185,000 Btu per hour, a 6-inch common vent may be used.

Example 5a: Common venting a draft-hood-equipped water heater with a fan-assisted furnace into a Type B vent.

In this case, a 35,000 Btu per hour input draft-hood-equipped water heater with a 4-inch-diameter draft hood *outlet*, 2 feet of connector rise, and 4 feet of horizontal length is to be common vented with a 100,000 Btu per hour fan-assisted furnace with a 4-inch-diameter flue collar, 3 feet of connector rise, and 6 feet of horizontal length. The common vent consists of a 30-foot height of Type B vent. What are the recommended vent diameters for each connector and the common vent? The installer would like to use a single-wall metal vent connector.

Solution: [Table 504.3(2)].

Water Heater Vent Connector Diameter. Since the water heater vent connector horizontal length of 4 feet is less than the maximum value listed in Section 504.3.2, the venting table values may be used without adjustments. Using the Vent Connector Capacity portion of Table 504.3(2), read down the Total Vent Height (*H*) column to 30 feet and read across the 2-foot Connector Rise (*R*) row to the first Btu per hour rating in the NAT Max column that is equal to or greater than the water heater input rating. The table shows that a 3-inch vent connector has a maximum input rating of 37,000

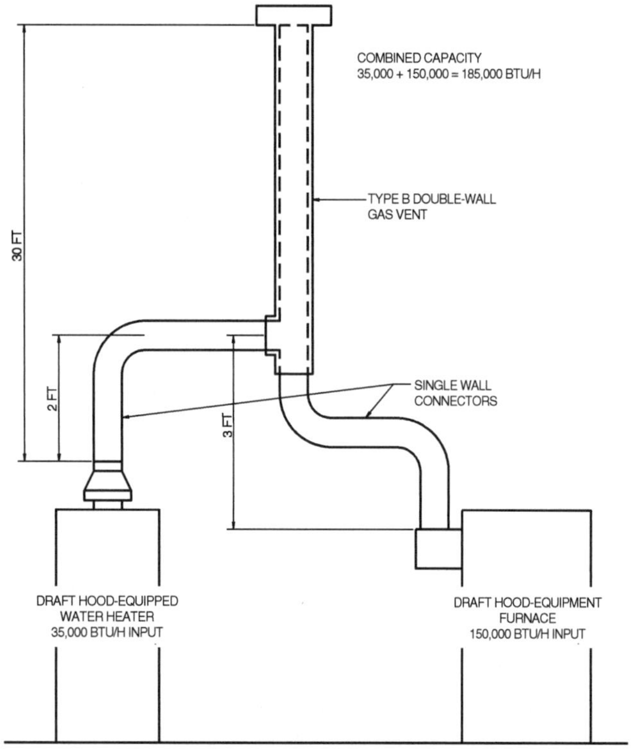

COMBINED CAPACITY
35,000 + 150,000 = 185,000 BTU/H
TYPE B DOUBLE-WALL GAS VENT
30 FT
2 FT
3 FT
SINGLE WALL CONNECTORS
DRAFT HOOD-EQUIPPED WATER HEATER 35,000 BTU/H INPUT
DRAFT HOOD-EQUIPMENT FURNACE 150,000 BTU/H INPUT

FIGURE B-17 (EXAMPLE 4)
COMMON VENTING TWO DRAFT-HOOD-EQUIPPED APPLIANCES

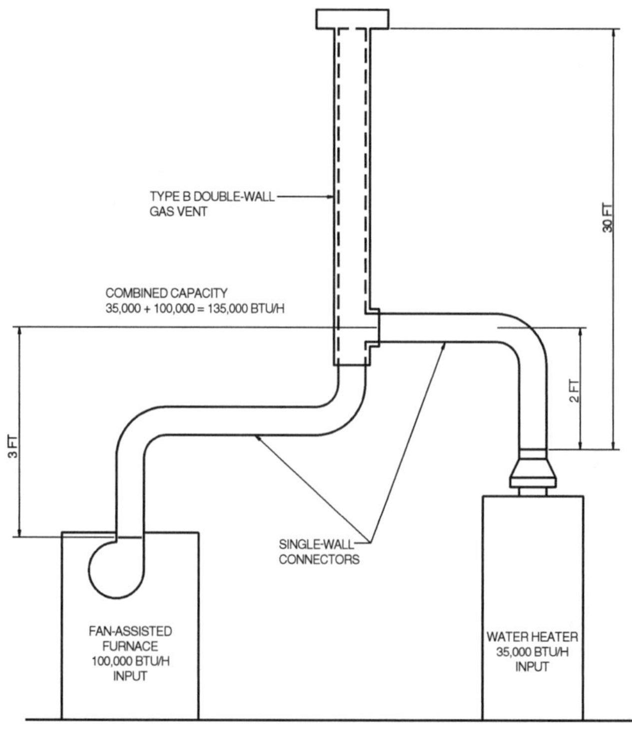

FIGURE B-18 (EXAMPLE 5A)
COMMON VENTING A DRAFT HOOD WITH A FAN-ASSISTED FURNACE INTO A TYPE B DOUBLE-WALL COMMON VENT

Btu per hour. Although this is greater than the water heater input rating, a 3-inch vent connector is prohibited by Section 504.3.21. A 4-inch vent connector has a maximum input rating of 67,000 Btu per hour and is equal to the draft hood *outlet* diameter. A 4-inch vent connector is selected. Since the water heater is equipped with a draft hood, there are no minimum input rating restrictions.

Furnace Vent Connector Diameter. Using the Vent Connector Capacity portion of Table 504.3(2), read down the Total Vent Height (*H*) column to 30 feet and across the 3-foot Connector Rise (*R*) row. Since the furnace has a fan-assisted combustion system, find the first FAN Max column with a Btu per hour rating greater than the furnace input rating. The 4-inch vent connector has a maximum input rating of 119,000 Btu per hour and a minimum input rating of 85,000 Btu per hour. The 100,000 Btu per hour furnace in this example falls within this range, so a 4-inch connector is adequate. Since the furnace vent connector horizontal length of 6 feet does not exceed the maximum value listed in Section 504.3.2, the venting table values may be used without adjustment. If the furnace had an input rating of 80,000 Btu per hour, then a Type B vent connector [see Table 504.3(1)] would be needed in order to meet the minimum capacity limit.

Common Vent Diameter. The total input to the common vent is 135,000 Btu per hour. Using the Common Vent Capacity portion of Table 504.3(2), read down the Total Vent Height (*H*) column to 30 feet and across this row to find the smallest vent diameter in the FAN + NAT column that has a Btu per hour rating equal to or greater than 135,000 Btu per hour. The 4-inch common vent has a capacity of 132,000 Btu per hour and the 5-inch common vent has a capacity of 202,000 Btu per hour. Therefore, the 5-inch common vent should be used in this example.

Summary. In this example, the installer may use a 4-inch-diameter, single-wall metal vent connector for the water heater and a 4-inch-diameter, single-wall metal vent connector for the furnace. The common vent should be a 5-inch-diameter Type B vent.

Example 5b: Common venting into a masonry chimney.

In this case, the water heater and fan-assisted furnace of Example 5a are to be common vented into a clay tile-lined masonry chimney with a 30-foot height. The chimney is not exposed to the outdoors below the roof line. The internal dimensions of the clay tile liner are nominally 8 inches by 12 inches. Assuming the same vent connector heights, laterals, and materials found in Example 5a, what are the recommended vent connector diameters, and is this an acceptable installation?

Solution:

Table 504.3(4) is used to size common venting installations involving single-wall connectors into masonry chimneys.

Water Heater Vent Connector Diameter. Using Table 504.3(4), Vent Connector Capacity, read down the Total Vent Height (*H*) column to 30 feet, and read across the 2-foot Connector Rise (*R*) row to the first Btu per hour rating in the NAT Max column that is equal to or greater than the water heater input rating. The table shows that a 3-inch vent connector has a maximum input of only 31,000 Btu per hour while a 4-inch vent connector has a maximum input of 57,000 Btu per hour. A 4-inch vent connector must therefore be used.

Furnace Vent Connector Diameter. Using the Vent Connector Capacity portion of Table 504.3(4), read down the Total Vent Height (*H*) column to 30 feet and across the 3-foot Connector Rise (*R*) row. Since the furnace has a fan-assisted combustion system, find the first FAN Max column with a Btu per hour rating greater than the furnace input rating. The 4-inch vent connector has a maximum input rating of 127,000 Btu per hour and a minimum input rating of 95,000 Btu per hour. The 100,000 Btu per hour furnace in this example falls within this range, so a 4-inch connector is adequate.

Masonry Chimney. From Table B-1, the equivalent area for a nominal liner size of 8 inches by 12 inches is 63.6 square inches. Using Table 504.3(4), Common Vent Capacity, read down the FAN + NAT column under the Minimum Internal Area of Chimney value of 63 to the row for 30-foot height to find a capacity value of 739,000 Btu per hour. The combined input rating of the furnace and water heater, 135,000 Btu per hour, is less than the table value, so this is an acceptable installation.

Section 504.3.17 requires the common vent area to be no greater than seven times the smallest *listed appliance* categorized vent area, flue collar area, or draft hood outlet area. Both appliances in this installation have 4-inch-diameter outlets. From Table B-1, the equivalent area for an inside diameter of 4 inches is 12.2 square inches. Seven times 12.2 equals 85.4, which is greater than 63.6, so this configuration is acceptable.

Example 5c: Common venting into an exterior masonry chimney.

In this case, the water heater and fan-assisted furnace of Examples 5a and 5b are to be common vented into an exterior masonry chimney. The chimney height, clay tile liner dimensions, and vent connector heights and laterals are the same as in Example 5b. This system is being installed in Charlotte, North Carolina. Does this exterior masonry chimney need to be relined? If so, what corrugated metallic liner size is recommended? What vent connector diameters are recommended?

Solution:

According to Section 504.3.20, Type B vent connectors are required to be used with exterior masonry chimneys. Use Tables 504.3(7a), (7b) to size FAN+NAT common venting installations involving Type-B double wall connectors into exterior masonry chimneys.

The local 99-percent winter design temperature needed to use Table 504.3(7b) can be found in the ASHRAE *Handbook of Fundamentals*. For Charlotte, North Carolina, this design temperature is 19°F.

Chimney Liner Requirement. As in Example 5b, use the 63 square inch Internal Area columns for this size clay tile liner. Read down the 63 square inch column of Table 504.3(7a) to the 30-foot height row to find that the combined *appliance* maximum input is 747,000 Btu per hour. The combined input rating of the appliances in this installation, 135,000 Btu per hour, is less than the maximum value, so this criterion is satisfied. Table 504.3(7b), at a 19°F design temperature, and at the same vent height and internal area used above, shows that the minimum allowable input rating of a space-heating appliance is 470,000 Btu per hour. The furnace input rating of 100,000 Btu per hour is less than this minimum value. So this criterion is not satisfied, and an alternative venting design needs to be used, such as a Type B vent shown in Example 5a or a *listed* chimney liner system shown in the remainder of the example.

According to Section 504.3.19, Table 504.3(1) or 504.3(2) is used for sizing corrugated metallic liners in masonry chimneys, with the maximum common vent capacities reduced by 20 percent. This example will be continued assuming Type B vent connectors.

Water Heater Vent Connector Diameter. Using Table 504.3(1), Vent Connector Capacity, read down the Total Vent Height (*H*) column to 30 feet, and read across the 2-foot Connector Rise (*R*) row to the first Btu/h rating in the NAT Max column that is equal to or greater than the water heater input rating. The table shows that a 3-inch vent connector has a maximum capacity of 39,000 Btu/h. Although this rating is greater than the water heater input rating, a 3-inch vent connector is prohibited by Section 504.3.21. A 4-inch vent connector has a maximum input rating of 70,000 Btu/h and is equal to the draft hood outlet diameter. A 4-inch vent connector is selected.

Furnace Vent Connector Diameter. Using Table 504.3(1), Vent Connector Capacity, read down the Vent Height (*H*) column to 30 feet, and read across the 3-foot Connector Rise (*R*) row to the first Btu per hour rating in the FAN Max column that is equal to or greater than the furnace input rating. The 100,000 Btu per hour furnace in this example falls within this range, so a 4-inch connector is adequate.

Chimney Liner Diameter. The total input to the common vent is 135,000 Btu per hour. Using the Common Vent Capacity Portion of Table 504.3(1), read down the Vent Height (*H*) column to 30 feet and across this row to find the smallest vent diameter in the FAN+NAT column that has a Btu per hour rating greater than 135,000 Btu per hour. The 4-inch common vent has a capacity of 138,000 Btu per hour. Reducing the maximum capacity by 20 percent (Section 504.3.19) results in a maximum capacity for a 4-inch corrugated liner of 110,000 Btu per hour, less than the total input of 135,000 Btu per hour. So a larger liner is needed. The 5-inch common vent capacity *listed* in Table 504.3(1) is 210,000 Btu per hour, and after reducing by 20 percent is 168,000 Btu per hour. Therefore, a 5-inch corrugated metal liner should be used in this example.

Single-Wall Connectors. Once it has been established that relining the chimney is necessary, Type B double-wall vent connectors are not specifically required. This example could be redone using Table 504.3(2) for single-wall vent connectors. For this case, the vent connector and liner diameters would be the same as found above with Type B double-wall connectors.

TABLE B-1
MASONRY CHIMNEY LINER DIMENSIONS WITH CIRCULAR EQUIVALENTS[a]

NOMINAL LINER SIZE (inches)	INSIDE DIMENSIONS OF LINER (inches)	INSIDE DIAMETER OR EQUIVALENT DIAMETER (inches)	EQUIVALENT AREA (square inches)
4 × 8	$2^1/_2 \times 6^1/_2$	4	12.2
		5	19.6
		6	28.3
		7	38.3
8 × 8	$6^3/_4 \times 6^3/_4$	7.4	42.7
		8	50.3
8 × 12	$6^1/_2 \times 10^1/_2$	9	63.6
		10	78.5
12 × 12	$9^3/_4 \times 9\ ^3/_4$	10.4	83.3
		11	95
12 × 16	$9^1/_2 \times 13^1/_2$	11.8	107.5
		12	113.0
		14	153.9
16 × 16	$13^1/_4 \times 13^1/_4$	14.5	162.9
		15	176.7
16 × 20	13 × 17	16.2	206.1
		18	254.4
20 × 20	$16^3/_4 \times 16^3/_4$	18.2	260.2
		20	314.1
20 × 24	$16^1/_2 \times 20^1/_2$	20.1	314.2
		22 ×	380.1
24 × 24	$20^1/_4 \times 20^1/_4$	22.1	380.1
		24	452.3
24 × 28	$20^1/_4 \times 20^1/_4$	24.1	456.2
28 × 28	$24^1/_4 \times 24^1/_4$	26.4	543.3
		27	572.5
30 × 30	$25^1/_2 \times 25^1/_2$	27.9	607
		30	706.8
30 × 36	$25^1/_2 \times 31^1/_2$	30.9	749.9
		33	855.3
36 × 36	$31^1/_2 \times 31^1/_2$	34.4	929.4
		36	1017.9

For SI: 1 inch = 25.4 mm, 1 square inch = 645.16 m^2.

a. Where liner sizes differ dimensionally from those shown in Table B-1, equivalent diameters may be determined from published tables for square and rectangular ducts of equivalent carrying capacity or by other engineering methods.

5°F
(-15°C)
-10°F
(-23°C)
17°F
(-8°C)
27°F
(-3°C)
37°F
(3°C)
-10°F
(-23°C)
-10°F
(-23°C)
5°F
(-15°C)
17°F
(-8°C)
27°F
(-3°C)
37°F
(3°C)

FIGURE B-19

APPENDIX C

EXIT TERMINALS OF MECHANICAL DRAFT AND DIRECT-VENT VENTING SYSTEMS

(This appendix is informative and is not part of the code. This appendix is an excerpt from the 2012 *International Fuel Gas Code,* coordinated with the section numbering of the *International Residential Code.*)

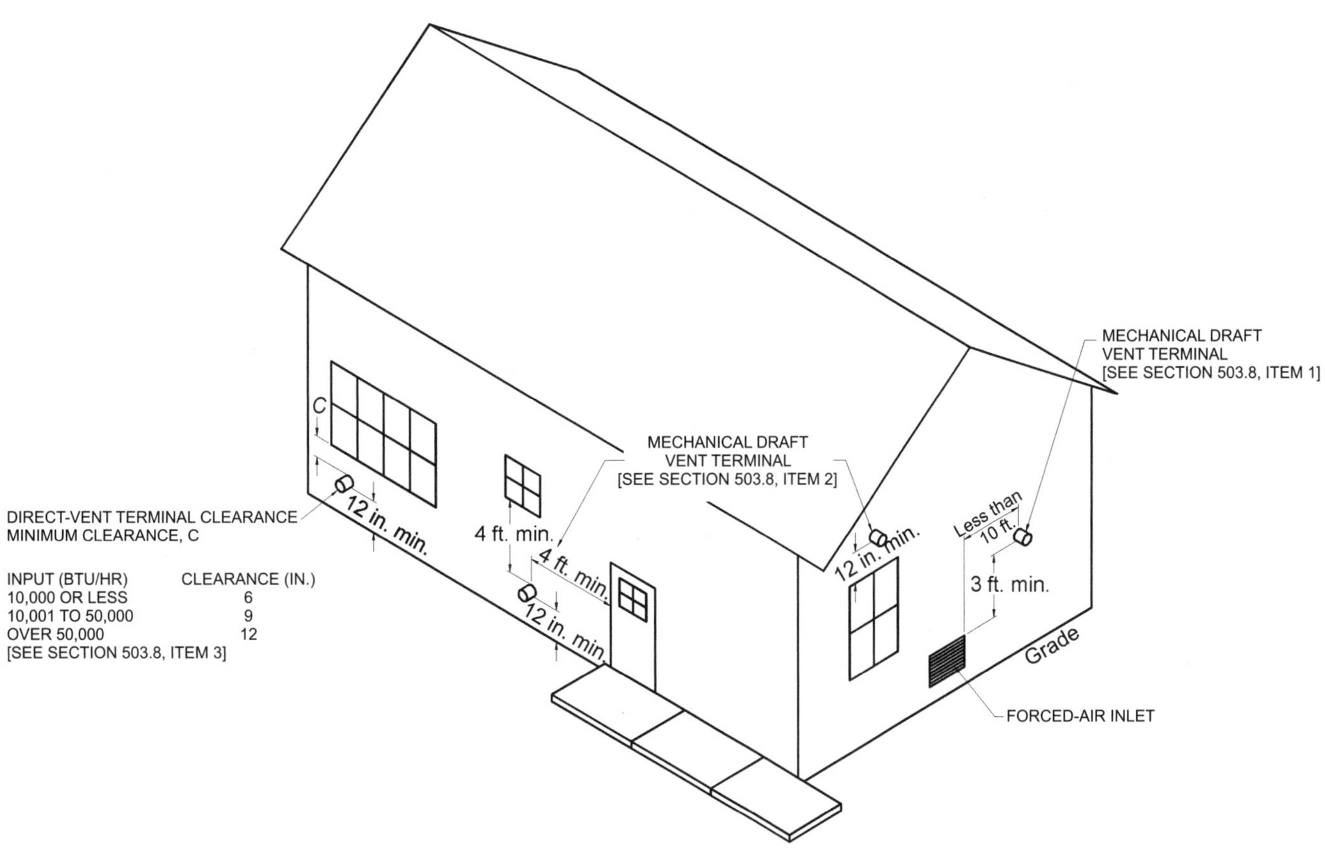

For SI: 1 inch = 25.4 mm, 1 foot = 304.8 mm, 1 British thermal unit per hour = 0.2931 W.

APPENDIX C
EXIT TERMINALS OF MECHANICAL DRAFT AND DIRECT-VENT VENTING SYSTEMS

APPENDIX D

RECOMMENDED PROCEDURE FOR SAFETY INSPECTION OF AN EXISTING APPLIANCE INSTALLATION

(This appendix is informative and is not part of the *code*. This appendix is an excerpt from the 2012 *International Fuel Gas Code,* coordinated with the section numbering of the *International Residential Code.*)

The following procedure is intended as a guide to aid in determining that an *appliance* is properly installed and is in a safe condition for continuing use.

This procedure is intended for cental furnace and boiler installations and may not be applicable to all installations.

(a) This procedure should be performed prior to any attempt at modification of the *appliance* or of the installation.

(b) If it is determined that there is a condition that could result in unsafe operation, shut off the appliance and advise the owner of the unsafe condition. The following steps should be followed in making the safety inspection:

1 Conduct a check for gas leakage. (See Section 406.6)

2. Visually inspect the venting system for proper size and horizontal pitch and determine there is no blockage or restriction, leakage, corrosion and other deficiencies that could cause an unsafe condition.

3 Shut off all gas to the *appliance* and shut off any other fuel-gas-burning *appliance* within the same room. **Use the shutoff valve in the supply line to each appliance.**

4. Inspect burners and crossovers for blockage and corrosion.

5. **Furnace installations:** Inspect the heat exchanger for cracks, openings or excessive corrosion.

6. **Boiler installations:** Inspect for evidence of water or combustion product leaks.

7. Close all building doors and windows and all doors between the space in which the *appliance* is located and other spaces of the building that can be closed. Turn on any clothes dryers. Turn on any exhaust fans, such as range hoods and bathroom exhausts, so they will operate at maximum speed. Do not operate a summer exhaust fan. Close *fireplace* dampers. If, after completing Steps 8 through 13, it is believed sufficient *combustion air* is not available, refer to Section 304 of this code.

8. Place the *appliance* being inspected in operation. **Follow the lighting instructions.** Adjust the thermostat so *appliance* will operate continuously.

9. Determine that the pilot, where provided, is burning properly and that the main burner ignition is satisfactory by interrupting and reestablishing the electrical supply to the *appliance* in any convenient manner. If the *appliance* is equipped with a continuous pilot, test all pilot safety devices to determine if they are operating properly by extinguishing the pilot when the main burner is off and determining, after 3 minutes, that the main burner gas does not flow upon a call for heat. If the *appliance* is not provided with a pilot, test for proper operation of the ignition system in accordance with the *appliance* manufacturer's lighting and operating instructions.

10. Visually determine that the main burner gas is burning properly (i.e., no floating, lifting or flashback). Adjust the primary air shutters as required. If the *appliance* is equipped with high and low flame controlling or flame modulation, check for proper main burner operation at low flame.

11. Test for spillage at the draft hood relief opening after 5 minutes of main burner operation. Use the flame of a match or candle or smoke.

12. Turn on all other fuel-gas-burning appliances within the same room so they will operate at their full inputs. **Follow lighting instructions for each appliance.**

13. Repeat Steps 10 and 11 on the *appliance* being inspected.

14. Return doors, windows, exhaust fans, *fireplace* dampers and any other fuel-gas-burning *appliance* to their previous conditions of use.

15. **Furnace installations:** Check both the limit control and the fan control for proper operation. Limit control operation can be checked by blocking the circulating air inlet or temporarily disconnecting the electrical supply to the blower motor and determining that the limit control acts to shut off the main burner gas.

16. **Boiler installations:** Verify that the water pumps are in operating condition. Test low water cutoffs, automatic feed controls, pressure and temperature limit controls and relief valves in accordance with the manufacturer's recommendations to determine that they are in operating condition.

APPENDIX E

MANUFACTURED HOUSING USED AS DWELLINGS

(The provisions contained in this appendix are not mandatory unless specifically referenced in the adopting ordinance.)

SECTION AE101 SCOPE

AE101.1 General. These provisions shall be applicable only to a *manufactured home* used as a single *dwelling unit* installed on privately owned (nonrental) lots and shall apply to the following:

1. Construction, *alteration* and repair of any foundation system which is necessary to provide for the installation of a *manufactured home* unit.
2. Construction, installation, *addition*, *alteration*, repair or maintenance of the building service *equipment* which is necessary for connecting *manufactured homes* to water, fuel, or power supplies and sewage systems.
3. *Alterations*, *additions* or repairs to existing *manufactured homes*. The construction, *alteration*, moving, demolition, repair and use of accessory buildings and structures, and their building service *equipment*, shall comply with the requirements of the codes adopted by this *jurisdiction*.

These provisions shall not be applicable to the design and construction of *manufactured homes* and shall not be deemed to authorize either modifications or *additions* to *manufactured homes* where otherwise prohibited.

Exception: In addition to these provisions, new and replacement *manufactured homes* to be located in flood hazard areas as established in Table R301.2(1) of the *International Residential Code* shall meet the applicable requirements of Section R322 of the *International Residential Code*.

SECTION AE102 APPLICATION TO EXISTING MANUFACTURED HOMES AND BUILDING SERVICE EQUIPMENT

AE102.1 General. *Manufactured homes* and their building service *equipment* to which *additions*, *alterations* or repairs are made shall comply with all the requirements of these provisions for new facilities, except as specifically provided in this section.

AE102.2 Additions, alterations or repairs. *Additions* made to a *manufactured home* shall conform to one of the following:

1. Be certified under the National Manufactured Housing Construction and Safety Standards Act of 1974 (42 U.S.C. Section 5401, et seq.).
2. Be designed and constructed to comply with the applicable provisions of the National Manufactured Housing Construction and Safety Standards Act of 1974 (42 U.S.C. Section 5401, et seq.).
3. Be designed and constructed in compliance with the code adopted by this *jurisdiction*.

Additions shall be structurally separated from the *manufactured home*.

Exception: A structural separation need not be provided when structural calculations are provided to justify the omission of such separation.

Alterations or repairs may be made to any *manufactured home* or to its building service *equipment* without requiring the existing *manufactured home* or its building service *equipment* to comply with all the requirements of these provisions, provided the *alteration* or repair conforms to that required for new construction, and provided further that no hazard to life, health or safety will be created by such *additions*, *alterations* or repairs.

Alterations or repairs to an existing *manufactured home*, which are nonstructural and do not adversely affect any structural member or any part of the building or structure having required fire protection, may be made with materials equivalent to those of which the *manufactured home* structure is constructed, subject to approval by the *building official*.

Exception: The installation or replacement of glass shall be required for new installations.

Minor *additions*, *alterations* and repairs to existing building service *equipment* installations may be made in accordance with the codes in effect at the time the original installation was made, subject to the approval of the *building official*, and provided such *additions*, *alterations* and repairs will not cause the existing building service *equipment* to become unsafe, insanitary or overloaded.

AE102.3 Existing installations. Building service *equipment* lawfully in existence at the time of the adoption of the applicable codes may have their use, maintenance or repair continued if the use, maintenance or repair is in accordance with the original design and no hazard to life, health or property has been created by such building service *equipment*.

AE102.4 Existing occupancy. *Manufactured homes*, which are in existence at the time of the adoption of these provisions, may have their existing use or occupancy continued if such use or occupancy was legal at the time of the adoption of these provisions, provided such continued use is not dangerous to life, health and safety.

The use or occupancy of any existing *manufactured home* shall not be changed unless evidence satisfactory to the *building official* is provided to show compliance with all applicable provisions of the codes adopted by this *jurisdiction*. Upon any change in use or occupancy, the *manufactured home* shall cease to be classified as such within the intent of these provisions.

AE102.5 Maintenance. All *manufactured homes* and their building service *equipment*, existing and new, and all parts thereof, shall be maintained in a safe and sanitary condition. All devices or safeguards which are required by applicable codes or by the *Manufactured Home* Standards shall be maintained in conformance to the code or standard under which it was installed. The owner or the owner's designated agent shall be responsible for the maintenance of *manufactured homes*, accessory buildings, structures and their building service *equipment*. To determine compliance with this section, the *building official* may cause any *manufactured home*, accessory building or structure to be reinspected.

AE102.6 Relocation. *Manufactured homes* which are to be relocated within this *jurisdiction* shall comply with these provisions.

SECTION AE201 DEFINITIONS

AE201.1 General. For the purpose of these provisions, certain abbreviations, terms, phrases, words and their derivatives shall be construed as defined or specified herein.

ACCESSORY BUILDING. Any building or structure, or portion thereto, located on the same property as a *manufactured home* which does not qualify as a *manufactured home* as defined herein.

BUILDING SERVICE EQUIPMENT. Refers to the plumbing, mechanical and electrical *equipment*, including piping, wiring, fixtures and other accessories which provide sanitation, lighting, heating, ventilation, cooling, fire protection and facilities essential for the habitable occupancy of a *manufactured home* or accessory building or structure for its designated use and occupancy.

MANUFACTURED HOME. A structure transportable in one or more sections which, in the traveling mode, is 8 body feet (2438 body mm) or more in width or 40 body feet (12 192 body mm) or more in length or, when erected on site, is 320 or more square feet (30 m^2), and which is built on a permanent chassis and designed to be used as a *dwelling* with or without a permanent foundation when connected to the required utilities, and includes the plumbing, heating, air-conditioning and electrical systems contained therein; except that such term shall include any structure which meets all the requirements of this paragraph, except the size requirements and with respect to which the manufacturer voluntarily files a certification required by the Secretary of the U.S. Department of Housing and Urban Development (HUD) and complies with the standards established under this title.

For mobile homes built prior to June 15, 1976, a *label* certifying compliance with the *Standard for Mobile Homes*, NFPA 501, ANSI 119.1, in effect at the time of manufacture, is required. For the purpose of these provisions, a mobile home shall be considered a *manufactured home*.

MANUFACTURED HOME INSTALLATION. Construction which is required for the installation of a *manufactured home*, including the construction of the foundation system, required structural connections thereto and the installation of on-site water, gas, electrical and sewer systems and connections thereto which are necessary for the normal operation of the *manufactured home*.

MANUFACTURED HOME STANDARDS. The *Manufactured Home* Construction and Safety Standards as promulgated by the HUD.

PRIVATELY OWNED (NONRENTAL) LOT. A parcel of real estate outside of a *manufactured home* rental community (park) where the land and the *manufactured home* to be installed thereon are held in common ownership.

SECTION AE301 PERMITS

AE301.1 Initial installation. A *manufactured home* shall not be installed on a foundation system, reinstalled or altered without first obtaining a *permit* from the *building official*. A separate *permit* shall be required for each *manufactured home* installation. When *approved* by the *building official*, such *permit* may include accessory buildings and structures, and their building service *equipment*, when the accessory buildings or structures will be constructed in conjunction with the *manufactured home* installation.

AE301.2 Additions, alterations and repairs to a manufactured home. A *permit* shall be obtained to alter, remodel, repair or add accessory buildings or structures to a *manufactured home* subsequent to its initial installation. *Permit* issuance and fees therefor shall be in conformance to the codes applicable to the type of work involved.

An *addition* made to a *manufactured home*, as defined in these provisions, shall comply with these provisions.

AE301.3 Accessory buildings. Except as provided in Section AE301.1, *permits* shall be required for all accessory buildings and structures, and their building service *equipment*. *Permit* issuance and fees therefor shall be in conformance the codes applicable to the types of work involved.

AE301.4 Exempted work. A *permit* shall not be required for the types of work specifically exempted by the applicable codes. Exemption from the *permit* requirements of any of said codes shall not be deemed to grant authorization for any work to be done in violation of the provisions of said codes or any other laws or ordinances of this *jurisdiction*.

SECTION AE302 APPLICATION FOR PERMIT

AE302.1 Application. To obtain a *manufactured home* installation *permit*, the applicant shall first file an application, in writing, on a form furnished by the *building official* for that purpose. At the option of the *building official*, every such application shall:

1. Identify and describe the work to be covered by the *permit* for which application is made.
2. Describe the land on which the proposed work is to be done by legal description, street address or similar description that will readily identify and definitely locate the proposed building or work.
3. Indicate the use or occupancy for which the proposed work is intended.
4. Be accompanied by plans, diagrams, computations and specifications, and other data as required in Section AE302.2.
5. Be accompanied by a soil investigation when required by Section AE502.2.
6. State the valuation of any new building or structure; or any *addition*, remodeling or *alteration* to an existing building.
7. Be signed by permittee, or permittee's authorized agent, who may be required to submit evidence to indicate such authority.
8. Give such other data and information as may be required by the *building official*.

AE302.2 Plans and specifications. Plans, engineering calculations, diagrams and other data as required by the *building official* shall be submitted in not less than two sets with each application for a *permit*. The *building official* may require plans, computations and specifications to be prepared and designed by an engineer or architect licensed by the state to practice as such.

Where no unusual site conditions exist, the *building official* may accept *approved* standard foundation plans and details in conjunction with the manufacturer's *approved* installation instructions without requiring the submittal of engineering calculations.

AE302.3 Information on plans and specifications. Plans and specifications shall be drawn to scale on substantial paper or cloth, and shall be of sufficient clarity to indicate the location, nature and extent of the work proposed and shown in detail that it will conform to the provisions of these provisions and all relevant laws, ordinances, rules and regulations. The *building official* shall determine what information is required on plans and specifications to ensure compliance.

SECTION AE303 PERMITS ISSUANCE

AE303.1 Issuance. The application, plans and specifications, and other data filed by an applicant for *permit* shall be reviewed by the *building official*. Such plans may be reviewed by other departments of this *jurisdiction* to verify compliance with any applicable laws under their *jurisdiction*. If the *building official* finds that the work described in an application for a *permit*, and the plans, specifications and other data filed therewith, conform to the requirements of these provisions, and other data filed therewith conform to the requirements of these provisions and other pertinent codes, laws and ordinances, and that the fees specified in Section AE304 have been paid, the *building official* shall issue a *permit* therefor to the applicant.

When the *building official* issues the *permit* where plans are required, the *building official* shall endorse in writing or stamp the plans and specifications *APPROVED*. Such *approved* plans and specifications shall not be changed, modified or altered without authorization from the *building official*, and all work shall be done in accordance with the *approved* plans.

AE303.2 Retention of plans. One set of *approved* plans and specifications shall be returned to the applicant and shall be kept on the site of the building or work at all times during which the work authorized thereby is in progress. One set of *approved* plans, specifications and computations shall be retained by the *building official* until final approval of the work.

AE303.3 Validity of permit. The issuance of a *permit* or approval of plans and specifications shall not be construed to be a *permit* for, or an approval of, any violation of any of the provisions of these provisions or other pertinent codes of any other ordinance of the *jurisdiction*. No *permit* presuming to give authority to violate or cancel these provisions shall be valid.

The issuance of a *permit* based on plans, specifications and other data shall not prevent the *building official* from thereafter requiring the correction of errors in said plans, specifications and other data, or from preventing building operations being carried on thereunder when in violation of these provisions or of any other ordinances of this *jurisdiction*.

AE303.4 Expiration. Every *permit* issued by the *building official* under these provisions shall expire by limitation and become null and void if the work authorized by such *permit* is not commenced within 180 days from the date of such *permit*, or if the work authorized by such *permit* is suspended or abandoned at any time after the work is commenced for a period of 180 days. Before such work can be recommenced, a new *permit* shall be first obtained, and the fee therefor shall be one-half the amount required for a new *permit* for such work, provided no changes have been made or will be made in the original plans and specifications for such work, and provided further that such suspension or abandonment has not exceeded one year. In order to renew action on a *permit* after expiration, the permittee shall pay a new full *permit* fee.

Any permittee holding an unexpired *permit* may apply for an extension of the time within which work may commence under that *permit* when the permittee is unable to commence work within the time required by this section for good and satisfactory reasons. The *building official* may extend the time for action by the permittee for a period not exceeding 180 days upon written request by the permittee showing that circumstances beyond the control of the permittee have prevented action from being taken. No *permit* shall be extended more than once.

AE303.5 Suspension or revocation. The *building official* may, in writing, suspend or revoke a *permit* issued under these provisions whenever the *permit* is issued in error or on the basis of incorrect information supplied, or in violation of any ordinance or regulation or any of these provisions.

SECTION AE304
FEES

AE304.1 Permit fees. The fee for each *manufactured home* installation *permit* shall be established by the *building official*.

When *permit* fees are to be based on the value or valuation of the work to be performed, the determination of value or valuation under these provisions shall be made by the *building official*. The value to be used shall be the total value of all work required for the *manufactured home* installation plus the total value of all work required for the construction of accessory buildings and structures for which the *permit* is issued, as well as all finish work, painting, roofing, electrical, plumbing, heating, air conditioning, elevators, fire-extinguishing systems and any other permanent *equipment* which is a part of the accessory building or structure. The value of the *manufactured home* itself shall not be included.

AE304.2 Plan review fees. When a plan or other data are required to be submitted by Section AE302.2, a plan review fee shall be paid at the time of submitting plans and specifications for review. Said plan review fee shall be as established by the *building official*. Where plans are incomplete or changed so as to require additional plan review, an additional plan review fee shall be charged at a rate as established by the *building official*.

AE304.3 Other provisions.

AE304.3.1 Expiration of plan review. Applications for which no *permit* is issued within 180 days following the date of application shall expire by limitation, and plans and other data submitted for review may thereafter be returned to the applicant or destroyed by the *building official*. The *building official* may extend the time for action by the applicant for a period not exceeding 180 days upon request by the applicant showing that circumstances beyond the control of the applicant have prevented action from being taken. No application shall be extended more than once. In order to renew action on an application after expiration, the applicant shall resubmit plans and pay a new plan review fee.

AE304.3.2 Investigation fees—work without a permit.

AE304.3.2.1 Investigation. Whenever any work for which a *permit* is required by these provisions has been commenced without first obtaining said *permit*, a special investigation shall be made before a *permit* may be issued for such work.

AE304.3.2.2 Fee. An investigation fee, in addition to the *permit* fee, shall be collected whether or not a *permit* is then or subsequently issued. The investigation fee shall be equal to the amount of the *permit* fee required. The minimum investigation fee shall be the same as the minimum fee established by the *building official*. The payment of such investigation fee shall not exempt any person from compliance with all other provisions of either these provisions or other pertinent codes or from any penalty prescribed by law.

AE304.3.3 Fee refunds.

AE304.3.3.1 Permit fee erroneously paid or collected. The *building official* may authorize the refunding of any fee paid hereunder which was erroneously paid or collected.

AE304.3.3.2 Permit fee paid when no work done. The *building official* may authorize the refunding of not more than 80 percent of the *permit* fee paid when no work has been done under a *permit* issued in accordance with these provisions.

AE304.3.3.3 Plan review fee. The *building official* may authorize the refunding of not more than 80 percent of the plan review fee paid when an application for a *permit* for which a plan review fee has been paid is withdrawn or canceled before any plan reviewing is done.

The *building official* shall not authorize the refunding of any fee paid, except upon written application by the original permittee not later than 180 days after the date of the fee payment.

SECTION AE305
INSPECTIONS

AE305.1 General. All construction or work for which a *manufactured home* installation *permit* is required shall be subject to inspection by the *building official*, and certain types of construction shall have continuous inspection by special inspectors as specified in Section AE306. A survey of the *lot* may be required by the *building official* to verify that the structure is located in accordance with the *approved* plans.

It shall be the duty of the *permit* applicant to cause the work to be accessible and exposed for inspection purposes. Neither the *building official* nor this *jurisdiction* shall be liable for expense entailed in the removal or replacement of any material required to allow inspection.

AE305.2 Inspection requests. It shall be the duty of the person doing the work authorized by a *manufactured home* installation *permit* to notify the *building official* that such work is ready for inspection. The *building official* may require that every request for inspection be filed at least one working day before such inspection is desired. Such request may be in writing or by telephone at the option of the *building official*.

It shall be the duty of the person requesting any inspections required, either by these provisions or other applicable codes, to provide access to and means for proper inspection of such work.

AE305.3 Inspection record card. Work requiring a *manufactured home* installation *permit* shall not be commenced until the *permit* holder or the *permit* holder's agent shall have posted an inspection record card in a conspicuous place on the premises and in such position as to allow the *building official* conveniently to make the required entries thereon regarding inspection of the work. This card shall be maintained in such position by the *permit* holder until final approval has been issued by the *building official*.

AE305.4 Approval required. Work shall not be done on any part of the *manufactured home* installation beyond the point indicated in each successive inspection without first obtaining the approval of the *building official*. Such approval shall be given only after an inspection has been made of each successive step in the construction as indicated by each of the inspections required in Section AE305.5. There shall be a final inspection and approval of the *manufactured home* installation, including connections to its building service *equipment*, when completed and ready for occupancy or use.

AE305.5 Required inspections.

AE305.5.1 Structural inspections for the manufactured home installation. Reinforcing steel or structural framework of any part of any *manufactured home* foundation system shall not be covered or concealed without first obtaining the approval of the *building official*. The *building official*, upon notification from the *permit* holder or the *permit* holder's agent, shall make the following inspections and shall either approve that portion of the construction as completed or shall notify the *permit* holder or the *permit* holder's agent wherein the same fails to comply with these provisions or other applicable codes:

1. Foundation inspection: To be made after excavations for footings are completed and any required reinforcing steel is in place. For concrete foundations, any required forms shall be in place prior to inspection. All materials for the foundation shall be on the job, except where concrete from a central mixing plant (commonly termed "transit mixed") is to be used, the concrete materials need not be on the job. Where the foundation is to be constructed of *approved* treated wood, additional framing inspections as required by the *building official* may be required.
2. Concrete slab or under-floor inspection: To be made after all in-slab or under-floor building service *equipment*, conduit, piping accessories and other ancillary *equipment* items are in place but before any concrete is poured or the *manufactured home* is installed.
3. Anchorage inspection: To be made after the *manufactured home* has been installed and permanently anchored.

AE305.5.2 Structural inspections for accessory building and structures. Inspections for accessory buildings and structures shall be made as set forth in this code.

AE305.5.3 Building service equipment inspections. All building service *equipment* which is required as a part of a *manufactured home* installation, including accessory buildings and structures authorized by the same *permit*, shall be inspected by the *building official*. Building service *equipment* shall be inspected and tested as required by the applicable codes. Such inspections and testing shall be limited to site construction and shall not include building service *equipment* which is a part of the *manufactured home* itself. No portion of any building service *equipment* intended to be concealed by any permanent portion of the construction shall be concealed until inspected and *approved*. Building service *equipment* shall not be connected to a water, fuel or power supply, or sewer system, until authorized by the *building official*.

AE305.5.4 Final inspection. When finish grading and the *manufactured home* installation, including the installation of all required building service *equipment*, is completed and the *manufactured home* is ready for occupancy, a final inspection shall be made.

AE305.6 Other inspections. In addition to the called inspections specified in Section AE305.5.4, the *building official* may make or require other inspections of any construction work to ascertain compliance with these provisions or other codes and laws which are enforced by the code enforcement agency.

SECTION AE306 SPECIAL INSPECTIONS

AE306.1 General. In addition to the inspections required by Section AE305, the *building official* may require the owner to employ a special inspector during construction of specific types of work as described in this code.

SECTION AE307 UTILITY SERVICE

AE307.1 General. Utility service shall not be provided to any building service *equipment* which is regulated by these provisions or other applicable codes, and for which a *manufactured home* installation *permit* is required by these provisions, until *approved* by the *building official*.

SECTION AE401 OCCUPANCY CLASSIFICATION

AE401.1 Manufactured homes. A *manufactured home* shall be limited in use to a single *dwelling unit*.

AE401.2 Accessory buildings. Accessory buildings shall be classified as to occupancy by the *building official* as set forth in this code.

SECTION AE402 LOCATION ON PROPERTY

AE402.1 General. *Manufactured homes* and accessory buildings shall be located on the property in accordance with applicable codes and ordinances of this *jurisdiction*.

SECTION AE501 DESIGN

AE501.1 General. A *manufactured home* shall be installed on a foundation system which is designed and constructed to sustain within the stress limitations specified in this code and all loads specified in this code.

Exception: When specifically authorized by the *building official*, foundation and anchorage systems which are constructed in accordance with the methods specified in Section AE600 of these provisions, or in the HUD, *Permanent Foundations for Manufactured Housing,* 1984 Edition, Draft, shall be deemed to meet the requirements of this appendix.

AE501.2 Manufacturer's installation instructions. The installation instructions as provided by the manufacturer of the *manufactured home* shall be used to determine permissible points of support for vertical loads and points of attachment for anchorage systems used to resist horizontal and uplift forces.

AE501.3 Rationality. Any system or method of construction to be used shall submit to a rational analysis in accordance with well-established principles of mechanics.

SECTION AE502 FOUNDATION SYSTEMS

AE502.1 General. Foundation systems designed and constructed in accordance with this section may be considered a permanent installation.

AE502.2 Soil classification. The classification of the soil at each *manufactured home* site shall be determined when required by the *building official*. The *building official* may require that the determination be made by an engineer or architect licensed by the state to conduct soil investigations.

The classification shall be based on observation and any necessary tests of the materials disclosed by borings or excavations made in appropriate locations. Additional studies may be necessary to evaluate soil strength, the effect of moisture variation on soil-bearing capacity, compressibility and expansiveness.

When required by the *building official*, the soil classification design-bearing capacity and lateral pressure shall be shown on the plans.

AE502.3 Footings and foundations. Footings and foundations, unless otherwise specifically provided, shall be constructed of materials specified by this code for the intended use and in all cases shall extend below the frost line. Footings of concrete and masonry shall be of solid material. Foundations supporting untreated wood shall extend at least 8 inches (203 mm) above the adjacent finish *grade*. Footings shall have a minimum depth below finished *grade* of 12 inches (305 mm) unless a greater depth is recommended by a foundation investigation.

Piers and bearing walls shall be supported on masonry or concrete foundations or piles, or other *approved* foundation systems which shall be of sufficient capacity to support all loads.

AE502.4 Foundation design. When a design is provided, the foundation system shall be designed in accordance with the applicable structural provisions of this code and shall be designed to minimize differential settlement. Where a design is not provided, the minimum foundation requirements shall be as set forth in this code.

AE502.5 Drainage. Provisions shall be made for the control and drainage of surface water away from the *manufactured home*.

AE502.6 Under-floor clearances—ventilation and access. A minimum clearance of 12 inches (305 mm) shall be maintained beneath the lowest member of the floor support framing system. Clearances from the bottom of wood floor joists or perimeter joists shall be as specified in this code.

Under-floor spaces shall be ventilated with openings as specified in this code. If combustion air for one or more heat-producing *appliance* is taken from within the under-floor spaces, ventilation shall be adequate for proper *appliance* operation.

Under-floor access openings shall be provided. Such openings shall be not less than 18 inches (457 mm) in any dimension and not less than 3 square feet (0.279 m^2) in area, and shall be located so that any water supply and sewer drain connections located under the *manufactured home* are accessible.

SECTION AE503 SKIRTING AND PERIMETER ENCLOSURES

AE503.1 Skirting and permanent perimeter enclosures. Skirting and permanent perimeter enclosures shall be installed only where specifically required by other laws or ordinances. Skirting, when installed, shall be of material suitable for exterior exposure and contact with the ground. Permanent perimeter enclosures shall be constructed of materials as required by this code for regular foundation construction.

Skirting shall be installed in accordance with the skirting manufacturer's installation instructions. Skirting shall be adequately secured to ensure stability, minimize vibration and susceptibility to wind damage, and compensate for possible frost heave.

AE503.2 Retaining walls. Where retaining walls are used as a permanent perimeter enclosure, they shall resist the lateral displacements of soil or other materials and shall conform to this code as specified for foundation walls. Retaining walls and foundation walls shall be constructed of *approved* treated wood, concrete, masonry or other *approved* materials or combination of materials as for foundations as specified in this code. Siding materials shall extend below the top of the exterior of the retaining or foundation wall, or the joint between the siding and enclosure wall shall be flashed in accordance with this code.

SECTION AE504 STRUCTURAL ADDITIONS

AE504.1 General. Accessory buildings shall not be structurally supported by or attached to a *manufactured home* unless engineering calculations are submitted to substantiate any proposed structural connection.

Exception: The *building official* may waive the submission of engineering calculations if it is found that the nature of the work applied for is such that engineering calculations are not necessary to show conformance to these provisions.

SECTION AE505
BUILDING SERVICE EQUIPMENT

AE505.1 General. The installation, *alteration*, repair, replacement, *addition* to or maintenance of the building service *equipment* within the *manufactured home* shall conform to regulations set forth in the *Manufactured Home* Standards. Such work which is located outside the *manufactured home* shall comply with the applicable codes adopted by this *jurisdiction*.

SECTION AE506
EXITS

AE506.1 Site development. Exterior stairways and ramps which provide egress to the public way shall comply with the applicable provisions of this code.

AE506.2 Accessory buildings. Every accessory building or portion thereof shall be provided with exits as required by this code.

SECTION AE507
OCCUPANCY, FIRE SAFETY AND ENERGY CONSERVATION STANDARDS

AE507.1 General. *Alterations* made to a *manufactured home* subsequent to its initial installation shall conform to the occupancy, fire safety and energy conservation requirements set forth in the *Manufactured Home* Standards.

SECTION AE600
SPECIAL REQUIREMENTS FOR FOUNDATION SYSTEMS

AE600.1 General. This section is applicable only where specifically authorized by the *building official*.

SECTION AE601
FOOTINGS AND FOUNDATIONS

AE601.1 General. The capacity of individual load-bearing piers and their footings shall be sufficient to sustain all loads specified in this code within the stress limitations specified in this code. Footings, unless otherwise *approved* by the *building official*, shall be placed level on firm, undisturbed soil or an engineered fill which is free of organic material, such as weeds and grasses. Where used, an engineered fill shall provide a minimum load-bearing capacity of not less than 1,000 pounds per square foot (48 kN/m^2). Continuous footings shall conform to the requirements of this code. Section AE502 of these provisions shall apply to footings and foundations constructed under the provisions of this section.

SECTION AE602
PIER CONSTRUCTION

AE602.1 General. Piers shall be designed and constructed to distribute loads evenly. Multiple-section homes may have concentrated roof loads which will require special consideration. Load-bearing piers may be constructed utilizing one of the following methods listed. Such piers shall be considered to resist only vertical forces acting in a downward direction. They shall not be considered as providing any resistance to horizontal loads induced by wind or earthquake forces.

1. A prefabricated load-bearing device that is listed and *labeled* for the intended use.
2. Mortar shall comply with ASTM C 270, Type M, S or N; this may consist of one part Portland cement, one-half part hydrated lime and four parts sand by volume. Lime shall not be used with plastic or waterproof cement.
3. A cast-in-place concrete pier with concrete having specified compressive strength at 28 days of 2,500 pounds per square inch (17 225 kPa).

Alternative materials and methods of construction may be used for piers which have been designed by an engineer or architect licensed by the state to practice as such.

Caps and leveling spacers may be used for leveling of the *manufactured home*. Spacing of piers shall be as specified in the manufacturer's installation instructions, if available, or by an *approved* designer.

SECTION AE603
HEIGHT OF PIERS

AE603.1 General. Piers constructed as indicated in Section AE602 may have heights as follows:

1. Except for corner piers, piers 36 inches (914 mm) or less in height may be constructed of masonry units, placed with cores or cells vertically. Piers shall be installed with their long dimension at right angles to the main frame member they support and shall have a minimum cross-sectional area of 128 square inches (82 560 mm^2). Piers shall be capped with minimum 4-inch (102 mm) *solid masonry* units or equivalent.
2. Piers between 36 and 80 inches (914 and 2032 mm) in height and all corner piers greater than 24 inches (610 mm) in height shall be at least 16 inches by 16 inches (406 mm by 406 mm) consisting of interlocking masonry units and shall be fully capped with minimum 4-inch (102 mm) *solid masonry* units or equivalent.
3. Piers greater than 80 inches (2032 mm) in height may be constructed in accordance with the provisions of Item 2, provided the piers shall be filled solid with grout and reinforced with four continuous No. 5 bars. One bar shall be placed in each corner cell of hollow

masonry unit piers or in each corner of the grouted space of piers constructed of *solid masonry* units.

4. Cast-in-place concrete piers meeting the same size and height limitations of Items 1, 2 and 3 may be substituted for piers constructed of masonry units.

SECTION AE604 ANCHORAGE INSTALLATIONS

AE604.1 Ground anchors. Ground anchors shall be designed and installed to transfer the anchoring loads to the ground. The load-carrying portion of the ground anchors shall be installed to the full depth called for by the manufacturer's installation instructions and shall extend below the established frost line into undisturbed soil.

Manufactured ground anchors shall be listed and installed in accordance with the terms of their listing and the anchor manufacturer's instructions, and shall include the means of attachment of ties meeting the requirements of Section AE605. Ground anchor manufacturer's installation instructions shall include the amount of preload required and load capacity in various types of soil. These instructions shall include tensioning adjustments which may be needed to prevent damage to the *manufactured home*, particularly damage that can be caused by frost heave. Each ground anchor shall be marked with the manufacturer's identification and listed model identification number which shall be visible after installation. Instructions shall accompany each listed ground anchor specifying the types of soil for which the anchor is suitable under the requirements of this section.

Each *approved* ground anchor, when installed, shall be capable of resisting an allowable working load at least equal to 3,150 pounds (14 kN) in the direction of the tie plus a 50-percent overload [4,725 pounds (21 kN) total] without failure. Failure shall be considered to have occurred when the anchor moves more than 2 inches (51 mm) at a load of 4,725 pounds (21 kN) in the direction of the tie installation. Those ground anchors which are designed to be installed so that loads on the anchor are other than direct withdrawal shall be designed and installed to resist an applied design load of 3,150 pounds (14 kN) at 40 to 50 degrees from vertical or within the angle limitations specified by the home manufacturer without displacing the tie end of the anchor more than 4 inches (102 mm) horizontally. Anchors designed for the connection of multiple ties shall be capable of resisting the combined working load and overload consistent with the intent expressed herein.

When it is proposed to use ground anchors and the *building official* has reason to believe that the soil characteristics at a given site are such as to render the use of ground anchors advisable, or when there is doubt regarding the ability of the ground anchors to obtain their listed capacity, the *building official* may require that a representative field installation be made at the site in question and tested to demonstrate ground-anchor capacity. The *building official* shall approve the test procedures.

AE604.2 Anchoring equipment. Anchoring *equipment*, when installed as a permanent installation, shall be capable of resisting all loads as specified within these provisions. When the stabilizing system is designed by an engineer or architect licensed by the state to practice as such, alternative designs may be used, providing the anchoring *equipment* to be used is capable of withstanding a load equal to 1.5 times the calculated load. All anchoring *equipment* shall be listed and *labeled* as being capable of meeting the requirements of these provisions. Anchors as specified in this code may be attached to the main frame of the *manufactured home* by an *approved* $^3/_{16}$-inch-thick (4.76 mm) slotted steel plate anchoring device. Other anchoring devices or methods meeting the requirements of these provisions may be permitted when *approved* by the *building official*.

Anchoring systems shall be so installed as to be permanent. Anchoring *equipment* shall be so designed to prevent self-disconnection with no hook ends used.

AE604.3 Resistance to weather deterioration. All anchoring *equipment*, tension devices and ties shall have a resistance to deterioration as required by this code.

AE604.4 Tensioning devices. Tensioning devices, such as turnbuckles or yoke-type fasteners, shall be ended with clevis or welded eyes.

SECTION AE605 TIES, MATERIALS AND INSTALLATION

AE605.1 General. Steel strapping, cable, chain or other *approved* materials shall be used for ties. All ties shall be fastened to ground anchors and drawn tight with turnbuckles or other adjustable tensioning devices or devices supplied with the ground anchor. Tie materials shall be capable of resisting an allowable working load of 3,150 pounds (14 kN) with no more than 2 percent elongation and shall withstand a 50-percent overload [4,750 pounds (21 kN)]. Ties shall comply with the weathering requirements of Section AE604.3. Ties shall connect the ground anchor and the main structural frame. Ties shall not connect to steel outrigger beams which fasten to and intersect the main structural frame unless specifically stated in the manufacturer's installation instructions. Connection of cable ties to main frame members shall be $^5/_8$-inch (15.9 mm) closed-eye bolts affixed to the frame member in an *approved* manner. Cable ends shall be secured with at least two U-bolt cable clamps with the "U" portion of the clamp installed on the short (dead) end of the cable to ensure strength equal to that required by this section.

Wood floor support systems shall be fixed to perimeter foundation walls in accordance with provisions of this code. The minimum number of ties required per side shall be sufficient to resist the wind load stated in this code. Ties shall be as evenly spaced as practicable along the length of the *manufactured home* with the distance from each end of the home and the tie nearest that end not exceeding 8 feet (2438 mm). When continuous straps are provided as vertical ties, such ties shall be positioned at rafters and studs. Where a vertical tie and diagonal tie are located at the same place, both ties may be connected to a single anchor, provided the anchor used is capable of carrying both loads. Multiple-section *manufactured homes* require diagonal ties only. Diagonal ties shall be installed on the exterior main frame and slope to the exterior at an angle of 40 to 50 degrees from the vertical or within the angle limitations specified by the home manufacturer. Vertical ties which are not continuous over the top of the *manufactured home* shall be attached to the main frame.

SECTION AE606
REFERENCED STANDARDS

ASTM C 270—04 Specification for Mortar for Unit Masonry. AE602

NFPA 501—03 Standard on Manufactured Housing AE201

APPENDIX F

RADON CONTROL METHODS

(The provisions contained in this appendix are not mandatory unless specifically referenced in the adopting ordinance.)

SECTION AF101 SCOPE

AF101.1 General. This appendix contains requirements for new construction in *jurisdictions* where radon-resistant construction is required.

Inclusion of this appendix by *jurisdictions* shall be determined through the use of locally available data or determination of Zone 1 designation in Figure AF101 and Table AF101(1).

SECTION AF102 DEFINITIONS

AF102.1 General. For the purpose of these requirements, the terms used shall be defined as follows:

DRAIN TILE LOOP. A continuous length of drain tile or perforated pipe extending around all or part of the internal or external perimeter of a *basement* or crawl space footing.

RADON GAS. A naturally occurring, chemically inert, radioactive gas that is not detectable by human senses. As a gas, it can move readily through particles of soil and rock, and can accumulate under the slabs and foundations of homes where it can easily enter into the living space through construction cracks and openings.

SOIL-GAS-RETARDER. A continuous membrane of 6-mil (0.15 mm) polyethylene or other equivalent material used to retard the flow of soil gases into a building.

SUBMEMBRANE DEPRESSURIZATION SYSTEM. A system designed to achieve lower submembrane air pressure relative to crawl space air pressure by use of a vent drawing air from beneath the soil-gas-retarder membrane

SUBSLAB DEPRESSURIZATION SYSTEM (Active). A system designed to achieve lower subslab air pressure relative to indoor air pressure by use of a fan-powered vent drawing air from beneath the slab.

SUBSLAB DEPRESSURIZATION SYSTEM (Passive). A system designed to achieve lower subslab air pressure relative to indoor air pressure by use of a vent pipe routed through the *conditioned space* of a building and connecting the subslab area with outdoor air, thereby relying on the convective flow of air upward in the vent to draw air from beneath the slab.

SECTION AF103 REQUIREMENTS

AF103.1 General. The following construction techniques are intended to resist radon entry and prepare the building for post-construction radon mitigation, if necessary (see Figure AF102). These techniques are required in areas where designated by the *jurisdiction*.

AF103.2 Subfloor preparation. A layer of gas-permeable material shall be placed under all concrete slabs and other floor systems that directly contact the ground and are within the walls of the living spaces of the building, to facilitate future installation of a subslab depressurization system, if needed. The gas-permeable layer shall consist of one of the following:

1. A uniform layer of clean aggregate, a minimum of 4 inches (102 mm) thick. The aggregate shall consist of material that will pass through a 2-inch (51 mm) sieve and be retained by a $^{1}/_{4}$-inch (6.4 mm) sieve.
2. A uniform layer of sand (native or fill), a minimum of 4 inches (102 mm) thick, overlain by a layer or strips of geotextile drainage matting designed to allow the lateral flow of soil gases.
3. Other materials, systems or floor designs with demonstrated capability to permit depressurization across the entire subfloor area.

AF103.3 Soil-gas-retarder. A minimum 6-mil (0.15 mm) [or 3-mil (0.075 mm) cross-laminated] polyethylene or equivalent flexible sheeting material shall be placed on top of the gas-permeable layer prior to casting the slab or placing the floor assembly to serve as a soil-gas-retarder by bridging any cracks that develop in the slab or floor assembly, and to prevent concrete from entering the void spaces in the aggregate base material. The sheeting shall cover the entire floor area with separate sections of sheeting lapped at least 12 inches (305 mm). The sheeting shall fit closely around any pipe, wire or other penetrations of the material. All punctures or tears in the material shall be sealed or covered with additional sheeting.

AF103.4 Entry routes. Potential radon entry routes shall be closed in accordance with Sections AF103.4.1 through AF103.4.10.

AF103.4.1 Floor openings. Openings around bathtubs, showers, water closets, pipes, wires or other objects that penetrate concrete slabs, or other floor assemblies, shall be filled with a polyurethane caulk or equivalent sealant applied in accordance with the manufacturer's recommendations.

AF103.4.2 Concrete joints. All control joints, isolation joints, construction joints, and any other joints in concrete slabs or between slabs and foundation walls shall be sealed with a caulk or sealant. Gaps and joints shall be cleared of loose material and filled with polyurethane caulk or other elastomeric sealant applied in accordance with the manufacturer's recommendations.

AF103.4.3 Condensate drains. Condensate drains shall be trapped or routed through nonperforated pipe to daylight.

AF103.4.4 Sumps. Sump pits open to soil or serving as the termination point for subslab or exterior drain tile loops shall be covered with a gasketed or otherwise sealed lid. Sumps used as the suction point in a subslab depressurization system shall have a lid designed to accommodate the vent pipe. Sumps used as a floor drain shall have a lid equipped with a trapped inlet.

AF103.4.5 Foundation walls. Hollow block masonry foundation walls shall be constructed with either a continuous course of *solid masonry*, one course of masonry grouted solid, or a solid concrete beam at or above finished ground surface to prevent the passage of air from the interior of the wall into the living space. Where a brick veneer or other masonry ledge is installed, the course immediately below that ledge shall be sealed. Joints, cracks or other openings around all penetrations of both exterior and interior surfaces of masonry block or wood foundation walls below the ground surface shall be filled with polyurethane caulk or equivalent sealant. Penetrations of concrete walls shall be filled.

AF103.4.6 Dampproofing. The exterior surfaces of portions of concrete and masonry block walls below the ground surface shall be dampproofed in accordance with Section R406.

AF103.4.7 Air-handling units. Air-handling units in crawl spaces shall be sealed to prevent air from being drawn into the unit.

Exception: Units with gasketed seams or units that are otherwise sealed by the manufacturer to prevent leakage.

AF103.4.8 Ducts. Ductwork passing through or beneath a slab shall be of seamless material unless the air-handling system is designed to maintain continuous positive pressure within such ducting. Joints in such ductwork shall be sealed to prevent air leakage.

Ductwork located in crawl spaces shall have all seams and joints sealed by closure systems in accordance with Section M1601.4.1.

AF103.4.9 Crawl space floors. Openings around all penetrations through floors above crawl spaces shall be caulked or otherwise filled to prevent air leakage.

AF103.4.10 Crawl space access. Access doors and other openings or penetrations between *basements* and adjoining crawl spaces shall be closed, gasketed or otherwise filled to prevent air leakage.

AF103.5 Passive submembrane depressurization system. In buildings with crawl space foundations, the following components of a passive submembrane depressurization system shall be installed during construction.

Exception: Buildings in which an *approved* mechanical crawl space ventilation system or other equivalent system is installed.

AF103.5.1 Ventilation. Crawl spaces shall be provided with vents to the exterior of the building. The minimum net area of ventilation openings shall comply with Section R408.1.

AF103.5.2 Soil-gas-retarder. The soil in crawl spaces shall be covered with a continuous layer of minimum 6-mil (0.15 mm) polyethylene soil-gas-retarder. The ground cover shall be lapped a minimum of 12 inches (305 mm) at joints and shall extend to all foundation walls enclosing the crawl space area.

AF103.5.3 Vent pipe. A plumbing tee or other *approved* connection shall be inserted horizontally beneath the sheeting and connected to a 3- or 4-inch-diameter (76 or 102 mm) fitting with a vertical vent pipe installed through the sheeting. The vent pipe shall be extended up through the building floors, and terminate at least 12 inches (305 mm) above the roof in a location at least 10 feet (3048 mm) away from any window or other opening into the *conditioned spaces* of the building that is less than 2 feet (610 mm) below the exhaust point, and 10 feet (3048 mm) from any window or other opening in adjoining or adjacent buildings.

AF103.6 Passive subslab depressurization system. In *basement* or slab-on-grade buildings, the following components of a passive subslab depressurization system shall be installed during construction.

AF103.6.1 Vent pipe. A minimum 3-inch-diameter (76 mm) ABS, PVC or equivalent gas-tight pipe shall be embedded vertically into the subslab aggregate or other permeable material before the slab is cast. A "T" fitting or equivalent method shall be used to ensure that the pipe opening remains within the subslab permeable material. Alternatively, the 3-inch (76 mm) pipe shall be inserted directly into an interior perimeter drain tile loop or through a sealed sump cover where the sump is exposed to the subslab aggregate or connected to it through a drainage system.

The pipe shall be extended up through the building floors, and terminate at least 12 inches (305 mm) above the surface of the roof in a location at least 10 feet (3048 mm) away from any window or other opening into the *conditioned spaces* of the building that is less than 2 feet (610 mm) below the exhaust point, and 10 feet (3048 mm) from any window or other opening in adjoining or adjacent buildings.

AF103.6.2 Multiple vent pipes. In buildings where interior footings or other barriers separate the subslab aggregate or other gas-permeable material, each area shall be fitted with an individual vent pipe. Vent pipes shall connect to a single vent that terminates above the roof or each individual vent pipe shall terminate separately above the roof.

AF103.7 Vent pipe drainage. All components of the radon vent pipe system shall be installed to provide positive drainage to the ground beneath the slab or soil-gas-retarder.

AF103.8 Vent pipe accessibility. Radon vent pipes shall be accessible for future fan installation through an *attic* or other area outside the *habitable space*.

Exception: The radon vent pipe need not be accessible in an *attic* space where an *approved* roof-top electrical supply is provided for future use.

AF103.9 Vent pipe identification. All exposed and visible interior radon vent pipes shall be identified with at least one *label* on each floor and in accessible *attics*. The *label* shall read: "Radon Reduction System."

AF103.10 Combination foundations. Combination *basement*/crawl space or slab-on-grade/crawl space foundations shall have separate radon vent pipes installed in each type of foundation area. Each radon vent pipe shall terminate above the roof or shall be connected to a single vent that terminates above the roof.

AF103.11 Building depressurization. Joints in air ducts and plenums in un*conditioned spaces* shall meet the requirements of Section M1601. Thermal envelope air infiltration requirements shall comply with the energy conservation provisions in Chapter 11. Fireblocking shall meet the requirements contained in Section R302.11.

AF103.12 Power source. To provide for future installation of an active submembrane or subslab depressurization system, an electrical circuit terminated in an *approved* box shall be installed during construction in the *attic* or other anticipated location of vent pipe fans. An electrical supply shall also be accessible in anticipated locations of system failure alarms.

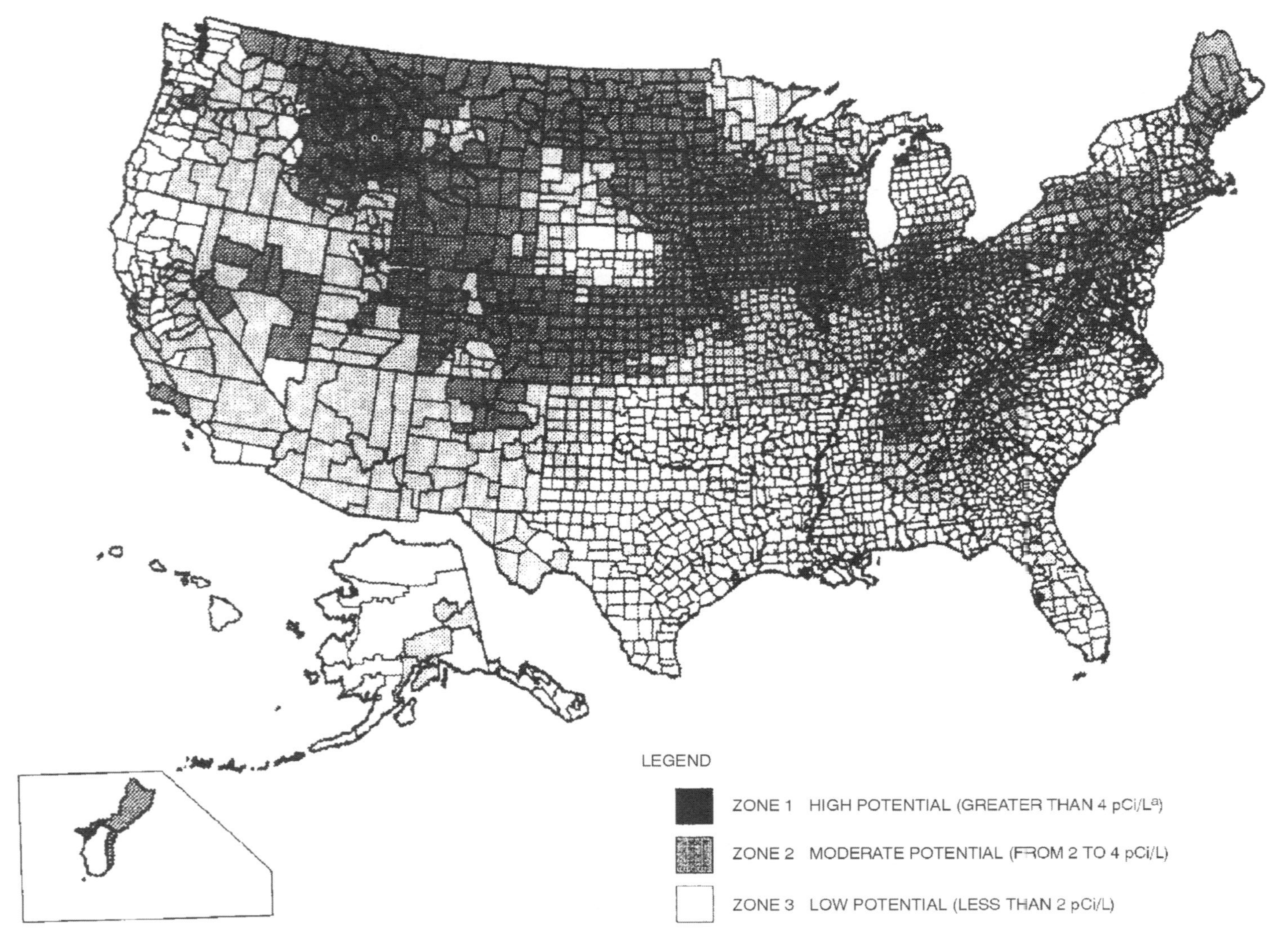

a. pCi/L standard for picocuries per liter of radon gas. The U.S. Environmental Protection Agency (EPA) recommends that all homes that measure 4 pCi/L and greater be mitigated.

The EPA and the U.S. Geological Survey have evaluated the radon potential in the United States and have developed a map of radon zones designed to assist building officials in deciding whether radon-resistant features are applicable in new construction.

The map assigns each of the 3,141 counties in the United States to one of three zones based on radon potential. Each zone designation reflects the average short-term radon measurement that can be expected to be measured in a building without the implementation of radon-control methods. The radon zone designation of highest priority is Zone 1. Table AF101 lists the Zone 1 counties illustrated on the map. More detailed information can be obtained from state-specific booklets (EPA-402-R-93-021 through 070) available through State Radon Offices or from EPA Regional Offices.

FIGURE AF101
EPA MAP OF RADON ZONES

TABLE AF101(1)
HIGH RADON-POTENTIAL (ZONE 1) COUNTIES[a]

ALABAMA
Calhoun
Clay
Cleburne
Colbert
Coosa
Franklin
Jackson
Lauderdale
Lawrence
Limestone
Madison
Morgan
Talladega

CALIFORNIA
Santa Barbara
Ventura

COLORADO
Adams
Arapahoe
Baca
Bent
Boulder
Chaffee
Cheyenne
Clear Creek
Crowley
Custer
Delta
Denver
Dolores
Douglas
El Paso
Elbert
Fremont
Garfield
Gilpin
Grand
Gunnison
Huerfano
Jackson
Jefferson
Kiowa
Kit Carson
Lake
Larimer
Las Animas
Lincoln
Logan
Mesa
Moffat
Montezuma
Montrose
Morgan
Otero
Ouray
Park
Phillips
Pitkin
Prowers
Pueblo
Rio Blanco
San Miguel
Summit
Teller
Washington
Weld
Yuma

CONNECTICUT
Fairfield
Middlesex
New Haven
New London

GEORGIA
Cobb
De Kalb
Fulton
Gwinnett

IDAHO
Benewah
Blaine
BoiseBonner
Boundary
Butte
Camas
Clark
Clearwater
Custer
Elmore
Fremont
Gooding
Idaho
Kootenai
Latah
Lemhi
Shoshone
Valley

ILLINOIS
Adams
Boone
Brown
Bureau
Calhoun
Carroll
Cass
Champaign
Coles
De Kalb
De Witt
Douglas
Edgar
Ford
Fulton
Greene
Grundy
Hancock
Henderson
Henry
Iroquois
Jersey
Jo Daviess
Kane
Kendall
Knox
La Salle
Lee
Livingston
Logan
Macon
Marshall
Mason
McDonough
McLean
Menard
Mercer
Morgan
Moultrie
Ogle
Peoria
Piatt
Pike
Putnam
Rock Island
Sangamon
Schuyler
Scott
Stark
Stephenson
Tazewell
Vermilion
Warren
Whiteside
Winnebago
Woodford

INDIANA
Adams
Allen
Bartholomew
Benton
Blackford
Boone
Carroll
Cass
Clark
Clinton
De Kalb
Decatur
Delaware
Elkhart
Fayette
Fountain
Fulton
Grant
Hamilton
Hancock
Harrison
Hendricks
Henry
Howard
Huntington
Jay
Jennings
Johnson
Kosciusko
LaGrange
Lawrence
Madison
Marion
Marshall
Miami
Monroe
Montgomery
Noble
Orange
Putnam
Randolph
Rush
Scott
Shelby
St. Joseph
Steuben
Tippecanoe
Tipton
Union
Vermillion
Wabash
Warren
Washington
Wayne
Wells
White
Whitley

IOWA
All Counties

KANSAS
Atchison
Barton
Brown
Cheyenne
Clay
Cloud
Decatur
Dickinson
Douglas
Ellis
Ellsworth
Finney
Ford
Geary
Gove
Graham
Grant
Gray
Greeley
Hamilton
Haskell
Hodgeman
Jackson
Jewell
Johnson
Kearny
Kingman
Kiowa
Lane
Leavenworth
Lincoln
Logan
Marion
Marshall
McPherson
Meade
Mitchell
Nemaha
Ness
Norton
Osborne
Ottawa
Pawnee
Phillips
Pottawatomie
Pratt
Rawlins
Republic
Rice
Riley
Rooks
Rush
Saline
Scott
Sheridan
Sherman
Smith
Stanton
Thomas
Trego
Wallace
Washington
Wichita
Wyandotte

KENTUCKY
Adair
Allen
Barren
Bourbon
Boyle
Bullitt
Casey
Clark
Cumberland
Fayette
Franklin
Green
Harrison
Hart
Jefferson
Jessamine
Lincoln
Marion
Mercer
Metcalfe
Monroe
Nelson
Pendleton
Pulaski
Robertson
Russell
Scott
Taylor
Warren
Woodford

MAINE
Androscoggin
Aroostook
Cumberland
Franklin
Hancock
Kennebec
Lincoln
Oxford
Penobscot
Piscataquis
Somerset
York

MARYLAND
Baltimore
Calvert
Carroll
Frederick
Harford
Howard
Montgomery
Washington

MASS.
Essex
Middlesex
Worcester

MICHIGAN
Branch
Calhoun
Cass
Hillsdale
Jackson
Kalamazoo
Lenawee
St. Joseph
Washtenaw

MINNESOTA
Becker
Big Stone
Blue Earth
Brown
Carver
Chippewa
Clay
Cottonwood
Dakota
Dodge
Douglas
Faribault
Fillmore
Freeborn
Goodhue
Grant
Hennepin
Houston
Hubbard
Jackson
Kanabec
Kandiyohi
Kittson
Lac Qui Parle
Le Sueur
Lincoln
Lyon
Mahnomen
Marshall
Martin
McLeod
Meeker
Mower
Murray
Nicollet
Nobles
Norman
Olmsted
Otter Tail
Pennington
Pipestone
Polk
Pope
Ramsey
Red Lake
Redwood
Renville
Rice
Rock
Roseau
Scott
Sherburne
Sibley
Stearns
Steele
Stevens
Swift
Todd
Traverse
Wabasha
Wadena
Waseca
Washington
Watonwan
Wilkin
Winona
Wright
Yellow Medicine

MISSOURI
Andrew
Atchison
Buchanan
Cass
Clay
Clinton
Holt
Iron
Jackson
Nodaway
Platte

MONTANA
Beaverhead
Big Horn
Blaine
Broadwater
Carbon
Carter
Cascade
Chouteau
Custer
Daniels
Dawson
Deer Lodge
Fallon
Fergus
Flathead
Gallatin
Garfield
Glacier
Granite
Hill
Jefferson
Judith Basin
Lake
Lewis and Clark
Madison
McCone
Meagher
Missoula
Park
Phillips
Pondera
Powder River
Powell
Prairie
Ravalli
Richland
Roosevelt
Rosebud
Sanders
Sheridan
Silver Bow
Stillwater
Teton
Toole
Valley
Wibaux
Yellowstone

(continued)

TABLE AF101(1)—continued
HIGH RADON-POTENTIAL (ZONE 1) COUNTIES[a]

NEBRASKA
Adams
Boone
Boyd
Burt
Butler
Cass
Cedar
Clay
Colfax
Cuming
Dakota
Dixon
Dodge
Douglas
Fillmore
Franklin
Frontier
Furnas
Gage
Gosper
Greeley
Hamilton
Harlan
Hayes
Hitchcock
Hurston
Jefferson
Johnson
Kearney
Knox
Lancaster
Madison
Nance
Nemaha
Nuckolls
Otoe
Pawnee
Phelps
Pierce
Platte
Polk
Red Willow
Richardson
Saline
Sarpy
Saunders
Seward
Stanton
Thayer
Washington
Wayne
Webster
York

NEVADA
Carson City
Douglas
Eureka
Lander
Lincoln
Lyon
Mineral
Pershing
White Pine

NEW HAMPSHIRE
Carroll

NEW JERSEY
Hunterdon
Mercer
Monmouth
Morris
Somerset
Sussex
Warren

NEW MEXICO
Bernalillo
Colfax
Mora
Rio Arriba
San Miguel
Santa Fe
Taos

NEW YORK
Albany
Allegany
Broome
Cattaraugus
Cayuga
Chautauqua
Chemung
Chenango
Columbia
Cortland
Delaware
Dutchess
Erie
Genesee
Greene
Livingston
Madison
Onondaga
Ontario
Orange
Otsego
Putnam
Rensselaer
Schoharie
Schuyler
Seneca
Steuben
Sullivan
Tioga
Tompkins
Ulster
Washington
Wyoming
Yates

N. CAROLINA
Alleghany
Buncombe
Cherokee
Henderson
Mitchell
Rockingham
Transylvania
Watauga

N. DAKOTA
All Counties

OHIO
Adams
Allen
Ashland
Auglaize
Belmont
Butler
Carroll
Champaign
Clark
Clinton
Columbiana
Coshocton
Crawford
Darke
Delaware
Fairfield
Fayette
Franklin
Greene
Guernsey
Hamilton
Hancock
Hardin
Harrison
Holmes
Huron
Jefferson
Knox
Licking
Logan
Madison
Marion
Mercer
Miami
Montgomery
Morrow
Muskingum
Perry
Pickaway
Pike
Preble
Richland
Ross
Seneca
Shelby
Stark
Summit
Tuscarawas
Union
Van Wert
Warren
Wayne
Wyandot

PENNSYLVANIA
Adams
Allegheny
Armstrong
Beaver
Bedford
Berks
Blair
Bradford
Bucks
Butler
Cameron
Carbon
Centre
Chester
Clarion
Clearfield
Clinton
Columbia
Cumberland
Dauphin
Delaware
Franklin
Fulton
Huntingdon
Indiana
Juniata
Lackawanna
Lancaster
Lebanon
Lehigh
Luzerne
Lycoming
Mifflin
Monroe
Montgomery
Montour
Northampton
Northumberland
Perry
Schuylkill
Snyder
Sullivan
Susquehanna
Tioga
Union
Venango
Westmoreland
Wyoming
York

RHODE ISLAND
Kent
Washington

S. CAROLINA
Greenville

S. DAKOTA
Aurora
Beadle
Bon Homme
Brookings
Brown
Brule
Buffalo
Campbell
Charles Mix
Clark
Clay
Codington
Corson
Davison
Day
Deuel
Douglas
Edmunds
Faulk
Grant
Hamlin
Hand
Hanson
Hughes
Hutchinson
Hyde
Jerauld
Kingsbury
Lake
Lincoln
Lyman
Marshall
McCook
McPherson
Miner
Minnehaha
Moody
Perkins
Potter
Roberts
Sanborn
Spink
Stanley
Sully
Turner
Union
Walworth
Yankton

TENNESEE
Anderson
Bedford
Blount
Bradley
Claiborne
Davidson
Giles
Grainger
Greene
Hamblen
Hancock
Hawkins
Hickman
Humphreys
Jackson
Jefferson
Knox
Lawrence
Lewis
Lincoln
Loudon
Marshall
Maury
McMinn
Meigs
Monroe
Moore
Perry
Roane
Rutherford
Smith
Sullivan
Trousdale
Union
Washington
Wayne
Williamson
Wilson

UTAH
Carbon
Duchesne
Grand
Piute
Sanpete
Sevier
Uintah

VIRGINIA
Alleghany
Amelia
Appomattox
Augusta
Bath
Bland
Botetourt
Bristol
Brunswick
Buckingham
Buena Vista
Campbell
Chesterfield
Clarke
Clifton Forge
Covington
Craig
Cumberland
Danville
Dinwiddie
Fairfax
Falls Church
Fluvanna
Frederick
Fredericksburg
Giles
Goochland
Harrisonburg
Henry
Highland
Lee
Lexington
Louisa
Martinsville
Montgomery
Nottoway
Orange
Page
Patrick
Pittsylvania
Powhatan
Pulaski
Radford
Roanoke
Rockbridge
Rockingham
Russell
Salem
Scott
Shenandoah
Smyth
Spotsylvania
Stafford
Staunton
Tazewell
Warren
Washington
Waynesboro
Winchester
Wythe

WASHINGTON
Clark
Ferry
Okanogan
Pend Oreille
Skamania
Spokane
Stevens

W. VIRGINIA
Berkeley
Brooke
Grant
Greenbrier
Hampshire
Hancock
Hardy
Jefferson
Marshall
Mercer
Mineral
Monongalia
Monroe
Morgan
Ohio
Pendleton
Pocahontas
Preston
Summers
Wetzel

WISCONSIN
Buffalo
Crawford
Dane
Dodge
Door
Fond du Lac
Grant
Green
Green Lake
Iowa
Jefferson
Lafayette
Langlade
Marathon
Menominee
Pepin
Pierce
Portage
Richland
Rock
Shawano
St. Croix
Vernon
Walworth
Washington
Waukesha
Waupaca
Wood

WYOMING
Albany
Big Horn
Campbell
Carbon
Converse
Crook
Fremont
Goshen
Hot Springs
Johnson
Laramie
Lincoln
Natrona
Niobrara
Park
Sheridan
Sublette
Sweetwater
Teton
Uinta
Washakie

a. The EPA recommends that this county listing be supplemented with other available State and local data to further understand the radon potential of a Zone 1 area.

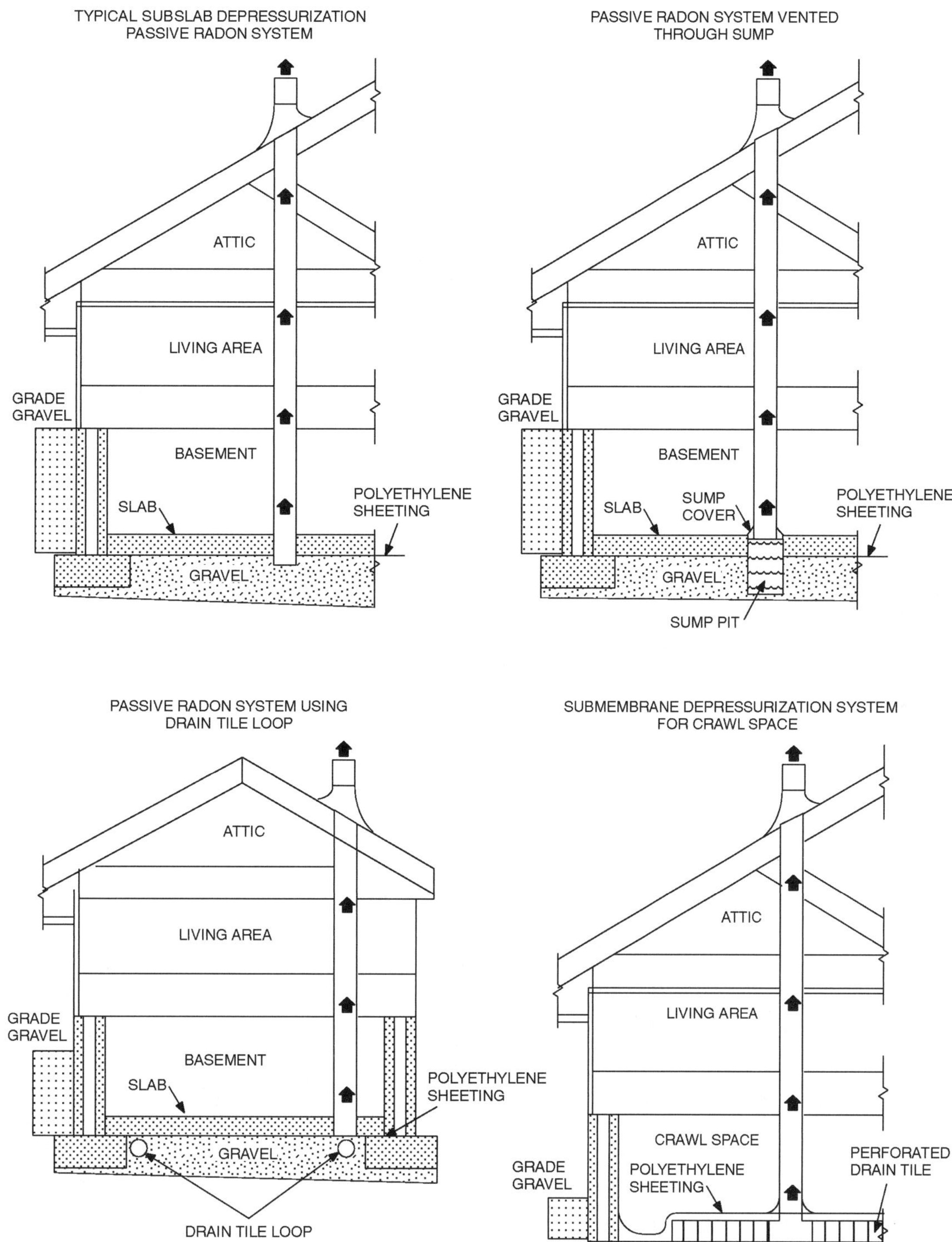

FIGURE AF102
RADON-RESISTANT CONSTRUCTION DETAILS FOR FOUR FOUNDATION TYPES

APPENDIX G

SWIMMING POOLS, SPAS AND HOT TUBS

(The provisions contained in this appendix are not mandatory unless specifically referenced in the adopting ordinance.)

SECTION AG101 GENERAL

AG101.1 General. The provisions of this appendix shall control the design and construction of swimming pools, spas and hot tubs installed in or on the *lot* of a one- or two-family dwelling.

AG101.2 Pools in flood hazard areas. Pools that are located in flood hazard areas established by Table R301.2(1), including above-ground pools, on-ground pools and in-ground pools that involve placement of fill, shall comply with Section AG101.2.1 or AG101.2.2.

Exception: Pools located in riverine flood hazard areas which are outside of designated floodways.

AG101.2.1 Pools located in designated floodways. Where pools are located in designated floodways, documentation shall be submitted to the *building official* which demonstrates that the construction of the pool will not increase the design flood elevation at any point within the *jurisdiction*.

AG101.2.2 Pools located where floodways have not been designated. Where pools are located where design flood elevations are specified but floodways have not been designated, the applicant shall provide a floodway analysis that demonstrates that the proposed pool will not increase the design flood elevation more than 1 foot (305 mm) at any point within the *jurisdiction*.

SECTION AG102 DEFINITIONS

AG102.1 General. For the purposes of these requirements, the terms used shall be defined as follows and as set forth in Chapter 2.

ABOVE-GROUND/ON-GROUND POOL. See "Swimming pool."

BARRIER. A fence, wall, building wall or combination thereof which completely surrounds the swimming pool and obstructs access to the swimming pool.

HOT TUB. See "Swimming pool."

IN-GROUND POOL. See "Swimming pool."

RESIDENTIAL. That which is situated on the premises of a detached one- or two-family dwelling, or a one-family *townhouse* not more than three stories in height.

SPA, NONPORTABLE. See "Swimming pool."

SPA, PORTABLE. A nonpermanent structure intended for recreational bathing, in which all controls, water-heating and water-circulating *equipment* are an integral part of the product.

SWIMMING POOL. Any structure intended for swimming or recreational bathing that contains water more than 24 inches (610 mm) deep. This includes in-ground, above-ground and on-ground swimming pools, hot tubs and spas.

SWIMMING POOL, INDOOR. A swimming pool which is totally contained within a structure and surrounded on all four sides by the walls of the enclosing structure.

SWIMMING POOL, OUTDOOR. Any swimming pool which is not an indoor pool.

SECTION AG103 SWIMMING POOLS

AG103.1 In-ground pools. In-ground pools shall be designed and constructed in compliance with ANSI/NSPI-5.

AG103.2 Above-ground and on-ground pools. Above-ground and on-ground pools shall be designed and constructed in compliance with ANSI/NSPI-4.

AG103.3 Pools in flood hazard areas. In flood hazard areas established by Table R301.2(1), pools in coastal high-hazard areas shall be designed and constructed in compliance with ASCE 24.

SECTION AG104 SPAS AND HOT TUBS

AG104.1 Permanently installed spas and hot tubs. Permanently installed spas and hot tubs shall be designed and constructed in compliance with ANSI/NSPI-3.

AG104.2 Portable spas and hot tubs. Portable spas and hot tubs shall be designed and constructed in compliance with ANSI/NSPI-6.

SECTION AG105 BARRIER REQUIREMENTS

AG105.1 Application. The provisions of this appendix shall control the design of barriers for residential swimming pools, spas and hot tubs. These design controls are intended to provide protection against potential drownings and near-drownings by restricting access to swimming pools, spas and hot tubs.

AG105.2 Outdoor swimming pool. An outdoor swimming pool, including an in-ground, above-ground or on-ground pool, hot tub or spa, shall be surrounded by a barrier which shall comply with the following:

1. The top of the barrier shall be at least 48 inches (1219 mm) above *grade* measured on the side of the barrier which faces away from the swimming pool. The maximum vertical clearance between grade and the bottom of the barrier shall be 2 inches (51 mm) measured on the side of the barrier which faces away from the swimming pool. Where the top of the pool structure is above grade, such as an above-ground pool, the barrier may be at ground level, such as the pool structure, or mounted on top of the pool structure. Where the barrier is mounted on top of the pool structure, the maximum vertical clearance between the top of the pool structure and the bottom of the barrier shall be 4 inches (102 mm).

2. Openings in the barrier shall not allow the passage of a 4-inch-diameter (102 mm) sphere.

3. Solid barriers which do not have openings, such as a masonry or stone wall, shall not contain indentations or protrusions, except for normal construction tolerances and tooled masonry joints.

4. Where the barrier is composed of horizontal and vertical members, and the distance between the tops of the horizontal members is less than 45 inches (1143 mm), the horizontal members shall be located on the swimming pool side of the fence. Spacing between vertical members shall not exceed $1^3/_4$ inches (44 mm) in width. Where there are decorative cutouts within vertical members, spacing within the cutouts shall not exceed $1^3/_4$ inches (44 mm) in width.

5. Where the barrier is composed of horizontal and vertical members, and the distance between the tops of the horizontal members is 45 inches (1143 mm) or more, spacing between vertical members shall not exceed 4 inches (102 mm). Where there are decorative cutouts within vertical members, spacing within the cutouts shall not exceed $1^3/_4$ inches (44 mm) in width.

6. Maximum mesh size for chain link fences shall be a $2^1/_4$-inch (57 mm) square, unless the fence has slats fastened at the top or the bottom which reduce the openings to not more than $1^3/_4$ inches (44 mm).

7. Where the barrier is composed of diagonal members, such as a lattice fence, the maximum opening formed by the diagonal members shall not be more than $1^3/_4$ inches (44 mm).

8. Access gates shall comply with the requirements of Items 1 through 7, and shall be equipped to accommodate a locking device. Pedestrian access gates shall open outward away from the pool, and shall be self-closing and have a self-latching device. Gates, other than pedestrian access gates, shall have a self-latching device. Where the release mechanism of the self-latching device is located less than 54 inches (1372 mm) from the bottom of the gate, the release mechanism and openings shall comply with the following:

   8.1. The release mechanism shall be located on the pool side of the gate at least 3 inches (76 mm) below the top of the gate; and

   8.2. The gate and barrier shall have no opening larger than $^1/_2$ inch (12.7 mm) within 18 inches (457 mm) of the release mechanism.

9. Where a wall of a *dwelling* serves as part of the barrier, one of the following conditions shall be met:

   9.1. The pool shall be equipped with a powered safety cover in compliance with ASTM F 1346;

   9.2. Doors with direct access to the pool through that wall shall be equipped with an alarm which produces an audible warning when the door and/or its screen, if present, are opened. The alarm shall be listed and *labeled* in accordance with UL 2017. The deactivation switch(es) shall be located at least 54 inches (1372 mm) above the threshold of the door; or

   9.3. Other means of protection, such as self-closing doors with self-latching devices, which are *approved* by the governing body, shall be acceptable as long as the degree of protection afforded is not less than the protection afforded by Item 9.1 or 9.2 described herein.

10. Where an above-ground pool structure is used as a barrier or where the barrier is mounted on top of the pool structure, and the means of access is a ladder or steps:

   10.1. The ladder or steps shall be capable of being secured, locked or removed to prevent access; or

   10.2. The ladder or steps shall be surrounded by a barrier which meets the requirements of Items 1 through 9. When the ladder or steps are secured, locked or removed, any opening created shall not allow the passage of a 4-inch-diameter (102 mm) sphere.

AG105.3 Indoor swimming pool. Walls surrounding an indoor swimming pool shall comply with Item 9 of Section AG105.2.

AG105.4 Prohibited locations. Barriers shall be located to prohibit permanent structures, *equipment* or similar objects from being used to climb them.

AG105.5 Barrier exceptions. Spas or hot tubs with a safety cover which comply with ASTM F 1346 shall be exempt from the provisions of this appendix.

SECTION AG106 ENTRAPMENT PROTECTION FOR SWIMMING POOL AND SPA SUCTION OUTLETS

AG106.1 General. Suction outlets shall be designed and installed in accordance with ANSI/APSP-7.

SECTION AG107
ABBREVIATIONS

AG107.1 General.

ANSI—American National Standards Institute
11 West 42nd Street
New York, NY 10036

APSP—Association of Pool and Spa Professionals
NSPI—National Spa and Pool Institute
2111 Eisenhower Avenue
Alexandria, VA 22314

ASCE—American Society of Civil Engineers
1801 Alexander Bell Drive
Reston, VA 98411-0700

ASTM—ASTM International
100 Barr Harbor Drive
West Conshohocken, PA 19428

UL—Underwriters Laboratories, Inc.
333 Pfingsten Road
Northbrook, IL 60062-2096

SECTION AG108
REFERENCED STANDARDS

AG108.1 General.

ANSI/NSP

ANSI/NSPI-3—99 Standard for Permanently Installed Residential Spas AG104.1

ANSI/NSPI-4—99 Standard for Above-ground/ On-ground Residential Swimming Pools AG103.2

ANSI/NSPI-5—03 Standard for Residential In-ground Swimming Pools. . AG103.1

ANSI/NSPI-6—99 Standard for Residential Portable Spas AG104.2

ANSI/APSP

ANSI/APSP-7—06 Standard for Suction Entrapment Avoidance in Swimming Pools, Wading Pools, Spas, Hot Tubs and Catch Basins AG106.1

ASCE

ASCE/SEI-24—05 Flood-resistant Design and Construction AG103.3

ASTM

ASTM F 1346—91 (2003) Performance Specification for Safety Covers and Labeling Requirements for All Covers for Swimming Pools Spas and Hot Tubs AG105.2, AG105.5

UL

UL 2017—2000 Standard for General-purpose Signaling Devices and Systems—with revisions through June 2004 . . . . AG105.2

APPENDIX H

PATIO COVERS

(The provisions contained in this appendix are not mandatory unless specifically referenced in the adopting ordinance.)

SECTION AH101 GENERAL

AH101.1 Scope. Patio covers shall conform to the requirements of Sections AH101 through AH105.

AH101.2 Permitted uses. Patio covers shall be permitted to be detached from or attached to *dwelling units*. Patio covers shall be used only for recreational, outdoor living purposes, and not as carports, garages, storage rooms or habitable rooms.

SECTION AH102 DEFINITION

AH102.1 General. The following word and term shall, for the purposes of this appendix, have the meaning shown herein.

PATIO COVER. A structure with open or glazed walls which is used for recreational, outdoor living purposes associated with a dwelling unit.

SECTION AH103 EXTERIOR WALLS AND OPENINGS

AH 103.1 Enclosure walls. Enclosure walls shall be permitted to be of any configuration, provided the open or glazed area of the longer wall and one additional wall is equal to at least 65 percent of the area below a minimum of 6 feet, 8 inches (2032 mm) of each wall, measured from the floor. Openings shall be permitted to be enclosed with the following:

1. Insect screening;
2. Approved translucent or transparent plastic not more than 0.125 inch (3.2 mm) in thickness;
3. Glass conforming to the provisions of Section R308; or
4. Any combination of the foregoing.

AH103.2 Light, ventilation and emergency egress. Exterior openings required for light and ventilation shall be permitted to open into a patio structure conforming to Section AH101, provided that the patio structure shall be unenclosed if such openings are serving as emergency egress or rescue openings from sleeping rooms. Where such exterior openings serve as an exit from the *dwelling unit*, the patio structure, unless unenclosed, shall be provided with exits conforming to the provisions of Section R310.

SECTION AH104 HEIGHT

AH104.1 Height. Patio covers are limited to one-story structures not exceeding 12 feet (3657 mm) in height.

SECTION AH105 STRUCTURAL PROVISIONS

AH105.1 Design loads. Patio covers shall be designed and constructed to sustain, within the stress limits of this code, all dead loads plus a minimum vertical live load of 10 pounds per square foot (0.48 kN/m^2), except that snow loads shall be used where such snow loads exceed this minimum. Such covers shall be designed to resist the minimum wind loads set forth in Section R301.2.1.

AH105.2 Footings. In areas with a frostline depth of zero as specified in Table R301.2(1), a patio cover shall be permitted to be supported on a slab-on-*grade* without footings, provided the slab conforms to the provisions of Section R506, is not less than 3.5 inches (89 mm) thick and the columns do not support live and dead loads in excess of 750 pounds (3.34 kN) per column.

SECTION AH106 SPECIAL PROVISIONS FOR ALUMINUM SCREEN ENCLOSURES IN HURRICANE-PRONE REGIONS

AH106.1 General. Screen enclosures in *hurricane-prone regions* shall be in accordance with the provisions of this Section.

AH106.1.1 Habitable spaces. Screen enclosures shall not be considered *habitable spaces*.

AH106.1.2 Minimum ceiling height. Screen enclosures shall have a ceiling height of not less than 7 feet (2134 mm).

AH106.2 Definition. The following word and term shall, for the purposes of this appendix, have the meaning shown herein.

SCREEN ENCLOSURE. A building or part thereof, in whole or in part self-supporting, and having walls of insect screening, and a roof of insect screening, plastic, aluminum or similar lightweight material.

AH106.3 Screen enclosures. Screen enclosures shall comply with Sections AH106.3.1 and AH106.3.2

AH106.3.1 Thickness. Actual wall thickness of extruded aluminum members shall be not less than 0.040 inch (1.02 mm).

AH106.3.2 Density. Screen density shall be a maximum of 20 threads per inch by 20 threads per inch mesh.

AH106.4 Design. The structural design of screen enclosures shall comply with Sections AH106.4.1 through AH106.4.4.

AH106.4.1 Wind load. Structural members supporting screen enclosures shall be designed to support the minimum wind loads given in Tables AH106.4(1) and AH106.4(2). Where any value is less than 10 pounds per square foot (psf) (0.479 kN/m^2) use 10 pounds per square feet (0.479 kN/m^2).

AH106.4.2 Deflection limit. For members supporting screen surfaces only, the total load deflection shall not exceed *l*/60. Screen surfaces shall be permitted to include a maximum of 25-percent solid flexible finishes.

AH106.4.3 Importance factor. The wind factor for screen enclosures shall be 0.77 in accordance with Section 6.5.5 of ASCE 7.

AH106.4.4 Roof live load. The minimum roof live load shall be 10 psf (0.479 kN/m^2).

AH106.5 Footings. In areas with a frost line depth of zero, a screen enclosure shall be permitted to be supported on a concrete slab-on-*grade* without footings, provided the slab conforms to the provisions of Section R506, is not less than $3^1/_2$ inches (89 mm) thick and the columns do not support loads in excess of 750 pounds (3.36 kN) per column.

TABLE AH106.4(2)
HEIGHT ADJUSTMENT FACTORS

MEAN	EXPOSURE	
Roof Height (feet)	**B**	**C**
15	1	0.86
20	1	0.92
25	1	0.96
30	1	1.00
35	1.05	1.03
40	1.09	1.06
45	1.12	1.09
50	1.16	1.11
55	1.19	1.14
60	1.22	1.16

For SI: 1 foot = 304.8 mm.

TABLE AH106.4(1)
DESIGN WIND PRESSURES FOR ALUMINUM SCREEN ENCLOSURE FRAMING WITH AN IMPORTANCE FACTOR OF 0.77[a, b, c]

LOAD CASE	WALL	Basic Wind Speed (mph)											
		100		110		120		130		140		150	
		Exposure Category Design Pressure (psf)											
		C	B	C	B	C	B	C	B	C	B	C	B
A[d]	Windward and leeward walls (flow thru) and windward wall (nonflow thru) *L/W* = 0-1	12	8	14	10	17	12	19	14	23	16	26	18
A[d]	Windward and leeward walls (flow thru) and windward wall (nonflow thru) *L/W* = 2	13	9	16	11	19	14	22	16	26	18	30	21
B[e]	Windward: Nongable roof	16	12	20	14	24	17	28	20	32	23	37	26
B[e]	Windward: Gable roof	22	16	27	19	32	23	38	27	44	31	50	36
	ROOF												
All[f]	Roof-screen	4	3	5	4	6	4	7	5	8	6	9	7
All[f]	Roof-solid	12	9	15	11	18	13	21	15	24	17	28	20

For SI: 1 mile per hour = 0.44 m/s, 1 pound per square foot = 0.0479 kPa, 1 foot = 304.8 mm.

a. Values have been reduced for 0.77 importance factor in accordance with Section AH106.4.3.

b. Minimum design pressure shall be 10 psf in accordance with Section AH106.4.1.

c. Loads are applicable to screen enclosures with a mean roof height of 30 feet or less. For screen enclosures of different heights, the pressures given shall be adjusted by multiplying the table pressure by the adjustment factor given in Table AH106.4(2).

d. For Load Case A flow thru condition, the pressure given shall be applied simultaneously to both the upwind and downwind screen walls acting in the same direction as the wind. The structure shall also be analyzed for wind coming from the opposite direction. For the nonflow thru condition, the screen enclosure wall shall be analyzed for the load applied acting toward the interior of the enclosure.

e. For Load Case B, the table pressure multiplied by the projected frontal area of the screen enclosure is the total drag force, including drag on screen surfaces parallel to the wind, which must be transmitted to the ground. Use Load Case A for members directly supporting the screen surface perpendicular to the wind. Load Case B loads shall be applied only to structural members which carry wind loads from more than one surface.

f. The roof structure shall be analyzed for the pressure given occurring both upward and downward.

APPENDIX I

PRIVATE SEWAGE DISPOSAL

(The provisions contained in this appendix are not mandatory unless specifically referenced in the adopting ordinance.)

SECTION AI101
GENERAL

AI101.1 Scope. Private sewage disposal systems shall conform to the *International Private Sewage Disposal Code.*

APPENDIX J

EXISTING BUILDINGS AND STRUCTURES

(The provisions contained in this appendix are not mandatory unless specifically referenced in the adopting ordinance.)

SECTION AJ101 PURPOSE AND INTENT

AJ101.1 General. The purpose of these provisions is to encourage the continued use or reuse of legally existing buildings and structures. These provisions are intended to permit work in existing buildings that is consistent with the purpose of this code. Compliance with these provisions shall be deemed to meet the requirements of this code.

AJ101.2 Classification of work. For purposes of this appendix, all work in existing buildings shall be classified into the categories of repair, renovation, *alteration* and reconstruction. Specific requirements are established for each category of work in these provisions.

AJ101.3 Multiple categories of work. Work of more than one category may be part of a single work project. All related work permitted within a 12-month period shall be considered a single work project. Where a project includes one category of work in one building area and another category of work in a separate and unrelated area of the building, each project area shall comply with the requirements of the respective category of work. Where a project with more than one category of work is performed in the same area or in related areas of the building, the project shall comply with the requirements of the more stringent category of work.

SECTION AJ102 COMPLIANCE

AJ102.1 General. Regardless of the category of work being performed, the work shall not cause the structure to become unsafe or adversely affect the performance of the building; shall not cause an existing mechanical or plumbing system to become unsafe, hazardous, insanitary or overloaded; and unless expressly permitted by these provisions, shall not make the building any less compliant with this code or to any previously *approved* alternative arrangements than it was before the work was undertaken.

AJ102.2 Requirements by category of work. Repairs shall conform to the requirements of Section AJ301. Renovations shall conform to the requirements of Section AJ401. *Alterations* shall conform to the requirements of Section AJ501 and the requirements for renovations. Reconstructions shall conform to the requirements of Section AJ601 and the requirements for *alterations* and renovations.

AJ102.3 Smoke detectors. Regardless of the category of work, smoke detectors shall be provided where required by Section R314.3.1.

AJ102.4 Replacement windows. Regardless of the category of work, when an existing window, including the sash and glazed portion, is replaced, the replacement window shall comply with the requirements of Chapter 11.

AJ102.5 Flood hazard areas. Work performed in existing buildings located in a flood hazard area as established by Table R301.2(1) shall be subject to the provisions of Section R105.3.1.1.

AJ102.6 Equivalent alternatives. These provisions are not intended to prevent the use of any alternative material, alternative design or alternative method of construction not specifically prescribed herein, provided any alternative has been deemed to be equivalent and its use authorized by the *building official*.

AJ102.7 Other alternatives. Where compliance with these provisions or with this code as required by these provisions is technically infeasible or would impose disproportionate costs because of structural, construction or dimensional difficulties, other alternatives may be accepted by the *building official*. These alternatives may include materials, design features and/or operational features.

AJ102.8 More restrictive requirements. Buildings or systems in compliance with the requirements of this code for new construction shall not be required to comply with any more restrictive requirement of these provisions.

AJ102.9 Features exceeding code requirements. Elements, components and systems of existing buildings with features that exceed the requirements of this code for new construction, and are not otherwise required as part of *approved* alternative arrangements or deemed by the *building official* to be required to balance other building elements not complying with this code for new construction, shall not be prevented by these provisions from being modified as long as they remain in compliance with the applicable requirements for new construction.

SECTION AJ103 PRELIMINARY MEETING

AJ103.1 General. If a building *permit* is required at the request of the prospective *permit* applicant, the *building official* or his designee shall meet with the prospective applicant to discuss plans for any proposed work under these provisions prior to the application for the *permit*. The purpose of this preliminary meeting is for the *building official* to gain an understanding of the prospective applicant's intentions for the proposed work, and to determine, together with the prospective applicant, the specific applicability of these provisions.

SECTION AJ104
EVALUATION OF AN EXISTING BUILDING

AJ104.1 General. The *building official* may require an existing building to be investigated and evaluated by a registered design professional in the case of proposed reconstruction of any portion of a building. The evaluation shall determine the existence of any potential nonconformities to these provisions, and shall provide a basis for determining the impact of the proposed changes on the performance of the building. The evaluation shall use the following sources of information, as applicable:

1. Available documentation of the existing building.
   - 1.1. Field surveys.
   - 1.2. Tests (nondestructive and destructive).
   - 1.3. Laboratory analysis.

Exception: Detached one- or two-family dwellings that are not irregular buildings under Section R301.2.2.2.5 and are not undergoing an extensive reconstruction shall not be required to be evaluated.

SECTION AJ105
PERMIT

AJ105.1 Identification of work area. The work area shall be clearly identified on all *permits* issued under these provisions.

SECTION AJ201
DEFINITIONS

AJ201.1 General. For purposes of this appendix, the terms used are defined as follows.

ALTERATION. The reconfiguration of any space; the *addition* or elimination of any door or window; the reconfiguration or extension of any system; or the installation of any additional *equipment*.

CATEGORIES OF WORK. The nature and extent of construction work undertaken in an existing building. The categories of work covered in this appendix, listed in increasing order of stringency of requirements, are repair, renovation, *alteration* and reconstruction.

DANGEROUS. Where the stresses in any member; the condition of the building, or any of its components or elements or attachments; or other condition that results in an overload exceeding 150 percent of the stress allowed for the member or material in this code.

EQUIPMENT OR FIXTURE. Any plumbing, heating, electrical, ventilating, air-conditioning, refrigerating and fire protection *equipment*; and elevators, dumb waiters, boilers, pressure vessels, and other mechanical facilities or installations that are related to building services.

LOAD-BEARING ELEMENT. Any column, girder, beam, joist, truss, rafter, wall, floor or roof sheathing that supports any vertical load in addition to its own weight, and/or any lateral load.

MATERIALS AND METHODS REQUIREMENTS. Those requirements in this code that specify material standards; details of installation and connection; joints; penetrations; and continuity of any element, component or system in the building. The required quantity, fire resistance, flame spread, acoustic or thermal performance, or other performance attribute is specifically excluded from materials and methods requirements.

RECONSTRUCTION. The reconfiguration of a space that affects an exit, a renovation and/or *alteration* when the work area is not permitted to be occupied because existing means-of-egress and fire protection systems, or their equivalent, are not in place or continuously maintained; and/or there are extensive *alterations* as defined in Section AJ501.3.

REHABILITATION. Any repair, renovation, *alteration* or reconstruction work undertaken in an existing building.

RENOVATION. The change, strengthening or *addition* of load-bearing elements; and/or the refinishing, replacement, bracing, strengthening, upgrading or extensive repair of existing materials, elements, components, *equipment* and/or fixtures. Renovation involves no reconfiguration of spaces. Interior and exterior painting are not considered refinishing for purposes of this definition, and are not renovation.

REPAIR. The patching, restoration and/or minor replacement of materials, elements, components, *equipment* and/or fixtures for the purposes of maintaining those materials, elements, components, *equipment* and/or fixtures in good or sound condition.

WORK AREA. That portion of a building affected by any renovation, *alteration* or reconstruction work as initially intended by the owner and indicated as such in the *permit*. Work area excludes other portions of the building where incidental work entailed by the intended work must be performed, and portions of the building where work not initially intended by the owner is specifically required by these provisions for a renovation, *alteration* or reconstruction.

SECTION AJ301
REPAIRS

AJ301.1 Materials. Except as otherwise required herein, work shall be done using like materials or materials permitted by this code for new construction.

AJ301.1.1 Hazardous materials. Hazardous materials no longer permitted, such as asbestos and lead-based paint, shall not be used.

AJ301.1.2 Plumbing materials and supplies. The following plumbing materials and supplies shall not be used:

1. All-purpose solvent cement, unless listed for the specific application;
2. Flexible traps and tailpieces, unless listed for the specific application; and
3. Solder having more than 0.2 percent lead in the repair of potable water systems.

AJ301.2 Water closets. When any water closet is replaced with a newly manufactured water closet, the replacement water closet shall comply with the requirements of Section P2903.2.

AJ301.3 Safety glazing. Replacement glazing in hazardous locations shall comply with the safety glazing requirements of Section R308.1.

AJ301.4 Electrical. Repair or replacement of existing electrical wiring and *equipment* undergoing repair with like material shall be permitted.

Exceptions:

1. Replacement of electrical receptacles shall comply with the requirements of Chapters 34 through 43.
2. Plug fuses of the Edison-base type shall be used for replacements only where there is no evidence of overfusing or tampering in accordance with the applicable requirements of Chapters 34 through 43.
3. For replacement of nongrounding-type receptacles with grounding-type receptacles and for branch circuits that do not have an *equipment* grounding conductor in the branch circuitry, the grounding conductor of a grounding-type receptacle outlet shall be permitted to be grounded to any accessible point on the grounding electrode system, or to any accessible point on the grounding electrode conductor, as allowed and described in Chapters 34 through 43.

SECTION AJ401 RENOVATIONS

AJ401.1 Materials and methods. The work shall comply with the materials and methods requirements of this code.

AJ401.2 Door and window dimensions. Minor reductions in the clear opening dimensions of replacement doors and windows that result from the use of different materials shall be allowed, whether or not they are permitted by this code.

AJ401.3 Interior finish. Wood paneling and textile wall coverings used as an interior finish shall comply with the flame spread requirements of Section R302.9.

AJ401.4 Structural. Unreinforced masonry buildings located in Seismic Design Category D_2 or E shall have parapet bracing and wall anchors installed at the roofline whenever a reroofing *permit* is issued. Such parapet bracing and wall anchors shall be of an *approved* design.

SECTION AJ501 ALTERATIONS

AJ501.1 Newly constructed elements. Newly constructed elements, components and systems shall comply with the requirements of this code.

Exceptions:

1. Openable windows may be added without requiring compliance with the light and ventilation requirements of Section R303.
2. Newly installed electrical *equipment* shall comply with the requirements of Section AJ501.5.

AJ501.2 Nonconformities. The work shall not increase the extent of noncompliance with the requirements of Section AJ601, or create nonconformity to those requirements which did not previously exist.

AJ501.3 Extensive alterations. When the total area of all the work areas included in an *alteration* exceeds 50 percent of the area of the *dwelling unit*, the work shall be considered a reconstruction and shall comply with the requirements of these provisions for reconstruction work.

Exception: Work areas in which the *alteration* work is exclusively plumbing, mechanical or electrical shall not be included in the computation of the total area of all work areas.

AJ501.4 Structural. The minimum design loads for the structure shall be the loads applicable at the time the building was constructed, provided that no dangerous condition is created. Structural elements that are uncovered during the course of the *alteration* and that are found to be unsound or dangerous shall be made to comply with the applicable requirements of this code.

AJ501.5 Electrical equipment and wiring.

AJ501.5.1 Materials and methods. Newly installed electrical *equipment* and wiring relating to work done in any work area shall comply with the materials and methods requirements of Chapters 34 through 43.

Exception: Electrical *equipment* and wiring in newly installed partitions and ceilings shall comply with all the applicable requirements of Chapters 34 through 43.

AJ501.5.2 Electrical service. Service to the *dwelling unit* shall be a minimum of 100 ampere, three-wire capacity and service *equipment* shall be dead front having no live parts exposed that could allow accidental contact. Type "S" fuses shall be installed when fused *equipment* is used.

Exception: Existing service of 60 ampere, three-wire capacity, and feeders of 30 ampere or larger two- or three-wire capacity shall be accepted if adequate for the electrical load being served.

AJ501.5.3 Additional electrical requirements. When the work area includes any of the following areas within a *dwelling unit*, the requirements of Sections AJ501.5.3.1 through AJ501.5.3.5 shall apply.

AJ501.5.3.1 Enclosed areas. Enclosed areas other than closets, kitchens, *basements*, garages, hallways, laundry areas and bathrooms shall have a minimum of two duplex receptacle outlets, or one duplex receptacle outlet and one ceiling- or wall-type lighting outlet.

AJ501.5.3.2 Kitchen and laundry areas. Kitchen areas shall have a minimum of two duplex receptacle outlets. Laundry areas shall have a minimum of one duplex receptacle outlet located near the laundry *equipment* and installed on an independent circuit.

AJ501.5.3.3 Ground-fault circuit-interruption. Ground-fault circuit-interruption shall be provided on newly installed receptacle outlets if required by Chapters 34 through 43.

AJ501.5.3.4 Lighting outlets. At least one lighting outlet shall be provided in every bathroom, hallway, stairway, attached garage and detached garage with electric power to illuminate outdoor entrances and exits, and in utility rooms and *basements* where these spaces are used for storage or contain *equipment* requiring service.

AJ501.5.3.5 Clearance. Clearance for electrical service *equipment* shall be provided in accordance with Chapters 34 through 43.

AJ501.6 Ventilation. All reconfigured spaces intended for occupancy and all spaces converted to habitable or occupiable space in any work area shall be provided with ventilation in accordance with Section R303.

AJ501.7 Ceiling height. *Habitable spaces* created in existing *basements* shall have ceiling heights of not less than 6 feet, 8 inches (2032 mm). Obstructions may project to within 6 feet, 4 inches (1930 mm) of the *basement* floor. Existing finished ceiling heights in nonhabitable spaces in *basements* shall not be reduced.

AJ501.8 Stairs.

AJ501.8.1 Stair width. Existing *basement* stairs and handrails not otherwise being altered or modified shall be permitted to maintain their current clear width at, above and below existing handrails.

AJ501.8.2 Stair headroom. Headroom height on existing *basement* stairs being altered or modified shall not be reduced below the existing stairway finished headroom. Existing *basement* stairs not otherwise being altered shall be permitted to maintain the current finished headroom.

AJ501.8.3 Stair landing. Landings serving existing *basement* stairs being altered or modified shall not be reduced below the existing stairway landing depth and width. Existing *basement* stairs not otherwise being altered shall be permitted to maintain the current landing depth and width.

SECTION AJ601
RECONSTRUCTION

AJ601.1 Stairways, handrails and guards.

AJ601.1.1 Stairways. Stairways within the work area shall be provided with illumination in accordance with Section R303.6.

AJ601.1.2 Handrails. Every required exit stairway that has four or more risers, is part of the means of egress for any work area, and is not provided with at least one handrail, or in which the existing handrails are judged to be in danger of collapsing, shall be provided with handrails designed and installed in accordance with Section R311 for the full length of the run of steps on at least one side.

AJ601.1.3 Guards. Every open portion of a stair, landing or balcony that is more than 30 inches (762 mm) above the floor or *grade* below, is part of the egress path for any work area, and does not have guards, or in which the existing guards are judged to be in danger of collapsing, shall be provided with guards designed and installed in accordance with Section R312.

AJ601.2 Wall and ceiling finish. The interior finish of walls and ceilings in any work area shall comply with the requirements of Section R302.9. Existing interior finish materials that do not comply with those requirements shall be removed or shall be treated with an *approved* fire-retardant coating in accordance with the manufacturer's instructions to secure compliance with the requirements of this section.

AJ601.3 Separation walls. Where the work area is in an attached *dwelling unit*, walls separating *dwelling units* that are not continuous from the foundation to the underside of the roof sheathing shall be constructed to provide a continuous fire separation using construction materials consistent with the existing wall or complying with the requirements for new structures. Performance of work shall be required only on the side of the wall of the *dwelling unit* that is part of the work area.

AJ601.4 Ceiling height. *Habitable spaces* created in existing *basements* shall be permitted to have ceiling heights of not less than 6 feet, 8 inches (2032 mm). Obstructions may project to within 6 feet, 4 inches (1930 mm) of the *basement* floor. Existing finished ceiling heights in nonhabitable spaces in *basements* shall not be reduced.

APPENDIX K

SOUND TRANSMISSION

(The provisions contained in this appendix are not mandatory unless specifically referenced in the adopting ordinance.)

SECTION AK101
GENERAL

AK101.1 General. Wall and floor-ceiling assemblies separating *dwelling units*, including those separating adjacent *townhouse* units, shall provide air-borne sound insulation for walls, and both air-borne and impact sound insulation for floor-ceiling assemblies.

SECTION AK102
AIR-BORNE SOUND

AK102.1 General. Air-borne sound insulation for wall and floor-ceiling assemblies shall meet a sound transmission class (STC) rating of 45 when tested in accordance with ASTM E 90. Penetrations or openings in construction assemblies for piping; electrical devices; recessed cabinets; bathtubs; soffits; or heating, ventilating or exhaust ducts shall be sealed, lined, insulated or otherwise treated to maintain the required ratings. *Dwelling unit* entrance doors, which share a common space, shall be tight fitting to the frame and sill.

AK102.1.1 Masonry. The sound transmission class of concrete masonry and clay masonry assemblies shall be calculated in accordance with TMS 0302 or determined through testing in accordance with ASTM E 90.

SECTION AK103
STRUCTURAL-BORNE SOUND

AK103.1 General. Floor/ceiling assemblies between *dwelling units*, or between a *dwelling unit* and a public or service area within a structure, shall have an impact insulation class (IIC) rating of not less than 45 when tested in accordance with ASTM E 492.

SECTION AK104
REFERENCED STANDARDS

ASTM

ASTM E 90—04 Test Method for Laboratory Measurement of Air-borne Sound Transmission Loss of Building Partitions and Elements AK102

ASTM E 492—04 Specification for Laboratory Measurement of Impact Sound Transmission through Floor-ceiling Assemblies Using the Tapping Machine. AK103

The Masonry Society

TMS 0302—07 Standard for Determining the Sound Transmission Class Rating for Masonry Walls AK102.1.1

APPENDIX L
PERMIT FEES

(The provisions contained in this appendix are not mandatory unless specifically referenced in the adopting ordinance.)

TOTAL VALUATION	FEE
$1 to $ 500	$24
$501 to $2,000	$24 for the first $500; plus $3 for each additional $100 or fraction thereof, up to and including $2,000
$2,001 to $40,000	$69 for the first $2,000; plus $11 for each additional $1,000 or fraction thereof, up to and including $40,000
$40,001 to $100,000	$487 for the first $40,000; plus $9 for each additional $1,000 or fraction thereof, up to and including $100,000
$100,001 to $500,000	$1,027 for the first $100,000; plus $7 for each additional $1,000 or fraction thereof, up to and including $500,000
$500,001 to $1,000,000	$3,827 for the first $500,000; plus $5 for each additional $1,000 or fraction thereof, up to and including $1,000,000
$1,000,001 to $5,000,000	$6,327 for the first $1,000,000; plus $3 for each additional $1,000 or fraction thereof, up to and including $5,000,000
$5,000,001 and over	$18,327 for the first $5,000,000; plus $1 for each additional $1,000 or fraction thereof

APPENDIX M
HOME DAY CARE—R-3 OCCUPANCY

(The provisions contained in this appendix are not mandatory unless specifically referenced in the adopting ordinance.)

SECTION AM101 GENERAL

AM101.1 General. This appendix shall apply to a home day care operated within a *dwelling*. It is to include buildings and structures occupied by persons of any age who receive custodial care for less than 24 hours by individuals other than parents or guardians or relatives by blood, marriage, or adoption, and in a place other than the home of the person cared for.

SECTION AM102 DEFINITION

EXIT ACCESS. That portion of a means-of-egress system that leads from any occupied point in a building or structure to an exit.

SECTION AM103 MEANS OF EGRESS

AM103.1 Exits required. If the occupant load of the residence is more than nine, including those who are residents, during the time of operation of the day care, two exits are required from the ground-level *story*. Two exits are required from a home day care operated in a *manufactured home* regardless of the occupant load. Exits shall comply with Section R311.

AM103.1.1 Exit access prohibited. An exit access from the area of day care operation shall not pass through bathrooms, bedrooms, closets, garages, fenced rear *yards* or similar areas.

Exception: An exit may discharge into a fenced *yard* if the gate or gates remain unlocked during day care hours. The gates may be locked if there is an area of refuge located within the fenced *yard* and more than 50 feet (15 240 mm) from the *dwelling*. The area of refuge shall be large enough to allow 5 square feet (0.5 m^2) per occupant.

AM103.1.2 Basements. If the *basement* of a *dwelling* is to be used in the day care operation, two exits are required from the *basement* regardless of the occupant load. One of the exits may pass through the *dwelling* and the other must lead directly to the exterior of the *dwelling*.

Exception: An emergency and escape window complying with Section R310 and which does not conflict with Section AM103.1.1 may be used as the second means of egress from a *basement*.

AM103.1.3 Yards. If the *yard* is to be used as part of the day care operation it shall be fenced.

AM103.1.3.1 Type of fence and hardware. The fence shall be of durable materials and be at least 6 feet (1529 mm) tall, completely enclosing the area used for the day care operations. Each opening shall be a gate or door equipped with a self-closing and self-latching device to be installed at a minimum of 5 feet (1528 mm) above the ground.

Exception: The door of any *dwelling* which forms part of the enclosure need not be equipped with self-closing and self-latching devices.

AM103.1.3.2 Construction of fence. Openings in the fence, wall or enclosure required by this section shall have intermediate rails or an ornamental pattern that do not allow a sphere 4 inches (102 mm) in diameter to pass through. In addition, the following criteria must be met:

1. The maximum vertical clearance between *grade* and the bottom of the fence, wall or enclosure shall be 2 inches (51 mm).
2. Solid walls or enclosures that do not have openings, such as masonry or stone walls, shall not contain indentations or protrusions, except for tooled masonry joints.
3. Maximum mesh size for chain link fences shall be $1^1/_4$ inches (32 mm) square, unless the fence has slats at the top or bottom which reduce the opening to no more than $1^3/_4$ inches (44 mm). The wire shall not be less than 9 gage [0.148 inch (3.8 mm)].

AM103.1.3.3 Decks. Decks that are more than 12 inches (305 mm) above *grade* shall have a guard in compliance with Section R312.

AM103.2 Width and height of an exit. The minimum width of a required exit is 36 inches (914 mm) with a net clear width of 32 inches (813 mm). The minimum height of a required exit is 6 feet, 8 inches (2032 mm).

AM103.3 Type of lock and latches for exits. Regardless of the occupant load served, exit doors shall be openable from the inside without the use of a key or any special knowledge or effort. When the occupant load is 10 or less, a night latch, dead bolt or security chain may be used, provided such devices are openable from the inside without the use of a key or tool, and mounted at a height not to exceed 48 inches (1219 mm) above the finished floor.

AM103.4 Landings. Landings for stairways and doors shall comply with Section R311, except that landings shall be required for the exterior side of a sliding door when a home day care is being operated in a Group R-3 occupancy.

SECTION AM104
SMOKE DETECTION

AM104.1 General. Smoke detectors shall be installed in *dwelling* units used for home day care operations. Detectors shall be installed in accordance with the approved manufacturer's instructions. If the current smoke detection system in the *dwelling* is not in compliance with the currently adopted code for smoke detection, it shall be upgraded to meet the currently adopted code requirements and Section AM103 before day care operations commence.

AM104.2 Power source. Required smoke detectors shall receive their primary power from the building wiring when that wiring is served from a commercial source and shall be equipped with a battery backup. The detector shall emit a signal when the batteries are low. Wiring shall be permanent and without a disconnecting switch other than those required for overcurrent protection. Required smoke detectors shall be interconnected so if one detector is activated, all detectors are activated.

AM104.3 Location. A detector shall be located in each bedroom and any room that is to be used as a sleeping room, and centrally located in the corridor, hallway or area giving access to each separate sleeping area. When the *dwelling* unit has more than one *story*, and in *dwellings* with *basements*, a detector shall be installed on each *story* and in the *basement*. In *dwelling* units where a *story* or *basement* is split into two or more levels, the smoke detector shall be installed on the upper level, except that when the lower level contains a sleeping area, a detector shall be installed on each level. When sleeping rooms are on the upper level, the detector shall be placed at the ceiling of the upper level in close proximity to the stairway. In *dwelling* units where the ceiling height of a room open to the hallway serving the bedrooms or sleeping areas exceed that of the hallway by 24 inches (610 mm) or more, smoke detectors shall be installed in the hallway and the adjacent room. Detectors shall sound an alarm audible in all sleeping areas of the *dwelling* unit in which they are located.

APPENDIX N

VENTING METHODS

(This appendix is informative and is not part of the code. This appendix provides examples of various of venting methods.)

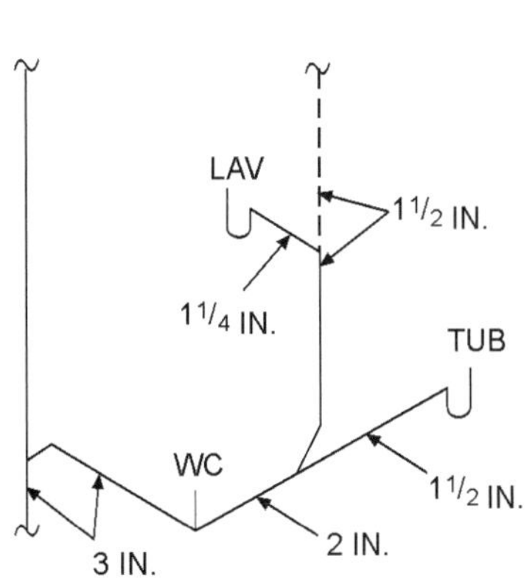

A. TYPICAL SINGLE-BATH ARRANGEMENT

1 1/2 IN.
1 1/4 IN.
WC
LAV
1 1/2 IN.
3 IN.

B. TYPICAL POWDER ROOM

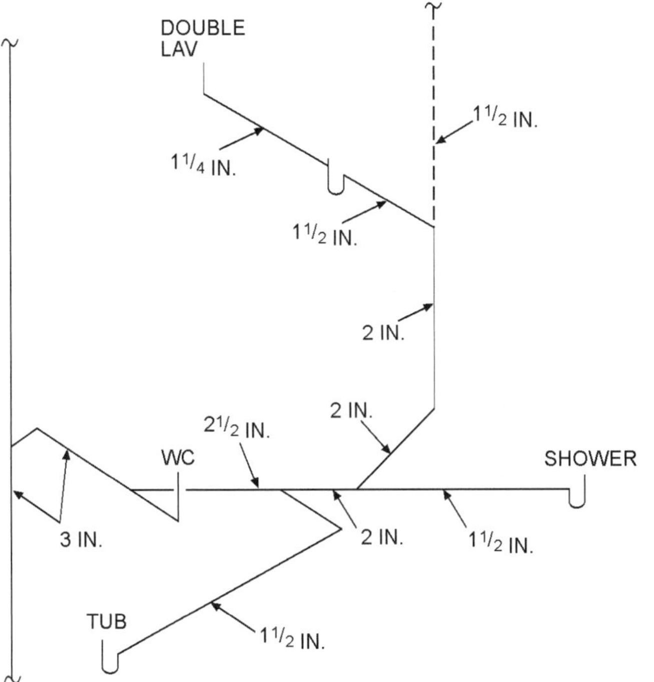

C. MORE ELABORATE SINGLE-BATH ARRANGEMENT

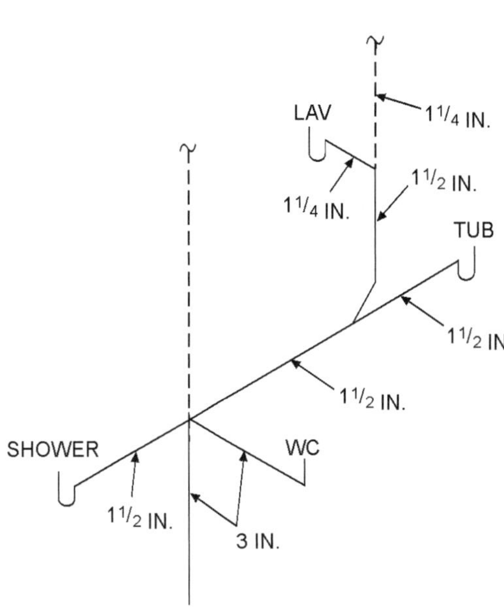

D. COMBINATION WET AND STACK VENTING WITH STACK FITTING

For SI: 1 inch = 25.4 mm.

FIGURE N1
TYPICAL SINGLE-BATH WET-VENT ARRANGEMENTS

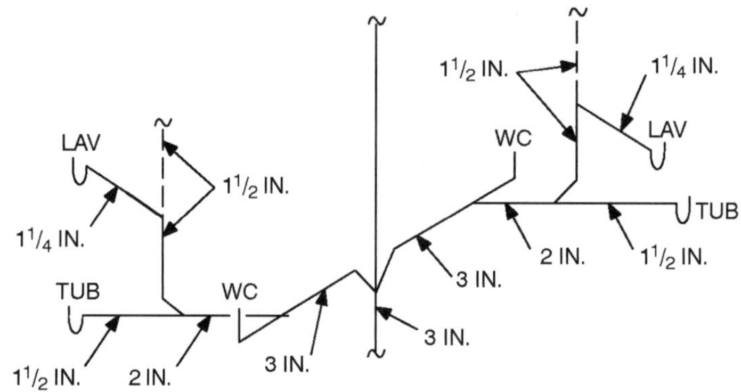

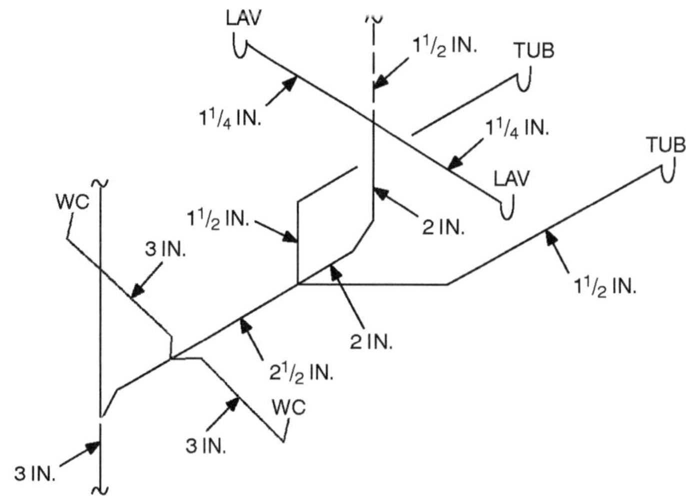

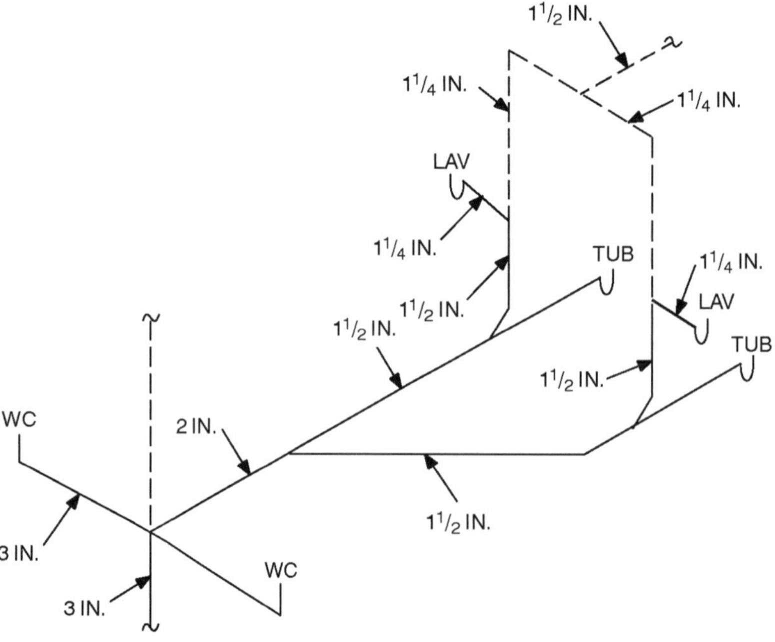

For SI: 1 inch = 25.4 mm.

FIGURE N2
TYPICAL DOUBLE-BATH WET-VENT ARRANGEMENTS

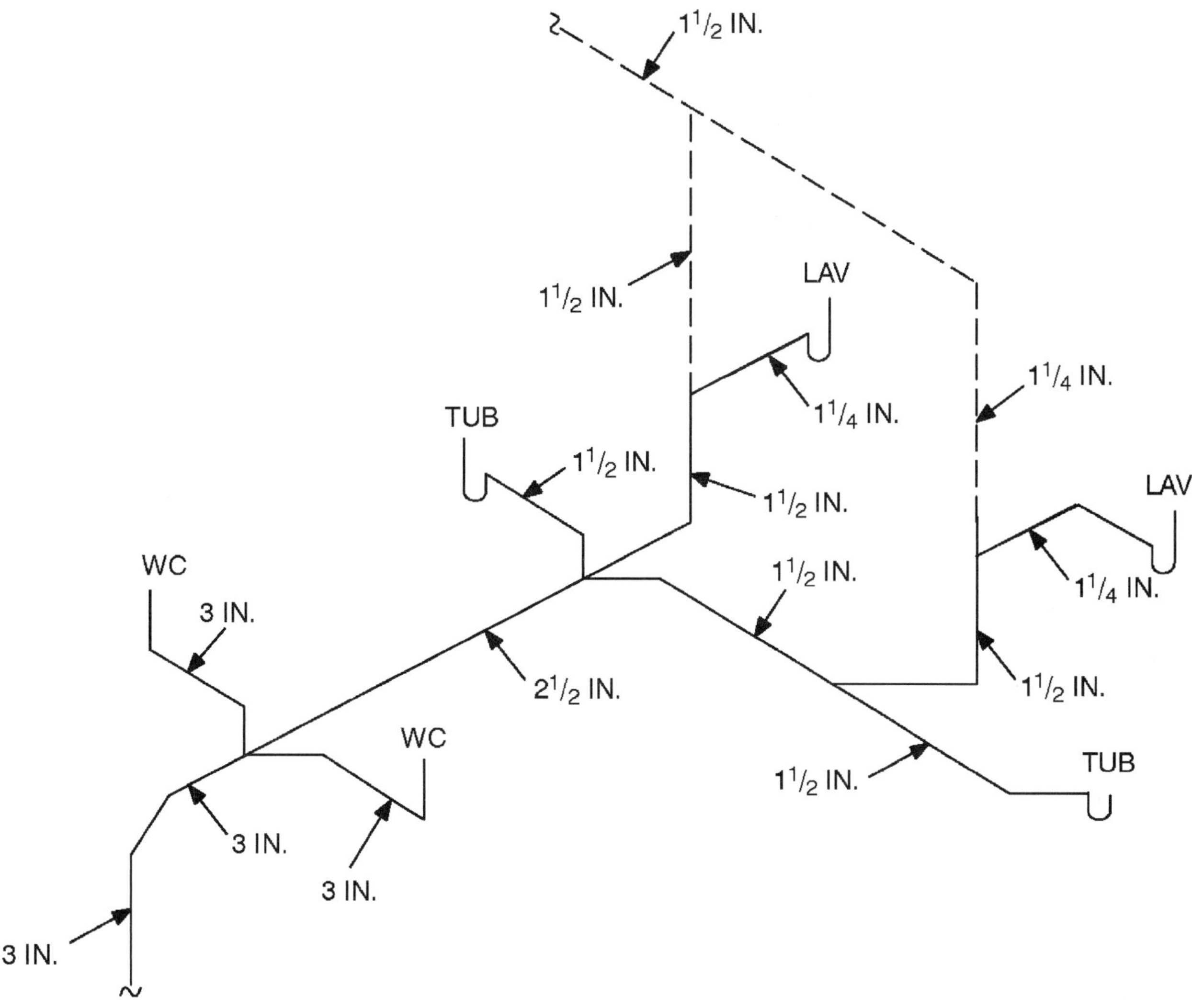

For SI: 1 inch = 25.4 mm.

FIGURE N3
TYPICAL HORIZONTAL WET VENTING

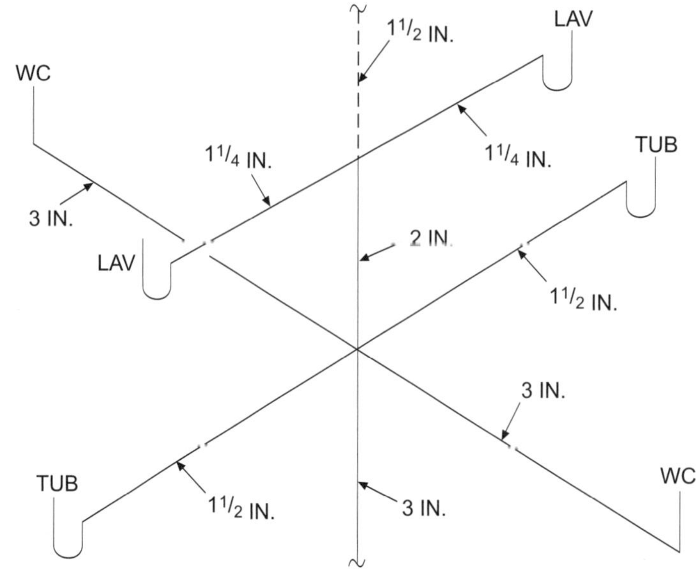

A. VERTICAL WET VENTING

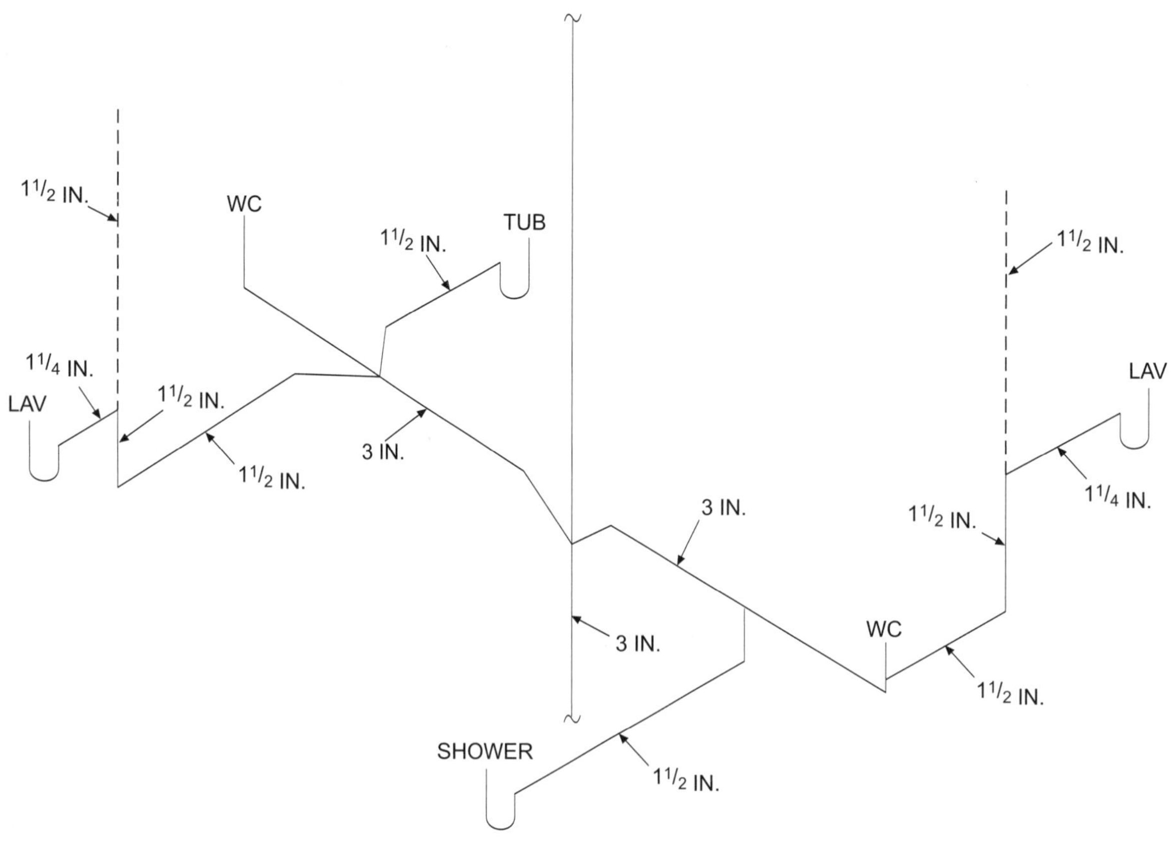

B. HORIZONTAL WET VENTING

For SI: 1 inch = 25.4 mm.

FIGURE N4
TYPICAL METHODS OF WET VENTING

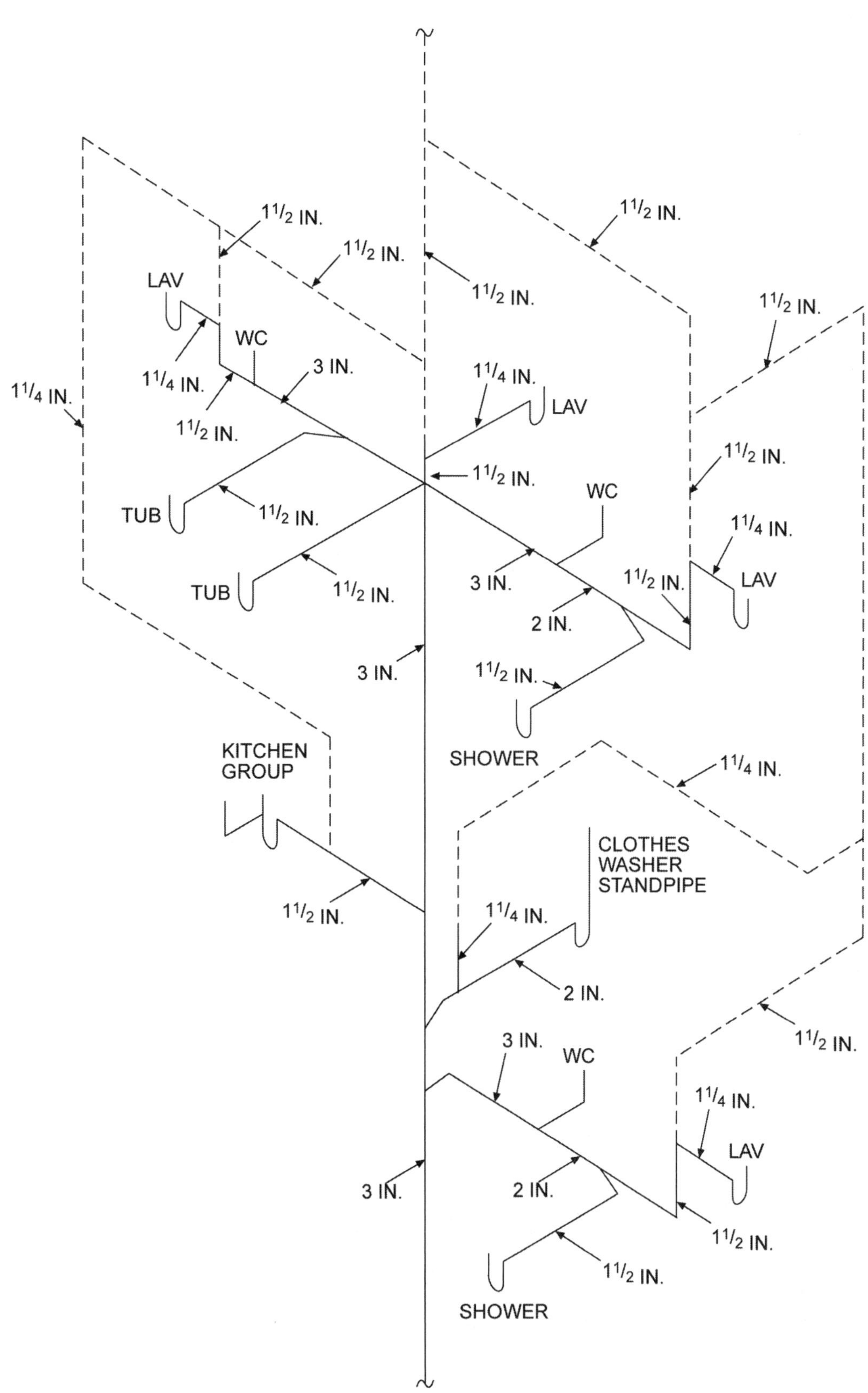

For SI: 1 inch = 25.4 mm.

FIGURE N5
SINGLE STACK SYSTEM FOR A TWO-STORY DWELLING

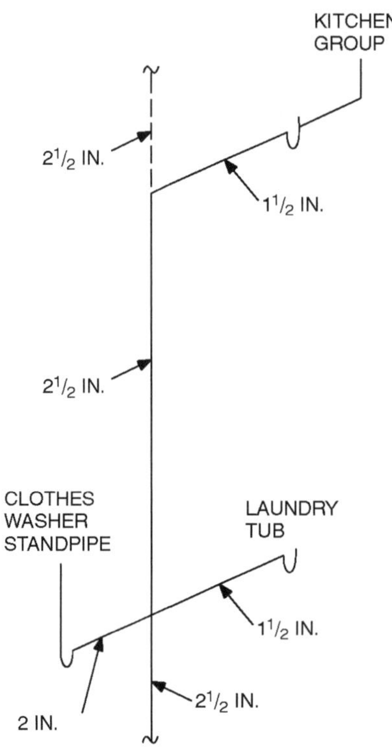

For SI: 1 inch = 25.4 mm.

FIGURE N6
WASTE STACK VENTING

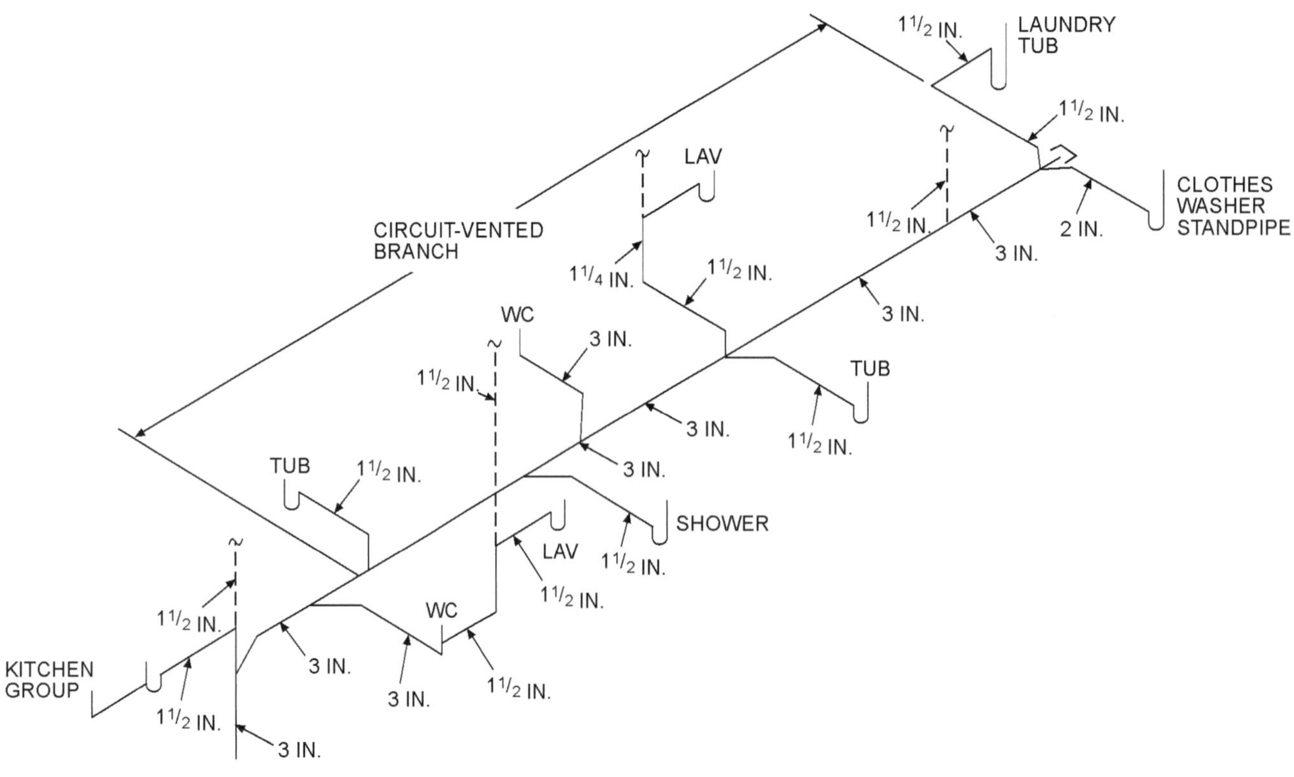

For SI: 1 inch = 25.4 mm.

FIGURE N7
CIRCUIT VENT WITH ADDITIONAL NONCIRCUIT VENTED BRANCH

APPENDIX O

AUTOMATIC VEHICULAR GATES

(The provisions contained in this appendix are not mandatory unless specifically referenced in the adopting ordinance.)

SECTION AO101
GENERAL

AO101.1 General. The provisions of this appendix shall control the design and construction of automatic vehicular gates installed on the lot of a one- or two-family dwelling.

SECTION AO102
DEFINITION

AO102.1 General. For the purposes of these requirements, the terms used shall be defined as follows and as set forth in Chapter 2.

VEHICULAR GATE. A gate that is intended for use at a vehicular entrance or exit to the lot of a one- or two-family dwelling, and that is not intended for use by pedestrian traffic.

SECTION AO103
AUTOMATIC VEHICULAR GATES

AO103.1 Vehicular gates intended for automation. Vehicular gates intended for automation shall be designed, constructed and installed to comply with the requirements of ASTM F 2200.

AO103.2 Vehicular gate openers. Vehicular gate openers, when provided, shall be listed in accordance with UL 325.

APPENDIX P

SIZING OF WATER PIPING SYSTEM

(The provisions contained in this appendix are not mandatory unless specifically referenced in the adopting ordinance.)

SECTION AP101 GENERAL

AP101.1 Scope.

AP101.1.1 This appendix outlines two procedures for sizing a water piping system (see Sections AP103.3 and AP201.1). The design procedures are based on the minimum static pressure available from the supply source, the head charges in the system caused by friction and elevation, and the rates of flow necessary for operation of various fixtures.

AP101.1.2 Because of the variable conditions encountered in hydraulic design, it is impractical to specify definite and detailed rules for sizing of the water piping system. Accordingly, other sizing or design methods conforming to good engineering practice standards are acceptable alternatives to those presented herein.

SECTION AP102 INFORMATION REQUIRED

AP102.1 Preliminary. Obtain the necessary information regarding the minimum daily static service pressure in the area where the building is to be located. If the building supply is to be metered, obtain information regarding friction loss relative to the rate of flow for meters in the range of sizes likely to be used. Friction loss data can be obtained from most manufacturers of water meters.

AP102.2 Demand load.

AP102.2.1 Estimate the supply demand of the building main and the principal branches and risers of the system by totaling the corresponding demand from the applicable part of Table AP103.3(3).

AP102.2.2 Estimate continuous supply demands, in gallons per minute (gpm) (L/m), for lawn sprinklers, air conditioners, etc., and add the sum to the total demand for fixtures. The result is the estimated supply demand for the building supply.

SECTION AP103 SELECTION OF PIPE SIZE

AP103.1 General. Decide from Table P2903.1 what is the desirable minimum residual pressure that should be maintained at the highest fixture in the supply system. If the highest group of fixtures contains flushometer valves, the pressure for the group should not be less than 15 pounds per square inch (psi) (103.4 kPa) flowing. For flush tank supplies, the available pressure should not be less than 8 psi (55.2 kPa) flowing, except blowout action fixtures must not be less than 25 psi (172.4 kPa) flowing.

AP103.2 Pipe sizing.

AP103.2.1 Pipe sizes can be selected according to the following procedure or by other design methods conforming to acceptable engineering practice and *approved* by the administrative authority. The sizes selected must not be less than the minimum required by this code.

AP103.2.2 Water pipe sizing procedures are based on a system of pressure requirements and losses, the sum of which must not exceed the minimum pressure available at the supply source. These pressures are as follows:

1. Pressure required at fixture to produce required flow. See Sections P2903.1 of this code and Section 604.5 of the *International Plumbing Code.*
2. Static pressure loss or gain (due to head) is computed at 0.433 psi per foot (9.8 kPa/m) of elevation change.

   Example: Assume that the highest fixture supply outlet is 20 feet (6096 mm) above or below the supply source. This produces a static pressure differential of 8.66 psi (59.8 kPa) loss [20 feet by 0.433 psi per foot (2096 mm by 9.8 kPa/m)].
3. Loss through water meter. The friction or pressure loss can be obtained from meter manufacturers.
4. Loss through taps in water main.
5. Loss through special devices, such as filters, softeners, backflow prevention devices and pressure regulators. These values must be obtained from the manufacturer.
6. Loss through valves and fittings. Losses for these items are calculated by converting to the *equivalent length* of piping and adding to the total pipe length.
7. Loss caused by pipe friction can be calculated when the pipe size, pipe length and flow through the pipe are known. With these three items, the friction loss can be determined. For piping flow charts not included, use manufacturers' tables and velocity recommendations.

Note: For all examples, the following metric conversions are applicable.

1 cubic foot per minute = 0.4719 L/s.

1 square foot = 0.0929 m^2.

1 degree = 0.0175 rad.

1 pound per square inch = 6.895 kPa.

1 inch = 25.4 mm.

1 foot = 304.8 mm.

1 gallon per minute = 3.785 L/m.

AP103.3 Segmented loss method. The size of water service mains, branch mains and risers by the segmented loss method, must be determined according to water supply demand [gpm (L/m)], available water pressure [psi (kPa)] and friction loss caused by the water meter and *developed length* of pipe [feet (m)], including the *equivalent length* of fittings. This design procedure is based on the following parameters:

1. The calculated friction loss through each length of pipe.
2. A system of pressure losses, the sum of which must not exceed the minimum pressure available at the street main or other source of supply.
3. Pipe sizing based on estimated peak demand, total pressure losses caused by difference in elevation, equipment, *developed length* and pressure required at the most remote fixture; loss through taps in water main; losses through fittings, filters, backflow prevention devices, valves and pipe friction.

Because of the variable conditions encountered in hydraulic design, it is impractical to specify definite and detailed rules for the sizing of the water piping system. Current sizing methods do not address the differences in the probability of use and flow characteristics of fixtures between types of occupancies. Creating an exact model of predicting the demand for a building is impossible and final studies assessing the impact of water conservation on demand are not yet complete. The following steps are necessary for the segmented loss method.

1. **Preliminary.** Obtain the necessary information regarding the minimum daily static service pressure in the area where the building is to be located. If the building supply is to be metered, obtain information regarding friction loss relative to the rate of flow for meters in the range of sizes to be used. Friction loss data can be obtained from manufacturers of water meters. Enough pressure must be available to overcome all system losses caused by friction and elevation so that plumbing fixtures operate properly. Section 604.6 of the *International Plumbing Code* requires that the water distribution system be designed for the minimum pressure available taking into consideration pressure fluctuations. The lowest pressure must be selected to guarantee a continuous, adequate supply of water. The lowest pressure in the public main usually occurs in the summer because of lawn sprinkling and supplying water for air-conditioning cooling towers. Future demands placed on the public main as a result of large growth or expansion should also be considered. The available pressure will decrease as additional loads are placed on the public system.
2. **Demand load.** Estimate the supply demand of the building main and the principal branches and risers of the system by totaling the corresponding demand from the applicable part of Table AP103.3(3). When estimating peak demand, sizing methods typically use water supply fixture units (w.s.f.u.) [see Table AP103.3(2)]. This numerical factor measures the load-producing effect of a single plumbing fixture of a given kind. The use of fixture units can be applied to a single basic probability curve (or table), found in the various sizing methods [see Table AP103.3(3)]. The fixture units are then converted into a gpm (L/m) flow rate for estimating demand.
   2.1. Estimate continuous supply demand in gpm (L/m) for lawn sprinklers, air conditioners, etc., and add the sum to the total demand for fixtures. The result is the estimated supply demand for the building supply. Fixture units cannot be applied to constant-use fixtures, such as hose bibbs, lawn sprinklers and air conditioners. These types of fixtures must be assigned the gpm (L/m) value.
3. **Selection of pipe size.** This water pipe sizing procedure is based on a system of pressure requirements and losses, the sum of which must not exceed the minimum pressure available at the supply source. These pressures are as follows:
   3.1. Pressure required at the fixture to produce required flow. See Section P2903.1 of this code and Section 604.5 of the *International Plumbing Code*.
   3.2. Static pressure loss or gain (because of head) is computed at 0.433 psi per foot (9.8 kPa/m) of elevation change.
   3.3. Loss through a water meter. The friction or pressure loss can be obtained from the manufacturer.
   3.4. Loss through taps in water main [see Table AP103.3(4)].
   3.5. Loss through special devices, such as filters, softeners, backflow prevention devices and pressure regulators. These values must be obtained from the manufacturers.
   3.6. Loss through valves and fittings [see Tables AP103.3(5) and AP103.3(6)]. Losses for these items are calculated by converting to the *equivalent length* of piping and adding to the total pipe length.
   3.7. Loss caused by pipe friction can be calculated when the pipe size, pipe length and flow through the pipe are known. With these three items, the friction loss can be determined using Figures AP103.3(2) through AP103.3(7). When using charts, use pipe inside diameters. For piping flow charts not included, use manufacturers' tables and velocity recommendations. Before attempting to size any water supply system, it is necessary to gather preliminary information which includes available pressure, piping material, select design velocity, elevation differences and *developed length* to the most remote fixture. The water supply system is divided into sections at major changes in eleva-

tion or where branches lead to fixture groups. The peak demand must be determined in each part of the hot and cold water supply system which includes the corresponding w.s.f.u. and conversion to gpm (L/m) flow rate to be expected through each section. Sizing methods require determination of the "most hydraulically remote" fixture to compute the pressure loss caused by pipe and fittings. The hydraulically remote fixture represents the most downstream fixture along the circuit of piping requiring the most available pressure to operate properly. Consideration must be given to all pressure demands and losses, such as friction caused by pipe, fittings and equipment; elevation; and the residual pressure required by Table P2903.1. The two most common and frequent complaints about water supply system operation are lack of adequate pressure and noise.

Problem: What size Type L copper water pipe, service and distribution will be required to serve a two-story factory building having on each floor, back-to-back, two toilet rooms each equipped with hot and cold water? The highest fixture is 21 feet above the street main, which is tapped with a 2-inch corporation cock at which point the minimum pressure is 55 psi. In the building basement, a 2-inch meter with a maximum pressure drop of 11 psi and 3-inch reduced pressure principle backflow preventer with a maximum pressure drop of 9 psi are to be installed. The system is shown in Figure AP103.3(1). To be determined are the pipe sizes for the service main, and the cold and hot water distribution pipes.

Solution: A tabular arrangement such as shown in Table AP103.3(1) should first be constructed. The steps to be followed are indicated by the tabular arrangement itself as they are in sequence, Columns 1 through 10 and Lines A through L.

Step 1

Columns 1 and 2: Divide the system into sections breaking at major changes in elevation or where branches lead to fixture groups. After Point B [see Figure AP103.3(1)], separate consideration will be given to the hot and cold water piping. Enter the sections to be considered in the service and cold water piping in Column 1 of the tabular arrangement. Column 1 of Table AP103.3(1) provides a line-by-line, recommended tabular arrangement for use in solving pipe sizing.

The objective in designing the water supply system is to ensure an adequate water supply and pressure to all fixtures and equipment. Column 2 provides the psi (kPa) to be considered separately from the minimum pressure available at the main. Losses to take into consideration are the following: the differences in elevations between the water supply source and the highest water supply outlet; meter pressure losses; the tap in main loss; special fixture devices, such as water softeners and backflow prevention devices; and the pressure required at the most remote fixture outlet.

The difference in elevation can result in an increase or decrease in available pressure at the main. Where the water supply outlet is located above the source, this results in a loss in the available pressure and is subtracted from the pressure at the water source. Where the highest water supply outlet is located below the water supply source, there will be an increase in pressure that is added to the available pressure of the water source.

Column 3: According to Table AP103.3(3), determine the gpm (L/m) of flow to be expected in each section of the system. These flows range from 28.6 to 108 gpm. Load values for fixtures must be determined as w.s.f.u. and then converted to a gpm rating to determine peak demand. When calculating peak demands, the w.s.f.u. are added and then converted to the gpm rating. For continuous flow fixtures, such as hose bibbs and lawn sprinkler systems, add the gpm demand to the intermittent demand of fixtures. For example, a total of 120 w.s.f.u. is converted to a demand of 48 gpm. Two hose bibbs × 5 gpm demand = 10 gpm. Total gpm rating = 48.0 gpm + 10 gpm = 58.0 gpm demand.

Step 2

Line A: Enter the minimum pressure available at the main source of supply in Column 2. This is 55 psi (379.2 kPa). The local water authorities generally keep records of pressures at different times of the day and year. The available pressure can also be checked from nearby buildings or from fire department hydrant checks.

Line B: Determine from Table P2903.1 the highest pressure required for the fixtures on the system, which is 15 psi (103.4 kPa), to operate a flushometer valve. The most remote fixture outlet is necessary to compute the pressure loss caused by pipe and fittings, and represents the most downstream fixture along the circuit of piping requiring the available pressure to operate properly as indicated by Table P2903.1.

Line C: Determine the pressure loss for the meter size given or assumed. The total water flow from the main through the service as determined in Step 1 will serve to aid in the meter selected. There are three common types of water meters; the pressure losses are determined by the American Water Works Association Standards for displacement type, compound type and turbine type. The maximum pressure loss of such devices takes into consideration the meter size, safe operating capacity [gpm (L/m)] and maximum rates for continuous operations [gpm (L/m)]. Typically, equipment imparts greater pressure losses than piping.

Line D: Select from Table AP103.3(4) and enter the pressure loss for the tap size given or assumed. The loss of pressure through taps and tees in psi (kPa) is based on the total gpm (L/m) flow rate and size of the tap.

Line E: Determine the difference in elevation between the main and source of supply and the highest fixture on the system. Multiply this figure, expressed in feet (mm), by 0.43 psi. Enter the resulting psi (kPa) loss on Line E. The difference in elevation between the water supply source and the highest water supply outlet has a significant impact on the sizing of the water supply system. The difference in elevation usually results in a loss in the available pressure because the water supply outlet is generally located above the water supply source. The loss is caused by the pressure required to lift the water to the outlet. The pressure loss is subtracted from the pressure at the water source. Where the highest water supply outlet is located below the water source, there will be an increase in pressure which is added to the available pressure of the water source.

Lines F, G and H: The pressure losses through filters, backflow prevention devices or other special fixtures must be obtained from the manufacturer or estimated and entered on these lines. Equipment, such as backflow prevention devices, check valves, water softeners, instantaneous, or tankless water heaters, filters and strainers, can impart a much greater pressure loss than the piping. The pressure losses can range from 8 to 30 psi.

Step 3

Line I: The sum of the pressure requirements and losses that affect the overall system (Lines B through H) is entered on this line. Summarizing the steps, all of the system losses are subtracted from the minimum water pressure. The remainder is the pressure available for friction, defined as the energy available to push the water through the pipes to each fixture. This force can be used as an average pressure loss, as long as the pressure available for friction is not exceeded. Saving a certain amount for available water supply pressures as an area incurs growth, or because of the aging of the pipe or equipment added to the system is recommended.

Step 4

Line J: Subtract Line I from Line A. This gives the pressure that remains available from overcoming friction losses in the system. This figure is a guide to the pipe size that is chosen for each section, incorporating the total friction losses to the most remote outlet (measured length is called *developed length*).

Exception: When the main is above the highest fixture, the resulting psi (kPa) must be considered a pressure gain (static head gain) and omitted from the sums of Lines B through H and added to Line J.

The maximum friction head loss that can be tolerated in the system during peak demand is the difference between the static pressure at the highest and most remote outlet at no-flow conditions and the minimum flow pressure required at that outlet. If the losses are within the required limits, every run of pipe will also be within the required friction head loss. Static pressure loss is at the most remote outlet in feet × 0.433 = loss in psi caused by elevation differences.

Step 5

Column 4: Enter the length of each section from the main to the most remote outlet (at Point E). Divide the water supply system into sections breaking at major changes in elevation or where branches lead to fixture groups.

Step 6

Column 5: When selecting a trial pipe size, the length from the water service or meter to the most remote fixture outlet must be measured to determine the *developed length*. However, in systems having a flushometer valve or temperature-controlled shower at the topmost floors, the *developed length* would be from the water meter to the most remote flushometer valve on the system. A rule of thumb is that size will become progressively smaller as the system extends farther from the main source of supply. Trial pipe size may be arrived at by the following formula:

Line J: (Pressure available to overcome pipe friction) × 100/*equivalent length* of run total *developed length* to most remote fixture × percentage factor of 1.5 (Note: a percentage factor is used only as an estimate for friction losses imposed for fittings for initial trial pipe size) = psi (average pressure drop per 100 feet of pipe).

For trial pipe size, see Figure AP103.3(3) (Type L copper) based on 2.77 psi and 108 gpm = $2^1/_2$ inches. To determine the *equivalent length* of run to the most remote outlet, the *developed length* is determined and added to the friction losses for fittings and valves. The *developed lengths* of the designated pipe sections are as follows:

A-B	54 feet
B-C	8 feet
C-D	13 feet
D-E	150 feet

Total *developed length* = 225 feet

The *equivalent length* of the friction loss in fittings and valves must be added to the *developed length* (most remote outlet). Where the size of fittings and valves is not known, the added friction loss should be approximated. A general rule that has been used is to add 50 percent of the *developed length* to allow for fittings and valves. For example, the *equivalent length* of run equals the *developed length* of run (225 feet × 1.5 = 338 feet). The total *equivalent length* of run for determining a trial pipe size is 338 feet.

Example: 9.36 (pressure available to overcome pipe friction) × 100/338 (*equivalent length* of run = 225 × 1.5) = 2.77 psi (average pressure drop per 100 feet of pipe).

Step 7

Column 6: Select from Table AP103.3(6) the *equivalent lengths* for the trial pipe size of fittings and valves on each pipe section. Enter the sum for each section in Column 6. (The number of fittings to be used in this example must be an estimate). The *equivalent length* of piping is the *developed length* plus the *equivalent lengths* of pipe corresponding to the friction head losses for fittings and valves. Where the size of fittings and valves is not known, the added friction head losses must be approximated. An estimate for this example is found in Table AP.1.

Step 8

Column 7: Add the figures from Columns 4 and 6, and enter in Column 7. Express the sum in hundreds of feet.

Step 9

Column 8: Select from Figure AP103.3(3) the friction loss per 100 feet of pipe for the gpm flow in a section (Column 3) and trial pipe size (Column 5). Maximum friction head loss per 100 feet is determined on the basis of the total pressure available for friction head loss and the longest *equivalent length* of run. The selection is based on the gpm demand, uniform friction head loss and maximum design velocity. Where the size indicated by the hydraulic table indicates a velocity in excess of the selected velocity, a size must be selected which produces the required velocity.

Step 10

Column 9: Multiply the figures in Columns 7 and 8 for each section and enter in Column 9.

Total friction loss is determined by multiplying the friction loss per 100 feet for each pipe section in the total *developed length* by the pressure loss in fittings expressed as *equivalent length* in feet (mm). Note: Section C-F should be considered in the total pipe friction losses only if greater loss occurs in Section C-F than in pipe Section D-E. Section C-F is not considered in the total *developed length*. Total friction loss in *equivalent length* is determined in Table AP.2.

Step 11

Line K: Enter the sum of the values in Column 9. The value is the total friction loss in *equivalent length* for each designated pipe section.

Step 12

Line L: Subtract Line J from Line K and enter in Column 10.

The result should always be a positive or plus figure. If it is not, repeat the operation using Columns 5, 6, 8 and 9 until a balance or near balance is obtained. If the difference between Lines J and K is a high positive number, it is an indication that the pipe sizes are too large and should be reduced, thus saving materials. In such a case, the operations using Columns 5, 6, 8 and 9 should be repeated.

The total friction losses are determined and subtracted from the pressure available to overcome pipe friction for the trial pipe size. This number is critical because it provides a guide to whether the pipe size selected is too large and the process should be repeated to obtain an economically designed system.

Answer: The final figures entered in Column 5 become the design pipe size for the respective sections. Repeating this operation a second time using the same sketch but considering the demand for hot water, it is possible to size the hot water distribution piping. This has been worked up as a part of the overall problem in the tabular arrangement used for sizing the service and water distribution piping. Note that consideration must be given to the pressure losses from the street main to the water heater (Section A-B) in determining the hot water pipe sizes.

TABLE AP.1

COLD WATER PIPE SECTION	FITTINGS/VALVES	PRESSURE LOSS EXPRESSED AS EQUIVALENT LENGTH OF TUBE (feet)	HOT WATER PIPE SECTION	FITTINGS/VALVES	PRESSURE LOSS EXPRESSED AS EQUIVALENT OF TUBE (feet)
A-B	3 – $2^1/_2$″ Gate valves	3	A-B	3 – $2^1/_2$″ Gate valves	3
	1 – $2^1/_2$″ Side branch tee	12	—	1 – $2^1/_2$″ Side branch tee	12
B-C	1 – $2^1/_2$″ Straight run tee	0.5	B-C	1 – 2″ Straight run tee	7
	—	—	—	1 – 2″ 90-degree ell	0.5
C-F	1 – $2^1/_2$″ Side branch tee	12	C-F	1 – $1^1/_2$″ Side branch tee	7
C-D	1 – $2^1/_2$″ 90-degree ell	7	C-D	1 – $^1/_2$″ 90-degree ell	4
D-E	1 – $2^1/_2$″ Side branch tee	12	D-E	1 – $1^1/_2$″ Side branch tee	7

For SI: 1 inch = 25.4 mm, 1 foot = 304.8 mm, 1 degree = 0.01745 rad.

TABLE AP.2

PIPE SECTIONS	FRICTION LOSS EQUIVALENT LENGTH (feet)	
	Cold Water	Hot Water
A-B	0.69 × 3.2 = 2.21	0.69 × 3.2 = 2.21
B-C	0.085 × 3.1 = 0.26	0.16 × 1.4 = 0.22
C-D	0.20 × 1.9 = 0.38	0.17 × 3.2 = 0.54
D-E	1.62 × 1.9 = 3.08	1.57 × 3.2 = 5.02
Total pipe friction losses (Line K)	5.93	7.99

For SI: 1 foot = 304.8 mm.

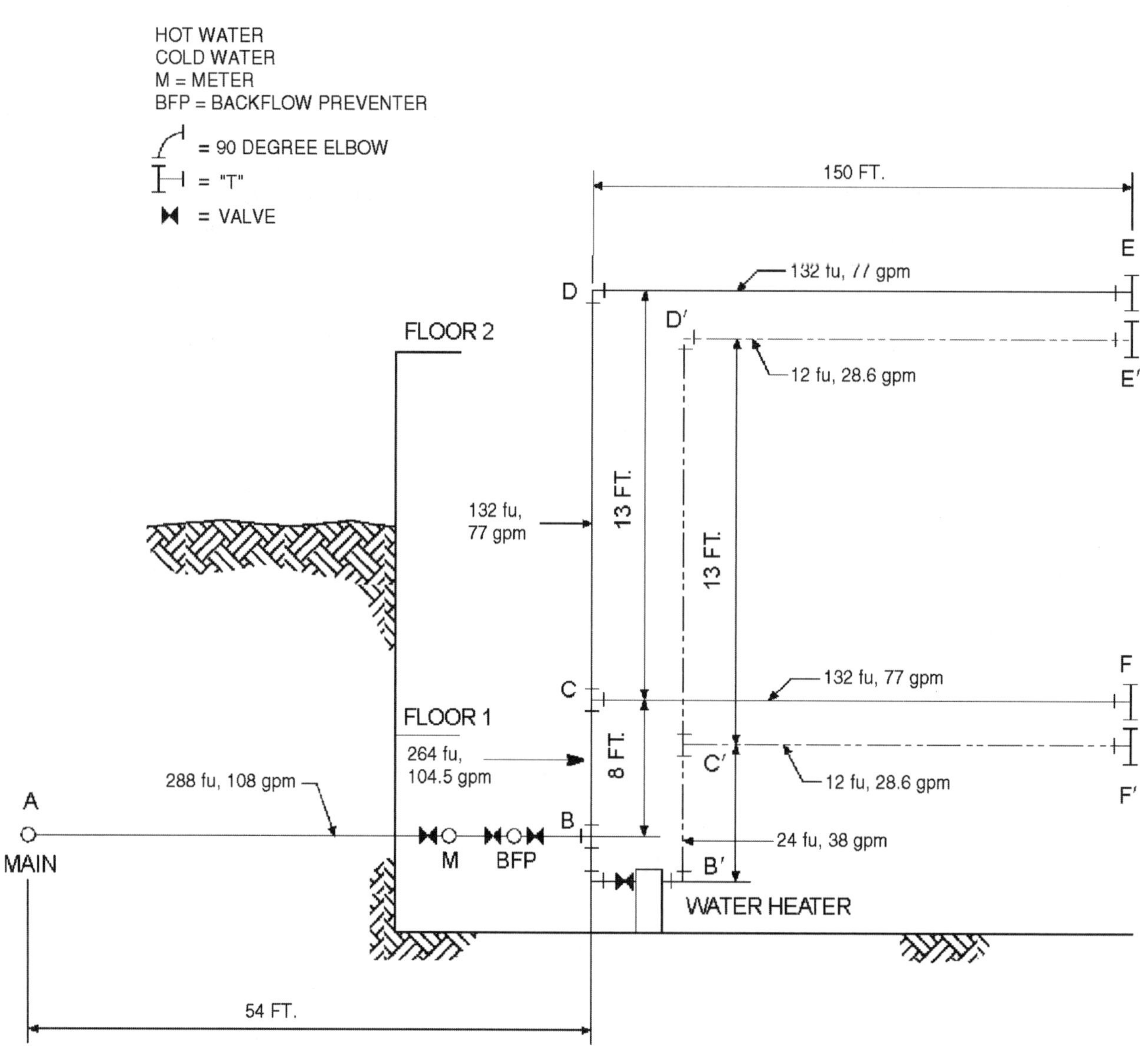

For SI: 1 foot = 304.8 mm, 1 gallon per minute = 3.785 L/m.

FIGURE AP103.3(1)
EXAMPLE—SIZING

TABLE AP103.3(1)
RECOMMENDED TABULAR ARRANGEMENT FOR USE IN SOLVING PIPE SIZING PROBLEMS

COLUMN	1	2		3	4	5	6	7	8	9	10
Line	Description		Pounds per square inch	Gallons per min through section	Length of section (feet)	Trial pipe size (inches)	Equivalent length of fittings and valves (feet)	Total equivalent length Columns 4 and 6 (100 feet)	Friction loss per 100 feet of trial size pipe (psi)	Friction loss in equivalent length Column 8 x Column 7 (psi)	Excess pressure over friction losses (psi)
A	Service and cold water distribution piping[a]	Minimum pressure available at main	55.00								
B		Highest pressure required at a fixture (see Table P2903.1)	15.00								
C		Meter loss 2″ meter	11.00								
D		Tap in main loss 2″ tap [see Table AP103.3(4)]	1.61								
E		Static head loss 21 ft × 0.43 psi/ft	9.03								
F		Special fixture loss backflow preventer	9.00								
G		Special fixture loss—Filter	0.00								
H		Special fixture loss—Other	0.00								
I		Total overall losses and requirements Sum of (Lines B through H)	45.64								
J		Pressure available to overcome pipe friction (Line A minus Line I)	9.36								
	DESIGNATION Pipe section (from diagram) Cold water distribution piping		FU	108.0	54		15.00	0.69	3.2	2.21	—
		A-B	.288	104.5	8	$2\frac{1}{2}$	0.5	0.85	3.1	0.26	—
		B-C	.264	77.0	13	$2\frac{1}{2}$	7.00	0.20	1.9	0.38	—
		C-D	.132	77.0	150	$2\frac{1}{2}$	12.00	1.62	1.9	3.08	—
		C-F[b]	.132	77.0	150	$2\frac{1}{2}$	12.00	1.62	1.9	3.08	—
		D-E[b]	.132			$2\frac{1}{2}$					
K	Total pipe friction losses (cold)			—	—	—	—	—	—	5.93	—
L	Difference (Line J minus Line K)			—	—	—	—	—	—	—	3.43
	Pipe section (from digram) Diagram Hot water Distribution Piping	A′B′	.288	108.0	54	$2\frac{1}{2}$	12.00	0.69	3.3	2.21	—
		B′C′	.24	38.0	8	2	7.5	0.16	1.4	0.22	—
		C′D′	.12	28.6	13	$1\frac{1}{2}$	4.0	0.17	3.2	0.54	—
		C′F′[b]	.12	28.6	150	$1\frac{1}{2}$	7.00	1.57	3.2	5.02	—
		D′E′[b]	.12	28.6	150	$1\frac{1}{2}$	7.00	1.57	3.2	5.02	—
K	Total pipe friction losses (hot)			—	—	—	—	—	—	7.99	—
L	Difference (Line J minus Line K)			—	—	—	—	—	—	—	1.37

For SI: 1 inch = 25.4 mm, 1 foot = 304.8 mm, 1 pounds per square inch = 6.895 kPa, 1 gallons per minute = 3.785 L/m.

a. To be considered as pressure gain for fixtures below main (to consider separately, omit from "I" and add to "J").

b. To consider separately, in Line K use Section C-F only if greater loss than Note a.

**TABLE AP103.3(2)
LOAD VALUES ASSIGNED TO FIXTURES[a]**

FIXTURE	OCCUPANCY	TYPE OF SUPPLY CONTROL	LOAD VALUES, IN WATER SUPPLY FIXTURE UNITS (w.s.f.u.)		
			Cold	Hot	Total
Bathroom group	Private	Flush tank	2.7	1.5	3.6
Bathroom group	Private	Flushometer valve	6.0	3.0	8.0
Bathtub	Private	Faucet	1.0	1.0	1.4
Bathtub	Public	Faucet	3.0	3.0	4.0
Bidet	Private	Faucet	1.5	1.5	2.0
Combination fixture	Private	Faucet	2.25	2.25	3.0
Dishwashing machine	Private	Automatic	—	1.4	1.4
Drinking fountain	Offices, etc.	$^3/_8$″ valve	0.25	—	0.25
Kitchen sink	Private	Faucet	1.0	1.0	1.4
Kitchen sink	Hotel, restaurant	Faucet	3.0	3.0	4.0
Laundry trays (1 to 3)	Private	Faucet	1.0	1.0	1.4
Lavatory	Private	Faucet	0.5	0.5	0.7
Lavatory	Public	Faucet	1.5	1.5	2.0
Service sink	Offices, etc.	Faucet	2.25	2.25	3.0
Shower head	Public	Mixing valve	3.0	3.0	4.0
Shower head	Private	Mixing valve	1.0	1.0	1.4
Urinal	Public	1″ flushometer valve	10.0	—	10.0
Urinal	Public	$^3/_4$″ flushometer valve	5.0	—	5.0
Urinal	Public	Flush tank	3.0	—	3.0
Washing machine (8 lb)	Private	Automatic	1.0	1.0	1.4
Washing machine (8 lb)	Public	Automatic	2.25	2.25	3.0
Washing machine (15 lb)	Public	Automatic	3.0	3.0	4.0
Water closet	Private	Flushometer valve	6.0	—	6.0
Water closet	Private	Flush tank	2.2	—	2.2
Water closet	Public	Flushometer valve	10.0	—	10.0
Water closet	Public	Flush tank	5.0	—	5.0
Water closet	Public or private	Flushometer tank	2.0	—	2.0

For SI: 1 inch = 25.4 mm, 1 pound = 0.454 kg.

a. For fixtures not listed, loads should be assumed by comparing the fixture to one listed using water in similar quantities and at similar rates. The assigned loads for fixtures with both hot and cold water supplies are given for separate hot and cold water loads, and for total load. The separate hot and cold water loads being three-fourths of the total load for the fixture in each case.

TABLE AP103.3(3)
TABLE FOR ESTIMATING DEMAND

SUPPLY SYSTEMS PREDOMINANTLY FOR FLUSH TANKS			SUPPLY SYSTEMS PREDOMINANTLY FOR FLUSHOMETERS		
Load	Demand		Load	Demand	
(w.s.f.u.)	(gpm	(cfm)	(w.s.f.u.)	(gpm)	(cfm)
1	3.0	0.04104	—	—	—
2	5.0	0.0684	—	—	—
3	6.5	0.86892	—	—	—
4	8.0	1.06944	—	—	—
5	9.4	1.256592	5	15.0	2.0052
6	10.7	1.430376	6	17.4	2.326032
7	11.8	1.577424	7	19.8	2.646364
8	12.8	1.711104	8	22.2	2.967696
9	13.7	1.831416	9	24.6	3.288528
10	14.6	1.951728	10	27.0	3.60936
11	15.4	2.058672	11	27.8	3.716304
12	16.0	2.13888	12	28.6	3.823248
13	16.5	2.20572	13	29.4	3.930192
14	17.0	2.27256	14	30.2	4.037136
15	17.5	2.3394	15	31.0	4.14408
16	18.0	2.90624	16	31.8	4.241024
17	18.4	2.459712	17	32.6	4.357968
18	18.8	2.513184	18	33.4	4.464912
19	19.2	2.566656	19	34.2	4.571856
20	19.6	2.620128	20	35.0	4.6788
25	21.5	2.87412	25	38.0	5.07984
30	23.3	3.114744	30	42.0	5.61356
35	24.9	3.328632	35	44.0	5.88192
40	26.3	3.515784	40	46.0	6.14928
45	27.7	3.702936	45	48.0	6.41664
50	29.1	3.890088	50	50.0	6.684
60	32.0	4.27776	60	54.0	7.21872
70	35.0	4.6788	70	58.0	7.75344
80	38.0	5.07984	80	61.2	8.181216
90	41.0	5.48088	90	64.3	8.595624
100	43.5	5.81508	100	67.5	9.0234
120	48.0	6.41664	120	73.0	9.75864
140	52.5	7.0182	140	77.0	10.29336
160	57.0	7.61976	160	81.0	10.82808
180	61.0	8.15448	180	85.5	11.42964
200	65.0	8.6892	200	90.0	12.0312
225	70.0	9.3576	225	95.5	12.76644

(continued)

TABLE AP103.3(3)—continued
TABLE FOR ESTIMATING DEMAND

SUPPLY SYSTEMS PREDOMINANTLY FOR FLUSH TANKS			SUPPLY SYSTEMS PREDOMINANTLY FOR FLUSHOMETERS		
Load	Demand		Load	Demand	
(w.s.f.u.)	(gpm)	(cfm)	(w.s.f.u.)	(gpm)	(cfm)
250	75.0	10.026	250	101.0	13.50168
275	80.0	10.6944	275	104.5	13.96956
300	85.0	11.3628	300	108.0	14.43744
400	105.0	14.0364	400	127.0	16.97736
500	124.0	16.57632	500	143.0	19.11624
750	170.0	22.7256	750	177.0	23.66136
1,000	208.0	27.80544	1,000	208.0	27.80544
1,250	239.0	31.94952	1,250	239.0	31.94952
1,500	269.0	35.95992	1,500	269.0	35.95992
1,750	297.0	39.70296	1,750	297.0	39.70296
2,000	325.0	43.446	2,000	325.0	43.446
2,500	380.0	50.7984	2,500	380.0	50.7984
3,000	433.0	57.88344	3,000	433.0	57.88344
4,000	535.0	70.182	4,000	525.0	70.182
5,000	593.0	79.27224	5,000	593.0	79.27224

For SI: 1 gallon per minute = 3.785 L/m, 1 cubic foot per minute = 0.000471 m^3/s.

TABLE AP103.3(4)
LOSS OF PRESSURE THROUGH TAPS AND TEES IN POUNDS PER SQUARE INCH (psi)

GALLONS PER MINUTE	SIZE OF TAP OR TEE (inches)						
	$^5/_8$	$^3/_4$	1	$1^1/_4$	$1^1/_2$	2	3
10	1.35	0.64	0.18	0.08	—	—	—
20	5.38	2.54	0.77	0.31	0.14	—	—
30	12.10	5.72	1.62	0.69	0.33	0.10	—
40	—	10.20	3.07	1.23	0.58	0.18	—
50	—	15.90	4.49	1.92	0.91	0.28	—
60	—	—	6.46	2.76	1.31	0.40	—
70	—	—	8.79	3.76	1.78	0.55	0.10
80	—	—	11.50	4.90	2.32	0.72	0.13
90	—	—	14.50	6.21	2.94	0.91	0.16
100	—	—	17.94	7.67	3.63	1.12	0.21
120	—	—	25.80	11.00	5.23	1.61	0.30
140	—	—	35.20	15.00	7.12	2.20	0.41
150	—	—	—	17.20	8.16	2.52	0.47
160	—	—	—	19.60	9.30	2.92	0.54
180	—	—	—	24.80	11.80	3.62	0.68
200	—	—	—	30.70	14.50	4.48	0.84
225	—	—	—	38.80	18.40	5.60	1.06
250	—	—	—	47.90	22.70	7.00	1.31
275	—	—	—	—	27.40	7.70	1.59
300	—	—	—	—	32.60	10.10	1.88

For SI: 1 inch = 25.4 mm, 1 pound per square inch = 6.895 kPa, 1 gallon per minute = 3.785 L/m.

TABLE AP103.3(5)
ALLOWANCE IN EQUIVALENT LENGTHS OF PIPE FOR FRICTION LOSS IN VALVES AND THREADED FITTINGS (feet)

FITTING OR VALVE	PIPE SIZE (inches)							
	1/2	3/4	1	1 1/4	1 1/2	2	2 1/2	3
45-degree elbow	1.2	1.5	1.8	2.4	3.0	4.0	5.0	6.0
90-degree elbow	2.0	2.5	3.0	4.0	5.0	7.0	8.0	10.0
Tee, run	0.6	0.8	0.9	1.2	1.5	2.0	2.5	3.0
Tee, branch	3.0	4.0	5.0	6.0	7.0	10.0	12.0	15.0
Gate valve	0.4	0.5	0.6	0.8	1.0	1.3	1.6	2.0
Balancing valve	0.8	1.1	1.5	1.9	2.2	3.0	3.7	4.5
Plug-type cock	0.8	1.1	1.5	1.9	2.2	3.0	3.7	4.5
Check valve, swing	5.6	8.4	11.2	14.0	16.8	22.4	28.0	33.6
Globe valve	15.0	20.0	25.0	35.0	45.0	55.0	65.0	80.0
Angle valve	8.0	12.0	15.0	18.0	22.0	28.0	34.0	40.0

For SI: 1 inch = 25.4 mm, 1 foot = 304.8 mm, 1 degree = 0.0175 rad.

TABLE AP103.3(6)
PRESSURE LOSS IN FITTINGS AND VALVES EXPRESSED AS EQUIVALENT LENGTH OF TUBE[a] (feet)

NOMINAL OR STANDARD SIZE (inches)	FITTINGS				Coupling	VALVES			
	Standard Ell		90-degree Tee			Ball	Gate	Butterfly	Check
	90 Degree	45 Degree	Side Branch	Straight Run					
3/8	0.5	—	1.5	—	—	—	—	—	1.5
1/2	1	0.5	2	—	—	—	—	—	2
5/8	1.5	0.5	2	—	—	—	—	—	2.5
3/4	2	0.5	3	—	—	—	—	—	3
1	2.5	1	4.5	—	—	0.5	—	—	4.5
1 1/4	3	1	5.5	0.5	0.5	0.5	—	—	5.5
1 1/2	4	1.5	7	0.5	0.5	0.5	—	—	6.5
2	5.5	2	9	0.5	0.5	0.5	0.5	7.5	9
2 1/2	7	2.5	12	0.5	0.5	—	1	10	11.5
3	9	3.5	15	1	1	—	1.5	15.5	14.5
3 1/2	9	3.5	14	1	1	—	2	—	12.5
4	12.5	5	21	1	1	—	2	16	18.5
5	16	6	27	1.5	1.5	—	3	11.5	23.5
6	19	7	34	2	2	—	3.5	13.5	26.5
8	29	11	50	3	3	—	5	12.5	39

For SI: 1 inch = 25.4 mm, 1 foot = 304.8 mm, 1 degree = 0.01745 rad.

a. Allowances are for streamlined soldered fittings and recessed threaded fittings. For threaded fittings, double the allowances shown in the table. The equivalent lengths presented in the table are based on a C factor of 150 in the Hazen-Williams friction loss formula. The lengths shown are rounded to the nearest half-foot.

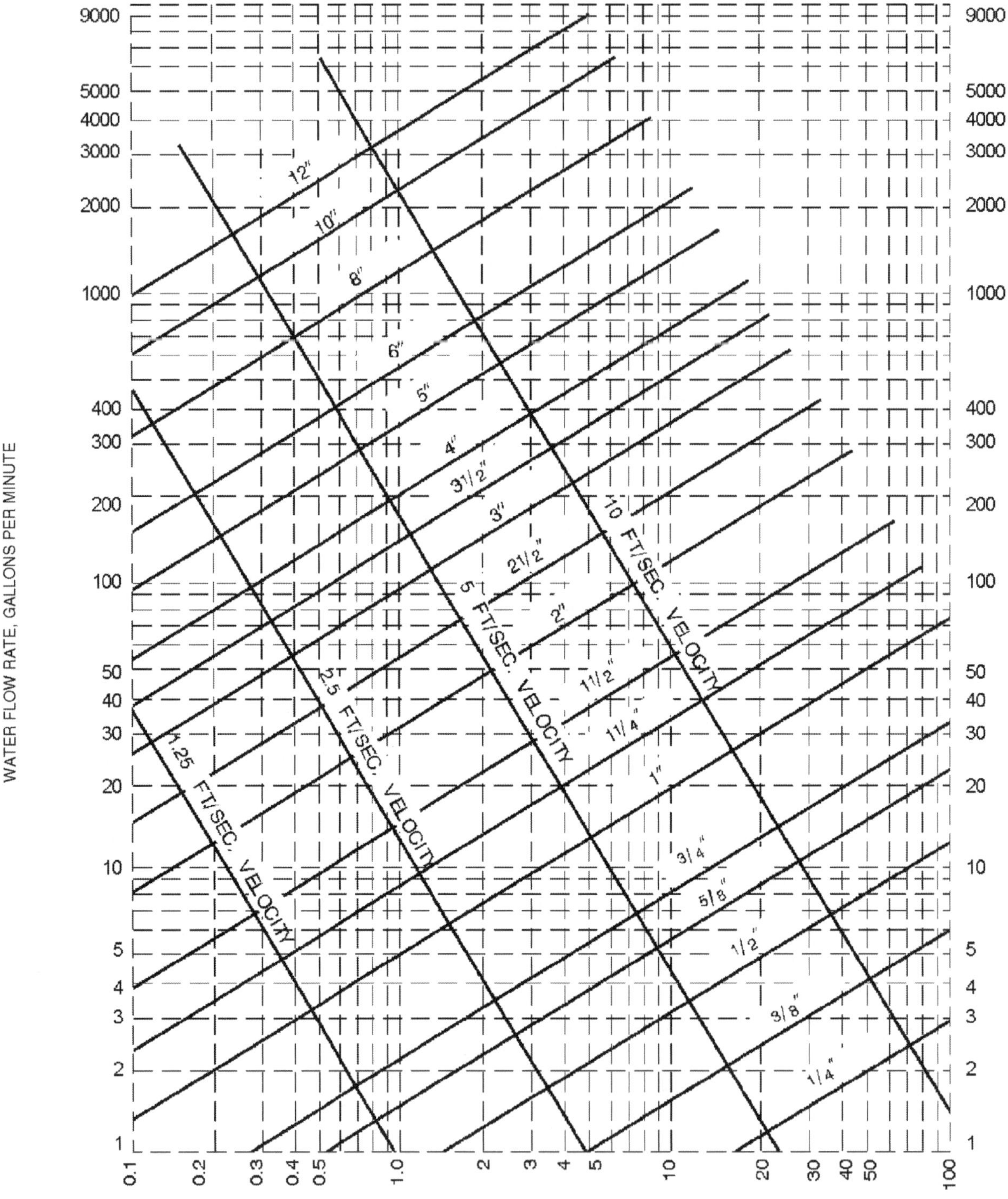

Note: Fluid velocities in excess of 5 to 8 feet per second are not usually recommended.

For SI: 1 inch = 25.4 mm, 1 foot = 304.8 mm, 1 gallon per minute = 3.785 L/m, 1 pounds per square inch = 6.895 kPa, 1 foot per second = 0.305 m/s.

a. This figure applies to smooth new copper tubing with recessed (streamline) soldered joints and to the actual sizes of types indicated on the diagram.

FIGURE AP103.3(2)
FRICTION LOSS IN SMOOTH PIPE[a]
(TYPE K, ASTM B 88 COPPER TUBING)

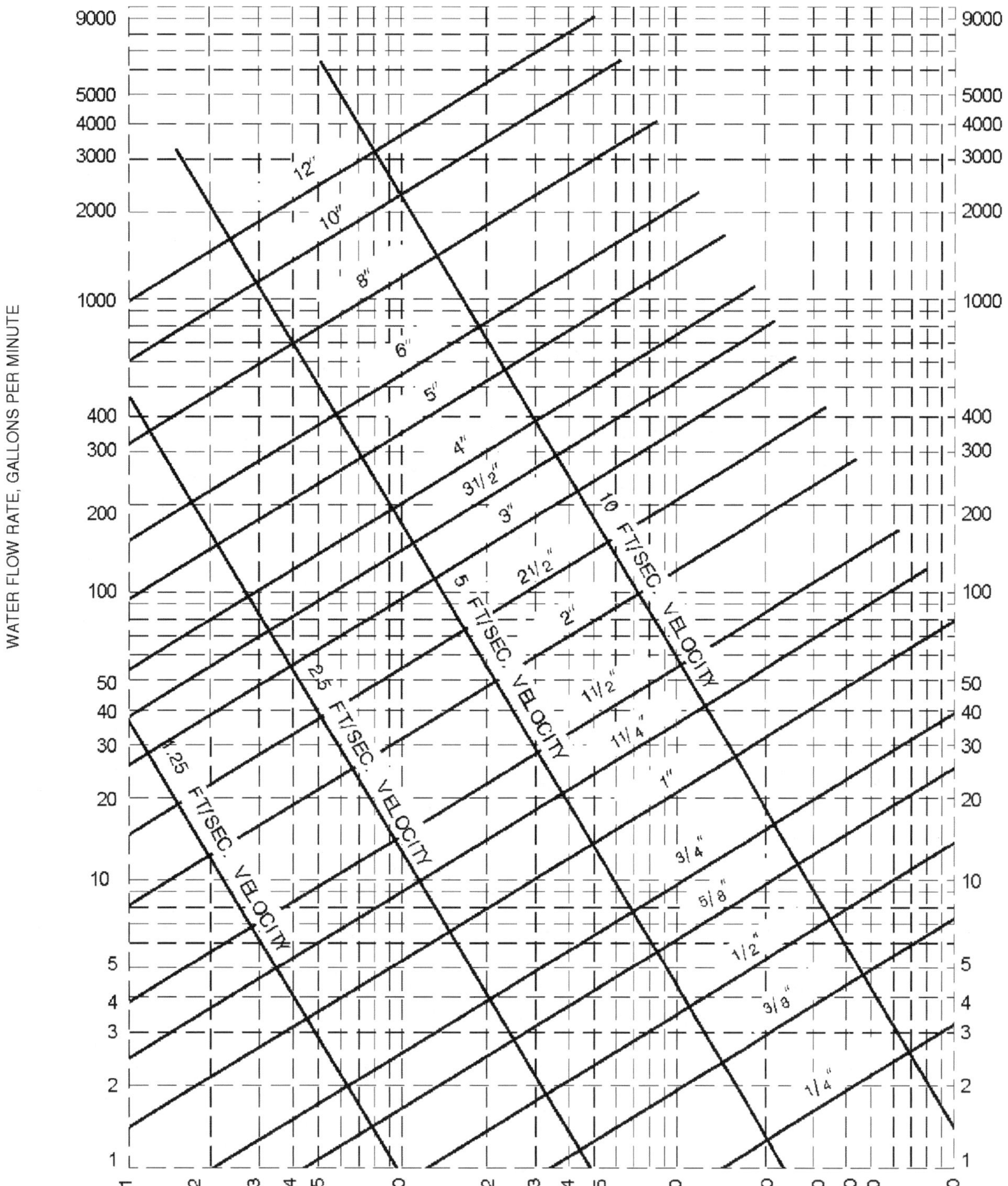

Note: Fluid velocities in excess of 5 to 8 feet per second are not usually recommended.

For SI: 1 inch = 25.4 mm, 1 foot = 304.8 mm, 1 gallon per minute = 3.785 L/m, 1 pounds per square inch = 6.895 kPa, 1 foot per second = 0.305 m/s.

a. This figure applies to smooth new copper tubing with recessed (streamline) soldered joints and to the actual sizes of types indicated on the diagram.

FIGURE AP103.3(3)
FRICTION LOSS IN SMOOTH PIPE[a]

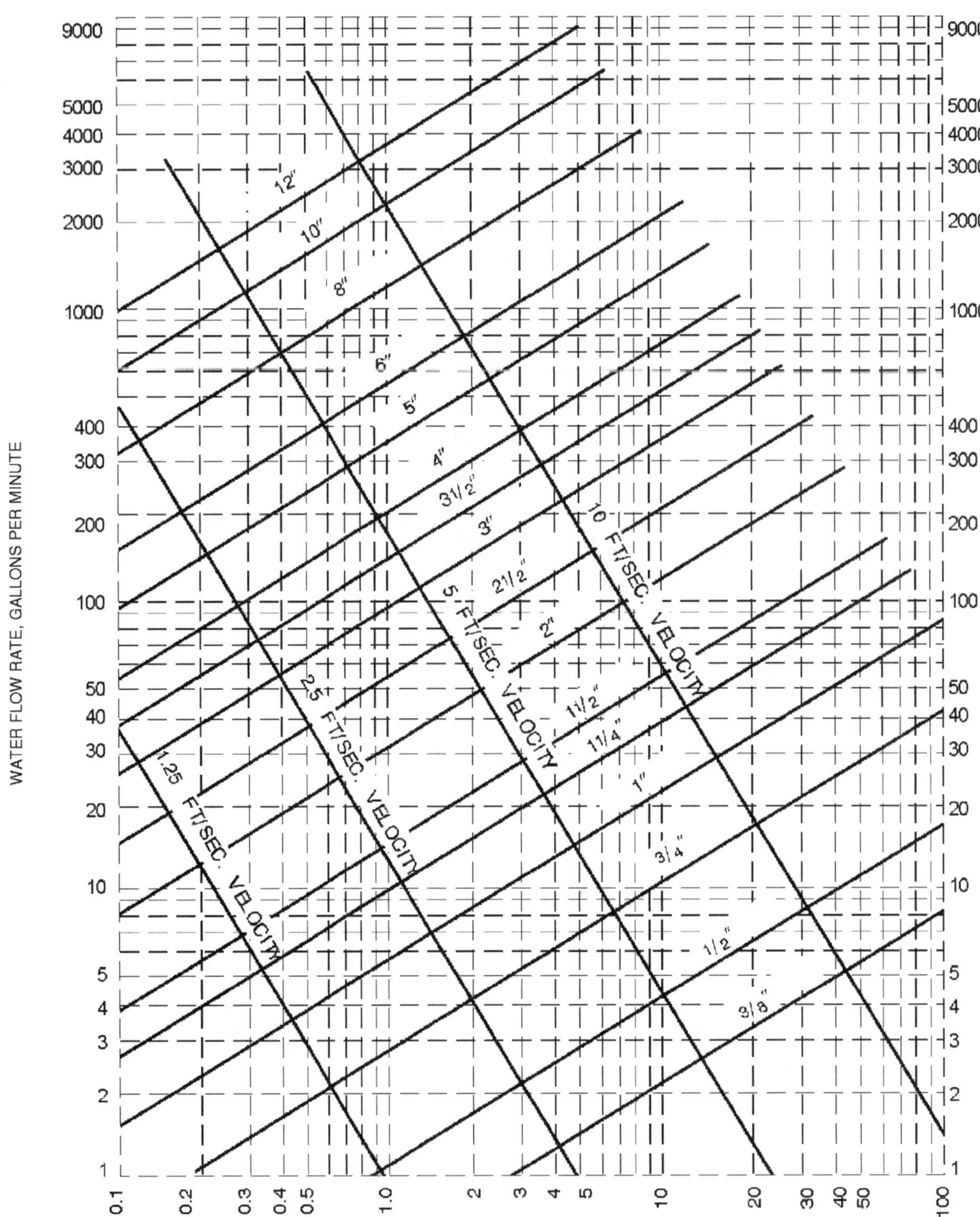

Note: Fluid velocities in excess of 5 to 8 feet per second are not usually recommended.

For SI: 1 inch = 25.4 mm, 1 foot = 304.8 mm, 1 gallon per minute = 3.785 L/m, 1 pounds per square inch = 6.895 kPa, 1 foot per second = 0.305 m/s.

a. This figure applies to smooth new copper tubing with recessed (streamline) soldered joints and to the actual sizes of types indicated on the diagram.

FIGURE AP103.3(4)
FRICTION LOSS IN SMOOTH PIPE[a]

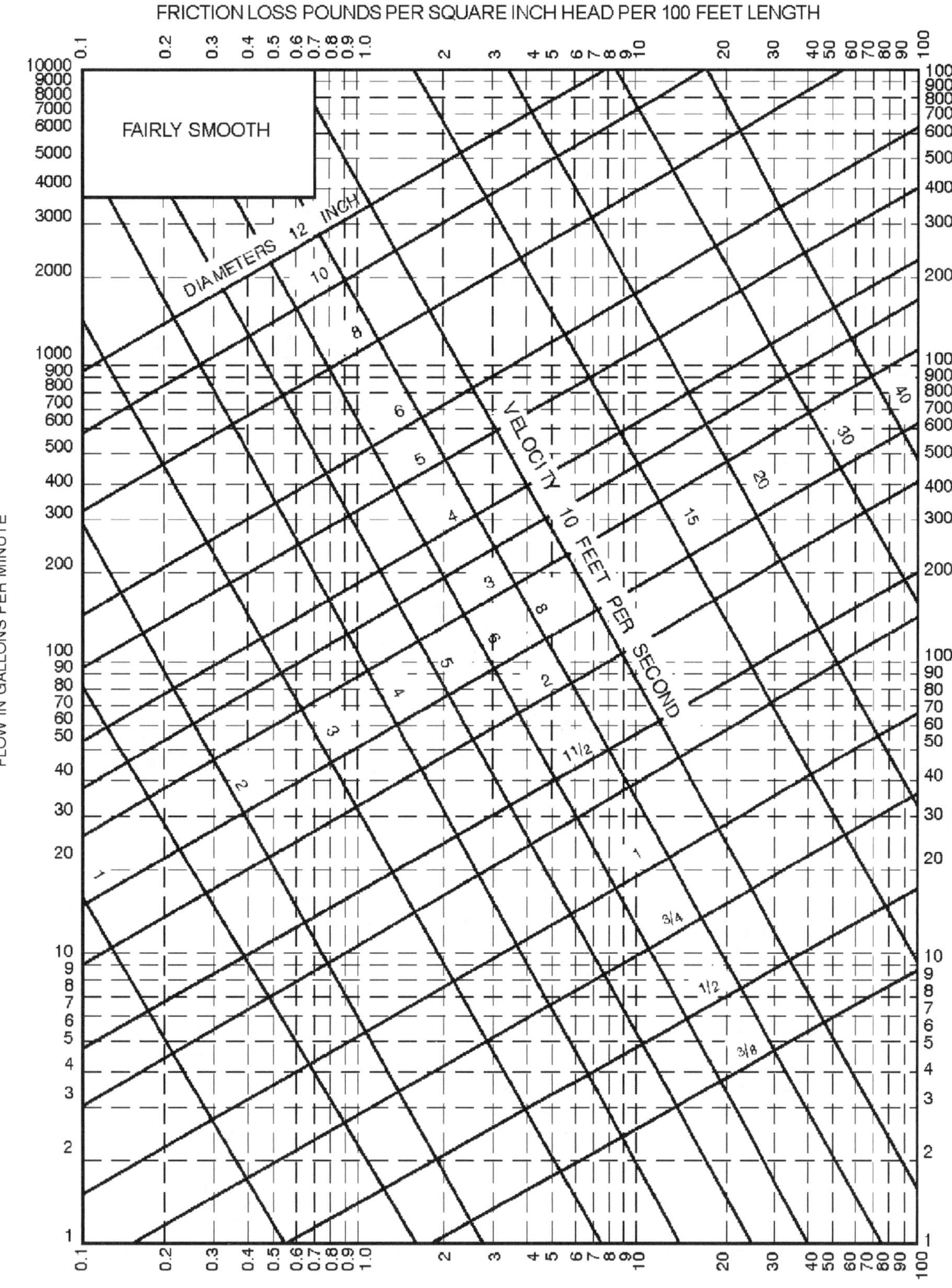

For SI: 1 inch = 25.4 mm, 1 foot = 304.8 mm, 1 gallon per minute = 3.785 L/m, 1 pounds per square inch = 6.895 kPa, 1 foot per second = 0.305 m/s.

a. This figure applies to smooth new steel (fairly smooth) pipe and to actual diameters of standard-weight pipe.

FIGURE AP103.3(5)
FRICTION LOSS IN FAIRLY ROUGH PIPE[a]

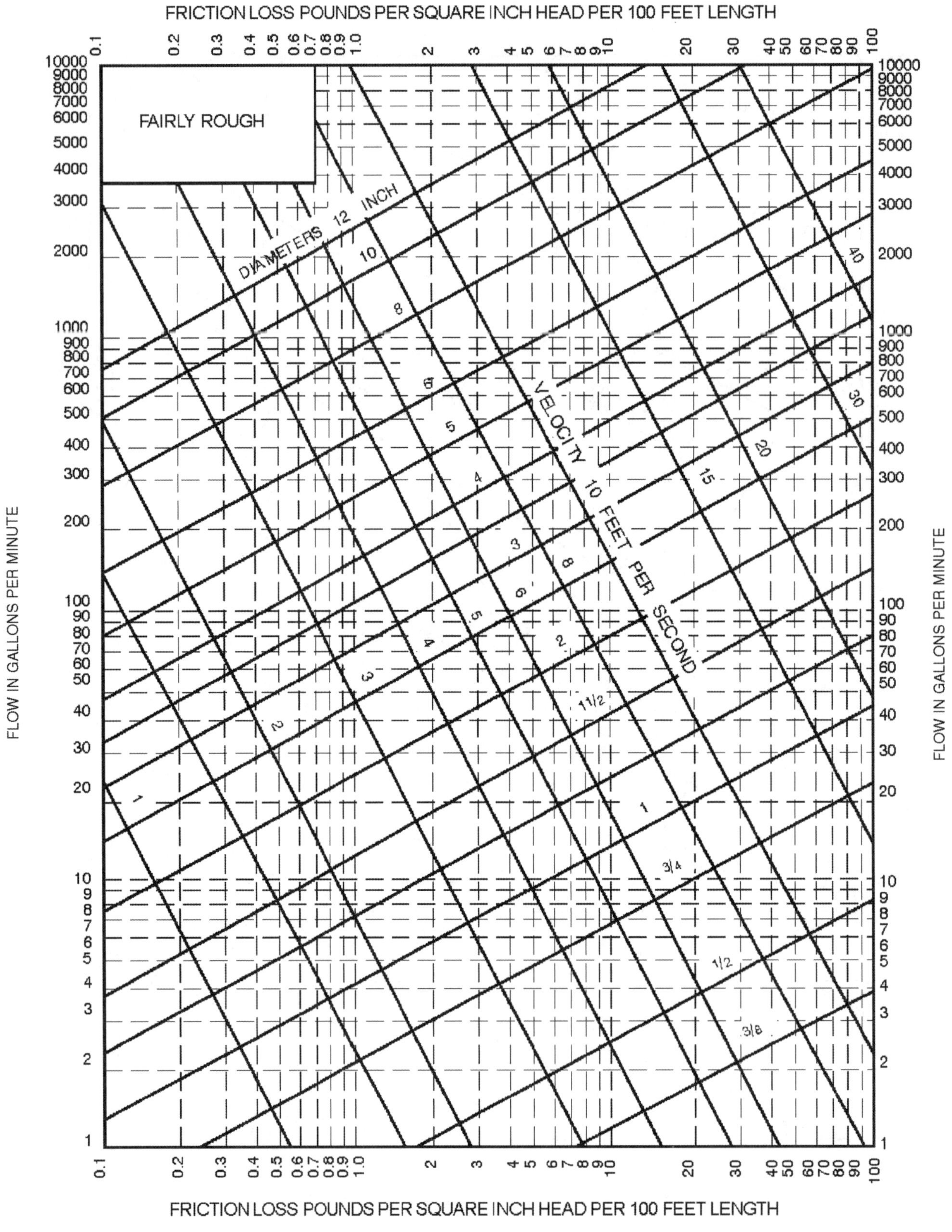

FRICTION LOSS POUNDS PER SQUARE INCH HEAD PER 100 FEET LENGTH

FAIRLY ROUGH

DIAMETERS 12 INCH

VELOCITY 10 FEET PER SECOND

FLOW IN GALLONS PER MINUTE

FRICTION LOSS POUNDS PER SQUARE INCH HEAD PER 100 FEET LENGTH

For SI: 1 inch = 25.4 mm, 1 foot = 304.8 mm, 1 gallon per minute = 3.785 L/m, 1 pounds per square inch = 6.895 kPa, 1 foot per second = 0.305 m/s.

a. This figure applies to fairly rough pipe and to actual diameters which, in general, will be less than the actual diameters of the new pipe of the same kind.

FIGURE AP103.3(6)
FRICTION LOSS IN FAIRLY ROUGH PIPE[a]

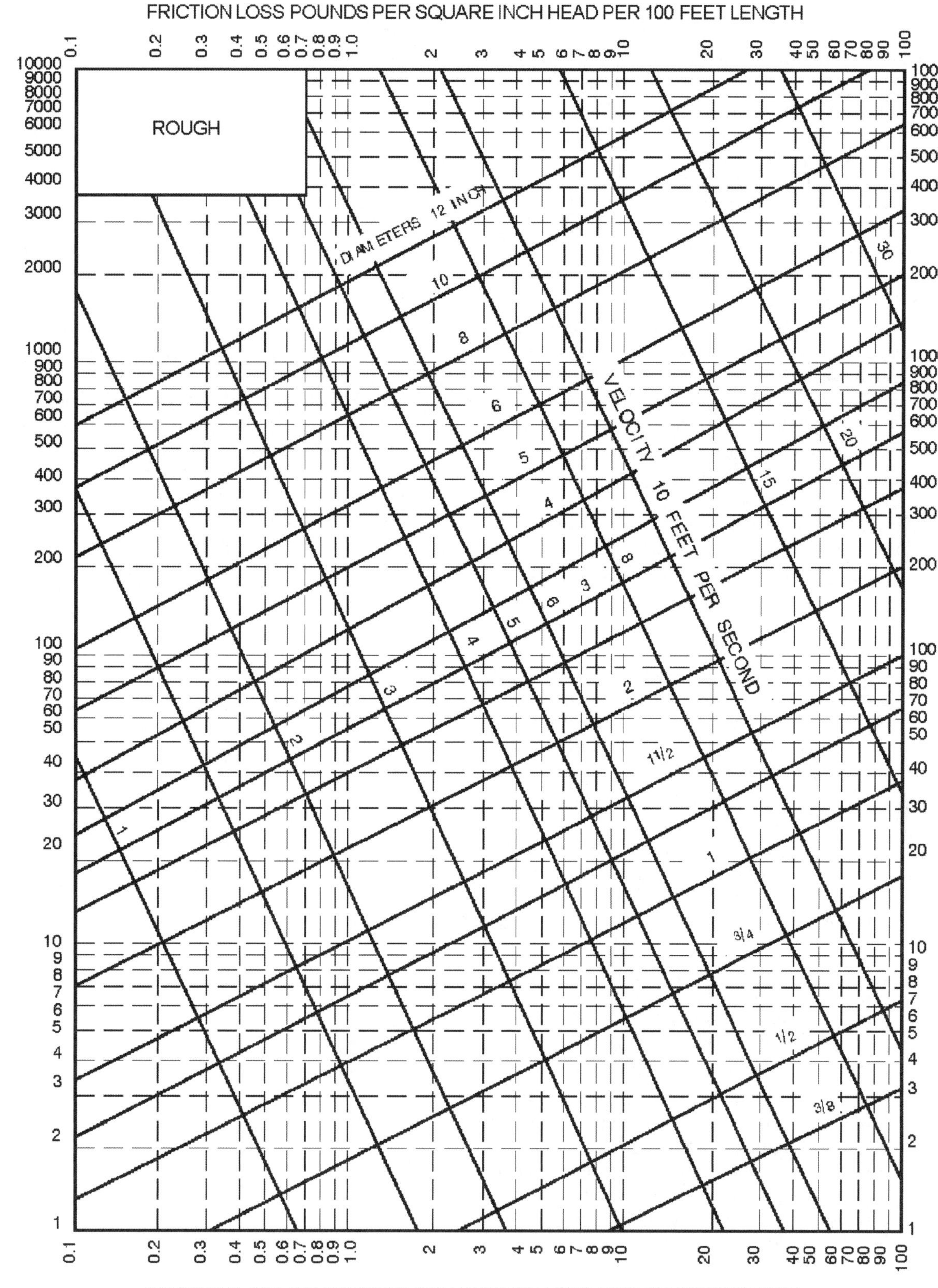

For SI: 1 inch = 25.4 mm, 1 foot = 304.8 mm, 1 gallon per minute = 3.785 L/m, 1 pounds per square inch = 6.895 kPa, 1 foot per second = 0.305 m/s.

a. This figure applies to very rough pipe and existing pipe, and to their actual diameters.

FIGURE AP103.3(7)
FRICTION LOSS IN FAIRLY ROUGH PIPE[a]

SECTION AP201
SELECTION OF PIPE SIZE

AP201.1 Size of water-service mains, branch mains and risers. The minimum size water service pipe shall be $^3/_4$ inch (19.1 mm). The size of water service mains, branch mains and risers shall be determined according to water supply demand [gpm (L/m)], available water pressure [psi (kPa)] and friction loss caused by the water meter and *developed length* of pipe [feet (m)], including the *equivalent length* of fittings. The size of each water distribution system shall be determined according to the procedure outlined in this section or by other design methods conforming to acceptable engineering practice and *approved* by the *building official*:

1. Supply load in the building water distribution system shall be determined by the total load on the pipe being sized, in terms of w.s.f.u., as shown in Table AP103.3(2). For fixtures not listed, choose a w.s.f.u. value of a fixture with similar flow characteristics.
2. Obtain the minimum daily static service pressure [psi (kPa)] available (as determined by the local water authority) at the water meter or other source of supply at the installation location. Adjust this minimum daily static pressure [psi (kPa)] for the following conditions:
   - 2.1. Determine the difference in elevation between the source of supply and the highest water supply outlet. Where the highest water supply outlet is located above the source of supply, deduct 0.5 psi (3.4 kPa) for each foot (0.3 m) of difference in elevation. Where the highest water supply outlet is located below the source of supply, add 0.5 psi (3.4 kPa) for each foot (0.3 m) of difference in elevation.
   - 2.2. Where a water pressure-reducing valve is installed in the water distribution system, the minimum daily static water pressure available is 80 percent of the minimum daily static water pressure at the source of supply or the set pressure downstream of the water pressure-reducing valve, whichever is smaller.
   - 2.3. Deduct all pressure losses caused by special equipment, such as a backflow preventer, water filter and water softener. Pressure loss data for each piece of equipment shall be obtained through the manufacturer of the device.
   - 2.4. Deduct the pressure in excess of 8 psi (55 kPa) resulting from the installation of the special plumbing fixture, such as temperature-controlled shower and flushometer tank water closet. Using the resulting minimum available pressure, find the corresponding pressure range in Table AP201.1.
3. The maximum *developed length* for water piping is the actual length of pipe between the source of supply and the most remote fixture, including either hot (through the water heater) or cold water branches multiplied by a factor of 1.2 to compensate for pressure loss through fittings. Select the appropriate column in Table AP201.1 equal to or greater than the calculated maximum *developed length*.
4. To determine the size of the water service pipe, meter and main distribution pipe to the building using the appropriate table, follow down the selected "maximum *developed length*" column to a fixture unit equal to or greater than the total installation demand calculated by using the "combined" w.s.f.u. column of Table AP201.1. Read the water service pipe and meter sizes in the first left-hand column and the main distribution pipe to the building in the second left-hand column on the same row.
5. To determine the size of each water distribution pipe, start at the most remote outlet on each branch (either hot or cold branch) and, working back toward the main distribution pipe to the building, add up the w.s.f.u. demand passing through each segment of the distribution system using the related hot or cold column of Table AP201.1. Knowing demand, the size of each segment shall be read from the second left-hand column of the same table and the maximum *developed length* column selected in Steps 1 and 2, under the same or next smaller size meter row. In no case does the size of any branch or main need to be larger that the size of the main distribution pipe to the building established in Step 4.

TABLE AP201.1
MINIMUM SIZE OF WATER METERS, MAINS AND DISTRIBUTION PIPING BASED ON WATER SUPPLY FIXTURE UNIT VALUES (w.s.f.u.)

METER AND SERVICE PIPE (inches)	DISTRIBUTION PIPE (inches)	MAXIMUM DEVELOPMENT LENGTH (feet)									
Pressure Range 30 to 39 psi		40	60	80	100	150	200	250	300	400	500
3/4	1/2[a]	2.5	2	1.5	1.5	1	1	0.5	0.5	0	0
3/4	3/4	9.5	7.5	6	5.5	4	3.5	3	2.5	2	1.5
3/4	1	32	25	20	16.5	11	9	7.8	6.5	5.5	4.5
1	1	32	32	27	21	13.5	10	8	7	5.5	5
3/4	1 1/4	32	32	32	32	30	24	20	17	13	10.5
1	1 1/4	80	80	70	61	45	34	27	22	16	12
1 1/2	1 1/4	80	80	80	75	54	40	31	25	17.5	13
1	1 1/2	87	87	87	87	84	73	64	56	45	36
1 1/2	1 1/2	151	151	151	151	117	92	79	69	54	43
2	1 1/2	151	151	151	151	128	99	83	72	56	45
1	2	87	87	87	87	87	87	87	87	87	86
1 1/2	2	275	275	275	275	258	223	196	174	144	122
2	2	365	365	365	365	318	266	229	201	160	134
2	2 1/2	533	533	533	533	533	495	448	409	353	311

METER AND SERVICE PIPE (inches)	DISTRIBUTION PIPE (inches)	MAXIMUM DEVELOPMENT LENGTH (feet)									
Pressure Range 40 to 49 psi		40	60	80	100	150	200	250	300	400	500
3/4	1/2[a]	3	2.5	2	1.5	1.5	1	1	0.5	0.5	0.5
3/4	3/4	9.5	9.5	8.5	7	5.5	4.5	3.5	3	2.5	2
3/4	1	32	32	32	26	18	13.5	10.5	9	7.5	6
1	1	32	32	32	32	21	15	11.5	9.5	7.5	6.5
3/4	1 1/4	32	32	32	32	32	32	32	27	21	16.5
1	1 1/4	80	80	80	80	65	52	42	35	26	20
1 1/2	1 1/4	80	80	80	80	75	59	48	39	28	21
1	1 1/2	87	87	87	87	87	87	87	78	65	55
1 1/2	1 1/2	151	151	151	151	151	130	109	93	75	63
2	1 1/2	151	151	151	151	151	139	115	98	77	64
1	2	87	87	87	87	87	87	87	87	87	87
1 1/2	2	275	275	275	275	275	275	264	238	198	169
2	2	365	365	365	365	365	349	304	270	220	185
2	2 1/2	533	533	533	533	533	533	533	528	456	403

(continued)

TABLE AP201.1—continued
MINIMUM SIZE OF WATER METERS, MAINS AND DISTRIBUTION PIPING BASED ON WATER SUPPLY FIXTURE UNIT VALUES (w.s.f.u.)

METER AND SERVICE PIPE (inches)	DISTRIBUTION PIPE (inches)	MAXIMUM DEVELOPMENT LENGTH (feet)									
Pressure Range 50 to 60 psi		40	60	80	100	150	200	250	300	400	500
3/4	1/2[a]	3	3	2.5	2	1.5	1	1	1	0.5	0.5
3/4	3/4	9.5	9.5	9.5	8.5	6.5	5	4.5	4	3	2.5
3/4	1	32	32	32	32	25	18.5	14.5	12	9.5	8
1	1	32	32	32	32	30	22	16.5	13	10	8
3/4	1 1/4	32	32	32	32	32	32	32	32	29	24
1	1 1/4	80	80	80	80	80	68	57	48	35	28
1 1/2	1 1/4	80	80	80	80	80	75	63	53	39	29
1	1 1/2	87	87	87	87	87	87	87	87	82	70
1 1/2	1 1/2	151	151	151	151	151	151	139	120	94	79
2	1 1/2	151	151	151	151	151	151	146	126	97	81
1	2	87	87	87	87	87	87	87	87	87	87
1 1/2	2	275	275	275	275	275	275	275	275	247	213
2	2	365	365	365	365	365	365	365	329	272	232
2	2 1/2	533	533	533	533	533	533	533	533	533	486

METER AND SERVICE PIPE (inches)	DISTRIBUTION PIPE (inches)	MAXIMUM DEVELOPMENT LENGTH (feet)									
Pressure Range Over 60		40	60	80	100	150	200	250	300	400	500
3/4	1/2[a]	3	3	3	2.5	2	1.5	1.5	1	1	0.5
3/4	3/4	9.5	9.5	9.5	9.5	7.5	6	5	4.5	3.5	3
3/4	1	32	32	32	32	32	24	19.5	15.5	11.5	9.5
1	1	32	32	32	32	32	28	28	17	12	9.5
3/4	1 1/4	32	32	32	32	32	32	32	32	32	30
1	1 1/4	80	80	80	80	80	80	69	60	46	36
1 1/2	1 1/4	80	80	80	80	80	80	76	65	50	38
1	1 1/2	87	87	87	87	87	87	87	87	87	84
1 1/2	1 1/2	151	151	151	151	151	151	151	144	114	94
2	1 1/2	151	151	151	151	151	151	151	151	118	97
1	2	87	87	87	87	87	87	87	87	87	87
1 1/2	2	275	275	275	275	275	275	275	275	275	252
2	2	365	368	368	368	368	368	368	368	318	273
2	2 1/2	533	533	533	533	533	533	533	533	533	533

For SI: 1 inch = 25.4, 1 foot = 304.8 mm, 1 pound per square inch = 6.895 kPa.

a. Minimum size for building supply is a 3/4-inch pipe.

APPENDIX Q

ICC *INTERNATIONAL RESIDENTIAL CODE* ELECTRICAL PROVISIONS/*NATIONAL ELECTRICAL CODE* CROSS REFERENCE

(This appendix is informative and is not part of the code. This table is a cross reference of Chapters 34 through 43 of this code, and the 2011 *National Electrical Code,* NFPA 70).

International Residential Code		*National Electrical Code*
CHAPTER 34	**GENERAL REQUIREMENTS**	
SECTION E3401	**GENERAL**	
E3401.1	Applicability	None
E3401.2	Scope	90.2
E3401.3	Not covered	90.2
E3401.4	Additions and alterations	None
SECTION E3402	**BUILDING STRUCTURE PROTECTION**	
E3402.1	Drilling and notching	None
E3402.2	Penetrations of fire-resistance-rated assemblies	300.21
E3402.3	Penetrations of firestops and draftstops	300.21
SECTION E3403	**INSPECTION AND APPROVAL**	
E3403.1	Approval	110.2
E3403.2	Inspection required	None
E3403.3	Listing and labeling	110.3
SECTION E3404	**GENERAL EQUIPMENT REQUIREMENTS**	
E3404.1	Voltages	110.4
E3404.2	Interrupting rating	110.9
E3404.3	Circuit characteristics	110.10
E3404.4	Enclosure types	
E3404.5	Protection of equipment	110.11
E3404.6	Unused openings	110.12(A)
E3404.7	Integrity of electrical equipment	110.12(B)
E3404.8	Mounting	110.13(A)
E3404.9	Energized parts guarded against accidental contact	110.27(A)
E3404.10	Prevent physical damage	110.27(B)
E3404.11	Equipment identification	110.21
E3404.12	Identification of disconnecting means	110.22
SECTION E3405	**EQUIPMENT LOCATION AND CLEARANCES**	
E3405.1	Working space and clearances	110.26
Figure E3405.1	Working space and clearances	110.26(A)
	Footnote 1	110.26(E)(1)(a)
	Footnote 2	110.26(A)(3)
	Footnote 3	110.26(B)
	Footnote 4	230.70(A); 240.24(D), (E) and (F); Note 5
	Footnote 5	110.26(D)
E3405.2	Working clearances for energized equipment and panelboards	110.26(A)(1), (2) and (3)

E3405.3	Dedicated panelboard space	110.26(F)(1)(a)
E3405.4	Location of working spaces and equipment	110.26(B); 230.70(A); and 240.24(D), (E) and (F)
E3405.5	Access and entrance to working space	110.26(C)(1)
E3405.6	Illumination	110.26(D)
SECTION E3406	**ELECTRICAL CONDUCTORS AND CONNECTIONS**	
E3406.1	General	Articles 110, 300 and 310
E3406.2	Conductor material	110.5
E3406.3	Minimum size of conductors	310.106(A)
E3406.4	Stranded conductors	310.106(C)
E3406.5	Individual conductor insulation	310.10(B) and (C); and 310.106(D)
E3406.6	Conductors in parallel	310.10(H)(2)
E3406.7	Conductors of the same circuit	300.3(B)
E3406.8	Aluminum and copper connections	110.14
E3406.9	Fine stranded conductors	110.14
Table E3406.9	Conductor stranding	Chapter 9, Table 10
E3406.10	Terminals	110.14(A)
E3406.11	Splices	110.14(B)
E3406.11.1	Continuity	300.13(A)
	Exception	300.13(A)
E3406.11.2	Device connections	250.148(B) and 300.13(B)
E3406.11.3	Length of conductor for splice or termination	300.14
E3406.12	Grounded conductor continuity	200.2(B)
SECTION E3407	**CONDUCTOR IDENTIFICATION**	
E3407.1	Grounded conductors	200.6(A) and (B); and 310.110(A)
E3407.2	Equipment grounding conductors	250.119 and 310.110(B)
E3407.3	Ungrounded conductors	200.7(C)(1) and 310.110(C), Exception
E3407.4	Identification of terminals	200.10
E3407.4.1	Device terminals	200.10(A)
E3407.4.2	Receptacles, plugs and connectors	200.10(B)
CHAPTER 35	**ELECTRICAL DEFINITIONS**	
SECTION 3501	**GENERAL**	Article 100, Definitions
CHAPTER 36	**SERVICES**	
SECTION E3601	**GENERAL SERVICES**	
E3601.1	Scope	230.1
E3601.2	Number of services	230.2
E3601.3	One building or other structure not to be supplied through another	230.3
E3601.4	Other conductors in raceway or cable	230.7
E3601.5	Raceway seal	230.8
E3601.6	Service disconnect required	230.70
E3601.6.1	Marking of service equipment and disconnects	230.66 and 230.70(B)
E3601.6.2	Service disconnect location	230.70(A) and 230.72(C)
E3601.7	Maximum number of disconnects	230.71(A)

SECTION E3602	**SERVICE SIZE AND RATING**	
E3602.1	Ampacity of ungrounded conductors	230.79(C) and (D)
E3602.2	Service load	220.82(A)
Table E3602.2	Minimum service load calculation	220.82(B) and (C)
E3602.2.1	Services under 100 amperes	None
E3602.3	Rating of service disconnect	230.79 and 230.80
E3602.4	Voltage rating	220.82(A)
SECTION E3603	**SERVICE, FEEDER AND GROUNDING ELECTRODE CONDUCTOR SIZING**	
E3603.1	Grounded and ungrounded service conductor size	310.15(B)(7)
Table E3603.1	Service conductor and grounding electrode conductor sizing	Table 310.15(B)(7) and Table 250.66
	Footnote a	250.64(E)
	Footnote b	250.64(B)
	Footnote c	250.64(B)
	Footnote d	250.66(A) and (B)
E3603.2	Ungrounded service conductors for accessory buildings and structures	230.79(D)
	Exception 1	230.42 and 230.79(A)
	Exception 2	230.42 and 230.79(B)
E3603.3	Overload protection	230.90
E3603.3.1	Ungrounded conductor	230.90(A)
	Exception	230.90(A), Exception 3
E3603.3.2	Not in grounded conductor	230.90(B)
E3603.3.3	Location	230.91
E3603.4	Grounding electrode conductor size	250.66
E3603.5	Temperature limitations	110.14(C)(1)
SECTION E3604	**OVERHEAD SERVICE AND SERVICE-ENTRANCE CONDUCTOR INSTALLATION**	
E3604.1	Clearances on buildings	230.9
Figure E3604.1	Clearances from building openings	230.9
E3604.2	Vertical clearances	230.24
E3604.2.1	Above roofs	230.24(A)
	Exception 1	230.24(A), Exception 1
	Exception 2	230.24(A), Exception 2
	Exception 3	230.24(A), Exception 3
	Exception 4	230.24(A), Exception 4
	Exception 5	230.24(A), Exception 5
Figure E3604.2.1	Clearances from roofs	230.24
E3604.2.2	Vertical clearance from grade	230.24(B)
	Item 1	230.24(B)(1)
	Item 2	230.24(B)(2)
	Item 3	230.24(B)(4)
E3604.3	Point of attachment	230.26
E3604.4	Means of attachment	230.27
E3604.5	Service masts as supports	230.28
E3604.6	Supports over buildings	230.29

SECTION E3605	**SERVICE-ENTRANCE CONDUCTORS**	
E3605.1	Insulation of service-entrance conductors	230.41
	Exception 1	230.41, Exception
	Exception 2	230.41, Exception
E3605.2	Wiring methods for services	230.43
E3605.3	Spliced conductors	230.46
E3605.4	Protection of underground service entrance conductors	230.50(A)
E3605.5	Protection of service cables against damage	230.50(B)
E3605.6	Direct sunlight exposure	310.10(D)
E3605.7	Mounting supports	230.51
E3605.8	Raceways to drain	230.53
E3605.9	Overhead service locations	230.54
E3605.9.1	Rain-tight service head	230.54(A)
E3605.9.2	Service cable, service head or gooseneck	230.54(B)
E3605.9.3	Service head location	230.54(C)
	Exception	230.54(C), Exception
E3605.9.4	Separately bushed openings	230.54(E)
E3605.9.5	Drip loops	230.54(F)
E3605.9.6	Conductor arrangement	230.54(G)
E3605.9.7	Secured	230.54(D)
SECTION E3606	**SERVICE EQUIPMENT—GENERAL**	
E3606.1	Service equipment enclosures	230.62
E3606.2	Working space	110.26
E3606.3	Available short-circuit current	110.9
E3606.4	Marking	230.66
SECTION E3607	**SYSTEM GROUNDING**	
E3607.1	System service ground	250.20(B)(1) and 250.24(A)
E3607.2	Location of grounding electrode conductor connection	250.24(A)(1) and (A)(5)
E3607.3	Buildings or structures supplied by feeder(s) or branch circuit(s)	250.32(A)
	Exception	250.32(A), Exception
E3607.3.1	Equipment grounding conductor	250.32(B) and Table 250.122
E3607.3.2	Grounded conductor	250.32(B)(1), Exception
E3607.4	Grounding electrode conductor	250.24(D)
E3607.5	Main bonding jumper	250.24(B)
E3607.6	Common grounding electrode	250.58
SECTION E3608	**GROUNDING ELECTRODE SYSTEM**	
E3608.1	Grounding electrode system	250.50
	Exception	250.50, Exception
E3608.1.1	Metal underground water pipe	250.52(A)(1)
E3608.1.1.1	Interior metal water piping	250.68(C)(1)
E3608.1.1.2	Installation	250.53(D) and 250.53(E)
E3608.1.2	Concrete-encased electrode	250.52(A)(3)
E3608.1.3	Ground rings	250.52(A)(4) and 250.53(F)
E3608.1.4	Rod and pipe electrodes	250.52(A)(5)

E3608.1.4.1	Installation	250.53(G)
E3608.1.5	Plate electrodes	250.52(A)(7) and 250.53(H)
E3608.1.6	Other listed electrodes	250.52(A)(6)
E3608.2	Bonding jumper	250.53(C)
E3608.3	Rod, pipe and plate electrode requirements	250.53(A), (B) and (E)
E3608.4	Supplemental electrode required	250.53(A)(2) and (A)(3)
E3608.5	Aluminum electrodes	250.52(B)(2)
E3608.6	Metal underground gas piping system	250.52(B)(1)
SECTION E3609	**BONDING**	
E3609.1	General	250.90
E3609.2	Bonding of service equipment	250.92(A)
E3609.3	Bonding to other systems	250.94
E3609.4	Method of bonding at the service	250.92(B)
E3609.4.1	Grounded service conductor	250.92(B)(1)
E3609.4.2	Threaded connections	250.92(B)(2)
E3609.4.3	Threadless couplings and connectors	250.92(B)(3)
E3609.4.4	Other devices	250.92(B)(4)
E3609.5	Sizing bonding jumper on supply side of service and main bonding jumper	250.28(D) and 250.102(C)
E3609.6	Metal water piping bonding	250.104(A)
E3609.7	Bonding other metal piping	250.104(B)
SECTION E3610	**GROUNDING ELECTRODE CONDUCTORS**	
E3610.1	Continuous	250.64(C) and (F)
E3610.2	Securing and protection against physical damage	250.64(A) and (B)
E3610.3	Enclosures for grounding electrode conductors	250.64(E)
E3610.4	Prohibited use	250.121
SECTION E3611	**GROUNDING ELECTRODE CONDUCTOR CONNECTION TO GROUNDING ELECTRODES**	
E3611.1	Methods of grounding connection to electrodes	250.70
E3611.2	Accessibility	250.68(A)
E3611.3	Effective grounding path	250.68(B)
E3611.4	Interior metal water piping	250.68(C)(1)
E3611.5	Protection of ground clamps and fittings	250.10
E3611.6	Clean surfaces	250.12
CHAPTER 37	**BRANCH-CIRCUIT AND FEEDER REQUIREMENTS**	
SECTION E3701	**GENERAL**	
E3701.1	Scope	None
E3701.2	Branch-circuit and feeder ampacity	210.19(A) and 215.2(A)(1)
E3701.3	Selection of ampacity	310.15(A)(2), Exception
E3701.4	Branch circuits with more than one receptacle	210.19(A)(2)
E3701.5	Multiwire branch circuits	210.4
E3701.5.1	Disconnecting means	210.4(B)
E3701.5.2	Grouping	210.4(D)

SECTION E3702	**BRANCH-CIRCUIT RATINGS**	
E3702.1	Branch-circuit voltage limitations	210.6(A), (B) and (C)
E3702.2	Branch-circuit ampere rating	210.3
E3702.3	Fifteen- and 20-ampere branch circuits	210.23(A)
E3702.4	Thirty-ampere branch circuits	210.23(B)
E3702.5	Branch circuits serving multiple loads or outlets	210.23(A)
E3702.6	Branch circuits serving a single motor	430.22(A)
E3702.7	Branch circuits serving motor-operated and combination loads	220.18(A)
E3702.8	Branch-circuit inductive lighting loads	220.18(B)
E3702.9	Branch-circuit load for ranges and cooking appliances	Table 220.55, Note 4
E3702.9.1	Minimum branch circuit for ranges	210.19(A)(3)
E3702.10	Branch circuits serving heating loads	422.13; and 424.3(A) and (B)
E3702.11	Branch circuits for air-conditioning and heat pump equipment	440.4(B), 440.35 and 440.62(A)
E3702.12	Branch circuits serving room air conditioners	440.62(A)
E3702.12.1	Where no other loads are supplied	440.62(B)
E3702.12.2	Where lighting units or other appliances are also supplied	440.62(C)
E3702.13	Branch-circuit requirement—summary	210.24 and 210.25
Table E3702.13	Branch-circuit requirement—summary	Table 210.24
SECTION E3703	**REQUIRED BRANCH CIRCUITS**	
E3703.1	Branch circuits for heating	422.12 and 422.12, Exceptions 1 and 2
E3703.2	Kitchen and dining area receptacles	210.52 (B)(1)
E3703.3	Laundry circuit	210.11(C)(2); and 210.23(A), Exception
E3703.4	Bathroom branch circuits	210.11(C)(3)
E3703.5	Number of branch circuits	210.11(A)
E3703.6	Branch-circuit load proportioning	210.11(B)
SECTION E3704	**FEEDER REQUIREMENTS**	
E3704.1	Conductor size	215.2 (A)(3) and (A)(4); 220.40; and Table 310.15(B)(7)
E3704.2	Feeder loads	220.40
Table E3704.2(1)	Feeder load calculation	Table 220.12, 220.14, Table 220.42(C), 220.50, 220.51, 220.52, 220.53, 220.54, 220.55 and 220.60
Table E3704.2(2)	Demand loads for electric ranges, wall-mounted ovens, counter-mounted cooking units and other cooking appliances over 13/4 kVA rating	Table 220.55
E3704.3	Feeder neutral load	220.61
E3704.4	Lighting and general use receptacle load	220.12, Table 220.12 and 220.14(J)
E3704.5	Ampacity and calculated loads	220.40
E3704.6	Equipment grounding conductor	215.6
SECTION E3705	**CONDUCTOR SIZING AND OVERCURRENT PROTECTION**	
E3705.1	General	310.15(A)
Table E3705.1	Allowable ampacities	Table 310.15(B)(16)
E3705.2	Correction factors for ambient temperatures	Table 310.15(B)(16)
Table E3705.2	Ambient temperature correction factors	Table 310.15(B)(2)(a)
E3705.3	Adjustment factor for conductor proximity	310.15(B)(3)(a)
Table E3705.3	Conductor proximity adjustment factors	Table 310.15(B)(3)(a)
E3705.4	Temperature limitations	110.14(C)

E3705.4.1	Conductors rated 60°C . . . 110.14(C)(1)(a)
E3705.4.2	Conductors rated 75°C . . . 110.14(C)(1)(b)
E3705.4.3	Separately installed pressure connectors . . . 110.14(C)(2)
E3705.4.4	Conductors of Type NM cable . . . 334.80 and 334.112
E3705.5	Overcurrent protection required . . . 240.4, 240.21 and Table 310.15(B)(6)
E3705.5.1	Cords . . . 240.5
E3705.5.2	Overcurrent devices of the next higher size . . . 240.4(B)
E3705.5.3	Small conductors . . . 240.4(D)
E3705.5.4	Air-conditioning and heat pump equipment . . . 240.4(G)
E3705.6	Fuses and fixed-trip circuit breakers . . . 240.6
E3705.7	Location of overcurrent devices in or on premises . . . 240.24(A), (C), (D), (E) and (F)
E3705.8	Ready access for occupants . . . 240.24(B)
E3705.9	Enclosures for overcurrent devices . . . 240.30(A) and (B)
SECTION E3706	**PANELBOARDS**
E3706.1	Panelboard rating . . . 408.30
E3706.2	Panelboard circuit identification . . . 408.4
E3706.3	Panelboard overcurrent protection . . . 408.36
E3706.4	Grounded conductor terminations . . . 408.41
E3706.5	Back-fed devices . . . 408.36(D)
CHAPTER 38	**WIRING METHODS**
SECTION E3801	**GENERAL REQUIREMENTS**
E3801.1	Scope . . . 300.1
E3801.2	Allowable wiring methods . . . 110.8 and 300.3(A)
Table E3801.2	Allowable wiring methods . . . None
E3801.3	Circuit conductors . . . 300.3(B)
E3801.4	Wiring method applications . . . Chapter 3 and 300.2
Table E3801.4	Allowable applications for wiring methods . . . Chapter 3 and 300.2
SECTION E3802	**ABOVE-GROUND INSTALLATION REQUIREMENTS**
E3802.1	Installation and support requirements . . . Chapter 3
Table E3802.1	General installation and support requirements for wiring methods . . . Chapter 3
E3802.2	Cables in accessible attics . . . 320.23 and 334.23
E3802.2.1	Across structural members . . . 320.23(A) and 334.23
E3802.2.2	Cable installed through or parallel to framing members . . . 320.23(B)
E3802.3	Exposed cable . . . 320.15 and 334.15
E3802.3.1	Surface installation . . . 334.15(A)
E3802.3.2	Protection from physical damage . . . 334.15(B)
E3802.3.3	Locations exposed to direct sunlight . . . 310.10(D)
E3802.4	In unfinished basements . . . 334.15(C)
E3802.5	Bends . . . 334.24 and 338.24
E3802.6	Raceways exposed to different temperatures . . . 300.7(A)
E3802.7	Raceways in wet locations above grade . . . 300.9

SECTION E3803	**UNDERGROUND INSTALLATION REQUIREMENTS**	
E3803.1	Minimum cover requirements	300.5(A)
Table E3803.1	Minimum cover requirements, burial in inches	Table 300.5
E3803.2	Warning ribbon	300.5(D)(3)
E3803.3	Protection from damage	300.5(D)
E3803.4	Splices and taps	300.5(E)
E3803.5	Backfill	300.5(F)
E3803.6	Raceway seals	300.5(G)
E3803.7	Bushing	300.5(H)
E3803.8	Single conductors	300.5(I)
E3803.9	Ground movement	300.5(J)
E3803.10	Wet locations	300.5(B)
E3803.11	Under buildings	300.5(C)
CHAPTER 39	**POWER AND LIGHTING DISTRIBUTION**	
SECTION E3901	**RECEPTACLE OUTLETS**	
E3901.1	General	210.50 and 210.52
E3901.2	General-purpose receptacle distribution	210.52(A)
E3901.2.1	Spacing	210.52(A)(2)
E3902.2.2	Wall space	210.52(A)(2)
E3901.2.3	Floor receptacles	210.52(A)(3)
E3901.2.4	Countertop receptacles	210.52(A)(4)
E3901.3	Small appliance receptacles	210.52(B)(1)
E3901.3.1	Other outlets prohibited	210.52(B)(2)
E3901.3.2	Limitations	210.52(B)(3)
E3901.4	Countertop receptacles	210.52(C)
E3901.4.1	Wall counter space	210.52(C)(1)
E3901.4.2	Island counter spaces	210.52(C)(2)
E3901.4.3	Peninsular counter space	210.52(C)(3)
E3901.4.4	Separate spaces	210.52(C)(4)
E3901.4.5	Receptacle outlet location	210.52(C)(5)
E3901.5	Appliance receptacle outlets	210.50(C)
E3901.6	Bathroom	210.52(D) and 406.5(E)
E3901.7	Outdoor outlets	210.52(E)
E3901.8	Laundry areas	210.52(F)
E3901.9	Basements and garages	210.52(G)
E3901.10	Hallways	210.52(H)
E3901.11	Foyers	210.52(I)
	HVAC outlet	210.63
SECTION E3902	**GROUND-FAULT AND ARC-FAULT CIRCUIT-INTERRUPTER PROTECTION**	
E3902.1	Bathroom receptacles	210.8(A)(1)
E3902.2	Garage and accessory building receptacles	210.8(A)(2)
E3902.3	Outdoor receptacles	210.8(A)(3)
E3902.4	Crawl space receptacles	210.8(A)(4)
E3902.5	Unfinished basement receptacles	210.8(A)(5)

E3902.6	Kitchen receptacles	210.8(A)(6)
E3902.7	Sink receptacles	210.8(A)(7) and 406.5(E)
E3902.8	Boathouse receptacles	210.8(A)(8)
E3902.9	Boat hoists	210.8(C)
E3902.10	Electrically heated floors	424.44(G)
E3902.11	Location of ground-fault circuit interrupters	210.8
E3902.12	Arc-fault circuit-interrupter protection	210.12(A)
E3902.13	Arc-fault circuit-interrupter protection for branch-circuit extensions or modifications	210.12(B)
SECTION E3903	**LIGHTING OUTLETS**	
E3903.1	General	210.70(A)
E3903.2	Habitable rooms	210.70(A)(1)
E3903.3	Additional locations	210.70(A)(2)
E3903.4	Storage or equipment spaces	210.70(A)(3)
SECTION E3904	**GENERAL INSTALLATION REQUIREMENTS**	
E3904.1	Electrical continuity of metal raceways and enclosures	300.10
E3904.2	Mechanical continuity—raceways and cables	300.12
E3904.3	Securing and supporting	300.11(A)
E3904.3.1	Prohibited means of support	300.11(C)
E3904.4	Raceways as means of support	300.11(B)
E3904.5	Raceway installations	300.18
E3904.6	Conduit and tubing fill	300.17; Chapter 9, Table 1; and Annex C
Tables E3904.6(1)–E3904.6(10)	Maximum number of conductors in conduit or tubing	300.17; and Chapter 9, Table 1, Note 1
E3904.7	Air handling—stud cavity and joist spaces	300.22(C), Exception
CHAPTER E3905	**BOXES, CONDUIT BODIES AND FITTINGS**	
E3905.1	Box, conduit body or fitting—where required	300.15
E3905.1.1	Equipment	300.15(B)
E3905.1.2	Protection	300.15(C)
E3905.1.3	Integral enclosure	300.15(E)
E3905.1.4	Fitting	300.15(F)
E3905.1.5	Buried conductors	300.15(G)
E3905.1.6	Luminaires	300.15(J)
E3905.2	Metal boxes	314.4
E3905.3	Nonmetallic boxes	314.3
E3905.3.1	Nonmetallic-sheathed cable and nonmetallic boxes	314.17(C)
E3905.3.2	Securing to box	314.17(C)
E3905.3.3	Conductor rating	314.17(C)
E3905.4	Minimum depth of outlet boxes	314.24
E3905.4.1	Outlet boxes without enclosed devices	
E3905.4.2	Utilization equipment	
E3905.5	Boxes enclosing flush-mounted devices	314.19
E3905.6	Boxes at luminaire (lighting fixture) outlets	314.27(A)(1) and(2)
E3905.6.1	Wall outlets	
E3905.6.2	Ceiling outlets	

E3905.7	Floor boxes	314.27(C)
E3905.8	Boxes at fan outlets	314.27(D)
E3905.9	Utilization equipment	314.27(E)
E3905.10	Conduit bodies, junction, pull and outlet boxes to be accessible	314.29
E3905.11	Damp or wet locations	314.15
E3905.12	Number of conductors in outlet, device, and junction boxes and conduit boxes	314.16
E3905.12.1	Box volume calculations	314.16(A)
Table E3905.12.1	Maximum number of conductors in metal boxes	Table 314.16
E3905.12.1.1	Standard boxes	314.16(A)(1)
E3905.12.1.2	Other boxes	314.16(A)(2)
E3905.12.2	Box fill calculations	314.16(B)
E3905.12.2.1	Conductor fill	Table 314.16 and 314.16(B)(1)
Table E3905.12.2.1	Volume allowance required per conductor	Table 314.16(B)
E3905.12.2.2	Clamp fill	314.16(B)(2)
E3905.12.2.3	Support fittings fill	314.16(B)(3)
E3905.12.2.4	Device or equipment fill	314.16(B)(4)
E3905.12.2.5	Equipment grounding conductor fill	314.16(B)(5)
E3905.12.3	Conduit bodies	314.16(C)(1)
E3905.12.3.1	Splices, taps or devices	314.16(C)(2)
E3905.12.3.2	Short radius conduit bodies	314.16(C)(3)
SECTION E3906	**INSTALLATION OF BOXES, CONDUIT BODIES AND FITTINGS**	
E3906.1	Conductors entering boxes, conduit bodies or fittings	314.17
E3906.1.1	Insulated fittings	300.4(G)
E3906.2	Openings	314.17(A)
E3906.3	Metal boxes, conduit bodies and fittings	314.17(B)
E3906.4	Unused openings	110.12(A)
E3906.5	In wall or ceiling	314.20
E3906.6	Noncombustible surfaces	314.21
E3906.7	Surface extensions	314.22
E3906.8	Supports	314.23
E3906.8.1	Surface mounting	314.23(A)
E3906.8.2	Structural mounting	314.23(B)
E3906.8.2.1	Nails and screws	314.23(B)(1)
E3906.8.2.2	Braces	314.23(B)(2)
E3906.8.3	Mounting in finished surfaces	314.23(C)
E3906.8.4	Raceway supported enclosures without devices or fixtures	314.23(E)
E3906.8.5	Raceway supported enclosures with devices or fixtures	314.23(F)
E3906.8.6	Enclosures in concrete or masonry	314.23(G)
E3906.9	Covers and canopies	314.25
E3906.10	Metal covers and plates	314.25(A)
E3906.11	Exposed combustible finish	314.25(B)
SECTION E3907	**CABINETS AND PANELBOARDS**	
E3907.1	Switch and overcurrent device enclosures	312.8
E3907.2	Damp or wet locations	312.2

E3907.3	Position in wall	312.3
E3907.4	Repairing noncombustible surfaces	312.4
E3907.5	Unused openings	110.12(A) and 408.7
E3907.6	Conductors entering cabinets	300.4(G) and 312.5(B)
E3907.7	Openings to be closed	312.5(A)
E3907.8	Cables	312.5(C)
SECTION E3908	**GROUNDING**	
E3908.1	Metal enclosures	250.86
E3908.2	Equipment fastened in place or connected by permanent wiring methods (fixed)	250.110(1), (2) and (3)
E3908.3	Specific equipment fastened in place or connected by permanent wiring methods	250.112(J), (L) and (M)
E3908.4	Effective ground-fault current path	250.4(A)(5)
E3908.5	Earth as a ground-fault current path	250.4(A)(5)
E3908.6	Load-side grounded conductor	250.24(A)(5)
E3908.7	Load-side equipment	250.142(B)
E3908.8	Types of equipment grounding conductors	250.118
E3908.8.1	Flexible metal conduit	250.118(5)
E3908.8.2	Liquid-tight flexible metal conduit	250.118(6)
E3908.8.3	Nonmetallic sheathed cable (Type NM)	334.108
E3908.9	Equipment fastened in place or connected by permanent wiring methods	250.134
E3908.10	Methods of equipment grounding	250 Part (VII)
E3908.11	Equipment grounding conductor installation	250.120
E3908.12	Equipment grounding conductor size	250.122(A) and (B)
Table E3908.12	Equipment grounding conductor sizing	Table 250.122
E3908.12.1	Multiple circuits	250.122(C)
E3908.13	Continuity and attachment of equipment grounding conductors to boxes	250.148
E3908.14	Connecting receptacle grounding terminal to box	250.146(A), (B) and (C)
E3908.15	Metal boxes	250.8 and 250.148(C)
E3908.16	Nonmetallic boxes	250.148(D)
E3908.17	Clean surfaces	250.12
E3908.18	Bonding other enclosures	250.96(A)
E3908.19	Size of equipment bonding jumper on load side of service	250.102(D) and 250.122
E3908.20	Installation—equipment bonding jumper	250.102(E)
E3908.20.1	Inside raceway or enclosure	
E3908.20.2	Outside raceway or enclosure	
E3908.20.3	Protection	
SECTION E3909	**FLEXIBLE CORDS**	
E3909.1	Where permitted	400.7 and 400.8
E3909.2	Loading and protection	240.4; 240.5(A) and 240.5(B)(1); 400.5; and 400.13
Table E3909.2	Maximum ampere load for flexible cords	Table 400.5(A)(1)
E3909.3	Splices	400.9
E3909.4	Attachment plugs	400.7(B)

CHAPTER 40	**DEVICES AND LIGHTING FIXTURES**
SECTION E4001	**SWITCHES**

IRC	Subject	NEC
E4001.1	Rating and application of snap switches	404.14(A)
E4001.2	CO/ALR snap switches	404.14(C)
E4001.3	Indicating	404.7
E4001.4	Time switches and similar devices	404.5
E4001.5	Grounding of enclosures	404.12
E4001.6	Access	404.8(A)
E4001.7	Damp or wet locations	404.4
E4001.8	Grounded conductors	404.2(B)
E4001.9	Switch connections	404.2(A)
E4001.10	Box mounted	404.10(B)
E4001.11	Snap switch faceplates	404.9(A)
E4001.11.1	Faceplate grounding	404.9(B)
E4001.12	Dimmer switches	404.14(E)
E4001.13	Multipole snap switches	404.8(C)
E4001.14	Cord- and plug-connected loads	401.14(F)
E4001.15	Switches controlling lighting loads	404.2(C)
SECTION E4002	**RECEPTACLES**	
E4002.1	Rating and type	406.3(A) and (B)
E4002.1.1	Single receptacle	210.21(B)(1)
E4002.1.2	Two or more receptacles	210.21(B)(3)
Table E4002.1.2	Receptacle ratings for various size multioutlet circuits	Table 210.21(B)(3)
E4002.2	Grounding type	406.4(A)
E4002.3	CO/ALR receptacles	406.3(C)
E4002.4	Faceplates	406.6(B)
E4002.5	Position of receptacle faces	406.5(D) and 406.6
E4002.6	Receptacles mounted in boxes	406.5(A) and 406.5(B)
E4002.7	Receptacles mounted on covers	406.5(C)
E4002.8	Damp locations	406.9(A)
E4002.9	15- and 20-ampere receptacles in wet locations	406.9(B)(1)
E4002.10	Other receptacles in wet locations	406.9(B)(2)
E4002.11	Bathtub and shower space	406.9(C)
E4002.12	Flush mounting with faceplate	406.9(E)
E4002.13	Exposed terminals	406.5(F)
E4002.14	Tamper-resistant receptacles	406.12
SECTION E4003	**LUMINAIRES**	
E4003.1	Energized parts	410.5
E4003.2	Luminaires near combustible material	410.11
E4003.3	Exposed conductive parts	410.42(A)
E4003.4	Screw-shell type	410.90
E4003.5	Recessed incandescent luminaires	410.115(C)
E4003.6	Thermal protection	410.130(E)(1)
E4003.7	High-intensity discharge luminaires	410.130(F)

E4003.8	Metal halide lamp containment	410.130(F)(5)
E4003.9	Wet or damp locations	410.10(A)
E4003.10	Lampholders in wet or damp locations	410.96
E4003.11	Bathtub and shower area	410.10(D)
E4003.12	Luminaires in clothes closets	410.2; and 410.16(A), (B), (C) and (D)
Figure E4003.12	Closet storage space	Figure 410.2
E4003.13	Luminaire wiring—general	410.48
E4003.13.1	Polarization of luminaires	410.50
E4003.13.2	Luminaires as raceways	410.64(A), (B) and (C)
SECTION E4004	**LUMINAIRE INSTALLATION**	
E4004.1	Outlet box covers	410.22
E4004.2	Combustible material at outlet boxes	410.23
E4004.3	Access	410.8
E4004.4	Supports	410.30(A)
E4004.5	Means of support	410.36(A)
E4004.6	Exposed components	410.136(A)
E4004.7	Combustible low-density cellulose fiberboard	410.136(B)
E4004.8	Recessed luminaire clearance	410.116(A)(1) and (2)
E4004.9	Recessed luminaire installation	410.116(B)
SECTION E4005	**TRACK LIGHTING**	
E4005.1	Installation	410.151(A) and (B)
E4005.2	Fittings	410.151(A) and (D)
E4005.3	Connected load	410.151(B)
E4005.4	Prohibited locations	410.151(C)
E4005.5	Fastening	410.154
E4005.6	Grounding	410.155(B)
CHAPTER 41	**APPLIANCE INSTALLATION**	
SECTION E4101	**GENERAL**	
E4101.1	Scope	422.1 and 424.1
E4101.2	Installation	110.3(B) and 422.17
E4101.3	Flexible cords	422.16
Table E4101.3	Flexible cord length	422.16(B)(1), (2) and (4)
E4101.4	Overcurrent protection	422.11
E4101.4.1	Single nonmotor-operated appliance	422.11(E)
E4101.5	Disconnecting means	422.30, 422.35 and 424.19
Table E4101.5	Disconnecting means	422.31(A), (B) and (C); 422.34; 422.35; 424.19; 424.20; and 440.14
E4101.6	Support of ceiling-suspended paddle fans	422.18
E4101.7	Snow-melting and deicing equipment protection	210.8(A)(3), Exception; and 426.28
CHAPTER 42	**SWIMMING POOLS**	
SECTION E4201	**GENERAL**	
E4201.1	Scope	680.1
E4201.2	Definitions	680.2

SECTION E4202	**WIRING METHODS FOR POOLS, SPAS, HOT TUBS AND HYDROMASSAGE BATH TUBS**
E4202.1	General . . . 680.7; 680.21(A); 680.23(B) and (F); 680.25(A); 680.40; 680.42; 680.43; and 680.70
E4202.2	Flexible cords . . 680.7(A) and (B); 680.22(B)(5); 680.23(B)(3); 680.42; and 680.43, Exception
E4202.3	Double-insulated pool pumps . . . 680.21(B)
SECTION E4203	**EQUIPMENT LOCATION AND CLEARANCES**
E4203.1	Receptacle outlets . . . 680.22(A)(5)
E4203.1.1	Location . . . 680.22(A)(1)
E4203.1.2	Where required . . . 680.22.(A)(3)
E4203.1.3	GFCI protection . . . 680.22(C)
E4203.1.4	Indoor locations . . . 680.43(A) and (A)(1)
E4203.1.5	Indoor GFCI protection . . . 680.43(A)(2)
E4203.2	Switching devices . . . 680.22(D), 680.43(C) and 680.72
E4203.3	Disconnecting means . . . 680.12
E4203.4	Luminaires and ceiling fans . . . 680.22(B)
E4203.4.1	Outdoor location . . . 680.22(B)(1)
E4203.4.2	Indoor location . . . 680.22(B)(2)
E4203.4.3	Existing lighting outlets and luminaires . . . 680.22(B)(3)
E4203.4.4	Indoor spas and hot tubs . . . 680.43(B)
E4203.4.5	GFCI protection in adjacent areas . . . 680.22(B)(4)
4203.5	Other outlets . . . 680.22(D)
E4203.6	Overhead conductor clearances . . . 680.8
Table E4203.6	Overhead conductor clearances . . . Table 680.8
E4203.7	Underground wiring . . . 680.10
Table E4203.7	Underground wiring . . . Table 680.10
SECTION E4204	**BONDING**
E4204.1	Performance . . . 680.26(A)
E4204.2	Bonded parts . . . 680.26(B); 680.26(E); 680.42(B); and 680.43(D)(4)
E4204.3	Pool water . . . 680.26(C)
E4204.4	Bonding of outdoor hot tubs and spas . . . 680.42 and 680.42(B)
E4204.5	Bonding of indoor hot tubs and spas . . . 680.43(D)
E4204.5.1	Methods . . . 680.43(E)
E4204.5.2	Connections . . . 680.26(B)
SECTION E4205	**GROUNDING**
E4205.1	Equipment to be grounded . . . 680.6
E4205.2	Luminaires and related equipment . . . 680.23(F)(2) and 680.24(F)
E4205.3	Nonmetallic conduit . . . 680.23(B)(2)(b)
E4205.4	Flexible cords . . . 680.23(B)(3)
E4205.5	Motors . . . 680.21(A)(1) and (4)
E4205.6	Feeders . . . 680.25(B) and (B)(1)
E4205.6.1	Separate buildings . . . 680.25(B)(2)
E4205.7	Cord-connected equipment . . . 680.7
E4205.8	Other equipment . . . Article 250, Parts V, VI and VII; and 680.6

SECTION E4206	**EQUIPMENT INSTALLATION**
E4206.1	Transformers . . . 680.23(A)(2)
E4206.2	Ground-fault circuit-interrupters . . . 680.5
E4206.3	Wiring on load side of ground-fault circuit-interrupters and transformers . . . 680.23(F)(3)
E4206.4	Underwater luminaires . . . 680.23(A)(1), (3) and (7)
E4206.4.1	Maximum voltage . . . 680.23(A)(4)
E4206.4.2	Luminaire location . . . 680.23(A)(5) and (6)
E4206.5	Wet-niche luminaires . . . 680.23(B)(1), (2), (4) and (5)
E4206.5.1	Servicing . . . 680.23(B)(6)
E4206.6	Dry-niche luminaires . . . 680.23(C)(1) and (2)
E4206.7	No-niche luminaires . . . 680.23(D)
E4206.8	Through-wall lighting assembly . . . 680.23(E)
E4206.9	Junction boxes and enclosures for transformers or ground-fault circuit-interrupters . . . 680.24(A) through (E)
E4206.9.1	Junction boxes . . . 680.24(A)(1) and (2)
E4206.9.2	Other enclosures . . . 680.24(B)(1) and (2)
E4206.9.3	Protection of junction boxes and enclosures . . . 680.24(C)
E4206.9.4	Grounding terminals . . . 680.24(D)
E4206.9.5	Strain relief . . . 680.24(E)
E4206.10	Underwater audio equipment . . . 680.27(A)
E4206.10.1	Speakers . . . 680.27(A)(1)
E4206.10.2	Wiring methods . . . 680.27(A)(2)
E4206.10.3	Forming shell and metal screen . . . 680.27(A)(3)
E4206.11	Electrically operated pool covers . . . 680.27(B)(1) and (2)
E4206.12	Electric pool water heaters . . . 680.9
E4206.13	Pool area heating . . . 680.27(C)
E4206.13.1	Unit heaters . . . 680.27(C)(1)
E4206.13.2	Permanently wired radiant heaters . . . 680.27(C)(2)
E4206.13.3	Radiant heating cables prohibited . . . 680.27(C)(3)
SECTION E4207	**STORABLE SWIMMING POOLS**
E4207.1	Pumps . . . 680.31
E4207.2	Ground-fault circuit-interrupters required . . . 680.32
E4207.3	Luminaires . . . 680.33
E4207.3.1	Within the low voltage contact limit . . . 680.33(A)
E4207.3.2	Over the low voltage contact limit . . . 680.33(B)
E4207.4	Receptacle locations . . . 680.34
SECTION E4208	**SPAS AND HOT TUBS**
E4208.1	Ground-fault circuit-interrupters . . . 680.44(A) and (B)
E4208.2	Electric water heaters . . . 680.9
E4208.3	Underwater audio equipment . . . 680.43(G)
E4208.4	Emergency switch for spas and hot tubs . . . 680.41
SECTION E4209	**HYDROMASSAGE BATHTUBS**
E4209.1	Ground-fault circuit-interrupters . . . 680.71
E4209.2	Other electric equipment . . . 680.72
E4209.3	Accessibility . . . 680.73

E4209.4	Bonding 680.74
CHAPTER 43	**CLASS 2 REMOTE-CONTROL, SIGNALING AND POWER-LIMITED CIRCUITS**
SECTION E4301	**GENERAL**
E4301.1	Scope 725.1
E4301.2	Definitions . . . . Article 100 and 725.2
SECTION E4302	**POWER SOURCES**
E4302.1	Power sources for Class 2 circuits . . . . 725.121(A)
E4302.2	Interconnection of power sources. . . . . 725.121(B)
SECTION E4303	**WIRING METHODS**
E4303.1	Wiring methods on supply side of Class 2 power source. 725.127
E4303.2	Wiring methods and materials on load side of Class 2 power source . . . . .725.130, 725.154, Table 725.154(G), 725.179
Table E4303.2	Cable uses and permitted substitutions. . . . . .725.154 and Table 725.154(G)
E4303.2.1	Type CL2P cables. . . . . 725.154(A) and 725.179(A)
E4303.2.2	Type CL2 cables. . . . . 725.179(C)
E4303.2.3	Type CL2X cables . . . . .725.154(E)(4) and 725.179(D)
E4303.2.4	Marking . . . . 725.179(L)
SECTION E4304	**INSTALLATION REQUIREMENTS**
E4304.1	Separation from other conductors. . . . . 725.136(A), (B), (C) and (D)
E4304.2	Other applications. . . . . .725.136(I)
E4304.3	Class 2 circuits with communication circuits. . . . . 725.139(D)
E4304.4	Class 2 cables with other circuit cables . . . . .725.139(E)(1) through (5)
E4304.5	Installation of conductors and cables . . . . .300.11(B)(2), 725.21, 725.24 and 725.143
E4304.6	Bushing. . . . . .725.3(J)

INDEX

A

ABSORPTION COOLING EQUIPMENT M1412
ACCESS
To appliances M1305
To attic R807
To crawl space R408.4
To equipment M1401.2
To floor furnace M1408.4
To plumbing connections P2704
To plumbing fixtures P2705
To whirlpool pumps P2720.1
ACCESSIBLE
Definition R202
Readily accessible, definition R202
ACCESSORY STRUCTURE
Definition R202
ADDRESS (SITE) R319
ADMINISTRATIVE . . . . Chapter 1
Authority R104
Entry R104.6
Inspections R109
Permits R105
Purpose R101.3
Violations R113
AIR
Combustion . . . . Chapter 17, G2407
Combustion air, definition R202
Return M1602
AIR CONDITIONERS
Branch circuits E3702.11
Room air conditioners E3702.12
ALLOWABLE SPANS
Of floor joists R502.3, R505.3.2
Of headers R602.7, R603.6, R613.10
Of rafters and ceiling joists R802.4, R802.5, R804.3.1, R804.3.2
ALTERNATIVE MATERIALS (see MATERIALS) R104.11
AMPACITY E3501
ANCHOR BOLTS R403.1.6
APPEAL
Board of R112.1
Right of R112
APPLIANCE
Access to M1305
Attic M1305.1.3
Clearance for M1306
Connectors, fuel-gas . . . . Chapter 24
Definition R202
Definition applied to electrical equipment E3501
Electrical appliance disconnection means E4101.5
Electrical appliance installation E4101
Equipment (general) . . . . Chapter 14
Floor furnace M1408
Flue area R1003.14
Fuel-burning . . . . Chapter 24
Heating and cooling . . . . Chapter 14
Installation M1307
Labeling M1303
Open-top broiler units M1505.1
Outdoor G2454
Ranges M1901
Room heaters M1410
Special fuel-burning equipment . . . . Chapter 19
Vented (decorative) . . . . Chapter 24
Wall furnace M1409
Warm-air furnace M1402
Water heaters . . . . Chapter 20, Chapter 24
APPLICATION
Plywood R703.3
APPROVAL
Mechanical M1302
APPROVED
Definition R202
Definition applied to electrical equipment E3501
ARC-FAULT CIRCUIT-INTERRUPTER E3902
AREA
Disposal, private sewage disposal P2602
Flue (appliances) R1003.14
Flue masonry R1003.15
ARMORED CABLE . . . . Table E3801.2
ASPHALT SHINGLES R905.2
ATTACHMENT PLUG (PLUG CAP) (CAP)
Definition E3501
ATTIC
Access R807
AUTOMATIC FIRE SPRINKLER SYSTEMS . . . R313

B

BACKFILL
For piping P2604
BACKFLOW, DRAINAGE
Definition R202
BACKWATER VALVE R202

BASEMENT WALL
Definition R202
BATH AND SHOWER SPACES R307
BATHROOM
Exhaust M1507.4
Group. . . . . R202, Table P3004.1
BATHTUB
Enclosure. P2713
Hydromassage. E4209
Whirlpool P2720
BEAM SUPPORTS. R606.14
BEARING
Of joists R502.6
BIDETS. P2721
BOILER
Definition R202
Requirements . . . . . Chapter 20
BONDING. E3609, E4204
Definition applied to electrical installations E3501
BONDING JUMPER. E3501
Bonding of service equipment E3609.2
Bonding to other systems E3609.3
Main bonding jumper E3607.5
Metal water piping bonding E3608.1.1
Sizing bonding jumpers E3609.5
BORED HOLES (see NOTCHING)
BOXES E3906
Nonmetallic boxes E3905.3
Support of boxes E3904.3
Where required E3905.1
BRANCH CIRCUIT. . . . . E3501, Chapter 37
Branch circuit ampacity E3701.2
Branch circuit ratings E3702
Branch circuits required. E3703
BUILDING
Definition R202
Drain, definition R202
Existing, definition R202
Sewer, definition R202
BUILDING OFFICIAL
Definition R202
Inspection and tests of fuel-gas piping . . . . . Chapter 24
BUILDING PLANNING
Automatic fire sprinkler systems R313
Carbon monoxide alarms. R315
Ceiling height. R305
Decay protection R317
Design criteria R301
Emergency escape R310
Exterior wall R302.1
Fire-resistant construction R302
Foam plastic. R316
Garages and carports R309
Glazing. R308
Guardrails. R312
Handrails R311.7.8, R311.8.3
Insulation R302.10
Landing R311.3, R311.3.1, R311.3.2, R311.5.1
Light, ventilation and heating R303
Means of egress. R311
Minimum room area R304
Planning . . . . . Chapter 3
Plumbing fixture clearances. R307
Radon protection . . . . . Appendix F
Ramps R311.8
Sanitation R306
Site address R319
Smoke alarms R314
Stairways R311.7
Storm shelters R323
Termite protection R318
BUILDING THERMAL ENVELOPE
Definition R202
BUILT-UP GIRDERS (see GIRDERS)
BUILT-UP ROOFING (see ROOFING)

C

CABINETS AND PANELBOARDS E3907
CAPACITY
Expansion tank. M2003.2
CARBON MONOXIDE ALARMS R315
CARPORT. R309.4
CEILING
Finishes R805
Height. R305
CEILING FANS. E4203.4
CENTRAL FURNACES (see FURNACES)
CHASES R606.7
CHIMNEYS
And fireplaces . . . . . Chapter 10
Caps. R1003.9.1
Clearance. R1003.18
Corbeling R1003.5
Crickets R1003.20
Design (masonry). R1003.1
Factory-built R1005
Fireblocking R1003.19
Flue area R1003.14, R1003.15
Flue lining. R100311

Load R1003.8
Masonry and factory built, size M1805
Multiple flue R1003.14
Rain caps 1003.9.3
Spark arrestors R1003.9.2
Termination R1003.9
Wall thickness R1003.10

CIRCUIT BREAKER
Definition E3401

CLASS 2 CIRCUITS
Class 2 remote-control, signaling and power-limited circuits . . . . Chapter 43

CLAY
Tiles R905.3

CLEANOUT
Definition R202
Drainage P3005.2
Masonry chimney R1003.17

CLEARANCE
Above cooking top M1901.1
For appliances M1306.1
For chimneys R1003.18
Reduction methods M1306.2
Vent connector M1803.3.4

CLEARANCES
Around electrical equipment E3405.1, E3604.1, E3604.2, E3604.3

CLOTHES CLOSETS
Lighting fixtures E4003.12

CLOTHES DRYERS . . . . M1502, Chapter 24

CLOTHES WASHING MACHINES P2718

COLUMNS R407

COMBUSTIBLE
Materials R202

COMBUSTION AIR
Air . . . . Chapter 17
Definition R202
Indoor M1702
Supply M1701.1

COMMON VENT
Definition R202
Requirements M1801.11

CONCRETE
Compressive Strength R402.2
Floors (on ground) R506
Tile (roof) R905.3
Weathering . . . . Figure R301.2(3), R402.2

CONCRETE-ENCASED ELECTRODE E3608.1.2

CONDUCTOR E3406
Ampacity E3705.1
Definition E3501
Identification E3407
Insulation E3406.5
Material E3406.2
Parallel E3406.6
Size E3406.3, E3704.1
Ungrounded E3603.1

CONDUIT BODY
Definition E3501

CONNECTION
Access to connections P2704
For fuel-burning appliances . . . . Chapter 24
For fuel-oil piping M2202
Joints P2904
Plumbing fixture P2601.2
To water supply P2902.1

CONNECTIONS
Aluminum E3406.8
Device E3406.10.2

CONNECTOR
Chimney and vent M1803
Vent, definition R202

CONNECTOR, PRESSURE (SOLDERLESS)
Definition E3501

CONSTRUCTION
Cavity wall masonry R608
Flood-resistant R322
Floors . . . . Chapter 5
Footings R403
Foundation material R402
Foundation walls R404
Foundations . . . . Chapter 4
Grouted masonry R609
Masonry R606, R607, R608, R609, R610
Reinforced hollow unit masonry R609.4
Roofs . . . . Chapter 8
Steel framing R505, R603, R804
Wood framing R502, R602, R802
Walls . . . . Chapter 6

CONTINUOUS LOAD
Definition E3501

CONTINUOUS WASTE
Definition R202

CONTROL
Devices . . . . Chapter 24

CONTROLS
For forced-air furnaces . . . . Chapter 24

COOKING UNIT, COUNTER-MOUNTED
Definition E3501

COOLING
Absorption cooling equipment M1412

Access to equipment M1401.2
Evaporative cooling equipment M1413
Installation M1401.1
Refrigeration cooling equipment M1404
Return air-supply source M1602
Screen inlet protection M1602.3
COPPER-CLAD ALUMINUM CONDUCTORS
Definition E3501
CORDS
Flexible E4101.3, E4202.2
COURT
Definition R202
COVER REQUIREMENTS E3803.1
COVERING
Exterior R703
Interior R702
Roof . . . . Chapter 9
Wall . . . . Chapter 7
CRAWL SPACE R408
CRITERIA
Design R301
CROSS CONNECTION
Definition R202

D

DAMPER, VOLUME
Definition R202
DECAY
Protection against R317
DECK
Supported by exterior wall R502.2.2
Wood/plastic composite boards. R502.1.7, R502.2.2.4
DECORATIVE APPLIANCES
Outdoor G2454
Vented . . . . Chapter 24
DEFINITIONS
Building R202
Electrical E3501
Mechanical system R202
Plumbing R202
DESIGN
Criteria R301
DIRECTIONAL
Fittings, plumbing. P2707
DISCONNECTING MEANS
Definition E3501
DISHWASHING MACHINES P2717
DOORS
Egress R311.2
Exterior. R311.3.2, R612
DRAFT HOOD
Definition R202
DRAFTSTOPPING R302.12, R502.12
DRAIN
Floor. P2719
Shower receptors. P2709
DRAINAGE
Cleanouts. P3005.2
Foundation. R405
Inspection and tests P2503
Storm drainage. Chapter 33
DRILLING AND NOTCHING (see NOTCHING)
DRIP LOOPS E3605.9.5
DRYERS
Domestic clothes M1502
DUCTS . . . . Chapter 16
Installation M1601.4
Insulation N1103.2, M1601.3
Material M1601.1
System, definition. R202
DWELLING
Definition R202, E3501
DWELLING UNIT
Definition R202, E3501
Separation R302.2, R302.3

E

ELECTRICAL
Appliance (labeling) M1303
Inspection. E3403
ELECTRICAL METALLIC TUBING . . . . Table E3904.6
ELECTRICAL NONMETALLIC TUBING . . . . Table E3904.6
ELECTRICAL RESISTANCE HEATERS
Baseboard convectors M1405
Duct heaters. M1407
Radiant heating M1406
ELECTRODES
Grounding E3608
EMERGENCY ESCAPE R202, R310
ENCLOSURE
Definition E3501
ENERGY CONSERVATION . . . . Chapter 11
ENTRY R104.6
EQUIPMENT
Definition applied to electrical equipment. E3501
General, mechanical . . . . Chapter 14
Heating and cooling . . . . Chapter 14

EXCAVATIONS
For appliance installation M1305.1.4.2
EXTERIOR
Covering R703
Insulation finish systems. R703.9
Lath. R703.6.1
Plaster R703.6
EXTERIOR WALL
Construction R302.1
Definition. R202
Fire-resistance rating R302.1

F

FACEPLATES. E4001.11, E4002.4
FACTORY BUILT
Chimneys R1005, M1805
Fireplace stoves R1005.3
Fireplaces R1004
FASTENING . . . . Table R602.3(1)
FAUCETS P2701
FEEDER
Ampacity. E3704.5
Conductor size E3704.1
Feeder neutral load E3704.3
Loads E3704.2
Requirements E3704
FENESTRATION
Definition. R202
FINISHES
Flame spread and smoke density. R302.9
For ceilings R805
Interior R302.9, R315, R702
FIRE-RESISTANT CONSTRUCTION R302
FIRE SPRINKLER SYSTEM
Inspections of P2904.8
Sizing of P2904.6
Sprinkler location P2904.1.1
FIREBLOCKING
Barrier between stories. R302.11, R602.8
Chimney R1003.19
Fireplace. R1001.12
FIREPLACES . . . . Chapter 10
Clearance R1001.11
Corbeling R1001.8
Factory-built R1004
Fireblocking R1001.12
Walls. R1001.5
FITTING
Definition applied to electrical installations E3501
FITTINGS
DWV piping. P3002.3
Prohibited joints P3003.2
Water supply. P2905.6
FIXTURE UNIT
Drainage, definition. R202
Unit valves P2903.9.3
Water supply, definition. R202
FIXTURES
Plumbing fixture, definition R202
Plumbing fixture, general . . . . Chapter 27
Trap seals P3201.2
FLAME SPREAD INDEX. R302.9, R302.10
FLASHING. R703.7.5, R703.8, R903.2, R905
FLEXIBLE CORDS E3909, E4101.3, E4202.2
FLEXIBLE METAL CONDUIT. . . . E3801.4, E3908.8.1
FLOOR FURNACE M1408
FLOORS
Concrete (on ground) R506
Steel framing. R505
Treated-wood (on ground) R504
Wood framing R502
FLUE
Area R1003.14, R1003.15
Lining R1003.11, R1003.12
Multiple R1003.13
FOAM PLASTICS R316
FOOD-WASTE GRINDERS. P2716
FOOTINGS R403
FOUNDATIONS. . . . . Chapter 4
Cripple walls R602.9, R602.10.9, R602.11.2
Frost protection. R403.1.4.1
Inspection R109.1.1
Walls. R404
FRAME
Inspection R109.1.4
FREEZE PROTECTION OF PLUMBING P2603.6
FUEL-BURNING APPLIANCES (see APPLIANCE, definition)
Identification . . . . Chapter 24
FUEL GAS. . . . . Chapter 24
FUEL OIL
Oil tanks M2201
Piping, fittings and connections M2202
Pumps and valves. M2204
FURNACES
Clearance of warm-air furnaces M1402.2
Exterior M1401.4
Floor (see Floor Furnace)
Wall (see Wall Furnace)
Warm-air. . . . . M1402, Chapter 24

FUSES E3705.6

G

GARAGES R309
GAS
Appliance labeling . . . . Chapter 24
GAS PIPING SYSTEM NOT TO BE USED AS GROUNDING ELECTRODE . . . . Chapter 24
GIRDERS R502.5
GLAZING R308
Aggregate R303.1
Protection of openings R301.2.1.2
GRADE
Definition R202
Of lumber R502.1, R602.1, R802.1
Plane, definition R202
Slope of piping, definition R202
GRAY WATER P3009
GROUND
Definition of electrical E3501
Floors (on ground) R504, R506
Joint connection P2905.8
GROUND-FAULT CIRCUIT-INTERRUPTER PROTECTION E3902, E4203, E4206, E4207.2, E4208
GROUND SOURCE HEAT PUMP M2105
GROUNDED
Definition E3501
GROUNDED CONDUCTOR
Definition E3501
Identification E3407
GROUNDED, EFFECTIVELY
Definition E3501
GROUNDING
Effective grounding path E3611.3
Of equipment E3908, E4205.1
GROUNDING CONDUCTOR
Definition E3501
Equipment E3501, E3607.3.1, E3908
GROUNDING ELECTRODE CONDUCTOR E3607.4
Connection E3607.2, E3611
Definition E3501
Size E3603.4
GROUNDING ELECTRODE SYSTEM E3608
GROUNDING ELECTRODES E3608, E3608.1
Resistance of E3608
GROUTED MASONRY (see MASONRY) R609
Reinforced R609.3
GUARDED
Definition applied to electrical equipment E3501
GUARDING OF ENERGIZED PARTS E3404.9
GUARDS R312
Definition R202
GYPSUM
Wallboard R702.3

H

HABITABLE SPACE
Definition R202
HALLWAYS R311.6
HANDRAILS R311.7.7, R311.8.3
Definition R202
HEADERS
SIP R613.10
Steel R603.6
Wood R602.7
HEARTH R1001.9
Extension R1001.9, R1001.10
HEATERS
Baseboard M1405
Central furnaces M1402
Duct M1407
Heat pumps M1403
Masonry R1002
Pool . . . . M2006, Chapter 24
Radiant M1406
Sauna . . . . Chapter 24
Unvented G2445
Vented room . . . . M1410, Chapter 24
Water . . . . Chapter 20
HEATING R303.8
HEATING EQUIPMENT . . . . Chapter 14
HEIGHT
Ceiling R305
HOLLOW-UNIT MASONRY (see MASONRY) R607.2.2.2, R608.1.1.2
Reinforced R609.4
HORIZONTAL
Pipe, definition R202
Vent length M1803.3.2
HOT TUBS . . . . E4201, Appendix G
HOT WATER
Definition R202
Distribution pipe rating P2905.5
Heaters . . . . M2005, Chapter 24
HYDROGEN GENERATING AND REFUELING M1307.4

HYDRONIC HEATING SYSTEMS
Baseboard convectors M2102
Boilers M2001
Expansion tanks M2003
Floor heating systems M2103
Operating and safety controls M2002
Piping systems installation M2101
Water heaters M2005

I

IDENTIFIED
Definition applied to electrical equipment E3501
INDIVIDUAL
Branch circuit, definition E3501
Sewage disposal systems R202, P2602
Water supply and sewage disposal systems P2602
INLET
To masonry chimneys R1003.16
INSPECTION
Card AE305.3
Excavation for piping P2604.1
Fuel-supply system . . . . . Chapter 24
Of plumbing system P2503
On site R109.1
INSTALLATION
Existing . . . . . Appendix J
Of appliances M1307
Of cooling and heating equipment M1401.1
Of ducts M1601.4
Of flue liners R1003.12
Of plumbing fixtures P2705
Of wall furnaces M1409.3
INSULATION R302.10
INTERIOR
Lath R702.2.3
Other finishes R702.5
Plaster R702.2
Wall covering . . . . . Chapter 7
INTERMEDIATE METAL CONDUIT E3801, E3904
INTERRUPTING RATING E3404.2
Definition E3501

J

JOINTS, PIPE
And connections P2905, P3003
Slip P2704.1, P3005.2.9
JOIST
Bearing R502.6, R606.14.1
JUMPERS
Bonding E3609
JUNCTION BOXES E3905, E4206.9.1

K

KITCHEN
Definition R202
KITCHEN RECEPTACLES E3703.2, E3901

L

L VENT TERMINATION M1804.2.4
LABELED
Definition R202
Definition applied to electrical equipment E3501
LABELING
Appliances M1303
Definition R202
LAMPHOLDERS E4003, E4004
LANDINGS R311.3, R311.3.1, R311.7.6, R311.8.2
LATERAL SUPPORT R502.7, R606.9, R610.5.2
LATH
Exterior R703.6.1
Interior R702.2.3
LAUNDRY CIRCUIT E3703.3
LAUNDRY TUBS P2715
LAVATORIES P2711
Clearances R307
Waste outlets P2711.3
LEAD
Bends and traps P3002.4.2
Caulked joints P3003.6.1
Flashing P3002.4.1
LIABILITY R104.8
LIGHT, VENTILATION AND HEATING R303
LIGHTING
Luminaire in clothes closets E4003.12
Luminaire installation E4004
Luminaries E4003, E4004, E4206
LIGHTING OUTLETS
Definition E3501
Required lighting outlets E3903
LINING
Flue R1003.11, R1003.12
LINTEL R606.10, R611.8, R703.7.3, R1001.7
LIQUID-TIGHT FLEXIBLE CONDUIT E3801.4, E3908.8.2

LISTED
Definition applied to electrical equipment E3501
LISTED AND LISTING
Definition applied to building and mechanical provisions R202
Mechanical appliances M1302.1
LOAD
Additional R1003.8
Roof R301.6
Seismic risk map . . . . Figure R301.2(2)
Snow load map . . . . Figure R301.2(5)
Wind speed map . . . . Figure R301.2(4)
LOADS
Branch circuit loads E3702, E3703.6
Dead load R301.4
Feeder load E3704.2
Feeder neutral load E3704.3
Horizontal load AE602.1
Live load R301.5
Service load E3602.2
LOADS, LIVE AND DEAD
Definition R202
LOCATION
Of furnaces M1408.3, M1409.2
LOCATION (DAMP) (DRY) (WET)
Definitions E3501
LUMBER
Grade R502.1, R602.1, R802.1
LUMINAIRE E4004
Clearances E4004.8

M

MACERATING TOILET P2723
MANUFACTURED HOME
Definition R202
Provisions . . . . Appendix E
MASONRY
Anchorage R606.11
Cavity wall R608
Chases R606.7
General R606
Grouted R609
Hollow unit R202, R607.2.2.2, R608.1.1.2
Inspection R109.1.4
Reinforced grouted .R609.3
Reinforced hollow unit R609.4
Seismic requirements R606.12
Solid, definition R202
Veneer R703.7
Veneer attachment .R703.4
MATERIALS
Alternative R104.11
Combustible R202, R1001.11, R1001.12, R1003.18, R1003.19
For ducts M1601
For fixture accessories P2702
For flue liners R1003.11
For fuel-supply systems . . . . Chapter 24
For hearth extension R1001.9
For siding R703.3
Hydronic pipe M2101, M2103, M2104
Plumbing pipe P2905, P3002
MECHANICAL
Inspection R109.1.2
System requirements . . . . Chapter 13
MEDIUM PRESSURE FUEL-GAS PIPING SYSTEM . . . . Chapter 24
MEMBRANE
Penetration R302.4.2
Polyethylene R504.2.2, R506.2.3
Water-resistive R703.2
Waterproofing R406.2
METAL
Ducts . . . . Chapter 16
Roof panels R905.10
Roof shingles R905.4
METAL-CLAD CABLE . . . . .Table E3701.2
METHODS
Water distribution pipe sizing . . . . Appendix P
MODIFICATIONS R104.10
MOISTURE CONTROL R601.3
MORTAR
Joints R607.2.1.1
MOTORS
Motor branch circuits E3702.6, E3702.7
MULTIPLE
Appliance venting systems M1801.11
Flues R1003.13
MULTIWIRE BRANCH CIRCUITS E3701.5

N

NONCOMBUSTIBLE MATERIAL
Definition R202
NONMETALLIC BOXES E3905.3, E3908.16
NONMETALLIC RIGID CONDUIT E3801.4
NONMETALLIC-SHEATHED CABLE E3705.4.4, E3801.4, E3905.3.1, E3908.8.3
NOTCHING
Electrical E3402.1

For mechanical systems M1308
For plumbing piping P2603.2
Steel joists R505.2.5, R505.3.5, R804.2.5, R804.3.4
Steel studs R603.2.5, R603.3.4
Wood joists R502.8, R802.7.1
Wood studs R602.6
Wood top plates R602.6.1

O

OCCUPIED SPACE
Definition R202
OIL
Piping and connections M2202
Supply pumps and valves M2204
Tanks M2201
OPEN-TOP GAS BROILER UNITS M1505.1
OPENING
Requirements, combustion air . . . . . Chapter 17, Chapter 24
Waterproofing of piping penetrations P2606.1
OUTLET
Definition . . . . . Chapter 24, E3501
OVEN, WALL-MOUNTED
Definition applied to electrical provisions E3501
OVERCURRENT
Definition E3501
OVERCURRENT PROTECTION E3705
OVERCURRENT-PROTECTION RATING . . . E3705.6
OVERLOAD
Definition E3501

P

PANELBOARD
Definition E3501
PANELBOARDS E3706
Clearance and dedicated space E3405.2, E3405.3
Headroom E3405.7
PARAPETS R307.2.2, R606.2.4
PARTICLEBOARD
Floor R503.3
Walls R605
PERMITS R105
PHOTOVOLTAIC SOLAR ENERGY SYSTEMS M2302
PIERS R606.6
Masonry R404.1.9
PIPE
Materials listing . . . . . Table M2101.1, P2608
Protection P2603
Standards, drain, waste and vent . . . . . Table P3002.1(1), Table P3002.1(2)
Standards, sewer . . . . . Table P3002.2
Standards, water supply P2905
PIPING
Connections and materials . . . Table M2101.1, P2905
Drain, waste and vent P3002.1
Fuel-gas size determination . . . . . Chapter 24
Protection P2603
Sizing methods, water distribution P2903.7
Support . . . . . M2101.9, Chapter 24, P2605
PLANNING
Building . . . . . Chapter 3
PLANS R106
PLASTER
Exterior R703.6
Interior R702.2
PLENUM
Definition R202
Definition, electrical installations E3501
PLUMBING
Fixture clearances R307
Fixtures . . . . . Chapter 27
Inspection R109.1.2
Materials P2702.2, P2905, P3002
Requirements and definitions R202
System, definition R202
Traps . . . . . Chapter 32
Vents . . . . . Chapter 31
PLYWOOD
Application R703.3
Materials, walls R604
POTABLE WATER
Definition R202
PRECAST CONCRETE
Footings R403.4
Foundation material R402.3.1
Foundation walls R404.5
PRESSURE, WATER SUPPLY
Maximum P2903.3.1
Minimum P2903.3
PRIVATE
Sewage disposal system . . . . . Appendix I
PROHIBITED
Receptors P2706.3
Return air sources M1602.2
Traps P3201.5
Valves below grade P2903.9.5

Water closets P2712.1

PROTECTION

Against decay and termites R319, R320

Against radon. . . . . Appendix F

Of backflow preventers P2902.6

Of ferrous gas piping . . . . . Chapter 24

Of potable water supply. P2902

PURLINS R802.5.1

PURPOSE. R101.3

R

***R*-VALUE**

Definition R202

RACEWAY

Definition E3501

Raceway installations E3904

Raceway seals. E3601.5, E3803.6

Raceways as means of support. E3904.4

RADON

Map . . . . . Appendix F

RAFTERS

Grade of lumber. R802.1

Spans. . . . . R802.5, Tables R802.5.1(1) – R802.5.1(8)

RAIN TIGHT

Definition applied to electrical provisions E3501

RAINPROOF

Definition applied to electrical provisions E3501

RAMPS. R311.8

RANGES

Branch circuits for ranges E3702.9

RANGES AND OVENS

Vertical clearance above cooking top M1901.1

READILY ACCESSIBLE

Definition R202

Definition, electrical installations E3501

RECEPTACLE

Definition E3501

RECEPTACLE OUTLET

Definition E3501

Required outlets. E3901

RECEPTACLES

Rating, type and installation. E4002

RECEPTORS

Plumbing fixtures and traps . . . . . Chapter 27

Shower receptor materials. P2709.2

Waste. P2706

RECESSED LUMINAIRES. E4003.5, E4003.12, E4004.8, E4004.9

REQUIREMENTS

Connections for fuel-burning appliances. . . . . Chapter 24

Return air M1602.1

RESISTANCE TO GROUND

Electrodes E3608.4

RESISTANT SIDING MATERIAL (see MATERIALS)

RETURN-AIR LIMITATION M1602.2

RIDGE BOARD. R802.3

RIGID METALLIC CONDUIT E3908

RIGID NONMETALLIC CONDUIT E3801.4

ROOF

Coverings. R905

Drainage. R903.4

Flashing R703.8, R903.2, R905

Steel framing R804

Wood framing. R802

ROOF-CEILING CONSTRUCTION (see CONSTRUCTION). . . . . Chapter 8

Wind uplift R802.11, R804.3.9

ROOFING

Built-up. R905.9

Liquid-applied coating R905.15

Modified bitumen R905.11

Sprayed polyurethane foam. R905.14

Thermoplastic single-ply R905.13

Thermoset single-ply R905.12

ROOM

Heaters, vented M1410

Minimum Sizes. R304

S

SANITATION R306

SEISMIC RISK MAP. . . . . Figure R301.2(2)

SEPTIC TANK

Definition R202

SERVICE

Definition E3501

SERVICE CABLE

Definition E3501

SERVICE CONDUCTORS

Definition E3501

Drip loops. E3605.9.5

Insulation E3605.1

Overload protection E3603.3

Rating of ungrounded service conductors E3602.1

Size E3603.1

SERVICE DISCONNECT

Location E3601.6.2

Marking of E3601.6.1

Maximum number of. E3601.7
Rating of E3602.3
Required E3601.6
SERVICE DROP
Clearances E3604.1, E3604.2
Definition. E3501
Point of attachment. E3604.3
SERVICE-ENTRANCE CONDUCTORS, OVERHEAD SYSTEM
Definition. E3501
SERVICE-ENTRANCE CONDUCTORS, UNDERGOUND SYSTEM
Definition. E3501
SERVICE EQUIPMENT
Definition. E3501
SERVICE LATERAL
Definition. E3501
SERVICE LOAD E3602.2
SERVICE POINT
Definition. E3501
SERVICE VOLTAGE RATING E3602.4
SERVICES. . . . . Chapter 36
SEWAGE
Disposal, private P2602
SEWER, BUILDING
Definition. R202
Size. P3005.4.2
Testing P2503.4
SHAKES
Wood R702.6, R703.5, R905.8
SHINGLE
Asphalt shingles R905.2
Metal. R905.4
Slate R905.6
Wood R905.7
SHOWER
Compartment R307.2
Receptor P2709
Stall dimensions P2708
Walls. P2710
SHUTOFF VALVE (see VALVES)
SIDING
Exterior coverings. R703
SINKS P2714
SITE
Address R319
Preparation. R408.5, R504.2, R506.2
SIZE
Of combustion air openings . . . . Chapter 17
Of rooms. R304
Of trap. P3201.7
SIZING METHODS
Water piping P2903.7
SKYLIGHTS. R308.6
SLATE SHINGLES R905.6
SMOKE ALARMS R314
SMOKE-DEVELOPED INDEX. R302.9, R302.10
SNOW LOAD MAP .R301.2(5)
SOLAR ENERGY SYSTEMS M2301
SPANS
Steel (allowable) R505.3.2, R804.3.2.1
Wood (allowable) R502.3, R802.5
SPAS E4208
SPLICES E3406.10
SPLICES AND TAPS E3803.4
STACK BOND R606.8
STAIRWAYS R311.7
STANDARDS. . . . . .Chapter 44, Appendix G
STEEL
Fireplace units. R1001.5.1
Floor construction R505
Roof-ceiling construction. R804
Walls R603
STORM SHELTERS R323
STORY
Definition. R202
STOVES
Factory-built fireplace M1414
STRUCTURAL AND PIPING PROTECTION . . .P2603
STRUCTURAL INSULATED PANEL (SIP). R613
STUDS
Wood. R602.2, R602.3
Spacing. R602.3.1
Steel R603.2, R603.3
SUPPLY
Combustion air . . . . Chapter 17, Chapter 24
Fuel systems. . . . . Chapter 24
Oil supply pumps and valves M2204
Required gas. . . . . Chapter 24
Return air M1602
Water . . . . Chapter 29
SUPPORT
Of decks R502.2.2
Of ducts. M1601.4.3
Of floor joists. R502.6, R505.3.2
Of masonry chimneys R1003.2, R1003.3, R1003.4
Of pipe . . . . .M2101.9, Chapter 24, P2605
SUPPORT REQUIREMENTS FOR WIRING METHODS. E3802.1
SWIMMING POOLS . . . . .Chapter 42, Appendix G

SWITCHES
Definition E3501
Rating and application E4001
SYSTEMS
Mechanical venting . . . . Chapter 18, Chapter 24
Plumbing, drainage, waste and venting . . . Chapter 25, Chapter 30, Chapter 31

T

TAIL PIECES P2703
TANK
For fuel oil-burning appliances M2201
Septic, definition R202
TEMPERATURE ISOLINES . . . . Figure R301.2(1)
TEMPERATURE LIMITATIONS E3705.4
TERMINALS E3406.9
TERMINATION
Of chimneys R1003.9
Of vents (general) M1804.2
TERMITES
Infestation probability map . . . . Figure R301.2(6)
Protection R318
TEST
For leaks in supply piping . . . . Chapter 24
Of building sewers P2503.4
Of plumbing systems P2503
Of shower liner P2503.6
THICKNESS
Of chimney walls R1003.10
TIES
Veneer R703.7.4
TILE
Shingles (clay and concrete) R905.3
TOWNHOUSE
Definition R202
Scope R101.2
Separation R302.2
TRACK LIGHTING E4005
TRAP . . . . Chapter 32
Arm, definition R202
TRUSSES
Steel R505.1.3, R804.3.7
Wood R502.11, R802.10
TYPE OF VENTING SYSTEMS REQUIRED . . . . . Chapter 18, Chapter 24

U

***U*-FACTOR**
Definition R202
UNDER FLOOR
Access to furnaces M1305.1.4
Plenums M1601.5
Space R408
UNDERGROUND INSTALLATION REQUIREMENTS
Duct systems M1601.1.2
Electrical cable E3803
Wiring E4103.6
UNVENTED ATTIC ASSEMBLIES R806.4
UTILIZATION EQUIPMENT
Definition E3401

V

VALVES
Backwater P3008
Meter . . . . Chapter 24
Oil-supply M2204
Relief, water heater P2803
Shutoff, fuel-gas . . . . Chapter 24
Shutoff, fuel-oil M2204.2
Water heaters P2903.9.2
Water supply P2903.9
VAPOR RETARDERS R601.3
Definition R202
VENEER
Masonry R703.7
VENT
B or BW vent . . . . Chapter 24
Definition . . . . Chapter 24
L vent . . . . M1804.2.4, Chapter 24
Plumbing system, definition R202
Termination M1804.2, M2203.5, P3103.5
VENTED
Decorative appliances . . . . Chapter 24
Floor furnaces M1408
Room heaters M1410
Wall furnaces M1409
VENTILATION
Bathroom and kitchen M1507
For hydrogen systems M1307.4
Roof R806
Under floor R408.1
Whole house M1507.3
VENTING (MECHANICAL)
Chimney and vent connectors M1803
Components M1802
General M1801
Required M1801, M2203.4

Systems . . . Chapter 18
VENTING (PLUMBING)
Air admittance valves . . . P3114
Circuit . . . P3110
Combination waste and vent system . . . P3111
Common . . . P3107
Connections and grades . . . P3104
Fixture . . . P3105
Individual . . . P3106
Island fixture . . . P3112
Pipe sizing . . . P3113
Stacks and stack vents . . . P3102
Systems . . . P3101
Terminals . . . P3103
Waste stack . . . P3109
Wet . . . P3108
VERTICAL
Clearances above cooking top . . . M1901.1
Pipe, definition . . . R202
VIOLATIONS
And penalties . . . R113
VOLTAGE
Definition . . . E3501
VOLTAGE RATING (SERVICES) . . . E3602.4
VOLTAGE TO GROUND
Definition . . . E3501
VOLTAGES . . . E3404.1

W

WALL FURNACE (see APPLIANCE)
General . . . M1409.1
Installation . . . M1409.3
Location . . . M1409.2
WALLBOARD
Gypsum . . . R702.3
WALLS
Bracing, steel . . . R603.9
Bracing, wood . . . R602.10
Construction . . . Chapter 6
Covering . . . Chapter 7
Cripple . . . R602.9
Deflection . . . R301.7
Exterior covering . . . R703
Finishes . . . R302.9, R702
Fireplace . . . R1001.5
Foundation . . . R404
Insulating concrete form . . . R611.3, R611.4, R611.5.3
Steel framing . . . R603
Structural insulated panels (SIP) . . . R613
Thickness, masonry chimneys . . . R1003.10
Wood framing . . . R602
WARM-AIR FURNACES . . . M1402
Access to . . . M1401.2
Clearance from . . . M1402.2
WASTE
Continuous, definition . . . R202
WATER
Distribution system, definition . . . R202
Heater, definition . . . R202
Heaters . . . Chapter 20, Chapter 24, Chapter 28
Individual supply and sewage disposal . . . P2602
Piping sizing methods . . . P2903
Supply and distribution . . . Chapter 29
Supply system, definition . . . R202
WATER CLOSET . . . R306.1
WATER TIGHT
Definition applied to electrical provisions . . . E3501
WATERPROOFING
And dampproofing . . . R406
Of openings through roofs and walls . . . P2606
WEATHERPROOF
Definition applied to electrical provisions . . . E3501
WHIRLPOOL BATHTUBS . . . P2720
WIND SPEED MAP . . . Figure R301.2(4)
WINDOW . . . R612
Fall prevention . . . R312.2
Opening limiting devices . . . R312.2.2
WIRING METHODS . . . Chapter 37
WOOD
Floor construction . . . R502
Foundation walls . . . R404.2
Roof-ceiling construction . . . R802
Shakes . . . R905.8
Shingles . . . R905.7
Trusses . . . R502.11, R802.10
Walls . . . R602
WORKING SPACE . . . E3606.2
Around electrical equipment . . . E3405.1, E3405.2
WORKMANSHIP, PLUMBING . . . P2607

Y

YARD
Definition . . . R202